Handbook of METALLOPROTEINS

Volume 3

Handbook of METALLOPROTEINS

Editors:

Albrecht Messerschmidt
Max-Planck-Institut für Biochemie, Martinsried bei München, Germany

Mirek Cygler
Biotechnology Research Institute, NRC, McGill University, Montreal, Quebec, Canada

and

Wolfram Bode
Max-Planck-Institut für Biochemie, Martinsried bei München, Germany

Series Editors:

Albrecht Messerschmidt
Robert Huber
Thomas Poulos
Karl Wieghardt

WILEY

Copyright © 2004 John Wiley & Sons, Ltd,
The Atrium,
Southern Gate,
Chichester,
West Sussex,
PO19 8SQ, England

Phone   (+44) 1243 779777
Email (for orders and customer service enquires): cs-books@wiley.co.uk
Visit our Home Page on www.wiley.co.uk or www.wiley.com

All Rights Reserved. No part of this publication may be reproduced, stored in a retrieval system or transmitted in any form or by any means, electronic, mechanical, photocopying, recording, scanning or otherwise, except under the terms of the Copyright Licensing Agency Ltd, 90 Tottenham Court Road, London, W1P 0LP, UK, without the permission in writing of the Publisher. Requests to the Publisher should be addressed to the Permissions Department John Wiley & Sons, Ltd, The Atrium, Southern Gate, Chichester, West Sussex, PO19 8SQ, England, or e-mailed to permreq@wiley.co.uk, or faxed to (44) 1243 770620.

This publication is designed to provide accurate and authoritative information in regard to the subject matter covered. It is sold on the understanding that the Publisher is not engaged in rendering professional services. If professional advice or other expert assistance is required, the services of a competent professional should be sought.

Other Wiley Editorial Offices

John Wiley & Sons, Inc. 111 River Street,
Hoboken, NJ 07030, USA

Jossey-Bass, 989 Market Street,
San Francisco, CA 94103-1741, USA

Wiley-VCH Verlag GmbH, Boschstr. 12,
D-69469 Weinheim, Germany

John Wiley & Sons Australia, Ltd, 33 Park Road,
Milton, Queensland, 4064, Australia

John Wiley & Sons (Asia) Pte Ltd, 2 Clementi Loop #02-01,
Jin Xing Distripark, Singapore 129809

John Wiley & Sons Canada Ltd, 22 Worcester Road,
Etobicoke, Ontario, Canada, M9W 1L1

Wiley also publishes its books in a variety of electronic formats. Some content that appears in print may not be available in electronic books.

Library of Congress Control Number: 2001026205

British Library Cataloguing in Publication Data

A catalogue record for this book is available from the British Library

ISBN 0-470-84984-3

Typeset in 9.5/12pt Sabon by Laserwords Private Limited, Chennai, India
Printed and bound in Spain by Grafos S.A., Barcelona, Spain.
This book is printed on acid-free paper responsibly manufactured from sustainable forestry in which at least two trees are planted for each one used for paper production.

Preface

The role of metal ions in biology has received serious and sustained attention and the field of bioinorganic chemistry has evolved into a flourishing scientific area at the beginning of the twenty-first century. The *Handbook of Metalloproteins* is a reflection of this development and serves the community with the annotation of the rapidly growing knowledge and data on metalloproteins. The first two volumes contain entries of metalloproteins of redox-active metals (iron, nickel, manganese, cobalt, molybdenum, tungsten, copper, vanadium). Volume 3 is dedicated to the proteins found in the redox-inactive ions of **zinc** and **calcium**. Both metal ions play key roles in many important biological processes. Most of the 58 contributions have been compiled in the same style as used in Volumes 1 and 2 with the 3D Structure of the protein on the first page of each contribution followed by several mandatory sections detailing the biological function, occurrence, amino acid sequence, spatial structure and functional properties of the protein. Furthermore, a detailed presentation and discussion of the metal site is given. The accession code in the Protein Data Bank (PDB-code) of the reported structure(s) is given in each contribution and a list of all cited PDB-codes is located at the end of the volume. Some contributions are reviews summarizing whole protein or domain families or dealing with special aspects, as the articles of D. Auld and W. Maret.

While there are more than 200 known 3D structures of zinc enzymes, this Handbook has focused on contributions for those enzymes most relevant in research today. This ranges from contributions devoted to zinc-containing proteases, which are of remarkable medical interest, to metallothioneins, a crucial part of a recently discovered system of proteins involved in zinc homeostasis. Another group of zinc proteins covered in the Handbook are the zinc finger domains, key elements in the molecular recognition of nucleic acid, proteins or lipids. The number of genes containing zinc finger domains exceeds 3% of about 32 000 identified human genes. The contributions on the different zinc finger domains underpin the importance of this class of zinc proteins.

Also covered extensively is the role of calcium ions as secondary messengers in controlling the many biological processes. The change of free Ca^{2+} concentration within the cell provides the possibility of signal transduction for a number of different cellular activities, such as muscle contraction, glycogen metabolism, cell growth and division, differentiation, development and apoptosis. Many of these functions are accomplished through the interaction of Ca^{2+} with specific proteins generating modulations of protein–protein interactions due to conformational changes of the Ca^{2+} receptors. Approximately a third of the Handbook has been devoted to dealing with Ca^{2+}-binding proteins or domains, covering all aspects of their biological functions.

The substantial contribution of Volume 3 to the scientific canon will hopefully be of great interest, not only for the readership of Volumes 1 and 2 (inorganic and bioinorganic chemists, biochemists, biophysicists, microbiologists, structural biologists) but also for any student or researcher involved with molecular medicine.

The Editors thank all the contributors for delivering such excellent articles, Martin Röthlisberger and David Hughes for their advice and support, and Sam Crowe for his splendid assistance throughout the project.

Albrecht Messerschmidt
Wolfram Bode
Mirek Cygler

Contents

Volume 3

ZINC

Oxidoreductases ... 3
Zn-dependent medium-chain dehydrogenases/reductases ... 5
Rob Meijers and Eila S Cedergren-Zeppezauer

Transferases ... 35
Protein prenyltransferases ... 37
Hong Zhang

Hydrolases: Acting on ester bonds ... 49
Nuclease P1 ... 51
Anne Volbeda, Christophe Romier and Dietrich Suck

5′-Nucleotidase ... 62
Norbert Sträter

E. coli alkaline phosphatase ... 71
Evan R Kantrowitz

Hydrolases: Acting on peptide bonds ... 83
Thermolysin ... 85
Brian W Matthews

Methionine aminopeptidase ... 95
Brian W Matthews

Neprilysin ... 104
Glenn E Dale and Christian Oefner

Astacin ... 116
Walter Stöcker and Irene Yiallouros

Matrix metalloproteinases ... 130
Wolfram Bode and Klaus Maskos

Serralysin ... 148
Ulrich Baumann

Leishmanolysin ... 157
Peter Metcalf and Robert Etges

Streptomyces albus G D-Ala-D-Ala carboxypeptidase ... 164
Paulette Charlier, Jean-Pierre Wery, Otto Dideberg and Jean-Marie Frère

Metallocarboxypeptidases ... 176
Josep Vendrell, Francesc X Aviles and Lloyd D Fricker

Pitrilysins/inverzincins ... 190
Klaus Maskos

Leucine aminopeptidase ... 199
Norbert Sträter and William N Lipscomb

Bacillus subtilis D-aminopeptidase DppA ... 208
Han Remaut, Colette Goffin, Jean-Marie Frère and Jozef Van Beeumen

Hydrolases: Acting on carbon–nitrogen bonds, other than peptide bonds ... 215
Metallo β-Lactamases ... 217
Osnat Herzberg and Paula MD Fitzgerald

Contents

GTP cyclohydrolase I 235
Herbert Nar

Lyases 247

Carbonic anhydrases (α-class) 249
David M Duda and Robert McKenna

Carbonic anhydrases (β-class) 264
Eiki Yamashita, Satoshi Mitsuhashi and Tomitake Tsukihara

Carbonic anhydrases (γ-class) 270
Caroline Kisker and Tina M Iverson

5-Aminolaevulinic acid dehydratase 283
Jonathan B Cooper and Peter T Erskine

6-Pyruvoyl-tetrahydropterin synthase 296
Herbert Nar

Zinc-fingers 305

Cys_2His_2 zinc finger proteins 307
John H Laity

Zinc modules in nuclear hormone receptors 324
Srikripa Devarakonda and Fraydoon Rastinejad

RING domain proteins 338
Cyril Dominguez, Gert E Folkers, Rolf Boelens

Zinc storage 351

Metallothioneins 353
Klaus Zangger and Ian M Armitage

Other zinc proteins 365

Insulin 367
G David Smith

LIM domain proteins 378
Georg Kontaxis, Klaus Bister and Robert Konrat

FYVE domain 390
Tatiana G Kutateladze and Michael Overduin

General aspects 401

Structural zinc sites 403
David S Auld

Cocatalytic zinc sites 416
David S Auld

Protein interface zinc sites: the role of zinc in the supramolecular assembly of proteins and in transient protein–protein interactions 432
Wolfgang Maret

CALCIUM 443

EF-hand Ca^{2+}-binding proteins 445

Calmodulin 447
Kyoko L Yap and Mitsuhiko Ikura

Troponin C 459
Stéphane M Gagné

Gating domain of calcium-activated potassium channel with calcium and calmodulin 471
Maria A Schumacher

Calpain 489
Peter L Davies, Robert L Campbell and Tudor Moldoveanu

Parvalbumin 501
Susumu Nakayama, Hiroshi Kawasaki, Robert H Kretsinger

Basement membrane protein BM-40 509
Erhard Hohenester and Rupert Timpl

PEFLINS: a family of penta EF-hand proteins 516
Miroslaw Cygler

3D structures of the calcium and zinc binding S100 proteins 529
Günter Fritz and Claus W Heizmann

EH domain 541
Michael Overduin and Mahadev Ravi Kiran

EGF-domains 551

Calcium-binding EGF-like domains 553
Emma J Boswell, Nyoman D Kurniawan and A Kristina Downing

GLA-domains 571

Gla-domain 573
Mark A Brown and Johan Stenflo

C2-like-domains 585

Membrane binding C2-like domains 587
Nuria Verdaguer, Senena Corbalán-García, Wendy F Ochoa, Juan Carmelo Gómez-Fernández and Ignacio Fita

C2-domain proteins involved in membrane traffic 599
Josep Rizo

Dockerin-domains 615

Dockerin domains 617
Brian F Volkman, Betsy L Lytle and J. H David Wu

Hemopexin domains 629

Hemopexin domains 631
F Xavier Gomis-Rüth

Annexins 647

Annexins: calcium binding proteins with unusual binding sites 649
Anja Rosengarth and Hartmut Luecke

Other Ca-proteins 665

Calcium pump (ATPase) of sarcoplasmic reticulum 667
Chikashi Toyoshima

Phospholipase A_2 677
Christian Betzel1, Tej P Singh, Dessislava Georgieva and Nicolay Genov

Calsequestrin 692
ChulHee Kang

C-type animal lectins 704
William I Weis

Structural calcium (trypsin, subtilisin) 718
Gary L Gilliland and Alexey Teplyakov

The superfamily of Cadherins: calcium-dependent cell adhesion receptors 731
Thomas Ahrens, Jörg Stetefeld, Daniel Häussinger and Jürgen Engel

Metal-dependent type II restriction endonucleases 742
Éva Scheuring Vanamee and Aneel K Aggarwal

List of Contributors 757

PDB-Code Listing 763

Index 779

Contents of Volumes 1 and 2 793

ZINC

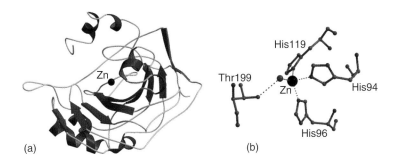

Oxidoreductases

Zn-dependent medium-chain dehydrogenases/reductases

Rob Meijers[†] and Eila S Cedergren-Zeppezauer[‡]

[†]Dana Farber Cancer Institute, Harvard Medical School, Boston, MA, USA
[‡]Biochemistry, Center for Chemistry and Chemical Engineering, Lund University, Lund, Sweden

FUNCTIONAL CLASS

Enzymes; medium-chain dehydrogenases/reductases (MDRs), alcohol dehydrogenases (ADHs) (EC 1.1.1.1; EC 1.2.1.1.) and polyol active enzymes ketose reductase (KR), sorbitol dehydrogenase (SDH), (EC 1.1.1.14), glucose dehydrogenase (GDH) (EC 1.1.1.47) (not to be mistaken for EC 1.1.99.17, which is not Zn-dependent). Distinct from short-chain dehydrogenases/reductases,[3] the MDRs (about 350–390 amino acids) belong to a large superfamily of proteins including the Zn-dependent ADHs[4] and several other Zn-dependent activities.[5]

These enzymes have one essential Zn^{2+} ion in the active site and frequently a second noncatalytic Zn site per subunit. They are dimers or tetramers and utilize either NADH or NADPH as electron carriers referred to as cofactors. The oxidation/reduction process is a two-electron/one-proton (hydride ion, H^-) transfer between substrates and cofactors. NAD/NADP is noncovalently bound to the protein

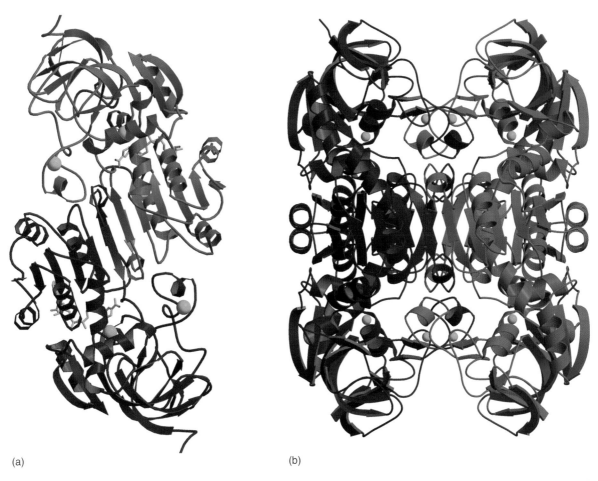

(a) (b)

3D Structure Ribbon representations of dimeric and tetrameric Zn-dependent NAD/NADP medium-chain dehydrogenases/reductases. Subunits differ in color and the green spheres are Zn ions. (a) The 'classical' liver alcohol dehydrogenase dimer (PDB code 1HET); (b) tetrameric ketose reductase (1E3J). If not otherwise mentioned, this and several of the pictures of 3D structures were prepared using MOLSCRIPT[1] and RASTER3D.[2]

and becomes indirectly linked to the metal upon binding.[6] The substrate becomes a direct ligand to the Zn^{2+} ion. The hydride-transfer reaction is dependent on the presence of a metal and zinc that has been exchanged for Co^{2+}, Cd^{2+}, or Ni^{2+} form catalytically active enzymes.[7] The compounds oxidized or reduced by MDRs vary considerably in structure ranging from small primary and secondary alcohols/aldehydes/ketones (aliphatic and aromatic) and sugars (sorbitol/fructose, glucose) to large hydrophobic substances like fatty acids, steroids, and retinoids.

OCCURRENCE

ADHs are widely distributed in nature and some MDR members are found in all species. Alcohol dehydrogenases are mostly described as cytosolic, but a mitochondrial form in yeast has been reported.[8] ADH and glucose dehydrogenase[9] from Archaea have been characterized.[10,11] ADHs (EC 1.1.1.2) isolated from thermophilic and mesophilic bacteria,[12] an Antarctic psychophile,[13] and the thermophilic eubacterial strains[14,15] indicate that alcohol dehydrogenase also arose in organisms existing under extreme conditions. A number of genes encoding for ADHs have been identified in *Escherichia coli*[16] among which classes I and III[17] of Zn-dependent enzymes are represented. (Classification is partly defined according to substrate specificity and enzyme characterization, treated in detail later.) The class III enzyme is regarded as the ancestral form, from which gene duplication has generated the class I type, an event estimated to have happened around 500 million years ago.[4] Plants have many genes encoding for ADH-related enzymes,[18] and similar to the prokaryotic case, these include the Zn-dependent proteins of classes I and III. Among mammalian ADHs, the major isozymes (ethanol and steroid-active) from horse liver are the most extensively examined. Human ADH forms are numerous and divided into five classes (Table 1),[19] among which class III is unique, not having ethanol activity, but defined as glutathione-dependent formaldehyde dehydrogenase (EC 1.2.1.1). Classes I, II, and IV share the ability to have both ethanol and retinol as substrates, and class IV enzymes are sometimes called retinol dehydrogenases. In addition, sorbitol dehydrogenase is present in human liver.[20]

Table 1 Old and new nomenclature for human ADH proteins and genes

Old nomenclature					
Protein	Class I α, β, γ	Class II π	Class III χ	Class IV σ or μ	Class V Class V
Genes	ADH1 ADH2 ADH3	ADH4	ADH5	ADH7	ADH6
New nomenclature applied in Figure 1					
Protein	ADH1s, A, B, C	ADH2	ADH3	ADH4	ADH5
Genes	The same but in italics.				

BIOLOGICAL FUNCTION

The physiological role of ADHs is not entirely clear, especially for the class II and class V ADHs.[21] The most persistent view is that they have a detoxifying function,[4,17] for instance, in the removal of consumed alcohol by class I ADH, abundant in human liver and present in most tissues. In the case of human glutathione-dependent formaldehyde dehydrogenase (FDH, class III ADH), toxic formaldehyde[22] is metabolized by spontaneous formation of a glutathione adduct (S-(hydroxymethyl)glutathione), which is the enzyme substrate.[23] In addition, S-nitrosoglutathione is reduced.[24] FDH is thus important in two cellular processes: formaldehyde detoxification and nitric oxide signaling.[25] A detoxifying effect has been suggested for ADH in *Sulfolobus solfataricus*.[26] It was observed that addition of sugars and substrates to the growth medium rapidly induced higher ADH activity. Since high temperatures and acidic growth conditions have a tendency to easily oxidize substrates to toxic compounds, a plausible role for the enzyme might be to reduce these substances.

In the butanol/isopropanol-producing *Clostridium beijerinckii*, the function of ADH (CbADH) apparently is to keep a balance in the concentrations of primary and secondary aldehydes/ketones and alcohols.[27] Glucose dehydrogenase (GDH) from the archaeon *Thermoplasma acidophilum* catalyzes the first step in a novel pathway for sugar metabolism.[28] Sorbitol is a metabolite derived either from glucose via reduction by aldose reductase or synthesized by reduction of fructose by ketose reductase (KR),[29] homologous to sorbitol dehydrogenase. High

Figure 1 (a) Unrooted, phylogenetic tree relating different ADH forms from a number of mammalian species. The initial alignment was made with Clustal W[65] with parameters to exclude gaps. Phylogenies were investigated using the neighbor-joining method. The tree was created with TreeView and numbers on branches represent the result of bootstrap analysis with 1000 replicates. Courtesy of Dr. J-O Höög. (b) Relationships of 116 MDR enzymes in an evolutionary tree generated using information from six genomes (see text). On the basis of the branching pattern the presence of eight families is suggested. The upper half of the graph shows the Zn-dependent dimeric and tetrameric alcohol and the polyol dehydrogenase groups treated in this chapter. (Reprinted from *Chem Biol Interact*, **143–144**, Jörnvall *et al.*, Multiplicity of Eukaryotic ADH and other MDR forms, 255–61, Copyright (2003), with permission from Elsevier).

Zn-dependent medium-chain dehydrogenases/reductases

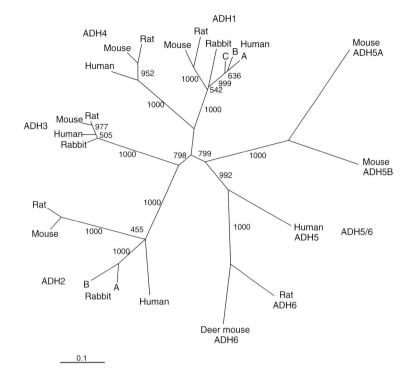

(a)

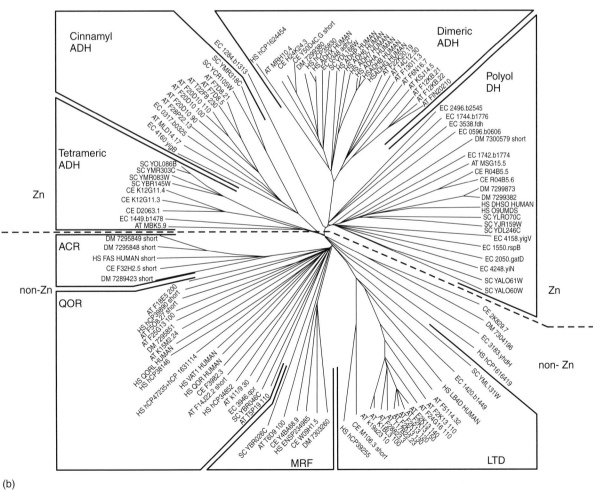

(b)

Figure 1

sorbitol levels in an organism can be either beneficial or harmful. In humans, sorbitol accumulation is responsible for several complications in diabetes patients.[30] On the contrary, for the agricultural pest *Bemisia argentifolii*, high sorbitol levels are critical for survival because it protects the insect against heat and osmotic stress.[31,32]

Among physiological substrates for class I ADH, 4-hydroxynonenal and aldehydes present during lipid peroxidation as well as steroids and ω-hydroxy fatty acids (reference 19 and references therein) can be mentioned. Retinoic acid regulates several cellular functions, particularly tissue differentiation and development.[33] Recently, the importance of ADH, class IV, in the synthesis of retinoic acid has been described.[34] Moreover, a connection has been suggested between the interference by ethanol in the retinol metabolism and the pathogenesis of fetal alcohol syndrome.[35] The class IV enzyme is otherwise the major stomach ADH involved in the first-pass metabolism of ethanol.[36]

AMINO ACID SEQUENCE INFORMATION

- *Equus caballus*, (horse liver), ADH class I, EE-isozyme, 374 amino acids (AA) determined from the protein. SWISSPROT Id code (SWP) P00327.[37]
- *Equus caballus*, (horse liver), ADH class I, SS-isozyme, 373 AA, determined from the protein SWP P00328.[38,39]
- *Homo sapiens*, (human liver), ADH class I, ADH1A or αα-isozyme, 374 AA determined from the cDNA sequence. SWP P07327.[40]
- *Homo sapiens*, (human liver), ADH class I, ADH1B*1 or $\beta_1\beta_1$-isozyme, 374 AA determined from the protein SWP P00325.[41]
- *Homo sapiens*, (human liver), ADH class I, ADH1C*1 or γγ-isozyme, 374 AA determined from protein (γ1 and 2 differ in two AA). SWP P00326.[42]
- *Homo sapiens*, (human liver) ADH class II, ππ-chain, 391 AA determined from protein in combination with cDNA. SWP P08319.[43]
- *Mus musculus*, (mouse liver), ADH class II, ADH2, 376 AA, determined from DNA.[44]
- *Homo sapiens*, (human liver), ADH class III, ADH3 or χχ-chain, glutathione-dependent formaldehyde dehydrogenase (FDH), 373 AA determined from protein. SWP P11766.[45]
- *Homo sapiens*, (human gastric) ADH class IV, ADH4 or σσ-chain, retinol converting dehydrogenase, 374 AA determined from protein. SWP P40394.[46]
- *Rana perezi*, (frog stomach), ADH class IV, 372 AA determined from the protein.[47]
- *Saccharomyces cerevisiae*, (bakers yeast), ADH1 and 2, classical fermentative isozyme, 374 AA determined from DNA. SWP P00330.[48,49]
- *Saccharomyces cerevisiae*, (intra-mitrochondrial location), ADH, 375 AA determined from mitochondrial DNA. SWP P07246.[8]
- *Entamoeba histolytica*, (parasite) ADH class I, 360 AA determined from DNA. SWP P35630.[50]
- *Acinetobacter calcoaceticus* (gram-negative bacterium), ADH class I, benzyl alcohol dehydrogenase, 371 AA determined from cDNA. SWP Q59096.[51]
- *Clostridium beijerinckii*, (mesophilic eubacterium), ADH class I, 351 AA determined from cDNA. SWP P25984.[52]
- *Thermoanaerobacter brockii*, (thermophilic eubacterium), ADH class I, 352 AA determined from cDNA. SWP P14941.[53]
- *Sulfolobus solfataricus*, (thermophilic archeon), ADH class I, 347 AA determined from cDNA. SWP P39462.[54]
- *Homo sapiens*, (human liver), sorbitol dehydrogenase, SDH, 356 AA determined from protein. SWP Q00796.[55]
- *Rattus norvegicus*, (rat liver) sorbitol dehydrogenase, SDH, 399 AA, sequenced from cDNA. SWP P27867.[56]
- *Bemisia argentifolii*, (whitefly), ketose reductase, KS, (sorbitol dehydrogenase), 352 AA determined from cDNA. SWP O96496.[32]
- *Thermoplasma acidophilum*, (thermophilic archaeon), glucose dehydrogenase, GDH, 360 AA determined from cDNA. SWP P13203.[9]

RELATIONSHIPS

Since such a large number of sequences are available for alcohol and polyol Zn-dependent MDRs, only those mentioned in the text are listed. Countless sequence comparisons have been made and it is recommended to use the Swissprot database to fetch the sequence data needed for further comparisons. Although the 3D Structures of Zn-dependent MDR are highly similar, the low sequence identity found sometimes could not immediately reveal such a close relationship. One example is glucose dehydrogenase from the archaeon *T. acidophilum*,[9] an early living form, and horse liver ADH.[57] *S. solfataricus* ADH (SsADH) is a second example sharing only 24% sequence identity to HLADH, 25% to yeast (YADH), and 24% to a bacterial secondary-alcohol dehydrogenase from *Thermoanaerobacter* (TbADH) (11 and references therein). Only structurally and functionally relevant residues have been conserved or conservatively substituted.[58] Various trends are observed between enzymes from distantly related organisms. Clostridium ADH (CbADH) has low sequence identity (27%) to HLADH[59] but a significant degree of identity (65%) to class I ADH from *E. histolytica*.[60] CbADH (mesophile) and TbADH (thermophile) are more closely related (75% sequence identity).[61] This latter number is similar to the sequence homology reported for class II and III ADHs compared to class I ADH of the human liver.[43]

Although the class I horse liver ADH and the human liver class III enzyme have different substrate preferences, they share 88% sequence identity.[62] Among the human isozymes (α, β, γ) (Figure 1(a)) an overall 93% identity is observed (reference 42 and references therein). The diversity among ADH enzymes in humans is large. Seven genes have been identified, mapped to chromosome 4,[63] and a new classification system has been suggested (shown in Table 1).[63,64]

In vertebrates, seven ADH classes have so far been distinguished on the basis of enzymatic and sequence characteristics.[64] Old and new classifications are mixed in the literature and in the sequence information listing above, and in Table 2, both the nomenclatures are mentioned in parallel in an attempt to clear the confusion. The phylogenetic relationship (Figure 1(b)) calculated for MDRs[5] using data from six totally sequenced genomes (*H. sapiens*, *Caenorhabditis elegans*, *Drosophila melanogaster*, *Arabidopsis thaliana*, *Saccharomyces cerevisiae* and *E. coli*) identifies more than 100 enzymes of which about 60 are Zn-dependent and divided into four

Table 2 List of Zn-dependent medium-chain dehydrogenases/reductases, their ligand complexes and mutants. Apo denotes unliganded enzymes in an 'open' protein conformation, holo is a cofactor complex in a 'semi-open' or 'closed' conformation. Wt is wild type. The latest Protein Data Bank (PDB) codes are from January 2003. The lowest resolution limit was selected to 2.9 Å

Origin	Enzyme	Complex dimers	Mutant	Resolution Å	PDB code	Reference
Equus caballus, liver EE-isozyme, class I	ADH1	Apoenzyme open conformation	Wt	2.4	8ADH, 1984	66
E. caballus, liver EE-isozyme, class I	ADH1	Holoenzyme NADH + DMSO	Wt	2.9	6ADH, 1984	67
				1.8	2OHX, 1993	68
E. caballus, liver EE-isozyme, class I	ADH1	Holoenzyme NADH + DMSO	Cu-enzyme	2.1	2OXI, 1993	69
E. caballus, liver EE-isozyme, class I	ADH1	Holoenzyme; abortive NADH + (MPD) complex	Wt	1.15	1HET, 2000	6
			Cd-enzyme	1.15	1HEU, 2000	6
E. caballus, liver EE-isozyme, class I	ADH1	Holoenzyme NAD + 2,3,4,5,6-pentafluorobenzyl alcohol	Wt	2.1	1HLD, 1993	70
E. caballus, liver EE-isozyme, class I	ADH1	Holoenzyme NADH + (1S,3R) 3-butylthiolane 1-oxide	Wt	2.0	1BTO, 1996	71
E. caballus, liver EE-isozyme, class I	ADH1	Holoenzyme NADH + (1S,3S) 3-butylthiolane 1-oxide	Wt	1.66	3BTO, 1996	71
E. caballus, liver EE-isozyme, class I	ADH1	Holoenzyme NADH + *N*-cyclohexyl formamide	Wt	2.5	1LDY, 1996	71
E. caballus, liver EE-isozyme, class I	ADH1	Holoenzyme NADH + *N*-formyl piperidine	Wt	2.5	1LDE, 1996	72
E. caballus, liver EE-isozyme, class I	ADH1	Holoenzyme NAD + 2,3-difluoro-benzyl alcohol	Wt	1.8	1MGO, 2002	73
E. caballus, liver EE-isozyme, class I	ADH1	Holoenzyme NAD + 2,3,4,5,6-pentafluorobenzyl alcohol	F93A	1.2	1MGO, 2002	73
Horse liver, EE-isozyme class I, expressed in *E. coli*	ADH1	Holoenzyme NAD + trifluoro ethanol	F93W	2.0	1AXE, 1997	74
Horse liver, EE-isozyme class I, expr. in *E. coli*	ADH1	Holoenzyme NAD + trifluoro ethanol	V203A	2.5	1AXG, 1997	74
Horse liver, EE-isozyme class I, expr. in *E. coli*	ADH1	Holoenzyme NAD + trifluoro ethanol	F93W; V203A	2.0	1A71, 1998	75

(*continued overleaf*)

Table 2 (Continued)

Origin	Enzyme	Complex dimers	Mutant	Resolution Å	PDB code	Reference
Horse liver, EE-isozyme class I, expr. in E. coli	ADH1	Apoconformation CPAD (NAD-analogue)	F93W; V203A	2.6	1A72, 1998	75
Horse liver, EE-isozyme class I, expr. in E. coli	ADH1	Apoconformation	G293A; P295T	2.8	1QLH, 1999	76
Horse liver, EE-isozyme class I, expr. in E. coli	ADH1	Apoconformation NAD trifluoroethanol	G293A; P295T	2.07	1QLJ, 1999	76
E. caballus, liver SS-isozyme, class I	ADH1	Holoenzyme NAD(H) + cholic acid	10 aa substitutions	1.54	1EE2, 2000	77
E. caballus, liver EE-isozyme, class I	ADH1	Apoconformation βTAD (NAD-analogue)	Wt	2.9	1ADF, 1993	78
E. caballus, liver EE-isozyme, class I	ADH1	Apoconformation βSAD (NAD-analogue)	Wt	2.7	1ADG, 1993	78
E. caballus, liver EE-isozyme, class I	ADH1	Holoenzyme CNAD(NAD-analogue)–ethanol	Wt	2.4	1ADB, 1993	79
E. caballus, liver EE-isozyme, class I	ADH1	Holoenzyme CPAD(NAD-analogue)–pentanol	Wt	2.7	1ADC, 1993	79
Horse liver, EE-isozyme, class I, expr. in E. coli	ADH1	Apoconformation NAD	V292S	2.0	1JU9, 2001	80
Cod liver, class I	ADH1	Mixed-conformation NAD	Wt	2.05	1CDO, 1995	81
Human liver, β1-isozyme, class I, expr. in E. coli	ADH1 B*1	Holoenzyme NAD(H)-cyclohexanol	Wt	2.5	1HDX, 1993	82
Human liver, β2-isozyme, class I, expr. in E. coli	ADH1 B*2	Holoenzyme NAD-4-iodopyrazole	Wt	2.5	1HDY, 1993	82
Human liver, β1-isozyme, class I, expr. in E. coli	ADH1 B*1	Holoenzyme NAD	R47G	2.45	1HDZ, 1993	82
Human liver, β1-isozyme, class I, expr. in E. coli	ADH1 B*1	Holoenzyme NAD-4-iodopyrazole	Wt	2.2	1DEH, 1995	41
Human liver, β3-isozyme, class I, expr. in E. coli	ADH1 B*3	Holoenzyme NAD-4-iodopyrazole	Wt	2.4	1HTB, 1995	41
Human liver, α-isozyme, class I expr. in E. coli	ADH1 A	Holoenzyme NAD(H)-(4-iodopyrazole)	Wt	2.5	1HSO, 2000	42
Human liver, β1-isozyme, class I, expr. in E. coli	ADH1 B*1	Holoenzyme NAD(H)	Wt	2.2	1HSZ, 2000	42
Human liver, γ2-isozyme, class I, expr. in E. coli	ADH1 C*2	Holoenzyme NAD(H)	Wt	2.0	1HTO, 2000	42
M. musculus (Mouse), class II, expr. in E. coli	ADH2	'Semiopen' conformation NADH	Wt	2.12	1E3E, 2000	83

Table 2 (*Continued*)

Origin	Enzyme	Complex dimers	Mutant	Resolution Å	PDB code	Reference
M. musculus (Mouse), class II, expr. in *E. coli*	ADH2	'Semiopen' conformation NADH + *N*-cyclohexyl-formamide	Wt	2.08	1E3I, 2000	83
M. musculus (Mouse), class II, expr. in *E. coli*	ADH2	'Semiopen' conformation NADH	P47H	2.5	1E3L, 2000	83
Human liver, χ-chain, class III, expr. in *E. coli*	ADH3 (FDH)	'Semiopen' conformation NAD(H)	Wt	2.7 2.2	1TEH, 1996 1PM0, 2003	45 84
Human liver, χ-chain, class III, expr. in *E. coli*	ADH3 (FDH)	Apoenzyme	Wt	2.0	1M6H, 2002	85
Human liver, χ-chain, class III expr. in *E. coli*	ADH3 (FDH)	'Semiopen' conformation 12-hydroxy-dodecanoic acid	Wt	2.3	1M6W, 2002	85
Human liver, χ-chain, class III expr. in *E. coli*	ADH3 (FDH)	'Semiopen' conformation NAD + dodecanoic acid	Wt	2.3	1MA0, 2002	85
Human liver, χ-chain, class III expr. in *E. coli*	ADH3 (FDH)	Mixed conformations NADH or NADH + S-(hydroxymethyl) glutathione	Wt	2.6	1MC5, 2002	86
Human gastric, σ-chain, class IV expr. in *E. coli*	ADH4 (retinol dehydrogenase)	Apoenzyme	Wt	2.5	1D1S, 1999	87
Human gastric, σ-chain, class IV, expr. in *E. coli*	ADH4 (retinol dehydrogenase)	Apoenzyme 4-iodopyrazole	L141M	2.4	1D1T, 1999	87
Tetrameric enzymes						
Pseudomonas putida, PFDH expr. in *E. coli*	Glutathione-independent	Holoenzyme NAD	Wt	1.65	1KOL, 2003	88
T. brockii, class I, expr. in *E. coli*	ADH (thermophile)	Holoenzyme NADP(H)	Wt	2.5	1YKF, 1996	89
C. beijerinckii, class I, expr. in *E. coli*	ADH (mesophile)	Apoenzyme	Wt	2.15	1PED, 1995	89
C. beijerinckii, class I, expr. in *E. coli*	ADH (mesophile)	Holoenzyme NADP(H)	Wt	2.05	1KEV, 1996	89
C. beijerinckii, class I, expr. in *E. coli*	ADH (mesophile)	Apoenzyme	Multimutant	1.97	1JQB, 2001	90
S. solfataricus, class I, expr. in *E. coli*	ADH, archaeon (thermophile)	Apoenzyme	Wt	1.85	1JVB, 2001	58
B. argentifolii, whitefly, expr. in *E. coli*	Ketose reductase	Apoenzyme	Wt	2.3	1E3J, 2000	29
T. acidophilum, culture	Glucose dehydrogenase	Apoenzyme	Wt	2.9	No PDB code	57
R. norvegicus (Rat), expr. in *E. coli*	Sorbitol dehydrogenase	Holoenzyme NADH	Wt	3	No PDB code	91
E. coli ADH class I	Ethanol induced	Holoenzyme, NAD(H)	Wt	2.0	No PDB code	92

DMSO is dimethylsulfoxide. NAD(H)/NADP(H) denotes unknown oxidation state of the cofactor. FDH is glutathione-dependent formaldehyde dehydrogenase. PFDH is glutathione-independent formaldehyde dehydrogenase.

groups; ≈10 are tetrameric ADHs, ≈20 are dimeric ADHs, 12 are cinnamyl ADHs, and more than 20 are polyol dehydrogenases. In spite of the large quantity of data available for the medium-chain dehydrogenases/reductases, new features and novel relationships are still being detected with most probably even more to come.

PROTEIN PRODUCTION, PURIFICATION, AND MOLECULAR CHARACTERIZATION

Before recombinant enzymes were more routinely prepared, ADHs were extracted from various organs among which horse liver has been the most common source first reported by Bonnichsen and Wassén in 1948.[93] In horse liver, two main isozymes are present – the ethanol-active 'EE'-ADH dimer and the ethanol/steroid-active 'SS'-enzyme.[94,95] The subunits E and S combine to form the hybrid ES-dimer. The effective separation and isolation of these three different molecular forms became a reason to modify early preparation procedures.[96] Ground tissue is extracted and centrifuged; the supernatant is fractionated with solid ammonium sulfate; and the precipitate is collected, resuspended, and dialyzed. After heat treatment at 52 °C and rapid cooling, the cleared supernate is applied to a DEAE-cellulose column. Isozymes can conveniently be separated on a phosphocellulose matrix by stepwise elution with phosphate buffer.[96] Finally, by utilizing the pH-dependent differences in solubility between the isozymes, crystallization in the presence of ethanol yields pure isozymes. Chromatography on CM- and DEAE-cellulose can resolve seven isozymes according to Lutstorf.[97] Affinity chromatography on AMP-Sepharose can be used to separate purified EE and SS isoenzymes.[98,99] They differ in 10 amino acid residues[38,39] resulting in clear differences in coenzyme affinity utilized to elute the adsorbed proteins. An optimal combination of the previously described preparation procedures for the large-scale isolation of horse LADH isoenzymes has been presented.[100] From 5 kg fresh horse liver, at least a gram of the EE-isoenzyme is obtained, up to 800 mg of the hybrid ES-dimer, and 20 to 400 mg of the less abundant steroid, metabolizing enzyme. In addition, isoforms were separated, one of which could be identified by mass spectrometry to be glycosylated.

Preparation of glucose dehydrogenase from *T. acidophilum*[57,101] or ADHs overexpressed in *E. coli* from other thermophilic organisms[11,90] generally include anion exchange chromatography applied to a crude extract followed by adsorption onto some kind of dye-affinity matrix (Matrex Gel Red/Blue A) or AMP-substituted gel. One such dye has been Cibacron Blue 3GA, a large aromatic chromophore, which readily accommodates into the large, adenine binding pocket of dehydrogenases. Crystallization, often used in the 'old days', is a good purification step also applied in the case of GDH.[57] In the preparation of GDH from a halophilic archaeon expressed in *E. coli*, the halophilicity of the protein was utilized in the purification procedure.[102] Taking advantage of the high solubility of the protein in high salt concentrations is suggested to be of general use in the purification of halophilic proteins expressed in mesophilic hosts. Ketose reductase[29] prepared from frozen whiteflies is fractionated after a Q-Sepharose column by ammonium sulfate precipitation whereafter affinity chromatography on Matrex Orange or Reactive Red is applied. KR is eluted using 0.2 to 2 mM NADPH depending on the dye-ligand column. The final step is a cation-exchange perfusion chromatography.[103] A rapid, one-step procedure using Zn-affinity chromatography has been useful in the preparation of tetrameric ADHs from baker's yeast,[104] but several other purification techniques have been described for YADH. Tetrameric *E. histolytica* ADH purifies well using the Zn-affinity gel but proved less successful in the isolation of dimeric ADH from rat liver.

The development of immobilized pyrazole derivatives for affinity chromatography was critical for the preparation of human liver ADH isoforms, which were otherwise difficult to resolve from a mixture.[105] Furthermore, the separation of three classes of human liver ADHs followed the protocol of Ditlow[106] until it became common practice to work with recombinant liver enzymes.[107] As described above, ion exchange and affinity chromatography generally comprise the initial phase followed by a final 'polishing' step. In the case of glutathione-dependent formaldehyde dehydrogenase (class III, FDH), initial purification on a Q-Sepharose column gave a final yield of 7 mg of enzyme per liter of bacterial culture.[23] Human class I, II, and III ADH expressed in *E. coli* are all purified to homogeneity in a three-step procedure.[108] The isolation of sorbitol dehydrogenase from human liver follows a multistep procedure involving HPLC on a cation-exchange column as the last purification step.[20]

Many ADHs are dimeric but *E. coli*,[92] yeast, and *E. histolytica* ADHs as well as ADHs isolated from thermophilic organisms are tetrameric.[11] A tetrameric glutathione-*independent* formaldehyde dehydrogenase (a nicitinoprotein) from *P. putida* (PFDH) has recently been reported.[88] Further tetrameric MDRs mentioned in this account are ketose reductase,[29] glucose dehydrogenase,[57,102] and sorbitol dehydrogenase.[20,91] An extensive analysis of the N-termi of 40 ADHs from various sources shows that they are all acetylated.[109] There are no disulfide bridges. The number of SH-groups varies considerably between structures, the maximum number being 14 per monomer (40 000 Da), some of which participate in metal binding.

METAL CONTENT AND COFACTORS

The metal content of Zn-dependent NAD/NADP MDR frequently is two metal ions per subunit. Exceptions are yeast ADH, sorbitol dehydrogenase,[20,91] and the two ADHs

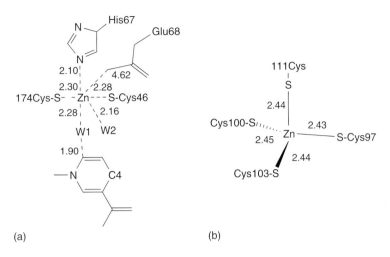

Figure 2 Schematic drawings of the two zinc coordination spheres in HLADH (PDB code 1HET). Metal–ligand bond distances are given in Å. (a) The catalytic Zn site in an enzyme-NADH complex; (b) the structural Zn site with four cysteine ligands in a tetrahedral geometry. Generated using ChemDraw.

from *T. brockii* and *C. beijerinckii*.[89] These enzymes lack the second Zn^{2+} ion, which is not essential for catalysis but has been ascribed a structural role. The zinc site is contained within a separate structural unit resembling iron–sulfur cluster structures. In the case of *S. solfataricus*, ADH thermostabilty was reduced when the second Zn ion was removed, indicating a structural role.[11,110] In many ADHs, the coordination sphere of the structural zinc site (Figure 2) consists of four cysteine residues[66] 97, 100, 103, and 111 (HLADH sequence numbering) arranged in a near-perfect tetrahedral geometry. In glucose dehydrogenase and SsADH, one of the cysteine ligands is either an aspartate (Cys111 = Asp115[57]) or a glutamate (Cys97 = Glu98[58]). The exchange for an acidic side chain within the metal binding loop has been suggested to be important to develop/maintain thermal stability of those enzymes. The bacterial, thermostable TbADH and CbADH lack the second metal and also the cluster of cysteines[89] inside the loop structure, which is conserved and serves as a contact area between subunits.

The active site zinc coordination sphere is much more diverse among the MDR. The geometry of the active site metal can differ and the coordination numbers vary between four to six, the highest coordination number exemplified by sorbitol dehydrogenase. The reference enzyme will again be HLADH, which has three protein ligands (Cys46, His67, and Cys174) and a water molecule bound to the metal in the apostructure.[66] The bond distances for a holoenzyme HLADH–NADH complex is shown in Figure 2 and this structure was chosen since the accuracy in the structure determination is among the highest available for ADHs, solved to 1.15 Å resolution.[6] H_2O (W1) or OH^- (W2) has been interpreted to bind in alternate positions to the Zn^{2+} ion linking the cofactor to the Zn-coordination sphere. Glu68 is shown in this complex at a second sphere-coordination distance, but this residue and the corresponding glutamate in other ADHs can move to become a metal–ligand (discussed in detail below).

NAD(H) or NADP(H) act as electron acceptors/donors and the stereo specificity is strictly A-side or pro-R[111] for Zn-dependent MDRs. Glucose dehydrogenase (GHD) from an archaeon is a rare example of dual cofactor usage with preference for NADP(H) over NAD(H).[57] An amphibian alcohol dehydrogenase (class IV like) has been reported to be NADP-dependent.[47] The hallmark of an NAD(H)-dependent MDR is that the adenosine ribose is firmly anchored to the binding site via hydrogen bonds to an aspartate, conserved among species. In the amphibian ADH, this residue corresponds to a glycine similar to the bacterial CbADH and TbADH, which are also NADP-dependent.[89] By site-directed mutation (E223A), yeast ADH was easily converted to an NADP-dependent enzyme. In GDH, with the ability to utilize both cofactor types, the aspartic acid side chain is substituted by an asparagine and thus allows for hydrogen bond interactions to the ribose. Close by is a histidine side chain, and the model building of NADP binding to GDH shows that the O2'-linked phosphate of the ribose can be accommodated in its neighborhood.[57] The adenosine phosphate charge is compensated by the presence of an arginine side chain in the bacterial Cb- and TbADHs and by two arginine residues in ketose reductase.[29] The ADH from *P. putida* (PFDH) is a nicotinoprotein, meaning that the cofactor NAD(H) is firmly and permanently bound. This is achieved by unusual and tight interactions between protein main-chain atoms and the adenosine part of the cofactor.[88]

ACTIVITY TEST

The activity assay performed for these enzymes rely on spectrophotometric measurements at a wavelength of

340 nm, where the absorption increases when NADH is a product of the reaction (oxidation of an R-OH group) and decreases when NADH is consumed (reduction of aldehydes or ketones). The original description for activity measurement of HLADH by Theorell and Bonnichsen from 1951[112] and later refined by Dalziel, 1957,[113] can still be followed. Reduction of NAD^+:

> 'Glycine-NaOH-buffer, pH 9.6, 3 ml final volume in a cuvette. Add 0.73 mg NAD^+ and 0.05 ml ethanol. 15–20 µg of pure HLADH is a suitable addition to initiate the reaction. Follow the absorbance change at 340 nm for 3 min. If the absorbance increase reaches 0.036 in 3 min per µg HLADH ($T = RT$) the turnover is about 140 mol NADH/mol ADH × min.'

A further observation made by Theorell and Bonnichsen, useful in enzyme analysis, is the shift in the absorption maximum of NADH to 325 nm upon binding to the enzyme.[112] This has been particularly important in X-ray crystallographic investigations of HLADH to establish the oxidation state of the bound cofactor in various crystalline complexes.[6,7,77,114]

Protein concentration is measured at 280 nm using 0.455 as the (extinction coefficient) absorptivity[115] or $18\,200\,M^{-1}\,cm^{-1}$, a value similar for all three horse liver isoenzymes.[116] Concentration of active sites of HLADH can be determined fluorimetrically by titration with NADH on the basis of the formation of a strong, ternary complex between the enzyme, isobutyramide, and the reduced cofactor.[117,118] Alternatively, spectral changes upon the formation of a different, strong, ternary complex between the enzyme, pyrazole, and NAD^+ can be followed and the number of active sites estimated.[119] However, these methods are not generally applicable since all ADHs do not form the ternary complexes. In a recent work on YADH, Leskovac et al.[120] determined enzyme concentration according to Hayes and Velick[121] and used a simple fluorometric method for the determination of the active site concentration based solely on the total concentration and fluorescence of ligand in the presence and in the absence of enzyme.[122] There are numerous variations of the standard activity assay for alcohol and polyol dehydrogenases found in the literature[20,23,44,58,90,101,103,108,123–125] owing to the differences in substrate preference for a particular enzyme, the pH optimum for the activity, or the temperature used, which is considerably higher for the thermophilic SsADH.[11]

HLADH can also act as an 'aldehyde dehydrogenase' and catalyzes a dismutation reaction. Aldehyde is converted in a sequential process to give two products, alcohol and acid, requiring two equivalents of cofactors resulting in a spectrophotometrically transparent mixture at certain substrate concentrations. Product analysis is therefore required to quantify this activity correctly, which is either made by gas chromatography/mass spectrometry[108] or by recording ^1H NMR spectra of a reaction mixture over time.[126,127]

An activity test, based on a color reaction, is a convenient way to trace ethanol-active isoenzymes in ADH preparations on native gels after electrophoresis.[11,128,129]

X-RAY STRUCTURES

Crystallization

Generally, it can be noticed that alcohol/polyol dehydrogenases crystallize readily. Various conditions have been found to include all categories of precipitating agents. Salts as crystallization agents are less common for this group of enzymes, but the two related enzymes ketose reductase and sorbitol dehydrogenase were precipitated from ammonium sulfate, tartrate,[29] or citrate.[91] The most commonly used precipitating agent is polyethylene glycol (PEG) ranging from low (400) to high (8000) molecular weights. For the steroid-converting isozyme (SS-ADH) from horse liver, a mixture of PEG400 and 8K was successfully used to promote increased thickness of crystals,[77] which otherwise grew as utterly thin plates not suitable for X-ray studies. Alcohol, often 2-methyl-2,4-pentanediol (MPD), has traditionally been used for the crystallization of numerous HLADH complexes, but ethanol has also been used particularly in cases in which inactive cofactor derivatives have been studied.[79] A mixture of PEG and 2-propanol was used for GDH[57] and SsADH,[58] both from thermophiles. The temperature for crystallization can vary from +4 °C to RT and the limitation in pH (from about pH 5.5 to 9.5) is mainly due to the instability of the proteins at extremely low pH values because of metal dissociation. Except for what is cited here, all other references to crystallization of MDRs are found in the Biological Macromolecular Crystallization Database, http://wwwbmcd.nist.gov:8080/bmcd/bmcd.html.[130] At least 25 different conditions were found.

Overall description of structures

Table 2 lists Zn-dependent NAD/NADP MDR, giving their origin of preparation, modern and old enzyme nomenclature for ADHs, as well as the Protein Data Bank (www.pdb.org) identification codes for deposited crystalline structures. Six more structures of ADHs are on hold in that database, including benzyl alcohol dehydrogenase from *A. calcoaceticus* characterized to be similar to the archetypal Zn-dependent HLADH.[51] Since horse liver ADH was the first 3D Structure in this enzyme group to be determined by X-ray crystallography,[66,67] structural comparisons are frequently made with HLADH as well as with sequence alignments.

Examination of the selected structures in Table 2 reveals that they essentially have the same polypeptide fold. The topology diagram in Figure 3 shows how strands and

Zn-dependent medium-chain dehydrogenases/reductases

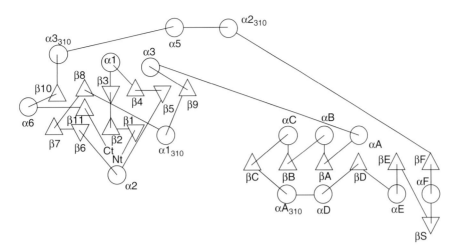

Figure 3 Topology diagram for the horse liver ADH monomer was generated with the program TOPS.[132] The residue numbers are given for each strand (S) and helix (H) defined according to the program STRIDE:[131] S-β1 from 7 to 14, S-β2 from 22 to 29; S-β3 from 35 to 45; H-α1, 47 to 55; S-β4, 69 to 77; S-β5, 88 to 92; H-α2, 101 to 105; S-β6, 130 to 133; S-β7, 135 to 137; S-β8, 149 to 154; H-3$_{10}$α1, 154 to 156; S-β9, 157 to 160; H-α3, 166 to 173; (beginning of Rossmann fold) H-αA, 175 to 185; S-βA, 194 to 199; H-αB, 202 to 214; S-βB, 218 to 223; H-αC, 229 to 236; S-βC, 239 to 242; H-3$_{10}$αA, 243 to 246; H-αD, 250 to 259; S-βD, 264 to 268; H-αE, 272 to 282; S-βE, 288 to 292; S-βS, 301 to 304; H-αF, 306 to 310; S-βF, 313 to 317; H-3$_{10}$α2, 319 to 321; (end of Rossmann fold) H-α5, 324 to 337; H-3$_{10}$α3, 343 to 345; S-β10, 346 to 352; H-α6, 355 to 364; S-β11, 369 to 374.

helices are organized taking the horse liver enzyme as a template, but this pattern is identical for the mammalian ADHs and corresponds closely to ketose reductase and to other tetrameric enzymes. Short 3$_{10}$-helices are included in the topology graph and the number of β-sheets in the catalytic domain is reduced to two (left part of the diagram) rather than three sheets originally defined for HLADH.[66] The secondary structure assignment was done with STRIDE.[131]

The subunit consists of two domains. The coenzyme binding domain is a six-stranded, parallel β-sheet surrounded by helices, frequently found in various NAD/NADP-dependent enzymes and called the *Rossmann-fold*.[133,134] The catalytic domain has a core composed of mainly two large β-sheets of antiparallel strands complemented by a few helices. This domain is classified as a partially open β-barrel and described as GRO-ES like.[135] A long, slightly bent helix (α3–αA in Figure 3) connects the two domains and serves as a hinge for domain motions in HLADH.[136] For CbADH, a hinge exists but has a different location.[89] The dimer is formed by contacts through the large hydrophobic surfaces at the C-terminal end of the coenzyme binding domains (Figure 4(a)). This association brings the β-strands in the Rossmann fold together and a continuous, 12-stranded β-sheet passes through the entire dimer (Figure 4(b)). It is around this symmetry-related core of the dimer that catalytic domain motions occur.

The molecule has a ligand-accessible surface and an opposite side impregnable to ligands. This latter surface has a large number of charged side chains (in some enzymes predominantly positively charged) forming the interaction area in tetrameric enzymes (3D Structure). These are thus defined as dimers of dimers.[29,57,58,88–92] Figure 4(c) shows that the dimers in the tetramer are arranged at an angle with respect to each other, an angle that can vary between the tetrameric enzymes. The difference between GDH and CbADH is around 30°.[89] The strength of interactions between dimers is also variable. Sorbitol dehydrogenase has only a few residues that make hydrogen bonds,[91] whereas the electrostatic interactions seem to play a particularly important role in the stabilization of the tetrameric structures from thermophiles.[58] Bogin *et al.*[90] discuss this aspect on the basis of a study of site-directed mutagenesis of CbADH (originating from a mesophilic bacterium) gaining thermal stability when new charged residues were introduced at the dimer–dimer interface.

The major structural differences between the MDRs are in loop connections between secondary structure elements of the cores of the two domains in the subunit, which are either cut or extended. These loop structure variations are related to monomer–monomer interactions, influencing the size and architecture of the substrate binding cavities, or directly connected with the oligomeric state. Figure 5(a) shows the locations of structural variations when tetrameric SsADH and dimeric HLADH are compared, which represents both categories of sequence alterations. In the comparison between ADH1 and mouse ADH2, evolutionary relatively closely related enzymes (Figure 1) show that the sequence differences are situated around the substrate binding area (Figure 5(b)).

Conformational variations

Zn-dependent MDRs are highly flexible molecules adapted for conformational variation. The assumption that hydride

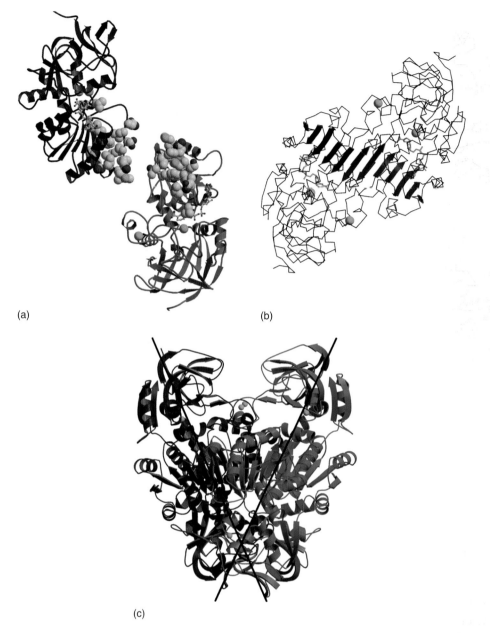

Figure 4 Molecular assembly. (a) ADH monomers (red and blue) have been separated to highlight the hydrophobic, complementary surfaces (space-filling residues) constituting the major subunit–subunit interaction area in the dimer (PDB code 1HET). The ball and stick model is coenzyme. (b) The dimers in all MDRs have a 12-stranded β-sheet (red arrows) passing through the dimer. Green spheres are Zn ions. Some enzymes have only an active site Zn/subunit, the metal ion closest to the cofactor (yellow) model. (c) Ketose reductase tetramer (PDB code 1E3J). Black lines indicate that the dimers are not packed parallel to each other but at an angle.

transfer between cofactor and substrate in HLADH requires a totally water-depleted enzyme interior originated from the interpretation of the two 3D structures[66,67] and theoretical calculations based on these structures.[137] HLADH undergoes large structural changes upon NADH binding but not when the nicotinamide is missing. This narrows coenzyme and substrate binding clefts excluding some of the water molecules present. Two conformation states were described – 'open'/'closed' or apo- and holoconformations. This transformation has been regarded as a necessary step for hydride transfer to occur.[75,138,139] Recently, these two proposed requirements for hydride transfer, namely, the total exclusion of water from the active site and the closed conformation have been challenged.[6,140]

The transformation between the open and closed conformations includes at least two steps. A loop motion is the prerequisite for 'total' closure, a 10° rigid–body rotation of the catalytic domain. Residues 292–298 (HLADH sequence numbering) are at the 'entrance' to the active site and are named the 'gating loop' by Tapia et al.[137] This loop moves away, allowing the catalytic domain to approach the coenzyme binding domain with which it

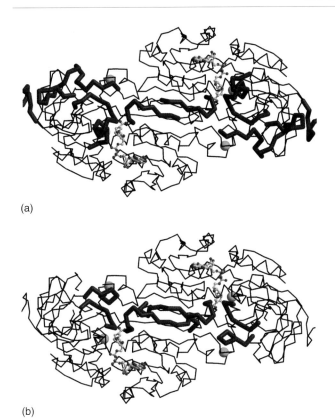

Figure 5 C$_\alpha$-traces of the HLADH dimer, in which variations in the sequence/structure are indicated (thick lines) when compared to other MDRs. (a) Tetrameric SsADH. Red denotes sequence parts, which are deleted or added in SsADH. Centered about the twofold axis is the sequence partaking in tetramer interactions. Functionally important amino acid differences are located around the active site Zn (green sphere at the cofactor binding area). (b) A similar comparison for dimeric, mouse liver ADH2. The significant changes are in three loop regions (V1-3 in reference 83) forming an active site accommodating large substrates. Green spheres are zinc ion positions.

finally makes contact. This is equal for both subunits as determined from numerous structural comparisons. Sometimes, crystal-packing effects[81] have been proposed to create conformational asymmetry in the dimer.

The pattern for loop and domain motions in HLADH is not a general feature of all Zn-dependent MDRs. It appears from the many structure determinations[45,74,75,78,80,81,83,85,86,89] that the angle for the domain rotations in these enzymes can take almost any value between 2 to 12° compared with that of an 'open' HLADH form. Apo-SsADH is an example for which the domain separation is even larger[58] than that observed for the 'open' conformation. Korkhin et al.[89] describe in detail the effects of domain motions in an NADP(H)-complex of the tetrameric CbADH structure. The difference in conformation between the apo- and holo-structure is on average small (2.5° domain rotation) and gives rise to the expansion of an internal cavity at the hydrophilic interface between dimers.

Regulation of conformation changes

Correct positioning of the nicotinamide is important. The domain closure mechanism seems to be highly sensitive to correct 'signaling' events. Several disparate observations can be connected to the inhibition of conformation changes in HLADH. Some coenzyme analogue molecules, substituted at the pyridine ring, have been found to interact with the enzyme in an unusual manner. In the βTAD and βSAD analogues, the pyridine ring is substituted by thiazole or selenazole rings.[78] These modifications prevent certain interactions in the active site crucial for the triggering of a conformation change. NADH binding in the presence of imidazole as a zinc ligand, substituting the Zn-bound water, forces the pyridine ring to bind with the B-side facing the active area (a *syn* conformation) and no gross structural change takes place.[141] (It should be noted that hydride transfer is A-side specific.) This resembles what is observed for the isosteric cofactor analogue CPAD[75] (the C2-position of the pyridine ring is a nitrogen atom) when bound to the double mutant V203A/F93W of HLADH. The underlying factor influencing the binding mode of the cofactor in the two latter cases is partially the same – steric interference. In the mutant, the bulkier tryptophan side chain hampers favorable contacts between the carboxamide group of the nicotinamide and protein main-chain atoms close to Trp93 including contacts to the carbonyl oxygen of 292, a 'gating loop' residue[75] (Figure 6). In the case of the imidazole-NADH complex, the imidazole inhibitor constitutes the steric hindrance. It interferes with the nicotinamide at the zinc site by making very close van der Waals contacts to the ring if positioned in its correct, productive orientation. An additional effect of ternary complex formation with imidazole-NADH is the Zn-ion shift, a motion toward the substrate channel[141] common to other enzymes and mentioned in coming sections. Thus, the carboxamide group interaction with a main-chain atom in the gating loop is part of the signaling path for domain closure, but how exactly this is linked to the metal center is still unclear.

The inactive NADH analogue 1,4,5,6-tetrahydro–NADH interacts quite differently with HLADH[143] when compared to a similar NADH complex. The electronic structure of the tetrahydropyridine ring does not permit H_2O/OH^- interaction at the 6-position, similar to a native, active dihydro-pyridine ring[6] shown in Figure 2. The tetrahydropyridine ring instead binds at a position remote from the active metal, blocking the motion of the 'gating loop' needed to allow for the 10° domain rotation. The structure of Cys46-carboxymethylated HLADH (Cys46 is a zinc ligand) in complex with NADH shows a different situation.[142] Here the nicotinamide of NADH is enclosed in the active site in a productive mode solely through the proper change in orientation of the flexible loop (Figure 6) induced by proper interactions with the carboxamide group atoms. No domain rotation takes place because of the

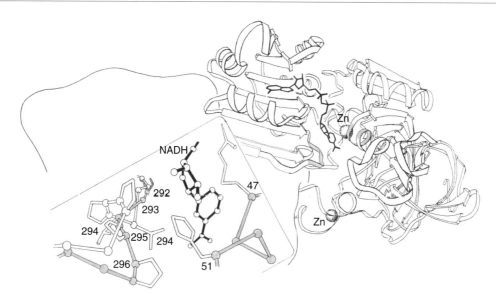

Figure 6 A ribbon drawing of the HLADH subunit. NADH is depicted in black. Residues 292–297, the 'gating' loop, are labeled in green. Inset: The loop (left) showing two positions. In the apoenzyme (green), Val294 is buried inside a hydrophobic subunit–subunit contact area. After a loop motion (a requirement for total domain closure), Val294 (open lines) approaches the pyridine ring of NADH and the nicotinamide of the cofactor is enclosed in the 'partially closed' conformation.[142] Mutations at positions G293A/P295T impair this motion.[76] The mutation of Val292 for serine (red side chain) might contribute to a shift in the conformation equilibrium.[80] (Artist: Dr Bo Furugren).

addition of the carboxymethyl group on the zinc ligand 46 protruding into the cleft between domains.

Is domain closure a requirement for hydride transfer? The NADH complex just described is called the *partially closed* conformation and the binding cleft between domains is not as narrow as in most closed complexes of HLADH. Nevertheless, the enzyme in these crystals as well as in other crystalline complexes is active in the open conformation for which substrate conversion was measured (reference 7, Chapter 29). The enzyme mutated in the gating loop (G293A/P295T) does not respond to the conformational change signals (the coenzyme-complex is open) and, from kinetic considerations, has been suggested to convert substrates in an open conformation.[76] This is opposite to what is proposed for the mutant V292S (Figure 6), which also crystallizes in an open conformation in the presence of the coenzyme.[80] The kinetic behavior of the single mutant, however, differs considerably from the double mutant, and hydride transfer was interpreted to take place in the closed enzyme after all. The structures of coenzyme complexes of ADHs belonging to classes I (CbADH), II, and III, found in 'semiopen' or 'half-closed' conformations[83,85,89] add strength to the assumption that there is no absolute requirement for a total closure of the structures for the hydride-transfer reaction to occur in all MDRs.[86]

Does the zinc environment influence domain motions? The examples given above suggest that there is a clear connection between the correct position of the nicotinamide and the signaling route for domain motion in HLADH. However, it is possible that the metal site could be a determinant actor in regulating conformation changes in MDR. In complexes of CbADH,[89] variations in the degree of domain closure correlate well with the flexibility of the Zn site including both metal–protein ligand reorganization and an actual shift of the metal atom position.

The notion that the state of the zinc coordination is connected to conformational changes seems to contradict the observations made earlier for active site zinc-depleted HLADH in complex with NADH. If the active site metal is removed, the enzyme becomes inactive. Calculated from the kinetic data given in the paper by Dietrich *et al.*,[144] NADH binding is by a factor >1000 stronger to enzyme from which Zn^{2+} has been removed as compared to native HLADH. The X-ray structure[145] showed that NADH was enclosed inside the protein in a binding mode resembling that of NADH bound to the closed structure and the stronger binding was explained as an effect of a more hydrophobic environment around the pyridine ring in a demetallized site.[145] The Zn^{2+} ion apparently is not required for the domain motions to go to completion in the NADH complex. Since the hinge axis for domain rotation passes through the zinc ligand Cys174 and the apo-HLADH structure can have large hinge fluctuations,[136] it cannot be excluded that in the horse liver enzyme also modulation of domain motions is governed by the state of the coordination sphere during a catalytic cycle.

Electrostatic interactions are a further trigger. There are two arginine side chains in HLADH, Arg47 and 369, interacting with the pyrophosphate bridge of NADH, both originating from the catalytic domain. Arg47 is a surface side chain. Arg369, on the contrary, is an interior residue making hydrogen bonds to Glu68, located close to the domain rotation axis and a potential zinc ligand (see Table 3). These latter residues are conserved in many ADHs. The sulfate ion competes for the pyrophosphate part of NAD, and Arg47 and Arg369 make similar interaction with sulfate as with NADH.[116] At high salt concentrations, a sulfate–enzyme complex is formed mimicking the closed-enzyme conformation although no coenzyme is present.[146] Both arginine side chains approach the anion from opposite directions and these electrostatic interactions are obviously strong enough to move catalytic domains in HLADH and clearly contribute to the trigger mechanism. In the light of all structural data available today, it appears that for a number of Zn-dependent MDRs the signaling events for conformation changes are ruled by a combination of factors and are more sophisticated than anticipated. For instance, not all ADHs have an arginine side chain at the position corresponding to the surface residue 47 in HLADH.[42,82,83]

Active site zinc geometry

Zn-dependent MDRs show diversity with respect to the active site zinc ligands, but they also have several features in common. (i) There is always one cysteine sulfur and a histidine nitrogen coordinating to the Zn^{2+} ion denoted 'Constant ligands' in Table 3. (ii) Irrespective of the type of side chain for the remaining ligands (cysteine, aspartic acid, or glutamic acid) the zinc ion is surrounded by negatively charged groups, 'Variable ligands' in Table 3. (iii) An arginine or a lysine side chain ('Neighboring residue' in Table 3) forms an ion pair with a glutamic acid side chain, here denoted the 'flexible Glu'. Figure 7 shows the coordination spheres for SsADH and ketose reductase and indicates the bonds between the charged residues. The 'flexible Glu' takes the position of a metal ligand in the apoenzyme form of these enzymes. The dissimilarities in the metal environment

Table 3 Metal ligands in Zn-dependent NAD/NADP medium-chain dehydrogenases/reductases. Cd-HLADH denotes cadmium-substituted ADH[6] highly similar to the native Zn-enzyme. SsADH from *S. solfataricus*, CbADH from *C. beijerinckii*. SDH is sorbitol dehydrogenase; GDH is a glucose dehydrogenase; KR is a ketose reductase; and PFDH is *P. putida* glutathione-independent formaldehyde dehydrogenase. In the column H_2O, S;I denotes a substrate or inhibitor-binding site. H_2O/(close) means bound to Zn/further than 3 Å away. The residue in brackets is not always a metal ligand

Enzyme/conformation	Constant ligands	Variable ligands	Neighboring residue	H_2O/S;I site	Comments
HLADH/Apo	Cys46; His67	Cys174, (Glu68)*	Arg369	H_2O	*Glu68 no Zn^{2+} ligand
HLADH/Holo	Cys46; His67	Cys174, (Glu68)	Arg369	H_2O/OH^-	Partially occupied H_2O site OH^- orientation toward NADH
					5-coordination an assumed intermediate state
Cd-HLADH/Holo	Cys46; His67	Cys174, Glu68	Arg369	H_2O/OH^-	Glu68 partially a metal ligand
$\beta_1\beta_1$ ADH1/Apo	Cys46; His67	Cys174, (Glu68)	Arg369	H_2O	3 protein ligands
ADH2/Holo	Cys46; His67	Cys178, (Glu68)	Arg371	H_2O (ligand?)	3 protein ligands
ADH3(FDH)/Apo	Cys44; His66	Cys173, (Glu67)	Arg368	H_2O/inhibitor	3 protein ligands
ADH3(FDH)/Holo	Cys44; His66	Cys173, Glu67	Arg368	H_2O at 3 Å	4 protein ligands, Glu67 included after reorganization. Zinc sphere inversion
ADH3(FDH)/ternary	Same as above with Glu67 bound or unbound depending on the complex.				Possible with 5-coordination
ADH4/Apo	Cys46; His67	Cys174, (Glu68)	Arg369	H_2O	3 protein ligands
SsADH1/Apo	Cys38; His68	Cys154, Glu69	Arg342	H_2O (close)	4 protein ligands, Glu69 included
CbADH1/Apo	Cys37; His59	Asp150, Glu60	Lys346	–	4 protein ligands, Glu60 included
CbADH1/Holo	Cys37; His59	Asp150, (Glu60)	Lys346	–	3 or 4 protein ligands. Glu60 detached from the metal in 3 subunits. Zinc movement
E. coli ADH	Cys, His	Cys, Glu	Arg	H_2O/(close)	3 or 4 protein ligands with Glu bound or unbound
PFDH	Cys46, His67	Asp169	Arg	H_2O	3 protein ligands
KR/Apo	Cys41; His66	(Glu152)*, Glu60	Lys344	H_2O	3 protein ligands. *Glu152 never a Zn^{2+} ligand
SDH/Holo	Cys37; His54	Glu150, Glu60	--?--	--?--	4 protein ligands. Possible with 6 ligands
GDH/Apo	Cys40; His67	Glu155, --?--	--?--	--?--	3 protein ligands. ? indicates no PDB file available

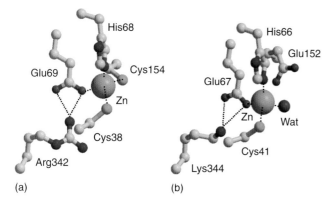

Figure 7 Zn-coordination in MDRs. (a) Apo-SsADH (class I) with four protein ligands to Zn^{2+} and no water molecule observed at bonding distance to the metal. Glu69 is hydrogen bonded to an arginine side chain (PDB code 1JVB). (b) Apo-KR (sorbitol dehydrogenase) with three protein ligands to Zn^{2+} and a water molecule. In this structure, the metal ligand Glu67 interacts with a lysine side chain (PDB code 1E3J). Glu152 is not a metal inner-sphere ligand but probably an adaptation to offer a favorable binding of a sugar substrate.

otherwise observed between them are most likely related to their different substrate preferences, alcohols versus sugar. Which structural changes and zinc coordination alterations may take place upon cofactor and substrate/inhibitor binding in SsADH and KR are not yet known.

In the apoconformation of CbADH, Zn^{2+} is coordinated to four protein ligands similar to SsADH. However, upon NADP(H) binding, it is observed that in three of the four subunits the 'flexible Glu' is no longer a metal ligand,[89] and hence this residue can alternate between metal-bound and unbound states. A recent report describing the structure of a tetrameric NAD-dependent ADH from *E. coli* shows the same tendency.[92] The most evident example that the glutamate takes alternative positions and that this action is functional is demonstrated for FDH, which is discussed below.[45,84,86] There are obviously secrets hidden behind the role of 'flexible Glu'.

The number of protein ligands to Zn^{2+} in MDRs including coordinating and neighboring water molecules are listed in Table 3. The metal geometry for MDRs is defined in the literature as 'distorted tetrahedral'. The largest deviation from a standard tetrahedral angle is shown for the sulfur ligands in many of the mammalian ADH enzymes. For HLADH, this distortion has been explained to be the result of repulsive forces (reference 7, Chapter 28). The S–Zn–S angle is obtuse and can vary considerably showing a maximum distortion from the ideal up to 25° in certain complexes. Sorbitol dehydrogenase is described to have octahedral geometry with four protein groups ligating to the metal and offering the sugar substrate two coordination sites[91] (Table 3). Whether the Zn^{2+}-ion in HLADH is four-[147] or five-coordinated[96] has been discussed ever since the structure was solved by X-ray methods. Strong arguments against five-coordination have been presented (reference 7, Chapter 28) on the basis of the examination of electron density maps of complexes of HLADH solved to medium resolution,[66,67] model building, and accessibility calculations using those structures. The narrow space at the coordination sphere in the presence of the cofactor in a closed enzyme conformation would bring a water molecule too close to the bulky sulfur ligands according to model building. A further argument has been introduced on the basis of theoretical considerations.[137] If hydride transfer in fact involves a hydride ion, this entity could react with water at the nicotinamide site leading to adverse side reactions (giving gaseous molecular hydrogen).

SPECTROSCOPY

Are metal-coordination changes essential? The novel structural information available today on MDR structures correlated to measurements done using other methods hopefully will advance the discussion about coordination number and intermediates formed during catalysis. Several authors claim that five-coordination ought to be an intermediate state during ADH catalysis. Interpretation of spectral changes in Co^{2+}-substituted HLADH has been one technique and the participation of metal-bound water in catalysis was not excluded.[140,148] Data obtained from ligand-binding studies to Cd-substituted HLADH using perturbed angular correlation of γ-rays (PAC) were explained in terms of 'flexible Glu' (residue 68 in HLADH) interaction to the metal increasing the coordination number.[149,150] Theoretical calculations indicate that Glu68 in HLADH coordinates intermittently to Cd^{2+}.[151]

Using rapid-scanning stopped-flow spectrophotometry, at least two intermediates have been detected during ternary complex interconversion in HLADH (reference 7, Chapter 35, references 152, 153). The presence of a negatively charged group interacting with Co^{2+} has been regarded essential in stabilizing the NAD^+ complex in the closed conformation.[154] Three different Co^{2+}-substituted enzymes among the MDRs have been examined with respect to their spectral shifts during catalysis – HLADH,[155–158] human class III ADH = FDH, and human liver sorbitol dehydrogenase.[159] The SDH shows unique spectral properties compared to the ADHs, which presumably derive from the differences in the protein residues (Table 3) coordinating to the metal. The mammalian class I and class III enzymes have exactly the same metal environment (Table 3) and the 'alkaline form' of the HLADH-cofactor spectrum resembles that of FDH,[159] but there are also slight differences observed in the spectral properties. Recent crystallographic data presented for FDH clearly indicate a great flexibility at the Zn^{2+} ion, which might oscillate between different coordination states. Apo-FDH has Cys44, His66, Cys173, and a water molecule (at the substrate site) as ligands (Figure 8(a)). In a binary NAD(H) complex,[84] Glu67 comes in as a

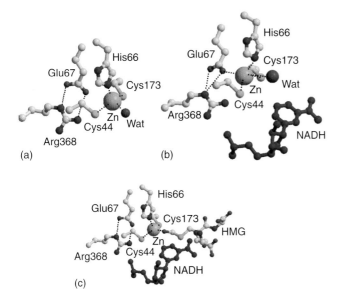

Figure 8 Variation in Zn-coordination in formaldehyde dehydrogenase. (a) Apoenzyme (PDB code 1M6H). Color codes: Red atoms are oxygen, blue are nitrogen, green are sulfur, and yellow are carbon; Zn, gray sphere. Dotted lines indicate bonding interactions. (b) Binary complex with NADH (PDB code 1MC5) for which only the NMN part is shown in blue. Wat is a water molecule at 3-Å distance. The Zn ion has moved 2.3 Å toward Glu67, which is now a ligand to the metal indicating coordination flexibility as discussed in the text. (c) HMG (exchanging water) is the substrate S-(hydroxymethyl)glutathione, which spontaneously forms to neutralize the poison formaldehyde in the liver. This substrate is oxidized in the presence of NAD$^+$ and the ternary complex shown here is thus an abortive complex with NADH (PDB code 1MC5). The Zn^{2+} ion in this complex is at a position comparable to apo-FDH and Glu67 is no longer a ligand.

ligand (Figure 8(b)), the water is still bound but at a longer distance, the metal has moved 2.3 Å toward the glutamate ligand, and the S$_{173}$-Zn bond length has increased. In a ternary complex with S-(hydroxymethyl)glutathione and NADH (Figure 8(c)), the substrate coordinates instead of water, the metal ion is brought back to its original position corresponding to the apoform, and the Glu67 is back to the same interaction with an arginine side chain as before.[86] Thus the 'flexible Glu' participates in metal–ligand reorganization and, interestingly, water is not totally excluded from the active site in FDH in the NADH complex.

The presence of a negatively charged group interacting with the metal in ADHs, predicted from spectral investigations on Co^{2+}-substituted enzymes, might be the 'flexible Glu', which in FDH apparently easily leaves or attaches to the metal-coordination sphere upon complex formation.[84,86] There are no crystallographic observations showing that Glu68 is a ligand to Zn^{2+} in HLADH. However, in a Cd^{2+}-substituted enzyme, which acts as a catalyst albeit with reduced efficiency,[160] the tendency for a similar interaction has been revealed for an NADH complex resolved to 1.15 Å.[161] In one of the subunits the OE1 oxygen of Glu68 is found at 3-Å distance from the metal, which is a motion 1.5 Å closer to the metal. This new glutamic acid site is only partially occupied (Figure 9). The electron density for the metal is enlarged and elongated into an ellipsoid. The larger radius of the Cd^{2+}-ion increases bond lengths to protein ligands by 0.25 Å as expected, but the rest of the active site is similar in configuration to the zinc enzyme and the ligands have moved marginally away from the metal to accommodate its increased size. On the basis of theoretical computations on the zinc center in HLADH, Ryde finds that Glu68 can coordinate and estimates the energy barrier between bound and unbound states to be low and therefore easily bypassed.[162] Both binding modes give stable structures and it is suggested that Glu68 coordination may facilitate ligand exchange reactions at the metal site. For the native HLADH, this situation has never been trapped using X-ray techniques.

Again, there are great variations among MDRs regarding the importance of the 'flexible Glu' on enzyme activity. Site-directed mutagenesis in yeast ADH showed that exchange of Glu68 for a glutamine reduces enzyme efficiency about a 100-fold compared to wt-YADH.[163] On the contrary, excluding the conserved, negatively charged group in TbADH (E60A) did not abolish activity (Table 2).[164]

Are snapshots of transition states possible to catch at the Zn-center? The most convincing data concerning the dynamics of Zn^{2+} coordination emerges from the time-resolved X-ray absorption spectroscopy measured on TbADH.[165] The fluctuations of Zn^{2+}-coordination number and local structural changes were followed during a catalytic cycle. The analysis started from an apoenzyme with tetrahedral coordination and the development of

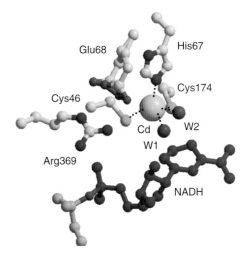

Figure 9 Reaction centers of HLADH. Active site of the Cd^{2+}-substituted HLADH–NADH complex showing Glu68 in double conformations.[161] The side-chain movement is 1.5 Å closer to Cd but the distance is still 0.5 Å longer than for a typical oxygen–metal bond. Atoms are color-coded as before. Cd, yellow sphere. NMN part of NADH in blue. W1, W2 denote alternate positions for water/hydroxide ion. Five-coordination is possible.

two new metal-coordination species within a time frame of 100 ms^{-1} were recorded. In the mechanistic model presented by Kleifeld et al., Glu60 and water are proposed to participate in exchange reactions. Strong support for this interpretation comes from crystallographic data on CbADH,[89] structurally identical to TbADH, because Glu60 was observed to have a changed position in the NADP(H) complex compared to the apoenzyme.

It seems unavoidable to regard the presence of water and high degree of flexibility as inherent properties of the metal site of MDRs for certain enzymes that are probably vital for efficient catalysis. The flexibility is manifested in various ways: (i) in variation of Zn to ligand bond lengths in FDH[84,86]; (ii) the 'flexible Glu' connecting to or disconnecting from the metal shown for quite different enzymes[86,89,92,161]; (iii) displacement of the metal atom itself[79,84,86,92,141] modifying the coordination sphere moderately. In addition, in HLADH, water/OH$^-$ is assumed to oscillate between two closely located sites (W1 and W2 in Figure 9) at the metal controlled by an essential serine residue.[6] This OH$^-$ group might be a further candidate accounting for the characteristic shifts observed in Co-ADH spectra assigned to be due to a negatively charged group.[154,157,159]

FUNCTIONAL ASPECTS

Within the scope of this chapter it is not possible to summarize all kinetic information on Zn-dependent NAD/NADP medium-chain dehydrogenases/reductases available in the literature. Therefore, only few examples selected mainly from the plentitude of data existing for ADHs, which have been studied most extensively, are given. Many references to the classical work on HLADH have been excluded but are found in reference 166 where Cook and Bertagnolli present a general description of the kinetics of pyridine nucleotide-dependent enzymes. Pettersson has given a comprehensive summary of the mechanistic aspects of HLADH in 1987.[167] To our knowledge, the most recent record on nicotinamide cofactor-dependent reactions is found in an article written by Clarke and Daffron[168] containing the cardinal references to some of the enzymes mentioned here. A further good source of information is the kinetic database http://www.brenda.uni-koeln.de/.

Sequential mechanisms: These enzymes use two substrates, and HLADH was the first for which the detailed kinetic mechanism for a two-substrate enzyme was presented.[94,169] Scheme 1 shows the general description of a Zn-dependent MDR-catalyzed reaction.

Substrate and cofactor must bind to form the ternary, central complex on the enzyme but the order in which the substrates arrive at the site is not always the same among the enzymes. Sometimes, binding of the cofactor is a first step. A general characteristic is a weaker binding of oxidized cofactor in comparison to reduced cofactor, and K_d values can differ by a factor of 100. The coenzyme product must bind weakly enough to maintain good rates of exchange in the steady state reaction to allow for the next turnover to start. As mentioned, an exception has been found.[88] The glutathione-*independent* formaldehyde dehydrogenase (PFDH in Table 2) is a nicotinoprotein from which NADP(H) never leaves the protein. An extra loop close to the adenosine moiety of NADP(H), not present in other ADHs, is thought to link the cofactor more strongly to the protein.

As soon as the correct geometry is achieved at the reaction center between the hydride donor and acceptor, hydride transfer to carbon and proton transport (H$^-$/H$^+$) can take place. In Scheme 2, this is shown for A-side stereospecific transfer of H$_R$ from NAD(P)H to an aldehyde molecule.

Steady state parameters

A sample of steady state kinetic parameters and substrate specificities for MDRs related to this review are found in Table 4 and in the following citations: SsADH,[11,110] *Moraxella* ADH[13] *B. stearothermophilus* ADH-hT,[14] human/mouse sorbitol dehydrogenase,[20,170,171] human ADH class III,[23,172] human/mouse ADH class IV,[34,87,173,174] human ADH class I isozymes,[42,175–179] human steroid-active ADH class I,[180] horse ADH isozymes,[72,73,76,77,80,94,96,116,158,167,181] glucose dehydrogenase,[101] CbADH/TbADH,[90,182] ketose reduc-

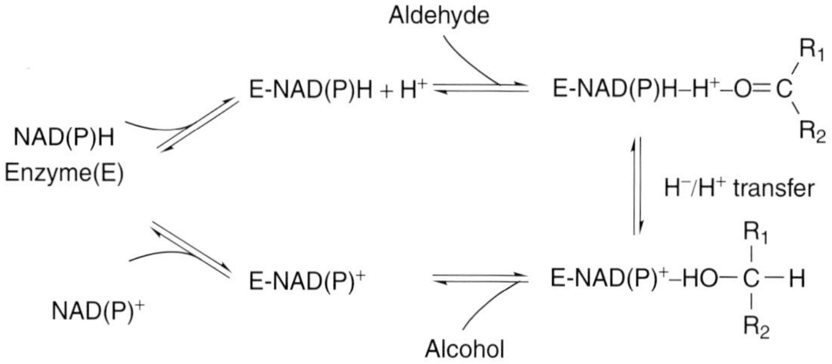

Scheme 1

Table 4 Summary of substrate specificities for mammalian ADHs. Data adapted from references 19,125

Mammalian ADHs	Ethanol	S-(hydroxymethyl) glutathione	Retinol	Aldehyde dismutation
ADH1	Yes	No	Yes	Yes
ADH2	Yes	No	Yes	Yes
ADH3	No	Yes	No	Not detectable
ADH4	Yes, but low	No	Yes, best	Not tested

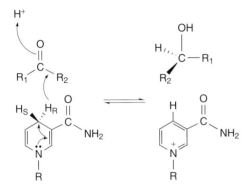

Scheme 2

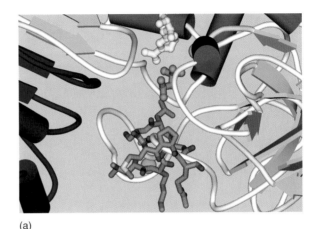

(a)

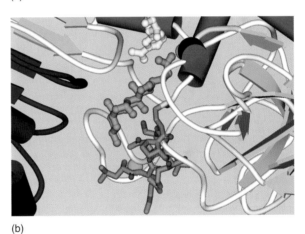

(b)

Figure 10 Substrate binding sites of ethanol- and steroid-converting HLADH. Helices and strands are cylinders and arrows, respectively, and a neighboring subunit is in black. Green spheres are Zn sites; yellow is the model part of NADH. (a) Methyl pentanediol (MPD) (orange) binding at a hydrophobic 'inner compartment' of the EE-ADH active site (PDB code 1HET). MPD is not a Zn-ligand. Side chains are shown only for the substrate, binding loop, residues 112–119; (b) Cholic acid (orange) binding to Zn, a potent inhibitor to SS-ADH (PDB code 1EE2). The difference in the substrate loop orientation compared to (a) gives a 50% larger binding volume.[77]

tase,[103] human ADH dismutase activity,[108] mouse/human ADH class II,[44,183] chick ADH class VII,[124] horse ADH dismutase activity,[126,127] benzyl-ADH,[51] *P. putida* ADH (FDH) dismutase activity,[184] retinoid dehydrogenases,[185] yeast: ADH class I,[120] *S. cerevisiae* ADH class VI,[186] *Galactocandida mastotermitis* xylitol-DH.[187]

Variable shape and size of the active site channel in MDRs. As mentioned initially, this enzyme group converts a great number of substrates with widely differing structures – small, large, aliphatic, aromatic, linear, branched, primary, secondary alcohols/aldehydes, ketones, and polyols. K_m and k_{cat} values show an enormous spread. No framework of rules has yet been established that correlate the diverse kinetic parameters with (i) the Zn-coordination status, (ii) the total enzyme charge, and (iii) water/no water participation in the chemical reaction.

However, a clear, general trend becomes apparent pointed out by Jörnvall *et al.*[188] The evolutionary strategy for MDRs has been to vary amino acids preferably at the reaction center and further away in the substrate binding area but otherwise maintain an overall subunit structure. Examining the many structures solved shows that the size and shape of the substrate channel is adapted for small or large substrates, hydrophilic or hydrophobic substances, modifications attained via few mutations, or significant deletions/additions of amino acids. A further opportunity to create multiplicity has been to vary the control mechanism for the protein conformation change and the oligomeric state. It may not be that easy to deduce any rules from the multifaceted pool of information available.

One example will be given, on the structural level, for how an adaptation evolved, changing from small to large substrates. The active sites of alcohol metabolizing

(EE)- and steroid-active (SS)-HLADH are compared in Figure 10(a) and (b).[77] Both isoenzymes can convert small substrates but SS-ADH binds considerably bulkier and more hydrophobic substrates. Ten amino acid substitutions distinguish the E- from the S-polypeptide (Figure 11(a)) and these mutations are mainly centered at an area around the 'substrate, binding loop' (residues 113–119). The crucial

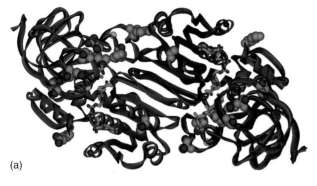

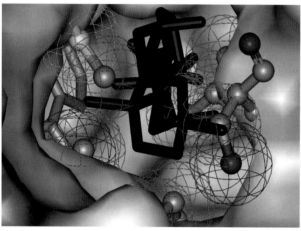

Figure 11 Comparison of EE- and SS-isozymes of HLADH. (a) Ribbon model, global view, with ligands and NAD(H) as stick models. Shown as space-filling models is the distribution and character of the 10 amino acid substitutions distinguishing the two isoenzymes (color code: Red, oxygen; blue, nitrogen; gray, carbon). The subunit of the steroid-active enzyme is in red superimposed onto the ethanol-active dimer (green). The SS-dimer differ by +6 charges compared to the EE-dimer. (b) The substrate channel. The steroid molecule (blue model) fitted into the substrate cavity of SS-ADH (gray surface). Yellow sphere at the bottom of the channel is the Zn^{2+} ion; the orange model (left) is the nicotinamide of NAD(H). Superimposed is the structure of the EE-isoenzyme showing how active site residues would impair steroid binding through steric interference to the ethanol-active enzyme. For Figures 10 and 11, the WebLab Viewer from MSI was used.

amino acid substitution is the deletion of an aspartate located in the middle of that loop causing a rearrangement such that the entrance to the reaction center (the Zn^{2+} ion) opens and widens (Figure 11(b)). The other substitutions either alter the charge of the enzyme influencing the kinetics[116,158] or the hydrophobicity and shape of the substrate channel.

Some ADHs have particularly broad substrate specificity, and HLADH probably beats the record for being tested for around 170 different compounds[189] and has frequently been utilized in enantio-selective synthesis.[190–193] SsADH and TbADH are more suitable for industrial purposes since they are stable at higher temperatures and survive in harsh organic solvents. Although HLADH is not as stable as the enzymes from the thermophilic organisms, greatly improved stability has been achieved by enclosing HLADH in reverse micelles.[194] Successful experiments are reported in which HLADH was active for several weeks in a microemulsion system in which cofactor generation and product separation was included in the process.

Kinetic models for selected enzymes

Ordered bi-bi mechanism. Many ADHs (class I, II, IV, and sorbitol dehydrogenase (although this is debated)) are described following an ordered reaction mechanism. This means a binary enzyme–cofactor complex forms, 'preparing' the protein to receive the second substrate for conversion. This 'isomerization' is the conformation change discussed in detail earlier, which encloses the cofactor deep inside the protein. Productive binding of the substrate is the direct coordination to the metal, first shown for an aldehyde substrate bound to HLADH.[143] It is reasonable to think that the essential process to activate a substrate for $NAD(P)^+$ reduction is similar for the Zn-dependent MDR enzymes. Because of binding to the Zn^{2+} ion and subsequent deprotonation of an alcoholic group, polarization of the substrate molecule in HLADH has been suggested to favor hydride transfer (167 and references therein). Binding studies have shown that all kinds of binary and ternary complexes can form and basically could proceed through a generalized random ordered mechanism. However, the steady state rate behavior establishes that a compulsory-order mechanism operates at concentration ranges where the Michaelis–Menten kinetics is obeyed. The equilibrium at pH 7 is 10^5 in favor of ethanol oxidation and NADH dissociation is rate-limiting. For larger secondary alcohols, the hydride-transfer efficiency decreases and could become rate-limiting.[169]

The textbook view concerning the HLADH mechanism is represented in Scheme 3 adapted from reference 167. The HLADH reaction is dependent on three pK_a values, which by different authors have been assigned to various groups: Zn-bound water, a histidine, or a lysine side chain. In Scheme 3, the activation of an alcohol substrate and pH-dependent $Zn-H_2O$ protonation/deprotonation steps are emphasized. NAD^+ binding (reaction 1) occurs to the protein in the $Zn-H_2O$ state and the bound water molecule is deprotonated (reaction 8). The alcohol substrate enters the site in neutral form but is stripped of a proton by the $Zn-OH^-$ ion (reaction 3). Water leaves the metal and an alcoholate ion is the metal-bound species polarized by the coordination to zinc promoting hydride delivery from the C1 carbon to NAD^+ (reaction 4).

Product release, opening of the coenzyme binding cleft, NADH dissociation, and the restoration of a water-bound Zn sphere (arriving at reaction 6) brings the enzyme to the

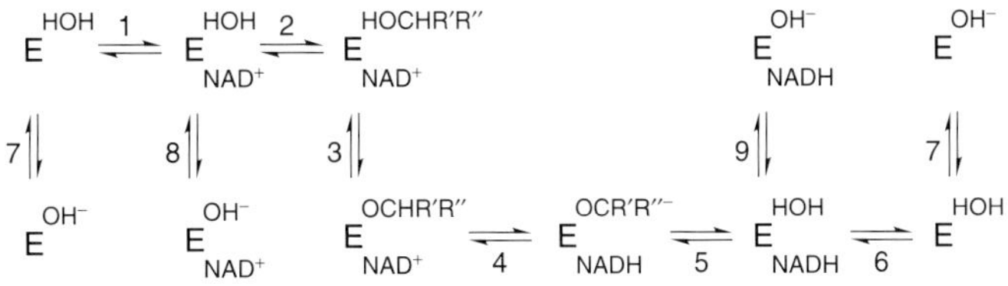

Scheme 3

[Scheme 3 diagram showing enzyme catalytic cycle with states:
E·HOH/NAD⁺ ⇌(1) E·HOH/NAD⁺ ⇌(2) E·HOCHR'R''/NAD⁺ ... E·OH⁻/NADH ... E·OH⁻
with vertical equilibria labeled 7, 8, 3, 9, 7 connecting to bottom row:
E·OH⁻/NAD⁺, E·OH⁻/NAD⁺, E·OCHR'R''/NAD⁺ ⇌(4) E·OCR'R''⁻/NADH ⇌(5) E·HOH/NADH ⇌(6) E·HOH]

starting point for a new catalytic cycle (reaction 1). This mechanistic model excludes five-coordination of the zinc based on the interpretation of the X-ray data on HLADH, suggesting that five-coordination was not possible; this was discussed in an earlier section.

Ping-Pong mechanism. The aldehyde dismutase activity has been analyzed for ADHs of class I[126,127] and II.[108] Alcohol oxidation differs fundamentally from aldehyde oxidation in that the addition of water is an extra step in acid formation (Scheme 4).

At certain aldehyde concentrations dismutation of aldehyde to equal amounts of alcohol and acid is observed. This mainly happens in a 'lag phase'[126] in the HLADH reaction during which no net NADH production can be detected (NADH increase/decrease is the normal method to follow activity) and is, therefore, notoriously overlooked. Aldehyde conversion by HLADH is a rapid process. Owing to the slowness with which NADH leaves the enzyme (the rate-limiting step), an aldehyde molecule has time to bind and alcohol is formed. The resulting NAD⁺ complex binds a new aldehyde molecule and acid is produced. This reaction follows a ping-pong mechanism. At the center of Scheme 5 (adapted from reference 126) is shown two molecular forms, the aldehyde, which is a substrate in the presence of NADH (k_{-1}) and its *gem*-diol, which is the form of the substrate that is oxidized to acid (k_2). This circumstance highly favors the presence of water at the active center of HLADH.

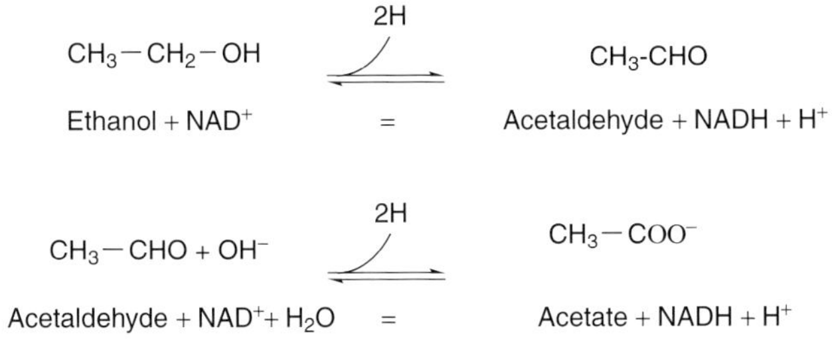

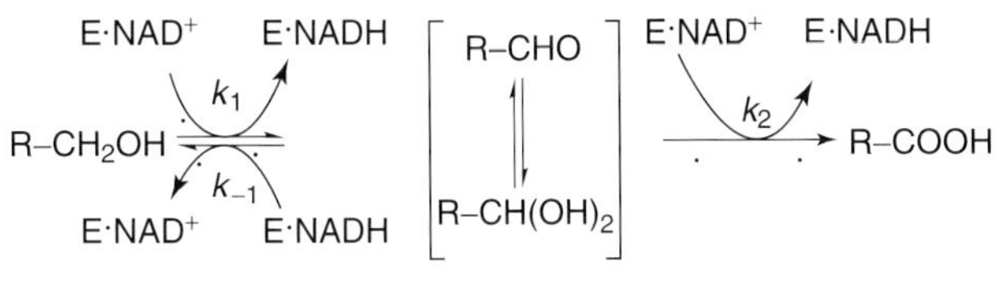

Random bi-bi mechanism. Liver ADH class III, FDH, is glutathione-dependent. Adduct formation between formaldehyde and glutathione is nonenzymatic and the major product is *S*-(hydroxymethyl)glutathione (HMG, see Figure 8(c)), a substrate of FDH, which does not convert ethanol.[85] Alcohol oxidation has been tested using HMG, which is a slow reaction ($k_{cat} = 63\ min^{-1}$). *S*-formylglutathione is unstable and cannot be used to measure the reverse reaction. Instead, 12-hydroxy dodecanoic acid (DA-OH in Figure 12) and its corresponding aldehyde, 12-oxododecanoic acid (12-DOOA) were used to collect kinetic parameters also for aldehyde reduction. Interpretation of product and dead-end inhibition patterns for the reduction of 12-DOOA resulted in the conclusion that FDH follows a random mechanism (Scheme 6, adapted from reference 85).

Binary complexes form between the enzyme and both the substrates independent of each other. Crystallographic evidence also strongly supports the random mechanism interpretation. One indication is the lack of total protein closure, which allows ligands to enter and leave without large changes in enzyme conformation. Figure 12 shows the alcohol complex (E·Alc in Scheme 6) with 12-DA-OH. Figure 8(b) is an example of E·NADH binding to a complex in which one subunit in the crystal binds the cofactor only, even when *S*-(hydroxymethyl) glutathione is present and the ternary abortive complex (Figure 9(c)) is formed in the second subunit.[86]

Mixed mechanisms depending on the substrate. ADH class IV can function as a retinol dehydrogenase.[125,173,174] The catalytic efficiency of ADH4 for ethanol/acetaldehyde as substrates is poor, but for substrates with longer carbon chains, the activity increased dramatically.[173] Within the human ADH family, ADH4 exhibits the highest catalytic efficiency for retinol oxidation. The enzyme conforms to the classical, ordered, sequential mechanism in the oxidation of ethanol and participates in ethanol metabolism in the stomach. An X-ray analysis[46] performed on an enzyme-NAD(H)-acetate complex shows high structural similarity to class I ADHs. Like HLADH, the enzyme is fully closed upon coenzyme binding, which is not the case for FDH mentioned above. In addition to a few other amino acid substitutions, the ADH4 substrate channel is enlarged by means of a deletion of residue 117 located in what is denoted the 'substrate loop' in Figure 10, which compares the ethanol- and steroid-converting isozymes from horse liver.

$$CH_3-CH_2-OH \xrightarrow{2H} CH_3\text{-}CHO$$
$$Ethanol + NAD^+ = Acetaldehyde + NADH + H^+$$

$$CH_3-CHO + OH^- \xrightarrow{2H} CH_3-COO^-$$
$$Acetaldehyde + NAD^+ + H_2O = Acetate + NADH + H^+$$

Scheme 4

[Scheme 5 diagram:
E·NAD⁺ → E·NADH (via k_1), R-CH₂OH, E·NAD⁺ ⇌ E·NADH (via k_{-1}); middle: R-CHO ⇌ R-CH(OH)₂; then E·NAD⁺ → E·NADH (via k_2) → R-COOH]

Scheme 5

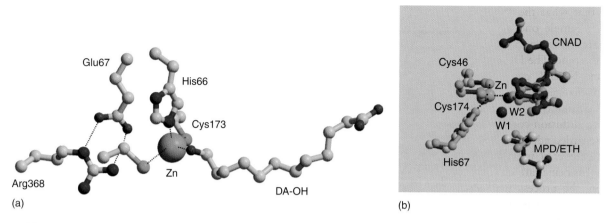

Figure 12 Various modes of binding of alcohols to formaldehyde dehydrogenase and HLADH. (a) Productive binding of an alcohol to the Zn^{2+} ion in a binary complex with FDH (PDB code 1M6W). 12-hydroxydodecanoic acid was used as a substrate to analyze the kinetic mechanism, which was found to be random – the complex forms independent of the presence of cofactor. (b) Nonproductive binding of alcohols to HLADH. The light blue model is the complex with a coenzyme analogue denoted CNAD (PDB code 1ADB). It is substituted with N at the C5 position of the pyridine ring and called a 'hydride' inhibitor.[79] N5 coordinates to Zn with a bond length of 2.19 Å, pulling the metal toward the ring. Ethanol (ETH) is not a metal ligand. Superimposed is an NADH complex (PDB code 1HET) (dark blue model, yellow Zn sphere and ligands). Red dots show the two possible H_2O/OH^- positions, W1 being the site for substrate binding. 2-methyl 2,4-pentanediol (MPD) binds at a similar position as ethanol remote from the active metal (compare Figure 10(a)).

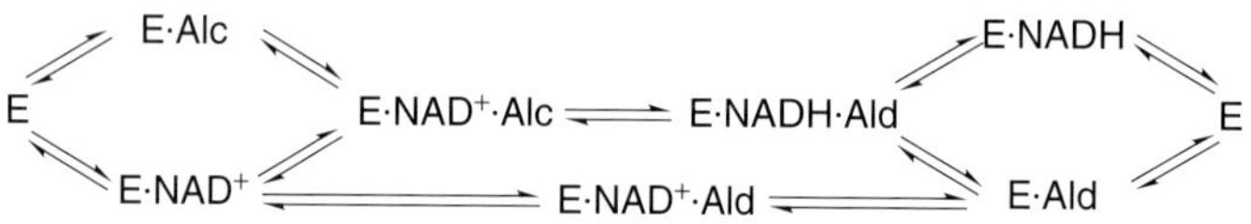

Scheme 6

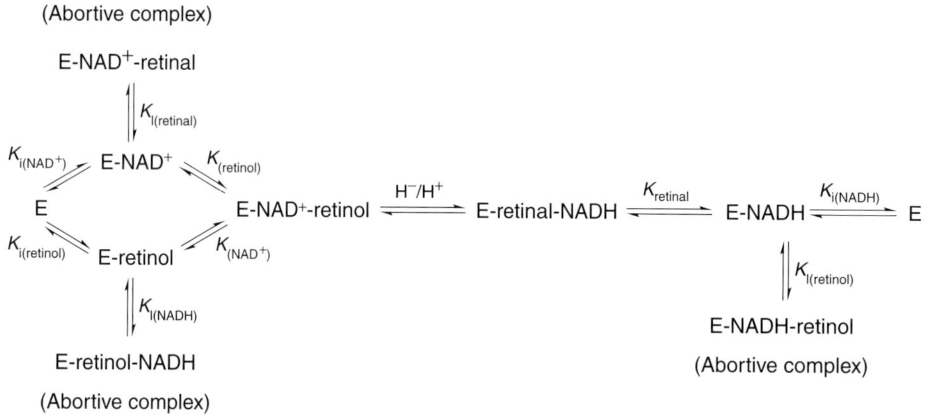

Scheme 7

A minimal reaction diagram for retinol dehydrogenase is shown in Scheme 7 (adapted from reference 174). A rapid equilibrium ordered mechanism with one dead-end ternary complex is deduced for retinal reduction (right-hand part of the diagram). For retinol oxidation, a symmetric, rapid equilibrium, random mechanism with two dead-end ternary complexes, is described. All-*trans*-retinol/retinal are suggested to be natural substrates to ADH4. A proposal has been made that retinol dehydrogenase is involved in retinoic acid metabolism in man and the enzyme may have an important physiological function. Retinoic acid is a hormonal ligand to a nuclear receptor regulating the signaling pathway for growth and development. A high concentration of ethanol interfering with retinol metabolism is believed to give rise to severe birth defects.[35]

TRANSIENT KINETICS

Theorell *et al.* were among the pioneers to study coenzyme binding to HLADH using stopped-flow techniques[195] and their comments in the paper of 1967 are slightly puzzling but indicate that what they had observed could not be explained by a simple one-step reaction. Binding of the adenosine part prior to the nicotinamide moiety was discussed at the time. The HLADH structures later clearly linked enzyme–NADH complex formation to protein conformation changes, but since the magnitude of the forward rate constant ($\approx 10^7$ M^{-1} s^{-1}) was close to that expected for a diffusion-controlled reaction, the binding process was regarded adequately described by a single step reaction.[196] Coates et al. implicated NAD$^+$ binding with an 'isomerization' as being rate-limiting for NAD$^+$ association when studying the process using pressure relaxation.[197] The term *isomerization* is frequently used to indicate a protein conformation-linked coenzyme interaction. Using stopped-flow experiments, Sekhar and Plapp measured a limiting rate for NAD$^+$ association of 500 s^{-1} and suggested a minimum two-step mechanism to explain that observation.[138] Furthermore, studies on human liver ADH mutants indicated a two-step mechanism[177] and rarely has the NADH binding process been treated as a reaction with more than two steps.

Greeves and Fink, however, used cryoenzymology to resolve the initial diffusion-controlled complexation from subsequent isomerization steps.[198] At temperatures below −30 °C, NADH binding produced biphasic changes in the protein fluorescence. A fast and a slow phase could be separated and the conditions used made it possible to measure at high coenzyme concentrations (up to 150 μM). The results were interpreted in terms of a three-step binding mechanism, two of which were related to enzyme conformation changes. The fact that NADH binding becomes more complicated at high coenzyme concentrations has been confirmed later also in measurements performed at 25 °C.[116,199]

Here, the rate constants for NAD$^+$/NADH binding to the EE- and SS-isozymes of HLADH are taken from reference 116. In addition to the mutations modifying the substrate binding area illustrated in Figure 11, these isozymes have a net charge difference of +6 charges/HLADH molecule. The rate constants were derived from fluorescence measurements performed under standardized conditions using a buffer system not interfering with coenzyme binding and avoiding ions found to be coenzyme-competitive inhibitors (sulfate and phosphate). The dead time of the stopped-flow spectrophotometer was 1.3 ms and from the transient kinetic measurements rates could be detected just beyond 1500 s^{-1}. Four constants accounting for a two-step association and a two-step dissociation of NAD$^+$ and NADH were determined in the low and medium concentration ranges according to a minimum reaction scheme including a conformational transition (Scheme 8). The data for NADH binding at higher concentrations could only be interpreted as connected to a further isomerization step.

The rate constant k_1 (Scheme 8) has a value of ≈ 5.5 M^{-1} s^{-1} $\times 10^7$ for the EE-isozyme and is similar for oxidized and reduced coenzyme (the values given for physiological pH). For SS-HLADH, k_1 is ≈ 2 M^{-1} s^{-1} $\times 10^8$, which approaches the magnitude of a diffusion-controlled rate. This first step does not result in fluorescence changes and has been related to the docking of the adenosine part of the coenzymes. The pH dependence of k_1 differs slightly, the pK_a value being 7.6 for the adenosine association of NAD$^+$ and 8.5 for NADH. Thus, electrostatic attraction/repulsion controls the docking process. The coenzyme encounters a protein in which the adenine pocket is easily accessible, E_oE_o in Scheme 8 where 'o' stands for open (conformation) and 'c' for closed.

The second step, governed by the rate constant k_2, (*NADH in Scheme 8) is related to nicotinamide association and subsequent isomerization (E_cE_c; k_3/k_{-3}). In this step, NAD$^+$ interaction is pH-dependent but NADH is pH-independent. The charge difference between isozymes has an influence upon NADH binding rates, but the dramatic differences in rate constants in this enzyme system are observed for oxidized and reduced coenzyme interacting with the EE-isozyme. When NADH has formed after NAD$^+$ reduction, the nicotinamide dissociation is slow, $k_{-2} = 20$ s^{-1}, suggesting that conformation changes may limit cofactor release. (No differences are noticed in this step between the isoenzymes.) After NAD$^+$ has formed at the active site, dissociation is, on the contrary, comparatively rapid ($k_{-2} \approx 430$ s^{-1}) and the very fast adenosine dissociation rate ($k_{-1} \approx 2000$ s^{-1}) presumably is favored by the open protein structure. Another fast reaction is associated with the enzyme–NAD$^+$–alcohol complex, which loses a proton rapidly, estimated to be >1000 s^{-1} and preceding hydride transfer.[139]

The kinetic mechanism of EE- and SS-HLADH appears to be identical. A substrate-specific comparison, however, results in differences of individual rate constants in the catalytic mechanism.[77,116,158] Since the coenzyme binding dynamics is a major determinant for substrate kinetics, changes in the coenzyme binding kinetics have a direct

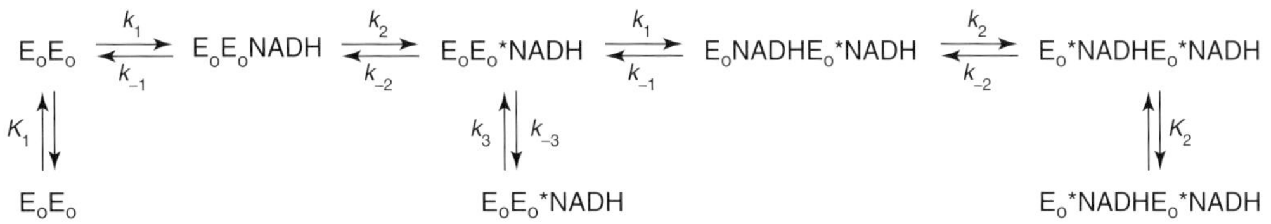

$$E_oE_o \underset{k_{-1}}{\overset{k_1}{\rightleftharpoons}} E_oE_oNADH \underset{k_{-2}}{\overset{k_2}{\rightleftharpoons}} E_oE_o*NADH \underset{k_{-1}}{\overset{k_1}{\rightleftharpoons}} E_oNADHE_o*NADH \underset{k_{-2}}{\overset{k_2}{\rightleftharpoons}} E_o*NADHE_o*NADH$$

$$K_1 \updownarrow \qquad\qquad\qquad k_3 \updownarrow k_{-3} \qquad\qquad\qquad K_2 \updownarrow$$

$$E_oE_o \qquad\qquad\qquad E_oE_o*NADH \qquad\qquad\qquad E_o*NADHE_o*NADH$$

Scheme 8

influence on the rates of substrate conversion. Differences observed in k_{cat} and K_m for primary alcohols are mainly due to the modified dissociation rate constants of NADH. Nicotinamide release is equally slow for both isoenzymes (SS-ADH $\approx 16\,s^{-1}$, pH 7) but the adenosine dissociation is slower for SS-HLADH ($\approx 70\,s^{-1}$, pH 7, and $\approx 330\,s^{-1}$ for EE-ADH). In summary: the nicotinamide part of oxidized cofactor leaves both isoenzymes with a similar rate ($k_{-2} \approx 430\,s^{-1}$) and the adenosine part extremely fast (maximum at pH7, $\approx 1000/2000\,s^{-1}$). In contrast, the nicotinamide part of NADH leaves both isoenzymes very slowly and SS-ADH shows the tendency to also dissociate adenosine slowly. A major charge difference between the two isozymes is an exchange of a Lys_{EE} for a Glu_{SS} at position 366 not far from the adenosine-binding crevice. The faster dissociation of the oxidized cofactor indicates that there may be differences in the degree of closure of the enzyme structure regulated by the oxidation state of the cofactor.

Changes of the substrate channel architecture in the SS-enzyme have a pronounced effect on the dynamics of substrate binding. The association rate constants for secondary alcohols are reduced by two orders of magnitude and those for ketones by one order of magnitude compared to the EE-isozyme, although the substrate channel of SS-ADH has a much larger volume (Figure 11). Whereas binding of small substrates to SS-HLADH is slow, binding of the bulky steroid substrates is three orders of magnitude faster.[77,116] It appears that docking of small substrates is slow because of the difficulties to bind in a productive ternary complex. Different conformations of the enzyme–substrate complex with nonproductive orientations might be possible because of a low energetic advantage of the productive conformation. Similarly, dissociation of small hydrophobic reaction products like ketones is slow since many possible interactions exist that make transfer to the bulk water less likely from the SS-isozyme active site, which has a wider and more hydrophobic substrate channel.

FURTHER FUNCTIONAL ASPECTS

Activation of coenzyme for hydride transfer

To understand the hydride-transfer mechanism, it is important to know precisely what conformation the pyridine ring in NADH assumes in an enzyme-bound state. Theoretical calculations on the nicotinamide in a simulated enzyme environment[200] showed ring distortion into a boat conformation promoting the departure of a hydride ion from the reduced ring. One of the functions of an NAD(P)-dependent enzyme is expected to be a perturbation of the nicotinamide to enhance the hydride-transfer rate. Structural modifications of the nicotinamide in enzyme–substrate complexes have been observed by NMR.[201] The typical UV-visible spectral shift accompanying NADH binding[112] to enzymes is another experimental indication. In earlier work, the activation of an alcohol substrate has been the focus in understanding the reaction mechanism. Rarely is the activation process for hydride ion delivery from NADH discussed.

A transition state intermediate of NADH in complex with HLADH has been published for both Zn- and Cd-HLADH.[6] These structures, which were solved to atomic resolution (≈ 1 Å), have revealed novel features not possible to be detected at medium resolution.[202] Most strikingly, the pyridine ring of NADH was puckered in a conformation that was predicted to enhance hydride transfer.[200] Two electron density peaks were present (Figure 13) at the active center, both at coordinating distances to the metal (either Zn or Cd). Each peak was half occupied and it was suggested that the two peaks together represented a water molecule or a hydroxide ion, alternating between two positions. The peak named OH^- in Figure 13(a) is at 1.90-Å distance to the nicotinamide of NADH (all Zn-ligand bond lengths are given in Figure 3). In the paradigm reaction mechanism (Scheme 3), a water molecule binds as a fourth ligand to the metal in the apo-HLADH structure,[66] and it is displaced from the active site.[167] A serine or a threonine side chain, present in active sites of MDRs, has been assigned the role of shuffling protons via the nicotinamide ribose and a histidine (or an equivalent residue) to bulk water. In the atomic resolution structure of the HLADH–NADH complex, Ser48 is found at hydrogen-bonding distance to both H_2O/OH^- sites and has space to oscillate between two orientations to interact with a proton-transport pathway and the nonprotein metal ligand. In this configuration, five-coordination of Zn is possible and does not lead to collisions between active site components.

To verify if an OH^- pyridine ring adduct could form, theoretical calculations were performed on a model system resembling the crystal structure.[6] Neither H_2O nor H_3O^+, added close to the pyridine ring, did generate a calculated model identical to the X-ray structure. Only when hydroxide ion was present did the theoretical and experimental models overlap. The distortion of the nicotinamide at the enzyme site, affected by the presence of a negatively charged oxygen ligand, influences three parameters: (i) the puckering of the ring (Figure 13(b)), (ii) a deviation from a standard carbon–carbon double bond length (1.33 Å) within the ring (C5–C6 = 1.41 Å), and (iii) accumulation of negative charge at the C4 atom was observed, which became more profound when Zn with ligands were included in the theoretical calculations. These modifications have been suggested to be an activation process of NADH[6] facilitating hydride transfer to an incoming substrate. Examination of various other NAD/NAD(P)-dependent dehydrogenases/reductases outside the MDR group shows that oxygen atoms are frequently located close to the nicotinamide (Meijers and Cedergren-Zeppezauer, unpublished results). When more atomic resolution structures have been

Zn-dependent medium-chain dehydrogenases/reductases

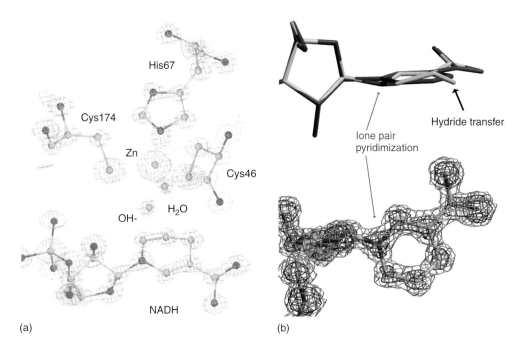

Figure 13 Electron density maps of HLADH at 1.1-Å resolution (PDB code 1HET). (a) A $2F_o - F_c$ electron density map, contoured at 2 and 3 σ-levels showing the active site Zn sphere and part of NADH. The density peak–denoted H_2O corresponds to the substrate binding site. OH^- is 1.9 Å from the C6 atom of the pyridine ring and 2.28 Å from Zn forming the link between metal and NADH. (b) Comparison between a planar nicotinamide superimposed onto the puckered, experimentally observed structure. The model was unrestrained and refined using SHELX97.[203]

solved, the similarities and differences in the NAD(P)H activation process between Zn-dependent and Zn-independent enzymes will be revealed.

Theoretical calculations as a further step to elucidate the hydride-transfer reaction. A current trend in LADH enzymology has been to approach the complex questions regarding the hydride-transfer reaction and the impact of enzyme dynamics on activity by comparing experimental data with sophisticated theoretical calculations. One area of interest has been to understand the basis behind nuclear quantum effects and if that is significant for the hydride-transfer reaction. Kinetic isotope-effect experiments have been performed using several mutants of HLADH to demonstrate the enhancement or attenuation of a quantum mechanical tunneling contribution to hydride transfer.[75] Hammes-Schiffer recently has given an excellent account on the calculations using HLADH and dihydrofolate reductase as models, and many references relevant to the field are found in that paper.[204]

Alcohol dehydrogenase returns to where it belongs. Other groups of Zn-enzymes are described as H_2O/OH^- dependent for catalysis changing between four- and five-coordination.[205] Alcohol dehydrogenase was so far considered an outsider within the family since water participation in catalysis was excluded (Scheme 3). As has been shown in this review, X-ray data indicate that water molecules can be detected in the active site in cofactor complexes. Figure 14 summarizes the X-ray data of the active site structures of glutathione-dependent formaldehyde dehydrogenase and

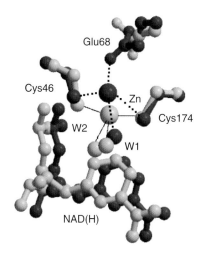

Figure 14 Superposition of the NAD(H) (unknown oxidation state) complex of FDH, red model, (PDB code 1PM0) onto the HLADH–NADH complex, yellow model, (PDB code 1HET). The active sites are viewed along the direction of the His67–N–Zn bond, omitting the histidine side chain for clarity. The magnitude of the Zn-ion movement in FDH toward Glu68 is clearly visible. The side chain of Cys46 changes conformation, whereas His67 and Cys174 are identical between the two structures. Five-coordination is possible. The red water dot in FDH is oriented in the same region as water/OH^- in HLADH (yellow dots).

HLADH in which these features are further confirmed. In both cases, water is found in the vicinity of NAD(H). In some studies, the limitations in resolution of the X-ray data prevent the detection of ligands with low occupancy. The

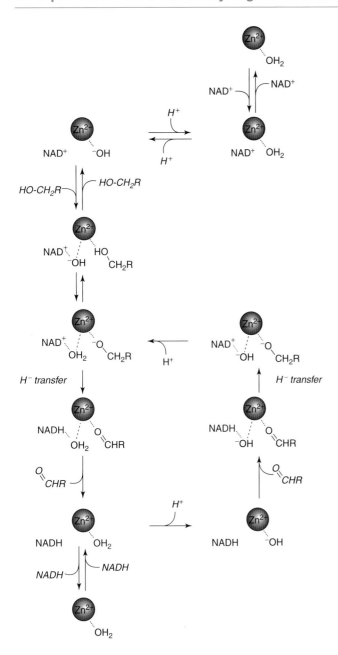

Figure 15 A reaction mechanism suggested for HLADH in which water is present at the active site during catalysis (reference 6). Dotted lines indicate Zn-ligand interactions. The longer dotted bonds are not related to bond length but are to distinguish the second water site from the substrate site. For the sake of clarity, the zinc ligands Cys46, Cys174, and His67 have been omitted. It is unknown at which steps Glu68 may enter the coordination sphere for HLADH.

common habit of studying strong Zn-binding inhibitors in complex with the coenzyme could hamper the detection of functional water binding since the inhibitor moderates flexibility at the metal-coordination sphere. The inhibitory effect may just be interference in the cofactor activation process requiring OH^- bound to the metal.

Figure 15 presents schematically a revised mechanistic proposal for HLADH.[6] The X-ray data does not allow prediction of proton-transport modes but numerous other experimental evidences support the release of a proton before alcoxide–hydride transfer.[139,167] It resembles Scheme 3 with the exception that H_2O/OH^- is present in one or the other form at all times and takes different positions at the coordination sphere over a catalytic cycle. Further development of time-resolved XAFS[165] and more detailed structures of MDRs will hopefully accumulate more evidence in support of the view that Zn-dependent MDR enzymes belong to the classical Zn-enzyme family.

REFERENCES

1. P Kraulis, *J Appl Crystallogr*, **24**, 946–50 (1991).
2. EA Merritt and DJ Bacon, Methods in enzymology, in CW Carter and RM Sweet (eds.), *Macromolecular Crystallography*, Vol. 277, Academic Press, San Diego, London, Boston, New York, Sidney, Tokyo, Toronto, pp 505–24 (1997).
3. H Jörnvall, JO Höög and B Persson, *FEBS Lett*, **445**, 261–64 (1999).
4. B Persson, JS Zigler Jr and H Jörnvall, *Eur J Biochem*, **226**, 15–22 (1994).
5. H Jörnvall, E Norling and B Persson, *Chem Biol Interact*, **143–144**, 255–61 (2003).
6. R Meijers, RJ Morris, HW Adolph, A Merli, VS Lamzin and ES Cedergren-Zeppezauer, *J Biol Chem*, **276**, 9316–21 (2001).
7. I Bertini, C Luchinat, W Maret and M Zeppezauer (eds.), *Progress in Inorganic Biochemistry and Biophysics*, Vol. 1, Zinc Enzymes, Birkhäuser, Boston-Basel-Stuttgart (1986).
8. ET Young and D Pilgrim, *Mol Cell Biol*, **5**, 3024–34 (1985).
9. JR Bright, D Byrom, MJ Danson, DW Hough and P Towner, *Eur J Biochem*, **211**, 549–54 (1993).
10. S Ammendola, CA Raia, C Caruso, L Camardella, S D'Auria, M De Rosa and M Rossi, *Biochemistry*, **31**, 12514–23 (1992).
11. CA Raia, A Giordano and M Rossi, Methods in enzymology, in MWW Adams and RM Kelly (eds.), *Hyperthermophilic Enzymes*, Vol. 331, Part B, Academic Press, San Diego, London, Boston, New York, Sidney, Tokyo, Toronto, pp 176–95 (2001).
12. Y Korkhin, F Frolow, O Bogin, M Peretz, AJ Kalb (Gilboa) and Y Burstein, *Acta Crystallogr, Sect D*, **52**, 882–86 (1996).
13. I Tsigos, K Velonia, I Smonou and V Bouriotis, *Eur J Biochem*, **254**, 356–62 (1998).
14. A Guagliardi, M Martino, I Iaccarino, M De Rosa, M Rossi and S Bartolucci, *Int J Biochem Cell Biol*, **28**, 239–46 (1996).
15. S D'Auria, F La Cara, F Nazzaro, N Vespa and M Rossi, *J Biochem (Tokyo)*, **120**, 498–506 (1996).
16. J Shafqat, JO Höög, L Hjelmqvist, UCT Oppermann, C Ibánez and H Jörnvall, *Eur J Biochem*, **263**, 305–11 (1999).
17. O Danielsson, S Atrian, T Luque, L Hjelmqvist, R Gonzáles-Duarte and H Jörnvall, *Proc Natl Acad Sci USA*, **91**, 4980–84 (1994).
18. ADH Working group, http://mbclserver.rutgers.edu/CPGN/AdhWeb/Adh.group.html.
19. H Jörnvall, JO Höög, B Persson and X Parés, *Pharmacology*, **61**, 184–91 (2000).
20. W Maret and DS Auld, *Biochemistry*, **27**, 1622–28 (1988).
21. CS Chen and A Yoshida, *Biochem Biophys Res Commun*, **181**, 743–47 (1991).
22. L Uotila and M Koivusalo, in D Dolphin, R Poulson and O Avramovic (eds.), *Coenzymes and Cofactors, Glutathione*.

23 PC Sanghani, CL Stone, BD Ray, EV Pindel, TD Hurley and WF Bosron, *Biochemistry*, **39**, 10720–29 (2000).
24 L Liu, A Hausladen, M Zeng, L Que, J Heitman and JS Stamler, *Nature*, **410**, 490–94 (2001).
25 L Deltour, MH Foglio and G Duester, *J Biol Chem*, **274**, 16796–801 (1999).
26 R Cannio, G Fiorentino, M Rossi and S Bartolucci, *FEMS Microbiol*, **170**, 31–39 (1999).
27 JS Chen, *FEMS Microbiol Rev*, **17**, 263–73 (1995).
28 MJ Danson, Central metabolism of the Archaea, in M Kates, DJ Kushner and AT Matheson (eds.), *Biochemistry of Archaea (Archaebacteria)*, Elsevier, The Netherlands, pp 1–21 (1993).
29 MJ Banfield, ME Salvucci, EN Baker and CA Smith, *J Mol Biol*, **306**, 239–50 (2001).
30 J Jeffery and H Jörnvall, *Adv Enzymol*, **61**, 47–106 (1988).
31 GR Wolfe, DL Hendrix and ME Salvucci, *J Insect Physiol*, **44**, 597–603 (1998).
32 GR Wolfe, CA Smith, DL Hendrix and ME Salvucci, *Insect Biochem Mol Biol*, **29**, 113–20 (1999).
33 JL Napoli, *Biochem Biophys Acta*, **1440**, 139–62 (1999).
34 CF Chou, CL Lai, YC Chang, G Duester and SJ Yin, *J Biol Chem*, **275**, 25209–16 (2002).
35 G Duester, *Alcohol Clin Exp Res*, **15**, 568–72 (1991).
36 PS Haber, T Gentry, KM Mak, AA Mirmiran-Yazdy, RJ Greenstein and CS Lieber, *Gastroenterology*, **111**, 863–70 (1996).
37 H Jörnvall, *Nature (London)*, **225**, 1133–40 (1970).
38 DH Park and BV Plapp, *J Biol Chem*, **26**, 296–302 (1991).
39 I Hubatsch, M Zeppezauer, D Waidelich and E Bayer, in H Weiner (ed.), *Enzymology and Molecular Biology of Carbonyl Metabolism*, Vol. 4, Plenum Press, New York, pp 451–55 (1993).
40 M Yasunami, I Kikuchi, D Sarapata and A Yoshida, *Genomics*, **7**, 152–58 (1990).
41 GJ Davis, WF Bosron, CL Stone, K Owusu-Dekyi and TD Hurley, *J Biol Chem*, **271**, 17057–61 (1996).
42 MS Niederhut, BJ Gibbons, S Perez-Miller and TD Hurley, *Protein Sci*, **10**, 697–706 (2001).
43 JO Höög, H von Bahr-Lindström, LO Heden, B Holmquist, K Larsson, J Hempel, BL Vallee and H Jörnvall, *Biochemistry*, **26**, 1926–32 (1987).
44 S Svensson, P Strömberg and JO Höög, *J Biol Chem*, **274**, 29712–19 (1999).
45 ZN Yang, WF Bosron and TD Hurley, *J Mol Biol*, **265**, 330–43 (1997).
46 P Xie, SH Parsons, DC Speckhard, WF Bosron and TD Hurley, *J Biol Chem*, **272**, 18558–63 (1997).
47 LM Peralba, E Cederlund, B Crosas, A Moreno, P Julià, SE Martinez, B Persson, J Farrés, X Parés and H Jörnvall, *J Biol Chem*, **274**, 26021–26 (1999).
48 JL Bennetzen and BD Hall, *J Biol Chem*, **81**, 4125–28 (1982).
49 DW Russell, M Smith, VM Williamson and ET Young, *J Biol Chem*, **258**, 2674–82 (1983).
50 A Kumar, P Shen, S Descoteaux, J Pohl, G Bailey and J Samuelson, *Proc Natl Acad Sci USA*, **89**, 10188–92 (1992).
51 DJ Gillooly, AGS Robertson and CA Fewson, *Biochem J*, **330**, 1375–81 (1998).
52 M Peretz, O Bogin, S Tel-Or, A Cohen, G Li, JS Chen and Y Burstein, *Anaerobe*, **3**, 259–70 (1997).
53 M Peretz and Y Burstein, *Biochemistry*, **28**, 6549–55 (1989).
54 R Cannio, G Fiorentino, P Carpinelli, M Rossi and S Bartolucci, *J Bacteriol*, **178**, 301–5 (1996).
55 C Karlsson, W Maret, DS Auld, JO Höög and H Jörnvall, *Eur J Biochem*, **186**, 543–50 (1989).
56 C Karlsson, H Jörnvall and JO Höög, *Eur J Biochem*, **198**, 761–65 (1989).
57 J John, SJ Crennell, DW Hough, MJ Danson and GL Taylor, *Structure*, **2**, 385–93 (1994).
58 L Esposito, F Sica, CA Raia, A Giordano, M Rossi, L Mazzarella and A Zagari, *J Mol Biol*, **318**, 463–77 (2002).
59 MF Reid and CA Fewson, *Crit Rev Microbiol*, **20**, 13–56 (1994).
60 HS Lo and RE Reeves, *Biochem J*, **171**, 225–30 (1978).
61 RJ Lamed and JG Zeikus, *Biochem J*, **195**, 183–90 (1981).
62 J Hempel, R Bühler, R Keiser, B Holmquist, C deZalenski, JP von Wartburg, BL Vallee and H Jörnvall, *Eur J Biochem*, **145**, 437–45 (1984).
63 HJ Edenberg, *Prog Nucleic Acid Res Mol Biol*, **64**, 295–314 (2000).
64 G Duester, J Farrés, MR Felder, RS Holmes, JO Höög, X Parés, BV Plapp, SJ Yin and H Jörnvall, *Biochem Pharm*, **58**, 389–95 (1999).
65 JD Thompson, DG Higgins and TJ Gibson, *Nucleic Acids Res*, **22**, 4673–80 (1994).
66 H Eklund, B Nordström, E Zeppezauer, G Söderlund, I Ohlsson, T Boiwe, BO Söderberg, O Tapia, CI Brändén and Å Åkesson, *J Mol Biol*, **102**, 27–59 (1976).
67 H Eklund, JP Samama, L Wallén, CI Brändén, Å Åkesson and AT Jones, *J Mol Biol*, **146**, 561–87 (1981).
68 S Al-Karadaghi, ES Cedergren-Zeppezauer, S Hovmöller, K Petratos, H Terry and KS Wilson, *Acta Crystallogr*, **D50**, 793–807 (1994).
69 S Al-Karadaghi, ES Cedergren-Zeppezauer, Z Dauter and KS Wilson, *Acta Crystallogr*, **D51**, 805–13 (1995).
70 S Ramaswamy, H Eklund and BV Plapp, *Biochemistry*, **33**, 5230–37 (1994).
71 H Cho, S Ramaswamy and BV Plapp, *Biochemistry*, **36**, 382–89 (1997).
72 S Ramaswamy, M Scholze and BV Plapp, *Biochemistry*, **36**, 3522–27 (1997).
73 JK Rubach and BV Plapp, *Biochemistry*, **41**, 15770–79 (2002).
74 BJ Bahnson, TD Colby, JK Chin, BM Goldstein and JP Klinman, *Proc Natl Acad Sci USA*, **94**, 12797–802 (1997).
75 TD Colby, BJ Bahnson, JK Chin, JP Klinman and BM Goldstein, *Biochemistry*, **37**, 9295–304 (1998).
76 S Ramaswamy, DH Park and BV Plapp, *Biochemistry*, **38**, 13951–59 (1999).
77 HW Adolph, P Zwart, R Meijers, I Hubatsch, M Kiefer, V Lamzin and ES Cedergren-Zeppezauer, *Biochemistry*, **39**, 12885–97 (2000).
78 H Li, WH Hallows, JS Punzi, VE Marquez, HL Carrell, KW Pankiewicz, KA Watanabe and BM Goldstein, *Biochemistry*, **33**, 23–32 (1994).
79 H Li, WH Hallows, JS Punzi, KW Pankiewicz, KA Watanabe and BM Goldstein, *Biochemistry*, **33**, 11734–44 (1994).
80 JK Rubach, S Ramaswamy and BV Plapp, *Biochemistry*, **40**, 12686–94 (2001).
81 S Ramaswamy, ME Ahmad, O Danielsson, H Jörnvall and H Eklund, *Protein Sci*, **5**, 663–71 (1996).
82 TD Hurley, WF Bosron, CL Stone and LM Amzel, *J Mol Biol*, **239**, 415–29 (1994).
83 S Svensson, JO Höög, G Schneider and T Sandalova, *J Biol Chem*, **302**, 441–58 (2000).

84 PC Sanghani, H Robinson, R Bennett-Lovsey, TD Hurley and WF Bosron, *Chem Biol Interact*, **143-144**, 195-200 (2003).

85 PC Sanghani, H Robinson, WF Bosron and TD Hurley, *Biochemistry*, **41**, 10778-86 (2002).

86 PC Sanghani, WF Bosron and TD Hurley, *Biochemistry*, **41**, 15189-94 (2002).

87 PT Xie and TD Hurley, *Protein Sci*, **8**, 2639-44 (1999).

88 N Tanaka, Y Kusakabe, K Ito, T Yoshimoto and KT Nakamura, *Chem Biol Interact*, **143-144**, 211-18 (2003).

89 Y Korkhin, AJ Kalb (Gilboa), M Peretz, O Bogin, Y Burstein and F Frolow, *J Mol Biol*, **278**, 967-981 (1998).

90 O Bogin, I Levin, Y Hacham, S Tel-Or, M Peretz, F Frolow and Y Burstein, *Protein Sci*, **11**, 2561-74 (2002).

91 K Johansson, M El-Ahmad, C Kaiser, H Jörnvall, H Eklund, JO Höög and S Ramaswamy, *Chem Biol Interact*, **130-132**, 351-58 (2001).

92 A Karlsson, M El-Ahmad, K Johansson, J Shafqat, H Jörnvall, H Eklund and S Ramaswamy, *Chem Biol Interact*, **143-144**, 239-45 (2003).

93 RK Bonnichsen and AM Wassén, *Arch Biochem Biophys*, **18**, 361-63 (1948).

94 R Pietruszko, HJ Ringold, TK Li, BL Valee, Å Åkesson and H Theorell, *Nature (London)*, **221**, 440-43 (1969).

95 R Pietruszko and H Theorell, *Arch Biochem Biophys*, **131**, 288-98 (1969).

96 RT Dworschack and BV Plapp, *Biochemistry*, **16**, 111-16 (1977).

97 UM Lutstorf, PM Schürch and JP von Wartburg, *Eur J Biochem*, **17**, 497-508 (1970).

98 L Andersson, H Jörnvall, Å Åkesson and K Mosbach, *Biochim Biophys Acta*, **364**, 1-8 (1974).

99 L Andersson, H Jörnvall and K Mosbach, *Anal Biochem*, **69**, 401-9 (1975).

100 I Hubatsch, P Maurer, D Engel and HW Adolph, *J Chromatogr, A*, **17**, 105-12 (1995).

101 LD Smith, N Budgen, SJ Bungard, MJ Danson and DW Hough, *Biochem J*, **261**, 973-77 (1989).

102 C Pire, J Esclapez, J Ferrer and MJ Bonete, *FEMS Microbiol Lett*, **200**, 221-27 (2001).

103 ME Salvucci, GR Wolfe and DL Hendrix, *Insect Biochem Mol Biol*, **28**, 357-63 (1998).

104 N Cabrera, P Rangel, R Hernández-Munoz and R Pérez-Montfort, *Protein Expr Purif*, **10**, 340-44 (1997).

105 LG Lange and BL Vallee, *Biochemistry*, **15**, 4681-86 (1976).

106 CC Ditlow, B Holmquist, M Morelock and BL Vallee, *Biochemistry*, **23**, 6363-68 (1984).

107 TD Hurley, HJ Edenberg and WF Bosron, *J Biol Chem*, **265**, 16366-72 (1990).

108 S Svensson, A Lundsjö, T Cronholm and JO Höög, *FEBS Lett*, **394**, 217-20 (1996).

109 L Hjelmqvist, M Hackett, J Shafqat, O Danielsson, J Iida, RC Hendrickson, H Michel, J Shabanowitz, DF Hunt and H Jörnvall, *FEBS Lett*, **367**, 237-40 (1995).

110 CA Raia, S D'Auria and M Rossi, *Biocatalysis*, **11**, 143-50 (1994).

111 FA Loewus, FH Westheimer and B Wennesland, *J Am Chem Soc*, **75**, 5018-23 (1953).

112 H Theorell and R Bonnichsen, *Acta Chem Scand*, **5**, 1105-26 (1951).

113 K Dalziel, *Acta Chem Scand*, **11**, 397-98 (1957).

114 E Bignetti, GL Rossi and ES Zeppezauer, *FEBS Lett*, **100**, 17-22 (1979).

115 R Bonnichsen, *Acta Chem Scand*, **4**, 715-21 (1950).

116 HW Adolph, M Kiefer and ES Cedergren-Zeppezauer, *Biochemistry*, **36**, 8743-54 (1997).

117 H Theorell and J McKinley-McKee, *Acta Chem Scand*, **15**, 1797-1810 (1961).

118 R Einarsson, L Widell and M Zeppezauer, *Anal Lett*, **9**, 815-23 (1976).

119 H Theorell and T Yonetani, *Biochem Z*, **338**, 537-53 (1963).

120 V Leskovac, S Trivic' and BM Anderson, *Eur J Biochem*, **264**, 840-47 (1999).

121 JH Hayes and SF Velick, *J Biol Chem*, **207**, 225-32 (1954).

122 V Leskovac, S Trivic' and M Pantelic', *Anal Biochem*, **214**, 431-34 (1993).

123 BV Plapp, *J Biol Chem*, **245**, 1727-35 (1970).

124 NY Kedishivili, WH Gough, EAG Chernoff, TD Hurley, CL Stone, KD Bowman, KM Popov, WF Bosron and YK Li, *J Biol Chem*, **272**, 7494-500 (1997).

125 SJ Svensson, Ph.D. Thesis, Karolinska Institutet, Stockholm, ISBN 91-628-3865-2 (1999).

126 GTM Henehan and MJ Oppenheimer, *Biochemistry*, **32**, 735-38 (1993).

127 GL Shearer, K Kim, KM Lee, CK Wang and BV Plapp, *Biochemistry*, **32**, 11186-94 (1993).

128 R Pietruszko and H Theorell, *Arch Biochem Biophys*, **131**, 288-98 (1969).

129 O Gabriel, in WB Jacoby (ed.), *Methods in Enzymology*, Vol. 22, Academic Press, New York, London, pp 578-604 (1971).

130 GL Gilliland, M Tung, DM Blakeslee and J Ladner, *Acta Crystallogr*, **D50**, 408-13 (1994).

131 D Frishman and P Argos, *Proteins*, **23**, 566-79 (1995).

132 D Gilbert, D Westhead, J Viksna and J Thornton, *Comput Chem*, **1**, 23-30 (2001).

133 MG Rossmann, D Moras and KW Olsen, *Nature (London)*, **250**, 194-97 (1974).

134 I Ohlsson, B Nordström and CI Brändén, *J Mol Biol*, **89**, 339-45 (1974).

135 SCOP, a data base that gives a fold description for each molecule stored in the PDB, http://scop.mrc-lmb.cam.ac.uk/scop/data/scop.b.c.eb.b.c.b.html.

136 F Colonna-Cesari, D Perahia, M Karplus, H Eklund, CI Brändén and O Tapia, *J Biol Chem*, **261**, 15273-80 (1986).

137 O Tapia, H Eklund and CI Brändén, Chap. 8, in G Náray-Szabó and K Simon (eds.), *Steric Aspects of Biomolecular Interactions*, CRC Press, Boca Raton, pp 159-80 (1986).

138 VC Sekhar and BV Plapp, *Biochemistry*, **27**, 5082-88 (1988).

139 VC Sekhar and BV Plapp, *Biochemistry*, **29**, 4289-95 (1990).

140 MW Makinen and MB Yim, *Proc Natl Acad Sci USA*, **78**, 6221-25 (1981).

141 ES Cedergren-Zeppezauer, *Biochemistry*, **22**, 5761-72 (1983).

142 ES Cedergren-Zeppezauer, I Andersson, S Ottonello and E Bignetti, *Biochemistry*, **24**, 4000-10 (1985).

143 ES Cedergren-Zeppezauer, JP Samama and H Eklund, *Biochemistry*, **21**, 4895-908 (1982).

144 H Dietrich, AKH MacGibbon, MF Dunn and M Zeppezauer, *Biochemistry*, **22**, 3432-38 (1983).

145 G Schneider, H Eklund, ES Cedergren-Zeppezauer and M Zeppezauer, *EMBO J*, **2**, 685-89 (1983).

146 E Johansson, N Aghajari, E Hjalmarsson, HW Adolph, Z Dauter and ES Cedergren-Zeppezauer, *Biochemistry* (2003); in preparation.
147 CI Brändén, in H Sund (ed.), *Pyridine Nucleotide-Dependent Dehydrogenases*, Walter de Gruyter & Co, pp 325–38 (1977).
148 MW Makinen, W Maret and MB Yim, *Proc Natl Acad Sci USA*, **80**, 2584–88 (1983).
149 R Bauer, HW Adolph, I Andersson, E Danielsson, G Formicka and M Zeppezauer, *Eur Biophys J*, **14**, 431–39 (1991).
150 L Hemmingsen, R Bauer, MJ Bjerrum, M Zeppezauer, HW Adolph, G Formicka and ES Cedergren-Zeppezauer, *Biochemistry*, **34**, 7145–53 (1995).
151 U Ryde and L Hemmingsen, *J Biol Inorg Chem*, **2**, 567–79 (1997).
152 MF Dunn, H Dietrich, AKH McGibbon, SC Koerber and M Zeppezauer, *Biochemistry*, **21**, 354–63 (1982).
153 M Gerber, M Zeppezauer and MF Dunn, *Inorg Chim Acta*, **79**, 161–64 (1983).
154 W Maret and M Zeppezauer, *Biochemistry*, **25**, 1584–88 (1986).
155 W Maret, I Andersson, H Dietrich, H Schneider-Bernlöhr, R Einarsson and M Zeppezauer, *Eur J Biochem*, **98**, 501–19 (1979).
156 C Sartorius, M Gerber, M Zeppezauer and MF Dunn, *Biochemistry*, **26**, 871–82 (1987).
157 W Maret and MW Makinen, *J Biol Chem*, **266**, 20636–44 (1991).
158 HW Adolph, P Maurer, H Bernlöhr-Schneider, C Sartorius and M Zeppezauer, *Eur J Biochem*, **201**, 615–25 (1991).
159 W Maret, *Biochemistry*, **28**, 9944–49 (1989).
160 M Zeppezauer, I Andersson, H Dietrich, M Gerber, W Maret, G Schneider and H Bernlöhr-Schneider, *J Mol Catal*, **23**, 377–87 (1984).
161 R Meijers, Ph.D. Thesis, Amsterdam University, Amsterdam (2001).
162 U Ryde, *Protein Sci*, **4**, 1124–32 (1995).
163 AJ Ganzhorn and BV Plapp, *J Biol Chem*, **263**, 5446–54 (1988).
164 O Kleifeld, SP Shi, R Zarivach, M Eisenstein and I Sagi, *Protein Sci*, **12**, 468–79 (2003a).
165 O Kleifeld, A Frenkel, JML Martin and I Sagi, *Nat Struct Biol*, **10**, 98–103 (2003b).
166 PF Cook and BL Bertagnolli, in D Dolphin, R Poulson and O Abramivic (eds.), *Pyridine Nucleotide Coenzymes*, Part A, John Wiley, New York, pp 405–47 (1987).
167 G Pettersson, *CRC Crit Rev Biochem*, **21**, 349–83 (1987).
168 AR Clarke and TR Daffron, in M Sinnott (ed.), *Comprehensive Biological Catalysis*, Vol. III, Academic Press, London, pp 1–78 (1998).
169 K Dalziel, in PD Boyer (ed.), *The Enzymes*, Vol. XI, Academic Press, New York, London, pp 1–60 (1975).
170 NG Leissing and ET McGuinness, *Int J Biochem*, **15**, 651–56 (1983).
171 P Guardina, MG Debiasi, M Derosa, A Gambacorta and V Buonocore, *Biochem J*, **236**, 517–22 (1986).
172 K Engeland, JO Höög, B Holmquist, M Estonius, H Jörnvall and BL Vallee, *Proc Natl Acad Sci USA*, **99**, 2491–94 (1993).
173 BV Plapp, JL Mitchell and KB Berst, *Chem Biol Interact*, **130-132**, 445–56 (2001).
174 SJ Yin, CF Chou, CL Lai, SL Lee and CL Han, *Chem Biol Interact*, **143-144**, 219–27 (2003).
175 T Ehrig, TD Hurley, HJ Edenberg and WF Bosron, *Biochemistry*, **30**, 1062–68 (1991).
176 CL Stone, MB Jipping, K Owusu-Dekyi, TD Hurley, TK Li and WF Bosron, *Biochemistry*, **38**, 5829–35 (1999).
177 CL Stone, WF Bosron and MF Dunn, *J Biol Chem*, **268**, 892–99 (1993).
178 WF Bosron, SJ Yin and TK Li, *Enzymology of Carbonyl Metabolism 2: Aldehyde Dehydrogenase, Aldo-Keto Reductase and Alcohol Dehydrogenase*, Alan R. Liss, Inc, New York, pp 193–206 (1985).
179 A Dubied, JP von Wartburg, DP Bolken and BV Plapp, *J Biol Chem*, **252**, 1464–70 (1977).
180 AJ McEvily, B Holmquist, DS Auld and BL Vallee, *Biochemistry*, **27**, 4284–88 (1988).
181 LA LeBrun and BV Plapp, *Biochemistry*, **38**, 12387–93 (1999).
182 EG Oestereicher, DA Pereira and BF Pinto, *J Biotechnol*, **46**, 23–31 (1996).
183 S Sellin, B Holmquist, B Mannervik and BL Vallee, *Biochemistry*, **30**, 2514–18 (1991).
184 NJ Oppenheimer, GTM Henehan, JA Heute-Perez and K Ito, in H Weiner (ed.), *Enzymology and Molecular Biology of Carbonyl Metabolism*, Vol. 6, Plenum Press, New York, pp 417–23 (1996).
185 G Duester, *Eur J Biochem*, **267**, 4315–24 (2000).
186 C Larroy, MR Fernández, E González, X Parés and JA Biosca, *Chem Biol Interact*, **143-144**, 229–38 (2003).
187 B Nidetzky, H Helmer, M Klimacek, R Lunzer and G Mayer, *Chem Biol Interact*, **143-144**, 533–42 (2003).
188 H Jörnvall, J Shafqat, M El-Ahmed, L Hjelmqvist, B Persson and O Danielsson, *Adv Exp Med Biol*, **414**, 281–89 (1996).
189 JB Jones and JF Beck, in J Jones, X Sih and M Perlman (eds.), *Techniques of Chemistry*, Vol. X, Part 1, Elsevier, pp 236–401 and Appendix B, pp 495–505 (1976).
190 AJ Irwin and JB Jones, *J Am Chem Soc*, **99**, 556–61 (1977).
191 IJ Jakovac, HB Goodbrand, KP Lok and JB Jones, *J Am Chem Soc*, **104**, 4659–65 (1982).
192 JA Haselgrave and JB Jones, *J Am Chem Soc*, **104**, 4666–71 (1982).
193 AJ Bridges, PS Raman, GSY Ng and JB Jones, *J Am Chem Soc*, **106**, 1461–67 (1984).
194 B Orlich, H Berger, M Lade and R Schomäcker, *Biotechnol Bioeng*, **70**, 638–46 (2000).
195 H Theorell, A Ehrenberg and C deZalinski, *Biochem Biophys Res Commun*, **27**, 304–14 (1967).
196 MC Detraglia, J Schmidt, MF Dunn and JT McFarland, *J Biol Chem*, **252**, 3493–500 (1977).
197 JH Coates, MJ Hardman, JD Shore and H Gutfreund, *FEBS Lett*, **84**, 25–28 (1977).
198 MA Greeves and AL Fink, *J Biol Chem*, **255**, 3248–50 (1980).
199 J Kovár and H Klukanova, *Biochim Biophys Acta*, **788**, 98–109 (1984).
200 O Almarsson and TC Bruice, *J Am Chem Soc*, **115**, 2125–38 (1993).
201 JR Burke and PA Frey, *Biochemistry*, **32**, 13220–30 (1993).
202 A Schmidt and VS Lamzin, *Curr Opin Struct Biol*, **12**, 698–703 (2002).
203 GM Sheldrick and TR Schneider, Methods in enzymology, in CW Carter and RM Sweet (eds.), *Macromolecular Crystallography*, Vol. 276, Academic Press, San Diego, London, Boston, New York, Sidney, Tokyo, Toronto, pp 319–43 (1997).
204 S Hammes-Schiffer, *Biochemistry*, **41**, 13335–43 (2002).
205 JE Coleman, *Curr Opin Chem Biol*, **2**, 222–34 (1998).

Transferases

Protein prenyltransferases

Hong Zhang
University of Texas, Southwestern Medical Center at Dallas, TX, USA

FUNCTIONAL CLASS

Enzyme; EC 2.5.1.; zinc metalloenzyme; Protein-cysteine-S:farnesyl diphosphate farnesyltransferase or protein-cysteine-S:geranylgeranyl diphosphate geranylgeranyltransferase.

Protein prenyltransferases (PPT) catalyze the formation of the thioether linkage between the C1 atom of the farnesyl or geranylgeranyl isoprenoid lipid and the cysteine residue at or near the carboxy-terminus of a protein acceptor. The protein prenyltransferase family contains three members: protein farnesyltransferase (PFT), protein geranylgeranyltransferase type I (PGGT-I), and protein geranylgeranyltransferase type II (PGGT-II).[1-3] PFT and PGGT-I are closely related and are collectively termed the CaaX protein prenyltransferases because they recognize a CaaX motif (C, cysteine; a, aliphatic amino acid; X, serine/methionine/alanine for PFT and leucine for PGGT-I) at the C-terminus of their protein substrate. The PGGT-II is also named RabGGT because it exclusively modifies the Rab family of small GTP-binding proteins that regulate intracellular vesicular trafficking. Distinct from the CaaX prenyltransferases, RabGGT catalyzes the transfer of two isoprenoid groups to the two cysteine residues of the C-terminus prenylation motif (-CC, -CCX, -CXC, or -CCXX) of Rab proteins.[4]

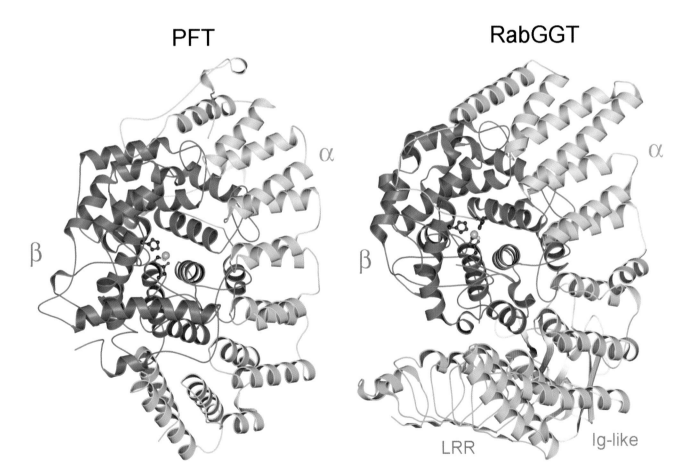

3D Structure Ribbon diagrams of rat protein farnesyltransferase (PFT, PDB code 1FT1) and rat Rab geranylgeranyltransferase (RabGGT, PDB code 1DCE). The PPTα repeat domain is colored cyan, the Ig-like domain yellow, the LRR domain green, and the β-subunit magenta. The Zn ion and the protein ligands are in ball-and-stick representation. The figures were prepared with MOLSCRIPT[5] and Gl_render (L. Esser, unpublished program).

OCCURRENCE

Protein prenyltransferases exist in most, if not all, eukaryotic organisms including animals, plants, yeast, fungi, and protozoan parasites, but not in prokaryotes.

BIOLOGICAL FUNCTION

Protein prenylation is an important posttranslational modification that involves the covalent attachment of either a 15-carbon farnesyl or 20-carbon geranylgeranyl isoprenoid to a conserved cysteine(s) of many cellular proteins. The protein substrates for PFT include Ras, yeast and fungi mating factors, nuclear lamin B, and several proteins involved in visual signal transduction such as transducin γ-subunit and rhodopsin kinase.[1,3,6,7] Known substrates for PGGT-I include most Ras-like small GTP-binding proteins such as Rho, Rac, as well as most γ-subunits of heterotrimeric G-proteins.[1,7] PGGT-II attaches geranylgeranyl groups to two C-terminal cysteines in a single family of Ras-related small GTPases, the Rab proteins (Ypt/Sec4 in yeast) that participate in the regulation of intracellular vesicle transporting.[3,8] Prenylated proteins are involved in a wide variety of cellular signaling activities and regulatory events. Attachment of the lipid promotes membrane association of these proteins and in some cases appears to mediate protein–protein interactions and is thus essential for the proper functioning of the modified proteins.[9,10] Inhibitors of PFT cause tumor regression in animals and are currently being evaluated in clinical trials for the treatment of human cancer.[11,12]

AMINO ACID SEQUENCE INFORMATION

Protein prenyltransferases (PFT, PGGT-I, and PGGT-II) are heterodimers composed of tightly associated α- and β-subunits. The α-subunits of PFT and PGGT-I are exchangeable,[13] but their β-subunits are distinct, sharing only 30% sequence identity. All three PPTs have been experimentally characterized in mammals and in yeast. PFTs from plants and *Trypanosomes* have also been cloned and their enzymatic properties have been analyzed. The PPTs clearly exist in *Caenorhabditis elegans*, *Drosophila melanogaster*, and in other eukaryotes whose complete genomes are sequenced, although their properties remain to be characterized experimentally. Crystal structures have been reported for the PFTs from rat[14–16] and human,[17] and for RabGGT from rat.[18] Rat PFT contains a 377–amino acid α-subunit (SWISS_PROT entry name PFTA_RAT) and a 437–amino acid β-subunit (PFTB_RAT). They share 97% sequence identity with the human enzyme (PFTA_HUMAN and PFTB_HUMAN). Rat RabGGT contains a 567–amino acid α-subunit (PGTA_RAT) and a 331–amino acid β-subunit (PGTB_RAT). The RabGGT α-subunit from higher eukaryotes is considerably larger than the one from PFT and GGT-I. In addition to the common PPTA repeat domain, it contains two more domains, namely, an immunoglobulin-like (Ig-like) domain and a leucine-rich repeat (LRR) domain (Figure 1).[18,19] The β-subunits of protein prenyltransferases are also composed of repeats termed PPTB repeat, which are divergent representatives of the QW motif-containing repeats found in squalene–hopene cyclases.[19,20] Yeast PGGT-II, encoded by genes *BET4* (α) and *BET2* (β), has a much smaller α-subunit (BET4_YEAST, 290 amino acids) and lacks the Ig-like and the LRR domain. On the other hand, *Trypanosoma cruzi* PFT has much larger α- and β-subunits (628 and 588 amino acids respectively), and their sequences (gi|18448721 and gi|18448723 respectively, in the nonredundant protein sequence database at NCBI, NIH, Bethesda, MD, USA) contain many insertions between the repeating units.[21]

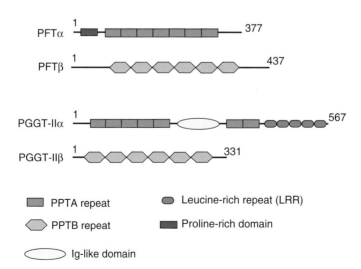

Figure 1 Domain organization of mammalian PFT and RabGGT.

PROTEIN PRODUCTION, PURIFICATION, AND MOLECULAR CHARACTERIZATION

Recombinant rat PFT was produced in both baculovirus/insect cell and *Escherichia coli* expression systems.[22–25] PFT expressed in the SF9 or Hi5 insect cells was purified by DEAE-Sephacel, Q-Sepharose HP and phenyl-Sepharose columns.[22,23] The *E. coli* expression of the heterodimeric PFT was achieved using a translationally coupled operon from the T7 promoter of the pET23a (Novagen) expression plasmid.[25] Purification of the *E. coli* expressed PFT is based on the procedure for the purification of Sf9 cell–expressed PFT with minor modifications.[25]

Properly folded RabGGT has also been expressed in both baculovirus/insect cell[26,27] and *E. coli* expression system.[28] The purification of the insect cell–expressed native RabGGT consists of Q-sepharose HP anion exchange and Superdex 200 size exclusion chromatography.[26] In order to obtain good diffraction-quality crystals, additional purification steps (Mono Q and phenyl-sepharose hydrophobic interaction chromatography) may be required.[18] The *E. coli* expression system contains two vectors harboring the α- and β-subunit of RabGGT respectively. A tandem 6xHis-GST tag and a tobacco etch virus (TEV) protease cleavage site were introduced at the N-terminus of the α-subunit, and both vectors were transformed into the *E. coli* expression strain. The expressed protein was purified by Ni-sepharose affinity chromatography. The 6His-GST tag was removed by TEV protease digestion to produce the native protein.[28]

METAL CONTENT

PPTs contain one Zn ion per heterodimer as was determined by the atomic absorption spectroscopy for PFT.[22] The presence of a single Zn ion in RabGGT was confirmed by the characteristic X-ray absorption at Zn K edge (1.2837 Å) from a RabGGT crystal and by the anomalous difference electron density peak (H. Zhang, unpublished observations). Neither PFT nor RabGGT binds any cofactors. The role of the Zn ion has been shown to be catalytical rather than structural.[29–31] Mg^{2+} is also required for optimal activity of PFT and RabGGT.[26,29] The optimal concentration of the Mg^{2+} required is high (∼5 mM), which indicates that Mg^{2+} is not directly bound to the enzyme. It was later shown that Mg^{2+} is likely to coordinate with the diphosphate group of farnesyl diphosphate (FPP) and may participate in the catalysis.[16]

ACTIVITY TESTS

Isotope-based assay

PPT activities were determined by quantifying the amount of ^3H transferred from [^3H]farnesyl diphosphate (FPP) or [^3H]geranylgeranyl diphosphate (GGPP) to the appropriate acceptor protein (CaaX peptide, Ras, Rab, etc.).[27,32] The ^3H labeled prenylated product is separated from the free form [^3H]FPP or [^3H]GGPP by acid precipitation and filtration through glass-fiber filters, after which the filter-bound reaction product is quantified in a scintillation counter.[27] The standard reaction mixture contains the following components in a final volume of 50 μl: 50 mM Tris-HCl, pH 7.7, 5 mM $MgCl_2$, 5 mM $ZnCl_2$, 20 mM KCl, 2 mM DTT (dithiothreitol), either 2 μM of [^3H]GGPP or [^3H]FPP (both typically at 3000 disintegrations/min (dpm/pmol)), appropriate protein or peptide substrate, and appropriate prenyltransferase. In the RabGGT assay mixture, 1 mM NP-40 and 30 ng Rab escort proteins (REP) are also needed. There have been various modifications of this assay,[33,34] and it can also be performed in a combinatorial manner.[35]

Continuous fluorescence assay for PFT and PGGT-I

Dansylated CaaX peptide, for example, dansyl-GCVLS, has a fluorescence emission maximum at 550 nm, which is shifted to 505 nm and enhanced 13-fold on farnesylation.[36] Thus, monitoring the change in fluorescence at 505 nm provides a method for observing the CaaX prenyltransferase reaction continuously. Dansylated peptides can be prepared by standard solid-phase synthesis methods on a peptide synthesizer.

HPLC-based assay for RabGGT

An HPLC-based prenylation assay that takes advantage of the different mobilities of un-, mono-, and doubly prenylated Rab on a C4-reversed phase column has been developed.[37] The reaction mixture, similar to that in the isotope-based assay, was quenched by adding 0.3% trifluoroacetic acid at defined time intervals and was applied to a C4 column. By using a semisynthetic dansylated Rab substrate generated by a protein ligation method,[38] the elution of Rab can also be monitored by its more sensitive fluorescence property even when its elution partially overlaps with other proteins.[37]

X-RAY STRUCTURES

Crystallization

Rat PFT crystals were grown at 17 °C from hanging drops containing a mixture of protein solution (4–16 mg mL^{-1} PFT in 20 mM KCl, 10 μM $ZnCl_2$, 10 mM DTT, 20 mM Tris-HCl, pH 7.7) and reservoir solution (13–15% PEG8000, 200 mM ammonium acetate, pH 7.0), equilibrated against the reservoir by vapor diffusion. The PFT crystals are hexagonal and belong to the space group

Protein prenyltransferases

Table 1 List of crystal structures of protein prenyltransferases

Protein	Species	Resolution (Å)	Ligands	S.G.[a]	Cell dimensions (Å)	PDB code	References
PFT	Rat	2.25	apo	$P6_5$	$a = b = 167.06, c = 97.90$	1FT1	14
	Rat	3.4	FPP	$P6_5$	$a = b = 166.66, c = 98.82$	1FT2	40
	Rat	2.75	FPP	$P6_1$	$a = b = 170.00, c = 68.86$	1FPP	41
	Rat	2.4	CIVM tetrapeptide + HFP[b]	$P6_1$	$a = b = 174.13, c = 69.71$	1QBQ	15
	Rat	2.0	11-mer K-Ras peptide + FII[c]	$P6_1$	$a = b = 170.92, c = 69.28$	1D8D	16
	Rat	3.0	−Zn, K-Ras peptide + FII[c]	$P6_1$	$a = b = 171.41, c = 69.66$	1D8E	16
	Rat	2.0	CVFM tetrapeptide + FPP	$P6_1$	$a = b = 171.23, c = 69.33$	1JCR	17
	Rat	2.2	FII[c]	$P6_1$	$a = b = 170.92, c = 69.43$	1JCS	17
	Rat	2.2	Farnesylated peptide	$P6_1$	$a = b = 171.20, c = 69.44$	1KZP	39
	Rat	2.1	Farnesylated peptide + FPP	$P6_1$	$a = b = 171.24, c = 69.36$	1KZO	39
	Human	2.3	FPP + inhibitor L-739,750[d]	$P6_1$	$a = b = 178.48, c = 64.84$	1JCQ	17
	Human	2.0	FPP + inhibitor U66[e]	$P6_1$	$a = b = 178.73, c = 64.55$	1LD7	42
	Human	1.8	FPP + inhibitor U49[f]	$P6_1$	$a = b = 178.13, c = 64.46$	1LD8	42
RabGGT	Rat	2.0	apo	P1	$a = 57.86, b = 77.44, c = 121.78$ $\alpha = 74.60°, \beta = 79.91°, \gamma = 67.86°$	1DCE	18
	Rat		REP + isoprenoid			1LTX[g]	43
PGGT-I	Rat		GGPP			1N4P[g]	44
	Rat		GGPP and peptide substrate			1N4Q[g]	44
	Rat		Prenylated peptide product			1N4R[g]	44
	Rat		GGPP + product			1N4S[g]	44

[a] Space group.
[b] α-hydroxyfarnesylphosphonic acid.
[c] [(3,7,11-trimethyl-dodeca-2,6,10-trienyloxycarbamoyl)-methyl]-phosphonic acid.
[d] 2(S)-{2(S)-[2(R)-amino-3-mercapto]propylamino-3(S)-methyl}pentyloxy-3-phenylpropionylmethionine sulfone.
[e] (20S)-19,20,22,23-tetrahydro-19-oxo-5H,21H-18,20-ethano-12,14-etheno-6,10-methenobenz[D]imodazo[4,3-L][1,6,9,13]oxatriaza-cyclonoadecosine-9-carbonitrile).
[f] (20S)-19,20,22,23-tetrahydro-19-oxo-5H-18,20-ethano-12,14-etheno-6,10-metheno-18H-benz[D]imodazo[4,3-K][1,6,9,12]oxatri-aza-cyclooctadecosine-9-carbonitrile).
[g] These coordinates were deposited but not yet released as of December 2002.

of $P6_5$ with unit cell dimensions $a = b = 167.06$ Å and $c = 97.9$ Å.[14] There is one heterodimeric PFT molecule in the asymmetric unit. The first PFT structure was solved at 2.25 Å resolution by the multiple isomorphous replacement (MIR) methods in 1997.[14] Since then, structures of the complexes of PFT with the lipid, with both lipid and peptide substrates, and with a series of inhibitors have been reported (Table 1). Additionally, the structures of a complex of PFT and prenylated peptide product as well as a complex that contains both farnesylated peptide product and an additional FPP have been reported recently[39] (Table 1). These structures provided important insights into the substrate binding and catalytic mechanisms of PFT.

Rat RabGGT crystals were grown by the hanging drop vapor diffusion method at 21°C. A 2-μl aliquot of 5 to 10 mg mL^{-1} RabGGT was mixed with an equal amount of reservoir solution (0.1 M Na acetate, pH 5.5, 0.25 M Mg acetate, 10 mM NaH$_2$PO$_4$, 6% ethylene glycol, and 18–21% PEG8000) and equilibrated against the reservoir.[18] The crystals tend to be twinned. The twinning could be eliminated by consecutive microseeding and macroseeding procedures (H. Zhang, unpublished results). RabGGT crystals are in the triclinic space group P1 with cell dimensions $a = 57.86$ Å, $b = 77.44$ Å, $c = 121.78$ Å, $\alpha = 74.60°$, $\beta = 79.91°$, and $\gamma = 67.89°$. There are two heterodimeric RabGGT molecules in the asymmetric unit. The structure of RabGGT was solved by the MIR method and has been refined to 2.0 Å resolution.[18]

Recently, the *in vitro* assembly and crystallization of RabGGT–REP and REP–Rab binary complexes, as well as a RabGGT–REP–Rab ternary complex, have been reported.[45–47] The structure of the RabGGT–REP complex has now been solved.[43] Progress has also been made in the structure determination of mammalian PGGT-I in complex with substrates and product.[44]

Overall description of the structure

PFT

The structure of rat PFT provided the first view of any protein prenyltransferases. It revealed a combination of two unusual domains: a crescent-shaped superhelical domain and an α/α barrel. The α-subunit of PFT is composed of 15 α-helices folded into a series of right-handed antiparallel coil-coils. These 'helical hairpins' are arranged

Protein prenyltransferases

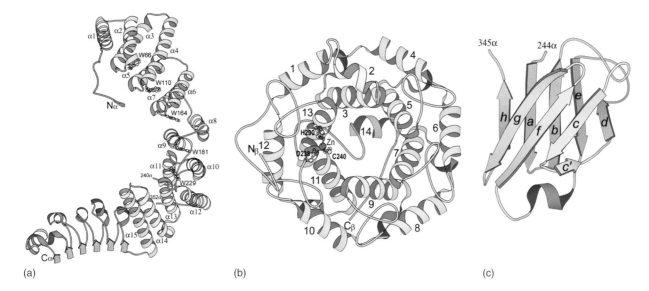

Figure 2 Ribbon diagrams illustrating the fold of individual domains in RabGGT. (a) The PPTA repeat helical domain (cyan) and the LRR domain (green) in the RabGGT α-subunit. The 15 helices are labeled from α1 to α15. (b) The β-subunit of RabGGT (magenta), with the Zn ion shown as a red sphere and the Zn–ligands Asp238$_\beta$, Cys240$_\beta$ and His290$_\beta$ in ball-and-stick representation. The helices are numbered 1 to 14. (c) The Ig-like domain of RabGGT α-subunit (yellow). The strands are labeled from *a* to *h*. The figures were prepared with MOLSCRIPT.[5]

in a double-layered right-handed superhelix resulting in a crescent-shaped subunit that envelops part of the β-subunit. The sequences of all PPT α-subunits contain a common domain composed of seven sequence repeats termed PPTA repeat. The seven PPTA sequence repeat motifs in PFT comprise the 2nd to 15th helices, with each repeating unit folded into an α-hairpin (Figure 2(a)). Similar right-handed α-superhelical folds have also been found in many other proteins such as lipovitellin,[48] bacterial muramidase,[49] and tetratricopeptide repeat (TPR) containing proteins.[50,51] An evolutionary relationship between PPTA repeats and TPR has been proposed.[52] In the β-subunit of PFT, 12 α-helices are folded into an α/α barrel, which is termed 'α/α toroid fold' in the SCOP (structure classification of proteins) protein structure classification database.[53] There are six helices lining the inside wall of the barrel and six helices located on the outside (Figure 2(b)). Each PPTB sequence repeat motif in the PFT β-subunit is also folded into a pair of helices and is connected with a characteristic glycine-rich loop containing the conserved QW motif.[20] The center of the α/α barrel forms a funnel-shaped pocket lined with mostly hydrophobic residues. The Zn ion is located at the opening of this pocket, marking the catalytic center of the enzyme. The bottom of this pocket is blocked by a loop, whereas the opposite end is open to the solvent. The PFT–substrate complex structures have revealed that both FPP and peptide substrates bind to this central cavity in the β-subunit.[15–17,40,41] Sequence analysis has suggested that PPT β-subunit is evolutionarily related to terpenoid cyclases and C3D compliment component protein.[19,54,55] In these related structures, the enzymatic active sites are also located in the central cleft.[54,55]

RabGGT

The α-subunit of RabGGT contains three distinct domains: the PPTA repeat domain that is closely related to the PFT α-subunit (with a root mean square deviation (rmsd) of 2.4 Å, and a sequence identity of 22%), an Ig-like domain, and an LRR domain (Figure 2). The Ig-like domain contains eight antiparallel β-strands folded into a β-sandwich with Greek-Key topology (Figure 2(c)). The LRR domain contains five typical LRR repeating units and is most similar to the U2A'LRR domain (rmsd 1.2 Å for 116 superimposed C$_\alpha$ atoms).[56] Neither the Ig-like domain nor the LRR domain makes contact with the β-subunit of RabGGT, and it appears that the catalytic activity of RabGGT does not require the presence of either one of these domains.[57] The exact function of these two additional domains in RabGGT remains unclear. Notably, both Ig-like and LRR domains are absent in yeast PGGT-II.

The β-subunit of RabGGT is very similar to that of PFT with an rmsd of 1.4 Å and a sequence identity of 30%. It is somewhat smaller than the β-subunit of PFT (331 residues in RabGGTβ versus 437 residues in PFTβ) and lacks the first α-helix and the C-terminal long loop of PFTβ.

Metal site geometry

A single Zn ion was located at the opening and near the edge of the central cavity of the PPT β-subunit α/α barrel. In rat PFT, this Zn ion is coordinated by the β-subunit residues Asp297$_\beta$, Cys299$_\beta$ from the N-terminus of helix 11$_\beta$, and His362$_\beta$ on helix 13$_\beta$ (Figure 3). These three

Protein prenyltransferases

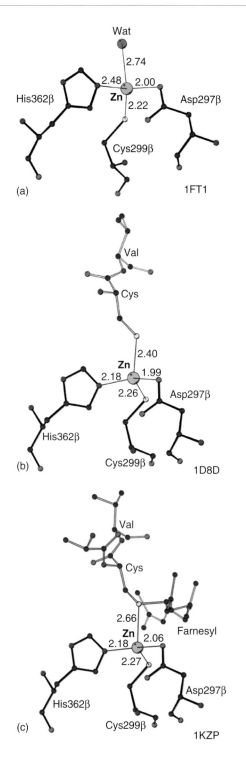

Figure 3 The Zn site geometry in (a) apo-PFT (PDB code 1FT1); (b) a PFT–substrate complex (PDB code 1D8D); and (c) PFT–product complex (PDB code 1KZP). The peptide substrate is colored cyan and the farnesyl moiety is colored green. The figures were prepared with MOLSCRIPT.[5]

Zn–ligand residues are invariant among all PPTs. In the apo-PFT structure, a well-ordered water molecule is also coordinated to the Zn, completing the tetracoordination of Zn (Figure 3(a)). In RabGGT, the three Zn–ligands from the β-subunit are $Asp238_\beta$, $Cys240_\beta$, and $His290_\beta$. However, the fourth Zn–ligand in RabGGT comes from a histidine residue ($His2_\alpha$) near the N-terminus of the α-subunit and it binds to the Zn in a self-inhibitory fashion.[18] In the PFT–substrate complexes, the peptide cysteine replaces the water molecule and directly coordinates to Zn through a thiolate linkage[15–17] (Figure 3(b)). Additionally, the thioether group of the prenylated peptide product remains coordinated to the Zn as well[39] (Figure 3(c)). These observations are consistent with the spectroscopic analysis of Co^{2+}-substituted PFT.[31] The bond distances between Zn and its ligands range from 1.99 to 2.74 Å, with the bond distance to the fourth ligand (either a water molecule or the thiol group of the peptide cysteine) being the longest. These observations, combined with spectroscopic and other biochemical studies, indicate that Zn ion plays a critical catalytic role by activating the protein substrate thiol for the nucleophilic attack on the C1 atom of FPP.[15–17,31,58]

Isoprene diphosphate binding sites

Substrate binding in PFT has been well characterized for both FPP and peptide substrates. The substrate binding pocket is primarily located in the β-subunit, with several residues from the α-subunit also being involved in FPP and peptide binding. The isoprene tail of FPP binds deeply in the active site cavity at the center of the β-subunit α/α barrel. Its diphosphate head binds to a positively charged cluster near the subunit interface. The positively charged side chains of residues $His248_\beta$, $Arg291_\beta$, $Lys294_\beta$, $Lys164_\alpha$, as well as the $Tyr300_\beta$ side-chain hydroxyl interact with the α- and β-phosphates of FPP (Figure 4). The hydrophobic isoprene tail of FPP contacts many hydrophobic residues that line the funnel-shaped substrate binding pocket, including $Trp102_\beta$, $Tyr154_\beta$, $Tyr205_\beta$, $Cys254_\beta$, $Tyr251_\beta$, $His248_\beta$, $Trp303_\beta$, and $Tyr166_\alpha$. The third isoprene unit packs against the aliphatic portion of $Arg202_\beta$ side chain as well as the Ile side chain of the CVIM tetrapeptide moiety in the reactant ternary complexes.[15–17] In one of the PFT–substrate complex structures determined in the presence of $MnSO_4$,[16] the Mn^{2+} ion was located in a position between the two phosphate groups of FPP, which indicates the potential binding site for Mg^{2+}. This observation is consistent with the proposed role for Mg^{2+} in stabilizing the pyrophosphate-leaving group during catalysis.

Although no complex structure has yet to be reported for RabGGT, comparison of the active site pockets revealed that two residues near the bottom of the active site cavity, $Trp102_\beta$ and $Tyr154_\beta$, in PFT are replaced by $Ser48_\beta$ and $Leu99_\beta$ in RabGGT respectively, which makes the substrate binding pocket of RabGGT both deeper and wider (Figure 5). Modeling of the GGPP molecule in the binding site indicates that the pocket in RabGGT should be

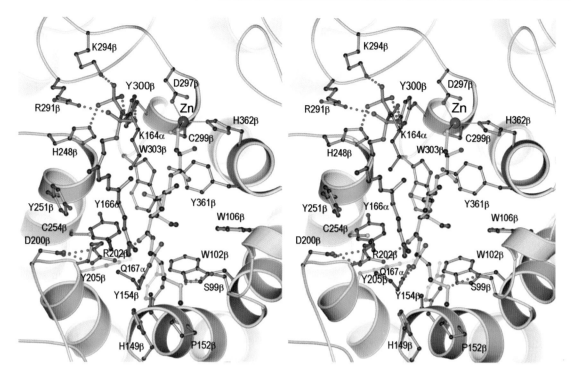

Figure 4 Stereo view of the substrate binding site and the active site of PFT. The CVIM peptide is colored green and FPP is colored orange. The Zn ion is shown as a purple sphere. Relevant protein side chains are shown in the ball-and-stick representation, with residues from PFT β-subunit colored cyan and residues from the α-subunit colored magenta. The hydrogen bonds are shown as dotted lines. This figure was prepared with MOLSCRIPT[5] and Gl_render (L. Esser, unpublished program).

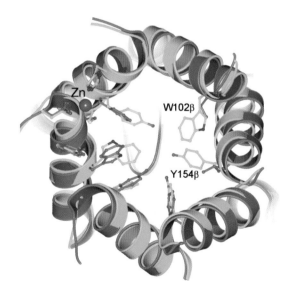

Figure 5 Superposition of the substrate binding sites of PFT (green) and RabGGT (magenta). The side chains of several relevant residues are shown in ball-and-stick representation. Residues W102$_\beta$ and Y154$_\beta$ of PFT are labeled; these residues correspond to S48$_\beta$ and L99$_\beta$ in RabGGT. This figure was prepared with MOLSCRIPT[5] and Gl_render (L. Esser, unpublished program).

able to accommodate the four isoprene units of GGPP while keeping the diphosphate head bound to the corresponding positively charged binding site.[18]

Peptide/protein substrate binding site

It has been shown that the C-terminal CaaX tetrapeptide of the CaaX prenyltransferase substrate is necessary and sufficient for effective prenylation by PFT and PGGT-I.[59–62] So far, a number of structures of PFT complexes containing CaaX peptides or peptidomimetic inhibitors and FPP or its nonhydrolyzable analogue have been reported (Table 1). Several PFT + peptide/FPP ternary complexes contain the tetrapeptide CVIM sequence,[15,16] a typical C-terminal CaaX motif in N- and K-Ras proteins. In these complex structures, the peptide adopts an extended conformation and occupies a large part of the active site cavity (Figure 4). The peptide interacts with atoms from both the enzyme and FPP. The Ile-to-Met peptide bond of the CVIM moiety packs against the aromatic ring of Tyr166$_\alpha$. A hydrogen bond is formed between the Ile carbonyl and the guanidinium group of the buried Arg202$_\beta$, which in turn forms a salt bridge with Asp200$_\beta$. The C-terminal Met side chain is nested in a narrow pocket defined by Trp102$_\beta$, His149$_\beta$, Ala151$_\beta$, and Pro152$_\beta$. The Ser99$_\beta$ hydroxyl appears to be hydrogen-bonded to the sulfur atom of the Met with a distance of 3.34 Å. The C-terminal carboxylate is hydrogen-bonded to Gln167$_\alpha$. The Ile side chain is sequestered in a pocket lined with the side chains of Trp102$_\beta$, Trp106$_\beta$, Tyr361$_\beta$, and the FPP isoprenoid tail. The following Val residue becomes partially exposed to the solvent. A wide variety of residues can be accommodated at this position in PFT protein and peptide substrates.[63]

The sulfur atom of the cysteine in the CaaX motif of the peptide is found directly coordinated to the zinc ion, replacing the bound solvent molecule in the apo-PFT structure. This coordination implicated Zn not only to participate directly in catalysis but also in orienting the bound peptide in a productive conformation. In a Zn-depleted PFT-peptide/FPP complex, the CVIM peptide is found to be in a nonproductive conformation with the cysteine side chain pointing away from the bound FPP.[16] It has been found that an additional stretch of polylysines upstream of the CaaX motif at the C-terminus of K-Ras4B increases the affinity of the peptide to PFT and that K-Ras is more resistant to the peptidomimetic inhibitors.[64] The crystal structure of PFT complexed with an FPP analogue and a 11-mer K-Ras4B peptide ($^+NH_3$-KKKSKTKCVIM-COO^-) showed that the polybasic region forms a type I β-turn and binds along the rim of the hydrophobic cavity in a region rich in negatively charged residues, including $Glu94_\beta$, $Asp91_\beta$, $Glu125_\alpha$, and $Glu161_\alpha$.[16] These additional favorable interactions may explain the enhanced affinity of PFT toward K-Ras4B.

RabGGT is unique in the protein prenyltransferase family. It does not recognize a short peptide containing the Rab C-terminal double cysteine motifs, nor does it recognize the Rab protein alone.[26,65] The geranylgeranylation of Rab requires the presence of Rab escort protein (Mr ~72 kDa in mammals).[66] A mechanism for the Rab prenylation cascade has been proposed: newly synthesized Rab binds REP and forms a stable Rab-REP complex with a dissociation constant of 0.15 to 0.4 μM.[65] RabGGT recognizes the Rab-REP complex as its protein substrate primarily through interactions with REP.[65] After prenylation, REP remains bound to the prenylated Rab and delivers Rab to its target membrane.[67] So far, no RabGGT-substrate complex structures have been reported, but progress has been made in the crystallization of RabGGT complexes with either REP or REP-Rab.[45,46]

PFT-product complexes

The structures of a PFT-farnesylated peptide product complex and a ternary complex containing both farnesylated peptide and an additional FPP have been reported recently.[39] In the product binary complex, the cysteine thiol of the thioether-linked product remains coordinated to the Zn ion, confirming spectroscopic studies.[31] Compared to the substrate complexes, the C1 atom of the first isoprene unit of FPP moved 5.7 Å and is now covalently linked to the cysteine sulfur (Figure 6). No significant conformational changes of the enzyme are observed. The peptide conformation in the product is essentially the same as that in the substrate complexes. Although a structural model for the transition state has been proposed, it is not clear what forces cause the large conformational changes of the lipid that bring the C1 atom close to the cysteine thiol of the protein substrate.

Kinetic studies have indicated that product release is slow for PFT and requires the addition of a substrate molecule.[33,68] This final step was captured in a ternary complex containing both farnesylated peptide product and an additional farnesyl diphosphate.[39] This exit complex structure revealed that binding of the additional FPP moved the farnesyl group of the farnesylated peptide to

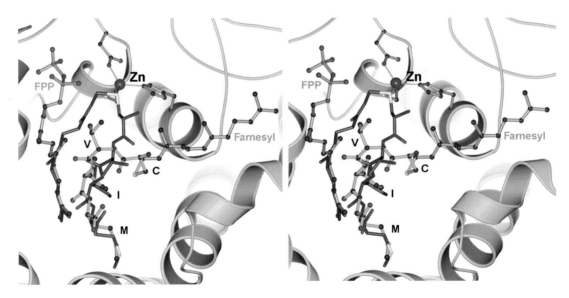

Figure 6 Stereo diagram of the farnesylated peptide product conformations at the PFT active site. The farnesylated peptide in the PFT-product binary complex (shown in black bonds) is superimposed onto the ternary PFT complex containing the product (colored green for the peptide portion and orange for the farnesyl moiety) and an additional FPP substrate (orange). The Zn ion and its ligands are shown in ball-and-stick representation. This figure was prepared with MOLSCRIPT[5] and Gl_render (L. Esser, unpublished program).

a new binding site, the so called 'exit groove' outside the central cavity, whereas the central cavity is now occupied by the additional FPP. This movement is accompanied by a conformational change in the Ca_1a_2X peptide. The two C-terminal residues of the Ca_1a_2X moiety (Ile-Met in this case) remain largely unchanged and occupy essentially the same binding location as those in the reactant and product complex. The Ca_1 residues of the Ca_1a_2X moiety (Cys-Val in this case), however, underwent drastic conformational changes (Figure 6). The stability of the exit complex and the observation that this complex persists even in the presence of high concentrations of FPP and dissociates only if excess peptide substrate is added would suggest that dissociation of the product from the enzyme may ultimately require binding of both FPP and unfarnesylated protein/peptide substrate.[39] This result has direct implications for the RabGGT double prenylation mechanism. A secondary geranylgeranyl binding site was proposed on the basis of the model of RabGGT complexed with a monoprenylated Rab peptide and an additional GGPP.[39] However, the detailed interactions of RabGGT with its lipid and protein substrates remain to be elucidated experimentally.

FUNCTIONAL ASPECTS

PFT: steady state and pre-steady state kinetics

Steady state kinetic and isotope-trapping analysis of PFT demonstrated a formally random but functionally ordered sequential mechanism for the enzyme,[33,69,70] whereby the preferred pathway for catalysis involves FPP binding first, followed by the peptide substrate, chemistry, and then product release (Figure 7). FPP binds to PFT with an effective dissociation rate constant of $0.013\,s^{-1}$ and an overall K_d of $2.8\,nM$.[33] The peptide reacted with PFT-FPP irreversibly with a second-order rate constant of $2.2 \times 10^5\,M^{-1}\,s^{-1}$.[33] The maximal rate constant for the formation of enzyme–bound prenylated product ($17\,s^{-1}$) is much larger than the overall K_{cat} ($0.06\,s^{-1}$), indicating that product release is the rate-limiting step in the reaction mechanism.[31,33] A detailed analysis of product release under single turnover condition led to the conclusion that the prenylated peptide product did not dissociate from the enzyme unless additional substrate was provided.[68] The rate for the product release in the presence of excess FPP was determined to be $0.05\,s^{-1}$.[68] The crystal structure of the PFT exit complex further captured a previously unknown reaction intermediate in which both product and FPP bind to the enzyme simultaneously, suggesting that ultimate release of the product would require binding of both isoprenoid and peptide substrate.[39]

RabGGT: interactions with REP or Rab–REP and double prenylation

The classic view of RabGGT prenylation assumes the initial formation of the Rab–REP binary complex, which subsequently binds to RabGGT loaded with GGPP. However, it has also been demonstrated that REP can associate with RabGGT in the absence of Rab and that this interaction is dramatically strengthened by the presence of phosphoisoprenoids such as GGPP.[71] The RabGGT binds REP with a K_d of $10\,nM$ in the presence of GGPP and a K_d in the micromolar range ($3.5-7.5\,\mu M$) in the absence of GGPP.[71] Additionally, REP interacts with Rab with a K_d value range from $1\,nM$ to $0.4\,\mu M$ depending on the different Rab proteins.[65,72] The REP–Rab complex binds to RabGGT with a K_d of $100\,nM$. In the presence of GGPP, the binding between RabGGT and REP–Rab is much tighter ($2\,nM$).[73] The interactions of RabGGT with phosphoisoprenoids are also characterized with steady state kinetics and transient kinetics using fluorescent isoprenoid derivatives.[74] GGPP binds to RabGGT with an affinity of $8 \pm 4\,nM$ while FPP binding is much weaker with an affinity of $60 \pm 8\,nM$.[74] The overall K_{cat} of the double prenylation reaction was determined to be 0.019 to $-0.029\,s^{-1}$.[65] More detailed studies using semisynthetic fluorescent prenylated Rab protein have shown that the rates for the first and second prenylation are $0.16\,s^{-1}$ and $0.039\,s^{-1}$ respectively, indicating that the second prenylation step is much slower and that most monoprenylated intermediates proceed to the second prenylation without dissociation from the enzyme.[37] Similar to PFT, binding of additional GGPP also weakens the affinity between the double prenylated product and the

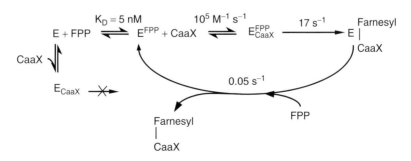

Figure 7 The PFT kinetic scheme (Adapted from Figure 1 of ref. 39).

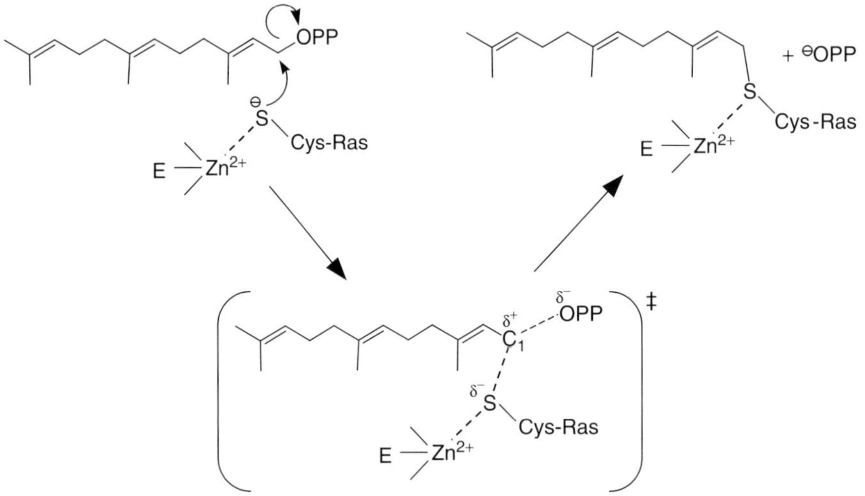

Figure 8 Proposed chemical mechanism for mammalian PFTs. The double dagger indicates the proposed transition state, which is associative and contains both carbocation and nucleophile characters (Adapted from Fig. 5 of ref. 7).

enzyme (K_d changes from 2 nM to 22 nM).[37] Thus, binding of the additional lipid facilitates the product release by RabGGT as well. Although it has been shown that the most C-terminal cysteine is preferred to be prenylated first,[37] the detailed mechanism of double prenylation, which presumably involves a conformational rearrangement of the monoprenylated intermediate, is not yet clear.

CATALYTIC MECHANISM

Detailed mechanistic studies have been carried out to elucidate the catalytic mechanism of PFT. Early work established the requirement for zinc and magnesium in the catalysis for PFT and RabGGT.[26,29] Both spectroscopic analysis of Co^{2+}-substituted PFT and structural analysis of PFT complexes have shown that Zn is directly coordinated to the cysteine thiol(ate) of the bound peptide substrate and to the thiol of the thioether-linked product.[15,16,31,39] The exact role of Zn^{2+} ion in the catalysis was further investigated by analysis of the pH dependence of interactions between peptide substrate and PFT.[58] These studies revealed that the pK_a of the cysteine thiol of the peptide substrate was lowered from 8.3 to 6.4 when it was bound to the PFT and FPP binary complex.[58] These findings support a mechanism by which Zn enhances the nucleophilicity of the cysteine thiol of the protein substrate CaaX motif for the attack at the C1 carbon of FPP. On the other hand, evidence for a carbocation character of the rate-contributing transition state has been obtained from the decrease in the rate constant for product formation when electron-withdrawing fluorines are substituted at the C3 methyl position of FPP.[75,76] Combining several lines of evidence, it has been suggested that PFT catalyzes protein farnesylation by an associative mechanism with an 'exploded' transition state where the metal-bound peptide/protein sulfur has a partial negative charge, the C1 of FPP has a partial positive charge, and the bridge oxygen between C1 and the α-phosphate of FPP has a partial negative charge[76] (Figure 8). Therefore, the reaction mechanism of PFT contains both nucleophilic and electrophilic components.

The general aspects of the chemical mechanism outlined above are likely to be shared among all PPTs, since the active site residues of all PPTs are highly conserved. Zn-mediated nucleophilic attack of a thiolate on alkyl groups has also been proposed as the catalytic mechanism for several other sulfur-alkylating metalloproteins, including methylcobalamide:coenzyme M methyltransferase, cobalamin-independent and cobalamin-dependent methionine synthase, betaine-homocystein methyltransferase, and the DNA repair protein Ada.[77,78] For all these enzymes, a metal-bound thiolate has been suggested as the nucleophile in the chemical step. However, the potential carbocation in these enzymes is significantly less stable than that in PFT.[78] The detailed catalytical mechanisms, especially the structural characterizations for each of these metalloenzymes, remain to be elucidated.

PFT INHIBITORS AND CANCER THERAPY

Of particular biomedical interest in the field of PPT is the observation that the farnesylation of Ras oncoproteins, which are associated with approximately 30% of all human cancers, is absolutely required for the transformation of cells to a tumorigenic state.[79] It has been shown that inhibition of PFT results in almost complete tumor regression in Ras transgenic mice without visible toxicity to the animal.[80] Therefore, the development of PFT inhibitors as novel anticancer agents has been one of the most active areas in anticancer research in the past 10 years.[81,82] Numerous potent PFT inhibitors have been developed, and their structure–activity relationships (SAR) have been the subject of extensive investigations.[42,83] These inhibitors can be separated into three different categories: (i) natural products; (ii) peptidomimetics and other CaaX-competitive inhibitors; and (iii) FPP mimetics or analogs and other FPP-competitive inhibitors. Surprisingly, PFT inhibitors do not exclusively target tumors expressing oncogenetic Ras protein.[84] It was later shown that some farnesylated non-Ras proteins, such as Rho family members, are also important for the support of cancer cell growth.[85] These data illustrate the point that the true target under investigation is PFT and not simply Ras. Currently, several PFT inhibitors have entered human clinical trials in different phases and some of them have shown partial responses.[12,79] PFT from *Trypanosoma brucei* has also been recognized as a drug target for combating infections that cause African sleeping sickness.[86,87] Developments of specific inhibitors for *T. brucei* PFT are currently underway.[88]

ACKNOWLEDGEMENTS

The author wishes to thank Kirill Alexandrov for sharing unpublished results before publication and for helpful comments, Mischa Machius and Nick Grishin for critical reading of the manuscript, and Hans Deisenhofer for support and encouragement.

REFERENCES

1. CA Omer and JB Gibbs, *Mol Microbiol*, **11**, 219–25 (1994).
2. FL Zhang and PJ Casey, *Annu Rev Biochem*, **65**, 241–69 (1996).
3. PJ Casey and MC Seabra, *J Biol Chem*, **271**, 5289–92 (1996).
4. CC Farnsworth, MC Seabra, LH Ericsson, MH Gelb and JA Glomset, *Proc Natl Acad Sci USA*, **91**, 11963–67 (1994).
5. PJ Kraulis, *J Appl Crystallogr*, **24**, 946–50 (1991).
6. PJ Casey, *Science*, **268**, 221–25 (1995).
7. HW Fu and PJ Casey, *Recent Prog Horm Res*, **54**, 315–42 (1999).
8. MC Seabra, EH Mules and AN Hume, *Trends Mol Med*, **8**, 23–30 (2002).
9. CJ Marshall, *Science*, **259**, 1865–66 (1993).
10. MC Seabra, *Cell Signal*, **10**, 167–72 (1998).
11. SR Johnston, *Lancet Oncol*, **2**, 18–26 (2001).
12. JE Karp, SH Kaufmann, AA Adjei, JE Lancet, JJ Wright and DW End, *Curr Opin Oncol*, **13**, 470–76 (2001).
13. MC Seabra, Y Reiss, PJ Casey, MS Brown and JL Goldstein, *Cell*, **65**, 429–34 (1991).
14. HW Park, SR Boduluri, JF Moomaw, PJ Casey and LS Beese, *Science*, **275**, 1800–4 (1997).
15. CL Strickland, WT Windsor, R Syto, L Wang, R Bond, Z Wu, J Schwartz, HV Le, LS Beese and PC Weber, *Biochemistry*, **37**, 16601–11 (1998).
16. SB Long, PJ Casey and LS Beese, *Struct Fold Des*, **8**, 209–22 (2000).
17. SB Long, PJ Hancock, AM Kral, HW Hellinga and LS Beese, *Proc Natl Acad Sci USA*, **98**, 12948–53 (2001).
18. H Zhang, MC Seabra and J Deisenhofer, *Struct Fold Des*, **8**, 241–51 (2000).
19. MA Andrade, CP Ponting, TJ Gibson and P Bork, *J Mol Biol*, **298**, 521–37 (2000).
20. K Poralla, A Hewelt, GD Prestwich, I Abe, I Reipen and G Sprenger, *Trends Biochem Sci*, **19**, 157–58 (1994).
21. FS Buckner, RT Eastman, JL Nepomuceno-Silva, EC Speelmon, PJ Myler, WC Van Voorhis and K Yokoyama, *Mol Biochem Parasitol*, **122**, 181–88 (2002).
22. WJ Chen, JF Moomaw, L Overton, TA Kost and PJ Casey, *J Biol Chem*, **268**, 9675–80 (1993).
23. JF Moomaw, FL Zhang and PJ Casey, *Methods Enzymol*, **250**, 12–21 (1995).
24. CA Omer, AM Kral, RE Diehl, GC Prendergast, S Powers, CM Allen, JB Gibbs and NE Kohl, *Biochemistry*, **32**, 5167–76 (1993).
25. KK Zimmerman, JD Scholten, CC Huang, CA Fierke and DJ Hupe, *Protein Expr Purif*, **14**, 395–402 (1998).
26. MC Seabra, JL Goldstein, TC Sudhof and MS Brown, *J Biol Chem*, **267**, 14497–503 (1992).
27. SA Armstrong, MS Brown, JL Goldstein and MC Seabra, *Methods Enzymol*, **257**, 30–41 (1995).
28. A Kalinin, NH Thoma, A Iakovenko, I Heinemann, E Rostkova, AT Constantinescu and K Alexandrov, *Protein Expr Purif*, **22**, 84–91 (2001).
29. Y Reiss, MS Brown and JL Goldstein, *J Biol Chem*, **267**, 6403–8 (1992).
30. Z Chen, JC Otto, MO Bergo, SG Young and PJ Casey, *J Biol Chem*, **275**, 41251–57 (2000).
31. CC Huang, PJ Casey and CA Fierke, *J Biol Chem*, **272**, 20–23 (1997).
32. MC Seabra and GL James, *Methods Mol Biol*, **84**, 251–60 (1998).
33. ES Furfine, JJ Leban, A Landavazo, JF Moomaw and PJ Casey, *Biochemistry*, **34**, 6857–62 (1995).
34. JA Thissen and PJ Casey, *Anal Biochem*, **243**, 80–85 (1996).
35. DL Pompliano, MD Schaber, SD Mosser, CA Omer, JA Shafer and JB Gibbs, *Biochemistry*, **32**, 8341–47 (1993).
36. PB Cassidy, JM Dolence and CD Poulter, *Methods Enzymol*, **250**, 30–43 (1995).
37. NH Thoma, A Niculae, RS Goody and K Alexandrov, *J Biol Chem*, **276**, 48631–36 (2001).
38. A Iakovenko, E Rostkova, E Merzlyak, AM Hillebrand, NH Thoma, RS Goody and K Alexandrov, *FEBS Lett*, **468**, 155–58 (2000).
39. SB Long, PJ Casey and LS Beese, *Nature*, **419**, 645–50 (2002).
40. SB Long, PJ Casey and LS Beese, *Biochemistry*, **37**, 9612–18 (1998).
41. P Dunten, U Kammlott, R Crowther, D Weber, R Palermo and J Birktoft, *Biochemistry*, **37**, 7907–12 (1998).
42. IM Bell, SN Gallicchio, M Abrams, LS Beese, DC Beshore, H Bhimnathwala, MJ Bogusky, CA Buser, JC Culberson, J Davide, M Ellis-Hutchings, C Fernandes, JB Gibbs, SL Graham, KA Hamilton, GD Hartman, DC Heimbrook, CF Homnick, HE Huber, JR Huff, K Kassahun, KS Koblan, NE Kohl, RB Lobell, JJ Lynch Jr, R Robinson, AD Rodrigues, JS Taylor, ES Walsh, TM Williams and CB Zartman, *J Med Chem*, **45**, 2388–409 (2002).
43. O Pylypenko, A Rak, R Reents, A Niculae, MD Cioaca, V Sidorovitch, E Bessolitsyna, NH Thoma, H Waldmann, I Schlichting, RS Goody and K Alexandrov, *Mol Cell*, **11**, 483–94 (2003).
44. JS Taylor, TS Reid and LS Beese, *Presented at ACA Annual Meeting*, Los Angeles, July 21–26 (2001).

45 A Rak, R Reents, O Pylypenko, A Niculae, V Sidorovitch, NH Thoma, H Waldmann, I Schlichting, RS Goody and K Alexandrov, *J Struct Biol*, **136**, 158–61 (2001).

46 A Rak, A Niculae, A Kalinin, NH Thoma, V Sidorovitch, RS Goody and K Alexandrov, *Protein Expr Purif*, **25**, 23–30 (2002).

47 A Rak, O Pylypenko, A Niculae, RS Goody and K Alexandrov, *J Struct Biol*, **141**, 93–5 (2003).

48 TA Anderson, DG Levitt and LJ Banaszak, *Structure*, **6**, 895–909 (1998).

49 AM Thunnissen, AJ Dijkstra, KH Kalk, HJ Rozeboom, H Engel, W Keck and BW Dijkstra, *Nature*, **367**, 750–53 (1994).

50 AK Das, PW Cohen and D Barford, *EMBO J*, **17**, 1192–99 (1998).

51 LM Rice and AT Brunger, *Mol Cell*, **4**, 85–95 (1999).

52 H Zhang and NV Grishin, *Protein Sci*, **8**, 1658–67 (1999).

53 TJ Hubbard, B Ailey, SE Brenner, AG Murzin and C Chothia, *Nucleic Acids Res*, **27**, 254–56 (1999).

54 KU Wendt, K Poralla and GE Schulz, *Science*, **277**, 1811–15 (1997).

55 B Nagar, RG Jones, RJ Diefenbach, DE Isenman and JM Rini, *Science*, **280**, 1277–81 (1998).

56 SR Price, PR Evans and K Nagai, *Nature*, **394**, 645–50 (1998).

57 B Dursina, NH Thoma, V Sidorovitch, A Niculae, A Iakovenko, A Rak, S Albert, AC Ceacareanu, R Kolling, C Herrmann, RS Goody and K Alexandrov, *Biochemistry*, **41**, 6805–16 (2002).

58 MJ Saderholm, KE Hightower and CA Fierke, *Biochemistry*, **39**, 12398–405 (2000).

59 Y Reiss, JL Goldstein, MC Seabra, PJ Casey and MS Brown, *Cell*, **62**, 81–88 (1990).

60 JL Goldstein, MS Brown, SJ Stradley, Y Reiss and LM Gierasch, *J Biol Chem*, **266**, 15575–78 (1991).

61 MS Brown, JL Goldstein, KJ Paris, JP Burnier and JC Marsters Jr, *Proc Natl Acad Sci USA*, **89**, 8313–16 (1992).

62 K Yokoyama, P McGeady and MH Gelb, *Biochemistry*, **34**, 1344–54 (1995).

63 Y Reiss, SJ Stradley, LM Gierasch, MS Brown and JL Goldstein, *Proc Natl Acad Sci USA*, **88**, 732–36 (1991).

64 GL James, JL Goldstein and MS Brown, *J Biol Chem*, **270**, 6221–26 (1995).

65 JS Anant, L Desnoyers, M Machius, B Demeler, JC Hansen, KD Westover, J Deisenhofer and MC Seabra, *Biochemistry*, **37**, 12559–68 (1998).

66 MC Seabra, MS Brown, CA Slaughter, TC Sudhof and JL Goldstein, *Cell*, **70**, 1049–57 (1992).

67 K Alexandrov, H Horiuchi, O Steele-Mortimer, MC Seabra and M Zerial, *EMBO J*, **13**, 5262–73 (1994).

68 WR Tschantz, ES Furfine and PJ Casey, *J Biol Chem*, **272**, 9989–93 (1997).

69 Y Reiss, MC Seabra, SA Armstrong, CA Slaughter, JL Goldstein and MS Brown, *J Biol Chem*, **266**, 10672–77 (1991).

70 DL Pompliano, E Rands, MD Schaber, SD Mosser, NJ Anthony and JB Gibbs, *Biochemistry*, **31**, 3800–7 (1992).

71 NH Thoma, A Iakovenko, RS Goody and K Alexandrov, *J Biol Chem*, **276**, 48637–43 (2001).

72 K Alexandrov, I Simon, A Iakovenko, B Holz, RS Goody and AJ Scheidig, *FEBS Lett*, **425**, 460–64 (1998).

73 NH Thoma, A Iakovenko, A Kalinin, H Waldmann, RS Goody and K Alexandrov, *Biochemistry*, **40**, 268–74 (2001).

74 NH Thoma, A Iakovenko, D Owen, AS Scheidig, H Waldmann, RS Goody and K Alexandrov, *Biochemistry*, **39**, 12043–52 (2000).

75 JM Dolence and CD Poulter, *Proc Natl Acad Sci USA*, **92**, 5008–11 (1995).

76 C Huang, KE Hightower and CA Fierke, *Biochemistry*, **39**, 2593–602 (2000).

77 RG Matthews and CW Goulding, *Curr Opin Chem Biol*, **1**, 332–39 (1997).

78 KE Hightower and CA Fierke, *Curr Opin Chem Biol*, **3**, 176–81 (1999).

79 AA Adjei, *J Natl Cancer Inst*, **93**, 1062–74 (2001).

80 NE Kohl, CA Omer, MW Conner, NJ Anthony, JP Davide, SJ deSolms, EA Giuliani, RP Gomez, SL Graham, K Hamilton, LK Handt, GD Hartman, KS Koblan, AM Kral, PJ Miller, SD Mosser, TJ O'Neill, E Rands, MD Schaber, JB Gibbs and A Oliff, *Nat Med*, **1**, 792–7 (1995).

81 SM Sebti and AD Hamilton, *Oncogene*, **19**, 6584–93 (2000).

82 RA Gibbs, TJ Zahn and JS Sebolt-Leopold, *Curr Med Chem*, **8**, 1437–65 (2001).

83 CL Strickland, PC Weber, WT Windsor, Z Wu, HV Le, MM Albanese, CS Alvarez, D Cesarz, J del Rosario, J Deskus, AK Mallams, FG Njoroge, JJ Piwinski, S Remiszewski, RR Rossman, AG Taveras, B Vibulbhan, RJ Doll, VM Girijavallabhan and AK Ganguly, *J Med Chem*, **42**, 2125–35 (1999).

84 L Sepp-Lorenzino, Z Ma, E Rands, NE Kohl, JB Gibbs, A Oliff and N Rosen, *Cancer Res*, **55**, 5302–9 (1995).

85 W Du, PF Lebowitz and GC Prendergast, *Mol Cell Biol*, **19**, 1831–40 (1999).

86 FS Buckner, K Yokoyama, L Nguyen, A Grewal, H Erdjument-Bromage, P Tempst, CL Strickland, L Xiao, WC Van Voorhis and MH Gelb, *J Biol Chem*, **275**, 21870–76 (2000).

87 K Yokoyama, P Trobridge, FS Buckner, J Scholten, KD Stuart, WC Van Voorhis and MH Gelb, *Mol Biochem Parasitol*, **94**, 87–97 (1998).

88 F Clerici, ML Gelmi, K Yokoyama, D Pocar, WC Van Voorhis, FS Buckner and MH Gelb, *Bioorg Med Chem Lett*, **12**, 2217–20 (2002).

Hydrolases: Acting on ester bonds

Nuclease P1

Anne Volbeda[†], Christophe Romier[‡] and Dietrich Suck[§]

[†]Institut de Biologie Structurale J.P. Ebel, CEA-CNRS-UJF, Grenoble, France
[‡]Institut de Génétique et de Biologie Moléculaire et Cellulaire, CNRS/INSERM/ULP, Illkirch, France
[§]European Molecular Biology Laboratory, Heidelberg, Germany

FUNCTIONAL CLASS

Enzyme; P1 nuclease; EC 3.1.30.1; P1 is a glycoprotein of 36 kDa that uses three zinc ions for both its phosphodiesterase and 3′-phosphomonoesterase enzymatic activities (Figure 1). It prefers single-stranded DNA and RNA substrates and hydrolyzes these completely, independent of the base sequence, to 5′-mononucleotides.[1–3] The phosphodiester bonds are cleaved with inversion of configuration.[4]

OCCURRENCE

Nuclease P1 is found in the mold *Penicillium citrinum*.[5] It belongs to a class of mostly secreted, extracellular nucleases that have been characterized in fungi,[6] plants,[7] protozoan parasites,[8] and bacteria.[9] Common features include the presence of a large extent of covalently bound carbohydrate, a strict dependence on zinc for activity, and pH optima that are in most cases in the acid range. Their molecular weights range between about 31 and 42 kDa.

BIOLOGICAL FUNCTION

P1 and related enzymes are widely used by molecular biologists as an analytical tool.[10–12] They completely hydrolyze single-stranded DNA and RNA substrates into 5′-mononucleotides. In addition, they may be used to detect non-A and non-B conformations in double-stranded DNA substrates, which, unlike regular A- and B-DNA structures, are susceptible to enzymatic cleavage.[13] Several physiological roles have been proposed. In trypanosomatid parasites, which are incapable of purine biosynthesis,

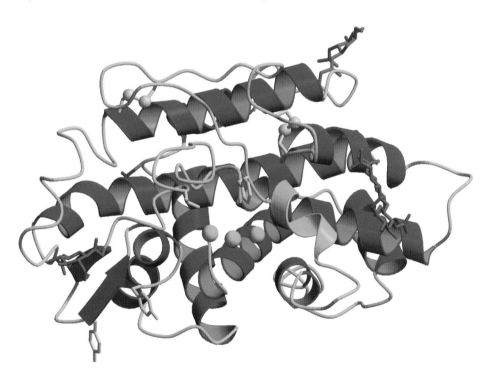

3D Structure   Schematic representation of the structure of nuclease P1. α-helices are depicted by blue ribbons, 3/10-helices by light-blue ribbons, and β-strands with violet arrows. Zinc ions are shown as light-green spheres, whereas two disulfide bonds are indicated in yellow. Four asparagine side chains constituting carbohydrate fixation sites and bound ordered sugar residues are shown in red. Selected residues of two mononucleotide binding sites are depicted in light gray. Coordinates are available from the Protein Data Bank,[37,38] entry 1AK0.[36] This figure, along with Figures 3 to 5 and 6(a), was prepared with programs MOLSCRIPT[39] and RASTER3D.[40]

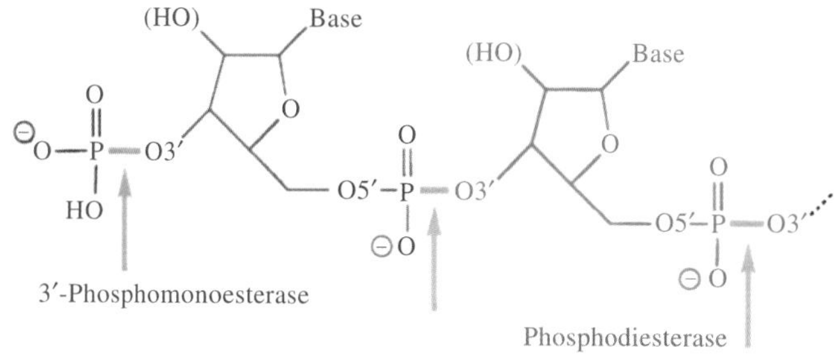

Nuclease P1

Figure 1 The two enzymatic activities of nuclease P1.

external membrane-bound P1-like nucleases are probably involved in the salvage of purines from host RNA and DNA.[8] In plants, nuclease I enzymes are thought to degrade RNA and single-stranded DNA substrates during several developmental processes, including senescence and programmed cell death.[14,15] The precise function of a homologous nuclease in celery, with a neutral pH optimum and a specificity for DNA mismatches, is still unclear.[16] A putative role in DNA damage repair has been suggested for a nuclease I enzyme that is associated with the endoplasmic reticulum in the parasite *Leishmania pifanoi*.[17]

AMINO ACID SEQUENCE INFORMATION

The following is an incomplete compilation of nucleases related to P1 (see also Figure 2). Many more homologous sequences may be found in rapidly growing databases such as EMBL[18] and GenBank.[19] In most, if not all, cases, a signal peptide seems to be encoded in the gene sequence before the N-terminal tryptophan residue.

Fungal enzymes:
Penicillium citrinum (P1), 270 amino acid residues (AA)[22]
Aspergillus oryzae (S1), 267 AA[23]

```
              W    GH      A                    L                      W  D  *           H*  *
P24289     WGALGHATVAYVAQHYVSP---EAASWAQGILGSSS------SSYLASIASWADEYRLTSAGKWSASLHFIDAEDN----
P24021     WGNLGHETVAYIAQSFVAS---STESFCQNILGDDS------TSYLANVATWADTYKYTDAGEFSKPYHFIDAQDN----
O81958     WGKEGHYMTCKIADGFLTS---EALTGVKALLPSWA------NGELAEVCSWADSQR---FRYRWSRSLHFADTPG----
Q9LL59     WSKEGHVMTCQIAQDLLEP---EAAHAVKMLLPDYA------NGNLSSLCVWPDQIRHWYKYRWTSSLHFIDTPDQ----
L35078     WWSKGHMSVALIAKRHMGASLVEKAELAAKVLSFSGPYP--KSPDMVQTAPWADDIK-TIGLKTLSTWHYITTPYY----
AF140355   WWSKGHMAVALIAQRHMSPTAVEKGNAAANVLCKTGPYP--LSPDMVQTASWADDIK-TIGLDTMSSWHFITTPYY----
O68530     WGQEGHAAVAEIAQHRLTS---SASDVVQRLLRAHLGLTGQQVVSMASIASWADDYR-ADGHKDTSNWHFVDIPLASLPG

                =                  =         =         /             L     H  GD QP H
P24289     -PPTNCNVDYERDCGSSG-----=CSISAIANYTQRVS----DSSLSSENHAEALRFLVHFIGDMTQPLHD--------EA
P24021     -PPQSCGVDYDRDCGSAG-----CSISAIQNYTNILL----ESPNGSE-ALNALKFVVHIIGDTHQPLHD--------EN
O81958     ----DCKFSYARDCHDTKGNKNVCVVGAINNYTAALQ----DSSSPFNPTESLMFLAHFVGDVHQPMHCG-------HV
Q9LL59     ----ACSFDYQRDCHDPHGGKDMCVAGAIQNFTSQLGHFRHGTSDRRYNMTEALLFLSHFMGDIHQPMHVG-------FT
L35078     -TDED-FTLDVSPVQTVN------VASVIPMLQTAIE----KPTANSDVIVQSLALLLHFMGDIHQPLHNVLFSNQYPE
AF140355   -PEGDTFRLSVSPVQAVN------VASVIPMLQSALQ----SKSATSEIIAQSLALLIHFMGDIHQPLHNANEFSTEYPT
O68530     GSSATTDYDAIRDCADDATYG-SCLLKALPAQEAILS----DATKDDESRWKALAFVIHLTGDLAQPLHCVQRVDG--SQ

            *  GGN              *    *  H   WD  *                /                 /
P24289     YAVGGNK-INVT---------FDGYHDNLHSDWDTYMPQKLIGGHALSDAESWAKTLVQNIESGNYTAQAIGWIKGDNI
P24021     LEAGGNG-IDVT---------YDGETTNLHHIWDTNMPEEAAGGYSLSVAKTYADLLTERIKTGTYSSKKDSWTDGIDI
O81958     DDLGGNT-IKLR---------WYRRKSNLHHVWDSDVITQTMKDFFDKDQDAMIESIQRNITD-DWSSEEKQWETCR--
Q9LL59     SDMGGNS-IDLR---------WFRHKSNLHHVWDREIILTAAADYHGKDMHSLLQDIQRNFTEGSWLQDVESWKEC---
L35078     SDLGGNKQLVVID--------SKGTKMLLHAYWDSMAEGKSGEDVPRPLSEADYDDLNNFADY-LEATYASTLTDK---
AF140355   SDLGGNKQTVIVD--------AAGTKMKLHAYWDSIAEGPSGSDMPRPLSADDYADLNTFVDY-LESTYASTLTDA---
O68530     KDQGGNTLTVTFNVTRPAPDNSTFRDFTTFHSVWDTDLITFKYYDWGLAAAEAEKLLPTLAADLLADDTPEKWLAEC---

                                  =              L     Y           G   L
P24289     SEPITTATRWASDANALVCTVVMPHGAAALQTGDLYPTYYDSVIDTIELQIAKGGYRLANWINEIHGSEIAK--------
P24021     KDPVSTSMIWAADANTYVCSTVLDDGLAYINSTDLSGEYYDKSQPVFEELIAKAGYRLAAWLDLIASQPS---------
O81958     SKTTTCAEKYAQESAVLACD----AYEGVEQDDTLGDEYYFKALPVVQKRLAQGGLRLAAILNRIFSGNGRLQSI-----
Q9LL59     DDISTCANKYAKESIKLACNW---GYKDVESGETLSDKYFNTRMPIVMKRIAQGGIRLSMILNRVLGSSADHSLA-----
L35078     EKNLVDTTEISKETFDLALKY---AYPGADNGATLSNEYKTNAKKISERQVLLAGYRLAKMLNTTLKSVSMDTILQGLKS
AF140355   EKTLLNATTISAETFDLAVEY---AYPGGDNGATLSATYKANAKRIAERQVLLGGYRLALMLNQTLRPVTMDAIQQGMKN
O68530     HR--QAEAAYQALPAGTPLKS-----DIGH-PVILDQAYFEKFHPVVTQQLALGGLHLAAELNEALKGGK----------
```

Figure 2 Multiple sequence alignment of selected Zn-containing nucleases homologous to P1. Swiss-Prot[20] (P24289, P24021, O81958, Q9LL59 and O68530) and GenBank[21] (L35078 and AF140355) accession numbers are given that correspond respectively to *Penicillium citrinum* P1 nuclease, *Aspergillus oryzae* S1 nuclease, *Hordeum vulgare* (barley) endonuclease, *Apium graveolens* (celery) CEL I mismatch endonuclease, *Mesorhizobium loti* endonuclease S1 homolog, *Leishmania donovani* 3′-nucleotidase, and *Crithidia luciliae* 3′-nucleotidase. The top lines indicate fully conserved residues, with zinc ligands underlined and bold letter-type used for residues in substrate binding pockets. In addition, the following special symbols are given for the P1 sequence: * for residues involved in nucleotide binding, = for residues forming SS-bonds, and / for carbohydrate binding residues.

Lentinus edodes (Le1), 290 AA[24]

Plant enzymes:
Hordeum vulgare (barley), 265 AA[14]
Zinnia elegans, 278 AA[14]
Hemerocallis sp. (daylily), 276 AA[25]
Arabidopsis thaliana, 277 AA[15]
Apium graveolens (celery), 274 AA[16]

Protozoan enzymes:
Leishmania donovani, 267 AA (without 125 AA N-terminal region)[8]
Leishmania pifanoi, 285 AA[17]
Crithidia lucilia, 352 AA[26]
Leishmania mexicana, 352 AA[27]

Bacterial enzyme:
Mesorhizobium loti, 288 AA[9]

PROTEIN PRODUCTION, PURIFICATION, AND MOLECULAR CHARACTERIZATION

P1 can be commercially purchased in a pure lyophilized form. The first specific studies of P1 were reported by Fujimoto and coworkers.[5] Initial estimates of its molecular weight, based on a number of methods such as gel filtration and SDS-polyacrylamide gel electrophoresis, varied from 42 to 50 kDa.[28] A subsequent amino acid sequence determination[22] indicated that the enzyme consists of 270 amino acid residues and a variable amount of bound sugars, providing molecular masses of the major species, as detected by mass spectrometry, of 35.7, 36.3, and 36.7 kDa. From these data, the carbohydrate content of P1 ranges from 18 to 20% of the total molecular mass. The sugar residues are bound at four asparagines and consist of mannose, glucose, and glucosamine residues in a molar ratio of 6:2:1.[28]

METAL CONTENT AND COFACTORS

On the basis of atomic absorption spectroscopic results, P1 was shown to contain three zinc ions per molecule,[28] in agreement with the first crystallographic analysis.[29] No other metals have been detected in significant amounts. Zinc is essential for catalytic activity, as the enzyme is completely inactivated upon treatment with EDTA (ethylene-diamine-tetra-acetate) but can be partially activated again upon subsequent addition of Zn^{2+}.[1] Similar properties are found for S1 nuclease, which shows 50% sequence identity with P1,[22] contains 3 zinc ions per molecule, and is also inactivated by EDTA treatment.[30]

ACTIVITY TESTS

Several methods have been employed for measuring the two activities of P1.[1,4,5,31] The 3′-phosphomonoesterase activity may be assayed by measuring the amount of liberated inorganic phosphate. This can be done spectrophotometrically using a detection buffer with 0.045% malachite green hydrochloride and 4.3% ammonium molybdate.[17,32,33] A method to determine the phosphodiesterase activity consists of incubating substrate solutions at 37 °C and pH 6 with a small amount of enzyme for different time intervals, each time stopping the reaction by adding ice-cold 0.2 M HCl in 20% ethanol.[4] The degree of hydrolysis may be obtained by integration of substrate and product peaks detected with anion-exchange HPLC (high-pressure liquid chromatography) or with other methods,[30] and kinetic parameters can be determined, for example, from the dependence of the obtained reciprocal reaction rate to the reciprocal substrate concentration.

SPECTROSCOPY

The UV absorption spectrum of a solution of P1 in 0.1 M acetate buffer shows a maximum at 281 nm, with a significant shoulder at about 290 nm. The extinction coefficient at 280 nm is reported to be 18.4,[34] in agreement with a high amount of aromatic residues. The molecular conformation of the enzyme has been investigated by optical rotatory dispersion (ORD) and circular dichroism (CD) spectroscopy. Both methods suggested an α-helix content of approximately 30%,[34] which corresponds to about half of the amount found in the crystal structure.[35] Partial removal of zinc ions by dialysis against a zinc-free solution resulted in a significant change in the CD spectrum, indicating a disruption of α-helix structure, whereas complete removal of zinc by treatment with EDTA gave rise to insolubility of the enzyme.[34] These results suggested that the presence of zinc is important not only for catalysis but also for maintaining the structural integrity of the enzyme.

X-RAY STRUCTURE OF NATIVE P1

Crystallization

The first crystallization experiments used commercially purchased P1 dissolved at a concentration of 10 mg mL^{-1} in 50 mM of sodium acetate buffer, pH 5.3. In a typical hanging drop setup, the protein solution contained, in addition, 1 mM zinc acetate and 30% saturated ammonium sulfate, which was equilibrated against a reservoir containing 65% ammonium sulfate in the same buffer.[29] Two different crystal forms were obtained under these conditions, which sometimes grew in the same drop (see also Table 1). Needle-shaped crystals turned out to be tetragonal with space group $P4_32_12$, whereas bipyramidal crystals were trigonal with space group $P3_221$.

Table 1 Crystal forms of P1 nuclease

Complex	AA[a]	Native	Native	AT[a]	ATTT[a]
Space group	$P4_32_12$	$P3_221$	$P2_12_12_1$	$P4_12_12$	$P2_12_12_1$
a (Å)	133.7	121.0	80.4	77.2	42.0
b (Å)	133.7	121.0	76.4	77.2	74.0
c (Å)	108.3	150.5	63.5	156.6	102.1
Resolution (Å)	2.7	3.0	2.2	2.2	1.8

[a] Further explained in the text.

A third, orthorhombic form was obtained using 16% (w/v) polyethylene glycol 4000 as the precipitating agent.[29]

More recently, when a significantly higher protein concentration was used, two other crystal forms were found during cocrystallization experiments with modified dithiophosporylated DNA oligonucleotides.[36] Hanging drops were formed by mixing 1 μl of a solution of 0.8 mM P1 and 1.6 mM DNA in 25 mM sodium acetate, pH 5.3, and 2 mM zinc chloride with 1 μL of reservoir solution. The latter contained 12 to 20% (w/v) polyethylene glycol 6000 in 25 mM sodium acetate, pH 5.3, either with or without 2 mM zinc chloride. Depending on the oligonucleotide used and on the zinc concentration, tetragonal ($P4_12_12$) or orthorhombic crystals were found. The first was obtained with modified AT and ATTT substrates, whereas the latter was obtained with modified ATAAAA and ATTT at a lower zinc concentration. In both forms, a zinc ion is involved in lattice contacts, whereas in the orthorhombic form the DNA substrate is bound between two symmetry-related molecules.[36]

Overall description of the structure

The structure of P1 was first solved for the tetragonal $P4_32_12$ crystal form, using multiple isomorphous replacement combined with anomalous scattering and solvent flattening.[35] Virtually identical folds were found subsequently for the other crystal forms. One molecule of the enzyme has the longest dimension of about 57 Å. The secondary structure consists mainly of helices that are arranged around the 31-residue-long C-terminal α-helix (3D Structure). Two disulfide bonds further stabilize the fold. Four carbohydrate attachment sites are found that involve asparagine side chains in Asn-X-Thr/Ser sequence motifs, and electron density for several ordered sugar residues has been observed. The three zinc ions of the active site are located at the bottom of a deep cleft, the walls of which are formed mainly by secondary structure elements, including a 7-residue-long 3/10 helix and a β-hairpin.

A sequence comparison of P1 with homologous nucleases shows that the sequences of the protozoan nucleases have diverged most (Figure 2). Nevertheless, a strict conservation of all zinc ligands (see below) is observed. Between the plant and fungal nucleases, both disulfide bonds and at least one of the four carbohydrate attachment sites appear to be conserved as well. The bacterial *Mesorhizobium loti* nuclease is less similar but may still have kept one of the two disulfide bonds. All sequenced nucleases contain at least two Asn-X-Thr/Ser sequence motifs that indicate potential sites for N-linked glycosylation.

The fold of P1 is remarkably similar to that of *Bacillus cereus* phospholipase C (PLC), another enzyme with an active site containing three zinc ions.[41] The similarity extends to about two-thirds of the structure of PLC: a superposition of 160 out of the 245 CA-carbons of PLC to P1 gives a root mean square discrepancy of 1.9 Å.[35] The sequence identity of the superimposed segments is only 16%. However, the coordination of the three zinc ions of P1 and PLC is almost identical, as discussed further in the text below. In accordance with this, both enzymes cleave a phosphodiester bond, present in a nucleic acid in the case of P1 and in a phospholipid in PLC. More recently, almost the same structure has been found in toxin α of *Clostridium perfringens*, which displays the same activity as PLC.[42] Although the sequence identity is rather low, the similarity in fold, zinc coordination, and function strongly suggests that these enzymes have evolved from a common ancestral phosphodiesterase.

Zinc site geometries

The active site of P1 contains three zinc ions that are distributed over a binuclear metal site (Zn1 and Zn3) and a nearby single metal site (Zn2). Zn1 and Zn3 are ≈3.4 Å apart, whereas Zn2 is at ≈6.0 Å from Zn1 and

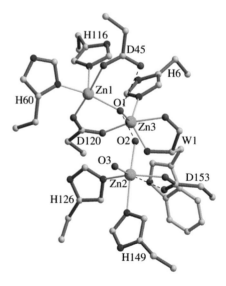

Figure 3 Zinc coordination in the active site of nuclease P1. Color code used: Zn–green, O–red, N–blue, and C–light gray. O1, O2, and O3 represent water/hydroxide ligands that are observed in the 2.2 Å resolution structure of an orthorhombic crystal form of P1.

Table 2 Zinc–zinc and zinc–ligand distances (Å) in native P1 nuclease

Atoms	d (Å)	Atoms	d (Å)	Atoms	d (Å)
Zn3–Zn1	3.4	Zn1–Zn2	6.0	Zn2–Zn3	4.9
Zn3–Trp1 N	2.1	Zn1–Asp45 OD1	2.5	Zn2 – His126 NE2	2.1
Zn3–Trp1 O	2.1	Zn1–His60 ND1	2.0	Zn2 – His149 NE2	2.1
Zn3–His6 NE2	2.0	Zn1–His116 NE2	2.0	Zn2–Asp153 OD2	1.9
Zn3–Asp120 OD1	2.2	Zn1–Asp120 OD2	2.0	Zn2–O2	1.9
Zn3–O1	1.9	Zn1–O1	1.9	Zn2–O3	2.2

≈4.9 Å from Zn3. A special feature is the binding of the N-terminal main chain nitrogen and the carbonyl oxygen of Trp1 to Zn3 (Figure 3). The polypeptide conformation is probably significantly different and perhaps not compatible with zinc binding before proteolytic cleavage of the signal peptide that is located before Trp1. Zn3 is further bound to the NE2-atom of His6 and to OD2 of Asp120, which is also bound to Zn1 by its OD1-atom. Other ligands of Zn1 are ND1 of His60, NE2 of His116, and OD1 of Asp45. Zn2 is coordinated by the NE2-atoms of His126 and His149 and by OD1 of Asp153.

Three additional water ligands are visible at 2.2 Å resolution for an orthorhombic crystal form of the enzyme.[35,43] The first, O1, occupies a bridging position between Zn1 and Zn3 of the binuclear site. The other two, O2 and O3, bind to Zn2. Almost the same zinc coordination is found in *Bacillus cereus* phospholipase C,[41] except for the substitution of Asp153 by a glutamate ligand. In summary, each of the zinc ions in the native enzyme has five ligands in a distorted trigonal bipyramidal arrangement, with metal–ligand distances varying between about 1.9 and 2.5 Å (Table 2).

FUNCTIONAL ASPECTS

Enzyme kinetics

Although P1 is not a sequence-specific nuclease, it does display some base and sugar preference. RNA substrates are, in general, more rapidly cleaved by P1 than DNA substrates, in contrast to S1 nuclease in which the inverse is found.[4,44] In addition, the pH optimum depends on the nature of the bases involved (Table 3). Considering the 3′-phosphomonoesterase activity, the base preference is G > A > C > U at the respective pH optima for ribonucleotides and C ≥ T > A ≥ G for deoxyribonucleotides. However, at pH values up to 6.5, the base preferences are almost identical for both kinds of substrates, becoming C ≥ U > A > G for ribonucleotides.[1] The K_m values of the various 3′-nucleotides are also pH-dependent, ranging between 70 and 600 µM at pH 5.3 and 37 °C, with the tightest binding interactions observed for purine bases. No phosphomonoesterase activity is observed with substrates such as (deoxy)ribose-3-phosphate and 5′-(deoxy)ribonucleotides, confirming the specificity of P1 for a phosphate group at the 3′-position and the requirement for the presence of a base moiety.

Cleavage of phosphodiester bonds by P1 takes place under acidic conditions (Table 3), with sharp pH optima. For example, at pH 4.0, no significant activity is observed for poly (A) and poly (C), whereas poly (U) is readily cleaved.[1] At pH 6.0, this situation is reversed, with low activity for poly (U) and high activity for poly (A) and poly (C). Another remarkable feature is the relatively low activity toward poly (G). RNA substrates are again more rapidly cleaved than denatured DNA, although the difference is less pronounced than for the 3′-phosphomonoesterase activity.

In a study of the susceptibility of various dinucleoside monophosphates to P1, maximal cleavage rates were found when an adenine base is in the 5′-position,[2] for example, V_{max} = 341 enzyme units for A-C and 48 units for G-C. Here, one unit is defined as the number of micromoles of substrate cleaved per milligram enzyme per minute at 37 °C and at pH 5.0. Activity increased upon addition of a terminal 5′-phosphate group, for example, V_{max} = 682 units for pA-C. In a more recent study of the susceptibility of 16 deoxydinucleoside monophosphates to P1, carried out at pH 5,[31] the observed base preference for the 5′-site was A ≥ C > G > T. For the 3′-site, the base preference was found to be C > G > A ≥ T with A, G, or T as the 5′-base, whereas it was A > T > G > C with C as the 5′-base. No significant cleavage activity of P1 was observed

Table 3 Enzymatic activity of nuclease P1 with respect to selected substrate molecules[1]

	Optimal pH	Hydrolysis rate[a]		Optimal pH	Hydrolysis rate[b]
3′-GMP	8.5	1440.0	Poly (I)	4.5	541.0
3′-AMP	7.2	1004.3	Poly (A)	6.0	490.1
3′-CMP	6.0	853.6	Poly (C)	6.0	305.8
3′-UMP	6.0	700.3	Poly (U)	4.0	390.0
3′-dGMP	5.0	14.0	Poly (G)	4.5	14.0
3′-dAMP	4.5	16.7	RNA[c]	6.0	336.0
3′-dCMP	4.5	49.6	Denatured DNA	5.3	218.4
3′-dTMP	4.5	33.6	Native DNA	5.3	1.1

[a] 3′-phosphomonoesterase activity.
[b] Phosphodiesterase activity, both defined as the number of micromoles of substrate cleaved per milligram enzyme per minute at 37 °C.
[c] Commercial low molecular weight RNA with a mean size of about 40 nucleotides.

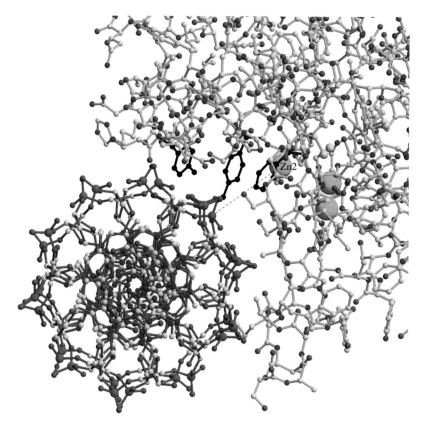

Figure 4 Docking of a 14-bp fragment of B-DNA into the P1 substrate binding cleft. Both molecules are shown as ball-and-stick models, using blue and red sticks for the protein and the DNA respectively. The same atom colors are used as in Figure 3, in addition to yellow for S and violet for P. The dashed line links Zn2 to the closest phosphate group of the docked DNA fragment. Three aromatic residues of P1 that are shown in black highlight the position of two mononucleotide binding sites.

for substrates with phosphodiester bonds other than those between subsequent 3' and 5' hydroxyl groups in oxy- and deoxynucleoside phosphates.[2] Additional studies of depurinated d-ApA substrates showed that the 5'-base of a dinucleotide is required for hydrolysis, whereas the 3'-base is not.[45]

Structural basis of single-strand specificity

P1 is a sequence a-specific nuclease in the sense that its activity is almost independent of the nucleotide sequence of its RNA and DNA substrates. However, it may be called structure-selective[46] as it recognizes certain structural features, including the presence of a single-strand-like conformation. Only low enzymatic activities are measured for native double-stranded DNA and RNA/DNA hybrids in A and B conformations, but cruciform structures and looped-out or otherwise stressed regions are readily attacked.[13] A simple explanation for these findings is provided by the observation that the P1 active site is located at the bottom of a cleft that is inaccessible for double-stranded DNA or RNA substrates in a regular A or B conformation. Assuming rigid conformations, a docking study shows that the closest distance of any B-DNA phosphate group to Zn2, the most exposed zinc ion, is about 7 Å (Figure 4). In addition, the crystallographic results discussed below show that substrate binding to P1 requires unpaired bases and these are not present in regular A- and B-DNA.

FUNCTIONAL DERIVATIVES

General remarks

In order to study enzyme–substrate interactions, complexes with P1 have been prepared either by soaking native crystals with uncleavable substrate analogs or by cocrystallizing the two. When one of the two nonbridging oxygen atoms of the phosphate group that is involved in the diester bond in RNA and DNA substrates is substituted by sulfur, providing a thiophosphorylated substrate analog, only the resulting S-diastereomer is cleaved, whereas the R-diastereomer is cleavage-resistant.[4,47] Likewise, dithiophosphorylated nucleotides[48] are completely cleavage-resistant.[36] In the following section, a selection of results with both kinds of substrate analogs is presented, an important point of

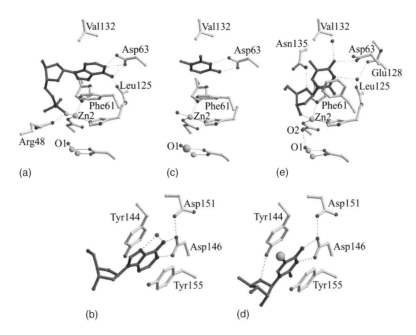

Figure 5 Crystallographically observed P1-substrate interactions. (a,b) binding of dAp(S) in dAp(S)A complex; (c,d) binding of Hg, C, and 5-Hg-Cytidine in 5Hg-CTP complex; (e) binding of dT in d[Ap(S$_2$)Tp(S$_2$)Tp(S$_2$)T] complex. Atoms are colored as in Figures 3 and 4, using, in addition, dark yellow for Hg. The bound nucleotides are shown with dark gray C-atoms and bonds.

discussion being the functional significance of the obtained complexes.

X-ray structure of a complex of P1 with the R-diastereomer of dAp(S)dA

Crystallographic analyses of substrate binding modes of P1 have first been reported at 3 Å resolution for a complex with the R-diastereomer of dAp(S)dA.[35] Reprocessing of the data at 2.7 Å resolution with XDS[49] and model refinement with PROLSQ[50] confirmed the presence of two binding sites at the surface of P1, separated by about 20 Å (3D Structure). Although no cleavage could be detected of this substrate analog upon exposure to P1 in solution, only one adenine base has been found in both binding sites (Figure 5(a) and (b)). The first of these is located close to the active site and involves a hydrophobic interaction of the adenine with the side chains of Phe61 and Val132, along with H-bonding to Asp63 and the carbonyl O-atom of Leu125. In the second nucleoside binding site, the adenine is stacked between the side chains of Tyr144 and Tyr155 and H-bonded to Asp146. Hydrogen bonding to Asp63 and Asp146 implies that either the protein side chain or the base is protonated. Moreover, in the case of the second binding site, Asp146 forms a pair of closely interacting carboxylates with Asp151, suggesting that the latter is protonated as well. Such protonation is compatible with the low pH of 5.3 used for crystallization and with the low pH optimum of P1.[35]

As far as the remainder of the dAp(S)dA substrate analog is concerned, density extending from N1 of the adenine base to Zn2 in the first binding site was assigned to a deoxyribose-5′-thiophosphate moiety that is only partially ordered, as it could not be satisfactorily refined with full occupancy (A. Volbeda, unpublished results). In the major refined conformation, the thiophosphate is bound by its S-atom to Zn2 and further stabilized by salt-bridge interactions with the guanidinium group of Arg48 (Figure 5(a)). In the second binding site, relatively weak electron density extending from the adenine base was assigned to about 50% occupied deoxyribose moiety. It cannot be excluded that in both sites the observed electron density represents a mixture of bound 5′- and 3′-nucleotides, with the adenine bound approximately in the same way and half of the dinucleotide not being detectable because of disorder.

X-ray structure of a 5-Hg-CTP complex of P1

An isomorphous derivative used for the structure solution of P1 was obtained by soaking a tetragonal P4$_3$2$_1$2 crystal with 5-mercury-cytidine triphosphate.[29,35] Two binding sites were found for the cytosine base, which correspond to the two adenine base binding sites discussed above (Figure 5(c) and (d)). Only in the second site, high electron density for a Hg ion bound at C5 of the cytosine was observed, whereas in the first site the bond between C5 and the Hg ion was cleaved, with the Hg ion apparently ending up at the position of Zn1. Upon refinement with PROLSQ[50] (A. Volbeda, unpublished results), all binding sites appeared to be only partially occupied, and no clear electron density was observed for the deoxyribose and phosphate moieties, suggesting disordered or highly mobile conformations. Like in the case of adenine binding, the cytosine is within H-bonding distance of an aspartic acid in both binding sites, suggesting that the latter, Asp63 and Asp146, are protonated.

Incompletely characterized functional derivatives

The binding of many other oligonucleotides to P1 has been investigated by crystal soaking experiments. In most cases, the two already known mononucleoside binding sites were reconfirmed. Possibly because of the binding of different bases to the same site in different P1 copies in the crystal, or because of an insufficient resolution of the X-ray data, the substrate electron densities were often difficult to interpret. An interesting result was obtained by a soaking experiment with inorganic phosphate, showing the binding of a phosphate anion to all three zinc ions. Likewise, in a complex of P1 and 5′-AMP, the phosphate moiety was found between the three zinc ions (A. Lahm and D. Suck, unpublished results).

A crystal of the apoenzyme was obtained by soaking tetragonal $P4_32_12$ crystals of P1 with EDTA.[29,35] All the three zinc ions were removed by this treatment. In contrast to solution studies, but probably due to crystal packing interactions, the overall structure of the apoenzyme did not change much. The obtained resolution of only 3.3 Å was insufficient to merit a detailed analysis of the modifications of the active site. However, the data obtained were very useful for phasing purposes by treating the crystal of the apoenzyme as an isomorphous derivative of the native state.[35]

X-ray structure of P1 in complex with dithiophosphorylated oligonucleotides

Owing to the ambiguity in interpreting the electron density for the dAp(S)dA thiodinucleotide at the active site, new oligonucleotides that contained both purine and pyrimidine moieties were synthesized. In order to further decrease the risk of digestion by P1, both nonbridging oxygen atoms of the phosphate groups were substituted by sulfurs. The following dithiophosphorylated nucleotides were used in cocrystallization studies:[36] $d[Ap(S_2)Tp(S_2)]$, $d[Ap(S_2)Tp(S_2)Tp(S_2)T]$, and $d[Ap(S_2)Tp(S_2)Ap(S_2)Ap(S_2)Ap(S_2)A]$ (called AT, ATTT, and ATAAAA hereafter). Various zinc concentrations in the crystallization droplets lead to crystals belonging either to tetragonal ($P4_12_12$) or orthorhombic ($P2_12_12_1$) space groups (Table 1). In the structures derived from the tetragonal crystals, only mononucleotides were observed at both binding sites, but with poor electron density at the site close to the zinc cluster, suggesting multiple binding orientations. Surprisingly, in the orthorhombic case the oligonucleotides link both binding sites but from symmetry-related molecules. Especially, data at 1.8 Å resolution could be collected on the P1/ATTT complex showing interpretable electron density for most of the tetranucleotide except for the two central bases. As in all other studied P1/dithiooligonucleotide complexes, an adenine is found in the remote nucleotide binding site, whereas the terminal thymidine is bound at the active site (Figure 5(e)). As for the P1/dAp(S)dA complex, no modified phosphate group is observed between the three zincs. Water molecules O1 (bridging Zn1 and Zn3) and O2 (bound to Zn2) are still present, but the most exposed water ligand of Zn2 (O3) is replaced by the O3′ oxygen of the terminal thymidine. An additional interaction with the protein is a H–bond between the ribose O4′-atom and the ND2-atom of Asn135. The latter residue is invariant in all the sequences of P1-like nucleases that have been determined so far (Figures 2 and 5(e)).

Base recognition and substrate specificity

Owing to their chemical differences, different binding modes necessarily apply for the various purines and pyrimidines. This may be partially compensated for by the intercalation of water molecules between the protein and the base (Figure 5). Apart from this, only little conformational changes are observed in the enzyme when it binds oligonucleotides. Two exceptions concern Arg48 and Phe61, which are both located close to the active site. Arg48 interacts with the thiophosphate of the bound dAp(S)dA substrate analog, but it is found in different conformations in other complexes, showing an intrinsic flexibility. The same is true for Phe61, which is found in significantly different orientations depending on whether an adenine or a thymine is stacked against it (Figure 5). Hence, the active site's nucleotide binding pocket may significantly adapt to the binding of different bases. This flexibility is most probably very important as, depending on the base bound, the (deoxy)ribose-phosphate moiety has to be correctly placed for enzymatic cleavage. Still, some base specificity is observed, and the pH optimum of P1 is also base-dependent, suggesting that not all the nucleotides bind with the same affinity or the same optimal conformation to the nuclease.

The binding site remote from the active site, which is composed of Tyr144, Tyr155, and Asp146, is unique to P1. It displays rather low flexibility in terms of adaptation to different substrates, but there is little evidence that this site is responsible for the difference in base specificity of P1. Other homologous enzymes that do not contain such a site also show different base specificity. Owing to its rather external position, it may be that this site is involved in unwinding double-stranded substrates, where a low flexibility might be advantageous. In agreement with this, P1 is significantly more active on double-stranded conformations than S1.[51,52]

A comparison of the sequences of homologous nucleases shows that the binding pocket close to the active site is quite well conserved (Figure 2). Conservative substitutions are observed in some sequences for Arg48 (into Lys), Phe61 (into Tyr), and Asp63 (into Asn or Thr), whereas Asn135 is part of an invariant GGN sequence. Gly133 is part of the base binding pocket, and its substitution with a larger residue may interfere with substrate binding. These observations suggest that all P1-like nucleases have similar substrate recognition and reaction mechanisms.

REACTION MECHANISM

Potter et al.[4,44] have shown that the stereochemical course of RNA/DNA hydrolysis by nucleases P1 and S1 proceeds with inversion of configuration at the phosphorus, which is compatible with a direct nucleophilic attack of an activated water molecule without the involvement of a covalent enzyme intermediate. On the basis of the P1/dAp(S)dA structure, which showed the location of the thiophosphate group between Zn2 and Arg48 (Figure 5(a)), it has first been proposed that one of the two waters (O2 or O3) coordinating Zn_2 could be activated into a hydroxide and acts as a nucleophile. After nucleophilic attack, Arg48 would stabilize the additional negative charge created in the resulting pentacovalent transition state. In this model, the binuclear zinc cluster (Zn1–Zn3) was thought to play mainly a structural role within P1, although an additional role in the activation of the attacking oxygen atom, along with Zn_2, was also considered.[35]

Other catalytic mechanisms have been proposed for P1 by taking into account studies of other enzymes with similar binuclear or trinuclear metal ion catalytic sites. One of them is known as the two–metal ion mechanism[53] and is based on the structure of a 3′,5′-exonuclease in complex with single-stranded DNA.[54] The catalytic site of this enzyme contains a binuclear center consisting of two divalent metals that can be Mn^{2+}, Mg^{2+}, or Zn^{2+}, situated at 3.9 Å from each other. In the complex with DNA, both metal ions coordinate one of the nonbridging oxygen atoms of a phosphate group. One of the metal ions activates a water molecule, assisted by a glutamate and a tyrosine side chain. The other facilitates the leaving of the 3′-O atom formed after cleavage of the bond opposite the attacking nucleophile in a pentacovalent transition state.

A similar two–metal ion mechanism has been postulated for alkaline phosphatase (AP), on the basis of enzyme/inhibitor structures.[55,56] This enzyme, like P1, contains a trinuclear center containing either Zn/Zn/Mg or Zn/Zn/Zn, with distances of 3.9, 4.9, and 7.1 Å between metals. Again a nonbridging oxygen atom of the phosphate group of an inhibitor complex bridges the two closest metal ions. However, in AP, the hydroxyl group of a serine presumably takes over the role of the water molecule in the 3′,5′-exonuclease as the attacking nucleophile. Therefore, the AP mechanism is thought to proceed through the formation of an enzyme–substrate covalent intermediate, which is cleaved in the next step. This explains why AP catalyzes the cleavage of phosphodiester bonds with retention of configuration, which is fundamentally different from what is found for P1.

Because, compared to P1, 3′,5′-exonuclease and AP have completely different structures and significantly different active sites and enzymatic properties, their catalytic mechanisms are likely to be significantly different as well. On the other hand, a similar mechanism would be expected for P1 and PLC, in view of their almost identical active site structures and highly related protein folds. Nevertheless, the structure of PLC in complex with a substrate analog having a nonbridging phosphate oxygen coordinated by both Zn1 and Zn3 suggested that a two–metal ion mechanism similar to the ones found for AP and 3′,5′-exonuclease might be correct.[57] This possibility was, however, weakened by the fact that no water was ideally placed for an in-line attack of the scissile P–O bond. Further molecular interaction energy and distance geometry calculations on this model gave rise to the proposal of a third mechanism[58] in which O1, the water/hydroxide ligand bridging Zn1 and Zn3, would in fact act as a nucleophile, its activation being assisted by Asp45 (P1 numbering). Several other binuclear zinc enzymes have been shown to act through zinc-activated bridging waters.[59,60] In the specific case of PLC, the negative charge of the leaving group resulting from the cleavage of the P–O bond opposite to the attacking nucleophile would be compensated by binding to Zn2.[58]

It may be argued that the thiophosphorylated nucleotides used for binding studies with P1 do not display functional binding modes to the active site. The presence of S may induce steric hindrance causing a different binding mode, although, presumably owing to this, the modified substrates are not cleaved. Steric hindrance would not be a problem for a complex between P1 and the 3′-end of one of the two reaction products (Figure 1), at least not at the level of the O3′ atom of the terminal deoxyribose. It appears that the ATTT dithiotetranucleotide found in orthorhombic crystals could be considered as such a partial product complex.[36] This would imply that the base 5′ to the scissile phosphate is bound to P1, in agreement with experiments from Weinfeld et al.[45] that showed that the bond between the phosphate and the ribose O3′-atom is only cleaved when a base is linked to the ribose. As already mentioned, the O3′-atom is bound to Zn2, just like the leaving group in the discussion of the PLC mechanism in the previous paragraph, replacing the O3 water ligand (Figure 3, 5(e)). Extending the DNA by one nucleotide beyond O3′ gives a model in which the phosphate sits between the three zinc ions, as found in some complexes (Figure 6). The phosphate O_R oxygen replaces water O2 and interacts with Zn2, whereas the O_S and O5′

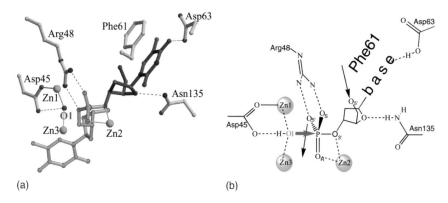

Figure 6 Reaction mechanism of P1, based on the structure of the ATTT productlike complex. (a) Productive enzyme–substrate complex modeled by extending the bound ATTT by one nucleotide (shown with light-blue bonds) in the 3′-direction. (b) Schematic picture of the proposed three–metal ion mechanism. Nucleophilic attack by O1 in-line with the P–O3′ bond produces a pentacovalent phosphate intermediate, which is stabilized by the interaction with Arg48. The subsequent cleavage of the P–O3′ bond with inversion of configuration of the phosphate is supported by Zn2.

oxygen atoms could form a salt bridge with Arg48.[36] In this model, the O1 ligand that bridges Zn1 and Zn3 is perfectly placed for an in-line attack at the phosphorus. Activation of a water molecule at the O1 position by both zinc ions and further by Asp45 would stabilize a hydroxide ion, which directs its lone pair toward the modeled phosphate group. This makes it a most attractive candidate for the attacking nucleophile. Involvement of both Zn1 and Zn3 also helps to explain the low pH optimum of P1.

A three–metal ion mechanism in which the Zn1–Zn3 pair is involved in the activation of the nucleophile and the more distant Zn2 ion stabilizes the negative charge of the leaving group is clearly in agreement with the model obtained by energy calculations for PLC. The absence of a suitable protein residue to protonate the leaving O3′ group suggests that the proton must come from the solvent. This may also partially explain the low pH optimum of P1. As discussed above, another reason for the low pH optimum is the involvement of aspartic acid residues, which need to be protonated for good binding interactions with the nucleotide bases. Considering the length of a P–S bond, which is longer than that of a P–O bond, and the larger Van der Waals radius of sulfur, this model also provides an explanation for the resistance of the R-diastereomers of thiophosphorylated nucleotides. Indeed, a bulkier atom at the O_R position creates steric hindrance with Zn2, which either prevents binding of the phosphate between the three zinc ions or repulses it slightly so that the O1 water cannot attack in-line anymore.

More recently, an almost identical mechanism was found for endonuclease IV, an enzyme that is involved in the repair of apurinic/apyrimidinic (abasic) DNA.[61] Although it has a completely different structure, displaying a TIM-barrel fold, it contains a trinuclear zinc cluster similar to the ones of P1 and PLC, with distances of 3.4, 4.7, and 5.4 Å between metal ions. Analysis at very high resolution (1.02 Å) showed the presence of three water/hydroxide ligands, one bridging between the two closest zinc ions and the two others bound to the remaining zinc. A 1.55-Å resolution analysis of the structure of the enzyme with a cocrystallized abasic DNA substrate analog showed this to be cleaved, with both parts of it still bound, forming therefore a full enzyme/product complex. As for the P1-ATTT complex, the O3′ oxygen of the ribose that was liberated after cleavage of the P–O3′ bond was bound to the more distal zinc ion. The cleaved phosphate itself was found between the three zinc ions, with one of its nonbridging oxygen atoms sitting between the dinuclear pair as would be expected after an in-line attack by the bridging water. As for P1 and PLC, this water may be activated by both zinc ions of the dinuclear pair, with the help of a H-bonded carboxylate group from the protein.

Very recently, an interesting example of bio-mimetic chemistry was reported with two model compounds having similar properties as P1.[62] Both contain a trinuclear metal center consisting of Zn/Zn/Cu in one case and of Zn/Zn/Zn in the other. Metal-to-metal distances are between 3.3 and 3.6 Å, and a water molecule bridges each metal pair. Both compounds display nuclease activity, but this is significantly higher for the one containing only zinc as metal. These results are consistent with the role of a bridging water molecule as the attacking nucleophile and further confirm the importance of having zinc in the active site.

REFERENCES

1. M Fujimoto, A Kunikaka and H Yoshino, *Agric Biol Chem*, **38**, 785–90 (1974).
2. M Fujimoto, A Kunikaka and H Yoshino, *Agric Biol Chem*, **38**, 1555–61 (1974).
3. M Fujimoto, A Kunikaka and H Yoshino, *Agric Biol Chem*, **38**, 2141–47 (1974).
4. BVL Potter, BA Connolly and F Eckstein, *Biochemistry*, **22**, 1369–77 (1983).

5   M Fujimoto, A Kunikaka and H Yoshino, *Agric Biol Chem*, **38**, 777–83 (1974).

6   T Ando, *Biochim Biophys Acta*, **114**, 158–68 (1966).

7   PH Johnson and M Laskowski Sr, *J Biol Chem*, **243**, 3421–24 (1968).

8   A Debrabant, M Gottlieb and DM Dwyer, *Mol Biochem Parasitol*, **71**, 51–63 (1995).

9   JT Sullivan and CW Ronson, *Proc Natl Acad Sci USA*, **95**, 5145–49 (1998).

10 K Shishido and T Ando, in SM Linn and RJ Roberts (eds.), *Nucleases*, Cold Spring Harbor Laboratory, Cold Spring Harbor, New York, pp 155–85 (1982).

11 MJ Fraser, RL Low, in SM Linn and RJ Roberts (eds.), *Nucleases Ed 2*, Cold Spring Harbor Laboratory, Cold Spring Harbor, New York, pp 171–207 (1993).

12 S Gite and V Shankar, *Crit Rev Microbiol*, **21**, 101–22 (1995).

13 DE Pulleyblank, M Glover, C Farah and DB Haniford, in RD Wells and SC Harvey (eds.), *Unusual DNA Structures*, Springer, Heidelberg, pp 23–44 (1988).

14 S Aoyagi, M Suguyama and H Fukuda, *FEBS Lett*, **429**, 134–38 (1998).

15 MA Pérez-Amador, ML Abler, EJ De Rocher, DM Thompson, A van Hoof, ND LeBrasseur, A Lers and PJ Green, *Plant Physiol*, **120**, 169–79 (2000).

16 B Yang, X Wen, NS Kodali, CA Oleykowski, CG Miller, J Kulinski, D Besack, JA Yeung, D Kowalski and AT Yeung, *Biochemistry*, **39**, 3533–41 (2000).

17 S Kar, L Soong, M Colmenares, K Goldsmith-Pestana and D McMahon-Pratt, *J Biol Chem*, **275**, 37789–97 (2000).

18 G Stoesser, W Baker, A van den Broek, E Camon, M Garcia-Pastor, C Kanz, T Kulikova, R Leinonen, Q Lin, V Lombard, R Lopez, N Redaschi, P Stoehr, MA Tuli, K Tzouvara and R Vaughan, *Nucleic Acids Res*, **30**, 21–26 (2002).

19 DA Benson, I Karsch-Mizrachi, DJ Lipman, J Ostell, BA Rapp and DL Wheeler, *Nucleic Acids Res*, **30**, 17–20 (2002).

20 Swiss-Prot Database, http://www.ebi.ac.uk/swissprot.

21 GenBank Database, http://www.ncbi.nlm.nih.gov.

22 K Maekawa, S Tsunasawa, G Dibo and F Sakiyama, *Eur J Biochem*, **200**, 651–61 (1991).

23 BR Lee, B Kitamoto, O Yamada and C Kumagai, *Appl Microbiol Biotechnol*, **44**, 425–31 (1995).

24 H Kobayashi, F Kumagai, T Itagaki, N Inokuchi, T Koyama, M Iwama, K Ohgi and M Irie, *Biosci Biotechnol Biochem*, **64**, 948–57 (2000).

25 T Panavas, A Pikula, PD Reid, B Rubinstein and EL Walker, *Plant Mol Biol*, **40**, 237–48 (1999).

26 M Yamage, A Debrabant and DM Dwyer, *J Biol Chem*, **275**, 36369–79 (2000).

27 WF Sopwith, A Debrabant, M Yamage, DM Dwyer and PA Bates, *Int J Parasitol*, **32**, 449–59 (2002).

28 M Fujimoto, A Kunikaka and H Yoshino, *Agric Biol Chem*, **39**, 1991–97 (1975).

29 A Lahm, A Volbeda and D Suck, *J Mol Biol*, **215**, 207–10 (1990).

30 K Shishido and N Habuka, *Biochem Biophys Acta*, **884**, 215–18 (1986).

31 HC Box, EE Budzinski, MS Evans, JB French and AE Maccubbin, *Biochim Biophys Acta*, **1161**, 291–94 (1993).

32 PA Lanzetta, LJ Alvarez, PS Reinach and AO Candia OA, *Anal Biochem*, **100**, 95–97 (1979).

33 GW Zlotnick and M Gottlieb, *Anal Biochem*, **153**, 121–25 (1986).

34 M Fujimoto, A Kunikaka and H Yoshino, *Agric Biol Chem*, **39**, 2145–48 (1975).

35 A Volbeda, A Lahm, F Sakiyama and D Suck, *EMBO J*, **10**, 1607–18 (1991).

36 C Romier, R Dominguez, A Lahm, O Dahl and D Suck, *Proteins*, **32**, 414–24 (1998).

37 HM Berman, J Westbrook, Z Feng, G Gilliland, TN Bhat, H Weissig, IN Shindyalov and PE Bourne, *Nucl Acids Res*, **28**, 235–42 (2000).

38 PDB: Protein Data Bank, http://www.rcsb.org/pdb.

39 PJ Kraulis, *J Appl Crystallogr*, **11**, 946–50 (1991).

40 EA Merrit and MEP Murphy, *Acta Crystallogr*, **D50**, 869–73 (1994).

41 E Hough, LK Hansen, B Birknes, K Jynge, S Hansen, A Hordvik, C Little, E Dodson and Z Derewenda, *Nature*, **338**, 357–60 (1989).

42 CE Naylor, JT Eaton, A Howells, N Justin, DS Moss, RW Titball and AK Basak, *Nature Struct Biol*, **5**, 738–46 (1998).

43 D Suck, R Dominguez, A Lahm and A Volbeda, *J Cell Biochem Supplement* **17C**, 154 (1993).

44 BVL Potter, PJ Romaniuk and F Eckstein, *J Biol Chem*, **258**, 1758–60 (1983).

45 M Weinfeld, M Liuzzi and MC Paterson, *Nucleic Acids Res*, **17**, 3735–45 (1989).

46 D Suck, *Biopolymers*, **44**, 405–21 (1997).

47 F Eckstein, *Annu Rev Biochem*, **54**, 367–402 (1985).

48 B Dahl, K Bjergårde, J Nielsen and O Dahl, *Tetrahedron Lett*, **31**, 3489–92 (1990).

49 W Kabsch, *J Appl Crystallogr*, **21**, 916–24 (1988).

50 WA Hendrickson and JH Konnert, in R Diamond, S Ramaseshan and D Venkatesan (eds.), *Computing in Crystallography*, Indian Institute of Science, Bangalore, India, pp 1301–23 (1980).

51 VM Vogt, *Eur J Biochem*, **33**, 192–200 (1973).

52 VM Vogt, *Methods Enzymol*, **65**, 248–55 (1980).

53 TA Steitz and JA Steitz, *Proc Natl Acad Sci USA*, **90**, 6498–6502 (1993).

54 LS Beese and TA Steitz, *EMBO J*, **10**, 25–33 (1991).

55 EE Kim and HW Wyckoff, *J Mol Biol*, **218**, 449–64 (1991).

56 JE Coleman, *Curr Opin Struct Biol*, **2**, 222–34 (1998).

57 S Hansen, E Hough, LA Svensson, Y-L Wong and SF Martin, *J Mol Biol*, **234**, 179–87 (1993).

58 S Sundell, S Hansen and E Hough, *Protein Eng*, **7**, 571–77 (1994).

59 DE Wilcox, *Chem Rev*, **96**, 2435–58 (1996).

60 N Sträter, WN Lipscomb, T Klabunde and B Krebs, *Angew Chem*, **108**, 2158–91 (1996).

61 DJ Hosfield, Y Guan, BJ Haas, RP Cunningham and JA Tainer, *Cell*, **98**, 397–408 (1999).

62 SR Korupoju, N Mangayarkarasi, PS Zacharias, J Mizuthani and H Nishibara, *Inorg Chem*, **41**, 4099–101 (2002).

5'-Nucleotidase

Norbert Sträter

Biotechnologisch-Biomedizinisches Zentrum, Universität Leipzig, Leipzig, Germany

FUNCTIONAL CLASS

Enzyme; 5'-Nucleotidase (5'-NT), E.C. 3.1.3.5., UDP-sugar hydrolase, E.C. 3.6.1.45. *Escherichia coli* 5'-NT displays UDP-sugar hydrolase activity, producing uridine monophosphate and glucose-1-phosphate from UDP-glucose, as well as 5'-nucleotidase activity, catalyzing the hydrolysis of phosphate esterified at carbon 5' of the ribose and deoxyribose groups of a nucleotide. Apparent K_m values for AMP and also for UDP-D-glucose are in the low micromolar range (1–30 μm). However, *E. coli* 5'-NT hydrolyzes all 5'-ribo- and 5'-deoxyribonucleotides (including di- and tri-phosphates) with preference for AMP.[1,2] 2'-, 3'-, or cyclic 2',3'-AMP, are no substrates. The unnatural chromogenic substrates *p*-nitrophenyl phosphate and bis(*p*-nitrophenyl) phosphate are readily hydrolyzed by *E. coli* 5'-NT. The related animal 5'-NTs are more specific. These ecto-enzymes are inhibited by ATP and ADP. *p*-Nitrophenylphosphate is not hydrolyzed by these enzymes. The pH optimum for the animal and bacterial enzymes is generally in the neutral or slightly alkaline range (pH 7–8), but depends on the substrate and metal cofactor.

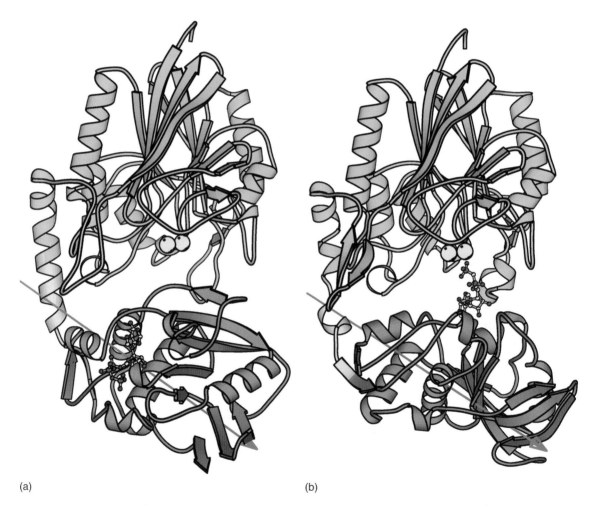

(a) (b)

3D Structure Fold of the (a) open and (b) closed form of 5'-NT. (a) In the inactive open form, the substrate ATP is bound to the substrate-specificity pocket formed by the C-terminal domain (red); (b) the structure of the closed conformation is shown with the ADP-analogue inhibitor α,β-methylene-ADP. The two conformers are related by a 96° domain rotation around the axis that is shown in blue. The bending residues, which enable the domain movement and do not rotate as rigid domains, are shown in yellow. This figure as well as Figures 1, 2, and 5 have been prepared using programs MOLSCRIPT[27] and RASTER3D.[28]

The extracellular bacterial or animal 5'-nucleotidases should not be confused with cytosolic 5'-nucleotidases, which are not related. For a more detailed account on the biochemical properties of 5'-NT, the reader is referred to a review article.[3]

OCCURRENCE

5'-NT occurs not only in bacteria but also in animals. The identity between the different bacterial species is about 60% and that between the bacterial and animal proteins is about 20%. Sequence signature motifs indicate that these enzymes belong to a large superfamily of distantly related metallophosphoesterases,[4,5] including purple acid phosphatases (PAPs) and the serine/threonine protein phosphatases PP-1 and PP-2B (calcineurin).

BIOLOGICAL FUNCTION

Escherichia coli 5'-NT is secreted into the periplasmic space by a 25-residue N-terminal signal peptide. In the periplasm, 5'-NT serves a nutritional role for the bacterium, by hydrolyzing external nucleotides to nucleosides and phosphate or extracellular UDP-glucose to uridine, glucose-1-phosphate, and phosphate for utilization by the cell.[1] In addition, *E. coli* 5'-NT can hydrolyze dinucleoside polyphosphates (Ap_2A, Ap_3A, Gp_3G, and Ap_4A).[6]

Related animal enzymes have a function in the hydrolysis of nucleotides, such as ATP, ADP, UTP, or diadenosine polyphosphates, that serve as extracellular signaling substances.[3] The hydrolysis of ATP is catalyzed by membrane-bound extracellular nucleotidases (ectonucleotidases), including ATPases, apyrases, and 5'-nucleotidases, which dephosphorylate ATP to ADP, AMP, and adenosine. It is generally assumed that these enzymes terminate the neurotransmitter actions of ATP in the brain and periphery. In addition, the surface-located 5'-NTs, which are glycosylated and sialylated, have also been implicated in cell–matrix and cell–cell interactions and in transmembrane signaling. In the case of human ecto-5'-NT (CD73), it has been shown that these functions are independent of the nucleotidase activity.[7]

AMINO ACID SEQUENCE INFORMATION

There are currently more than 50 sequences in the Swiss-Prot Database, which are related to *E. coli* 5'-nucleotidase. Referenced here are the sequences that are not derived from genome sequencing projects. Most of these enzymes have been expressed and partially characterized. The sequences are ordered by their similarity to the *E. coli* sequence.

Escherichia coli, 550 amino acid residues (AA), USHA_ECOLI[8]

Salmonella typhimurium, 550 AA, USHA_SALTY, 95% identity to USHA_ECOLI for 524 aligned amino acids (AA)[9]

Vibrio parahaemolyticus, 560 AA, 5NTD_VIBPA, 63% for 525 AA[10]

Boophilus microplus (Cattle tick), 580 AA, 5NTD_BOOMI, 26% for 567 AA[11]

Lutzomyia longipalpis (sand fly), 572 AA, 5NTD_LUTLO, 25% for 564 AA[12]

Homo sapiens (human placenta), 574 AA, 5NTD_HUMAN, 25% for 543 AA[13]

Bos taurus (bovine liver), 574 AA, 5NTD_BOVIN, 24% for 549 AA[14]

Rattus norvegicus (rat), 576 AA, 5NTD_RAT, 23% for 547 AA[15]

Mus musculus (mouse), 576 AA, 5NTD_MOUSE, 23% for 547 AA[16]

Discopyge ommata (Electric ray), 577 AA, 5NTD_DISOM, 26% for 550 AA[17]

PROTEIN PRODUCTION, PURIFICATION, AND MOLECULAR CHARACTERIZATION

5'-NT was purified and characterized not only from *E. coli* but also from other *Enterobacteriaceae*.[1,2,18–20] The purified enzyme of *E. coli* hydrolyzed all 5'-ribo and deoxyribonucleotides, but showed optimal activity for 5-AMP. The enzyme is inactive against all sugar phosphates except for UDP-glucose, which is a good substrate. For the crystallographic studies described in this article, a homologous overexpression system was used and a purification procedure has been described.[21] It consists of an ammonium sulfate precipitation, a gel filtration chromatography, an ion exchange chromatography, and a second gel filtration step. The enzyme can also be His_6-tagged at the C-terminus for an affinity purification step (unpublished results).

E. coli 5'-NT is a monomeric protein of 525 amino acids (58.1 kDa). The N-terminal 25 residues are cleaved off after the enzyme has been exported to the periplasmic space. A disulfide bridge is present between Cys258 and Cys275. The related animal 5'-NTs are dimeric proteins and contain several glycosylation sites and disulfide bridges. These proteins are attached to the cell membrane via a glycosyl phosphatidylinositol (GPI) anchor.[3]

An intracellular protein inhibitor for 5'-NT is present in *E. coli* and has been partially purified.[1,22] It has been proposed that the inhibitor controls the activity of 5'-NT in the bacterial cell so that the enzyme does not disrupt the nucleic acid metabolism of the cell.

METAL CONTENT AND COFACTORS

When *E. coli* cells are grown in the presence of radioactive ^{65}Zn, the 5'-NT released after osmotic shock was shown to be radioactive.[18] This indicates that E. coli 5'-NT contains at least one tightly bound zinc ion. However, it is unclear if only one or both metal binding sites in the active site are occupied with Zn^{2+}. Inactive metal-free 5'-NT is significantly activated by Mg^{2+}, Co^{2+}, and Mn^{2+}, whereas Zn^{2+} alone does not activate the enzyme.[1,2]

For the crystal structure analyses, the enzyme was purified in the metal-free apoform. This enzyme form is stable for weeks at 4 °C. For crystallization or kinetic assays, the appropriate metal ions are added and they bind quickly to both metal binding sites. However, for full occupancy of both binding sites, millimolar concentrations of Mn^{2+} are required.

ACTIVITY TESTS

The activity of 5'-NT is most conveniently assayed by the hydrolysis of bis-p-nitrophenylphosphate to the yellow product p-nitrophenol ($\lambda = 405$ nm). The more specific activities toward nucleotides can be assayed by detection of the liberated phosphate from AMP or other nucleotides.[23,24] For the animal ectonucleotidases, K_m values for AMP of 1 to 50 µm have been determined.[3] The Michaelis constants for the hydrolysis of AMP as well as of UDP-glucose by the bacterial 5'-NTs are in a similar range of 1 to 30 µM.

X-RAY STRUCTURE OF NATIVE 5'-NT

Crystallization

Four different crystal forms of 5'-NT have been described.[21,25,26] In two crystal forms, I and II, the protein is present in an open conformation, whereas crystal forms III and IV contain 5'-NT in a closed conformation. The two conformations differ in the relative orientation of the two protein domains. Table 1 gives an overview of the crystal forms that have been obtained so far.

Protein fold

5'-NT has a monomeric structure with dimensions of ~80 × 45 × 55 Å (3D Structure).[21] The protein consists of two domains (25–342 and 362–550) connected by a long α-helix (343–361). The larger N-terminal domain has a four-layered structure of the composition α/β-β-β-α, including two sandwiched mixed β-sheets. This domain binds the two catalytic metal ions and is related to a superfamily of dinuclear metallophosphatases. A typical feature of this domain are the two sandwiched βαβαβ motifs, where the parallel β-strands are a part of the two sandwiched layers of β-sheets. The C-terminal domain has a four-layered structure of the composition α/β-β-α-β. A five-stranded β-sheet forms the core of this domain. No folds homologous to the C-terminal domain are currently known in the PDB.

At the time when the X-ray structure of 5'-NT was determined, it was unclear if an open, inactive conformation of the enzyme was present in both crystal forms (I and II) used for the study.[21] Only when crystal forms III and IV were obtained, did it become clear that the orientation of the two protein domains differs by a rotation of as much as 96° between the active closed conformation and the open conformation. The substrate ATP binds to the C-terminal domain of the enzyme in the open form at a distance of about 25 Å away from the dimetal catalytic site (3D Structure). In the closed conformation, however, the substrate analogue inhibitor α,β-methylene-ADP binds close to the two metal ions (3D Structure). In both complex structures, the adenosine part of the substrate/inhibitor binds to the same region of the C-terminal domain. In the closed conformation, the adenosine moiety is rotated with the substrate-specificity pocket of the C-terminal domain close to the dimetal center of the N-terminal domain.

Dimetal site

In the active site of the open conformation in crystal form I, two five-coordinated zinc ions are present at a distance of 3.3 Å. The crystal structures of the closed enzyme forms were determined in the presence of Mn^{2+}-ions and either the products adenosine and phosphate or the substrate

Table 1 Crystal forms and X-ray structures of E. coli 5'-NT

Crystal form	Inhibitor	Domain rotation[a]	Space group[b]	d (Å)[c]	PDB ID	References
I	–	96.1°	$P4_12_12$	1.7	1USH	21
I–ATP	ATP	96.1°	$P4_12_12$	1.7	1HP1	25
II	–	a:91.2°, b:90.3°	$P2_12_12_1$	2.2	2USH	21
III	Adenosine	a:2.8°, b:3.9°	$P2_12_12_1$	2.1	1HO5	25,26
IV	AMPCP	a:7.0°, b:8.4°, c:0°, d:5.4°	P1	1.85	1HPU	25,26

[a] Relative to molecule c of crystal form IV, the letters a–d refer to the chain identifier of the molecule in the asymmetric unit.
[b] With the following cell constants: I: $a = 83.6$ Å, $c = 181.6$ Å; II: $a = 93.2$ Å, $b = 116.3$ Å, $c = 132.8$ Å; III: $a = 69.9$ Å, $b = 75.7$ Å, $c = 221.7$ Å; IV: $a = 89.6$ Å, $b = 90.1$ Å, $c = 96.4$ Å ($\alpha = 110.7°$, $\beta = 106.4°$, $\gamma = 107.7°$).
[c] Crystallographic resolution.

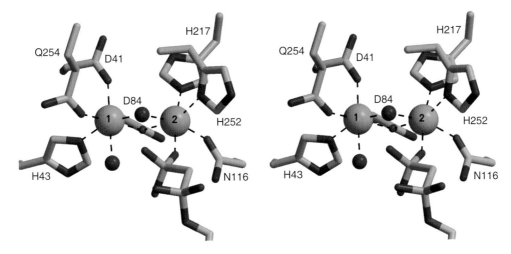

Figure 1 Structure of the dimetal center complexed with the β-methylene-phosphate group of AMPCP. The metal–ligand distances are listed in Table 2.

Table 2 Metal–ligand distances for the 5′-NT-AMPCP complex

Metal	Ligand atom	Distance (Å)
Mn1	OD2 Asp41	2.15
	NE2 His43	2.20
	OD2 Asp84	2.34
	OE1 Gln254	2.17
	OH2 WAT 412	2.16
	OH2 WAT 1137	2.43
Mn2	OD2 Asp84	2.27
	OD1 Asn116	2.17
	NE2 His217	2.24
	ND1 His252	2.27
	OH2 WAT 412	2.27
	O1B AMPCP	2.18
	Mn1	3.47

analogue inhibitor α,β-methylene-ADP (AMPCP). In these structures, both metal ions are octahedrally coordinated (Figure 1). The binding site 2 has a higher affinity to bind manganese ions compared to binding site 1, which was found to be not occupied or only partially occupied in several structures despite the presence of concentrations of Mn^{2+} of more than 1 mM.[25] Mn2 is coordinated by two histidines, a carboxylate, and by a carboxamide side chain. It also binds the terminal phosphate group of the substrate.

In the complex structure with the substrate analogue, two water molecules are coordinated to the metal ions and might serve as nucleophiles: One water molecule (or hydroxide ion) bridges the two metal ions symmetrically, whereas the second one is terminally coordinated to Mn1.

COMPLEX STRUCTURES

X-ray structures of 5′-NT in complex with the substrate analogue AMPCP and of the product complex with adenosine and phosphate provide insight into the catalytic mechanism. The substrate analogue binds at the interface between both domains (Figure 2). The specificity pocket for nucleotides is mainly formed by the C-terminal domain, whereas residues of the N-terminal domain provide most of the catalytic residues, in particular, the dimetal center and His117. Arg419, which is in direct contact to the β-phosphate group, may also contribute to the transition-state stabilization. In addition to the dimetal center and Arg419, Arg375 and Arg379 probably contribute to the distinct positive potential at the active site, which might be necessary for binding of the substrate phosphate groups and for transition-state stabilization.

A hydrophobic stacking interaction with Phe429 and Phe498 is the most prominent feature of the binding mode of the adenine ring to the enzyme (Figure 2). In addition, there are several hydrogen-bonding interactions between the adenine ring or the ribose ring and residues of the C-terminal domain. The three arginines (375, 379, and 410) mainly bind the α-phosphate group of the substrate.

The binding mode of adenosine to the substrate specificity pocket of 5′-NT is very similar to that of AMPCP. A difference is seen in the three arginine bases, which are not in hydrogen-bonding distance to adenosine or to the phosphate ion. The phosphate ion in the product complex binds in an almost identical binding mode as does the β-phosphate group of AMPCP.

CATALYTIC MECHANISM

On the basis of the binding modes of the substrate analogue and the products, a catalytic mechanism for 5′-NT has been proposed (Figure 3). The terminal phosphate group of the substrate binds with one oxygen atom to the site 2 metal ion, as seen in the structure of 5′-NT with AMPCP. Arg410 and His117 bind and polarize

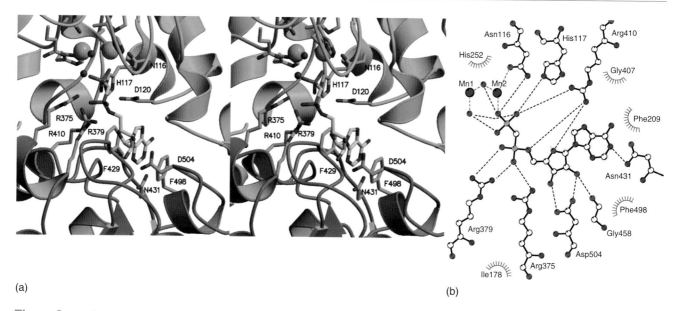

Figure 2 Binding mode of AMPCP at the interface between the two protein domains: (a) Stereo representation. The ribbon representation is colored green for the N-terminal domain, red for the C-terminal domain and yellow for the bending residues (b) Schematic view of the interaction between the inhibitor AMPCP and 5′-NT.

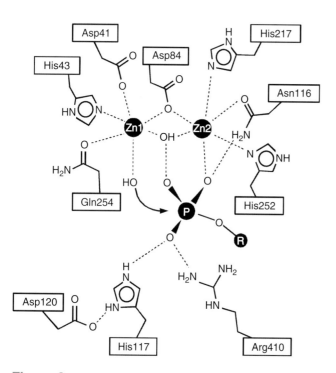

Figure 3 Proposed structure of the Michaelis complex for 5′-NT catalysis. The phosphoryl group is directly transferred from the leaving group to the nucleophile without the formation of a covalent intermediate.

the phosphate group for attack of the nucleophile. The nucleophilic water molecule is supposed to be the terminal water ligand to the site 1 metal ion. In the structure with AMPCP, this water ligand is located at a distance of 3.2 Å to the electrophilic phosphorus atom. The angle between this water, the phosphorus atom, and the leaving group is 155°, which deviates by 25° from a perfect inline arrangement.

The transition state is stabilized by the two metal ions, Arg410 and His117. For protonation of the leaving alcohol group, no suitable general acid is appropriately positioned. It is therefore assumed that a water molecule provides a proton for the leaving group. After the leaving group has left, the phosphate group remains bound bidentately bridging the two metal ions. This postulated primary product binding mode differs from the complex as seen in the structure with adenosine and phosphate, where the phosphate group is bound similar to the substrate phosphate group. After an inline phosphoryl transfer reaction with inversion of the configuration around the phosphorus atom, the binding mode of the phosphate group should be different compared to the substrate. Furthermore, such a bridging binding mode for the phosphate ion has been observed in crystal structures of the purple acid phosphatases (PAPs) and serine/threonine protein phosphatases (PPs). The proposed mechanism for 5′-NT resembles that suggested for the PAPs based on structural and many spectroscopic data.[29] This includes the direct transfer of the phosphoryl group from the substrate to water under inversion of the configuration at the phosphorus atom.[30]

Alternative possibilities for the catalytic mechanism have been proposed for the related enzymes of the calcineurin superfamily (see below) in which the metal-bridging water molecule is the nucleophile and/or the substrates phosphate group binds in a bridging mode to both metal ions as observed for some product complexes with phosphate.

COMPARISON TO THE ANIMAL 5'-NUCLEOTIDASES

The core catalytic residues (the metal ligands, His117, and Asp120) are conserved between *E. coli* 5'-NT and the animal nucleotidases with the exception of Gln254, which is an asparagine in the animal enzymes.[21] In the 5'-NTs from *Haemophilus influenzae*, *Boophilus microplus*, and *Archaeglobus fulgidus*, a histidine is present at this position, similar to the site 1 ligand sphere in the PAPs. The cystine bridge between Cys258 and Cys275 appears to be present in the bacterial enzymes but not in the animal ones.

The substrate specificity pocket, in particular, Arg379, Arg410, Gly407, Phe429, Gly458, and Asp504, is well conserved between the bacterial 5'-NT and the animal counterparts. Phe498 is replaced by a tyrosine in several enzymes, which should be a good replacement for the hydrophobic stacking interaction with the adenine ring. In addition, an analysis of the positions of disulfide bridges and glycosylation sites strongly indicates that the core structures of both domains of 5'-NT are similar between the bacterial and animal enzymes.

COMPARISON TO THE CALCINEURIN SUPERFAMILY OF METALLOPHOSPHATASES

5'-NT belongs to a large superfamily of phosphatases with a dinuclear metal center in the active site. A large variety of substrates are hydrolyzed by these enzymes, including phosphoproteins, nucleotides, as well as RNA and DNA. Crystal structures have been determined for several enzymes as summarized in Table 3.

The active sites of 5'-NT, the serine/threonine protein phosphatases (PPs), and the PAPs are similar, but they also have interesting differences (Figures 4 and 5). The conserved metal ligands and the catalytic histidine (His117) superimpose in identical conformations. Interestingly, the aspartate which is hydrogen bonded to the catalytic histidine is in all of these enzymes derived from different (nonhomologous) sequence positions. This aspartate is not present in the animal PAPs. A mutant of bacteriophage λPPase in which this aspartate was mutated to an asparagine showed a 400-fold reduced k_{cat} with little change in K_m.[41]

Whereas all ligands to metal site 2 (the high-affinity metal binding site in 5'-NT) are strictly conserved, some variation is seen in the ligands coordinated to the site 1 metal ion. In the purple acid phosphatases, the binding site 1 is occupied by a Fe(III) ion. In addition to the two metal ions and the histidine, additional residues are present in the active sites of these enzymes. These residues, which are presumably involved predominantly in substrate binding but possibly also in catalysis, are different between the enzyme families.

The catalytic residues of the active sites superimpose well as shown in Figure 5 for λPP and 5'-NT. This includes the metal-bridging water molecule and the water molecule terminally coordinated to the site 1 metal ion. The binding mode of a sulfate ion to λPP is similar to that of the β-phosphate group of AMPCP to the site 1 metal ion.[25,40] These similarities indicate that the enzymes of the calcineurin superfamily might also share a common catalytic mechanism.

DOMAIN MOVEMENT

Escherichia coli 5'-NT shows a unique type of hinge-bending domain rotation in which the C-terminal domain rotates approximately around its center (3D Structure). As a result, the residues at the domain interface (where the active site is located) move predominantly along the interface. This type of movement is different from the classical closure motion, where a cleft between the domains opens up (or closes) with the movement and the residues at the domain interface move mainly perpendicular to the interface.

A typical feature of a hinge-bending domain rotation is that the rotation axis passes through the bending residues, where the rotational transition between the two connected rigid domains occur. In 5'-NT residues 352–364 are the bending residues. In the open conformation, these residues are part of the long α-helix connecting the two domains. In the closed conformation, the hydrogen-bonding pattern of this helix is broken in the region of residues Lys355 and

Table 3 Structurally known enzymes of the calcineurin superfamily

Enzyme	Organism	Resolution (Å)	Inhibitor	PDB ID	References
5'-NT	*E. coli*	1.85	AMPCP (1HPU)	1HPU	25
PAP	Red kidney bean	2.65	Native PO_4^{3-}	2KBP 4KBP	31,32
PAP	Rat	2.2	SO_4^{2-} PO_4^{3-}	1QHW 1QFC	33,34
PAP	Pig	1.55	PO_4^{3-}	1UTE	35
PP-1	Rabbit	2.1	PO_4^{3-}	1FJM	36
PP-1	Human	2.5	WO_4^{3-}	–	37
PP-2B	Human	2.1	Native	1AUI	38
PP-2B	Bovine	2.5	PO_4^{3-}	1TCO	39
λPP	Bacteriophage λ	2.15	SO_4^{2-}	1G5B	40

5'-Nucleotidase

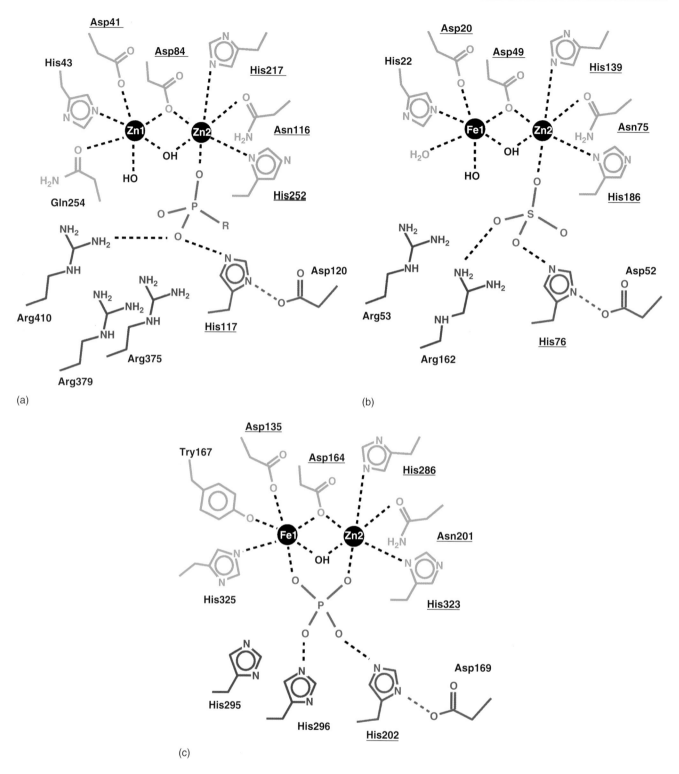

Figure 4 Comparison of the active site structures of (a) *E. coli* 5'-NT; (b) bacteriophage λ-protein phosphatase; (c) kidney bean purple acid phosphatase. The conserved residues are marked by underlined residue labels. Also shown is the binding mode of the β-phosphate group of AMPCP to 5'-NT and of a sulfate anion to λPP.

Gly356. These are also the residues where the principal main-chain torsion angle changes take place.

Using the four crystal forms, nine independent protein conformers could be observed. Three conformers are in the open conformation (all from crystal forms I and II) and six conformers are in the closed conformation. The open conformers differ by an interdomain rotation angle of up to 10.4°, whereas the closed conformers differ by up to

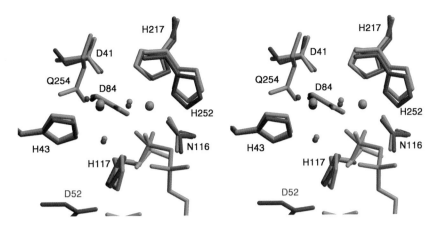

Figure 5 Superposition of the active site structures of AMPCP (blue) bound to 5'-NT (green) with λPP (red) and the bound sulfate ion (yellow). The labels refer to 5'-NT except for D52.

8.4°. The change between conformers of the closed conformation and between conformers of the open conformation all involve movements where the residues move along the domain interface.

The influence of the domain movement on the enzyme function remains to be experimentally shown. It appears likely that the domain movement is necessary for substrate binding and product leaving. However, it is unclear if the large 96° movement is necessary for efficient catalysis or if a smaller flexibility within the closed conformation is sufficient. The domain movement is facilitated by little direct protein–protein contacts between the two domains. Instead, a layer of water molecules is present between the domains at the domain interface.

REFERENCES

1. L Glaser, A Melo and R Paul, J Biol Chem, **242**, 1944–54 (1967).
2. HC Neu, J Biol Chem, **242**, 3896–904 (1967).
3. H Zimmermann, Biochem J, **285**, 345–65 (1992).
4. EV Koonin, Protein Sci, **3**, 356–58 (1994).
5. N Sträter, WN Lipscomb, T Klabunde and B Krebs, Angew Chem, Int Ed Engl, **35**, 2024–55 (1996).
6. A Ruiz, C Hurtado, JM Ribeiro, A Sillero and MAG Sillero, J Bacteriol, **171**, 6703–9 (1989).
7. W Gutensohn, R Resta, Y Misumi, Y Ikehara and LF Thompson, Cell Immunol, **161**, 213–17 (1995).
8. DM Burns and IR Beacham, Nucleic Acids Res, **14**, 4325–42 (1986).
9. DM Burns and IR Beacham, J Mol Biol, **192**, 163–75 (1986).
10. Y Tamao, K Noguchi, Y Sakai-Tomita, H Hama, T Shimamoto, H Kanazawa, M Tsuda and T Tsuchiya, J Biochem, **109**, 24–29 (1991).
11. N Liyou, S Hamilton, C Elvin and P Willadsen, Insect Mol Biol, **8**, 257–66 (1999).
12. JMC Ribeiro, ED Rowton and R Charlab, Insect Biochem Mol Biol, **30**, 279–85 (2000).
13. Y Misumi, S Ogata, K Ohkubo, S Hirose and Y Ikehara, Eur J Biochem, **191**, 563–69 (1990).
14. K Suzuki, Y Furukawa, H Tamura, N Ejiri, R Suematsu, R Taguchi, S Nakamura, Y Suzuki and H Ikezawa, J Biochem, **113**, 607–13 (1993).
15. Y Misumi, S Ogata, S Hirose and Y Ikehara, J Biol Chem, **265**, 2178–83 (1990).
16. R Resta, SW Hooker, KR Hansen, AB Laurent, JL Park, MR Blackburn, TB Knudsen and LF Thompson, Gene, **133**, 171–77 (1993).
17. W Volknandt, M Vogel, J Pevsner, Y Misumi, Y Ikehara and H Zimmermann, Eur J Biochem, **202**, 855–61 (1991).
18. HF Dvorak and LA Heppel, J Biol Chem, **243**, 2647–53 (1968).
19. IR Beacham and MS Wilson, Arch Biochem Biophys, **218**, 603–8 (1982).
20. HC Neu, Biochemistry, **7**, 3766–73 (1968).
21. T Knöfel and N Sträter, Nat Struct Biol, **6**, 448–53 (1999).
22. HC Neu, J Biol Chem, **242**, 3905–11 (1967).
23. PA Lanzetta, LJ Alvarez, PS Reinach and OA Candia, Anal Biochem, **100**, 95–97 (1979).
24. PP van Veldhoven and GP Mannaerts, Anal Biochem **161**, 45–48 (1987).
25. T Knöfel and N Sträter, J Mol Biol, **309**, 239–54 (2001).
26. T Knöfel and N Sträter, J Mol Biol, **309**, 255–66 (2001).
27. PJ Kraulis, J Appl Crystallogr, **24**, 946–50 (1991).
28. EA Meritt and MEP Murphy, Acta Crystallogr, Sect D, **50**, 869–73 (1994).
29. T Klabunde and B Krebs, Struct Bonding, **89**, 177–98 (1997).
30. EG Mueller, MW Crowder, BA Averill and JR Knowles, J Am Chem Soc **115**, 2974–75 (1993).
31. N Sträter, T Klabunde, P Tucker, H Witzel and B Krebs, Science, **268**, 1489–92 (1995).
32. T Klabunde, N Sträter, R Fröhlich, H Witzel and B Krebs, J Mol Biol **259**, 737–48 (1996).
33. Y Lindqvist, E Johansson, H Kaija, P Vihko and G Schneider, J Mol Biol, **291**, 135–47 (1999).
34. J Uppenberg, F Lindqvist, C Svensson, B Ek-Rylander and G Andersson, J Mol Biol, **290**, 201–11 (1999).
35. LW Guddat, AS McAlpine, D Hume, S Hamilton, J de Jersey and JL Martin, Structure, **7**, 757–67 (1999).
36. J Goldberg, H Huang, Y Kwon, P Greengard, AC Nairn and J Kuriyan, Nature, **376**, 745–53 (1995).

37 M-P Egloff, PTW Cohen, P Reinemer and D Barford, *J Mol Biol*, **254**, 942–59 (1995).

38 CR Kissinger, HE Parge, DR Knighton, CT Lewis and E Villafranca, et al., *Nature*, **378**, 641–44 (1995).

39 JP Griffith, JL Kim, EE Kim, MD Sintchak, JA Thomson, MJ Fitzgibbon, MA Fleming, PR Caron, K Hsiao and MA Navia, *Cell*, **82**, 507–22 (1995).

40 WC Voegtli, DJ White, NJ Reiter, F Rusnak and AC Rosenzweig, *Biochemistry*, **39**, 15365–74 (2000).

41 S Zhuo, JC Clemens, RL Stone and JE Dixon, *J Biol Chem*, **269**, 26234–38 (1994).

E. coli alkaline phosphatase

Evan R Kantrowitz

Department of Chemistry, Boston College, Merkert Chemistry Center, Chestnut Hill, MA, USA

FUNCTIONAL CLASS

Enzyme: alkaline phosphatase; EC 3.1.3.1; a zinc, magnesium metalloenzyme; known as phosphomonoesterase, alkaline phosphomonoesterase, and alkaline phosphatase (AP).

AP from *Escherichia coli* catalyzes the hydrolysis of phosphomonoesters from a wide variety of substrates. The pH optimum of the *E. coli* enzyme is 8.0. AP can also transfer the phosphate group from the substrate to an acceptor alcohol via a transphosphorylation reaction. Alcohols such as Tris and ethanolamine are often used as the acceptor alcohols *in vitro*.

OCCURRENCE

In *E. coli* cells, AP is found in the periplasmic space between the inner and outer membranes. A comparison of the amino acid sequence of the mature protein (449 amino acids)[1] isolated from the periplasmic space to the amino acid sequence derived from the nucleotide sequence[2] reveals that the mature protein undergoes a processing step in which a 22-amino acid peptide is removed from the N-terminus. This predominantly hydrophobic peptide is a signal peptide that directs the export of the polypeptide chain to the periplasm.[3] Structurally and functionally related APs have been found in organisms ranging from bacteria to mammals.

AMINO ACID SEQUENCE INFORMATION

Escherichia coli, accession number NP_414917, precursor, 471 amino acid residues,[2] mature protein, 449 amino acids.[1] The SMART (Simple Modular Architecture Research Tool, http://smart.embl-heidelberg.de) program has identified 120 sequences (alkPPc) from bacterial, archaeal, and eukaryotic sources that are homologues of *E. coli* alkaline phosphatase.

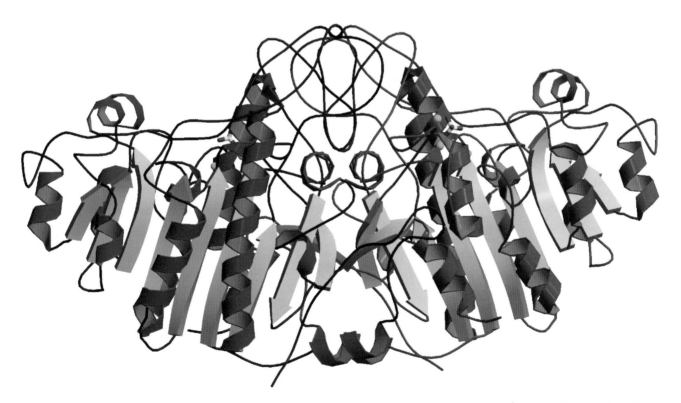

3D Structure Schematic representation of the 3D Structure of dimeric *E. coli* AP based on the 1.75 Å structure[16] with P$_i$ bound (PDB code 1ED8). The monomers are shown in blue and red. Zn$_1$ and Zn$_2$ are shown in turquoise and Mg is green. Prepared with the program MOLSCRIPT[21] and RASTER3D.[22]

BIOLOGICAL FUNCTION

In *E. coli*, the main biological function of AP is to supply inorganic phosphate (P_i). The level of expression of AP is regulated by the cellular levels of P_i, being repressed in the presence of P_i and activated in the absence of P_i.[4,5] When cells are grown in media containing high levels of P_i, low levels of AP are observed, whereas if cells are grown in media depleted of P_i, high levels of AP are observed.

PROTEIN PRODUCTION, PURIFICATION, AND MOLECULAR CHARACTERIZATION

The standard expression system from *E. coli* AP is based on the strain SM547 [Δ(*phoA-proC*), *phoR*, tsx::Tn5, Δlac, *galK*, *galU*, *leu*, str^r]. This strain has a deletion in the *phoA* gene, which codes for AP, as well as a defect in the *phoR* gene, a regulatory gene for the *pho* regulon. Expression of the wild-type or mutant AP is achieved by the transformation of SM547 with a plasmid carrying the wild-type (pPhoA) or mutant *phoA* gene, with its natural promoter. Growth of SM547 harboring pPhoA in rich media results in the accumulation of AP in the periplasmic space of the *E. coli* cells.[6] Osmotic shock is then used to break the outer cell membrane and to release the proteins from the periplasmic space while most of the cellular proteins and nucleic acids are retained in spheroplasts.[7] Standard purification involves ammonium sulfate precipitation and anion-exchange chromatography.[8] For higher purity, an additional hydrophobic interaction chromatography step is added.[9]

E. coli AP exists in a dimeric quaternary structure in the periplasmic space. The signal sequence at the N-terminal of AP directs the translocation of the protein to the periplasmic space.[10] Following passage through the inner membrane, the signal sequence is removed[10,11] and the monomers acquire metals and undergo dimerization. The acquisition of metals by the monomer is critical for the dimerization of the enzyme.

METAL CONTENT

Each monomer of the dimeric AP has three metal binding sites identified as M1, M2, and M3. The M1 and M2 sites have a high affinity for zinc, while the M3 site has a high affinity for Mg. The metal content of AP is dependent upon metal availability and, therefore, under some circumstances the actual metal content may be different from the expected two zinc, one magnesium per active site, for example, if *E. coli* cells are grown in media depleted of metals, a nonmetal containing apo form of AP is obtained.[12] Upon addition of metals, this apoenzyme formed *in vivo* is converted to a form identical to the normal metal containing AP. If *E. coli* cells are grown in media depleted of Zn^{2+}, but with high concentrations of other metals such as Co^{2+} or Cd^{2+}, the AP isolated contains Co^{2+} or Cd^{2+} respectively.[12]

ACTIVITY TEST

The standard assay for AP activity, developed by Garen and Levinthal,[13] involves the use of *p*-nitrophenylphosphate. Under basic conditions the release of *p*-nitrophenolate can be easily measured spectrophotometrically at 410 nm. If the reaction is carried out in the presence of a phosphate acceptor such as 1.0 M Tris, the activity measured is the sum of the hydrolysis and transphosphorylation reactions. Hydrolysis alone is measured in the absence of a phosphate acceptor in a buffer such as 4-morpholinepropanesulfonic acid (MOPS). A unit is defined as 1 μmol *p*-nitrophenylphosphate hydrolyzed per minute.

X-RAY STRUCTURE OF NATIVE AP

Crystallization

Although the ability of AP to crystallize was known as far back as 1964,[14] the crystallization conditions that resulted in the formation of crystals of sufficient quality for X-ray structure determination were developed in the Wyckoff Laboratory at Yale University.[15] The highest quality crystals of wild-type AP were obtained by vapor diffusion using hanging drops.[16] The enzyme solution, at approximately 30 mg mL^{-1}, is first dialyzed against a 20% saturated solution of $(NH_4)_2SO_4$ in 100 mM Tris, 10 mM $MgCl_2$, 0.01 mM $ZnCl_2$ at pH 9.5. Crystals are formed in reservoirs with the ammonium sulfate concentration between 39 and 43% saturation. Before data collection, the crystals are normally transferred into a stabilizing solution containing 55 to 65% saturated $(NH_4)_2SO_4$, 100 mM Tris, 10 mM $MgCl_2$, 1 mM $ZnCl_2$, 2 mM P_i at pH 7.5.

The structure of AP

The X-ray structure of AP was solved by Wyckoff and colleagues at Yale University.[15,17,18] They found that the quality of the crystals of AP was improved dramatically when P_i was added to the stabilizing solution. These higher quality crystals resulted in the first high-resolution structure of AP to 2.0 Å resolution.[19] Improved structures of AP, both in the presence and absence of P_i, have recently been determined to 1.75 Å resolution.[16] The high quality of these data allowed the final stages of the refinement to be carried out with SHELXL-97[20] and provided the opportunity to perform an anisotropic refinement of the active site residues, providing detailed information about the mobility of these residues.

E. coli alkaline phosphatase

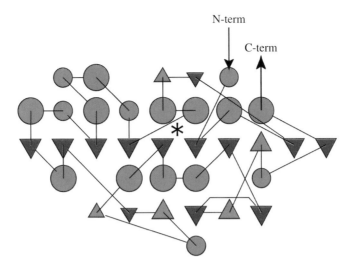

Figure 1 A topology diagram of AP monomer. Circles represent α-helices and triangles represent β-sheet. The red and green β-sheets are oriented in opposite directions. The asterisk represents the approximate position of the active site. Prepared using the topology of protein structure server (TOPS) at http://www.tops.leeds.ac.uk.

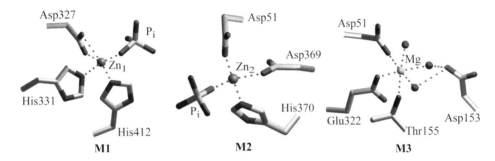

Figure 2 The three metal binding sites in AP based on the 1.75 Å structure[16] with P_i bound (PDB code 1ED8). For the *E. coli* enzyme, M1 and M2 are zinc sites and M3 is a magnesium site. The three water molecules in the M3 site are not labeled.

Shown in the 3D Structure is a ribbon plot of the AP dimer shown with the metals (two Zn and one Mg) and P_i bound in each active site. The structure is that of a symmetric dimer (100 Å × 50 Å × 50 Å). The active sites are approximately 30 Å apart and are not shared across the dimer interface. As seen in the topology diagram of a monomer (Figure 1), the basic structure is of the α/β type. The core of the structure is composed of a 10-strand β-sheet with all but one of the strands running parallel, surrounded on both sides by α-helices. Additional layers of α-helices and β-sheet comprise the remainder of the structure.

Metal sites

Plocke *et al.*[23] showed in 1962 that AP contained at least one zinc atom per monomer. Later studies utilizing nuclear magnetic resonance (NMR)[24] and X-ray crystallography[17] conclusively established that there were three metal binding sites per monomer of AP. In the presence of both Zn^{2+} and Mg^{2+}, two of the sites are filled by Zn^{2+} and one by Mg^{2+}. Under appropriate conditions, other metals can be substituted for Zn^{2+} and Mg^{2+} such as Cd^{2+} and Co^{2+}.[25] Shown in Figure 2 are the three metal sites in each active site of AP as observed by X-ray crystallography.[19,26] The M1 and M2 sites normally bind Zn^{2+} and are therefore also referred to as the Zn_1 and Zn_2 sites. The M3 site normally binds Mg^{2+} and is therefore also referred to as the Mg site. Given in Table 1 is a list of the bond distances between the ligands and the metal ions in each of the metal sites.

The Zn_1 site

The Zn_1 site, which has the highest affinity for Zn, binds the Zn^{2+} in a tetragonal pyramidal geometry. The Zn^{2+} is ligated to the NE nitrogens of His331 and His412, one oxygen of the phosphate, and both the carboxylate oxygens of Asp327, which acts as a bidentate ligand. The two phosphate oxygens act as a bridge between the Zn_1 and Zn_2 sites.

E. coli alkaline phosphatase

Table 1 Interactions to the three metal sites in AP[a]

Site	Ligand	Atom	Distance A subunit (Å)	Distance B subunit (Å)
Zn_1	Asp327	OD1	2.01	2.06
Zn_1	Asp327	OD2	2.45	2.41
Zn_1	His331	NE2	2.20	2.26
Zn_1	His412	NE2	1.93	2.06
Zn_1	PO_4^{2-}	O2	1.84	1.82
Zn_2	Asp51	OD1	1.93	2.00
Zn_2	Asp369	OD1	2.04	2.01
Zn_2	His370	NE2	2.01	2.15
Zn_2	PO_4^{2-}	O3	1.86	1.89
Mg	Asp51	OD2	2.10	2.09
Mg	Thr155	OG1	2.15	2.23
Mg	Glu322	OE2	2.02	1.91
Mg	Water$_1$	O	2.23	1.98
Mg	Water$_2$	O	2.15	2.13
Mg	Water$_3$	O	2.14	2.27

[a] These distances were obtained from the 1.75 Å structure of the enzyme with P$_i$ bound[16] (PDB code 1ED8).

The Zn$_2$ site

The Zn$_2$ site, which has a lower affinity for Zn^{2+} than the Zn$_1$ site, binds the Zn^{2+} in a distorted tetrahedral geometry. The Zn^{2+} is ligated to the NE2 nitrogen of His370, one oxygen of the phosphate, and the carboxylate oxygens of Asp51 and Asp369. The carboxylate of Asp51 acts as a bridge between the Zn$_2$ and Mg sites.

The Mg site

Of the three metal sites, the Mg site has the lowest affinity for Zn^{2+} and the highest affinity for Mg^{2+}. The Mg^{2+} binds with octahedral geometry and is ligated to the carboxylate oxygens of Asp51 and Glu322 as well as to the hydroxyl oxygen of Thr155. The other three coordination sites are completed by oxygens from water molecules. Two of these water molecules are held in place by additional hydrogen bonding interactions with the carboxylate of Asp153. As indicated above, the carboxylate of Asp51 acts as a bridge between the Zn$_2$ and Mg sites.

The active site

The active site of AP is a shallow pocket on the surface of the enzyme; this pocket is barely large enough to bind the phosphate portion of the substrate, which provides a structural explanation for the lack of substrate specificity found with AP. Because of the presence of the three metals ions as well as the positively charged side chains such as Arg166, the active site pocket is extremely electropositive. Shown in Figure 3 is a surface representation of an AP

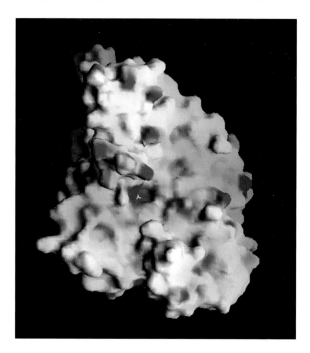

Figure 3 A surface representative of one monomer of AP. Mapped on the surface is the electrostatic potential that was calculated in the absence of P$_i$. P$_i$ has been added to show the position of the active site. Prepared with the program GRASP.[27]

monomer onto which the electrostatic potential has been mapped. The highly electropositive nature of the active site pocket is a critical component for the attraction of the substrate to the enzyme.

Shown in Figure 4 is the active site of AP with the product P$_i$ bound.[16,19] This structure represents the noncovalent enzyme–phosphate complex (E•P$_i$). The spatial arrangement of the three metal sites is easily observed in this representation. The P$_i$ is bound in the active site by interactions with Zn$_1$, Zn$_2$, the guanidinium group of Arg166, and a water-mediated interaction with Lys328. Arg166, which is important for the binding of both the P$_i$ and the substrates, is held in the proper position by a salt-link to Asp101 and a water-mediated interaction with the carboxylate of Asp153. The carboxylate of Asp153 is also involved in hydrogen bonding interactions with two water molecules that are ligands to Mg and in a salt-link to Asp328.

FUNCTIONAL ASPECTS

Kinetic scheme

In 1961, Schwartz and Lipmann[28] showed that a serine residue was phosphorylated during the catalytic reaction creating a phosphoseryl intermediate (E–P). In the accepted kinetic scheme of AP (Scheme 1),[29] Ser102 is transiently phosphorylated giving a covalent E–P.[28] This E–P complex can either be hydrolyzed to give a noncovalent

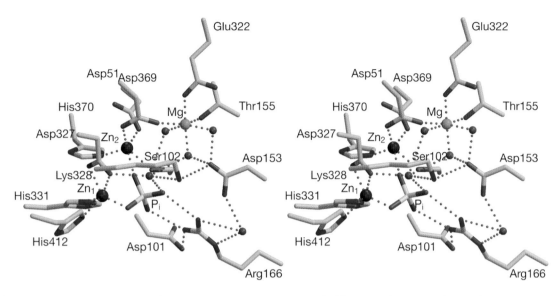

Figure 4 Stereoview of the active site of AP with P_i bound, the $E \cdot P_i$ complex[16] (PDB code 1ED8). The P_i is bound to Zn_1 and Zn_2 and has a bidentate interaction with the guanidinium group of Arg166. This figure also shows the relative position of the three metal binding sites in relationship to the phosphate binding site and the active site nucleophile, Ser102. The carboxylate of Asp153 forms hydrogen bonds to two of the water molecules (red spheres) that are ligands to the Mg^{2+}. The position of Asp153 is stabilized by a salt-link with Lys328. The binding of the P_i, and presumably of the substrate as well, is enhanced by a water-mediated interaction between Lys328 and the P_i. The critical position of Arg166 is stabilized by a water-mediated interaction to Asp153 and a salt-link with Asp101. Prepared with the program MOLSCRIPT[21] and RASTER3D.[22]

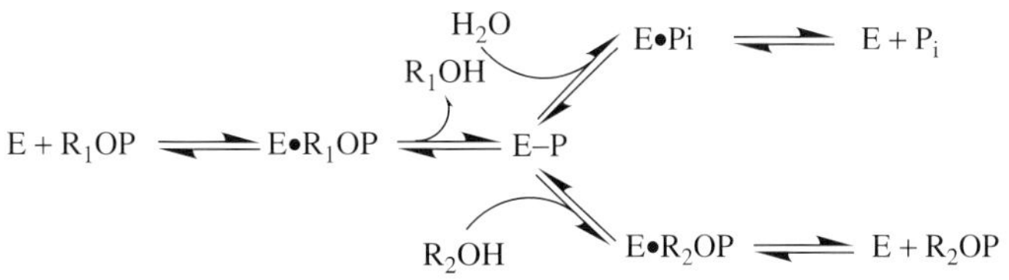

Scheme 1

enzyme–phosphate complex ($E \cdot P_i$) or it can also undergo a transphosphorylation reaction where the P_i is transferred to a phosphate acceptor (R_2OH) such as Tris or ethanolamine.[30–32] Consistent with the double displacement mechanism, the overall reaction proceeds with retention of the configuration.[33]

^{31}P NMR has been extremely useful in studying the kinetics of the AP reaction, since E–P and $E \cdot P_i$ can easily be distinguished.[24] ^{31}P NMR experiments have shown that the hydrolysis of the covalent E–P complex is rate-limiting under acidic conditions, whereas P_i dissociation from the noncovalent $E \cdot P_i$ complex is rate-limiting under basic conditions.[24,34] The enhanced rate observed in the presence of a phosphate acceptor is due to the decomposition of E–P in two ways – by hydrolysis and by transphosphorylation. The change in the rate-determining step from hydrolysis of the covalent E–P complex to P_i dissociation from the noncovalent $E \cdot P_i$ complex coincides with the sigmoidal pH versus activity profile. This profile reflects the ionization state of the hydroxide coordinated to Zn_1 that acts as the nucleophile in the hydrolysis of E–P. When the metals in AP are replaced with Cd^{2+}, there is a dramatic decrease in reaction rate and an alteration in the pH profile of the enzyme,[35] which correlates with the expected shift in pK_a of a hydroxyl coordinated to Cd^{2+} compared with Zn^{2+}, providing convincing evidence that the Zn-coordinated hydroxide controls the rate-limiting step in the AP mechanism.

Steady state kinetics has been used extensively to characterize both the wild-type and the mutant versions of AP. Table 2 summarizes the kinetic parameters for AP.

Pre–steady state kinetics

As would be expected for a reaction that proceeds via a covalent enzyme intermediate, burst kinetics are observed in the pre–steady state reaction of AP.[36,37] Analysis of the pre–steady state kinetics was hindered for some time until it was realized that during the normal purification of

E. coli alkaline phosphatase

Table 2 Kinetic parameters for wild-type AP[a]

k_{cat}[b] (s^{-1})	K_m (μM)	k_{cat}/K_m (M^{-1} s^{-1})	Phosphate acceptor
44.4 (±0.5)	9.4 (±0.1)	4.7 × 10^6	No[c]
80.5 (±1.5)	21.1 (±0.3)	3.8 × 10^6	Yes[d]

[a] Assays were performed at 25 °C with the use of p-nitrophenyl phosphate as substrate.
[b] The k_{cat} values are calculated from the V_{max} by use of a dimer molecular weight of 94 000. The k_{cat} per active site would be half of the value indicated.
[c] 0.1 M MOPS buffer, pH 8.0. The ionic strength was adjusted to 0.557 with NaCl.
[d] 1.0 M Tris buffer, pH 8.0.

the enzyme some residual P_i remained bound in the active site.[38]

Molecular complementation

A variety of early experiments using hybrid enzymes[39,40] led to the observation that certain combinations of mutant versions of AP yielded hybrid enzymes (heterodimers) that were more active than would be expected on the basis of the activity of the homodimeric parental enzymes.[41–44] This intragenic complementation is most often observed in enzymes that have the active site composed of residues contributed from two adjacent chains, a shared active site. However, structural studies of AP show that the active sites are not shared. Thus, E. coli AP belongs to a rare group of enzymes that exhibit intragenic complementation without shared active sites. In this case, intragenic complementation would only be expected to occur if the combination of mutants in the heterodimer enhanced preexisting cooperative interactions between the subunits.

Using polymerase chain reaction (PCR) and sequence analysis, the amino acid substitutions responsible for intragenic complementation in a number of E. coli strains were determined to be at amino acid positions that contribute side chains involved in metal binding, not at the subunit interface as might be expected.[45] To investigate these alterations in more detail, heterodimeric forms of AP were isolated in pure form, which became possible by altering the overall charge of one subunit by replacement of the C-terminal lysine residue with three aspartic acid residues. This modification had no effect on the kinetic properties of the enzyme. An increase in k_{cat} as well as a large reduction in K_m was observed for certain combinations of mutant enzymes. These results suggest that the structural assembly of E. coli AP into the dimer induces cooperative interactions between the monomers necessary for the establishment of the functional form of the holoenzyme.[45] The absence of cooperative interactions may also be responsible for the putative lack of activity of E. coli AP monomers.

CRYSTALLOGRAPHY OF REACTION INTERMEDIATES

General remarks

Using X-ray crystallographic techniques, the structures of various steps in the catalytic mechanism of AP have been captured including (i) the free enzyme in the absence of P_i, (ii) the E–P covalent complex, (iii) the enzyme bound with a transition state analog, and (iv) the E•P_i noncovalent complex. These structures, discussed below, have been useful both for the confirmation of certain aspects of the proposed catalytic mechanism[19] as well as for extending the mechanism to include a role for the magnesium.[16]

X-ray structure of AP in the absence of P_i

The initial structure of the enzyme in the absence of P_i was determined at low resolution >3Å because the crystals were of poor diffraction quality.[19] Recently, a new crystallization strategy was used to obtain crystals of AP without P_i that diffract to high resolution (<1.8 Å).[16] As seen in Figure 5(a), in the structure of AP without P_i, the active site contains three water molecules in place of P_i. The hydroxyl group of Ser102 coordinates Zn_2. The average distance between the Oγ of Ser102 and the Zn_2 of 1.91 Å suggests that the hydroxyl of Ser102 is deprotonated. A water molecule bridges the Oγ oxygen atom of Ser102 and one of the water ligands to the Mg^{2+}, while another water molecule bridges between the Oγ oxygen atom of Ser102 and Zn_1. This water molecule is the first in a short chain of three active site water molecules. The second water molecule in this chain forms a hydrogen bond to a guanidinium nitrogen (NH1) of Arg166. These water molecules are relatively close to each other, possibly suggesting that at least one is deprotonated and exists as a hydroxide ion. The side chain of Ser102 is disordered in the structure of AP without P_i, and the position of the side chain of Ser102 correlates with the metal bound in the M3 site (Mg 60%, Zn 40%). The strong correlation between the metal ion in the M3 site and the Ser102 side chain conformation in this structure suggests that Mg is required for deprotonation of the hydroxyl group of the nucleophile.

Proton abstraction from the Ser102 hydroxyl group requires a general base. One of the water molecules coordinated to Mg^{2+} is in close proximity to Zn_2 (4.7 Å) and the side chain hydroxyl of Ser102 (3.1 Å). On the basis of the short oxygen-to-magnesium distance, this water molecule may be coordinated as a hydroxide ion. A Mg-coordinated hydroxide ion could function as a general base

E. coli alkaline phosphatase

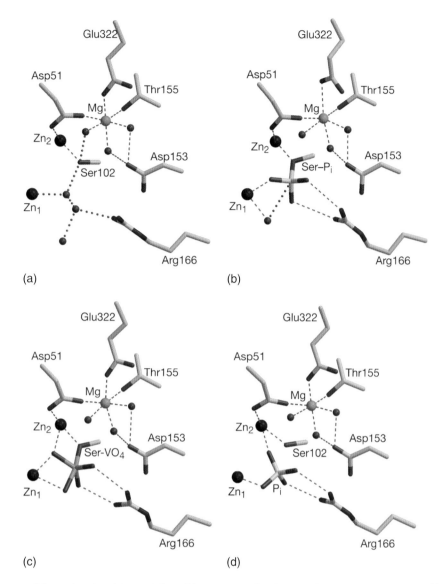

Figure 5 The structure of the active site of AP. (a) The wild-type AP in the absence of substrates and products[16] (PDB code 1ED9). The hydrogen-bonded network of water molecules in the active site is also shown. (b) The H331Q mutant AP[46] that stabilizes the phosphoseryl intermediate (Ser-P_i) (PDB code 1HJK). A water molecule coordinated to Zn_2 is in perfect position for the apical attack on the phosphorus. (c) Wild-type enzyme in complex with vanadate that mimics the structure of the trigonal bipyramidal transition state[47] (Ser-VO_4) (PDB code 1B8J). (d) The wild-type AP in the presence of P_i[16] (PDB code 1ED8). Hydrogens are omitted for clarity. Water molecules are represented as red spheres. Prepared with the program MOLSCRIPT[21] and RASTER3D.[22]

and accept the proton from the Oγ of the Ser102 side chain. Following this proton transfer, the Oγ oxygen of Ser102 is stabilized by coordination to Zn_2.

X-ray structure of the covalent enzyme–phosphate complex

The structure of the E–P covalent complex was determined using a mutant AP (H331Q) in the presence of excess P_i.[46] In the presence of P_i, the enzyme exists in a pH-dependent equilibrium between the E•P_i and E–P forms. For a wild-type enzyme, the equilibrium significantly favors the E–P form at pH 5.5. In the case of the mutant enzyme, the relatively free energies of the E•P_i and E–P forms is modified slightly such that the E–P form of the mutant predominates at pH 7.5. Thus, by using site-specific mutagenesis, the E–P intermediate was stabilized, and direct observation of this covalent form became possible.

The active site of the H331Q structure differed significantly from the wild-type structure. In the wild-type E•P_i structure (see Figures 4 and 5(d)), P_i is held in the active site pocket through interactions with Zn_1, Zn_2, Arg166, and a variety of water-mediated interactions, while P_i is not close enough to Ser102 to form a covalent bond. In the H331Q structure, the phosphorus and the Oγ oxygen of Ser102 are within covalent bonding distance (1.6 Å) (see Figure 5(b)). In this structure, one of the oxygens of the phosphoserine is coordinated to Zn_1, while another oxygen is coordinated to Zn_2 (see Figure 5(b)). The stereochemistry of the phosphorus in the H331Q structure, corresponding to the E–P complex, is inverted when compared to the stereochemistry of the phosphorus in the wild-type structure (E•P_i), as would be expected after one step of the double in-line displacement mechanism (compare Figure 5(b) and (d)).

In the X-ray structure of the H331Q enzyme, a water molecule is coordinated to Zn_1, which places the water in the ideal position for apical attack on the phosphorus (see Figure 5(b)). Furthermore, this water molecule is in direct line with the oxygen of the phosphoserine, an optimal position for in-line displacement of the P_i. Here, for the first time, using a mutant AP, the reaction is observed at the instant in which an activated water molecule is poised to attack the phosphoseryl intermediate, thus providing details of enzyme catalysis by AP on the molecular level. The proximity and altered pK_a of the bound water, due to coordination to Zn_2, contributes to the catalytic rate acceleration.

X-ray structure of a model for the transition state

Vanadate, which is isostructural and isoelectronic with P_i, is known to be a potent inhibitor of AP.[48,49] This potent inhibition of the enzyme by vanadate is due to the ability of vanadium to become five-coordinate and perhaps mimic the transition state of the reaction. In the X-ray crystal structure of the enzyme•vanadate complex, refined at 1.9 Å resolution, the vanadium ion is clearly bound in the active site of AP with five-coordinate geometry.[47] The structure (Figure 5(c)) shows a trigonal bipyramidal geometry for vanadate when bound in its complex with AP. On the basis of bond lengths, the Oγ of Ser102 is covalently bound to the vanadium center. The covalent bond distance of this axial interaction is the same as the distance of the three covalent equatorial vanadium–oxygen bonds. In contrast, the interaction between the vanadium atom and the oxygen of the axial hydroxyl ligand is slightly longer (av. 1.90 Å). The close proximity of this oxygen ligand to the vanadium atom indicates significant bond formation at this axial position. In comparison, the distance between the Zn_1 ligand and the oxygen atom of the coordinated hydroxyl group is a much longer interaction (av. 2.40 Å). The equatorial plane of the complex is defined by the three oxygen atoms spaced 120° apart. The guanidinium group of Arg166 forms two strong hydrogen bonds to two equatorial oxygen atoms of vanadate. The third oxygen atom in the equatorial plane is positioned to form equally strong interactions with both Zn^{2+} ions.

The trigonal bipyramidal complex of vanadate in the active site of AP provides details about the formation and breakdown of the axial coordination of the phosphorus atom as the covalently bound species is converted into P_i. The AP–vanadate complex is positioned more closely to the Ser102 nucleophile; in fact, Ser102 is rotated toward vanadate and assumes the same orientation observed in the covalent phosphoseryl intermediate of the E–P complex (Figure 5(b)). On the basis of the respective structures of AP with P_i and vanadate, the Oγ oxygen of Ser102 can be presumed to move approximately 0.8 Å from the noncovalent complex to form the transition state. In the AP–vanadate complex, the longer interaction between the coordinated hydroxyl occupying the second axial position and vanadium indicates a weaker complex in this direction. The strength of this axial interaction is most probably influenced by Zn_1. The weak but evident interaction between Zn_1 and this axial oxygen atom is consistent with the role of zinc in activating a water molecule for attack on the phosphoseryl intermediate. In fact, the axial hydroxyl group of vanadate is nearly superimposable with a phosphate oxygen atom in the structure of the noncovalent E•P_i complex (Figure 4).

In this unique manner, through the acquisition of structural snapshots, the pathway of the nucleophilic water molecule can be traced using the available X-ray structures of AP shown in Figure 5. Initially, it is fully bound by Zn_1 (2.2 Å) in the E–P intermediate (Figure 5(b)). In the AP–vanadate structure, the Zn_1–OH distance increases (2.4 Å) with significant bonding between this oxygen atom and the vanadium center (Figure 5(c)). In the E•P_i complex, the oxygen atom is covalently attached to the phosphorus atom to form P_i (Figure 5(d)). By examination of these X-ray crystal structures, the axial bonding properties of the AP–vanadate complex are clearly found to be intermediate between those of the E–P intermediate and the E•P_i complex. The structure of the AP–vanadate complex shows that only minor changes in the positions of Ser102 and Arg166 are required in the catalytic reaction and that the active site arrangement is optimized for stabilization of a trigonal bipyramidal species.

The bond distances of the vanadate ion, although not identical with those of phosphate, provide an accurate picture of the crucial structural elements involved in transition state stabilization. Furthermore, the increased bond distances of about 0.2 Å in the vanadate ion compared with the phosphate ion simulate the bond formation and the breakage that takes place in the transition state of the phosphatase reaction. From the AP–vanadate crystal structure, it is evident that both Zn^{2+} ions and Arg166 clearly play an important role in stabilizing the transition state. Zn_1 stabilizes the five-coordinate intermediate while also activating a water molecule for the subsequent hydrolysis. The structure of the AP–vanadate complex serves as a direct and convincing model for the five-coordinate trigonal bipyramidal transition state in the catalytic reaction of AP. The stabilization of the transition

E. coli alkaline phosphatase

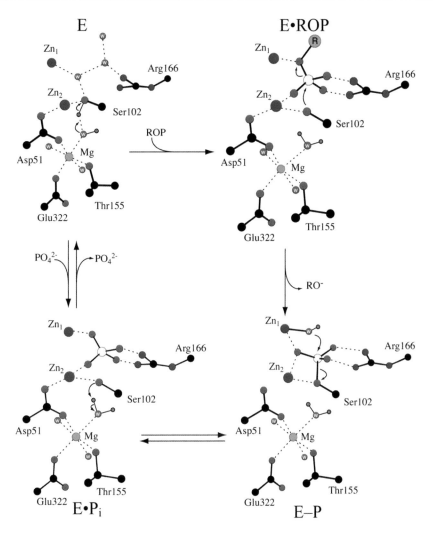

Figure 6 The catalytic mechanism of AP.[16] The steps in the mechanisms are shown from the free enzyme (E), to the enzyme–substrate complex (E•ROP), to the covalent enzyme–phosphate intermediate (E–P), and to the noncovalent enzyme–phosphate complex (E•P$_i$). Hydrogen atoms have been included only at catalytically relevant sites. For clarity, most ligands to Zn$_1$ and Zn$_2$ have been omitted.

state by the enzyme provides a mechanism for the observed catalytic rate acceleration by AP.

CATALYTIC MECHANISM

The structures of the AP determined have not only provided the necessary details to propose a revised mechanism for the AP reaction but also to determine the role that Mg^{2+} plays in catalysis. The catalytic mechanism for AP[16] shown in Figure 6 builds upon the one originally proposed by Kim and Wyckoff.[19] In the free enzyme (E, Figure 6, top left), three water molecules fill the active site, and the Ser102 hydroxyl participates in a hydrogen bond with the Mg-coordinated hydroxide ion. Upon binding of the phosphomonoester (ROP) to form the Michaelis enzyme–substrate complex (E•ROP, Figure 6, top right), the Ser102 Oγ becomes fully deprotonated for nucleophilic attack with the concomitant transfer of the proton to the Mg-coordinated hydroxide to form a Mg-coordinated water molecule. Coordination to Zn$_2$ stabilizes the Ser102 Oγ in its nucleophilic state. In the first in-line displacement, the activated hydroxyl of Ser102 attacks the phosphorus center of the substrate in the enzyme–substrate complex (E•ROP) to form a covalent phosphoseryl intermediate (E–P, Figure 6, bottom right). Zn$_1$ participates in this step by coordinating the bridging oxygen of the substrate and facilitating the departure of the alcohol leaving group (RO$^-$). In the second in-line displacement step, a nucleophilic hydroxide coordinated to Zn$_1$ attacks the phosphorous, hydrolyzing the covalent phosphoseryl intermediate to form the noncovalent enzyme–phosphate product complex (E•P$_i$, Figure 6, bottom left) and regenerating the nucleophilic Ser102. Zn$_1$ lowers the pK_a of the coordinated water molecule to form the nucleophilic hydroxide ion, while the Mg-coordinated water molecule acts as a general acid to reprotonate the Oγ of Ser102. Protonation of Ser102 may facilitate departure of the P$_i$

product from the noncovalent E•P$_i$ complex. Alternatively, the Mg-coordinated water may directly protonate the P$_i$ for its release. The release of P$_i$ from the E•P$_i$ complex to give the free enzyme (E, Figure 6, top left) may also be facilitated by the increased mobility of the Arg166 side chain.

RELATIONSHIP BETWEEN THE *E. coli* AND MAMMALIAN APs

Although the amino acid sequences of AP from bacteria such as *E. coli* and from mammals is 25 to 30% conserved, there are some important functional differences between the bacterial and mammalian enzymes. The bacterial enzymes, using *E. coli* AP as a model, are much more heat-stable than the mammalian enzymes, while the mammalian enzymes exhibit 20- to 30-fold higher catalytic activity as well as a shift in the pH of optimal activity toward higher pH. Some mammalian alkaline phosphatases also require relatively high concentrations of magnesium in order to achieve maximal activity.[50,51]

A comparison of the structure of *E. coli* AP[16,19] to that of the human placental AP[52] reveals that the overall structure is quite similar. Most of the gross differences between the bacterial and mammalian enzymes involve insertions and or deletions of loop regions. In the active site region, Ser102 is conserved along with Arg166, as well as the metal binding ligands except for residue 155, which is threonine in bacteria and serine in mammals. Close to the active site there are two noticeable differences. In *E. coli*, AP positions 153 and 328 are aspartic acid and lysine respectively (see Figure 4), whereas in the mammalian enzymes histidine is observed at both these positions.

In order to determine whether either or both of these residues were responsible for the enhanced catalytic activity of the mammalian enzyme, site-specific mutagenesis was used to replace either 153 and 328 or both with histidine in the *E. coli* AP.[53] Under optimal conditions, the K328H, D153H, and D153H/K328H mutant enzymes all exhibited higher catalytic activity than the wild-type *E. coli* AP. The D153H/K328H double mutant had sixfold higher catalytic activity along with significantly lower affinity for both the substrate and the product P$_i$. Since the release of P$_i$ from the noncovalent E•P$_i$ complex is the slow step in the mechanism, the enhanced catalytic activity is due to the reduced affinity of the mutant enzyme for P$_i$. These results also provide evidence for the importance of the water-mediated interaction between Lys328 and P$_i$ for the binding of P$_i$ in the active site of AP (see Figure 4).

MUTANTS OF AP WITH ENHANCED CATALYTIC ACTIVITY

The known enhanced catalytic activity of mammalian APs compared to the *E. coli* enzyme has resulted in a number of attempts to create more active versions of the *E. coli* enzyme. One of the first was the modification of Asp101 (see Figure 4), a residue conserved in nearly all AP sequences. The D101A mutant exhibited a threefold increase in specific activity, compared to the wild-type enzyme under optimal conditions.[54] The most likely explanation of this enhanced activity is based on a reduced affinity of the enzyme for P$_i$, which makes the product release faster. Another mutation at this same position, D101S produced an even better enzyme with a 35-fold increase in specific activity over wild-type under optimal conditions, and a sixfold increase in the k_{cat}/K_m ratio.[55] X-ray structure determination of the D101S enzyme revealed that the alterations induced by the mutation affected primarily the side chain of Arg166.[56] The loss of the hydrogen bonding interaction between the carboxylate of Asp101 and guanidinium of Arg166 (see Figure 4) allows the side chain of Arg166 to become more flexible. This enhanced flexibility of Arg166 is responsible for enhanced P$_i$ release and the ensuing enhanced catalytic activity.[56]

The salt-link between Asp153 and Lys328 also has an influence on catalytic activity. As mentioned above, these residues are both histidine in mammalian APs. When they are replaced with histidine in *E. coli* AP, enzymes with enhanced catalytic activity are obtained.[53,57] When Asp153 is replaced by Glycine, the resulting mutant enzyme has a fivefold higher specific activity but no change in K_m compared to the wild-type AP.[58] Again, in this case the increase in specific activity has been attributed to faster P$_i$ release resulting from lower P$_i$ affinity. By combining the D153G with other mutations, even more active APs have been isolated,[59] for example, the D153G/D330N mutant had a specific activity 40-fold higher than wild-type AP under optimal conditions.[59] Thus, with appropriate changes in residues near the active site, significantly enhanced catalytic activity of AP can be achieved. The lower catalytic activity of *E. coli* AP relative to the APs from higher organisms may be due to the ability of *E. coli* to easily and quickly turn on the expression of the enzyme itself. Thus, there is no evolutionary pressure on *E. coli* to evolve a more active AP.

METAL SPECIFICITY OF APs FROM DIFFERENT SPECIES

Although *E. coli* AP is most active with Zn^{2+} and Mg^{2+}, this is not true for all APs; for example, both the *Bacillus subtilis* APs[60] and the *Thermotoga maritima* AP[9] are most active with Co^{2+}, while the *Halobacterium halobium* AP requires Mn^{2+}.[61] Since the amino acids that are ligands to the metals are, for the most part, conserved in the APs with known sequences, metal specificity is governed by factors other than the exact nature of the metal ligands. Factors that may play a role in metal selectivity include the exact geometry of the metal site, the nature of side chains near the site, conformational flexibility, as well as the overall electrostatic field at the binding site.

The *T. maritima* AP[9] is approximately eightfold more active with Co^{2+} than with Zn^{2+}, and the enzyme exhibits enhanced affinity for Co^{2+} relative to Zn^{2+}. In order to understand the altered metal specificity of the *T. maritima* AP, an analysis of conserved/nonconserved residues was carried out moving outward from the center of the active site. In the three enzymes that have been shown to require Co^{2+} for activity, the *T. maritima* AP and the two *B. subtilis* APs, position 328 is a tryptophan rather than the aspartic acid or histidine observed in other APs.[9] In order to determine whether the tryptophan substitution at position 328 was responsible for the altered metal specificity, the K328W mutation was introduced into *E. coli* AP.[9] Under optimal conditions, the *E. coli* K328W AP was almost twice as active with Co^{2+} than with Zn^{2+}, and the Co^{2+} K328W mutant AP exhibited approximately threefold higher specific activity than the wild-type *E. coli* AP. Thus, the tryptophan substitution at position 328 is at least partially responsible for the altered metal specificity of some APs.

E. coli AP AS A MODEL FOR OTHER ALKALINE PHOSPHATASES

On the basis of the similarities in amino acid sequence and three-dimensional structures, the *E. coli* enzyme has become an excellent model for all APs. The catalytic mechanism of the *E. coli* enzyme is expected to be extremely similar if not identical for all APs.

ACKNOWLEDGEMENT

The work reported here and performed in my laboratory was supported by grant GM42833 from the National Institute of Health. I wish to thank Dr. S. C. Pastra-Landis for critically reading the manuscript.

REFERENCES

1. RA Bradshaw, F Cancedda, LH Ericsson, PA Newman, SP Piccoli, K Schlesinger and KA Walsh, *Proc Natl Acad Sci USA*, **78**, 3473–77 (1981).
2. CN Chang, W-J Kuang and EY Chen, *Gene*, **44**, 121–25 (1986).
3. H Inouye and J Beckwith, *Proc Natl Acad Sci USA*, **74**, 1440–44 (1977).
4. T Horiuchi, S Horiuchi and D Mizuno, *Nature (London)*, **183**, 1529–30 (1959).
5. A Torriani, *Biochim Biophys Acta*, **38**, 460–469 (1960).
6. MH Malamy and BL Horecker, *Biochemistry*, **3**, 1889–93 (1994).
7. HC Neu and LA Heppel, *J Biol Chem*, **240**, 3685–92 (1965).
8. A Chaidaroglou, JD Brezinski, SA Middleton and ER Kantrowitz, *Biochemistry*, **27**, 8338–43 (1988).
9. CL Wojciechowski and ER Kantrowitz, *J Biol Chem*, **277**, 50476–481 (2002).
10. H Inouye, C Pratt, J Beckwith and A Torrini, *J Mol Biol*, **110**, 75–87 (1977).
11. CN Chang, H Inouye, P Model and J Beckwith, *J Bacteriol*, **142**, 726–28. (1980).
12. MI Harris and JE Coleman, *J Biol Chem*, **243**, 5063–73 (1968).
13. A Garen and C Levinthal, *Biochim Biophys Acta*, **38**, 470–83 (1960).
14. MH Malamy and BL Horecker, *Biochemistry*, **3**, 1893–97 (1994).
15. JM Sowadski, BA Foster and HW Wyckoff, *J Mol Biol*, **150**, 245–72 (1981).
16. B Stec, KM Holtz and ER Kantrowitz, *J Mol Biol*, **299**, 1303–11 (2000).
17. JM Sowadski, MD Handschumacher, HMK Murthy, C Kundrot and HW Wyckoff, *J Mol Biol*, **170**, 575–81 (1983).
18. JM Sowadski, MD Handschumacher, HMK Murthy, BA Foster and HW Wyckoff, *J Mol Biol*, **186**, 417–33 (1985).
19. EE Kim and HW Wyckoff, *J Mol Biol*, **218**, 449–64 (1991).
20. GM Sheldrick and TR Schneider, *Methods Enzymol*, **277**, 319–43 (1997).
21. PJ Kraulis, *J Appl Crystallogr*, **24**, 946–50 (1991).
22. EA Merritt and DJ Bacon, *Methods Enzymol*, **277**, 505–24 (1997).
23. DJ Plocke and BL Vallee, *Biochemistry*, **1**, 1039–43 (1962).
24. P Gettins and JE Coleman, *J Biol Chem*, **258**, 408–16 (1983).
25. JE Coleman, KI Nakamura and JF Chlebowski, *J Biol Chem*, **258**, 386–95 (1983).
26. B Stec, MJ Hehir, C Brennan, M Nolte and ER Kantrowitz, *J Mol Biol*, **277**, 647–62 (1998).
27. A Nicholls, KA Sharp and B Honig, *Proteins*, **11**, 281–96 (1991).
28. JH Schwartz and F Lipmann, *Proc Natl Acad Sci USA*, **47**, 1996–2005 (1961).
29. IB Wilson and J Dayan, *Biochemistry*, **4**, 645–49 (1965).
30. TW Reid and IB Wilson, in Boyer, P. (ed.), *E. coli alkaline phosphatase, The Enzymes*, 3rd edn, Vol. 4 Academic Press, New York, pp 373–415 (1971).
31. IB Wilson, J Dayan and K Cyr, *J Biol Chem*, **239**, 4182–85 (1964).
32. J Dayan and IB Wilson, *Biochim Biophys Acta*, **81**, 620–23 (1964).
33. SR Jones, LA Kindman and JR Knowles, *Nature (London)*, **275**, 564–65 (1978).
34. WE Hull, SE Halford, H Gutfreund and BD Sykes, *Biochemistry*, **15**, 1547–61 (1976).
35. P Gettins and JE Coleman, *J Biol Chem*, **258**, 396–407 (1983).
36. DR Trentham and H Gutfreund, *Biochem J*, **106**, 455–60 (1968).
37. HN Fernley and PG Walker, *Biochem J*, **111**, 187–94. (1969).
38. W Bloch and MJ Schlesinger, *J Biol Chem*, **248**, 5794–5805 (1973).
39. C Levinthal, ER Signer and K Fetherolf, *Proc Natl Acad Sci USA*, **48**, 1230–37 (1962).
40. MJ Schlesinger and K Barrett, *J Biol Chem*, **240**, 4284–92 (1965).

41. DP Fan, MJ Schlesinger, A Torriani, KJ Barrett and C Levinthal, *J Mol Biol*, **15**, 32–48 (1966).
42. A Garen and S Garen, *J Mol Biol*, **7**, 13–22 (1963).
43. MJ Schlesinger and C Levinthal, *J Mol Biol*, **7**, 1–12 (1963).
44. MJ Schlesinger, A Torrini and C Levinthal, *Cold Spring Harbor Symp Quant Biol*, **28**, 539–42 (1963).
45. MJ Hehir, JE Murphy and ER Kantrowitz, *J Mol Biol*, **304**, 645–56 (2000).
46. JE Murphy, B Stec, L Ma and ER Kantrowitz, *Nat Struct Biol*, **4**, 618–21 (1997).
47. KM Holtz, B Stec and ER Kantrowitz, *J Biol Chem*, **274**, 8351–54 (1999).
48. DC Crans, A Keramidas and C Drouza, *Phosphorus, Sulfur Silicon*, **109–110**, 245–48 (1996).
49. V Lopez, T Stevens and RN Lindquist, *Arch Biochem Biophys*, **175**, 31–38 (1976).
50. C Brunel and G Cathala, *Biochim Biophys Acta*, **309**, 104–15 (1973).
51. G Cathala, C Brunel, D Chappelet-Tordo and M Lazdunski, *J Biol Chem*, **250**, 6046–53 (1975).
52. MH Le Du, T Stigbrand, MJ Taussig, A Menez and EA Stura, *J Biol Chem*, **276**, 9158–65. (2001).
53. JE Murphy, TT Tibbitts and ER Kantrowitz, *J Mol Biol*, **253**, 604–17 (1995).
54. A Chaidaroglou and ER Kantrowitz, *Protein Eng*, **3**, 127–32 (1989).
55. W Mandecki, MA Shallcross, J Sowadski and S Tomazic-Allen, *Protein Eng*, **4**, 801–4 (1991).
56. L Chen, D Neidhart, WM Kohlbrenner, W Mandecki, S Bell, J Sowadski and C Abad-Zapatero, *Protein Eng*, **5**, 605–10 (1992).
57. X Xu and ER Kantrowitz, *Biochemistry*, **30**, 7789–96 (1991).
58. CG Dealwis, L Chen, C Brennan, W Mandecki and C Abad-Zapatero, *Protein Eng*, **8**, 865–71 (1995).
59. BH Muller, C Lamoure, MH Le Du, L Cattolico, E Lajeunesse, F Lemaitre, A Pearson, F Ducancel, A Menez and JC Boulain, *Chem Bio Chem*, **2**, 517–23 (2001).
60. FM Hulett, EE Kim, C Bookstein, NV Kapp, CW Edward and HW Wyckoff, *J Biol Chem*, **266**, 1077–84 (1991).
61. ML Bonet, FI Llorca and E Cadenas, *Biochem Mol Biol Int*, **34**, 1109–20 (1994).

Hydrolases: Acting on peptide bonds

Thermolysin

Brian W Matthews

Howard Hughes Medical Institute, University of Oregon, Eugene, OR, USA

FUNCTIONAL CLASS

Enzyme: thermolysin (TLN) (E.C.3.4.24.4); a metalloprotease with a single zinc ion in the active site. The enzyme also binds four calcium ions that contribute to thermostability.

Thermolysin is an endopeptidase of molecular weight 34 600 Da isolated from the thermophile *Bacillus thermoproteolyticus*.[1] It preferentially cleaves peptides with a bulky, hydrophobic residue (phenylalanine or leucine) in the R'_1 position.[2-4] As noted below, thermolysin is a member of a very large family and the different family members display a broad range of specificities.

OCCURRENCE

Thermolysin is the prototypical member of the so-called 'M4' or 'thermolysin' family of secreted eubacterial endopeptidases defined by Rawlings and Barrett.[5] All members of the family contain the HEXXH sequence motif in which the two histidines coordinate the zinc ion and the

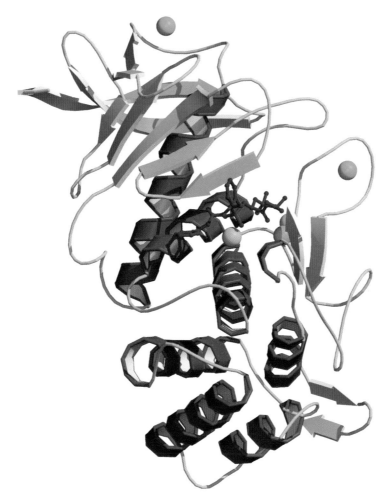

3D Structure Ribbon drawing showing the overall structure of thermolysin. The view is toward the active site with the zinc ion shown in light blue. A bound inhibitor (ZF^{P}LA) is shown in dark blue. Four calcium ions are shown as green spheres. The amino-terminal domain is located in the top half of the figure. It includes an extended β-sheet that terminates across the 'top' of the substrate binding site and contributes to substrate binding. The carboxy-terminal domain, located in the bottom half of the figure, is predominantly α-helical. [Figure, based on PDB file 4TMH, courtesy of Doug Juers was prepared with MOLSCRIPT[17] and RASTER3D.[18]]

glutamic acid is critical for catalysis (see below). Family members include enzymes from the pathogens *Legionella*, *Listeria*, *Pseudomonas*, and *Vibrio*.[5]

Proteins that are structurally and functionally related to thermolysin, and presumably evolved from a common precursor, extend well beyond the M4 family. Rawlings and Barrett[5] define the 'HEXXH + E Metallopeptidase Clan (MA).' This includes families M4 (thermolysin), M5 (mycolysin), M13 (neprilysin), M1 (membrane alanyl aminopeptidase), and M2 (peptidyl-dipeptidase A). All members of these families are clearly related to thermolysin. Rawlings and Barrett[5] also define the 'HEXXH + H Metallopeptidase Clan (MB).' This includes families M12 (astacin and reprolysin), M10 (interstitial collagenase, serralysin, and matrixins), M11 (autolysin), and M7 (*Streptomyces* small neutral protease). Members of the latter clan have also been defined as the metzincins[6] and have been shown to have significant structural relationship to thermolysin.

BIOLOGICAL FUNCTION

Thermolysin and its immediate family members are secreted eubacterial peptidases. As such, the presumed biological function of the enzyme is to degrade ambient peptides that can be utilized by the host. More distantly related relatives are involved in a broad range of digestive activities. They also are involved in the processing of biologically active peptides. Perhaps the best-known example is that of the angiotensin-converting enzyme for which the structure has recently been determined.[7]

AMINO ACID SEQUENCE INFORMATION

The amino acid sequence of thermolysin was determined by Titani *et al.*[8] by classical methods. It was also supported by the parallel X-ray structure determination.[9,10] The active site sequences of the zinc endopeptidases most closely related to thermolysin are given by Rawlings and Barrett.[5]

NUCLEOTIDE SEQUENCE

Although the sequence of the thermolysin gene itself seems not to have been reported, an almost identical relative has been described. Kubo and Imanaka[11] determined the sequence of the gene *nprM* for the highly thermostable neutral protease of *Bacillus stearothermophilus* MK232. The inferred amino acid sequence of the extracellular form of this protease is identical to that of thermolysin except for the two amino acid substitutions Asp37 to Asn37 and Glu119 to Gln119. Both the thermal stability and the specific activity of the protease encoded by gene *nprM* are somewhat higher than that of thermolysin.[11]

PROTEIN PRODUCTION, PURIFICATION, AND MOLECULAR CHARACTERIZATION

Thermolysin was initially isolated and purified by Endo.[1] The host organism is held on a proprietary basis and the purified enzyme is available commercially through Calbiochem (San Diego, CA). The enzyme can be recrystallized by dissolving at high pH and reducing the pH to neutrality.

METAL CONTENT

Thermolysin binds a single zinc ion that is essential for catalysis.[12,13] The structure of the protein[9,10,14,15] showed the zinc ion to be liganded tetrahedrally by His142, His146, Glu166, and a water molecule.

X-RAY STRUCTURE

Large X-ray quality crystals of thermolysin were initially obtained by dissolving the protein to very high concentrations (about 400 mg ml^{-1}) in 45% dimethylsulfoxide and then allowing water vapor to slowly diffuse into the solution of protein.[9] Even though the total volume of the solution increased, the solubility of the protein sharply decreased and resulted in multiple crystals, often intergrown.

A more economical method has subsequently been developed[16] and is as follows.

100 mg of 3× recrystallized thermolysin (Calbiochem) is dissolved in 1 mL of 50 mM 2-N-morpholinoethanesulfonic acid (MES), pH 6.0, 45% (v/v) dimethylsulfoxide (DMSO) by gently rocking at room temperature for 1 h. Insoluble material is removed by centrifugation and the clarified supernatant is used for crystallization at room temperature. 2 µL of protein solution is mixed with 2 µL of 50 mM MES, pH 6.0, 0–2.2 M NaCl, 0.04 M zinc acetate, and 45% DMSO in sitting drops over a well containing 0.5 mL of 30 to 50% saturated ammonium sulfate. With this procedure, large hexagonal crystals grew in 1 to 3 days in drops initially containing 1.0 to 1.1 M NaCl, but no zinc acetate, over 35% ammonium sulfate. These have space group $P6_122$ and are the same as those used for the original structure determination.[9] Large tetragonal crystals grew in 4 to 5 days in drops initially containing 0.7 to 0.9 M NaCl, 0.4 M zinc acetate over wells containing 30% ammonium sulfate.

Overall description of the structure

The overall fold of the molecule is shown as the 3D Structure. It is distinctly bilobal with the active site cleft formed at the junction of the two domains. There is an α-helix (residues 137–151) that is at the bottom of this cleft and contains the HEXXH motif (His142-Glu143-Leu144-Thr145-His146).

It was assumed that the structure, as initially determined,[9,14] was that of the free enzyme. Subsequent high-resolution refinement revealed the presence of a presumed dipeptide, valine–lysine, bound in the active site.[15] Recently, the structure of thermolysin has been determined in another crystal form that appears to be free of bound substrate or product.[16] In this crystal form, there is a hinge-bending movement such that the two domains rotate by 5°, opening the active site cleft (Figure 1). On the basis of prior comparison of the structures of several proteases related to thermolysin, Holland *et al.*[15] had proposed that hinge bending might play a role in their catalytic mechanism. Members of the family with substrate analogs bound in their active sites were observed to display 'closed' conformations, while those that had unoccupied active sites were seen to have conformations that were more 'open'. Further evidence for hinge-bending motion has come from molecular dynamics simulations[19] and glycine-to-alanine substitutions in the thermolysin-like protease from *Bacillus stearothermophilus*.[20]

Taken together, these results suggest that thermolysin and its relatives undergo hinge-bending motion during catalysis, closing when substrates are bound and opening to release products. The hinge-bending motion appears to be essentially rigid-body in character, although there are three side chains in the hinge region (Met120, Glu143, and Leu144) that undergo distinct changes in conformation in going from the closed to the open form. These same residues also undergo related changes when thermolysin crystals are exposed to cadmium, manganese, or excess zinc.[21]

STRUCTURAL BASIS FOR THERMOSTABILITY

Thermolysin was the first protein from a thermophilic organism for which the three-dimensional structure was determined. As such, there was considerable interest in identifying any feature of the structure that might be responsible for the enhanced stability of the molecule. In fact, there were none and it was concluded that the enhanced stability of thermostable proteins relative to thermolabile ones cannot be attributed to a common determinant such as metal ion or hydrophobic stabilization. Rather, the stability of unusually stable proteins was suggested to be due to rather subtle differences in hydrophobic character, metal binding, hydrogen bonding, ionic interactions, or a combination of all of these.[22]

During the subsequent years, a very large number of structures from thermophiles have been described. Also, the basis for protein stability has been tested by many mutagenesis experiments. We believe that the conclusions based on the thermolysin structure still apply. In different situations, many different types of interaction can contribute to the stability of a given protein. These interactions can include, but need not be limited to, hydrophobic interactions, van der Waals interactions, hydrogen bonding, electrostatic interactions, metal binding, disulfide bridges, and ligand binding. By combining multiple favorable interactions, the stability of a given protein can be gradually increased.[23,24] This is, no doubt, what happens during evolution. There is no need for a unique type of interaction to make a protein thermostable.

INHIBITOR COMPLEXES

The first three-dimensional structure of a thermolysin-inhibitor complex was that with phosphoramidon.[25,26] This is a potent, naturally occurring inhibitor. Its high affinity of binding can be attributed in part to the leucine–tryptophan moiety, which mimics peptide binding in the P'_1 and P'_2 subsites, and to the phosphoramide portion, which mimics the putative tetrahedral transition state during catalysis.

On the basis of these principles, a broad series of related inhibitors has been developed (Table 1; Figure 2). These have illustrated the modes of binding of a broad spectrum of inhibitors to the zinc peptidases. BAG, for example

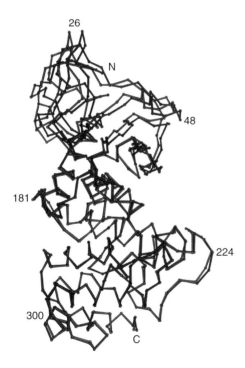

Figure 1 Superposition of the backbone of the 'closed' (blue) and 'open' (red) forms of thermolysin. The structures are superimposed on the basis of the C-terminal domains that are in the lower half of the figure (from Hausrath, AC and Matthews BW, (2002) Thermolysin in the absence of substrate has an open conformation. *Acta Crystallogr*, **D58**, 1002–1007. Figure 1. Reproduced by permission of the International Union of Crystallography).[16]

Table 1 Some inhibitors and ligands of thermolysin analyzed crystallographically

Inhibitor[a]	Abbreviated name	K_i (μM)	PDB code	References
β-(phenylpropionyl)-L-phenylalanine	βPPP	1600	–	13,29
Carbobenzoxy-L-phenylalanine	Z-Phe	510	–	29,30
L-valyl-L-tryptophan	Val–Trp		3TMN	31
N-((α-L-rhamnopyranosyloxy)hydroxy-phosphinyl)-L-Leu-L-Trp	Phosphoramidon	0.028	1TLP	25,32
N-phosphoryl-L-leucinamide	P-Leu-NH$_2$	21.3	2TMN	26,33
N-[[1-[(phenylmethoxycarbonyl)amino]-2-phenylethyl]methoxy-phosphinyl]-L-Leu-L-Ala-methyl ester (Cbz-Phe^P-L-Leu-L-Ala)	ZF^{P}LA	0.00068	4TMN	34,35
Cbz-Gly^P-L-Leu-L-Leu	ZG^{P}LL	0.0091	5TMN	35,36
Cbz-Gly^P(O)-L-Leu-L-Leu	ZG^P(O)LL	9.0	6TMN	36,37
Constrained analog of ZG^{P}LL	(S,S)-1	0.004	–	38
(2-benzyl-3-mercaptopropanoyl)-L-alanylglycinamide	BAG	0.75	–	39,40
L-benzylsuccinic acid	BZSA	3800	1HYT	41–43
N-(1-carboxy-3-phenylpropyl)-L-Leu-L-Trp	CLT	0.05	1TMN	28,44
N-[1-carboxy-3-((N'-carbobenzoxy-L-phenylalanyl)amino)propyl]-L-Leu-L-Trp	CCPALT	0.011	–	45
L-leucine hydroxamic acid	Leu-NHOH	190	4TLN	46,47
HONH-(benzylmalonyl)-L-Ala-Gly-p-nitroanilide	HONH-BAGN	0.43	5TLN	46,47
N-(chloroacetyl)-DL-N-hydroxyleucine methyl ester	CHME	Covalent	7TLN	48,49

[a] The inhibitors are illustrated in Figure 2.

(Figure 2), represents the binding of mercaptans in which the sulfur atom coordinates directly with the zinc ion. As another example, CLT is related to enalapril, the well-known inhibitor of the angiotensin-converting enzyme.[27] Its binding illustrates how the N-carboxymethyl group ligands the zinc ion.[28]

Selected pairs of matched inhibitors also have provided insights into the kinetics and thermodynamics of binding. For example, the inhibitor ZF^{P}LA binds much more slowly than its counterpart ZG^{P}LA.[34] This difference can be attributed to the restriction of rotational motion imposed by leucine relative to glycine at the P_1 subsite. In order for these inhibitors to bind to the enzyme or a water molecule, it first needs to exit the active site cleft. It appears that the more flexible glycine-containing inhibitor can enter the active site cleft as the water molecule leaves. The leucine-containing analog, however, is more rigid and in effect has to wait until the water molecule has departed the active site.

Comparison of inhibitors of the form ZG^P(X)LL where X can be NH, O, or CH_2 also illustrates the factors that influence the energetics of binding. For example, the inhibitor ZG^P(O)LL has an affinity of binding K_i = 9000 nM, whereas its counterpart, ZG^P(NH)LL, has K_i = 9.1 nM.[36] The two compounds bind to thermolysin in modes that are virtually identical.[37] The only difference is that the NH group in the latter compound makes a 3.0-Å hydrogen bond to the carbonyl oxygen of Ala113, whereas the ester oxygen, ZG^P(O)LL, is not capable of making such a hydrogen bond. The intrinsic binding energy of this interaction can therefore be assigned as 4.1 kcal mol^{-1} in agreement with theoretical estimates.[50,51] Interestingly, the compound in which the NH group is replaced by CH_2 binds with almost equal affinity even though it cannot make an analogous hydrogen bond.[52] These results can be rationalized by considering the desolvation of the inhibitor that accompanies binding to the protein.[51,52] The compound with the NH group loses a hydrogen bond owing

Figure 2 Schematic diagram showing the interactions between thermolysin and a number of inhibitors that have been analyzed crystallographically. Presumed hydrogen bonds are shown as dotted lines and interactions with the zinc as broken lines. The interactions shown at the top of the figure are those that are *assumed* to occur for the tetrahedral transition state of an extended polypeptide substrate. The bond to be cleaved is indicated by an arrowhead. This figure is to be read in conjunction with Table 1, which provides references and additional information (from Matthews, BW (1988) Structural basis of the action of thermolysin and related zinc peptidases. *Acct Chem Res*, **21**, 333–340. Figure 2. Reproduced by permission of the American Chemical Society).[45]

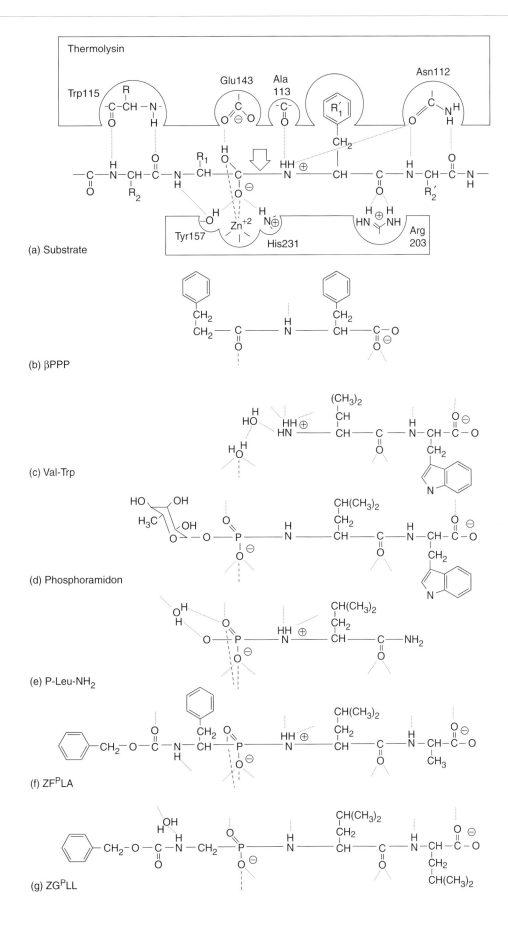

Figure 2

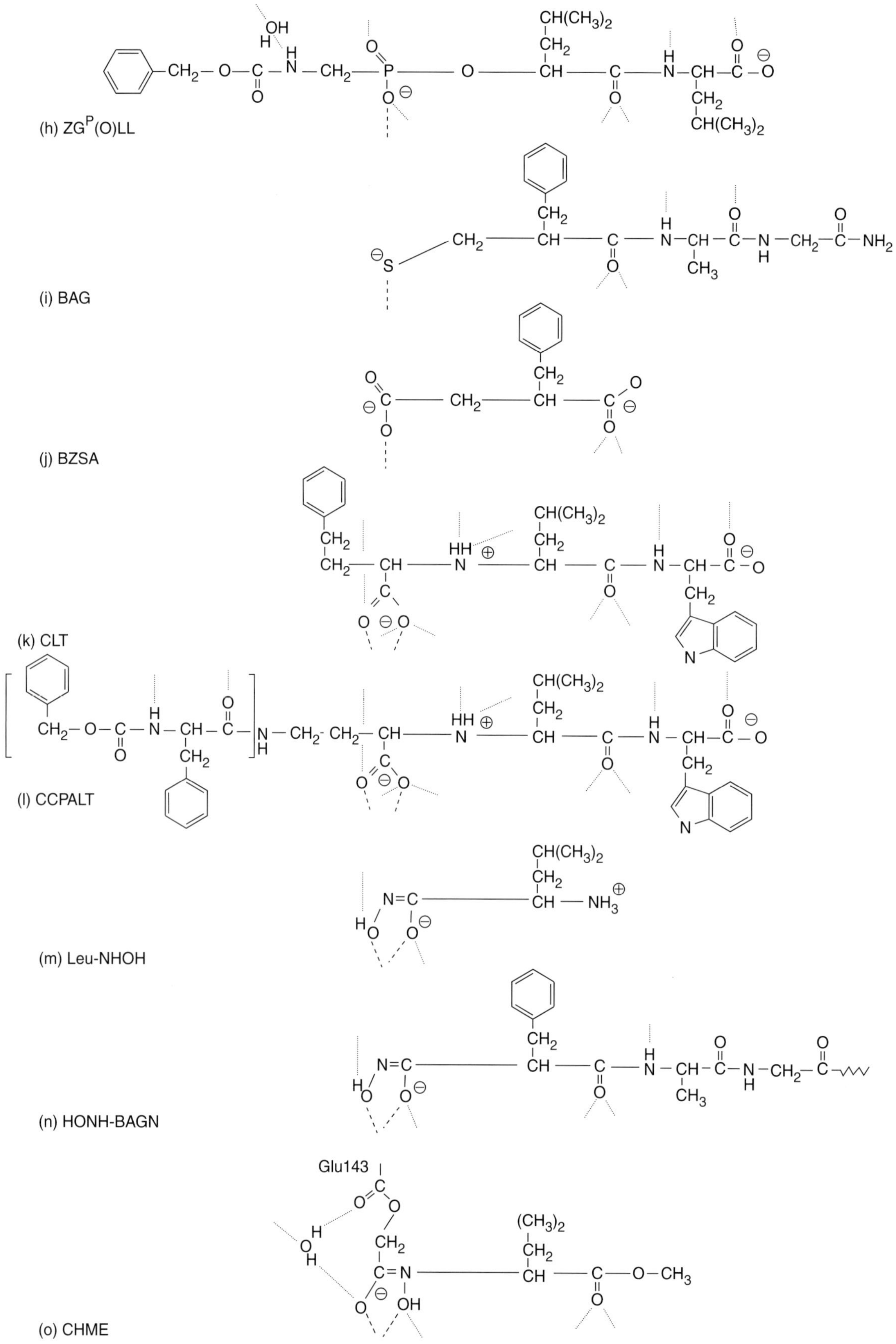

Thermolysin

(h) ZG^P(O)LL

(i) BAG

(j) BZSA

(k) CLT

(l) CCPALT

(m) Leu-NHOH

(n) HONH-BAGN

(o) CHME

Figure 2 (*Continued*)

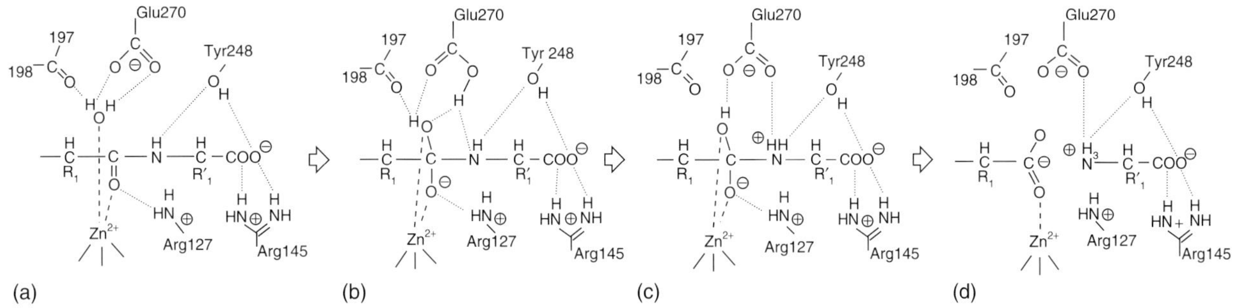

Figure 3 Mechanism of action of thermolysin as proposed by Hangauer et al.[54]

to desolvation but makes a hydrogen bond to the protein. The compound with the ester oxygen also loses a hydrogen bond owing to solvation, but does not hydrogen bond to the protein. This unsatisfied hydrogen-bonding potential weakens binding. The CH_2-containing inhibitor does not hydrogen-bond to the enzyme but neither does it lose a hydrogen bond to solvent. Thus, it retains high affinity.

In yet another study, it was shown that the rigidification of a substrate, if accomplished without introducing steric clashes with the enzyme, can decrease the entropy of binding and thus increase affinity.[38]

CATALYTIC ACTIVITY OF THERMOLYSIN

Measurement of activity

The measurement of activity is normally based on the hydrolysis of 3-(2-furylacryloyl)-L-glycyl-L-leucine amide or related substrates.[53] Hydrolysis results in a decrease in absorption at 345 nm.

Mechanism of catalysis

The currently accepted mechanism of action of thermolysin was first proposed by Hangauer et al.[54] and has become the prototype not only for thermolysin family members but a number of other metallopeptidases including carboxypeptidase A[28,55] and methionine aminopeptidase.[56]

The proposed mechanism is illustrated in Figure 3. The key stereochemical features are illustrated by the observed mode of binding of ZF^PLA (Figure 4) in which the tetrahedral phosphoramide group is presumed to mimic the transition state. In the free enzyme, a water molecule is bound to the active site Zn^{2+} ion. As the substrate approaches this water molecule (or hydroxide ion), it remains bound to the zinc but is displaced toward Glu143, which helps promote its nucleophilic attack on the scissile peptide bond (Figure 3), leading to a tetrahedral intermediate. This intermediate is stabilized by hydrogen bonding interactions with Tyr157 and His231 (Figure 2). The proton that is accepted by Glu143 from the attacking

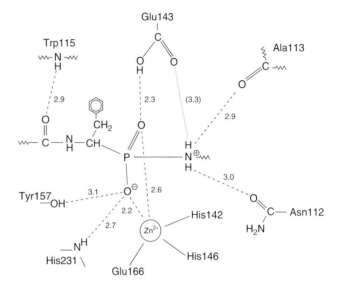

Figure 4 Schematic diagram showing the hydrogen bonding interactions (broken lines) in the complex of ZF^PLA with thermolysin.[35] Hydrogen bonds are shown as broken lines.

water molecule is transferred to the leaving nitrogen. In the transition state, the doubly protonated tetrahedral nitrogen of the scissile bond is stabilized by hydrogen bonds to the side-chain oxygen of Asn112 and the backbone carbonyl of Ala113. In the ZF^PLA complex, for example, these distances are 3.1 Å and 2.7 Å (Figure 4). Evidence that the zinc coordination can change to five-coordinate as in the presumed transition state is provided not only by the phosphoramide-containing inhibitors but also by other inhibitors such as CLT and the hydroxamate HONH-BAGN (Figure 2).

THE ROLE OF DIFFERENT METALS IN CATALYSIS

The zinc ion at the active site of thermolysin is bound in approximate tetrahedral coordination by His142, His146, and Glu166 plus a solvent molecule, the coordination distances being summarized in Table 2. The precise coordination geometry of Glu166 seems to be somewhat sensitive to the pH.

Thermolysin

Table 2 Ligand distances for different metals bound in the active site of thermolysin[a]

Ligand	Distance to substituted metal (Å)					Native	
	Co^{2+}	Cd^{2+}	Mn^{2+}	Fe^{2+}	Excess Zn^{2+} Zn(1)	Zn^{2+}[b]	Zn^{2+}[c]
Glu166 OE2	2.2	1.9	2.1	1.9	2.1	1.9	2.4
Glu166 OE1	2.7	3.0	3.0	2.9	2.7	2.8	2.2
His142 NE2	2.0	2.0	2.1	1.9	1.9	1.9	2.0
His146 NE2	2.1	2.1	2.1	2.0	2.0	2.0	2.0
Wat231[d]	2.2	2.3	2.2	2.2	3.2 (Zn2)	2.2	2.4
Wat143[d]	2.4	2.3	2.3	2.6	1.9	2.2	
Glu143$_A$ OE1[e]		2.5					
Distance Wat231:Wat143[d]	2.6	2.5	2.3	2.0	NA	NA	2.0
Effective coordination	5	6	5	4	4	4	4

[a] The distances quoted are those within 3.0 Å. PDB codes for the crystal structures are 1LNA, 1LNB, 1LNC, 1LND, 1LNE, and 1LNF (from reference 21).
[b] Data at pH 7.2 (from reference 15).
[c] Data at pH 6.8 (from reference 21).
[d] Wat231 and Wat143 are thought to be either two distinct solvent molecules (for the Co^{2+} and Cd^{2+} substitutions) or alternative sites occupied by a single water (for the Mn^{2+}, Fe^{2+}, and Zn^{2+} substitutions).
[e] Subscript 'A' denotes alternative side-chain conformation.
NA: Not applicable.

By exposing the crystals to a combination of EDTA and 1,10-phenanthroline, it is possible to remove the zinc ion from the crystalline enzyme.[9] The three-dimensional structure of the apoenzyme[22] is very similar to the native protein, showing that the zinc ion is not required to maintain the three-dimensional structure.

Removal of zinc in solution is not straightforward because chelating agents used to bind the zinc also tend to remove calcium. This destabilizes the enzyme and renders it very susceptible to autoproteolysis.[13,57,58] Notwithstanding these difficulties, Holmquist and Vallee[13,59] showed that removal of zinc yielded an inactive enzyme and that varying levels of esterase and peptidase activity could be regained by substitution with transition metals. Zn^{2+}, Co^{2+}, and Mn^{2+}, when added in stoichiometric amounts, restored 100, 200, and 10% of the activity of the native enzyme toward furylacryloylglycyl-L-leucyl amide (FAGLA). In concentrations up to 1 mM, these metals retained or restored activity to some extent. Fe^{2+}, in high molar excess, restored about 60% of the native activity. Zn^{2+} in excess of the amount required for catalytic activity actually inhibited the enzyme. The transition metals Mg^{2+}, Cr^{2+}, Ni^{2+}, Cu^{2+}, Mo^{2+}, Pb^{2+}, Hg^{2+}, Cd^{2+}, Nd^{2+}, Pr^{2+}, and Dy^{2+} all failed to restore activity above 2% that of the native enzyme.

To investigate the structural basis for these differences, the structures of a series of metal-substituted thermolysin crystals were determined by Holland et al.[21] The structure of the Co^{2+}-substituted enzyme is very similar to that of wild-type except that two solvent molecules are liganded to the metal at positions that are thought to be occupied by the two oxygens of the hydrated scissile peptide in the transition state. Thus, the enhanced activity toward some substrates of the cobalt, relative to the zinc-substituted enzyme, may be due to enhanced stabilization of the transition state. The ability of Zn^{2+} and Co^{2+} to accept tetrahedral coordination in the Michaelis complex, as well as the fivefold coordination in the transition state, may also contribute to their effectiveness in catalysis. The Cd^{2+}- and Mn^{2+}-substituted thermolysins display conformational changes that disrupt the active site to varying degrees, and could explain the associated reduction of activity. The conformational changes involve not only the essential catalytic residue, Glu143, but also concerted side-chain rotations in the adjacent residues Met120 and Leu144. Some of these side-chain movements are similar to adjustments that have been observed previously in association with the 'hinge-bending' motion that is presumed to occur during catalysis by the zinc endoproteases.

In the presence of excess zinc, a second zinc ion is observed to bind at His231 within 3.2 Å of the zinc bound to native thermolysin, explaining the inhibitory effect.[21]

USE OF THERMOLYSIN IN PEPTIDE COUPLING REACTIONS

The low-calorie sweetener aspartame is widely used, with about 10 000 metric tons being produced per year. In 1979, Isowa et al.[60] showed that the chiral aspartame precursor Z-L-Asp-L-Phe-OMe could be produced from the thermolysin-promoted condensation of Z-L-Asp and L-Phe-OMe. The results were confirmed by Wayne and Fruton[61] and Nakanishi et al.[62,63] The rate of reaction is, however, slow. St. Clair and Navia[64] and Persichetti et al.[65] have shown that cross-linked crystals of thermolysin are stable in mixtures of water with various organic solvents such as dimethylformamide, tetrahydrofuran, or acetone. By using such conditions they are attempting to develop feasible alternatives to the traditional methods of peptide coupling.

THERMOLYSIN AS A MODEL FOR OTHER HEXXH FAMILY MEMBERS

Because of the relation of thermolysin to a number of physiologically important enzymes, it has often been used to model the binding of inhibitors of these proteins. In general, the anticipated similarities have been supported by subsequent experimental studies. In a small number of cases, however, possible differences have been reported. These are discussed briefly below.

In 2000, Hanson and Stevens[66] described the complex of the 38-residue inhibitor synaptobrevin II with Botulinum Neurotoxin Type B protease. In the proposed complex, the mechanism of peptide cleavage was suggested to be related to that of thermolysin but the model for the bound peptide had the opposite orientation. Rupp and Segelke[67] have questioned the experimental evidence for the presence of the peptide on the grounds that the crystallographic B-factors are very high (averaging 128 Å^2) and that the stereochemistry for most of the 36 modeled residues are at variance with expected stereochemistry. Therefore, the precise alignment of this peptide remains something of an open question. Another complex of the Botulinum Neurotoxin Type B protease has also been described[68] but has subsequently been withdrawn.[69]

Recently, Pannifer et al.[70] reported the crystal structure of the anthrax lethal factor, a thermolysin-related protease, in complex with a fragment of its substrate. The authors state that their model for the bound peptide lies in the direction expected for a thermolysin family member, but the coordinates are for a peptide with the opposite alignment. Paul A. Bartlett, Dale E. Tronrud, B.W.M, and Bernhard Rupp (unpublished) have noted that the average value of the B-factors for the modeled peptide is 168 Å^2. Also, the proposed protein–ligand interactions are very tenuous, with only three contacts less than 3.0 Å for the entire 16-residue peptide. Thus, the experimental evidence supporting the model for the bound peptide is not compelling and its alignment remains uncertain.

ACKNOWLEDGEMENTS

Many members of my group have made invaluable contributions to the structural studies of thermolysin described herein. Indeed, the work would not have been possible without them. I am also grateful to Doug Juers for help with preparing the figures and tables. The work was supported in part by NIH grant GM20066.

REFERENCES

1. S Endo, *J Ferment Technol (Tokyo)*, **40**, 346–53 (1962).
2. H Matsubara, R Sasaki, A Singer and TH Jukes, *Arch Biochem Biophys*, **115**, 324–31 (1965).
3. J Feder and JM Schuck, *Biochemistry*, **9**, 2784–91 (1970).
4. K Morihara and H Tsuzuki, *Eur J Biochem*, **15**, 374–80 (1970).
5. ND Rawlings and AJ Barrett, *Methods Enzymol*, **248**, 183–228 (1995).
6. W Stöcker, F Grams, U Baumann, P Reinemer, F-X Gomis-Rüth, DB McKay and W Bode, *Protein Sci*, **4**, 823–40 (1995).
7. NM Hooper and AJ Turner, *Nat Struct Biol*, **10**, 155–57 (2003).
8. K Titani, MA Hermodson, LH Ericsson, KA Walsh and H Neurath, *Nat New Biol*, **238**, 35–37 (1972).
9. BW Matthews, JN Jansonius, PM Colman, BP Schoenborn and D Dupourque, *Nat New Biol*, **238**, 37–41 (1972).
10. BW Matthews, PM Colman, JN Jansonius, K Titani, KA Walsh and H Neurath, *Nat New Biol*, **238**, 41–43 (1972).
11. M Kubo and T Imanaka, *J Gen Microbiol*, **134**, 1883–92 (1988).
12. SA Latt, B Holmquist and BL Vallee, *Biochem Biophys Res Commun*, **37**, 333–39 (1969).
13. B Holmquist and BL Vallee, *J Biol Chem*, **249**, 4601–7 (1974).
14. MA Holmes and BW Matthews, *J Mol Biol*, **160**, 623–29 (1982).
15. DR Holland, DE Tronrud, HW Pley, KM Flaherty, W Stark, JN Jansonius, DB McKay and BW Matthews, *Biochemistry*, **31**, 11310–16 (1992).
16. AC Hausrath and BW Matthews, *Acta Crystallogr*, **D58**, 1002–7 (2002).
17. P Kraulis, *J Appl Crystallogr*, **24**, 946–50 (1991).
18. E Merritt and MEP Murphy, *Acta Crystallogr*, **D50**, 869–73 (1994).
19. DM van Aalten, A Amadei, AB Linssen, VG Eijsink, G Vriend and HJC Berendsen, *Proteins*, **22**, 45–54 (1995).
20. OR Veltman, VGH Eijsink, G Vriend, A de Kreij, G Venema and B Van den Berg, *Biochemistry*, **37**, 5305–11 (1998).
21. DR Holland, AC Hausrath, D Juers and BW Matthews, *Protein Sci*, **4**, 1955–65 (1995).
22. BW Matthews, LH Weaver and WR Kester, *J Biol Chem*, **249**, 8030–44 (1974).
23. L Serrano, AG Day and AR Fersht, *J Mol Biol*, **233**, 305–12 (1993).
24. X-J Zhang, WA Baase, BK Shoichet, KP Wilson and BW Matthews, *Protein Eng*, **8**, 1017–22 (1995).
25. LH Weaver, WR Kester and BW Matthews, *J Mol Biol*, **114**, 119–32 (1977).
26. DE Tronrud, AF Monzingo and BW Matthews, *Eur J Biochem*, **157**, 261–68 (1986).
27. AA Patchett, E Harris, EW Tristham, MJ Wyvratt, MT Wu, D Taub, ER Peterson, TJ Ikeler, J ten Broeke, LG Payne, DL Ondeyka, ED Thorsett, WJ Greenlee, NS Lohr, RD Hoffsommer, H Joshua, WV Ruyle, JW Rothrock, SD Aster, AL Maycock, FM Robinson and R Hirschmann, *Nature*, **288**, 280–83 (1980).
28. AF Monzingo and BW Matthews, *Biochemistry*, **23**, 5724–29 (1984).
29. WR Kester and BW Matthews, *Biochemistry*, **16**, 2506–16 (1977).
30. J Feder, N Auferheide and BS Wildi, in H. Zuber (ed.), *Enzymes and Proteins from Thermophilic Microorganisms*, Birkhauser Verlag, Basel, Switzerland, pp 41–54 (1976).
31. HM Holden and BW Matthews, *J Biol Chem*, **263**, 3256–60 (1987).

32. H Suda, T Aoyagi, T Takeuchi and J Umezawa, *J Antiobiot*, **26**, 621–23 (1973).
33. C-M Kam, N Nishino and JC Powers, *Biochemistry*, **18**, 3032–38 (1979).
34. PA Bartlett and CK Marlowe, *Biochemistry*, **26**, 8553–61 (1987).
35. HM Holden, DE Tronrud, AF Monzingo, LH Weaver and BW Matthews, *Biochemistry*, **26**, 8542–53 (1987).
36. PA Bartlett and CK Marlowe, *Science*, **235**, 569–71 (1987).
37. DE Tronrud, HM Holden and BW Matthews, *Science*, **235**, 571–74 (1987).
38. BP Morgan, DR Holland, BW Matthews and PA Bartlett, *J Am Chem Soc*, **116**, 3251–60 (1994).
39. N Nishino and JC Powers, *Biochemistry*, **18**, 4340–47 (1979).
40. AF Monzingo and BW Matthews, *Biochemistry*, **21**, 3390–94 (1982).
41. LD Byers and R Wolfenden, *J Biol Chem*, **247**, 606–8 (1972).
42. MC Bolognesi and BW Matthews, *J Biol Chem*, **254**, 634–39 (1979).
43. AC Hausrath and BW Matthews, *J Biol Chem*, **269**, 18839–42 (1994).
44. AL Maycock, DM DeSousa, LG Payne, J ten Broeke, MT Wu and AA Patchett, *Biochem Biophys Res Commun*, **102**, 963–69 (1981).
45. BW Matthews, *Acct Chem Res*, **21**, 333–40 (1988).
46. N Nishino and JC Powers, *Biochemistry*, **17**, 2846–50 (1978).
47. MA Holmes and BW Matthews, *Biochemistry*, **20**, 6912–20 (1981).
48. D Rasnick and JC Powers, *Biochemistry*, **17**, 4363–69 (1978).
49. MA Holmes, DE Tronrud and BW Matthews, *Biochemistry*, **22**, 236–40 (1983).
50. PA Bash, PA Kollman, U Singh, FK Brown and R Langridge, *Science*, **235**, 574–76 (1987).
51. KM Mertz and PK Kollman, *J Am Chem Soc*, **111**, 5649–58 (1989).
52. BP Morgan, JM Scholtz, MD Ballinger, ID Zipkin and PA Bartlett, *J Am Chem Soc*, **113**, 297–307 (1991).
53. J Feder, *Biochem Biophys Res Commun*, **32**, 326–32 (1968).
54. DG Hangauer, AF Monzingo and BW Matthews, *Biochemistry*, **23**, 5730–41 (1984).
55. DW Christianson and WN Lipscomb, *Acc Chem Res*, **22**, 62–69 (1989).
56. WT Lowther and BW Matthews, *Chem Rev*, **102**, 4581–607 (2002).
57. H Drucker and SL Borchers, *Arch Biochem Biophys*, **147**, 242–48 (1971).
58. J Feder, LR Garrett and DS Wildi, *Biochemistry*, **10**, 4552–56 (1971).
59. B Holmquist and BL Vallee, *Biochemistry*, **15**, 101–7 (1976).
60. Y Isowa, M Ohmori, T Ichikawa, K Mori, Y Nonaka, K-I Kihara, K Oyama, H Satoh and S Nishimura, *Tetra Lett*, **28**, 2611–12 (1979).
61. SI Wayne and JS Fruton, *Proc Natl Acad Sci U S A*, **80**, 3241–44 (1983).
62. K Nakanishi, T Kamikubo and R Matsuno, *Bio/Technology*, **3**, 459–64 (1985).
63. K Nakanishi, Y Kimura and R Matsuno, *Eur J Biochem*, **161**, 541–49 (1986).
64. NL St. Clair and MA Navia, *J Am Chem Soc*, **114**, 7314–16 (1992).
65. RA Persichetti, NL St. Clair, JP Griffith, MA Navia and AL Margolin, *J Am Chem Soc*, **117**, 2732–37 (1995).
66. MA Hanson and RC Stevens, *Nat Struct Biol*, **7**, 687–92 (2000).
67. B Rupp and B Segelke, *Nat Struct Biol*, **8**, 663–64 (2001).
68. MA Hanson, TK Oost, C Sukonpan, DH Rich and RC Stevens, *J Am Chem Soc*, **122**, 11268–69 (2000).
69. MA Hanson, TK Oost, C Sukonpan, DH Rich and RC Stevens, *J Am Chem Soc*, **124**, 10248 (2002).
70. AD Pannifer, TY Wong, R Schwarzenbacher, M Renatus, C Petosa, J Bienkowska, DB Lacy, RJ Collier, S Park, SH Leppla, P Hanna and RC Liddington, *Nature*, **414**, 229–33 (2001).

Methionine aminopeptidase

Brian W Matthews

Howard Hughes Medical Institute, Institute of Molecular Biology, University of Oregon, Eugene OR, USA

FUNCTIONAL CLASS

Enzyme: methionine aminopeptidase (MetAP) (E.C.3.4.11); a metalloprotease generally described as having two cobalt ions in the active site, although the number and identity of metals bound *in vivo* is in dispute.

MetAPs catalyze the nonprocessive removal of the N-terminal initiator methionine from nascent polypeptide chains (Table 1).[1–3] The enzyme also removes N-terminal methionine from polypeptides. The *Escherichia coli* enzyme is the prototypical member of the family and has a preference for cleavage next to short amino acids.[4]

OCCURRENCE

Gene knockouts in *E. coli* and *Salmonella typhimurium*[5,6] are lethal, suggesting that the enzyme is essential. This observation, coupled with the role of the enzyme in processing newly synthesized polypeptide chains[1,2] suggest that its occurrence may be widespread. The enzyme from a number of organisms including *E. coli*,[4] yeast,[7] human,[8] *S. typhimurium*,[9] pig,[10] and *Pyrococcus furiosis*[11] has been extensively characterized.

On the basis of sequence and structure comparisons, the enzyme has been divided into two classes. The type II

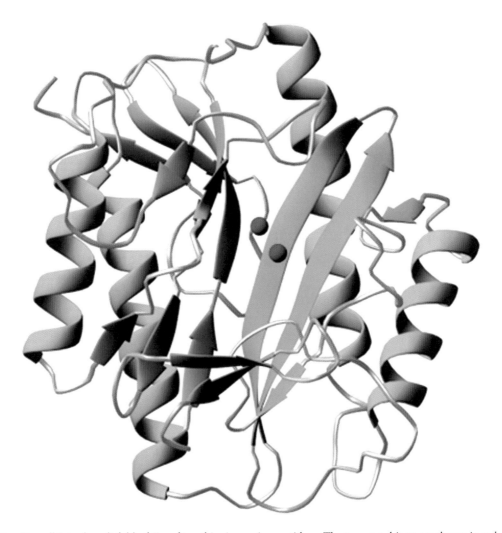

3D Structure   Overall 'pita-bread' fold of *E. coli* methionine aminopeptidase. The two metal ions are shown in red. Figure courtesy of Todd Lowther. PDB Code 2MAT.

Table 1 Specificities of enzymes with structures similar to methionine aminopeptidase

Enzyme	Substrate
Methionine aminopeptidase	H-Met-↓-Xaa-Yaa-...
Aminopeptidase P	H-Xaa-↓-Pro-Yaa-...
Prolidase	H-Xaa-↓-Pro-OH
Creatinase	H_2N-C(=NH)-N(CH_3)-CH_2-COOH

Table 2 Sequences of methionine aminopeptidases

Source	Number of amino acids	References
Escherichia coli	264	4
Pyrococcus furiosis	295	11
Salmonella typhimurium	263	15–17
Bacillus subtilis	248	17–19
Saccharomyces cerevisiae	387	17
Homo sapiens, type 1	394	20
Homo sapiens, type 2	478	8,21
Synechocystis sp.	253	22
Synechocystis sp.	274	22
Synechocystis sp.	305	22
Saccharomyces cerevisiae, type I	387	17
Saccharomyces cerevisiae	421	23,24
Haemophilus influenzae	268	25
Mycoplasma genitalium	248	26
Mycoplasma pneumoniae	248	27
Helicobacter pylori	253	28
Methanococcus jannaschii	294	29
Methanothermus fervidus	188	30
Rattus norvegicus	478	8,31

enzymes are differentiated from type I by the addition of a helical subdomain of about 60 residues inserted within the C-terminal domain. The function of this extra segment is not known. *Escherichia coli*, for example, has only type I MetAP, but yeast and human have both the type I and type II enzymes.[8]

BIOLOGICAL FUNCTION

The presumed key biological role of the methionine aminopeptidases, at least in prokaryotes, is to cleave the N-terminal methionine from nascent polypeptide chains. The enzyme is not thought to act as a generalized protease, for example, in digestion. At the same time, the presence of two types of MetAP in yeast and human, and presumably in other eukaryotes, suggests that the enzyme may have additional roles. In this context, it was striking and unexpected to find that the type II MetAP in humans is the site of action of fumagillin and related drugs.[12–14] These drugs are in clinical trials for the control of cancer. Their presumed role is to inhibit the growth of blood vessels that are required for tumor development. This observation suggests that one role of type II MetAP in humans is to cleave a protein or peptide that participates in angiogenesis.

AMINO ACID SEQUENCE INFORMATION

The number of amino acids in a series of methionine aminopeptidases and the references to their sequences are given in Table 2. In general, the sequences were determined by translation of the cDNA sequence.

PROTEIN PRODUCTION, PURIFICATION, AND MOLECULAR CHARACTERIZATION

The MetAP from *E. coli* was originally cloned, overproduced, and purified as described by Ben-Bassat et al.[4] An improved expression plasmid has been developed by Lowther et al.[32] A 4-L fermentation culture of *E. coli* cells containing the expression plasmid was grown in Luria-Bertani broth with kanamycin ($100\,mg\,L^{-1}$) at $37\,°C$.

The expression was induced by the addition of isopropyl β-D-thiogalactoside to 1 mM at 1.0 OD_{600} for 3 h at $25\,°C$. The cells were lysed by French Press in 100 mL of +T/G buffer [50 mM HEPES, pH 7.9/10% glycerol/0.1% Triton X-100/0.5 M KCl/40 µg mL^{-1} DNase/1 mM $MgCl_2$/15 mM methionine/5 mM imidazole/2 Complete/EDTA-free (Boehringer Mannheim) inhibitor tablets] and centrifuged at $40\,000 \times g$ for 45 min. The supernatant was loaded onto a 10-mL nitrilotriacetic acid–agarose column (Qiagen) equilibrated with +T/G buffer. After washing with +T/G and −T/G buffer (+T/G buffer without glycerol, Triton X-100, and inhibitor cocktail), MetAP was eluted with −T/G buffer containing 60 mM imidazole directly into 1 mL of 500 mM EDTA, pH 8.0. Additional EDTA was added, if necessary, to give a final concentration of 5 mM. After dialysis at $4\,°C$ against 25 mM HEPES buffer, pH 7.9, 150 mM KCl, 15 mM methionine, the poly-histidine tail was removed by incubation of 100 to 200 mg of MetAP with no more than 0.25 units mg^{-1} of biotinylated thrombin (Novagen) at $15\,°C$ for 18 to 20 h. The biotinylated thrombin was eliminated by treatment with excess streptavidin agarose (Novagen) prewashed with −T/G buffer. Passage of the protein through another nitrilotriacetic acid–agarose column equilibrated with −T/G resulted in His-tag-free protein that was subsequently loaded onto a Superdex 75 Hi-load, prep-grade 16/60 gel filtration column (Pharmacia) equilibrated with 25 mM HEPES, pH 6.8, 25 mM K_2SO_4, 100 mM NaCl, 1 mM $CoCl_2$, 15 mM methionine. Protein concentrations were determined by absorption at 280 nm with the extinction coefficient of $16\,350\,M^{-1}\,cm^{-1}$ calculated by using the Genetics Computer Group program PEPTIDESORT. Typical yields were 125 to 200 mg L^{-1} of culture.

Characterization of other representative MetAPs are as follows: *S. typhimurium*,[9,15] yeast,[7,17] *P. furiosis*,[11] porcine liver,[10] human.[33]

METAL CONTENT

Methionine aminopeptidase is a metalloenzyme. Early activity studies[4] suggested that it was maximally stimulated by cobalt, although the identity of the metal ion(s) bound *in vivo* was not firmly established. The crystal structure of the *E. coli* enzyme[34] (see below) showed it to have two metal ions in the active site, both apparently cobalt.

MetAP activity is observed not only in the presence of Co^{2+}; depending on the source, Zn^{2+}, Mn^{2+}, or Ni^{2+} can also be effective. On the basis of the observation that the Zn^{2+}-substituted yeast enzyme was as active as the Co^{2+}-enzyme and that only Zn^{2+}-mediated activity was retained in the presence of reduced glutathione, Walker[35,36] proposed that Zn^{2+} was the *in vivo* metal ion. Holz and coworkers, however,[37–39] have presented data showing that the addition of Zn^{2+} to the *E. coli* enzyme is inhibitory. Zn^{2+} has also been shown to inhibit human MetAP.[40]

On the basis of measurements of the metal content of bacterial extracts, D'Souza and Holz[38] have argued that the cofactor of MetAP in the cell is Fe^{2+}. D'Souza and Holz found that if metal ions are completely removed from *E. coli* MetAP, the enzyme is maximally stimulated on readdition of a single equivalent of Co^{2+} or Fe^{2+}. The addition of excess metal ions (>50 equivalents) resulted in a loss of activity. In contrast, the addition of more than one equivalent of Co^{2+} to human MetAP leads to optimal activity.[40]

Additional work is required to resolve these differences. What seems clear is that the MetAPs have a strong and a weak metal binding site, and that occupancy of the stronger site is sufficient to confer activity.[41,42] The identity of the metal bound *in vivo* has not been definitively established and may be dictated by the concentration and availability of metal ions in the cell.

SPECTROSCOPY

The ^{1}H NMR spectrum of reconstituted *E. coli* MetAP containing one equivalent of Co^{2+} suggested that the tight metal binding discussed in the previous section occurs in Site 1[39] (see below). An EXAFS study also confirmed the presence of a mononuclear metal center with a histidine ligand.[39] To date, such mononuclear binding has not been seen crystallographically.

X-RAY STRUCTURE

Crystallization

The first crystals of *E. coli* MetAP were in space group $P2_1$ with unit cell dimensions $a = 39.0\,\text{Å}$, $b = 61.7\,\text{Å}$, $c = 54.5\,\text{Å}$, and $\beta = 107.3°$, and they permitted the determination of the structure at 2.4-Å resolution.[34,43]

Improved crystals were obtained using protein expressed and purified as above. These crystals were grown at room temperature by vapor diffusion in 20 to 30 μL sitting drops. The protein, 12 mg mL^{-1}, was in storage buffer containing 48.8 mM *N*-octanoyl sucrose (Calbiochem-Novabiochem Corp., La Jolla, CA). This is mixed with an equal volume of the well solution containing 24 to 25% PEG 4000, 0.1 M HEPES, pH 7.0 to 7.2 and fresh 2 mM $CoCl_2$. The crystals have cell dimensions $a = 39.3\,\text{Å}$, $b = 67.7\,\text{Å}$, $c = 48.9\,\text{Å}$, and $\beta = 111.2°$ and permit the resolution to be increased to 1.9 Å (PDB code 2MAT).[44]

Overall description of the structure

Escherichia coli MetAP is a monomer with a 'pita-bread' fold (3D Structure, Figure 1) similar to that originally

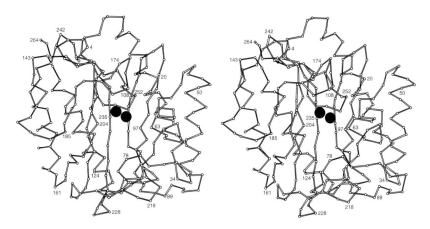

Figure 1 Stereo diagram showing the three-dimensional structure of *E. coli* MetAP. The direction of view is similar to that in the 3D Structure, and the two cobalt ions in the active site are shown as solid spheres. PDB Code 1MAT. Figure drawn with ORTEP.[45]

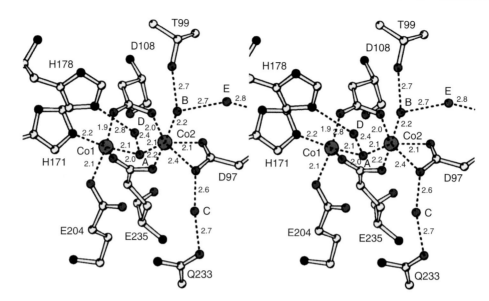

Figure 2 Stereo diagram showing the metal center of *E. coli* MetAP. Oxygen atoms are colored red; nitrogen, blue; carbon, yellow; and cobalt, cyan. Water molecules seen bound in the crystalline enzyme are labeled A through E.[32] PDB Code 2MAT. Figure drawn with BOBSCRIPT.[50]

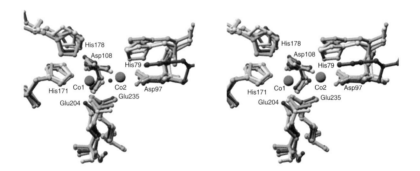

Figure 3 Comparison of the dinuclear centers of *E. coli* MetAP (red),[32] *P. furiosis* MetAP (cyan),[48] human MetAP (yellow)[33] and aminopeptidase P[51] (after WT Lowther and BW Matthews, *Chem Rev*, **102**, 4581–607 (2002)).[41] The numbering is for the *E. coli* enzyme. PDB codes 2MAT, 1XGS, 1BN5, and 1A16, respectively. Figure drawn with MOLSCRIPT and RASTER3D.[52]

described for creatinase[46–48] (Table 1). The overall shape of the molecule is cylindrical with a radius of about 22 Å and height of 25 Å. The central part of the molecule is made up, in large part, of an eight-stranded β-sheet that is folded on itself to 'wrap around' the two active site cobalt ions. It is this folded sheet that is also described as having a 'pita-bread' shape. The 'outside' of the β-sheet, that is, the side distal to the active site, is enclosed by four α-helices plus a region of extended structure made up of the C-terminal region (residues 242–264).

The overall structure also includes two domains (residues 11–116 and 120–241) that have the same topology and very similar three-dimensional structures. These two domains are related by a rotation of 174°, and the rotation axis passes within 4 Å of the midpoint of the two cobalt ions. In other words, the backbone fold (but not the detailed structure) has approximate two-fold symmetry. The degree of sequence similarity between the two structurally related

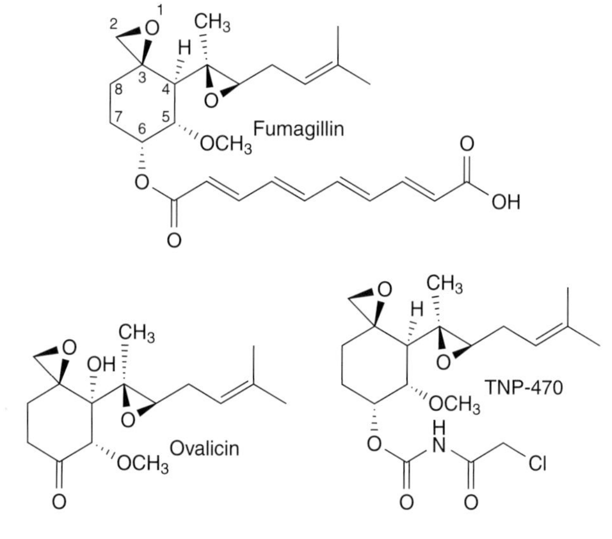

Figure 4 The antiangiogenesis compounds fumagillin, ovalicin, and TNP-470. The epoxide at C3 is the site of covalent attack.

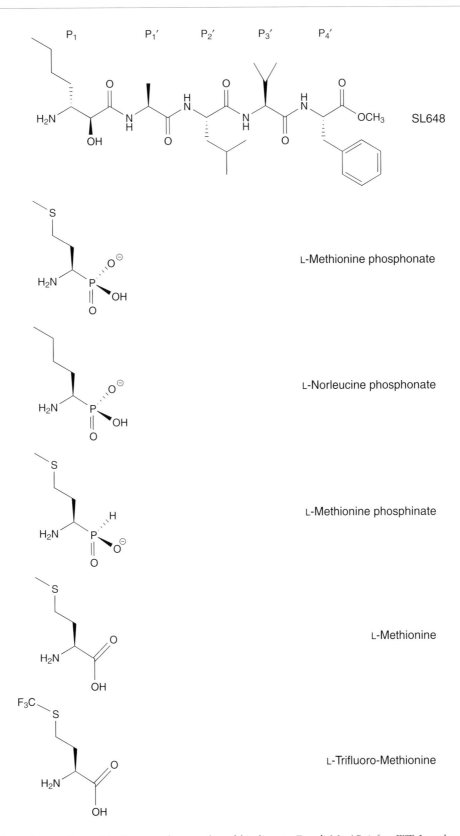

Figure 5 Ligands used crystallographically to analyze modes of binding to *E. coli* MetAP (after WT Lowther and BW Matthews, *Chem Rev*, **102**, 4581–607 (2002)).[41] PDB codes of the enzyme–ligand complexes are SL648 (3MAT), L-methionine phosphonate (1C23), L-norleucine phosphonate (IC27), L-methionine phosphinate (1C24), L-methionine (1C21), L-trifluoro-methionine (1C22).

domains is low, although it seems very likely that the molecule could have arisen by an ancestral gene duplication.[34]

The structures of *P. furiosis* and human MetAPs are similar to the *E. coli* enzyme although these type II enzymes are distinguished by having an additional helical subdomain of approximately 60 residues inserted within the C-terminal domain.[33,49]

Geometry of the dinuclear metal site

The dinuclear cobalt binding site seen in the X-ray structure of *E. coli* MetAP (Figure 2) has as ligands two glutamic acids, two aspartic acids, a histidine, a water molecule that bridges between Co1 and Co2, and another water molecule that is bound just by Co2. The two cobalt ions are 3.2 Å apart.[44] The coordination geometry of Co1 is a distorted trigonal bipyrimidal, while that of Co2 is a distorted octahedral.

Strikingly similar dinuclear metal centers have also been seen in the crystal structures of *P. furiosis* MetAP,[49] human MetAP,[33] and aminopeptidase P[51] (Table 1, Figure 3).

INHIBITOR COMPLEXES

Covalent complexes: fumagillin

Fumagillin and related expoxides (Figure 4) are in clinical tests for the suppression of tumor growth.[53,54] Their site of action has been shown to be the type 2 human MetAP.[12-14] Although more reactive toward the type 2 enzymes, fumagillin will inhibit *E. coli* MetAP, a type 1 enzyme, and has been shown to form a covalent adduct with His79.[32]

Direct evidence for the nature of the complex of fumagillin with human MetAP-2 has come from a crystal structure of the complex (PDB code 1BOA).[33]

Noncovalent complexes

Crystal structures of a series of complexes of *E. coli* MetAP with different ligands (Figures 5 and 6) have helped clarify the likely mode of binding of substrates and the transition state.

SL648 ((3R)-amino-(2S)-hydroxyheptanoyl-L-Ala-L-Leu-L-Val-L-Phe-OMe, Figure 5) is an analogue of bestatin, adapted for the substrate specificity of MetAP (the norleucine at P_1 serves as an analogue of the terminal methionine and avoids the possibility of oxidation to methionine sulfoxide). It inhibits *E. coli* MetAP with an IC50 of 5 μM.[55] The complex of SL648 with the enzyme[32] (Figure 6) helps explain the known substrate specificity. In particular, the norleucine side chain occupies a narrow hydrophobic pocket that would accommodate methionine but not the more bulky side chains of, for example, leucine or phenylalanine. Also, the alanine at the P_1 position occupies a shallow subsite that would not obviously accommodate larger side chains.

Phosphonate and phosphinate analogues of the reaction product methionine (Figure 5) mimic the putative tetrahedral transition-state during catalysis. Two phosphonates and one phosphinate bind to the metal center with similar N-terminal and metal-bridging interactions[56] (Figure 7). The loop containing His79 moves toward the metal center, allowing His179 to make a hydrogen bond to the tetrahedral moiety of the inhibitor.

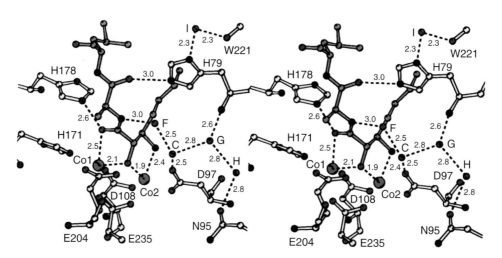

Figure 6 Stereo diagram showing the mode of binding of SL648 (see Figure 5) to *E. coli* MetAP. The backbone of the inhibitor is shown in red and that of the protein in yellow. Atoms are colored as in Figure 2. The valine–phenylalanine moiety at the C-terminus of the inhibitor is not seen in the electron-density maps and is presumably disordered owing to weak interactions with the enzyme (from WT Lowther, DA McMillen, AM Orville and BW Matthews, *Proc Natl Acad Sci USA*, 95, 12153–57 (1998)).[32] PDB Code 3MAT.

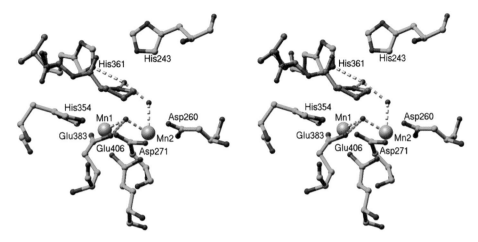

Figure 7 Stereo view showing the binding of L-methionine phosphonate to *E. coli* MetAP (from WT Lowther and BW Matthews, *Chem Rev*, **102**, 4581–607 (2002),[41] and WT Lowther, Y Zhang, PB Sampson, JF Honek and BW Matthews, *Biochemistry*, **38**, 14810–19 (1999)).[56] PDB Code 1C23.

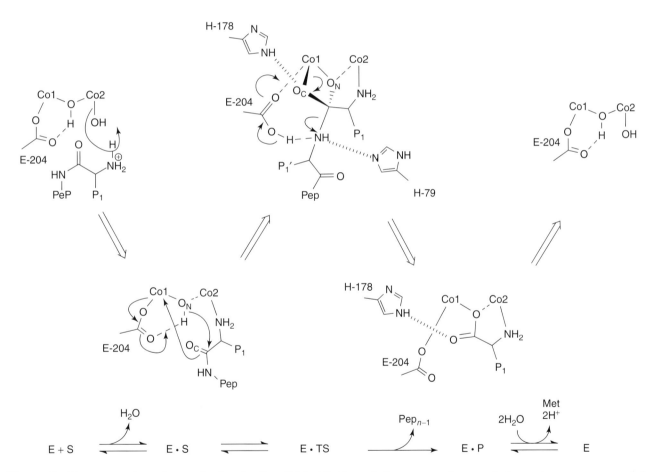

Figure 8 The proposed reaction mechanism for *E. coli* MetAP. O_N and O_C correspond, respectively, to the oxygen atom of the nucleophile and the carbonyl oxygen of the scissile bond. Dashed and solid lines indicate interactions with the ligands or metal ions; hashed lines indicate hydrogen bonds (from WT Lowther, Y Zhang, PB Sampson, JF Honek and BW Matthews, *Biochemistry*, **38**, 14810–19 (1999)).[56]

CATALYTIC ACTIVITY OF MetAP

Measurement of activity

The measurement of the catalytic activity is typically based on the hydrolysis of methionine or norleucine-containing peptides.[32,38] In the system proposed by Lowther et al.,[32] for example, the release of the N-terminal norleucine from the substrate Nle-Ale-Ala-Glu-Glu is measured by derivatization with AQC (6-aminoquinolyl-N-hydroxysuccinimidylcarbamate).

Mechanism of catalysis

On the basis of the crystallographic analyses described above, together with analysis of various mutant enzymes, the mechanism of action outlined in Figure 8 has been proposed.[41,56,57]

The key feature of the proposed mechanism is that Glu204, in concert with the metal center, promotes the nucleophilic attack of a metal-bound water or hydroxide ion on the carbonyl carbon of the substrate to form a tetrahedral noncovalently bound intermediate. The proton, which is accepted by Glu204, is shuttled to the leaving nitrogen in a manner similar to that originally proposed for thermolysin and carboxypeptidase A.[58–61]

The significant residual activity of the mutant H178A of *E. coli* MetAP suggests that the function of this residue is to help stabilize the transition state, but is not essential for activity.

Recently, Holz et al.[42] have proposed that a similar mechanism of action can apply to MetAP with a single metal ion in the active site.

ACKNOWLEDGEMENTS

I am greatly indebted to a number of individuals in my group, in particular, to Todd Lowther and Steve Roderick, for outstanding contributions to the structure–function analysis of *E. coli* methionine aminopeptidase. The work was supported in part by NIH grant GM20066.

REFERENCES

1. A Ben-Bassat and K Bauer, *Nature*, **326**, 315 (1987).
2. A Taylor, *FASEB J*, **7**, 290–98 (1993).
3. RA Bradshaw, WW Brickey and KW Walker, *Trends Biochem Sci*, **23**, 263–67 (1998).
4. A Ben-Bassat, K Bauer, SY Chang, K Myambo, A Boosman and S Chang, *J Bacteriol*, **169**, 751–57 (1987).
5. SY Chang, EC McGary and S Chang, *J Bacteriol*, **171**, 4071–72 (1989).
6. CG Miller, AM Kukral, JL Miller and NR Movva, *J Bacteriol*, **171**, 5215–17 (1989).
7. Y-H Chang, U Teichert and JA Smith, *J Biol Chem*, **265**, 19892–97 (1990).
8. SM Arfin, RL Kendall, L Hall, LH Weaver, AE Stewart, BW Matthews and RA Bradshaw, *Proc Natl Acad Sci USA*, **92**, 7714–18 (1995).
9. CG Miller, KL Strauch, AM Kukral, JL Miller, PT Wingfield, GJ Mazzei, RC Werlen, P Graber and NR Movva, *Proc Natl Acad Sci USA*, **84**, 2718–22 (1987).
10. RL Kendall and RA Bradshaw, *J Biol Chem*, **267**, 20667–73 (1992).
11. S Tsunasawa, Y Izu, M Miyagi and I Kato, *J Biochem*, **122**, 843–50 (1997).
12. EC Griffith, Z Su, BE Turk, S Chen, YH Chang, Z Wu, K Biemann and JO Liu, *Chem Biol*, **4**, 461–71 (1997).
13. EC Griffith, S Zhuang, S Niwayama, CA Ramsay, YH Chang and JO Liu, *Proc Natl Acad Sci USA*, **95**, 15183–88 (1998).
14. N Sin, L Meng, MQ Wang, JJ Wen, WG Bornmann and CM Crews, *Proc Natl Acad Sci USA*, **94**, 6099–103 (1997).
15. PT Wingfield, P Graber, G Turcatti, NR Movva, M Pelletier, S Craig, K Rose and CG Miller, *Eur J Biochem*, **180**, 23–32 (1989).
16. NR Movva, D Semon, C Meyer, E Kawashima, P Wingfield, JL Miller and CG Miller, *Mol Gen Genet*, **223**, 345–48 (1990).
17. Y-H Chang, U Teichert and JA Smith, *J Biol Chem*, **267**, 8007–11 (1992).
18. K Nakamura, A Nakamura, H Takamatsu, H Yoshikawa and K Yamane, *J Biochem*, **107**, 603–7 (1990).
19. JW Suh, SA Boylan, SH Oh and CW Price, *Gene*, **168**, 17–23 (1996).
20. T Nagase, N Miyajima, A Tanaka, T Sazuka, N Seki, S Sato, S Tabata, K Ishikawa, Y Kawarabayasi and H Kotani, *DNA Res*, **2**, 37–43 (1995).
21. X Li and Y-H Chang, *Biochim Biophys Acta*, **1260**, 333–36 (1995).
22. T Kaneko, A Tanaka, S Sato, H Kotani, T Sazuka, N Miyajima, M Sugiura and S Tabata, *DNA Res*, **2**, 153–66 (1995).
23. X Li and Y-H Chang, *Proc Natl Acad Sci USA*, **92**, 12357–61 (1995).
24. B Obermaier, J Gassenhuber, E Piravandi and H Domdey, *Yeast*, **11**, 1103–12 (1995).
25. RD Fleischmann, MD Adams, O White, RA Clayton, EF Kirkness, AR Kerlavage, CJ Bult, J-F Tomb, BA Dougherty, JM Merrick, K McKenney, G Sutton, W FitzHugh, C Fields, JD Gocayne, J Scott, R Shirley, L-I Liu, A Glodek, JM Kelley, JF Weidman, CA Phillips, T Spriggs, E Hedblom, MD Cotton, TR Utterback, MC Hanna, DT Nguyen, DM Saudek, RC Brandon, LD Fine, JL Fritchman, JL Fuhrmann, NSM Geoghagen, CL Gnehm, LA McDonald, KV Small, CM Fraser, HO Smith and JC Venter, *Science*, **269**, 496–512 (1995).
26. CM Fraser, JD Gocayne, O White, MD Adams, RA Clayton, RD Fleischmann, CJ Bult, AR Kerlavage, G Sutton, JM Kelley, JL Fritchman, JF Weidman, KV Small, M Sandusky, J Fuhrmann, D Nguyen, TR Utterback, DM Saudek, CA Phillips, JM Merrick, J-F Tomb, BA Dougherty, KF Bott, P-C Hu, TS Lucier, SN Peterson, HO Smith, CA Hutchison III and JC Venter, *Science*, **270**, 397–403 (1995).
27. H Hilbert, R Himmelreich, H Plagens and R Herrmann, *Nucleic Acids Res*, **24**, 628–39 (1996).
28. J-F Tomb, O White, AR Kerlavage, RA Clayton, GG Sutton, RD Fleischmann, KA Ketchum, HP Klenk, S Gill, BA Dougherty, K Nelson, J Quackenbush, L Zhou, EF Kirkness, S Peterson, B Loftus, D Richardson, R Dodson, HG Khalak, A Glodek, K McKenney, LM Fitzgerald, N Lee, MD Adams, EK Hickey,

DE Berg, JD Gocayne, TR Utterback, JD Peterson, JM Kelley, MD Cotton, JM Weidman, C Fujii, C Bowman, L Watthey, E Wallin, WS Hayes, M Borodovsky, PD Karp, HO Smith, CM Fraser and JC Venter, *Nature*, **388**, 539–47 (1997).

29. CJ Bult, O White, GJ Olsen, L Zhou, RD Fleischmann, G Sutton, JA Blake, LM FitzGerald, RA Clayton, JD Gocayne, AR Kerlavage, BA Dougherty, J-F Tomb, MD Adams, CI Reich, R Overbeek, EF Kirkness, KG Weinstock, JM Merrick, A Glodek, JL Scott, NSM Geoghagen, JF Weidman, JL Fuhrmann, D Nguyen, TR Utterback, JM Kelley, JD Peterson, PW Sadow, MC Hanna, MD Cotton, KM Roberts, MA Hurst, BP Kaine, M Borodovsky, H-P Klenk, CM Fraser, HO Smith, CR Woese and JC Venter, *Science*, **273**, 1058–73 (1996).

30. ES Haas, CJ Daniels and JN Reeve, *Gene*, **77**, 253–63 (1989).

31. S Wu, S Gupta, N Chatterjee, RE Hileman, TG Kinzy, ND Denslow, WC Merrick, D Chakrabarti, JC Osterman and NK Gupta, *J Biol Chem*, **268**, 10796–801 (1993).

32. WT Lowther, DA McMillen, AM Orville and BW Matthews, *Proc Natl Acad Sci USA*, **95**, 12153–57 (1998).

33. S Liu, J Widom, CW Kemp, CM Crews and J Clardy, *Science*, **282**, 1324–27 (1998).

34. SL Roderick and BW Matthews, *Biochemistry*, **32**, 3907–12 (1993).

35. KW Walker and RA Bradshaw, *Protein Sci*, **7**, 2684–87 (1998).

36. KW Walker and RA Bradshaw, *J Biol Chem*, **274**, 13403–9 (1999).

37. VM D'Souza, B Bennett, AJ Copik and RC Holz, *Biochemistry*, **39**, 3817–26 (2000).

38. VM D'Souza and RC Holz, *Biochemistry*, **38**, 11079–85 (1999).

39. NJ Cosper, VM D'Souza, RA Scott and RC Holz, *Biochemistry*, **40**, 13302–9 (2001).

40. G Yang, RB Kirkpatrick, T Ho, GF Zhang, PH Liang, KO Johanson, DJ Casper, ML Doyle, JP Marino Jr, SK Thompson, W Chen, DG Tew and TD Meek, *Biochemistry*, **40**, 10645–54 (2001).

41. WT Lowther and BW Matthews, *Chem Rev*, **102**, 4581–607 (2002).

42. A Copik, SI Swierczek, WT Lowther, VM D'Souza, BW Matthews and RC Holz, *Biochemistry*, **42**, 6283–92 (2003).

43. SL Roderick and BW Matthews, *J Biol Chem*, **263**, 16531 (1988).

44. WT Lowther, AM Orville, DT Madden, S Lim, DH Rich and BW Matthews, *Biochemistry*, **38**, 7678–88 (1999).

45. CK Johnson, ORTEP: A Fortran Thermal-Ellipsoid Plot Program for Crystal Structure Illustrations, ORNL-3794 Revised, Oak Ridge National Laboratory, Oak Ridge, Tennessee (1965).

46. M Coll, SH Knof, Y Ohga, A Messerschmidt, R Huber, H Moellering, L Russmann and G Schumacher, *J Mol Biol*, **214**, 597–610 (1990).

47. HW Hoeffken, SH Knof, PA Bartlett, R Huber, H Moellering and G Schumacher, *J Mol Biol*, **204**, 417–33 (1988).

48. JF Bazan, LH Weaver, SL Roderick, R Huber and BW Matthews, *Proc Natl Acad Sci USA*, **91**, 2473–77 (1994).

49. TH Tahirov, H Oki, T Tsukihara, K Ogasahara, K Yutani, K Ogata, Y Izu, S Tsunasawa and I Kato, *J Mol Biol*, **284**, 101–24 (1998).

50. J Esnouf, *J Mol Graph*, **15**, 132–38 (1997).

51. MC Wilce, CS Bond, NE Dixon, HC Freeman, JM Guss, PE Lilley and JA Wilce, *Proc Natl Acad Sci USA*, **95**, 3472–77 (1998).

52. EA Merritt and DJ Bacon, *Methods Enzymol*, **277**, 505–24 (1997).

53. D Ingber, T Fujita, S Kishimoto, K Sudo, T Kanamaru, H Brem and J Folkman, *Nature*, **348**, 555–57 (1990).

54. EJ Corey, A Guzman-Perez and MC Noe, *J Am Chem Soc*, **116**, 12109–10 (1994).

55. SJ Keding, NA Dales, S Lim, D Beaulieu and DH Rich, *Synth Commun*, **28**, 4463–70 (1998).

56. WT Lowther, Y Zhang, PB Sampson, JF Honek and BW Matthews, *Biochemistry*, **38**, 14810–19 (1999).

57. WT Lowther and BW Matthews, *Biochim Biophys Acta*, **1477**, 157–67 (2000).

58. AF Monzingo and BW Matthews, *Biochemistry*, **23**, 5724–29 (1984).

59. DG Hangauer, AF Monzingo and BW Matthews, *Biochemistry*, **23**, 5730–41 (1984).

60. BW Matthews, *Acc Chem Res*, **21**, 333–40 (1988).

61. DW Christianson and W Lipscomb, *Acc Chem Res*, **22**, 62–69 (1989).

Neprilysin

Glenn E Dale and Christian Oefner
Morphochem AG, Basel, Switzerland

FUNCTIONAL CLASS

Enzyme; Neprilysin (also known as NEP, neutral endopeptidase, enkephalinase, CALLA, CD10, EC 3.4.24.11) is a type II integral membrane zinc-dependent endopeptidase belonging to the M13 subfamily. The members of the family are all glycoproteins with small N-terminal cytoplasmic segments and large C-terminal extracellular domains. Neprilysin is essentially an oligopeptidase hydrolyzing peptides up to about 40 amino acids in length.

OCCURRENCE

NEP is widely, although not ubiquitously, distributed in mammalian cells. In peripheral tissues, NEP is particularly abundant in membranes of the brush border epithelial cells of kidney and intestine, the lymph nodes and placenta; it is found at lower concentrations in lung, testis, prostate, fibroblasts, neutrophils, chondrocytes in articular cartilage, exocrine glands, and various epithelial and endocrine cells.[1] In the central nervous system (CNS), NEP is discretely distributed in the brain with the highest concentrations in the choroid plexus, substantia nigra, caudate cumbens,

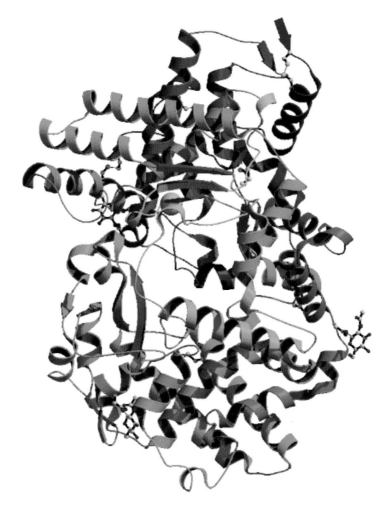

3D Structure   Schematic presentation of the soluble extracellular domain of human NEP showing the monomeric structure with the catalytically active zinc ion (yellow) in the center of the enzyme, PDB code: 1DMT. Sugar moieties and disulfide bridges are indicated. Prepared with the programs MOLSCRIPT[71] and RASTER3D.[72]

globus pallidus, olfactory tubercle, nucleus accumbens, and the substantia gelatinosa of the spinal cord.[2]

BIOLOGICAL FUNCTION

NEP

NEP was originally extracted and purified from the brush border of a rabbit kidney as a peptidase capable of hydrolyzing insulin B chain.[3] As it is widely distributed in mammalian tissues, it is involved in the inactivation of a variety of signaling peptides.[2,4,5] NEP has been implicated in the regulation of opioid peptide action through the degradation of endogenously released enkephalins.[6] Moreover, NEP is involved in the physiological degradation of the peptides modulating blood pressure, such as the cardiac hormone atrial natriuretic peptide (ANP), bradykinin, and endothelin. More recently, NEP has been implicated in the degradation of amyloid β peptide (Aβ1-42), the primary pathogenic agent in Alzheimer's disease.[7,8] Owing to the central role of NEP in hormone regulation, there is great interest in the identification of novel inhibitors of NEP for application in clinical treatment.[9–11] Potent inhibitors of NEP have been synthesized, which produce a pharmacological response through increase in opioid or vasoactive peptide levels, indicating their therapeutic potential as novel analgesics or antihypertensive agents.[12–16]

In recent years, several mammalian homologues of NEP have been described, including the endothelin-converting enzyme (ECE-1),[17] ECE-2,[18] the erythrocyte surface antigen KELL,[19] the phosphate-regulating gene (*PEX*) on the X chromosome,[20] SEP, soluble secreted endopeptidase,[21,22] and the recently identified damage-induced neuronal endopeptidase (DINE)/X-converting enzyme (XCE).[23,24] Moreover, NEP homologues have been found in simpler organisms, including prokaryotes,[25,26] with over 20 NEP-like genes present in *Caenorhabditis elegans* and *Drosophila melanogaster*.[27,28] The overall homology between NEP and the M13 family members ranges from 55 to 25% but increases if only the 250 C-terminal residues are considered, as can be seen in Figure 1. This degree of similarity is sufficient to indicate that these proteins share a common origin and a similar fold.

Other members of the mammalian NEP family

Only NEP and ECE-1 were discovered on the basis of their activities. All other members were identified through their sequence similarities and only subsequently was their enzymatic activity detected. The endothelins are potent vasoconstrictor peptides and are synthesized as preproendothelins, converted to inactive big endothelin, for which further processing is required to generate the active hormone, endothelin.[29,30] ECE-1 cleaves the tryptophan–valine bond in big endothelin-1, thereby performing the final and rate-limiting step in the conversion of preproendothelin to endothelin-1.[31] Although it has been shown that ECE-2 can convert big endothelin to endothelin, it is unclear whether it performs this function *in vivo*.[18] The KELL protein constitutes a major antigen on human erythrocytes and is clinically important because incompatibility to KELL antigens can cause hemolytic reactions to blood transfusions as well as erythroblastosis in newborn children.[32] Until recently, KELL was a member of the M13 family by homology only. It has now been shown that KELL preferentially cleaves big endothelin-3 to endothelin-3.[33] The human *PEX* gene was identified from studies of patients with X-linked hypophosphatemic rickets.[20] The function of PEX is unknown; current information suggests that impaired renal phosphate conservation in X-linked hypophosphatemia is due to the failure of PEX to either degrade an undefined phosphaturic factor or activate a novel phosphate-conserving hormone. XCE was originally identified by homology cloning from human brain cDNA.[23] There is no known substrate for the enzyme but targeted deletion in mouse have shown it to be essential and it appears to be critical in control of respiration.[34] Finally, SEP, a soluble secreted endopeptidase, was cloned from ECE-1 knockout mouse embryos; however, very little is known of its localization or physiological roles. The activity and specificity of SEP most closely resembles NEP and it is inhibited by both phosphoramidon and thiorphan.[21,22]

AMINO ACID SEQUENCE INFORMATION

- *Homo sapiens* (Human), 742 amino acid residues (AA), translation of cDNA sequence.[35]

- *Rattus norvegicus* (Rat), 742 amino acid residues (AA), translation of cDNA sequence.[36]

- *Oryctolagus cuniculus* (Rabbit), 750 amino acid residues (AA), translation of cDNA sequence.[37]

- *Mus musculus* (Mouse), 750 amino acid residues (AA), translation of cDNA sequence.[38]

- *Caenorhabditis elegans*, over twenty NEP-like genes.[27]

- *Drosophila melanogaster*, (Fruit Fly) over twenty NEP-like genes.[28]

PROTEIN PRODUCTION, PURIFICATION, AND MOLECULAR CHARACTERIZATION

NEP has been purified from various tissues, such as kidney,[39–42] pituitary,[43] brain,[44,45] and intestine,[46] using various detergents or lipophilic solvents to solubilize the membrane-bound protein. The initial purifications followed a conventional scheme consisting of a solubilization step followed by subsequent use of anion

Neprilysin

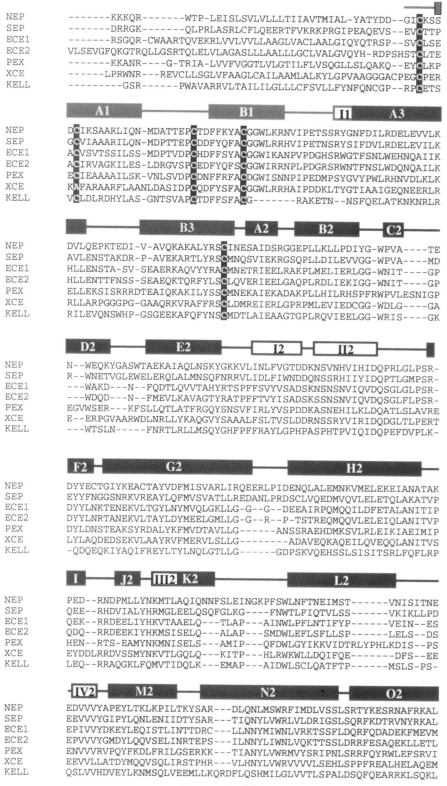

Figure 1 (a)

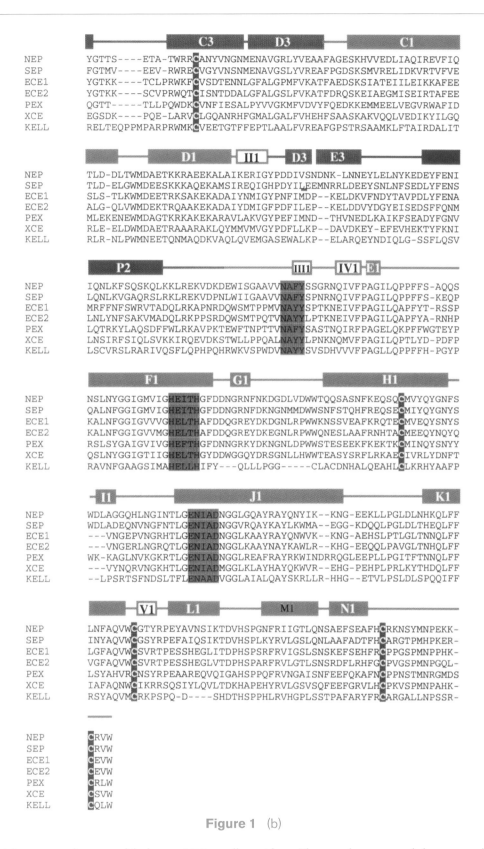

Figure 1 (b)

Figure 1 Multiple sequence alignment of the human M13 metallopeptidases. The secondary structural elements are those derived from the NEP structure. The α-helices (filled rectangles), β-strands (open rectangles) are drawn above the sequence. The color indicates the catalytic domain (red), the smaller domain 2 (blue), and the three interdomain linker fragments (cyan). The conserved motifs for zinc ligation and catalysis are shaded in green and the conserved cysteine residues are indicated.

exchange, hydroxylapatite, and gel filtration column chromatography. The purification of the native enzyme has subsequently been greatly improved by the use of immunoaffinity chromatography and improved solubilizing agents.[47] More recently, after the cloning of the cDNA coding for the complete primary structure of NEP, the first recombinant proteins were produced. The large-scale expression of enzymatically active full-length NEP has been achieved in mammalian cells.[48,49] The soluble ectodomain has been produced, expressed, and purified from various sources including COS-1 cells,[50] baculovirus-infected insect cells,[51] *Schizosaccharomyces pombe*,[52] as well as *Pichia pastoris*.[53]

Molecular cloning of NEP revealed it to be a type II integral membrane protein consisting of a short NH2-cytoplasmic domain, a transmembrane region of 22 hydrophobic residues, with the bulk of the protein, including the active site, facing the extracellular space.[34–36] The molecular weight of NEP ranges from about 85 000 to 100 000 Da depending on the tissue source, the variation being attributable to the differences in glycosylation.[45] Rabbit NEP contains five potential N-linked glycosylation sites and the human and rat NEP contain six. The human enzyme has 12 cysteine residues, 10 of which are conserved among the other members of the family. The extracellular cysteines are involved in interchain disulphide bridges that are important for maintenance of structure and activity.[54] In most species, NEP appears to exist as a noncovalent homodimer when inserted in the membrane; however, the recombinant soluble ectodomain exists as a monomer in solution.[52,53]

The catalytic properties of NEP resemble those of a group of zinc-dependent bacterial endopeptidases of which thermolysin (TLN) is the best characterized. Early structure activity studies showed that TLN and NEP cleave the peptide bond of their substrates at the amino side of hydrophobic residues and that both enzymes are inhibited by the same type of molecules, such as phosphoramidon and thiorphan, with the same stereochemical dependence.[55–57] Although there is very limited homology between the primary sequences of mammalian and bacterial enzymes, they are characterized by two consensus sequences HExxH and ExxA/GD, involved in zinc binding and catalysis.[58] Hydrophobic cluster analysis suggested important similarities in their active sites, and putative residues involved in zinc coordination, catalysis, and substrate binding have been identified in NEP by site-directed mutagenesis experiments.[59–64]

METAL CONTENT

NEP contains one molecule of zinc/subunit and no other metal ions are known to be associated with the protein.

ACTIVITY TEST

There are numerous sensitive fluorometric and colorimetric assays available for NEP based on a wide variety of substrates. Methods include measuring the tripeptide [^{3}H] Tyr-Gly-Gly or that formed by cleavage of the glycine–phenylalanine bond of [^{3}H] Enkephalin.[65] Additionally, internally quenched fluorogenic substrates such as the commercially available dansyl-D-Ala-Gly-Phe(pNO2)-Gly allows continuous recording of NEP activity.[66] A particularly convenient and sensitive two-stage method that can be easily adapted to microtiter plates involves the substrate Suc-Ala-Ala-Leu-NHPhNO2 in the presence of *Schizosaccharomyces griseus* aminopeptidase as coupling enzyme.[67] The release of Leu-NHPhNO2 by NEP followed by the release of *p*-nitroaniline by aminopeptidase can be measured at 405 nm. NEP exhibits a pH optimum of 6.0.[3] It is inhibited by zinc-chelating agents as well as a number of inhibitors of which phosphoramidon and thiorphan are the best described. Mercapto-, phosphonamidate-, carboxyalkyl-hydroxamate, and *N*-formyl hydroxylamine inhibitors have all been reported.[2,68]

Table 1 Structures of the competitive NEP inhibitors phosphoramidon (I), thiorphan (II), and CGS 31447 (III)

X-RAY STRUCTURE OF THE SOLUBLE EXTRACELLULAR DOMAIN OF HUMAN NEP (sNEP)

Crystallization

The first crystals of *human* sNEP, which led to the determination of its three-dimensional structure by X-ray crystallography, were grown under the conditions published by Dale et al.,[53] using Endo-F1 glycosidase–treated enzyme at a concentration of 20 mg mL^{-1}. A binary complex with the inhibitor phosphoramidon [N-(α-L-rhamnopyranosyl-oxyhydroxy-phosphinyl)-L-leucyl-L-tryptophan], a metabolite produced by *Streptomyces tanashiensis*, was formed to obtain crystals of the inhibited protein by vapor diffusion. The experimental conditions were 25% PEG (polyethylene glycol) 3350, 200 mM ammonium sulfate, and 100 mM bis–Tris at pH 7.5 as well as 1 mM inhibitor. The crystals belong to the trigonal space group P3$_2$21 with $a = b = 107.6$ Å and $c = 112.8$ Å. They contain one molecule per asymmetric unit and diffract to a resolution of 2.0 Å. The structure has been solved by the multiple isomorphous replacement method measuring all diffraction intensities from flash frozen crystals at 120 K using a cryoprotectant solution corresponding to the reservoir conditions containing 30% (v/v) glycerol.[69] Diffraction data were also collected for other potent and selective inhibitors (Table 1), which are the neutral endopeptidase inhibitor thiorphan and the dual action NEP/ECE inhibitor CGS 31447, giving insight into zinc ligation and subsite specificity of the enzyme. The structure of *human* NEP complexed with thiorphan has been refined to 2.5 Å resolution, and that of the complex with CGS 31447 to 2.8 Å.

Overall description of the structure

The structure of the soluble extracellular domain of human sNEP, comprising residues D52 to W749 is shown schematically in the 3D structure and Figure 2. The monomeric ectodomain, which consists of 699 residues, has an ellipsoidal shape with approximate dimensions

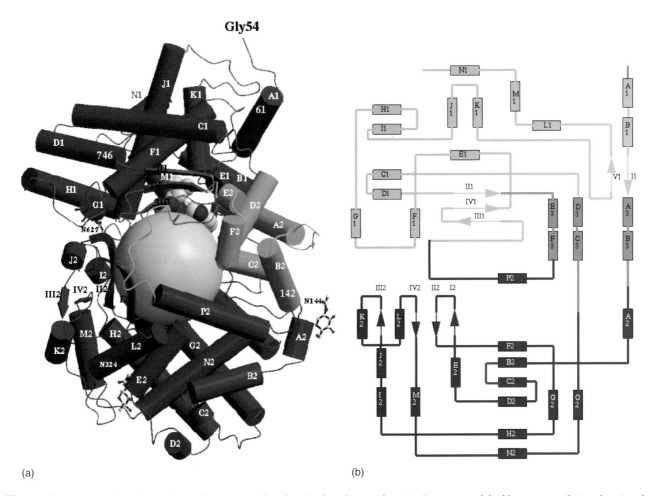

Figure 2 Ribbon plot and topology of sNEP complexed with phosphoramidon. (a) Cartoon model of human neprilysin, showing the catalytic domain 1 (violet), the smaller domain 2 (blue) and the four interdomain linker fragments (cyan). The transition-state mimetic phosphoramidon is shown in CPK representation. Glycosylation sites and disulfide bonds are indicated. The figure was produced with MOLSCRIPT[71] and RASTER3D[72] (b) Topological diagram of sNEP colored as (a). Arrows denote strands and cylinders indicate helices.

Neprilysin

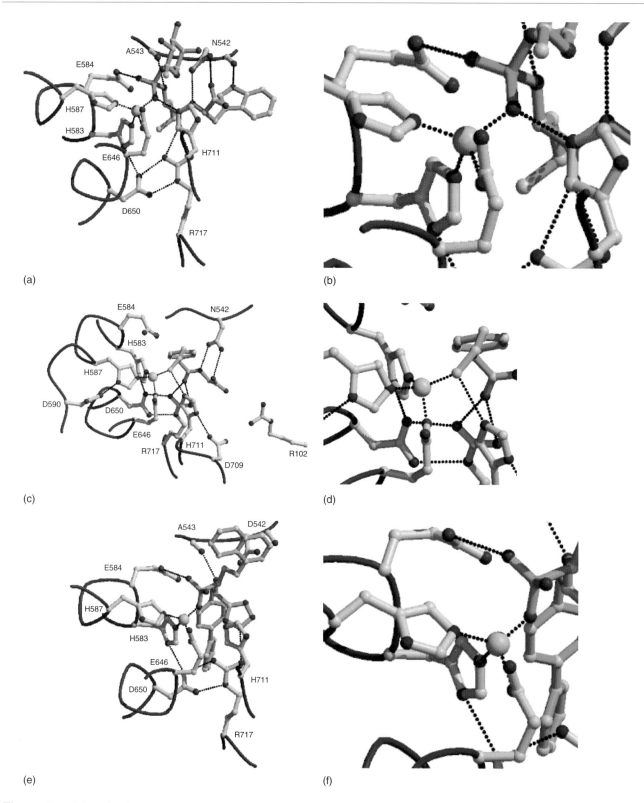

Figure 3 Inhibitor binding mode. The inhibitors (a) phosphoramidon; (b) thiorphan; and (c) CGS 31447 are shown in green. The zinc ion is indicated as yellow sphere, H-bonds are presented as dotted lines. A close-up of the zinc-binding geometry for each inhibitor is shown in the right panel.

Table 2 Geometry of the Zn^{2+} binding site

Ligands	Distance to Zn^{2+} (Å)	Angles (degree)		
		His583 (NE2)	His587 (NE2)	Glu646 (OE1)
Phosphoramidon				O1P
His583NE2	2.03		101	94	116
His587NE2	2.01			109	128
Glu646OE1	1.96				104
PHO-O1P	1.93				
Thiorphan					S10
His583NE2	1.88		101	106	124
His587NE2	1.94			96	123
Glu646OE1	1.87				104
S10	2.12				
CGS 31447					O19
His583NE2	2.12		100	89	109
His587NE2	1.96			124	124
Glu646OE1	1.96				114
O19	1.70				

$95 \times 55 \times 55$ Å and is built up by three domains. Two largely α-helical domains, which can be considered as two lobes connected by intertwining polypeptide segments, are bridged by four interdomain linker fragments. They form together a central, spherical cavity with a diameter of approximately 20 Å, bearing the active site of the enzyme. The largest N-terminal domain (domain I) comprises 321 residues, including the residues involved in zinc ligation and catalysis. These belong to the highly conserved residues of the consensus sequences ^{583}HExxH^{687} and ^{646}ExxxD^{650} and the residues of the ^{542}NAFY^{545} motif.[58,70] They are part of the zinc-binding helix F1, the helix J1, which forms the base of the active site and β-strand III1, which acts as a cap, holding the substrate or inhibitor respectively (Figure 2(b)). The smaller second domain (domain II), which comprises 286 residues, restricts the access of the active site region together with the linker fragment and functions as a molecular sieve. The size of the natural substrates is limited to about 3000 Da, which is the main difference between the mammalian metallopeptidases of the neprilysin family and the bacterial proteases.

The monomeric enzyme structure exhibits six disulfide bridges. Five are conserved among the other members of the mammalian membrane metallopeptidase family. Four are located within the catalytic domain I, one is found within the interdomain linker fragments and the nonconserved disulfide bridge is present within domain II. N-Glycosylation on Asn144, Asn304, and Asn627 is confirmed by electron density consistent with a single N-acetyl-glucosamine moiety.

Zinc binding site geometry

A stereo diagram of the catalytically active zinc binding site is shown in Figure 3; distances and angles of the approximately tetrahedral arrangement are listed for the various complexes in Table 2. The data observed for phosphoramidon are compared with the data obtained for the mercapto inhibitor thiorphan and the phosphonic acid inhibitor CGS 31447, which inhibit human NEP with an IC$_{50}$ of 1.9 nM and 5 nM respectively.[70,73] The mononuclear zinc binding site is located in the catalytic domain I and its coordination involves – in all three complexes – the side chains of the three conserved amino acid residues His583 (NE2), His587 (NE2), and Glu646 (OE1) with interatomic distances of 2.03 Å, 2.01 Å, and 1.96 Å respectively. These residues are located around the zinc atom such that a catalytic water molecule would complete a distorted tetrahedral coordination. Within the ligand complex, the zinc ion is inaccessible to the solvent and close to the inner surface of the protein, which is structurally most closely related to other zinc metalloproteases such as TLN, bacterial metalloelastase, and the neutral protease from *Bacillus cereus*. Tridentate combinations of histidine and glutamic or aspartic acid have proved to be characteristic of catalytic function by providing coordination sites open for water and/or substrate complexes and their transition-state intermediates.[74] In the complex with phosphoramidon, the inhibitor is involved in a monodentate recognition with the zinc ion involving a single oxygen (O1P) of its tetrahedral N-phosphoryl moiety. The interatomic distance to the metal ion is 1.93 Å. The second oxygen of the N-phosphoryl moiety, O2P, is hydrogen bonded to OE1 of the catalytic residue Glu584 at a distance of 2.73 Å. The atom O2P replaces the catalytic water molecule, which is polarized by the acidic amino acid side chain during catalysis.[60] The structure was solved by using X-ray crystallography data at 2.1-Å resolution.

The first potent, synthetic NEP inhibitor, thiorphan (N-[(R,S)-(3-mercapto-2-benzyl-propanoyl)]-glycine ($K_i = $

4 nM), demonstrates the energetically greater importance of the zinc-coordinating thiol group than the stereochemically dependent van der Waals interactions governing subsite recognition for small molecule binding.[75,76] The thiol moiety of the competitive and selective endopeptidase inhibitor acts as a monodentate ligand for the zinc ion. The thiol metal distance is 2.12 Å as revealed by the complexed structure of human NEP solved and refined to 2.5-Å resolution. The peptidic backbone of the inhibitor is involved in the same intermolecular interactions as observed for phosphoramidon.

Zinc recognition of the phosphonic acid inhibitor CGS 31447 is very similar to that observed for phosphoramidon. A single oxygen (O19) of the terminal phosphate moiety is involved in a monodentate metal recognition with an intermolecular bonding distance of 1.70 Å. The second oxygen is hydrogen bonded to the side chain of Glu584 (3.05 Å) and the third oxygen of the phosphonate points toward the boundary of the S_1 subsite.

Subsite specificity

Extensive studies of enkephalin analogues and dipeptides have shown that the specificity of NEP is essentially ensured by the S'_1 subsite, which interacts primarily with aromatic or large hydrophobic residues. The S'_2 subsite has poor specificity, although there is increased affinity with the presence of a carboxyl group owing to favorable

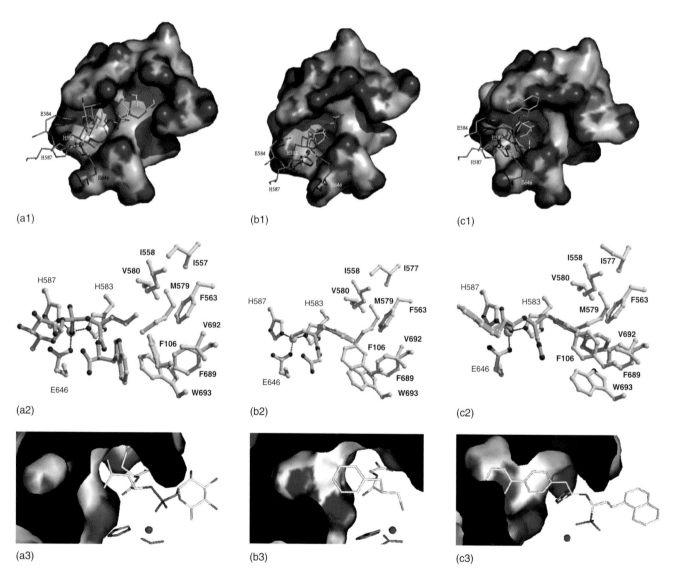

Figure 4 Primed subsites of human NEP. (a) The inhibitors phosphoramidon; (b) thiorphan; and (c) CGS 31447 are shown together (first row) with the molecular surface (second row) with the zinc binding site and the hydrophobic S'_1 subsite presented in a ball and stick model and (third row) in a close-up of the S'_1 binding pocket. Pictures of the first and last row have been produced with PyMOL.[80] Pictures of the second row have been produced with MOLSCRIPT[71] and RASTER3D.[72] The pictures within each row are presented in the same orientation; between rows they are rotated by 90° to the left (row 2) or 90° to the right (row 3).

interactions with Arg717. There is a minor stabilizing role of the S_1 subsite for small molecule binding in NEP as expected from the slight increase in affinity of inhibitors with a P_1 side chain.[77–79] In the bound conformation of phosphoramidon, the rhamnose moiety of the P1 residue is mainly exposed to solvent, leaving 80% of the inhibitor surface in direct contact with the enzyme. This is also true for the naphthyl substituent of CGS 31447, which is also slightly interacting with the side chain of Phe544.

The absence of this stringent specificity is in contrast with that observed for the S'_1 subsite. The S'_1 residues interact preferentially with aromatic or large hydrophobic moieties and involve Phe106, Ile558, Phe563, Met579, Val580, Val692, and Trp693. For the various inhibitor complexes, the primed subsite recognition varies significantly as shown in Figure 4. This can most clearly be seen for the dual action ECE/NEP inhibitor CGS 31447. Its P'_1 biphenyl substituent nearly doubles the volume of the subsite to approximately 420Å^3 when compared to the binding of the smallest L-leucyl substituent of phosphoramidon. This is mainly achieved by the indole movement of Trp693 toward the surface, which is accompanied by an adaptation on Asn105 and Phe106.

Reduced specificity has been observed for the S'_2 subsite, which is large and extending to the solvent across a linker fragment toward the side chains of Arg102, Asp107, and Arg110. In all three observed cases, the terminal carboxylate of phosphoramidon and thiorphan as well as the tetrazole moiety of CGS 31447 point toward the basic side chains of Arg102 and Arg110. They are, however, not directly involved in intermolecular interactions. The extended P'_2 substituent of phosphoramidon leads to a decreased volume of the S'_1 subsite when compared to the unsubstituted P'_2 residue of thiorphan and the CGS compound. This is a direct consequence of the β-indole interaction with the side chain of Phe106, which affects the side-chain position of Trp693. Interestingly, no residue of domain II is involved in active center or subsite formation.

X-ray structure of the apoenzyme

Presently, no structural information is available for the apoenzyme, which would give insight into the natural apo-zinc-binding geometry. All structural analyses of untreated crystal material carried out so far, using difference Fourier techniques and crystallographic refinement, revealed an enzyme active site inhibited by an unknown small molecule, which most likely comes from the yeast peptone medium.

Structural similarities with thermolysin

In the absence of structural information for any member of the neprilysin family, the observed three-dimensional fold for the TNL family has been used to generate a model of the active site region of NEP for the design of new NEP inhibitors.[70,77–79] Figure 5 shows the superposition of the active site region of human NEP and thermolysin (PDB code 1TLP) as observed in a complex with phosphoramidon. As proposed by site-directed mutagenesis, all residues involved in zinc ligation, enzyme reaction, and hydrophilic interactions with the inhibitor are conserved in function and provide a conserved framework for substrate hydrolysis.

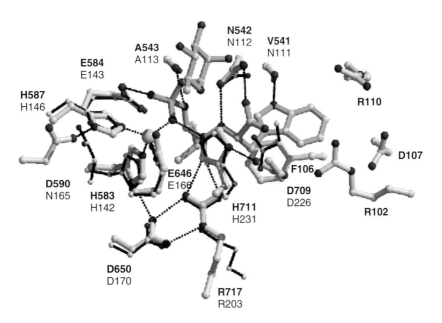

Figure 5 Superposition of the active site residues between human NEP (white) and thermolysin (black). The zinc atom is shown as yellow sphere. Phosphoramidon, as observed in the structure of NEP, is shown in green. H-bonds are indicated by dotted lines. The neprilysin residue numbering is given in bold.

The S_1 subsite, which has a minor stabilizing role for NEP, involves Tyr157 in thermolysin. This residue has no counterpart in the NEP structure, and its side-chain hydroxyl is involved in stabilizing the hydrated substrate species.[58] The primed subsites, however, which are important for the specificity of NEP, differ extensively between both enzymes, and there is no counterpart for the S'_2 pocket in TLN, which is mainly formed by charged residues in the human enzyme.

REFERENCES

1. AJ Kenny, SL Stephenson and AJ Turner, in AJ Kenny and AJ Turner (eds.), *In Mammalian Ectoenzymes*, Elsevier Science Publisher, Amsterdam, The Netherlands, pp 169–210 (1987).
2. BP Roques, F Noble, V Dauge, M-C Fournie-Zaluski and A Beaumont, *Pharmacol Rev*, **45**, 87–146 (1993).
3. MA Kerr and AJ Kenny, *Biochem J*, **137**, 489–95 (1974).
4. EG Erdos and RA Skidgel, *FASEB J*, **3**, 145–51 (1989).
5. AJ Turner and K Tanzawa, *FASEB J*, **11**, 355–64 (1997).
6. BP Roques, M-C Fournie-Zaluski, D Florentin, G Waksman, A Sassi, P Chaillet and J Costentin, *Nature*, **288**, 286–88 (1980).
7. N Iwata, S Tsubaki, Y Takaki, K Watanabe, M Sekiguchi, E Hosoki, M Kawashima-Morishima, HJ Lee, E Hama, Z Sekine-Aizawa and TC Saido, *Nat Med*, **6**, 143–50 (2000).
8. K Shiratoni, S Tsubaki, N Iwata, Y Takaki, W Harigaya, K Maruyama, S Kiryu-Seo, H Kiyama, H Iwata, T Tomita, T Iwatsubo and TC Saido, *J Biol Chem*, **276**, 21895–901 (2001).
9. AJ Kenny and SL Stephenson, *FEBS Lett*, **232**, 1–8 (1988).
10. I Pham, AI El-Amrani, M-C Fournie-Zaluski, P Corvol, BP Roques and JB Micheal, *J Pharmacol Exp Ther* **265**, 1339–47 (1993).
11. ZA Abassi, JE Tate, E Golombe and HR Keiser, *Hypertension*, **20**, 89–95 (1992).
12. JC Burnett, *J Hypertens*, **17**, 37–43 (1999).
13. H Chen, F Noble, P Coric, M-C Fournie-Zaluski and BP Roques, *Proc Natl Acad Sci USA*, **95**, 12028–33 (1998).
14. JA Robl, CQ Sun, J Stevenson, DE Ryono, LM Simpkins, MP Cimarusti, T Dejneka, WA Slusarchyk, S Chao, L Stratton, RN Misra, MS Bednarz, MM Asaad, HS Cheung and BE, PL Smith, PD Mathers, M Fox, TR Schaeffer, AA Seymour and NC Tripodo, *J Med Chem*, **40**, 1570–77 (1997).
15. JA Robl, R Sulsky, E Sieber-McMaster, DE Ryono, MP Cimarusti, LM Simpkins, DS Karanewsky, DS Chao, MM Asaad, AA Seymour, M Fox, PL Smith and NC Tripodo, *J Med Chem*, **42**, 305–11 (1999).
16. BP Roques and A Beaumont, *Trends Pharmacol Sci*, **11**, 245–49 (1990).
17. K Shimada, M Takahashi and K Tanzawa, *J Biol Chem*, **269**, 18275–278 (1994).
18. N Emoto and M Yanagisawa, *J Biol Chem*, **270**, 15262–68 (1995).
19. S. Lee, ED Zambas, WL Marsh and CM Redman, *Proc Natl Acad Sci USA*, **88**, 6353–57 (1991).
20. The HYP Consortium, *Nat Genet*, **11**, 130–36 (1995).
21. K Ikeda, N Emoto, SB Raharjo, Y Nurhantari, K Saiki, M Yokoyama and M Matsuo, *J Biol Chem*, **274**, 32469–77 (1999).
22. G Ghaddar, AF Ruchon, M Carpentier, M Marcinkiewicz, NG Seidah, P Crine, L Degroseillers and G Boileau, *Biochem J*, **347**, 565–70 (2000).
23. O Valdenaire, JG Richards, RLM Faull and A Schweizer, *Mol Brain Res*, **64**, 211–21 (1999).
24. S Kiryu-Seo, M Sasaki, H Yokohama, S Nakagomi, T Hiazama, S Aoki, K Wada and H Kiyama, *Proc Natl Acad Sci USA*, **97**, 4345–50 (2000).
25. I Mierau, PST Tan, AJ Haandrikman, J Kok, KJ Leenhouts, WN Konings and G Venema, *J Bacteriol*, **175**, 2087–96 (1993).
26. S Awano, T Ansai, H Mochizuki, K Tanzawa, AJ Turner and T Takehara, *FEBS Lett* **460**, 139–44 (1999).
27. C. elegans Sequencing Consortium, *Science*, **282**, 2012–18 (1998).
28. EW Myers, GG Sutton, AL Delcher, IM Dew, DP Fasulo, MJ Flanigan, SA Kravitz, CM Moberry, KHJ Reinert and KA Reminton, *Science*, **287**, 2196–204 (2000).
29. M Yanasigawa, H Kurihara, S Kimura, Y Tomobe, Y Kobiashi, Y Mitsui, Y, Yazaki, K Goto and T Masaki, *Nature*, **332**, 411–15 (1988).
30. FE Stranchen and DJ Webb, *Emerg Drugs*, **3**, 95–112 (1998).
31. AJ Turner and LJ Murphy, *Biochem Pharmacol*, **51**, 91–102 (1996).
32. S Lee, D Russo and C Redman, *Transfusion Med Rev*, **14**, 93–103 (2000).
33. S Lee, M Lin, A Mele, Y Cao, J Farmar, D Russo and C Redman, *Blood*, **94**, 1440–50 (1999).
34. A Schweizer, O Valdenaire, A Koester, Y Lang, G Schmitt, B Lenz, H Bluetmann and J Rohrer, *J Biol Chem*, **274**, 20450–56 (1999).
35. B Malfroy, W-J Kuang, PH Seeburg, AJ Mason and PR Schofield, *FEBS Lett*, **229**, 206–10 (1988).
36. B Malfroy, PR Schofield, W-J Kuang, PH Seeburg, AJ Mason and WJ Henzel, *Biochem Biophys Res Commun*, **144**, 59–66 (1987).
37. A Devault, C Lazure, C Nault, H LeMoual, NG Seidah, M Chretian, P Kahn, J Powell, J Mallet, A Beaumont and BP Roques, *EMBO J*, **6**, 1317–22 (1987).
38. CY Chen, G Salles, MF Seldin, AE Kister, EL Reinherz and MA Shipp, *J Immunol*, **148**, 2187–825 (1992).
39. MA Kerr and AJ Kenny, *Biochem J*, **137**, 477–88 (1974).
40. M Ishida, M Ogawa, G Kosaki, T Mega and T Ikenaka, *J Biochem*, **94**, 17–24 (1983).
41. B Malfroy and J-C Schwartz, *J Biol Chem*, **259**, 14365–70 (1984).
42. JT Gafford, RA Skidgel, EG Erdos and LB Hersh, *Biochemistry*, **22**, 3265–71 (1983).
43. J Almenoff, S Wilk and M Orlowski, *J Neurochem*, **42**, 151–57 (1981).
44. C Gorenstein and SH Snyder, *Proc R Soc London B, Biol Sci*, **210**, 123–32 (1980).
45. JM Relton, NS GEE, R Matsas, AJ Turner and AJ Kenny, *Biochem J*, **215**, 519–23 (1983).
46. IS Fulcher, MF Chaplin and AJ Kenny, *Biochem J*, **215**, 317–23 (1983).
47. M Aubry, A Berteloot, A Beaumont, BP Roques and P Crine, *Biochem Cell Biol*, **65**, 398–404 (1987).
48. CM Gorman, D Gies, PR Schofield, H Kado-Fong and B Malfroz, *J Cell Biochem*, **39**, 277–84 (1989).
49. G Lemay, M Zollinger, G Waksman, BP Roques, P Crine and G Boileau, *Biochem J*, **267**, 447–52 (1990).
50. G Lemay, G Waksman, BP Roques, P Crine and G Boileau, *J Biol Chem*, **264**, 15620–23 (1989).
51. F Fossiez, G Lemay, N Labonte, F Parmentier-Lesage, G Boileau and P Crine, *Biochem J*, **284**, 53–59 (1992).

52. H Beaulieu, A Elagoz, P Crine and LA Rokeach, *Biochem J*, **340**, 813–19 (1999).
53. GE Dale, B D'Arcy, C Yuvanyama, B Wipf, C Oefner and A D'Arcy, *Acta Crystallogr, Sect D Biol Crystallogr*, **56**, 894–97 (2000).
54. LT Tam, S Engelbrecht, JM Talent, RW Gracy and EG Erdos, *Biochem Biophys Res Commun*, **133**, 1187–92 (1985).
55. T Benchetrit, M-C Fournie-Zaluski and BP Roques, *Biochem Biophys Res Commun*, **147**, 1034–40 (1987).
56. IS Fulcher, R Matsas, AJ Turner and AJ Kenny, *Biochem J*, **203**, 519–22 (1982).
57. LB Hersh and K Morihara, *J Biol Chem*, **261**, 6433–37 (1986).
58. B Matthews, *Acc Chem Res*, **21**, 333–40 (1988).
59. T Benchetrit, V Bissery, JP Mornon, A Devault, P Crine and BP Roques, *Biochemistry*, **27**, 592–97 (1988).
60. A Devault, N Sales, C Nault, A Beaumont, BP Roques, P Crine and G Boileau, *FEBS Lett*, **231**, 54–58 (1988).
61. N Dion, H LeMoual, P Crine and G Boileau, *FEBS Lett*, **318**, 301–4 (1993).
62. N Dion, H LeMoual, M-C Fournie-Zaluski, BP Roques, P Crine and G Boileau, *Biochem J*, **311**, 623–27 (1995).
63. H Le Moual, A Devault, BP Roques, P Crine and G Boileau, *J Biol Chem*, **266**, 15670–74 (1991).
64. H Le Moual, N Dion, BP Roques, P Crine and G Boileau, *Eur J Biochem*, **221**, 475–80 (1994).
65. Z Vogel and M Altstein, *FEBS Lett*, **80**, 332–36 (1977).
66. D Florentin, A Sassi and BP Roques, *Anal Biochem*, **141**, 62–69 (1984).
67. FE Indig, M Pecht, N Trainin, Y Burstein and S Blumberg, *Biochem J*, **278**, 891–94 (1989).
68. JA Robl, LM Simpkins and MM Asaad, *Bioorg Med Lett*, **10**, 257–60 (2000).
69. C Oefner, A D'Arcy, M Hennig, FK Winkler and GE Dale, *J Mol Biol*, **296**, 341–49 (2000).
70. G Tiraboschi, N Jullian, V Thery, S Antonczak, MC Fournie-Zaluski and BP Roques, *Protein Eng*, **12**, 141–49 (1999).
71. P Kraulis, *J Appl Crystallogr*, **26**, 946–50 (1991).
72. EA Merrirr and MEP Murphy, *Acta Crystallogr*, **D50**, 869–73 (1994).
73. S De Lombaert, L Blanchard, LB Stamford, J Tan, EM Wallace, Y Satoh, J Fitt, D Hoyer, D Simonsbergen, J Moliterni, N Marcopoulos, P Savage, M Chou, AJ Trapani and AY Jeng, *J Med Chem*, **43**, 488–504 (2000).
74. BL Vallee and DS Aild, *Biochemistry*, **29**, 5647–59 (1990).
75. R Bouboutou, G Waksmanm, J Devin, MC Fournie-Zaluski and BP Roques, *Life Sci*, **35**, 1023–30 (1984).
76. MC Fournie-Zaluski, A Coulaud, R Bouboutou, P Chaillet, J Devin, G Waksman, J Costentin and BP Roques, *J Med Chem*, **28**, 1158–69 (1985).
77. JF Gaucher, M Selkti, G Tiraboschi, T Prange, BP Roques, A Tomas and MC Fournie-Zaluski, *Biochemistry*, **38**, 12569–76 (1999).
78. GM Ksander, R de Jesus, A Yuan, RD Ghai, A Trapani, C McMartin and R Bohacek, *J Med Chem*, **40**, 495–505 (1997).
79. GM Ksander, R de Jesus, A Yuan, RD Ghai, A Trapani, C McMartin and R Bohacek, *J Med Chem*, **40**, 506–14 (1997).
80. WL DeLano, *The PyMOL Use's Manual*, DeLano Scientific, San Carlos, CA (2001).

Astacin

Walter Stöcker and Irene Yiallouros
Institut für Zoophysiologie, Westfälische Wilhelms-Universität Münster, Germany

FUNCTIONAL CLASS

Enzyme; mono-zinc peptide bond hydrolase; IUBMB: EC 3.4.24.21; MEROPS classification: clan MA(M); family M12A; peptidase M12.001; MEROPS link: M12.001; CAS registry: 143179-21-9.

OCCURRENCE

Astacin was first purified from the digestive juice of the European noble crayfish *Astacus astacus* L. (i.e. *Astacus fluviatilis* Fabr.) by Pfleiderer *et al.* (1967).[1–3]

Since then, astacin-like digestive proteolytic enzymes have been observed in several decapod species (Metazoa, Arthropoda), including the Camchatka crab *Chionoecetes opilio*[4,5] and the American freshwater crayfish *Procambarus* sp.[6]

Astacin is the prototype of a family of extracellular zinc peptidases termed the 'astacin family'[7–9] and of the metzincin superfamily, which combines the MMPs, the bacterial serralysins, the adamalysins/reprolysins/ADAMs (a disintegrin and metalloprotease), the pappalysins, and other zinc proteases.[10–13] All astacins are secreted into the extracellular space or they stay membrane bound at the cell

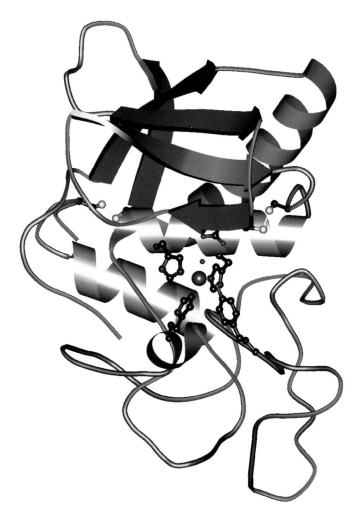

3D Structure Schematic representation of the structure of astacin, PDB code 1AST,[40,41] prepared with the program MOLSCRIPT.[42] Details shown are the catalytic zinc ion (gray sphere), the zinc ligands, His92, His96, His102, Tyr149, Sol300 (red sphere), the disulfide forming cysteines, and the catalytically important Glu93.

surface. They are found within the animal kingdom and also in bacteria. Examples are the bone morphogenetic protein 1 (BMP-1), which is identical with the procollagen C-proteinase (i.e. BMP-1, bone morphogenetic protein),[14–16] tolloid from *Drosophila*[17], and the membrane protease meprin, which is anchored to or secreted from the surface of epithelial cells.[18–20]

BIOLOGICAL FUNCTION

The physiological function of crayfish astacin is the digestion of food proteins.[21] Crayfish digestive proteases depend on the capability to cleave proteins in their native conformation, since the crayfish stomach does not provide an acidic denaturing environment. Astacin is able to degrade the triple helix of type I collagen, gelatin, and other native proteins at neutral pH.[22] Most other astacins do not have a digestive function, but are involved in developmental processes. *Drosophila* tolloid and related mammalian enzymes are involved in dorso-ventral patterning by cleaving antagonists of transforming growth factor β-like molecules. Others like BMP1 catalyze the limited proteolysis of procollagens and a variety of other matrix protein precursors, and some astacins are specialized for the cleavage of egg envelopes during embryonic hatching.[18,20]

AMINO ACID SEQUENCE INFORMATION

The astacin gene, obtained from an *A. astacus* genomic library, spans 2616 bp; it includes 5 exons, 4 introns, and a TATA-box in the promoter region[23] (EMBL genomic database: X95684). As deduced from the cDNA, *Astacus* astacin is translated as a pre-pro-protein comprising 251 residues including a signal sequence of 15 and a pro-domain of 34 residues respectively[24,25] (EMBL cDNA database: AJ242595; SwissProt: P07584). The only other primary structure information available for a crustacean digestive astacin protease is an amino-terminal sequence of 20 residues obtained from the Camchatka crab *C. opilio*[4,5] (SwissProt: P34156).

The activated form of astacin from the crayfish digestive tract is a single domain protein. Most of the other members of the astacin family have a multimodular structure including one astacin-like catalytic domain, which exhibits at least about 30% sequence identity with the crayfish proteinase (Figure 1).[9,18]

BIOSYNTHESIS AND PRO-ASTACIN ACTIVATION

Crayfish astacin is synthesized in the midgut gland as a pre-pro-enzyme. Like other astacins, it requires proteolytic removal of the amino-terminal pro-peptide to gain full catalytic activity.[31] The proenzyme is present only transiently after biosynthesis within the ducts leading from the hepatopancreas to the stomach, where the enzyme is stored extracellularly as an active protease, which is remarkably stable against self-digestion.[21,32] In these traits, the crayfish digestive proteinases differ drastically from the mammalian pancreatic enzymes, which are stored inside the cells as zymogen granules.[33]

Pro-astacin activation follows a two-step mechanism.[31] In the first step, the crayfish trypsin cleaves within the pro-peptide, resulting in a premature enzyme that carries two additional amino-terminal residues. These are cleaved off by astacin itself in the second step. In the absence of a trypsin-like activity, the heterologously expressed and folded pro-astacin is capable of slow intramolecular autoproteolytic activation *in vitro*, yielding the precisely trimmed mature amino-terminus.[31]

In other astacins, zymogen activation can be directly catalyzed by trypsin-like serine proteases as in the case of meprin.[34] However, there are subtle differences between meprin α- and β-subunits, which assemble to oligomers of different sizes. Meprin α forms homo-oligomers with molecular masses of up to 900 kDa, whereas meprin β homo-oligomers are maximally dimers.[35,36] This varying oligomeric arrangement has consequences for the activation of pro-meprin in various tissues. Human pro-meprin α can be activated by trypsin inside the gut and by plasmin outside the gut.[37] This is not the case for human pro-meprin β, which is activated by trypsin, but not by the larger protease plasmin, presumably because of the steric hindrance in the pro-meprin β-dimer.[36] This might explain the observation that outside the intestine, as in the kidney brush border membrane, meprin oligomers have been observed consisting of active α- but latent (inactive) β-subunits.[18] Other astacin-proteases can be activated on the secretory pathway by furin-like pro-hormone convertases as observed in BMP-1/tolloid-like astacins.[38]

PROTEIN PRODUCTION, PURIFICATION, AND MOLECULAR CHARACTERIZATION

Astacin can be isolated from the digestive juice of the crayfish *A. astacus*.[3,7,22] After the posttranslational removal of the signal- and pro-sequence and the two carboxy-terminal residues, the covalent structure of mature active astacin comprises a single chain of 200 residues, corresponding to an M_r of 22614. Two disulfide bridges cross-link Cys42–Cys198 and Cys64–Cys84.[25]

The recombinant mature enzyme, the proenzyme, and several mutant variants of both have been expressed in bacteria and folded into the native conformation from dissolved inclusion bodies.[24,31,39]

```
                     .        10         .        20         .        30
astacin      : AAILGDEYLWSGGVIPYTFAG-VSGADQSAILSGM :  34
meprin_alpha : NGLRDPNTRWT-FPIPYILADNLGLNAKGAILYAF :  34
meprin_beta  : NSIIGEKYRWP-HTIPYVLEDSLEMNAKGVILNAF :  34
BMP1         : AATSRPERVWPDGVIPFVIGGNFTGSQRAVFRQAM :  35
tolloid      : AVTVRKERTWDYGVIPYEIDTIFSGAHKALFKQAM :  35
                            sssssssss   hhhhhhhhhhh

                   .        40         .        50         .        60         .
astacin      : QELEEKTCIREVPRTTE--SDYVEIFTSGSGCWSY :  67
meprin_alpha : EMFRLKSCVDEKPYEGE--SSYIIFQQ-FDGCWSE :  66
meprin_beta  : ERYRLKTCIDEKPWAGE--TNYISVFK-GSGCWSS :  66
BMP1         : RHWEKHTCVTELERTDE--DSYIVFTYRPCGCCSY :  68
tolloid      : RHWENFTCIKEVERDPNLHANYIYFTVKNCGCCSF :  70
               hhhhhhhsssssss           ssssss    sssss

                 70         .        80         .        90         .       100
astacin      : VGRI-SGAQQVSLQANGCVYHGTIIHELMHAIGFY : 101
meprin_alpha : VGDQ-HVGQNISIGQ-GCAYKAIIEHEILHALGFY :  99
meprin_beta  : VGNRRVGKQELSIGAN-CDRIATVQHEFLHALGFW : 100
BMP1         : VGRRGGPQAISIGKN-CDKFGIVVHELGHVVGFW  : 102
tolloid      : LGKNGNGRQPISIGRN-CEKFGIIIHELGHTIGFH : 104
               ss    sssssssss          hhhhhhhhhhhhhh

                *  .       110         .       120         .       130
astacin      : HEHTRMDRDNYVTINYQNVDPSMTSNED-IDT-YS : 134
meprin_alpha : HEQSRTDRDDYVNIWWDQILSGYQHNFDTYDDSLI : 134
meprin_beta  : HEQSRSDRDDYVRIMWDRILSGREHNFNTYSDDIS : 135
BMP1         : HEHTRPDRDRHVSIVRENIQPGQEYNFLKMEPQEV : 137
tolloid      : HEHARGDRDKHIVINKGNIMRGQEYNFDVLSPEEV : 139
                                                    hhhhhh

                   .       140         .       150         .       160         .
astacin      : RYVGEDYQYYSIMHYGKYSFSIQWGVLETIVPLQN : 169
meprin_alpha : TDLNTPYDYESLMHYQPFSEN-KNASVPTITAKIP : 168
meprin_beta  : DSLNVPYDYTSVMHYSKTAFQ--NGTEPTIVTRIS : 168
BMP1         : ESLGETYDFDSIMHYARNTFS-RGIFLDTIVPKYE : 171
tolloid      : DLPLLPYDLNSIMHYAKNSFS-KSPYLDTITPIGI : 173

                170         .       180         .       190         .       200
astacin      : GIDLT--DPYDKAHMLQTDANQINNLYTNECSL-- : 200
meprin_alpha : EFNSI--IG-QRLDFSAIDLERLNRMY--NCTTTH : 198
meprin_beta  : DFEDV--IG-QRMDFSDSDLLKLNQLY--NCSSSL : 198
BMP1         : VNGVKPPIG-QRTRLSKGDIAQARKLY--KCPACG : 203
tolloid      : PPGTHLELG-QRKRLSRGDIVQANLLY--KCASCG : 205
                                                 hhhhhhhhhhhhhh
```

Figure 1 Sequence alignment of the catalytic domains of crayfish astacin, human meprin α, human meprin β, human BMP1 (bone morphogenetic protein, i.e. procollagen C-proteinase), and *Drosophila* tolloid. SwissProt accession numbers are P07584 (astacin),[24,25] Q16820 (meprin α),[26] Q16819 (meprin β),[27] P13497 (BMP1),[28] and P25723 (tolloid)[17] respectively. The alignment was obtained with CLUSTALX[29] and rearranged with GENEDOC[30] according to a structural overlay of the proteins onto the astacin structure.[9] The numbers on top correspond to the numbering of astacin, whereas the numbers on the right label correspond to the individual sequences. The secondary structure elements are indicated in the respective bottom lines. Color code: red, zinc ligands; blue, catalytic Glu93; green, methionine of the Met-turn; yellow, cysteine; black, other strictly conserved residues; gray, less conserved positions.

METAL CONTENT AND COFACTORS

Astacin preparations purified from the natural source have one single zinc per catalytic domain as determined by atomic absorption spectrometry; other metals or cofactors were not detectable.[43] Other astacins have bound calcium ions in their C-terminal EGF-like domains.

ACTIVITY TEST

The pH-optimum of astacin ranges between pH 6 and pH 8 depending on the substrate.[7] Astacin activity is assayed with denatured casein, with synthetic nitroanilides like Succinyl-Ala-Ala-Ala-NHPhNO$_2$ (STANA), or with more sensitive fluorescent substrates.[22,44–46] The most convenient are the quenched fluorescent substrates like Dansyl-Pro-Lys-Arg+Ala-Pro-Trp-Val (Dansyl = 5-(dimethylamino)naphthalene-1-sulfonyl)[44] Abz-Arg-Pro-Ile-Phe+Ser-Pro-Npa-Arg (Abz = 4-aminobenzoyl; Npa = 4-nitrophenylalanine),[45] which are the best substrates out of a series of tubulin-derived heptapeptides and bradykinin-derived octapeptides respectively.

CLEAVAGE SPECIFICITY

The inspection of 71 cleavage sites in the denatured chains of α- and β-tubulin revealed a high preference of astacin for alanine (or another small uncharged residue) in P'_1, proline in P'_2 and P_3, hydrophobic residues in P'_3 and P'_4, and lysine, arginine, asparagine, tyrosine in P_1 and P_2.[22,44,47,48]

INHIBITORS

Astacin and other members of the astacin family are not inhibited by TIMP1, TIMP2, and TIMP4 (tissue inhibitors of metalloproteinases, unpublished results) or by phosphoramidon. The most potent natural inhibitor known so far is α$_2$-macroglobulin.[49]

Efficient synthetic inhibitors for astacin are phosphinic transition state analogue pseudopeptides like Fmoc-Pro-Lys-Phe-Ψ(PO$_2$CH$_2$)-Ala-Pro-Leu-Val-OH (Fmoc = 9-fluorenylmethyloxycarbonyl) (K_i = 42 nM; k_{on} = 96.8 M^{-1}s^{-1}; k_{off} = 4.1 × 10^{-6} s^{-1}).[50,51] The rate constants k_{on} and k_{off} of these inhibitors are indicative of their extremely slow binding behavior.

Metal chelating inhibitors include 1,10-phenanthroline, EDTA, dipicolinic acid, 8-hydroxyquinoline-5-sulfonic acid, 2,2'-bipyridyl, amino acid hydroxamates, and thiol compounds.[43,52,53] The peptide hydroxamate Pro-Leu-Gly-NHOH (K_i = 16 μM)[50] is used for affinity purification of recombinant astacin.[24]

X-RAY CRYSTAL STRUCTURE OF NATIVE ASTACIN

Crystallization

The protein material was prepared from the digestive juice from *A. astacus* L. Of the three bands visible on polyacrylamide gels in the absence of SDS and reducing agents, the most prominent upper band (astacin a) was purified and used for crystallization.[7] Crystals of the mature form of astacin, dissolved in 1.0 mM MES/NaOH, pH 7.0 were grown by the hanging drop method in a final solution of 1.0 M (NH$_4$)$_2$SO$_4$ buffered to pH 7.0 by 0.1 M sodium phosphate.[40,41] The trigonal crystals had the space group P3$_1$21 with $a = b = 61.96$ Å and $c = 98.52$ Å and a single monomer per asymmetric unit. The X-ray structure was analyzed via the multiple isomorphous replacement technique using one native, one apo, and six heavy atom derivative crystals. The final resolution was refined to 1.8 Å.[41]

Overall description of the structure

The polypeptide chains are arranged in the crystals as monomers with threefold symmetry. Astacin is a kidney-shaped molecule with dimensions of 52 Å × 37 Å × 45 Å. A deep cleft subdivides the protein into two subdomains.[40,41] The amino-terminal, upper part (see 3D Structure and Figure 2) is made up of a twisted five-stranded ß-pleated sheet formed by four parallel strands (strands I, II, III, and V), one antiparallel strand (strand IV), and two long α-helices (helices A and B) (Figure 2). The lower, carboxy-terminal subdomain is folded more irregularly except for a short 3^{10}-helix (helix C) and a long α-helix (helix D), which is fixed to the amino-domain via the disulfide bond Cys42–Cys198. The other disulfide bond (Cys64–Cys84) clamps ß-strand IV to the loop between strand V and helix B. The sequential arrangement of secondary structure elements except for the short 3^{10}-helix (helix C) is depicted in Figure 3.

The catalytic zinc resides at the bottom of the cleft between the two subdomains in the center of the molecule. The metal is coordinated by three histidine residues, a tyrosine residue, and a water molecule, which is hydrogen bonded by Glu93 and Tyr149, resulting in a trigonal bipyramidal coordination sphere with His96NE2 and the phenolic oxygen of Tyr149 at the pyramid's vertices and His92NE2 and His102NE2 and the water molecule Sol300 forming the central trigonal plane (Figure 4, Table 1). His92 and His96 are part of the central helix B separated by a single helix-turn (Figure 4). They are located at distances of 2.0 Å and 2.2 Å removed from the metal. The third zinc ligand, His102 (2.0 Å), follows only six residues upstream of the second (Table 1). This is facilitated by an invariant glycine residue, Gly99, which terminates helix

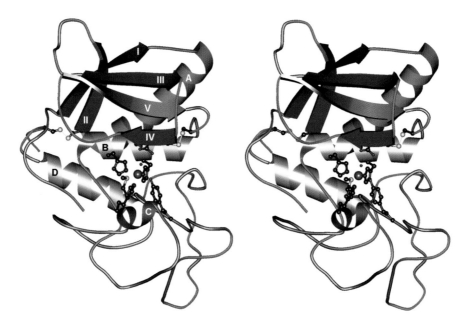

Figure 2 Schematic stereo representation of the structure of astacin with the assignment of secondary structure elements.[40,41] Helices are labeled with capital letters, ß-strands with Roman numbering. The catalytic zinc (gray sphere), the zinc ligands (His92, His96, His102, Tyr149, Sol300 as a red sphere), Met147 of the Met-turn and the disulfide bonds (yellow) are shown. Produced with the program MOLSCRIPT.[42]

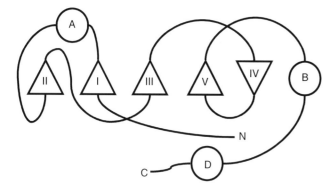

Figure 3 Topology diagram showing the sequential arrangement of secondary structure elements in astacin.[41] Each strand is shown by a triangle whose apex points up or down depending on whether the strand is viewed from the amino- or the carboxy-terminus. α-Helices are represented as circles. The short 3^{10}-helix C has been omitted.

B and induces a sharp bend in the chain (Figures 1, 2, and 4). The fourth and the fifth zinc ligands are supplied by the hydroxyl oxygen of Tyr149 and a water molecule (Sol300) at distances of 2.5 Å and 2.0 Å removed from the metal, respectively (Figure 4).

The three imidazole zinc ligands and the catalytically important glutamic acid are combined in the consensus motif **HEXXHXXGXXH** (one letter code, X is any amino acid residue), seen in all astacins (Figures 1 and 2). This is also a typical distinguishing feature for the metzincin superfamily of zinc peptidases. In astacin, the peptide chain leaves the zinc binding active site and folds into a mostly irregular coiled structure in the carboxy-terminal domain before it comes close to the zinc site again by forming another key structure of the metzincins, the so-called Met-turn with the underlying consensus sequence SXMHY within the astacin family (Figures 1 and 2). This conserved methionine containing 1,4-ß-turn provides the fifth zinc ligand, Tyr149. The tyrosine has only been observed in the astacins and serralysins, but not in other metzincins as, for example, the MMPs and ADAMs. This different arrangement in the active sites might be a reason why the astacins are not inhibited by TIMPs, in contrast to the MMPs and ADAMs. Four strands of the ß-pleated sheet and long helices A and B of the amino-terminal subdomain of the metzincin superfamily exhibit topological similarity to the archetypical zinc-proteinase thermolysin.[41] This includes helix B, whose counterpart in thermolysin carries two of the zinc-binding histidines in a more general consensus motif **HEXXH**, shared by the majority of metalloproteinases of the zincin-group.[10,54,55]

A unique structural feature of astacin is a water-filled cavity in the lower subdomain, which harbors the two amino-terminal residues Ala1 and Ala2 (Figure 5).[9,40,41] Ala1 is connected with the active site region by a water-mediated salt bridge between its ammonium group and the carboxylate of Glu103, the direct neighbor of the zinc ligand His102. A similar situation has been observed in the trypsin-like serine proteinases (EC 3.4.21.4), whose activation requires the removal of an amino-terminal pro-peptide from the inactive zymogen, resulting in the formation of a salt bridge between the new amino-terminus and Asp194, the neighbor of the nucleophilic serine residue.[56] The

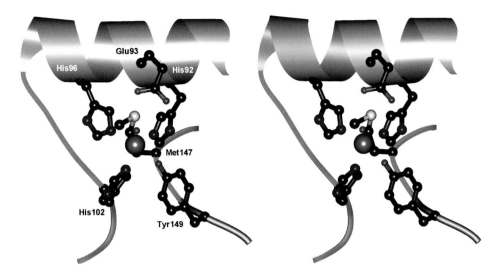

Figure 4 Close-up stereo view of the active site in native astacin with the bipyramidally coordinated zinc ion,[41] drawn by MOLSCRIPT.[42] The water Sol300, hydrogen bonded to the zinc ion, to Glu93 and Tyr149, is coplanar with His102 and His92. His96 and Tyr149 form the tips of the vertical axis of the bipyramid. Also shown is Met147 of the Met-turn located beneath the zinc site. For distances and angles refer Table 1 and Figure 10.

concomitant conformational rearrangement increases the activity by several orders of magnitude.[56] In astacin, the arrangement of the amino-terminus suggested a similar activation mechanism, because the amino-terminally extended proenzyme would not be able to form a salt bridge as seen in the mature protease.[40,41] However, site-directed mutagenesis of Glu103 in astacin[31] and the corresponding residue in mouse meprin α[34] did not essentially alter the activity of the protease, but rendered the enzyme extremely unstable. In the astacins, the conserved salt bridge appears to be indispensable for the structural integrity of the protein rather than for catalytic function.

FUNCTIONAL ASPECTS

Subsite mapping of astacin with synthetic substrates

The extended binding site of astacin requires up to four residues on either side of the cleavage spot to become saturated. Hence, the specificity constants, k_{cat}/K_m, for the release of nitroaniline from substrates of the general structure Suc-Ala$_n$-pNA ($n = 2,3,5$) and Ala$_n$-pNA ($n = 1,2,3$) increase with the number of alanine residues.[7] However, the nitroanilides fill the substrate binding region only in part and, therefore, the k_{cat}/K_m value, even of the longest of these compounds, does not exceed 5.6×10^4 M^{-1} cm^{-1}.

Fluorescent, N-dansylated oligopeptides based on the tubulin cleavage pattern are turned over much faster than the nitroanilides, reaching k_{cat}/k_m values of up to 1.4 * 10^6 M^{-1} s^{-1} at 25 °C.[44] Lysine and arginine in positions P_1 and P_2 yield high-turnover substrates, and in P_3 the enzyme prefers Pro > Val > Leu > Ala > Gly (P_1, P_2, P_3,..., and P'_1, P'_2, P'_3,..., designate the residues in the substrate amino-terminally and carboxy-terminally to the cleavage spot, respectively; S_1, S_2, S_3,..., and S'_1, S'_2, S'_3,..., are the corresponding binding sites of the enzyme.[48] The substitution of lysine by glycine in P_2 causes a 330-fold increase in K_m, demonstrating the importance of this interaction for substrate binding. The presence of a small side chain in P'_1 is absolutely essential, since the activity drops by 30 000-fold if alanine is replaced by leucine. Also crucial is a proline in P'_2, whose substitution by serine decreases k_{cat} 50-fold, whereas K^m stays about the same. As indicated by the tubulin pattern, astacin accepts hydrophobic side chains including tryptophan in position P'_3, which has been convenient for the design of a sensitive quenched fluorescent assay (see ACTIVITY TEST).[44] The investigation of substrates varying in length revealed that an optimal substrate for astacin comprises seven or more amino acid residues and minimally requires about five, which is in accordance with the three-dimensional structure.

Structural basis of the cleavage specificity of astacins

The cleavage specificities of astacin and other astacins like meprin or the procollagen C-proteinase show some similarity. In comparing astacin and meprin α, it becomes obvious that both enzymes prefer extended substrates with bulky residues in P_1 and P_2 and proline residues in P_3 and P'_2 respectively. However, they differ markedly in their preference for residues in P'_1, where astacin tolerates only small uncharged residues, but meprin α also cleaves substrates with arginine, phenylalanine, or lysine residues in this position.[45,57] The reason for this difference is the replacement of Pro176 for glycine and

Astacin

Table 1 Geometry of the metal centers in zinc(II)-, copper(II)-, cobalt(II)-, nickel(II)- and mercury(II)-astacin.[41] Reproduced by permission of The American Society for Biochemistry and Molecular Biology

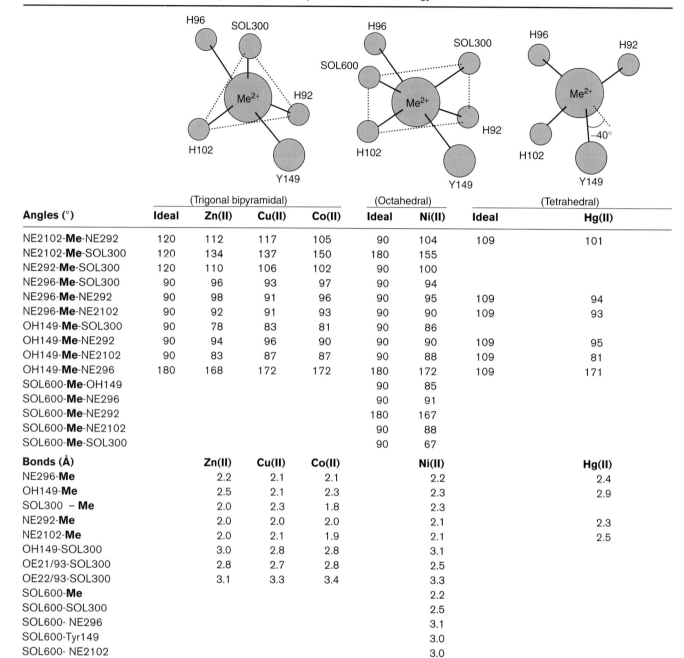

Angles (°)	(Trigonal bipyramidal)				(Octahedral)		(Tetrahedral)	
	Ideal	Zn(II)	Cu(II)	Co(II)	Ideal	Ni(II)	Ideal	Hg(II)
NE2102-**Me**-NE292	120	112	117	105	90	104	109	101
NE2102-**Me**-SOL300	120	134	137	150	180	155		
NE292-**Me**-SOL300	120	110	106	102	90	100		
NE296-**Me**-SOL300	90	96	93	97	90	94		
NE296-**Me**-NE292	90	98	91	96	90	95	109	94
NE296-**Me**-NE2102	90	92	91	93	90	90	109	93
OH149-**Me**-SOL300	90	78	83	81	90	86		
OH149-**Me**-NE292	90	94	96	90	90	90	109	95
OH149-**Me**-NE2102	90	83	87	87	90	88	109	81
OH149-**Me**-NE296	180	168	172	172	180	172	109	171
SOL600-**Me**-OH149					90	85		
SOL600-**Me**-NE296					90	91		
SOL600-**Me**-NE292					180	167		
SOL600-**Me**-NE2102					90	88		
SOL600-**Me**-SOL300					90	67		
Bonds (Å)		Zn(II)	Cu(II)	Co(II)		Ni(II)		Hg(II)
NE296-**Me**		2.2	2.1	2.1		2.2		2.4
OH149-**Me**		2.5	2.1	2.3		2.3		2.9
SOL300 – **Me**		2.0	2.3	1.8		2.3		
NE292-**Me**		2.0	2.0	2.0		2.1		2.3
NE2102-**Me**		2.0	2.1	1.9		2.1		2.5
OH149-SOL300		3.0	2.8	2.8		3.1		
OE21/93-SOL300		2.8	2.7	2.8		2.5		
OE22/93-SOL300		3.1	3.3	3.4		3.3		
SOL600-**Me**						2.2		
SOL600-SOL300						2.5		
SOL600- NE296						3.1		
SOL600-Tyr149						3.0		
SOL600- NE2102						3.0		

the concomitant deletion of Tyr177, which results in a more open configuration of the S'_1 subsite in meprin α. The upper border of the S'_1/S'_2 subsite is formed by Cys64 and Thr89 in both astacin and meprin α. However, at the floor of this subsite, Asp175 of astacin is replaced by isoleucine or leucine in meprin and other astacins. Binding of aromatic side chains like tryptophan in S'_3 of astacin might be promoted by Tyr177 and/or Trp158, which are deleted and replaced by Asn158 in meprin, respectively.

In meprin β, the S'_1 subsite is formed in part by Arg85, which is presumably responsible for the preference of acidic side chains in the corresponding substrate position P'_1.[57] In the procollagen C-proteinase and tolloid-like astacins, a second basic residue is presumably involved in the formation of the S'_1 subsite, which would explain the preference of these astacins for acidic side chains in the P'_1 and P'_2 positions of their substrates.[58]

Taken together, there are significant differences in the primed subsites of astacin compared to those of the other members of the astacin family, whereas the nonprimed regions appear to be more similar. Interestingly, this seems to be a general feature valid for the metzincin superfamily

Astacin

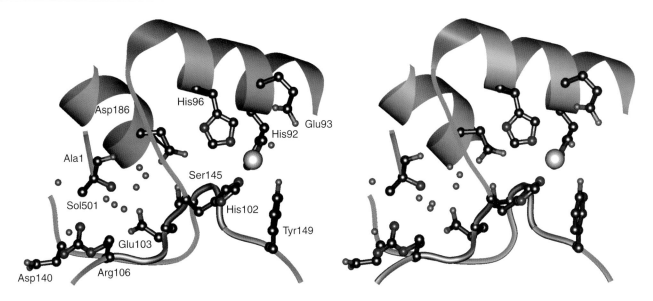

Figure 5 Buried salt bridge between the amino terminus of astacin and Glu103 in the active site,[40,41] drawn with MOLSCRIPT in stereo view.[42] Eight water molecules, shown as red spheres, are caged in a deep pocket. One of these, labelled as Sol501, is engaged in the salt bridge linking the ammonium group of Ala1 to the carboxylate of Glu103, the neighbor of zinc ligand His102. Also shown is Asp186 of helix D, which is strictly conserved within the astacin family and which is also found in other metzincins as in the MMPs. It is part of a hydrogen-bonding network linking the N-terminus with the Met-turn region via Ser145.

as a whole.[11,13] This is exemplified by the short nonprimed side inhibitor PLG-NHOH, which originally was designed as an inhibitor of interstitial collagenase (MMP1)[59] and which turned out to be more effective against astacin[50] and meprin.[60]

COMPLEX STRUCTURES

Structure of astacin complexed with Pro-Leu-Gly-hydroxamate

Astacin binds the hydroxamate-based inhibitor PLG-NHOH with a K_i value of 16 μM.[50] Crystals of the complex of astacin with this inhibitor were obtained by soaking crystals of mature astacin in a solution of 4.0 mM PLG-NHOH in 1.5 M $(NH_4)_2SO_4$, buffered to pH 7.0 by 0.1 M sodium phosphate. The structure was solved to a final resolution of 1.86 Å. In this complex, the hydroxamate inhibitor chelates the metal with its hydroxamic acid moiety and forms two backbone hydrogen bonds with the edge strand IV of the amino-terminal β-sheet (Figure 6). This only antiparallel β-strand of the β-sheet is part of the upper rim of the active site cleft. Incoming substrates and inhibitors align with strand IV, thereby extending the β-sheet by one strand. An important role for substrate recognition in astacin is attributed to the nichelike S_3

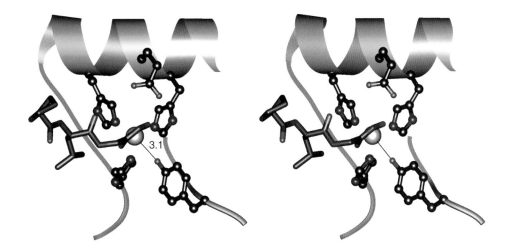

Figure 6 Stereo ribbon plot of the complex of astacin with PLG-NHOH,[50] drawn with MOLSCRIPT;[42] PDB code 1QJJ. The inhibitor chelates the zinc with its C-terminal hydroxamic acid moiety and binds to the nonprimed subsites of the active site cleft (see text). The distance (Å) between the zinc and the phenolic group of TYR149 is indicated.

subsite, which is formed by the side chains of Trp65 and Tyr67 of the edge strand. This niche is optimized for the binding of the proline in the P_3-position of PLG-NHOH (Figure 6).

Structure of astacin complexed with a transition state analog phosphinic peptide

The complex of astacin with the phosphinic pseudopeptide Cbz-PKFΨ(PO_2CH_2)AP-OCH_3 (Cbz = carbobenzoxy, i.e. benzoyl-oxycarbonyl) ($K_i = 14\,\mu M$)[51] was likewise obtained by soaking native crystals in a 4.0 mM solution of the inhibitor.[50] The structure was solved to a resolution of 2.14 Å. In the complex, the phosphinyl group chelates the zinc ion, which thus adopts the geometry of a tetrahedrally coordinated carbon during peptide bond cleavage (Figure 7). The nonprimed residues of the inhibitor are hydrogen bonded via backbone carbonyl and amide groups as seen in the PLG-NHOH structure, and the proline in P_3 slots as well into the niche of the S_3 subsite. However, there are no further backbone interactions possible although the phosphinic peptide spans five rather than three subsites as compared to PLG-NHOH (Figures 6, 7, and 8). The reason might be that in contrast to other metzincins, astacin does not provide backbone hydrogen-bonding interactions in the primed side region of the substrate binding cleft.[12] Furthermore, a possible hydrogen bond between the nitrogen of the scissile bond and the carbonyl oxygen of Cys64 in the edge strand cannot be formed in the inhibitor complex, since the amide group is replaced by a methylene moiety in the phosphinic peptide (Figure 8).[50] However, this additional hydrogen bond would most likely be formed during normal enzyme–substrate interactions (Figure 8).

Upon binding of the transition state analogue inhibitor, the tyrosine side chain moves away from the zinc into a position about 5.0 Å from the metal and becomes hydrogen bonded with the PO_2 group that mimics a water-attacked peptide bond (Figure 7).[50] This unique 'tyrosine switch' is a special feature of the astacin-like proteinases[12] and also of the serralysins.[61]

Catalytic mechanism of astacin

A model for the action of zinc peptidases implies that the metal-bound water is polarized between the zinc(II) ion, acting as a Lewis acid, and the glutamic acid residue, acting as the general base, for nucleophilic attack of the peptide bond (Figure 9).[62–64]

In astacin, Glu93 seemed to play the part of the general base during catalysis. However, the catalytic water is bound not only by Glu93 and the zinc ion but also by Tyr149 (Figure 4, Table 1). Therefore, one could imagine that Tyr149 rather than Glu93 might function as the base, as suggested for corresponding residues in the serralysins.[65]

In order to assess the roles of the tyrosine–zinc ligand in catalysis, it was replaced by site-directed mutagenesis with a phenylalanine residue.[39] The Y149F mutant retained a residual activity of 2.5% compared to the recombinant wild-type astacin. Hence, the tyrosine is obviously not indispensable for catalysis, but it seems to stabilize the transition state by hydrogen bonding. In conclusion, astacin's Tyr149 appears to have a similar function as His231 of thermolysin[63,66] or Arg127 of carboxpeptidase A.[64] On the other hand, mutagenesis of Glu93 in astacin to alanine resulted in a completely inactive enzyme.[39] This suggests that Glu93 of astacin acts as a general base during catalysis in analogy to Glu143 of thermolysin[64] or Glu270 of carboxypeptidase A (Figure 9).[62,64]

Glu93 of astacin most likely acts as a proton shuttle during catalysis; the location of one carboxyl oxygen of Glu93 in astacin close to the upper phosphinyl oxygen (at 2.9 Å distance) is in agreement with this proposal (Figure 9). It can be presumed that Glu93 is protonated, giving a hydrogen bond to the upper phosphinyl oxygen, since the pK_a of the phosphinyl group (1.8) is lower than the pK_a of a glutamic acid (about pH 4.5).[50]

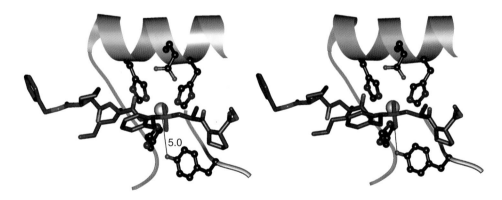

Figure 7 Ribbon plot in stereo view of astacin with CBZ-PKFΨ(PO_2CH_2)AP-OCH_3,[50] drawn with MOLSCRIPT;[42] PDB code 1QJI. The distance (Å) between the zinc and the phenolic group of Tyr149 is indicated.

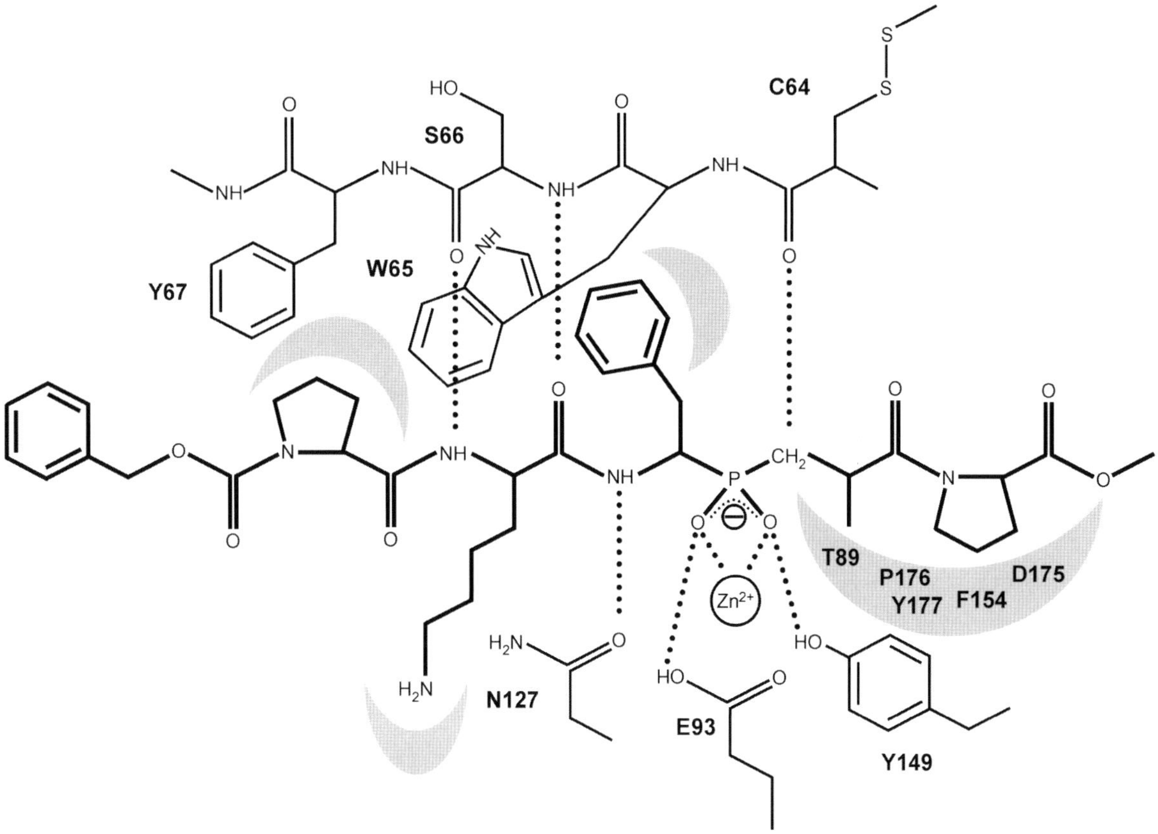

Figure 8 Scheme of the substrate binding region with bound CBZ-PKFΨ(PO₂CH₂)AP-OCH₃.[50] Reproduced by permission of Nature Structural Biology.

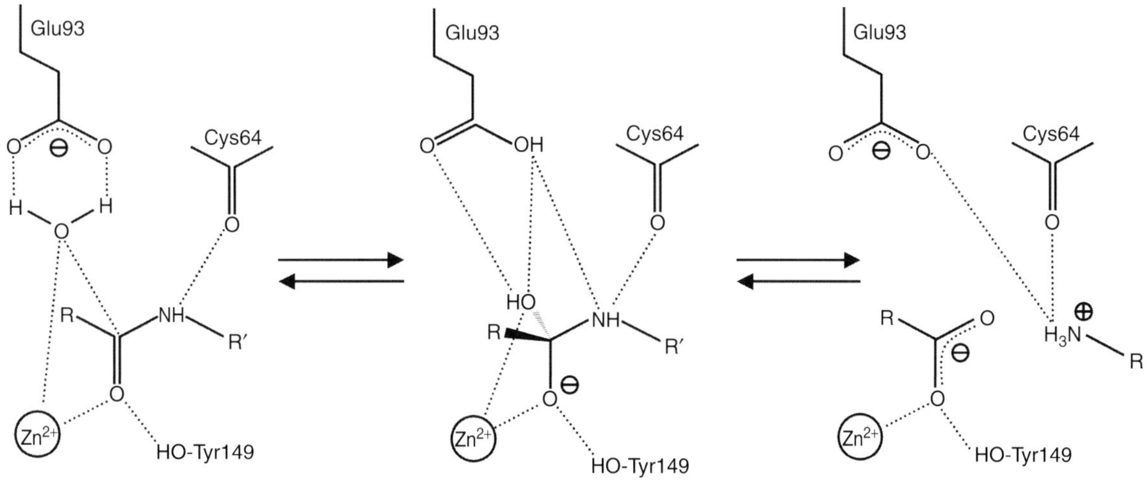

Figure 9 Catalytic mechanism of astacin.[50] Reproduced by permission of Nature Structural Biology.

STRUCTURES OF METAL DERIVATIVES OF ASTACINS

Removal of the metal

The inactivation of astacin by chelators like 1,10-phenanthroline has been monitored with the STANA-assay either instantaneously or after 1 h preincubation of enzyme and inhibitor.[43] The observed mechanism implies that in the first (fast) step, two chelator molecules bind into the active site in a cooperative manner.[43,52] In the second (slow) step, 1,10-phenanthroline removes the metal and forms higher bis- and tris-1,10-phenanthroline-zinc complexes in solution. At neutral pH, the inactivation by 1,10-phenanthroline is much faster than by EDTA.[43] The half-life of metal dissociation at pH 7.8

Astacin

in the absence of a chelator was calculated to be 40 days.[43]

Apo-astacin is produced by dialysis of the enzyme dissolved in 50 mM HEPES/NaOH-buffer, pH 8.0 for four days at 4 °C versus four changes of 100 mL of the same buffer containing 10 mM 1,10-phenanthroline.[43] The resulting apo-astacin contains less than 0.007 gram-atoms of zinc per mole of protein, and exhibits about 3% of the catalytic activity of zinc-astacin when assayed with 1-μM enzyme, which reflects the background of zinc.[43]

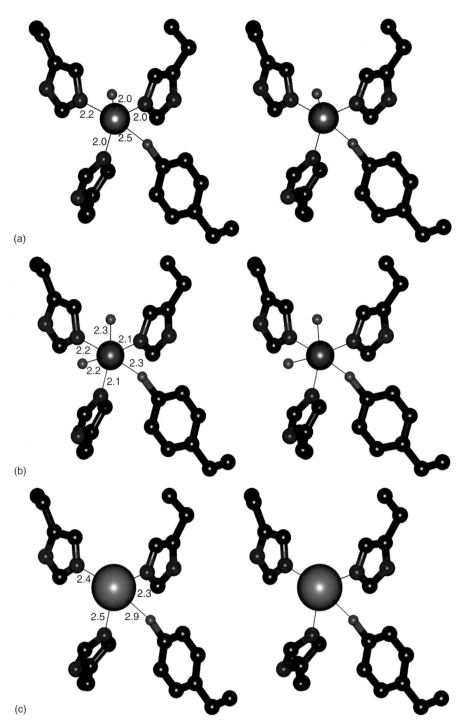

Figure 10 (a) Metal coordination of zinc (II)-, copper(II)-, and cobalt(II)-astacin; (b) nickel(II)-astacin; (c) mercury(II)-astacin, drawn with MOLSCRIPT.[42] PDB codes: 1IAA zinc replaced with copper; 1IAB zinc replaced with cobalt; 1IAC zinc replaced with mercury; 1IAE zinc replaced with nickel.[67] The zinc, cobalt, and copper derivatives have almost identical coordination spheres of their respective metal sites. In the nickel enzyme, a second solvent molecule (Sol600), in addition to Sol300 has entered the central plane (see also Table 1). In the mercury enzyme, no water at all is detectable at the metal ion. The metal–ligand distances are indicated in Å.

Preparation of metal-substituted astacin derivatives

Stoichiometric amounts of zinc(II) ions to apo-astacin fully restore its catalytic activity. Also, copper(II) and cobalt(II) ions[67] as well as manganese(II) ions, and iron(II) ions (unpublished results), but not mercury(II), nickel(II), or calcium(II) (among others) reactivate the apoenzyme.[43,67] On the basis of their specificity constants (k_{cat}/K_m), cobalt(II)- and copper(II)-astacin display 140 and 37% activity of the zinc enzyme, respectively, whereas mercury(II)-astacin and nickel(II)-astacin are almost inactive.[67]

X-ray crystal structures of astacin metal derivatives

Cobalt(II)-, copper(II)-, nickel(II)-, and mercury(II)-astacin were produced by dialysis of the apoenzyme against solutions of the respective metal chlorides and crystallized under the same conditions as the zinc enzyme.[40,41,67] Apo-astacin crystals were obtained by soaking zinc-astacin crystals in 8 mM 1,10-phenanthroline.[40,41]

The X-ray crystal structures of native zinc(II)-, copper(II)-, cobalt(II)-, mercury(II)-, nickel(II)- and apo-astacin were solved to resolutions of 1.8 Å, 1.90 Å, 1.79 Å, 2.10 Å, 1.83 Å, and 2.30 Å, respectively.[40,41,67] They exhibit an identical overall framework, but the modes of metal coordination in their active sites differ (Figure 10).

The active compounds copper(II)- and cobalt(II)-astacin contain trigonal bipyramidally coordinated metal as is seen in the zinc enzyme (Figure 10, Table 1).[67] By contrast, in the inactive mercury astacin, the catalytically important water ligand is missing, leaving a tetra-coordinated metal, and in the nickel derivative, an additional water ligand has entered the central plane, expanding it from trigonal to tetragonal, which results in an octahedral ligand geometry (Figure 10, Table 1).[67] The nucleophilicity of the general base should be reduced in this case. Hence, both the inactive derivatives apparently lack a catalytic water molecule for the attack of the scissile peptide bond.

Spectroscopy of metal derivatives of astacin

Cobalt(II)- and copper(II)-astacin were obtained by titration of the zinc-free apoenzyme at pH 8.0.[43,67] The electron paramagnetic resonance spectrum of copper(II)-astacin is indicative of 'normal' axial type II copper values. Albeit, the spectrum reveals a fine structure indicating some rhombic distortion being consistent with the crystallographically determined copper-coordination sphere (see below), which exhibits an almost ideal trigonal bipyramidal geometry.

The cobalt enzyme exhibits an absorption maximum at 514 nm ($\varepsilon = 77\,M^{-1}\,cm^{-1}$) with shoulders at 505 nm and 550 nm. The spectrum resembles the corresponding spectra of cobalt(II)-thermolysin (EC 3.4.24.27) and cobalt(II)-carboxypeptidase A (EC 3.4.17.1).[43] On the other hand, the spectrum of copper(II)-astacin exhibits a completely different behavior as compared to copper(II)-thermolysin or copper(II)-carboxypeptidase. It shows an intense absorption band at 445 nm ($\varepsilon = 1900\,M^{-1}\,cm^{-1}$) and an additional band at 325 nm ($\varepsilon = 1600\,M^{-1}\,cm^{-1}$) (Figure 11).[67]

The copper(II)-astacin UV–vis spectrum resembles those of copper(II)-transferrin, copper(II)-lactoferrin, and copper(II)-conalbumin, which, unlike thermolysin or carboxypeptidase, contain tyrosine-metal ligands and whose respective spectral properties have been interpreted

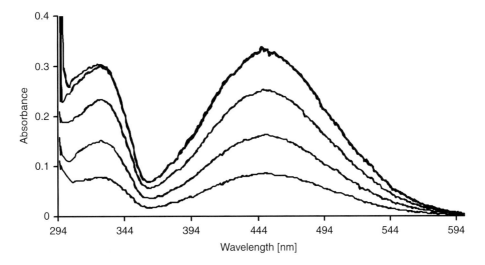

Figure 11 UV–vis spectrum of copper (II) astacin.[67] Apo-astacin (0.2 mM in 50 mM HEPES-NaOH-buffer, pH 8.0) was titrated with increasing amounts of CuSO$_4$. Saturation was reached at a molar ratio of 1:1. Reproduced by permission of The American Society for Biochemistry and Molecular Biology. F-X Gomis-Rüth, F Grams, I Yiallouros, H Nar, U Küsthardt, R Zwilling, W Bode and W Stöcker *J Biol Chem* **269**, 17111–17 (1994).

as phenol-metal charge transfer bands.[67] Interestingly, the intense charge transfer band at 445 nm is abolished upon inhibitor binding, which spectroscopically visualizes the 'tyrosine switch' during catalysis.[6]

Recently, a copper(II)-acetate complex of the heteroscorpionate ligand (2-hydroxyl-3-t-butyl-methylphenyl)bis(3,5-dimethylpyrazolyl)methane has been synthesized as a model of the trigonal bipyramidal zinc coordination seen in astacin and serralysin.[68] This complex reproduces some spectroscopic features of astacin including the charge transfer interaction between the phenolic oxygen and the metal.

REFERENCES

1. G Pfleiderer, R Zwilling and HH Sonneborn, *Hoppe-Seyler's Z Physiol Chem*, **348**, 1319–31 (1967).
2. R Zwilling, H Dorsam, HJ Torff and J Rödl, *FEBS Lett*, **127**, 75–78 (1981).
3. R Zwilling and H Neurath, *Methods Enzymol*, **80**, 633–44 (1981).
4. OA Klimova, SI Borukhov, NI Solovyeva, TO Balaevskaya and A Strongin, *Biochem Biophys Res Commun*, **166**, 1411–20 (1990).
5. OA Klimova, V Vedishcheva Iu and A Strongin, *Dokl Akad Nauk SSSR*, **317**, 482–84 (1991).
6. HI Park and LJ Ming, *J Inorg Biochem*, **72**, 57–62 (1998).
7. W Stöcker, B Sauer and R Zwilling, *Hoppe-Seyler Biol Chem*, **372**, 385–92 (1991a).
8. E Dumermuth, EE Sterchi, WP Jiang, RL Wolz, JS Bond, AV Flannery and RJ Beynon, *J Biol Chem*, **266**, 21381–85 (1991).
9. W Stöcker, F-X Gomis-Rüth, W Bode and R Zwilling, *Eur J Biochem*, **214**, 215–31 (1993).
10. W Bode, F-X Gomis-Rüth and W Stöcker, *FEBS Lett*, **331**, 134–40 (1993).
11. W Stöcker, F Grams, U Baumann, P Reinemer, F-X Gomis-Rüth, DB McKay and W Bode, *Protein Sci*, **4**, 823–40 (1995).
12. W Stöcker and W Bode, *Curr Opin Struct Biol*, **5**, 383–90 (1995).
13. F-X Gomis-Rüth, *Mol Biotechnol*, **24**, 157–202 (2003).
14. E Kessler, K Takahara, L Biniaminov, M Brusel and DS Greenspan, *Science*, **271**, 360–62 (1996).
15. SW Li, AL Sieron, A Fertala, Y Hojima, WV Arnold and DJ Prockop, *Proc Natl Acad Sci USA*, **93**, 5127–30 (1996).
16. N Suzuki, PA Labosky, Y Furuta, L Hargett, R Dunn, AB Fogo, K Takahara, DM Peters, DS Greenspan and BL Hogan, *Development*, **122**, 3587–95 (1996).
17. MJ Shimell, EL Ferguson, SR Childs and MB O'Connor, *Cell*, **67**, 469–81 (1991).
18. JS Bond and RJ Beynon, *Protein Sci*, **4**, 1247–61 (1995).
19. W Stöcker and I Yiallouros, in AJ Barrett, ND Rawlings and JF Woessner Jr (eds.), *Handbook of Proteolytic Enzymes*, Academic Press, San Diego, in press (2003).
20. F Möhrlen, JS Bond and W Stöcker, in AJ Barret, ND Rawlings and FJ Woessner Jr (eds.), *Handbook of Proteolytic Enzymes*, Academic Press, San Diego, in press (2003).
21. G Vogt, W Stöcker, V Storch and R Zwilling, *Histochemistry*, **91**, 373–81 (1989).
22. W Stöcker and R Zwilling, *Methods Enzymol*, **248**, 305–25 (1995).
23. G Geier, E Jacob, W Stöcker and R Zwilling, *Arch Biochem Biophys*, **337**, 300–7 (1997).
24. S Reyda, E Jacob, R Zwilling and W Stöcker, *Biochem J*, **344**, 851–57 (1999).
25. K Titani, HJ Torff, S Hormel, S Kumar, KA Walsh, J Rödl, H Neurath and R Zwilling, *Biochemistry*, **26**, 222–26 (1987).
26. E Dumermuth, JA Eldering, J Grünberg, W Jiang and EE Sterchi, *FEBS Lett*, **335**, 367–75 (1993).
27. JA Eldering, J Grünberg, D Hahn, HJ Croes, JA Fransen and EE Sterchi, *Eur J Biochem*, **247**, 920–32 (1997).
28. JM Wozney, V Rosen, AJ Celeste, LM Mitsock, MJ Whitters, RW Kriz, RM Hewick and EA Wang, *Science*, **242**, 1528–34 (1988).
29. JD Thompson, TJ Gibson, F Plewniak, F Jeanmougin and DG Higgins, *Nucleic Acids Res*, **25**, 4876–82 (1997).
30. KB Nicholas, HB Nicholas and DW Deerfield, *EMBNEW.NEWS* **4**, 14 (see also http://www.psc.edu/biomed/genedoc) (1997).
31. I Yiallouros, R Kappelhoff, O Schilling, F Wegmann, MW Helms, A Auge, G Brachtendorf, E Große Berkhoff, B Beermann, HJ Hinz, S König, J Peter-Katalinic and W Stöcker, *J Mol Biol*, **324**, 237–46 (2002).
32. F Möhrlen, S Baus, A Gruber, HR Rackwitz, M Schnölzer, G Vogt and R Zwilling, *Eur J Biochem*, **268**, 2540–46 (2001).
33. GE Palade, *Science*, **189**, 347–58 (1975).
34. GD Johnson and JS Bond, *J Biol Chem*, **272**, 28126–32 (1997).
35. GP Bertenshaw, MT Norcum and JS Bond, *J Biol Chem*, **278**, 2522–32 (2003).
36. C Becker, M-N Kruse, KA Slotty, D Köhler, JR Harris, S Rösmann, EE Sterchi and W Stöcker, *Biol Chem*, **384**, 825–31 (2003).
37. S Rösmann, D Hahn, D Lottaz, M-N Kruse, W Stöcker and EE Sterchi, *J Biol Chem*, **277**, 40650–58 (2002).
38. M Leighton and KE Kadler, *J Biol Chem*, **278**, 18478–84 (2003).
39. I Yiallouros, E Große Berkhoff and W Stöcker, *FEBS Lett*, **484**, 224–28 (2000).
40. W Bode, F-X Gomis-Rüth, R Huber, R Zwilling and W Stöcker, *Nature*, **358**, 164–67 (1992).
41. F-X Gomis-Rüth, W Stöcker, R Huber, R Zwilling and W Bode, *J Mol Biol*, **229**, 945–68 (1993).
42. PJ Kraulis, *J Appl Crystallogr*, **24**, 946–50 (1991).
43. W Stöcker, RL Wolz, R Zwilling, DJ Strydom and DS Auld, *Biochemistry*, **27**, 5026–32 (1988).
44. W Stöcker, M Ng and DS Auld, *Biochemistry*, **29**, 10418–25 (1990).
45. RL Wolz, *Arch Biochem Biophys*, **310**, 144–51 (1994).
46. RL Wolz and JS Bond, *Methods Enzymol*, **248**, 325–45 (1995).
47. E Krauhs, H Dörsam, M Little, R Zwilling and H Ponstingl, *Anal Biochem*, **119**, 153–57 (1982).
48. I Schechter and A Berger, *Biochem Biophys Res Commun*, **27**, 157–62 (1967).
49. W Stöcker, S Breit, L Sottrup-Jensen and R Zwilling, *Comp Biochem Physiol, B*, **98**, 501–9 (1991b).
50. F Grams, V Dive, A Yiotakis, I Yiallouros, S Vassiliou, R Zwilling, W Bode and W Stöcker, *Nat Struct Biol*, **3**, 671–75 (1996).
51. I Yiallouros, S Vassiliou, A Yiotakis, R Zwilling, W Stöcker and V Dive, *Biochem J*, **331**, 375–79 (1998).

52 RL Wolz and R Zwilling, *J Inorg Biochem*, **35**, 157–67 (1989).

53 RL Wolz, C Zeggaf, W Stöcker and R Zwilling, *Arch Biochem Biophys*, **281**, 275–81 (1990).

54 W Jiang and JS Bond, *FEBS Lett*, **312**, 110–14 (1992).

55 NM Hooper, *FEBS Lett*, **354**, 1–6 (1994).

56 R Huber and W Bode, *Acc Chem Res*, **11**, 114–22 (1978).

57 GP Bertenshaw, BE Turk, SJ Hubbard, GL Matters, JE Bylander, JM Crisman, LC Cantley and JS Bond, *J Biol Chem*, **276**, 13248–55 (2001).

58 N Hartigan, L Garrigue-Antar and KE Kadler, *J Biol Chem*, **278**, 18045–49 (2003).

59 WM Moore and CA Spilburg, *Biochemistry*, **25**, 5189–95 (1986).

60 D Köhler, M-N Kruse, W Stöcker and EE Sterchi, *FEBS Lett*, **465**, 2–7 (2000).

61 T Hege and U Baumann, *J Mol Biol*, **314**, 187–93 (2001).

62 DS Auld and BL Vallee, in A Neuberger and K Brocklehurst (eds.), *Hydrolytic Enzymes*, Elsevier, Amsterdam, pp 201–255 (1987).

63 BW Matthews, *Acc Chem Res*, **21**, 333–40 (1988).

64 DW Christianson and WN Lipscomb, *Acc Chem Res*, **22**, 62–69 (1989).

65 WL Mock and J Yao, *Biochemistry*, **36**, 4949–58 (1997).

66 A Beaumont, MJ O'Donohue, N Paredes, N Rousselet, M Assicot, C Bohuon, MC Fournie-Zaluski and BP Roques, *J Biol Chem*, **270**, 16803–8 (1995).

67 F-X Gomis-Rüth, F Grams, I Yiallouros, H Nar, U Küsthardt, R Zwilling, W Bode and W Stöcker, *J Biol Chem*, **269**, 17111–17 (1994).

68 CR Warthen and CJ Carrano, *J Inorg Biochem*, **94**, 197–99 (2003).

Matrix metalloproteinases

Wolfram Bode and Klaus Maskos
Max-Planck-Institut für Biochemie, Abteilung für Strukturforschung, Martinsried, Germany

FUNCTIONAL CLASS

Hydrolytic enzymes; mononuclear zinc endopeptidases; matrix metalloproteinases (MMPs), matrixins, EC 3.4.24.-, according to the MEROPS database[1] grouped together in the MMP or matrixin subfamily A of the metalloproteinase M10 family in clan MB; contain, besides other domains, a metzincin-like catalytic domain.[2–4] According to their (presumed) functions or membrane anchoring, these MMPs have been subgrouped as the 'collagenases', 'stromelysins', 'gelatinases', 'membrane-type MMPs (MT-MMPs)', and so on. The first MMP was detected as a metamorphogenic 'activity' in involuting tadpole tails in 1962.[5,6] The MMPs cleave internal peptide bonds, preferentially before leucine/isoleucine residues. The collagenases (in particular, MMPs 1 and 8, but also 2, 13, 14 and 18), for example, exhibit a special triple-helicase activity to start cleavage of interstitial (types I–III) collagens. Some

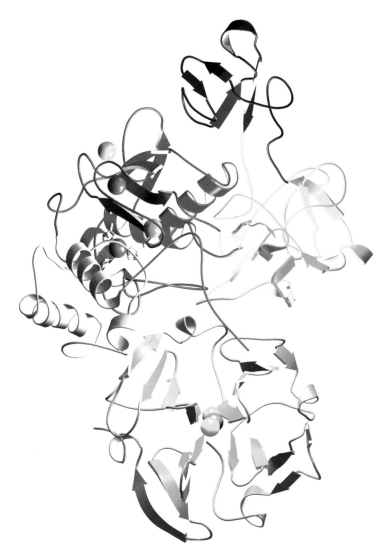

3D Structure Ribbon representation of the progelatinase A/proMMP-2 structure (PDB code 1CKX), shown together with the catalytic and the structural zinc ions (pink spheres), two calcium ions (yellow spheres) binding to the catalytic domain, and two calcium ions and an intercostal chlorine ion (green sphere) binding to the tunnel of the hemopexin-like domain.[8] Figure made with MOLSCRIPT[102] and RASTER3D.[103]

MMPs recognize their specific substrates through exosite interaction.[7]

OCCURENCE

MMPs have been identified in mammals including man, mouse, rat, rabbit, pig, cattle, and horse; in other vertebrates such as frogs; and in invertebrates such as *Drosophila, Caenorhabditis elegans*, sea urchin, and hydra, and in green algae and plants (soybean, *Arabidopsis thaliana*), but not in prokaryotes.[9–11] In mammals, the MMPs are either stored in the specific granules of polymorphonuclear leukocytes (MMP-8 and MMP-9), or transported as active enzymes to the cellular membrane, or secreted as latent pro-enzymes. MMP-11, MMP-27, and the MT-MMPs are activated during secretion by furin-like serine proteinases, while the other MMPs are activated in the extracellular space by plasmin and other trypsin-like serine proteinases, or by already activated MMPs.[12] The soluble MMPs can degrade their connective tissue substrates also at a distance from the cells, while the MT-MMPs and some cell membrane–localized secreted MMPs act at or near the cell surface.

BIOLOGICAL FUNCTION

MMPs are collectively able, *in vitro* and *in vivo*, to degrade all kinds of extracellular matrix (ECM) protein components such as interstitial and basement-membrane collagens, proteoglycans, fibronectin, elastin, aggrecan, and laminin, and are thus implicated in the connective tissue remodeling processes associated with embryonic development, angiogenesis, tissue growth, wound healing, and so on. Many of these MMPs, however, are (also) involved in the shedding and release of latent growth factors, growth factor binding proteins, cytokines, and cell surface receptors; in the activation of proMMPs and other pro-proteinases; and in the inactivation of proteinase and angiogenesis inhibitors, thus participating in diverse physiological processes.[11,13,14] Only a few MMPs are expressed at detectable levels in normal, healthy resting tissues, while most MMPs are only expressed in growing cultures, diseased or inflamed tissues, and activated cells.[15]

Normally, the degenerative potential of the MMPs is mainly held in check by the nonspecific inhibitor α2-macroglobulin and by the co-secreted or ECM-anchored specific tissue inhibitors of metalloproteinases (TIMPs) 1 to 4.[16–19] Disruption of this MMP–TIMP balance can result in pathologies such as rheumatoid and osteoarthritis, atherosclerosis, heart failure, fibrosis, pulmonary emphysema, tumor growth, cell invasion, and metastasis.[20–25]

AMINO ACID SEQUENCE INFORMATION ON HUMAN MMPs

MMPs are expressed and secreted as single-chain latent proMMPs, consisting of several domains (Figure 1). Individual vertebrate MMPs have often been named according to their presumed substrates. Normally, they are classified by sequential numbers, which run from MMP-1 to MMP-28, omitting numbers 4 to 6 (Table 1).

Twenty-four human MMP genes encode 23 different active MMPs, of which 22 have been well characterized:[7,26–29]

- MMP-1 (SwissProt accession code: P03956); precursor amino acid residues AA 1–469 (including the signal peptide).
- MMP-2 (P08253); precursor AA 1–660 (including three fibronectin-type II domains).
- MMP-3 (P08254); precursor AA 1–477.
- MMP-7 (P09237); AA 1–267 (no hemopexin domain).
- MMP-8 (P22894); AA 1–467.
- MMP-9 (P14780); AA 1–707 (including three fibronectin-type II domains).
- MMP-10 (P09238); AA 1–476
- MMP-11 (P24347); AA 1–488.
- MMP-12 (P39900); AA 1–470.
- MMP-13 (P45452); AA 1–471.
- MMP-14 (P50281); AA 1–582 (including a hemopexin, a transmembrane, and a cytoplasmic domain).
- MMP-15 (P51511); AA 1–669 (including a hemopexin, a transmembrane, and a cytoplasmic domain).
- MMP-16 (P51512); AA 1–607 (including a hemopexin, a transmembrane, and a cytoplasmic domain).
- MMP-17 (Q9ULZ9); AA 1–606 (before GPI attachment).
- MMP-19 (Q99542); AA 1–508.
- MMP-20 (O60882); AA 1–483.
- MMP-23; AA 1–390.
- MMP-24 (SP: Q9Y5R2); AA 1–645 (including a hemopexin, a transmembrane, and a cytoplasmic domain).
- MMP-25 (SP: Q9NPA2; TrEMBL code: Q96TE2); AA 1–562 (before GPI attachment).
- MMP-26 (SP: Q9NRE1); AA 1–261 (no hemopexin domain).
- MMP-27 (SP: Q99542; TR: Q9H306);
- MMP-28 (SP: Q9H239); AA 1–520.

All of these MMPs are synthesized with an N-terminal signal peptide, which is removed upon insertion into the endoplasmic reticulum yielding the latent proenzymes. All human proMMPs have in common an N-terminal ~80 amino acid residue prodomain (PRO) and an adjacent characteristic catalytic domain (CAT), which, except for the two gelatinases A and B with additional fibronectin type II domains (FN) inserts, consists of ~175 amino acid residues (Figure 1). In MMP-23[27,29] and probably

Matrix metalloproteinases

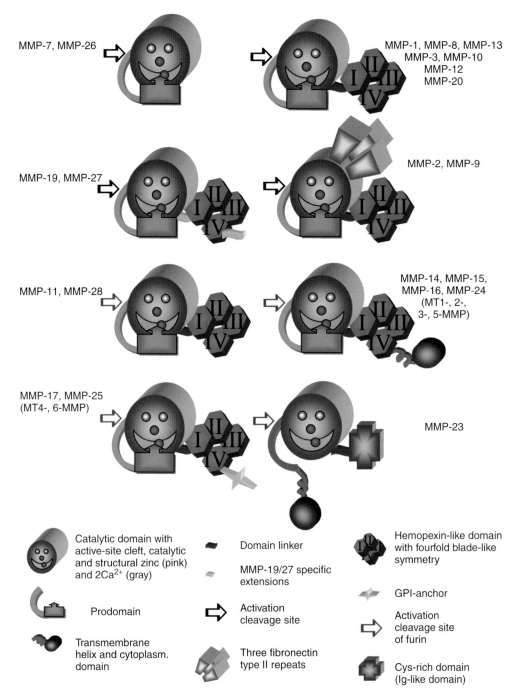

Figure 1 Schematic representation of the domain structure of the human MMPs, grouped according to furin-like and non-furin-like activation cleavage sites and different domains.[19]

also in the structurally related human MMPs 21/22,[26] the N-terminal part of the PRO domain seems to contain a transmembrane (TM) domain rendering these proteinases transmembrane-type-II proteins.[29] All other human MMP PRO domains except that of proMMP-26 contain a 'cysteine switch'[30] PRCXXPD consensus sequence, whose cysteine blocks the catalytic zinc in the latent pro-form. The linker between the PRO and the CAT domains is susceptible to proteolytic activation cleavage; one-third of the classified MMPs exhibit an alkaline R-X-(R/K)-R consensus cleavage site, rendering them activatable by furin and related proconvertases[29,31,32] (see Figure 1). Except the two matrilysins, MMP-7 and MMP-26, the human MMPs exhibit additional C-terminal domains. MMP-23 and human MMP-21/22 are unique in possessing a Cys/Pro-rich 'IL-1-type-II receptor'-like[26] or 'Ig'-like C2-type domain.[29] In all other human MMPs, the CAT domain is covalently connected through an up to 70-residue

Table 1 Classification of the matrix metalloproteinase (human, if not stated differently)

MMP number	EC	Trivial names
MMP-1	3.4.24.7	Collagenase-1, Fibroblast collagenase
MMP-2	3.4.24.24	Gelatinase A, TBE-1, 72 kDa gelatinase, type IV collagenase
MMP-3	3.4.24.17	Stromelysin-1, Transin-1, SL-1
MMP-7	3.4.24.23	Matrilysin (−1), Pump-1 protease, Uterine MP, Matrin
MMP-8	3.4.24.34	Collagenase-2, Neutrophil (PMNL) collagenase
MMP-9	3.4.24.35	Gelatinase B, GELB, 92 kDa gelatinase, type IV collagenase
MMP-10	3.4.24.22	Stromelysin-2, Transin-2, SL-2
MMP-11	3.4.24.-	Stromelysin-3, ST3, SL-3
MMP-12	3.4.24.65	Macrophage metalloelastase, HME
MMP-13	3.4.24.-	Collagenase-3
MMP-14	3.4.24.-	MT1-MMP, Membrane-type-1 MMP
MMP-15	3.4.24.-	MT2-MMP, Membrane-type-2 MMP
MMP-16	3.4.24.-	MT3-MMP, Membrane-type-3 MMP
MMP-17	3.4.24.-	MT4-MMP, Membrane-type-4 MMP
MMP-18	3.4.24.-	Collagenase-4 (*Xenopus*), xCol4
MMP-19	3.4.24.-	Stromelysin-4 (MMP-18), RASI-1
MMP-20	3.4.24.-	Enamelysin, Enamel metalloproteinase
MMP-21		XMMP (*Xenopus*)
MMP-22		CMMP (*Chicken*)
MMP-23		CA-MMP (Cysteine array), MIFR, Femalysin
MMP-24	3.4.24.-	MT5-MMP
MMP-25	3.4.24.-	MT6-MMP, Leukolysin
MMP-26	3.4.24.-	Endometase, Matrilysin-2
MMP-27		–
MMP-28	3.4.24.-	Epilysin

Pro-rich hinge to an ∼195-residue C-terminal hemopexin-like (PEX) domain. In the MT-MMPs, the polypeptide chains continue to end in additional C-terminal tails, which either include a transmembrane helix (TM) and terminate in a short cytoplasmic domain (CYT),[31] or whose termini act as a glycosil phosphatidyl inositol (GPI) membrane–anchoring signal, which gets replaced by the GPI membrane anchor.[33]

Of 11 MMPs, crystal structures are available at least for their catalytic domains[19] (see Figure 4). Figure 2 shows the sequences of their catalytic domains aligned according to topological equivalencies.[34]

PROTEIN PRODUCTION, PURIFICATION, AND MOLECULAR CHARACTERIZATION

Forty years ago, Gross and colleagues detected a metamorphogenic activity in tadpole tails,[5] which later was identified as an interstitial MMP. Since then, several natural (glycosylated) MMPs have been isolated from tissues and from cultured cells, mainly in their latent forms. Detailed protocols for the purification of proMMP-1, proMMP-2, and proMMP-3 from cultured human fibroblasts, of MMP-7 from human rectal carcinoma cells, of human proMMP-8 and MMP-8 from human neutrophils, of proMMP-9 from leukemia or fibrosarcoma cells, of human proMMP-10 from human keratinocytes, of MMP-12 from peritoneal macrophages have been described by Shimokawa and Nagase[38] and others.[39–43] These MMPs are mostly stored as latent enzymes at −80 °C, and are activated either by 1 mM 4-aminophenylmercuric acetate (APMA) or by activated MMP-3. Some of the active MMPs are not stable in solution, but undergo rapid autodegradation. Active MMP stock solutions can be frozen in small aliquots at −80 °C; however, frequent freezing–thawing should be avoided.[44]

Several expression systems have been established to make recombinant MMPs, including mammalian cells, insect cells, yeast (*Saccharomyces cerevisiae* and *Pichia pastoris*), and bacteria (*Escherichia coli*). Detailed protocols for cloning, culturing, transfection, expression, and purification of various MMPs are given by Yeow *et al.*[45] and Butler *et al.*,[46] by Vallon and Angel,[47] by Doyle,[48] and by Windsor and Steele.[49] It should be noted that the pro-peptide is (e.g. upon *E. coli* expression) not essential for proper folding.[50,51]

ACTIVITY AND INHIBITION TESTS

The proteolytic activity of the MMPs can be determined in various ways, for example, by measuring the fragments released from the (radiolabeled) macromolecular substrates such as collagen, casein, or Azocoll,[39,52] or by gelatin zymography, if the purification of an MMP is followed, or by utilizing chromogenic or fluorogenic peptides, if highly purified enzymes are studied.[53,54] Particularly convenient are the quenched fluorescent peptide substrates, which allow to monitor the fluorescence relieve of a fluorescent donor quenched in the intact peptide by a nearby (contact-quenched) or a more distant (resonance energy transfer quenched) group upon separation by proteolytic cleavage.[53,55] Donor/acceptor pairs frequently used are, for example, tryptophan (Trp)/N-2, 4-dinitrophenyl

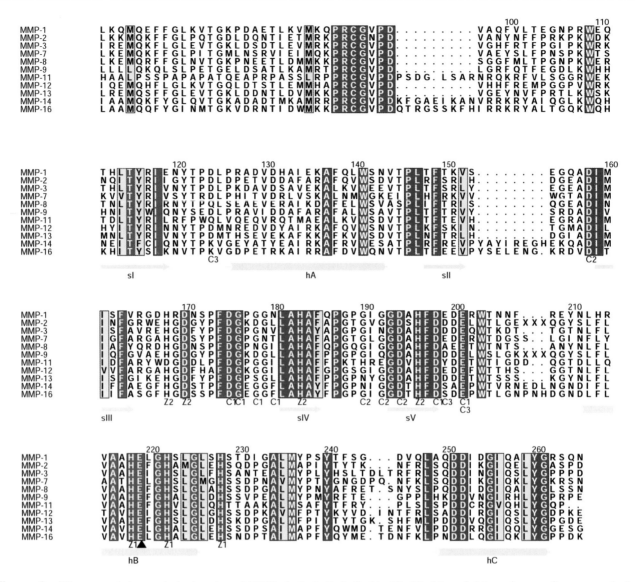

Figure 2 Alignment of the catalytic domains of MMPs 1, 2, 3, 7, 8, 9, 11, 12, 13, 14, and 16, made according to topological equivalencies.[34] The residue numbering is that of pre-proMMP-1.[35,36] The location and extent of the α-helices and β-strands are represented by cylinders and arrows; strictly or strongly conserved residues are shown with a red and a yellow background respectively; and special symbols denote residues involved in main and side-chain interactions with the catalytic (Z1) and the structural zinc (Z2), the first (C1), second (C2), and third calcium ion (C3) of the catalytic domain. Both gelatinases possess additional 175- and 176-residue FN inserts in the gap between sV and hB, marked as xxx. Figure made with ALSCRIPT.[37]

(Dnp), (7-methoxycoumarin-4-yl)acetyl (Mca)/Dnp, 5-[(2-aminoethyl)amino]-naphthalin-1-sulfonic acid (Edans)/4-(4-dimethylaminophenylazo)benzoic acid (Dabcyl).

These synthetic substrates were initially designed according to the protein sequences surrounding MMP cleavage sites.[56–58] Systematic studies with oligopeptides showed that in most cases the occupancy of the enzyme subsites S3 to S3′ (with P1, P2, P3 etc. and P1′, P2′, P3′ etc. indicating the residues in N- and C-terminal directions of the scissile peptide bond of a bound peptide substrate (analogue), and S1, S2, S3, etc. and S1′, S2′, S3′ etc. the opposite binding sites on the enzyme[59]) is necessary for rapid hydrolysis,[60–62] in agreement with the strict endopeptidase nature of MMPs. Although the hydrolysis of some natural substrates can be extremely slow (with e.g. turnover numbers for the fibroblast collagenase of less than one cleavage per minute in type I collagen at 30 °C), the specificity constant k_{cat}/K_m can reach values beyond 10^6 M^{-1} s^{-1} for optimized peptide substrates.[63] In contrast to many kinetic data measured with protein substrates, the hydrolysis data obtained with peptide substrates do not depend much on the presence of the C-terminal hemopexin-like domains. By combining favorable interactions at the different subsites, hepta- to octapeptide substrates have been developed, which somewhat allow to discriminate between distinct MMPs. A list of useful fluorescence-quenching substrates for various MMPs, convenient assay conditions, and their synthesis is given by Fields.[55]

The pH-k_{cat}/K_m profile of matrilysin from pH 4.0 (in the presence of some extra zinc to suppress zinc release) to pH 10 shows a broad bell-shaped curve, with a pH-optimum from 5.0 to 9.0 and flanks governed by ionizing groups with pK_a values around 4.5 and 9.5,[51] while stromelysin-1 exhibits a more narrow profile, with a peak activity at pH 6.0 and an alkaline shoulder.[64] Ca^{2+} concentrations above 1 mM are usually required to maintain full enzymatic activity around neutral pH, while the addition of extra Zn^{2+} seems to be necessary only at quite low pH values. Almost complete inhibition is observed with 10 mM EDTA or with 5 mM 1,10-phenanthroline (OP). Dialysis of MMP-7 in the presence of OP and $CaCl_2$ removes only the catalytic zinc ion; this mono-zinc enzyme can be (partially) reactivated by the addition of zinc, manganese, nickel, and cobalt respectively.[51] The OP inhibition is biphasic, with the first step characterized by an enzyme-OP complex, followed by the release of Zn^{2+}-OP and the formation of a Zn^{2+}-$(OP)_3$ complex.[51,65]

Most small molecule MMP inhibitors (MMPIs) designed so far bind to the catalytic site. In most MMPIs, a peptidyl or peptido-mimetic moiety based on the known substrate specificities is combined with a zinc chelating group such as a hydroxamate, carboxylate, N-carboxyalkyl, thiol, mercaptoalkyl, sulfodiimine, phosphonoalkyl, phosphinic, or phosphonic acid group. A typical example is the hydroxamic acid inhibitor batimastat (BB-94; 4-(N-hydroxyamino)-2(R)-isobutyl-3(S)-[(2-thienylthiomethyl)succinyl]-L-phenylalanine-N-methylamide), a broad-spectrum inhibitor with low nanomolar activity against MMPs, which for a long time was considered as the MMPI 'gold standard' (see the batimastat-MMP-8 structure[66]). With the zinc ligand positioned at the cleavage site, the peptide moiety has been extended to the N-terminal end (left-hand-side inhibitors), the C-terminal end (right-hand-side inhibitors), or in both directions of the substrate.[67] Besides, a number of nonpeptide inhibitors and nonpeptide natural product MMP inhibitors have been investigated.[68–70] The rate constants for MMP association with synthetic as well as protein inhibitors are often determined by the analysis of the progress curves of the hydrolysis of fluorescent substrates.[71]

X-RAY AND NMR STRUCTURES OF MMPs

Crystallization

Several MMPs, mainly their isolated catalytic domains, have been crystallized, in most cases in the presence of one or the other synthetic inhibitor (see Table 2). The Phe79 and the Met80 crystal form of the recombinant MMP-8 catalytic domain, for example, that is used for the structure determination of various synthetic inhibitors,[66,72–75] is crystallized by mixing 1 μL of the 10 mg mL^{-1} protein solution (containing 5 mM $CaCl_2$, 100 mM NaCl, 3 mM MES-NaOH, 0.02% NaN_3, pH 6.0), 3 μL of a PEG solution (10% PEG 6000, 0.2 M MES-NaOH, pH 6.0), and 1 μL of a 100 mM inhibitor solution, followed by vapor diffusion against a reservoir buffer of 1.0 M potassium phosphate, pH 6.0.[72]

Since early 1994, a large number of MMP-related structures (Table 2), determined by NMR and, in particular, by X-ray techniques, have been published.[17,19] The first X-ray crystal structures, determined with the help of 'classical' heavy metal derivatives, were made of the catalytic (CAT) domains (blocked by various synthetic inhibitors) of human fibroblast collagenase-1/MMP-1[76–79] and human neutrophil collagenase-2/MMP-8.[72,73,78] At the same time, an NMR structure of the catalytic domain of stromelysin-1/MMP-3[80] became available. These CAT structures were later complemented by additional X-ray and NMR structures of the CAT domains of MMP-1,[81] matrilysin-1/MMP-7,[82] of MMP-3 with synthetic inhibitors[83–86] and with TIMP-1,[87] of MMP-8,[66,74] and of human MT1-MMP in complex with TIMP-2.[88] In 1995, the first X-ray crystal structure of an MMP proform, the C-terminally truncated pro-stromelysin-1, was published,[83,86] and the first structure of a mature full-length MMP, namely, that of porcine fibroblast collagenase-1/MMP-1,[89] was described. At that time, structures of the isolated hemopexin-like domains (PEX) from human gelatinase A/MMP-2[90,91] and from collagenase-3/MMP-13[92] were also reported. In 1999, the formidable structure of full-length proMMP-2/progelatinase A[8] appeared, accompanied by the fibronectin type II (FN)-deleted CAT domain of MMP-2.[93] More recently followed the X-ray structures of the CAT domains of human[94] and mouse collagenase-3/MMP-13,[95] mouse stromelysin-3/MMP-11,[96] human macrophage metalloelastase/MMP-12,[50,97] and human MMP-16.[34] Quite recently, the FN-truncated CAT domain[98] and the full-length pro-CAT domain[99] of recombinant human MMP-9, the progelatinase A/TIMP-2 complex,[100] and the PEX-9 domain of MMP-9[101] were added as new members to the MMP structure zoo. At the end of 2002, more than 80 MMP-related structures were deposited in the PDB, almost all of them in complex with synthetic inhibitors.[19,68–70] Some of the early representative structures are listed in Table 2.

Catalytic domain

The catalytic domains of MMPs (Figure 3) exhibit the shape of an oblate ellipsoid. In the 'standard' orientation, which, here as well as in most other MMP papers, is preferred for the display of the CAT domains, a small active site cleft notched into the flat ellipsoid surface extends horizontally across the domain surface to bind peptide substrates from left to right. This cleft, harboring the 'catalytic zinc' separates the smaller 'lower subdomain' from the larger 'upper subdomain'.

Table 2 Selection of the 3D structures of MMPs currently deposited in the PDB

MMP	Domains	Inhibitor*	Method	PDB code	Reference
MMP-1	CAT	CA	X-RAY	1CGL	76
MMP-1	CAT	N-term	X-RAY	1CGE,CGF	81
MMP-1	CAT	HY	X-RAY	2TCL	77
MMP-1	CAT	HY	X-RAY	1HFC	79
MMP-1	CAT + PEXpig	HY	X-RAY	1FBL	89
MMP-2	PEX	–	X-RAY	1GEN	90
MMP-2	PEX	–	X-RAY	1RTG	91
MMP-2	PRO + CAT + FN + PEX	–	X-RAY	1CK7	8
MMP-2	CAT-FN	HY	X-RAY	1QIB	93
MMP-2	PRO + CAT + FN + PEX + TIMP-2	–	X-RAY	1GXD	100
MMP-3	CAT	CA	NMR	2SRT	80
MMP-3	CAT	CA	X-RAY	1SLN, 1HFS	83
MMP-3	PRO + CAT	–	X-RAY	1SLM	83
MMP-3	CAT	HY	NMR	1UMT, 1UMS	85
MMP-3	CAT + TIMP-1	TIMP-1	X-RAY	1UEA	87
MMP-7	CAT	CL	X-RAY	1MMP, 1MMQ, 1MMR	82
MMP-8	CAT	HY	X-RAY	1JAP	72
MMP-8	CAT	HY	X-RAY	1JAN	73
MMP-8	CAT	TH,HY	X-RAY	1JAO, 1JAQ	74
MMP-8	CAT	HY	X-RAY	1MMB	66
MMP-8	CAT	HY	X-RAY	1MNC	78
MMP-9	CAT-FN	HY	X-RAY	1GKC, 1GKD	98
MMP-9	CAT + FN + PRO	–	X-RAY	1L6J	99
MMP-9	PEX	–	X-RAY	1ITV	101
MMP-11	CATmouse	PI	X-RAY	1HV5	96
MMP-12	CAT	HY	X-RAY	1JK3	50
MMP-12	CAT	HY	X-RAY	1JIZ	97
MMP-13	PEX	–	X-RAY	1PEX	92
MMP-13	CAT	HY	X-RAY	456C, 830C	94
MMP-13	CATmouse	NP	X-RAY	1CXV	95
MMP-14	CAT + TIMP-2	TIMP-2	X-RAY	1BUV, 1BQQ	88
MMP-16	CAT	HY	X-RAY	–	34

*CA: carboxyalkyl; HY: hydroxamate; CL: carboxylate; PI: phosphinate; TH: thiol.

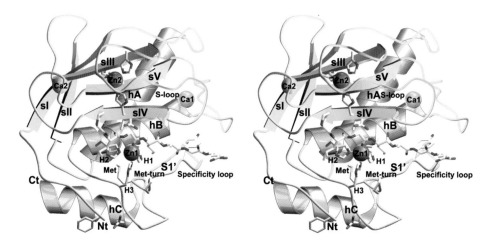

Figure 3 Ribbon structure of the superactive MMP-8 catalytic domain (PDB code 1JAN)[73] shown in standard orientation together with a modeled heptapeptide substrate productively bound in the active site cleft.[74] Strands (yellow) and helices (red) are labeled with Roman numerals and letters respectively, and the Met-turn, the specificity loop, and the S1' specificity pocket are emphasized. The catalytic and the structural zinc (center and top) and the two flanking calcium ions are shown as pink and yellow spheres, and the three histidine residues (H1, H2, and H3) ligating the catalytic zinc, the catalytic glutamic acid in between, the four ligands of the structural zinc, the characteristic methionine (Met) residue of the Met-turn, the proline and the tyrosine side chains of the S1' wall-forming segment, and the N-terminal phenylalanine and the first aspartic acid of the aspartic acid pair forming the surface-located salt bridge are shown with all nonhydrogen side-chain atoms. Figure made with MOLSCRIPT[102] and RASTER3D.[103]

The upper subdomain (Figure 3) formed by the first three-quarters of the polypeptide chain consists of a five-stranded β-pleated sheet flanked by three surface loops on its convex side and by two long regular α-helices on its concave side embracing a large hydrophobic core. The polypeptide chain starts on the molecular surface of the lower subdomain, where its N-terminal Phe100 ammonium group, if correctly tailored, can become engaged in a surface-located salt bridge with Asp251 (see Figure 3), causing the 'superactivity' phenomenon (see reference 73). The chain turns to the upper domain, passes β-strand sI, the amphipathic α-helix hA, and β-strands sII, sIII, sIV, and sV, before entering the 'active site helix' hB (for the nomenclature, see Figure 3). In the classical MMPs, strands sII and sIII are connected by a relatively short loop bridging sI; in the membrane-type MT-MMPs 14, 15, 16, and 24, however, this loop is expanded into the spur-like, solvent-exposed 'MT-specific loop'[88] (see Figure 4). In all MMPs, strands sIII and sIV are linked via an 'S-shaped double loop', which is clamped via the 'structural zinc' and the first of two to three bound calcium ions to the β-sheet. This S-loop extends into the cleft-sided 'bulge' continuing in the antiparallel 'edge strand' sIV. This bulge-edge segment is of prime importance for the binding of peptidic substrates as well as for peptido-mimetic and nonpeptidic (e.g. heterocyclic) inhibitors (Figures 3, 4, and 10). The sIV-sV connecting loop sandwiches together with the sII-sIII bridge the second bound calcium. After strand sV, the chain (except in both gelatinases, where it turns toward the 180-residue FN domains, see Figure 11) passes the large open sV-hB loop, before entering the active site helix hB. This helix provides the first two histidine ligands of the catalytic zinc (His218 and His222) and the 'catalytic' Glu219 in between (see Figure 1), representing the N-terminal part of the 'zinc binding consensus sequence' HEXXHXXGXXH characteristic of the metzincin superfamily.[2,3]

This active site helix abruptly stops at Gly225, where the peptide chain bends down and descends to present the third zinc liganding histidine, His228, and runs through a wide right-handed spiral terminating in the 1, 4-tight 'Met-turn' (Figure 3). The chain then turns back to the molecular surface to an almost invariant Pro238, forms with a conserved Pro238-Xaa-Tyr240 segment (the 'S1' wall-forming segment') the outer wall of the S1' pocket, and runs through the wide 'specificity loop' of slightly variable length and conformation, before it passes the C-terminal α-helix hC, which ends with the conserved Tyr260-Gly261 residue pair.

The overall structures of all MMP CAT domains known so far are very similar (see Figure 4). Larger main-chain differences occur (i) in the N-terminal segment up to Pro107 (depending on the length of the N-terminus and the intactness of the N-terminus-involved salt bridge); (ii) in the sII-sIII bridge (with the elongated and exposed 'MT loops' of the MT-MMPs 1, 2, 3, and 5); (iii) in the specificity loop folding around the most important S1' pocket; and (iv) in the sV-hB loop. In the pro-gelatinases A and B, this usually short loop segment between Leu205 and Tyr210 (proMMP-1 nomenclature, see Figure 1) is replaced by 179 and 180 residue inserts respectively. These inserts form a large cloverleaf-like FN domain[8,99] consisting of three tandem copies of fibronectin-type-II-like (FN) modules (see Figure 11). The three FN domains are packed together, leaving a deep cleft, which levels off to the five-stranded surface sheet of the CAT domain, that is, not in line with the active site cleft.

Metal site geometry

Besides the catalytic zinc (Figure 5), all MMP catalytic domains possess another zinc ion, the structural zinc

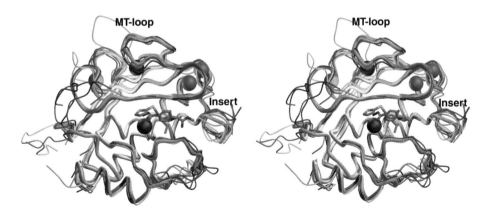

Figure 4 Superposition of the CAT domains of superactive MMP-8 (1JAN) (thick yellow rope), MMP-2 (1QIB) and MMP-9 (1GKC) (only conserved parts, see Figure 11), MMP-1 (1CGL), MMP-3 (1UEA), MMP-7 (1MMP), MMP-11 (1HV5), MMP-12 (1JK3), MMP-13 (456C), and MMP-14 (1BUV, thin green rope), together with batimastat[66] shown in standard orientation. The specific MT-loop of the MT-MMPs and the insertion sites for the FN domains in both gelatinases are marked. The catalytic and the structural zinc (center and top), and the three calcium ions visible in MMP-12 are shown as pink and orange spheres. Figure made with MOLSCRIPT[102] and RASTER3D.[103]

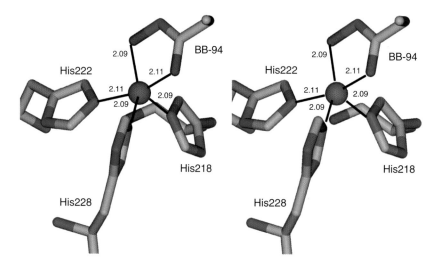

Figure 5 Stereo view of the catalytic zinc (pink sphere) of the MMP-12–batimastat complex (PDB code 1JK3),[50] approximately shown in standard orientation. Besides the three liganding histidine–imidazole ligands and part of the catalytic glutamic acid side chain, the hydroxamic acid group of the bound batimastat inhibitor (BB-94) is shown, and the five Zn-N/O contacts are given together with the distances. Figure made with DS ViewerPro.[107]

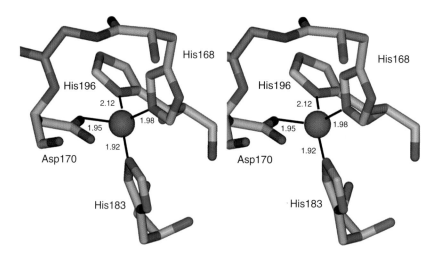

Figure 6 Stereo view of the structural zinc (pink sphere) of MMP-12,[50] with one of the alternative conformations, approximately shown in standard orientation. The three histidine imidazole and the aspartic acid–carboxylate ligands are shown, and the four Zn-N/O contacts are given together with the distances. Figure made with DS ViewerPro.[107]

(Figure 6), and one (MMP-11, MMP-13), two (MMP-8, MMP-2; MMP-14, MMP-16), or three (MMP-1, MMP-3, MMP-7, MMP-9, MMP-12) bound calcium ions (Figures 4, 7, 8, and 9). The catalytic zinc, situated at the bottom of the active site cleft, is in the hydroxamate-inhibited MMPs (e.g. in the 1.1 Å batimastat complex with MMP-12[50]), penta-coordinated in the manner of an almost exact squared pyramid, with the two hydroxamate oxygen atoms and the NE2 atoms of His218 and His222 forming a squared base, and the His228 NE2 placed at the tip (Figure 5). In the MMP-12 structure,[50] the distances to all five ligands are within 2.1 (±0.1) Å, in agreement with the minimal Zn–N distances of 2.0 Å for penta-coordinated zinc clusters in small molecules.[104] In an MMP-8 complex with a right-side-binding peptidyl-thiol compound,[74] the sulfur atom forms, together with the three imidazole nitrogens, a nearly exact tetrahedron. In unliganded MMPs, the sulfur atom is (as in the related metzincins astacin[105] and adamalysin II[106]) replaced by a fixed water molecule, which simultaneously is in hydrogen bond distance to the carboxylate group of the catalytic Glu219.

The structural zinc (Figure 6) is coordinated in an almost ideal tetrahedral manner by the two NE2 and one ND1 imidazole atoms of His168, His183, and His196 (Figure 1), and by the Oδ1 atom of the Asp170 carboxylate group, respectively. The Zn-ligand distances (1.92 to 2.12 Å[50]) are just above the minimal values derived for tetra-coordinated Zn–O (1.91 Å) and Zn–N (1.92 Å) bonds in small molecules.[104] In the 1.1-Å MMP-12 structure,[50] the His183 imidazole ring adopts two alternative conformations, and the high temperature factor

Matrix metalloproteinases

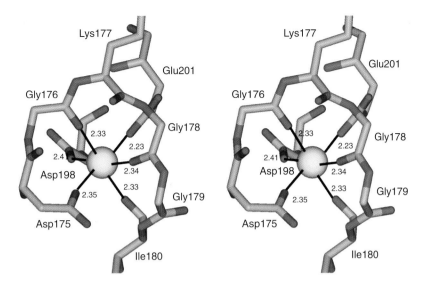

Figure 7 Stereo view of the first bound calcium ion (yellow sphere) of MMP-12,[50] shown approximately in standard orientation. The three asparagine and glutamic acid carboxylate and three carbonyl oxygen ligands are shown, and the four Ca–O contacts are given together with the distances. Figure made with DS ViewerPro.[107]

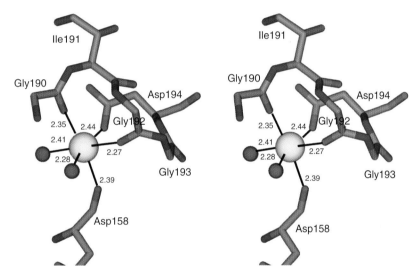

Figure 8 Stereo view of the second bound calcium ion of MMP-12,[50] shown approximately in standard orientation. Figure made with DS ViewerPro.[107]

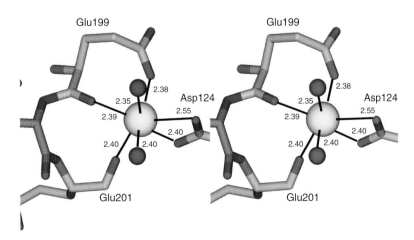

Figure 9 Stereo view of the third bound calcium ion of MMP-12,[50] shown approximately in standard orientation. Figure made with DS ViewerPro.[107]

anisotropy indicates two slightly different positions. This structural zinc ion, completely buried in the protein matrix, seems to be bound much tighter than the catalytic zinc.[51,72]

Dialysis of matrilysin against 1,10-phenanthroline in the presence of 5 mM $CaCl_2$ yields an inactive 'apoenzyme' with about one zinc ion per mole. Zn^{2+} can restore most of the hydrolytic activity, and Mn^{2+}, Ni^{2+}, and Co^{2+} can restore partial activity, while this apoenzyme remains inactive in the presence of Cd^{2+} or Cu^{2+}.[51]

The first (probably most tightly bound) calcium ion (Figure 7), which packs the second part of the S-shaped double loop against the side-chain carboxylate groups of an invariant aspartic acid–glutamic acid couple protruding from strand sV, is coordinated by the carbonyl oxygen atoms of Gly176 and of residues 178 and 180, and by one carboxylate oxygen atom of the conserved residues Asp175, Asp198, and Glu201 in a nearly ideal octahedral manner. The three Ca-carbonyl oxygen bonds are (in the 1.09 Å MMP-12 structure[50]) limited to 2.33 to 2.34 Å, while the Ca distances to the (unidentate) carboxylate oxygen atoms vary from 2.23 to 2.41 Å. The second calcium ion (Figure 8), sandwiched between the sIV-sV loop and sIII, is octahedrally coordinated by three carbonyl groups, a bulk solvent and one carboxylate oxygen of the invariant Asp194. The loop immediately following sV encircles the third calcium (Figures 4, 9) and coordinates it through the two carbonyl groups of residues 199 and 201, together with two fixed water molecules and the carboxylate oxygens of Glu199 and Asp124, which occupy the corners of a pentagonal bipyramid. The presence of Asp124 seems to correlate with this third calcium site (see Figure 1).

Substrate binding and specificity

Bounded at the upper rim by the bulge-edge segment and at the lower side by the third zinc-liganding imidazole and the S1' wall-forming segment, the active site cleft of all MMPs is relatively flat at the left ('nonprimed') side, but is notched into the molecular surface at the catalytic zinc and to the right ('primed') side, leveling-off again to the surface further to the right. The S1' specificity pocket (Figures 3, 4) invaginates immediately to the right of the catalytic zinc, and varies, in size and shape, considerably among the various MMPs. Of the synthetic inhibitors determined in complex with an MMP, only a very few bind to the left-hand-side 'nonprimed' subsites ('left-side inhibitor') or across the active site, while the majority of synthetic (nonpeptidic or peptidic) inhibitors studied so far interact with the primed right-hand-side subsites ('right-side inhibitors'), inserting in the narrow primed-site cleft created between the (antiparallel) bulge-edge segment and the (parallel) S1' wall-forming segment of the cognate MMP, forming a number of inter-main-chain hydrogen bonds (see Figure 10). When linked to a

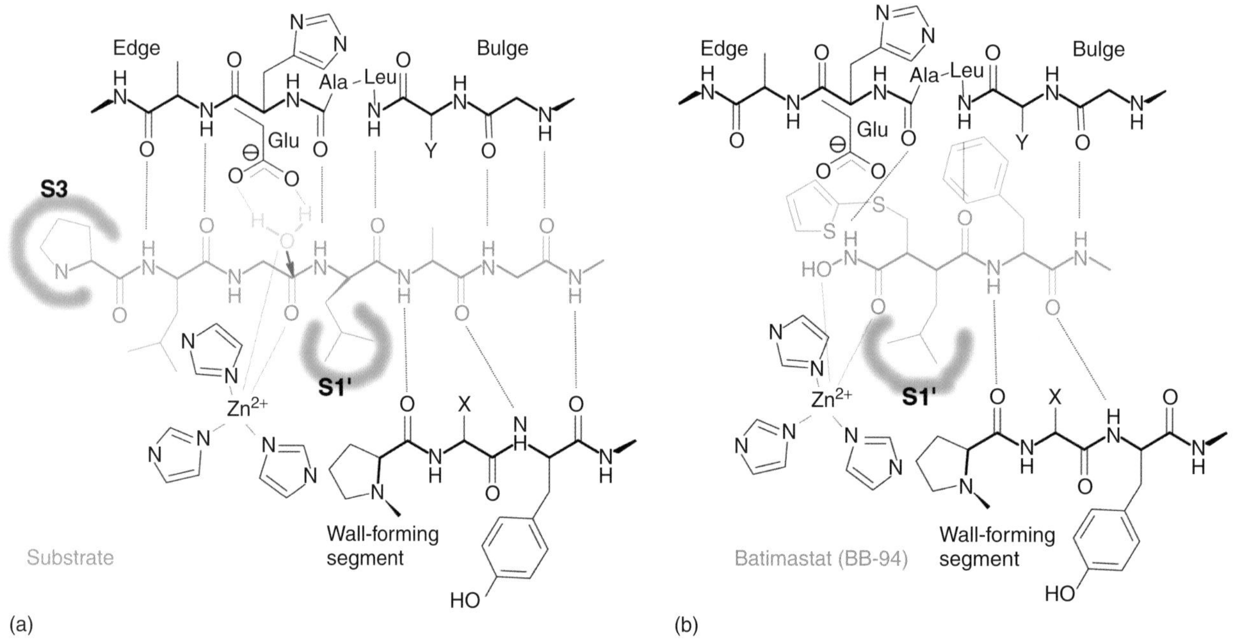

Figure 10 Peptide substrate, inhibitor interaction, and specificity.[17] Schematic drawing of the putative encounter complex between (a) modeled Pro-Leu-Gly-Leu-Ala-Gly-amide hexapeptide substrate (left)[74] and (b) batimastat (right),[66] interacting with the MMP active site. The substrate polypeptide chain and the inhibitor peptide mimic (cream-faced connections) run antiparallel to the bulge-edge strand (top, black connections) and parallel to the S1' wall-forming segment (bottom, black), forming several inter-main-chain hydrogen bonds (dashed black lines). The substrate makes dominant hydrophobic interactions, via its P1' and P3 side chains, with the S1' pocket and the S3 subsite (highlighted by green troughs). In the enzyme–substrate encounter, the catalytic water (activated by the catalytic zinc) is suitably placed and is activated to attack the carbonyl group of the scissile P1–P1' peptide bond.

peptidic moiety, only an L-configured P1'-like side chain can extend into the hydrophobic bottleneck of the S1' pocket.[74] This P1'–S1' interaction is the main determinant for the affinity of most inhibitors and for the cleavage position of peptide substrates.

Depending in particular on the length and character of residue 214 harbored in the N-terminal part of the active site helix hB (see Figure 2), the size of the S1' pocket differs considerably among the MMPs. In MMP-1 and MMP-7, for example, the side chains of Arg214 and Tyr214 extend into the S1' opening, limiting it to a size and shape still compatible with the accomodation of medium-sized P1' residues, but less for very large side chains, in agreement with peptide cleavage studies on model peptides.[60–62] It has been shown, however, that the Arg214 side chain can swing out of its 'normal' site, thus also allowing binding of synthetic inhibitors with larger P1' side chains.[94] The smaller Leu214 residues of MMPs 2, 3, 9, 12, 13, and 14 do not bar the internal S1' 'pore', which, in these MMPs, extends right through the molecule to the lower surface, that is more like a long solvent-filled 'tube'. In spite of a small Leu214 residue, however, the S1' pocket of MMP-8 is of medium size and closed at the bottom, because of the Arg243 side chain extending into the S1' space from the specificity loop.[72] In MMP-9, the Arg424 (corresponding to MMP-1 residue 241) side chain also can close the end of the pocket.[98] In uninhibited MMP-3, the phenolic side chain of the Pro-X239-Tyr S1' wall-forming segment has been observed to cover the S1' entrance, while it becomes displaced by the P1'-like group of a binding inhibitor.[108]

Other important substrate determinants of MMPs are the S3, S2, S1, and S2' subsites[109–113] (Figure 10). The P3 residue (in collagen cleavage sites, always a proline residue) nestles into the mainly hydrophobic S3 pocket. The polarity and extension of the S2 subsite, in most MMPs a shallow depression extending on top of the imidazole ring of the third zinc-liganding His228, are particularly determined by the nature of residue 227, which is quite variable (see Figure 2). Longer side chains of P1 residues are placed in the surface groove lined by the His183 side chain of the edge strand and the medium-sized residue 180 (marked with a 'Y' in Figure 10), which varies from serine via isoleucine/leucine to phenylalanine. P2' side chains will become squeezed between the bulge rim and the side chain of the middle residue of the Pro-X239-Tyr wall-forming segment, with residue 239 (marked with an 'X' in Figure 10) in the MMPs 14, 15, and 16 and MMP-11 represented by an exposed phenylalanine.[88] Further to the right, the molecular surface again has a hydrophobic/polar depression, which could accomodate P3' side chains of differing natures. Some MMPs acquire their substrate specificity also through substrate-binding exosites located outside the active site cleft.[7]

Zymogen structures

ProMMP structures have so far been known for the C-terminally truncated proMMPs 3[83] and 9,[99] and for full-length proMMP-2[8,100] (Figure 11). The PRO domain exhibits an egglike shape, attached with its rounded-off

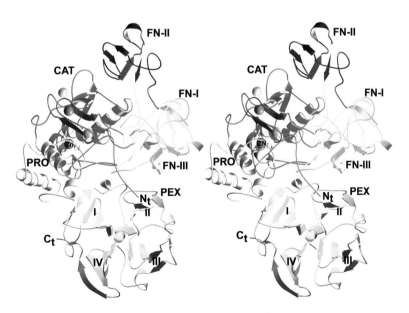

Figure 11 Stereo ribbon representation of pro-MMP-2 (PDB code 1CK7),[8] consisting of (from the N_t- to the C_t-terminus) the pro-domain (PRO, brown ribbon, in front); the conserved part of the catalytic domain (CAT, left upper red ribbon, enveloping the catalytic and the structural zinc, (pink spheres), and two calcium ions,(orange spheres)), with the three fibronectin-type-II-domains (FN-I (orange), FN-II (dark blue) and FN-III (green)) inserted in the sV-hB loop; and the hemopexin-like domain (PEX, lower light blue ribbon comprising blades I, II, III, and IV and containing up to three ions in its central channel) connected via a hinge segment. The molecule is shown such that its CAT domain is in standard orientation. Figure made with MOLSCRIPT[102] and RASTER3D.[103]

side to the active site of the catalytic domain. It essentially consists of three mutually perpendicular α-helices, connected by relatively flexible and proteolysis-susceptible loops, and an almost invariant Pro90-Arg-Cys-Gly-Val-Pro-Asp96 'cysteine switch' loop (Figure 2). This latter peptide segment runs through the active site cleft in a direction opposite to the bound substrates and further extends into the CAT domain. The cysteine side chain, coordinates, together with the three histidine ligands, the catalytic zinc.

The proteolytic activation of these 'classical' proMMPs seems to proceed via a stepwise mechanism: some early cleavages occurring in the flexible, exposed helix1–helix2 loop might destabilize this domain, exposing other (downstream) cleavage sites and weakening and finally breaking the Cys-catalytic zinc bond, eventually leading to a liberation and flexibilization of the Xaa99-Phe/Tyr100 activation cleavage peptide bond (as originally predicted by the 'cysteine switch hypothesis'[30]), allowing the maturation cleavage by another proteinase. If the cleavage occurs exactly before the Phe/Tyr100 N-terminus (in the classical MMPs), the activation cleavage is accompanied by a substantial rearrangement of the N-terminal 100–107 segment, with the N-terminal residue moving about 17 Å to form the above-mentioned surface-located salt bridge with the carboxylate of Asp250 of the helix C-based Asp251/Asp252 pair.[73]

Other domains

Except for MMPs 7, 26, and 23, all vertebrate/human MMPs are expressed with a C-terminal PEX domain (see Figure 1). Some of these PEX domains have been shown to be involved in substrate recognition and to confer substrate specificity, most dramatically in the collagenase subfamily, where the capability to cleave native triple-helical collagen is associated with the covalently bound PEX domain.[7,75,113] The PEX domains exhibit the structure of a four-bladed β-propeller of pseudo fourfold symmetry[89–92] (see Figure 11). The polypeptide chain is essentially organized in four β-sheets (blades) I to IV, which are arranged almost symmetrically around a central axis in consecutive order. Each propeller blade is twisted and made up of four antiparallel β-strands connected in a W-like topology. The first innermost strands in all four blades enter the propeller at one site and run almost parallel to one another along the propeller axis forming a central funnel-shaped tunnel, which opens slightly toward the exit and often accomodates some ions (Figure 11). The C-terminus of the blade IV helix is clamped to the entering strand of blade I via a single disulfide bridge, rigidifying the whole domain.

Chimeric constructs made to define the structural features encoding the triple-helicase specificity of collagenases have shown that the CAT, the hinge, as well as the PEX domain possess important determinants, and that all three must be suitably arranged to act in concert to confer helicase specificity. The Glu(respectively Asn)209-Tyr-Asn-Leu segments of MMPs 1 and 8, preceding the active site helix and located in the corridor connecting the CAT and the PEX domains, seem to form a critical exosite for triple-helical collagen substrates.[75,114] Intriguingly, Tyr209 in MMPs 1 and 8 is preceded by a cis-bond, distinguishing these more powerful type I and II collagenases from other collagen-degrading MMPs such as MMP-2, MMP-13, or MMP-14.[75] Triple-helical collagen substrates might thus bind into the active site cleft of the cognate MMP under partial unwinding from P3 to P3′, bend at the 209 exosite, and interact with blade II of the PEX domain.[115] Other potential triple-helicase mechanisms have recently been discussed in detail by Overall.[7]

Structures of MMP–TIMP complexes

The wedge-shaped TIMP molecules consist of an N-terminal segment, an all-β-structure N-terminal part, an all-helical center, and a C-terminal β-turn structure.[87] The N-terminal half, consisting essentially of a closed five-stranded β-barrel[116] and the C-terminal half of the polypeptide chain form two opposing subdomains (Figure 12). The TIMP edge is formed by five sequentially separate chain segments, namely, the extended N-terminal segment Cys1-Pro5 and two flanking loops on either side provided by the N- and the C-terminal parts. The particularly remarkable features of TIMP-2 are the quite elongated sA-sB β-hairpin loop and the much longer, negatively charged flexible C-terminal tail.[88]

In inhibitory complexes with MMPs, the wedge-shaped TIMPs bind with their edge into the entire length of the active site cleft of their cognate MMPs,[87,88] under some rigidification of the participating loops.[118–120] The first five TIMP residues Cys1 to Pro5 bind to the MMP active site cleft in a quasi-substrate-like manner that is similar to the manner in which P1, P1′, P2′, P3′, and P4′ peptide substrate residues insert between the bulge and the S1′ wall-forming segment, forming five intermolecular inter-main-chain hydrogen bonds. Cys1 is located directly above the catalytic zinc, with its N-terminal α-amino nitrogen and its carbonyl oxygen coordinating the catalytic zinc together with the three imidazole rings from the cognate MMP. The threonine/serine side chain of the second TIMP residue extends, similar to the side chain of a P1′ peptide substrate residue, into the S1′ pocket of the cognate MMP, without filling this pocket properly.[16]

CATALYTIC MECHANISM

By combining the structural information obtained from the many peptide inhibitor–MMP complexes available,

Matrix metalloproteinases

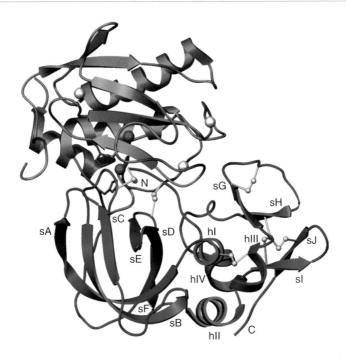

Figure 12 The MMP–TIMP complex formed between the MMP-3 catalytic domain (top, orange ribbon) and TIMP-1 (bottom, blue ribbon) (PDB code 1UEA),[87] with the CAT domain rotated around a horizontal axis for 90° compared with the standard orientation. Strands and helices of TIMP-1 are labeled, and the two zinc and three calcium ions are represented as pink and silver spheres. The wedge-shaped inhibitor binds with its edge made of six segments into the entire active site cleft of the cognate MMP-3, which, in this view, is directed toward the bottom. The N-terminal Cys1 of TIMP-2 is located on top of the catalytic zinc, ligating it through the amino and the carbonyl group, and the first five N-terminal residues bind to the active site in a substrate-like manner. Figure made with SETOR.[117]

a contiguous peptide substrate can be constructed, indicating the probable binding geometry of productive substrate-MMP encounter complexes[74] (Figures 3, 10). This substrate polypeptide chain is aligned in an extended manner to the continuous bulge-edge segment, under the formation of an antiparallel two-stranded β-pleated sheet, which on the right-hand side is expanded into a three-stranded mixed parallel–antiparallel β-sheet, owing to the additional alignment of the S1′ wall-forming segment. A bound peptide substrate (such as the hepta/hexapeptides shown in Figures 3 and 10) can form five and two inter-main-chain hydrogen bonds to both crossing-over MMP segments respectively. Similar to the reaction mechanism previously suggested for the more distantly related zinc endopeptidase thermolysin,[121] the MMP-catalyzed cleavage of the scissile peptide bond could proceed via a general-base mechanism[74] (Figure 13(a)): The P1–P1′ scissile peptide bond is appropriately placed and strongly polarized through interactions with the Ala182 carbonyl group and the catalytic zinc; the water molecule found in native MMP structures (Bode and Reinemer, unpublished results) as the fourth zinc ligand (at an equivalent position as found for free astacin[105] and adamalysin[106]) will be squeezed between this carbonyl and the nearby Glu219 carboxylate group. Owing to its interaction with the catalytic zinc, this water molecule is strongly acidified to transfer a proton to the Glu219 carboxylate group and to turn into a more nucleophilic hydroxyl ion, and is properly oriented to attack with its second lone pair electrons the electrophilic carbonyl carbon of the scissile peptide bond, under the formation of and transition through a tetrahedral hemiketal intermediate with a penta-coordinated catalytic zinc. In contrast to thermolysin[121] or astacin,[122] the MMPs do not possess any proteinaceous electrophilic group, which, besides the catalytic zinc, could further stabilize the oxyanion of the hemiketal intermediate. The quite positively charged and hence more stabilizing catalytic zinc might obviate the need for additional electrophils in MMPs.[68] As the equivalent carboxylate in thermolysin, the catalytic Glu219 carboxylate group might act as a proton shuttle, transferring one proton from the zinc-coordinated 'catalytic water' to the leaving amino group, followed by a break of the scissile peptide bond and collapse into products (Figure 13(a)).

The acidic pK_a around 4.3 of an ionizing group controlling the catalytic activity of matrilysin at low pH[51] has been assigned to the carboxylate group of the Glu219 equivalent. The low but detectable catalytic activities of matrilysin mutants, where this glutamic acid had been replaced by aspartic acid, cysteine, glutamine, and alanine residues, show that this glutamic acid residue plays an important role in catalysis but might not be essential, and the relatively unaltered pH-k_{cat}/K_m profiles question the pK_a assignment to its carboxylate group.[124] The pK_a of a water bound to a zinc ion ligated by three imidazole

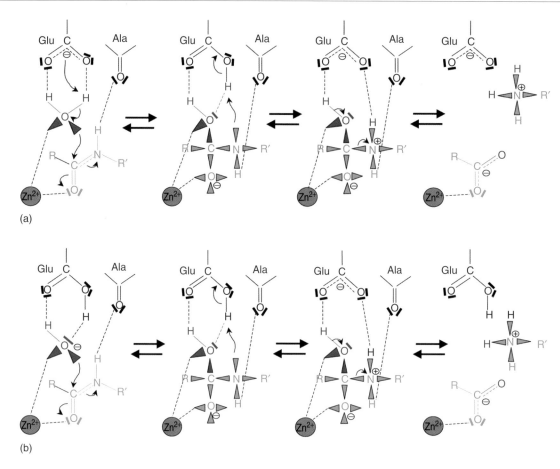

Figure 13 (a) Proposed general-base mechanism for the MMPs,[74] derived in analogy to the mechanisms proposed for thermolysin[121] and astacin.[123] Part of the peptide substrate, with the P1-(R) and the P1'-residue (R') flanking the scissile peptide bond, is shown in green, the 'catalytic' Glu219 carboxylate group and the Ala182 carbonyl group of the cognate MMP in black, the attacking water in red, and the catalytic zinc as a pink sphere. (b) Alternative general acid mechanism, starting with a Zn^{2+}-hydroxyl pair and a Glu219 carboxylic acid group.

groups is quite low. On the other hand, the carboxylate group of this Glu219 might rather possess an abnormally high pK_a, because of its hydrophobic environment and the lack of any charge-stabilizing hydrogen bond donor.[72] Therefore, alternatively, around neutral pH the Glu219 carboxylic acid group might be neutral/protonated, coexisting with the hydroxyl ion and acting as a general acid catalyst,[124] which stabilizes the negatively charged hemiketal formed by nucleophilic attack by the hydroxyl ion, shuttling its proton to the leaving group nitrogen (Figure 13(b)).

ACKNOWLEDGEMENT

We are indebted to a number of colleagues in Martinsried (R. Huber, P. Reinemer, F. Grams, F.-X. Gomis-Rüth, R. Lang, H. Brandstetter) and from abroad (H. Tschesche, G. Murphy, H. Nagase, K. Brew), who essentially helped to elucidate the MMP and TIMP structures, and to D. Jozic for her help in preparing some of the figures and the cover figure. The financial support by the SFB469, the Fonds der Chemischen Industrie and by the EU projects HPRN-CT-2002-00181, QLK3-CT-2002-02136, and SPINE contract QLG2-CT-2002-00988 is kindly acknowledged.

REFERENCES

1. AJ Barrett, N.D. Rawlings and JF Woessner Jr (ed.), *Handbook of Proteolytic Enzymes*, Academic Press, London (1998).
2. W Bode, F-X Gomis-Rüth and W Stöcker, *FEBS Lett*, **331**, 134–40 (1993).
3. W Stöcker, F Grams, U Baumann, P Reinemer, FX Gomis-Rüth, DB McKay and W Bode, *Protein Sci*, **4**, 823–40 (1995).
4. FX Gomis-Rüth, *Mol Biotechnol*, **24**, 157–202 (2003).
5. J Gross and CM Lapiere, *Proc Natl Acad Sci USA*, **48**, 1014–22 (1962).
6. CE Brinckerhoff and LM Matrisian, *Nat Rev Mol Biol*, **3**, 207–14 (2002).
7. CM Overall, *Mol Biotechnol*, **22**, 51–86 (2002).
8. E Morgunova, A Tuuttila, U Bergmann, M Isupov, Y Lindqvist, G Schneider and K Tryggvason, *Science*, **284**, 1667–70 (1999).
9. I Massova, LP Kotra, R Fridman and S Mobashery, *FASEB J*, **12**, 1075–95 (1998).

10. JF Woessner, in IM Clark (ed.), *Methods in Molecular Biology Vol. 151. Matrix Metalloproteinase Protocols*, Humana Press, Totowa, NJ, pp 1–23 (2000).
11. MD Sternlicht and Z Werb, *Ann Rev Cell Dev Biol*, **17**, 463–516 (2001).
12. V-M Kähäri and U Saarialho-Kere, *Ann Med*, **31**, 34–45 (1999).
13. G Murphy and J Gavrilovic, *Curr Opin Cell Biol*, **11**, 614–21 (1999).
14. H Nagase and JF Woessner Jr, *J Biol Chem*, **274**, 21491–94 (1999).
15. J Lohi, CL Wilson, JD Roby and WC Parks, *J Biol Chem*, **276**, 10134–44 (2001).
16. K Brew, D Dinakar Pandian and H Nagase, *Biochim Biophys Acta*, **1477**, 267–83 (2000).
17. W Bode and K Maskos, in IM Clark (ed.), *Methods in Molecular Biology Vol. 151. Matrix Metalloproteinase Protocols*, Humana Press, Totowa, NJ, pp 45–77 (2000).
18. AH Baker, DR Edwards and G Murphy, *J Cell Sci*, **115**, 3719–27 (2002).
19. K Maskos and W Bode, *Mol Biotechnol*, **20**, (2003); in press.
20. WG Stetler-Stevenson, *J Clin Invest*, **103**, 1237–41 (1999).
21. LJ McCawley and LM Matrisian, *Mol Med Today*, **6**, 149–56 (2000).
22. CE Brinckerhoff, JL Rutter and U Benbow, *Clin Cancer Res*, **6**, 4823–30 (2000).
23. G Bergers, R Brekken, G McMahon, TH Vu, T Itoh, K Tamaki, K Tanzawa, P Thorpe, S Itohara, Z Werb and D Hanahan, *Nat Cell Biol*, **2**, 737–44 (2000).
24. HD Foda and S Zucker, *Drug Discovery Today*, **6**, 478–82 (2001).
25. WC Parks and SD Shapiro, *Respir Res*, **2**, 10–19 (2001).
26. R Gururajan, J Grenet, JM Lahti and VJ Kidd, *Genomics*, **52**, 101–6 (1998).
27. G Velasco, AM Pendas, A Fueyo, V Knauper, G Murphy and C Lopez-Otin, *J Biol Chem*, **274**, 4570–76 (1999).
28. RPT Somerville, SA Oblander and SS Apte, *Genome Biol*, **4**, 216-1–216-11 (2003).
29. D Pei, T Kang and H Qi, *J Biol Chem*, **275**, 33988–97 (2000).
30. HE VanWart and H Birkedal-Hansen, *Proc Natl Acad Sci USA*, **87**, 5578–82 (1990).
31. H Sato, T Takino, Y Okada, J Cao, A Shinagawa, E Yamamoto and M Seiki, *Nature*, **370**, 61–65 (1994).
32. D Pei and SJ Weiss, *Nature*, **375**, 244–47 (1995).
33. Y Itoh, M Kajita, H Kinoh, H Mori, A Okada and M Seiki, *J Biol Chem*, **274**, 34260–66 (1999).
34. R Lang, M Braun, NE Sounni, JM Foidart, A Noel, W Bode, F Frankenne and K Maskos (2003); Submitted for publication.
35. SE Whitham, G Murphy, P Angel, HJ Rahmsdorf, B Smith, A Lyons, TJR Harris, JJ Reynolds, P Herrlich and AJP Docherty, *Biochem J*, **240**, 913–16 (1986).
36. GI Goldberg, SM Wilhelm, A Kronberger, EA Bauer, GA Grant and AZ Eisen, *J Biol Chem*, **261**, 6600–5 (1986).
37. GJ Barton, *Protein Eng*, **6**, 37–40 (1993).
38. K-I Shimokawa and H Nagase, in IM Clark (ed.), *Methods in Molecular Biology Vol. 151. Matrix Metalloproteinase Protocols*, Humana Press, Totowa, NJ, pp 275–304 (2000).
39. M Dioszegi, P Cannon and HE Van Wart, *Methods Enzymol*, **248**, 413–31 (1995).
40. H Tschesche, *Methods Enzymol*, **248**, 431–49 (1995).
41. H Nagase, *Methods Enzymol*, **248**, 449–70 (1995).
42. G Murphy and T Crabbe, *Methods Enzymol*, **248**, 470–84 (1995).
43. JF Woessner Jr, *Methods Enzymol*, **248**, 485–495 (1995).
44. V Knäuper and G Murphy, in IM Clark (ed.), *Methods in Molecular Biology Vol. 151. Matrix Metalloproteinase Protocols*, Humana Press, Totowa, NJ, pp 377–87 (2000).
45. KM-L Yeow, BW Phillips, PP Beaudry, KJ Leco, G Murphy and DR Edwards, in IM Clark (ed.), *Methods in Molecular Biology Vol. 151. Matrix Metalloproteinase Protocols*, Humana Press, Totowa, NJ, pp 181–189 (2000).
46. MJ Butler, M-P d'Ortho and SJ Atkinson, in IM Clark (ed.), *Methods in Molecular Biology Vol. 151. Matrix Metalloproteinase Protocols*, Humana Press, Totowa, NJ, pp 239–255 (2000).
47. R Vallon and P Angel, in IM Clark (ed.), *Methods in Molecular Biology Vol. 151. Matrix Metalloproteinase Protocols*, Humana Press, Totowa, NJ, pp 207–18 (2000).
48. GA Doyle, in IM Clark (ed.), *Methods in Molecular Biology Vol. 151. Matrix Metalloproteinase Protocols*, Humana Press, Totowa, NJ, pp 219–38 (2000).
49. LJ Windsor and DL Steele, in IM Clark (ed.), *Methods in Molecular Biology Vol. 151. Matrix Metalloproteinase Protocols*, Humana Press, Totowa, NJ, pp 191–205 (2000).
50. R Lang, A Kocourek, M Braun, H Tschesche, R Huber, W Bode and K Maskos, *J Mol Biol*, **312**, 731–42 (2001).
51. J Cha, MV Pedersen and DS Auld, *Biochemistry*, **35**, 15831–38 (1996).
52. TE Cawston, P Koshy and AD Rowan, in IM Clark (ed.), *Methods in Molecular Biology Vol. 151. Matrix Metalloproteinase Protocols*, Humana Press, Totowa, NJ, pp 389–97 (2000).
53. CG Knight, *Methods Enzymol*, **248**, 18–34 (1995).
54. MS Stack and RD Gray, *J Biol Chem*, **264**, 4277–81 (1989).
55. GB Fields, in IM Clark (ed.), *Methods in Molecular Biology Vol. 151. Matrix Metalloproteinase Protocols*, Humana Press, Totowa, NJ, pp 495–518 (2000).
56. Y Nagai, Y Masui and S Sakakibara, *Biochim Biophys Acta*, **445**, 521–24 (1976).
57. RD Gray and HH Saneii, *Anal Biochem*, **120**, 339–46 (1982).
58. H Nagase and GB Fields, *Biopolymers*, **40**, 399–416 (1996).
59. I Schechter and A Berger, *Biochem Biophys Res Commun*, **27**, 157–62 (1967).
60. S Netzel-Arnett, G Fields, H Birkedal-Hansen and HE Van Wart, *J Biol Chem*, **266**, 6747–55 (1991).
61. L Niedzwiecki, J Teahan, RK Harrison and RL Stein, *Biochemistry*, **31**, 12618–23 (1992).
62. S Netzel-Arnett, QX Sang, WGI Moore, M Navre, B Birkedal-Hansen and HE VanWart, *Biochemistry*, **32**, 6427–32 (1993).
63. A Mucha, P Cuniasse, R Kannan, F Beau, A Yiotakis, P Basset and V Dive, *J Biol Chem*, **273**, 2763–68 (1998).
64. RK Harrison, B Chang, L Niedzwiecki and RL Stein, *Biochemistry*, **31**, 10757–62 (1992).
65. EB Springman, H Nagase, H Birkedal-Hansen and HE VanWart, *Biochemistry*, **34**, 15713–20 (1995).
66. F Grams, M Crimmin, L Hinnes, P Huxley, M Pieper, H Tschesche and W Bode, *Biochemistry*, **34**, 14012–20 (1995).
67. WH Johnson, NA Roberts and N Borkakoti, *J Enzyme Inhibition*, **2**, 1–22 (1987).
68. A Zask, JI Levin, LM Killar and JS Skotnicki, *Curr Pharm Des*, **2**, 624–61 (1996).
69. KM Bottomley, WH Johnson and DS Walter, *J Enzyme Inhibition*, **13**, 79–101 (1998).

70. RP Becket and M Whittaker, *Exp Opin Ther Patents*, **8**, 259–82 (1998).
71. F Willenbrock, T Crabbe, PM Slocombe, CW Sutton, AJ Docherty, MI Cockett, M O'Shea, K Brocklehurst, IR Phillips and G Murphy, *Biochemistry*, **32**, 4330–37 (1993).
72. W Bode, P Reinemer, R Huber, T Kleine, S Schnierer and H Tschesche, *EMBO J*, **13**, 1263–69 (1994).
73. P Reinemer, F Grams, R Huber, T Kleine, S Schnierer, M Pieper, H Tschesche and W Bode, *FEBS Lett*, **338**, 227–33 (1994).
74. F Grams, P Reinemer, JC Powers, T Kleine, M Pieper, H Tschesche, R Huber and W Bode, *Eur J Biochem*, **228**, 830–41 (1995).
75. H Brandstetter, F Grams, D Glitz, A Lang, R Huber, W Bode, H-W Krell and RA Engh, *J Biol Chem*, **276**, 17405–12 (2001).
76. B Lovejoy, A Cleasby, AM Hassell, K Longley, MA Luther, D Weigl, G McGeehan, AB McElroy, D Drewry, MH Lambert and SR Jordan, *Science*, **263**, 375–77 (1994).
77. N Borkakoti, FK Winkler, DH Williams, A D'Arcy, MJ Broadhurst, PA Brown, WH Johnson and EJ Murray, *Nat Struct Biol*, **1**, 106–10 (1994).
78. T Stams, JC Spurlino, DL Smith, RC Wahl, TF Ho, MW Qoronfleh, TM Banks and B Rubin, *Nat Struct Biol*, **1**, 119–23 (1994).
79. JC Spurlino, AM Smallwood, DD Carlton, TM Banks, KJ Vavra, JS Johnson, ER Cook, J Falvo, RC Wahl, TA Pulvino, JJ Wendoloski and DL Smith, *Proteins: Struct Funct Genet*, **19**, 98–109 (1994).
80. PR Gooley, JF O'Connell, Al Marcy, GC Cuca, SP Salowe, BL Bush, JD Hermes, CK Esser, WK Hagmann, JP Springer and BA Johnson, *Nat Struct Biol*, **1**, 111–118 (1994).
81. B Lovejoy, AM Hassell, MA Luther, D Weigl and SR Jordan, *Biochemistry*, **33**, 8207–17 (1994).
82. MF Browner, WW Smith and AL Castelhano, *Biochemistry*, **34**, 6602–10 (1995).
83. JW Becker, Al Marcy, LL Rokosz, MG Axel, JJ Burbaum, PMD Fitzgerald, PM Cameron, CK Esser, WK Hagmann, JD Hermes and JP Springer, *Protein Sci*, **4**, 1966–76 (1995).
84. V Dhanaraj, Q-Z Ye, LL Johnson, DJ Hupe, DF Ortwine, JB Dunbar, JR Rubin, A Pavlovsky, C Humblet and TL Blundell, *Structure*, **4**, 375–86 (1996).
85. SR vanDoren, AV Kurochkin, W Hu, QZ Ye, LL Johnson, DJ Hupe and ER Zuiderweg, *Protein Sci*, **4**, 2487–98 (1995).
86. DR Wetmore and KD Hardman, *Biochemistry*, **35**, 6549–58 (1996).
87. FX Gomis-Rüth, K Maskos, M Betz, A Bergner, R Huber, K Suzuki, N Yoshida, H Nagase, K Brew, GP Bourenkov, H Bartunik and W Bode, *Nature*, **389**, 77–81 (1997).
88. C Fernandez-Catalan, W Bode, R Huber, D Turk, JJ Calvete, A Lichte, H Tschesche and K Maskos, *EMBO J*, **17**, 5238–48 (1998).
89. J-Y Li, P Brick, MC O'Hare, T Skarzynski, LF Lloyd, VA Curry, IM Clark, HF Bigg, BL Hazleman, TE Cawston and DM Blow, *Structure*, **3**, 541–49 (1995).
90. A Libson, A Gittis, I Collier, B Marmer, G Goldberg and EE Lattman, *Nat Struct Biol*, **2**, 938–42 (1995).
91. U Gohlke, F-X Gomis-Rüth, T Crabbe, G Murphy, AJP Docherty and W Bode, *FEBS Lett*, **378**, 126–30 (1996).
92. F-X Gomis-Rüth, U Gohlke, M Betz, V Knäuper, G Murphy, C Lopez-Otin and W Bode, *J Mol Biol*, **264**, 556–66 (1996).
93. V Dhanaraj, MG Williams, Q-Z Ye, F Molina, LL Johnson, DF Ortwine, A Pavlovsky, R Rubin, RW Skeean, AD White, C Humblet, DJ Hupe and TL Blundell, *Croat Chem Acta*, **72**, 575–91 (1999).
94. B Lovejoy, AR Welch, S Carr, C Luong, C Broka, RT Hendricks, JA Campbell, KAM Walker, R Martin, H Van Wart and MF Browner, *Nat Struct Biol*, **6**, 217–21 (1999).
95. I Botos, E Meyer, SM Swanson, V Lemaitre, Y Eeckhout and EF Meyer, *J Mol Biol*, **292**, 837–44 (1999).
96. AL Gall, M Ruff, R Kannan, P Cuniasse, A Yiotakis, V Dive, MC Rio, P Basset and D Moras, *J Mol Biol*, **307**, 577–86 (2001).
97. H Nar, K Werle, MM Bauer, H Dollinger and B Jung, *J Mol Biol*, **312**, 743–51 (2001).
98. S Rowsell, P Hawtin, CA Minshull, H Jepson, SM Brockbank, DG Barratt, AM Slater, WL McPheat, D Waterson, AM Henney and RA Pauptit, *J Mol Biol*, **319**, 173–81 (2002).
99. PA Elkins, YS Ho, WW Smith, CA Janson, KJ D'Alessio, MS McQueney, MD Cummings and AM Romanic, *Acta Crystallogr, Sect D: Biol Crystallogr*, **58**, 1182–92 (2002).
100. E Morgunova, A Tuuttila, U Bergmann and K Tryggvason, *Proc Natl Acad Sci USA*, **99**, 7414–19 (2002).
101. H Cha, E Kopetzki, R Huber, M Lanzendorfer and H Brandstetter, *J Mol Biol*, **320**, 1065–79 (2002).
102. PJ Kraulis, *J Appl Crystallogr*, **24**, 946–50 (1991).
103. EA Merritt and DJ Bacon, *Methods Enzymol*, **277**, 505–27 (1997).
104. JP Glusker, M Lewis, and M Rossi, *Crystal Structure Analysis for Chemists and Biologists*, VCH Verlagsgesellschaft GmbH, Weinheim, Germany (1994).
105. W Bode, F-X Gomis-Rüth, R Huber, R Zwilling and W Stöcker, *Nature*, **358**, 164–66 (1992).
106. F-X Gomis-Rüth, LF Kress and W Bode, *EMBO J*, **12**, 4151–57 (1993).
107. DS ViewerPro, Acelrys Inc, San Diego (2002).
108. L Chen, TJ Rydel, F Gu, CM Dunaway, S Pikul, KM Dunham and BL Barnett, *J Mol Biol*, **293**, 545–57 (1999).
109. BE Turk, LL Huang, ET Piro and LC Cantley, *Nature Biotech.*, **19**, 661–7 (2001).
110. El Chen, SJ Kridel, EW Howard, W Li, A Godzik and JW Smith, *J Biol Chem*, **277**, 4485–91 (2002).
111. SJ Kridel, E Chen, LP Kotra, EW Howard, S Mobashery and JW Smith, *J Biol Chem*, **276**, 20572–78 (2001).
112. SJ Deng, DM Bickett, JL Mitchell, MH Lambert, RK Blackburn, HL Carter III, J Neugebauer, G Pahel, MP Weiner and ML Moss, *J Biol Chem*, **275**, 31422–27 (2000).
113. G Murphy and V Knäuper, *Matrix Biol*, **15**, 511–518 (1997).
114. L Chung, K Shimokawa, D Dinakar Pandian, F Grams, GB Fields and H Nagase, *J Biol Chem*, **275**, 29610–17 (2000).
115. J Ottl, D Gabriel, G Murphy, V Knauper, Y Tominaga, H Nagase, M Kröger, H Tschesche, W Bode and L Moroder, *Chem Biol*, **7**, 119–32 (2000).
116. RA Williamson, G Martorell, MD Carr, G Murphy, AJ Docherty, RB Freedman and J Feeney, *Biochemistry*, **33**, 11745–59 (1994).
117. SV Evans, *J Mol Graph*, **11**, 134–38 (1993).
118. FW Muskett, TA Frenkiel, J Feeney, RB Freedman, MD Carr and R Williamson, *J Biol Chem*, **273**, 21736–43 (1998).

119 A Tuuttila, E Morgunova, U Bergmann, Y Lindqvist, K Maskos, C Fernandez-Catalan, W Bode, K Tryggvason and G Schneider, *J Mol Biol*, **284**, 1133–40 (1998).

120 B Wu, S Arumugam, G Gao, G Le, V Semenchenko, W Huang, K Brew and SR VanDoren, *J Mol Biol*, **295**, 257–68 (2000).

121 BW Matthews, *Acc Chem Res*, **21**, 333–40 (1988).

122 I Yiallouros, E Große-Berkhoff and W Stöcker, *FEBS Lett*, **484**, 224–28 (2000).

123 F Grams, V Dive, A Yiotakis, I Yiallouros, S Vassiliou, R Zwilling, W Bode and W Stöcker, *Nat Struct Biol*, **3**, 671–75 (1996).

124 J Cha and DS Auld, *Biochemistry*, **36**, 16019–24 (1997).

Serralysin

Ulrich Baumann

Departement für Chemie und Biochemie, University of Berne, Switzerland

FUNCTIONAL CLASS

Enzyme; Peptidase subclass EC 3.4.24, serralysin *Serratia marcescens* EC 3.4.24.40; metalloendoproteinase Clan MA(M) ≫ Family 10 ≫ Subfamily B ≫ Peptidase M10.051 (MEROPS Release 6.0, http://merops.sanger.ac.uk/). Serralysins generally catalyze the hydrolysis of peptide bonds with a hydrophobic residue in P1' and P2' position (nomenclature according to Reference 1). The hydrolysis rate decreases markedly with substrate chain lengths less than 4 to 6 amino acids. However, the precise substrate specificity may vary between the individual family members.

OCCURRENCE

Occurs in bacteria, enterobacteria. Serralysins are 50-kDa extracellular proteases secreted by gram-negative bacteria such as *Serratia* sp., *Pseudomonas* or *Erwinia* sp.

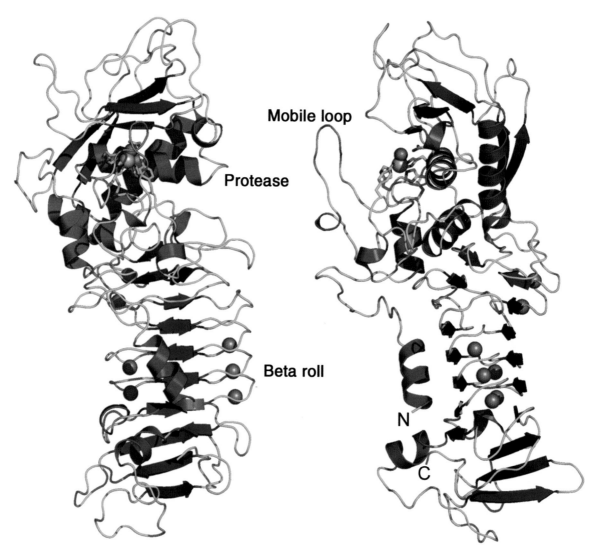

3D Structure Two orthogonal views of mature SMP (1SAT). The active site residues are shown as sticks. Metal ions are in gray. The mobile loop is drawn in orange. All figures were created with PYMOL (http://www.pymol.org).

148   HANDBOOK OF METALLOPROTEINS

BIOLOGICAL FUNCTION

The main function is probably bacterial nutrition. Serralysins from *S. marcescens* or *Pseudomonas aeruginosa* are considered to act as virulence factors by cleaving host-defense molecules such as immunoglobulins and interleukins or cause shedding of the host cell-surface proteins.[2–8]

AMINO ACID SEQUENCE INFORMATION

More than 32 sequences have been described from the following: bacteria (*Anabaena* sp., *Bacteroides fragilis*, *Caulobacter crescentus*, *Erwinia chrysanthemi*, *E. amylovora*, *Escherichia freundii*, *Nostoc punctiforme*, *Pectobacterium carotovorum*, *Prochlorococcus marinus*, *Proteus mirabilis*, *Pseudomonas aeruginosa*, *P. fluorescens*, *P. brassicacearum*, *P. fragi*, *P. tolaasii*, *P.* sp. 'TAC II 18', *Serratia marcescens*, *S. piscatram* (sp. E-15), *Sinorhizobium meliloti*, *Rhodobacter capsulatus*, *R. sphaeroides*, *Yersinia enterocolitica*, *Y. pestis*, *Y. pseudotuberculosis*, *Y. ruckeri*).

The sequence entries of the serralysins described more closely in the text are APR: S26699 (GenBank); PrtC: PRTC_ERWCH, accession number P16317 (SwissProt); SMP: SMP_SERMA, accession number Q06517 (SwissProt).

PROTEIN PRODUCTION, PURIFICATION, AND MOLECULAR CHARACTERIZATION

Serralysin can be isolated from the medium supernatant of bacterial cultures.[9] Purification procedures usually consist of ion exchange chromatography (diethylaminoethyl (DEAE) cellulose, High Q) and gel filtration. Recombinant serralysins have been produced mostly as extracellular proteins in *Escherichia coli* utilizing the native transport machinery of the original host, which is constitutively expressed from a separate plasmid.[10,11,22]

The serralysins possess an M_r of about 50 kDa. The N-terminal 220 amino acids harbor the active site that is characterized by the sequence fingerprint HEXXHXXGXXHP, where the first 10 amino acids represent the typical *metzincin* fingerprint[12,13] and the proline residue is characteristic of the serralysin family within the metzincin clan. The three histidines in this motif are zinc ligands (the fourth ligand is the catalytic water molecule) and the glutamic acid acts as the catalytic base (see below). The C-terminal half of the polypeptide chain contains six tandem repeats (residues ~330–390) of a calcium-binding glycine-rich sequence motif GGXGXDX(L/I/F)X, which is characteristic of the RTX-toxin family.[14–16] The last 50 C-terminal residues contain the secretion signal that is recognized by an ABC (ATP binding cassette) transporter secretion system.[17–19] Secretion of the inactive zymogen occurs, as for all members of the RTX-toxin family, in one step from the cytosol to the extracellular medium following the so-called *E. coli* hemolysin pathway. The zymogens are then activated autocatalytically in the presence of divalent cations in the external medium by cleavage of the N-terminal ~20-residue propeptide.

A conspicuous feature of the primary sequence of serralysins, and RTX toxins in general, is the absence of cysteine residues and hence the absence of disulfide bonds.

METAL CONTENT AND COFACTORS

There are seven to eight calcium ions and one zinc ion bound per molecule. No other cofactors are present.

ACTIVITY TEST

Activity is assayed in typical protease assays, for example, employing azocasein, resurofin casein, thiolester substrates, *para*-nitroanilides or fluorogenic substrates.[9,11,20,21] While the azocasein or resurofin casein assay is rather general, the substrate must be optimized for the particular serralysin in the case of the other methods. *Para*-nitroanilides are easy to handle but offer only little sensitivity due to the general importance of the P1′ residue for peptide bond hydrolysis by serralysins. Milk agar plates can be used to detect secreted protease.[19,22] As all metalloproteases, serralysins can be inhibited by mM concentrations of chelators, for example, ethylenediaminetetraacetate (EDTA) or *ortho*-phenanthroline. Hydroxamic acid derivatives are powerful inhibitors as well. Additionally, there are proteinaceous inhibitors (see below).

SPECTROSCOPY

Since there are only Ca^{2+} and Zn^{2+} as metal ions and no other cofactors, no characteristic absorption bands or other spectroscopic properties can be observed. For astacin, a homologous metalloprotease from the metzincin clan, the catalytic zinc was replaced by cobalt, nickel, mercury, and copper.[23] A characteristic band at 445 nm was interpreted to originate from charge-transfer transitions between the copper and Tyr149, a more distant fifth zinc ligand in the free astacin, which has its counterpart in the serralysins (Tyr216 in SMP). For spectroscopic studies on serralysin derivatives, see below.

X-RAY STRUCTURES

The first serralysin of which the X-ray structure was determined in 1993 was the alkaline protease from *Ps.*

Table 1 Serralysin structures in the Protein Data Bank (December 2002)

PDB ID	Space group	a	b	c	Alpha	Beta	Gamma	Resolution (Å)	Year	References
1SRP	$P2_12_12_1$	109.14	150.89	42.64	90.0	90.0	90.0	2.00	1993	25
1KAP	$P6_5$	106.90	106.90	97.00	90.0	90.0	120.0	1.64	1993	24
1SAT	$P2_12_12_1$	151.00	109.20	42.6	90.0	90.0	90.0	1.75	1994	26
1AF0	$P2_12_12_1$	151.00	109.20	42.60	90.0	90.0	90.0	1.80	1996	Unpub
1SMP	$P4_3$	108.83	108.83	87.95	90.0	90.0	90.0	2.30	1995	27
1AKL	$P2_12_12_1$	77.16	176.69	51.12	90.0	90.0	90.0	2.00	1995	28
1GO7	$P3_121$	102.70	102.70	121.20	90.0	90.0	120.0	2.10	2001	29
1GO8	$P3_121$	102.02	102.02	121.30	90.0	90.0	120.0	2.00	2001	29
1K7G	$P3_121$	101.89	101.89	122.01	90.0	90.0	120.0	2.00	2001	11
1K7I	$P3_121$	101.97	101.97	122.30	90.0	90.0	120.0	1.59	2001	11
1K7Q	$P3_121$	102.40	102.40	121.75	90.0	90.0	120.0	1.80	2001	11
1JIW	$P2_12_12$	75.63	118.43	91.99	90.0	90.0	90.0	1.74	2001	30

1SRP, 1SAT: *S. marcescens* 50-kDa protease; 1KAP, 1AKL: *Ps. aeruginosa* alkaline protease; 1AF0: *S. marcescens* 50-kDa protease hydroxamate inhibitor complex; 1SMP: *S. marcescens* 50-kDa protease – *E. chrysanthemi* inhibitor (proteinaceous) complex; 1GO7: *E. chrysanthemi* M226C-E189K double mutant; 1GO8: *E. chrysanthemi* M226L mutant; 1K7G: PrtC *E. chrysanthemi*; 1K7I: PrtC *E. chrysanthemi* Y228F mutant; 1K7Q: *E. chrysanthemi* E189A mutant; 1JIW: *Ps. aeruginosa* alkaline protease – alkaline protease inhibitor (proteinaceous) complex.

aeruginosa.[24] Since then, about a dozen serralysin structures have been published. Table 1 summarizes the current entries in the Protein Data Bank.

Overall description of the structure

Mature serralysins are elongated molecules (3D Structure) with approximate dimensions of $100 Å \times 35 Å \times 30 Å$. The structure is divided into two domains, the N-terminal protease domain and a C-terminal domain that binds calcium.

The protease domain

The N-terminal domain exhibits the typical metzincin fold consisting of a twisted β-sheet and three α-helices. The N-terminus extends down to the C-terminal domain and is in close proximity to the C-Terminus, approximately 50 Å away from the active site (3D Structure). Thus, proenzyme activation clearly requires some conformational rearrangement. The active site helix, which carries the first two zinc ligands His176 (numbering as in serralysin from *S. marcescens*, SMP) and His180 of the consensus motif (176)HEXXHXXGXXH(186)P, is covered by a five-stranded, predominantly parallel β-sheet. The *outmost* β-strand is antiparallel to the rest of the sheet and forms one edge of the active site cleft (therefore being called *edge strand*). The coordination of the catalytically active zinc ion is similar to that found in other metzincins.[12] The three histidines of the consensus sequence form three ligands; the coordination sphere is completed by a water molecule in the free proteases, which is the nucleophilic agent attacking the peptide bond. Furthermore, a more distant ligand is provided by the phenolic hydroxyl group of Tyr216 as was shown in the structure of the free metalloprotease from *S. marcescens* (SMP). In APR (PDB code 1KAP), the situation was initially different: a peptide was found bound in the active site cleft in the crystal structure, which probably originated from partial autodigestion during the crystallization process. This peptide, corresponding to residues P3, P2, and P1, binds along the S-sites (nomenclature according to reference 1) of the protease and coordinates to the zinc *via* its C-terminal carboxyl group, which acts as a bidentate ligand. The tyrosine moves away from the zinc and donates a hydrogen bond to one of the carboxylic oxygen of the peptide while the other is hydrogen bonded to the glutamic acid of the HEXXH motif. The peptide is bound to the edge of the β-sheet in an antiparallel manner with the characteristic backbone-to-backbone hydrogen bonds. On the other side, it is anchored to the side chain of Asn191 (SMP numbering, Asn203 in SMP) by two hydrogen bonds. An intriguing feature is the loop 191–196, which closes the active site when a ligand is bound. In the structure of free SMP, this loop is disordered. Also, in APR, this segment becomes disordered upon peptide removal. Thus, the serralysins show an induced-fit mechanism in which this loop in the ligand-free state is highly flexible and adopts a well-ordered conformation upon peptide binding. Tyr216 acts here as a switch: in the free protease it may be weakly bound to the zinc, in the substrate/inhibitor/product complexes its phenolic hydroxyl group donates a hydrogen bond to the carbonyl oxygen of the P1 residue and accepts a hydrogen bond from the amide backbone residue 193 (alanine in APR and SMP and PrtC) and anchors the loop 191 to 196 in this way. Tyr216 stabilizes in this way the developing partial negative charge on the carbonyl oxygen in the transition state and contributes in this way to catalysis (see below).

The C-terminal domain

The C-terminal domain of the serralyins forms an extended, mostly parallel, 21-stranded β-sandwich. Eight calcium ions are found within this domain in APR and seven in SMP/PrtC. This domain can formally be further subdivided into the interface domain, the parallel β-roll, and the secretion-signal domain.

The interface domain mediates the contacts between the proteolytic domain and the C-terminal part of the molecule. This interface is surprisingly hydrophilic with 12 internal water molecules mediating the contacts. These water molecules are conserved in all three crystal structures described here despite the different crystallization conditions, crystal forms, and the different molecular species. Three salt bridges, each formed by an arginine–aspartate pair, help stabilize the contacts between the proteolytic domain and the β-sandwich. Two calcium ions are found within the interface domain. They bind between the loops connecting parallel strands and are sevenfold coordinated (see below).

The central part of the C-terminal domain is built by the glycine-rich sequence repeats mentioned above (residues 345–386). It can be described as a parallel β-sandwich in which the connections between the strands are such that a right-handed helix of parallel β-strands is made. Each turn of this β-helix or 'parallel β-roll' is made by two sequence repeats; in other words, there are 18 residues per turn. The pitch of the helix is approximately 4.8 Å. The first six residues of each sequence motif, that is, GGXGXD, build the loops ('calcium turns') connecting the strands of the two sheets of the sandwich, and the remaining three residues form a short β-strand. The glycine-rich tandem repeats are a hallmark of the RTX toxins (Repeats in ToXins). The number of the repeats is variable and depends on the size of the protein: serralysins (M_r 50 kDa) possess 6, E. coli α-hemolysin (112 kDa) 16 and cyclolysin from B. pertussis (178 kDa) has about 50 of these repeats.

The central core of the sandwich is made by hydrophobic residues, especially by interdigitating leucines at position 8 of the sequence motif. The top and the bottom of the sandwich contain hydrophilic cavities that are occupied by calcium ions; these will be described in more detail below. The requirement for the individual amino acids in the glycine-rich repeats can be understood: Gly22 and Gly44 have backbone-dihedral angles that are disfavored for nonglycine residues, and a Cβ on the Gly1 would make a close contact to the carboxyl oxygen of Asp6. Position 8 is ideally a leucine as mentioned above, while residues 3 and 5 are generally small and hydrophilic (serine, asparagine). The 'parallel β-roll' is a highly regular structure, the short strands exhibit only a little twist, and the side chains of consecutive strands are stacked.

The secretion-signal domain is folded into a mixed parallel–antiparallel β-sandwich with rather irregular connections. It is the most flexible part of the molecule and is easily distorted by crystal contacts. The essential secretion signal is located approximately within the last 50 to 100 residues. This part does not show a significant sequence homology between the different proteins secreted by the hemolysin pathway with the exception of the last four residues, which have been shown to be important for the secretion process. A consensus sequence of the terminal 4 amino acids has been defined D/EUUU-COOH, where the 'U' is an unpolar amino acid, for example, valine, isoleucine or leucine.[31]

Metal site geometries

Catalytic site (Zn^{2+})

The only available apo-structures of a wild-type serralysin are those of S. marcescens metalloprotease (PDB entries 1SAT, 1SRP)[25,32] and one crystal form of APR (PDB code 1AKL)[28]. The zinc site in the unliganded enzyme is depicted in Figure 1. The zinc is approximately tetrahedrally coordinated by the NE2 atoms of the three histidines (distances between 1.9 and 2.1 Å, Table 2) and the catalytic water molecule/hydroxide ion (distance 1.9 to 2.2 Å). In the free serralysins, a fifth, more distant (2.8 Å) zinc ligand

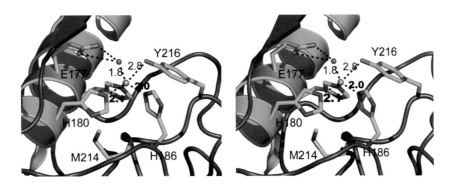

Figure 1 Catalytic center of SMP (PDB code 1SAT). The distances to the zinc are as indicated in Table 2.

Table 2 Zinc geometry in SMP

Partner	Distance (Å)	Angle (deg)
H176 – Zn^{2+}	2.07	
H180 – Zn^{2+}	2.13	
H186 – Zn^{2+}	1.99	
Zn^2 – HOH	1.84	
Y216 – Zn^{2+}	2.77	
H176 – Zn^{2+} – H180		101.6
H180 – Zn^{2+} – H186		102.4
H186 – Zn^{2+} – H176		114.7
H186 – Zn^{2+} – Y216		89.2
H180 – Zn^{2+} – Y216		167.9
H176 – Zn^{2+} – Y216		76.2
HOH – Zn^{2+} – Y216		71.3
H176 – Zn^{2+} – HOH		107.4
H180 – Zn^{2+} – HOH		98.5
H186 – Zn^{2+} – HOH		127.3

is Tyr216, which is located in the so-called met-turn. Upon substrate binding, this tyrosine swings away from the zinc and coordinates to the carbonyl oxygen of the cleavable peptide bond. Since there is no structure at true atomic resolution available, metal–ligand distances have to be interpreted with caution. The free Serratia protease was solved in the same unit cell twice independently and refined using different force fields (PDB entries 1SAT and 1SRP). The distances between Zn^{2+} and NE2 atoms are remarkably similar in both structures; the only exception is the coordinating water molecule that is 1.84 and 2.10 Å apart from the zinc in entries 1SAT and 1SRP respectively. The average distance found in tridentate tetrahedral model complex compounds is about 1.85 Å for the hydroxo complexes.[33]

Ca^{2+} sites

Hepta-coordinated Ca^{2+}: Two such sites are found in the intermediate domain that connects the N-terminal metzincin and parallel β-roll domains (Figure 2). These calcium ions are located next to each other in loops between β-strands and are bridged by an aspartic acid, similar to those Ca^{2+} ions that are located in the parallel β-roll domain (see below). However, they show the sevenfold pentagonal-bipyramidal coordination

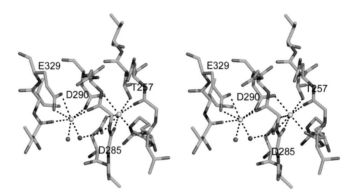

Figure 2 Hepta-coordinated calcium ions in the interface domain.

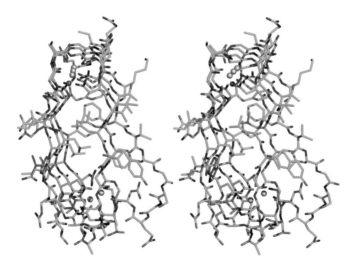

Figure 3 The parallel β-roll viewed along the helix axis.

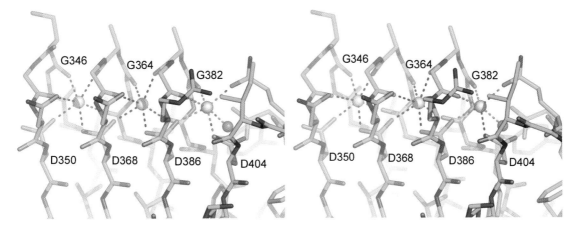

Figure 4 Hexa-coordinated calcium ions in the parallel β-roll domain. The periodicity of the β-roll is 18 residues per turn. Ca^{2+} ions bridge the aspartic acids from two GGXGXDXLX motifs.

that is found most frequently for calcium in proteins, whereby the oxygen ligands stem entirely from the protein for the more buried Ca1. Ca2, which is more at the edge of the β-sheet, has two water molecules as ligands. The $O–Ca^{2+}$ distances vary from 2.3 to 2.6 Å.

Hexa-coordinated Ca^{2+} ions: These Ca^{2+} ions are bound to the loops in the β-roll motif that is formed by the tandem GGXGXDXLX motifs (Figures 3 and 4). The calcium ions bind within the turns connecting the β-strands and are octahedrally coordinated. Each turn forms two half-sites with the first loop contributing carbonyls of Gly2, Gly4, and one carboxyl oxygen of Asp6 (*anti*-lone pair). The second loop, which is 18 residues further toward the C-terminus, contributes carbonyls of Gly1″, X3″, and a carboxyl oxygen of Asp6″ (*syn*-lone pair). The calcium ions are generally completely internal and have very slow exchange rates, with the exception of those located at the edges of the 'parallel β-roll', which have water molecules filling up their coordination sphere.

COMPLEX STRUCTURES

Serralysin has been crystallized in the presence of Cbz-Leu-Ala-NHOH (PDB code 1AF0, Baumann, unpublished results). Figure 5 shows the binding mode. The Zn^{2+} is now fivefold, coordinated (three histidines and the hydroxamic acid chelator) with an approximate trigonal-bipyramidal geometry. The inhibitor's main chain runs antiparallel to the edge strand of the β-sheet and forms the characteristic hydrogen bonds to the main chain of Ala136. Similar binding modes have been observed for peptide complexes of APR.

Frequently, an additional gene for a periplasmic 10-kDa inhibitor is found on the same operon containing the protease gene and the three genes encoding the transport machinery.[34] Its physiological function is probably to inactivate protease molecules that enter the periplasm through leaky outer membranes or get liberated during the secretion process. The reported inhibition constants K_i vary widely from $\mu M^{34,35}$ to $pM.^{21}$ It seems to be likely that the low-affinity constants are at least partially artifacts due to inhomogeneous N-termini.

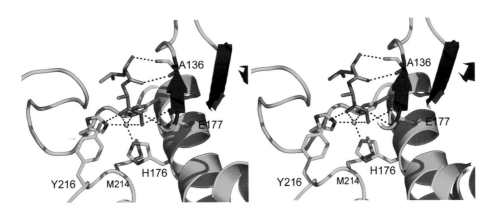

Figure 5 SMP-Cbz-Leu-Ala-NHOH complex (PDB code 1AF0). The inhibitor binds as antiparallel β-strand to the edge strand of the β-sheet. The hydroxamic acid acts as bidentate ligand and replaces the catalytic water molecule.

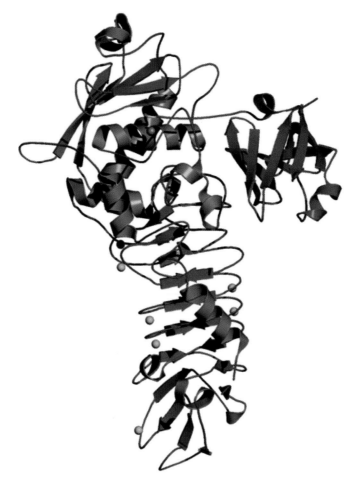

Figure 6 Cartoon representation of the APR–APRin complex (PDB code 1JIW). This is the cognate complex between the alkaline protease (red and blue) and its inhibitor (magenta) from *Ps. aeruginosa*. Zinc is shown in yellow, calcium ions in cyan.

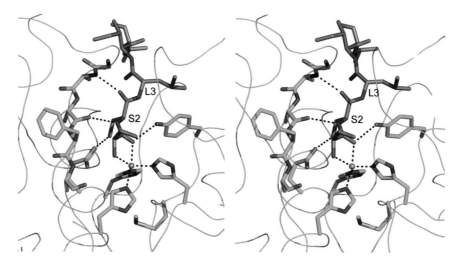

Figure 7 Insertion of the N-terminal trunk into the active site. Interactions between the amino-terminal residue and the Zn^{2+} and H-bond patterns are indicated.

While the hydroxamic acid inhibitor binds in a way that mimics binding of substrates to the S-sites of the protease, those naturally occurring proteinaceous inhibitors block access to the active site by occupying the S'-sites (Figure 6) *via* the N-terminal trunk (Figure 7), as shown in the example of the complex between APR and its cognate inhibitor APRin[30] or the complex between the inhibitor from *E. chrysanthemi* and SMP.[27] The inhibitor folds into a compact eight-stranded antiparallel β-barrel with up–down topology. The major contacts are made by the five N-terminal residues that insert into the active site cleft. The amino-terminal residue chelates to the Zn^{2+} with its carbonyl oxygen and amino group, replacing the catalytic water molecule. The second residue, an invariant serine, occupies the S1' pocket and forms a hydrogen bond to the glutamic acid of the HEXXH motif. Leu3 of the inhibitor fits nicely into a pocket formed by residues Tyr216, Trp217, Tyr169, and Ala192 of the protease and the turn 64–67 of the inhibitor. Further hydrogen bonds are formed by the main chain of the inhibitor's Leu3 and the edge strand of the protease. The complete hydrogen bonding pattern of the N-terminal residues and the zinc chelation by the amino-terminus and the carbonyl oxygen of the first amino acid are completely conserved in the nonhomologous tissue matrix metalloprotease inhibitor family (TIMP)–MMP complexes.[36]

The importance of the N-terminal five residues for the tight binding ($K_i = 4$ pM) in the APR–APR in complex has been underlined by site-directed mutagenesis.[21]

FUNCTIONAL ASPECTS

Reaction mechanism

In general, Serralysins hydrolyze polypeptides preferentially at small-to-medium-sized hydrophobic amino acids in P1' position. The reaction mechanism was initially formulated analogous to the one proposed for thermolysin.[37,38] This mechanism has the following key features: (i) acidification of the zinc-bound water molecule, enhancing its nucleophilicity; (ii) the glutamic acid of the HEXXH motif acts as catalytic base; (iii) enhanced polarization of the C=O group of the cleavable peptide bond by coordination to the zinc, and (iv) stabilization of the transition state by the distant zinc-ligand Tyr216. Later, a reversed mechanism was proposed in which Tyr216 acts as catalytic base.[39] This mechanism provides no convincing role for the conserved glutamic acid of the HEXXEH motif. Furthermore, Tyr216 is not conserved amongst the metzincins; it only occurs in serralysins and astacins. Site-directed mutagenesis studies on PrtC from E. chrysanthemi have provided clear evidence that the glutamic acid (Glu177) is the catalytic base and Tyr216 stabilizes the transition state by forming an H-bond to the former carbonyl oxygen in the tetrahedral adduct.[11] An analogous study on the crayfish protease astacin led to the same results.[20]

Conserved methionine

The methionine is located 20 to 40 amino acids downstream of the HEXXH motif and lies in a region that is not conserved between the various metzincin subfamilies. It was first identified by Bode et al.[12] who gave the name 'metzincin' to the whole protease clan. The methionine resides under the base of the pyramid of the three histidines that coordinate the zinc. The SD sulfur atom of the methionine and the catalytic zinc are about 6 Å apart, too far for any direct interaction. Earlier studies reported the replacement of the methionine by seleno-methionine in two matrix metalloproteases with slightly different results: in one case no change in catalytic activity was noticed;[40] in the other an increase of the K_m value by a factor of 2 was reported.[41] A recent publication[29] on PrtC from the serralysin family reports the directed mutagenesis of M226 (prtC numbering) versus alanine, isoleucine, and leucine. For the wild-type enzyme, the K_{cat} and K_m values are $0.023 \, s^{-1}$ and 3.1 mM using AlaAlaAla-para-nitroanilide as assay. While the alanine and isoleucine mutants show a decrease in K_{cat} as well as an increase in K_m, the leucine mutant exhibits a moderate decrease in K_{cat} by about 30%. The crystal-structure analysis revealed slight distortions in the zinc geometry. A further inactive double mutant M226C/E189K, where the glutamic acid 189 from the HEXXH motif was changed to a lysine, showed an even further increased rearrangement of the zinc-binding histidines.

Metal exchange

A systematic study by Park and Ming[42] has been reported recently. Mn^{2+}, Co^{2+}, Ni^{2+}, Cu^{2+}, and Cd^{2+} derivatives were produced. Of those, the cobalt-, copper-, and nickel-substituted enzymes were more active against benzoyl-arg-para-nitroanilide than the original zinc-containing protease. The cobalt and copper derivatives, especially, had an increased activity by a factor of 33 and 26 respectively. However, in a caseinolytic assay, the zinc enzyme had the highest activity.

The electronic spectrum of Co^{2+}-serralysin shows a maximum at 506 nm ($78 \, M^{-1} \, cm^{-1}$) and shoulders at 470 and 530 nm, which are assigned to the d–d transitions of the Co^{2+} center. From the similarity with the Co^{2+}-astacin spectrum, a trigonal-bipyramidal geometry can be inferred. Cu^{2+}-serralysin shows three absorption bands at 725 nm ($128 \, M^{-1} \, cm^{-1}$, d–d transition), at 320 nm ($1070 \, M^{-1} \, cm^{-1}$), and 450 nm ($1283 \, M^{-1} \, cm^{-1}$). The latter, more intense absorptions are assigned to tyrosinate-to-copper charge-transfer transitions.

REFERENCES

1   I Schechter and A Berger, *Biochem Biophys Res Commun*, **27**, 157–62 (1967).

2   JA Hobden, *DNA Cell Biol*, **21**, 391–96 (2002).

3   AM Firoved and V Deretic, *J Bacteriol*, **185**, 1071–81 (2003).

4   C Coker, CA Poore, X Li and HL Mobley, *Microbes Infect*, **2**, 1497–505 (2000).

5. CM Pillar, LD Hazlett and JA Hobden, *Curr Eye Res*, **21**, 730–39 (2000).
6. PS Estrellas Jr, LG Alionte and JA Hobden, *Curr Eye Res*, **20**, 157–65 (2000).
7. P Vollmer, I Walev, S Rose-John and S Bhakdi, *Infect Immun*, **64**, 3646–51 (1996).
8. JL Kadurugamuwa and TJ Beveridge, *J Bacteriol*, **177**, 3998–4008 (1995).
9. H Maeda and K Morihara, *Methods Enzymol*, **248**, 395–413 (1995).
10. Y Suh and MJ Benedik, *J Bacteriol*, **174**, 2361–66 (1992).
11. T Hege and U Baumann, *J Mol Biol*, **314**, 187–93 (2001).
12. W Bode, FX Gomis-Ruth and W Stocker, *FEBS Lett*, **331**, 134–40 (1993).
13. W Bode, F Grams, P Reinemer, FX Gomis-Ruth, U Baumann, DB McKay and W Stocker, *Adv Exp Med Biol*, **389**, 1–11 (1996).
14. RA Welch, *Curr Top Microbiol Immunol*, **257**, 85–111 (2001).
15. MB Chancellor, *Urology*, **57**, 106–7 (2001).
16. ET Lally, RB Hill, IR Kieba and J Korostoff, *Trends Microbiol*, **7**, 356–61 (1999).
17. P Delepelaire and C Wandersman, *Mol Microbiol*, **5**, 2427–34 (1991).
18. P Delepelaire, *J Biol Chem*, **269**, 27952–57 (1994).
19. P Delepelaire and C Wandersman, *J Biol Chem*, **264**, 9083–89 (1989).
20. I Yiallouros, E Grosse Berkhoff and W Stocker, *FEBS Lett*, **484**, 224–28 (2000).
21. RE Feltzer, RD Gray, WL Dean and WM Pierce Jr, *J Biol Chem*, **275**, 21002–9 (2000).
22. P Delepelaire and C Wandersman, *J Biol Chem*, **265**, 17118–25 (1990).
23. FX Gomis-Ruth, F Grams, I Yiallouros, H Nar, M Kusthardt, R Zwilling, W Bode and W Stocker, *J Biol Chem*, **269**, 17111–17 (1994).
24. U Baumann, S Wu, KM Flaherty and DB McKay, *EMBO J*, **12**, 3357–64 (1993).
25. K Hamada, Y Hata, Y Katsuya, H Hiramatsu, T Fujiwara and Y Katsube, *J Biochem (Tokyo)*, **119**, 844–51 (1996).
26. U Baumann, *J Mol Biol*, **242**, 244–51 (1994).
27. U Baumann, M Bauer, S Letoffe, P Delepelaire and C Wandersman, *J Mol Biol*, **248**, 653–61 (1995).
28. H Miyatake, Y Hata, T Fujii, K Hamada, K Morihara and Y Katsube, *J Biochem (Tokyo)*, **118**, 474–9 (1995).
29. T Hege and U Baumann, *J Mol Biol*, **314**, 181–86 (2001).
30. T Hege, RE Feltzer, RD Gray and U Baumann, *J Biol Chem*, **9**, 35087–92 (2001).
31. JM Ghigo and C Wandersman, *J Biol Chem*, **269**, 8979–85 (1994).
32. U Baumann, *J Mol Biol*, **242**, 244–51 (1994).
33. CE MacBeth, BS Hammes, VG Young Jr and AS Borovik, *Inorg Chem*, **40**, 4733–41 (2001).
34. S Letoffe, P Delepelaire and C Wandersman, *Mol Microbiol*, **3**, 79–86 (1989).
35. KH Bae, IC Kim, KS Kim, YC Shin and SM Byun, *Arch Biochem Biophys*, **352**, 37–43 (1998).
36. FX Gomis-Ruth, K Maskos, M Betz, A Berger, R Huber, K Suzuki, N Yoshida, H Nagase, K Brew, GP Bourenkov, H Bartunik and W Bode, *Nature*, **389**, 77–81 (1997).
37. HM Holden and BW Matthews, *J Biol Chem*, **263**, 3256–60 (1988).
38. DE Tronrud, SL Roderick and BW Matthews, *Matrix Suppl*, **1**, 107–11 (1992).
39. WL Mock and J Yao, *Biochemistry*, **36**, 4949–58 (1997).
40. MW Qoronfleh, TF Ho, PG Brake, TM Banks, TA Pulvino, KF Vavra, F Falvo and RB Cicarelli, *J Biotechnol*, **39**, 119–28 (1995).
41. M Pieper, M Betz, N Budisa, FX Gomis-Ruth, W Bode and H Tschesche, *J Protein Chem*, **16**, 637–50 (1997).
42. HI Park and LJ Ming, *J Biol Inorg Chem*, **7**, 600–10 (2002).

Leishmanolysin

Peter Metcalf[†] and Robert Etges[‡]

[†]School of Biological Sciences, University of Auckland, New Zealand
[‡]IBFB Pharma GmbH, Leipzig, Germany

FUNCTIONAL CLASS

Enzyme; zinc proteinase, EC 3.4.24.36; metzincin subclass neutral metalloproteinase, MEROPS peptidase family M8. Leishmanolysin was previously termed gp63, promastigote surface proteinase (PSP) or major surface proteinase (MSP). Leishmanolysin is a major polymorphic surface glycoprotein antigen of *Leishmania* promastigotes.

OCCURRENCE

Leishmanolysin is the major surface glycoprotein antigen of the infective form of *Leishmania*, a widespread pathogenic human parasitic protozoan carried by sand flies.[1] *Leishmania* alternates between the promastigote form found in the sandfly vector and the amastigote form that occurs within macrophages of mammalian hosts, in particular,

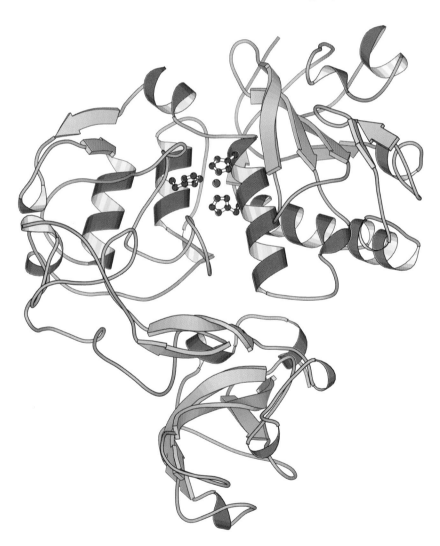

3D Structure Ribbon diagram depicting the atomic structure of the zinc proteinase leishmanolysin, PDB code 1LML. The structure consists of three domains. The active site zinc atom (purple sphere) is positioned between two histidines in the N-terminal catalytic domain (top right) and a third histidine in the central domain (top left). The cysteine-rich C-terminal domain is shown at the bottom of the figure. Prepared with the program MOLSCRIPT.[28]

the macrophages from the granulomatous skin lesions that are the most common symptoms of human infection. Leishmanolysin is a major component of the outer membrane of *Leishmania* promastigotes,[2] attached by a C-terminal glycophosphatidylinositol (GPI) anchor.[3] In *L. major*, there are $\sim 5 \times 10^5$ leishmanolysin molecules per promastigote cell.[2] Leishmanolysin homologs occur in all known *Leishmania* species, and also in the protozoan parasite *Trypanosoma*. Soluble intracellular forms of leishmanolysin, occurring in much smaller amounts than the surface antigen form, have been described within the amastigote lysosomal compartment.[4,5] More distant homologs have been identified in *Crithidia fasciculata* and in *Herpetomonas samuelpessoai*, and recently in the genomes of humans and mice, *Drosophila*, *Dictyostelium*, and *Caenorhabditis elegans*, and in the bacterium *Deinococcus radiodurans*, suggesting that leishmanolysin homologs, like other metzincin class metalloproteinases, may occur in most branches of life.

BIOLOGICAL FUNCTION

The function of leishmanolysin in *Leishmania* has been the subject of active research since the protein was first identified.[6] Early work established that leishmanolysin was the major surface glycoprotein antigen of *L. major* promastigotes,[7] together with lipophosphoglycans (LPG) and glycoinositolphospholipid (GIPL).[8] Leishmanolysin is a nonspecific neutral metalloproteinase, active *in situ* on promastigotes.[9] Since leishmanolysin forms a significant part of the promastigote cell surface, it is thought most likely to play a role in the insect or in the early stages of mammalian infection. Expression is much reduced in amastigotes, the form of the parasite that occurs in mammalian host macrophages in chronic leishmaniasis. Recently, viable *L. major*, with the 20-kb region containing all seven leishmanolysin genes deleted, has been used to investigate leishmanolysin function.[10] Growth in sand flies was normal, but the symptoms of mouse infection were delayed and promastigotes were more sensitive to complement-mediated lysis. These effects could be reversed by expression of a cloned leishmanolysin gene in the knockout strain. Thus, *L. major* leishmanolysin appears to be involved in the defence against complement-mediated attack early in the infection. Leishmanolysin polymorphism is thought to be due to mammalian host-selection pressure,[11] which is consistent with this idea. Leishmanolysin may limit complement-mediated lysis by inactivating the bound C3b[12,13] and enhance infectivity by binding to macrophage integrins.[14] The substrates of leishmanolysin *in vivo* have not been determined. Very little is known about the function of leishmanolysin homologs in other species.

AMINO ACID SEQUENCE INFORMATION

L. major leishmanolysin (SWISSPROT[15] code GP63_LEIMA) is synthesized as an inactive precursor protein of 602 amino acid residues. The mature active proteinase that occurs on the surface of *L. major* promastigotes has 478 residues and includes three N-linked carbohydrate groups, the C-terminal GPI anchor, and a single zinc atom bound at the active site. Limited proteolytic processing involves the cleavage of an N-terminal signal peptide and a regulatory propeptide, resulting in the mature polypeptide chain starting at Val100. At the C-terminus, a hydrophobic sequence of 25 amino acid residues is replaced with a GPI structure, which anchors the protein to the promastigote outer membrane. In leishmanolysin from *L. major*, the anchor is attached to the C-terminal N577 by an ethanolamine phosphate bridge. The chemical structure of the *L. major* GPI anchor has been determined.[16] The HEXXH sequence motif, characteristic of zinc proteinases, occurs at residue 264 in the N-terminal catalytic domain. The two motif histidines are ligands for the active site zinc atom. The third histidine zinc ligand is in the central domain of the molecule at residue 334 and the metzincin-defining methionine is at residue 345. There are 18 cysteines, 12 of which occur in the C-terminal cysteine-rich domain, and all cysteines form disulfide bonds. The 18 cysteines, the active site HEXXH motif, the third histidine zinc ligand, and the metzincin methionine are conserved in leishmanolysin homologs. There are carbohydrate attachment sites at Asn300 in the central domain and Asn407 and Asn534 in the C-terminal domain. Neither the *L. major* C-terminal GPI attachment site at Asn577 nor the carbohydrate attachment sites are conserved, even in other *Leishmania* species. Amino acid sequences and multiple sequence alignments for leishmanolysin homologs are listed under the MEROPS database[17] peptidase family M8.

PROTEIN PRODUCTION, PURIFICATION, AND MOLECULAR CHARACTERIZATION

Amphiphilic detergent-solubilized leishmanolysin can be purified in $\sim$40-mg yields from membrane preparations derived from 30-L ($\sim 2 \times 10^{12}$ cells) stationary-phase cultures of the lipophosphoglycan (LPG)-deficient *L. major* strain LRC-L119.[2] The time required to grow sufficient promastigotes ($\sim 2 \times 10^{12}$ cells, 1 week per 5-L batch, six batches pooled) is an important factor in protein production. The GPI anchor is enzymatically removed from the detergent-solubilized protein by incubation with phosphatidylinositol-specific phospholipase C (PI-PLC). Soluble protein with the diacylglycerol lipid part of the GPI anchor cleaved is extracted by two cycles of Triton X-114 phase separation and the detergent-free protein further purified by MonoQ anion exchange chromatography. Enzymatically active fractions are collected. The yield of soluble protein is primarily limited by the low (~20%) yield of the PI-PLC cleavage step of the purification protocol. The soluble protein is stored at 4 °C, is stable, and active for long periods.

Characterization

Leishmanolysin prepared by the method described in the previous section shows two closely spaced bands at ~55 kDa on SDS-PAGE, which are also observed by matrix-assisted laser desorption (MALDI) mass spectrometry.[18] Both PNGase F and Endo H deglycosylation produced a single band at ~52 kDa showing that the doublet was a result of glycosylation. The deglycosylated protein was proteolytically active. The GPI cross-reacting determinant (CRD) epitope[19] was present in the crystalline protein used for atomic structure determination.[18] Isoelectric focusing gels revealed a series of bands in the pI range 4.6 to 5.0.

ACTIVITY AND INHIBITION TESTS

The protease activity of leishmanolysin is most easily determined in solution with an azocasein assay.[9] Azocasein is digested by the isolated enzyme as well as by intact *Leishmania* promastigotes. More recently, cleavage of the synthetic peptide substrate LIAY/LKKAT has been analyzed by high-pressure liquid chromatography (HPLC) separation of the cleavage products.[20] Radioiodination of the tyrosine residue does not affect cleavage, allowing rapid analysis of the reaction by thin-layer chromatography.[21] An internally quenched fluorogenic peptide was described by Bouvier *et al*.[22] Leishmanolysin exhibits an anomalous behavior on casein micelles, presumably due to the presence of the GPI anchor: the amphiphilic form rapidly clears a 1.5% powdered milk solution, whereas the phospholipase-cleaved hydrophilic form of the enzyme slowly causes the casein micelles to precipitate.[2] The milk assay may be conveniently conducted in microtiter plates and analyzed by nephelometry.

Alternatively, leishmanolysin activity can be analyzed by zymographic methods in which protein substrates such as fibrinogen or gelatin are copolymerized in polyacrylamide gels. Leishmanolysin is electrophoresed in the presence of sodium dodecylsulfate without prior heating or reduction of the enzyme. Activity is revealed as a clear band on a Coomassie Blue–stained substrate protein background.[2]

Leishmanolysin is inhibited by zinc chelators like 1,10-phenanthroline and bathophenanthroline at millimolar concentrations, but not by ethylenediaminetetraacetic acid (EDTA). Bouvier *et al*.[20] identified a hydroxamate dipeptide able to inhibit *L. major* leishmanolysin with a $K_i =$ 17 μM. More recently, Bangs *et al*.[23] screened a series of peptidomimetic inhibitors designed for mammalian matrix metalloproteinases (MMPs) and identified several with IC_{50} values in the low micromolar range. Despite the similarity between leishmanolysin and mammalian collagenases in terms of active site structure and inhibitor profile, the enzyme does not cleave native collagen type 1. Data for leishmanolysin substrates is tabulated at the BRENDA.[24]

X-RAY STRUCTURE

Crystallization, structure determination

The crystallization of leishmanolysin[18] proved to be time-consuming and difficult, with results differing markedly between protein preparations for unknown reasons. Monoclinic and tetragonal crystal forms were grown by the standard vapor-diffusion hanging-drop method using the precipitant methyl pentanediol (MPD) at approximately pH 9. Both crystal forms were obtained under similar conditions. Protein was recycled in some cases from hanging drops and macroseeding methods were eventually used to obtain large crystals suitable for X-ray crystallography. Monoclinic crystals (space group C2, $a = 107.2$ Å, $b = 90.6$ Å, $c = 70.6$ Å, $\beta = 110.6°$) grown using macroseeds added to the drop after 2 days, diffract to at least 1.9 Å with synchrotron radiation. A single tetragonal crystal (space group $P4_12_12$, $a = b = 63.6$ Å, $c = 251.4$ Å) that diffracted to 2.7 Å with synchrotron radiation was grown using recycled protein without seeding. Data was collected with nonfrozen crystals and little radiation decay was evident. The structure was originally determined with data from native and trimethyl lead-acetate derivatized crystals in both crystal forms, using cross-crystal averaging and density modification methods.[25] Each crystal form contains a single molecule in the asymmetric unit. The monoclinic form structure coordinates and X-ray data have PDB[26] code 1LML. The structure has been subsequently determined using automatic methods with only the monoclinic form data.[27] The lower resolution tetragonal form structure is not significantly different.

Overall description of the structure

Leishmanolysin is a compact molecule containing predominantly β-sheet secondary structure with overall dimensions $45 \times 50 \times 70$ Å. The primary sequence folds into three consecutive domains (see Figure 1).

The N-terminal domain has a fold similar to the catalytic modules of zinc proteinases,[29] with a catalytic zinc atom and an active site helix containing the two histidines of the zinc proteinase sequence motif HEXXH. The following two domains have novel folds. There are

Leishmanolysin

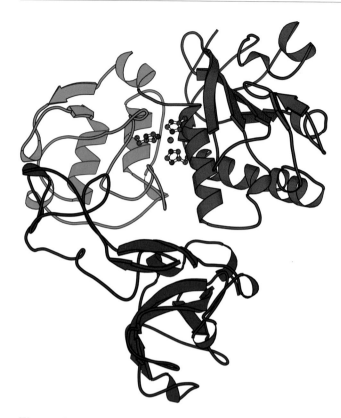

Figure 1 Schematic representation of leishmanolysin showing the three domains (red, catalytic domain; green, central domain; blue C-terminal domain) and the active site. Produced using the program MOLSCRIPT.[28]

deep surface indentations at domain interfaces, which join the active site cleft between the N-terminal domain and the central domain of the molecule. The calculated surface charge distribution shows a large region of negative charge surrounding the active site cleft, but charge distribution features suggesting the orientation of the molecule on the promastigote membrane are not evident.

N-terminal domain

The N-terminal domain of leishmanolysin shares the basic topology of zinc proteinase catalytic domains, which consists of two helices (A and B) packed against one side of a five-stranded twisted β-sheet. The HEXXH motif is on the second helix B, and the two histidine side chains are ligands of the active site zinc atom. The N-terminal domain of leishmanolysin has these basic features (Figure 2) and also includes 40- and 35-residue inserted 'flaps' that cover the side of the β-sheet opposite the helices (Figure 3). The 40-residue 'flap' replaces the second strand of the central sheet in the conserved topology. There are two disulphide bonds, one within the 35-residue 'flap' and the other linking the 40-residue flap to strand

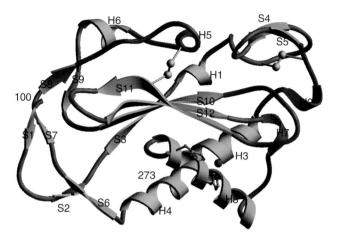

Figure 2 Schematic representation of the N-terminal catalytic domain of leishmanolysin showing the secondary structure elements, the active site zinc, and the two histidine side chains from the active site helix H8 that bind the zinc atom. The figure was made using the program RIBBONS.[30]

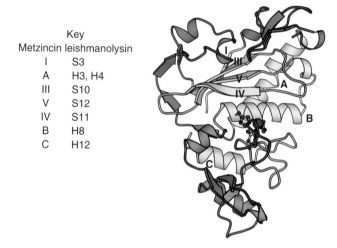

Figure 3 Schematic diagram showing the metzincin characteristic secondary structure elements in leishmanolysin. The key lists the metzincin elements (helices A, B, C; strands I, III–V), and the corresponding leishmanolysin helices and strands are numbered as in Reference 25. The 40- and 35-residue insertions in the catalytic domain are colored green and orange. The blue region is the 62-residue insertion in the central domain shown in Figure 5. Generated using MOLSCRIPT.[28]

S11 of the central sheet, which borders the active site cleft.

Central domain

Residues 274–391 of leishmanolysin fold into a compact domain (Figure 4) with antiparallel helices H11 and H12 forming the core of the domain together with the antiparallel sheet formed by strands S13, S14, and S15. Residues in the long loop joining H11 and H12 form part of the active site. Strand S16 forms an antiparallel sheet with strand S6 forming the main connection between the N-terminal and central domains. A single disulphide bond connects the C-terminus of the domain to the beginning of the long loop between H11 and H12.

The central domain contains structural features identifying leishmanolysin as a metzincin class zinc proteinase.[29]

These are a tight turn including a conserved methionine at the base of the active site and a following helix C, which is roughly parallel to the active site helix B.[31] In leishmanolysin, the metzincin tight turn consists of residues (Asp342–Ala348) and the metzincin helix C is H12. Leishmanolysin is unusual because in other metzincins the third zinc–histidine ligand occurs in an extended HEXXHXXGXXH active site motif that is six residues after the first two ligands. In leishmanolysin, an insert of 65 residues separate the second and third ligands, comprising about half of the central domain (Figure 5).

C-terminal domain

The elongated C-terminal domain of leishmanolysin consists of mainly antiparallel β-strand and random coil structure with only minor helical contributions (Figure 6).

The domain contains six disulphide bonds. The N-terminal part of the domain contains three two-stranded antiparallel sheets (S17,S18; S20,S21; S19, S22). These are connected via two short helices (H16, H17) to an antiparallel six-stranded β-sheet folded into a sandwich structure (S23–28) and this is in turn attached via two short helices (H19, H20) to the GPI anchor. The GPI anchor, although present in the crystals, was not visible in the electron density map.

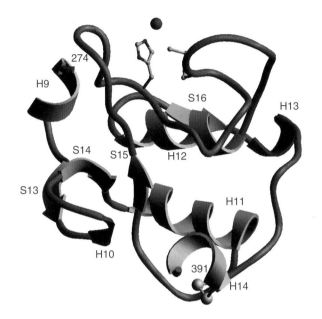

Figure 4 Schematic diagram showing the central domain of leishmanolysin, the active site zinc atom, and the side chains of two central domain active site residues H334 and M345 near the zinc atom. Figure prepared using RIBBONS.[30]

The zinc site

The geometry of leishmanolysin active site residues in the immediate vicinity of the catalytic zinc atom is similar to

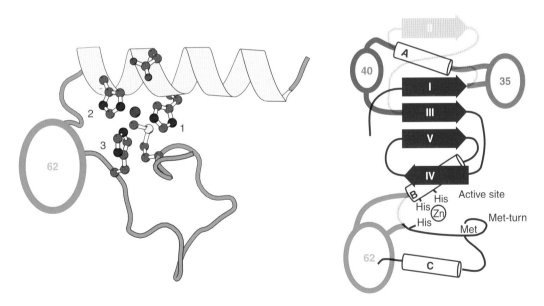

Figure 5 Schematic diagrams showing the 62-residue leishmanolysin insertion between the second and third active site zinc ligands His268 and His334. The diagram on the left shows the active site region, with the side chains of the HEXXH motif residues, the third zinc ligand H334, and the metzincin characteristic methionine (Met345) shown in ball and stick representation. The three histidine zinc ligands are labeled. The diagram on the right is a cartoon representation of metzincin topology[29] showing the location of the 62-residue insertion. This region is shown also in Figure 3. Prepared using MOLSCRIPT.[28]

Leishmanolysin

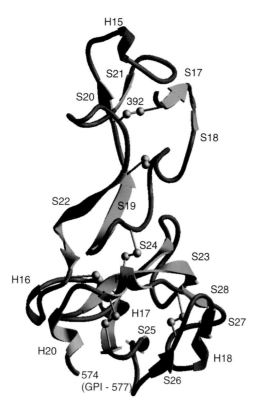

Figure 6 Schematic diagram showing the C-terminal cysteine-rich domain of leishmanolysin. Figure prepared using RIBBONS.[30]

the three histidine ligands: a water molecule held between the zinc and the oxygens of the conserved glutamic acid in the zinc proteinase motif HEXXH. Astacin is unusual in having a fifth zinc ligand – a tyrosine phenolic oxygen. The water molecule zinc ligand is thought to play the key role in proteolysis, being the nucleophile that attacks the substrate peptide bond. Electron density at the expected position of the catalytic water molecule zinc ligand is absent in the leishmanolysin electron density map, and the glutamate side chain is twisted away from the usual position it adopts in other zinc proteinases. The leishmanolysin active site zinc appears to be penta-cooordinated (approximately trigonal bipyramidal), with two ligands to a region of connected unidentified density modeled as glycine. In the model, one zinc ligand is the glycine amino nitrogen that is near the position normally occupied by the catalytic water in zinc proteinase structures (Figure 7).

A glycine carboxyl oxygen forms the fifth zinc ligand and is located near the position of the phenolic oxygen of the tyrosine that forms the fifth zinc ligand in astacin. Similar N–O zinc coordination has been observed in a complex of the metzincin human MMP 3/stromelysin-1 with the protein inhibitor TIMP-1, involving the amino nitrogen and carbonyl oxygen of the first residue of the inhibitor.[34]

The similarity of leishmanolysin with metzincin class zinc proteinases extends beyond the immediate vicinity of the active site zinc atom. When the metzincin human neutrophil collagenase (MMP-8) (PDB code 1JAP[35]) is aligned to leishmanolysin, 89 corresponding carbon alpha positions in the active site regions of the two molecules match with an rms difference of 1.8 Å, 53 in the leishmanolysin N-terminal domain, and 36 in the central domain. Leishmanolysin and neutrophil collagenase share residues near the active site in addition to the conserved metzincin motif residues, and both molecules have large S1′ substrate pockets. Collagenase inhibitors in the structural alignment fit reasonably well to the modeled leishmanolysin surface,

that of other zinc proteinases. The zinc atom is coordinated to the ε-nitrogen atoms of His264, His268, and His334 at distances 2.18, 2.18, and 2.12 Å respectively, similar to values observed in other zinc proteinase structures (2.0–2.2 Å).[29] The arrangement of the three zinc ligand histidine side chains (His264, His268, His334) and the underlying Met-turn methionine (Met345) are very similar to those for metzincin class zinc proteinases, such as astacin[32] and collagenase.[33] Structures of zinc proteinases without inhibitors have a fourth zinc ligand in addition to

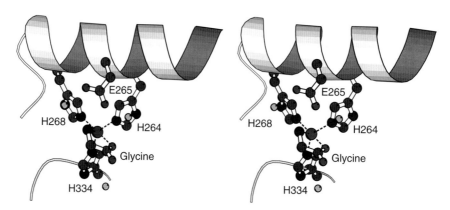

Figure 7 Stereo diagram showing the active site of leishmanolysin. The active site zinc atom is shown with five ligands: εN atoms from His264, His268, and His334, and the amino nitrogen and carboxyl oxygen from a glycine molecule modeled into observed electron density. The His264, His268, and His334 ligand distances are 2.18, 2.18, and 2.12 Å respectively. Prepared using MOLSCRIPT.[28]

and have been recently shown to inhibit leishmanolysin *in vitro*.[23]

REFERENCES

1. DH Molyneux and RW Ashford, *The Biology of Trypanosoma and Leishmania, Parasites of Man and Domestic Animals*, Taylor & Francis, London (1983).
2. J Bouvier, P Schneider and R Etges, *Methods Enzymol*, **248**, 614–33 (1995).
3. BR Voth, BL Kelly, PB Joshi, AC Ivens and WR McMaster, *Mol Biochem Parasitol*, **93**, 31–41 (1998).
4. TO Frommel, LL Button, Y Fujikura and WR McMaster, *Mol Biochem Parasitol*, **38**, 25–32 (1990).
5. E Medina-Acosta, RE Karess, H Schwartz and DG Russell, *Mol Biochem Parasitol*, **37**, 263–73 (1989).
6. M Klemba and DE Goldberg, *Annu Rev Biochem*, **71**, 275–305 (2002).
7. J Bouvier, RJ Etges and C Bordier, *J Biol Chem*, **260**, 15504–9 (1985).
8. MJ McConville and MA Ferguson, *Biochem J*, **294**, 305–24 (1993).
9. R Etges, J Bouvier and C Bordier, *J Biol Chem*, **261**, 9098–101 (1986).
10. PB Joshi, BL Kelly, S Kamhawi, DL Sacks and WR McMaster, *Mol Biochem Parasitol*, **120**, 33–40 (2002).
11. K Victoir and JC Dujardin, *Trends Parasitol*, **18**, 81–85 (2002).
12. DG Russell and SD Wright, *J Exp Med*, **168**, 279–92 (1988).
13. A Brittingham, CJ Morrison, WR McMaster, BS McGwire, KP Chang and DM Mosser, *J Immunol*, **155**, 3102–11 (1995).
14. A Brittingham, G Chen, BS McGwire, KP Chang and DM Mosser, *Infect Immun*, **67**, 4477–84 (1999).
15. B Boeckmann, A Bairoch, R Apweiler, MC Blatter, A Estreicher, E Gasteiger, MJ Martin, K Michoud, C O'Donovan, I Phan, S Pilbout and M Schneider, *Nucleic Acids Res*, **31**, 365–70 (2003).
16. P Schneider, MAJ Ferguson, MJ McConville, A Mehlert, SW Homans and C Bordier, *J Biol Chem*, **265**, 16955–64 (1990).
17. ND Rawlings, E O'Brien and AJ Barrett, *Nucleic Acids Res*, **30**, 343–46 (2002).
18. E Schlagenhauf, R Etges and P Metcalf, *Proteins*, **22**, 58–66 (1995).
19. C Bordier, RJ Etges, J Ward, MJ Turner and ML Cardoso de Almeida, *Proc Natl Acad Sci USA*, **83**, 5988–91 (1986).
20. J Bouvier, P Schneider, R Etges and C Bordier, *Biochemistry*, **29**, 10113–19 (1990).
21. P Schneider and TA Glaser, *Mol Biochem Parasitol*, **58**, 277–82 (1993).
22. J Bouvier, P Schneider and B Malcolm, *Exp Parasitol*, **76**, 146–55 (1993).
23. JD Bangs, DA Ransom, M Nimick, G Christie and NM Hooper, *Mol Biochem Parasitol*, **114**, 111–17 (2001).
24. I Schomburg, A Chang and D Schomburg, *Nucleic Acids Res*, **30**, 47–49 (2002).
25. E Schlagenhauf, R Etges and P Metcalf, *Structure*, **6**, 1035–46 (1998).
26. HM Berman, J Westbrook, Z Feng, G Gilliland, TN Bhat, H Weissig, IN Shindyalov and PE Bourne, *Nucleic Acids Res*, **28**, 235–42 (2000).
27. A Perrakis, R Morris and VS Lamzin, *Nat Struct Biol*, **6**, 458–63 (1999).
28. PJ Kraulis, *J Appl Crystallogr*, **24**, 946–50 (1991).
29. W Stöcker and W Bode, *Curr Opin Struct Biol*, **5**, 383–90 (1995).
30. M Carson, *J Appl Crystallogr*, **24**, 958–61 (1991).
31. W Stöcker, F Grams, U Baumann, P Reinemer, FX Gomis-Rüth, DB McKay and W Bode, *Protein Sci*, **4**, 823–40 (1995).
32. W Bode, FX Gomis-Rüth, R Huber, R Zwilling and W Stöcker, *Nature (London)*, **358**, 164–67 (1992).
33. W Bode, R Reinemer, R Huber, T Kleine, S Schnierer and H Tschesche, *EMBO J*, **13**, 1263–69 (1994).
34. F Gomis-Ruth, K Maskos, M Betz, A Bergner, R Huber, K Suzuki, N Yoshida, H Nagase, K Brew, GP Bourenkov, H Bartunik and W Bode, *Nature*, **389**, 77–81 (1997).

Streptomyces albus G D-Ala-D-Ala carboxypeptidase

Paulette Charlier[†], Jean-Pierre Wery[‡], Otto Dideberg[§] and Jean-Marie Frère[†]

[†]Centre d'Ingénierie des Protéines, Université de Liège, Institut de Chimie B6, Sart Tilman, Belgium
[‡]Concurrent Pharmaceuticals, Inc., West Center Office Drive, Fort Washington, PA, USA
[§]LCM, Institut de Biologie Structurale, Rue Jules Horowitz, Grenoble-Cedex1, France

FUNCTIONAL CLASS

Enzyme: Zn^{2+}-D-Ala-D-Ala carboxypeptidase (DDC), EC 3.4.17.14, a metalloprotein containing one Zn^{2+} ion per molecule and related to VanX and VanY, components of the vancomycin resistance systems in streptococci.

The zinc D-Ala-D-Ala carboxypeptidase hydrolyzes the C-terminal peptide bond of peptides of general structure R-D-Ala-D-Xaa. It was first called an 'endopeptidase' because it could solubilize the peptidoglycans of various bacteria[1] by cleaving their peptide cross-links; but it was later recognized as a strict carboxypeptidase since the hydrolyzed bond was always in α of a free carboxylic group[2] (Figure 1).

OCCURRENCE

The enzyme is secreted into the culture medium by *Streptomyces albus G*, which is probably a *Streptomyces griseus* strain.

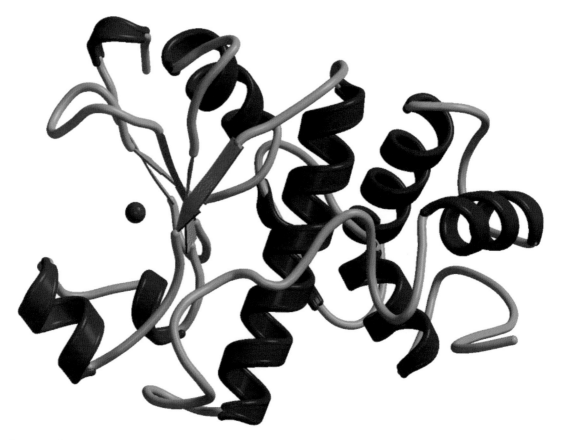

3D Structure Schematic representation of the zinc D-Ala-D-Ala carboxypeptidase, PDB code: 1LBU. Produced with the programs MOLSCRIPT[3] and RASTER3D.[4]

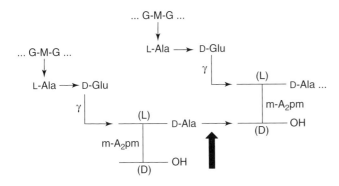

Figure 1 Structure of the peptide cross-link hydrolyzed by the zinc D-Ala-D-Ala-carboxypeptidase. The heavy arrow identifies the hydrolyzed peptide bond. G = N-acetylglucosamine, M = N-acetylmuramic acid, m-A$_2$pm = meso-diaminopimelic acid.

BIOLOGICAL FUNCTION AND SUBSTRATE SPECIFICITY

The best substrate described so far is N$^\alpha$ N$^\varepsilon$-(acetyl)$_2$-L-Lys-D-Ala-D-Ala ($k_{cat} = 3\,s^{-1}$, $K_m = 0.33\,mM$). With the tripeptides, the activity decreases when the length of the side chain of the N-terminal residue decreases (less than 1% with Ac-L-Ala-D-Ala-D-Ala and nearly zero with Ac-Gly-D-Ala-D-Ala). The amino groups of L-Lys must be blocked. In the second position, D-Ala can only be replaced by glycine, but at the expense of a 50-fold increase in K_m while D-Leu and L-Ala result in a complete loss of activity. Similarly, no hydrolysis of Ac$_2$-L-Lys-D-Ala-L-Ala is observed, but the C-terminal residue can be replaced by Gly, D-Lys, D-Leu, D-Glu yielding substrates hydrolyzed with 10 to 100% of the best catalytic efficiency.[2,5] The lytic activity of the enzyme and its extracellular location suggest that it might be used by *Streptomyces* for fighting competitors in its ecological niche, since the enzyme does not hydrolyze the *Streptomyces* peptidoglycan.

AMINO ACID SEQUENCE INFORMATION

Streptomyces albus G, Swissprot P00733, PIR A00913, EMBLX55794 (cDNA)

The sequence of the 213-residue ($M_r = 22\,156$) mature, secreted protein was first determined by chemical degradation[6] and confirmed by sequencing the corresponding gene, the only modifications being that the N-terminal residue was aspartic acid instead of asparagine and an alanine was inserted after Pro67. The gene encodes a 255-residue precursor containing a 41- or 42-residue typical signal peptide. The cleavage site (Leu-Asp) is somewhat unexpected, and it is possible that the signal peptidase acts on the preceding alanine–leucine bond and that the leucine residue is removed by an aminopeptidase, since aspartic acid was clearly established as the N-terminal residue of the mature protein.

The protein most similar to the D-Ala-D-Ala carboxypeptidase from *S. albus* G (DDC) is the product of the *pdca* gene from *Micrococcus xanthus*, which is also a D-D-carboxypeptidase.[7] Both proteins share an overall amino acid identity of 25%.

DDC is also related to the VanX, VanY, and VanXY proteins of vancomycin-resistant enterococci.[8] These proteins regulate the synthesis of new resistant peptidoglycan precursors and the elimination of wild-type sensitive peptidoglycan precursors. VanX acts selectively as a D-Ala-D-Ala dipeptidase allowing only D-Ala-D-Lac to accumulate and become incorporated into bacterial cell wall intermediates in the presence of vancomycin.[9] It is associated with VanS, VanR, and VanH, and is required for resistance against vancomycin in the VanA and VanB classes of enterococci. VanY has a D-D-carboxypeptidase activity. It removes D-Ala from peptidoglycan precursors ending in D-Ala-D-Ala. It is not required for resistance.[10] In the VanC-mediated resistance, a VanXYc protein has both D-Ala-D-Ala dipeptidase and carboxypeptidase activities.[11] VanX-like related proteins have been found in several other bacterial strains, including gram-negative bacteria. Since, in these later cases, the outer membrane prevents access of vancomycin to its target, these VanX-like proteins must have roles other than mediation of vancomycin resistance.[12] Sequence alignments of these enzymes with DDC highlight the lack of similarity within the N-terminal sequence up to residue 83, while the C-terminal sequence is more conserved.

PROTEIN PRODUCTION, PURIFICATION, AND MOLECULAR CHARACTERIZATION

Low amounts of protein were produced by growing the original strain[11] ($0.5\,mg\,L^{-1}$), but the culture conditions were not optimized. Higher yields ($10\,mg\,L^{-1}$) are obtained after cloning of the gene on a multicopy plasmid in *Streptomyces lividans*. Purification involves separation of the culture supernatant, adsorption on a cation exchanger at pH 4.0, negative chromatography on DEAE-cellulose, followed by chromatographies on CM-cellulose, Sephadex G-100, and CM-Sepharose[13] (yield: 25%).

The protein is monomeric. The presence of a smaller form detected by polyacrylamide gel electrophoresis[11] could later be attributed to incomplete reduction of one of the disulfide bridges. Each monomer contains three such bridges, extending between residues 3–81, 94–142 and 170–211.[6,14]

METAL CONTENT

Proton-induced X-ray emission (PIXE) revealed the presence of 0.95 to 1.04 Zn^{2+} ion per 22.000 M_r protein molecule.[15] Interestingly, no zinc salt had been added to the buffers during the purification procedure. This indicated a high affinity for Zn^{2+} ions since the enzyme appeared to be able to scavenge the metal from buffers prepared without specific treatment for removal of divalent cations. The association constant of $2 \times 10^{14}\,M^{-1}$ reported in the same paper[15] (and determined on the basis of competition experiments between enzyme and ethylenediaminetetraacetic acid (EDTA) for a limited amount of Zn^{2+} ions) is probably strongly overestimated since no account was taken of the nonspecific binding of EDTA to the enzyme and of the presence of contaminating Zn^{2+} ions in the buffers. The active enzyme could be reconstituted by adding Co^{2+} or Fe^{2+} ions to the apoenzyme (but not with Ni^{2+}, Cd^{2+} or Mg^{2+}). Binding of Co^{2+} to the apoenzyme was confirmed by PIXE experiments.[15]

In the native enzyme, the Zn^{2+} ligands are His154, Asp161 and His197 (see below).

ACTIVITY TEST

Enzyme activity is routinely determined by monitoring the initial rate of D-alanine release from Ac_2-L-Lys-D-Ala-D-Ala at 37 °C with the help of a discontinuous colorimetric assay involving oxidation of D-alanine by D-amino acid oxidase and the utilization of the produced hydrogen peroxide to oxidize o-dianisidine or 2,2′-Azinobis(3-ethylbenzothiazoline-6-sulfonic acid) (ABTS) in the presence of horse-radish peroxidase.[16]

SPECTROSCOPY

The enzyme exhibits a typical protein near-UV spectrum with a maximum at 280 nm. The maximum of the fluorescence emission spectrum is at 350 nm (excitation at 285 nm).[13]

X-RAY STRUCTURE

Crystallization and structure determination

Crystals of the zinc D-Ala-D-Ala carboxypeptidase were grown from 12% PEG (polyethylene glycol) 6000, 10 mM NaN_3 and 5 mM $MgCl_2$ in 50 mM Tris-HCl buffer at pH 8.0. They belong to the monoclinic space group $P2_1$ with unit cell parameters $a = 51.08$ Å, $b = 49.70$ Å, $c = 38.65$ Å, and $\beta = 100.6°$ and one molecule per asymmetric unit. The structure was solved by multiple isomorphous replacement (MIR) using a 2.5-Å native data set and three heavy atom derivatives. The preliminary X-ray structure was published in 1982.[17] Higher resolution data have been collected on in-house systems and the structure has been refined to 1.8 Å (PDB accession code: 1LBU).

Overall structure

The molecule is an ellipsoid with dimensions $48 \times 34 \times 28$ Å, consisting of two globular domains where the N-terminal and the C-terminal ends are pulled apart. The two domains are connected by a single link. The small N-terminal domain has an all-α-helices structure (α1, α2, α3) and the largest C-terminal domain belongs to the α + β-type secondary structure (α4, β1, α5, α6, β2, α7, β3, β4, β5, α8). The mixed five-stranded β-sheet forms the core of the C-terminal domain and the lining of one side of the catalytic cavity. At the interface of the two domains lies the longest α-helix (α4) from Ala105 to Ala125 (Figure 2). The structure is stabilized by the three disulfide bridges.

The N-terminal domain

The all-α N-terminal domain, from residue Asp1 to Pro83, exhibits a typical three-helix bundle fold, characteristic of the peptidoglycan binding domain (PGBD). It possesses an elongated cavity of 59 Å^3 whose likely function would be substrate recognition. The three helices and their connecting loops show amino acid sequence segments that are also found in a number of bacterial wall lytic enzymes and bacterial cell-surface-associated proteins.[18] These multimodular proteins possess, in addition to the catalytic domain, another domain that is frequently at or near the N- or C-terminal ends of the polypeptide chain, and that contains conserved sequence segments, as shown in Table 1. The substrate recognition and binding sites would be defined by the strictly conserved aspartic acid (Asp39 in DDC), glutamine (Gln54 in DDC), and threonine (Thr69 in DDC) residues and several hydrophobic residues along the cavity defining motifs. The different enzymes exhibit different functions and activities against the bacterial wall peptidoglycan (peptidase, amidase), and the variations in the sequence binding motifs could be responsible for the substrate specificities.

More surprisingly, a similar observation is made of another family of enzymes, the human matrix metalloproteinases (MMPs). MMPs are also multidomain enzymes that degrade the extracellular matrix of the basal membrane, and that are believed to play a role in pathological states such as tumor invasion and arthritis. The whole structure of human MMP9 has been recently determined.[19] It consists of an N-terminal prodomain, a zinc binding catalytic domain, and three fibronectin type III domains (Figure 3(a)). The prodomain is closely associated with the catalytic domain and consists of a cluster of three helices that lie roughly perpendicular to each other, as observed in the N-terminal domain of the zinc D-Ala-D-Ala carboxypeptidase (Figure 3(b)). If a structural similarity is undoubtedly observed, sequence similarities with the peptidoglycan binding motif of bacterial enzymes is less evident. As highlighted in Table 1, sequence identity and isology

Streptomyces albus G D-Ala-D-Ala carboxypeptidase

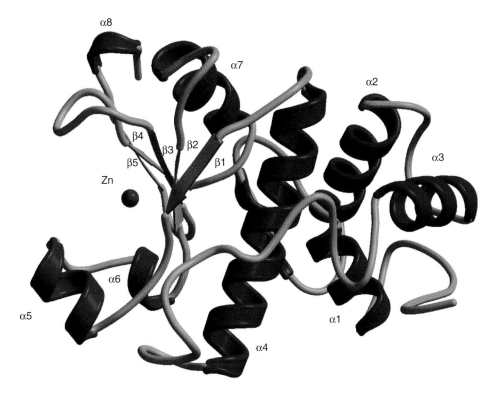

Figure 2 Schematic representation of the structure of the zinc D-Ala-D-Ala carboxypeptidase with assignment of the secondary structure elements. Produced with the programs MOLSCRIPT[3] and RASTER3D.[4]

between MMPs and PGBDs are restricted to the glutamine, the threonine, and a few hydrophobic residues in helices 2 and 3. The root mean square deviation (rmsd) between the Cα of those identical and equivalent residues is quite low as shown in Table 2, and their side chains create two hydrophobic zones that define the putative cavity involved in substrate recognition (Figure 3(c)).

The C-terminal domain

The C-terminal domain, from residue Val83 to Ile213, belongs to the α + β class. The core of the domain is formed by a mixed five-stranded β-sheet with the usual left-handed twist and is bordered in the back by the largest α4 helix. There are only two structural homologs to the C-terminal domain of the DDC of *S. albus* G: the N-terminal fragment of the murine SHH, a signaling protein involved in natal morphogenesis and formation of embryonic patterning centers (PDB code: 1VHH)[20] and the VanX D-Ala-D-Ala dipeptidase of *Enterococcus faecium* (VAX), a protein involved in the vancomycin resistance systems in streptococci (no PDB code available, coordinates kindly provided by Pr. C. Park).[21] The three proteins share 17% sequence identity and show large areas of structural similarity around the region of the β-sheet and the cavities identified as the catalytic sites of the three proteins. The C-terminal domain of DCC overlaps with SHH on 89 Cα atoms with an rmsd value of 2.3 Å (Figure 4(a)) and with VAX on 91 Cα atoms with an rmsd value of 2.6 Å (Figure 4(b)), while SHH overlaps with VAX on 99 Cα atoms with an rmsd value of 1.8 Å.

The active site

The active site of DCC is held within a channel-shaped cleft of 486 Å^3 on the surface of the C-terminal domain, and is delimited in the back by the β-sheet and on the sides by two jawlike segments: from Asn134 to Gly157, including the α5 and α6 helices with their connecting loop; and from Leu185 to Thr196, the loop connecting β3 to β4 strands. The zinc ion occupies a roughly central position in the active site and is coordinated by the His154 NE2, the Asp161 OD1 and the His197 ND1 residues, situated respectively on helix α6, strand β2, and strand β3, the fourth ligand being a structural water molecule (Wat234) (Figure 5). Table 3 shows the zinc–ligand bond distances. The zinc coordination is almost tetrahedral and the three ligands have quite stable conformations as indicated by their low temperature factors. They are stabilized by additional interactions: the second carboxyl oxygen OD2 of Asp161 is hydrogen-bonded to a water molecule Wat306, and the second imidazole nitrogens of the histidine ligands, ND1 of His154 and NE2 of His197, are hydrogen-bonded respectively to the carbonyl oxygen of Gly136 and the carboxyl oxygen OD1 of Glu183. In addition to

Table 1 Sequence alignments showing conserved sequence segments between the N- and C-terminal domains of the zinc D-Ala-D-Ala carboxypeptidase from *S. albus G* (DCC) and the following proteins: the *pdca* gene product (pdca) from *Myxococcus xanthus*, the spore cortex-lytic enzymes (SCLE) from *Bacillus cereus* and *Clostridium perfringens*, the *N*-acetylmuramoyl-L-alanine amidases from *Bacillus subtilis* (CWLL) and *Bacillus licheniformis* (CWLX), the 1,4-beta-*N*-acetylmuramidase (LYC) from *Clostridium acetobutylicum*, the cell wall lytic enzyme (LYTE) from *Bacillus halodurans*, the sporulation protein (SPOII2) from *Thermoanaerobacter tengcongensis*, the human collagenases, the human gelatinase A, the human stromelysin, the human matrix metalloproteinase (MMPase), the D-Ala-D-Ala dipeptidase (VanX) and the D-D-carboxypeptidase (VanY) from *Enterococcus faecium*, the D-D-dipeptidase/D-D-carboxypeptidase (VanXYc) from *Enterococcus gallinarum* and the N-terminal fragment of the murine sonic hedgehog (SHH). The numbers indicate the first residue numbers in the aligned sequence segments. Several well-conserved residues are highlighted in red in the consensus sequences, and the zinc ligands are marked with an asterisk

```
N-terminal domain        Helix 1                      Helix 2                    Helix 3
                         ---------                    ---------------            ----------
DDC_S.albus  (1LBU)   14 SSGEAVRQLQIRVAG //       38 IDGQFGPATKAAVQRFQSAYGLAADGIAGPATFNKIYQL //
pdcA_M.xanthus        26 GARGAVTQLQNKLRA //       49 SDGVFGPKTQSAVKAFQQSRGLVADGIVGPKTWDKLGIN //
SCLE_B.cereus          1 MRQKAIFKIAVLLAF //       64 VDGVFGWGTYWALRNFQEKFGLPVDGLAGAKTKQMLVKA //
SCLE_C.perfringens   337 YSGEPVRVIQEQLNA //      364 VDGKYGPKTREAVKTFQKIFNLPQTGEVDYATWYKISDV //
CWLL_B.subtilis      200 ASGSQVKALQKRLIA //      320 IDGYYGPKTANAVKRFQLMGHLSADGIYGSDTKAKLKTL //
CWLX_B.lichenif      199 MSGSHVKKLQTRLVA //      314 IDGYYGMKTANAVKRFQLMYGLGADGIYGPKTKAKMLSL //
LYC_C.acetobutyl     197 GGDDNIKAIQQDLNI //      285 VDGTFGSGTKAKVAAWQSNQGLMADGVVGSATWSKLLDE //
LYTE_B.halodurans    214 SRGDAVRDLQSKLKD //      236 IDGIFGAGTTTAVREFQRKMGLTVDGVAGPQTLNALHVN //
SPOIID2_T.tengcong   233 MRSPEVRKLQENLNR //      279 PDGIFGFKTQNAVVQFQKANGLLADGIVGPATQKVLLQK //
Collagenase H. (1EAK) 15 DKELAVQYLNTFYGC //       31 KESCNLFVLKDTLKKMQKFFGLPQTGDLDQNTIETMRKP //
Gelatinase A H.(1CK7) 17 DKELAVQYLNTFYGC //       33 KESCNLFVLKDTLKKMQKFFGLPQTGDLDQNTIETMRKP //
Collagenase H. (1GXD) 17 PKTDKELAVQYLNTF //       33 KESCNLFVLKDTLKKMQKFFGLPQTGDLDQNTIETMRKP //
Stromelysin H. (1SLM) 13 SMNLVQKYLENYYDL //       35 VRRKDSGPVVKKIREMQKFLGLEVTGKLDSDTLEVMRKP //
MMPase H.      (1L6J) 22 DRQLAEEYLYRYGYT //       40 EMRGESKSLGPALLLLQKQLSLPETGELDSATLKAMRTP //

Consensus PGBD           -----V--LQ--L--              -DG-FG--T---V--FQ---GL--DGI-G---T---L---
                              I IA V/I                    Y      L   W   N  TVD      I/M
Consensus MMP            -------YL------              --------L---L--MQKF-GL--TGD-D--T---MRKP
                                AV                            V    I  LQS   E/K        T

C-terminal domain        Helix Strand                 Strand
                         ----- ------                 ------
DDC_S.albus  (1LBU)  152 SRHMYGHAADLGAGSQGFCA //  191 GHNDHTHVAGGD //
pdcA_M.xanthus       255 SNHQGGIAVDVNTGGTGTST //  291 VPSEPWHWEY //
VanX_E.faecium       114 SSHSRGSAIDLTLYRLDTGE //  178 YSLEWWHYVLRD //
VanY_E.faecium       161 SEHNSGLSLDVGSSLTKMER //  210 IQYEPWHIRYVG //
VanXYc_E.gallinarum   93 SEHQIGLAIDVGLKKQEDDD //  150 ISYEPWHFRYVG //
Murine SHH   (1VHH)  139 SLHYEGRAVDITTSDRDRSK //  177 ESKAHIHCSVKA //

Consensus                S-H*--G-A-D*                 -H*-
```

Table 2 Root mean square deviations between the Cαs of conserved motifs in helices 2 and 3 of the N-terminal domain of the zinc D-Ala-D-Ala carboxypeptidase from *S. albus G* (DCC) and the human matrix metalloproteinase MMP9

DDC	RMSD (Å)	MMP9
(1LBU, residues in purple)		(1L6J, residues in cyan)
Val50	0.63	Leu74
Phe53	0.28	Met77
Gln54	0.09	Gln78
Gly58	1.27	Gly82
Leu59	1.06	Leu83
Asp62	1.09	Thr86
Gly63	1.24	Gly87
Thr69	0.34	Thr93
Ile73	1.22	Met9

Table 3 Zinc–ligand distances in zinc D-Ala-D-Ala carboxypeptidase from *S. albus G* (DCC), the VanX D-Ala-D-Ala dipeptidase of *E. faecium* in complex with the substrate D-Ala-D-Ala (VAX) and the N-terminal fragment of the murine sonic hedgehog (SHH)

Zinc		DCC	VanX	SHH
His NE2	Residue no.	154	116	141
	Distance (Å)	2.16	2.42	2.06
Asp OD1	Residue no.	161	123	148
	Distance (Å)	1.98	2.43	1.97
His ND1	Residue no.	197	184	183
	Distance (Å)	2.15	3.17	2.08
Fourth ligand	Residue no.	Wat234	O carbonyl D-Ala	Wat201
	Distance (Å)	2.12	2.48	2.05

the zinc-bound Wat234, 10 additional water molecules were identified within the active site. Four of them are defined within a sphere of 5-Å diameter centered on the zinc ion (Wat271, Wat306, Wat307 and Wat312) and form a dense hydrogen-bonding network. The zinc-bound hydroxyl ion forms hydrogen bonds with Wat307 and Wat312. Wat306 forms hydrogen bonds with Asp161 OD1 and Arg138 NH1. The only acidic residue buried in the vicinity of the active site is Asp194. The Asp194 OD2 forms a hydrogen bond with the His195 ND1. The dyad Asp194-His195 has a crucial role in the catalytic mechanism hypothesis.

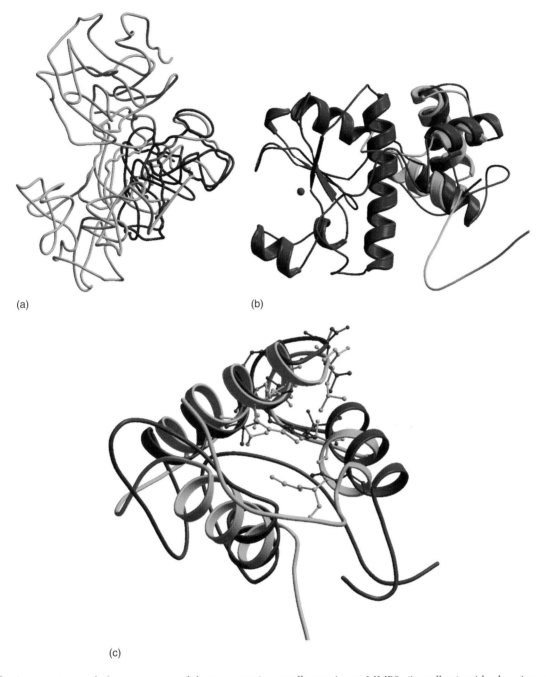

Figure 3 Superposition of the structures of human matrix metalloproteinase MMP9 (in yellow) with the zinc D-Ala-D-Ala carboxypeptidase DCC (in red): (a) the whole structures; (b) the N-terminal prodomain of MMP9 with the whole DCC; (c) the clusters of the 3 α-helices in both MMP9 and DCC, showing the conserved residues in helices 2 and 3. Produced with the programs MOLSCRIPT[3] and RASTER3D.[4]

Sequence alignment of the C-terminal domain of DDC with the N-terminal fragment of the murine SHH, the VanX and VanY proteins of *E. faecium*, the VanXYc protein of *E. gallinarum*, and the *pdca* gene product of *M. xanthus*, highlights a consensus sequence in the regions containing the active site residues. The motif SxH*xxGxAxD* (aa* being ligands of the zinc atom) is typical of this class of zinc hydrolases (Table1).

Beyond the overall structural similarity between the C-terminal domain of DCC and VAX and SHH, several residues within the active site, including those of the above motif, are structurally conserved. Of course, the zinc ligands of DCC (His154, Asp161 and His197) are mirrored in VAX (His116, Asp123 and His184) and SHH (His141, Asp148 and His183) (Table 3). Ser152 of DCC is also conserved in both VAX (Ser114) and SHH (Ser139). His195 of DCC is equivalent to the indole ring of Trp183

Streptomyces albus G D-Ala-D-Ala carboxypeptidase

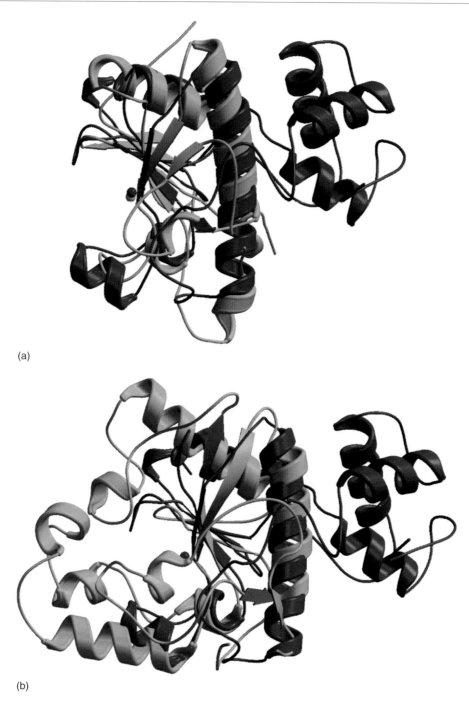

(a)

(b)

Figure 4 (a) Superposition of the structures of the Zinc D-Ala-D-Ala carboxypeptidase DCC, in red, with the murine sonic hedgehog (SHH) and (b) the VanX D-Ala-D-Ala dipeptidase VAX, both in yellow. Produced with the programs MOLSCRIPT[3] and RASTER3D.[4]

in VAX and is conserved in SHH (His181). Two other catalytic residues have their counterparts: His192 (Glu181 in VAX and Glu177 in SHH) and Arg138 (Arg71 in VAX and His135 in SHH). Figures 6(a) and 6(b) show the active sites of SHH and VAX and Figures 7(a) and 7(b) their respective superposition with DCC.

In comparison with the widely opened and large active site of DCC, the active sites of the VAX and SHH enzymes are more compact. The SHH active site is placed within a long channel of about 308 Å^3 along one face of the protein, while the VAX active site is entirely contained within a small cavity of about 150 Å^3. The configuration of the different active sites is consistent with the function of each protein. The proteolytic activity of SHH is probably to cleave itself.[20] The VAX protein only hydrolyzes dipeptides that contain D-amino acids. Owing to steric constraints, L residues would make inappropriate contacts, and only a small side chain (D-Ala or D-Ser) is allowed in position 1,

Streptomyces albus G D-Ala-D-Ala carboxypeptidase

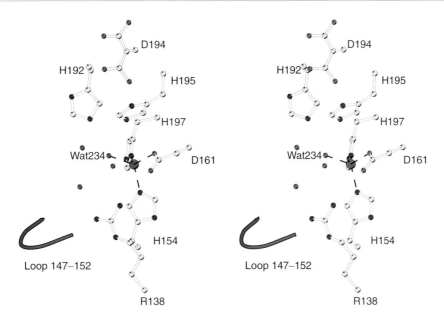

Figure 5 Stereo view of the active site of DCC, showing the zinc atom (in brown), the three zinc ligands, the catalytic residues (CPK colored) and water molecules (in red), and the loop defining the active site entrance (in purple). Produced with the program MOLSCRIPT.[3]

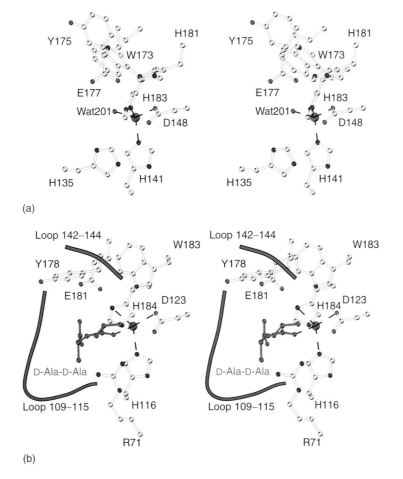

Figure 6 Stereo views of the active sites of (a) SHH; (b) VAX, showing the zinc atom (in brown), the three zinc ligands, the catalytic residues (CPK colored) and water molecule (in red), and the D-Ala-D-Ala substrate of VAX (in red). The orientation for both SHH and VAX is the same as for DCC in Figure 5. The active site of SHH is more open than that of DCC. By contrast, two large loops (in purple) reduce the accessibility of the VAX active site and narrow the cavity. Produced with the program MOLSCRIPT.[3]

Streptomyces albus G D-Ala-D-Ala carboxypeptidase

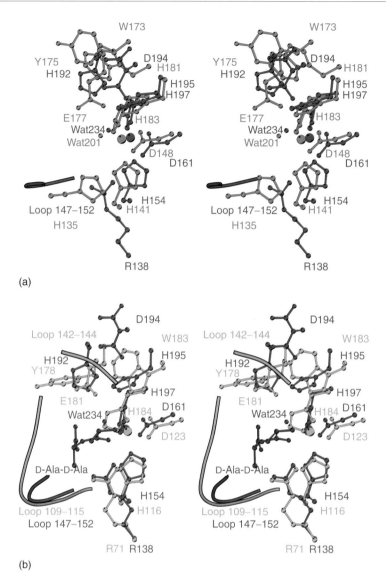

Figure 7 Stereo views of the superposition of the active sites of DCC (in purple) with (a) SHH (in green) and (b) VAX (in orange). Produced with the program MOLSCRIPT.[3]

while a larger D residue is allowed in position 2.[21] The DCC function, on the other hand, requires an easily accessible active site, which is provided by a long and large channel at the edge of the C-terminal domain.

CATALYTIC MECHANISM

The D-Ala-D-Ala carboxypeptidase of S. albus G acts as a lytic agent for bacterial walls where R-D-Ala-D-Xaa peptide linkages serve to cross-link the peptidoglycan subunits.[1] It hydrolyzes with high efficiency a variety of C-terminal R-D-Ala-D-peptide bonds irrespective of the structure of the side chain of the terminal D-amino acid residue.[2,5] As explained above, the examination of the kinetic parameters of the reaction involving a series of peptide analogs (standard reaction: $N^\alpha\ N^\varepsilon$-(acetyl)$_2$-L-Lys^1-D-Ala^2-D-Ala^3 + H$_2$O → $N^\alpha\ N^\varepsilon$-(acetyl)$_2$-L-Lys^1-D-Ala^2 + D-Ala^3)

has shown that substrate activity requires the occurrence of a long side chain at position 1, strictly depends on a D-alanine at position 2, and accommodates, although at the expense of the catalytic efficiency, glycine or a D-amino acid other than D-alanine at position 3. An L-alanine at the C-terminal position abolishes substrate activity.

The size of the groove on the enzyme surface and the positions of the catalytically important groups and other side chains suggest a mode of binding for the standard substrate $N^\alpha\ N^\varepsilon$-(acetyl)$_2$-L-Lys-D-Ala-D-Ala. The spatial requirements for an efficient positioning of the substrate for catalysis may be summarized as follows: (i) the carbonyl carbon atom of the scissile amide bond must approach the zinc ion; (ii) the terminal carboxylate group must make appropriate electrostatic interactions with the protein; (iii) the leaving D-amino acid in C-terminal position must be positioned in a large cavity that

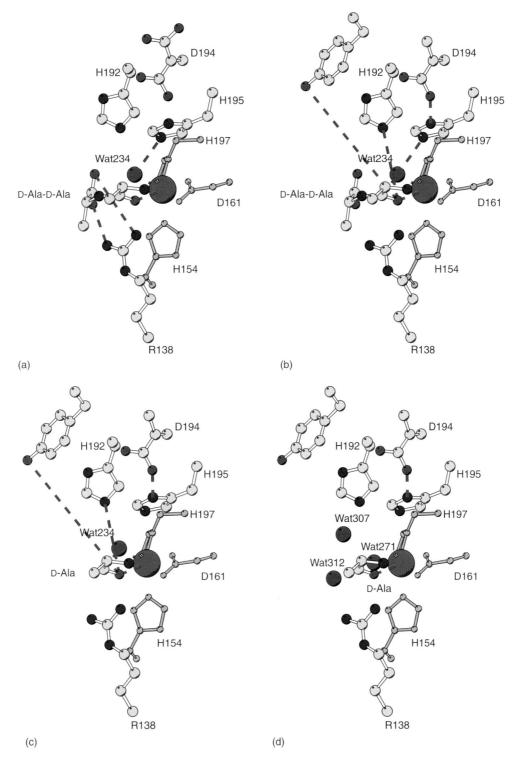

Figure 8 Proposed catalytic mechanism of DCC. The analog substrate D-Ala-D-Ala has been docked by analogy with the structure of the VAX complex (by courtesy of Dr. C. Park). The catalytic residues and the analog substrate are CPK colored, the zinc atom is in brown and the catalytic water molecules are in red. The three zinc ligands are colored orange and the red dashed lines represent the different interactions. Produced with the program MOLSCRIPT.[3]

can accommodate larger side chains in a D configuration; (iv) the D-alanine in position 2, is much more restrictive and should be placed in a smaller cavity; and (v) the long side chain in L configuration (or the remaining disaccharide – L-Ala-D-Glu-L-A$_2$ pm) must be positioned in an open channel making appropriate interactions with the protein. On the basis of these restrictions, a hypothesis can be proposed for a probable catalytic mechanism. The first step consists of an electrostatic interaction between the Arg138 NH1 and NH2 moieties and the substrate terminal carboxylate. The carbonyl group of the scissile amide bond orientates toward the zinc ion, forming a 'fifth' ligand, while the fourth ligand, water molecule Wat234, is shifted to the His195 NE2 (Figure 8(a)). The proton transfer from Wat234 to the scissile amide bond nitrogen is facilitated by Asp194, which makes an H-bond with the His195 ND1. The so-formed tetrahedral intermediate is stabilized by hydrogen bonds between the carbonyl oxygen atom and the NE2 nitrogen atom of His192 and the hydroxyl group of Tyr189 (Figure 8(b)). A second proton transfer from Wat234 to the nitrogen atom, catalyzed by His195, induces the amide bond rupture (Figure 8(c)). The restoration of the lone electron pair of the nitrogen atom NE2 of His195 induces the rotation of the carbonyl, and the release of the remaining substrate is performed by a neighboring water molecule that becomes the new fourth ligand of the zinc ion (Figure 8(d)).

Previous work has outlined the role of an acidic residue, aspartic or glutamic acid, as the catalytic base that activates the zinc-bound water molecule of zinc peptidases to attack the amide carbonyl carbon to form the tetrahedral intermediate, followed by the proton back-donation to the scissile amide nitrogen and the cleavage of the peptide.[22] In the VanX D-Ala-D-Ala dipeptidase, the catalytic water is found to be hydrogen-bonded to Glu181 at a distance of 2.9 Å,[21] and in the N-terminal fragment of the murine SHH, it is found to be hydrogen-bonded to Glu177 at a distance of 2.53 Å, confirming their putative roles as the catalytic bases. Unless we assume large-scale conformational changes of the protein, residue Asp194 in the D-Ala-D-Ala carboxypeptidase of *S. albus* G is too far away to play a direct role in the catalytic mechanism. When the active sites of the three proteins are superimposed, Asp194 of DCC is at a distance of more than 7.5 Å from the VAX Glu181 and SHH Glu177. On the other hand, the catalytic water in DCC is at H-bond distances of His192 NE2 (3.17 Å) and His195 NE2 (3.48 Å), suggesting that those residues may be more likely to play the role of the catalytic base, as described above.

INHIBITORS

Classical β-lactam compounds, which readily inactivate active site serine DD-peptidases, are very sluggish inactivators of the zinc D-Ala-D-Ala carboxypeptidase. The second-order rate constants for the inactivation reaction[23] range between 5×10^{-3} and $5 \times 10^{-2} s^{-1}$. Reactivation of the enzyme upon elimination of the excess of β-lactam occurs at an even slower rate (10^{-6} to $10^{-5} s^{-1}$). Interestingly, a radioactive adduct could be isolated by gel filtration after reaction of ^{14}C benzylpenicillin with the holoenzyme, but not with the apoenzyme. The first reversible step of the interaction appeared to be noncompetitive with cephalothin and cephalosporin C, and competitive with cephaloglycin and 7-[2-(*p*-iodo-phenyl)-acetamido] cephalosporanate. Upon soaking of the crystals in a solution of the latter compound, the X-ray data indicated that the β-lactam was bound near the zinc ion. By contrast, cephalothin and cephalosporin C destroyed the crystal lattice. 6-β-iodopenicillanic acid inactivated the enzyme according to a branched kinetic pathway. The inactivation phenomenon was characterized by a second-order rate constant of about $1 M^{-1} s^{-1}$. After diffusion of the compound in the crystal, a density corresponding to the inactivator was found near the zinc ion and His192, but the iodine atom was clearly eliminated. This suggested alkylation of His192 by the inactivator acting as an analog of iodoacetamide. It seems impossible to fit a penicillin or a cephalosporin in the same way as the peptidic substrate. Those inhibitors are probably stabilized in another way with much less efficiency in the catalytic cavity, before being hydrolyzed by nucleophilic attack performed by a solvent water molecule. This is probably the cause of the resistance of the zinc D-Ala-D-Ala carboxypeptidase of *S. albus* G to β-lactam antibiotics.

Various peptide analogs of the substrates were found to act as inhibitors of the enzyme with K_i values (0.25–0.5 mM) similar to the K_m for the best substrate.[5] They all contained D-residues: Ac-D-Ala-D-Glu, Suc$_2$-L-Lys-D-Ala-D-Glu (where suc = succinyl) and Ac$_2$-L-Lys-D-Glu-D-Ala. However, the best competitive inhibitors were 2- and 3-mercaptopropionate (K_i = 50 and 5 nM respectively) and, surprisingly, both L- and D-3-mercaptoisobutyrate (K_i = 60 and 10 nM respectively).[23] Other thiol derivatives were less active in the µM range (acetyl-cysteine, D-cysteine, D-penicillamine, formylpenicillamine, and L-captopril). Succinic acid monohydroxamate exhibited a similar inhibitory activity. All these compounds might behave as bidentate ligands of the Zn^{2+} ion.

REFERENCES

1   J-M Ghuysen, *Bacteriol Rev*, **32**, 425–64 (1968).

2   M Leyh-Bouille, J-M Ghuysen, R Bonaly, M Nieto, HR Perkins, KH Schleifer and O Kandler, *Biochemistry*, **9**, 2961–71 (1970).

3   P Kraulis, *J Appl Crystallogr*, **24**, 946–50 (1991).

4   EA Merritt and MEP Murphy, *Acta Crystallogr*, **D50**, 868–73 (1994).

5   M Nieto, HR Perkins, M Leyh-Bouille, J-M Frère and J-M Ghuysen, *Biochem*, **131**, 163–71 (1973).

6   B Joris, J Van Beeumen, F Casagrande, Ch Gerday, J-M Frère and J-M Ghuysen, *Eur J Biochem*, **130**, 53–69 (1983).

7   Y Kimura, Y Takashima, Y Tokumasu and M Sato, *J Bacteriol*, **181**, 4696–99 (1999).

8   CA Arias, P Courvalin and PE Reynolds, *Antimicrob Agents Chemother*, **44**, 1660–66 (2000).

9   PE Reynolds, F Depardieu, M Dutka-Malen, M Arthur and P Courvalin, *Mol Microbiol*, **13**, 1065–107 (1994).

10   M Arthur, PE Reynolds and P Courvalin, *Trends Microbiol*, **4**, 401–7 (1996).

11. PE Reynolds, CA Arias and P Courvalin, *Mol Microbiol*, **34**, 341–49 (1999).
12. F Hilbert, F Garcia del Portillo and EA Groisman, *J Bacteriol*, **181**, 2158–65 (1999).
13. C Duez, J-M Frère, F Geurts, J-M Ghuysen, L Dierickx and L Delcambe, *Biochem J*, **175**, 793–800 (1978).
14. C Duez, B Lakaye, S Houba, J Dusart and J-M Ghuysen, *FEMS Microbiol Lett*, **71**, 215–20 (1990).
15. O Dideberg, B Joris, J-M Frère, J-M Ghuysen, G Weber, R Robaye, J-M Delbrouck and I Roelandts, *FEBS Lett*, **117**, 215–18 (1980).
16. J-M Frère, M Leyh-Bouille, J-M Ghuysen, M Nieto and HR Perkins, in L Lorand (ed.), *Methods in Enzymology*, Vol. XLV, *Proteolytic Enzymes*, Part B, Academic Press, 610–35 (1976).
17. O Dideberg, P Charlier, G Dive, B Joris, J-M Frère and J-M Ghuysen, *Nature*, **299**, 469–70 (1982).
18. J-M Ghuysen, J Lamotte-Brasseur, B Joris and GD Shockman, *FEBS Lett*, **342**, 23–28 (1994).
19. PA Elkins, YS Ho, WW Smith, CA Janson, KJ D'Alessio, MS McQueney, MD Cummings and AM Romanic, *Acta Crystallogr*, **D58**, 1182–92 (2002).
20. TM Hall, JA Porter, PA Beachy and DJ Leahy, *Nature*, **378**, 212–16 (1995).
21. DE Bussiere, SD Pratt, L Katz, JM Severin, T Holzman and CH Park, *Mol Cell*, **2**, 75–84 (1998).
22. WL Mock and M Askamawati, *Biochem J*, **302**, 57–68 (1994).
23. P Charlier, O Dideberg, JC Jamoulle, J-M Frère, J-M Ghuysen, Dive G and J Lamotte-Brasseur, *Biochem J*, **219**, 763–72 (1984).

Metallocarboxypeptidases

Josep Vendrell[†], Francesc X Aviles[†] and Lloyd D Fricker[‡]

[†]Departament de Bioquímica i Biologia Molecular and Institut de Biotecnologia i Biomedicina, Universitat Autònoma de Barcelona, Spain

[‡]Department of Molecular Pharmacology, Albert Einstein College of Medicine, NY, USA

FUNCTIONAL CLASS

Enzyme; metallopeptidases; EC 3.4.17.-; contain one catalytic zinc atom per molecule; some members bind other metals such as Ca^{2+}; known as carboxypeptidases (CP)

Metallocarboxypeptidases (CP) catalyze the removal of C-terminal amino acids from proteins and/or peptides. The different members of the CP family perform different functions, typically differing in their specificity for basic versus hydrophobic C-terminal residues. Four classes may be distinguished according to their specificity: (i) for aromatic/hydrophobic residues (A-like); (ii) for basic residues (B-like); (iii) for acidic residues (based on active site modeling); and (iv) inactive toward standard CP substrates (discussed below). Some CPs from lower organisms have broad specificities that are both A- and B-like.[1] By comparing the amino acid sequences and crystal structures, the CPs can be divided into two subfamilies that have been named after the first two members of each family that were identified.[2] One is the A/B subfamily, whose members are generally produced as proenzymes, contain an approximately 300-residue CP catalytic domain, and have

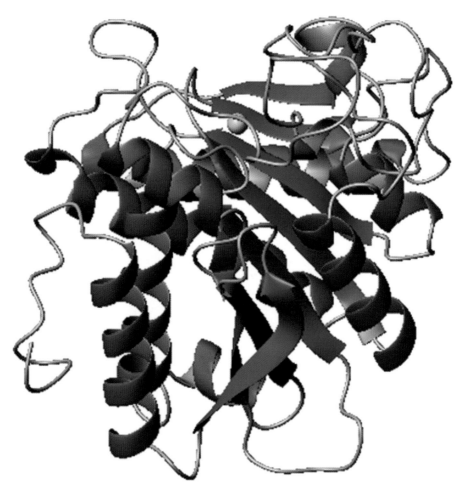

3D Structure Schematic representation of the 3D Structure of bovine carboxypeptidase A, the prototype enzyme for metallocarboxypeptidases, showing the zinc atom. PDB code: 5CPA. Prepared with the program MolMol.[69]

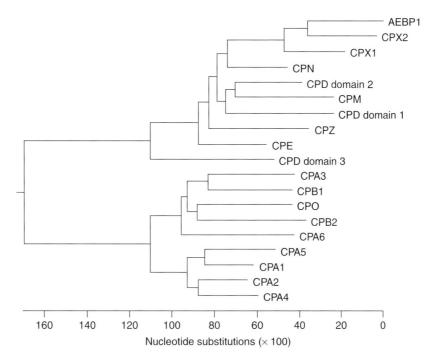

Figure 1 Phylogenetic tree of human metalloCP. The amino acid sequences were aligned using the ClustalV (PAM250) method with the Lasergene program (DNASTAR, Inc., Madison, WI, USA). Sequences were obtained from translation of the following GenBank nucleotide sequences: AEBP1 (NM_001129), CPX2 (XM_058409), CPX1 (NM_019609), CPN (NM_001308), CPD (NM_001304), CPM (XM_006768), CPZ (NM_003652), CPE (NM_001873), CPA3/mast cell CPA (NM_001870), CPB1 (NM_001871), CPO (NM_173077), CPB2/CPU (NM_001872), CPA6 (NM_020361), CPA5 (AF384667), CPA1 (NM_001868), CPA2 (NM_001869), and CPA4 (AF095719). The length of each branch shows the relative distance of each family member from the average of all other family members, taking into account the amino acid sequence matches, and the number and length of the gaps introduced for optimal alignment.

the greatest amino acid sequence identity to the exocrine pancreatic enzymes CPA and CPB. The proenzymes of A/B subfamily members are known as procarboxypeptidases (PCPs). Their pro-region, usually called 'activation segment', is approximately 95 residues long and is placed at the N-terminus of the enzyme. It folds in a globular independent unit known as 'activation domain', which is linked to the enzyme through a connecting segment. The other subfamily is the N/E group whose members are not produced as inactive proenzymes. Instead of the pro-domain found in members of the A/B subfamily, members of the N/E subfamily contain an approximately 80-residue region following the 300-residue CP domain; this extra 80-residue region has structural homology to transthyretin.[3] Figure 1 shows a phylogenetic tree of human metalloCP that illustrates the variety of these enzymes in mammals and their distribution into subfamilies.

OCCURRENCE

CPs are found throughout the animal kingdom, with at least one member in every animal species whose genome has been sequenced to date. CPs are also found in plants (*Arabidopsis thaliana*) and some bacteria and unicellular organisms (such as *Streptomyces capreolus*, *Streptomyces griseus*, *Streptomyces coelicolor*, *Thermoactinomyces vulgaris*, *Sulfolobus solfataricus*, *Metarhizium anisopliae*, *Staphylococcus epidermidis*, *Porphyromonas gingivalis*, *Bacillus subtilis*, *Bacillus sphaericus*, and others). Altogether, there are 17 members in the human genome, most of which have been expressed and characterized. Within mammals, at least one CP is present in all tissues or fluids that have been examined. However, most CPs have fairly restricted patterns of tissue distribution, reflecting a specific physiological function.

BIOLOGICAL FUNCTION

In mammals, CPs perform a variety of functions ranging from degradation to biosynthesis, depending on the peptidase. The enzymes produced by the exocrine pancreas and secreted into the digestive tract (CPA1, CPA2, and CPB) function in the breakdown of peptides in food;[4–6] these peptides are largely generated by the action of endopeptidases such as trypsin and chymotrypsin. Mast cell CPA (which was recently renamed CPA3) presumably functions in anaphylactic and inflammatory processes and cleaves peptides and proteins following the action of chymase.[7] CPU (also known as CPB2, CPR, plasma CPB, and TAFI) circulates in plasma and is activated by the thrombin-thrombomodulin complex.[8] Once activated, CPU controls the rate of fibrinolysis by cleaving C-terminal lysine residues that serve as plasminogen binding sites. There are four additional members of the A/B subfamily that have been identified, but only one of these (CPA5) has been expressed and shown to encode an active enzyme.[2] On the basis of modeling studies of the active sites, the others are predicted to encode active CPs. One of these, named CPO, appears to encode an enzyme with a specificity for C-terminal acidic residues.[2]

Several members of the N/E subfamily of CPs function in the processing of peptide hormones and neurotransmitters. CPE is a neuroendocrine-specific enzyme that is primarily involved in the intracellular production of most neuroendocrine peptides.[9,10] CPE removes C-terminal lysine and arginine residues from intermediates formed by the action of prohormone convertases 1 and 2 on the peptide precursor.[11] CPD also contributes to the processing of peptides, although it is very broadly distributed throughout the body and primarily functions in the *trans*-Golgi network and constitutive secretory pathway; putative substrates include growth factors and receptors that are produced from larger precursors by cleavage at basic amino acid–containing sites.[9,12] CPM is attached to a wide variety of cells via a glycosylphosphatidylinositol linkage and functions in the extracellular processing of peptides and proteins.[13] CPN circulates in plasma together with an 83-kDa binding protein, and removes basic residues from the C-terminus of various blood proteins/peptides.[14] CPZ is present in the extracellular matrix, with a broad distribution during embryogenesis and a more restricted pattern in adult tissues.[15] Although the precise function of CPZ is not known, it is likely that this enzyme cleaves intermediates generated by the various matrix endopeptidases.

Interestingly, all animal species examined express proteins that are members of the CP family, based on amino acid–sequence conservation, but which lack one or more critical active site or substrate-binding residues. For example, 3 of the 17 human members of this family (CPX-1, CPX-2, and AEBP-1) are missing key catalytic and/or substrate-binding residues and do not have significant activity toward standard CP substrates.[15] The function of the inactive animal CP-like proteins is not yet known. It is possible that they cleave other types of substrates. Alternatively, because some of them lack residues that are important for catalytic activity, they may function as binding proteins.

Nonmammalian animals also contain CPs of both the A/B and the N/E subfamilies. For example, the *Drosophila melanogaster* genome has 2 N/E-like CPs (one of which is very similar to mammalian CPD)[16] and over a dozen A/B-like CPs. The *Drosophila* CPD homolog appears to function in the processing of growth factors and/or growth factor receptors based on the phenotype of animals with mutations within the CPD gene.[17]

The yeast genome (*Saccharomyces cerevisiae*) contains a single member of the CP family (ECM14/YHR132C). However, this protein is missing the catalytically important Glu270 equivalent (discussed below), and thus may not encode an active enzyme. Despite this putative lack of CP function, the protein is thought to function in the organization and/or biogenesis of the cell wall, based on the hypersensitivity of a mutant to a cell surface perturbing agent.[18]

A CP-like protein in *B. sphaericus* (gamma-D-glutamyl-(L)-meso-diaminopimelate peptidase 1) is missing several important substrate-binding residues.[19] Despite having considerable amino acid sequence similarity to other members of the CP family, gamma-D-glutamyl-(L)-meso-diaminopimelate peptidase 1 is not a functional CP, but instead cleaves a peptide bond of a peptidoglycan fragment.[19]

The function of CPs in plants and in bacteria and other unicellular organisms has not been reported.

AMINO ACID SEQUENCE INFORMATION

Subfamily A/B (Merops database: Clan MC; subfamily M14A)

- Carboxypeptidase A1; *Bos taurus* (bovine), precursor (pre-proenzyme), 419 amino acid residues (AA), translation of cDNA sequence[20] (Swiss-Prot: P00730)
- Carboxypeptidase A1; *Homo sapiens*, precursor (pre-proenzyme), 419 AA, translation of cDNA sequence[21] (Swiss-Prot: P15085)
- Carboxypeptidase A; *Helicoverpa armigera* (cotton bollworm), precursor (pre-proenzyme), 433 AA, translation of cDNA sequence[22] (TrEMBL O97389)
- Carboxypeptidase A2; *Homo sapiens*, precursor (pre-proenzyme), 417 AA, translation of cDNA sequence[23] (Swiss-Prot: P48052)
- Carboxypeptidase B; *Homo sapiens*, precursor (pre-proenzyme), 417 AA, translation of cDNA sequence[24] (Swiss-Prot: P15086)
- Carboxypeptidase B; *Bos taurus* (bovine), sequence of active enzyme, 306 AA[25] (Swiss-Prot: P00732)
- Carboxypeptidase U; *Mus musculus* (mouse) precursor, 422 AA, translation of cDNA sequence[26] (TrEMBL: Q9JHH6)
- Carboxypeptidase T; *Thermoactinomyces vulgaris*, precursor (pre-proenzyme), 424 AA, translation of cDNA sequence[27] (Swiss-Prot: P29068)

Subfamily N/E (Merops database: Clan MC; subfamily M14B)

- Carboxypeptidase D domain II; *Anas platyrhynchos* (duck), 380 AA, internal sequence of metallocarboxypeptidase D multipeptidase, translation of cDNA sequence[28] (Merops M14.016)

- Carboxypeptidase N catalytic chain; *Homo sapiens*, precursor, 458 AA, translation of cDNA sequence[29] (Swiss-Prot: P15169)
- Carboxypeptidase E; *Bos taurus*, precursor, 434 AA, translation of cDNA sequence[30] (Swiss-Prot: P04836)
- Carboxypeptidase M; *Homo sapiens*, precursor, 443 AA, translation of cDNA sequence[31] (Swiss-Prot: P14384)
- Carboxypeptidase Z; *Homo sapiens*, precursor, 641 AA, translation of cDNA sequence[32] (TrEMBL: O00520)

Besides those described above, over 100 entries for metallocarboxypeptidases can be found in the Swiss-Prot or MEROPS databases.

PROTEIN PRODUCTION, PURIFICATION, AND MOLECULAR CHARACTERIZATION

Pancreatic CPs, either in the enzyme or proenzyme forms, have been isolated from the pancreas of several species including human, bovine, porcine, and others.[33–36] A stable enzyme can be purified from pancreatic acetone powder through successive steps of salt precipitation and anion exchange, DEAE-based chromatographies.[36,37] This method also allows for the purification of the intact proenzyme forms. Internal proteolysis and degradation may occur if the selected source is pancreatic secretion.[38] The purified enzymes are stable for years in the precipitated or frozen states. The methylotrophic yeast *Pichia pastoris* has been used to overexpress pancreatic pro-CPs at high yield. Using the yeast α-mating factor signal for secretion, the proteins are recovered from the culture medium and purified by hydrophobic and anion exchange chromatographies.[39,40] Other nonpancreatic members of the A/B subfamily have been obtained from plasma[41] or mast cells.[42] Plasma CPU is isolated from plasma by differential precipitation, followed by affinity chromatography on a monoclonal antibody-affinity column, followed by two additional chromatographic steps,[43] or by anion and gel-filtration chromatographies, and a final step of affinity chromatography on a plasminogen-Sepharose column.[44] Baby hamster kidney cell lines transfected with a plasmid containing CPU cDNA express a recombinant protein that can be purified from the medium through chromatographic steps on antibody-Sepharose and DEAE-Sepharose columns.[45] CPA3 can be isolated from mast cells by an affinity chromatography method, using the potato carboxypeptidase inhibitor (PCI).[42,46]

Members of the N/E subfamily of CPs have been expressed in insect cells using the baculovirus system and in *P. pastoris*.[8,47,48] Purification is readily achieved on the substrate affinity resin *p*-aminobenzoyl-Arg Sepharose.[49] For CPE and CPZ, the bound protein is eluted by shifting the pH from acidic (5.5–6) to basic (8–8.5) conditions. CPD, N, and M require competition with an active site-directed reagent (i.e. arginine or an arginine-based inhibitor) to elute the bound protein. This single-step purification results in homogeneous preparations of CPE, D, M, and N when starting with material from an overexpression system. CPZ can be expressed with the baculovirus system but the resulting protein cannot be extracted from the cells; thus, mammalian cell culture systems have been used to express the protein.[50] Owing to the lower levels of expression, additional chromatography on ion exchange or heparin agarose is required for homogeneous CPZ.[50] CPX-1 and CPX-2 have not been successfully purified to homogeneity from any expression system. AEBP-1 was expressed in bacteria as a His6-fusion protein and purified on a metal-chelate column.[51] Although the authors claimed that the resulting protein was active, the amount of activity was extremely low and may have resulted from a contaminating protease. Furthermore, attempts to repeat the enzyme assay in another laboratory using the same reagents and procedure were unsuccessful (LD Fricker, unpublished results).

Pancreatic carboxypeptidases contain one to three disulfide bridges and no glycosylation sites. Two other pancreatic-like forms, human plasma CPU and *H. armigera* CPA contain respectively three and one disulfide bridges in the catalytic domain and four glycosylation sites in the pro-region in the former case and one glycosylation site in the catalytic region in the latter.[52] The precise number of disulfide bridges in the members of the N/E family has been experimentally determined only in a few cases. Thus, the central repeat of CPD, CPD-2, contains one disulfide bridge and three N-linked glycosylation sites; CPE contains two to three disulfide bridges, depending on the species, and two N-linked glycosylation sites.[53]

Whereas all members of the A/B subfamily contain a single carboxypeptidase domain, all members of the N/E family contain additional domains that probably serve to target the protein. Some of these domains, such as the C-terminal regions of CPE, CPM, and CPN, are small. CPE contains an amphipathic helix that allows the protein to bind peripherally to membranes within the acidic pH of the secretory vesicle.[54] CPM contains a sequence for the addition of glycosylphosphatidylinositol, which, in turn anchors the protein to the extracellular membrane.[55] CPN binds to an 83-kDa binding protein as a multimeric 280-kDa complex in plasma.[14] Other members of the N/E family contain larger and/or multiple domains. CPD contains three CP-like domains followed by a transmembrane domain and a 60-residue cytosolic tail.[56–58] Of the three CP-like domains, the first two are enzymatically active while the third is missing key catalytic residues and does not cleave standard CP substrates.[48,59] CPZ contains domains that target the protein to the extracellular matrix, which are presumably located within the C-terminal region.[60] In addition, this protein contains a cysteine-rich region that has amino acid sequence similarity to the Wnt protein-binding region of a variety of proteins that function in growth and development;[60] Wnt derives from Wg and int, both referring to chromosomal loci involved in critical aspects of early embryonic development. CPX-1, CPX-2, and AEBP-1 contain a domain that has amino acid sequence similarity to the carbohydrate binding domain of the slime mold protein discoidin.[51,61,62] The function of this domain is not known.

Most A/B metalloCPs are monomeric, but several of them have been reported to form stable oligomeric structures in the proenzyme form with zymogens of serine proteases.[63] The oligomerization seems to be species-dependent, since ternary complexes involving PCPA1 are found in ruminants (ox, goat, camel), and binary complexes involving PCPA1 and A2 in nonruminants (human and rat).

METAL CONTENT

All of the enzymatically active CPs bind one atom of Zn^{2+} at the active site.[64,65] The Zn coordination at the active center of CPA is shown in Figure 2. Members of the CP family that are inactive against standard CP substrates lack some of the cation-binding ligands and may therefore not be able to bind the metal. CPA can bind a second zinc ion, which perturbs the active site and renders the enzyme inactive.[66] In addition to zinc, CPT from T. vulgaris binds Ca^{2+} through four calcium binding sites, a fact that could account for the high thermostability of the protein.[67] CPE was reported to bind Ca^{2+}, with the binding having a small, pH-dependent effect on enzyme activity and stability, suggesting that calcium may play a role in the regulation of the enzyme activity.[68]

ACTIVITY TEST

A large number of substrates have been employed to assay for the various CPs. Simple spectrophotometric assays employ either the hippuryl group (benzoyl-Gly) or the furylacryloyl group (FA) and monitor an increase in absorption at 256 nm (hippuryl substrates) or a decrease in absorption at 336 nm (FA substrates) upon cleavage.[4–7,14] Depending on the attached amino acid, these substrates are either selective for A-like cleavages (hippuryl-Phe, FA-Phe-Phe) or B-like cleavages (hippuryl-Lys, hippuryl-Arg, FA-Ala-Lys, FA-Ala-Arg). Carbobenzoxy derivatives, arazoformyl peptide surrogates,[70] and other compounds are also used as substrates. Although the enzymatically active members of the N/E subfamily will cleave the hippuryl and FA substrates with C-terminal lysine or arginine residues, most of these CPs greatly prefer longer substrates.[8,12] For CPE, tripeptides are cleaved much more rapidly than dipeptides, and the standard substrate is dansyl-Phe-Ala-Arg.[71] Unlike the hippuryl- or FA-containing substrates, the dansyl-Phe-Ala-Arg does not undergo a spectral shift upon cleavage of the C-terminal bond. Instead, the standard assay uses chloroform to extract the product (the substrate remains in the acidified aqueous phase). The assay can be performed in glass test tubes; if enough chloroform is added to the tube to push the aqueous phase above the path of the light beam, the phases do not need to be mechanically separated prior to reading in a fluorimeter.[71]

X-RAY STRUCTURE

Crystallization

The first crystal structures of CPs were obtained for bovine CPA by Lipscomb et al.[64,72] The protein, prepared from pancreatic acetone powder, gave monoclinic crystals that belonged to the space group $P2_1$ with $a = 51.6$ Å, $b = 60.27$ Å, $c = 47.25$ Å (1 Å = 0.1 nm) in lithium chloride (0.25 to 0.35 M), pH 7.5. The crystals diffracted to a resolution of 1.54 Å. The high-resolution structure of bovine CPA (entry 5CPA in the PDB) has been used in molecular replacement procedures for the derivation of most of the X-ray structures of related CPs. The 3D structures of four additional active enzymes have been solved: bovine CPB,[73] rat CPA2,[74] T. vulgaris CPT,[67] and the second

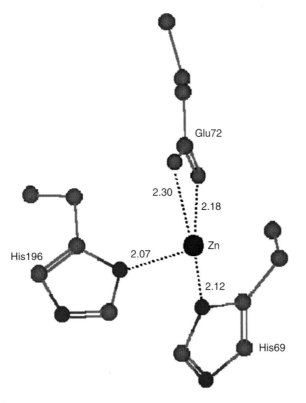

Figure 2 Drawing of the Zn coordination at the active center of CPA. The displayed bond distances are expressed in angstroms (Å). Figure made with MolMol.[69]

domain of duck CPD.[3] Other CP structures have been obtained from PCPs, the precursor forms of the enzyme: porcine PCPB (1NSA),[75] porcine PCPA1 (1PCA),[76] bovine PCPA1 (1PYT) (in complex with two serine-proteinase precursors),[77] human PCPA2 (1AYE),[78] *H. armigera* PCPA (1JQG),[52] and human PCPB (1KWM).[79]

Overall description of CP structures

The tertiary folding of CPs corresponds to the α/β hydrolase fold and is formed by a central mixed parallel/antiparallel 8-strand β-sheet, with a 120° twist between the first and the last strand, over which 8 α-helices pack on both sides to form a globular molecule (see 3D Structure). The active center is located in a cavity formed by parallel strands of the internal β-sheet, two helices and a nonregular extension that partially covers it. The active site Zn^{2+} as well as the peptide substrates are held at the active center by residues that are mainly located in turns or in loops connecting those secondary structures and protruding from them. This catalytically obligatory zinc ion is penta-coordinated to the ND1 atoms of His69 and His196, the OE1 and OE2 atoms of Glu72, and to a water molecule in a slightly distorted tetrahedral manner, with one of the tetrahedron vertices occupied by the two glutamate oxygen ligands[80] (the numbering system corresponds to bovine CPA and will be used throughout).

Many structural studies with bovine CPA involving the use of inhibitors, substrate or transition-state analogs, reaction-coordinate analogs, slowly hydrolyzed substrates,[80] as well as site-directed mutagenesis studies[81,82] have led to the identification of the residues involved in the catalytic machinery and the binding of substrates. These residues can be ascribed to several subsites: S1' (including Arg145, Try248, and Asn144, involved in the anchoring and neutralization of the C-terminal carboxylate of the substrate), S1 (including Arg127 and Glu270, involved in the polarization of the carbonyl of the peptide to be cleaved, and in the proton exchange respectively), and the secondary sites S2 (Arg71, Tyr198, Ser197, Ser199, and Tyr248). Besides, subsites S3 and S4 have been suggested for extended substrates.[83,84] The active site architecture and the relative positions of the residues involved in substrate binding are common to all the known structures with the exception of Tyr248. In the presence of substrates, inhibitors, or anions that occupy the specificity pocket, Tyr248 rotates about its χ1 torsion angle in a movement that spans about 13 Å and fixes it in the 'down' position, closing the access to the active site, as observed in the structures of porcine PCPA1, human PCPA2, and CPD-2, among others. The long-range movement of Tyr248 and of its associated loop has been used as an example of the induced fit between enzyme and substrate.[80]

The varied specificities of different CPs are largely due to the character of the residue at position 255 that is an isoleucine or an aspartic acid in enzymes with A-type or B-type activities respectively. In CPD-2, this role is carried out by Asp192, at a position equivalent to Asp255 in CPB.[3] Species that are evolutionarily distant from mammals may display a less restricted specificity and thus have variations in this common scheme, as is the case in *H. armigera* CPA, which has a serine at position 255.[52]

The single disulfide bridge (Cys138–Cys161) observed in bovine CPA is also found in most members of the A/B subfamily with the exception of CPT. However, A2 forms display an additional bond between residues 210 and 244, and the B-forms display two additional bonds between residues 66–79 and 152–166. One disulfide bridge between Cys230 and Cys275 (numbering system of CPD-2) is observed in the only available structure for the N/E subfamily. No S–S bridges are found in the activation domain of PCPs.

Only members of the N/E subfamily contain an additional C-terminal domain of about 80 amino acids. Figure 3 displays its position relative to the enzyme domain. This domain displays a rodlike shape with dimensions 25 × 25 × 40 Å, smaller than those of the enzyme moieties of about 50 × 50 × 40 Å,[3] and folds in a 7-stranded β-barrel that shares topological similarity with transthyretin, a plasma protein associated with several forms of amyloidosis.[85] The transthyretin-like C-terminal domain displays two N-linked glycosylation sites, and one additional site is found at the enzyme domain of CPD-2. Besides CPD-2, only plasma PCPU (also known as TAFI) is predicted to support glycosylation in its activation domain.[79]

Besides the structures directly deduced from X-ray diffraction studies, some CP sequences for which the growth

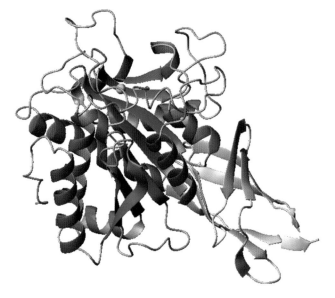

Figure 3 Schematic representation of the structure of CPD domain 2. The enzyme domain is shown in the same colors and orientation as the initial 3D Structure, whereas the C-terminal, transthyretin-like domain is shown in gray.

of crystals has proved difficult have been modeled. Among N/E members, the model structures of domains 1 and 3 of duck carboxypeptidase D and human CPE were built on the basis of the structure of CPD domain 2, using a method of comparative modeling by satisfaction of spatial constraints.[86] The generated models predict slightly different shapes in the access to the active site, implying some structural selection for proteins or peptides. They also confirm the prediction of CPD-3 being an inactive domain because of the absence of catalytically indispensable residues at the expected positions. Modeling of CPU,[79] a thrombin-activatable fibrinolysis inhibitor, besides showing that its fold is closer to B- than to A-forms, has also given some clues about the intrinsic instability of the active enzyme in plasma.[45] Also, three further members of the A/B subfamily, namely, PCPA5, PCPA6, and CPO, were modeled following similar procedures.[2] On the basis of this model, CPO appears to represent a novel CP that is specific for acidic C-terminal residues.

Comparison among CP forms

All metallocarboxypeptidases share the same global fold in their enzyme moieties, although greater similarities are found in closer members of the family. For instance, bovine and porcine CPA have an rmsd of 0.38 Å, and the corresponding value for the porcine and human forms of PCPB is 0.53 Å, with the main differences residing in surface loops.[76,79] In contrast, superposition of the 303 common Cα atoms of bovine CPA and CPT gives an average deviation of 2.1 Å.[67] In the latter case, the deviation arises from the different positions of some secondary structure elements, from insertions in surface loops, and from the different orientations of the N- and C-termini.

At the active site of human CPB, although nearly identical to that observed in porcine CPB or bovine CPA, His69 displays two alternate conformations, with approximately equal occupancies.[79] This induces several changes in other close residues, which, in the published structure of the human B zymogen, make new interactions with residues at the activation domain. These rearrangements also bring about a side-chain flip of Tyr248 that displays the 'down' position, representing the only case in which this observation is made for an enzyme with an empty specificity pocket.

Most members of the N/E subfamily of CPs function in the processing of biologically active peptides, whereas most members of the A/B subfamily act in the digestion of dietary protein. These functional differences are also reflected in the 3D structures. The CPD-2 catalytic domain displays topological similarity with both CPA/CPB and CPT, although the similarity is less than that observed between CPA and CPT. The most important differences are observed in three chain segments forming the funnel-like access to the active site cleft,[3] where insertions and deletions shape an entrance to the active site of CPD-2 that is clearly distinguishable from the various members of the A/B subfamily. This difference suggests that CPD has a distinct selectivity toward larger substrates, which is supported by experimental evidence showing that this enzyme is not inhibited by potato or leech carboxypeptidase inhibitors.

Zymogen structure

The first PCP to have its 3D Structure determined was PCPB. Parallel X-ray diffraction and NMR studies described respectively, the fold of the complete proenzyme and of the isolated activation domain,[75,87] serving as one of the first examples that demonstrated the coincidence of high-resolution structures obtained by both methods.[88] The nonenzymatic moiety of the zymogen consists of a globular activation domain of about 80 residues and a partially helical connecting region of about 15 residues that spans to the N-terminus of the enzyme. The helical portion of the connecting segment is followed by a loosely arranged loop that contains the primary targets for tryptic activation (Figure 4). The activation domain folds as an open sandwich with a four-stranded antiparallel β-sheet on one side and two antiparallel α-helices on top, defining a hydrophobic core that itself maintains the domain stable in solution, without the contribution of any disulfide bridge. It interacts with the enzyme at the side of the β-sheet,

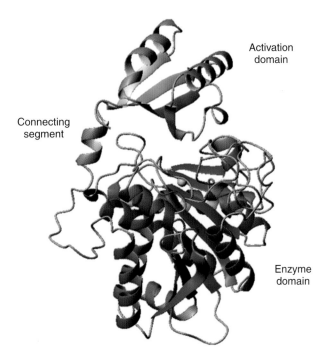

Figure 4 Schematic representation of the structure of procarboxypeptidase B. The enzyme domain is shown in the same colors and orientation as the initial 3D Structure, whereas the pro-region (containing the activation domain and the connecting segment) is shown in yellow and gray. Figure made with MolMol.[69]

particularly through the second strand and the loop that links it to the third strand, shielding the preformed active site and establishing specific interactions with residues that are important for substrate recognition. The above-mentioned loop contains a *cis*-pro bond and a 3_{10}-helix turn that have been found to be responsible for the complete lack of intrinsic activity of the zymogens of B-forms. The structure of the CP moiety in PCPB superimposes without any remarkable difference to that of the isolated enzyme, showing that the mechanism of activation does not imply structural rearrangements of the active site as is the case in serine proteases.[89]

All of the activation segment structures deduced so far share the general topology described above for PCPB, with minor differences between the various forms. The connecting segment that links the activation domain to the enzyme moiety folds partially as an α-helix, which is two turns long in the B proenzymes, four turns long in A1 forms, and five turns long in A2 forms. In all the cases, however, the primary activation target is fully solvent-accessible. Although the globular activation domain shields the preformed active center, none of the residues involved in subsite S1 is directly affected by the shielding, nor are S1′ residues except in the case of the B-forms (see further on). In contrast, subsites S2, S3, and S4, involved in the binding of extended substrates, are blocked by specific contacts with activation domain residues.

The more important differences between activation domains from different forms are observed at the region that covers the entrance to the active site of the enzyme: the *cis*-pro bond and the 3_{10} helical turn observed in the two B-forms for which a structure has been released are absent from all the A-forms studied so far. Moreover, sequence alignments predict that this is a characteristic shared by all proteins with a known sequence. The effect of this conformation in B proenzymes is that Asp41 in the activation segment is positioned to form a salt bridge with Arg145 in the enzyme moiety, a residue directly involved in the anchoring of the C-terminal carboxylate of the enzyme. This interaction is likely to be responsible for the absence of any intrinsic activity in B zymogens. The A-forms, which lack both the *cis*-pro bond and the 3_{10} helical turn show some residual activity against small synthetic substrates.[75]

Complex structures

Structural studies with CPs complexed with substrate analogs or inhibitors have been of much help in the definition of their catalytic mechanism and mode of substrate binding.[80] In most cases, however, the low–molecular weight compounds that have been used can afford no information about extended substrate binding. Two complex structures are available with natural protein inhibitors/substrates of CPs, thus providing a means of circumventing the problem mentioned and also of understanding the mechanism of action of protein inhibitors on proteases.[90,91]

In the crystal structure of the complex between bovine CPA and the 39-residue PCI, the four C-terminal residues of the inhibitor are bound to the active site groove of the enzyme, defining subsites S1′ to S3.[90] The inhibitor acts primarily as a substrate since its C-terminal glycine is cleaved, but it remains trapped at the active site pocket while the rest of the molecule remains bound to CPA through interactions with the C-terminal tail and also with a secondary contact region in which five residues of PCI participate through polar and nonpolar but not ionic interactions.[90]

Leech CP inhibitor (LCI) is also a small (66-residue) monodomain protein inhibitor of A/B subfamily CPs, whose structure is also stabilized by disulfide bridges (4 in this case, as compared to 3 in the case of PCI). However, LCI is richer in secondary structure elements as it contains a five-stranded antiparallel β-sheet and a short helix. All cysteine involved in S–S bridges are located within the regular secondary structure. A primary interaction site and a secondary contact region can also be defined in this case. In the former, the occupancy of the active site groove is very similar to that of PCI, with the C-terminal residues Pro63 to Glu66 occupying subsites S3 to S1′. LCI also acts as a substrate, Glu66 is cleaved off and trapped in the active site, and Val65 becomes the new C-terminal of LCI and is coordinated to the Zn atom.[91] The secondary contact region is larger and differs from the corresponding region in PCI, involving some hydrogen bonds between main chain atoms from the enzyme. This may explain the stronger interaction of LCI as measured by K_i constants.[39]

CATALYTIC MECHANISM

Most crystallographic data support the promoted-water pathway[80] for the hydrolysis of peptides by CPA. In this model, the zinc ion of CPA is a classical electrophylic catalyst that provides electrostatic stabilization for the negatively charged intermediates generated during the hydrolysis[80] and promotes a water molecule as a potent nucleophile[92,93] to attack the scissile peptide bond of the substrate, leading to a tetrahedral transition state (Figure 5). The nucleophilicity of the metal-bound water is enhanced by a hydrogen bond with Glu270, which acts as the general acid–base catalyst. While Glu270 abstracts a proton from the water molecule, Arg127 polarizes the peptide bond prior to hydrolysis and also stabilizes the tetrahedral transition state.[80] As commented above, the zinc binding residues are His69, Glu72 and His196, with the catalytic water molecule also affording a coordination bond. From crystallographic studies, amino acids Asn144, Arg145, Tyr248 in S1′ and Arg127 and Glu270 in S1 have been defined as important for substrate binding and

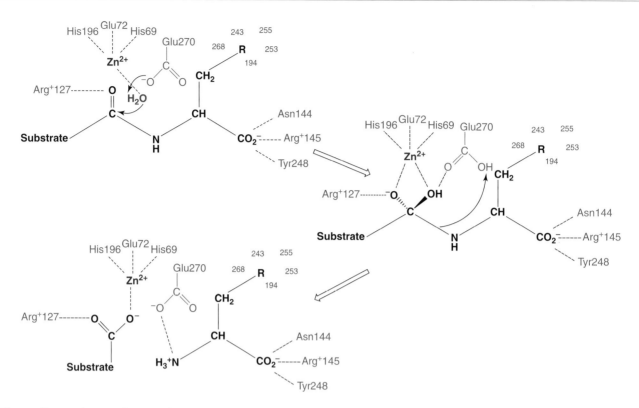

Figure 5 Catalytic mechanism of CPA (adapted from reference 79). Substrate atoms are in black, and some active site residues and residues involved in the specificity pocket are in red and purple respectively.

catalysis and are shared by all enzymatically active members of the metalloCP family. The terminal carboxylate group of the peptide substrate is fixed by Asn144, Arg145, and Tyr248, while the carbonyl group of the scissile peptide bond becomes positioned near Glu270, Arg127, and zinc. The differences are restricted to the positions that define subsites S2, S3, and S4, and to the amino acids that define the specificity for a C-terminal side chain. The amino acid residue at position 255 appears to be the main determinant of specificity. It is an isoleucine in CPA1 and CPA2 and an aspartic acid in CPB, allowing for the binding of hydrophobic and basic residues respectively. The enzymatically active members of the N/E subfamily do not use the residue in the position equivalent to residue 255 of bovine CPA to provide the specificity for basic C-terminal residues.[3] This residue is either a glutamine (CPE, CPD domains 1 and 2, CPM, CPN) or serine (CPZ). On the basis of the crystal structure,[3,86] the role of binding to the positive charge of the substrate's side chain in CPD domain 2 is carried out by Asp192, at a position equivalent to Gly207 in CPA. The use of small synthetic inhibitors has also been useful in defining the active center of the CPs from the N/E subfamily, as can be appreciated in Figure 6, where the active site of CPD domain 2 in complex with the guanidinoethylmercaptosuccinic acid (GEMSA) inhibitor is shown. Because Asp192 is conserved in all members of the N/E subfamily, it is likely to perform the same function in all of them.

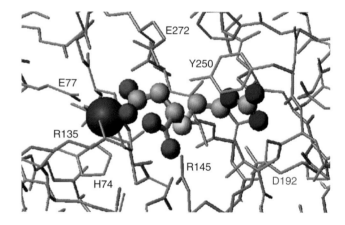

Figure 6 Structure of the GEMSA inhibitor-complexed active site cleft of CPD domain 2 where the two carboxylate groups of the inhibitor coordinate the catalytic zinc ion. Some residues are numbered and Asp192 is marked in red. Tyr250 (equivalent to Tyr248 in CPA) is in the 'down' conformation. His74, Glu77, and Glu272 are equivalent to His69, Glu72, and Glu270 in CPA. Figure made with MolMol.[69]

The other residues that line the specificity pocket also show a higher variability among the various members of the CP family. For instance, the replacement of Ile243 by glycine in the B-forms contributes to the creation of a polar environment. Leu203 and Thr268 in A1 are replaced by methionine and alanine respectively, in A2,[78] rendering a binding pocket more capable of accommodating a bulky substrate side chain. The presence of a serine at position 255 makes the specificity pocket of *H. armigera* CPA smaller and more polar.[52]

FUNCTIONAL ASPECTS

Specificity

The various CPs differ in their optimal substrates. For those enzymes produced in the same tissues, the substrate preferences tend to be complementary. For example, CPA1 prefers aliphatic and smaller aromatic residues such as phenylalanine and tyrosine, while CPA2 prefers the bulkier tryptophan residue[4,5] (see Table 1 for characteristic kinetic constants of members of the A/B subfamily). Mechanistically, this difference can be largely attributed to the amino acid differences in the side-chain binding pockets of the two enzymes commented upon in the previous section. The dramatic difference in specificity between CPB-like and CPA-like enzymes has been attributed to the residue in position 255. In the A-like enzymes, the presence of a hydrophobic residue in this position (isoleucine, valine, leucine, methionine) produces the specificity for cleavage of aliphatic/aromatic residues, while the presence of an acidic residue (aspartic acid) in CPB and CPU produces the specificity for cleavage of basic residues. The bacterial enzymes that can cleave either hydrophobic or charged C-terminal residues have a small chain residue (such as threonine) in the position comparable to residue 255 in bovine CPA. An aspartic acid residue plays the same role in N/E subfamily members in a structurally similar, although sequentially distinct, position.

The various members of the N/E subfamily that are enzymatically active all cleave C-terminal basic residues. CPE shows a slight preference for arginine over lysine, as do CPM and the first domain of CPD.[8,13,48] In contrast, the second domain of CPD and CPN prefer lysine to arginine[14,48] (Table 2). The substrate specificity of CPZ has not been examined. CPM and CPN are maximally active at neutral pH, while CPE is maximally active at pH 5–6.[8,13,14] This preference reflects the environment in which the enzyme functions; CPM is extracellular, CPN circulates in plasma, and CPE is enriched in peptide-containing secretory vesicles.[8,13,14] The two domains of CPD that are active as CP have distinct pH optima; the first domain is optimal at neutral pH while the second domain is optimal at slightly acidic pH values.[48] This presumably reflects the broad range of pH encountered by CPD during its movement from the *trans*-Golgi network to the cell surface and back.[9]

Zymogens and activation

Pancreatic and pancreatic-like carboxypeptidases are synthesized as inactive zymogens known as procarboxypeptidases, which are peculiar among pancreatic proenzymes in that their pro-peptides are remarkably long: they may vary from 94 to 96 residues, which accounts for about one-fourth of the molecule. The pro-region or activation segments, besides maintaining the enzyme inactive in the presecretional phase, may also act as a postactivation mechanism of activity control[100,101] and also constitute important elements for the proper folding of the enzyme moiety. Thus, it has been observed that the presence of this piece greatly facilitates the heterologous expression of the proenzymes,[102] with the pro-region probably assisting in the folding of pancreatic carboxypeptidases *in vivo*.[40]

The sequence identity between the pro-region of digestive metalloCPs is lower than for the enzymes and varies from 50 to 20%. However, an overall identity in folding topology is observed in all the proteins studied. The ~80-residue globular core of the pro-region of PCPs, or activation domain, covers the active site of the enzymes, shielding the S2, S3, and S4 subsites. The activation domain is an independent folding unit with no disulphide bridges and constitutes a good model for studies on protein folding and redesign. The activation domain of human PCPA2 has been extensively studied by site-directed mutagenesis with the aim of understanding its folding process and stability, and to get information about the structural determinants for the improvement of its inhibitory action. The initially high stability of the protein can be further increased by

Table 1 Kinetic constants for peptide substrate hydrolysis by members of the A/B subfamily of metallocarboxypeptidases

Substrate	CPA1[a]			CPA2[b]			CPB[c]		
	K_m (μM)	k_{cat} (s^{-1})	k_{cat}/K_m (μM^{-1} s^{-1})	K_m (μM)	k_{cat} (s^{-1})	k_{cat}/K_m (μM^{-1} s^{-1})	K_m (μM)	k_{cat} (s^{-1})	k_{cat}/K_m (μM^{-1} s^{-1})
Cbz-Gly-Gly-Phe	172	131.5	0.76	372	58.3	0.15			
Cbz-Gly-Gly-Tyr	102	56.3	0.55	125	70	0.56			
Cbz-Gly-Gly-Trp	NM	NM		146	90.3	0.62			
Hippuryl-Arg							180	61	0.34

Cbz: carbobenzoxy; Hippuryl: N-benzoylglycine; NM: nonmeasurable.
[a] Bovine CPA1; kinetic data taken from reference 94.
[b] Human CPA2; kinetic data taken from reference 95.
[c] Human CPB; kinetic data taken from reference 96.

Table 2 Kinetic constants for peptide substrate hydrolysis by members of the N/E subfamily of metallocarboxypeptidases

Substrate	CPE[a]			CPD1[b]			CPD2[b]			CPM[c]			CPN[d]		
	K_m (μM)	k_{cat} (s^{-1})	k_{cat}/K_m (μM^{-1} s^{-1})	K_m (μM)	k_{cat} (s^{-1})	k_{cat}/K_m (μM^{-1} s^{-1})	K_m (μM)	k_{cat} (s^{-1})	k_{cat}/K_m (μM^{-1} s^{-1})	K_m (μM)	k_{cat} (s^{-1})	k_{cat}/K_m (μM^{-1} s^{-1})	K_m (μM)	k_{cat} (s^{-1})	k_{cat}/K_m (μM^{-1} s^{-1})
Dansyl-Phe-Ala-Arg	34	13	0.38	15	5.1	0.35									
YGGFLR (Leu-Enk-R)				23	2.16	0.09	25	11	0.41	63	1.77	0.028	57	6.25	0.11
YGGFMR (Met-Enk-R)							95.4	1.57	0.016	46	15.6	0.34	49	17	0.34
YGGFMK (Met-Enk-K)										375	11.1	0.030	216	103	0.48

Dansyl: 5-dimethylaminonaphthalene-1-sulphonyl; Enk: enkephalin.
[a] Bovine CPE; kinetic data taken from reference 97.
[b] Duck CPD; kinetic data taken from reference 48.
[c] Human CPM; kinetic data taken from reference 98.
[d] Human CPN; kinetic data taken from reference 99.

the improvement of helix propensities through site-directed mutagenesis.[103]

CPs can be inhibited as long as the activation domain of the pro-region is kept in place. Two regions of the pro-regions are clearly responsible for the interaction with the enzyme: the globular activation domain and the connecting region that covalently links both globular moieties. The contact areas of both moieties with the enzyme domain are necessary, and none of them are sufficient to maintain the zymogen state of PCPs, as has been shown by site-directed mutagenesis studies on the B form[40] and by modeling and theoretical approaches.[24] In the PCPs of some species, the whole activation segment of the A1 forms seems to behave as a strong inhibitor of the enzyme after being cut from it, whilst this is not the case with those of the A2 and B-forms.[101]

Protein inhibitors of CPs

Unlike endoproteases, for which numerous examples of protein inhibitors have been reported, natural polypeptides that specifically inhibit metalloCPs have few examples described. Two of these proteins have been crystallized in complex with a CP. The PCI inhibitor isolated from potato, a 39-residue protein, is among the smallest globular proteins described. PCI belongs to the cystine-knot or T-knot super-family of proteins, named for their particular pattern of disulfide bridges. From the 27-residue globular core of PCI protrudes a 7-residue N-terminal and 5-residue C-terminal tail. No functional role has so far been assigned to the N-terminal tail (residues 1–6), whose conformation seems to be undefined. The C-tail docks into the active site of the carboxypeptidase, leading to a stopper-like inhibition mechanism. Site-directed mutants of PCI and computer simulation analysis have allowed the identification of the key residues for PCI–CPA interaction in the PCI tail and the estimation of the energy-related contribution of their chemical groups to the binding to CPA.[104,105]

The LCI CP inhibitor isolated from leeches displays a structure fairly different from that of PCI, although the inhibitory motif is identical in both proteins.[91] In this case, the position of the heterologous protein inhibitor LCI relative to the enzyme in the complex is similar to that adopted both by PCI and its own activation segment, which acts as the autologous inhibitor (Figure 7). The homology between the C-terminal tails of LCI and PCI represents a clear example of convergent evolution dictated by the target protease. Considering that some enzymes of the A/B subfamily are not involved in digestive processes but are involved in other physiologically relevant ones, knowledge about the control mechanism of such enzymes by these protein inhibitors may be used as a starting point for the structure-based design of drugs that promote or interfere with these mechanisms. One of the best K_i values displayed by LCI is against CPU, a plasma enzyme that establishes a link between the coagulation and fibrinolysis processes and functions as a fibrinolysis inhibitor. Thus, LCI, or drugs developed on the basis of its structural determinants, could act as fibrinolysis enhancers. CP inhibitors could also find commercial use in protecting plants from predators. For example, PCI behaves as a potent inhibitor of a CPA from

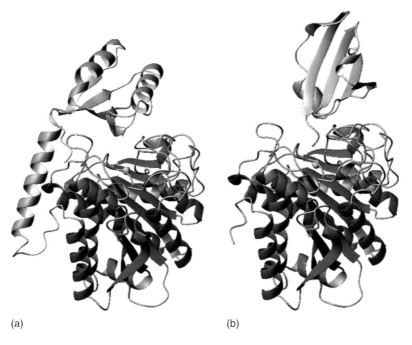

Figure 7 Comparison of the structures of autologous and heterologous inhibitors of CPs. (a) The structure of PCPA2, which contains its own pro-region. (b) The complex between CPA2 and LCI. The enzyme domain is shown in the same colors and orientation as the initial 3D Structure, whereas the pro-region and LCI are shown in yellow and gray respectively. Figure made with MolMol.[69]

H. armigera, a devastating pest of many crop plants. These are two examples of the potentiality of carboxypeptidase inhibitors in biotechnological and biomedical applications.

ACKNOWLEDGEMENTS

The authors wish to thank Juan Cedano (Institut de Biotecnologia i Biomedicina, UAB) for his expertise and help in the elaboration of the figures.

REFERENCES

1. VM Stepanov, in AJ Barrett, ND Rawlings and JF Woessner (eds.), *Handbook of Proteolytic Enzymes*, Academic Press, San Diego, pp 1336–38 (1998).
2. S Wei, S Segura, J Vendrell, FX Aviles, E Lanoue, R Day, Y Feng and LD Fricker, *J Biol Chem*, **277**, 14954–64 (2002).
3. FX Gomis-Rüth, V Companys, Y Qian, LD Fricker, J Vendrell, FX Aviles and M Coll, *EMBO J*, **18**, 5817–26 (1999).
4. DS Auld, in AJ Barrett, ND Rawlings and JF Woessner (eds.), *Handbook of Proteolytic Enzymes*, Academic Press, San Diego, pp 1321–26 (1998a).
5. DS Auld, in AJ Barrett, ND Rawlings and JF Woessner (eds.), *Handbook of Proteolytic Enzymes*, Academic Press, San Diego, pp 1326–28 (1998b).
6. FX Aviles and J Vendrell, in AJ Barrett, ND Rawlings and JF Woessner (eds.), *Handbook of Proteolytic Enzymes*, Academic Press, San Diego, pp 1333–35 (1998).
7. EB Springman, in AJ Barrett, ND Rawlings and JF Woessner (eds.), *Handbook of Proteolytic Enzymes*, Academic Press, San Diego, pp 1330–33 (1998).
8. DF Hendriks, in AJ Barrett, ND Rawlings and JF Woessner (eds.), *Handbook of Proteolytic Enzymes*, Academic Press, San Diego, pp 1328–30 (1998).
9. LD Fricker, in AJ Barrett, ND Rawlings and JF Woessner (eds.), *Handbook of Proteolytic Enzymes*, Academic Press, San Diego, pp 1341–44 (1998a).
10. LD Fricker, in RE Dalbey and DS Sigman (eds.), *The Enzymes, Volume 23: Co- and Posttranslational Proteolysis of Proteins*, 3rd edn, Academic Press, San Diego, pp 421–52 (2002).
11. LD Fricker and EH Leiter, *Trends Biochem Sci*, **24**, 390–93 (1999).
12. LD Fricker, in AJ Barrett, ND Rawlings and JF Woessner (eds.), *Handbook of Proteolytic Enzymes*, Academic Press, San Diego, pp 1349–51 (1998b).
13. RA Skidgel, in AJ Barrett, ND Rawlings and JF Woessner (eds.), *Handbook of Proteolytic Enzymes*, Academic Press, San Diego, pp 1347–49 (1998).
14. RA Skidgel and EG Erdos, in AJ Barrett, ND Rawlings and JF Woessner (eds.), *Handbook of Proteolytic Enzymes*, Academic Press, San Diego, pp 1344–47 (1998).
15. SE Reznik and LD Fricker, *Cell Mol Life Sci*, **58**, 1790–804 (2001).
16. G Sidyelyeva and LD Fricker, *J Biol Chem*, **277**, 49613–20 (2002).
17. SHJ Settle, MM Green and KC Burtis, *Proc Natl Acad Sci USA*, **92**, 9470–74 (1995).
18. G Giaever et al., *Nature*, **418**, 387–91 (2002).
19. M Guinand, in AJ Barrett, ND Rawlings and JF Woessner (eds.), *Handbook of Proteolytic Enzymes*, Academic Press, San Diego, pp 1338–40 (1998).
20. I le Hueerou, P Guilloteau, R Toullec, A Puigserver and C Wicker, *Biochem Biophys Res Commun*, **175**, 110–16 (1991).
21. L Catasús, V Villegas, R Pascual, FX Avilés, C Wicker-Planquart and A Puigserver, *Biochem J*, **287**, 299–303 (1992).
22. DP Bown, HS Wilkinson and JA Gatehouse, *Insect Biochem Mol Biol*, **28**, 739–49 (1998).
23. L Catasús, J Vendrell, FX Avilés, S Carreira, A Puigserver and M Billeter, *J Biol Chem*, **270**, 6651–57 (1995).
24. P Aloy, L Catasús, V Villegas, D Reverter, J Vendrell and FX Avilés, *Biol Chem*, **379**, 149–55 (1998).
25. K Titani, LH Ericsson, KA Walsh and H Neurath, *Proc Natl Acad Sci USA*, **72**, 1666–70 (1975).
26. PF Marx, GTM Wagenaar, A Reijerkerk, MJ Tiekstra, AGSH van Rossum, MFGB Gebbink and JCM Meijers, *Thromb Haemost*, **83**, 297–303 (2000).
27. SV Smulevitch, AL Osterman, OV Galperina, MV Matz, OP Zagnitko, RM Kadyrov, IA Tsaplina, NV Grishin, GG Chestukhinaand and VM Stepanov, *FEBS Lett*, **291**, 75–78 (1991).
28. K Kuroki, F Eng, T Ishikawa, C Turck, F Harada and D Ganem, *J Biol Chem*, **270**, 15022–28 (1995).
29. W Gebhard, M Schube and M Eulitz, *Eur J Biochem*, **178**, 603–7 (1989).
30. LD Fricker, CJ Evans, FS Esch and E Herbert, *Nature*, **323**, 461–64 (1986).
31. F Tan, SJ Chan, DF Steiner, JW Schilling and RA Skidgel, *J Biol Chem*, **264**, 13165–70 (1989).
32. L Song and LD Fricker, *J Biol Chem*, **272**, 10543–50 (1997).
33. AG Lacko and H Neurath, *Biochemistry*, **9**, 4680–90 (1970).
34. GR Reeck, KA Walsh and H Neurath, *Biochemistry*, **10**, 4690–98 (1971).
35. FJ Burgos, R Pascual, J Vendrell, CM Cuchillo and FX Avilés, *J Chromatogr*, **479**, 27–37 (1989).
36. R Pascual, FJ Burgos, M Salvà, F Soriano, E Méndez and FX Avilés, *Eur J Biochem*, **179**, 609–16 (1989).
37. JE Folk, KA Piez, WR Carroll and JA Gladner, *J Biol Chem*, **235**, 2272–77 (1960).
38. JW Brodrick, MC Geokas and C Largman, *Biochim Biophys Acta*, **97**, 39–43 (1976).
39. D Reverter, S Ventura, V Villegas, J Vendrell and FX Avilés *J Biol Chem*, **273**, 3535–41 (1998a).
40. S Ventura, V Villegas, J Sterner, J Larson, J Vendrell, CL Hershberger and FX Avilés, *J Biol Chem*, **274**, 19925–33 (1999).
41. DL Eaton, BE Malloy, SP Tsai, W Henzel and D Drayna, *J Biol Chem*, **266**, 21833–38 (1991).
42. DS Reynolds, RL Stevens, DS Gurley, WS Lane, KF Austen and WE Serafin, *J Biol Chem*, **264**, 20094–99 (1989).
43. AK Tan and DL Eaton, *Biochemistry*, **34**, 5811–16 (1995).
44. L Bajzar, R Manuel and ME Nesheim, *J Biol Chem*, **270**, 14477–84 (1995).
45. MB Boffa, R Bell, WK Stevens and ME Nesheim, *J Biol Chem*, **275**, 12868–78 (2000).
46. MT Everitt and H Neurath, *FEBS Lett*, **110**, 292–96 (1980).
47. O Varlamov, EH Leiter and LD Fricker, *J Biol Chem*, **271**, 13981–86 (1996).
48. EG Novikova, FJ Eng, L Yan, Y Qian and LD Fricker, *J Biol Chem*, **274**, 28887–92 (1999).

49 THJ Plummer and MY Hurwitz, *J Biol Chem*, **253**, 3907–12 (1978).

50 EG Novikova and LD Fricker, *Biochem Biophys Res Commun*, **256**, 564–68 (1999).

51 GP He, A Muise, AW Li and HS Ro, *Nature*, **378**, 92–96 (1995).

52 E Estebanez-Perpina, A Bayes, J Vendrell, MA Jongsma, DP Bown, JA Gatehouse, R Huber, W Bode, FX Aviles and D Reverter, *J Mol Biol*, **313**, 629–38 (2001).

53 LD Fricker and L Devi, in FX Aviles (ed.), *Innovations in Proteases and their Inhibitors*, Walter de Gruyter, Berlin, pp 259–77 (1993).

54 LD Fricker, B Das and RH Angeletti, *J Biol Chem*, **265**, 2476–82 (1990).

55 RA Skidgel, F Tan, PA Deddish and X Li, *Biomed Biochim Acta*, **50**, 815–20 (1991).

56 K Kuroki, F Eng, T Ishikawa, C Turck, F Harada and D Ganem, *J Biol Chem*, **270**, 15022–28 (1995).

57 X Xin, O Varlamov, R Day, W Dong, MM Bridgett, EH Leiter and LD Fricker, *DNA Cell Biol*, **16**, 897–909 (1997).

58 F Tan, M Rehli, SW Krause and RA Skidgel, *Biochem J*, **327**, 81–87 (1997).

59 SR Nalamachu, L Song and LD Fricker, *J Biol Chem*, **269**, 11192–95 (1994).

60 FJ Eng, EG Novikova, K Kuroki, D Ganem and LD Fricker, *J Biol Chem*, **273**, 8382–88 (1998).

61 EG Novikova, SE Reznik, O Varlamov and LD Fricker, *J Biol Chem*, **275**, 4865–70 (2000).

62 Y Lei, X Xin, D Morgan, JE Pintar and LD Fricker, *DNA Cell Biol*, **18**, 175–85 (1999).

63 X Xin, R Day, W Dong, Y Lei and LD Fricker, *DNA Cell Biol*, **17**, 897–909 (1998).

64 FX Gomis-Rüth, M Gómez, J Vendrell, S Ventura, W Bode, R Huber and FX Avilés, *J Mol Biol*, **269**, 861–80 (1997).

65 FA Quiocho and WN Lipscomb, *Adv Protein Chem*, **25**, 1–78 (1971).

66 RA Skidgel, in NM Hooper (ed.), *Zinc Metalloproteases in Health and Disease*, Taylor & Francis, London, pp 241–83 (1996).

67 M Gomez-Ortiz, FX Gomis-Ruth, R Huber and FX Aviles, *FEBS Lett*, **400**, 336–40 (1997).

68 A Teplyakov, K Polyakov, G Obmolova, B Strokopytov, I Kuranova, A Osterman, N Grishin, S Smulevitch, O Zagnitko, O Galperina, M Matz and V Stepanov *Eur J Biochem*, **208**, 281–88 (1992).

69 R Koradi, M Billeter and K Wüthrich, *J Mol Graphics*, **14**, 51–55 (1996).

70 WL Mock, Y Liu and DJ Stanford, *Anal Biochem*, **239**, 218–22 (1996).

71 LD Fricker, *Methods Neurosci*, **23**, 237–50 (1995).

72 DC Rees, M Lewis and WN Lipscomb, *J Mol Biol*, **168**, 367–87 (1983).

73 MF Schmid and JR Herriot, *J Mol Biol*, **103**, 175–90 (1976).

74 Z Famming, B Kobe, C-B Stewart, WJ Rutter and EJ Goldsmith, *J Biol Chem*, **266**, 24606–12 (1991).

75 M Coll, A Guasch, FX Aviles and R Huber, *EMBO J*, **10**, 1–9 (1991).

76 A Guasch, M Coll, FX Aviles and R Huber, *J Mol Biol*, **224**, 141–57 (1992).

77 FX Gomis-Rüth, M Gomez, W Bode, R Huber and FX Aviles, *EMBO J*, **14**, 4387–94 (1995).

78 I Garcia-Saez, D Reverter, J Vendrell, FX Aviles and M Coll, *EMBO J*, **16**, 6906–13 (1997).

79 PJB Pereira, S Segura-Martin, B Oliva, BC Ferrer-Orta, FX Aviles, M Coll, FX Gomis-Rüth and J Vendrell, *J Mol Biol*, **321**, 537–47 (2002).

80 DW Christianson and WN Lipscomb, *Acc Chem Res*, **22**, 62–69 (1989).

81 SJ Gardell, CS Craik, D Hilvert, M Urdea and WJ Rutter, *Nature*, **317**, 551–55 (1985).

82 MA Phillips, AP Kaplan, WJ Rutter and PA Bartlett, *Biochemistry*, **31**, 959–63 (1992).

83 N Abramowitz and I Schechter, *Isr J Chem*, **12**, 543–55 (1974).

84 FX Aviles, J Vendrell, A Guasch, M Coll and R Huber, *Eur J Biochem*, **211**, 381–89 (1993).

85 CCF Blake and SL Oatley, *Nature*, **268**, 115–20 (1977).

86 P Aloy, V Companys, J Vendrell, FX Aviles, LD Fricker, M Coll and FX Gomis-Rüth, *J Biol Chem*, **276**, 16177–84 (2001).

87 J Vendrell, M Billeter, G Wider, FX Avilés and K Wüthrich, *EMBO J*, **10**, 11–15 (1991).

88 M Billeter, J Vendrell, M Coll, A Guasch, FX Avilés, R Huber and K Wüthrich, *J Biomol NMR*, **2**, 1–10 (1992).

89 R Huber and W Bode, *Acc Chem Res*, **11**, 114–22 (1978).

90 DC Rees and WN Lipscomb, *J Mol Biol*, **160**, 475–98 (1982).

91 D Reverter, C Fernández-Catalán, R Baumgartner, R Pfänder, R Huber, W Bode, J Vendrell, TA Holak and FX Avilés, *Nat Struct Biol*, **7**, 322–28 (2000).

92 TH Fife and TJ Przystas, *J Am Chem Soc*, **108**, 4631–36 (1986).

93 S Alvarez-Santos, A Gonzalez-Lafont, JM Lluch, B Oliva and FX Aviles, *New J Chem*, **22**, 319–25 (1998).

94 SJ Gardell, CS Craik, E Clauser, E Goldsmith, CB Stewart, M Graf and W Rutter, *J Biol Chem*, **263**, 17929–36 (1988).

95 D Reverter, I Garcia-Sáez, L Catasús, J Vendrell, M Coll and FX Aviles, *FEBS Lett*, **420**, 7–10 (1997).

96 M Edge, C Forder, J Hennam, I Lee, D Tonge, I Hardern, J Fitton, K Eckersley, S East, A Shufflebotham, D Blakey and A Slater, *Protein Eng*, **11**, 1229–34 (1998).

97 LD Fricker and L Devi, *Anal Biochem*, **184**, 21–27 (1990).

98 RA Skidgel, RM Davis and F Tan, *J Biol Chem*, **264**, 2236–41 (1989).

99 RA Skidgel, AR Johnson and EG Erdöss, *Biochem Pharmacol*, **33**, 3471–78 (1984).

100 J Vendrell, M Billeter, G Wider, FX Avilés and K Wüthrich, *EMBO J*, **10**, 11–15 (1991).

101 J Vendrell, E Querol and FX Aviles, *Biochim Biophys Acta*, **1477**, 284–98 (2000).

102 MA Phillips and WJ Rutter, *Biochemistry*, **35**, 6771–76 (1996).

103 V Villegas, AR Viguera, FX Avilés and L Serrano, *Struct Fold Des*, **1**, 29–34 (1996).

104 MA Molina, C Marino, B Oliva, FX Aviles and E Querol, *J Biol Chem*, **269**, 21467–72 (1994).

105 G Venhudova, F Canals, E Querol and FX Aviles, *J Biol Chem*, **276**, 11683–90 (2001).

Pitrilysins/inverzincins

Klaus Maskos
Max-Planck-Institut für Biochemie, Martinsried bei München, Germany

FUNCTIONAL CLASS

Enzyme, EC 3.4.24.55, endopeptidase, M16, inverzincins.[1]

OCCURRENCE

Pitrilysin, the first member to be discovered, was described as a peptide-cleaving activity in the periplasm of *Escherichia coli*.[2] Relatives are found not only in a variety of bacteria, but also in yeast, plants, and animals. Members of the mitochondrial processing peptidase subfamily are present not only as part of the cytochrome c reductase complex, but also in soluble form in mitochondria.

AMINO ACID SEQUENCE INFORMATION

- *Pitrilysin subfamily (precursors)*
  Pitrilysin from *E. coli*: P05458, 962 AA[3]
  Insulin-degrading enzyme from *Homo sapiens*: P14735, 1018 AA[4]
  N-arginine dibasic convertase from *H. sapiens*: O43847, 1150 AA[5]
  PQQF, coenzyme PQQ synthesis protein F, from *Klebsiella pneumoniae*: P27508, 761 AA[6]
  AXL1 from *Saccharomyces cerevisiae*: P40851, 1208 AA[7]

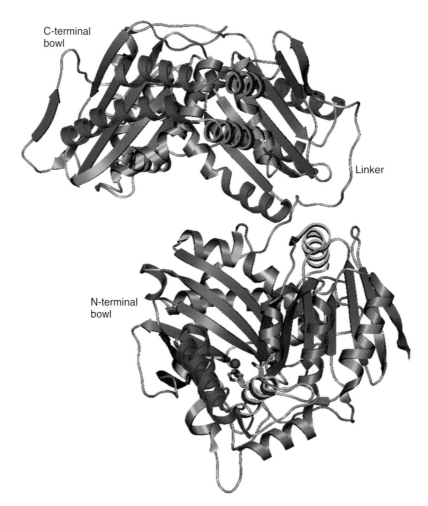

3D Structure Ribbon model of pitrilysin (PDB: 1Q2L). The active site zinc is shown in pink, active site residues are indicated as stick models. Figure was made with SETOR.[73]

CPE, chloroplast processing enzyme (SPP), from *Arabidopsis thaliana*: O48870, 1265 AA[8]
Human metalloendoprotease 1: O95204, 1038 AA[9]
Falcilysin: Q9U7N7, 1193 AA[10]

- *Mitochondrial processing peptidase subfamily*
  β-subunits:
  bovine core I: P31800[11–13]
  Saccharomyces cerevisiae core I: P07256[14]
  Saccharomyces cerevisiae β-MPP: P10507[15]
  Solanum tuberosum β-MPP: Q41440[16]
  Neurospora crassa β-MPP: P11913[17]
  α-subunits:
  bovine core II: P23004[11–13]
  Saccharomyces cerevisiae core II: P07257[14]
  Saccharomyces cerevisiae α-MPP: P11914[18]
  Solanum tuberosum α-MPP: P29677[16]
  Neurospora crassa α-MPP: P23955[19]

BIOLOGICAL FUNCTIONS

Several potential biological functions of pitrilysin have already been discussed in the literature. Since the cytoplasmic membrane contains only transport systems for small oligopeptides and amino acids,[20] the original purpose of pitrilysin in the periplasmic space of bacteria seems to be the processing of nutritional peptides in combination with other periplasmic proteinases and aminopeptidases and/or the general turnover of proteins. Correctly folded proteins are not among the substrates of pitrilysin. However, pitrilysin is known to take part in the degradation of misfolded recombinant or fusion proteins in the periplasm of *E. coli*,[21] which seem to be recognized as substrates only when they expose unfolded or improperly folded areas. Furthermore, owing to the cleavage site specificity of pitrilysin near hydrophobic and positively charged sites, the degradation of polyanions might also group it with detoxifying enzymes.

PROTEIN PRODUCTION, PURIFICATION, AND MOLECULAR CHARACTERIZATION

In 1979, Cheng and Zipser purified and described an endopeptidase that cleaved β-galactosidase fragments, which they named protease III.[2] Three years later, Swamy and Goldberg[22] could prove the identity between their insulin B-chain cleaving activity of the periplasm of *E. coli*, which they had termed *protease Pi*, and *protease III*. In 1992, the IUBMB renamed the enzyme as *pitrilysin* since it is the product of the *ptr* gene in *E. coli*.[23]

Pitrilysin lacks the consensus sequence HExxH described for most metalloendopeptidases, and only in 1992, when Becker and Roth cloned pitrilysin into a modified Tacterm vector, the zinc binding site was identified via mutational analysis.[24] The consensus sequence turned out to be present at the N-terminus of the 107 kDa protein in the inversed orientation of the classical motif ($H^{88}xxEH^{92}$). In 1993, Becker and Roth could prove that E169 is the third zinc ligand, while E162 only modulates the activity.[25] In sum, the full active site sequence of inverzincins reads $HxxEHx_{(74-76)}E$.

Pitrilysin degrades several fragments of β-galactosidase of less than 7000 Da. So far, the highest observed degradation rate is observed for insulin B-chain, which is cleaved at position 16 (...Leu-Tyr^{16}*Leu-Val...). Anastasi *et al.*[26] found several other oligopeptides to be among the substrates such as secretin, various forms of the vasoactive intestinal peptide (VIP 1–28, 10–28, 16–28), thyrocalcitonin, substance P, angiotensinogen, glucagons, and luteinizing-releasing hormone (LRH). However, VIP 1–12, angiotensin I and II, dynorphin-A (1–13 and 1–8), insulin-A chain, and bradykinin were not cleaved in this assay, which argues against a general oligopeptidase activity. There seem to be further restrictions to substrate selection such as a minimum peptide length. At the moment, the smallest substrate is LRH, which contains 10 amino acids. Furthermore, none of these peptides was cleaved less than three amino acids from either N- or C-terminus. In case of glucagon and insulin, the ability to cleave them depends on the substrate concentration. In fact, they are processed effectively at a concentration of 1 μM, while they are not processed or are significantly slower processed at 50 μM concentration.[26] A possible explanation might be the inherent tendency of these peptides to form oligomers, which are not readily available to the active site.

The affinity for several substrates with regular tertiary structure was additionally proved by cross-linking experiments with insulin, insulin-like growth factors I and II (IGF-I and -II), relaxin, and bombyxin II.[27]

ACTIVITY TESTS AND INHIBITION CHARACTERISTICS

The most convenient assay for pitrilysin activity is the degradation of the oxidized insulin B-chain, while assays with radiolabeled insulin and a coupled insulin receptor-binding assay have also been performed.[27,28] Additionally, Anastasi *et al.*[26] have synthesized a quenched fluorescent-peptide substrate based on the cleavage sites of VIP 18–25 (A^{18}-V-K-K-Y-L-N-S^{25}). The fluorescent peptide is called QF27 (Mca-Nle-A^{18}-V-K-K-Y-L-N-S^{25}-K(Dnp)-L-D-k) and is cleaved at position 22 and 24.

The metal-depleted enzyme could be reactivated by 0.1 mM Zn^{2+} (96%), 0.5 mM Co^{2+} (75%), 0.5 mM Mn^{2+} (52%), and 1 mM Ca^{2+} (100%), while 0.5 mM Cu^{2+}, Cd^{2+}, or Mg^{2+} did not activate the enzyme nor did it interfere with reactivation by Zn^{2+}. However, inhibition tests performed with the native enzyme and QF27 by Anastasi *et al.*[26] showed that 1 mM Zn^{2+}, Mn^{2+}, Co^{2+}, and Ca^{2+} inhibited the activity by 90, 32, 15, and 9% respectively.

The lack of inhibition by α_2-macroglobulin as determined for pitrilysin is a general property of oligopeptidases and is used to distinguish them from endoproteases. Ninety-three and seventy-five percent inhibition of pitrilysin could be obtained with 5 mM DTT or 2 mM of the dipeptide hydroxamic acid Zinkov, respectively. The weak inhibition by Zinkov contrasts with findings on thermolysin, which is inhibited much more potently ($K_d = 0.48\,\mu M$). Interestingly, the antibiotic bacitracin inhibits pitrilysin in a zinc-dependent manner, indicating that a zinc–bacitracin complex might allow tight binding, since 1.6 zinc atoms per bacitracin molecule improved inhibition more than tenfold to 13 μM. Fifty to one hundred micrometers of other metalloprotease inhibitors like captopril, thiorphan, bestatin, amastatin, several cinnamonyl-derived inhibitors, and 0.3 μM tissue inhibitor of matrix metalloproteases-1 did not inhibit pitrilysin.[26]

X-RAY STRUCTURES

Pitrilysin exhibits a typical α/β fold and forms two similarly sized bowl-like moieties embracing a central cavity.[29] The two bowls face each other, forming an opening angle of about 54° resembling a pacman (3D Structure and Figure 1). These bowls are connected via a 28–amino acid linker peptide that bridges about 60 Å. Most interbowl contacts are found in the central linker region (Leu500 to Leu513) and are mediated via this linker peptide. Only four amino acids of each bowl participate in direct contacts. A sequence alignment shows about 12% sequence identity and about 20% similarity between both bowls, which share a similar secondary-structure topology.

Domain topology

The two bowls of the molecule are further subdivided into two similar folding units, which are composed of ~240 amino acids each. Each domain essentially consists of a central 6- or 7-stranded β-sheet, which together form central elements of the inside surface of each bowl and participate in the domain contacts, which lead to bowl formation. The outer surface of each bowl is covered by α-helices, which shield the β-sheets from bulk solvent.

Interestingly, all four domains show a similar overall fold, while no significant sequence homology can be detected between them. Superposition of all four domains shows high similarity in the center of the domains, while more distal secondary-structure elements vary in length and arrangement, such as the upper half of the β-sheet and α-helix h5 (Figures 1 and 2), both involved in the intrabowl domain contacts.

Catalytic domain

A seven-stranded antiparallel β-sheet dominates the upper half of the catalytic (first) domain (Figure 2). Only strands s1 to s3 appear in consecutive order, while strand s7

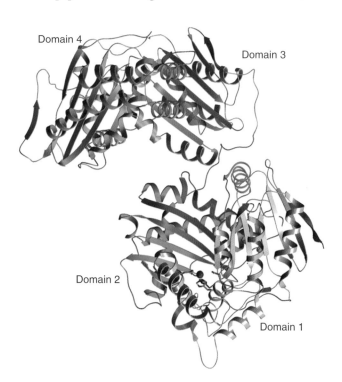

Figure 1 Ribbon model of pitrilysin indicating the location of the four domains. Domains 1, 2, 3, and 4 are shown in gray, green, yellow, and blue respectively. Figure was made with SETOR.[73]

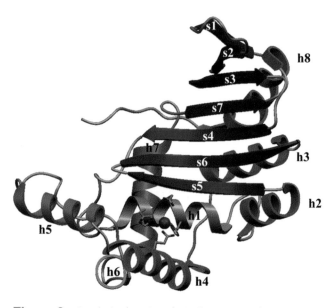

Figure 2 Catalytic domain of pitrilysin. Secondary-structure elements are numbered in their consecutive order. The active site zinc is represented by a pink sphere; active site residues are shown as stick models. Figure was made with SETOR.[73]

is inserted between strands s3 and s4 and strand s6 intercalates between strands s4 and s5. Additionally, strand s7 breaks the anti-parallel order of the strands. After strand s4, the polypeptide chain turns downwards and enters helix h1, the active site helix, which contains two zinc-coordinating histidines, His88 and His92, and the catalytic glutamate Glu91. Together with the succeeding helix h2, both are located below the following strand s5. Strand s5 forms the upper boundary of the active site and is supposed to provide hydrogen-bonding partners for polypeptide substrates. After strand s5, the chain forms a tight turn into strand s6 and proceeds into helix h3, which covers the back of strands s5 and s6 together with helix h7. Helix h4 contains two further active site residues, the activity-modulating Glu162 and the zinc-binding glutamate Glu169.[25] Helix h5 protrudes from the domain center and nestles toward the β-sheet of the second domain. Strand 3 of the second domain embraces helix h5 and forms several contacts (Figure 1). With helices h6 and h7 the chain turns up again, completes the β-sheet (strand 7), and enters helix h8, which is located at the back of the upper half of the β-sheet. After helix h8, the chain traverses to the following domain via an extended linker of ~20 amino acids.

PROPOSED CATALYTIC MECHANISM

Pitrilysin-catalyzed peptide cleavage is supposed to proceed via a general-base mechanism as proposed for thermolysin.[30] Thus, an incoming substrate is bound next to the water molecule that interacts with both the zinc ion and the carboxyl group of Glu91 (Figure 3). The zinc ion will acidify the water, so that a proton could be pushed over to the carboxylate group of the catalytic Glu91 producing a nucleophilic hydroxyl ion that attacks the carbonyl carbon of the peptide bond. Additionally, the protonated Glu91 is expected to shuttle its proton to the amide nitrogen of the attacked peptide bond of the substrate. Thus, a tetrahedral intermediate is formed, with a tetrahedral carbonic group and a tetrahedral secondary amino group. Upon shuttling of another proton to the leaving group nitrogen and the C–N break, the tetrahedral intermediate decomposes and the products are released.

Asn119 and Ala120, both part of strand s5 and absolutely conserved in this family and several other metalloprotease families, are in appropriate distance to be involved in substrate binding and catalysis. In short, Ala120 O and Asn119 OD1 may interact with the amino group of P1′, while Asn119 ND2 could additionally be properly oriented to form a hydrogen bond to the carbonyl group of the P1′ residue. No residue of pitrilysin is located in an appropriate position to contribute directly to oxyanion stabilization, which is probably mainly performed by the zinc ion and several solvent molecules that are found in the vicinity of the active site. Thr122, which forms a hydrogen bond to the carboxylate group of the active site glutamate (Glu91), is part of a hydrogen-bonding network including the conserved Thr127, a water molecule, and the conserved Tyr226. This hydrogen bond network is expected to modulate the pK-properties of the active site glutamate.

ZINC-BINDING ARRANGEMENT AND COMPARISON WITH THE CLASSICAL ZINC-BINDING MOTIF

The classical metalloprotease consensus sequence HExxH as represented by the gluzincins and metzincins is inversed in the inverzincin family.[1] Superposition of the catalytical zincs of a matrix metalloprotease[31] and pitrilysin shows that both right-handed active site helices have a similar

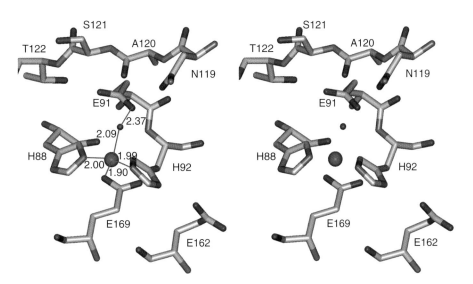

Figure 3 Stereo view of the active site of pitrilysin shown with zinc–ligand distances in Å. The catalytic water molecule is represented by a red ball. Figure was made with DS ViewerPro.[74]

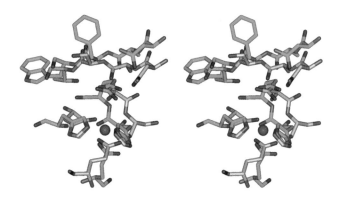

Figure 4 Stereo view of a superposition of the active site of thermolysin (green, PDB: 1KJO) and pitrilysin (gray). Figure was made with DS ViewerPro.[74]

length but are displaced by about half of a turn and extend in opposite direction. These helices show also conserved residues in the left half on the active site helix (HExxH(S/A)LG in MMPs and GL(S/A)HxxEH in pitrilysin and other pitrilysin-like enzymes). Strand s5 of pitrilysin forms the upper edge of the active site and extends in the same direction as in metzincins or gluzincins, probably providing main-chain hydrogen-bonding groups to substrates or inhibitors. In fact, glucincins as exemplified by thermolysin do not share all details of the active site helix but coincide with a second helix that contributes to the active site. This helix provides the third zinc binding residue, which is also a glutamate as in inverzincins (Figure 4).[32]

OTHER PITRILYSIN-LIKE INVERZINCIN FAMILY MEMBERS

Insulin-degrading enzyme

Mammalian insulin-degrading enzymes (insulysin, IDE) are among the closest relatives of pitrilysin, with 26% identity and 47% similarity and three regions of more than 50% identity.[4] Human IDE and pitrilysin share major substrate binding and processing activities[27] such as the high affinity for insulin ($K_m = <0.1\,\mu M$ and $1\,\mu M$, respectively). Monoclonal antibodies against IDE allowed demonstrating that IDE is responsible for most of the insulin-degrading activity in cells or cell lysates.[33,34] Nevertheless, it is still a matter of debate if IDE is the only relevant insulin-degrading activity, and in which cellular compartment the degradation may happen. IDE possesses a peroxisomal targeting sequence (Ala/Ser-Lys-Leu) and functions as a homodimer of ~220 kDa, which is probably cross-linked via a disulfide bridge. A variety of physiologically important polypeptides are processed by IDE such as insulin,[35] glucagon,[36] atrial natriuretic factor,[37] transforming growth factor alpha,[38] β-endorphin,[39] and β-amyloid protein.[40,41] IDE, like pitrilysin, is not a general oligopeptidase and does not act on many other peptides such as somatostatin, bradykinin, vasopressin, and platelet-derived growth factor. The cleavage sites show little or no similarity to each other, although the preference for basic and large hydrophobic residues on the carboxyl side of the cleavage site is again observed. Since there are no significant sequence similarities among substrates, it is speculated that secondary or tertiary structural elements may be the critical determinants for specificity.[42] Additionally, IDE has been implicated in the degradation of modified proteins like oxidatively damaged hemoglobin[43] and signal peptides such as the leader peptide of thiolase.[44] Recent results additionally support the view that IDE plays an important role in regulating the levels of Aβ monomers, although it is ineffective in controlling the progressive accumulation of insoluble Aβ.[45] IDE degrades both Aβ40 and Aβ42 at multiple sites,[46–48] and the relevance of this processing in vivo has recently been indicated by the genetic linkage between Alzheimer's disease and the IDE gene locus on chromosome 10.[49,50] Interestingly, several IDE substrates, which are also pitrilysin substrates, have the proposed ability to form amyloidic aggregates.[51]

Additionally, regulatory functions of IDE such as the influence on proteasome function via insulin-dependent displacement of IDE from a proteasome/IDE complex have been described. Dissociation of IDE from the complex leads to an accumulation of ubiquitinoylated proteins in vivo and in vitro.[52] Furthermore, IDE has been implicated in development, cellular differentiation, and steroid-mediated signaling.[53]

Nardilysin

Nardilysin (N-arginine dibasic convertase, NRD convertase) is found in the cytosol as well as at the cell surface, and in the extracellular matrix. It is described as a metalloendopeptidase cleaving upstream of dibasic amino acid pairs. An outstanding difference to other family members is a 90–amino acid insertion at the very N-terminus of the enzyme, which is supposed to form a unique acidic domain called DAC domain.[54] At the cell surface, it binds the heparin-binding epidermal growth factor–like growth factor (HB-EGF) and enhances HB-EGF-induced cell migration, potentiating its physiological effect even in the absence of its proteolytic activity. Additionally, HB-EGF was found to bind to the DAC domain and probably simultaneously to the active site in a calcium-dependent manner that leads to a potent inhibition of nardilysin.[54] In the cytosol or the extracellular space, nardilysin is supposed to function as a proprotein convertase similar to furin-like proteases.[55]

Axl1

Yeast harbors a relative of pitrilysin, called Axl1, which is involved in two independent processes, namely, prohormone processing and bud-site selection.[56] Processing of a-factor prior to its release is necessary for efficient cell fusion during mating.[57] Axl1 is expressed only in haploid yeast and leads to axial cell division via complex formation with several other budding factors.[58] Expression of Axl1 in diploids results in conversion of their bipolar budding pattern to an axial pattern,[58] while glucose depletion and concomitant decline in Axl1 level in haploid strains leads to unipolar-distal budding, underscoring the central position of Axl1 in bud-site selection.[59]

Chloroplast processing peptidase

Chloroplast processing peptidase (CPE or SPP) is involved in the maturation of proteins involved in central processes such as photosynthesis, fatty acid synthesis, and amino acid synthesis during their import into chloroplasts. The targeting sequences show wide size variability, are enriched in serine and threonine residues, and possess a few acidic residues. There is no clear consensus sequence to be seen at the cleavage sites of imported proteins; only a weakly conserved motif (I/V)-X-(A/C)*A was suggested.[60] However, purified SPP cleaved most of its predicted substrate preproteins after a basic residue.[61,62] A concerted action with other chloroplast proteases could explain the obvious discrepancy and would allow obtaining the final N-termini observed *in vivo*.[63]

SPP has a molecular weight of about 145 kDa, has a pH optimum near pH 9, and possesses a large C-terminal region of unknown function not found in other pitrilysin-like enzymes. Antisense RNA experiments were used to decrease the level of SPP and led to drastic effects in chloroplast biogenesis suggesting its necessity for efficient protein import into chloroplasts.[64]

MEMBERS OF THE MITOCHONDRIAL PROCESSING PEPTIDASE SUBFAMILY

The second subfamily within the inverzincins, the mitochondrial processing peptidase-like proteins, are found as soluble proteins in the matrix of mitochondria and as part of the cytochrome c reductase complex (bc1 complex) with full or very limited proteolytic activity depending on the species under consideration (see below). Mitochondrial processing peptidases (MPPs) function as heterodimers with approximately 50 kDa α- and β-subunits, which are called Core II and I in bc1 complexes respectively. Only the β-subunit shows catalytic activity against mitochondrial targeting signal sequences, which can be processed only in the presence of the α-subunit.

MPPs recognize a wide variety of mitochondrial precursor proteins, which are recognized by the mitochondrial import machinery, and cleave off the signal peptide at a single site.[65] The newly formed N-terminus, in some cases, becomes further truncated (eight amino acids) by the mitochondrial intermediate peptidase. Signal sequences of proteins imported into mitochondria are characterized by a high content of basic amino acids and are predicted to form α-helical amphipathic structures. In most cases, MPP recognizes signal sequences via so-called distal and proximal arginines. These arginines, which appear in position P2 or P3 (proximal) and 4 to 10 amino acids closer to the N-terminus (distal), and an aromatic or large hydrophobic residue in P1′ are believed to be critical determinants for the specific cleavage of substrates.

X-RAY STRUCTURES OF THE MPP SUBFAMILY

The first X-ray structures were solved for the core proteins of cattle,[11] chicken,[13] and yeast,[14] which are part of cytochrome c reductase complexes. The corresponding proteins are expected to serve mostly a structural role in these complexes.

More recently, X-ray structures of yeast MPP and an inactive variant in the presence and absence of two signal peptides have been determined.[66]

The core proteins

Core proteins are thought to be relics of processing enzymes, which have lost their catalytic activity during evolution.[67] The structures of Core I and II resemble the domain- and secondary-structure arrangement of the N- and C-terminal bowl of pitrilysin, respectively, albeit with several deviations such as bowl volume and absence of a few secondary-structure elements especially in the C-terminal bowl. This increasing dissimilarity along the polypeptide chain is well illustrated in sequence alignments, which show the highest identity in the first (catalytic) domain. Furthermore, the Core proteins form a ball-like structure with a small crack remaining, in contrast to the open arrangement of the building blocks of pitrilysin, which resemble a pacman.

The unexpected presence of the presequence of the iron–sulfur protein (subunit IX of the bovine bc1 complex,[12]) within a cavity formed by Core I/II indicated the possibility to bind a signal peptide along the edge of a β-sheet in the inner surface of such a ball-like self-compartmentalizing protein. The situation is different in *S. cerevisiae*, where no such peptide is found in the complex.[14] This discrepancy may be explained by the observation that the corresponding signal sequence of the iron–sulfur protein is removed in two steps in *S. cerevisiae* before it is further processed by a yet unidentified protease.

However, subunit IX is bound in a nonproductive orientation in the bovine Core I/II complex with the C-terminal end forming a two-stranded β-sheet motif about 25 Å away from the predicted active site with

Pitrilysins/inverzincins

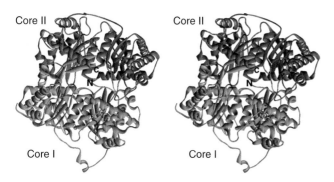

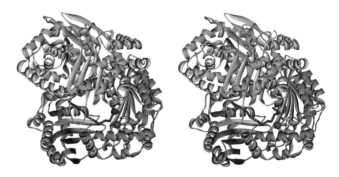

Figure 5 Stereo view of the bovine Core I/II complex (green and orange, respectively) with the signal sequence of subunit IX (blue) of the bc$_1$ complex bound to Core II. The N- and C-terminus of subunit IX are indicated (PDB: 1BGY). Residues contributing to the 'incomplete' zinc binding consensus sequence of Core I are shown. Figure was made with DS ViewerPro.[74]

Figure 6 Stereo view of yeast αβ-MPP (β-MPP mutant E73Q; α- and β-MPP shown in gray and green respectively) with the signal sequence of COX IV (red) bound to β-MPP along the edge strands of the β-sheets of β-MPP (yellow, PDB: 1HR8). Figure was made with DS ViewerPro.[74]

the modified active site motif YxxEH(x)$_{76}$H (Figure 5).[12] Interestingly, a recent study with recombinant bovine Core proteins showed that the iron–sulfur protein precursor is indeed processed by Core I and that the resulting peptide immediately blocks the active site after the cleavage. Thus, the structure of the bovine Core proteins shows an example of perfect 1:1 end product inhibition. Unexpectedly, the catalytic activity is not influenced by the mutation of Tyr57 to histidine (as in soluble MPP), phenylalanine or, tryptophan but reduces considerably for the threonine mutant, which indicates that a bulky hydrophobic residue is needed for this catalysis rather than a free electron pair for improving zinc binding.[68] Thus, the catalytic procedure is expected to vary from the one described above for pitrilysin.

Yeast MPP

The large cavity formed by the α- and β-subunit is decorated with many negatively charged residues, and it is therefore well suited to accommodate and interact with positively charged residues that form important elements for the sequence of mitochondrial matrix targeting signal sequences.[66] Recent studies support the view that the electrostatic interaction of targeting signals by MPP is an important determinant for the recognition of substrates.[69]

An important contribution to catalysis was expected from a highly conserved glycine-rich loop in α-MPP (residues 284–301), which is located near the active site of the enzyme, because it was observed before that the binding and catalysis of the substrate is diminished upon deletion of this loop (Figure 6).[70] However, the weak electron density for this loop indicates its high flexibility, and the proximal location to the active site in combination with the absence of a direct interaction in the active site, even in the presence of a signal sequence, suggests a role in the recognition of the whole substrate protein and not only a signal peptide.

SUBSTRATE BINDING SITES

The E73Q substitution in β-MPP allowed the cocrystallization of the signal peptide of yeast cytochrome c oxidase subunit IV (COX IV, residues 2–25: LSLRQS^{7}IRFFKPATRT*LC^{19}SSRYLL).[66] Residues 7–19 are observed in electron density, while the others are disordered. Two β-strands of MPP help to bind and orient the peptide. One interaction site is formed between residues 16–18 (P2–P1′) and the edge strand of the β-sheet of domain 1 in β-MPP, while the N-terminal part is loosely arranged along the β-strand formed by residues 322–329 (Figure 7).

NMR experiments had shown that residues 4–11 of the COX IV signal peptide form an α-helical structure in a micellar environment while the C-terminal part of the signal sequence was unstructured.[71] However, these residues bind in an elongated manner in the complex with MPP, while the carbonyl oxygen of residue 17 (the cleavage site) points toward the active site zinc.

The proximal P2 arginine forms a salt bridge with Glu160β (the suffix β indicates residues of β-MPP) in a shallow pocket decorated with Glu160β and Asp164β, both highly conserved in β-MPPs. A bulky hydrophobic residue is often found in the P1′ site of substrates, and in the case of COX, a leucine is located near the conserved Phe77β.

The second cocrystallized signal peptide of yeast malate dehydrogenase (MDH, residues 2–17: LSRVAKRA^9*FSSTVANP) was obviously cleaved during crystallization with the E73Q mutant of β-MPP and shows one of the C-terminal oxygens of Ala9 in position of the water molecule in the native enzyme, while residues 10–17 are not detected. Both peptides, COX IV and MDH, are positioned within the β-subunit of MPP in a similar manner. The binding of the MDH signal peptide along the edge strand of the β-sheet contrasts with the predicted α-helical structure for its residues 4–12. Thus, sequence analysis and biophysical studies of both signal peptides suggest that they should adopt an α-helical structure, which was disproved

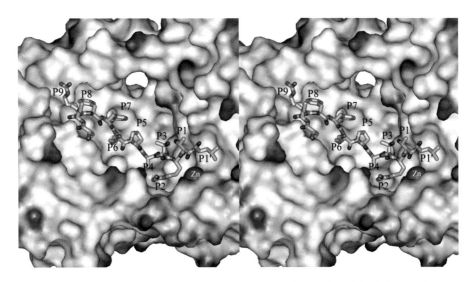

Figure 7 Stereo view of a surface representation of yeast β-MPP E73Q with a stick model of the signal sequence of COX IV (PDB: 1HR8). Substrate binding sites are indicated by the respective numbering. Figure was made with DS ViewerPro.[74]

by the X-ray structure results. Interestingly, NMR data of the signal peptide of rat aldehyde dehydrogenase (residues 12–22) clearly demonstrate the importance of forming an amphiphilic α-helix.[72] This helix is recognized by the mitochondrial import machinery via complex formation with the hydrophobic side of the helix and a hydrophobic patch in a groove of TOM20, a general mitochondrial import receptor. Thus, while the hydrophobic groove of TOM20 favors the α-helical structure of the signal peptide, the local hydrophilic environment in the self-compartmentalizing MPP with the exposed β-strands may disfavor the α-helix and leads to binding of peptides along the exposed edge strands of the β-sheets of domain 1 and 2. Since the binding along β-strands is rather sequence independent, the right cleavage site is mostly defined by the so-called proximal (P3 or P2) and distal arginines together with the large hydrophobic residue in P1'. Since both MPP subunits present four edge strands into the lumen, even much longer peptides could find suitable interaction sites. In summary, signal peptides may have evolved in a way to allow the formation of an amphiphilic α-helix or a β-strand in a context-dependent manner.

REFERENCES

1   NA Hooper, *FEBS Lett*, **354**, 1–6 (1994).
2   YS Cheng and D Zipser, *J Biol Chem*, **254**, 4698–706 (1979).
3   PW Finch, RE Wilson, K Brown, ID Hickson and PT Emmerson, *Nucleic Acid Res*, **14**, 7695–703 (1986).
4   JA Affholter, VA Fried and RA Roth, *Science*, **242**, 1415–18 (1988).
5   P Fumagalli, M Accarino, A Egeo, P Scartezzini, G Rappezzo, A Pizutti, V Avvantaggiato, A Simeone, G Arrigo, O Zuffardi, S Ottolenghi and R Taramelli, *Genomics*, **47**, 238–45 (1988).
6   JJM Meulenberg, E Sellink, NH Riegman and PW Postma, *Mol Gen Genet*, **232**, 284–94 (1992).
7   A Fujita, C Oka, Y Arikawa, T Katagai, A Tonouchi, S Kuhara and Y Misumi, *Nature*, **372**, 567–70 (1994).
8   S Richter and GK Lamppa, *Proc Natl Acad Sci USA*, **95**, 7463–68 (1998).
9   N Mzhavia, YL Berman, Y Qian, L Yan and LA Devi, *DNA Cell Biol*, **18**, 369–80 (1999).
10 KK Eggleson, KL Duffin and DE Goldberg, *J Biol Chem*, **274**, 32411–17 (1999).
11 D Xia, CA Yu, H Kim, JZ Xia, AM Kachurin, L Zhang, L Yu and J Deisenhofer, *Science*, **277**, 60–66 (1997).
12 S Iwata, JW Lee, K Okada, LK Lee, M Iwata, B Rasmussen, TA Link, S Ramaswamy and BK Jap, *Science*, **281**, 64–71 (1998).
13 Z Zhang, L Huang, VM Shulmeister, Y-I Chi, KK Kim, L-W Hung, AR Crofts, EA Berry and S-H Kim, *Nature*, **392**, 677–84 (1998).
14 C Hunte, J Koepke, C Lange, T Roßmanith and H Michel, *Structure*, **8**, 669–84 (2000).
15 C Witte, RE Jensen, MP Yaffe and G Schatz, *EMBO J*, **7**, 1439–47 (1988).
16 HP Braun, M Emmermann, V Kruft and UK Schmitz, *EMBO J*, **11**, 3219–27 (1992).
17 G Hawlitschek, H Schneider, B Schmidt, M Tropschug, F-U Hartl and W Neupert, *Cell*, **53**, 795–806 (1988).
18 RA Pollock, F-U Hartl, MY Cheng, J Ostrermann, A Horwich and W Neupert, *EMBO J*, **7**, 3493–500 (1988).
19 H Schneider, M Arretz, E Wachter and W Neupert, *J Biol Chem*, **265**, 9881–87 (1990).
20 JC Andews, TC Blevins and SA Short, *J Bacteriol*, **165**, 428–33 (1986).
21 F Baneyx and G Giorgiou, *J Bacteriol*, **173**, 2696–703 (1991).
22 KH Swamy and AL Goldberg, *J Bacteriol*, **149**, 1027–33 (1982).
23 CC Dykstra, D Prasher and SR Kushner, *J Bacteriol*, **157**, 21–27 (1984).
24 AB Becker and RA Roth, *Proc Natl Acad Sci USA*, **89**, 3835–39 (1992).
25 AB Becker and RA Roth, *Biochem J*, **292**, 137–42 (1993).
26 A Anastasi, CG Knight and AJ Barrett, *Biochem J*, **290**, 601–7 (1993).

27. L Ding, AB Becker, A Suzuki and RA Roth, *J Biol Chem*, **267**, 2414–20 (1992).
28. KHS Swamy and AL Goldberg, *Nature*, **292**, 652–54 (1981).
29. K Maskos, C Fernandez-Catalan, GP Bourenkov, D Jozic and W Bode (2003); unpublished.
30. BW Matthews, *Acc Chem Res*, **21**, 333–40 (1988).
31. W Bode, FX Gomis-Ruth and W Stocker, *FEBS Lett*, **331**, 143–40 (1993).
32. KS Makarova and NV Grishin, *Protein Sci*, **8**, 2537–40 (1999).
33. K Shii and RA Roth, *Proc Natl Acad Sci USA*, **83**, 4147–51 (1986).
34. K Shii, K Yokono, S Baba and RA Roth, *Diabetes*, **35**, 675–83 (1986).
35. WC Duckworth, FG Hamel, DE Peavy, JJ Liepnieks, MP Ryan, MA Hermodson and BH Frank, *J Biol Chem*, **263**, 1826–33 (1988).
36. RJ Kirschner and AL Goldberg, *J Biol Chem*, **258**, 967–76 (1983).
37. D Müller, C Schulze, H Baumeister, F Buck and D Richter, *Biochemistry*, **31**, 11138–43 (1992).
38. FG Hamel, BD Gehm, MR Rosner and WC Duckworth, *Biochim Biophys Acta*, **1338**, 207–14 (1997).
39. A Safavi, BC Miller, L Cottam and LB Hersh, *Biochemistry*, **35**, 14318–25 (1996).
40. WQ Qiu, DM Walsh, Z Ye, K Vekrellis, J Zhang, MB Podlisny, MR Rosner, A Safavi, LB Hersh and DJ Selkoe, *J Biol Chem*, **273**, 32730–38 (1998).
41. JR McDermott and AM Gibson, *Neurochem Res*, **22**, 49–56 (1997).
42. IV Kurochkin, *TIBS*, **26**, 421–25 (2001).
43. JM Fagan and L Waxman, *J Biol Chem*, **267**, 23016–22 (1992).
44. F Authier, JJM Bergeron, W-J Ou, RA Rachubinski, BI Posner and PA Walton, *Proc Natl Acad Sci USA*, **92**, 3859–63 (1995).
45. K Vekrellis, Z Ye, WQ Qiu, D Walsh, D Hartley, V Chesneau, MR Rosner and DJ Selkoe, *J Neurosci*, **20**, 1657–65 (2000).
46. V Chesneau, K Vekrellis, MR Rosner and DJ Selkoe, *Biochem J*, **351**, 509–16 (2000).
47. G Evin and A Weidemann, *Peptides*, **23**, 1285–97 (2002).
48. JA Carson and AJ Turner, *J Neurochem*, **81**, 1–8 (2002).
49. L Bertram, D Blacker, K Mullin, D Keeney, J Jones, S Basu, S Yhu, MG McInnis, RC Go, K Vekrellis, DJ Selkoe, AJ Saunders and RE Tanzi, *Science*, **290**, 2302–3 (2000).
50. N Ertekin-Taner, N Graff-Radford, LH Younkin, C Echman, J Adamson, DJ Schaid, J Blangero, M Hutton and SG Younkin, *Science*, **290**, 2303–4 (2000).
51. IV Kurochkin, *TIBS*, **26**, 421–25 (1998).
52. RG Bennett, FG Hamel and WC Duckworth, *Endocrinology*, **141**, 2508–17 (2000).
53. WC Duckworth, RG Bennet and FG Hamel, *Endocrinol Rev*, **19**, 608–24 (1998).
54. V Hospital, E Nishi, M Klagsbrun, P Cohen, NG Seidah and A Prat, *Biochem J*, **367**, 229–38 (2002).
55. NG Seidah and A Prat, *Essays Biochem*, **38**, 79–94 (2002).
56. N Adames, K Blundell, MN Ashby and C Boone, *Science*, **270**, 464–67 (1995).
57. V Brizzio, AE Gammie, S Nijbroek, S Michaelis and MD Rose, *J Cell Biol*, **135**, 1727–39 (1996).
58. M Lord, F Inose, T Hiroko, T Hata, A Fujita and J Chant, *Curr Biol*, **12**, 1347–52 (2002).
59. PJ Cullen and GF Sprague Jr, *Mol Biol Cell*, **13**, 2990–3004 (2002).
60. Y Gavel and G von Heijne, *Protein Eng*, **4**, 33–37 (1990).
61. S Richter and GK Lamppa, *Proc Natl Acad Sci USA*, **95**, 7463–68 (1998).
62. S Richter and GK Lamppa, *J Biol Chem*, **277**, 43888–94 (2002).
63. O Emanuelsson, H Nielsen and G von Heijne, *Protein Sci*, **8**, 978–84 (1999).
64. J Wan, D Bringloe and GK Lamppa, *Plant J*, **15**, 459–68 (1998).
65. O Gakh, P Cavadini and G Isaya, *Biochim Biophys Acta*, **1592**, 63–77 (2002).
66. AB Taylor, BS Smith, S Kitada, K Kojima, H Miyaura, Z Otwinowski, A Ito and J Deisenhofer, *Structure*, **9**, 615–25 (2001).
67. HP Braun and UK Schmitz, *TIBS*, **20**, 171–75 (1995).
68. K Deng, SK Shenoy, S-C Tso, L Yu and C-A Yu, *J Biol Chem*, **276**, 6499–505 (2001).
69. S Kitada and A Ito, *J Biochem*, **129**, 155–61 (2001).
70. Y Nagao, S Kitada, K Kojima, K Toh, S Kuhara, T Ogishima and A Ito, *J Biol Chem*, **275**, 34552–56 (2000).
71. T Endo, I Shimada, D Roise and F Inagaki, *J Biochem*, **106**, 396–400 (1989).
72. Y Abe, T Shodai, T Muto, K Mihara, H Torii, S Nishikawa, T Endo and D Kohda, *Cell*, **100**, 551–60 (2000).
73. SV Evans, *J. Mol. Graph.*, **11**, 134–38 (1993).
74. DS ViewerPro 5.0 from www.accelrys.com/dstudio/ds_viewer.

Leucine aminopeptidase

Norbert Sträter[†] and William N Lipscomb[‡]

[†]Biotechnologisch-Biomedizinisches Zentrum der Universität Leipzig, Leipzig, Germany
[‡]Department of Chemistry and Chemical Biology, Harvard University, Cambridge, USA

FUNCTIONAL CLASS

Enzyme; leucine aminopeptidase (LAP), leucyl aminopeptidase, E.C. 3.4.11.1. LAP belongs to the peptidase clan F, family M17. The enzyme activity of an aminopeptidase that cleaved leucylglycine about 20 times faster than glycylglycine was first described in 1929 in extracts of pig intestinal mucosa.[1] Although it is now known that LAP hydrolyzes a wide variety of peptides and amides, this name has persisted. However, LAPs prefer substrates bearing hydrophobic N-terminal amino acid residues.[2]

OCCURRENCE

LAPs are cytosolic aminopeptidases found in animals, plants, and bacteria. The homology between the LAPs

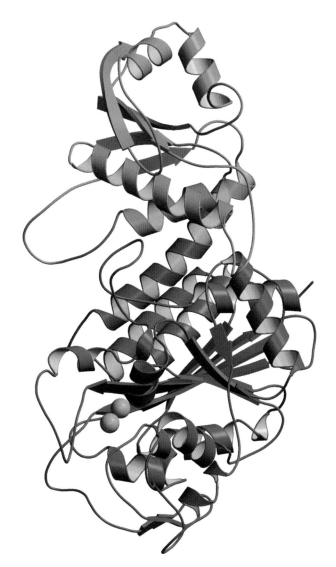

3D Structure View of the blLAP protomer structure (PDB id 1LAM). The catalytic C-terminal domain is colored in red and the N-terminal domain in green. The two zinc ions are shown in yellow. This figure has been prepared using programs MOLSCRIPT[38] and RASTER3D.[39]

from different kingdoms is about 30 to 40% for the whole protein and higher in the catalytic C-terminal domain.

BIOLOGICAL FUNCTION

LAP has different tissue-specific physiological roles in the processing and degradation of peptides. Human LAP catalyzes the postproteasomal trimming of the N-terminus of antigenic peptides for presentation on major histocompatibility complex (MHC) class I molecules.[3] Most MHC-presented peptides seem to be trimmed at the N-terminus of the larger proteasome products by aminopeptidase hydrolysis.[4] In addition to its function in promoting proteasomal cleavage, interferon-γ also induces LAP.[5] Altered LAP activity has also been implicated in several pathological conditions, including human eye cataracts,[6] or is used as a marker in other clinical assays such as prenatal diagnosis of cystic fibrosis.[7]

The related bacterial aminopeptidases have functions as DNA binding proteins. PepA is involved in site-specific DNA recombination[8] and in the transcriptional regulation of the operon that encodes the enzyme carbamoylphosphate synthetase.[9] The functions are independent of the aminopeptidase active site, as has been demonstrated by site-directed mutations that impair the aminopeptidase activity.[9,10]

AMINO ACID SEQUENCE INFORMATION

There are currently more than 50 sequences in the Swiss-Prot Database for leucine aminopeptidases. Referenced here are the sequences for the two best-studied enzymes:

- *Bos taurus* (Bovine), 487 amino acid residues (AA), from nucleic acid[11] and peptide sequence,[12] AMPL_BOVIN.

- *Escherichia coli*, 503 AA, from nucleic acid,[8] AMPA_ECOLI.

PROTEIN PRODUCTION, PURIFICATION, AND MOLECULAR CHARACTERIZATION

LAP has been isolated and characterized from bovine lens,[13] porcine kidney[14] and tomato leaves.[15] The enzyme from bovine lens (blLAP) is best considered for our review. For purification of blLAP, fresh cattle lenses were dissolved in 0.85% sodium chloride solution. After precipitations by zinc sulfate (6 mM) and the application of heat (54 °C), the protein is obtained by ammonium sulfate precipitation and size exclusion chromatography. The production of recombinant animal LAPs has not been reported yet.

The bacterial homologue PepA from *E. coli* has been produced by homologous overexpression.[10] Following a fractionating ammonium sulfate precipitation, the low solubility of PepA under low salt concentrations is used to purify the protein, followed by size exclusion chromatography. PepA contains no disulfide bonds.

blLAP is a homohexamer (324 kDa per monomer). The protomer of 487 amino acid residues contains 7 cysteine residues, but no disulfide bond. Porcine kidney LAP also forms hexamers and has a molecular weight of 326 kDa per monomer. No glycosylation sites have been reported.

METAL CONTENT AND COFACTORS

The metal dependency of catalysis has been well studied for blLAP. Atomic absorption spectroscopy indicated the presence of two Zn ions per protomer.[16] Removal of the zinc by dialysis results in an inactive, metal-free enzyme. The enzyme activity is restored by stoichiometric readdition of Zn^{2+}, but not by other divalent metal ions. However, the addition of Mg^{2+} or Mn^{2+} to native di-Zn^{2+}-blLAP activates the enzyme by exchange of one of its zinc ions.[17,18] However, since the enzyme binds zinc much more strongly, it has been concluded that the physiological state of the enzyme is the zinc–zinc form rather than the Zn–Mg or Zn–Mn form.[16] The readily exchangeable site (site 1, Zn1) can bind Zn^{2+}, Mn^{2+}, Mg^{2+}, and Co^{2+}.[13,16,18] By structure determination of the Mg^{2+}–Zn^{2+} form, it was shown that Zn-488 in the X-ray structure corresponds to the readily exchangeable site.[19] The zinc ion in the tight binding site (site 2) can be exchanged against Co^{2+} only when both binding sites are unoccupied.[13] The dissociation constants for site 1 and site 2 metals in blLAP are 10^{-2}–10^{-3} M and 10^{-9}–10^{-11} M, respectively. It is now clear that metal substitution in both binding sites significantly affects both K_m and k_{cat}.[13] Thus, both metals participate in substrate binding and transition-state stabilization.

ACTIVITY AND INHIBITION TESTS

The enzyme activity is most conveniently determined by the hydrolysis of L-leucine-*p*-nitroanilide following the appearance of the yellow product *p*-nitroanilide ($\Delta\varepsilon_{405}$ = 9900 M^{-1}), or by detecting the cleavage of L-leucine amide ($\Delta\varepsilon_{238}$ = $-14.3\,M^{-1}\,cm^{-1}$). The Zn^{2+}/Zn^{2+} form has a k_{cat} of 0.7 s^{-1} and a K_m of 6 mM for the hydrolysis of L-leucine-*p*-nitroanilide.[13]

The hydrolysis of dipeptides or larger peptides can be measured by the microtitration method[20] or by other assays for the released amino acids.

LAP is inhibited by the naturally occurring inhibitors bestatin (Figure 1) and amastatin with K_i values of 1.3 nM (cattle lens enzyme) and 0.2 µM (pig kidney LAP) respectively.[21,22] Compared to a true peptide substrate, these inhibitors have an extra carbon atom that bears a hydroxyl group. However, bestatin does not specifically inhibit LAP but also inhibits other aminopeptidases. The natural inhibitor shows low toxicity in humans and has been used to explore immune

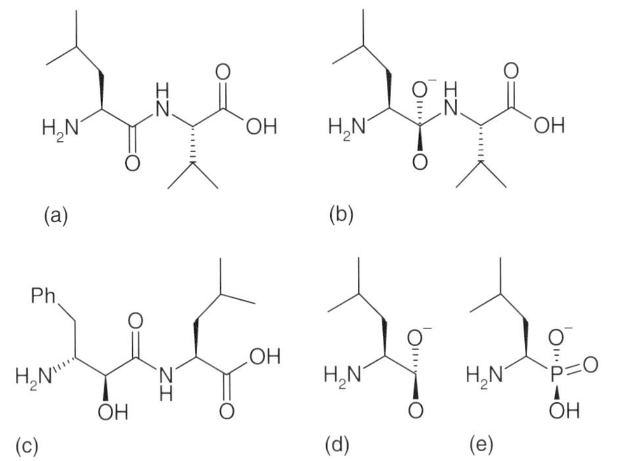

Figure 1 Chemical structures of various peptide inhibitors of LAP: (a) A dipeptide leucine–valine substrate, (b) the presumed *gem*-diol transition state, (c) bestatin, (d) leucinal, and (e) leucine phosphonic acid.

response, retardation of tumor invasion and growth, HIV viral entry, and degradation of peptides and proteins.[23,24] Further, synthetic inhibitors are available, which resemble the *gem*-diol intermediate after the attack of a water nucleophile on the peptide bond, and thus act as transition-state analogues (Figure 1). These include aminophosphonates,[25,26] aminoaldehydes,[27] chloromethanes,[28] and peptides containing a ketomethylene amide bond replacement.[29] Also, amino acid hydroxamates[30] and L-leucylthiol[31] are strong inhibitors of LAP.

X-RAY STRUCTURE OF NATIVE LAP

Crystallization

Crystals of bovine LAP were obtained back in 1937 in Sumners laboratory in crystalline preparations of bovine liver catalase.[32] The crystals were shaped like American footballs, and since the identity of the protein was not recognized at that time, the crystallized protein was called 'football protein.' Much later it became clear that this protein is blLAP.[33] The particular shape of these crystals was due to impure blLAP preparations. Pure blLAP forms hexagonal rod-shaped crystals that are obtained in the presence of $ZnSO_4$, NaCl and 2-methyl-2,4-pentanediol (MPD) at pH 7.8 (Tris).

The crystals belong to space group $P6_322$ with unit cell parameters $a = 132$ Å and $c = 122$ Å.[34] Data sets to 1.6 Å resolution have been obtained using this crystal form. The asymmetric unit contains one protomer. The homohexamer of 32 symmetry is generated by the crystallographic symmetry operators.

Crystals of *E. coli* PepA are obtained by equilibrating the protein in a buffer containing high salt concentration (KCl) at pH 8.0 (Tris) against a buffer containing low salt concentration.[35] These crystals have the shape of trigonal needles or rods and belong to space group $P3_2$ with $a = 178$ Å and $c = 244$ Å and diffract to 2.5-Å resolution. The asymmetric unit contains two hexamers.

Protein fold

The LAP monomer has a mixed $\alpha + \beta$ structure and consists of an N-terminal domain (150 amino acid residues) and a catalytic C-terminal domain (327 residues).[36,37] The N-terminal domain folds into a five-stranded β-sheet sandwiched between four α-helices (3D Structure). A long helix (residues 151 to 172) connects the N-terminal domain with the C-terminal domain. The C-terminal domain is dominated by a central, eight-membered, mixed β-sheet that is sandwiched between groups of α-helices. The two zinc ions and the active site are located within the C-terminal domain at the edge of the central eight-stranded β-sheet.

Hexamer structure

LAP has a characteristic hexameric structure of 32-symmetry, which is triangular in shape when viewed down the threefold molecular axis (Figure 2). The triangle edge length is about 115 Å, and the thickness along the trimer axis is 90 Å. The catalytic C-terminal domains are located near the threefold axis, whereas the N-terminal domains form the corners of the triangle. The upper and lower trimers are related by twofold molecular axes, which are perpendicular to the threefold axis.

A characteristic feature of the hexamer structure is the presence of a large disk-shaped solvent cavity of 15-Å radius and 10-Å height. The catalytic sites are located at the edge of the cavity (Figure 2). Access to this cavity is provided by three channels that are at the twofold molecular axis, and at the interface between two N-terminal and two C-terminal domains. Such a compartmentalization of the catalytic region is not very common for peptidases, but is well known for the proteasome and for some other proteases. Consequently, access to the active site of LAP is limited to molecules with a maximal diameter of about 7 Å. Thus, the compartmentalization of the active site ensures that LAP processes small peptides but cannot attack proteins.

A comparison of blLAP and PepA shows that the C-terminal domains of the two proteins superimpose well and are very similar. Also, the overall hexamer structure is well conserved between the two related proteins, including the presence of the large central solvent cavity and the orientation of subunits within the hexamer. However, the two domains differ in their relative orientation by a rotation of 19° and a translation of 2.9 Å when the two structures are superimposed. The central sheet of the N-terminal domain contains six strands in PepA compared to five in LAP.

Leucine aminopeptidase

Figure 2 View of the LAP hexamer structure along the threefold molecular axis. The upper trimer is colored in red and the lower trimer in gray. The zinc ions, which are located at the edge of a large solvent cavity in the center of the hexamer, are shown in yellow. This figure has been prepared using programs MOLSCRIPT[38] and RASTER3D.[39]

Dimetal site

The active site of the native protein contains two five-coordinated zinc ions that are 3.0 Å apart (Figure 3, Table 1).[36,37,40] One water molecule bridges the two metal ions symmetrically. Furthermore, the carboxylate side chains of Asp273, Asp255, Asp332 and Glu344, the amino group of Lys250, and the backbone carbonyl group of Asp332 coordinate the dimetal center. Because of the highly basic property of the amino group of a lysine side chain, the coordination of lysine side chains is relatively rare in metalloproteins. Interestingly, the metal ion that is

Table 1 Metal–ligand distances in unliganded blLAP

Metal	Ligand atom	Distance (Å)
Zn1	OD2 Asp255	2.12
	O Asp255	2.12
	OD1 Asp332	1.98
	OE1 Glu334	2.02
	OH2 WAT 236	2.01
Zn2	NZ Lys250	2.17
	OD2 Asp255	2.60
	OD2 Asp273	1.98
	OE2 Glu334	2.02
	OH2 WAT 236	1.95
	Zn1	3.02

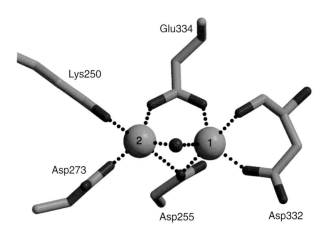

Figure 3 Coordination sphere of the two catalytic zinc ions. The metal–ligand distances are listed in Table 1. This figure has been prepared using programs MOLSCRIPT[38] and RASTER3D.[39]

coordinated to Lys250 also binds the terminal amino group of the substrate peptide.

By structure analysis of the Mg^{2+}-Zn^{2+}-form, it was possible to show that the metal ion bound to Lys250 (Zn2, Zn-489 in the X-ray structure) is the tightly bound metal ion.[19]

Activation by bicarbonate

The 1.6-Å electron density map of native LAP revealed the presence of a triangular planar molecule of XO_3 geometry near the Arg336 side chain, which was proposed to be a carbonate or bicarbonate ion (Figure 4(a)).[40] Since no activation of blLAP by bicarbonate ions was found, it was assumed that the binding of bicarbonate in the catalytic site was an artifact of the crystallization conditions. Later, when

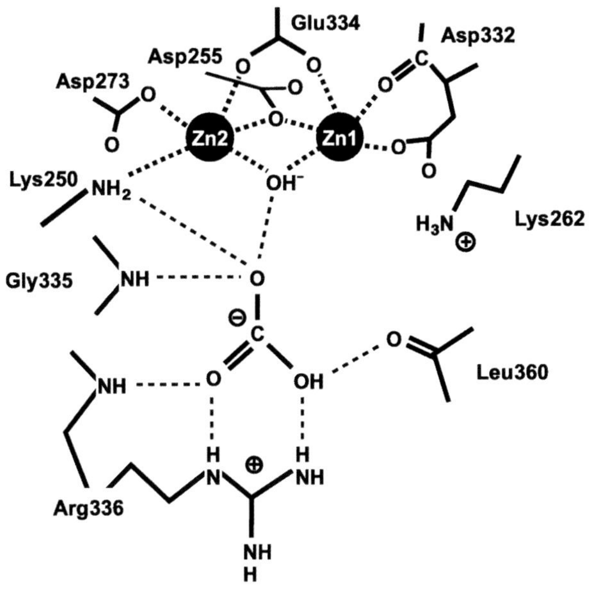

Figure 4 Schematic view (a) of the active site pocket of the native enzyme and (b) of the binding of the transition-state analogue inhibitor L-leucinal.

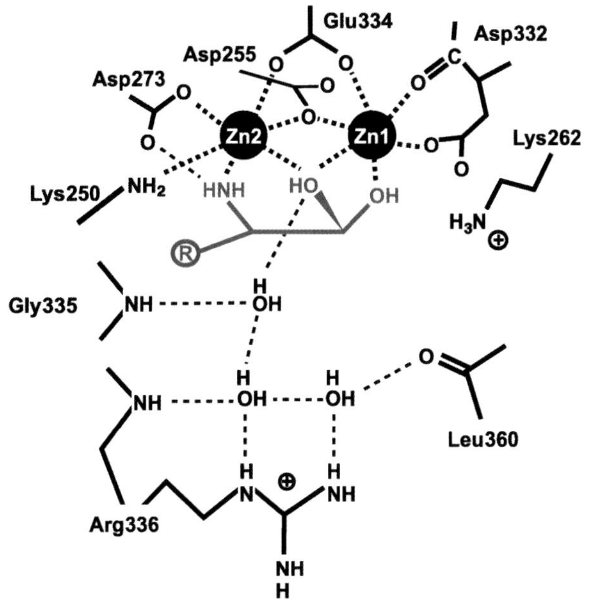

the X-ray structure of *E. coli* PepA was determined, a bicarbonate ion was found at the same location as in LAP.[35] Furthermore, PepA is activated about 10-fold by bicarbonate ions with an apparent association constant of $K_a = 0.2$ mM.[41] Thus, the bicarbonate ion most likely has a functional role in the catalytic mechanism of these enzymes. The binding of the bicarbonate ion is facilitated by a favorable hydrogen-bonding environment near Arg336 and by the distinct positive electrostatic potential of the binding site.

The absence of any activation of blLAP by bicarbonate ions might have one of two reasons: (i) the presence or absence of a bicarbonate ion next to Arg336 has no influence on the enzyme activity or (ii) the bicarbonate ion is so tightly bound that it cannot be removed by degassing the buffer, in contrast to the situation in PepA.

COMPLEX STRUCTURES

Table 2 gives an overview of the native and complex structures determined for blLAP and *E. coli* PepA. No significant conformational changes occur on inhibitor binding. The complex structure with L-leucinal (L-leucine aldehyde) probably resembles the binding mode of the presumed *gem*-diolate transition state most closely, since it binds as a hydrated *gem*-diol (Figures 4(b) and 5).[40] One hydroxyl group bridges the metal ions at the position in which the metal-coordinated water molecule is bound in the native structure and the second OH group is coordinated to Zn1. The terminal amino group is coordinated to Zn2. Thus, both metal ions are six-coordinated in a distorted octahedral geometry in the structure of the complex.

L-leucine phosphonic acid binds in a similar mode to LAP as does L-leucinal.[42] In both structures of these complexes, the bicarbonate ion is replaced by three water molecules. Two of these water molecules are in close proximity of about 2.3 to 2.5 Å distance between their oxygen atoms. It has been proposed that one of the water molecules is deprotonated to form a bihydroxide ion (a water molecule hydrogen-bonded to a hydroxide ion) near the distinctly positive potential of Arg336.[40,41]

Table 2 X-ray structures determined for blLAP and *E. coli* PepA

Protein	Inhibitor	Space group	Resolution (Å)	PDB id	References
blLAP	–	P6₃22	1.6	1LAM	40
blLAP	Leucinal	P6₃22	1.9	1LAN	40
blLAP	LeuP[a]	P321	1.65	1LCP	42
blLAP	Bestatin	P6₃22	2.25	–	37,43,44
blLAP	Amastatin	P6₃22	2.4	1BLL	45
PepA	–	P32	2.5	1GYT	35,41

[a] L-leucine phosphonic acid.

Leucine aminopeptidase

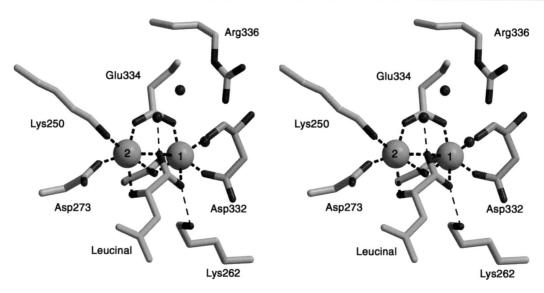

Figure 5 Binding mode of L-leucinal to the active center. The carbon atoms of the inhibitor are colored green, whereas those of the protein residues are shown in gray. The two zinc ions are depicted in yellow. The carbonate ion in the structure of the unliganded enzyme replaces the three water molecules next to Arg336. This figure has been prepared using programs MOLSCRIPT[38] and RASTER3D.[39]

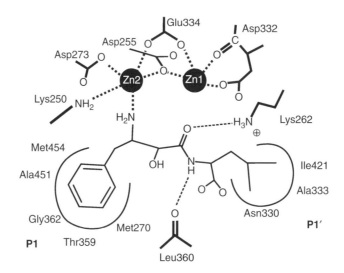

Figure 6 Schematic view of the binding mode of bestatin.

The structures in complex with the natural inhibitors bestatin (dipeptide analogue inhibitor)[37,43,44] and amastatin (tetrapeptide analogue)[45] are important because they define the substrate specificity pockets (Figure 6). The P1 pocket that binds the N-terminal amino acid (nomenclature by Schechter and Berger[46]) is formed by amino acids Met270, Thr359, Gly362, Ala451, and Met454. The P1′ binding pocket is formed by amino acids Asn330, Ala333, and Ile421. Beyond the P1′ portion, the bound amastatin is near the solvent cavity at the center of the LAP hexamer and there are few enzyme–inhibitor interactions for the P2′ and P3′ portions of amastatin.

Both natural inhibitors contain an additional hydroxymethyl group between the C_α-atom and the peptide carbonyl group. Thus, it is difficult to decide if they inhibit by acting as substrate or transition-state analogues. The additional hydroxyl group binds such that it bridges the two metal ions as does one of the *gem*-diol groups in leucinal. The carbonyl group is not coordinated to the metal ions, but it is hydrogen-bonded to Lys262.

FUNCTIONAL ASPECTS

Steady state kinetics and mutational analysis

Pig kidney LAP is maximally active between pH 9 and 9.5 and cleaves a variety of amino acid amides, dipeptides, and other compounds.[14,47,48] Compounds that have L-leucyl residues in the N-terminal position are the preferred substrates, but all tested substances that have an N-terminal L-amino acid (or glycine) are hydrolyzed. Some characteristic kinetic data are as follows: L-Leu-Gly: $K_m = 8.4\,\text{mM}$, $k_{cat} = 1670\,\text{s}^{-1}$ for the Mg^{2+}-activated enzyme, L-Leu-Phe: $K_m = 0.06\,\text{mM}$, $k_{cat} = 270\,\text{s}^{-1}$ without activator (probably Zn–Zn form) and L-Gly-Phe: $K_m = 1.0\,\text{mM}$, $k_{cat} = 0.5\,\text{s}^{-1}$ without activator (Reference 49 and references therein). For the chromogenic substrate L-leucine-p-nitroanilide, Michaelis-Menten kinetic constants of $K_m = 4.1\,\text{mM}$, $k_{cat} = 0.003\,\text{mmol}\,\text{s}^{-1}\,\text{mg}^{-1}$ and of $K_m = 2.0\,\text{mM}$, $k_{cat} = 0.04\,\text{mmol}\,\text{s}^{-1}\,\text{mg}^{-1}$ have been reported for the Zn–Zn and Mg–Zn forms respectively.[18] Peptides with a proline residue in P1' (Xaa–Pro-) are not cleaved by LAP. Esters are cleaved at about 10% of the rate of the corresponding amides. A comparison of the activities of cattle lens and pig kidney LAP on amino acid amides, aminoacyl-β-naphthylamides, and aminoacyl-p-nitroanilides showed that the general specificities of the two enzymes are very similar.[50]

Studies by site-directed mutagenesis are available for E. coli PepA.[41] The mutation of Arg336 (residue numbering of blLAP; the correct residue numbers of active site residues for PepA are obtained by adding 20) showed that this residue contributes to catalysis, but is not essential. Arg336 binds the bicarbonate ion that activates PepA about 10-fold. The mutants R336A and R336K have approximately the same rate constant as the wild-type enzyme in the absence of activation by bicarbonate ions. This result indicates that the role of Arg336 is indeed to bind an exogenous bicarbonate ion in the positively charged pocket.

Mutation of Lys262 to an alanine reduces the catalytic activity as well as the substrate binding affinity such that the second-order rate constant (k_{cat}/K_m) is reduced about 10 000-fold. These results indicate that this lysine side chain is involved in substrate binding as well as in transition-state stabilization. Mn^{2+}-activated PepA has a K_m of 0.23 mM and a k_{cat} of $3.1\,\text{s}^{-1}$ for the hydrolysis of L-leucine-p-nitroanilide.[41]

CATALYTIC MECHANISM

In the proposed catalytic mechanism (Figure 7), Zn2 binds the terminal amino group of the substrate and Zn1 polarizes the carbonyl group for attack of the nucleophile. The zinc-bridging water molecule is activated by the two metal ions by a reduction of its pK_a such that it can be easily deprotonated to the better nucleophilic hydroxide ion.[41] The bicarbonate ions accept a proton from the water molecule and thus function as a general base similar to the carboxylic acid side chains in the mono–zinc peptidases, such as carboxypeptidase A[51] and thermolysin.[52] However, in these enzymes, mutations of the general base residue have a much larger impact on the catalytic activity, compared to the contribution of the bicarbonate ion (about 10-fold activation) in the two Zn^{2+} enzymes. For example, mutations of Glu270 in carboxypeptidase A radically lowers enzyme activity.[53]

Binding of the nucleophile to two metal ions (instead of one) might reduce the nucleophilicity of the hydroxide ion more than necessary (e.g. for deprotonation at pH ∼ 7). However, the type and charge of the zinc coordinating ligands as well as the dielectric constant and electrostatic potential of the protein environment will also influence the pK_a and nucleophilicity of the water nucleophile. It is also possible that the bridging water ligand is a hydroxide ion that is deprotonated to an oxide ion on nucleophilic attack.

The gem-diolate transition state is stabilized by coordination to the two metal ions such that one gem-diol group bridges the two zinc ions symmetrically and the second gem-diol group is coordinated to Zn1. In addition, Lys262 is hydrogen-bonded to the gem-diol OH, which is terminally coordinated. The bicarbonate ion shuttles the proton from the nucleophile to the leaving amino group.

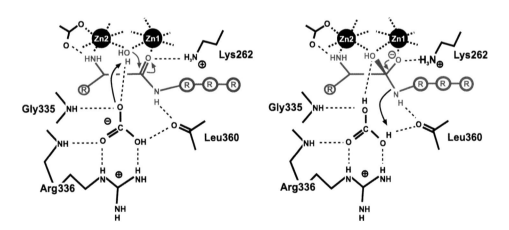

Figure 7 Proposed catalytic mechanism. The mechanism is based on the structures of the unliganded enzyme and the structure in complex with transition-state analogue inhibitors (Figure 4). Both metal ions are involved in substrate binding and stabilization of the transition state. The nucleophile is a water molecule or a hydroxide ion bound to both metal ions.

COMPARISON WITH RELATED STRUCTURES

LAP and PepA are as members of the metallohydrolase family 17 (M17) related to other members of the zinc hydrolase superfamily.[54–56] These include families M14 (carboxypeptidases A, A2, B, D, T-cpA, cpA2, cpB, cpD, and cpT) and the metallopeptidase H clan (*Aeromonas proteolytica* aminopeptidase, *Streptomyces griseus* aminopeptidase, carboxypeptidase G2). These proteins are derived from a common evolutionary ancestor protease and share a similar structural scaffold consisting of eight β-strands and six α-helices. The active sites are located at the C-terminal end of the central four parallel β-strands.

However, when the different families are superimposed on the basis of the conserved fold, the zinc binding sites and the active sites do not superimpose well.[54] The metal positions as well as the metal-ligand spheres have changed during the course of evolution. These changes are mainly due to the donation of one or more zinc ligands from different regions of the backbone rather than to movements of the protein backbone in the vicinity of the zinc-ligating residues. However, when the di-zinc enzymes of the metallopeptidase H clan and of the M17 family are superimposed, one of the two zinc ions is conserved between the two families (Zn1 of *Aeromonas p.* Aminopeptidase with Zn489 of LAP). Carboxypeptidase A and LAP do not share any homologous zinc binding sites in this superposition.

Alternatively, the active sites can also be superimposed on the basis of the local active site structure (the two metal ions, the nucleophile, the substrate binding pocket and the general base). Such a superposition stresses the evolution of enzyme active sites to structures that appear to converge at similar solutions for a particular catalytic mechanism. In the case of the zinc hydrolase superfamily, both convergent and divergent evolution has resulted in structures with similar scaffolds and apparently similar catalytic mechanism, but the local active site structures are only partially conserved and in the case of the dimetal sites, the roles of the two ions appear to have interchanged during evolution. For a more detailed analysis of the relationships among these enzymes, the reader is referred to an article by Wouters and Husain.[54]

PROTEIN/DNA INTERACTION

PepA serves as a DNA binding protein in Xer site-specific DNA recombination.[8] PepA and the arginine repressor (ArgR) serve as accessory proteins, ensuring that recombination is exclusively intramolecular. In addition to these two proteins, the homologous recombinases XerC and XerD are necessary for the recombination reaction. PepA binds to about 160 bp of accessory sequences flanking the recombination sites. The DNA binding sites of PepA are not known, but it has been proposed that the DNA binds along a large groove that runs from the lower trimer face across the twofold molecular axis to the upper trimer face.[35] LAP has no known DNA binding function.

REFERENCES

1   K Linderstrøm-Lang, *Z Physiol Chem*, **182**, 151–74 (1929).
2   EL Smith and DH Spackman, *J Biol Chem*, **212**, 271–99 (1955).
3   J Beninga, KL Rock and AL Goldberg, *J Biol Chem*, **273**, 18734–42 (1998).
4   P Cascio, C Hilton, AF Kisselev, KL Rock and AL Goldberg, *EMBO J*, **20**, 2357–66 (2001).
5   CA Harris, B Hunte, MR Krauss, A Taylor, LP Epstein, *J Biol Chem*, **267**, 6865–69 (1992).
6   A Taylor, T Surgenor, DKR Thomson, RJ Graham and HC Oettgen, *Exp Eye Res*, **38**, 217–29 (1984).
7   GJ Buffone, JE Spence, SD Fernbach, MR Curry, WE O'Brien and AL Beaudet, *Clin Chem*, **34**, 933–37 (1988).
8   CJ Stirling, SD Colloms, JF Collins, G Szatmari and DJ Sherratt, *EMBO J*, **8**, 1623–27 (1989).
9   D Charlier, G Hassanzadeh, A Kholti, D Gigot, A Pierard and N Glansdorff, *J Mol Biol*, **250**, 392–406 (1995).
10 R McCulloch, ME Burke and DJ Sherratt, *Mol Microbiol*, **12**, 241–51 (1994).
11 BP Wallner, C Hession, R Tizard, AZ Frey, A Zuliani, C Mura, J Jahngen-Hodge and A Taylor, *Biochemistry*, **32**, 9296–301 (1993).
12 HT Cuypers, LAH van Loon-Klaassen, WTM Vree Egberts, WW de Jong and H Bloemendal, *J Biol Chem*, **257**, 7077–85 (1982).
13 MP Allen, AH Yamada and FH Carpenter, *Biochemistry*, **22**, 3778–83 (1983).
14 DH Spackman, EL Smith and DM Brown, *J Biol Chem*, **212**, 255–69 (1955).
15 YQ Gu, FM Holzer and LL Walling, *Eur J Biochem*, **263**, 726–35 (1999).
16 FH Carpenter and JM Vahl, *J Biol Chem*, **248**, 294–304 (1973).
17 MJ Johnson, GH Johnson and WH Peterson, *J Biol Chem*, **116**, 515–26 (1936).
18 GA Thompson and FH Carpenter, *J Biol Chem*, **251**, 53–60 (1976).
19 H Kim and WN Lipscomb, *Proc Natl Acad Sci USA*, **90**, 5006–10 (1993).
20 W Grassmann and W Heyde, *Z Physiol Chem*, **183**, 32–8 (1929).
21 DH Rich, BJ Moon and S Harbeson, *J Med Chem*, **27**, 417–22 (1984).
22 A Taylor, CZ Peltier, FJ Torre and N Hakamian, *Biochemistry*, **32**, 784–90 (1993).
23 OA Scornik and V Botbol, *Curr Drug Metab*, **2**, 67–85 (2001).
24 G Pulido-Cejudo, B Conway, P Proulx, R Brown and CA Izaguirre, *Antiviral Res*, **36**, 167–77 (1997).
25 B Lejczak, P Kafarski and J Zygmunt, *Biochemistry*, **28**, 3549–55 (1989).
26 PP Giannousis and PA Bartlett, *J Med Chem*, **30**, 1603–09 (1987).
27 L Andersson, J MacNeela and R Wolfenden, *Biochemistry*, **24**, 330–33 (1985).

28 PL Birch, HA El-Obeid and M Akhtar, *Arch Biochem Biophys*, **148**, 447–51 (1972).
29 SL Harbeson and DH Rich, *J Med Chem*, **32**, 1378–92 (1989).
30 WWC Chan, P Dennis, W Demmer and K Brand, *J Biol Chem*, **257**, 7955–57 (1982).
31 WWC Chan, *Biochem Biophys Res Commun*, **116**, 297–302 (1983).
32 AL Dounce and PZ Allen, *Trends Biochem Sci*, **13**, 317–20 (1988).
33 AL Dounce and PZ Allen, *Arch Biochem Biophys*, **257**, 13–16 (1987).
34 MM Chernaya, MK Nurbekov and AA Fedorov, *Biofizika*, **30**, 700–1 (1985).
35 N Sträter, DJ Sherratt and SD Colloms, *EMBO J*, **18**, 4513–22 (1999a).
36 SK Burley, PR David, A Taylor and WN Lipscomb, *Proc Natl Acad Sci USA*, **87**, 6878–82 (1990).
37 SK Burley, PR David, RM Sweet, A Taylor and WN Lipscomb, *J Mol Biol*, **224**, 113–40 (1992).
38 PJ Kraulis, *J Appl Crystallogr*, **24**, 946–50 (1991).
39 EA Meritt and MEP Murphy, *Acta Crystallogr, Sect D*, **50**, 869–73 (1994).
40 N Sträter and WN Lipscomb, *Biochemistry*, **34**, 14792–800 (1995).
41 N Sträter, L Sun, ER Kantrowitz and WN Lipscomb, *Proc Natl Acad Sci USA*, **96**, 11151–55 (1999).
42 N Sträter and WN Lipscomb, *Biochemistry*, **34**, 9200–10 (1995).
43 SK Burley, PR David and WN Lipscomb, *Proc Natl Acad Sci USA*, **88**, 6916–20 (1991).
44 H Kim, SK Burley and WN Lipscomb, *J Mol Biol*, **230**, 722–24 (1993).
45 H Kim and WN Lipscomb, *Biochemistry*, **32**, 8465–78 (1993).
46 I. Schechter and A Berger, *Biochem Biophys Res Commun*, **27**, 157–62 (1967).
47 EL Smith and DH Spackman, *J Biol Chem*, **212**, 271–99 (1955).
48 RJ Delange and EL Smith, in PD Boyer (ed.), *The Enzymes*, 3rd edn, Vol. III, Academic Press, London, pp 81–118 (1971).
49 H Hanson and M Frohne, *Methods Enzymol*, **45**, 504–21 (1976).
50 H Hanson, D Glässer, M Ludewig, HG Mannsfeldt, M John and H Nesvadba, *Z Physiol Chem*, **348**, 689–704 (1967).
51 DW Christianson and WN Lipscomb, *Acc Chem Res*, **22**, 62–69 (1989).
52 BW Matthews, *Acc Chem Res*, **21**, 333–40 (1988).
53 SJ Gardell, *Abstract Meeting Physico-Chemical Approaches for the Analysis of Biological Catalysis*, Florence, Italy, June 16–20 (1986).
54 MA Wouters and A Husain, *J Mol Biol*, **314**, 1191–207 (2001).
55 PJ Artymiuk, HM Grindley, JE Park, DW Rice and P Willett, *FEBS Lett*, **303**, 48–52 (1992).
56 KS Makarova and NV Grishin, *J Mol Biol*, **292**, 11–17 (1999).

Bacillus subtilis D-aminopeptidase DppA

Han Remaut[†], Colette Goffin[‡], Jean-Marie Frère[‡] and Jozef Van Beeumen[†]

[†]Laboratory of Protein Biochemistry and Protein Engineering, Gent University, K.L. Ledeganckstraat, Belgium
[‡]Centre for Protein Engineering, Liège University, Institut de Chimie B6, Sart Tilman, Belgium

FUNCTIONAL CLASS

Enzyme; D-aminopeptidase; EC 3.4.11; a D-stereospecific dinuclear zinc metalloprotease.

OCCURRENCE

DppA was first detected in *Bacillus pumilus* during a functional search for enzymes that were able to hydrolyze D-alanyl-para-nitroanilide (D-Ala-*p*Na), then it was detected in other *Bacillus* strains, and finally it has been purified from *Bacillus subtilis* strain 168.[1] An N-terminal sequence fragment was determined that identified the enzyme as the product of the *dppA* gene. This gene, previously called *dciAA*, constitutes the first open reading frame (ORF) of the dipeptide permease operon (*dpp*, previously *dciA*).[2] Dpp is under negative control of both the general stationary-phase regulator AbrB and a nutrient-sensing regulator CodY.[3–6] As a result, *dppA* is expressed only during the early stationary phase. In the case of *B. subtilis*, the *dpp* operon is expressed shortly after induction of sporulation (*dppA* transcripts are detected 8 min after decoyinine-induced sporulation).[2] In *B. subtilis*, DppA can be found in supernatants of ageing cultures. However, the absence of a signal sequence indicates that the enzyme is more likely cytoplasmic and that extracellular DppA activity results from lysis of the cells.

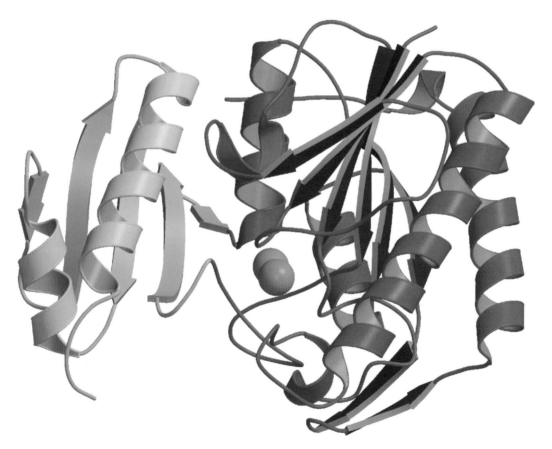

3D Structure Schematic representation of the monomer structure of DppA plus two zinc ions, PDB code 1HI9. Prepared with programs MOLSCRIPT[15] and RASTER3D.[16]

BIOLOGICAL FUNCTION

DppA hydrolyzes D-alanyl-D-alanine (D-Ala-D-Ala), and is expressed during early stationary phase as part of a dipeptide permease operon. Though unknown, its physiological role is probably an adaptation to nutrient deficiency. D-Alanine, the product resulting from (D-Ala)$_2$ hydrolysis, could be oxidized to carbon dioxide and acetate by the actions of D-alanine dehydrogenase and pyruvate oxidase, and could thus be used as backup metabolic fuel under starvation conditions. Such a role has been proposed for the VanX homologue (DdpX) in *Escherichia coli*, a D-Ala-D-Ala dipeptidase, which is also part of a dipeptide transport operon expressed at the end of the logarithmic growth phase.[7,8] In *E. coli*, the increased direct cross-linking of *meso*-diaminopimelic acid residues in the peptidoglycan peptides during stationary phase could provide a periplasmic pool of (D-Ala)$_2$, which could be internalized by the *dpp* encoded dipeptide permease.[7] DppA, however, does not share any significant sequence similarity with *E. coli* DdpX.

One reference suggests a possible role of DppA in extrachromosomal maintenance of some shuttle plasmids in *Bacillus methanolicus*.[9] Working on genetic transformation systems in this bacterium, Cue *et al.* identified the presence of a *dpp* fragment in a locus required for extrachromosomal stability of the plasmids.[9] This *dpp* fragment contains the full-length *B. methanolicus dppA* gene (76% amino acid identity with *B. subtilis* DppA) and the beginning of the *dppB* gene. However, the minimal fragment required for replication has not been identified yet, so it remains to be determined whether the *dppA* expression product is involved in this process.

AMINO ACID SEQUENCE INFORMATION

ORF's encoding putative DppA-like enzymes were identified in 22 microorganisms for which the determination of the genome sequences is completed or is in progress. Homologues exist in hyper-thermophilic Archaea (*Pyrococcus* (#3); *Aeropyrum* (#1)), in gram-negative bacteria (*Bordetella* (#3); *Burkholderia* (#3); *Erwinia* (#1); *Ralstonia* (#2); *Mesorhizobium* (#1)) and in gram-positive bacteria (*Bacillus* (#2); *Deinococcus* (#1); *Enterococcus* (#1); *Clostridium* (#1); *Listeria* (#2); *Desulfitobacterium* (#1); *Streptomyces* (#1)); the number of species is indicated in parentheses. All these sequences have a conserved SXDXEG motif located near the N-terminus (residues 6–11 in *B. subtilis*, D8 and E10 are zinc-coordinating residues).

Representative sequences are

- *Bacillus subtilis* 168, 274 amino acid residues (AA), translated ORF in completed genome sequence[10]
- *Pyrococcus abyssi*, 278 AA, translated ORF in completed genome sequence[11]
- *Ralstonia solanacearum*, 271 AA, translated ORF in completed genome sequence[12]
- *Listeria innocua*, 272 AA, translated ORF in completed genome sequence[13]

PROTEIN PRODUCTION, PURIFICATION, AND MOLECULAR CHARACTERIZATION

Initially, the enzyme was purified from the supernatants of ageing cultures of *B. subtilis*. However, much higher yields of protein were obtained after the *dppA* gene was cloned in the expression plasmid pET22bK^r and overexpressed in *E. coli* BL21 (DE3) cells.[1] For the recombinant protein, purification involves an initial crude fractionation on a Q-Sepharose ion exchange column at pH 6.5. Subsequent steps involve a Q-Sepharose column at pH 8.5, a Superdex-200 gel filtration column and a polishing step on a mono-Q column at pH 8.5.

The M_r value determined by electrospray mass spectrometry (ESMS) was $30\,152.7 \pm 6$, in good agreement with that calculated from the cloned gene (30 158.7). The recombinant protein differs from the sequence in the DDBJ/EMBL/Genbank databases (accession number P26902) in that Arg190 is replaced by an alanine. The specific activity of the enzyme overproduced in *E. coli* is similar to that of the *Bacillus* purified enzyme. The isoelectric pH was estimated to be 5.0 (calculated value is 5.12), whatever the way of production.[1] No disulfide bridges are present in DppA and the enzyme does not undergo post-translational modification, although SDS Page and ESMS analysis revealed that about 5% of the purified enzyme was cleaved after Ser61.

On the basis of its behavior upon molecular-sieve filtration, the enzyme was first thought to be an octamer,[1] but the crystallographic data clearly indicate a decameric structure (see below).[14]

METAL CONTENT AND COFACTORS

Recombinant *B. subtilis* DppA, purified from *E. coli* has two zinc ions per 30.150 M_r monomer, assessed using the anomalous signal in the X-ray data collected on a single crystal at 1.2832-Å wavelength (the zinc K-edge absorption is located at 1.2837 Å). No other cofactors are present in the molecule.

ACTIVITY TEST

Enzyme activity is determined from the initial rate of *p*-nitroaniline release from D-Ala-*p*Na by monitoring the increase in absorbance at 405 nm ($\varepsilon = 11\,500\,M^{-1}\,cm^{-1}$). One activity unit (IU) corresponds to the release of one µmol min^{-1} of *p*-nitroaniline.

X-RAY STRUCTURE OF NATIVE DppA

Crystallization

Recombinant DppA was crystallized in at least three different crystal forms.[14] All crystal forms were grown under very similar conditions. Orthorhombic crystals, used for structure determination, were grown from 18% (W/V) PEG (polyethylene glycol) 6000, 100 mM Tris, pH 8.5, 5 mM NaN_3, 5 mM $MgCl_2$ and 5 mM NaCl. The stock protein solution contained 18 mg mL^{-1} enzyme in 10 mM Tris, pH 8.5, and 5 mM NaN_3. All reported crystals were obtained by sitting drop vapor diffusion (1:1 ratio of protein solution to precipitant solution) at 294 K. Frequently tetragonal and monoclinic crystals were obtained under the same conditions depending on subtle differences in growth temperature or in the $MgCl_2$ and NaCl concentrations. However, these latter crystals were of poor diffraction quality.

The crystals used for structure determination were in the $C222_1$ space group, with cell dimensions $a = 145.11$ Å, $b = 165.68$ Å and $c = 109.92$ Å. These crystals have five monomers per asymmetric unit, corresponding to half of the decameric DppA molecule. The anomalous signal from the zinc cofactors (5 times 2) was used for phasing the X-ray data in a three-wavelength multiple anomalous diffraction (MAD) experiment, and the structure was refined to 2.4-Å resolution.[14]

Overall description of the structure

DppA is organized as a barrel-shaped decamer with D52 point group symmetry and outer dimensions of approximately $110 \times 110 \times 85$ Å^3. Monomeric DppA (274 residues), shown in the representation of the 3D Structure and Figure 1, consist of two domains. The N-terminal domain, which forms the core of the monomeric enzyme, encompasses the zinc-coordinating residues, and is primarily responsible for building up the active site. The much smaller C-terminal domain is mainly involved in the quaternary organization of the protein. Together, the N- and C-terminal domains give rise to a kidney-shaped monomeric subunit. A large active site cleft is formed in the concave joint between both domains.

The N-terminal domain (residues 1–119 and 132–213) has a Rossmann-like α/β/α-fold formed by a central six-stranded parallel β-sheet in the order 3-2-1-4-7-8, flanked on both sides by two α-helices (Figures 1 and 2). A dinuclear zinc site is located near the C-terminal edge of this β-sheet. Two additional β-strands, numbered 9 and 10, are located between strand 8 and helix F and are important for pentamerization of the monomers. The antiparallel strands 9 and 10 of one subunit align with, and extend, the six-stranded β-sheet of a preceding subunit giving rise to a ring-shaped pentamer with fivefold symmetry (Figure 3).

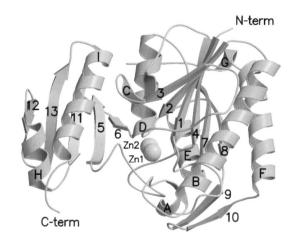

Figure 1 Schematic representation of the monomer structure of DppA with assignment of the secondary structure elements. Produced with the programs MOLSCRIPT[15] and RASTER3D.[16]

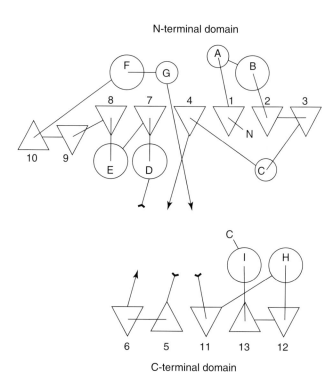

Figure 2 Topology diagram of the monomer structure of DppA with assignment of the secondary structure elements. The upper and lower diagrams represent the N-terminal and C-terminal domain respectively.

The C-terminal domain (residues 12–131 and 214–274) folds into a five-stranded antiparallel β-sheet, flanked on one side by two α-helices (Figures 1 and 2). The C-terminal domains stand up from the annular plane formed by the N-terminal domains within a pentamer. Two such pentameric units link with each other to form the barrel-shaped decamer (Figure 3). They interact by hydrophobic and electrostatic contacts between the C-terminal domains.

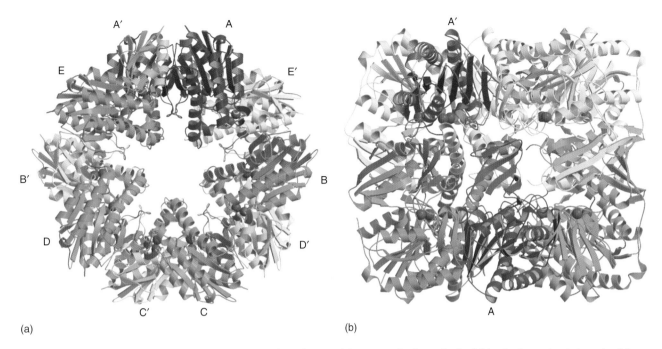

Figure 3 Representation of the DppA decamer viewed (a) along and (b) perpendicular to its fivefold axis. One subunit in each of the two pentamers is differentiated by color in order to visualize the subunit interactions in the complex. Prepared with programs MOLSCRIPT[15] and RASTER3D.[16]

Quaternary structure

DppA is a dimer of pentamers, forming a barrel-shaped complex. Two opposite conical-shaped channels give access to a central 50-Å wide cavity (Figure 4). Ten active sites are located inside this cavity where they proceed in a zigzag course along the inner wall of the compartment. They are therefore effectively shielded from the protein exterior. Free access to this active site cavity is controlled by the two entrance gates formed by the central channel crossing the decamer. The diameter of the entrance channels is about 50 Å at the surface of the complex and narrows to about 15 Å at their apex, where they open into the catalytic chamber.

The general architecture of the complex strongly resembles that found in 'self-compartmentalizing' proteases such as the 20S proteasomes,[18] HslV,[19] ClpP,[20] Lon,[21] Bleomycin hydrolase[22] and Tricorn.[23] Though unrelated in sequence or tertiary structure, all these proteases form complexes of stacked rings of subunits (3, 6, or 7 subunits per ring) that form a barrel-shaped molecule in which the active sites are sequestered inside a central 'active site chamber'.

Zinc site geometries and active site structure

DppA contains two Zn^{2+} ions per monomer, which are located close to each other in the N-terminal domain, at the C-terminal edge of inner strands 1 and 4 of the six-stranded parallel β-sheet. Both zinc(II) ions are bound in a

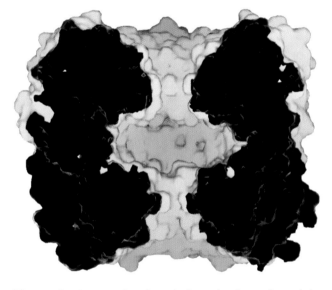

Figure 4 Cross-section through the molecular surface of the DppA decamer. The architecture of the complex gives rise to an inner 'active site chamber' that is accessed via two 15-Å wide channels. Prepared with the program GRASP.[17]

tridentate coordination site. They give rise to a (μ-aqua)(μ-carboxylato)di-zinc(II) core with terminal glutamate and histidine residues coordinating each metal ion, resulting in symmetric, tetrahedral coordination spheres for both metal ions (Figure 5). The first zinc, Zn(1), is coordinated by Asp8 (OD1), Glu10 (OE1), and His60 (ND1), and the second zinc, Zn(2), by the residues Asp8 (OD2), His104 (NE2), and Glu133 (OE1). All five residues responsible

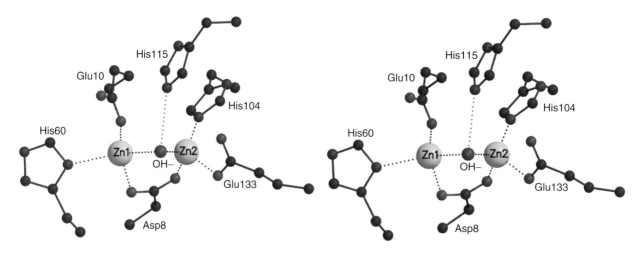

Figure 5 Stereo drawing of the dinuclear zinc(II) site in the monomeric DppA subunit. Bond distances are given in Table 1. Figure prepared with the program MOLSCRIPT.[15]

for coordinating the zinc ions are part of the N-terminal domain. The bridging H_2O or OH^- ligand is hydrogen bonded to His115 (NE2). Table 1 displays the binding distances in the catalytic site.

In the DppA X-ray structure, both zinc sites have different occupancies. From the anomalous data measured at the zinc K-edge absorption peak (1.2832-Å wavelength), it was calculated that the average occupancy of the Zn(1) site was about half of that in the Zn(2) site. This corresponds well with the significant shorter coordination distances for Zn(2) compared to that for Zn(1) (Table 1). Together these data indicate that the dissociation constant for the Zn(2) site is smaller than that for the Zn(1) site. A difference in binding constants for both ions in dinuclear metal centers has also been observed in other zinc(II) aminopeptidases, such as *Aeromonas proteolytica* (AAP)[24] and *Streptomyces griseus* aminopeptidase[25] and bovine lens leucine aminopeptidase (blLAP).[26] Moreover, the mononuclear form of AAP retains ~80% of its activity.[27]

FUNCTIONAL ASPECTS

Specific activity and specificity

The hydrolysis of 120 μM D-Ala-*p*Na remained strictly first order throughout complete hydrolysis time courses, from which a k_{cat}/K_m value of $100\,000 \pm 10\,000\,M^{-1}\,s^{-1}$ was deduced. With 1 mM D-Ala-*p*Na, at 30 °C in 50 mM potassium phosphate, pH 8.0, and 40 μM $ZnSO_4$, the specific activity of the enzyme was 80 IU mg^{-1}.[1] D-Leu-*p*Na and D-Phe-*p*Na are hydrolyzed with much lower efficiencies.

For oligopeptides, the rate of D-alanine release from D-Ala-D-Ala, $(D-Ala)_3$, $(D-Ala)_4$, D-Ala-Gly, and D-Ala-Gly-Gly indicated that D-Ala-D-Ala and D-Ala-Gly-Gly were the best substrates, with turnover numbers of about 5 min^{-1} (at 8 mM substrate concentration). DppA behaves as a strict D-aminopeptidase. No hydrolysis was observed with oligopeptides constituted of L-derivatives or L-Ala-D-Ala after a 24-h incubation at 30 °C with 12 μM pure DppA.[1]

Catalytic mechanism

Though the primary and tertiary structures of the proteins are unrelated, the DppA active site is remarkably similar to those of AAP and blLAP, for which a detailed catalytic mechanism has been proposed.[27,28] The zinc coordination geometry in DppA is very similar to that in AAP, in particular (Figure 6). In both enzymes, the active site contains a (μ-aqua)(μ-carboxylato)di-zinc(II) core with terminal carboxylate and histidine residues ligating each metal ion. In AAP, a nonmetal coordinating glutamate (Glu151) forms a hydrogen bond with the bridging OH^-

Table 1 Bond distances in the dinuclear zinc center of DppA. Bond distances are given for each of the five subunits (A to E) in the asymmetric unit of the *B. subtilis* DppA crystal structure

Atom1	Atom2	Distance (Å) in DppA subunits				
		A	B	C	D	E
Zn2	D8 OD2	2.02	1.98	1.95	2.05	2.07
	H104 NE2	1.91	1.97	1.96	1.99	1.99
	E133 OE1	1.98	1.95	1.97	1.87	1.88
	OH^-	2.10	2.13	2.10	2.07	2.01
Zn1	D8 OD1	2.29	2.31	2.29	2.33	2.32
	E10 OE1	2.29	2.31	2.38	2.28	2.26
	H60 ND2	2.52	2.52	2.46	2.55	2.47
	OH^-	2.55	2.60	2.59	2.49	2.51
Zn1	Zn2	3.14	3.16	3.21	3.18	3.23
H115 NE2	OH^-	2.94	2.89	2.83	2.85	2.94

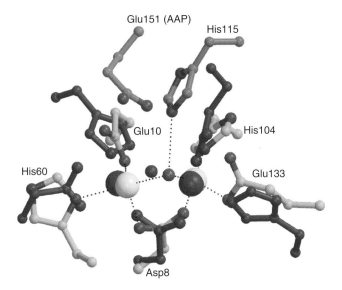

Figure 6 Superposition of the zinc coordination sites of *B. subtilis* DppA and *A. proteolytica* aminopeptidase (PDB code 1AMP). Residues in DppA are indicated in atom color, the possible acid/base catalyst, His115, in darker orange. Residues in AAP are colored blue; the acid/base catalyst, Glu151 is colored in light blue. Residue names are given for DppA. Figure prepared with programs MOLSCRIPT[15] and RASTER3D.[16]

or H_2O ligand, which serves as the nucleophile during attack of the scissile peptide bond.[27] This glutamate assists in deprotonation of the nucleophile and stabilization of the *gem*-diolate transition state intermediate. In DppA, this acid/base catalyst is probably a histidine residue (His115). Given the high degree of similarity between the cocatalytic zinc(II) sites of DppA and AAP, the catalytic mechanism of the former is expected to be similar to that proposed for the latter.[27]

Compartmentalization

In established examples of self-compartmentalizing proteases, this assembly is believed to be a protective measure acting as a 'molecular sieve', only allowing small peptides to enter the active site chamber and excluding larger polypeptide substrates from unwanted hydrolysis.[29] A number of the compartmentalizing proteases have further evolved to cooperate with ATPase subunits that associate with the complex near the entrance to the active site chamber. These modular protease complexes, like the 26S proteasome, act on protein targets in an energy-dependent way and are the main effectors in the cellular protein-breakdown pathways. The ATPase units are believed to be involved in substrate recognition, in its unfolding and translocation into the active site chamber, as well as in gating of the channels giving access to this chamber.[29,30] At present, there is no evidence that the DppA complex might associate with ATPase subunits. The reasons for the self-compartmentalizing architecture in DppA are as yet not understood. Indeed, the absence of cytoplasmic metabolites or peptides, other than D-Ala-D-Ala, with D-amino acids at their N-terminus seems to contradict the hypothesis that compartmentalization in DppA is needed as a protective measure to prevent unwanted hydrolysis of larger substrates.

REFERENCES

1. A Cheggour, L Fanuel, C Duez, B Joris, F Bouillenne, B Devreese, G Van Driessche, J Van Beeumen, JM Frère and C Goffin, *Mol Microbiol*, **37**, 1–12 (2000).
2. C Mathiopoulos, JP Mueller, FJ Slack, CG Murphy, S Patankar, G Bukusoglu and AL Sonenshein, *Mol Microbiol*, **5**, 1903–13 (1991).
3. FJ Slack, JP Mueller, MA Strauch, C Mathiopoulos and AL Sosenhein, *Mol Microbiol*, **5**, 1915–25 (1991).
4. MA Strauch, *Prog Nucleic Acid Res Mol Biol*, **46**, 121–53 (1993).
5. FJ Slack, P Serror, E Joyce and AL Sonenshein, *Mol Microbiol*, **15**, 689–702 (1995).
6. P Serror and AL Sosenhein, *Mol Microbiol*, **20**, 843–52 (1996).
7. IAD Lessard, SD Pratt, DG McCafferty, DE Bussiere, C Hutchins, BL Wanner, L Katz and CT Walsh, *Chem Biol* **5**, 489–504 (1998).
8. IAD Lessard and CT Walsh, *Proc Natl Acad Sci USA*, **96**, 11028–32 (1999).
9. D Cue, H Lam, RL Dillingham, RS Hanson and MC Flickinger, *Appl Environ Microbiol*, **63**, 1406–20 (1997).
10. F Kunst and 149 co-authors, DDBJ/EBML/Genbank databases, accession code P26902.
11. R Heilig, DDBJ/EMBL/Genbank databases, accession code Q9V149.
12. M Salanoubat and 27 co-authors, DDBJ/EMBL/Genbank databases, accession code Q8XZL1.
13. P Glaser and 54 co-authors, DDBJ/EMBL/Genbank databases, accession code Q92FA5.
14. H Remaut, C Bompard-Gilles, C Goffin, JM Frère and J Van Beeumen, *Nat Struct Biol*, **8**, 674–78 (2001).
15. P Kraulis, *J Appl Crystallogr*, **24**, 946–50 (1991).
16. EA Merritt and MEP Murphy, *Acta Crystallogr*, **D50**, 869–73 (1994).
17. A Nicholls, KA Sharp and B Honig, *Proteins*, **11**, 281–96 (1991).
18. J Lowe, D Stock, B Jap, P Zwickl, W Baumeister and R Huber, *Science*, **268**, 533–39 (1995).
19. M Bochtler, L Ditzel, M Groll and R Huber, *Proc Natl Acad Sci USA*, **94**, 6070–74 (1997).
20. J Wang, JA Hartling and JM Flanagan, *Cell*, **91**, 447–56 (1997).
21. H Stahlberg, E Kutejova, K Suda, B Wolpensinger, A Lustig, G Schatz, A Engel and CK Suzuki, *Proc Natl Acad Sci USA*, **96**, 6787–90 (1999).
22. L Joshua-Tor, HE Xu, SA Johnston, DC Rees, *Science*, **269**, 945–50 (1995).
23. H Brandstetter, JS Kim, M Groll and R Huber, *Nature*, **414**, 466–69 (2001).
24. B Chevrier, C Schalk, H D'Orchymont, JM Rondeau, D Moras and C Tarnus, *Struct Fold Des*, **2**, 283–91 (1994).

25. HM Greenblatt, O Almog, B Maras, A Spungin-Bialik, D Barra, S Blumberg and G Shoham, *J Mol Biol*, **265**, 620–36 (1997).
26. H Kim and WN Lipscomb, *Proc Natl Acad Sci USA*, **90**, 5006–10 (1993).
27. RC Holz, *Coord Chem Rev*, **232**, 5–26 (2002).
28. N Strater, L Sun, ER Kantrowitz and WN Lipscomb, *Proc Natl Acad Sci USA*, **96**, 11151–55 (1999).
29. A Lupas, JM Flanagan, T Tamura and W Baumeister, *Trends Biochem Sci*, **22**, 399–404 (1997).
30. CN Larsen and D Finley, *Cell*, **91**, 431–34 (1997).

Hydrolases: Acting on carbon–nitrogen bonds, other than peptide bonds

Metallo β-Lactamases

Osnat Herzberg[†] and Paula MD Fitzgerald[‡]

[†]Center for Advanced Research in Biotechnology, University of Maryland Biotechnology Institute, MD, USA
[‡]Merck Research Laboratories, NJ, USA

FUNCTIONAL CLASS

Enzyme; metallo-β-lactamase; zinc-dependent β-lactamase; class B β-lactamase; EC 3.5.2.6. Metallo-β-lactamase catalyzes the hydrolysis of β-lactam compounds by cleaving the β-lactam C—N bond (Scheme 1). It confers bacterial resistance to β-lactam antibiotics such as penicillin and cephalosporin. There are four β-lactamase sequence families, termed class A, B, C, and D. The class B enzymes depend on the presence of zinc for their activity. Most family members exhibit a broad-spectrum substrate profile, acting not only on penicillin and cephalosporin antibiotics but also on carbapenem antibiotics as well as some mechanism-based inhibitors of the class A β-lactamases. The class B β-lactamases are divided into three sequence-based subclasses called B1, B2, and B3.[1]

OCCURRENCE

Metallo-β-lactamases occur in the periplasm or the extracellular space of growing bacteria. The first metallo-β-lactamase was characterized in *Bacillus cereus*.[2] The list of identified metallo-β-lactamases has grown rapidly and, at the time of writing, includes enzymes from 28 different

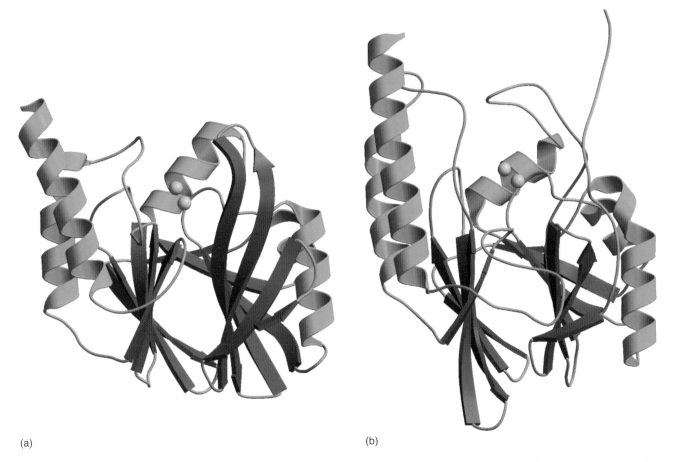

(a) (b)

3D Structure The overall fold of metallo-β-lactamases. (a) A subclass B1 metallo-β-lactamase exemplified with a structure of the CcrA enzyme from *B. fragilis* (PDB entry 1A7T);[55] (b) a subclass B3 metallo-β-lactamase, exemplified with a structure of the L1 enzyme from *S. maltophilia* (PDB entry 1SML).[31] The coloring scheme is red for β strands, green for α helices, and yellow for zinc atoms. Prepared with the programs MOLSCRIPT[61] and RASTER3D.[62,63]

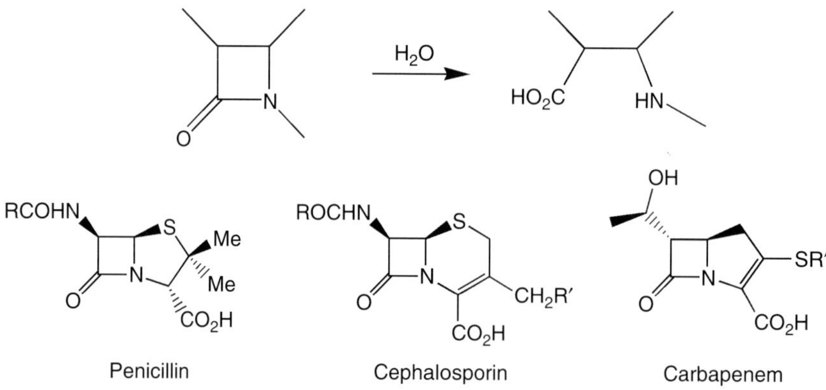

Metallo β-Lactamases

Scheme 1

species of bacteria. Moreover, some organisms code for several sequence variants, reflecting a rapid evolutionary process. The earliest metallo-β-lactamase enzymes characterized were encoded by nontransferable chromosomal genes, but, subsequently, metallo-β-lactamase genes carried on transferable genetic material (plasmid or integron) were detected in Japan, where genes coding for IMP-variant enzymes in *Pseudomonas aeruginosa*, *Serratia marcescens*, and *Klebsiella pneumoniae* have been reported.[3-5] More recently, integron-encoded VIM-variant enzymes from *P. aeruginosa* have been identified in Italy and France.[6,7] In December of 2002, a VIM-1 metallo-β-lactamase was identified in *Escherichia coli* in Greece.[8] It is interesting to note that in 1996–1998, the first outbreak of VIM-1 resistance in *P. aeruginosa* occurred in Greece.[9] Many new IMP and VIM variants have now been identified in various bacteria worldwide and the numbers are growing rapidly. The potential for lateral gene transfer, the broad-spectrum substrate profile, and the lack of clinically useful inhibitors raise the concern that bacteria that have acquired resistance to β-lactam antibiotics via metallo-β-lactamases will ultimately pose a greater clinical threat than bacteria with resistance mediated by the class A β-lactamases.[10,11]

BIOLOGICAL FUNCTION

β-Lactam antibiotics interfere with bacterial cell wall biosynthesis and repair by inhibiting cell enzymes involved in the formation of the peptidoglycan matrix. The antibiotic compounds inhibit the step in which a D-Ala-D-Ala peptide is cleaved – the step just prior to the transpeptidation reaction that links polysaccharide chains. As a defense against these antibiotics, some bacteria have evolved the ability to produce β-lactamases, periplasmic enzymes that hydrolyze β-lactam compounds before they reach their target enzymes.

AMINO ACID SEQUENCE INFORMATION

Sequence analysis has shown that the metallo-β-lactamases belong to a superfamily of enzymes that perform a wide range of biochemical reactions. A common feature of the chemistry of many superfamily members is the action on an ester linkage with involvement of a negative charge. In addition to metallo-β-lactamase, the superfamily includes glyoxylase II, aryl sulfatase, cytidine monophosphate-N-acetyl neuraminic acid hydrolase, cAMP phosphodiesterase, alkylsulfatase, rubredoxin oxygen:oxidoreductase, teichoic acid phosphorylcholine esterase, insecticide hydrolases, a polyadenylation factor involved in mRNA processing, phnP, an enzyme involved in alkylphosphonate uptake, romA, an enzyme associated with multidrug resistance, enzymes involved in DNA interstrand cross-link repair, proteins involved in DNA uptake for natural genetic transformation, and others.[12,13]

Currently, the nonredundant database contains over 50 sequences of confirmed metallo-β-lactamases, in addition to homologues from genome scale sequencing that may or may not correspond to metallo-β-lactamases. The proteins vary in length between approximately 240 and 310 amino acid residues and contain signal peptides of 17 to 30 amino acids. The sequences have been organized into subclasses B1, B2, and B3, and a standard numbering scheme based on sequence alignment has been developed by a metallo-β-lactamase working group.[1] In the following discussion, residues are numbered according to this scheme. Where a specific crystal structure is under discussion, the residue numbers used in the original publication and in the coordinates deposited in the Protein Data Bank are provided in parenthesis. The first residue in the standard numbering scheme corresponds to residue number 1 of the signal peptide sequence of the largest enzyme in the family, variant L1 (subclass B3) from *Stenotrophomonas maltophilia*.[14]

The key markers along the polypeptide chains are the residues that serve as ligands to the two active site zinc ions, which are referred to as Zn1 and Zn2. The differences between the three subclasses can be seen in the zinc ligands listed in Table 1. Subclass B2 contains an asparagine residue postulated to coordinate to Zn1 at position 116 instead of the histidine residue seen in all other metallo-β-lactamases; the remaining zinc ligands are the same as in subclass B1.

Table 1 Consensus numbering of zinc ligands in class B β-lactamases

Sub-class	Zn1 ligands			Zn2 ligands		
B1	His116	His118	His196	Asp120	Cys221	His263
B2	Asn116	His118	His196	Asp120	Cys221	His263
B3	His/Gln116	His118	His196	Asp120	His121	His263

In all known subclass B3 enzymes, the second ligand of Zn2 is His121 instead of the Cys221 seen in the subclass B1 and B2 enzymes; the residue equivalent to Cys221 in the B3 enzymes is Ser221, which plays only an indirect role in metal binding. In some subclass B3 enzymes, His116 is replaced by Gln116.

PROTEIN PRODUCTION, PURIFICATION, AND MOLECULAR CHARACTERIZATION

Early work on the enzyme from *B. cereus* predated recombinant DNA technology and enzyme samples were obtained from *B. cereus* cell cultures.[15] Currently, all metallo-β-lactamases are produced in recombinant form by expression in *E. coli*. The following reports describe the production and purification of enzyme variants for which three-dimensional structures are available: Subclass B1: *B. cereus*,[16–18] *Bacteriodes fragilis*,[19–23] *P. aeruginosa* (variant IMP-1);[24] subclass B3: *S. maltophilia* (variant L1),[25] *Fluoribacter gormanii*, (variant FEZ-1).[26] So far, there are no structures of subclass B2 metallo-β-lactamases, although production and purification of the subclass B2 enzyme CphA from *Aeromonas maltophilia* have been reported.[27]

In solution, most metallo-β-lactamases exist as monomers of $M_r \sim 25\,000$ Da. The exception is the *S. maltophilia* L1 enzyme (subclass B3), which is a tetramer of $M_r \sim 29\,000$ Da subunits.[28–31] FEZ-1, the second subclass B3 enzyme to be studied crystallographically, is a monomer.[26]

The metallo-β-lactamases are inhibited by metal chelators such as ethyethylenediaminetetraacetic acid (EDTA), dipicolinic acid, and 1,10–o–phenanthroline in contrast to the serine β-lactamases,[2,6,7,15,20,26,30,32–37] and they are not inhibited by mechanism-based inhibitors of the class A β-lactamases (on the contrary, the metallo-β-lactamases degrade some of these inhibitors).

METAL CONTENT AND COFACTORS

In vivo, metallo-β-lactamases function in the presence of Zn(II), although the enzymes are also active in the presence of Cd(II) and Co(II).[15,19,20] All metallo-β-lactamases have the potential to bind two Zn(II) ions, presumably forming a binuclear center. The *B. cereus*, *B. fragilis*, and *S. maltophilia* L1 enzymes are active also in the mono-zinc form, albeit with lower activity.[18,25,38] Both zinc metals bind to the *B. fragilis* enzyme with binding affinity lower than $10\,\mu M$.[38,39] The most recent equilibrium dialysis studies of zinc binding to the enzyme from *B. cereus* found an equilibrium constant of $0.3\,\mu M$ for a single Zn(II) site and $3\,\mu M$ for two metal sites.[18] Zinc binding was also studied by fluorescence spectroscopy using the chromophoric chelator Mag-fura-2; this study reported a dissociation constant of $0.62\,nM$ for the mono-zinc form and $1.50\,\mu M$ for the binuclear form.[40] This study also measured the affinity of the *B. cereus* metallo-β-lactamase for Cd(II) ($8.3\,nM$ (mono-cadmium) and $5.9\,\mu M$ (di-cadmium)) and for Co(II) ($0.093\,\mu M$ and $66.7\,\mu M$). It is worth noting that affinity values from earlier experiments were carried out either in the presence of salts that could chelate the metal or at low pH that could lead to protonation of active site groups essential for metal binding.[41] Binding data using the fluorescence of Mag-fura-2 are also available for the subclass B1 enzyme from *C. meningosepticum* NCTC10585. Dissociation constants in this case were $5.1\,nM$ (mono-zinc) and $7\,nM$ (di-zinc).[41] Dissociation constants for the subclass B3 metallo-β-lactamase L1 from *S. maltophilia* are $2.6\,nM$ (mono-zinc) and $6\,nM$ (di-zinc).[41]

In contrast to the subclass B1 and B3 enzymes, the binding of a second Zn(II) ion to the subclass B2 enzymes is inhibitory. The dissociation contact of the first Zn(II) bound to the subclass B2 enzyme from *Aeromonas hydrophila* is lower than $20\,nM$, but the addition of a second Zn(II) inhibits the enzyme activity with a K_i of $46\,\mu M$.[42] The dissociation constants using Mag-fura-2 as a spectroscopic probe are $7\,pM$ (mono-zinc) and $50\,\mu M$ (di-zinc).[41] The addition of a second Zn(II) ion to the subclass B2 AER 14M enzyme from *Aeromonas sobria* inhibits the enzyme activity with an IC_{50} of $8\,\mu M$.[43]

Binding of zinc in the presence of substrate (imipenem or nitrocefin) was investigated for enzymes from each subclass (*B. cereus* 569/H/9 and *C. meningosepticum* NCTC10585 for subclass B1, *A. hydrophila* for subclass B2, *S. maltophilia* L1 for subclass B3); the presence of substrate was found to increase the affinity for a first zinc ion to an approximately picomolar dissociation constant.[41] In this study, the dissociation constant for a second zinc ion in the presence of substrate was found to be 4 to 6 orders of magnitude higher than that for the first, with the precise affinity depending on the substrate. Although the concentration of free zinc in the periplasm and in the extracellular space is unknown, the authors argue that it is likely to be lower than the nanomolar concentration of the dissociation constant found in the absence of substrate and therefore speculate that the metallo-β-lactamases may be in the inactive apostate in the periplasm. They also suggest that despite the fact that the enzymes included in the study have higher catalytic activity in the binuclear form, they may function in the mono-zinc state under physiological conditions.

Unique to the *B. fragilis* enzyme is the presence of one monovalent cation, which was interpreted as a sodium ion on the basis of the presence of sodium in the crystallization experiment and on the analysis of coordination geometry and interaction distances in the crystal structure.[39] Metallo-β-lactamase molecules do not contain any cofactors.

ACTIVITY TEST

In vivo, the presence of metallo-β-lactamases is detected by the ability of bacteria to grow in media containing β-lactam antibiotics, in particular carbapenem compounds. The β-lactam susceptibility is expressed in terms of a minimum inhibitory concentration (MIC). Metallo-β-lactamase is assumed if the crude extract exhibits carbapenemase activity and if that activity is inhibited by EDTA.

In vitro, the rate of β-lactam hydrolysis is followed spectroscopically by monitoring absorbance changes arising from the opening of the β-lactam ring. The wavelength and variation in extinction coefficient, $\Delta\varepsilon$, are characteristics of each β-lactam compound. The absorbance changes are lower for penicillin compounds (for example $\Delta\varepsilon_{252} = -940\,M^{-1}\,cm^{-1}$ for benzylpenicillin and $\Delta\varepsilon_{235} = -820\,M^{-1}\,cm^{-1}$ for ampicillin) and higher for cephalosporin and carbapenem compounds (for example, $\Delta\varepsilon_{260} = -10\,700\,M^{-1}\,cm^{-1}$ for cephaloridine, $\Delta\varepsilon_{262} = -7250\,M^{-1}\,cm^{-1}$ for cefotaxime, and $\Delta\varepsilon_{300} = -9000\,M^{-1}\,cm^{-1}$ for imipenem). Chromogenic β-lactams are not used clinically, but they are useful for monitoring activity because of the high absorbance changes at wavelengths at which the protein does not absorb light (for example, $\Delta\varepsilon_{500} = 15\,900\,M^{-1}\,cm^{-1}$ for nitrocefin).

SPECTROSCOPY

Direct observation of zinc binding is possible using X-radiation tuned to the wavelength of the zinc K-absorption edge, an experiment that is carried out at a synchrotron facility. Indeed, the crystal structure determination of the *B. fragilis* metallo-β-lactamase exploits this edge, which reveals a zinc binuclear center.[39] More recently, extended X-ray absorption fine structure (EXAFS) measurements were carried out on the enzyme from *B. cereus*.[18,40] The conclusion from these studies was that metal binding in the mono-zinc form is distributed between the two binding sites, which would imply similar Zn(II) affinity to the two sites, and that the various zinc-ligand distances are slightly shorter (by ∼0.1 Å) than those reported in the crystal structures. Similarly, EXAFS data for the enzyme from *A. hydrophila* were also rationalized as arising from a single Zn(II) ion partitioned between the two binding sites.[44]

Since zinc ions are silent to most spectroscopic techniques, many spectroscopic studies have used metal-substituted forms of the enzyme. Co(II) binding can be monitored by electron paramagnetic resonance (EPR) spectroscopy. The low-temperature EPR spectra of the Co-β-lactamases from *B. cereus* and *B. fragilis* are very similar in shape and g values.[38,45] Using this technique, Crowder *et al.* found 1.9 mole of Co(II) per mole of the *B. fragilis* enzyme, which is in agreement with atomic absorption measurements. They also found that the signal is temperature-dependent and disappears at temperatures greater than 30 K. Analysis of the power saturation properties at different temperatures was interpreted as indicating that the two cobalt ions are both high-spin and not spin-coupled, and that one or both are five- or six-coordinate. Bicknell *et al.*[38,45] interpreted low-temperature EPR spectra of the *B. cereus* enzyme by analogy to the 1:1 adduct between Co(II)-substituted carbonic anhydrase B and azide, and suggested that one penta-coordinated Co(II) is bound to the *B. cereus* enzyme.

When protein refolded from inclusion bodies was studied, the UV-visible difference spectrum between the Co(II)-substituted and the Zn(II)-containing *B. fragilis* enzyme was characterized by an intense absorption at 320 nm and several broad weak peaks between 550 and 650 nm.[38] The 320-nm band is characteristic of a single ligand-to-metal charge-transfer transition from sulfur to Co(II). When the protein was expressed in a soluble form, the difference spectrum showed a 340-nm band corresponding to the sulfur-to-Co(II) charge-transfer transition and, in addition, four Co(II) d–d transition bands at 510, 548, 615, and 635 nm.[23] The pattern of the d–d transition bands was attributed to the existence of two distinct Co(II) binding sites, one with distorted tetrahedral and one with trigonal bipyramidal coordination geometry, consistent with what is seen in the crystal structure of the zinc-bound enzyme. Spectra of the Co(II)-substituted *B. cereus* enzyme were similar to that of the enzyme from *B. fragilis*.[15,40,45,46] Spectra obtained by titration of the mono Zn-enzyme with Co(II) showed the sulfur to Co(II) charge-transfer band, consistent with the zinc ion binding in the Zn1 site, and the cobalt ion binding in the Zn2 site, the only site with a cysteine ligand to the metal (Table 1).[47]

The Cd(II)-substituted metallo-β-lactamase from *B. cereus* was studied by perturbed angular correlation (PAC) of γ-rays.[48] Two nuclear quadrupole interactions were assigned to the binding of two interacting cadmium ions. PAC measurement of the Cys221Ala mutant enabled assignment of the two Cd(II) sites and showed that Cd(II) binds preferentially to the cysteine-containing Zn2 site. In the hybrid Zn(II)Cd(II) enzyme, Cd(II) occupies the cysteine-containing Zn2 site and Zn(II) occupies the three-hisdine-containing Zn1 site.

An early ^{1}H NMR spectroscopy study of the Cd-β-lactamase from *B. cereus* implicated three histidine residues as ligands of the high-affinity metal ion and a fourth histidine as a ligand of the low-affinity metal ion.[49] This study is consistent with the crystal structure but is in conflict with the EXAFS and PAC experiments, which were interpreted as showing that the first zinc ion that is bound is distributed between the two sites. More recently, a ^{1}H, ^{15}N, and ^{113}Cd NMR study showed that none of the imidazole protons that serve as metal ligands is affected until one equivalent of Cd(II) is titrated into the protein, at which point all of the imidazole proton shifts seen with two Cd(II) equivalents appear.[50] The ^{113}Cd NMR data show that one resonance is observed when Cd(II) is titrated up to a ratio of one Cd(II) ion per molecule and that this single resonance disappears and is replaced by two new resonances as the ratio of Cd(II) to enzyme is increased from one to two. The ^{113}Cd data at Cd(II) to enzyme ratios less than one disagree with the PAC spectra, which exhibit two equally populated coordination geometries under the same experimental conditions. To account for the seemingly conflicting results, the authors suggested that the single bound Cd(II) exchanges sites in a timescale between the time regimes of the PAC and NMR experiments (100 ns vs. 0.01 ms), so that only an average spectrum is detected by NMR.

X-RAY STRUCTURES

Crystallization

A wide variety of conditions have been employed in the crystallization of metallo-β-lactamases, and it is beyond the scope of this review to summarize the details of each experiment. Thus, one example is provided for the crystallization conditions of a B1 metallo-β-lactamase and one for a B3 metallo-β-lactamase.

Crystals of the B1 enzyme from the QMCN3 *B. fragilis* isolate (CcrA gene cloned with the 17-residue amino-terminal signal peptide omitted) were obtained at room temperature by vapor diffusion in hanging drops.[39] The protein drops were prepared by mixing equal volumes of 5 to 8 mg mL^{-1} protein solution (in 10 mM HEPES, pH 7.0, and 10 μM ZnCl$_2$) and reservoir solution containing 26 to 28% PEG (polyethylene glycol) 2000, 100 mM HEPES, pH 7.0. The crystals belong to the tetragonal space group P4$_3$2$_1$2 with unit cell dimensions $a = b = 78.3$ Å, $c = 140.9$ Å. There are two molecules in the asymmetric unit and the solvent constitutes 40% of the cell.

Crystals of the B3 enzyme from *S. maltophilia* L1 (with the 21-residue amino-terminal signal peptide removed post-translationally) were obtained at 4 °C by vapor diffusion in hanging drops.[31] The protein drops were prepared by mixing equal volumes of 16 mg mL^{-1} protein solution (in 10 mM Tris-HCl, pH 7.0, 100 mM NaCl, 5 mM ZnSO$_4$, 1 mM β-mercaptoethanol) and reservoir solution containing 2.0 M (NH$_4$)$_2$SO$_4$, 100 mM HEPES, pH 7.75, 1.5% (v/v), and PEG400. The crystals belong to the hexagonal space group P6$_4$22 with unit cell dimensions $a = b = 105$ Å, $c = 98$ Å.

Since reaction mechanisms have been proposed on the basis of some of these structure determinations, it is of interest to note the pH at which those structures were determined. For the early 2.5-Å resolution *B. cereus* structure,[51] the pH was 5.6 (10 mM cacodylate and 25 mM sodium citrate), and the same buffer was used for the subsequent higher-resolution structure.[52] An independent structure determination of the *B. cereus* enzyme in a trigonal space group[53] was based on crystals grown at pH 4.5 (Tris-HCl) and stabilized at pH 5.2. For comparison, the *B. fragilis* structure that is the basis for a mechanistic proposal was determined from crystals grown at pH 7.0 (HEPES).[39] Crystals of the L1 enzyme form *S. maltophilia*[31] were grown at pH 7.75 (HEPES), and crystals of the FEZ-1 enzyme from *F. gormanii*[54] were grown at pH 6.0 (cacodylate).

Density that was interpreted as bicarbonate was observed bound to the active site of the 1.85-Å structure of the *B. cereus* enzyme, but no mention of bicarbonate is made in the reported crystallization conditions for that structure.[52] Morpholineethanesulfone (MES) was found bound to the active site of the *B. fragilis* enzyme,[55] which was crystallized from a solution containing 100 mM MES. Sulfate ions were found in the active site of the FEZ-1 enzyme,[54] which was crystallized from a solution containing 200 mM ammonium sulfate.

Overall structure of metallo-β-lactamases

Structures have been reported for three members of the B1 metallo-β-lactamase subclass (the BcII enzyme from *B. cereus*,[51–53] the CcrA and Cfia enzymes from *B. fragilis*,[39,55–58] and the IMP-1 enzyme from *P. aeruginosa*[59,60]) and for two members of the B3 subclass (the L1 enzyme from *S. maltophilia*[31] and FEZ-1 from *F. gormanii*[54]). All these enzymes share a fold that can be described as an αβ/βα sandwich. Two β-sheets are packed against one another to form the core of the enzyme, while the two α-helices are packed on the surface of each sheet (see 3D Structure). A fifth α-helix is located topologically between the last two β-strands of the N-terminal sheet; this helix and the two β-strands form a structural bridge to the C-terminal domain. The N-terminal half of the polypeptide chain comprises the first αβ unit and the C-terminal half forms the second, with the zinc cluster and active site positioned in a channel at the top of the interface between the two domains. The topology of the C-terminal αβ layer (ββββαβα) is also present at the center of the N-terminal αβ layer, leading to the speculation that a gene-duplication event may have occurred during the evolution of the fold.[51] It has been noted, however,[39] that the two helices in the

C-terminal αβ layer are antiparallel, while the two helices in the N-terminal layer are parallel.

When the first crystal structure of a metallo-β-lactamase was determined,[51] this fold had not previously been observed and it was thought for some time that the fold was unique to the metallo-β-lactamase family of enzymes. However, subsequent structure determinations of glyoxylase II[64] and rubredoxin oxygen:oxidoreductase,[65] members of the metallo-β-lactamase sequence superfamily,[12,13] revealed that these proteins possess the same fold.

The 3D Structure shows the overall folds of the subclass B1 (CcrA from *B. fragilis*) and B3 (L1 from *S. maltophilia*) metallo-β-lactamases. While the overall topologies are clearly the same, significant differences do exist. The flap (residues 58–68), a flexible secondary structure element that has been shown to interact directly with bound inhibitors in the subclass B1 enzymes,[55,56,58–60] has no structural analog in the subclass B3 enzymes. Both the L1 and FEZ-1 structures possess a disulfide bond, C256–C290 (C218–C248), that is not present in the subclass B1 structures. The extended strand at the N-terminus of the L1 enzyme, which mediates some of the interactions involved in tetramerization,[31] replaces the first β-strand in the N-terminal β-sheet of the monomeric subclass B1 enzymes, with the result that the N-terminal β-sheet of L1 is one strand shorter than the equivalent sheet in the subclass B1 enzymes. In the monomeric FEZ-1 enzyme, the N-terminal β-sheet is also one strand shorter, but the N-terminal β-strand is not extended. FEZ-1 contains an α-helical insertion at the edge of the C-terminal β-sheet.[54]

A striking feature of all metallo-β-lactamase structures determined to date is a severely strained main chain conformation at position 84, a residue that is either aspartic acid or asparagine in all subclass B1 and B3 enzymes. This residue does not participate directly in the zinc coordination, but it does form part of the second shell of structure around the zinc cluster, and it has been speculated[39] that the strained conformation of the main chain at this position helps to establish the proper geometry for the zinc ligands. The strained conformation is in turn stabilized by the polar interactions that bridge the side chain of Asp/Asn84 to the carbonyl oxygen of residue 69, but the nature of the bridge varies widely. In the *B. cereus* structure, the bridging atom comes from the side chain of Arg121(91), in IMP-1 it comes from the side chain of Lys68(34), and in the *B. fragilis* structure it is provided by a buried sodium ion. In the L1 structure, the bridging position is occupied by a water molecule, and in the FEZ-1 structure, the interaction between the Asn84(84) side chain and the carbonyl oxygen of Ala69(69) is direct. Interestingly, the amino acid at position 84 in the subclass B2 enzymes is glycine, so a very different structural mechanism for stabilizing the active site must occur in these proteins.

Zinc coordination

The active sites of most metallo-β-lactamases with known structure contain two zinc atoms positioned approximately 3.5 Å apart. The zinc atoms are labeled Zn1 and Zn2 by convention. The exception is the original structure determination of the enzyme from *B. cereus* at moderate resolution (2.5 Å), in which only one zinc ion occupying the Zn1 site, was observed.[51] This is not, however, a true mono-zinc enzyme structure, as, when the diffraction data were extended to 1.85-Å resolution, a second, partially occupied, zinc atom was observed in the Zn2 site.[52,53] In the subclass B1 enzymes (Figure 1), Zn1 is tetra-coordinate, with three ligands contributed by the side chains of histidine residues and one by solvent, in a tetrahedral arrangement. Zn2 is penta-coordinate with trigonal bipyramidal geometry, with three ligands from protein (histidine, aspartic acid, and cysteine) and two ligands from solvent. One of the solvent atoms coordinates to both zinc atoms, and the highly positively charged environment argues that in fact it is a hydroxide. The side chain of Asp120 interacts with

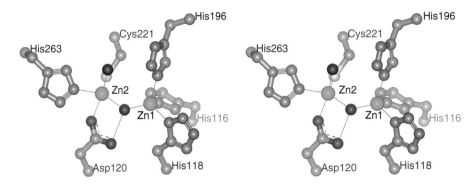

Figure 1 A stereo pair illustrating the zinc ligation of subclass B1 metallo-β-lactamases. The structure illustrated is native *B. fragilis* enzyme (PDB entry 1ZNB).[39] The atom coloring scheme is blue for nitrogen, red for oxygen, yellow for sulfur, green for zinc, and gray for carbon. The residue-numbering scheme is the standardized version of Reference (1). Prepared with ViewerPro 4.2, Accelrys Inc., San Diego, CA, USA.

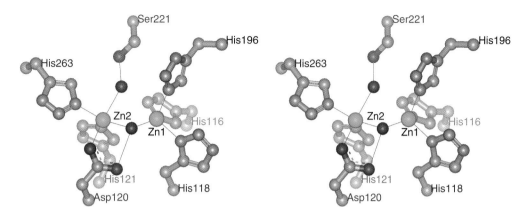

Figure 2 A stereo pair illustrating the zinc ligation of subclass B3 metallo-β-lactamases. The structure illustrated is the L1 enzyme from *S. maltophilia* (PDB entry 1SML).[31] The atom coloring scheme is blue for nitrogen, red for oxygen, yellow for sulfur, green for zinc, and gray for carbon. The residue-numbering scheme is the standardized version of Reference (1). For Figures 2 and 3, the *B. fragilis* and *S. maltophilia* structures were aligned and the same orientation was used for both figures, to illustrate the similarities and differences in the zinc coordination. Prepared with ViewerPro 4.2, Accelrys Inc., San Diego, CA, USA.

both Zn2 and this shared hydroxide, and is thus itself a bridging element. The second solvent atom interacts with Zn2 but not with Zn1; given its position in the trigonal bipyramidal coordination of Zn2, this is referred to as the apical water (the second apical position is occupied by the side chain of Asp120).

While the coordination of Zn1 is similar in the subclass B1 and B3 enzymes, the coordination of Zn2 is not. The coordination geometry of Zn2 in the subclass B3 structures remains trigonal bipyramidal, but the arrangement of the ligands has changed (Figure 2). The planar ligands His263 and the shared hydroxide are the same as those found in the subclass B1 structures, but the third planar ligand is provided by His121, dramatically changing the orientation of the plane. The apical position is still occupied by a water molecule, but this water lies at the position of the sulfur atom of the Cys221 ligand in the subclass B1 structures. Position 221 in the subclass B3 enzymes is serine rather than cysteine; in the L1 enzyme, Ser221 makes only an indirect interaction with Zn2 through the apical water, and in the FEZ-1 enzyme, the interaction is even more remote, involving two bridging water molecules.

The position of the shared hydroxide (W1 in Table 2) is somewhat closer to Zn1 than to Zn2 in most structure determinations of metallo-β-lactamases (0.1–0.2 Å). Fabiane *et al.*, studying the di-zinc *B. cereus* enzyme,[53] noted that the ligand to zinc distances in the Zn2 site were longer than in Zn1 and described the arrangement of ligands in the Zn2 site as 'looser' than in the Zn1 site. These longer distances may be attributed to the low pH of both the *B. cereus* structure determinations, which may have led to protonation of some active site side chains. As can be seen in Table 2, the difference between coordination distances for the two sites is not observed in most metallo-β-lactamase crystal structures; such differences as do exist fall within the expected coordinate error for the structure determinations.

Structural variability within the metallo-β-lactamase subclasses

Figure 3 shows a comparison of representative structures from the three subclass B1 enzymes. Of particular relevance to efforts at drug design is the difference in the conformation of the loop (residues 222–234) that forms the left-hand side of the active site (as shown in Figure 3). This loop contains residues Lys223 and Asn233, which have been shown to interact directly with bound ligands in crystal structures of inhibited complexes of subclass B1 metallo-β-lactamases.[55,56,58–60] The sequence of the IMP-1 enzyme is three residues shorter than the sequences of the *B. cereus* and *B. fragilis* enzymes in this loop, with the result that this binding loop is narrower and positioned further away from the zinc cluster in the IMP-1 structure. The figure also shows a significant difference in the position of the flap, but the indicated positions are a consequence of the ligands that were bound in the structures illustrated, or of a stabilization of the mobile flap by crystal packing interactions, not of an intrinsic difference in the positions of the flap between the three enzymes. Variations in the position of flap of equivalent magnitude have been observed between structures of the *B. fragilis* enzyme complexed with inhibitors of different chemical structure.[55,56,58]

Figure 4 shows a comparison of the overall folds of the L1[31] and FEZ-1[54] subclass B3 enzymes. The structural differences between the two proteins are greater than those observed when comparing the three subclass B1 enzymes (Figure 3). Most significantly, the L1 enzyme possesses an N-terminal extension, with residues in that extension mediating many of the interactions that stabilize the L1 tetramer. Although neither structure possesses the flap found in the subclass B1 enzymes, there is an extended loop following the third helix that might play an equivalent role in closing down over bound substrates or inhibitors.

Metallo β-Lactamases

Table 2 Key interatomic distances in the active site Zn cluster

Structure	Distances to Zn1					Distances to Zn2				
	H116	H118	H196	W1	Zn2	W1	D120	C221	H263	W2
B. cereus										
Native (1BC2)	2.3	2.0	2.0	1.9	3.9	2.5	2.8	1.9	2.6	2.6
Bicarbonate (1BVT)	2.2	2.2	2.2	2.1	3.8	3.7	2.3	2.4	2.2	<u>2.9</u>
B. fragilis										
Native (1ZNB)	2.1	2.1	2.1	1.9	3.5	2.1	2.3	2.3	2.1	2.3
Native (2BMI)	2.1	2.1	2.1	1.8	3.4	1.9	2.2	2.4	2.0	2.2
MES (1A7T)	2.1	2.1	2.1	1.0	3.5	2.2	2.3	2.2	2.2	2.3
Tetrazole (1A8T)	2.3	2.1	2.1	n.a.	2.9	n.a.	2.2	2.3	2.1	<u>2.4</u>
Tricyclic (1KR3)	2.1	2.4	2.4	n.a.	3.4	n.a.	3.0	2.1	3.0	<u>2.6</u>
IMP-1										
Mercapto (1DD6)	2.2	2.2	2.2	<u>2.2</u>	3.6	<u>2.4</u>	2.1	2.4	2.3	n.a.
Succinate (1JJE)	2.1	2.1	2.0	<u>2.0</u>	3.6	<u>2.1</u>	2.0	2.3	2.2	<u>2.3</u>
L1									H121	
Native (1SML)	2.0	2.1	2.1	1.9	3.5	2.1	2.1	2.0	2.1	2.4
FEZ-1										
Sulfate (1K07)	2.1	2.1	2.1	2.1	3.7	2.1	2.1	2.2	2.0	2.0

The PDB code for each structure is given in parenthesis. All distances are in Angstroms. W1 is the shared hydroxide and W2 is the apical water. n.a. indicates that the structure does not have an atom in the corresponding position. An underlined value indicates that the distance was measured to an atom in a bound ligand that occupies the same position as a water molecule in the unliganded structure. Structures that involve active site mutants or chemical modifications of active site residues are not included in the table. Where structures contain more than one molecule per asymmetric unit, only the first is included in the analysis. Where more than one coordinate set has been deposited for a given structure, the most recent deposition was used (therefore, the 2.5-Å mono-zinc *B. cereus* structure, which was superceded by a 1.85-Å di-zinc structure, is not included in the table).

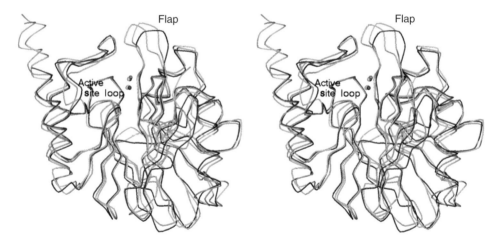

Figure 3 A stereo pair comparing the alpha carbon backbones of the subclass B1 metallo-β-lactamases. Shown in red is the enzyme from *B. cereus* (PDB entry 1BVT),[52] in cyan is the enzyme from *B. fragilis* (PDB entry 1A7T),[55] and in yellow is the IMP-1 enzyme from *P. aeruginosa* (PDB entry 1JJE).[60] The positions of the zinc atoms are shown as spheres of the appropriate color. The active site loop, which is three residues shorter in the IMP-1 enzyme than in the other two, is labeled. Prepared with ViewerPro 4.2, Accelrys Inc., San Diego, CA, USA.

Ligand binding studies

Although several classes of inhibitors of metallo-β-lactamases have been reported,[60,66–71] crystal structures of inhibited complexes have been obtained for representatives of only a few of these structural classes. There have not yet been three-dimensional structures of the subclass B1 *B. cereus* enzyme with inhibitors bound, although the 1.85-Å refinement[39,52] of the structure revealed density that was interpreted as a bicarbonate molecule bound to the Zn cluster. Three structures have been reported for complexes of small molecule inhibitors with the subclass B1

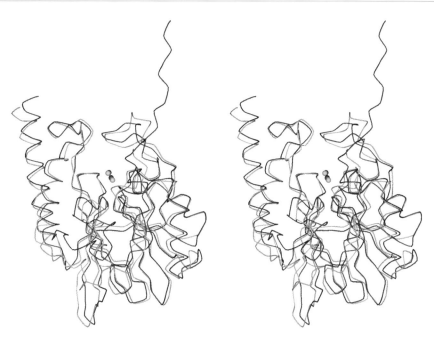

Figure 4 A stereo pair comparing the alpha carbon backbones of the subclass B3 metallo-β-lactamases. Shown in red is L1 from *S. maltophilia* (PDB entry 1SML)[31] and in cyan is FEZ-1 from *F. gormanii* (PDB entry 1K07).[54] The positions of the zinc atoms are shown as spheres of the appropriate color. Prepared with ViewerPro 4.2, Accelrys Inc., San Diego, CA, USA.

metallo-beta-lactamase from *B. fragilis*. In the first of these, the bound inhibitor was MES, a component of the crystallization buffer. This complex revealed that Lys224(161) and Asn233(167) provide crucial polar interactions with the bound ligand; these residues are conserved in most subclass B1 and B2 enzymes, and crucial roles have been assigned to them in the proposed reaction mechanisms for these enzymes (see below). Significantly, the subclass B3 residue equivalent to Lys224 is a serine in L1 and a glycine in FEZ-1, and the residues equivalent to Asn233 (Asn196 in L1 and Asn233 in FEZ) are positioned ~16 Å from the enzyme active site; thus, the subclass B3 residues are unlikely to play equivalent roles in substrate binding or catalysis.[31,54] A consequence of the adventitious binding of the buffer component in the MES complex was the ordering of the residues in the flap, residues that are either poorly ordered[39,52] or ordered via packing interactions with a symmetry related molecule[72] in the structures of the *B. fragilis* enzyme determined without bound ligand.

Two further complexes with the *B. fragilis* enzyme have been determined, one with a biphenyl tetrazole inhibitor[56] and one with a tricyclic natural product.[58] Both complexes confirmed the importance of the active site residues Lys224(161) and Asn233(167) in coordinating the bound ligand through polar interactions, and the involvement of the hydrophobic flap residues Val61(25), Trp64(28), and Val67(31) in providing nonpolar interactions with the portion of the ligand distant from the zinc cluster.

It is not clear whether the most effective clinical strategy would be the development of a broad-spectrum inhibitor that targets all metallo-β-lactamases or a narrow spectrum inhibitor targeted at the IMP or VIM enzymes, which seem to be emerging as a major clinical threat. Complexes with inhibitors that exemplify both approaches have been reported with the IMP-1 enzyme. A complex with a broad-spectrum mercaptocarboxylate inhibitor (IC_{50} 90 vs. IMP-1) is illustrated in Figure 5.[59] The sulfhydryl group of the inhibitor assumes the position of the shared hydroxide and the apical water is displaced, with the result that the coordination of Zn2 is tetrahedral in this complex. The carboxylate group of the inhibitor forms polar interactions with the side chain of Lys224(161) and the main chain of Asn233(167), while the carbonyl group forms a solvent-mediated interaction with the side chain of Asn233(167). The benzyl substituent is tucked into a pocket at the base of the flap, a binding site made accessible by the positioning of the side chain of Phe87(51) away from the active site.

The structure of a complex of the IMP-1 enzyme with a narrow spectrum biaryl succinic acid inhibitor (IC_{50} 3.7 nM) is shown in Figure 6.[60] In this complex, one oxygen atom of the rightmost carboxylate group (as shown in the figure) assumes the position of the shared hydroxide, while one oxygen of the leftmost carboxylate assumes the position of the apical water. The remaining oxygen atoms of the succinate core of the inhibitor form polar interactions with the side chain of Lys224(161) and the main chain and side chain of Asn233(167). In this complex, the side chain of Phe87(51) is oriented toward the base of the flap, and its closest interaction with the inhibitor is 4.1 Å. Overall, the succinic acid inhibitor has a more rigid core than the mercaptocarboxylate inhibitor and it makes more direct polar and nonpolar interactions, which correlates with its

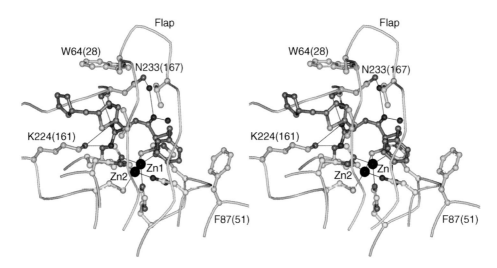

Figure 5 Binding of a mercaptocarboxylate inhibitor to the active site of IMP-1 (PDB entry 1DD6).[59] The coloring scheme is blue for nitrogen, red for oxygen, gold for sulfur, and black for zinc. Carbon atoms in the inhibitor are colored orange-red, carbon atoms of residues that coordinate the zinc atoms are colored yellow, carbon atoms of key active site polar residues are colored green, and carbon atoms of the hydrophobic residues that interact with the inhibitor are colored cyan. Prepared with ViewerPro 4.2, Accelrys Inc., San Diego, CA, USA.

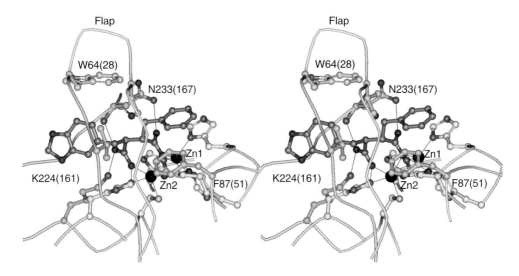

Figure 6 A stereo pair illustrating a diaryl succinic acid inhibitor bound to the active site of IMP-1 (PDB entry 1JJE).[60] The coloring scheme is blue for nitrogen, red for oxygen, gold for sulfur, and black for zinc. Carbon atoms in the inhibitor are colored orange-red, carbon atoms of residues that coordinate the zinc atoms are colored yellow, carbon atoms of key active site polar residues are colored green, and carbon atoms of the hydrophobic residues that interact with the inhibitor are colored cyan. Prepared with ViewerPro 4.2, Accelrys Inc., San Diego, CA, USA.

greater potency (IC_{50} of 3.7 vs. 90). However, the very precision of the complementarity between the succinic acid inhibitor and the IMP-1 active site may be the cause of the relatively poor efficacy of this inhibitor against metallo-β-lactamases from other species.

The subclass B3 FEZ-1 structure has been determined both in an unliganded state and in complex with D-captopril.[54] However, the occupancy of the bound inhibitor is low and only a portion of the molecule could be fit to reliable electron density. D-captopril makes no direct interactions with the Zn cluster; instead, there is a sulfate ion that does.

STRUCTURAL STUDIES BY NMR

Complete backbone NMR resonance assignments have been made for the metallo-β-lactamase from *B. fragilis* in the presence and absence of the inhibitor 3-[2′-(S)-benzyl-3′-mercaptopropanoyl]-4-(S)-carboxy-5,5-dimethylthiazolidine.[73] The secondary structure assignment is in agreement with that derived from the crystal structure. ^{15}N NMR relaxation measurements of complexes with and without bound inhibitors show that the flap becomes more ordered upon inhibitor binding, in agreement with the results seen in crystal structure analysis.[74]

FUNCTIONAL ASPECTS

Steady state kinetics

Metallo-β-lactamase variants are characterized by their substrate profiles, which span a wide range of catalytic efficiencies.[17,75,76] For example, the enzymes from Aeromonas species have the narrowest specificity profiles, while the *B. fragilis* enzyme exhibits the broadest substrate profile. Numerous steady state kinetic parameters are reported in the literature, and the reader may refer to the following publications for steady state kinetic profiles of specific enzymes: *B. fragilis*,[20,21] *P. aeruginosa* IMP-1,[24] *P. aeruginosa* VIM-1 and VIM-2,[37] *A. hydrophila*,[17] *A. sobria*,[43,77] *S. maltophilia*,[25] *C. meningosepticum* GOB1,[78] BlaB,[79] *F. gormanii* FEZ-1,[26] and *C. crescentus* Mb11b,[35,80] also termed CAU-1.[35,80]

Cryoenzymology studies of the metallo-β-lactamase from *B. cereus* at −30 °C revealed that the hydrolysis of benzylpenicillin by the Co(II)-enzyme and nitrocefin by the Zn(II)-enzyme proceed through a branched pathway.[81] For benzylpenicillin, burst kinetics was observed prior to reaching steady state, and the size of the burst was greater than the enzyme concentration. The branching step was attributed to a change in conformation of the enzyme–substrate complex. For nitrocefin, the progress curves were triphasic, with two transients preceding the linear steady state phase.

The pH dependence and kinetic solvent isotope effects of the hydrolysis of benzylpenicillin and cephaloridine by the *B. cereus* mono-zinc enzyme were studied at steady state conditions.[82] Both substrates showed characteristic bell-shaped curves for the logarithmic pH-dependent change of K_{cat}/K_m. The slope of the acidic part of the curve was 2.0 and not the usual 1.0. This was attributed to two acidic groups with pK_a values of 5.6 controlling activity. On the basic side, the data supported one ionizing group with a pK_a of 9.5. While the residues have not been identified experimentally, the authors proposed that the shared hydroxide and the carboxylate side chain of Asp120 are the groups with the low pK_a values.

For the metallo-β-lactamase from *B. fragilis*, both K_{cat} and K_m for nitrocefin hydrolysis remain flat between pH 5.25 and 10.[23] Thus, for this binuclear zinc enzyme, no ionizable group within this range of pH appears to be involved in catalysis. As proposed on the basis of the structural studies,[82] the shared hydroxide could serve as the nucleophilic moiety with a pK_a below 5.25.

Transient kinetics

Rapid-scanning stopped-flow spectroscopy was used to study benzylpenicillin hydrolysis by the Co(II)-substituted *B. cereus* enzyme in aqueous solution at 3 to 20 °C.[45] A branched kinetic pathway was observed, with an additional transient intermediate occurring prior to the first intermediate seen in cryoenzymology studies.[81]

Rapid-scanning and single-wavelength stopped-flow methods revealed the accumulation of an enzyme-bound intermediate during turnover of nitrocefin by the binuclear metallo-β-lactamase from *B. fragilis*.[83] Kinetic rate constants for a minimum kinetic mechanism were fitted, showing that in this case the pathway is linear rather than branched. According to the proposed mechanism, four steps are involved:

$$E + S \rightleftharpoons ES \longrightarrow EI \longrightarrow EP \rightleftharpoons E + P$$

The Michaelis complex, ES, is formed during the first step. Next, the intermediate, EI, is formed. Breakdown of the intermediate to yield the enzyme–product complex, EP, occurs during the third step, and the fourth step corresponds to product release. The rate-limiting step is the breakdown of the intermediate, and the red-shifted absorbance of this intermediate was attributed to an anionic species with the negative charge located on the nitrogen of the cleaved β-lactam C—N bond.

Similar results to those reported for the *B. fragilis* enzyme were also seen with the L1 enzyme from *S. maltophilia*.[84] However, in a more recent publication, pre–steady state tryptophan fluorescence experiments were carried out with nitrocefin, as well as with the cephalosporin cefaclor and the carbapenem meropenem.[85] In contrast to nitrocefin hydrolysis, the fluorescence quenching curves of the cefaclor and meropenem reactions indicated the presence of an additional reaction step corresponding to rearrangement of the ES complex into an intermediate (ES*) after which a second intermediate with an intact C—N bond was formed (EI_2). The rate-limiting step was the cleavage of the β-lactam C—N bond of EI_2. The authors concluded that the mechanism of nitrocefin hydrolysis differs from that of the hydrolysis of the majority of β-lactam substrates.

Site-directed mutagenesis

Mutagenesis studies probe the role that specific residues play in metal binding and catalysis. Such studies have been carried out on the metallo-β-lactamases from *B. fragilis*, *B. cereus* and *S. maltophilia*. Table 3 provides kinetic data, monitoring hydrolysis of nitrocefin, benzylpenicillin, and, whenever available, imipenem, for selected mutants that probe key active site residues. The reader should refer to the referenced publications for detailed discussion of the impact of a wide range of mutations, as well as kinetics using other substrates.[17,20,38,83,86–89] In general, most of the residues probed by mutagenesis were those that serve as metal ligands, those that provide the framework for orienting metal ligands, or those thought to interact with substrates. These replacements affect zinc binding and substrate kinetics to a varying degree, which is not

Metallo β-Lactamases

Table 3 Kinetic parameters of wild-type and mutant metallo-β-lactamases

Mutant	Substrate	B. fragilis				B. cereus				S. maltophilia			
		k_{cat} (s^{-1})	K_m μM	Zn[b]	References[a]	k_{cat} (s^{-1})	K_m μM	Zn[b]	References[a]	k_{cat} (s^{-1})	K_m μM	Zn[b]	References[a]
Wt	Nitrocefin	154.3	5.9	2	1	43	9	1.4	4	41	4	1.9	5
	Benzylpenicillin	94	25	2	2	230	170		4	600	38		5
	Imipenem	74	140		2	>100	>1000		6	370	57		5
H116N	Benzylpenicillin	0.28	150	2	2								
	Imipenem	0.029	180		2								
H116S	Nitrocefin					44	180	1.0	4				
	Benzylpenicillin					47	530		4				
H118S	Nitrocefin					5	220	0.95	4				
	Benzylpenicillin					36	2300		4				
D120V	Nitrocefin	0.024	31.0	0.4	1								
D120N	Benzylpenicillin	40	510	1&2 mix	2	3.5	440	1.4	4				
	Imipenem	3.9	14		2								
	Nitrocefin			1.6	2	2.2	133		4				
D120C	Nitrocefin	1.9	39		3								
	Benzylpenicillin	3.3	321	1.7	3								
D120S	Nitrocefin	<0.003	ND		3								
	Benzylpenicillin	<0.59	ND		3								
H196S	Nitrocefin					0.2	160	1.0	4				
	Benzylpenicillin					7	5700						
C221S[c]	Nitrocefin	0.014	30.4	1.9	1	1.3; 63	8; 21	1.2; 1.7	4				
	Benzylpenicillin	0.65	170	1	2	15; 200	475; 775		4				
	Imipenem	0.05	380		2								
C221A[c]	Nitrocefin					0.8; >80	10; na	1; 2	7				
	Benzylpenicillin					1.8; 194	405; 2040		7				
K224R	Benzylpenicillin	620	420	1&2 mix	2								
	Imipenem	110	700	2	2								

Metallo β-Lactamases

Mutant	Substrate						
K224A	Nitrocefin	99	4.8				3
	Benzylpenicillin	174	1005	1.5			3
K224E	Nitrocefin	100	15				3
	Benzylpenicillin	121	3941				3
S224A	Nitrocefin				48	4	1.8	5
	Benzylpenicillin				580	70		5
	Imipenem				100	29		5
S224K	Nitrocefin				480	14	1.0	5
	Benzylpenicillin				124	44		5
	Imipenem				14	60		5
S224D	Nitrocefin				2.3	11	1.7	5
	Benzylpenicillin				42	1600		5
	Imipenem				17	42		5
N233D	Benzylpenicillin	790	930	2	440	60	1.5	5
	Imipenem	17	570	2	158	71		5
	Nitrocefin	51	5.1	3	21	9		5
N233L	Nitrocefin	47	21	1.4	62	7	1.8	5
	Benzylpenicillin	224	756	3	184	33		5
	Imipenem			3	250	36		5
N233Y	Nitrocefin		6		17		1.0	4
H263S	Nitrocefin		1.3		365		1.0	4
	Benzylpenicillin		>14		>6000			

[a] References for Table 3: 1:(38); 2. Data for TAL3636 strain:(86); 3:(87); 4. For the 569/H/9 strain:(88); for imipenem, only K_{cat}/K_m values were provided, and they are between one to two orders of magnitude lower for the mutants compared with wild-type enzyme; 5:(89); 6. Data for the 5/B/6 strain:(17); 7. For the 569/H/9 strain:(18).

[b] Zinc content from reference 2 corresponds to the mass spectrometry results, which is capable of distinguishing between mono- and di-zinc species. Mass spectrometry is also a more reliable method to assess the metal binding than atomic absorption.

[c] For the C221S and C221A metallo-β-lactamase from *B. cereus*, kinetic parameters are provided for the mono- and di-zinc enzyme.

always possible to rationalize. For example, replacements of Cys221 or Asp120 by a serine have the largest impact on the activity of the *B. fragilis* enzyme. However, replacement of Asp120 by asparagine is not detrimental to the activity of either *B. fragilis* or *B. cereus* enzymes, which raises doubts as to whether this residue serves as a general base (see the section on Catalytic Mechanism). In fact, none of the changes to key active site residues abolishes β-lactam hydrolysis, implying that no single mutation eliminates the nucleophilic machinery completely.

To date, the only active site mutant to be studied crystallographically is C221S; structures have been reported for this mutant for both the *B. fragilis*[90] and *B. cereus*[91] enzymes. Consistent with the mass spectrometry results, a single zinc ion is observed bound to the *B. fragilis* enzyme. This zinc ion occupies the Zn1 position and forms a distorted tetrahedral coordination to the three histidine residues and to a water molecule. The same result was found with the *B. cereus* enzyme.

CATALYTIC MECHANISM

The sequence fingerprint for a binuclear zinc center is present in all metallo-β-lactamases, although the mono-zinc form is also active. The mono- and di-zinc states have different electrostatic environments and different constraints on the orientations and flexibilities of active site residues. Therefore, the catalytic mechanism of the two states must be considered separately.

The structure-based mechanism of the mono–zinc-β-lactamase is based on the crystal structure of the *B. cereus* metallo-β-lactamase, which was originally thought to be occupied by a single zinc ion based on a 2.5-Å diffraction data, but found to also contain a second, partially occupied, zinc atom when the resolution was extended to 1.85 Å. As proposed by Carfi *et al.*,[51] the main features of this mechanism resemble those of zinc peptidases such as carboxypeptidase and thermolysin.[92,93] Figure 7(a) illustrates this mechanism schematically. In the mono-zinc active site, the zinc ion has presumably tetrahedral coordination with three ligands provided by the protein and the fourth provided by a postulated water molecule. In turn, the postulated water molecule interacts with Asp120, which is proposed to serve as a general base during the reaction. The zinc ion and Asp120 polarize the water molecule, and the resulting hydroxide attacks the incoming β-lactam carbonyl carbon atom. A β-lactam oxyanion is formed, stabilized by the zinc ion and possibly by the side-chain amide group of Asn233 (although that was not proposed by the authors). As a result, the zinc ion coordination changes

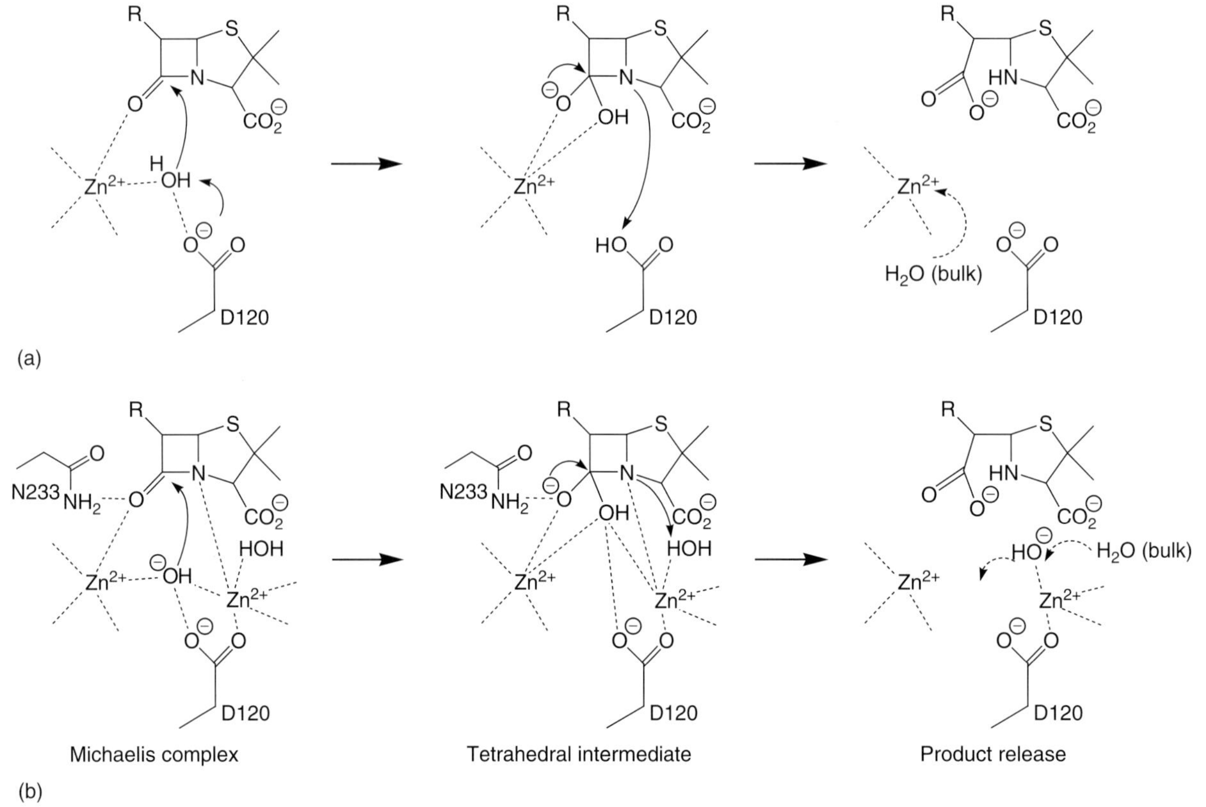

Figure 7 Proposed catalytic mechanisms for metallo-β-lactamase. A penicillin molecule is used. Similar reactions could be drawn for other types of β-lactam compounds. For clarity, only the entities that are directly involved in the chemical process are shown and residues that are involved in substrate binding are not included. (a) The mono-zinc enzyme; (b) the di-zinc enzyme.

from tetrahedral to trigonal bipyramidal. Following the formation of a tetrahedral intermediate at the carbonyl carbon, the β-lactam C—N bond is cleaved as the proton accepted by Asp120 is transferred to the nitrogen atom.

In contrast to the above proposal, the pH-activity dependence of the *B. cereus* enzyme led Bounaga *et al.* to propose that the hydroxide is supported solely by the zinc ion, and that a proton transfer from this hydroxide to Asp120 yields a dianionic tetrahedral intermediate.[82] The significant activity of the D120N mutant (Table 3) is inconsistent with Asp120 playing the role of sole general base that is a feature of both proposals.

The role of Cys221 and His263 in the catalytic mechanism of the mono-zinc enzyme is unclear. If Cys221 interacts with the hydrolytic water molecule and has a low pK_a value, it could enhance the nucleophilicity of the water molecule. That is consistent with the larger impact that mutations of Cys221 have on the activity of the mono-zinc enzyme compared with the di-zinc enzyme (Table 3). His263 could help orient the side chain of Asp120.

A proposal for the catalytic mechanism of the binuclear zinc enzyme was first articulated for the *B. fragilis* metallo-β-lactamase.[39] This mechanism was based on models of Michaelis complexes of the enzyme with β-lactam antibiotics and on an analogy to the proposed mechanisms of other hydrolases with a binuclear metal center. The mechanism is illustrated in Figure 7(b). For the binuclear zinc enzyme, the solvent molecule shared by the two zinc ions is likely to exist as a hydroxide. Its interaction with the carboxylate group of Asp120 may help modulate the strength of the hydroxide–zinc interaction. This interaction also assures that the lone pair electrons of the oxygen are oriented appropriately for a nucleophilic attack on the β-lactam carbonyl carbon atom. Substrate binding is further mediated by the interaction of the carboxylate group with Lys224 and by interaction of the aromatic side-chain substituent with a hydrophobic pocket formed by Ile87 and the flap residues Ala59, Ile61, and Val67. As the hydroxide attacks the β-lactam carbonyl carbon, an oxyanion tetrahedral transition state is formed at the β-lactam carbonyl, stabilized by Zn1 and the amide group of the side chain of Asn233. Although not discussed by Concha *et al.*,[39] the implication of this transition state is that the coordination of Zn1 changes from tetrahedral to trigonal bipyramidal and that of Zn2 changes from trigonal bipyramidal to octahedral. This was later pointed out by Ullah *et al.* in their proposal for the mechanism of the L1 enzyme from *S. maltophilia*.[31]

The reaction proceeds by cleavage of the C—N bond and proton transfer to the amide nitrogen of the β-lactam bond. On the basis of the models of bound substrates, the apical water molecule that coordinates to Zn2 is appropriately positioned to donate this proton. Once the apical water is converted into a hydroxide, it moves to occupy the vacated nucleophilic hydroxide site, as the tetrahedral transition state decomposes and the degraded substrate diffuses away from the active site. A water molecule from bulk solvent is then recruited to the apical position, restoring the pentacoordination of Zn2.

A somewhat modified catalytic mechanism was proposed on the basis of kinetic experiments that showed accumulation of an intermediate with absorbance at 665 nm during hydrolysis of nitrocefin.[83] The absorbance was attributed to a cleaved anionic β-lactam containing a negatively charged nitrogen atom. According to this mechanism, substrate binds so that the β-lactam nitrogen atom coordinates directly to Zn2 by either displacing the apical water molecule, or by changing the coordination from trigonal bipyramidal to octahedral. Following a nucleophilic attack by the hydroxide, cleavage of the β-lactam bond yields a deprotonated nitrogen that is stabilized by interaction with Zn2. Next, a water molecule from bulk solvent protonates the negatively charged nitrogen, leading to a breakdown of the intermediate and product formation. While displacement of the apical water molecule by the β-lactam nitrogen appears chemically feasible, modeling for the current review shows that it leads to steric clashes between the substrate and protein. To avoid clashes, the enzyme must undergo significant conformational adjustments, including perturbations of the key active site residues His263 and Lys224. Note that the mechanism proposed by Concha *et al.*[39] does not exclude formation of an anionic intermediate. This is simply a matter of which event happens first, C—N bond cleavage or nitrogen protonation. In addition, the highly conjugated nitrocefin system may stabilize a negatively charged nitrogen, but this conjugated system is not a property of most other β-lactam substrates.

Another catalytic mechanism for the binuclear center metallo-β-lactamase was proposed on the basis of the di-zinc structure of the *B. cereus* enzyme.[53] The pH of the structure determination is reported to be pH 5.2, although the coordinates deposited in the PDB (entry code 1BC2) quote pH 4.5. Either pH is below the pK_a of postulated catalytic groups and one at which the enzyme has considerably reduced activity.[82] Indeed, the structure revealed unusual Zn2-ligand distances (1BC2 in Table 2). The hydroxide in this structure is positioned 1.9 Å away from Zn1 and 2.5 Å away from Zn2. Fabiane *et al.*[53] proposed that the hydroxide mounts a nucleophilic attack on the substrate, in accord with the proposed mechanisms for other enzymes with a binuclear zinc center. In their model of bound substrate, the apical water at Zn2 is replaced by the carboxylate moiety of the β-lactam substrate and the lone pair of the β-lactam nitrogen interacts with Zn2. As with the mono-zinc mechanism, they propose that the side-chain carboxylate of Asp120 plays the role of general base. It accepts a proton from the hydroxide, which leads to a dianionic species, which then transfers the proton to the β-lactam nitrogen atom. Here again, the proposed mechanism is inconsistent with the fact that the D120N mutant enzyme exhibits considerable enzymatic activity (Table 3).

The catalytic mechanism of the subclass B3 metallo-β-lactamases is somewhat different from that of the subclass B1 enzymes because of the different coordination of Zn2. The proposed mechanism is based on modeling and Monte-Carlo simulations of substrate binding to the structure of the L1 enzyme from *S. maltophilia*.[31] It is supported by fast kinetic experiments with nitrocefin, which reveal the same 665-nm intermediate as was seen with the *B fragilis* enzyme.[84] Substrate binding changes the coordination state of Zn2 from trigonal bipyramidal to octahedral and the coordination state of Zn1 from tetrahedral to trigonal bipyramidal. The β-lactam carboxylate group interacts with Ser224 (equivalent to Lys224 in the subclass B1 and B2 enzymes). The nucleophilic attack is mounted by the hydroxide, but in this case the proton is delivered to the nitrogen by the water molecule (W2 in Table 2) that bridges Zn2 to Ser221 (equivalent to Cys221 in the subclass B1 and B2 enzymes). To accomplish this, Ullah *et al.* proposed that the formation of the oxyanion transition state is accompanied by a loss of coordination to Zn1 by the carbonyl oxygen, as the geometry about the carbonyl carbon alters from planar to tetrahedral and W2 moves to coordinate to Zn1 while remaining coordinated to Zn2. Upon proton transfer, the newly formed hydroxide is therefore appropriately positioned to replace the original hydroxide and a water molecule from bulk solvent is recruited to occupy the vacated position between Zn2 and Ser221 as the product is released.

Finally, no structures have yet been determined for a representative member of the subclass B2 metallo-β-lactamases. It is expected that the catalytic mechanism of these enzymes will differ somewhat from the mechanisms of the B1 and B3 subclasses. These differences should account for the narrow substrate profile and for the fact that the binding of a second zinc ion inhibits the activity of the B2 enzymes.

REFERENCES

1. M Galleni, J Lamotte-Brasseur, GM Rossolini, J Spencer, O Dideberg and JM Frere, *Antimicrob Agents Chemother*, **45**, 660–63 (2001).
2. LD Sabath and EP Abraham, *Biochem J*, **98**, 11C–13C (1966).
3. M Watanabe, S Iyobe, M Inoue and S Mitsuhashi, *Antimicrob Agents Chemother*, **35**, 147–51 (1991).
4. H Ito, Y Arakawa, S Ohsuka, R Wacharotayankun, N Kato and M Ohta, *Antimicrob Agents Chemother*, **39**, 824–29 (1995).
5. H Yamaguchi, M Nukaga and T Sawai, GeneBank, accession code D29636, gi:473726.
6. L Lauretti, ML Riccio, A Mazzariol, G Cornaglia, G Amicosante, R Fontana and GM Rossolini, *Antimicrob Agents Chemother*, **43**, 1584–90 (1999).
7. L Poirel, T Naas, D Nicolas, L Collet, S Bellais, JD Cavallo and P Nordmann, *Antimicrob Agents Chemother*, **44**, 891–97 (2000).
8. IK Neonakis, EV Scoulica and YJ Tselentis, DDBJ/EMBL/Genbank databases, accession code AAN77714, gi:26514699.
9. A Tsakris, S Pournaras, N Woodford, MF Palepou, GS Babini, J Douboyas and DM Livermore, *J Clin Microbiol*, **38**, 1290–92 (2000).
10. K Bush, *Curr Pharm Des*, **5**, 839–45 (1999).
11. DM Livermore and N Woodford, *Curr Opin Microbiol*, **3**, 489–95 (2000).
12. L Aravind, *In Silico Biol*, **1**, 69–91 (1999).
13. H Daiyasu, K Osaka, Y Ishino and H Toh, *FEBS Lett*, **503**, 1–6 (2001).
14. TR Walsh, L Hall, SJ Assinder, WW Nichols, SJ Cartwright, AP MacGowan and PM Bennett, *Biochim Biophys Acta*, **1218**, 199–201 (1994).
15. RB Davies and EP Abraham, *Biochem J*, **143**, 129–35 (1974).
16. RW Shaw, SD Clark, NP Hilliard and JG Harman, *Protein Expr Purif*, **2**, 151–57 (1991).
17. A Felici and G Amicosante, *Antimicrob Agents Chemother*, **39**, 192–99 (1995).
18. R Paul-Soto, R Bauer, JM Frere, M Galleni, W Meyer-Klaucke, H Nolting, GM Rossolini, D de Seny, M Hernandez-Valladares, M Zeppezauer, HW Adolph, *J Biol Chem*, **274**, 13242–49 (1999).
19. K Bandoh, Y Muto, K Watanabe, N Katoh and K Ueno, *Antimicrob Agents Chemother*, **35**, 371–72 (1991).
20. Y Yang, BA Rasmussen and K Bush, *Antimicrob Agents Chemother*, **36**, 1155–57 (1992).
21. BA Rasmussen, Y Yang, N Jacobus and K Bush, *Antimicrob Agents Chemother*, **38**, 2116–20 (1994).
22. JH Toney, JK Wu, KM Overbye, CM Thompson and DL Pompliano, *Protein Expr Purif*, **9**, 355–62 (1997).
23. Z Wang and SJ Benkovic, *J Biol Chem*, **273**, 22402–8 (1998).
24. N Laraki, N Franceschini, GM Rossolini, P Santucci, C Meunier, E de Pauw, G Amicosante, JM Frere and M Galleni, *Antimicrob Agents Chemother*, **43**, 902–6 (1999).
25. MW Crowder, TR Walsh, L Banovic, M Pettit and J Spencer, *Antimicrob Agents Chemother*, **42**, 921–26 (1998).
26. PS Mercuri, F Bouillenne, L Boschi, J Lamotte-Brasseur, G Amicosante, B Devreese, J van Beeumen, JM Frere, GM Rossolini and M Galleni, *Antimicrob Agents Chemother*, **45**, 1254–62 (2001).
27. M Hernandez Villadares, M Galleni, JM Frere, A Felici, M Perilli, N Franceschini, GM Rossolini, A Oratore and G Amicosante, *Microb Drug Resist*, **2**, 253–56 (1996).
28. R Bicknell, EL Emanuel, J Gagnon and SG Waley, *Biochem J*, **229**, 791–97 (1985).
29. Y Saino, F Kobayashi, M Inoue and S Mitsuhashi, *Antimicrob Agents Chemother*, **22**, 564–70 (1982).
30. R Paton, RS Miles and SG Amyes, *Antimicrob Agents Chemother*, **38**, 2143–49 (1994).
31. JH Ullah, TR Walsh, IA Taylor, DC Emery, CS Verma, SJ Gamblin and J Spencer, *J Mol Biol*, **284**, 125–36 (1998).
32. E Osono, Y Arakawa, R Wacharotayankun, M Ohta, T Horii, H Ito, F Yoshimura and N Kato, *Antimicrob Agents Chemother*, **38**, 71–78 (1994).
33. A Carfi, R-P Soto, E Duée, M Galleni, J-M Frère and O Dideberg, PDB Entry 1BMI. (1997).
34. K Marumo, A Takeda, Y Nakamura and K Nakaya, *Microbiol Immunol*, **39**, 27–33 (1995).
35. JD Docquier, F Pantanella, F Giuliani, MC Thaller, G Amicosante, M Galleni, JM Frere, K Bush and GM Rossolini, *Antimicrob Agents Chemother*, **46**, 1823–30 (2002).

36 G Lombardi, F Luzzaro, JD Docquier, ML Riccio, M Perilli, A Coli, G Amicosante, GM Rossolini and A Toniolo, *J Clin Microbiol*, **40**, 4051–55 (2002).

37 JD Docquier, J Lamotte-Brasseur, M Galleni, G Amicosante, JM Frere and GM Rossolini, *J Antimicrob Chemother*, **51**, 257–66 (2003).

38 MW Crowder, Z Wang, SL Franklin, EP Zovinka and SJ Benkovic, *Biochemistry*, **35**, 12126–32 (1996).

39 NO Concha, BA Rasmussen, K Bush and O Herzberg, *Structure*, **4**, 823–36 (1996).

40 D de Seny, U Heinz, S Wommer, M Kiefer, W Meyer-Klaucke, M Galleni, JM Frere, R Bauer and HW Adolph, *J Biol Chem*, **276**, 45065–78 (2001).

41 S Wommer, S Rival, U Heinz, M Galleni, JM Frere, N Franceschini, G Amicosante, B Rasmussen, R Bauer and HW Adolph, *J Biol Chem*, **277**, 24142–47 (2002).

42 M Hernandez Valladares, A Felici, G Weber, HW Adolph, M Zeppezauer, GM Rossolini, G Amicosante, JM Frere and M Galleni, *Biochemistry*, **36**, 11534–41 (1997).

43 Y Yang and K Bush, *FEMS Microbiol Lett*, **137**, 193–200 (1996).

44 M Hernandez Valladares, M Kiefer, U Heinz, RP Soto, W Meyer-Klaucke, HF Nolting, M Zeppezauer, M Galleni, JM Frere, GM Rossolini, G Amicosante and HW Adolph, *FEBS Lett*, **467**, 221–25 (2000).

45 R Bicknell, A Schaffer, SG Waley and DS Auld, *Biochemistry*, **25**, 7208–15 (1986).

46 GS Baldwin, A Galdes, HA Hill, SG Waley and EP Abraham, *J Inorg Biochem*, **13**, 189–204 (1980).

47 EG Orellano, JE Girardini, JA Cricco, EA Ceccarelli and AJ Vila, *Biochemistry*, **37**, 10173–80 (1998).

48 R Paul-Soto, M Zeppezauer, HW Adolph, M Galleni, JM Frere, A Carfi, O Dideberg, J Wouters, L Hemmingsen and R Bauer, *Biochemistry*, **38**, 16500–6 (1999).

49 GS Baldwin, A Galdes, HA Hill, BE Smith, SG Waley and EP Abraham, *Biochem J*, **175**, 441–47 (1978).

50 L Hemmingsen, C Damblon, J Antony, N Jensen, H Adolph, S Wommer, G Roberts and R Bauer, *J Am Chem Soc*, **123**(42), 10329–35 (2001).

51 A Carfi, S Pares, E Duée, M Galleni, C Duez, JM Frère and O Dideberg, *EMBO J*, **14**, 4914–21 (1995).

52 A Carfi, E Duée, M Galleni, J-M Frère and O Dideberg, *Acta Crystallogr*, **D54**, 313–23 (1998).

53 SM Fabiane, MK Sohi, T Wan, DJ Payne, JH Bateson, T Mitchell and BJ Sutton, *Biochemistry*, **37**, 12404–11 (1998).

54 I Garcia-Saez, PS Mercuri, C Papamicael, R Kahn, JM Frere, M Galleni, GM Rossolini and O Dideberg, *J Mol Biol*, **325**, 651–60 (2003).

55 PMD Fitzgerald, JK Wu and JH Toney, *Biochemistry*, **37**, 6791–800 (1998).

56 JH Toney, PMD Fitzgerald, N Grover-Sharma, SH Olson, WJ May, JG Sundelof, DE Vanderwall, KA Cleary, SK Grant, JK Wu, JW Kozarich, DL Pompliano and GG Hammond, *Chem Biol*, **5**, 185–96 (1998).

57 NO Concha, BA Rasmussen, K Bush and O Herzberg, *Protein Sci*, **6**, 2671–76 (1997).

58 DJ Payne, JA Hueso-Rodriguez, H Boyd, NO Concha, CA Janson, M Gilpin, JH Bateson, C Cheever, NL Niconovich, S Pearson, S Rittenhouse, D Tew, E Diez, P Perez, J De La Fuente, M Rees and A Rivera-Sagredo, *Antimicrob Agents Chemother*, **46**, 1880–6 (2002).

59 NO Concha, CA Janson, P Rowling, S Pearson, CA Cheever, BP Clarke, C Lewis, M Galleni, JM Frere, DJ Payne, JH Bateson and SS Abdel-Meguid, *Biochemistry*, **39**, 4288–98 (2000).

60 JH Toney, GG Hammond, PM Fitzgerald, N Sharma, JM Balkovec, GP Rouen, SH Olson, ML Hammond, ML Greenlee and YD Gao, *J Biol Chem*, **276**, 31913–18 (2001).

61 P Kraullis, *J Appl Crystallogr*, **24**, 946–50 (1991).

62 DJ Bacon and WF Anderson, *J Mol Graph*, **6**, 219–20 (1988).

63 EA Merrit and MEP Murphy, *Acta Crystallogr*, **D50**, 869–73 (1994).

64 AD Cameron, M Ridderstrom, B Olin and B Mannervik, *Struct Fold Des*, **7**, 1067–78 (1999).

65 C Frazao, G Silva, CM Gomes, P Matias, R Coelho, L Sieker, S Macedo, MY Liu, S Oliveira, M Teixeira, AV Xavier, C Rodrigues-Pousada, MA Carrondo and J Le Gall, *Nat Struct Biol*, **7**, 1041–45 (2000).

66 DJ Payne, W Du and JH Bateson, *Expert Opin Investig Drugs*, **9**, 247–61 (2000).

67 R Nagano, Y Adachi, T Hashizume and H Morishima, *J Antimicrob Chemother*, **45**, 271–76 (2000).

68 MI Quiroga, N Franceschini, GM Rossolini, G Gutkind, G Bonfiglio, L Franchino and G Amicosante, *Chemotherapy*, **46**, 177–83 (2000).

69 C Mollard, C Moali, C Papamicael, C Damblon, S Vessilier, G Amicosante, CJ Schofield, M Galleni, JM Frere and GC Roberts, *J Biol Chem*, **276**, 45015–23 (2001).

70 S Bounaga, M Galleni, AP Laws and MI Page, *Bioorg Med Chem*, **9**, 503–10 (2001).

71 S Siemann, DP Evanoff, L Marrone, AJ Clarke, T Viswanatha and GI Dmitrienko, *Antimicrob Agents Chemother*, **46**, 2450–57 (2002).

72 A Carfi, E Duée, R Paul-Soto, M Galleni, J-M Frère and O Dideberg, *Acta Crystallogr*, **D54**, 47–57 (1998).

73 SD Scrofani, J Chung, JJ Huntley, SJ Benkovic, PE Wright and HJ Dyson, *Biochemistry*, **38**, 14507–14 (1999).

74 JJ Huntley, SD Scrofani, MJ Osborne, PE Wright and HJ Dyson, *Biochemistry*, **39**, 13356–64 (2000).

75 BA Rasmussen and K Bush, *Antimicrob Agents Chemother*, **41**, 223–32 (1997).

76 A Felici, G Amicosante, A Oratore, R Storm, P Ledent, B Joris, L Fanuel and J-M Frère, *Biochem J*, **291**, 151–55 (1993).

77 TR Walsh, S Gamblin, DC Emery, AP MacGowan and PM Bennett, *J Antimicrob Chemother*, **37**, 423–31 (1996).

78 S Bellais, D Aubert, T Naas and P Nordmann, *Antimicrob Agents Chemother*, **44**, 1878–86 (2000).

79 S Vessillier, JD Docquier, S Rival, JM Frere, M Galleni, G Amicosante, GM Rossolini and N Franceschini, *Antimicrob Agents Chemother*, **46**, 1921–27 (2002).

80 AM Simm, CS Higgins, ST Pullan, MB Avison, P Niumsup, O Erdozain, PM Bennett and TR Walsh, *FEBS Lett*, **509**, 350–54 (2001).

81 R Bicknell and SG Waley, *Biochemistry*, **24**, 6876–87 (1985).

82 S Bounaga, AP Laws, M Galleni and MI Page, *Biochem J*, **331**(Pt 3), 703–11 (1998).

83 Z Wang, W Fast and SJ Benkovic, *Biochemistry*, **38**, 10013–23 (1999).

84 S McManus-Munoz and MW Crowder, *Biochemistry*, **38**, 1547–53 (1999).

85 J Spencer, AR Clarke and TR Walsh, *J Biol Chem*, **276**, 33638–44 (2001).

86 Y Yang, D Keeney, X Tang, N Canfield and BA Rasmussen, *J Biol Chem*, **274**, 15706–11 (1999).

87 MP Yanchak, RA Taylor and MW Crowder, *Biochemistry*, **39**, 11330–39 (2000).

88 D Seny, C Prosperi-Meys, C Bebrone, GM Rossolini, MI Page, P Noel, JM Frere and M Galleni, *Biochem J*, **363**, 687–96 (2002).

89 AL Carenbauer, JD Garrity, G Periyannan, RB Yates and MW Crowder, *BMC Biochem*, **3**, 4 (2002), Epub 2002 Feb 13.

90 Z Li, BA Rasmussen and O Herzberg, *Protein Sci*, **8**, 249–52 (1999).

91 L Chantalat, E Duee, M Galleni, JM Frere and O Dideberg, *Protein Sci*, **9**, 1402–6 (2000).

92 WN Lipscomb, *Annu Rev Biochem*, **52**, 17–34 (1983).

93 BW Matthews, *Acc Chem Res*, **21**, 333–40 (1988).

GTP cyclohydrolase I

Herbert Nar

Boehringer Ingelheim Pharma, Biberach, Germany

FUNCTIONAL CLASS

Enzyme; GTP Cyclohydrolase I (GTP-CH-I), hydrolase, EC 3.5.4.16; zinc-dependent enzyme.

GTP Cyclohydrolase I (GTP-CH-I) is a homodecameric protein complex of approximately 250 kDa molecular weight.[1,2] GTP-CH-I catalyzes the conversion of guanosine triphosphate (GTP) to dihydroneopterin triphosphate (H_2NTP), the committing step in the biosynthesis of pteridines. The proposed complex reaction involves the hydrolytic opening of the imidazole ring of GTP and the formation of a formamidopyrimidine intermediate. Release of formate, an Amadori rearrangement of the ribose, and closure of the dihydropyrazine ring affords the product. The turnover rate under steady state conditions is 0.05 s^{-1}/subunit[3] and the K_m for the substrate GTP is 0.02 μM.[4] Catalysis is independent of any cofactor other than zinc.

OCCURRENCE

Occurs across all species. GTP-CH-I in higher species is ubiquitously expressed. Regulation of enzymatic activity in various cell and tissue types may occur on transcriptional and posttranscriptional level as well as via feedback inhibition mediated by tetrahydrobiopterin-dependent (BH_4-dependent) binding of a regulatory protein, GTP cyclohydrolase feedback regulatory protein (GFRP).[5]

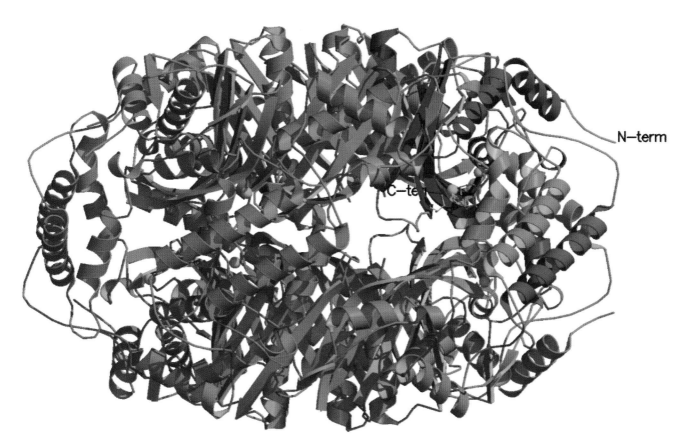

3D Structure Ribbon representation of the functional decameric form of *E. coli* GTP-CH-I (PDB code 1FBX). One subunit is shown in color with the constituents of its metal binding site displayed. The 10 metal binding sites are indicated by the green spheres. The subunit folds into an α + β structure with a four-stranded antiparallel β-sheet surrounded by loops and helices. The association of five subunits along their β-sheets results in the formation of a 20-stranded β-barrel that forms the heart of the GTP-CH-I pentamer. The functional complex is a dimer of two such pentamers.

As shown by immunohistochemical staining with anti–GTP-CH-I antibodies, the expression pattern in various tissues and cell types is highly specific, and colocalization was generally found with aromatic amino acid hydroxylases.[5] The studies revealed a cytosolic and nuclear localization for GTP-CH-I in specific but various cell types.

BIOLOGICAL FUNCTION

GTP-CH-I catalyzes the biosynthesis of pteridines from GTP (Scheme 1).[6,7] Pteridines serve as cofactors for a variety of enzyme-catalyzed reactions. Specifically, tetrahydrofolate (in plants, bacteria, and eukaryotic organisms) and tetrahydromethanopterin (in archaea) mediate the transfer of one-carbon fragments, tetrahydrobiopterin (in animals) is implicated in the hydroxylation of aromatic amino acids and the formation of nitric oxide in animals,[8,9] and molybdopterin is required as a cofactor by a variety of redox enzymes, for example, xanthine dehydrogenase. The metabolic roles of these cofactors have been reviewed repeatedly.[5,10–12]

In human newborns, lack of tetrahydrobiopterin leads to hyperphenylalaninemia and a deficiency of the biogenic amine neurotransmitters dopamine and serotonin, and if it is untreated, results in severe progressive mental retardation.[9,13–16] One less common form of BH_4 deficiency is due to the absence of GTP-CH-I activity.[5,12,15]

The bacterial enzymes involved in the biosynthesis of tetrahydrofolate represent important anti-infective drug targets. The antimicrobial activity of sulfonamides and trimethoprim are based on the inhibition of dihydropteroate synthase and dihydrofolate reductase. The growing resistance of human pathogens against all major anti-infective drugs in current use stimulates the exploration of novel drug targets, among which GTP-CH-I represents an interesting candidate.[17]

AMINO ACID SEQUENCE INFORMATION

The primary structure of GTP-CH-I from a variety of organisms has been determined either by protein sequencing from native protein material isolated from bacterial, insect, or mammalian species or by expression-cloning efforts.[4,18–24] Since the late 1990s, genome projects contribute to a growing number of annotated GTP-CH-I sequences.

Here is a list of some GTP-CH-I representatives of bacteria, eukaryota, and archaea:

- GCH1_BACSU, P19465, *Bacillus subtilis*, GTP cyclohydrolase I, 190 AA residues[25]
- GCH1_HUMAN, P30793, *Homo sapiens*, GTP cyclohydrolase I, 250 AA residues[26]
- GCH1_RAT, P22288, *Rattus norvegicus*, GTP cyclohydrolase I precursor, 241 AA residues[23]
- GCH1_DROME, P48596, *Drosophila melanogaster*, GTP cyclohydrolase I, 324 AA residues[27]
- GCH1_HAEIN, P43866, *Haemophilus influenzae*, GTP cyclohydrolase I, 218 AA residues[28]
- GCH1_ECOLI, P27511, *Escherichia coli*, GTP cyclohydrolase I, 221 AA residues[21]
- GCH1_CAEEL, Q19980, *Caenorhabditis elegans*, probable GTP cyclohydrolase I, 223 AA residues (McMurray, A., Submitted APR-1996 to the EMBL GenBank DDBJ databases)
- GCH1_SYNP7, Q54769, *Synechococcus* sp. PCC 7942, GTP cyclohydrolase I, 213 AA residues (Phung, L.T., Haselkorn, R., Genes encoding the alpha subunit of carboxyltransferase of the acetyl-CoA carboxylase complex and GTP cyclohydrolase I from cyanobacterium *Synechococcus* sp. PCC 7942. Submitted MAY-1996 to the EMBL GenBank DDBJ databases)
- GCH1_SULTO, Q971G9, *Aeropyrum pernix* K1, GTP cyclohydrolase I, 215 AA residues[29]

The sequence homology within the complete GTP-CH-I family suggests that the evolution of the protein has been relatively conservative. For instance, the C-terminal domain of approximately 120 residues is especially conserved exhibiting a 60% identity between the *E. coli* and the *human* enzymes. Greater sequence variability is found at the N-termini of known sequences, which also differ in length. The sequence conservation of the C-terminal domains that

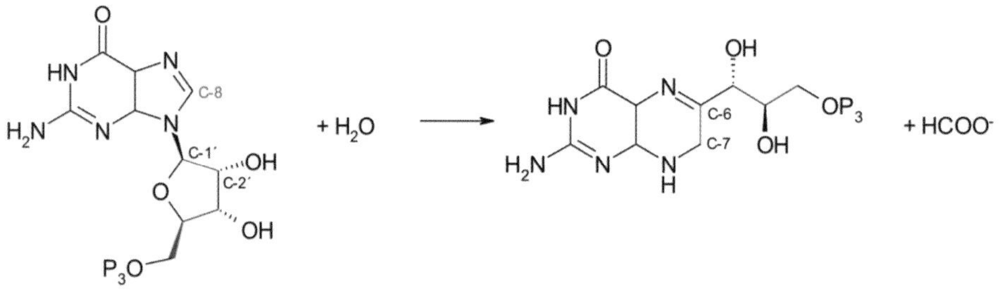

Scheme 1 Reaction catalyzed by GTP-CH-I. The substrate GTP is converted to dihydroneopterin triphosphate. The carbon atom C-8 of the imidazole ring portion of the heterocycle is released as formate during the reaction. Carbon atoms C-1' and C-2' are incorporated as atoms C-6 and C-7 into the pterin ring, whereas carbon atoms C-3' to C-5' form the side chain of the product. This complex reaction is catalyzed by a catalytic machinery of the enzyme involving zinc as the only cofactor.

are responsible for oligomerization of the protomers also suggests that the quaternary structures of GTP-CH-I are very similar across species. All residues involved in metal binding or implicated in catalysis are strictly conserved (see below).

PROTEIN PRODUCTION, PURIFICATION, AND MOLECULAR CHARACTERIZATION

GTP-CH-I from a variety of species has been isolated or recombinantly expressed and characterized. Native material was isolated from *E. coli* and several other microorganisms, drosophila, rat, and human liver.[4,18–20,22,24] Purification of the proteins was typically done by anion exchange chromatography, GTP-affinity chromatography, and gel-permeation chromatography techniques.[30–33]

The protein is relatively heat-stable with a half-life of 7 min at 82 °C. Isolation and partial purification of *E. coli* GTP-CH-I was first reported by Burg and Brown in 1968.[6] Early electron microscopic work by the same group[4] demonstrated that the enzyme is an oligomer forming a torus-shaped three-dimensional assembly. Electron microscopy on freeze-etched crystal surfaces unequivocally showed that the active enzymatic particle exhibits fivefold particle symmetry strongly suggesting a homodecameric structure.[1] These suggestions were confirmed ultimately by the resolution of the crystal structure of the *E. coli* enzyme.[2,34]

Protein biochemical work shows that the oligomer reversibly dissociates to dimers in 0.3 M potassium chloride.[4] The oligomerization apparently leads to cooperativeness in the case of the rat enzyme[35] but not for the *E. coli* enzyme.[4] Known competitive inhibitors of GTP-CH-I are guanosine tetraphosphate ($K_i = 0.13\,\mu M$), dGTP ($K_i = 0.24\,\mu M$), GDP ($K_i = 1.5\,\mu M$), and ATP ($K_i = 0.24\,\mu M$).[4]

GTP-CH-I is subject to genetic regulation and other regulatory mechanisms. The enzyme is expressed upon stimulation of various cytokines.[36–38] Multiple mRNA forms have been detected in human, rat, and drosophila.[39,40]

Posttranslational processing involves cleavage of the N-terminal 11 amino acids as observed for the rat liver enzyme and protein phosphorylation.[22,23] The role of the N-terminal processing is unknown; however, phosphorylation has been shown to modulate enzyme activity. One serine residue (Ser167 in the rat and mouse sequences), which is conserved in the proteins of higher species and which is exposed at the protein surface, has been proposed to be a potential target site for protein kinase C.

Regulation of GTP-CH-I activity is mediated by its substrate GTP, the pathway end product BH_4, and phenylalanine. Intracellular levels of GTP modulate GTP-CH-I activity by changes of enzyme kinetics due to cooperative binding.[35] BH_4 and phenylalanine modulate enzymic activity via GFRP binding to GTP-CH-I. GFRP mediates end product feedback inhibition by BH_4 and stimulates GTP-CH-I activity at higher levels of phenylalanine.[41,42] GFRP is a homopentamer of 9.5 kDa subunits. Recently, the structure of GFRP alone[43] and of the phenylalanine-induced stimulatory complex of GFRP with rat GTP-CH-I has been determined,[44,45] and it shows an assembly with two pentamers of GFRP bound to the outer faces of the GTP-CH-I decamer.

METAL CONTENT AND COFACTORS

Initially, it was assumed that GTP-CH-I was a cofactor-independent enzyme. The first X-ray structure of the *E. coli* protein revealed an active site that contained an unusual disulfide bridge in its center.[2,34,46] The existence of a catalytic zinc ion at the active site of GTP-CH-I was discovered during the crystallographic analysis of the human protein that had been isolated without the use of chelating agents. In the same study, it was shown by crystallographic and analytic techniques that the *E. coli* protein also contains stoichiometric amounts of one zinc ion per protomer.[47] The *E. coli* protein originally used for the initial structure analysis had been extensively treated with ethylenediaminetetraacetic acid (EDTA), which completely depletes the enzyme of zinc. During protein purification and crystallization, the two metal coordinating cysteines were oxidized to the cystine, a process that was observed also for the zinc protein cytidine deaminase, which has a similar zinc binding site.[48–50]

Apart from zinc, the enzyme apparently does not require any cofactor. Magnesium ions are not needed for charge compensation of the triphosphate moiety of the substrate because of the presence of a cluster of basic residues at the rim of the active site pocket.

ACTIVITY TEST

GTP-CH-I activity is determined by UV detection of the reaction product H_2NTP at 330 nm. Assay mixtures typically contain 100 mM Tris hydrochloride, pH 8.5, 100 mM KCl, 2.5 mM EDTA, 0.2 mM GTP, and protein. The samples are incubated at 37 °C, and absorbance at 330 nm is recorded. The concentration of H_2NTP is estimated using an absorption coefficient of $\varepsilon\,(330\,nm) = 6300\,M^{-1}$.[32]

X-RAY STRUCTURE OF *E. COLI* GTP-CH-I

The first crystal structure of a GTP-CH-I was obtained from the *E. coli* enzyme.[2] More recently, the structures of various point mutants of the *E. coli* enzyme,[17] of the N-terminally truncated human protein,[47] and of the rat GTP-CH-I in complex with the regulatory protein GFRP were determined.[45]

Crystallization

N-terminally truncated human GTP-CH-I (ΔN42) was crystallized from 1.0 M ammonium sulfate, 3.2% isopropanol, 2 mM DTT, pH 7.5. Crystals are of the space group $P6_522$, $a = b = 115.1$ Å, $c = 387.3$ Å with a pentamer in the asymmetric unit.

The rat GTP-CH-I/GFRP complex crystallized from 24% 2-methyl-2,4-pentanediol, 50 mM potassium chloride, 5 mM phenylalanine, 75 mM Tris/HCl, pH 7.5 in space group $P2_1$ with unit cell constants $a = 123.2$ Å, $b = 111.4$ Å, $c = 125.8$ Å, $\beta = 97.7°$. The asymmetric unit contains 10 subunits each of GTP-CH-I and GFRP.

Crystals of *E. coli* GTP-CH-I are obtained at room temperature using vapor diffusion under the following conditions.

(a) 0.1 M sodium Mes (pH 6.0), containing 0.2 M sodium acetate and 3 mM sodium azide.

(b) 0.1 M Mops (pH 7.0), with 10% (w/v) polyethylene glycol 6000 and 0.1 M ammonium sulfate.

(c) 0.1 M Tris hydrochloride (pH 8.5), containing 0.2 M ammonium dihydrogen phosphate and 50% (v/v) MPD.

(d) 0.5 M sodium citrate, pH 7.4

Crystals obtained using the two first crystallization conditions belong to the space group $C222_1$ with 15 monomers in the asymmetric unit and lattice constant $a = 315.9$ Å, $b = 220.6$ Å, $c = 131.4$ Å. Crystals obtained under condition (c) belong to the space group $P4_32_12$ with 10 monomers in the asymmetric unit and lattice constants $a = b = 123.9$ Å, $c = 388.6$ Å. Condition (d) yields crystals of the monoclinic space group $P2_1$ with unit cell constants $a = 204.2$ Å, $b = 210.4$ Å, $c = 71.8$ Å, $\beta = 95.8°$ containing 20 monomers per asymmetric unit.

The initial structure analysis of GTP-CH-I was done using the monoclinic crystal form of the *E. coli* protein. Crystal packing considerations led to the conclusion that there are two decamers in the asymmetric unit. Information about the packing arrangement was independently obtained by electron microscopy on the same crystal form[1] and showed the relative arrangement of the decamer particles most prominently in the crystal ab plane projection. This picture revealed a view of the fivefold particle axes that are roughly perpendicular to the crystal plane. The crystal structure was solved by single isomorphous replacement and electron density averaging techniques. The refined model consists of residues 3 to 217 of the 221-residue sequence of *E. coli* GTP-CH-I.

Overall description of the structure (PDB code 1FBX)

Monomer structure

The GTP-CH-I monomer folds into a structure with a predominantly α-helical N-terminal part (3D Structure). An antiparallel α-helix pair, composed of α-helix h2 (residues 32 to 49; numbering of secondary structural elements according to secondary structure–sequence assignments in Figure 1) and α-helix h3 (residues 62 to 72), is remote from the main body of the molecule. The compact C-terminal domain (residues 95 to 217) comprises a sequential four-stranded antiparallel β-sheet (β-strands b2, b3, b6, b7) including a 45-residue insertion between β-strands b3 and b6, which contains two antiparallel α-helices (h4 and h5) layered on one side of the β-sheet. The C-terminus is formed by a three-turn α-helix, positioned on the other side of the β-sheet. The N-terminal α-helix h1 (residues 5 to 17) lies on top of α-helices h4 and h5.

Pentamer structure

Five monomers associate in a highly symmetrical fashion along their β-sheets to form a 20-stranded antiparallel β-barrel with a diameter of 35 Å. The N-terminal β-strand b2 of one monomer thereby forms a hydrogen bond ladder with the C-terminal β-strand b7 of the neighboring monomer (Figure 2). Within the pentamer, further interactions are made via two extended loop regions on the latter monomer (residues 85 to 95 and 130 to 140) that interlock with a groove in the former. The β-barrel is occupied by the C-terminal helices from each monomer, which form a parallel α-helix bundle with a diameter of about 15 Å between the helix centers. The pentamer has the overall shape of a crab with five legs composed of the N-terminal α-helix pairs of the individual monomers.

Decamer structure

The GTP-CH-I decamer is formed by face-to-face association of two pentamers. The legs of one pentamer clasp the body of the other, that is, the antiparallel α-helices h2 and h3 on one monomer are intertwined with those of another monomer, wedged into the cleft between these and the C-terminal domain. Since the arrangement of the two five-fold symmetric pentamers follows a twofold symmetry, the decamer has perfect D_5 particle symmetry and is toroidal with an approximate height of 65 Å and a diameter of 100 Å. It encloses a cavity of dimensions $30 \times 30 \times 15$ Å, which is accessible through the pores formed by the five α-helix bundles in the center of the pentamers, but has no openings at the decamer equator. This cavity is partially occupied by the four C-terminal residues, which are not

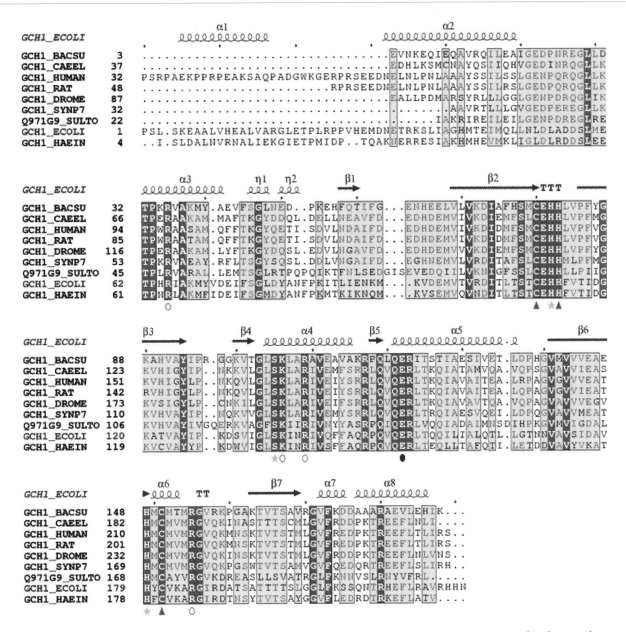

Figure 1 Sequence alignment of representatives of the GTP-CH-I family. The chosen representatives correspond to the named sequences in the text. All residues implicated in metal binding (▲), catalysis (★), and substrate recognition (●) are totally conserved throughout the species. Secondary structural elements are symbolized as arrows and spirals on top of each block.

defined in the electron density and are probably disordered.

Comparison of *E. coli* GTP-CH-I to the rat and human enzymes (PDB codes 1IS8, 1FB1)

The tertiary and quaternary structures of the human and rat enzymes are very similar to the *E. coli* enzyme structure.[45,47] The root mean square deviation (rmsd) of human and rat GTP-CH-I subunits to the *E. coli* subunit is 1.6 Å. Relative to the *E. coli* protein, the mammalian enzymes exhibit extended N-terminal sequences. The crystal structures, however, do not give any indication as to their structures, because the N-termini are either truncated during expression cloning or disordered.[45,47]

Active site location and zinc binding site

The first indication of the location of the active sites of GTP-CH-I came from topological analyses of the protein structure and a comparison with the structurally related enzyme 6-pyruvoyl-tetrahydropterin synthase.[51] An unequivocal proof was provided by a cocrystal structure of *E. coli* GTP-CH-I with the substrate analog dGTP.[34]

The active site of GTP-CH-I is located at the interface of three subunits (two from one pentamer and one from the

GTP cyclohydrolase I

Figure 2 Top view of the GTP-CH-I structure along the fivefold particle axis. This figure, as well as Figures (3–6), were prepared with MOLSCRIPT (Per J. Kraulis, *J. Appl. Cryst.* (1991) **24**, 946–950).

other pentamer). Owing to the 52-point group symmetry of the homodecameric enzyme complex, there are 10 equivalent active sites per functional unit.

The Zn(II) binding site is located at the center of the cavity. The metal is bound to the two cysteine residues CysA110 (nomenclature used for numbering amino acids: all residue numbers refer to the *E. coli* protein sequence; residue numbers are preceded by A and A' for subunits from one pentamer and B for residues of the subunit on the other pentamer) and CysA181 as well as to ND1 of HisA113 and a water molecule in the resting state of the enzyme (Figure 3).[17]

The coordination environment of zinc is distorted tetrahedral. During catalysis, the coordination state of the metal probably changes transiently to a pentacoordinated state with the formyl group of an intermediate and a water molecule bound to zinc in addition to the protein side chain groups. For the human enzyme, a coordinated isopropanol molecule from the crystallization buffer was identified as a fourth ligand to the zinc (see Table 1 below).

The active site surrounding loops formed by residues 109 to 113, 150 to 153, and 179 to 181 are structurally stabilized by extensive short-range hydrogen bond interactions and are, furthermore, mutually linked by the zinc binding site, the salt bridge Glu111-Arg153, and the

Table 1 Zinc–ligand bond distances in GTP-CH-I in (Å), *E. coli* sequence residue numbering

	E. coli enzyme (1FBX)	Rat enzyme (1IS8)	Human enzyme (1FB1)
ZN-ND1 His113	2.1	2.2	2.2
ZN-SG Cys110	2.3	2.2	2.2
ZN-SG Cys181	2.2	2.3	2.2
ZN-O wat	2.4	2.2	
ZN-O isoprop			2.7

hydrogen bond interaction Gln151-His179. This results in a strictly confined conformational space available to residues involved in substrate recognition and catalysis and explains why the protein structure is hardly changed even when the metal is depleted from the protein or when its binding site is mutated.[17,34]

Substrate binding

The protein acceptor site for the pyrimidine portion of the purine ring system is found at the bottom of the 10-Å deep active site cavity (Figure 4). GluA152 forms a salt

GTP cyclohydrolase I

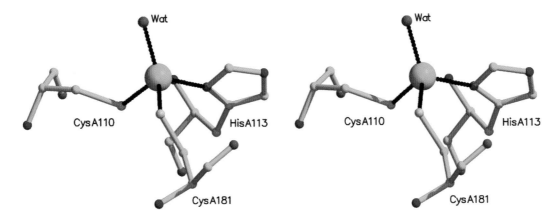

Figure 3 Stereo view of the zinc binding site of the resting form of the enzyme. The metal ion is coordinated by two cysteine thiol groups, ND1 of HisA113 and a water molecule in a tetrahedral fashion.

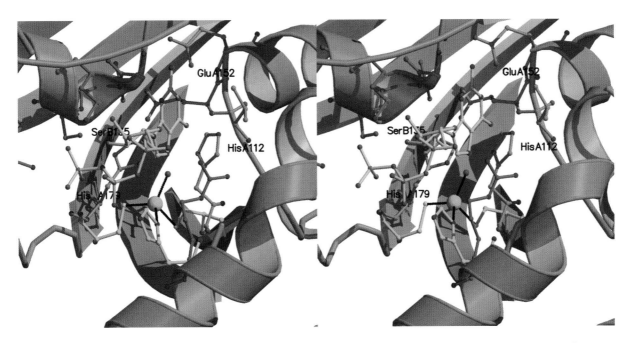

Figure 4 Stereo view of a model of substrate bound to the active site of the enzyme. This model is based on several experimental structures of dGTP bound to the zinc-depleted wild-type enzyme,[2,34] GTP bound to various inactive mutants of E. coli GTP-CH-I and uncomplexed wild-type enzyme.[17,47] GTP (light blue carbon atoms) fits snugly in the binding pocket enclosed by amino acid residues of three adjacent GTP-CH-I subunits (orange and green, subunit A; magenta, subunit B; and blue, subunit A'). Substrate recognition is facilitated by the interaction of the pyrimidine ring of GTP with GluA152 deep in the pocket. The imidazole of the substrate is in the vicinity of the zinc-bound water molecule.

bridge with the guanidine moiety. The preceding peptide bond between ValA150 and GlnA151 forms a hydrogen bond with its amide hydrogen to the C-4 oxo group of the purine. Further, a hydrogen bond is formed between the N-2 of the nucleobase and the carbonyl oxygen of residue IleA'132. This peculiar recognition pattern is specific for guanine and explains the selectivity of the enzyme of GTP over ATP.[4] The inner wall of the pocket that lies parallel to the pyrimidine ring plane is lined by residues ValA150 and LeuA'134 that create a suitably hydrophobic environment. The imidazole portion, in particular, C-8 of the purine ring is in the vicinity of the zinc binding site and is close to HisA179. There is space left on both sides of its ring plane that might be occupied by solvent during catalysis. A cluster of basic residues, ArgA'185, LysA'136, ArgA'139, ArgB65, and LysB68, at the pocket entrance binds the triphosphate group of GTP. Since these residues provide for complete charge compensation, Mg(II)-assisted binding to the protein, as found in other nucleoside triphosphate binding proteins, is neither necessary nor realized in GTP-CH-I. The above residues that are in contact with the substrate are strictly conserved across enzyme sequences of all species with the exception of Lys68 (Figure 1).

GTP cyclohydrolase I

GTP-CH-I/GPFR Complex

The crystal structures of the GFRP alone and of the GFRP in complex with rat GTP-CH-I were recently determined.[43,45] GFRP is a pentamer of 9.5 kDa subunits whose 6-stranded β-sheet secondary structural elements form a symmetrical five-membered ring similar to β-propellers. The complex structure shows that the GTP-CH-I decamer is sandwiched by two GFRP pentamers with 5 phenylalanine molecules buried inside the interface between GFRP and GTP-CH-I thereby enhancing the binding of the protein oligomers (Figure 5). While this structure suggests that the phenylalanine-induced complex formation induces an active state of the rat enzyme that is normally present in the unregulated bacterial enzyme, it remains to be elucidated how BH_4-mediated inactivation of GTP-CH-I activity is conferred by GFRP.

Comparison of GTP-CH-I to 6-pyruvoyl tetrahydropterin synthase (PTPS, PDB code 1GTQ)

The structure of the C-terminal domain of GTP-CH-I (*E. coli* protein residues 97 to 200) is topologically identical to PTPS (Figure 6).[51] In the latter, this single domain comprises 140 amino acid residues. A structural alignment of identically positioned amino acids in the secondary structural elements common to both protein subunits results in an rmsd of Cα-atom positions of 1.9 Å for 71 atoms.[2] Significant sequence homology between the protein families cannot be recognized even after topological alignment, with the exception of a few residues that are involved in purine and pterin ring binding. The structural similarity of the two protein subunits extends beyond the level of tertiary structure. In PTPS, hydrogen bonding between three four-stranded β-sheets of neighboring monomers leads to an assembly with a 12-stranded antiparallel β-barrel at its core. The same mode of association is used in GTP-CH-I for the formation of the 20-stranded β-barrel.

Structures of other enzymes involved in pterin metabolism, epimerase, and 7,8-dihydroneopterin aldolase as well as urate oxidase exhibit similar monomer topologies to PTPS and GTP-CH-I and associate to tetramers or octamers via an identical association mode to form 16-stranded β-barrels.[52–55]

FUNCTIONAL ASPECTS AND CATALYTIC MECHANISM

The formation of pterins by ring expansion of guanosine was first suggested by Weygand *et al.*[56] on the basis of

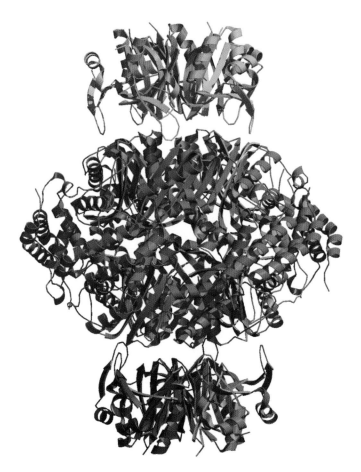

Figure 5 Ribbon representation of the GTP-CH-I/GFRP complex. GFRP pentamers (colored) sandwich the GTP-CH-I decamer (gray).

GTP cyclohydrolase I

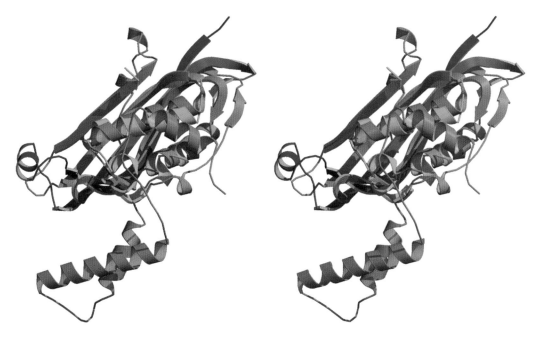

Figure 6 Stereo view of the superposition of GTP-CH-I with 6-pyruvoyltetrahydropterin synthase (PTPS). The secondary structural elements comprising the C-terminal domain of GTP-CH-I (gray) and the entire single-domain subunit of PTPS (blue) superimpose very well. Insertions and deletions of loop regions confer protein-specific zinc binding site residues and catalytic apparati.

in vivo studies using ^{14}C-labeled precursors. Subsequent studies by Brown and Burg[6] and by Shiota *et al.*[7] showed that the first committed step in the biosynthesis of tetrahydrofolate and BH$_4$ is catalyzed by the enzyme GTP-CH-I.

A hypothetical mechanism for GTP cyclohydrolase I was suggested by the same authors and implicates the hydrolytic opening of the imidazole ring of GTP, an Amadori rearrangement of the ribose moiety and pyrazin ring closure. In this mechanism, C-8 of GTP is released as formate, carbon atoms C-1′ and C-2′ of the ribose moiety are utilized for the formation of the dihydropyrazine ring, and carbon atoms C-3′–C-5′ of the ribose moiety afford the pterin ring atom 6 side chain of dihydroneopterin triphosphate (Scheme 1).

The reaction formally requires the removal of hydrogen from C-2′ of GTP followed by the reintroduction of hydrogen, which becomes ultimately located at C-7 of the product.

Experimental evidence for an Amadori rearrangement preceding the formation of the dihydropterin system has been obtained recently by deuterium tracer experiments monitored by NMR spectroscopy, which showed that a hydrogen atom is incorporated into the pro-7R position of dihydroneopterin triphosphate from solvent water.[32]

Role of the zinc ion in catalysis

The hydrolytic opening of the imidazole ring of the nucleobase (Scheme 2) requires an activated nucleophile attacking C-8 of the heterocycle. Two consecutive nucleophilic attacks of water result in the release of C-8 as formate.

The cocrystal structures of GTP-CH-I with dGTP and of inactive mutants of the enzyme with GTP have shown that this reaction takes place in the immediate vicinity of the metal ion binding site. Therefore, it is assumed that the zinc ion functions primarily as a Lewis acid that activates water molecules for hydrolysis of the GTP imidazole and for hydrolysis of the formyl group of intermediate 2. During catalysis, water, the formyl containing intermediate, and formyl itself may coordinate to zinc, necessitating a transient increase in the zinc coordination number from four to five, much in analogy to the proposed mechanism of zinc proteinases.[57–59] Residue HisA112 is bound to the zinc-coordinating water molecule in the resting state of the enzyme and appears to be required for its proper positioning for nucleophilic attack. Protonation events accompanying the imidazole hydrolysis probably involve residues HisA112 and HisA179, which are placed favorably.

The following reaction steps comprise the Amadori-type isomerization of the carbohydrate moiety and the pterin ring closure condensation step (Scheme 3). These reactions require acid–base chemistry only and may take place independently of the zinc ion. Again, residue HisA112 may be implicated for protonation of the ribose ring oxygen atom facilitating the transient formation of the Schiff base intermediate 6. SerA′135 may be implicated in the proton abstraction from C-2′. It

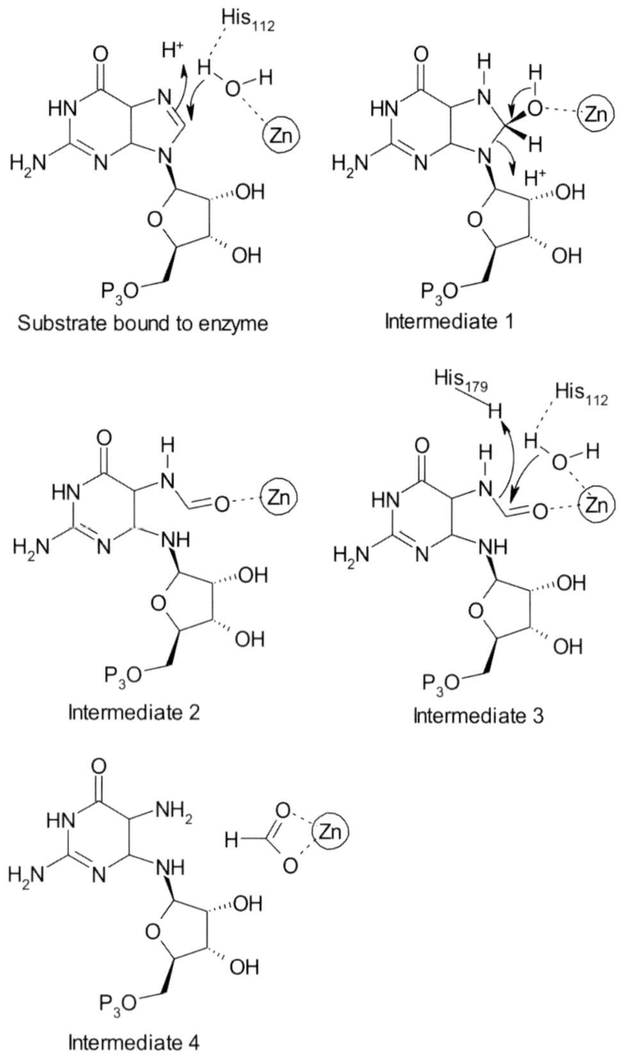

GTP cyclohydrolase I

Scheme 2 Putative reaction mechanism of the purine ring hydrolysis reaction catalyzed by the active site GTP-CH-I and involving the zinc ion. GTP is hydrolyzed in two steps via three intermediates (1–3) resulting in the formation of intermediate 4, a diaminopyridine derivative. Hydrolysis is initiated by the nucleophilic attack of a zinc-bound water molecule on C-8 of the purine ring. The formyl derivative (intermediate 2) has been shown to be the product of HisA179 mutants of GTP-CH-I.[33] A second nucleophilic attack on this intermediate and amide bond breakage leads to the release of formate.

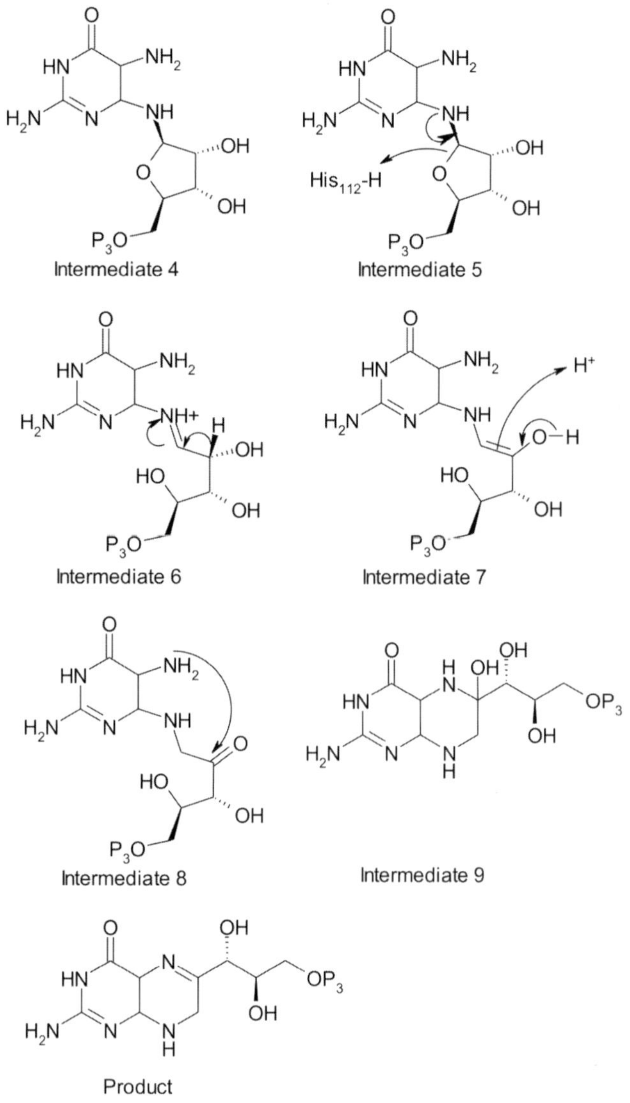

Scheme 3 Putative reaction mechanism of the carbohydrate isomerization and product formation reaction steps. Protonation of the ring oxygen of the ribose leads to the Schiff base intermediate 6. Proton abstraction from C-2' initiates the isomerization of the ribose leading to the C-2' carbonyl intermediate 8, which then probably spontaneously forms the pterin ring system.

has been suggested that the basicity of this residue is increased by the interaction with the γ-phosphate of GTP.[34]

Kinetic studies have recently shown that the velocity of the ring-opening reaction exceeds that of product formation by roughly one order of magnitude,[3,60,61] the hydrolysis of the formamide bond being in turn faster than the ring opening. Therefore, the hydrolytic reactions involving zinc catalysis are rapid compared to the isomerization of the carbohydrate moiety.

REFERENCES

1   W Meining, A Bacher, L Bachmann, C Schmid, S Weinkauf, R Huber and H Nar, *J Mol Biol*, **253**, 208–18 (1995).
2   H Nar, R Huber, W Meining, C Schmid, S Weinkauf and A Bacher, *Structure*, **3**, 459–66 (1995).
3   N Schramek, A Bracher and A Bacher, *J Biol Chem*, **276**, 2622–26 (2001).
4   JJ Yim and GM Brown, *J Biol Chem*, **251**, 5087–94 (1976).
5   B Thony, G Auerbach and N Blau, *Biochem J*, **347**(Pt 1), 1–16 (2000).
6   AW Burg and GM Brown, *J Biol Chem*, **243**, 2349–58 (1968).
7   T Shiota, MP Palumbo and L Tsai, *J Biol Chem*, **242**, 1961–69 (1967).

8   MA Marletta, *Adv Exp Med Biol*, **338**, 281–84 (1993).

9   CR Scriver, RC Eisensmith, SL Woo and S Kaufman, *Annu Rev Genet*, **28**, 141–65 (1994).

10 JT Keltjens and GD Vogels, *Biofactors*, **1**, 95–103 (1988).

11 C Kisker, H Schindelin and DC Rees, *Annu Rev Biochem*, **66**, 233–67 (1997).

12 CA Nichol, GK Smith and DS Duch, *Annu Rev Biochem*, **54**, 729–64 (1985).

13 S Kaufman, S Berlow, GK Summer, S Milstien, JD Schulman, S Orloff, S Spielberg and S Pueschel, *N Engl J Med*, **299**, 673–79 (1978).

14 N Blau and JL Dhondt, *Adv Exp Med Biol*, **338**, 255–61 (1993).

15 A Niederwieser, N Blau, M Wang, P Joller, M Atares and J Cardesa-Garcia, *Eur J Pediatr*, **141**, 208–14 (1984).

16 CR Scriver, CL Clow, P Kaplan and A Niederwieser, *Hum Genet*, **77**, 168–71 (1987).

17 J Rebelo, G Auerbach, G Bader, A Bracher, H Nar, C Hosl, N Schramek, J Kaiser, A Bacher, R Huber and M Fischer, *J Mol Biol*, **326**, 503–16 (2003).

18 G Schoedon, U Redweik and HC Curtius, *Eur J Biochem*, **178**, 627–34 (1989).

19 EP Weisberg and JM O'Donnell, *J Biol Chem*, **261**, 1453–58 (1986).

20 G Katzenmeier, C Schmid and A Bacher, *FEMS Microbiol Lett*, **54**, 231–34 (1990).

21 G Katzenmeier, C Schmid, J Kellermann, F Lottspeich and A Bacher, *Biol Chem Hoppe Seyler*, **372**, 991–97 (1991).

22 K Hatakeyama, T Harada, S Suzuki, Y Watanabe and H Kagamiyama, *J Biol Chem*, **264**, 21660–64 (1989).

23 K Hatakeyama, Y Inoue, T Harada and H Kagamiyama, *J Biol Chem*, **266**, 765–69 (1991).

24 N Blau and A Niederwieser, *J Clin Chem Clin Biochem*, **23**, 169–76 (1985).

25 A De Saizieu, P Vankan and AP van Loon, *Biochem J*, **306**(Pt 2), 371–77 (1995).

26 A Togari, H Ichinose, S Matsumoto, K Fujita and T Nagatsu, *Biochem Biophys Res Commun*, **187**, 359–65 (1992).

27 JR McLean, S Krishnakumar and JM O'Donnell, *J Biol Chem*, **268**, 27191–97 (1993).

28 RD Fleischmann, MD Adams, O White, RA Clayton, EF Kirkness, AR Kerlavage, CJ Bult, JF Tomb, BA Dougherty and JM Merrick, *Science*, **269**, 496–512 (1995).

29 Y Kawarabayasi, Y Hino, H Horikawa, S Yamazaki, Y Haikawa, K Jin-no, M Takahashi, M Sekine, S Baba, A Ankai, H Kosugi, A Hosoyama, S Fukui, Y Nagai, K Nishijima, H Nakazawa, M Takamiya, S Masuda, T Funahashi, T Tanaka, Y Kudoh, J Yamazaki, N Kushida, A Oguchi and H Kikuchi, *DNA Res*, **6**, 83–101, 145–52 (1999).

30 C Schmid, R Ladenstein, H Luecke, R Huber and A Bacher, *J Mol Biol*, **226**, 1279–81 (1992).

31 C Schmid, W Meining, S Weinkauf, L Bachmann, H Ritz, S Eberhardt, W Gimbel, T Werner, HW Lahm and H Nar, *Adv Exp Med Biol*, **338**, 157–62 (1993).

32 A Bracher, W Eisenreich, N Schramek, H Ritz, E Gotze, A Herrmann, M Gutlich and A Bacher, *J Biol Chem*, **273**, 28132–41 (1998).

33 A Bracher, M Fischer, W Eisenreich, H Ritz, N Schramek, P Boyle, P Gentili, R Huber, H Nar, G Auerbach and A Bacher, *J Biol Chem*, **274**, 16727–35 (1999).

34 H Nar, R Huber, G Auerbach, M Fischer, C Hosl, H Ritz, A Bracher, W Meining, S Eberhardt and A Bacher, *Proc Natl Acad Sci USA*, **92**, 12120–25 (1995).

35 G Schoedon, U Redweik, G Frank, RG Cotton and N Blau, *Eur J Biochem*, **210**, 561–68 (1992).

36 S Milstien, S Kaufman and N Sakai, *J Inherit Metab Dis*, **16**, 975–81 (1993).

37 C Lapize, C Pluss, ER Werner, A Huwiler and J Pfeilschifter, *Biochem Biophys Res Commun*, **251**, 802–5 (1998).

38 C Hesslinger, E Kremmer, L Hultner, M Ueffing and I Ziegler, *J Biol Chem*, **273**, 21616–22 (1998).

39 A Togari, M Arai, M Mogi, A Kondo and T Nagatsu, *FEBS Lett*, **428**, 212–16 (1998).

40 M Gutlich, K Schott, T Werner, A Bacher and I Ziegler, *Biochim Biophys Acta*, **1171**, 133–40 (1992).

41 T Harada, H Kagamiyama and K Hatakeyama, *Science*, **260**, 1507–10 (1993).

42 S Milstien, H Jaffe, D Kowlessur and TI Bonner, *J Biol Chem*, **271**, 19743–51 (1996).

43 G Bader, S Schiffmann, A Herrmann, M Fischer, M Gutlich, G Auerbach, T Ploom, A Bacher, R Huber and T Lemm, *J Mol Biol*, **312**, 1051–57 (2001).

44 N Maita, K Okada, S Hirotsu, K Hatakeyama and T Hakoshima, *Acta Crystallogr, Sect D Biol Crystallogr*, **57**, 1153–56 (2001).

45 N Maita, K Okada, K Hatakeyama and T Hakoshima, *Proc Natl Acad Sci USA*, **99**, 1212–17 (2002).

46 H Nar, R Huber, W Meining, A Bracher, M Fischer, C Hosl, H Ritz, C Schmid, S Weinkauf and A Bacher, *Biochem Soc Trans*, **24**, 37S (1996).

47 G Auerbach, A Herrmann, A Bracher, G Bader, M Gutlich, M Fischer, M Neukamm, M Garrido-Franco, J Richardson, H Nar, R Huber and A Bacher, *Proc Natl Acad Sci USA*, **97**, 13567–72 (2000).

48 L Betts, S Xiang, SA Short, R Wolfenden and CW Carter Jr, *J Mol Biol*, **235**, 635–56 (1994).

49 S Xiang, SA Short, R Wolfenden and CW Carter Jr, *Biochemistry*, **34**, 4516–23 (1995).

50 S Xiang, SA Short, R Wolfenden and CW Carter Jr, *Biochemistry*, **35**, 1335–41 (1996).

51 H Nar, R Huber, CW Heizmann, B Thony and D Burgisser, *EMBO J*, **13**, 1255–62 (1994).

52 T Ploom, C Haussmann, P Hof, S Steinbacher, A Bacher, J Richardson and R Huber, *Structure Fold Des*, **7**, 509–16 (1999).

53 M Hennig, A D'Arcy, IC Hampele, MG Page, C Oefner and GE Dale, *Nat Struct Biol*, **5**, 357–62 (1998).

54 N Colloc'h, M el Hajji, B Bachet, G L'Hermite, M Schiltz, T Prange, B Castro and JP Mornon, *Nat Struct Biol*, **4**, 947–52 (1997).

55 N Colloc'h, A Poupon and JP Mornon, *Proteins*, **39**, 142–54 (2000).

56 F Weygand, H Simon, G Dahms, M Waldschmidt, HJ Schliep and H Wacker, *Angew Chem*, **73**, 402–7 (1961).

57 DG Hangauer, AF Monzingo and BW Matthews, *Biochemistry*, **23**, 5730–41 (1984).

58 DW Christianson, PR David and WN Lipscomb, *Proc Natl Acad Sci USA*, **84**, 1512–15 (1987).

59 HM Holden and BW Matthews, *J Biol Chem*, **263**, 3256–60 (1988).

60 A Bracher, N Schramek and A Bacher, *Biochemistry*, **40**, 7896–902 (2001).

61 N Schramek, A Bracher, M Fischer, G Auerbach, H Nar, R Huber and A Bacher, *J Mol Biol*, **316**, 829–37 (2002).

Lyases

Carbonic anhydrases (α-class)

David M Duda and Robert McKenna
Department of Biochemistry and Molecular Biology, University of Florida, Gainesville, FL, USA

FUNCTIONAL CLASS

Enzyme; Carbonic anhydrase (CA): carbonate hydro-lyase, EC 4.2.1.1.; zinc metalloenzyme. CA catalyzes the interconversion of carbon dioxide to bicarbonate. The α-class of CA (and the CA domains in more complex isoforms) is monomeric with a molecular weight of approximately 30 kDa. At present, 14 isoforms of the α-CA are known.[1,2] The activity of CO_2 hydration ranges from the maximal rate of $10^6 \, s^{-1}$ for most efficient CA isozyme II to $10^3 \, s^{-1}$ for isozyme III. Several other isoforms (VIII, X, and XI) appear to entirely lack CO_2-hydration activity, most probably because of amino acid substitutions in one or more of the three histidine residues required to bind the active-site zinc ion.

OCCURRENCE

There are three broad classes α, β, and γ of CA, divided into three genetically unrelated families, namely, animal, plant, and bacterial CAs, respectively. There is no amino acid homology between classes,[3] although there is some overlap of the occurrence of these genes. At present, it is believed that only the α-genes are present in vertebrate organisms. The tissue distribution of α-CA seems to be ubiquitous with cytoplasmic, transmembrane, mitochondrial, and secreted forms.

The mammalian α-class of CAs is composed of 14 known isozymes (CA I–XIV) with varying tissue distributions and catalytic activity.[1,2] CA I is the major, soluble,

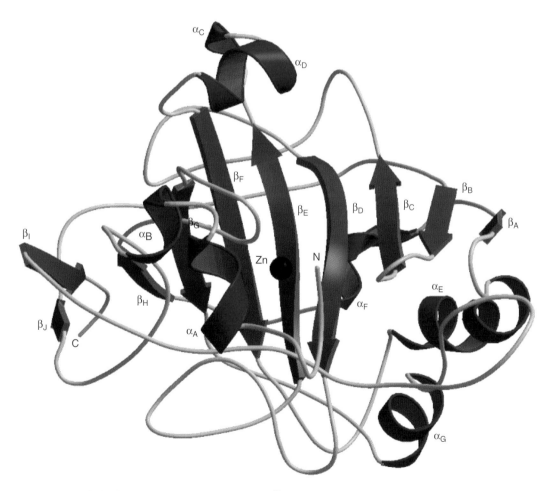

3D Structure Tertiary fold of hCA II (PDB accession 1MOO).[89] Secondary structure is colored as follows: α-helices (blue), β-strands (red), and coil (gray). The zinc ion is shown as a black sphere. Figure created using Bobscript and Raster3D.[94,95]

nonhemoglobin (12 mg CA I/gram of hemoglobin) protein in erythrocytes. In addition to erythrocytes, CA I is expressed in the epithelium of the large intestine, adipose tissue, sweat glands, and corneal epithelium.[4,5] The physiological function of CA I is unclear but it has been proposed that it serves as a backup system for CA II.[6]

CA II is the most extensively studied isozyme of CA and has a wide tissue distribution and can be found in some cells of virtually every tissue type with large amounts located in red blood cells where it is involved in respiration.

CA III is primarily found in red skeletal muscle cells where it is the major soluble protein.[6] CA III is distinguished from other α-class isozymes by its lower catalytic activity[7] and its resistance to most sulfonamide inhibitors.[8] The function of CA III is unclear but it may play a role in facilitated diffusion of CO_2 to the tissue capillaries.[9] It has also been shown that adipocytes have a high concentration of CA III,[10] although its function is not well understood. CA III is also expressed at lower levels in other tissues, including salivary glands, smooth muscle cells in the uterus, red cells, prostate, lung, colon, kidney, and testis.[5]

CA IV is membrane-bound via a glycosyl phosphatidylinositol group linked to residue Ser284 on the C-termius of the enzyme. Although CA IV was originally purified from Bovine lung,[11,12] it has also been found in epithelial cells in the nephron as well as the plasma face of endothelial cells of capillary beds, the plasma membrane of the lower gastrointestinal tract, and brush border membranes of rat kidney.[4,13]

CA V is found predominately in mammalian liver mitochondria. It has been proposed that CA V is responsible for providing bicarbonate for use in the gluconeogenesis and ureagenesis pathways.[14]

CA VI is the only known isozyme of this family to be secreted[15–17] and has been detected in saliva.[18] The proposed function of this isozyme is in the protection of teeth[19] and the neutralization of excess acid in the esophageal and gastric epithelium.[20]

The gene for human CA VII was initially identified by screening a genomic library using a mouse CA II cDNA clone as a probe and is thought to be cytosolic.[21]

CA VIII is a CA-related protein (CA-RP) and was originally isolated as a mouse brain cDNA and has no CA activity.[22] This protein is expressed in Purkinje cells and the human form is 98% identical to the mouse form, although no function has yet been assigned.[23]

CA IX is an integral membrane protein consisting of an N-terminal signal peptide, an extracellular domain, a transmembrane spanning region, and an intracellular C-terminal domain. The carbonic anhydrase domain is extracellular and consists of 257 amino acids, which bear a 36% sequence identity to hCA II.[24] CA IX shows an increased expression in some colorectal tumors and may be involved in pathogenesis of the disease.[25]

CA isozymes X and XI are both CA-RPs that show specific expression of mRNAs in the adult human brain.[26–28]

CA XII is a transmembrane glycoprotein whose extracellular domain retains CA activity. It has been found that CA XII mRNAs are overexpressed in renal cell cancers[29] and in a lung cancer cell line.[30] Fujikawa–Adachi isolated the first full-length cDNA of CA XIV.[31] Northern blot analysis has shown that CA XIV mRNA is expressed in adult human heart, brain, liver, and skeletal muscle.[27]

BIOLOGICAL FUNCTION

CA II with its wide tissue distribution has varied physiological roles throughout the body. In addition to catalyzing the reversible hydration of CO_2 and HCO_3^- for respiration, CA II also acidifies urine by eliminating H^+ in the renal tubules and collecting ducts of the kidney, provides H^+ necessary for the bone resorption function of osteoclasts, produces HCO_3^- for use in pyrimidine biosynthesis, supplies H^+ and maintains pH balance in the choroid plexus for cerebro spinal fluid (CSF) formation, involved in saliva production by producing HCO_3^- for acinar and ductal cells, provides H^+ to gastral parietal cells for stomach acidification, and provides HCO_3^- to liver epithelial duct cells for bile production and to epithelial duct cells of the pancreas for pancreatic juice formation. A specific clinical phenotype has been noted for CA II deficiency – osteopetrosis and renal tubular acidosis, which in some cases is accompanied by mental retardation.[4]

Koester[32] determined that CA III exhibits tyrosine phosphatase second messenger activity that depends on glutathiolation of a surface cysteine residue (Cys186).[33] It has been suggested that the redox state of the cell may control CA III indirectly through the glutathione level. It has also been suggested that the two free thiol groups on the surface of CA III (Cys181 and 186) may be involved in scavenging oxygen radicals in skeletal muscle.[34] A recent study by Raisanen[35] indicated that cells overexpressing CA III are significantly more resistant to H_2O_2-induced apoptosis.

CATALYTIC MECHANISM

CA catalyzes the interconversion of CO_2 to HCO_3^- in a two-stage ping-pong reaction. Human CA II (hCA II) catalyzes the reversible hydration of CO_2 in two distinct half reactions.[6,36] The first step of the reaction involves the trapping of the CO_2 substrate within a putative hydrophobic pocket.[37] The CO_2 displaces a water molecule, 'the deep water', in the active site by associating with the amide nitrogen of Thr199 in a hydrogen-bonding interaction prior to a nucleophilic attack on the substrate carbon to form bicarbonate (Figure 1). The bicarbonate is

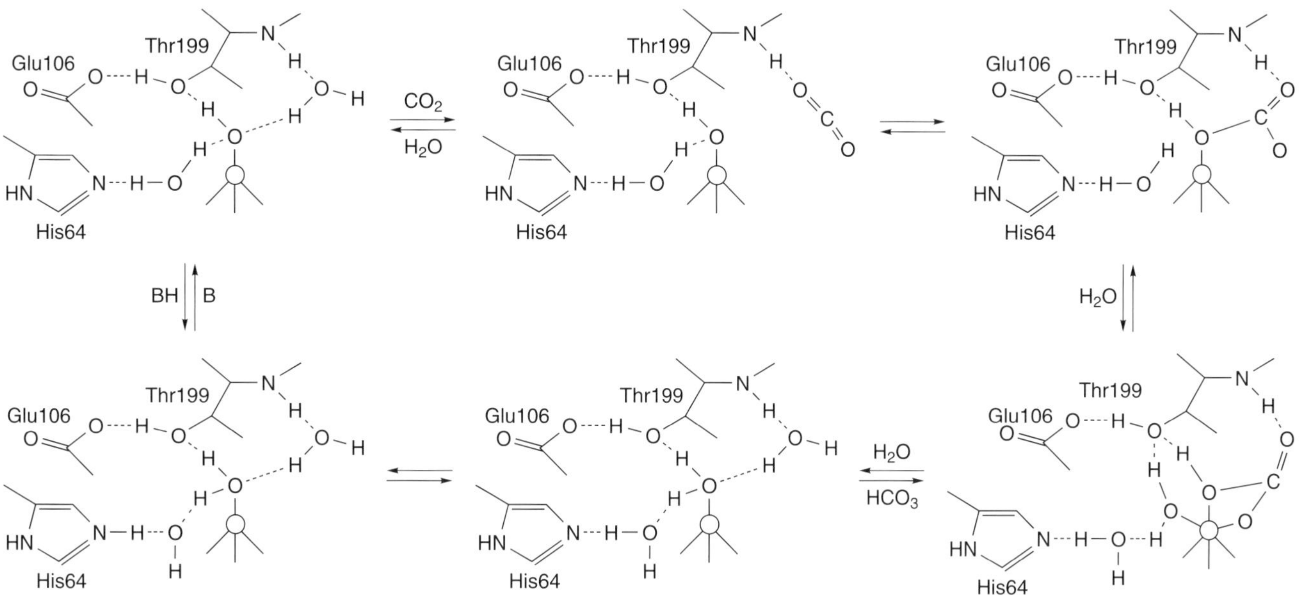

Figure 1 Schematic representation of the catalytic mechanism of carbonic anhydrase.

then displaced from the zinc ion by an active-site water molecule, concluding the first-half reaction (Equation (1)).

$$CO_2 + EZnOH^- \rightleftharpoons EZnHCO_3^- \xrightarrow{H_2O} EZnH_2O + HCO_3^- \quad (1)$$

$$EZnH_2O + B \rightleftharpoons EZnOH^- + BH^+ \quad (2)$$

The second-half reaction involves the transfer of a proton from the zinc-bound water molecule to residue His64 through a chain of hydrogen-bonded water molecules.[36,38] This intramolecular proton transfer is followed by an intermolecular proton transfer from His64 to bulk solvent (B) of the system. This second step regenerates the zinc-bound hydroxyl group, allowing for another round of catalysis to proceed (Equation (2), Figure 1).

AMINO ACID SEQUENCE INFORMATION

There have been at least 14 α-CA isozymes (CA I–XIV) that have been identified.[1,2] The information below gives details of the *homosapien* isoforms, except for CA XIII, which is given for *Mus musculus*.

- CA I (*Homo sapiens*), 261 amino acid residues (AA), based on cDNA.[39]
- CA II (*Homo sapiens*), 260 AA, based on cDNA.[40]
- CA III (*Homo sapiens*), 260 AA, based on cDNA.[41]

Table 1 Amino acid sequence identity between isozymes of the α-class of carbonic anhydrase

	I	II	III	IV	V	VI	VII	VIII	IX	X	XI	XII	XIII	XIV
I	–	60.5	54.4	28.8	43.7	30.9	50.6	37.9	31.3	29.1	27.4	29.8	62.1	32.6
II	158	–	58.9	32.4	49.4	32.5	55.9	40.1	31.4	30.6	31.4	28.4	60.9	36.0
III	142	153	–	30.2	43.3	32.8	52.9	36.0	29.5	27.6	27.0	27.3	59.0	33.0
IV	78	88	82	–	27.9	31.3	33.1	29.6	29.4	24.3	23.4	26.0	30.9	34.2
V	114	129	113	76	–	28.6	47.9	34.9	29	27.3	30.2	24.3	47.1	32.6
VI	83	87	88	85	77	–	35.3	32.5	41	25.8	23.9	35.4	33.5	41.4
VII	132	146	138	90	125	95	–	40.4	34.6	30.2	28.5	31.6	54	36.3
VIII	103	109	98	82	94	89	110	–	31.8	28	29.8	24.8	38.6	29.4
IX	85	85	80	80	79	110	94	88	–	24.7	26.2	31.2	32.7	41.9
X	80	84	76	67	75	72	83	79	68	–	60.6	22.1	28.4	24.6
XI	75	86	74	65	83	66	78	84	72	166	–	21.4	28.1	26.7
XII	79	75	72	70	64	93	84	67	83	61	59	–	30.5	37.8
XIII	162	159	154	84	123	90	141	105	89	78	77	81	–	35.1
XIV	86	82	88	93	87	110	97	80	113	68	74	99	93	–

(Top right) Percent sequence identity between isozymes. (Bottom left) Number of conserved amino acid residues between isozymes.

Carbonic anhydrases (α-class)

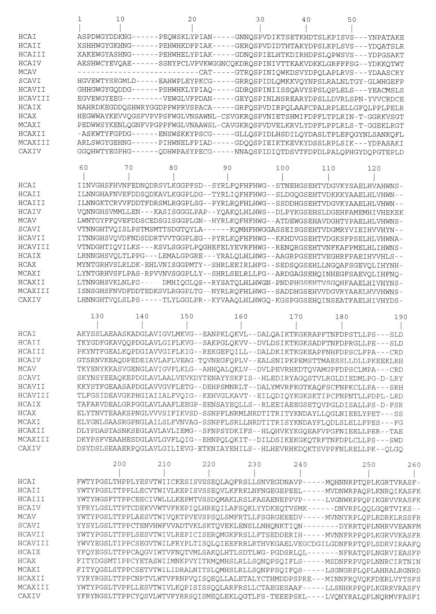

Figure 2 Multiple pairwise sequence alignment of the 14 isozymes of the α-class of CA. Amino acid numbering is according to carbonic anhydrase II. Figure created using CLUSTALW.[52]

- CA IV (*Homo sapiens*), 312 AA, based on cDNA.[42]
- CA V (*Homo sapiens*), 305 AA, based on cDNA.[43]
- CA VI (*Homo sapiens*), 308 AA, based on cDNA.[44]
- CA VII (*Homo sapiens*), 264 AA, based on cDNA.[21]
- CA VIII (*Homo sapiens*), 289 AA, based on cDNA.[23]
- CA IX (*Homo sapiens*), 459 AA, based on cDNA.[45]
- CA X (*Homo sapiens*), 328 AA, based on cDNA.[46]
- CA XI (*Homo sapiens*), 267 AA, based on cDNA.[47]
- CA XII (*Homo sapiens*), 354 AA, based on cDNA.[48]
- CA XIII (*Mus musculus*), 262 AA, based on cDNA.[49]
- CA XIV (*Homo sapiens*), 337 AA, based on cDNA.[31]

Comparative sequence analysis of CA isozymes I to VII from various species have been used to determine the evolution of these isozymes.[5,50,51] From this analysis, it was determined that isozymes IV and VI are the most ancient forms. CA V and VII diverged at an intermediate stage, and the most recent isozymes are CA I, II, and III. Figure 2 shows a pairwise, multiple-sequence alignment of the 14 isozymes of the α-class of CA. Twenty-three or 4.7% of the aligned amino acids show complete conservation, while an additional 32 or 6.6% are strongly similar. The percent identity of isozymes I to XIV have been determined (Table 1).

PROTEIN PRODUCTION, PURIFICATION, AND MOLECULAR CHARACTERIZATION

Discovery of the first enzyme in the CA family occurred over 60 years ago when it was isolated from bovine erythrocytes.[53,54] CA II is purified by affinity chromatography utilizing the CA inhibitor p-aminomethylbenzenesulfonamide (pAMBS).[55] Recombinant CA II has been produced using the pET vector system and has allowed for the expression of large quantities of enzyme in *Escherichia coli* strain BL21(DE3)pLysS.[56]

CA III is resistant to sulfonamide inhibition and therefore cannot be purified by affinity chromatography utilizing pAMBS.[8] CA III with a C-terminal hexahistidine tag has recently been constructed for ease of purification.[57] The purification procedure for CA III involved affinity chromatography using a nickel chelating column and anion exchange chromatography.

The α-class of CAs has a molecular mass of approximately 30 000 Da and exists as monomers. Two cysteine residues (Cys183 and 188) are present on the surface of CA III (Figure 2), which have made crystallographic studies difficult because of solubility problems.[57] A disulfide bridge (Cys23 and Cys203) has been reported in human CA IV that is not present in the other isozymes[58] (Figure 2). It is thought that these disulfide linkages are responsible for the increased stability of CA IV against heat and sodium dodecyl sulfate (SDS).[11] CA IV is also membrane anchored by a phosphatidylinositol linkage to the C-terminal Ser266.

METAL CONTENT AND COFACTOR

CAs have a single zinc ion tetrahedrally coordinated in the active site by three histidine residues (His94, 96, and 119) and a bound water molecule, which serves as the fourth ligand. The zinc ion is very rigidly coordinated and has an estimated dissociation constant of $1.4 \times 10^{-6}\,s^{-1}$ for hCA II corresponding to a half-life of about five days.[59]

Several structures of hCA II with metal substitutions have been reported by Hakansson.[60] Crystals of the apoenzyme can be formed by soaking the crystals in a stabilizing solution in the presence of dipicolinic acid-NaOH for several days.[61,62] The apoenzymes are inactive but the activity can be fully restored with the addition of zinc ions.[6] Cobalt-substituted CA II is the only metal-substituted form of the enzyme that retains activity and binds with a tetrahedral coordination geometry. Mn and Ni substitutions result in octahedral coordination and Cu substitutions result in trigonal bipyramidal coordination[60,63]

ACTIVITY TEST

Several methods exist for monitoring the activity of CA. Initial rates of CO_2 hydration can be measured through stopped-flow spectrophotometry using the pH-indicator method.[64,65] Oxygen-18 exchange kinetics can also be used to measure activity. This method is based on the exchange of ^{18}O between ^{12}C and ^{13}C-containing species of CO_2 and water that occurs because of the hydration–dehydration reaction of CA.[66,67]

Table 2 CO_2 hydration and proton-transfer rates for α-class carbonic anhydrases

Isozyme	k_{cat}/K_M ($M^{-1}\,s^{-1}$)	k_{cat} (s^{-1})	References
I	5.0×10^7	2.0×10^5	64
II	1.5×10^8	1.4×10^6	64
III	3.0×10^5	1.0×10^4	70
IV	5.0×10^7	1.1×10^6	71
V	3.0×10^7	3.0×10^5	72
VI	1.6×10^7	7×10^4	71
VII	7.6×10^7	9.4×10^5	73
VIII	NA	NA	74
IX	5.5×10^7	3.8×10^5	75
X	NA	NA	74
XI	NA	NA	74
XII	3.4×10^7	4.0×10^5	2
XIII	Active[b]	Active[b]	76
XIV	3.9 ± 0.35[a]	[a]	31

[a] The catalytic activity of CA XIV was determined from sonicated COS-7 cells.[31]
[b] Isozyme XIII is presumed to be active on the basis of the translated cDNA sequence.[76]

Enzyme activity can also be determined using the compound p-nitrophenyl acetate. CA has been shown to catalyze the hydrolysis of a wide range of synthetic ester substrates.[68] The ester bond of p-nitrophenyl acetate is hydrolyzed by CA, yielding p-nitrophenolate and acetate. p-nitrophenolate is intensely yellow at 400 nm and can easily be detected by spectrophotometric methods. Utilizing this method to measure the change in absorbance over time will yield the specific activity of the enzyme.[69]

The parameter k_{cat} reflects the reactions involved in intra and intermolecular proton transfer, while the parameter k_{cat}/K_M reflects the steps involved in CO_2/HCO_3^- interconversion. CA II is the most efficient isozyme of the α-class with a k_{cat}/K_M 1.5×10^8 ($M^{-1}\,s^{-1}$) and k_{cat} of 1.4×10^6 (s^{-1}). CA III is the least efficient isozyme with k_{cat}/K_M 3.0×10^5 ($M^{-1}\,s^{-1}$) and k_{cat} of 1.0×10^4 (s^{-1}) (Table 2).

SPECTROSCOPY

Spectroscopy has been very useful in analyzing the interactions of CA with various inhibitors and substrates. The Co^{2+}-substituted form of CA has been shown to retain catalytic activity and shows electronic spectra that are sensitive to the metal ion environment.[7,77,78] This sensitivity has allowed for a spectroscopic investigation of the inhibition/enhancement mechanisms of various CA inhibitors/activators.[79] Briganti[80,81] reported the electronic spectra of Co^{2+}-substituted CA with the catalytic activators histamine and phenylalanine. The spectrum of Co^{2+}-hCA II with histamine was shown to differ from Co^{2+}-hCA II in the absence of histamine. The observed spectrum in the presence of histamine did not resemble the spectra for other known anionic or sulfonamide inhibitors, indicating that histamine most likely is not interacting with the metal ion. The spectra did resemble that seen for Co^{2+}-hCA II with phenol, which is the only reported competitive inhibitor of CO_2.[82] The crystal structure of Co^{2+}-hCA II with phenol indicated that the inhibitor hydrogen bonds with the metal-bound solvent and the amide nitrogen of Thr199.[83] This result indicates that histamine binds to the active site but not directly to the metal ion.

X-RAY STRUCTURE OF NATIVE CA

Crystallization

Preliminary crystallization studies of CA II were performed by Strandberg[84] and later reproduced by Tilander.[85] The initial crystallization method involved the dialysis of 0.5 to 2.0 mL of a 1% protein solution in 50 mM Tris-HCl pH 8.5 against 20 to 100 mL of ammonium sulfate. The optimum ammonium sulfate concentration was determined to be between 1.75 and 2.5 M. These conditions produced plates that have a tendency to aggregate. It was found that the use of methylmercuric acetate (MMA) leads to the formation of parallellepipedon-shaped crystals that grow bigger and show less tendency to aggregate. These conditions, both in the presence and absence of MMA, resulted in crystals belonging to the $P2_1$ space group with unit cell parameters $a = 43.1$, $b = 42.1$, $c = 73.6$ Å, $\beta = 104.6°$ and two molecules per unit cell. The preliminary three-dimensional structure of hCA II was published by Liljas[86] and was later crystallized using the hanging drop method[87] (2.4 M ammonium sulfate, 50 mM Tris-HCl (pH 8.5), and 1 mM $HgCl_2$) and reported at 2.0 Å resolution by Eriksson.[38] The highest resolution structure of wild-type hCA II reported to date is 1.54 Å, published by Hakansson.[88] More recently, a structure of the His64 to Ala variant of hCA II in complex with the activator 4-methylimidazole has been reported to be 1.05 Å.[89] This represents the highest resolution structure of any CA reported thus far.

Overall description of the structure

CA II is a mixed α/β, globular protein (3D Structure), which is nearly spherical and has approximate dimensions of $5 \times 4 \times 4$ nm^3.[6] These enzymes can be considered one-domain proteins with the possible exception of the amino-terminal region (approximately 24 amino acids), which are more loosely connected to the rest of the molecule. An aromatic cluster at the amino terminus of the enzyme consisting of residues Trp5, Tyr7, Trp16, and Phe20 has been suggested to anchor this region to the rest of the enzyme.[6] It has also been shown that the removal of the amino-terminal region does not result in loss of protein stability or enzyme activity.[90] The central structural motif of hCA II can be described as a 10-stranded (βA-βJ) twisted β-sheet, which is flanked by seven α-helices (αA-αG). The twisted β-sheet structure is antiparallel with the exception of two pairs of parallel strands (3D Structure). Several mutants have been identified that affect the stability of CA. Replacement of Ser29 with Cys results in a near total loss of stability.[91] It has been shown that mutations at Gly196 result in insoluble aggregates during expression in *E. coli*.[92] The replacement of Trp209 with either Ser or Gly residues has a major effect on stability.[92,93]

Zinc site geometries

The catalytic active site is characterized by a conical cleft that is approximately 15 Å deep with a zinc ion residing deep in the interior (Figure 3(a)). The zinc ion is located at the bottom of the conical active-site cleft and is tetrahedrally coordinated by three histidine ligands (His94, His96, and His119) and a bound hydroxyl group (Figure 3(b)). The zinc-ligand distances are all 2.1 Å including the zinc-bound water molecule.[88] It has been proposed by Cox et al.[96] that there is a hierarchy of zinc ligands in the active site (that function as distinct shells

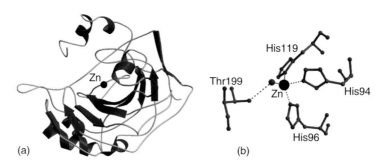

Figure 3 Zinc ion coordination. (a) Looking into the active-site cavity. Secondary structure is colored as follows: α-helices (blue), β-strands (red), and coil (gray). The zinc ion is shown as a black sphere. (b) Close-up view of the active site showing the histidine ligands (His94, 96, and 119) as well as Thr199 and the zinc-associated water molecule. Figure created using Bobscript and Raster3D.[94,95]

of residues to stabilize the zinc ion). The first shell, or direct zinc ligands, are the three histidine residues His94, His96, and His119. The second shell, or indirect ligands, stabilize the direct ligands and help position them for zinc ion coordination. Residue Gln92 stabilizes His94, Glu117 stabilizes His119, and the backbone carbonyl oxygen of Asp244 stabilizes His96, while residue Thr199 hydrogen bonds with the zinc-bound hydroxyl ion (Figure 3(b)). Finally, a third shell of stabilization was proposed of a cluster of aromatic residues (Phe93, Phe95, and Trp97) that anchor the β-strand βF that contains His94 and His96. The second-shell residue Thr199 also plays an important role in catalysis. The zinc-bound hydroxyl ion donates a hydrogen bond to the hydroxyl side chain of Thr199, which in turn donates a hydrogen bond to the carboxyl side chain of Glu106. This interaction with Thr199 serves to orient the zinc-bound hydroxyl ion for optimal nucleophilic attack on the CO_2 molecule. Thr199 also serves to stabilize the transition state of the reaction through a hydrogen bond and serves to destabilize the bicarbonate ion product.[97] Thr199 is said to have a 'gatekeeper function' in the catalytic reaction by selecting only protonated molecules to interact with the zinc ion. The hydrogen bond–accepting ability of Thr199 hydroxyl side chain enables this selection (Figure 1).

Proton transfer

The rate-limiting step in catalysis by carbonic anhydrase is an intramolecular proton transfer (Equation 2) that occurs from the zinc-bound water molecule to a specific residue in the active site capable of accepting protons. This residue has been identified as His64 in CA II.[7,65,98] The three-dimensional structure of hCA II shows His64 located in the active-site cavity with its imidazole ring in the 'in' conformation, approximately 7 Å from the zinc ion.[38] The position of His64 and its distance from the zinc ion suggests that proton transfer proceeds through intervening hydrogen-bonded water bridges (Figure 4). The mechanism of proton translocation through this hydrogen-bonded water bridge, or proton wire, is stepwise, where a proton is transferred from the zinc-bound water molecule to an adjacent hydrogen-bonded water molecule. This water molecule then, in turn, transfers a different proton to another hydrogen-bonded water molecule, which, in turn, transfers a different proton to His64 in a mechanism known as *Grotthus diffusion*.[99,100] Water is a poor acceptor of protons and a maximal turnover of $10^3 \, s^{-1}$ would be expected if the catalytic mechanism had to rely entirely on direct proton transfer to bulk water.[101] An observation that established the function of His64 in the proton-transport pathway was the reduction in catalysis by a site-specific mutant of hCA II in which His64 is replaced by Ala (H64A hCA II), a residue that does not support proton transport. The turnover number for

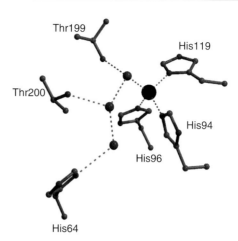

Figure 4 Proton pathway. Hydrogen-bonded water molecule (red spheres) network between the zinc ion (black sphere) and His64 (gray stick) in the active site of hCA II (PDB accession 2CBA).[88] Figure created using Bobscript and Raster3D.[94,95]

catalysis, k_{cat}, of CO_2 hydration by H64A hCA II was decreased approximately 10-fold compared to the wild-type enzyme.[98]

CA III has a lysine residue at position 64 (Figure 2), which has been shown to be pH-independent, indicating that this residue does not effectively participate in proton transfer.[102] Replacement of Lys64 with His64 by mutagenesis results in a 10-fold enhancement of CO_2 hydration and a pH profile indicative of a predominant, active proton shuttle group in the active site.[70]

Previous structural studies have demonstrated that the size of the residue at position 65 is important for efficient proton-shuttling activity of the adjacent His64.[103,104] Amino acid substitutions at position 65 with residues bulkier than a threonine residue were found to disrupt the hydrogen-bonded solvent network and result in a reduction of proton transfer (k_{cat}) up to 30-fold.

The mitochondrial CA V has a tyrosine at position 64. In addition, position 65 is occupied by a bulky phenylalanine side chain. It has been proposed that Phe65 restricts the normally free motion of His64 in Y64H CA V, resulting in only small increases in proton transfer.[105,106] The double mutant of Y64H and F65A for CA V results in a 100-fold increase in proton-transfer capability, which indicates that free motion of the proton shuttle is necessary for efficient catalysis in CA V.

Mercury binding site

Crystallization of hCA II in the presence of organomercury compounds has been shown to enhance crystal quality.[85] Previous crystallographic analysis of H64A hCA II has revealed that a mercury ion binds to the side chain thiol group of Cys206 (Figure 5).[107] This interaction caused the

Carbonic anhydrases (α-class)

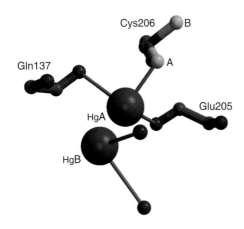

Figure 5 Binding site of mercury on the surface of hCA II (PDB accession 1MOO).[89] Mercury ion is shown as a gray sphere. Alternate conformational positions of both the mercury ion and the side chain of cysteine 206 are indicated as 'A' and 'B'. Figure created using Bobscript and Raster3D.[94,95]

Cys206 (χ_1) in h64A HCA II to change conformation from that of the wild-type isozyme structure by 86°,[88,107] enabling the Sγ sulfur of Cys206 and a water molecule to ligand the Hg^{2+} ion. Other electrostatic interactions with the mercury ion include that with the carbonyl oxygen of Gln137, Glu205, and a second water molecule.

A more recent crystallographic study indicated that the mercury ion was occupying two spatial positions with respect to the Sγ sulfur of Cys206, which also was clearly seen with two distinct positions.[89] The mercury ion had approximate occupancies of three-fourth and one-fourth for positions A and B respectively. The significantly higher occupancy values seen for position A indicate a tendency toward the bound state between the mercury ion and Cys206 (Figure 5).

FUNCTIONAL ASPECTS

Steady state kinetics

Kinetic analysis of the catalyzed reaction of CA in the steady state and at chemical equilibrium has led to a model of its mechanism involving two ionizing groups in the active site with pK_a values near 7.[7,65] One of these ionizable groups corresponds to the zinc-bound water molecule that ionizes to a hydroxyl ion. This group is involved in the interconversion of CO_2 and HCO_3^-. The other ionizable group is involved in intramolecular proton transfer and has been identified as His64.[98] The most catalytically efficient form of CA, isozyme II, shows k_{cat}/K_M values of 120 μM^{-1} s^{-1} and k_{cat} values of 1000 m s^{-1}.[65] Intramolecular proton transfer is defined as the transfer of a proton from the zinc-bound water molecule to His64, resulting in the ionization of the water molecule regenerating the catalytically active hydroxyl group. It has been shown that intramolecular proton transfer is rate-limiting at high buffer concentrations. At low buffer concentrations, intermolecular proton transfer, the transfer of protons from His64 to buffer in solution, is rate-limiting.[7]

FUNCTIONAL DERIVATIVES

General remarks

It has been proposed that the catalytic substrate CO_2 interacts with a hydrophobic pocket that is approximately 3 to 4 Å from the zinc ion. The dissociation constant for this interaction has been estimated to be 100 mM.[108] Molecular dynamic studies have indicated that the hydrophobic pocket is composed of Val121, Val143, Leu198, Val207, and Trp209 (Figure 2).[109,110] It was found that CO_2-hydration activity decreases with the size of the residue at position 143.[111] Replacements of Val143 with Phe and Tyr residues almost completely block the hydrophobic pocket.[112] Lindahl[113] reported the crystal structure of the cyanate ion (NCO^-) in complex with CA. NCO^- is isoelectronic with CO_2 and was found to be near the hydrophobic pocket and able to hydrogen bond the amide group of Thr199. Carbon dioxide–hydration activity does not decrease in a similar way to NCO^- affinity as the size of residue 143 is increased.[111] Merz[114] proposed that a hydrogen-bond network exists between the OH^- of Thr199, Glu106, and the zinc-bound OH^- that orients a lone pair of electrons of the zinc-bound OH^- group to react with CO_2 (Figure 1). Disruption of this network by replacing Thr199 with Ala causes a 100-fold decrease in k_{cat}/K_M for CO_2 hydration.[115,116] Another effect of this hydrogen-bonding system is that the OH^- group of Thr199 must accept a hydrogen bond from the zinc-bound solvent. Because of this interaction, Thr199 is known as the 'gatekeeper' and requires that protonated molecules interact with the zinc ion. CA inhibitors having a protonated ligand atom such as HSO_3^- and HS^- bind the zinc ion, replacing the zinc-bound solvent group.[117,118]

Carbonic anhydrase inhibitors

Inorganic and organic anions have been useful in studying the properties of the metal center in carbonic anhydrase. Most monovalent anions inhibit CA with varying affinities. These inhibitors bind to the metal ion and disrupt the coordination of the zinc-OH^- group that disrupts the catalytic activity of the enzyme.[119] Spectrophotometric analysis of Co^{2+}-substituted CA suggests that some anions have the ability to replace the zinc-OH^- group, while others displace this group forming a pentacoordinated metal-binding sphere.[77] The main determining factor for whether the OH^- group is replaced or displaced is the gatekeeper function of Thr199 formed through the active-site hydrogen-bond network.[110,120] Because of this hydrogen bond accepting function of Thr199, anionic inhibitors possessing a protonated group can bind directly to the zinc ion substituting for the OH^- group. This class of inhibitors can maintain a hydrogen bond with Thr199 without distortion of the tetrahedral metal site geometry.

Fully ionized anionic ligands bind the metal ion and displace the zinc-bound OH^- group that maintains its hydrogen bond to Thr199. SCN^-,[38] formate, and acetate ions[88,121] are examples of this class of anionic inhibitor.

Another group of anionic inhibitors that replaces the zinc-bound OH^- group but does not interact with Thr199 has also been identified. Among this class of inhibitors is azide and the halogen ions Br^- and I^-.[122,123] Azide is a potent inhibitor of CA and binds directly to the zinc ion and forms a hydrogen bond with the amide nitrogen of Thr199.[124]

Aromatic and heterocyclic sulfonamides of the form R-SO_2NH_2 or R-$SO_2NH(OH)$ are selective and powerful CA inhibitors, which bind to the zinc ion as anions through the sulfonamide group nitrogen atom[125] (Figure 6). Several crystallographic studies of CA-sulfonamide complexes have been performed, all showing similar interactions; the ionized sulfonamide group NH binds to the metal ion and satisfies the gate-keeper function of Thr199 by donating a hydrogen bond.[6] Another of the sulfonamide oxygens interacts with the backbone nitrogen of Thr199, displacing the deep water. The most common sulfonamide inhibitor in clinical use is acetazolamide (DIAMOX), which shows strong inhibition of CA II with a Ki value of $0.01\,\mu M$.[126] This inhibitor has been used for a wide range of ailments including congestive heart failure, for its diuretic properties, and altitude sickness.[127,128] Structural analysis of acetazolamide in complex with CA revealed the binding interactions of this compound.[129] The thiadiazole ring is in van der Waals contact with Val121, Leu198, and Thr200, and the carbonyl oxygen of the amido group shares a hydrogen-bond interaction with the side chain amide of Gln92. The methyl group was shown to interact with the side chain of Phe131.[129] Another clinically relevant sulfonamide inhibitor is dorzolamide (TRUSOPT), which is applied topically to the eye in the treatment of glaucoma. Dorzolamide reduces the production rate of aqueous humor, which is involved in the increased pressure on the eye in glaucoma, by disrupting the ability of CA to produce HCO_3^- ions in the anterior segments of the eye.

Metal substitutions

An extensive crystallographic analysis of wild-type hCA II with various metal substitutions for the zinc ion was performed by Hakansson.[60] Cobalt-substituted CA retains approximately 50% of the CO_2-hydration activity seen for the wild-type enzyme. The activities are negligible for cupric, ferrous, nickel, and manganese substitutions

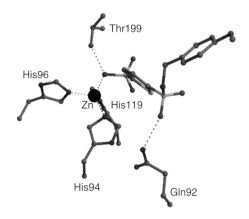

Figure 6 Inhibitory interaction of N-[(4-methoxyphenyl)methyl]-2,5-thiophenedisulfonamide with hCA II (PDB accession 1BN4).[130] Figure created using Bobscript and Raster3D.[94,95]

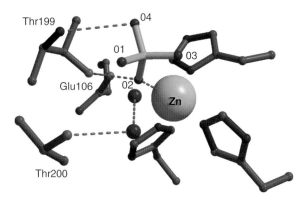

Figure 7 The binding of a sulfate ion to hCA II at pH 5.0 (unpublished result). Figure created using Bobscript and Raster3D.[94,95]

even though the pK, charge, and size are similar.[131–133] The structure of the cobalt-substituted enzyme at pH 7.8 retains tetrahedral coordination of the zinc ion. At pH 6.0, however, a SO_4^{2-} ion binds directly to the metal ion and displaces the metal-bound OH^- group (Figure 7). This SO_4^{2-} ion also displaces the deep water and hydrogen bonds to the backbone amide of Thr199. The copper-substituted form of the enzyme assumes a pentacoordinated geometry for the metal ion with the additional ligand coming from the diatomic oxygen that has displaced the deep-water molecule.[60] In this coordination, all of the ligands are in a plane with His94, occupying an apical position. As in the low pH cobalt-substituted enzyme, a nickel substitution has a SO_4^{2-} ion bound to the metal ion. The SO_4^{2-} ion is located more toward the hydrophilic side of the active site but still retains a hydrogen bond with the backbone N of Thr199. The metal coordination for this substitution is described as hexacoordinated (all ligand angles are close to 90 or 180°) with an additional water molecule and SO_4^{2-} oxygen interacting with the nickel ion.[60] The manganese-substituted enzyme also contains a metal-bound SO_4^{2-} ion, and the coordination is described as octahedral.[60] Two SO_4^{2-} oxygens and a water molecule provide the additional coordination of the manganese ion.

pH effects

A crystallographic study of hCA II at pH 5.7, 6.5, and 8.5 revealed structural stability even at low pH.[134] An overall rms coordinate error of 0.2 Å was estimated between the three structures, indicating no major pH-dependent variations. The active-site hydrogen-bonded solvent network and residues also exhibited no major pH-dependent conformational changes with the exception of the proton-transfer residue His64. It was noted that as the pH decreased, the position of His64 increasingly occupied a position oriented away from the zinc ion in the active site. This alternate conformation was termed the *out conformation* and it predominated at pH 5.7. The positions of the two pH-dependent conformations ('in' and 'out') of His64 are related by a 64° rotation about $\chi 1$ (Figure 8).[134]

The experimentally determined pK_a of His64 is 7.1[135] and therefore the imidazole side chain of this residue must be fully charged at pH 5.7. This conformational shift toward the 'out' conformation could be a result of electrostatic repulsion between the positively charged imidazole group and the positive charge of the zinc-coordination polyhedron.[134]

The site-specific mutant T200H of hCA II at pH 8.0 induces a similar, but more extreme, conformational shift of His64 away from the active site.[136] In this mutant, His64 rotates 105° away from the 'in' conformation. This mutant exhibits normal CO_2 hydration and proton-transfer kinetics.[136] This result indicates that the kinetic activity of CA is not dependent on the specific position of the proton-transfer group.

Carbonic anhydrase activators

Structural identification of the binding site for a CA activator came in 1997 with the binary complex of

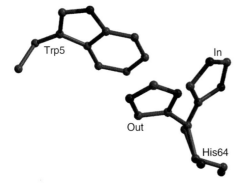

Figure 8 Alternate conformational states of His64 in hCA II (PDB accession 2CBA).[88] Histidine 64 is shown with the 'in' and 'out' conformations, gray sticks). Figure created using Bobscript and Raster3D.[94,95]

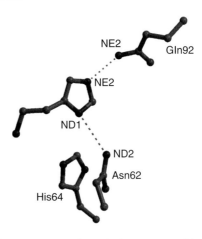

Figure 9 Binary complex of the activator histamine with carbonic anhydrase (PDB accession 1AVN).[80] Figure created using Bobscript and Raster3D.[94,95]

histamine and hCA II.[80] In this structure, the histamine molecule was found to bind at the entrance of the active-site cavity (Figure 9). The ND1 and NE2 atoms of the imidazole ring of histamine were found to hydrogen bond with the side chains of Asn62 and Gln92 as well as an additional interaction with a water molecule. The terminal aliphatic amino group of histamine was not engaged in any interactions and extended away from the cavity and into the solvent. The electron density for the imidazole ring had some rotational disorder, but clear alternate positions could not be identified and the estimated occupancy for the ring was about one-third.[80] The usual conformational disorder seen for the side chain of His64[134,137] was not evident in the histamine complex structure, as the side chain of His64 was oriented toward the active site. A hydrogen-bonded solvent bridge, consisting of two water molecules, was also evident, linking the zinc-bound water to the imidazole moiety of histamine. A similar bridge, corresponding to the wild-type solvent bridge, was observed between the zinc-bound water and His64.[88] The existence of two potential proton-transfer pathways in the active site of the enzyme gives an explanation of the enhancement of catalytic activity observed in the presence of histamine.[80]

Another structure of hCA II in complex with an amino acid activator phenylalanine shared a similar binding site.[81] The phenylalanine bound at the mouth of the active-site cavity with residue His64 found to be well ordered and oriented into the active-site cavity. A hydrogen bond (2.4 Å) was observed between the carboxylate group of phenylalanine and the NE2 atom of His64. The phenylalanine amino group hydrogen bonds to an active-site water molecule, forming a bridge between the activator and the zinc ion.[81]

Activation data for many other amines and amino acid activators (including phenylethylamine, dopamine, noradrenaline, adrenaline, isoprotenerole, 2-pyridyl-ethylamine, serotonin, 4-hydroxyphenylalanine, 3,4-hydroxyphenylalanine, 4-fluorophenylalanine, 3-amino-4-hydroxyphenylalanine, 4-aminophenylalanine, histidine, and tryptophan) have been reported and reviewed.[138]

Chemical rescue

Removal of the predominant proton shuttle residue (His64) in hCA II by site-directed mutagenesis results in a 10-fold reduction in proton-transfer activity. This decrease in catalysis has been shown to be rescued in a saturable manner by the addition of exogenous proton acceptor/donors in solution, such as imidazole and its derivatives.[98] It has been an intriguing observation that the chemical rescue of H64A hCA II by small exogenous compounds is substantial, with catalysis at saturation levels of additives approaching that of hCA II.[98]

The catalytic properties of H64A hCA II with and without the exogenous proton acceptor/donor 4-methylimidazole (4-MI) have recently been compared with hCA II.[107] The effect of 4-MI binding on catalysis was measured by ^{18}O exchange at chemical equilibrium. The enhancement of activity caused by the addition of 4-MI to H64A hCA II was saturable with a maximum rate that approached that of the wild-type enzyme. The crystal structure of H64A hCA II complexed with 4-MI shows that this chemical proton rescuer bound to the enzyme in close proximity to the vacant space in the active site, previously occupied by the imidazole ring of His64 in the enzyme. Specifically, the 4-MI molecule π-stacks with the indole ring of Trp5 (Figure 10).

It was initially thought that the chemical rescue occurs due to the positional substitution of 4-MI for His64. The functionally equivalent nitrogens, N1 and N3, of 4-MI in comparison with ND1 and NE2 of His64, are 13.44 and 12.01 Å from the zinc ion, respectively. The 4-MI is located close to the 'out' conformation of His64 in the structure of hCA II[88] (Figure 10).

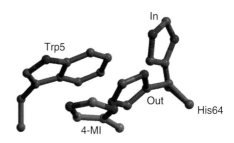

Figure 10 The primary binding site of the carbonic anhydrase activator 4-methylimidazole to hCA II (PDB accession 1G0E).[107] Figure created using Bobscript and Raster3D.[94,95]

Many plausible pathways for proton transfer from the zinc-bound water to the bulk solvent can be assigned along different water molecules in the active-site cavity of the structures of wild-type hCA II,[88] H64A hCA II and H64A hCA II in complex with 4-MI.[107] However, the number of molecules utilized differs. In the wild-type structure,[88] three water molecules form the network from the zinc ion to the ND1 nitrogen of His64 in the 'in' conformation. However, in the H64A hCA II soaked with 4-MI, four water molecules form this network to the N3 nitrogen of the 4-MI.[107]

Subsequent kinetic analysis of the site-specific mutant W5A hCA II in the presence of 4-MI by An[139] using ^{18}O-exchange methods and stopped-flow spectrophotometry have indicated that 4-MI is not actively involved in chemical rescue of proton transfer from the primary binding site with Trp5, although the W5A mutant of hCA III in the presence of 4-MI showed a complete loss of chemical rescue activity, indicating that Trp5 is the primary, active binding site for 4-MI in hCA III. The current explanation of these results is that the level of activation seen for wild-type hCA III is 10^4 s^{-1} (10^3 s^{-1} in the absence of 4-MI) at 200 mM 4-MI and that this level of activation would be undetectable in H64A hCA II, which has a proton-transfer rate of 10^5 s^{-1} in the presence of 4-MI[139] (Figure 11). These results may indicate that 4-MI binds to Trp5 in both H64A hCA II and hCA III and has the capacity to transfer protons, but other more predominant binding sites exist for H64A hCA

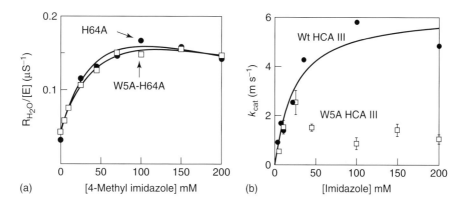

Figure 11 Chemical rescue of carbonic anhydrase by 4-methylimidazole (4-MI).[139] (a) Chemical rescue of H64A and H64A/W5A carbonic anhydrase II by 4-MI. (b) Chemical rescue of wild-type and W5A carbonic anhydrase III.

Carbonic anhydrases (α-class)

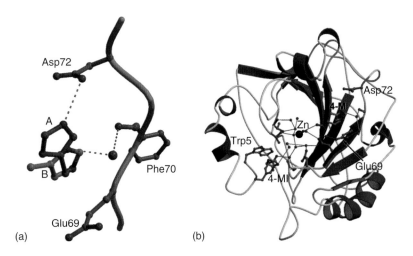

Figure 12 The interaction of the activator 4-methylimidazole (4-MI) with hCA II (PDB accession 1MOO).[89] Figure created using Bobscript and Raster3D.[94,95] (a) Binding site for 4-methylimidazole in a secondary position between Asp72 and Glu69 (gray sticks). The alternate conformational positions of 4-MI are labeled 'A' (green) and 'B' (orange). (b) Representation of the hydrogen-bonded water structure (red spheres) in the active site with both positions of 4-MI indicated (orange sticks).

II that mask the chemical rescue capability of 4-MI bound to Trp5.[139]

More recent high-resolution structural analysis of H64A hCA II in complex with 4-MI at 1.05 Å resolution has confirmed the π-stacking interaction between 4-MI and Trp5[89] and the identification of a second 'weaker' binding site for 4-MI on the opposite side of the active-site cavity from the primary position near Trp5. In this position, the 4-MI molecule was modeled in two distinct conformations with an approximate occupancy of two-fifth and three-fifth for the A and B positions respectively (Figure 12(a)). In this position, the 4-MI molecule is bound in a pocket formed by the side chains of Glu69, Ile91, and Asp72.

Similar interactions in the binding pocket were seen for 4-MI in the B-conformation.[89] The 4-MI ND1 and NE2 nitrogens are slightly further from the zinc ion in this secondary binding position[89] (Figure 12(b)) with the ND1 and NE2 approximately 14.0 Å from the zinc ion. In the B-conformation, ND1 and NE2 are further from the zinc ion with respective distances of 15.2 Å and 16.6 Å.

CONCLUDING REMARKS

Proton transfer is an important process underlying many fundamental biological systems. This process occurs in systems as diverse as the bacterial reaction center, which catalyzes light-induced, proton-coupled electron transfer, resulting in a proton gradient that ultimately drives ATP synthesis. The cytochrome complexes of the electron-transport chain also harness the free energy of electron transport to transfer protons across the mitochondrial inner membrane from the matrix to the intermembrane space, generating a proton gradient. This proton gradient is further utilized by F_1F_0 ATPase that transfers the protons through its membrane-embedded F_0 subunit to generate the energy needed for ATP synthesis.

A detailed understanding of how the process of proton transfer is mediated in a biological environment will aid in the overall knowledge of these systems. CA is an ideal model system for the study of biological proton transfer owing to its simple, monomeric, and less complex form.

REFERENCES

1 S Parkkila, in WR Chegwidden, ND Carter and YH Edwards (eds.), *Carbonic Anhydrases New Horizons: An Overview of the Distribution and Function of Carbonic Anhydrase in Mammals*, Birkhauser Verlag Basel, Switzerland, pp 80–93 (2000).

2 B Ulmasov, A Waheed, GN Shah, JH Grubb, WS Sly, CK Tu and DN Silverman, *Proc Natl Acad Sci USA*, **97**, 14212–17 (2000).

3 D Hewett-Emmett and RE Tashian, *Mol Phylogenet Evol*, **5**, 50–77 (1996).

4 WS Sly and PY Hu, *Annu Rev Biochem*, **64**, 375–401 (1995).

5 RE Tashian, *Adv Genet*, **30**, 321–56 (1992).

6 S Lindskog, *Pharmacol Ther*, **74**, 1–20 (1997).

7 DN Silverman and S Lindskog, *Acc Chem Res*, **21**, 30–36 (1988).

8 G Sanyal, ER Swenson, NI Pessah and TH Maren, *Mol Pharmacol*, **22**, 211–20 (1982).

9 DA Riley, S Ellis and J Bain, *Histochem Cytochem*, **30**, 1275–88 (1982).

10 SS Spicer, ZH Ge, RE Tashian, DJ Hazen-Marten and BA Schulte, *Am J Anat*, **187**, 60–89 (1990).

11 PL Whitney and TV Briggle, *J Biol Chem*, **257**, 12056–59 (1982).

12 RE Fleming, EC Crouch, CA Ruzicka and WS Sly, *Am J Physiol*, **265**, L627–35 (1993).

13 D Brown, XL Zhu and WS Sly, *Proc Natl Acad Sci USA*, **87**, 7457–61 (1990).

14. SJ Dodgson, in SJ Dodgson, RE Tashian, G Gros and ND Carter (eds.), *The Carbonic Anhydrases: Cellular Physiology and Molecular Genetics*, Plenum, New York, pp 267–305 (1991).
15. RT Fernley, RD Wright and JP Coghlan, *FEBS Lett*, **105**, 299–302 (1979).
16. JB Feldstein and DN Silverman, *J Biol Chem*, **259**, 5447–53 (1984).
17. H Murakami and WS Sly, *J Biol Chem*, **262**, 1382–88 (1987).
18. S Parkkila, K Kaunisto, L Rajaniemi, T Kumpulainen, K Jokinen and HJ Rajaniemi, *Histochem Cytochem*, **38**, 941–47 (1990).
19. J Kivela, S Parkkila, A-K Parkkila, J Leinonen and H Rajaniemi, *J Physiol*, **520**, 315–20 (1999).
20. S Parkkila, A-K Parkkila, J Lehtola, A Reinila, H-J Sodervik, M Rannisto and H Rajaniemi, *Dig Dis Sci*, **42**, 1013–19 (1997).
21. JC Montgomery, PJ Venta, RL Eddy, YS Fukushima, TB Shows and RE Tashian, *Genomics*, **11**, 835–48 (1991).
22. K Kato, *FEBS Lett*, **271**, 137–40 (1990).
23. LA Skaggs, NC Bergenhem, PJ Venta and RE Tashian, *Gene*, **126**, 291–92 (1993).
24. R Opavsky, S Pastorekova, S Zelnik, A Gibadulinova, EJ Standbridge, J Zevada, R Kettmann and J Pastorek, *Genomics*, **33**, 480–87 (1996).
25. J Saarnio, S Parkkila, AK Parkkila, K Haukipuro, S Pastorekova, J Pastorek, MI Kairaluoma and TJ Karttunen *Am J Pathol*, **153**, 279–85 (1998).
26. K Taniuchi, I Nishimori, T Takeuchi, K Fujikawa-Adachi, Y Ohtsuki and S Onishi, *Neuroscience*, **112**, 93–99 (2002).
27. K Fujikawa-Adachi, I Nishimori, T Taguchi, K Yuri and S Onishi, *Biochim Biophys Acta*, **1431**, 518–24 (1999).
28. N Okamoto, K Fujikawa-Adachi, I Nishimori, K Taniuchi and S Onishi, *Biochim Biophys Acta*, **1518**, 311–16 (2001).
29. O Tureci, U Sahin, E Vollmer, S Siemer, E Gottert, G Seitz, AN Parkkila, GN Shah, JH Grubb and M Pfreundschuh, *Proc Natl Acad Sci USA*, **95**, 7608–13 (1998).
30. RM Torczynski and AP Bollon, US Patent 5,589,579 (1996).
31. K Fujikawa-Adachi, I Nishimori, T Taguchi and S Onishi, *Genomics*, **61**, 74–81 (1999).
32. MK Koester, LM Pullan and EA Noltman, *Arch Biochem Biophys*, **211**, 632–42 (1981).
33. E Cabiscol and RL Levine, *Proc Natl Acad Sci USA*, **93**, 4170–74 (1996).
34. E Cabiscol and RL Levine, *J Biol Chem*, **270**, 14742–47 (1995).
35. SR Raisanen, P Lehenkari, M Tasanen, P Rahkila, PL Harkonen and HK Vaananen, *FASEB J*, **13**, 513–22 (1999).
36. DW Christianson and CA Fierke, *Acc Chem Res*, **29**, 331–39 (1996).
37. S Lindskog DN Silverman, in WR Chegwidden, ND Carter and YH Edwards (eds.), *The Carbonic Anhydrases New Horizons: The catalytic Mechanism of Mammalian Carbonic Anhydrases*, Birkhauser Verlag Basel, Switzerland, pp 175–95 (2000).
38. AE Eriksson, TA Jones and A Liljas, *Proteins: Struct, Funct, Genet*, **4**, 274–82 (1988).
39. JH Barlow, N Lowe, YH Edwards and PH Butterworth, *Nucleic Acids Res*, **15**, 2386 (1987).
40. JC Montgomery, PJ Venta, RE Tashian and D Hewett-Emmett, *Nucleic Acids Res*, **15**, 4687 (1987).
41. JC Lloyd, H Isenberg, DA Hopkinson and YH Edwards, *Ann Hum Genet*, **49**, 241–51 (1985).
42. T Okuyama, S Sato, XL Zhu, A Waheed and WS Sly, *Proc Natl Acad Sci USA*, **89**, 1315–19 (1992).
43. Y Nagao, JS Platero, A Waheed and WS Sly, *Proc Natl Acad Sci USA*, **90**, 7623–27 (1993).
44. P Aldred, P Fu, G Barrett, JD Penschow, RD Wright, JP Coghlan and RT Fernley, *Biochemistry*, **30**, 569–75 (1991).
45. J Pastorek, S Pastorekova, I Callebaut, J Mornon, V Zelnik, R Opavsky, M Zatovicova, S Liao, D Portetelle, EJ Standbridge, J Zavada and A Burny, *Oncogene*, **9**, 2877–88 (1994).
46. JJ Kleiderlein, PE Nisson, J Jessee, WB Li, KG Becker, ML Derby, CA Ross and RL Margolis, *Hum Genet*, **103**, 666–73 (1998).
47. DA Lovejoy, D Hewett-Emmett, CA Porter, D Cepoi, A Sheffield, WW Vale and RE Tashian, *Genomics*, **54**, 484–93 (1998).
48. SV Ivanov, I kuzmin, MH Wei, S Pack, L Geil, BE Johnson, EJ Standbridge and MI Lerman, *Proc Natl Acad Sci USA*, **95**, 12596–601 (1998).
49. D Hewett-Emmett, in WR Chegwidden, ND Carter and YH Edwards (eds.), *The Carbonic Anhydrases New Horizons: Evolution and Distribution of the Carbonic Anhydrase Gene Families*, Birkhauser Verlag Basel, Switzerland, pp 29–76 (2000).
50. RE Tashian, *BioEssays*, **10**, 186–92 (1989).
51. D Hewett-Emmett RE Tashian, in SJ Dodgson, RE Tashian, G Gros and ND Carter (eds.), *The Carbonic Anhydrases: Cellular Physiology and Molecular Genetics*, Plenum, New York, pp 15–32 (1991).
52. EA Merritt and MEP Murphy, *Acta Crystallogr*, **D50**, 869–73 (1995).
53. NU Meldrum and FJW Roughton, *J Physiol*, **80**, 113–42 (1933).
54. WC Stadie and H O'Brien, *J Biol Chem*, **103**, 521–29 (1933).
55. RG Kalifah, DJ Strader, SH Bryant and SM Gibson, *Biochemistry*, **16**, 2241–47 (1977).
56. SM Tanhauser, DA Jewell, CK Tu, DN Silverman and PJ Laipis, *Gene*, **117**, 113–17 (1992).
57. DM Duda, C Yoshioka, L Govindasamy, H An, CK Tu, DN Silverman and R McKenna, *Acta Crystallogr*, **D58**, 849–52 (2002).
58. T Stams, SK Nair, T Okuyama, A Waheed, WS Sly and DW Christianson, *Proc Natl Acad Sci USA*, **93**, 13589–94 (1996).
59. LL Keifer, SA Paterno and CA Fierke, *J Am Chem Soc*, **117**, 6831–37 (1995).
60. K Hakansson, A Wehnert and A Liljas, *Acta Crystallogr*, **D50**, 93–100 (1994).
61. S Lindskog and BG Malmstrom, *J Biol Chem*, **237**, 1129–37 (1962).
62. JB Hunt, MJ Rhee and CB Storm, *Anal Biochem*, **79**, 614–17 (1977).
63. K Hakansson and A Wehnert, *J Mol Biol*, **228**, 1212–18 (1992).
64. RG Kalifah, *J Biol Chem*, **246**, 2561–73 (1971).
65. H Steiner, B-H Jonsson and S Lindskog, *Eur J Biochem*, **59**, 253–59 (1975).
66. DN Silverman, CK Tu, S Lindskog and GC Wynns, *J Am Chem Soc*, **101**, 6734–40 (1979).
67. DN Silverman, *Methods Enzymol*, **87**, 732–52 (1982).
68. Y Pocker and S Sarkanen, *Adv Enzymol*, **47**, 149–274 (1978).
69. JA Verpoorte, S Mehta and JT Edsall, *J Biol Chem*, **242**, 4221–29 (1967).
70. DA Jewell, CK Tu, SR Parananwithana, SM Tanhauser, PV LoGrasso, PJ Laipis and DN Silverman, *Biochemistry*, **30**, 1484–90 (1991).
71. TT Baird, A Waheed, T Okuyama, WS Sly and CA Fierke, *Biochemistry*, **36**, 2669–78 (1997).
72. RW Heck, SM Tanhauser, R Manda, CK Tu, PJ Laipis and DN Silverman, *J Biol Chem*, **269**, 24742–46 (1994).

73. JN Earnhardt, M Qian, CK Tu, MM Lakkis, NCH Bergenhem, PJ Laipis, RE Tashian and DN Silverman, *Biochemistry*, **37**, 10837–45 (1998).
74. RE Tashian, in WR Chegwidden, ND Carter and YH Edwards (eds.), *Carbonic Anhydrases New Horizons: Carbonic anhydrase (CA)-Related Proteins (CA-RPs), and Transmembrane Proteins with CA or CA-RP Domains*, Birkhauser Verlag Basel, Switzerland, pp 105–20 (2000).
75. T Wingo, CK Tu, PJ Laipis and DN Silverman, *Biochem Biophys Res Comm*, **288**, 666–69 (2001).
76. WR Chegwidden and N Carter, in WR Chegwidden, ND Carter and YH Edwards (eds.), *Carbonic Anhydrases New Horizons: Introduction to the Carbonic Anhydrases*, Birkhuaser Verlag Basel, Switzerland, pp 13–28 (2000).
77. I Bertini, G Canti, C Luchinat and A Scozzafava, *J Am Chem Soc*, **100**, 4873–77 (1978).
78. I Bertini, C Luchinat and A Scozzafava, *Bioinorg Chem*, **9**, 93–100 (1978).
79. I Bertini, C Luchinat and A Scozzafava, *Struct Bonding*, **48**, 45–92 (1982).
80. F Briganti, S Mangani, P Orioli, A Scozzafava, G Vernaglione and CT Supuran, *Biochemistry*, **36**, 10384–92 (1997).
81. F Briganti, V Iaconi, S Mangani, P Orioli, A Scozzafava, G Vernaglione and CT Supuran, *Inorg Chim Acta*, **275**, 295–300 (1998).
82. I Simonsson, BH Jonsson and S Lindskog, *Biochem Biophys Res Commun*, **108**, 1406–12 (1982).
83. SK Nair, PA Ludwig and DW Christianson, *J Am Chem Soc*, **116**, 3659–60 (1994).
84. B Strandberg, B Tilander, K Fridborg, S Lindskog and PO Nyman, *J Mol Biol*, **5**, 583–584 (1962).
85. B Tilander, B Strandberg and K Fridborg, *J Mol Biol*, **12**, 740–60 (1965).
86. L Liljas, KK Kannan, P-C Bergsten, I Waara, K Fridborg, B Strandberg, U Carlbom, L Jarup, S Lovgren and M Petef, *Nat N Biol*, **235**, 131–37 (1972).
87. A McPherson, *Preparation and Analysis of Protein Crystals*, John Wiley & Sons, New York (1982).
88. K Hakansson, M Carlsson, LA Svensson and A Liljas, *J Mol Biol*, **227**, 1192–204 (1992b).
89. DM Duda, L Govindasamy, M Agbandje-McKenna, CK Tu, DN Silverman and R McKenna, *Acta Crystallogr*, **59**, 93–104 (2003); accepted for publication.
90. G Aronsson, L-G Martensson, U Carlsson and B-H Jonsson, *Biochemistry*, **34**, 2153–62 (1995).
91. L-G Martensson, B-H Jonsson, M Andersson, A Kihlgren, N Bergenhem and U Carlsson, *Biochim Biophys Acta*, **1118**, 179–86 (1992).
92. JF Krebs and CA Fierke, *J Biol Chem*, **268**, 948–54 (1993).
93. L-G Martensson, P Jonasson, P-O Freskgard, M Svensson, U Carlsson and B-H Jonsson, *Biochemistry*, **34**, 1011–21 (1995).
94. JD Thompson, DG Higgins and TJ Gibson, *Nucleic Acids Res*, **22**, 4673–80 (1994).
95. RM Esnouf, *J Mol Graphics Modell*, **15**, 132–34 (1997).
96. JD Cox, JA Hunt, KM Compher, CA Fierke and DW Christianson, *Biochemistry*, **39**, 13687–94 (2000).
97. DW Christianson and JD Cox, *Annu Rev Biochem*, **68**, 33–57 (1999).
98. CK Tu and DN Silverman, *Biochemistry*, **28**, 7913–18 (1989).
99. N Agmon, *Chem Phys Lett*, **244**, 456–62 (1995).
100. J Lobaugh and GA Voth, *J Chem Phys*, **104**, 2056–69 (1996).
101. M Eigen, *Angew Chem, Int Ed Engl*, **3**, 1–72 (1964).
102. PV LoGrasso, CK Tu, DA Jewell, GC Wynns, PJ Laipis and DN Silverman, *Biochemistry*, **30**, 8463–70 (1991).
103. JE Jackman, KM Merz and CA Fierke, *Biochemistry*, **35**, 16421–28 (1996).
104. LR Scolnick and DW Christianson, *Biochemistry*, **35**, 16429–34 (1996).
105. RW Heck, PA Boriack-Sjodin, MZ Qian, CK Tu, DW Christianson, PJ Laipis and DN Silverman, *Biochemistry*, **35**, 11605–11 (1996).
106. PA Boriack-Sjodin, RW Heck, PJ Laipis, DN Silverman and DW Christianson, *Proc Natl Acad Sci USA*, **92**, 10949–53 (1995).
107. DM Duda, CK Tu, M Qian, PJ Laipis, M Agbandje-McKenna, DN Silverman and R McKenna, *Biochemistry*, **40**, 1741–48 (2001).
108. JF Krebs, F Rana, RA Dluhy and CA Fierke, *Biochemistry*, **32**, 4496–505 (1993b).
109. J-Y Liang and WN Lipscomb, *Proc Natl Acad Sci USA*, **87**, 3675–79 (1990).
110. KM Merz, *J Am Chem Soc*, **113**, 406–11 (1991).
111. CA Fierke, TL Calderone and JF Krebs, *Biochemistry*, **30**, 11054–63 (1991).
112. RS Alexander, SK Nair and DW Christianson, *Biochemistry*, **30**, 11064–72 (1991).
113. M Lindahl, LA Svensson and A Liljas, *Proteins: Struct, Funct, Genet*, **15**, 177–82 (1993).
114. KM Merz, *J Mol Biol*, **214**, 799–802 (1990).
115. JF Krebs, JA Ippolito, DW Christianson and CA Fierke, *J Biol Chem*, **268**, 27458–66 (1993).
116. Z Liang, Y Xue, G Behraven, B-H Jonsson and S Lindskog, *Eur J Biochem*, **211**, 821–27 (1993).
117. K Hakansson, M Carlsson, LA Svensson and A Liljas, *J Mol Biol*, **227**, 1192–1204 (1992).
118. S Mangani and A Liljas, *Eur J Biochem*, **210**, 867–71 (1992).
119. I Bertini and C Luchinat, *Acc Chem Res*, **16**, 272–79 (1983).
120. A Liljas, K Hakansson, BH Jonsson and Y Xue, *Eur J Biochem*, **219**, 1–10 (1994).
121. K Hakansson, C Briand, V Zaitsev, Y Xue and A Liljas, *Acta Crystallogr*, **D50**, 101–4 (1994).
122. BM Jonsson, K Hakansson and A Liljas, *FEBS Lett*, **322**, 186–90 (1993).
123. V Kumar, KK Kannan and P Sathyamurthi, *Acta Crystallogr*, **D50**, 731–38 (1994).
124. SK Nair and DW Christianson, *Eur J Biochem*, **213**, 507–15 (1993).
125. S Lindskog PJ Wistrand, in M Sandler and HJ Smith (eds.), *Design of Enzyme Inhibitors as Drugs*, Oxford University Press, Oxford, pp 698–723 (1988).
126. TH Maren and CW Conroy, *J Biol Chem*, **268**, 26233–39 (1993).
127. TH Maren, *Ann NY Acad Sci*, **429**, 10–17 (1984).
128. JR Sutton, CS Houston, AL Mansell, MD McFadden, PM Hackett, JRA Rigg and ACP Powles, *N Engl J Med*, **301**, 1329–331 (1979).
129. J Vidgren, A Liljas and NPC Walker, *Int J Biol Macromol*, **12**, 342–44 (1990).
130. A Boriack-Sjodin, S Zeitlin, H Chen, L Crenshaw, S Gross, A Dantanrayana, P Delgado, EA May, T Dean and DW Christianson, *Protein Sci*, **7**, 2483–89 (1998).
131. JE Coleman, *Nature*, **214**, 193–94 (1967).

132 S Lindskog and PO Nyman, *Biochim Biophys Acta*, **85**, 462–74 (1964).
133 A Thorslund and S Lindskog, *Eur J Biochem*, **3**, 117–23 (1967).
134 SK Nair and DW Christianson, *J Am Chem Soc*, **113**, 9455–58 (1991).
135 ID Campbell, S Lindskog and AI White, *J Mol Biol*, **98**, 597–614 (1975).
136 JF Krebs, CA Fierke, RS Alexander and DW Christianson, *Biochemistry*, **30**, 9153–60 (1991).
137 GM Smith, RS Alexander, DW Christianson, BM McKeever, GS Ponticello, JM Springer, WC Randall, JJ Baldwin and CN Habecker, *Protein Sci*, **3**, 118–25 (1994).
138 CT Supuran and I Puscas, in I Puscas (ed.), *Carbonic Anhydrase and Modulation of Physiologic and Pathologic Processes in the Organism*, Helicon, Timisoara, pp 113–45 (1994).
139 H An, CK Tu, DM Duda, I Montanez-Clemente, K Math, PJ Laipis, R McKenna and DN Silverman, *Biochemistry*, **41**, 3235–42 (2002).

Carbonic anhydrases (β-class)

Eiki Yamashita[†], Satoshi Mitsuhashi[‡] and Tomitake Tsukihara[†]

[†]Institute for Protein Research, Osaka University, Osaka, Japan
[‡]Kyowa Hakko Kogyo Co., Ltd, Tokyo, Japan

FUNCTIONAL CLASS

Enzyme; carbonate hydrolyase; EC 4.2.1.1; known as carbonic anhydrase.

Carbonic anhydrase (CA) catalyzes the reversible hydration of CO_2 to HCO_3^- in aqueous solution.

$$CO_2 + H_2O \longleftrightarrow H^+ + HCO_3^-$$

All animal CAs characterized thus far belong to a single gene family referred to as the α-CAs. In the early 1990s, cDNA sequences encoding several chloroplast CAs from leaves of higher plants were determined and could not be aligned with sequences of α-CAs.[1] The second gene family was referred to as the β-carbonic anhydrases (β-CA). At present, all CAs are divided into three distinct classes (α-CAs, β-CAs, and γ-CAs) that have little sequence homology to one another and are thought to have originated independently and evolved convergently.[2] These three CA families are interesting examples of convergent evolution of the same catalytic function.

OCCURRENCE

β-CA is found in species from three domains of life, with representatives in eubacteria including *Escherichia coli*,[3]

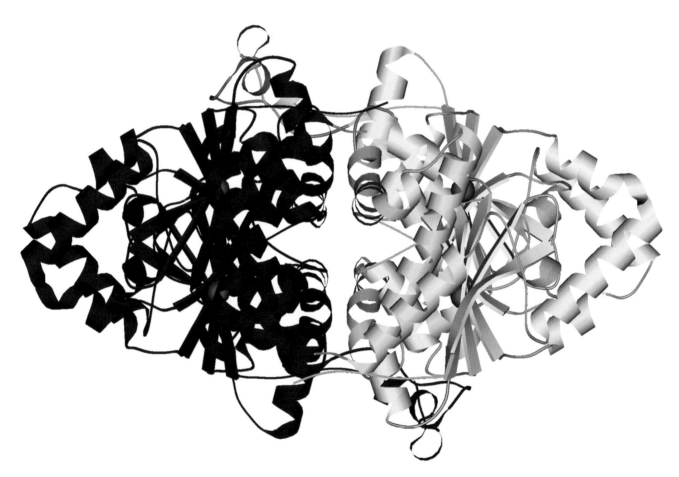

3D Structure Overall structure of *P. purpureum* β-CA dimer looking down along the twofold axis. Two monomers are colored in yellow and blue, respectively. Red spheres are zinc ions. The figure and subsequent molecular structures were drawn by the program MOLSCRIPT.[24] PDB code: 1DDZ.

in archaea including *Methanobacterium thermoautotrophicum*,[4] and in a variety of higher plants and algae. By searching available sequence databases in 1999, Smith *et al.* identified 51 ORFs having significant homology and identity to the β-CA from *M. thermoautotrophicum*.[5]

In higher plants, occurrence of β-CAs is dependent on the method of fixing carbon dioxide. C4 plants have a specialized anatomy and a specialized biochemistry for collecting CO_2 in the mesophyll and concentrating it in the bundle sheath cells. C4 plants use phosphoenolpyruvate (PEP) carboxylase for initial CO_2 fixation in the mesophyll cells and produce a four-carbon organic acid that is transported to the bundle sheath cells. In C4 plants, β-CA is located predominantly in the cytoplasm of mesophyll cells. Normal plants are called C3 plants because the first product of carbon dioxide fixation is 3-phosphoglycerate. In C3 plants, β-CA is an abundant protein, representing 0.5 to 2% of total soluble leaf protein, and is predominantly located in the stroma of chloroplasts. In algae, β-CAs are located in the cytosol of the unicellular green alga *Coccomyxa*[6] and associated with the cytoplasmic periphery of the plasma membrane in the red alga *Porphyridium purpureum*.[7]

BIOLOGICAL FUNCTION

β-CA plays an essential role in photosynthesis, by concentrating CO_2 in the proximity of ribulose bisphosphate carboxylase/oxygenase for CO_2 fixation.[8,9] In C4 plants, β-CA converts CO_2 to HCO_3^-, which can then serve as the substrate for phosphoenolpyruvate carboxylase. Little is known about the physiological roles performed by β-CA in nonphotosynthetic organisms. In *M. thermoautotrophicum*, the physiological role of β-CA may be to provide HCO_3^- to enzymes important in CO_2 fixation pathways of microbes.

AMINO ACID SEQUENCE INFORMATION

β-Carbonic anhydrase

- *Porphyridium purpureum*, 496 AA, translation of cDNA.[10]
- *Pisum sativum*, 221 AA, translation of cDNA.[11]
- *Methanobacterium thermoautotrophicum*, 170 AA, translation of DNA sequence.[4]
- *Escherichia coli*, 220 AA, translation of DNA sequence.[3]

PROTEIN PRODUCTION, PURIFICATION, AND MOLECULAR CHARACTERIZATION

β-CAs from *P. purpureum*, *P. sativum*, *M. thermoautotrophicum*, and *E. coli* were all cloned and expressed in *E. coli*. Their X-ray crystal structures have been determined.[12–15] The purification procedure for *P. purpureum* β-CA consists of the preparation of cell extracts, followed by affinity chromatography on *p*-aminomethylbenzene sulfonamide conjugated Sepharose 6B, and anion-exchange chromatography on Q-Sepharose.[10]

β-CAs have been found in different oligomeric states ranging from dimers to tetramers to octamers.[16] In solution, *P. purpureum* β-CA exists as a homodimer with a molecular weight of about 110 kDa. The β-CA monomer (55 kDa) has two domains, each of which is equivalent to that of other β-CAs (25–30 kDa) and which exhibit approximately 70% identity with one another.[10] Thus, it was suggested that the *P. purpureum* β-CA gene was formed by duplication and fusion of a primordial β-CA gene.

METAL CONTENT AND COFACTORS

Both β-CAs from *M. thermoautotrophicum* and *E. coli* have one zinc atom per monomer, while β-CA from *P. purpureum* contains two zinc atoms per monomer.

ACTIVITY TEST

CA activity was measured as the time needed for pH to change from 8.3 to 7.3 (t) after addition of 2-mL CO_2-saturated water to 12 mM sodium 5,5-diethylbarbiturate-HCl buffer (pH 8.3) containing an enzyme solution and 5 mM NaCl (final concentration in a total volume of 5 mL). The reaction was carried out at $2\,°C$. An enzyme activity unit was calculated using the equation, unit = $(t_0 - t)/t$, in which t_0 is the time required for the pH change using buffer without enzyme.[10]

The CA activity was also measured by pH/dye indicator method.[17]

SPECTROSCOPY

X-ray absorption spectra have been recorded for spinach β-CA.[18,19] The models that best fit the data have average Zn–N(O) distances of 1.99 to 2.06 Å, average Zn–S distances of 2.31 to 2.32 Å, and a total coordination number of 4 to 6. Analysis of the extended fine structure (EXAFS) suggests that the coordination sphere of Zn in spinach CA must have one or more sulfur ligands, in contrast to animal α-CAs, which have only nitrogen and oxygen ligands. Recently, EXAFS results obtained using β-CA from *M. thermoautotrophicum* indicated that the active zinc site of the enzyme is coordinated by two sulfur and two O/N ligands, with the possibility that one of the O/N ligands is derived from histidine and the other from water.[20]

X-RAY STRUCTURES OF β-CA

Crystallization and structure determination

Four crystal structures of β-CAs from *P. purpureum* (PDB code: 1DDZ), *P. sativum* (PDB code: 1EKJ), *M. thermoautotrophicum* (PDB code: 1G5C), and *E. coli* (PDB code: 1I6P) have been determined so far. The crystals from *P. purpureum* were grown by a vapor-diffusion method, using a 30 mg mL^{-1} protein solution containing 20 mM NaCl, 20 mM Tris-HCl, pH 8.5, and a precipitant solution containing 24% polyethylene glycol 4000, 300 mM ammonium sulfate, and 50 mM sodium cacodylate-HCl, pH 6.75 at 20 °C. They belong to the monoclinic space group P2$_1$ with cell dimensions of $a = 63.8$ Å, $b = 113.9$ Å, $c = 73.8$ Å, and β 104.1°.[21] The structure was determined at 2.2 Å resolution by multiwavelength anomalous dispersion (MAD) using the signal from the intrinsic zinc atoms. The electron density map calculated with the MAD phases was refined by solvent flattening and noncrystallographic symmetry averaging methods. The refined map was of sufficient quality to build an initial model. The model was refined to an R-factor of 0.208 and an Rfree of 0.274. The rms deviation of bond lengths and angles for the final model was 0.011 Å and 2.3° respectively.[12]

P. sativum β-CA was crystallized by a vapor-diffusion method against 16% polyethylene glycol 4000, 400 mM ammonium acetate, 50 mM dithiothreitol, and 100 mM sodium citrate pH 5.0 at 4 °C. The crystals were orthorhombic, with the space group C222 having cell dimensions of $a = 136.9$ Å, $b = 143.3$ Å, and $c = 202.1$ Å.[22] The structure was determined by MAD using the signal from the intrinsic zinc atoms combined with the noncrystallographic symmetry averaging and phase extension. The model was refined to an R-factor of 0.227 and an Rfree of 0.252 at 1.93 Å resolution. The rms deviation of bond lengths and angles for the final model was 0.012 Å and 1.5° respectively.[13]

M. thermoautotrophicum β-CA was crystallized by a vapor-diffusion method against 35% ethanol, 12% 2-methyl-2,4-pentanediol, 50 mM calcium acetate, and 100 mM HEPES, pH 7.5 at 22 °C. The crystals belong to the orthorhombic space group P2$_1$2$_1$2$_1$ with cell dimensions of $a = 54.9$ Å, $b = 113.2$ Å, and $c = 156.2$ Å.[14] The structure was determined by MAD using the signal from the intrinsic zinc atoms combined with the noncrystallographic symmetry averaging and phase extension. The model was refined to an R-factor of 0.211 and an Rfree of 0.246 at 2.1 Å resolution. The rms deviation of bond lengths and angles for the final model was 0.013 Å and 1.7° respectively.[14]

Three different crystal forms were obtained for *E. coli* β-CA. Two (forms 1 and 2) were used for crystal structure analysis. Both crystals were obtained by a vapor-diffusion method. Form 1 crystals belonging to the space group P4$_2$2$_1$2, with $a = b = 68.54$ Å and $c = 85.88$ Å, were grown using 1.8 to 1.6 M ammonium sulfate and 0.1 M MES (2-morpholinoethanesulfonic acid), pH 6.3 at 20 °C. Form 2 crystals of space group P4$_3$22, with $a = b = 81.24$ Å and $c = 162.14$ Å, were grown by adding 4% polyethylene glycol 400 to the form 1 condition.[23] The structure in the form 2 crystal was determined by MAD using the signal from the intrinsic zinc and selenium atoms. The structure in the form 1 crystal was solved using molecular replacement, with a model of the form 2 structure as a search model. The model in the form 1 crystal was refined to an R-factor of 0.177 and an Rfree of 0.203 at 2.0 Å resolution. The rms deviations of bond lengths and angles for the final model in the form 1 crystal was 0.004 Å and 1.1° respectively.[15]

Overall description of structure

The monomeric structure of *P. purpureum* β-CA is depicted by a ribbon drawing of the main chain (Figure 1).[12] The N-terminal half from amino acids 86 to 309 and the C-terminal half from amino acids 340 to 563 are highly homologous to one another, both in their amino acid sequences and in their main chain folds. Thus, two homologous motifs are repeated in the *P. purpureum* β-CA polypeptide.[10] Each motif is composed of two components: an α/β domain and an N-terminal arm. The α/β domain

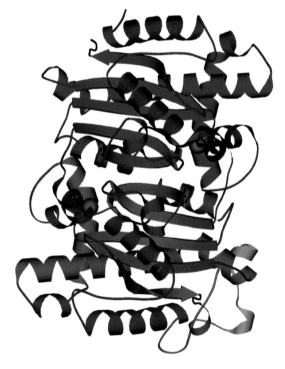

Figure 1 Ribbon drawing of *P. purpureum* β-CA monomer. The N- and C-terminal halves colored in green and light blue respectively exhibit an identical fold. Produced using the program MOLSCRIPT.[24]

Carbonic anhydrases (β-class)

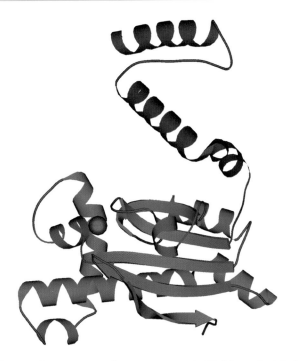

Figure 2 Structure of *P. purpureum* β-CA N-terminal half. Produced using the program MOLSCRIPT.[24]

consists of α-β-α units exhibiting a Rossman fold with a four-stranded parallel β-sheet core ordered 2-1-3-4 (7-6-8-9 at C-terminal half) and an antiparallel β-strand at the C-terminal side (Figure 2). The N-terminal arm runs on the surface of the α/β domain formed by the opposite motif of the same monomer. The two motifs in a monomer are related by a pseudo twofold axis centered between the β2 strand of the N-terminal half and the β7 strand of the C-terminal half. An asymmetric unit cell of *P. purpureum* β-CA crystal contains a dimer of two identical subunits, each with a molecular mass of 55 KDa (3D Structure). The dimer has approximate dimensions of $90 \times 70 \times 60 \, \text{Å}^3$. The two monomers are related by a twofold axis perpendicular to the pseudo twofold axis in the monomer. Consequently, the dimer has a pseudo 222 symmetry.

E. coli β-CA and *P. sativum* β-CA are composed of a tetramer and an octamer respectively (Figures 3(a) and (b)).[13,15] The dimer of *P. purpureum* β-CA resembles a tetramer of *E. coli* β-CA. A dimer of *P. sativum* β-CA, which resembles the monomer of *P. purpureum* β-CA, is the basic building block of the octamer. The octamer has 222 symmetry with the basic block. This oligomerization is mediated by the β5 strand of the C-terminus.[12]

Zinc site geometries

In the *P. purpureum* β-CA structure, the two zinc atoms of the monomer are located in two clefts on both sides of the monomeric molecule. They are located at the C-terminal ends of the parallel β-sheets, as often found in α/β enzymes with Rossmann folds. One of the catalytic zinc atoms is coordinated in a tetrahedral manner with the SG atom of Cys149, the OD2 of Asp151, the NE2 of His205, and the SG of Cys208 in the N-terminal half (Figure 4). The other is coordinated with equivalent atoms of Cys403, Asp405, His459, and Cys462 in the C-terminal half. *E. coli* β-CA structure shows nearly the same zinc-coordination scheme. The zinc atom at the active site of any β-CA is coordinated by two cysteines and one histidine (Table 1). The fourth ligand of the zinc atoms in the three different β-CAs is not a specific amino acid residue (Table 1). In *P. sativum* β-CA, the fourth ligand is a carboxyl group of acetate that was included in the crystallization buffer,[13] while

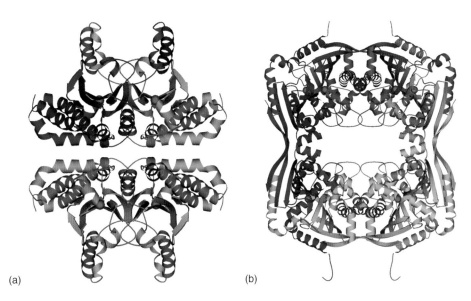

Figure 3 Subunit arrangements of (a) *P. sativum* β-CA tetramer (PDB code: 1EKJ) and (b) *E. coli* β-CA octamer (PDB code: 1I6P). Each subunit in the molecule is shown in individual color. Produced using the program MOLSCRIPT.[24]

Carbonic anhydrases (β-class)

in *M. thermoautotrophicum* β-CA, the ligand is a water molecule.[14]

FUNCTIONAL ASPECTS

In photosynthetic CO_2 fixation, CO_2 is the substrate for ribulose-1,5-biphosphate carboxylase/oxygenase. However, HCO_3^- is the dominant species in alkaline chloroplast stroma, and the rate of spontaneous interconversion between CO_2 and HCO_3^- is insufficient to cope with metabolic demand. Steady state kinetics of β-CAs from higher plants show high catalytic efficiency at high pH with k_{cat} values between 10^5 and $10^6 \, s^{-1}$ per subunit and k_{cat}/K_m values between 10^7 and $10^8 \, M^{-1} s^{-1}$.[25,26] It has been suggested that CA plays an essential role in the fixation of CO_2 in the Calvin–Benson cycle. β-CA is thought to be involved in the CO_2-concentrating mechanism, which maintains a favorable CO_2 level at the carboxylation site and compensates for the relatively lower affinity of algal ribulose-1,5-biphosphate carboxylase/oxygenase for CO_2.[9,27]

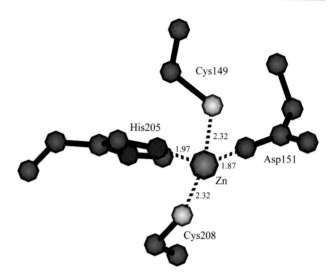

Figure 4 Coordination structure of a zinc ion at an active site of *P. purpureum* β-CA. The zinc ion is coordinated by two sulfur atoms of cysteines, a nitrogen atom of a histidine, and an oxygen atom of aspartate in a distorted tetrahedral or a distorted trigonal pyramidal configuration. The other zinc ion in the monomer is similar to the present zinc in its coordination structure. The figure was produced using the program MOLSCRIPT.[24] PDB code: 1DDZ.

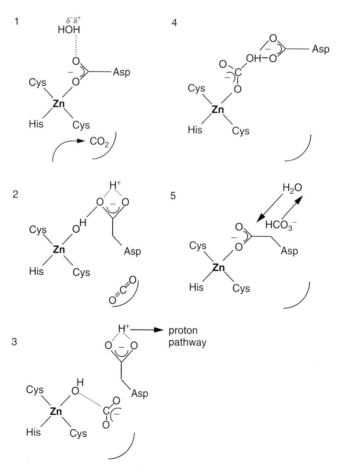

Figure 5 Proposed CO_2 hydration mechanism for *P. purpureum* β-CA. See text for the explanation of each step.

Table 1 Zinc-coordination bond distances in the β-CAs

PDB code	1DDZ	1I6P	1EKJ	1G5C
Zn–S$_{Cys}$	2.32	2.34	2.29	2.40
	2.33	2.37	2.30	2.35
Zn–N$_{His}$	1.97	2.21	2.05	2.03
Zn–O	1.87	2.10	2.81	2.03
(Fourth ligands)	(Asp)	(Asp)	(Acetate)	(Water)

1DDZ: *P. purpureum*; 1I6P: *E. coli*; 1EKJ: *P. sativum*; *M. thermoautotrophicum*.

CATALYTIC MECHANISM

A possible catalytic mechanism of β-CA, based on the structure of *P. purpureum* CA shown in Figure 5, is proposed.[12] The substrate of CO_2 is accommodated in the hydrophobic pocket beside the zinc-active center. The zinc-bound aspartate functions as a base to accept a proton from a hydrogen-bonded water molecule and yields a nucleophilic hydroxide (Figure 5, step 1). Consequently, the protonated aspartate is released from the zinc and the resulting nucleophilic hydroxide moves toward and binds the zinc (Figure 5, step 2). In the next step, the hydroxide attacks the CO_2 molecule to generate zinc-bound HCO_3^-. The proton is then transferred from the protonated aspartate to the bulk solvent or buffer, possibly through one of the hydrogen-bonded pathways immediately surrounding the zinc ligands (Figure 5, step 3). Then, the zinc-bound HCO_3^- is replaced with a deprotonated Asp, releasing the HCO_3^- and leaving a zinc-bound Asp (Figure 5, step 4). Finally, a water molecule binds OD1 of the zinc-bound Asp to regenerate the initial stage (Figure 5, step 5).

Following the solution of the structure of the *P. purpureum* CA, several β-CA structures have recently been determined. It should be noted that in the structure of *M. thermoautotrophicum* β-CA, the active site zinc is coordinated by protein ligands Cys32, His87, and Cys90, with the tetrahedral coordination completed by a water molecule.[14] A zinc hydroxide mechanism has also been proposed for the *M. thermoautotrophicum* β-CA. Studies on structure–function relationship will be necessary to clarify the catalytic mechanism of β-CAs.

REFERENCES

1. TW Fawcett, JA Browse, M Volokita and SG Bartlett, *J Biol Chem*, **265**, 5414–17 (1990).
2. D Hewett-Emmett and RE Tashian, *Mol Phylogenet Evol*, **5**, 50–77 (1996).
3. MB Guilloton, JJ Korte, AF Lamblin, JA Fuchs and PM Anderson, *J Biol Chem*, **267**, 3731–34 (1992).
4. KS Smith and JG Ferry, *J Bacteriol*, **181**, 6247–53 (1999).
5. KS Smith, C Jakubzick, TS Whittam and JG Ferry, *Proc Natl Acad Sci USA*, **96**, 15184–89 (1999).
6. T Hiltonen, J Karlsson, K Palmqvist, AK Clarke and G Samuelsson, *Planta*, **195**, 345–51 (1995).
7. S Mitsuhashi, M Kawachi, N Kurano and S Miyachi, *Phycologia*, **40**, 319–23 (2001).
8. JV Moroney, A Somanchi, *Plant Physiol*, **119**, 9–16 (1999).
9. MR Badger and GD Price, *Annu Rev Plant Physiol Plant Mol Biol*, **45**, 369–92 (1994).
10. S Mitsuhashi and S Miyachi, *J Biol Chem*, **271**, 28703–9 (1996).
11. N Majeau and JR Coleman, *Plant Physiol*, **95**, 264–68 (1991).
12. S Mitsuhashi, T Mizushima, E Yamashita, M Yamamoto, T Kumasaka, H Moriyama, T Ueki, S Miyachi and T Tsukihara, *J Biol Chem*, **275**, 5521–26 (2000).
13. MS Kimber and EF Pai, *EMBO J*, **19**, 1407–18 (2000).
14. P Strop, KS Smith, TM Iverson, JG Ferry and DC Rees, *J Biol Chem*, **276**, 10299–305 (2001).
15. JD Cronk, JA Endrizzi, MR Cronk, JW O'neill and KY Zhang, *Protein Sci*, **10**, 911–22 (2001).
16. BC Tripp, K Smith, JG Ferry, *J Biol Chem*, **276**, 48615–18 (2001).
17. RG Khalifah, *J Biol Chem*, **246**, 2561–73 (1971).
18. MH Bracey, J Christiansen, P Tovar, SP Cramer and SG Bartlett, *Biochemistry*, **33**, 13126–31 (1994).
19. RS Rowlett, MR Chance, MD Wirt, DE Sidelinger, JR Royal, M Woodroffe, YF Wang, RP Saha and MG Lam, *Biochemistry*, **33**, 13967–76 (1994).
20. KS Smith, NJ Cosper, C Stalhandske, RA Scott, JG Ferry, *J Bacteriol*, **182**, 6605–13 (2000).
21. S Mitsuhashi, T Mizushima, E Yamashita, S Miyachi and T Tsukihara, *Acta Crystallogr D Biol Crystallogr*, **56**, 210–11 (2000).
22. MS Kimber, JR Coleman and EF Pai, *Acta Crystallogr D Biol Crystallogr*, **56**, 927–29 (2000).
23. JD Cronk, JW O'Neill, MR Cronk, JA Endrizzi, KY Zhang, *Acta Crystallogr D Biol Crystallogr*, **56**, 1176–79 (2000).
24. PJ Kraulis, *J Appl Crystallogr*, **24**, 946–50 (1991).
25. IM Johansson and C Forsman, *Eur J Biochem*, **218**, 439–46 (1993).
26. RS Rowlett, MR Chance, MD Wirt, DE Sidelinger, JR Royal, M Woodroffe, YF Wang, RP Saha and MG Lam, *Biochemistry*, **33**, 13967–76 (1994).
27. K Aizawa and S Miyachi, *FEMS Microbiol Rev*, **39**, 215–33 (1986).

Carbonic anhydrases (γ-class)

Caroline Kisker[†] and Tina M Iverson[‡]

[†]Department of Pharmacological Sciences, Center for Structural Biology, State University of New York at Stony Brook, Stony Brook, NY, USA

[‡]Imperial College of Science, Technology and Medicine, Division of Biomedical Sciences, Wolfson Laboratories, London, UK

FUNCTIONAL CLASS

Enzyme; lyase; EC 4.2.1.1; a trimeric carbonic anhydrase containing one zinc per monomer, known as Cam.

Three distinct classes of carbonic anhydrases are currently recognized: α, β, and γ. These classes have no detectable sequence or structural similarity, but all coordinate a Zn at the active site and have been proposed to catalyze the interconversion of CO_2 and HCO_3^- by a 'zinc hydroxide' mechanism. The α-class is generally thought of as mammalian, even though prokaryotic α-class carbonic anhydrases have recently been identified.[1,2] The β-class is most often found in plants where it plays a role in the Calvin cycle, but it is also present in algae, bacteria, and archaea.[3] The only characterized member of the γ-class is the secreted carbonic anhydrase, Cam, from the moderately thermophilic methanogenic archaeon *Methanosarcina thermophila*.[4] During acetate catabolism, Cam catalyzes the interconversion of CO_2 and HCO_3^- outside the cell.[5] Cam is a trimeric enzyme[6] with the active site located between monomers, and three active sites per trimer (3D Structure). The fold of each monomer is that of

3D Structure Schematic representation of the γ-class carbonic anhydrase trimer (PDB code 1QRG), with the monomers in magenta, green, and blue. The three orange spheres in between the monomers represent the active-site Zn ions. This figure as well as Figures 1, 5, 8–11 were prepared with the programs MOLSCRIPT[7] and RASTER3D.[8]

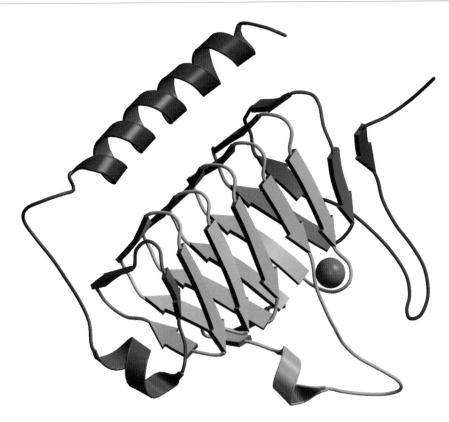

Figure 1 Ribbon representation of the γ-class carbonic anhydrase monomer highlighting the fold of the left-handed β-helix (PDB code 1QRG). The amino-terminus is colored blue, and the chain can be traced by following the color gradient to the carboxy terminus in red. The zinc is shown as an orange sphere.

a left-handed β-helix (Figure 1) with each turn connected by left-handed crossovers.

OCCURRENCE

γ-Class carbonic anhydrases are predominantly found in the archaea, but have sequence homologs in bacteria and eukarya.[9] The function of Cam homologs from other organisms is unknown, although they could act in a similar capacity.

BIOLOGICAL FUNCTION

Methanogenic archaea living in strictly anaerobic environments[10] can derive all their energy from the methanogenesis of a very limited number of substrates. These substrates are restricted to acetate, carbon dioxide, hydrogen, formate, methanol, and methylamines.[11] During acetate fermentation, M. thermophila obtains energy for growth by cleaving the C–C bond of acetate and reducing the methyl group to CH_4, while oxidizing the carbonyl group to CO_2.[12] The activity of a secreted carbonic anhydrase (Cam) increases during acetate catabolism,[13,14] and it has been proposed that CO_2 is converted to HCO_3^- outside the cell to facilitate either the uptake of acetate or the removal of CO_2 from the cytoplasm.

The aceticlastic pathway (Figure 2) is best understood in M. thermophila, where the fermentation occurs in six enzymatic steps. In the first enzymatic step, acetate kinase (Ack) phosphorylates the carboxyl group of acetate,[15,16] consuming one molecule of ATP. In the second enzymatic step, phosphoacetate is transferred to coenzyme A (CoA) by phosphotransacetylase (Pta), yielding acetyl-CoA.[17] The key enzyme in acetate catabolism catalyzes the third enzymatic step and is the five-subunit, multifunctional carbon monoxide dehydrogenase (CODH). CODH cleaves the C–C and C–S bonds of acetyl-CoA,[18,19] oxidizes the carboxyl group to CO_2, and transfers the methyl group to tetrahydrosarcinapterin[20] (H_4SPT). The fourth enzymatic step is catalyzed by an integral-membrane methyltransferase[21] (Tase) and involves the transfer of the methyl group from CH_3–H_4SPT to coenzyme M (CoM) to form CH_3–S–CoM. The activated CH_3–S–CoM can now be reduced in the fifth enzymatic step catalyzed by methyl-CoM reductase (Mer), an enzyme common to all methanogenic pathways.[22] Electrons are supplied for this reaction by heterodisulfide reductase[23] (Hdr) and coenzyme B (CoB). The final enzymatic step involves the conversion of CO_2, a by-product of the CODH reaction in step 3, to HCO_3^- outside the cell and is catalyzed by the trimeric γ-class carbonic anhydrase, Cam.[4]

Energy-yielding steps occur at two points of the acetate fermentation pathway. The first is during the oxidation

Carbonic anhydrases (γ-class)

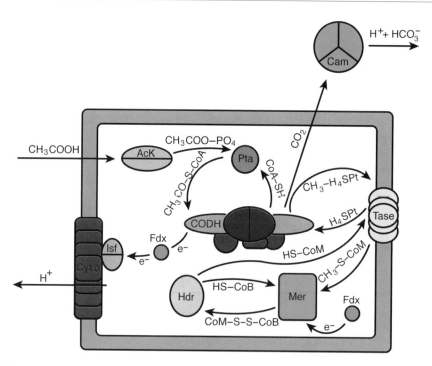

Figure 2 Schematic description of the aceticlastic pathway in *M. thermophila*, where the fermentation occurs in six enzymatic steps. Acetate kinase (Ack) phosphorylates the carboxyl group of acetate, which is then transferred to CoA by phosphotransacetylase (Pta) yielding acetyl-CoA. Carbon monoxide dehydrogenase (CODH) cleaves the C–C and C–S bonds of acetyl-CoA, oxidizes the carbonyl group to CO_2, and transfers the methyl group to tetrahydrosarcinapterin (H_4SPT) and concomitantly transfers electrons through a type-II ferredoxin (Fdx) and iron–sulfur flavoprotein (Isf) to a membrane-bound, proton-pumping cytochrome *b* (Cyt *b*) complex. Methyltransferase (Tase) is responsible for the transfer of the methyl group from CH_3–H_4SPT to CoM to form CH_3–S–CoM. The latter is reduced in the fifth enzymatic step catalyzed by methyl-CoM reductase (Mer). Electrons are supplied for this reaction by heterodisulfide reductase (Hdr) and CoB. The final enzymatic step involves the conversion of CO_2, a by-product of the CODH reaction, to HCO_3^- outside of the cell and is catalyzed by the trimeric γ-class carbonic anhydrase (Cam).

of the carboxyl group of acetate by CODH. Electrons derived from this reaction are shuttled to an integral-membrane cytochrome *b* complex by ferredoxin.[24] The cytochrome *b* complex may act as a proton pump to form a transmembrane proton gradient.[25] The second energy-yielding step occurs when the transfer of the methyl group from H_4SPT to CoM by the integral-membrane methyltransferase is coupled to the pumping of Na^+ ions across the membrane. Both the transmembrane proton and sodium gradients are used to drive ATP synthesis.

AMINO ACID SEQUENCE INFORMATION

Sequence similarity between Cam and other open reading frames in sequenced genomes (Figure 3) as well as Western blotting analysis combined with assays of cell lysates for carbonic anhydrase activity predicts that γ-class carbonic anhydrases are widely distributed throughout the archaea and additionally have homologs in bacteria and eukarya.[9] All of the putative γ-class carbonic anhydrase sequences predict a mature protein with three absolutely conserved histidine residues. These residues are known to be the Zn-ligands in Cam. The abundance of genome sequencing projects has discovered three proteins with better than 70% sequence similarity, but no biochemical analysis has been performed with these gene products. With greater than 60% sequence similarity, the protein CcmM from the cyanobacterium *T. elongatus*,[26] which participates in a CO_2-concentrating mechanism of the organism,[27] is the most closely related protein to Cam that has a described physiological role. The involvement of CcmM in a CO_2-concentrating pathway suggests that it could function as a carbonic anhydrase; however, functional studies have not been performed on purified CcmM or any purified homologs of Cam to show that they actually exhibit carbonic anhydrase activity.

The gene encoding Cam in *M. thermophila*[4] is preceded by a 34–amino acid signal sequence that targets the protein for secretion, which is followed by a cleavage site. Each monomer of the mature, posttranslationally cleaved Cam contains 213 amino acids yielding a molecular mass of 23 kD.

PROTEIN PRODUCTION, PURIFICATION, AND MOLECULAR CHARACTERIZATION

Although Cam is maximally expressed during acetate catabolism, the expression levels are very low even during

Carbonic anhydrases (γ-class)

```
                         10           20           30           40
Cam   Q E I T V D E F S N I R E N P V T P W N P E P S A P V D P T A Y P Q A S V I G E V T I G
CcmM  . . M A V Q S . . . . . . . . Y A A P P T A E P L A P T A Y H S F S N L G D V R E K
Fbp   F E F V K S E E F R L V S R H R T L M N V F D K A P I V D K T A I G G K V I Q
Ara   M E R T L . . . . . . T T V S Y A F E G L I P V H P T A F H P S A V I G D V H I G
Eco                                                                      I N E

                         50           60           70           80
Cam   A N V M V S P M A S I R S D E G M P I F F V G D R S N V Q D D G V V L H A L E T I N E
CcmM  D Y V V H A P G T S I R A D E V P T P F H I G D S R T N I Q D D G V V H G L Q . . . .
Fbp   A N V S H F V P G P Y A V C V L R A D E V D A D G G M Q P I V S A N S N Q D D G V I H S K S . . . .
Ara   R G S S I W Y G C V L G G D Y G . . T V Q A G A N Q D D G V I M H G Y T . . . .
Eco   A G V Y I G P L A S . . . . . . . . . R L I Q A G A N Q D D G C I M H G Y T . . . .

                         90          100          110          120          130
Cam   E G E P I E D N I V E V D G K E Y A V Y I G N N V S L A H Q S I V H G P A A V I G D D T F I G G M
CcmM  . . . . Q G R V I G D D G Q E Y S V W I G D N V S I T H M A L I H G P A Y V I G D D G C F F G G F
Fbp   . . . . . . . . . . . . . L S G K V H P T I G D T I A H R S I V H G P A P E T V E D E T F I G G M
Ara   . . . . . . . . . L S G . L S G K V H P T I G E N G H S V T I G D T I H G A L V H G P A Y I G D D G C F G G F
Eco                             D T D T I V G E N V S V A H R S I V H G P A Y V I G D D G D E T F I G G M
                                                                       R D A L V G M

                        140          150          160          170
Cam   Q A F V F K . S K R . . . G A N N C V L E P R S A A I G . V T I P D G R Y I P . A G M V T S Q A E
CcmM  R S T V F N . . A R . . . G A G C V M M H A V L G Q D . V E P P G K Y V P P . S G M V T Q A D
Fbp   N S V L F N . C R . . . G G D G C V V R H N A V A G A L . . C D L T R I P S G E V W G G N P A R F L R K L T
Ara   G A T L L D G V V . . . G E K H G M V A A G A L . V D G . T R Q N V R Q N P A F Y V P . S T E R F L R K L T
Eco   N S V I M D G A V I G E E S I V A A M S F V K A G F H G E K R Q L L M G . T P A R A V R S V S

                        180          190          200          210
Cam   A D K L P E V T D . D Y A Y S H T N E A V V Y V N V H . . . L A E G Y K E T S . . . . .
CcmM  A D R L P N V E . E S D I H F A Q H V V G I N E A L L S G Y Q C A E N I A C I A P I R N E L Q
Fbp   L A S M P R V S Q S A T N Y S N L A Q A H V V G I N E A L L S G Y Q C A E N I A C I A P I R N E L Q
Ara   D E E I A F I S Q S A T N Y S N L A Q A H V A G E A E N A K P L N V I E F K V L R K K H A L K D E
Eco   L N T K E Y Q D L V G R C H A S . L H E T Q P L R Q M E E N P R L
```

Figure 3 Multiple sequence alignment of putative proteins sharing significant identity with Cam. Identical residues are highlighted in yellow, type conserved aliphatic residues pointing into the interior of the β-helix are highlighted in cyan. Insertions are denoted as periods and every 20th residue of Cam is labeled. Abbreviations are CcmM for carbon dioxide concentrating mechanism protein from *Thermosynechococcus elongates* (NP_681734), Fbp, a hypothetical protein from *Pseudomonas aeruginosa* (NP_254227), Ara from *Arabidopsis thaliana* (AAF79435), and Eco from *Escherichia coli* is possibly involved in the synthesis of the cofactor for carnitine racemase and dehydratase (NP_285730). The first 50 and the last 35 residues of Ara as well as the last 10 and 443 residues of Eco and CcmM were omitted in this sequence alignment.

acetate growth conditions. The first purification of Cam[4] from acetate-grown *M. thermophila* thus required multiple repetitions of ion exchange, hydrophobic, gel filtration, and affinity chromatography. After each purification step, the fractions that exhibited carbonic anhydrase activity were pooled. A DEAE (diethyl aminoethyl) ion-exchange column eluted with a NaCl gradient was followed by phenyl-sepharose chromatography eluted with a decreasing $(NH_4)_2SO_4$ gradient and phenyl-superose column eluted with a step gradient. These first three columns were repeated three to four times. The pooled fractions were subjected to Mono Q anion exchange chromatography eluted with a gradient of KCl and followed by gel filtration chromatography. Finally, affinity chromatography was performed using a *p*-aminomethylbenzenesulfonamide-agarose column and eluted with a step gradient of sodium perchlorate. Protein yields from this purification scheme were low, and one protein preparation would yield only tens of micrograms of protein. Heterologous expression of Cam in *E. coli*[5] resulted in milligram quantities of protein and allowed for a much simpler purification protocol involving a single ion-exchange step on a Q-sepharose column eluted with an increasing linear gradient of NaCl and one phenyl-sepharose column eluted with a decreasing linear gradient of $(NH_4)_2SO_4$. Size-exclusion chromatography under native conditions estimates the molecular mass as 84 kD, suggesting that Cam is either a trimer or a tetramer. Analytical ultracentrifugation analysis[28] suggests that Cam forms both a trimer and a hexamer. The two oligomeric species are in equilibrium in solution, but with the trimer as the predominant species. Both the trimeric and the hexameric states were later observed in crystal structures determined from different crystal forms.[6,29]

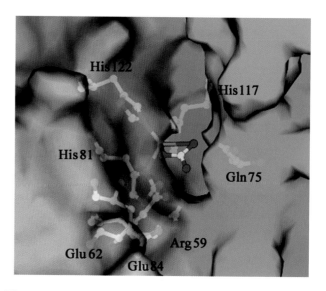

Figure 4 Surface representation of the active site of Cam. The surface is rendered transparent and colored in cyan for one and gray for the other monomer. Residues from both monomers close to the active site Zn are shown in ball-and-stick representation. This figure was prepared with SPOCK.[31]

METAL CONTENT AND COFACTORS

Low yields of enzyme purified from *M. thermophila* precluded metal quantitation on the native protein; however, the enzyme heterologously expressed in *E. coli* contains three metal ions per trimer and 0.6 molecules of Zn per monomer of Cam.[5] Heterologously produced Cam denatured in guanidine-HCl and depleted of metal using the chelator dipicolinic acid can be reconstituted in the presence of $ZnSO_4$ to achieve an incorporation of one molecule of Zn per monomer. The Zn-reconstituted enzyme exhibits kinetic characteristics that suggest that zinc acts as the active site nucleophile,[30] as it does in both the α- and β-class carbonic anhydrases. Zinc coordination by Cam requires three histidine ligands (Figures 3 and 4) at the interface of two monomers with two of the ligands provided by one monomer and the third ligand contributed by the second monomer. Two water molecules complete the zinc coordination sphere (Figure 5, Table 1).

No additional cofactors are known to be present in the enzyme.

In Cam, replacement of zinc by cobalt[30] (Co–Cam) or iron[32] (Fe–Cam) increases the enzymatic activity, whereas substitution by any other metal reduces the catalytic activity to undetectable levels. Co–Cam exhibits an increased number of water ligands and longer metal–ligand distances.[29,30] Replacing Zn by Co in human CAII has the opposite effect compared to Cam[33] and reduces the activity of CAII by 50% compared to the wild-type protein.[34]

ACTIVITY TEST

Activity of the γ–carbonic anhydrase is determined at room temperature using a modification of the electrometric method of Wilbur and Anderson[35] in which the time required for a pH change is measured. The enzyme is mixed with a solution containing 20 mM veronal*H_2SO_4, pH 8.3, and 1 μM zinc sulfate. The reaction is started by the addition of CO_2-saturated water and the time required for a pH decrease from pH 7.85 to pH 7 is recorded. Units of activity are calculated through $U = (t_0 - t)/t$, where t_0 is the time required for the pH change using boiled enzyme or buffer without the enzyme.

SPECTROSCOPIC PROPERTIES

All spectroscopic studies of Cam have been performed on the heterologously produced enzyme with artificially

Carbonic anhydrases (γ-class)

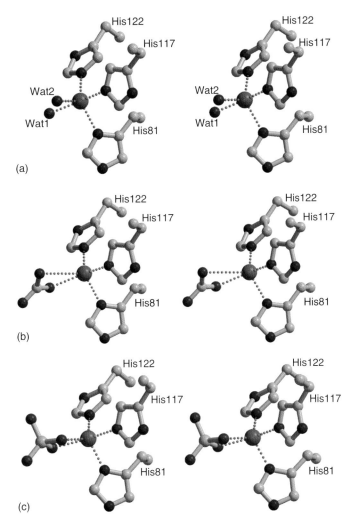

Figure 5 Stereo view of the active site coordination of Cam. (a) The active site Zn exhibits distorted trigonal bipyramidal geometry with two water molecules and three histidine side chains coordinating the metal (PDB code 1QRG). (b) Coordination of Cam bound to bicarbonate (PDB code 1QRL). Bicarbonate binds in a bidentate fashion, replacing both water molecules. (c) Coordination of Cam bound to sulfate (PDB code 11QRM). Binding of the sulfate molecule replaces both water molecules and retains the distorted trigonal bipyramidal coordination seen in (a).

Table 1 Active site geometry of Zn–Cam

Atom	Distance to zinc (Å)	X–Zn–W2 (deg)	X–Zn–81 (deg)	X–Zn–117 (deg)	X–Zn–122 (deg)
Wat 1	2.14	71.9	92.8	161.4	93.1
Wat 2	2.13		121.4	90.1	119.1
His81	2.15			118.0	92.3
His117	2.23				100.3
His122	2.06				
Average	2.14				

incorporated metal.[30] Spectroscopic studies have thus far focused on Cam and Co–Cam since these both have been known to exhibit carbonic anhydrase activity for several years. Fe–Cam, which also exhibits activity, is currently being characterized.

Optical spectroscopy

Zn(II) does not display absorption in the visible range, and indeed protein solutions containing Cam appear colorless. Co(II) has a maximal absorption at a wavelength of 527 nm. Co–Cam solutions, therefore, appear as light pinkish, but with a low extinction coefficient ($\varepsilon = 40\,M^{-1}\,cm^{-1}$).

EXAFS spectroscopy

Cam and Co–Cam have both been investigated by extended X-ray absorption fine structure (EXAFS) spectroscopy. For Cam-containing Zn, the spectra show a coordination by three imidazole ligands, as predicted from the sequence and structural analyses as well as two or three N/O ligands assumed to be water molecules.[30] The average ligand distance for the imidazole was calculated to be 2.06 Å, most closely resembling that of a five-coordinate Zn, which was later observed in the high-resolution crystal structure.[29] The metal–ligand distances determined by EXAFS closely agree with those later assigned in the crystal structures.

Co–Cam exhibits an EXAFS spectrum consistent with protein ligation by three imidazole ligands and two or three N/O ligands.[30] A four-coordinate Co was ruled out because of the absence of a transition at 7720 eV. The spectra suggested slightly longer coordination distances of 2.09 Å for Co than for Zn. Six-coordination for Co in Co–Cam was independently confirmed in the crystal structure.[29]

X-RAY STRUCTURE

Crystallization

The crystallization of Cam has been reported in four distinct crystal forms. 2.0 M $(NH_4)_2SO_4$, 3% PEG 400, and 0.1 M Sodium Cacodylate, pH 6.5 resulted in the formation of tetragonal crystals belonging to the space group $P4_32_12$ with $a = b = 71.84$ Å and $c = 333.51$ Å that diffracted to 2.8 Å resolution.[6] These tetragonal crystals were used for the initial structure determination by the method of multiple isomorphous replacement. A second distinct crystal form that grew from 30% PEG 400, 0.2 M $MgCl_2$, and 0.1 M HEPES pH 7.5 was simultaneously reported that belonged to space group $P2_12_12_1$, with $a = 67.22$ Å, $b = 70.40$ Å, and $c = 311.07$ Å but diffracted to lower resolution.[6] Two other crystal forms were later reported that grew from 3 to 5% PEG 8000 and from 0.1 to 0.5 M $(NH_4)_2SO_4$.[29] The first belonged to the C-centered orthorhombic space group $C222_1$, with $a = 87.4$ Å, $b = 166.4$ Å, $c = 108.7$ Å and the crystals diffracted to 2.5 Å resolution, while the second crystal form diffracted to 1.5 Å resolution and belonged to the cubic space group $P2_13$ with $a = 82.64$ Å. The cubic crystal form was solved by molecular replacement and was used for the detailed characterization of the enzyme.

Description of the polypeptide fold

Each monomer of Cam displays an unusual left-handed β-helix fold (3D structure and Figure 1) topped by a short α-helix and a C-terminal α-helix, which is positioned antiparallel to the axis of the β-helix.[6] When the structure of Cam was first reported, the only other protein known to contain this unusual fold was uridine diphosphate (UDP) N-acetylglucosamine O-acyltransferase.[36] Since then, numerous other proteins have been reported that contain a left-handed β-helix fold (Table 2), and it appears that this fold may represent a stable scaffolding for many types of enzymatic reactions. The basic architecture of the β-helix is an equilateral triangle with three short β-strands of nearly equal length.[6] Another unusual feature of the β-sheets in this left-handed helix is the absence of a twist, which is reflected in the pronounced tendency of the main chain dihedral angles to adopt values of $\phi = -\psi$. The sequence determinants of the left-handed β-helix fold include an internal repeat of branched hydrophobic residues (Figure 6) that point into the center of the helix and stack in adjacent turns of the helix and a very low number of glycine residues. This led to the proposal that side chain stacking interactions helped to dictate β-helix formation.

The minimal oligomeric association of Cam is a trimer, with three monomers arranged around a threefold axis of symmetry (3D Structure). In the hexameric enzyme observed in the tetragonal crystal form,[6] the monomers pack with 32-point group symmetry (Figure 7), with the N-terminal 9 residues of the 6 monomers forming 3 antiparallel β-sheets, each containing 2 strands. These N-terminal residues are poorly ordered when the enzyme forms a trimer in the cubic crystal form.[29]

Active site coordination

As predicted by EXAFS spectroscopy, the active site Zn is coordinated by three histidine ligands and two solvent molecules in trigonal bipyramidal geometry (Figures 4 and 5). The Zn is located at the interface between two monomers with two histidine ligands from one monomer, while the third originates from the second monomer. Interestingly, the stereochemistry of metal coordination by the histidines is remarkably conserved between α-class carbonic anhydrases and γ-class carbonic anhydrases. Two of the histidine ligands coordinate via the Nε2 of the imidazole group, while the third utilizes the Nδ1

Table 2 Structures containing a left-handed β-helix fold

PDB code	Protein	rmsd (Å)	% Identity	Organism	Reference
1THJ	γ-Carbonic anhydrase	0.0	100	Methanosarcina thermophila	6,29
1LXA	UDP-N-acetylglucosamine O-acyltransferase	2.1	15	Escherichia coli	36
1FWY	UDP-N-acetylglucosamine pyrophosphorylase	2.3	11	Escherichia coli	37
1EWW,1M8N	Antifreeze protein	2.3	12	Choristoneura fumiferana	38,39
3TDT	Tetrahydrodipicolinate N-succinyltransferase	2.9	12	Mycobacterium bovis	40
1KQA	Galactoside N-acetyltransferase	3.8	21	Escherichia coli	41
2XAT	Xenobiotic acetyltransferase	3.8	17	Pseudomonas aeruginosa	42

```
              β 1          β 2         β 3
              *  *        *   *        *  *
      25    A P V I D P T A Y I D P Q A S V I G

      43    E V T I G A N V M V S P M A S I R S D E G M

      65    P I F V G D R S N V Q D G V V L H A L E T I N ...

     106    A V Y I G N N V S L A H Q S Q V H G

     124    P A A V G D D T F I G M Q A F V F

     141    K S K V G N N C V L E P R S A A I

     158    G V T I P D G R Y I P A G M V V T S Q A E ...
```

Figure 6 Sequence alignment of the internal repeat of Cam showing residues pointing to the interior of the β-helix by a *, type conserved aliphatic residues in blue, the Zn-ligands in yellow, and β1–β3 are the three β-strands in each turn of the β-helix.

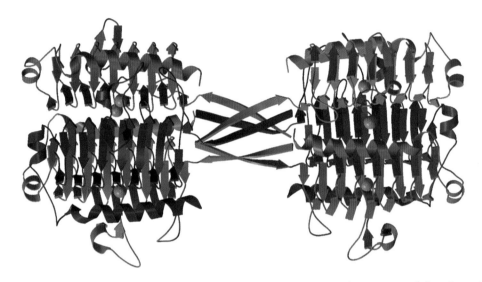

Figure 7 The Cam hexamer viewed along one of its twofold axes of symmetry. Each monomer of the trimers is shown in green, magenta, and blue, respectively. Formation of the hexamer is due to the formation of three antiparallel, two-stranded β-sheets at the N-terminus of each monomer.

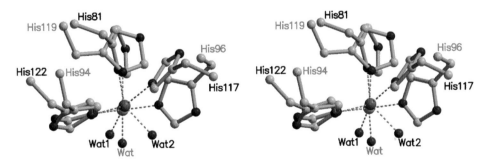

Figure 8 Active site superposition of the Zn ions and the coordinating histidines of Cam and CAII (PDB code 2CBA). The side chains of Cam are shown in gray and the side chains of CAII are shown in green with their labels in black and green, respectively.

atom. A superposition of the active-site residues from α-class and γ-class carbonic anhydrases results in a root-mean-square deviation (rmsd) of 0.76 Å (Figure 8). This amazing example of convergent evolution of Zn ligation occurs in the presence of entirely distinct folds, different oligomeric states, different numbers of solvent ligands, and coordination geometry for Zn. Other residues near the Cam active site closely superimpose with residues in α-class carbonic anhydrases and may be their catalytic counterparts.

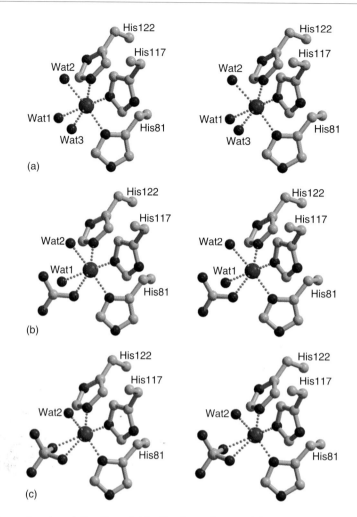

Figure 9 Stereo view of the active site of Co–Cam. (a) In Co–Cam, the metal ligands exhibit distorted octahedral geometry with three water molecules and three histidine imidazoles coordinating the metal (PDB code 1QQO). (b) Coordination of Co–Cam bound to bicarbonate (PDB Code 1QRE). Bicarbonate binds in a monodentate fashion, replacing one water molecule (Wat3). (c) Coordination of Co–Cam bound to sulfate (PDB code 1QRF). Sulfate binds in a bidentate fashion and replaces two water molecules, Wat1 and Wat3.

Co(II) incorporation yields a hexa-coordinated octahedral metal site with three histidine side chains and three solvent molecules as ligands (Figure 9). One of these solvent molecules, Wat1, occupies the same position as one of the solvent ligands of Zn in the native Cam structure (Figures 5 and 9). It is likely that this conserved solvent position represents the position of the catalytic hydroxide involved in enzyme turnover. In α-class carbonic anhydrases, the number of metal ligands has been postulated to be important in determining the rate of catalysis.[43] When Zn is substituted by other metals that lead to an increase in the number of ligands, enzyme activity decreases to undetectable levels. The α-class carbonic anhydrases normally contain a tetrahedrally coordinated zinc at the active site ligated by the side chains of three histidine residues and one solvent molecule. Interestingly, the active sites of Cam and Co–Cam most closely resemble the metal coordination observed in the catalytically incompetent Ni- and Mn- substituted α-class carbonic anhydrases respectively. The fact that Cam activity is elevated when the exchange of the active site metal leads to an increase of the number of ligands was unexpected, considering that α- and γ-class carbonic anhydrases have been proposed to catalyze the hydration of CO_2 by an essentially identical mechanism.

Cam and Co–Cam have additionally been cocrystallized with the substrate HCO_3^- and the mild activator SO_4^{2-}. The binding modes observed for these two anions differ for the two metal-substituted enzymes (Figures 5 and 9). Interestingly, binding of these two molecules to the different metal substituted active sites replaces different metal-ligating water molecules. This simple alteration of ligation positions suggests a deviation from the mechanism proposed for the α-class carbonic anhydrases since α-class carbonic anhydrases utilize a single binding position during enzyme turnover. Zn–Cam retains a penta-coordinated distorted trigonal bipyramidal geometry with either bound ligand. Although Co–Cam remains hexa-coordinated with either ligand, the geometry is significantly distorted in the Co–Cam structure with bound SO_4^{2-}.

FUNCTIONAL ASPECTS

Enzyme activity

The kinetic properties of Cam have been investigated using a modified electrometric method of Wilbur and Anderson or stopped-flow spectroscopy and pH monitoring. Cam exhibits a maximal turnover number of $6 \times 10^4\,s^{-1}$, which approaches the activity of 1×10^6 observed in the fastest α-class enzyme, human carbonic anhydrase II. The catalytic activity of Cam is inhibited by similar chemical compounds as the α-class, albeit to a lesser extent for all of the inhibitors tested (Table 3). Cam activity is pH-dependent, and the pK_a values relevant to enzyme turnover have been determined by both steady state analysis and ^{18}O exchange. There is one ionization event relevant to k_{cat} with a pK_a between 6.5 and 6.8 and two relevant pK_a values affecting k_{cat}/K_m during enzymatic turnover, one between 6.7 and 6.9 and the other between 8.2 and 8.4. One of these ionizations, with a pK_a of 8.5 appears to be involved in the rate-determining step of the enzyme.

The kinetic characterization of Cam suggests that the enzyme mechanism may be similar to that of α-class carbonic anhydrases, which includes a rate-determining step involving a side chain ionization. Since the α-class carbonic anhydrases have been extensively studied for decades, the interpretation of the biochemical studies of Cam can be greatly enhanced by a comparison to this class. α-Class carbonic anhydrases catalyze the hydration of CO_2 by a zinc hydroxide mechanism that occurs in two half reactions (Equations (1) and (2)). In the first half reaction, the active form of the enzyme contains a Zn-bound hydroxide that performs a nucleophilic attack on a CO_2 molecule, thus converting it to HCO_3^-. This first half reaction is aided by the 'gatekeeper' residue Thr199, which orients the CO_2 molecule for attack. The second half reaction is the rate-limiting step of the enzyme and involves the regeneration of the active site Zn-OH by ionizing a Zn-bound H_2O molecule and the transfer of the proton to bulk solvent.

$$E-Zn^{2+}-OH + CO_2 \longleftrightarrow E-Zn^{2+}-HCO_3^- \quad (1a)$$

$$E-Zn^{2+}-HCO_3^- + H_2O \longleftrightarrow E-Zn^{2+}-H_2O + HCO_3^- \quad (1b)$$

$$E-Zn^{2+}-H_2O \longleftrightarrow {}^+H-E-Zn^{2+}-OH^- \quad (2a)$$

$${}^+H-E-Zn^{2+}-OH^- + B \longleftrightarrow E-Zn^{2+}-OH^- + BH^+ \quad (2b)$$

(1a) and (1b): First half reaction, interconversion of CO_2 and HCO_3^- by α-class carbonic anhydrases.

(2a) and (2b): Second half reaction, rate-determining intramolecular and intermolecular proton-transfer steps. B represents the buffer in the bulk solvent.

In vitro, intramolecular proton transfer (Equation (2a)) is rate-determining at high buffer concentrations, while intermolecular proton transfer (Equation (2b)) is rate-determining at low buffer concentrations. In α-class carbonic anhydrases, the second half reaction is aided by the presence of an intermediate proton-shuttling residue, which is His64 in human carbonic anhydrase II.[48,49] Mutational and crystallographic analyses strongly indicate that the analogous residue to His64 is Glu84 in Cam.[29,50] In both cases the side chains adopt multiple discrete conformations in their respective crystal structures.[29,51] In addition, site-directed mutagenesis replacing His64 in human carbonic anhydrase II or Glu84 in Cam by nonionizable residues results in large decreases in k_{cat}/K_m with the enzyme being rescuable by the addition of imidazole to the assay[50] (Figure 10).

Differences in the kinetic behaviors of these enzymes suggest that modifications of the α-class carbonic anhydrase mechanism may be required to fully understand the mechanism of Cam. Some of the α-class carbonic anhydrases exhibit esterase activity; however, assays for Cam activity using *p*-nitrophenylacetate as the substrate showed no detectable turnover.[4] Since these two classes of carbonic anhydrases exhibit no sequence or structural similarity, it is not unexpected that catalysis could proceed by a slightly

Table 3 Comparison of inhibition constants for α- and γ-class carbonic anhydrases

Inhibitor	IC$_{50}$ α-CA	Reference	IC$_{50}$ Cam	Reference	IC$_{50}$ Heterologously produced cam	Reference
Acetazolamide	1.5×10^{-8}	44	$40\,000 \times 10^{-8}$	4	$57\,000 \times 10^{-8}$	5
Sulfanilamide	7.6×10^{-6}	44	$>5000 \times 10^{-6}$	4	$>6000 \times 10^{-6}$	5
Azide	3.4×10^{-3}	45	3.4×10^{-3}	4	3.1×10^{-3}	5
Cyanide	4.8×10^{-5}	46	31×10^{-5}	4	27×10^{-5}	5
Cyanate	2.9×10^{-5}	47	28×10^{-5}	4	17×10^{-5}	5
Iodide	2.3×10^{-2}	47	24×10^{-2}	4,5	ND	5

Carbonic anhydrases (γ-class)

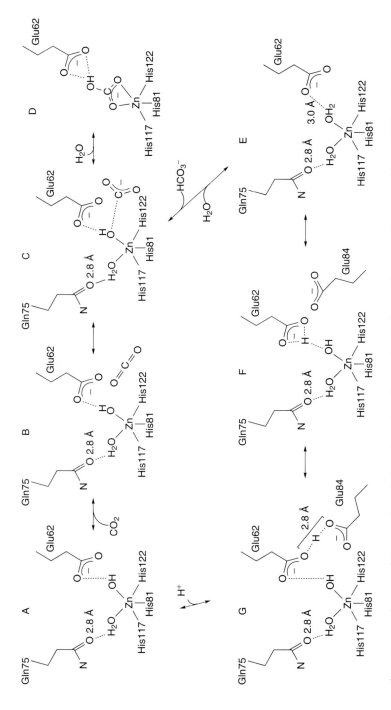

Figure 10 Proposed reaction mechanism of Zn-coordinated Cam. Steps A to E comprise the first half reaction (Equations (1a and 1b)), while F and G constitute the second half reaction (Equations (2a and 2b)). (a) Zn^{2+} is coordinated by one water molecule and one hydroxide ion at the beginning of the first half reaction cycle. (b) CO_2 enters the active site along the hydrophobic pocket. (c) CO_2 is attacked by the Zn-bound hydroxide. (d) The bicarbonate may have several stable binding modes. (e) The first half reaction ends with the exchange of bicarbonate and a water molecule form the solvent. (f) The second half reaction begins with the deprotonation of a zinc-bound water molecule, with the proton being transferred to Glu62. During this process, the side chain of Glu84 swings in so that it can accept the proton. (g) The proton is passed from Glu62 to Glu84. With the subsequent transfer of the proton to the solvent, the second half reaction is complete and state A is regenerated.

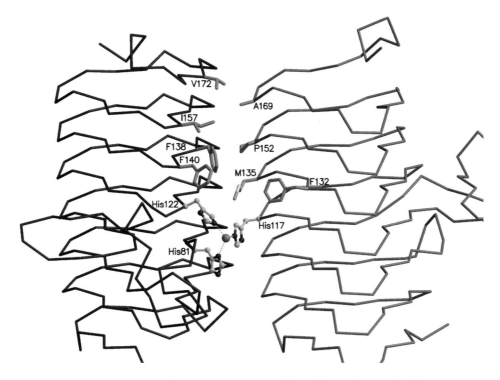

Figure 11 Cα trace of two Cam monomers shown in magenta and blue. The active site histidines are shown in ball-and-stick representation and hydrophobic residues indicating the cleft to the active site are shown in all-bonds representation.

different manner. In human carbonic anhydrase II, both the side chain and amide nitrogen of Thr199 as well as a special water molecule known as the deep water are essential for the reaction mechanism. It is unclear if any residues perform similar functions in Cam. Additionally, mutational analysis of the absolutely conserved active site Arg59 of Cam shows a remarkable decrease in both k_{cat} and k_{cat}/K_m, suggesting that this residue is involved in the CO_2-hydration step (Equation (1a)) of enzyme turnover.[52] No analogous step in the mechanism is known to exist for the α-class carbonic anhydrases. It remains unclear how the CO_2-hydration step occurs in Cam. Additionally, there are other absolutely conserved active site residues in Cam that differ in identity to those in α-class carbonic anhydrases such as Gln75 and Asp61, which await further characterization.

Kinetic analyses indicate that the catalytic mechanism of Cam (Figure 10) should resemble that determined for α-class carbonic anhydrases, utilizing a zinc hydroxide mechanism with proton transfer as the rate-limiting step. CO_2 is believed to enter the active site through a hydrophobic funnel/cleft (Figure 11) in both α- and γ-class carbonic anhydrases. This involves the side chains of Val121, Val143, Leu198, and Trp209 in human carbonic anhydrase II. In Cam, the active site is located between two monomers and the hyrophobic cleft comprises Leu83, Phe132, Met135, Phe138, Phe140, Ile157 and Val172. This funnel in both enzymes ends at the active site metal. A Zn-bound hydroxyl acts as the nucleophile attacking CO_2, thus forming a C−OH bond. Kinetic analyses of mutagenic variants of Arg59 suggest that this residue is involved in the CO_2-hydration step; however, it is unclear in what capacity. HCO_3^- is released from the active site and is replaced by a water molecule on the zinc. The rate-limiting step of catalysis involves deprotonation of the Zn-bound water molecule. The proton is first transferred to Glu62 and subsequently transferred to the intermediate proton-shuttling residue Glu84 before being transferred to bulk solvent. Hydrogen bonding of these two residues to each other and to the oxygen groups of HCO_3^- in the unliganded and HCO_3^- bound high-resolution crystal structures, respectively, indicate that the pK_a values of the side chains are altered by their protein environment.

REFERENCES

1. D Hewett-Emmett and R Tashian, *Mol Phylogenet Evol*, **5**, 50–77 (1996).
2. KS Smith, C Jakubzick, TS Whittam and JG Ferry, *Proc Natl Acad Sci USA*, **96**, 15184–89 (1999).
3. E Soltes-Rak, ME Mulligan and JR Coleman, *J Bacteriol*, **179**, 769–74 (1997).
4. BE Alber and JG Ferry, *Proc Natl Acad Sci USA*, **91**, 6909–13 (1994).
5. BE Alber and JG Ferry, *J Bacteriol*, **178**(11), 3270–74 (1996).
6. C Kisker, H Schindelin, BE Alber, JG Ferry and D Rees, *EMBO J*, **15**(10), 2323–30 (1996).
7. PJ Kraulis, *J Appl Crystallogr*, **24**, 946–50 (1991).
8. EA Merritt and MEP Murphy, *Acta Crystallogr*, **D50**, 869–73 (1994).

9. KS Smith, C Jakubzick, TS Whittam and JG Ferry, *Proc Natl Acad Sci USA*, **96**(26), 15184–89 (1999).
10. W Whitman, D Coleman and W Wiebe, *Proc Natl Acad Sci USA*, **95**, 6578–83 (1998).
11. S Zinder, in J Ferry (ed.), *Methanogenesis*, Chapman & Hall, New York, pp 128–206 (1993).
12. J Ferry, *J Bacteriol*, **174**(17), 5489–95 (1992).
13. M Karrasch, M Bott and R Thauer, *Arch Microbiol*, **151**, 137–42 (1989).
14. P Jablonski, A DiMarco, T Bobik, M Cabell and J Ferry, *J Bacteriol*, **172**(3), 1271–75 (1990).
15. D Aceti and J Ferry, *J Biol Chem*, **263**, 15444–88 (1988).
16. R Fisher and R Thauer, *FEBS Lett*, **228**, 249–53 (1988).
17. L Lundie and J Ferry, *J Biol Chem*, **264**, 18392–96 (1989).
18. D Abbanat and J Ferry, *Proc Nat Acad Sci USA*, **88**, 3272–76 (1991).
19. S Raybuck, S Ramer, D Abbanat, J Peters, W Orme-Johnson, J Ferry and C Walsh, *J Bacteriol*, **173**, 929–32 (1991).
20. D Grahame, *J Biol Chem*, **266**, 22227–33 (1991).
21. H Al-Mahrouq, S Carper and JJ Lancaster, *FEBS Lett*, **207**, 262–65 (1986).
22. P Jablonski and J Ferry, *J Bacteriol*, **173**, 2481–87 (1991).
23. R Hedderich, A Berkessel and R Thauer, *Eur J Biochem*, **193**, 255–61 (1990).
24. R Fisher and R Thauer, *FEBS Lett*, **269**, 368–72 (1990).
25. D Mountfort, *Biochem Biophys Res Commun*, **85**, 1346–50 (1978).
26. Y Nakamura, T Kaneko, S Sato, M Ikeuchi, H Katoh, S Sasamoto, A Watanabe, M Iriguchi, K Kawashima, T Kimura, Y Kishida, C Kiyokawa, M Kohara, M Matsumoto, A Matsuno, N Nakazaki, S Shimpo, M Sugimoto, C Takeuchi, M Yamada and S Tabata, *DNA Res*, **9**(4), 123–30 (2002).
27. G Price, S Howitt, K Harrison and M Badger, *J Bacteriol*, **175**(10), 2871–79 (1993).
28. J Behlke and O Ristau, *Eur Biophys J*, **25**, 325–32 (1997).
29. TM Iverson, BE Alber, C Kisker, JG Ferry and DC Rees, *Biochemistry*, **39**(31), 9222–31 (2000).
30. BE Alber, CM Colangelo, J Dong, CMV Stålhandske, TT Baird, C Tu, CA Fierke, DN Silverman, RA Scott and JG Ferry, *Biochemistry*, **38**, 13119–28 (1999).
31. JA Christopher, *Spock: The Structural Properties Observation and Calculation Kit* (Program Manual), The Center for Macromolecular Design, Texas A&M University, College Station, TX.
32. BC Tripp, K Smith and JG Ferry, *J Biol Chem*, **276**(52), 48615–18 (2001).
33. K Håkansson, A Wehnert and A Liljas, *Acta Crystallogr*, **D50**, 93–100 (1994).
34. J Hunt, M Rhee and C Storm, *Anal Biochem*, **79**, 614–17 (1977).
35. KM Wilbur and NG Anderson, *J Biol Chem*, **176**, 147–54 (1948).
36. CRH Raetz and SL Roderick, *Science*, **270**, 997–1000 (1995).
37. K Brown, F Pompeo, S Dixon, D Mengin-Lecreulx, C Cambillau and Y Bourne, *EMBO J*, **18**, 4096–107 (1999).
38. D Doucet, MG Tyshenko, MJ Kuiper, SP Graether, BD Sykes, AJ Daugulis, PL Davies and VK Walker, *Eur J Biochem*, **277**(36), 33349–52 (2000).
39. EK Leinala, PL Davies, D Doucet, MG Tyshenko, VK Walker and Z Jia, *J Biol Chem*, **277**(36), 33349–52 (2002).
40. TW Beaman, JS Blanchard and SL Roderick, *Biochemistry*, **37**(29), 10363–69 (1998).
41. XG Wang, LR Olsen and SL Roderick, *Structure*, **10**(4), 581–88 (2002).
42. TW Beaman, M Sugantino and SL Roderick, *Biochemistry*, **37**(19), 6689–96 (1998).
43. K Håkansson and WA, *J Mol Biol*, **228**, 1212–18 (1992).
44. G Sanyal, NI Pessah and TH Maren, *Biochim Biophys Acta*, **657**(1), 128–37 (1981).
45. L Tibell, C Forsman, I Simonsson and S Lindskog, *Biochim Biophys Acta*, **789**(3), 302–10 (1984).
46. G Sanyal, ER Swenson, NI Pessah and TH Maren, *Mol Pharmacol*, **22**(1), 211–20 (1982).
47. TH Maren and EO Couto, *Arch Biochem Biophys*, **196**(2), 501–10 (1979).
48. CK Tu, DN Silverman, C Forsman, BH Jonsson and S Lindskog, *Biochemistry*, **28**(19), 7913–18 (1989).
49. S Taoka, CK Tu, KA Kistler and DN Silverman, *J Biol Chem*, **269**(27), 17988–92 (1994).
50. B Tripp and J Ferry, *Biochemistry*, **39**, 9232–40 (2000).
51. J Krebs, C Fierke, R Alexander and D Christianson, *Biochemistry*, **30**(38), 9153–60 (1991).
52. BC Tripp, C Tu and JG Ferry, *Biochemistry*, **41**, 669–78 (2002).

5-Aminolaevulinic acid dehydratase

Jonathan B Cooper and Peter T Erskine
School of Biological Sciences, University of Southampton, UK

FUNCTIONAL CLASS

Enzyme: 5-Aminolaevulinic acid dehydratase (ALAD, E.C.4.2.1.24), also referred to as porphobilinogen synthase, catalyzes one of the initial steps in the biosynthesis of tetrapyrroles involving the condensation of two 5-aminolaevulinic acid (ALA) molecules to form the pyrrole porphobilinogen (PBG) (Scheme 1).[1-5] Single-turnover experiments have shown that the first substrate molecule

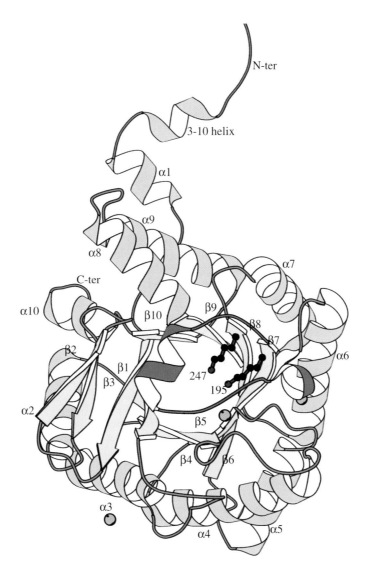

3D Structure Shows a ribbon diagram of the fold of the *E. coli* ALAD monomer (PDB code 1B4E). The dominant feature of the tertiary structure is the closed $(\alpha/\beta)_8$ or TIM-barrel with the active site located in a pronounced cavity (facing reader). Helical segments are shown in lilac, strands in yellow, and the connecting loop regions in cyan. The active site flap (197–220) is colored magenta and the enzyme's N-terminal arm region (residues 1–30), which forms extensive quaternary contacts, is shown above the barrel. The two adjacent lysines implicated in catalysis (195 and 247) are shown in blue. Lys247 forms a Schiff base link with P-site ALA. Green spheres indicate the locations of the zinc binding sites. The active site zinc ion held by three cysteines can be seen close to the two lysines in the center of the TIM-barrel. The zinc ion on the right-hand side (β-site) is involved in intersubunit contacts and that on the lower left is involved in crystal contacts. This figure was prepared using the program BOBSCRIPT.[30]

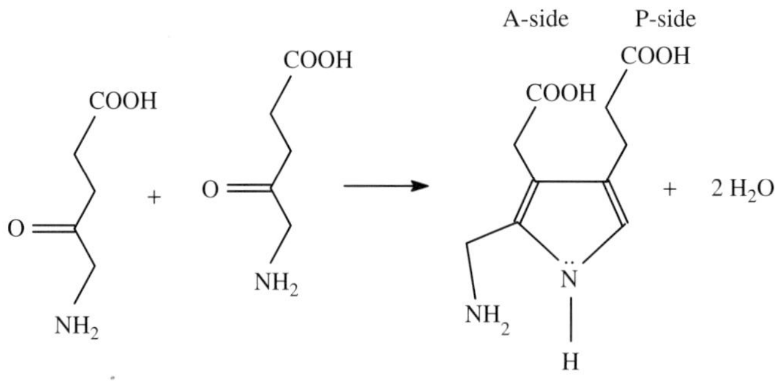

Scheme 1 The Knorr-type condensation reaction catalyzed by 5-aminolaevulinic acid dehydratase (ALAD) indicating the A- and P-sides of the product porphobilinogen.

to bind to the enzyme ultimately forms the 'propionate' half of the product PBG, whilst the second substrate molecule forms the 'acetate' half of PBG.[6] This has led to the widely used terminology of the 'P' and 'A' binding sites in the enzyme that bind the substrates forming the propionate and acetate halves of the product, respectively.

OCCURRENCE

5-aminolaevulinic acid dehydratases (ALADs) have been purified from a variety of sources including bovine liver,[7] human erythrocytes,[8] bacteria such as *Escherichia coli*,[9] and plants such as spinach.[10,11] There are differences between the ALAD enzymes in terms of their metal requirements, kinetic parameters, pH dependence, inactivation by inhibitors, and susceptibility to oxidation. There is evidence that all ALADs require a metal ion for activity with animal enzymes using zinc but with plant enzymes requiring magnesium. Representatives of both classes exist in bacteria.

BIOLOGICAL FUNCTION

The enzyme plays an essential role in tetrapyrrole biosynthesis by catalyzing the formation of the pyrrole porphobilinogen (PBG) from two molecules of 5-aminolaevulinic acid. Four porphobilinogen molecules are subsequently condensed in a reaction catalyzed by porphobilinogen deaminase to form the linear tetrapyrrole, preuroporphyrinogen, which is cyclized and rearranged by uroporphyrinogen synthase to give uroporphyrinogen III, the first macrocyclic tetrapyrrole in the pathway.[1–5] The three steps from 5-aminolaevulinic acid (ALA) to uroporphyrinogen III are common to the biosynthesis of heme, chlorophyll, cobalamins, and all other tetrapyrroles.

The hereditary deficiency of ALAD in humans is associated with the genetic disease Doss or ALAD porphyria,[12] a disease with severe neurological symptoms due to the accumulation of 5-aminolaevulinate, which structurally resembles the neurotransmitter γ-aminobutyric acid (GABA), and may have pharmacologically significant properties.[13] The zinc ions in ALAD are displaced upon addition of lead, a potent inhibitor of the zinc-dependent enzymes.[1–5] Inhibition of ALAD by lead ions is one of the major manifestations of acute lead poisoning, which often leads to neurological and psychotic disturbances. It has been shown that the population has a very wide range of ALAD activity with 2% of individuals having less than 50% of the normal level.[14]

Elevated levels of ALA are also associated with the hereditary disease type I tyrosinaemia,[15–17] which is thought to stem from the accumulation of succinylacetone, a breakdown product of tyrosine and a potent inhibitor of ALAD. Hereditary tyrosinaemia type I is a rare and lethal disease that affects sufferers in early life. The disease is characterized principally by raised plasma tyrosine, liver cirrhosis, and renal tubule problems. The course of tyrosinaemia type I can vary from fulminating liver failure within a few months of birth to a more slowly developing form with a later onset that progresses to liver malignancy within the first decades of life.[16] Patients suffer from neurological crises, and it has been found that peripheral axons of long nerves are subject to degeneration and secondary demyelination.[17] Large amounts of succinylacetoacetate and succinylacetone are excreted by patients with type I tyrosinaemia. Succinylacetone is an effective inhibitor of ALAD having a K_i of 0.8 mM for the *E. coli* enzyme,[18] and it is thought to be responsible for many of the symptoms of this disease.

AMINO ACID SEQUENCE INFORMATION

The protein/cDNA/gene sequences are now available for ALADs from dozens of species. All ALADs share a high degree of amino acid sequence identity, contain about 350 residues per subunit (although the plant enzymes are extended by about 100 residues at the N-terminus) and are usually octameric.[1–5] Studies on bovine and human ALAD showed that catalysis proceeds by the formation of a Schiff base link at the P-site between the 4-keto-group of substrate and an invariant lysine residue equivalent to Lys247 in

E. coli ALAD.[19] Later structural studies established the importance of a second invariant lysine residue (195 in *E. coli* ALAD).[20] Comparison of ALAD primary sequences reveals a strong degree of similarity that is most notable in the vicinity of these two active site lysine residues. Another conserved region of the ALAD sequence has been implicated in metal ion binding at the active site.[21]

PROTEIN PRODUCTION, PURIFICATION, AND MOLECULAR CHARACTERIZATION

Whilst the human and bovine ALAD enzymes have been purified in high yield from blood and liver where the enzyme is present in high abundance,[7,8] the majority of recent studies have involved the production of recombinant enzyme in *E. coli*.[18] In the majority of cases, purification is by conventional chromatographic techniques involving ion exchange and gel filtration, with the presence of reducing agents such as dithiothreitol (DTT) or β-mercaptoethanol being essential to prevent oxidation of the enzyme's metal-binding thiol residues. Gel-filtration studies have shown the majority of ALAD enzymes to be octameric, although there are reports that the plant enzyme is hexameric.[1] Further evidence that the enzyme forms octamers was provided by electron microscopy (EM) and solution scattering studies[22,23] and was finally confirmed by X-ray structural studies.[20]

METAL CONTENT AND COFACTORS

Most of the zinc-dependent ALADs can bind two zinc ions per subunit and these sites have been classified variously as A and B or α and β,[9,24,25] where the metal at the α-site is essential for catalytic activity and the metal at the β-site is thought to have more of a structural role. The catalytic α-site is now known to consist of three conserved zinc-binding cysteine residues. In contrast, ALADs from plants and some prokaryotes lack several of these characteristic cysteines and have a requirement for magnesium ions. For example, two of the conserved cysteines are replaced by aspartates, which are more appropriate for coordination of the essential Mg^{2+} ions.[21] Accordingly, the magnesium-dependent enzymes are less sensitive to oxidation than the zinc-dependent ALADs.

ACTIVITY TEST

The enzyme activity can be assayed spectrophotometrically by the use of Ehrlich's reagent,[26] which forms a colored adduct with the product porphobilinogen. The absorbance of this adduct is measured at 555 nm where its molar absorption coefficient is $6.02 \times 10^4 \, M^{-1} \, cm^{-1}$.

SPECTROSCOPY

The nature of the metal ligands in bovine ALAD has been studied using extended X-ray absorption fine structure (EXAFS), which suggested that one of the enzyme's zinc ions has four cysteine ligands, whereas the other is coordinated by more polar ligands.[24] Subsequent structural studies established that the catalytic zinc ion is coordinated tetrahedrally by only three cysteines, allowing one ligand position to be occupied by a solvent molecule that is likely to be displaced by a bound substrate during the reaction.[20]

X-RAY STRUCTURE

A number of high-resolution X-ray crystal structures have been determined recently, (PDB codes 1AW5, 1B4E, 1B4K),[20,27,28] confirming that the subunits of the enzyme associate to form compact octamers. The monomers of the ALAD octamer are organized with 422 or D_4 point group symmetry with their active sites oriented toward the solvent region. The octamer of *E. coli* enzyme (PDB code 1B4E), which will be treated as the archetypal ALAD for the remainder of this chapter, has overall dimensions of $104 \times 104 \times 83 \, Å^3$ consistent with those deduced from EM and solution scattering studies of the bovine enzyme.[22,23] Each subunit adopts the $(\alpha/\beta)_8$ or TIM-barrel fold with an N-terminal arm approximately 30 residues in length, which forms extensive intersubunit interactions. The TIM-barrel fold is shared by a number of aldolases, a class of enzymes that are functionally related to ALAD. In ALAD, the most extensive of the quaternary contacts are those between pairs of monomers that associate with their arms wrapped around each other to form compact dimers. These dimers further associate to form the octamer that possesses a solvent-filled channel of 15 to 20 Å diameter passing through the center.

Structure of the monomer

The first 30 amino acids of *E. coli* ALAD form an extended armlike structure pointing away from the compact α/β domain (3D Structure). The arm, which is of variable length in different ALADs, includes a region of distorted 3_{10}-helix (residues 8–12) followed by an α-helix (residues 15–21) denoted α1. The remaining tertiary structure of the ALAD monomer is dominated by the $(\alpha/\beta)_8$ or TIM-barrel formed by an eight-membered cylindrical β-sheet surrounded by eight α-helices. In all enzymes adopting the $(\alpha/\beta)_8$ fold, the active site is located in an opening formed by the loops connecting the C-terminal ends of the parallel β-strands in the barrel with their ensuing α-helical segments.[29] In *E. coli* ALAD, the TIM-barrel is formed from the following elements of secondary structure from N- to C-terminus: β1, α2, β4, α3, β5, α4, β6, α5, β7, α6, β8, α7, β9, α9, β10, and α10 as shown in the 3D Structure.[27] The loop regions between these α- and β-segments are elaborated extensively at the active site end of the barrel where the regular alternation of helices and strands is broken by the insertion of several extra secondary structure elements. One example of this is the β-hairpin, which lies between β1 and α2. Here, the hydrogen bonds made by the strands of this hairpin (β2 and β3) mean that they are part of the main β-sheet of the molecule and provide an extension to it at the active site end of the barrel. Another elaboration on the basic TIM-barrel fold is the loop covering the active site (197–220), which includes a region of α-helix involving residues 202 to 208 (in *E. coli*) denoted α_{act}. There was very little electron density for this region of the yeast ALAD molecule when crystallized in the absence of an inhibitor,[20] but it is very well defined in the *E. coli* ALAD structure with the competitive inhibitor laevulinic acid bound. Finally, there is also a helix (α8) that lies between β9 and α9 in the primary sequence. This helix is important for quaternary interactions about a twofold axis within the dimer and will be discussed later.

The longest stretches of conserved sequence in ALADs are all involved in maintaining the structure of the active site, which is formed mainly by the loop regions at the exposed end of the β-barrel. Several of these loops make quaternary contacts with neighboring subunits, implying that they may play a role in transmitting conformational change throughout the octamer. The residues in the loop covering the active site (197–220 in *E. coli* ALAD) are also very strongly conserved, indicating that they have an important role in substrate binding during catalysis.

Structure of the dimer

The N-terminal arm of ALAD (residues 1–30 in *E. coli*) adopts a conformation in which it partly wraps around the $(\alpha/\beta)_8$ barrel domain of a neighboring monomer to form a dimer (Figure 1(a)). These interactions occur between monomers related by an intervening twofold axis so as to generate dimers resembling the number 69. The $(\alpha/\beta)_8$ domains of each monomer interact extensively about the twofold axis so that the two active sites are pointing approximately perpendicular to one another. This interaction is common to all ALADs that have been analyzed structurally and is in accord with cross-linking studies of bovine ALAD,[23] which suggested that the strongest oligomeric interaction between ALAD monomers is dimer formation. The N-terminus of the arm region of each monomer interacts with helices α4 and α5 of the second subunit in the dimer. The arm region itself adopts helical conformations between residues 8 to 12 and 15 to 21 (in *E. coli* ALAD) and these regions interact with helix α6 of the adjacent subunit in the dimer.

The principal barrel–barrel contacts in the dimer involve the helical segments α7, α8, and α9. Helix α8 is approximately perpendicular to the twofold axis of the dimer and pairs in an antiparallel manner with the equivalent helix in the neighboring monomer. Helix α8 appears to be an elaboration of the classical $(\alpha/\beta)_8$ barrel, as mentioned above, perhaps to allow extensive nonpolar intersubunit contacts within the ALAD dimers. Helix α7 interacts with the equivalent helix in the neighboring monomer in a parallel manner as they are roughly aligned with the intervening twofold axis. Residues in helix α7 and α9 form several intra- and intermolecular electrostatic interactions involving conserved or invariant residues indicating that these largely buried ionic interactions may be important for conserving the quaternary structures of all ALADs.

Structure of the octamer

The dimers in the ALAD octamer have the gross appearance of ellipsoids with their long axes inclined slightly with respect to the fourfold symmetry axis. The overall appearance of the octamer is shown in Figure 1(b), where the dimers are colored separately. All dimers within the octamer interact with each other in an identical manner, and these interactions are mediated principally by the arm regions that are on the surface of each dimer. In the dimer–dimer contacts, the two helical segments in the arm region (residues 8–12 and 15–21 in *E. coli* ALAD) associate with one end of a β-barrel in the neighboring dimer. This end of the β-barrel is essentially 'capped' by the arm region of the neighboring subunit, leaving the other end of the barrel, which forms the active site, exposed on the surface of the octamer.

In *E. coli* ALAD, a number of intersubunit interactions are mediated by sulfate anions (present in the crystallization conditions), which suggests that interactions with more physiological dibasic anions, such as phosphate, may play an important role in stabilizing the octamer *in vivo*. The apparent role of dibasic anions in the structure may partly explain the effect that the phosphate buffer has on prolonging the activity of the *E. coli* enzyme during storage.[9]

The active site

The loops at the exposed end of the β-barrel form a pronounced cavity, the base of which is dominated by two spatially adjacent lysine side chains numbered 195 and 247 in *E. coli* ALAD (Figure 2). Both these residues occur at the C-terminal ends of adjacent β-strands in the barrel, namely β7 and β8. The side chains of these two lysines emerge from a hydrophobic pocket formed by the side chains of several tyrosines and other highly conserved residues. The active site also contains a number of invariant polar residues such as Glu123, Ser273, Asp118, and Ser165, whose putative roles in substrate binding and catalysis are discussed later. Lys247, which is known to form a Schiff base to the P-site substrate molecule, has a more hydrophobic environment than its neighbor (Lys195).

The substrate binding cleft in ALAD has a number of aromatic residues, which are predominantly tyrosines whose hydroxyl groups point toward the active site; these aromatic groups are almost totally invariant. The two phenylalanine side chains at positions 36 and 79 appear to form a hydrophobic patch against which nonpolar residues in the helical segment (α_{act}) of the loop covering the active site are packed. The hydroxyl groups of tyrosines 270 and 312 are in hydrogen-bonding contact and are close to the side chain $-NH_2$ group of Lys247. As will be discussed later, these tyrosine residues are likely to have an important role in substrate binding since they interact with the inhibitor, laevulinic acid, covalently bound to Lys247. The $-NH_2$ group of the other lysine, Lys195, is approximately 5 Å from a well-defined zinc

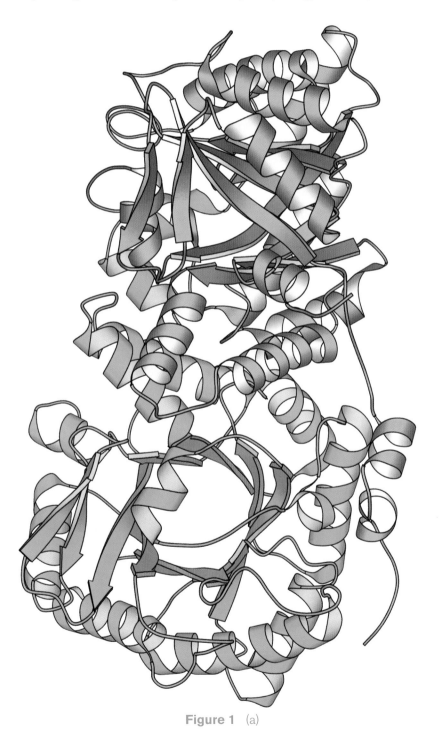

Figure 1 (a)

5-Aminolaevulinic acid dehydratase

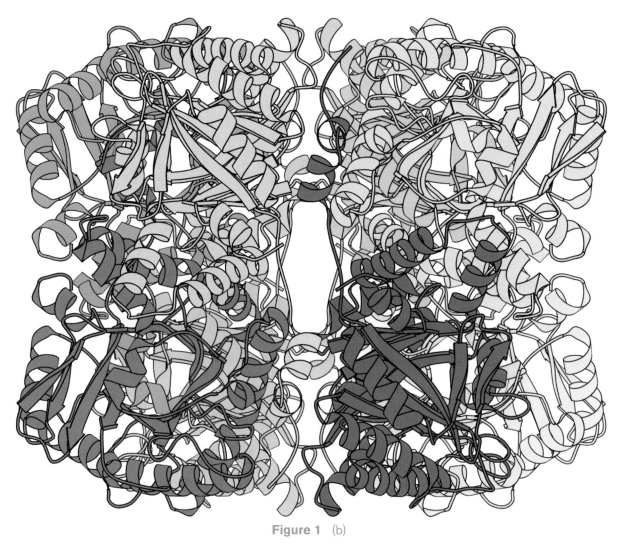

Figure 1 (b)

Figure 1 (a) Shows the formation of *E. coli* ALAD dimers. The two monomers (colored differently) associate extensively about an intervening twofold axis, and their arm regions wrap around the TIM-barrel domain of the neighboring subunit. (b) Shows the organization of the *E. coli* ALAD octamer. The dimers (colored differently) are oriented with their long axes slightly inclined with respect to the central fourfold axis of the octamer (vertical). Interactions between adjacent dimers are mediated principally by the arm regions. For clarity, the subunits within each dimer have been given slightly different colors. This figure was prepared using the program BOBSCRIPT.[30]

ion, which is held by three cysteine side chains and a solvent molecule.

The active site metal ion (α-site)

The cysteine residues numbered 120, 122, and 130 in *E. coli* ALAD are located in the loop connecting β5 and α4 (3D Structure). This region of the molecule was previously identified as a potential metal-binding region from primary sequence comparisons, since it possesses a number of likely zinc binding cysteine residues that are absent in the magnesium-binding plant ALADs.[21] Two of the above three cysteine residues (122 and 130) are replaced by aspartate in plant ALADs and Cys120 becomes Ala. The three cysteine side chains in *E. coli* ALAD along with a water molecule coordinate the zinc ion with tetrahedral geometry. The cysteines coordinate the zinc ion with a mean Zn–S distance of 2.2 Å, and the water ligand has a Zn–O distance of 1.7 Å in the *E. coli* ALAD structure. There is good electron density for this solvent ligand, which lies between the metal ion and the two active site lysines. The zinc ion is close to the side chain of the invariant Ser165, which itself hydrogen bonds to the carboxyl group of an invariant aspartate (Asp118). These two residues interact indirectly with the zinc ion through a network of hydrogen-bonded water molecules. The proximity of these interacting groups to the Schiff base lysine indicates that they most probably participate in substrate binding or catalysis. It is very unusual to find an active site zinc ion coordinated by three cysteine ligands since zinc shows a preponderance of imidazole and carboxyl ligands at the active sites of other enzymes such as the functionally related metalloaldolases.

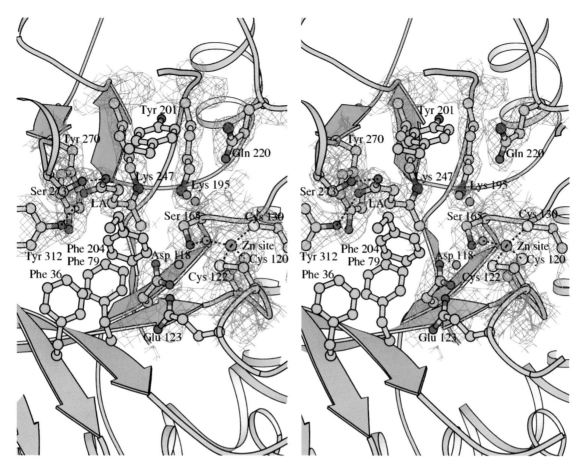

Figure 2 A stereoview of the active site of *E. coli* ALAD showing the two lysines implicated in catalysis (247 and 195) in the top center. Lys247 has formed a Schiff base link to the inhibitor laevulinic acid (LA), which is hydrogen-bonded to Ser273 and Tyr312 (center left). The zinc binding site consists of cysteines 120, 122, and 130 and a solvent molecule that coordinates the metal ion (shown on the right). The $2F_o–F_c$ electron density at 2.0 Å resolution (contoured at 1.2 σ) is shown for selected residues as purple lines, and solvent molecules are colored blue. Phe204 (shown below the laevulinic acid moiety) is notable since it was disordered in the native yeast ALAD structure, that is, its interaction with the inhibitor may contribute to ordering of the active site flap (197–220). The feature of electron density containing two solvent molecules just above Ser165 and in front of Lys195 may originate from a disordered laevulinic acid residue at the A-site. This figure was prepared using the program BOBSCRIPT.[30]

The structure of magnesium-dependent ALAD from the pathogenic bacterium *Pseudomonas aeruginosa* has been solved with laevulinic acid bound,[28] (PDB code 1B4K) revealing that this octameric enzyme effectively consists of four dimers. The monomers of each dimer differ from each other by having 'open' and 'closed' active site pockets. In the 'closed' subunit, the active site is shielded by the active site loop, which is in a well-defined conformation covering the bound laevulinic acid residue. In the 'open' subunit, the active site loop is partially disordered although laevulinic acid appears to be bound. The most surprising finding is that no metal ions could be detected in the active site of either subunit. However, a hydrated magnesium ion was found at the second metal binding site (β-site) of the 'closed' subunit, which is not a conserved feature of ALADs (see below). In the 'open' subunit, conformational differences in the side chains of amino acids preclude magnesium from binding to the β-site.

The allosteric metal binding site (β-site)

Adjacent to the side chain of Lys195 in *E. coli* ALAD is a water-filled pocket that lies between the TIM-barrel domain and the N-terminal arm of the neighboring subunit in the dimer. This pocket contains a zinc ion, which is coordinated octahedrally by one side chain oxygen of Glu232 (in helix $α_6$) and five solvent molecules, presumed to be waters (Figure 3). The glutamate side chain coordinates the zinc with a Zn–O distance of 2.2 Å and the waters have a mean Zn–O distance of 2.0 Å. These water molecules are hydrogen-bonded to surrounding side chain atoms of Asp169, Gln171, and Asp236. There have been many studies of the metal ion requirements of ALAD, and there is experimental evidence for the enzyme having at least two distinct metal binding sites that contribute to its catalytic function.[24,25] While the closest side chain atom of Lys195 is less than 8 Å from this zinc ion, the lysine's ε–NH_2 group

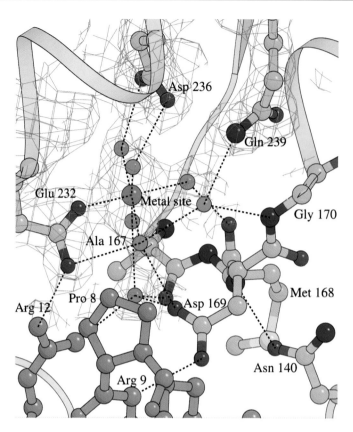

Figure 3 The allosteric metal binding site (β-site) of *E. coli* ALAD occupied by a penta-hydrated zinc ion bound to Glu232. This glutamate side chain and five water molecules coordinate a zinc ion with approximate octahedral geometry. This site is at the subunit interface in the octamer. The lilac arginine and proline side chains (bottom left) originate from the N-terminal arm region of the adjacent monomer. The $2F_o-F_c$ electron density at 2.0 Å resolution (contoured at 1.2 σ) is shown for selected residues as purple lines. This figure was prepared using the program BOBSCRIPT.[30]

is over 11 Å from the metal ion. Hence, it is difficult to envisage a direct catalytic role for this metal site. Nonetheless, it would be expected to contribute significantly to the stability of the octamer since the arm residue Arg12 (from the neighboring subunit) forms a salt bridge with the residue coordinating the metal ion (Glu232). This would appear to be the most likely site at which Mg^{2+} could bind to this enzyme. The apparent activating properties of magnesium on *E. coli* ALAD may therefore stem from an effect on the active site flap, which passes close to this metal ion, or an effect on the enzyme structure close to the active site. In *E. coli* ALAD, it has been shown that magnesium activates the enzyme even at optimal zinc concentrations.[9,18] However, this effect is not observed with yeast ALAD, and in the latter enzyme the β-site is disrupted by an arginine side chain (Arg251), which forms a five-membered salt-bridge interaction involving Asp180, Arg185, Arg251, Asp252, and Asp183. In *E. coli* ALAD, several of these residues are replaced by smaller side chains, thereby creating space for the second metal binding site. The absence of this site (β-site) in yeast ALAD may account for the failure of magnesium ions to activate ALAD from this species at optimal zinc ion concentrations.[18] It is difficult to rationalize this intriguing difference between ALAD enzymes.

In the *P. aeruginosa* enzyme, the subunits within each dimer differ significantly at the catalytic and metal binding sites. Only in the 'closed' monomer is magnesium bound to the enzyme at the β-site. The hydrated magnesium ion at this site interacts with an arginine side chain (Arg181). In contrast, in the 'open' subunit, the absence of bound magnesium causes this arginine to adopt a different conformation in which it points toward the catalytic site and forms a salt bridge with one of the putative active site metal ligands, Asp139, thereby pulling the side chain of this residue away from the active site. Thus, it is proposed that the side chain of Arg181 acts as a conformational switch coupling the two metal binding sites. When the β-site is occupied by a metal ion, Arg181 will be in a position that allows Asp139 to fulfill its putative metal-binding role at the catalytic α-site. Thus, occupation of the β-site leads to an increase in enzyme activity. However, it should be emphasized that the role of a metal ion at the active site of the *P. aeruginosa* enzyme has been questioned since electron paramagnetic resonance (EPR) and nuclear magnetic resonance (NMR) experiments using Mn^{2+} and ^{13}C-labeled PBG appear to exclude the involvement of a metal ion in substrate binding.[31]

FUNCTIONAL DERIVATIVES AND ASPECTS

Definition of the P-site

The X-ray structure of the competitive inhibitor laevulinic acid bound to the P-site of the enzyme has been analyzed for ALADs from a number of species. The inhibitor makes a Schiff base to Lys247 and its carboxyl group forms three hydrogen bonds, one with the side chain of Tyr312 and two with Ser273 involving its hydroxyl and main chain nitrogen (Figure 2). These two residues are invariant in ALADs, as are most of the residues that interact with the hydrophobic moiety of the laevulinic acid. One important aromatic residue belonging to the loop covering the active site is Phe204. This residue packs against the laevulinic acid methylene groups and shields them from the solvent. The active site loop was completely invisible in the native yeast ALAD structure, presumably because of its flexibility in the absence of substrate or inhibitor. Hence, the hydrophobic interaction between P-site laevulinic acid or ALA and Phe204 may be crucially important for ordering the remaining loop residues so that they can function in binding the second substrate.

The structures of yeast ALAD complexed with substrate (ALA) and three inhibitors: laevulinic acid, succinylacetone (SA), and 4-keto-5-aminolaevulinic acid (KAH) have been solved at high resolution (Scheme 2) (PDB codes 1H7O, 1H7N, 1H7R, and 1H7P, respectively).[32] The structure of the complex with ALA was solved with data from crystals of the enzyme that had been purified solely by ammonium sulfate fractionation. Thus, it would appear that ALA remains bound to the enzyme during initial extraction and purification, and it was speculated that ALA would normally be lost during the later stages of enzyme purification, most probably during gel filtration. All of these ligands form a Schiff base link with Lys263 (equivalent to Lys247 in *E. coli*) at the catalytic center, with the exception of KAH, which appears to be linked covalently to the lysine but has become trapped as a carbinolamine intermediate.[32]

The substrate ALA differs from the inhibitor laevulinic acid by the possession of a C-5 amino group, which interacts with the side chains of Asp131 and Ser179. These residues, which are equivalent to Asp118 and Ser165 in *E. coli* ALAD, are invariant, implying that they have an important role in catalysis. Both ALA and laevulinic acid bind to yeast ALAD in two slightly different conformations,[32] but they interact with the same set of residues in each case.

All of the inhibitors induce a significant ordering of the flap covering the active site. Succinylacetone appears to be unique by inducing a number of conformational changes in loops covering the active site, which may be important for understanding the cooperative properties of ALAD enzymes. The part of the active site flap most affected by conformational change upon SA binding is in the region of amino acids 225 to 235. These residues come close to the side chain of Lys210 (Lys195 in *E. coli*) at the catalytic center of the enzyme. The tip of this side chain appeared to move about 2 Å from its position in the native enzyme

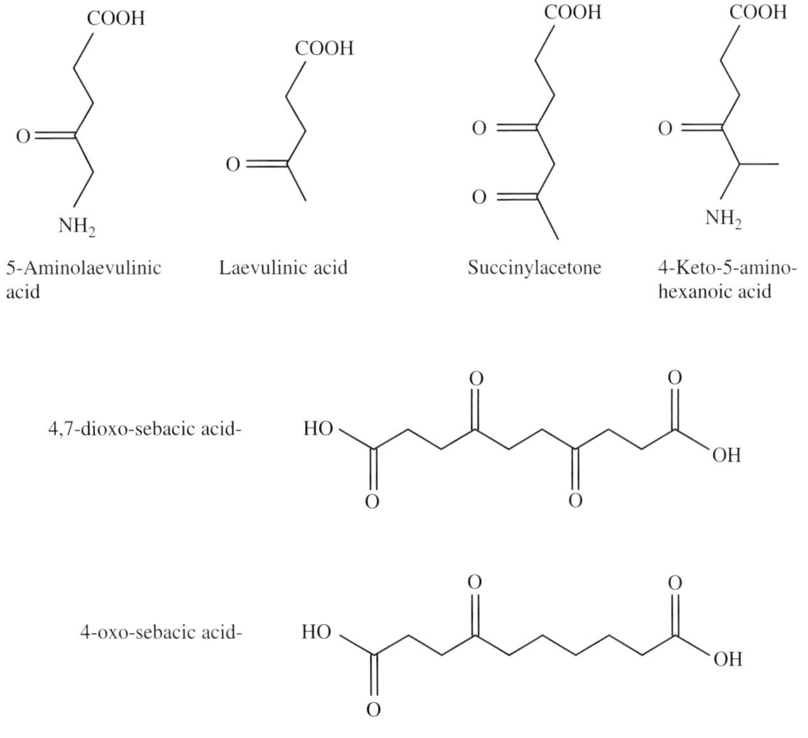

Scheme 2 The formulae of substrate ALA together with various inhibitors of ALAD.

upon SA binding but was largely unaffected by the binding of the other ligands. Thus, it has been suggested that the large conformational changes in the active site flap, which occur upon SA binding, are a knock-on effect of a very local conformational change in Lys210 caused by SA binding to Lys263. The active site flap is also in contact with residues in the N-terminal arm region of a neighboring monomer. These residues are sandwiched between the active site flap and the base of a TIM barrel within an adjacent dimer. This suggests that conformational change in an active site in one dimer can be communicated to neighboring dimers via this contact point. The effects these conformational changes would have on the active sites of neighboring subunits are difficult to predict since the active sites in ALAD are far apart. It is reasonable to expect that changes in the conformation of the active site flap due to substrate binding in one subunit could transmit a conformational change to neighboring monomers via the subunit contacts described above. Subtle changes in relative subunit orientation may affect the flexibility or conformation of the active site flap since it forms part of the subunit interface. Thus, substrate or inhibitor binding to one subunit may affect the catalytic turnover of neighboring monomers.

An intriguing result was obtained with 4-keto-5-amino-hexanoic acid, which seems to form a stable carbinolamine intermediate with Lys263.[32] It appears to define the structure of an intermediate of Schiff base formation, which the substrate forms upon binding to the P-site of the enzyme. The intermediate seems to be held by hydrogen bonds between its 5-amino group and the side chains of Tyr207 and Asp131, and between its carbinolamine −OH group and the side chain of Lys210.

Definition of the A-site

Kinetic studies of ALAD have suggested that the K_m of the P-site is lower than that of the A-site (4.6 μM as against 66 μM for the E. coli enzyme).[33,34] This correlates with the finding that many substrate analogues studied up till now bind predominantly in the P-site and the adjacent A-site appears to be occupied only by water molecules, and that a number of the side chains forming this site are disordered. ALADs typically have turnover numbers (k_{cat}) of around 1.0 per second at their optimal pH values, which are usually in the pH range of 8.0 to 9.0.[18]

The interactions that an ALA molecule could make at the A-site of the enzyme can be predicted with reasonable confidence from the laevulinic acid complexes analyzed so far (Figure 2). Adjacent to the P-site laevulinic acid is a solvent-filled pocket lined by the following highly conserved residues: Asp118, Ser165, Lys195, Arg205, Arg216, Gln220, and the zinc–sulphur cluster involving cysteines 120, 122 and 130 (E. coli numbering). There is good electron density for a solvent molecule bound datively to the zinc ion and this is within H-bonding distance of the amino group of P-side ALA. It has been suggested that this water molecule may be a zinc-bound hydroxide that abstracts a proton from C-3 of A-side ALA during formation of the C–C bond, which eventually links both substrates. In addition, an electrostatic link may exist between the C-5 amino group of P-side substrate and the zinc-bound hydroxide. This putative hydroxide forms an H–bond with Ser165 and forms an indirect H–bond with Asp118 via a water molecule. Both these residues have been implicated in the mechanism by site-directed mutagenesis studies,[35] and their proximity to the C-5 amino group of P-side ALA indicates that they have a role in the catalytic mechanism. Further evidence that this may be the A-site of the enzyme is provided by the presence of an unusual elongated feature of electron density in the E. coli structure, which might represent an additional laevulinic acid molecule bound with low occupancy (Figure 2).

It was suggested that dicarboxylic acids of appropriate length may be able to cross-link the carboxylic acid binding groups associated with the A and P substrate binding sites.[36,37] The structures of ALAD from E. coli and yeast have been analyzed in complex with the 10 carbon-chain irreversible dicarboxylic acid inhibitors 4-oxosebacic acid and 4,7-dioxosebacic acid (formulae are shown in Scheme 2) (PDB codes 1I8J, 1L6S, 1L6Y, 1EB3, 1GJP).[38–40] Both inhibitors bind by forming a Schiff base link with Lys247 (E. coli numbering) at the active site. The most intriguing result of these studies is the novel finding that 4,7-dioxosebacic acid forms a second Schiff base with the enzyme involving Lys195. It has been known for many years that P-side substrate forms a Schiff base (with Lys247) but prior to work on the dicarboxylic acid inhibitors there has been no evidence that binding of A-side substrate involves formation of a Schiff base with the enzyme.

In the 4,7-dioxosebacic acid complexes analyzed, it is notable that hydrogen bonds are formed between the A-site carboxyl group and the side chain of Gln220 as well as the guanidinium groups of Arg205 and Arg216 (in E. coli numbering) (Figure 4). Hence, these arginine residues are very likely to form a salt bridge with the carboxylate of A-side substrate. These arginines (along with Gln220) are strongly conserved residues. Interestingly, 4,7-dioxosebacic acid was found to have higher potency for the zinc-requiring ALADs than for the magnesium-requiring enzymes.[39]

The structure of the substrate analog 5-fluorolaevulinic acid bound to a mutant of P. aeruginosa ALAD has been determined at high resolution (PDB code 1GZG).[41] The mutation D139N converts one of the putative metal ligands into an asparagine residue that is less likely to coordinate metal ions. The same residue has been implicated in coupling the two metal binding sites (catalytic and allosteric; see above). In this remarkable complex, two inhibitor moieties are bound to the enzyme via Schiff base linkages with the two invariant lysines (K205 and K260) and a metal ion (probably sodium) is coordinated by the C-5 fluoro-groups of the two inhibitor molecules. Intriguingly,

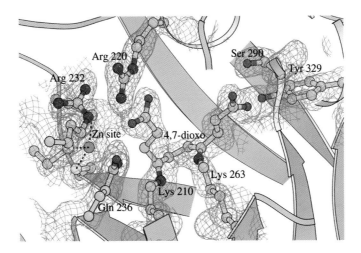

Figure 4 The refined electron density map for the inhibitor 4,7-dioxosebacic acid bound to yeast ALAD solved at 1.75 Å resolution and contoured at 1.5 rms. The P-site is on the right-hand side, and the zinc-binding cysteines, as well as the conserved arginines, which form the A-site, are on the left. This figure was prepared using the program BOBSCRIPT.[30]

the other putative metal ligand Asp131 is involved in binding the metal ion, as are the invariant Ser175 (165 in *E. coli*) and Asp127 (118 in *E. coli*) residues. These findings are consistent with the idea that whilst the native enzyme may not bind a metal ion at the active site, a metal binding site is formed when the substrates bind. These findings may be applicable to all the plant-like ALAD enzymes, and *in vivo*, the metal ion involved is more likely to be potassium than sodium.

CATALYTIC MECHANISM

The synthesis of PBG from two molecules of ALA requires the formation of a C–N bond and a C–C bond with the loss of two water molecules. Single-turnover experiments established that the substrate moiety to bind first forms a Schiff base link with the enzyme at the P-site.[6] It has been suggested that the enzyme may function by forming an additional Schiff base linking the A-site substrate to the enzyme.[37] This was proposed prior to determination of the X-ray structure, which showed that the side chain of the Schiff base lysine (Lys247) is spatially adjacent to the side chain of another invariant lysine (Lys195). Further evidence for a double Schiff base mechanism comes from recent structural studies in which both the invariant active site lysines have been observed forming Schiff bases with bound inhibitor molecule(s).[38–41]

In addition to the Schiff base(s) formed between the enzyme and substrate, another Schiff base is formed between the two substrate moieties. This involves the C-4 atom of the A-side substrate and the amino group of the P-side substrate, eventually forming the C–N bond linking both ALA molecules in the product. The C–C bond formed between the substrates results from the nucleophilic attack by a stabilized carbanion at the C-3 position of the A-side substrate, on the carbonyl carbon of P-side substrate. The latter would be rendered highly electropositive by its Schiff base link to the enzyme.

The ε-NH$_2$ group of Lys247 would have to be neutral for it to act as a nucleophile and to condense with the first ALA molecule to bind to the enzyme. Of the two lysines at the active site, Lys195 is more likely to be positively charged since its environment is more polar than that of Lys247. A positive charge on Lys195 would serve to lower the pK_a of Lys247 and make it a more effective nucleophile for Schiff base formation with P-side ALA. The Schiff base formed by Lys247 and ALA could then become protonated and act in a similar way on the pK_a of Lys195 allowing it to nucleophilically attack the A-side ALA molecule. The central region of the A-side ALA would be in a position to interact with the zinc ion and its associated water molecule or hydroxide ion. These groups could facilitate the deprotonation reactions and assist stabilization of the C-3 carbanion prior to the aldol condensation, which forms the C–C bond linking the two ALA molecules. The C–N bond results from the formation of an intersubstrate Schiff base.

The major issue of discussion in the field has been the order in which the bonds linking the two ALA molecules are formed, that is, whether C–C bond formation occurs before or after C–N bond formation. The formation of a Schiff base between the A-side ALA moiety and Lys195 would further help to labilize the C-3 hydrogen atoms of A-side ALA yielding a bound enamine (see Scheme 3). This could lead to nucleophilic attack of the enamine in the A-site on the C-4 of P-side ALA, thus forming the first C–C bond linking the substrates. The intersubstrate C–N bond could then form. Whilst it is possible that the C–N bond could form first, thus yielding a Schiff base linking both substrates, this would imply that the Schiff base between A-side ALA and Lys195 observed in recent studies[38–41] has no role in the mechanism other than perhaps to anchor the A-side substrate prior to catalysis. If instead it is assumed that the

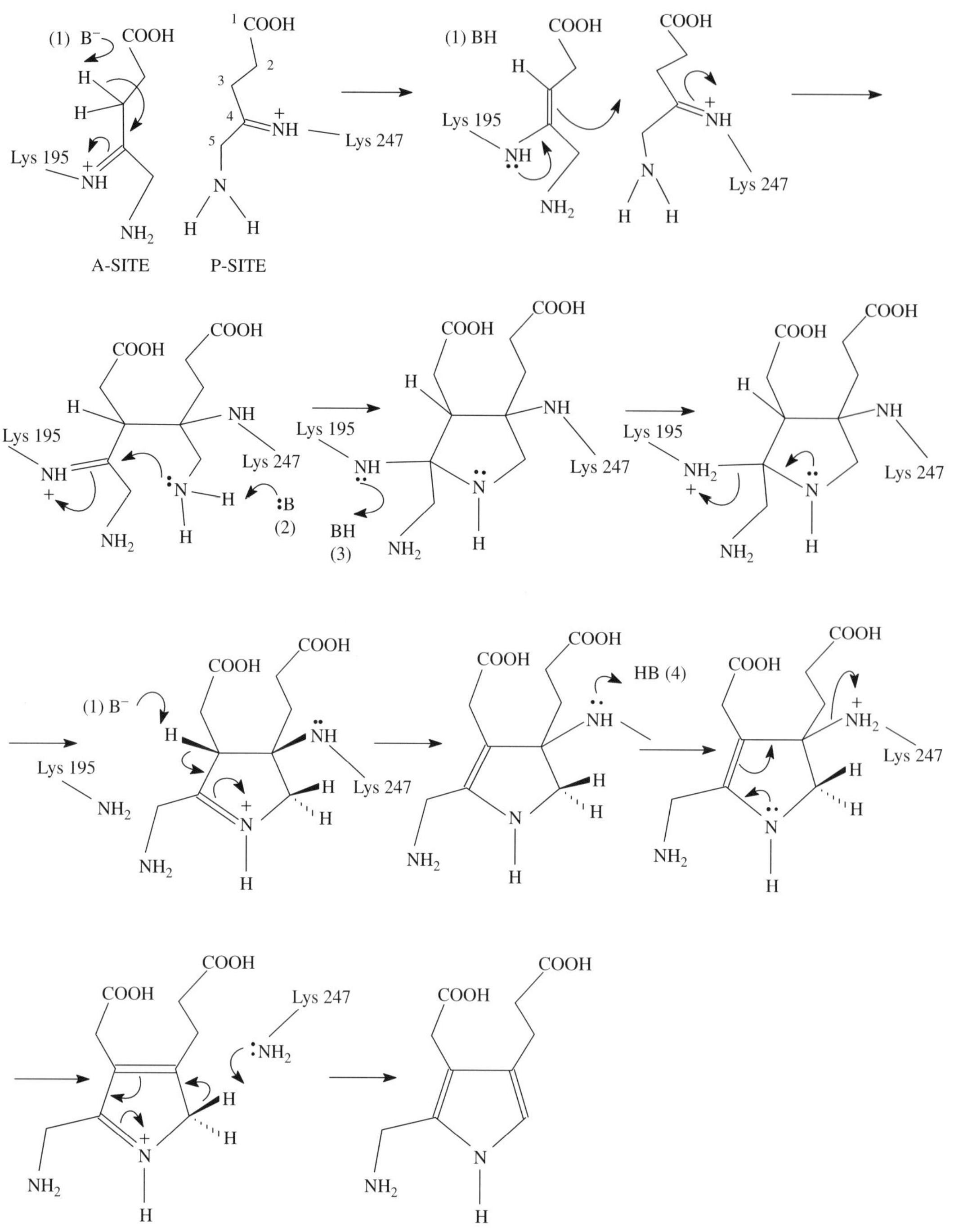

Scheme 3 A proposed reaction mechanism of ALAD catalysis. Note that both ALA molecules are bound to the enzyme by Schiff bases to lysines 195 and 247 (*E. coli* ALAD numbering).

Schiff base linking A-side ALA with Lys195 has a catalytic function (rather than just a passive binding role), then it is likely that intersubstrate C–C bond formation would occur before the C–N bond formation, as shown in Scheme 3.

The final deprotonation shown in Scheme 3 involves the C-5 of the P-side ALA. This step, which has been shown to be stereospecific for the *pro-R* hydrogen,[42] could be catalyzed by Lys247 in view of its proximity. In the reaction, a base (B(2) in Scheme 3) assists the deprotonation of the N-5 of P-side ALA. It is possible that the two proximal active site residues Asp118 and Ser165 may be involved in this process and site-directed mutagenesis[35] implicates both these residues in the mechanism.

REFERENCES

1   PM Jordan, *New Compr Biochem*, **19**, 1–65 (1991).
2   PM Jordan, *Curr Opin Struct Biol*, **4**, 902–11 (1994).

3. MJ Warren and AI Scott, *TIBS*, **15**, 486–91 (1990).
4. EK Jaffe, *J Bioenerg Biomembr*, **27**, 169–79 (1995).
5. EK Jaffe, *Acta Crystallogr*, **D56**, 115–28 (2000).
6. PM Jordan and PNB Gibbs, *Biochem J*, **227**, 1015–20 (1985).
7. K Gibson, A Neuberger and JJ Scott, *Biochem J*, **61**, 618–29 (1955).
8. PM Anderson and RJ Desnick, *J Biol Chem*, **254**, 6924–30 (1979).
9. P Spencer and P Jordan, *Biochem J*, **290**, 279–87 (1993).
10. HAW Schneider and W Liedgens, *Z Naturforsch*, **36C**, 44–50 (1981).
11. W Liedgens, C Lutz and HAW Schneider, *Eur J Biochem*, **135**, 75–79 (1983).
12. M Doss, R Von-Tieperman, J Schneider and H Schmid, *Klin Wochenschr*, **57**, 1123–27 (1979).
13. MJW Brennan and RC Cantrill, *Nature (Lond)*, **280**, 514–15 (1979).
14. S Thunell, L Holmberg and J Lundgren, *J Clin Chem Clin Biochem*, **25**, 5–14 (1987).
15. G Mitchell, J Larochelle, M Lambert, J Michaud, A Grenier, H Ogier, M Gauthier, J Lacroix, M Vanasse, A Larbrisseau, K Paradis, A Weber, Y Lefevre, S Melancon and L Dallaire, *N Engl J Med*, **322**, 432–37 (1990).
16. S Lindstedt, E Holme, EA Lock, O Hjalmarson and B Strandvik, *Lancet*, **340**, 813–17 (1992).
17. B Lindblad, S Lindstedt and G Steen, *Proc Natl Acad Sci USA*, **74**, 4641–45 (1977).
18. N Senior, PG Thomas, JB Cooper, SP Wood, PT Erskine, PM Shoolingin-Jordan and MJ Warren, *Biochem J*, **320**, 401–12 (1996).
19. PNB Gibbs and PM Jordan, *Biochem J*, **236**, 447–51 (1986).
20. PT Erskine, N Senior, S Awan, R Lambert, G Lewis, IJ Tickle, M Sarwar, P Spencer, P Thomas, MJ Warren, PM Shoolingin-Jordan, SP Wood and JB Cooper, *Nat Struct Biol*, **4**, 1025–31 (1997).
21. QF Boese, AJ Spano, J Li and MP Timko, *J Biol Chem*, **266**, 17060–66 (1991).
22. W Wu, D Shemin, KE Richards and RC Williams, *Proc Natl Acad Sci USA*, **71**, 1767–70 (1974).
23. I Pilz, E Schwarz, M Vuga and D Beyersmann, *Biol Chem Hoppe-Seyler*, **369**, 1099–103 (1988).
24. A Dent, D Beyersmann, C Block and SS Hasnain, *Biochemistry*, **29**, 7822–28 (1990).
25. LW Mitchell and EK Jaffe, *Arch Biochem Biophys*, **300**, 169–77 (1993).
26. D Mauzerall and S Granick, *J Biol Chem*, **219**, 435–46 (1956).
27. PT Erskine, E Norton, JB Cooper, R Lambert, A Coker, G Lewis, P Spencer, M Sarwar, SP Wood, MJ Warren and PM Shoolingin-Jordan, *Biochemistry*, **38**, 4266–76 (1999).
28. N Frankenberg, PT Erskine, JB Cooper, PM Shoolingin-Jordan, D Jahn and DW Heinz, *J Mol Biol*, **289**, 591–602 (1999).
29. C Branden and J Tooze, *Introduction to Protein Structure*, Garland Publishing, New York (1991).
30. R Esnouf, *J Mol Graphics Modell*, **15**, 132–34 (1997).
31. N Frankenberg, D Jahn and EK Jaffe, *Biochemistry*, **38**, 13976–82 (1999).
32. PT Erskine, R Newbold, AA Brindley, SP Wood, PM Shoolingin-Jordan, MJ Warren and JB Cooper, *J Mol Biol*, **312**, 133–41 (2001).
33. C Jarret, F Stauffer, ME Henz, M Marty, RM Luond, J Bobalova, P Schurmann and R Neier, *Chem Biol*, **7**, 185–96 (2000).
34. R Neier, *J Heterocycl Chem*, **37**, 487–508 (2000).
35. PM Shoolingin-Jordan, P Spencer, M Sarwar, PT Erskine, KM Cheung, JB Cooper and EB Norton, *Biochem Soc Trans*, **30**, 584–90 (2002).
36. F Stauffer, E Zizzari, C Engeloch-Jarret, J-P Faurite, J Babalova and R Neier, *ChemBioIChem*, **2**, 343–54 (2001).
37. R Neier, *Adv Nitrogen Heterocycles*, **2**, 35–146 (1996).
38. PT Erskine, L Coates, R Newbold, AA Brindley, F Stauffer, SP Wood, MJ Warren, JB Cooper, PM Shoolingin-Jordan and R Neier, *FEBS Lett*, **503**, 196–200 (2001).
39. J Kervinen, EK Jaffe, F Stauffer, R Neier, A Wlodawer and A Zdanov, *Biochemistry*, **40**, 8227–36 (2001).
40. EK Jaffe, J Kervinen, J Martin, F Stauffer, R Neier, A Wlodawer and A Zdanov, *J Biol Chem*, **277**, 19792–99 (2002).
41. F Frere, W-D Schubert, F Stauffer, N Frankenberg, R Neier, D Jahn and DW Heinz, *J Mol Biol*, **320**, 237–47 (2002).
42. AG Chaudhry and PM Jordan, *Biochem Soc Trans*, **4**, 760–61 (1976).

6-Pyruvoyl-tetrahydropterin synthase

Herbert Nar

Boehringer Ingelheim Pharma, Biberach, Germany

FUNCTIONAL CLASS

Enzyme; 6-pyruvoyl-tetrahydropterin synthase (PTPS), lyase, EC 4.6.1.10; zinc-dependent enzyme.

PTPS is a heat-stable, homohexameric protein complex of approximately 90 kDa molecular weight.[1] PTPS catalyzes the conversion of dihydroneopterin triphosphate (H_2NTP) to 6-pyruvoyl-tetrahydropterin, the second of three

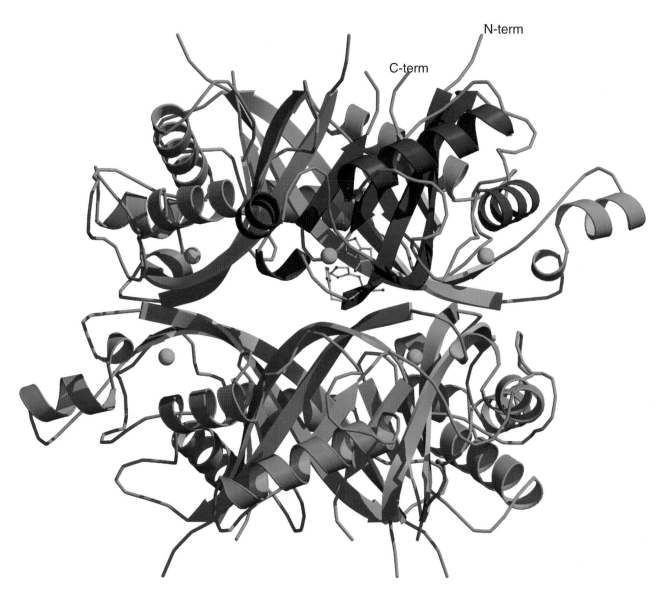

3D Structure   Ribbon representation of the functional hexameric form of PTPS (PDB code 1B6Z). One subunit is shown in colors with the constituents of its metal binding site displayed. The six metal binding sites are indicated by the green spheres. The subunit folds into an $\alpha + \beta$ structure with a four-stranded antiparallel β-sheet surrounded by loops and helices. The association of three subunits along their β-sheets results in the formation of a 12-stranded β-barrel that forms the heart of the PTPS trimer. The functional complex is a dimer of two such trimers. They are associated by stacking of two dyad-related β-sheet regions whose strands run perpendicular to each other.

enzymatic steps in the synthesis of tetrahydrobiopterin (BH$_4$) from guanosine 5′-triphosphate (GTP). The reaction evolves in two steps involving an internal redox transfer and triphosphate elimination. The K_m for the substrate is around 10 μM.[2] The binding of the substrate is Mg(II) dependent.[3]

OCCURRENCE

Occurs across all species. PTPS protein from several species has been purified and characterized. PTPS is considered to be constitutively expressed and subject to posttranslational modifications in higher species.[4] Immunohistochemical studies revealed a cytosolic and nuclear localization of PTPS.[5]

BIOLOGICAL FUNCTION

In vertebrates, PTPS serves as the enzyme catalyzing the second of three enzymatic steps in the synthesis of BH$_4$ from GTP (Scheme 1). BH$_4$ functions as the natural cofactor for the aromatic amino acid hydroxylases, phenylalanine-4-hydroxylase, tyrosine-3-hydroxylase, and tryptophan-5-hydroxylase, the latter two being key enzymes in the biosynthesis of biogenic amines. In addition to the hydroxylation of aromatic amino acids,[6] BH$_4$ serves as the cofactor for nitric oxide synthase[7] and glyceryl-ether monooxygenase.[8]

Lack of BH$_4$ in newborns leads to hyperphenylalaninemia and a deficiency of the biogenic amine neurotransmitters, dopamine, and serotonin, and if untreated, results in severe progressive mental retardation.[6,9–12] The most common form of BH$_4$ deficiency is due to the autosomal recessively inherited absence of PTPS activity. Changes in BH$_4$ metabolism have been observed in different neurological disorders such as Parkinson's disease, Alzheimer's disease, dystonia, and depression.[5]

Glycosidic forms of BH$_4$ have been found in cyanobacteria and *Chlorobium* sp.[13,14] In most prokaryotes, no function for BH$_4$ has yet been detected. Therefore, the role of the bacterial PTPS orthologs must be connected to a hitherto unknown biochemical pathway. Recently, for the recombinantly expressed *Escherichia coli* PTPS, in addition to the established function of the production of 6-pyruvoyl-tetrahydropterin from dihydroneopterin triphosphate, the catalysis of the conversion of sepiapterin to 7,8-dihydropterin was demonstrated.[15] Whether this activity is of relevance for the *in vivo* role of bacterial PTPS orthologs remains to be elucidated.

AMINO ACID SEQUENCE INFORMATION

Amino acid sequence information for PTPS was derived from native protein material isolated from human, rat, and salmon,[16–18] from recombinant cloning of human and rat enzymes[18–20] and from the genome projects that revealed orthologs of PTPS in all cellular species (Figure 1).

Here is a list of some PTPS representatives of bacteria, eukaryota, and archaea:

- O31676, *Bacillus subtilis*, ykvk protein, 149 AA residues[21]
- YB90_HAEIN, *Haemophilus influenzae*, hypothetical protein hi1190, 143 AA residues[22]
- PTPS_ECOLI, *Escherichia coli*, putative 6-pyruvoyl-tetrahydrobiopterin synthase, 121 AA residues[23]
- Q54774, *Synechococcus* sp. (strain pcc 7942), hypothetical 14.6 kDa protein, 128 AA residues
- Q9WXP5, *Thermotoga maritima*, 6-pyruvoyl-tetrahydrobiopterin synthase, putative, 120 AA residues[24]
- PTPS_CAEEL, *Caenorhabditis elegans*, putative 6-pyruvoyl-tetrahydrobiopterin synthase, 126 AA residues (submitted to GenBank DB by Fulton & Wohldmann, 1997)
- PTPS_DROME, *Drosophila melanogaster* (fruit fly), 6-pyruvoyl-tetrahydropterin synthase, 168 AA residues[25]
- PTPS_HUMAN, *Homo sapiens* (human), 6-pyruvoyl-tetrahydrobiopterin synthase, 145 AA residues[20]
- PTPS_RAT, *Rattus norvegicus* (rat), 6-pyruvoyl-tetrahydrobiopterin synthase precursor, 144 AA residues[18]
- Y440_ARCFU, *Archaeoglobus fulgidus*, hypothetical protein af0440, 115 AA residues[26]

PROTEIN PRODUCTION, PURIFICATION, AND MOLECULAR CHARACTERIZATION

PTPS from a few species has been isolated or recombinantly expressed.[18,19] Native protein was isolated from human liver and pituitary gland, rat liver, and salmon.[16,27,28] Purification of the recombinant protein species was done

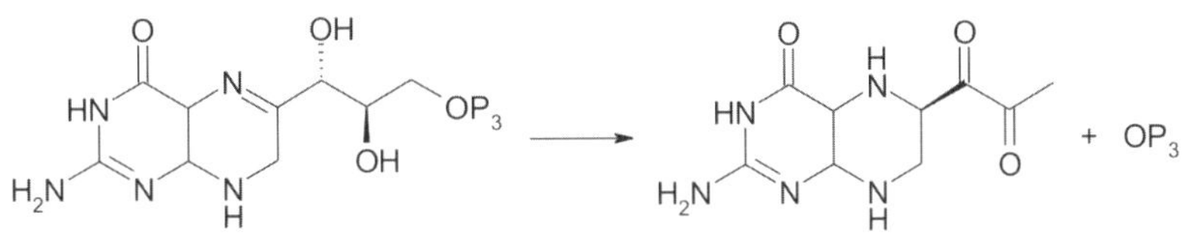

Scheme 1 Reaction catalyzed by PTPS. The substrate dihydroneopterin triphosphate undergoes triphosphate elimination and an internal redox transfer to yield 6-pyruvoyl-tetrahydropterin. This complex reaction is catalyzed by a catalytic machinery of the enzyme including a zinc ion, a nucleophile, and a catalytic triad.

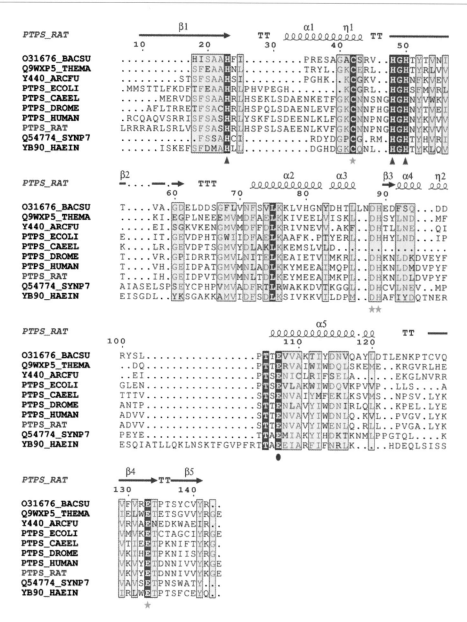

Figure 1 Sequence alignment of representatives of the PTPS family. The chosen representatives correspond to the named sequences in the text. All residues implicated in metal binding (▲), catalysis (★), and substrate recognition (●) are totally conserved throughout the species.

by affinity chromatography[1,15] and by generic protein chromatography steps including gel filtration and heat treatment (due to the extraordinary heat stability of the enzyme). Most detailed studies, including structural work, were performed on the rat liver enzyme.[3,29] A comparison of the isolated and recombinantly expressed forms of this enzyme showed that the mature form of the protein lacks four N-terminal amino acids.[1] The enzymatically active species turned out to be a hexamer of identical subunits that are noncovalently associated in a dimer of trimers (see the section 'X-ray Structure of Rat Liver PTPS').[29] The only cysteine residue in the rat PTPS sequence is implicated in catalysis since covalent modifications and mutagenesis yield inactive protein.[1,3]

Several point mutations were characterized in order to elucidate their role in catalysis or in hyperphenylalaninemia.[3,4] Mutations of residues involved in zinc binding (His23, His48, and His50 rat sequence numbers) inevitably lead to complete loss of activity. Residues involved in catalysis (see below) His89, Cys42, and Glu133 may not be altered, since the mutants tested show either a complete or a dramatic loss of enzymatic activity. All mentioned residues are fully conserved in the PTPS family (Figure 1).

Residues involved in inherited hyperphenylalaninemia are R16C, R25Q, P87L, and a deletion mutant ΔV57 (human PTPS sequence numbers). Further, it was shown that PTPS undergoes phosphorylation and that it requires not-yet-identified posttranslational modifications for its *in vivo* function.[4]

METAL CONTENT AND COFACTORS

Initially, PTPS was described as a Mg(II)-dependent enzyme, and the low amounts of native protein isolated from various species did not allow the detection of other metal cofactors. The first X-ray crystal structure of the recombinantly expressed rat enzyme showed that in the putative active site a typical transition metal binding site was present and was occupied by an atom much richer in electron than Mg(II).[29] Based on this finding, a detailed study of the metal content of heterologously expressed protein and the dependence of its enzymatic activity from the nature of bound metal was initiated.[3] It turned out that the rat protein, as isolated from *E. coli*, contained one zinc ion per subunit and that zinc in combination with the cofactor Mg(II) exhibited the largest activation of the apoenzyme. Co(II) is able to substitute for Zn(II) in the binding site reactivating the apoenzyme to roughly 60% of the holoenzyme activity. The apoprotein can be produced from the holoenzyme by treatment with a mixture of the chelating agents ethylenediaminetetraacetic acid (EDTA) and *o*-phenanthroline.

ACTIVITY TEST

PTPS activity is measured on the basis of the observation that BH_4 is produced from H_2NTP by PTPS and sepiapterin reductase in the presence of Mg(II) and NADPH.[3] The standard mixture contains the following components: 100 mM Tris-HCl (pH 7.4), 8 mM $MgCl_2$, 10 mM dithioerythritol (DTE), 3-m units sepiapterin reductase, 60 mM H_2NTP, and the appropriate amount of PTPS in a final volume of 100 mL. After incubation at 37 °C for 30 min, the reaction is terminated by adding 30 mL of 30% (v/v) trichloroacetic acid. The produced BH_4 is oxidized to biopterin by adding 5 mg MnO_2 within 10 min at room temperature. After centrifugation, an aliquot of the supernatant is injected into the high-performance liquid chromatography (HPLC) system and biopterin is measured by a fluorometric detector at 350/450 nm. One unit PTPS is defined as the amount that catalyzes the production of 1 mmol BH_4/min. For the determination of the K_m value, the following amounts of substrate and PTPS are usually used in the incubation mixture: 2 to 64 mM H_2NTP with 2 ng PTPS. The k_{cat} values are measured by incubating 60 mM H_2NTP with 4 ng PTPS.[2]

X-RAY STRUCTURE OF RAT LIVER PTPS

Crystallization

PTPS is crystallized at 4 °C employing the vapor diffusion technique[2,29,30]. Hanging droplets are made by mixing 5 μL protein solution (protein concentration 6 mg mL^{-1} in 1 mM Mes/NaOH pH 7.4) and 1.7 μL precipitating buffer. The reservoir buffer consists of 1.3 M ammonium sulfate, 0.1 M Tris-HCl (pH 9.0), 2 mM $MgCl_2$.[2,29] Crystals belong to the trigonal space group P321 with unit cell constants $a = 121$ Å, $b = 61$ Å. Two subunits form the asymmetric unit. The functional hexameric form of the enzyme thereby occupies a special position on a crystallographic triad axis.

Overall description of the structure (PDB code 1GTQ)

Monomer structure

The recombinant rat liver PTPS used for crystallization comprises residues 7–144, which fold into a compact, single-domain α + β structure. A sequential, four-stranded, antiparallel β-sheet is composed of residues 8–24, 48–60, 128–133, and 137–142. Layered onto one side of this β-sheet are two antiparallel α-helices comprising residues 72–88 and 106–120 in a sequence segment inserted between β-strands 2 and 3. The helices are connected through a polypeptide stretch folded into a sequence of α-helical turns and a short tripeptide (89–91), which adopts a β-strand conformation and extends the β-sheet structure by hydrogen bonding to residues 20–24 of β-strand 1. Between strands 1 and 2, there is a 25-residue insert containing a two-turn α-helix (residues 32–38), which lies below the two longer helices on the same face of the β-sheet. Strands 3 and 4 are connected via an α-helical turn made by residues 134–136. The domain has an ellipsoidal shape with dimensions 60 × 24 × 18 Å. Its topology belongs to a family of structures characterized by a packing of a layer of β-sheet against a layer of α-helices, both enclosing a hydrophobic core.

Trimer structure

Three monomers of PTPS assemble into a trimer by tight hydrogen bonding between the N- and C-terminal β-strands of adjacent monomers. A 12-stranded antiparallel β-barrel structure is thereby formed, surrounded by a ring of α-helices. The trimer is disc shaped with 60 Å diameter and a height of 30 Å.

The monomers are held together not only by perfect antiparallel β-strand interactions between residues 11–15 on the N-terminal and residues 136–141 on the C-terminal strands but are also interlocked via additional interactions between the loop regions 62–72 on one and 102–111 on the neighboring subunit. The β-strands of the barrel are tilted by 30° with respect to the barrel axis. The barrel is conically shaped with the smaller opening on the side on which the monomer chain termini are located. It encloses a hydrophilic pore of 6 to 12 Å diameter. An unusual cluster of basic and aromatic residues accumulates

6-Pyruvoyl-tetrahydropterin synthase

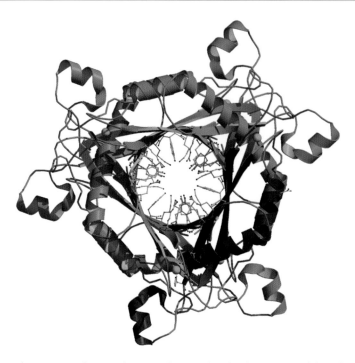

Figure 2 Top view of the PTPS hexamer, with one reference subunit colored. The interior of the β-barrel is formed by the side chains of residues Arg12, Lys53, Lys128, Lys130, and Lys142, as well as Tyr127 and Tyr132 from each subunit, which stretch radially into the pore. The diameter of the pore is 6 Å at the upper end where the chain termini are located and it enlarges to 12 Å at the bottom owing to the conical shape of the barrel.

in the barrel interior. Four lysines, one arginine, and two tyrosine residues from each subunit stretch radially into the pore whose wall is complemented by eight additional hydrophobic or uncharged polar side chains (Figure 2).

Hexamer structure

Two trimers arrange themselves in a head-to-head fashion to form the PTPS hexamer, the functional complex. The contact region between trimers is formed by the 'horizontal' part of the β-sheet of two monomers (residues 20–24, 48–51, and 89–91). The β-strands of the two subunits are <4 Å apart and run almost perpendicular to each other. The hexamer arrangement is stabilized by heterologous, rotationally symmetrical hydrogen bonds between subunits of each trimer. The overall dimensions of the PTPS hexamer are 60 × 60 × 60 Å. As a consequence of the conical shape of the β-barrel of both trimers, the hexamer encloses a large solvent-filled cavity of dimensions 20 × 20 × 15 Å, which opens up not only to the barrel pores but also equatorially to the hexamer surroundings.

Zinc binding site

The PTPS active site is a 12-Å deep cavity that is located on the interface of three monomers, two from one trimer (nomenclature used for numbering amino acids: A and A′)

and one subunit from the other trimer (B). The Zn(II) binding site is located at the center of the cavity. The transition metal is bound to the NE2-atoms of the three histidine residues HisA23, HisA48, and HisA50 (Table 1). In a high-resolution structure of the native protein, a water molecule was identified as the fourth ligand that completes the tetrahedral coordination environment of the metal ion (PDB code 1B6Z). Juxtaposed above the zinc ion are two catalytically important residues GluA133 and CysA42 (Figure 3(a)), which are close to the transition metal at distances of 3.5 to 4.5 Å, but not coordinating to it.

As shown by a complex structure of the catalytically incompetent Cys42Ala mutant of rat PTPS with substrate (PDB code 1B66), the water ligand is replaced by the substrate upon initiation of the catalyzed reaction. Binding of both C1′-OH and C2′-OH groups of the dihydroneopterin triphosphate side chain to the metal

Table 1 Zinc–ligand bond distances in PTPS in (Å)

	Resting state	Dihydroneopterin triphosphate complex
Zn-NE2 His23	2.3	2.3
Zn-NE2 His48	2.3	2.3
Zn-NE2 His50	2.3	2.3
Zn-O water	2.4	
Zn-O1′H		2.4
Zn-O2′H		2.4

6-Pyruvoyl-tetrahydropterin synthase

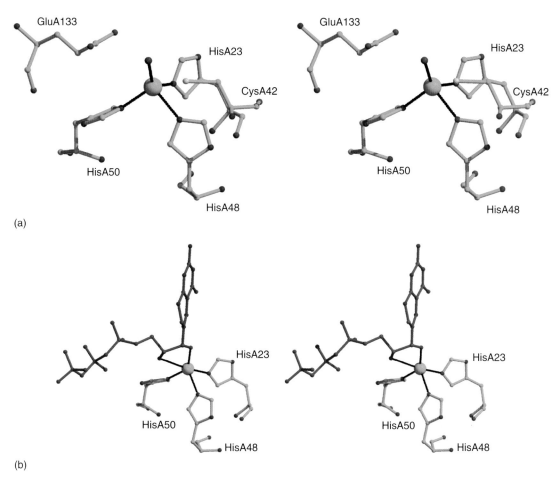

Figure 3 (a) Stereo view of the zinc binding site of the resting form of the enzyme. The zinc ion is bound to the protein via the NE2 atoms of the three histidine residues HisA23, HisA48, and HisA50. The fourth ligand is a water molecule. Also shown are two catalytically important residues, GluA133 and CysA42, which are in close proximity but not coordinating to the metal. (b) Stereo view of the substrate-bound form of the enzyme (PDB code 1B66) with the triphosphate moiety that is not defined in the experimental electron density modeled. The side chain of dihydroneopterin triphosphate directs its C1'-OH and C2'-OH groups toward the zinc ion, replacing the water molecule as a zinc ligand. Thereby, a pentavalent coordination environment is formed with the coordinating ligand atoms at 2.3 to 2.4-Å distance to the metal ion.

yields a distorted trigonal bipyramidal coordination environment, which is most common for catalytic zinc ions (Figure 3(b)).

Both the extension of the coordination environment as well as the proximity of Lewis acids implicated in catalysis are features that are found similarly in other zinc-dependent enzymes, most notably in the metzincin family.[31]

FUNCTIONAL ASPECTS AND CATALYTIC MECHANISM

The catalytic triad

PTPS contains an intersubunit catalytic triad motif composed of the amino acid residues CysA42, HisB89, and AspB88 (Figure 4).[3] The catalytic triad in PTPS is involved in the abstraction of protons from the substrate side-chain carbons with CysA42 acting as the proton acceptor.

Other known catalytic triads or tetrads are involved in the hydrolysis of amide or ester bonds or in peptide cis–trans isomerizations.[32,33] The charge relay system of AspB88 and HisB89 activate the nucleophile CysA42 for proton abstraction. The distance between the CysA42 SG atom and HisB89 NE2 atom is 3.9 Å in the wild-type structure. The histidine is polarized by a hydrogen bond between the ND1 and the carboxylate group of AspB88 (distance 3.3 Å).

Substrate recognition

The atomic nature of the enzyme–substrate interaction was initially derived from the X-ray structure of the uncomplexed wild-type enzyme[29] and modeling studies.[3] Later, it was confirmed by a cocrystal structure of a complex of the catalytically incompetent Cys42Ala mutant of rat PTPS with substrate.

6-Pyruvoyl-tetrahydropterin synthase

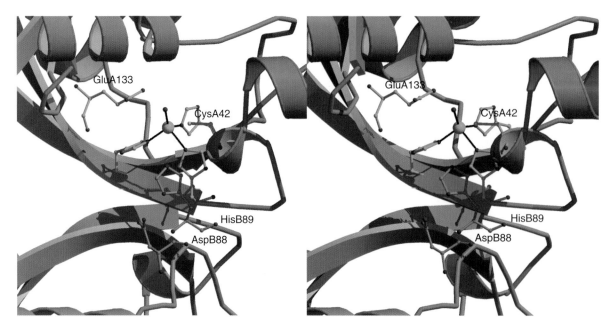

Figure 4 Stereo view of the zinc binding site of the resting state of PTPS in the catalytic environment (residue color scheme: reference monomer A, green carbons; monomer B, yellow). CysA42 of reference subunit A is involved in an intersubunit catalytic triad with residues AspB88 and HisB89 of a neighboring subunit B (yellow carbon atoms) yielding a nucleophile thiolate.

At the bottom of the pocket containing the active site, GluA107 and the peptide bond between ThrA105 and ThrA106 are positioned such as to serve as an acceptor site for the pyrimidine ring of the substrate that is analogous to the purine acceptor site in G-proteins.[34] The walls of the pocket are lined with apolar residues of different subunits (LeuA25, PheA39, MetA'68, and LeuA'72), which flank the heterocyclic ring system of the substrate.

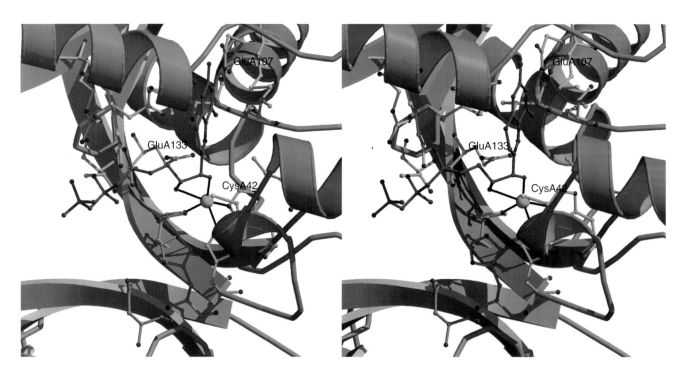

Figure 5 Stereo view of the active site of PTPS with bound substrate (residue color scheme: reference monomer A, green carbons; monomer A', light blue; monomer B, yellow; substrate, magenta). The key residue implicated in substrate binding is GluA107 found at the bottom of the 12-Å deep pocket. Metal coordination positions the side-chain protons of the substrate close to the thiolate of CysA42. The side-chain hydroxyls as well as the ring carbon atom C6 are in proximity to GluA133 that probably is the key player in the stereospecific reduction of the N5–C6 double bond.

This positioning of the dihydroneopterin moiety puts the region of the substrate converted by the enzyme in close proximity to the Zn(II) ion binding site and the catalytic residues GluA133 and CysA42. The substrate binds with its vicinal hydroxyl functions to the vacant coordination sites of Zn(II). Coordination of the two hydroxyls at position C1′ and C2′ of the substrate's side chain with the Zn(II) ion brings the C1′ and the C2′ proton close to the CysA42 thiol group of the active center (Figure 5).

In the crystal structure, the triphosphate group is not defined in the electron density.[2] Modeling of this moiety onto the crystal structure[3] shows that the triphosphate is probably located at the periphery of the active site in which it could be bound by ArgA′15, LysA′76, and by Mg(II) for which there are several possible protein ligands nearby (e.g. ThrA134, AspB88, or SerB20).

Reaction mechanism

On the basis of structural features and kinetics, a hypothetical reaction mechanism can be devised,[2,3] assuming that the reaction starts with proton abstraction from C1′ (Scheme 2). The nucleophilicity of CysA42 Sγ is increased by HisB89 and AspB88. Furthermore, the Zn(II) ion, located at 4.5-Å distance to the Cys Sγ atom, via electrostatic effects leading to a lowering of the CysA42 pK_a, should also contribute to the stabilization of the thiolate Sγ. In the first step, the activated CysA42 abstracts a proton from C1′, leading to intermediate 1. This intermediate 1 is stabilized by conjugation to the pterin ring and by electrostatic interaction with Zn(II). The next step involves the stereospecific protonation of C6, probably catalyzed by GluA133 to produce a 6-R stereoisomer followed by the formation of the C1′ keto group (intermediate 2). GluA133 may thereby serve as a proton shuttle accepting the proton of the C1′ hydroxyl group and transferring it to C6. Prior to the next step, CysA42 has to be deprotonated possibly again via action of the catalytic triad charge relay system. In intermediate 2, the pK_a of C2′ should be lower than that in the substrate because it is now in α-position to the keto group. This should facilitate proton abstraction followed by triphosphate elimination yielding the intermediate 3, which undergoes a keto–enol tautomerization to the product. The stabilizing coordination of the C1′ keto and C2′ hydroxyl groups by Zn(II) ion prevents C1′–C2′ bond cleavage as a side reaction. This is the main reaction in 7,8-dihydroneopterin aldolase.[35] Key elements of the catalytic apparatus in the reaction mechanism of PTPS thus appear to be GluA107 for specific recognition of the pterin moiety, CysA42 as the general base for proton abstraction from C1′ and C2′, and GluA133 as the proton shuttle responsible for the stereospecific protonation of the pterin ring C6 atom, and the Zn(II) ion, which correctly positions the substrate, polarizes the carbon–oxygen bonds for facile proton release from the substrate, and prevents cleavage of the C1′–C2′ bond.[2]

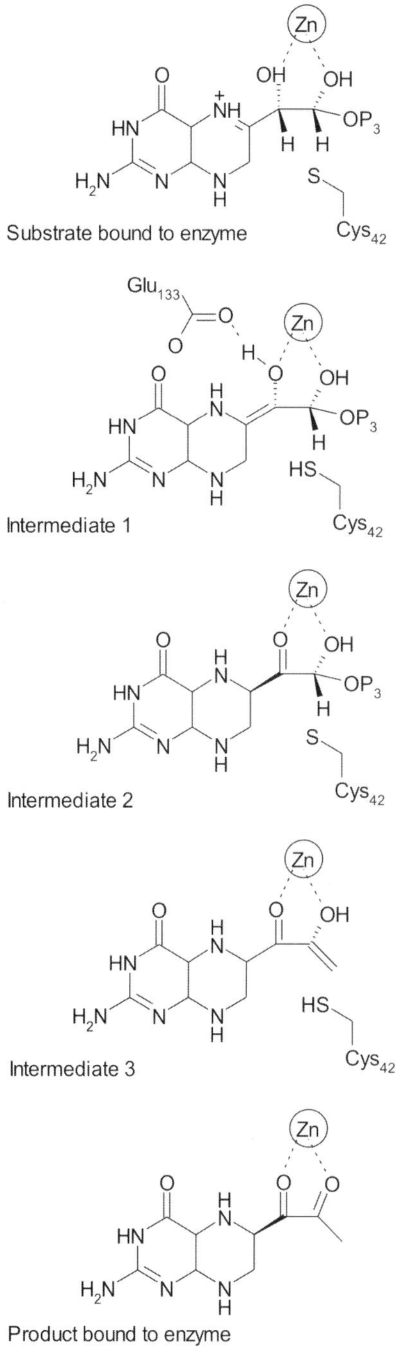

Scheme 2 Putative reaction mechanism of PTPS according to kinetic and structural data. The substrate is converted to the product via three intermediates. The conversion is initiated by reduction of the dihydroneopterin ring system followed by triphosphate elimination.

REFERENCES

1 DM Burgisser, B Thony, U Redweik, P Hunziker, CW Heizmann and N Blau, *Eur J Biochem*, **219**(1–2), 497–502 (1994).

2. T Ploom, B Thony, J Yim, S Lee, H Nar, W Leimbacher, J Richardson, R Huber and G Auerbach, *J Mol Biol*, **286**(3), 851–60 (1999).
3. DM Burgisser, B Thony, U Redweik, D Hess, CW Heizmann, R Huber and H Nar, *J Mol Biol*, **253**(2), 358–69 (1995).
4. T Oppliger, B Thony, H Nar, D Burgisser, R Huber, CW Heizmann and N Blau, *J Biol Chem*, **270**(49), 29498–506 (1995).
5. B Thony, G Auerbach and N Blau, *Biochem J*, **347**(Pt 1), 1–16 (2000).
6. CR Scriver, RC Eisensmith, SL Woo and S Kaufman, *Annu Rev Genet*, **28**, 141–65 (1994).
7. MA Marletta, *Adv Exp Med Biol*, **338**, 281–84 (1993).
8. S Kaufman, RJ Pollock, GK Summer, AK Das and AK Hajra, *Biochim Biophys Acta*, **1040**(1), 19–27 (1990).
9. CR Scriver, CL Clow, P Kaplan and A Niederwieser, *Hum Genet*, **77**(2), 168–71 (1987).
10. S Kaufman, S Berlow, GK Summer, S Milstien, JD Schulman, S Orloff, S Spielberg and S Pueschel, *N Engl J Med*, **299**(13), 673–79 (1978).
11. A Niederwieser, N Blau, M Wang, P Joller, M Atares and J Cardesa-Garcia, *Eur J Pediatr*, **141**(4), 208–14 (1984).
12. N Blau and JL Dhondt, *Adv Exp Med Biol*, **338**, 255–61 (1993).
13. HJ Chung, YA Kim, YJ Kim, YK Choi, YK Hwang and YS Park, *Biochim Biophys Acta*, **1524**(2-3), 183–88 (2000).
14. SH Cho, JU Na, H Youn, CS Hwang, CH Lee and SO Kang, *Biochim Biophys Acta*, **1379**(1), 53–60 (1998).
15. HJ Woo, YK Hwang, YJ Kim, JY Kang, YK Choi, CG Kim and YS Park, *FEBS Lett*, **523**(1-3), 234–38 (2002).
16. T Hasler and HC Curtius, *Eur J Biochem*, **180**(1), 205–11 (1989).
17. S Takikawa, HC Curtius, U Redweik and S Ghisla, *Biochem Biophys Res Commun*, **134**(2), 646–51 (1986).
18. Y Inoue, Y Kawasaki, T Harada, K Hatakeyama and H Kagamiyama, *J Biol Chem*, **266**(31), 20791–96 (1991).
19. B Thony, W Leimbacher, D Burgisser and CW Heizmann, *Biochem Biophys Res Commun*, **189**(3), 1437–43 (1992).
20. B Thony, W Leimbacher, N Blau, CW Heizmann and D Burgisser, *Adv Exp Med Biol*, **338**, 187–90 (1993).
21. F Kunst, N Ogasawara, I Moszer, AM Albertini, G Alloni, V Azevedo, MG Bertero, P Bessieres, A Bolotin, S Borchert, R Borriss, L Boursier, A Brans, M Braun, SC Brignell, S Bron, S Brouillet, CV Bruschi, B Caldwell, V Capuano, NM Carter, SK Choi, JJ Codani, IF Connerton and A Danchin, *Nature*, **390**(6657), 249–56 (1997).
22. RD Fleischmann, MD Adams, O White, RA Clayton, EF Kirkness, AR Kerlavage, CJ Bult, JF Tomb, BA Dougherty and JM Merrick, *Science*, **269**(5223), 496–512 (1995).
23. FR Blattner, G Plunkett III, CA Bloch, NT Perna, V Burland, M Riley, J Collado-Vides, JD Glasner, CK Rode, GF Mayhew, J Gregor, NW Davis, HA Kirkpatrick, MA Goeden, DJ Rose, B Mau and Y Shao, *Science*, **277**(5331), 1453–74 (1997).
24. KE Nelson, RA Clayton, SR Gill, ML Gwinn, RJ Dodson, DH Haft, EK Hickey, JD Peterson, WC Nelson, KA Ketchum, L McDonald, TR Utterback, JA Malck, KD Linher, MM Garrett, AM Stewart, MD Cotton, MS Pratt, CA Phillips, D Richardson, J Heidelberg, GG Sutton, RD Fleischmann, JA Eisen and CM Fraser, *Nature*, **399**(6734), 323–29 (1999).
25. N Kim, J Kim, D Park, C Rosen, D Dorsett and J Yim, *Genetics*, **142**(4), 1157–68 (1996).
26. HP Klenk, RA Clayton, JF Tomb, O White, KE Nelson, KA Ketchum, RJ Dodson, M Gwinn, EK Hickey, JD Peterson, DL Richardson, AR Kerlavage, DE Graham, NC Kyrpides, RD Fleischmann, J Quackenbush, NH Lee, GG Sutton, S Gill, EF Kirkness, BA Dougherty, K McKenney, MD Adams, B Loftus and JC Venter, *Nature*, **390**(6658), 364–70 (1997).
27. CR Hauer, W Leimbacher, P Hunziker, F Neuheiser, N Blau and CW Heizmann, *Biochem Biophys Res Commun*, **182**(2), 953–59 (1992).
28. S Takikawa, HC Curtius, U Redweik, W Leimbacher and S Ghisla, *Eur J Biochem*, **161**(2), 295–302 (1986).
29. H Nar, R Huber, CW Heizmann, B Thony and D Burgisser, *EMBO J*, **13**(6), 1255–62 (1994).
30. G Auerbach and H Nar, *Biol Chem*, **378**(3-4), 185–92 (1997).
31. W Stocker and W Bode, *Curr Opin Struct Biol*, **5**(3), 383–90 (1995).
32. W Bode and R Huber, *Biomed Biochim Acta*, **50**(4-6), 437–46 (1991).
33. R Ranganathan, KP Lu, T Hunter and JP Noel, *Cell*, **89**(6), 875–86 (1997).
34. HR Bourne, DA Sanders and F McCormick, *Nature*, **349**(6305), 117–27 (1991).
35. C Haussmann, F Rohdich, E Schmidt, A Bacher and G Richter, *J Biol Chem*, **273**(28), 17418–24 (1998).

Zinc-fingers

Cys₂His₂ zinc finger proteins

John H Laity

Division of Cell Biology & Biophysics, School of Biological Sciences, University of Missouri-Kansas City, Kansas City, MO, USA

FUNCTIONAL CLASS

Transcription factor; Cys_2His_2 (C2H2) zinc finger motifs are a class of transcription regulatory proteins that control a wide range of cellular processes such as differentiation, development, and tumor suppression. They are one of the major structural motifs involved in eukaryotic nucleic acid binding.[1-7] Functional specificity of individual C2H2 zinc finger–containing transcription factors is achieved through both the transactivation and C2H2 DNA binding domains. Each individual finger contains 22 to 30 amino acids (AAs) and recognizes a specific DNA sequence spanning 3 to 5 bp.[7] Tandem repeats of the C2H2 motif connected by highly conserved peptide linker sequences are found in most transcription factors of this class. Repetition of the zinc finger domain is often accompanied by altered DNA-site preferences at different fingers. Thus, transcriptional control of individual genes within a large gene pool is achieved through sequence-specific DNA recognition and high-affinity binding by a number of small repetitive C2H2 domains. Functions of the C2H2 zinc finger motif extend beyond DNA binding to protein–RNA and protein–protein interactions. The number of repeated C2H2 motifs found in a single protein ranges from 1 up to 37. Each finger contains one Zn(II) atom coordinated by two cysteines and two histidines at the N-terminal and C-terminal regions of the protein respectively. The general sequence is of the form X_2-Cys-X_{2-4}-Cys-X_{9-12}-His-X_{2-6}-His-L, where X is any amino acid, and L is the peptide linker connecting adjacent fingers.

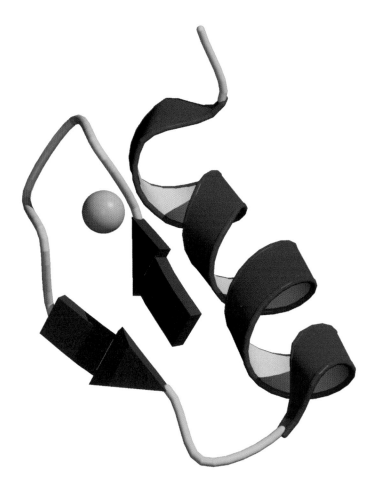

3D Structure   Schematic representation of the solution structure of the C2H2 zinc finger from Xfin (PDB code 1ZNF; model 1).

OCCURRENCE

The C2H2 zinc finger domain was first identified in the *Xenopus laevis* transcription factor TFIIIA,[1] and shortly thereafter in the *Krüppel* segmentation gene of *Drosophila melanogaster*.[8] Genes containing the C2H2 motif have been located in eukaryotic, eukaryotic virus, and prokaryotic phyla. The largest numbers of known sequences to date are from humans (*Homo sapiens*), mouse (*Mus musculus*), African clawed frog (*Xenopus laevis*), yeast (*Saccharomyces cerevisiae*), and the fruit fly (*Drosophila melanogaster*). Approximately 30 proteins from plants containing the C2H2 motif have also been identified.[9] The C2H2 zinc finger Ros protein, which regulates the expression of several virulence genes, was identified in the plant tumor–inducing gram-negative bacterium *Argobacterium tumefaciens*.[10] Homologues of Ros have recently been identified in other bacterial organisms.[11] The bacterial C2H2 Ros protein and Ros homologues contain nine residues between the second cysteine and first histidine, while most eukaryotic motifs contain 12 residues. The point of evolutionary divergence of bacterial and eukaryotic C2H2 zinc fingers is still unclear.

Cellular localization of proteins containing the C2H2 motif is dependent on the protein with which the C2H2 domain is associated, although transcription factor proteins that encompass the majority of functionally identified C2H2 zinc fingers are usually located in the DNA-rich domains of the nucleus. Expression and subcellular localization patterns of many C2H2-containing transcription factors are tissue-specific, and in some cases depend on the growth stage of the organism. For example, in the +KTS and −KTS isoforms of the human proto-oncogene Wilms' tumor suppressor protein (WT1),[12,13] subnuclear localization of the C2H2 zinc finger proteins depends on the nucleic acid binding target. The +KTS isoform of WT1 is an RNA binding protein that preferentially colocalizes with pre-mRNA splicing domains.[14] By contrast, the −KTS isoform of WT1 functions as a classic transcriptional regulatory protein and colocalizes primarily with the DNA-rich domains of the nucleus.[15]

BIOLOGICAL FUNCTION

The major role of C2H2 zinc fingers, which is also the best characterized, appears to be sequence-specific binding to DNA for regulation of gene expression. The number of C2H2 proteins in the vertebrate genomes is myriad. Well over 100 such proteins are in the human brain alone.[16] The known functions and ubiquitous presence of C2H2 regulatory proteins in humans make these proteins potential candidates for many genetic disorders such as neoplasia that may affect cellular proliferation and differentiation. Considerable effort has been devoted to understanding the molecular basis for zinc finger–DNA specificity, through structure and function studies,[7,17–34] and through protein design and phage display approaches.[35–47]

Three distinct groups of C2H2 proteins have been proposed: (i) triple-fingered (tC2H2), (ii) multi-adjacent-fingered (maC2H2), and (iii) separated-paired-fingered (spC2H2).[48] The triple-fingered group comprises three-fingered repeating C2H2 motifs connected mostly by short, conserved peptide linkers with the canonical sequence TGEKP. The extensively studied Egr1 (Zif268) protein for which the first tC2H2 structure bound to DNA was determined is included in this class.[7,27,49] Egr1 seems to have evolved for efficient binding to the GC-rich cognate DNA sequence 5′-GCGTGGGCG-3′ as evidenced by the three residue–DNA base contacts per finger observed in the X-ray crystal structure (described in detail later). The other family of triple-fingered proteins known as Krüppel-like factors (KLF) differ from the Egr1 type of motif at finger 1.[8] In the KLF protein Sp1, finger 1 contributes weakly to DNA binding through only one residue.[50] However, Sp1 binds the primary strand of DNA–RNA heteroduplexes and also interacts with a different class of zinc finger (C2C2), GATA-1, through interactions involving all three fingers of Sp1.[51] The data available for KLF-type three-finger C2H2 proteins suggest that they are capable of binding more than one ligand at a time such as protein, DNA, RNA, or a DNA–RNA hybrid, but not simultaneously since the binding sites overlap.

Proteins of the maC2H2 group have four or more fingers separated by short peptide linkers, which are not necessarily of the TGEKP consensus if the adjacent fingers are not involved in high-affinity binding to the major groove of DNA. The maC2H2 domains are capable of multiligand binding and, in some cases, simultaneously. The maC2H2 *Xenopus* TFIIIA protein is required for correct transcriptional initiation of the 5SRNA gene by RNA polymerase III.[2,52] The observation that large quantities of TFIIIA are also found in the immature ovaries of *Xenopus* in a 7S complex with the 5SRNA transcript can be explained by the differing roles of the nine tandem C2H2 finger repeats.[1,53] Fingers 1–3 are responsible for sequence-specific binding to the C-block of the 5SRNA gene,[18–22,25] fingers 4–7 are required for binding the 5SRNA transcript,[23,54–58] and fingers 8 and 9 are required for transcriptional activation.[58,59] WT1 is a four-finger maC2H2 variant with an alternate splice site that inserts (+KTS) or omits (−KTS) a Lys-Ser-Thr tripeptide in the conserved TG(KTS)EKP linker between fingers 3 and 4. Fingers 2–4 of WT1 are highly homologous (64% sequence identity) to the Egr1 protein and bind to DNA with higher affinity than the four-finger domain of WT1.[60,61] The KTS insertion in the linker between fingers 3 and 4 of the +KTS isoform of WT1 increases the linker flexibility,[34] disrupts the C-cap hydrogen bonding structure between the linker and the C-terminal portion of the α-helix in finger 3,[33,62] and abrogates binding of the fourth finger to the DNA.[34] While the biological function of the +KTS isoform of WT1 is unknown, preferential binding to the mouse *Igf-2* gene mRNA transcript by +KTS was observed, in which finger 1 was more important for RNA binding than finger 4.[61]

The zap1 protein from *Saccharomyces cerevisiae*, which regulates expression of genes that encode for a number of zinc transporter proteins, is a C2H2 transcription factor responsible for sensing intracellular Zn(II) levels.[63] There are 7 maC2H2 motifs in zap1. One of the transactivation domains (ADII) encompasses fingers 1 and 2, and is separated by 65 amino acids from the DNA binding domain consisting of fingers 3–7.[63–66]

The third group categorized as spC2H2 zinc fingers occur as pairs that are widely spaced from each other by long polypeptide sequences. Little is known about this group of zinc fingers. The human PRDII-BF1 protein has two pairs of C2H2 fingers separated by 1630 residues, and both pairs recognize the same DNA sequence.[67] The role of this protein is not understood, although hypothetical interactions between distant recognition sites on the same chromosome or, alternatively, interchromosomal linking have been suggested.

AMINO ACID SEQUENCE INFORMATION

More than 700 human genes encoding C2H2 zinc fingers have been identified, making them the most abundant protein motif in the human genome.[68,69] As of December 2002, more than 3000 C2H2 domains from over 600 protein sequences are in the Swiss-Prot database.[70] The C2H2 motif is estimated to comprise as much as 1% of total mammalian protein.[71] Examples from representative phyla and well-studied proteins are given below:

- *Homo sapiens* (Human), WT1, 449 AAs, 4 C2H2 fingers (116 AA), accession number P19544 in the Swiss-Prot database.[12,13]
- *Homo sapiens* (Human), Zfy, 801 AA, 13 C2H2 fingers (374 AA), accession number P08048 in the Swiss-Prot database.[72,73]
- *Mus musculus* (Mouse), Egr1, 533 AA, 3 C2H2 fingers (81 AA), accession number P08046 in the Swiss-Prot database.[49]
- *Tribolium castaneum* (Red flour beetle) HB, 524 AA, 6 C2H2 fingers (163 AA), accession number Q01791 in the Swiss-Prot database.[74]
- *Xenopus laevis* (African clawed frog), TFIIIA, 366 AA, 9 C2H2 fingers (265 AA), accession number P03001 in the Swiss-Prot database.[75]
- *Xenopus laevis* (African clawed frog), Xfin, 1350 AA, 37 C2H2 fingers (1190 AA), accession number P08045 in the Swiss-Prot database.[76]
- *Saccharomyces cerevisiae* (Bakers yeast), Zap1, 880 AA, 7 C2H2 fingers (268 AA), accession number P47043 in the Swiss-Prot database.[77]
- *Saccharomyces cerevisiae* (Bakers yeast), ADR1, 1323 AA, 2 C2H2 fingers (52 AA), accession number P47043 in the Swiss-Prot database.[78]
- *Caenorhabditis elegans*, Sdc-3, 2150 AA, 2 C2H2 fingers (63 AA), accession number P34706 in the Swiss-Prot database.[79]
- *Drosophila melanogaster* (Fruit fly), Kr, 466 AA, 5 C2H2 fingers, (133 AA), accession number P07247 in the Swiss-Prot database.[8]
- *Nasonia vitripennis* (Parasitic wasp), Reverse transcriptase, 1025 AA, 1 C2H2 finger, (24 AA), accession number Q03278 in the Swiss-Prot database.[80]
- Lymphocytic chloriomeningitis virus (strain Pasteur), Z, 51 AA, 1 C2H2 finger (22 AA), accession number P19326 in the Swiss-Prot database.[81]
- Lymphocytic chloriomeningitis virus (strain Traub), Z, 60 AA, 1 C2H2 finger (22 AA), accession number P19325 in the Swiss-Prot database.[81]
- *Argobacterium radiobacter*, Ros, 142 AA, 1 C2H2 fingers (22 AA), accession number Q04152 in the Swiss-Prot database.[10]
- *Argobacterium radiobacter*, RosAR, 142 AA, 1 C2H2 fingers (22 AA), accession number P55324 in the Swiss-Prot database.[82]
- *Rhizobium* sp., MucR, 143 AA, 1 C2H2 fingers (22 AA), accession number P55363 in the Swiss-Prot database.[83]

PROTEIN PRODUCTION, PURIFICATION, AND MOLECULAR CHARACTERIZATION

The earliest purification of a C2H2 domain prior to publication of any gene sequence information involved isolation of the nine C2H2 zinc fingers of TFIIIA transcription factor as an intact 7S particle containing the protein in complex with the 5SRNA transcript from immature *Xenopus laevis* oocytes.[84] The 7S particle was purified using high-speed centrifugation, followed by fractionation on a gel filtration column and a precipitation step using calcium chloride. The yield was 0.5 to 1 mg of pure 7S particle per ovary. The stability of this particle in the absence of chelating agents such as ethylenediaminetetraacetic acid (EDTA) and dithiothreitol (DTT) suggested that a metal was involved in the binding of the protein to RNA.[2] Models of the nine repeating peptide units for Xenopus TFIIIA with two cysteine and two histidine Zn(II) binding ligands for each unit were subsequently predicted from the published cDNA sequence of TFIIIA;[75] limited proteolysis experiments on the 7S particle using trypsin, chymotrypsin, elastase, and papain; and the measured zinc content extended X-ray absorption fine structure (EXAFS) studies described in the next section.[1,2,5,85]

Preparations of isolated single C2H2 zinc finger domains were prepared by solid-phase peptide synthesis followed by reversed-phase high-performance liquid chromatography (HPLC) using a C4 or C18 column and 0.05 to 0.1% trifluoroacetic acid (TFA) with a linear water-CH_3CN gradient.[6,86–88] Multifingered C2H2 domains were produced through recombinant bacterial expression. Purification was done with reversed-phase chromatography, as described above; gel filtration; cation exchange;[62,89,90] heparin sepharose;[31,91] DNA affinity chromatography; or a combination of these methods.[7,28,29,92,93] Proteins for heteronuclear NMR study were prepared through recombinant expression in E. coli in minimal media using ^{15}N- and ^{13}C-labeled nitrogen and carbon sources respectively.[29,94] Selected polypeptides from an isolated C2H2 finger and larger multifinger complexes with DNA were selectively and uniformly deuterated, respectively, at nonlabile positions in the proteins to resolve issues of sensitivity and spectral crowding for NMR studies.[32,62,95]

METAL CONTENT AND COFACTORS

Analysis by atomic absorption spectroscopy of the intact 7S particles containing the nine-finger TFIIIA protein bound to 5SRNA revealed 7 to 11 atoms of Zn(II) per mole of complex.[1] Subsequent EXAFS spectra of the 7S particle showed that the coordination sphere of the Zn(II) contains two cysteine and two histidine residues.[96] Co(II)-substitution studies clearly showed that the metal binding site of the C2H2 TFIIIA finger 2 was tetrahedral.[97] Intense d-d transitions ($\varepsilon = 400\,M^{-1}\,cm^{-1}$) at 635 nm ruled out higher coordination numbers for the metal site. Charge transfer sites at 310 nm and 340 nm corroborated the earlier EXAFS studies that identified cysteines as ligands for metal binding. Competition experiments showed that Zn(II) bound to the metal site of TFIIIA finger 2 with higher affinity than Co(II) did.[97] Circular dichroism spectroscopy of isolated TFIIIA finger 2 and ADR1 zinc finger domains also demonstrated that the C2H2 motif folded in the presence of Zn(II).[86,97]

NMR SOLUTION STRUCTURES

The number of NMR and X-ray crystal structures of C2H2 zinc finger proteins found in the PDB (Protein Data Bank) number is 20 for single and multifinger polypeptides free in solution and 22 for C2H2 zinc fingers bound to DNA. Therefore, only some of these structures will be considered in the following sections. Since understanding the structural basis for C2H2 finger-DNA recognition equates to elucidating the 'function' for this class of protein domain, detailed descriptions of structural features will be presented for selected isolated finger and finger-DNA motifs. Structures were chosen either because they were the first of their type (free or bound to DNA) or because they contained new features that were judged to best add to the understanding of C2H2 structure and function within the scope of the 'Handbook of Metalloproteins'. A list of all known C2H2 zinc finger polypeptides with a corresponding PDB entry code is summarized in Table 1. It should be noted that, to date, no structures of C2H2 proteins bound to other proteins or RNA have been published, although it is anticipated that structures of these complexes will be reported in the near future.

Zinc finger 31 from the Xfin *Xenopus laevis* transcription factor was the first high-resolution structure of a C2H2 motif (3D Structure).[6] This and all subsequent structures of free (non-DNA bound) C2H2 zinc finger motifs were determined by NMR spectroscopy. Prior NMR studies of ADR1 zinc fingers from *Saccharomyces cerevisiae* indicated the qualitative C2H2 zinc finger fold[86] and corroborated previous predictions of the overall structure.[5,85]

Xfin finger 31

Residues 1–10 form a β-hairpin encompassing the two cysteine ligands that coordinate the Zn(II) ion (Figure 1). A well-defined helix extends from Glu12 to Lys24 (the number immediately following the three-letter amino acid code hereafter refers to the residue position from the N-terminal starting position of the finger). This helix contains both of the histidine ligands that complete the Zn(II) coordination sphere. The Zn(II) atom is buried in the interior core of the protein. The overall structure of the protein is compared to a 'right hand', in which the Zn(II) atom and the C-terminal residue are at the base of the hand, and the amino terminus extends like a thumb towards the fingertip (N-terminal residues of the helix and second β-sheet).[6] The N- and C-terminal residues are at opposite ends of the domain, and their respective Cα atoms are separated by 17.3 ± 1.7 Å (average from 37 structures).

A number of residues form a hydrophobic core in the interior of the protein. These include Phe10, Ala15, Leu16, His19, Gln20, and Val22. The structure of the C2H2 hydrophobic core will be discussed in more detail later with the description of the ZFY-6 and ZFY-'swap',[95,100,101] and SWI5 NMR structures.[94] The family of 10 lowest energy NMR structures from Xfin all had NH → CO hydrogen bonds from Tyr1 → Phe10 and from Phe10 → Tyr1. Additional internal hydrogen bonds from Leu5 → Cys3 and/or from Cys6 → Cys3 were also observed in all 10 NMR structures. Collectively, this hydrogen-bonding pattern delineated the short antiparallel β-sheet and β-hairpin structures. The α-helical region of the Xfin zinc finger was defined by hydrogen bonds between Leu16 → Glu12, Ser17 → Lys13, Arg18 → Ser14, His19 → Ala15, and, in half of the 10 structures, an additional Gln20 → Leu16 hydrogen bond. A 3_{10}-helical structure was evident from the $NH_i \rightarrow CO_{(i-3)}$ hydrogen bonds in the C-terminal residues (Figure 1). The bending of the peptide backbone between histidine ligands to accommodate the tetrahedral coordination to the Zn(II) ion is the cause of this alternative helical structure. The residues with 3_{10} hydrogen bonds were Arg21 → Arg18 (9 structures) and Val22 → His19

Table 1 Summary of published structures of free and DNA-bound Cys$_2$His$_2$-zinc finger motifs

Protein	Source	Finger number(s)	PDB code	NMR/X-ray	Rmsd/resol.[a] (Å)	Reference(s)
Free proteins						
X*fin*	Frog	31	1ZNF	NMR	0.81	6
MBP1	Human					
30 residues		2	3ZNF, 4ZNF	NMR	0.40	
57 residues		1–2	1BBO	NMR	0.32(1), 0.33(2)	98
SWI5	Yeast					
70 residues		1–2	1ZFD	NMR	1.41(1), 0.61(2)	99
45 residues		1	1NCS	NMR	0.39 (T20-H56)	94
ZFY-6	Human					
30 residues		6	5ZNF	NMR	0.6	95,100
Y10K, S12F		6	7ZNF	NMR	~0.6	101
Y10F		6	1KLR	NMR	0.31–0.83[b]	102
Y10L		6	1KLF	NMR	0.44–0.98[b]	102
ADR1	Yeast					
29 residues		1	1ARD	NMR	0.36	103
H118A		1	1ARE	NMR	0.36	104
H118Y		1	1ARF	NMR	0.49	104
PAPA		2	1PAA	NMR	0.21 (Y132-H155)	105
BBA1[c]	Synthetic	1	1HCW	NMR	0.90 (Y1-A22)	39
SP1	Human					
31 residues		2	1SP2	NMR	0.43 (F3-T28)	106
29 residues		3	1SP1	NMR	0.43 (F3-Q26)	106
FSD-1[c]	Synthetic	1	1FSD, 1FSV	NMR	0.54 (Y3-K26)	42
FSD-EY[c]	Synthetic	1	1FME	NMR	0.40 (Y3-K26)	45
ATF-2	Human	1	1BHI	NMR	0.32	90
B3YY1	Human	3	1ZNM	NMR	0.34	90
DNA-bound proteins						
Zif268	Mouse					
90 residues		1–3	1ZAA	X-ray	2.1	7
90 residues		1–3	1AAY	X-ray	1.6	27
DSNR(GACC)[d]		1–3	1A1F	X-ray	2.1	107
DSNR(GCGT)[d]		1–3	1A1G	X-ray	1.9	107
QGSR(GCAC)[d]		1–3	1A1H	X-ray	1.6	107
RADR(GCAC)[d]		1–3	1A1I	X-ray	1.6	107
RADR(GCGT)[d]		1–3	1A1J	X-ray	2.0	107
RADR(GACC)[d]		1–3	1A1K	X-ray	1.9	107
RDER(GCAC)[d]		1–3	1A1L	X-ray	2.3	107
(TATA)zf[d]		1–3	1G2F	X-ray	2.0	46
(TATA)zf*[d]		1–3	1G2D	X-ray	2.2	46
D20A(GCG)[d]		1–3	1JK1	X-ray	1.9	108
D20A(GCT)[d]		1–3	1JK2	X-ray	1.9	108
Dimer[e]		(1–2)$_2$	1F2I	X-ray	2.35	109
GLI	Human	1–5	2GLI	X-ray	2.6	17
Tramtrack	Fruit fly	1–2	2DRP	X-ray	2.8	24
YY1	Human	1–4	1UBD	X-ray	2.5	28
MEY	Synthetic	1–3	1MEY	X-ray	2.2	38

(*continued overleaf*)

Table 1 (continued)

Protein	Source	Finger number(s)	PDB code	NMR/X-ray	Rmsd/resol.[a] (Å)	Reference(s)
TFIIIA	Frog					
92 residues		1–3	1TF3	NMR	0.64	29
190 residues		1–6	1TF6	X-ray	3.1	91
GAGA	Fruit fly	1	1YUI, 1YUJ	NMR	0.42 (P24-F58)	30
ADR1-DBD	Yeast	1–2	2ADR	NMR	3.02(1), 3.17(2)	32

[a] Root-mean-square-distributions (rmsd) between reported members of the ensemble are presented for NMR-derived structures, and reported crystallographic diffraction resolution is given for X-ray crystal structures. Individual rmsd values are reported for each finger (in parentheses) in the NMR structures of the free proteins and the ADR1-DBD DNA complex due to the flexibility of peptide linkers connecting fingers.

[b] Rmsd range given since individual values have been reported for β-hairpin, 'finger tip', and helix regions.

[c] Synthetic peptides designed to adopt ββα fold without bound zinc.

[d] Engineered variants of Zif268 bound to different DNA sites (in parentheses).

[e] Dimer of Zif268 fingers 1 and 2 with N-terminal 15-residue peptide 'extension' bound to symmetric DNA site.

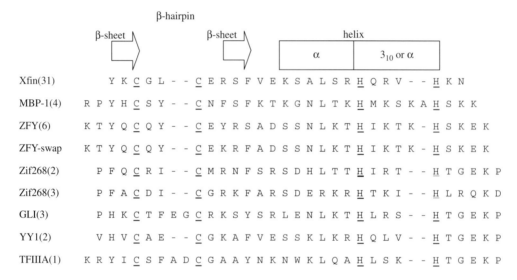

Figure 1 Amino acid sequence alignment of selected C2H2 zinc finger motifs showing the metal binding cysteine and histidine ligands (underlined). Proteins are listed by name followed by finger number in parentheses. Secondary structural elements (open arrows: β-hairpin structure; open rectangles: α-/3_{10}-helix) are shown above the alignment.

(10 structures). A bifurcated hydrogen bond from both His23 → Gln20 and Lys24 → Gln20 was also observed in 8 of the 10 structures.[6]

Zn(II) binding site

Each zinc finger domain binds a single Zn(II) ion with tetrahedral coordination involving two cysteines and two histidines. Both histidines coordinate the Zn(II) atom at the NE2 position in both the free and the DNA-bound state.[6,7] Figure 2 shows the metal coordination site and bond distances and Table 2 lists relevant bond angles for the metal–ligand coordination sphere from the solution structure of the human enhancer binding protein C-terminal C2H2 zinc finger (amino acid sequence shown in Figure 1).[87] With the exception of SG (Cys8)–Zn(II)–NE2 (His27), the bond angles of the tetrahedron are very close to the perfect values of 109.5°. A similar coordination geometry was observed for the HIV *gag* protein p55,[110] in which one Zn(II) ion coordinates to three cysteines and one histidine (C3H). In p55, 5 bond angles average $111.7 \pm 3.9°$, and the bond angle involving the first cysteine and the histidine is $96.5 \pm 0.53°$. The same approximate bond geometry is also observed for the noncatalytic zinc atom in the closed form of the C4 Zn(II) coordination site from horse liver alcohol dehydrogenase, where five bonds in the tetrahedron average 111.4 ± 7.4 and the sixth angle is 97°.[111] The Zn(II) ion is required for structural stability of most C2H2 domains, although novel apo variants of the motif have been designed that adopt the canonical fold.[39,42,45]

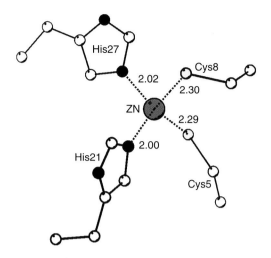

Figure 2 Drawing of the tetrahedral Zn(II) metal site from the average solution structure of the C-terminal C2H2 zinc finger of human enhancer binding protein (MBP-1; PDB code 4ZNF).[87] Figure made with the MOLSCRIPT[112] program.

Table 2 Zinc coordination geometry for C-terminal C2H2 zinc finger of MBP-2[87]

Bonds	Angle (degrees)
SG (Cys5)–Zn(II)–SG(Cys8)	111.8 ± 0.25
SG (Cys5)–Zn(II)–NE2(His21)	111.3 ± 0.31
SG (Cys5)–Zn(II)–NE2(His27)	111.1 ± 0.58
SG (Cys8)–Zn(II)–NE2(His21)	110.4 ± 0.17
SG (Cys8)–Zn(II)–NE2(His27)	97.9 ± 1.31
NE2(His21)–Zn(II)–NE2(His27)	111.8 ± 0.49

Hydrophobic packing in ZFY finger 6

ZFY is a sex-related C2H2-containing protein that differs from the general class by the presence of a two-finger repeat. Odd-numbered fingers of ZFY contain the 'consensus' aromatic residue phenylalanine at position 12 of the zinc finger (see ZFY-swap in Figure 1), whereas even-numbered fingers have an aromatic residue at position 10, which has been proposed to alter the DNA-binding surface of the domain (see ZFY finger 6 in Figure 1).[7,88,95,100,101,113] Both finger variants exhibit equivalent thermodynamic stability, although the internal packing arrangement of the hydrophobic residues is markedly different (Figure 3).[88,95,100,101,113] In the consensus ZFY-swap finger shown in Figure 3(a), the Phe12 residue packs primarily against the highly conserved Leu18 and makes an edge-to-face contact with the aromatic ring of His21. In stark contrast, the phenolic ring of Tyr10 in the ZFY-6 nonconsensus type finger is nearly parallel to the His21 immidazole ring and interacts in a different manner with Leu18 (Figure 3(b)).[101] This different packing arrangement for Tyr10 in ZFY-6 contributes to a contiguous region of the protein surface that was shown to contact DNA in the consensus Zif268 C2H2 zinc finger–DNA complex (described in detail in a later section).[7,95,100] Interestingly, an arginine residue at position 10 in all three fingers of Zif268 contacts the phosphate backbone of DNA. It is proposed that the two-finger repeat in ZFY, which is evolutionarily conserved, encodes a weakly polar switch between distinct and likewise evolutionarily conserved DNA surfaces.[101]

A recent NMR and thermodynamic study involving wild-type and mutant ZFY-6 fingers, in which a phenylalanine-to-leucine substitution was made at residue position 10, demonstrated that the highly conserved Phe10 residue is not necessary for high-affinity Zn(II) binding.[102] However, the relative enthalpic and entropic contributions to Zn(II) binding in the two zinc fingers were significantly different. The enthalpy of Zn(II) binding was comparatively favorable in the mutant C2H2 domain, but the entropy was correspondingly disfavored. Similarly, amide proton/deuterium exchange rates of the mutant domain were considerably faster throughout, suggesting accelerated segmental

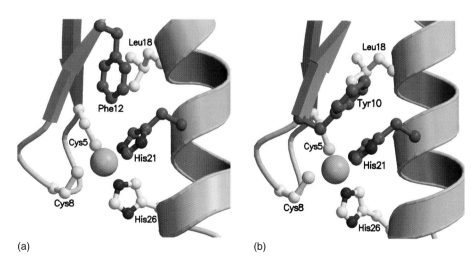

Figure 3 Schematic showing the differences in hydrophobic packing interactions in the (a) ZFY-swap (PDB code: 5ZNF[95,100]) and (b) ZFY-6 C2H2 zinc fingers (PDB codes: 7ZNF[101]). Figure made with the MOLSCRIPT[112] and RASTER3D[114] programs.

unfolding and folding and greater accessibility to water in the internal space of the finger. The similar Zn(II) binding stability of the Phe10 to Leu10 mutant is attributed to a 'rescuing' of the Zn(II) affinity by allowing water into the cavity of the protein created by the void where the Phe10 ring resides in the wild-type finger. This water is proposed to contribute to enthalpic stability through enhanced water–water and water–protein hydrogen bonding, but to destabilize the protein through a loss of entropy from ordering of the water molecules in the interior of the finger. It is further hypothesized that the driving force for conservation of phenylalanine at residue 10 is that it acts as an aromatic anchor that excludes water from the core of the zinc finger, thus enabling it to act as a rigid scaffold that is necessary to present a surface to the DNA that facilitates highly specific binding to cognate nucleotide bases.[102]

SWI5 finger 1

SWI5 is a transcription factor from *Saccharomyces cerevisiae* containing three C2H2 zinc finger motifs that is involved in regulating mating-type switching.[115] Structural determination of the N-terminal finger revealed a new 15-residue accessory domain that adds an extra β-sheet and small helix to the otherwise conserved structure of the finger (Figure 4).[94] The functional significance of the 14 amino acids preceding the N-terminus of finger 1 was originally inferred from the observation that deleting this small peptide region was sufficient to inactivate SWI5 *in vivo*.[116] The overall hydrogen-bonding pattern of the canonical C2H2 SWI5 finger 1 is very similar to that observed for other domains that do not contain additional structural elements.[6,87,95,98,103,106] A network of hydrophobic interactions in SWI5 finger 1 that extends from the canonical finger to the interface between the two helices stabilizes the short N-terminal helical segment (Figure 4(b)). Specifically, the rings of His56, Phe36, and Pro37 from the canonical finger 1 motif are packed together with Tyr23 from the N-terminal helix. Phe32, Phe43, and Ile49 link the antiparallel β-sheet to the N-terminus of the finger helix, which, in turn, links to the histidine ligands and on through to the accessory N-terminal helix.[94] Residues Ile20, Tyr23, and Val24 in the N-terminal helix interact with Ile53, Gln54, and Leu57 in the helix of the SWI5 finger 1 to further stabilize the core of the extended zinc finger domain. With the exception of the conservative phenylalanine-to-tyrosine substitution at residue position 36, all of these residues in the hydrophobic network of SWI5 finger 1 and in the N-terminal accessory domain are conserved in another yeast transcription factor ACE2, suggesting that these accessory domains are involved in DNA binding.[94] An earlier NMR solution structure of SWI5 fingers 1 and 2, which lacked the N-terminal accessory helical domain, revealed essentially no interfinger residue contacts, suggesting that the two C-terminal fingers

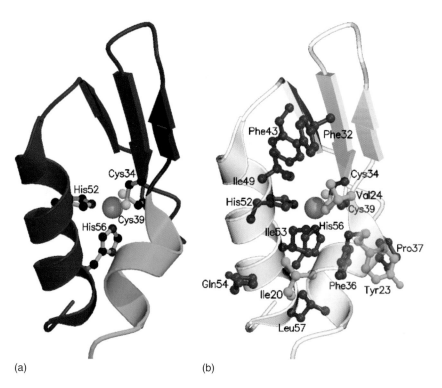

Figure 4 Cartoon model of SWI5 zinc finger 1 showing (a) the consensus C2H2 finger (red), accessory domain helix (green), and extra β-sheet (blue); cysteine and histidine Zn(II) ligands are also shown and (b) hydrophobic residues that form an extended network linking the accessory helical domain to the consensus finger. Residue side-chain colors match those of the secondary structures shown in (a) from which they originate (PDB code 1NCS[94]). Figure made with the MOLSCRIPT[112] and RASTER3D[114] programs.

of SWI5 do not affect the interactions of finger 1 with the accessory helical domain.[99]

Four hydrogen bonds connect the β-sheet from the accessory domain with the second β-sheet (first β-sheet of the canonical finger), while two hydrogen bonds connect the two sheets of the C2H2 finger. The accessory β-sheet and the first canonical β-sheet are connected through a type 3:3 β-turn at residues Pro28-Lys30, while the two canonical β-sheets are connected through a type II β-turn at Phe36-Cys39.[94,117] The extra β-sheet of the accessory domain contributes no detectable residue interactions to the hydrophobic core of the protein. However, this third β-sheet is required for the folding of the SWI5 finger 1.[92] The N-terminal helix seems to play a role in DNA binding, since addition of this polypeptide region to the three-finger SWI5 polypeptide resulted in a 16-fold increase in DNA-binding affinity.[94]

ATF-2

A recent NMR solution structure of the C-terminal transactivation domain of activating transcription factor 2 (ATF-2) reveals a C2H2 zinc finger motif in the N-subdomain[90,118] and a highly flexible and disordered C-subdomain. The overall structure of the zinc finger is nearly identical to previously determined structures of the C2H2 motif.[6,87,95,98–101,103] However, in ATF-2, the surface of a DNA-binding C2H2 zinc finger, which is normally positively charged (usually rich in arginines and lysines) is rather neutral, with some hydrophobic amino acids that form a patch on the surface.[90] This patch is formed by a leucine, an alanine, and a valine at positions 4, 5, and 6 of the helix respectively along with another leucine in the first β-sheet and a methionine C-terminal to the end of the helix (where the linker connecting multiple finger domains is normally located).[90] This atypical surface for the C2H2 finger of ATF-2 is consistent with a possible functional role involving protein–protein interactions, such as that recently observed for the Ikaros transcription factor that uses two C2H2 zinc fingers to form a homodimer interface.[119]

X-RAY AND NMR STRUCTURES OF DNA-BOUND C2H2 REPEATS

Zif268

The molecular basis for site-specific recognition of DNA by three tandem C2H2 motifs was first demonstrated from the 2.1-Å X-ray crystal structure of Zif268 bound to the 5′-GCGTGGGCG-3′ DNA sequence.[7] This X-ray structure was later refined to 1.6 Å.[27] Structural changes in DNA that accompany binding by Zif268 and other C2H2 proteins will be discussed in a later section. The overall structure of the protein forms a C-shape or semicircular arrangement of the three fingers along the major groove of the DNA (Figure 5). The structure of each finger is virtually unchanged from that reported for the unbound domains.[6,87,90,95,98–101,103] There is one interdomain contact each between fingers 1 and 2 and fingers 2 and 3 in which the backbone NH group of the conserved arginine at position 9 of the helix donates a hydrogen bond to the backbone carbonyl of the serine at position −2 relative to the start of the helix. There is periodicity with respect to adjacent fingers, with a rotation of approximately 96° or 3 × 32° per finger around the DNA axis.[7] The helices of the fingers are tilted at a steeper angle (45°) than the overall rise of the plane of the base pairs (32°). The 2.1 Å structure showed that each finger uses the N-terminal portion of its helix in a similar fashion to make specific residue–base contacts within adjacent 3-bp subsites in the major groove of the DNA. The orientation of the protein with respect to the DNA is antiparallel; that is, finger 1 makes contact with the 3-bp subsite at the 3′ end (coding strand) of the DNA duplex.

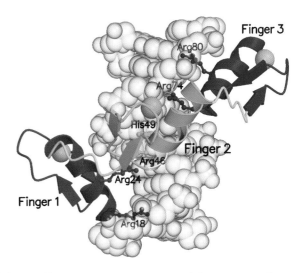

Figure 5 Schematic representation of the interaction between the three C2H2 zinc fingers of Zif268 and the cognate DNA.[7,27] Fingers 1, 2, and 3 are displayed in red, green, and blue respectively; coordinating Zn(II) ions in each finger are represented as gray spheres; and conserved linkers that connect fingers 1 and 2 and 2 and 3 are displayed in yellow. Residue side chains that make base-specific contacts are colored according to the finger they originate from. The DNA is depicted with a CPK model (PDB code 1AAY). Figure made with the MOLSCRIPT[112] and RASTER3D[114] programs.

An especially intricate and therefore specific residue–base interaction occurs in all three fingers involving the arginine and aspartate residues at positions −1 and 2, relative to the start of the helix (Figure 6(a)). The arginine NH1 and NH2 groups form hydrogen bonds with the N7 and O6 atoms respectively of the guanine base on the coding strand at position 3 of each finger's base pair subsite. Both side-chain oxygens of the aspartate residue (OD1 and OD2) at position 2 of the helices hydrogen bond

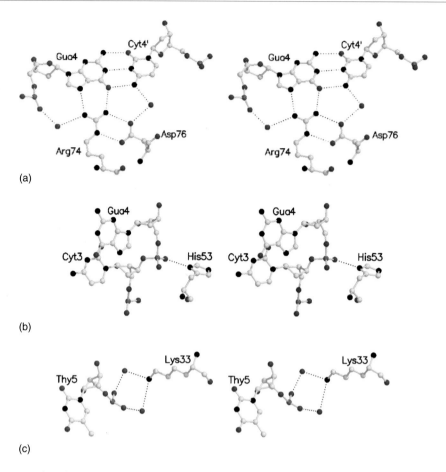

Figure 6 Stereo drawing of (a) the Arg74-base contact to the Gua4-Cyt4' base pair, (b) the His53 contact to the phosphate backbone, and (c) the water-mediated interaction between Lys33 and the phosphate backbone in the Zif268 zinc finger–DNA complex. Reprinted from *Structure*, 4, M Elrod-Erickson, MA Rould, L Nekludova and CO Pabo, Zif268 protein-DNA complex refined at 1.6 Å: a model system for understanding zinc finger-DNA interactions, 1171–80 (1996), with permission from Elsevier Science.[27] Figure made with the MOLSCRIPT program.[112]

to the NE and NH2 groups of the arginine respectively, stabilizing the position of the arginine side chain for optimal interaction with the guanine base. Additional water-mediated hydrogen bonds from the NH1 group of arginine to the phosphate backbone on the coding strand, and from the OD2 atom of aspartate to the N4 of cytosine on the noncoding strand, were uncovered unambiguously from the 1.6-Å structure.[27] Fingers 1 and 3 also form identical residue–base contacts at helix position 6 involving arginine NH1 and NH2 groups contacting the guanine N7 and O6 atoms respectively. These contacts are the same as those observed for the arginine at the −1 helix position. The refined 1.6-Å structure identified additional water-mediated contacts between the arginine NH2 groups at helix position 6 of fingers 1 and 3, and the neighboring guanine O6 atom on the noncoding strand.[27] That both of the specific residue–base interactions of finger 1 and 3 are identical is not surprising, given that the GCG DNA subsites for these fingers are also identical. Finger 2, which recognizes a different DNA subsite (TGG), has an alternative residue–base contact involving the histidine at position 3 of the helix (this histidine is not a ligand of the Zn(II) ion). The NE2 of this histidine forms a hydrogen bond with the N7 of the guanine in the middle of the 3-bp subsite in finger 2.[7] Thus, each finger in Zif268 makes two highly specific residue–base contacts, which form a basis for site-specific recognition of DNA and concomitant function of the C2H2 domains.

A number of residue contacts to the phosphate backbone of the DNA also contribute to the stability of the complex. The arginines at the second residue position after the second cysteine ligand (position 1 of the second β-sheet; Figure 1), and the first histidine ligand at position 7 of the helices (Figure 6(b)) make phosphate backbone contacts in all three fingers, with the exception of the histidine–phosphate backbone contact in finger 3, which adopts an alternate water-mediated hydrogen bond due to the truncated DNA sequence.[27] Remarkably, the specificity of the histidine ND1-phosphate contacts is dictated by the Zn(II) metal, which positions the immidazole ring favorably for this interaction through coordination by histidine at its NE2 group. Both of these residue–phosphate contacts are highly conserved, both in terms of amino acid–type preference at the respective finger positions and in terms of the

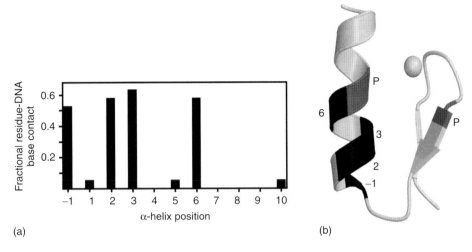

Figure 7 (a) Summary of fractional residue contacts as a function of the position from the N-terminus of the α-helix in C2H2 zinc finger–DNA complexes. Fractional contacts were calculated as the total number of fingers considered, divided by the total number of fingers with residues at a given helix position that contact a DNA base. Data from published structures of protein–DNA structures: 1AAY,[27] 2GLI,[17] 2DRP,[24] 1UBD,[28] 1TF3,[29] 1YUI,[30] and 2ADR.[32] (b) The residue positions that most often contact the DNA bases (black) and phosphate backbone (gray) are mapped onto a cartoon model of a C2H2 zinc finger (PDB code 1UBD; finger 2). Figure made with the MOLSCRIPT[112] and RASTER3D[114] programs.

frequency of these interactions in other reported zinc finger complexes with DNA.[17,24,28,29] A summary of C2H2 zinc finger positions for residue–base and residue–phosphate backbone contacts from all reported structures is shown in Figure 7. The 1.6-Å structure also identified a number of water-mediated hydrogen bonds from amino acid side chains to the phosphate backbone in the helices, and in the conserved linkers connecting adjacent fingers. The water-mediated hydrogen bonds between the lysine residues in position 4 of the conserved linkers connecting fingers 1 and 2 (Figure 6(c)) and fingers 3 and 4 in Zif268 may help explain its conservation in many C2H2 linker sequences.[8]

Other C2H2 finger–DNA structures

GLI

The X-ray structure of the five C2H2 zinc finger domains from the human glioblastoma oncogene (GLI) protein bound to a 22-bp DNA site was determined by Pabo and coworkers shortly after they published the Zif268–DNA complex structure.[17] Several new features of protein–DNA recognition were uncovered from this structure, although fingers 2 to 5 are bound inside the major groove of DNA in an orientation similar to fingers 1 to 3 of Zif268. Remarkably, finger 1 made extensive protein–protein contacts with finger 2 but no DNA contacts. The precise way in which the fingers contacted the DNA was different when compared to the Zif268 complex. Fingers 2–4 all made contact with DNA, but fingers 4 and 5 accounted for all but one of the residue–base contacts with the DNA. The finger 4 and 5 subsites were each 4-bp long compared to 3 bp in Zif268, and these sites were separated by 1 bp that made no contact with the protein.[17] It is important to note that the 9-bp region of the DNA site that was contacted by fingers 4 and 5 is the only region that is conserved in all known GLI binding sites[120] and that the choice of binding site used for the study was based on *in vitro* selection,[120] so the biological significance of the interactions described is not clear. However, the lack of specificity for DNA of fingers 1–3 suggest that they may have an alternate role in protein–protein or protein–RNA interactions as was observed for TFIIIA.[23,54–59] The residues that made contact with the DNA bases were tyrosine at helix position 2 of finger 2, alanine, serine, aspartate, and lysine at helix positions 1, 2, 3, and 6 of finger 4, and aspartate, serine, serine, arginine, and lysine at helix positions −1, 2, 3, 5, and 6 of finger 5. Notably, the contacts were roughly evenly divided between the coding and noncoding strands, which is in stark contrast to the Zif268 complex, in which all residue–base contacts were with the coding strand.[7] Residue–phosphate backbone contacts similar to those observed in the Zif268 structure were also detected in GLI. The most significant structural difference in the finger structures from those of Zif268 was linked to the 4-residue spacing between histidine ligands in 1, 4, and 5 of GLI compared to 3 residues for Zif268. In these fingers, the helix terminated one residue before the second histidine ligand and the backbone continued on with a wider turn. Additional structural effects of this spacing could not be resolved in the 2.6-Å GLI structure.

Tramtrack

The 2.8-Å X-ray structure of the two-finger domain bound to DNA from the *Drosophila* development gene regulatory protein tramtrack (TTK) was the first structure of C2H2 domains complexed with an AT-rich DNA sequence.[96] Although most aspects of the overall structure were similar to those of the Zif268– and GLI–DNA complexes, there were some notable previously unseen features. First, there was a third β-sheet N-terminal to finger 1, which was analogous to that observed in SWI5 finger 1 (Figure 4).[94] The additional β-sheet does not contact the DNA directly, but is required for DNA binding by TTK.[89] This observation suggests that this extra β-sheet is required for the proper folding of finger 1, as is the case for the corresponding finger of SWI5.[92] Each finger contacts 3-bp subsites that overlap 1 bp (5-bp site for both fingers). Residue positions 1, 2, and 6 contact DNA bases in finger 1, while positions −1, 2, and 3 make base-specific contacts in finger 2.[24] Various residue–phosphate contacts are present in the TTK complex, four of which are located in the N-terminal β-sheet to the helix. There is a 20° bend in the DNA at the finger 1 binding site, which has been seen to varying extents in other protein–DNA complexes such as those involving LEF-1.[121] The linker connecting the fingers is not of the canonical TGEKP sequence and is reported to be quite flexible, as evidenced from its high average temperature factor.[24]

YY1

The X-ray crystal structure of the Ying-Yang protein (YY1) bound to the AAVP5 promotor was determined at 2.5 Å.[28] The overall finger structure, orientations, and residue–phosphate backbone contacts are very similar to the previously described complexes.[7,17,24] All four YY1 fingers bind the major groove of DNA. Fingers 2–4 make multiple residue–base contacts to the DNA, and finger 1 makes one contact at helix position 6. Residue–base contact positions in finger 2 are −1, 2, 3, and 6 from the start of the helix, while contacts at helix positions −1, 2, and 3 occur in both fingers 3 and 4. With the exception of one threonine–thymine contact in finger 4, all the bases downstream of the transcriptional initiation site on the noncoding strand are accessible for strand separation (YY1 makes no additional contacts and the strand is solvent accessible), while neither strand on the upstream side of the promotor is accessible because YY1 makes numerous base contacts with both strands. YY1 promotes unidirectional initiation of mRNA production from two adjacent sites within the adeno-associated virus P5 promoter (containing AAVP5). Thus, a model in which YY1 serves as a mediator of unidirectional RNA strand synthesis was proposed from the X-ray crystal structure.[28]

TFIIIA

An NMR solution structure of a C2H2 zinc finger–DNA complex was determined for the first three fingers of TFIIIA (Zf1-3) bound to a 15-bp region of the internal control region (ICR) from the 5S RNA gene.[29] Zf1-3 is responsible for the high-specificity DNA binding of the nine-finger domain.[18–22,25] A number of new characteristics of C2H2 zinc finger binding and DNA recognition were revealed from the NMR solution structure of Zf1-3.[29] Although fingers 2 and 3 are bound to the major groove in an orientation similar to that observed previously,[7,17,24,28] finger 1 is oriented more 'across' the major groove than parallel to it. This alternate orientation is largely a consequence of substantial residue interactions between fingers 1 and 2. The amount of buried surface area between fingers 1 and 2 was determined to be 120 Å^2, compared to 80 Å^2 for the corresponding interface between fingers 2 and 3.[29] Each finger contacts a 4- to 5-bp subsite of DNA. Residue–base contacts at helix positions −1, 2, and 3 (finger 1); 2 and 6 (finger 2); and 3, 6 and 10 (finger 3) were identified. Contacts spanning the entire length of the finger such as those of Zf1-3 finger 3 had not been observed previously,[7,17,24,28] and may help explain the diversity of DNA sites recognized by C2H2 zinc fingers. The NMR side-chain resonances corresponding to the lysines at positions −1 and 3 of finger 1 were severely broadened, suggesting conformational fluctuations in the μs–ms timescale, with the possibility of making more than one base contact within the span of this fluctuation. While the corresponding resonances of lysine at position 6 of finger 3 were not broadened, there were NOE constraints assigned from this side chain to both strands of the DNA, which leads to multiple conformations for this side chain in the ensemble of solution structures. This 'dynamic disordering' of the lysine side chains in Zf1-3 is proposed to decrease the entropic cost of DNA binding.[29,122]

A 3.1-Å X-ray structure of fingers 1–6 of TFIIIA, reported later, is in general agreement with the high-resolution NMR structure, although the orientation of finger 1 is somewhat different.[91] Packing between adjacent fingers is greatly reduced by an increased spacing between fingers 1 and 2 in the X-ray structure. The origin of the differences between the two structures is not clear, although recent structural refinement of the Zf1-3 complex with DNA using NMR-derived residual dipolar coupling data corroborate the orientation of finger 1 that was derived in the original NMR solution structure.[29,123] The majority of residue–base contacts occur in fingers 1–3 in the X-ray structure. The differences in specific residue–base contacts from those reported in the NMR structure are located in finger 3 (residue positions −1, 2 and 10 for X-ray versus 3, 6, and 10 for NMR). Residue–base contacts at the 3 and 6 positions of the helix are also observed for finger 5 in the 3.1 Å structure, which corresponds to the IE element of the 5SRNA gene. These protein–DNA contact points are consistent with previously reported biochemical studies that mapped the binding sites within the ICR of the 5SRNA gene for all nine fingers of TFIIIA.[23,25,124] Fingers 4 and 6, which do not bind the major groove in the X-ray structure, are reported to traverse the minor groove and make a few residue contacts to the phosphate backbone. The finger 4–6 region is 'anchored' to the DNA by residue–base and residue–phosphate backbone contacts in finger 5.[91] Overall, the fingers 4–6 reportedly form a continuous 'platform-like' surface on the DNA that is proposed as a possible docking point for other components of the transcription complex.[52]

GAGA factor

The molecular basis for sequence-specific high-affinity DNA binding by the antirepressor GAGA factor from *Drosophila melanogaster* has been demonstrated from the NMR solution structure and is represented schematically in Figure 8.[30] The GAGA factor is the only known single-finger C2H2 DNA binding domain (the ATF-2 C2H2 finger is in the transactivation domain). High-affinity binding to DNA, which is normally insufficient with a single C2H2 zinc finger because of the small number of residue–base contacts,[125,126] is achieved through an N-terminal extension that contains two basic regions (BR1 and BR2). The recognition sequence 5'-GAGAG-3' is contacted by arginine,[93] asparagine, and arginine residues at positions 2, 3, and 6 of the zinc finger helix respectively. Additional residue–base contacts in the major groove of the DNA involving an arginine at position 3 of a small 5-residue helix formed in BR2, and surprising residue–base contacts by arginine and lysine residues from the unstructured BR1 region to the minor groove of the DNA, confer the additional necessary binding specificity and affinity to augment the residue–base contacts from the single finger. A number of additional hydrophobic interactions involving residues from the zinc finger, and residue–phosphate backbone contacts in the finger and the small helix in BR2, supplement the binding of the GAGA factor polypeptide to the DNA cognate. The structure of the N-terminal extension of the GAGA factor is reminiscent, in some structural respects, to that observed in the first finger of SWI5.[94] However, in SWI5 the domain extensions are intimately connected with the consensus finger through an extensive network of residue side-chain interactions (Figure 4), which is not the case for the GAGA factor.[30] The orientation of the single GAGA factor zinc finger in the major groove of DNA is essentially identical to that of finger 2 from the TTK-DNA crystal structure.[24,30] Remarkably, the positioning of the small helix in BR2 approximately superimposes onto that of the helix of finger 1 from TTK. The orientation and residue–base minor groove contacts of the BR1 region, however, are unprecedented in previously studied C2H2 zinc finger complexes with DNA.[7,17,24,28,38] The general mode of DNA binding by the GAGA factor through a helix and loop interacting in the major groove, followed by a polypeptide extension into the minor groove that makes specific residue–base contacts, is generally similar to the mode of interaction seen in the GATA-1 complex with DNA.[127] In the case of GATA-1, the peptide extension into the minor groove is on the C-terminal side of a CCCC zinc-binding motif that is structurally related to the glucocorticoid receptor.[128] Simultaneous high-affinity major- and minor-groove DNA interactions could potentially be an advantage in one of the putative functions of the GAGA factor. Repression of transcription by linker histones H1 and H5 is alleviated by the GAGA factor,[129] possibly by displacement of the linker histones from the nucleosome core. It is possible that the winged helix-turn-helix domain of the linker histones may be displaced by the GAGA factor because they have a lower affinity for major and minor groove binding to DNA.[30,130]

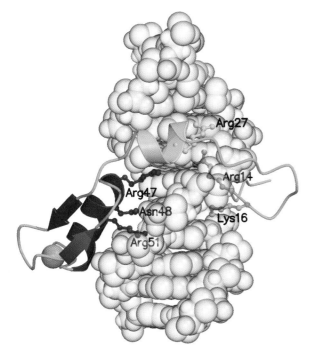

Figure 8 Schematic representation of the specific GAGA factor bound to DNA.[30] The β-sheets and helix of the canonical C2H2 finger are colored blue and red respectively, and the basic region helix (BR2) is colored yellow. All regions of the polypeptide that lack regular secondary structure, including the N-terminal basic region (BR1) that contacts both the major and minor grooves of the DNA, are colored green. The coordinating Zn(II) ion is represented as a gray sphere. Residue side chains that make base-specific contacts are colored according to the region of the polypeptide from which they originate. The DNA is depicted by means of a CPK model (PDB code 1YUJ; average structure). Figure made with the MOLSCRIPT[112] and RASTER3D[114] programs.

ADR1

The NMR structure of another unusual DNA-bound C2H2 zinc finger motif that contains an additional 20-residue proximal accessory region (PAR) has recently been solved.[32] The ADR1 transcription factor from

Saccharomyces cerevisiae regulates genes governing carbon source metabolism and binds to an upstream activation sequence, UAS1, which contains two symmetric and opposed binding sites.[131–134] NMR chemical shift, NOE, hydrogen/deuterium exchange, and paramagnetic Co(II)-substitution studies show that the PAR region N-terminal to the ADR1 zinc finger domain is largely unstructured in the absence of UAS1, but upon binding UAS1 adopts an individual compact domain structure consisting of three antiparallel β-sheets that contacts A-T base pairs in the major groove.[31,32,135] An NMR-based dynamics study also demonstrates a 'disorder to order' transition for the PAR upon binding UAS1.[31] It should be noted that a structure of ADR1 bound to UAS1 has been deposited in the PDB (code 2ADR), although, as of this writing, the coordinates for both the PAR and the DNA in the complex are not contained in the file. However, a number of NOE constraints are reported between the PAR and the DNA that led to the model of the ADR1-PAR structure bound to UAS1.[32] The PAR appears to be essential for ADR1 function, as no detectable binding to UAS1 is observed for the two C2H2 zinc fingers of ADR1 alone. The two fingers of ADR1 do not appear to make contact with each other, and the linker connecting them has been shown to have random coil hydrogen/deuterium exchange rates in the presence and absence of DNA.[31] This 'isolation' of each finger and of the PAR domain in the UAS1 complex is in stark contrast to the importance of finger–finger packing interactions for DNA-binding observed for other zinc finger complexes.[7,29] In the two-finger domain with PAR bound to UAS1, residue positions −1, 3, and 6 make contacts to DNA bases in finger 1, while only residue positions −1 and 2 contact DNA bases in finger 2. In contrast to the GAGA factor, the PAR domain is believed to make primarily nonspecific residue contacts to the phosphate backbone of DNA, thus adding necessary binding affinity to the complex, but leaving most of the DNA-site recognition function to the two canonical C2H2 zinc fingers.[30,32] The only putative residue–base contact between PAR and DNA is a hydrophobic interaction involving a leucine and a thymidine base methyl group that may account for the slight A-T preference of the ADR1 PAR.[32] It appears that a number of novel strategies have evolved for C2H2 zinc finger transcription factors to achieve high DNA-binding affinity and specificity with only one- or two-finger motifs that even extends to minor groove recognition in at least one case.[24,30,32,94]

Linker structure

A highly conserved TGEKP linker sequence connects approximately 70% of repeating C2H2 zinc finger domains (see Figure 1 for examples).[33,71,136] For most tandem-repeating zinc finger domains, these peptide regions remain flexible in the unbound state. However, upon binding DNA, the linkers adopt a highly ordered conformation that stabilizes DNA binding.[7,29,33,34,62,122] In Figure 9(a) the TGEKP linker that connects fingers 1 and 2 of Zf1-3 is superimposed for the 10 lowest energy structures (X. Liao and P.E. Wright, unpublished results), and it is quite clear that there is considerable conformational heterogeneity in this linker region of the otherwise well-ordered three-finger C2H2 domain. Upon binding to DNA, the linkers adopt nearly identical conformations in the 10 lowest energy structures of the Zf1-3–DNA complex (Figure 9(b)).[29] The remarkable similarity in the TGEKP linker structure from a number of otherwise unrelated C2H2 zinc finger–DNA complexes are shown in Figure 9(c). Several factors contribute to linker ordering: (i) a DNA-induced capping motif involving the glycine at position 2 of the linker results in the formation of one (α_L) or two (Schellman) hydrogen bonds at the C-terminal region of the helix in the preceding finger;[7,33,137] (ii) the lysine at position 4 of the linker contacts the phosphate backbone in a number of the finger–DNA complexes;[17,28,29] and (iii) the linker structure serves to orient adjacent fingers

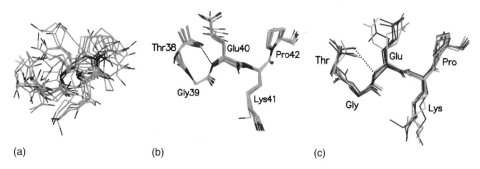

Figure 9 Alignment of the backbone structures of the TGEKP linkers from (a) the 10 lowest energy structures of free TFIIIA fingers 1–3 (X. Liao & P.E. Wright, unpublished results), (b) the 10 lowest energy structures of Zf1-3 bound to DNA[29] and (c) an alignment of the backbone structures from conserved TGEKP linkers in protein–DNA complexes from PDB codes 1MEY linker 1–2 (blue),[38] 1MEY linker 2–3 (cyan), 1AAY linker 2–3 (magenta),[7,27] 2GLI linker 2–3 (orange),[17] 2GLI linker 3–4 (yellow), and 1UBD linker 1–2 (green) onto the 1TF3 linker 1–2 (red) from the average NMR structure.[28,29] Reprinted from *J Mol Biol*, **295**, JH Laity, HJ Dyson and PE Wright, DNA-induced alpha-helix capping in conserved linker sequences is a determinant of binding affinity in Cys2-His2 zinc fingers, 719–27, (2000) with permission from Elsevier Science.[33] Figure made with the NAB,[138] MOLSCRIPT,[112] and RASTER3D[114] programs.

along the major groove to facilitate optimal residue–base and interfinger contacts.[29] A model has been proposed in which the linker flexibility necessarily allows interdomain flexibility of the individual fingers to search the DNA for the correct sequence.[34] Once the correct sequence is encountered transiently, the linkers 'snaplock' into an ordered conformation that facilitates the DNA recognition process and concomitantly stabilizes binding.

DNA structure

Comparisons of DNA conformations in the Zif268,[7] GLI,[17] and Tramtack[24] C2H2 zinc finger–DNA complexes[89] with those of the trp,[139] glucocorticoid,[127] MetJ,[140] and engrailed[141] non-C2H2 zinc finger proteins–DNA complexes reveal an enlarged (deeper and wider) major groove in the DNA for all of these complexes that was coined B_{eg}-DNA (where eg means enlarged major groove).[26] B_{eg}-DNA has some characteristics that more resemble A-DNA, such as displacement and inclination, but many other features such as δ-angle and groove width appear more B-DNA-like. An altered DNA conformation is necessary for C2H2 zinc finger binding to facilitate proper alignment of the protein residues with the DNA bases simultaneously in all bound fingers.[7,27] A structure for DNA, which is distinctly different from that of either A- or B-DNA, is proposed in a variety of protein–DNA complexes in which the major groove is both wide and deep, but the minor groove remains fully accessible.[26] A number of other complexes such as α2 and GCN4 have DNA conformations intermediate between B_{eg}-DNA and B-DNA.[142,143] One fundamental question that remains unanswered is whether these altered DNA conformations exist in the absence of bound protein, therefore making the major groove more accessible for protein binding, or whether they are an induced structural consequence of major groove binding by a variety of DNA binding proteins.[26] Studies of DNA binding by the three-finger Sp1 C2H2 zinc finger domain suggest that the DNA sites recognized by Sp1 are B-like, and that they become unwound upon zinc finger binding.[144]

SELECTION AND DESIGN OF ZINC FINGERS

The small DNA subsites recognized by individual C2H2 zinc fingers provide a versatile framework for designing zinc finger combinations with novel DNA recognition sequences. Over the past 10 years, considerable attention using modeling, phage display, and sequence comparisons has been devoted to creating repeating zinc finger motifs with altered specificity in one or more fingers, and to creating new zinc finger proteins that target a preselected DNA site.[35,145,37,146,36,147,107,40,41,43,46] A detailed discussion of this approach is outside the scope of this book.

However, it is worth mentioning the design of novel C2H2 zinc finger proteins, since they have tremendous potential for medicinal applications as gene therapy agents. Initially, design strategies assumed that each 3-bp DNA subsite recognized by an individual finger was independent, but more recent selection studies using phage display followed by analysis of the resulting X-ray structures of the protein–DNA complexes uncovered many contacts within each finger that occur outside these subsites. Several strategies have evolved to overcome these 'context dependent' interactions,[46,107–109,148] including a sequential approach in which protein domain specificity is altered in a segmental manner, one finger–DNA subsite at a time.[41]

REFERENCES

1 J Miller, AD McLachlan and A Klug, *EMBO J*, **4**, 1609–14 (1985).

2 A Klug and D Rhodes, *Trends Biochem Sci*, **12**, 464–69 (1987).

3 RM Evans and SM Hollenberg, *Cell*, **52**, 1–3 (1988).

4 F Payre and A Vincent, *FEBS Lett*, **234**, 245–50 (1988).

5 JM Berg, *Proc Natl Acad Sci USA*, **85**, 99–102 (1988).

6 MS Lee, GP Gippert, KV Soman, DA Case and PE Wright, *Science*, **245**, 635–37 (1989).

7 NP Pavletich and CO Pabo, *Science*, **252**, 809–17 (1991).

8 R Schuh, W Aicher, U Gaul, S Cote, A Preiss, D Maier, E Seifert, U Nauber, C Schroder and R Kemler, *Cell*, **47**, 1025–32 (1986).

9 H Takatsuji, *Plant Mol Biol*, **39**, 1073–78 (1999).

10 AY Chou, J Archdeacon and CI Kado, *Proc Natl Acad Sci USA*, **95**, 5293–98 (1998).

11 N Bouhouche, M Syvanen and CI Kado, *Trends Microbiol*, **8**, 77–81 (2000).

12 KM Call, T Glaser, CY Ito, AJ Buckler, J Pelletier, DA Haber, EA Rose, A Kral, H Yeger and WH Lewis, *Cell*, **60**, 509–20 (1990).

13 M Gessler, A Poustka, W Cavenee, RL Neve, SH Orkin and GA Bruns, *Nature*, **343**, 774–78 (1990).

14 SH Larsson, JP Charlieu, K Miyagawa, D Engelkamp, M Rassoulzadegan, A Ross, F Cuzin, H van, V and ND Hastie, *Cell*, **81**, 391–401 (1995).

15 M Little, G Holmes and P Walsh, *BioEssays*, **21**, 191–202 (1999).

16 KG Becker, JW Nagle, RD Canning, WE Biddison, K Ozato and PD Drew, *Hum Mol Genet*, **4**, 685–91 (1995).

17 NP Pavletich and CO Pabo, *Science*, **261**, 1701–7 (1993).

18 QM You, N Veldhoen, F Baudin and PJ Romaniuk, *Biochemistry*, **30**, 2495–2500 (1991).

19 KR Clemens, X Liao, V Wolf, PE Wright and JM Gottesfeld, *Proc Natl Acad Sci USA*, **89**, 10822–26 (1992).

20 XB Liao, KR Clemens, L Tennant, PE Wright and JM Gottesfeld, *J Mol Biol*, **223**, 857–71 (1992).

21 JJ Hayes and KR Clemens, *Biochemistry*, **31**, 11600–5 (1992).

22 N Veldhoen, Q You, DR Setzer and PJ Romaniuk, *Biochemistry*, **33**, 7568–75 (1994).

23 KR Clemens, V Wolf, SJ McBryant, P Zhang, X Liao, PE Wright and JM Gottesfeld, *Science*, **260**, 530–33 (1993).

24 L Fairall, JW Schwabe, L Chapman, JT Finch and D Rhodes, *Nature*, **366**, 483–87 (1993).

25. KR Clemens, P Zhang, X Liao, SJ McBryant, PE Wright and JM Gottesfeld, *J Mol Biol*, **244**, 23–35 (1994).
26. L Nekludova and CO Pabo, *Proc Natl Acad Sci USA*, **91**, 6948–52 (1994).
27. M Elrod-Erickson, MA Rould, L Nekludova and CO Pabo, *Structure*, **4**, 1171–80 (1996).
28. HB Houbaviy, A Usheva, T Shenk and SK Burley, *Proc Natl Acad Sci USA*, **93**, 13577–82 (1996).
29. DS Wuttke, MP Foster, DA Case, JM Gottesfeld and PE Wright, *J Mol Biol*, **273**, 183–206 (1997).
30. JG Omichinski, PV Pedone, G Felsenfeld, AM Gronenborn and GM Clore, *Nat Struct Biol*, **4**, 122–32 (1997).
31. DE Hyre and RE Klevit, *J Mol Biol*, **279**, 929–43 (1998).
32. PM Bowers, LE Schaufler and RE Klevit, *Nat Struct Biol*, **6**, 478–85 (1999).
33. JH Laity, HJ Dyson and PE Wright, *J Mol Biol*, **295**, 719–27 (2000).
34. JH Laity, HJ Dyson and PE Wright, *Proc Natl Acad Sci USA*, **97**, 11932–35 (2000).
35. JR Desjarlais and JM Berg, *Proc Natl Acad Sci USA*, **89**, 7345–49 (1992).
36. Y Choo and A Klug, *Proc Natl Acad Sci USA*, **91**, 11163–67 (1994).
37. JR Desjarlais and JM Berg, *Proc Natl Acad Sci USA*, **91**, 11099–103 (1994).
38. CA Kim and JM Berg, *Nat Struct Biol*, **3**, 940–45 (1996).
39. MD Struthers, RP Cheng and B Imperiali, *Science*, **271**, 342–45 (1996).
40. Y Choo and A Klug, *Curr Opin Struct Biol*, **7**, 117–125 (1997).
41. HA Greisman and CO Pabo, *Science*, **275**, 657–61 (1997).
42. BI Dahiyat and SL Mayo, *Science*, **278**, 82–87 (1997).
43. DJ Segal, B Dreier, RR Beerli and CF Barbas III, *Proc Natl Acad Sci USA*, **96**, 2758–63 (1999).
44. A Klug, *J Mol Biol*, **293**, 215–18 (1999).
45. CA Sarisky and SL Mayo, *J Mol Biol*, **307**, 1411–18 (2001).
46. SA Wolfe, RA Grant, M Elrod-Erickson and CO Pabo, *Structure (Camb)*, **9**, 717–23 (2001).
47. SA Wolfe, HA Greisman, EI Ramm and CO Pabo, *J Mol Biol*, **285**, 1917–34 (1999).
48. S Iuchi, *Cell Mol Life Sci*, **58**, 625–35 (2001).
49. VP Sukhatme, XM Cao, LC Chang, CH Tsai-Morris, D Stamenkovich, PC Ferreira, DR Cohen, SA Edwards, TB Shows and T Curran, *Cell*, **53**, 37–43 (1988).
50. M Yokono, N Saegusa, K Matsushita and Y Sugiura, *Biochemistry*, **37**, 6824–32 (1998).
51. M Merika and SH Orkin, *Mol Cell Biol*, **15**, 2437–47 (1995).
52. DD Brown and MS Schlissel, *Cell*, **42**, 759–67 (1985).
53. B Picard and M Wegnez, *Proc Natl Acad Sci USA*, **76**, 241–45 (1979).
54. MK Darby and KE Joho, *Mol Cell Biol*, **12**, 3155–64 (1992).
55. O Theunissen, F Rudt, U Guddat, H Mentzel and T Pieler, *Cell*, **71**, 679–90 (1992).
56. DF Bogenhagen, *Mol Cell Biol*, **13**, 5149–58 (1993).
57. DR Setzer, SR Menezes, S Del Rio, VS Hung and G Subramanyan, *RNA*, **2**, 1254–69 (1996).
58. KE Vrana, ME Churchill, TD Tullius and DD Brown, *Mol Cell Biol*, **8**, 1684–96 (1988).
59. JF Smith, J Hawkins, RE Leonard and JS Hanas, *Nucleic Acids Res*, **19**, 6871–76 (1991).
60. H Nakagama, G Heinrich, J Pelletier and DE Housman, *Mol Cell Biol*, **15**, 1489–98 (1995).
61. A Caricasole, A Duarte, SH Larsson, ND Hastie, M Little, G Holmes, I Todorov and A Ward, *Proc Natl Acad Sci USA*, **93**, 7562–66 (1996).
62. JH Laity, J Chung, HJ Dyson and PE Wright, *Biochemistry*, **39**, 5341–48 (2000).
63. H Zhao and DJ Eide, *Mol Cell Biol*, **17**, 5044–52 (1997).
64. H Zhao, E Butler, J Rodgers, T Spizzo, S Duesterhoeft and D Eide, *J Biol Chem*, **273**, 28713–20 (1998).
65. A Bird, MV Evans-Galea, E Blankman, H Zhao, H Luo, DR Winge and DJ Eide, *J Biol Chem*, **275**, 16160–66 (2000).
66. AJ Bird, H Zhao, H Luo, LT Jensen, C Srinivasan, M Evans-Galea, DR Winge and DJ Eide, *EMBO J*, **19**, 3704–13 (2000).
67. CM Fan and T Maniatis, *Genes Dev*, **4**, 29–42 (1990).
68. ES Lander, LM Linton, B Birren, C Nusbaum, MC Zody, et al., *Nature*, **409**, 860–921 (2001).
69. JC Venter, MD Adams, EW Myers, PW Li, RJ Mural, et al., *Science*, **291**, 1304–51 (2001).
70. Swiss Institute of Bioinformatics, CMU – Rue Michel-Servet 1 1211 Genève 4, Switzerland (2003). Available on the World Wide Web: *http://us.expasy.org/*.
71. JM Hoovers, M Mannens, R John, J Bliek, H van, V, DJ Porteous, NJ Leschot, A Westerveld and PF Little, *Genomics*, **12**, 254–63 (1992).
72. MS Palmer, P Berta, AH Sinclair, B Pym and PN Goodfellow, *Proc Natl Acad Sci USA*, **87**, 1681–85 (1990).
73. YF Lau and KM Chan, *Am J Hum Genet*, **45**, 942–52 (1989).
74. C Wolff, R Sommer, R Schroder, G Glaser and D Tautz, *Development*, **121**, 4227–36 (1995).
75. AM Ginsberg, BO King and RG Roeder, *Cell*, **39**, 479–89 (1984).
76. A Altaba, H Perry-O'Keefe and DA Melton, *EMBO J*, **6**, 3065–70 (1987).
77. TM Pohl and G Ajinovic, EMBL/Swiss Prot, accession number P47043.
78. TA Hartshorne, H Blumberg and ET Young, *Nature*, **320**, 283–87 (1986).
79. RD Klein and BJ Meyer, *Cell*, **72**, 349–64 (1993).
80. WD Burke, DG Eickbush, Y Xiong, J Jakubczak and TH Eickbush, *Mol Biol Evol*, **10**, 163–85 (1993).
81. MS Salvato and EM Shimomaye, *Virology*, **173**, 1–10 (1989).
82. G Brightwell, H Hussain, A Tiburtius, KH Yeoman and AW Johnston, *Mol Plant Microbe Interact*, **8**, 747–54 (1995).
83. C Freiberg, R Fellay, A Bairoch, WJ Broughton, A Rosenthal and X Perret, *Nature*, **387**, 394–401 (1997).
84. J Miller, L Fairall and D Rhodes, *Nucleic Acids Res*, **17**, 9185–92 (1989).
85. TJ Gibson, JP Postma, RS Brown and P Argos, *Protein Eng*, **2**, 209–18 (1988).
86. G Parraga, SJ Horvath, A Eisen, WE Taylor, L Hood, ET Young and RE Klevit, *Science*, **241**, 1489–92 (1988).
87. JG Omichinski, GM Clore, E Appella, K Sakaguchi and AM Gronenborn, *Biochemistry*, **29**, 9324–34 (1990).
88. MA Weiss, KA Mason, CE Dahl and HT Keutmann, *Biochemistry*, **29**, 5660–64 (1990).
89. L Fairall, SD Harrison, AA Travers and D Rhodes, *J Mol Biol*, **226**, 349–66 (1992).
90. A Nagadoi, K Nakazawa, H Uda, K Okuno, T Maekawa, S Ishii and Y Nishimura, *J Mol Biol*, **287**, 593–607 (1999).

91 RT Nolte, RM Conlin, SC Harrison and RS Brown, *Proc Natl Acad Sci USA*, **95**, 2938–43 (1998).
92 Y Nakaseko, D Neuhaus, A Klug and D Rhodes, *J Mol Biol*, **228**, 619–36 (1992).
93 PV Pedone, R Ghirlando, GM Clore, AM Gronenborn, G Felsenfeld and JG Omichinski, *Proc Natl Acad Sci USA*, **93**, 2822–26 (1996).
94 RN Dutnall, D Neuhaus and D Rhodes, *Structure*, **4**, 599–611 (1996).
95 M Kochoyan, HT Keutmann and MA Weiss, *Biochemistry*, **30**, 7063–72 (1991).
96 GP Diakun, L Fairall and A Klug, *Nature*, **324**, 698–99 (1986).
97 AD Frankel, JM Berg and CO Pabo, *Proc Natl Acad Sci USA*, **84**, 4841–45 (1987).
98 JG Omichinski, GM Clore, M Robien, K Sakaguchi, E Appella and AM Gronenborn, *Biochemistry*, **31**, 3907–17 (1992).
99 D Neuhaus, Y Nakaseko, JW Schwabe and A Klug, *J Mol Biol*, **228**, 637–51 (1992).
100 M Kochoyan, TF Havel, DT Nguyen, CE Dahl, HT Keutmann and MA Weiss, *Biochemistry*, **30**, 3371–86 (1991).
101 M Kochoyan, HT Keutmann and MA Weiss, *Proc Natl Acad Sci USA*, **88**, 8455–59 (1991).
102 MJ Lachenmann, JE Ladbury, NB Phillips, N Narayana, X Qian and MA Weiss, *J Mol Biol*, **316**, 969–89 (2002).
103 RE Klevit, JR Herriott and SJ Horvath, *Proteins*, **7**, 215–26 (1990).
104 RC Hoffman, SJ Horvath and RE Klevit, *Protein Sci*, **2**, 951–65 (1993).
105 BE Bernstein, RC Hoffman, S Horvath, JR Herriott and RE Klevit, *Biochemistry*, **33**, 4460–70 (1994).
106 VA Narayan, RW Kriwacki and JP Caradonna, *J Biol Chem*, **272**, 7801–9 (1997).
107 M Elrod-Erickson, TE Benson and CO Pabo, *Structure*, **6**, 451–64 (1998).
108 JC Miller and CO Pabo, *J Mol Biol*, **313**, 309–15 (2001).
109 BS Wang and CO Pabo, *Proc Natl Acad Sci USA*, **96**, 9568–73 (1999).
110 MF Summers, TL South, B Kim and DR Hare, *Biochemistry*, **29**, 329–40 (1990).
111 H Eklund and C-I Branden, in TG Spiro (ed.), *Zinc Enzymes*, John Wiley, New York (1983).
112 P Kraulis, *J Appl Crystallogr*, **24**, 946–50 (1991).
113 MA Weiss and HT Keutmann, *Biochemistry*, **29**, 9808–13 (1990).
114 EA Merritt and MEP Murphy, *Acta Crystallogr*, **D50**, 869–73 (1994).
115 DJ Stillman, AT Bankier, A Seddon, EG Groenhout and KA Nasmyth, *EMBO J*, **7**, 485–94 (1988).
116 G Tebb, T Moll, C Dowzer and K Nasmyth, *Genes Dev*, **7**, 517–28 (1993).
117 BL Sibanda, TL Blundell and JM Thornton, *J Mol Biol*, **206**, 759–77 (1989).
118 T Maekawa, H Sakura, C Kanei-Ishii, T Sudo, T Yoshimura, J Fujisawa, M Yoshida and S Ishii, *EMBO J*, **8**, 2023–28 (1989).
119 L Sun, A Liu and K Georgopoulos, *EMBO J*, **15**, 5358–69 (1996).
120 KW Kinzler and B Vogelstein, *Mol Cell Biol*, **10**, 634–42 (1990).
121 JJ Love, X Li, DA Case, K Giese, R Grosschedl and PE Wright, *Nature*, **376**, 791–95 (1995).
122 MP Foster, DS Wuttke, I Radhakrishnan, DA Case, JM Gottesfeld and PE Wright, *Nat Struct Biol*, **4**, 605–8 (1997).
123 V Tsui, L Zhu, TH Huang, PE Wright and DA Case, *J Biomol NMR*, **16**, 9–21 (2000).
124 KR Clemens, X Liao, V Wolf, PE Wright and JM Gottesfeld, *Proc Natl Acad Sci USA*, **89**, 10822–26 (1992).
125 A Klug and JW Schwabe, *FASEB J*, **9**, 597–604 (1995).
126 JM Berg and Y Shi, *Science*, **271**, 1081–85 (1996).
127 JG Omichinski, GM Clore, O Schaad, G Felsenfeld, C Trainor, E Appella, SJ Stahl and AM Gronenborn, *Science*, **261**, 438–46 (1993).
128 BF Luisi, WX Xu, Z Otwinowski, LP Freedman, KR Yamamoto and PB Sigler, *Nature*, **352**, 497–505 (1991).
129 GE Croston, LA Kerrigan, LM Lira, DR Marshak and JT Kadonaga, *Science*, **251**, 643–49 (1991).
130 H Granok, BA Leibovitch, CD Shaffer and SC Elgin, *Curr Biol*, **5**, 238–41 (1995).
131 SK Thukral, ML Morrison and ET Young, *Mol Cell Biol*, **12**, 2784–92 (1992).
132 SK Thukral, ML Morrison and ET Young, *Proc Natl Acad Sci USA*, **88**, 9188–92 (1991).
133 SK Thukral, A Eisen and ET Young, *Mol Cell Biol*, **11**, 1566–77 (1991).
134 C Cheng, N Kacherovsky, KM Dombek, S Camier, SK Thukral, E Rhim and ET Young, *Mol Cell Biol*, **14**, 3842–52 (1994).
135 M Schmiedeskamp and RE Klevit, *Biochemistry*, **36**, 14003–11 (1997).
136 GH Jacobs, *EMBO J*, **11**, 4507–17 (1992).
137 R Aurora, R Srinivasan and GD Rose, *Science*, **264**, 1126–30 (1994).
138 TJ Macke and DA Case, in NB Leontis and J Santa Lucia (eds.), *Molecular Modeling of Nucleic Acids: Modeling Unusual Nucleic Acid Structures*, American Chemical Society, Washington, DC, pp 379–93 (1998).
139 Z Otwinowski, RW Schevitz, RG Zhang, CL Lawson, A Joachimiak, RQ Marmorstein, BF Luisi and PB Sigler, *Nature*, **335**, 321–29 (1988).
140 WS Somers and SE Phillips, *Nature*, **359**, 387–93 (1992).
141 CR Kissinger, BS Liu, E Martin-Blanco, TB Kornberg and CO Pabo, *Cell*, **63**, 579–90 (1990).
142 C Wolberger, AK Vershon, B Liu, AD Johnson and CO Pabo, *Cell*, **67**, 517–28 (1991).
143 TE Ellenberger, CJ Brandl, K Struhl and SC Harrison, *Cell*, **71**, 1223–37 (1992).
144 Y Shi and JM Berg, *Biochemistry*, **35**, 3845–48 (1996).
145 AC Jamieson, SH Kim and JA Wells, *Biochemistry*, **33**, 5689–95 (1994).
146 EJ Rebar and CO Pabo, *Science*, **263**, 671–73 (1994).
147 AC Jamieson, H Wang and SH Kim, *Proc Natl Acad Sci USA*, **93**, 12834–39 (1996).
148 BS Wang, RA Grant and CO Pabo, *Nat Struct Biol*, **8**, 589–93 (2001).

Zinc modules in nuclear hormone receptors

Srikripa Devarakonda[†,‡] *and Fraydoon Rastinejad*[†,‡,§]

[†]Department of Pharmacology, University of Virginia, Charlottesville, VA, USA
[‡]Interdisciplinary Program in Biophysics, University of Virginia, Charlottesville, VA, USA
[§]Department of Biochemistry and Molecular Genetics, University of Virginia, Charlottesville, VA, USA

FUNCTIONAL CLASS

Transcription factor; nuclear hormone receptor; zinc finger protein; nucleic acid binding protein. Nuclear hormone receptors form a class of ligand-activated proteins that, when bound to specific gene-regulatory sequences representing DNA response elements (DREs), serve as on–off switches for transcription. The recognition of DREs takes place via the zinc finger motifs of these receptors.[1]

OCCURRENCE

Nuclear hormone receptors are found in a variety of higher eukaryotes[2–5] and their sequences have been conserved through evolution. In humans, these receptor proteins have been detected in nearly all major tissues and organs, including metabolic organs, reproductive organs, and the skin.[6–9] At the cellular level, the receptors can be found both in the cytoplasm as well as in the nucleus. The intracellular location of the nuclear receptors depends on the dynamic equilibrium between nuclear-cytoplasmic export and import.[10–12] Most nonsteroid receptors appear to be localized in the nucleus even in the absence of ligands, whereas the steroid receptors (except the estrogen receptor) appear cytoplasmically localized in the absence of ligands, localizing to the nucleus when liganded.[1,13,14]

BIOLOGICAL FUNCTION

Nuclear receptors are key mediators involved in signal transduction. They regulate the transcription of genes and control the development and differentiation of skin, bone and behavioral centers in the brain, as well as the continual regulation of reproductive tissues.[6–9] They are also responsible for mediating the physiological action of a variety of hormones such as steroid and thyroid hormones and the retinoids.[15,16] More recently, several

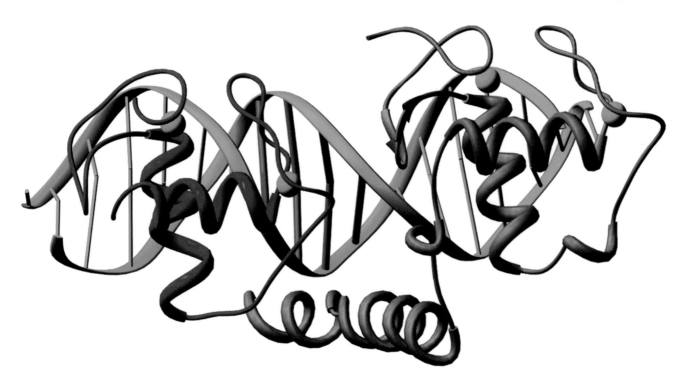

3D Structure   Schematic representation of the heterodimeric assembly of RXR and TR on DR4 DNA, showing the position of zinc ions. PDB code is 1NLL. The figure was prepared using RIBBONS.[81]

orphan receptors such as the liver X receptor (LXR) and the bile acid receptor Farnesoid X Receptor (FXR) have been shown to be sensors for hepatic lipid and cholesterol levels.

Hormone-activated nuclear receptors play crucial roles in cell growth, morphogenesis, and differentiation. Vitamin D receptor (VDR) plays a central role in calcium and phosphate homeostasis. Mutations associated with this receptor lead to vitamin D resistance and rickets.[17] Peroxisome proliferator-activated receptors (PPAR) induce differentiation of preadipocyte cell lines.[18] Thyroid hormone receptor (TR) is responsible for the actions of T3 and T4 hormones, and plays a key role in the development of the central nervous system (CNS), control of basal metabolism, and the proper functioning of the hypothalamus-pituitary-thyroid axis.[19] The retinoic acid receptors, (RAR) and retinoid X receptor (RXR), are responsible for the development of proper limb formation, as well as for the patterning and growth of embryos. These receptors are also required in eye morphogenesis and for the maintenance of the differentiated epithelia.[20] As a result, RAR and RXR are the drug targets for both skin disorders and for epithelial cancers.

In addition to the ligand-activated receptors, there are numerous orphan receptors in this family of transcription factors, which have as yet no known ligands. Examples include the steroidogenic factor-1 (SF-1) receptor, which controls the differentiation of human male genital apparatus and is critical for the development and function of the hypothalamic-pituitary-gonadal axis.[21] RevErb, an orphan receptor, acts as a transcriptional repressor in muscle and adipocyte differentiation and nervous system development in mice.[22-24] Another orphan receptor, the nerve growth factor–induced-B (NGFI-B), is involved in T-cell-mediated apoptosis, as well as neuronal differentiation and function.[25-27]

AMINO ACID SEQUENCE INFORMATION

As mentioned earlier, nuclear hormone receptors are found in a wide variety of species. Table 1 provides a list of receptors from five different organisms. While all

Table 1   Amino acid sequence information, including the SWISSPROT accession numbers and the number of amino acids for nuclear receptors from five different species. All known receptors from *H. sapiens* have been listed

Species	Gene	Amino acids	SWISSPROT ACC #
Caenorhabditis elegans	CNR3	553	P41828
	CNR14	534	P41830
	CSR1	579	Q17370
Drosophila melanogaster	Ecdysone receptor	878	P34021
	Fushi tarazu-F1	1198	Q9VVS8
	Ultraspiracle	508	P20153
Gallus gallus	Estrogen (α)	589	P06212
	Progesterone receptor	786	P07812
	Retinoic acid (β)	455	P22448
Mus musculus	Androgen receptor	899	P19091
	Farnesoid X receptor	488	Q60641
	Thyroid hormone (α)	461	P37242
Homo sapien Classical receptors	Glucocorticoid	777	P04150
	Mineralocorticoid	984	P08235
	Progesterone	933	P06401
	Estrogen (α, β)	595	P03372
		530	Q92731
	Androgen	919	P10275
	Thyroid hormone (α, β)	490	P10827
		461	P10828
		476	P37243
	Vitamin D	427	P11473
	Retinoic acid (α, β, γ)	462	P10276
		455	P10826
		454	P13631
		443	P22932
Orphan receptors	CAR	348	Q14994
	COUP (α, β)	423	P10589
		414	P24468

(continued overleaf)

Table 1 (continued)

Species	Gene	Amino acids	SWISSPROT ACC #
	DAX	470	P51843
	ERR (α, β, γ)	519	P11474
		500	O95718
		458	O75454
	FXR	476	Q96RI1
	GCNF1	480	Q15406
	HNF4 (α, γ)	465	P41235
		408	Q14541
	LRH1	541	O00482
	LXR (α, β)	447	Q13133
		461	P55055
	NGFI-B (α, β, γ)	598	P22736
		598	P43354
		626	Q92570
	PPAR (α, β, γ)	468	Q07869
		441	Q03181
		505	P37231
	PXR	434	O75469
	RetinaXR	410	Q9Y5X4
	RevErb (α, β)	614	P20393
		579	Q14995
	ROR (α, β, γ)	556	P35398
		459	Q92753
		560	P51449
	RXR (α, β, γ)	462	P19793
		533	P28702
		463	P48443
	SF1 (α)	461	Q13285
	SHP	257	Q15466
	Tlx	385	Q9Y466
	TR2 (α, β)	483	P13056
		596	P49116

known receptors in *Homo sapiens* have been listed, this is not an exhaustive list for the receptors found in the other organisms.

MOLECULAR CHARACTERIZATION

Modular structure of nuclear receptors

Nuclear hormone receptors are modular in structure, with separate domains responsible for distinct functions.[28] Figure 1 shows the overall architecture of these receptors. Domains C and E are evolutionarily the most conserved in sequence among the family members. In isolation, these domains have also been structurally well characterized.[29–34] The N-terminal region is the least conserved part of the receptor, being variably sized with no examples of three-dimensional structures to date. In some receptors, this region harbors a transcriptional activation function (AF-1), allowing the receptor to activate transcription even in the absence of a bound ligand.[1,28,35] Domain C, the DNA binding domain (DBD), is responsible for the receptor's selective binding to DREs. This domain also forms a DNA-dependent dimerization surface for most receptors. The carboxy-terminal E region, the ligand binding domain (LBD), binds to lipophilic molecules and also forms a second dimerization surface for certain receptors. The region F in most receptors is associated with a secondary activation function (AF-2), which regulates the transcriptional activity of the receptor in a strictly ligand-dependent manner. Region D, which links the DBD to the LBD, is referred to as the receptor-hinge region. This region, like the A/B region, is both variable in size and in sequence within the family. In some of the structures reviewed in this chapter, the hinge region has been visualized to play a significant role in the receptor's DNA target selectivity.[36–38]

The DBD is the most conserved region within the nuclear hormone receptors.[1,28,35] This conserved region covers a stretch of 66 residues referred to as the 'core DBD'. Figure 2(a) shows sequence alignment of the core DBDs (top) and of the adjacent hinge regions (bottom) of various receptors. A common numbering scheme has been adopted in order to provide a simplified way to compare the many receptor DBDs.[28] In this scheme, the numbering starts at

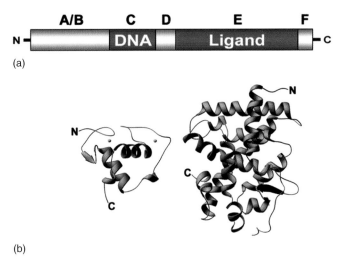

Figure 1 (a) A schematic representation of the modular structure of nuclear receptors. (b) On the left is a ribbon diagram of TR-DBD, and on the right is a ribbon diagram of TR-LBD. These structures have not been visualized together in a single protein, but rather solved independently.

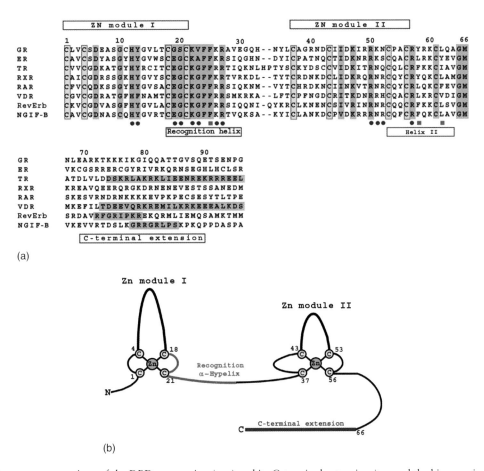

Figure 2 (a) Sequence comparison of the DBD core region (top) and its C-terminal extension (toward the hinge region) of eight nuclear receptors. A common numbering scheme used throughout the chapter starts with the first conserved cysteine being marked position 1. In the core region, residues belonging to the two zinc modules/fingers and the recognition helix are indicated. The conserved cysteines involved in zinc coordination are shaded in yellow. The residues shaded in orange are conserved in most of the nuclear receptors. The blue circles indicate residues that are involved in DNA binding and red squares indicate residues that contribute to the hydrophobic core. In the CTE region, residues shaded in pink are responsible for spacer discrimination and those shaded in green are responsible for recognition of 5′ extensions of the half-sites. (b) Schematic representation of the DBD showing the two zinc modules. The cysteines involved in zinc coordination are indicated in yellow along with the zinc ions shown in green. The portion that forms the recognition helix is in red and the CTE is in blue.

the first conserved cysteine residue in the core sequence and ends in the conserved methionine terminating the DBD core. The residues beyond this core have also been shown to play an important role in the function of the DBD despite their lower degree of conservation.[39–41]

The zinc finger motifs

Each nuclear hormone receptor contains two zinc ions per molecule, as first identified by atomic absorption spectroscopy and extended X-ray absorption fine structure (EXAFS) spectroscopy. A comparison of the protein EXAFS spectrum with that of a model compound containing a tetrahedral zinc:4-sulfur coordination complex revealed that there are two zinc ions in the receptor DBD and these are coordinated in a rigid tetrahedral geometry by four sulfur atoms.[42]

Importantly, removal of zinc from the protein abolishes the DNA binding function of the receptor. Though cadmium does substitute for zinc in terms of restoring DNA binding, inductively coupled plasma mass spectrometry on purified glucocorticoid receptor (GR) shows that the native receptor contains zinc and not significant levels of cadmium.[42] Proteolytic digestion of the metallo- and apoprotein forms of GR indicate that presence of the metal confers protease resistance to the DNA binding domain. These results indicate that metal coordination is essential for proper folding of the protein into its native and active structure.[42]

Figure 2(b) shows the coordination scheme for the two zinc ions within the nuclear receptor DNA binding domains.[42] Each of the two zinc ions is coordinated by sulfur atoms from four absolutely conserved cysteines in the core DBD (shaded in yellow in Figure 2(a)). This results in the formation of two finger-like motifs folded around the zinc ions. These two modules, referred to as zinc subdomains or zinc fingers, do not act in DNA binding independently, but rather as a single scaffold for DNA binding.

DNA response-element discrimination

Nuclear receptors regulate genes through binding sites located in gene promoter elements.[1,28,35,42] Despite the large number of nonsteroid nuclear receptors, there is a single-consensus hexad sequence that these receptors recognize and bind to: 5′-PuGGTCA-3′.[1] Characteristics of the variations in this hexad sequence that result in a repertoire of DREs for the many nuclear receptors have been described below. Figure 3 shows the various classes into which nuclear receptors can be classified on the basis of the response element they recognize.

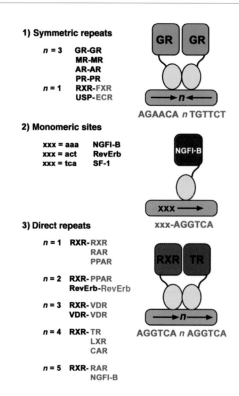

Figure 3 Classification of nuclear receptors based on the types of response elements that they recognize. The first class of receptors binds to symmetric or palindromic repeats, the second class binds to direct repeats, and the third class binds to extended half-sites as monomers.

Sequence of the response element Two distinct classes of response elements have emerged from mutations in the core sequence. The steroid receptors (GR, mineralocorticoid receptor (MR), androgen receptor (AR), progesterone receptor (PR)) have been shown to recognize the hexad sequence 5′-AGAACA-3′[43–45] and all of the nonsteroid receptors and also estrogen receptor (ER) have been shown to recognize 5′-AGGTCA-3′.[46–50] In addition, within each class there could be mutations of the DRE that lead to added specificity of the response element.

Topology of receptor binding Receptors have been shown to be organized on their DREs as monomers, homodimers, or heterodimers. Dimerization of receptors allows binding of extended regions that consist of two side-by-side hexad sequences, and thus increases the specificity of receptors for their cognate DNA sites.[1] The two hexad elements can be oriented in a head-to-tail fashion (direct repeats or 'DR' orientation) or a head-to-head fashion (palindromic repeats). Steroid receptors form homodimers and bind strictly to palindromic repeats (which are inverted or 'IR' orientated). Most nonsteroid receptors bind to direct repeats as homo- or heterodimers. Some orphan receptors bind to monomeric hexad sequences, which are extended on the 5′ side with additional specificity sequences.[32–34,36–41]

Table 2 Response elements of some nuclear receptors, including the consensus sequences

DRE	DRE sequence	Receptor
Half-site	AAAAGGTCA	NGFI-B[41]
DR1	AGGTCANAGGTCA	RXR dimer[33]
DR2	AGGTCANNAGGTCA	RevErb dimer[32,40]
DR3	AGGTCANNNAGGTCA	VDR dimer[34]
DR4	AGGTCANNNNAGGTCA	RXR and LXR[51]
DR5	AGGTCANNNNNAGGTCA	RXR and RAR[38]
IR1	AGGTCANTGACCT	RXR and FXR[52]
IR3	AGAACANNNTGTTCT	GR dimer[36]

In cases in which the DBDs bind as dimers to their cognate response elements, another aspect of the response element that is responsible for generating site selectivity is the size of the spacing between repeats. The length of the spacer is thought to play a more significant role than the sequence.[32–34,36–40,48] Table 2 gives a list of some of the response elements and their sequences.

PROTEIN PRODUCTION AND PURIFICATION

Nuclear hormone receptors have been cloned and identified using nuclear-receptor-specific antibodies and later with degenerate oligonucleotides on the basis of the peptide sequences of the known receptors.[53–56] Using antibodies, a few receptors were initially isolated, purified, and sequenced. The degenerate oligonucleotides were used to screen cDNA libraries for nuclear receptor sequences, which were then expressed in cells to obtain the proteins.[53,54] The remarkable sequence conservation in the DBD region has proven useful for cloning the complementary DNA sequences encoding various nuclear receptors. The vast majority of the members of this superfamily have been cloned using this strategy.[57–62]

The full-length proteins as well as their isolated domains, such as the DBDs, have also been cloned into various expression systems and overproduced in *Escherichia coli*. Most of these clones have employed either the pGEX or the pET vector systems,[38–40] as these vectors provide convenient N-terminal affinity tags that assist protein purification. The DBDs are mostly expressed and purified from the soluble fraction of the cell lysate using affinity chromatography. The pGEX expression system (Pharmacia) allows GST-fusion protein production and subsequent purification using glutathione-sepharose affinity chromatography. The glutathione S transferase (GST) tag can subsequently be removed by thrombin or other protease digestion.

The pET vectors (Novagen) allow expression of receptor fragments as histidine-tagged fusions, which can then be purified using nickel affinity chromatography. The histidine-tag can also be removed using a suitable protease. In both cases the DBD is finally purified using cation exchange resin to obtain pure and homogeneous protein.

The samples are analyzed using standard SDS-gel electrophoresis. DBDs are monomers in solution in the absence of DNA, and need to be maintained in a reducing environment due to their high content of oxidation-sensitive cysteines. While the above-described procedures involving affinity tags are the more commonly used procedures for obtaining purified DBDs, various other methods have been and can be employed for this purpose in the case of receptors that are expressed without tags. For example, the GR-DBD has been purified using ammonium sulfate precipitation,[36,42] VDR-DBD has been purified from inclusion bodies and refolded to the native state,[34] and NGFI-B-DBD has been purified using ion exchange chromatography.[41]

X-RAY AND NMR STRUCTURE

The successful overexpression and purification of the DBDs have made their structural studies feasible. The various structures determined to date by two-dimensional nuclear magnetic resonance (NMR) and X-ray diffraction studies have been summarized in Table 3. The list includes NMR structures of the GR, ER, RAR, RXRDBDs[63–67] in the absence of DNA. Assemblies of DBD dimers on DNA have significantly larger molecular weights as compared to free DBD. Hence, structures of all the complexes to date have been determined using X-ray diffraction. Crystal structures listed include those of GR, ER, RevErb, RXR and VDR homodimers bound to DNA,[32–34,36,37,40] of RXR in a heterodimeric DBD complex with TR and also with RAR on DNA[38,39] and the monomeric complex of NGFI-B-DBD on DNA.[41] In addition to these crystal structures, there is also a structure of a GR/ER hybrid DBD on a noncognate response element[68] and that of ER dimer bound to a nonconsensus response element.[69]

Table 3 X-ray and NMR structures solved to date including resolution of the X-ray structures

#	Receptor	Method	Resolution (Å)	PDB code
1	GR[63,64]	NMR	–	1RGD
2	ER[65]	NMR	–	1HCP
3	RAR[66]	NMR	–	1HRA
4	RXR[67]	NMR	–	1RXR
5	(GR)$_2$ + DNA[36]	X-ray	2.90	1GLU
6	(ER)$_2$ + DNA[37]	X-ray	2.40	1HCQ
7	RXR-TR + DNA[39]	X-ray	1.90	2NLL
8	E/Tr-Gr + DNA[68]	X-ray	1.90	1LAT
9	EREvariant + DNA[69]	X-ray	2.60	Not available
10	RevErb + DNA[32,40]	X-ray	2.30	1A6Y, 1GA5
11	NGFI-B + DNA[41]	X-ray	2.70	1CIT
12	(RXR)$_2$ + DNA[33]	X-ray	2.10	1BY4
13	RXR-RAR + DNA[38]	X-ray	1.70	1DSZ
14	VDR + DNA[34]	X-ray	2.70	1KB2

These structural studies have essentially created a database of DBD structures and have aided in understanding the basis of DRE selection and nuclear receptor dimerization. A detailed analysis of these structures helps in understanding how it is possible for relatively few hexad sequences to be placed in a multitude of geometric arrangements such that there is adequate specificity and diversity created in the repertoire of nuclear receptor binding sites.

Crystallization

Sample purity is known to be a critical factor for crystallization and hence samples used for this purpose should be as pure and homogeneous as possible.[70] DBDs of very high purity can be obtained using one of the purification techniques mentioned in the previous section. In order to obtain the corresponding response elements for formation of a protein–DNA complex, complementary strands of the DRE are synthesized using solid-phase phosphoramidite method.[71] This is the preferred technique for DNA synthesis as it couples high yields with easily purified products. The oligonucleotides thus obtained can be purified using reverse-phase high-performance liquid chromatography (HPLC) chromatography and annealed in order to obtain the double-stranded DNA targets. It is necessary to purify the DNA oligomers to homogeneity as impurities can have deleterious effects on cocrystallization with DBD.[70,72] The DBDs and DNA are then combined in stoichiometric ratio in order to obtain the complex for crystallization.

Table 4 lists the conditions in which various DBD/DNA complexes have been successfully crystallized. This list includes the sequence of the oligonucleotide used, temperature of crystallization and the protein:DNA molar ratio. Crystallization of the complexes under these conditions resulted in high-quality crystals, which were then used to obtain the necessary diffraction data for structure solution.

Global structure

Figure 4 shows the overall fold of the TR-DBD in the RXR-TR structure.[39] In both RXR and TR, each zinc is coordinated in a rigid tetrahedral geometry by the sulfurs from the conserved cysteines.[28,42] The two cysteine-coordinated zinc modules in each subunit differ in the chirality they adopt with respect to the zinc. Zinc module I in both RXR and TR adopts an S configuration, and zinc module II adopts an R configuration.[28] Figure 5(a) shows

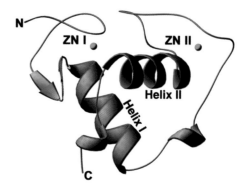

Figure 4 Ribbon diagram of a typical DBD depicting the two Zn modules (ZN I and ZN II) and the two helices that are arranged perpendicular to each other (Helix I and Helix II).

Table 4 Crystallization conditions including temperature, response-element sequence and protein:DNA ratio (P:D) used in crystallization

#	Receptor	Oligonucleotide	Crystallization condition	P:D
1	(GR)$_2$	CC**AGAACA**CAG**TGACCT**G GTCTTGTGTCACTGGACC	8% 2-methyl-2,4-pentanediol (MPD), 24 mM Sodium cacodylate pH 6.0, 0.2 M NaCl, 1.8 mM Spermine, 2 µM ZnCl$_2$, 8 °C[36]	2:1
2	(ER)$_2$	CC**AGGTCA**CAG**TGACCT**G GTCCAGTGTCACTGGACC	10% MPD, 20 mM 2-(N-morpholino) ethanesulfonic acid (MES) pH 6.0, 30–80 mM NaCl, 2–8 mM CaCl$_2$, 1.8 mM Spermine, 2 µM ZnCl$_2$, 20 °C[37]	2.2:1
3	RXR-TR	CC**AGGTCA** TTT**CAGGTCA**G GTCCAGTAAAGTCCAGTCC	40% Polyethyleneglycol (PEG) 3350, 25 mM Imidazole pH 7.0, 400 mM NH$_4$Cl, 10 mM DTT, 5 mM MgCl$_2$, 17 °C[39]	2:1
4	RevErb	CAACT**AGGTCA**CT**AGGTCA**G GTTGATCCAGTGATCCAGTC	20–25% PEG 8000, Tris pH 7.5, 400 mM NaCl, 5 mM MgCl$_2$, 17 °C[32]	2:1
5	NGFI-B	CCCGAA**AGGTCA**TGCG GGCTTTCCAGTACGCC	29–30% PEG 4000, 50 mM 3-(N-morpholino) propanesulfonic acid (MOPS) pH 7.0, 250 mM NH$_4$Cl, 5 mM DTT, 27 °C[41]	1:1.2
6	(RXR)$_2$	T**AGGTCA**A**AGGTCA**G ATCCAGTTCCAGTC	18–23% Peg 3350, 25 mM Tris pH 7.5, 400 mM NH$_4$Cl, 5 mM MgCl$_2$, 8 °C[33]	2:1
7	RXR-RAR	T**AGGTCA**A**AGGTCA**G ATCCAGTTCCAGTC	18–23% Peg 3350, 25 mM Tris pH 7.5, 400 mM NH$_4$Cl, 5 mM MgCl$_2$, 8 °C[38]	2:1
8	(VDR)$_2$	CAC**AGGTCA**CGA**AGGTCA** GTGTCCAGTGCTTCCAGT	4–6% PEG 4000, 50 mM MES pH 5.6, 5 mM MgCl$_2$, 10% glycerol, 10 mM DTT, 18 °C[34]	2:1

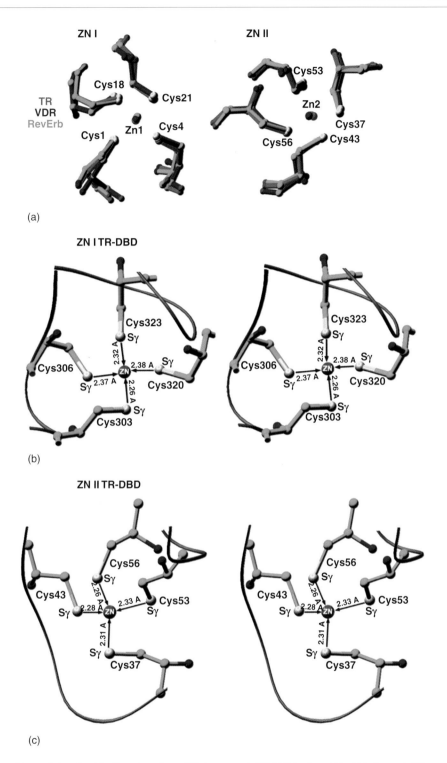

Figure 5 (a) Superposition of the zinc binding sites from TR (magenta), VDR (blue) and RevErb (green). The cysteines are numbered using the scheme shown in Figure 2(a). (b) and (c) show stereo images of the two zinc modules, with distances from the zinc to the cysteines indicated. The numbering scheme is the same as in Figure 2(a) and the distances are for TR from the RXR-TR structure. The average Zn–sulfur distance is 2.32 Å.

a superposition of the zinc binding sites of three nuclear receptors, TR, VDR and RevErb, suggesting that the zinc coordination is likely to occur in this identical manner for all members of the family. Figure 5 ((b) and (c)) also shows stereo views of the individual zinc modules.

While indicating the distances between the sulfur and the zinc ions, the figures also show the chiralities of zinc coordination.

In addition to the two zinc modules, the DBD also consists of two amphipathic alpha helices. The first helix

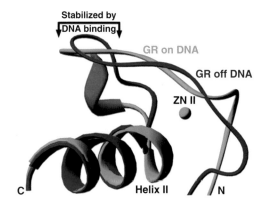

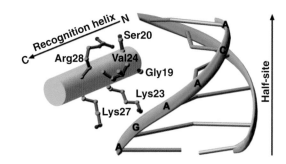

Figure 6 Superposition of Zn module II and Helix II of GR with (light blue) and without bound DNA (dark blue). In the absence of DNA, some of the residues in ZN II are very flexible. They become more ordered when bound to DNA, as can be seen from the formation of the helix, indicated in red.

Figure 7 The recognition helix recognizes the major groove of the half-site. The P-box residues are colored in blue and other residues in the helix that interact with DNA are also shown.

encompasses residues 18 to 30 and the second helix encompasses residues 53 to 64.[28] These helices each follow the two zinc modules and pack at right angles to each other, crossing almost at their midpoints.[1] The two helices bury a cluster of hydrophobic side chains between them, giving rise to the central core of the molecule.[28,35] The region between the first zinc module and the first helix also has hydrophobic residues that form a loop and interact with the hydrophobic core of the protein.[28,35] The charged face of helix I forms sequence specific major groove contacts at the half-site and hence is referred to as the recognition helix.[1,28,35–37] Figure 2(a) indicates the amino acids that comprise the two zinc modules, in addition to the residues that form the recognition helix and those that make contacts with DNA.

Comparison of GR-DBD bound to DNA (X-ray) with the free GR-DBD (NMR) in solution structure reveals their overall similarity. Differences are seen in the zinc module II region (Figure 6). In the absence of DNA, the residues in this region are flexible (dark blue), but become more ordered upon binding to DNA (light blue).[36,63,64] This reconfiguration is indicated by the formation of a short helix (red) when bound to DNA.

DNA recognition

Sequence-based discrimination

The overall shape of the DBD bound to DNA is very similar to the one without DNA.[36,63,64] The DNA structure in the response elements is in the B-form without significant distortion.[32–34,36–41] The amino-terminal region of the DBD helps to orient the residues that make DNA phosphate contacts. The recognition helix provides all the base-specific contacts in the major groove of DNA. Ordered water molecules also extend the protein-DNA interfaces, by bridging specific interactions between protein side chains and DNA functional groups.[35] Three residues (residues 19, 20 and 23 in Figure 2(a)) in the recognition helix of the receptors, referred to as the P-box, are responsible for half-site discrimination.[73–75] Receptors with a GS..V in their P-box recognize the 5'-AGAACA-3 half-site, while receptors that typically contain EG..G residues at the corresponding positions recognize 5'-AGGTCA-3 half-sites.[1,28,35,36,38] Figure 7 shows the recognition helix of GR making specific contacts with the half-site.

In addition to these specific contacts made by the P-box residues, there are a number of nonspecific interactions between basic side chains of the DBD and the DNA-phosphodiester backbone, as indicated in Figure 7. The DBD surface buried along the DNA maintains a positive electrostatic potential that stabilizes its association with the negatively charged backbone of the DNA.

Topology-based discrimination

Although the recognition helix determines the half-site specificity of a receptor, it does not play a role in discrimination of target-site symmetry. As mentioned earlier, DRE selectivity is derived from the geometry associated with a response element with respect to how the two half-sites are arranged relative to each other (Figure 3, Table 2). In order to recognize the geometry of these bipartite structures, receptors bind cooperatively as homo- or heterodimers (Figure 8).[28,35,76] In cases in which the receptors bind to the response elements as monomers, they rely on novel DNA binding regions positioned adjacent to their core DBD for extending their monomeric interaction (Figures 8, 9, and 10(b)).[32,40,41]

Dimerization

Dimerization between DBDs is a general mechanism used by the nuclear receptors to increase binding site affinity, specificity, and diversity.[1] In the absence of DNA, the protein–protein interactions do not form very readily and the interface is too small and not correctly bridged

Zinc modules in nuclear hormone receptors

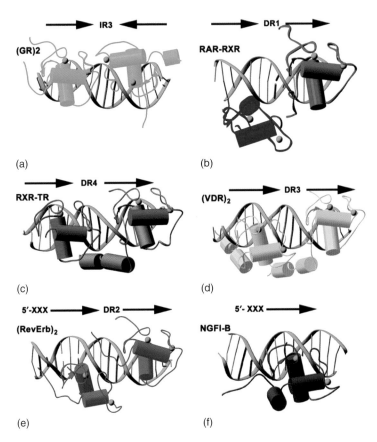

Figure 8 Structures of complexes between nuclear receptor DBDs and their cognate response elements. (a) GR homodimer bound to a palindromic response element (IR3), (b) and (c) heterodimers of RXR with RAR and TR, bound to DR1 and DR4 respectively, (d) VDR homodimer bound to a cognate DR3 element, (e) RevErb homodimer bound to cognate response element, which includes the 5′ extension site and (f) NGFI-B bound to its extended monomeric site.

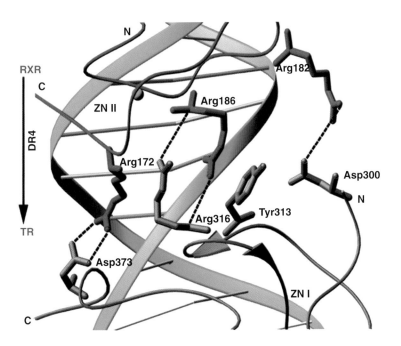

Figure 9 Residues involved in the formation of the dimer interface between RXR and TR.

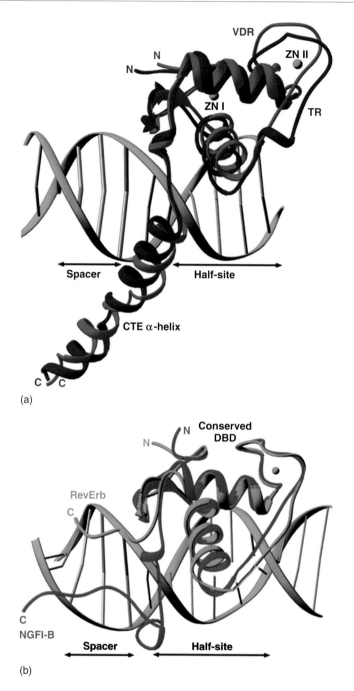

Figure 10 (a) Superposition of TR and VDR bound to a consensus site. The CTE forms an extended helix that plays a role in spacer discrimination. Also shown is the spacer between the half-sites. (b) Superposition of the upstream monomer from the RevErb structure and the NGFI-B monomer bound to a consensus site with the 5′ extension. The CTEs of both the proteins bind to the minor groove and recognize the corresponding 5′ extension sequence. Figures 1 and 4–10 have been made using RIBBONS.[81] Superposition of various structures in Figures 5(a), 6, 10(a) and 10(b) was done using CCP4.[82]

by DNA. Therefore, in the absence of dimerization, the affinity of a single receptor to an isolated response element can be reduced by as much as 100-fold in comparison to the affinity associated with the joint binding of two DBDs. Dimerization of DBDs on DNA not only increases the surface area of interaction between the two proteins but also leads to cooperative enhancement of DNA affinity.[35]

Crystallographic examination of various complexes has illustrated the detailed mechanisms for joint cooperative binding of receptor DBDs on bipartite response elements (Figure 8).[32–34,36–40] The GR crystal structure shows the basis for the formation of a GR homodimer on a symmetric response element.[36] Owing to the symmetric nature of the response element, the GR dimer is formed in a head-to-head fashion as shown in Figure 8. Various studies performed on GR have shown the importance of the spacing between the half-sites in response-element recognition. Improper insertion of an extra base pair between binding sites forces the recognition helix of one subunit out of alignment with its half-site, disabling the expected contacts at the price of preserving the dimerization interface.[36] The other steroid receptors (MR, AR, PR) have been shown to recognize the same response element as GR.[77–79] However, studies done to date have not explained how these receptors discriminate their target sites. It is possible that they utilize differences in the regions flanking the half-sites or mutations in the consensus half-site itself to achieve this discrimination. In addition, they could also rely on other configurations of the half-sites *in vivo* that have yet to be characterized.[77]

Figure 8 shows that all the dimeric complexes on DRs form in a head-to-tail fashion distinct from the symmetric organization seen in GR. There are two elements that stand out as recognition features in the direct-repeat structures: the spacer between the repeats and the polarity of the heterodimers. DR sequences are typically the targets for the subset of nuclear receptors that dimerize with RXR (Figure 3). Four such heterodimeric assemblies have been studied to date.[33,38,39,76] In each case, the spacer orients the two DBDs in the dimer in such a way as to create a unique dimerization interface between RXR and its partner. Figure 9 shows the dimer interface in the RXR-TR structure straddling the minor groove of the spacer. Any change in this spacer has serious geometric consequences, displacing and reorienting the interacting subunits, leading to a disruption of the favorable dimerization interface.[35,39] On the basis of the specific response element, RXR changes its conformation to make the diverse range of protein–protein interactions required for all partners. This, in part, could explain the preferential binding of RXR to the 5′ site in most of the dimeric assemblies on DRs. One exception known to date is the RXR-RAR dimer on DR1 in which RAR occupies the 5′ site.[38] On the other hand, there is biochemical evidence that shows that on a DR5, RXR binds to the upstream site and RAR to the downstream site,[80] indicating that the polarity of this dimer is response-element–dependent. These findings together indicate that the interactions between RXR and its partners are highly dependent on the geometry of their DR binding sites.[35]

Receptor hinge in spacer discrimination

Sequence alignment of the DBD region indicates that despite the high degree of sequence identity in the core DBD, there is a substantial degree of sequence divergence in the C-terminal extension (CTE) of the DBD that is within the hinge regions (Figure 2(a)). Some of the DBDs have been crystallized with part or most of their hinge region[32,34,39,41] and thus offer an opportunity to examine the role of this poorly conserved region. The structure of each CTE determined to date has been unique, in accordance with its poor sequence conservation.

In the RXR-TR and the VDR homodimer structures (Figure 10(a)), the CTE has been shown to form an elongated α-helix that projects across the minor groove. A comparison of the VDR and TR structures would seem to predict the structural homology seen crystallographically. Although both the VDR and TR contain helices in the CTEs of their DBDs, these helices project in somewhat different directions. Modeling studies with TR have shown that this region plays a significant role in determining the spacer length and thus acts as a molecular ruler for the discrimination of the inter-half-site spacing in a DR.[39] In preventing the dimeric assembly onto a misspaced response element, the CTE appears to ensure the formation of a favorable dimerization interface between the two partners. In TR, the CTE helix also makes extensive contacts with DNA in the minor groove and along the phosphodiester backbone. Therefore, this region also serves to increase the affinity of the protein to DNA.[34]

Receptor hinge in DNA binding

Though dimerization is the most common way of achieving higher DNA affinity and target-site selectivity,[28,35] there are some monomeric nuclear receptors that recognize response elements with a single copy of the half-site.[41] Biochemical studies have suggested that unique flanking sequences preceding the half-site play an important role in determining response-element specificity for SF-1, RevErb, NGFI-B and ROR (RAR related orphan receptor).[32,83] Figure 3 shows the elements of the response element that are responsible in determining the specificity. Structural studies of RevErb and NGFI-B have provided detailed insight into how these DBDs use the 5′ flanking regions of the core site to discriminate the response elements.[32,40,41]

RevErb DBD complexed with DNA was studied as a homodimer, although this receptor is very efficient also at binding to an extended half-site as a monomer.[32,40] Since the 5′ flanking region of the response element had the appropriate extension (ACT) associated with RevErb's monomeric binding site, it was possible to infer how the monomeric RevErb would use this region of DNA for efficient binding. The crystal structures revealed that the CTE of RevErb makes direct and sequence-specific contacts with the half-site 5′ extension, all of which are confined to the minor groove and backbone regions of the DNA. The structure of a monomeric complex of NGFI-B bound to DNA later provided a second view of how this recognition is mediated. In this case also, it was seen that the CTE region of the DBD forms extensive contacts with the minor groove recognizing the nucleotides AAA of the 5′ flanking site, as opposed to ACT in the case of RevErb. Figure 2(a) shows the residues in the C-terminal regions of RevErb and NGFI-B that are responsible for discriminating between the 5′ flanking regions of their cognate response elements. Figure 10(b) is a superposition of the RevErb and the NGFI-B monomers bound to the extended half-site showing how the CTEs project across the minor groove and recognize the 5′ flanking sites.

COMPARISON WITH RELATED STRUCTURES

Zinc binding is associated with a diverse set of protein motifs characterized by mini-domains that are DNA binding transcription factors. It has been estimated that nearly 1% of the human genome encodes for zinc-containing proteins, making this a ubiquitous motif in biology. Single zinc binding modules in transcription factors could be broadly categorized into distinct classes on the basis of a signature amino acid sequence: C_2H_2, C_4, where C and H represent the amino acids cysteine and histidine respectively. The repeated nature of this invariant sequence in the protein results in the formation of anywhere between 2 to 37 repeats of the zinc binding motif.[84] DNA recognition by zinc binding proteins occurs primarily due to an interaction between the complementary surfaces of a protein α-helix and the DNA major groove. The nuclear hormone receptors belong to the C_4 class, with a signature sequence of $CX_2CX_{13}CX_2CX_{14}CX_5CX_9CX_2C$. Importantly, distinct sets of four cysteine clusters coordinate two distinct zinc ions in each nuclear receptor DBD. Unlike many other zinc finger proteins, these two fingers are not the result of a duplication event in evolution and represent complementary surfaces within a single fold.[28] While zinc binding motifs from the classical zinc finger proteins act independent of each other when bound to DNA, the two fingers from nuclear receptors pack onto each other and form a compact globular domain, with each zinc module playing a distinct role in DNA recognition and binding.

REFERENCES

1. H Gronemeyer and V Laudet, *Protein Profile*, **2**, 1171–308 (1995).
2. DB Lubahn, DR Joseph, M Sar, J Tan, HN Higgs, RE Larson, FS French and EM Wilson, *Science*, **240**, 327–30 (1988).
3. WW He, LK Fischer, S Sun, DL Bilhartz, XP Zhu, CY Young, DB Kelley and DJ Tindall, *Biochem Biophys Res Commun*, **171**, 697–704 (1990).
4. A Krust, S Green, P Argos, V Kumar, P Walter, JM Bornert and P Chambon, *EMBO J*, **5**, 891–97 (1986).
5. MR Koelle, WS Talbot, WA Segraves, MT Bender, P Cherbas and DD Hogness, *Cell*, **67**, 59–77 (1991).
6. JA Tan, DR Joseph, VE Quarmby, DB Lubahn, M Sar, FS French and EM Wilson, *Mol Endocrinol*, **2**, 1276–85 (1988).
7. S Green, P Walter, V Kumar, A Krust, JM Bornert, P Argos and P Chambon, *Nature*, **320**, 134–39 (1986).
8. DP McDonnell, DJ Mangeldorf, JW Pike, MR Haussler and BW O'Malley, *Science*, **235**, 1214–17 (1987).
9. A Zelent, A Krust, M Petkovich, P Kastner and P Chambon, *Nature*, **339**, 714–17 (1989).
10. A Guichon Mantel, H Loosefelt, P Lescop, S Christin Maitre, M Perrot Applanat and E Milgrom, *J Steroid Biochem Mol Biol*, **41**, 209–15 (1992).
11. M Perrot Applanat, A Guichon Mantel and E Milgrom, *Cancer Surv*, **14**, 5–30 (1992).
12. JF Savouret, M Perrot Applanat, P Lescop, A Guichon Mantel and E Milgrom, *Steroid Biochem*, **24**, 19–24 (1986).
13. WJ King and GL Greene, *Nature*, **307**, 745–47 (1984).
14. KH Lin, MC Willingham, CM Liang and SY Cheng, *Endrocrinology*, **128**, 2601–9 (1991).
15. H Gronemeyer, *FASEB J*, **6**, 2524–29 (1992).
16. H Gronemeyer, *J Recept Res*, **13**, 667–91 (1993).
17. MR Hughes, PJ Malloy, DG Keiback, RA Kesterson, JW Pike, D Feldman and BW O'Malley, *Science*, **242**, 1701–5 (1988).
18. A Ijpenberg, E Jeannin, W Wahli and B Desvergne, *J Biol Chem*, **272**, 20108–17 (1997).
19. D Forrest, *Semin Cancer Biol*, **5**, 167–76 (1994).
20. LM De Luca, *FASEB J*, **5**, 2924–33 (1991).
21. HA Ingraham, DS Lala, Y Ikeola, X Luo, WH Shen, MW Nachtigal, R Abbud, JH Nilson and KL Parker, *Genes Dev*, **8**, 2302–12 (1994).
22. E Bonnelye, JM Vanacker, X Desbians, A Begue, D Stehelin and V Laudet, *Cell Growth Differ*, **5**, 1357–65 (1994).
23. A Chawla and MA Lazar, *J Biol Chem*, **268**, 16265–69 (1993).
24. M Downes, AJ Carozzi and GE Muscat, *Mol Endocrinol*, **9**, 1666–78 (1995).
25. ZG Liu, SW Smith, KA McLaughlin, LM Schwartz and BA Orborne, *Nature*, **367**, 281–84 (1994).
26. S Bandoh, T Tsukada, K Maruyama, N Okhura and K Yamaguchi, *J Neuroendocrinol*, **9**, 3–8 (1997).
27. PA Crawford, Y Sadovsky, K Woodson, SL Lee and J Milbrandt, *Mol Cell Biol*, **15**, 4331–416 (1995).
28. F Rastinejad, in LP Freedman (ed.), *The Molecular Biology of Steroid and Nuclear Hormone Receptors*, Birkhauser, Boston, MA, pp 105–31 (1997).
29. W Bourguet, M Ruff, P Chambon, H Gronemeyer and D Moras, *Nature*, **375**, 377–82 (1995).
30. RL Wagner, JW Apriletti, ME McGrath, BL West, JD Baxter and RJ Fletterick, *Nature*, **378**, 690–97 (1995).
31. D Moras and H Gronemeyer, *Curr Opin Cell Biol*, **10**, 384–91 (1998).
32. Q Zhao, S Khorasanizadeh, Y Miyoshi, MA Lazar and F Rastinejad, *Mol Cell*, **1**, 849–61 (1998).
33. Q Zhao, SA Chasse, S Devarakonda, ML Sierk, B Ahvazi and F Rastinejad, *J Mol Biol*, **296**, 509–20 (2000).
34. PL Shaffer and DT Gewirth, *EMBO J*, **21**, 2242–52 (2002).
35. S Khorasanizadeh and F Rastinejad, *Trends Biochem Sci*, **26**, 383–90 (2001).
36. BF Luisi, WX Xu, Z Otwisnowski, LP Freedman, KR Yamamoto and PB Sigler, *Nature*, **352**, 497–505 (1991).
37. JWR Schwabe, L Chapman, JT Finch and D Rhodes, *Cell*, **75**, 567–78 (1993).
38. F Rastinejad, T Wagner, Q Zhao and S Khorasanizadeh, *EMBO J*, **19**, 1045–54 (2000).
39. F Rastinejad, T Perlmann, RM Evans and PB Sigler, *Nature*, **375**, 203–11 (1995).

40 ML Sierk, Q Zhao and F Rastinejad, *Biochemistry*, **40**, 12833–43 (2001).
41 G Meinke and PB Sigler, *Nat Struct Biol*, **6**, 471–77 (1999).
42 LP Freedman, BF Luisi, Z Richard Korszun, R Basavappa, PB Sigler and KR Yamamoto, *Nature*, **334**, 543–46 (1988).
43 M Beato, *Cell*, **56**, 335–44 (1989).
44 RM Evans, *Science*, **240**, 889–95 (1988).
45 CK Glass, *Endocrinol Rev*, **15**, 391–407 (1994).
46 DJ Mangelsdorf, C Thummel, M Beato, P Herrlich, G Schutz, K Umesono, B Blumberg, P Kastner, M Mark, P Chambon and RM Evans, *Cell*, **83**, 835–39 (1995).
47 DJ Mangelsdorf and RM Evans, *Cell*, **83**, 841–50 (1995).
48 S Mader, JY Chen, Z Chen, J White, P Chambon and H Gronemeyer, *EMBO J*, **12**, 5029–41 (1993).
49 C Zechel, XQ Shen, JY Chen, ZP Chen, P Chambon and H Gronemeyer, *EMBO J*, **13**, 1425–33 (1994).
50 T Perlmann, PN Rangarajan, K Umesono and RM Evans, *Genes Dev*, **7**, 1411–22 (1993).
51 PJ Willy, K Umesono, ES Ong, RM Evans, RA Heyman and DJ Mangelsdorf, *Genes Dev*, **9**, 1033–45 (1995).
52 BA Laffitte, HR Kast, CM Nguyen, AM Zavacki, DD Moore and PA Edwards, *J Biol Chem*, **275**, 10638–47 (2000).
53 S Green, P Watter, V Kumar, A Krust, JM Bornert, P Argos and P Chambon, *Nature*, **320**, 134–39 (1986).
54 GL Greene, P Gilna, M Waterfield, A Baker, Y Hort and J Shine, *Science*, **231**, 1150–54 (1986).
55 P Walter, S Green, G Greene, A Krust, JM Bornert, JM Jeltsch, A Staub, E Jensen, G Scrace and M Waterfield, *Proc Natl Acad Sci USA*, **82**, 7889–93 (1985).
56 SM Hollenberg, C Weinberger, ES Ong, G Cerelli, A Oro, R Lebo, EB Thompson, MG Rosenfeld and RM Evans, *Nature*, **318**, 535–641 (1985).
57 V Giguere, ES Ong, P Segui and RM Evans, *Nature*, **330**, 624–29 (1987).
58 M Petkovich, NJ Brand, A Krust and P Chambon, *Nature*, **330**, 444–50 (1987).
59 DJ Mangelsdorf, ES Ong, JA Dyck and RM Evans, *Nature*, **345**, 224–29 (1990).
60 G Feigl, M Gram and O Pongs, *Nucleic Acids Res*, **17**, 7167–78 (1989).
61 M Rothe, U Nauber and H Jackle, *EMBO J*, **8**, 3087–94 (1989).
62 I Issemann and S Greene, *Nature*, **347**, 645–50 (1990).
63 T Hard, E Kellenbach, R Boelens, BA Maler, K Dahlman, LP Freedman, J Carlstedt Duke, KR Yamamoto, JA Gustafsson and R Kaptein, *Science*, **249**, 157–60 (1990).
64 MA van Tilborg, AM Bonvin, K Hard, AL Davis, B Maler, R Boelens, KR Yamamoto and R Kaptein, *J Mol Biol*, **247**, 689–700 (1995).
65 JW Schwabe, D Neuhaus and D Rhodes, *Nature*, **348**, 458–61 (1990).
66 M Katahira, RMA Knegtel, R Boelens, D Eib, JG Schilthuis, RA van der Saag and R Kaptein, *Biochemistry*, **31**, 6474–80 (1992).
67 MS Lee, SA Kliewer, J Provencal, PE Wright and RM Evans, *Science*, **260**, 1117–21 (1993).
68 DT Gewirth and PB Sigler, *Nat Struct Biol*, **2**, 386–94 (1995).
69 JW Schwabe, L Chapman and D Rhodes, *Structure*, **3**, 201–13 (1995).
70 DG Brown and PS Freemont, in C Jones, B Mulloy and M Sanderson (ed.), *Methods in Molecular Biology: Crystallographic Methods and Protocols*, 56, Humana Press, Totowa, NJ, pp 293–318 (1996).
71 MH Carruthers, *Science*, **230**, 281–85 (1985).
72 SR Jordan, TV Whitcombe, JM Berg and CO Pabo, *Science*, **230**, 1383–85 (1985).
73 M Danielson, L Hinck and GM Ringold, *Cell*, **57**, 1131–38 (1989).
74 S Mader, V Kumar, H de Verneuil and P Chambon, *Nature*, **338**, 271–74 (1989).
75 K Umesono and RM Evans, *Cell*, **57**, 1139–46 (1989).
76 F Rastinejad, *Curr Opin Struct Biol*, **11**, 33–38 (2001).
77 E Schoenmakers, G Verrijdt, B Peeters, G Verhoeven, W Rombauts and F Claessens, *J Biol Chem*, **275**, 12290–97 (2000).
78 BA Lierberman, BJ Dona, DP Edwards and SK Nordeen, *Mol Endocrinol*, **7**, 515–27 (1993).
79 D Pearce and KR Yamamoto, *Science*, **259**, 1161–65 (1993).
80 R Kurokawa, J DiRenzo, M Boehm, J Sugarman, B Gloss, MG Rosenfeld, RA Heyman and CK Glass, *Nature*, **371**, 528–31 (1994).
81 M Carson, in CW Carter and RM Sweet (eds.), *Methods in Enzymology: Macromolecular Crystallography*, Academic Press, Boston, MA, pp 493–505 (1997).
82 Collaborative Computational Project, Number 4, *Acta Crystallogr, Sect D*, **50**, 760–763 (1994).
83 TE Wilson, TJ Fahrner and J Milbrandt, *Mol Cell Biol*, **13**, 5794–804 (1993).
84 A Klug and JWR Schwabe, *FASEB J*, **9**, 597–604 (1995).

RING domain proteins

Cyril Dominguez, Gert E Folkers and Rolf Boelens

Department of NMR Spectroscopy, Bijvoet Center for Biomolecular Research, Utrecht University, Utrecht, The Netherlands

FUNCTIONAL CLASS

Eukaryotic zinc binding motif: this motif is related to the zinc finger motif present in many transcription factors[1,2], and was found initially in the really interesting new gene (RING1) and therefore named RING finger.[3] Subsequently, this motif was detected in a number of unrelated proteins.[3,4]

Initially, the function of RING domains was not clear, although they were known to mediate protein–protein interactions and to be involved in a range of cellular processes, including development, oncogenesis, apoptosis, and viral replication.[5,6] By 1999, the function of the RING domain was clarified, with the observation that the RING domain of c-Cbl mediates a protein–protein interaction with proteins known to be involved in the protein ubiquitination and 26S proteasome degradation pathways.[7–9] Thereafter, a similar function was deduced for a number of RING proteins.[10]

OCCURRENCE

RING domains of RING finger proteins are one of the most common zinc binding motifs in eukaryotes.[11] There have been more than 380 RING motifs identified in the human genome and the number of sequences in all eukaryotes corresponding to a RING motif is currently more than 1200 (http://www.sanger.ac.uk/Software/Pfam).[12] RING finger–containing proteins can be found in a large variety of different species ranging from yeast to human, including double-strand DNA viruses, and in all kind of cells or tissues[13]. RING proteins are not found in bacteria, which testifies to their unique role in the ubiquitination pathway that is absent in prokaryotes.

BIOLOGICAL FUNCTION

RING domains have been shown to mediate a crucial step in protein degradation. Figure 1 shows how a ubiquitin

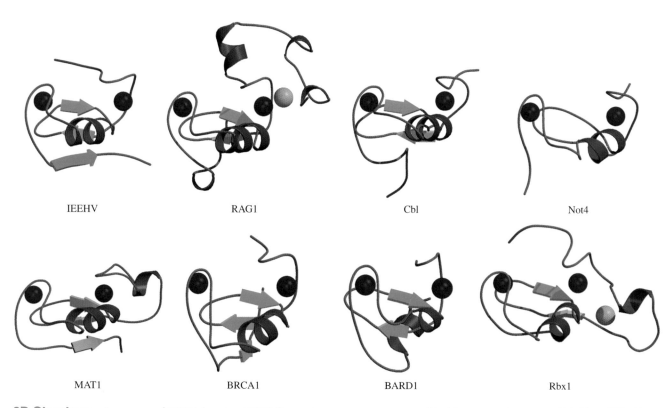

3D Structure Structures of RING domains: RING finger zinc atoms are represented in blue, extra zinc atoms (RAG1 and Rbx1) are represented in pink. The figures have been generated with the program MOLSCRIPT[47] and RASTER3D.[54]

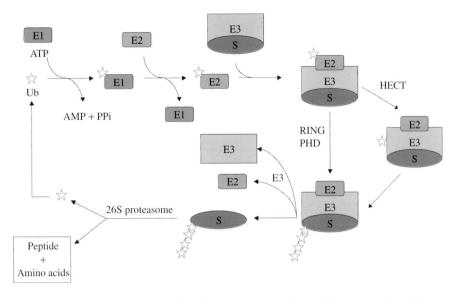

Figure 1 The ubiquitination pathway. Ubiquitin molecules are represented as yellow stars, ubiquitin-activating enzyme as E1, ubiquitin-conjugating enzyme as E2, ubiquitin ligase as E3, and substrate as S.

chain can be covalently linked to protein substrates.[6,13–15] These polyubiquitinated substrates will then be recognized by the 26S proteasome and degraded by proteolysis.[16–18] Three enzymes are involved in the ubiquitination pathway. Firstly, an E1 or ubiquitin-activating enzyme forms a thiol ester with the carboxyl terminal group of the small protein ubiquitin at position Gly76. The ubiquitin is then transferred to the ubiquitin-conjugating enzyme (E2). Finally, ubiquitin ligase (E3) transfers the ubiquitin from E2 to the target protein promoting the ubiquitination of the substrate. At least three different classes of E3 ligases have been found that mediate substrate ubiquitination. These E3 enzymes differ in the domain that recognizes the E2 enzymes, which can be a RING, a plant homeodomain (PHD), or a Homologous to E6AP COOH terminus (HECT) domain.[17–19] In case of the HECT type E3 ligases, the small ubiquitin protein is captured from the E2, and covalently bound to the E3, before it is transferred to the substrate.[20] In contrast, in the case of RING- or PHD-containing protein ligases, no evidence for a stable E3-ubiquitin intermediate exists; therefore, the E3 ligases that contain a RING or a PHD domain are thought to have only a role in bringing the E2 and the substrate together. In all cases, the interaction between the E2 and E3 enzymes is mainly accomplished by the HECT, PHD or RING domains.

It is now clear that protein degradation is an important feature in the regulation of cellular processes. Many vital functions like signaling, growth, transcription, and DNA repair are all regulated by the 26S proteasome.[18] Aberrant expression caused by deregulated protein degradation leads to severe diseases. Frequently, RING finger proteins have been related to such diseases (see below).[21–23]

Recently, another role for ubiquitination has been suggested. Whereas polyubiquitination is generally the marker on a substrate that will be degraded by the proteasome, a monoubiquitination of a protein can have a regulatory function as, for example, in signal transduction, transcription regulation, chromatin remodeling, and DNA repair.[24–28] This new role suggests that the ubiquitination pathway is not only involved in degradation but might also regulate protein function. This protein modification can, like phosphorylation, acetylation, and methylation, regulate protein activity in which the ubiquitin moiety acts as a molecular switch in many important cellular processes.

AMINO ACID SEQUENCE INFORMATION AND TOPOLOGY

The only common feature of RING finger proteins is the presence of the cysteine-rich sequence motif that has similarities with zinc fingers. This consensus RING motif, also called the C3HC4 motif, can be defined as a unique pattern of cysteine and histidine residues at defined positions in a peptide sequence, which is Cys-X_2-Cys-X_{9-39}-Cys-X_{1-3}-His-X_{2-3}-Cys-X_2-Cys-X_{4-48}-Cys-X_2-Cys, and where X can be any amino acid. It was suggested that the eight cysteine and histidine residues form a binding site for two zinc atoms. Since these transition metals can also be coordinated by other residues, it can be expected that variations on the RING consensus sequence are possible. Correspondingly, a second large group of RING finger proteins, the RING-H2 family, has been identified, in which the cysteine at position 4 is replaced by a histidine.[5] Also, a third consensus sequence for RING domains was found, which consists of the C4C4 motif in which all the zinc-ligating residues correspond to cysteines.[29] In all three cases, the size of RING domains is approximately 70 amino acids (Figure 2), and representation of all three families have been shown to fold

RING domain proteins

```
IEEHV:     1            MATVAERCPICLEDPSN----YSMALPCLHAF--CYVCITRWIRQNPT---CPLCKVPVESVVHTIESDSEF      63
PML:      49            EEEFQFLRCQQCQAEAKC-----PKLLPCLHTL--CSGCLEASGMQ------CPICQAPWPLGADTPAL      104
RAG1:    281            PAHFVKSISCQICEHILAD-----PVETSCKHLF--CRICILRCLKVMGS---CPSCRYPCFPTDLESP      340
Cbl:     376            STFQLCKICAENDKD-----VKIEPCGHLM--CTSCLTSWQESEGQG--CPFCRCEIKGTEPIVVDPF      434
Not4:      1  MSRSPDAKEDPVECPLCMEPLEI---DDINFFPCTCGYQICRFCWHRIRTDENGL--CPACRKPYPEDPAVYKPLSQEELQRI   78
MAT1:      1            MDDQGCPRCKTTKYRNPSLKLMVNVCGHTL--CESCVDLLFVRGAGN--CPECGTPLRKSNFRVQLFED   65
BRCA1:    21            ILECPICLELIKE-----PVSTKCDHIF--CKFCMLKLLNQKKGPSQCPLCKNDITKRSLQESTRFS   80
BARD1:    49            LLRCSRCTNILRE----PVCLGGCEHIF--CSNCVSDCIGTG-----CPVCYTPAWIQDLKINRQLDSMI  105
```

Figure 2 Sequence alignment of RING domains. Zinc ligands are highlighted in yellow. Residues involved in the binding to an E2 enzyme (c-Cbl and CNOT4) are underlined. Secondary structure elements are based on the structure of IEEHV.

with a characteristic topology around the zinc ions. From sequence comparisons, it was proposed that the RING domain family also includes the U-box, a domain that is also involved in the ubiquitination pathway.[30] However, this domain does not bind zinc atoms, but models predict a similar structure,[31] although structure determination is required to confirm that this domain indeed adopts a similar fold.

ACTIVITY TEST

The involvement of RING domains in the ubiquitination pathway is generally tested using an *in vitro* assay.[32] The most common method is the so-called in-solution ubiquitination essay[9,33,34] that uses ubiquitin, the E1, E2, and E3 enzymes, ATP and an ATP-regenerating system in a reaction mixture. In cases in which the protein substrate for the ubiquitination is known, it is also added to the reaction. Ubiquitination will lead to an accumulation of high molecular weight polyubiquitin adducts bound to the substrate, or in the case of a substrate-independent assay, autoubiquitination of the E3 ligase.[9,10,35] The characteristic ladder corresponding to multiple ubiquitin adducts is usually detected on polyacrylamide gel using western blot analysis (with antiubiquitin or antisubstrate antibodies) or autoradiography (with radiolabeled ubiquitin).

PROTEIN PRODUCTION, PURIFICATION, AND MOLECULAR CHARACTERIZATION

RING domains are frequently expressed and purified as Glutathione S-transferase (GST)-RING fusion proteins.[36-38] However, protocols using fusion constructs with the maltose binding protein (MBP) or a His$_6$-Tag have also been described.[29,39] The recombinant RING proteins are overexpressed in *Escherichia coli*. Since zinc is needed for proper folding, ZnCl$_2$ is added to the bacteria during protein production. The GST-RING fusion proteins are purified using Glutathione affinity chromatography, after which the GST part is cleaved off by proteases. The RING domains or proteins are then further purified by standard methodology such as ion exchange and/or gel filtration chromatography. Buffers used for the purification should contain ZnCl$_2$, preferably a low concentration of phosphate, EDTA should be absent, and the pH should not be lower than 6 to 6.5 in order to prevent the protonation of the zinc-coordinating residues and the subsequent loss of the metals, which generally leads to unfolding and precipitation of the RING domain.

METAL CONTENT

The RING domains comprise eight metal-binding ligands and bind two zinc(II) ions with each metal ion in a tetrahedral coordination. In a C3HC4 RING domain, one zinc ion is bound to four cysteines, and the other to three cysteines and a histidine, whereas in a RING-H2 domain both zinc ions are coordinated to three cysteines and a histidine. The metal content of RING domain proteins has been deduced by optical spectroscopy[3] and by atomic absorption spectroscopy.[40,41]

ZINC SITE GEOMETRIES

The arrangement of the coordinating residues around the zinc atoms in the RING domain has often been erroneously ascribed to two tandemly arranged zinc finger domains of the CCCH and CCCC type. However, the zinc ligation topology of RING fingers is quite distinct and is referred to as a 'cross-brace' motif (Figure 3). In this motif, the first pair of ligands (Cys1 and Cys2) together with the third pair (Cys5 and Cys6) coordinates with one zinc atom, and the second (Cys3 and His4) and fourth (Cys7 and Cys8) pairs bind the second zinc atom. In contrast, in the classical zinc finger proteins, the first four coordinating residues would bind to one zinc atom and the next four to the next zinc

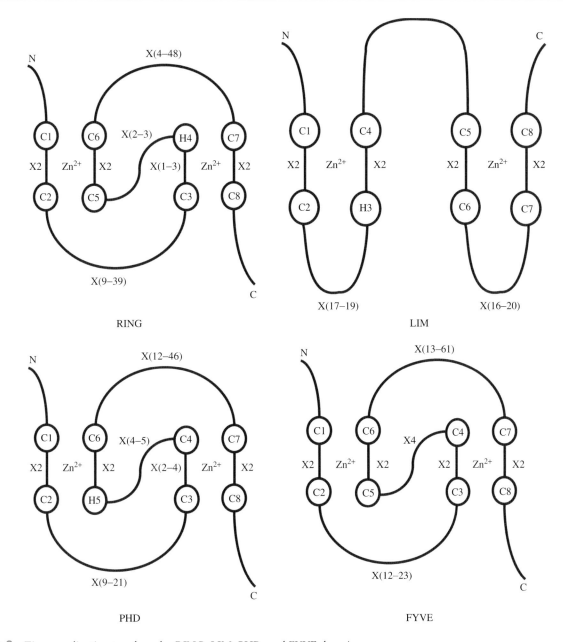

Figure 3 Zinc coordination topology for RING, LIM, PHD, and FYVE domains.

atom. Sometimes, these sequential zinc fingers form independent stable folds connected by flexible peptide segments as found in several transcription factors. In other cases, a stable fold requires the close packing of both fingers,[2] as is the case in the DNA binding domain of the steroid hormone receptor and the LIM domain[42] (Figure 3). However, in the case of the 'cross-brace' RING motif, it is clear that both zinc atoms are required for an intact structure.

This 'cross-brace' motif is not unique for RING fingers, but can also be found in the PHD[43] and in the FYVE (Fab1p, YOTB, Vac1p, EEA1) domain,[44,45] a domain that was originally observed in the Fab1p (formation of aploid and binucleate cells), YOTB (Hypothetical *Caenorhabditis elegans* protein ZK632.12 in chromosome III.), Vac1p (also known as VSP19 involved in vacuolar segregation) and early endosome antigen 1 proteins (EEA1) (Figure 3).

The zinc ions of the RING domains are coordinated to the sulfur of the cysteines. The coordination to the histidine imidazole ring is unusual. Commonly in zinc fingers, the zinc coordinates to the NE2 of the imidazole group, but in the RING domains, coordination to the ND1 atom was demonstrated.[39,46] This was explained by the close spacing between the ligands 3 and 4, which is conserved among all C3HC4 RING finger proteins, whereas, in most zinc finger proteins, a more relaxed two-residue spacing is found. In the recombination-activating gene protein (RAG)1 structure,[39] the distances between the cysteine SG and the zinc atoms are between 2.23 and 2.36 Å, while

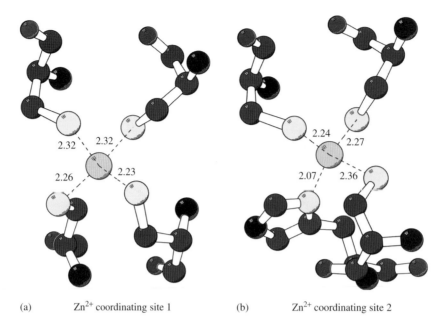

Figure 4 Zinc-coordinating site geometry for the C4 site (site I) (a) and the C3H site (site II) (b) of the RAG1 RING domain. Zinc atoms are represented in gray; cysteine SG and histidine ND1 atoms are represented in yellow. Distances in Å between the cysteine SG or histidine ND1 and the zinc ion are displayed. The figures have been generated with the program MOLSCRIPT.[47]

the distance between the histidine ND1 and the zinc atom is 2.07 Å (Figure 4).

SPECTROSCOPIC AND METAL-BINDING PROPERTIES OF RING FINGER DOMAINS

There are only a few spectroscopic techniques that enable the direct observation of zinc ions. In principle, zinc-extended X-ray absorption fine structure spectroscopy (EXAFS) allows the direct study of the metal-binding properties of zinc proteins[48] but this has not been applied thus far for RING fingers. The spectroscopic properties of RING domains can be investigated indirectly using a variety of spectroscopic techniques due to the fact that one can substitute the zinc ion by other metals such as cobalt(II) or cadmium(II).[3] The replacement of zinc by cobalt allows the observation of the metal site by UV–vis and fluorescence spectroscopy. Two studies showed that the C3HC4 RING domains of breast cancer type 1 (BRCA1) and hdm2 (human double minute 2 protein) bind the two cobalt atoms in a sequential manner, where the C4 site (site I) has a higher affinity than the C3H site (site II), which has been ascribed to an intrinsic differential stability of the two sites for the metal binding.[49,50] An alternate explanation, however, could be that this differential stability is directly related to a lower affinity of the histidine imidazole group for the cobalt ion, decreasing the affinity of the second site with respect to that of the first. Similarly, zinc can be replaced by cadmium and this allows the analysis of the coordination site by NMR. Hanzawa *et al.* (unpublished data) studied the zinc–cadmium exchange by NMR titration experiments using the RING domain of CNOT4 (negative on TATA), which has an unusual C4C4 motif as confirmed by ^{113}Cd–^{1}H HSQC experiments. In this case, it was found that the first site exchanges the zinc before the second site. These experiments demonstrate that the lower stability of the second metal binding site of the RING domain is not a general property of all RING domains.

X-RAY AND NMR STRUCTURES OF RING FINGER DOMAINS

First structural information about RING domains came from the analysis of the NMR chemical shifts of the RING domain of the immediate early EHV-1 (IEEHV-1) protein from equine herpesvirus.[40] Shortly thereafter, the three-dimensional structure of immediate early equine herpesvirus (IEEHV) was solved by the same group[46] (Table 1). Since then, a number of structures of other RING domains have been solved by NMR and X-ray crystallography, both in the free state[29,36,39,51] or in complexes with other proteins[37,38,52] (Table 1).

Overall description of the structure

Comparison of the nine RING domain structures shows that all RING fingers adopt a similar fold, but significant differences are present. All RING fingers display the so-called 'cross-brace' motif for the zinc ligation. The overall

Table 1 Ring finger protein structures deposited in the PDB. Represented are the PDB code, the name of the protein, the type of RING, the technique used (X-ray or NMR), the resolution of the structure (X-ray) in Å or the backbone rmsd from mean structure (NMR) and the function of each RING

PDB code	Protein name	Type of RING	Technique	Res/rmsd	Function
1CHC	IEEHV	C3HC4	NMR	0.55	Regulation of the equine herpesvirus gene expression
1BOR	PML	C3HC4	NMR	0.88	Cellular defense mechanism
1RMD	RAG1	C3HC4	X-ray	2.1	Assembly of antibody and T-cell receptor genes
1E4U	CNOT4	C4C4	NMR	0.58	Part of a global regulator of RNA polymerase II transcription
1G25	Mat1	C3HC4	NMR	0.67	Subunit of the human transcription/DNA repair factor TFIIH
1JM7	BRCA1	C3HC4	NMR	0.87	DNA repair and transcriptional regulation
1JM7	BARD1	C3HC4	NMR	0.95	DNA repair and transcriptional regulation
1FBV	c-Cbl	C3HC4	X-ray	2.9	Negative regulator of tyrosine kinase-coupled receptors
1LDD/1LDJ	Rbx1	RING-H2	X-ray	3	Part of the SCF E3 ligase complex

structure is characterized by a βαβ fold, but the number of β-strands and the exact position of the β-sheet differ among the structures when analyzed with a DSSP algorithm[53] (3D Structure).

A common feature of all C3HC4 structures is the 14-Å distance between the two zinc atoms of the RING motif. In all structures, a hydrophobic cluster, which stabilizes the ternary structure, is found. The main differences observed among the nine structures relate to the presence and length of the secondary structure elements.

Structures of the free RING domains

The NMR structure of the IEEHV (Table 1) has a cross-brace coordination and a secondary structure that is composed of 3 β-strands and a central α-helix (3D Structure). The first loop (residues 1 to 19) contains the first pair of zinc-ligating residues (Cys8 and Cys11). This loop is followed by two small β-strands (residues 19 to 21 and 26 to 28) connected by a short turn containing the second pair of zinc-ligating residues (Cys24 and His26). The second β-strand is followed by an α-helix (residues 32 to 40) with the third pair of zinc-ligating residues (Cys29 and Cys32) between both secondary structure elements. The C-terminal part of the structure consists of a long loop (residues 33 to 63) containing the fourth pair of zinc-ligating residues (Cys43 and Cys46) and a third short β-strand (residues 53 to 55).

The C_3HC_4 RING domain of the promyelocytic leukemia (PML) protooncoprotein was the second RING structure that was determined by NMR[51] (Table 1). Although the metal ligation is similar to IEEHV and the distance between the two zinc atoms is conserved, the ternary structure is quite different from that of IEEHV, and the protein does not possess the central α-helix.

The first X-ray structure of a RING domain was published two years thereafter[39] (Table 1). The structure of the dimerization domain of the protein RAG1 consists of an N-terminal RING finger domain and a C-terminal zinc finger. The structure of the RING finger domain of RAG1 is highly similar to that of IEEHV (3D Structure). A superposition of 40 of the 44 Cα atoms of the RAG1 RING finger onto the corresponding Cα atoms of IEEHV finger results in an rmsd of 1.65 Å. The highly conserved hydrophobic core residues Phe309 and Ile314 corresponding to Phe28 and Ile33 residues in the IEEHV adopt identical conformations in the two structures. However, in the structure of RAG1, the third β-strand is missing.

The structure of a C4C4 type RING domain of the CNOT4 protein from the human CCR4-NOT transcription complex has been solved using NMR[29] (Table 1). Despite the fact that the sequence of the N-terminal domain of CNOT4 did not show the consensus C3HC4 RING motif, the fold of the domain and the zinc coordination topology of CNOT4 resembles that of the RING proteins (3D Structure). The structure of CNOT4 RING domain consists of three loops and an α-helix between the second and third loop. No regular β-strand could be detected in the structure using the Kabsch–Sander algorithm as implemented in the DSSP program. However, despite the differences in secondary structure elements that are probably due to the presence of many prolines in the CNOT4 sequence, the overall structure of CNOT4 is similar to the structure of IEEHV (Cα rmsd: 1.7 Å) and RAG1 (Cα rmsd: 1.6 Å). The metal binding site was investigated by replacing the zinc atoms by ^{113}Cadmium. The ^{113}Cd–^{1}H HSQC spectrum showed the presence of two metal binding sites. The cysteines 14, 17, 38, and 41 coordinate one metal ion and the cysteines 31, 33, 53, and 56 bind the second metal ion, in both cases via the SG atoms, confirming the cross-brace motif of the CNOT4 RING domain. The three loops in CNOT4 are stabilized by the coordination with the zinc ions and by hydrophobic interactions in a similar manner as for the C3HC4 RING finger structures. Interestingly, the distance between the two zinc atoms is slightly longer (15 Å) than the conserved value of 14 Å as found in the C3HC4 RING finger structures of IEEHV and RAG1. This difference is possible owing to the difference in residue spacing between the ligands 4 and 5 of CNOT4 versus those of the C3HC4 RINGs (Figure 2).

Next to CNOT4, three other structures of C3HC4 RING domain were also solved in 2001. One corresponds to the NMR structure of the RING domain of the human transcription factor TFIIH MAT1 (ménage a trois) subunit[36] (Table 1). The structure adopts the ββαβ fold that is typical of RING domains. The core of the domain consists of a three-stranded antiparallel β-sheet packed along a two-turn α-helix (3D Structure). The other two structures correspond to the NMR structure of the complex between two RING domains, the BRCA1-BARD1 heterodimeric RING–RING complex[52] (Table 1). The BRCA1 RING domain is characterized by a short three-stranded antiparallel β-sheet, two large zinc binding loops, and a central α-helix (3D Structure). The BRCA1-associated RING domain protein 1 (BARD1) is structurally homologous but lacks the central helix between the third and fourth pair of zinc ligands (3D Structure). The dimerization interface of BRCA1 and BARD1 does not involve the RING motifs directly as discussed below, and the two structures closely resemble those of the free RING domains.

Structures of related domains

A number of domains, such as the PHD domain found in E3 ligases and the FYVE domain found in proteins involved in the membrane recruitment of cytosolic proteins, show the same zinc cross-brace ligation motif as the RING domain.[43,45] The LIM domain found in transcription factors shows a more conventional sequential ligation pattern.[42]

The LIM domain is commonly found in transcription factors, like many zinc fingers. The domain is involved in protein–protein interaction with other transcription accessory factors. Six structures of LIM domains have been solved thus far, all of them by NMR.[55–60] A comparison shows that the LIM domain is less compact than the core structure of the RING domain, probably because of the difference in the zinc ligation topology (Figure 5). The conserved central helix in RING fingers is not present in LIM domains. However, the fold of the LIM domain resembles that of part of the RING domain. A superposition of the structures for the residues around the first zinc (site I) and the two first β-strands leads to a backbone rmsd of

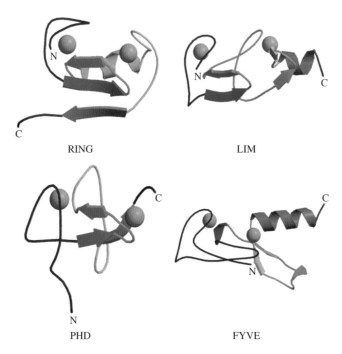

Figure 5 Comparison between the structures of the RING domain of the IEEHV protein, the LIM domain of the CRP2 protein (PDB:1A7I), the PHD domain of the Kap1 protein (PDB:1FPO), and the FYVE domain of the VSP27P protein (PDB:1VFY). The figures have been generated with the program MOLSCRIPT[47] and RASTER3D.[54]

3 Å between the CRP2 LIM domain and the IEEHV RING finger protein.

Initially, the function of proteins containing a PHD domain was unclear, although there was a general consensus that the PHD domains are involved in protein–protein interactions. More recently, PHD-containing proteins have been shown to function in the ubiquitination pathway as E3 ligases. The PHD domain in these proteins has the same function as the RING and the HECT domain, and is involved in the interaction with the E2 ubiquitin-conjugating enzyme.[19] The structure of the PHD domains from the human Williams–Beuren syndrome transcription factor (WSTF) and from the Kap-1 transcriptional repressor have been solved by NMR.[61,62] The zinc coordination topology of the PHD domain is similar to that of the RING finger domain, but it turns out that the ternary fold is quite different (Figure 5). The structure of the PHD domain consists of an N-terminal loop followed by an antiparallel β-sheet, a second loop, and a third β-strand. The site I zinc ligand pairs are located in the N-terminal loop (first pair) and at the beginning of the second loop (third pair). The site II zinc ligand pairs are located in the turn between the two first β-strands (second pair) and at the end of the third β-strand (fourth pair). The arrangement of the site I zinc ligands, as well as the two first β-strands are similar to that of the RING domain (a superposition of these regions between the Kap-1 PHD domain and the IEEHV RING domain leads to a backbone rmsd of 1.3 Å).[62] However, the position of the second zinc with respect to the first one is different between the two domains and the conserved α-helix found in the RING domain is not present in the PHD structures.

Also the cross-brace motif found in the FYVE domain is very similar to that of the RING motif. Since the FYVE domain is found in proteins involved in the membrane recruitment of cytosolic proteins and binds to phosphatidylinositol 3-phosphate (PI3P) located in membranes,[45] there is clearly no functional resemblance with the RING and the PHD domains, involved in ubiquitination. Four structures of FYVE domains have been solved by X-ray[44,63,64] and NMR.[65] The FYVE domain is composed of two central β-strands and a C-terminal α-helix (Figure 5). Despite the similarity in zinc coordination topology with the RING and PHD domains, there is no structural relationship between the FYVE and these two domains. The secondary structure elements of the FYVE domain are located at totally different positions as compared to both RING and PHD domains.

Structures of the RING domains in complex with other proteins

Ubiquitination consists of the covalent attachment of ubiquitin molecules to protein substrates. The conjugation of ubiquitin to substrates is accomplished by an enzymatic cascade involving three enzymes (see biological function). How these enzymes recognize each other and what are the molecular basis of these interactions are key questions to understand the specific mode of action of the ubiquitination pathway. Detailed molecular insights into the E2–E3 interaction came from the crystal structure of E6AP, an E3 protein ligase containing a HECT domain, bound to UbcH7 (human ubiquitin-conjugating enzyme 7) an E2 enzyme.[66] One year later, the crystal structure of the RING finger protein c-Cbl (CAS-BR-M murine ecotropic retroviral transforming sequence homolog) bound to UbcH7[37] extended our knowledge to the RING-E2 interaction mechanism (Figure 6). E3 protein ligases represent a diverse family of enzymes that can either be single proteins or large protein complexes. The SCF (Skp1, Cullin, F-box protein) complexes, which consist of four different proteins, are one of the largest E3 protein ligase complexes. One of the proteins of the SCF complex contains a RING domain that is required for the E2 recognition. How the E3 complex spatially arranges the transfer of ubiquitin from the E2 to the substrate is an important question toward the understanding of the molecular basis of ubiquitination. The crystal structure of the SCF complex composed of the cullin gene family member 1 (Cul1), the RING box protein 1 (Rbx1), the suppressor of kinetochore protein 1 (Skp1), and the F-box domain of the suppressor of kinetochore protein 2 (Fbox^{skp2}) provides considerable insights into this issue[38] (Figure 7).

The c-Cbl-Ubch7 complex

The structure of the full length UbcH7 in complex with the N-terminal half of the 100 kDa protein c-Cbl has been solved by X-ray crystallography. The N-terminal half of c-Cbl corresponds to a tyrosine kinase binding (TKB) domain (residues 47 to 344), a linker sequence (residues 345 to 380) and a C3HC4 RING domain (residues 381 to 434).[37] The structure of the RING domain in the complex is similar to that of the free RING domains. The backbone rmsd between c-Cbl and RAG1 for the 40 core residues is 1.9 Å, similar to the rmsd values found between all RING domains[37] (3D Structure). The structure of the complex shows how the RING domain of the E3 ligase interacts with the E2 conjugating enzyme. The interaction involves the two well-conserved zinc-chelating loops and the α-helix of the RING domain (Figure 6). The interaction is stabilized mainly by hydrophobic contacts in which the aromatic side chain of Phe63 of UbcH7 plays a crucial role. This residue makes close van der Waals contacts with Ile383 in the loop1 and Trp408, Ser407 and Ser411 in the α-helix of the c-Cbl RING domain. Other important hydrophobic contacts are between Pro97 and Ala98 of Ubch7 and Ile383, Trp408, Pro417, and Phe418 of c-Cbl. In addition, the charged residues Glu366 and Glu369 of the linker region of c-Cbl and Arg5 and Arg15 of UbcH7 make

RING domain proteins

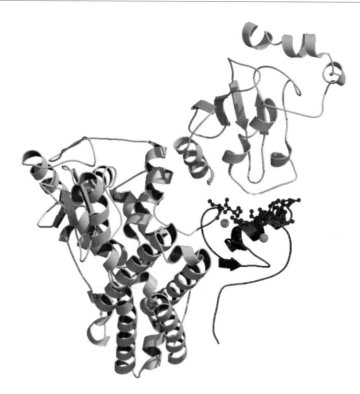

Figure 6 The c-Cbl–UbcH7 complex. The RING domain of c-Cbl is represented in red, the zinc atoms in cyan, and the residues important for the interaction (Ile383, Cys384, Cys404, Ser407, Trp408, Ser411, Pro417, Phe418 and Arg420) in blue. UbcH7 is represented in yellow. The figure has been generated with the program MOLSCRIPT[47] and RASTER3D.[54]

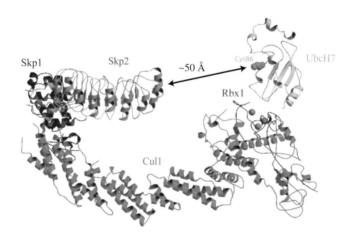

Figure 7 The SCF ubiquitin ligase complex. The X-ray complex consists of the Rbx1 RING (red), the Cul1 (green), the Skp1 (blue) proteins, and the Skp2 F-box domain (purple). The UbcH7 structure (yellow) has been docked on the basis of the c-Cbl-UbcH7 complex[37] as well as the full Skp2 structure (purple) on the basis of the Skp1–Skp2 complex.[67] The black arrow indicates the 50-Å gap between the active site of the E2 (Cys86) and the tip of Skp2. The figure has been generated with the program MOLSCRIPT[47] and RASTER3D.[54]

electrostatic contacts. A comparison of the structure of the c-Cbl-UbcH7 complex and the E6AP-UbcH7 complex reveals considerable similarities in the interaction surface of UbcH7 with these two structurally unrelated E3 ligases. The binding of UbcH7 with both proteins involves the same set of residues, even though, in the RING and the HECT domains, the structural elements and residues that are recognized by UbcH7 are different. The importance of the loops and α-helix of the RING domain for the interaction with the E2 enzyme are underlined by the NMR studies on the CNOT4-UbcH5B complex.[35] The NMR studies showed that the regions of the CNOT4 RING domain involved in the interaction with UbcH5B are similar to the ones of c-Cbl in the interaction with UbcH7. However, the types of amino acids involved in the interaction are different, which possibly explains the differences in specificity in the E2–E3 interaction. Whereas in c-Cbl RING domain, the residues Ser407, Trp408, and Ser411 are involved in the interaction with UbcH7, in CNOT4, the corresponding residues Arg44, Ile45, and Asp48 are involved in the interaction with UbcH5B (Figure 2).

The SCF ubiquitin ligase complex

Recently, a second structure of a RING domain-containing protein complex has been solved by X-ray[38] (Figure 7). This complex consists of the RING-H2 protein Rbx1 in association with Cul1, Skp1, and Fbox[skp2] and corresponds to the SCF ubiquitin ligase complex. In this case, the complex provides information about the mechanism of ubiquitin transfer by an E3 ligase. The proteins Rbx1 and Cul1 form the catalytic core of this complex. Rbx1 is responsible for the E2 recruitment. Cul1 consists of a long α-helical structure composed of three cullin repeats and binds Skp1 via its N-terminus and Rbx1 via its C-terminus. The F-box protein Skp2 binds the substrate. Skp1 is an adapter between the F-box protein and Cul1. The protein Rbx1 has a RING-H2 domain that adopts the same fold as that of the canonical RING motif. The RING-H2 domain of Rbx1 is stabilized by two zinc ions, but possess a 20-residue insertion between the first and the second pair of zinc ligand. This insertion contains three additional zinc ligands and together with a fourth zinc ligand from the RING motif, these residues form a new zinc binding site (3D Structure). The docking of the SCF complex onto the c-Cbl-UbcH7 complex[37] on one side and the Skp1–Skp2 complex[67] on the other (Figure 7) shows that the binding site of Rbx1 for an E2 enzyme is not affected by this additional zinc finger motif and that the active site Cys86 of UbcH7 and the tip of the F-box protein Skp2 are on the same side of the SCF complex. The distance between the E2 enzyme and the substrate binding site of the F-box protein is about 50 Å. This distance is suitable for the insertion of the substrate, p27, between Skp2 and UbcH7 and for the positioning of p27 at a suitable topology for its ubiquitination.

Another interaction involving RING proteins that has been observed is the direct interaction between two RING domains in the dimeric BRCA1–BARD1 complex.[52] In this heterodimeric complex, the complex formation is mainly governed by extensive interactions involving hydrophobic residues of a four-helix bundle formed between the two proteins, while there are only few direct contacts between the two RING domains.

MEDICAL ASPECT OF RING FINGER PROTEINS

Many human diseases can be related to a malfunctioning in the regulation of gene expression. As ubiquitination is a key step in the protein degradation pathway, it is not surprising that alterations of this process leads to severe diseases. Since it has been shown that many RING domain proteins function as E3 ligases, there has been a rapid increase in the number of articles that link RING domains with disease. In this review, we will focus on only three RING domain proteins, two involved in cancer and one involved in Parkinson's disease, for which the mode of action is now well defined. The interested reader is referred to recent reviews[21–23,68] for more details.

The BRCA1 protein

The gene for the putative tumor suppressor BRCA1 was first cloned in 1990.[69] BRCA1 encodes a protein of 220 kDa (1863 amino acids) that has a highly conserved amino terminal RING finger and an acidic carboxyl terminal domain characteristic of many transcription factors. BRCA1 has been shown to be involved in several important cellular functions including DNA repair, regulation of transcription, cell cycle control, and ubiquitination.[70] During the last decade, researchers have been able to confidently link mutations in the *BRCA1* gene to familial breast and ovarian cancers.[71,72] It was found that mutations in the *BRCA1* gene are responsible for about 5% of the total of these cancers, but it also became clear that a woman who inherits one mutant allele of BRCA1 from either her mother or father has currently a >80% risk of developing breast cancer during her life. It has been estimated that between 1/500 and 1/800 women carry a mutation in their *BRCA1* gene.[73] More than 600 different mutations predisposing to a high risk of cancer have been identified in the *BRCA1* gene, some of them in the RING domain.[10,74] In particular, mutations of the ligand cysteines 39[75], 61,[76] and 64[77] that probably disrupt the integrity of the RING structure underscore the importance of the RING for functionality of BRCA1.

The p53-Mdm2 complex

Exposure to ionizing radiation or ultraviolet light damages the DNA and leads, in the worst case, to cellular transformation. For maintaining genomic stability, the cell is equipped with repair systems that recognize the DNA damages, initiate repair or direct death of the damaged cells. One of the proteins playing a central part in damage recognition, signal transduction, initiation of apoptosis, and repair is the tumor suppressor protein p53.[78,79] It accumulates to high levels after DNA damage and this increase in abundance, presumably in combination with activating modifications, leads to cell cycle arrest to allow repair of the DNA before the next round of replication or it implements cell death by activating transcription of proapoptotic target genes. In normal cells, p53 is degraded by the 26S proteasome. The multiubiquitin chain is attached to p53 by the oncoprotein Mdm2 (murine double minute chromosome clone number 2).[78] The central domain of Mdm2 is constitutively phosphorylated and this phosphorylation is essential for p53 degradation. Mdm2 possesses a carboxyl terminal RING domain[80] that promotes the ubiquitination,[81] and is necessary for the nuclear exclusion[82,83] and the degradation of p53. It has been estimated that 50% of human cancers are due to mutations of the p53 protein and some of these mutations prevent the Mdm2 mediated degradation.[78]

The Parkin protein

Autosomal recessive juvenile parkinsonism (AR-JP) is one of the most common forms of familial Parkinson's disease and is characterized by selective and massive loss of dopaminergic neurons, leading to a deficiency of dopamine supplies, and absence of Lewy bodies, cytoplasmic inclusions consisting of insoluble protein aggregates. The clinical features of the disease consist of resting tremor, cogwheel rigidity, bradykinesia, and postural instability.[84] It is the second most frequent neurodegenerative disorder after Alzheimer's disease. The causative gene of AR-JP is the Parkin gene that encodes a protein of 52 kDa (465 amino acids) that is composed of an amino terminal ubiquitin-like domain and a carboxyl terminal RING-IBR-RING motif. The RING-IBR-RING motif corresponds to two RING domains separated by an additional cysteine/histidine rich domain termed in between RING (IBR) finger.[85] The RING-IBR-RING domain is necessary and sufficient for binding UbcH7.[84] Parkin is involved in protein degradation as an E3 ubiquitin ligase[86] suggesting that dysfunction of the ubiquitin–proteasome pathway plays a role in Parkinson's disease. It has been shown that Parkin can ubiquitinate itself, thereby promoting its own degradation.[87] Furthermore, Parkin also induces the degradation of the synaptic vesicle-associated protein, cell division control related protein 1 (CDCrel-1). Therefore, mutations in Parkin could lead to an accumulation of CDCrel-1 in the brain promoting an inhibition of the release of dopamine.[87] Some of the Parkin mutations that are found in the Parkinson's disease involve residues within the RING domains.[88] Concerning the RING2, a mutation of the sixth cysteine ligand may disrupt the integrity of the structure. Interestingly, mutations were also found within the RING-1 that could, on the basis of sequence homology with c-Cbl, be involved in the E2 recognition. Further studies should elucidate the mechanism underlying the relationship between the AR-JP form of the familial Parkinson's disease and the deficiency of the ubiquitin ligase activity of Parkin.

CONCLUDING REMARKS

For many years, the eukaryotic protein degradation pathway by the 26S proteasome, and therefore by the ubiquitination pathway, was considered as a simple protein removal system of the cells. However, it is now clear that the ubiquitination pathway is involved in many regulatory processes. The observation that the RING motif is a key player in the ubiquitination pathway as being an E3 ligase not only shed light on the function of RING finger-containing proteins but also provided an explanation for the diverse biological processes that the RING finger proteins participate in. In the last few years, a large body of evidence demonstrated that the ubiquitination pathway mediated by RING domains is an indispensable regulatory process of the cells. This is further underscored by the observation that many diseases, such as certain cancers, are due to mutations in RING domains that prevent an efficient ubiquitination and degradation process.

The general dogma, ubiquitination leads to protein degradation by the proteasome, was recently challenged by the observation that (mono)ubiquitination functions as a protein modification that, in the same manner as phosphorylation, acetylation, or methylation, regulates cellular location and activity of many proteins involved in transcription, chromatin remodeling or DNA repair. This new twist raises many questions with respect to the function and reaction mechanism of monoubiquitination versus polyubiquitination and the role of RING fingers herein. Recent structures of protein complexes of E2-conjugating enzymes and E3 ligases provided clear insight into the reaction mechanism of the ubiquitin transfer but many questions remain unanswered. Determination of structures of protein complexes containing ubiquitin-conjugating enzymes and ubiquitin ligases together with specific substrates will provide detailed insight into the reaction mechanism of ubiquitin transfer from E2 to the substrates. Such structures could further explain the observed E2–E3 specificities and might clarify the role of ubiquitination in protein activation or destruction and the involvement of RING domains in these processes.

ACKNOWLEDGEMENT

The authors are grateful to Sandrine Jayne for critical reading of the manuscript. The financial support of the Center for Biomedical Genetics is also acknowledged.

REFERENCES

1. JWR Schwabe and A Klug, *Nat Struct Biol*, **1**, 345–49 (1994).
2. GE Folkers, H Hanzawa and R Boelens, in H Sigel (ed.), *Handbooks on Metalloproteins*, Marcel Dekker, New York, Basel, pp 961–1000 (2001).
3. R Lovering, IM Hanson, KLB Borden, S Martin, NJ O'Reilly, GI Evan, D Rahman, DJC Pappin, J Trowsdale and PS Freemont, *Proc Natl Acad Sci USA*, **90**, 2112–16 (1993).
4. PS Freemont, IM Hanson and J Trowsdale, *Cell*, **64**, 483–84 (1991).

5   KLB Borden and PS Freemont, *Curr Opin Struct Biol*, **6**, 395–401 (1996).

6   KLB Borden, *J Mol Biol*, **295**, 1103–12 (2000).

7   H Waterman, G Levkowitz, I Alroy and Y Yarden, *J Biol Chem*, **274**, 22151–54 (1999).

8   M Yokouchi, T Kondo, A Houghton, M Bartkiewicz, WC Horne, H Zhang, A Yoshimura and R Baron, *J Biol Chem*, **274**, 31707–12 (1999).

9   CAP Joazeiro, SS Wing, H Huang, JD Leverson, T Hunter and YC Liu, *Science*, **286**, 309–12 (1999).

10 KL Lorick, JP Jensen, S Fang, AM Ong, S Hatakeyama and AM Weissman, *Proc Natl Acad Sci USA*, **96**, 11364–69 (1999).

11 AJ Saurin, KLB Borden, MN Boddy and PS Freemont, *Trends Biochem Sci*, **21**, 208–14 (1996).

12 A Bateman, E Birney, L Cerruti, R Durbin, L Etwiller, SR Eddy, S Griffiths-Jones, KL Howe, M Marshall and EL Sonnhammer, *Nucleic Acids Res*, **30**, 276–80 (2002).

13 PS Freemont, *Curr Biol*, **10**, R84–R87 (2000).

14 CAP Joazeiro and AM Weissman, *Cell*, **102**, 549–52 (2000).

15 PK Jackson, AG Eldridge, E Freed, L Furstenthal, JY Hsu, BK Kaiser and JDR Reimann, *Trends Cell Biol*, **10**, 429–39 (2000).

16 AM Weissman, *Nat Rev Mol Cell Biol*, **2**, 169–78 (2001).

17 CM Pickart, *Annu Rev Biochem*, **70**, 503–33 (2001).

18 MH Glickman and A Ciechanover, *Physiol Rev*, **82**, 373–428 (2002).

19 L Coscoy and D Ganem, *Trends Cell Biol*, **13**, 7–12 (2003).

20 M Scheffner, U Nuber and JM Huibregtse, *Nature*, **373**, 81–83 (1995).

21 S Fang, KL Lorick, JP Jensen and AM Weissman, *Semin Cancer Biol*, **13**, 5–14 (2003).

22 D Michael and M Oren, *Semin Cancer Biol*, **13**, 49–58 (2003).

23 K Shtiegman and Y Yarden, *Semin Cancer Biol*, **13**, 29–40 (2003).

24 L Deng, C Wang, E Spencer, L Yang, A Braun, J You, C Slaughter, C Pickart and ZJ Chen, *Cell*, **103**, 351–61 (2000).

25 C Wang, L Deng, M Hong, GR Akkaraju, J Inoue and ZJ Chen, *Nature*, **412**, 346–51 (2001).

26 P Kaiser, K Flick, C Wittenberg and SI Reed, *Cell*, **102**, 303–14 (2000).

27 L Kuras, A Rouillon, T Lee, R Barbey, M Tyers and D Thomas, *Mol Cell*, **10**, 69–80 (2002).

28 DL Mallery, CJ Vandenberg and K Hiom, *EMBO J*, **21**, 6755–62 (2002).

29 H Hanzawa, MJ de Ruwe, TK Albert, PC van Der Vliet, HTM Timmers and R Boelens, *J Biol Chem*, **276**, 10185–90 (2001).

30 AP VanDemark and CP Hill, *Curr Opin Struct Biol*, **12**, 822–30 (2002).

31 L Aravind and EV Koonin, *Curr Biol*, **10**, R132–34 (2000).

32 T Belz, AD Pham, C Beisel, N Anders, J Bogin, S Kwozynski and F Sauer, *Methods*, **26**, 233–44 (2002).

33 M Koegl, T Hoppe, S Schlenker, HD Ulrich, TU Mayer and S Jentsch, *Cell*, **96**, 635–44 (1999).

34 R Swanson, M Locher and M Hochstrasser, *Genes Dev*, **15**, 2660–74 (2001).

35 TK Albert, H Hanzawa, YIA Legtenberg, MJ de Ruwe, FAJ van den Heuvel, MA Collart, R Boelens and HTM Timmers, *EMBO J*, **21**, 355–64 (2002).

36 V Gervais, D Busso, E Wasielewski, A Poterszman, JM Egly, JC Thierry and B Kieffer, *J Biol Chem*, **276**, 7457–64 (2001).

37 N Zheng, P Wang, PD Jeffrey and NP Pavletich, *Cell*, **102**, 533–39 (2000).

38 N Zheng, BA Schulman, L Song, JJ Miller, PD Jeffrey, P Wang, C Chu, DM Koepp, SJ Elledge, M Pagano, RC Conaway, JW Conaway, JW Harper and NP Pavletich, *Nature*, **416**, 703–9 (2002).

39 SF Bellon, KK Rodgers, DG Schatz, JE Coleman and TA Steitz, *Nat Struct Biol*, **4**, 586–91 (1997).

40 RD Everett, P Barlow, A Milner, B Luisi, A Orr, G Hope and D Lyon, *J Mol Biol*, **234**, 1038–47 (1993).

41 AG von Arnim and XW Deng, *J Biol Chem*, **268**, 19626–31 (1993).

42 IB Dawid, JJ Breen and R Toyama, *Trends Genet*, **14**, 156–62 (1998).

43 R Aasland, TJ Gibson and AF Stewart, *Trends Biochem Sci*, **20**, 56–59 (1995).

44 S Misra and JH Hurley, *Cell*, **97**, 657–66 (1999).

45 H Stenmark, R Aasland and PC Driscoll, *FEBS Lett*, **513**, 77–84 (2002).

46 PN Barlow, B Luisi, A Milner, M Elliott and R Everett, *J Mol Biol*, **237**, 201–11 (1994).

47 PJ Kraulis, *J Appl Crystallogr*, **24**, 946–50 (1991).

48 CD Garner, SS Hasain, I Bremner and J Bordas, *J Inorg Biochem*, **16**, 253–56 (1982).

49 PC Roehm and JM Berg, *Biochemistry*, **36**, 10240–45 (1997).

50 Z Lai, DA Freedman, AJ Levine and GL McLendon, *Biochemistry*, **37**, 17005–15 (1998).

51 KLB Borden, MN Boddy, J Lally, NJ O'Reilly, S Martin, K Howe, E Solomon and PS Freemont, *EMBO J*, **14**, 1532–41 (1995).

52 PS Brzovic, P Rajagopal, DW Hoyt, MC King and RE Klevit, *Nat Struct Biol*, **8**, 833–37 (2001).

53 W Kabsch and C Sander, *Biopolymers*, **22**, 2577–637 (1983).

54 EA Merrit and MEP Murphy, *Acta Crystallogr*, **D50**, 869–73 (1994).

55 GC Perez-Alvarado, C Miles, JW Michelsen, HA Louis, DR Winge, MC Beckerle and MF Summers, *Nat Struct Biol*, **1**, 388–98 (1994).

56 GC Perez-Alvarado, JL Kosa, HA Louis, MC Beckerle, DR Winge and MF Summers, *J Mol Biol*, **257**, 153–74 (1996).

57 R Konrat, R Weiskirchen, B Krautler and K Bister, *J Biol Chem*, **272**, 12001–7 (1997).

58 X Yao, GC Perez-Alvarado, HA Louis, P Pomies, C Hatt, MF Summers and MC Beckerle, *Biochemistry*, **38**, 5701–13 (1999).

59 A Velyvis, Y Yang, C Wu and J Qin, *J Biol Chem*, **276**, 4932–39 (2001).

60 A Hammarstrom, KD Berndt, R Sillard, K Adermann and G Otting, *Biochemistry*, **35**, 12723–32 (1996).

61 J Pascual, M Martinez-Yamout, HJ Dyson and PE Wright, *J Mol Biol*, **304**, 723–29 (2000).

62 AD Capili, DC Schultz, IF Rauscher and KLB Borden, *EMBO J*, **20**, 165–77 (2001).

63 Y Mao, A Nickitenko, X Duan, TE Lloyd, MN Wu, H Bellen and FA Quiocho, *Cell*, **100**, 447–56 (2000).

64 JJ Dumas, E Merithew, E Sudharshan, D Rajamani, S Hayes, D Lawe, S Corvera and DG Lambright, *Mol Cell*, **8**, 947–58 (2001).

65 T Kutateladze and M Overduin, *Science*, **291**, 1793–96 (2001).

66 L Huang, E Kinnucan, G Wang, S Beaudenon, PM Howley, JM Huibregtse and NP Pavletich, *Science*, **286**, 1321–26 (1999).

67. BA Schulman, AC Carrano, PD Jeffrey, Z Bowen, ERE Kinnucan, MS Finnin, SJ Elledge, JW Harper, M Pagano and NP Pavletich, *Nature*, **408**, 381–86 (2000).
68. Y Mizuno, N Hattori, H Mori, T Suzuki and K Tanaka, *Curr Opin Neurol*, **14**, 477–82 (2001).
69. JM Hall, MK Lee, B Newman, JE Morrow, LA Anderson, B Huey and MC King, *Science*, **250**, 1684–89 (1990).
70. P Kerr and A Ashworth, *Curr Biol*, **11**, R668–76 (2001).
71. Y Miki, J Swensen, D Shattuck-Eidens, PA Futreal, K Harshman, S Tautigian, Q Liu, C Cochran, LM Bennet, W Ding, R Bell, J Rosenthal, C Hussey, T Tran, M Mc Clure, C Frye, T Hattier, R Phelps, A Haugen-Strano, H Katcher, K Yakumo, Z Gholami, D Shaffer, S Stone, S Bayer, C Wray, R Bogden, P Dayananth, J Ward, P Tonin, S Narod, PK Bristow, FH Norris, L Helvering, P Morrisson, P Rosteck, M Lai, JC Barrett, C Lewis, S Neuhausen, L Cannon-Albright, D Goldgar, R Wiseman, A Kambs and MH Skolnick, *Science*, **266**, 66–71 (1994).
72. D Ford, DF Easton, DT Bishop, SA Narod and DE Goldgar, *Lancet*, **343**, 692–95 (1994).
73. D Ford, DF Easton and J Peto, *Am J Hum Genet*, **57**, 1457–62 (1995).
74. O Serova, M Montagna, D Torchard, SA Narod, P Tonin, B Sylla, HT Lynch, J Feunteun and GM Lenoir, *Am J Hum Genet*, **58**, 42–51 (1996).
75. M Santarosa, A Viel, R Dolcetti, D Crivellari, MD Magri, MA Pizzichetta, MG Tibiletti, A Gallo, S Tumolo, L Del Tin and M Boiocchi, *Int J Cancer*, **78**, 581–86 (1998).
76. LS Friedman, EA Ostermeyer, CI Szabo, P Dowd, ED Lynch, SE Rowell and MC King, *Nat Genet*, **8**, 399–404 (1994).
77. LH Castilla, FJ Couch, MR Erdos, KF Hoskins, K Calzone, JE Garber, J Boyd, MB Lubin, ML Deshano, LC Brody, FS Collins and BL Weber, *Nat Genet*, **8**, 387–91 (1994).
78. D Michael and M Oren, *Curr Opin Genet Dev*, **12**, 53–59 (2002).
79. AJ Levine, *Cell*, **88**, 323–31 (1997).
80. MN Boddy, PS Freemont and KLB Borden, *Trends Biochem Sci*, **19**, 198–99 (1994).
81. S Fang, JP Jensen, RL Ludwig, KH Vousden and AM Weissman, *J Biol Chem*, **275**, 8945–51 (2000).
82. SD Boyd, KY Tsai and T Jacks, *Nat Cell Biol*, **2**, 563–68 (2000).
83. RK Geyer, ZK Yu and CG Maki, *Nat Cell Biol*, **2**, 569–73 (2000).
84. K Tanaka, T Suzuki, T Chiba, H Shimura, N Hattori and Y Mizuno, *J Mol Med*, **79**, 482–94 (2001).
85. T Kitada, S Asakawa, N Hattori, H Matsumine, Y Yamamura, S Minoshima, M Yokochi, Y Mizuno and N Shimizu, *Nature*, **392**, 605–8 (1998).
86. H Shimura, N Hattori, S Kubo, Y Mizuno, S Asakawa, S Minoshima, N Shimizu, K Iwai, T Chiba, K Tanaka and T Suzuki, *Nat Genet*, **25**, 302–5 (2000).
87. Y Zhang, J Gao, KKK Chung, H Huang, VL Dawson and TM Dawson, *Proc Natl Acad Sci USA*, **97**, 13354–59 (2000).
88. A West, M Periquet, S Lincoln, CB Lucking, D Nicholl, V Bonifati, N Rawal, T Gasser, E Lohmann, JF Deleuze, D Maraganore, A Levey, N Wood, A Durr, J Hardy, A Brice and M Farrer, *Am J Med Genet*, **114**, 584–91 (2002).

Zinc storage

Metallothioneins

Klaus Zangger[†] and Ian M Armitage[‡]

[†]Institute of Chemistry, Organic and Bioorganic Chemistry, University of Graz, Graz, Austria
[‡]Department of Biochemistry, Molecular Biology and Biophysics, University of Minnesota, Minneapolis, MN, USA

FUNCTIONAL CLASS

Metal binding protein; a small cysteine-rich protein containing seven zinc or cadmium atoms per molecule; MT3 = growth-inhibitory factor.

Metallothioneins (MTs) are a class of small (6–7 kDa) intracellular cysteine-rich (∼33% of the amino acids are cysteines) proteins with the highest-known metal content after ferritins.[1,2] Binding of both essential (Cu^+ and Zn^{2+}) and nonessential (Cd^{2+}, Ag^+ and Hg^{2+}) metals in MTs has a high thermodynamic but low kinetic stability. Thus, metal binding is very tight, but there is facile metal exchange with other proteins.[3–5] MTs, which, by definition are *'polypeptides resembling equine renal metallothionein in several of their features'*, are classified on the basis of their amino acid sequence and structural characteristics into three classes (I, II, and III) with class I describing proteins in which the locations of cysteine residues closely resemble equine renal MT. All the MT proteins to be described in this contribution belong to class I, except the cyanobacterial Smta, a member of class II, in which the locations of cysteines are only distantly related to equine renal MT. The atypical, nontranslationally synthesized metal-thiolate polypeptides are contained in class III and will not be described here.

In class I, there are currently four known isoforms of metallothionein, which are called MT1, 2, 3, and 4.[6] By far the most studied systems are isoforms 1 and 2, which show a similar pattern as far as locations of occurrence and metal-binding characteristics are concerned. The binding of the divalent Zn^{2+} and Cd^{2+} metal ions occurs in two separate domains in vertebrate isoforms MT1 and MT2.[7] The N-terminal β-domain, comprising residues 1–30 binds 3 metals, whereas 4 metals are found in the C-terminal α-domain (residues 31–61/62 or 68).[7] The metal cluster arrangement is inverted in sea urchin MTA,[8] in which four metals are bound in the N-terminal domain and three in the C-terminal cluster. Only six metals are bound in two 3-metal clusters in the crustacean MT1 from the blue crab *Callinectes sapidus*[9] and the American lobster

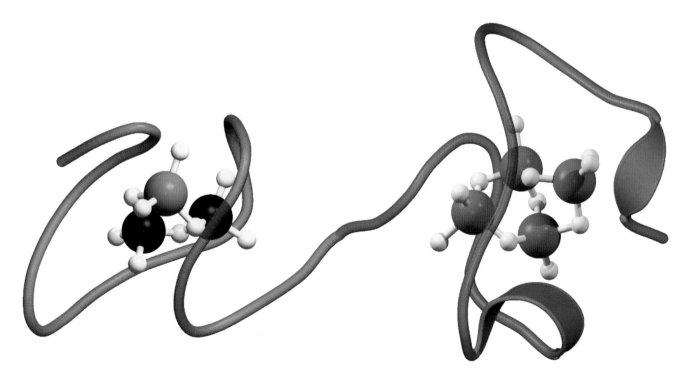

3D Structure Ribbon diagram of the X-ray structure of rat liver MT2 showing the metal-sulfur clusters with zinc atoms drawn as black and cadmium atoms as gray spheres, PDB code: 4MT2. Prepared with the programs MOLSCRIPT[38] and RASTER3D.[39]

Homarus americanus.[10,11] The MT from cyanobacterial Smta contains only one 4-metal cluster.[12,13] MT1 and MT2 are thought to function in the homeostasis of zinc and copper[1,2,14] as well as in the detoxification of heavy metals.[15,16] Metal ions, especially the ones bound in the β-domain, can be released by oxidized glutathione,[17,18] reactive oxygen species, (ROS)[19–21] and nitric oxide.[22–25] The isoform MT3 acts as a growth-inhibitory factor (GIF) of neurons.[26,27] Not much is known about MT4, which can be found in various stratified tissues.[28]

OCCURRENCE

MTs are ubiquitous proteins, found in animals, higher plants, eukaryotic organisms, and some prokaryotes.[1,2] In mammals, the two major metallothionein isoforms, MT1 and MT2, are most abundant in parenchymatous tissues, that is, in liver, kidney, pancreas, and intestines,[29–32] but their occurrence and biosynthesis have been documented in many tissues and cell types. The absolute amounts of MT1 and MT2 vary considerably with age, state of development, dietary status, and other not yet completely understood factors. MT3 can be found only in the brain[26] and MT4 is most abundant in stratified squamous epithelia.[28]

BIOLOGICAL FUNCTION

Despite the fact that MTs have been known for over 40 years, their functional role(s) still remain somewhat elusive,[33] although evidence has been accumulating for various proposed roles of MTs. They are thought to function biologically as intracellular distributors and mediators of the metals they bind,[1,2,14] and they seem to play a fundamental role in heavy metal detoxification.[15,16] The facile metal exchange involving the β-domain is thought to play a part in metal homeostasis, while the tighter metal-ion binding in the C-terminal α-domain has a role in the detoxification of heavy metals. The role of MTs in heavy metal detoxification, which has been shown by MT knockout studies in mice[34–36] is, however, very likely *not* the evolutionarily conserved function that propagated this protein, but rather just a property of these cysteine-rich proteins. MTs have also been shown to protect against ROS[19–21] and they scavenge nitrogen oxide (NO).[22–25] Because of the latter effect, it has been proposed that MTs have a role in regulating the cytotoxic activity of NO at inflammatory sites.[25] In contrast to the ubiquitous isoforms MT1 and MT2, the brain-specific member of the MT family of proteins, MT3, inhibits the growth and survival of neurons. MT3 was discovered in the central nervous system during research aimed at understanding the pathogenesis of Alzheimer's disease (AD).[26] Brain extracts from AD patients stimulated the survival and growth of neurons to a greater degree than regular brain extracts did, owing to the loss of a growth-inhibitory factor, which was later found to be a member of the MT family.[37] Not much is known about the function of MT4, but it is believed that it is involved in pumping excess zinc out of cells and plays a special role in regulating zinc metabolism during differentiation of stratified epithelia.[28]

AMINO ACID SEQUENCE INFORMATION

Owing to the high number of different known MT sequences (currently about 200 sequences http://www.expasy.ch/cgi-bin/lists?metallo.txt) only the amino acid sequences for zinc or cadmium MTs with known 3D Structure are given here:

- *Homo sapiens* MT2 (human), 61 AA, Swiss-Prot entry MT2_HUMAN, accession# P02795[40]
- *Oryctolagus cuniculus* MT2a (rabbit), 61 AA, Swiss-Prot entry MT2A_RABIT, accession# P18055[41]
- *Rattus rattus* MT2 (rat), 61 AA, Swiss-Prot entry MT2_RAT, accession# P04355[42]
- *Mus musculus* MT1 (mouse), 61 AA, Swiss-Prot entry MT1_MOUSE, accession# P02802[43]
- *Mus musculus* MT3 (mouse), 68 AA, Swiss-Prot entry MT3_MOUSE, accession# P28184[44]
- *Strongylocentrotus purpuratus* MTA (sea urchin), 64 AA, Swiss-Prot entry MTA_STRPU, accession# P04734[45]
- *Callinectes sapidus* MT1 (blue crab), 59 AA, Swiss-Prot entry MT1_CALSI, accession# P55949[46]

```
                                                                                                Homology
human MT2            MDPN-CSCAAG-DSCTCA-GSCKCK-ECKCTSCKKSCCSCCPVGCAKCA-QGCICKG-----ASDKCSCCA-    100.0
rabbit MT2           MDPN-CSCAAAGDSCTCA-NSCTCK-ACKCTSCKKSCCSCCPPGCAKCA-QGCICKG-----ASDKCSCCA-     91.9
rat MT2              MDPN-CSCATD-GSCSCA-GSCKCK-QCKCTSCKKSCCSCCPVGCAKCS-QGCICKE-----ASDKCSCCA-     88.5
mouse MT1            MDPN-CSCSTG-GSCTCT-SSCACK-NCKCTSCKKSCCSCCPVGCSKCA-QGCVCKG-----AADKCTCCA-     82.0
mouse MT3            MDPETCPCPTG-GSCTCS-DKCKCK-GCKCTNCKKSCCSCCPAGCEKCA-KDCVCKGEEGAKAEAEKCSCCQ-    61.8
sea urchin Mta       MPDVKCVCCKEGKECACFGQDCCKT-GECCKDG--TCCGICTNAACKCA-NGCKCGSG-----CSCTEGNCAC     37.3
blue crab MT1        -MPG--PCCND--KCVCQEGGCKA--GCQCTSCRCSPCQKCTSGC-KCA-TKEECSKT-----CTKPCSCCPK     43.8
lobster MT1          --PG--PCCKD--KCECAEGGCKT--GCKCTSCRCAPCEKCTSGC-KCP-SKDECAKT-----CSKPCSCCPT     43.8
cyanobacterial Smta  TSTTLVKCACE--PCLCNVDPSKAIDRNGLYYCSEACADGHTGGSKGCGHTGCNCHG--------------     18.5
```

Figure 1 Sequence alignment of the structurally characterized zinc/cadmium binding class I and class II MTs. The highly conserved cysteine residues are shown in red, all other cysteines in green. The sequence homology is calculated relative to human MT2 with the program CLUSTALW.[49]

- *Homarus americanus* MT1 (lobster), 58 AA, Swiss-Prot entry MT1_HOMAM, accession# P29499[47]
- *Synechococcus* sp. Smta (cyanobacterial), 55 AA, Swiss-Prot entry MT_SYNP7, accession# P30331[48]

A sequence alignment of all these MTs is shown in Figure 1 together with the relative homology compared to human MT2. The highest homology can be found, as expected, for the mammalian MT isoforms 1 and 2, whereas the cyanobacterial Smta, a member of class II MTs, shows only 18.5% homology to human MT2.

PROTEIN PRODUCTION, PURIFICATION, AND MOLECULAR CHARACTERIZATION

Traditionally, MTs were extracted from natural sources following the administration of metal ions (Cd^{2+}, Zn^{2+}, Ag^+, Cu^+) over a period of several days.[50,51] A typical protocol for cadmium, which is the strongest inducer of MTs, includes nine injections three times a week of 0.1 M $CdCl_2$ in 0.15 M NaCl at a dose of 1 mg/kg body weight.[51] Despite the use of cadmium for induction in vertebrates, MT isolated from these sources never contains more than 5 mol Cd/mol MT, the rest being zinc. After sacrificing the animal, MT is obtained from the liver by ethanol precipitation and purified, first on a Sephadex G-50 gel-filtration column (150 × 5 cm) and subsequently on a DEAE-cellulose ion-exchange column (40 × 3 cm).[50] It should be noted that since MTs contain no aromatic amino acids, the absorption at 280 nm is very low. Instead, the charge-transfer bands for Cd–S and Zn–S at 250 and 231 nm respectively[52] can be used to observe the elution of MT-containing fractions. As mentioned above, MT isolated from natural sources never contain more than 5 cadmium atoms per MT molecule, and to obtain Cd_7-MT, the protein needs to be reconstituted, as described by Vašak.[53] This procedure involves first the removal of bound metals by lowering the pH to 1, followed by passing the protein solution over a gel-filtration column. After the addition of approximately 8 mol equivalents of either cadmium or zinc, the pH is raised to 8.5 with 0.5 M Tris base and the excess metal is removed with Chelex-100. Today, MT is most often obtained through recombinant techniques. A typical expression and purification scheme involves expression of MT in *Escherichia coli* cells containing the corresponding metallothionein gene. Metal salts (e.g. $ZnCl_2$) are added after induction with isopropyl-β-D-thiogalactapyranoside (IPTG).[54] The protein is purified first in a *Sephadex* G75 gel-filtration column, and the collected metal-containing fractions are loaded onto a DE-32 anion-exchange column and eluted with a tris-HCl gradient.[55] The combined MT-containing fractions are concentrated in an *Amicon* apparatus using a YM-3 membrane under nitrogen pressure. The ^{113}Cd-enriched protein can then be obtained by the mass action exchange of Zn through the addition of an excess of $^{113}CdCl_2$, incubation for several hours, and subsequent removal of the excess metal ions; or by the preparation of the apoprotein followed by reconstitution with ^{113}Cd.[53]

METAL CONTENT AND COFACTORS

For mammalian MTs, a series of well-characterized complexes have been described, including Zn_7-MT, Cd_7-MT, Cu_{12}-MT, Hg_7-MT, Hg_{18}-MT, Ag_{12}-MT, and Ag_{17}-MT.[56] However, in this contribution, only Zn_x-MT and its isomorphously substituted NMR-active counterpart Cd_x-MT are described. MTs of vertebrates contain seven divalent metal ions bound in two domains.[7] The N-terminal β-domain binds three metals in a M_3S_9 cluster, whereas four metals are arranged in a M_4S_{11} cluster in the C-terminal α-domain. As an example, the metal-cysteine connectivities of mouse MT1[57] are shown in Figure 2. The same pattern is found for all other mammalian MTs described to date.[58–61] Each metal is coordinated to four cysteine sulfur atoms and each sulfur atom binds to either one (terminal) or two (bridging) metal ions. In sea urchin MT, the orientation of the metal clusters is reversed.[8] A four-metal Zn_4Cys_{11} cluster is found in the N-terminal domain and three metals are bound in the C-terminal Zn_3Cys_9 cluster. Accordingly, in this particular MT, the nomenclature has been reversed; the α-domain here is the N-terminal domain, while the β-domain is at the C-terminus. The six bound metals in both blue crab and lobster Zn_6-MT are found in two Zn_3Cys_9 clusters in separate protein domains.

The naturally bound Zn^{2+} or Cu^+ ions can be partially replaced by administering cadmium or silver solutions to the living animals, followed by the subsequent isolation of the protein. Mixed Cd/Zn-MT obtained by this procedure contains approximately 5 cadmium and 2 zinc atoms per MT molecule, where the cadmium is preferentially bound in the C-terminal and zinc in the N-terminal domain.[58] Homogeneous Cd_7-MT can be obtained from recombinant methods when the growth media contains cadmium or by

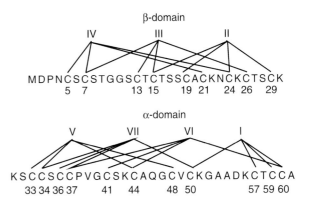

Figure 2 Metal-cysteine connectivities in mouse MT1. The numbering of the metals corresponds to the increasing field in the ^{113}Cd NMR spectrum. Zinc is coordinated in exactly the same way.

the preparation of the apo-MT from natural MT by the removal of all metals at low pH and reconstitution with cadmium.[53]

NMR SPECTROSCOPY

All but one of the metallothionein structures known to date were obtained by nuclear magnetic resonance (NMR) spectroscopy. Not only can the backbone structure be determined with NMR-derived proton–proton distances through nuclear Overhauser effect (NOE) experiments,[62] but the NMR-active, spin 1/2 nuclei cadmium-111/113 or silver-107/109 can also be used to replace the native metals zinc and copper-(I) and provide cysteine-metal connectivities through the metal-cysteine β-proton scalar coupling.[63]

The spin of 5/2 for ^{67}Zn and its inherent quadrupole moment have prevented similar NMR studies from being carried out on zinc in metallothioneins. However, the NMR-active cadmium isotopes are isomorphic replacements for zinc due to their similar ionic radii and coordination geometry.[64] A typical ^{113}Cd NMR spectrum of human-[Cd$_7$]-metallothionein2[60] is shown in Figure 3. One resonance can be found for each cadmium atom, with resonances belonging to the β-domain (Cadmium II, III, and IV) showing reduced intensity, which has been attributed to the increased mobility in this domain. The cadmium-cysteine connectivities can be found through ^{1}H-^{113}Cd-HMQC or heteronuclear single quantum coherence (HSQC) type experiments,[63] where the three–bond scalar coupling between cadmium and the cysteine β-protons is used to establish these correlations. Even more information about

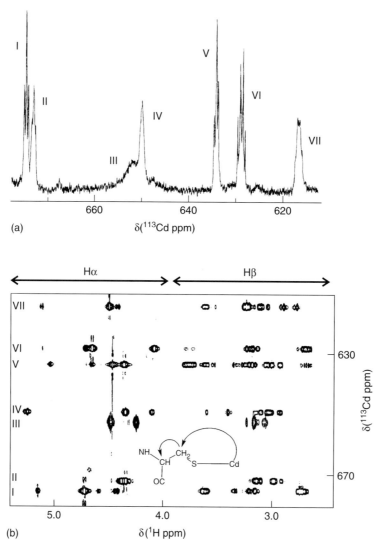

Figure 3 One-dimensional ^{113}Cd-NMR spectrum (a) and two-dimensional ^{113}Cd-^{1}H-Relayed heteronuclear multiple quantum coherence (HMQC) of human-[^{113}Cd$_7$]-MT2. Owing to the additional relay step (insert in b) correlations are observed from each cadmium atom not only to the cysteine β-protons but also to the α-protons. Reprinted from *J Mol Biol*, **214**, BA Messerle, A Schäffer, M Vašák, JH Kägi and K Wüthrich, Three-dimensional structure of human [^{113}Cd$_7$] metallothionein-2 in solution determined by nuclear magnetic resonance spectroscopy, 765–79 (1990). Reproduced with permission from Elsevier Science.[60]

the cadmium-cysteine connectivities can be obtained from ^{1}H-^{113}Cd Relayed-HMQC or HSQC type experiments,[65] which provide correlations for each cadmium and the β- and α-protons of the coordinated cysteine (Figure 3b).

Structure calculation of metallothioneins from NMR data

The tertiary structure of MTs is dictated by the wrapping of the polypeptide chain around the metal clusters and is largely devoid of β-sheet and α-helix secondary structural elements. As a consequence, the number of proton–proton NOEs is rather low, especially in the less well defined β-domain as shown in Figure 4, which displays the number of NOEs per residue for mouse MT1.[57]

Therefore, it is essential to obtain additional constraints for the solution structure determination, which has been established from the scalar coupling between the cysteine β-protons and specifically bound cadmium or silver metal ions. Using these connectivities, the complete Cd-S clusters are defined during the structure calculation with the information about the cadmium–sulfur bond lengths as well as Cd-S-Cd, S-Cd-S and CysCβ-S-Cd angles taken from X-ray structures of model complexes[66,67] and from the X-ray structure of rat liver MT2.[58] In the NMR-derived MT structures to date, the number of proton–proton NOEs used for the structure calculation is higher for the α-domain, which has been assumed to be because of the increased mobility in the β-domain. Since, in all solution structures of MTs published so far, no interdomain NOEs have been found, the structures of the two domains have been calculated separately. This is why the *relative orientation* of the two domains in MTs could only be obtained from the X-ray structure. While historically, the low number of NOEs made the determination of MT structures dependent upon the determination of metal-cysteine connectivities, Bertini *et al.*[68] recently showed that with the increased number of NOEs obtained at 800 MHz, the structure of the peptide backbone of yeast Cu$_7$-MT could be obtained from proton–proton NOEs alone.

OTHER SPECTROSCOPIC TECHNIQUES

NMR spectroscopy has been used extensively from the mid 1970s to characterize MTs, since the use of UV spectroscopy provided only limited information. This was due in part to the complete absence of aromatic amino acids in MTs, resulting in minimal 280-nm absorption[52] and in part to the overlapping nature of the multiple cadmium–sulfur or zinc–sulfur charge-transfer transitions, which are responsible for pronounced absorption at around 250 nm and 231 nm respectively. Nevertheless, the elution of these metal-containing MTs from chromatography columns is still routinely monitored by their absorbance at 231 or 250 nm. Additionally, these metal-thiolate transitions allow for the monitoring of metal binding to apo-MT where the absorption profiles below 300 nm, obtained by adding a cadmium solution to MT, show a family of closely similar spectra with increasing intensity around 250 nm as a function of Cd concentration.[69] In the UV spectrum, a 6-nm red shift of the lowest unresolved absorption band was observed when more than three equivalents of cadmium had been added. This was paralleled by a changeover in the circular dichroism (CD) appearance from a broad monophasic (positive ellipticity bands near 220 and 240 nm) to a biphasic (positive ellipticity at 224 and 260 nm and negative ellipticity at 240 nm) spectrum.[69] Both features can be attributed to the presence of individual Cd-thiolate ligands at low concentration and to Cd-thiolate clusters when the number of metals is higher than three.

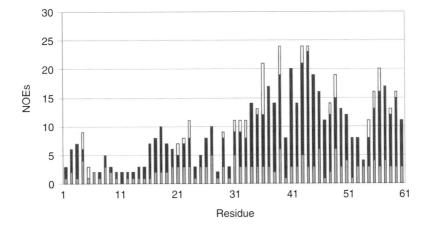

Figure 4 Number of NOEs per residue in mouse MT1. Intraresidue, short (i–j < 5) and long range (i–j ≥ 5) NOEs are depicted as gray, black, and white bars respectively. Reprinted from *Protein Sci* 8, K Zangger, G Öz, JD Otvos and IM Armitage, Three-dimensional solution structure of mouse [Cd$_7$]-metallothionein-1 by homonuclear and heteronuclear NMR spectroscopy, 2630–38 (1999). Adapted with permission from Elsevier Science.[57]

Metal binding in metalloproteins in general, and in MTs in particular, can also be investigated using electrospray ionization mass spectrometry ESI-MS.[70,71] This method offers a soft ionization method that enables the detection of the entire protein and allows the metal-to-protein stoichiometries to be determined. Yu et al.[71] were the first to analyze MTs by ESI-MS and, using this method, Gehrig et al.[70] found that the prominent peak in the holo-human MT2 spectrum corresponded to the Zn_7 form.

X-RAY STRUCTURE OF RAT LIVER MT2

Crystallization

The only MT structure solved by X-ray crystallography has been rat liver Cd_5Zn_2-MT2.[58] The MT2 used for crystallization was isolated from rat liver and had a metal composition of 5 mol Cd^{2+} and 2 mol Zn^{2+} per mol of MT.[72] The amino acid sequence consisted of 61 residues, including 20 cysteine residues, and contained N-acetyl-methionine at the N-terminus. Lyophilized samples were used to make a solution of $10\,mg\,mL^{-1}$ in MT2 containing 1.0 M sodium formate and 0.2 M phosphate at pH 7.5. The solution was equilibrated in 5- to 10-μL volumes against 1.0-mL volumes of 5.0 M sodium formate, pH 7.5, using vapor diffusion and hanging drops. Repetitive seeding was used to produce large single crystals. The crystals are tetragonal with unit cell constants $a = b = 30.9$ Å, $c = 120.4$ Å with 1 molecule per asymmetric unit. The space group is $P4_32_12$. The crystal packing reveals intimate association of molecules about the diagonal twofold axes and trapped phosphate and sodium ions of crystallization.

Overall description of the structure

The X-ray structure of rat liver MT2 has been refined at 2.0-Å resolution. It shows a dumbbell-shaped molecule with the seven metal ions (a mixture of zinc and cadmium) bound in two separate domains (Figure 5).

In the N-terminal domain, two zinc and one cadmium atom are bound in a M_3S_9 cluster, with each metal being tetrahedrally coordinated by four cysteines. The C-terminal domain binds four cadmium atoms in a M_4S_{11} cluster, again with each metal being tetrahedrally held in place by four cysteine sulfur ligands. The cadmium atoms and the bridging sulfur ligands are arranged in a six-membered ring in a slightly distorted boat conformation in the three-metal cluster and two fused six-membered rings in a distorted boat–boat conformation in the four-metal domain. The metal-sulfur distances found in the crystal structure of rat liver MT2 are given in Table 1.

The protein backbone basically wraps around these metal clusters. The polypeptide chain is folded right-handed around the N-terminal Cd_3S_9 cluster and left-handed around the C-terminal Cd_4S_{11} cadmium-cysteine cluster. Like all other structurally characterized MTs, the crystal structure of rat MT2 shows only a few short sequences of regular secondary structural elements. In most mammalian MTs, two short stretches of a 3_{10} helix can be found around residues 42–45 and 58–60 and several turns are spread along the whole backbone. The polypeptide fold can be described as a series of loops of variable length, with both ends tied to a metal atom.

SOLUTION NMR STRUCTURES OF MTs

Rabbit, rat, human MT2, and mouse MT1

These MTs are treated together since only minor differences have been found in their three-dimensional structures. A least-square fitted superposition of the individual N- and C-terminal domains of these MTs together with the crystal structure of rat MT2 is shown in Figure 6.

The pairwise rmsd of the backbone atoms is less than 2 Å for all proteins shown. The first solution structure of an

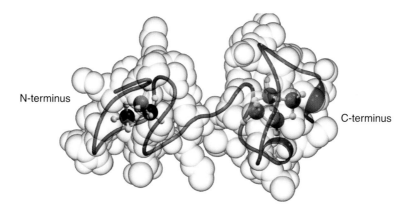

Figure 5 Ribbon diagram of the crystal structure of rat MT2 (PDB code: 4MT2), showing the metal-cysteine clusters as ball-and-stick models (sulfur:yellow; cadmium:gray; zinc:black) and the whole protein excluding protons as a transparent cpk drawing. The N-terminal β-domain is represented by the red and the C-terminal α-domain by the blue ribbon. Prepared with the programs MOLSCRIPT[38] and RASTER3D.[39]

Table 1 Cadmium–sulfur and zinc–sulfur distances in the crystal structure of rat liver Cd_5Zn_2-MT2

Metal[a]	Cysteine[b]	Distance (Å)
Cd-I	50	2.57
	57	2.50
	59	2.51
	60	2.50
Zn-II	15	2.48
	19	2.30
	24	2.41
	29	2.41
Zn-III	7	2.33
	13	2.37
	15	2.37
	26	2.37
Cd-IV	5	2.48
	7	2.54
	21	2.49
	24	2.54
Cd-V	33	2.55
	34	2.45
	44	2.62
	48	2.47
Cd-VI	37	2.52
	41	2.48
	44	2.51
	60	2.60
Cd-VII	34	2.51
	36	2.49
	37	2.44
	50	2.50

[a] The metal atoms are indicated by the roman numeral of the corresponding cadmium signal in the ^{113}Cd NMR spectra, even for the zinc atoms (see Figures 2 and 3) with metals I, V, VI, and VII belonging to the C-terminal domain and II, III, and IV to the N-terminal cluster.

[b] The ligands are denoted by the amino acid residue number in rat liver MT2.

MT to be determined was rabbit MT2a,[59] which was also the first *correct* structure obtained for any MT, since the first published MT crystal structure[74] was later found to be flawed. Discrepancies between this first published *crystal* structure of rat liver MT2 and the NMR structure of rabbit MT2a and later rat MT2[61] paved the way for renewed attempts to determine the correct crystal structure of rat MT2.[58] The solution structures of all mammalian MT1[57] and MT2[59–61] isoforms determined to date are, within their range of accuracy, identical to the crystal structure and therefore will not be further discussed. By inspection of the NOE pattern in rabbit MT2a, Wagner et al.[75] found a new secondary structure element called a 'half-turn', which can be thought of as a type II turn where the angle ϕ_3 is rotated from $+90°$ to $-90°$. Half-turns have so far only been found in MTs, and are very likely formed because

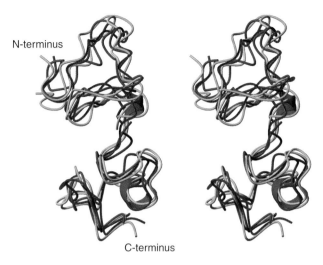

Figure 6 Stereo view of a least-square superposition of all structurally characterized mammalian MT isoforms 1 and 2. The relative domain orientation is known only for the crystal structure of rat MT2 (black ribbon, PDB code: 4MT2), and onto this structure, all others are fitted. The rabbit MT2 (PDB codes: 1MRB, 2MRB), human MT2 (PDB codes: 1MHU, 2MHU), rat MT2 (NMR structure, PDB code: 1MRT, 2MRT), and mouse MT1 (PDB codes: 1DFT, 1DFS) are shown in cyan, red, blue, and green respectively. For easier visualization, the metal-cysteine clusters have been omitted. Prepared with the program MOLMOL.[73]

of large constraints on the protein backbone by the large number of metal binding sites. The half-turns found are located between residues 3–5, 5–7, 10–12, 19–21, 24–26, 34–36, and 48–50.

As mentioned above, in all solution structures of MTs, the relative orientation of the two domains could not be determined and this is why the domains were calculated separately. Since the calculation of the solution structures from NMR data does not lead to *one specific* structure but to a *whole bundle* of structures that fit the measured data, the accuracy of the determined structure can be evaluated by the rmsd of atom positions. It has been found that the longer the loop connecting two metals, the less well defined its structure is. An overlay of 10 accepted structures for the N- and C-terminal domains of mouse MT1[57] (Figure 7) clearly shows the higher structural flexibility in the N-terminal domain, which is a consequence of the lower number of NOEs in this domain (see Figure 4). The identity of MTs with zinc and cadmium bound has been confirmed through the equivalence of the 2D nuclear Overhauser effect spectroscopy (NOESY) spectra of human Zn_7-MT and Cd_7-MT.[64] It is further corroborated by the very high degree of similarity between the crystal structure of rat Cd_5Zn_2-MT2[58] and the solution structure of rat Cd_7-MT2.[61]

The 3D solution structure of mouse MT1,[57] the only structurally characterized mammalian MT1 isoform, shows basically the same backbone fold and metal cluster arrangement as all the mammalian MT2s. There are,

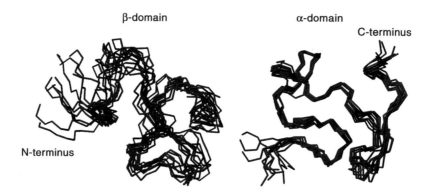

Figure 7 Stereo view of a least-square superposition of the backbone traces of 10 lowest energy structures of the N-terminal (PDB code: 1DFT) and C-terminal (PDB code: 1DFS) domains of mouse MT1. The average rmsd of backbone atoms in residues 5–25 is 1.9 Å in the β-domain and for residues 35–55 it is 0.5 Å in the α-domain. Prepared with the program MOLMOL.[73]

however, certain indications of increased mobility in the MT1 N-terminal domain. These include a lower number of NOEs per residue, increased metal–metal exchange rates, and the absence of slowly exchanging amide protons in the N-terminal domain.

Mouse MT3

Compared to the mammalian isoforms MT1 and MT2, the brain-specific MT3[76] consists of 68 amino acids, whose most remarkable differences are an acidic 6 amino acid insertion in the C-terminus and a Cys-Pro-Cys-Pro tetrapeptide at the N-terminus. The latter was shown to be responsible for the growth-inhibitory action of MT3.[77] Owing to the flexible nature of the N-terminal β-domain, its NMR solution structure could not be determined.[76] The C-terminal α-domain structure (Figure 8) is very similar to the previously mentioned C-terminal MT domains, with the acidic insertion found as a rather flexible loop, which, it has been suggested, might be involved in protein–protein interactions.

Sea urchin MTA

Sea urchin MTA consists of 65 amino acids and binds 7 zinc or cadmium atoms in 2 domains.[8] However, there is a marked difference from mammalian MTs in that the N-terminal domain contains the M_4Cys_{11} cluster while the M_3S_9 cluster is located in the C-terminal domain (Figure 9). Therefore, the nomenclature α- and β-domain is reversed when compared to the previously mentioned MTs. The overall arrangement of the polypeptide chain around the metal-sulfur clusters in the α-and β-domains in sea urchin MTA and mammalian MT1 and MT2 are similar, although the cysteine sequence spacing is quite different.

As for the other MTs, the β- (or C-terminal) domain in sea urchin MTA is more flexible than the α- (or N-terminal) domain, which has been further characterized by ^{15}N relaxation measurements.[8] Differences with the mammalian MTs can be found in the metal-bridged sulfur conformation in the four-metal cluster. In sea urchin MTA, one of the two fused six-membered rings is in boat conformation, while the other is in a chair form. Additionally, in the β-domain the protein backbone is wrapped left-handedly around the M_3S_9 cluster.

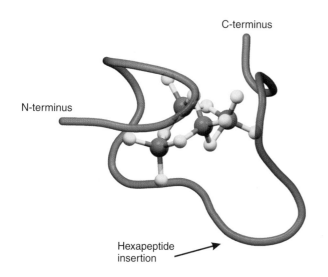

Figure 8 Ribbon diagram of the C-terminal α-domain of mouse MT3, PDB code: 1JI9. The metal-cysteine cluster is drawn as a ball-and-stick model. Prepared with MOLSCRIPT[38] and RASTER3D.[39]

Blue crab and lobster MT1

These two crustacean MTs[9–11] bind six metals in two M_3S_9 clusters in separate domains. The metal-cysteine connectivities, as well as the overall polypeptide fold, are identical (Figure 10).

The N-terminal β-domain shows a high resemblance to the N-terminal domain of vertebrate MT1 and 2 with the protein backbone being right-handedly wrapped around

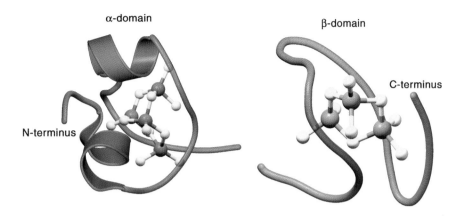

Figure 9 Ribbon diagram of the N-terminal α- (PDB code: 1QJK) and the C-terminal β-domain (PDB code: 1QJL) of sea urchin MTA. The metal-cysteine cluster is drawn as a ball-and-stick model. Prepared with MOLSCRIPT[38] and RASTER3D.[39]

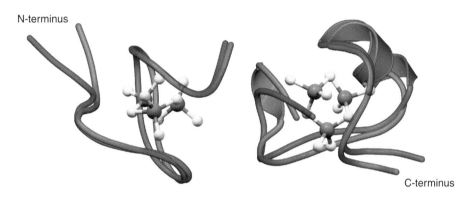

Figure 10 Least-square superposition of the individual domains of blue crab (blue ribbon, PDB codes: 1DMC,1DME) and lobster (red ribbon PDB codes: 1J5L,1J5M) MT1. For easier visualization, the metal-cysteine cluster is drawn as a ball-and-stick model only for blue crab MT1. Prepared with MOLSCRIPT[38] and RASTER3D.[39]

the metal-sulfur cluster. However, the N-terminal metal cluster in blue crab[9] and lobster MT1[10,11] is less exposed to the solvent. It is noteworthy that while 8 out of 9 cysteine residues in crustacean MTs occupy homologous positions compared to mammalian MT, 3 of the 12 Cd-Cys connectivities are different. This indicates that there is a multitude of possible M_3S_9 clusters that can be formed from a sequence of 30 amino acids containing 9 cysteines.

The C-terminal domain is half right-handed and half left-handed. The α-domain of these crustacean MTs is more compact than the α-domains of mammalian MTs, due, most likely, to the mixed handedness of this domain and the presence of only three metals. There is also evidence of a short α-helix in this domain in blue crab MT1[9] extending from Lys42 to Thr48. Differences in flexibilities between the N- and C-terminal domains in blue crab and lobster MT1 are rather minor, based on well-defined structures with low rmsds obtained for both domains.

Cyanobacterial Smta

The MT from the cyanobacterium *Synechococcus* PCC 7942[12,13] is presently the only structurally characterized zinc-/cadmium-containing class II prokaryotic metallothionein. Both its amino acid sequence and its three-dimensional structure show rather large differences from the eukaryotic MTs presented above (Figure 11).

The percentage of cysteine in the primary sequence (ca. 20%), is low compared to other MTs (ca. 33%). It contains nine cysteines and three histidine residues and binds four divalent metals. This is achieved by forming a $M_4Cys_9His_2$ cluster. Most noteworthy, and unusual for MTs, is the presence of several well-defined secondary structural elements. An α-helix was found from Glu34-Gly39 and four short antiparallel β-strands were formed by residues Val7-Lys8, Asn17-Val18, Ile24-Asp25, and Tyr30-Tyr31. A remarkable similarity has been found between the arrangement of one cadmium and the associated α-helix and antiparallel β-strands with the zinc site, and secondary structural elements in the C-terminal zinc finger of the eukaryotic DNA binding protein GATA-1[78] as well as zinc finger 2 of LIM domains.[79] Although the positions of the metal ligands (cysteines and histidines) in this prokaryotic MT are not very similar to all previously mentioned ones, the conformation of the

Metallothioneins

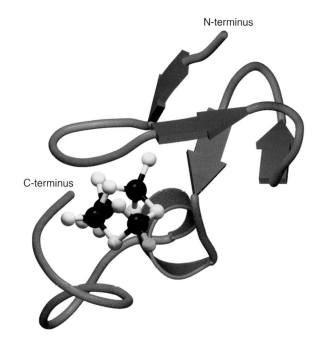

Figure 11 Ribbon diagram of the cyanobacterial Smta that belongs to metallothionein class II (PDB code: 1JJD). The $M_4S_9N_2$ cluster is drawn as a ball-and-stick model. Prepared with MOLSCRIPT[38] and RASTER3D.[39]

four-metal cluster is very similar to the mammalian four-metal domains, showing two fused six-membered rings in distorted boat–boat conformation. Two of the terminal cysteine ligands are replaced by histidines in Smta.

FUNCTIONAL ASPECTS

Metal binding and exchange

Mammalian MTs bind seven equivalents of divalent metals in two domains quite tightly. The dissociation constant for zinc at pH = 7 is about 10^{-11} to $10^{-12}\,M^{-1}$ with Cd(II), Cu(I) and Ag(I) binding several orders of magnitude more tightly.[5] The divalent metals zinc and cadmium bind preferentially in the C-terminal four-metal cluster, whereas silver and copper(I) exhibit the opposite preference. When a mixture of cadmium and zinc is available, cadmium is found preferentially in the C-terminal domain and zinc in the looser N-terminal domain.[3] It is remarkable that despite the fact that metal binding in MTs is rather tight, the exchange of metals bound in MTs is quite rapid when free metal ions or other metal binding proteins are present in solution. Apparently, the solvent–exposed cysteine residues can be easily attacked by incoming metals, which facilitate the dechelation of bound metal. Adding free cadmium to Zn_7-MT results in a complex mixture of Zn_xCd_y-MT whose composition is different from the Zn/Cd-MT formed naturally *in vivo*.[3] However, by mixing Zn-MT and Cd-MT, the natural distribution of Zn and Cd between the two clusters is obtained.[3] Therefore, it was concluded that metal exchange *in vivo* occurs through intermolecular metal exchange between MT molecules, rather than by the exchange of free metal with MT. This intermolecular metal exchange involving MTs is catalyzed by excess metal ions, which are assumed to form transient bonds with thiolate ligands in both exchanging molecules (Figure 12).

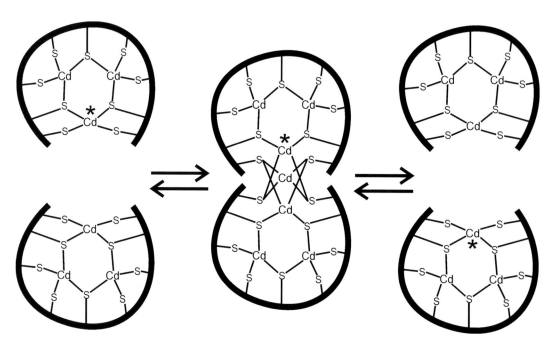

Figure 12 Proposed model for the intermolecular metal exchange in metallothioneins in the presence of excess metal ion. An asterisk is used to indicate the path of the exchanging metal atom. Reprinted from *J Inorg Biochem*, 88, K Zangger and IM Armitage, Dynamics of interdomain and intermolecular interactions in mammalian metallothioneins, 135–43 (2002). Reproduced with permission from Elsevier Science.[80]

Intramolecular metal exchange is much faster in the N-terminal domain, where it occurs approximately on the order of 0.2 to $2\,s^{-1}$ as monitored by ^{113}Cd-NMR.[5] The differential mobilities of metals in the two clusters of vertebrate MTs have been assumed to be responsible for the dual role of MTs, where the zinc binding in the flexible N-terminal domain has been associated with zinc distribution to other proteins, and the tight metal binding in the C-terminal domain is responsible for the heavy metal detoxification.

Modulation of metal release

A number of compounds have been found that modulate the metal release from metallothioneins. It has been shown that the zinc release into solution, or exchange to other proteins, is enhanced by adding oxidized glutathione to a MT solution.[17,18] In contrast, reduced glutathione decreases the already slow metal release. This finding is especially interesting because it offers the possibility of controlling the metal release of redox inert zinc in MTs through the redox state of glutathione, which might be the underlying principle of how MTs regulate the homeostasis of essential metals like zinc and copper(I). Recently, it was found that nitric oxide also releases zinc and cadmium from MTs[22–24,81] and we later determined that only the metal from the N-terminal domain is released from mouse [Cd$_7$]-MT1.[25] Since both metallothionein and the nitric oxide–producing enzyme 'inducible' nitric oxide synthase (iNOS), are induced by α-tumor necrosis factor, lipopolysaccharides, and interleukin-1, and since the zinc released from MT *in vivo* suppresses further expression of iNOS, we proposed that MT plays a role at inflammatory sites by modulating the pathotoxic action of NO.[25]

REFERENCES

1. JH Kägi and Y Kojima, *Exs*, **52**, 25–61 (1987).
2. JH Kägi and A Schäffer, *Biochemistry*, **27**, 8509–15 (1988).
3. DG Nettesheim, HR Engeseth and JD Otvos, *Biochemistry*, **24**, 6744–51 (1985).
4. JD Otvos, HR Engeseth, DG Nettesheim and CR Hilt, *Exs*, **52**, 171–78 (1987).
5. JD Otvos, X Liu, H Li, G Shen and M Basti, in KT Suzuki, N Imura and M Kimura (eds.), *Metallothionein III*, Birkhäuser Verlag, Basel/Switzerland, pp 57–74 (1993).
6. M Vašak and DW Hasler, *Curr Opin Chem Biol*, **4**, 177–83 (2000).
7. JD Otvos and IM Armitage, *Proc Natl Acad Sci USA*, **77**, 7094–98 (1980).
8. R Riek, B Precheur, Y Wang, EA Mackay, G Wider, P Güntert, A Liu, JH Kägi and K Wüthrich, *J Mol Biol*, **291**, 417–28 (1999).
9. SS Narula, M Brouwer, Y Hua and IM Armitage, *Biochemistry*, **34**, 620–31 (1995).
10. Z Zhu, EF DeRose, GP Mullen, DH Petering and CF Shaw III, *Biochemistry*, **33**, 8858–65 (1994).
11. A Munoz, DH Petering and CF Shaw III, *Inorg Chem*, **39**, 6114–23 (2000).
12. MJ Daniels, JS Turner-Cavet, R Selkirk, H Sun, JA Parkinson, PJ Sadler and NJ Robinson, *J Biol Chem*, **273**, 22957–61 (1998).
13. CA Blindauer, MD Harrison, JA Parkinson, AK Robinson, JS Cavet, NJ Robinson and PJ Sadler, *Proc Natl Acad Sci USA*, **98**, 9593–98 (2001).
14. TY Li, AJ Kraker, CF Shaw and DH Petering, *Proc Natl Acad Sci USA*, **77**, 6334–38 (1980).
15. DH Hamer, *Annu Rev Biochem*, **55**, 913–51 (1986).
16. MG Cherian, SB Howell, N Imura, CD Klaassen, J Koropatnick, JS Lazo and MP Waalkes, *Toxicol Appl Pharmacol*, **126**, 1–5 (1994).
17. LJ Jiang, W Maret and BL Vallee, *Proc Natl Acad Sci USA*, **95**, 3483–88 (1998).
18. C Jacob, W Maret and BL Vallee, *Proc Natl Acad Sci USA*, **95**, 3489–94 (1998).
19. M Sato and I Bremner, *Free Radical Biol Med*, **14**, 325–37 (1993).
20. PJ Thornalley and M Vašak, *Biochim Biophys Acta*, **827**, 36–44 (1985).
21. W Maret, *Neurochem Int*, **27**, 111–17 (1995).
22. CT Aravindakumar, J Ceulemans and M De Ley, *Biochem J*, **344 Pt 1**, 253–58 (1999).
23. KD Kröncke, K Fehsel, T Schmidt, FT Zenke, I Dasting, JR Wesener, H Bettermann, KD Breunig and V Kolb-Bachofen, *Biochem Biophys Res Commun*, **200**, 1105–10 (1994).
24. LL Pearce, RE Gandley, W Han, K Wasserloos, M Stitt, AJ Kanai, MK McLaughlin, BR Pitt and ES Levitan, *Proc Natl Acad Sci USA*, **97**, 477–82 (2000).
25. K Zangger, G Öz, E Haslinger, O Kunert and IM Armitage, *FASEB J*, **15**, 1303–5 (2001).
26. Y Uchida, Y Ihara and M Tomonaga, *Biochem Biophys Res Commun*, **150**, 1263–67 (1988).
27. Y Uchida and M Tomonaga, *Brain Res*, **481**, 190–93 (1989).
28. CJ Quaife, SD Findley, JC Erickson, GJ Froelick, EJ Kelly, BP Zambrowicz and RD Palmiter, *Biochemistry*, **33**, 7250–59 (1994).
29. M Karin and HR Herschman, *Eur J Biochem*, **107**, 395–401 (1980).
30. I Bremner and BW Young, *Biochem J*, **157**, 517–20 (1976).
31. S Ohi, G Cardenosa, R Pine and PC Huang, *J Biol Chem*, **256**, 2180–84 (1981).
32. KS Squibb, RJ Cousins and SL Feldman, *Biochem J*, **164**, 223–28 (1977).
33. RD Palmiter, *Proc Natl Acad Sci USA*, **95**, 8428–30 (1998).
34. J Liu, Y Liu, SS Habeebu and CD Klaassen, *Toxicol Appl Pharmacol*, **159**, 98–108 (1999).
35. T Abe, O Yamamoto, S Gotoh, Y Yan, N Todaka and K Higashi, *Arch Biochem Biophys*, **382**, 81–88 (2000).
36. CP Yao, JW Allen, LA Mutkus, SB Xu, KH Tan and M Aschner, *Brain Res*, **855**, 32–38 (2000).
37. Y Uchida, K Takio, K Titani, Y Ihara and M Tomonaga, *Neuron*, **7**, 337–47 (1991).
38. PJ Kraulis, *J Appl Crystallogr*, **24**, 946–50 (1991).
39. EA Merritt and DJ Bacon, *Methods Enzymol*, **277**, 505–24 (1997).
40. M Karin and RI Richards, *Nature*, **299**, 797–802 (1982).
41. PE Hunziker, *Methods Enzymol*, **205**, 421–26 (1991).

42 DR Winge, KB Nielson, RD Zeikus and WR Gray, *J Biol Chem*, **259**, 11419–25 (1984).

43 N Glanville, DM Durnam and RD Palmiter, *Nature*, **292**, 267–69 (1981).

44 RD Palmiter, SD Findley, TE Whitmore and DM Durnam, *Proc Natl Acad Sci USA*, **89**, 6333–37 (1992).

45 P Harlow, E Watkins, RD Thornton and M Nemer, *Mol Cell Biol*, **9**, 5445–55 (1989).

46 M Brouwer, J Enghild, T Hoexum-Brouwer, I Thogersen and A Truncali, *Biochem J*, **311**, 617–22 (1995).

47 M Brouwer, DR Winge and WR Gray, *J Inorg Biochem*, **35**, 289–303 (1989).

48 JW Huckle, AP Morby, JS Turner and NJ Robinson, *Mol Microbiol*, **7**, 177–87 (1993).

49 WR Pearson and DJ Lipman, *Proc Natl Acad Sci USA*, **85**, 2444–48 (1988).

50 M Vašak, *Methods Enzymol*, **205**, 41–44 (1991).

51 M Vašak, *Methods Enzymol*, **205**, 39–41 (1991).

52 M Vasak, JH Kagi and HA Hill, *Biochemistry*, **20**, 2852–66 (1981).

53 M Vašak, *Methods Enzymol*, **205**, 452–58 (1991).

54 JC Erickson, AK Sewell, LT Jensen, DR Winge and RD Palmiter, *Brain Res*, **649**, 297–304 (1994).

55 H Li and JD Otvos, *Biochemistry*, **35**, 13929–36 (1996).

56 MJ Stillman, D Thomas, C Trevithick, X Guo and M Siu, *J Inorg Biochem*, **79**, 11–19 (2000).

57 K Zangger, G Öz, JD Otvos and IM Armitage, *Protein Sci*, **8**, 2630–38 (1999).

58 AH Robbins, DE McRee, M Williamson, SA Collett, NH Xuong, WF Furey, BC Wang and CD Stout, *J Mol Biol*, **221**, 1269–93 (1991).

59 A Arseniev, P Schultze, E Wörgötter, W Braun, G Wagner, M Vašak, JH Kägi and K Wüthrich, *J Mol Biol*, **201**, 637–57 (1988).

60 BA Messerle, A Schäffer, M Vašak, JH Kägi and K Wüthrich, *J Mol Biol*, **214**, 765–79 (1990).

61 P Schultze, E Wörgötter, W Braun, G Wagner, M Vašak, JH Kägi and K Wüthrich, *J Mol Biol*, **203**, 251–68 (1988).

62 K Wüthrich, *NMR of Proteins and Nucleic Acids*, Wiley, New York (1986).

63 D Live, IM Armitage, DC Dalgarno and D Cowburn, *J Am Chem Soc*, **107**, 1775–77 (1985).

64 BA Messerle, A Schäffer, M Vašak, JH Kägi and K Wüthrich, *J Mol Biol*, **225**, 433–43 (1992).

65 MH Frey, G Wagner, M Vašak, OW Sørensen, D Neuhaus, E Wörgötter, JHR Kägi, RR Ernst and K Wüthrich, *J Am Chem Soc*, **107**, 6847–51 (1985).

66 AD Watson, CP Rao, JR Dorfman and RH Holm, *Inorg Chem*, **24**, 2820–26 (1985).

67 S Lacelle, WC Steven, DM Kurtz, JW Richardson and RA Jacobson, *Inorg Chem*, **23**, 930–55 (1984).

68 I Bertini, HJ Hartmann, T Klein, G Liu, C Luchinat and U Weser, *Eur J Biochem*, **267**, 1008–18 (2000).

69 H Willner, M Vašak and JH Kägi, *Biochemistry*, **26**, 6287–92 (1987).

70 PM Gehrig, C You, R Dallinger, C Gruber, M Brouwer, JH Kägi and PE Hunziker, *Protein Sci*, **9**, 395–402 (2000).

71 X Yu, M Wojciechowski and C Fenselau, *Anal Chem*, **65**, 1355–59 (1993).

72 KA Melis, DC Carter, CD Stout and DR Winge, *J Biol Chem*, **258**, 6255–57 (1983).

73 R Koradi, M Billeter and K Wüthrich, *J Mol Graphics*, **14**, 51–55 (1996).

74 WF Furey, AH Robbins, LL Clancy, DR Winge, BC Wang and CD Stout, *Science*, **231**, 704–10 (1986).

75 G Wagner, D Neuhaus, E Wörgötter, M Vašak, JH Kägi and K Wüthrich, *J Mol Biol*, **187**, 131–35 (1986).

76 G Öz, K Zangger and IM Armitage, *Biochemistry*, **40**, 11433–41 (2001).

77 AK Sewell, LT Jensen, JC Erickson, RD Palmiter and DR Winge, *Biochemistry*, **34**, 4740–47 (1995).

78 JG Omichinski, GM Clore, O Schaad, G Felsenfeld, C Trainor, E Appella, SJ Stahl and AM Gronenborn, *Science*, **261**, 438–46 (1993).

79 A Velyvis, Y Yang, C Wu and J Qin, *J Biol Chem*, **276**, 4932–39 (2001).

80 K Zangger and IM Armitage, *J Inorg Biochem*, **88**, 135–43 (2002).

81 RR Misra, JF Hochadel, GT Smith, JC Cook, MP Waalkes and DA Wink, *Chem Res Toxicol*, **9**, 326–32 (1996).

Other zinc proteins

Insulin

G David Smith

The Hospital for Sick Children, Toronto, Ontario Hauptman-Woodward Medical Research Institute Inc., Buffalo, NY, USA

FUNCTIONAL CLASS

Protein hormone; stored in the pancreas as a zinc-containing hexamer.

OCCURRENCE

Insulin is found in all vertebrates and is synthesized as a single-chain precursor (preproinsulin) in the β-cells of the pancreas from a single gene.[1] Following the enzymatic removal of an N-terminal signal sequence, proinsulin folds spontaneously and forms the correct disulfide bonds.[2] In humans, proinsulin contains 86 residues: a 30-residue B-chain, a 35-residue connecting peptide (C-peptide), and, finally, a 21-residue A-chain. In the presence of zinc ions, six proinsulin molecules assemble with the connecting peptide on the surface of the hexamer. Owing to the presence of charged residues within the C-peptide segment, proinsulin is quite soluble, but proteolytic cleavage[3] of the C-peptide from the complex results in a zinc–insulin hexamer $[Zn_{1/3}(AB)_2Zn_{1/3}]_3$ with reduced solubility. Insulin is then stored in granules as an insoluble semicrystalline precipitate until required. Following secretion of the hexamers directly into the blood stream where zinc concentrations are very low, the complex dissociates into dimers and finally into the biologically active monomers.

BIOLOGICAL FUNCTION

Unlike most metalloproteins where the metal ion plays a key role in catalysis, the function of zinc when bound to the insulin hexamer is to stabilize the complex. In the absence of zinc, the hexamer freely dissociates into dimers, and thus zinc can be thought of as the molecular glue that

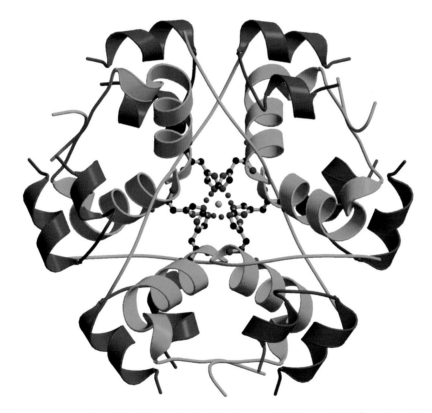

3D Structure   A ribbon representation of the T_6 porcine insulin hexamer, PDB entry 4INS,[4] as viewed down the crystallographic threefold axis. A-chains are colored red, B-chains green, bonds within the HisB10 side chains cyan, water molecules coordinated to zinc red, and the zinc ion gray. Since both zinc ions lie on the threefold axis, only the upper one can be seen in the illustration. Prepared with the programs MOLSCRIPT[5] and RASTER3D.[6]

holds the hexamer together with a zinc binding constant of approximately 10^6 M^{-1}.[7] Following the dissociation of insulin dimers into monomers, the monomer circulates freely in the blood, where it interacts with its receptor, a 480 kDa $(\alpha\beta)_2$ protein.[8] Binding of an insulin monomer to the extracellular domain of its receptor (α-subunit) results in autophosphorylation of specific tyrosine residues within the intracellular domain (β-subunit), initiating a signal transduction cascade that ultimately results in glucose absorption and metabolism.[9]

AMINO ACID SEQUENCE INFORMATION

- *Bos taurus* (bovine), preproinsulin, 105 AA, P01317.[10]
- *Homo sapiens* (human), preproinsulin, 110 AA, P01308.[10]
- *Sus scrofa* (pig), preproinsulin, 108 AA, P01315.[10]

PROTEIN PRODUCTION AND PURIFICATION

Today, nearly all commercial insulins are produced biosynthetically from proinsulin. Insulin was originally extracted and purified from pancreatic tissue, which was minced and extracted with acidified alcohol. Concentration of the extract followed by the addition of sodium chloride at a pH of 2 to 3 produced a salt cake. Following the dissolution of the salt cake in acidified water, insulin will precipitate at a pH of 5 as amorphous particles that can then be isolated and dried.[11]

X-RAY STRUCTURES

Insulin allosterism

The structure of hexameric porcine insulin was first reported by Adams *et al.*[12] and was called *2Zn insulin* since there are two zinc ions bound to each hexamer. An exhaustive description of the 2Zn insulin structure at 1.5 Å resolution and its biological implications were published in 1988.[4] An analysis of the 2Zn-insulin structure showed that a crystallographic threefold axis generates the hexamer from an insulin dimer. The two independent monomers that make up the dimer have nearly identical conformations and are related by a local twofold axis.

Each insulin monomer consists of an A- and B-chain, linked by two interchain disulfide bonds between residues A7–B7 and A20–B19; an additional intrachain disulfide bond exists between residues A6 and A11. The conformation of the A-chain consists of two α-helical strands connected by a central extended segment of polypeptide chain, while the B-chain conformation contains a central α-helical segment flanked on both sides by the polypeptide chain in extended conformations. Dimer stability is enhanced by four hydrogen bonds in a short section of antiparallel β-sheet involving B-chain residues 24–26 from both monomers. The two zinc ions within the hexamer are separated by approximately 16 Å and lie on the crystallographic threefold axis. Each zinc ion is octahedrally coordinated by three symmetry related HisB10 NE2 atoms and by three water molecules.

The structure of the hexamer does explain why insulins lacking histidine at B10 are unable to form hexamers. The structure of hagfish insulin, which, among other substitutions has HisB10 replaced by Asp, has been determined and shows that the hormone crystallizes as a dimer even though the conformation of the monomers are nearly identical to that of porcine insulin.[13] Thus, the presence of histidine at B10 is critical for zinc binding and subsequent hexamer formation. However, the presence of a histidine residue at B10 does not ensure hexamer formation. The removal of the last five residues of the B-chain produces an analogue known as *des-penta insulin*, which exists only as a monomer because of the removal of the dimer-forming residues.[14,15]

In 2Zn-insulin hexamers, the first eight residues of each B-chain are in an extended conformation and stretch from the central portion of the hexamer to the surface near the equator. The addition of 1 M sodium chloride to the crystallizing media induces a conformational change in residues B4–B8 from extended to α-helical in one trimer; this form was named *4Zn insulin* since there are four potential zinc binding sites in each hexamer.[16] The addition of phenol to the crystallizing media generates a hexamer in which the first eight residues in all six B-chains adopt α-helical conformations.[17] Additional spectroscopic[18–21] and crystallographic[22–25] studies have subsequently shown that the insulin hexamer is an allosteric complex. In order to avoid confusion in the naming of these hexamers, a nomenclature has been adopted for which T and R refer to an extended or α-helical conformation in the first eight residues of the B-chain, respectively;[26] R^f refers to an α-helical conformation for residues B4 through B8.[27] Thus, the original 2Zn insulin hexamer is now known as T_6 *insulin*, 4Zn insulin is referred to as $T_3R_3^f$, and the hexamer grown in the presence of phenol is called R_6.

Accompanying the T $\rightarrow$ R transition within the insulin hexamer is a change in the coordination of the zinc ions. In the T-state trimer in T_6 hexamers, each of the two zinc ions lies on the surface on opposite sides of the hexamer and is coordinated by three HisB10 residues and three water molecules. Following the T $\rightarrow$ R transition and the conformational change of the B-chains from extended to α-helical, a narrow channel is produced that extends 12 Å from the surface of the hexamer to the zinc binding HisB10 residues. In this arrangement, the zinc ion adopts tetrahedral coordination since there is no longer space in the narrow channel for octahedral coordination.

At present, there are 34 zinc-containing insulin structures in the Protein Data Bank.[28] One entry was eliminated from this review owing to unreasonably long zinc–ligand distances. Of the remaining 33 entries, 6, 16, and 11 are T_6, $T_3R_3^f$, and R_6 respectively. An additional structure of bovine insulin[29] not yet in the database has been included by the author.

T_6 structures

Crystallization T_6 insulin hexamers are typically grown from a solution consisting of 5 mg mL^{-1} insulin, 0.01 M HCl, 0.007 M zinc acetate, 0.05 M sodium citrate, and 17% acetone. The components are added in the order given, the pH is raised to approximately 8.5 by the addition of sodium hydroxide to ensure complete dissolution, and finally, the solution is back-titrated with hydrochloric acid to a pH of approximately 6.3 or until the solution becomes slightly turbid.[11] The solution is then warmed to 50 °C, placed in a dewar, and allowed to cool slowly to room temperature over a period of three to four days. Colorless rhomboid-shaped crystals of approximately 0.4 to 0.6 mm on an edge typically result. The species (mutation), PDB codes, space group, and resolution of the data for the T_6 hexameric insulin structures are listed in Table 1.

Description of structure As described above, the two crystallographically independent monomers are related by a local twofold axis, and the hexamer is generated by the crystallographic threefold axis. Thus, the T_6 hexamer has the approximate symmetry of point group D_3. The A-chain conformation consists of a central extended segment flanked by two α-helical segments. Residues 1–8 of the B-chain adopt an extended conformation followed by a continuous α-helix from B9 to B19. The transition from an extended conformation to that of an α-helical one takes place at GlyB8 where this residue adopts the conformation of a D-amino acid

Table 1 T_6 hexameric insulin structures

Species (mutation), PDB code	Space group	Resolution (Å)
Human; 1MSO[30]	R3	1.00
Porcine; 4INS[4]	R3	1.50
Porcine; 3INS[31]	R3	1.50[a]
Des-(B28-B30)[b]; 1HTV[32]	P2$_1$2$_1$2$_1$	1.90
Porcine (B13E → Q); 1IZB[33]	R3	2.00
Bovine[29]	R3	2.25
Des-(B1) Bovine; 2INS[34]	R3	2.50

[a] Joint refinement using 1.50 Å X-ray and 2.20 Å neutron data.
[b] In the absence of residue B30, it is not possible to distinguish between human and porcine insulin.

with φ/ψ torsion angles of approximately +55°/−135°. The C-terminus of the B-chain (B21–B30) is in an extended β-strand conformation, and the formation of an antiparallel β-pleated sheet with its counterpart in an adjacent monomer produces the dimer. In this arrangement, three symmetry related HisB10 residues lie near the central crystallographic threefold axis with PheB1 lying near the surface of the hexamer at the dimer–dimer interface. At 120 K the N-terminus of one B-chain undergoes a conformational change that displaces PheB1 by 7.9 Å,[30] but this displacement has no effect upon the binding of zinc ion. The hexameric T_6 insulin structure is illustrated in the 3D Structure.

Zinc geometry In space group R3, each zinc ion is coordinated to an insulin trimer that is generated from one monomer by the crystallographic threefold axis. In the T_6 hexamer, each zinc ion is octahedrally coordinated by NE2 of three symmetry related HisB10 residues and by three symmetry related water molecules shown in Figure 1. Owing to the threefold symmetry in space group R3, there is only one independent Zn–NE2 bond distance and one independent Zn–O bond distance for each zinc ion. In entry 1HTV, des-(B28–B30), the space group is P2$_1$2$_1$2$_1$ and the asymmetric unit consists of a complete hexamer; as a result of the truncation of the C-terminus of the B-chains, monomer–monomer interactions are weakened and therefore only a single zinc ion is bound to the hexamer. In the seven T_6 hexamer entries in the PDB, Zn–NE2 distances range from 1.94 to 2.29 Å with a mean Zn–NE2 distance of 2.12 (σ = 0.08) Å. In one entry (1IZB), a zinc–oxygen distance is unreasonably long (2.68 Å) and no water is associated with the second zinc ion. In a second entry (1HTV), zinc–oxygen distances range from 2.58 to 2.93 Å. Excluding these Zn–O distances of the above two entries from the calculation of the mean results in a mean Zn–O distance of 2.25 (σ = 0.07) Å.

$T_3R_3^f$ structures

Crystallization Crystals of this hexameric form of insulin can be grown under several sets of conditions. The first structure of $T_3R_3^f$ porcine insulin was determined from crystals grown using the T_6 insulin recipe in which the crystallizing media was supplemented with 1 M NaCl.[11,16] Other salts can also be used to effect the T → R transition, the most potent being potassium thiocyanate. Subsequent crystallization experiments have shown that the minimum potassium thiocyanate or sodium chloride concentrations necessary to produce $T_3R_3^f$ crystals is 0.02 M or 0.4 M respectively.[35] The structures of both human and porcine $T_3R_3^f$ hexameric insulin have been reported for crystals grown in the presence of 0.75 M NaCl[22] and

Insulin

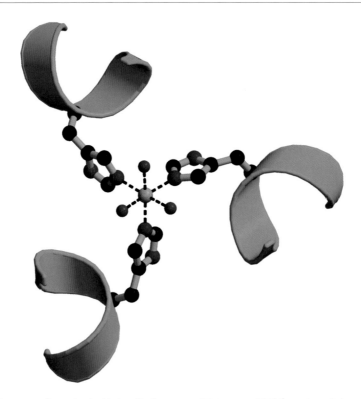

Figure 1 Octahedral coordination of zinc in the T_6 insulin hexamer, PDB entry 4INS,[4] as viewed down the crystallographic threefold axis. The B-chain segments are colored green, coordinating water molecules red, the zinc ion gray, and the bonds of the HisB10 side chains cyan. Prepared with the programs MOLSCRIPT[5] and RASTER3D.[6]

0.211 M KSCN[24] respectively. The addition of phenolic derivatives such as methylparaben, p-hydroxybenzamide, or 4′-hydroxyacetanalide to the $T_3R_3^f$ crystallization recipe also yields crystals in which three molecules of the phenolic derivative are bound to each insulin hexamer.[23,24,36] In the presence of very low concentrations of phenol (5 mM), $T_3R_3^f$ crystals are obtained[24] rather than R_6. $T_3R_3^f$ hexameric crystals of (ProB28 → Lys; LysB28 → Pro) human insulin were obtained in the presence of phenol but with no added sodium chloride.[27] Although the addition of salts can be used to grow $T_3R_3^f$ insulin crystals, higher concentrations of salt alone are unable to drive the T → R transition to completion to produce an R_6 hexamer. $T_3R_3^f$ hexameric structures retrieved from the PDB[28] are listed in Table 2.

Description of structures The T_3 trimer is constructed from three T-state monomers related by the crystallographic threefold axis and is similar to the T-state trimers observed in T_6 hexamers. The second trimer consists of three crystallographically related R^f-state monomers in which the B1–B3 segment is in an extended conformation, while B4–B19 is a continuous α-helix. The $T_3R_3^f$ insulin hexamer is illustrated in Figure 2. As a result of the change in conformation of three of the N-termini of the B-chains, the point group symmetry of the hexamer is reduced from D_3 to C_3. The transformation from T to R^f

Table 2 $T_3R_3^f$ hexameric insulin structures

Species (mutation), PDB code	Space group	Resolution (Å)
Human; 1G7A[37]	R3	1.20
Human; 1G7B[37]	R3	1.30
Human; 1BEN[36]	R3	1.40
Porcine; 1ZNI[16]	R3	1.50
Human; 1TRZ[22]	R3	1.60
Porcine; 2TCI[24]	R3	1.80
Human; 1TYL[23]	R3	1.90
Human; 1TYM[23]	R3	1.90
Porcine; 3MTH[24]	R3	1.90
Human (A8T → Diaminobutyric acid); 1J73[38]	R3	2.00
KB29-GA1 Cross-link[a]; 6INS[39]	R3	2.00
Human (B28P → K; B29K → P); 1LPH[27]	R3	2.30
Porcine; 1MPJ[24]	R3	2.30
Human (B5H → Y); 1QJO[40]	R3	2.40
Human (A8T → K); 1JCA[38]	R3	2.50
Human (A2I → Alloisoleucine); 1LW8[41]	R3	2.50

[a] In the absence of residue B30, it is not possible to distinguish between human and porcine insulin.

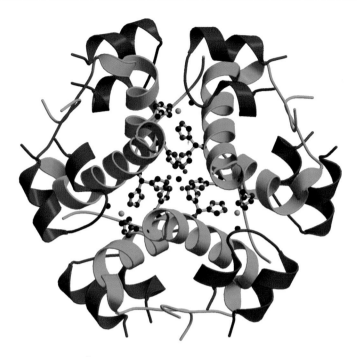

Figure 2 Ribbon representation of the $T_3R_3^f$ insulin hexamer, PDB entry 1ZNI,[16] viewed along the crystallographic threefold axis toward the R-state trimer. A-chains are colored red and B-chains are colored green. Histidine side chains are drawn with cyan colored bonds. The HisB10 residue is disordered and can form bonds to either the zinc ion (gray) on the threefold axis or to the zinc ions (gray) in the off-axial sites. Water molecules coordinating the T-state zinc ion are colored red. The chloride ion, which obscures both the T- and R-state zinc ions, is colored dark gray. Prepared with the programs MOLSCRIPT[5] and RASTER3D.[6]

is accompanied by a conformational change of GlyB8 from a 'D' conformation to an α-helical one ($\phi/\psi = -67°/-36°$). Owing to the displacement of the N-termini of the three B-chains, three elliptical cavities are produced between symmetry related R^f-state monomers in the vicinity of the displaced T-state LeuB6 side chain. Depending upon the crystallization conditions, this cavity can contain water, an off-axial zinc ion, or a phenolic derivative.

Frozen $T_3R_3^f$ insulin crystals exhibit a phase change in which the c-axis of the rhombohedral unit cell is doubled[37,42] because of a rotation of one dimer by approximately 9° relative to a second c-translationally related dimer. In the first hexamer, a zinc ion and two chloride ions are found to occupy the off-axial binding site, but a disordered glycerol molecule and several water molecules are found in this site in the second hexamer. Two additional and partially occupied zinc ions are observed at the interface between independent dimers.

Zinc geometries The results from early studies on $T_3R_3^f$ porcine insulin showed that the side chain of the R^f-state HisB10 adopts two discrete conformations.[16,37] In approximately half of the hexamers, the HisB10 side chain is directed toward the crystallographic threefold axis, while in the remainder of the hexamers this side chain is directed toward the off-axial binding site. Within this off-axial binding site, a zinc ion is tetrahedrally coordinated by two chloride ions, NE2 from HisB5 of an adjacent R^f-state monomer, and NE2 of the second orientation of the HisB10 side chain. In the other half of the hexamers in the crystal, a zinc ion lies on the crystallographic threefold axis where it is tetrahedrally coordinated by three symmetry related HisB10 side chains in the second alternate orientation and a chloride ion that also lies on the threefold axis. Owing to the crystallographic threefold axis, any given R_3^f trimer binds either a single axial zinc ion or three off-axial zinc ions. Mixed TR^f trimers have never been observed, a result of severe steric clashes between the N-termini of T- and R^f-state B-chains. The coordination of the axial and off-axial zinc ions is illustrated in Figure 3.

In subsequent studies (2TCI and 1TRZ), the HisB10 residue is not disordered and the off-axial sites contained only water molecules.[22,24] The off-axial site in those structures that were crystallized in the presence of either phenol, methylparaben, p-hydroxybenzamide, or 4'-hydroxyacetanalide is occupied by the phenolic compound in which the phenolic hydroxyl group forms hydrogen bonds to the carbonyl oxygen atom of CysA6 and the amide nitrogen of CysA11 of the R^f-state monomer.[43]

The binding of zinc ion in T-state trimers is similar to that in the T_6 structures. However, in four of the $T_3R_3^f$ structures that have been deposited in the PDB (1TRZ, 1G7A, 1G7B, 2TCI), the zinc ion coordinated to the T-state trimer is found to have dual coordination:[22,24,37] half of the hexamers contain a zinc ion with water molecules completing the octahedral coordination sphere, while in

Insulin

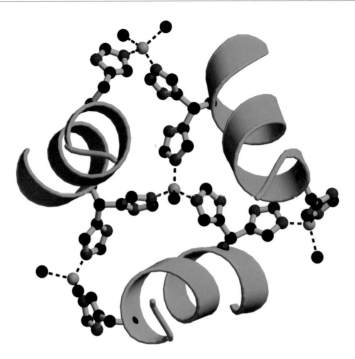

Figure 3 A close-up view of the zinc binding sites in the R^f-state trimer in PDB entry, 1ZNI,[16] with the crystallographic threefold axis rotated 15° from perpendicular to the plane of the figure. The ribbons representing residues B3–B12 are colored green, bonds within the imidazole ring cyan, zinc ions gray, and chloride ions dark gray. Prepared with the programs MOLSCRIPT[5] and RASTER3D.[6]

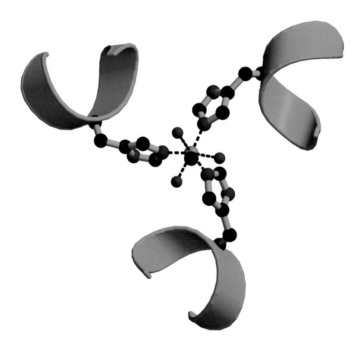

Figure 4 A close-up view of the dual coordination in the T-state trimer in PDB entry 1TRZ,[22] with the crystallographic threefold axis rotated 18° from the perpendicular to the plane of the figure. B-chain segments are colored green, bonds within the imidazole ring are colored cyan, zinc ions gray, and chloride ions dark gray. One-half of the T-state zinc ions within the crystal adopt octahedral coordination, while the other half adopts tetrahedral coordination. Prepared with the programs MOLSCRIPT[5] and RASTER3D.[6]

the remaining hexamers the coordination is tetrahedral with the chloride ion acting as the fourth ligand. The dual coordination of zinc in these four structures is illustrated in Figure 4.

The collection of $T_3R_3^f$ hexameric structures in the PDB presents somewhat of a problem in attempting to calculate mean distances and standard deviations. PDB entries were not included in the calculation of mean values for those structures in which the coordination sphere of the zinc ion is incomplete or in which dual coordination exists around a zinc ion. In cases in which there is dual coordination, the refined position of the zinc represents an average position from both tetrahedral and octahedral geometry and will make Zn–NE2 bond distances an average of the two geometries.

There are only three examples of octahedral geometry around zinc in the T-state trimer, but one structure (1TYL) has Zn–NE2 and Zn–O bonds that are considerably longer than expected. The mean Zn–NE2 and Zn–O distances in the remaining two structures (1BEN and 1ZNI) are 2.08 and 2.28 Å.

A thiocyanate ion completes the tetrahedral coordination sphere of an R^f-state zinc ion in one entry in the PDB (2TCI), but the Zn–N distance is somewhat shorter (1.69 Å) than would be expected for a Zn–N bond.

There are a total of three structures in the PDB that have zinc in the off-axial binding site, 1G7A, 1G7B, and 1ZNI. For the off-axial zinc ions, the mean Zn–NE2 bond distance was calculated to be 2.00 ($\sigma = 0.08$) Å and the mean Zn–Cl distance was 2.25 ($\sigma = 0.03$) Å. The tetrahedral geometry around an axial zinc ion [Zn(NE2)$_3$Cl] gives mean Zn–NE2 and Zn–Cl distances for all $T_3R_3^f$ entries of 2.03 ($\sigma = 0.06$) and 2.23 ($\sigma = 0.10$) Å respectively. In the cryofrozen $T_3R_3^f$ crystals, two additional zinc ions with partial occupancies are observed between the two independent dimers, as illustrated in Figure 5. The first zinc ion is tetrahedrally coordinated by NE2 of one of the orientations of a disordered T-state HisB5 residue (mean distance, 1.92 Å), NE2 of an R^f-state HisB5 (mean distance, 1.88 Å), and two water molecules (mean distances, 2.33, 2.48 Å). The second zinc ion is coordinated to ND1 of the second orientation of the disordered T-state HisB5 side chain (mean distance, 1.85 Å) and three water molecules (mean distances, 1.79, 2.16, and 2.16 Å). Because of the severe disorder within these two sites, bond distances involving the two zinc ions are perturbed from their expected values.

R_6 structures

Crystallization To date, crystals containing R_6 insulin hexamers have only been obtained when phenol, *m*-cresol, or resorcinol are present in the crystallizing media. The crystallization of R_6 hexamers follows a similar procedure as described for the $T_3R_3^f$ hexamers, except that the crystallizing media contains phenol, *m*-cresol, or resorcinol at a concentration of 50 to 100 mM. To facilitate dissolution, the phenolic compounds are usually dissolved in either acetone or ethanol. In the presence of 1 M sodium chloride and at a pH of 6.7, a mixture of rhombohedral and monoclinic crystals are obtained; at a pH of 8.5, only rhombohedral crystals are produced. In the absence of sodium chloride and at a pH of 6.5, only monoclinic crystals are obtained.[44] Table 3 lists the relevant R_6 hexameric insulin structures.

Table 3 R_6 hexameric insulin structures

Species (mutation), PDB code	Space group	Resolution (Å)
Human (B28P → D); 1ZEH[45]	R3	1.50
Human (B28P → D); 1ZEG[45]	R3	1.60
Human; 1EV3[44]	R3	1.78
KB29-tetradecanoyl, des-(B30)[a]; 1XDA[46]	R3	1.80
Human; 1EV6[44]	P2$_1$	1.90
Human; 1EVR[44]	P2$_1$	1.90
Human (B28P → D; TB30-GA1 cross-link); 1ZEI[45]	P2$_1$	1.90
Human (B5H → Y); 1QIZ[40]	P2$_1$	2.00
Porcine; 7INS[47]	P4$_3$2$_1$2	2.00
Porcine; 1ZNJ[17]	P2$_1$	2.00
Human (B5H → Y); 1QIY[40]	P2$_1$	2.30

[a] In the absence of residue B30, it is not possible to distinguish between human and porcine insulin.

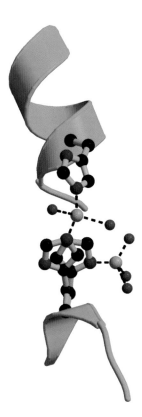

Figure 5 The interdimer zinc binding sites in PDB entries, 1G7A and 1G7B,[37] as viewed nearly perpendicular to the crystallographic threefold axis. B-chain segments are colored green, zinc ions gray, coordinating water molecules red, and bonds within the imidazole rings cyan. The lower disordered histidine residue is His5B of a T-state monomer, while the upper histidine side chain is His5B of an R^f-state monomer. Prepared with the programs MOLSCRIPT[5] and RASTER3D.[6]

Insulin

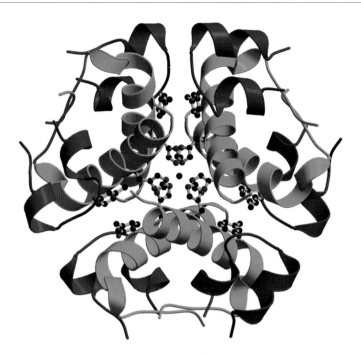

Figure 6 Schematic drawing of the R_6 hexamer of PDB entry 1EV6[44] as viewed along the crystallographic threefold axis. A-chains are colored red, B-chains are green, bonds within the histidine imidazole rings are colored cyan, chloride ions are dark gray, and *m*-cresol molecules are yellow. Note that the *m*-cresol molecules occupy nearly the same position as do the off-axial zinc ions illustrated in Figure 2. Prepared with the programs MOLSCRIPT[5] and RASTER3D.[6]

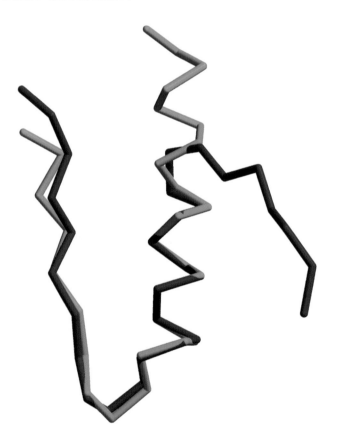

Figure 7 Cα trace illustrating the differences in conformation between the T-state (red) and R-state (green) B-chains. The B-chain of monomer 1 of PDB entry 4INS was superimposed upon monomer 1 of PDB entry 1EV6, minimizing the displacements of main chain atoms in residues B11 through B19. Following the minimization, the mean displacement was calculated to be 0.170 Å. Prepared with the programs MOLSCRIPT[5] and RASTER3D.[6]

Description of structures Owing to the conformational transition from T to R, the N-termini of all six B-chains exist in an α-helical conformation, producing a continuous α-helix from PheB1 to CysB19. The R_6 hexamer is illustrated in Figure 6. With the exception of PDB entry 7INS, both zinc ions in each hexamer are tetrahedrally coordinated by three NE2 HisB10 atoms and one chloride ion; in 7INS, a water molecule is the fourth ligand. In all cases, the off-axial site is occupied by the phenolic derivative where the phenolic hydroxyl group forms hydrogen bonds to the carbonyl oxygen of CysA6 and the nitrogen of CysA11. As was the case in the T_6 hexamers, the conformations of all six monomers in the R_6 hexamer have nearly identical conformations, and pairs of monomers are related by a local twofold axis, resulting again in the symmetry of point group D_3. In the rhombohedral space group R3, the hexamer is perfectly threefold symmetric and the asymmetric unit consists of an insulin dimer; in the monoclinic cases (space group $P2_1$), the hexamer is pseudo-threefold symmetric as the entire hexamer is contained in the asymmetric unit; in the tetragonal case (PDB entry 7INS, space group $P4_32_12$), the hexamer is generated by the action of a crystallographic twofold axis on one R_3 trimer. Packing of the hexamers within the unit cell is quite different. In the rhombohedral space group R3, the threefold axis of all hexamers are parallel, but in the monoclinic space group $P2_1$, the local threefold axis of one hexamer is nearly perpendicular to that of an adjacent hexamer.

Zinc geometries Differences in conformation between the T- and R-state B-chains, illustrated in Figure 7, result in a displacement of the PheB1 residue of approximately 30 Å. Because of the transition from an extended to an α-helical conformation in all six B-chain N-termini, the zinc ion can only assume tetrahedral coordination as shown in Figure 8. As in the R^f-state trimers in the $T_3R_3^f$ hexamers, each zinc ion is coordinated by three HisB10 NE2 atoms and either a chloride ion or a water molecule. There are a total of 10 R_6 insulin hexamer entries in the PDB in which a chloride ion completes the tetrahedral coordination sphere of zinc. Zn–NE2 distances range from 1.64 to 2.21 Å with a mean value of 2.01 (σ = 0.10) Å, while Zn–Cl distances range from 2.14 to 2.39 Å with a mean of 2.23 (σ = 0.07) Å. Entry 7INS crystallizes in a tetragonal space group and has a single R-state trimer in the asymmetric unit. In this structure, the three independent Zn–NE2 bond distances range from 1.74 to 2.18 Å (mean value of 1.96 Å) and the Zn–O bond distance is 2.16 Å.

In a 2.5-Å resolution structure of a rhombohedral R_6 hexameric structure[25] complexed with phenol, the electron density in the vicinity of one of the zinc ions suggested the presence of a threefold disordered phenol molecule

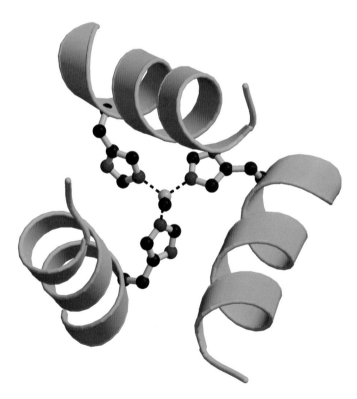

Figure 8 Close-up view of the tetrahedral coordination of zinc in one of the R-state trimers in the R_6 insulin hexamer (1EV6[44]), viewed nearly along the local threefold axis. B-chain segments are illustrated from PheB1 to ValB12 and are colored green. The bonds within the imidazole rings of HisB10 are colored cyan, the zinc ion gray, and the chloride ion dark gray. Prepared with the programs MOLSCRIPT[5] and RASTER3D.[6]

coordinated to the zinc ion. No such density was present near the other zinc ion. Although no attempt was made to refine the position of the phenol, the electron density and the diameter of the channel was consistent with the presence of a phenol molecule with its hydroxyl group coordinated to the zinc ion at a distance of approximately 2.10 Å.

Summary

Mean $Zn-NE2$ and $Zn-Cl$ bond distances and their standard deviations are summarized in Table 4. The standard deviations are for the most part large and would at first suggest that differences in bond distances between tetrahedral and octahedral geometry are not significant. A major contributing factor to the relatively large standard deviations is the fact that only three structures were determined at a resolution of 1.2 Å or better, and the average resolution of the 34 structures is approximately 1.9 Å. The resolution of none of the structures exceeds 1.0 Å. Other contributors to the lack of precision in the bond distances include the method of refinement, the temperature of the experiment, disorder, and the care taken in performing the refinements. In spite of the lack of significance, a trend does exist in the bond distances. For octahedral geometry, the $Zn-NE2$ bond distances are approximately 2.10 Å, while for tetrahedral geometry these distances are 2.00 Å. In contrast, $Zn-Cl$ distances are approximately 2.25 Å, regardless of the number of ligands to zinc.

Table 4 Mean zinc–ligand bond distances and standard deviations

Bond type	Distance (Å)	Standard deviation (Å)
T_6 hexamers, octahedral		
Zn–NE2	2.12	0.08
Zn–O	2.25	0.07
$T_3R_3^f$ T-state, octahedral		
Zn–NE2	2.08	a
Zn–Cl	2.28	a
$T_3R_3^f$ R^f-state off-axial, tetrahedral		
Zn–NE2	2.00	0.08
sZn–Cl	2.25	0.03
$T_3R_3^f$ R^f-state axial, tetrahedral		
Zn–NE2	2.03	0.06
Zn–Cl	2.23	0.10
R_6 hexamers, tetrahedral		
Zn–NE2	2.01	0.10
Zn–Cl	2.23	0.07

[a] Standard deviations are not reported since the mean was calculated from only two observations.

ACKNOWLEDGEMENT

This work was supported in part by NIH grant GM56829.

REFERENCES

1 SJ Chan, P Keim and DF Steiner, *Proc Natl Acad Sci USA*, **73**, 1964–68 (1976).

2 DF Steiner and JL Clark, *Proc Natl Acad Sci USA*, **57**, 473–80 (1968).

3 HW Davidson, CJ Rhodes and JC Hutton, *Nature*, **333**, 93–96 (1988).

4 EN Baker, TL Blundell, JF Cutfield, SM Cutfield, EJ Dodson, GG Dodson, DC Hodgkin, RE Hubbard, NW Isaacs, CD Reynolds, K Sakabe, N Sakabe and NM Vijayan, *Philos Trans R Soc London, Ser B*, **319**, 369–456 (1988).

5 P Kraulis, *J Appl Crystallogr*, **24**, 946–50 (1991).

6 EA Merritt and MEP Murphy, *Acta Crystallogr*, **D50**, 869–73 (1994).

7 J Goldman and F Carpenter, *Biochemistry*, **13**, 4566–74 (1976).

8 A Ullrich and J Schlessinger, *Cell*, **61**, 203–12 (1990).

9 MR White and R Kahn, *J Biol Chem*, **269**, 1–4 (1994).

10 The ExPASY Proteome WWW Server, Swiss Institute of Bioinformatics, 1, Rue Michel Servet, 1211 Geneva 4, Switzerland, http://www.expasy.ch.

11 J Schlichtkrull, *Insulin Crystals*, Ejnar Munksgaard, Munksgaard, Copenhagen (1958).

12 MJ Adams, TL Blundell, EJ Dodson, GG Dodson, M Vijayan, EN Baker, MM Harding, DC Hodgkin, B Rimmer and S Sheat, *Nature*, **224**, 491–95 (1969).

13 JF Cutfield, SM Cutfield, EJ Dodson, GG Dodson, SF Emdin and CD Reynolds, *J Mol Biol*, **132**, 85–100 (1979).

14 RC Bi, Z Dauter, E Dodson, G Dodson, F Gordiano, R Hubbard and C Reynolds, *Biopolymers*, **23**, 391–95 (1984).

15 J-C Diao, Z-L Wan, W-R Chang and D-C Liang, *Acta Crystallogr*, **D53**, 507–12 (1997).

16 G Bentley, E Dodson, G Dodson, D Hodgkin and D Mercola, *Nature*, **261**, 166–68 (1976).

17 U Derewenda, Z Derewenda, EJ Dodson, GG Dodson, C Reynolds, K Sparks, GD Smith and DC Swenson, *Nature*, **338**, 594–96 (1989).

18 P Krüger, G Gilge, Y Çabuk and A Wollmer, *Biol Chem Hoppe-Seyler*, **371**, 669–73 (1990).

19 WE Choi, ML Brader, V Aguilar, NC Kaarsholm, MF Dunn, *Biochemistry*, **32**, 11638–45 (1993).

20 PS Brović, WE Choi, D Borchardt, NC Kaarsholm, MF Dunn, *Biochemistry*, **33**, 13057–69 (1994).

21 CR Bloom, WE Choi, PS Brović, JJ Huang, NC Kaarsholm and MF Dunn, *J Mol Biol*, **245**, 324–330 (1995).

22 E Ciszak and GD Smith, *Biochemistry*, **33**, 1512–17 (1994).

23 GD Smith and E Ciszak, *Proc Natl Acad Sci USA*, **91**, 8851–55 (1994).

24 JL Whittingham, S Chaudhuri, E Dodson, PCE Moody and GG Dodson, *Biochemistry*, **34**, 15553–63 (1995).

25 GD Smith and GG Dodson, *Proteins: Struct, Funct, Genet*, **14**, 401–8 (1992).

26 NC Kaarsholm, H Ko and MF Dunn, *Biochemistry*, **28**, 4427–35 (1989).

27 E Ciszak, JM Beals, BH Frank, JC Baker, ND Carter and GD Smith, *Structure*, **3**, 615–22 (1995).

28. HM Berman, J Westbrook, Z Feng, G Gilliland, TN Bhat, H Weissig, IN Shindyalov and PE Bourne, *Nucleic Acids Res*, **28**, 235–42 (2000).
29. GD Smith, WA Pangborn and RH Blessing, unpublished results.
30. GD Smith, WA Pangborn and RH Blessing, *Acta Crystallogr*, **D59**, 474–82 (2003).
31. A Wlodawer, H Savage and GG Dodson, *Acta Crystallogr*, **B45**, 99–107 (1989).
32. J Ye, W Chang and D Liang, *Biochim Biophys Acta*, **1547**, 18–25 (2001).
33. GA Bentley, J Brange, Z Derewenda, EJ Dodson, GG Dodson, J Markussen, AJ Wilkinson, A Wollmer and B Xiao, *J Mol Biol*, **228**, 1163–76 (1992).
34. GD Smith, WL Duax, EJ Dodson, GG Dodson, RAG DeGraaf and CD Reynolds, *Acta Crystallogr*, **B38**, 3028–32 (1982).
35. RAG DeGraaf, A Lewit-Bentley and SP Tolley, in G Dodson, JP Glusker and D Sayre (eds.), *Structural Studies on Molecules of Biological Interest: A Volume in Honour of Professor Dorothy Hodgkin*, Clarendon Press, Oxford, pp 547–56 (1981).
36. GD Smith, E Ciszak and WA Pangborn, *Protein Sci*, **5**, 1502–11 (1996).
37. GD Smith, WA Pangborn and RH Blessing, *Acta Crystallogr*, **D57**, 1091–1100 (2001).
38. MA Weiss, Z Wan, M Zhao, Y-C Chu, SH Nakagawa, GH Burke, W Jia, R Hellmich and P Katsoyannis, *J Mol Biol*, **315**, 103–11 (2002).
39. U Derewenda, Z Derewenda, EJ Dodson, GG Dodson, X Bing, *J Mol Biol*, **220**, 425–33 (1991).
40. L Tang, JL Whittingham, CS Verma, LSD Caves and GG Dodson, *Biochemistry*, **38**, 12041–51 (1999).
41. Z-L Wan, B Xu, C Chu, P Katsoyannis and MA Weiss, unpublished results.
42. RB Von Dreele, PW Stephens, GD Smith, and RH Blessing, *Acta Crystallogr*, **D56**, 1549–53 (2000).
43. GD Smith, *J Mol Struct*, **469**, 71–80 (1998).
44. GD Smith, E Ciszak, LA Magrum, WA Pangborn and RH Blessing, *Acta Crystallogr*, **D56**, 1541–48 (2000).
45. JL Whittingham, DJ Edwards, AA Antson, JM Clarkson and GG Dodson, *Biochemistry*, **37**, 11516–23 (1998).
46. JL Whittingham, S Havelund and I Jonassen, *Biochemistry*, **36**, 2826–31 (1997).
47. P Balschmidt, FB Hansen, EJ Dodson, GG Dodson and F Korber, *Acta Crystallogr*, **B47**, 975–86 (1991).

LIM domain proteins

Georg Kontaxis[†], Klaus Bister[‡] and Robert Konrat[†]

[†]Institute of Theoretical Chemistry and Molecular Structural Biology, University of Vienna, Austria
[‡]Institute of Biochemistry, University of Innsbruck, Austria

FUNCTIONAL CLASS

Protein-binding module; the LIM domain proteins do not form a functional protein family but rather represent a variety of dissimilar proteins, which share a common sequence motif. Although originally recognized as a distinct protein motif within LIM-homeodomain (LIM-HD) proteins, the LIM protein family now comprises LIM proteins that lack homeodomains (HD), either being composed almost entirely of LIM domains or containing additional functional domains of various types (e.g. kinase domains). The LIM domain contains a cysteine-rich motif that was first identified in the protein products of three genes: *lin-11* and *mec-3* from *Caenorhabditis elegans* and *ISL1* from rat.[1,2] The LIM-HD proteins are encoded by a subfamily of homeobox genes in which the various homeodomains show a high degree of similarity. Homeodomains found in LIM-HD proteins to date are POU domain (named after the proteins Pit, Oct, and Unc), paired homeobox (PAX), and homeobox (HOX) homeodomain proteins.[3] Another distinct group of LIM domain proteins are the LIM-only proteins (LMO), which can be further subdivided into predominantly nuclear LIM-only

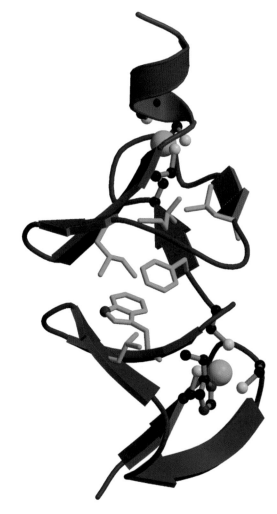

3D Structure Schematic representation of the solution structure of the LIM domain protein quail CRP2(LIM2) also showing the zinc ions, PDB code 1QLI. Prepared with the program MOLSCRIPT.[49]

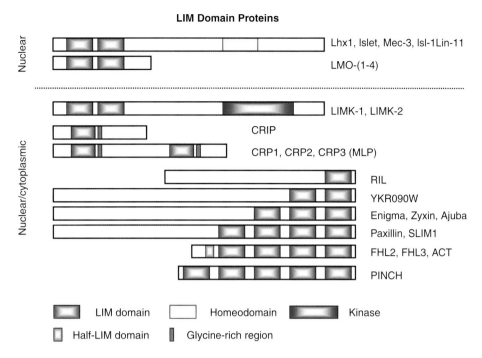

Figure 1 The topology of LIM domain proteins. LIM proteins exist as LIM-only proteins (LMO, CRP, FHL subfamilies) or contain additional functional domains (DNA-binding homeodomains or kinase domains).

proteins (LMO; subfamily members LMO1–LMO4) and LIM-only proteins that display a dual cellular localization behavior (e.g. cysteine-rich proteins (CRP), particularly interesting new cysteine–histidine protein (PINCH), Zyxin, Ajuba, lipoma preferred partner (LPP), Trip6, and Four-and-a-Half LIM domain (FHL) proteins). In addition to their LIM domains, family members of the latter group sometimes carry additional functional protein domains at their N-termini.[3] Finally, LIM domains are also fused to protein kinase domains (LIMK) and additional distinct protein motifs (e.g. PDZ domains, which are modular 70–90 amino acid domains that are named after the three proteins in which this domain was first described, namely, PSD-95, Dlg, and ZO-1 (PDZ)). These proteins also show a dual cellular localization and comprise two tandemly arrayed LIM domains analogous to LMO proteins (LMO1–4) and LIM-HD proteins respectively. An overview of the domain topologies of LIM domain proteins is given in Figure 1.

OCCURRENCE

Occur in eukaryotes; they are predominantly found in humans and animals, but have also been identified in yeast and plants. At the cellular level, LIM domain proteins occur both in the cytoplasm and in the nucleus.

BIOLOGICAL FUNCTION

LIM domain proteins are implicated in fundamental developmental processes, and the functional hallmark of all LIM domains is presumably the mediation of protein–protein interactions. LIM domain–containing transcription factors (LIM-HD proteins) are involved in general cell fate determination, for example, tissue patterning and differentiation, and, to date, substantial experimental evidence has been provided for their prominent involvement in neural patterning.[3] The disruption of several LIM-HD genes has demonstrated their importance in the development of neuronal lineages. The presence of the LIM domain distinguishes the LIM-HD proteins from other homeodomain proteins, which lack an additional protein interaction module (apart from the HOX cluster proteins). LIM-HD proteins interact with a diverse set of other transcription factors (see the section on Protein–protein Interactions). Thus, by means of their LIM domain protein-binding capacities, LIM-HD proteins have acquired a unique potential for regulating transcriptional activities in a highly sophisticated manner. For example, it was recently shown that the LIM-HD encoding genes *lim3* and *islet* constitute a combinatorial code that generates distinct motorneuron identities.[4] The functional hallmark of LIM-HD proteins thus is the formation of multiorder, homomeric or heteromeric, transcriptional regulator complexes that allow for tissue-specific control of developmental processes.[3]

LIM-only proteins also have important roles in development and in the regulation of cell differentiation and proliferation. Nuclear LMOs with their tandemly arrayed LIM domains act as adaptor molecules arranging functional transcription factor complexes. Furthermore, LMOs have also been shown to be involved in oncogenesis.[5] The nuclear LIM-only proteins LMO-1 and LMO-2 were identified on the basis of their genomic localization at translocation breakpoints in T-cell leukemia and on the abnormal expression of LMO-1 and LMO-2 in thymus and T cells.[6] LMO-2 is essential for embryonic hematopoiesis.[7] LMO-2 was also demonstrated to be an obligatory regulator of neo-vascularization of tumors and therefore proposed to constitute a suitable drug target in cancer therapy.[8] While little is known about LMO-3, which was discovered solely on the basis of sequence similarity, LMO-4 is a negative regulator of mammary epithelial development and is involved in breast oncogenesis.[9] However, the molecular mechanism by which LMO-4 exerts its oncogenic function is still elusive and awaits the identification of the target genes for LMO-4-containing transcriptional complexes.

The primarily cytoplasmic CRP proteins (CRP1, CRP2, CRP3) are associated with actin filaments and myofibers. All three isoforms are capable of interacting with the cytoskeletal proteins α-actinin and zyxin *in vitro*.[10] Owing to their expression patterns, the three family members may perform similar functions by contributing to cell differentiation via effects on cytoarchitecture. For the CRP family member CRP3 (also called muscle LIM protein (MLP), a nuclear localization has been observed, and a recent report links CRP3 to an enhanced MyoD activity, an essential muscle-specific bHLH transcription factor.[11]

The Zyxin subfamily of LIM proteins (e.g. zyxin, LPP, Trip6, Ajuba) is characterized by the additional presence of proline-rich regions. While the cellular function is largely unknown, a common feature of these family members is the localization to focal adhesions, to the actin cytoskeleton, and to cell–cell contact sites in epithelial cells.[12–14] In fibroblasts, Zyxin and Trip6 were found to affect cell motility.[15,16] Interestingly, they contain nuclear export signals, which allow for shuttling between the nucleus and the cytoplasm.[13,14] While the functional role of nuclear localization of Zyxin and Trip6 is unknown, Ajuba plays a role in growth control and differentiation.[14] Recently, it was shown that at cadherin adhesive complexes Ajuba interacts with α-catenin, which is required for the efficient recruitment of Ajuba to cell junctions and, as such, influences the formation or stabilization of cadherin-mediated cell–cell adhesion.[17] The LIM-containing LPP is also localized at sites of cell adhesion and transiently in the nucleus. Interestingly, in various benign and malignant tumors, LPP is mutated, which leads to a permanent nuclear localization.[18] PINCH is a recently identified adaptor protein, which is also involved in cell adhesion, growth, and differentiation.[19] It interacts with a membrane-proximal integrin-linked kinase (ILK), which itself colocalizes with β1-integrin in the focal adhesion plaques.[20,21] Members of the four-and-a-half LIM domain (FHL) protein family were found to act as transcriptional coactivators. The FHL family member DRAL/FHL2 (down-regulated in rhabdomyosarcoma LIM domain protein) augments transcriptional repression via physical interaction with the sequence-specific repressor promyelocytic leukemia zinc finger (PLZF).[22] Additionally, it also enhances the transcriptional activity of the androgen receptor.[23] The FHL protein ACT (activator of CREM in testis) specifically associates with CREM, and its expression level is also synchronized with the expression level of CREM. It significantly stimulates CREM activity and also displays an intrinsic activation function. Also, a tissue-specific coactivator activity was proposed for ACT.[24] Another FHL family member for which transcriptional coactivator activity was found is the isoform FHL1 (also known As KyoT2). It interacts with the DNA-binding transcription factor RBP-J and negatively regulates transcription.[25] The heart-specific FHL2 protein was shown to interact with human DNA-binding nuclear protein, hNP220, and seems to be particularly relevant for heart muscle differentiation and heart phenotype integrity.[26]

It thus seems tenable that the FHL protein family constitutes a homogeneous set of transcriptional modulator proteins possessing a dual activator–repressor function. Although little is known about the detailed molecular function of LIM kinase proteins in intact cells, experimental data suggest a central cytoplasmic role in the organization of the actin cytoskeleton and in the regulation of cell motility and morphogenesis. For example, LIMK-1 participates in Rac-mediated actin cytoskeletal reorganization, presumably by phosphorylation of cofilin,[27] and is itself phosphorylated by the small guanosine triphosphate Rho-associated kinase ROCK.[28]

PROTEIN–PROTEIN INTERACTIONS

Specific interactions between LIM-HD or LIM-only proteins and other DNA-binding proteins have been demonstrated. It was shown that in erythroid cells, LMO-2 cooperates with GATA-1, TAL1, E2A, and Ldb1/NLI in DNA binding.[29] The LIM-HD protein Lmx1 and the basic helix-loop-helix (bHLH) protein E47/Pan-1 activate the insulin promoter in transfected fibroblasts.[30] Other examples of the assembly of LIM domain proteins into nuclear complexes that regulate gene expression are the association of the LIM-HD protein Lim-3 with the POU-domain protein Pit-1[31] and the activation of transcription by the POU-type homeodomain protein UNC-86 and the LIM-HD protein Mec-3 in the nematode *Caenorhabditis elegans*.[32,33] The Nuclear LIM interactor (NLI or Ldb1) was identified in the nuclei of developing embryonic neuronal cells.[34] It binds to LIM-HD and LMO proteins, including LMO-2 and the related protein LMO-1. Other examples of specific LIM-HD binding partners are Pan-1,[35] CLIM-1,[36] and Chip.[37] Additionally, the participation of LIM-HD proteins in functional transcription complexes, however, is more complex. For example, the transcriptional activity of Lim-3 is enhanced by the HD protein P-Otx but only in the presence of CLIM-1,[36] and it has been shown that the structural integrity of the ternary complex is LIM-dependent.

The cytoplasmic CRP subfamily member CRP1 protein was shown to colocalize with zyxin *in vivo*,[38] and *in vitro* studies indicate that zyxin-CRP1 interactions are mediated by the N-terminal LIM domain,[39] thereby demonstrating that a solitary LIM domain in a CRP protein defines a discrete functional unit that can mediate protein–protein interactions. The amino-terminal LIM1 domain of CRP1 localizes along the actin stress fibers and associates with the cytoskeleton by direct interactions with the actin binding protein, α-actinin.[40] Additionally, specific nuclear interactions of CRP3 with the myogenic basic helix-loop-helix (bHLH) transcription factor MyoD were reported to enhance the DNA-binding activity of the MyoD-E47 transcription factor complex.[12] The FHL family member DRAL/FHL2 interacts with the promyelocytic leukemia zinc finger protein (PLZF).[22] Another FHL2 protein interacts with the insulin-like growth factor (IGF) binding protein IGFBP-5. The heart-specific FHL2 protein was shown to interact with human DNA-binding nuclear protein, hNP220.[26]

Again, these data support the notion that proteins comprising multiple LIM domains may function as adaptor molecules arranging two or more protein constituents into nuclear transcription or cytoskeletal complexes, respectively. Nevertheless, there is evidence that protein recognition by LIM domain proteins may require more than a single LIM domain. For example, the interaction between DRAL/FHL2 and PLZF required the full complement of LIM domains (only the N-terminal half LIM domain could be deleted).[22] To date, however, no structural information of a LIM domain protein complex is available. NMR chemical shift mapping data obtained from PINCH(LIM1) have identified regions that are important for binding[19] and which suggest a highly variable LIM domain protein recognition motif.

AMINO ACID SEQUENCE INFORMATION

The LIM motif is basically composed of two zinc finger structures separated by a two-amino acid spacer and conforms to the consensus sequence $CX_2CX_{16-23}HX_2C X_2CX_{16-21}CX_2(C/H/D)$.[41,42] The conserved amino acid sequence and a sketch of the domain topology are shown in Figure 2. Amino acid sequences for a variety of LIM domain proteins can be found in the SWISS-PROT database. There are currently 119 entries.

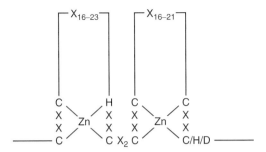

Figure 2 Amino acid sequence and zinc-binding modules in LIM domains. Schematic diagram of the N-terminal CCHC and C-terminal CCCC zinc-binding modules in LIM domain proteins.

PROTEIN EXPRESSION, PURIFICATION, AND MOLECULAR CHARACTERIZATION

The details of recombinant LIM domain protein expression have been reported.[43–46] Briefly, a pET expression vector is typically used to direct the synthesis of the LIM domain protein. Uniform ^{13}C and ^{15}N labeling, as needed for NMR studies, is achieved by growing a 1:100 dilution of a 20-mL bacterial culture prepared overnight in ZB-media [5 g of NaCl and 10 g of NZ-Amine (Aldag, Hamburg, Germany) in 1 L of H_2O] in minimal medium [4.8 g Na_2HPO_4, 3 g KH_2PO_4, 0.5 g NaCl, 1 g $^{15}NH_4Cl$ (Cambridge Isotope Laboratories, CIL) in 1 L of H_2O] supplemented with 20 mL of an 18% (w/v) [^{13}C]-D-glucose (CIL) solution, 2 mL of 1 M $MgSO_4$ solution, 4 mL of a 0.01 M $ZnSO_4$ solution, and ampicillin and chloramphenicol to final concentrations of 100 μg mL^{-1} and 25 μg mL^{-1} respectively. After reaching an optical density of approximately 0.5 at 600 nm, cells are induced to express the $^{13}C/^{15}N$-labeled LIM domain proteins by the addition of 0.5 mL of a 1M solution of isopropyl-β-D-thiogalactoside (IPTG) to a final concentration of 0.5 mM, and the incubation is continued for 5 h at 37 °C. The cells are collected by centrifugation and resuspended in 20 mL of ice-cold buffer.[45,46] Bacteria are lysed by a freeze and thaw cycle, and the cell lysate is cleared by centrifugation. The supernatant containing the soluble protein fraction is then fractionated by CM-52 cation exchange chromatography and subsequent gel-filtration chromatography.[45,46] The purity of recombinant LIM domain proteins is analyzed by gel electrophoresis, and the structural integrity is verified by amino-terminal sequencing and mass spectrometry.

NMR STRUCTURE

The solution structures of several LIM proteins have been determined by NMR spectroscopy,[43–48] including CRIP,[44] the N- and C-terminal LIM domains (commonly referred to as LIM1 and LIM2) of chicken CRP1[43] and of quail CRP2,[45,46] PINCH(LIM1),[19] and the N-terminal zinc finger of Lasp-1 LIM.[47] All proteins showed good

NMR signal dispersion (characteristic of predominantly β-stranded proteins) and their solution structures could be determined by either homonuclear ^{1}H or heteronuclear ^{1}H-^{15}N NMR spectroscopy. The LIM domain proteins CRP1 and CRP2 were later refined using triple resonance ^{1}H-^{15}N-^{13}C NMR spectroscopy. LIM domain proteins share a highly conserved and very modular fold whose topology and structure are outlined in Figures 1 and 3.

It was shown in the structural study of proteins with multiple LIM domains (CRP1 and CRP2 with two LIM domains each) that the two LIM domains are independent structural and functional modules with an apparently unstructured linker region of ~50 aa connecting them.[50,51] Their structures and dynamic properties in the full-length protein were essentially identical to the isolated domains. Their structural independence and separation supports speculations about the biological role of CRP proteins as adaptor and/or linker in protein–protein interactions.

One LIM domain comprises two zinc finger modules in a tandem arrangement, each consisting of two short antiparallel β-sheets followed by a short piece of α-helix. The two modules closely pack together through hydrophobic interactions. Although one LIM domain always contains a pair of zinc fingers, it has been shown that one zinc binding module is capable of forming an independently folded entity.[47] Each LIM domain coordinates two Zn(II) ions that are bound independently in so-called CCHC and CCCC modules (labeling according to the metal coordinating amino acids). Despite a close structural similarity between the C-terminal CCCC zinc finger module and the GATA-1 transcription factor,[52] there is no conclusive evidence for DNA-binding capabilities of LIM domain proteins.

Overall description of secondary and tertiary structure

The LIM domain fold consists of two well-defined subdomains, the N-terminal CCHC and the C-terminal CCCC zinc binding modules, which pack together through a distinct hydrophobic interface made up of conserved hydrophobic and aromatic residues, respectively. A schematic ribbon drawing is shown in Figure 4.

The LIM domain is predominantly made up of β-sheets with a short C-terminal α-helix. The alignment of secondary structure elements followed unambiguously from the analysis of characteristic strong sequential HA(i-1) to HN(i)

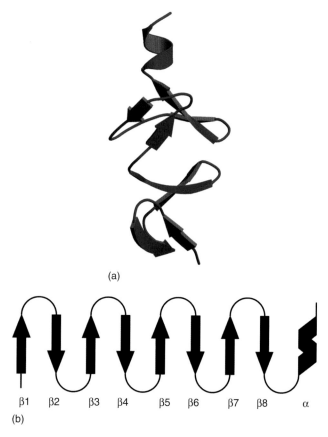

Figure 3 Solution structure of the LIM domain protein CRP2(LIM2). (a) Ribbon diagram of a selected representative structure from the calculated set for CRP2(LIM2) (residues 118–174). The figure was compiled with the program MOLSCRIPT.[49] (b) Topology diagram of the LIM domain. β-strands and α-helices are depicted by arrows and helical symbols respectively.

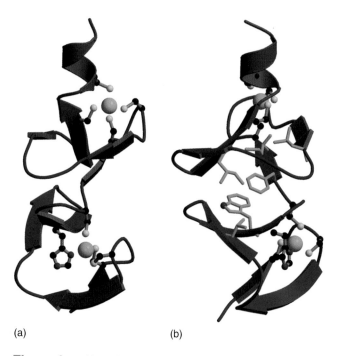

Figure 4 Ribbon drawing of CRP2(LIM2). (a) and (b) Orthogonal views of a selected representative (PDB code 1QLI) structure from the calculated NMR ensemble, including the zinc binding cysteine and histidine residues respectively. The zinc ions are represented as spheres. (b) Side chains of the hydrophobic core residues are shown.

NOEs in the predominantly β-stranded substructures and strong sequential HN(i) to HN(i+1) NOEs in loop/turn regions, between residues in opposite β-strands and between adjacent residues in the C-terminal α-helix respectively. Additional information on the location of secondary structure elements came from characteristic secondary chemical shifts (i.e. shift differences between experimental NMR chemical shifts and random coil values). Both the N-terminal CCHC and the C-terminal CCCC zinc binding modules of the LIM domain are characterized by a similar arrangement of secondary structure elements. They are arranged in the order βI–IV, and α-helix, where β-sheets I and II are oriented orthogonal to each other and comprise the CCHC zinc binding module. Similarly, the orthogonal β-sheets III and IV, together with the α-helix, form the C-terminal CCCC subdomain. The first two β-strands (β1 and β2) of the first antiparallel β-sheet (βI) are connected by a characteristic 'Rubredoxin knuckle'–type turn,[53,54] which immediately follows the first coordinating cysteine residue and also contains the second coordinating cysteine. After an extended and flexible loop, the second β-sheet (βII) with strands β3 and β4 (connected by a tight reverse-type turn) terminates at the coordinating histidine residue. This linker region appears to be conformationally flexible in all LIM domain proteins without any noticeable hydrogen bonding. The second β-sheet (βII) is better and more tightly defined than the previous one. A tight turn (possibly of the 3_{10} helix type), which extends to the last cysteine residue of the first CCHC module, leads to the second CCCC module, whose tertiary structure is very similar to the first module. Again, the first two (of four) coordinating cysteine residues are part of a loose β-sheet (βIII, strands β5 and β6) and a 'Rubredoxin knuckle' connecting its two strands. An extended loop and a tight β-sheet (βIV, strands β7 and β8) with a reverse turn lead up to the third cysteine. After a turn, a short α-helix starts just at the last coordinating cysteine residue and extends for two turns.

Metal content and coordination

The metal binding properties of LIM domains were exemplified for the C-terminal LIM2 domain of the CRP family member chicken CRP1.[55,56] From spectroscopic studies of Co(II)- and Cd(II)-substituted cCRP1, it was concluded that each metal ion is tetrahedrally coordinated with predominantly cysteinyl sulfurs. For Cd-CRP1, an ultraviolet transition consistent with cysteine(II) ligand-to-metal charge transfer was observed. A similar UV transition band was found for CRP1 in which Zn(II) was replaced with Cu(I). To get more details about the coordination geometry, denatured apo-cCRP1 was also reconstituted with Co(II), yielding blue-green protein samples. The electronic spectrum of Co(II)-CRP1 was dominated by d–d transitions with maxima at 620 nm [$\varepsilon = 513\,M^{-1}\,cm^{-1}$ per Co(II)], 701 nm [$\varepsilon = 662\,M^{-1}\,cm^{-1}$], and 740 nm [$\varepsilon = 621\,M^{-1}\,cm^{-1}$], which are typical values for spin-allowed transitions of distorted tetrahedral Co(II) complexes. The band at 740 nm is indicative of thiolate coordination.[57] Co(II)-thiolate ligation is additionally indicated by charge-transfer transitions in the near ultraviolet spectral region (band at 340–360 nm).[55] Additional information came from ^{113}Cd-NMR studies. Chemical shift values of 646 and 707 ppm (relative to 1M Cd(ClO$_4$)$_2$) were observed for the metal ions in the two coordination sites of the LIM domain.

Structural refinement of zinc binding sites and assignment of Rd-knuckle type

In recent years, novel NMR constraints based on cross-correlated NMR spin relaxation have been introduced into biomolecular NMR spectroscopy. In addition to conventional NMR constraints (e.g. nuclear Overhauser enhancements (NOE) or scalar coupling constants)[58] cross-correlated NMR spin relaxation[59–62] was employed for the structural refinement of ^{13}C,^{15}N-labeled CRP2(LIM2). It was demonstrated that the application of ^{1}H(i)-^{15}N(i)-^{13}C'(i) and ^{13}Cα(i)-^{1}Hα(i)-^{13}C'(i) dipole-chemical shift anisotropy and relaxation interferences substantially improves the accuracy and precision of the derived structures, particularly for regions located within the zinc-binding modules of the LIM domains of CRP2(LIM2).[63] In general, the relaxation interference–optimized backbone dihedral angles for all the residues in the zinc binding sites of CRP2(LIM2) are in better agreement with the corresponding residues of other zinc finger proteins.[63] For example, in the first turns of both the N-terminal CCHC (Cys120 to Gly124) and the C-terminal CCCC zinc finger (Cys147 to Gly151), only minor structural deviations occur and the geometry is typical for a rubredoxin 'full knuckle.' In both turns, the protein backbone displays extended conformations, which is a requirement for a 'full knuckle' structure. The geometries of the second halves of both zinc-binding modules are, however, distinctly different. The second turn of the CCCC binding site is partially incorporated in the fold-terminating α-helix, which precludes the formation of an extended backbone structure. This zinc binding site is thus structurally comparable to a rubredoxin 'half knuckle.' The slight backbone dihedral angle deviations for the turn-terminating residue (Tyr172) are presumably due to additional side-chain interactions governing the orientation of the C-terminal α-helix. In contrast, the second turn of the N-terminal CCHC zinc-binding module does not resemble either of the two rubredoxin knuckle types. This may be due to the specific steric demands of the two-residue linker connecting the CCHC and CCCC zinc coordination sites respectively. In particular, the local backbone geometry at residue Phe145, which immediately follows the CCHC zinc binding site, is very different when compared with classical knuckle types.

This is most likely due to the fact that the side chain of Phe145 is involved in hydrophobic interactions that stabilize the global fold of the LIM domain of CRP2(LIM2) and determine the relative orientations of the two zinc-binding subdomains. The relevance of these interactions is emphasized by the fact that aromatic amino acids are conserved at this position in the CRP family of LIM domain proteins.[64] Furthermore, residues Lys142, Asn143, and Cys144 undergo conformational reorientational processes in the ms–μs timescale, presumably as a result of unfavorable sterical strain in this part of the polypeptide chain. The surprising observation that residues that are part of the zinc coordination site also display conformation mobility had also been previously made for the zinc finger proteins Xfin[65] and E. coli Ada.[66]

Bispheric coordinative structuring of zinc binding sites

In LIM domain proteins (as well as in other zinc finger proteins), histidine and cysteine side chains are the characteristic ligands for the Zn(II)-ion. On the basis of the structural study of a mutant protein (e.g. CRP2(LIM2)E155G),[67] we have characterized an extended H-bonding network in the coordination polyhedron around the Zn(II)-ions in the two zinc finger units of CRP2(LIM2). This report was also the first experimental demonstration of the protonation states of the zinc-coordinating ligands in a natural zinc finger protein. In addition to the direct 'inner sphere coordination' of the zinc ion by thiolate anions and/or histidine nitrogens, the metal gives rise to an extended network of H-bonds in the coordination polyhedron of the central ion. This indirect or 'outer sphere coordination' is important in a zinc finger protein in which the Zn(II) centers are assigned a structural role. In catalytically active zinc finger proteins, this indirect metal coordination has been recognized as an important determinant of the catalytic activity, presumably by modulating the basicity and/or nucleophilicity of the coordinating ligands.[68,69] The term 'bispheric coordination' was coined to describe these effects, which were studied in great detail in CRP2(LIM2) and were verified by point mutational analysis.[67] The structures of the N-terminal

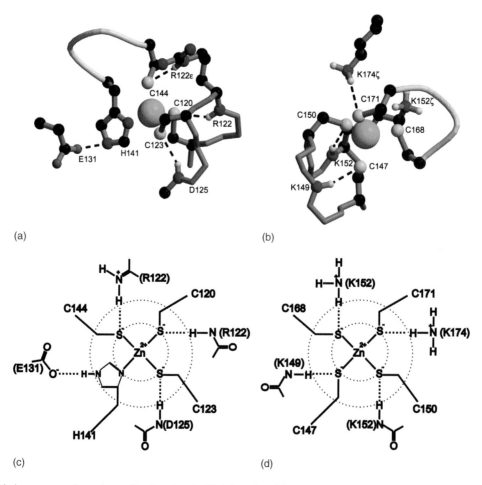

Figure 5 Detailed geometry of metal coordination sites in LIM domains. (Top) Drawing of the zinc-coordinating residues in the CCHC (a) and CCCC (b) zinc-binding modules of CRP2(LIM2) respectively. (Bottom) Schematic drawing of the inner and outer coordination spheres and their associated H-bonding networks around the two zinc centers, CCHC (c) and CCCC (d). Zinc ligand bond distances are Zn-SG: 2.30 Å, Zn-ND1: 2.00 Å.[45]

CCHC and the C-terminal CCCC zinc finger modules are illustrated in Figure 5 in greater detail.

Within the 'Rubredoxin knuckles' (C_i-X_2-C_{i+3}-X_2) the backbone amide of residue (i + 2) forms a H-bond to CysSG of residue i. Analogously, there is an additional H-bond between the HN of residue (i + 5) and SG of residue (i + 3). Furthermore, the residue (i + 4) assumes a backbone conformation with a positive ϕ angle. Interestingly, within the CCHC module the histidine residue coordinates the zinc through the ND1 atom, as has been confirmed by the observation of NOEs between HisHE2 and the surrounding residues. The NOE between HisHD2 and HB/G of a glutamate in its vicinity (Glu131 in the case of CRP2(LIM2)), and similar interactions found in CRIP and CRP1(LIM2), prove the existence of a H-bond between HE2 and the carboxylate group of the glutamate residue. It is therefore suggested that this interaction is important in defining the structure of the CCHC module. The H-bond donor for the remaining cysteine was found to be the HE of an arginine residue (Arg122), whose backbone HN is already the donor for a H-bond to Cys110. It therefore 'bridges' the zinc finger and provides additional structural stabilization. Proof of this interaction comes from the fact that mutating this residue, as in CRP2(LIM2)R122A, results in significant shift changes not only in the vicinity of the mutation but also at Cys144, the proposed acceptor. An analogous H-bonding pattern can be deduced for the CCCC module (Figure 5). Again, the HN(i + 2) and (i + 5) form H-bonds to the SG of the two coordinating cysteine residues and the residues involved assume the backbone angles characteristic of the 'Rubredoxin knuckle' motif. As donors for the remaining two cysteine residues, the protonated side-chain amino groups of two lysine residues could be identified. The HZ/NZ group of Lys152, whose backbone HN already forms a H-bond to Cys150, interacts with Cys168 SG. Lys152, therefore, bridges and stabilizes the CCCC module. A similar interaction could be deduced for Cys171 SG and the side chain of Lys174, which is already part of the short C-terminal helix and possibly helps to orient the helix relative to the remainder of the protein by 'tethering' it to the zinc finger module.

Structural and dynamical features of individual LIM domains

Despite high sequence identity >50% between various LIM domains, there is considerable variability in the relative orientations of the two zinc-binding modules, and typical pairwise coordinate rmsds are 2 to 3 Å for the backbone atoms. However, a separate comparison of the two zinc fingers yields a higher structural similarity. In particular, in proteins with two LIM domains (CRP1 and CRP2), there is a pronounced difference between the N-terminal LIM1 and the C-terminal LIM2 domains, and in the N-terminal domain, the CCHC and the CCCC zinc fingers appear to be less tightly packed against each other, in part, because the side chains of hydrophobic residues in the protein core are, on average, shorter in LIM1 than in LIM2, and thus hydrophobic packing interactions are less pronounced.[46] The dynamic nature of this hydrophobic interface is also reflected in backbone ^{15}N relaxation, as a number of residues whose side chains are buried in the protein core display internal dynamics in an intermediate ns timescale, which may, in turn, reflect as a time-dependent (re)orientation of the two zinc fingers.[46] The study of the point mutant CRP2(LIM2)R122A revealed that, as a consequence of the disruption of the 'outer sphere' H-bonding network by the amino acid replacement, the hydrophobic core had to rearrange itself. This resulted in a significantly different relative orientation[48] as well as a somewhat increased collective intramodular dynamic behavior, as detected by cross-correlated spin relaxation. This effect is illustrated in Figure 6, where the different relative domain orientations are illustrated either as a result of enhanced flexibility and/or (re)packing in the protein core for CRP2(LIM1), CRP2(LIM2), and CRP2(LIM2) R122A. The relative orientations of the two zinc fingers

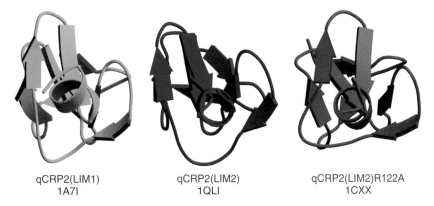

qCRP2(LIM1)   qCRP2(LIM2)   qCRP2(LIM2)R122A
1A7I   1QLI   1CXX

Figure 6 Intradomain variability in LIM domains. Schematic representation showing the change in the relative orientations of the two CCHC and CCCC zinc binding subdomains for the LIM domain proteins CRP2(LIM1) (left, green, PDB code 1A7I), CRP2(LIM2) (middle, blue, PDB code 1QLI), and the mutant protein CRP2(LIM2)R122A (right, red, PDB code 1CXX).

thus reflect a delicate balance of hydrophobic packing and dynamic disorder, and this could have implications for the physiological function of LIM domains, such as 'fine tuning' the interaction with a potential protein-binding partner.

MOLECULAR DYNAMICS OF LIM DOMAINS

Anisotropy of overall tumbling

As relaxation parameters are sensitive to anisotropic overall tumbling, ^{15}N NMR relaxation data were used to determine the rotational diffusion tensor of the CRP2(LIM2) protein. This was accomplished by using the local diffusion approximation.[70] This method analyzes local diffusion coefficients derived from relaxation rates in order to determine the diffusion tensor, assuming isotropic, axially symmetric, and asymmetric diffusion.[70] The fit yielded F-values for the axial model between 2.6 and 7.04 for the individual structures of an NMR-derived bundle and 5.30 for the mean structure. The values obtained for $D_\parallel/D_\perp$ were 1.35 for the mean structure and 1.36, as the mean value, for the individual structures. The angle θ between the principal axis of the inertia tensor and the axial diffusion tensor is 41° on average (38° for the mean structure). The significantly different relaxation rates for some residues could be explained with the anisotropy of the overall tumbling motion. For example, residues Cys120 and Lys174 have reduced R_1 and elevated R_2 values respectively. The N–H bond vectors for these residues are oriented preferentially along the symmetry axis of the diffusion tensor and are thus relaxed mainly by slower motions about the perpendicular axes. The relaxation derived values for the rotational diffusion tensor of CRP2(LIM2) clearly underscores the nonspherical shape of the LIM domain fold.

Intramolecular dynamics

Analysis of ^{15}N relaxation data of CRP2(LIM2) revealed a good correlation between backbone dynamics and secondary structure in LIM domains.[45] Residues in loop regions consistently exhibited significantly different ^{15}N relaxation rates compared with residues located in well-defined secondary structure elements. In addition to local backbone dynamics, ^{15}N relaxation data also indicated significant conformational flexibility for the linker region (Asn143–Arg146) connecting the two rigid zinc-binding subdomains. These findings suggested the existence of conformational flexibility involving extended structural segments of the LIM domain. Analysis of the highly refined structural ensemble of CRP2(LIM2)[71] with the program DynDom[72] additionally corroborated the existence of this intramolecular mobility in LIM domains and suggested that the two zinc-binding subdomains constitute, to a large extent, rigid substructures and that the relative orientations of the two subdomains define a significant degree of freedom.

Collective motions in LIM domains

A more detailed picture of the intramolecular dynamics of LIM domains was obtained from ^{13}C,^{15}N-labeled CRP2(LIM2) for which the backbone dynamics of residues located in the structured part of the LIM domain was investigated by means of novel proton-detected ^{13}C'(i-1)-^{15}N(i) and ^{15}N(i)-^{1}H^{N}(i) multiple-quantum relaxation respectively.[71,73] Rather uniform relaxation rates were found for most of the backbone positions, but significant deviations were found for residues located in the loop region connecting the two β-strands β6 and β7. Interestingly, the different relative contributions of ^{1}H and ^{15}N spin relaxation at sites Leu154, Thr158, and Leu159 were interpreted assuming fast conformational exchange between different conformational states. Reversible opening and closing of hydrogen bonds at these sites transiently forms open, unprotected conformational substates, which subsequently can undergo conformational reorientations. The backbone amide group of Thr158 exhibited only very weak hydrogen bonding and thus easily undergoes a conformational exchange process, presumably between two states. In contrast, Leu154 and Leu159 form stronger hydrogen bonds and thus show a more complicated relaxation behavior.[73] This differential conformational flexibility is intriguing, given that residues in this loop segment are conserved in the CRP protein family and may thus be of functional relevance, for example, in fine-tuning of intermolecular interactions along the protein–protein binding interface with authentic CRP binding partners.

Recently, a more elaborate dynamical study of CRP2(LIM2) revealed the surprising and unexpected presence of cross-correlated exchange processes for a number of residues that are located in well-defined secondary structure elements.[71] Significant intramolecular motional dynamics in the ms–μs timescale were observed for residues that are connected via an extended hydrogen-bonding network, indicating that CRP2(LIM2) undergoes long-range concerted motions. A thorough comparison with structural data of the mutant protein CRP2(LIM2)R122A, where the hydrogen-bonding network around the N-terminal zinc-binding module was altered, revealed that regions undergoing motional dynamics correlate with residues that are structurally altered in the mutant protein. These data thus provided a unique correlation between protein dynamic fluctuations and structural reorientations upon modification of hydrogen-bonding patterns relevant to the overall protein fold. The existence of such multiple conformational substates in a LIM domain indicates that allosteric regulation of reversible LIM domain protein interactions may be triggered by shifting the populations of these microstates.

FUNCTIONAL ASPECTS

An interesting functional feature of LIM domain proteins is represented by their dual protein binding functionalities. For example, LIM-HD proteins support the DNA-binding activity by enhancing protein–protein interactions, and thus allow cooperatively regulated tissue-specific promoter binding. They also negatively regulate LIM-HD activity, presumably by preventing homeodomain association with DNA. This LIM-HD activity can be additionally tuned by interactions with other proteins. Specifically, there is evidence that the LIM domains in LIM-HD proteins interact intramolecularly with the homeodomain to impair its DNA binding, thereby providing a means to regulate the transcriptional activity of the LIM-HD protein. For example, the mouse Isl-1 HD binds to the TAAT consensus binding sequence. In the full-length Isl-1 protein, this DNA-binding capability is lost, but it can be restored by removal of the Isl-1 LIM domains or by disruption of the global fold stabilizing zinc finger structure.[74] Similarly, LIM domains inhibit in vitro binding of Lim-3 to the αGSU and βTSH promoters,[31] and of Mec-3 to the mec-3 promoter.[32] Again, removal or disruption of the LIM domains restores promoter binding. The nature of LIM inhibition of DNA binding by HDs is still unknown and could result either from direct intramolecular interaction between the LIM domain and the homeodomain or indirectly from association with a repressor protein. In both cases, interference might be relieved by binding to another protein, such as Ldb1, a LIM binding protein that has been demonstrated to increase LIM1 homeodomain association with specific promoters.[75] The involvement of LIM domain proteins in oligomeric protein complexes requires the controlled modulation of the protein-binding affinity of a LIM domain. It thus seems plausible that LIM domains exhibit an inherent conformational flexibility that can be used to tune the intermolecular interactions, reminiscent of allosteric regulation. A first insight into how protein binding by LIM domains may be fine-regulated was provided by a structural and dynamical study of a mutant CRP protein, CRP2(LIM2)R122A. In this mutant protein, the fold-relevant hydrogen-bonding pattern around the two zinc coordination centers was modified by substituting alanine for the hydrogen bond donor arginine, Arg122. It was observed that this disruption of hydrogen bonding (see Figures 5 and 7) resulted in a subtle structural reorientation of the LIM domain (Figure 6). Specifically, the relative orientation of the two zinc-binding modules in CRP2(LIM2)R122A was altered. Although the various LIM domains of the CRP protein family display a nearly identical global fold, they differ in terms of the relative orientations of the CCHC and CCCC subdomains. This difference in orientation leads to a widening of the hydrophobic core region with an extended hydrophobic groove, and hence to a different in exposure to and accessibility of residues located at the interface of the CCHC and CCCC subdomains. In view of the pronounced binding specificities of LIM domains, it was thus proposed that this structural feature might provide a mechanism to modulate the protein-binding capacity of the LIM domain. Furthermore, in a subsequent NMR spectroscopic study, it was shown by cross-correlated NMR relaxation experiments that regions having significant conformational flexibility in the ms–μs timescale correlate with residues that are structurally altered in the mutant protein CRP2(LIM2)R122A, and that these residues are part of an extended hydrogen-bonding network connecting the two zinc binding sites (Figure 7).

These data indicate the presence of long-range collective motions in the LIM domain, consistent with the notion of conformational substates, for example, energetically comparable protein conformations that interconvert because of a small energy barrier. Presumably this conformational plasticity is important for the fine-tuning of essential molecular interactions, leading to the assembly of functionally important protein complexes.

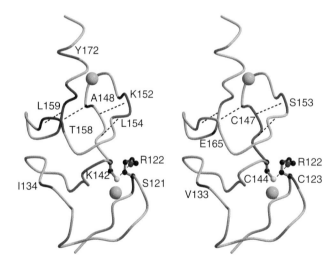

Figure 7 Long-range collective motions in LIM domains. Backbone trace showing the correlation between cross-correlated ms–μs timescale protein backbone motions in CRP2(LIM2) and structural changes caused by the disruption of the zinc finger hydrogen bond network in CRP2(LIM2)R122A. (Left) Residues for which significant conformational flexibility was observed by NMR spectroscopy. (Right) Location of chemical shift changes of $^1H^N$ and ^{15}N (indicating structural changes) introduced by the point mutation (R122A). Hydrogen bonds are indicated by dashed lines. The figure was produced with the program MOLSCRIPT.[49] The side chains of the residues of the mutation site (R122A) and the zinc-coordinating cysteine (Cys144) are shown. The (outer coordination sphere, see text) hydrogen bond between Arg122 and Cys144 is indicated. The zinc ions are shown as spheres.

REFERENCES

1. G Freyd, SK Kim and HR Horvitz, *Nature*, **344**, 876–79 (1990).
2. O Karlsson, S Thor, T Norberg, H Ohlsson and T Edlund, *Nature*, **344**, 879–82 (1990).
3. O Hobert and H Westphal, *Trend Genet*, **16**, 75–83 (2000).
4. S Thor, SGE Andersson, A Tomlinson and JB Thomas, *Nature*, **397**, 76–80 (1999).
5. TH Rabbits, *Genes Dev*, **12**, 2651–57 (1998).
6. I Bach, *Mech Dev*, **91**, 5–17 (2000).
7. AJ Warren, WH Colledge, MB Carlton, MJ Evans, AJ Smith and TH Rabbitts, *Cell*, **78**, 45–57 (1994).
8. Y Yamada, R Pannell, A Forster and TH Rabbitts, *Oncogene*, **21**, 1309–15 (2002).
9. JE Visvader, D Venter, K Hahm, M Santamaria, EYM Sum, L O'Reilly, D White, R Williams, J Armes and GJ Lindeman, *Proc Natl Acad Sci USA*, **98**, 14451–57 (2001).
10. HA Louis, JD Pino, KL Schmeichel, P Pomiès and MC Beckerle, *J Biol Chem*, **272**, 27484–91 (1997).
11. Y Kong, MJ Flick, AJ Kudla and SF Konieczny, *Mol Cell Biol*, **17**, 4750–60 (1997).
12. AW Crawford and MC Beckerle, *J Biol Chem*, **266**, 5847–53 (1991).
13. MMR Petit, J Fradelizi, RM Golsteyn, TAY Ayoubi, B Menichi, D Louvard, WJM Van de Ven and E Friederich, *Mol Biol Cell*, **11**, 1117–29 (2000).
14. J Kanungo, SJ Pratt, H Marie and GD Longmore, *Mol Biol Cell*, **11**, 3299–13 (2000).
15. J Yi, S Kloeker, CC Jensen, S Bockholt, H Honda, H Hirai and MC Beckerle, *J Biol Chem*, **277**, 9580–89 (2002).
16. BE Drees, KM Andrews and MC Beckerle, *J Cell Biol*, **147**, 1549–59 (1999).
17. H Marie, SJ Pratt, M Betson, H Eppler, J Kittler, L Meek, S Moss, S Troyanovsky, D Attwell, GD Longmore and VMM Braga, *J Biol Chem*, **278**, 1220–8 (2003).
18. MMR Petit, SMP Meulemans and WJM Van de Ven, *J Biol Chem*, **278**, 2157–68 (2003).
19. A Velyvis, Y Yang, C Wu and J Qin, *J Biol Chem*, **276**, 4932–39 (2001).
20. Y Tu, F Li, S Goicoechea and C Wu, *Mol Cell Biol*, **19**, 2425–34 (1999).
21. F Li, Y Zhang and C Wu, *J Cell Sci*, **112**, 4589–99 (1999).
22. P McLoughlin, E Ehler, G Carlile, JD Licht and BW Schäfer, *J Biol Chem*, **277**, 37045–53 (2002).
23. JM Muller, U Isele, E Metzger, A Rempel, M Moser, A Pscherer, T Breyer, C Holubarsch, R Buettner and T Honjo, *EMBO J*, **19**, 359–69 (2000).
24. GM Fimia, D De Cesare and P Sassone-Corsi, *Nature*, **398**, 165–69 (1999).
25. Y Taniguchi, T Furukawa, T Tun, H Han and T Honjo, *Mol Cell Biol*, **18**, 644–54 (1998).
26. EK Ng, KK Chan, CH Wong, SK Tsui, SM Ngai, SM Lee, M Kotaka, CY Lee, MM Waye and KP Fung, *J Cell Biochem*, **84**, 556–66 (2002).
27. N Yang, O Higuchi, K Ohashi, K Nagata, A Wada, K Kangawa, E Nishida and K Mizuno, *Nature*, **393**, 809–12 (1998).
28. M Maekawa, T Ishizaki, S Boku, N Watanabe, A Fujita, A Iwamatsu, T Obinata, K Ohashi, K Mizuno and S Narumiya, *Science*, **285**, 895–98 (1999).
29. IA Wadman, H Osada, H GG Grütz, AD Agulnick, H Westphal, A Forster and TH Rabbits, *EMBO J*, **16**, 3145–57 (1997).
30. JD Johnson, W Zhang, A Rudnick, WJ Rutter and MS German, *Mol Cell Biol*, **17**, 3488–96 (1997).
31. I Bach, SJ Rhodes II, RV Pearse, T Heinzel, B Gloss, KM Scully, PE Sawchenko and MG Rosenfeld, *Proc Natl Acad Sci USA*, **92**, 2720–24 (1995).
32. D Xue, Y Tu and YM Chalfie, *Science*, **261**, 1324–28 (1993).
33. S Lichtsteiner and R Tijan, *EMBO J*, **14**, 3937–45 (1995).
34. LW Jurata, DA Kenny and DAGN Gill, *Proc Natl Acad Sci USA*, **93**, 11693–98 (1996).
35. MS German, J Wang, RB Chadwick and WJ Rutter, *Genes Dev*, **6**, 2165–76 (1992).
36. I Bach, C Carrière, HP Ostendorff, B Andersen and B MG Rosenfeld, *Genes Dev*, **11**, 1370–80 (1997).
37. P Morcillo, C Rosen, MK Baylies and D Dorsett, *Genes Dev*, **11**, 2729–40 (1997).
38. I Sadler, AW Crawford, JW Michelsen and MC Beckerle, *J Cell Biol*, **119**, 1573–87 (1992).
39. KL Schmeichel and MC Beckerle, *Cell*, **79**, 211–19 (1994).
40. P Pomiès, HA Louis and MC Beckerle, *J Cell Biol*, **139**, 157–68 (1997).
41. I Sánchez-Garcia and TH Rabbitts, *Trends Genet*, **10**, 315–20 (1994).
42. IB Dawid, JJ Breen and R Toyama, *Trends Genet*, **14**, 156–62 (1998).
43. GC Pérez-Alvarado, C Miles, JW Michelsen, HA Louis, DR Winge, MC Beckerle and MF Summers, *Nat Struct Biol*, **1**, 388–98 (1994).
44. GC Pérez-Alvarado, JL Kosa, HA Louis, MC Beckerle, DR Winge and MF Summers, *J Mol Biol*, **257**, 153–74 (1996).
45. R Konrat, R Weiskirchen, B Kräutler and K Bister, *J Biol Chem*, **272**, 12001–7 (1997).
46. G Kontaxis, R Konrat, B Kräutler, R Weiskirchen and K Bister, *Biochemistry*, **37**, 7127–34 (1998).
47. A Hammarström, KD Berndt, R Sillard, K Adermann and G Otting, *Biochemistry*, **35**, 12723–32 (1996).
48. K Kloiber, R Weiskirchen, B Kräutler, K Bister and R Konrat, *J Mol Biol*, **292**, 893–908 (1999).
49. P Kraulis, *J Appl Crystallogr*, **24**, 946–50 (1991).
50. R Konrat, B Kräutler, R Weiskirchen and K Bister, *J Biol Chem*, **273**, 23233–40 (1998).
51. X Yao, GC Pérez-Alvarado, HA Louis, P Pomiès, C Hatt, MF Summers and MC Beckerle, *Biochemistry*, **38**, 5701–13 (1999).
52. JG Omichinski, GM Clore, O Schaad, G Felsenfeld, C Trainor, E Appella, SJ Stahl and AM Gronenborn, *Science*, **261**, 438–46 (1993).
53. MW Day, BT Hsu, L Joshua-Tor, J-B Park, ZH Zhou, MWW Adams and DC Rees, *Protein Sci*, **1**, 1494–1507 (1992).
54. L Fairall, JWR Schwabe, L Chapman, JT Finch and D Rhodes, *Nature*, **366**, 483–87 (1993).
55. JW Michelsen, KL Schmeichel, MC Beckerle and DR Winge, *Proc Natl Acad Sci USA*, **90**, 4404–8 (1993).
56. JW Michelsen, AK Sewell, HA Louis, JI Olsen, DR Davis, DR Winge and MC Beckerle, *J Biol Chem*, **269**, 11108–13 (1994).
57. I Bertini and C Luchinat, *Adv Inorg Biochem*, **6**, 72–111 (1994).
58. K Wüthrich, *NMR of Proteins and Nucleic Acids*, Wiley, New York (1986).
59. B Reif, M Hennig and C Griesinger, *Science*, **276**, 1230–33 (1997).

60 D Yang, R Konrat and LE Kay, *J Am Chem Soc*, **119**, 11938–40 (1997).
61 K Kloiber and R Konrat, *J Biomol NMR*, **18**, 33–42 (2000).
62 K Kloiber and R Konrat, *J Am Chem Soc*, **122**, 12033–34 (2000).
63 K Kloiber, W Schüler and R Konrat, *J Biomol NMR*, **19**, 347–54 (2001).
64 R Weiskirchen, JD Pino, T Macalma, K Bister and MC Beckerle, *J Biol Chem*, **270**, 28946–54 (1995).
65 AG Palmer III, M Rance and PE Wright, *J Am Chem Soc*, **113**, 4371–80 (1991).
66 J Habazettl, LC Myers, F Yuan, GL Verdine and G Wagner, *Biochemistry*, **35**, 9335–48 (1996).
67 R Konrat, R Weiskirchen, K Bister and B Kräutler, *J Am Chem Soc*, **120**, 7127–29 (1998).
68 WN Lipscomb and N Sträter, *Chem Rev*, **96**, 2375–2433 (1996).
69 DW Christianson and CA Fierke, *Acc Chem Res*, **29**, 331–39 (1996).
70 R Brüschweiler, X Liao and PE Wright, *Science*, **268**, 886–89 (1995).
71 W Schüler, K Kloiber, T Matt, K Bister and R Konrat, *Biochemistry*, **40**, 9596–9604 (2001).
72 S Hayward and HJC Berendsen, *Proteins*, **30**, 144–54 (1998).
73 K Kloiber and R Konrat, *J Biomol NMR*, **18**, 33–42 (2000).
74 I Sánchez-Garcia, H Osada, A Forster and TH Rabbits, *EMBO J*, **12**, 4243–50 (1993).
75 AD Agulnick, M Taira, JJ Breen, T Tanaka, IB Dawid and H Westphal, *Nature*, **384**, 270–72 (1996).

FYVE domain

Tatiana G Kutateladze and Michael Overduin
Department of Pharmacology, University of Colorado Health Sciences Center, Denver, CO, USA

FUNCTIONAL CLASS

Membrane interaction module; binds phosphatidylinositol 3-phosphate; stabilized by the coordination of two zinc ions; known as the FYVE domain.

OCCURRENCE

The FYVE domain is a 65-residue module that binds phosphatidylinositol 3-phosphate [PtdIns(3)P], a phospholipid present in endocytic membranes.[1–3] The structures of three FYVE domains have been solved, providing insights into the mechanisms of PtdIns(3)P ligation, membrane insertion, and zinc coordination.[4–8] The FYVE domain is found in many eukaryotic proteins and is named after the initials of four: Fab1, YOTB, Vac1, and EEA1.[9] Its sequence is defined by several motifs (Figure 1). Eight cysteine residues are absolutely conserved in human FYVE domains and they coordinate zinc ions. Three sequences that directly contact phosphatidylinositol 3-phosphate (PtdIns(3)P) are also highly conserved: the central R+HHCRxCG motif (where + and x refer to basic and any amino acid residue, respectively) and flanking N-terminal WxxD and C-terminal RVC motifs. The zinc coordination pattern of the FYVE domain is similar to that of the RING finger domain, but the two domains are functionally unrelated.[10]

The family of FYVE-domain-containing proteins is extensive, particularly in higher eukaryotes, and is involved in a variety of cellular processes including endocytosis,

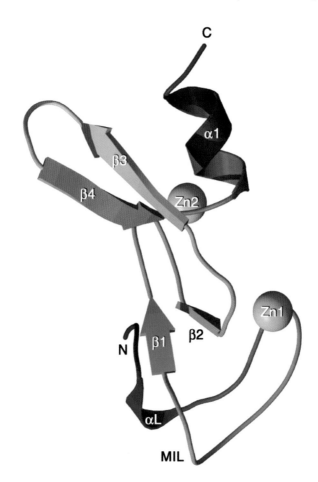

3D Structure Schematic representation of the solution structure of the FYVE domain of the human EEA1 protein including two zinc ions. Prepared with MOLSCRIPT[37] and RASTER3D.[38] (PDB code: 1HYI).

FYVE domain

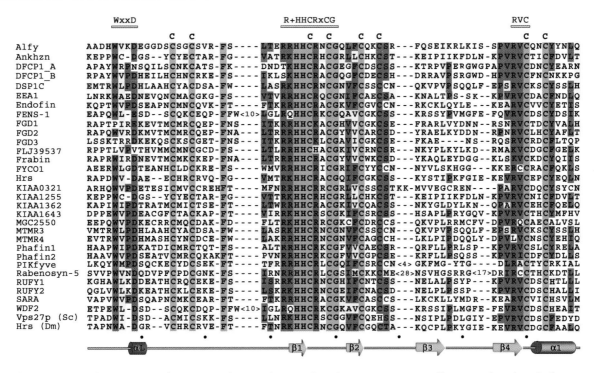

Figure 1 Alignment of the amino acid sequences of FYVE domains from human proteins as well as two others for which structures are available: *Drosophila* Hrs and *Saccharomyces cerevisiae* Vps27p. The PtdIns(3)P and zinc-binding residues are indicated above by orange lines and 'c' respectively, and the positions of the α-helix, linker helix (Lα), and β-strands are indicated below.

signal transduction, and cytoskeletal organization. We identify 29 *Homo sapiens* genes that encode proteins with one or more FYVE domains bearing the canonical motifs. In contrast, FYVE domains are found in only five *Saccharomyces cerevisiae* proteins. The *Arabidopsis thaliana*, *Caenorhabditis elegans*, and *Drosophila melanogaster* genomes encode 12, 13, and 14 proteins with canonical FYVE domains respectively. The alignment of the human FYVE domain sequences along with those of *Drosophila* Hrs and *S. cerevisiae* Vps27p, for which structures have been determined, is shown in Figure 1. For comparison, the residues that ligate zinc and PtdIns(3)P are indicated. Adjacent modules that are found in this diverse family of proteins include phosphoinositide binding domains such as pleckstrin homology (PH) domains, protein interaction modules such as coiled coil regions, VHS domains, WD40 repeats, and catalytic domains including phosphoinositide kinases and phosphatases.

BIOLOGICAL FUNCTION

Proteins that possess FYVE domains often localize to endosomes. These acidic compartments are derived from the plasma membrane and dissociate receptor–ligand complexes. The limiting and intralumenal membranes of early and multivesicular endosomes contain the highest intracellular concentrations of PtdIns(3)P, although a nuclear pool has also been identified.[11] This phosphoinositide comprises about 0.25% of the inositol-containing lipids in mammalian cells and is maintained at local concentrations of about 200 μM.[12] The endosomal PtdIns(3)P pool is targeted by the FYVE domains of mammalian proteins, including the early endosome antigen 1 (EEA1) and its relative, Rabenosyn-5. These proteins, in conjunction with the Rab5 GTPase and SNARE proteins, are involved in the regulated fusion of endocytic membranes.[13–15] Similarly, the fusion of vesicles derived from the plasma membrane and Golgi with the yeast vacuole is mediated by Vac1p.[16] Another FYVE-domain-containing protein Hrs[17] and its yeast homolog (Vps27p)[18] also mediate endocytic protein trafficking. Accordingly, FYVE domains are usually inferred to target host proteins to PtdIns(3)P in endosomal membranes.

There are several exceptions to this rule. For example, DFCP1 contains two FYVE domains and binds PtdIns(3)P, yet localizes to the Golgi and endoplasmic reticulum.[19,20] The murine Lz-fyve protein interacts with transcription factors and is found in the nucleus during embryogenic development.[21] Fgd1 is linked to a human disease, faciogenital dysplasia,[22] and associates with actin filaments, the Golgi apparatus, and the plasma membrane.[23,24] Along with PH and Dbl homology domains, Fgd1 contains an atypical FYVE domain, which is missing a WxxD motif and binds both PtdIns(3)P and PtdIns(5)P.[25] Furthermore, its close relative, Frabin, is primarily localized to the plasma membrane, perhaps also as a function of its PH domain.[26]

Similar to their noncatalytic relatives, several enzymes are localized to vesicular membranes by their FYVE domains.

For example, PIKfyve is a phosphoinositide kinase that regulates membrane invagination.[27] Its FYVE domain plays a key role in targeting endosomes and controlling their size and vacuolation.[28] The homologous yeast kinase, Fab1, regulates vacuole homeostasis and vesicle formation.[29,30] Myotubularin-related protein 3 (MTMR3) and MTMR4 possess FYVE domains and function as dual-specificity phosphatases that can act on PtdIns(3)P.[31] Intriguingly, these proteins use PtdIns(3)P as ligand and substrate, suggesting competitive binding. Alternatively, the FYVE domains may recruit and orient these enzymes toward their substrates. Other enzymes such as the Pib1p ubiquitin ligase also contain FYVE domains that drive their endosomal and vacuolar localization.[32]

In a new wrinkle in the cellular fabric of signal transduction pathways, some FYVE domains sequester signaling complexes inside the cell. For example, the Smad anchor for receptor activation (SARA) protein is targeted by its FYVE domain to endosomal PtdIns(3)P[20] where it interacts with the activated transforming growth factor β-receptor. This intracellular complex recruits and phosphorylates Smad proteins that translocate to the nucleus to control transcription.[33]

Several proteins contain modules that are closely related to FYVE domains. Six *A. thaliana* proteins are distinguished by KRHNCYxCG binding sequences, in which noncanonical residues are underlined. These unusual motifs may contribute to the broader phosphoinositide specificity of some of these proteins.[34,35] FYVE-domain-like modules are also found in several human proteins including FLJ14840, Momo, MyRIP[36] and Sakura, as well as a hypothetical human protein with multiple membrane-spanning elements. These variant FYVE domains contain the eight critical cysteine residues but are missing WxxD motifs and have divergent basic elements in place of the canonical motifs.

AMINO ACID SEQUENCE INFORMATION

The following human proteins contain FYVE domains. (The relevant sequences are shown in Figure 1.)

- Alfy, also known as WDF3, residues 3447–3514
- Ankhzn, residues 1097–1161
- DFCP1_A, similar to TAFF1, residues 591–659 and 708–775
- DSP1C, residues 1112–1179
- EEA1, residues 1344–1409
- Endofin, residues 740–805
- FENS-1, also known as WDF1, residues 277–352
- FGD1, residues 723–790
- FGD2, residues 451–518
- FGD3, residues 525–588
- FLJ39537, residues 100–166
- Frabin, residues 552–619
- FYCO1, also known as FCCD1, residues 1166–1231
- Hrs, residues 156–220
- KIAA0321, residues 808–875
- KIAA1255, residues 1163–1227
- KIAA1362, residues 547–6134
- KIAA1643 residues 916–983
- MGC2550, residues 37–104
- MTMR3, also known as DSP1A, residues 1075–1142
- MTMR4, residues 1087–1149
- Phafin1, residues 145–212
- Phafin2, residues 145–212
- PIKfyve, residues 151–229
- Rabenosyn, residues 150–260
- RUFY1, also known as EIP1, residues 527–592
- RUFY2, also known as Rabip4, residues 533–598
- SARA, also known as MADI, residues 692–758
- WDF2, residues 277–352

PROTEIN PRODUCTION, PURIFICATION, AND MOLECULAR CHARACTERIZATION

The FYVE domains of three proteins have been characterized at atomic resolution. In particular, three-dimensional structures have been solved for the FYVE domains of human EEA1,[4,5,8] yeast Vps27p,[6] and *Drosophila* Hrs[7] (see 3D Structure and Figure 2). Each of these FYVE domain constructs was expressed in *Escherichia coli* with attached glutathione *S*-transferase[4–6] or chitin binding domains,[7] which were removed prior to structural characterization, or with uncleaved His$_6$ tags.[8] Two zinc ions are constitutively bound by all FYVE domains. Monomeric FYVE domains are able to bind PtdIns(3)P, although dimerization may influence their membrane affinity.[4,7,8] No posttranslational modifications of the FYVE domain have been demonstrated to occur *in vivo*.

ACTIVITY AND INHIBITION TESTS

The PtdIns(3)P specificity of several FYVE domains was discovered by *in vitro* liposome binding experiments and substantiated by the protein's subcellular localization *in vivo*.[1–3] Furthermore, the association of FYVE domain proteins with PtdIns(3)P-containing liposomes was abolished by removal of zinc with chelators or by mutation of the residues that ligate lipid or metal. The localization of various FYVE-domain-containing proteins to endosomes and yeast vacuoles was shown by fluorescence microscopy. This subcellular recruitment was blocked by elimination of the kinase that produces PtdIns(3)P, mutations of conserved FYVE domain residues, or by the presence of kinase inhibitors such as wortmannin or LY294002. Thus, the vesicular localization of these proteins was attributed to the PtdIns(3)P interactions of their FYVE domains.

Following these initial discoveries, a plethora of assays has been used to measure the *in vitro* binding properties

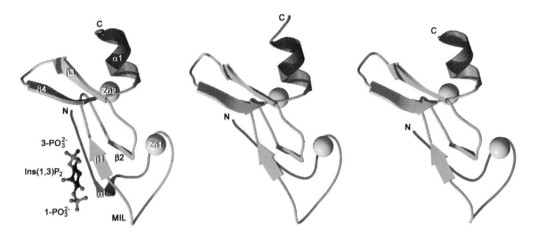

Figure 2 Schematic representation of the crystal structures of the FYVE domains of EEA1, Hrs, and Vps27p from left to right, PDB codes: 1JOC, 1DVP and 1VFY respectively. The backbone ribbon is colored from the blue N-terminus to the red C-terminus. The secondary structure units are labeled for EEA1, and the inositol 1,3-bisphosphate is depicted. The zinc ions are shown in yellow. Prepared with MOLSCRIPT[37] and RASTER3D.[38]

of FYVE domains. On the basis of tryptophan fluorescence and NMR studies, it has been shown that these domains possess a micromolar affinity for soluble forms of PtdIns(3)P based on tryptophan fluorescence and nuclear magnetic resonance (NMR) spectroscopy studies.[4,5,8] A nanomolar-range affinity of various FYVE domains for PtdIns(3)P in monolayers and bilayers is measured using liposome binding experiments,[25,28,39,40] surface plasmon resonance (SPR),[11,19,40] and surface pressure studies.[41] In addition, the phospholipid specificities of FYVE domains have been studied by dot-blot procedures.[20,25,35] Dimerization of a variety of FYVE domain proteins has been investigated by analytical ultracentrifugation,[4,25] pulsed field gradient NMR,[4] and gel filtration.[6] The results of these assays support a common PtdIns(3)P binding mode despite some differences in membrane affinities and dimerization propensities.

X-RAY AND NMR STRUCTURES

NMR spectroscopy

The solution structure of the FYVE domain of human EEA1 has been characterized by multidimensional ^{1}H/^{13}C/^{15}N NMR spectroscopy by Kutateladze et al.[4,5] Spectra were typically collected in 20 mM Tris (pH 6.7), 200 mM KCl, and 20 mM DTT, with varying amounts of ligands. Unfolded states were obtained by removing zinc with ethylenediaminetetraacetic acid (EDTA) or N,N,N',N'-tetrakis(2-pyridylmethyl)ethylenediamine (TPEN) chelators, and the refolded states were then obtained by adding zinc back to the unfolded protein. In addition, four functional states have been compared: (i) the ligand-free state, (ii) dibutanoyl PtdIns(3)P-bound, (iii) dodecylphosphocholine (DPC) micelle-associated, and (iv) ternary complex with PtdIns(3)P and DPC micelles.

Crystallization

Three crystal structures of FYVE domains have been solved by multiwavelength anomalous dispersion phasing. The first was the *S. cerevisiae* Vps27p FYVE domain, which was crystallized by vapor diffusion by Misra and Hurley.[6] The protein solution contained 20 mM Tris-HCl (pH 7.5), 50 mM NaCl, and 10 mM DTT. The well solution consisted of 100 mM sodium acetate (pH 4.6), 200 mM ammonium acetate, and 15% PEG 4000. The crystals were in the space group P1 with $a = 24.1$ Å, $b = 26.5$ Å, $c = 31.8$ Å, $\alpha = 112.0°$, $\beta = 92.5°$, $\gamma = 105.7°$ and included one protein molecule per asymmetric subunit. This structure has been refined at 1.15-Å resolution and has an R-value of 17.4% and an R_{free} of 18.1%.

Crystals of an N-terminal 219-residue fragment of the *Drosophila* Hrs protein that encompasses the tandem VHS and FYVE domains were grown by vapor diffusion by Mao et al.[7] The protein solution contained 50 mM citrate (pH 5.5) and 1 mM DTT. The well solution contained 100 mM HEPES (pH 7.4), 5 mM DTT, and 15% PEG 10000. The crystals were in the space group C2 with $a = 116.7$ Å, $b = 69.7$ Å, $c = 41.8$ Å, $\beta = 94.77°$ and included one protein molecule per asymmetric subunit. The Hrs structure has been refined at 2-Å resolution and has an R-value of 21.5% and an R_{free} of 25.4%.

A C-terminal fragment of human EEA1 containing a portion of the coiled coil region and the FYVE domain was crystallized by Dumas et al.[8] Crystals were microseeded into hanging drops containing 1.5 mM inositol 1,3-bisphosphate, 50 mM HEPES buffer (pH 7.0), 60 mM ammonium acetate, and 11% PEG 4000. The well solution contained 100 mM HEPES (pH 7.4), 5 mM DTT, and 15% PEG 10000. The crystals were in the space group $P2_12_12_1$ with $a = 36.5$ Å, $b = 85.1$ Å, $c = 88.0$ Å and included one molecule per asymmetric subunit. The EEA1 crystal structure has been refined at 2.2-Å resolution and has an R-value of 22.1% and an R_{free} of 28.1%.

Overall description of the structures

Three-dimensional structures of the EEA1, Hrs, and Vps27p FYVE domains have been solved by X-ray crystallography and NMR spectroscopy. All are similar. They share a pair of antiparallel β-hairpins followed by a C-terminal α-helix, as shown in the 3D solution Structure and Figure 2. The functionally critical first β strand (β1) spans three residues of the R+HHCRxCG motif, and pairs with the β2 strand that links the two zinc clusters. The β1 strand is preceded by an exposed hydrophobic loop that inserts into the membrane. The β3–β4 hairpin loop is the most heterogeneous sequence element and extends away from the domain and binding site. Another heterogeneous element follows the N-terminal WxxD motif; in EEA1 it forms an α-helical turn, and in Hrs and Vps27p it adopts a nonstandard strand. A three-residue deletion here in the Hrs/Vps27p subfamily of FYVE domains preserves the position of the WxxD motif side chains within the ligand binding site. The structural conservation in the active site suggests a common binding mechanism for FYVE domains with only subtle functional or regulatory variations.

Zinc coordination

The fold of the FYVE domain is stabilized by tetrahedral coordination of two zinc ions. The metals are bound by four CxxC motifs in a cross-braced topology. The first ion (Zn1) is bound by the first and third CxxC motifs, while Zn2 is coordinated by the second and fourth CxxC motifs. In Vps27p, the Zn1 ion is ligated by cysteines 176, 179, and 200, as well as His203, which replaces the usual fourth cysteine (Figure 3).[6] The Zn2 site of Vps27 is canonical, consisting of cysteines 192, 195, 222, and 225. All the residues that coordinate the zinc ions are in α-helical conformations, except for the second pair of Zn1 binding residues, which are in 3/10 conformations.[6]

The zinc coordination sites undergird the PtdIns(3)P binding site. As a result, the coordinating cysteine residues experience significant chemical shift changes upon PtdIns(3)P docking. The resonances of the first and third cysteines that ligate Zn2 (EEA1 residues 1373 and 1401, respectively) are the most perturbed, consistent with their positions in the conserved PtdIns(3)P binding motifs.[4,5] In contrast, the third cysteine (EEA1 residue 1381) that ligates Zn1 experiences the largest chemical shift perturbations upon micelle interaction, reflecting its proximity to the membrane insertion loop (MIL).

Zinc coordination is required for the stability of the FYVE domain. Mutation of any of the zinc-coordinating residues abolishes the function of the domain, presumably

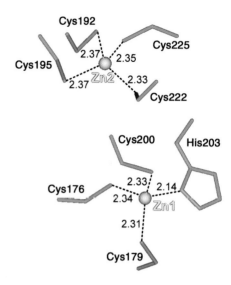

Figure 3 Drawings of the two zinc coordination sites of the FYVE domain of Vps27p. Side chains of the zinc chelating residues are depicted as sticks and colored green. The zinc ions are shown as yellow spheres. The bond distances are displayed (PDB code 1VFY). Prepared with InsightII.

by destabilizing the structure.[1,2,19,32,40] The addition of excess zinc chelators causes the dispersed resonances of the EEA1 FYVE domain to collapse to the random coil region,[5] indicating conversion to an unfolded or molten-globule state. Moreover, removal of only one equivalent of zinc leads to the presence of two sets of NMR signals that correspond to the resonances of the folded and unfolded states of the protein. No partially folded states with only one zinc site occupied are detected, suggesting cooperative zinc ligation and folding. Reintroduction of zinc leads to restoration of the NMR signals of the folded protein, showing that unfolding is reversible.

Comparison to other structures

Zinc finger proteins are encoded by a large portion of eukaryotic genomes.[42] While many of these domains bind nucleic acids, FYVE domains are uniquely adapted to bind membranes. Structurally, they bear similarity to the rabphilin zinc binding domain.[4,6,43] Despite minimal sequence conservation outside the zinc-coordinating residues, the bulk of the structure superimposes well, with a root mean square deviation of 1.2 Å when compared to the Vps27p structure.[6] However, major differences in the N-terminus, ligand binding residues, and β3–β4 hairpin loop preclude any functional similarities between the two domains.

LIPID BINDING MECHANISM

Structures of the EEA1 FYVE domain complexed with dibutanoyl PtdIns(3)P[5] and inositol 1,3-bisphosphate[8] have been solved. The NMR and X-ray studies provide complementary insights into the lipid binding mechanism. For simplicity, the binding site can be subdivided into four pockets that accommodate the 3-phosphate, 1-phosphate, unphosphorylated side of the inositol ring, and acyl chain, respectively. The respective regions are described below for EEA1.

Specific recognition of the 3-phosphate of PtdIns(3)P is the most crucial function of the FYVE domain. This recognition is directly mediated by the His1371 and Arg1374 residues that are underlined in the R+HHCRxCG motif. Critical hydrogen bonding and electrostatic interactions are provided by His1371's imidazole ring and Arg1374's guanidino group (Figure 4).[8] Accordingly, rotation becomes restricted about the Arg1374 Nε-Cζ bond, as evidenced by the pronounced separation of its NHη resonances in the complex.[5] In addition, the Arg1374 NHε resonances are shifted downfield dramatically, presumably due to the inductive effect of the bidentate hydrogen bonds that hold the 3-phosphate. The resonances of the Arg1399 NHε group are also shifted upon PtdIns(3)P ligation, consistent with its water-mediated hydrogen bond to the 3-phosphate, as seen in the crystal.[8] Together, these three side chains constitute the distal basic cluster, so-called because of its more elevated position in the plane of the membrane (Figure 5). All the corresponding groups of Vps27p and Hrs are positioned similarly, indicating a conserved mode of 3-phosphate ligation.

The 1-phosphate sits deeper in the bilayer than the 3-phosphate and is recognized by two adjacent basic residues that comprise the proximal basic cluster. In EEA1, these conserved residues are represented by Arg1369 and Arg1370. A bidentate pair of hydrogen bonds between Arg1369 and the 1-phosphate is evident in the crystal structure (Figure 4)[8] and is supported by a dramatic downfield shift of the corresponding NHε resonance.[5] The role of Arg1370 is less straightforward. Its backbone carboxyl group forms a water-mediated contact with the 1-phosphate. However, its side chain is oriented away from the bound inositol 1,3-bisphosphate and points instead toward the N-terminal hinge, suggesting a fixed structural role.[8] However, the rotation about the Arg1370's Cζ-Nε bond becomes more restricted in the PtdIns(3)P complex, as evidenced by the ligand-induced separation of its NHη NMR signals.[5] This fact suggests that either the engagement of the Arg1370 side chain by the N-terminal hinge is reinforced by ligand binding or this side chain moves to ligate another PtdIns(3)P moiety such as the glycerol group or 1-phosphate. The glycerol group was not present in the crystal, and its absence would alter the 1-phosphate pK_a. Another possibility is that, in the context of a membrane, the Arg1370 side chain contacts the head groups of the surrounding phospholipids.

Recognition of the unphosphorylated face of the PtdIns(3)P ring is crucial for stereospecificity and exclusion of alternately phosphorylated phosphoinositides. This recognition is mediated by Asp1351 and His1372. The 4- and 5-hydroxyls of the inositol ring are hydrogen bonded to the His1372 inositol ring.[8] The hydroxyls at the 5 and 6 positions are contacted by the Asp1351 side chain in the conserved N-terminal WxxD motif. These interactions induce significant conformational changes, as inferred from several observations. The Trp1348 and Asp1351 residues are conformationally heterogeneous in the three crystal structures.[8] More importantly, when PtdIns(3)P is ligated, the Trp1348 and Ala1349 residues pack closer against His1372, Arg1374, Gly1377, Asn1378, and Ile1379 in the binding site, as evidenced by changes in nuclear Overhauser effect patterns.[5] This induced fit hinges on the GluValGln1355 sequence, which exhibits changes in φ angles upon ligation that are consistent with a backbone extension.

The precise orientations of the acyl chains and glycerol moiety of PtdIns(3)P are unknown. The dibutanoyl chains of PtdIns(3)P are unstructured in the NMR studies of the EEA1 FYVE domain, although they are predicted to lie alongside the exposed hydrophobic loop that inserts into micelles.[5]

MEMBRANE INSERTION

The site of EEA1 penetration into membranes has been mapped by NMR using micelles.[5] The FYVE domain inserts the FSVTV sequence (EEA1 residues 1364–1368) into DPC micelles, which mimic the predominant phospholipid in mammalian membranes. The resonances of these residues experience line broadening and chemical shift changes upon micelle interaction, and therefore this motif constitutes

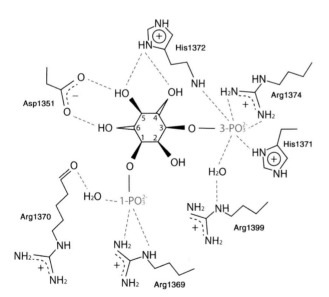

Figure 4 Drawing of the inositol 3-phosphate coordination pattern of the FYVE domain of the EEA1 protein (PDB: code 1JOC). Intermolecular hydrogen bonds are shown as blue dotted lines.

FYVE domain

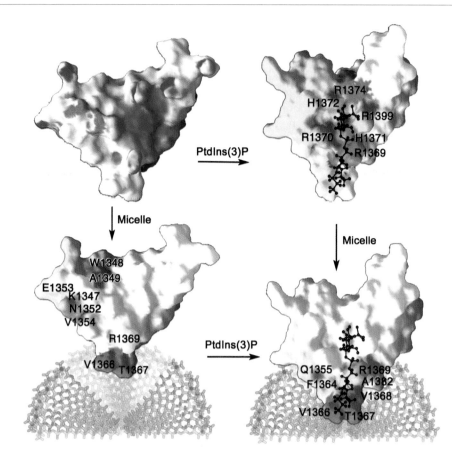

Figure 5 Proposed membrane docking mechanism by the monomeric FYVE domain of the EEA1 protein. The electrostatic surface potential of the lipid-free FYVE domain is colored blue and red for positive and negative charges respectively (upper left). The PtdIns(3)P binding pocket (upper right), nonspecific micelle association site (lower left), and MIL (lower right) are identified by the extent of chemical shift changes in ^{13}C, ^{15}N, and nonexchangeable ^{1}H atoms, induced by the addition of PtdIns(3)P and/or DPC. Residues whose resonances show large, medium, small, and negligible perturbations are colored red, orange, yellow, and white respectively. Reprinted with permission from T Kutateladze and M Overduin, Structural mechanisms of endosome docking by the FYVE domain, *Science*, **291**, 1793–96 (2001). Copyright 2001 American Association for the Advancement of Science.[5]

the membrane insertion loop or MIL (Figure 5). Val1366 and Thr1367 are located at the most exposed tip of the MIL, and are positioned to penetrate deep into the micelle's hydrophobic interior. Other residues that are located nearby, including Gln1355 and Ala1382, may contact the interfacial zone of the membrane as shown by micelle-induced chemical shift changes. Resonances of residues that ligate the 1-phosphate such as Arg1369 are also altered by micelle insertion, suggesting rearrangements in PtdIns(3)P coordination in a membrane. In addition, stabilizing hydrophobic interactions of the PtdIns(3)P acyl chains, MIL, and membrane can be envisaged. Together, these interactions serve to enhance the affinity of the FYVE domain for PtdIns(3)P in the context of a membrane environment by several orders of magnitude.[4,5,40]

In addition to targeting PtdIns(3)P-containing membranes, FYVE domains can also associate weakly with phospholipid surfaces that are free of the ligand.[5] Structurally, these nonspecific interactions are also mediated by the MIL. However, they are less extensive, involving only millimolar-range affinities, shorter residency times, induced chemical shift perturbations, and little membrane penetration.[5,41] Interestingly, nonspecific micelle binding appears to slightly open rather than close the PtdIns(3)P site, that is, the hinge and WxxD motif residues experience chemical shift changes that are opposite to those seen upon PtdIns(3)P ligation. This suggests that nonspecific membrane association serves not only to concentrate the FYVE domain near its ligand in the two-dimensional plane of the membrane, but may also promote PtdIns(3)P binding by favoring the open state of the domain.

The hydrophobic nature of the MIL is conserved among the FYVE domains. Consequently, FYVE domains are likely to share the synergy of binding PtdIns(3)P and insertion into membranes. The Hrs and Vps27 proteins contain FTFTN and FSLLN sequences in place of the FSVTV sequence of EEA1, and these exposed loops occupy similar conformations (Figure 2). Moreover, the effects of mutations of the MIL residues of EEA1, Hrs, and Vps27p support a similar role of this element in membrane association. In particular, the subcellular localization of the EEA1 protein is lost when the Val1366 and

Thr1367 residues are replaced with glycine or glutamine residues.[4] Similarly, the residency time of the Vps27p FYVE domain on lipid monolayers is compromised by mutating the corresponding Leu185 and Leu186 to alanine residues, leading to a sevenfold reduction in affinity for the membrane-bound PtdIns(3)P.[41] Mutation of the corresponding Hrs residue, Phe173, to an alanine reduces the PtdIns(3)P affinity by a factor of 20, again by reducing the monolayer penetration and residency of its FYVE domain.[41] Despite the MIL conservation, outliers do exist. The FYVE domains of FENS-1 and WDF2 have insertions of 10 extra residues in their MILs, and the FENS-1 FYVE domain binds particularly tightly to membranes.[19] Furthermore, these two FYVE domains are distinguished by RQHHCRKCG sequences that differ from the canonical motif at the underlined positions,[5,19] suggesting a variation to the canonical phospholipid binding mechanism.

MULTIPLE VALENCY

The membrane-binding activity of a protein may be profoundly influenced by multivalent phospholipid interactions. For example, the affinity of tandem FYVE domains for PtdIns(3)P-containing bilayers can be substantially enhanced, as shown by SPR studies of designed pairs of linked FYVE domains.[11] Moreover, some proteins contain a tandem pair of FYVE domains. For example, DFCP1 binds PtdIns(3)P selectively and interacts tightly with the membrane. Interestingly, its tandem FYVE domains may not be sufficient for subcellular membrane targeting.[19] Some proteins contain a PH domain in addition to their FYVE domain, providing a second potential phosphoinositide recognition element.

FYVE domain proteins can also form dimers with the help of adjacent coiled coil regions. For example, the coiled coil region of EEA1 forms a parallel homodimer that juxtaposes the two FYVE domains.[44] When attached to this coiled coil, other cytosolic FYVE domains become localized to endosomes.[2] Coiled coils are also predicted next to the FYVE domains of human proteins including FYCO1, Hrs, MTMR3, MTMR4, Rabenosyn-5, and RUFY2. Consequently, parallel dimerization may represent a common means of boosting membrane affinity by juxtaposing FYVE domains. It is also possible that the coiled coils of Hrs[45] and EEA1[14,15] may interact with syntaxin partners to form SNARE assemblies that fuse membranes.

The intimate nature of part of EEA1's parallel coiled coil interface, which spans residues 281–1346, has been revealed by X-ray crystallography.[8] This structure also identifies a 1200 Å^2 interface between the two FYVE domains that was devoid of hydrogen bonding and electrostatic interactions (Figure 6). The dimer interface involves the positioning of Trp1348 residue of the WxxD motif against the variable SerSerLys1396 sequence in the β3–β4 loop.[8] Also, Pro1392 is placed near the

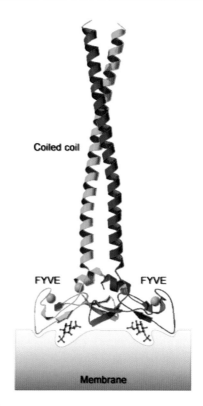

Figure 6 Schematic representation of the dimer structure of EEA1 coiled coil and FYVE domain (PDB: code 1JOC). Prepared with MOLSCRIPT[37] and RASTER3D.[38]

GlnCysGly1375 sequence of the β1–β2 loop. Together with the more extensive coiled coil interaction, this juxtaposes the two FYVE domains of EEA1 to simultaneously ligate two PtdIns(3)P molecules in a bilayer (Figure 6). In this model, the MIL also penetrates into the bilayer, although the angle of insertion differs by ∼45° from that measured for micelle insertion by NMR.[5] Consequently, the X-ray model also predicts bilayer contacts by the tandem lysine residues in the β3–β4 loop–interactions that are not supported by NMR (Figure 5).[5] Given the conformational freedom available in the FYVE domain interface and linkers, it is conceivable that the FYVE domains may swing into a closer, more parallel orientation when inserted together into a membrane. Alternatively, multiple membrane-docked conformers may exist, for example, to reshape the bilayer surface during membrane fusion events.

A distinct dimer involving the *Drosophila* Hrs FYVE domain has also been observed by crystallography.[7] Antiparallel packing between the Hrs FYVE domains buries 1920 Å^2 in this interface, including the functionally important N-terminal tryptophan and the second histidine in β1. Because of partial occlusion of the binding site, a different PtdIns(3)P coordination and membrane docking mode is proposed. This putative Hrs interface could explain the weak dimerization of the VHS and FYVE tandem, which is detected by dynamic light scattering.[7] However, analytical ultracentrifugation studies have shown that the homologous human Hrs FYVE domain does not dimerize, and that it binds PtdIns(3)P-containing vesicles tightly as a monomer.[25] Thus, the physiological relevance of the *Drosophila* Hrs dimer interface and its implications for PtdIns(3)P ligation remain unclear.

Despite the ability of some FYVE domain proteins to dimerize, it is apparent that isolated FYVE domains can target proteins to endosomes *in vivo*. For example, both full-length FENS-1 and its isolated FYVE domain localize to endosomes, and they bind comparably to PtdIns(3)P.[19] Similarly, the endosomal localization of Endofin mirrors the distribution of its isolated FYVE domain to cytoplasmic vesicles.[46] Also, the localization of EEA1 and Rabenosyn-5 to early endosomes is similar to that of truncated versions that contain primarily their FYVE domains.[1,47] Thus, dimerization of FYVE domain proteins does not appear to be required for physiological membrane targeting.

FUNCTIONAL ASPECTS

Tight PtdIns(3)P binding by FYVE domains depends on the presence of a membrane environment. The FYVE domain of EEA1 binds to soluble dibutanoyl PtdIns(3)P with an affinity of $130\,\mu M$.[8] However, a 50 nM affinity is measured when long-chain PtdIns(3)P is presented in a liposome.[1,3,11,40] Thus, the protein binds to PtdIns(3)P in bilayers about 260 times more tightly than to soluble PtdIns(3)P. A comparable degree of stabilization by bilayer insertion is suggested by the millimolar-range affinity of the FYVE domain for DPC micelles.[5] In addition, the two FYVE domains of the EEA1 dimer bind soluble PtdIns(3)P independently and without cooperativity, suggesting, at most, an additive increase in affinity by the dimer for membrane-bound PtdIns(3)P.[8]

Other FYVE domains bind PtdIns(3)P with roughly comparable affinities. The FENS-1 protein exhibits a 48 nM affinity for liposomes containing PtdIns(3)P, and similar binding is displayed by its isolated FYVE domain.[19] The interaction of PIKfyve with PtdIns(3)P-containing liposomes involves an affinity of at least 550 nM, and is mediated by its FYVE domain.[28] The FYVE domains of *Drosophila* Hrs and Vps27p have affinities of 42 and 75 nM respectively for monolayers that contain PtdIns(3)P.[41] The FYVE domain of human Hrs binds PtdIns(3)P-containing vesicles with an affinity of about $2.5\,\mu M$,[25] and this interaction is only weakly inhibited by the inositol 1,3-bisphosphate. Consequently, bilayer insertion by the monomeric Hrs FYVE domain strongly stabilizes the membrane interaction in a manner similar to EEA1.

The critical roles of the PtdIns(3)P binding residues have been tested by alanine substitution experiments. Mutation of the conserved basic or histidine residues that contact PtdIns(3)P abolishes or reduces the affinity for PtdIns(3)P-containing membranes and yields proteins with diffuse cytosolic distributions.[40,41] Mutations of the conserved tryptophan or aspartic acid residues of the WxxD motif also compromise PtdIns(3)P affinity and lead to cytosolic localization.[1,8,40] Similar effects are seen when the MIL residues are mutated.[4,41] Thus, the FYVE domain docking to PtdIns(3)P-enriched membranes involves a substantial network of interactions that span the phosphoinositide head group site, N-terminal mobile element, and MIL residues.

REGULATION OF ACTIVITY

FYVE domains may be regulated at several levels. The concentration of PtdIns(3)P in the cell is increased by the activity of phosphatidylinositol 3-kinases and decreased by PtdIns(3)P 5-kinases[27,29,48] and MTMR phosphatases.[31] The endocytic PtdIns(3)P pool is hydrolyzed in lysosomal compartments.[49] In addition, local pools of PtdIns(3)P may be stabilized and trafficked by proteins that contain FYVE and Phox domains.[50]

The actions of FYVE domains may be modulated by protein interactions. For example, the EEA1 dimer may be dissociated by proteins such as calmodulin or Rab5 that bind to elements in the coiled coil region.[51] Alternatively, Rab5 may bind to the membrane-associated EEA1 dimer to regulate homotypic endosome fusion.[52] Proteins may also bind directly to FYVE domains. For example, the Src homology domains of Etk bind to the FYVE domain of RUFY1.[53] However, it remains to be resolved why most FYVE domain proteins target endocytic pools of PtdIns(3)P, and not, for example, the nuclear accumulations. Moreover, the mechanisms whereby proteins such as DFCP1, Frabin, or Lz-fyve become distributed to the Golgi,[19,20] plasma membrane[26] or nucleus,[21] respectively, remain to be explained, as do the roles of their FYVE domains there.

REFERENCES

1. CG Burd and SD Emr, *Mol Cell*, **2**, 157–62 (1998).
2. JM Gaullier, A Simonsen, A D'Arrigo, B Bremnes, H Stenmark and R Aasland, *Nature*, **394**, 432–33 (1998).
3. V Patki, DC Lawe, S Corvera, JV Virbasius and A Chawla, *Nature*, **394**, 433–34 (1998).
4. TG Kutateladze, KD Ogburn, WT Watson, T de Beer, SD Emr, CG Burd and M Overduin, *Mol Cell*, **3**, 805–11 (1999).
5. T Kutateladze and M Overduin, *Science*, **291**, 1793–96 (2001).
6. S Misra and JH Hurley, *Cell*, **97**, 657–66 (1999).
7. Y Mao, A Nickitenko, X Duan, TE Lloyd, MN Wu, H Bellen and FA Quiocho, *Cell*, **100**, 447–56 (2000).
8. JJ Dumas, E Merithew, E Sudharshan, D Rajamani, S Hayes, D Lawe, S Corvera and DG Lambright, *Mol Cell*, **8**, 947–58 (2001).
9. H Stenmark, R Aasland, BH Toh and A D'Arrigo, *J Biol Chem*, **271**, 24048–54 (1996).
10. PN Barlow, B Luisi, A Milner, M Elliott and R Everett, *J Mol Biol*, **237**, 201–11 (1994).

11 DJ Gillooly, IC Morrow, M Lindsay, R Gould, NJ Bryant, JM Gaullier, RG Parton and H Stenmark, *EMBO J*, **19**, 4577–88 (2000).

12 H Stenmark and DJ Gillooly, *Semin Cell Dev Biol*, **12**, 193–99 (2001).

13 A Simonsen, R Lippe, S Christoforidis, JM Gaullier, A Brech, J Callaghan, BH Toh, C Murphy, M Zerial and H Stenmark, *Nature*, **394**, 494–98 (1998).

14 A Simonsen, JM Gaullier, A D'Arrigo and H Stenmark, *J Biol Chem*, **274**, 28857–60 (1999).

15 HM McBride, V Rybin, C Murphy, A Giner, R Teasdale and M Zerial, *Cell*, **98**, 377–86 (1999).

16 MR Peterson, CG Burd and SD Emr, *Curr Biol*, **9**, 159–62 (1999).

17 M Komada and N Kitamura, *Mol Cell Biol*, **15**, 6213–21 (1995).

18 RC Piper, AA Cooper, H Yang and TH Stevens, *J Cell Biol*, **131**, 603–17 (1995).

19 SH Ridley, N Ktistakis, K Davidson, KE Anderson, M Manifava, CD Ellson, P Lipp, M Bootman, J Coadwell, A Nazarian, H Erdjument-Bromage, P Tempst, MA Cooper, JW Thuring, ZY Lim, AB Holmes, LR Stephens and PT Hawkins, *J Cell Sci*, **114**, 3991–4000 (2001).

20 PC Cheung, L Trinkle-Mulcahy, P Cohen and JM Lucocq, *Biochem J*, **355**, 113–21 (2001).

21 JC Dunkelberg and A Gutierrez-Hartmann, *DNA Cell Biol*, **20**, 403–12 (2001).

22 NG Pasteris, A Cadle, LJ Logie, ME Porteous, CE Schwartz, RE Stevenson, TW Glover, RS Wilroy and JL Gorski, *Cell*, **79**, 669–78 (1994).

23 K Nagata, M Driessens, N Lamarche, JL Gorski and A Hall, *J Biol Chem*, **273**, 15453–57 (1998).

24 MF Olson, NG Pasteris, JL Gorski and A Hall, *Curr Biol*, **6**, 1628–33 (1996).

25 VG Sankaran, DE Klein, MM Sachdeva and MA Lemmon, *Biochemistry*, **40**, 8581–87 (2001).

26 H Obaishi, H Nakanishi, K Mandai, K Satoh, A Satoh, K Takahashi, M Miyahara, H Nishioka, K Takaishi and Y Takai, *J Biol Chem*, **273**, 18697–700 (1998).

27 OC Ikonomov, D Sbrissa and A Shisheva, *J Biol Chem*, **276**, 26141–47 (2001).

28 D Sbrissa, OC Ikonomov and A Shisheva, *J Biol Chem*, **277**, 6073–79 (2002).

29 FT Cooke, SK Dove, RK McEwen, G Painter, AB Holmes, MN Hall, RH Michell and PJ Parker, *Curr Biol*, **8**, 1219–22 (1998).

30 G Odorizzi, M Babst and SD Emr, *Cell*, **95**, 847–58 (1998).

31 GS Taylor, T Maehama and JE Dixon, *Proc Natl Acad Sci USA*, **97**, 8910–15 (2000).

32 ME Shin, KD Ogburn, OA Varban, PM Gilbert and CG Burd, *J Biol Chem*, **276**, 41388–93 (2001).

33 T Tsukazaki, TA Chiang, AF Davison, L Attisano and JL Wrana, *Cell*, **95**, 779–91 (1998).

34 B Heras and BK Drobak, *J Exp Bot*, **53**, 565–67 (2002).

35 RB Jensen, T La Cour, J Albrethsen, M Nielsen and K Skriver, *Biochem J*, **359**, 165–73 (2001).

36 A El-Amraoui, JS Schonn, P Kussel-Andermann, S Blanchard, C Desnos, JP Henry, U Wolfrum, F Darchen and C Petit, *EMBO Rep*, **3**, 463–70 (2002).

37 P Kraulis, *J Appl Crystallogr*, **24**, 946–50 (1991).

38 EA Merritt and DJ Bacon, *Methods Enzymol*, **277**, 505–24 (1997).

39 A Hayakawa and N Kitamura, *J Biol Chem*, **275**, 29636–42 (2000).

40 JM Gaullier, E Ronning, DJ Gillooly and H Stenmark, *J Biol Chem*, **275**, 24595–600 (2000).

41 RV Stahelin, F Long, K Diraviyam, KS Bruzik, D Murray and W Cho, *J Biol Chem*, **277**, 26379–88 (2002).

42 JH Laity, BM Lee and PE Wright, *Curr Opin Struct Biol*, **11**, 39–46 (2001).

43 C Ostermeier and AT Brunger, *Cell*, **96**, 363–74 (1999).

44 J Callaghan, A Simonsen, JM Gaullier, BH Toh and H Stenmark, *Biochem J*, **338**, 539–43 (1999).

45 AJ Bean, R Seifert, YA Chen, R Sacks and RH Scheller, *Nature*, **385**, 826–29 (1997).

46 LF Seet and W Hong, *J Biol Chem*, **276**, 42445–54 (2001).

47 E Nielsen, S Christoforidis, S Uttenweiler-Joseph, M Miaczynska, F Dewitte, M Wilm, B Hoflack and M Zerial, *J Cell Biol*, **151**, 601–12 (2000).

48 JD Gary, AE Wurmser, CJ Bonangelino, LS Weisman and SD Emr, *J Cell Biol*, **143**, 65–79 (1998).

49 AE Wurmser and SD Emr, *EMBO J*, **17**, 4930–42 (1998).

50 TK Sato, M Overduin and SD Emr, *Science*, **294**, 1881–85 (2001).

51 IG Mills, S Urbe and MJ Clague, *J Cell Sci*, **114**, 1959–65 (2001).

52 DC Lawe, A Chawla, E Merithew, J Dumas, W Carrington, K Fogarty, L Lifshitz, R Tuft, D Lambright and S Corvera, *J Biol Chem*, **277**, 8611–17 (2002).

53 J Yang, O Kim, J Wu and Y Qiu, *J Biol Chem*, **277**, 30219–26 (2002).

General aspects

Structural zinc sites

David S Auld

Center for Biochemical and Biophysical Sciences and Medicine and Department of Pathology, Harvard Medical School, Cambridge, MA, USA

The early structural studies of metalloenzymes led to the suggestion that metal atoms that have a catalytic role would be composed of ligands from different parts of the polypeptide chain, whereas ligands to metal atoms in structural sites would be close together. We now know that both types of zinc sites generally contain at least two ligands that are separated by a short spacer.[1–3] In addition, while many structural zinc sites are composed of ligands from a relatively short sequence in the protein, this is not always the case as might be anticipated from inspection of disulfide structural elements that often come from vastly separated regions of the protein primary structure.

The role of structural zinc sites is probably twofold. They maintain the structure of the protein in the immediate vicinity of the metal site. In this manner, it may influence enzymatic activity by providing active site residues involved in catalysis and/or affecting the chemical environment of catalytic groups through interaction with amino acids originating from within its metal-binding spacers. However, in many cases, these metal sites also affect the overall stability of the protein as judged by temperature and pH criteria.

GENERAL FEATURES

While zinc can readily form four, five, and six coordinate complexes, structural zinc sites in proteins have four protein ligands and no bound water molecule. This type of coordination is also observed in many of the *Protein Interface* zinc sites.[2,3] The first zinc enzymes recognized to have a structural zinc site were horse liver alcohol dehydrogenase, ADH,[4] and the regulatory subunit of *Escherichia coli* aspartate carbamoyltransferase.[5] In both cases the zinc is bound to four cysteines in a relatively short linear sequence of 15 to 33 amino acids, respectively. There are today over seven dozen structural zinc sites with representatives in five of the six enzyme classes defined by the International Union of Biochemists. The ligands to these metal sites are provided by protein sequences as small as 10 amino acids and as large as 210 amino acids, but the majority fall within the range of 25 to 50 amino acids (Table 1). There is nearly always one short spacer separating two amino acids in these sites. The ligands are frequently supplied from within or just before or after a β-sheet secondary structure (Table 1). Several ligands are supplied by a short amino acid sequence of three to six amino acids (loop) between two peptide regions of regular (β-sheet or α-helical) secondary structure. The majority of these loop regions are composed of two β-sheets. While the sulfur of a cysteine is by far the preferred ligand for such sites, it is also found in combination with an imidazole nitrogen of histidine, and/or a carboxylate oxygen from aspartic acid or glutamic acid (Figure 1). The second most prevalent ligand is histidine, which is often found in combination with cysteine.

The presence of histidine and/or cysteine in the same zinc binding site frequently gets some form of a 'zinc finger' nomenclature. On the basis of only 'loose' criteria, some of the sites listed here are still referred to as zinc finger domains,[85] even if they do not fit the criteria of having DNA binding properties and/or the ligand nature (2His/2Cys), sequence length (~30 amino acids), and presence of both α-helix and β-sheets in the zinc binding site associated with a classical zinc finger.[86] Several reviews exist on the expansion of this nomenclature in the area of zinc protein sites interacting with nucleic acids.[87–89] The expansion thus far deals mainly with the type of secondary structure present in combination with cysteine and histidine ligands. However, even if one considers only the four main amino acid ligand donors, cysteine, histidine, aspartic acid, and glutamic acid, there are 22 permutations of a zinc bound to these 4 ligands, 8 of which have been observed in metalloenzymes (Figure 1, Table 1). Several other features of the site can also be important to the resulting influence of the zinc ion. These include the type of secondary structure, outer shell interactions, how many short spacers occur, and the order of the ligand donors. This is why one continues

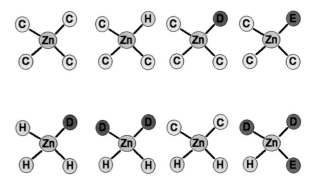

Figure 1 Schematic of the type of zinc sites observed thus far in terms of the type of ligand bound. The one-letter codes C, D, E, and H are given for the amino acids cysteine, aspartic acid, glutamic acid, and histidine respectively.

Structural zinc sites

Table 1 Structural zinc sites[a]

Enzyme	PDB#	L$_1$	X	L$_2$	Y	L$_3$	Z	L$_4$	References
				CLASS I: OXIDOREDUCTASES					
Alcohol dehydrogenase family									
Horse EE	8ADH, 3BTO	Cys	2	Cys$_{b\alpha}$	2	Cys$_\alpha$	7	Cys	6
Cod	1CDO	Cys	2	Cys$_{b\alpha}$	2	Cys$_\alpha$	7	Cys	7
Mouse Class II	1E3E	Cys	2	Cys$_{b\alpha}$	2	Cys$_\alpha$	7	Cys	8
Human Class I									
$\alpha\alpha$	1HSO	Cys	2	Cys$_{b\alpha}$	2	Cys$_\alpha$	7	Cys	9
$\beta_1\beta_1$	1HDZ	Cys	2	Cys$_{b\alpha}$	2	Cys$_\alpha$	7	Cys	10
$\beta_2\beta_2$	1HDY	Cys	2	Cys$_{b\alpha}$	2	Cys$_\alpha$	7	Cys	11
$\beta_3\beta_3$	1DEH	Cys	2	Cys$_{b\alpha}$	2	Cys$_\alpha$	7	Cys	12
$\gamma\gamma$	1HT0	Cys	2	Cys$_{b\alpha}$	2	Cys$_\alpha$	7	Cys	9
Human Class III									
$\chi\chi$	1TEH	Cys	2	Cys$_{b\alpha}$	2	Cys$_\alpha$	7	Cys	13
or glutathione-dependent formaldehyde dehydrogenase	1M6H	Cys	2	Cys$_{b\alpha}$	2	Cys$_\alpha$	7	Cys	14
Human Class IV									
$\sigma\sigma$	1AGN	Cys	2	Cys$_{b\alpha}$	2	Cys$_\alpha$	7	Cys	15
Silverleaf whitefly ketose reductase	1E3J	Cys$_L$	2	Cys$_{b\alpha}$	2	Cys$_\alpha$	7	Cys	16
Sulfolobus solfataricus ADH	1JVB	Glu$_L$	2	Cys$_{b\alpha}$	2	Cys$_\alpha$	7	Cys	17
Thermoplasma acidophilum Glucose dehydrogenase		Cys	2	Cys	7	Cys	5	Asp$_{b\beta}$	18
Bovine heart cytochrome c oxidase	1OCC	Cys$_\beta$	1	Cys$_{2a\beta}$	19	Cys$_\beta$	2	Cys$_L$	19
				CLASS II: TRANSFERASES					
Aspartate carbamoyltransferase	1AT1	Cys	4	Cys$_{2b\alpha}$	23	Cys$_\beta$	2	Cys$_L$	20
Human Burton's tyrosine kinase	1BTK	His$_{a\beta}$	10	Cys$_{a\beta}$	0	Cys$_{2a\beta}$	9	Cys$_{b\beta}$	21
Bacillus stearothermophilus adenylate kinase	1ZIP	Cys$_\beta$	2	Cys$_L$	16	Cys$_\beta$	2	Cys$_L$	22
Protein kinase family									
Rat protein kinase C-α	1TBN	Cys$_\beta$	2	Cys$_L$	21	His$_{a\beta}$	2	Cys$_{b\alpha}$	23,24 NMR
		His$_{b\beta}$	29	Cys$_\beta$	2	Cys$_L$	15	Cys	
Rabbit C-α Cys-rich domain		Cys$_\beta$	2	Cys$_L$	21	His$_\beta$	2	Cys$_\alpha$	25 NMR
		His$_{b\beta}$	29	Cys$_\beta$	2	Cys$_{b\beta}$	15	Cys	
Mouse C-δ activator binding domain	1PTQ	Cys$_\beta$	2	Cys$_L$	21	His$_{a\beta}$	2	Cys$_\alpha$	26
		His$_{b\beta}$	29	Cys$_\beta$	2	Cys$_L$	15	Cys	
Human Raf-1	1FAR	Cys$_\beta$	2	Cys$_L$	17	His$_{a\beta}$	2	Cys	27 NMR
		His	25	Cys$_\beta$	2	Cys$_L$	15	Cys	
Human casein kinase, Truncated dimerCK2β	1QF8	Cys$_\beta$	4	Cys$_L$	22	Cys$_\beta$	2	Cys$_L$	28
		Cys$_\beta$	4	Cys$_L$	22	Cys$_\beta$	2	Cys$_L$	
Human casein kinase, Holoenzyme CK2	1JWH	Cys$_\beta$	4	Cys	22	Cys$_\beta$	2	Cys$_L$	29
		Cys$_\beta$	4	Cys$_L$	22	Cys$_{a\beta}$	2	Cys	
Yeast RNA polymerase II subunit A	1K83	Cys	2	Cys	16	Cys	2	His	30
		Cys	2	Cys	37	Cys$_\beta$	18	Cys$_L$	
Yeast RNA polymerase II subunit B	1K83	Cys$_\beta$	2	Cys$_{2a\beta}$	15	Cys$_{a\beta}$	2	Cys	
Yeast RNA polymerase II subunit C	1K83	Cys	1	Cys	3	Cys	2	Cys$_{2b\beta}$	
Yeast RNA polymerase II subunit I	1K83	Cys	2	Cys	17	Cys$_\beta$	2	Cys$_L$	
		Cys	2	Cys	24	Cys$_\beta$	2	Cys$_L$	
Yeast RNA polymerase II subunit J	1K83	Cys$_\beta$	2	Cys$_L$	34	Cys$_\alpha$	0	Cys$_\alpha$	
Yeast RNA polymerase II subunit L	1K83	Cys	2	Cys	13	Cys	2	Cys (CO)	
Thermococcus celer RNA polymerase II RPB9	1QYP	Cys$_\beta$	2	Cys$_L$	24	Cys$_\beta$	2	Cys$_L$	31 NMR
Methanobacterium thermautotrophicum RNA polymerase subunit RPB10	1EF4	Cys	2	Cys	33	Cys$_\alpha$	0	Cys$_\alpha$	32 NMR
Thermus aquaticus RNA polymerase	1HQM	Cys	81	Cys	6	Cys	2	Cys	33,34
B. stearothermophilus DNA primase	1D0Q	Cys$_{2a\beta}$	2	His	17	Cys$_\beta$	2	Cys$_L$	35

Table 1 (continued)

Enzyme	PDB#	L$_1$	X	L$_2$	Y	L$_3$	Z	L$_4$	References
Pyrococcus furiosus DNA primase	1G71	Cys	3	His	5	Cys$_{b\alpha}$	6	Cys$_\alpha$	36
E. coli DNA polymerase III δ′-subunit	1A5T	Cys$_{a\alpha}$	8	Cys	2	Cys$_{b\alpha}$	2	Cys$_\alpha$	37
E. coli DNA polymerase III γ-subunit	1JR3	Cys	8	Cys	2	Cys$_{b\alpha}$	2	Cys$_\alpha$	38
Human O-6-alkylguanine-DNA alkyltransferase	1EH6	Cys	18	Cys$_\beta$	4	His$_\beta$	55	His	39
Galactose-1-phosphate uridylyltransferase	1GUP	Cys	2	Cys	59	His	48	His$_{b\beta}$	40,41
CLASS III: HYDROLASES									
Matrix metalloproteinase family									
Human fibroblast collagenase (MMP-1)	1CGL	His	1	Asp	12	His$_\beta$	12	His$_\beta$	42,43
Human fibroblast collagenase (MMP-1)	3AYK	His	1	Asp	12	His$_\beta$	12	His$_\beta$	43,44 NMR
Human progelatinase 72 kDa (MMP-2)	1CK7	His	1	Asp	12	His$_\beta$	12	His$_\beta$	45
Human stromelysin-1 (MMP-3)	2SRT, 1BM6	His	1	Asp	12	His$_\beta$	12	His$_\beta$	46–48 NMR
Human stromelysin-1 (MMP-3)	1B3D	His	1	Asp	12	His$_\beta$	12	His$_\beta$	49,50
Human prostromelysin-1 (MMP-3)	1SLM	His	1	Asp	12	His$_\beta$	12	His$_\beta$	51
Human matrilysin (MMP-7)	1MMP	His	1	Asp	12	His$_\beta$	12	His$_\beta$	52
Human neutrophil collagenase (MMP-8)	1KBC	His	1	Asp	12	His$_\beta$	12	His$_\beta$	53,54
Human gelatinase 92 kDa (MMP-9)	1GKC	His	1	Asp	12	His$_\beta$	12	His$_\beta$	55
Human macrophage elastase (MMP-12)	1JIZ	His	1	Asp	12	His$_\beta$	12	His$_\beta$	56
Human collagenase-3 (MMP-13)	830C	His	1	Asp	12	His$_\beta$	12	His$_\beta$	57
Mouse collagenase-3 (MMP-13)	1CXV	His	1	Asp	12	His$_\beta$	12	His$_\beta$	58
Human membrane Mt1 (MMP-14)	1BQQ	His	1	Asp	12	His$_\beta$	12	His$_\beta$	59
B. stearothermophilus L1 lipase	1KU0	Asp$_\alpha$	19	His$_\alpha$	5	His$_L$	150	Asp$_{2b\alpha}$	60
Physarum polycephalum endonuclease I-PpoI	1A73, 1EVX	Cys$_\beta$	58	Cys$_{a\beta}$	4	Cys$_\beta$	4	His$_\beta$	61,62
		Cys$_{2a\alpha}$	6	Cys	1	His	3	Cys	
Bacteriophage T4 endonuclease VII	1EN7, 1ESD	Cys$_\beta$	2	Cys$_L$	31	Cys$_L$	2	Cys$_\alpha$	63,64
E. coli DNA mismatch endonuclease	1VSR	Cys$_\alpha$	4	Cys$_{2a\alpha}$	1	Cys	43	Cys$_\alpha$	65
Formamidopyrimidine-DNA glycosylase MutM (Fpg) family									
Thermus thermophilus HB8 MutM	1EE8	Cys$_\beta$	2	Cys$_L$	16	Cys$_{a\beta}$	2	Cys	66
Lactococcus lactis Fpg	1KFV	Cys$_\beta$	2	Cys$_L$	16	Cys$_{a\beta}$	2	Cys	67
E. coli Fpg	1K82	Cys$_\beta$	2	Cys$_L$	16	Cys$_{a\beta}$	2	Cys	68
B. stearothermophilus MutM	1L1T	Cys$_\beta$	2	Cys$_L$	16	Cys$_{a\beta}$	2	Cys	69
E. coli Endonuclease VIII (Nei)	1K3W	Cys$_\beta$	2	Cys$_L$	16	Cys$_{a\beta}$	2	Cys	70
Human picornavirus endoprotease 2A	2HRV	Cys	1	Cys$_{b\beta}$	57	Cys$_\beta$	1	His$_L$	71
CLASS V: ISOMERASES									
E. coli Rhamnose isomerase	1DE5	Glu$_\beta$	32	Asp$_\beta$	26	His$_\beta$	39	Asp$_\beta$	72
CLASS VI: LIGASES									
tRNA synthetase family									
E. coli MetRS	1QQT, 1F4L	Cys$_\beta$	2	Cys$_L$	9	Cys$_\beta$	2	Cys$_L$	73 NMR 74,75
T. thermophilus HB8 MetRS	1A8H	Cys$_\beta$	2	Cys$_L$	13	Cys	2	His	76
T. thermophilus HB8 ProRS	1HC7	Cys$_L$	4	Cys$_\alpha$	25	Cys$_\beta$	2	Cys$_L$	77,78
T. thermophilus IleRS	1ILE	Cys$_\beta$	2	Cys$_L$	37	Cys$_\beta$	1	Cys$_L$	79
		Cys$_\beta$	2	Cys$_L$	204	Cys$_\beta$	2	Cys$_L$	
T. thermophilus ValRS	1GAX	Cys$_\beta$	2	Cys$_L$	17	Cys	2	Cys	80
		Cys$_\beta$	2	Cys$_L$	164	Cys$_\beta$	2	Cys$_L$	
E. coli IleRS	1QU2	Cys$_\beta$	2	Cys$_L$	16	Cys$_{b\alpha}$	2	Cys$_\alpha$	81
Staphylococcus aureus IleRS	1FFY	Cys$_\beta$	7	Cys$_L$	6	Cys$_{b\alpha}$	2	Cys$_\alpha$	81
HIV-1 integrase	1WJA	His	3	His$_\alpha$	23	Cys$_L$	2	Cys$_\alpha$	82 NMR
HIV-2 integrase	1E0E	His$_\alpha$	3	His$_L$	23	Cys$_L$	2	Cys$_\alpha$	83 NMR
Thermus filiformis DNA ligase	1DGS	Cys$_\beta$	2	Cys$_L$	12	Cys$_\beta$	4	Cys	84

[a] The amino acid spacers between consecutive ligands are X, Y, and Z. The subscripts α, β refer to the α or 3$_{10}$ helix and β-sheet structure respectively, which supply the ligand. The letter subscript L denotes an amino acid sequence of ≤6 residues between 2 structural elements. The subscripts a and b indicate that the ligand is either one (or two) residues after or before the secondary structural element. CO is a carboxyl group.

to see the word 'novel' associated with new sites because of the difficulty in classifying the site in terms of the relatively small number of proposed names such as zinc-box, -bundle, -cluster, -finger, -knuckle, -ribbon, -ring, and so on. The one feature they have in common is the presence of four protein ligands and no coordinated water molecule.[1,2]

CLASS I: OXIDOREDUCTASES

Dehydrogenase family

The alcohol and polyol dehydrogenase family contains medium-sized, oligomeric enzymes, most of which contain zinc.[90] Many of these enzymes contain both a catalytic and a structural zinc site.[2] The crystal structure of horse liver ADH shows that the subunits of the dimeric enzyme can be further divided into two domains that are separated by a cleft containing a deep pocket that accommodates the substrate and the nicotinamide moiety of the coenzyme that is crucial for catalysis.[91] One of the domains is the catalytic domain that contains both types of zinc sites (separated by 19.3 Å), while the other domain is involved in binding the coenzyme, NADH. The structural zinc ion is bound to four sulfur atoms of cysteines in a short sequence with spacers of 2, 2, and 7 amino acids between cysteine ligands. The amino acid segment of these enzymes (residues 93 to 128) is not organized in any particular secondary structure except for a short helix of four amino acids that starts after the second cysteine ligand and contains the third one (Table 1). This zinc site is highly conserved in both the dimeric and tetrameric ADH.[1] On the other hand, neither the mammalian sorbitol dehydrogenases nor the bacterial NADP(H)-dependent ADHs have this motif fully conserved and they generally have no structural zinc. However, the recent X-ray structure of the tetrameric NADP(H)-dependent sorbitol dehydrogenase from silverleaf whitefly, *Bemisia argentifolii*,[16] contains a structural zinc site that is very similar to that found in the dimeric ADHs (Table 1).

The structural zinc site in the mammalian ADHs is part of a lobe that projects out of the catalytic domain.[91] The zinc atom is not accessible to solution and the lobe makes few interactions with the remaining part of the subunit. The exact function of this zinc site is not known but the nearness of the site to the subunit interface has led to the suggestion that it may be important to protein refolding. Removal of metal from this site in the tetrameric yeast ADH leads to decreased protein stability consistent with a structural role.[92] The presence of the metal site near the surface of the protein, however, indicates that it may have other, as yet unidentified, roles.

Sites containing cysteine and aspartic acid/glutamic acid

Structural zinc sites that contain a carboxylic acid containing amino acid residue, for example, glutamic acid or aspartic acid in combination with cysteine residues have recently been reported in archaeon dehydrogenases. The archaea kingdom contains organisms distinct from bacteria and eukaryotes that are of interest because they function in extreme environments such as volcanic areas and sulfur springs.[93] The catalytic domain of the glucose dehydrogenase from the thermophilic archaeon *T. acidophilum* shows significant structural homology to the catalytic domain of horse liver alcohol dehydrogenase.[18] It possesses a structural zinc site that is composed of one aspartic acid and three cysteine ligands (Table 1).

The alcohol dehydrogenase from the hyperthermophilic sulfur-dependent archaeon *S. solfataricus*, SsADH, contains a structural zinc site composed of one glutamic acid and three cysteine ligands.[17] The enzymatic activity of SsADH is still increasing at 95 °C and the half-life for activity is 5 h at 70 °C,[94] stimulating interest in understanding the structural reasons for the stability. Removal of the structural zinc site with chelators results in a concomitant loss of structural stability.[95] SsADH is an NADH-dependent homotetrameric enzyme containing both structural and catalytic zinc sites in each subunit.[17] The catalytic zinc site is quite similar to the mammalian ADHs,[2] including an interaction with the glutamic acid residue next to the histidine ligand in some of the crystal forms. The structural zinc site differs in one regard – the replacement of the first cysteine ligand by a glutamic acid ligand. The presence of the glutamic acid with its network of H-bonds is considered to impart a higher rigidity to the structural lobe supplying the ligands to this site.[17] However, replacement of the equivalent cysteine residue with a glutamic acid in mammalian Class III χχ ADH results in very low yields of an unstable enzyme that does not bind zinc.[96] The results of mutation of any of the cysteine residues to alanine or serine indicate that the translated, mutated proteins lacking the zinc-stabilized local fold are subject to rapid degradation. These results have led to the suggestion that the zinc site in the mammalian enzymes may be involved in the protein folding, while the glutamic acid–containing site in the archaeon enzymes may be important to tetramer stabilization.[17]

Cytochrome c oxidase

Cytochrome c oxidase is the terminal enzyme in the respiratory chain, a process that reduces molecular oxygen to water with the electrons from cytochrome c, coupled with pumping protons from the matrix side of the mitochondrial membrane toward the cytosolic side. The crystal structure of the bovine heart mitochondrion cytochrome c oxidase reveals a dimeric enzyme, each containing 13 different subunits.[97] Two heme iron, two copper, one magnesium, and one zinc site are found in the protein. The carboxyl terminal domain of subunit Vb (full length 98 amino acids) contains a bound zinc with the properties of a structural zinc site. The amino acid sequence between amino acids

54 and 94 contains four short β-sheets that are involved in providing the cysteine ligands (Table 1). The zinc site is far removed from the inner membrane side of the protein that contains the other metal sites. The function of the site is unknown at present.

CLASS II: TRANSFERASES

Aspartate carbamoyl transferase

One of the first structural zinc sites was identified in the regulatory subunit of *E. coli* aspartate carbamoyl transferase.[5] The zinc is bound tetrahedrally to four cysteine residues separated by 4, 23, and 2 amino acids. The zinc atom holds together two peptide loops that form part of the interface between the regulatory and catalytic subunits. Zinc probably influences the local conformation and structure of the regulatory subunit, thus fine-tuning its interaction with the catalytic subunit.

Burton's tyrosine kinase

This cytoplasmic protein tyrosine kinase is an enzyme that is involved in maturation of B-cells. The structure of the N-terminal part of the enzyme reveals a pleckstrin homology (PH) domain followed by a Burton's tyrosine kinase, Btk, motif.[21] The Btk motif is a target for mutations causing X-linked agammaglobulinaemia, XLA, in man. The globular core of this motif packs against the β-strands 5–7 of the PH domain. The structure is described as a long loop that folds back upon itself and is held together by the zinc binding site. The zinc is bound in a distorted tetrahedral geometry to the ND1 nitrogen of His143, and the sulfurs of Cys154, Cys155, and Cys165, which are conserved in enzymes having the Btk motif. All of the ligands come from amino acids residing just after or before very short β-sheets (Table 1).

Adenylate kinase

The zinc binding site in *B. stearothermophilus* adenylate kinase was first investigated since it had a sequence similar to zinc fingers.[22] However, the structure shows that the role of the zinc is probably structural. It is bound to four cysteine residues in the lid domain. The zinc site replaces an intricate network of hydrogen bonds that hold the lid together in the structures of gram-negative bacteria and all of the eukaryotic large-form adenylate kinases. Gram-positive bacteria, on the other hand, have the potential of forming this type of zinc site. However, sequence alignment and mutagenic studies of the adenylate kinase from *Bacillus subtilis* suggest that the fourth cysteine is replaced by an aspartic acid residue in this structural zinc site.[98] On the basis of the knowledge gained on these systems, a structural zinc binding site equivalent to the *B. stearothermophilus* site was engineered into the *E. coli* adenylate kinase that does not normally bind zinc.[99] The incorporation of the zinc site into the *E. coli* enzyme improved its thermal stability while not adversely affecting its catalytic properties in agreement with a structural role for this zinc site.

Protein kinase C family

This group of enzymes contains an interwoven zinc site. Thus, one zinc site is formed from residues residing within the two long spacers of the other zinc site (Figure 2). These enzymes are serine/threonine protein kinases that depend on phospholipids and diacyl glycerol and are known to play crucial roles in intracellular signal transduction events elicited by various extracellular stimuli such as growth factors, hormones, and neurotransmitters.[100] Rat brain protein kinase C, PKC, is a zinc metalloenzyme in which zinc is bound within the lipid binding domain.[100] NMR studies indicate that the regulatory domain of rat PKC contains two cysteine-rich, independent subdomains in which two zinc atoms each are bound in a C_3H coordination.[23,24] These studies are in agreement with an XAFS study[101] that favored 3S/1N ligand sites without bridging zinc sites. The same type of site is observed in the crystal structure of the cysteine-rich activator domain of mouse PKCδ.[26] The metal ligands from the two zinc sites are interwoven. The ligands in the outer Zn site are His231, Cys261, Cys264, and Cys280, while those in the inner zinc site are Cys244, Cys247, His269, and Cys272 (Figure 2). Thus, the C2C ligation spacer of the inner zinc site is contained within the first 29 amino acid spacer of the outer zinc site, while the H2C ligation spacer of the inner zinc site resides within the last 15 amino acid spacer of the outer zinc site. This interwoven type of zinc site has been

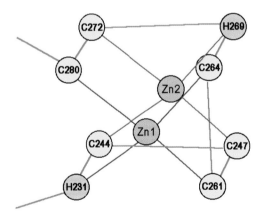

Figure 2 Interwoven zinc sites observed in the protein kinases. The schematic is based on the zinc sites in the Cys2 activator binding domain of mouse protein kinase Cδ.[26] The peptide backbone is shown in green and the coordination bonds for the inner and outer zinc sites are given in red and blue respectively.

observed in the solution structures of human Raf-1 protein kinase[27] and rabbit PKC-α[25] (Table 1). The zinc ligands are conserved in C1 domains of protein kinases encompassing a wide range of organisms.[102]

Casein kinase II, CK2

Protein kinase CK2 is considered to be an unspecific eukaryotic protein kinase because of its ability to phosphorylate numerous proteins on serine/threonine and tyrosine and using either ATP or GTP as phosphoryl donors.[29] The crystal structures of the human truncated β-subunit,[28] and the holoenzyme containing two C-terminal truncated catalytic α-subunits and two regulatory β-subunits,[29] have been reported. Zinc binds to four cysteine residues, separated by 2, 22, and 2 amino acids in the regulatory subunit. The ligands are positioned within, between, or before β-sheets (Table 1). The truncated regulatory β-subunit forms a dimer with a number of protein interface interactions coming from amino acids originating from within the zinc binding peptide.[28] The holoenzyme has the shape of a butterfly in which the regulatory subunits form the dimer interface through the zinc sites and the catalytic subunits do not interact with one another. The zinc sites are unusual in the respect that the second cysteine ligand has a short coordination bond length of 2.0 to 2.2 Å, while the fourth cysteine sulfur is 2.7 to 2.9 Å from the zinc.

DNA-dependent RNA polymerases (RNAP)

The crystal structure of the 10 subunit yeast (*Saccharomyces cerevisiae*) RNA polymerase II (4181 amino acids) has been determined at 2.8-Å resolution.[30] Eight zinc binding sites are found in subunits A(2), B, C, I(2), J, and L (Table 1). No two sites are identical. They are formed from as small a linear amino acid sequence as 10 or as large as 61. Nearly all the sites contain four cysteine ligands. However, one has a histidine ligand and one apparently uses the carbonyl oxygen of a Cys (2.55 Å) instead of the S (3.73 Å) as the ligand. Ligands generally come from within, or just before or after β-sheets. However, in one case, an α-helix supplies two ligands with no spacer between. The function of these sites is, as yet, unknown.

NMR structures of the zinc sites in peptides from subunits RPB9 and RPB10 of *T. celer*[31] and *M. thermautotrophicum*[32] RNA polymerase II, respectively, have been reported. These zinc sites correspond to the spacing characteristics of those found in subunits I and J of yeast RNA II polymerase (Table 1). The zinc site in the peptide from the RPB10 subunit of *M. thermautotrophicum*[32] is called a *zinc bundle*. The zinc is bound tetrahedrally to the cysteine sulfurs of the $CX_2CX_{33}CC$ sequence motif and stabilizes the overall structure of the oligopeptide. Removal of the zinc by EDTA treatment leads to loss in 222-nm circular dichroism intensity indicative of the unfolding of the peptide. A CX_2CX_nCC motif is found in eukaryotic, archaeal, and viral RNA polymerases. Site-directed mutagenesis of the cysteine ligands in this motif of the ABC10β protein, a conserved subunit shared by all three yeast RNA polymerases, shows that it is essential for yeast growth.[103]

Both the zinc binding sites found in subunit I in yeast RNA polymerase II have been predicted on the basis of alignment of sequences of archaea and eukaryotic RNA polymerases.[31] Eukaryotic RNA polymerase II subunit 9 (RPB9) regulates start-site selection and elongation arrest. The NMR structure of the polypeptide corresponding to residues 58–110 of subunit 9, containing the zinc-binding motif $CX_2CX_{24}CX_2C$, shows that the ligands to the zinc come from within or from loops between β-sheets (Table 1). This structure, termed a zinc ribbon, is also found in the human transcription factor IIS, TFIIS,[104] and the archaeal TFIIB.[105]

The *T. aquaticus* RNA polymerase has a core structure of five subunits ($α_2ββ'ω$). The β' subunit contains a zinc binding site not readily predicted because of a long spacer of 81 amino acids.[33,34] However, the cysteine ligands are conserved in prokaryotes but not in eukaryotes. The zinc–sulfur ligand distances are quite unusual, being 3.13, 1.88, 2.11, and 2.65 Å. The location of the site suggests that it plays a structural role in the folding of the subunit.[33] The amino-terminal zinc binding domain may also contact the DNA backbone.[106]

DNA primases catalyze the synthesis of short oligoribonucleotides on single-stranded DNA. These short strands of RNA then serve as primers for DNA replication by DNA polymerases. The primases are multidomain enzymes containing a zinc binding domain that is responsible for template recognition, a polymerase domain, and a domain that interacts with the replicative helicase.[35] The properties of the zinc binding sites in the primases from the bacteria *B. stearothermophilus*,[35] and the hyperthermophilic archaeon *Pyrococcus furiosus*,[36] differ markedly in the spacing characteristics and type of secondary structure involved in supplying the ligands (Table 1). The primase from *B. stearothermophilus* is referred to as a member of the zinc ribbon family, while the *P. furiosus* primase is described as a novel zinc knuckle.

DNA transferase enzymes

The crystal structure of the processivity clamp loader γ-complex of *E. coli* DNA polymerase III has been reported recently.[38] The γ-complex, consisting of a minimum of three kinds of subunits (γ, δ', and δ), loads the sliding clamp β onto the DNA in bacterial replication systems. The crystal structure reveals a pentameric arrangement of subunits of the form δ'γγγδ. The structure suggests a mechanism in which an element in δ that interacts with β is either hidden by δ' in a closed protein state or exposed in an open state as is observed in the crystal structure. Essentially, the same type of zinc binding site occurs in the δ′- and γ-subunits but not in the δ-subunit.[37,38] While the zinc site uses cysteine ligands as do many of the other Class II enzymes discussed above, this site depends on α-helices to supply the ligands (Table 1). The function of the zinc site is unknown. However, all the zinc sites are in the N-terminal region of each subunit and are part of the outer collar of the clamp holder, far away from the DNA, suggesting that the sites may play some structural role.

The human O^6-alkylguanine-DNA alkyl transferase (AGT) directly reverses endogenous alkylation at the O^6-position of guanine and thus confers resistance to alkylation chemotherapies.[39] The crystal structure reveals a zinc binding site that is quite different from other members of this class of enzymes (Table 1). The zinc site bridges three strands of β-sheets and the coil immediately preceding a domain-spanning helix. The zinc is on an active site face but 20 Å from the cysteine involved in the alkylation process. The role of the zinc is likely to be involved in stabilizing the interface between the N- and C-terminal domains of the protein.

Galactose-1-phosphate uridylyltransferase

This enzyme plays an important role in the metabolism of galactose, a major constituent of dairy products. The uridylyltransferase from E. coli is a homodimer containing one equivalent each of iron and one zinc ion per monomer of 348 amino acids.[40] The zinc and iron are 8 and 30 Å respectively from the active site. Mutagenesis studies have indicated that zinc is essential for the activity while iron is not.[41] The zinc site has one short spacer and two very long ones, spanning about one-third of the protein (Table 1). A combination of kinetics and site-directed mutagenesis studies has suggested that His166 is critical to the catalysis of this enzyme. This finding may explain why the zinc is important for this activity since one of the zinc ligands is His164. The triad His164Pro165His166 is highly conserved in these enzymes. The backbone carbonyl of His166 forms a H-bond with the ND1proton of His166. The zinc site, while likely important to overall protein folding, also influences local conformation, particularly by maintaining the proper alignment of His166 for catalysis as well as influencing its acid/base properties.

CLASS III: HYDROLASES

There are a number of examples in this class of enzymes in which the structural zinc site may indirectly affect function. Thus, one or more amino acid residues within the active site may be provided by the amino acid spacers between zinc ligands (Figure 3). The side chain of these amino acids may be involved in substrate binding, bond cleavage, or in modulating the chemical environment of the active site. In addition, other active site residues are often provided by the peptide sequence N-terminal to the structural zinc site.

Matrix metalloproteases

There are only a few examples in which aspartic acid and glutamic acid in combination with histidine residues are found (Table 1). Metalloproteases frequently use either disulfides or calcium ions to aid in stabilizing the structure of the enzyme. However, the matrix metalloproteinase class of enzymes contains a noncysteine zinc binding site that has the characteristics of a structural zinc site, that is, four ligands, and no metal-bound water. Early metal analyses of the matrix metalloproteinases (MMPs) indicated that these enzymes contained a zinc site in addition to the catalytic one proposed for the entire astacin superfamily.[107] This site contains three histidine residues and an aspartic acid residue in a linear sequence spanning 28 amino acids (Table 1).[108,109] The site is highly conserved in the MMPs, occurring in the great majority of sequence-identified MMPs (125/138 sequences in MEROPS 1.7 data set)[110] and has a signature of $HX_1DX_{12}HX_{12}H$. As is probably the case for many structural metal sites, this site may be indirectly affecting the activity of the enzyme. Although this zinc is separated by 12 Å from the catalytic zinc site, the amino acids adjacent to the third and fourth ligands (histidine) to the 'structural' zinc site provide a number of hydrophobic residues that border the catalytic glutamic acid residue. These regions contain Leu-Ala and Ala-Phe/Tyr sequences on the two sides of one histidine and an alanine and phenylalanine adjacent to the other histidine. These residues could provide a hydrophobic environment for the catalytic glutamic acid carboxylate group of the MMPs and thus raise its pK_a, allowing it to

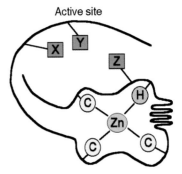

Figure 3 Schematic of a structural zinc binding site that provides active site residues from amino acids located within the metal coordination spacers. Other active site residues are often provided by the peptide sequence N-terminal to the structural zinc site. The symbol Z can represent more than one amino acid. The cysteine and histidine ligands can also vary for the different enzymes (Table 1).

L1 lipase

The thermoalkylophilic lipase from *B. stearothermophilus* L1 contains a zinc binding site in an extra domain that accounts for the larger molecular size for these lipases in comparison to other microbial lipases.[60] This extra domain and the ligands to zinc are conserved in other members of the lipase family I.5. The zinc is tetrahedrally coordinated to one oxygen of the side chain carboxylates of Asp61 and Asp238 and the NE2 nitrogens of His81 and His87 (Table 1). Incubation of the enzyme with the zinc chelator, N,N,N',N' tetrakis(2-pyridyl-methylenediamine) (TPEN) lowers the thermotolerance of the enzyme by 20 °C, suggesting that zinc is important to the thermostability of the enzyme.[60] The extra domain may also play a role in the regulation of the lid opening of these enzymes since the region of the zinc binding site makes tight contacts with a long loop that is extended from the C-terminus of the lid helix.

Endonucleases

Three endonucleases have structural zinc sites that have little in common (Table 1). Homing endonucleases recognize 14 to 40-bp DNA targets.[61] They have been grouped into four families on the basis of conserved sequence motifs, one of which is for zinc binding (histidine–cysteine box). The enzyme from *P. polycephalum* is a homodimer containing two zinc sites per monomer of 324 amino acids. The two zinc sites originate in the N-terminus of the protein-spanning residues 41 to 138. The first zinc site is formed from Cys41, Cys100, Cys105, and His110 supplied by three β-sheets, while the second zinc site uses Cys125, Cys132, His134, and Cys138 as ligands that are supplied from a largely unstructured region of the protein except for one short α-helix (Table 1). The two zinc ions are separated by 15.3 Å. Both sites contain three cysteine and one histidine ligand, although the order of ligand coordination is different. The second zinc site may play a role in stabilizing the dimer interface since the indole rings of Trp125 from each monomer are stacked against each other.[62] Neither site is involved in DNA binding since both are 15 to 17 Å from the substrate complex. However, the active site His98 amino acid is one residue away from the second cysteine ligand in the first zinc site.

Phage 4 endonuclease VII, a homodimer (157 amino acid per monomer), recognizes a diverse group of DNA substrates ranging from branch chain to single base pair mismatches.[63] The dimer is formed by the monomers binding to each other in the reverse direction (N to C). The zinc site appears to stabilize the protein fold by placing two helices from each monomer in juxtaposition to each other. Mutation of the outer cysteine ligands to serine results in loss of zinc binding and enzyme activity, while the analogous mutations of the inner cysteine ligands, one at a time, still binds zinc to some degree.[112]

The *E. coli* very short patch endonuclease, Vsr, plays a crucial role in the repair of TG mismatched base pairs that are generated by the spontaneous degradation of methylated cytidines.[65] The enzyme contains a zinc site on the C-terminus end of the protein. An α-helix is involved in supplying the first two ligands, while a β-sheet supplies the last ligand (Table 1). The zinc is believed to be involved in stabilizing protein structure. The catalytic Mn atom is 12 Å removed from the zinc site.

Formamidopyrimidine-DNA glycosylase family

The bacterial formamidopyrimidine-DNA glycosylase (alias Fpg or MutM) is a bifunctional base repair enzyme (DNA glycosylase/AP lyase) that removes a wide range of oxidized purines from oxidatively damaged DNA.[67] The crystal structures of the Fpg (MutM) enzymes from *T. thermophilus*,[66] *L. lactis*,[67] *E. coli*,[68] *B. stearothermophilus*,[69] and the sequence similar-endonuclease VIII (Nei) from *E. coli*,[70] all share the same type of structural zinc binding site at the end of the C-terminus of the proteins. This zinc site is composed of four cysteine sulfur ligands (Table 1). The first three cysteines are supplied by two β-sheets. The zinc site influences the function of the enzyme through an amino acid located within its metal-binding spacer. The guanidinium group of an arginine residue from within the large central coordination spacer makes two H-bonds with the p^{-1} and p^0 phosphates of the pyrimidine cleavage site of the substrate. In addition, several other binding interactions come from the N-terminal side of the zinc binding site.

Rhinovirus-2 proteinase

The human 2A proteinase from the common cold (rhinovirus) is structurally related to the cysteine proteases.[71] Its active site contains a characteristic catalytic AspHisCys triad. It also contains a structural zinc binding site in the N-terminus of the protein composed of Cys52, Cys54, Cys112, and His114 as ligand donors (Table 1). Two β-sheets supply the last three amino acids. The site is believed to be structural since it is not accessible to other ligands and is unlikely to have any functional role in RNA binding. Chelation and site-directed mutagenesis studies indicate that zinc is essential for the formation of an enzymatically active form of the enzyme.[113,114] The zinc-depleted enzyme shows a mostly unchanged secondary structure as determined by CD spectroscopy, but not a fully denatured random coil as obtained by guanidinium hydrochloride.[115] The zinc may therefore be stabilizing local conformation of the protein that is critical for activity.

This zinc site also probably influences enzymatic function through its effect on local conformation. In this case, the crucial catalytic residue Cys106 is supplied from within the large central coordination spacer and the other members of the triad, His14 and Asp35, come from the N-terminal side of the zinc binding site.

CLASS V: ISOMERASES

Rhamnose isomerase

The crystal structure of *E. coli* rhamnose isomerase[72] shows zinc bound to ligands in the amino acid sequence Glu (OE1)-X_{31}-Asp (OD1)-X_{26}-His (ND1)-X_{39}-Asp (OD1) (Table 1). This is one of the few zinc sites where no short spacer is found. In the presence of a substrate, a second 'catalytic' manganese binding site is found in close proximity to this zinc binding site. The catalytic site appears to be occupied only when the substrate is present, suggesting that the true substrate is the Mn-bound form of it. Two very similar metal binding sites are observed in *Streptomyces rubiginous* xylose isomerase,[116] although the overall identity to rhamnose isomerase is only 13%.[72] Manganese is identified as binding to both metal sites in xylose isomerase.[116] In this case, the structural site is made up of two aspartic acid and two glutamic acid ligands and no histidine ligands. It may be that a site composed of multiple glutamic acid and aspartic acid residues only is too flexible for zinc and leads to weak binding constants due to fast dissociation rates of the zinc from such sites. The usual presence of Ca^{2+} and Mg^{2+} in such acidic ligand sites correlates with the weak binding constants of such ions for protein binding sites and the fast dissociation rate constants for such metals.

The ligand-binding characteristics of Zn2 in *E. coli* endonuclease IV are very similar to those found in the structural zinc binding site in *E. coli* rhamnose isomerase (Table 1).[117] In addition, a protein that may play a role in inositol catabolism has a structural homology to *E. coli* endonuclease IV and xylose isomerase.[118] This protein also has a single zinc binding site that is very similar to that of the structural zinc site found in rhamnose isomerase (Table 1). The zinc is bound tetrahedrally to one carboxylate oxygen of Glu142, Asp174, and Glu246 and the ND1 nitrogen of His200 (PDB # 1I6N, note that ligand Glu142 is given as His177 in the paper). All of the ligands are provided by different β-sheets as in the rhamnose isomerase.

CLASS VI: LIGASES

tRNA synthetase family

Amino acyl-tRNA synthases play a crucial role in the protein translation apparatus by ensuring the proper esterification of tRNA with a particular amino acid. A few of these enzymes contain zinc binding sites. Three-dimensional structures of these enzymes have revealed a number of different types of zinc binding sites in this family of enzymes (Table 1). In nearly all cases, four cysteine amino acids tetrahedrally coordinate the zinc. Several of them have a binding motif of the type, $CX_2CX_nCX_2C$, where n ranges from 9 to 204 amino acids. The cysteine ligands are supplied from within a β-sheet or from a short loop between two β-sheets in nearly all of these sites. The varying size of the sites suggests that some zinc sites may be important to the overall fold of the protein, while many probably influence local protein conformation. Recent mutagenesis studies show that the cytoplasmic yeast MetRS[119] has a zinc binding site with the coordination properties of the *E. coli* MetRS (Table 1).[73,74]

HIV integrase

The human immunodeficiency virus (HIV) integrase is the enzyme responsible for the insertion of a DNA copy of the viral genome into the host DNA, an essential step in the replication cycle of the virus.[82] The N-terminal domain of the integrase contains a zinc binding site similar to those seen for zinc fingers. It is conserved in all the integrases. In this case, both the cysteine and histidine ligands are supplied by three α-helices (Table 1). The function of the site is unknown.

DNA ligase

The NAD^+-dependent DNA ligase from *T. filiformis* is a member of a superfamily that includes mammalian DNA ligases.[84] These enzymes catalyze the joining of breaks in double-stranded DNA during DNA replication, repair, and recombination. They are multidomain proteins. Subdomain 3A contains a zinc binding site composed of four closely spaced cysteine residues that bind a zinc ion (Table 1). The cysteine residues are strictly conserved in bacterial ligases. This site is said to structurally resemble the first of two zinc sites in the DNA binding domains of steroid/nuclear hormone receptors[87] that use a structural motif C-terminal to the zinc site to bind to the DNA recognition site. This zinc site is proposed to make a direct interaction with the nicked DNA as well as provide structural support to subdomain 3b and domain 4.[84]

Structural zinc sites

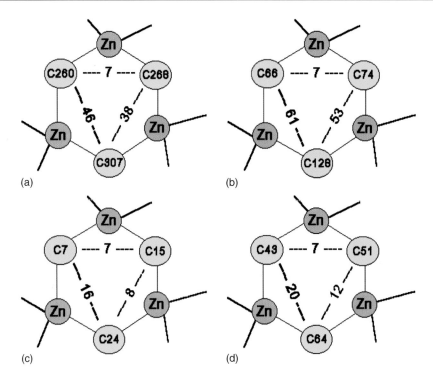

Figure 4 Properties of some zinc cluster sites observed in zinc enzymes and metallothionein. The bridging cysteine residues as well as their spacing are identified for (a) *S. pombe*. Lysine methyltransferase Clr4,[123] (b) *N. crassa* histone H3 methyltransferase DIM-5,[122] (c) β-domain of human metallothionein-2,[125] and (d) β-domain of sea urchin metallothionein MTA.[127] The PDB numbers are respectively: 1MVH, 1ML9, 1MHU, and 1QJL.

ZINC CLUSTERS

Class II: histone lysine methyl transferases

Histones undergo extensive posttranslational modifications such as acetylation, phosphorylation, and methylation on their N-terminal peptides that extend out of the nucleosome. Such modifications have been proposed to be part of a histone code that is recognized by cellular factors that influence gene expression.[120] Some of the SET (Su(var), Enhancer of zeste, Trithorax) domain proteins,[121] in combination with a cofactor S-adensoyl-L-methionine (SAM), are enzymes that methylate lysine 9 of histone H3. These enzymes contain minimally a histone methyltransferase (HMTase) catalytic domain that includes the conserved SET domain flanked by cysteine-rich pre- and post-SET motifs.[122,123] A zinc cluster-type binding site, Zn_3Cys_9, is believed to be important to the structural stability of the pre-SET domains of the histone lysine methyl transferases, DIM-5,[122] (the gene product of *Neurospora crassa* defective in methylation gene, *dim-5*,[124]) and the *Schizosaccharomyces pombe* cryptic loci regulator 4, Clr4.[123]

These sites have several features in common with the β-domain three metal clusters of eukaryote metallothioneins. All of these clusters can be viewed as simple hexagons in two dimensions (Figure 4). The hexagon is composed of alternating zinc ions and bridging cysteine ligands. Each zinc has two other nonbridging cysteine sulfurs ligated to it. In each case, there is a short spacer of 7 amino acids between two of the bridging cysteine ligands. However, the spacers to the third bridging cysteine are much larger for the histone methyltransferases. This is due to large inserts of 28 and 46 amino acids between the fifth and sixth cysteines in the Clr4 and DIM-5 HMTases respectively, compared to 1 and 3 amino acid spacers for the human,[125] mouse,[126] and sea urchin,[127] metallothioneins respectively. The spacing characteristics of the last four cysteine ligands in the HMTases is highly conserved compared to that of the first five cysteine ligands. A search using the motif [YFIK]ECX$_3$CXCX$_3$CX$_2$R leads to only 46 proteins, many of which are suspected to be HMTases (Auld, unpublished). In these cases, the majority of the putative large spacers between the fifth and sixth cysteine ligands are between 24 and 29 amino acids, while two are 46 and 66 amino acids. The resulting variation in the putative spacing of the first five cysteine residues is given by the subscripts in the sequence: $CX_{1-2}CX_{2-6,8}CX_{3-7}CX_{0-1}C$.

REFERENCES

1 BL Vallee and DS Auld, *Biochemistry*, **29**, 5647–59 (1990).
2 DS Auld, *BioMetals*, **14**, 271–313 (2001).
3 DS Auld, in I Bertini, A Sigel and H Sigel (eds.), *Handbook on Metalloproteins*, Marcel Dekker, New York, pp 881–959 (2001).

4   H Eklund, B Nordstrom, E Zeppezauer, G Soderlund, I Ohlsson, T Boiwe and CI Branden, *FEBS Lett*, **44**, 200–4 (1974).

5   RB Honzatko, JL Crawford, HL Monaco, JE Ladner, BF Ewards, DR Evans, SG Warren, DC Wiley, RC Ladner and WN Lipscomb, *J Mol Biol*, **160**, 219–63 (1982).

6   H Eklund, JP Samma, L Wallen, CI Branden, A Akeson and TA Jones, *J Mol Biol*, **146**, 561–87 (1981).

7   S Ramaswamy, M el Ahmad, O Danielsson, H Jornvall and H Eklund, *Protein Sci*, **5**, 663–71 (1996).

8   S Svensson, JO Hoog, G Schneider and T Sandalova, *J Mol Biol*, **302**, 441–53 (2000).

9   MS Niederhut, BJ Gibbons, S Perez-Miller and TD Hurley, *Protein Sci*, **10**, 697–706 (2001).

10 TD Hurley, WF Bosron, JA Hamilton and LM Amzel, *Proc Natl Acad Sci USA*, **88**, 8149–53 (1991).

11 TD Hurley, WF Bosron, CL Stone and LM Amzel, *J Mol Biol*, **239**, 415–29 (1994).

12 GJ Davis, WF Bosron, CL Stone, K Owusu-Dekyi and TD Hurley, *J Biol Chem*, **271**, 17057–61 (1996).

13 ZN Yang, WF Bosron and TD Hurley, *J Mol Biol*, **265**, 330–43 (1997).

14 PC Sanghani, H Robinson, WF Bosron and TD Hurley, *Biochemistry*, **41**, 10778–86 (2002).

15 P Xie, SH Parsons, DC Speckhard, WF Bosron and TD Hurley, *J Biol Chem*, **272**, 18558–63 (1997).

16 MJ Banfield, ME Salvucci, EN Baker and CA Smith, *J Mol Biol*, **306**, 239–50 (2001).

17 L Esposito, F Sica, CA Raia, A Giordano, M Rossi, L Mazzarella and A Zagari, *J Mol Biol*, **318**, 463–77 (2002).

18 J John, SJ Crennell, DW Hough, MJ Danson and GL Taylor, *Structure*, **2**, 385–93 (1994).

19 S Karlin, ZY Zhu and KD Karlin, *Biochemistry*, **37**, 17726–34 (1998).

20 JE Gouaux, RC Stevens and WN Lipscomb, *Biochemistry*, **29**, 7702–15 (1990).

21 M Hyvonen and M Saraste, *EMBO J*, **16**, 3396–404 (1997).

22 MB Berry and GN Phillips Jr, *Proteins*, **32**, 276–88 (1998).

23 U Hommel, M Zurini and M Luyten, *Nat Struct Biol*, **1**, 383–87 (1994).

24 RX Xu, T Pawelczyk, TH Xia and SC Brown, *Biochemistry*, **36**, 10709–17 (1997).

25 S Ichikawa, H Hatanaka, Y Takeuchi, S Ohno and F Inagaki, *J Biochem (Tokyo)*, **117**, 566–74 (1995).

26 G Zhang, MG Kazanietz, PM Blumberg and JH Hurley, *Cell*, **81**, 917–24 (1995).

27 HR Mott, JW Carpenter, S Zhong, S Ghosh, RM Bell and SL Campbell, *Proc Natl Acad Sci USA*, **93**, 8312–17 (1996).

28 L Chantalat, D Leroy, O Filhol, A Nueda, MJ Benitez, EM Chambaz, C Cochet and O Dideberg, *EMBO J*, **18**, 2930–40 (1999).

29 K Niefind, B Guerra, I Ermakowa and OG Issinger, *EMBO J*, **20**, 5320–31 (2001).

30 DA Bushnell, P Cramer and RD Kornberg, *Proc Natl Acad Sci USA*, **99**, 1218–22 (2002).

31 B Wang, DN Jones, BP Kaine and MA Weiss, *Structure*, **6**, 555–69 (1998).

32 CD Mackereth, CH Arrowsmith, AM Edwards and LP McIntosh, *Proc Natl Acad Sci USA*, **97**, 6316–21 (2000).

33 G Zhang, EA Campbell, L Minakhin, C Richter, K Severinov and SA Darst, *Cell*, **98**, 811–24 (1999).

34 L Minakhin, S Bhagat, A Brunning, EA Campbell, SA Darst, RH Ebright and K Severinov, *Proc Natl Acad Sci USA*, **98**, 892–97 (2001).

35 H Pan and DB Wigley, *Structure*, **8**, 231–39 (2000).

36 MA Augustin, R Huber and JT Kaiser, *Nat Struct Biol*, **8**, 57–61 (2001).

37 B Guenther, R Onrust, A Sali, M O'Donnell and J Kuriyan, *Cell*, **91**, 335–45 (1997).

38 D Jeruzalmi, M O'Donnell and J Kuriyan, *Cell*, **106**, 429–41 (2001).

39 DS Daniels, CD Mol, AS Arvai, S Kanugula, AE Pegg and JA Tainer, *EMBO J*, **19**, 1719–30 (2000).

40 JB Thoden, FJ Ruzicka, PA Frey, I Rayment and HM Holden, *Biochemistry*, **36**, 1212–22 (1997).

41 S Geeganage and PA Frey, *Biochemistry*, **38**, 13398–406 (1999).

42 B Lovejoy, A Cleasby, AM Hassell, K Longley, MA Luther, D Weigl, G McGeehan, AB McElroy, D Drewry, MH Lambert and SR Jordan, *Science*, **263**, 375–77 (1994).

43 FJ Moy, PK Chanda, S Cosmi, MR Pisano, C Urbano, J Wilhelm and R Powers, *Biochemistry*, **37**, 1495–504 (1998).

44 FJ Moy, PK Chanda, JM Chen, S Cosmi, W Edris, JS Skotnicki, J Wilhelm and R Powers, *Biochemistry*, **38**, 7085–96 (1999).

45 E Morgunova, A Tuuttila, U Bergmann, M Isupov, Y Lindqvist, G Schneider and K Tryggvason, *Science*, **284**, 1667–70 (1999).

46 PR Gooley, JF O'Connell, AI Marcy, GC Cuca, SP Salowe, BL Bush, JD Hermes, CK Esser, WK Hagmann, JP Springer and BA Johnson, *Nat Struct Biol*, **1**, 111–18 (1994).

47 SR Van Doren, AV Kurochkin, W Hu, QZ Ye, LL Johnson, DJ Hupe and ER Zuiderweg, *Protein Sci*, **4**, 2487–98 (1995).

48 YC Li, X Zhang, R Melton, V Ganu and NC Gonnella, *Biochemistry*, **37**, 14048–56 (1998).

49 V Dhanaraj, QZ Ye, LL Johnson, DJ Hupe, DF Ortwine, JB Dunbar Jr, JR Rubin, A Pavlovsky, C Humblet and TL Blundell, *Structure*, **4**, 375–86 (1996).

50 L Chen, TJ Rydel, F Gu, CM Dunaway, S Pikul, KM Dunham and BL Barnett, *J Mol Biol*, **293**, 545–57 (1999).

51 JW Becker, AI Marcy, LL Rokosz, MG Axel, JJ Burbaum, PM Fitzgerald, PM Cameron, CK Esser, WK Hagmann, JD Hermes and JP Springer, *Protein Sci*, **4**, 1966–76 (1995).

52 MF Browner, WW Smith and AL Castelhano, *Biochemistry*, **34**, 6602–10 (1995).

53 W Bode, P Reinemer, R Huber, T Kleine, S Schnierer and H Tschesche, *EMBO J*, **13**, 1263–69 (1994).

54 M Betz, P Huxley, SJ Davies, Y Mushtaq, M Pieper, H Tschesche, W Bode and FX Gomis-Ruth, *Eur J Biochem*, **247**, 356–63 (1997).

55 S Rowsell, P Hawtin, CA Minshull, H Jepson, SM Brockbank, DG Barratt, AM Slater, WL McPheat, D Waterson, AM Henney and RA Pauptit, *J Mol Biol*, **319**, 173–81 (2002).

56 H Nar, K Werle, MM Bauer, H Dollinger and B Jung, *J Mol Biol*, **312**, 743–51 (2001).

57 B Lovejoy, AR Welch, S Carr, C Luong, C Broka, RT Hendricks, JA Campbell, KA Walker, R Martin, H Van Wart and MF Browner, *Nat Struct Biol*, **6**, 217–21 (1999).

58 I Botos, E Meyer, SM Swanson, V Lemaitre, Y Eeckhout and EF Meyer, *J Mol Biol*, **292**, 837–44 (1999).

59 C Fernandez-Catalan, W Bode, R Huber, D Turk, JJ Calvete, A Lichte, H Tschesche and K Maskos, *EMBO J*, **17**, 5238–48 (1998).

60 ST Jeong, HK Kim, SJ Kim, SW Chi, JG Pan, TK Oh and SE Ryu, *J Biol Chem*, **277**, 17041–47 (2002).

61. EA Galburt, MS Chadsey, MS Jurica, BS Chevalier, D Erho, W Tang, RJ Monnat Jr and BL Stoddard, *J Mol Biol*, **300**, 877–87 (2000).
62. KE Flick, MS Jurica, RJ Monnat Jr and BL Stoddard, *Nature*, **394**, 96–101 (1998).
63. H Raaijmakers, O Vix, I Toro, S Golz, B Kemper and D Suck, *EMBO J*, **18**, 1447–58 (1999).
64. H Raaijmakers, I Toro, R Birkenbihl, B Kemper and D Suck, *J Mol Biol*, **308**, 311–23 (2001).
65. SE Tsutakawa, T Muto, T Kawate, H Jingami, N Kunishima, M Ariyoshi, D Kohda, M Nakagawa and K Morikawa, *Mol Cell*, **3**, 621–28 (1999).
66. M Sugahara, T Mikawa, T Kumasaka, M Yamamoto, R Kato, K Fukuyama, Y Inoue and S Kuramitsu, *EMBO J*, **19**, 3857–69 (2000).
67. L Serre, K Pereira de Jesus, S Boiteux, C Zelwer and B Castaing, *EMBO J*, **21**, 2854–65 (2002).
68. R Gilboa, DO Zharkov, G Golan, AS Fernandes, SE Gerchman, E Matz, JH Kycia, AP Grollman and G Shoham, *J Biol Chem*, **277**, 19811–16 (2002).
69. JC Fromme and GL Verdine, *Nat Struct Biol*, **9**, 544–52 (2002).
70. DO Zharkov, G Golan, R Gilboa, AS Fernandes, SE Gerchman, JH Kycia, RA Rieger, AP Grollman and G Shoham, *EMBO J*, **21**, 789–800 (2002).
71. JF Petersen, MM Cherney, HD Liebig, T Skern, E Kuechler and MN James, *EMBO J*, **18**, 5463–75 (1999).
72. IP Korndorfer, WD Fessner and BW Matthews, *J Mol Biol*, **300**, 917–33 (2000).
73. D Fourmy, F Dardel and S Blanquet, *J Mol Biol*, **231**, 1078–89 (1993).
74. Y Mechulam, E Schmitt, L Maveyraud, C Zelwer, O Nureki, S Yokoyama, M Konno and S Blanquet, *J Mol Biol*, **294**, 1287–97 (1999).
75. L Serre, G Verdon, T Choinowski, N Hervouet, JL Risler and C Zelwer, *J Mol Biol*, **306**, 863–76 (2001).
76. I Sugiura, O Nureki, Y Ugaji-Yoshikawa, S Kuwabara, A Shimada, M Tateno, B Lorber, R Giege, D Moras, S Yokoyama and M Konno, *Struct Fold Des*, **8**, 197–208 (2000).
77. A Yaremchuk, S Cusack and M Tukalo, *EMBO J*, **19**, 4745–58 (2000).
78. A Yaremchuk, M Tukalo, M Grotli and S Cusack, *J Mol Biol*, **309**, 989–1002 (2001).
79. O Nureki, DG Vassylyev, M Tateno, A Shimada, T Nakama, S Fukai, M Konno, TL Hendrickson, P Schimmel and S Yokoyama, *Science*, **280**, 578–82 (1998).
80. S Fukai, O Nureki, S Sekine, A Shimada, J Tao, DG Vassylyev and S Yokoyama, *Cell*, **103**, 793–803 (2000).
81. LF Silvian, J Wang and TA Steitz, *Science*, **285**, 1074–77 (1999).
82. M Cai, R Zheng, M Caffrey, R Craigie, GM Clore and AM Gronenborn, *Nat Struct Biol*, **4**, 567–77 (1997).
83. AP Eijkelenboom, FM van den Ent, R Wechselberger, RH Plasterk, R Kaptein and R Boelens, *J Biomol NMR*, **18**, 119–28 (2000).
84. JY Lee, C Chang, HK Song, J Moon, JK Yang, HK Kim, ST Kwon and SW Suh, *EMBO J*, **19**, 1119–29 (2000).
85. A Klug and D Rhodes, *Trends Biochem Sci*, **12**, 464–69 (1987).
86. JM Berg, *Science*, **232**, 485–87 (1986).
87. A Klug and JW Schwabe, *FASEB J*, **9**, 597–604 (1995).
88. JM Berg and Y Shi, *Science*, **271**, 1081–85 (1996).
89. JP Mackay and M Crossley, *Trends Biochem Sci*, **23**, 1–4 (1998).
90. H Jornvall, O Danielsson, L Hjelmqvist, B Persson and J Shafqat, *Adv Exp Med Biol*, **372**, 281–94 (1995).
91. H Eklund and CI Branden, *The Role of Zinc in Alcohol Dehydrogenase*, John Wiley & Sons, New York, pp. 123–52 (1983).
92. E Magonet, P Hayen, D Delforge, E Delaive and J Remacle, *Biochem J*, **287**(Pt 2), 361–65 (1992).
93. JR Brown and WF Doolittle, *Microbiol Mol Biol Rev*, **61**, 456–502 (1997).
94. R Rella, CA Raia, M Pensa, FM Pisani, A Gambacorta, M De Rosa and M Rossi, *Eur J Biochem*, **167**, 475–79 (1987).
95. S Ammendola, CA Raia, C Caruso, L Camardella, S D'Auria, M De Rosa and M Rossi, *Biochemistry*, **31**, 12514–23 (1992).
96. J Jelokova, C Karlsson, M Estonius, H Jornvall and JO Hoog, *Eur J Biochem*, **225**, 1015–19 (1994).
97. T Tsukihara, H Aoyama, E Yamashita, T Tomizaki, H Yamaguchi, K Shinzawa-Itoh, R Nakashima, R Yaono and S Yoshikawa, *Science*, **272**, 1136–44 (1996).
98. V Perrier, WK Surewicz, P Glaser, L Martineau, CT Craescu, H Fabian, HH Mantsch, O Barzu and AM Gilles, *Biochemistry*, **33**, 9960–67 (1994).
99. V Perrier, S Burlacu-Miron, S Bourgeois, WK Surewicz and AM Gilles, *J Biol Chem*, **273**, 19097–101 (1998).
100. AF Quest, J Bloomenthal, ES Bardes and RM Bell, *J Biol Chem*, **267**, 10193–97 (1992).
101. SR Hubbard, WR Bishop, P Kirschmeier, SJ George, SP Cramer and WA Hendrickson, *Science*, **254**, 1776–1779 (1991).
102. JH Hurley, AC Newton, PJ Parker, PM Blumberg and Y Nishizuka, *Protein Sci*, **6**, 477–80 (1997).
103. O Gadal, GV Shpakovski and P Thuriaux, *J Biol Chem*, **274**, 8421–27 (1999).
104. X Qian, C Jeon, H Yoon, K Agarwal and MA Weiss, *Nature*, **365**, 277–79 (1993).
105. W Zhu, Q Zeng, CM Colangelo, M Lewis, MF Summers and RA Scott, *Nat Struct Biol*, **3**, 122–24 (1996).
106. KS Murakami, S Masuda, EA Campbell, O Muzzin and SA Darst, *Science*, **296**, 1285–90 (2002).
107. D Soler, T Nomizu, WE Brown, M Chen, QZ Ye, HE Van Wart and DS Auld, *Biochem Biophys Res Commun*, **201**, 917–23 (1994).
108. D Soler, T Nomizu, WE Brown, Y Shibata and DS Auld, *J Protein Chem*, **14**, 511–20 (1995).
109. QA Sang and DA Douglas, *J Protein Chem*, **15**, 137–60 (1996).
110. AJ Barrett and ND Rawlings, http://www.merops.co.uk (2001).
111. DS Auld, *Structure Bonding*, **89**, 29–50 (1997).
112. MJ Giraud-Panis, DR Duckett and DM Lilley, *J Mol Biol*, **252**, 596–610 (1995).
113. W Sommergruber, G Casari, F Fessl, J Seipelt and T Skern, *Virology*, **204**, 815–18 (1994).
114. W Sommergruber, J Seipelt, F Fessl, T Skern, HD Liebig and G Casari, *Virology*, **234**, 203–14 (1997).
115. T Voss, R Meyer and W Sommergruber, *Protein Sci*, **4**, 2526–31 (1995).
116. M Whitlow, AJ Howard, BC Finzel, TL Poulos, E Winborne and GL Gilliland, *Proteins: Struct, Funct, Genet*, **9**, 153–73 (1991).
117. DS Auld, in A Messerschmidt, W Bode and M Cygler (eds.), *The Handbook of Metalloproteins*, Vol. 3, John Wiley & Sons, Chichester, pp. 416–431 (2003).
118. RG Zhang, I Dementieva, N Duke, F Collart, E Quaite-Randall, R Alkire, L Dieckman, N Maltsev, O Korolev and A Joachimiak, *Proteins*, **48**, 423–26 (2002).

119 B Senger, L Despons, P Walter, H Jakubowski and F Fasiolo, *J Mol Biol*, **311**, 205–16 (2001).
120 T Jenuwein and CD Allis, *Science*, **293**, 1074–80 (2001).
121 TO Yeates, *Cell*, **111**, 5–7 (2002).
122 X Zhang, H Tamaru, S Khan, J Horton, L Keefe, E Selker and X Cheng, *Cell*, **111**, 117–27 (2002).
123 J Min, X Zhang, X Cheng, SI Grewal and RM Xu, *Nat Struct Biol*, **9**, 828–32 (2002).
124 H Tamaru and EU Selker, *Nature*, **414**, 277–83 (2001).
125 BA Messerle, A Schaffer, M Vasak, JH Kagi and K Wuthrich, *J Mol Biol*, **214**, 765–79 (1990).
126 K Zangger, G Oz, JD Otvos and IM Armitage, *Protein Sci*, **8**, 2630–38 (1999).
127 R Riek, B Precheur, Y Wang, EA Mackay, G Wider, P Guntert, A Liu, JH Kagi and K Wuthrich, *J Mol Biol*, **291**, 417–28 (1999).

Cocatalytic zinc sites

David S Auld

Center for Biochemical and Biophysical Sciences and Medicine and Department of Pathology, Harvard Medical School, Cambridge, MA, USA

Cocatalytic zinc sites occur in enzymes in which two or three metals are closely grouped to bring about catalysis. There are over four dozen representatives of this type of zinc site with the great majority belonging to the Class III hydrolases (Table 1). The fact that many zinc enzymes required more than one metal for full activity was known for many years. However, the location of these metals relative to one another was often in debate. Prior to structural analyses, these metal atoms were often referred to as a combination of a catalytic zinc site and a 'modulating' or 'regulatory' zinc.[1] Catalytic zinc sites, in general, contain three protein ligands, a metal-bound water molecule, and one short and one long spacer.[2,3] The zinc frequently can be removed reversibly without a major effect on the structure of the protein. Cocatalytic metal sites, on the other hand, are often formed from ligands that extend over a large portion of the protein structure. It is, therefore, not unreasonable that enzymes containing two or more metals have been described as having one catalytic metal and a second metal that has structural and/or regulatory effects on activity. Thus far, three-dimensional structures of these enzymes show that the metals are always in close proximity.

GENERAL FEATURES

The alkaline phosphatase family contains three metals in the active site while the rest of the cocatalytic zinc sites contain two metals (Table 1). Some of these sites contain metals such as copper, iron, and magnesium in combination with zinc. Combinations of Zn/Mg are seen in alkaline phosphatase and lens aminopeptidase; Fe(III)/Zn in the purple acid phosphatase family and Cu(II)/Zn in the superoxide dismutase, SOD, family (Table 1).

A novel feature of these sites is the bridging of two of the metal sites by a side chain moiety of a single amino acid residue, usually aspartic acid, and sometimes a water molecule.[6,9] In principle, any sp^2 center containing two nucleophilic atoms should have a bridging potential. Thus, the ring nitrogens of the imidazole group of histidine and the carboxylate oxygens of aspartic acid, glutamic acid, or of a carboxylated lysine, $LysCO_2^-$, have been found to bridge such sites (Figure 1). Such an interaction would, of course, require the metals to be in close proximity to each other. The distance between the metals in these sites depends on the bridging amino acid. In the case of aspartic acid, glutamic acid or $LysCO_2^-$ carboxylate, it is generally between 3 and 4 Å (Table 1). In the case of a histidine imidazole group, the distance increases to about 6 Å.

Aspartic acid and histidine predominate as ligands in cocatalytic zinc sites where the frequency is Asp $\simeq$ His > glutamic acid. These sites also occasionally contain unusual zinc ligands such as amide carbonyls provided by asparagine, glutamine, and the peptide backbone; hydroxyl groups from serine, threonine, and tyrosine, and the amine nitrogen of lysine or the N-terminal amino group of the protein. Perhaps remarkable is the absence of cysteine ligands with the possible exception of the β-lactamases (see below). The ligands to cocatalytic zinc sites often come from nearly the entire length of the protein. The metals in these sites may therefore be important to the overall fold of the protein as well as to the catalytic function. The secondary structure of the protein plays a major role in providing the ligands to these sites. Some of the families use almost exclusively β-sheets to provide the ligands while other families use all α-helices. In many cases, ligands are provided by amino acids residing one or two residues before or after a β-sheet or an α-helix. The zinc ions are often penta-coordinate and arranged in a trigonal bipyramidal geometry.

The bridging amino acids and H_2O probably have critical roles in catalysis. Thus, their dissociation from either metal atom during catalysis will change the charge on the metal promoting its action as a Lewis acid or allowing interaction with an electronegative atom of the substrate. In addition, a released bridging ligand can participate transiently in the reaction as a nucleophile or general acid/base catalyst. The flexibility of the arm supplying the bridging ligands (e.g. one carbon, C, for aspartic acid and histidine, 2 C for glutamic acid or 5 C/N for $LysCO_2^-$) would be expected to influence the stability and reactivity of the two metal sites. Thus, it can be envisioned for cocatalytic sites in hydrolytic enzymes that substrate binding involves one zinc site acting as a template for substrate binding while the other zinc provides hydroxide for nucleophilic attack on the sp^2 center of the ester or amide bond of the substrate. In the next step the roles of the metals can be reversed. In this manner, the metal atoms and their associated ligands play specific roles in each step of the reaction that works to bring about catalysis. The ligands in these sites, in particular the histidines, are often involved in further hydrogen-bonding

Table 1 Cocatalytic zinc sites[a]

Enzyme	PDB#	Metal	R,Å	L$_1$	X	L$_2$	Y	L$_3$	Z	L$_4$	L$_5$	References
						CLASS I: OXIDOREDUCTASES						

Superoxide dismutase family

Enzyme	PDB#	Metal	R,Å	L$_1$	X	L$_2$	Y	L$_3$	Z	L$_4$	L$_5$	References
Bovine	2SOD	Cu(II)	6.2	His$_\beta$	1	His$_\beta$	14	***His**$_\beta$*	56	His$_\beta$ (C)	Solv	4
		Zn		***His**$_\beta$*	7	His	8	His$_{2b\beta}$	2	Asp$_{2b\beta}$ (C)		
Human	1SPD	Cu(II)	5.5	His$_\beta$	1	His$_\beta$	14	***His**$_\beta$*	56	His$_\beta$ (C)		5
		Zn		***His**$_\beta$*	7	His	8	His$_{2b\beta}$	2	Asp$_{2b\beta}$ (C)		
Frog (*Xenopus laevis*)	1XSO	Cu(II)	6.0	His$_\beta$	1	His$_\beta$	14	***His**$_\beta$*	56	His$_\beta$ (C)	Solv	6
		Zn		***His**$_\beta$*	7	His	8	His$_{2b\beta}$	2	Asp$_{2b\beta}$ (C)		
Spinach (*Spinacia oleracea*)	1SRD	Cu(II)	6.0	His$_\beta$	1	His$_\beta$	14	***His**$_\beta$*	56	His$_\beta$ (C)		7
		Zn		***His**$_\beta$*	7	His	8	His$_{2b\beta}$	2	Asp$_\beta$ (C)		
Yeast (*S. cerevisiae*)	1SDY	Cu(II)	6.1	His$_\beta$	1	His$_\beta$	14	***His**$_\beta$*	56	His$_\beta$ (C)	Solv	8
		Zn		***His**$_\beta$*	7	His	8	His$_{2b\beta}$	2	Asp$_\beta$ (C)		
Yeast (*Candida albicans*)	1YSO	Cu(I)	6.5	His$_\beta$	1	His$_\beta$	14	His$_\beta$	56	His$_\beta$ (C)	Solv	9
		Zn		His$_\beta$	7	His	8	His$_{2b\beta}$	2	Asp$_\beta$ (C)		
Escherichia coli	1ESO	Cu(II)	6.5	His$_\beta$	1	His$_\beta$	14	***His**$_\beta$*	56	His$_\beta$ (C)		10
		Zn		***His**$_\beta$*	7	His	8	His$_{2b\beta}$	2	Asp$_\beta$ (C)		
Salmonella typhimurium	1EQW	Cu(II)	6.5	His$_\beta$	1	His$_\beta$	14	***His**$_\beta$*	56	His$_\beta$ (C)		11
		Zn		***His**$_\beta$*	7	His	8	His$_{2b\beta}$	2	Asp$_\beta$ (C)		
Photobacterium leiognathi	1YAI	Cu(II)	6.2	His$_\beta$	1	His$_\beta$	22	***His**$_\beta$*	54	His$_\beta$ (C)		12
		Zn		***His**$_\beta$*	8	His	8	His$_{2b\beta}$	2	Asp$_\beta$ (C)		
Brucella abortus		Cu(II)		His	1	His	22	***His***	54	His (C)		13
		Zn		***His***	8	His	7	His (C)	–	?		
Actinobacillus pleuropneumoniae	2APS	Cu(II)	6.4	His$_\beta$	1	His$_\beta$	22	***His**$_\beta$*	55	His$_\beta$ (C)		14
		Zn		***His**$_\beta$*	8	His	8	His$_{2b\beta}$	2	Asp$_\beta$ (C)		
						CLASS III: HYDROLASES						

Phosphatase family

Enzyme	PDB#	Metal	R,Å	L$_1$	X	L$_2$	Y	L$_3$	Z	L$_4$	L$_5$	References
E. coli	1ALK	Zn1	4.0	Asp$_\alpha$	3	His$_{a\alpha}$	80	His (C)	–	H$_2$O		15,16
		Zn2	4.6	His	0	Asp$_{2a\beta}$	266	Ser$_{b\alpha}$	50	***Asp**$_\beta$* (N)		
		Mg		***Asp**$_\beta$*	103	Thr$_\alpha$	166	Glu$_\beta$ (C)	–	H$_2$O		
Human	1EW2	Zn1	4.0	Asp$_\alpha$	3	His$_\alpha$	121	His (C)	–	H$_2$O		17
		Zn2	4.8	His$_L$	0	Asp$_L$	264	Ser$_{b\alpha}$	49	***Asp**$_\beta$* (N)		
		Mg		***Asp**$_\beta$*	112	Ser$_\alpha$	155	Glu$_\beta$ (C)		H$_2$O		
Shrimp	1KW7	Zn1	4.5	Asp$_\alpha$	3	His$_{a\alpha}$	121	His (C)	–	H$_2$O		18
		Zn2	4.5	His$_L$	0	Asp$_L$	269	Ser$_{b\alpha}$	48	***Asp**$_\beta$* (N)		
		Zn3		***Asp**$_\beta$*	111	His$_{b\alpha}$	1	Thr$_\alpha$	158	Glu$_\beta$ (C)	H$_2$O	
E. coli D153H	2ANH	Zn1	4.1	Asp$_\alpha$	3	His$_{a\alpha}$	80	His (C)	–	H$_2$O		19
		Zn2	4.7	His	0	Asp$_{2a\beta}$	269	Ser$_{b\alpha}$	48	***Asp**$_\beta$* (N)		
		Zn3		***Asp**$_\beta$*	101	His$_{b\alpha}$	1	Thr$_\alpha$	166	Glu$_\beta$ (C)		
E. coli D153H, D330N	1KHK	Zn1	4.4	Asp$_\alpha$	3	His$_{a\alpha}$	80	His (C)	–	H$_2$O		20
		Zn2	4.7	His$_L$	0	Asp$_L$	269	Ser$_{b\alpha}$	48	***Asp**$_\beta$* (N)		
		Mg		***Asp**$_\beta$*	103	Thr$_\alpha$	166	Glu$_\beta$ (C)		H$_2$O		

(*continued overleaf*)

Cocatalytic zinc sites

Table 1 (continued)

Enzyme	PDB#	Metal	R,Å	L_1	X	L_2	Y	L_3	Z	L_4	L_5	References
B. cereus phospholipase C1	1AH7	Zn1	6.1	Glu_α	3	His_α	13	His_α (N)	–	H_2O		21
		Zn2	3.6	**$Asp_{a\alpha}$**	3	His_α	48	His	13	$Asp_{a\alpha}$ (N)	H_2O	
		Zn3		$Trp_{b\beta}$	12	His_α	107	**$Asp_{a\alpha}$** (C)	–	**H_2O**		
Clostridium perfringens α-Toxin												
Avian strain (SWCP)	1KHO	Zn1	6.3	Glu_α	3	His_α	11	His_α (N)	–	H_2O		22
		Zn2	3.3	**Asp_α**	3	His_α	57	His	11	Asp (N)		
		Zn3		Trp	10	His_α	118	**Asp_α** (C)	–			
Bovine strain CER89L43, open	1CA1	Cd1	5.5	Glu_α	3	His_α	11	His_α (N)	–	H_2O		23
		Cd2	3.5	**Asp_α**	3	His_α	57	His	11	$Asp_{a\alpha}$ (N)	H_2O	
		Zn3		Trp	10	His_α	118	**Asp_α** (C)	–			
Bovine strain CER89L43, closed	1GYG	Zn2	3.4	**Asp_α**	3	His_α	57	His	11	$Asp_{a\alpha}$ (N)		24
		Zn3		Trp	10	His_α	118	**Asp_α** (C)	–			
Penicillium citrinum P1 nuclease	1AK0	Zn1	5.9	$Asp_{a\alpha}$	3	His_α	12	His (N)	–	H_2O		25
		Zn2	3.7	**Asp_α**	3	His_α	55	$His_{a\alpha}$	14	Asp_α (N)		
		Zn3		Trp	4	His_α	113	**Asp_α** (C)	–	H_2O		
E. coli Endonuclease IV	1QTW	Zn1	4.7	His_β	3	$Asp_{b\beta}$	46	His_α (N)	–	H_2O		26
		Zn2	3.4	**Glu_β**	33	Asp_β	36	His_β	44	$Glu_{a\beta}$ (C)	H_2O	
		Zn3		**Glu_β**	35	$His_{a\beta}$	39	His_β (N)	–	**H_2O**		

Phosphotriesterase family

Enzyme	PDB#	Metal	R,Å	L_1	X	L_2	Y	L_3	Z	L_4	L_5	References
Pseudomonas diminuta	1EYW	Zn1	3.5	$His_{a\beta}$	1	His	111	**$Lys_\beta\ CO_2^-$**	131	$Asp_{2a\beta}$ (C)	**OH**	27
		Zn2		**$Lys_\beta\ CO_2^-$**	31	His_β	28	$His_{a\beta}$ (C)	–	**OH**		
Escherichia coli	1BP6	Zn1	3.4	$His_{2a\beta}$	1	His	110	**$Glu_{b\beta}$**	117	$Asp_{2a\beta}$ (C)	**Unk**	28
		Zn2		**$Glu_{b\beta}$**	32	His_β	27	$His_{a\beta}$ (C)	–	**Unk**		

Purple acid phosphatase family

Enzyme	PDB#	Metal	R,Å	L_1	X	L_2	Y	L_3	Z	L_4	L_5	References
Kidney bean	1KBP	Fe(III)	3.3	Asp_β	28	**Asp_β**	2	$Tyr_{2a\beta}$	157	$His_{2a\beta}$ (C)		29,30
		Zn		**Asp_β**	36	$Asn_{b\alpha}$	84	$His_{a\beta}$	36	$His_{2a\beta}$ (C)	H_2O	
Rat	1QFC	Fe(III)	3.1	$Asp_{2a\beta}$	37	**$Asp_{2a\beta}$**	2	Tyr	167	$His_{b\beta}$ (C)	H_2O	31,32
		Fe(II)		**$Asp_{2a\beta}$**	38	$Asn_{b\alpha}$	94	$His_{a\beta}$	34	$His_{2a\beta}$ (C)	H_2O	
Porcine (uteroferrin)	1UTE	Fe(III)	3.3	$Asp_{2a\beta}$	37	**$Asp_{2a\beta}$**	2	Tyr	167	$His_{b\beta}$ (C)	**OH**	33
		Fe(II)		**$Asp_{2a\beta}$**	38	Asn_α	94	$His_{a\beta}$	34	$His_{2a\beta}$ (C)	**OH**	
Human protein phosphatase I		Fe(III)	3.5–4.0	Asp	1	His	25	**Asp**	179	Tyr (C)	**H_2O**	34
		Mn		**Asp**	31	Asn	48	His	74	His (C)	**H_2O**	
Human brain calcineurin	1AUI	Fe(III)	3.1	Asp	1	His	25	**$Asp_{2a\beta}$** (C)		H_2O	**H_2O**	35,36
		Zn		**$Asp_{2a\beta}$**	31	Asn	48	$His_{a\beta}$	81	$His_{2a\beta}$ (C)	**H_2O**	
E. coli UDP-sugar hydrolase	1USH, 1HP1	Zn1	3.3	$Asp_{2a\beta}$	1	His	40	**$Asp_{2a\beta}$**	169	Gln_L (C)	**H_2O**	37,38
		Zn2		**$Asp_{2a\beta}$**	31	$Asn_{2a\beta}$	100	His_β	34	$His_{a\beta}$ (C)	**H_2O**	

Amidohydrolase family

Enzyme	PDB#	Metal	R,Å	L_1	X	L_2	Y	L_3	Z	L_4	L_5	References
Escherichia coli dihydroorotase	1J79	Zn1	3.5	His	1	His	93	**$Lys_\beta\ CO_2^-$**	147	Asp (C)	**OH**	39
		Zn2		**$Lys_\beta\ CO_2^-$**	36	His_β	37	His_β (C)		**OH**		
Arthobacter aurescens L-hydantoise	1GKR	Zn1	3.6	His	1	His	84	**$Lys_\beta\ CO_2^-$**	164	$Asp_{2a\beta}$ (C)	**OH**	40
		Zn2		**$Lys_\beta\ CO_2^-$**	35	His_β	55	$His_{a\beta}$ (C)		**OH**		

Table 1 (continued)

Enzyme	PDB#	Metal	R,Å	L$_1$	X	L$_2$	Y	L$_3$	Z	L$_4$	L$_5$	References
Thermus sp. D-hydantoise	1GKP	Zn1 Zn2	3.6	His **Lys$_\beta$ CO$_2^-$**	1 32	His His$_\beta$	88 55	**Lys$_\beta$ CO$_2^-$** His$_{a\beta}$ (C)	164	Asp$_{2a\beta}$ (C) –	**OH** **OH**	41
Bacillus stearothermophilus D-hydantoinase	1K1D	Zn1 Zn2	3.0	His **Lys$_\beta$ CO$_2^-$**	1 32	His His$_\beta$	89 55	**Lys$_\beta$ CO$_2^-$** His$_\beta$ (C)	164	Asp (C) –	**OH** **OH**	42
Burkholderia picketti D-hydantoinase	1NFG	Zn1 Zn2	3.0	His$_{2a\beta}$ **Lys$_\beta$ CO$_2^-$**	1 32	His His$_\beta$	88 55	**Lys$_\beta$ CO$_2^-$** His$_{a\beta}$ (C)	164	Asp (C) –	**OH** **OH**	43
Escherichia coli Isoaspartyl dipeptidase	1ONW	Zn1 Zn2	3.4	His$_\beta$ **Lys$_\beta$ CO$_2^-$**	1 38	His His$_\beta$	91 28	**Lys$_\beta$ CO$_2^-$** His$_{a\beta}$ (C)	162	Asp (C) –	**H$_2$O** **H$_2$O**	44
Aminopeptidase family												
Bovine lens	1BLL	Zn1 Zn2	2.9	**Glu$_{b\alpha}$** Lys$_\beta$	1 4	Asp **Asp$_{2a\beta}$**	76 17	**Asp$_{2a\beta}$** (N) Asp$_{2b\alpha}$	– 60	**H$_2$O** **Glu$_{b\alpha}$** (C)	**H$_2$O**	45
Escherichia coli PepA	1GYT	Zn1 Zn2	3.0	**Glu$_{b\alpha}$** Lys$_\beta$	1 4	Asp **Asp$_{2a\beta}$**	76 17	**Asp$_{2a\beta}$** (N) **Glu$_{b\alpha}$** (C)	–	**H$_2$O** –		46
Aeromonas proteolytica	1AMP, 1LOK	Zn1 Zn2	3.5	**Asp$_{2b\alpha}$** His$_\beta$	34 19	Glu **Asp$_{2b\alpha}$**	103 61	His (C) Asp$_{a\beta}$ (C)	– –	**H$_2$O** **H$_2$O**		47,48
Streptomyces griseus	1XJO, 1QQ9	Zn1 Zn2	3.6	**Asp$_{2b\alpha}$** His$_\beta$	34 11	Glu **Asp$_{2b\alpha}$**	114 62	His (C) Asp$_{a\beta}$ (C)	– –	**H$_2$O** **H$_2$O**		49,50
Methionine aminopeptidases												
Escherichia coli methionine-1	1MAT	Co1 Co2	2.9	Asp$_\beta$ **Asp$_\beta$**	10 62	**Asp$_\beta$** His$_\beta$	126 32	**Glu$_{b\beta}$** (C) Glu$_{a\beta}$	– 30	**Glu$_{b\beta}$** (C)		51,52
Pyrococcus furiosus methionine-2	1XGM	Co1 Co2	2.8	Asp$_\beta$ **Asp$_\beta$**	10 59	**Asp$_\beta$** His$_\beta$	186 33	**Glu$_\beta$** (C) Glu$_{a\beta}$	– 92	**H$_2$O** **Glu$_\beta$** (C)	**H$_2$O**	53
Human methionine-2	1B59	Co1 Co2	3.1	Asp$_\beta$ **Asp$_{a\beta}$**	10 68	**Asp$_{a\beta}$** His$_\beta$	196 32	**Glu$_\beta$** (C) Glu$_\beta$	– 94	**H$_2$O** **Glu$_\beta$** (C)	**H$_2$O**	54
Escherichia coli proline	1AZ9	Mn1 Mn2	3.3	Asp$_\beta$ **Asp$_\beta$**	10 82	**Asp$_\beta$** His$_\beta$	134 28	**Glu$_\beta$** (C) Glu$_\beta$	– 22	**H$_2$O** **Glu$_\beta$** (C)	**H$_2$O** H$_2$O	55
Other peptidases												
Pseudomonas sp. CPD G$_2$	1CG2	Zn1 Zn2	3.3	**Asp$_{2b\alpha}$** His$_\beta$	34 28	Glu **Asp$_{2b\alpha}$**	208 58	His (C) Glu$_\beta$ (C)	– –	**H$_2$O** **H$_2$O**		56
Lactobacillus delbrueckii PEPV	1LFW	Zn1 Zn2	3.8	**Asp$_{2b\alpha}$** His$_\beta$	34 31	Glu **Asp$_{2b\alpha}$**	284 57	His (C) Asp$_{b\beta}$ (C)	– –	**H$_2$O** **H$_2$O**		57
S. typhimurium Peptidase T	1FNO	Zn1 Zn2	3.4	**Asp$_{2b\alpha}$** His$_\beta$	33 61	Glu **Asp$_{2b\alpha}$**	204 55	His (C) Asp$_{a\beta}$ (C)	– –			58
Bacillus subtilis Dppa	1HI9	Zn1 Zn2	3.2	**Asp$_\beta$** **Asp$_\beta$**	1 95	Glu$_{2a\beta}$ His$_{2a\beta}$	49 28	His$_{a\beta}$ (C) Glu$_\alpha$ (C)		**H$_2$O** **H$_2$O**		59
Human renal dipeptidase	1ITU	Zn1 Zn2	3.3	His$_{2a\beta}$ **Glu$_\beta$**	1 72	Asp$_{2b\alpha}$ His$_L$	102 20	**Glu$_\beta$** (C) His$_{2a\beta}$		H$_2$O H$_2$O		60
Human glyoxalase II	1QH3, 1QH5	Zn1 Zn2	3.3	His$_{2a\beta}$ Asp	1 0	His His	53 74	His **Asp$_L$**	23 38	**Asp$_L$** (C) His$_\beta$ (C)	**H$_2$O** **H$_2$O**	61

(continued overleaf)

Cocatalytic zinc sites

Table 1 (continued)

Enzyme	PDB#	Metal	R,Å	L_1	X	L_2	Y	L_3	Z	L_4	L_5	References
β-lactamase family												
Bacillus cereus	1BMC	Zn1		His$_{2\alpha\beta}$	1	His	60	His (C)	–	H_2O		62
Bacillus cereus	1BC2	Zn1	3.8–4.4	His$_{2\alpha\beta}$	1	His$_L$	60	His (C)	–	H_2O		63,64
		Zn2		Asp$_\alpha$	77	Cys$_{2\alpha\beta}$	41	His (C)	–	H_2O	H_2O	
Bacteroides fragilis	1ZNB	Zn1	3.5	His	1	His	60	His (C)	–	***H$_2$O***		65,66
		Zn2		Asp	77	Cys$_{2\alpha\beta}$	41	His (C)	–	***H$_2$O***	H_2O	
Stenotrophomonas maltophilia	1SML	Zn1	3.5	His	1	His	73	His$_L$ (C)	–	***H$_2$O***		67
		Zn2		Asp	0	His	135	His (C)	–	***H$_2$O***	H_2O	
Pseudomonas aeruginosa	1DD6	Zn1	3.6	His$_{2\alpha\beta}$	1	His	59	His (C)	–			68
		Zn2		Asp	76	Cys$_{2\alpha\beta}$	38	His (C)	–	H_2O		

[a] The amino acid spacer between ligands L_1 and L_2 is X; that between L_3 and the nearest ligand L_1 or L_2 is Y, and that between L_3 and L_4 is Z. The symbols N and C indicate that L_3 is located on the amino (N) or the carboxyl (C) side of L_2 respectively. The subscripts α, β refer to the α- or 3_{10} helix and β-sheet structure respectively that supply the ligand. The letter subscript L denotes an amino acid loop sequence of ≤6 residues between two structural elements. The subscripts *na* and *nb* indicate the ligand that is either one or two ($n = 2$) residues after or before the secondary structural element. The amino acid residue, which bridges the two metal sites, is shown in *italic bold face*. When the symbol H_2O is given, it may represent from one to three metal-bound water molecules. R is the distance between the metal atoms in Angstroms. Distances are for between consecutive metals.

interactions with other amino acids. These interactions should effect the charge on the metal and the stability of the metal complex thus fine-tuning the catalysis and the stability of the metal sites.[70,71]

CLASS I: OXIDOREDUCTASE

Superoxide dismutase family

Superoxide dismutase (SOD) plays a critical role in the physiological control of oxygen radicals by catalyzing the dismutation of the superoxide anion into molecular oxygen and hydrogen peroxide.[72] SOD is a small dimeric enzyme in all eukaryotic species (2 × 16 kDa), with both subunits containing a cocatalytic site containing copper and zinc.[73] The three-dimensional structure of this enzyme has been determined in both the crystalline and solution states (Table 1; see also reference 73). The zinc site is composed of three histidine residues coordinated tetrahedrally by their ND1 nitrogens, unusual for zinc sites, and the OD1 oxygen of Asp81. However, the site shows a strong distortion toward a trigonal pyramid with the buried Asp81 as the apex.[4] The copper site, bridged to the zinc site through His61, has three histidine residues coordinated by their NE2 nitrogens and one histidine by its ND1 (His44) in an uneven tetrahedral distortion of a square plane.[4] A fifth exposed axial coordination site is filled by a solvent molecule. The majority of the ligands are supplied by amino acids residing within β-sheets (Table 1). The structure of the Cu-free form of human SOD in solution shows that the copper ligands are in a conformation close to that observed in the holoenzyme indicating that the copper binding site is preorganized.[74] The role of zinc in the SOD family is generally considered supportive to that of the copper ion, which undergoes oxidation/reduction during catalysis. However, zinc may be important to substrate specificity. Thus, the zinc-deficient SOD has been proposed to participate in both sporadic and familial amyotrophic lateral sclerosis by an oxidative mechanism involving nitric oxide.[75]

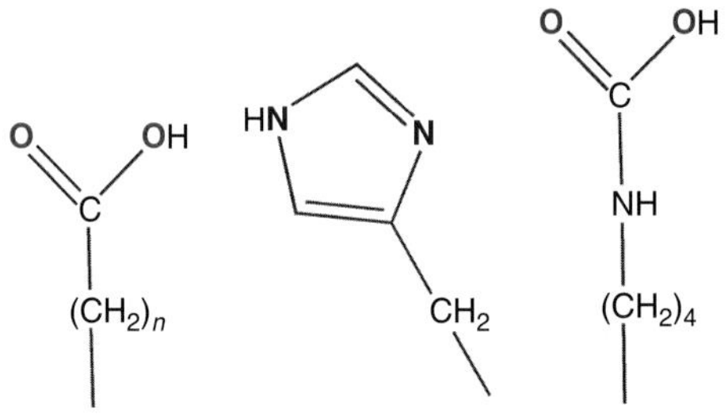

Figure 1 The amino acids that provide bridging ligands (carboxyl group or imidazole) to cocatalytic zinc sites are aspartic acid, glutamic acid, histidine and the carboxylated lysine.

CLASS II: HYDROLASE

Phosphatase family

Several of the zinc enzymes that catalyze phosphomonoester hydrolysis have cocatalytic zinc sites containing three metal atoms in close proximity (Table 1).[3,70] Two of the metal

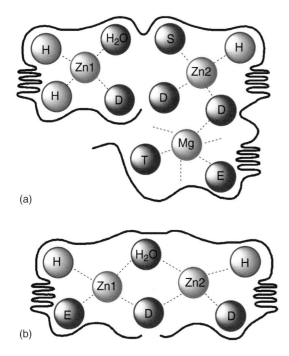

Figure 2 (a) Schematics of three metal cocatalytic sites represented by *E. coli* alkaline phosphatase;[15] (b) two metal cocatalytic sites represented by *A. proteolytica* amino peptidase.[47] The one letter codes D, E, and H are given for the amino acids aspartic acid, glutamic acid, and histidine respectively.

sites are linked by a bridging amino acid while the third zinc site has the properties of a catalytic zinc site, that is, three protein ligands and a metal-bound water molecule (Figure 2). *E. coli* alkaline phosphatase is the best-studied representative of this group. It has a cocatalytic zinc site in both of its subunits composed of two zinc atoms and one magnesium that form a nonequilateral triangle with the metals as the apices.[15] The metal ions are separated by 3.99 Å (Zn1 to Zn2), 4.64 Å (Zn2 to Mg) and 6.88 Å (Zn1 to Mg). The ligands to these metals and the adjacent amino acids are highly conserved for a large family made up of representatives from bacteria, yeast, and mammalian sources.[69] While β-sheets exclusively supply the ligands to the zinc and copper ions in SOD, the cocatalytic sites of the alkaline phosphatase class rely heavily on α-helices as ligand donors (Table 1).

One zinc site in alkaline phosphatase has the properties of a catalytic zinc site being formed from two ligands, Asp327 (both carboxylate oxygens) and His331 (NE2) supplied from a short α-helix, a third protein ligand His412 (NE2) supplied by a β-strand, and a water molecule in the free state.[15] In the phosphate-inhibited state, the water is displaced by one of the phosphate oxygens. The second zinc, Zn2, is coordinated by the imidazole of His370 (NE2), and a single carboxylate oxygen of both Asp369 and the bridging Asp51. The oxygen of either Ser102 in the free state or a phosphate oxygen in the inhibited state completes the coordination. This is the first zinc site in which a reactant amino acid in catalysis, Ser102, is found to be a ligand to a metal (Zn2) in the resting state. The serine is believed to be bound as an alkoxide ion since the oxygen–Zn distance is 1.91 Å.[76] The pK_a of the serine is estimated to be as low as 5.5.[77] A hydroxide bound to the Mg maybe responsible for aiding in deprotonating the zinc-bound serine hydroxyl. The Mg coordination is a slightly distorted octahedron.[15] The second carboxylate oxygen of Asp51, one of the carboxylate oxygens of Glu322, the hydroxyl group of Thr155 and three water molecules form the octahedron (Table 1).

The alkaline phosphatase from human placenta has a cocatalytic site closely similar to that of the *E. coli* enzyme.[17] The only difference is the replacement of a threonine hydroxyl by a serine hydroxyl as a ligand to the Mg site (Table 1). However, the catalytic activity responses of the human placental and *E. coli* phosphatases toward the mutation of ligands to the first metal site (Zn1) can be markedly different. Thus mutation of the aspartic acid ligand to an alanine has a negligible effect on the activity (k_{cat}/K_m) of the placental enzyme toward hydrolysis of *p*-nitrophenyl phosphate,[78] while the corresponding mutation in the *E. coli* enzyme reduces activity 10^7-fold.[79] However, the mutation Asp327Asn in the *E. coli* enzyme reduces the k_{cat}/K_m value by only 20-fold at a maximal zinc concentration. The reason for the marked sensitivity to the ligand change is not known. However, while the identity of the ligands is quite high in the two enzymes, other amino acids are more variable in their active sites. The *E. coli* enzyme with the Asp327Ala mutation may therefore be more susceptible to the binding of inhibitory anions or cations or movement of neighboring amino acid side chains.

The shrimp phosphatase containing three zinc ions in its cocatalytic site is isolated.[18] In this case, His149 (NE2) binds to the third zinc ion. The corresponding imidazole nitrogen of His153 of the placental enzyme is nearly 5 Å removed from the bound Mg ion. The equivalent residue in the *E. coli* enzyme is Asp153, which salt bridges Lys328 and thus does not bind the Mg ion.[19] The mutant *E. coli* enzyme Asp153His displays many of the features characteristic of the placental enzyme including the low activity in the absence of magnesium and a time-dependent enhancement of activity in its presence. The X-ray structure indicates that the octahedral Mg binding site of the wild-type enzyme has been converted to a tetrahedral Zn binding site in the mutant enzyme with the mutant His153 now binding the zinc (Table 1).[19,80] The same results are found for the double mutant Asp153His and Asp330Asn for the three zinc cocatalytic site of the *E. coli* enzyme.[20] The two zinc and one magnesium cocatalytic site of this mutant was also examined. In this case, the Mg site is octahedral with the metal bound by one carboxylate oxygen of Asp51 and Glu322, the hydroxyl of Thr155, and three water molecules (Table 1). The increased activity in the presence of Mg is believed to be due to the displacement of the third Zn by magnesium with a concomitant change in the coordination site. The denaturation kinetics of the apo *E. coli* enzyme in the presence of Zn alone or with Zn and Mg suggests Mg maybe important to the stability of the enzyme.[81]

Mutation of the bridging Asp51 ligand to asparagine has profound effect on the activity and structure of the cocatalytic site of the *E. coli* enzyme.[82] At pH8, the activity of the mutant enzyme is about 1% of the wild type. At pH9 and in the presence of Mg, the activity returns in a time-dependent fashion. The X-ray structure indicates that the low pH, low magnesium form of the enzyme no longer has a third metal binding site. The activation by Mg at high pH in this case is postulated to be due to the binding of Mg to the Zn2 site. Under these conditions the third metal site is vacant.

X-ray crystallographic, NMR, and kinetic studies on the Cd and Co substituted enzymes have aided in deciphering the mechanism of action of the *E. coli* enzyme (Figure 3).[15,83,84] The rate-determining step is strongly pH dependent. In the alkaline pH region, the release of the noncovalently bound product phosphate (E•P_i → E + P_i) is rate-limiting, while in the acidic pH region the breakdown of the covalent phosphoryl intermediate (E–P → E•P_i) is postulated to be rate-limiting. Ser102, a ligand to Zn2, is the nucleophile in the first step of the reaction.

The breakdown of the serine phosphoryl intermediate, E–P, is believed to be through a zinc-bound water/hydroxide on the catalytic zinc in the proposed mechanism (Figure 3). The enzyme•vanadate complex has been proposed to mimic the transition state complex.[85] The vanadate ion is bound in a trigonal bipyramidal geometry with the active site Ser102 and the water molecule in opposite apical positions. The equatorial oxygens are stabilized by interaction with the guanidinium group of Arg166. Mutation of the catalytic zinc ligand, His331Gln yields an enzyme in which the covalent phosphate intermediate can

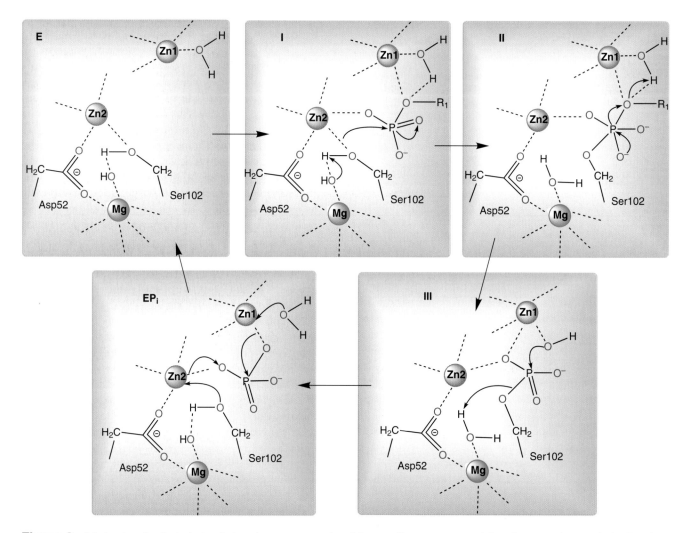

Figure 3 Mechanism for hydrolysis of phosphate esters catalyzed by metalloenzymes containing three metal cocatalytic zinc sites. The schematic is based on the studies performed on alkaline phosphatase. Steps **I** to **III** show the conversion of an initial ES complex into the covalent E–P complex depicted in **III**. The E–P complex is then converted into the noncovalent E•P_i complex. The metal-bound water on Zn1 is used in steps **II** and **III** as a specific acid/base catalyst. Since some of the nonphysiological phosphate esters used in mechanistic studies have activated leaving groups, protonation of RO$^-$ is not necessary in these cases at pH values well above the pK_a of the leaving group.

be observed in the crystal structure.[86] The structure shows that the zinc-bound water on the catalytic zinc is in a position for apical attack on the Ser102 phosphoryl bond.

In the E·P$_i$ complex, the phosphate ion is coordinated to both Zn1 and Zn2 and to two of its oxygens to the guanidinium group of Arg166. The phosphate is further hydrogen-bonded to the amide of Ser102 and a water molecule that is coordinated to the Mg.[15] Mutation of Ser102 to glycine, alanine or cysteine decreases activity 10^4- to 10^5-fold with only the cysteine mutant having an effect on the position of the phosphate.[16]

Extended phosphatase family

Several other enzymes that catalyze the hydrolysis of phosphomono- and phosphodi-esters also have three metal cocatalytic sites (Table 1). In the following discussion, I have used a nomenclature for numbering the metals that follows the numbering used for alkaline phosphatase and the definitions for catalytic and bridging amino acid sites. Thus, Zn1 is defined as the nonbridging, catalytic-like zinc site. Zn2 and Zn3 refer to the bridging amino acid site. The phospholipase C from *B. cereus*,[21] cleaves membrane phospholipids liberating the polar head and leaving behind diacylglycerol. The three zinc ions form the apices of a nonequilateral triangle. The catalytic-like Zn1 ion is 4.58 and 6.12 Å from the two zinc ions that are linked by a bridging carboxylate group, restricting the distance between these two metals to 3.59 Å. An α-helix extending from residue 141 to 150 provides both His142 (NE2) and Glu146 (carboxylate O) as ligands to the catalytic Zn1 site.[21] A second α-helix provides the third ligand His128 (NE2). Two water molecules complete a penta-coordinate site. Asp122 and a water molecule bridge the other two zinc ions, Zn2 and Zn3. Zn2 also binds to His118 (NE2), His69 (ND1), and to one carboxylate oxygen of Asp55. Zn3 is bound by the N-terminal amino group and peptide carbonyl of Trp1, the NE2 nitrogen of His14, and the bridging water molecule and an oxygen of the carboxylate group of Asp122. All three zinc sites are penta-coordinate with α-helices providing nearly all the ligands. A phosphonate inhibitor binds to the three metals by displacing the water molecules bound to Zn1 with one phosphonyl oxygen while the other phosphonyl oxygen replaces the bridging water between Zn2 and Zn3.[87]

The alpha-toxin from *C. perfringens* is a phospholipase C that possesses hemolytic and toxic properties. It is composed of an N-terminal domain of 246 amino acids that contains the phospholipase activity and a C-terminal 124-residue domain that contributes to the toxin properties.[88,89] The cocatalytic metal site for this enzyme is very similar to that found for the nontoxic *B. cereus* PLC (Table 1). Crystal structures of a bovine strain, CER89L43, in an open three metal form,[23] and a closed two zinc form,[24] have been reported (Table 1). The open form was prepared in the presence of 50 mM Cd ions, resulting in Cd binding to the catalytic-like site (Cd1, Table 1) that is closest to the surface of the protein.[23] A second Cd and a Zn bind to the other two metal sites. Nearly all the ligands are supplied by α-helices (Table 1). The zinc is bound by the N-terminal amino and peptide carbonyl groups of Trp1, His11 (NE2) and one carboxylate oxygen of the bridging ligand Asp130. The second carboxylate oxygen of Asp130, His126 (NE2), His68 (ND1), one carboxylate oxygen of Asp56 and a water molecule bind to Cd2. Cd1 binds to the NE2 nitrogens of His136 and His148, both oxygens of the carboxylate of Glu152, and a water molecule.

The closed form of the enzyme, lacking a metal at the catalytic-like metal site, is prepared by growing crystals under the mildly acidic conditions of pH 4.6.[24] The two zinc ions are in the identical positions found for Cd2 and Zn3 of the open enzyme form (Table 1). These metals probably stabilize the structure of the N-terminal domain of the enzyme since the metal ligands come from the first 55% of the sequence of the protein.

The structure of an avian strain, SWCP, contains a three zinc cocatalytic site,[22] with ligands identical to that found in the open bovine strain (Table 1). Close inspection of the site reveals some slight differences. The distances between the catalytic-like Zn1 site and Zn2 is 6.3 Å, slightly longer than is usually observed in these sites (Table 1). This may in part be due to the fact that His11 and His126 are binding by an ND1 nitrogen rather than an NE2 nitrogen seen in the bovine forms while the reverse is found for His68 which also has an unusually long bond of 3.4 Å to Zn2 (PDB# 1KHO). In addition, the α-helix that occurs just before Asp56 is missing in the avian form of the enzyme. Different solvent conditions are used to prepare these crystals, which may be influencing not only the number of metals bound but also the type of coordination metal spheres formed.

The P1 nuclease from *P. citrinum*, a glycoprotein, has a cocatalytic zinc site,[25] very similar to those of the phospholipase C enzymes (Table 1). In this case, the enzyme catalyzes the cleavage of bond between the 3′-hydroxyl and 5′-phosphoryl group of adjacent nucleotides, in particular, conformations of double and single stranded nucleic acids.

The *E. coli* endonuclease IV has a three zinc cocatalytic site that is formed from a different secondary structure compared to others in this group (Table 1). This enzyme is a phosphodiesterase that cleaves the DNA backbone between a purinic and a pyrimidinic site.[26] While all of the phospholipase C-like enzymes use a number of α-helices to supply the ligands to the zinc ions, endonuclease IV relies on β-sheets (Table 1). The residues that provide the ligands span nearly 60% of the protein sequence. Two of the zinc ions, Zn2 and Zn3, are partially buried. The ligands to these metals emanate from six separate strands of a β-barrel.[26] The metals are bridged by a water molecule and the carboxylate of Glu145. The Zn2 site has a trigonal bipyramidal geometry with carboxylate oxygens of Glu145 and Glu261 as axial ligands and a carboxylate oxygen of

Cocatalytic zinc sites

Asp179, the ND1 nitrogen of His216 and a water molecule as the equatorial ligands. Zn3 is tetrahedrally coordinated to His69 (NE2), His109 (ND1), and two bridging oxygens from a water molecule and the bridging carboxylate of Glu145. The catalytic zinc, Zn1, is coordinated to the NE2 nitrogens of His182 and His231, both carboxylate oxygens of Asp229, and a water molecule in a distorted tetrahedron.

A protein that may play a role in inositol catabolism has a structural homology to *E. coli* endonuclease IV.[90] This protein has a single zinc binding site that is very similar to that of the Zn2 site found in the endonuclease IV enzyme. The zinc is bound tetrahedrally to one carboxylate oxygen of Glu142, Asp174, and Glu246 and the ND1 nitrogen of His200 (PDB# 1I6N, note ligand Glu142 is given as His177 in the reference). All of the ligands are provided by different β-sheets as in endonuclease IV (Table 1). The characteristics of the structural zinc binding site in the *E. coli* rhamnose isomerase is also very similar to that found in the Zn2 site of endonuclease IV.[91,92]

Phosphotriesterase family

The soil dwelling bacterium *P. diminuta* contains a phosphotriesterase (PTE) that catalyzes the hydrolysis of organophosphate nerve agents such as sarin, soman, and VX.[93] This enzyme contains a two zinc cocatalytic site that uses a carboxylated lysine, $LysCO_2^-$, and a hydroxide (or H_2O) as bridging ligands (Table 1). The more buried zinc, Zn1, is coordinated by the NE2 nitrogens of His55 and His57, a carboxylate oxygen from Asp301 and $Lys169CO_2^-$, and a hydroxide ion in a trigonal bipyramidal geometry in a number of phosphonate complexes.[27,94] The more solvent exposed Zn2 is coordinated by His201 (ND1), His230 (NE2), the second oxygen of the bridging $Lys169CO_2^-$, and a bridging hydroxide in a trigonal bipyramidal geometry if the inhibitor phosphonate oxygen binds to the zinc. The geometry becomes more tetrahedral when the phosphonate oxygen of an inhibitor is further removed from the metal coordination sphere. Since the cadmium and manganese enzymes are also catalytically active, the crystalline structures of the Cd/Cd, Zn/Cd, Mn/Mn PTEs have been determined.[95] In each case, the more buried metal remains in a trigonal bipyramidal geometry, while the coordination geometry of the exposed metal is octahedral for Cd and Mn and tetrahedral for Zn. The octahedral geometry is achieved by adding two additional water molecules with little disruption in the position of the other ligands.

It has been suggested that the PTE from *P. diminuta* maybe an example of natural evolution of a new type of enzyme activity in response to environmental changes.[28] The *E. coli* enzyme PTE homology protein (PHP), was examined on the basis of its sequence similarity to PTE. The protein has a similar α/β barrel fold with a cocatalytic zinc site in a cleft at the C-terminal end of the barrel.

The two zinc ions are bound in a nearly identical fashion to that found in PTE with the exception of a glutamic acid side chain carboxylate replacing the $LysCO_2^-$ as the bridging ligand (Table 1). The enzyme differs from PTE in that it does not catalyze the hydrolysis of nonspecific phosphotriesters.

Purple acid phosphatase

The purple acid phosphatases (PAP) are a group of nonspecific phosphomonoesterases that have been found in animal, plant, and fungal sources.[96] The characteristic purple color of this subclass of acid phosphatases comes from a tyrosine phenolate-Fe(III) charge-transfer transition at 560 nm. The presence of Fe(III) is universally found in these enzymes. The 35-kDa mammalian PAP, or tartrate-resistant acid phosphatases (TRAP), contain a Fe(III)–Fe(II) iron center,[31] in contrast to the Fe(III)-Zn(II) center found in the 110 kDa kidney bean enzyme.[29] Nevertheless, despite the low sequence homology, monomeric versus dimeric structural nature, and different metal types, the mammalian and plant purple acid phosphatases have an almost identical fold of the active site domain.[97] The serine/threonine human protein phosphatase 1 and calcineurin also contain a very similar cocatalytic Fe(III)-M(II) (where M is Zn or Mn) site to that of the PAPs.[34,35] In this case, there is no Tyr–Fe(III) interaction but the general ligand nature, the distance between metals, and the presence of a bridging aspartic acid residue, is common to all of these enzymes (Table 1). The ligands are generally provided by a secondary structure of β-sheets in marked contrast to that found in the alkaline phosphatase family. A mechanism has been proposed for the PAPs in which the phosphate ester oxygen binds to the Zn(II) or Fe(II) site and the phosphate bond undergoes nucleophilic attack by a Fe(III)-coordinated hydroxide ion.[96,97]

UDP-sugar hydrolase

The *E. coli* 5′-nucleotidase is a zinc-containing enzyme that displays UDP-sugar hydrolase as well as 5′-nucleotidase activity.[37] The structure of the enzyme has been obtained with either zinc or manganese in the metal sites.[37,38] When occupied by zinc, both sites are trigonal bipyramidal. A water molecule and one oxygen of Asp84 bridges both sites (Table 1). It is not clear if both metal sites are essential for the activity of the enzyme.

Amidohydrolase family

Examination of the enzyme architecture of urease, phophotriesterase and adenosine deaminase has led to the formation of an amidohydrolase superfamily.[98] Several recent crystal structures are in agreement with this classification. Dihydroorotase (DHO) is a zinc metalloenzyme that catalyzes the cyclization of carbamoyl-L-aspartate to L-dihydroorotate in the biosynthetic pathway for the assembly of pyrimidine nucleotides.[39] Each subunit of the dimer contains a 'TIM' barrel motif with eight strands of parallel β-sheet flanked on the outer surface by α-helices. The active site contains two zinc ions that are bridged by a carboxylated Lys102 and a water or hydroxide ion. The more buried zinc, Zn1, is bound in a trigonal bipyramidal geometry with the NE2 nitrogens of His16 and His18, and the bridging hydroxide/water in the equatorial positions and one oxygen of the carboxylates of Asp250 and Lys102CO_2^- as the axial ligands.[39] The more solvent exposed Zn2 is coordinated by the NE2 nitrogens of His139 and His177, the second oxygen of the bridging Lys102CO_2^-, and the bridging hydroxide in a distorted tetrahedral geometry. All of the ligands are provided by β-sheets of the TIM barrel (Table 1). Several secondary interactions between neighboring amino acids and the histidine ligands also occur that could be involved in fine-tuning the chemical reaction.[71] These include hydrogen bonds from the carboxylates of Glu141 and Glu176 to the nonmetal binding nitrogens of the imidazole rings of His139 and His16 respectively. In addition, the side chain amide carbonyl of Asn44 hydrogen bonds the nonbonding nitrogen of His18.

The overall geometry and ligand nature of this site is remarkably similar to that of the cocatalytic zinc site for the *P. diminuta* phosphotriesterase (Table 1). In addition, the L-hydantoinase or dihydropyrimidinase from *A. aurescens*,[40] and the D-hydantoinases from *thermus* sp.,[41] *B. stearothermophilus*,[42] and *Burkholderia picketti*[43] contain cocatalytic zinc sites that are very similar to those of the dihydroorotase and phosphotriesterase enzymes (Table 1). The dihydropyrimidinases belong to a class of cyclic amidases whose physiological function is the hydrolysis of dihydropyrimidines as the second step in the reductive catabolism of pyrimidines.[40] These enzymes are used commercially to produce optically pure L- and D-amino acids from racemic hydantoins. The recent structural determination of the isoaspartyl dipeptidase from *E. coli* indicates that it also belongs to this family of enzymes (reference 44; see also Table 1). This enzyme converts β L-isoaspartyl linkages to normal L-aspartyl residues.

Aminopeptidase family

Aminopeptidases containing cocatalytic zinc sites catalyze the hydrolysis of a wide variety of N-terminal peptides and amino acid derivatives. These enzymes are widely distributed in bacteria, yeast, and in plant and animal sources. The structures of several aminopeptidases containing cocatalytic zinc sites have been reported (Table 1). The ligand nature and spacing characteristics allow some subgrouping of the zinc sites found in these enzymes. Thus, the bovine lens,[45] and the *E. coli* PepA,[46] form one group, the *A. proteolytica*,[47] and *S. griseus*,[49] form a second set and the methionine,[51,53,54] and proline[55] aminopeptidases a third group. Lastly, the cocatalytic sites of the *L. delbrueckii* D-aminopeptidase,[57] and the *Pseudomonas* sp. carboxypeptidase G_2 are quite similar.[56]

The most extensive steady state and pre–steady state kinetic, spectral, and structural studies have been performed on bovine lens leucine aminopeptidase (BLAP), *A. proteolytica* aminopeptidase (AAP) and *S. griseus* aminopeptidase (SGAP). The zinc binding site of the hexameric BLAP is composed of two zinc atoms separated by 2.91 Å.[45] Zn1 is defined as the fast exchange, weak binding site in which Mg can substitute for Zn.[99] It is coordinated to a carboxylate oxygen of Asp255, a carboxylate oxygen and backbone carbonyl of Asp332, and one of the bridging carboxylate oxygens of Glu334. Zn2, is defined as the tight-binding, slow exchange site. Zn2 is coordinated to a carboxylate oxygen of Asp255, a carboxylate oxygen of Asp273, the side chain ε-amino group nitrogen of Lys250, and a bridging carboxylate oxygen of Glu334. No bound water molecules are observed in the free enzyme or in any of the inhibitor complexes of these crystals and there are no nucleophilic amino acids in the active site, which are in a position for direct attack on the scissile carbonyl bond of the substrate.[99,100] A general base catalyzed mechanism is still favored using the zinc ligand Asp255 as the general base acting on an active site water molecule and Lys262 as the general acid catalyst.[45,99,101,102] It has been proposed that Zn2 is involved in substrate binding while Zn1 along with Arg336 are the electrophiles that polarize the scissile carbonyl bond.[102] *E. coli* alkaline phosphatase serves as a precedent for a cocatalytic site containing a ligand that is involved in catalysis (see above discussion) by forming a phosphorylated intermediate of Ser102.

Alternative mechanisms have been suggested on the basis of the results of inhibitor studies using the transition state analogs, L-leucinephosphonic acid, and the *gem*-diolate analog L-leucinal.[30,100] In the L-leucinephosphonic acid complex, one of the phosphorous oxygens bridges both zinc ions, while the second phosphorous oxygen is closer to Zn2. In the L-leucinal complex, the amino terminal nitrogen is bound to Zn2 and one of the *gem*-diolate oxygens is bound to Zn1, while the other bridges both metals. In the course of these studies, a new crystal form was found in which a water molecule does bridge the two metal sites in the inhibitor-free form.[103] These results led to the proposal for a modified mechanism in which a metal-bound hydroxide is the nucleophile and the carbonyl of the substrate is polarized by both zinc ions acting as Lewis acid catalysts. These studies also observed a bicarbonate or carbonate ion bound near Arg336 in the inhibitor-free form of the enzyme.[103] This led to the discovery of bicarbonate acting as a general base catalyst,[104] in the structurally similar aminopeptidase PepA. The bicarbonate is believed to be bound to Arg336 (BLAP numbering) since no activation is seen in the lysine, alanine, methionine, and glutamic acid mutants. These studies also indicate Arg336 is not essential to catalysis. The bicarbonate is proposed to facilitate proton transfer from the zinc-bridging water nucleophile to the peptide-leaving group. The mutant enzyme Lys282Ala reduces catalytic activity 10 000-fold, suggesting that it may be important in stabilizing the transition state.

The structure of *A. proteolytica* aminopeptidase has been determined in its free state,[47] and in complexes with a hydroxamate inhibitor,[101] and butaneboronic acid.[105] Zn1 is coordinated to both carboxylate oxygens of Glu152, the NE2 of His256, and a bridging water and a bridging carboxylate oxygen of Asp117.[47] Zn2, is bound to both carboxylate oxygens of Asp179, the NE2 of His97, and a bridging water molecule and a carboxylate oxygen of Asp117. The structure of SGAP has been determined in its free state[49] and in complexes with the L-amino acid products L-Phe,[106] L-Leu, and L-Met.[50] SGAP and AAP have very similar cocatalytic zinc sites (Table 1). Both enzymes have a histidine and an aspartic acid ligand bound to Zn1 and a histidine and a glutamine ligand bound to Zn2. The zinc ions are separated by 3.5 to 3.6 Å. They also have an aspartic acid and a water/hydroxide that coordinate both metals. In each case, a nearby potential acid/base catalyst is found; Glu151 for AAP and Glu131 for SGAP.[106]

The results of a p-iodo-D-phenylalanine hydroxamate inhibitor complex with AAP have led to a proposed mechanism for the AAP catalyzed hydrolysis of peptides.[101] This structure shows that a single oxygen of the inhibitor bridges the two metals after displacing the bridging water molecule. Both the hydroxyl and carbonyl oxygen atoms of the hydroxamate bind to Zn1, while only the hydroxyl oxygen, albeit at a shorter distance, binds to Zn2. In the complex, the distance between the zinc increases to 3.7 Å. In addition, the carboxylate oxygen of Glu151 hydrogen bonds the hydroxylamine nitrogen in a manner seen in several of the matrix metalloproteinases[107,108] and thermolysin.[109] The similarity of these structures led the authors to propose Glu151 as the general base catalyst in AAP.[101]

The results of electron paramagnetic resonance (EPR) studies on the Co(II)Zn and ZnCo(II) enzymes suggest the competitive inhibitor 1-butaneboronic acid binds to the first site only in the heterosubstituted AAPs.[110] The X-ray structure of the complex of the inhibitor with the di-zinc enzyme support this conclusion reasonably well.[105] The inhibitor displaces the bridging water resulting in a complex with both the oxygens of the boronic acid 2.5 Å and 2.7 Å from Zn1 and the first oxygen 3.0 Å from Zn2. This structure has been said to be intermediate between a substrate Michaelis complex and the transition state.

The binding of product amino acids, phenylalanine, leucine, and methionine to SGAP has been examined by X-ray crystallography.[50] These inhibitors bind in very similar ways by displacing the bridging water molecule. The carboxylate group bridges the two zinc ions. In the leucine complex, one oxygen of the carboxylate is 2.22 Å from Zn1 while the second oxygen forms a short (2.09 Å) and long bond (2.68 Å) with Zn2 and a long H-bond to the potential acid/base catalyst in the reaction Glu131 (2.81 Å) and the OH of Tyr246 (2.90 Å). The amine group forms H-bonds with carboxylate oxygens of Glu131 (2.90 Å), Asp160 (2.86 Å), and the backbone carbonyl of Arg202 (2.50 Å). This complex is similar to the phenylalanine hydroxamate complex of AAP.[101] Tyr246 has no counterpart in AAP. A potential product amine site also exists in AAP.[106] This complex must represent the last step in the hydrolytic reaction.

The importance of the zinc-bound water to catalysis has been indicated by studies of fluoride ion inhibition.[111] Fluoride inhibits in an uncompetitive fashion, which is weakened, as a group in the ES complex ionizes with a pK_a of 7.0. The zinc-bound bridging water molecule is proposed to be the group with this pK_a on the basis of the belief that substrate binding should break the bridging water bond and allow fluoride to displace it. Such uncompetitive inhibition by chloride and bromide has been observed for thermolysin catalysis.[112,113]

These studies allow the formulation of a potential mechanism for the cocatalytic aminopeptidase enzymes that is similar in nature to the one for the zinc peptidases containing catalytic zinc sites with two histidine residues and one glutamic acid residue as ligands (Figure 4).[71] In this mechanism, the scissile peptide carbonyl disrupts the bridging water of the cocatalytic zinc site and interacts with Zn1 in the first step. Glu151/Glu131 could act as a general base first by removing a proton from the Zn2-bound water. In the next step, the Zn2-bound hydroxide could attack the carbonyl of the scissile peptide bond to form a tetrahedral intermediate. In the last step, the protonated form of Glu131/Glu151 could act as a general acid catalyst by donating a proton to the scissile amide nitrogen allowing the tetrahedral intermediate to collapse to the product complex (Figure 4).

Methionyl aminopeptidases

The methionyl aminopeptidases (MetAP) family play a critical role in protein synthesis since they remove the N-terminal amino acid methionine from newly formed proteins.[114] The importance of their role in the formation of functional proteins has led to the identification of human MetAP as a molecular target for antiangiogenesis and anticancer drugs. This class was initially divided into two classes through analyses of their sequences. The X-ray structures of several of these enzymes reveal the presence of a potential cocatalytic metal site.[51,53,54,115] While these structures were of the di-cobalt and di-manganese enzymes (Table 1), several divalent metals (Co, Zn, Mn, Ni, and Fe) have been reported to activate these

Cocatalytic zinc sites

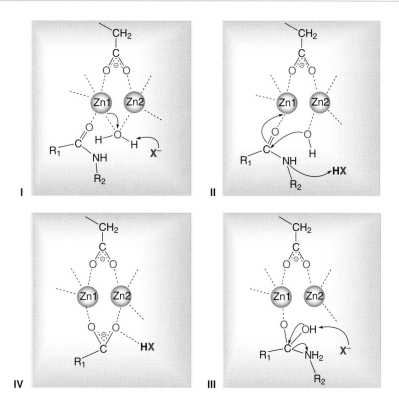

Figure 4 Schematic for a proposed mechanism of peptide hydrolysis for two metal cocatalytic zinc sites.[71]

enzymes.[115–119] As these studies have progressed, the Fe(II) ion has been suggested to be the physiological metal for *E. coli* MetAP.[119,120] Recent extended X-ray absorption fine structure studies of the Fe(II) and Co(II) enzymes have in fact suggested that the active form of the enzyme needs only one metal.[121]

Other peptidases

The catalytic domain of the exopeptidase *Pseudomonas* sp. carboxypeptidase G$_2$, CPDG2, shows a similarity to a number of the mono- and di-zinc exopeptidases that contain catalytic and cocatalytic zinc sites respectively.[56] The mono-zinc carboxypeptidases, CPDA, CPDB, and CPDT (M14 class of metallopeptidases[122]), and the cocatalytic peptidases *A. proteolytica* aminopeptidase, AMP (M28), bovine lens aminopeptidase, LAP (M17) and CPDG2 (M20), all contain eight-stranded β-sheets and six (CPDA, CPDB, CPDT) or seven (CPDG2, AMP, and LAP) structurally equivalent helices. In all these enzymes, the zinc sites are found at the C-terminal side of four central parallel β-strands.[56]

The cocatalytic zinc site of the *L. delbrueckii* D-aminopeptidase, PepV,[57] is also quite similar to that of CPDG2 (Table 1). The nature, spacing, and backbone support for the ligands is very similar. In both cases, a bridging aspartic acid residue resides two amino acids after an α-helix of 17 amino acids. A bridging water molecule is also seen in both structures. The one difference is the use of glutamic acid as a ligand to the Zn2 site in PepV in place of an aspartic acid in CPDG2. The difference in the distance between the two zinc ions is probably due to the type of inhibitor complex observed for PepV.[57] The cocatalytic zinc site of the *Salmonella typhimurium* peptidase T or tripeptidase[58] is also quite similar to those of PepV and CPDG2 (Table 1). No zinc-bound water molecule is found; the closest water being 3.1 Å from Zn1.

Two other metalloprotease structures have been reported that contain unique cocatalytic zinc sites (Table 1). The *B. subtilis* D-specific aminopeptidase (DppA) preferring D-Ala-D-Ala as a substrate, has a unique tertiary structure consisting of a barrel shaped decamer.[59] The core domain (an α/6 parallel β, 2 antiparallel β/α fold) and the cocatalytic zinc site are different from other known metalloproteases. The zinc ions are located between strands 1 and 4 near the C-terminal of the central β-strand. Zn1 is bound to one oxygen of the carboxylate of bridging Asp8, a carboxylate oxygen of Glu10 and His60 (ND1), while Zn2 is bound to the other carboxylate oxygen of Asp8, His104 (NE2), and a carboxylate oxygen of Glu133. A bridging water molecule completes the tetrahedral coordination of both zinc ions.

The human renal dipeptidase, a membrane-bound glycoprotein, is involved in the hydrolytic metabolism of penem and carbapenem β-lactam antibiotics.[60] Both zinc ions are bound in a distorted trigonal bipyramid. The zinc ions are bridged by the carboxylate oxygens of Glu125. Each zinc binds two water molecules. A carboxylate oxygen of

Cocatalytic zinc sites

Asp20 and His22 (NE2) complete the coordination of Zn1, while the NE2 nitrogens of His198 and His219 complete the coordination of Zn2 (Table 1). The binding of the inhibitor cilastatin yields a structure in which only one bridging water is found.[60] The other ligands remain the same and one carboxylate oxygen of the inhibitor binds to Zn2. The resulting coordination spheres are a trigonal bipyramid for Zn1 and a square pyramid for Zn2.

Glyoxalase II

Glyoxalase II, a thiolesterase, is part of a two enzyme pathway that converts toxic 2-oxoaldehydes (such as methylglyoxal) to the corresponding 2-hydroxycarboxylic acids using glutathione as the coenzyme.[123] The enzyme contains an N-terminal domain that has a βαβαβ-fold resembling that found in the whole structure of β-lactamases.[61] The N-terminal sheet of the metallo β-lactamases contains two extra strands at the N-terminus. Protein database searches using a zinc binding motif, T/SHXHXD, found in the β-lactamases was used to predict the ligands to glyoxalase II and the prokaryotic aryl sulfatases.[124] This study correctly predicted several of the metal ligands in the cocatalytic zinc site of glyoxalase II. However, the conservation of some of the putative zinc ligands as well as the adjacent amino acid residues was not good in some of the selections.[70] As a result, the choice of a cysteine ligand for the glyoxalase II enzyme was incorrect and the bridging amino acid, Asp134 or His59 ligands were not predicted.

The two metals are bridged by a single oxygen of the carboxylate of Asp134 and a water molecule. Zn1 is also coordinated to the imidazole nitrogens of His54 (NE2), His56 (ND1), and His173 (NE2), while Zn2 additionally binds His59 (NE2) and His173 (NE2). When the structure is determined in the presence of the slowly turned over substrate S-(N-hydroxy-N-bromophenylcarbamoyl) glutathione the carbonyl group of the substrate replaces the water bound to Zn1 and the sulfur is 3.3 Å removed from Zn2 in subunit B. In subunit A, only glutathione is bound with its sulfur now 2.8 Å from Zn2 (PDB# 1QH5). The crystal structure was performed on an enzyme that contained 1.5 mol of zinc and 0.7 mol of iron.[61] Cytosolic glyoxalase II from *Arabidopsis thaliana* was originally shown to be dependent on two moles of zinc for activity.[125] However, recent studies suggest that this enzyme is a zinc/iron enzyme.[126] The difference in the results of the two studies is believed to be due to an inadvertent mutation, R248W in the former enzyme that causes the enzyme to prefer zinc over iron.

β-Lactamase family

The interest in β-lactamases comes from their ability to catalyze the hydrolysis of the amide bond in β-lactam antibiotics.[127] These enzymes have been grouped into four classes based on sequence comparisons. Three of these classes are serine type hydrolases while class B is a metalloenzyme. The first reported β-lactamase structure was on the *B. cereus* enzyme from crystals grown at pH 5.6. This enzyme contained one zinc site.[62] The zinc is bound to the NE2 nitrogens of His84 and His149 and the ND1 nitrogen of His88 and a water molecule in a distorted tetrahedron (Table 1) – features typical of catalytic zinc sites. However, the structure of the *B. fragilis* enzyme showed that it contained two zinc sites.[65] This enzyme was crystallized at pH 7.0 in a 10-μM ZnCl$_2$ HEPES buffer. The structure contained the catalytic-like zinc site seen in the *B. cereus* enzyme (Zn1, Table 1) and a second zinc, Zn2, bound to an oxygen of the carboxylate of Asp90, the sulfur of Cys168, the NE2 nitrogen of His210 (numbering according to *B. cereus* enzyme), and two water molecules in a trigonal bipyramidal geometry. One of the water molecules bridges the two zinc ions. This structure was followed by those for the β-lactamases from *S. maltophilia*[67] and *Pseudomonas aeruginosa*,[68] both of which contain two bound zincs in the active site. In addition, a structure for the *B. cereus* enzyme that contains two bound zinc ions was also reported.[63,64] In all of these structures, the features of the Zn1 site remains the same. However, the features of Zn2 vary considerably (Table 1).

It is not yet clear if this is a true cocatalytic zinc site. If so, it will be the first cocatalytic zinc site in which there is no bridging amino acid ligand. A water molecule does bridge the metals in some of the structures. A bridging water molecule is not observed in the *Ps. aeruginosa* structure at 3.1 Å but is assumed to be there.[68] No shared water molecule is observed in the *B. cereus* enzyme that binds two zinc ions.[63] The distance between the metals can vary from 3.9 to 4.4 Å, while that for the Zn1 to its bound water remains constant at 1.9 Å.[64] It should also be noted that zinc hydroxide bridges are inhibitory in some zinc proteases.[70,128–130] The ligand nature of the second zinc site in the β-lactamases is not conserved in the few enzymes that have been examined. Thus, the *S. maltophilia* β-lactamase has no cysteine in the second zinc site.[67]

The importance of the second zinc site to catalytic activity is still not clear. The mono-zinc *B. cereus* enzyme is active and the *A. hydrophila* AE06 enzyme is inhibited by Zn with a K_i of 46 μM,[131] while the catalytic zinc binds to the enzyme with a dissociation constant lower than 20 nM. The kinetic parameters for the mono- and di-zinc *B. fragilis* enzyme differ only slightly for four substrates.[132] Mutation of Cys168 to serine in the *B. fragilis* enzyme eliminates the second zinc site.[133] The k_{cat} values for this enzyme is reduced 140- to 1500-fold, while the k_{cat}/K_m values are reduced 970- to 3700-fold dependent on the substrate used.[134] This reduction in activity could be due to the importance of the second zinc to catalysis or the loss of a residue that plays a role in the transition state in catalysis. The mutant Cys168Ser *B. cereus* enzyme can bind a second zinc weakly.[135] Changes in the interaction of Asp90 and Arg91 in the mono-zinc mutant enzyme make it difficult to interpret the activity effects.[136] The determination of metal dissociation constants for the enzymes from *B.cereus*, *Chryseobacterium mengosepticum*, *Aeromonas hydrophila*, and *S. maltophilia* in the absence and presence of substrates suggest that the second zinc binding site may exist only at nonphysiologic high zinc concentrations.[137]

Thus, to summarize, all the β-lactamases are dependent on the presence of one zinc that has the characteristics of a catalytic zinc site. The functional importance of the second zinc site is in doubt for many of the enzymes in this class.

REFERENCES

1. BL Vallee and DS Auld, *Matrix Suppl*, **1**, 5–19 (1992).
2. BL Vallee and DS Auld, *Acc Chem Res*, **26**, 543–51 (1993).
3. DS Auld, *BioMetals*, **14**, 271–313 (2001).
4. JA Tainer, ED Getzoff, KM Beem, JS Richardson and DC Richardson, *J Mol Biol*, **160**, 181–217 (1982).
5. HE Parge, RA Hallewell and JA Tainer, *Proc Natl Acad Sci USA*, **89**, 6109–13 (1992).
6. KD Carugo, A Battistoni, MT Carri, F Polticelli, A Desideri, G Rotilio, A Coda and M Bolognesi, *FEBS Lett*, **349**, 93–98 (1994).
7. Y Kitagawa, N Tanaka, Y Hata, M Kusunoki, GP Lee, Y Katsube, K Asada, S Aibara and Y Morita, *J Biochem (Tokyo)*, **109**, 477–85 (1991).
8. K Djinovic, G Gatti, A Coda, L Antolini, G Pelosi, A Desideri, M Falconi, F Marmocchi, G Rotilio and M Bolognesi, *J Mol Biol*, **225**, 791–809 (1992).
9. NL Ogihara, HE Parge, PJ Hart, MS Weiss, JJ Goto, BR Crane, J Tsang, K Slater, JA Roe, JS Valentine, D Eisenberg and JA Tainer, *Biochemistry*, **35**, 2316–21 (1996).
10. A Pesce, C Capasso, A Battistoni, S Folcarelli, G Rotilio, A Desideri and M Bolognesi, *J Mol Biol*, **274**, 408–20 (1997).
11. A Pesce, A Battistoni, ME Stroppolo, F Polizio, M Nardini, JS Kroll, PR Langford, P O'Neill, M Sette, A Desideri and M Bolognesi, *J Mol Biol*, **302**, 465–78 (2000).
12. Y Bourne, SM Redford, HM Steinman, JR Lepock, JA Tainer and ED Getzoff, *Proc Natl Acad Sci USA*, **93**, 12774–79 (1996).
13. YL Chen, S Park, RW Thornburg, LB Tabatabai and A Kintanar, *Biochemistry*, **34**, 12265–75 (1995).
14. KT Forest, PR Langford, JS Kroll and ED Getzoff, *J Mol Biol*, **296**, 145–53 (2000).
15. EE Kim and HW Wyckoff, *J Mol Biol*, **218**, 449–64 (1991).
16. B Stec, MJ Hehir, C Brennan, M Nolte and ER Kantrowitz, *J Mol Biol*, **277**, 647–62 (1998).
17. MH Le Du, T Stigbrand, MJ Taussig, A Menez and EA Stura, *J Biol Chem*, **276**, 9158–65 (2001).
18. M de Backer, S McSweeney, HB Rasmussen, BW Riise, P Lindley and E Hough, *J Mol Biol*, **318**, 1265–74 (2002).
19. JE Murphy, TT Tibbitts and ER Kantrowitz, *J Mol Biol*, **253**, 604–17 (1995).
20. MH Le Du, C Lamoure, BH Muller, OV Bulgakov, E Lajeunesse, A Menez and JC Boulain, *J Mol Biol*, **316**, 941–53 (2002).
21. E Hough, LK Hansen, B Birknes, K Jynge, S Hansen, A Hordvik, C Little, E Dodson and Z Derewenda, *Nature*, **338**, 357–60 (1989).
22. N Justin, N Walker, HL Bullifent, G Songer, DM Bueschel, H Jost, C Naylor, J Miller, DS Moss, RW Titball and AK Basak, *Biochemistry*, **41**, 6253–62 (2002).
23. CE Naylor, JT Eaton, A Howells, N Justin, DS Moss, RW Titball and AK Basak, *Nat Struct Biol*, **5**, 738–46 (1998).
24. JT Eaton, CE Naylor, AM Howells, DS Moss, RW Titball and AK Basak, *J Mol Biol*, **319**, 275–81 (2002).
25. A Volbeda, A Lahm, F Sakiyama, and D Suck, *EMBO J*, **10**, 1607–18 (1991).
26. DJ Hosfield, Y Guan, BJ Haas, RP Cunningham and JA Tainer, *Cell*, **98**, 397–408 (1999).
27. MM Benning, SB Hong, FM Raushel and HM Holden, *J Biol Chem*, **275**, 30556–60 (2000).
28. JL Buchbinder, RC Stephenson, MJ Dresser, JW Pitera, TS Scanlan and RJ Fletterick, *Biochemistry*, **37**, 5096–106 (1998).
29. T Klabunde, N Strater, R Frohlich, H Witzel and B Krebs, *J Mol Biol*, **259**, 737–48 (1996).
30. N Strater, T Klabunde, P Tucker, H Witzel and B Krebs, *Science*, **268**, 1489–92 (1995).
31. J Uppenberg, F Lindqvist, C Svensson, B Ek-Rylander and G Andersson, *J Mol Biol*, **290**, 201–11 (1999).
32. Y Lindqvist, E Johansson, H Kaija, P Vihko and G Schneider, *J Mol Biol*, **291**, 135–47 (1999).
33. LW Guddat, AS McAlpine, D Hume, S Hamilton, J de Jersey and JL Martin, *Struct Fold Des*, **7**, 757–67 (1999).
34. MP Egloff, PT Cohen, P Reinemer and D Barford, *J Mol Biol*, **254**, 942–59 (1995).
35. CR Kissinger, HE Parge, DR Knighton, CT Lewis, LA Pelletier, A Tempczyk, VJ Kalish, KD Tucker, RE Showalter, EW Moomaw, LN Gastinel, N Habuka, X Chen, F Maldonado, JE Baker, R Bacquet and VJ E, *Nature*, **378**, 641–44 (1995).
36. JP Griffith, JL Kim, EE Kim, MD Sintchak, JA Thomson, MJ Fitzgibbon, MA Fleming, PR Caron, K Hsiao and MA Navia, *Cell*, **82**, 507–22 (1995).
37. T Knofel and N Strater, *J Mol Biol*, **309**, 239–54 (2001).
38. T Knofel and N Strater, *Nat Struct Biol*, **6**, 448–53 (1999).
39. JB Thoden, GN Phillips Jr, TM Neal, FM Raushel and HM Holden, *Biochemistry*, **40**, 6989–97 (2001).
40. J Abendroth, K Niefind, O May, M Siemann, C Syldatk and D Schomburg, *Biochemistry*, **41**, 8589–97 (2002).
41. J Abendroth, K Niefind and D Schomburg, *J Mol Biol*, **320**, 143–56 (2002).
42. YH Cheon, HS Kim, KH Han, J Abendroth, K Niefind, D Schomburg, J Wang and Y Kim, *Biochemistry*, **41**, 9410–17 (2002).
43. Z Xu, Y Liu, Y Yang, W Jiang, E Arnold and J Ding, *J Bacteriol*, **185**, 4038–49 (2003).
44. JB Thoden, R Marti-Arbona, FM Raushel and HM Holden, *Biochemistry*, **42**, 4874–82 (2003).
45. SK Burley, PR David, RM Sweet, A Taylor and WN Lipscomb, *J Mol Biol*, **224**, 113–40 (1992).
46. N Strater, DJ Sherratt and SD Colloms, *EMBO J*, **18**, 4513–22 (1999).
47. B Chevrier, C Schalk, H D'Orchymont, JM Rondeau, D Moras and C Tarnus, *Structure*, **2**, 283–91 (1994).
48. W Desmarais, D Bienvenue, K Bzymek, R Holz, G Petsko and D Ringe, *Structure (Camb)*, **10**, 1063–72 (2002).
49. HM Greenblatt, O Almog, B Maras, A Spungin-Bialik, D Barra, S Blumberg and G Shoham, *J Mol Biol*, **265**, 620–36 (1997).

50. R Gilboa, HM Greenblatt, M Perach, A Spungin-Bialik, U Lessel, G Wohlfahrt, D Schomburg, S Blumberg and G Shoham, *Acta Crystallogr, Sect D Biol Crystallogr*, **56**, 551–58 (2000).
51. SL Roderick and BW Matthews, *Biochemistry*, **32**, 3907–12 (1993).
52. WT Lowther, Y Zhang, PB Sampson, JF Honek and BW Matthews, *Biochemistry*, **38**, 14810–19 (1999).
53. TH Tahirov, H Oki, T Tsukihara, K Ogasahara, K Yutani, CP Libeu, Y Izu, S Tsunasawa and I Kato, *J Struct Biol*, **121**, 68–72 (1998).
54. S Liu, J Widom, CW Kemp, CM Crews and J Clardy, *Science*, **282**, 1324–27 (1998).
55. MC Wilce, CS Bond, NE Dixon, HC Freeman, JM Guss, PE Lilley and JA Wilce, *Proc Natl Acad Sci USA*, **95**, 3472–77 (1998).
56. S Rowsell, RA Pauptit, AD Tucker, RG Melton, DM Blow and P Brick, *Structure*, **5**, 337–47 (1997).
57. D Jozic, G Bourenkow, H Bartunik, H Scholze, V Dive, B Henrich, R Huber, W Bode and K Maskos, *Structure (Camb)*, **10**, 1097–106 (2002).
58. K Hakansson and CG Miller, *Eur J Biochem*, **269**, 443–50 (2002).
59. H Remaut, C Bompard-Gilles, C Goffin, JM Frere and J Van Beeumen, *Nat Struct Biol*, **8**, 674–78 (2001).
60. Y Nitanai, Y Satow, H Adachi and M Tsujimoto, *J Mol Biol*, **321**, 177–84 (2002).
61. AD Cameron, M Ridderstrom, B Olin and B Mannervik, *Struct Fold Des*, **7**, 1067–78 (1999).
62. A Carfi, S Pares, E Duee, M Galleni, C Duez, JM Frere and O Dideberg, *EMBO J*, **14**, 4914–21 (1995).
63. A Carfi, E Duee, M Galleni, JM Frere and O Dideberg, *Acta Crystallogr, Sect D Biol Crystallogr*, **54**, 313–23 (1998).
64. SM Fabiane, MK Sohi, T Wan, DJ Payne, JH Bateson, T Mitchell and BJ Sutton, *Biochemistry*, **37**, 12404–11 (1998).
65. NO Concha, BA Rasmussen, K Bush and O Herzberg, *Structure*, **4**, 823–36 (1996).
66. A Carfi, E Duee, R Paul-Soto, M Galleni, JM Frere and O Dideberg, *Acta Crystallogr, Sect D Biol Crystallogr*, **54**, 45–57 (1998).
67. JH Ullah, TR Walsh, IA Taylor, DC Emery, CS Verma, SJ Gamblin and J Spencer, *J Mol Biol*, **284**, 125–36 (1998).
68. NO Concha, CA Janson, P Rowling, S Pearson, CA Cheever, BP Clarke, C Lewis, M Galleni, JM Frere, DJ Payne, JH Bateson and SS Abdel-Meguid, *Biochemistry*, **39**, 4288–98 (2000).
69. BL Vallee and DS Auld, *Biochemistry*, **32**, 6493–500 (1993).
70. DS Auld, in I Bertini, A Sigel and H Sigel (eds.), *Handbook on Metalloproteins*, Marcel Dekker, New York, pp 881–959 (2001).
71. DS Auld, in AJ Barrett, ND Rawlings and JF Woessner (eds.), *Handbook of Proteolytic Enzymes*, Academic Press, London, pp XXX–XXX (2003).
72. JV Bannister, WH Bannister and G Rotilio, *CRC Crit Rev Biochem*, **22**, 111–80 (1987).
73. D Bordo, A Pesce, M Bolognesi, ME Stroppolo, M Falconi and A Desideri, in A Messerschmidt, R Huber, K Poulos and K Wieghardt (eds.), *Handbook of Metalloproteins*, Vol. 2, Wiley, Chichester pp 752–67 (2001).
74. L Banci, I Bertini, F Cantini, M D'Onofrio and MS Viezzoli, *Protein Sci*, **11**, 2479–92 (2002).
75. AG Estevez, JP Crow, JB Sampson, C Reiter, Y Zhuang, GJ Richardson, MM Tarpey, L Barbeito and JS Beckman, *Science*, **286**, 2498–500 (1999).
76. B Stec, KM Holtz and ER Kantrowitz, *J Mol Biol*, **299**, 1303–11 (2000).
77. PJ O'Brien and D Herschlag, *Biochemistry*, **41**, 3207–25 (2002).
78. A Kozlenkov, T Manes, MF Hoylaerts and JL Millan, *J Biol Chem*, **277**, 22992–99 (2002).
79. X Xu and ER Kantrowitz, *J Biol Chem*, **267**, 16244–51 (1992).
80. JE Murphy, X Xu and ER Kantrowitz, *J Biol Chem*, **268**, 21497–500 (1993).
81. E Dirnbach, DG Steel and A Gafni, *Biochemistry*, **40**, 11219–26 (2001).
82. TT Tibbitts, JE Murphy and ER Kantrowitz, *J Mol Biol*, **257**, 700–15 (1996).
83. BL Vallee and A Galdes, *Adv Enzymol Relat Areas Mol Biol*, **56**, 283–430 (1984).
84. JE Coleman, *Annu Rev Biophys Biomol Struct*, **21**, 441–83 (1992).
85. KM Holtz, B Stec and ER Kantrowitz, *J Biol Chem*, **274**, 8351–834 (1999).
86. JE Murphy, B Stec, L Ma and ER Kantrowitz, *Nat Struct Biol*, **4**, 618–22 (1997).
87. S Hansen, LK Hansen and E Hough, *J Mol Biol*, **231**, 870–76 (1993).
88. CE Naylor, M Jepson, DT Crane, RW Titball, J Miller, AK Basak and B Bolgiano, *J Mol Biol*, **294**, 757–70 (1999).
89. M Jepson and R Titball, *Microbes Infect*, **2**, 1277–84 (2000).
90. RG Zhang, I Dementieva, N Duke, F Collart, E Quaite-Randall, R Alkire, L Dieckman, N Maltsev, O Korolev and A Joachimiak, *Proteins*, **48**, 423–26 (2002).
91. IP Korndorfer, WD Fessner and BW Matthews, *J Mol Biol*, **300**, 917–33 (2000).
92. DS Auld, in A Messerschmidt, W Bode and M Cygler (eds.), *The Handbook of Metalloproteins*, Vol. 3, John Wiley & Sons, Chichester, pp xxx–xxx (2003).
93. SR Caldwell and FM Raushel, *Appl Biochem Biotechnol*, **31**, 59–73 (1991).
94. JL Vanhooke, MM Benning, FM Raushel and HM Holden, *Biochemistry*, **35**, 6020–25 (1996).
95. MM Benning, H Shim, FM Raushel and HM Holden, *Biochemistry*, **40**, 2712–22 (2001).
96. T Klabunde and B Krebs, *Struct Bonding*, **89**, 177–98 (1997).
97. A Vogel, F Spener and B Krebs, in A Messerschmidt, R Huber, K Poulos and K Wieghardt (eds.), *Handbook of Metalloproteins*, Vol. 2, Wiley, Chichester, pp 1284–300 (2001).
98. L Holm and C Sander, *Proteins: Struct, Funct, Genet*, **28**, 72–82 (1997).
99. H Kim and WN Lipscomb, *Adv Enzymol Relat Areas Mol Biol*, **68**, 153–213 (1994).
100. N Strater and WN Lipscomb, *Biochemistry*, **34**, 9200–10 (1995).
101. B Chevrier, H D'Orchymont, C Schalk, C Tarnus and D Moras, *Eur J Biochem*, **237**, 393–98 (1996).
102. H Kim and WN Lipscomb, *Biochemistry*, **32**, 8465–78 (1993).
103. N Strater and WN Lipscomb, *Biochemistry*, **34**, 14792–800 (1995).
104. N Strater, L Sun, ER Kantrowitz and WN Lipscomb, *Proc Natl Acad Sci USA*, **96**, 11151–55 (1999).
105. CC De Paola, B Bennett, RC Holz, D Ringe and GA Petsko, *Biochemistry*, **38**, 9048–53 (1999).
106. R Gilboa, A Spungin-Bialik, G Wohlfahrt, D Schomburg, S Blumberg and G Shoham, *Proteins*, **44**, 490–504 (2001).
107. W Bode, P Reinemer, R Huber, T Kleine, S Schnierer and H Tschesche, *EMBO J*, **13**, 1263–69 (1994).

108 MF Browner, WW Smith and AL Castelhano, *Biochemistry*, **34**, 6602–10 (1995).
109 MA Holmes and BW Matthews, *Biochemistry*, **20**, 6912–20 (1981).
110 B Bennett and RC Holz, *Biochemistry*, **36**, 9837–46 (1997).
111 G Chen, T Edwards, VM D'Souza and RC Holz, *Biochemistry*, **36**, 4278–86 (1997).
112 S Li, *Biochemical Sciences*, Harvard University, Boston, pp 1–61 (1982).
113 JJ Yang, DR Artis and HE Van Wart, *Biochemistry*, **33**, 6516–23 (1994).
114 WT Lowther and BW Matthews, *Biochim Biophys Acta*, **1477**, 157–67 (2000).
115 WT Lowther, AM Orville, DT Madden, S Lim, DH Rich and BW Matthews, *Biochemistry*, **38**, 7678–88 (1999).
116 X Li and YH Chang, *Biochem Biophys Res Commun*, **227**, 152–59 (1996).
117 KW Walker and RA Bradshaw, *Protein Sci*, **7**, 2684–87 (1998).
118 L Meng, S Ruebush, VM D'Souza, AJ Copik, S Tsunasawa and RC Holz, *Biochemistry*, **41**, 7199–208 (2002).
119 VM D'Souza, B Bennett, AJ Copik and RC Holz, *Biochemistry*, **39**, 3817–26 (2000).
120 VM D'Souza and RC Holz, *Biochemistry*, **38**, 11079–85 (1999).
121 NJ Cosper, VM D'Souza, RA Scott and RC Holz, *Biochemistry*, **40**, 13302–9 (2001).
122 AJ Barrett and ND Rawlings, http://www.merops.co.uk (2001).
123 PJ Thornalley, *Mol Aspects Med*, **14**, 287–371 (1993).
124 S Melino, C Capo, B Dragani, A Aceto and R Petruzzelli, *Trends Biochem Sci*, **23**, 381–82 (1998).
125 MW Crowder, MK Maiti, L Banovic and CA Makaroff, *FEBS Lett*, **418**, 351–54 (1997).
126 TM Zang, DA Hollman, PA Crawford, MW Crowder and CA Makaroff, *J Biol Chem*, **276**, 4788–95 (2001).
127 SG Waley, *Sci Prog*, **72**, 579–97 (1988).
128 KS Larsen and DS Auld, *Biochemistry*, **30**, 2613–18 (1991).
129 JT Bukrinsky, MJ Bjerrum and A Kadziola, *Biochemistry*, **37**, 16555–64 (1998).
130 M Gomez-Ortiz, FX Gomis-Ruth, R Huber and FX Aviles, *FEBS Lett*, **400**, 336–40 (1997).
131 M Hernandez Valladares, A Felici, G Weber, HW Adolph, M Zeppezauer, GM Rossolini, G Amicosante, JM Frere and M Galleni, *Biochemistry*, **36**, 11534–41 (1997).
132 R Paul-Soto, M Hernandez-Valladares, M Galleni, R Bauer, M Zeppezauer, JM Frere and HW Adolph, *FEBS Lett*, **438**, 137–40 (1998).
133 Z Li, BA Rasmussen and O Herzberg, *Protein Sci*, **8**, 249–52 (1999).
134 Y Yang, D Keeney, X Tang, N Canfield and BA Rasmussen, *J Biol Chem*, **274**, 15706–11 (1999).
135 R Paul-Soto, R Bauer, JM Frere, M Galleni, W Meyer-Klaucke, H Nolting, GM Rossolini, D de Seny, M Hernandez-Valladares, M Zeppezauer and HW Adolph, *J Biol Chem*, **274**, 13242–49 (1999).
136 L Chantalat, E Duee, M Galleni, JM Frere and O Dideberg, *Protein Sci*, **9**, 1402–6 (2000).
137 S Wommer, S Rival, U Heinz, M Galleni, JM Frere, N Franceschini, G Amicosante, B Rasmussen, R Bauer and HW Adolph, *J Biol Chem*, **277**, 24142–47 (2002).

Protein interface zinc sites: the role of zinc in the supramolecular assembly of proteins and in transient protein–protein interactions

Wolfgang Maret

Center for Biochemical and Biophysical Sciences and Medicine, Harvard Medical School, Cambridge, MA, USA

INTRODUCTION

Zinc manifests its biological essentiality and functional versatility in numerous zinc proteins.[1] In addition to its function as a catalytic metal ion in zinc enzymes, zinc serves a structural function in proteins, either by stabilizing their tertiary structure or by organizing domains for the interaction with other proteins, nucleic acids, or lipids. The focus of this chapter is on yet another function of zinc, namely, its role in establishing protein quaternary structure when different polypeptide chains supply its ligands. Such sites, for which the term 'protein interface zinc sites' has been introduced,[2,3] have many of the general characteristics of zinc binding sites in proteins. Thus, with few exceptions, protein ligands of zinc are the oxygen atom(s) of glutamate or aspartate, the nitrogen atoms of histidine, and the sulfur atom of cysteine. Four ligands with a tetrahedral coordination geometry of zinc are encountered most frequently, but the coordination sphere can be extended to five and six ligands, in particular, with donor atoms other than sulfur. Two protein ligands are minimal to ensure a nucleus for zinc binding. In zinc enzymes, a short 'spacer' of amino acids often separates these two ligands of zinc, while the spacing between the second and third, and occasionally a fourth, ligand is longer.[4] The recognition of these common features defined metal-binding signatures that can be used to predict zinc binding sites in protein sequences.[5] In fact, mining eukaryotic genomes for zinc-binding motifs led to the discovery of more than 1000 putative zinc proteins. The success of this approach is largely based on the relative ease of recognizing zinc finger motifs whose four zinc ligands are usually closely spaced in the primary structure. A zinc binding site might be predicted from pattern searches, if the monomer supplies three ligands, but an additional ligand supplied by another protein subunit cannot be identified. The predictive capacity is very poor for sites with only two ligands, in particular if the spacer is long, and it fails altogether if the ligands stem from two different molecules, which are all characteristics of some protein interface zinc sites. Consequently, genome databases cannot be searched for protein interface zinc sites in the same manner as for other types of zinc sites. The difficulty in predicting or detecting protein interface zinc sites possibly signifies that the majority of interactions have eluded our attention. Protein interface zinc sites and other sites such as inhibitory zinc sites[6] substantially increase the number and potential functions of zinc sites in proteins. Thus, zinc proteomes based on known motifs underestimate the number and significance of zinc–protein interactions.[7]

The following overview of a seemingly heterogeneous set of zinc binding sites in protein crystal structures suggests important roles of interface zinc sites in organizing higher-order protein structure. Why is there such a role for zinc? The association of protein protomers into oligomers is believed to enlarge the repertoire of structures and functions during evolution.[8] An evolutionary record for this process is the 'hydrophilic effect', that is, the occurrence of polar and charged residues at otherwise hydrophobic interfaces of oligomers.[9] These residues are thought to be a vestige from a state in which the monomer interacted with the solvent. In fact, a tight binding metal ion can employ polar, metal-binding ligands very efficiently, offering additional opportunities to control the oligomeric state of proteins and increasing the level of complexity in homologous and heterologous protein–protein interactions. In contrast to the other redox-inert cations Mg^{2+} and Ca^{2+}, the stereochemically flexible zinc ion is a borderline Lewis acid and therefore particularly well suited for this task. Regulation of the availability of zinc itself, the choice of ligands, which determines the coordination chemistry of the zinc site, and the role of additional nonprotein ligands as effectors are all levels at which control can be exerted over the dynamics of molecular recognition in protein–protein interactions.

ZINC CROSS-LINKS AMONG FOUR POLYPEPTIDE CHAINS

The only example in this category (Figure 1(a)) is the squash trypsin inhibitor, which crystallizes as a 4:1 protein–zinc complex (1LUO).[10] The ligands of the zinc ion are four oxygen atoms of glutamates (Glu19), resulting in a tetrahedral coordination (coordination type: ZnO_4). Presumably, two of the ligands are protonated, that is, they are glutamic acid, because the oxygen atoms of a sulfate

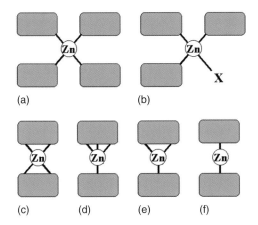

Figure 1 Binding modes of protein interface zinc sites. (a) Zinc cross-links four polypeptide chains. (b) Zinc cross-links three polypeptide chains. X is a nonprotein ligand. In (a) and (b), coordination numbers 5 and 6 have not been considered (c–f) Zinc cross-links two polypeptide chains with coordination number 4 (2:2, (c) or 3:1, (d)), or 3 (2:1, (e)), and 2 (1:1, (f)).

ion are in hydrogen-bonding distance to the oxygen atoms of two of the glutamates and there is a precedent of zinc being bound to both carboxylic and carboxylate groups in a small molecule crystal. Induction of crystal formation appears to be the sole function of zinc in this example.

ZINC CROSS-LINKS AMONG THREE POLYPEPTIDE CHAINS

There are three examples of 3:1 protein–zinc complexes (Figure 1(b)). Coordination numbers of zinc range from four to six.

A homotrimer of apoptosis-inducing ligand 2 (Apo2L/Trail), a cytokine and a member of the tumor necrosis factor (TNF) family of proteins, is formed by interaction of a zinc ion with the sulfur donor atoms of three cysteines (Cys230), each from a different molecule (1DG6).[11] The fourth ligand is a solvent molecule and completes a tetrahedral coordination of the zinc ion (ZnS_3X, where X is a putative chloride ion). Decreased biological activity upon removal of the zinc ion suggests that it is required for the structure and function of this protein in solution and that its presence in the crystal structure is not an artifact of crystallization.

The insulin hexamer is a stable storage form of the insulin monomer, which is the biologically active form. In the hexamer, zinc bridges three monomers. The ligands are the nitrogen donor atoms of His10 from three β-chains. During the conformational transition from the T_6 to the R_6 state, the coordination of zinc changes from octahedral (coordination type: ZnN_3X_3; six ligands: three histidines and three water molecules) to tetrahedral (coordination type: ZnN_3X; four ligands: three histidines and a solvent/solute molecule).[12,13] Anions and phenolic compounds that bind to the nonprotein coordination site of zinc influence the stability of the hexamer through allosteric ligand interactions.[14] This finding illustrates how nonprotein ligands of interface zinc sites can affect the thermodynamics and kinetics of protein–protein interactions.

In avian pancreatic polypeptide, the coordination geometry of zinc is trigonal bipyramidal (1PPT).[15] The five ligands are the N-terminal amine nitrogen and carbonyl oxygen of Gly1, the amide nitrogen of Asn23, a nitrogen atom of His34 and a water molecule (coordination type: ZnN_3OX). Each of these three amino acids stems from a different molecule. Asparagine and histidine are not present at corresponding positions in the primary structure of the bovine ortholog. It was speculated that pancreatic polypeptide-producing cells might form storage granules containing the crystalline hormone and that zinc might organize the structure of the protein in a manner similar to that in crystals grown in the laboratory.

ZINC CROSS-LINKS BETWEEN TWO POLYPEPTIDE CHAINS

In this group, which is by far the largest one, the coordination number of zinc is even more variable and ranges from two to five, including solvent molecule(s) as ligands. When zinc at the interface of two polypeptide chains has a complement of four ligands, the two polypeptide chains provide either two ligands each (2:2, Figure 1(c)) or one provides three and the other one (3:1, Figure 1(d)). However, there are also instances when zinc has apparently only a total of three (2:1, Figure 1(e)) and even only two (1:1, Figure 1(f)) protein ligands.

Oxygen and nitrogen donor atoms

Superantigens (Sags) are toxic proteins from viral or bacterial sources. Sags form complexes with both MHC-II (major histocompatibility complex class II) molecules and T-cell receptors. The concomitant T-cell activation is orders of magnitude stronger than that elicited by conventional antigens and, therefore, can lead to, for example, toxic shock, food poisoning, or scarlet fever. The Sags are not processed to small peptides in antigen-presenting cells and they do not bind to the peptide binding site of the MHC-II molecule. Instead, they form a brace that links the MHC-II molecule and the T-cell receptor. The best-characterized toxins are those from *Staphylococcus aureus* (staphylococcal enterotoxins, SEs) and *Streptococcus pyogenes* (streptococcal pyrogenic exotoxins, SPEs).[16] They are globular proteins of 22 to 29 kDa, belong to three subfamilies, and share a two-domain architecture. Zinc binding sites are not present in all Sags. When present, three coordination types have been observed. A zinc-dependent homodimerization of Sags can involve two different zinc binding sites, resulting in one or two cross-links. The third type is a zinc binding site in a Sag molecule where the fourth ligand is quite variable and can stem from an MHC-II molecule. Three different zinc sites in Sags are located at the bottom of the cleft between the two domains (site 1), on the surface of the so-called β-grasp motif (C-terminal domain) (site 2), or on the edge of the β-barrel small domain surface (site 3).[17]

The following zinc binding sites (coordination types: ZnN_2O, ZnN_3O, ZnN_2O_2) are involved in intermolecular interactions (Table 1). In the Sag SEC2, the ligands are Asp83, His118, His122 from one and Asp9 from a neighboring molecule in the crystals (1STE).[18] Zinc is also bound to SEC2 in solution.

In SED, two zinc ions are at the dimer interface.[19] In both cases, the ligands are Asp182, His220, Asp222 from one molecule and His218 from a second. An unusual feature of a second site in SED is a lysine ligand. The site contains two ligands from each molecule with an intervening spacer of three amino acids, that is, His114 and Lys118 from one and His13 and Glu17 from the other.[19] Zinc does not participate in the interaction of SED with the MHC-II α-chain (Figure 2(a)),[16] but it has been suggested that the second zinc ion could interact with the β-chain.[20]

In SEA, the zinc binding site corresponds to the first zinc site in SED. However, the first ligand of zinc in the primary structure of SEA is a histidine and not an aspartate, that is, His187, His225, and Asp227.[21] The fourth ligand in SEA is variable. It is His61 from a neighboring molecule in one molecule of the asymmetric unit and a water molecule in the other. No evidence for dimerization in solution was found. The dissociation constant of the protein–zinc complex is 0.3 μM. In SEA, there is a low-affinity Sag-binding on the α-chain of MHC-II ($K_d = 1 \times 10^{-5}$ M) and a high-affinity, zinc-dependent binding site on the β-chain

Table 1 Zinc binding sites engaged in the homo-dimerization of superantigens and the heterodimerization of superantigens with MHC-II[a]

Superantigen	Zinc site	Zinc ligands
Subfamily I		
SEA	2	His187, His225, Asp227, **His61 (or His81 from MHC-II)**
SED	2	Asp182, His220, Asp222, **His218**
	1	His114, Lys118, **His13**, **Glu17**
SEH	2	His206, Asp208, **His81 from MHC-II**
Subfamily II		
SEC2	1	Asp83, His118, His122, **Asp9**
Subfamily III		
SPEC	2	His167, His201, Asp203, **His81 (or His81 from MHC-II)**
	3	His35, Glu54, **His35**, **Glu54**

[a] Ligands from another molecule are shown in bold.

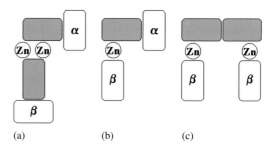

Figure 2 Binding modes of interface zinc sites in the interaction between superantigens or between superantigens and MHC-II molecules (adapted from Reference 16). (a) Two zinc ions bridge the dimer interface of two Sag molecules in complex with the α-chains of MHC-II (e.g. SED). (b) A zinc ion bridges a Sag molecule and the β-chain of MHC-II (e.g. SEA/SEC2). (c) Zinc ions bridge Sag molecules and the β-chain of MHC-II with zinc-independent dimerization of the Sag molecules (e.g. SPEC). Open structures: MHC-II; filled structures: Sag.

($K_d = 1 \times 10^{-7}$ M). The zinc ion is believed to bridge the Sag molecule and the MHC-II β-chain with His81 as the ligand from MHC-II instead of His61 from a second SEA molecule (Figure 2(b)).[21]

In the crystal structure of dimeric SPEC, no electron density could be ascribed to a bound metal ion (1AN8).[22] However, when SPEC crystals were soaked in 50 mM zinc acetate, three zinc binding sites were found. The first is a bridging zinc ion bound by His167, His201, and Asp203 from one molecule and His81 from the other. It is analogous to the interface zinc site in SEA. The second is also a bridging zinc ion and it is coordinated by His35 and Glu54 and by the same pair of ligands from a symmetry-related molecule. The third site was deemed to be an adventitious zinc binding site as it contains only the two protein ligands Asp26 and His68 and two water molecules. The dimerization of SPEC occurs at pH values above 6 and is not affected by ethylenediaminetetraacetic acid (EDTA). Hence, zinc is not essential for, but rather stabilizes dimer formation. In contrast, EDTA abolishes the interaction of SPEC with MHC-II. In an attempt to characterize this interaction, a complex HLA-DR2a/MBP/SPEC – where MBP is a peptide from myelin basic protein – was crystallized.[23] Instead of the symmetry-related Sag molecule, the β-chain of HLA now provides His81 as the fourth ligand of the bridging zinc. Possibly, there is dimerization as shown in Figure 2(c).[16]

When crystals of SEH were soaked in 100 mM zinc chloride, two zinc ions were sandwiched in the dimer (1EWC).[24] Each molecule binds one zinc ion, which is bound by only two protein ligands, that is, His206 and Asp208, and two water molecules (Figure 3(a) and (b)). His206 forms a hydrogen bond to Ser205, and additional hydrogen bonds stabilize the zinc coordination environment. SEH is a monomer in solution. Thus, strictly speaking, these zinc sites are located at a protein interface, but they are not interface zinc sites, since the ligands stem from one

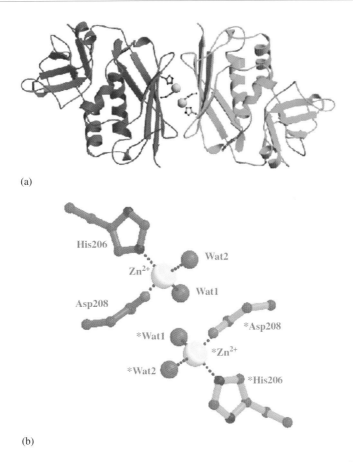

Figure 3 Zinc coordination in the crystal structure of staphylococcal enterotoxin H (SHE; PDB reference: 1EWC). (a) Structure of the SEH dimer with two zinc ions bound at the dimer interface of two SEH molecules. (b) Coordination of the two zinc ions with His206, Asp208, and two water molecules (Reprinted from *J Mol Biol*, 302, M Håkansson, K Petersson, H Nilsson, G Forsberg, P Björk, P Antonsson and LA Svensson, The Crystal Structure of Staphylococcal Enterotoxin H: Implications for Binding Properties to MHC Class II and TcR Molecules, 527–37 (2000) with permission from Elsevier[24]).

polypeptide chain. In the complex of SEH with HLA-DR1, the MHC-II molecule binds to the SEH surface that would otherwise serve as the dimerization interface with a second SEH molecule and provides H81 from its β-chain as a third protein ligand to one zinc ion.[25] In the 2.6 Å resolution structure, low occupancy in the fourth coordination position precluded the assignment of a water molecule as a possible ligand. The structure of the SEH dimer demonstrates that zinc can be bound by only two protein ligands. Therefore, the question arises whether zinc binding sites such as the third site in the structure of SPEC should be dismissed as adventitious. In fact, crystallization in the presence of rather high concentrations of zinc might reveal potential interface zinc sites.

The motif HExxH in SEC (or SPEA) is reminiscent of the signature of the zinc binding site in some metalloproteinases. Furthermore, the presence of a water molecule as a ligand is another characteristic of catalytic zinc sites in enzymes. Conceivably, an additional amino acid could block the catalytic activity of these sites in a manner observed in matrix metalloproteinases, where a thiol group contributed by a cysteine blocks the enzyme in the zymogen form.[26] To date, no evidence exists that any of the protein interface zinc sites discussed here is catalytic rather than structural. (Catalytic zinc sites with ligands from different subunits occur in glyoxalase I and in the γ-class of carbonic anhydrase. Zinc binding sites in these proteins are discussed elsewhere.[27,28]) These sites mediate zinc-dependent protein–protein interactions, and, to use a quotidian analogy, they work like snap buttons in garments.

Zinc binding sites of Sags epitomize the general difficulties encountered in establishing the number of interface zinc sites and in determining their functions. The physiological significance of an interface zinc site may be revealed only when the complex of a protein with its correct biological partner is investigated. Of course, such studies require prior identification of the partner. In the absence of the partner, zinc may coordinate differently or may be either present in substoichiometric amounts or absent altogether owing to a lack of sufficient stabilization.

Equally important, the effect of zinc on protein function also needs to be established *in vivo*. Although the total zinc concentration in blood is about 100 μM, the available concentration of 'free' zinc is orders of magnitude lower in the picomolar range.[29] Similarly, low values have been determined for cellular 'free' zinc.[30–32] Because some of the zinc binding sites in protein crystals were detected when millimolar concentrations of zinc were present in the mother liquor, it remains uncertain whether these sites will bind zinc under physiological conditions. Evaluation of the binding constants of interface zinc sites, determination of the extent to which zinc can be lowered in the mother liquor without compromising saturation of these sites with zinc, and biological assays in solution as a function of the zinc concentration would all appear necessary in order to determine whether zinc might be available for these sites *in vivo*. Interactions with zinc should be detectable at rather low concentrations of zinc, since zinc dissociation constants of zinc proteins are quite low, that is, in the picomolar range. It may be that in addition to physiological zinc sites, crystallographic studies have revealed sites in Sags or other proteins where zinc has a pharmacological action. Indeed, zinc enhances endotoxin activity. Therefore, therapeutic zinc supplementation may pose potential risks that have to be carefully weighed against beneficial effects such as improved immune responses when serum zinc concentrations are elevated during pharmacological zinc supplementation.[33]

The examples discussed demonstrate a biological role of protein interface zinc sites in immune functions and host–pathogen interactions and raise the possibility that such zinc-mediated interactions are important for yet additional immune functions. Therefore, a comment on the significance of zinc for immune functions seems to be pertinent. Effects of zinc on the immune system have been known for a long time, but few, if any, molecular targets had been identified. Retrospectively, it is perhaps becoming clear from this discussion why this is so. These intermolecular interactions are not easy to detect and only the pursuit of crystal structures has provided us with insights on where zinc might be important.

Zinc-dependent oligomerization such as discussed for Apo2L/Trail and Sags also occurs in class I interferons (IFN-α and -β). Zinc is bound to His121 from one molecule and to His93 and His97 of another molecule at the dimer interface of IFN-β (coordination type: ZnN_3X, where X = water). Secondary interactions maintain the structure of this site, which has only three protein ligands. A hydrogen bond to Glu43 orients the conformation of His121. There is also a hydrogen bond between His97 and Gln94 (1AUI).[34] Hydrogen bonds between histidine and oxygen atoms outside the primary coordination sphere of zinc are typical for zinc binding sites in proteins,[35] and many interface zinc sites follow this rule. In their discussion of the structure of IFN-β at 2.2 Å resolution, Karpusas *et al.* point out that zinc was not added during crystallization,[34] but that traces of zinc may have originated from the zinc-chelate chromatography employed in purifying the protein. In contrast, IFN-α2B was crystallized in the presence of 40 mM zinc acetate.[36] The structure of this protein at 2.9 Å resolution shows Glu41 and Glu42 and the corresponding residues from a symmetry-related protein molecule as ligands of a zinc ion (coordination type: ZnO_4). The physiological significance of these interactions in IFNs is unclear since evidence for zinc-dependent dimer formation in solution is lacking. Multimerization of killer cell Ig-like receptors (KIR) in the presence of zinc through the putative zinc-binding motif HEGVH and the interaction of KIRs with class I MHC ligands indicate similar roles of zinc in other systems.[37]

Several members of the S100 family of proteins (A2, A3, A5, A6, A7, A9, A12, S100B) bind zinc with concomitant modulation of their affinity for calcium. 3D structures revealed either a zinc binding site or an organization of ligands in such a manner that a zinc binding site is preformed. In the A7 protein (psoriasin), a single zinc ion bridges the homodimer (3PRS).[38] The ligands are His17 and Asp24 from one monomer and His86 and His90 from the other (coordination type: ZnN_3O). Sequence homologies suggest that similar zinc binding sites exist in the A9 and A12 proteins. The binding of two zinc ions to the A12 dimer induces the high-affinity binding of two calcium ions.[39] The structure of the A12 protein, a member of the calgranulin family, in the two space groups R3 and $P2_1$ revealed possible zinc binding sites (1GQM).[40] In the R3 structure, the putative zinc binding site is in a conformation close to that seen in the A7 isoform although the site does not contain zinc. In the hexamer (of dimers), seven subunits are in the zinc-bound conformation, four in the zinc-free with Glu25 turned away, whereas no electron density was detected for the side chain of Glu25 in one subunit.

The binding of four zinc ions to the A2 protein antagonizes the binding of calcium.[41] The A5 dimer binds two zinc ions in addition to four copper ions.[42] Four zinc ions bind to the dimer of the A3 protein (1KSO).[43] One putative zinc binding site involves Cys83, Cys86, and His87 at the C-terminus. The S100B dimer binds zinc with nanomolar affinity and interacts with the IQGAP1 protein in a zinc-dependent manner.[44] Potentially, the situation in S100 proteins is analogous to that in Sags, where zinc participates in dimerization and/or interactions with another protein.

Nerve growth factor (NGF) exists in an inactive 7S complex that comprises two α-subunits, two β-subunits, two γ-subunits, and between one and two zinc ions.[45] The analysis of murine β-NGF in two different space groups, $P2_12_12_1$ and C2, revealed two zinc binding sites (1BTG).[46] The C2 form contains monomers A, B, C. A binuclear zinc site is located between monomers A and C. One zinc ion is bound by His84 and Asp105 from monomer C and by two water molecules. Asp105 is also a ligand of a second zinc ion, which is the only interface zinc ion, since it interacts with Asp94 from monomer A in addition to two water molecules. The occupancy of these zinc sites is less than 1 in crystals of the other space group. In a structure of the

7S complex, zinc bridges the α- and γ-subunits (1SGF).[47] However, zinc ions were not found in the β-subunit. The ligands are Glu75 and His82 of the α-subunit and His217 and Glu222 of the γ-subunit (coordination type: ZnN_2O_2).

A coordination of the type ZnN_2O_2 with a pair of oxygen and nitrogen ligands on each molecule is also found in three other protein interface zinc sites. In dimers of mouse survivin, a member of the inhibitor of apoptosis proteins (IAPs), a zinc ion bridges the dimer interface with Glu76 and His80 and the corresponding symmetry-related residues from the other subunit as ligands (1F3H).[48] His80 is in hydrogen-bonding distance to the main chain carbonyl of Glu65 of the other molecule. Zinc was not added to the purification, dialysis, or crystallization buffer. The human ortholog dimerizes without zinc.

Zinc can have a function in modulating protein–receptor interactions. It enhances the binding of human growth hormone to the human prolactin receptor about 10^4-fold. The ligands from the receptor are Asp217 and His218 and those from the hormone are His18 and Glu174 (1BP3).[49]

The archaeal cytochrome P450 from *Sulfolobus solfataricus* binds a metal ion at the dimer interface, putatively assigned as a zinc ion (1F4T).[50] The ligands are rather far apart in the primary structure. They are Glu139 and His178 and their counterparts from a symmetry-related molecule (coordination type: ZnN_2O_2). This zinc site does not seem to be an artifact of crystal lattice formation, because zinc was not included in the crystallization buffers.

Zinc inhibits the activity of tonin, a serine protease. When crystals were grown with zinc in the mother liquor, a zinc binding site of the coordination type ZnN_3O comprises His57, which is the catalytic histidine, His97, and His99 from one subunit and Glu148 from another subunit (1TON).[51]

The glucose-specific phosphocarrier protein IIA^{Glc} of the *Escherichia coli* phosphotransferase system binds a zinc ion with His75 and His90 and with Glu478 of glycerol kinase (1GLC).[52] In the absence of glycerol kinase, Glu148 of another IIA^{Glc} molecule is a ligand (1F3Z);[53] a water molecule is a fourth ligand in both instances (coordination type: ZnN_2OX). The affinity of IIA^{Glc} for zinc is utilized for zinc-promoted association of the proteins, complementing the relatively weak protein–protein interactions. The interaction is thought to be important for recognition of the many targets that IIA^{Glc} regulates.[54]

Sulfur donor atoms

The nuclease Mre11, which repairs DNA double-strand breaks, is associated with the ATPase Rad50 yielding a heterotetramer (A_2B_2)(1L8D).[55] A zinc dimerization motif, which is conserved from bacteria to humans and in which each molecule contributes a CX_2C binding site with the ligands Cys444 and Cys447 (coordination type: ZnS_4), is found in the crystal structure of the central region of the Rad50 coiled coil (Figure 4(a)). Two main chain hydrogen bonds, a small hydrophobic interface, and a salt bridge stabilize the zinc-mediated interaction between the two chains. The authors introduced the term 'zinc hook' for the ligand motif in the region between the antiparallel coiled coils and the interlocking of the coiled coils by zinc (Figure 4(b)). The molecular architecture determined by zinc provides a molecular basis for a

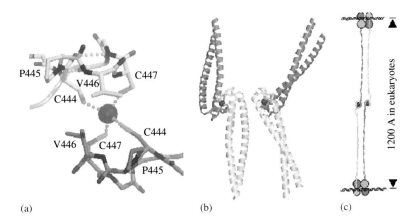

Figure 4 The zinc hook motif (PDB reference: 1L8D). (a) Interface zinc site in the dimer of Rad50. Each chain (yellow and orange) contributes two cysteines (green) for the tetrahedral coordination of a zinc ion (magenta). Two main-chain hydrogen bonds stabilize the interaction (white dots). (b) Zinc-bound coiled coil dimer of Rad50. The Rad50 molecules form an antiparallel coiled coil around a hook-shaped region of 14 amino acids containing the CX_2C motif. The assembly on the right is rotated by 90° relative to the one on the left. (c) The A_2B_2 heterodimer of Mre11 and Rad50 in the complex with DNA. The Mre11 dimer (blue spheres) binds to the coiled coils of two Rad50 molecules adjacent to the ATPase domain (red/yellow spheres). The DNA-binding heads are linked to those of the sister chromatid strand by the zinc hook (zinc in magenta), which is located in the central region of the coiled coil. Reproduced by permission of Nature Publishing Group from K-P Hopfner, L Craig, G Moncalian, RA Zinkel, T Usui, BAL Owen, A Karcher, B Henderson, J-L Bodmer, CT McMurray, JP Carney, JHJ Petrini and JA Tainer, The Rad50 zinc-hook is a structure joining Mre11 complexes in DNA recombination and repair *Nature*, **418** 562–66 (2002).[55]

fundamental biological function. The assembled proteins have the length and conformational properties to link sister chromatids in homologous recombination (Figure 4(c)) and DNA ends in nonhomologous end-joining. In this manner, a metal-mediated bridging between two DNA-binding heads is achieved, suggesting a zinc-linked zipper for ATP-dependent cross-linking.

In endothelial nitric oxide synthase, a zinc ion bridges the dimer. Two of the four cysteine ligands stem from each subunit. The strict conservation of the CX_4C motif in endothelial, inducible, and neuronal nitric oxide synthase suggests that this zinc bridge is present in all mammalian forms of the enzyme (1NSE).[56] (The zinc ion is absent in nitric oxide synthase from *Bacillus subtilis*,[57] however.) The ligands are Cys96 and Cys101 from each subunit (coordination type: ZnS_4). Further stabilization of the site is achieved by hydrogen bonds from the cysteine ligands to the backbone amides of Leu102 and Gly103 and a hydrogen bond from the amide of Cys101 to the carbonyl of Asn468. A total of eight hydrogen bonds at the dimer interface are gained upon binding of zinc.[58] Redox reactions at the zinc center control the activity of the enzyme as peroxynitrite releases zinc from this site.[59]

The *E. coli* colicin E3 immunity protein (IM3) dimerizes with a bridging zinc ion between Cys47 of two molecules (coordination type: ZnS_2).[60] The crystal structure at 1.8 Å resolution (3EIP) did not identify any additional ligand. Hence, zinc binds with two ligands only, and, remarkably, each supplied by a different molecule. Zinc was not added during purification or crystallization. Because Cys47 of IM3 is essential for inhibiting the ribonucleolytic activity of colicin E3, the zinc-dependent dimerization may be a way of protecting IM3 from proteolysis and keeping the protein in an inactive state.

A zinc-binding motif $HX_5CX_{20}CC$ is highly conserved in 62 sequences of the Shab, Shaw, Shal, but not the Shaker subfamily of voltage-gated K^+ channels.[61] In the Shaw T1 (tetramerization domain) structure (3KVT), a zinc ion is bound by His85, Cys102, and Cys103 from one monomer and Cys81 from another monomer (coordination type: $ZnNS_3$). The zinc ion is assumed to be important for the electrophysiology and regulation of the channel. Removal and addition of zinc reversibly switches the T1 domain between a monomeric and a tetrameric state in solution.[62]

Five examples where high-resolution crystal structures are not yet available will be presented, because they either demonstrate similar principles in other proteins or suggest yet other binding modes and functions in interface zinc sites with sulfur ligands.

Zinc is essential for the association between the tyrosine kinase Lck (p56^{lck}) and the T-cell coreceptors CD4 or CD8a.[63,64] This intracellular interaction is necessary for lymphocyte development and activation. The ligands from the cytoplasmic tail of the receptor are Cys420 and Cys423, while those of the tyrosine kinase are Cys20 and Cys23.[64] Zinc has been termed a molecular 'adhesive' for this interaction.[63] Zinc tetrathiolate cross-links appear to be more common as they have been proposed for the homodimerization of the HIV Tat protein[65] and the *S. aureus* CadC repressor.[66]

Plant cellulose synthase proteins form the CesA complex that catalyzes glucan polymerization and assembles into microfibrils, so-called rosettes, in the plasma membrane. Dimerization of cotton fiber cellulose synthase is brought about by oxidation of the zinc binding domains. Thus, in this protein, zinc binding maintains the monomeric state, while its release initiates dimerization, a functional potential opposite to that of all the zinc-induced oligomerizations discussed here.[67]

Two zinc-binding motifs of the type CX_2CXGXG are present in the AT protein of *B. subtilis* (AT = anti-TRAP, TRAP = trp RNA binding attenuation protein). The stoichiometry of zinc binding is one per monomer. Studies in solution indicate that the hexamer is composed of two trimers, in which three zinc ions bridge three polypeptide chains, such that each of the two motifs interacts with a different chain (Figure 5).[68]

The significance of zinc/sulfur interactions

The categorization of zinc binding sites into those with oxygen and nitrogen ligands on one hand and those with only cysteine ligands on the other likely will turn out to be not as exclusive as conveyed here, because some zinc proteins have mixed-ligand coordination environments. A case in point is the interface zinc site in K^+ channels with its cysteine and histidine ligands. To a certain degree, a preference for a particular ligand of zinc depends on the location of the zinc protein in the cell. Extracellular space and cytosol are different redox environments. Accordingly, extracellular proteins have disulfides while cytosolic ones usually do not, because disulfides are not stable under the more reducing conditions of the cytosol. In cytosolic proteins, *redox-inert* zinc ions can serve as a bridge between the sulfhydryl groups of cysteine side chains, however. In bringing thiolate ligands together, zinc determines their thiol/disulfide redox potential. Hence, these zinc/thiolate coordination environments are not inert. Through coupling with corresponding biological redox pairs, they can undergo oxidoreduction, because the *redox-active* sulfur ligand confers redox activity on the zinc–sulfur bond.[69] In

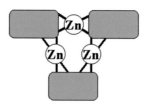

Figure 5 Zinc bridges postulated for the AT protein of *B. subtilis*.[68]

contrast to other thiol/disulfide equilibria, redox reactions at zinc/thiolate sites are linked to the binding and release of zinc ions. Since both the availability of zinc and redox reactions are tightly controlled, zinc and the redox state affect the interaction of proteins with zinc/thiolate cross-links. Thus, the propensity of sulfur versus oxygen and histidine ligands is critical for the coordination dynamics of the zinc site. Indeed, the examples of nitric oxide and cellulose synthase demonstrate the interplay between zinc availability and redox state in organizing macromolecular assemblies and modulating their biological function.

FUNCTIONS OF INTERFACE ZINC SITES

To evaluate the significance of a zinc binding site in a protein, it is necessary to determine whether the metal is lost or introduced during purification. In this regard, bacterial expression systems require special scrutiny as they are prone to yield proteins with a nonnative metal content. The evaluation should continue with examining conditions for crystallization. Much as the addition of chelating agents such as EDTA or reducing agents such as dithiothreitol (DTT), which, incidentally, is also a strong chelating agent, may lead to the loss of a metal ion, the addition of zinc or spurious zinc in the crystallization medium may also influence the metal composition of the protein. For instance, some commercial formulations for crystallization of proteins such as Crystal Screen™ (Hampton Research, Laguna Hills, CA) contain zinc, which, owing to its strong interactions with proteins, redox-inert nature, and flexibility in both the choice of ligands and coordination number, is particularly well suited for protein crystallization ever since it was introduced for the crystallization of insulin[70] and for the fractionation of plasma proteins.[71]

Finally, there is the question of how zinc is identified in the crystal structure. The most direct way is a metal analysis of the protein. Remarkably, analyses of interface zinc sites could yield nonintegral stoichiometries when expressed per monomer. Thus, theoretical values are 0.5 (one zinc per dimer), 0.33 (one zinc per trimer), and 0.25 (one zinc per tetramer) respectively. Methods of identifying the metal in the crystal structure include the use of the bond-valence parameters of Brese and O'Keeffe[72] and calculation of anomalous difference electron density maps from data collected at the wavelength having the highest expected anomalous signal for zinc. Even if some interface zinc sites are considered to be merely crystallization artifacts, zinc might actually have a biological function in crystallization of the protein if it is stored in a crystalline form *in vivo*. Regardless of whether some of the interface zinc sites discussed here are artifacts of crystallization, the evidence is now overwhelming that protein interface zinc sites represent a new principle of protein structure and new functions of zinc in biology. This role complements that of structural zinc ions organizing small protein domains for protein–protein interaction. In particular, the role of the zinc-containing RING domain, which is the third most abundant class of zinc binding domains, in the self-assembly of supramolecular structures such as cellular dots, speckles, and bodies, promises to open a related frontier.[73] The authors summarize the implication of their work as follows: '... this is the first example of a small domain increasing its specific activity and the activities of its partners through high-order supramolecular self-assembly' or '... is the first exposition of the functional linkage between assembly and partner activity and control of biological activity by supramolecular assembly'. In the same manner, zinc-dependent oligomerization through interface zinc sites is important for constructing scaffolds for macromolecular structures, thereby creating new catalytic, structural, or regulatory functions. Moreover, dynamic properties have evolved through the use of ligand-centered chemistry in coordination environments with sulfur ligands. Because a homeostatic system presides over the availability of zinc, it will be interesting to learn how interface zinc sites obtain their zinc, how zinc is removed, and what effects zinc deficiency and other perturbations of zinc homeostasis will have on the functions of these sites.

NOTE ADDED IN PROOF

The coordination type ZnNO2X has been observed in the crystal structure of the C-terminal domain of the RAP74 subunit of human transcription factor IIF (1I27).[74] Glu503 and 517 are the ligands from one protomer. Another protomer contributes His512. A water molecule (hydroxide ion) completes the tetrahedral arrangement of the zinc ion. The conservation of these ligands in higher eukaryotes suggests a physiological role of zinc in addition to making a major contribution to lattice packing.

ACKNOWLEDGEMENTS

I thank Dr. James F. Riordan for valuable advice, Dr. K. Ravi Acharya for helpful discussions regarding superantigens, and Dr. H. Haase for his help in preparing the figures.

REFERENCES

1. BL Vallee and KH Falchuk, *Phys Rev*, **73**, 79–118 (1993).
2. DS Auld, *BioMetals*, **14**, 271–313 (2001).
3. DS Auld, in I Bertini, A Sigel and H Sigel (eds.), *Handbook on Metalloproteins*, Marcel Dekker, New York, pp 881–959 (2001).
4. BL Vallee and DS Auld, *FEBS Lett*, **257**, 138–40 (1989).
5. BL Vallee and DS Auld, *Biochemistry*, **29**, 5647–59 (1990).
6. W Maret, C Jacob, BL Vallee and EH Fischer, *Proc Natl Acad Sci USA*, **96**, 1939–40 (1999).

7   W Maret, *BioMetals*, **14**, 187–90 (2001).
8   G D'Alessio, *Eur J Biochem*, **269**, 3122–30 (2002).
9   G D'Alessio, *Prog Biophys Mol Biol*, **72**, 271–98 (1999).
10 R Thaimattam, E Tykarska, A Bierzynski, GM Sheldrick and M Jaskolski, *Acta Crystallogr*, **D58**, 1448–61 (2002).
11 SG Hymowitz, MP O'Connell, MH Ultsch, A Hurst, K Totpal, A Ashkenazi, AM de Vos and RF Kelley, *Biochemistry*, **39**, 633–40 (2000).
12 EN Baker, TL Blundell, JF Cutfield, SM Cutfield, EJ Dodson, GG Dodson, DC Hodgkin, RE Hubbard, NW Isaacs, CD Reynolds, K Sakabe, N Sakabe and NM Vijayan, *Philos Trans R Soc London, Ser B*, **319**, 369–456 (1988).
13 U Derewenda, Z Derewenda, EJ Dodson, GG Dodson, CD Reynolds, GD Smith, C Sparks and D Swenson, *Nature*, **338**, 594–96 (1989).
14 S Rahuel-Clermont, CA French, NC Kaarsholm and MF Dunn, *Biochemistry*, **36**, 5837–45 (1997).
15 TL Blundell, JE Pitts, IJ Tickle, SP Wood and C-W Wu, *Proc Natl Acad Sci USA*, **78**, 4175–79 (1981).
16 AC Papageorgiou and KR Acharya, *Trends Microbiol*, **8**, 369–75 (2000).
17 Y-I Chi, I Sadler, LM Jablonski, SD Callantine, CF Deobald, CV Stauffacher and GA Bohach, *J Biol Chem*, **277**, 22839–46 (2002).
18 AC Papageorgiou, KR Acharya, R Shapiro, EF Passalacqua, RD Brehm and HS Tranter, *Structure*, **3**, 769–79 (1995).
19 M Sundström, L Abrahamsén, P Antonsson, K Mehindate, W Mourad and M Dohlsten, *EMBO J*, **15**, 6832–40 (1996).
20 R Al-Daccak, K Mehindate, F Damdoumi, P Etongue-Mayer, H Nilsson, P Antonsson, M Sundström, M Dohlsten, RP Sekaly and W Mourad, *J Immunol*, **160**, 225–32 (1998).
21 M Sundström, D Hallén, A Svensson, E Schad, M Dohlsten and L Abrahamsén, *J Biol Chem*, **271**, 32212–16 (1996).
22 A Roussel, BF Anderson, HM Baker, JD Fraser and EN Baker, *Nat Struct Biol*, **4**, 635–43 (1997).
23 YL Li, HM Li, N Dimasi, JK McCormick, R Martin, P Schuck, PM Schlievert and RA Mariuzza, *Immunity*, **14**, 93–104 (2001).
24 M Håkansson, K Petersson, H Nilsson, G Forsberg, P Björk, P Antonsson and LA Svensson, *J Mol Biol*, **302**, 527–37 (2000).
25 K Petersson, M Håkansson, H Nilsson, G Forsberg, LA Svensson, A Liljas and B Walse, *EMBO J*, **20**, 3306–12 (2001).
26 EB Springman, EL Angleton, H Birkedahl-Hansen and HE Van Wart, *Proc Natl Acad Sci USA*, **87**, 364–68 (1990).
27 AD Cameron, B Olin, M Ridderström, B Mannervik and TA Jones, *EMBO J*, **16**, 3386–95 (1997).
28 TM Iverson, BE Alber, C Kisker, JG Ferry and DC Rees, *Biochemistry*, **39**, 9222–31 (2000).
29 P Zhang and JC Allen, *J Nutr*, **125**, 1904–10 (1995).
30 EJ Peck Jr and WJ Ray Jr, *J Biol Chem*, **246**, 1160–67 (1971).
31 TJB Simons, *J Membr Biol*, **123**, 63–71 (1991).
32 D Atar, PH Backx, MM Appel, WD Gao and E Marban, *J Biol Chem*, **270**, 2473–77 (1995).
33 C Driessen, K Hirv, H Kirchner and L Rink, *Immunology*, **84**, 272–77 (1995).
34 M Karpusas, M Nolte, CB Benton, W Meier, WN Lipscomb and S Goelz, *Proc Natl Acad Sci USA*, **94**, 11813–18 (1997).
35 IL Alberts, K Nadassy and SJ Wodak, *Protein Sci*, **7**, 1700–16 (1998).
36 R Radhakrishnan, LJ Walter, A Hruza, P Reichert, PP Trotta, TL Nagabhushan and MR Walter, *Structure*, **4**, 1453–63 (1996).
37 M Valés-Gómez, RA Erskine, MP Deacon, JL Strominger and HT Reyburn, *Proc Natl Acad Sci USA*, **98**, 1734–39 (2001).
38 DE Brodersen, J Nyborg, M Kjeldgaard, *Biochemistry*, **38**, 1695–1704 (1999).
39 EC Dell'Angelica, CH Schleicher and JA Santome, *J Biol Chem*, **269**, 28929–36 (1994).
40 OV Moroz, AA Antson, EJ Dodson, HJ Burrell, SJ Grist, RM Lloyd, NJ Maitland, GG Dodson, KS Wilson, E Lukanidin and IB Bronstein, *Acta Crystallogr*, **D58**, 407–13 (2002).
41 C Franz, I Durussel, JA Cox, BW Schäfer and CW Heizmann, *J Biol Chem*, **273**, 18826–34 (1998).
42 BW Schäfer, J-M Fritschy, P Murmann, H Troxler, I Durussel, CW Heizmann and JA Cox, *J Biol Chem*, **275**, 30623–30 (2000).
43 G Fritz, PRE Mittl, M Vasak, MG Grütter and CW Heizmann, *J Biol Chem*, **277**, 33092–98 (2002).
44 GO Mbele, JC Deloulme, BJ Gentil, C Delphin, M Ferro, J Garin, M Takahashi and J Baudier, *J Biol Chem*, **277**, 49998–50007 (2002).
45 M Young and MJ Koroly, *Biochemistry*, **19**, 5316–21 (1980).
46 DR Holland, LS Cousens, W Meng and BW Matthews, *J Mol Biol*, **239**, 385–400 (1994).
47 B Bax, TL Blundell, J Murray-Rust and NQ McDonald, *Structure*, **5**, 1275–85 (1997).
48 SW Muchmore, J Chen, C Jacob, D Zakula, ED Matayoshi, W Wu, H Zhang, F Li, SC Ng and DC Altieri, *Mol Cell*, **6**, 173–82 (2000).
49 W Somers, M Ultsch, AM de Vos and AA Kossiakoff, *Nature*, **372**, 478–81 (1994).
50 JK Yano, LS Koo, DJ Schuller, H Li, PR Ortiz de Montellano and TL Poulos, *J Biol Chem*, **275**, 31086–92 (2000).
51 M Fujinaga and MN James, *J Mol Biol*, **195**, 373–96 (1987).
52 MD Feese, DW Pettigrew, ND Meadow, S Roseman and SJ Remington, *Proc Natl Acad Sci USA*, **91**, 3544–48 (1994).
53 MD Feese, L Comolli, ND Meadow, S Roseman and SJ Remington, *Biochemistry*, **36**, 16087–96 (1997).
54 CK Holtman, AC Pawlyk, N Meadow, S Roseman and DW Pettigrew, *Biochemistry*, **40**, 14302–8 (2001).
55 K-P Hopfner, L Craig, G Moncalian, RA Zinkel, T Usui, BAL Owen, A Karcher, B Henderson, J-L Bodmer, CT McMurray, JP Carney, JHJ Petrini and JA Tainer, *Nature*, **418**, 562–66 (2002).
56 CS Raman, H Li, P Martásek, V Král, BSS Masters and TL Poulos, *Cell*, **95**, 939–50 (1998).
57 K Pant, AM Bilwes, S Adak, DJ Stuehr and BR Crane, *Biochemistry*, **41**, 11071–79 (2002).
58 H Li, CS Raman, CB Glaser, E Blasko, TA Young, JF Parkinson, M Whitlow and TL Poulos, *J Biol Chem*, **274**, 21276–84 (1999).
59 M-H Zou, C Shi and RA Cohen, *J Clin Invest*, **109**, 817–26 (2002).
60 C Li, D Zhao, A Djebli and M Shoham, *Struct Fold Des*, **7**, 1365–72 (1999).
61 KA Bixby, MH Nanao, NV Shen, A Kreusch, H Bellamy, PJ Pfaffinger and S Choe, *Nat Struct Biol*, **6**, 38–43 (1999).
62 AW Jahng, C Strang, D Kaiser, T Pollard, P Pfaffinger and S Choe, *J Biol Chem*, **277**, 47885–90 (2002).
63 RS Lin, C Rodriguez, A Veillette and HF Lodish, *J Biol Chem*, **273**, 32878–82 (1998).
64 M Huse, MJ Eck and SC Harrison, *J Biol Chem*, **273**, 18729–33 (1998).

65 AD Frankel, DS Bredt and CO Pabo, *Science*, **240**, 70–73 (1988).

66 MD Wong, Y-F Lin and BP Rosen, *J Biol Chem*, **277**, 40930–36 (2002).

67 I Kurek, Y Kawagoe, D Jacob-Wilk, M Doblin and D Delmer, *Proc Natl Acad Sci USA*, **99**, 11109–14 (2002).

68 A Valbuzzi and C Yanofsky, *J Biol Chem*, **277**, 48574–78 (2002).

69 W Maret and BL Vallee, *Proc Natl Acad Sci USA*, **95**, 3478–82 (1998).

70 DA Scott, *Biochem J*, **28**, 1592–602 (1934).

71 AA Green and WL Hughes, *Methods Enzymol*, **1**, 67–90 (1955).

72 NE Brese and M O'Keeffe, *Acta Crystallogr*, **B47**, 192–97 (1991).

73 A Kentsis, RE Gordon and KLB Borden, *Proc Natl Acad Sci USA*, **99**, 15404–9 (2002).

74 K Kamada, J De Angelis, RG Roeder and SK Burley, *Proc Natl Acad Sci USA*, **98**, 3115–20 (2001).

CALCIUM

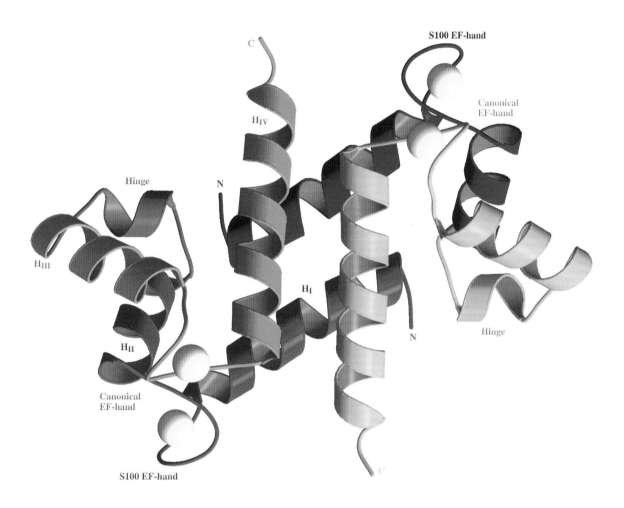

EF-hand Ca²⁺-binding proteins

Calmodulin

Kyoko L Yap and Mitsuhiko Ikura

Department of Medical Biophysics, University of Toronto and Division of Molecular & Structural Biology, Ontario Cancer Institute, Canada

FUNCTIONAL CLASS

Calcium binding protein; activator (less commonly, inactivator) of numerous proteins by binding and modulating their structures, typically in a Ca^{2+}-dependent manner; commonly abbreviated as CaM.

OCCURRENCE

CaM is a ubiquitously expressed protein found in most, if not all, eukaryotes. Concentration of CaM is tissue-, organ- and cell-specific, with high expression in the brain and reproductive organs.[1] In mammals, CaM protein and mRNA are 5 to 15 times more abundant in the central nervous system than in other tissues.[2] In plants, the expression of CaM is dynamically modulated and can be induced by external stimuli.[3] Intracellular concentration of free (unliganded, unbound) CaM is dependent on transient intracellular calcium concentration; below 2 μM free Ca^{2+}, no Ca^{2+}-CaM is detectable in the cell.[4] The total concentration of CaM binding proteins is approximately twofold higher than that of CaM,[5] and therefore high-affinity targets are efficiently activated throughout the cell, while low-affinity targets are efficiently activated only when local free Ca^{2+}-CaM concentration can be enhanced.[4]

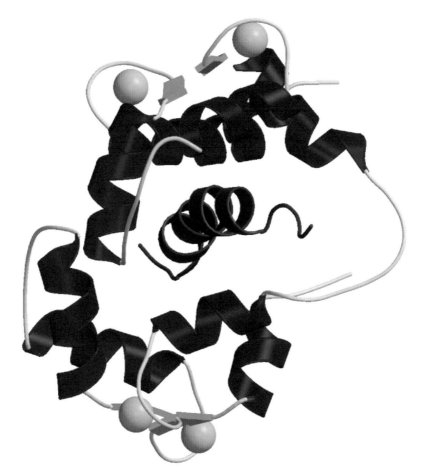

3D Structure Schematic representation of the NMR structure of Ca^{2+}-CaM in complex with a peptide from skeletal muscle MLCK, PDB code 2BBM. CaM is colored blue, target peptide red, and calcium ions yellow. Prepared with programs MOLSCRIPT[133] and RASTER3D.[134]

Calmodulin

BIOLOGICAL FUNCTION

CaM is not itself an enzyme and it does not catalyze chemical reactions alone. Rather, CaM functions as a regulator of enzymes and various intracellular and membrane proteins via direct binding to those target proteins.[6,7] As one of the major Ca^{2+} 'sensor' proteins in the cell, it responds to an increase in the concentration of intracellular calcium from the resting state by undergoing a large conformational change upon binding Ca^{2+} ions, enabling it to interact with and, as a result, activate its targets. In many cases, the CaM binding domain on the target overlaps or is proximate to an autoinhibitory domain (AID) that maintains enzyme inactivity at resting levels of Ca^{2+}. CaM binding to this target displaces the AID to render the enzyme active (see below, Calmodulin Binding Proteins).

AMINO ACID SEQUENCE INFORMATION

Representatives >40% identity with human CaM shown below:

- Animal: Human, mouse, rat, rabbit, bovine, chicken, duck, *Xenopus laevis*, *Arbacia punctulata*, salmon, ricefish, electric ray (P02593, P99014, P70667, Q61379, Q61380), 148 residues; Electric eel (P02594, Q90496), 148 residues, 99.3% identity with human form; *Drosophila melanogaster* (P07181, Q9V3T4), 148 residues, 98.0% identity; Scallop (P02595), 148 residues, 96.6% identity.
- Plant: Tomato (P27161), 148 residues, 91.2% identity; Rice (P29612), 148 residues, 90.5% identity; *Petunia hybrida* isoform 1 (P27162, P34792), 148 residues, 89.9% identity; *Arabidopsis thaliana* isoforms 2, 3, and 5 (P25069), 148 residues, 89.2% identity.
- Fungi: *Phytophthora infestans* (P27165), 148 residues, 94.6% identity; *Neurospora crassa* (Q02052, P40907), 148 residues, 85.1% identity; *Emericella (Aspergillus) nidulans* (P19533), 148 residues, 84.5% identity; *Schizosaccharomyces pombe* (P05933), 150 residues; 74.0% identity; *Saccharomyces cerevisiae* (P06787), 147 residues, 58.4% identity.
- Protists: *Tetrahymena pyriformis* (P02598), 148 residues, 91.9% identity; *Plasmodium falciparum* (P24044), 149 residues, 88.6% identity; *Paramecium tetraurelia* (P07463), 148 residues, 88.5% identity; *Dictyostelium discoideum* (P02599), 151 residues, 87.4% identity; *Chlamydomonas reinhardtii* (P04352), 162 residues, 80.3% identity.

PROTEIN PRODUCTION, PURIFICATION, AND MOLECULAR CHARACTERIZATION

CaM was first described in 1970 as an activator for cyclic nucleotide phosphodiesterase (PDE).[8,9] In 1981, a partial CaM cDNA clone was isolated from the electric eel and sequenced.[10] Soon after, the efficient purification of CaM by phenyl-sepharose affinity chromatography was reported,[11] as CaM binds to phenyl- and octyl-sepharose quantitatively in the presence of micromolar Ca^{2+}, and elution occurs with the addition of ethyleneglycol-*bis*(2aminoethyl)N, N, N', N',-tetraacetic acid (EGTA), ethylene glycol, low–ionic-strength buffer, or nonionic detergent. In the years that followed, cDNA and genomic clones for other species were isolated and characterized.[12–17] CaM has been purified from organs that express the protein abundantly, such as brain and testes,[18] while most biochemical and molecular biology studies with recombinant CaM employ overexpression of a mammalian CaM gene in *Escherichia coli*[19,20] and purification with phenyl-sepharose and/or diethyl-aminoethyl (DEAE) columns.

CaM, CaM isoforms, and highly related CaM-like proteins in different species vary in molecular weight (15–22 kDa) and sequence length (~135–200 residues). It is a monomeric, globular, and typically soluble protein amenable to biochemical and biophysical studies. In certain species such as *Drosophila*, the protein is encoded by a single gene, but other species contain numerous CaM or CaM-like genes.[17] In recent years, several highly homologous (>95% identity) plant CaM isoforms have been isolated. These isoforms appear to have specificity for certain targets and are differentially expressed.[21]

Among prokaryotes, there are analogous proteins in bacteria and cyanobacteria containing Ca^{2+} binding sites and possessing the ability to stimulate known CaM targets such as PDE, Ca^{2+}-ATPase, and myosin light chain kinase (MLCK).[22] Studies involving prokaryote crude extracts have demonstrated the presence of antigens to antimammalian calmodulin antibodies, detection of the complementarity to restriction fragments via probes for mammalian CaM, and inhibition of target activation by CaM inhibitors.[22,23] However, because of low sequence homology between these bacterial proteins and eukaryotic CaM, they are normally not considered to be part of the CaM family.

CaM contains a number of well-conserved hydrophobic residues and, in most species and isoforms, does not contain cysteines. The protein has a relatively high proportion of acidic residues, resulting in an isoelectric point between 3.9 and 4.3.[22] Perhaps not coincidentally, many CaM binding sequences are highly basic in nature. CaM also contains an unusually high number of methionine residues, which are believed to confer promiscuous binding ability owing to the flexible nature of the side chains.[24] These residues comprise part of the hydrophobic, target-interacting surface that is exposed when CaM is bound to Ca^{2+}.

Native CaM possesses a number of posttranslational modifications, including acetylation of the N-terminal A1, trimethylation of K115 in mammalian CaM, dimethylation of K13 in *Paramecium* CaM, and phosphorylation at various threonine, serine, and tyrosine residues.[25] However, the significance of these modifications *in vivo* is still unclear. *In vitro* studies have shown that phosphocalmodulin in place of native CaM binds some targets with different binding affinities and even affects subsequent target kinase activities.[26]

The structure of CaM consists of two highly homologous domains, each containing two Ca^{2+} binding helix-loop-helix (EF-hand) motifs. All EF-hands undergo a change ('closed' to 'open') in helix–helix orientation upon binding Ca^{2+}. A three-residue β-strand precedes the second helix of the motif, and a β-sheet is formed between the strands of each pair of EF-hands by means of a hydrogen bond between the middle residues. The linker between the two domains is flexible[27,28] and enables domain–domain rearrangement to adapt to different target proteins. Consequently, high-resolution structures of CaM, both in the absence and presence of Ca^{2+}, have demonstrated that the relative orientation of one domain to the other is not fixed.

X-RAY STRUCTURE OF CALCIUM-BOUND CALMODULIN

The 3.0-Å crystal structure of mammalian Ca^{2+}-CaM was solved in 1986[29] (Figure 1) with several refined structures to follow,[30–32] thereby verifying that CaM is composed of two structurally homologous domains as had been predicted from the structure of troponin C.[33] The two helices of the four EF-hand motifs are in an identical conformation (with interhelical angles between the helix pairs ranging from 92° to 107°), each motif coordinating a Ca^{2+} ion in the same manner as that observed for parvalbumin, for which the EF-hand was first described.[34] The two domains are separated by a long central α-helix such that the molecule resembles a dumbbell about 65 Å long. This interdomain helix, which has a slight kink at the center, is in fact composed of the second helix of the second EF-hand and the first helix of the third EF-hand.

Crystallization conditions[35] are as follows: Lyophilized CaM is dissolved in 0.004 M $CaCl_2$ solution to 10 mg mL^{-1} and subjected to the hanging drop method using a reservoir solution of 45 to 60% 2-methyl-2,4-pentanediol (MPD) in 0.05 M cacodylate buffer at pH ~5.5. Crystals are obtained after 4 to 7 days at 4 °C. Seeding is usually required. The crystals are stable indefinitely at 4 °C but can be kept at room temperature in a stabilizing solution of 75% MPD in cacodylate buffer. The crystals belong to triclinic space group P1 with unit cell dimensions $a = 29.84$ Å, $b = 53.72$ Å, $c = 24.94$ Å, $α = 93.3$ Å, $β = 97.45$ Å, $γ = 88.77$ Å with one molecule per unit cell.

Metal ion coordination: The two calcium ions in each domain are separated by ~11.3 Å and are coordinated to main-chain oxygen atoms and oxygen atoms of acidic side chains. In the heavy-atom derivative, sites I and II were occupied by Pb^{2+}, indicating that sites III and IV were probably high-affinity sites, thereby confirming the findings of earlier Ca^{2+} binding and nuclear magnetic resonance (NMR) studies (see also Calcium Binding and Coordination, below).

A higher resolution structure (1.0 Å) solved 12 years later[36] showed that considerable disorder was found 'at every accessible length-scale' but particularly in residues within the interdomain linker and those comprising the hydrophobic binding pockets. The overall disorder led to the suggestion that CaM in the crystalline state occupies a discrete set of conformational states and, in solution, might also sample a quasi-continuous range of conformations.[37]

The 1.5-Å crystal structure of Ca^{2+}-bound human CaM-like protein (43% identity)[38] shows strong similarity in backbone conformation, with a marked difference in central helix orientation such that one domain is displaced by 30° with respect to the homologous domain in previous crystal structures of CaM.

NMR STRUCTURE OF CALCIUM-BOUND CALMODULIN

Recently, Bax and coworkers solved a high-resolution solution structure of Ca^{2+}-CaM with the use of residual dipolar couplings, showing that the two domains remain flexible despite conformational change upon Ca^{2+} binding[39] (Figure 2). The tertiary structure of the C-terminal domain is similar to that found in the crystalline state, but the interhelical angles of the N-terminal domain EF-hands are smaller and approximate the semi-open conformation observed for the EF-hands of the myosin light chains.[40] Interestingly, while plasticity of the backbone was demonstrated, hydrophobic, target-binding side chains

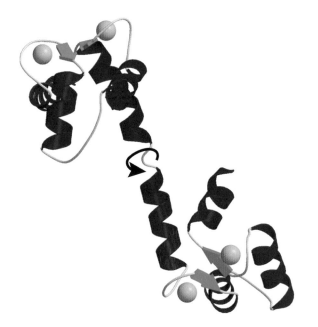

Figure 1 Schematic representation of the 2.2-Å crystal structure of Ca^{2+}-CaM, PDB code 3CLN.[30] The arrow indicates flexibility of the linker in solution.

were locked in the same rotameric states as found in complex with targets.

Solution conditions are as follows: purified CaM is prepared in 250 µL of 95% (v/v) H_2O/5% D_2O, pH 7.0. The buffer contains 1 mM CaM, 16 mM $CaCl_2$, 100 mM KCl. The aligned sample for residual dipolar coupling measurements contains 15 or 18 mg mL^{-1} filamentous phage Pf1 in addition to the buffer above.

NMR studies on S. cerevisiae Ca^{2+}-CaM indicate that, unlike mammalian CaM, the two domains interact with each other via their hydrophobic surfaces, suggestive of a collapsed state,[41] which may explain its decreased ability to efficiently bind and activate targets of vertebrate Ca^{2+}-CaM.

NMR STRUCTURE OF CALCIUM-FREE CALMODULIN

In 1995, three groups independently solved the solution structure of apo(Ca^{2+}-free)-CaM, a compact molecule with two globular domains connected by a highly flexible linker[42–44] (Figure 3). The NMR structure of apo-CaM

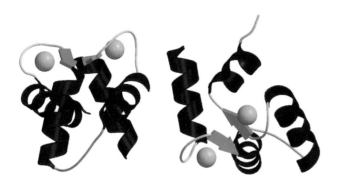

Figure 2 Schematic representation of the NMR structures of the N-terminal (left) and C-terminal (right) domains of Ca^{2+}-CaM, PDB codes 1J7O and 1J7P model 1 respectively.[39]

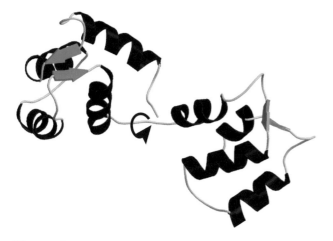

Figure 3 Schematic representation of the NMR structure of Ca^{2+}-free CaM, PDB code 1DMO model 11.[42] The arrow indicates flexibility of the linker in solution.

from S. cerevisiae is similarly flexible.[45] As indicated by earlier small-angle X-ray scattering studies[46,47] and in later molecular dynamics simulations,[48,49] the apo structure is more compact than the Ca^{2+}-bound protein. Because of the unfixed domain–domain orientation, an ensemble of calculated structures can only be superimposed over one of the domains, with the second domain occupying a 'mushroom cap'–like space. Considerable backbone flexibility is also observed in the loop regions within the EF-hands. The structure clearly illustrates the Ca^{2+}-induced conformational change undergone by CaM, particularly the EF-hand helix orientation. The interhelical angle difference between CaM in calcium-free and calcium-bound forms for all four EF-hands is approximately 40°, which is very similar to those observed for the N-terminal EF-hands of troponin C.[50] The secondary structure of Ca^{2+}-free CaM is essentially conserved from the Ca^{2+}-bound structure, although circular dichroism measurements indicate a lower degree of α-helicity in the apo structure as compared to the Ca^{2+}-bound structure.[51] This is suggested to be reflective of the higher flexibility of the apo molecule or rearrangement of the existing EF-hand helices.[52]

STRUCTURES OF CALCIUM-BOUND CALMODULIN IN COMPLEX WITH TARGET PEPTIDES OR PROTEINS

Although no structure of a full-length kinase bound to CaM has been determined to date, the structures of Ca^{2+}-CaM in complex with peptides derived from the CaM binding domains of CaM-dependent kinases I (CaMKI)[53] and II (CaMKII),[54] CaM-dependent kinase kinase (CaMKK),[55,56] and MLCK[57,58] have been solved by crystallography and NMR. These peptides are typically unstructured in the absence of CaM and form an α-helical structure when bound to CaM. The structures are generally quite similar, with the two domains of CaM wrapping around the helical peptide (Figure 4). The CaM–CaMKK peptide complex differs from the others in that the orientation of the peptide is in the opposite direction and there is additional interaction with the unstructured tail following the helix.

The target peptide is characterized by two bulky hydrophobic residues spaced 8, 12, or 14 residues apart, known as hydrophobic anchors.[59,60] Each anchor interacts with the hydrophobic pocket of one CaM domain. Most of the CaM residues mediating contact with the target peptide comprise the hydrophobic patch and include alanines A15 and A88, leucines L18, L32, L39, L112, L116, isoleucines I27, I52, I100, I125, aromatics F12, F19, F68, F89, F92, valines V35, V55, V108, V121, V136, and eight of the nine methionines M36, M51, M71, M72, M109, M124, M144, M145. One anchor binds one pocket of each domain.

Two recently solved crystal structures of Ca^{2+}-CaM–peptide complexes demonstrate that other targets interact in a similar manner to the CaM-dependent

Calmodulin

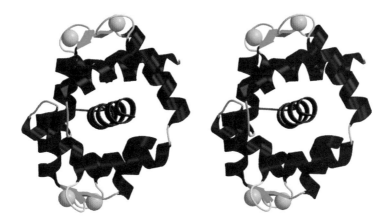

Figure 4 Stereo view of the 1.7-Å crystal structure of Ca^{2+}-CaM in complex with a peptide from CaMKI, PDB code 1MXE.[53]

kinases: CaM in complex with a peptide from endothelial nitric oxide synthase[61] illustrates an interaction similar to that of the CaM–MLCKp complex, while the interaction between CaM and a peptide from myristoylated alanine-rich C kinase substrate (MARCKS)[62] results in a similar CaM conformation, despite an elongated peptide structure containing only a single-turn helix.

The solution structure of Ca^{2+}-bound CaM in complex with the N-terminal half of the CaM binding domain of the plasma membrane Ca^{2+} pump (C20W) shows that peptide interaction involves only the C-domain.[63] The N-domain retains a conformation like that of the target-free, Ca^{2+}-CaM. The C20W peptide forms an α-helix as in previous kinase peptide complexes, but lacks the second hydrophobic anchor. Although both domains are probably involved in binding the full-length CaM binding domain (peptide C24W), only the C-domain of CaM is necessary to activate the pump.[64]

The model of CaM–target interaction as two domains wrapped around a single, amphipathic α-helix remained until the crystal structure of CaM bound to the gating domain of the Ca^{2+}-activated K^+ channel (SK2) was solved[65] (Figure 5(a)). This was the first structure to show an alternate binding stoichiometry of CaM to target molecule – an SK2 dimer binds two molecules of CaM. Interestingly, calcium binds only to the N-terminal domain of CaM, which has lower affinity for Ca^{2+}, although both CaM domains interact with SK2. In the absence of Ca^{2+}, the constitutively formed CaM–SK2 complex is believed to be monomeric, such that binding of Ca^{2+} induces conformational change in CaM, dimerization of the CaM binding domain, and subsequent rotation of the adjacent pore helices to open the channel.

The crystal structure of CaM in complex with the anthrax adenylyl cyclase (aAC) demonstrates a much more extensive CaM interaction with a multidomain target[66] (Figure 5(b)). Here, only the C-domain of CaM is bound to Ca^{2+}. aAC consists of three domains of which the first two (C_A and C_B) form the catalytic core; the active site resides at the interface of the domains. A third 'helical' domain forms a positive surface that is amenable to binding CaM, which is acidic. Upon CaM binding – which, from mutational studies, is believed to be a multistep process – the helical domain is displaced from C_A, leading to stabilization of a

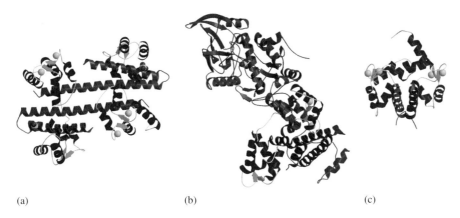

(a)   (b)   (c)

Figure 5 (a) Schematic representation of the 1.6-Å crystal structure of half-Ca^{2+}-saturated CaM in complex with SK2, PDB code 1G4Y.[65] (b) Schematic representation of the 2.95-Å crystal structure of half-Ca^{2+}-saturated CaM in complex with aAC, PDB code 1K93.[66] (c) Schematic representation of the NMR structure of Ca^{2+}-CaM in complex with a dimeric peptide from GAD, PDB code 1NWD.[67]

region near the nucleotide substrate and enabling activation of aAC.

Most recently, the solution structure of CaM in complex with a dimeric peptide from plant glutamate decarboxylase (GAD) has illustrated another unusual binding stoichiometry[67] (Figure 5(c)). CaM binding to GAD is believed to relieve autoinhibition and induce dimerization, subsequently activating the enzyme. However, the CaM–GAD dimer may be a subunit of a larger oligomeric complex (see Induced Dimerization below).

Despite the unique stoichiometry and secondary and tertiary structures of the targets bound to CaM in recently determined structures, CaM in all complexes display a remarkably conserved tertiary backbone (excluding the Ca^{2+}-free domains of CaM in complex with SK2 or aAC), with the primary difference being the orientation of the N- and C-domains with respect to each other. This multiple-target binding ability of CaM should be attributed to the flexible nature of the linker as well as to the flexibility of the side chains comprising the hydrophobic patch exposed upon Ca^{2+} binding, as demonstrated in recent structures of Ca^{2+}-CaM.

TARGET INTERACTION WITH CALCIUM-FREE CALMODULIN

A number of proteins have been identified as targets that bind to apo-CaM. Often the CaM binding domain is an IQ motif (IQXXGRXXXR)[68] such as those found in myosins. CaM can bind some of these proteins in both a Ca^{2+}-dependent and Ca^{2+}-independent manner, and sometimes the target-binding domains for these interactions overlap. In other cases, only Ca^{2+}-free CaM binds the target, or has lower affinity for targets in the presence of Ca^{2+}. It is believed that one role of apo-CaM binding proteins is to sequester CaM away from activating CaM-dependent enzymes or to inhibit association of CaM with these enzymes.[69] No structure of Ca^{2+}-free CaM bound to a target or target peptide has yet been determined, although it is likely that the tertiary conformation would differ from that in a Ca^{2+}-CaM complex, and CaM residues mediating the apo-CaM interaction are not entirely the same as those involved in Ca^{2+}-dependent interactions.[70]

CALCIUM BINDING AND COORDINATION

When fully saturated, CaM binds four Ca^{2+} ions via its four EF-hands. Its affinity for Ca^{2+} is within the range of intracellular Ca^{2+} concentrations, varying from $K_d = 5 \times 10^{-7}$ M to 5×10^{-6} M, enabling it to detect and respond to physiological calcium concentration changes in the cell.[5] The C-terminal domain has a three- to fivefold higher affinity for Ca^{2+} than the N-terminal domain. The Hill coefficient is 1.33, indicating that Ca^{2+} binding is cooperative.[71] Cooperativity between the two adjacent Ca^{2+} binding sites is supported by the small change in interhelical angles between adjacent helices (e.g. helices A and D, B and C) undergone by Ca^{2+} binding to CaM.[42] Ca^{2+} binding to the N- or C-domains of CaM may also play a dual activation/inactivation role, as for the P/Q-type Ca^{2+} channels (see Constitutive Binding, below).

The coordination of the calcium ion within the EF-hand is strongly conserved among the EF-hand superfamily of proteins.[72] As in CaM, EF-hands typically occur in pairs. The canonical EF-hand sequence (followed by the EF-hand sequences of mammalian CaM) is denoted by

```
En**nn**nX*Y*ZGyIx**zn**nn**n
EFKEAFSLFDKDGDGTITTKELGTVMRSL
ELQDMINEVDADGNGTIDFPEFLTMMARK
EIREAFRVFDKDGNGYISAAELRHVMTNL
EVDEMIREANIDGDGQVNYEEFVQMMTAK
```

where E, G, and I indicate very conserved glutamate, glycine, and isoleucine residues respectively; n indicates a conserved hydrophobic residue that interacts with those of its partner EF-hand; and asterisks indicate variable hydrophilic or other residues. The underlined sections indicate the two helices, and the three residues immediately preceding the second helix comprise the β-strand. In vertebrate CaM (and similarly for other species), six residues are involved in coordinating calcium in a pentagonal bipyramidal manner: the side chain oxygen atoms of aspartate or asparagine residues at positions X, Y, and Z, a bridging water molecule from the polar residue at x, the carbonyl oxygen of the polar residue at y, and two side chain oxygen atoms of the glutamate at z. In high-resolution crystal structures of CaM[31,32,36] and CaM–target complexes,[53,56,65] distances between the coordinating oxygen atoms and the calcium ion range from 2.3 to 4.5 Å with shorter distances on average (2.3 to 2.7 Å) observed between the calcium ion and coordinating oxygens from the latter half of the EF-hand loop sequence. Although the arrangement can vary slightly for EF-hands with insertions and amino acid replacements, the coordination will usually involve some combination of oxygen atoms from carbonyl and carboxylate groups.

CALMODULIN BINDING PROTEINS

Despite a significant number of structural studies that have been published on CaM and its complex with various targets, there remains much to learn about the details of the mechanism of activation, not to mention characterization of the other hundreds of CaM target proteins that have been identified over the past two decades.[60] These targets include numerous ion transporters and receptors, proteins involved in cell division, cellular and viral signal peptides, glycoproteins and membrane-associated proteins, and a

Table 1 Calmodulin target proteins
Calcium channels
• L, N, P/Q, R-type
• Transient receptor potential (TRP) proteins and TRP-like (TRPL) proteins
• Ryanodine receptor (Ca^{2+} release channel)
Other channels
• Voltage-dependent sodium channel
• Calcium-activated potassium channel
• Cyclic nucleotide gated olfactory channels
Ion exchangers, receptors, and pumps
• Calcium ATPase, plasma membrane calcium pump
• Na^+/H^+ exchanger
• N-methyl-D-aspartic acid (NMDA) receptor
• IP3 receptor
• Metabotropic glutamate receptor
• Dopamine receptor
Kinases and phosphatases
• CaM-dependent kinases
• Titin kinase
• Calcineurin
• Rhodopsin protein phosphatase
Other enzymes
• Nitric oxide synthases
• Adenylyl cyclases
• Glutamate decarboxylase
Others
• Myosins and other muscle-associated proteins
• Glycoproteins and membrane-associated proteins
• Cell division proteins
• Transcription factors
• Cellular and viral signal peptides
• Heat shock proteins

large number of other protein kinases (Table 1). On the surface, it would seem that the majority among the collection of target proteins are functionally unrelated. In addition, the CaM binding sequences of these targets can be categorized only according to rather general consensus motifs, and there are a great many that fail to conform to any detectable motif. Perhaps this reflects a potentially huge variation in activation or inactivation mechanisms employed by different CaM target proteins.[73–75] Three-dimensional structures and complementary biochemical and biophysical studies have helped understand some such mechanisms, and a general classification based on the specific role of CaM binding has been proposed (Table 2). As we learn more about these mechanisms, it becomes clear that often CaM may be involved in multiple roles for particular interactions.

Reversible CaM binding

Most target proteins become associated with CaM upon increase in intracellular Ca^{2+} concentration, conferring Ca^{2+} dependence onto the enzyme's functionality. In these cases, when Ca^{2+} levels fall back to the resting state, CaM dissociates from the target.

Autoinhibitory domain release

Many Ca^{2+}-CaM-regulated enzymes are activated by CaM binding to a region within or proximate to an autoinhibitory domain (AID). As illustrated in the crystal structure of CaM-dependent kinase I (CaMKI),[76] the AID contains a pseudo-substrate that remains in the active site while the enzyme is inactive, to which CaM binding induces removal of the AID leading to activation. The earliest CaM–target complex structures showed that peptides derived from the AID of different kinases were α-helical, a property induced by CaM binding. Several well-known CaM target enzymes are activated in this manner, including CaMKI, CaMKII, CaMKIV, MLCK, titin kinase, and calcineurin, and the autoinhibitory/CaM binding region is often near or at the C-terminus of the enzyme. The CaM binding sequences of many of these enzymes have given rise to several of the motif classes defined by the positions of the hydrophobic anchor residues that interact with CaM.[59] These remain the most easily identifiable motifs in a sequence-based homology plus biophysical parameter (e.g. α-helical propensity, hydrophobicity) similarity search.[60]

Conversely, identification of such a motif may not indicate that the mode of activation is of AID removal. While domain displacement may be a means to activation for these kinases, CaM binding can also induce conformational changes in its targets to build an active site or precipitate oligomerization (see Induced Dimerization below), to enable protein–protein interaction, DNA or membrane binding, or transport to another location in the cell.

Active site remodeling

The recent structure of CaM bound to aAC[66] highlights a novel role of binding in which the active site of the enzyme is built from separate, solvent-exposed regions brought together by CaM (active site remodeling, or ASR). As an enzyme from a bacterium that does not possess native CaM but requires it to be lethal, this is a particularly clever way to remain inactive until it encounters the appropriate host.

The relative lack of structural information available for full-length CaM targets has not enabled the identification of other targets that undergo ASR upon binding; this, however, is almost certainly soon to change with the current explosion in genomics and proteomics research. As an example of a 'discontinuous' CaM binding domain (one of an increasing number being identified in recent years), it is practically impossible to define a consensus motif with which other sequences can be compared. Even the standard method of searching for sequences that would

Table 2 Roles of calmodulin binding to targets

Target protein	CaM association with target	CaM action	Binding sequence description	References
CaM kinases including myosin light chain kinase, titin kinase	Reversible	Activation by autoinhibitory domain release after phosphorylation, trapped CaM keeps kinase partially active	C-terminal domain overlaps or adjacent to inhibitory domain containing pseudo-substrate: two hydrophobic anchors	76–80
Glutamate decarboxylase	Reversible?	Activation by autoinhibitory domain release and dimerization	C-terminal domain overlaps inhibitory domain; no homologous motif known	67,81–83
Anthrax adenylyl cyclase	Reversible	Activation by active site remodeling	Discontinuous and extensive	66
Phosphorylase kinase	Constitutive	Activation by autoinhibitory subunit release	Discontinuous and extensive	84–89
L, P/Q-type Ca^{2+} channel	Constitutive	Channel facilitation/inhibition possibly mediated by Ca^{2+} binding to each CaM domain	Two binding sites separated by ~50 residues: IQ motif binds apo/Ca^{2+}; another Ca^{2+} only? (P/Q: site is homologous to adenylyl cyclase 8)	90–97
Small conductance Ca^{2+}-activated K^+ channel	Constitutive	Anchoring and dimerization possibly mediated by SK2 binding to individual CaM domains	Discontinuous, but one of three regions is classic 1 to 14	65,98–100
Myosins	Constitutive?	Activation of adjacent motor domain by induced conformational change	Tandem IQ motifs that bind both apo and Ca^{2+}-CaM	40,68,101

form basic, amphipathic α-helices[102] would probably fail to identify all binding regions. The most convincing method of determining an incidence of ASR would probably be the elucidation of atomic resolution structures of CaM-free and CaM-bound target proteins. It remains to be seen whether other targets currently known to contain discontinuous CaM binding sequences belong to this class.

Constitutive binding

Primarily on the basis of truncation mutagenesis studies, a number of target proteins, including the voltage-dependent sodium channel[103] and the ryanodine receptor,[104–107] have been identified as having both apo- and Ca^{2+}-CaM binding sites that either overlap or are enclosed one within another. The advantage of this structural arrangement is that localization of CaM to this region of the target in response to changes in local Ca^{2+} concentration are unnecessary, such that CaM may remain constitutively associated or tethered to the target.

Several CaM-dependent proteins have been shown to bind and respond to CaM in the absence and presence of calcium, with CaM interaction occurring at sites that are one hundred or more residues apart. In these cases, it is unclear whether CaM remains associated to its target independent of Ca^{2+} signaling. These targets may require either the apo- or Ca^{2+}-bound form of CaM for different functions, or the binding sites may in fact be adjacent when folded to simulate an arrangement like that of overlapping sites.

Most classes of myosins, the actin motor proteins, possess a regulatory 'neck' domain that is adjacent to the motor domain and contains a variable number of tandem light-chain binding IQ motifs. In many myosins, the regulating light chain is Ca^{2+}-free CaM.[101] Binding to the IQ motif, which is also found in some Ca^{2+} channels and neuronal proteins, is known to be a predominantly Ca^{2+}-independent association,[68] and in fact for some targets, such as neuromodulin, the affinity of CaM for the target is lower in the presence of Ca^{2+}.[108] In the absence of Ca^{2+}, up to three molecules have been shown to associate with the four IQ motifs of myosin 1c;[109,110] but with elevated levels of Ca^{2+}, one CaM molecule still remains bound to the IQ region. Binding of Ca^{2+} has been shown to inhibit motility,[109] but it is unclear whether high Ca^{2+} concentrations can remove CaM from the IQ region *in vivo*.[68] Interestingly, a study of CaM binding to a double IQ motif from an unconventional myosin suggested that a change in CaM-IQ stoichiometry was more responsible for Ca^{2+}-dependent motility regulation rather than a decrease in affinity.[111]

The model for voltage-gated Ca^{2+} channels, such as the L[90–93] and P/Q-types,[94,95] assumes that CaM is constitutively bound near the channel pore,[96] such that it may rapidly bind Ca^{2+} ions allowed through by depolarization.[97] Ca^{2+} binding to CaM would then permit facilitation (promotion) or inactivation (inhibition) of the channel. This is accompanied by the fascinating observation that the two highly homologous domains of CaM may behave differently from each other in the same target interaction: in a recent study of the P/Q-type Ca^{2+} channels,[95] Ca^{2+} binding to the N-domain initiates channel inactivation, while binding to the C-domain induces channel facilitation. This is quite possibly the same mechanism for other voltage-dependent calcium channels.

Induced dimerization

Several studies have suggested or shown that CaM binding can induce dimerization of its target, resulting in enzyme or channel activation. In the structure of the SK2 potassium channel, two molecules of CaM bind a dimer of the SK2 gating domain, but the channel is completely active in a tetrameric 4:4 complex.[65] As described earlier, the CaM–SK2 interaction is also constitutive,[112] and as with the P/Q-type Ca^{2+} channels, the individual domains of CaM play different roles in SK2 association: the apo C-domain anchors CaM to the channel, while the Ca^{2+}-bound N-domain is involved in dimerization.

The plant form of GAD also appears to undergo dimerization upon CaM binding to its C-terminus.[67,81] Studies have also suggested that the CaM binding domain is also an AID.[82,83] It is unclear whether the CaM-free enzyme is monomeric, since GAD in other species appears to be a constitutive dimer via a conserved catalytic domain upstream of the plant-only CaM binding domain. The 1:2 CaM:GAD complex may therefore be composed of a unit within a larger oligomer that is fully active. In contrast, CaMKII appears to be a large oligomer independent of AID-binding CaM,[113] but traps CaM after autophosphorylation to keep itself partially active.[114]

CALMODULIN AS A DRUG TARGET

CaM in some cases is more highly expressed in cancer cells compared to those in normal cells, and cancer cell growth has been correlated with Ca^{2+}-CaM levels.[115–117] Likely, this is a result of its role in cell proliferation, and CaM antagonists such as trifluoperazine (TFP) and N-(6-aminohexyl)-5-chloro-1-naphthalenesulfonamide (W-7) inhibit cell proliferation by arresting the cell cycle at the G_1/S boundary, and, at initiation and termination of mitosis.[118,119] Additionally, CaM may be involved in stimulation of the effect of insulin and growth factors on glycolysis, thereby contributing to adenosine 5′ triphosphate (ATP) energy for cell growth.[120] Application of the CaM antagonist clotrimazole has shown that this effect of insulin and growth factors can be inhibited, resulting in cell death within three hours of treatment. Treatments involving combinations of antagonists, such as tamoxifen with TFP or W-7 have been shown to be effective in rapidly inducing apoptosis in breast cancer cells.[121] Design of a CaM antagonist has few basic requirements – an aromatic group, which interacts with the hydrophobic surface of Ca^{2+}-CaM and a cationic group to interact with the acidic side chains of CaM.[122] The number of antagonists actively used in *in vitro* and *in vivo* studies today probably number in the hundreds.

The structures of CaM in complex with three different antagonists, TFP,[123–125] W-7,[126] and N-(3,3-diphenylpropyl)-N′-[1-R(3,4-*bis*-butoxyphenyl)-ethyl]-propylenediamine, or AAA[127] have been solved by both crystallography and NMR, and the tertiary structures of each of the N- and C-domains of CaM are very similar to those of Ca^{2+}-CaM bound to target peptides and proteins. Although there appear to be conflicting stoichiometries for the crystal structures of CaM bound to TFP or AAA, the domain–domain orientation is practically identical for these complexes. The differing stoichiometry may be due

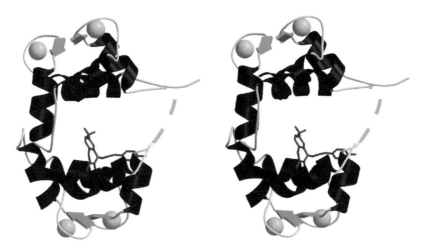

Figure 6 Stereo view of the 2.45-Å crystal structure of Ca^{2+}-CaM in complex with one molecule of TFP (in magenta), PDB code 1CTR.[123] Residues 75–80 are absent from the structure (dashed lines).

to nonspecific binding of the small molecule. Small-angle X-ray scattering analysis of the CaM-W-7 complex[128] and crystallographic studies of the CaM-TFP complex indicate that drug binding can induce an interaction between the two domains of CaM, leading to a globular complex structure (Figure 6). Binding of different drug molecules involves many of the same hydrophobic residues that are involved in target protein or peptide binding, but the precise coordination is specific to each drug and its stoichiometry. These structural studies suggest that the more elongated shape of CaM found in the Ca^{2+}-bound state is energetically unstable[129] and can be easily collapsed into a more globular structure upon complex formation with its binding partners, including small organic compounds such as TFP, W-7, and AAA.

Although CaM antagonists have a large number of desirable therapeutic effects such as anti-estrogen, antipsychotic, and antidepressant properties,[130,131] and have proven to be useful as pharmacological tools, further studies are needed for their use in medicine. As CaM is the primary receptor for Ca^{2+} signals and the activator of hundreds of crucial enzymes, cell cycle proteins and ion-regulating membrane proteins, the inhibition of CaM using general antagonists is not a practical therapy. However, it is possible to develop antagonists that selectively inhibit certain CaM–target interactions, for example, by synthesizing covalent adducts of CaM and classical antagonists.[132] As long as technologies to improve selective inhibition continue to be developed, CaM should remain of interest as a potential drug target.

ACKNOWLEDGEMENT

This work was supported by grants from the Canadian Institutes of Health Research.

REFERENCES

1. D Chin and AR Means, *Trends Cell Biol*, **10**, 322–28 (2000).
2. A Palfi, E Kortvely, E Fekete, B Kovacs, S Varszegi and K Gulya, *Life Sci*, **70**, 2829–55 (2002).
3. WA Snedden and H Fromm, *Trends Plant Sci*, **3**, 299–304 (1998).
4. A Persechini and B Cronk, *J Biol Chem*, **274**, 6827–30 (1999).
5. A Persechini and PM Stemmer, *Trends Cardiovasc Med*, **12**, 32–37 (2002).
6. A Crivici and M Ikura, *Annu Rev Biophys Biomol Struct*, **24**, 85–116 (1995).
7. MR Nelson and WJ Chazin, in LJ van Eldik and DM Watterson (eds.), *Calmodulin and Signal Transduction*, Academic Press, New York, pp 17–64 (1998).
8. WY Cheung, *Biochem Biophys Res Commun*, **38**, 533–38 (1970).
9. S Kakiuchi and R Yamazaki, *Biochem Biophys Res Commun*, **41**, 1104–10 (1970).
10. RP Munjaal, T Chandra, SL Woo, JR Dedman and AR Means, *Proc Natl Acad Sci USA*, **78**, 2330–34 (1981).
11. R Gopalakrishna and WB Anderson, *Biochem Biophys Res Commun*, **104**, 830–36 (1982).
12. L Lagace, T Chandra, SL Woo and AR Means, *J Biol Chem*, **258**, 1684–88 (1983).
13. JA Putkey, KF Ts'ui, T Tanaka, L Lagace, JP Stein, EC Lai and AR Means, *J Biol Chem*, **258**, 11864–70 (1983).
14. YH Chien and IB Dawid, *Mol Cell Biol*, **4**, 507–13 (1984).
15. H Goldhagen and M Clarke, *Mol Cell Biol*, **6**, 1851–54 (1986).
16. MK Yamanaka, JA Saugstad, O Hanson-Painton, BJ McCarthy and SL Tobin, *Nucleic Acids Res*, **15**, 3335–48 (1987).
17. CA West, JV Bannister, BA Levine and RN Perham, *Protein Eng*, **2**, 307–311 (1988).
18. J Haiech and JP Capony, in MP Thompson (ed.), *Calcium-Binding Proteins*, CRC Press, Boca Raton, pp 11–20 (1988).
19. AR Shatzman and M Rosenberg, *Methods Enzymol*, **152**, 661–73 (1987).
20. M Ikura, D Marion, LE Kay, H Shih, M Krinks, CB Klee and A Bax, *Biochem Pharmacol*, **40**, 153–60 (1990).
21. S Luan, J Kudla, M Rodriguez-Concepcion, S Yalovsky and W Gruissem, *Plant Cell*, **14**, S389–S400 (2002).
22. LA Onek and RJ Smith, *J Gen Microbiol*, **138**, 1039–49 (1992).
23. K Yang, *J Mol Microbiol Biotechnol*, **3**, 457–59 (2001).
24. HJ Vogel and M Zhang, *Mol Cell Biochem*, **149-150**, 3–15 (1995).
25. G Benaim and A Villalobo, *Eur J Biochem*, **269**, 3619–31 (2002).
26. M Quadroni, EL L'Hostis, C Corti, I Myagkikh, I Durussel, J Cox, P James and E Carafoli, *Biochemistry*, **37**, 6523–32 (1998).
27. M Ikura, S Spera, G Barbato, LE Kay, M Krinks and A Bax, *Biochemistry*, **30**, 9216–28 (1991).
28. G Barbato, M Ikura, LE Kay, RW Pastor and A Bax, *Biochemistry*, **31**, 5269–78 (1992).
29. YS Babu, JS Sack, TJ Greenhough, CE Bugg, AR Means and WJ Cook, *Nature*, **315**, 37–40 (1985).
30. YS Babu, CE Bugg and WJ Cook, *J Mol Biol*, **204**, 191–204 (1988).
31. R Chattopadhyaya, WE Meador, AR Means and FA Quiocho, *J Mol Biol*, **228**, 1177–92 (1992).
32. ST Rao, S Wu, KA Satyshur, KY Ling, C Kung and M Sundaralingam, *Protein Sci*, **2**, 436–47 (1993).
33. O Herzberg and MN James, *Nature*, **313**, 653–59 (1985).
34. RH Kretsinger and CE Nockolds, *J Biol Chem*, **248**, 3313–26 (1973).
35. YS Babu, CE Bugg and WJ Cook, *Methods Enzymol*, **139**, 632–42 (1987).
36. MA Wilson and AT Brunger, *J Mol Biol*, **301**, 1237–56 (2000).
37. J Evenas, S Forsen, A Malmendal and M Akke, *J Mol Biol*, **289**, 603–17 (1999).
38. BG Han, M Han, H Sui, P Yaswen, PJ Walian and BK Jap, *FEBS Lett*, **521**, 24–30 (2002).
39. JJ Chou, S Li, CB Klee and A Bax, *Nat Struct Biol*, **8**, 990–97 (2001).
40. A Houdusse and C Cohen, *Proc Natl Acad Sci USA*, **92**, 10644–47 (1995).
41. SY Lee and RE Klevit, *Biochemistry*, **39**, 4225–30 (2000).
42. M Zhang, T Tanaka and M Ikura, *Nat Struct Biol*, **2**, 758–67 (1995).
43. H Kuboniwa, N Tjandra, S Grzesiek, H Ren, CB Klee and A Bax, *Nat Struct Biol*, **2**, 768–76 (1995).

44 BE Finn, J Evenas, T Drakenberg, JP Waltho, E Thulin and S Forsen, *Nat Struct Biol*, **2**, 777–83 (1995).

45 H Ishida, K Nakashima, Y Kumaki, M Nakata, K Hikichi and M Yazawa, *Biochemistry*, **41**, 15536–42 (2002).

46 BA Seaton, JF Head, DM Engelman and FM Richards, *Biochemistry*, **24**, 6740–43 (1985).

47 DB Heidorn and J Trewhella, *Biochemistry*, **27**, 909–15 (1988).

48 C Yang and K Kuczera, *J Biomol Struct Dyn*, **19**, 801–19 (2002).

49 Y Komeiji, Y Ueno and M Uebayasi, *FEBS Lett*, **521**, 133–39 (2002).

50 KL Yap, JB Ames, MB Swindells and M Ikura, *Proteins*, **37**, 499–507 (1999).

51 SR Martin and PM Bayley, *Biochem J*, **238**, 485–90 (1986).

52 M Zhang and T Yuan, *Biochem Cell Biol*, **76**, 313–23 (1998).

53 JA Clapperton, SR Martin, SJ Smerdon, SJ Gamblin and PM Bayley, *Biochemistry*, **41**, 14669–79 (2002).

54 WE Meador, AR Means and FA Quiocho, *Science*, **262**, 1718–21 (1993).

55 M Osawa, H Tokumitsu, MB Swindells, H Kurihara, M Orita, T Shibanuma, T Furuya and M Ikura, *Nat Struct Biol*, **6**, 819–24 (1999).

56 H Kurokawa, M Osawa, H Kurihara, N Katayama, H Tokumitsu, MB Swindells, M Kainosho and M Ikura, *J Mol Biol*, **312**, 59–68 (2001).

57 M Ikura, GM Clore, AM Gronenborn, G Zhu, CB Klee and A Bax, *Science*, **256**, 632–38 (1992).

58 WE Meador, AR Means and FA Quiocho, *Science*, **257**, 1251–55 (1992).

59 AR Rhoads and F Friedberg, *FASEB J*, **11**, 331–40 (1997).

60 KL Yap, J Kim, K Truong, M Sherman, T Yuan and M Ikura, *J Struct Funct Genom*, **1**, 8–14 (2000).

61 M Aoyagi, AS Arvai, JA Tainer and ED Getzoff, *EMBO J*, **22**, 766–75 (2003).

62 E Yamauchi, T Nakatsu, M Matsubara, H Kato and H Taniguchi, *Nat Struct Biol*, **10**, 226–31 (2003).

63 B Elshorst, M Hennig, H Forsterling, A Diener, M Maurer, P Schulte, H Schwalbe, C Griesinger, J Krebs, H Schmid, T Vorherr and E Carafoli, *Biochemistry*, **38**, 12320–32 (1999).

64 D Guerini, J Krebs and E Carafoli, *J Biol Chem*, **259**, 15172–77 (1984).

65 MA Schumacher, AF Rivard, HP Bachinger and JP Adelman, *Nature*, **410**, 1120–24 (2001).

66 CL Drum, SZ Yan, J Bard, YQ Shen, D Lu, S Soelaiman, Z Grabarek, A Bohm and WJ Tang, *Nature*, **415**, 396–402 (2002).

67 KL Yap, T Yuan, TK Mal, HJ Vogel and M Ikura, *J Mol Biol*, **328**, 193–204 (2003).

68 M Bahler and A Rhoads, *FEBS Lett*, **513**, 107–13 (2002).

69 JR Slemmon, B Feng and JA Erhardt, *Mol Neurobiol*, **22**, 99–113 (2000).

70 JL Urbauer, JH Short, LK Dow and AJ Wand, *Biochemistry*, **34**, 8099–8109 (1995).

71 TH Crouch and CB Klee, *Biochemistry*, **19**, 3692–98 (1980).

72 H Kawasaki and RH Kretsinger, *Protein Profile*, **1**, 343–517 (1994).

73 OB Peersen, TS Madsen and JJ Falke, *Protein Sci*, **6**, 794–807 (1997).

74 KP Hoeflich and M Ikura, *Cell*, **108**, 739–42 (2002).

75 SW Vetter and E Leclerc, *Eur J Biochem*, **270**, 404–14 (2003).

76 J Goldberg, AC Nairn and J Kuriyan, *Cell*, **84**, 875–87 (1996).

77 DR Knighton, RB Pearson, JM Sowadski, AR Means, LF Ten Eyck, SS Taylor and BE Kemp, *Science*, **258**, 130–35 (1992).

78 H Yokokura, MR Picciotto, AC Nairn and H Hidaka, *J Biol Chem*, **270**, 23851–59 (1995).

79 G Zhi, SM Abdullah and JT Stull, *J Biol Chem*, **273**, 8951–57 (1998).

80 P Amodeo, MA Castiglione Morelli, G Strazzullo, P Fucile, M Gautel and A Motta, *J Mol Biol*, **306**, 81–95 (2001).

81 T Yuan and HJ Vogel, *J Biol Chem*, **273**, 30328–35 (1998).

82 G Baum, S Lev-Yadun, Y Fridmann, T Arazi, H Katsnelson, M Zik and H Fromm, *EMBO J*, **15**, 2988–96 (1996).

83 WA Snedden, N Koutsia, G Baum and H Fromm, *J Biol Chem*, **271**, 4148–53 (1996).

84 C Picton, CB Klee and P Cohen, *Eur J Biochem*, **111**, 553–61 (1980).

85 KF Chan and DJ Graves, *J Biol Chem*, **257**, 5956–61 (1982).

86 M Dasgupta, T Honeycutt and DK Blumenthal, *J Biol Chem*, **264**, 17156–63 (1989).

87 P Newsholme, KL Angelos and DA Walsh, *J Biol Chem*, **267**, 810–18 (1992).

88 LM Heilmeyer Jr, AM Gerschinski, HE Meyer and HP Jennissen, *Mol Cell Biochem*, **127-128**, 19–30 (1993).

89 NA Rice, OW Nadeau, Q Yang and GM Carlson, *J Biol Chem*, **277**, 14681–87 (2002).

90 RD Zuhlke, GS Pitt, K Deisseroth, RW Tsien and H Reuter, *Nature*, **399**, 159–62 (1999).

91 C Romanin, R Gamsjaeger, H Kahr, D Schaufler, O Carlson, DR Abernethy and NM Soldatov, *FEBS Lett*, **487**, 301–6 (2000).

92 RD Zuhlke, GS Pitt, RW Tsien and H Reuter, *J Biol Chem*, **275**, 21121–29 (2000).

93 GS Pitt, RD Zuhlke, A Hudmon, H Schulman, H Reuter and RW Tsien, *J Biol Chem*, **276**, 30794–30802 (2001).

94 A Lee, ST Wong, D Gallagher, B Li, DR Storm, T Scheuer and WA Catterall, *Nature*, **399**, 155–59 (1999).

95 CD DeMaria, TW Soong, BA Alseikhan, RS Alvania and DT Yue, *Nature*, **411**, 484–89 (2001).

96 SR Ikeda, *Science*, **294**, 318–19 (2001).

97 MD Ehlers and GJ Augustine, *Nature*, **399**, 105–8 (1999).

98 BM Zhang, V Kohli, R Adachi, JA Lopez, MM Udden and R Sullivan, *Biochemistry*, **40**, 3189–95 (2001).

99 H Wen and IB Levitan, *J Neurosci*, **22**, 7991–8001 (2002).

100 E Yus-Najera, I Santana-Castro and A Villarroel, *J Biol Chem*, **277**, 28545–53 (2002).

101 JR Sellers, *Biochim Biophys Acta*, **1496**, 3–22 (2000).

102 KT O'Neil and WF DeGrado, *Trends Biochem Sci*, **15**, 59–64 (1990).

103 M Mori, T Konno, T Ozawa, M Murata, K Imoto and K Nagayama, *Biochemistry*, **39**, 1316–23 (2000).

104 CP Moore, G Rodney, JZ Zhang, L Santacruz-Toloza, G Strasburg and SL Hamilton, *Biochemistry*, **38**, 8532–37 (1999).

105 N Yamaguchi, C Xin and G Meissner, *J Biol Chem*, **276**, 22579–85 (2001).

106 M Samso and T Wagenknecht, *J Biol Chem*, **277**, 1349–53 (2002).

107 H Zhang, JZ Zhang, CI Danila and SL Hamilton, *J Biol Chem*, **278**, 8348–55 (2003).

108 ED Apel, MF Byford, D Au, KA Walsh and DR Storm, *Biochemistry*, **29**, 2330–35 (1990).

109 T Zhu, K Beckingham and M Ikebe, *J Biol Chem*, **273**, 20481–86 (1998).
110 PG Gillespie and JL Cyr, *BMC Biochemistry*, **3**, 31–47 (2002).
111 SR Martin and PM Bayley, *Protein Sci*, **11**, 2909–23 (2002).
112 JE Keen, R Khawaled, DL Farrens, T Neelands, A Rivard, CT Bond, A Janowsky, B Fakler, JP Adelman and J Maylie, *J Neurosci*, **19**, 8830–38 (1999).
113 EP Morris and K Torok, *J Mol Biol*, **308**, 1–8 (2001).
114 SI Singla, A Hudmon, JM Goldberg, JL Smith and H Schulman, *J Biol Chem*, **276**, 29353–60 (2001).
115 JW Wei, HP Morris and RA Hickie, *Cancer Res*, **42**, 2571–74 (1982).
116 JL Chien and JR Warren, *Int J Pancreatol*, **3**, 113–27 (1988).
117 GX Liu, HF Sheng and S Wu, *Br J Cancer*, **73**, 899–901 (1996).
118 CD Rasmussen and AR Means, *Trends Neurosci*, **12**, 433–38 (1989).
119 N Takuwa, W Zhou and Y Takuwa, *Cell Signal*, **7**, 93–104 (1995).
120 J Penso and R Beitner, *Eur J Pharmacol*, **342**, 113–17 (1998).
121 OS Frankfurt, EV Sugarbaker, JA Robb and L Villa, *Cancer Lett*, **97**, 149–54 (1995).
122 TT Sakai and NR Krishna, *Bioorg Med Chem*, **7**, 1559–65 (1999).
123 WJ Cook, LJ Walter and MR Walter, *Biochemistry*, **33**, 15259–65 (1994).
124 M Vandonselaar, RA Hickie, JW Quail and LT Delbaere, *Nat Struct Biol*, **1**, 795–801 (1994).
125 BG Vertessy, V Harmat, Z Bocskei, G Naray-Szabo, F Orosz and J Ovadi, *Biochemistry*, **37**, 15300–10 (1998).
126 M Osawa, MB Swindells, J Tanikawa, T Tanaka, T Mase, T Furuya and M Ikura, *J Mol Biol*, **276**, 165–76 (1998).
127 V Harmat, Z Bocskei, G Naray-Szabo, I Bata, AS Csutor, I Hermecz, P Aranyi, B Szabo, K Liliom, BG Vertessy and J Ovadi, *J Mol Biol*, **297**, 747–55 (2000).
128 M Osawa, S Kuwamoto, Y Izumi, KL Yap, M Ikura, T Shibanuma, H Yokokura, H Hidaka and N Matsushima, *FEBS Lett*, **442**, 173–77 (1999).
129 AL Lee, SA Kinnear and AJ Wand, *Nat Struct Biol*, **7**, 72–77 (2000).
130 WC Prozialeck and B Weiss, *J Pharmacol Exp Ther*, **222**, 509–16 (1982).
131 R Beitner, *Mol Genet Metab*, **64**, 161–68 (1998).
132 SP Zhang, WC Prozialeck and B Weiss, *Mol Pharmacol*, **38**, 698–704 (1990).
133 PJ Kraulis, *J Appl Crystallogr*, **24**, 946–50 (1991).
134 EA Merritt and DJ Bacon, *Methods Enzymol*, **277**, 505–24 (1997).

Troponin C

Stéphane M Gagné
Université Laval, Québec, Canada

FUNCTIONAL CLASS

EF-hand calcium binding protein; muscle-contraction protein; calmodulin-like protein.

Troponin C (TnC) is a regulatory Ca^{2+}-dependent protein that triggers muscle contraction upon binding of calcium. The skeletal isoform of TnC (sTnC) has four calcium binding sites: two low-affinity sites located in the N-terminal domain that bind calcium specifically and two high-affinity sites located in the C-terminal domain that can bind both Ca^{2+} and Mg^{2+}. The cardiac isoform of TnC (cTnC) has three calcium binding sites – one in the N-terminal domain and two in the C-terminal domain.

OCCURRENCE

Occurs in all muscle cells. The skeletal isoform is found in fast skeletal muscle fibers. The cardiac isoform is found in slow skeletal and cardiac muscle fibers.

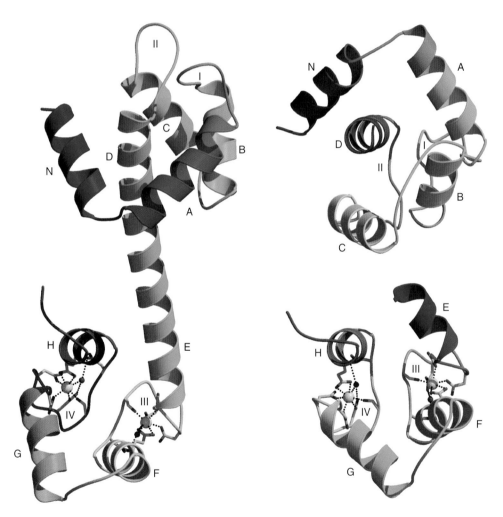

3D Structure 3D Structure of turkey skeletal muscle troponin C with the C-domain in the calcium-bound form and the N-domain in the apoform (PDB ID: 5TNC). Helices are labeled N, A, B, C, and D for the N-domain, and E, F, G, and H for the C-domain. Calcium binding sites are labeled I, II, III, and IV. The whole molecule is shown on the left, and the two domains are shown in a similar orientation on the right. It should be noted that a portion of the central helix is flexible in solution. The figure was made with MOLSCRIPT[1] and rendered with RASTER3D.[2]

Troponin C

BIOLOGICAL FUNCTION

Troponin, which is located on the filament of vertebrate striated muscles, regulates muscle contraction. Troponin consists of three subunits: troponin I (TnI, inhibitory subunit of myosin–actin interaction), troponin T (TnT, tropomyosin binding subunit), and TnC (Ca^{2+}-binding subunit).

In resting muscle, calcium ions are actively pumped out of muscle fibers, concentrating the calcium ions within the spaces of the sarcoplasmic reticulum. The concentration of free calcium in muscle fiber in the resting state is in the neighborhood of 10^{-7} M, and the concentration of magnesium is about 10^{-3} M. Consequently, in the resting state, the high-affinity sites in the C-domain of TnC are filled, and the low-affinity sites in the N-domain are in the apostate. The 3D Structure shown is therefore related to the conformation of TnC in relaxed muscle. Upon neural stimulation, calcium is released from the sarcoplasmic reticulum to the cytoplasm (sarcoplasm) of the muscle fiber. This calcium influx ($>10^{-6}$ M) causes the binding of calcium to the regulatory domain of TnC, inducing a conformational change, and triggering a sequence of events that ultimately leads to muscle contraction. The primary role of TnI is to inhibit the formation of the actomyosin complex. As TnC binds calcium, the TnC–TnI interactions change, and the interactions between TnI and actin weaken. The troponin–tropomyosin complex is believed to be relocated following the calcium-induced structural change in TnC, allowing the formation of the actomyosin complex and ultimately leading to the power stroke.

AMINO ACID SEQUENCE INFORMATION

In 1963, Ebashi discovered a new protein factor that was necessary for the calcium sensitivity of actomyosin.[3] This protein system was called 'native tropomyosin', but it was soon found that this system was a complex of tropomyosin and a new globular protein, which was named troponin.[4] Hartshorne and Mueller[5] were the first to observe that troponin was made of subunits. They obtained two fractions, which they named troponin A and troponin B. Troponin A, which also carried the name 'EGTA sensitizing factor', was found to be required for calcium control of ATP hydrolysis by myosin.[6,7] Greaser and Gergely[8,9] later resolved troponin into three subunits and named them troponin T, troponin I, and troponin C, the latter corresponding to the troponin A observed previously.

Thirty-nine sequences of TnC have been sequenced so far. It is worth noting that the majority of the differences between cardiac and skeletal TnC are located around site I. The average sequence length for TnC is 157 amino acids (AA). The number in parenthesis refers to the Swiss-Prot or TrEMBL accession number.

- *Anguilla anguilla* (European freshwater eel), skeletal muscle, 160 AA (Swiss-Prot P81660)[10]
- *Balanus nubilis* (Giant barnacle), isoform 1, 158 AA (Swiss-Prot P21797)[11]
- *Balanus nubilis* (Giant barnacle), isoform 2, 151 AA, (Swiss-Prot P21798)[11]
- *Brachydanio rerio* (Zebra fish) (Danio rerio), fast skeletal muscle, 160 AA (TrEMBL Q9I8U8)[12]
- *Branchiostoma floridae* (Florida lancelet), 164 AA (TrEMBL P90687)[13]
- *Branchiostoma lanceolatum* (Common lancelet), 163 AA (Swiss-Prot P80322)[14]
- *Caenorhabditis elegans*, isoform 2, 160 AA (Swiss-Prot Q09665)[15]
- *Chlamys nipponensis akazara* (Akazara scallop), adductor muscle, 153 AA (TrEMBL Q27428)[16]
- *Coturnix coturnix japonica* (Japanese quail), slow skeletal and cardiac muscles, 161 AA (Swiss-Prot P05936)[17]
- *Drosophila melanogaster* (Fruit fly), isoform 1, 154 AA (Swiss-Prot P47947)[18]
- *Drosophila melanogaster* (Fruit fly), isoform 2, 155 AA (Swiss-Prot P47948)[18]
- *Drosophila melanogaster* (Fruit fly), isoform 3, 155 AA (Swiss-Prot P47949)[18]
- *Gallus gallus* (Chicken), skeletal muscle, 162 AA (Swiss-Prot P02588)[19,20]
- *Gallus gallus* (Chicken), slow skeletal and cardiac muscles, 161 AA (Swiss-Prot P09860)[21,22]
- *Halocynthia roretzi* (Sea squirt), 158 AA (TrEMBL P92193, TrEMBL P92206)[23]
- *Halocynthia roretzi* (Sea squirt), body wall muscle, 155 AA (Swiss-Prot P06706)[24]
- *Homarus americanus* (American lobster), isoform 1, 150 AA (Swiss-Prot P29289)[25]
- *Homarus americanus* (American lobster), isoform 2A, 150 AA (Swiss-Prot P29290)[25]
- *Homarus americanus* (American lobster), isoform 2B, 150 AA (Swiss-Prot P29291)[25]
- *Homo sapiens* (Human), *Bos taurus* (Bovine), *Sus scrofa* (Pig), slow skeletal and cardiac muscles, 161 AA (Swiss-Prot P02590)[26-28]
- *Homo sapiens* (Human), skeletal muscle, 159 AA (Swiss-Prot P02585)[29]
- *Lampetra japonica* (Japanese lamprey), 162 and 167 AA (TrEMBL O42137, TrEMBL O42136)[13]
- *Meleagris gallopavo* (Common turkey), skeletal muscle, 162 AA (Swiss-Prot P10246)[20]
- *Mus musculus* (Mouse), skeletal muscle, 159 AA (Swiss-Prot P20801)[30]
- *Mus musculus* (Mouse), slow skeletal and cardiac muscles, 161 AA (Swiss-Prot P19123)[31]
- *Oryctolagus cuniculus* (Rabbit), skeletal muscle, 159 AA (Swiss-Prot P02586)[32-34]
- *Oryctolagus cuniculus* (Rabbit), slow skeletal and cardiac muscles, 161 AA (Swiss-Prot P02591)[35]

- *Patinopecten yessoensis* (Ezo giant scallop), striated adductor muscle, 151–153 AA (Swiss-Prot P35622, TrEMBL Q9U8L7, TrEMBL Q9U8L8)[36,37]
- *Perinereis vancaurica* tetradentata (Sandworm), 152 AA (TrEMBL Q9GN70)[38]
- *Pontastacus leptodactylus* (Narrow-fingered crayfish), isotype alpha, 150 AA (Swiss-Prot P06707)[39]
- *Pontastacus leptodactylus* (Narrow-fingered crayfish), isotype gamma, 150 AA (Swiss-Prot P06708)[39]
- *Protopterus dolloi* (Lungfish), skeletal muscle (Fragment), 52 AA (Swiss-Prot P81074)[40]
- *Rana esculenta* (Edible frog), skeletal muscle, 162 AA (Swiss-Prot P02589)[41]
- *Sus scrofa* (Pig), skeletal muscle, 159 AA (Swiss-Prot P02587)[42]
- *Tachypleus tridentatus* (Japanese horseshoe crab), skeletal muscle, 153 AA (Swiss-Prot P15159)[43]
- *Todarodes pacificus* (Japanese flying squid), mantle muscle, 147 AA (Swiss-Prot Q9BLG0)[44]
- *Xenopus laevis* (African clawed frog), cardiac, 161 AA (TrEMBL O12998)[13]
- *Xenopus laevis* (African clawed frog), fast skeletal alpha, 163 AA (TrEMBL O12996)[13]
- *Xenopus laevis* (African clawed frog), fast skeletal beta, 163 AA (TrEMBL O12997)[13]

PROTEIN PRODUCTION, PURIFICATION, AND MOLECULAR CHARACTERIZATION

Recombinant skeletal TnC (sTnC), cardiac TnC (cTnC), and their respective N- and C-domains (sNTnC, sCTnC, cNTnC, and cCTnC) can be overexpressed in very high yield in *E. coli*.[45–51] The purification typically consists of extraction from the harvested cells with a French press, ion exchange and Sephadex chromatographies.[20] TnC is highly soluble in water and therefore well suited for NMR studies, although forms of sTnC where calcium is bound to the N-terminal domain tend to weakly dimerize.[45]

X-RAY AND NMR STRUCTURES

Overall description of the structure from the first X-ray structures

The overall fold of TnC was first described by the crystal structures of both chicken and turkey skeletal TnC that were solved in 1985[52,53] and refined to 2.0 Å in 1988.[54,55] As shown in the 3D Structure, these 65% α-helical structures have two globular domains, each containing two calcium binding sites, connected by a single 31-residue-long central helix. The N-domain has five helices (N, A, B, C, and D), whereas the C-domain has four helices (E, F, G, and H). The two calcium binding loops of each domain are coupled via a short β-sheet. In these structures, sites III and IV in the C-domain are occupied by calcium, whereas the regulatory sites (I and II) in the N-domain are in the apostate. This structure will therefore be referred to as sTnC•Ca$_2$. All four calcium binding sites show the helix-loop-helix motif termed the *EF-hand motif*.[56] Although both domains are highly homologous, the helix packing is different in the two domains. The structure of the calcium-bound C-domain was found to be very similar to other homologous calcium binding proteins. However, the apo N-domain was considerably different, mainly in terms of interhelical angles (Table 1).

Calcium binding sites

Skeletal TnC has four calcium binding sites: two with high affinity (sites III and IV) and two with low affinity (sites I and II). In cardiac muscle, there is only one low-affinity site (site II), as site I is defunct.[59] The high-affinity sites are believed to be always occupied by either calcium ($K_{Ca} = 2 \times 10^7$ M^{-1}) or magnesium ($K_{Mg} = 2 \times 10^3$ M^{-1}) under physiological conditions, and have primarily a structural role. The low-affinity sites are calcium specific and assume a regulatory role in muscle contraction. Sequences for the calcium binding loops of sTnC and cTnC are shown in Figure 1.

In the literature, controversial results may be found regarding the binding affinity and the cooperativity of the low-affinity sites. Some groups report cooperative coupling between sites I and II, while others propose that the binding process is sequential. The published binding affinity varies between 10^4 M^{-1} and 10^5 M^{-1}. However, a detailed NMR study on sNTnC has quantitatively determined the binding constants in sites I and II.[48] A calcium titration, which was monitored by 2D ^{15}N, ^{1}H-HMQC experiments, has revealed a stepwise binding for sNTnC with affinities (K_{Ca}) of 6.3×10^4 and 5.9×10^5 M^{-1}.

Calcium binding sites in TnC belong to the well-characterized EF-hand calcium binding motifs. A typical EF-hand consists of an α-helix (residues 1–10), a 12-residue site that binds calcium (residues 10–21), and a second α-helix (residues 19–21). Calcium ions are coordinated by seven oxygen ions that are furnished by side chains, peptide backbone carbonyls, and bridging water molecules. The coordination geometry is a pentagonal bipyramid, as illustrated in Figure 2. EF-hands are typically paired and connected in space through a short antiparallel β-sheet.

HMJ model for the calcium-induced structural change

By comparing the crystal structure of the apo N-domain of sTnC•Ca$_2$ with its homologous calcium-bound C-domain (see 3D Structure), Herzberg *et al.* proposed the Herzberg, Moult and James (HMJ) model for the conformational change that occurs in the N-domain of TnC upon calcium binding.[61] Using molecular modeling and the assumption that the calcium-filled N-domain would adopt a conformation similar to the calcium-filled C-domain,

Troponin C

Table 1 Interhelical angles in troponin C

Structure	PDB ID	Interhelical angle			
		A/B or E/F		C/D or G/H	
(a) Skeletal TnC, N-domain, apo					
N-domain of sTnC•Ca$_2$	5TNC	139		146	
TR$_1$C•Ca$_0$	a	135 ± 15	(94–159)	133 ± 25	(82–170)
sNTnC•Ca$_0$	1TNP	127 ± 3	(120–135)	124 ± 4	(115–132)
(b) Skeletal TnC, N-domain, Ca$_2$					
sNTnC•Ca$_2$ (HMJ model)	b	101		112	
sNTnC•Ca$_2$ (NMR)	1TNQ	85 ± 4	(74–92)	68 ± 5	(58–81)
sNTnC•Ca$_2$ (X-ray)	1AVS	97		83	
sTnC•Ca$_4$ (X-ray)	1TN4	104		97	
sTnC•Ca$_4$ (X-ray)	1TCF	95		89	
E41A-sNTnC•Ca$_2$	1SMG	131 ± 4	(124–138)	131 ± 5	(120–138)
N-domain of sTnC•Ca$_4$ (NMR)	1TNW	81 ± 5	(74–92)	78 ± 7	(67–89)
(c) Skeletal TnC, C-domain, Ca$_2$					
C-domain of sTnC•Ca$_2$	5TNC	107		109	
IV-IV•Ca$_2$	c			~90[c]	
III-III•Ca$_2$	1CTD	118 ± 7	(101–129)		
III-IV•Ca$_2$	1PON	103 ± 5	(92–115)	110 ± 10	(86–129)
C-domain of sTnC•Ca$_4$	1TNW	89 ± 6	(81–104)	104 ± 7	(91–117)
(d) Cardiac TnC					
cNTnC•Ca$_0$	1SPY	139 ± 3	(132–146)	128 ± 5	(116–139)
cNTnC•Ca$_1$	1AP4	135 ± 4	(129–146)	118 ± 3	(111–125)
N-domain of cTnC•Ca$_3$	2CTN	144 ± 3	(137–149)	110 ± 4	(102–118)
C-domain of cTnC•Ca$_3$	3CTN	111 ± 4	(101–120)	115 ± 3	(110–121)

Note: The axis orientation for an α-helix was defined by two points, where the first point is the average of the first 11 backbone atoms and the second point is the average of the last 11 backbone atoms. The program used to calculate the angles was IHA (Gagné) and is available upon request. Helix definitions are as defined in reference 57.
[a] Provided by BD Sykes.
[b] Provided by MNG James.
[c] Value quoted from reference 58 because PDB coordinates are not available.

```
              21
Site I        EFKAAFDMF  DADGGGDISTKE  LGTVMRMLG
              57
Skeletal  Site II   ELDAIIEEV  DEDGSGTIDFEE  FLVMMVRQM
TnC           97
Site III      ELANCFRIF  DKNADGFIDIEE  LGEILRATG
              133
Site IV       DIEDLMKDS  DKNNDGRIDFDE  FLKMMEGVQ

              19
(Site I)      EFKAAFDIF  VLGAEDGCISTKE  LGKVMRMLG
Cardiac       56
TnC  Site II  ELQEMIDEV  DEDGSGTVDFDE  FLVMMVRCM
              96
Site III      ELSDLFRMF  DKNADGYIDLEE  LKIMLQATG
              132
Site IV       DIEELMKDG  DKNNDGRIDYDE  FLEFMKGVE
```

Figure 1 Sequences of helix-loop-helix calcium binding motifs in chicken skeletal and cardiac TnC. Calcium binding loops are underlined (including site I of cTnC, which is defunct). Residues directly involved in calcium coordination are indicated in bold. Residues involved in calcium coordination only via a water molecule are indicated in italic and bold. In site I of skeletal TnC, the missing ligand is replaced by a water molecule that is hydrogen-bonded to D30 and D36.[60] The last glutamic acid residue in each calcium binding loop contributes two coordinating oxygens.

they deduced a model structure of sTnC in the Ca$_4$ state (sTnC$_{HMJ}$•Ca$_4$). The major conformational change in their model is the movement of the B-C pair of helices away from the A-D pair, thereby exposing a hydrophobic patch. This model was the only structural representation of the calcium-saturated form of the regulatory domain of TnC for nearly 10 years.

List of structures solved by NMR and X-ray crystallography

There are now several structures of TnC. Many structures of skeletal and cardiac TnC have been solved by NMR and X-ray, both for isolated domains and for intact protein. More recently, several structures of TnC complexes have also been solved. Since the C-domain of TnC is unstructured in the absence of calcium, there are no structures of the C-domain in the apoform. A list of the structures that have been solved to date is presented below.

Skeletal TnC, N-domain in the apoform

- sTnC + 2Ca^{2+}, turkey, X-ray, 1988, 5TNC[54]
- sTnC + 2Ca^{2+}, chicken, X-ray, 1987, 4TNC[55]
- sTnC + 2Ca^{2+}, chicken, X-ray, 1993, 1TOP[62]

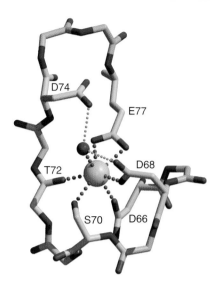

Figure 2 Calcium coordination site in TnC. Site II of sTnC is shown (PDB ID: 1AVS). Carbons of residues involved in calcium coordination are shown in yellow, all oxygens are shown in red, all nitrogens are shown in blue, the coordinating water is shown in red, the calcium ion is shown in white. Dotted orange lines indicate calcium coordination and dotted cyan lines indicate hydrogen bonds with the coordinating water. Calcium-liganding distances are 2.26 Å (Asp66 OD1), 2.34 Å (Asp68 OD1), 2.59 Å (Ser70 OG), 2.59 Å (Thr72 O), 2.37 Å (Glu77 OE1), 2.43 Å (Glu77 OE2), and 2.46 Å (water).

- sTnC + $2Cd^{2+}$, chicken, X-ray, 1996, 1NCX[63]
- sTnC + $2Mn^{2+}$, chicken, X-ray, 1996, 1NCY[63]
- sTnC + $2Tb^{2+}$, chicken, X-ray, 1996, 1NCZ[63]
- sNTnC + $0Ca^{2+}$, chicken, NMR, 40 structures, 1995, 1TNP[64]
- sNTnC + $0Ca^{2+}$ (4 °C), chicken, NMR, 40 structures, 1999, 1SKT[65]
- sNTnC + $0Ca^{2+}$ (4 °C), chicken, NMR, average structure, 1999, 1ZAC[65]
- sNTnC + $0Ca^{2+}$ (TR1C fragment), turkey, NMR, 1 structure, 1993 1TRF[66]

Skeletal TnC, N-domain in the calcium-bound form

- sTnC + $4Ca^{2+}$, chicken, NMR, 23 structures, 1995, 1TNW[67]
- sTnC + $4Ca^{2+}$, chicken, NMR, average structure, 1995, 1TNX[67]
- sTnC + $4Ca^{2+}$, rabbit, X-ray, 1997, 2TN4[49]
- sTnC + $4Ca^{2+}$, rabbit, X-ray, 1998, 1TCF[68]
- sTnC + $4Ca^{2+}$, rabbit, X-ray, 1997, 1TN4[49]
- sNTnC + $2Ca^{2+}$, chicken, NMR, 40 structures, 1995, 1TNQ[64]
- sNTnC + $2Ca^{2+}$, chicken, X-ray, 1997, 1AVS[60]
- sNTnC + $2Ca^{2+}$ (E41A mutant), chicken, NMR, 40 structures, 1997, 1SMG[69]

Skeletal TnC, peptides

- III-III sTnC homodimer + $2Ca^{2+}$, chicken, NMR, average structure, 1992, 1CTA[70]
- III-III sTnC homodimer + $2Ca^{2+}$, chicken, NMR, 7 structures, 1992, 1CTD[70]
- III-IV sTnC heterodimer + $2Ca^{2+}$, chicken, NMR, 42 structures, 1996, 1PON[71]

Cardiac TnC, N-domain apo

- cNTnC + $0Ca^{2+}$, human, NMR, 40 structures, 1997, 1SPY[72]

Cardiac TnC, N-domain in the calcium-bound form

- cTnC + $3Ca^{2+}$, chicken, NMR average structure, 1997, 1AJ4[51]
- cTnC + $3Ca^{2+}$ (N-domain portion), chicken, NMR, 30 structures, 1997, 2CTN[51]
- cTnC + $3Ca^{2+}$ (C-domain portion), chicken, NMR, 30 structures, 1997, 3CTN[51]
- cNTnC + $1Ca^{2+}$, human, NMR, 40 structures, 1997, 1AP4[72]

NMR structures of sTnC peptides

The first TnC structures to be solved by NMR were of peptides corresponding to the calcium binding sites of the C-domain of sTnC.[73] In 1991, the structure of a 39–amino acid proteolytic fragment of rabbit sTnC was solved by Kay et al.[58] This fragment includes residues 121–159 of rabbit sTnC. The helix-loop-helix motif corresponds to the fourth calcium binding site in sTnC and dimerizes upon calcium binding. In 1992, Shaw et al.[70] solved the structure of a 34–amino acid synthetic peptide corresponding to the third calcium binding site of chicken sTnC (residues 93–126). As for the IV-IV homodimer, calcium binding induces dimerization of this peptide to form the so-called III-III homodimer. Shaw and Sykes[71] also solved the solution structure of another calcium-bound synthetic dimer; this one is of a synthetic III-IV heterodimer (residues 93–126 and 129–162 of chicken sTnC). Structurally, the III-IV heterodimer is very similar to the wild-type. Unlike the homodimers that have differences in interhelical angles of +10° (III-III) or −20° (IV-IV) with the wild-type, the III-IV heterodimer has an identical degree of opening compared to the crystal structure of the wild-type (Table 1).

NMR structure of the proteolytic fragment TR1C in the apostate

In 1994, Findlay et al.[66] published the solution structure of TR_1C•apo, a proteolytic fragment of turkey skeletal

TnC (residues 12–87). The fragment corresponds to the N-domain of sTnC minus the N-terminal helix. Although helix C is well defined by itself in the TR$_1$C•apo structure, its orientation relative to the other helices is not. It was not clear at the time whether helix C really possessed structural heterogeneity in solution. On the basis of a more recent analysis of the nuclear Overhauser effect (NOE) data (Sykes, unpublished data), it is now believed that the lack of definition of helix C was a consequence of overlapping NOE peaks, which could not be unambiguously assigned from homonuclear NMR spectra.

First structure of the regulatory domain of TnC in the calcium-bound state

The first structure of a regulatory domain of TnC in the calcium-bound state was solved by NMR in 1995, 10 years after the first X-ray structure of TnC.[64] The study involved the determination of the structures of sNTnC•apo and sNTnC•Ca$_2$ by NMR. The resulting structures are shown in Figure 3. Both structures consist of five helices (N, A, B, C, and D) and two calcium binding loops (site I and II) connected by a short β-strand, as was determined in the NMR study of the secondary structure[47] and in the crystal structures of the apo N-domain.[54,55] The number of NOE violations (>0.2 Å) per structure is 0.30 and 0.07 and reflects the agreement between the structures generated and the experimental data. The analysis of the rmsd statistics indicates the following: (i) the five helices are individually well defined (rmsd < 0.3 Å), and their relative orientations are also well defined but to a lesser extent (overall rmsd ~0.7 Å); (ii) the calcium binding sites are better defined in sNTnC•Ca$_2$ compared to sNTnC•apo, mostly owing to the inclusion of restraints to the calcium ligands. It is interesting to note that although there is a two-fold difference in distance restraints for the two states of sNTnC,[64] the rmsd numbers are similar. One potential explanation for this observation is that most of the additional information in the sNTnC•Ca$_2$ data comes from intraresidue and short-range NOEs. Also, it is likely that an open structure, like sNTnC•Ca$_2$, requires more NOEs than a more compact structure, like sNTnC•apo, to achieve the same definition. Nevertheless, the quality of the structures of sNTnC in the apo and calcium forms was adequate to provide an experimental understanding of calcium-induced structural change (Figure 3).

NMR structure of sTnC in the Ca$_4$ state

The solution structure of whole sTnC•Ca$_4$ was published shortly after the structures of the N-domain.[67] This structure determination was done in the presence of TFE in order to eliminate calcium-induced dimerization.[45] One of the major conclusions of this structure determination was the confirmation that the central helix is flexible and thus the relative orientations of the two domains is not fixed in solution. Comparison of the NMR structure of the C-domain of sTnC•Ca$_4$ with the corresponding domain in the crystal structure of sTnC•Ca$_2$ reveals that the NMR structure is slightly more open. Although the G/H interhelical angle is the same in both structures, the angle between helix E and helix F is more open by 16° in the NMR structure. Comparison of the NMR structure

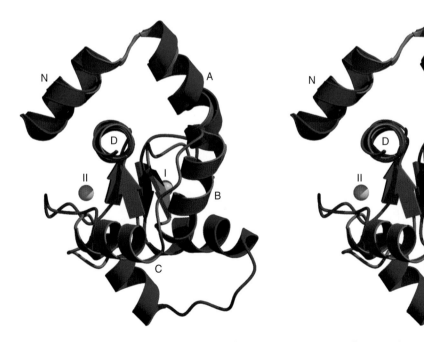

Figure 3 Stereo view of the calcium-induced structural change in the N-domain of sTnC. The apo structure (PDB ID: 1TOP) is shown in red and the calcium-bound structure (PDB ID: 1TNQ) is shown in blue.

Troponin C

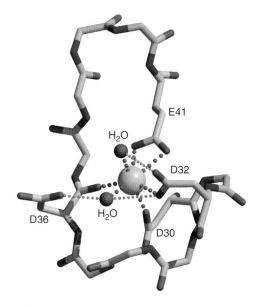

Figure 4 Calcium coordination site in site I of skeletal TnC (PDB ID: 1AVS). Carbons of residues involved in calcium coordination are shown in yellow, all oxygens are shown in red, all nitrogens are shown in blue, the two coordinating water molecules are shown in red, and the calcium ion is shown in white. Dotted orange lines indicate calcium coordination and dotted cyan lines indicate hydrogen bonds with the coordinating water molecules.

of the N-domain of sTnC·Ca$_4$ with the NMR structure of sNTnC·Ca$_2$ shows a near-identical level of opening.

The A/B and C/D interhelical angles are the same within experimental error, although the average C/D angle is more open by 10° in the structure of sNTnC·Ca$_2$ (Table 1). This structural similarity confirms that the N-domain behaves the same structurally in sNTnC as in intact sTnC, and that structural studies of the isolated domain are valid. The similarity also suggests that neither partial dimerization nor TFE affects the level of opening. However, there is the possibility that both conditions actually affect the conformation of the N-domain in a similar fashion.

X-ray structure of sNTnC in the Ca$_2$ state

Soon after the NMR structure of sNTnC·Ca$_2$, the crystal structure of chicken sNTnC·Ca$_2$ was reported by Strynadka et al.[60] The paper reported very detailed features of the calcium binding loops, which were not as well defined in the NMR structures. The X-ray structure of sNTnC·Ca$_2$ is somewhat less open than the average sNTnC·Ca$_2$ and sTnC·Ca$_4$ NMR structures, although the crystal structure is at one end of the range of interhelical angles covered by the NMR structures (Table 1). This structure also provided the first detailed description of calcium binding in site I of sTnC (Figure 4), which is different from sites II-IV owing to a sequence in the calcium binding site that is not typical.

X-ray structures of skeletal rabbit TnC in the Ca$_4$ state

In 1997, the crystal structure of rabbit sTnC·Ca$_4$ was solved by Cohen et al.[49] The structure of rabbit sTnC·Ca$_4$ was solved in two different crystal forms.[49] A very interesting result was that the two crystal forms gave a significantly different level of opening in the structure of the calcium-bound C-domain.

As shown in Table 1, the N-domain in these crystal structures of rabbit sTnC·Ca$_4$ is more closed than the other structures of sTnC in the calcium-bound state (NMR structure of sNTnC·Ca$_2$, NMR structure of sTnC·Ca$_4$, and X-ray structure of sNTnC·Ca$_2$). At the time, it was not clear why this was so. One possibility was that this was simply a result of sequence differences between skeletal rabbit and skeletal avian TnC.

Soon after, the group of Phillips et al. also solved the X-ray structure of rabbit sTnC·Ca$_4$,[68] the same protein that was solved by Cohen et al. Following a familiar trend, this structure, in terms of opening, was different from others solved to date. The Phillips structure is more open than the Cohen structures and more closed than the X-ray and NMR structures of sNTnC·Ca$_2$ (Table 1).

Early evidences of conformational changes

The first evidence of the calcium-induced conformational changes in sTnC was reported in 1972 by Murray and Kay.[74] They used CD spectroscopy to demonstrate that the binding of calcium induced a conformational change in sTnC (or, more accurately, the ancestor of troponin C, troponin A). Over the years, ^{1}H NMR spectroscopy has been used extensively to characterize the structural changes of the two domains of sTnC. These studies have shown that binding of calcium to the high-affinity sites (C-domain) induces changes in the protein fold, and that binding to the low-affinity sites (N-domain) leads mainly to changes among hydrophobic side chains.[75-80] Laser Raman spectroscopy was used, and it revealed a large increase in α-helical content due to calcium binding to the C-domain sites, and none associated with the N-domain.[81] Other studies, using CD and fluorescence, reported large spectroscopic changes associated with calcium binding to the C-domain, and small changes when the low-affinity sites were filled.[82-85] These studies reported a large increase of negative ellipticity when sites III and IV were filled, consistent with the NMR results. The estimates of the magnitude of the calcium-induced far-UV CD ellipticity changes attributable to the N-domain transition were more subtle, less clearly defined, and somewhat contradictory. A more recent characterization of the calcium-induced spectroscopic changes of the recombinant sNTnC clearly indicated a significant increase in the negative far-UV CD ellipticity.[47] These data suggested

Troponin C

that the binding of calcium to the N-domain induced a significant increase in α-helix content. However, a 1994 NMR study demonstrated[47] that binding of calcium to sNTnC does not affect the secondary structure content, and that the spectral changes seen by CD are most likely because of the reorientation of the α-helices relative to each other.[47]

Calcium-induced structural change in the regulatory domain of sTnC

As was concluded from early NMR studies,[47] no change in regular secondary structure is observed upon calcium binding. Except for the straightening of helix B,[47] there is no increase in α-helical content upon calcium binding, therefore reinforcing the fact that the increase in negative ellipticity observed by circular dichroism is not due to a change in secondary structure but due to a change in tertiary structure.[47] This change is depicted in Figure 3, where helices N, A, and D of the apo and calcium-saturated forms of sNTnC are superimposed, thus showing the movement of helices B and C when calcium is bound. This opening of the structure is quantified by measuring the difference in interhelical angle between the apo and the calcium structure (Table 1).

The relative orientation of helices N, A, and D is not affected by calcium binding, whereas the A-B and C-D interhelical angles show significant changes. The concept of a calcium-induced opening of the structure has similarities with the HMJ model that was predicted in 1986.[61] The magnitude of the movement of helices B and C is characterized by a large change in interhelical distance, resulting in displacements of 12 Å for the C-terminal end of helix B, 16 Å for the N-terminal end of helix C, and 19 Å for the center of the B-C linker.

Although the changes in interhelical angle quantify the opening of the structure, it is also of interest to ask what is actually exposed. The entity that is exposed is depicted in Figure 5. The changes in interhelical angles result in the exposure of an extensive hydrophobic patch that is partially buried in the apoform. The calcium-induced change in accessible surface area (ASA) for the hydrophobic residues is $+500 \text{Å}^2$. The binding of calcium to the regulatory domain can therefore be viewed as a switch that turns on a 'sticky' hydrophobic patch.

In order to obtain an insight into the mechanism of regulation within sTnC, that is, the coupling between calcium binding and subsequent structural change, the structure of a calcium-bound mutant of sNTnC was studied – E41A-sNTnC•Ca$_2$. In this mutant, the bidentate ligand to the calcium in site I (E41) was removed. The initial goal was to explore the contribution of site I to the calcium-induced structural change by making site I defunct. The calcium titration of E41A-sNTnC revealed that despite removal of the bidentate ligand, site I was still able to bind calcium although with a diminished affinity ($K_d = 1-2$ mM).

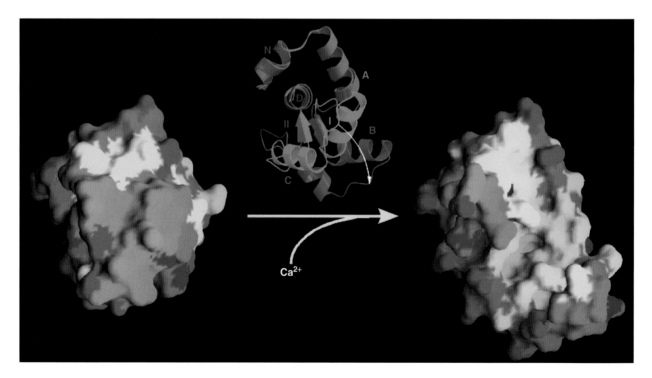

Figure 5 Exposure of the hydrophobic patch in the N-domain of sTnC upon calcium binding. The hydrophobic surface is shown in yellow, negative charges are shown in red, and positive charges are shown in blue. A ribbon representation of the structural change is also shown. (PDB ID: 1TNP and 1TNQ).

Troponin C

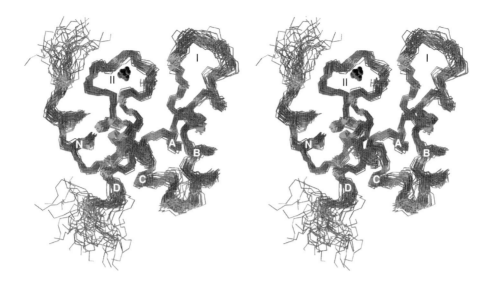

Figure 6 Solution structure of E41A-NTnC•Ca$_2$. The backbone (N, C, Cα) of the family of 40 structures is shown in 'rods' representations, and the site II Ca^{2+} position in each of these structures is shown by a small sphere. Although Ca^{2+} is present in solution in site I, there is no Ca^{2+} shown here because none was used in the calculations, as the coordination state of site I was not known *a priori*. The five helices are labeled N, A, B, C, and D. The two Ca^{2+} binding loops are labeled I and II. (PDB ID: 1SMG).

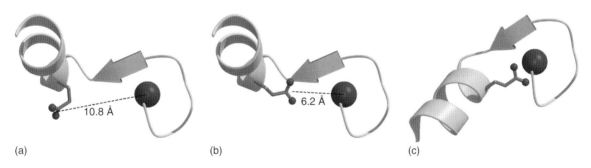

Figure 7 Glu41 cannot coordinate the Ca^{2+} ion when the domain is in the closed form. The backbone of loop I and helix B are shown in light gray, the side chain of Glu41 is shown in the dark gray stick representation, the position of the Ca^{2+} ion is shown by the large sphere, and the coordination oxygens of Glu41 are shown by small spheres. The indicated distance is between the Ca^{2+} ion and the center of the two coordinating oxygens. (a) Glu41 in closed form (NTnC•apo). The kink at residue Glu41 is clearly visible. (b) Same as in (a) but with the side chain χ$_1$ and χ$_2$ angles of Glu41 modified in order to minimize the distance of the ligands to the Ca^{2+} ion. The coordinating oxygens are still 6.2 Å away from the Ca^{2+} ion, 3.8 Å short of the coordinating distance (2.4 Å). (c) Glu41 in NTnC•Ca$_2$.

The structure of E41A-sNTnC•Ca$_2$ remains closed upon calcium binding (Figure 6), indicating that the linkage between calcium binding and the induced conformational change has been broken. This provides a snapshot of sTnC between the 'off' and the 'on' state. Comparison of the helix packing in E41A-sNTnC•Ca$_2$ with sNTnC•apo and sNTnC•Ca$_2$ reveals valuable information. The interhelical angles of E41A-NTnC•Ca$_2$ are like the apoforms rather than the Ca$_2$ form. Therefore, the hydrophobic patch is not exposed upon calcium binding unless the bidentate ligand in site I (E41) is involved in calcium coordination. Although several factors contribute to the triggering mechanism, the opening of the sNTnC structure is ultimately dependent, in a temporal and energy-balance sense, on one amino acid – Glu41 (Figure 7).[57,69]

NMR structures of cardiac TnC

The three-dimensional structure of chicken recombinant cTnC (C35S, C84S mutant) in the calcium-saturated state was solved by NMR.[51] The overall structure of cTnC•Ca$_3$ consists of two separate N- and C-globular domains, connected by a flexible linker, like the calcium-saturated structures of sTnC and calmodulin. The most striking feature of cTnC•Ca$_3$ is that the calcium-bound N-terminal regulatory domain is partially closed, resulting in significantly less exposed hydrophobic surfaces than those found in calcium-bound sTnC, a direct consequence of the defunct calcium binding site I in cTnC. The helix packing in the calcium-bound N-domain of cTnC•Ca$_3$ is similar to that in the apo N-domain of sTnC. With no calcium bound in site I, the A/B helices run almost antiparallel to each other, exhibiting the typical A/B interhelical angle found in the apostate of sTnC (Table 1). On the other hand, with calcium bound at site II, helix C is in fact in a partially open conformation, but does not open up to the extent seen in sTnC (Table 1). These observations demonstrate that the inability of site I to bind calcium results in a more compact conformation of helix B and helix C than is observed in sTnC. The structure of the C-domain of cTnC•Ca_3 is similar to the one in sTnC as indicated by the comparable interhelical angle of the two EF-hands (Table 1).

The structures of the N-domain of human recombinant cTnC in both the apo and the calcium-saturated states were also solved by NMR[72] (Figure 8). This pair of structures defines the calcium-induced structural transition in the regulatory domain of cTnC. The B/C structural unit in cNTnC moves only slightly away from the NAD structural unit (Figure 8). This movement is only a small fraction of the analogous change in the skeletal isoform. The global fold of cNTnC•apo is very similar to that of sNTnC•apo. The interhelical angles of cNTnC•apo are comparable to those of sNTnC•apo (Table 1), indicative of a closed conformation for the apostate of the regulatory domain in both isoforms. The structure of cNTnC•Ca_1 differs significantly from that of sNTnC•Ca_2, but displays similar interhelical angles to those in cTnC•Ca_3, indicating that both cNTnC and cTnC remain in a predominantly closed conformation upon calcium binding. This evidence demonstrates that separation of the two domains in cTnC has little structural effect on the regulatory domain.

The structural differences between cardiac and skeletal TnC[51,72] can be explained by what is in fact the most striking functional difference between the two proteins, namely, that site I in cTnC is inactive owing to an insertion and substitutions of key ligands (Figure 1). The structures confirm what was originally proposed on the basis of the structure of E41A-sNTnC•Ca_2; in cTnC there is no calcium at site I to pull the Glu40 (cTnC equivalent of Glu41 in sTnC), and opening of the structure is not observed upon calcium binding.

Structures of skeletal TnC complexes

The structures of three complexes of sTnC with various fragments of TnI are currently available, and a list of these structures is provided below.

- sNTnC + 2Ca^{2+} + TnI(96–148), chicken, NMR, 29 structures, 1998, 1BLQ[86]
- sCTnC + 2Ca^{2+} + TnI(1–40), chicken, NMR, 30 structures, 2001, 1JC2[87]
- sTnC + 2Ca^{2+} + TnI(1–47), rabbit, 1998, X-ray, 1A2X[88]
- rhodamine-labeled sNTnC mutant + 2Ca^{2+} + TnI (115–131), 1NPQ (currently on hold)

Structures of cardiac TnC complexes

The following structures have been solved for complexes between cardiac TnC and fragments of TnI, as well as with other compounds. Unlike the closed structures of the calcium-bound N-domain of cTnC, the structures of this domain in complex with TnI or drugs are in the open form (Figure 9). Given below is a list of the structures that have been determined to date for cardiac TnC complexes.

- cTnC + 3Ca^{2+} + bepridil, chicken, X-ray, 2000, 1DTL[89]

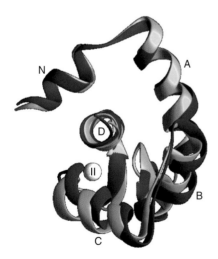

Figure 8 Calcium-induced structural change in the N-domain of cardiac TnC. The apostructure is shown in yellow (PDB ID: 1SPY) and the calcium-bound structure is shown in blue (PDB ID: 1AP4). Calcium binding site I is defunct in cardiac TnC, and therefore, only one calcium binds to the N-domain. Consequently, the binding of calcium does not induce an opening of the structure as observed in skeletal TnC.

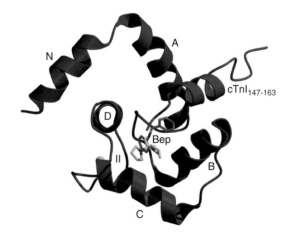

Figure 9 Solution structure of the cNTnC•Ca_2cTnI$_{147-163}$•bepridil complex. The N-domain of cTnC is shown in green, the cTnI$_{147-163}$ peptide is shown in pink, and bepridil is shown in the stick representation. Unlike the uncomplexed calcium-bound structure of cNTnC, this structure is found in the open form. (PDB ID: 1LXF).

- cCTnC + 2Ca^{2+} + NTnI, chicken, NMR, 20 structures, 1999, FI5[90]
- cCTnC + 2Ca^{2+} + drug Emd 57033, human, NMR, 30 structures, 2001, 1IH0[91]
- cCTnC + 3Ca^{2+} + TnI, chicken, NMR, average structure, 2002, 1LA0[92]
- cNTnC + 1Ca^{2+} + TnI(147–163) + bepridil, human, NMR, 30 structures, 2002, 1LXF[93]
- cNTnC + 1Ca^{2+} + TnI(147–163), human, NMR, 40 structures, 1999, 1MXL[94]
- 46 kDa domain of human cardiac troponin complex, X-ray, 1J1D (currently on hold)
- 52 kDa domain of human cardiac troponin complex, X-ray, 1J1E (currently on hold)
- cCTnC + 2Ca^{2+} + TnI (inhibitory), human, NMR, 1OZS (currently on hold)

REFERENCES

1. PJ Kraulis, *J Appl Crystallogr*, **24**, 946–50 (1991).
2. EA Merritt and DJ Bacon, *Methods Enzymol*, **277**, 505–24 (1997).
3. S Ebashi, *Nature*, **200**, 1000–10 (1963).
4. S Ebashi and A Kodama, *J Biochem (Tokyo)*, **58**, 107–8 (1965).
5. DJ Hartshorne and H Mueller, *Biochem Biophys Res Commun*, **31**, 647–53 (1968).
6. MC Schaub and SV Perry, *Biochem J*, **115**, 993–1004 (1969).
7. MC Schaub, SV Perry and W Hacker, *Biochem J*, **126**, 237–49 (1972).
8. ML Greaser and J Gergely, *J Biol Chem*, **248**, 2125–33 (1973).
9. ML Greaser and J Gergely, *J Biol Chem*, **246**, 4226–33 (1971).
10. JM Francois, C Gerday, FG Prendergast and JD Potter, *J Muscle Res Cell Motil*, **14**, 585–93 (1993).
11. JH Collins, JL Theibert, JM Francois, CC Ashley and JD Potter, *Biochemistry*, **30**, 702–7 (1991).
12. Y Xu, J He, X Wang, TM Lim and Z Gong, *Dev Dyn*, **219**, 201–15 (2000).
13. HJ Yuasa, JA Cox and T Takagi, *J Biochem (Tokyo)*, **123**, 1180–90 (1998).
14. T Takagi, T Petrova, M Comte, T Kuster, CW Heizmann and JA Cox, *Eur J Biochem*, **221**, 537–46 (1994).
15. H Terami, BD Williams, S Kitamura, Y Sakube, S Matsumoto, S Doi, T Obinata and H Kagawa, *J Cell Biol*, **146**, 193–202 (1999).
16. T Ojima, H Tanaka and K Nishita, *Arch Biochem Biophys*, **311**, 272–76 (1994).
17. PC Maisonpierre, KE Hastings and CP Emerson Jr, *Methods Enzymol*, **139**, 326–37 (1987).
18. C Fyrberg, H Parker, B Hutchison and E Fyrberg, *Biochem Genet*, **32**, 119–35 (1994).
19. JM Wilkinson, *FEBS Lett*, **70**, 254–56 (1976).
20. K Golosinska, JR Pearlstone, T Borgford, K Oikawa, CM Kay, MR Carpenter and LB Smillie, *J Biol Chem*, **266**, 15797–809 (1991).
21. JA Putkey, SL Carroll and AR Means, *Mol Cell Biol*, **7**, 1549–53 (1987).
22. N Toyota, Y Shimada and D Bader, *Circ Res*, **65**, 1241–46 (1989).
23. HJ Yuasa, S Sato, H Yamamoto and T Takagi, *J Biochem (Tokyo)*, **121**, 671–76 (1997).
24. T Takagi and K Konishi, *J Biochem (Tokyo)*, **94**, 1753–60 (1983).
25. L Garone, JL Theibert, A Miegel, Y Maeda, C Murphy and JH Collins, *Arch Biochem Biophys*, **291**, 89–91 (1991).
26. A Roher, N Lieska and W Spitz, *Muscle Nerve*, **9**, 73–77 (1986).
27. T Kobayashi, T Takagi, K Konishi, S Morimoto and I Ohtsuki, *J Biochem (Tokyo)*, **106**, 55–59 (1989).
28. JP van Eerd and K Takahshi, *Biochemistry*, **15**, 1171–80 (1976).
29. AE Romero-Herrera, O Castillo and H Lehmann, *J Mol Evol*, **8**, 251–70 (1976).
30. MS Parmacek, AR Bengur, AJ Vora and JM Leiden, *J Biol Chem*, **265**, 15970–76 (1990).
31. MS Parmacek and JM Leiden, *J Biol Chem*, **264**, 13217–25 (1989).
32. AS Zot, JD Potter and WL Strauss, *J Biol Chem*, **262**, 15418–21 (1987).
33. Q Chen, J Taljanidisz, S Sarkar, T Tao and J Gergely, *FEBS Lett*, **228**, 22–6 (1988).
34. JH Collins, ML Greaser, JD Potter and MJ Horn, *J Biol Chem*, **252**, 6456–62 (1977).
35. JM Wilkinson, *Eur J Biochem*, **103**, 179–88 (1980).
36. K Nishita, H Tanaka and T Ojima, *J Biol Chem*, **269**, 3464–68 (1994).
37. HJ Yuasa and T Takagi, *Gene*, **245**, 275–81 (2000).
38. HJ Yuasa and T Takagi, *Gene*, **268**, 17–22 (2001).
39. T Kobayashi, T Takagi, K Konishi and W Wnuk, *J Biol Chem*, **264**, 18247–59 (1989).
40. JM Francois, A Altintas and C Gerday, *Comp Biochem Physiol, B: Biochem Mol Biol*, **117**, 589–98 (1997).
41. JP van Eerd, JP Capony, C Ferraz and JF Pechere, *Eur J Biochem*, **91**, 231–42 (1978).
42. PA Lorkin and H Lehmann, *FEBS Lett*, **153**, 81–87 (1983).
43. T Kobayashi, O Kagami, T Takagi and K Konishi, *J Biochem (Tokyo)*, **105**, 823–28 (1989).
44. T Ojima, T Ohta and K Nishita, *Comp Biochem Physiol, B: Biochem Mol Biol*, **129**, 787–96 (2001).
45. CM Slupsky, CM Kay, FC Reinach, LB Smillie and BD Sykes, *Biochemistry*, **34**, 7365–75 (1995).
46. MX Li, M Chandra, JR Pearlstone, KI Racher, G Trigo-Gonzalez, T Borgford, CM Kay and LB Smillie, *Biochemistry*, **33**, 917–25 (1994).
47. SM Gagné, S Tsuda, MX Li, M Chandra, LB Smillie and BD Sykes, *Protein Sci*, **3**, 1961–74 (1994).
48. MX Li, SM Gagné, S Tsuda, CM Kay, LB Smillie and BD Sykes, *Biochemistry*, **34**, 8330–40 (1995).
49. A Houdusse, ML Love, R Dominguez, Z Grabarek and C Cohen, *Structure*, **5**, 1695–1711 (1997).
50. MX Li, SM Gagné, L Spyracopoulos, CP Kloks, G Audette, M Chandra, RJ Solaro, LB Smillie and BD Sykes, *Biochemistry*, **36**, 12519–25 (1997).
51. SK Sia, MX Li, L Spyracopoulos, SM Gagné, W Liu, JA Putkey and BD Sykes, *J Biol Chem*, **272**, 18216–21 (1997).
52. O Herzberg and MN James, *Nature*, **313**, 653–59 (1985).
53. M Sundaralingam, R Bergstrom, G Strasburg, ST Rao, P Roychowdhury, M Greaser and BC Wang, *Science*, **227**, 945–48 (1985).
54. O Herzberg and MN James, *J Mol Biol*, **203**, 761–79 (1988).
55. KA Satyshur, ST Rao, D Pyzalska, W Drendel, M Greaser and M Sundaralingam, *J Biol Chem*, **263**, 1628–47 (1988).

56. RH Kretsinger, *CRC Crit Rev Biochem*, **8**, 119–174 (1980).
57. SM Gagné, MX Li, RT McKay and BD Sykes, *Biochem Cell Biol*, **76**, 302–12 (1998).
58. LE Kay, JD Forman-Kay, WD McCubbin and CM Kay, *Biochemistry*, **30**, 4323–33 (1991).
59. JP van Eerd and K Takahashi, *Biochem Biophys Res Commun*, **64**, 122–27 (1975).
60. NC Strynadka, M Cherney, AR Sielecki, MX Li, LB Smillie and MN James, *J Mol Biol*, **273**, 238–55 (1997).
61. O Herzberg, J Moult and MN James, *J Biol Chem*, **261**, 2638–44 (1986).
62. KA Satyshur, D Pyzalska, M Greaser, ST Rao and M Sundaralingam, *Acta Crystallogr, Sect D: Biol Crystallogr*, **50**, 40–49 (1994).
63. ST Rao, KA Satyshur, ML Greaser and M Sundaralingam, *Acta Crystallogr, Sect D: Biol Crystallogr*, **52**, 916–22 (1996).
64. SM Gagné, S Tsuda, MX Li, LB Smillie and BD Sykes, *Nat Struct Biol*, **2**, 784–89 (1995).
65. S Tsuda, A Miura, SM Gagné, L Spyracopoulos and BD Sykes, *Biochemistry*, **38**, 5693–700 (1999).
66. WA Findlay, FD Sonnichsen and BD Sykes, *J Biol Chem*, **269**, 6773–78 (1994).
67. CM Slupsky and BD Sykes, *Biochemistry*, **34**, 15953–64 (1995).
68. J Soman, T Tao and GN Phillips Jr, *Proteins*, **37**, 510–11 (1999).
69. SM Gagné, MX Li and BD Sykes, *Biochemistry*, **36**, 4386–92 (1997).
70. GS Shaw, RS Hodges and BD Sykes, *Biochemistry*, **31**, 9572–80 (1992).
71. GS Shaw and BD Sykes, *Biochemistry*, **35**, 7429–38 (1996).
72. L Spyracopoulos, MX Li, SK Sia, SM Gagné, M Chandra, RJ Solaro and BD Sykes, *Biochemistry*, **36**, 12138–46 (1997).
73. GS Shaw, RS Hodges and BD Sykes, *Science*, **249**, 280–83 (1990).
74. AC Murray and CM Kay, *Biochemistry*, **11**, 2622–27 (1972).
75. JS Evans, BA Levine, PC Leavis, J Gergely, Z Grabarek and W Drabikowski, *Biochim Biophys Acta*, **623**, 10–20 (1980).
76. BA Levine, DM Coffman and JM Thornton, *J Mol Biol*, **115**, 743–60 (1977).
77. BA Levine, JM Thornton, R Fernandes, CM Kelly and D Mercola, *Biochim Biophys Acta*, **535**, 11–24 (1978).
78. KB Seamon, DJ Hartshorne and Bothner-By AA, *Biochemistry*, **16**, 4039–46 (1977).
79. S Tsuda, Y Hasegawa, M Yoshida, K Yagi and K Hikichi, *Biochemistry*, **27**, 4120–26 (1988).
80. S Tsuda, K Ogura, Y Hasegawa, K Yagi and K Hikichi, *Biochemistry*, **29**, 4951–58 (1990).
81. EB Carew, PC Leavis, HE Stanley and J Gergely, *Biophys J*, **30**, 351–58 (1980).
82. PC Leavis, SS Rosenfeld, J Gergely, Z Grabarek and W Drabikowski, *J Biol Chem*, **253**, 5452–59 (1978).
83. B Nagy and J Gergely, *J Biol Chem*, **254**, 12732–37 (1979).
84. JD Johnson and JD Potter, *J Biol Chem*, **253**, 3775–77 (1978).
85. MT Hincke, WD McCubbin and CM Kay, *Can J Biochem*, **56**, 384–95 (1978).
86. RT McKay, JR Pearlstone, DC Corson, SM Gagné, LB Smillie and BD Sykes, *Biochemistry*, **37**, 12419–30 (1998).
87. P Mercier, L Spyracopoulos and BD Sykes, *Biochemistry*, **40**, 10063–77 (2001).
88. DG Vassylyev, S Takeda, S Wakatsuki, K Maeda and Y Maeda, *Proc Natl Acad Sci USA*, **95**, 4847–52 (1998).
89. Y Li, ML Love, JA Putkey and C Cohen, *Proc Natl Acad Sci USA*, **97**, 5140–45 (2000).
90. GM Gasmi-Seabrook, JW Howarth, N Finley, E Abusamhadneh, V Gaponenko, RM Brito, RJ Solaro and PR Rosevear, *Biochemistry*, **38**, 8313–22 (1999).
91. X Wang, MX Li, L Spyracopoulos, N Beier, M Chandra, RJ Solaro and BD Sykes, *J Biol Chem*, **276**, 25456–66 (2001).
92. A Dvoretsky, EM Abusamhadneh, JW Howarth and PR Rosevear, *J Biol Chem*, **277**, 38565–70 (2002).
93. X Wang, MX Li and BD Sykes, *J Biol Chem*, **277**, 31124–33 (2002).
94. MX Li, L Spyracopoulos and BD Sykes, *Biochemistry*, **38**, 8289–98 (1999).

Gating domain of calcium-activated potassium channel with calcium and calmodulin

Maria A Schumacher

Department of Biochemistry and Molecular Biology, Oregon Health and Science University, Portland, OR, USA

FUNCTIONAL CLASS

Potassium channel (K+ channel); Small Conductance Ca^{2+}-activated K+ channel (SK channel). SK channels are coassembled complexes of the pore-forming SK subunit (termed the α-subunit) and the Ca^{2+}-binding protein calmodulin (CaM), the β-subunit. Three SK α-channel subunits have been cloned – SK1, SK2, and

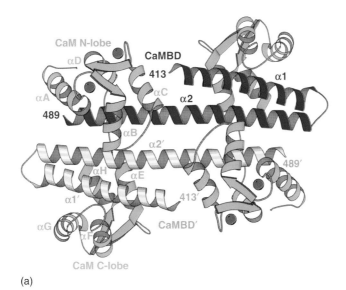

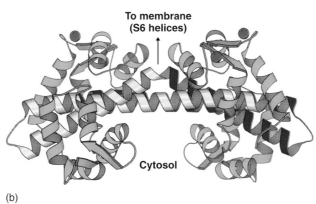

3D Structure The three-dimensional structure of the rSK2 CaMBD in complex with Ca^{2+}/CaM.[55] (a) Ribbon diagram of the Ca^{2+}/CaM/CaMBD dimeric complex. CaMBD subunits are represented as blue and yellow ribbons. CaM molecules are shown as green ribbons and the Ca^{2+} ions are shown as red spheres. Secondary structural elements and the first and last observed residues of the CaMBD molecules are labeled. (b) View in which (a) is rotated by 90° to show the orientation of the complex relative to the membrane. The arrow indicates the positions of the first observed CaMBD residues that are linked to the S6 pore α-helices. (Figure produced with the program MOLSCRIPT.[76] PDB code: 1G4Y). Reprinted with permission from M. A. Schumacher, A. F Rivard, H. P. Bachinger and J. P. Adelman, *Nature*, **410** 1120–24, Copyright 2001 *Nature* Publishing Group (http://www.nature.com/).

SK3.[1] CaM is constitutively associated with the SK α-pore-forming subunit.[2] SK channels are K$^+$-selective and voltage-independent, and are activated by increases in the intracellular levels of Ca^{2+} that occur during an action potential.[3–6] SK channels function to limit the firing frequency during a train of action potentials and are, therefore, required for normal neurotransmission.

Ca^{2+}-binding protein (CaBP): calmodulin (CaM). Highly conserved eukaryotic protein that functions as the primary regulator of intracellular Ca^{2+} signaling.

The following description involving occurrence and biological function refers to the SK channel complex, that is, the α-pore-forming SK channel and its constitutively bound β-subunit, CaM.

OCCURRENCE

SK channels

SK channel α-subunits and CaM have been identified only in eukaryotes. The three SK channel subtypes were first isolated from rat brain (rSK1, rSK2, and rSK3).[1] Subsequently, SK channels were also cloned in trout (TSK2 and TSK3), chick (cSK2), *Caenorhabditis elegans* (NSK), mouse (mSK1), and human (hSK1, hSK2, and hSK3).[7–12] The developmental distribution of trout SK channels has been well studied.[7] As ascertained by mRNA levels, both TSK2 and TSK3 are initially expressed in the trout brain upon hatching. In the mature trout, both TSK2 and TSK3 continue to be expressed widely in the brain, including the forebrain, tectum, brainstem, and cerebellum. Outside the nervous system, TSK2 and TSK3 are expressed in the muscle and the liver.[7]

Mammalian SK channels also appear to exhibit partially overlapping expression patterns.[1] In mice, Ro *et al.* found that mSK1 channels are expressed in brain and liver and, to a lesser extent, the heart, kidney, testes, and colon.[10] mSK2 channels are also expressed in brain, liver, and colon with less expression observed in kidney and testes. mSK3 channels, on the other hand, are ubiquitously expressed.[10] A detailed study of SK expression patterns in the mammalian brain was carried out by Stocker and Pedarzani, who found that both SK1 and SK2 are expressed in the neocortex, CA1–3 layers of the hippocampus, and the reticularis thalami.[13] SK3 channels are expressed in the supraoptic nucleus, and both SK2 and SK3 channels are expressed in the inferior olivary nucleus.[13] Interestingly, in rat, SK2 channels were found to be expressed in immature Purkinje cells,[14] which provided the first evidence that SK channels might play specific roles in neural development in higher eukaryotes.[14]

Calmodulin (CaM)

The genomes of all examined eukaryotes contain at least one CaM encoding gene. In fact, CaM is one of the most highly conserved eukaryotic proteins; all vertebrate CaMs are identical (see references 15 and 16 for reviews). In the lower metazoans, the *Paramecium* CaM is 94% similar to the mammalian CaM while the *Aspergillis* CaM is 92% similar[17]. All mammalian species have three CaM genes that are 80% identical in nucleotide sequence but, remarkably, encode proteins with 100% amino acid identity.[18] CaM is a small (17 kDa), acidic (pI = 4), and highly abundant protein. It accounts for 0.5% of all brain protein and is found at a concentration of 1 to 10 µM in typical eukaryotic cells. Within the cell, approximately half of the CaM is associated with membranes and the rest is distributed between the cytosol and the nucleus.[18,19] Consistent with this ubiquitous cellular distribution, CaM activates over 50 types of enzymes and channels. The list of target proteins that are activated by CaM will no doubt increase as studies on this key Ca^{2+}-binding protein continues. Given its high-sequence conservation and ubiquitous nature in eukaryotes, it is not surprising that CaM is essential. This was demonstrated in *Saccharomyces cerevisiae* in which a CaM knockout produced nonviable cells.[20,21] Similarly, *Drosophila* CaM null mutants die after the first larval stage, when the maternal CaM is no longer available.[22]

BIOLOGICAL FUNCTION

SK channels belong to a very large family of essential membrane proteins, the K$^+$ channels. The first K$^+$ channel to be cloned was the Drosophila *Shaker* channel, in 1987.[23] Since then, over 50 K$^+$ channel subunits have been cloned and it is predicted that there are over 100 unique K$^+$ channels in *Chaenorabditis elegans* alone.[24] Sequence comparisons of all K$^+$ channels reveal a single region of conservation among these proteins, the so-called P-loop region (for pore forming). This region consists of two transmembrane (TM) helices that are connected by a 'pore loop' (H-P-loop-H) in which the P-loop contains the functionally critical, conserved signature motif, GY(F)G. Before any structural information was available for a K$^+$ channel, it was demonstrated that four of these regions combine to form the K$^+$-selective pore in the functional tetrameric K$^+$ channel.[25] Structurally, K$^+$ channels can be grouped into three main families on the basis of the number of predicted TM helices: the 2TM, one-pore family, as exemplified by the bacterial KcsA protein; the 6TM, one-pore family; and the 4TM, two-pore family that includes inward rectifier K$^+$ channels. In the 2TM family, four subunits combine to form the pore, while within the 6TM family the last 2TM helices form the pore and in the 4TM family the H-P-loop-H motif is repeated two times and these proteins dimerize to form functional channels.[26]

K⁺ channels are responsible for the depolarization stage of the action potential. Indeed, fast electrical signaling in excitable cells is made possible by the slow homeostatic mechanisms that maintain a cellular environment with high [Na⁺] in the blood and extracellular space and high [K⁺] in the cytosol. Opening a Na⁺ channel creates a more positive intracellular voltage. The resulting flow of K⁺ ions down the concentration gradient out of a cell, via K⁺ channels, restores the negative voltage of the cell. The various types of K⁺ channels, which include voltage-activated, Ca^{2+}-activated and ligand-activated K⁺ channels, play distinct roles in shaping the repolarization kinetics of a given cell.

When the action potential decays in neurons, it leads to membrane repolarization and an increase in the intracellular concentration of Ca^{2+}. The result is a biphasic afterhyperpolarization (AHP) caused by Ca^{2+}-activated K⁺ channels. Ca^{2+}-activated K⁺ channels were first categorized on the basis of their characteristic single-channel conductances as 'big' (BK channels, 100–250 pS), 'intermediate' (IK channels, 20–50 pS), and 'small' (SK channels, 5–15 pS).[27] This classification was established before these channels were cloned. The fast phase of the AHP is due to BK channels, while the slow AHP or sAHP was assigned to SK channels. Prior to the cloning of SK channels, two types of sAHPs had been characterized that were distinguished by their pharmacological properties, second-messenger regulation, and kinetics.[3,28,29] Pharmacologically, the two sAHPs are characterized by their sensitivity to the bee-venom toxin, apamin (see below). The generation of these sAHPs, or long-lasting hyperpolarization, by SK channels in excitable cells limits burst frequency because during a train of action potentials the sAHP becomes longer lasting until the cell is unable to reach the action potential threshold. This adaptation, termed spike-frequency adaptation, is essential for normal neurotransmission as it protects against the harmful effects of continuous tetanic activity.[3,4,30–32]

Structure/function: K⁺ channels and shared gating region

The two fundamental features of K⁺ channels are K⁺-selective permeation and gating. K⁺ channels are at least 100 times more permeable to K⁺ ions than Na⁺ ions. The structural basis for this fundamental K⁺-selectivity feature was revealed in the structure of the *Streptomyces lividans* KcsA K⁺ channel.[33–35] KcsA is a 2TM K⁺ channel and the structure revealed that the four subunits, each with a TM1-P-loop-TM2 structure, combine to form the active tetramer. The structure resembles an inverted teepee in which each of the TM1 helices in the tetramer face the lipid on the outside and the four TM2 helices combine to create the lining of the inner pore. In the tetramer, each of the four highly conserved signature motifs form a selectivity filter in which the carbonyl oxygens of the conserved GY(F)G residues create a binding site that is exquisitely specific for dehydrated K⁺ ions.[33–35] On the basis of the sequence conservation of this K⁺ channel signature motif, a similar selectivity filter is predicted to exist in all K⁺ channels, suggesting a conserved K⁺-selective permeation mechanism for these channels.

In contrast to selective permeation, the diverse array of K⁺-channel subtypes are very distinctive in terms of what motivates them to open and close, that is, their gating mechanisms. Indeed, the diversity of gating cues, ranging from voltage to a variety of metabolic ligands, might reflect inherently different gating mechanisms. However, electron paramagnetic resonance(EPR) spectroscopy studies on the *S. lividans* KcsA channel and several studies on voltage and ligand-dependent K⁺ channels indicate that TM6 (TM2 in KcsA), which forms the channel pore, also functions as the gate.[36–45] These data suggest that although K⁺ channels respond to diverse metabolic signals to activate gating, all may gate by a conserved mechanism involving the pore-lining helix as the gate (TM2 in 2TM family and TM6 in the 6TM family).[36–45] This is further supported by the fact that most of the identified gating domains of K⁺ channels, which are intracellular, are attached immediately C-terminal to this pore-lining transmembrane helix. Therefore, signaling through these domains, via a conformational change, either chemo-mechanical or electro-mechanical, can be transmitted directly to the TM6 pore/gate to effect its structure.

SK channels: Ca²⁺-activated gating

Small conductance Ca^{2+}-activated K⁺ channels (SK) are voltage-independent channels that are gated solely by increases in the levels of intracellular Ca^{2+}, such as that occurs during an action potential. Interestingly, in many cell types expressing SK channels, L- or N-type Ca^{2+} channels have been shown to be found in proximity, indicating a direct functional coupling of the signal of increases in intracellular [Ca^{2+}] and gating by SK channels.[46,47] In addition to the three cloned SK channels, a related K⁺ channel with an intermediate conductance (IK) has been cloned recently and is more homologous to the SK channels than any other K⁺ channel. IK channels are also voltage-independent and gated solely by increases in intracellular Ca^{2+}.[48] The overall architecture of SK channels is consistent with the 6TM family of K⁺ channels, whereby the N- and C-termini reside in the cytosol. The primary sequences of these channels, however, place them in a distinct branch of the 6TM potassium channel family. Within the SK proteins themselves, there is marked sequence conservation with the exception of the extreme N- and C-terminal extensions, which vary not only in sequence but also in length and contain multiple potential phosphorylation sites for a variety of kinases. Heterologous expression studies demonstrated that all three

SK channel isoforms display the biophysical properties that are associated with the properties originally attributed to 'SK channels'.[1] All SK channels are gated by submicromolar concentrations of Ca^{2+} with half-maximal activation at 0.3 to $0.5 \times 10^{-6}\,M\,Ca^{2+}$. Further, macroscopic and single-channel recordings showed that these channels have unit conductances of 10 pS in symmetrical 120 mM potassium concentrations.[1]

The sequences of SK channels themselves have no Ca^{2+} binding motifs, and mutation of all the acidic residues within the C-terminal cytosolic portion of the channel had no effect on gating[2]. The basis for the Ca^{2+} activation of SK channels was eventually found to be mediated through constitutively associated CaM.[2] The CaM binding region was localized to a highly conserved cytosolic region of the protein just C-terminal to TM6, the pore-lining helix. This region was subsequently termed the 'Calmodulin binding domain' (CaMBD). CaM is an EF hand–containing CaBP, which functions as the main Ca^{2+} sensor in eukaryotic cells.[13,14,49] It contains four EF hands, two in its N-terminal lobe (N-lobe), EF1 and EF2, and two in its C-terminal lobe (C-lobe), EF3 and EF4. These four Ca^{2+} binding sites in CaM have a reasonably high affinity for Ca^{2+} ($K_a \sim 10^{-6}\,M^{-1}$), which is optimally tuned to respond to the signal generated by a transient increase in Ca^{2+} concentration such as that occurs during an action potential. Specifically, influx of Ca^{2+} through Ca^{2+} channels produces an increase in intracellular Ca^{2+} from the basal level of $10^{-7}\,M$ to $10^{-5}\,M$. Thus, only a small amount of CaM will be complexed with Ca^{2+} in the resting state of cells, while transient influx of Ca^{2+} will result in rapid CaM target activation. For example, this shift in intracellular Ca^{2+} concentration would result in the half activation of SK channels, which, as mentioned, occurs in the range of around $0.5 \times 10^{-6}\,M$ $[Ca^{2+}]$.

Ca^{2+} binding within each CaM lobe (i.e. EF1 and EF2 of the N-lobe and EF3 and EF4 of the C-lobe) is linked and likely cooperative. However, the complexities associated with measuring binding constants of a system containing four similar binding sites dispersed over two domains makes data interpretation difficult and thus the mechanism of Ca^{2+} binding by CaM remains a topic of considerable debate. Nonetheless, data support the notion that CaM most likely does bind Ca^{2+} in some cooperative manner. This cooperativity would be expected to be critical in Ca^{2+} signaling as it provides a tightly controlled all-or-nothing response to changes in Ca^{2+} concentration permitting a tight separation between activated and deactivated states.

Remarkably, in SK channels, it was demonstrated that the CaM N-lobe and C-lobes appear to mediate different functions of SK gating. Specifically, two-hybrid experiments revealed that the Ca^{2+}-independent, that is, constitutive, interaction of CaM with the SK α-subunit CaMBD requires only the C-lobe of CaM, while the N-lobe mediates the Ca^{2+}-dependent gating response.[48] Even more interesting, Ca^{2+} needs to bind only the EF hands within the CaM N-lobe to trigger gating. Indeed, the Ca^{2+}-coordinating residues within the CaM C-lobe can be mutated, destroying Ca^{2+} binding, without any effect on gating.[50] Thus, these data suggest that the different domains of CaM act independently and carry out distinct functions, a situation distinct from the well-studied CaM-activated kinases. This chapter describes the structural analysis of the SK2 CaMBD gating domain in complex with Ca^{2+}/CaM and what this structure reveals about the mechanism of SK channel gating.

AMINO ACID SEQUENCE INFORMATION

SK channels

- *Oncorhynchus mykiss* (Trout) SK2, 545 AA, translation of mRNA sequence[7]
- *Oncorhynchus mykiss* (Trout) SK3, 752 AA, translation of mRNA sequence[7]
- *Rattus norvegicus* (Rat) SK1, 536 AA, translation of mRNA sequence[1]
- *Rattus norvegicus* (Rat) SK2, 580 AA, translation of mRNA sequence[1]
- *Rattus norvegicus* (Rat) SK3, 732 AA, translation of mRNA sequence[1]
- *Mus musculus* (Mouse) SK1, 580 AA, translation of mRNA sequence[10]
- *Mus musculus* (Mouse) SK2, 574 AA, translation of mRNA sequence[10]
- *Mus musculus* (Mouse) SK3, 731 AA, translation of mRNA sequence[10]
- *Homo sapiens* (human) SK1, 561 AA, translation of mRNA sequence[8]
- *Homo sapiens* (human) SK2, 579 AA, translation of mRNA sequence[8]
- *Homo sapiens* (human) SK3, 736 AA, translation of mRNA sequence[8]
- *Gallus gallus* (chick) SK2, 553 AA, translation of mRNA sequence (unpublished)
- *Sus scrofa* (Pig) SK3, 724 AA, translation of mRNA sequence[11]
- *C. elegans* (worm) NSK, 471 AA, translation of mRNA sequence[12]

PROTEIN PRODUCTION, PURIFICATION, AND MOLECULAR CHARACTERIZATION

The rSK2 calmodulin binding domain (CaMBD)

For crystallization, a synthetic gene (in which the codons were optimized for protein expression in *Escherichia coli*) encoding the CaMBD of rat SK2 (rSK2) residues 395 to 490 was constructed.[50] A C-terminal hexa-histidine tag encoding sequence was added and the synthetic gene was subcloned into pET33b and expressed in *E. coli* BL-21(DE3). The sequence of the CaMBD (including the His-tag) used for expression is

MGRKELTKAEKHVHNFMMDTQLTKRVKNAAANVL RETWLIYKNTKLVKKIDHAKVRKHQRKFLQAIHQLR SVKMEQRKLNDQANTLVDLAKTQLEHHHHHH

The majority of the expressed protein is found in inclusion bodies, thus requiring resolubilization by sodium N-lauroylsarcosinate (Sarkosyl) prior to Ni^{2+}-NTA affinity purification. For solubilization, the pellet is dissolved in 20 mM Tris pH 8.0, 500 mM NaCl, and 1.5% sarkosyl. Following solubilization, the solution is dialyzed extensively to remove the Sarkosyl before loading onto the Ni^{2+}-NTA column. The protein is eluted by a gradient of 80 to 200 mM Imidazole. The pure fractions are dialyzed into 100 mM NaCl buffered at pH 7.0.

Calmodulin (CaM)

For structural studies, rat CaM, rCaM, was cloned into pET23b, expressed in BL-21(DE3), and purified on a low-substitution phenyl sepharose column. Specifically, pelleted cells are resuspended in 500 mM NaCl, 20 mM Tris pH 8.0, 10 mM $CaCl_2$ and lysed by French press. The lysate is loaded onto the low-substitution phenyl sepharose column equilibrated with the reconstitution buffer. As the Ca^{2+} loaded form of CaM has exposed hydrophobic domains, the protein is retained on the column and, following extensive washing, can be eluted with 500 mM NaCl, 20 mM Tris pH 8.0, 10 mM ethylene glycol tetraacetic acid (EGTA). The pure CaM protein is dialyzed into a 100 mM NaCl solution buffered at pH 7.0.

Complex formation

To form the rCaM/rSK2(CaMBD) complex, equimolar ratios of protein are combined by slowly adding the CaMBD to the CaM solution. The resulting complex is then dialyzed into the appropriate buffer. For crystallization of the Ca^{2+} bound form, that buffer is 10 mM Tris pH 7.5, 50 mM NaCl, and 10 mM $CaCl_2$.

METAL CONTENT AND COFACTORS

Functional SK channels are coassembled complexes consisting of the α-channel-forming SK subunit and a tightly and constitutively associated molecule of CaM. CaM is bound to the region of the SK channel immediately C-terminal to the pore-forming TM6 (S6) helix. Only 2 moles of Ca^{2+} per CaM/SK channel complex are required to bind for activity. Briefly, for channel opening to occur, two Ca^{2+} ions must bind the N-lobe of CaM; the Ca^{2+} coordinating residues of the C-lobes of CaM can be mutated without effect on gating.[50] The important role that CaM plays as both a subunit and a regulator of SK channel function is not unprecedented. In fact, a surprising array of ion channels have been demonstrated to use CaM as either constitutive or dissociable Ca^{2+}-sensing subunit. These channels are too numerous to list here, but include the voltage-gated Ca^{2+} channels, various non–voltage-regulated Ca^{2+} channels, ligand-gated channels, tryptophan family channels, and Ca^{2+} induced Ca^{2+} release channels found in specific organelles (reviewed in reference 51).

The utilization of CaM as a channel subunit was first established for the Na^+ channel.[52,53] In this study, Saimi *et al.* isolated a group of behavioral mutants in *Paramecium* called *pantophobiacs*, so named because they overreact even to such general stimuli as culture media.[52] Genetic analyses showed that *pantophobiac* is allelic to *fast-2*, which is a group of mutants with opposite behavior. Both behaviors were eventually linked to Ca^{2+}-regulated Na^+ channels. Subsequently, it was found that the *pantophobiac*s and *fast-2* mutants are localized to mutations in the CaM C-lobe and N-lobe respectively, both of which affect Na^+ channel function.[53] Those studies not only set the stage for future studies showing the utilization of CaM in channel function but also predicted an important hypothesis concerning CaM. This hypothesis, called 'functional bipartism' by Saimi and Kung,[51] refers to the ability of the CaM N- and C-lobes to act as entities capable of performing quite distinct functional roles.

SPECTROSCOPIC STUDIES

Two types of spectroscopic studies have been carried out on the CaM/CaMBD complex; fluorescence emission measurements,[50] which monitored the structural consequence of Ca^{+2} binding to the complex and nuclear magnetic resonance (NMR) studies,[54] which examined the structure of the rSK2 CaMBD in the absence of CaM. The CaMBD region contains a single tryptophan residue, W432, making it amenable to fluorescence emission studies. These studies revealed that in the absence of CaM there is an observed fluorescence emission maximum at 346 nm, indicating that W432 is in a solvent-exposed environment in its CaM-free state. This fluorescence maximum is shifted to 329 nm upon addition of apoCaM, indicating that the CaMBD undergoes a structural change in which W432 is moved into a hydrophobic environment. Subsequent addition of 100 µM Ca^{2+} resulted in a large decrease in the fluorescence, which was accompanied by a further shift in the emission maximum to 326 nm. These latter results were interpreted to indicate that W432 remains in a hydrophobic environment upon Ca^{2+} addition and that the CaMBD remains complexed with CaM under these conditions. Thus, the information drawn from fluorescence studies is that CaM binding to the CaMBD induces a conformational change in this domain of the SK channel and addition of Ca^{2+} to this complex further alters its conformation.[50] These studies also indicate that in the absence of CaM the CaMBD is likely mostly unfolded.

NMR studies of the same CaMBD fragment, later used for cocrystallization studies with Ca^{2+}/CaM, showed that, aside from a highly distorted helical region from residues 423–427, the CaMBD is completely disordered in the absence of CaM, consistent with the fluorescence studies. Interestingly, prior to these NMR studies, the crystal structure had revealed that residues 423–427 lie within a part of the CaMBD that mediates Ca^{2+}-independent interactions with CaM, that is, the region of the CaMBD that is presumably responsible for the constitutive interaction of the CaMBD with CaM (see below). These findings lead the authors to suggest that this short helical region might serve as a scaffold for the initial binding of apoCaM during complex formation.[54] Presumably, however, the CaMBD region of SK channels are immediately complexed with CaM upon channel folding and, therefore, it is possible that chaperones or other proteins may also be involved in this folding/complex formation. Thus, the *in vivo* mechanism of SK channel complex formation is yet unknown.

X-RAY STRUCTURE OF THE RAT SK2 CALMODULIN BINDING DOMAIN–CALCIUM–CALMODULIN COMPLEX

Crystallization and structure determination

Crystals of the purified Ca^{2+}/CaM/CaMBD complex were grown via the hanging drop vapor diffusion method by mixing a 1 mM protein complex 1:1 with the reservoir solution (1.25 M Li_2SO_4, 0.5 M $(NH_4)_2SO_4$, 0.1 M citrate pH 5.6) and equilibrated against 1 mL of the reservoir solution at room temperature. The crystals are monoclinic, space group C2, with $a = 77.6$ Å, $b = 66.3$ Å, $c = 64.6$ Å, $\beta = 93.9°$ and contain one CaMBD subunit, one CaM subunit, and two Ca^{2+} ions in the crystallographic asymmetric unit (ASU). The structure was originally solved by data collected at room temperature by multiple isomorphous replacement (MIR) using $PbCl_2$ and selenomethionine-substituted CaMBD as derivatives.[55] Difference Patterson maps revealed the presence of two lead sites per ASU, which were later found to have replaced the Ca^{2+} ions present only in the CaM N-lobe. The room temperature structure was refined to 2.40-Å resolution, and subsequently conditions were obtained for cryopreservation of the crystals by dipping them in the crystallization solution supplemented with 25% glycerol. Although causing no significant structural alterations, except small shifts in side chains, this cryopreservation enhances the diffraction quality of the crystals permitting a 1.60-Å resolution data set to be collected. The final structure was refined to an R_{work} of 22.8% and an R_{free} of 25.2% using all data to 1.60-Å resolution.

Overall description of the complex structure

The structure of the CaMBD from rat SK2 (residues 395–490) complexed to Ca^{2+}/CaM includes CaM residues 1 to 147, the nonhelical CaM linker region, two Ca^{2+} ions that are bound in the CaM N-lobe, and CaMBD residues 413 to 489 (3D Structure). No density was observed for CaMBD residues 395 to 412, which links the CaMBD to S6 (TM6), the pore-lining helix of the channel. CaM contains eight α-helices, αA–αH; αA: residues 5–20, αB: residues 28–39, αC: residues 44–55, αD: residues 66–74, αE: residues 81–91, αF: residues 102–110, αG: residues 117–129, αH: residues 138–147. The CaMBD consists of two long α-helices α1 (residue 413–440) and α2 (residues 446–489) that are connected by a loop (residues 441–445). CaM is a member of the Ca^{2+}-binding protein superfamily (CaBPs), the members of which contain Ca^{2+}-binding EF hand motifs that are characterized by a signature helix-loop-helix (HLH) motif. The term 'EF hand' was first coined in 1973 by Kretsinger to describe parvalbumin, the first CaBP structure that was solved.[56] The basic structural/functional unit of CaBPs are pairs of EF hands, and the EF hand elements are usually packed in a parallel manner, that is, face to face. This feature permits the formation of a stabile globular domain and provides a means of cooperativity between adjacent EF hands. CaM, which is monomeric, contains two EF hand–containing lobes, the N-lobe and the C-lobe. A linker region that is structurally and conformationally flexible connects the N- and C-lobes of CaM. Importantly, this region can adopt coil or helical conformations. In the Ca^{2+}/CaM/CaMBD complex, the linker is composed of residues 75 to 80.

The crystallographic ASU contains a single Ca^{2+}/CaM/CaMBD complex and the crystal symmetry creates a clear dimeric complex. In the CaMBD, this dimer is formed by the side-by-side antiparallel interaction of helices α2 and α2′, where the prime indicates the other subunit. This CaMBD α2/α2′ interaction buries ~ 400 Å^2 of the surface of each monomer. No other CaMBD:CaMBD contacts, even side-chain interactions, are observed in the crystal. Two CaM molecules are bound to the CaMBD dimer such that each CaM molecule grips an end of the dimer (3D Structure). In this interaction, each molecule of CaM contacts both subunits of the CaMBD dimer, forming a highly elongated complex with dimensions of ~ 80 Å $\times 54$ Å $\times 50$ Å. The surface area buried upon complex formation is striking. Indeed, CaM nearly completely engulfs the CaMBD dimer, burying over 80% of the latter's surface area. Consistent with this extensive interface, the Ca^{2+}/CaM/CaMBD complex is stabilized by an extensive number of contacts (>140 with distance ≤4.1 Å) (Figure 1). In addition to overall electrostatic complementarity between the positive CaMBD (pI = 10.5; charge = +14.7) and acidic CaM surfaces, CaM is anchored onto the CaMBD by several hydrogen bond 'tethers' primarily involving residues from the CaM linker and two CaM loops (Figures 1 and 2(a–c)). These interactions anchor the CaM onto the CaMBD; however, the specific contact surface between the two proteins is mediated by hydrophobic interactions.

The structure of the Ca^{2+}/CaM/CaMBD complex was unique compared to any of the previously described structures of CaM/peptide complexes (see below), as in this interaction CaM binds three α-helices instead of one, and the N-lobe and C-lobe of each CaM molecule contact different CaMBD monomers. Perhaps more remarkable was the finding that Ca^{2+} is not bound in the C-lobe EF hands, thus providing the first view of any peptide or protein bound to a calcium-free (decalcified) form of CaM (3D Structure). However, unlike the C-lobe, the N-lobe clearly contains two Ca^{2+} ions (is calcified). Thus, the two lobes effectively carry out independent functions; the C-lobe binds a portion of the CaMBD to form the constitutive interaction, which cannot be disrupted by the application of Ca^{2+}, CaM inhibitors, or mild denaturants,[50] while the N-lobe, which can bind Ca^{2+}, undergoes a conformational change in response to Ca^{2+} binding that must somehow drive gating. Importantly, these structural findings are consistent with data obtained from *in vivo* studies on full-length SK channels that demonstrate that SK channel gating depends only on Ca^{2+} binding to EF hands 1 and 2 in the CaM N-lobe. Thus, the Ca^{2+}/CaM/CaMBD structure details both Ca^{2+}-dependent and Ca^{2+}-independent CaM interactions in a single complex and provides the first visualization of CaM functional bipartism.

Ca^{2+} coordination: calcified and noncalcified CaM lobes, functional bipartition

Structures of Ca^{2+}/CaM complexes from several organisms have been determined.[59–63] In addition, there are numerous structures available for EF hand–containing CaBPs.[64–67] Therefore, the Ca^{2+} coordination of EF hand proteins such as CaM is extremely well described. EF hand proteins are characterized by a distinct sequence homology found within the Ca^{2+}-binding loops that are 12 residues in length. The consensus consists of a conserved pattern of Ca^{2+}-coordinating acidic residues as well as residues that play a key role in maintaining the unique stereochemistry of the loop, most notably a conserved glycine residue at 'position 6'.[64–66] With the exception of the glycine at position 6, the residues at positions 1, an aspartate, and 12, a glutamate, are the most conserved, and are involved in direct Ca^{2+} coordination. Of these residues, the glutamate at position 12 appears to play a central role in Ca^{2+} coordination and perhaps signaling as it provides the only bidentate interaction with Ca^{2+} (see below).

The most interesting feature of the Ca^{2+}/CaM/CaMBD complex structure in terms of Ca^{2+} binding, is the finding that only one lobe, the N-lobe, is calcified, while the C-lobe is Ca^{2+} free (Figure 3(a) and (b)). The Ca^{2+} coordination within each of the N-lobe EF hands in the Ca^{2+}/CaM/CaMBD structure is shown in Figure 3(a). As observed in other Ca^{2+}/CaM structures, the coordination is best described as a pentagonal bipyramidal Ca^{2+} ligation. In EF1, D20 (conserved position 1, in EF hand nomenclature), D22 (position 3), D24 (position 5), the carbonyl of T26 (position 7) and E31 (conserved position 12) coordinate the Ca^{2+} along with a single water molecule, W1 (Figure 3(a)). Similarly, in EF2, D56 (position 1), D58 (position 3), N60 (position 5), the carbonyl of T62 (position 7), E67 (position 12), and water, W2, coordinate the Ca^{2+}.

As revealed in electron-density maps calculated for the high-resolution Ca^{2+}/CaM/CaMBD structure, the CaM C-lobes in the structure are clearly Ca^{2+}-free (Figure 4). Instead, clear density in these lobes is observed for three water molecules that fill the expanded Ca^{2+} binding sites within EF3 and EF4 (Figure 4). Superimposition of the Cα-atoms of residues 81 to 146 of the CaMBD complexed CaM C-lobe onto those of the Ca^{2+}/CaM C-lobe[62] results in an rmsd of 2.80 Å, in contrast to the rmsd of 0.58 Å obtained from a similar overlay of the Cα-atoms of residues 10 to 67 of the corresponding N-lobes (Figure 5). Thus, as expected, the calcified N-lobe in the Ca^{2+}/CaM/CaMBD complex takes the Ca^{2+}-bound conformation, while the C-lobe is structurally distinct, perhaps retaining the Ca^{2+}-free or apo structure. Indeed, comparison of the structure of the C-lobe from the Ca^{2+}/CaM/CaMBD complex to those from NMR apoCaM structures[68–70] reveals the same overall structural trends entailing a reorientation of the helical pairs αG/αF with respect to αE/αH and the subsequent repacking of the hydrophobic core, which leads to the creation of a more compact lobe compared to the Ca^{2+}-bound form. Further, superimposition of the C-lobe Cα atoms of residues 81 to 146 of the CaMBD-bound C-lobe onto those of the NMR apo C-lobe structure[68–70] results in an rmsd of ~1.7 Å, confirming their strong similarity.

Consistent with the more apo-like conformation of the CaMBD-bound C-lobe, there is a severe disruption of the C-lobe EF hand Ca^{2+} coordination spheres that is underscored by the large increase in the Cα-to-Cα distances of Asn97 and Glu104 (EF3), and Asp133 and Glu140 (EF4), which are 11.3 Å and 11.4 Å respectively, in the Ca^{2+}-bound conformation compared to 13.6 Å and 14.3 Å in the CaMBD complex (Luzzati coordinate error for the Ca^{2+}/CaM/CaMBD complex = 0.23 Å). Therefore, the side

Gating domain of calcium-activated potassium channel with calcium and calmodulin

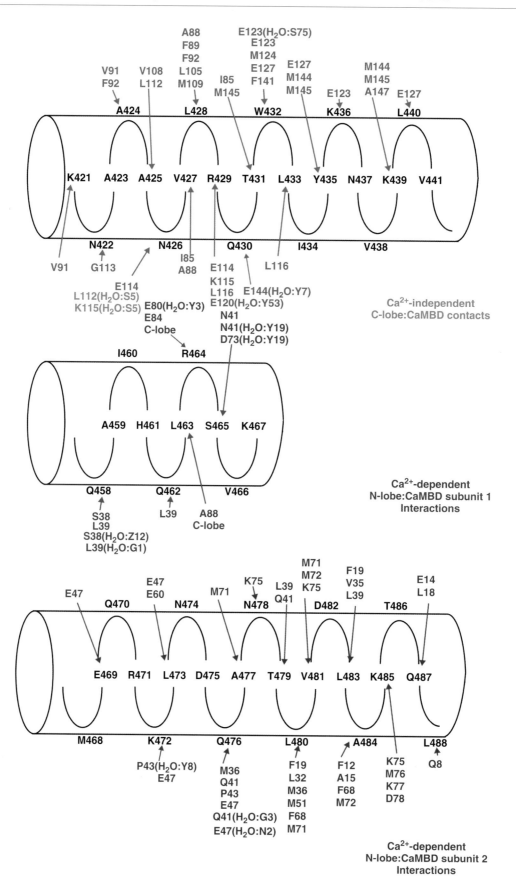

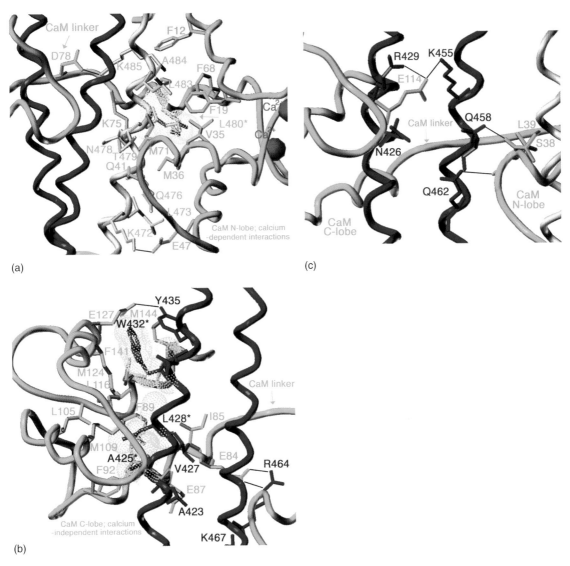

Figure 2 CaM/CaMBD interactions. (a) Ca^{2+}-dependent interactions between the CaMBD and the CaM N-lobe. CaMBD subunits are colored blue and yellow; CaM is colored green. Hydrogen bonding interactions are shown as black lines. The van der Waals surface (yellow) of the key hydrophobic prong, L480, is shown. (b) Ca^{2+}-independent interactions between the CaM C-lobe and the CaMBD. The van der Waals surface (cyan) is shown for the three important 'prongs' that interact with this lobe, A425, L428, and W432. (c) Tethering contacts between the CaMBD and CaM. Of note are the interactions between the CaM N-lobe and the second CaMBD dimer (blue), where the contacts to the first subunit are shown in (a). These figures and Figures 3(a) and (b), 5(a) and (b), 6(a), and 7 were produced with Swiss PDW Viewer[57] and rendered with POV-RAY.[58] Reprinted with permission from M. A. Schumacher, A. F Rivard, H. P. Bachinger and J. P. Adelman, *Nature*, **410** 1120–24, Copyright 2001 *Nature* Publishing Group (http://www.nature.com/).

Figure 1 Schematic diagram of the CaM/CaMBD interactions. Shown are all contacts ≤ 4.1 Å. The CaMBD is represented as helical cylinders and contacts from CaM, indicated by arrows, are shown in color. Water-mediated contacts are also represented with bridging waters indicated in parentheses by number. The top schematic shows the Ca^{2+}-independent C-lobe interactions to the CaMBD. CaM residues mediating these contacts, and the arrows that point from each CaM residue to the corresponding CaMBD residue, are colored green. The CaM N-lobe makes contacts with both CaMBD subunits in the dimer. These contacts are shown in the middle and bottom schematic where the Ca^{2+}-dependent N-lobe:CaMBD subunit 1 contacts are colored purple and the Ca^{2+}-dependent N-lobe:CaMBD subunit 2 contacts are colored red.

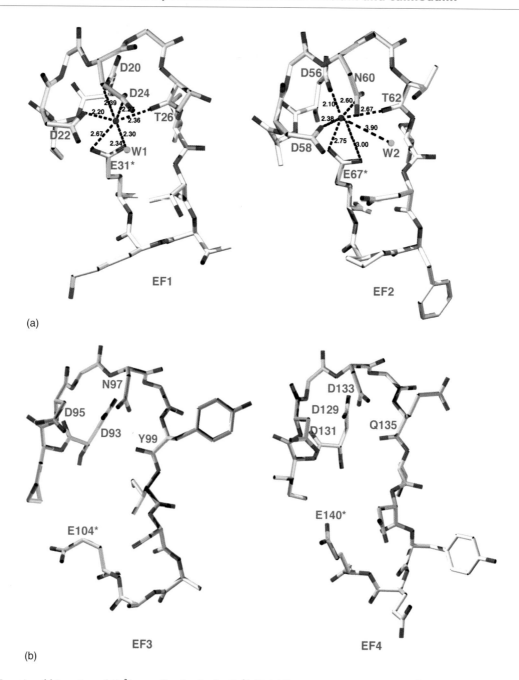

Figure 3 Functional bipartism of Ca^{2+} coordination in the Ca^{2+}/CaM/CaMBD complex. In the Ca^{2+}/CaM/CaMBD complex, only the N-lobes of CaM are calcified. (a) The Ca^{2+} coordination of the N-lobe EF hand, EF1 and EF2, loops, with coordinating distances shown. (b) The conformation of the C-lobe Ca^{2+}-coordinating residues of EF3 and EF4. Note the dramatic expansion of the Ca^{2+}-coordination loops, in which the key bifurcating residues, E104 and E140 (asterisks), are effectively removed from the Ca^{2+} coordination sphere. The atoms of each residue are represented as sticks with carbon, nitrogen, oxygen, and Ca^{2+} shown in white, blue, red, and dark orange respectively. The Ca^{2+}-coordinating water molecules, W1 and W2, in the N-lobe EF hands are colored light blue.

chains of Glu104 and Glu140 (at positions 12 in EF3 and EF4 respectively), which play key roles in Ca^{2+} binding as they provide bifurcated contacts to the ions, are effectively removed from their Ca^{2+}-binding positions (Figure 5). Indeed, this structure suggests that these residues likely play the pivotal role in the Ca^{2+}-induced CaM conformational changes, as, upon Ca^{2+} binding, they must move into the Ca^{2+} coordination sphere, a conformational change that would lever their attached helices inward. The reverse process would account for the helical reorientation that occurs upon Ca^{2+} release.

Ca^{2+}-dependent interactions: the gating trigger

In the Ca^{2+}-dependent CaMBD interaction, CaM N-lobe residues Phe12, Phe19, Val35, Met36, Leu39, Phe68,

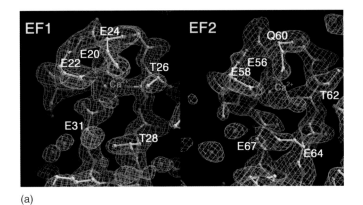

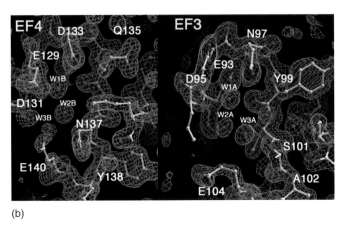

Figure 4 $2F_o-F_c$ composite omit maps (blue mesh) contoured at 1.5 σ around the EF hand region of the Ca^{2+}-coordinating loops of EF1 and EF2 (a), of the N-lobe and EF3 and EF4 (b) of the C-lobe. The model is superimposed and represented as sticks with carbon, nitrogen, oxygen, and Ca^{2+} shown in yellow, blue, red, and orange respectively. This figure was prepared using O.[71]

Met71, and Met72 create a hydrophobic patch for interaction with CaMBD residues Leu480, Leu483, and Ala484 (Figure 2(a)). This interaction is comparable to other CaM-Ca^{2+}-peptide structures[72–75] in that it contains a 'hydrophobic anchor point', namely the side chain of Leu480, which reaches into the N-lobe core. However, a feature of the N-lobe:CaMBD interaction not previously observed in other CaM/peptide structures is that each N-lobe interacts with two CaMBD subunits; in addition to the main N-lobe:CaMBD interaction involving the Leu480 prong, a short CaM N-lobe loop (residues Gly40 to Pro43) reaches around to engage residues from the other subunit of the CaMBD dimer (Figure 2(a) and (c)). This N-lobe loop along with a loop from the C-lobe composed of residues Asn111 to Leu116 encases the 'front' of the CaMBD dimer (3D Structure, Figure 2(c)). The side chain of Glu114, from the C-lobe loop, plays an important role in anchoring the CaM C-lobe onto the CaMBD (Figure 2(c)). In other Ca^{2+}/CaM/peptide structures in which CaM binds and encases a short peptide, these two loops are brought into proximity such that they interact with each other and not the target peptide. In contrast, CaM complexed to the CaMBD adopts an elongated conformation by the extension of its linker region. Notably, this extended structure is essential in enabling a single CaM to wrap around three CaMBD α-helices.

Ca^{2+}-independent interactions: the constitutive, tethering interaction

The CaMBD:C-lobe complex represents a novel type of CaM interaction. This results from the apo-like conformation adopted by the CaM C-lobe in the complex, which leads to the rearrangement of the helical pairs αG/αF compared to αE/αH and the subsequent repacking of the hydrophobic core. A striking feature of this repacking is the separation of the principal hydrophobic binding pocket found in the Ca^{2+}-bound form, normally used for binding residues of peptides or proteins, into several smaller and discrete pockets (Figure 6(a) and (b)). Thus, instead of the typical one-pronged interaction, the CaMBD complements these discrete pockets with the three hydrophobic side chains, Ala425, Leu428, and Trp432 (Figures 2(b) and 6(b)). Trp432 is located at the 'top' of the binding site and its side chain is stacked between the aliphatic portions of the side chains of residues Glu127, Met124, Phe141, and Met144. Unlike Trp432, Leu428 is completely buried and plugs a hydrophobic hole located at the center of

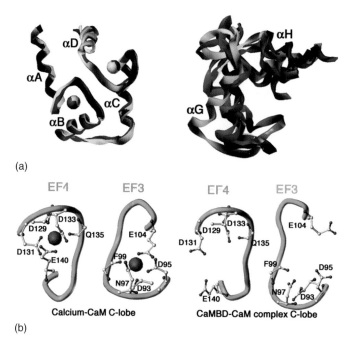

Figure 5 Structure of the two CaM lobes in the Ca^{2+}/CaM/CaMBD complex. (a) Shown in the left panel is the Cα superimposition of N-lobe atoms from residues 10 to 67 from the Ca^{2+}/CaM/CaMBD complex (blue) onto those of the Ca^{2+}/CaM complex (yellow). Ca^{2+} ions are illustrated as spheres. Shown on the right is the Cα superimposition of the C-lobe atoms of residues 81–146 of the Ca^{2+}/CaM/CaMBD complex (blue) and the NMR apoCaM C-lobe structure (magenta) onto those of the Ca^{2+}/CaM complex (yellow). (b) Comparison of the C-lobe EF hand regions of the Ca^{2+}/CaM/CaMBD complex with the Ca^{2+}/CaM complex. Ca^{2+}-coordinating residues are shown as ball and stick and colored white, red, and blue for carbon, oxygen, and nitrogen respectively. Note the large displacements of key Ca^{2+} coordinating residues D95, D131, and especially E104 and E140, out of the Ca^{2+}-coordinating sphere. Reprinted with permission from M. A. Schumacher, A. F Rivard, H. P. Bachinger and J. P. Adelman, Nature, **410** 1120–24, Copyright 2001 Nature Publishing Group (http://www.nature.com/).

the CaM C-lobe formed by Phe89, Phe92, Leu105, and Met109 (Figure 2(b)). Importantly, Phe89 is recruited from its buried position (in the Ca^{2+}-bound form of CaM[62]) into the CaMBD binding pocket by the repacking of the C-lobe core. In this reorganization, Phe89 switches stacking partners from Phe141 to Tyr138 (Figure 6(a)). The side chain of Ala425 is critical in binding to the third pocket as it fills the small cleft that is formed by atoms of the C-lobe backbone and the side chains of Val91 and Phe92. The large number of interactions and the relatively low average B values, 29.3 Å^2, for residues of the C-lobe compared with that of 35.4 Å^2 for all residues in the complex, indicate a very stabile CaMBD:C-lobe complex. Indeed, functional analyses[50] reveal that, in the absence of Ca^{2+}, the CaMBD and CaM C-lobe form an extremely strong interaction such that the conformation of the C-lobe, once bound to the CaMBD, cannot be switched to the Ca^{2+}-loaded configuration.

FUNCTIONAL ASPECTS

Channel inhibition/activation studies

Several lines of data suggest that SK channels play important roles in multiple physiological processes including excitability, the sleep–wake cycle, learning and memory, and digestive system activity.[78–82] Interestingly, a recent study demonstrated a potential link between SK3 channels and schizophrenia.[83] Thus, numerous channel blockers have been identified and used not only as tools in exploring various experimental SK channel models but have been implicated for use as potential chemotherapeutics. As mentioned, one hallmark feature of SK channels is their sensitivities to the 18-residue bee-venom toxin apamin. Pharmacologically, two types of sAHPs, attributable to SK channels can be distinguished. The first type of sAHP is sensitive to apamin, is maximally activated following an action potential, and decays over several hundred milliseconds.[28–30,84] The second type of sAHP is apamin-insensitive, displays an increasing phase after an action potential, and decays over several seconds. These two types of sAHPs are frequently found in the same cell.

The pharmacological profiles observed for sAHPs are reflected by the three cloned SK isoforms. Specifically, SK1 channels are not blocked by apamin, while SK2 and SK3 channels are sensitive to the toxin with IC$_{50}$ values of 60 pM and 1 nM, respectively.[1,84] In situ hybridization studies in rat brain revealed that the distribution of SK2 and SK3 channels correspond to the distribution of radiolabeled apamin binding sites, in which sAHP

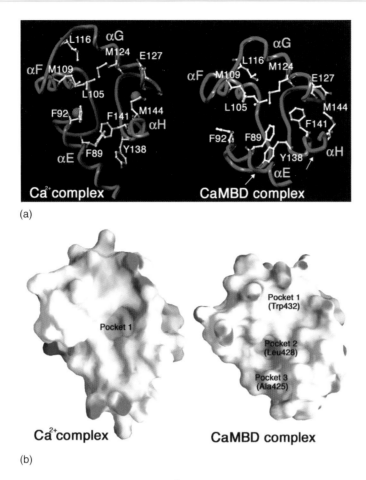

Figure 6 Repacking of the C-lobe hydrophobic core in the Ca^{2+}/CaM/CaMBD complex. (a) Comparison of the C-lobe hydrophobic core of the Ca^{2+}/CaM complex (left) and the Ca^{2+}/CaM/CaMBD complex (right). Using αG as a reference (similarly placed in both structures), note the large reorganization of the lobe in the Ca^{2+}/CaM/CaMBD, which leads to the creation of three smaller hydrophobic pockets. Upon this repacking, F89, which is buried in the Ca^{2+}/CaM structure, is translocated into the core for target protein binding. (b) Surface representation of the structures in (a), contoured for electrostatic potential (red, negative; blue, positive) to show the surface of the pockets. Figure 6(b) was made using GRASP.[77] Reprinted with permission from M. A. Schumacher, A. F Rivard, H. P. Bachinger and J. P. Adelman, *Nature*, **410** 1120–24, Copyright 2001 *Nature* Publishing Group (http://www.nature.com/).

sensitivity to apamin has been noted. Moreover, SK1 channels are expressed in regions where apamin-insensitive sAHPs have been documented. Apamin blocks channel function by binding not to the cytosolic gating domain but to an extracellular region of the pore. Interestingly, the residues that comprise this region of the pore are highly conserved in all SK channels. In fact, there are only three residues, corresponding to K328, E330, and H357 in the apamin-insensitive SK1 isoform that are not conserved. These residues are Q, D, and N and Q, D and H in the apamin-sensitive channels, SK2 and SK3, respectively. Mutational analyses revealed that substitution of SK1 residues E330 and H357 to the corresponding residues in the SK2 isoform conferred apamin sensitivity in an additive manner with the double SK1 E330D/H357N mutant displaying apamin sensitivity nearly identical to that of SK2 channels.[84] Substitution of SK1 residue K328 to Q, however, had no effect indicating that only two SK residues, corresponding to SK1 residues 330 and 357, are responsible for apamin sensitivity. SK3 channels contain only one of these two residues, that is, the D but not the N, thus explaining the intermediate apamin sensitivity observed for these channels.[84] Interestingly, coexpression of SK1 and SK2 isotypes leads to the formation of channels with intermediate sensitivity, suggesting that SK channels can heterooligomerize.[84]

In addition to apamin, which produces a long-lasting channel block, several nonpeptidic SK channel blockers including (+) tubocurarine, bicuculline salts, and gallamine have been identified. Like apamin, these blocking agents lack specificity and affect other targets such as the $GABA_A$, nicotinic, or muscarinic receptors.[85–89] More recently, methyl-laudanosine, which has an IC_{50} value of 4 mM for SK channels, has been demonstrated to be specific for SK channels and, unlike apamin, is reversible.[90] Thus, these advantages present methyl-laudanosine as a potential tool in examining SK function and, potentially, in the treatment of central nervous disorders.

Several SK channel blockers have been identified; however, few activators have been characterized molecularly. One compound, 1-ethyl-2-benzimidazolinone (EBIO), was recently shown to activate SK channels by shifting the Ca^{2+} concentration response relationship such that Ca^{2+} concentrations in the nanomolar range are sufficient to elicit robust activity.[91] As a result, SK channel deactivation is slowed by over 10-fold by EBIO application. EBIO does not function by activating channel opening, rather it stabilizes the open state by an unknown mechanism.[91] Because the open state is formed when Ca^{2+} binds CaM and induces a conformational change in the CaMBD, it has been postulated that EBIO functions to stabilize the interaction between the SK channel CaMBD and CaM in the Ca^{2+} complexed form. However, direct binding of EBIO to the CaMBD has not yet been demonstrated.

Neurotransmitter regulation and kinetics of SK channel subtypes

In addition to distinctive sensitivities to apamin and other toxins, SK channel subtypes also display distinct second-messenger regulation and kinetics. Specifically, unlike apamin-sensitive channels, apamin-insensitive channels are regulated by neurotransmitter-induced second messengers. Noradrenaline, dopamine, serotonin, histamine, glutamate, acetylcholine, and several neuropeptides have been shown specifically to suppress apamin-insensitive SK channels.[92–94] This modulation of the apamin-insensitive sAHP is a key mechanism involved in controlling the excitatory functional state of the brain by setting the overall level of forebrain neuron excitation. It has been suggested that such regulation may be important in regulating the sleep–wake cycle, arousal, attention, and several cognitive functions.[32]

The apamin-sensitive channels (SK2 and SK3) also differ from apamin-insensitive SK1 channels in their kinetics; apamin-sensitive channels have distinctly faster kinetics than apamin-insensitive channels.[93,95] In well-studied cells such as hippocampal neurons, the apamin-sensitive sAHP is maximal after an action potential and decays with a half-time on the order of hundreds of milliseconds. By contrast, apamin-insensitive AHPs, best characterized in hippocampal pyramidal neurons, exhibit a rising phase and subsequently decay over the time course of several seconds. This difference might reflect different gating kinetics of these channels. However, steady state Ca^{2+} gating is essentially identical for the cloned SK2, SK3, and SK1 channels, which all gate very rapidly with activation-time constants <10 ms. Thus, it is thought that the vastly different kinetics displayed by these channels must be indicative of different rates of Ca^{2+} exposure and, therefore, distinctive subcellular localizations of SK channel subtypes.

Comparison with other structures

The vast number of important cellular targets regulated by interactions with CaM spurred early attempts to identify consensus CaM binding motifs in these proteins.[96] However, it was found that the CaM binding regions of several of the initially well-characterized CaM target proteins such as kinases and phosphatases, which bind in a Ca^{2+}-dependent manner, show weak sequence homology. Nonetheless, most of these Ca^{2+}-dependent CaM binding sites share specific characteristics such as an overall net positive charge, moderate hydrophobicity, and a propensity to form a helical structure. Comparison of several such sites reveal that they can be roughly placed into two groups – one group consists of the so-called 1-8-14 motif and the second the 1-5-10 motif, where the numbers represent the positions of hydrophobic residues.[96]

Subsequent to these findings, the structures of several Ca^{2+}/CaM/target peptide complexes were solved including CaM in complex with fragments from myosin light chain kinase (MLCK), CaM kinase II-α (CaMKIIα), and CaM kinase kinase.[72–75] These structures define the canonical Ca^{2+}/CaM–peptide interaction as defined by the 1-5-10 and 1-8-14 motifs and reveal a similar bent conformation of CaM, which, in these cases, binds a single peptide α-helix, encasing the α-helix between its two calcified lobes (Figure 7). These structures demonstrated that the hydrophobic residues at the numbered positions of the peptides, especially the end positions, form key anchor contact points with CaM. These structures also unveiled the remarkable flexibility inherent in the CaM linker region, which allows it to wrap its N- and C-lobes around different target proteins in a highly adaptable way (Figure 7).

In addition to these Ca^{2+}-dependent CaM binding motifs, a well-characterized Ca^{2+}-independent CaM binding motif found in proteins such as neuromodulin and neurogranin is the so-called IQ motif, which has the consensus IQXXXRGXXXR.[96] However, there are no structures yet available for a CaM/IQ peptide complex and in fact, prior to the structure determination of the Ca^{2+}/CaM/CaMBD complex, there were no structures of any complex of noncalcified CaM bound to a peptide or protein. Furthermore, the Ca^{2+}/CaM/CaMBD represents the first structure of CaM bound to a protein domain. In this case, CaM does not bind an IQ motif or any of the Ca^{2+}-dependent 1-8-14 and 1-5-10 motifs. Instead CaM binds to noncontiguous regions of the SK2 CaMBD such that each CaM molecule encases three CaMBD α-helices, again, made possible by CaM linker region flexibility. Following the structure determination of the Ca^{2+}/CaM/CaMBD complex, the structure of the edema factor, a CaM-activated adenylyl cyclase from *Bacillus anthracis*, was resolved in complex with CaM, thus revealing the first CaM/protein complex.[97] Interestingly, enzymatic activation of edema factor requires that Ca^{2+} binds only to the CaM C-lobe and in the structure only the

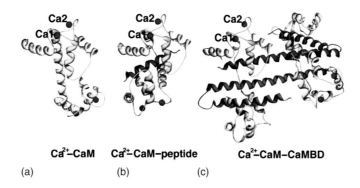

Figure 7 Comparison of CaM structures demonstrating CaM flexibility. (a) The structure of Ca^{2+}/CaM[62] (PDB code, 1CLL). (b) Ca^{2+}/CaM complexed with a peptide from smooth muscle myosin light chain kinase[74] (PDB code, 1CDL). This structure is representative of the canonical Ca^{2+}/CaM/peptide interaction. (c) The Ca^{2+}/CaM/CaMBD complex,[55] highlighting the difference between this complex and the canonical Ca^{2+}/CaM/peptide interaction. In the CaM/CaMBD complex, CaM wraps around three α-helices, two from one subunit and one from the other, thus stabilizing the CaMBD dimer. For reference, the CaM N-lobes (with Ca^{2+} ions colored red and labeled Ca1 and Ca2) are similarly placed in all three structures. Note, the dramatically distinct conformations of the CaM linkers and distinct conformations of the CaM C-lobes in the three structures.

C-lobe is calcified, providing yet another example of the functional bipartism of the CaM lobes.

Proposed gating mechanism for SK channels

Although it has become established that gating in K$^+$ channels is mediated through the TM6 helices either by voltage, as sensed by the TM4 voltage sensor, or via a ligand binding to a gating domain, the precise molecular details of the physical gate remain obscure. For example, EPR studies on the KcsA channel indicate that the physical gate is located just C-terminal to the TM6 helices.[34] Similarly, for the Shaker Kv channel, the substituted cysteine accessibility method (SCAM) and cross-linking experiments suggest that the TM6 bundle crossing near the TM6 C-terminus also forms the physical gate. Interestingly, the recent high-resolution structures of KcsA in the presence and absence of K$^+$ indicate that structural changes in the selectivity filter, are coupled to the large gating rearrangements in the activation gate.[32,33] These latter studies are particularly relevant as studies on the cyclic nucleotide gated (CNG) channels suggest that the selectivity filter itself may form the physical gate and that the signal of ligand binding is somehow transferred to the selectivity filter.[98]

The structure of the Ca^{2+}/CaM/CaMBD complex was a surprise as it revealed a dimer rather than a tetramer, as found for K$^+$ channel pore domains. It should be pointed out that it is not uncommon for a multidomain protein to have distinct oligomerization states within each of its domains. A particularly relevant example is the tetrameric LacI repressor, which contains two separate dimeric DNA binding domains attached via a four helix bundle tetramerization domain.[99] To characterize in solution the oligomeric state of the CaM/CaMBD complex further biophysically, dynamic light scattering (DLS) and sedimentation equilibrium studies were performed. Interestingly, DLS studies on the CaM/CaMBD complex (molecular weight of monomer complex = 29 kDa) indicate that it is monomeric in the absence of Ca^{2+} (30 kDa) and dimeric (61 kDa) in the presence of Ca^{2+}.[55] This was confirmed by equilibrium sedimentation analyses (29 kDa and 54 kDa in the absence and presence of Ca^{2+} respectively).[55] Because the intracellular CaMBD of SK channels resides immediately adjacent to TM6 such that structural changes within or between CaMBD molecules could be directly transmitted to elicit gating, these combined data suggest a possible gating mechanism in which Ca^{2+} binding to each CaM N-lobe exposes a hydrophobic patch and thus allows it to then interact with an adjacent CaMBD monomer. As each N-lobe on adjacent monomers engage the C-terminal region of the other CaMBD, a rotary force would be created between them and transmitted to the attached TM6 pore helices in the gate region. In this model, which was termed the 'chemo-mechanical model',[55] two CaMBD dimers would serve as mechanical levers to drive the opening of the physical gate of the channel (Figure 8).

A critical question remains – where is the physical gate of SK channels located? To address this question SCAM studies on TM6 were carried out to elucidate those residues in SK channels that comprise the physical gate.[100] These studies suggest that the physical gate of SK channels is not formed by residues within the cytosolic end of TM6, but by residues in or near the selectivity filter.[100] Thus, the location of the physical gate of SK channels appears to be more comparable to CNG channels than to KcsA and voltage-gated K$^+$ channels. While the structure of the KcsA channel was solved in its closed state, the recent structure of a second bacterial 2TM K$^+$ channel, the *Methanobacterium thermoautotrophicum* MthK channel, was determined in its open state.[101,102] In addition to the pore, this structure included a proposed gating domain. Although previously uncharacterized, the authors demonstrated that this channel can be gated by

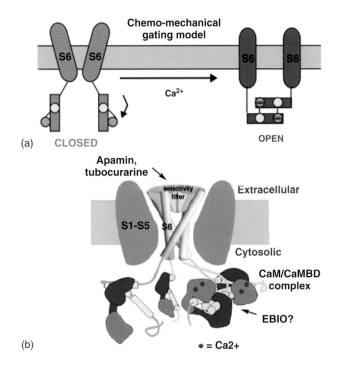

Figure 8 Gating mechanism of SK channels. (a) Cartoon showing the proposed chemo-mechanical gating model derived from the Ca^{2+}/CaM/CaMBD complex structure and biophysical studies. Shown is a cutaway view of two of the four subunits of the channel (the other two would be in the foreground). Only the S6 helices of the SK channel, the CaMBD, the linker from the CaMBD to S6 and CaM are shown for clarity. The CaM C- and N-lobes are colored yellow and orange respectively. Crucial to this model are biophysical studies revealing that the CaM/CaMBD complex is monomeric in the absence of Ca^{2+} and dimeric in the presence of Ca^{2+}. In this model, a rotational force (exaggerated in the picture) would be created with the formation of the dimeric complex, when the CaM N-lobe grabs an adjacent CaMBD molecule on a nearby α-subunit. This force would ultimately drive a rotation between the attached S6 helices to open the gate. (b) Cartoon model of the SK channel structure showing the location of the CaMBD relative to the pore S6 (TM6) helices. The two CaMBD subunits on the left represent the Ca^{2+}-free, monomeric form, while the two subunits on the right indicate how binding of Ca^{2+} ions (red spheres) to CaM results in CaMBD dimerization. Also shown are the binding sites for SK pore inhibitors, as represented by apamin and tubocurarine, and the SK activator, EBIO, which is postulated to bind to the CaMBD region and stabilize its Ca^{2+}-dependent interaction with CaM. Reprinted with permission from M. A. Schumacher, A. F Rivard, H. P. Bachinger and J. P. Adelman, Nature, 410 1120–24, G. Yellen, Nature, 419, 35–42 Copyright 2001 Nature Publishing Group (http://www.nature.com/).

increases in the $[Ca^{2+}]$, which bind the regulator of K^+ conductance (RCK) gating domains, just C-terminal to the TM6 helices. The structure revealed that the C-terminal ends of the TM6 helices are rotated wide open. Interestingly, the pivot point or the location of the physical gate was found to coincide with a glycine residue that is found within the membrane embedded part of TM6, close to the selectivity filter. The finding is important as sequence comparisons of the pore-lining residues of all K^+ channels reveal that a glycine at this position is almost absolutely conserved, suggesting that the physical gate may be conserved among K^+ channels. This finding may also explain discrepancies in the studies in which the physical gate of K^+ channels appear to reside either at the TM6 C-terminal cytosolic region or the selectivity filter as the glycine is located between these regions. Clearly, the elucidation of the exact location of the physical gate and the specific gating mechanism of SK channels, and indeed, other K^+ channels, will require further structural studies.

REFERENCES

1. M Köhler, B Hirschberg, CT Bond, JM Kinzie, NV Marrion, J Maylie and JP Adelman, Science, **273**, 1709–14 (1998).
2. X-M Xia, B Fakler, A Rivard, G Wayman, T Johnson-Pais, JE Keen, T Ishii, B Hirschberg, CT Bond, S Lutsenko, J Maylie and JP Adelman, Nature, **395**, 503–7 (1998).
3. AL Blatz and KL Magelby, Nature, **323**, 718–20 (1996).
4. P Sah, TINS, **4**, 150–4 (1996).
5. P Sah and P Davies, Clin Exp Pharmacol Physiol, **27**, 657–63 (2000).
6. P Pedarzani, J Moshbacher, AF Rivard, LA Cingolani, D Oliver, M Stocker, JP Adelman and B Fakler, J Biol Chem, **276**, 9762–69 (2001).
7. F Panofen, T Piwowarski and G Jeserich, Mol Brain Res, **101**, 1–11 (2002).
8. R Desai, A Peretz, H Idelson, P Lazarovici and B Attali, J Biol Chem, **275**, 39954–63 (2000).
9. JM Vernier, Ann Embryol Morphol, **2**, 495–520 (1969).
10. S Ro, WJ Hatton, SD Koh and B Horowitz, Am J Physiol Gastrointest Liver Physiol, **281**, G964–73 (2001).
11. MP Burnham, R Bychkov, M Feletou, GR Richards, PM Vanhoutte, AH Weston and G Edwards, Br J Pharmacol, **135**, 1133–43 (2002).
12. The C. elegans Sequencing Consortium, Science, **282**, 2012–18 (1998).
13. M Stocker and P Pedarzani, Mol Cell Neurosci, **15**, 476–93 (2000).
14. LA Cingolani, M Gymnopoulos, A Boccaccio, M Stocker and P Pedarzani, J Neurosci, **11**, 4456–67 (2002).
15. P James, T Vorherr and E Carafoli, Trends Biochem, **20**, 38–42 (1995).
16. D Chin and AR Means, Trends Cell Biol, **10**, 322–28 (2000).
17. C Kung, RR Preston, ME Maley, KY Ling, JA Kanabrocki, BR Seavey and Y Saimi, Cell Calcium, **13**, 413–25 (1992).
18. SL Toutenhoofd and EE Strehler, Cell Calcium, **28**, 83–96 (2000).
19. L Santella and E Carafoli, FASEB J, **11**, 1091–1109 (1997).
20. TN Davies and J Thorner, Proc Natl Acad Sci USA, **86**, 7909–13 (1989).
21. JR Geiser, D van Tuinen, SE Brockerhoff, MM Neff and TN Davies, Cell, **65**, 949–59 (1991).
22. RG Heiman, RC Atkinson, BF Andruss, C Bolduc, GE Kovalick and K Beckingham, Proc Natl Acad Sci USA, **93**, 2420–25 (1996).

23. DM Papazian, TL Schwarz, BL Tempel, YN Jan and LY Jan, *Science*, **237**, 749–53 (1987).
24. CC Shieh, M Coghlan, JP Sullivan and M Gopalakrishnan, *Pharmacol Rev*, **52**, 227–94 (2000).
25. RL Nakamura, JA Anderson and RF Gaber, *J Biol Chem*, **272**, 1011–18 (1997).
26. F Lesage and M Lazdunski, *J Physiol Renal Physiol*, **279**, F793–F801 (2000).
27. P Sah and EM McLachlan, *Neuron*, **7**, 257–64 (1991).
28. DG Lang and AK Ritchie, *J Physiol*, **425**, 117–32 (1990).
29. TM Ishii, J Maylie and JP Adelman, *J Biol Chem*, **272**, 23195–200 (1997).
30. DV Madison and RA Nicholl, *J Physiol*, **354**, 319–31 (1984).
31. B Lancaster and PR Adams, *J Neurophys*, **55**, 1268–82 (1986).
32. B Hille, *Ionic Channels of Excitable Membranes*, Sinauer Associates, Sunderland, MA (1992).
33. DA Doyle, JH Morais-Cabral, RA Pfuetzner, A Kuo, JM Gulbis, SL Cohen, BT Chait and R MacKinnon, *Science*, **280**, 69–77 (1998).
34. Y Zhou, JH Morais-Cabral, A Kaufman and R MacKinnon, *Nature*, **414**, 43–48 (2001).
35. JH Morais-Cabral, Y Zhou and R MacKinnon, *Science*, **280**, 37–42 (1998).
36. E Perozo, DM Cortes and LG Cuello, *Science*, **285**, 73–78 (1999).
37. ER Liman, P Hess, F Weaver and G Koren, *Nature*, **535**, 752–56 (1991).
38. E Loots and EY Isacoff, *J Gen Physiol*, **112**, 377–89 (1998).
39. M Holmgren, KS Shin and G Yellen, *Neuron*, **21**, 617–21 (1998).
40. G Yellen, *Q Rev Biophys*, **31**, 239–303 (1998).
41. KS Glauner, LM Mannussu, CS Gandhi and EY Isacoff, *Nature*, **402**, 813–17 (1999).
42. T Lu, B Nguyen, X Zhang and J Yang, *Neuron*, **22**, 571–80 (1999).
43. E Loots and EY Isacoff, *J Gen Physiol*, **116**, 623–36 (2000).
44. CS Gandhi, E Loots and EY Isacoff, *Neuron*, **27**, 673–84 (2000).
45. D del Camino, M Holmgren, Y Liu and G Yellen, *Nature*, **403**, 321–25 (2000).
46. RP Robitaille, ML Garcia, GJ Kaczorowski and MP Charlton, *Neuron*, **11**, 645–55 (1993).
47. NV Marrion and SJ Tavalin, *Nature*, **395**, 900–5 (1998).
48. M Hay and DL Kunze, *Neurosci Lett*, **167**, 179–82 (1994).
49. MR Nelson and WJ Chazin, in LJ van Eldik and DM Watterson (eds.), *Calmodulin and Signal Transduction*, Academic Press, New York, pp 17–58 (1998).
50. JE Keen, R Khawaled, DL Farrens, T Neelands, A Rivard, CT Bond, A Janowsky, B Fakler and JP Adelman, *J Neurosci*, **19**, 8830–38 (1999).
51. Y Saimi and C Kung, *Annu Rev Physiol*, **64**, 289–311 (2002).
52. Y Saimi, RD Hinrichsen, M Forte and C Kung, *Proc Natl Acad Sci USA*, **80**, 5112–16 (1983).
53. RD Hinrichsen, A Burgess-Cassler, BC Soltvedt, T Hennessey and C Kung, *Science*, **232**, 503–6 (1986).
54. R Wissmann, W Bildl, H Neumann, AF Rivard, N Klocker, D Weitz, U Schulte, JP Adelman and B Fakler, *J Biol Chem*, **277**, 4558–64 (2002).
55. MA Schumacher, AF Rivard, HP Bachinger and JP Adelman, *Nature*, **410**, 1120–24 (2001).
56. RH Kretsinger and CE Nockolds, *J Biol Chem*, **248**, 3313–26 (1973).
57. N Guex and MC Peitsch, *Electrophoresis*, **18**, 2714–23 (1997).
58. POV-Ray, Persistence of Vision Raytracer, Version 3.1 (www.povray.org).
59. YS Babu, JS Sack, TJ Greenhough, CE Bugg, AR Means and WJ Cook, *Nature*, **315**, 37–40 (1995).
60. C Ban, B Ramakrishnan, KY Ling, C Kung and M Sundaralingam, *Acta Crystallogr, Sect D*, **50**, 50–63 (1994).
61. G Barbato, M Ikura, LE Kay, RW Pastor and A Bax, *Biochemistry*, **31**, 5269–78 (1992).
62. R Chattopadhyaya, WE Meador, AR Means and FA Quiocho, *J Mol Biol*, **228**, 1177–92 (1992).
63. MA Wilson and AT Brunger, *J Mol Biol*, **301**, 1237–56 (2000).
64. NCJ Strynadka and MNG James, *Annu Rev Biochem*, **58**, 951–98 (1989).
65. H Kawasaki and RH Kretsinger, *Protein Profile*, **2**, 297–490 (1995).
66. M Ikura, *Trends Biochem Sci*, **21**, 14–17 (1996).
67. KL Yap, JB Ames, MB Swindells and M Ikura, *Proteins*, **37**, 499–507 (1999).
68. M Zhang, T Tanaka and M Ikura, *Nat Struct Biol*, **2**, 758–67 (1995).
69. H Kuboniwa, N Tjandra, S Grzesiek, H Ren, CB Klee and A Bax, *Nat Struct Biol*, **2**, 768–76 (1995).
70. BE Finn, J Evenas, T Drakenberg, JP Waltho, E Thulin and S Forsen, *Nat Struct Biol*, **2**, 777–83 (1995).
71. TA Jones, J-Y Zou, SW Cowan and M Kjeldgaard, *Acta Crystallogr*, **A47**, 110–19 (1991).
72. M Ikura, GM Clore, AM Gronenborn, G Shu, CB Klee and A Bax, *Science*, **256**, 632–38 (1992).
73. M Ikura, G Barbato, CB Klee and A Bax, *Cell Calcium*, **13**, 391–400 (1992).
74. WE Meador, AR Means and FA Quiocho, *Science*, **257**, 1251–55 (1992).
75. M Osawa, H Tokumitsu, MB Swindells, H Kurihara, M Orita, T Shibanuma, T Furuya and M Ikura, *Nat Struct Biol*, **6**, 819–24 (1999).
76. PJ Kraulis, *J Appl Crystallogr*, **24**, 946–50 (1991).
77. A Nicholls, K Sharp and BH Honig, *Proteins: Struct, Funct, Genet*, **11**, 281–96 (1991).
78. TJ McCown and GR Breese, *Eur J Pharmacol*, **187**, 49–58 (1990).
79. G Gandolfo, H Schweitz, M Landunski and C Gottesmann, *Brain Res*, **736**, 344–47 (1996).
80. C Messier, C Mourre, B Bontempi, J Sif, M Lazdunski and C Destrade, *Brain Res*, **551**, 322–26 (1991).
81. CC Heurteauz, C Messier, C Destrade and M Landunski, *Mol Brain Res*, **3**, 17–22 (1993).
82. M Hugues, H Schmid, G Romey, D Duval, C Frelin and M Lazdunski, *EMBO J*, **9**, 1039–42 (1982).
83. H Tomita, VG Shakkottai, GA Gutman, GA Sun, WE Bunney, MD Cahalan, KG Chandy and JJ Gargus, *Mol Psychiatry* **8**, 524–535 (2003).
84. TM Ishii, J Maylie and JP Adelman, *J Biol Chem*, **272**, 23195–200 (1997).
85. NH Lee and EE el-Fakahany, *J Pharmacol Exp Ther*, **256**, 2762–69 (1991).
86. PM Dunn, DC Benton, J Rosa Campos, CR Ganellin and DH Jenkinson, *Br J Pharmacol*, **117**, 35–42 (1996).

87 SW Seutin and SW Johnson, *Trends Pharmacol Sci*, **105**, 268–70 (1999).
88 J Scuvee-Moreau, J-F Liegeois, L Massotte and V Seutin, *J Pharmacol Exp Ther*, **302**, 1176–83 (2002).
89 D Strobaek, TD Jorgensen, P Christophersen, PK Ahring and S Olensen, *Br J Pharmacol*, **129**, 991–99 (2002).
90 J Scuvee-Moreau, J-F Liegeois, L Massotte and V Seutin, *J Pharmacol Exp Ther*, **302**, 1176–83 (2002).
91 P Pedarzani, J Mosbacher, A Rivard, LA Cingolani, D Oliver, M Stocker, JP Adelman and B Fakler, *J Biol Chem*, **276**, 9762–69 (2001).
92 RA Nicholl, *Science*, **241**, 545–51 (1988).
93 N Gorelova and PB Reiner, *J Neurophysiol*, **75**, 695–706 (1996).
94 JF Storm, *Prog Br Res*, **83**, 161–87 (1990).
95 SH Chandler, C-F Hsaio, T Ingue and LJ Goldberg, *J Neurophysiol*, **71**, 129–45 (1994).
96 AR Rhoads and F Friedberg, *FASEB J*, **11**, 331–39 (1997).
97 CL Drum, S-Z Yan, J Bard, Y-Q Shen, D Lu, S Soelalman, Z Grabarek, A Bohm and W-J Tang, *Nature*, **415**, 396–401 (2002).
98 GF Flynn and WN Zagotta, *Neuron*, **30**, 689–98 (2001).
99 M Lewis, G Chang, NC Horton, MA Kercher, MA Schumacher, RG Brennan and P Lu, *Science*, **266**, 763–70 (1994).
100 A Bruening-Wright, MA Schumacher, JP Adelman and J Maylie, *J Neurosci*, **22**, 6499–506 (2002).
101 Y Jiang, A Lee, J Chen, M Cadene, BT Chait and R MacKinnon, *Nature*, **417**, 515–522 (2002).
102 Y Jiang, A Lee, J Chen, M Cadene, BT Chait and R MacKinnon, *Nature*, **417**, 523–26 (2002).

Calpain

Peter L Davies, Robert L Campbell and Tudor Moldoveanu

Department of Biochemistry, Queen's University, Kingston, Ontario, Canada

FUNCTIONAL CLASS

Enzyme; Ca^{2+}-dependent neutral cysteine protease; EC 3.4.22.17; a papain-like protease linked to other domains.

m-Calpain[1] (also known as calpain 2) catalyzes the limited proteolysis of numerous substrates inside the cell in response to Ca^{2+} signaling. It also undergoes extensive Ca^{2+}-dependent autoproteolysis. m-Calpain is a heterodimer of a large (80 kDa) catalytic subunit bound to a small (28 kDa) regulatory subunit. It is one of 14 paralogs in the human genome that have a papain-like protease core[2,3] in common. (http://ag.arizona.edu/calpains/)

OCCURRENCE

Diverse members of the calpain family of proteins are found in organisms ranging from mammals to bacteria.[4]

The heterodimeric m-calpain and its closely related isoform μ-calpain (calpain 1) are recognizable in mammals where they are found in virtually all cell types. Most of the other mammalian calpains are considered to be tissue-specific, and many of these are monomeric without an obvious requirement for the regulatory subunit that is present in m- and μ-calpains. Calpains are intracellular and nonlysosomal, but may have transient Ca^{2+}-dependent associations with membranes.[5,6] Some isoforms show evidence of nuclear localization.

BIOLOGICAL FUNCTION

The physiological role of the ubiquitous calpains is thought to be the catalysis of limited, discrete proteolysis in response to highly localized, transient elevations in Ca^{2+} ion concentrations.[7,8] Among the many

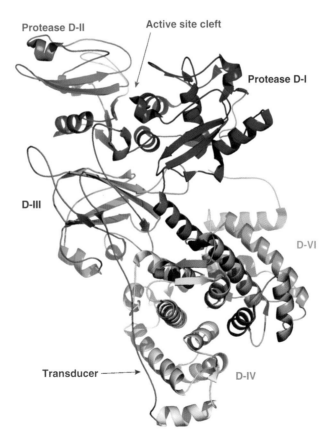

3D Structure Schematic representation of the heterodimeric human m-calpain in the absence of Ca^{2+}. Individual domains are color-coded and labeled. The anchor helix (residues 1–19 of the large subunit) is colored red. Domain V is not visible. PDB code: 1KFU. This figure and Figures 1 through 7 were prepared with the program PyMOL.[62]

well-documented target proteins are spectrin (fodrin), talin, ADP-ribosyltransferase, tau, and focal adhesion kinase.[9-11] These proteolytic events are those required for normal cellular processes like enzyme activation, receptor modulation, cytoskeleton remodeling, cell proliferation, and differentiation.[12-16] The Ca^{2+} concentrations needed for *in vitro* half-maximal activation of μ- and m-calpain (5–50 and 250–500 μM, respectively) are approximately four orders of magnitude greater than the resting intracellular Ca^{2+} concentrations (nM). Nevertheless, it is generally acknowledged that Ca^{2+} concentrations in microenvironments close to Ca^{2+} channels connecting to the extracellular medium or internal Ca^{2+} stores are high enough to begin the activation process. (Note that the transport of 100 Ca^{2+} ions into 1% of the volume of a mitochondrion ($\sim 0.01\,\mu^3$) would elevate the Ca^{2+} concentration by 10 to 20 μM.) In cytoplasm, where protein occupies most of the volume, the effective Ca^{2+} concentration increase resulting from such an influx could be several fold higher. These transient 'sparks', 'puffs', or 'waves' of Ca^{2+} ions are rapidly dissipated by diffusion and active transport out of the cytoplasm causing calpain activity to cease.

AMINO ACID SEQUENCE INFORMATION

- *Homo sapiens* (human), large subunit calpain 1 has 714 amino acid residues (AA), accession number (#): X04366.[17]
- Human large subunit calpain 2 has 700 AA, #: NM_001748.[18]
- Human small subunit calpain 4 has 275 AA, #: NM_001749.[19]
- *Rattus norvegicus* (rat), large subunit calpain 1 has 713 AA, #: U53858.[20]
- Rat large subunit calpain 2 has 700 AA, #: L09120.[21]
- Rat small subunit calpain 4 has 275 AA, #: U10861.[22]

NOMENCLATURE

Calpain 1 (CAPN1) is a recent designation for μ-calpain, which is a heterodimer of the 80-kDa large subunit (product of the *capn1* gene) and the 28-kDa small subunit (product of the *capn4* gene) Calpain 2 (CAPN2) is a recent designation for m-calpain, which is a heterodimer of a different 80-kDa large subunit (product of the *capn2* gene) and the same 28-kDa small subunit (product of the *capn4* gene).

PROTEIN PRODUCTION, PURIFICATION, AND MOLECULAR CHARACTERIZATION

The heterodimeric calpains 1 and 2 have been isolated from a variety of mammalian tissues (skeletal muscle, erythrocytes, platelets, kidney, placenta) in adequate quantities for characterization.[23] Some tissues are richer in one isoform than the other. Yields are of the order of milligrams from a kilogram of tissue. Extractions are typically done in the presence of Ca^{2+} chelating agents to prevent autoproteolysis, and reducing agents to prevent oxidation of cysteines, including the active site cysteine. A conventional purification sequence starts with anion exchange chromatography (DEAE-cellulose), which can separate μ-calpain from the more tightly bound m-calpain. An additional complication is the presence of the natural inhibitor, calpastatin.[24] Although this requires Ca^{2+} to bind to the calpains, it must be separated from them in the early steps of the purification scheme because it can mask calpain activity during the enzyme assay. Further purification typically involves hydrophobic, size exclusion, and additional ion exchange chromatography steps. Calpain 2 can be concentrated by ammonium sulfate precipitation.

The calpains and their subunits and the domains for which structures have been obtained have all been produced as recombinant proteins in *E. coli* or animal cells. In *E. coli*, the two subunits of rat m-calpain have been expressed either from compatible plasmids maintained under the selection of two different antibiotics, or from a bi-cistronic expression vector.[25] The ability to overproduce a single isoform in the absence of the other (and of calpastatin) is a big advantage. Production of recombinant calpains has enabled a His-tag to be introduced at the C-terminal end of the large subunit to facilitate purification. Also, the mutation of the active site cysteine to serine eliminates the danger of autoproteolysis during purification and when attempting crystallization in the presence of Ca^{2+}.

Proteolysis seems to be the main barrier to homogeneity. During purification from natural sources, there is often some loss of the anchor peptide at the N-terminal end of the large subunit and cleavage of domain V at the N-terminal end of the small subunit. This might result from autoproteolysis during purification as well as some prior activation in the cell. The unstructured glycine-rich domain V is susceptible to proteolysis in *E. coli* even when calpain activity is eliminated by the active site mutation (C105S in rat m-calpain). For this reason, domain V is typically omitted from the small subunit construct.

METAL CONTENT AND COFACTORS

Direct measurements of Ca^{2+} binding to the whole enzyme (calpain 1) can be complicated by Ca^{2+}-induced conformational changes that lead to precipitation and subunit dissociation.[26] However, Michetti *et al.* were able to use equilibrium dialysis to detect the binding of 8 Ca^{2+} to calpain 2 from human erythrocytes.[27] Several of these sites can be accounted for by Ca^{2+} binding to the EF-hand domains as detected by filter assays with $^{45}Ca^{2+}$.[28] Ca^{2+} binding to the penta-EF-hand (PEF) domain VI, which self-associates as a homodimer,[29] has been defined by X-ray crystallography.[30,31] Of the five EF-hands in domain VI, the fifth is used for dimerization, the fourth binds Ca^{2+} with difficulty, being occupied at >50 mM Ca^{2+} but not at 1 mM, and the other three sites are occupied at 1 mM Ca^{2+}. By inference, the same is true for the homologous PEF domain IV, and this has been substantiated by EF-hand mutagenesis studies.[32] Ca^{2+} also binds to the protease core that consists of domains I + II[33] and to domain III.[34] Two non–EF-hand Ca^{2+} sites are occupied in the core at >0.1 mM, one each in domains I and II.[33] The number and location of Ca^{2+} ions bound to the C_2-like domain III has not been defined, but the highly acidic loop (residues 392–402 of m-calpain) is one possible binding site. If at least one Ca^{2+} ion is bound there, this would add up to a total of 11 Ca^{2+} binding sites per molecule of calpain. Since neither of the fourth EF-hand sites are likely to be occupied under activation conditions, the total number of bound Ca^{2+} could be very close to the estimate of Michetti *et al.*[27]

ACTIVITY TEST

A commonly used fixed-point assay for calpain is the Ca^{2+}-dependent release of acid-soluble peptides from a denatured substrate like casein.[35] Denaturation of the substrate facilitates cleavage of peptide bonds, which might otherwise be protected by secondary and tertiary structures. Fluorescently labeling the casein and measuring release of peptides by fluorescence increases the sensitivity of the assay.[36] A semiquantitative assay is gel zymography in which casein is uniformly incorporated into a native polyacrylamide gel during polymerization.[37,38] After electrophoresis of samples containing calpain, the gel is incubated in the presence of Ca^{2+} and enzyme activity is revealed after Coomassie blue staining as a clear zone. One advantage of this method is that the main isoforms, μ- and m- are well separated during electrophoresis. Real-time assays have been developed around the hydrolysis of fluorogenic peptide substrates[39] like Succinyl-Leu-Tyr-MCA.[40] The poor dissociation constants ($K_m \sim 0.2$ mM), inner filter effect, and low solubility can limit their range of linearity and render absolute kinetic constants unreliable. However, the assay is particularly useful for comparing activities of mutants. A number of *in vivo* methods have been developed for assaying calpain activity in cells. One example is the use of antibodies to detect a calpain-specific cleavage site generated in the cytoskeletal protein spectrin.[41] There are many variations on these themes.

SPECTROSCOPY

The binding of Ca^{2+} to calpain causes substantial alterations in intrinsic tryptophan fluorescence that reflects conformational changes associated with enzyme activation. Intrinsic tryptophan fluorescence has been especially informative for studying activation in the protease core in which W298 (in μ-calpain) undergoes a radical change in environment during realignment of the active site cleft.[33]

X-RAY STRUCTURE OF HUMAN APO-m-CALPAIN

Crystallization of recombinant protein

A full-length version of human m-calpain (including domain V) has been crystallized in the Ca^{2+}-free state.[42] This enzyme was made in insect cells using a baculovirus expression system.[43] For crystallization, the protein was concentrated to 14 mg mL^{-1} in 10 mM Tris-HCl (pH 7.5), 50 mM NaCl, 1 mM ethylenediaminetetraacetic acid (EDTA), and 1 mM dithiothreitol (DTT). Under initial screening conditions, small crystals were sometimes obtained in 0.1 M HEPES/NaOH (pH 7.5), 20% polyethyleneglycol (PEG) 10 000. Larger crystals were prepared by macro-seeding in 0.1 M HEPES/NaOH (pH 7.7), 15% PEG 10 000, 2.2% 2-propanol mixed in a 3:1 ratio by volume with 1 M guanidinium chloride. They were of two monoclinic forms in space group $P2_1$. Form I crystals were rhombic platelets with maximum dimensions of $1.0 \times 1.0 \times 0.1$ mm and unit-cell parameters $a = 65$ Å, $b = 133$ Å, $c = 76$ Å. Form II crystals had platelet-like to prism-like morphology with maximum dimensions of $1.0 \times 0.2 \times 0.1$ mm and unit-cell parameters $a = 52$ Å, $b = 170$ Å, $c = 64$ Å. PDB ID codes for crystal forms II and I are 1KFU (update of 1DKV) and 1KFX, respectively.

Overall description of the structure

Both the N- and C-termini of the large subunit make contact with domain VI of the small subunit (3D Structure). Thus, the domains of the heterodimer are curved around to form an oval disk of dimensions $100 \times 60 \times 50$ Å. Domain VI is a penta-EF-hand domain (PEF), the founding member of this family.[44] The C-terminal contact to the large subunit is through the fifth EF-hand pairing to the fifth EF-hand of the homologous PEF domain in the large subunit, domain IV. This heterodimeric PEF structure is very similar to that observed for the homodimer of domain VI.[30,31] Domain VI's other contact to the large subunit is to the latter's N-terminal 19-residue α-helix (now referred to as an anchor peptide[34]) that binds to a cavity in the small subunit. One of the key contacts appears to be a salt bridge between Lys7 in the anchor peptide and Asp154 in domain IV.[45,46] The anchor peptide connects to the papain-like protease core that is made up of domains I and II (Figure 1). (Note that Strobl *et al.*[47] refer to the protease core as a bilobular structure composed of subdomains IIa and IIb and the anchor peptide as domain I). Of the two protease core domains, domain I has diverged the most from its papain equivalent. Although its hydrophobic

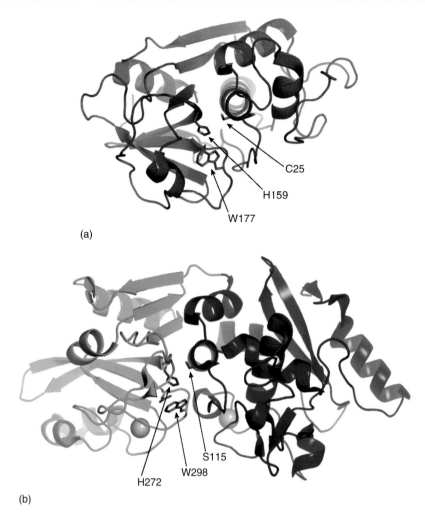

Figure 1 (a) Papain structure with key residues of the active site marked by arrows. PDB code: 9PAP. Yellow indicates SH of C25. (b) Similar orientation of μ-calpain protease core. Blue indicates domain I, cyan specifies domain II and gold spheres represent Ca^{2+}. Red indicates OH of the inactivating S115 mutation.

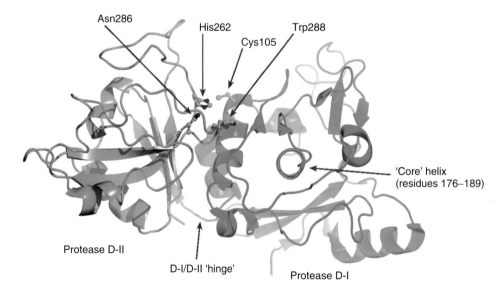

Figure 2 The protease core (domains I and II) of human m-calpain oriented to display the core helix of domain I and the hinge between domains I and II.[48] Side chains of the principal active site residues are displayed and labeled.

core containing the central α-helix (residues 176–189) is papain-like, it contains an additional 80-residue, compact N-terminal region. The α-helix containing the active site cysteine (C105) at one end is shorter in calpain than the equivalent helix in other members of the papain family. The junction between the core protease domains (D-I/D-II hinge) is at the G209–G210 sequence (Figure 2). In domain II, the N-terminal helix and the two antiparallel β-sheets overlap well with those of papain, although there are slight differences in topology as well as an extra helix in the C-terminal region (Figure 1).

Following the protease core is a C_2-like domain (III) characterized by a sandwich of two β-sheets, each of which has four antiparallel β-strands (Figure 3). This domain has a very acidic loop ($E_{392}EEDEDDEDGE_{402}$) that makes two salt linkages to basic residues in domain II. C_2-like domains have been characterized in phospholipase C and protein kinase C, where they are associated with Ca^{2+}-mediated binding to phospholipids.[49] The domain III of calpain has been likened to the first C_2 domain of synaptotagmin,[48,50,51] a protein involved in membrane trafficking. The connection between domain III and the C-terminal PEF domain is via an extended 15-residue linker that runs the length of domain IV and forms one short β-sheet (residues 516–518) with three residues on the surface of that domain.

X-RAY STRUCTURE OF RAT APO-m-CALPAIN

Crystallization of recombinant protein

Crystals of Ca^{2+}-free rat m-calpain made in *E. coli* were initially obtained from a 10 mg mL^{-1} protein solution in 0.1 M MES (2[N-morpholino]ethanesulfonic acid) (pH 6.5), 0.05 M NaCl, 10–12% PEG 6000 by the hanging drop method.[52,53] This version of calpain has a His-tag at the C-terminus of the large subunit, lacks the glycine-rich domain V of the small subunit, and has the active site cysteine replaced by serine (C105S) as a precaution against autolysis during crystallization. The enzyme was concentrated in 0.05 M Tris-HCl (pH 7.6), 2 mM EDTA and 0.01% sodium azide and was used fresh or stored at 203 K after being flash-frozen in liquid N_2. The crystals grew in the triclinic space group P1, but were typically too small for data collection. A few larger crystals (0.2 × 0.2 × 0.2 mm) diffracted to 2.6–3.0 Å resolution with unit-cell parameters $a = 58$ Å, $b = 80$ Å, $c = 81$ Å and a solvent content of ~60%. The only reducing agent present under these conditions was residual 2-mercaptoethanol from the enzyme preparation. Subsequently, a second crystal form, $P2_1$, was obtained after streamlining the enzyme preparation, including reducing agents and 0.01% (w/v) sodium azide in all buffers, and adding 10 mM DTT in the enzyme storage buffer and drop solution. The $P2_1$ crystals, which grew to

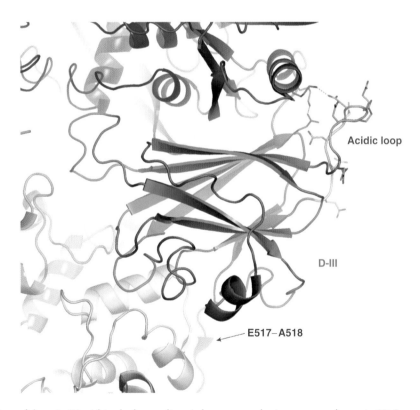

Figure 3 Close-up view of domain III within the heterodimeric human m-calpain structure shown in 3D Structure. The acidic loop that forms salt bridges to domain II is colored red. Also shown is the β-sheet interaction between the transducer peptide and domain IV.

a size of $1.0 \times 0.3 \times 0.3$ mm, gave better diffraction data and had unit-cell parameters $a = 52$ Å, $b = 158$ Å, $c = 65$ Å and a solvent content of 45%.
PDB ID code 1DFO.

Comparison of the rat and human m-calpain structures

Using the form II crystals, which have similar symmetry parameters as the rat $P2_1$ crystals, the human m-calpain structure was solved at 2.8-Å resolution and was refined to a more complete model than the rat structure. As a result, some loops (particularly residues T245 to K260 and Q290 to N321 in domain II), for which electron density could not be traced in the rat structure, are seen in the human enzyme (PDB ID code 1KFU). Interestingly, the form I crystals of human m-calpain gave a structure (PDB ID code 1KFX) that had electron density missing in the same six regions as the rat structure.[45] Although the glycine-rich domain V is present in the human m-calpain structure, most of it (up to Thr85) cannot be seen. In other details, the human and rat calpain structures are essentially the same.

Misalignment of the active site residues in the apo-m-calpains

Aside from defining the domain structure and redrawing domain boundaries, the structure of calpain explained why the apoenzyme is inactive. In essence, the two core domains are held too far apart and are slightly rotated from their juxtaposition in calpain. The active site Cys105 in domain I is 10.5 Å away from His262 in domain II rather than 3.7 Å apart as seen for equivalent residues in papain. Moreover, Trp288 protrudes into the active site cleft like a wedge and is not positioned to shield the hydrogen bond between His262 and Asn286 from the solvent (Figures 2 and 4).

FUNCTIONAL DERIVATIVES

X-ray structure of rat Ca^{2+}-bound μ-calpain protease core

m-Calpain undergoes substantial conformational changes on binding Ca^{2+} as revealed by its much greater susceptibility to proteolysis.[54] Hot spots for cleavage of the heterodimer appear in domain III and the anchor peptide, whereas domains IV and VI, and the protease core seem to be stabilized by Ca^{2+}. Attempts to produce and crystallize a Ca^{2+}-bound form of intact rat m-calpain have been hampered by aggregation and precipitation, and by dissociation of the small subunit.[15] However, release of the calpain protease core by proteolysis and its stabilization by Ca^{2+} provided additional targets for structural analysis.

The μ-calpain protease core from residues 29 to 356, inclusive, with a C-terminal His-tag and conversion of the active site cysteine (Cys115) to serine was produced in *E. coli*.[33] (Note that the numbering of these residues is offset by 10 because μ-calpain has a longer anchor peptide region.) The core enzyme (also referred to as μ-minicalpain or μI–II) was concentrated and stored frozen in 10 mM HEPES-NaOH (pH 7.6), 10 mM DTT. It was crystallized at 12.5 mg mL^{-1} in the presence of Ca^{2+} by the hanging drop vapor diffusion method. The well solution was 1.5 M NaCl, 0.1 M MES-NaOH (pH 6.0), 2% PEG 6000, 15% glycerol and 10 mM $CaCl_2$. The drops, which contained equal volumes of storage buffer and well solution, yielded blade-shaped crystals of maximum dimensions $1.0 \times 0.2 \times 0.03$ mm, belonging to the C_2 space group, after a few days. Their unit-cell parameters were $a = 149$ Å, $b = 41$ Å, $c = 132$ Å.
PDB ID code 1KXR.

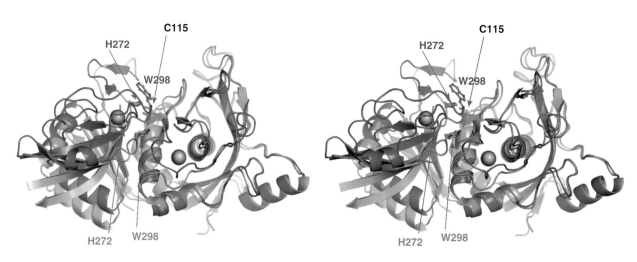

Figure 4 Stereo view of the protease core of Ca^{2+}-bound rat μ-calpain (blue) superimposed on that of calcium-free rat m-calpain (orange). The bound Ca^{2+} atoms are drawn as gold spheres. Twin arrows indicate the movement of key residues in response to Ca^{2+} binding.

Comparison of the apo- and Ca²⁺-bound protease core structures

The Ca^{2+}-bound enzyme core has dimensions of $70 \times 40 \times 35\,\text{Å}$. It has two Ca^{2+} ions bound, one to each domain (Figure 4). It differs from the core regions of the rat and human apo-m-calpains in the locations of loops that move to coordinate the Ca^{2+}s into two non–EF-hand sites, and in the juxtaposition of domains I and II, which is similar to that seen in papain. The tryptophan (W298) wedge is removed from the cleft and lies flat on a hydrophobic surface in domain II generated during formation of Ca^{2+} binding site 2. The realignment of these two domains narrows the active site cleft. As a result, the active site Cys105 in domain I is 3.7 Å away from His262 in domain II as seen for equivalent residues in papain. Indeed, the side chains of all the catalytic residues of Ca^{2+}-bound μI–II can be superimposed over the equivalent residues of papain.[33]

Ca²⁺ coordination at non–EF-hand sites in the protease core

Site 1 in domain I is formed by two peptide loops that directly (or indirectly via water molecule stabilization) donate eight coordinations to the Ca^{2+} (Figure 5(a), Table 1). Two of these come from the carboxyl group of E185, which lies in a fixed loop connected to the core α-helix (residues 186–199 in μ-calpain). Four coordinations

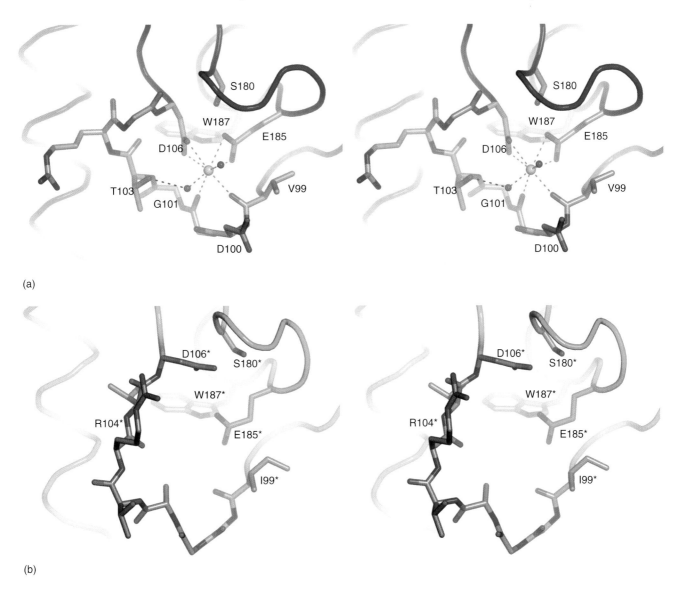

Figure 5 Ca^{2+} binding site 1 (domain I) in (a) Ca^{2+}-bound rat μ-calpain and (b) Ca^{2+}-free rat m-calpain. The Ca^{2+} atom is drawn as a gold sphere and the waters as red spheres. Hydrogen bonds are drawn as dashed lines. To facilitate comparison of the two structures, the residues in the apo-m-calpain site 1 have been renumbered (by the addition of 10 residues) to match those in μ-calpain. An asterisk next to the number signifies this.

Calpain

Table 1 Calcium-ligand distances

Domain I		Domain II	
Atom	Distance (Å)	Atom	Distance (Å)
99 VAL O	2.37	302 GLU OE1	2.48
101 GLY O	2.43	302 GLU OE2	2.41
106 ASP OD1	2.54	309 ASP OD1	2.22
106 ASP OD2	2.46	309 ASP OD2	3.16
185 GLU OE1	2.65	329 MET O	2.48
185 GLU OE2	2.50	331 ASP OD1	2.48
86 HOH O	2.29	333 GLU O	2.26
87 HOH O	2.55	16 HOH O	2.37

are supplied by a flexible loop with well-defined density that closes in on the Ca^{2+} (Figure 5(a) and (b)). This loop precedes the helix that contains the active site cysteine (C115). In it, two coordinations come from the D106 side chain and two come from the backbone carbonyl oxygens of V99 and G101. Two tightly bound water molecules (WAT1 and WAT2) supply the remaining two coordinating oxygen atoms, the former being stabilized by a hydrogen bond interaction provided by the T103 side chain.

In domain II, seven O atoms arranged in a typical pentagonal bipyramidal arrangement coordinate the site 2 Ca^{2+} (Figure 6(a), Table 1). Again, two peptide loops are involved, but both undergo considerable movement to bind

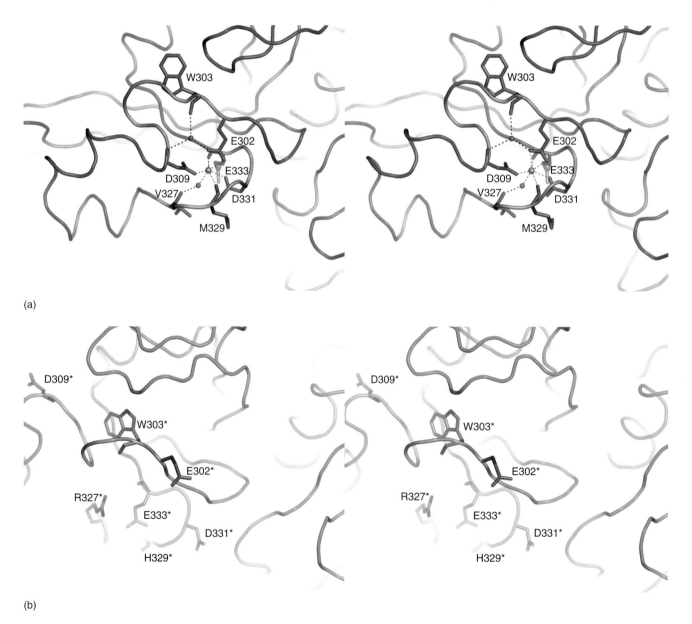

(a)

(b)

Figure 6 Ca^{2+} binding site 2 (domain II) in (a) Ca^{2+}-bound rat μ-calpain and (b) Ca^{2+}-free human m-calpain. The Ca^{2+} atom is drawn as a gold sphere and the waters as red spheres. Hydrogen bonds are drawn as dashed lines. To facilitate comparison of the two structures, the peptide backbone of the flexible loops is marked in pink, and residues are renumbered as described in the legend to Figure 6.

the Ca^{2+} (Figure 6(a) and (b)). One loop directly supplies three coordinations, two from the backbone O atoms of E333 and M329, and one from the D331 side chain. A fourth coordination is derived from a water molecule (WAT3) held by the D331 side chain and the backbone carbonyl oxygen of V327. The other loop provides three side-chain coordinations, two from E302 and one from D309. This loop is stabilized by a water molecule (WAT4) held by the E302 side chain, the backbone oxygen of W303 and the backbone nitrogen of D309.

Conformational changes induced by Ca^{2+} via the R104–E333 double salt bridge

The two Ca^{2+} ions are 17 Å apart in different domains but are linked by a double salt bridge between R104 and E333 that forms as a result of, and in concert with, the conformational changes induced by Ca^{2+} binding (Figure 7). R104 is part of the flexible loop in domain I that rotates as its neighboring residues (V99, G101, and D106) move in to coordinate Ca^{2+} at site 1. On the basis of a comparison with the m-calpain apostructure, the carboxyl group of D106 moves ~4.5 Å to coordinate the Ca^{2+}, causing ~8-Å shift in the guanido group of R104. E333 is on one of the two loops that move to coordinate the Ca^{2+} in site 2. Indeed, its side chain occupies the Ca^{2+} site in the apoform and must be rotated through 180° such that its backbone O atom coordinates the incoming Ca^{2+} and its side chain extends to salt bridge to R104. This salt bridge appears to be the structural basis for the cooperativity observed in Ca^{2+} binding to the protease core that results in enzyme activation.[33] On the basis of a comparison with the m-calpain apostructure, the carboxyl group of D309 moves ~11 Å to coordinate the Ca^{2+}, and the carboxyl group of E333 moves ~22 Å to form the double salt bridge with the guanido group of R104.

X-ray structure of rat Ca^{2+}-bound m-calpain protease core

The protease core of m-calpain from residues 19 to 346, inclusive, (analogous to the μ-calpain protease core) was produced in E. coli, but had barely any enzyme activity in comparison to the μ-core.[55] The structural explanation for this observation was revealed by X-ray crystallography. The m-calpain protease core (otherwise referred to as m-minicalpain or mI–II) at 15 mg mL^{-1} was crystallized in the presence of Ca^{2+} by the hanging drop vapor diffusion method. The well solution was 0.1 M sodium acetate (pH 5.5), 10% PEG 6000, and 30 mM $CaCl_2$. The drops, which contained equal volumes of protein stock solution and well solution, yielded rectangular plate-shaped crystals of maximum dimensions 0.7 × 0.7 × 0.07 mm, belonging to the $P2_1$ space group, after a few days. There were two molecules per asymmetric unit and their unit-cell parameters were $a = 63$ Å, $b = 81$ Å, $c = 75$ Å. In the 1.9-Å X-ray structure, the Trp106 side chain, the neighbor of the catalytic Cys105, protrudes into the active site cleft in such a way as to sterically hinder substrate binding, whereas, in μI–II, the equivalent Trp (116) is part of the domain I hydrophobic core. The protrusion of Trp106 in m-minicalpain can be traced back through the rearrangement of the domain I hydrophobic core to the collapse of one coil in α-helix 7. The metastability of this helix has been linked to the presence of a glycine residue at this site in this

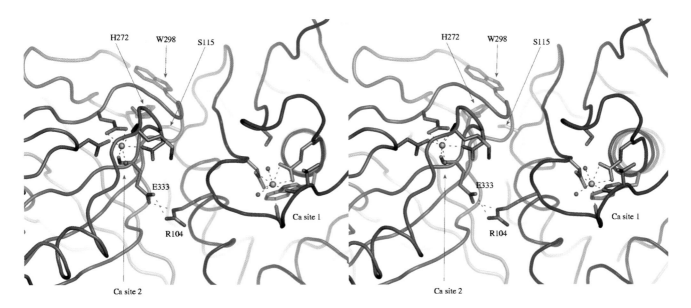

Figure 7 The salt bridge linking the two Ca^{2+} binding sites in domains I and II in the Ca^{2+}-bound form of rat μ-calpain. Ca^{2+} atoms are drawn as gold spheres and waters as red spheres. Hydrogen bonds are drawn as dashed lines.

and some other isoforms, which is an alanine in μ-calpain and in other isoforms. This inactivation can be reversed when adjacent domains are in place to help support the core structure.

Ca^{2+} binding sites outside the calpain core

The long-established Ca^{2+}-binding sites in m-calpain are those of the homologous PEF domains IV and VI at the C-termini of the two subunits. Although the fifth EF-hand in each PEF domain is used for dimerization and is Ca^{2+}-free, the other four EF-hands can bind Ca^{2+} in or near their loop region. The chapter by M. Cygler (see the chapter PEFLINS by M. Cygler on PEF proteins in this volume) provides a detailed description of these PEF domains and their EF-hand structures. The other potential extra-core Ca^{2+}-binding site(s) lies in domain III. By analogy with other C2-like domains, it is likely that one or more Ca^{2+} ions are bound to the acidic loop that lies adjacent to the core. There is some biochemical evidence that domain III binds Ca^{2+},[34] but no structural verification is currently available. If so, then all five domains shown in the 3D Structure can bind Ca^{2+}, which together comprise a remarkably varied set of more than 10 sites.[33]

Overall activation mechanism

In the heterodimeric calpains, only two of the 10+ Ca^{2+} binding sites are directly involved in the cooperative realignment of the active site residues. This is the basic, fundamental mechanism of activation. Consistent with this there are calpain isoforms and orthologs that consist only of the two core protease domains.[4] It seems likely that other Ca^{2+} binding domains have been added to the protease core to refine regulation of the enzyme by Ca^{2+}. This possibility was illustrated when the protease core of m-calpain was substituted by the μ-calpain core in the context of m-calpain. The μ-calpain core and whole m-calpain require very different half-maximal Ca^{2+} concentrations of 5 to 50 μM and 250 to 350 μM, respectively.[56] The hybrid μ/m-calpain had an intermediate Ca^{2+} requirement for half-maximal activity of 120 μM, as if the ancillary domains were restricting activation of the μ-calpain core.[22] A similar conclusion can be drawn from experiments in which EF-hands in the PEF domains are inactivated by mutation. Accumulative EF-hand knockouts require higher Ca^{2+} concentrations for activation.[21] This is entirely in accordance with deductions that were made about the apo-m-calpain crystal structure. The two domains of the protease core appear to be restrained by contacts with neighboring domains. Domain I is tethered through contact of the anchor helix with the small subunit, and through direct interactions with the basic loop of Domain III (residues 413–425). Domain II makes electrostatic interactions with domain III, which in turn is connected via the extended linker to domain IV (Figure 3).

Several mechanisms have been proposed for the release of constraints on either side of the protease core, and these mechanisms are not mutually exclusive. Farthest away from the core are the PEF domains. A comparison of the apo[31] and Ca^{2+}-bound domain VI homodimers[30,31] shows there is very little conformational change induced by Ca^{2+} binding to six moderate affinity sites in EF-hands 1, 2, and 3, and two low affinity sites near EF-hand 4. This is likely to be true for domain IV also. Although the PEF domains were originally referred to as calmodulin-like,[57] this is a misnomer as far as the calpain activation mechanism is concerned. Aside from the presence of paired EF-hands, there is little in common. The PEF domains lack the long central helix of calmodulin that can fold to promote radical structural change (see the chapter PEFLINS by M. Cygler on PEF proteins in this volume). It is possible that the small conformational change resulting from Ca^{2+} binding to the PEF domains could be transduced, via the linker polypeptide, to domain III. A more immediate release mechanism might result from the considerable conformational changes in domain III that occur on addition of Ca^{2+}. These generate presumptive flexible regions that are readily cleaved by exogenous proteases like trypsin and chymotrypsin,[54] and by autoproteolysis.[58,59]

On the other side of the enzyme, another potential release mechanism is the dissociation of the anchor helix from the small subunit. This could follow from disruption of the salt bridge interaction between Lys7 in the anchor helix and Asp154 in the small subunit as a result of Ca^{2+} binding to the nearby EF-hand 2.[46,47,60] The anchor helix is fairly rigid in the apo-m-calpain structures and resistant to proteolysis, but becomes susceptible to tryptic and chymotryptic cleavage on addition of Ca^{2+}. This is consistent with its release from the binding site on the small subunit. The anchor helix is also cleaved by autoproteolysis, which is conditional on the addition of Ca^{2+}. One other potential release mechanism is the dissociation of the small subunit. Although this is not a precondition for calpain activation, subunit dissociation does occur under certain conditions of Ca^{2+} loading and should be at least as effective as anchor-peptide release in removing constraints on the protease domain.

The increase in enzyme activity of μ- and m-calpains as a function of Ca^{2+} concentration is sigmoidal, suggesting that the overall process of activation is cooperative. These two isoforms have very different [Ca^{2+}]$_{50}$ of 20 to 50 μM and 250 to 350 μM respectively, but share the same small subunit. It has been argued that the Ca^{2+} requirement differences must lie in the large subunit, but it has not been possible to identify a specific sequence or domain that confers the Ca^{2+} requirement.[61]

CONCLUDING REMARKS

A combination of sequence alignments and structural studies support the concept of a calcium-activatable papain being at the heart of all calpains. Onto this underlying core structure, additional domains have been added in eukaryotes to further refine control of these intracellular proteases by calcium. For example, the calpain-specific calcium-dependent inhibitor, calpastatin, binds to calpain through the penta-EF-hand domains. Other domains have been implicated in membrane/phospholipid binding that may help activate calpain or lower its calcium requirement for activation. However, there is still much to learn about the mechanisms of regulation by calcium. There is also much to learn about the other calpain isoforms with different combinations of ancillary domains, their substrates, inhibitors, and mechanisms of activation and regulation.

REFERENCES

1. T Murachi, K Tanaka, M Hatanaka and T Murakami, *Adv Enzyme Regul*, **19**, 407–24 (1980).
2. TN Dear and T Boehm, *Gene*, **274**, 245–52 (2001).
3. Y Huang and KK Wang, *Trends Mol Med*, **7**, 355–62 (2001).
4. H Sorimachi and K Suzuki, *J Biochem (Tokyo)*, **129**, 653–64 (2001).
5. S Gil-Parrado, O Popp, TA Knoch, S Zahler, F Bestvater, M Felgentrager, A Holloschi, A Fernandez-Montalvan, EA Auerswald, H Fritz, P Fuentes-Prior, W Machleidt and E Spiess, *J Biol Chem*, **278**, 16336–46 (2003).
6. M Molinari and E Carafoli, *J Membr Biol*, **156**, 1–8 (1997).
7. H Sorimachi, S Ishiura and K Suzuki, *Biochem J*, **328**, 721–32 (1997).
8. E Carafoli and M Molinari, *Biochem Biophys Res Commun*, **247**, 193–203 (1998).
9. KK Wang, *Trends Neurosci*, **23**, 20–26 (2000).
10. N Dourdin, AK Bhatt, P Dutt, PA Greer, JS Arthur, JS Elce and A Huttenlocher, *J Biol Chem*, **276**, 48382–88 (2001).
11. NO Carragher, MA Westhoff, D Riley, DA Potter, P Dutt, JS Elce, PA Greer and MC Frame, *Mol Cell Biol*, **22**, 257–69 (2002).
12. NO Carragher and MC Frame, *Int J Biochem Cell Biol*, **34**, 1539–43 (2002).
13. W Zhang, RD Lane and RL Mellgren, *J Biol Chem*, **271**, 18825–30 (1996).
14. JE Fox, *Thromb Haemost*, **82**, 385–91 (1999).
15. A Glading, DA Lauffenburger and A Wells, *Trends Cell Biol*, **12**, 46–54 (2002).
16. BJ Perrin and A Huttenlocher, *Int J Biochem Cell Biol*, **34**, 722–25 (2002).
17. K Aoki, S Imajoh, S Ohno, Y Emori, M Koike, G Kosaki and K Suzuki, *FEBS Lett*, **205**, 313–17 (1986).
18. S Imajoh, K Aoki, S Ohno, Y Emori, H Kawasaki, H Sugihara and K Suzuki, *Biochemistry*, **27**, 8122–28 (1988).
19. S Ohno, Y Emori and K Suzuki, *Nucleic Acids Res*, **14**, 5559 (1986).
20. H Sorimachi, S Amano, S Ishiura and K Suzuki, *Biochim Biophys Acta*, **1309**, 37–41 (1996).
21. CI DeLuca, PL Davies, JA Samis and JS Elce, *Biochim Biophys Acta*, **1216**, 81–93 (1993).
22. K Graham-Siegenthaler, S Gauthier, PL Davies and JS Elce, *J Biol Chem*, **269**, 30457–60 (1994).
23. VF Thompson and DE Goll, *Methods Mol Biol*, **144**, 3–16 (2000).
24. M Maki, H Ma, E Takano, Y Adachi, WJ Lee, M Hatanaka and T Murachi, *Biomed Biochim Acta*, **50**, 509–16 (1991).
25. JS Elce, C Hegadorn, S Gauthier, JW Vince and PL Davies, *Protein Eng*, **8**, 843–48 (1995).
26. GP Pal, JS Elce and Z Jia, *J Biol Chem*, **276**, 47233–38 (2001).
27. M Michetti, F Salamino, R Minafra, E Melloni and S Pontremoli, *Biochem J*, **325**, 721–26 (1997).
28. Y Minami, Y Emori, S Imajoh-Ohmi, H Kawasaki and K Suzuki, *J Biochem (Tokyo)*, **104**, 927–33 (1988).
29. H Blanchard, Y Li, M Cygler, CM Kay, J Simon, C Arthur, PL Davies and JS Elce, *Protein Sci*, **5**, 535–37 (1996).
30. GD Lin, D Chattopadhyay, M Maki, KK Wang, M Carson, L Jin, PW Yuen, E Takano, M Hatanaka, LJ DeLucas and SV Narayana, *Nat Struct Biol*, **4**, 539–47 (1997).
31. H Blanchard, P Grochulski, Y Li, JS Arthur, PL Davies, JS Elce and M Cygler, *Nat Struct Biol*, **4**, 532–38 (1997).
32. P Dutt, JS Arthur, P Grochulski, M Cygler and JS Elce, *Biochem J*, **348**, 37–43 (2000).
33. T Moldoveanu, CM Hosfield, D Lim, JS Elce, Z Jia and PL Davies, *Cell*, **108**, 649–60 (2002).
34. P Tompa, Y Emori, H Sorimachi, K Suzuki and P Friedrich, *Biochem Biophys Res Commun*, **280**, 1333–39 (2001).
35. L Waxman, *Methods Enzymol*, **80**, 664–80 (1981).
36. SS Twining, *Anal Biochem*, **143**, 30–34 (1984).
37. KJ Raser, A Posner and KK Wang, *Arch Biochem Biophys*, **319**, 211–16 (1995).
38. JS Arthur and DL Mykles, *Methods Mol Biol*, **144**, 109–16 (2000).
39. T Sasaki, T Kikuchi, N Yumoto, N Yoshimura and T Murachi, *J Biol Chem*, **259**, 12489–94 (1984).
40. AH Schmaier, HN Bradford, D Lundberg, A Farber and RW Colman, *Blood*, **75**, 1273–81 (1990).
41. GY Huh, SB Glantz, S Je, JS Morrow and JH Kim, *Neurosci Lett*, **316**, 41–44 (2001).
42. H Masumoto, K Nakagawa, S Irie, H Sorimachi, K Suzuki, GP Bourenkov, H Bartunik, C Fernandez-Catalan, W Bode and S Strobl, *Acta Crystallogr, D*, **56**, 73–75 (2000).
43. H Masumoto, T Yoshizawa, H Sorimachi, T Nishino, S Ishiura and K Suzuki, *J Biochem (Tokyo)*, **124**, 957–61 (1998).
44. RH Kretsinger, *Nat Struct Biol*, **4**, 514–16 (1997).
45. D Reverter, M Braun, C Fernandez-Catalan, S Strobl, H Sorimachi and W Bode, *Biol Chem*, **383**, 1415–22 (2002).
46. K Nakagawa, H Masumoto, H Sorimachi and K Suzuki, *J Biochem (Tokyo)*, **130**, 605–11 (2001).
47. S Strobl, C Fernandez-Catalan, M Braun, R Huber, H Masumoto, K Nakagawa, A Irie, H Sorimachi, G Bourenkow, H Bartunik, K Suzuki and W Bode, *Proc Natl Acad Sci USA*, **97**, 588–92 (2000).
48. CM Hosfield, Ph.D., Queen's University, Kingston (2001).
49. J Rizo and TC Sudhof, *J Biol Chem*, **273**, 15879–82 (1998).
50. L Pallanck, *Trends Neurosci*, **26**, 2–4 (2003).
51. RB Sutton, BA Davletov, AM Berghuis, TC Sudhof and SR Sprang, *Cell*, **80**, 929–38 (1995).
52. CM Hosfield, JS Elce, PL Davies and Z Jia, *EMBO J*, **18**, 6880–89 (1999).

53 CM Hosfield, Q Ye, JS Arthur, C Hegadorn, DE Croall, JS Elce and Z Jia, *Acta Crystallogr, D*, **55**, 1484–86 (1999).

54 T Moldoveanu, CM Hosfield, Z Jia, JS Elce and PL Davies, *Biochim Biophys Acta*, **1545**, 245–54 (2001).

55 T Moldoveanu, CM Hosfield, D Lim, Z Jia and PL Davies, *Nat Struct Biol*, **10**, 371–78 (2003).

56 DE Croall and GN DeMartino, *Biochim Biophys Acta*, **788**, 348–55 (1984).

57 S Ohno, Y Emori, S Imajoh, H Kawasaki, M Kisaragi and K Suzuki, *Nature*, **312**, 566–70 (1984).

58 T Nishimura and DE Goll, *J Biol Chem*, **266**, 11842–50 (1991).

59 C Crawford, NR Brown and AC Willis, *Biochem J*, **296**, 135–42 (1993).

60 D Reverter, H Sorimachi and W Bode, *Trends Cardiovasc Med*, **11**, 222–29 (2001).

61 P Dutt, CN Spriggs, PL Davies, Z Jia and JS Elce, *Biochem J*, **367**, 263–69 (2002).

62 WL DeLano, http://www.pymol.org (2002).

Parvalbumin

Susumu Nakayama[†], *Hiroshi Kawasaki*[‡] *and Robert H Kretsinger*[§]

[†]Department of Biochemistry, Nagasaki University Graduate School of Biomedical Sciences, Sakamoto, Nagasaki, Japan
[‡]Division of Plant Genetic Engineering, Kihara Institute for Biological Research, Yokohama City University, Yokohama, Japan
[§]Department of Biology, University of Virginia, Charlottesville, VA, USA

OVERVIEW AND HISTORY OF PARVALBUMIN AND HOMOLOGOUS EF-HAND PROTEINS

Parvalbumin is one of the 77 known subfamilies in the homolog family of EF-hand containing proteins. Most of these have been demonstrated or inferred to be involved in one of two interrelated functions. Some, like parvalbumin, help shape the spatial and temporal distribution of calcium functioning as a secondary messenger in the cytosol. Others, such as the archetypical calmodulin, transduce the information of that calcium signal into a change in conformation of a structural protein or change in activity of an enzyme. Five of the 77 EF-hand subfamilies are found in bacteria and one is found in a virus (Table 1). Bacteria extrude calcium and maintain nanomolar cytosolic concentrations, as do eukaryotes; however, there is little evidence that bacteria use calcium as a secondary messenger.

Parvalbumin is considered from three distinct perspectives. It is best understood in terms of its evolution and comparison with other EF-hand proteins.[1] Second, there is strong evidence that parvalbumin functions as a calcium buffer and/or transporter within the cytosols of many cells. The mechanism of this buffering can be related to its structure. Its characteristic intracellular and extracellular distributions, especially in nervous tissue and avian thymus, indicate that it may have additional, yet to be identified functions. Finally, parvalbumin is a small, stable protein. It can readily be extracted and/or expressed in quantity and purified. The crystal structures of several isoforms and mutants are available. It is an appropriate and convenient subject for many physical studies including calcium binding.

Parvalbumin was first isolated by Henrotte[3] and colleagues[4] and characterized as the 'myogen II' fraction of albumins extracted from carp muscle. He noted that it is quite acidic by electrophoretic mobility; it has high phenylalanine content with little or no tyrosine and tryptophan; it has sedimentation coefficient, $S_{20} = 1.44\,S$; it does not precipitate after heating at $50\,°C$; and it crystallizes in the 90 to 100% saturated ammonium sulfate cut. He also recognized that it is not present in rabbit muscle but that a similar fraction is present in extracts from the muscle of plaice and of frog. Konosu *et al.*[5] determined the amino acid compositions of the three main components of the myogen 1.5 S fraction and confirmed that they do not contain tyrosine or tryptophan. Pechère[6] proposed that all of these low molecular weight albumins are homologous and that they be called *parv*(small)*albumins*. Kretsinger[7] and colleagues[8] determined the first crystal structure – carp parvalbumin, fraction 3 – and coined the term 'EF-hand'. The structure consists of three helix-loop-helix domains; these were initially called AB, CD, and EF. Alpha-helices E and F are most canonical and were adopted as the prototype. The E-helix, EF-loop about the Ca^{2+} ion, and the F-helix resemble the extended forefinger, clenched middle finger, and extended thumb of the right hand – hence the moniker 'EF-hand' (Figure 1). Inspired by the close superimposibility of the EF-hand and the CD-hand, Kretsinger[7] compared the amino acid sequences[9] of the CD and the EF regions and noted their obvious similarity. Further comparison with the AB sequence indicated weak similarity, with a deletion of two residues in the AB region (Figure 2(a)). He proposed that parvalbumin had originated by gene triplication and fusion. More recent analyses support this interpretation; however, when one deals with domains of only 30 residues, the statistical significance of alignments is weaker and the distinction between homologs and analogs is more tenuous.

PARVALBUMIN WITHIN THE EF-HAND HOMOLOG FAMILY

Seventy-three of the 77 known subfamilies of EF-hand proteins, (Table 1[11]) contain 2, 4, 6, 8, or 12 EF-hands; these EF-hands always occur in adjacent pairs and appear to have evolved as pairs – hence our comparison of the structures of pairs (Figure 3). Most *odd*-numbered domains cluster together in phylogenic analysis as do the *even*-numbered ones. The most parsimonious interpretation is that duplication and fusion of an *ur-EF-hand* gene generated an *odd* and an *even* precursor and that all, certainly most, pairs of EF-hands evolved from this original, precursor pair. Calpain, sorcin, and probably URE3-BP (members of the penta-EF-hand subfamilies), each have

Table 1 The 77 known EF-hand homolog subfamilies are described in several groups: *CTER*, those that are congruent with calmodulin, troponin C, essential, and regulatory light chains; *CPV*, those that are congruent with calcineurin B, p22, and visinin (recoverin); *Pairings*, those closely related between or among themselves but not closely related to other subfamilies; Self, those (some of) whose EF-hands are most closely related to other EF-hands within the same subfamily; and *Miscellaneous*, those whose domains do not show a strong and consistent pattern of similarity with other EF-hand subfamilies. The first domain of (pre)parvalbumin is inferred to have been deleted; hence, its domains are listed as 2(AB), 3(CD) and 4(EF). In congruent subfamilies, all of the EF-hands 1 (or n) resemble one another more closely than they resemble other EF-hands within their own protein. For 26 subfamilies, only one sequence is available; this is indicated by inclusion of genus name in parentheses or as part of the name of that subfamily. *APFP* refers to whether the protein is found in Animals, Plants, Fungi, and/or Protists. Five subfamilies are found in prokaryotes (bact) and one is found in a virus. The *Function/Struct* columns indicate whether a function is known ('+') or not ('?') and whether a crystal structure is available ('X') or not ('?'). The symbol '***' before the first EF-hand column or after the last column indicates a protein with the non-EF-hand domain(s) to either the N-side or C-side of the EF-hands. For P26 and for EP15 '***' indicates another domain between EF-hands 2 and 3 and EF-hands 4 and 5. Thirty-four, including the nine enzymes, of the 77 subfamilies are heterochimeric. For AEQ and PPTS the '*–*' indicates that domain is not recognizable as an EF-hand by analysis of its sequence; however, its proximity to an otherwise unpaired EF-hand suggests that it may be an EF-hand and is inferred not to bind calcium. For each EF-hand, it is indicated whether calcium binding is observed as (or inferred from sequence) '+' or not '–'. For some EF-hands, there are instances of both binding and not binding calcium '±'. There are four examples – 'a, b, c, and d' – of noncanonical EF-hand loops that bind calcium. Some loops inferred not to bind calcium may provide additional examples of noncanonical calcium binding loops. EF12 has 12 and LPS has 8 EF-hands; for ease of formatting, EF-hands 7 and on are listed under EF-hands 1 and on. Of the 77 distinct EF-hand proteins, 56 have been found in animals. The functions of only 27 of the 77 are known. Nine of these are enzymes and have been demonstrated or inferred to be activated by the binding of calcium. Many, but certainly not all, of the remaining 50 function in the information–transduction pathway summarized for calcium modulated proteins, such as calmodulin

Name	Animal Plant Fungi Protist	Func Struct	1 (7	2 8	3 9	4 10	5 11	6 12)
CTER								
CAM	Calmodulin	APFP	+X	+	+	+	±	
TNC	Troponin C	A...	+X	±	+	±	+	
ELC	Essential light chain, myosin	A.F.	+X	a/–	±	±	±	
RLC	Regulatory light chain	A.FP	+X	+	–	–	–	
TPNV	Troponin, nonvertebrate	A...	+?	–	+	±	+	
CLAT	CAM-like leaf (*Arabidopsis*)	P..	??	+	+	+	+***	
SQUD	Squidulin (*Loligo*)	A...	+?	+	+	+	+	
CDC	CDC31 and caltractin	APFP	??	+	±	±	+	
CAL	Cal1 (*Caenorhabditis*)	A...	??	+	+	+	+	
CAST	CAST	.P..	??	***+	–	+	+	
CPV								
CLNB	Calcineurin B	A.F.	+X	+	+	+	+	
P22	p22	A...	??	+	–	+	+	
VIS	Visinin and recoverin	A...	+X	–	+	+	–	
CALS	Calsenilin (*Homo*)	A...	??	***–	–	+	+	
DREM DRE antag. modul. (*Homo*)		A...	+?	***–/?	+/?	+	+	
CMPK CAM dep prot kinase (*Lilium*)		P..	+?	***+	+	+	+	
SOS3	Ca sens homo (*Arabidopsis*)	.P..	+?	–	–	–	–	
Pairings								
RTC	Reticulocalbin (*Mus*)	A...	??	+	?/+	?/+	+	+	+
SCF	DNA supercoil fact (*Bombyx*)	A...	??	***+	–	+	+	+	+
CALP	Calpain	A...	+X	***+	+	–	–		
SORC	Sorcin/grancyclin	A.F.	+X	***+	+	–	–		
S100	S100	A...	?X	b/–	+				
ICBP	Intestinal Ca binding protein	A...	?X	b/–	+				
HYFL	Trichohylin profilag	A...	??	b/?	+***				
DGK	Diacylglycerol kinase	A...	+?	***+	+***				
NUBN	Nucleobindin and NEFA	A...	??	***+	+***				
CRGP	CAM rel gene product (*Homo*)	A...	??	+	+				

Table 1	(Continued)								
Name	Animal Plant Fungi Protist	Func Struct	1 (7	2 8	3 9	4 10	5 11	6 12)	
ACTN	α-actinin	A.F.	+ X	***±	±				
FDRN	α-spectrin and α-fodrin	A...	+ ?	***−	−				
GPD	Glycerol-P-dehydrogenase	A...	+ ?	***−	+				
AIF1	Allograft inflammatory factor	A...	? ?	***+	+/?				
BM40	Osteonectin, SPARK	A...	? X	***c	+				
QR1	QR1 and SC1	A...	? ?	−	+				
Self									
EF12	Ca binding prot of nematodes	A...	? ?	***+ +	+ +	+ +	− +	+ +	+ +
LPS	*L. pictus* SPEC resemble prot	A...	? ?	+ +	+ −	+	+	+	+
CLBN	Calbindin 28 kDa, calretinin	A...	? ?	+	±	+	+	+	−
EP15	EP15	A...	+ X	***−	− ***	−	+***	+	−***
TCBP	Tetrahymena CaBP	...P	? ?	−	+	+	+		
P26	p260lf (*Rana*)	A...	? ?	−	−***	−			
PLC	Phospholipase C	A.F.	+ X	***−	−	−	−***		
CBP	CBP1, CBP2 (*Dictyostelium*)	..F.	? ?	+	+	+	+		
Miscellaneous									
PFS	Surface protein (*Plasmodium*)	...P	? ?	***+	+	+	+	+	−***
CLSM	Calsymin	bact	? ?	+	−	+	−	+	−
UEBP	URE3-BP	A...	+ ?	+	+	+	−?	−?	
CDPK	Ca-dependent protein kinase	.P..	+ ?	***+	+	+	+		
PFPK	Protein kinase (*Plasmodium*)	...P	+ ?	***+	+	+	+		
SPEC	*Strongylocentrotus* CaBP	A...	? ?	±	+	+	±		
TPP	p24 thyroid protein	A...	? ?	+	+	+	?		
1F8	1F8 and TB17 and calflagin	...P	? ?	+	+	+	+		
SARC	Sarcoplasm Ca bind prot	A...	? X	+	±	+	±		
AEQ	Aequorin and luciferin BP	A...	+ X	+	*−*	+	+		
PPTS	Protein phosphatase	A...	+ ?	***−	*−*	+	+		
H32	HRA32 (*Phaseolus*)	.P..	? ?	+	+	+	+		
EFH5	EFH5	...P	? ?	−	±	±	−		
CVP	Ca vector prot (*Branch.*)	A...	? X	−	−	+	+		
PMAT	Memb. assoc. (*Arabidopsis*)	.P..	? ?	***+	+	+	+		
LAV	LAV1 (*Physarum*)	..F.	? ?	***+	+	+	+		
CMSE	CaBP (*Saccharopolyspora*)	bact	? ?	±	±	±	±		
MSV	MSV097 (*Entomopoxvirinae*)	virus	? ?	+	+				
PARV	Parvalbumin	A...	? X	del	−	+	+		
BCBP	Brain calcium binding protein	A...	? ?	±	−	+	+		
CSCJ	S. coelicolor CBP	bact	? ?	+	+	−	+		
DYSN	Dystrophin	A...	+ X	***−	−	−	−		
FIMB	Fimbrin	A.F.	+ ?	±	±***				
GRP	Ras guan releasing prot (*Rattus*)	A...	? ?	***+/?	+/?***				
PKD	PKD2L/polycystin	A...	? ?	***−	+***				
RYR	Ryanodine receptor/Ca release	A...	+ ?	***−	+***				
CBL	Proto-oncogene Cbl	A...	+ X	***−	d***				
CIB	Ca and integrin binding protein	A...	? ?	***+	+				
SENS	Calsensin (*Haemopsis*)	A...	? ?	+	+				
GRV	Groovin (*Drosophila*)	A...	? ?	+	+				
BET4	Calcium bind. pollen allergen	.P..	? X	+	+				
CSCD	S. coelicolor CBP	bact	? ?	+	+				
CBCC	C. crescentus CBP	bact	? ?	+	+				
ACHE	Acetylcholine esterase	A...	+ X	***−	−				
NCAB	Neuronal CaBP	A...	? ?	+	−***				
SWPN	Swiprosin	A...	? ?	***−	+***				

Parvalbumin

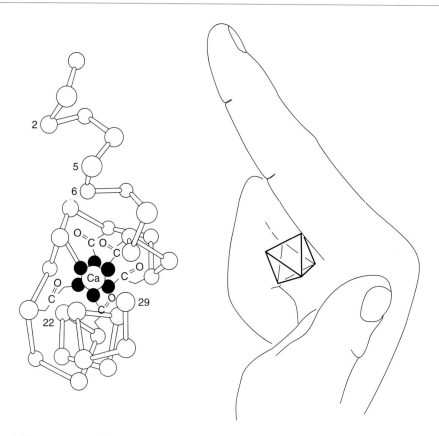

Figure 1 The EF-hand. The canonical EF-hand consists of α-helix E, (forefinger, residues 1–10), a loop around the Ca^{2+} ion (clenched middle finger, 10–21), and α-helix F (thumb, 19–29). Residue 1 is often glutamic acid; the insides of the helices (palmer surfaces) usually have hydrophobic residues that contact the insides of the other EF-hand of the pair. The side chains of five residues, approximating the vertices of an octahedron (X, residue 10; Y, 12; Z, 14; −X, 18; and −Z, 21), provide oxygen atoms to coordinate Ca^{2+}; residue 16 at −Y bonds to Ca^{2+} with its carbonyl oxygen. The positions of these ligands within the loop are often referred to as 1, 3, 5, 7, 9, and 12. The Ca^{2+} ion is actually seven coordinate in a pentagonal bipyramid with major axis, X, −X. There are five oxygen atoms in the Y, Z plane, since the −Z ligand (residue 21, usually glutamic acid) coordinates Ca^{2+} with both oxygen atoms of its carboxylate group. A glycine at 15 permits a tight bend; residue 17 has a hydrophobic side chain that attaches the loop to the hydrophobic core of the pair of EF-hands. Several variations of this canonical calcium coordination scheme have been inferred from amino acid sequence and confirmed in crystal structures of other EF-hand proteins. Nearly a third of all known EF-hands do not bind calcium; those with no indels (insertions or deletions), have a nonoxygen containing side chain substituted at positions 10, 12, 14, 18, or 21. Other EF-hands have indels; most notable is EF-hand 1 of the S-100 subfamily (Table 1) in which several carbonyl oxygens, instead of side-chain oxygens, coordinate Ca^{2+}. When trying to identify new EF-hands in protein sequences, it is much more reliable to search the entire sequence database with BLAST or FASTA for similarity to known EF-hands than using the heuristic of the preceding paragraph. Scores of proteins, some of which actually do bind calcium, have been referred to as containing EF-hand(like) domains, for example, in bacteria,[2] even though the similarity of the sequence or of the structure is not statistically significant. There exist many calcium-binding domains; the term 'EF-hand' should be reserved for those domains where homology is strongly supported.

five EF-hands; the fifth pairs with the fifth domain of another penta-EF-hand protein. Only parvalbumin has three domains; the unique orphan, AB-domain, does not pair with another domain. Parvalbumin does not normally oligomerize; however, as discussed below, the 24-kDa allergen of cod[12] and the CPV3 parvalbumin of chicken thymus[13] may be exceptions.

FUNCTIONAL CHARACTERISTICS OF PARVALBUMIN

Parvalbumins were initially observed in high concentration in amphibia and fish and their presence was suggested to be related to an aquatic existence. As more tissues were examined, it was realized that parvalbumin is present in especially high concentration in white, or fast, muscle. Several groups[16] proposed that parvalbumin is involved in spatio-temporal buffering. This is related to the different coordinations of calcium and magnesium by oxygen.

In proteins, the Ca^{2+} ion, of atomic radius 0.99 Å, is usually bound by seven oxygen atoms in an approximately pentagonal bipyramid conformation (Figure 2(b)); the oxygens have a bit of lateral flexibility. The Mg^{2+} ion, radius 0.65 Å, is usually coordinated by six oxygen atoms at the vertices of an octahedron; these oxygens are in tight van der Waals contact with one another.

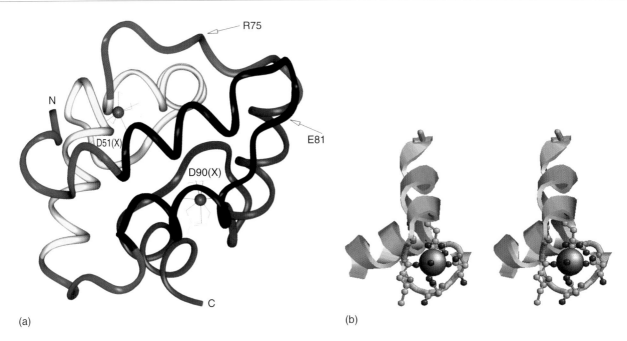

Figure 2 Trace of the main chain of parvalbumin and coordination of the Ca^{2+} ion. (a) parvalbumin is viewed down the approximate twofold axis that relates the CD and the EF domains; the two calcium-binding loops are far from the viewer. The main chain of the AB-hand is dark blue; CD is yellow; EF is green. The linkers, AB/CD and CD/EF, as well as the eight residues preceding helix A are light blue. The residues at the +X vertices of calcium coordination, Asp51 and Asp90 in carp and most parvalbumins, are indicated. Note that the Ca^{2+} ions are seven coordinate. Both the glutamic acids at −Zs coordinate with both oxygens of their carboxylate groups. The hydrophobic patch, near the viewer, formed by the pair of EF-hands, CD and EF, is covered by α-helices A and B, an EF-hand with a deletion of two residues. The hydrogen bonds between invariant Arg75 and Glu81 are inferred to stabilize the CD and EF pair. As seen in Figure 3(a), the pair of EF-hands (#s 3 and 4 of Table 1) in the calci form is in a more open conformation as seen in the third and fourth EF-hands of calci calmodulin. (b) The coordination of the Ca^{2+} ion of the EF domain of parvalbumin (*Esox lucius*, PDB 3PAL) is viewed in stereo. With the conventions described in the legend to Figure 1, the +X-axis points away from the viewer; +Y points to the right; and +Z points down. Helix F is vertical with the C terminus at the top. In the EF domain, the residue at −X (#18 in domain numbering) is glycine; the Ca^{2+} ion is coordinated by water. The X-axis corresponds to the unique axis of the pentagonal bipyramid. There are five oxygen atoms from four residues in the Y, Z plane; glutamic acid (#21) coordinates with both oxygens of its carboxylate group. The residue at −Y (#16) coordinates the Ca^{2+} ion with its carbonyl oxygen. Oxygens are colored red; nitrogen, blue; sulfur, yellow; backbone trace, green; and Ca^{2+} ion dark gray with atomic radius 1.7 Å, as drawn with RASMOL.[10]

The dissociation constant is the ratio of the off rate to the on rate: K_d (M) = k_{off} (s^{-1})/k_{on} (M^{-1}s^{-1}). The rate-limiting dehydration of $Ca(H_2O)_7^{2+}$ is fast, $\sim 10^{8.0}$ s^{-1}, while that of $Mg(H_2O)_6^{2+}$ is slow, $\sim 10^{4.6}$ s^{-1}. This reflects loose pentagonal bipyramidal versus tight octahedral packing. These rates are extremely important for modeling the flux of Ca^{2+} ions through the cytosol and the attendant binding of proteins. The increase in affinity of most proteins for calcium relative to magnesium derives primarily from this difference in k_{on} (for example, Table 2).

It is intriguing that most EF-hand proteins bind calcium $\sim 10\,000$ times more strongly than they bind magnesium.

Table 2

	K_d(M)	k_{off} (s^{-1})	k_{on} (M^{-1}s^{-1})
Parvalbumin/Ca^{2+}	$10^{-8.0}$	$10^{0.0}$	$10^{8.0}$
Parvalbumin/Mg^{2+}	$10^{-4.1}$	$10^{0.5}$	$10^{4.6}$
Troponin C (1 and 2)/Ca^{2+}	$10^{-6.5}$	$10^{1.5}$	$10^{8.0}$
Troponin C (1 and 2)/Mg^{2+}	$10^{-2.3}$	$10^{2.3}$	$10^{4.6}$

This surely reflects natural selection for function and not the inherent affinities of the Ca^{2+} and Mg^{2+} ions for oxygen ligands. The cytosolic concentration of the free Ca^{2+} ion is $\sim 10^{-7.2}$ M and that of Mg^{2+} is $\sim 10^{-2.8}$; in contrast both pCa$_{out}$ and pMg$_{out}$ are ~ 2.8. This means that a cytosolic protein, such as parvalbumin, that binds calcium with high affinity, for example, pK_d(Ca^{2+}) ~ 8.0 will also bind magnesium with relatively high affinity, for example, pK_d(Mg^{2+}) ~ 4.1 and, in the resting cell, will be in the magnesium state. In contrast, lower affinity sites, such as EF-hands 1 and 2 of troponin C have lower affinities for divalent cations, pK_d(Ca^{2+}) ~ 6.5, and pK_d(Mg^{2+}) ~ 2.3 and are apo in the resting state. This leads to the counterintuitive situation in which a pulse of messenger calcium binds first to the weaker 'calcium-specific' aposites in domains 1 and 2 of troponin C. Only after the Mg^{2+} ion diffuses off the strong site of parvalbumin can the Ca^{2+} ion, whose concentration during the pulse approaches $10^{-6.0}$ M, bind to that apo-site. Parvalbumin, with higher affinity for divalent cations, binds calcium after the weaker troponin C and thereby helps relax the muscle and prevents

Parvalbumin

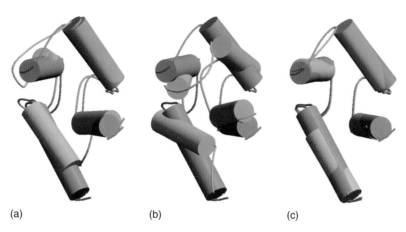

Figure 3 Conformations of pairs of EF-hands. Calmodulin consists of two pairs (1 and 2, and 3 and 4) of EF-hands. A flexible tether, seven-residues long, connects helix F2 to helix E3. In the crystal structure, F2, the tether, and E3 form a single continuous helix. Calmodulin is widely distributed and interacts with at least 30 different target proteins. Further, its four EF-hands are all canonical; hence, it serves as a reference point for many evaluations and comparisons. (a) The backbone of the pair calmodulin-3 and 4 in its dicalci form is shown as a cylinder when it is α-helical and as a strand elsewhere. It is viewed down its approximate twofold axis and is shown in light salmon color. As in Figures 1 and 2, the calcium-binding loops are far from the viewer. Dicalciparvalbumin-CD and EF is superimposed on dicalcicalmodulin-3 and 4; their conformations are very similar. The orientation and color of parvalbumin-CD and EF (yellow, light blue, and green) is the same as in Figure 2. (b) Apocalmodulin-3 and 4 (yellow, blue, and green) is superimposed on dicalcicalmodulin-3 and 4 (salmon); the binding of calcium opens the cleft between the two helices of each EF-hand and between the two EF-hands of the pair. This is supposedly a necessary step permitting calmodulin to interact with its numerous targets. The structure of apoparvalbumin is not known. (c) Dicalcicalmodulin-3 and 4 (yellow, light blue, and green) as complexed with the target peptide from myosin light-chain kinase (not shown) is superimposed on dicalcicalmodulin-3 and 4 (salmon). Both the 3 and 4 and the 1 and 2 pairs of calmodulin undergo little additional change in conformation to bind their targets. The binding of a target peptide facilitates binding of calcium and vice versa. There is no evidence of parvalbumins interacting with targets. Figures were drawn with MOLSCRIPT[14] and with RASTER3D.[15]

tetany. Myocytes do not normally express parvalbumin; rat myocytes, expressing a transferred parvalbumin gene, have an increased relaxation rate.[17] One infers that parvalbumin performs a similar buffering function in tissues other than muscle, helping to shape the calcium pulse(s).[18] As has been proposed for the intestinal calcium binding protein (ICBP) subfamily,[19] parvalbumins may also be involved in the cytosolic transport of calcium by facilitated diffusion.[20] Those sites with a lower affinity for divalent cations are inappropriately referred to as calcium-specific even though the ratio of affinities for calcium and for magnesium are similar to that of the high-affinity sites.

Most organisms have several isoforms of parvalbumin. When these are all represented in a large dendrogram, there are two distinct branches, α and β.[11] The crystal structures of β and α parvalbumins are very similar. There is no single characteristic that distinguishes the two; it is best to do an entire cladistic analysis, considering each residue to be a character. Most αs have pI ~5.2 and 109 residues; most βs have pI ~4.6 and 108 residues with the terminal helix F being one residue shorter.

The α-branch contains those parvalbumins commonly found in mammals and in the white (fast) muscle of many fish. Distinctions between functions and cellular distributions of α and β have not been noted in fish or in amphibia. However, in birds and mammals, the α-form(s) seem to be generally distributed. In mammals, neurons immunoreactive to parvalbumin, in contrast to those for calbindin and calretinin, are more prevalent in the middle and lower layers of the cortex.[21] Comparison of paired-pulse stimulations at GABAergic (gamma-aminobutyric acid) synapses between interneurons and Purkinje cells in parvalbumin. +/+ and −/− knockout mice indicate that parvalbumin modulates short-term synaptic plasticity.[22]

In contrast, the β-form(s) seem to have more specialized distributions and/or functions in birds and mammals. The avian β-parvalbumins, thymic hormone, and CPV3 are found in the cortex and stroma of the thymus of chickens and are inferred to be involved in the stimulus of antibody production.[23] Oncomodulin was first isolated from rat hepatomas.[24] Subsequently, it has been identified in the placenta and in the cochlear outer hair cells of the organ of corti of rats, so far the only normal adult tissue of mammals.[25]

About 2% of the general population has an allergic reaction to fish. Aas and Jebsen[26] first described the principle allergen, DS 22, of cod (*Gadus callarias*) and noted its similarity to parvalbumins. Parvalbumins, both α and β, (allergen M) from dozens of fishes and amphibia[27] have been shown to cross-react with IgE immunoglobulins from people showing class I allergies. The high immunogenicity and cross-reactivity is inferred to be related to the thermostability and resistance to proteolysis of parvalbumin; this explains its ability to survive cooking and passage through the gastrointestinal tract. The immunoreactivity of Gad c I (previously allergen M of the Baltic cod, *G. callarias*) is reduced by treatment with periodate; this implies that this parvalbumin is glycosylated.[28] Such posttranslational modifications have not been reported from numerous other studies of this nominally cytosolic protein.[29] An allergenic dimer, Gad m I of $M_r = \sim 24\,000$ from Atlantic cod (*Gadus morhua*), of parvalbumin is not converted to a monomer under conditions that should reduce disulfide bonds,[12] even though its DNA encodes a monomer. One might speculate that it results from domain swapping, as observed for Phl p 7[30] of the BET4 subfamily (Table 1).

PHYSICAL CHARACTERISTICS OF PARVALBUMIN

Parvalbumins are acidic, pIs ~5.2 for αs and ~4.6 for βs. They are thermostable, Tms ~46 °C for the dicalci α and ~54 °C for β. They are quite resistant to proteolysis *in vitro*. Most lack tryptophan, tyrosine, or cysteine; hence, they can be mutated and specifically labeled. However, a β-parvalbumin, CPV3 from chicken, contains cysteines 18 and 72; these may be cross-linked *in vitro*, to form oligomers.[23] Parvalbumins are an attractive system for studying the complementary questions of how the distribution of ligands affects the affinities of the protein for calcium and for other divalent cations as well as for lanthanides and conversely of the effects of calcium binding on the stability and conformation of the protein.

There have been reports both of positive and of negative cooperativity in binding calcium for both parvalbumin and calmodulin. In either case, the effect is rather small; the Hill coefficient is, within experimental error, ~1.0. Numerous studies have evaluated the effects on thermostability and fluorescence of natural (tryptophan) or introduced markers of binding of various cations.

Evaluation of the observed or inferred calcium-binding domains of the multiple isoforms of the 77 proteins summarized in Table 1 reveals from one to five carboxylate groups in the primary coordination sphere of the Ca^{2+} ion. There may be a bridging water molecule between some of these ligands, often at the −Y vertex, and the Ca^{2+} ion. For only a few parvalbumins and other EF-hand proteins is the affinity for calcium well determined. Further, since parvalbumin binds two Ca^{2+} ions, it is difficult to assign a $K_d(Ca^{2+})$ to a specific site. Reid and Hodges[31] surveyed the distribution of amino acid side chains coordinating calcium in parvalbumin and in troponin C and proposed that the highest affinities are obtained when two carboxylates are at the +X and −X vertices and two carboxylates are at the +Z and −Z vertices, as is often observed for parvalbumin (Figure 2(a)). This generalization does not necessarily apply to all EF-hand proteins. For example, both the S55D and the G98D mutants of rat (β) oncomodulin increase the number of carboxylates from four to five and increase the affinity for calcium.[32] The E101D (−Z vertex of the EF domain) mutant of eel parvalbumin binds Mg^{2+} ten times more strongly and Ca^{2+} a hundred times less strongly than the wild type.[33]

Lanthanum and the 14 lanthanides, as well as cadmium, are often substituted for calcium since calcium is spectrally silent and the lanthanides have a range of useful spectral properties. Also, they are often substituted for calcium as heavy atom derivatives in crystallographic studies. At most calcium binding sites, EF-hand or other, lanthanides are bound with 10 to 1000 times higher affinity; however, this varies with the site. As for calcium and magnesium, the affinity for lanthanides depends not only on the ligands but also on the conformation of the entire protein. One can acquire a great deal of empirical data regarding the affinities of various EF-hand sites and their mutants; however, we cannot predict affinity for a divalent cation based just on the five side chains at X, Y, Z, −X, and −Z vertices.

We strongly emphasize two points that are often overlooked in these experiments. First, one must consider the change in free energy, both enthalpic and entropic contributions, of the entire system – complete amino acid sequence plus solvent – when comparing metallo- and apo forms.[34] Considering only the ligands that coordinate calcium will not do. Second, high affinity and/or selectivity for calcium or thermostability are not necessarily what was selected by evolution. Some EF-hands, such as the AB-hand of parvalbumin, do not bind calcium; they are certainly not 'nonfunctional'.

As noted, all EF-hands, excepting the AB-hand of parvalbumin occur in pairs, either with the adjacent EF-hand of that protein or with another fifth domain of another penta-EF-hand protein. Much of the stability, divalent-ion affinity, and character of EF-hand proteins derive from the conformations of their pairs of hands.[34] Dicalcicalmodulin-3 and 4 is taken as the reference point and is viewed down its approximate twofold rotation axis. Dicalciparvalbumin-CD and EF is superimposed (Figure 3(a)); the two pairs of domains are very similar. In Figure 3(b), apocalmodulin-3 and 4 is superimposed on dicalcicalmodulin-3 and 4; there is a significant narrowing of the cleft between the *odd* and the *even* domains upon loss of calcium and of the cleft between helix E and helix F of both members of the pair. The more open structure associated with the binding of calcium is similar to the conformations of pairs of EF-hands observed with the binding of target peptides, or mimics, by calmodulin, troponin C, and myosin light chains (Figure 3(c)). That is, the binding of calcium and of the target to the exposed hydrophobic cleft is synergistic. Consistent with these observations, the structure of dicalcides1-37 rat parvalbumin that lacks the AB-domain is more closed and has over a hundredfold reduced affinity for calcium.[35] It is not known whether parvalbumin plays any function other than buffering calcium pulses. Perhaps the unique AB-hand covers this pudendum to assure that parvalbumin does *not* interact with targets meant for calmodulin and other signal-transducing EF-hand proteins.

REFERENCES

1. A Lewit-Bentley and S Réty, *Curr Opin Struct Biol*, **10**, 637–43 (2000).
2. J Michies, C Xi, J Verhaert and J Vanderleyden, *Trends Microbiol*, **10**, 87–93 (2002).
3. JG Henrotte, *Nature*, **176**, 1221 (1955).
4. JG Henrotte and G Hamoir, *Acta Chem Scand*, **12**, 351–52 (1958).
5. S Konosu, G Hamoir and J-F Pechère, *Biochem J*, **96**, 98–112 (1965).
6. J-F Pechère, *Comp Biochem Physiol*, **24**, 289–95 (1968).
7. RH Kretsinger, *Nat New Biol*, **240**, 85–88 (1972).
8. CE Nockolds, RH Kretsinger, CJ Coffee and RA Bradshaw, *Proc Natl Acad Sci*, **69**, 581–84 (1972).
9. CJ Coffee and R Bradshaw, *J Biol Chem*, **248**, 3302–12 (1973).
10. RA Sayle and EJ Milner White, *Trends Biochem Sci*, **20**, 374–76 (1995).
11. S Nakayama, H Kawasaki and RH Kretsinger, *Top Biol Inorg Chem*, **3**, 29–58 (2000).
12. S Das Dores, C Chopin, C Villaume, J Fleurence and JL Guéant, *Allergy*, **57**(Suppl. 72), 79–83 (2002).
13. MT Henzl, HM Zhao and CT Saez, *FEBS Lett*, **375**, 137–42 (1995).
14. PJ Kraulis, *J Appl Crystallogr*, **24**, 946–50 (1991).
15. EA Merritt and DJ Bacon, *Methods Enzymol*, **277**, 505–24 (1997).
16. J-M Gilles, DB Thomason, J LeFevre and RH Kretsinger, *J Muscle Res Cell Motil*, **3**, 377–98 (l982).
17. P Coutu and JM Metzger, *Biophys J*, **82**, 2565–79 (2002).
18. LM John, M Mosquera-Caro, P Camacho and JD Lechleiter, *J Physiol (London)*, **35**, 3–16 (2001).
19. RH Kretsinger, JE Mann, JG Simmonds, Model of facilitated diffusion of calcium by the intestinal calcium binding protein, in AW Norman (ed.), *Proceedings of the Fifth Workshop on Vitamin D*, Walter de Gruyter & Co., New York, pp. 233–48 (1982).
20. DW Maughan and RE Godt, *J Muscle Res Cell*, **20**, 199–209 (1999).
21. PR Hof, H Glezer, F Conde, RA Flagg, MB Rubin, EA Nimchinshky and DMV Weisenhorn, *J Chem Neuroanat*, **16**, 77–116 (1999).
22. O Caillard, H Moreno, B Schwaller, I Llano, MR Celio and A Marty, *Proc Natl Acad Sci*, **97**, 13372–77 (2000).
23. RC Hapak, CM Stanley and MT Henzl, *Exp Cell Res*, **222**, 234–45 (1996).
24. JP MacManus, *Cancer Res*, **39**, 3000–5 (1979).
25. N Sakaguchi, MT Henzl, I Thalmann, R Thalmann and BA Schulte, *J Histochem Cytochem*, **46**, 29–39 (1998).
26. K Aas and JW Jebsen, *Int Arch Allergy*, **32**, 1–20 (1967).
27. C Hilger, F Grigioni, L Thill, L Mertens and F Hentges, *Allergy*, **57**, 1053–58 (2002).
28. A Bugajska-Schretter, L Elfman, T Fuchs, S Kapiotis, H Rumpold, R Valenta and S Spitzauer, *J Allergy Clin Immunol*, **101**, 67–74 (1998).
29. UG Föhr, BR Weber, M Müntener, W Staudenmann, GJ Hughes, F Frutiger, D Banville, BW Schäfer and CW Heizmann, *Eur J Biochem*, **215**, 719–27 (1993).
30. P Verdino, K Westritschnig, R Valenta and W Keller, *EMBO J*, **21**, 5007–16 (2002).
31. RE Reid and RS Hodges, *J Theor Biol*, **84**, 401–44 (1980).
32. MT Henzl, WG Wycoff, JD Larson and JJ Likos, *Protein Sci*, **11**, 158–73 (2002).
33. MS Cates, MB Berry, EL Ho, Q Li, JD Potter and GN Phillips, *Struct Fold Des*, **7**, 1269–78 (1999).
34. MR Nelson, E Thulin, PA Fagan, S Forsén and WJ Chazin, *Protein Sci*, **11**, 198–205 (2002).
35. M Thépaut, M-P Strub, A Cavé, J-L Banères, MW Berchtold, C Dumas and A Padilla, *Proteins: Struct, Funct, Genet*, **45**, 117–28 (2001).

Basement membrane protein BM-40

Erhard Hohenester[†] and Rupert Timpl[‡]

[†]Department of Biological Sciences, Imperial College, London, UK
[‡]Max-Planck-Institut für Biochemie, Martinsried, Germany

FUNCTIONAL CLASS

Basement membrane protein of M_r 40 000 Da (BM-40),[1] also known as osteonectin[2] or SPARC (for secreted protein, acidic, and rich in cysteine),[3] is a calcium-binding counteradhesive protein secreted by many different cell types. BM-40 is a component of several extracellular matrices. It primarily appears to regulate cell–matrix interactions, but may also contribute to the matrix structure.

OCCURRENCE

BM-40 is one of the few extracellular matrix proteins that are found in all metazoa whose genomes have been sequenced to date.[4] BM-40 expression is regulated spatially and temporally during vertebrate development.[5–7] In the mouse embryo, the protein is initially localized to the heart primordia, somites, and extraembryonic membranes, but BM-40 expression becomes more widespread at later stages, with high levels in many organs, developing bone, epithelia, and endothelia. In the adult animal, the expression of BM-40 is more restricted and usually associated with tissues undergoing repair or remodeling, such as gut and bone. BM-40 is also highly expressed during wound healing and by certain invasive tumors.[6]

BIOLOGICAL FUNCTION

The biological function of BM-40 is still unclear. *In vitro* BM-40 binds to hydroxyapatite, several types of collagen, vitronectin, thrombospondin-1, platelet-derived growth factor (PDGF) and vascular endothelial growth

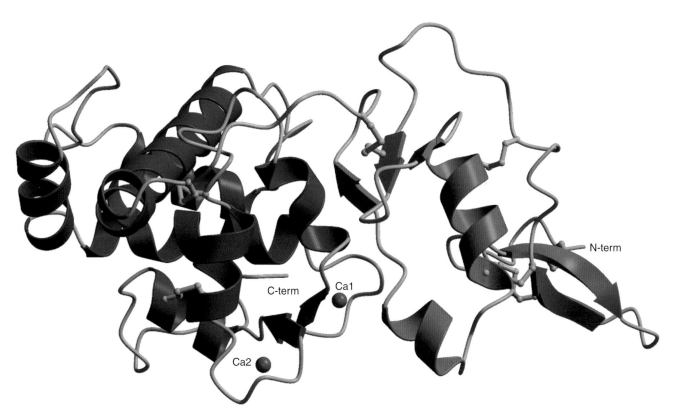

3D Structure Schematic representation of the structure of the BM-40 FS–EC domain pair containing two calcium ions, PDB code 1BMO.[35] The FS and EC domains are in green and blue respectively. Disulfide bridges are shown as yellow sticks. Calcium ions are shown as pink spheres. Made with BOBSCRIPT[38] and RASTER3D.[39]

factor (VEGF), endothelial cells and platelets.[5-7] BM-40 appears to regulate cellular phenotypes and has been described as an antiproliferative and counteradhesive protein.[5-7] Studies in invertebrate model systems have demonstrated critical roles in embryogenesis.[8,9] Ablation of BM-40 in the mouse does not perturb embryogenesis, but results in early-onset cataractogenesis.[10,11] Other phenotypic abnormalities include osteopenia, accelerated dermal wound healing, and altered collagen fibrils.[7]

AMINO ACID SEQUENCE INFORMATION

The SWISS-PROT/TrEMBL database contains BM-40 sequences from 12 organisms. A representative selection is given here.

- *Homo sapiens* (human), precursor, 303 amino acid residues (AA), SWISS-PROT/TrEMBL accession code P09486.[12-14]
- *Mus musculus* (mouse), precursor, 302 AA, accession code P07214.[3,12,15]
- *Xenopus laevis* (African clawed frog), precursor, 300 AA, accession code P36278.[16]
- *Drosophila melanogaster* (fruit fly), precursor, 304 AA, accession code O97365.[17]
- *Caenorhabditis elegans*, precursor, 264 AA, accession code P34714.[18]

PROTEIN PRODUCTION, PURIFICATION, AND MOLECULAR CHARACTERIZATION

BM-40 has been extracted from bovine and human bone using 4 M guanidinium hydrochloride, 0.5 M ethylenediaminetetraacetic acid (EDTA)[2,19] and, less harshly, from the mouse Engelbreth–Holm–Swarm tumor using Tris-buffered saline, 10 mM EDTA.[15] BM-40 has also been purified from the culture medium of bovine aortic endothelial cells[3] and mouse teratocarcinoma-derived PSY-2 cells,[20] as well as from the supernatant of thrombin-activated platelets.[21] Recombinant BM-40 has been obtained by stable expression in human embryonic kidney 293 cells and fibrosarcoma HT 1080 cells.[22] Several BM-40 fragments and mutants for functional and structural studies have been obtained by either stable or episomal expression in 293 cells.[23-25] Expression in *Escherichia coli* has been reported, but it is not clear whether this material is natively folded.[26] Finally, BM-40 has been expressed in a baculovirus system.[27] Purification of BM-40 is usually achieved by a combination of anion exchange and gel filtration chromatography.[15,22]

BM-40 is a monomeric protein of calculated M_r 32 700 Da (human protein). On denaturing polyacrylamide gel electrophoresis, the recombinant protein migrates as a band of $M_r \approx 40\,000$ Da due to glycosylation at Asn99 (the numbering scheme refers to the mature human protein). The glycan of BM-40 varies according to the tissue the protein is extracted from: platelet BM-40 (and presumably 293 cell-derived BM-40) possesses a complex-type structure, whereas bone BM-40 carries a high mannose-type structure.[28] BM-40 purified from tissues, and not the recombinant protein, is proteolytically nicked at the Leu198–Leu199 peptide bond.[15] This processing increases collagen binding and occurs within various tissues, as shown by cleavage-specific neopeptide antibodies.[29,30]

BM-40 is a modular protein composed of three distinct domains: an acidic segment (1–52), a follistatin-like (FS) domain (53–137), and an α-helical, calcium binding domain (138–286), termed EC domain (for extracellular calcium binding).[24,31] BM-40 contains a total of seven disulfide bridges, five in the FS domain and two in the EC domain.

METAL CONTENT

Calcium binding to BM-40 has been monitored by circular dichroism (CD) spectroscopy and is accompanied by a large increase in α-helix content.[31,32] The acidic N-terminal domain binds several calcium ions with low affinity.[32] The EC domain was initially reported to bind a single calcium ion with high affinity,[23,24,31] but subsequent structure analysis revealed the presence of a second calcium ion.[33] A detailed analysis, employing the EGTA (ethylene glycol tetraacetic acid)/Ca-EGTA buffer system to accurately adjust calcium concentrations and monitoring calcium binding by changes in intrinsic protein fluorescence, showed that BM-40 binds two calcium ions with macroscopic equilibrium dissociation constants of 490 and 57 nM.[25] Calcium binding is highly cooperative and influenced by the interaction between the FS and EC domains.[25]

ACTIVITY TESTS

There is no established activity test for BM-40, but various *in vitro* activities have been used to assess the structural integrity of tissue-derived or recombinant BM-40. These include calcium-dependent collagen binding[21,28,34] and the inhibition of endothelial cell spreading and DNA synthesis.[26]

X-RAY STRUCTURE

Crystallization

Full-length BM-40 has not been crystallized to date. Crystals of the BM-40 EC domain were grown using 0.7 M sodium–potassium tartrate, 0.1 M sodium HEPES buffer (pH 7.5) as precipitant.[24] The space group of the crystals is $P2_12_12_1$ with $a = 54.7$ Å, $b = 55.2$ Å, $c = 75.5$ Å (1 Å = 0.1 nm), with one molecule in the asymmetric unit; the structure has been determined at 2.0-Å resolution.[33] Crystals of the glycosylated FS–EC domain pair of human BM-40 were grown using 14% polyethylene glycol (PEG) 4000, 0.15 M sodium acetate, 0.1 M sodium HEPES buffer (pH 7.5) as precipitant.[35] The space group of the crystals is P2$_1$ with $a = 70.8$ Å, $b = 56.4$ Å, $c = 111.6$ Å, $\beta = 104.4$ Å, with two molecules in the asymmetric unit; the structure has been determined at 3.1-Å resolution.[35] Crystals of a BM-40 FS–EC mutant with increased affinity for collagen were grown using 12–14% PEG4000, 10% 2-propanol, 0.1 M sodium HEPES buffer (pH 7.5) as precipitant.[36] The space group of the crystals is P2$_1$2$_1$2$_1$ with $a = 51.9$ Å, $b = 88.1$ Å, $c = 153.7$ Å, with two molecules in the asymmetric unit; the structure has been determined at 2.8-Å resolution.[36]

Overall description of the structure

The BM-40 FS–EC structure consists of two distinct domains (3D Structure).[35] The elongated FS domain (53–137) is composed of two rather independent subdomains, a twisted β-hairpin (54–77) and a pair of α-helices connected to a small three-stranded antiparallel β-sheet (78–137). The latter subdomain is similar to proteinase inhibitors of the Kazal family,[37] but BM-40 does not appear to have any proteinase inhibitor activity.[35] The five disulfide bridges in the FS domain are linked 55–66, 60–76, 78–113, 84–106, and 95–132. The glycosylation site at Asn99 is located in the turn between β3 and β4. Two N-acetylglucosamine moieties are defined by the electron density map and interact with the side chain of Tyr134.

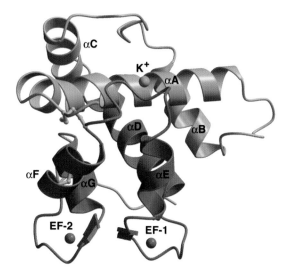

Figure 1 Schematic representation of the structure of the BM-40 EC domain.[33] Disulfide bridges and metal ions are shown as yellow sticks and spheres, respectively. The EF-hand pair is in orange. Secondary structure elements are labeled. Made with BOBSCRIPT[38] and RASTER3D.[39]

The EC domain (138–286) folds into a compact, α-helical structure (Figure 1). A long N-terminal helix αA is connected to a pair of calcium-binding EF-hands[40–43] by two helices, αB and αC, and three surface loops. The first EF-hand motif, EF-1, is made up of helices αD and αE, with the latter helix containing a proline residue, Pro237. The second motif, EF-2, is made up of helices αF and αG. Both EF-hands contain a calcium ion bound with apparently full occupancy. The tight association of helix αA and the EF-hand pair resembles the binding mode of amphipathic target helices to calmodulin.[33] The two disulfide bridges in the EC domain are linked 138–248 and 256–272. There are no glycosylation sites in the EC domain.

The FS and EC domains of BM-40 interact through a predominantly polar interface that buries 1100 Å^2 of solvent-accessible surface. The interface involves strand β5 of the FS domain and the EF-hand pair of the EC domain, in particular helix αE of EF-1.

Structure of the EF-hand calcium binding sites

The overall paired arrangement of EF-1 and EF-2 is very similar to canonical EF-hand pairs in intracellular proteins.[41,43] The two EF-hands interact intimately through hydrophobic contacts between the helices and a short antiparallel β-sheet formed by the two calcium-binding loops. Both calcium ions are bound with pentagonal bipyramidal coordination, with a water molecule at one apex. The detailed arrangement of the residues coordinating one of the metal ions is unusual, however (Figure 2).

Calcium ion 1 (Ca1) bound to EF-1 is coordinated by the side chains of Asp222 (position X of the EF-hand consensus; metal–ligand distance, 2.14 Å),[41,43] Asp227 (Z; 2.41 Å), and Glu234 (−Z; bidentate interaction; 2.55 Å and 2.58 Å), the peptide carbonyl groups of Pro225 (Y; 2.38 Å), and Tyr229 (−Y; 2.44 Å), and a water molecule (−X; 2.18 Å). The average Ca1–ligand distance is 2.38 Å. The coordination by a peptide carbonyl group at position Y is unusual (typical EF-hands use an acidic or polar amino acid side chain) and results from a one-residue insertion between X and Y in BM-40, which is accommodated by the looping out of residues 223–224 and a cis-peptide bond between His224 and Pro225.[33] Several residues in EF-1 (His224, His232, Thr233) participate in the interface to the FS domain.[35]

Calcium ion 2 (Ca2) bound to EF-2 is coordinated by the side chains of Asp257 (X; 2.29 Å), Asp259 (Y; 2.37 Å), Asp261 (Z; 2.31 Å), and Glu268 (−Z; bidentate; 2.36 Å and 2.74 Å), the peptide carbonyl group of Tyr263 (−Y; 2.43 Å), and a water molecule (−X; 2.42 Å). The average Ca1–ligand distance is 2.42 Å. The Cys256–Cys272 disulfide bridge links helices F and G of EF-2, replacing a hydrophobic contact in canonical EF-hand motifs.

Basement membrane protein BM-40

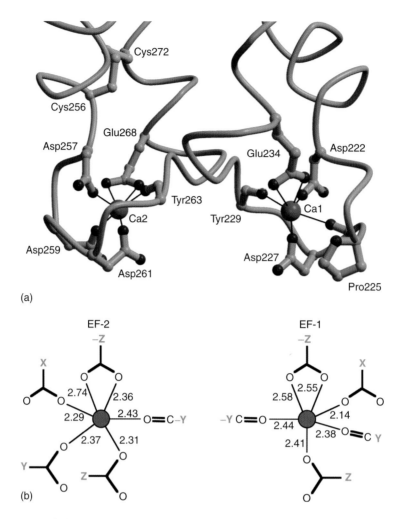

Figure 2 (a) Detailed structure of the EF-hand calcium-binding loops in the BM-40 EC domain. Calcium ions are shown as pink spheres. Residues coordinating the calcium ions are shown in atomic detail and are labeled. Made with BOBSCRIPT[38] and RASTER3D.[39] (b) Schematic drawing of the calcium coordination in the BM-40 EC domain, showing the calcium–ligand distances in Å and the EF-hand nomenclature (brown).[41,43]

Structure of a third metal binding site

In the crystal structure of the EC domain, a third metal ion was observed bound to the tip of the loop connecting EF-1 and EF-2.[33] This metal ion is coordinated by the side chain of Glu246, the peptide carbonyl groups of Pro241 and Ile243, and three water molecules (Figure 3). The coordination geometry is octahedral, with an average metal–ligand distance of 2.8 Å. The metal ion was assigned as potassium, consistent with the high potassium concentration in the crystallization buffer and its replacement by rubidium in a soaking experiment.[33] In the crystal structure of a BM-40 FS–EC mutant with increased affinity for collagen,[36] a metal ion was observed in the same position. However, two of the three metal ions were replaced by amino acid ligands from another BM-40 molecule in the crystal, and the average metal–ligand distance was reduced to 2.3 Å, more characteristic of a calcium ion. The functional significance (if any) of the third calcium ion in BM-40 is unknown, but it has been speculated that the ion may be involved in calcium-dependent collagen binding.[36]

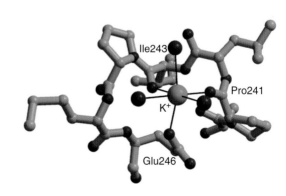

Figure 3 Detailed structure around the putative potassium ion (green) in the BM-40 EC domain. Made with BOBSCRIPT[38] and RASTER3D.[39]

FUNCTIONAL ASPECTS

Energetics of calcium binding

The family of EF-hand calcium binding proteins mainly consists of intracellular proteins involved in calcium signaling or buffering in the cytosol, whose affinities for calcium are tuned to the changes in calcium concentration upon activation (0.1 μM in the resting cell compared with 10 μM in the activated cell). The free calcium concentration in the extracellular space is high (1.2 mM) and constant, most likely ensuring full occupation of both EF-hands in BM-40 at all times.[44]

In a detailed study of calcium binding to BM-40, macroscopic equilibrium dissociation constants of 490 and 57 nM were determined for full-length BM-40.[25] The higher affinity for the second equivalent is an indication of the cooperative nature of calcium binding. The recombinant FS–EC domain pair has very similar calcium-binding characteristics. In contrast, in the isolated EC domain of BM-40, the macroscopic equilibrium dissociation constants are 11 750 and 56 nM, demonstrating a significant stabilizing influence of the FS domain on the EF-hand pair of BM-40.[25] This is consistent with the BM-40 FS–EC structure, in which the variant EF-hand 1 is intimately involved in the domain interface.[35] Mutation of key EF-hand residues had dramatic effects, abolishing calcium binding to the mutated site and, hence, all cooperativity.[25]

The structural basis for the large increase in α-helix content upon calcium binding to BM-40 detected by CD spectroscopy is not known.[24,31,32] Calcium-depleted BM-40 is more susceptible to proteolysis, but the unfolding by guanidinium hydrochloride is not affected by the removal of calcium.[32]

Influence of calcium binding on the biological activities of BM-40

Binding of BM-40 to triple–helical collagen is the only biological activity of BM-40 that has been conclusively shown to be affected by calcium. In the presence of 2 mM CaCl$_2$, full-length BM-40[23,34] and the recombinant EC domain[24] bind to collagen IV, as well as to the fibrillar collagens I, II, III, and V, whereas no binding is observed in 10 mM EDTA. It is assumed that the large conformational change upon calcium removal eliminates a discontinuous collagen binding site in the EC domain of BM-40.[36] The dissociation constants of collagen bound to BM-40 have been determined using a surface plasmon resonance assay and are 3 to 7 μM and 10 to 20 μM for tissue-derived BM-40 and the recombinant EC domain respectively.[29] Limited digestion with various matrix metalloproteinases (MMP-2, MMP-3, MMP-7, MMP-13) leads to a 10- to 20-fold increase in collagen affinity, which can be attributed to a single proteolytic nick of the Glu197–Leu198 peptide bond.[29] A similar increase in collagen affinity can be achieved by recombinant deletion of helix αC (residues Val196–Phe203).[36] The crystal structure of the FS–EC domain pair of this activated BM-40 mutant demonstrates significant structural rearrangements, exposing regions on the αA helix that are buried in the wild-type structure.[36] On the basis of this observation, a total of 17 single and two double mutants were used to map the collagen binding site on the EC domain. Five residues were shown to be critical: Arg149 and Asn156 on the αA helix; and Leu242, Met245, and Glu246 on the αE–αF loop connecting the two EF-hand motifs.[36] The collagen binding site in the EC domain defined by mutagenesis is adjacent to the glycosylation site in the FS domain (Figure 4). Conflicting data exist on the influence of the glycan on collagen binding.[36,45]

The EC domain of BM-40 has also been shown to bind with high affinity (dissociation constant 5–10 nM) to PDGF AA and BB.[46] This binding was not dependent on calcium and not affected by mutations affecting collagen binding. The biological role of PDGF binding to BM-40 is not known. An earlier report indicated that BM-40 might inhibit the binding of PDGF to its receptor,[47] but this was not observed by Göhring et al.[46]

A synthetic peptide spanning EF-hand 2 of BM-40 (residues Thr255–Gly274), termed 4.2, has been used extensively in in vitro studies addressing BM-40 function.

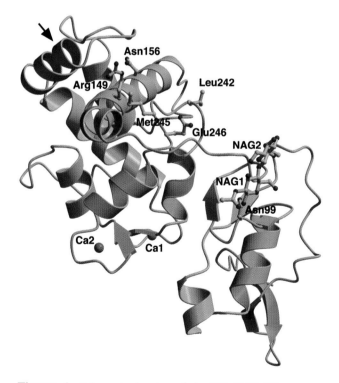

Figure 4 Schematic drawing of the BM-40 FS–EC structure showing the collagen binding site and the asparagine-linked glycan. Residues in the EC domain implicated by mutagenesis in collagen binding[36] are shown in atomic detail, as are the two N-acetylglucosamine moieties attached to Asn99. The arrow marks the peptide bond cleaved by matrix metalloproteinases. Made with BOBSCRIPT[38] and RASTER3D.[39]

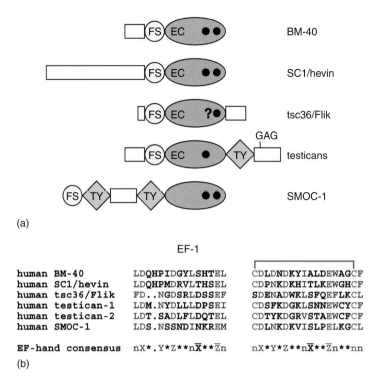

Figure 5 The BM-40 protein family. (a) Domain organization of human proteins containing an EC domain. FS, follistatin-like domain; EC, extracellular calcium binding domain; thyroglobulin-like (TY) domain; glycosaminoglycan (GAG) modification. Open rectangles represent domains of unknown structure. In the EC domain, black circles indicate calcium ions; (b) Alignment of EF-hand sequences.

Since the original observation that this sequence modulates cell shape,[48] peptide 4.2 has been reported to bind to cells, thereby inhibiting cell spreading and proliferation.[5–7] Recently, peptide 4.2 has been shown to affect morphogenesis in X. laevis, providing the first evidence of its counteradhesive activity in vivo.[49] Because peptide 4.2 contains all the calcium-binding residues of EF-2, as well as the cysteines that form the circularizing disulfide bridge, it has always been assumed that it binds calcium with high affinity. Indeed, treatment with chelating agents[48] or replacement of key residues[50] abrogates the in vitro effects of peptide 4.2. Maurer et al. studied a slightly longer peptide (Thr250–Ile275) and found that it binds calcium with apparent dissociation constants ranging from 0.3 to 3.8 mM, depending on the peptide concentration; this was shown to be the result of calcium-dependent peptide dimerization.[24] Thus, at the conditions used in most in vitro assays (typically 10–100 μM peptide, 1–2 mM calcium),[50] peptide 4.2 is likely to act as a homodimer. It would be important to confirm the counteradhesive activity of peptide 4.2 using site-directed mutagenesis of correctly folded BM-40, but such experiments have not been reported to date.

THE BM-40 PROTEIN FAMILY

A number of vertebrate proteins contain an EC domain homologous to the one in BM-40. In mammals, the BM-40 protein family includes the close BM-40 homolog SC1/hevin;[51] tsc36/Flik;[52] the proteoglycans testican-1,[53] testican-2,[54] and testican-3;[55] and SMOC-1[56] (Figure 5). The biological functions of these proteins are largely unknown. Calcium binding has been demonstrated experimentally for testican-1,[57] testican-2,[54] and SMOC-1.[56] Interestingly, the EC domains of testicans appear to bind only one calcium ion with modest affinity (30–70 μM dissociation constant). This is consistent with the EF-hand sequences, which show a canonical calcium-binding loop in EF-1 and an inactivating replacement in EF-2 (position Y, which normally is an acidic residue, is occupied by an aromatic residue).

REFERENCES

1. M Dziadek, M Paulsson, M Aumailley and R Timpl, Eur J Biochem, **161**, 455–64 (1986).
2. JD Termine, HK Kleinman, SW Whitson, KM Conn, ML McGarvey and GR Martin, Cell, **26**, 99–105 (1981).
3. IJ Mason, A Taylor, JG Williams, H Sage and BLM Hogan, EMBO J, **5**, 1465–72 (1986).
4. RO Hynes and Q Zhao, J Cell Biol, **150**, F89–F95 (2000).
5. Q Yan and EH Sage, J Histochem Cytochem, **47**, 1495–505 (1999).
6. RA Brekken and EH Sage, Matrix Biol, **19**, 569–80 (2000).
7. AD Bradshaw and EH Sage, J Clin Invest, **107**, 1049–54 (2001).

8 L Purcell, J Guia-Gray, S Scanga and M Ringuette, *J Exp Zool*, **265**, 153–64 (1993).
9 MC Fitzgerald and JE Schwarzbauer, *Curr Biol*, **8**, 1285–88 (1998).
10 DT Gilmour, GJ Lyon, MB Carlton, JR Sanes, MJ Cunningham, JR Anderson, BL Hogan, MJ Evans and WH Colledge, *EMBO J*, **17**, 1860–70 (1998).
11 K Norose, JI Clark, NA Syed, A Basu, S Heber-Katz, EH Sage and CC Howe, *Invest Opthalmol Vis Sci*, **39**, 2674–80 (1998).
12 B Lankat-Buttgereit, K Mann, R Deutzmann, R Timpl and T Krieg, *FEBS Lett*, **236**, 352–56 (1988).
13 A Swaroop, BLM Hogan and U Francke, *Genomics*, **2**, 37–47 (1988).
14 XC Villarreal, KG Mann and GL Long, *Biochemistry*, **28**, 6483–91 (1989).
15 K Mann, R Deutzmann, M Paulsson and R Timpl, *FEBS Lett*, **218**, 167–72 (1987).
16 S Damjanovski, F Liu and MJ Ringuette, *Biochem J*, **281**, 513–17 (1992).
17 MD Adams, JC Venter, et al., *Science*, **287**, 2185–95 (2000).
18 JE Schwarzbauer and CS Spencer, *Mol Biol Cell*, **4**, 941–52 (1993).
19 RW Romberg, PG Werness, P Lollar, BL Biggs and KG Mann, *J Biol Chem*, **260**, 2728–36 (1985).
20 H Sage, B Vernon, SE Funk, EA Everitt and J Agnello, *J Cell Biol*, **109**, 341–56 (1989).
21 RJ Kelm Jr and KG Mann, *Blood*, **75**, 1105–13 (1990).
22 R Nischt, J Pottgieser, T Krieg, U Mayer, M Aumailley and R Timpl, *Eur J Biochem*, **200**, 529–36 (1991).
23 J Pottgiesser, P Maurer, U Mayer, R Nischt, K Mann, R Timpl, T Krieg and J Engel, *J Mol Biol*, **238**, 563–74 (1994).
24 P Maurer, C Hohenadl, E Hohenester, W Göhring, R Timpl and J Engel, *J Mol Biol*, **253**, 347–57 (1995).
25 E Busch, E Hohenester, R Timpl, M Paulsson and P Maurer, *J Biol Chem*, **275**, 25508–15 (2000).
26 JA Bassuk, LP Braun, K Motamed, F Baneyx and EH Sage, *Int J Biochem Cell Biol*, **28**, 1031–43 (1996).
27 AD Bradshaw, JA Bassuk, A Francki and EH Sage, *Mol Cell Biol Res Commun*, **3**, 345–51 (2000).
28 RJ Kelm and KG Mann, *J Biol Chem*, **266**, 9632–39 (1991).
29 T Sasaki, W Göhring, K Mann, P Maurer, E Hohenester, V Knäuper, G Murphy and R Timpl, *J Biol Chem*, **272**, 9237–43 (1997).
30 T Sasaki, N Miosge and R Timpl, *Matrix Biol*, **18**, 499–508 (1999).
31 J Engel, W Taylor, M Paulsson, H Sage and B Hogan, *Biochemistry*, **26**, 6958–65 (1987).
32 P Maurer, U Mayer, M Bruch, P Jenö, K Mann, R Landwehr, J Engel and R Timpl, *Eur J Biochem*, **205**, 233–40 (1992).
33 E Hohenester, P Maurer, C Hohenadl, R Timpl, JN Jansonius and J Engel, *Nat Struct Biol*, **3**, 67–73 (1996).
34 U Mayer, M Aumailley, K Mann, R Timpl and J Engel, *Eur J Biochem*, **198**, 141–50 (1991).
35 E Hohenester, P Maurer and R Timpl, *EMBO J*, **16**, 3778–86 (1997).
36 T Sasaki, E Hohenester, W Göhring and R Timpl, *EMBO J*, **17**, 1625–34 (1998).
37 W Bode and R Huber, *Eur J Biochem*, **204**, 433–51 (1991).
38 RM Esnouf, *J Mol Graph*, **15**, 132–34 (1997).
39 EA Merritt and DJ Bacon, *Methods Enzymol*, **277**, 505–24 (1997).
40 RH Kretsinger and CE Nockolds, *J Biol Chem*, **248**, 3313–26 (1973).
41 NCJ Strynadka and MNG James, *Annu Rev Biochem*, **58**, 951–98 (1989).
42 S Nakayama and RH Kretsinger, *Annu Rev Biophys Biomol Struct*, **23**, 473–507 (1994).
43 A Lewit-Bentley and S Rety, *Curr Opin Struct Biol*, **10**, 637–43 (2000).
44 P Maurer and E Hohenester, *Matrix Biol*, **15**, 569–80 (1997).
45 R-L Xie and GL Long, *J Biol Chem*, **270**, 23212–17 (1995).
46 W Göhring, T Sasaki, CH Heldin and R Timpl, *Eur J Biochem*, **255**, 60–66 (1998).
47 EW Raines, TF Lane, ML Iruela-Arispe, R Ross and EH Sage, *Proc Natl Acad Sci USA*, **89**, 1281–85 (1992).
48 TF Lane and EH Sage, *J Cell Biol*, **111**, 3065–76 (1990).
49 M-H Huynh, EH Sage and M Ringuette, *Dev Growth Differ*, **41**, 407–18 (1999).
50 EH Sage, JA Bassuk, JC Yost, MJ Folkman and TF Lane, *J Cell Biochem*, **57**, 127–40 (1995).
51 IG Johnston, T Paladino, JW Gurd and IR Brown, *Neuron*, **2**, 165–76 (1990).
52 M Shibanuma, J Mashimo, A Mita, T Kuroki and K Nose, *Eur J Biochem*, **217**, 13–19 (1993).
53 PM Alliel, J-P Perin and FJ Bonnet, *Eur J Biochem*, **214**, 347–50 (1993).
54 C Vannahme, S Schübel, M Herud, S Gösling, H Hülsmann, M Paulsson, U Hartmann and P Maurer, *J Neurochem*, **73**, 12–20 (1999).
55 U Hartmann and P Maurer, *Matrix Biol*, **20**, 23–35 (2001).
56 C Vannahme, N Smyth, N Miosge, S Gösling, C Frie, M Paulsson, P Maurer and U Hartmann, *J Biol Chem*, **277**, 37977–86 (2002).
57 E Kohfeldt, P Maurer, C Vannahme and R Timpl, *FEBS Lett*, **414**, 557–61 (1997).

PEFLINS: a family of penta EF-hand proteins

Miroslaw Cygler

Biotechnology Research Institute, National Research Council of Canada, Montreal, Quebec, Canada

FUNCTIONAL CLASS

Penta-EF hand proteins (PEFLINs) are regulatory Ca^{2+}-binding proteins. They belong to the EF-hand superfamily and are characterized by five EF-hand motifs. Usually, the Ca^{2+}-binding domain is preceded by a Gly/Pro-rich sequence. PEFLINs associate with membranes in a calcium-dependent manner and form homo- and heterodimers.

OCCURRENCE

Penta EF-hand Ca^{2+}-binding proteins, with the exception of the calpain large chain, are usually relatively small proteins, approximately 200 to 250 residues long. The Ca^{2+}-binding domain (~170 residues) is associated with a 20- to 100-residue long N terminal Pro/Gly rich region, which in the Ca^{2+}-bound state leads to exposure of a hydrophobic surface and association with membranes. These proteins play regulatory roles, and a number of partners have recently been identified for several of them.[1-5] They are found predominantly in higher eukaryotes with cellular distribution varying from broad (ALG-2) to more specific (grancalcin). Recently, sequences homologous to that of ALG-2 have been found in slime mould *Dictyostelium discoideum*,[6] suggesting a wider occurrence of this protein in nature.

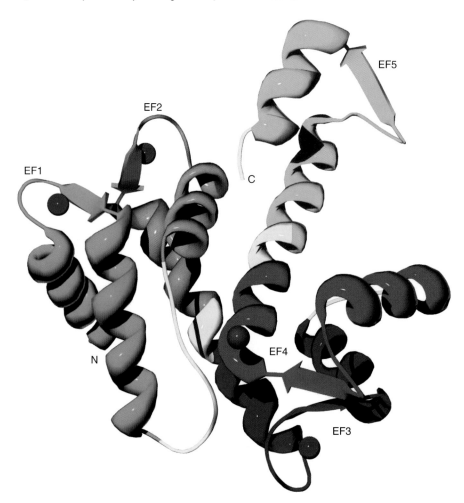

3D Structure Domain VI of calpain (PDB code 1DVI) with four bound Ca^{2+} ions. EF1–5 are painted green, orange, blue, magenta, and cyan, respectively. The connections between EF-hands are marked in yellow. Figure prepared with sPDBviewer[63] and PovRay (www.povray.org).

BIOLOGICAL FUNCTION

This ~170-residue-long domain was first identified in the calpain's heavy[7] and light[8] chains. The sequences analysis showed low level of similarity to calmodulin, with four potential calcium-binding EF-hands discerned in the sequence. The three-dimensional structure of domain VI of calpain (light chain) showed that there are five, rather than the predicted four, EF-hand motifs.[9,10] In subsequent years, this domain was identified in several other proteins, namely, sorcin, grancalcin, ALG-2, and peflin,[11] and the name PEFLINS was coined for the entire family. Members of this family have by themselves no enzymatic activity but appear to have diverse functions regulated through the common feature of calcium binding.

Calpain

Calpain is a ubiquitous calcium-activated cysteine protease composed of the heavy (~700 residues) and the light (270 residues) polypeptide chains.[12] Each chain contains a Ca^{2+}-binding domain at the C-terminus that is essential for the formation of heterodimers.[13–15] A large number of calpain isoforms have been found, the two principal ones being μ- and m-calpains.[16–19] These differ in their large, catalytic subunit while associating with the same light subunit, and have different *in vitro* Ca^{2+} binding properties with affinities ranging from low to high micromolar. These proteases coexist with a calpain-specific endogenous inhibitor protein called calpastatin.[16,20] Calpains are postulated to play a role in many cellular events and a large number of potential substrates have been identified *in vitro*. Despite this, their detailed biological role and the distinct functions of different isoforms are far from being fully understood.[21]

Sorcin

Sorcin (<u>so</u>luble <u>r</u>esistance-related <u>c</u>alcium binding prote<u>in</u>) is a 21.6-kDa protein and was initially isolated from the cytosol of multidrug-resistant cancer cells where it was co-expressed with *P*-glycoprotein.[22] Cloning of this gene overexpressed in the multidrug-resistant Chinese hamster ovary cell line showed that it is homologous to Ca^{2+}-binding domains of calpain, displaying ~35% sequence identity. Sorcin is also expressed in a range of normal tissues including skeletal muscle, heart, and brain, but at lower levels.[23] The biological role of sorcin is not yet fully understood. It has been shown that in the presence of calcium sorcin undergoes translocation from the cytosol to membranes, where it binds to target proteins.[24] Among proteins identified as sorcin targets are the ryanodine receptor, a Ca^{2+}-release channel responsible for Ca^{2+} release from the sarcoplasmic reticulum,[23] the pore-forming subunit of the voltage-dependent L-type Ca^{2+} channel,[25] annexin VII,[4] and presenilin 2.[26] A recent report suggests a role of sorcin in the inhibition of calcium release and modulation of excitation–contraction coupling in the heart through control of the ryanodine receptor activity.[27]

Grancalcin

Grancalcin is a cytosolic Ca^{2+}-binding protein, specifically associated with cells originating in the bone marrow. It is found predominantly in neutrophils and monocytes, and in relatively small amounts in lymphocytes.[28,29] In the presence of physiological concentrations of Ca^{2+} in human neutrophils, the grancalcin binds reversibly to secretory vesicles and plasma membranes and might, therefore, play a role in the regulation of granule-membrane fusion and degranulation.[30] L-plastin, a leukocyte-specific, actin-bundling protein, was identified some time ago as a grancalcin-binding partner.[3] Most recently, a strong interaction between grancalcin and sorcin was established.[31] A grancalcin-deficient mouse showed that the absence of grancalcin affected neither the generation of mature neutrophils in the bone marrow nor neutrophil or macrophage recruitment in response to the sterile inflammatory stimulus and that its role, if any, in the control of leukocyte activation may be a subtle one.[32]

ALG-2

Apoptosis-linked gene-2 protein ALG-2 is a 22-kDa Ca^{2+}-binding protein, highly conserved and ubiquitously expressed with proapoptotic function.[33] Inhibition of ALG-2 blocks apoptosis that is triggered by cell surface receptor Fas (CD95) and other stimuli in a variety of cell lines. It was thus concluded that ALG-2 acts downstream of caspase-3 or independently of this enzyme.[34] Several proteins that interact with ALG-2 in a calcium-dependent manner have been discovered in the last few years including AIP/Alix,[1,2] peflin,[35] annexin VII,[36] annexin XI[37] and an RNA-binding protein,[36] indicating possible involvement of ALG-2 not only in apoptosis but also in other processes. However, ALG-2-deficient mice show normal viability, and the development and differentiation of their immune system seems unimpaired and the experiments with the T cells obtained from these mice demonstrated that ALG-2 deficiency failed to block apoptosis induced by a variety of stimuli, indicating that ALG-2 is not essential or that another protein compensates for it in these mice.[38] Nevertheless, several reports describe increased levels of ALG-2 in cancerous and metastatic tissues.[35,36,39]

Peflin

By searching a human DNA database of expressed sequence tags (ESTs), Kitaura *et al.*[40] identified a partial sequence with significant similarity to ALG-2 and subsequently cloned the full-length cDNA. This protein has a long hydrophobic N-terminal segment of ~100 residues with several repeats of a Gly/Pro-rich nanopeptide sequence. Peflin is ubiquitously expressed in animal tissues,[40] and like other proteins from this family, it translocates to membranes in the presence of Ca^{2+}. The formation of heterodimers between peflin and ALG-2 was shown,[35] but the significance of this association is not clear. Indeed, the biological function of peflin is not presently known.

AMINO ACID SEQUENCE INFORMATION

Gene coding for PEFLIN proteins have been identified in a number of eukaryotic species. The most widely distributed is ALG-2, which has been found in protists, plants, fungi, nematodes, fruit fly, and mammals. The remaining proteins have been found to date only in higher eukaryotes. As many as 14 mammalian isoforms have been described for calpain large chain, containing the Ca^{2+}-binding domain at the C-terminus.[19] Representative examples of amino acid sequences for various PEFLINs are referenced below in Table 1.

The sequence conservation of ALG-2 proteins during evolution is quite remarkable. Between the human and *Drosophila melanogaster* proteins there is 41% sequence identity. Two proteins homologous to ALG-2 have been identified in the slime mould *D. discoideum* and they share ~28% sequence identity with the human protein. The level of similarity between the Ca^{2+}-binding domains of different PEFLINs from the same species is also rather high and, for example, in the human genome varies between 24 and 56% (Table 2).

Maki *et al.*[44] subdivided PEFLINs into two groups on the basis of the length of the first EF-hand (EF1) and analyzed the intron structure of their genes. ALG-2 and peflin have the EF1 of canonical length (group I), while calpains, sorcin, and grancalcin have one residue deletion in EF1 compared to the canonical EF-hand.[45]

Table 1 References to sequence data for representative PEFLINs

Name	Source	Length (AA)	NCBI GI number	References
Peflin	Human	284	6 912 582	40
Grancalcin	Human	217	6 912 388	28
ALG-2	Human	191	7 019 485	33
Sorcin	Human	198	267 021	41
Calpain small subunit	Human	268	67 665	42
Calpain large subunit isoform 1	Human	714	67 663	43

Table 2 Percent identity between human PEFLINs (excluding N-terminal extensions)

% Identity	ALG-2	Peflin	Sorcin	Grancalcin	Calpain dVI	Calpain dIV
ALG-2	100	41	36	35	30	24
Peflin	41	100	35	34	29	26
Sorcin	36	35	100	56	35	33
Grancalcin	35	34	56	100	32	33
Calpain dVI	30	29	35	32	100	50
Calpain dIV	24	26	33	33	50	100

PROTEIN PRODUCTION, PURIFICATION, AND MOLECULAR CHARACTERIZATION

To date, all penta EF-hand proteins or their calcium-binding domains, except peflin, were successfully overexpressed in *Escherichia coli* in fully active forms. Peflin has been expressed in HEK293 cells.

Calpain dVI

Various calpain isoforms have been successfully expressed.[46] Here we restrict description to the Ca^{2+}-binding domain (dVI) of the rat calpain small subunit, which provided the first structural view of PEFLIN proteins. The expression of this domain was accomplished in *E. coli* BL21(DE3).[46] The cloned domain corresponded to a protein of 184 residues with a calculated molecular mass of 21 257 Da. The predicted amino acid sequence is 91% identical to that of the corresponding portion of human calpain small subunit. The dVI protein was purified on a DEAE-Sephacel column and eluted from the column at about 125 mM NaCl.

Grancalcin

Recombinant grancalcin was expressed in *E. coli* as a glutathione S-transferase (GST) fusion. The supernatant from the cell lysate was loaded onto a glutathione Sepharose column and incubated with thrombin. Cleaved grancalcin was eluted from the column and loaded on a phenyl–Sepharose column in the presence of calcium, which increased its hydrophobicity. After washing, grancalcin was eluted by a low ionic strength buffer in the presence of ethylene glycol-O,O'-bis-[2-amino-ethyl]-N,N,N',N',-tetraacetic acid (EGTA).

Sorcin

Full-length sorcin and its Ca^{2+}-binding domain (SCBD, residues 33–198) were expressed in *E. coli* BL21(DE3).[47] The expressed full-length recombinant protein or the SCBD was extracted into the supernatant by centrifugation and purified by ion-exchange chromatography using a MonoQ FPLC column or a DEAE-5PW HPLC column with a 0 to 0.5 M NaCl gradient.

ALG-2

This cDNA was obtained after reverse transcription of mouse liver RNA and expressed in *E. coli* BL21(DE3) after induction with 1 mM isopropyl-beta-D-thiogalactopyranoside (IPTG). The protein was purified by Ca^{2+}-induced precipitation from total bacterial extracts followed by anion exchange chromatography and elution using a 0 to 0.5 M NaCl gradient.[48]

Peflin

Various peflin constructs were expressed in the HEK293 cell line as part of the investigations of its interaction with ALG-2.[49]

ACTIVITY AND INHIBITION TESTS

The PEFLINS bind from one to four Ca^{2+} ions. For some of them, the stoichiometry and binding constants have been determined experimentally. Berchtold *et al.*[48] measured Ca^{2+} binding by ALG-2. They identified two forms of ALG-2 differing by an insertion of only two residues, yet displaying different binding affinities for calcium. Addition of calcium in the absence of detergent leads to protein aggregation and precipitation during titration. In the presence of the nonionic detergent Triton X-100 (1%), this precipitation is prevented, suggesting an involvement of hydrophobic interactions in the Ca^{2+} induced precipitation. Indeed, ALG-2 exhibits a Ca^{2+}-dependent exposure of a hydrophobic surface as monitored with a fluorescent hydrophobicity probe.[50] Each of the two isoforms show two high-affinity Ca^{2+}-binding sites with the Ca^{2+} concentration at half saturation ($[Ca^{2+}]_{0.5}$) of 1.2 to 3.1 μM and a low-affinity site with $[Ca^{2+}]_{0.5}$ of 300 μM, measured in the presence of 0.5% Tween. The presence of detergent increases the affinity of the high-affinity sites 5 to 10 times. *In vivo*, the binding of calcium induces association with membranes.

Binding of calcium to grancalcin was studied by flow dialysis.[3] This analysis indicated the existence of two Ca^{2+}-binding sites per monomer. In the absence of a detergent, the protein precipitated rapidly upon addition of calcium. The measured $[Ca^{2+}]_{0.5}$ was 83 μM. In the presence of 25 mM octyl-β-glucopyranoside, the $[Ca^{2+}]_{0.5}$ decreased to 25 μM. The truncated forms of grancalcin devoid of the Gly/Pro-rich N-terminal segment were less susceptible to precipitation with calcium but had a somewhat higher $[Ca^{2+}]_{0.5}$ value (35–50 μM).[3]

Sorcin was also shown to bind two Ca^{2+} per monomer, based on indirect fluorescence titration experiments carried out in 0.1 M Tris-HCl at pH 7.5 in the presence of 2, 2-({2-[bis (carboxymethyl)-amino]methyl}phenoxy)-6-methoxy[bis(carboxymethyl)amino]quinoline.[51] The value $[Ca^{2+}]_{0.5}$ estimated from data shown by Zamparelli *et al.*[52] is ~30 μM.

Ca^{2+} binding to μ-calpain from human erythrocytes has been thoroughly investigated, both with the full-length calpain and with the isolated domains using dialysis with radioactive $^{45}Ca^{2+}$. Four Ca^{2+} ions were thought to bind each subunit (chain) with a K_d of 15 μM for the heavy chain and 25 μM for the light chain.[53] A comprehensive study on the effect of mutations disrupting Ca^{2+} binding on the activity of m-calpain showed that the activation of the enzyme is affected the most by the mutations of the EF3, less by mutations in EF2 and EF1, and the least by mutations in EF4.[54]

X-RAY STRUCTURE

The first three-dimensional structure of a protein from the PEFLIN family, that of calpain domain VI (dVI), was determined by crystallography in 1997.[9,10] Within the next few years the structures of grancalcin, ALG-2, and sorcin were also determined (Table 3). The structure of the full-length heterodimeric calpain was also solved and is discussed elsewhere in this volume (**Calpain**). That leaves peflin as the remaining target for structure determination.

Table 3 Presently known three-dimensional structures of PEFLINs

Protein	Resolution	PDB code	Reference
Rat Ca^{2+}-free calpain dVI	2.3 Å	1AJ5	9
Rat Ca^{2+}-bound calpain dVI	2.3 Å	1DVI	9
Bovine Ca^{2+}-bound calpain dVI	1.9 Å	1ALV, 1ALW	10
Human Ca^{2+}-free m-calpain	2.5 Å	1DKV	55
Human m-calpain form II	2.5 Å	1KFU	55
Human m-calpain form I	3.15 Å	1KFX	55
Rat m-calpain	2.6 Å	1DF0	56
Sorcin	2.2 Å	1GJY, 1JUO	57,58
Grancalcin	1.7 Å	1F4O, 1F4Q, 1K94, 1K95	59,60
ALG-2	2.3 Å	1HQV	61

Crystallization

PEFLINs contain, in addition to the Ca^{2+}-binding domain, an N-terminal hydrophobic Gly/Pro-rich extension. In most cases, presence of this extension caused difficulties in crystallization of the protein and crystals could only be obtained with a recombinant protein devoid of this extension. In cases in which the extension, or part of it, was present in the construct, it was not observed in the crystal due to disorder. Some of the crystallization conditions are given below.

Rat calpain domain VI

The construct contained residues 98–270 of the light chain. Crystals were obtained in the presence and absence of calcium. Apo-dVI (space group $C222_1$, $a = 69.4$ Å, $b = 73.9$ Å, $c = 157.4$ Å) crystallized from 1.6 M $(NH_4)_2SO_4$ at pH 6.8; high Ca^{2+}-dVI (space group $C222_1$, $a = 75.0$ Å, $b = 87.4$ Å, $c = 118.3$ Å) from 17% PEG 8000, 50 mM cacodylate pH 6.5, 10% glycerol, 200 mM $CaCl_2$; and low Ca^{2+}-dVI (space group $P2_12_12_1$ $a = 50.3$ Å, $b = 56.5$ Å, $c = 141.3$ Å) in 14% PEG 8 K, 50 mM cacodylate, pH 6.5, 1 mM $CaCl_2$.

Sorcin

Crystals of full-length sorcin were obtained in the presence of 0.8 M $(NH_4)_2SO_4$ and 100 mM MES buffer pH 5.8 but diffracted only to 4.5 Å.[47] Crystals of SCBD grew only in the absence of Ca^{2+}.[47] The tetragonal crystals grew from 0.8 M $(NH_4)_2SO_4$ in 100 mM MES buffer pH 5.6, 1 mM DTT and 1 mM NaN_3 (space group $P4_21_2$, $a = 103.3$, $c = 79.2$ Å). They diffracted to 2.7-Å resolution. The monoclinic crystals (space group C2, $a = 130.9$, $b = 103.8$, $c = 78.65$ Å, $\beta = 118.0°$) grew at 0.67 M $(NH_4)_2SO_4$, pH 5.4 and diffracted to 2.1-Å resolution.

ALG-2

Attempts to crystallize full-length protein were unsuccessful. Therefore, ALG-2 was proteolyzed with elastase. N-terminal sequencing showed that proteolyzed ALG-2 started at residue Ala21. Following cleavage the protein was dialyzed and concentrated to 6 mg mL^{-1}. Crystals of des1-20ALG-2 grew by vapor diffusion method only in the presence of Ca^{2+}.[61] These tetragonal crystals (space group $P4_122$, $a = 72.1$ and $c = 91.5$ Å) were obtained at 0.1 M sodium cacodylate pH 6.5, 100 mM $CaCl_2$, 12 to 15% PEG 8000.

Grancalcin

Ca^{2+}-free grancalcin was obtained in two forms.[62] The first monoclinic form (space group $P2_1$, $a = 48.4$, $b = 81.1$, $c = 46.6$ Å, $\beta = 111.3°$) crystallized from 0.1 M sodium acetate acetic acid pH 5.6, 0.1 M NaCl, and 1 M $NH_4H_2PO_4$. The second monoclinic form (space group C2, $a = 97.0$, $b = 51.9$, $c = 75.9$ Å, $\beta = 108.5°$) was obtained from 0.1 M NaAc pH 4.5, 0.8 M $(NH_4)_2SO_4$.

Crystals of Ca^{2+}-bound grancalcin were obtained from 15% poly(ethylene glycol) 20 000 (PEG20000), 15% (w/v) glycerol, 50 mM N-[2-hydroxyethyl]piperazine-N'-[2-ethanesulfonic acid (HEPES) pH 6.8 and 5 mM $CaCl_2$.[59] They were monoclinic, space group C2, $a = 97.4$, $b = 50.3$, $c = 77.6$ Å, $\beta = 108.2°$. Their cell parameters are very similar to those of the second Ca^{2+}-free form. Attempts to grow these crystals at higher Ca^{2+} concentrations (10–50 mM) led to poorly diffracting crystals.

Overall structure

The significant level of amino acid sequence similarity results, as expected, in highly similar three-dimensional structures among the penta EF-hand proteins. Their crystal structures reveal that the molecules associate into homodimers in an identical manner in all crystal forms. The only example of the three-dimensional structure of a heterodimer is that of intact calpains where domains IV and VI form a heterodimer.[55,56]

The crystal structure of domain VI of calpain, the first PEFLIN to have its three-dimensional structure determined, showed a compact monomer, predominantly α-helical and containing, surprisingly, five EF-hand, super-secondary structural elements (3D structure), rather than four predicted from its amino acid sequence, arranged in a novel fold.[9,10] Most of the known EF-hand proteins contain an even number of EF-hand motifs. Apart from the PEFLIN family, the only other proteins with an odd number of EF-hands are parvalbumins with three EF-hands (see **Parvalbumin**, this volume). Unlike parvalbumins, PEFLINs form dimers through the association of unpaired EF-hands. Therefore, in the PEFLIN protein dimer, all EF-hands are assembled into pairs. The general features of PEFLINs will be described using calpain dVI as an example (3D structure).

The first, EF1 (α1-L1-α2), and the second, EF2 (α3-L2-α4), pack together with loops L1 and L2 forming a characteristic short antiparallel β-sheet.[64] These two EF-hand motifs are interconnected by a segment longer than usually observed in other EF-hand proteins. A six-turn-long helix, α4, leads into the third EF-hand motif, EF3 (α4-L3-α5), and can therefore be considered a fusion of the F-helix from EF2 and the E-helix from EF3. EF3 packs against EF4 (α6-L4-α7) with the loops L3 and L4 forming a short antiparallel β-sheet giving rise to the second pair of EF-hands. This arrangement of two pairs of EF-hands separated by a long α-helix shows a superficial resemblance to calmodulin and troponin C, where the two pairs of EF-hands form distinct lobes.[65,66] However, the disposition of the two pairs of EF-hands in PEFLINs is quite different and allows for cross-communication between the two pairs. A second long helix, α7, leads into the EF5 motif (α7–L5–α8), which extends away from the rest of the molecule with the C-terminal helix α8 wrapping back toward the centre and ending close to EF2. This fifth EF-hand is unpaired in the monomer, but plays a crucial role in dimerization. Like helix α4, the helix α7 correspond to helix F from EF4 fused to helix E from EF5 into one long helix (3D structure).

Ca^{2+}-binding motifs

The proteins from the PEFLIN family contain five EF-hand motifs (Figure 1). However, it is clear from sequence analysis, Ca^{2+}-binding studies, and determined structures that some of these EF-hands in each protein have lost their ability to bind Ca^{2+} ions. The structure of calpain domain VI showed that three EF-hands, EF1–3, bind Ca^{2+} ions, while EF5 is empty, even at a very high concentration of calcium.[9,10] A Ca^{2+} ion was also observed in the EF4 but its binding is atypical (see below) and it is not clear if it plays a physiological role (3D Structure). In the structure of grancalcin, two Ca^{2+} ions were observed, at EF1 and EF3.[59] Finally, the structure of ALG-2 showed three bound Ca^{2+} ions at EF1, EF3 and, surprisingly, at EF5.[61] No structural data on Ca^{2+} binding to sorcin exist as the molecule was crystallized only in the absence of calcium. However, spectroscopic data indicate that two Ca^{2+} ions may bind to each monomer, EF1 and EF2 being the most likely candidates as their amino acid sequences are the closest to the canonical EF-hand.[51]

The Ca^{2+} ion is usually liganded by seven oxygen atoms that form a pentagonal bipyramid. Following the nomenclature introduced by Kretsinger (see reference 67), the ligand positions are marked as X, Y, Z, −Y, −Z (Figure 2). In the canonical EF-hand, the ligands marked X, Y, and Z are side-chain oxygens, −Y is a backbone carbonyl oxygen, and −Z provides two oxygens from the side chain of a glutamate. The seventh ligand in position −X is a water molecule. The naming of these positions follows a right-handed coordinate system with the center at the Ca^{2+} ion and the axes along the vectors toward the corresponding X, Y, and Z oxygens.

The EF-hands that appear functional (i.e. bind Ca^{2+}) in most PEFLINs are EF1 and EF3. The EF2 is functional in calpain domains (and possibly in sorcin), while EF5 is capable of binding Ca^{2+} only in ALG-2. EF4 varies the most in length, lacks the −Z glutamate side chain, and, in general, does not bind calcium. The various EF-hands are discussed in detail below. The typical Ca^{2+}-ligand distances observed in various PEFLINs are shown in Table 4.

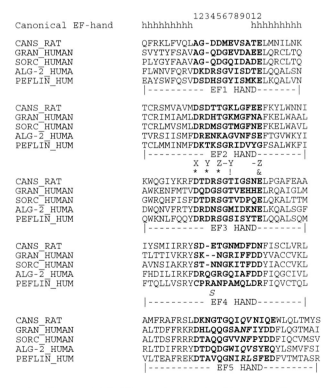

```
                           123456789012
Canonical EF-hand          hhhhhhhhh           hhhhhhhh

CANS_RAT                   QFRKLFVQLAG-DDMEVSATELMNILNK
GRAN_HUMAN                 SVYTYFSAVAG-QDGEVDAEELQRCLTQ
SORC_HUMAN                 PLYGYFAAVAG-QDGQIDADELQRCLTQ
ALG-2_HUMA                 FLWNVFQRVDKDRSGVISDTELQQALSN
PEFLIN_HUM                 EAYSWFQSVDSDHSGYISMKELKQALVN
                           |---------- EF1 HAND -------|

CANS_RAT                   TCRSMVAVMDSDTTGKLGFEEFKYLWNNI
GRAN_HUMAN                 TCRIMIAMLDRDHTGKMGFNAFKELWAAL
SORC_HUMAN                 TCRLMVSMLDRDMSGTMGFNEFKELWAVL
ALG-2_HUMA                 TVRSIISMFDRENKAGVNFSEFTGVWKYI
PEFLIN_HUM                 TCLMMINMFDKTKSGRIDVYGFSALWKFI
                           |---------- EF2 HAND -------|
                                    X Y Z-Y    -Z
                                    * * * !    &

CANS_RAT                   KWQGIYKRFDTDRSGTIGSNELPGAFEAA
GRAN_HUMAN                 AWKENFMTVDQDGSGTVEHHELRQAIGLM
SORC_HUMAN                 GWRQHFISFDTDRSGTVDPQELQKALTTM
ALG-2_HUMA                 DWQNVFRTYDRDNSGMIDKNELKQALSGF
PEFLIN_HUM                 QWKNLFQQYDRDRSGSISYTELQQALSQM
                           |---------- EF3 HAND -------|

CANS_RAT                   IYSMIIRRYSD-ETGNMDFDNFISCLVRL
GRAN_HUMAN                 TLTTIVKRYSK--NGRIFFDDYVACCVKL
SORC_HUMAN                 AVNSIAKRYST-NNGKITFDDYIACCVKL
ALG-2_HUMA                 FHDILIRKFDRQGRGQIAFDDFIQGCIVL
PEFLIN_HUM                 FTQLLVSRYCPRANPAMQLDRFIQVCTQL
                                  S
                           |---------- EF4 HAND -------|

CANS_RAT                   AMFRAFRSLDKNGTGQIQVNIQEWLQLTMYS
GRAN_HUMAN                 ALTDFFRKRDHLQQGSANFIYDDFLQGTMAI
SORC_HUMAN                 ALTDSFRRRDTAQQGVVNFPYDDFIQCVMSV
ALG-2_HUMA                 RLTDIFRRYDTDQDGWIQVSYEQYLSMVFSI
PEFLIN_HUM                 VLTEAFREKDTAVQGNIRLSFEDFVTMTASR
                           |---------- EF5 HAND -------|
```

Figure 1 The alignment of five EF-hands in PEFLINs. The numbering of residues according to the canonical EF-hand is shown at the top. Positions of Ca^{2+} ligands are marked with letters indicating their position. The symbols '*', '!' and '&' mark residues with side chain, main chain, and two side-chain ligands, respectively.

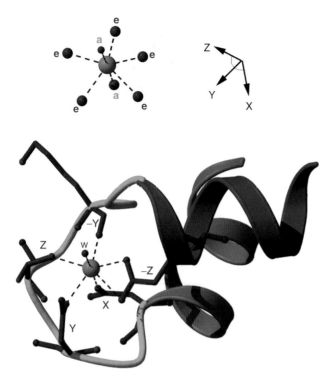

Figure 2 The example of a canonical EF-hand with the ligand positions marked as X, Y, Z, −Y, and −Z. Symbols 'a' and 'e' mark equatorial and axial positions of the pentagonal bipyramid. Figure prepared with MOLSCRIPT[68] and RASTER3D.[69]

EF1

The EF1 hand motifs fall into two categories. Those of ALG-2 and peflin follow the canonical pattern and indeed, the EF1 of ALG-2 binds Ca^{2+} in the expected way.[61] The Ca^{2+} ion displays pentagonal bipyramidal coordination. The protein provides six of the ligands, while the seventh is a water molecule in an axial position. The liganding oxygens come from the side chains of Asp36, Asp38, Ser40, and Glu47 (both side-chain oxygens), the carbonyl of Val42, and a water molecule, and are 2.2 to 2.6 Å from the Ca^{2+} ion (Figure 3(a)).

On the other hand, the EF1 of calpain domain VI, grancalcin, and likely sorcin bind Ca^{2+} in a different way, observed only in this protein subfamily (Figure 3(b)). In these proteins, the loop between helices E and F is one residue shorter than the canonical loop and lacks the oxygen-containing side chains in positions X and Y (Figure 1). The side chains in positions X and Y (Leu110 and Gly112) are not capable of providing ligands to the Ca^{2+} ion. Instead, the N-terminal segment of L1 adopts a conformation that brings the main-chain carbonyl oxygen of Ala111 into position to coordinate the Ca^{2+} and substitutes the coordination normally provided by the side chain at the X position. Pentagonal–bipyramidal coordination is completed by a unique arrangement consisting of the main-chain carbonyl of Glu116, the side

Table 4 Ca^{2+}–ligand distances observed in PEFLINs. When two molecules are present in the asymmetric unit, there are two equivalent distances shown

Ca^{2+}	dVI (1DVI)	ALG-2 (1HQV)	Grancalcin (1K94)
EF1	2.25/2.2 (O^{Ala111})	2.3 ($OD1^{Asp36}$)	2.3 (O^{Ala62})
	2.5/2.4 ($OD1^{Asp114}$)	2.2 ($OD2^{Asp38}$)	2.4 ($OD1^{Asp65}$)
	2.3/2.25 (O^{Glu116})	2.6 (OG^{Ser40})	2.3 (O^{Glu67})
	2.5/2.6 ($OE1^{Glu121}$)	2.4 (O^{Val42})	2.55 ($OE1^{Glu72}$)
	2.5/2.6 ($OE2^{Glu121}$)	2.6 ($OE1^{Glu47}$)	2.4 ($OE2^{Glu72}$)
	2.4/2.4 (Wat1)	2.5 ($OE2^{Glu47}$)	2.4 (Wat1)
	2.5/2.5 (Wat2)	2.2 (Wat)	2.4 (Wat2)
EF2	2.4/2.5 ($OD1^{Asp154}$)		
	2.4/2.45 ($OD1^{Asp156}$)		
	2.4/2.5 ($OG1^{Thr158}$)		
	2.2/2.2 (O^{Lys160})		
	2.5/2.4 ($OE1^{Glu165}$)		
	2.9/2.8 ($OE2^{Glu165}$)		
	2.5/2.5 (Wat)		
EF3	2.4/2.3 ($OD1^{Asp184}$)	2.3 ($OD1^{Asp103}$)	2.4/2.3 ($OD2^{Asp132}$)
	2.45/2.4 ($OD1^{Asp186}$)	2.5 ($OD2^{Asp105}$)	2.4/2.4 ($OD1^{Asp134}$)
	2.3/2.4 (OG^{Ser188})	2.6 (OG^{Ser107})	2.4/2.4 (OG^{Ser136})
	2.6/2.5 ($OG1^{Thr190}$)	2.3 (O^{Met109})	2.4/2.4 ($OG1^{Thr138}$)
	2.4/2.4 ($OE1^{Glu195}$)	2.4 ($OE1^{Glu114}$)	2.4/2.3 ($OE2^{Glu140}$)
	2.85/2.7 ($OE2^{Glu195}$)	2.2 ($OE2^{Glu114}$)	2.4/2.4 ($OE1^{Glu143}$)
	2.2/2.3 (Wat)	2.6 (Wat)	2.6/2.6 ($OE2^{Glu143}$)
EF4	2.3/2.4 ($OD1^{Asp139}$)		
	2.55/2.7 ($OD1^{Asp227}$)		
	2.5/2.6 ($OD2^{Asp227}$)		
	2.7/2.6 ($OD1^{Asp229}$)		
	2.6/2.6 ($OD2^{Asp229}$)		
	2.3/2.3 ($OD1^{Asn230}$)		
	2.3/2.4 (Wat1)		
	2.65/2.5 (Wat2)		
EF5		2.3 ($OD1^{Asp169}$)	
		2.1 ($OD1^{Asp171}$)	
		2.2 ($OD1^{Asp173}$)	
		2.4 (O^{Trp175})	
		2.3 (Wat1)	
		2.7 (Wat2)	

chains of Asp114, Glu121 (bidentate), and two water molecules, one in a planar and one in an axial position (Figure 3(b)). A glycine residue is highly conserved in the center of the calcium-binding loop in the EF-hand motifs. This residue assumes a $\phi/\psi = +/+$ conformation possible for glycine but generally unfavorable for amino acids with side chains. Indeed, in sorcin and grancalcin, this position is occupied by a glycine; however, in calpain domains a Met115 is at this position. Nevertheless, this methionine adopts a less favorable conformation with main-chain torsion angles $\phi/\psi = +60^0/+20^0$. The coordination of Ca^{2+} in calpain is different from the other noncanonical variants observed in S100 proteins,[70] molluscan myosin essential light chain,[71] and BM40.[72]

EF1 undergoes a conformational reorganization upon binding calcium. Comparison of Ca^{2+}-bound and apoforms

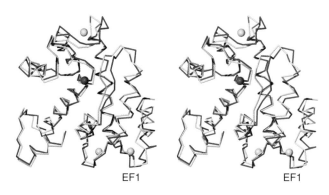

Figure 4 The stereo view of the superposition of calcium-bound (yellow, 1DVI) and calcium-free (blue, 1AJ5) domain VI of calpain showing relatively small rearrangements caused by Ca^{2+} binding. Ca^{2+} bound to EF4 is painted in dark orange. Figure prepared with sPDBviewer[63] and PovRay (www.povray.org).

EF3

The EF3 supersecondary motif conforms fully to the canonical EF-hand motif. The hallmark of the EF-hand is the 12-residue long loop between two helices, which provides coordination for Ca^{2+}. A typical EF3, here taken from calpain domain VI, with Kretsinger's nomenclature is shown in Figure 2.

EF4

In the Ca^{2+}-dVI crystal obtained in the presence of 200 mM concentration of calcium, a Ca^{2+} ion was found also in EF4. This ion binds in an atypical way, at the C-terminal end of L4, and is coordinated by eight oxygen atoms that include two water molecules (Figure 5). The side chains in the Ca^{2+} coordination sphere are Asp227 (bidentate), Asp229 (bidentate), Asn230 from helix α7 and Asp139, distant in linear sequence, from the loop connecting EF1 and EF2. This EF4 Ca^{2+}-binding site is probably not physiologically relevant as in crystals grown at 1 mM concentration of calcium only EF1–3 were occupied.[9]

EF5

The loop between helices E and F in the EF5 of all identified PEFLINs has an insertion of two residues as compared to the canonical EF-hand. These two helices E5 and F5 are nearly antiparallel. The insertion into the EF5 loop shifts the usual −Z glutamate Ca^{2+} ligand, located at the beginning of the F-helix, further away from the canonical Ca^{2+}-binding site located in the N-terminal part of the loop. The replacement of the −Z glutamate in a Ca^{2+}-binding EF-hand in calpain[54] or ALG-2[33,74] results in a loss of Ca^{2+} binding. Therefore, the lack of Ca^{2+} binding to the EF5 of calpain dVI and grancalcin was readily explained from the structure. The function of this EF-hand was thought to be solely to facilitate dimer formation.

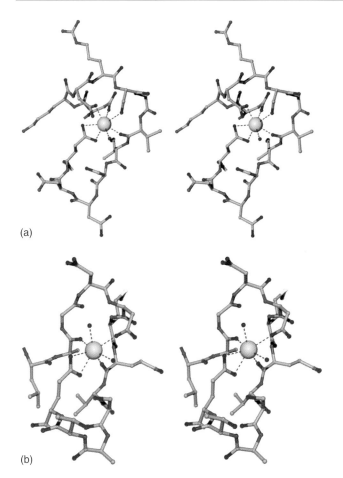

Figure 3 The stereo view of EF1 of (a) ALG-2 conforming to the canonical binding mode (1HQV); (b) calpain domain VI with a noncanonical binding mode (1DVI). The carbon, nitrogen, oxygen, and calcium atoms are colored gray, blue, red, and orange, respectively. Green dashed lines mark Ca^{2+}-ligand distances. Figure prepared with sPDBviewer[63] and PovRay (www.povray.org).

of calpain dVI showed that the carbonyl group of Ala111 points outside and flips toward the center of the loop upon Ca^{2+} binding. This is associated with a change in conformation of the N-terminal segment of the loop[9] (Figure 4).

EF2

This EF-hand has been shown experimentally to bind Ca^{2+} only in calpain dVI.[9,73] Grancalcin and peflin are unlikely to bind Ca^{2+} in this EF-hand, as they are missing the glutamate residue at position −Z. Site-directed mutagenesis of calpain dVI domain showed that the replacement of −Z glutamate in each of EF1,2,3 led to the disruption of Ca^{2+} binding in each case.[54] The structure of ALG-2 also showed no Ca^{2+} bound to EF2. On the other hand, the sequence of EF2 of sorcin conforms to the canonical pattern and most likely binds calcium. It appears that binding of Ca^{2+} ion to EF2 of calpain dVI does not induce a conformational change (Figure 4).[9]

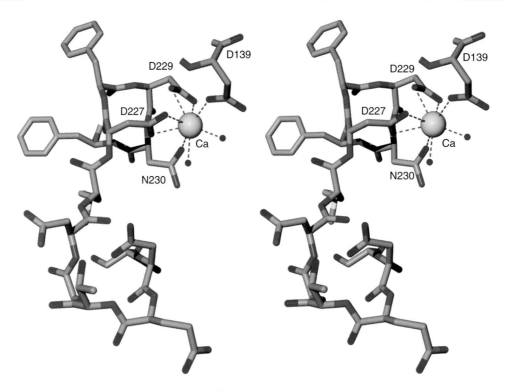

Figure 5 The EF4 of calpain domain VI with bound Ca^{2+} (1DVI). The ion binds in an unusual way with one ligand provided by Asp139 (green) from the loop between EF1 and EF2 outside this EF-hand. Ca^{2+} is painted orange. Figure prepared with sPDBviewer[63] and PovRay (www.povray.org).

Unexpectedly, the structure of ALG-2 showed that EF5 can also bind a Ca^{2+} ion, and therefore this EF-hand could potentially play a role in a Ca^{2+}-directed regulatory function. The Ca^{2+} binds to EF5 in a canonical location at the N-terminal part of the loop and is coordinated by six oxygen ligands, four of which come from canonical positions within EF5: Asp169 (axial), Asp171, Asp173, and the main-chain carbonyl group of Trp175 (Figure 6). A water molecule occupies the second axial position. There is no acidic side chain in a close-enough proximity to provide the two additional ligands. Instead, a second water molecule occupies the position midway between the two usual positions of the −Z glutamate side-chain oxygens. The access of this water molecule to the Ca^{2+} ion is possibly due to the larger size of the EF5 loop. However, the affinity for Ca^{2+} at this site is rather low, measured at 300 μM by flow dialysis.[48] Comparison of the EF5 sequences of PEFLINs shows that, of the three aspartate ligands in ALG-2, only Asp169 is conserved.[60] Therefore, Ca^{2+} binding by EF5 in ALG-2 may be an exception rather than the rule and may not be directly related to the Ca^{2+} sensory function of PEFLINs.

Dimer formation

All PEFLINs form homo- or heterodimers. Heterodimerization is essential for calpains where the association of the heavy and light subunits occurs through interaction of the Ca^{2+}-binding domains. The ability to form heterodimers was also reported between ALG-2 and peflin[35] and between sorcin and grancalcin.[31] The two PEFLIN molecules associate into a dimer through the packing of EF5 hands from both monomers (Figure 7). In calpain dVI, grancalcin, and sorcin, an area of 1800 to 2000 Å^2 of each monomer is buried upon dimerization, which corresponds to ∼20% of the total surface of each monomer. The interface is formed predominantly from residues on the sides of helices α3, α6, α7, and α8 with the major contribution from the two EF5 helices (α7 and α8). These are located in the center of the interface and form an antiparallel four-helix bundle. The L5 loops combine into a short antiparallel two-stranded β-sheet completing a pairwise arrangement of these two EF-hands. The residues at the interface are of mixed character, with approximately half of them being charged or polar and the other half being hydrophobic. Many hydrogen bonds participate in the interface.

The ALG-2 dimer has a distinctive 'V' shape, rather different from that of calpain, grancalcin, and sorcin (Figure 7(b)). The difference in the overall shape is due to the different position of EF1–2 relative to the dimer interface in ALG-2 and the other three proteins and has been attributed to binding of a hydrophobic peptide.[61] Dimerization of ALG-2 buries a surface area of 1440 Å^2 on each monomer, corresponding to burying approximately 13% of the total surface of each monomer.

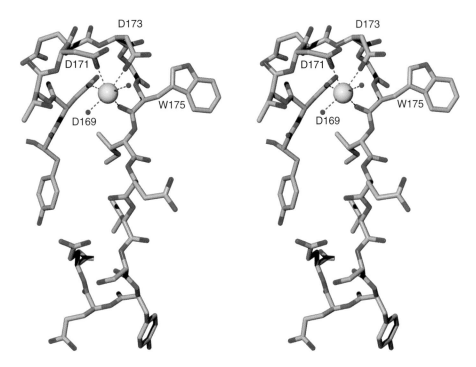

Figure 6 Binding of Ca^{2+} to EF5 of ALG-2 despite the lack of $-Z$ glutamate (1HQV). The open C-terminal part of the EF-hand loop permits a water molecule (small red ball) to settle in a location normally taken by the oxygens of $-Z$ glutamate. Figure prepared with sPDBviewer[63] and PovRay (www.povray.org).

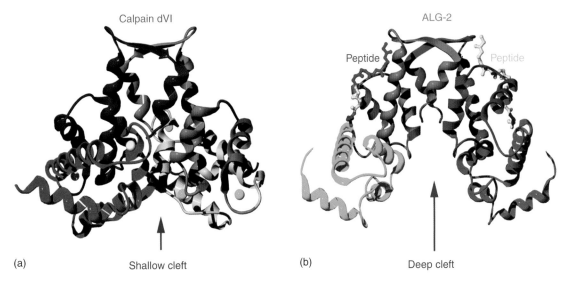

Figure 7 Structure of a PEFLIN protein dimer: (a) calpain dVI (1DVI); (b) ALG-2 (1HQV). Figure prepared with sPDBviewer[63] and PovRay (www.povray.org).

This is substantially less than the dimerization interface of grancalcin, calpain, and sorcin, but still an indication of strong interaction between the monomers. As in these other proteins, the interactions involve hydrophobic contacts and hydrogen bonds.

Binding site for hydrophobic peptide

The structure of Ca^{2+}-loaded des1-20ALG-2 showed unexpectedly strong electron density features that were interpreted as two copies of a hydrophobic decapeptide bound to the ALG-2 dimer.[61] The source of this peptide was thought to be the N-terminal Gly/Pro-rich part of ALG-2 that was removed from the wild-type protein by elastase digestion. This decapeptide is found on the surface of the protein, bound within a largely hydrophobic cleft formed between helix α7 and the loop connecting EF3 and EF4 (Figure 8). Its N-terminus is anchored near the C-terminal end of helix α7, while its C-terminus is buttressed against

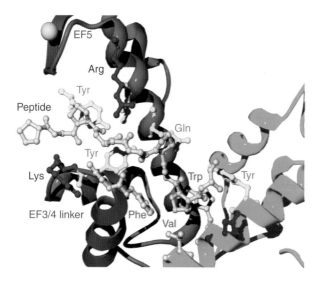

Figure 8 Binding of a hydrophobic peptide to ALG-2 (1HQV). Close-up of the peptide binding site. Figure prepared with sPDBviewer[63] and PovRay (www.povray.org).

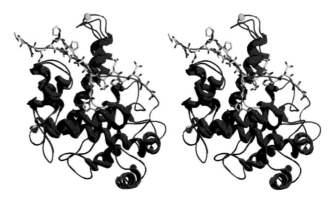

Figure 9 Superposition of ALG-2 (1HQV) and the calpain domain IV (1DF0) showing similar paths taken by the peptide bound to ALG-2 and the segment connecting dIV to domain III. The hydrophobic C-terminal end of the ALG-2-bound peptide enters a hydrophobic pocket created by a more open conformation of ALG-2 as compared to calpain. The connecting segment in calpain is hydrophilic and follows along the surface of this 'closed' conformation. Figure prepared with sPDBviewer[63] and PovRay (www.povray.org).

helix α4 of EF2 (residues 91–98). The two ends of the peptide bind to hydrophobic pockets on the des1-20ALG-2 surface, while its center is exposed on the surface. The first hydrophobic pocket is formed by Tyr124, the aliphatic chain of Arg166, Thr162, and with a contribution of Tyr180 from the second monomer. This pocket is occupied by the N-terminal sequence Gly-Pro-Gly of the decapeptide. The second hydrophobic pocket is created by Tyr91, Asp94 (backbone), Trp95, Val98 and Leu119, Phe122, Gly123 and Leu158 and is filled by the C-terminal Gly/Pro of the decapeptide. Aromatic residues form more than half of all residues lining these hydrophobic sites. The interaction of the peptide with des1-20ALG-2 also involves some polar contacts and hydrogen bonds: Arg166 forms hydrogen bonds to Gly11 and Gly12. Thus, the peptide is bound tightly to ALG-2 in a way resembling the binding of SH3 domains to their Pro-rich targets.[75]

Comparison of the des1-20ALG-2 structure with that of grancalcin, calpain, and sorcin suggests that binding of such hydrophobic peptide requires movement of EF1-2 (relative to the C-terminal half and dimer interface).[61] The peptide binds to the surface of ALG-2 with its C-terminal portion inserted into a hydrophobic pocket that is inaccessible in the conformations observed in the structures of other PEFLINs. When the peptide dissociates, a reverse movement of EF1-2 to bury the hydrophobic pockets would be expected. Measurements of 2-p-toluidinylnaphthalene-6-sulfonate (TNS) fluorescence upon Ca^{2+} binding to ALG-2 indicate some rearrangement in the molecule, which increases its hydrophobic surface,[50] consistent with this notion. Unfortunately, the structure of ALG-2 has not yet been determined in the absence of the peptide or without Ca^{2+}.

The structure of intact calpain in the absence of Ca^{2+} provides support to the notion that this part of the surface of PEFLIN proteins is utilized for binding to other partners. The path taken by the peptide in ALG-2 is partially occupied in calpain by an extended polypeptide chain that connects domain III and the penta EF-hand domain IV (Figure 9). However, this part of the calpain chain is hydrophilic and does not require the exposure of the hydrophobic patch described above. In addition, the N-terminus of the heavy chain is docked at the calcium-binding domain VI of the light chain in a similar general location to that occupied by the peptide in ALG-2.[56] On the basis of these observations, one could hypothesize on the mechanism of sensing Ca^{2+} and transmitting the signal by PEFLINs. According to this model, binding of Ca^{2+} alone causes only small local conformational rearrangements. Subsequent binding of a hydrophobic peptide (e.g. from the N-terminus of PEFLINs or from a binding partner) in the presence of Ca^{2+} induces a concomitant larger conformational change.[61] This mechanism is conceptually similar to calmodulin action.[76,77] The two lobes of calmodulin are connected by a long and rather flexible polypeptide, which partially assumes a helical conformation. The binding of Ca^{2+} induces initial reorganization of the EF-hands and exposes hydrophobic cavities. In the presence of calmodulin-binding peptide, the linker connecting the lobes unfolds permitting them to bind the peptide.

REFERENCES

1. M Missotten, A Nichols, K Rieger and R Sadoul, *Cell Death Differ*, **6**, 124–29 (1999).
2. P Vito, L Pellegrini, C Guiet and LD Adamio, *J Biol Chem*, **274**, 1533–40 (1999).

3. K Lollike, AH Johnsen, I Durussel, N Borregaard and JA Cox, *J Biol Chem*, **276**, 17762–69 (2001).
4. AM Brownawell and CE Creutz, *J Biol Chem*, **272**, 22182–90 (1997).
5. D Verzili, C Zamparelli, B Mattei, AA Noegel and E Chiancone, *FEBS Lett*, **471**, 197–200 (2000).
6. S Ohkouchi, K Nishio, M Maeda, K Hitomi, H Adachi and M Maki, *J Biochem (Tokyo)*, **130**, 207–15 (2001).
7. S Ohno, Y Emori, S Imajoh, H Kawasaki, M Kisaragi and K Suzuki, *Nature*, **312**, 566–70 (1984).
8. T Sakihama, H Kakidani, K Zenita, N Yumoto, T Kikuchi, T Sasaki, R Kannagi, S Nakanishi, M Ohmori and K Takio, *Proc Natl Acad Sci USA*, **82**, 6075–79 (1985).
9. H Blanchard, P Grochulski, Y Li, JSC Arthur, PL Davies, JS Elce and M Cygler, *Nat Struct Biol*, **4**, 532–38 (1997).
10. GD Lin, D Chattopadhyay, M Maki, KK Wang, M Carson, L Jin, PW Yuen, E Takano, M Hatanaka, LJ DeLucas and SV Narayana, *Nat Struct Biol*, **4**, 539–47 (1997).
11. M Maki, SV Narayana and K Hitomi, *Biochem J*, **328**, 718–20 (1997).
12. DE Goll, VF Thompson, H Li, W Wei and J Cong, *Physiol Rev*, **83**, 731–801 (2003).
13. T Kikuchi, N Yumoto, T Sasaki and T Murachi, *Arch Biochem Biophys*, **234**, 639–45 (1984).
14. DE Croall and GN DeMartino, *Physiol Rev*, **71**, 813–47 (1991).
15. H Sorimachi, TC Saido and K Suzuki, *FEBS Lett*, **343**, 1–5 (1994).
16. T Murachi, *Biochem Soc Symp*, **49**, 149–67 (1984).
17. M Molinari and E Carafoli, *J Membr Biol*, **156**, 1–8 (1997).
18. H Sorimachi, S Ishiura and K Suzuki, *Biochem J*, **328**, 721–32 (1997).
19. Y Huang and KK Wang, *Trends Mol Med*, **7**, 355–62 (2001).
20. T Murachi, K Tanaka, M Hatanaka and T Murakami, *Adv Enzyme Regul*, **19**, 407–24 (1980).
21. BJ Perrin and A Huttenlocher, *Int J Biochem Cell Biol*, **34**, 722–25 (2002).
22. AM Van der Bliek, MB Meyers, JL Biedler, E Hes and P Borst, *EMBO J*, **5**, 3201–8 (1986).
23. MB Meyers, VM Pickel, SS Sheu, VK Sharma, KW Scotto and GI Fishman, *J Biol Chem*, **270**, 26411–18 (1995).
24. MB Meyers, C Zamparelli, D Verzili, AP Dicker, TJ Blanck and E Chiancone, *FEBS Lett*, **357**, 230–34 (1995).
25. MB Meyers, TS Puri, AJ Chien, T Gao, PH Hsu, MM Hosey and GI Fishman, *J Biol Chem*, **273**, 18930–35 (1998).
26. E Pack-Chung, MB Meyers, WP Pettingell, RD Moir, AM Brownawell, I Cheng, RE Tanzi and TW Kim, *J Biol Chem*, **275**, 14440–45 (2000).
27. EF Farrell, AM Gomez, A Antaramian and HH Valdivia, *J Biol Chem*, **278**, 34660–66 (2003).
28. A Boyhan, CM Casimir, JK French, CG Teahan and AW Segal, *J Biol Chem*, **267**, 2928–33 (1992).
29. CG Teahan, NF Totty and AW Segal, *Biochem J*, **286**, 549–54 (1992).
30. K Lollike, O Sorensen, JR Bundgaard, AW Segal, A Boyhan and N Borregaard, *J Immunol Methods*, **185**, 1–8 (1995).
31. C Hansen, S Tarabykina, JM la Cour, K Lollike and MW Berchtold, *FEBS Lett*, **545**, 151–54 (2003).
32. J Roes, BK Choi, D Power, P Xu and AW Segal, *Mol Cell Biol*, **23**, 826–30 (2003).
33. P Vito, E Lacana and LD Adamio, *Science*, **271**, 521–25 (1996).
34. E Lacana, JK Ganjei, P Vito and LD Adamio, *J Immunol*, **158**, 5129–35 (1997).
35. Y Kitaura, S Matsumoto, H Satoh, K Hitomi and M Maki, *J Biol Chem*, **276**, 14053–58 (2001).
36. J Krebs, P Saremaslani and R Caduff, *Biochim Biophys Acta*, **1600**, 68–73 (2002).
37. H Satoh, Y Nakano, H Shibata and M Maki, *Biochim Biophys Acta*, **1600**, 61–67 (2002).
38. IK Jang, R Hu, E Lacana, L D'Adamio and H Gu, *Mol Cell Biol*, **22**, 4094–100 (2002).
39. J Krebs and R Klemenz, *Biochim Biophys Acta*, **1498**, 153–61 (2000).
40. Y Kitaura, M Watanabe, H Satoh, T Kawai, K Hitomi and M Maki, *Biochem Biophys Res Commun*, **263**, 68–75 (1999).
41. SL Wang, MF Tam, YS Ho, SH Pai and MC Kao, *Biochim Biophys Acta*, **1260**, 285–93 (1995).
42. S Ohno, Y Emori and K Suzuki, *Nucleic Acids Res*, **14**, 5559 (1986).
43. K Aoki, S Imajoh, S Ohno, Y Emori, M Koike, G Kosaki and K Suzuki, *FEBS Lett*, **205**, 313–17 (1986).
44. M Maki, Y Kitaura, H Satoh, S Ohkouchi and H Shibata, *Biochim Biophys Acta*, **1600**, 51–60 (2002).
45. RH Kretsinger, *Nat Struct Biol*, **4**, 514–16 (1997).
46. K Graham-Siegenthaler, S Gauthier, PL Davies and JS Elce, *J Biol Chem*, **269**, 30457–60 (1994).
47. V Nastopoulos, A Ilari, G Colotti, C Zamparelli, D Verzili, E Chiancone and D Tsernoglou, *Acta Crystallogr*, **D57**, 862–64 (2001).
48. S Tarabykina, AL Moller, I Durussel, J Cox and MW Berchtold, *J Biol Chem*, **275**, 10514–18 (2000).
49. Y Kitaura, H Satoh, H Takahashi, H Shibata and M Maki, *Arch Biochem Biophys*, **399**, 12–18 (2002).
50. M Maki, K Yamaguchi, Y Kitaura, H Satoh and K Hitomi, *J Biochem (Tokyo)*, **124**, 1170–77 (1998).
51. C Zamparelli, A Ilari, D Verzili, L Giangiacomo, G Colotti, S Pascarella and E Chiancone, *Biochemistry*, **39**, 658–66 (2000).
52. C Zamparelli, A Ilari, D Verzili, P Vecchini and E Chiancone, *FEBS Lett*, **409**, 1–6 (1997).
53. M Michetti, F Salamino, R Minafra, E Melloni and S Pontremoli, *Biochem J*, **325**(Pt 3), 721–26 (1997).
54. P Dutt, JS Arthur, P Grochulski, M Cygler and JS Elce, *Biochem J*, **348**, 37–43 (2000).
55. S Strobl, C Fernandez-Catalan, M Braun, R Huber, H Masumoto, K Nakagawa, A Irie, H Sorimachi, G Bourenkow, H Bartunik, K Suzuki and W Bode, *Proc Natl Acad Sci USA*, **97**, 588–92 (2000).
56. CM Hosfield, JS Elce, PL Davies and Z Jia, *EMBO J*, **18**, 6880–89 (1999).
57. A Ilari, KA Johnson, V Nastopoulos, D Verzili, C Zamparelli, G Colotti, D Tsernoglou and E Chiancone, *J Mol Biol*, **317**, 447–58 (2002).
58. X Xie, MD Dwyer, L Swenson, MH Parker and MC Botfield, *Protein Sci*, **10**, 2419–2425 (2001).
59. J Jia, N Borregaard, K Lollike and M Cygler, *Acta Crystallogr*, **D57**, 1843–49 (2001).
60. J Jia, Q Han, N Borregaard, K Lollike and M Cygler, *J Mol Biol*, **300**, 1273–83 (2000).
61. J Jia, S Tarabykina, C Hansen, MW Berchtold and M Cygler, *Structure*, **9**, 267–75 (2001).

62 Q Han, J Jia, Y Li, K Lollike and M Cygler, *Acta Crystallogr*, **D56**, 772–74 (2000).

63 N Guex and MC Peitsch, *Electrophoresis*, **18**, 2714–23 (1997).

64 H Kawasaki and RH Kretsinger, *Protein Profile*, **2**, 297–490 (1995).

65 YS Babu, CE Bugg and WJ Cook, *J Mol Biol*, **204**, 191–204 (1988).

66 O Herzberg and MN James, *J Mol Biol*, **203**, 761–79 (1988).

67 S Nakayama and RH Kretsinger, *Annu Rev Biophys Biomol Struct*, **23**, 473–507 (1994).

68 PJ Kraulis, *J Appl Crystallogr*, **24**, 946–50 (1991).

69 EA Merritt and DJ Bacon, *Methods Enzymol*, **277**, 505–24 (1997).

70 BW Schafer and CW Heizmann, *Trends Biochem Sci*, **21**, 134–40 (1996).

71 I Rayment, WR Rypniewski, K Schmidt-Base, R Smith, DR Tomchick, MM Benning, DA Winkelmann, G Wesenberg and HM Holden, *Science*, **261**, 50–58 (1993).

72 E Hohenester, P Maurer, C Hohenadl, R Timpl, JN Jansonius and J Engel, *Nat Struct Biol*, **3**, 67–73 (1996).

73 M Cygler, P Grochulski and H Blanchard, *Methods Mol Biol*, **172**, 243–60 (2002).

74 KW Lo, Q Zhang, M Li and M Zhang, *Biochemistry*, **38**, 7498–508 (1999).

75 BK Kay, MP Williamson and M Sudol, *FASEB J*, **14**, 231–41 (2000).

76 BE Finn and S Forsen, *Structure*, **3**, 7–11 (1995).

77 M Zhang and T Yuan, *Biochem Cell Biol*, **76**, 313–23 (1998).

3D structures of the calcium and zinc binding S100 proteins

Günter Fritz[†,‡] and Claus W Heizmann[‡]

[†]Mathematisch-Naturwissenschaftliche Sektion, Fachbereich Biologie, Universität Konstanz, Konstanz, Germany
[‡]Department of Pediatrics, Division of Clinical Chemistry and Biochemistry, University of Zürich, Zürich, Switzerland

FUNCTIONAL CLASS

Signaling molecule, S100 proteins; a protein family containing two EF-hand calcium binding motifs regulating different cellular processes; several members of the family bind zinc and copper with high affinity.

OCCURRENCE

S100 proteins are found exclusively in vertebrates. Proteins have been found in lungfish (*Lepidosiren paradoxus*), frog (*Rana catesbeiana*), chicken (*Gallus gallus*), mouse (*Mus musculus*), rat (*Rattus norvegicus*), pig (*Sus scrofa*), cow (*Bos taurus*), and human (*Homo sapiens*). Expression profiles indicate that the more complex vertebrates such as mammalia contain a larger number of S100 proteins, pointing toward specified functions of S100 proteins in the development of the organism. The individual members of the S100 family show a tissue- and cell type-specific expression pattern. Common to all organisms are high expression levels of S100 proteins in the brain and the heart.

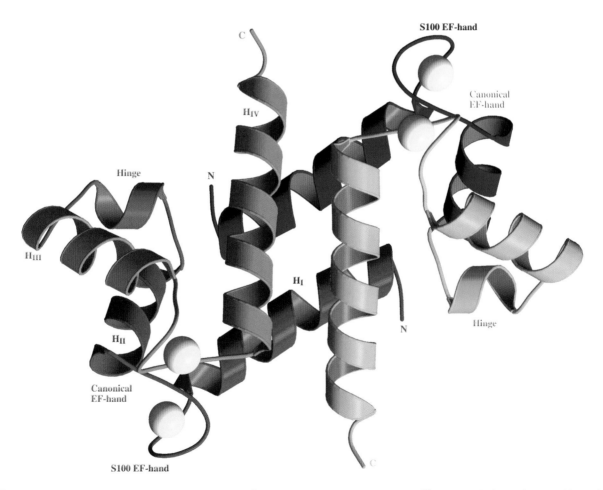

3D Structure Schematic representation of human Ca^{2+}-loaded S100A6 (PDB code 1K96)[23] showing the homodimer and bound Ca^{2+} ions (yellow). The figure was made with MOLSCRIPT[55] and RASTER3D.[56]

BIOLOGICAL FUNCTION

Calcium (Ca^{2+}) functions as a messenger regulating a great variety of cellular processes.[1] A major part of the molecular Ca^{2+}-signaling network is represented by the family of Ca^{2+}-binding proteins characterized by the EF-hand structural motif.[2] S100 proteins represent the largest subgroup within this family and have received increasing attention in recent years due to their cell- and tissue-specific expression and their involvement in widely different processes such as cell cycle regulation, cell growth, cell differentiation, and mobility. The family of EF-hand proteins can be divided into two primary classes: Ca^{2+} sensors, which transduce the Ca^{2+} signals, and Ca^{2+} signal modulators, which are involved in uptake, transport, and buffering of the Ca^{2+} signal. Most members of the S100 family belong to the class of Ca^{2+} sensors. Upon Ca^{2+}-binding they undergo a large conformational change that exposes a hydrophobic surface responsible for target protein binding and regulation. Only calbindin acts as a Ca^{2+} buffer sequestering Ca^{2+} in the cytoplasm after a raise in concentration, whereby the Ca^{2+}-induced conformational change is small and does not lead to the exposure of a hydrophobic patch.[3] Unlike other EF-hand proteins, many S100 proteins bind Zn^{2+} with high affinity in binding sites distinct from the EF-hand Ca^{2+}-binding sites. In some cases, it has been proposed that Zn^{2+} instead of Ca^{2+} regulates the biological function. Furthermore, several S100 proteins are secreted into the extracellular space,[4] where they act in cytokine-like manner. The individual members seem to utilize distinct pathways (ER-Golgi route, tubulin or actin dependent) for their

Table 1 Biological functions of S100 proteins

Name	Previous symbols/synonyms	Function
S100A1	S100α	Regulation of cell motility, muscle contraction, phosphorylation, Ca^{2+} release channel, transcription
S100A2	S100L, CaN19	Tumor suppression, nuclear functions, chemotaxis
S100A3	S100E	Hair shaft formation, tumor suppression, secretion, and extracellular functions
S100A4	CAPL, p9ka, pEL98, mts 1, metastatin, calvasculin, murine placental calcium protein, Fsp1	Regulation of cell motility, secretion, and extracellular functions, angiogenesis
S100A5	S100D	Ca^{2+}-, Zn^{2+}-, and Cu^{2+}-binding protein in the central nervous system (CNS); unknown function
S100A6	Calcyclin, CACY, 2A9, PRA, CaBP, 5B10	Regulation of insulin release, prolactation secretion, Ca^{2+} homeostasis, tumor progression
S100A7	Psoriasin, PSOR1, BDA11, CAAF2	S100A7-fatty acid binding protein complex regulates differentiation of keratinocytes
S100A8	Calgranulin A, CAGA, CFAg, MRP8, p8, MAC387, 60B8Ag, L1Ag, CP-10, MIF, NIF	S100A8/A9 heterodimer: Chemotactic activities, adhesion of neutrophils, myeloid cell differentiation, apoptosis, fatty acid metabolism
S100A9	Calgranulin B, CAGB, CFAg, MRP-14, p14, MAC 387, 60B8Ag, L1Ag, MIF, NIF	S100A8/A9 heterodimer; see above
S100A10	Calpactin light chain, CAL12, CLP11, p11, p10, 42C	Inhibition of phospholipase A2, neurotransmitter release, regulates membrane traffic in connection with annexin II, and ion currents
S100A11	Calgizzarin, S100C	Organization of early endosomes, inhibition of annexin I function, regulation of phosphorylation, physiological role in keratinocyte cornified envelope
S100A12	Calgranulin C, p6, CAAF1, CGRP, cornea-associated antigen	Host–parasite interaction, differentiation of squamous epithelial cells and extracellular functions
S100A13		Regulation of FGF-1 and synaptotagmin-1 release
S100A14		Unknown
S100A11P		Almost identical to S100A11, probably similar function
S100B	S100β, neurite extension factor, calcium-dependent guanylate cyclase activator protein (CD-GCAP)	Cell motility, proliferation, inhibition of phosphorylation, inhibition of microtubule assembly, transcription, regulation of nuclear kinase, extracellular functions, e.g. neurite extension; activation of rod outer segment guanylate cyclase
S100P		Function in the placenta
S100Z		Function in spleen and leukocytes
Calbindin	9k CALB3, CaBP9k, I CaBP	Ca^{2+} buffer and Ca^{2+} transport

For references see 8,9.

translocation to the extracellular space.[5] The extracellular action of S100B and S100A12 is mediated via binding to the receptor for advanced glycation endproduct (RAGE) – a multiligand member of the immunoglobulin superfamily.[6,7] For details on biological functions of S100 proteins, see also Table 1.

PROTEIN PRODUCTION, PURIFICATION, AND MOLECULAR CHARACTERIZATION

In general, the expression levels of signaling molecules in the cell are low compared to metabolic proteins, making it rather difficult to isolate S100 proteins from their sources. A few S100 proteins are expressed at high levels in the brain and heart tissue, for example, S100B, which represents about 0.5% of total cellular brain protein. In fact, the first S100 proteins were isolated from bovine brain where the proteins had been found soluble at 100 (S100) percent of ammonium sulfate.[10] Almost all S100 proteins are cloned into expression vectors for high expression in *Escherichia coli*. The different recombinant proteins are purified by ammonium sulfate precipitation, Ca^{2+}-dependent hydrophobic interaction, size exclusion, or anion exchange chromatography. S100 proteins have a size of 10 to 12 kDa and form homo- and heterodimers in solution. Only one protein of the S100 family, calbindin, occurs as a monomer. This difference in the structure is also reflected in calbindin's function as Ca^{2+}-buffer versus the signaling activity of the other S100 proteins. The dimerization plane of S100 proteins is composed of strictly conserved hydrophobic residues (highlighted in green in Figure 1), which are missing in the case of calbindin.

METAL-BINDING PROPERTIES

Generally, the dimeric S100 proteins bind four Ca^{2+} per dimer ($K_d = 20–500\,\mu M$) with strong positive cooperativity. Like many other EF-hand proteins, S100 proteins are able to bind Mg^{2+} into their EF-hand sites, but the affinities for Mg^{2+} are rather low ($K_d = 1–125\,mM$).[11,12] Therefore, Mg^{2+} at physiological concentrations (0.5 mM) has only a minor effect on the Ca^{2+}-binding.[11,13–15] Besides Ca^{2+}, a number of S100 proteins bind Zn^{2+} with a wide range of affinity ($K_d = 4\,nM$ to $2\,mM$).[16] Among the S100 proteins, S100A3 displays by far the highest affinity for Zn^{2+} ($K_d = 4\,nM$) and, interestingly, the lowest affinity for Ca^{2+} ($K_d = \sim 20\,mM$),[17] implying that S100A3 functions rather as a Zn^{2+}-signaling protein instead of a Ca^{2+}-signaling protein. Spectroscopic studies[17] and the crystal structure of metal-free S100A3[18] allowed the identification of one preformed Zn^{2+} binding site (distinct from the EF-hand) in the C-terminus of each subunit in which the Zn^{2+}-ion is coordinated by one histidine and three cysteine residues. Recently, Cu^{2+} was identified as another metal ion that binds to S100 proteins. Binding of four copper ions per homodimer was reported for S100B ($K_d = 0.46\,\mu M$)[19] and S100A5 ($K_d = 4\,\mu M$).[20]

X-RAY AND NMR STRUCTURES

Ca^{2+}-free state versus Ca^{2+}-loaded state

The amount of detailed structural data for S100 proteins is growing rapidly with structures determined by X-ray crystallography and NMR (see Table 2). For several S100 proteins, the 3D structures of the Ca^{2+}-free and of the Ca^{2+}-loaded state are known and they reveal the atomic details of the Ca^{2+}-dependent conformational changes. The structures determined so far show an architecture typical for all S100 proteins. Except for calbindin, the S100 proteins are organized as a tight homodimer. Each S100 subunit is composed of two helix-loop-helix Ca^{2+}-binding domains connected by a central hinge region. The C-terminal EF-hand contains the so-called 'canonical' Ca^{2+}-binding motif, common to all EF-hand proteins. This Ca^{2+}-binding loop has a typical sequence signature with a length of 12 amino acids (Figure 1) and is flanked by helix H_{III} and H_{IV} (Figure 2(a)). The N-terminal EF-hand is different from the canonical EF-hand motif and is characteristic for the S100 proteins. This EF-hand has a 14–amino acid consensus-sequence motif flanked by helix H_I and H_{II} and is called the 'S100-specific' or 'pseudo EF-hand' (Figure 1). In both EF-hands, the Ca^{2+}-ion is coordinated in a pentagonal bipyramidal configuration (Figure 3(b) and (c)). The six residues involved in the binding are denoted by X, Y, Z, −Y, −X and −Z. The invariant glutamate or aspartate at position −Z provides a carboxylate group for bidentate coordination of the Ca^{2+}-ion (Figures 1, 3(b) and (c)).

The dimer interface of the S100 homodimer is built up by helix H_I of the N-terminal S100-specific EF-hand and helix H_{IV} of the C-terminal canonical EF-hand of each subunit, forming a stable four-helix bundle (3D Structure). The large dimer interface ($\sim 2500\,Å^2$) is highly hydrophobic (Figure 4), driving the high affinity dimerization at even picomolar concentrations.[57] The hydrophobic residues of the interface are partially missing in calbindin, and mutational studies have shown that loss of these residues causes monomerization.[58]

Upon Ca^{2+}-binding, the typical EF-hand of the S100 protein undergoes a conformational change. In the Ca^{2+}-free conformation, the helices that flank the EF-hand are arranged in an antiparallel orientation, the so-called closed conformation. Ca^{2+}-binding causes a movement of the helices toward a perpendicular orientation (Figure 2(b)). This Ca^{2+}-loaded open conformation is characterized by the exposure of a concave hydrophobic surface that is required for target recognition.[60]

3D structures of the calcium and zinc binding S100 proteins

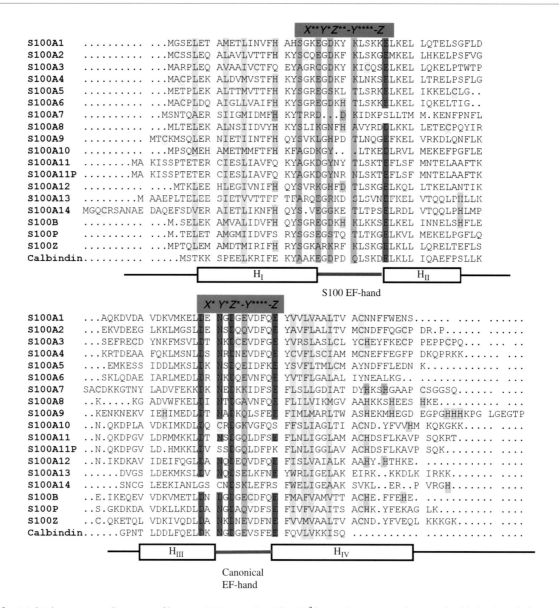

Figure 1 Multiple sequence alignment of human S100 proteins. The Ca^{2+}-coordinating residues are highlighted in dark red (side chain coordination), and in light red (backbone oxygen coordination). The Zn^{2+}-coordinating residues in S100A7 and the conserved binding site in S100A12 are highlighted in magenta. Hydrophobic residues that are essential for dimerization are highlighted in green. Residues that are putative Zn^{2+} ligands in other S100 proteins are highlighted in yellow (cysteine) and blue (histidine) respectively.

N-terminal S100-specific EF-hand

The N-terminal EF-hands of calbindin, S100B, and S100A6 exhibit only small conformational changes upon Ca^{2+}-binding.[24–27,38,40,43,61,62] Helix H_{II} undergoes a small rearrangement in direction of the Ca^{2+}-binding site with practically no change in the interhelical angle of helix H_I and H_{II} (Figure 2(a)). The crystal structure of Ca^{2+}-free S100A3[18] and S100A6[23] revealed two tightly bound water molecules in the N-terminal EF-hand, which stabilize the Ca^{2+}-free N-terminal EF-hand in a Ca^{2+}-loaded-like conformation. One water molecule is located exactly at the position of the subsequently bound Ca^{2+}-ion and can form various hydrogen bonds to the protein, which resemble the coordination of the Ca^{2+}-ion (Figure 3(a) and (b)). The Ca^{2+}-binding site is mainly formed by backbone oxygen atoms in the loop and the carboxylate group of a glutamate residue located on the start of helix H_{II}. In the Ca^{2+}-free structures, this carboxylate does not provide a direct hydrogen bond to the firmly bound water molecule; however, the second water molecule bridges the glutamate side chain oxygens with the first central water molecule. Since the affinity of the N-terminal EF-hand toward Mg^{2+} is rather low,[11] the water filled EF-hand most probably represents the resting state of the Ca^{2+}-free protein *in vivo*. The binding Ca^{2+}-ion displaces the two water molecules and becomes coordinated by the glutamate side chain. This strong

Table 2 Amino acid sequence information and available structures

Name	Chromosomal location (HUM)	SwissProt accession number	Available structures PDB code (technique): description	References
S100A1	1q21	P23297 (HUM) P02639 (BOV) P56565 (MOUS) P35467 (RAT)	1K2H (NMR): apo	21
S100A2	1q21	P29034 (HUM) P10462 (BOV)		
S100A3	1q21	P33764 (HUM) P56566 (MOUS)	1KSO (xtal): apo	18
S100A4	1q21	P26447 (HUM) P35466 (BOV) P07091 (MOUS) P05942 (RAT)	1M31 (NMR): apo	22
S100A5	1q21	P33763 (HUM) O88945 (MOUS)		
S100A6	1q21	P06703 (HUM) Q98953 (CHICK) O77691 (HORS) P14069 (MOUS) P30801 (RABIT) P05964 (RAT)	1K8U, 1K9P (xtal): apo; 1K96, 1K9K (xtal): 4 Ca^{2+} 1CNP, 2CNP, (NMR): apo; 1A03, 1JWD (NMR) 4 Ca^{2+}	23 24–27
S100A7	1q21	P31151 (HUM) Q28050 (BOV)	1PSR (xtal): 2 Ho^{3+}; 2PSR (xtal): 2 Zn^{2+}, 2 Ca^{2+}; 3PSR (xtal): 2 Ca^{2+}	28,29
S100A8	1q21	P05109 (HUM) P28782 (BOV) P27005 (MOUS) P50115 (RAT)	1MR8 (xtal): 4 Ca^{2+}	30
S100A9	1q21	P06702 ((HUM) P28783 (BOV) P31725 (MOUS) P50117 (RABIT) P50116 (RAT)	1IRJ	31
S100A10	1q21	P08206 (HUM) P27003 (CHICK) P08207 (MOUS) P04163 (PIG) P05943 (RAT)	1A4P (xtal) apo; 1BT6 (xtal): apo, annexin II peptide	32
S100A11	1q21	P31949 (HUM) P24479 (CHICK) P50543 (MOUS) P31950 (PIG) P24480 (RABIT)	1QLS (xtal): 4 Ca^{2+} annexin I peptide	33
S100A12	1q21	P80511 (HUM) P79105 (BOV) P80310 (PIG) O77791 (RABIT)	1E8A, (xtal) 4 Ca^{2+}; 1GQM (xtal): hexamer 18 Ca^{2+}	34,35
S100A13	1q21	Q99584 (HUM) P79342 (BOV) P97352 (MOUS)		
S100A14	1q21	Q9HCY8 (HUM)		

(continued overleaf)

Table 2 (continued)

Name	Chromosomal location (HUM)	SwissProt accession number	Available structures PDB code (technique): description	References
S100A11P	7q22-q31.1	O60417 (HUM)		
S100B	21q22	P04271 (HUM)	1UWO (NMR): 4 Ca^{2+}; 1MQ1 (NMR) 4 Ca^{2+}, peptide; 1CFP (NMR): apo; 1MHO (xtal): 4 Ca^{2+}	36,37
		P02638 (BOV)		38,39
		P50114 (MOUS)	1SYM (NMR): apo; 1B4C (NMR): apo; 1QLK (NMR): 4 Ca^{2+}; 1DT7 (NMR) 4Ca^{2+}, p53 peptide; 1MWN (NMR) 4 Ca^{2+}, peptide	
		P04631 (RAT)		40–45
S100P	4p16	P25815 (HUM)	1J55 (xtal): 4 Ca^{2+}	46
S100Z	5q12-q13	Q8WXG8 (HUM)		
Calbindin	Xp22	P29377 (HUM) P51964 (CHICK) P97816 (MOUS) P02632 (PIG) P02634 (RAT)		
		P02633 (BOV)	1CLB (NMR): apo (P43G); 2BCB (NMR): 2 Ca^{2+} (mutant P43G); 4ICB (xtal) 2 Ca^{2+}; 1IG5 (xtal): 1 Mg^{2+}; 1IGV (xtal): 1 Mn^{2+}; 1BOD (NMR): 2Ca^{2+} (mutant canonical EF1); 1BOC (NMR) 2 Ca^{2+} (mutant, A15D, P20G); 1CDN1 (NMR): 1Cd^{2+} (mutant P43G); 1D1O (NMR): 2 Ca^{2+} (mutant N56A); 1B1G (NMR) 2 Ca^{2+}, 1KCY (NMR): apo (mutant F36G, P43G)	3,11,25,47–54

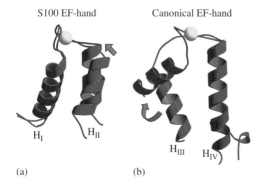

Figure 2 Conformational change in EF-hands of S100 proteins. The S100 specific EF-hand is depicted in (a), the canonical EF-hand in (b). The Ca^{2+}-free protein is shown in blue, the Ca^{2+}-loaded form in red. Ca^{2+} ions are shown as yellow spheres. The coordinates were taken from the crystal structures of human S100A6 (PDB codes 1K8U, 1K96).[23] The figures were made with MOLSCRIPT[55] and RASTER3D.[56]

bidentate coordination drives the short movement of helix H_{II} (Figure 2(a)).

C-terminal canonical EF-hand

The C-terminal canonical EF-hand represents the target interaction sites of the Ca^{2+} sensor S100 proteins. The EF-hand is characterized by a sequence of 12 residues with the pattern X * Y * Z *–Y *–X * *–Z, where X, Y, Z, –X, –Y, and –Z represent the ligands that participate in Ca^{2+} coordination and the stars represent the intervening residues (Figures 1 and 3(c)). In the Ca^{2+}-free state, the helices H_{III} and H_{IV} flanking the EF-hand loop adopt an antiparallel conformation (Figure 2(b)) similar to the EF-hands in Ca^{2+}-free calmodulin.[63] In the Ca^{2+}-loaded state of S100 proteins, there is a large change in the orientation of the helix H_{III} (Figure 2(b)), whereas the helix H_{IV} that is engaged in the dimer interface does not move. This movement of helix H_{III} results in a perpendicular orientation of helices H_{III} and H_{IV}, changing the interhelical angle by approximately 90°. The Ca^{2+} induced conformational change opens the structure and exposes a hydrophobic cleft formed by residues of the hinge region, helix H_{III}, and the C-terminal loop region. This Ca^{2+}-induced conformation of S100 proteins is strikingly different from that of calmodulin or troponin C, where helix H_{IV} reorients upon Ca^{2+}-binding. Additional structural changes upon Ca^{2+}-binding are observed in helix H_{IV}.[23,36,43] In the Ca^{2+}-loaded structures of S100B and S100A6, helix H_{IV} is two turns longer than in the Ca^{2+}-free structures. This leads to a reorientation of several residues in the C-terminus, which might influence the target recognition.

An open Ca^{2+}-loaded-like conformation of S100 proteins was observed in the structure of Ca^{2+}-free

3D structures of the calcium and zinc binding S100 proteins

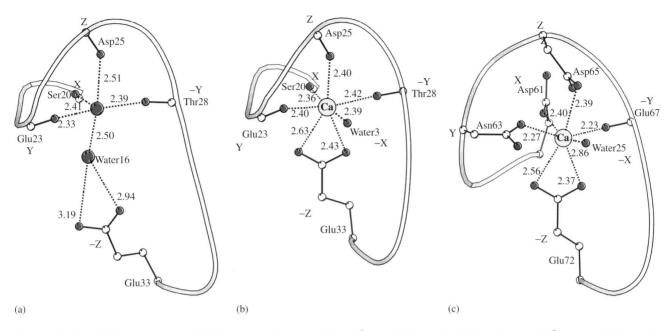

Figure 3 The Ca^{2+} binding sites of S100 proteins. Drawing of the Ca^{2+}-free S100-specific EF-hand (a), the Ca^{2+}-loaded S100-specific EF-hand (b), and of the Ca^{2+}-loaded canonical EF-hand (c). The coordinates were taken from the crystal structures of human S100A6 (PDB codes 1K8U, 1K96).[23] The figures were prepared with MOLSCRIPT.[55]

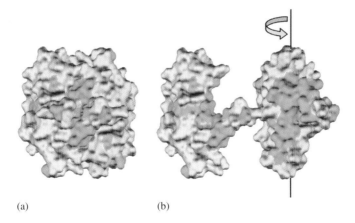

Figure 4 The hydrophobic dimer interface of S100 proteins. Surface representation of human S100A3 (PDB code 1KSO).[18] (a) The intact homodimer is shown, whereby the subunits are shown in blue and red; hydrophobic residues are shown in green. (b) The red subunit is rotated by 90° now showing the hydrophobic dimer interface. The figure was prepared with WebLab Viewer.[59]

S100A10. Although S100A10 is not able to bind Ca^{2+}, it mimics the Ca^{2+}-loaded conformation.[32] This enables Ca^{2+}-free S100A10 to interact with its target molecule annexin II in a way very similar to the binding of Ca^{2+}-loaded S100A11 to annexin I.[33]

ZN^{2+} BINDING SITE

Although a number of S100 proteins bind Zn^{2+} with high affinity, only the structure of Ca^{2+}/Zn^{2+}-loaded S100A7 is available so far.[29] S100A7 binds two Zn^{2+} ions per homodimer. The Zn^{2+}-binding sites are formed by both subunits and are located close to the dimer interface (Figure 5(a)). The coordination sphere involves two histidines from one subunit and another histidine and an aspartate from the second subunit (Figure 5(b)). The pattern that is formed by these residues in the primary sequence is fully conserved in S100A9 and S100A12 (Figure 1), suggesting that these two proteins coordinate Zn^{2+} in similar binding sites. In the case of S100A12, it was suggested that this site also binds Cu^{2+}.[34]

LARGER ASSEMBLIES

Although the structures determined so far confirmed the dimeric organization of S100 proteins, larger assemblies

3D structures of the calcium and zinc binding S100 proteins

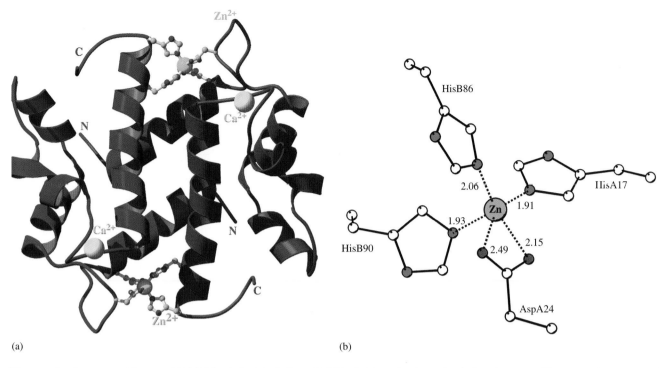

(a) (b)

Figure 5 Structure of human S100A7 loaded with Ca^{2+} and Zn^{2+}. (a) S100A7 dimer loaded with two Ca^{2+} ions in the canonical EF-hands and two Zn^{2+} ions close to the dimer interface. (b) Illustration of one Zn^{2+} site in S100A7. (PDB code 2PSR).[29] The figure was made with MOLSCRIPT[55] and RASTER3D.[56]

were observed under certain conditions. Recently, the structure of hexameric Ca^{2+}-loaded S100A12 was reported.[34] The hexamer is composed of three S100A12 dimers, whereby six additionally bound Ca^{2+} ions stabilize the contact between two adjacent dimers (Figure 6). Two Ca^{2+} ions at each interface are coordinated by residues from both dimers. The formation of the hexamer in solution can be driven by an increase of the Ca^{2+} concentration. It was proposed that such large assemblies are formed when S100A12 is secreted into extracellular space and is required for receptor-binding and activation by receptor multimerization.[34]

STRUCTURES OF COMPLEXES

Detailed information on the interaction mode of S100 proteins with their target proteins was obtained from five different structures of complexes with peptide fragments of target proteins. All structures consist of one S100 dimer that grasps two peptides in the hydrophobic cleft of each subunit. A characteristic of the S100-target complexes is the apparently diverse binding mode of the different peptides to the hydrophobic cavity of S100 proteins. So far the structures of S100A10 in complex with annexin II (PDB code 1BT6)[32] and of S100A11 with annexin I (PDB code 1QLS)[33] were determined by X-ray crystallography, as well as the structures of S100B with peptides derived from p53 (PDB code 1DT7),[44] and with a synthetic peptide TRTK-12

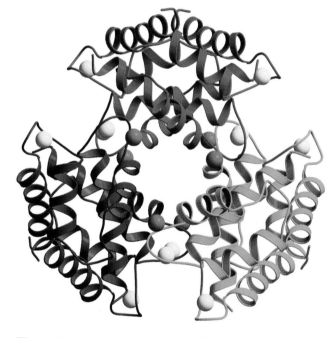

Figure 6 Hexameric structure of Ca^{2+}-loaded S100A12. Three S100A12 dimers (shown in green, blue, and magenta) form a hexamer. The Ca^{2+} ions bound to the EF-hands are shown as bright yellow spheres. At the hexamer-forming interface, six additional Ca^{2+} ions are located, (orange spheres). The figure was made with MOLSCRIPT[55] and RASTER3D.[56]

(PDB codes 1MQ1 and 1MWN)[37,45] derived from random bacteriophage screenings.

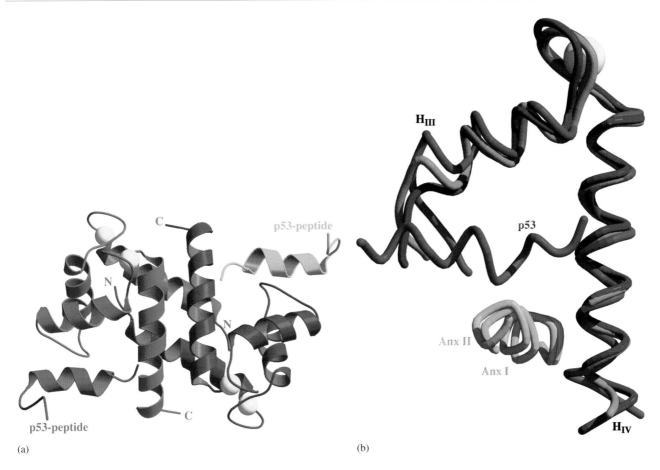

Figure 7 Target binding to S100 proteins. (a) Complex of Ca^{2+}-loaded human S100B dimer with two peptides derived from p53 (PDB code 1DT7). (b) Structural alignment of human S100A10-annexin II (PDB code 1BT6),[32] S100A11-annexin I (PDB code 1QLS),[33] and S100B-p53 (PDB code 1DT7)[44] complexes.

The complexes of S100A10 and S100A11 with the annexin peptides revealed that both subunits are involved in the binding of each peptide. The binding of the annexin peptides with the protein is mediated by hydrophobic and polar interactions, whereby the required residues are located on the hinge region and H_{IV} of one subunit and on helix H'_I of the second subunit. In contrast to the annexin peptides, each p53 peptide interacts with residues of the hinge region, helix H_{III}, and helix H_{IV} of one S100B subunit, with the peptide rotated by about 90°. The different binding mode is evident from a structural alignment of the three protein–peptide complexes (Figure 7). The structure of the S100B-TRTK-12 peptide complex is still a matter of debate. Two independent research groups determined independently the structure of S100B with TRTK-12 and obtained different results. Whereas one structure has the bound peptide in a partially α-helical conformation,[45] the second structure shows the peptide in a random coil conformation.[37]

FUNCTIONAL ASPECTS

S100 proteins generally are multifunctional signaling proteins and are involved in the regulation of diverse cellular functions such as cell differentiation, cell motility, transcription, and cell cycle progression.[8,9] This wide range of activities is achieved by the specificity of the more than 20 different S100 proteins and by their regulation through Ca^{2+}- and Zn^{2+}-binding. The ongoing identification of new target proteins is the basis for the molecular understanding of the S100 protein–regulated cellular processes.

Regulation by Ca^{2+}

S100 proteins regulate different enzymes, Ca^{2+} homeostasis, cytoskeletal organization, and transcription in a Ca^{2+}-dependent manner. Detailed information is available in recent reviews.[8,9,16,64,65]

Regulation by Zn^{2+}

Many S100 proteins bind Zn^{2+} besides Ca^{2+} with high affinity. However, most of the reported Zn^{2+} dissociation constants are in the μM range, which stands in contrast to the low nM concentrations of free Zn^{2+} (2 to 10 nM) in the cytoplasm. Owing to this low concentration, binding of Zn^{2+} to most S100 proteins does not occur in the cytoplasm, as shown for S100B and S100A6.[18] However, Zn^{2+}-binding might occur in some vesicles that contain Zn^{2+} in μM to mM concentrations, or in the extracellular space in which the Zn^{2+} concentrations are in the μM range.[66] An example for a Ca^{2+}-independent but Zn^{2+}-dependent target protein recognition is the interaction of S100B with tau protein.[67]

Copper-ion binding

Copper-ion binding to S100B might have a neuroprotective function. Bovine S100B that is highly expressed in the brain can efficiently sequester copper ions and hereby suppress copper-ion–induced cell damage.[19,68]

Extracellular function

Besides their intracellular functions, several S100 proteins such as S100B, S100A8/A9, S100A4 and S100A12 are secreted and act in a cytokine-like manner. For example, the S100A8/A9 heterodimer acts as a chemotactic molecule in an inflammation,[69] whereas S100B exhibits neurotrophic activity[7] and S100A4 has angiogenic effects.[70] One way of mediating these extracellular activities was recently identified: several S100 proteins bind to the extracellular domain of the receptor RAGE and thereby activate an intracellular signal cascade including MAP-kinase and NFκB.[6,7]

S100 domains participating in protein–protein interaction

It is interesting to note that several proteins involved in differentiation and signaling contain domains with high similarity to S100 proteins. These S100-like domains serve as protein–protein interaction modules. Generally, two distinct modes of protein–protein interaction should be considered. Proteins containing an S100-like domain could utilize the S100 target recognition site (Figure 7) for interaction. Other proteins that contain an S100-like domain might interact via the hydrophobic surface area of the domain that corresponds to the dimer interface in dimeric S100 proteins (Figure 4). Protein–protein interaction via the target recognition site of S100 domains occurs in proteins containing the so-called Eps15 homology (EH) motif that consists of three S100-like domains.[71,72] These domains bind proteins containing specific sequence motifs composed of hydrophobic amino acid residues. It was shown that a peptide derived from these sequence motifs binds to residues located on the corresponding helices H_{III} and H_{IV} of S100 proteins.[72] It appears that the EH domain utilizes the S100 target recognition site in a Ca^{2+}-independent manner for specific protein–protein interaction.

Multiple sequence alignments of S100 proteins with the N-terminal domains of trichohyalin (SwissProt Accession No. Q07283) and profilaggrin (SwissProt Accession No. Q01212) involved in the differentiation of the mammalian epidermis[73] as well as a recently identified esophagus specific protein[74] (GenBank Identification No. 7706635) revealed a high homology of these domains to S100 proteins. The sequence alignment data suggest that the N-terminal domains adopt a fold very similar to the S100 proteins. Most interestingly, the hydrophobic residues required for dimerization in S100 proteins (Figure 1) are strictly conserved in these S100-like domains. Therefore, these S100 domains presumably have a hydrophobic surface area similar to the hydrophobic dimerization plane of S100 proteins (Figure 4). Experimental support for this hypothesis comes from yeast two hybrid experiments,[75–79] which show that the dimerization plane of single S100 subunits indeed functions as a highly specific protein–protein interaction site. These experiments in which protein fusions with S100 protein subunits are used in the search for target proteins fused to another protein, always revealed the S100 protein itself as a binding partner, whereby the interaction of the two fused S100 subunits in yeast two hybrid experiments is driven by the dimerization plane.

Involvement in diseases

The central role of S100 proteins in cellular signaling is evident from their close association with several human diseases such as rheumatoid arthritis, acute inflammatory lesions, cardiomyopathy, Alzheimer's disease, and cancer (Table 3). Further elucidation of the S100 protein and protein-target structures in the future will be the basis for specific inhibition of their extra- and intracellular pathologies.

Table 3 Involvement of S100 proteins in diseases

Proteins	Association with diseases	Reference[a]
S100B	Alzheimer's disease	80
	Down syndrome	
S100A1	Cardiomyopathy	81
S100A2	Cancer, Tumor suppression	82
S100A4	Cancer, Metastasis	70
S100A6	Cancer	83
S100A7	Psoriasis	84
S100A8/A9	Inflammation	85
	Wound healing	86
S100A11	Cancer	87
S100A12	Inflammation	88
S100P	Cancer	89

[a] Selected references, for further references see 8.

ACKNOWLEDGEMENT

This work was supported in part by the Wilhelm Sander-Stiftung (to G.F.), by the NCCR on Neural Plasticity and Repair supported by the Swiss National Science Foundation (to C.W.H.), and by a grant of the Nachwuchsförderungskommission of the University of Zürich (to G.F.).

REFERENCES

1. MJ Berridge, P Lipp and MD Bootman, *Nat Rev Mol Cell Biol*, **1**, 11–21 (2000).
2. H Kawasaki, S Nakayama and RH Kretsinger, *Biometals*, **11**, 277–95 (1998).
3. NJ Skelton, J Kördel, M Akke, S Forsen and WJ Chazin, *Nat Struct Biol*, **1**, 239–45 (1994).
4. GE Davey, P Murmann and CW Heizmann, *J Biol Chem*, **276**, 30819–26 (2001).
5. HL Hsieh, BW Schäfer, JA Cox and CW Heizmann, *J Cell Sci*, **115**, 3149–58 (2002).
6. MA Hofmann, S Drury, C Fu, W Qu, A Taguchi, Y Lu, C Avila, N Kambham, A Bierhaus, P Nawroth, MF Neurath, T Slattery, D Beach, J McClary, M Nagashima, J Morser, D Stern and AM Schmidt, *Cell*, **97**, 889–901 (1999).
7. HJ Huttunen, J Kuja-Panula, G Sorci, AL Agnelletti, R Donato and H Rauvala, *J Biol Chem*, **275**, 40096–105 (2000).
8. CW Heizmann, BW Schäfer and G Fritz, in R Bradshaw and E Dennis (eds.), *Handbook of Cell Signaling*, Elsevier Science, San Diego (2003); in press.
9. CW Heizmann, G Fritz and BW Schäfer, *Front Biosci*, **7**, D1356–68 (2002).
10. BW Moore, *Biochem Biophys Res Commun*, **19**, 739–44 (1965).
11. M Andersson, A Malmendal, S Linse, I Ivarsson, S Forsen and LA Svensson, *Protein Sci*, **6**, 1139–47 (1997).
12. AV Gribenko and GI Makhatadze, *J Mol Biol*, **283**, 679–94 (1998).
13. S Matsuda, *J Biochem (Tokyo)*, **104**, 989–90 (1988).
14. M Pedrocchi, BW Schäfer, I Durussel, JA Cox and CW Heizmann, *Biochemistry*, **33**, 6732–38 (1994).
15. K Ridinger, BW Schäfer, I Durussel, JA Cox and CW Heizmann, *J Biol Chem*, **275**, 8686–94 (2000).
16. CW Heizmann and JA Cox, *Biometals*, **11**, 383–97 (1998).
17. G Fritz, CW Heizmann and PM Kroneck, *Biochim Biophys Acta*, **1448**, 264–76 (1998).
18. G Fritz, PR Mittl, M Vasak, MG Grütter and CW Heizmann, *J Biol Chem*, **277**, 33092–98 (2002).
19. T Nishikawa, IS Lee, N Shiraishi, T Ishikawa, Y Ohta and M Nishikimi, *J Biol Chem*, **272**, 23037230–41 (1997).
20. BW Schäfer, JM Fritschy, P Murmann, H Troxler, I Durussel, CW Heizmann and JA Cox, *J Biol Chem*, **275**, 30623–30 (2000).
21. RR Rustandi, DM Baldisseri, KG Inman, P Nizner, SM Hamilton, A Landar, DB Zimmer and DJ Weber, *Biochemistry*, **41**, 788–96 (2002).
22. KM Vallely, RR Rustandi, KC Ellis, O Varlamova, AR Bresnick and DJ Weber, *Biochemistry*, **41**, 12670–80 (2002).
23. LR Otterbein, J Kordowska, C Witte-Hoffmann, CL Wang and R Dominguez, *Structure (Camb)*, **10**, 557–67 (2002).
24. BC Potts, J Smith, M Akke, TJ Macke, K Okazaki, H Hidaka, DA Case and WJ Chazin, *Nat Struct Biol*, **2**, 790–96 (1995).
25. M Sastry, RR Ketcham, O Crescenzi, C Weber, MJ Lubienski, H Hidaka and WJ Chazin, *Structure*, **6**, 223–31 (1998).
26. L Mäler, BC Potts and WJ Chazin, *J Biomol NMR*, **13**, 233–47 (1999).
27. L Mäler, M Sastry and WJ Chazin, *J Mol Biol*, **317**, 279–90 (2002).
28. DE Brodersen, M Etzerodt, P Madsen, JE Celis, HC Thogersen, J Nyborg and M Kjeldgaard, *Structure*, **6**, 477–89 (1998).
29. DE Brodersen, J Nyborg and M Kjeldgaard, *Biochemistry*, **38**, 1695–1704 (1999).
30. K Ishikawa, A Nakagawa, I Tanaka, M Suzuki and J Nishihira, *Acta Crystallogr, Sect D, Biol Crystallogr*, **56**, 559–66 (2000).
31. H Itou, M Yao, I Fujita, N Watanabe, M Suzuki, J Nishihira and I Tanaka, *J Mol Biol*, **316**, 265–76 (2002).
32. S Rety, J Sopkova, M Renouard, D Osterloh, V Gerke, S Tabaries, F Russo-Marie and A Lewit-Bentley, *Nat Struct Biol*, **6**, 89–95 (1999).
33. S Rety, D Osterloh, JP Arie, S Tabaries, J Seeman, F Russo-Marie, V Gerke and A Lewit-Bentley, *Struct Fold Des*, **8**, 175–84 (2000).
34. OV Moroz, AA Antson, EJ Dodson, HJ Burrell, SJ Grist, RM Lloyd, NJ Maitland, GG Dodson, KS Wilson, E Lukanidin and IB Bronstein, *Acta Crystallogr, Sect D, Biol Crystallogr*, **58**, 407–13 (2002).
35. OV Moroz, AA Antson, GN Murshudov, NJ Maitland, GG Dodson, KS Wilson, I Skibshoj, EM Lukanidin and IB Bronstein, *Acta Crystallogr, Sect D, Biol Crystallogr*, **57**, 20–29 (2001).
36. SP Smith and GS Shaw, *Structure*, **6**, 211–22 (1998).
37. KA McClintock and GS Shaw, *J Biol Chem*, **278**, 6251–7 (2003).
38. PM Kilby, LJ Van Eldik and GC Roberts, *Structure*, **4**, 1041–52 (1996).
39. H Matsumura, T Shiba, T Inoue, S Harada and Y Kai, *Structure*, **6**, 233–41 (1998).
40. AC Drohat, JC Amburgey, F Abildgaard, MR Starich, D Baldisseri and DJ Weber, *Biochemistry*, **35**, 11577–88 (1996).
41. AC Drohat, DM Baldisseri, RR Rustandi and DJ Weber, *Biochemistry*, **37**, 2729–40 (1998).
42. AC Drohat, DM Baldisseri, RR Rustandi and DJ Weber, *Biochemistry*, **37**, 2729–40 (1998).
43. AC Drohat, N Tjandra, DM Baldisseri and DJ Weber, *Protein Sci*, **8**, 800–9 (1999).
44. RR Rustandi, DM Baldisseri and DJ Weber, *Nat Struct Biol*, **7**, 570–74 (2000).
45. KG Inman, R Yang, RR Rustandi, KE Miller, DM Baldisseri and DJ Weber, *J Mol Biol*, **324**, 1003–14 (2002).
46. H Zhang, G Wang, Y Ding, Z Wang, R Barraclough, PS Rudland, DG Fernig and Z Rao, *J Mol Biol*, **325**, 785–94 (2003).
47. C Johansson, M Ullner and T Drakenberg, *Biochemistry*, **32**, 8429–38 (1993).
48. J Kördel, NJ Skelton, M Akke and WJ Chazin, *J Mol Biol*, **231**, 711–34 (1993).
49. M Akke, S Forsen and WJ Chazin, *J Mol Biol*, **252**, 102–21 (1995).
50. NJ Skelton, J Kördel and WJ Chazin, *J Mol Biol*, **249**, 441–62 (1995).
51. J Kördel, DA Pearlman and WJ Chazin, *J Biomol NMR*, **10**, 231–43 (1997).
52. BB Kragelund, M Jonsson, G Bifulco, WJ Chazin, H Nilsson, BE Finn and S Linse, *Biochemistry*, **37**, 8926–37 (1998).

53. L Mäler, J Blankenship, M Rance and WJ Chazin, *Nat Struct Biol*, **7**, 245–50 (2000).
54. MR Nelson, E Thulin, PA Fagan, S Forsen and WJ Chazin, *Protein Sci*, **11**, 198–205 (2002).
55. PJ Kraulis, *J Appl Crystallogr*, **24**, 946–50 (1991).
56. EA Merritt and DJ Bacon, *Methods Enzymol*, **277**, 505–24 (1997).
57. AC Drohat, E Nenortas, D Beckett and DJ Weber, *Protein Sci*, **6**, 1577–82 (1997).
58. S Tarabykina, DJ Scott, P Herzyk, TJ Hill, JR Tame, M Kriajevska, D Lafitte, PJ Derrick, GG Dodson, NJ Maitland, EM Lukanidin and IB Bronstein, *J Biol Chem*, **276**, 24212–22 (2001).
59. Molecular Simulations Inc (MSI), Cambridge, UK, WebLab Viewer 3.20 (1999).
60. MR Nelson and WJ Chazin, *Biometals*, **11**, 297–318 (1998).
61. BC Potts, G Carlstrom, K Okazaki, H Hidaka and WJ Chazin, *Protein Sci*, **5**, 2162–74 (1996).
62. L Mäler, BC Potts and WJ Chazin, *J Biomol NMR*, **13**, 233–47 (1999).
63. H Ishida, K Nakashima, Y Kumaki, M Nakata, K Hikichi and M Yazawa, *Biochemistry*, **41**, 15536–42 (2002).
64. CW Heizmann, *Methods Mol Biol*, **172**, 69–80 (2002).
65. R Donato, *Int J Biochem Cell Biol*, **33**, 637–68 (2001).
66. DD Perrin and AE Watt, *Biochim Biophys Acta*, **230**, 96–104 (1971).
67. WH Yu and PE Fraser, *J Neurosci*, **21**, 2240–46 (2001).
68. N Shiraishi and M Nishikimi, *Arch Biochem Biophys*, **357**, 225–30 (1998).
69. RA Newton and N Hogg, *J Immunol*, **160**, 1427–35 (1998).
70. N Ambartsumian, J Klingelhofer, M Grigorian, C Christensen, M Kriajevska, E Tulchinsky, G Georgiev, V Berezin, E Bock, J Rygaard, R Cao, Y Cao and E Lukanidin, *Oncogene*, **20**, 4685–95 (2001).
71. S Koshiba, T Kigawa, J Iwahara, A Kikuchi and S Yokoyama, *FEBS Lett*, **442**, 138–142 (1999).
72. B Whitehead, M Tessari, A Carotenuto, PM van Bergen en Henegouwen and GW Vuister, *Biochemistry*, **38**, 11271–77 (1999).
73. RB Presland and BA Dale, *Crit Rev Oral Biol Med*, **11**, 383–408 (2000).
74. Z Xu, MR Wang, X Xu, Y Cai, YL Han, KM Wu, J Wang, BS Chen, XQ Wang and M Wu, *Genomics*, **69**, 322–30 (2000).
75. M Koltzscher and V Gerke, *Biochemistry*, **39**, 9533–39 (2000).
76. S Tarabykina, M Kriajevska, DJ Scott, TJ Hill, D Lafitte, PJ Derrick, GG Dodson, E Lukanidin and I Bronstein, *FEBS Lett*, **475**, 187–91 (2000).
77. H Hiyama, M Yokoi, C Masutani, K Sugasawa, T Maekawa, K Tanaka, JH Hoeijmakers and F Hanaoka, *J Biol Chem*, **274**, 28019–25 (1999).
78. Q Yang, D O'Hanlon, CW Heizmann and A Marks, *Exp Cell Res*, **246**, 501–9 (1999).
79. C Propper, X Huang, J Roth, C Sorg and W Nacken, *J Biol Chem*, **274**, 183–88 (1999).
80. WS Griffin, JG Sheng, JE McKenzie, MC Royston, SM Gentleman, RA Brumback, LC Cork, MR Del Bigio, GW Roberts and RE Mrak, *Neurobiol Aging*, **19**, 401–5 (1998).
81. R Kiewitz, C Acklin, E Minder, PR Huber, BW Schäfer and CW Heizmann, *Biochem Biophys Res Commun*, **274**, 865–71 (2000).
82. R Wicki, C Franz, FA Scholl, CW Heizmann and BW Schäfer, *Cell Calcium*, **22**, 243–54 (1997).
83. HB Guo, B Stoffel-Wagner, T Bierwirth, J Mezger and D Klingmuller, *Eur J Cancer*, **31A**, 1898–1902 (1995).
84. J Hitomi, K Maruyama, Y Kikuchi, K Nagasaki and K Yamaguchi, *Biochem Biophys Res Commun*, **228**, 757–63 (1996).
85. PA Hessian, J Edgeworth and N Hogg, *J Leukocyte Biol*, **53**, 197–204 (1993).
86. IS Thorey, J Roth, J Regenbogen, JP Halle, M Bittner, T Vogl, S Kaesler, P Bugnon, B Reitmaier, S Durka, A Graf, M Wockner, N Rieger, A Konstantinow, E Wolf, A Goppelt and S Werner, *J Biol Chem*, **276**, 35818–25 (2001).
87. M Tanaka, K Adzuma, M Iwami, K Yoshimoto, Y Monden and M Itakura, *Cancer Lett*, **89**, 195–200 (1995).
88. Z Yang, T Tao, MJ Raftery, P Youssef, N Di Girolamo and CL Geczy, *J Leukocyte Biol*, **69**, 986–94 (2001).
89. ID Guerreiro Da Silva, YF Hu, IH Russo, X Ao, AM Salicioni, X Yang and J Russo, *Int J Oncol*, **16**, 231–40 (2000).

EH domain

Michael Overduin and Mahadev Ravi Kiran
Department of Pharmacology, University of Colorado Health Sciences Center, Denver, CO, USA

FUNCTIONAL CLASS

Protein interaction module; contains two EF hands capable of coordinating calcium; binds to proteins containing Asn-Pro-Phe (NPF) sequences and related sequences; known as the Eps15 homology (EH) domain.

OCCURRENCE

The EH domain is a 100-residue motif that consists of two EF hands followed by a C-terminal proline-rich element (3D Structure). The EF hand is a helix-loop-helix motif that typically binds calcium tightly for structural stability or with an affinity in the physiological range for regulatory purposes. The EH domain is either found singly or in a series of two or three tandem repeats in a variety of eukaryotic proteins.[1] The function of the EH domain is the specific recognition of proteins containing NPF motifs or related sequences.[2,3] The structures, calcium coordination, and peptide binding modes of five monomeric EH domains have recently been elucidated, providing mechanistic insights into their contributions to endocytosis, vesicle transport, and signal transduction.

The EH domain is one of about 66 folds built from pairs of EF hands, the most common calcium binding motif.[6] It is

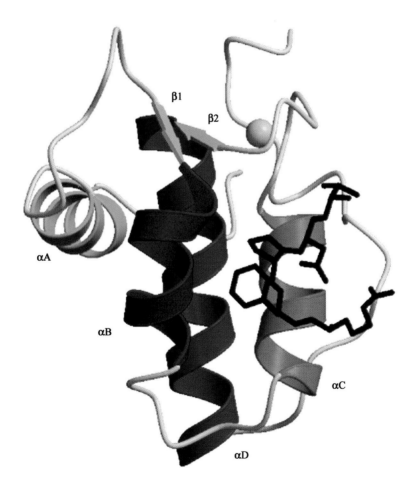

3D Structure Schematic representation of the monomer structure of the EH domain bound to a calcium ion and the TNPFR ligand residues. αA is shown in green, αB in blue, αC in magenta, and αD in red. The two short β-strands are shown in cyan. The calcium ion is shown in yellow and the bound peptide is represented as black sticks. The various secondary structural elements are also labeled. Prepared with MOLSCRIPT[4] and RASTER3D.[5] PDB code: 1F8H.

EH domain

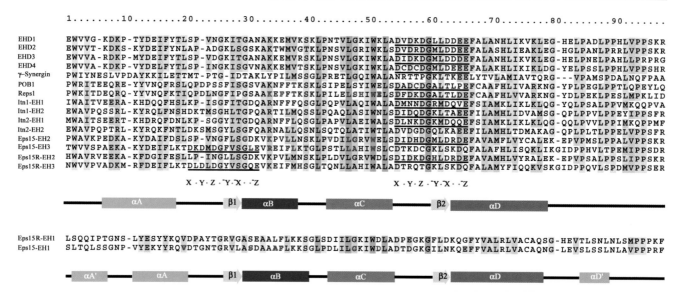

Figure 1 Alignment of the amino acid sequences of EH domains from various human proteins, with the structurally variant EH$_1$ sequences depicted as a separate subgroup. The positions of α-helices and β-strands are indicated. The calcium-binding EF hand motifs are underlined and the calcium-coordinating residues are marked. Residues determined to be involved in peptide binding are shown in red. Absolutely conserved residues are shown in a green background, similar residues are shown in cyan, and residue positions with more than 60% conservation are shown in a yellow background. The alignment was generated using the CLUSTALW[18] program and the coloring was done using the BOXSHADE[19] program in the SDSC Workbench.[20]

named after Eps15, where it was first discovered as a set of three repeats.[1,7] Most EH domains share between 25 and 40% sequence identity. The most highly conserved motif is the functionally important LPxxϕLπxIWxLxD sequence, where x, ϕ, and π refer, respectively, to any, hydrophobic and small residues. Other hallmarks of the EH domain sequence include a calcium-binding loop that is usually found in the second EF hand and a 12-residue proline-rich element at the C-terminus.

Proteins that contain one or more EH domains are common in eukaryotic proteomes, but have not been detected in archeae or bacteria. There are 11 *Homo sapien* proteins that contain EH domains. Seven have a single EH domain: EHD1,[8] EHD2, EHD3, EHD4,[9] γ-synergin,[10] POB1,[11] and Reps1.[12] There are two EH domains in Intersectin-1 and its isoform Intersectin-2.[13,14] Three EH domains are present in Eps15 and its close relative, Eps15R.[1,7] The amino acid sequences of these domains are aligned in Figure 1, where their calcium-binding loops are underlined. In addition, five *Saccharomyces cerevisiae* proteins, which are generally involved in endocytosis and actin cytoskeletal organization, contain from one to three EH domains.[3,15–17] The *Arabidopsis thaliana*, *Caenorhabditis elegans*, and *Drosophila melanogaster* genomes encode 3, 5, and 5 EH domain-containing proteins, respectively.

BIOLOGICAL FUNCTION

The protein-binding functions of EH domains were discovered by filter binding[7] and phage display studies.[2,3] Three types of ligands have been identified. Most EH domains bind NPF motifs, and thus these are defined as class I peptides. Multiple copies of these motifs are found in ligand proteins including Epsin, Hrb, Numb, and Synaptojanin, suggesting multivalent interactions.[2,21,22] Class II ligand peptides contain Phe-Trp (FW), Trp-Trp (WW), or Ser-Trp-Gly (SWG) motifs. Examples of class II motifs are found in the mannose 6-phosphate receptor[23] and are recognized by the third EH domain of Eps15. Class III peptides contain a His-Ser/Thr-Phe motif, which are bound, for example, by the N-terminal EH domain of the End3 protein of *S. cerevisiae*. In addition, some EH domain interactions may not require these motifs. For example, the EH domain of End3p binds the C-terminal region of Sla1p which does not contain canonical binding motifs.[24] Also, the EH domains of Eps15, POB1, and Reps1 may interact with the C-terminal DPF repeats in Eps15.[25–27]

The occurrence of EH domains alongside other protein–protein interaction modules gives rise to the possibility of forming complex networks of proteins. These assemblies function in various cellular processes including actin cytoskeletal organization, endocytosis, exocytosis, protein sorting, signal transduction, and nucleocytosolic transport.[28,29] For example, both Eps15 and Eps15R interact with adaptor protein 2 complex and Epsin, and localize to clathrin-coated pits and endocytic vesicles through which activated receptors are internalized.[22] Two other EH domain–containing proteins, POB1 and Reps1, help regulate the actin cytoskeleton and interact with RalBP1, a GTPase-activating protein.[11,12]

The EHD1 protein localizes to endocytic vesicles and Golgi apparatus,[8] and its paralog EHD4 is unique among EH proteins, being an extracellular matrix protein.[30]

Oligomerization is a recurring theme among the EH domain–containing proteins and is often mediated by neighboring regions. For example, the coiled coil region of Eps15 mediates the formation of parallel homodimer, which may further assemble into antiparallel tetramers.[26,31] In addition, heteromeric interaction of Eps15 with Eps15R[32] and Intersectin[14] have been described. Intersectin dimerization is presumably also mediated by its coiled coil region and may serve to orient its multiple src homology 3, Dbl homology, pleckstrin homology, and C2 domains. In addition to its EH domains, the Reps1 protein contains a coiled coil region, and it interacts with Rab11-FIP2, a protein that contains a C2 domain, coiled coil, and NPF motifs.[33] These homo- and hetero-oligomers and tandem sets of binding domains could provide opportunities for multivalent and cooperative docking to proteins and membranes.

AMINO ACID SEQUENCE INFORMATION

The following human proteins contain one, two, or three EH domains and are listed with their domain boundaries and GenBank accession numbers.[34] Their sequences are shown in the alignment in Figure 1.

- EHD1, also known as Testilin, Eme1, or Past1, amino acid residues 438–531 (aad45866)
- EHD2, residues 443–536 (aaf40470)
- EHD3, residues 438–531 (aaf40471)
- EHD4, also known as Past2, residues 382–475 (aaf40472)
- Eps15, residues 8–103 (EH_1), 121–215 (EH_2), 217–313 (EH_3), (aaa52101)
- Eps15R, residues 8–103 (EH_1), 120–214 (EH_2), 268–364 (EH_3), (NP_067058)
- γ-synergin, residues 291–383 (aad49732)
- Intersectin, also known as Dap160, EHSH, or Ese1, residues 14–108 (EH_1), 214–309 (EH_2), (aac78611)
- Intersectin-2, also known as Ese2, residues 14–108 (EH_1), 236–331 (EH_2), (aaf59903)
- POB1, also known as Reps2, residues 136–231 (aac02901)
- Reps1, residues 226–321 (np_114128)

PROTEIN PRODUCTION, PURIFICATION, AND MOLECULAR CHARACTERIZATION

Five EH domains have been characterized at atomic resolution. In particular, solution structures have been determined for the EH domains of human POB1,[35] mouse Reps1 (which is identical to the human sequence),[25] the first EH domain of mouse Eps15 (EH_1),[36] and the second[37,38] and third[39] EH domains (EH_2 and EH_3) of human Eps15. Each of these EH domains was ^{13}C and ^{15}N labeled in *Escherichia coli* and purified as His_6 tag or glutathione S-transferase fusions that were cleaved prior to structural characterization. In addition, the tendency of the Eps15 EH_3 domain to dimerize was eliminated by mutation of the exposed Cys274 to a serine residue.[39] Post-translational modifications of cytoplasmic EH domains have not been demonstrated to occur *in vivo*. However, the αA helix of the Eps15 EH_1 domain can be tyrosine phosphorylated *in vitro*.[36]

METAL CONTENT AND COFACTORS

One calcium ion is bound by most EH domains. Calcium has been shown to be important for the stability of the POB1 structure, as its removal causes aggregation or denaturation.[35] Also, calcium removal destabilizes the Eps15 EH_3 domain, which can be refolded in the presence of calcium or a large excess of the weaker magnesium ligand.[39] However, calcium is not essential for the function of some EH domains. For example, the Eps15 EH_1 domain binds NPF motif peptides in the absence of metals, and its interaction with calcium is too weak to be of physiological significance.[36] Moreover, the third EH domain of Eps15 can bind NPF and FW motif peptides in the magnesium-bound state, albeit more weakly than in the calcium-bound state.[39]

ACTIVITY AND INHIBITION TESTS

Quantitative estimates of the peptide binding activities of individual EH domains have been obtained by NMR spectroscopy and surface plasmon resonance (SPR) detection. The NMR experiments involve adding a peptide stepwise to an isotope labeled EH domain protein in aqueous buffer. Progressive changes in ^{1}H, ^{13}C, or ^{15}N chemical shifts of the protein are then monitored as a function of ligand concentration. The peptide ligands typically bind in the fast-exchange regime of the NMR timescale, allowing the affinities and off-rates to be estimated. The SPR experiments involve immobilizing a ligand peptide to a sensor chip. The response represents the amount of injected protein that binds the immobilized peptide, and is measured as a function of injected protein concentration. In the case of the monovalent EH domain interactions, these binding curves have been used to estimate the equilibrium binding affinities.

The consensus of these studies is that the peptide interactions of isolated EH domain involve affinities in the micromolar range.[25,38,39] The fast on and off rates of the interactions are consistent with the relatively exposed binding site and intrinsic β-turn propensity of the peptide ligands.[38] In addition, cyclic inhibitors have been designed to reinforce the β-turn conformation of the NPF motif peptide and can constitute high-affinity competitive ligands.[25,38]

NMR SPECTROSCOPY

Multidimensional NMR spectra of the $^{13}C/^{15}N$-labeled Eps15 EH_1 domain were collected by Whitehead et al.[36] The conditions used for structural characterization and peptide binding included 100 mM KH_2PO_4 (pH 5.2) and 100 mM NaCl. Spectra of calcium-occupied states were obtained at pH 7.0, since calcium ligation was reduced at lower pH values.

Heteronuclear NMR spectra of the Eps15 EH_2 domain have been collected by de Beer and colleagues.[37,38] The spectra were typically collected in 20 mM Tris (pH 7.0), 2 mM $CaCl_2$, 100 mM KCl, and 100 μM DTT. Spectra of calcium-bound and free states were compared by removing calcium by ethylenediaminetetraacetic acid (EDTA) treatment. In addition, interactions of several linear and cyclic peptides were compared (Table 1). Spectra of the Eps15 EH_3 domain were collected under similar conditions by Enmon et al.[39] Calcium-bound and free states of the EH_3 domain were generated, first using EDTA to remove calcium, and then refolding the protein from a guanidinium hydrochloride solution by slow dilution in the presence of excess magnesium. Folding of the domain is reversible, but requires the binding of divalent cations.[39] The binding sites

Table 1 Peptide affinities of EH domains

EH domain	Peptide	Source	Average K_d	Conditions	References
Eps15 EH_1	SSST**NPF**L	Rab/Hrb	590 μM	100 mM KH_2PO_4, pH 5.2, 100 mM NaCl	36
Eps15 EH_2	PTGSSST**NPF**L	Rab/Hrb	560 μM	20 mM Tris, pH 7.0, 2 mM $CaCl_2$, 100 mM KCl	37,38
Eps15 EH_2	PTGSSST**NPF**R	Rab/Hrb	<560 μM	20 mM Tris, pH 7.0, 2 mM $CaCl_2$, 100 mM KCl	
Eps15 EH_3	PTGSSST**NPF**L	Rab/Hrb	<560 μM	20 mM Tris, pH 6.5, 2 mM $CaCl_2$, 100 mM KCl	39
Eps15 EH_3	RTAAPG**NPF**RVQ	Synaptojanin	high μM[a]	20 mM Tris, pH 6.5, 2 mM $CaCl_2$, 100 mM KCl	
Eps15 EH_3	DSTPGQVA**FW**	Phage display	mid μM[a]	20 mM Tris, pH 6.5, 2 mM $CaCl_2$, 100 mM KCl	
Eps15 EH_3	DMEQFPHLA**FW**QDL	Mannose receptor	high μM[a]	20 mM Tris, pH 6.5, 2 mM $CaCl_2$, 100 mM KCl	
Reps1 EH	YEST**NPF**TAKF	Rab11–FIP2 (2)	46 μM	10 mM imidazole, pH 6.7, 2 mM $CaCl_2$, 10 mM NaCl	25,33
Reps1 EH	PRKK**NPF**EESS	Rab11–FIP2 (1)	423 μM	10 mM imidazole, pH 6.8, 2 mM $CaCl_2$, 10 mM NaCl	
Reps1 EH	IPDS**NPF**DATA	Rab11–FIP2 (3)	7 mM	10 mM imidazole, pH 6.8, 2 mM $CaCl_2$, 10 mM NaCl	
Mutant peptide ligands					
Reps1 EH	YESTD**PF**TAKF	Rab11–FIP2 (2)[b]	530 μM	10 mM imidazole, pH 6.7, 2 mM $CaCl_2$, 10 mM NaCl	25,33
Reps1 EH	YEST**NP**YTAKF	Rab11–FIP2 (2)[b]	1080 μM	10 mM imidazole, pH 6.7, 2 mM $CaCl_2$, 10 mM NaCl	
Magnesium-saturated peptides					
Eps15 EH_3	PTGSSST**NPF**L	Rab/Hrb	μM–mM[a]	20 mM PIPES, pH 7.2, 160 mM $MgCl_2$, 100 mM KCl	39
Eps15 EH_3	DSTPGQVA**FW**	Phage display	μM–mM[a]	20 mM PIPES, pH 7.2, 160 mM $MgCl_2$, 100 mM KCl	
Cyclic peptides					
Eps15 EH_2	SSDC^c**TNPF**RSC^{c}WRS	Phage display	12 μM	20 mM Tris, pH 7.0, 2 mM $CaCl_2$, 100 mM KCl	38
Reps1 EH	EYEC^c**TNPF**TAKC^c	Rab11–FIP2	~65 μM	10 mM imidazole, pH 6.7, 2 mM $CaCl_2$, 10 mM NaCl	25

[a] Estimate only.
[b] Contains a substitution.
[c] Cystine.

of the EH_3 domain for class I and II peptides were also compared.

The solution structure of the Reps1 EH domain was characterized by Kim et al.[25] The NMR spectra were obtained in 10 mM imidazole (pH 6.7), 2 mM $CaCl_2$, 10 mM NaCl, and 5 mM DTT. The interactions of the Reps1 EH domain with a variety of linear and cyclic peptide ligands were examined (Table 1). Also, the NMR structure of the POB1 EH domain was solved by Koshiba et al. in 20 mM Tris (pH 7.5), 5–10 mM $CaCl_2$, 50 mM KCl, and 5 mM DTT.[35] Spectra of calcium-free state of the POB1 EH domain were compared by removing calcium by EDTA treatment.

OVERALL DESCRIPTION OF THE STRUCTURE

The first structure of an EH domain was determined in 1998 by heteronuclear NMR spectroscopy,[37] and it has been confirmed by solution structures of four other EH domains.[25,35,36,39] Figure 2 depicts all the structures of the peptide-free EH domains determined to date, along with the calcium ions. Figure 3 shows the structure of EH_2 domain of Eps15 in a stereo representation. The EH domain consists of four α-helices (αA, αB, αC, and αD) and two short β-strands (β1 and β2) that form a pair of EF hands. The helices are tightly packed, with the long αD helix running roughly parallel to αB and being the most buried. The two β-strands form a small antiparallel sheet that links the EF hand loops. Both EF hands are capable of ligating a single calcium ion, although only one site is typically functional, the other containing incompatible basic residues, short deletions, or prolines in key calcium-coordinating positions. The conserved proline-rich element follows the last helix and is oriented near the N-terminus of most EH domains. A relatively shallow and hydrophobic pocket binds NPF sequences and is located between the αB and αC helices, about 10 Å from the calcium ion as seen in the EH_2 structure.[38]

VARIATIONS ON THE CONSERVED FOLD

Essentially the same fold is observed in all EH structures with only small differences. The Eps15 EH_1 domain

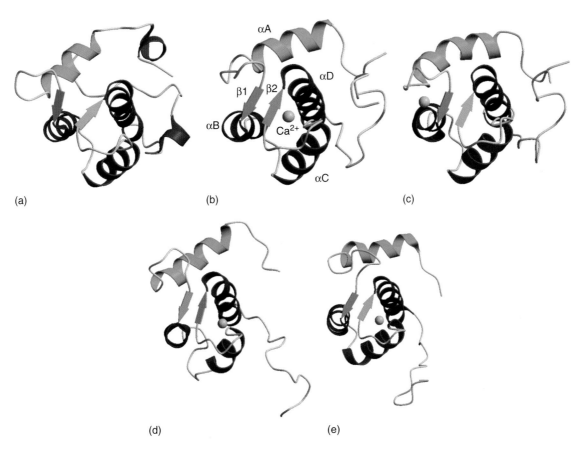

Figure 2 Schematic representation of the peptide-free structures of the (a) Eps15 EH_1, (b) EH_2, and (c) EH_3 domains and (d) the EH domains of POB1 and (e) Reps1 with bound calcium ions. The secondary structural elements are colored as in Figure 1 and labeled in the Eps15 EH_2 structure. The two additional α-helices of EH_1 are shown in brown. All the structures were superimposed using the backbone heavy atoms of the αD helix. Prepared with MOLSCRIPT[4] and RASTER3D.[5] PDB codes: 1QJT, 1EH2, 1C07, 1IQ3, and 1FI6, respectively.

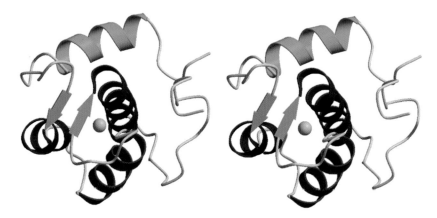

Figure 3 Stereo representation of the calcium-bound, peptide-free structure of the Eps15 EH_2 domain. The secondary structural elements are colored as in the 3D Structure. Prepared with MOLSCRIPT[4] and RASTER3D.[5] PDB code: 1EH2.

contains two inserts (Figure 1) that form additional α-helices, one before αA and the other after αD (Figure 2).[36] In addition, the register of the helical packing in the first, noncanonical EF hand of EH_1 is altered, and the N- and C-termini are oriented toward opposite sides of the domain. Despite these structural differences, EH_1 still prefers classical NPF ligands.[3] The additional N- and C-terminal helices are also likely to exist in the Eps15R EH_1 domain (Figure 1), and neither EH_1 domain is likely to bind calcium with much affinity. Thus, they appear to constitute a calcium-free subgroup of EH domains that may be structurally divergent. When all the structures are compared, some conformational variability is seen in the calcium-free EF hand loops, and the αA helix of the Reps1 EH domain is displaced relative to its position in the other structures.[25]

COMPARISON WITH OTHER EF HAND STRUCTURES

The classical EH domain fold, as exemplified by the EH_2, EH_3, POB1, and Reps1 structures, is most similar to that of calbindin D9k, a monomeric EF hand protein of the S100 class.[40,41] One distinction between the two folds is that the EH domain's buried αD helix packs against αC in a more antiparallel orientation. The EH_1 subfamily also resembles the S100 class of proteins, but again differs in its interhelical angles.[36] Another unique feature of EH domains is the proline-rich C-termini that presumably contributes to the spatial orientation of adjacent domains. Functional differences also exist between classical EF hands and those found in EH domains. For example, EH domains contain a unique NPF-specific protein binding site between the αB and αC helices that is not found in other EF hand proteins.

CALCIUM AFFINITY

Calcium is bound tightly by EH domains containing a classical EF hand. In fact, the affinities are so high that they are nontrivial to measure. For example, calcium is difficult to remove from the Eps15 EH_2 domain, even by treating it with an excess of EDTA, which has a 20 picomolar affinity for calcium.[37] The calcium-bound and apo states of this domain exist in slow exchange on the NMR timescale, indicative of a slow off-rate and tight binding. Moreover, calcium binding induces local rather than global conformational changes based on the chemical shift perturbations induced by calcium removal.

CALCIUM COORDINATION SITE

One calcium ion is bound by each of the structurally characterized EH domains other than EH_1. The metal occupancy is seen by local chemical shift perturbations that are induced by the addition or removal of calcium[36,37] or by its replacement by magnesium.[39] In most EH domains calcium is ligated by the EF hand motif formed by the αC and αD helices. However, EH_3 binds calcium in its first EF hand, which contains a kink in the helix αA (Figure 2) and a narrower peptide-binding groove.[39] Owing to the absence of high-resolution X-ray crystal structures, precise metal–ligand bond distances are not available for EH domains.

All the high-affinity calcium binding sites of the EH domain structures are canonical. They are therefore predicted to coordinate calcium in a pentagonal bipyramidal arrangement. In such cases, the two helices are separated by a 12-residue loop, which consists of the motif $X \bullet Y \bullet ZG^- Y \bullet^- X \bullet \bullet^- Z$, where X, Y, Z, ^-X, ^-Y, and ^-Z represent the residues that serve as calcium ligands, G is a glycine, and a dot represents any residue. There are aspartic acid ligands at the X, Y, and Z positions, and a glutamic acid ligand at ^-Z in the second EF hand of the Eps15 EH_2, POB1 and Reps1 EH domains (Figure 4(a)). The same pattern is found in the first EF hand of the Eps15 EH_3 domain, which also binds calcium tightly (Figure 4(b)).

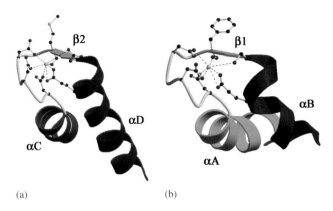

Figure 4 Drawings of the calcium coordination sites in the second and first EF hands of the (a) Eps15 EH$_2$ and (b) EH$_3$ domains respectively. Heavy atoms of residues predicted to coordinate calcium are shown in ball and stick representation in CPK colors. The calcium ion is depicted as a small yellow sphere. For the purpose of clarity, the atomic radii are not drawn to scale. The two fragments were superimposed using the backbone heavy atoms of the calcium-binding loop and an adjacent turn of each of the flanking helices. Prepared with MOLSCRIPT[4] and RASTER3D.[5] PDB codes: 1C07 and 1EH2, respectively.

Each of the residues of a canonical EF hand has a distinct, conserved role that has been defined by high-resolution crystal structures of other proteins.[42,43]

The canonical EF hands of EH domains always begin with an aspartic acid at the X position. The pair of side-chain oxygens initiates the calcium coordination sphere and receives a hydrogen bond from the conserved glycine in the center of the loop. The previous α-helix terminates immediately prior to this residue.

The second residue of EF hand loops is variable, and its side chain is exposed to the solvent. This residue is found between the X and Y positions and is often a hydrophobic residue in the calcium-binding EF hands of EH domains. The noncanonical EF hands of EH domains are found to contain a proline or polar residue in this position.

An aspartic acid is usually found to bind calcium at the Y position of the EH domain loops, although an asparagine can also provide the necessary direct coordination through its side-chain carbonyl. In addition, its main-chain amide can hydrogen bond to the absolutely conserved glutamine at the $^-$Z position. The amide of the variable, fourth residue (which lies between the Y and Z positions) forms a hydrogen bond to the aspartic acid at the X position of the canonical EF hands.

The Z position of calcium-binding EF hands of EH domains is essentially always an aspartic acid, although a serine or threonine is occasionally found in this position. However, those EF hands that contain aspartic acid residues at the X, Y, and Z positions have particularly high calcium stabilization. This is consistent with the high calcium affinity suggested for the EH$_2$ and EH$_3$ domains, which also contain this aspartic acid triplet.[37–39]

An invariant glycine is found between the Z and $^-$Y positions in the canonical EF hands of EH domains. This glycine is needed to redirect the backbone to complete the calcium coordination sphere. A strong hydrogen bond to the X position results in a dramatic downfield shift of the glycine amide ^{1}H resonance. This shift is a signature of EF hand proteins, and is evident in the NMR spectra of all of the characterized EH domains.[25,35,37,39]

The variable seventh residue at the $^-$Y position of EH domains can help coordinate calcium through its backbone carbonyl oxygen. Its side chain is exposed to solvent, and its main chain initiates a short β-strand, which pairs with that of the juxtaposed EF hand. The residue between $^-$Y and $^-$X is usually a leucine; it forms the center of the β-strand, with its long aliphatic side chain packing into the hydrophobic core.

The ninth residue at the $^-$X position is usually occupied by an aspartic acid or threonine in the canonical EF hands of EH domains. This residue engages the calcium through water-mediated hydrogen bonds. In addition, its side chain stabilizes the first turn of the EF hand's second helix by hydrogen bonding to the glutamine at the $^-$Z position.

The tenth position of the first and second EF hand loops of EH domains are occupied by a glycine and variable residues respectively. This residue initiates the second helix of the EF hand motif. The eleventh position is occupied by a variable or acidic residue in the first and second EF hands of EH domains respectively, providing conformational and electrostatic stabilization of the second helix.

The glutamine side chain at the twelfth, $^-$Z position is required for stable calcium ligation. In the canonical EF hands of the EH domains, its side-chain oxygens are predicted to provide bidentate ligation of calcium, thus completing the pentagonal bipyramidal sphere of coordination and ensuring metal specificity. This glutamine residue is often substituted in the noncanonical EF hands of EH domains.

The noncanonical EF hands of EH domains bind calcium very weakly or not at all. This can be attributed to the occurrence of incompatible residues at the usual calcium-coordinating positions. For example, the first EF hand in the POB1 and Reps1 EH and Eps15 EH$_2$ domains contains a valine at $^-$Z, a proline at $^-$X, and a lysine at $^-$Z respectively. The second EF hand of the Eps15 EH$_1$ and EH$_3$ domains have a lysine residue at the Z and Y positions respectively. In each case, the substitution of a residue that provides a crucial oxygen ligand would be expected to repel calcium, as is confirmed by the lack of physiologically relevant calcium binding in each of these sites. The first EF hand of the Eps15 EH$_1$ domain is actually a pseudo-EF hand that consists of 14 rather than 12 residues.[36,44] However, its affinity for calcium is compromised by the presence of a leucine residue in which a side-chain oxygen ligand would be expected. Thus, from the wealth of information about EF hand motifs,[42] the occupancies (although not

EH domain

affinities) of most EH domains can now be inferred from the sequence alone.

PEPTIDE AFFINITY

The affinities of isolated EH domains for peptide ligands have been estimated to be in the micromolar range. The three EH domains of Eps15 all bind comparably to the C-terminal NPF sequence of the human Hrb protein, also known as Rab, with affinities estimated to be around 500 μM (Table 1).[36,37,39] Interestingly, the Eps15 EH_3 domain, which has specificity for class II ligands, can interact even more strongly with some class I ligands than the EH_2 domain.[39]

The affinity of the Reps1 EH domain for the three NPF motifs of the Rab11-FIP2 protein have been measured and have been found to range from 46 μM to 7 mM.[25,33] The affinity is highest for the second NPF motif; the others are negatively charged and may be repelled by the acidic binding pocket of Reps1. The on-rate appears to be near the diffusion-controlled limit.[25] Cyclization of the linear NPF peptide through a disulfide bond was observed to reduce the off-rate from $1800 s^{-1}$ by a factor of 4.[25]

The peptide affinities of EH domains can be selectively abolished by mutation of the absolutely conserved tryptophan in the binding site to an alanine residue. Substituting a tyrosine here only reduces peptide affinity by a factor of 3.[37] Affinities are also reduced when residues that line the pocket are mutated, indicating a cooperative array of interactions that determine specificity.[25,38,39]

PEPTIDE BINDING MECHANISM

Two structures of complexes have been solved by NMR, both of the Eps15 EH_2 domain bound to 10-residue NPF motif peptides.[38] Complementing this work, the binding sites and affinities of the Reps1 EH, Eps15 EH_1 and EH_3 domains for a variety of class I and II peptides have also been characterized. Together, these studies define the conserved mechanism, whereby short β-turn peptides insert into a shallow hydrophobic groove between the αB and αC helices (Figure 5). For simplicity, the binding mechanism can be subdivided into the elements that accommodate the N-terminal, asparagine, proline, phenylalanine, and C-terminal residues, which are numbered as the −1, 0, +1, +2, and +3 positions respectively.

Phage display studies had originally suggested that EH domains may possess some specificity for small polar residues immediately N-terminal to the NPF motif.[3] However, no contacts were observed between the EH_2 domain and the serine or threonine residues that preceded the NPF motif of the Hrb protein in the NMR-derived structures. Rather, these ligand residues were found to be relatively unstructured and were positioned near the side

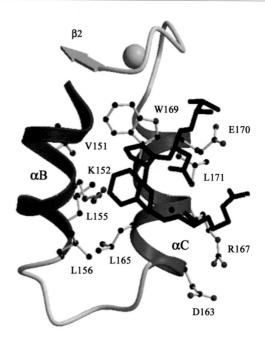

Figure 5 Drawing of the TNPFR ligand residues bound between the αB and αC helices of the EH_2 domain of Eps15. The side-chain heavy atoms of interacting residues are shown as ball and stick models in CPK colors and are also labeled. The threonine residue in the peptide ligand is included for the purpose of illustration and is poorly defined in the ensemble of NMR structures. Prepared with MOLSCRIPT[4] and RASTER3D.[5] PDB code: 1F8H.

chain of the variable fourth residue of the second EF hand loop.[38] This suggests that the steric complementarity and stabilizing interactions between these two residues could enhance ligand affinity.

The recognition of the asparagine residue at the 0 position of the NPF motif is critical for the selectivity of most EH domains. The asparagine residue helps stabilize the conformation of the bound ligand exhibiting several intramolecular contacts with the residues at the +2 and +3 positions, and its side-chain oxygen hydrogen bonds to the +2 and +3 backbone amides. Similar turns are commonly stabilized by aspartic acid or asparagine side chains in unrelated protein structures, although only in EH ligands are these so-called Asx-Pro turns also recognized through protein interactions.[38,45] In the EH_2 ligand, the asparagine side chain interacts with Trp169, Glu170, and Leu171, which are found in the LPxxϕLπxIWxLxD motif. This orientation places its side-chain amide within hydrogen bonding distance of the Gly166 and Arg167 carbonyls. The conservation of these residues suggests that other EH domains are likely to bind an asparagine side chain of a peptide ligand in a similar way. However, the acidic aspartic acid side chain of a DPF motif would be unable to participate in this way and could be repelled by the acidic group of the nearby Glu170 of the EH_2 domain.[38] Alternatively, the electronegative binding pocket of some EH domains may repel an acidic aspartic acid side chain, as is suggested by 12-fold lower affinity of the Reps1 EH domain for DPF over NPF sequences (Table 1).[25]

The proline residue at the +1 position stabilizes the turn of class I ligands and makes contacts with conserved elements in the pocket.[38] In particular, this proline residue lies near the Gly148, Val151, Lys152, and Leu155 residues in αB, as well as Trp169 in αC (Figure 5). The proline carbonyl is positioned to the hydrogen bond at the Lys152 side chain. This interaction is functionally critical. Mutation of the corresponding arginine of the EH_3 domain to a lysine or alanine reduces the peptide affinity,[39] and mutation of the comparable lysine of Reps1 is also debilitating.[25]

The phenylalanine residue of the NPF motif is most deeply buried in the binding groove. Its side chain provides the bulk of the 64 intermolecular contacts identified from nuclear Overhauser effects.[38] Direct contacts are observed to Val151, Lys152, Leu155, Leu156, and Leu165 (Figure 5). On the basis of NMR studies of EH_3, the conserved aromatic residue of the other classes of EH ligands is predicted to occupy this binding pocket similarly.[39] It is the hydrophobic character of this interdigitation that is the key, with the mutation of Leu155 in EH_2 to a phenylalanine only slightly affecting ligand affinity.[33] Moreover, the reciprocal phenylalanine to leucine mutation in EH_3 had only moderate effects on its ligand specificity.[39] Thus, this interaction may contribute more significantly to binding affinity than to specificity. Indeed, a peptide in which this phenylalanine is switched to a tyrosine binds in a similar way to the Reps1 EH domain, albeit 23 times more weakly (Table 1).[25]

Specificity is provided by the +3 residue that lies immediately C-terminal to the NPF motif.[3] This residue helps stabilize the conformation of the ligand, and its side chain contacts variable groups on the EH domain surface. In particular, its backbone amide forms a hydrogen bond with the asparagine at the zero position. Hydrophobic interactions between the +3 side chain and Val162 in αC favor an aliphatic +3 ligand, while electrostatic interaction with Asp163 and Glu170 favor a +3 basic residue here. The latter interactions explain the preference of EH_2 domain for NPFR over NPFL sequences. Other EH_2 residues positioned near the +3 residue of the ligand are Arg167 and Met203. Together these five residues of the EH domain help determine the specificity for class I ligands.

The cyclization of ligand peptides provides a rational basis for inhibitor design, as was first demonstrated by Kay and colleagues using Intersectin.[13] It was subsequently shown that a peptide containing a disulfide cross-link, SSDC*TNPFRSC*WRS, (cystines are denoted with an asterisk), preforms a type I β-turn about the central Pro-Phe, such that its cysteine side chains are juxtaposed. Consequently, this cyclic peptide binds to the EH_2 domain with an affinity of 12 μM, which is an order of magnitude stronger than the linear peptides.[38] Another cyclic peptide, EYEC*TNPFTAKC*, proved to be a poorer ligand of the Reps1 EH domain.[25] However, its cysteine residues may be less ideally positioned to form a comparable intramolecular cross-link, or the cross-link may be unable to stabilize a type I β-turn.

REGULATION OF ACTIVITY

Phosphorylation is thought to regulate proteins that contain EH domains. In particular, Eps15, Reps1, and POB are phosphorylated by the epidermal growth factor (EGF) receptor kinase upon EGF stimulation. The Lck kinase is able to phosphorylate one of the partially exposed tyrosine residues in the αA helix of the EH_1 domain.[36] This phosphorylation may be important for the monoubiquitination and turnover of Eps15 in response to EGF stimulation.[46] Dimerization has been proposed as a mechanism for increasing the valency of EH domain interactions,[26,31] although the regulatory mechanisms are unclear. Lastly, disulfide bond formation has been noted in the EHD4 protein, possibly involving its EH domain, and may serve to alter its extracellular activity.[30]

REFERENCES

1. WT Wong, C Schumacher, AE Salcini, A Romano, P Castagnino, PG Pelicci and P Di Fiore, *Proc Natl Acad Sci USA*, **92**, 9530–34 (1995).

2. AE Salcini, S Confalonieri, M Doria, E Santolini, E Tassi, O Minenkova, G Cesareni, PG Pelicci and PP Di Fiore, *Genes Dev*, **11**, 2239–49 (1997).

3. S Paoluzi, L Castagnoli, I Lauro, AE Salcini, L Coda, S Fre, S Confalonieri, PG Pelicci, PP Di Fiore and G Cesareni, *EMBO J*, **17**, 6541–50 (1998).

4. P Kraulis, *J Appl Crystallogr*, **24**, 946–50 (1991).

5. EA Merritt and MEP Murphy, *Acta Crystallogr*, **D50**, 869–73 (1994).

6. S Nakayama, H Kawasaki and RH Kretsinger, *Top Biol Inorg Chem*, **3**, 29–58 (2000).

7. F Fazioli, L Minichiello, B Matoskova, WT Wong and PP Di Fiore, *Mol Cell Biol*, **13**, 5814–28 (1993).

8. L Mintz, E Galperin, M Pasmanik-Chor, S Tulzinsky, Y Bromberg, CA Kozak, A Joyner, A Fein and M Horowitz, *Genomics*, **59**, 66–76 (1999).

9. U Pohl, JS Smith, I Tachibana, K Ueki, HK Lee, S Ramaswamy, Q Wu, HW Mohrenweiser, RB Jenkins and DN Louis, *Genomics*, **63**, 255–62 (2000).

10. LJ Page, PJ Sowerby, WW Lui and MS Robinson, *J Cell Biol*, **146**, 993–1004 (1999).

11. M Ikeda, O Ishida, T Hinoi, S Kishida and A Kikuchi, *J Biol Chem*, **273**, 814–21 (1998).

12. A Yamaguchi, T Urano, T Goi and LA Feig, *J Biol Chem*, **272**, 31230–34 (1997).

13. M Yamabhai, NG Hoffman, NL Hardison, PS McPherson, L Castagnoli, G Cesareni and BK Kay, *J Biol Chem*, **273**, 31401–7 (1998).

14. AS Sengar, W Wang, J Bishay, S Cohen and SE Egan, *EMBO J*, **18**, 1159–71 (1999).

15. PP Di Fiore, PG Pelicci and A Sorkin, *Trends Biochem Sci*, **22**, 411–13 (1997).

16 B Wendland, JM McCaffery, Q Xiao and SD Emr, *J Cell Biol*, **135**, 1485–500 (1996).

17 B Gagny, A Wiederkehr, P Dumoulin, B Winsor, H Riezman and R Haguenauer-Tsapis, *J Cell Sci*, **113**(Pt 18), 3309–19 (2000).

18 JD Thompson, DG Higgins and TJ Gibson, *Nucleic Acids Res*, **22**, 4673–80 (1994).

19 K Hofmann and MD Baron, Boxshade Version 3.3.1.

20 S Subramaniam, *Proteins* **32**, 1–2 (1998).

21 C Haffner, K Takei, H Chen, N Ringstad, A Hudson, MH Butler, AE Salcini, PP Di Fiore and P De Camilli, *FEBS Lett*, **419**, 175–80 (1997).

22 H Chen, S Fre, VI Slepnev, MR Capua, K Takei, MH Butler, PP Di Fiore and P De Camilli, *Nature*, **394**, 793–97 (1998).

23 A Schweizer, S Kornfeld and J Rohrer, *Proc Natl Acad Sci USA*, **94**, 14471–76 (1997).

24 HY Tang, J Xu and M Cai, *Mol Cell Biol*, **20**, 12–25 (2000).

25 S Kim, DN Cullis, LA Feig and JD Baleja, *Biochemistry*, **40**, 6776–85 (2001).

26 P Cupers, E ter Haar, W Boll and T Kirchhausen, *J Biol Chem*, **272**, 33430–34 (1997).

27 S Nakashima, K Morinaka, S Koyama, M Ikeda, M Kishida, K Okawa, A Iwamatsu, S Kishida and A Kikuchi, *EMBO J*, **18**, 3629–42 (1999).

28 E Santolini, AE Salcini, BK Kay, M Yamabhai and PP Di Fiore, *Exp Cell Res*, **253**, 186–209 (1999).

29 S Confalonieri and PP Di Fiore, *FEBS Lett*, **513**, 24–29 (2002).

30 HJ Kuo, NT Tran, SA Clary, NP Morris and RW Glanville, *J Biol Chem*, **276**, 43103–10 (2001).

31 F Tebar, S Confalonieri, RE Carter, PP Di Fiore and A Sorkin, *J Biol Chem*, **272**, 15413–18 (1997).

32 L Coda, AE Salcini, S Confalonieri, G Pelicci, T Sorkina, A Sorkin, PG Pelicci and PP Di Fiore, *J Biol Chem*, **273**, 3003–12 (1998).

33 DN Cullis, B Philip, JD Baleja and LA Feig, *J Biol Chem* **277**, 49158–66 (2002).

34 DA Benson, I Karsch-Mizrachi, DJ Lipman, J Ostell, BA Rapp and DL Wheeler, *Nucleic Acids Res*, **30**, 17–20 (2002).

35 S Koshiba, T Kigawa, J Iwahara, A Kikuchi and S Yokoyama, *FEBS Lett*, **442**, 138–42 (1999).

36 B Whitehead, M Tessari, A Carotenuto, PM van Bergen en Henegouwen and GW Vuister, *Biochemistry*, **38**, 11271–77 (1999).

37 T de Beer, R Carter, K Lobel, A Sorkin and M Overduin, *Science*, **281**, 1357–60 (1998).

38 T de Beer, AN Hoofnagle, JL Enmon, RC Bowers, M Yamabhai, BK Kay and M Overduin, *Nat Struct Biol*, **7**, 1018–22 (2000).

39 JL Enmon, T de Beer and M Overduin, *Biochemistry*, **39**, 4309–19 (2000).

40 LA Svensson, E Thulin and S Forsen, *J Mol Biol*, **223**, 601–6 (1992).

41 R Donato, *Int J Biochem Cell Biol*, **33**, 637–68 (2001).

42 NC Strynadka and MN James, *Annu Rev Biochem*, **58**, 951–98 (1989).

43 A Lewit-Bentley and S Rety, *Curr Opin Struct Biol*, **10**, 637–43 (2000).

44 CW Heizmann and W Hunziker, *Trends Biochem Sci*, **16**, 98–103 (1991).

45 DR Wilson and BB Finlay, *Protein Eng*, **10**, 519–29 (1997).

46 S van Delft, R Govers, GJ Strous, AJ Verkleij and PM van Bergen en Henegouwen, *J Biol Chem*, **272**, 14013–16 (1997).

EGF-domains

Calcium-binding EGF-like domains

Emma J Boswell, Nyoman D Kurniawan and A Kristina Downing
Biochemistry Department, University of Oxford, Oxford, UK

FUNCTIONAL CLASS

Most extracellular proteins in higher organisms are assembled from a selection of protein domains or modules.[1] Epidermal growth factor–like domains are extremely widespread, and of these domains a subset contains a consensus sequence associated with calcium binding.[2,3] Calcium-binding epidermal growth factor–like domains (cbEGFs) are found in a variety of extracellular proteins of diverse functions, including cell development, cholesterol metabolism, connective tissue fiber formation, blood coagulation, fibrinolysis, and in the complement system.[1]

OCCURRENCE

The EGF domain is a structural motif of approximately 45 amino acids and is one of the most common types of domain found in extracellular eukaryotic proteins (see review[4]). It appears to have evolved from a common ancestor through gene duplication, exon shuffling, and point mutations.[5]

The Nell-1 and Nell-2 proteins, expressed in neural tissues, contain 6 EGF domains (3 of which are cbEGFs), and although a portion is secreted, there is also a high proportion retained in the cytoplasm, which may constitute rare examples of intracellular EGF-containing proteins.[6] The domains mainly occur in animal proteins, although EGF and cbEGF domains have been reported in a number of plant protein kinases and a plant vacuolar sorting receptor.[7,8]

According to InterPro (http://www.ebi.ac.uk/interpro/index.html), 2057 proteins are listed as containing EGF domains (accession no: IPR000561) and almost 35% (719) of those contain cbEGF domains (accession no: IPR001881). Although there are occurrences of a single cbEGF adjoining various other domains, in a large number of proteins (approximately 85%) the cbEGF domain appears either as a pair with a non-calcium-binding EGF domain or in a string of domains composed of cbEGFs and EGFs (see Figure 1). Fibrillin-1 contains the highest occurrence of cbEGFs in a single protein, with 43 cbEGFs and 4 EGFs. The longest uninterrupted string of cbEGFs in this protein comprises 12 domains.

BIOLOGICAL FUNCTION

Three biological roles of EGF domains have been identified to date. EGF domains may be used as spacer units to

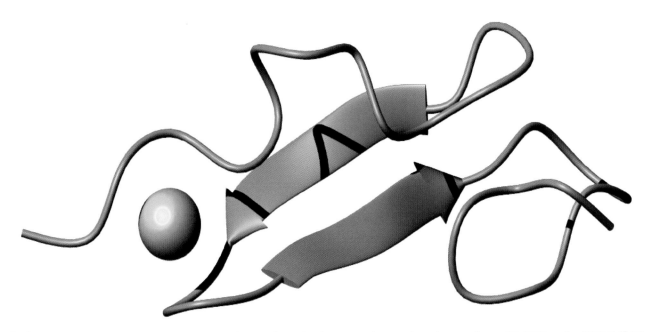

3D Structure Schematic ribbon drawing of the calcium-binding EGF domain from human factor IX (PDB code: 1EDM).[49] The calcium ion is rendered in gold. This figure was created using MOLMOL (2000).[53]

Calcium-binding EGF-like domains

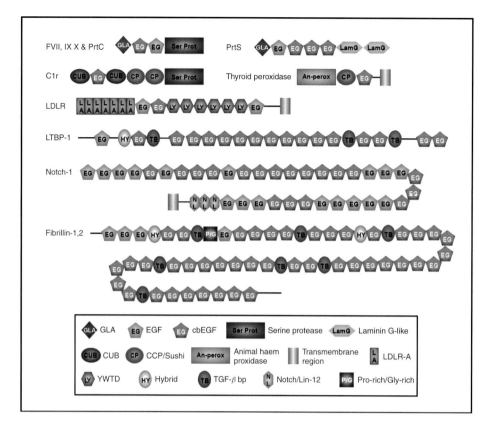

Figure 1 Schematic representation of the modular architectures of a representative set of human cbEGF domain–containing proteins. For clarity, only the extracellular portions of the thyroid peroxidase, low-density lipoprotein receptor (LDLR), and Notch-1 are shown, and the activation domains of FVII, IX, X, and protein C (which lie between the EGF and serine protease domains) are omitted. Domain symbols and abbreviations are according to Bork et al.[1]

provide the correct distance between other domains. This is exhibited in the blood coagulation factors, where two EGF domains are needed to position the active site of the protease domain at a distance from the biological membrane that is commensurate with cofactor interaction and substrate activation.[9–11] Secondly, EGF domains, and more specifically cbEGFs, can play a stabilizing role within a protein, as clearly seen in the case of fibrillin-1, where the binding of calcium to cbEGF domains enhances the rigidity of the protein.[12] EGF domains can also be directly involved in protein–protein interactions, which if involving cbEGFs, can be calcium-dependent. These functions will be discussed in greater detail in the following sections.

AMINO ACID SEQUENCE INFORMATION

Each EGF domain comprises ~40 to 45 amino acids, containing 6 signature conserved cysteines, which form three disulphide bonds, pairing in a 1–3, 2–4, 5–6 pattern.[1] A specific subset of these domains contains a consensus sequence associated with calcium binding: [DEQN]-x-[DEQN]$_2$-C-x_{3-14}-C-x_{3-7}-C-x-[DN]*-x_4-[FY]-x-C, where * indicates possible β-hydroxylation.[2,3,13] Figure 2(a) shows a multiple alignment of a selection of cbEGFs highlighting their sequence variability. The residues in the consensus sequence provide calcium ligands and maintain the geometry of the calcium-binding site (see Figure 2(b)).

Analysis of the EGF–AB pair from the LDLR showed that both domains bound a calcium ion, even though only domain B possesses the complete calcium-binding consensus sequence. The non-canonical sequence in domain A (GTNE-C-x_6-C-x_3-C-x-D-x_4-Y-x-C) deviates from the classical consensus sequence only in the first ligand residue. It has been suggested that this ligand is substituted by the threonine hydroxyl.[15]

PROTEIN PRODUCTION, PURIFICATION, AND MOLECULAR CHARACTERIZATION

Expression and synthesis

A large variety of methods have been used in the production of cbEGF domains and proteins containing cbEGF domains. One of the major problems associated with the production of the cbEGF is the correct

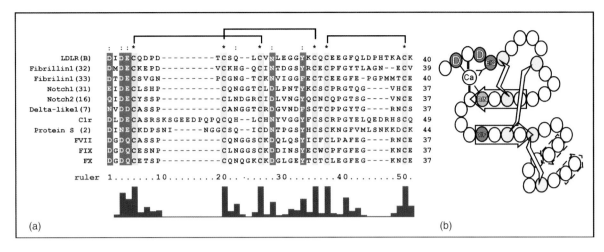

Figure 2 (a) A multiple sequence alignment of a selection of cbEGF domains. The cysteines that form the consensus sequence 1–3, 2–4 and 5–6 disulphide bonds are highlighted in yellow, and the lines indicate the pairing. The conserved calcium-binding consensus sequence is highlighted in red and the putatively β-hydroxylated Asn/Asp residues are highlighted in blue. The alignment was produced using ClustalX (1.8),[14] using the Gonnet250 protein weight matrix, with the gap opening and extension parameters set to 10 and 2.0 respectively. The bottom bars show the residue similarity as calculated according to the weight matrix. (b) A schematic illustration of the primary and secondary structure of the cbEGF domain. The calcium coordination by the carboxyl side chains is shown with single and branched lines and the coordination of the backbone carbonyl with thick lines. The solid and dashed arrows show the positions of the major and minor β-hairpin respectively. The color scheme in this diagram corresponds to that used in the alignment.

pairing of the disulphide bonds within the domain. The following references have been selected to highlight the different methodologies used, but by no means form a comprehensive review of proteins that have been produced.

Single EGF domains are of an amenable size for chemical synthesis. However, correct formation of the disulphide pairing requires a further step, either utilizing a refolding protocol or through directional disulphide formation. Both fibrillin-1 cbEGF 13 and the cbEGF domain of factor VII were chemically synthesized for nuclear magnetic resonance (NMR) studies and refolded via reoxidation of a fully reduced thiol precursor using cysteine:cystine redox conditions.[16,17] A similar approach was used with the C1r cbEGF domain using a refolding mixture of oxidized and reduced glutathione.[18] A novel approach to the production of the cbEGF from factor VII was presented by Husbyn and Cuthbertson[19] with a one-pot regioselective synthesis of an Asn^{57}β-Asp analog. The combination of t-butyl and either acetamidomethyl (Acm) or 4-methylbenzyl (MBn) protected cysteines in a peptide introduced orthogonality to the disulphide-folding process, controlled simply by varying the temperature in the one-pot reaction, and allowed the disulphides to be formed in 1–3, 5–6, 2–4 order.

Single cbEGF domains have also been expressed recombinantly. The cbEGF domain of factor VII was expressed in *Escherichia coli*, using a periplasmic secretion system,[20,21] while the cbEGF domain of factor IX has been expressed in *Saccharomyces cerevisiae* and refolded using cysteine:cystine redox conditions.[16] A similar refolding system has been used for pairs and triplets of cbEGF domains expressed in *E.coli*, in the cases of fibrillin-1 12–13;[22] fibrillin-1 13–14;[16] fibrillin-1 12–14;[22] fibrillin-1 32–33;[23] fibrillin-1 TB6-cbEGF32;[24] LDLR-AB,[15,25] and protein S 3–4.[26]

Baculovirus expression has also been used in the production of recombinant cbEGF domains, for example, in the cases of the CUB-EGF domain pair from C1r[27] and the Gla-cbEGF of factor IX.[28] The advantage of both baculovirus and mammalian expression systems is the secretion of the protein with disulphide bonds formed, negating the need for refolding of the domain.

Recombinant full-length proteins or extracellular portions of transmembrane proteins containing cbEGF domains are often expressed using a mammalian expression system, ensuring correct post-translational modifications.[29–32]

An alternative approach to the production of recombinant full-length samples is the purification of proteins from natural sources. In the case of the blood coagulation factors and complement proteins, the proteins can be purified directly from plasma.[33,34] Individual domains have also been isolated from proteolytic digests of the full-length protein.[35–37] These approaches to protein production have largely been replaced with the use of recombinant proteins.

Post-translational modifications

Three types of unusual post-translational modifications, including Asn/Asp β-hydroxylation and two forms of O-glycosylation, can occur in cbEGF domains, although the occurrence and level of modification is variable between proteins and between domains within a single protein. Two types of O-linked glycosylation have been found in EGF domains: O-linked fucose and O-linked glucose saccharides – both occurring as monosaccharide and oligosaccharide species.[38] In combination with the β-hydroxylation of Asn/Asp residues, each of the modifications has been associated with a putative consensus sequence[39,40] and occurs between different pairs of cysteine residues in the domain sequence. The consensus sequence for β-hydroxylation forms part of the calcium-binding consensus sequence, while the consensus sequence for O-linked glucose modifications occurs with greater frequency in cbEGFs than the O-linked fucose consensus sequence. It is possible for all three modifications to be present on a single domain, as in the case of factor IX cbEGF.[41]

O-linked glucose modifications have been associated with the putative consensus sequence C_1-x-S-x-P-C_2, sited between the first and second conserved cysteines of the domain, and have been characterized in the cbEGFs of factors VII and IX.[42] Mutation of Ser[52] of factor VII to alanine showed that the glycosyl moieties linked to this residue do not contribute to the catalytic efficiency of factor VIIa toward its substrates, factors IX and X, or to the association with its cofactor, tissue factor (TF).[43]

The putative consensus sequence established for O-linked fucose modifications is located between the second and third conserved cysteines of the EGF domain (C_2-X-X-G-G-S/T-C_3).[40] Replacement of the two glycine residues for alanines in this sequence in the factor VII cbEGF, either individually or jointly, resulted in multiple species of the domain with differing disulphide pairings. However, only the molecules with the correct disulphide pairing could be fucosylated.[20] A detailed investigation of the O-fucosylation of Ser[60] in factor VII by Kao et al.[21] showed that while the overall structure of the non-fucosylated form was very similar to the fucosylated form, even in the region of the fucosylation site, the calcium affinity is decreased twofold in the non-fucosylated form. The fucosylation site occurs in the context of a type I' β-turn, and it has been suggested that this β-turn structure is essential for recognition of O-fucosyltransferase.

A functional role for O-fucose glycans on the Notch protein has been established. Fringe is an O-fucose β1,3-N-acetylglucosminyltransferase that extends the O-fucose moieties on Notch by the addition of N-acetylglucosamine as well as also modifying Notch ligands (see review[44]). Specifically, Fringe has been shown to inhibit Serrate-mediated activation of Notch in dorsal *Drosophila* wing cells, while stimulating Delta-mediated activation of Notch in ventral cells at the border. The observation of similar Notch activation in *Drosophila* eye and leg, as well as chick wing, firmly establishes Fringe as a modulator of the Notch signaling pathway. Sequence analysis leads to the interesting observation that proteins only containing the O-linked fucose glycosylation consensus sequence have a developmental function (Neurocan, Serrate, and Slit). Apart from factors VII and IX and fibropellin (a sea-urchin extracellular matrix protein), proteins containing the O-linked fucose consensus sequence in combination with the O-linked glucose consensus sequence are also associated with a developmental function (Crumbs, Delta, Jagged, and Notch).

β-hydroxylation of Asn/Asp residues is linked with the consensus sequence C_3-x-D/N*-x_4-F/Y-x-C_4-x-C_5 between the third and fourth conserved cysteines. Although cbEGF

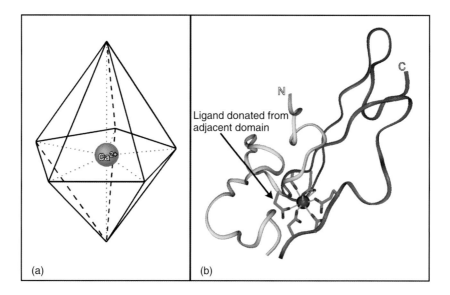

Figure 3 (a) Schematic representation of the pentagonal bipyramidal coordination of calcium found in cbEGF domains. (b) Schematic illustration of the calcium-binding site of human factor IX cbEGF, determined from the X-ray crystal structure (PDB code 1EDM).[49] The average Ca^{2+}-O distance is 2.3 Å and the calcium-binding ligand generally donated by a water molecule is replaced by a ligand from an adjacent domain (shown in yellow). This figure and figures 4 to 7 were created using MOLMOL (2000).[53]

domains contain the consensus sequence, it is also present in a number of non-calcium-binding EGFs. The presence of this sequence does not necessarily guarantee hydroxylation. Indeed, the site is often only partially hydroxylated or even not at all.[45,46] It was initially suggested that the β-hydroxyl group could serve as a calcium ligand. However, neither the calcium-binding properties nor the tertiary structures of cbEGF domains from factors IX and X were observed to change significantly in the presence or absence of this post-translational modification.[47] High-resolution structural studies have shown that the β-hydroxyl group points away from the calcium-binding site.[48,49] To date, the biological role for this modification remains elusive. It has been postulated that the aspartyl-β-hydroxylase may modify signaling through the Notch pathway by direct hydroxylation of Notch receptors and/or ligands, all of which contain multiple cbEGF domains.[50] Aspartyl-β-hydroxylase null mice display multiple developmental defects, with some abnormalities resembling the perturbations observed with mutants of the Jagged-2 (Serrate-2) pathway.[51]

METAL CONTENT

Each cbEGF domain binds a single calcium ion. In many Ca^{2+}-regulated proteins, calcium ions are typically sevenfold coordinated by oxygen atoms arranged in a pentagonal bipyramidal geometry (see Figure 3(a)). The oxygen atoms may be provided either by backbone carbonyl groups, side chain carboxylates, carboxyamides and hydroxyl groups, or by a water molecule. The mean Ca^{2+}-ligand distance is typically ~2.4 Å.[52] The specific ligand geometry of the cbEGF binding site has been determined through the crystal structures of the domains from factors VII and IX and follows the standard pentagonal bipyramidal arrangement.[10,49]

The packing of two domains in the crystal asymmetric unit of the cbEGF domain from factor IX leads to the donation of a side chain oxygen ligand by the adjacent domain (see Figure 3(b)). It has been suggested from these observations that calcium can directly mediate intramolecular domain–domain interactions.[49] It is also possible that this type of calcium ligation may play a general role in protein–protein interactions.[48,54]

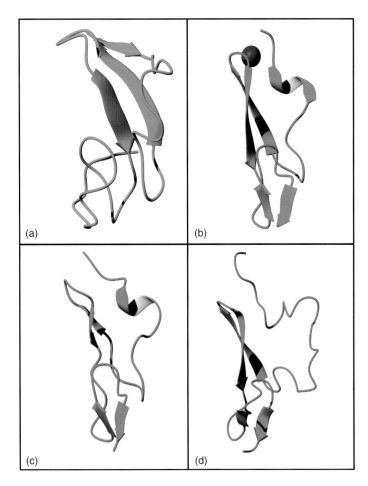

Figure 4 Schematic ribbon drawings comparing a non–calcium-binding EGF (cyan) and the apo and holo forms of calcium-binding EGFs (green): (a) Murine EGF (PDB code: 3EGF),[55] (b) human factor VII cbEGF holo form (PDB code: 1DAN),[10] (c) human factor VII cbEGF apo form (PDB code: 1BF9),[17] and (d) human C1r cbEGF apo form (PDB code: 1APQ).[18] The calcium ion is rendered as a red sphere.

The second [DEQN] residue of the calcium-binding consensus sequence (see above) is not a calcium ligand, and it only plays an indirect role in binding the calcium ion. In the factor IX cbEGF crystal structure, the carboxylate side chain forms an Asx turn with the backbone amide nitrogen of the i+2 cysteine residue, while also playing a structural role of maintaining the binding site geometry through hydrophobic packing interactions.[49]

X-RAY OR NMR STRUCTURE

Structures of single domains

The EGF domain fold is principally characterized by a major double-stranded β-hairpin that includes the third and fourth cysteine residues, although a short N-terminal region of α-helix and a second shorter β-hairpin near the C-terminus are also often observed. A comparison of the structure of epidermal growth factor and the holo and apo forms of the factor VII EGF in Figure 4(a) to (c) illustrates the relative spatial dispositions of these secondary structural elements. Detailed comparison of these three structures demonstrates that no major structural rearrangement of the domain is associated with the incorporation of a calcium-binding site. In addition, only minor structural changes are associated with calcium ligation.

The structures of a number of cbEGF domains have been determined in isolation or as part of larger constructs (Table 1). The structures solved are from proteins associated with diverse biological functions, including blood coagulation, connective tissue homeostasis, and cholesterol uptake. Nevertheless, a comparison of these structures based on the coordinates of the EGF 'core' region, which comprises a constant number of residues including the major β-hairpin,[54] has highlighted their similarity (backbone RMSDs ranging from 0.22–1.6 Å).[56] However, the sequences and structures of regions outside the 'core' are more variable and may be important for functional specificity. For example, the loop between the first and second cysteine residues in the C1r EGF domain is particularly long, as shown in Figure 4(d). This loop has been speculated to play a role in protein–protein interactions, although since the conformation of the loop is unaffected by calcium binding, these interactions are not likely to be calcium-dependent.[18]

The cbEGF domain has a smaller hydrophobic core than would be expected for a domain of this size, and its structure is principally stabilized by the three consensus disulphide bonds and calcium ligation. In particular, calcium binding plays an important role in mediating the flexible properties of the domain, as has been shown via the analysis of NMR backbone dynamics data.[64,65] Calcium ligation correlates with stabilization of the N-terminal region of the domain in the vicinity of the calcium-binding site. Furthermore, the amplitudes of slower timescale (μs-ms) motions of cysteine residues, which may be associated with disulphide bond rotamerization, are attenuated upon calcium binding. These results indicate

Table 1 Structures of cbEGF-like domains

PDB Entry	Description	Method	Apo/Holo	Reference
1BF9	N-terminal cbEGF from human factor VII	NMR	Apo	17
1F7M	N-terminal cbEGF from human factor VII	NMR	Holo	21
1FFM	N-terminal cbEGF from human factor VII fucosylated at Ser60	NMR	Holo	21
1DAN	Human factor VIIa complexed with tissue factor	X-ray (2.0 Å)	Holo	10
1FAK	Human factor VIIa complexed with tissue factor and inhibited with a Bpti mutant	X-ray (2.1 Å)	Holo	57
1IXA	N-terminal cbEGF from human factor IX	NMR	Apo	58
1EDM	N-terminal cbEGF from human factor IX	X-ray (1.5 Å)	Holo	49
1PFX	Porcine factor IXa	X-ray (3.0 Å)	Apo	11
1APO	N-terminal cbEGF from bovine factor X	NMR	Apo	59
1CCF	N-terminal cbEGF from bovine factor X	NMR	Holo	48
1WHE	Gla-cbEGF from human factor X	NMR	Apo	60
1WHF	Gla-cbEGF from human factor X	NMR	1Ca[a]	60
1AUT	Human activated protein C – Gla-domainless	X-ray (2.8 Å)	Apo	61
1DX5	Human thrombin complexed with EGFs 4–5 and cbEGF6 from thrombomodulin	X-ray (2.3 Å)	Holo	62
1APQ	Human C1r cbEGF	NMR	Apo[b]	18
1HJ7	cbEGF A–B pair from human LDL receptor	NMR	Holo	63
1I0U	cbEGF A–B pair from human LDL receptor	NMR	Holo	25
1EMN	cbEGF32–33 from human fibrillin-1	NMR	Holo	54

[a] Calcium is bound to the cbEGF domain but not to the Gla domain.
[b] Reference 18 compares the structures of both the apo and holo forms.

Calcium-binding EGF-like domains

that calcium binding to the domain imparts a global structural stabilization.

Comparison with related structures

Interestingly, in the absence of a calcium-binding site, alternative modifications of primary, secondary, and tertiary structure may improve the stability of the domain structure. For example, the structure of EGF encompasses an additional region of β-strand near the N-terminus that hydrogen bonds to the first strand of the major β-hairpin (Figure 4(a)). The integrin and laminin EGF domains (I-EGFs and L-EGFs) are more distantly related homologs that contain N- and C-terminal extensions respectively.[66,67] As highlighted in Figure 5, each of these EGF variants encompasses a canonical EGF domain sequence, as defined by the consensus disulphide-bonding pattern, and an additional pair of disulphide-bonded cysteines.

Structures of paired domains

Further insight into the role of calcium binding to EGF domains has been provided by the structures of EGF module pairs. A comparison of EGF–EGF, cbEGF–EGF, and cbEGF–cbEGF structures is presented in Figure 6. In the structure of the non-calcium-binding EGF pair from *Plasmodium falciparum* merozite surface protein-1, the two domains adopt a U-shaped fold, with a large hydrophobic interface. The two domains in the cbEGF–EGF pair taken from the X-ray structure of apo clotting factor IXa are arranged at an angle of 110°, which is stabilized by an interdomain salt bridge.[11] In contrast, a tandem pair of cbEGF domains from fibrillin-1 manifests an elongated, rod-like arrangement of the two domains, which is stabilized by hydrophobic packing interactions and calcium binding by the C-terminal domain. Multiple sequence alignment methods were used in the prediction that all tandem cbEGF pairs from fibrillin and many other functionally distinct proteins would adopt the same pairwise domain arrangement.[54] This hypothesis has been supported by structures of another pair of cbEGF domains from fibrillin-1 as well as a cbEGF domain pair from the low-density lipoprotein receptor.[63] NMR relaxation studies of wild-type and mutant domain pairs from fibrillin-1 also support a role for calcium in maintaining the stability of pairwise domain linkages.[64]

It is important to note that not all tandem cbEGF domain pairs will necessarily adopt the extended arrangement observed for the fibrillin-1 and LDLR structures. Multiple-sequence alignment data indicates that an alternative mode of pairwise domain assembly may exist that might resemble the crystal packing interactions observed for the N-terminal cbEGF domain from factor IX.[54] In this study, it was observed that two domains in the crystal asymmetric unit adopt a V-shaped arrangement, with their N- and C-termini

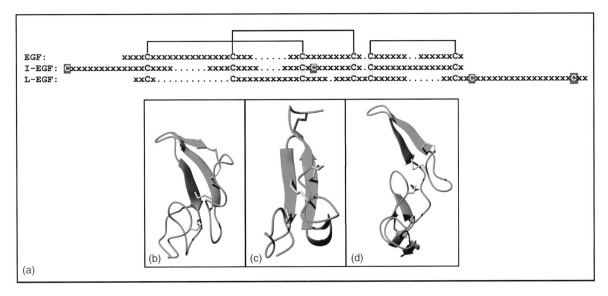

Figure 5 (a) Alignment of conserved cysteines in EGF, I-EGF, and L-EGF sequences. Each letter represents a residue position in a multiple sequence alignment. Residue insertions between the three alignments are shown by '.', and have been positioned randomly. The EGF alignment is taken from Figure 2. The I-EGF alignment is from Beglova et al.[66] and the L-EGF spacing is from the Pfam laminin EGF entry (http://www.sanger.ac.uk/Software/Pfam/). The cysteines that join in a 1–3, 2–4, 5–6 pattern are shown in yellow and joined by lines. The additional cysteines that disulphide bond in the I-EGF and L-EGF sequences are highlighted in red. Schematic ribbon drawings of EGF domains: (b) Murine EGF (PDB code 3EGF),[55] (c) Human I-EGF (PDB code: 1L3Y),[66] and (d) Murine L-EGF (PDB code: 1KLO).[67] The representations have the same orientation based on the superimposition of the conserved cysteine residues (disulphide bonds shown in yellow) and the variant disulphide bonds are shown in red.

Calcium-binding EGF-like domains

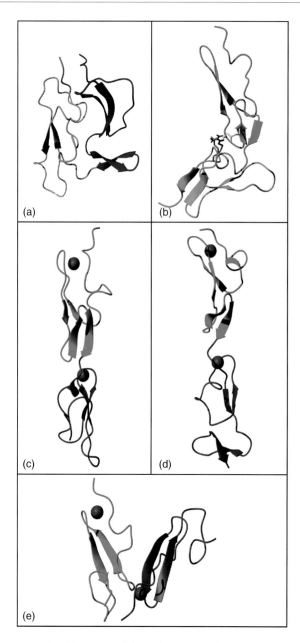

Figure 6 Schematic ribbon drawings of EGF pairs from (a) *Plasmodium falciparum* merozite surface plasmodium-1 (EGF–EGF; PDB code: 1CEJ),[68] (b) porcine factor IXa (cbEGF–EGF; PDB code: 1PFX),[11] (c) human fibrillin-1 32–33 (cbEGF–cbEGF; PDB code: 1EMN),[54] (d) human LDLR-AB (cbEGF–cbEGF; PDB code: 1IOU),[25] and (e) human factor IX cbEGF crystal packing (PDB code: 1EDM).[49] Calcium atoms are rendered as red spheres. In (b), the residues forming a salt bridge between the two domains are highlighted.

calcium, in that these types of domain pairs have also been observed to adopt extended arrangements.[66,67] This observation lends support to the theory that rod-like arrangements of domains may be required for specific macromolecular biophysical properties, which relate to function.

Complex structures

Only two structures of complexes containing cbEGF domains have been reported to date, and these are shown in Figure 7. The major double-stranded β-sheet of the cbEGF domain from factor VII packs against the interface of two fibronectin type-III modules of soluble TF (sTF).[10] This association stabilizes the structure of factor VIIa mainly via specific cbEGF domain contacts that bury 823 Å^2 of surface area. In contrast, the structure of sTF is virtually unaffected at the binding interface. The binding involves a mixture of polar and nonpolar interactions, and the face of the cbEGF domain that is involved does not have unusual surface properties. Since both O-glycosylated sites on the cbEGF domain are solvent exposed in the complex, this type of post-translational modification does not appear to play an important role in recognition. Calcium-binding data suggest that fucosylation may affect the interaction of the two molecules by influencing cbEGF affinity.[21] In this case, increased stability of the cbEGF domain associated with enhanced calcium binding may decrease the entropic penalty associated with complex formation.

In the structure of thrombin complexed with thrombomodulin EGF domains 4 to 6, the calcium containing junction between EGF domains 5 and 6 packs against the thrombin molecule, whose structure is virtually unaffected by the interaction.[62] Interestingly, EGF5 and cbEGF6 pack in an extended conformation, not unlike that of the fibrillin-1 or LDLR cbEGF domain pairs, in spite of a divergent disulphide-bonding pattern for EGF5 (1–2, 3–4, 5–6).[69] Specific contacts between Tyr^{413} in EGF5 and the end of the major double-stranded β-hairpin in cbEGF6 mimic intermodular hydrophobic packing interactions observed in the cbEGF domain pairs. The calcium atom bound by cbEGF6 plays a key role in stabilizing the relative domain orientation and in complex formation.[70] As in the factor VIIa–sTF complex, a mixture of polar and nonpolar interactions mediate macromolecular recognition; however, in this case there is a cluster of hydrophobic residues situated at the EGF5–6 junction that forms the core of the ~900 Å^2 interaction surface. The secondary structure regions that are involved in complex formation comprise the C-terminal minor β-sheet of EGF5 and the N-terminal α-helix of cbEGF6. The observation that different cbEGF secondary structure regions are involved in complex formation in the two complexes described suggests that distinct

in close proximity and a calcium ligand donated by an Asn residue from the adjacent domain.[49] As noted above, more distantly related EGF homologs, I-EGFs and L-EGFs, have extended sequences and an additional disulphide bond, which may play a role in stabilizing the domain structure in the absence of calcium. A further analogy exists between the role of these structural elements and

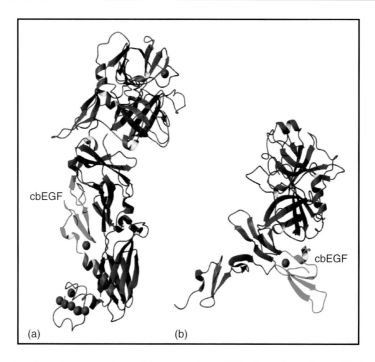

Figure 7 Schematic ribbon drawing of two complexes: (a) human factor VII·soluble tissue factor (PDB code: 1DAN)[10] and (b) human thrombomodulin(EGF4–6)·thrombin (PDB code: 1DX5).[62] Both tissue factor and thrombin are colored blue, while the cbEGF-containing proteins (factor VII and thrombomodulin) are shown in red, with the cbEGF domain highlighted in green. The calcium ions are rendered as red spheres.

functional patches have evolved on the surfaces of cbEGF domains and that it will be difficult to predict interaction surfaces on the basis of known features of macromolecular recognition.

FUNCTIONAL ASPECTS

Class I and class II domain pairs

Previous analysis of 250 sequences of EGF–cbEGF and cbEGF–cbEGF pairs extracted from the Swissprot database showed that they could be divided into two distinct classes (I and II), based only on the number of linking residues between the domains (one or two, respectively).[54] This analysis can now be extended to over 1000 sequences extracted from the Swissprot and Trembl databases (using the same search criteria). Of the sequences analyzed, 55% follow the class I domain linkage. Table 2 shows the proteins containing these pairs divided into the two classes.

In general, proteins containing multiple tandem repeats of the consensus EGF–cbEGF sequence incorporate only pairs of one class, implying that domain–domain interactions are usually homogenous within proteins. Protein S and Gas-6 are exceptions to that rule. Both contain four EGF domains (EGF_1-$cbEGF_2$-$cbEGF_3$-$cbEGF_4$) with the 1–2 pair being class II and 2–3 and 3–4 pairs being class I.[71,72] The long isoform of nephronectin contains 5 EGF domains (EGF_1-$cbEGF_2$-EGF_3-$cbEGF_4$-$cbEGF_5$) and also contains a mix of class I and class II pairs (1–2 = class II; 3–4 and 4–5 = class I).[73] Interestingly, the short isoform of the protein does not contain the complete consensus sequence for EGF_1 and therefore lacks the class II pairing of domains.

It is possibly easier to dissect differences found between the two classes of domain pairs by studying the sequence properties of individual domains that occur in the context of pairs and longer strings (see Figure 8). Sequence analysis of the two classes of pairs revealed that the glycine residue between the third and fourth conserved cysteine residues in (C_3-x-[DN]-x-x-G-x-[FY]-x-C_4), involved in domain–domain packing, forms part of the consensus sequence for class I domains, but not for class II.[54] This residue is only conserved in cbEGFs with a classical calcium-binding consensus sequence. As shown in Figure 8, it is present in protein S cbEGF2–3 and LDLR–cbEGF B, all of which form the class I linkage. Interestingly, LDLR–cbEGF A possesses a non-canonical calcium-binding consensus sequence and does not have the conserved glycine residue.

In class II domains, a conserved carboxylate/carboxyamide side chain is observed in the position of the calcium-binding ligand donated from the N- to the C-terminal domain in the crystal structure of the factor IX cbEGF (see Figure 3(b)), between the second and third

Calcium-binding EGF-like domains

Table 2 Classification of proteins containing EGF–cbEGF/cbEGF–cbEGF pairs

Class I: one interdomain residue	Class II: two interdomain residues
ApoE receptor 2 (human, mouse)	Bet (mouse)
CD 93 (human, mouse, rat)	Crumbs (*Drosophila*)
CD 97 (human, bovine, mouse, pig)	Crumbs homolog (human, mouse)
Comp (human, rat)	Cubilin (human, dog, rat)
Creld-1 (human)	Delta (*Drosophila*)
EGF precursor (human, cat, dog, mouse, pig, rat)	Delta-like (human, mouse, rat, sea urchin)
EMR-1 (human, mouse)	DNER (human, mouse)
EMR-2 (human)	FAT homolog (human, rat)
EMR-3 (human)	Fibropellin-1 (sea urchin)
EMR-4/Fire (mouse)	Fibropellin-3 (sea urchin)
Endosialin (mouse)	Fibrosurfin (sea urchin)
Fibulin-1 (human, *Caenorhabditis elegans*, chicken, mouse, zebrafish)	Gas-6 (human, mouse, rat)
Fibulin-2 (human, mouse)	GLP-1 (*C. elegans*)
Fibulin-3 (human, rat)	HR Notch (sea squirt)
Fibulin-4 (human, Chinese hamster, mouse)	Jagged-1 (human, mouse, rat, zebrafish)
Fibulin-5 (human, mouse, rat)	Jagged-2 (human, mouse, rat, zebrafish)
Fibulin-6 (human)	Jagged-3 (zebrafish)
Fibrillin (jellyfish)	Lin-12 (*C. elegans*)
Fibrillin-1 (human, bovine, mouse, pig, rat)	Nephronectin-long form (mouse)
Fibrillin-2 (human, chicken, mouse, rat)	Neurocan core protein (human, chicken, mouse, rat)
Fibrillin-3 (human)	Notch (*Drosophila*, sea urchin, Xenopus)
Gas-6 (human, mouse, rat)	Notch-1 (human, mouse, rat, zebrafish)
LDL receptor (human, Chinese hamster, mouse, pig, rabbit, rat, shark, Xenopus)	Notch-2 (human, fugu, mouse, rat)
Lipophorin (locust)	Notch-3 (human, mouse, rat)
LRP (human, *C. elegans*, chicken, mouse)	Notch-4 (human, mouse)
LTBP-1 (human, mouse, rat, Xenopus)	Polydom (mouse)
LTBP-2 (human, bovine, mouse, rat)	Protein S (human, bovine, macaque, mouse, pig, rabbit, rat)
LTBP-3 (human, green monkey, mouse)	Serrate (*Drosophila*)
LTBP-4 (human, mouse)	SLC (Notch homolog) (greenbottle fly)
MBP-1 (human)	Slit (Human, *Drosophila*, mouse, rat, Xenopus)
Megalin (human, rat)	Slit-2 (human, mouse)
Microneme-4 (slime mould, *Eimeria tanella*)	Slit-3 (human)
Nel (chicken)	Versican core protein (human, bovine, chicken, macaque, mouse, rat)
Nell-1 (human, rat)	Xerl (Xenopus)
Nell-2 (human, mouse, rat)	
Nephronectin-long form (mouse)	
Nidogen/entactin (human, chicken, mouse)	
Nidogen-2/entactin-2 (human, mouse)	
Poem (nephronectin-short form) (mouse)	
Protein S (human, bovine, macaque, mouse, pig, rabbit, rat)	
Scube-1 (mouse)	
Thrombomodulin (human, bovine, mouse, rat)	
Thrombospondin-1 (human, bovine, mouse, Xenopus)	
Thrombospondin-2 (human, bovine, chicken, mouse)	
Thrombospondin-3 (human, mouse, Xenopus)	
Thrombospondin-4 (human, rat, Xenopus)	
Uromodulin (human, bovine, mouse, rat)	
Vitellogenin receptor (cockroach, *Drosophila*, trout, Xenopus)	
VLDL receptor (human, chicken, mouse, rabbit, rat)	
Wall associated kinases (1,2,4) (Arabidopsis, rice)	

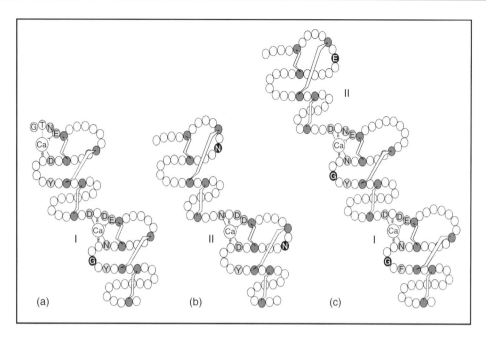

Figure 8 Schematic illustration of class I and class II EGF·cbEGF/cbEGF·cbEGF domain pairs. (a) Class I pair from human LDLR (cbEGF-AB), (b) class II pair from human Delta-like-1 (EGF6–cbEGF7), and (c) class II–class I triplet from human protein S (EGF1-cbEGF2-cbEGF3). Amino acid residues are shown as *circles*. Conserved cysteines, forming disulphide bonds, are represented in mid-gray. Residues of the calcium-binding consensus sequence are highlighted in light gray. Conserved residues associated with either class I or class II pairs are colored in black.

conserved cysteines. It is not exclusive to cbEGF domains, as highlighted in Figure 8 for Delta-like-1 EGF6–cbEGF7 and protein S cbEGF1. It is possible therefore that in class II pairs, a calcium-binding ligand will be donated from the N-terminal domain to the C-terminal calcium-binding site (see above).

Another noticeable difference between the two classes of domains is the difference in the length of the loop between the fifth and sixth conserved cysteines, with class II domains possessing a shorter loop than class I domains. The difference in the length of the loop may be important in determining different pairwise domain interactions in the two classes.

For proteins with non-homologous domain pairs (protein S, Gas-6, and nephronectin-long-isoform), the initial domain pair is formed from a class II domain–class I domain pairing. This is regardless of whether the C-terminal domain forms the N-terminal domain of the next pair. Structural information is required to fully determine the differences between the classes of domains and their linkages. At present, structures have only been determined for class I pairs (LDLR-AB[25,63] and fibrillin-1 32–33[54]).

Calcium binding

Calcium binding to an isolated cbEGF domain was first measured in Protein C through the use of a calcium-sensitive electrode.[36] Since then, a number of methods have been employed to determine the calcium dissociation constants for cbEGFs, including 1D and 2D NMR, chromophoric chelator methods, intrinsic protein fluorescence, and equilibrium dialysis. The accuracy and sensitivity of the techniques is discussed in the review by Downing *et al.*[56] Table 3 lists the current range of calcium dissociation constants determined for cbEGF domains.

Calcium dissociation constants measured for isolated or N-terminal cbEGFs are in the range of 1 to 10 mM K_d, although it is possible to explain differences of up to one order of magnitude through varying sample conditions. It is also possible to correlate calcium affinity in part to the side chain ligands to the calcium ion. Analysis of wild-type and mutant versions of the factor IX cbEGF, which created various combinations of the carboxyl and carboxyamide side chains presenting ligands, showed that the calcium affinity could vary as much as 20-fold.[3]

A number of calcium dissociation constants have been determined for specific cbEGF domains having both the N-terminus free and being N-terminally linked to an adjacent domain (see Figure 9). Linkage with an N-terminal cbEGF appears to have the potential to support higher affinity binding in the C-terminal domain. Knott *et al.*[23] measured the calcium affinities of a cbEGF pair from fibrillin-1 through NMR and provided evidence that calcium plays a major role in defining the interdomain linkage in multiple repeats of cbEGFs. The stabilization of calcium-binding sites by the presence of both N-terminally (N–C stabilization) and C-terminally (C–N) linked domains has

Table 3 Calcium dissociation constants determined for cbEGF domains[a]

Domain(s)	K_d(s)	pH	I	Method	Reference
EGF precursor cbEGF2	3.5 mM	7.5	0.1	Fluorescence	74
C1r cbEGF	10 mM	6.6	0	1D-NMR	75
C1r/C1s CUB–cbEGF	32–38 µM	7.4	0.145	Equilibrium dialysis	76
LDLR cbEGF-AB	**A**:~50 µM; **B**:10–20 µM	7.5	0.15	Chromophoric chelator/2D-NMR	15
Factor IX cbEGF	~250 µM	7.4	0	1D-NMR	77
Factor IX cbEGF	1.8 mM	7.5	0.15	1D-NMR	13
HsFactor IX cbEGF	1 mM	7.4	0.15	Fluorescence	78
Factor IX Gla-cbEGF[b]	60 µM	7.5	0.1	Fluorescence	37
Factor IX Gla-cbEGF[b]	160 µM	7.4	0.1	Fluorescence	28
Factor X cbEGF	2.7 mM	7.4	0.15	1D-NMR	35
Factor X Gla-cbEGF[b]	120 µM	7.5	0.1	Fluorescence	79
Protein C cbEGF	100 µM	7.5	0.1	Calcium-sensitive electrode	36
Protein S cbEGF3	6.1 mM	7.5	0.15	1D-NMR	80
Protein S cbEGF4	8.6 mM	7.5	0.15	1D-NMR	80
Protein S cbEGF2–3	**a**:15 µM; **b**: ≥10 µM	7.5	0	2D-NMR	81
Protein S cbEGF3–4	**3**:4.8 mM **4**:1 µM	7.5	0.15	Chromophoric chelator/2D-NMR	26
Protein S cbEGF1–3	**a**: ≥100 µM; **b**: ≥100 µM	7.5	0.15	Chromophoric chelator	81
Protein S cbEGF2–4	**a**:1 µM; **b**: ≥10 µM; **c**: ≥10 µM	7.5	0.15	Chromophoric chelator	81
Protein S EGF1–cbEGF2–4	**a**:20 nM; **b**:2 µM; **c**:5 µM	7.5	0.15	Chromophoric chelator	81
Thrombomodulin EGF4–6	2.5 µM	7.5	0.1	Equilibrium dialysis	70
Notch cbEGF11–12	**a**:34 µM; **b**:>0.25 mM	7.5	0.15	Chromophoric chelator	82
Notch cbEGF10–13	**a**:3.1 µM; **b**:160 µM; **c**:>250 µM	7.5	0.15	Chromophoric chelator	82
Fibrillin-1 cbEGF12–13	**12**:1.6 mM; **13**: ≤30 µM	6.5	0.15	2D-NMR/fluorescence	83
Fibrillin-1 cbEGF13	2–3 mM	6.5	0.15	2D-NMR	16
Fibrillin-1 cbEGF13–14	**13**:3 mM; **14**: ≤100 µM	6.5	0.15	2D-NMR	16
Fibrillin-1 cbEGF22–23	~400 µM (n = 2)	7.5	0.15	Equilibrium dialysis	84
Fibrillin-1 cbEGF22–31	~500 µM (n = 10)	7.5	0.15	Equilibrium dialysis	84
Fibrillin-1 cbEGF25–31	250 µM (n = 6 − 7)	7.5	0.17	Equilibrium dialysis	46
Fibrillin-1 cbEGF32	4.3 mM	7.4	0.15	1D-NMR	85
Fibrillin-1 TB6-cbEGF32	1.6 mM	6.5	0.15	2D-NMR	24
Fibrillin-1 cbEGF32–33	**32**:9.2 mM; **33**: 350 µM	6.5	0.15	2D-NMR	23

[a] For studies in which the calcium dissociation constants were measured under varying conditions of pH and ionic strength, the values corresponding to the most physiological conditions are reported.

[b] In these samples, the Gla residues were decarboxylated to Glu so the calcium binding to the cbEGF was exclusively monitored.

been shown through studies on cbEGF pairs and triplets from Protein S.[81] Analysis of a cbEGF domain pair and quadruplet from Notch-1 proposed that C to N stabilization can result in a more than 10-fold increase in calcium affinity.[82] Although both N-terminal and C-terminal domain linkage can be associated with an increase in calcium affinity, it has been observed that the magnitude of the affinity enhancement depends on the extent of the pairwise domain interactions and the flexibility of the interdomain linkage, which is also dependent on whether the linkage occurs between homologous domains.[83] In the case of cbEGF13 from fibrillin-1, linkage with cbEGF12 is associated with at least a 70-fold increase in calcium affinity over the isolated cbEGF13 domain. In contrast, the N-terminal linkage of the non-homologous TB6 domain to cbEGF32 results in an increase in calcium affinity of ~2.7-fold,[24] but has been shown to possess a more flexible linker than the cbEGF32–33 pair.[64,65] It is possible that interdomain calcium ligation (as proposed from the crystal packing of factor IX cbEGF) may also increase cbEGF domain affinity, although no structure corroborating this has been reported to date.

There has been a range of methods, other than NMR and X-ray crystallography, employed in elucidating the structural effect of calcium binding to proteins containing cbEGFs, especially with fibrillin-1 in the context of forming microfibrils. Through velocity sedimentation experiments, Reinhardt et al.[86] demonstrated a shortening of fibrillin by ~25% and an increase in width of ~13 to 17% on the removal of calcium, further suggesting that the interdomain regions are more rigid in the presence of calcium. This result was extended to microfibrils with the use of rotary shadowing, with a large alteration in microfibril structure and increase in flexibility observed upon the removal of calcium.[12] Wess et al.[87] used X-ray diffraction and scanning transmission electron microscopy to show the importance of calcium, not only in the molecular packing of individual microfibrils, but also in higher-order hydrated microfibrillar arrays such as those occurring in vivo. Use of the micro-needle technique showed that calcium depletion

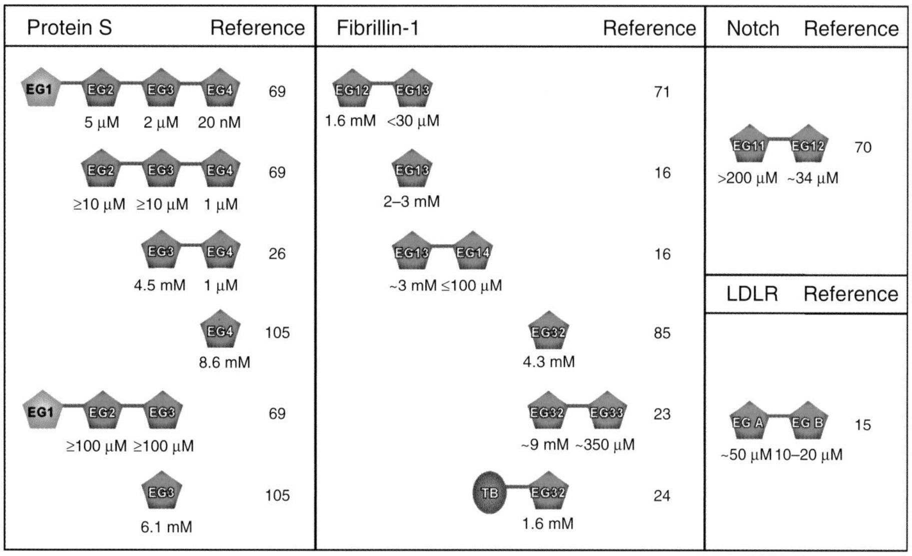

Figure 9 Schematic representations of the calcium dissociation constants for the human recombinant cbEGF domains from protein S, fibrillin-1, Notch, and LDLR. The dissociation constants were obtained in the presence of 0.15M NaCl by titration with chromophoric chelators (Quin-2, 5,5'-Br$_2$BAPTA, and 5-NBAPTA) or by NMR-monitored titrations. Domain symbols are according to Bork et al.[1]

had a dramatic effect on the rest length and reduced the stiffness of fibrillin-rich microfibrils, clearly highlighting the significance of calcium for the mechanical function of the microfibrils.[88] Gel filtration and intrinsic Tyr fluorescence provided evidence of CUB–EGF pair adopting a more compact structure in the presence of calcium.[27]

Interaction studies

Calcium-binding EGF domains have been directly implicated in numerous protein–protein interactions. The interaction between factor VII and TF has been studied in detail, aided by the elucidation of the crystal structure of the complex (see above). Binding of calcium to the cbEGF is partly responsible for the demonstrated calcium dependence of the factor VIIa·TF interaction,[89,90] even though neither the calcium ion nor the side chains involved in calcium coordination are in direct contact with TF. Mutational analysis of residues within the calcium-binding consensus sequence suggests that the binding of calcium restricts flexibility and orientates the Gla domain in relation to the cbEGF domain, which are requirements for optimal binding with TF.[91]

The binding between CD55 and CD97 is mediated exclusively by interactions of the SCR (CD55) and EGF (CD97) domains located at the N-terminus of each protein.[92] CD97 occurs in several splice forms, which vary the number of EGF domains encoded. The strongest interaction with CD55 is found with the 1–2–5 splice form, which was 5 times stronger than the 1–2–3–4–5 splice form. Both EGF domains 2 and 5 of CD97 possess the calcium-binding consensus sequence and the interaction with CD55 is calcium-dependent. EMR2, an EGF-TM7 protein, shares high sequence homology with CD97, differing by only three amino acids within EGFs 1–2–5.[93] These differences do not disrupt the calcium-binding consensus sequence but do substantially reduce the level of interaction with CD55 by at least an order of magnitude.[92] The EGF·cbEGF and cbEGF·cbEGF pairs of the 1–2–5 splice forms of both proteins are of class I configuration. It is suggested that calcium binding to the 1–2–5 splice form of CD97 is important in maintaining an overall rod-like structure of the three domains and could explain the reduction in binding seen with the 1–2–3–4–5 splice form. Although stability of the structure is obviously an important feature for the interaction, the amino acid differences between the 1–2–5 forms of CD97 and EMR2 show that it is not the sole determinant of high-affinity binding.

The interaction between C1r and C1s is calcium-dependent and the CUB–cbEGF pair of C1r has been determined as the minimal segment required for high-affinity calcium-binding and calcium-dependent interaction with C1s.[27] It is thought that calcium binding to the C1r CUB–cbEGF pair induces a more compact structure (see above) and provides the appropriate conformation and ligands required for interaction with the homologous CUB–cbEGF pair of C1s.

Post-translational modifications to cbEGF domains can affect protein–protein interactions. As stated above, control of the glycosylation state of the EGF domains of Notch, by selective expression of Fringe, is used to regulate the interactions of Notch with its ligands.[44] In contrast, the work on the CD97–CD55 interaction demonstrated that glycosylation of extracellular domains is not necessarily required to modulate protein–protein interactions.[92]

Calcium-binding EGF-like domains

A number of cbEGF-containing proteins also have RGD sequences present, either within the cbEGF or in neighboring domains. RGD sequences are recognized by many integrins and are associated with cell adhesion activities of the protein.[94] So far only LTBP1 and uromodulin possess an RGD within a cbEGF domain, located between the second and third conserved cysteines for LTBP1 and between the fifth and sixth conserved cysteines for uromodulin. In both cases, the sequences are positioned in a β-turn (based on the factor IX cbEGF crystal structure[49]), which agrees with the finding that in 80% of proteins with known three-dimensional structures and not having a documented cell adhesion function, the sequence is part of a loop. A novel protein was engineered by inserting the GRGDS motif of fibronectin into the loop between the first and second conserved cysteines of the cbEGF domain of C1r.[95] The insertion of the RGD sequence had no significant effect on the structure of the domain but was shown to induce cell adhesion via integrin receptors, although in the fibronectin/α5β1 system targeted in the study the cbEGF–RGD domain alone was not sufficient to yield optimal receptor–ligand interaction and cellular signaling transduction.

Mutations and diseases

Numerous inherited diseases have been associated with amino acid changes in proteins. Several diseases are linked to mutations within proteins containing cbEGF domains, which affect blood coagulation, cholesterol metabolism, connective tissue, cardiovascular, visual, and nervous systems. Table 4 summarizes only those diseases in which mutations occur within the cbEGF domains.

Deletion of an entire cbEGF domain can have drastic consequences. In the case of factor VII, deletion of the cbEGF results in a homozygous neonatally lethal form of factor VII deficiency.[113] Exon-skipping mutations in fibrillin-1, which delete single cbEGF domains, are often associated with the more severe forms of Marfan syndrome and in one case with a neonatally lethal disorder.[114] In these cases, normal protein production is observed but with a severe to less severe reduction in the extracellular matrix depositation.

On the basis of their predicted consequences, point mutations within cbEGF domains can be classified into three groups. Those that either replace a conserved cysteine involved in a disulphide bond or introduce an additional cysteine are most probably associated with domain misfolding. An example of this group is found in factor X with a Cys^{81}-Tyr mutation resulting in type I factor X deficiency with reduced protein levels.[100]

A second group of mutations affect the calcium-binding consensus sequence and are thought to result in reduced calcium affinity and greater interdomain linker flexibility.[85,115] In the case of fibrillin-1, those mutations affecting the calcium-binding consensus sequence constitute a significant proportion (~10%) of the Marfan syndrome–causing mutations found to date.[101]

Prediction of the consequences of the last group of mutations is not as straightforward, and therefore these should be considered on an individual basis. The effects of a selection of mutations in this group have been explained by structural studies. For example, analysis of the Arg^{79}-Gln mutation of factor VII, which causes factor VII deficiency, showed that while activation of the variant factor VII was identical to that of wild-type, FVII-R79Q has a reduced affinity for TF.[116] The structure of factor VIIa complexed to soluble TF showed that Arg^{79} is engaged in a salt bridge with Glu^{24} of soluble TF.[10] A Glu^{78}-Lys mutation in factor

Table 4 Heritable genetic disorders associated with mutations to cbEGF domains[a]

Protein(s)	Disease(s)	OMIM#	Reference
Crumbs homolog	Retinitis pigmentosa 12	600105	96
	Leber congenital amaurosis	604210	97
Factor VII	Factor VII deficiency	227500	98
Factor IX	Hemophilia B	306900	99
Factor X	Factor X deficiency	227600	100
Fibrillin-1	Marfan Syndrome	154700	101
Fibrillin-2	Contractual arachnodactyly, congenital	121050	102
Fibulin-3	ML/DHRD	601548/126600	103
Fibulin-5	Cutis Laxa	(604580)	104
Jagged-1	Alagille syndrome	118450/601920	105
LDLR	Familial hypercholesterolemia	143890	106
LTBP-2	Marfan-like	602091	107
Notch-3	CADASIL	125310/600276	108
Protein C	Protein C deficiency	176860	109
Protein S	Protein S deficiency	176880	110
Thyroid peroxidase	Thyroid peroxidase deficiency	606765	111

[a] Information on the genetic diseases was obtained from Human Gene Mutation Database at the Institute of Medical Genetics in Cardiff (http://archive.uwcm.ac.uk/uwcm/mg/hgmd0.html).[112]

IX results in hemophilia B. The study of the structure of porcine FIX and functional analysis of mutations to Glu[78] and Arg[94] proposed the existence of a salt bridge between the two EGF domains, important for maintaining the integrity of the factor VIII light-chain binding site.[11,117]

In familial hypercholesterolemia, mutations in the low-density lipoprotein receptor are categorized into five classes depending on their functional effect. A number of mutations have been detected along the length of the cbEGF–AB pair, comprising all three categories of cbEGF mutations and which, through molecular and cell biological characterization, can be classified functionally.[118] This creates another level by which the effect of the mutation can be determined, increasing the existing knowledge of the structure–function relationship. The LDLR cbEGF-AB mutations have been biochemically characterized into either of two classes, with no distinction between the cbEGF mutation category and familial hypercholesterolemia class. More detailed analysis, both structurally and functionally, is required to determine if a genotype–phenotype relationship can be derived for the three types of cbEGF mutations associated with familial hypercholesterolemia.

Proteolytic susceptibility

Calcium bound to cbEGFs plays a key role in stabilizing fibrillin-1 microfibril architecture by restricting conformational flexibility between cbEGF domains.[54,65] It has also been suggested that calcium plays another role, that of protecting cbEGFs from proteolysis in fibrillin-1[119] or fibulin-1 and -2.[120]

The structural and functional consequences were investigated for two mutations of fibrillin-1, one (N548I) leads to the classical form of Marfan syndrome, the other (E1073K) to the 'neonatal' form, and both mutations affect critical residues within cbEGFs directly involved in calcium binding.[32] Extensive protease-degradation assays of large recombinant polypeptides with a variety of proteases showed that both N548I and E1073K polypeptides were much more susceptible to proteolytic cleavage than the wild-type ones in the presence of calcium, while being equally susceptible in the absence of calcium.

Changes in proteolytic susceptibility associated with mutations have also been studied in multidomain constructs from fibrillin-1 in order to probe the extent of their effects. Two mutations that result in Marfan syndrome and affect calcium ligands in adjacent domains were analyzed.[121] The N2144S mutation was investigated in a TB6–cbEGF32 pair and a cbEGF32–33 pair and a N2183S mutation in the cbEGF32–33 pair. Only the N2183S mutation in cbEGF33 resulted in increased proteolytic susceptibility, consistent with calcium binding to cbEGF33 stabilizing the interdomain linkage between cbEGF32 and cbEGF33. The N2144S mutation in cbEGF32 had no discernible effect on proteolytic susceptibility in either the TB6–cbEGF32 or cbEGF32–33 pair, indicating that the domain interactions between TB6 and cbEGF32 are calcium-independent, unlike the interactions between cbEGF32–33. This demonstrates that the structural consequences of calcium-binding mutations in fibrillin-1 cbEGF domains can be influenced by the domain context.

A separate analysis involved the Marfan syndrome G1127S mutation in cbEGF13 of fibrillin-1, which does not affect a calcium-binding ligand.[22] Study of both the cbEGF12–13 pair and cbEGF12–14 triplet showed that the presence of the mutation introduced an additional endoproteinase Glu-C cleavage site in cbEGF13 in both constructs and two additional tryptic sites in cbEGF13 in the triplet. The position and calcium-dependent properties of the cleavage sites in the flanking domains (cbEGFs 12 and 14) were identical between the wild-type and mutant forms, indicating that the effect of the mutation was localized to cbEGF13.

In contrast, the investigation with the N548I and E1073K fibrillin-1 polypeptides (see above) showed, in addition to the short range effects surrounding the mutation site, the presence of proteolytically sensitive sites a distance away from the mutation site. Longer-range stabilizing effects have been postulated on the basis of the calcium-binding properties of Notch-1 cbEGF domains.[82] The tandemly repeated cbEGFs of fibrillin-1 are class I linked, while those of Notch-1 are class II linked,[54] suggesting that transmission of longer-range effects is not dependent on the length of the linker between cbEGF repeats. Comparison of (i) protease digestion profiles, (ii) calcium dependence of proteolysis, and (iii) calcium-binding properties can therefore be used to predict the short *versus* the longer-range consequences of various mutations.

CONCLUSIONS

EGF domains form one of the most abundant extracellular protein modules. A particular subset has the ability to bind calcium (cbEGF) and is found in a wide variety of proteins of diverse function. Although there are occurrences of single cbEGFs forming part of the modular structure of proteins, the domains are most commonly found tandemly linked by one or two residues. Extensive biophysical characterization has elucidated a number of properties for cbEGFs in terms of their structure, post-translational modifications, calcium binding, protein–protein interactions, and roles in numerous genetic disorders. In this review, we have focused on selected research that illustrates the three main functional roles of this type of domain, as a spacer unit, in structural stabilization, and in protein–protein interactions. We have also highlighted research demonstrating that cbEGF domains can undergo three unusual post-translational modifications, which can have a functional role. In general, calcium binding appears to play a general role in restricting the conformational flexibility of interdomain linkages and/or orienting domain pairs as well as reducing the proteolytic susceptibility of the domain.

REFERENCES

1. P Bork, AK Downing, B Kieffer and ID Campbell, *Q Rev Biophys*, **29**, 119–67 (1996).
2. DJ Rees, IM Jones, PA Handford, SJ Walter, MP Esnouf, KJ Smith and GG Brownlee, *EMBO J*, **7**, 2053–61 (1988).
3. PA Handford, M Mayhew, M Baron, PR Winship, ID Campbell and GG Brownlee, *Nature*, **351**, 164–67 (1991).
4. ID Campbell and P Bork, *Curr Opin Struct Biol*, **3**, 385–92 (1993).
5. L Patthy, *Curr Opin Struct Biol*, **1**, 351–61 (1991).
6. S Kuroda, M Oyasu, M Kawakami, N Kanayama, K Tanizawa, N Saito, T Abe, S Matsuhashi and K Ting, *Biochem Biophys Res Commun*, **265**, 79–86 (1999).
7. ZH He, I Cheeseman, D He and BD Kohorn, *Plant Mol Biol*, **39**, 1189–96 (1999).
8. T Shimada, M Kuroyanagi, M Nishimura and I Hara Nishimura, *Plant Cell Physiol*, **38**, 1414–20 (1997).
9. EJ Husten, CT Esmon and AE Johnson, *J Biol Chem*, **262**, 12953–61 (1987).
10. DW Banner, A D'Arcy, C Chene, FK Winkler, A Guha, WH Konigsberg, Y Nemerson and D Kirchhofer, *Nature*, **380**, 41–46 (1996).
11. H Brandstetter, M Bauer, R Huber, P Lollar and W Bode, *Proc Natl Acad Sci USA*, **92**, 9796–800 (1995).
12. CM Cardy and PA Handford, *J Mol Biol*, **276**, 855–60 (1998).
13. M Mayhew, P Handford, M Baron, AG Tse, ID Campbell and GG Brownlee, *Protein Eng*, **5**, 489–94 (1992).
14. JD Thompson, TJ Gibson, F Plewniak, F Jeanmougin and DG Higgins, *Nucleic Acids Res*, **25**, 4876–82 (1997).
15. S Malby, R Pickering, S Saha, R Smallridge, S Linse and AK Downing, *Biochemistry*, **40**, 2555–63 (2001).
16. P Whiteman, AK Downing, R Smallridge, PR Winship and PA Handford, *J Biol Chem*, **273**, 7807–13 (1998).
17. A Muranyi, BE Finn, GP Gippert, S Forsen, J Stenflo and T Drakenberg, *Biochemistry*, **37**, 10605–15 (1998).
18. B Bersch, JF Hernandez, D Marion and GJ Arlaud, *Biochemistry*, **37**, 1204–14 (1998).
19. M Husbyn and A Cuthbertson, *J Pept Res*, **60**, 121–27 (2002).
20. Y Wang and MW Spellman, *J Biol Chem*, **273**, 8112–18 (1998).
21. YH Kao, GF Lee, Y Wang, MA Starovasnik, RF Kelley, MW Spellman and L Lerner, *Biochemistry*, **38**, 7097–110 (1999).
22. P Whiteman, RS Smallridge, V Knott, JJ Cordle, AK Downing and PA Handford, *J Biol Chem*, **276**, 17156–62 (2001).
23. V Knott, AK Downing, CM Cardy and P Handford, *J Mol Biol*, **255**, 22–27 (1996).
24. S Kettle, X Yuan, G Grundy, V Knott, AK Downing and PA Handford, *J Mol Biol*, **285**, 1277–87 (1999).
25. ND Kurniawan, K Aliabadizadeh, IM Brereton, PA Kroon and R Smith, *J Mol Biol*, **311**, 341–56 (2001).
26. Y Stenberg, A Muranyi, C Steen, E Thulin, T Drakenberg and J Stenflo, *J Mol Biol*, **293**, 653–65 (1999).
27. NM Thielens, K Enrie, M Lacroix, M Jaquinod, JF Hernandez, AF Esser and GJ Arlaud, *J Biol Chem*, **274**, 9149–59 (1999).
28. KE Persson, J Astermark, I Bjork and J Stenflo, *FEBS Lett*, **421**, 100–4 (1998).
29. M Hyytiainen, J Taipale, CH Heldin and J Keski Oja, *J Biol Chem*, **273**, 20669–76 (1998).
30. G Kemball Cook, I Garner, Y Imanaka, T Nishimura, DP O'Brien, EG Tuddenham and JH McVey, *Gene*, **139**, 275–9 (1994).
31. D Zhong, MS Bajaj, AE Schmidt and SP Bajaj, *J Biol Chem*, **277**, 3622–31 (2002).
32. DP Reinhardt, RN Ono, H Notbohm, PK Muller, HP Bachinger and LY Sakai, *J Biol Chem*, **275**, 12339–45 (2000).
33. SB Yan, *J Mol Recognit*, **9**, 211–18 (1996).
34. GJ Arlaud, RB Sim, AM Duplaa and MG Colomb, *Mol Immunol*, **16**, 445–50 (1979).
35. E Persson, M Selander, S Linse, T Drakenberg, AK Ohlin and J Stenflo, *J Biol Chem*, **264**, 16897–904 (1989).
36. AK Ohlin, S Linse and J Stenflo, *J Biol Chem*, **263**, 7411–17 (1988).
37. J Astermark, I Bjork, AK Ohlin and J Stenflo, *J Biol Chem*, **266**, 2430–37 (1991).
38. DJ Moloney, LH Shair, FM Lu, J Xia, R Locke, KL Matta and RS Haltiwanger, *J Biol Chem*, **275**, 9604–11 (2000).
39. J Stenflo, A Lundwall and B Dahlback, *Proc Natl Acad Sci USA*, **84**, 368–72 (1987).
40. RJ Harris and MW Spellman, *Glycobiology*, **3**, 219–24 (1993).
41. RJ Harris, H van Halbeek, J Glushka, LJ Basa, VT Ling, KJ Smith and MW Spellman, *Biochemistry*, **32**, 6539–47 (1993).
42. H Nishimura, S Kawabata, W Kisiel, S Hase, T Ikenaka, T Takao, Y Shimonishi and S Iwanaga, *J Biol Chem*, **264**, 20320–25 (1989).
43. S Bjoern, DC Foster, L Thim, FC Wiberg, M Christensen, Y Komiyama, AH Pedersen and W Kisiel, *J Biol Chem*, **266**, 11051–57 (1991).
44. RS Haltiwanger and P Stanley, *Biochim Biophys Acta*, **1573**, 328–35 (2002).
45. L Thim, S Bjoern, M Christensen, EM Nicolaisen, T Lund Hansen, AH Pedersen and U Hedner, *Biochemistry*, **27**, 7785–93 (1988).
46. RW Glanville, RQ Qian, DW McClure and CL Maslen, *J Biol Chem*, **269**, 26630–34 (1994).
47. MS Sunnerhagen, E Persson, I Dahlqvist, T Drakenberg, J Stenflo, M Mayhew, M Robin, P Handford, JW Tilley, ID Campbell and GG Brownlee, *J Biol Chem*, **268**, 23339–44 (1993).
48. M Selander Sunnerhagen, M Ullner, E Persson, O Teleman, J Stenflo and T Drakenberg, *J Biol Chem*, **267**, 19642–49 (1992).
49. Z Rao, P Handford, M Mayhew, V Knott, GG Brownlee and D Stuart, *Cell*, **82**, 131–41 (1995).
50. DD Monkovic, WJ VanDusen, CJ Petroski, VM Garsky, MK Sardana, P Zavodszky, AM Stern and PA Friedman, *Biochem Biophys Res Commun*, **189**, 233–41 (1992).
51. JE Dinchuk, RJ Focht, JA Kelley, NL Henderson, NI Zolotarjova, R Wynn, NT Neff, J Link, RM Huber, TC Burn, MJ Rupar, MR Cunningham, BH Selling, J Ma, AA Stern, GF Hollis, RB Stein and PA Friedman, *J Biol Chem*, **277**, 12970–77 (2002).
52. P Maurer, E Hohenester and J Engel, *Curr Opin Cell Biol*, **8**, 609–17 (1996).
53. R Koradi, M Billeter and K Wuthrich, *J Mol Graph*, **14**, 29–32, 51–55 (1996).
54. AK Downing, V Knott, JM Werner, CM Cardy, ID Campbell and PA Handford, *Cell*, **85**, 597–605 (1996).

55. GT Montelione, K Wuthrich, AW Burgess, EC Nice, G Wagner, KD Gibson and HA Scheraga, *Biochemistry*, **31**, 236–49 (1992).
56. AK Downing, PA Handford and ID Campbell, in E Carafoli and J Krebs (eds.), *Calcium Homeostasis*, Vol. 3, Springer-Verlag, Heidelberg, pp 83–99 (2000).
57. E Zhang, R St Charles and A Tulinsky, *J Mol Biol*, **285**, 2089–104 (1999).
58. M Baron, DG Norman, TS Harvey, PA Handford, M Mayhew, AG Tse, GG Brownlee and ID Campbell, *Protein Sci*, **1**, 81–90 (1992).
59. M Ullner, M Selander, E Persson, J Stenflo, T Drakenberg and O Teleman, *Biochemistry*, **31**, 5974–83 (1992).
60. M Sunnerhagen, GA Olah, J Stenflo, S Forsen, T Drakenberg and J Trewhella, *Biochemistry*, **35**, 11547–59 (1996).
61. T Mather, V Oganessyan, P Hof, R Huber, S Foundling, C Esmon and W Bode, *EMBO J*, **15**, 6822–31 (1996).
62. P Fuentes Prior, Y Iwanaga, R Huber, R Pagila, G Rumennik, M Seto, J Morser, DR Light and W Bode, *Nature*, **404**, 518–25 (2000).
63. S Saha, J Boyd, JM Werner, V Knott, PA Handford, ID Campbell and AK Downing, *Structure (Camb)*, **9**, 451–56 (2001).
64. X Yuan, JM Werner, J Lack, V Knott, PA Handford, ID Campbell and AK Downing, *J Mol Biol*, **316**, 113–25 (2002).
65. JM Werner, V Knott, PA Handford, ID Campbell and AK Downing, *J Mol Biol*, **296**, 1065–78 (2000).
66. N Beglova, SC Blacklow, J Takagi and TA Springer, *Nat Struct Biol*, **9**, 282–87 (2002).
67. J Stetefeld, U Mayer, R Timpl and R Huber, *J Mol Biol*, **257**, 644–57 (1996).
68. WD Morgan, B Birdsall, TA Frenkiel, MG Gradwell, PA Burghaus, SE Syed, C Uthaipibull, AA Holder and J Feeney, *J Mol Biol*, **289**, 113–22 (1999).
69. BA Sampoli Benitez, MJ Hunter, DP Meininger and EA Komives, *J Mol Biol*, **273**, 913–26 (1997).
70. DR Light, CB Glaser, M Betts, E Blasko, E Campbell, JH Clarke, M McCaman, K McLean, M Nagashima, JF Parkinson, G Rumennik, T Young and J Morser, *Eur J Biochem*, **262**, 522–33 (1999).
71. B Dahlback, A Lundwall and J Stenflo, *Proc Natl Acad Sci USA*, **83**, 4199–203 (1986).
72. G Manfioletti, C Brancolini, G Avanzi and C Schneider, *Mol Cell Biol*, **13**, 4976–85 (1993).
73. R Brandenberger, A Schmidt, J Linton, D Wang, C Backus, S Denda, U Muller and LF Reichardt, *J Cell Biol*, **154**, 447–58 (2001).
74. C Valcarce, I Bjork and J Stenflo, *Eur J Biochem*, **260**, 200–207 (1999).
75. JF Hernandez, B Bersch, Y Petillot, J Gagnon and GJ Arlaud, *J Pept Res*, **49**, 221–31 (1997).
76. NM Thielens, CA Aude, MB Lacroix, J Gagnon and GJ Arlaud, *J Biol Chem*, **265**, 14469–75 (1990).
77. PA Handford, M Baron, M Mayhew, A Willis, T Beesley, GG Brownlee and ID Campbell, *EMBO J*, **9**, 475–80 (1990).
78. PE Hughes, G Morgan, EK Rooney, GG Brownlee and P Handford, *J Biol Chem*, **268**, 17727–33 (1993).
79. C Valcarce, M Selander Sunnerhagen, AM Tamlitz, T Drakenberg, I Bjork and J Stenflo, *J Biol Chem*, **268**, 26673–78 (1993).
80. Y Stenberg, K Julenius, I Dahlqvist, T Drakenberg and J Stenflo, *Eur J Biochem*, **248**, 163–70 (1997).
81. Y Stenberg, S Linse, T Drakenberg and J Stenflo, *J Biol Chem*, **272**, 23255–60 (1997).
82. MD Rand, A Lindblom, J Carlson, BO Villoutreix and J Stenflo, *Protein Sci*, **6**, 2059–71 (1997).
83. RS Smallridge, P Whiteman, K Doering, PA Handford and AK Downing, *J Mol Biol*, **286**, 661–68 (1999).
84. DP Reinhardt, DR Keene, GM Corson, E Poschl, HP Bachinger, JE Gambee and LY Sakai, *J Mol Biol*, **258**, 104–16 (1996).
85. P Handford, AK Downing, Z Rao, DR Hewett, BC Sykes and CM Kielty, *J Biol Chem*, **270**, 6751–56 (1995).
86. DP Reinhardt, DE Mechling, BA Boswell, DR Keene, LY Sakai and HP Bachinger, *J Biol Chem*, **272**, 7368–73 (1997).
87. TJ Wess, PP Purslow, MJ Sherratt, J Ashworth, CA Shuttleworth and CM Kielty, *J Cell Biol*, **141**, 829–37 (1998).
88. TA Eriksen, DM Wright, PP Purslow and VC Duance, *Proteins*, **45**, 90–95 (2001).
89. E Persson, OH Olsen, A Ostergaard and LS Nielsen, *J Biol Chem*, **272**, 19919–24 (1997).
90. M Osterlund, R Owenius, E Persson, M Lindgren, U Carlsson, PO Freskgard and M Svensson, *Eur J Biochem*, **267**, 6204–11 (2000).
91. CR Kelly, CD Dickinson and W Ruf, *J Biol Chem*, **272**, 17467–72 (1997).
92. HH Lin, M Stacey, C Saxby, V Knott, Y Chaudhry, D Evans, S Gordon, AJ McKnight, P Handford and S Lea, *J Biol Chem*, **276**, 24160–69 (2001).
93. HH Lin, M Stacey, J Hamann, S Gordon and AJ McKnight, *Genomics*, **67**, 188–200 (2000).
94. E Ruoslahti and MD Pierschbacher, *Science*, **238**, 491–97 (1987).
95. F Vella, JF Hernandez, A Molla, MR Block and GJ Arlaud, *J Pept Res*, **54**, 415–26 (1999).
96. AI den Hollander, JB ten Brink, YJ de Kok, S van Soest, LI van den Born, MA van Driel, DJ van de Pol, AM Payne, SS Bhattacharya, U Kellner, CB Hoyng, A Westerveld, HG Brunner, EM Bleeker Wagemakers, AF Deutman, JR Heckenlively, FP Cremers and AA Bergen, *Nat Genet*, **23**, 217–21 (1999).
97. AJ Lotery, SG Jacobson, GA Fishman, RG Weleber, AB Fulton, P Namperumalsamy, E Heon, AV Levin, S Grover, JR Rosenow, KK Kopp, VC Sheffield and EM Stone, *Arch Ophthalmol*, **119**, 415–20 (2001).
98. JH McVey, E Boswell, AD Mumford, G Kemball Cook and EG Tuddenham, *Hum Mutat*, **17**, 3–17 (2001).
99. F Giannelli, PM Green, SS Sommer, M Poon, M Ludwig, R Schwaab, PH Reitsma, M Goossens, A Yoshioka, MS Figueiredo and GG Brownlee, *Nucleic Acids Res*, **26**, 265–68 (1998).
100. F Peyvandi, M Menegatti, E Santagostino, S Akhavan, J Uprichard, DJ Perry, SJ Perkins and PM Mannucci, *Br J Haematol*, **117**, 685–92 (2002).
101. G Collod Beroud, C Beroud, L Ades, C Black, M Boxer, DJ Brock, KJ Holman, A de Paepe, U Francke, U Grau, C Hayward, HG Klein, W Liu, L Nuytinck, L Peltonen, AB Alvarez Perez, T Rantamaki, C Junien and C Boileau, *Nucleic Acids Res*, **26**, 229–33 (1998).
102. PA Gupta, EA Putnam, SG Carmical, I Kaitila, B Steinmann, A Child, C Danesino, K Metcalfe, SA Berry, E Chen, CV Delorme, MK Thong, LC Ades and DM Milewicz, *Hum Mutat*, **19**, 39–48 (2002).
103. EM Stone, AJ Lotery, FL Munier, E Heon, B Piguet, RH Guymer, K Vandenburgh, P Cousin, D Nishimura, RE Swiderski,

G Silvestri, DA Mackey, GS Hageman, AC Bird, VC Sheffield and DF Schorderet, *Nat Genet*, **22**, 199–202 (1999).

104 B Loeys, L Van Maldergem, G Mortier, P Coucke, S Gerniers, JM Naeyaert and A De Paepe, *Hum Mol Genet*, **11**, 2113–18 (2002).

105 C Crosnier, C Driancourt, N Raynaud, S Dhorne Pollet, N Pollet, O Bernard, M Hadchouel and M Meunier Rotival, *Gastroenterology*, **116**, 1141–48 (1999).

106 L Villeger, M Abifadel, D Allard, JP Rabes, R Thiart, MJ Kotze, C Beroud, C Junien, C Boileau and M Varret, *Hum Mutat*, **20**, 81–87 (2002).

107 KR Mathews and M Godfrey, *Am J Hum Genet*, **61**, 228 (1997).

108 A Joutel, K Vahedi, C Corpechot, A Troesch, H Chabriat, C Vayssiere, C Cruaud, J Maciazek, J Weissenbach, MG Bousser, JF Bach and E Tournier Lasserve, *Lancet*, **350**, 1511–15 (1997).

109 PH Reitsma, *Thromb Haemostasis*, **78**, 344–50 (1997).

110 D Borgel, S Gandrille and M Aiach, *Thromb Haemostasis*, **78**, 351–56 (1997).

111 H Bikker, T Vulsma, F Baas and JJ de Vijlder, *Hum Mutat*, **6**, 9–16 (1995).

112 M Krawczak and DN Cooper, *Trends Genet*, **13**, 121–22 (1997).

113 JH McVey, EJ Boswell, O Takamiya, G Tamagnini, V Valente, T Fidalgo, M Layton and EG Tuddenham, *Blood*, **92**, 920–26 (1998).

114 W Liu, C Qian, K Comeau, T Brenn, H Furthmayr and U Francke, *Hum Mol Genet*, **5**, 1581–87 (1996).

115 DM McCord, DM Monroe, KJ Smith and HR Roberts, *J Biol Chem*, **265**, 10250–54 (1990).

116 DP O'Brien, G Kemball Cook, AM Hutchinson, DM Martin, DJ Johnson, PG Byfield, O Takamiya, EG Tuddenham and JH McVey, *Biochemistry*, **33**, 14162–69 (1994).

117 OD Christophe, PJ Lenting, JA Kolkman, GG Brownlee and K Mertens, *J Biol Chem*, **273**, 222–27 (1998).

118 HH Hobbs, MS Brown and JL Goldstein, *Hum Mutat*, **1**, 445–66 (1992).

119 DP Reinhardt, RN Ono and LY Sakai, *J Biol Chem*, **272**, 1231–36 (1997).

120 T Sasaki, K Mann, G Murphy, ML Chu and R Timpl, *Eur J Biochem*, **240**, 427–34 (1996).

121 AJ McGettrick, V Knott, A Willis and PA Handford, *Hum Mol Genet*, **9**, 1987–94 (2000).

GLA-domains

Gla-domain

Mark A Brown and Johan Stenflo
Lund University, University Hospital, Malmö, Sweden

FUNCTIONAL CLASS

Metal binding domain found at the N-terminus of several vitamin K–dependent proteins.

DEFINITION

The term 'Gla-domain' is used to describe the γ-carboxyglutamic acid (Gla)-containing region of certain vitamin K–dependent proteins such as prothrombin and factor IX. The domain comprises the first 45 or so amino acids at the N-terminus of these proteins and contains 9–13 Gla residues that can bind divalent metal cations in a cooperative manner, the most physiologically important being Ca^{2+}. Binding of Ca^{2+} by the Gla residues induces a conformational transition in the domain that allows it to bind to phospholipid-containing membranes, a requisite step for the biological function of the proteins. Gla residues are also found in two proteins of bone and the extracellular matrix (bone Gla protein or osteocalcin, and matrix Gla protein), and in neurotoxic peptides synthesized by molluscs of the genus *Conus*, but these polypeptides are not considered to have a Gla-domain as such, and will not be discussed here.

OCCURRENCE

Gla was first identified in 1974,[1] and since then 12 proteins have been shown to contain Gla-domains (Figure 1). Factors VII, IX, and X, prothrombin, and proteins C, S, and

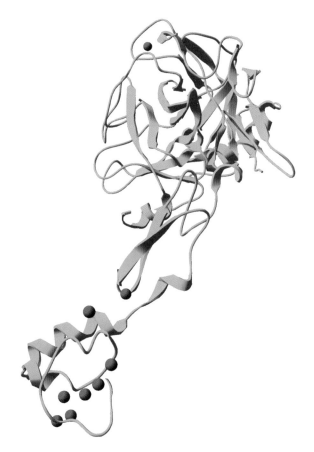

3D Structure Schematic representation of the structure of human factor VIIa. The Gla-domain with seven Ca^{2+} ions bound to it is visible at the bottom of the molecule, PDB code: 1DAN. Prepared with the program Swiss-PdbViewer 2.7 (available online at http://www.expasy.ch/spdbv/mainpage.htm) and rendered with QuickDraw™ 3D (Apple Computer, Inc.).

Gla-domain

	ω-loop	disulfide loop		hydrophobic stack region
	+1 10	20	30	40
Factor VII	ANA-FLγγLRPGSLγRγCKγγQCSFγγARγIFKDAγRTKLFWISYS
Factor IX	YNSGKLγγFVQGNLγRγCMγγKCSFγγARγVFγNTγRTTγFWKQYV
Factor X	ANS-FLγγMKKGHLγRγCMγγTCSYγγARγVFγDSDKTNγFWNKYK
Prothrombin	ANT-FLγγVRKGNLγRγCVγγTCSYγγAFγALγSSTATDVFWAKYT
Protein C	ANS-FLγγLRHSSLγRγCIγγICDFγγAKγIFQNVDDTLAFWSKHV
Protein S	ANS-LLγγTKQGNLγRγCIγγLCNKγγARγVFγNDPγTDYFYPKYL
Gas6	AFQ-VFγγAKQGHLγRγCVγγLCSRγγARγVFγNDPγTDYFYPRYL
Protein Z	AGSYLLγγLFγGNLγKγCYγγICVYγγARγVFγNγVVTDγFWRRYK
PRGP1	ANG-FFEEIRQGNIERECKEEFCTFEEAREAFENNEKTKEFWSTYT
PRGP2	ANHWDLELLTPGNLERECLEERCSWEEAREYFEDNTLTERFWESYI
TMG3	ANE-FLEELRQGTIERECMEEICSYEEVKEVFENKEKTMEFWKGYP
TMG4	LLYNRFDLELFTPGNLERECNEELCNYEEAREIFVDEDKTIAFWQEYS

Figure 1 The primary structure of the Gla-domains of human vitamin K–dependent proteins. Positions known to contain Gla residues are depicted as γ. The location of the Gla residues in the PRGP and TMG proteins is uncertain. Amino acids are numbered according to the sequence of mature factor VII. Residues identical in five or more of the sequences are boxed and shaded, and those that form the ω-loop, disulfide loop, and hydrophobic stack region are indicated.

Z are synthesized primarily in the liver, although synthesis of protein S and prothrombin also occurs in extrahepatic tissues.[2,3] Gas6, proline-rich Gla protein (PRGP) 1 and 2, and transmembrane Gla protein (TMG) 3 and 4 are synthesized in diverse tissues and cell types.[4,5] The proteins are expressed throughout fetal development, childhood, and adulthood.

Apart from the PRGP and TMG proteins, which are membrane-bound (with the Gla-domain probably extracellular), the proteins are secreted into the blood and extracellular fluids. Fragment 1 of prothrombin, an N-terminal fragment containing the Gla-domain, is found in considerable amounts in urine.[6] The venom of some snakes contains large amounts of an activated factor X-like protein with a Gla-domain.[7,8] Proteins containing a Gla-domain have been identified in many mammals, birds, reptiles, amphibians, and fish. All such proteins found to date appear to be homologues of the human vitamin K–dependent proteins. The Gla-domain appears early in the vertebrate lineage, being found in prothrombin from hagfish, but it has not been found in invertebrates.

BIOLOGICAL FUNCTION

The negatively charged dicarboxylate side chain of Gla allows the amino acid to chelate divalent cations. Physiologically the most important of these is Ca^{2+}. Whereas the affinity of a single Gla residue for Ca^{2+} is very low ($K_d \sim$ 30 mM for malonate), the presence of multiple Gla residues enables the Gla-domain to bind the metal at several sites with positive cooperativity.[9] Thus, the average K_d for Ca^{2+} binding at these sites is ~0.3 to 0.7 mM and the sites are presumed to be essentially saturated at the concentration of free Ca^{2+} in extracellular fluids, that is, ~1.2 mM.[10,11] Binding of Ca^{2+} to the Gla-domain is crucial for the biological function of vitamin K–dependent proteins because it induces a marked conformational rearrangement of the domain that allows it to bind to biological membranes, principally those upon which negatively charged phospholipids (especially phosphatidylserine) are exposed.[12–14] This greatly enhances the interaction of the proteins with their membrane-bound cofactors and substrates by concentrating the various components on a suitable biological surface where they are more likely to encounter one another than in solution. Naturally occurring point mutations of the Gla residues may severely diminish the biological activity of the affected protein. For example, factor IX Seattle 3 (Gla27Lys) and factor IX Oxford b2 (Gla7Ala) cause hemophilia B. However, systematic mutation of the Gla residues by recombinant methods has revealed that the importance of individual Gla residues is variable and while some substitutions may greatly impair Ca^{2+} binding and biological activity, others have no major effect.[15,16] This has been best examined for protein C, in which Gla residues at positions 7, 16, 20, and 26 were found to be indispensable for its biological activity, whereas those at positions 6, 14, and 19 were relatively unimportant.[15]

Ca^{2+}-induced folding appears to occur in two stages.[17,18] Binding of the first two or three Ca^{2+} ions causes a primary conformational transition that is considered to be nonspecific because it can be induced by various ions. This reorients and buries the 'hydrophobic stack' region of the Gla-domain (residues 40/41–45/46). A second transition, which is inducible specifically by Ca^{2+} (and to a lesser extent by Sr^{2+}) and required for full biological activity, occurs in the N-terminal part of the Gla-domain. The structural rearrangement sequesters several of the Gla residues and Ca^{2+} ions inside the domain and forces the side chains of hydrophobic residues located near the N-terminus into the solvent.[19–23] The Ca^{2+}-induced conformation is stabilized somewhat by interactions with the protein region adjacent to the Gla-domain[23–25] and may be modulated during membrane binding by the phospholipid composition of the membrane.[26]

Historically, the mode of interaction of the Gla-domain with cell membranes was believed to be mainly electrostatic in nature, with Ca^{2+} acting as a bridging ion between Gla residues and the negatively charged headgroups of phospholipids.[27] More recently, analyses of the three-dimensional structures of the Ca^{2+}-loaded and metal-free Gla-domain[19–23,28] and site-directed mutagenesis experiments[29,30] suggest a model in which hydrophobic residues located in the 'ω-loop' near the N-terminus (Phe4, Leu5, and Leu8 in human factor VII) are crucial for membrane interaction, and also for modulating the affinity of individual proteins for the membrane (affinities vary by more than 100-fold despite the highly conserved sequence of the Gla-domain).[31] This model also anticipates a role for electrostatic interactions in modulating membrane binding and affinity, although the nature of these interactions is poorly understood. An alternative model proposes that ion pairing occurs between a single phospholipid headgroup and metal ions contained within a pore in the Gla-domain.[31]

In addition to its role in membrane binding, the Gla-domain contributes to certain protein–protein interactions. Examples include the interactions between protein C and the endothelial cell protein C receptor,[32] factor IX and collagen IV,[33] and factor VIIa and its cofactor, the tissue factor.[34] In the latter case, mutagenesis experiments suggest that the Gla-domain of factor VIIa may modulate docking of the tissue factor during substrate binding.[35]

Most of the proteins that contain a Gla-domain play important roles in hemostasis, either as serine proteases or as cofactors. Gas6 is a ligand for the Axl/Tyro3 group of transmembrane receptors. Factors VII, IX, X, prothrombin, and Gas6 have procoagulant functions, whereas proteins C, S, and Z have anticoagulant roles. The function(s) of the PRGP and TMG proteins have yet to be defined. The presence of prothrombin fragment 1 in urine and renal calculi,[6] and the demonstrated synthesis of prothrombin by kidney cells,[3] suggests that its Gla-domain may also function as an inhibitor of urolithiasis.

BIOSYNTHESIS

The nascent vitamin K–dependent proteins are synthesized as single-chain precursors with an N-terminal signal peptide that directs their transport to the secretory pathway of the cell, and is cleaved off prior to the synthesis of Gla. Located between the signal peptide and the Gla-domain is a propeptide comprising 18 to 28 amino acids that marks the proteins as substrates for γ-glutamyl carboxylase, a resident enzyme of the endoplasmic reticulum.[36] PRGP1 and TMG3 are exceptions in that they lack a signal peptide preceding the propeptide.[4,5] The propeptide not only mediates binding of the substrate to the γ-carboxylase but also directly activates the enzyme. It is removed by proteolytic cleavage in the Golgi apparatus prior to secretion of the proteins.[37] Mutations that disrupt the propeptide cleavage site lead to the secretion of molecules that retain the propeptide or have one or more extra amino acids at the N-terminus. This prevents the Gla-domain from folding into a membrane-binding conformation and such proteins are biologically inactive. Depending on the protein affected, this can result in either a bleeding tendency or a hypercoagulable state.[38,39]

The biosynthesis of Gla is coupled to a vitamin K redox cycle[40] (Figure 2). In a reaction that requires CO_2, O_2, and the reduced (dihydroquinone) form of vitamin K as cofactors, the γ-carboxylase replaces a proton on the γ-carbon

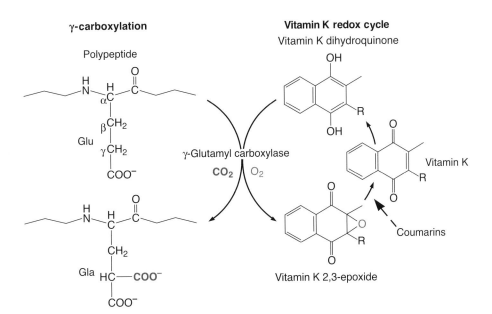

Figure 2 Biosynthesis of Gla is coupled to a vitamin K redox cycle. A CO_2 molecule is added to the side chain of specific glutamyl residues within the Gla-domain by γ-glutamyl carboxylase. During the reaction, vitamin K is converted to an epoxide, which is recycled to an active dihydroquinone by a two-step reductive process that is sensitive to coumarin compounds such as warfarin.

Gla-domain

of a glutamic acid residue with a CO_2 molecule. Only the 9–13 glutamic acid residues located within the Gla-domain serve as targets for the carboxylase. Although the mechanism for proton abstraction is unclear, it is assumed that the carboxylase converts vitamin K dihydroquinone to a peroxide intermediate that reacts further to form a basic dialkoxide capable of abstracting the proton.[41] The dialkoxide then collapses to a 2,3-epoxide that is reduced back to the dihydroquinone in two steps. Vitamin K antagonists such as warfarin, which is commonly used in anticoagulant therapy, inhibit the enzyme that catalyzes the first reductive step (vitamin K epoxide reductase). As a consequence, partially and noncarboxylated forms of the vitamin K–dependent proteins accumulate in the rough endoplasmic reticulum and their plasma concentrations decrease.[42,43] Moreover, many of the circulating molecules are only partially γ-carboxylated and therefore biologically inactive. Because humans cannot synthesize vitamin K, dietary deficiency of the vitamin produces a similar effect.

PRIMARY STRUCTURE

The Gla-domains of vitamin K–dependent proteins from vertebrates exhibit a highly conserved primary structure and intron/exon organization, indicating that they were recruited from a common ancestral sequence by gene duplication.[44] The Gla-domain of the mature proteins is a single functional unit encoded by two exons. One exon encodes the propeptide and the first 37–40 residues of the Gla-domain, whereas the other encodes the 8 or so C-terminal residues containing the hydrophobic stack region. The 2 exons are joined by a type 0-splice junction, whereas their outer splice junctions are both of type 1. Thus, a genetic unit encoding three functionally linked peptides (the propeptide, and the Gla-containing and hydrophobic stack regions of the Gla-domain) appears to have been shuffled among the vitamin K–dependent proteins.[43] Three tandem pairs of Gla residues are conserved, as are the Gla residues corresponding to positions 14, 16, and 29 in factor VII (Figure 1). Two cysteinyl residues are conserved among the proteins and across species and the disulfide bond formed between them is a crucial structural element of the Gla-domain. Mutations that disrupt this disulfide bridge severely affect the function of the proteins.[45]

AMINO ACID SEQUENCE INFORMATION

- *Homo sapiens* factor II (prothrombin), precursor, 622 amino acid residues (AA), translation of cDNA sequence,[46] mature 579 AA, signal peptide/propeptide, 43 AA, Gla-domain, 45 AA (+1–45).
- *Homo sapiens* (human) factor VII, precursor, 466 AA, translation of cDNA sequence,[47] mature, 406 AA, signal peptide/propeptide, 60 AA, Gla-domain, 45 AA (+1–45).
- *Homo sapiens* factor IX, precursor, 454–461 AA, translation of cDNA sequence,[48] mature, 415 AA, signal peptide, 21–28 AA, propeptide, 18 AA, Gla-domain, 46 AA (+1–46).
- *Homo sapiens* factor X, precursor, 488 AA, translation of genomic DNA sequence,[49] mature, 445 AA (light chain, 139 AA, heavy chain, 306 AA), signal peptide/propeptide, 40 AA, Gla-domain, 45 AA (+1–45 of light chain).
- *Homo sapiens* protein C, precursor, 461 AA, translation of cDNA sequence,[50] mature, 417 AA (light chain, 155 AA, heavy chain, 262 AA), signal peptide, 18 AA, propeptide, 24 AA, Gla-domain, 45 AA (+1–45 of light chain).
- *Homo sapiens* protein S, precursor, 676 AA, translation of cDNA sequence,[51] mature, 635 AA, signal peptide/propeptide, 41 AA, Gla-domain, 45 AA (+1–45).
- *Homo sapiens* protein Z, precursor, 400 AA, translation of cDNA sequence,[52] mature, 360 AA, signal peptide/propeptide, 40 AA, Gla-domain, 46 AA (+1–46).
- *Homo sapiens* Gas6, precursor, 678 AA, translation of cDNA sequence,[53] mature, 630 AA, signal peptide/propeptide, 48 AA, Gla-domain, 45 AA, (+1–45).
- *Homo sapiens* PRGP1, precursor, 218 AA, translation of cDNA sequence,[4] mature (predicted), 198 AA, signal peptide (none predicted), propeptide (predicted), 20 AA, Gla-domain, 45 AA, (+1–45).
- *Homo sapiens* PRGP2, precursor, 202 AA, translation of cDNA sequence,[4] mature (predicted), 153 AA, signal peptide/propeptide (predicted), 49 AA, Gla-domain, 46 AA, (+1–46).
- *Homo sapiens* TMG3, precursor, 231 AA, translation of cDNA sequence,[5] mature (predicted), 212 AA, signal peptide (none predicted), propeptide (predicted), 19 AA, Gla-domain, 45 AA, (+1–45).
- *Homo sapiens* TMG4, precursor, 226 AA, translation of cDNA sequence,[5] mature (predicted), 177 AA, signal peptide/propeptide (predicted), 49 AA, Gla-domain, 48 AA, (+1–48).

The sequences of homologues of several of the above proteins have been determined for other vertebrates. These are available from online databases such as GenBank and SWISS-PROT.

PROTEIN PRODUCTION, PURIFICATION, AND MOLECULAR CHARACTERIZATION

Methods to purify Gla-domain-containing proteins from blood plasma have traditionally incorporated an initial precipitation step in which the Gla-domain is adsorbed to an insoluble barium salt, such as barium citrate (formed by the addition of barium chloride to citrate-containing plasma). This is often followed by ammonium sulfate precipitation and anion-exchange chromatography. To obtain a homogeneous preparation, chromatography on heparin/benzamidine-conjugated resins is usually necessary.[54,55] Immunoaffinity chromatography with polyclonal or monoclonal antibodies raised against the protein of interest has also been utilized with success.[56] Recently, monoclonal antibodies have been developed that bind specifically to Gla residues and are therefore useful for purifying a range of Gla-containing proteins.[8] Several of the above antibodies are available from commercial suppliers. Recombinant Gla-domain-containing proteins have been expressed in mammalian cell lines[57–59] and in the milk of transgenic animals.[60] Under these conditions, some vitamin K–dependent proteins are subject to inefficient γ-carboxylation and undergo aberrant proteolytic processing.[59–61] Immunoaffinity chromatography using conformation-dependent antibodies[56] or chromatography on hydroxyapatite[16,61] can resolve the fully γ-carboxylated protein from noncarboxylated molecules. In the case of factor X, substitution of the native propeptide with that from prothrombin improves the extent of γ-carboxylation[61] and the efficiency of propeptide removal can be increased by site-directed mutagenesis of the cleavage site.[59]

The Gla-domain can be liberated from the parent molecule by treatment with chymotrypsin, and purified by chromatographic methods.[24,25] Alternatively, it can be chemically synthesized in large quantities, which permits selective isotopic labeling for NMR studies. The synthetic Gla-domain has also been ligated to other synthetic protein domains.[62]

A wide variety of methods has been developed for the qualitative or quantitative identification of Gla in proteins, including specific stains, isotopic labeling, mass spectrometry, and immunochemical methods.[11] Many of these methods are limited by a requirement for purified proteins or highly enriched preparations. Gla residues are not detected during routine protein sequence analyses that employ Edman degradation chemistry. Consequently, methyl esterification of the Gla residues prior to sequencing is commonly employed to allow their identification.[63] Amino acid analysis of an alkaline hydrolysate may be used for quantitative analysis of the Gla content of a protein.[64]

METAL CONTENT

The molar content and composition of metals bound to a Gla-domain *in vivo* is unknown. Studies have therefore investigated binding of various metals *in vitro*. A wide variety of divalent and trivalent metal ions may be bound by the Gla-domain but except for Ca^{2+} and Mg^{2+}, binding of these ions has not been shown to have physiological relevance. In all cases, Ca^{2+} binding is a requisite for full biological activity, and in the case of factor IX, binding of Mg^{2+} ions at sites distinct from those occupied by Ca^{2+} appears to induce an additional conformational transition in the Gla-domain and augments the activity of the protein.[65] Measurements of the number of metal ions bound by a Gla-domain, most often performed by dialysis methods, have given quite variable results.[66] Buffer composition, ionic strength, pH, and protein concentration-dependent dimerization have pronounced effects on these measurements.

In the crystal structure of Ca^{2+}-loaded bovine prothrombin fragment 1, seven Ca^{2+} ions are bound to the Gla-domain.[19] This number is consistent with the results of some equilibrium dialysis experiments, which indicate that 5–7 Ca^{2+} ions are bound per molecule,[67,68] although it should be noted that in this fragment, one Ca^{2+}-binding site resides outside the Gla-domain. Binding at three or four of the Gla-domain sites is cooperative, whereas the others are noninteracting. In the Sr^{2+}–loaded crystal structure, eight Sr^{2+} ions are ligated to the Gla-domain,[69] which concurs with the nine sites indicated by equilibrium dialysis assays.[70]

Seven Ca^{2+} ions are bound to the Gla-domain in the crystal structure of the factor VIIa–tissue factor complex,[34] while membrane filtration experiments and subsequent studies have suggested that only around four Ca^{2+}-binding sites reside in the Gla-domain.[71] Likewise, seven Ca^{2+} ions are bound to the Gla-domain in the crystal structures of the bovine factor X Gla-domain (bound to an inhibitor from snake venom)[72] and the human protein C Gla-domain (in a complex with the endothelial cell protein C receptor).[73] Analysis of the NMR structure of the human factor IX Gla-domain suggests that binding of up to nine Ca^{2+} ions could be accommodated.[74]

ASSAYS OF METAL BINDING

The structural transition of the Gla-domain induced by metal ion binding can be monitored by many methods, including rocket immunoelectrophoresis,[75] agarose gel electrophoresis,[76] circular dichroism,[25,77,78] intrinsic fluorescence assays,[24–27,68,78,79] electron spin resonance (ESR) spectroscopy,[80] NMR spectroscopy,[78,81] differential scanning calorimetry,[25,82] Fourier-transform infrared spectroscopy,[83] and immunoassays using conformation-specific antibodies.[56] Metal ion binding has also been studied directly by dialysis methods[9,67,68] and Ca^{2+}-selective electrode titrations.[84]

Most commonly, fluorescence methods have been employed. The metal-induced conformational transition alters the interaction of the disulfide loop region with the hydrophobic stack region. Consequently, the indole ring of Trp41/42 becomes buried and reoriented adjacent to the disulfide loop and the resulting disulfide–π electron interaction causes quenching of its intrinsic fluorescence.[19,21,26,79]

Gla-domain

ASSAYS OF MEMBRANE BINDING

A variety of methods have been employed to study the dynamics of the interaction between the Gla-domain and phospholipid-containing membranes. These include light-scattering assays,[68,85] differential scanning calorimetry,[82] Fourier-transform infrared spectroscopy,[83] and fluorescence energy transfer measurements.[30,86] In general, membrane binding by the isolated Gla-domain exhibits similar behavior to that of the domain in the parent protein, although this is not observed in all cases.[84]

SPECTROSCOPY

The structural features and arrangement of the metal binding sites in the Gla-domain have been investigated by X-ray crystallographic,[19,34,69,72,73,87] solution NMR,[20–23,28] and molecular dynamics modeling methods.[74,88,89] Comparison of the NMR and crystal structures of the metal-free and metal-bound Gla-domain of various proteins has revealed the same global fold and structural features and indicated that the conformational transition induced by Ca^{2+} binding is a conserved process, even for domains with somewhat different primary structures.

X-RAY STRUCTURES OF THE METAL-BOUND Gla-DOMAIN

Three-dimensional structures of the metal-free forms of bovine prothrombin fragment 1 (residues 1–156)[90–92] and porcine factor IX[93] have been determined by X-ray crystallography (highest resolution, 2.25 Å). Most of the apo-Gla-domain is disordered, implying a considerable degree of flexibility in the polypeptide backbone. The only defined structural element is an α-helix formed by residues 36–47, which encompasses the hydrophobic stack region. The side chains of hydrophobic residues in this region (Phe41, Trp42, and Tyr45) are positioned in a stacked arrangement and, notably, are solvent-exposed in the apo structures.

By contrast, Ca^{2+}-loaded bovine prothrombin fragment 1 produces a highly ordered crystal structure that has been solved at a resolution of 2.2 Å.[19] Seven Ca^{2+} ions are bound to the Gla-domain, almost exclusively by the side chains of Gla residues, and several are buried away from the solvent (Figure 3). With the exception of Gla33, which is not observed in the structure, all of the Gla residues point toward the interior of the domain. Three α-helices are present (9 turns in total), comprising residues 14–17, 25–31, and 36–47, and a tight loop is observed between the disulfide-bonded residues, Cys18 and Cys23. In contrast to the metal-free structure, the side chains of Phe41, Trp42, and Tyr45 are stacked inside the domain adjacent to the disulfide loop, and Tyr45 is hydrogen bonded to Cys18. This arrangement explains the quenching of intrinsic tryptophan fluorescence that occurs upon metal binding. The N-terminal region (residues 1–12) is folded to form the so-called ω-loop, so that Ala1 is completely buried and the side chains of Phe5, Leu6, and Val9 project into the solvent. The α-amino group of Ala1 forms hydrogen-bonding interactions with the carboxyl groups of Gla residues 17, 21, and 27, and with Pro22 in the disulfide loop. It may also

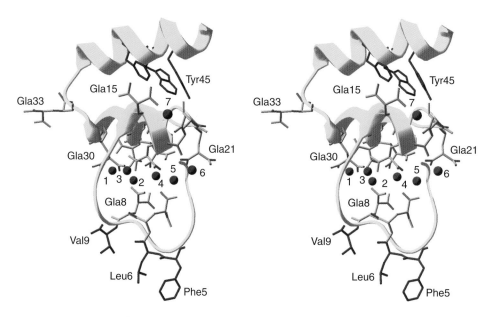

Figure 3 Stereo view diagram of the Ca^{2+}-loaded Gla-domain of bovine prothrombin. The seven bound Ca^{2+} ions are depicted as red spheres (numbered) and the side chains of the Gla residues are shown in light blue, some of which have been labeled. The C-terminal α-helix and the side chains of aromatic residues in the hydrophobic stack region (Phe41, Trp42, and Tyr45) are visible at the top of the figure. The exposed hydrophobic side chains of Phe5, Leu6, and Val9 in the ω-loop are proposed to constitute a membrane binding site. The model was drawn from the coordinates in PDB file 2 PF2.[19]

interact with Ca4 and Ca5. These interactions appear to preserve the ω-loop in the conformation required for membrane binding. Thus, Ca^{2+} binding to the Gla-domain seems to serve a primarily structural purpose, inducing folding of the domain, connecting Gla residues that are remote in the primary structure, and providing the energy required to force hydrophobic side chains into the solvent.

Pro22 is in the *trans* conformation in the crystal structure. However, it has been reported that this proline undergoes a *trans* to *cis* isomerization in the presence of Ca^{2+} ions, which has been postulated to be important for membrane binding.[94] Recent molecular dynamics simulations indicate that isomerization of Pro22 would give rise to only minimal structural changes.[88] Also, prothrombin from several other species, including humans, does not contain proline at this position.

All of the Gla residues observed in the crystal structure (Gla33 was not resolved) are involved in binding Ca^{2+} ions. The seven Ca^{2+} ions are coordinated by the carboxylate oxygen atoms of the Gla residues and by water molecules in a complex network of distorted polyhedral arrangements that somewhat resemble Ca^{2+}-malonate complexes[19,95] (Figure 4). The O^δ atom of Asn2 is also a ligand for Ca4. A cluster of six Ca^{2+} ions is seen with Ca1–Ca5 bound in an approximately linear array within the folded domain. Ca1 bridges Gla26 and Gla30, whereas Ca^{2+} ions 2–5 each contact three or more Gla residues. Interion distances in the linear array are: Ca1–Ca2, 4.4 Å; Ca2–Ca3, 4.0 Å; Ca3–Ca4, 3.8 Å; Ca4–Ca5, 3.8 Å. Ca6, which bridges Gla20 and Gla21, is separated from Ca5 of the linear array by 5.4 Å. Ca7 bridges Gla15 and Gla20 and is separated from Ca6 by a distance of 8.5 Å. Three of the Ca^{2+} ions (Ca1, Ca6, and Ca7) are partially exposed to the solvent. Details of the Ca^{2+}–oxygen network observed in the crystal structure are given in Table 1.

A crystal structure of strontium-bound bovine prothrombin fragment 1 has been determined at a resolution of 2.5 Å.[69] The conformation of the Gla-domain induced by this metal strongly resembles the Ca^{2+}-loaded structure, which may explain the ability of Sr^{2+} to partially support the biological activity of prothrombin in assays *in vitro*.[13] A total of eight Sr^{2+} ions are bound to the Gla-domain. Five are buried within the domain in an approximately linear array similar to that observed for Ca^{2+}, with interion distances of around 4.0 Å; the other three are solvent-exposed. One is bound near Gla30 at a site not occupied in the Ca^{2+}-prothrombin fragment 1 crystal structure. Binding of Sr^{2+} shows similarities to that observed with Ca^{2+}, but differs in lacking coordination with H_2O molecules (Table 1).

Crystal structures of the Gla-domain in complex with other proteins have also been determined. The 2.0-Å crystal structure of human factor VIIa in complex with the tissue factor indicates that the Ca^{2+}-loaded Gla-domain of factor VIIa has a very similar structure to that of

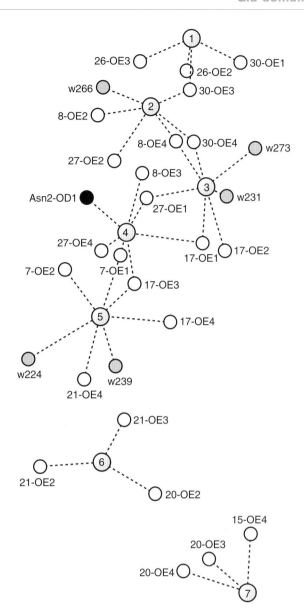

Figure 4 The Ca^{2+} ion–O atom network in the Gla-domain of bovine prothrombin. Ligands constitute O^ε atoms contributed by the carboxyl groups of Gla residues (empty circles) and H_2O molecules (cross-hatched circles), and the O^δ atom of Asn2 (filled circle). The figure was drawn from the coordinates in PDB file 2PF2.[19]

Ca^{2+}-prothrombin fragment 1.[34] Seven Ca^{2+} ions are bound to the Gla-domain, six of them in a linear array, and within the ω-loop the side chains of Phe4, Leu5, and Leu8 protrude into the solvent. Notably, the Gla-domain forms several contacts with the tissue factor over an interface of some 400 Å^2, most of them of a hydrophobic nature and involving the hydrophobic stack region.

A crystal structure of the Gla-domain of human protein C bound to the endothelial cell protein C receptor (EPCR) has been solved at a resolution of 1.55 Å.[73] Formation of this complex is Ca^{2+}-dependent. Electron density was observed only for residues 1–33 of the Gla-domain but

Gla-domain

Table 1 Coordination of Ca^{2+} and Sr^{2+} by the Gla-domain of bovine prothrombin fragment 1

Ion No.	Coord. No. Ca^{2+} [Sr^{2+}]	Type of linkage[a] Ca^{2+} [Sr^{2+}]	Ligand/O atom	d(Å) Ca^{2+} [Sr^{2+}]
1	4 [4]	mal_2 [mal_2]	Gla26 OE2	2.4 [2.7]
			Gla26 OE3	2.5 [2.7]
			Gla30 OE1	2.4 [2.2]
			Gla30 OE3	2.5[b] [2.6][b]
2	6 [4]	uni/bi/mal/H_2O [uni_2/bi]	Gla8 OE2	2.7[b] [3.5]
			Gla8 OE4	2.7 [–]
			Gla27 OE2	2.8 [2.9]
			Gla30 OE3	2.4[b] [2.6][b]
			Gla30 OE4	2.5[b] [2.7][b]
			O_{water}266	2.7 [–]
3	7 [4]	uni_3/bi/$(H_2O)_2$ [bi/mal]	Gla8 OE4	2.4[b] [–]
			Gla17 OE1	2.5[b] [2.7][b]
			Gla17 OE2	2.7 [3.2]
			Gla27 OE1	2.7[b] [3.0][b]
			Gla30 OE4	2.6[b] [2.8][b]
			O_{water}231	2.4 [–]
			O_{water}273	2.4 [–]
4	7 [6]	uni_3/mal_2/O^δ 1 [uni/mal_2]	Gla7 OE1	2.8[b] [3.2][b]
			Gla8 OE3	2.7 [–]
			Gla17 OE1	3.1[b] [3.3][b]
			Gla17 OE3	2.0[b] [2.3][b]
			Gla27 OE1	2.6[b] [2.7][b]
			Gla27 OE4	2.6 [2.8]
			Asn2 OD1	2.1 [3.0]
5	7 [6]	uni/bi_2/$(H_2O)_2$ [bi_3]	Gla7 OE1	2.5[b] [2.7][b]
			Gla7 OE2	2.5 [2.9]
			Gla17 OE3	2.5[b] [2.6][b]
			Gla17 OE4	2.9 [3.4]
			Gla21 OE3	– [3.2][b]
			Gla21 OE4	2.8 [2.7]
			O_{water}224	3.2 [–]
			O_{water}239	2.1 [–]
6	3 [3]	uni/mal [uni/mal]	Gla20 OE2	2.6 [2.9]
			Gla21 OE1	– [2.9]
			Gla21 OE2	2.7 [–]
			Gla21 OE3	2.4 [2.3]
7	3 [3]	uni/bi [uni/bi]	Gla15 OE4	2.8 [2.6]
			Gla20 OE3	2.4 [2.7]
			Gla20 OE4	2.8 [2.4]
8	– [3]	–[uni/CO/H_2O]	Gla30 O	– [3.1]
			Gla30 OE2	– [3.0]
			O_{water}199	– [2.2]
Average				2.6 [2.8]

Source: Data taken from M Soriano-Garcia, K Padmanabhan, AM de Vos and A Tulinsky, *Biochemistry*, **31**, 2554–66 (1992),[19] TP Seshadri, E Skrzypczak-Jankun, M Yin and A Tulinsky, *Biochemistry*, **33**, 1087–92 (1994),[69] and PDB files 2PF2 and 2SPT.

[a] Uni, bi, and mal denote unidentate, bidentate, and malonate-type linkages, respectively.
[b] Atom participates in an μ-oxo bond between metal ions.

folding in this region, and the coordination of seven Ca^{2+} ions, is akin to that of other Gla-domain–Ca^{2+} structures. The most important interactions with EPCR appear to be hydrophobic bonds involving Phe4 and Leu8 in the ω-loop, indicating that in protein C, this loop has an extra function in addition to its role in membrane binding.

A 2.3-Å-resolution crystal structure of the Gla-domain of bovine factor X bound to a protein isolated from snake venom has been reported.[72] The snake toxin (factor IX/X binding protein) inhibits factor Xa by binding and masking the Gla-domain, thereby precluding its interaction with biological membranes. Seven Ca^{2+} ions are bound to the

Gla-domain and its structure is much like that of the other Gla-domain–Ca^{2+} crystal structures. A Ca^{2+}-binding site not previously observed is formed by Gla35 and Gla39 in the C-terminal helix region and involves four Ca–oxygen interactions. One or both of the Gla residues forming this site are missing from most of the vitamin K–dependent proteins, but it is present in factor IX (Figure 1).

NMR STRUCTURES OF THE METAL-FREE AND METAL-BOUND Gla-DOMAIN

Solution NMR structures of the metal-free Gla-domain have been determined for human factor IX[20] and human factor X.[22,23] In the case of factor IX, a synthetic peptide (residues 1–47) was used, whereas for factor X, a fragment comprising the Gla- and adjacent EGF1-domains was employed. An NMR structure of the Ca^{2+}-loaded factor IX Gla-domain (residues 1–47) has also been determined,[21] although it should be noted that denaturants (urea and guanidine) were required to maintain the solubility of the Ca^{2+}–peptide complex.

The isolated Gla-domain is subject to a considerable degree of mobility in the absence of Ca^{2+}, but some defined structural elements are present, which are connected by a flexible polypeptide backbone. A loop near the N-terminus (residues 6–9), the disulfide region loop (residues 18–23), and a C-terminal α-helix (residues 35–46) are observed in the structure of the apo-factor IX Gla-domain.[20] Corresponding elements are seen in the apo-factor X Gla–EGF1 structure, as well as two additional, well-defined α-helices (residues 13–18 and 24–30).[22,23] In fact, when attached to the adjacent (Ca^{2+}-loaded) EGF1-domain, the apo-Gla-domain of factor X displays a similar global fold and most of the structural elements observed for the Ca^{2+}-loaded Gla-domain, although the structure is more mobile (Figure 5). This indicates that EGF1 exerts a stabilizing effect on the factor X Gla-domain. Moreover, Ca^{2+}-binding to EGF1 significantly affects the conformation and orientation of the apo-Gla-domain, especially in the C-terminal stack region, and appears to lock the hinge region between the Gla- and EGF1-domains in a more rigid conformation.[23] A similar influence of the EGF1-domain on the Gla-domain of factor VII is predicted on the basis of computer-modeling experiments.[89] When compared to the Ca^{2+}-loaded crystal structures, the most notable features of the metal-free Gla-domain are that the side chains of the Gla residues face the solvent, those of the hydrophobic residues in the ω-loop (Phe4, Leu5, and Val8 in factor X) are clustered and face the interior of the structure, and the N-terminus is mobile and does not form contacts with other parts of the Gla-domain.

In the presence of Ca^{2+}, the NMR structure of the factor IX Gla-domain largely resembles the crystal structure of

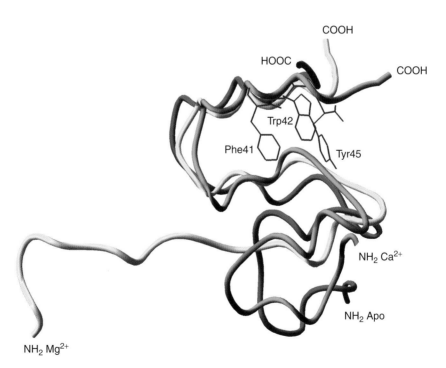

Figure 5 Solution NMR structures of the Gla-domain. Ribbon diagrams of the Ca^{2+}-loaded[21] (green) and Mg^{2+}-loaded[28] (yellow) Gla-domain of human factor IX (residues 1–47) have been superimposed on the apo form of the Gla–EGF1 fragment of human factor X[22] (red; only the Gla-domain is shown). In the absence of Ca^{2+}, residues 1–11 remain mobile and do not assume the ω-loop required for membrane binding. The side chains of aromatic residues in the hydrophobic stack region are displayed for the Ca^{2+}-loaded structure. For the factor IX models, the energy-minimized NMR structures in PDB files 1MGX and 1CF1 were used, and for the factor X model, the first structure contained in PDB file 1WHE.

the Ca^{2+}-loaded prothrombin Gla-domain.[19,28] The Ca^{2+}-loaded domain adopts a compact tertiary structure with significantly higher α-helical content than in its absence (four to five additional helical turns) and the interhelical regions form structured loops and turns not observed in the absence of the metal ion (Figure 5).[28,74] The C-terminal three-quarters of the Ca^{2+}-loaded domain comprises a globular core with α-helices formed by residues 14–17, 25–32, and 35–46. The side chains of Phe41, Trp42, and Tyr45 within the hydrophobic stack region are oriented toward the interior in a stacked arrangement and interact with the disulfide loop region, an interaction not observed in the metal-free NMR structures. N-terminal residues 1–12 form the ω-loop, which is positioned adjacent to the globular core, as in the prothrombin Gla-domain crystal structure. This loop, absent in the metal-free form, is closed by contacts formed between Tyr1 and residues 21–24. In direct contrast to the metal-free structures, the side chains of most of the Gla residues are oriented inwards, except for those of the three C-terminal Gla residues.

Binding of Mg^{2+} and certain other metal ions induces a stable conformational transition in the Gla-domain. However, only Ca^{2+} (and to a lesser extent Sr^{2+}) can produce a conformer capable of binding effectively to phospholipid membranes.[13] The structural basis for these differences was defined by the determination of the solution NMR structure of the Mg^{2+}-loaded Gla-domain of factor IX.[28] Comparison to the Ca^{2+}-loaded NMR structure[21] revealed that Ca^{2+} and Mg^{2+} elicit a similar conformational change in much of the Gla-domain (residues 14–21 and 24–47) but residues 1–11 remain mobile in the presence of Mg^{2+} and do not assume the ω-loop (Figure 5). Thus, a properly structured ω-loop, with its hydrophobic side chains oriented outward (residues Leu6, Phe9, and Val10 in factor IX), seems to constitute at least part of the interface required for binding to phospholipid membranes.

REFERENCES

1. J Stenflo, P Fernlund, W Egan and P Roepstorff, *Proc Natl Acad Sci USA*, **71**, 2730–33 (1974).
2. LM Stenberg, E Nilsson, O Ljungberg, J Stenflo and MA Brown, *Biochem Biophys Res Commun*, **283**, 454–59 (2001).
3. LM Stenberg, MA Brown, E Nilsson, O Ljungberg and J Stenflo, *Biochem Biophys Res Commun*, **280**, 1036–41 (2001).
4. JD Kulman, JE Harris, BA Haldeman and EW Davie, *Proc Natl Acad Sci USA*, **94**, 9058–62 (1997).
5. JD Kulman, JE Harris, L Xie and EW Davie, *Proc Natl Acad Sci USA*, **98**, 1370–75 (2001).
6. AM Stapleton, CJ Dawson, PK Grover, A Hohmann, R Comacchio, V Boswarva, Y Tang and RL Ryall, *Kidney Int*, **49**, 880–8 (1996).
7. JS Joseph, MCM Chung, K Jeyaseelan and RM Kini, *Blood*, **94**, 621–31 (1999).
8. MA Brown, LM Stenberg, U Persson and J Stenflo, *J Biol Chem*, **275**, 19795–802 (2000).
9. J Stenflo and P-O Ganrot, *Biochem Biophys Res Commun*, **50**, 98–104 (1973).
10. J Stenflo, *Crit Rev Eukaryot Gene Expr*, **9**, 59–88 (1999).
11. LM Stenberg, MA Brown and J Stenflo, in T Creighton (ed.), *Encyclopedia of Molecular Medicine*, Vol. 2, John Wiley & Sons, New York, pp 1367–70 (2001).
12. SN Gitel, WG Owen, CT Esmon and CM Jackson, *Proc Natl Acad Sci USA*, **70**, 1344–48 (1973).
13. GL Nelsestuen, M Broderius and G Martin, *J Biol Chem*, **251**, 6886–93 (1976).
14. J Stenflo and JW Suttie, *Annu Rev Biochem*, **46**, 157–72 (1977).
15. L Zhang, A Jhingan and FJ Castellino, *Blood*, **80**, 942–52 (1992).
16. PJ Larson, RM Camire, D Wong, NC Fasano, DM Monroe, PB Tracy and KA High, *Biochemistry*, **37**, 5029–38 (1998).
17. HA Liebman, SA Limentani, BC Furie and B Furie, *Proc Natl Acad Sci USA*, **82**, 3879–83 (1985).
18. HA Liebman, *Eur J Biochem*, **212**, 339–45 (1993).
19. M Soriano-Garcia, K Padmanabhan, AM de Vos and A Tulinsky, *Biochemistry*, **31**, 2554–66 (1992).
20. SJ Freedman, BC Furie, B Furie and JD Baleja, *J Biol Chem*, **270**, 7980–7 (1995).
21. SJ Freedman, BC Furie, B Furie and JD Baleja, *Biochemistry*, **34**, 12126–37 (1995).
22. M Sunnerhagen, S Forsén, AM Hoffrén, T Drakenberg, O Teleman and J Stenflo, *Nat Struct Biol*, **2**, 504–9 (1995).
23. M Sunnerhagen, GA Olah, J Stenflo, S Forsén, T Drakenberg and J Trewhella, *Biochemistry*, **35**, 11547–59 (1996).
24. E Persson, I Björk and J Stenflo, *J Biol Chem*, **266**, 2444–52 (1991).
25. A Vysotchin, LV Medved and KC Ingham, *J Biol Chem*, **268**, 8436–46 (1993).
26. M Hof, *Biochim Biophys Acta*, **1388**, 143–53 (1998).
27. GL Nelsestuen, *J Biol Chem*, **251**, 5648–56 (1976).
28. SJ Freedman, MD Blostein, JD Baleja, M Jacobs, BC Furie and B Furie, *J Biol Chem*, **271**, 16227–36 (1996).
29. L Zhang and FJ Castellino, *J Biol Chem*, **269**, 3590–95 (1994).
30. LA Falls, BC Furie, M Jacobs, B Furie and AC Rigby, *J Biol Chem*, **276**, 23895–902 (2001).
31. JF McDonald, AM Shah, RA Schwalbe, W Kisiel, B Dahlbäck and GL Nelsestuen, *Biochemistry*, **36**, 5120–7 (1997).
32. LM Regan, JS Mollica, AR Rezaie and CT Esmon, *J Biol Chem*, **272**, 26279–84 (1997).
33. AS Wolberg, DW Stafford and DA Erie, *J Biol Chem*, **272**, 16717–20 (1997).
34. DW Banner, A D'Arcy, C Chène, FK Winkler, A Guha, WH Konigsberg, Y Nemerson and D Kirchhofer, *Nature*, **380**, 41–6 (1996).
35. S Dittmar, W Ruf and TS Edgington, *Biochem J*, **321**, 787–93 (1997).
36. B Furie, BA Bouchard and BC Furie, *Blood*, **93**, 1798–808 (1999).
37. C Stanton, R Taylor and R Wallin, *Biochem J*, **277**, 59–65 (1991).
38. J Ware, DL Duiguid, HA Liebman, M-J Rabiet, CK Kasper, BC Furie, B Furie and DW Stafford, *J Biol Chem*, **264**, 11401–6 (1989).
39. B Lind, AH Johnsen and S Thorsen, *Blood*, **89**, 2807–16 (1997).
40. JW Suttie, *FASEB J*, **7**, 445–52 (1993).

41. P Dowd, S-W Ham, S Naganathan and R Hershline, *Annu Rev Nutr*, **15**, 419–40 (1995).
42. C Stanton and R Wallin, *Biochem J*, **284**, 25–31 (1992).
43. J Stenflo and B Dahlbäck, in G Stamatoyannopoulos, PW Majerus, RM Perlmutter and H Varmus (eds.), *The Molecular Basis of Blood Diseases*, 3rd edn, W. B. Saunders, Philadelphia, pp 579–613 (2001).
44. RTA MacGillivray, DE Cool, MR Fung, ER Guinto, ML Koschinsky and BA Van Oost, in JK Setlow (ed.), *Genetic Engineering: Principles and Methods*, Plenum Press, New York, pp 265–330 (1993).
45. EG Wojcik, P Simioni, Mvd Berg, A Girolami and RM Bertina, *Thromb Haemost*, **75**, 70–75 (1996).
46. RTA MacGillivray, DM Irwin, ER Guinto and JC Stone, *Ann N Y Acad Sci*, **485**, 73–79 (1986).
47. FS Hagen, CL Gray, P O'Hara, FJ Grant, GC Saari, RG Woodbury, CE Hart, M Insley, W Kisiel, K Kurachi and EW Davie, *Proc Natl Acad Sci USA*, **83**, 2412–16 (1986).
48. K Kurachi and EW Davie, *Proc Natl Acad Sci USA*, **79**, 6461–64 (1982).
49. SP Leytus, DC Foster, K Kurachi and EW Davie, *Biochemistry*, **25**, 5098–102 (1986).
50. RJ Beckmann, RJ Schmidt, RF Santerre, J Plutzky, GR Crabtree and GL Long, *Nucleic Acids Res*, **13**, 5233–47 (1985).
51. HK Ploos van Amstel, AL van der Zanden, PH Reitsma and RM Bertina, *FEBS Lett*, **222**, 186–90 (1987).
52. A Ichinose, H Takeya, E Espling, S Iwanaga, W Kisiel and EW Davie, *Biochem Biophys Res Commun*, **172**, 1139–44 (1990).
53. G Manfioletti, C Brancolini, G Avanzi and C Schneider, *Mol Cell Biol*, **13**, 4976–85 (1993).
54. CM Jackson, TF Johnson and DJ Hanahan, *Biochemistry*, **7**, 4492–505 (1968).
55. RG Di Scipio, MA Hermodson, SG Yates and EW Davie, *Biochemistry*, **16**, 698–706 (1977).
56. SB Yan, *J Mol Recognit*, **9**, 211–18 (1996).
57. L Thim, S Bjoern, M Christensen, EM Nicolaisen, T Lund-Hansen, AH Pedersen and U Hedner, *Biochemistry*, **27**, 7785–93 (1988).
58. KL Berkner, *Methods Enzymol*, **222**, 450–77 (1993).
59. AE Rudolph, MP Mullane, R Porche-Sorbet and JP Miletich, *Protein Expr Purif*, **10**, 373–78 (1997).
60. KE Van Cott, SP Butler, CG Russell, A Subramanian, H Lubon, FC Gwazdauskas, J Knight, WN Drohan and WH Velander, *Genet Anal*, **15**, 155–60 (1999).
61. RM Camire, PJ Larson, DW Stafford and KA High, *Biochemistry*, **39**, 14322–29 (2000).
62. TM Hackeng, JA Fernández, PE Dawson, SBH Kent and JH Griffin, *Proc Natl Acad Sci USA*, **97**, 14074–78 (2000).
63. JR Cairns, MK Williamson and PA Price, *Anal Biochem*, **199**, 93–97 (1991).
64. P Fernlund, J Stenflo, P Roepstorff and J Thomsen, *J Biol Chem*, **250**, 6125–33 (1975).
65. F Sekiya, M Yoshida, T Yamashita and T Morita, *J Biol Chem*, **271**, 8541–44 (1996).
66. CM Jackson, in JW Suttie (ed.), *Current Advances in Vitamin K Research*, Elsevier Science Publishing, New York, pp 305–24 (1988).
67. DW Deerfield II, DL Olson, P Berkowitz, PA Byrd, KA Koehler, LG Pedersen and RG Hiskey, *J Biol Chem*, **262**, 4017–23 (1987).
68. GA Zapata, P Berkowitz, CM Noyes, JS Pollock, DW Deerfield II, LG Pedersen and RG Hiskey, *J Biol Chem*, **263**, 8150–56 (1988).
69. TP Seshadri, E Skrzypczak-Jankun, M Yin and A Tulinsky, *Biochemistry*, **33**, 1087–92 (1994).
70. NW Huh, P Berkowitz, RG Hiskey and LG Pedersen, *Anal Biochem*, **198**, 391–93 (1991).
71. AK Sabharwal, JJ Birktoft, J Gorka, P Wildgoose, LC Petersen and SP Bajaj, *J Biol Chem*, **270**, 15523–30 (1995).
72. H Mizuno, Z Fujimoto, M Koizumi, H Kano, H Atoda and T Morita, *J Mol Biol*, **289**, 103–12 (1999).
73. V Oganesyan, N Oganesyan, S Terzyan, D Qu, Z Dauter, NL Esmon and CT Esmon, *J Biol Chem*, **277**, 24851–54 (2002).
74. L Li, TA Darden, SJ Freedman, BC Furie, B Furie, JD Baleja, H Smith, RG Hiskey and LG Pedersen, *Biochemistry*, **36**, 2132–38 (1997).
75. J Stenflo, *Acta Chem Scand*, **24**, 3762–63 (1970).
76. J Stenflo and P-O Ganrot, *J Biol Chem*, **247**, 8160–66 (1972).
77. I Björk and J Stenflo, *FEBS Lett*, **32**, 343–46 (1973).
78. HC Marsh, PJ Robertson, ME Scott, KA Koehler and RG Hiskey, *J Biol Chem*, **254**, 10268–75 (1979).
79. A Häfner, F Merola, G Duportail, R Hutterer, FW Schneider and M Hof, *Biopolymers*, **57**, 226–34 (2000).
80. SP Bajaj, T Nowak and FJ Castellino, *J Biol Chem*, **251**, 6294–99 (1976).
81. BC Furie, M Blumenstein and B Furie, *J Biol Chem*, **254**, 12521–30 (1979).
82. BR Lentz, C-M Zhou and JR Wu, *Biochemistry*, **33**, 5460–68 (1994).
83. JR Wu and BR Lentz, *Biophys J*, **60**, 70–80 (1991).
84. TL Colpitts and FJ Castellino, *Biochemistry*, **33**, 3501–8 (1994).
85. L Zhang and FJ Castellino, *J Biol Chem*, **268**, 12040–45 (1993).
86. M Jacobs, SJ Freedman, BC Furie and B Furie, *J Biol Chem*, **269**, 25494–501 (1994).
87. M Soriano-Garcia, CH Park, A Tulinsky, KG Ravichandran and E Skrzypczak-Jankun, *Biochemistry*, **28**, 6805–10 (1989).
88. L Perera, TA Darden, LG Pedersen and G Lee, *Biochemistry*, **37**, 10920–27 (1998).
89. L Perera, TA Darden and LG Pedersen, *Biophys J*, **77**, 99–113 (1999).
90. CH Park and A Tulinsky, *Biochemistry*, **25**, 3977–82 (1986).
91. A Tulinsky and CH Park, in JW Suttie (ed.), *Current Advances in Vitamin K Research*, Elsevier Science Publishing, New York, pp 295–304 (1988).
92. TPJ Seshadri, A Tulinsky, E Skrzypczak-Jankun and CH Park, *J Mol Biol*, **220**, 481–94 (1991).
93. H Brandstetter, M Bauer, R Huber, P Lollar and W Bode, *Proc Natl Acad Sci USA*, **92**, 9796–800 (1995).
94. TC Evans Jr and GL Nelsestuen, *Biochemistry*, **35**, 8210–15 (1996).
95. Y Yokomori and DJ Hodgson, *Inorg Chem*, **27**, 2008–11 (1988).

C2-like-domains

Membrane binding C2-like domains

*Nuria Verdaguer[†], Senena Corbalán-García[‡], Wendy F Ochoa[†],
Juan Carmelo Gómez-Fernández[‡] and Ignacio Fita[†]*

[†]Departamento de Biología Estructural, IBMB (C.S.I.C.), Jordi Girona, Barcelona, Spain
[‡]Departament de Bioquímica y Biología Molecular (A), Facultad de Veterinaria, Universidad de Murcia, Murcia, Spain

FUNCTIONAL CLASS

C2-domains are independently folded modules, of about 130 residues, found in a large and diverse set of eukaryotic proteins.[1,2] Many of the best-characterized C2-domains are found in proteins involved in membrane trafficking and fusion, such as synaptotagmins or rabphilin, and in signal transduction, such as phospholipases A2 (cPLA2) and C (PLC) or the protein kinase C isoforms (PKCs).

C2-like domains are structurally defined as all beta protein members of the C2-domain superfamily of calcium/lipid binding domains (CaLB) (SCOP: http//scop.mrc-lmb.cam.ac.uk/scop/). The superfamily includes two families: (i) the synaptotagmin-like variants, also referred to as the *S family* or the type I topology and (ii) the PLC-like variants, also known as the *P family* or the type II topology. When the structure of the second C2-domain (C2B-domain) from rabphilin was

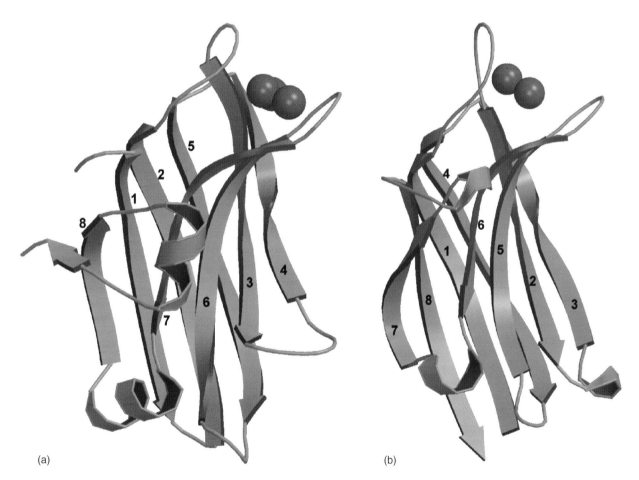

(a) (b)

3D Structure Ribbon diagrams of the C2-domain structures from (a) protein kinase Cα (PKCα) with PDB code 1DSY and (b) phospholipase C (PLC-δ1) with PDB code 1DJY, as representative members of the P and S families (topologies I and II) respectively. The two different topologies result from a circular permutation of the β-strands that leaves the N- and C-termini either at the top or at the bottom of the β-sandwich. Calcium ions found at the top of the β-sandwich are explicitly shown as solid spheres.

determined, a consistent local alternative in the C2-domain topologies became apparent (see below and reference 3). Consequently, it was proposed that the two CaLB families could be further divided into subfamilies A and B. In this classification, topology IA includes C2-domains from the first C2-domain of synaptotagmins (C2A-domain), while the C2B-domains of synaptotagmins, rabphilin, PKCα and β corresponds to topology IB. In turn, topology IIA includes the C2-domains of PLC-δ1 and cPLA2, while topology IIB is represented by the C2-domain from the novel PKCδ.

Despite their definition as calcium-dependent domains, a growing number of proteins containing C2-domains, such as the novel and atypical PKCs,[4] the PKC-related kinases (PRKs),[5,6] the tumor-suppressor PTEN/MMAC1,[7,8] and phosphoinositide 3 kinases (PI3Ks),[9] have now been found to be calcium independent. C2 domains have been the subjects of intensive research, and a number of reviews dealing with the physiology, the biochemistry, and the structure of C2-domains are available.[10–14] The present work will focus on the wealth of functional and structural information, which has been obtained at an exponential pace, on the membrane binding C2-domains that require calcium.

OCCURRENCE

C2-domains were first described as the second of the four conserved functional domains found in the α-, β- and γ-isoforms of mammalian calcium-dependent PKCs.[15–19] C2-domains have now been found to be widely distributed among eukaryotes but essentially absent in the prokaryotic world. The only exception reported to date is the α-toxin protein from *Clostridium perfringens*, which contains a C-terminal domain with a high structural homology to the eukaryotic C2-domains.[20] A sequence search shows 730 C2-domains containing proteins that can be grouped according to the similarity of their domain organization.[21] Some of these proteins contain a single C2-domain, for example, classical and novel PKCs, whereas some others include two consecutive C2-domains, for example, synaptotagmins, and a few, like Munc13 or dysferlin, have multiple C2-domains separated by sequences of varied size.

BIOLOGICAL FUNCTION

C2-domains are found in proteins that, as indicated above, participate in many different biological functions. The main role of C2-domains, among membrane-binding calcium-regulated proteins that contain this type of domain, is to act as the calcium-activated membrane-targeting motif.[11,12] In many cases, for example, in classical PKCs and cPLA2, the C2-domains bind multiple Ca^{2+} ions, then dock to specific intracellular membranes where a different protein domain, a catalytic domain, exerts its function.[12] In other cases, for example in synaptotagmins, proteins can be constitutively anchored to membranes throughout an N-terminal transmembrane region, and the two C2-domains act as calcium sensors.[22,23] The most relevant biological functions of calcium-binding C2-domain-containing proteins can be summarized as follows:

(1) *Protein phosphorylation*: For example, proteins from the PKC family or the catalytic subunit of the yeast cAMP-dependent kinase.[24]

(2) *Lipid modification*: There are a number of enzymes that generate or inactivate lipid-derived second messengers such as plant phospholipase D,[25] yeast phosphatidylserine decarboxylase (PSD2),[26] 5-Lipoxygenase,[27,28] or the PLC and cPLA2 protein families.[29–33]

(3) *Vesicular transport*: There are some proteins involved in neuronal vesicular trafficking such as the synaptotagmin family.[22,23] Other C2-domain-containing proteins that also participate in the neurotransmitter release regulation include phorbol ester/DOG binding protein Munc-13,[34] Double C2 protein (DOC2),[35] B/K protein,[36,37] the Rab3-interacting protein rabphilin3A,[38] and Piccolo/aczonin.[39,40]

Nonneuronal vesicular trafficking also includes C2-domain-containing proteins such as the synaptotagmin-like family.[41]

(4) *Small GTPase regulation*: Only in GAP1m and in RasGAP-activating-like proteins the corresponding C2-domain sequences show four of the aspartate residues that are conserved in calcium-binding C2-domains (see below and reference 42), although their regulation by calcium is still under study.[43]

(5) *Ubiquitination*: This is the case of members from the large family of HECT domain–containing ubiquitin ligases.[44,45]

(6) *Miscellaneous*: Many other C2-domain-containing proteins with different functions have been described, for example, *copines*,[46,47] *intersectins*,[48] *calpains*,[49] *perforins* (also known as pore-forming protein),[50] *Ferlins*,[51] and toll-interacting protein.[52]

AMINO ACID SEQUENCE INFORMATION

Several hundred C2-domain sequences are, as indicated above, now available.[21,53] Sequence alignment can be established between all of them when the first strand in the C2-domain with topology I is compared with the eighth strand in C2-domain with topology II. These strands occupy equivalent positions in the tertiary structure due to the cyclic relationship of both topologies (see

Membrane binding C2-like domains

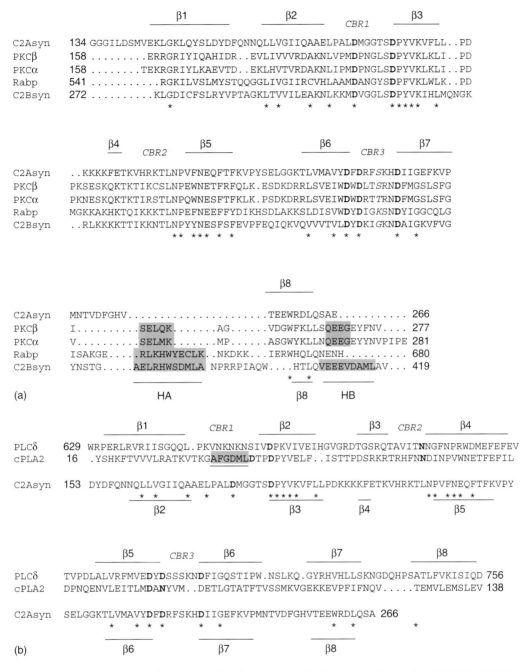

Figure 1 (a) Sequence alignment of the five structurally characterized C2-domains with topology I: PKCβ (1A25), PKCα (1DSY), rabphilin (3RPB), and domains C2A (1RSY) and C2B (1K5W) from synaptotagmins; (b) sequence alignment of the two structurally characterized C2-domains with topology II: PLC-δ1 (1DJY), cPLA2 (1BCI). The names in brackets correspond to the PDB codes of the structures used. The sequence of the C2A domain from synaptotagmins is also included as a reference in order to show the circular correspondence between the two topologies. The highly conserved aspartic residues, which participate in the coordination of the calcium ions, are represented in bold characters. Yellow boxes correspond to the β-strands where structural similarity is well preserved. Blue boxes represent the α-helical structures. The 'consensus' residues present in more than 50% of the sequences used for the structural alignment are indicated as stars at the bottom of the alignment.

below and Figure 1). Tandem C2-domains exhibit either topology, but with both members belonging to the same topological family. Exon boundaries in cPLA2 (following Ile11 and Val138) are located at the ends of the C2-domain, according to the view that the C2-domain is a functional module that has been inserted into a wide array of proteins.[11,54]

Sequence alignments of C2-domains, using structural information, reveal about 26 positions that exhibit greater than 50% identity, the 'consensus' residues (Figure 1).

Many other 'conserved' positions also display a high side-chain homology and, together with the consensus positions, are mostly situated in the β-strands where a clear pattern of alternating polar and nonpolar residues is often apparent in the strands that are not lying at the edges of the two β-sheets.[11,55] Proline and glycine residues are concentrated in the connections between strands, where sometimes large insertions are observed likely imparting protein-specific functions. Among the C2-domains that bind calcium, there is also a very high conservation of the residues that are implicated in the coordination of Ca^{2+} ions, in particular, five aspartates (see below and Figures 1 and 2). However, some variability is found at one or more of these highly conserved positions, which would allow specialization of the different C2-domains, presumably to provide optimized calcium-binding parameters or docking interactions for different biological functions.[11] Among C2-domains with topology I, there is another conserved region formed by a cluster of basic residues, mostly lysines, from strands β3 and β4, which has been involved in negatively charged phospholipid binding.[56-59] Other functions, such as protein–protein interactions, have also been attributed to this region, for example, between PKCβ and its anchoring protein receptor for activated C-kinases (RACK).[60]

Prior to the structure determination of the C2B-domain of rabphilin, sequence alignments had erroneously predicted the sequence corresponding to the last β-strand of the β-sandwich (see below and reference 3). This error,

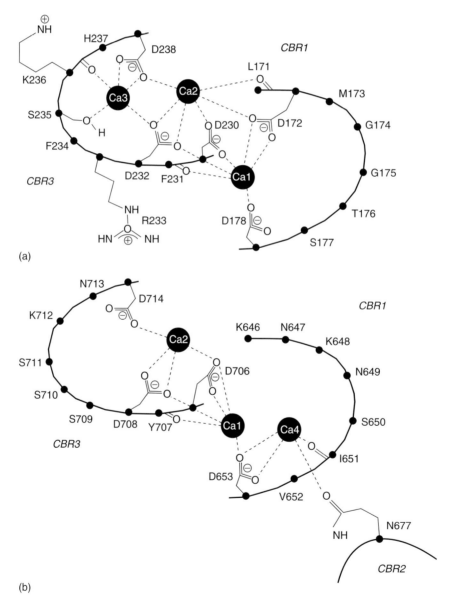

Figure 2 Diagram summarizing the Ca^{2+}-binding modes of the C2-domains with topology I (a) and II (b). In (a), the amino acid sequence used is based on the C2A-domain of synaptotagmin I, and the model is based on the results discussed in detail by Ubach et al.[3] In (b), the amino acid sequence used is based on the C2-domain of PLCδ1.[82]

significant for the taxonomy of C2-domains, was primarily due to the assumption that the conserved tryptophan residue of C2A-domains corresponded to Trp658 in the C2B-domain of rabphilin. Instead, the tryptophan previously thought to be part of the C-terminal extension of the C2B-domain, Trp672 in rabphilin, is the one that occupies the position of the conserved tryptophan in the C2A-domains. It is interesting to note that both tryptophans, as well as their neighboring histidines, are absolutely conserved in C2B-domains.[3] Finally, it is worth mentioning that C2-domain sequences of a given protein isolated from different species present significantly less divergence than is observed between functionally distinct proteins.

PROTEIN PRODUCTION, PURIFICATION, AND MOLECULAR CHARACTERIZATION

C2-domains can, in general, be obtained as recombinant proteins expressed in *Escherichia coli* in the soluble fraction of the cells.[55,61] However, the C2-domain from cPLA2 appeared only in the insoluble fraction from where it was solubilized using a refolding protocol.[62] The C2A-domain of synaptotagmin, the first C2-domain structurally studied, was obtained by expression in *E. coli* BL21 (DE3) strain and purified through affinity chromatography on GST-agarose beads followed by a cleavage with thrombin and purification through gel filtration on Superdex columns.[55,63] The C2-domain of PKCα has been expressed in the same strain of *E. coli* cells fused to a 6His tag and purified by affinity to a Ni-NTA agarose column. The tag was cleaved with thrombin and removed by a second incubation with Ni-NTA resin.[61,64]

METAL CONTENT AND COFACTORS

A variable number of Ca^{2+} ions are the only metal ions regularly found in membrane-binding C2-domains that require calcium. Different phospholipids can also bind specifically to different C2-domains at one or more locations, which in turn can alter the affinity of the domain by calcium (see below).

ACTIVITY TEST

C2-domains do not perform catalytic activities by themselves. However, the calcium and phospholipid binding properties of the membrane-binding C2-domains that require calcium can be measured by a number of different techniques.

(1) Calcium binding to C2-domains can be evaluated by equilibrium dialysis and isothermal titration microcalorimetry.[65–67] NMR is also a very powerful technique that enables us to study calcium binding by analysis of the ^{1}H-^{15}N heteronuclear single quantum coherence (HSQC) spectra, which shows one cross-peak for each nonproline residue in the sequence.[63] Direct visualization of the number, disposition, and occupancy of Ca^{2+} ions bound to a C2-domain can, of course, be achieved by X-ray crystallography of the domain complexed with calcium (see below).

A number of less direct techniques to evaluate calcium binding to C2-domains have also been used. For example, it has been proved that calcium binding can be monitored by tryptophan fluorescence. This is likely because, upon calcium addition, the fluorescence intensity of the one or two tryptophan residues that are very often located in the vicinity of the calcium-binding region decreases.[68]

(2) Phospholipid binding to C2-domains is mainly analyzed by three methodologies: fluorescence resonance energy transfer (FRET), ultracentrifugation, and affinity binding. (i) FRET, the most frequently used, is based on the quenching of tryptophans owing to a protein-to-membrane fluorescence resonance energy transfer.[65,69] These experiments can be performed in steady state conditions, but additional information about the kinetics of binding and dissociation can be obtained with stopped-flow methods.[68,70–72] The penetration depth of the fluorophore into the bilayers can be calculated according to the parallax analysis.[73,74] Alternatively, an estimation of membrane approximation and penetration can be obtained by site-directed spin labeling and EPR power saturation[74,75] or by monolayer surface pressure.[76] (ii) Lipids resuspended in a buffer-containing sucrose produce, on extrusion, a homogenous mixture of large unilamellar vesicles. Ultracentrifugation allows the separation of the lipid vesicles with bound protein from the free protein in solution.[77] (iii) The protein construct of glutathione S-transferase (GST) fused to the C2-domain, immobilized in resin matrices, is used for affinity binding assays with radioactively labeled liposomes of defined phospholipid composition.[78,79] Care has to be taken because in some cases the GST-C2-domain fusion proteins are not further purified and some contradictory results have been reported in the literature.[80]

X-RAY AND NMR STRUCTURES OF MEMBRANE-BINDING C2-DOMAINS

Current data on the structures and interactions of calcium-binding C2-domains are derived from a growing number of crystallographic studies. In particular, for C2-domains with topology I, the data are from synaptotagmin I-III[55,81] and from classical PKCs α[58,64] and β.[82] In the case of

Membrane binding C2-like domains

C2-domains with topology II, the structures available to date are from PLCδ1[83] and cPLA2.[62,84] NMR structure determinations have also made a major contribution, especially for the C2A-domain of synaptotagmin I[64] and for the C2B-domains of rabphilin and synaptotagmin I.[3,85]

Crystallization of C2-domains with topology I

The first crystals of the calcium-free form of the C2A-domain from synaptotagmin I were obtained with the hanging drop vapor diffusion technique at 21 °C using Li_2SO_4 as a precipitant at pH 7.2.[55] They diffracted beyond 1.7-Å resolution. Calcium was then introduced by soaking the calcium-free crystals in a solution containing 0.1 mM $CaCl_2$ for 24 h. The crystals were extremely sensitive to Ca^{2+} ions and cracked or dissolved if exposed to concentrations above 0.1 mM.

Alternatively, in many other cases, crystals of the calcium-bound form were obtained by including calcium in the crystallization buffer. This was so in the case of the PKCα-C2-domain[58,64] and the PKCβ-C2 domain[82] among others.

Crystallization of C2-domains with topology II

A recombinant fragment of the PLCδ1 that included the C2-domain as well as the catalytic and a calmodulin-like EF-hand domains, was crystallized in a cubic space group $F4_132$ with unit cell dimensions $a = 398$ Å and two monomers per asymmetric unit with 70% solvent content.[83] Crystals were obtained at 12 °C, pH 4.75, in hanging drops with protein at $10\,mg\,mL^{-1}$ and 1.4 M ammonium phosphate plus 0.1% w/v of detergent CHAPSO (3-(3-cholamidopropyl) dimethylammonio-2-hydroxy-1-propane-sulphonate). Crystals from complexes of PLCδ1 with D-myo-inositol-1,4,5-trisphosphate ($InsP_3$) were also obtained by soaking the PLCδ1 crystals with $InsP_3$ and calcium chloride. Crystals, both from the full-length cPLA2[84] and from its C2-domain,[62] which present a type II topology, have also been reported.

Overall description of C2-domain structures

All C2-domains share a common overall fold: a single compact Greek-key motif organized as an eight-stranded antiparallel β-sandwich consisting of a pair of four-stranded β-sheets – the A and B sheets – with a long axis of about 50 Å. Connections between β-strands emerge from the top and bottom of the domain according to the two distinct topologies of C2-domains. In topology I (the S family), the first β-strand occupies the same structural position as the eighth β-strand of topology II (the P family), which shifts in a circular permutation the order of homologous strands in the primary structure (Figures 1 and 3). Structural comparison of the PKCβ-C2-domain and the synaptotagmin I-C2A-domain, both with type I topologies, gives a root mean square deviation (rmsd) of 0.91 Å for 108 equivalent residues. In turn, comparison of the PKCβ-C2-domain with the type II topology C2-domains from PLA2 and PLCδ, gives rmsd values of 0.82 Å and 0.95 Å for 79 and 87 equivalent residues respectively. Therefore, permutation of β-strands seems to have little impact on structural conservation. The N- and C-ends are situated at the top or at the bottom of the β-sandwich depending on whether the C2-domain topology is I or II (Figure 3). These may play a functional role by conferring different orientations to the C2-domain relative to the rest of the protein, according to the different topologies.

The taxonomy of C2-domains can be further subdivided, as mentioned above, into the A and B subtypes or subfamilies. The main feature of the B subtype is the presence of an α-helix in the strand connections at the bottom of the β-sandwich, which is absent in the A subtype (Figure 3).

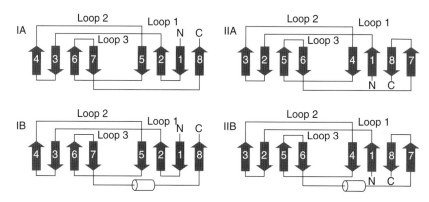

Figure 3 Representation of secondary elements in C2-domain structures. Short helices were omitted for simplicity. Topologies I and II, related to each other by a circular permutation of the β-strands, can be further subdivided into subtypes A and B. The subdivision would be mainly characterized by the presence in the B subtype of a well-defined helix in the connections at the bottom of the β-sandwich (see in the text and reference 3).

In the C2B-domains, from rabphilin, synaptotagmins, and classical PKCs, belonging to topology I and subtype B, the α-helix is situated in the connection between β-strands 7 and 8 packed against the interface between the two β-sheets and spans about 5–10 residues.[3,64,82,85]

β-sheets of C2-domains are distinctly concave (as seen in the 3D Structure). In the C2-domain from classical PKCs α and β, the curvature of the sheet is accentuated likely because of the β3-β4 connection twist that, in turn, appears to be induced by the presence of the cisPro-Asp-transPro motif. In the C2A-domain from synaptotagmin I, four β-bulges were found embedded within the β-sandwich. Three of the β-bulges, which span the A-sheet strands β1, β2, and β5, appear to be present also in the calcium binding C2-domains with topology II.[82] Together, the three bulges impose an archlike bend upon the three participating β-strands in the plane of the sheet. The global effect of this distortion is apparently to position the β2-β3 connection, in C2-domains with topology I, relative to the opposing β6-β7 connection such that the two loops, correctly juxtaposed, can build the calcium binding pocket (see below and Figure 3). C2-domains with topology I also contain a classic (+) β-bulge at the amino-terminus of the β6-β7 connection, which appears to contribute again to the proper orientation of this connection in the calcium binding site.

The greatest variability among C2-domain structures is found in the N-terminal region and in the strand connections. The three connections at the top of the β-sandwich, also named *calcium-binding regions* (CBRs), provide all the residues that participate in calcium coordination. CBR1, which, in C2-domains with topology I, corresponds to the β2-β3 connection, presents significant differences among C2-domain structures though it is well conserved between classical PKCs α and β (see below and reference 64). In C2-domains with topology II, CBR1s correspond to the β1-β2 connections and contain a short α-helix in cPLA2 that is absent in PLC-δ1. The organization of CBR2 and of CBR3, which, in C2-domains with topology I, correspond respectively to the β4-β5 and β6-β7 connections, appears to be highly conserved among C2-domain structures with the same topology. At the bottom of the β-sandwich, large variations, even with differences larger than 3 Å for the equivalent residues of the closely related C2-domains from PKCs α and β, are seen for the β3-β4 connection, while β5-β6 is less variable and often contains a short helix. Finally, the β7-β8 connection is, in members from the B subtype, a long helix, which is absent in the C2-domain structures belonging to the A subtype. Helices are often seen at the C2-domain ends, in particular at the N- and C-termini of PKCs α and β and at the C-termini of the C2B-domain of synaptotagmin I. However, these helices often fall outside the C2-domain homology region and correspond, in fact, to part of the linkers with other protein domains.[64,82,85]

Proteins containing C2-domains can experience large conformational changes upon calcium binding, often involving changes in the relative disposition of domains.[86] Calcium binding can also induce long-range rearrangements inside the C2-domains, as demonstrated for piccolo/aczonin[40] and perhaps also in classical PKCs,[75] where it has not yet been possible to crystallize the calcium-free forms. Calcium binding was found to induce a conformational change in the CBR1 of PLC-δ1, moving this loop apart from CBR3.[87] However, in the C2A-domain of synaptotagmin I, calcium binding appears to cause only some rigidity within the side chains of residues from the CBRs.[88] Finally, it is worth mentioning that no important rearrangements in the PKCα C2-domain with calcium were observed when binding to the short-chain lipids DCPS, DAPS, or DCPA (see below and reference 58).

Calcium binding sites of C2-domains

In C2-domains, four distinct calcium binding sites (Ca1 to Ca4) have been identified inside the calcium binding motif defined by the CBRs at the top end of the β-sandwich (3D Structure and Figures 2, 4, and 5). These calcium binding sites are also named as sites I (Ca4), II (Ca1), III (Ca2), and IV (Ca3) according to the nomenclature introduced with PLC-δ1.[89] Not all the calcium binding sites may be occupied in any one species; in fact, C2-domains with topology I have never been reported to use site Ca4. The C2B-domain from rabphilin and the C2-domain from cPLA2 seem to bind only two calcium ions at sites Ca1/Ca2 and Ca1/Ca4, respectively, whereas sites Ca1/Ca2/Ca3 have been observed to be occupied in the C2A-domains from synaptotagmins and in the C2-domain from PKCs α and β (Figures 2(a) and 4). The PLC-δ1 C2-domain is likely to bind calcium ions at three sites (Ca1, Ca2, and Ca4) or perhaps even at all four sites[88] (Figure 2(a) and (b)). Large differences in the calcium-binding affinities between the different calcium binding sites, both among the several calcium binding sites in one domain and among C2-domains, together with cooperative effects versus specific phospholipids, allow a versatile tuning of the interactions of C2-domains with membranes.[66]

Adjacent binding of several calcium ions to the C2-domains is possible because of the clustering of the five highly conserved aspartic residues from CBR1 and CBR3.[89] These residues conform to a bipartite calcium binding site where Ca1 and Ca2 are related to each other by a pseudo-dyad symmetry axis (Figures 2 and 4).[63,89] Three of the aspartic residues (187, 246, and 248 in PKC α) coordinate simultaneously with the calcium ions at sites Ca1 and Ca2. The aspartic residue situated on the pseudo-dyad axis (246 in PKC α) contributes one carboxylate oxygen to each of the Ca^{2+} ions, while the other two aspartic residues (187 and 248 in PKC α) are related to each other by the pseudo-dyad axis and show bidentate interactions with calcium at sites Ca1 and Ca2, respectively. Each of these two aspartic residues still participates in another coordination bond with Ca1 (Asp248) and with Ca2 (Asp187) (Figures 2(a) and 4).

Membrane binding C2-like domains

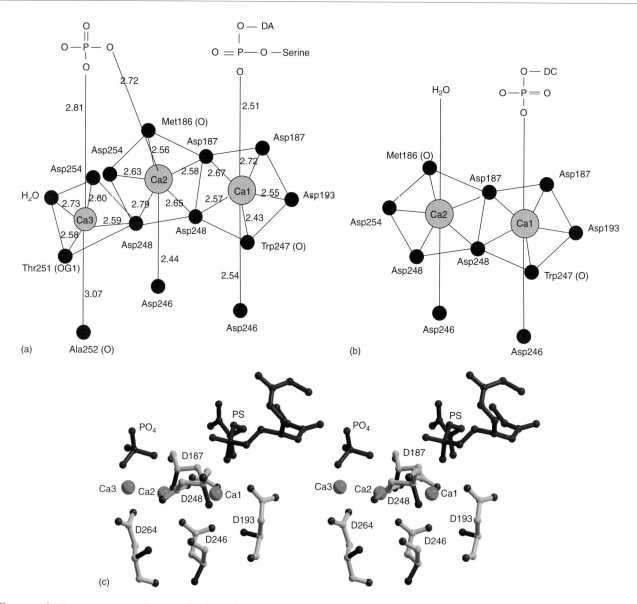

Figure 4 Coordination schemes of the calcium ions in the ternary complexes of the C2-domain from PKCα with (a) 1,2-diacetyl-sn-phosphatidyl-L-serine (DAPS) and (b) 1,2-dicaproyl-sn-phosphatidic acid (DCPA). Coordination distances, in angstroms, are explicitly indicated for the DAPS complex, where calcium site Ca3 presents a high occupancy by calcium. (c) Stereo image of the DAPS complex showing explicitly the disposition of the five conserved aspartic residues and the three calcium ions (represented as solid spheres). The ligand molecules of phosphate and DAPS are also shown (blue).

The remaining two conserved aspartic residues contribute a ligand oxygen to the coordination sphere of calcium at sites Ca1 (Asp193) and Ca2 (Asp254) and interact, often forming bidentate bonds, with sites Ca3 (Asp254) and Ca4 (Asp193). The main-chain oxygen atoms of two residues (Met186 and Trp247 in PKCα or Leu171 and Phe231 in the C2A-domain of synaptotagmin I) participate in the coordination of calcium at sites Ca2 and Ca1 respectively. For calcium at site Ca3, ligands from the protein also include one side-chain and one main-chain oxygen atoms from two CBR3 residues (Thr251 and Arg252 in PKC α or Ser235 and Lys236 in the C2A-domain of synaptotagmin I). In turn, the residue corresponding to Ser235 in the C2A-domain of synaptotagmin I is, respectively, a glycine or a lysine (pointing away from the calcium binding pocket) in the C2B-domains from synaptotagmin and rabphilin, which has been proposed to be the cause for no calcium binding at site Ca3 by these domains.[3,85] In C2-domains with topology II, a side-chain oxygen atom from a CBR2 residue (Asn65 in PLA2 and Asn677 in PLCδ1) provides a coordination bond for calcium at site Ca4 (Figure 2(b)). The coordination sphere at site Ca4 also includes oxygen atoms from the side chain and main chain of two CBR1 residues (Asp653 and Ile651 in PLC-δ1, respectively). Therefore, CBR1 and CBR3 provide, in general, six of the coordination bonds for sites Ca1 and

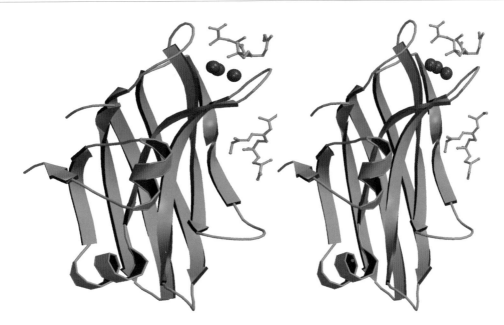

Figure 5 Stereo images of the overall structure of the C2-domain of PKCα bound to DAPS. Calcium ions and ligand molecules are explicitly represented. The second lipid binding site found in the domain is situated in the proximity of strands β3 and β4 (see in the text).

Ca2. However, only CBR3 participates in the coordination bonds of site Ca3, while CBR1 and CBR2 participate in the coordination bonds of Ca4 (Figure 2). CBR3 in cPLA2 is shorter, and the side chains of Asn95 and Asp99, which replace two highly conserved aspartic residues that coordinate Ca2 and Ca3, point outside the calcium binding pocket. The impossibility of forming these coordination bonds seems to explain the absence of calcium binding at

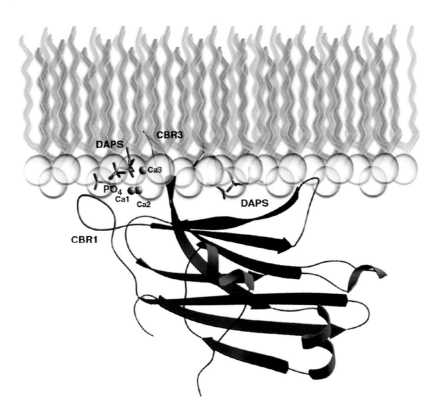

Figure 6 Proposed docking of the C2-domain from PKCs onto model membranes. In the resulting model, only the central part of CBR3 from the C2-domain is inserted into the lipid bilayer. The phosphate and the head groups from the two DAPS molecules found in the structure of the DAPS-C2-domain complex could correspond to three polar heads on the surface of the anionic membranes, which would define a feasible anchorage plane in the C2-domain structure.

sites Ca2/Ca3 in the C2-domain from cPLA2 (Figure 2(a) and (b)).

Water molecules and phosphate groups, often belonging to phospholipids, complete the coordination spheres of the calcium ions bound to C2-domains. In the structures of the ternary complexes of PKCα-C2-Ca^{2+} with DCPS, DAPS, or DCPA, calcium ions at sites Ca1 and Ca2 were always present with full occupancy and hepta-coordination. In contrast, site Ca3 was occupied by a hexacoordinated calcium ion only in the structure of the DAPS complex (Figure 4). In these complexes, the nature and disposition of the seventh ligand at sites Ca1 and Ca2 departs from the pseudo-symmetric organization of the calcium binding region. At site Ca1, the seventh ligand was an oxygen atom from the phosphoryl group of either the DCPS, DAPS, or DCPA molecule in the corresponding complexes. At site Ca2, the seventh ligand was mainly a water molecule, though a phosphate group was found in the complex with DAPS (Figure 4). In the latter complex, the fifth and sixth ligands at site Ca3 were a water molecule and the phosphate group that also interacts with Ca2 (Figure 4). In PKCβ, the organization of these ligands differs from that of PKCα in that the seventh coordination position in the former was empty for Ca1 and occupied by the side chain of a glutamic residue from a neighbor molecule for Ca2.[82]

FUNCTIONAL ASPECTS

Activation and docking mechanisms

For synaptotagmins and PKCs, it has been proposed, according to the structural and biochemical information, that the binding of the calcium ions to the C2-domain could trigger the interaction with negatively charged phospholipids at the membrane surface. Additional interactions, mainly between residues located in CBR3 and the phospholipids, would enable the domain to partially penetrate into the membrane.[14] In the case of synaptotagmins, they are Ca^{2+} sensors with distinct Ca^{2+}-binding properties, which in turn lead to synaptic or endocrine exocytosis.[22] However, in other enzymes like phospholipases or protein kinases, the Ca^{2+}/phospholipid-membrane interaction leads to the activation of these enzymes by proximity to their specific targets.[12] The finding of a second phospholipid binding site in the vicinity of the lysine-rich cluster of the C2-domains from classical PKCs (Figure 5) appears to define a structurally feasible anchorage plane of these domains onto anionic membranes[58,59] (Figure 6).

Some of the structural features of the C2-domain can show their full relevance only in the context of the different conformations that the corresponding multidomain molecules can adopt (Figure 7). For example, in PKCs, experimental data and modeling support the possibility of docking the C1A domain into the concave face

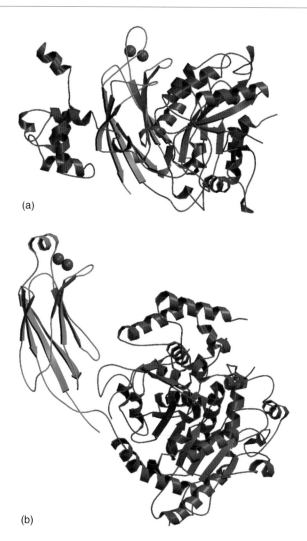

Figure 7 Ribbon representation of the multidomain structures (C2-domains in green) from (a) PLC-δ1 (PDB code 1DJY) and (b) cPLA2 (PDB code 1CJY). Calcium ions found in the C2-domains are explicitly represented as solid spheres. The accessibility of the C2 domain is affected by the disposition of the remaining protein regions (see in the text).

of the C2 domain, leading to a contact between the DAG binding site, in the C1A domain, and the CBR3 region, in the C2 domain. This would explain the inaccessibility of the DAG binding site found experimentally.[86,90] Calcium, when present, would trigger the binding of the C2 domain onto the anionic membranes and also a conformational change that would facilitate both the DAG binding pocket and the lysine-rich cluster to become accessible for further interactions.[86]

CONCLUSIONS

Calcium-binding C2-domains are remarkable modules present in a wide variety of proteins, in particular, related with membrane trafficking and signal transduction.

Taken together, the structural information defines adjacent calcium binding sites within a broad acidic cleft that contains highly conserved acidic residues, which provide most of the oxygen atoms that coordinate with the calcium ions. Coordination distances ranging, in general, from 2.3 to 2.6 Å are only slightly larger than the calcium–oxygen distances observed in EF-hand proteins, such as calmodulin, with a higher affinity (K_d of 10^{-5} M to 10^{-7} M).[89] The available C2-domain structures confirm that the number of Ca^{2+} ions as well as the number of phosphate and phospholipid molecules bound to the domain varies. The interplay between the binding of calcium ions and phosphates or phospholipid molecules in the C2-domain is strongly supported by the specificity and spatial organization of the binding sites in the domain and by the variable occupancies of ligands found in the different crystal structures. These structural results, together with the wealth of biochemical and biological information available, allow envisioning of an exquisite and versatile tuning of activity at the level of the binding affinity of the C2-domain to membranes.

REFERENCES

1. JH Hurley and S Misra, *Annu Rev Biophys Biomol Struct*, **29**, 49–79 (2000).
2. EA Nalefski, MA Wisner, JZ Chen, SR Sprang, M Fukuda, K Mikoshiba and JJ Falke, *Biochemistry*, **40**, 3089–100 (2001).
3. J Ubach, J Garcia, MP Nittler, TC Sudhof and J Rizo, *Nat Cell Biol*, **1**, 106–12 (1999).
4. AC Newton, *Chem Rev*, **101**, 2353–64 (2001).
5. H Mukai and Y Ono, *Biochem Biophys Res Commun*, **199**, 897–904 (1994).
6. RH Palmer, J Hidden and PJ Parker, *Eur J Biochem*, **227**, 344–51 (1995).
7. PA Steck, MA Pershouse, SA Jasser, WK Yung, H Lin, AH Ligon, LA Langford, ML Baumgard, T Hattier, T Davis, C Frye, R Hu, B Swedlund, DH Teng and SV Tavtigian, *Nat Genet Apr*, **15**, 356–62 (1997).
8. JO Lee, H Yang, MM Georgescu, A Di Cristofano, T Maehama, Y Shi, JE Dixon, P Pandolfi and NP Pavletich, *Cell*, **99**, 323–34 (1999).
9. EH Walker, O Perisic, C Ried, L Stephens and RL Williams, *Nature*, **402**, 313–20 (1999).
10. CP Ponting and PJ Parker, *Protein Sci*, **5**, 162–66 (1996).
11. EA Nalefski and JJ Falke, *Protein Sci*, **5**, 2375–90 (1996).
12. J Rizo and TC Sudhof, *J Biol Chem*, **273**, 15879–82 (1998).
13. M Katan and VL Allen, *FEBS Lett*, **452**, 36–40 (1999).
14. W Cho, *J Biol Chem*, **275**, 32407–10 (2001).
15. L Coussens, PJ Parker, L Rhee, TL Yang-Feng, E Chen, MD Waterfiel, U Francke and A Ullrich, *Science*, **233**, 859–66 (1986).
16. JL Knopf, MH Lee, LA Sultzman, RW Kriz, CR Loomis, RM Hewick and RB Bell, *Cell*, **46**, 491–502 (1986).
17. Y Ono, T Kurokawa, K Kawahara, O Nishimura, R Marumoto, K Igarashi, Y Sugino, U Kikkawa, K Ogita and Y Nishizuka, *FEBS Lett*, **203**, 111–15 (1986a).
18. Y Ono, T Kurokawa, T Fujii, K Kawahara, K Igarashi, U Kikkawa, K Ogita and Y Nishizuka, *FEBS Lett*, **206**, 347–52 (1986b).
19. PJ Parker, L Coussens, N Totty, L Rhee, S Young, E Chen, S Stabel, MD Waterfield and A Ullrich, *Science*, **233**, 853–59 (1986).
20. CE Nailor, JT Eaton, A Howells, N Justin, DS Moss, RW Titball and AK Basak, *Nat Struct Biol*, **5**, 738–46 (1998).
21. A Bateman, E Birney, L Cerruti, R Durbin, L Etwiller, SR Eddy, S Griffiths-Jones, KL Howe, M Marshall and EL Sonnhammer, *Nucleic Acids Res*, **30**, 276–80 (2002).
22. TC Südhof, *J Biol Chem*, **277**, 7629–32 (2002).
23. WC Tucker and ER Chapman, *Biochem J*, **366**, 1–13 (2002).
24. Y Nishizuka, *FASEB J*, **9**, 484–96 (1995).
25. L Zheng, R Krishnamoorthi, M Zolkiewski and X Wang, *J Biol Chem*, **275**, 19700–6 (2000).
26. PJ Trotter, J Pedretti, R Yates and DR Voelker, *J Biol Chem*, **270**, 6071–80 (1995).
27. XS Chen, TA Naumann, U Kurre, NA Jenkins, NG Copeland and CD Funk, *J Biol Chem*, **270**, 17993–99 (1995).
28. T Hammarberg, P Provost, B Persson and O Radmark, *J Biol Chem*, **275**, 38787–93 (2000).
29. RL Williams, *Biochim Biophys Acta*, **1441**, 255–67 (1999).
30. MJ Rebecchi and SN Pentyala, *Physiol Rev*, **80**, 1291–35 (2000).
31. SG Rhee, *Annu Rev Biochem*, **70**, 281–312 (2001).
32. DA Six and EA Dennis, *Biochim Biophys Acta*, **1488**, 1–19 (2000).
33. A Dessen, *Structure*, **8**, R15–R22 (2000).
34. N Brose, K Hofmann, Y Hata and TC Sudhof, *J Biol Chem*, **270**, 25273–80 (1995).
35. A Naito, S Orita, A Wanaka, T Sasaki, G Sakaguchi, M Maeda, H Igarashi, M Tohyama and Y Takai, *Brain Res Mol Brain Res*, **44**, 198–204 (1997).
36. OJ Kwon, H Gainer, S Wray and H Chin, *FEBS Lett*, **378**, 135–39 (1996).
37. M Fukuda and K Mikoshiba, *Biochem J*, **360**, 441–48 (2001).
38. H Shirataki, K Kaibuchi, T Sakoda, S Kishida, T Yamaguchi, K Wada, M Miyazaki and Y Takai, *Mol Cell Biol*, **13**, 2061–68 (1993).
39. C Cases-Langhoff, B Voss, AM Garner, U Appeltauer, K Takei, S Kindler, RW Veh, P De Camilli, ED Gundelfinger and CC Garner, *Eur J Cell Biol*, **69**, 214–23 (1996).
40. SH Gerber, J Garcia, J Rizo and TC Sudhof, *EMBO J*, **20**, 1605–19 (2001).
41. TS Kuroda, M Fukuda, H Ariga and K Mikoshiba, *J Biol Chem*, **277**, 9212–18 (2002).
42. M Fukuda and K Mikoshiba, *J Biol Chem*, **271**, 18838–42 (1996).
43. JA Koehler and MF Moran, *Cell Growth Differ*, **12**, 551–61 (2001).
44. JM Huibregtse, M Scheffner, S Beaudenon and PM Howley, *Proc Natl Acad Sci USA*, **92**, 2563–67 (1995).
45. S Kumar, KF Harvey, M Kinoshita, NG Copeland, M Noda and NA Jenkins, *Genomics*, **40**, 435–43 (1997).
46. CE Creutz, JL Tomsig, SL Snyder, MC Gautier, F Skouri, J Beisson and J Cohen, *J Biol Chem*, **273**, 1393–402 (1998).
47. JL Tomsig, SL Snyder and CE Creutz, *J Biol Chem*, **278**, 10048–54 (2003).
48. M Guipponi, HS Scott, H Chen, A Schebesta, C Rossier and SE Antonarakis, *Genomics*, **53**, 369–76 (1998).

49. BJ Perrin and A Huttenlocher, *Int J Biochem Cell Biol*, **34**, 722–25 (2002).
50. JA Trapani and MJ Smyth, *Nat Rev Immunol*, **2**, 735–47 (2002).
51. DB Davis, AJ Delmonte, CT Ly and EM McNally, *Hum Mol Genet*, **9**, 217–26 (2002).
52. F Volpe, J Clatworthy, A Kaptein, B Maschera, AM Griffin and K Ray, *FEBS Lett*, **419**, 41–44 (1997).
53. AG Murzin, SE Brenner, T Hubbard and C Chothia, *J Mol Biol*, **247**, 536–40 (1995).
54. JD Clark, AR Schievella, EA Nalefski and LL Lin, *J Lipid Mediators Cell signal*, **12**, 83–118 (1995).
55. RB Sutton, BA Davletov, AM Berghuis, TC Südhof and SR Sprang, *Cell*, **80**, 929–38 (1995).
56. M Fukuda, T Kojima, M Aruga, M Niinobe and K Mikoshiba, *J Biol Chem*, **270**, 26523–27 (1995).
57. K Ibata, M Fukuda and K Mikoshiba, *J Biol Chem*, **273**, 12267–73 (1998).
58. WF Ochoa, S Corbalan-Garcia, R Eritja, JA Rodriguez-Alfaro, JC Gómez-Fernández, I Fita and N Verdaguer, *J Mol Biol*, **320**, 277–91 (2002).
59. S Corbalan-Garcia, J Garcia-Garcia, JA Rodriguez-Alfaro and JC Gomez-Fernandez, *J Biol Chem*, **278**, 4972–80 (2003).
60. D Ron, J Luo and D Mochly-Rosen, *J Biol Chem*, **270**, 24180–87 (1995).
61. J Garcia-Garcia, S Corbalan-Garcia and JC Gomez-Fernandez, *Biochemistry*, **38**, 9667–75 (1999).
62. O Perisic, S Fong, DE Lynch, M Bycroft and RL Williams, *J Biol Chem*, **273**, 1596–604 (1998).
63. X Shao, BA Davletov, RB Sutton, TC Südhof and J Rizo, *Science*, **273**, 248–51 (1996).
64. N Verdaguer, S Corbalan-Garcia, WF Ochoa, I Fita and JC Gomez-Fernandez, *EMBO J*, **18**, 6329–38 (1999).
65. MD Bazzi and GL Nelsestuen, *Biochemistry*, **29**, 7624–31 (1990).
66. JA Grobler and JH Hurley, *Biochemistry*, **37**, 5020–28 (1998).
67. G-Y Xu, T McDonagh, H-A Yu, EA Nalefski, JD Clark and DA Cumming, *J Mol Biol*, **280**, 485–500 (1998).
68. EA Nalefski, MM Slazas and JJ Falke, *Biochemistry*, **36**, 12011–18 (1997).
69. N Brose, AG Petrenko, TC Sudhof and R Jahn, *Science*, **256**, 1021–25 (1992).
70. EA Nalefski, MA Wisner, JZ Chen, SR Sprang, M Fukuda, K Mikoshiba and JJ Falke, *Biochemistry*, **40**, 3089–100 (2001).
71. SC Kohout, S Corbalan-Garcia, A Torrecillas, JC Gomez-Fernandez and JJ Falke, *Biochemistry*, **41**, 11411–24 (2002).
72. J Bai, CA Earles, JL Lewis and ER Chapman, *J Biol Chem*, **275**, 25427–35 (2000).
73. J Bai, P Wang and ER Chapman, *Proc Natl Acad Sci USA*, **99**, 1665–70 (2002).
74. A Ball, R Nielsen, MH Gelb and BH Robinson, *Proc Natl Acad Sci USA*, **96**, 6637–42 (1999).
75. SC Kohout, S Corbalan-Garcia, JC Gomez-Fernandez and JJ Falke, *Biochemistry*, **42**, 1254–65 (2003).
76. RV Stahelin and W Cho, *Biochem J*, **359**, 679–85 (2001).
77. M Rebecchi, A Peterson and S McLaughlin, *Biochemistry*, **31**, 12742–47 (1992).
78. BA Davletov and TC Südhof, *J Biol Chem*, **268**, 26386–90 (1993).
79. X Zhang, J Rizo and TC Südhof, *Biochemistry*, **37**, 12395–403 (1998).
80. J Ubach, Y Lao, I Fernandez, D Arac, TC Sudhof and J Rizo, *Biochemistry*, **40**, 5854–60 (2001).
81. RB Sutton, JA Ernst and AT Brunger, *J Cell Biol*, **147**, 589–98 (1999).
82. RB Sutton and SR Sprang, *Structure*, **6**, 1395–405 (1998).
83. L-O Essen, O Perisic, R Cheung, M Katan and RL Williams, *Nature*, **380**, 595–602 (1996).
84. A Dessen, J Tang, H Schmidt, M Stahl, JD Clark, J Seehra and WS Somers, *Cell*, **97**, 349–60 (1999).
85. I Fernandez, D Arac, J Ubach, SH Gerber, OH Shin, Y Gao, RGW Anderson, TC Sudhof and J Rizo, *Neuron*, **32**, 1057–69 (2001).
86. SR Bolsover, JC Gomez-Fernandez and S Corbalan-Garcia, *J Biol Chem*, **78**, 10282–90 (2003).
87. JA Grobler, LO Essen, RL Williams and JH Hurley, *Nat Struct Biol*, **3**, 788–95 (1996).
88. X Shao, I Fernandez, TC Sudhof and J Rizo, *Biochemistry*, **37**, 16106–15 (1998).
89. LO Essen, O Perisic, DE Lynch, M Katan and RL Williams, *Biochemistry*, **36**, 2753–62 (1997).
90. E Oancea and T Meyer, *Cell*, **9599**, 307–18 (1998).

C2-domain proteins involved in membrane traffic

Josep Rizo

University of Texas Southwestern Medical Center at Dallas, TX, USA

FUNCTIONAL CLASS

Ca^{2+}- and phospholipid-binding protein modules.

C$_2$-domains are widespread protein modules of 120–150 amino acid residues.[1,2] More than 100 C$_2$-domains are present in the human genome. Ca^{2+} binding is the most defining activity of C$_2$-domains. They are the second-most abundant Ca^{2+}-binding modules in nature next to protein domains containing the EF-hand motif. Many C$_2$-domains bind phospholipids in a Ca^{2+}-dependent manner. However, some C$_2$-domains do not bind Ca^{2+}, and some bind phospholipids constitutively.

C$_2$-domains are particularly abundant in proteins involved in signal transduction, which are described in a separate chapter of this volume (see **Membrane binding C2-like domains**), and in proteins that function in membrane traffic, which are discussed in this chapter. While proteins in the former category typically contain only one C$_2$-domain, those involved in membrane traffic usually contain multiple C$_2$-domains that are commonly referred to as C$_2$A-domain,

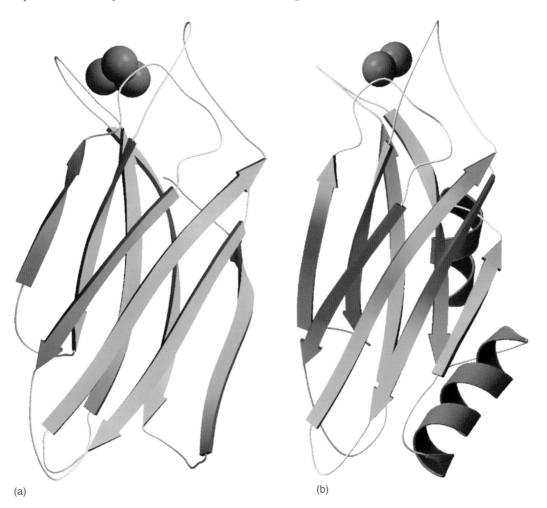

(a) (b)

3D Structure Ribbon diagrams of the three-dimensional structures of the Ca^{2+}-bound synaptotagmin 1 C$_2$A-domain (a) and C$_2$B-domain (b). The bound Ca^{2+} ions are displayed as orange spheres. PDB codes: 1BYN and 1K5W. Prepared with the programs MOLSCRIPT[99] and RASTER3D.[100]

C_2B-domain, and so on, according to their order in the sequence from N- to C-terminus. Two tandem C_2-domains that are separated by a short sequence are present in the C-terminus of many of these proteins, including those from the synaptotagmin family,[3] rabphilin,[4] Doc1/2[5] and B/K.[6] Other membrane traffic proteins such as munc13s,[7,8] RIMs,[9] piccolo/aczonin[10,11] and dysferlins[12,13] contain multiple C_2-domains that are separated by large sequences. This chapter will focus primarily on the synaptic vesicle protein synaptotagmin 1,[14] since this is the most extensively characterized among these proteins. However, studies of the structure and Ca^{2+}-binding properties of C_2-domains from additional membrane traffic proteins, including other synaptotagmin isoforms, rabphilin, and piccolo, will also be summarized to give an overview of the range of properties exhibited by these important Ca^{2+}-binding modules.

OCCURRENCE

Synaptotagmins constitute a large protein family with at least 15 human isoforms and about half as many isoforms in invertebrates [reviewed in 3]. Most synaptotagmins are neuronspecific proteins, but some isoforms are also expressed in nonneuronal cells. Synaptotagmin 1, the first isoform identified, is highly conserved from invertebrates to mammals. For instance, the amino acid sequences of the two C_2-domains from rat and drosophila synaptotagmin 1 exhibit a 78% identity, and are practically 100% identical among mammals.[15] Synaptotagmin 1 and the closely related synaptotagmin 2 are the most abundant synaptotagmins in mammals. Both these isoforms are localized to synaptic vesicles and secretory granules and are expressed in the brain in a largely nonoverlapping pattern.[14–17] In contrast, synaptotagmins 3, 6, and 7 are present in the plasma membrane, while the localization of other synaptotagmin isoforms is still under investigation [reviewed in reference 3].

Studies of rabphilin and piccolo have focused on a single isoform of each of these proteins. Rabphilin is conserved from invertebrates to mammals and is expressed in neurons and neuroendocrine cells in which it is localized in presynaptic terminals shuttling between the cytoplasm and synaptic vesicles together with the small GTPase Rab3A [reviewed in reference 18]. Piccolo is found in vertebrates but not in invertebrates, and is localized in the cytomatrix that forms the active zone of presynaptic terminals [reviewed in reference 19].

BIOLOGICAL FUNCTION

Synaptotagmin 1 is widely believed to act as the major Ca^{2+} sensor in the release of neurotransmitters by Ca^{2+}-evoked synaptic vesicle exocytosis.[3,20] Synaptotagmin 2 probably acts as an alternate Ca^{2+} sensor performing the same function as synaptotagmin 1, given the similarity of

Figure 1 Schematic representation of the domain structures of rat synaptotagmin 1 and rabphilin. The transmembrane region of synaptotagmin 1 is labeled TM, and the zinc-finger domain of rabphilin as ZF.

their sequences and their largely nonoverlapping expression patterns. The functions of other synaptotagmins are less clear, but it is likely that at least some of them act as complementary exocytotic Ca^{2+} sensors with a hierarchy of Ca^{2+} affinities.[21] The Ca^{2+}-sensing functions of synaptotagmins are executed by their tandem C_2-domains, which form most of their cytoplasmic region (exemplified by the domain diagram of synaptotagmin 1 shown in Figure 1).

Rabphilin was identified by virtue of its interaction with Rab3A, a small GTPase from the Rab family that plays a regulatory role in neurotransmitter release [reviewed in reference 18]. This interaction is mediated by an N-terminal Zn^{2+}-finger domain of rabphilin, while the tandem C_2-domains are located at the rabphilin C-terminus (Figure 1). Hence, rabphilin could act as an effector of Rab3A with a Ca^{2+}-sensing function that may be regulated by the GTP/GDP cycle of Rab3A. However, deletion of rabphilin does not lead to the synaptic defects observed in the absence of Rab3A.[22] Additional genetic experiments suggested that rabphilin potentiates the function of soluble N-ethylmaleimide sensitive factor attachment protein receptors (SNAREs),[23] which play a central role in membrane fusion,[24] but further research will be required to establish the precise function of rabphilin.

Piccolo is a large (>500 kDa) multidomain protein that exhibits extensive alternative splicing and is believed to play a scaffolding role by helping in organizing the structure of the active zone [reviewed in reference 19]. The most abundant splice forms contain one C-terminal C_2-domain (the C_2A-domain) that exhibits unusual Ca^{2+}-binding properties and could be involved in Ca^{2+}-dependent processes that regulate the efficiency of neurotransmitter release.[25] Some piccolo splice variants contain an additional C-terminal C_2-domain (the C_2B-domain) that is unlikely to bind Ca^{2+}.

AMINO ACID SEQUENCE INFORMATION

- Synaptotagmin 1, *Rattus norvegicus* (rat), 421 amino acid residues (AA), SWISS-PROT P21707.[14]
- Synaptotagmin 1, *Homo Sapiens* (human), 422 AA, SWISS-PROT P21579.[15]
- Synaptotagmin 1, *Drosophila Melanogaster* (fruit fly), 474 AA, SWISS-PROT P21521.[15]
- Synaptotagmin 2, *Rattus norvegicus* (rat), 422 AA, SWISS-PROT P29101.[17]

- Synaptotagmin 3, *Rattus norvegicus* (rat), 588 AA, SWISS-PROT P40748.[26]
- Synaptotagmin 4, *Rattus norvegicus* (rat), 425 AA, SWISS-PROT P50232.[16]
- Synaptotagmin 7, *Rattus norvegicus* (rat), 403 AA, SWISS-PROT Q62747.[27]
- Synaptotagmin 11, *Rattus norvegicus* (rat), 430 AA, SWISS-PROT O08835.[28]
- Rabphilin, *Rattus norvegicus* (rat), 684 AA, SWISS-PROT P47709.[29]
- Aczonin/Piccolo, *Rattus norvegicus* (rat), 5085 AA, SWISS-PROT Q9JKS6.[11]

PROTEIN PRODUCTION, PURIFICATION, AND MOLECULAR CHARACTERIZATION

Native synaptotagmin 1 can be purified from detergent extracts of rat brain using several fractionation and chromatographic steps.[30] However, most structural and biochemical studies of synaptotagmin 1, as well as of other synaptotagmin isoforms, rabphilin, and piccolo, have used recombinant fragments spanning the region or domain of interest. The C_2-domains from synaptotagmins, rabphilin, and piccolo can be obtained in large (milligram) quantities through bacterial expression as glutathione S-transferase fusion proteins, isolation from cell supernatants by affinity chromatography on glutathione-agarose, thrombin cleavage, and final purification by gel filtration and/or ion exchange chromatographies.[25,31–34] The structures of these C_2-domains have been analyzed by NMR spectroscopy and X-ray crystallography.[25,31–33,35–39] NMR spectroscopy has been particularly useful in studying the Ca^{2+}-binding properties of these C_2-domains because it allows the measurement of dissociation constants, and because their low intrinsic Ca^{2+} affinities hinder cocrystallization and/or saturation of the crystals with Ca^{2+}. Other biochemical properties of these C_2-domains, particularly phospholipid binding, have been studied with a range of techniques, most notably pull-down experiments with GST-fusions immobilized on glutathione-agarose.[40]

X-RAY AND NMR STRUCTURES

Crystallization and solution conditions for NMR structure determination

The C_2A-domain of rat synaptotagmin 1 (residues 140–265) was crystallized in 2.0 M Li_2SO_4, 100 mM BES (pH 7.2) at 21 °C.[31] The crystals were monoclinic, space group $P2_1$ with $a = 42.3$ Å, $b = 38.9$ Å, and $c = 44.7$ Å, and had one molecule per asymmetric unit. The NMR structures of the Ca^{2+}-free and Ca^{2+}-bound synaptotagmin 1 C_2A-domain (same fragment) were solved with the C_2A-domain (1.1 mM) dissolved in 40 mM sodium acetate (pH 5.0), 100 mM NaCl, and 0.2 mM EGTA, or 30 mM $CaCl_2$.[37] Resonance assignments were also obtained in 40 mM Tris (pH 7.4), 100 mM NaCl to determine its intrinsic Ca^{2+} affinities and Ca^{2+}-binding mode.[35,36] The NMR structure of the Ca^{2+}-bound rat rabphilin C_2B-domain (residues 524–684) was solved with samples of 0.7 to 1.2 mM C_2B-domain dissolved in 20 mM sodium acetate (pH 6.1), 150 mM NaCl, and 6 mM $CaCl_2$.[33] The NMR structure of the Ca^{2+}-bound rat synaptotagmin 1 C_2B-domain (residues 271–421) was solved with 1.3 mM samples of the C_2B-domain dissolved in 50 mM Mes (pH 6.3), 150 mM NaCl, and 20 mM $CaCl_2$.[38] The C_2AB-fragment of mouse synaptotagmin 3 (residues 259–569) was crystallized in 1.5 M $MgCl_2$, 100 mM MES (pH 6.5) at 20 °C.[39] The crystals were hexagonal, space group $P6_222$, with $a = 126$ Å, $b = 126$ Å, and $c = 118$ Å, and contained one molecule per asymmetric unit.

Structure of the synaptotagmin 1 C_2A-domain

The three-dimensional structure of the Ca^{2+}-free synaptotagmin 1 C_2A-domain determined by X-ray crystallography[31] provided the first structure at atomic resolution of a C_2-domain and revealed the general architecture of these widespread protein modules. Figure 2(a) shows a ribbon diagram of the structure, and a stereo diagram of the backbone is shown in Figure 3(a). The structure consists of a Greek-key β-sandwich formed by two four-stranded antiparallel β-sheets. Four β-bulges constitute a distinctive feature of the β-sandwich. One bulge is in the middle of strand 2 and two overlapping bulges are in the contiguous strand 5, yielding an unusual network of interstrand hydrogen bonds. The bulges induce an arch-like distortion in the sheet, which, as a result, has a largely convex surface. The opposite β-sheet is formed by two β-hairpins and has a concave surface. In this sheet, a fourth β-bulge causes a sharp bend at the top of strand 7 (according to the orientation of Figure 2(a)). Loops connecting the β-strands emerge at the top and the bottom of the β-sandwich. A short α-helix is present in the bottom loop connecting strands 5 and 6. This short helix has been observed in some C_2-domains and not in others, but note that such a short structural element may or may not be assigned as a helix by standard secondary structure analysis programs depending on small changes in dihedral angles and/or hydrogen bonding patterns. Hence, the presence or absence of this helix in the ribbon diagrams of published structures of C_2-domains (e.g. those in Figure 2) does not necessarily reflect a substantial structural difference.

Diffusion of 0.1 mM $CaCl_2$ into the crystals of the synaptotagmin 1 C_2A-domain allowed visualization of one bound Ca^{2+} ion, but the crystals cracked or dissolved at higher Ca^{2+} concentrations,[31] probably because of precipitation of $CaSO_4$. Studies in solution by NMR spectroscopy allowed determination of the complete Ca^{2+}-binding mode

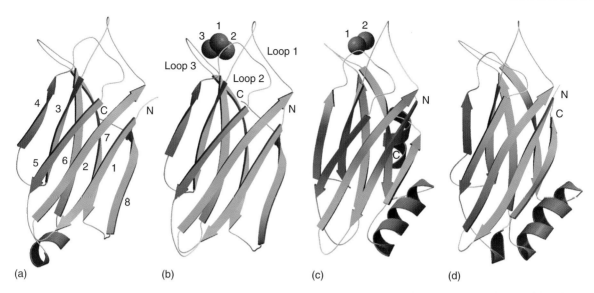

Figure 2 Ribbon diagrams of the structures of the synaptotagmins 1 C_2A-domain in the absence (a) and in the presence (b) of Ca^{2+}, of the synaptotagmin 1 C_2B-domain (c), and of the rabphilin C_2B-domain. The β-strands are numbered from 1 to 8 in (a), and the same numbering applies to the other three structures. The Ca^{2+} ions bound to the synaptotagmin 1 C_2A-domain (b) and C_2B-domain (c) are represented by orange spheres. The Ca^{2+}-binding loops are labeled loops 1 to 3 in (b). N and C indicate the N- and C-termini of each structure. PDB codes: 1RSY (a), 1BYN (b), 1K5W (c), and 3RPB (d). Prepared with the programs MOLSCRIPT[99] and RASTER3D.[100]

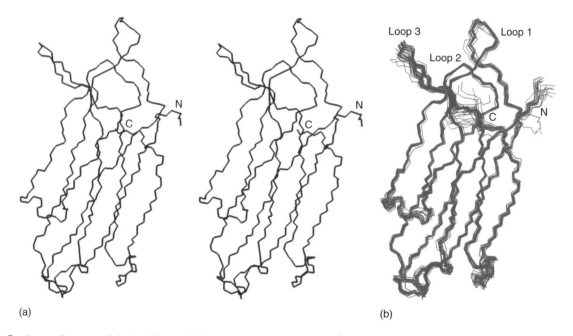

Figure 3 Stereo diagram of the backbone of the X-ray structure of the Ca^{2+}-free synaptotagmin 1 C_2A-domain (a), and backbone superposition of this structure (orange) with the ensemble of 20 NMR structures calculated for the Ca^{2+}-bound synaptotagmin 1 C_2A-domain (blue) (b). PDB codes 1RSY and 1BYN. Prepared with the program InsightII (MSI, San Diego, California).

of the synaptotagmin 1 C_2A-domain and of its Ca^{2+}-saturated three-dimensional structure (Figure 2(b)).[35–37] As described below in more detail, Ca^{2+} binds exclusively to the top loops of the C_2A-domain (loops 1–3). Surprisingly, no substantial differences were observed between the Ca^{2+}-bound solution structure and the Ca^{2+}-free X-ray structure (Figure 3(b)), except for changes in the rotameric states of the Ca^{2+} ligands. The NMR analysis in solution showed that the top loops are more flexible in the absence of Ca^{2+} than upon Ca^{2+} binding,[37] which correlates with the observation of decreased temperature factors in the C_2A-domain crystals upon partial Ca^{2+} binding.[31] Thus, Ca^{2+} binding to the synaptotagmin 1 C_2A-domain stabilizes its structure, but does not induce a change from a well-defined structure to another. These observations contrast with the typical behavior of EF-hand proteins, which usually exhibit major Ca^{2+}-induced structural changes that expose hydrophobic surfaces involved in binding to target molecules (see **Calmodulin** in this volume). The lack of substantial Ca^{2+}-induced conformational changes observed for the synaptotagmin 1 C_2A-domain is shared by most C_2-domains that have been studied in the absence and presence of Ca^{2+},[33,38,41] suggesting that C_2-domains generally function by a mechanism that is different from that commonly used by EF-hand proteins (see below).

A final point to note about the structure of the synaptotagmin 1 C_2A-domain is that its β-strand topology is different from that of the phospholipase C-δ1 (PLC-δ1) C_2-domain, the second C_2-domain whose structure was elucidated.[42] The two different topologies are referred to as topology I (for the synaptotagmin 1 C_2A-domain) and topology II (PLC-δ1 C_2-domain), and result from a circular permutation of the β-strands whereby the C-terminal strand from topology II occupies the same position in the β-sandwich as the N-terminal strand of topology I (Figure 4). Sequence analyses showed that C_2-domains can generally be aligned to one topology or the other,[2] suggesting that additional topologies resulting from other circular permutations of β-strands are unlikely. Topology II appears to be more abundant than topology I, but all known tandem C_2-domains involved in membrane traffic are predicted to adopt topology I, which has been confirmed experimentally by the structures determined so far.[33,38,39] The topological differences are important because they dictate the orientation of the contiguous domains in the protein sequence, thus influencing, for instance, the manner in which the tandem C_2-domains of synaptotagmin can cooperate with each other (see below).

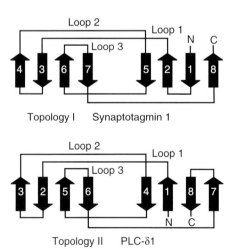

Figure 4 Topology diagrams of the synaptotagmin 1 C_2A-domain (above) and the PLC-δ1 C_2-domain (below). The N- and C-termini are indicated by N and C, respectively.

Structures of the rabphilin and synaptotagmin C_2B-domains

The sequences of the C_2A- and C_2B-domains of proteins with tandem C_2-domains exhibit evolutionarily conserved differences that suggest functional differentiation.[15] While initial sequence analyses suggested that the structures of C_2B-domains should be analogous to the structure of the synaptotagmin 1 C_2A-domain, the solution structure of the Ca^{2+}-bound rabphilin C_2B-domain determined by NMR spectroscopy (Figure 2(d)) revealed an important difference.[33] The overall structure of the rabphilin C_2B-domain also consists of an eight-stranded β-sandwich with topology I, and the NMR analysis showed that Ca^{2+} also binds at the top of the β-sandwich without inducing major conformational changes. However, a distinctive feature of the rabphilin C_2B-domain is an α-helix in the bottom loop connecting strands 7 and 8 (Figure 2(d)). Sequence comparisons indicated that this helix is present in all C_2B-domains from proteins with tandem C_2-domains, but not in their C_2A-domains.[33] In particular, a HW sequence in the center of this helix and a WH sequence in the C-terminal strand 8 constitute a signature that is conserved in all the C_2B-domains of these proteins. The two histidine residues from these sequences form an internal hydrogen bond that likely stabilizes the structure in this region of the domain. The fact that this α-helix is located at the bottom of the C_2B-domain, opposite the Ca^{2+}-binding region, suggests that this α-helix may mediate Ca^{2+}-independent activities that are specific to C_2B-domains. This proposal was supported by the observation that a 20-residue peptide containing the bottom α-helix of rabphilin has a selective inhibitory effect on the Ca^{2+}-independent component of secretion in chromaffin cells.[43] Altogether, these observations led to the notion that C_2-domains may act as Janus-faced modules with a Ca^{2+}-dependent top surface and a Ca^{2+}-independent bottom surface.[33]

The X-ray structure of a synaptotagmin 3 fragment containing the C_2A- and C_2B-domains in the absence of Ca^{2+}[39] and the solution structure of the Ca^{2+}-bound synaptotagmin 1 C_2B-domain[38] supported the prediction that all C_2B-domains from tandem C_2-domain proteins have similar structures. Thus, both the synaptotagmin 3 and synaptotagmin 1 C_2B-domains contain the bottom α-helix and the internal hydrogen bond between the characteristic histidines of the bottom helix and strand 8. However, the synaptotagmin 1 contains an additional α-helix at the very C-terminus that packs on the side of the β-sandwich (Figure 2(c)). This α-helix is also evolutionarily conserved in synaptotagmin 1s from different species, but is only shared by two other mammalian isoforms, synaptotagmins 2 and 8, suggesting that it may participate in interactions that are specific to this subset of synaptotagmins.[38] Because of the presence of two α-helices in the synaptotagmin 1 C_2B-domain that are not present in the C_2A-domain and were not predicted in initial sequence alignments, the

C_2B-domain is larger than it was initially thought to be and extends to the very C-terminus of the synaptotagmin 1 sequence. NMR analysis of the synaptotagmin 1 C_2B-domain also revealed that it binds Ca^{2+} through the top loops without undergoing substantial conformational changes, as observed for the C_2A-domain.[38]

Ca^{2+}-binding mode of the synaptotagmin 1 C_2A-domain

Studies of the Ca^{2+}-binding mode of the synaptotagmin 1 C_2A-domain by NMR spectroscopy helped in establishing how C_2-domains bind Ca^{2+} in general, and led to the development of new methodology to determine the location of Ca^{2+}-binding sites in proteins.[35,36] This methodology made extensive use of ^1H-^{15}N heteronuclear single quantum correlation (HSQC) spectra. These spectra contain one cross-peak for each nonproline residue in the molecule, and the positions of these cross-peaks are extremely sensitive to the chemical environment. Thus, ^1H-^{15}N HSQC spectra can be viewed as protein fingerprints and changes in these fingerprints due to binding to a ligand can be used to locate the ligand-binding site, as well as to measure the binding affinity.

Figure 5(a) illustrates the changes in the ^1H-^{15}N HSQC spectrum of the synaptotagmin 1 C_2A-domain caused by Ca^{2+} binding. Despite the absence of substantial conformational changes, some cross-peaks exhibit large Ca^{2+}-induced shifts that can be attributed to electrostatic changes, slight structural rearrangements, and Ca^{2+} coordination in some cases. All the Ca^{2+}-shifted cross-peaks correspond to residues from or around the top loops of the domain, showing that Ca^{2+} binding is restricted to this region of the molecule.[35] A Ca^{2+} titration revealed three components in the changes observed for some of the cross-peaks (illustrated for the cross-peak corresponding to D238 NH in Figure 6(a)), demonstrating the presence of three Ca^{2+}-binding sites.[36] The locations of these sites (referred to as sites Ca1, Ca2, and Ca3; see Figure 7(a)) were inferred from examination of the chemical shift changes, and they were confirmed by the effects that single-point mutations had on the individual components of the titration. For instance, mutation of S235 to alanine led to the loss of the third component (Figure 6(b)) since the S235 side chain forms part of the Ca3 site (Figure 7(a)). The locations of the sites were also confirmed by monitoring the broadening of ^1H-^{15}N HSQC cross-peaks upon addition of Mn^{2+}, a divalent metal ion that can substitute for Ca^{2+} in binding to the synaptotagmin 1 C_2A-domain and that causes broadening of resonances in a distance-dependent manner due to its paramagnetic nature.[36] The final Ca^{2+}-binding mode (Figures 7(a) and 8(a)) was obtained through progressive introduction of ligand/Ca^{2+}-ion restraints, based on the experimental data, into the structure calculations of the synaptotagmin 1 C_2A-domain.[37]

The three Ca^{2+} ions bind to the synaptotagmin 1 C_2A-domain in a tight cluster at the tip of the β-sandwich (Figures 2(b), 7(a) and 8(a)). One serine and five aspartate side chains, as well as three backbone carbonyl groups, participate in Ca^{2+} coordination. All these ligands are located in loops 1 and 3, but an amide group from loop 2 (K200 NH) also participates indirectly in binding by forming a hydrogen bond with the D178 side chain, thus helping in positioning the carboxylate group in the

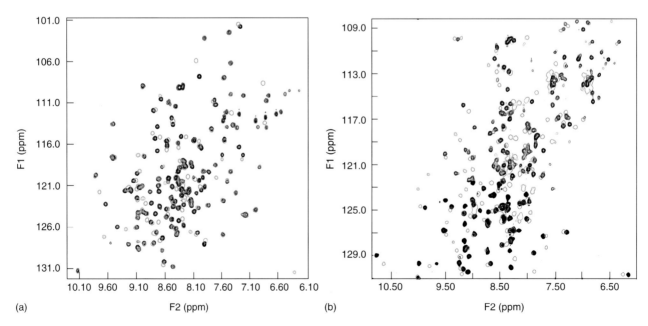

Figure 5 ^1H-^{15}N HSQC spectra of the synaptotagmin 1 C_2A-domain (a) and the piccolo C_2A-domain (b) in the absence (black contours) and in the presence of Ca^{2+} (red contours).

C2-domain proteins involved in membrane traffic

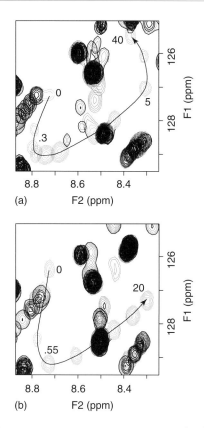

Figure 6 Expansions of superpositions of ^{1}H-^{15}N HSQC spectra of the wild type (a) and S235A mutant (b) synaptotagmin 1 C$_2$A-domain acquired at different Ca^{2+} concentrations. The cross-peak corresponding to D238 NH is highlighted in red and its progression through the Ca^{2+} titration is emphasized by arrows. The Ca^{2+} concentrations (in millimolar units) are indicated for a few cross-peaks.

proper orientation. The Ca^{2+}-binding mode of the synaptotagmin 1 C$_2$A-domain is analogous to that determined for the PKCβ C$_2$-domain by X-ray crystallography,[44] as can be observed by comparing the ligands that form sites Ca1 to Ca3 in both domains (Table 1). Only the ligand/Ca^{2+}

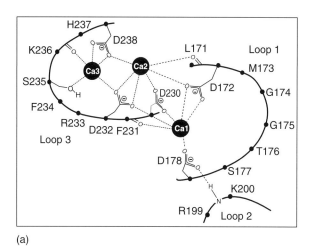

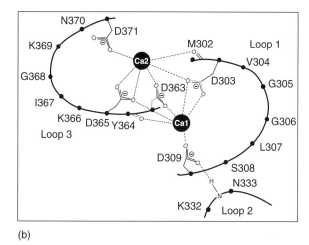

Figure 7 Schematic representations of the Ca^{2+}-binding modes of the synaptotagmin 1 C$_2$A-domain (a) and the synaptotagmin 1 C$_2$B-domain (b). The bound Ca^{2+} ions are represented by black spheres labeled Ca1 to Ca3. Only the side chains and backbone groups involved in Ca^{2+} binding are shown. The originally proposed nomenclature for the Ca^{2+}-binding sites has been used,[35] but note that different numbering is often used in the literature.

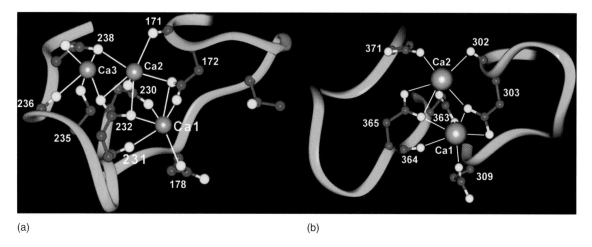

Figure 8 Three-dimensional views of the Ca^{2+}-binding sites of the synaptotagmin 1 C$_2$A-domain (a) and C$_2$B-domain (b). The residue numbers of the Ca^{2+} ligands are indicated. Prepared with the program InsightII (MSI, San Diego, Calfornia).

Table 1 Ca^{2+} ligands of the synaptotagmin 1 C$_2$A-domain and the PKCβ C$_2$-domain[a]

Ca^{2+}	Syt1 C$_2$A	PKCβ C$_2$	Distance
Ca1	D172 OD1	D187 OD1	2.62
Ca1	D172 OD2	D187 OD2	2.47
Ca1	D178 OD1	D193 OD1	2.32
Ca1	D230 OD1	D246 OD1	2.45
Ca1	F231 O	W247 O	2.34
Ca1	D232 OD1	D248 OD1	2.35
Ca2	L171 O	M186 O	2.64
Ca2	D172 OD1	D187 OD1	2.37
Ca2	D230 OD2	D246 OD2	2.42
Ca2	D232 OD1	D248 OD1	2.62
Ca2	D232 OD2	D248 OD2	2.51
Ca2	D238 OD2	D254 OD2	2.44
Ca3	D232 OD2	D248 OD2	2.49
Ca3	S235 OG	S251 OG	2.67
Ca3	K236 O	R252 O	2.52
Ca3	D238 OD1	D254 OD1	2.52
Ca3	D238 OD2	D254 OD2	3.38

[a] Ligands for the three Ca^{2+} bindings sites (Ca1–Ca3) are listed for the synaptotagmin 1 C$_2$A-domain, PDB code 1BYN (Syt1 C$_2$A), and the PKCβ C$_2$-domain, PDB code 1A25 (PKCβ C$_2$). The Ca^{2+}/ligand distances were measured from the crystal structure of the PKCβ C$_2$-domain.[44]

distances in the PKCβ C$_2$-domain are listed in Table 1 since the indirect nature of the information used to derive that Ca^{2+}-binding mode of the synaptotagmin 1 C$_2$A-domain by NMR spectroscopy does not allow accurate measurement of ligand/Ca^{2+} distances. The coordination spheres of the three Ca^{2+} ions bound to the synaptotagmin 1 C$_2$A-domain are incomplete (Figures 7(a) and 8(a)). Consequently, the intrinsic Ca^{2+} affinities of these sites measured from the ^{1}H-^{15}N HSQC Ca^{2+} titrations (Figure 6(a)) are weak [K_d 54 mM, 530 mM and >10 mM for sites Ca1, Ca2, and Ca3, respectively[20,36]]. Note that the lack of cooperativity indicated by the differences in binding constants is consistent with the absence of conformational changes upon Ca^{2+} binding. In contrast, Ca^{2+}-dependent phospholipid binding to the C$_2$A-domain is cooperative and occurs with much higher Ca^{2+} affinity (see below).

The C$_2$-motif

The five aspartates involved in Ca^{2+} binding to the synaptotagmin 1 C$_2$A-domain are conserved in many C$_2$-domains. This observation led to the proposal that the bipartite Ca^{2+} binding motif formed by sites Ca1 and Ca2, which was named the C$_2$-motif (Figure 9), is widespread in nature. This proposal was later supported experimentally by numerous structural studies of other Ca^{2+}-binding C$_2$-domains involved in membrane traffic or signal transduction.[33,38,41,44,45] However, additional Ca^{2+}-binding sites may exist in C$_2$-domains, such as site Ca3 in the synaptotagmin 1 C$_2$A-domain itself, or an additional site formed by ligands from loops 1 and 2 that is present in the PLC-δ1 and cPLA$_2$ C$_2$-domains (see **Membrane binding C2-like domains**).[41,46,47] It is important to note that the C$_2$-motif is very different from the EF-hand. Thus, while the EF-hand binds one single Ca^{2+} ion through contiguous sequences that form a helix-loop-helix structure, the C$_2$-motif forms two close Ca^{2+}-binding sites at the tip of a β-sandwich through residues that are distant in the sequence. Thus, the characteristic signature of the C$_2$-motif includes two groups of aspartate residues separated by approximately sixty residues; the first group contains two aspartates separated by five residues (X1 and X7, Figure 9), while the second group contains three aspartates separated by one and five residues (Y1, Y3, and Y9). Three of these aspartates (X1, Y1, and Y3) act as multidentate ligands that participate in Ca^{2+} binding to both site Ca1 and site Ca2 and form a 'triangle' with one binding site on each side. Note the pseudosymmetry of the motif, with a C$_2$-axis that goes through Y1 and relates Ca1 to Ca2, X1 to Y3, and X7 to Y9.[35]

Ca^{2+}-binding modes of other C$_2$-domains involved in membrane traffic

Analysis of the synaptotagmin 1 C$_2$B-domain by analogous NMR methods to those used for the C$_2$A-domain showed that the C$_2$B-domain binds two Ca^{2+} ions at sites Ca1 and Ca2 (Figures 7(b) and 8(b)), as predicted by the fact that it contains the five aspartate residues characteristic of the C$_2$-motif, but it has a glycine (G368) in the position homologous to S235 of the C$_2$A-domain.[38] In fact, mutation of G368 to serine induced binding of a third Ca^{2+} ion to site Ca3, indicating that the sequence of the C$_2$B-domain 'is designed' to bind two Ca^{2+} ions rather than three. Similarly, the rabphilin C$_2$B-domain was shown to bind two Ca^{2+}

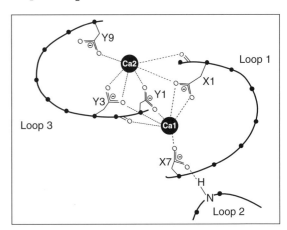

Figure 9 Schematic representation of the C$_2$-motif. The two aspartate Ca^{2+} ligands from loop 1 are labeled X1 and X7, and those from loop 3 are labeled Y1, Y3, and Y9.

ions as expected from sequence alignments.[33] However, the rabphilin C_2B-domain exhibits higher intrinsic Ca^{2+} affinities [K_d 7–11 μM;[33]] than the synaptotagmin 1 C_2B-domain [K_d 300–600 μM[38]]. Indeed, the intrinsic Ca^{2+} affinities of both the synaptotagmin 1 C_2A- and C_2B-domain are significantly lower than those of most C_2-domains studied so far. Although the apparent Ca^{2+} affinities of C_2-domains in the presence of phospholipids are more likely to be physiologically relevant (see below), the low intrinsic Ca^{2+} affinities of the synaptotagmin 1 C_2-domains bode well for a role in triggering neurotransmitter release, which is induced at significantly higher Ca^{2+} concentrations[48,49] than most intracellular biological processes regulated by Ca^{2+}. The total number of Ca^{2+} ions bound between the synaptotagmin 1 C_2A- and C_2B-domain (five) also correlates with the number of receptor Ca^{2+}-binding sites (five) that best fit the Ca^{2+}-dependence of neurotransmitter release in fast central synapses.[48,49] However, it is worth noting that this could be just a coincidence since synaptotagmin 1 is likely to function as a dimer or multimer (see below), and the apparent cooperativities observed in electrophysiological measurements may yield a low estimate of the number of Ca^{2+}-binding sites involved in triggering release.

In addition to synaptotagmin 1, intrinsic Ca^{2+} binding has been studied for two other synaptotagmin isoforms, synaptotagmins 4 and 11. The C_2A-domains of both these isoforms lack one of the central aspartate residues of the C_2-motif (Y1; see Figure 9) and, consequently, no Ca^{2+} binding to these C_2-domains was observed.[28] Mutation of the residue in the Y1 position to aspartate restored Ca^{2+} binding to both the synaptotagmin 4 and 11 C_2A-domains. These observations led to the proposal that evolutionary pressure may have inactivated Ca^{2+} binding to these C_2-domains, which presumably thus perform a Ca^{2+}-independent function.[28] It should be noted that the structure of the fragment containing the C_2A- and C_2B-domain of synaptotagmin 3[39] is often presented in the literature with four bound 'divalent cations', three bound to the C_2A-domain and one to the C_2B-domain, which might suggest that the C_2B-domain binds only one Ca^{2+} ion. However, these four ions are in fact Mg^{2+} ions that are electrostatically bound rather than fully coordinated by the protein, and Ca^{2+}-binding to synaptotagmin 3 was not analyzed in this study.[39] Overall, the available data indicate that the Ca^{2+}-binding sites present in C_2-domains can be generally predicted on the basis of the Ca^{2+}-binding sites found originally in the synaptotagmin 1 C_2A-domain[31,35] and the PLC-δ1 C_2-domain,[41] although exceptions may exist (see the example of synaptotagmin 6 below).

FUNCTIONAL ASPECTS

C_2-domains as multifunctional protein modules

The three-dimensional structures of C_2-domains involved in membrane traffic discussed above and those of C_2-domains from signal transduction proteins (MET050) generally share a similar β-sandwich, whereas they exhibit some variability in the loops connecting the β-strands. In all Ca^{2+}-binding C_2-domains studied, Ca^{2+} binding is restricted to the top loops. The picture that emerges from these observations is that the β-sandwich constitutes a scaffold for sequences at the top and the bottom, which may have conserved features and thus perform similar functions, or may form distinct structural elements that mediate specific roles. This stable scaffold also allows clustering of several negatively charged residues at the top that can bind Ca^{2+} without the need for conformational changes. The most common property of C_2-domains is Ca^{2+}-dependent phospholipid binding,[30,35,40,50,51] and this activity has been shown to be critical for Ca^{2+} regulation of C_2-domain containing proteins in several cases.[52–54] However, other Ca^{2+}-dependent and Ca^{2+}-independent interactions have been described for C_2-domains, including constitutive phospholipid binding, and some C_2-domains do not bind Ca^{2+} [reviewed in 1, 2]. Hence, C_2-domains are versatile protein modules with common and diverse functions. The results on synaptotagmin 1 and other proteins discussed below have brought insights into both specific functional aspects of these proteins and properties that are likely shared by many C_2-domains.

Synaptotagmin 1/SNARE interactions: the C_2A-domain as an electrostatic switch

The finding that synaptotagmin 1 is an abundant synaptic vesicle protein that binds phospholipids in a Ca^{2+}-dependent manner gave a first insight that this protein may act as a Ca^{2+}-sensor in neurotransmitter release.[14,30] This proposal was supported by genetic experiments showing a severe decrease in Ca^{2+}-triggered neurotransmitter release in the absence of synaptotagmin 1 in *Caenorhabditis elegans*[55] and in drosophila.[56,57] In mice, where a major, fast component, and a minor, slower component of release can be distinguished, deletion of synaptotagmin 1 abolished the fast component, whereas slow release was unaffected.[58] This observation showed that synaptotagmin 1 is critical for synchronous release but not for the asynchronous component. These and other results [reviewed in reference 59] spurred intense research to demonstrate that it indeed constitutes the major Ca^{2+} sensor in release and to understand the mechanism of action of synaptotagmin 1. In this context, the interactions of synaptotagmin 1 with SNARE proteins have attracted much attention.

The synaptic vesicle protein synaptobrevin and the plasma membrane proteins syntaxin and SNAP-25, which are collectively known as SNAREs, play a key role in neurotransmitter release and have homologues in most types of intracellular membrane fusion, suggesting that they constitute central components of a conserved membrane fusion machinery [reviewed in reference 24]. The three synaptic

C2-domain proteins involved in membrane traffic

SNAREs form a tight complex, known as the core complex, that brings the synaptic and plasma membranes together, which may cause membrane fusion.[60,61] Hence, synaptotagmin 1/SNARE interactions are interesting because they may couple Ca^{2+} sensing with membrane fusion. Synaptotagmin 1 and syntaxin were first found to bind constitutively,[62] and the isolated synaptotagmin 1 C_2A-domain was later shown to bind to syntaxin in a Ca^{2+}-dependent manner.[27] Many additional studies have described Ca^{2+}-dependent and Ca^{2+}-independent interactions between different synaptotagmin 1 fragments containing the C_2A-domain, the C_2B-domain, or both, and syntaxin, SNAP-25, syntaxin/SNAP-25 complexes or the core complex.[63–74] While some of these studies have provided evidence supporting the physiological relevance of these interactions, it is still unclear to what extent they arise from the promiscuity of both synaptotagmin 1 and the SNAREs in binding experiments performed *in vitro*.

Regardless of the potential relevance of synaptotagmin/SNARE interactions, analysis of the Ca^{2+}-dependent interaction between the synaptotagmin 1 C_2A-domain and syntaxin by NMR spectroscopy[34] provided critical insights to understand the nature of this interaction and, more generally, the mechanism of action of Ca^{2+}-dependent C_2-domains. The NMR studies showed that the C_2A-domain binds to syntaxin through the region surrounding the Ca^{2+}-binding sites. In the absence of Ca^{2+}, this region is characterized by a strongly negative electrostatic potential in the center owing to the five aspartate residues that form the C_2-motif, and a ring of positive electrostatic potential caused by several basic residues (Figures 10(a) and (c)). Upon Ca^{2+} binding, the electrostatic potential becomes highly positive (Figure 10(b)). Since syntaxin was known to be highly acidic, these observations suggested that electrostatic interactions play a critical role in mediating the Ca^{2+}-dependent binding of the C_2A-domain to syntaxin, and led to an electrostatic switch model of neurotransmitter release that attempted to explain the fact that release is triggered very fast by Ca^{2+} influx into a presynaptic terminal (<0.5 ms), yet it does not occur in the absence of Ca^{2+}.[34] According to this model, electrostatic repulsion between the synaptotagmin 1 C_2A-domain and syntaxin might prevent release before Ca^{2+} influx, whereas, upon Ca^{2+} binding to synaptotagmin 1, repulsion becomes attraction, activating release. An attractive feature of this model is that the absence of conformational changes in the Ca^{2+} receptor may facilitate the high speed of release.

As mentioned above, the lack of Ca^{2+}-dependent conformational changes that is common in C_2-domains contrasts with the fact that the Ca^{2+}-dependent interactions of EF-hand proteins usually involve Ca^{2+}-induced conformational changes, suggesting that C_2-domains generally function by a different mechanism. The NMR study of the synaptotagmin 1 C_2A-domain/syntaxin interaction provided an explanation for the different mechanisms of action of EF-hand proteins and C_2-domains, showing how changes in electrostatic potential, rather than in conformation, can induce Ca^{2+}-dependent interactions in C_2-domains. This mechanism is made possible by the ability of C_2-domains to bind multiple Ca^{2+} ions in a tight cluster

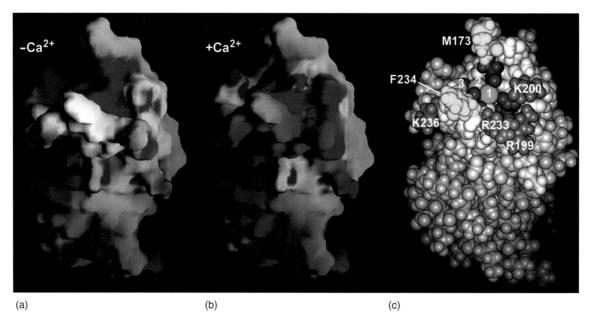

Figure 10 Electrostatic potential of the synaptotagmin 1 C_2A-domain in the absence (a) and in the presence of Ca^{2+} (b), and space filling model of its structure (c). In (a) and (b), negative electrostatic potential is represented in red and positive potential in blue. In (c), the bound Ca^{2+} ions are colored in green, the aspartate ligands in red, the basic residues surrounding the Ca^{2+}-binding sites in blue, and the hydrophobic residues exposed at the tips of loops 1 and 3 in yellow. Prepared with programs Grasp[101] and InsightII (MSI, San Diego, California).

at the tip of a β-sandwich (Figure 2). Clearly, other types of forces can also contribute to a greater or lesser extent to the Ca^{2+}-dependent interactions of C_2-domains but there is now little doubt that electrostatic forces play a substantial role in most cases,[75] as shown also by the studies of Ca^{2+}-dependent phospholipid binding summarized below. The importance of electrostatic interactions for Ca^{2+}-dependent binding of C_2-domains to target molecules can also explain the promiscuity of some C_2-domains. For instance, the synaptotagmin 1 C_2A-domain was shown to bind to a negatively charged surface of the N-terminal region of syntaxin,[76] but it also binds to the C-terminal sequence involved in core complex formation,[63,64,73] which is also highly acidic. The observation of additional Ca^{2+}-dependent interactions of synaptotagmin 1 with SNAP-25 and SNARE complexes can also be explained by their highly acidic nature. Thus, while the concept that the C_2-domains of synaptotagmin 1 act as electrostatic switches remains attractive, the main challenge is to identify their true physiological target(s).

Synaptotagmin 1 as a 'phospholipid-binding machine'

As mentioned above, Ca^{2+}-dependent phospholipid binding was the first Ca^{2+}-dependent activity identified for synaptotagmin 1.[30] This activity was initially mapped to the C_2A-domain and was shown to require negatively charged phospholipids without specificity for a particular headgroup.[40] These observations emphasized the importance of electrostatic interactions for phospholipid binding, and the NMR analysis of the C_2A-domain/syntaxin interaction suggested that phospholipid binding might involve a similar mechanism. This notion was confirmed by the observation that mutations of acidic residues involved in Ca^{2+} binding or basic residues around the Ca^{2+}-binding sites disrupt phospholipid binding.[77] On the other hand, Ca^{2+}-dependent phospholipid binding to the C_2A-domain occurs with high Ca^{2+} cooperativity and with a much higher apparent Ca^{2+} affinity [EC_{50} ca. 10 µM;[20,78]] than intrinsic Ca^{2+} binding (see above). This increased affinity correlates with that expected for the Ca^{2+}-sensor in neurotransmitter release[48,49] and is likely to arise because the phospholipid head groups complete the coordinate spheres of the Ca^{2+} ions bound to the C_2A-domain,[77] which has been demonstrated by crystallographic studies of the PKCα C_2-domain.[45] In addition, the observation of two exposed hydrophobic groups (M173 and F234) at the tip of loops 1 and 3 of the synaptotagmin 1 C_2A-domain (Figure 10(c)) suggested that hydrophobic interactions resulting from insertion of these residues into the lipid bilayer could also contribute to binding,[77] which was confirmed by fluorescence experiments[68,79] and mutational analysis.[80] Altogether, these observations suggest a model whereby phospholipids bind to the top of the synaptotagmin 1 C_2A-domain in a Ca^{2+}-dependent manner by a combination of electrostatic interactions, Ca^{2+} coordination by phospholipid head groups, and hydrophobic forces.[77] Definition of the membrane-bound orientation and position of the C_2A-domain by EPR experiments has largely supported this model.[81] This binding mode is likely to be shared by many C_2-domains, although the relative contributions from different types of forces may differ. For instance, the $cPLA_2$ C_2-domain binds preferentially to neutral phospholipids in an interaction dominated by hydrophobic forces[82] in which Ca^{2+} binding probably acts to neutralize negative charges. It is also interesting to note that the synaptotagmin 1 C_2A-domain has a particularly fast kinetics of binding/unbinding to Ca^{2+}/phospholipids,[68] which bodes well for a role in triggering fast neurotransmitter release.

Initial studies of the synaptotagmin 1 C_2B-domain using GST-pulldown experiments did not reveal Ca^{2+}-dependent phospholipid binding,[83,84] but acidic bacterial contaminants in the GST-fusion proteins used might have interfered with these assays,[32] and elucidation of the C_2B-domain three-dimensional structure did not reveal any significant differences with respect to the C_2A-domain that would justify the different behavior of the two C_2-domain.[38] Indeed, experiments performed in solution with highly purified isolated C_2B-domain revealed Ca^{2+}-dependent phospholipid binding with similar Ca^{2+}-dependence to that observed for the C_2A-domain.[38] This critical result showed that the two C_2-domains of synaptotagmin 1 can cooperate in a common activity – phospholipid binding. Both C_2-domains are linked by a short sequence. Although the two C_2-domains have been suggested to interact intramolecularly in a Ca^{2+}-dependent manner,[85] this result could not be confirmed by NMR experiments, which showed that the two C_2-domains are flexibly linked in the absence and presence of Ca^{2+} (J. Ubach and J. Rizo, unpublished). On the other hand, the topology of the synaptotagmin 1 C_2-domains (topology I) dictates that their N- and C-terminal ends emerge at the top of the β-sandwich (Figures 2 and 4). Because of these features, the Ca^{2+}-binding sites of both C_2-domains can be easily oriented next to each other so that they can cooperate in phospholipid binding (Figure 11),[38] which has been confirmed by the finding that a fragment containing both C_2-domains of synaptotagmin 1 (C_2AB-fragment) binds phospholipids with a higher apparent Ca^{2+} affinity than the individual C_2-domains.[86] These observations have been critical to rationalize genetic experiments of synaptotagmin 1 function (see below) and have suggested that synaptotagmin 1 may act as a 'phospholipid-binding machine'.[38]

Function of synaptotagmin 1

The characterization of the Ca^{2+}-dependent interactions of the synaptotagmin 1 C_2-domains summarized above

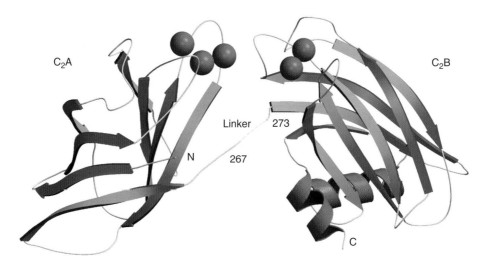

Figure 11 Ribbon diagrams of the two synaptotagmin 1 C_2-domains illustrating how they can easily orient their Ca^{2+}-binding surfaces close to each other to cooperate in phospholipid binding. The N-terminus of the C_2A-domain is labeled N and the C-terminus of the C_2B-domain is labeled C. The C-terminus of the C_2A-domain and the N-terminus of the C_2B-domain are labeled with their corresponding residue numbers, and the short linker between them is represented by a dashed yellow line. PDB codes 1BYN and 1K5W. Prepared with programs MOLSCRIPT[99] and RASTER3D.[100]

allowed the design of single-point mutations to test whether synaptotagmin 1 acts as a Ca^{2+} sensor in release by genetic and electrophysiological experiments. Particularly useful for this purpose was an R233Q mutation in the C_2A-domain. Because R233 participates in phospholipid binding and the lipids in turn help to bind Ca^{2+}, the R233Q mutation leads to a two- to threefold decrease in the apparent Ca^{2+} affinity of synaptotagmin 1 in Ca^{2+}-dependent phospholipid-binding assays.[20] Importantly, introduction of this mutation in the endogenous synaptotagmin 1 gene in mice leads to a corresponding shift in the Ca^{2+}-dependence of neurotransmitter release.[20] It is important to note that this mutation only represents a change of a few atoms in an exposed side chain of a 421-residue protein without affecting its structure, and hence is much less likely to cause indirect effects than removal of the whole protein. Hence, the observation that such a small change results in a substantial shift in the Ca^{2+}-dependence of neurotransmitter release provides very strong evidence supporting the proposed role for synaptotagmin 1 as the major Ca^{2+} sensor in release.

Surprisingly, analogous experiments with individual point mutations in two of the Ca^{2+} ligands of the synaptotagmin 1 C_2A-domain (D232N and D238N) cause little effect on neurotransmitter release even though these mutations severely disrupt Ca^{2+}-dependent phospholipid binding to the isolated C_2A-domain in vitro.[86] This apparent paradox was solved by examination of Ca^{2+}-dependent phospholipid binding to the synaptotagmin 1 C_2AB-fragment, which was unaffected by the same mutations.[86] Because phospholipid binding involves multiple interactions contributed by the two C_2-domains, it is not surprising that the mutations have a much stronger effect on binding to the isolated C_2A-domain than to the C_2AB-fragment. Note that, in drosophila, mutation of the aspartate corresponding to D178 of rat synaptotagmin 1 to asparagine (called D2N mutation) also disrupted phospholipid binding to the isolated C_2A-domain, whereas expression of synaptotagmin 1 bearing this mutation rescued neurotransmitter release in synaptotagmin 1 nulls.[87] However, neurotransmitter release was decreased in the D2N rescue experiments compared to rescues with wild-type synaptotagmin 1, in correlation with the observation that the D2N mutation only disrupts phospholipid binding partially when tested with the C_2AB fragment. On the other hand, mutations in the Ca^{2+}-binding sites of the drosophila synaptotagmin 1 C_2B-domain severely decreased Ca^{2+}-dependent phospholipid binding in vitro and Ca^{2+}-triggered neurotransmitter release in vivo.[88]

These genetic experiments have further supported the role of synaptotagmin 1 as the major Ca^{2+} sensor that triggers neurotransmitter release and have suggested that Ca^{2+}-dependent phospholipid binding constitutes its most physiologically relevant activity. This conclusion has also been supported by the observation of a strong correlation between the apparent Ca^{2+} affinity of synaptotagmin C_2-domain variants in lipid binding and their ability to inhibit secretion in PC12 cells, while SNARE binding exhibited no clear correlation.[89] Competition experiments have shown that Ca^{2+}-dependent binding of the synaptotagmin 1 C_2AB-fragment to the core complex is incompatible with phospholipid binding, and the phospholipids displace the core complex bound to the C_2AB-fragment.[90] Furthermore, Sr^{2+} can support neurotransmitter release and phospholipid binding to synaptotagmin 1, but not synaptotagmin 1/SNARE interactions.[74] It is currently unclear how Ca^{2+}-dependent phospholipid binding to synaptotagmin 1 triggers neurotransmitter release. It has been suggested that binding of synaptotagmin 1 to the plasma membrane helps the synaptic vesicles to approach the plasma membrane, facilitating core complex formation and membrane fusion.[91] However, this model seems incompatible with the high speed of neurotransmitter release, and alternative models envisage that the core complex is assembled before Ca^{2+} influx, perhaps with the help of complexin, a protein that binds tightly to the core complex [reviewed in reference 24]. In these models, the membranes may already be close before excitation, or even hemifused. Ca^{2+}-dependent binding of synaptotagmin 1 to the lipids may then exert membrane tension, which may be necessary to complete membrane merger and opening of the fusion pore.

While these are plausible models of synaptotagmin 1 function, the stronger disruption of neurotransmitter release caused by mutations in the Ca^{2+}-binding sites of the C_2B-domain,[88] compared to the effects of C_2A-domain mutations,[86,87] suggest that an additional Ca^{2+}-dependent activity associated with the C_2B-domain may also contribute to triggering release. In this context, the C_2B-domain has been implicated in Ca^{2+}-dependent dimerization of synaptotagmin 1.[92,93] While it was later found that recombinant C_2B-domain does not dimerize in solution,[32,85] a recent study has shown the formation of synaptotagmin 1 oligomers through the C_2B-domain upon Ca^{2+}-dependent phospholipid binding.[94] Thus, it is possible that the action of synaptotagmin 1 on the membranes during neurotransmitter release is aided by oligomerization. It should be noted that the C_2B-domain has also been implicated in a number of Ca^{2+}-independent interactions, including binding to the clathrin adaptor protein AP-2,[95] Ca^{2+} channels,[67] and inositol polyphosphates.[96] The significance of some of these interactions needs to be further studied since they all seem to involve a highly basic region of the C_2B-domain that has also been implicated in dimerization,[97] and has a high tendency to bind acidic bacterial contaminants,[32] but several lines of evidence have supported a role for the synaptotagmin 1/AP-2 interaction in endocytosis [reviewed in reference 98]. Overall, the results summarized above indicate that the synaptotagmin 1 C_2-domains act as multifunctional modules that may couple exocytosis to endocytosis.

Function of other C_2-domains involved in membrane traffic

Although the functions of other synaptotagmins are less understood than that of synaptotagmins 1 and 2, the abundance of synaptotagmin isoforms in mammals suggests that nature has taken advantage of the versatility of C_2-domains to generate a family of proteins with related but differentiated roles. Thus, studies of different synaptotagmins have revealed shared and distinct biochemical properties [reviewed in reference 3]. For instance, the C_2A-domains of synaptotagmins that contain the C_2-motif (e.g. synaptotagmins 3, 5, and 7) generally bind phospholipids and syntaxin in a Ca^{2+}-dependent manner, while these activities are not observed for those in which the C_2-motif is not complete (e.g. synaptotagmins 4, 8, and 11).[27,28] However, no Ca^{2+}-dependent phospholipid binding was observed for the synaptotagmin 6 C_2A-domain, even though it contains all five aspartate residues that form the C_2-motif,[27] suggesting that perhaps some structural feature may have rendered this C_2-domain Ca^{2+} independent. On the other hand, the apparent Ca^{2+} affinities of synaptotagmins 3 and 7, which have been localized to the plasma membrane, are significantly higher than those of synaptotagmin 1.[21] These observations suggest that the plasma membrane synaptotagmins may mediate the asynchronous component of neurotransmitter release, which is associated to a Ca^{2+} sensor with a higher affinity than that of the synchronous Ca^{2+} sensor, and that synaptotagmins provide a hierarchy of exocytotic Ca^{2+} sensors with different Ca^{2+} affinities.[3,21] In addition, at least some synaptotagmin isoforms are likely to play Ca^{2+}-independent roles.

The functions of the C_2-domains from other proteins involved in membrane traffic such as rabphilin, Doc2s, B/K, munc13s, RIMs, dysferlins, and piccolo are also less understood than those of the synaptotagmin 1 C_2-domains, and three-dimensional structural information has only been described for the rabphilin C_2B-domain (see above). However, a recent study has shown that the Piccolo C_2A-domain has an unusual behavior that emphasizes the range of properties of C_2-domains.[25] The sequence of this C_2-domain is similar to those of the synaptotagmin C_2A-domains, including all residues involved in Ca^{2+} binding, but contains a unique nine-residue sequence in a predicted bottom loop (connecting strands 3 and 4). Comparison of the ^1H-^{15}N HSQC spectra of the piccolo C_2A-domain in the absence and presence of Ca^{2+} (Figure 5(b)) revealed widespread Ca^{2+}-induced changes that contrast with the observation of changes in only a subset of cross-peaks observed for the synaptotagmin 1 and rabphilin C_2-domains (e.g. Figure 5(a)). This and other results showed that, unlike most C_2-domains, the piccolo C_2A-domain undergoes an overall conformational change upon Ca^{2+} binding.[25] In addition, the piccolo C_2A-domain was shown to dimerize as a function of Ca^{2+} and to have substantially lower intrinsic and apparent Ca^{2+} affinities than those of the synaptotagmin 1 C_2A-domain. These affinities increased upon mutation of residues in the unique bottom sequence, suggesting that this sequence induces a conformation incompatible with Ca^{2+} binding. Altogether, these observations suggested that the piccolo C_2A-domain may play a role in presynaptic plasticity processes when Ca^{2+} concentrations accumulate as a result of repetitive stimulation.[25] The unusual properties of the piccolo C_2A-domain show that the range of mechanisms of action of C_2-domains is more diverse than what it was previously thought to be. Future studies of other C_2-domains are likely to further emphasize this versatility.

REFERENCES

1. J Rizo and TC Sudhof, *J Biol Chem*, **273**, 15879–82 (1998).
2. EA Nalefski and I Falke, *J Protein Sci*, **5**, 2375–90 (1996).
3. TC Sudhof, *J Biol Chem*, **277**, 7629–32 (2002).
4. H Shirataki, K Kaibuchi, T Sakoda, S Kishida, T Yamaguchi, K Wada, M Miyazaki and Y Takai, *Mol Cell Biol*, **13**, 2061–68 (1993).
5. S Orita, T Sasaki, A Naito, R Komuro, T Ohtsuka, M Maeda, H Suzuki, H Igarashi and Y Takai, *Biochem Biophys Res Commun*, **206**, 439–48 (1995).
6. OJ Kwon, H Gainer, S Wray and H Chin, *FEBS Lett*, **378**, 135–39 (1996).
7. IN Maruyama and S Brenner, *Proc Natl Acad Sci USA*, **88**, 5729–33 (1991).
8. N Brose, K Hofmann, Y Hata and TC Sudhof, *J Biol Chem*, **270**, 25273–80 (1995).
9. Y Wang, M Okamoto, F Schmitz, K Hofmann and TC Sudhof, *Nature*, **388**, 593–98 (1997).
10. X Wang, M Kibschull, MM Laue, B Lichte, E Petrasch-Parwez and MW Kilimann, *J Cell Biol*, **147**, 151–62 (1999).
11. SD Fenster, WJ Chung, R Zhai, C Cases-Langhoff, B Voss, AM Garner, U Kaempf, S Kindler, ED Gundelfinger and CC Garner, *Neuron*, **25**, 203–14 (2000).
12. J Liu, M Aoki, I Illa, C Wu, M Fardeau, C Angelini, C Serrano, JA Urtizberea, F Hentati, MB Hamida, S Bohlega, EJ Culper, AA Amato, K Bossie, J Oeltjen, K Bejaoui, D McKenna-Yasek, BA Hosler, E Schurr, K Arahata, PJ de Jong and RH Brown Jr, *Nat Genet*, **20**, 31–36 (1998).
13. R Bashir, S Britton, T Strachan, S Keers, E Vafiadaki, M Lako, I Richard, S Marchand, N Bourg, Z Argov, M Sadeh, I Mahjneh, G Marconi, MR Passos-Bueno, ES Moreira, M Zatz, JS Beckmann and K Bushby, *Nat Genet*, **20**, 37–42 (1998).
14. MS Perin, VA Fried, GA Mignery, R Jahn and TC Sudhof, *Nature*, **345**, 260–63 (1990).
15. MS Perin, PA Johnston, T Ozcelik, R Jahn, U Francke and TC Sudhof, *J Biol Chem*, **266**, 615–22 (1991).
16. B Ullrich, C Li, JZ Zhang, H McMahon, RG Anderson, M Geppert and TC Sudhof, *Neuron*, **13**, 1281–91 (1994).
17. M Geppert, BT Archer III and TC Sudhof, *J Biol Chem*, **266**, 13548–52 (1991).
18. M Geppert and TC Sudhof, *Annu Rev Neurosci*, **21**, 75–95 (1998).
19. CC Garner, S Kindler and ED Gundelfinger, *Curr Opin Neurobiol*, **10**, 321–27 (2000).
20. R Fernandez-Chacon, A Konigstorfer, SH Gerber, J Garcia, MF Matos, CF Stevens, N Brose, J Rizo, C Rosenmund and TC Sudhof, *Nature*, **410**, 41–49 (2001).
21. S Sugita, OH Shin, W Han, Y Lao and TC Sudhof, *EMBO J*, **21**, 270–80 (2002).
22. OM Schluter, E Schnell, M Verhage, T Tzonopoulos, RA Nicoll, R Janz, RC Malenka, M Geppert and TC Sudhof, *J Neurosci*, **19**, 5834–46 (1999).
23. J Staunton, B Ganetzky and ML Nonet, *J Neurosci*, **21**, 9255–64 (2001).
24. J Rizo and TC Sudhof, *Nat Rev Neurosci*, **3**, 641–53 (2002).
25. SH Gerber, J Garcia, J Rizo and TC Sudhof, *EMBO J*, **20**, 1605–19 (2001).
26. M Mizuta, N Inagaki, Y Nemoto, S Matsukura, M Takahashi and S Seino, *J Biol Chem*, **269**, 11675–78 (1994).
27. C Li, B Ullrich, JZ Zhang, RG Anderson, N Brose and TC Sudhof, *Nature*, **375**, 594–99 (1995).
28. C von Poser, K Ichtchenko, X Shao, J Rizo and TC Sudhof, *J Biol Chem*, **272**, 14314–19 (1997).
29. C Li, K Takei, M Geppert, L Daniell, K Stenius, ER Chapman, R Jahn, P De Camilli and TC Sudhof, *Neuron*, **13**, 885–98 (1994).
30. N Brose, AG Petrenko, TC Sudhof and R Jahn, *Science*, **256**, 1021–25 (1992).
31. RB Sutton, BA Davletov, AM Berghuis, TC Sudhof and SR Sprang, *Cell*, **80**, 929–38 (1995).
32. J Ubach, Y Lao, I Fernandez, D Arac, TC Sudhof and J Rizo, *Biochemistry*, **40**, 5854–60 (2001).
33. J Ubach, J Garcia, MP Nittler, TC Sudhof and J Rizo, *Nat Cell Biol*, **1**, 106–12 (1999).
34. X Shao, C Li, I Fernandez, X Zhang, TC Sudhof and J Rizo, *Neuron*, **18**, 133–42 (1997).
35. X Shao, BA Davletov, RB Sutton, TC Sudhof and J Rizo, *Science*, **273**, 248–51 (1996).
36. J Ubach, X Zhang, X Shao, TC Sudhof and J Rizo, *EMBO J*, **17**, 3921–30 (1998).
37. X Shao, I Fernandez, TC Sudhof and J Rizo, *Biochemistry*, **37**, 16106–15 (1998).
38. I Fernandez, D Arac, J Ubach, SH Gerber, O Shin, Y Gao, RG Anderson, TC Sudhof and J Rizo, *Neuron*, **32**, 1057–69 (2001).
39. RB Sutton, JA Ernst and AT Brunger, *J Cell Biol*, **147**, 589–98 (1999).
40. BA Davletov and TC Sudhof, *J Biol Chem*, **268**, 26386–90 (1993).
41. LO Essen, O Perisic, DE Lynch, M Katan and RL Williams, *Biochemistry*, **36**, 2753–62 (1997).
42. LO Essen, O Perisic, R Cheung, M Katan and RL Williams, *Nature*, **380**, 595–602 (1996).
43. SH Chung, WJ Song, K Kim, JJ Bednarski, J Chen, GD Prestwich and RW Holz, *J Biol Chem*, **273**, 10240–48 (1998).
44. RB Sutton and SR Sprang, *Structure*, **6**, 1395–1405 (1998).
45. N Verdaguer, S Corbalan-Garcia, WF Ochoa, I Fita and JC Gomez-Fernandez, *EMBO J*, **18**, 6329–38 (1999).
46. O Perisic, S Fong, DE Lynch, M Bycroft and RL Williams, *J Biol Chem*, **273**, 1596–1604 (1998).
47. JA Grobler, LO Essen, RL Williams and JH Hurley, *Nat Struct Biol*, **3**, 788–95 (1996).
48. JH Bollmann, B Sakmann and JG Borst, *Science*, **289**, 953–57 (2000).
49. R Schneggenburger and E Neher, *Nature*, **406**, 889–93 (2000).
50. PJ Plant, H Yeger, O Staub, P Howard and D Rotin, *J Biol Chem*, **272**, 32329–36 (1997).
51. EA Nalefski, LA Sultzman, DM Martin, RW Kriz, PS Towler, JL Knopf and JD Clark, *J Biol Chem*, **269**, 18239–49 (1994).
52. R Uellner, MJ Zvelebil, J Hopkins, J Jones, LK MacDougall, BP Morgan, E Podack, MD Waterfield and GM Griffiths, *EMBO J*, **16**, 7287–96 (1997).
53. EA Nalefski and JJ Falke, *Biochemistry*, **37**, 17642–50 (1998).

54. AS Edwards and AC Newton, *Biochemistry*, **36**, 15615–23 (1997).
55. ML Nonet, K Grundahl, BJ Meyer and JB Rand, *Cell*, **73**, 1291–1305 (1993).
56. A DiAntonio, KD Parfitt and TL Schwarz, *Cell*, **73**, 1281–90 (1993).
57. JT Littleton, M Stern, K Schulze, M Perin and HJ Bellen, *Cell*, **74**, 1125–34 (1993).
58. M Geppert, Y Goda, RE Hammer, C Li, TW Rosahl, CF Stevens and TC Sudhof, *Cell*, **79**, 717–27 (1994).
59. TC Sudhof and J Rizo, *Neuron*, **17**, 379–88 (1996).
60. PI Hanson, R Roth, H Morisaki, R Jahn and JE Heuser, *Cell*, **90**, 523–35 (1997).
61. T Weber, BV Zemelman, JA McNew, B Westermann, M Gmachl, F Parlati, TH Sollner and JE Rothman, *Cell*, **92**, 759–72 (1998).
62. MK Bennett, N Calakos and RH Scheller, *Science*, **257**, 255–59 (1992).
63. ER Chapman, PI Hanson, S An and R Jahn, *J Biol Chem*, **270**, 23667–71 (1995).
64. Y Kee and RH Scheller, *J Neurosci*, **16**, 1975–81 (1996).
65. T Sollner, MK Bennett, SW Whiteheart, RH Scheller and JE Rothman, *Cell*, **75**, 409–18 (1993).
66. G Schiavo, G Stenbeck, JE Rothman and TH Sollner, *Proc Natl Acad Sci USA*, **94**, 997–1001 (1997).
67. ZH Sheng, CT Yokoyama and WA Catterall, *Proc Natl Acad Sci USA*, **94**, 5405–10 (1997).
68. AF Davis, J Bai, D Fasshauer, MJ Wolowick, JL Lewis and ER Chapman, *Neuron*, **24**, 363–76 (1999).
69. RR Gerona, EC Larsen, JA Kowalchyk and TF Martin, *J Biol Chem*, **275**, 6328–36 (2000).
70. X Zhang, MJ Kim-Miller, M Fukuda, JA Kowalchyk and TF Martin, *Neuron*, **34**, 599–611 (2002).
71. E Chieregatti, JW Witkin and G Baldini, *Traffic*, **3**, 496–511 (2002).
72. C Rickman and B Davletov, *J Biol Chem*, **278**, 5501–4 (2003).
73. MF Matos, J Rizo and TC Sudhof, *Eur J Cell Biol*, **79**, 377–82 (2000).
74. OH Shin, JS Rhee, J Tang, S Sugita, C Rosenmund and TC Sudhof, *Neuron*, **37**, 99–108 (2003).
75. D Murray and B Honig, *Mol Cell*, **9**, 145–54 (2002).
76. I Fernandez, J Ubach, I Dulubova, X Zhang, TC Sudhof and J Rizo, *Cell*, **94**, 841–49 (1998).
77. X Zhang, J Rizo and TC Sudhof, *Biochemistry*, **37**, 12395–403 (1998).
78. B Davletov, JM Sontag, Y Hata, AG Petrenko, EM Fykse, R Jahn and TC Sudhof, *J Biol Chem*, **268**, 6816–22 (1993).
79. ER Chapman and AF Davis, *J Biol Chem*, **273**, 13995–14001 (1998).
80. SH Gerber, J Rizo and TC Sudhof, *Diabetes*, **51**(Suppl. 1), S12–S18 (2002).
81. AA Frazier, CR Roller, JJ Havelka, A Hinderliter and DS Cafiso, *Biochemistry*, **42**, 96–105 (2003).
82. B Davletov, O Perisic and RL Williams, *J Biol Chem*, **273**, 19093–96 (1998).
83. G Schiavo, QM Gu, GD Prestwich, TH Sollner and JE Rothman, *Proc Natl Acad Sci USA*, **93**, 13327–32 (1996).
84. J Bai, CA Earles, JL Lewis and ER Chapman, *J Biol Chem*, **275**, 25427–35 (2000).
85. RA Garcia, CE Forde and HA Godwin, *Proc Natl Acad Sci USA*, **97**, 5883–88 (2000).
86. R Fernandez-Chacon, OH Shin, A Konigstorfer, MF Matos, AC Meyer, J Garcia, SH Gerber, J Rizo, TC Sudhof and C Rosenmund, *J Neurosci*, **22**, 8438–46 (2002).
87. IM Robinson, R Ranjan and TL Schwarz, *Nature*, **418**, 336–40 (2002).
88. JM Mackler, JA Drummond, CA Loewen, IM Robinson and NE Reist, *Nature*, **418**, 340–44 (2002).
89. OH Shin, J Rizo and TC Sudhof, *Nat Neurosci*, **5**, 649–56 (2002).
90. D Arac, T Murphy and J Rizo, *Biochemistry*, **42**, 2774–80 (2003).
91. K Hu, J Carroll, S Fedorovich, C Rickman, A Sukhodub and B Davletov, *Nature*, **415**, 646–50 (2002).
92. S Sugita, Y Hata and TC Sudhof, *J Biol Chem*, **271**, 1262–65 (1996).
93. ER Chapman, S An, JM Edwardson and R Jahn, *J Biol Chem*, **271**, 5844–49 (1996).
94. Y Wu, Y He, J Bai, SR Ji, WC Tucker, ER Chapman and SF Sui, *Proc Natl Acad Sci USA*, **100**, 2082–87 (2003).
95. JZ Zhang, BA Davletov, TC Sudhof and RG Anderson, *Cell*, **78**, 751–60 (1994).
96. M Fukuda, M Aruga, M Niinobe, S Aimoto and K Mikoshiba, *J Biol Chem*, **269**, 29206–11 (1994).
97. ER Chapman, RC Desai, AF Davis and CK Tornehl, *J Biol Chem*, **273**, 32966–72 (1998).
98. ER Chapman, *Nat Rev Mol Cell Biol*, **3**, 498–508 (2002).
99. PJ Kraulis, *J Appl Crystallogr*, **24**, 946–50 (1991).
100. EA Merritt and MEP Murphy, *Acta Crystallogr*, **50**, 869–73 (1994).
101. A Nicholls, KA Sharp and B Honig, *Proteins*, **11**, 281–96 (1991).

Dockerin-domains

Dockerin domains

Brian F Volkman[†], Betsy L Lytle[†,‡] and J H David Wu[‡]

[†]Department of Biochemistry, Medical College of Wisconsin, Milwaukee, WI, USA
[‡]Department of Chemical Engineering, University of Rochester, Rochester, NY, USA

FUNCTIONAL CLASS

Dockerin domains are small protein modules that mediate assembly of cellulosomes, large multienzyme complexes that degrade crystalline cellulose and other polymeric substances associated with plant biomass,[2] such as hemicellulose (Figure 1). Dockerins possess no known intrinsic enzymatic activity, but occur as independently folding domains in various subunits of the cellulosome. By virtue of high-affinity interactions with protein domains called cohesins, dockerin domains link a myriad of different polypeptide chains to construct an extracellular complex of over 2 MDa total molecular weight. Dockerin domains have been identified, along with the cohesins, as conserved homologous sequence elements of the proteins that make up the cellulosome scaffold (scaffoldin) and enzymatic subunits (glycosyl hydrolases). Most of the glycosyl hydrolases contain a type I dockerin domain C-terminal to the catalytic domain, which binds type I cohesins found in the scaffold. Some scaffoldins (categorized as class I)[3] contain a different type of dockerin, type II, used for anchoring the complex to the cell wall via interaction with complementary type II cohesins of cell surface proteins. Bacterial dockerin

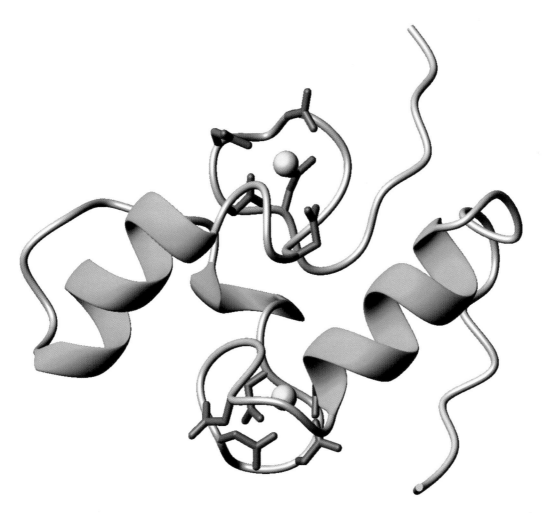

3D Structure   Schematic representation of the type I dockerin domain structure with calcium ions and coordinating side chains from each calcium-binding loop shown. PDB code: 1DAQ. All structural representations were generated with the program MOLMOL.[1]

Dockerin domains

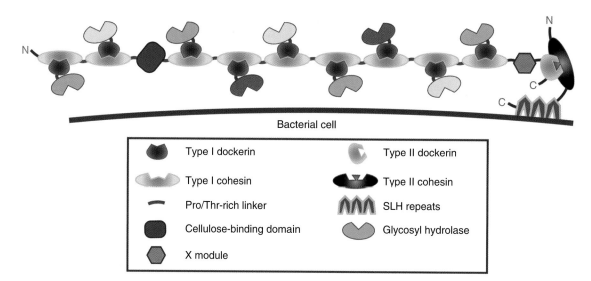

Figure 1 A schematic diagram of the *Clostridium thermocellum* cellulosome. The type I dockerins mediate attachment of the catalytic subunits to the scaffoldin, which is composed of nine cohesins, a cellulose-binding domain (CBD), a hydrophilic domain of unknown function (X), and a type II dockerin. The scaffoldin likewise binds through its type II dockerin domain to a type II cohesin–containing protein on the bacterial cell surface that is thought to anchor the complex through a series of three surface-layer homology (SLH) domains.

domains bind Ca^{2+} through a pair of homologous 22-residue sequences that bear close resemblance to EF-hand consensus sequences. However, the dockerin tertiary structure described below represents a distinct fold in the calcium-binding protein superfamily. In the absence of Ca^{2+}, these domains lack a stable tertiary fold and are thus unable to bind cohesin. Three-dimensional structural information is available for only one example of a bacterial dockerin domain (PDB code: 1DAV), but conservation of key hydrophobic core and Ca^{2+}-binding residues suggests that other family members will adopt a similar calcium-dependent conformation. While the crystal structures of three different cohesin domains (PDB accession 1ANU,[4] 1AOH,[5] and 1G1K[6]) define a conserved β-sandwich fold, no dockerin–cohesin complex structures have been reported, and the details of this high-affinity protein–protein interaction remain to be elucidated.

OCCURRENCE

Dockerin domains were first found in the subunits of the cellulosome from the thermophilic and cellulolytic anaerobe *Clostridium thermocellum*,[7] Cellulosomes have also been discovered in other clostridia, including *C. cellulovorans*,[8] *C. cellulolyticum*,[9] *C. papyrosolvens*,[10] *C. josui*,[11] *C. acetobutylicum*,[12] *Bacteroides cellulosolvens*,[13] and *Acetivibrio cellulolyticus*.[3] In addition, there is evidence for cellulosomes or cellulosome-like structures in other anaerobic bacteria[14] such as *Ruminococcus flavefaciens*[15] and *R. albus*;[16,17] in the aerobic bacteria *Vibrio* sp.;[18] and in the anaerobic fungi, *Neocallimastix patriciarum*,[19,20] *Piromyces*,[21] and *Orpinomyces*.[22]

Interestingly, while a small protein domain from *Piromyces equi* with dockerin-like activity has been characterized,[21] its sequence and three-dimensional structure are completely different from that of the bacterial dockerin and it does not bind calcium.[23]

To date, the Pfam (Protein Families) database (www.sanger.ac.uk/Software/Pfam/index.shtml) lists more than 80 proteins containing the dockerin domain, produced by the microorganisms, including many of those mentioned above, *Clostridium perfringens*,[24,25] *Deinococcus radiodurans*,[26] *Rickettsia conorii*,[27] and archaea including *Halobacterium* sp.,[28] *Methanosarcina mazei*,[29] and *Archaeoglobus fulgidus*.[30,31] The ongoing microbial genome-sequencing efforts will undoubtedly reveal additional family members, some of which may play novel functional roles distinct from cellulosome assembly. BLAST search of the draft genome sequence of *C. thermocellum* identified more than 35 dockerin-containing proteins encoded by the genome, (MJ Newcomb, DP Russell, R Zagursky and JHD Wu, unpublished results) consistent with the Pfam listing (33 entries) and indicating the diversity of dockerin-containing proteins even within one microorganism.

BIOLOGICAL FUNCTION

The cellulosome exists as an extracellular supramolecular complex that comprises proteins synthesized within the bacterial cell. Assembly of the active multimeric enzyme is therefore complicated by the requirements for protein secretion and regulation of the quaternary structure assembly. Dockerin domains function as modules that link numerous protein components through specific recognition of cohesin receptor domains. In addition, as discussed below, dockerin domains may also serve to regulate cellulosome assembly through a calcium-dependent mechanism.

The majority of dockerin domains (type I in most of the known clostridial enzymes) are responsible for incorporating their associated glycosyl hydrolases into the bacterial cellulosome via interactions with a cohesin domain (type I) contained in a scaffoldin (Figure 1). These dockerins can typically bind nonselectively to any of the cohesins existing in tandem within the scaffoldin,[32] but not the type II cohesin. In studies involving *C. thermocellum*, it has been determined that the type II dockerin domain cannot bind to the scaffoldin cohesin domains, but instead is responsible for anchoring the scaffoldin to cell surface proteins via complementary type II cohesin domains.[33,34] In *C. thermocellum* cell surface proteins, SdbA, Orf2p, and OlpB contain one, two, and four type II cohesins, respectively.[33,34] Another cell surface protein, OlpA, contains a type I cohesin, which binds cellulosomal enzymes directly via their type I dockerins.[35]

Characterization of cellulosomes from other bacteria suggests that enzyme dockerins can display distinct specificities, unlike the apparent nonselective binding of type I dockerins to the various scaffoldin cohesins within each *Clostridium* species.[32] For example, the dockerin sequences of *R. flavefaciens* enzymes display a high degree of divergence among themselves.[15] At least two enzymes do not appear to bind to the ScaA scaffoldin, but instead may bind to another unidentified scaffoldin or other cellulosomal component. This expands the potential complexity and diversity of cellulosomal organization beyond the current model based on the *Clostridium* species.[36]

AMINO ACID SEQUENCE INFORMATION

Sequences of dockerin domains have been obtained through molecular cloning of scaffoldins and hydrolases from various species mentioned above. Like the cohesins, the 60–70 amino acid sequence of the dockerins is highly conserved (Figure 2). It is made up of two 22-residue sequence repeats, separated by a linker of 9–16 residues. Mutagenesis experiments[37] and homology modeling of a cohesin–dockerin complex[6] suggest that specific recognition of a cohesin domain may require a binding site composed of residues from only one of the two repeats. However, both dockerin subdomains are required for cohesin binding,[38] presumably to preserve correct folding. Within each duplicated sequence is a segment with sequence similarity to the 12-residue Ca^{2+}-binding loop of the EF-hand motif.[39] As illustrated in Figure 2, residues that coordinate Ca^{2+} through their side chain functional groups are conserved in loop positions 1, 3, 5, 9, and 12 of nearly all dockerins. Hydrophobic amino acids consistently occupy a number of other positions within the repeated sequence elements, some of which are also conserved in EF-hand proteins.

One exception is the dockerin from the *Ruminococcus* scaffoldin, which has a normal EF-hand motif in the first sequence repeat and a second putative Ca^{2+}-binding loop that more closely resembles the 'pseudo EF-hand'[41,42] (Figure 2). Pseudo EF-hand sequences differ from the canonical EF-hand in that they consist of 14 rather than 12 residues and bind Ca^{2+} through four carbonyl groups from the polypeptide backbone and two side-chain ligands. In

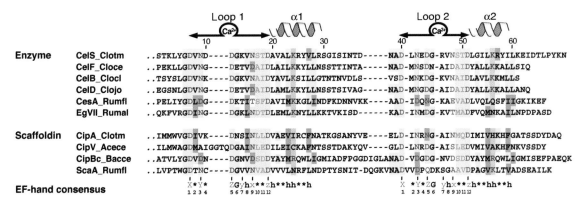

Figure 2 Sequence alignment of dockerin domains from hydrolases and scaffoldins of various cellulolytic bacteria. The name of the cellulosomal protein is followed by the organism: Clotm, *C. thermocellum*; Cloce, *C. cellulolyticum*; Clocl, *C. cellulovorans*; Clojo, *C. josui*; Rumfl, *R. flavefaciens*; Rumal, *R. albus*; Acece, *A. cellulolyticus*; Bacce, *B. cellulosolvens*. The SWISS-PROT or GenBank accession numbers are as follows: P38686 (CelS_Clotm), P37698 (CelF_Cloce), P28621 (CelB_Clocl), O82831 (CelD_Clojo), Q9L3K3 (CesA_Rumfl), Q9LCC4 (EgVII_Rumal), Q06851 (CipA_Clotm), AAF06064 (CipV_Acece), Q9FDJ9 (CipBc_Bacce), and AJ278969 (ScaA_Rumfl). Identical amino acids in at least five of the sequences are shaded yellow and conservative substitutions are shaded green. Residues in loop positions 10 and 11 are highlighted as sequence elements that may confer binding specificity between dockerin subfamilies or within a species or bacterial family. Type I/type II discrimination may also involve Lys24 and Lys56, highlighted to show the high level of conservation among the type I dockerin domains. In the canonical EF-hand, the Ca^{2+} ion is coordinated by one oxygen atom or bridging water molecule (x) of the side chains of residues at positions X, Y, Z, x, a backbone carbonyl oxygen at position y, and both side chain oxygens at position z.[40] Hydrophobic and variable residues in the EF-hand consensus repeats are indicated by h and * respectively. The dockerin residues with side chains thought to coordinate calcium ions are highlighted in red.

addition, the *A. cellulolyticus* scaffoldin contains a four-residue insert in the first loop (Figure 2). A definitive description of the Ca^{2+} coordination geometry for such divergent dockerin sequences must await their further structural characterization.

Most of these sequence elements are conserved for their roles in stabilization of the dockerin fold, though some may be components of a cohesin recognition surface shared across types and species. Amino acids that enable selective recognition of type I and type II cohesins must therefore reside in other variable sequence positions. Analysis of dockerin multiple sequence alignments has led to speculation about specific residues that may serve as these specificity determinants. In particular, divergence in the dipeptide sequence corresponding to residues 10 and 11 in each Ca^{2+}-binding loop is thought to confer type I/II and/or species specificity on the basis of the examination of the dockerins of *C. thermocellum* and *C. cellulolyticum*.[43–45] Similarly, basic residues found in the α-helices of type I (but not type II) clostridial dockerins might interact with acidic groups that are highly conserved only in the type I cohesins.[45]

PROTEIN PRODUCTION, PURIFICATION, AND MOLECULAR CHARACTERIZATION

The CelS subunit of the cellulosome from the thermophilic bacteria *C. thermocellum* is the most abundant catalytic subunit of the cellulosome and is the first exoglucanase, or processive cellulase, subunit of the cellulosome to be characterized.[46,47] As the dockerin domain from this enzyme is the only one for which tertiary structure has been reported, we focus here on its molecular characterization.

The CelS dockerin domain (Ct-Doc; previously called DS or DSCelS)[38,48] consisting of amino acids 673 (Ser) to 741 (Asn) has been cloned and expressed in *Escherichia coli*. To facilitate structural analysis of Ct-Doc by NMR spectroscopy, ^{15}N-labeled protein has been produced by expression on minimal medium containing ^{15}N ammonium chloride as the sole nitrogen source.[48] Purification to homogeneity has been achieved by chitin affinity chromatography followed by size-exclusion chromatography and the purity assessed by SDS-PAGE.[48] Metal binding activity, cohesin binding activity, oligomeric state, and tertiary structure have been characterized as described below.

METAL CONTENT AND COFACTORS

Metal content of dockerin-containing glycosyl hydrolases has not been assessed on protein purified from natural sources. However, sequence analysis and cohesin binding,[43] direct measurements of Ca^{2+} binding,[49] and NMR studies of Ca^{2+}-dependent folding,[48] all indicate that recombinant bacterial dockerin domains bind two Ca^{2+} ions, one in each of the 12-residue loops with homology to EF-hand metal binding sites. No other metal or cofactor binding activity is known for the dockerin domains. EDTA-inhibition of the crude cellulase preparation from *C. thermocellum*, which is reversed by Ca^{2+} and partially reversed by Mg^{2+},[50] may be a result of the loss of structure-stabilizing Ca^{2+} ions from the dockerin.

ACTIVITY TEST

The catalytic subunits of the cellulosome bind to cohesin domains via a type I dockerin domain predominantly found at their C termini.[51] Scaffoldins from *C. thermocellum*, *A. cellulolyticus*, *B. cellulosolvens*, and *R. flavefaciens* contain a type II dockerin domain at the C-terminus. Binding studies that demonstrated the specificity of type I versus type II interactions and the requirement for Ca^{2+} have been performed using nondenaturing gel shift or affinity blotting assays. Similarly, it has been shown that, while type I dockerins of *C. thermocellum* bind promiscuously to type I cohesins of the same species, this cross-reactivity does not extend to cohesin domains from different species.[32,43] Surface plasmon resonance (SPR) and isothermal titration calorimetry have also been used for quantitative measurements of dockerin–cohesin binding kinetics and thermodynamics, as well as comparative analyses of the dockerin mutants and chimeras.[37,49,52]

METAL-DEPENDENT FOLDING AND OLIGOMERIC STATE OF THE DOCKERIN DOMAIN

On the basis of its small size (~8 kDa), the CelS dockerin domain from *C. thermocellum* (Ct-Doc) is a suitable candidate for structural analysis by NMR spectroscopy. Surprisingly, a 2D HSQC spectrum of ^{15}N-labeled Ct-Doc shows minimal chemical shift dispersion, suggesting the lack of a stable tertiary structure (Figure 3, upper left). As has been observed for a few members of the EF-hand family of Ca^{2+}-binding proteins,[53–55] Ct-Doc is a completely disordered polypeptide in the absence of the metal cofactor. However, dramatic changes in the NMR spectrum have been observed with the addition of Ca^{2+} (Figure 3). New signals appear and increase in intensity as a function of Ca^{2+} concentration, while the intensities of signals observed in the initial spectrum gradually diminish. The well-dispersed spectrum observed at the final titration point ($[Ca^{2+}] = 14.8$ mM) is consistent with the formation of a well-defined folded structure.[48]

The binding reaction occurs in the slow-exchange regime on the NMR chemical shift timescale, as evidenced by the appearance of distinct resonances for the Ca^{2+}-loaded (folded) and Ca^{2+}-free (unfolded) Ct-Doc species during the course of the titration. Since the amino acid sequence contains two Ca^{2+}-binding loops, we compared plots of the

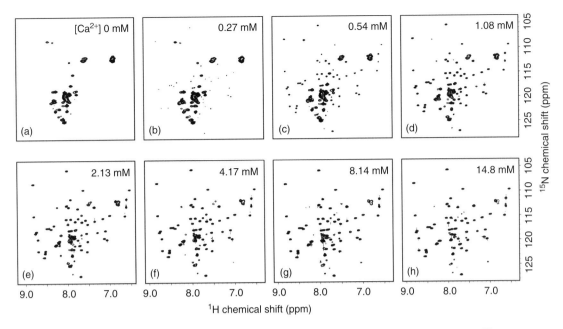

Figure 3 Ca^{2+}-dependent folding of the dockerin domain. 2D ^{15}N-^{1}H HSQC spectra were acquired on ^{15}N-labeled Ct-Doc in the presence of increasing amounts of Ca^{2+}. Total concentrations of Ca^{2+} (mM) were as follows: (a) 0.00, (b) 0.27, (c) 0.54, (d) 1.08, (e) 2.13, (f) 4.17, (g) 8.14, and (h) 14.8. The lack of chemical shift dispersion without added calcium clearly indicates the absence of a unique folded tertiary structure for the dockerin. Reprinted from *Arch Biochem Biophys*, 379, BL Lytle, BF Volkman, WM Westler and JH Wu, Secondary structure and calcium induced folding of the clostridium thermocellum dockerin domain determined by NMR spectroscopy, 237–44, Copyright (2000), with permission from Elsevier.

HSQC peak heights as a function of Ca^{2+} concentration and saw a similar progression for all residues, suggesting that metal binding to both sites and folding of the dockerin domain are cooperatively coupled. Consistent with this hypothesis, nonlinear fitting to the titration data using NMR intensities representing the fraction of folded protein resulted in a Hill coefficient of ~1.9, and K_d values in the 1–20 μM range. (BL Lytle, WM Westler and JHD Wu, unpublished results.)

While dockerin–cohesin binding occurs with very high affinity ($K_d < 10^{-9}$ M), in the absence of cohesin the dockerin domain binds Ca^{2+} with relatively low affinity. Because the cohesin–dockerin interaction is dependent on the Ca^{2+}-induced folding of the dockerin domain, local concentrations of Ca^{2+} may regulate this protein–protein interaction and, by extension, cellulosome assembly. Cytoplasmic Ca^{2+} concentrations of ~1 μM are probably too low to promote dockerin domain folding. In this model for type I dockerin function, only in an elevated Ca^{2+} environment outside the cell will this novel metal-dependent domain adopt its active, folded structure and be able to anchor its associated catalytic domain(s) to a type I cohesin in the scaffoldin protein.

Because the dockerins display sequence homology to the EF-hand family, some of which form dimers, it is important to establish the oligomeric state of the folded dockerin domain. NMR pulsed-field gradient (PFG) self-diffusion measurements[56] have been performed on samples of Ct-Doc (MW = 7947 Da) and human ubiquitin as a reference

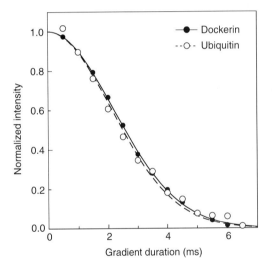

Figure 4 Self-diffusion of monomeric Ca^{2+}-loaded dockerin at 35 °C. Intensities from the 1D ^1H spectrum of Ct-Doc and human ubiquitin are plotted as a function of the total gradient duration in a PFG diffusion experiment, and self-diffusion coefficients (D_s) were obtained from nonlinear fitting.[56] With molecular weights of ~8 kDa, measured D_s values of 2.22×10^{-6} cm^2 s^{-1} and of 2.24×10^{-6} cm^2 s^{-1} for the monomeric protein ubiquitin and Ct-Doc respectively provide strong evidence that the dockerin domain is also monomeric in solution.

molecule, since it is a well-characterized monomeric protein of similar size (MW = 8565 Da) (Figure 4). Self-diffusion coefficient (D_s) values of 2.22×10^{-6} cm^2 s^{-1} and of 2.24×10^{-6} cm^2 s^{-1} have been obtained for ubiquitin and the dockerin domain, respectively, and from the very close agreement it has been concluded that Ct-Doc is monomeric under the conditions used for NMR.[45]

NMR STRUCTURE OF THE TYPE I DOCKERIN FROM CelS

Complete ^{1}H and ^{15}N chemical shift assignments for folded, Ca^{2+}-loaded Ct-Doc have been deduced from 2D and 3D NMR data acquired on unlabeled and ^{15}N-labeled protein samples at 55 °C and pH 6.[48] Peaks in the 2D ^{1}H-^{15}N HSQC spectrum provide a unique identifier for each residue in the protein reflected in the combination of backbone amide ^{1}H and ^{15}N chemical shifts (Figure 5). The three-dimensional structure of Ct-Doc has been determined by restrained molecular dynamics calculations employing 838 interproton distance constraints derived from crosspeaks assigned in the 2D and 3D NOESY spectra[45] and 12 distance constraints linking the Ca^{2+} ions with their coordinating ligands from each metal-binding loop. The NMR structure is represented by a bundle of 20 conformers (PDB code: 1DAV) calculated using the same input data and random starting geometries (Figure 6(a)), and a representative structure (PDB code: 1DAQ) derived from the average coordinate positions of the bundle (Figure 6(b)).

The structure consists of a pair of loop-helix motifs connected by a 10-residue linker (Figure 6). The two loop-helix elements are related by a twofold pseudosymmetry axis, reflected not only in the backbone conformation but also in the packing of symmetry-related side chains (Figure 6(b)). The first helix (α1) is a typical α-helix, extending from Ala20 to Arg29 as indicated by $(i, i + 4)$ hydrogen bonds. The second helix (α2) extends from Asp51 to Ile59, with a kink at Gly53 due to a nonhelical ψ backbone dihedral angle. This distortion has been anticipated in the prediction of helical residues by chemical shift index[57,58] analysis.[48] The angle between the two α-helices is 157°. The only other element of regular secondary structure is a turn of 3$_{10}$ helix composed of residues Thr36 to Asn39 in the linker.

Two distinct clusters of hydrophobic side chains stabilize the Ct-Doc tertiary structure (Figure 6(c)). Hydrophobic contacts within these clusters are indicated by NOE contacts involving the side chains. The first cluster is made up of Tyr6, Gly7, Val9, Leu23, Tyr26, Val27, Ile32, and Ile34 and the second consists of Leu41, Ile54, Tyr58, Ile59, Ile63, and Leu66. In addition, the side chains

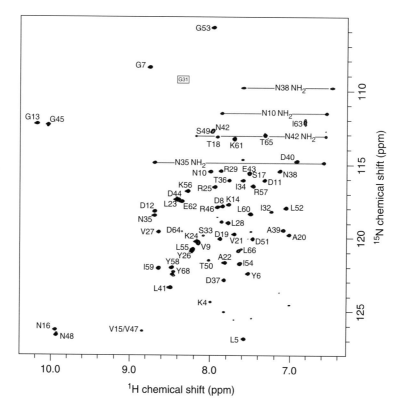

Figure 5 750 MHz ^{15}N-^{1}H HSQC spectrum of the ^{15}N-labeled Ct-Doc domain in 90% H$_2$O/10% D$_2$O, 100 mM KCl, 20 mM CaCl$_2$, pH 6 and 55 °C. All amide ^{1}H/^{15}N correlations are labeled according to their primary sequence positions. The side chain amide resonances of asparagine residues are connected by horizontal lines and labeled. The position of Gly31 is marked based on the resonances observed in the 3D TOCSY and NOESY spectra. Reprinted from *Arch Biochem Biophys*, 379, BL Lytle, BF Volkman, WM Westler and JH Wu, Secondary structure and calcium induced folding of the clostridium thermocellum dockerin domain determined by NMR spectroscopy, 237–44, Copyright (2000), with permission from Elsevier.

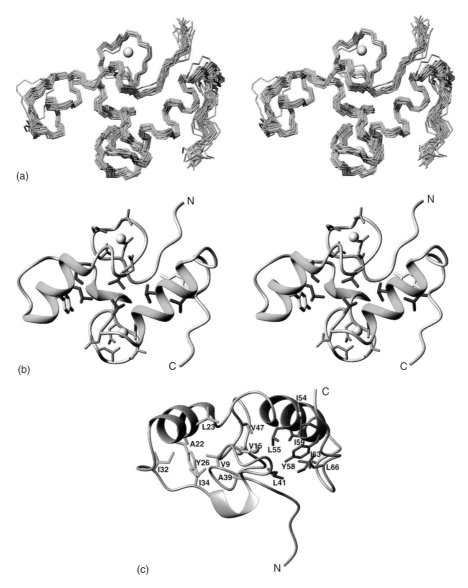

Figure 6 NMR structure of the dockerin domain from the *C. thermocellum* cellobiohydrolase CelS. (a) Stereo view of the backbone atoms (N, C$^\alpha$, and C′) of the selected 20 Ct-Doc structures. The structures are superimposed against the mean structure using residues 5–29 and 32–66. The α-helices are shown in cyan and the turn of 3$_{10}$ helix is highlighted green. Ca^{2+} ions are shown as yellow spheres. (b) Stereo view of a ribbon diagram of the energy-minimized averaged Ct-Doc structure in the same orientation as (a). Pairs of side chains from corresponding locations in the two homologous sequence repeats (Figure 2) are arranged symmetrically in the dockerin structure: Val15/Val47 (magenta), Leu23/Leu55 (red), Tyr26/Tyr58 (yellow), and Ile34/Leu66 (blue). Calcium-binding side chains from each loop (green) are also shown. (c) Side view of Ct-Doc with side chains comprising the two hydrophobic clusters shown in red and green and the conserved Val15 and Val47 (loop position 8) shown in yellow. Reprinted from *J Mol Biol*, 307, BL Lytle, BF Volkman, WM Westler, MP Heckman and JH Wu, Solution structure of a type I dockerin domain, a novel prokaryotic, extracellular calcium binding domain, 745–53, Copyright (2001), with permission from Elsevier.

of Val15 and Val47 (position 8 of the calcium-binding loops) are poised to interact with a pair of conserved hydrophobic residues in each helix (Ala20, Leu23, Leu52, Leu55). While the side chain of Ala20 in α1, equivalent to Leu52 in α2, does not contact the valine side chains directly, it is replaced in most other dockerins by a valine, leucine, isoleucine, or tryptophan (Figure 2) that could be expected to contribute to the stabilization of this apolar cluster. Additional stabilization of the structure appears to be provided by hydrogen bonds between the amide proton of Gly7 and the side-chain amide of Asn38 and between the amide proton of Asn35 and the carbonyl oxygen of Val9.

Consistent with the structural results, NMR measurements of ^{15}N relaxation to monitor internal mobility of the polypeptide backbone[59] showed that the dockerin domain is well ordered throughout the structure, with two exceptions. Amino-terminal residues preceding the first Ca^{2+}-binding loop show gradually increasing generalized

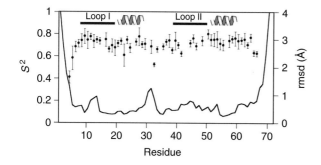

Figure 7 Backbone dynamics of Ct-Doc. Generalized order parameters derived from amide ^{15}N relaxation parameters (black circles; left vertical axis) and backbone (N, C$^\alpha$, and C') atomic rmsd values (solid line; right vertical axis) are plotted as a function of amino acid residue. S^2 values were obtained using the program Modelfree4.1.[60] The locations of the calcium-binding loops and α-helices are indicated.

order parameter (S^2) values, consistent with fraying of the structure reflected in atomic rmsd values for the ensemble of NMR structures (Figure 7). Another localized region of enhanced mobility is observed in the linker. Residues 30–32 are poorly defined in the structure, as reflected in the high local rmsd values, and ^{15}N relaxation data confirm the presence of increased disorder in this region, particularly for the NH of Ile 32 ($S^2 = 0.52$).

Ct-Doc is the only bacterial dockerin domain for which the three-dimensional structure is known, but on the basis of the high degree of sequence conservation (Figure 2) it may be anticipated that other dockerin domains will adopt the same overall fold. In the absence of Ca^{2+}, Ct-Doc is unstructured and fails to bind cohesin, but it remains to be seen whether this is true of all bacterial dockerins.

COMPARISONS WITH EF-HAND DOMAINS

As noted above, the dockerin calcium-binding subdomains differ from the typical EF-hand motif in that the E-helix preceding each calcium-binding loop is missing. Comparisons of the Ct-Doc structure with EF-hand domains show that the dockerin domain is clearly distinct from this well-studied family of Ca^{2+}-binding proteins. A Dali search of the Protein Data Bank identified no structural homologs to Ct-Doc, indicating that this paired loop-helix motif represents a novel fold. Upon comparison to the C-terminal domain of cardiac troponin C, which displays the typical EF-hand fold (Figure 8), it is seen that the calcium-binding loops of Ct-Doc are somewhat buried within the dockerin structure, while they project above the helices of troponin C. The orientation of the Ct-Doc helices is also clearly different from EF-hand proteins. The angle between the helices is 157°, whereas the angles between the equivalent helices in the calcium-bound forms of calcyclin, calbindin D$_{9k}$, and the N- and C-terminal domains of calmodulin are 36°, 33°, 45°, and 37° respectively.[61,62]

In fact, in the EF-hand Calcium-Binding Proteins Data Library (www.structbio.vanderbilt.edu/cabp_database) the largest interhelical angle reported between helices 2 and 4 of functional EF-hand domains from 15 different proteins is 92°.

Comparisons of the backbone structures of the Ct-Doc loops with canonical EF-hand loops suggests that the Ct-Doc loops adopt an unusual conformation, despite the fact that the sequences are highly homologous to those of the canonical EF-hand loops. While superposition of all 12 residues of either Ct-Doc loop on representative EF-hand Ca^{2+}-binding loops produces poor agreement (rmsd > 2), superposition of either the first 5 or last 4 residues in the loops shows that the local geometry is consistent with the conserved loop structure (rmsd 0.3–0.7) as shown in Figure 9 for Ct-Doc loop 1. In both loops the position 7 residue (Lys14 and Arg46) has an unusual dihedral angle, which causes a bend in the loop. In addition, the dihedral angles of both position 6 glycine residues deviate from the typical values of +90° and 0° for φ and ψ respectively.

The atypical loop conformations may be the result of the absence of the short, antiparallel β-sheet interaction found between the position 7–9 residues of all paired EF-hand domains studied to date (Figure 8(b)). While chemical shifts suggest that residues 7 and 8 of both loops might adopt β-sheet conformations, no such interactions have been observed in the dockerin structure. Degeneracy of the Val15 and Val47 (position 8) chemical shifts might have masked NOE contacts connecting the two loops; however, structures calculated with added H-bond restraints between the backbone carbonyl and amide groups of the position 8 valine residues are clearly inconsistent with the experimental constraints. Therefore, we have concluded that a β-sheet interaction between the two Ct-Doc loops is not present.[45]

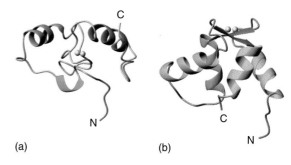

Figure 8 Dockerin domains are distinct from the EF-hand family of proteins. Comparison of the topology of (a) Ct-Doc and (b) C-terminal domain of cardiac muscle troponin C (PDB code 3CTN). The two structures are oriented similarly with respect to the axis defined by the two Ca^{2+} ions shown in yellow. Reprinted from *J Mol Biol*, **307**, BL Lytle, BF Volkman, WM Westler, MP Heckman and JH Wu, Solution structure of a type I dockerin domain, a novel prokaryotic, extracellular calcium binding domain, 745–53, Copyright (2001), with permission from Elsevier.

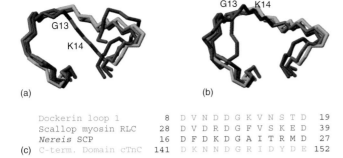

```
                          G13  K14
Dockerin loop 1       8   D V N D D G K V N S T D   19
Scallop myosin RLC   28   D V D R D G F V S K E D   39
Nereis SCP           16   D F D K D G A I T R M D   27
C-term. Domain cTnC 141   D K N N D G R I D Y D E  152
```
(a) (b) (c)

Figure 9 Comparison of the first Ct-Doc calcium-binding loop 1 (red) with other EF-hand loops. *Nereis* sarcoplasmic Ca^{2+}-binding protein (blue; PDB accession code 2SCP), C-terminal domain of cardiac troponin C (green; PDB accession code 3CTN), and regulatory light chain (RLC) of scallop myosin (magenta; PDB accession code 1WDC). The scallop RLC loop has Mg^{2+} bound to it, while the other loops are bound with Ca^{2+}. The Ct-Doc backbone atoms have been superimposed against the other three loops to yield a best fit for (a) residues 8–12 (average pairwise rmsd of 0.50 Å) and (b) residues 16–19 (average pairwise rmsd of 0.36 Å). Superposition of the other three loops utilized the backbones of all 12 residues. Superposition of all residues of the Ct-Doc loop 1 yielded an average pairwise rmsd of 2.34 Å. (c) Amino acid sequences of the loops used in the comparison.

In addition to the missing β-sheet interaction, Ct-Doc also lacks the conserved interhelix hydrophobic interactions present in most Ca^{2+}-bound EF-hand proteins. The strongest of these interactions normally occurs between helices A and D (preceding EF-loop I and following EF-loop II, respectively) with a packing angle for these helices of about 120°.[63] Obviously, the lack of an equivalent helix A in Ct-Doc prevents the formation of these contacts. The only conserved EF-hand interactions that remain in the Ct-Doc structure involve the position 8 valine residues, as mentioned above. It has been noted that these side chains form the nucleus of the hydrophobic core in EF-hand proteins.[63]

FUNCTIONAL DETERMINANTS OF DOCKERIN–COHESIN INTERACTIONS

Dockerin–cohesin binding selectivity has been observed on the basis of both species and functional specificity. Correct cellulosome assembly relies on discrimination by dockerin domains between cohesin domains from the scaffoldin and the cell surface anchoring proteins. As described above, sequence comparisons can be used to determine which residues might encode species or type I/type II binding specificity. Knowledge of the three-dimensional Ct-Doc structure enables the identification of residues conserved for structural purposes (Ca^{2+}-binding or hydrophobic core interactions), leaving a relatively small subset of surface-exposed residues that are likely to participate in cohesin recognition and discrimination.

Residues at positions 10 and 11 in the Ca^{2+}-binding loop have been predicted to serve as recognition determinants in the cohesin–dockerin interaction, on the basis of a series of observations. An ST dipeptide is preserved at this location in both loops of all the type I dockerins of *C. thermocellum*, and replaced with LL and MQ sequences in the loops of the type II dockerin of the CipA scaffoldin.[44] Likewise, dissimilarity at positions 10 and 11 for dockerin domains from *C. thermocellum* (ST) and the mesophile *C. cellulolyticum* (AI) prevents the recognition of cohesin domains between the two species.[43] With the discovery of other dockerin sequences containing an AI dipeptide at this position, it seems possible that dockerin and cohesin domains may cross-react between these mesophilic species, but not with the domains from the thermophile *C. thermocellum* (Figure 2). In contrast, dissimilarity at the 10–11 dipeptide for dockerins of two different glycosyl hydrolase components of the *C. cellulovorans* cellulosome may explain their ability to discriminate between individual cohesin domains within the CbpA scaffoldin.[64] Sequences from other bacterial species are also divergent at these sequence positions, with the *R. flavefaciens* type I dockerin sequences displaying a similar variability as described above.[36]

Figure 10(b) shows a molecular surface representation of Ct-Doc, which highlights a Y-shaped, solvent-accessible hydrophobic patch made up of residues Ala20, Val21, Leu23, Val27, Leu28, Val47, and Leu52. This region is surrounded by the loop position 10 and 11 residues Ser17, Thr18, Ser49, and Thr50. At either end of the hydrophobic patch are a number of conserved basic residues, including Lys24 and Lys56, from α1 and α2 respectively. These lysine residues each occur in a position of the helix normally occupied in EF-hand proteins by a hydrophobic residue; lysine is an extremely rare residue in this position, occurring only once out of 567 EF-hand sequences.[65] Lys24 and Lys56 are highly conserved among type I but not type II dockerins. Thus, these residues may be involved in electrostatic interactions specific to type I cohesin binding.

To gain more insight into the potential cohesin interaction surface, the backbone ^{1}H and ^{15}N resonances have been assigned for Ct-Doc in complex with cohesin 3 from CipA, the scaffoldin subunit of the *C. thermocellum* cellulosome. Comparison of chemical shift values for free and bound Ct-Doc reveals significant backbone ^{1}H and ^{15}N chemical shift changes for residues Asn10, Asp19, Val21, Val27, Arg29, Ile32, Thr50, Gly53, and Leu60. In addition, residues Leu23, Arg25, Tyr26, Leu28, Ser33, Asn48, Asp51, and Glu62 displayed moderate chemical shift changes. Additional perturbed residues are indicated by the disappearance (Thr18 and Ser49) or appearance (Ser30) of crosspeaks in the spectrum of the complex. Most of the perturbed residues (Figure 10(c)) are located in or adjacent to the hydrophobic surface (Figure 10(b)), suggesting that this region is indeed involved in the binding

Dockerin domains

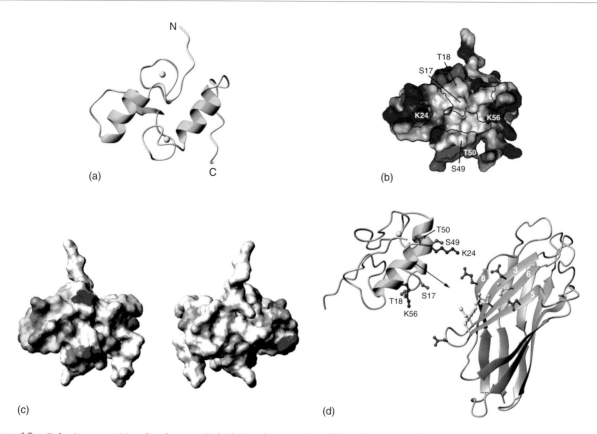

Figure 10 Cohesin recognition by the type I dockerin domain. (a) Ribbon diagram and molecular surface representations of the putative cohesin-binding surface of Ct-Doc shown from identical views according to (b) electrostatic potential (negative and positive regions are red and blue, respectively; the conserved serine, threonine, and lysine residues are labeled and the hydrophobic residues are outlined) or (c) chemical shift perturbations (residues with combined ^{1}H and ^{15}N chemical shift changes >140 Hz are colored orange). (d) Postulated docking of Ct-Doc to the 8,3,6,5 β-sheet of cohesin 7. The conserved serine and threonine side chains of Ct-Doc are shown in violet and the conserved lysine side chains are shown in blue. The side chains of solvent-accessible hydrophobic (yellow) and acidic (red) residues on the 8,3,6,5 β-sheet of cohesin are also shown. (a), (b), and (d) reprinted from *J Mol Biol*, **307**, BL Lytle, BF Volkman, WM Westler, MP Heckman and JH Wu, Solution structure of a type I dockerin domain, a novel prokaryotic, extracellular calcium binding domain, 745–53, Copyright (2001), with permission from Elsevier.

of cohesin. Few chemical shift changes mapped to the back face of Ct-Doc. It should be noted that these chemical shift changes can result from either direct contacts with the protein or from conformational changes induced as a result of binding, as both processes will alter the nuclear chemical environment.

The results of cohesin–dockerin docking calculations using the Ct-Doc NMR structure and the X-ray crystal structure of cohesin 7 from the CipA scaffoldin are also consistent with the proposed binding surface. In roughly 70% of the predicted complexes, the binding interface contains the conserved residues Ser17, Thr18, and Lys24. This binding site has been predicted in 6 of the 10 lowest energy models, and four of these solutions also involve a conserved hydrophobic region of the cohesin, the four-stranded 8,3,6,5 β-sheet. This face is involved in crystallographic dimer formation for both cohesin 7 and cohesin 2 and has previously been suggested as a possible dockerin-binding surface.[4,5] In addition to the conserved hydrophobic residues, this region has significant negative electrostatic potential associated with it.[4] We therefore suggested that the association of the cellulolytic subunits with the scaffoldin occurs through a combination of extensive hydrophobic contacts and complementary interactions involving the 8,3,6,5 β-sheet of the type I cohesin and a series of conserved hydrophobic and basic residues of the type I dockerin (Figure 10(d)).[45] It is also possible that the dockerin undergoes a conformational change upon binding, which could further expose hydrophobic side chains.

Results from recent mutagenesis studies are consistent with the model based on NMR shift perturbations and docking calculations. Substitution of acidic and hydrophobic residues on the 8,3,6,5 surface of the seventh cohesin domain from *C. thermocellum* scaffoldin CipA eliminates binding to the CelD dockerin domain,[37] thus reaffirming the potential role for conserved basic residues of the dockerin. Residues 10 and 11 of the Ca^{2+}-loops have been likewise shown to be essential components of the binding interface in the same dockerin–cohesin complex.[37]

IMPLICATIONS FOR SCAFFOLDING OF ENGINEERED MULTIENZYME COMPLEXES

The cellulosome represents an excellent example of an efficient, self-assembling molecular machine with high specific activity. Future structural studies of dockerin and cohesin interactions will provide insights into the mechanism of cellulosome assembly, enabling the design of novel enzyme complexes. This concept has already been demonstrated with scaffoldins engineered to contain cohesin domains with distinct dockerin specificities, enabling the assembly of arbitrary mixtures of enzymes tethered to the appropriate dockerin domains.[66] Protein-engineering efforts that borrow from the modular concepts discovered in the cellulosomes have the potential for broad applications in a variety of industrial settings that will benefit from improved catalysts.

Other future research directions may involve the development of 'designer' dockerin and cohesin domains. Once the critical residues in the binding interface are identified, it will be possible to vary the specificity of these domains, thus allowing the assembly of highly complex mixtures of enzymes on a single scaffold.[67] The cohesin–dockerin interaction itself may also have useful applications in biotechnology. With a sub-nanomolar affinity that depends on Ca^{2+}, this protein heterocomplex could provide an alternative to other affinity methods for analysis of protein interactions with the potential benefit of controlled binding through metal chelation by EDTA treatment, for example.[49] Thus, one can imagine numerous potential uses for this reversible interaction.

REFERENCES

1. R Koradi, M Billeter and K Wüthrich, *J Mol Graphics*, **14**, 51–55 (1996).
2. WH Schwarz, *Appl Microbiol Biotechnol*, **56**, 634–49 (2001).
3. SY Ding, EA Bayer, D Steiner, Y Shoham and R Lamed, *J Bacteriol*, **181**, 6720–29 (1999).
4. LJ Shimon, EA Bayer, E Morag, R Lamed, S Yaron, Y Shoham and F Frolow, *Structure*, **5**, 381–90 (1997).
5. GA Tavares, P Beguin and PM Alzari, *J Mol Biol*, **273**, 701–13 (1997).
6. S Spinelli, HP Fierobe, A Belaich, JP Belaich, B Henrissat and C Cambillau, *J Mol Biol*, **304**, 189–200 (2000).
7. EA Bayer, R Kenig and R Lamed, *J Bacteriol*, **156**, 818–27 (1983).
8. RH Doi, M Goldstein, S Hashida, JS Park and M Takagi, *Crit Rev Microbiol*, **20**, 87–93 (1994).
9. JP Belaich, C Tardif, A Belaich and C Gaudin, *J Biotechnol*, **57**, 3–14 (1997).
10. M Pohlschroder, SB Leschine and E Canale-Parola, *J Bacteriol*, **176**, 70–76 (1994).
11. M Kakiuchi, A Isui, K Suzuki, T Fujino, E Fujino, T Kimura, S Karita, K Sakka and K Ohmiya, *J Bacteriol*, **180**, 4303–8 (1998).
12. F Sabathe, A Belaich and P Soucaille, *FEMS Microbiol Lett*, **217**, 15–22 (2002).
13. C Lin, JW Urbance and DA Stahl, *FEMS Microbiol Lett*, **124**, 151–55 (1994).
14. EA Bayer, H Chanzy, R Lamed and Y Shoham, *Curr Opin Struct Biol*, **8**, 548–57 (1998).
15. V Aurilia, JC Martin, SI McCrae, KP Scott, MT Rincon and HJ Flint, *Microbiology*, **146**, 1391–97 (2000).
16. H Ohara, S Karita, T Kimura, K Sakka and K Ohmiya, *Biosci Biotechnol Biochem*, **64**, 254–60 (2000).
17. H Ohara, J Noguchi, S Karita, T Kimura, K Sakka and K Ohmiya, *Biosci Biotechnol Biochem*, **64**, 80–88 (2000).
18. Y Tamaru, T Araki, T Morishita, T Kimura, K Sakka and K Ohmiya, *J Ferment Bioeng*, **83**, 2 (1997).
19. HJ Gilbert, GP Hazlewood, JI Laurie, CG Orpin and GP Xue, *Mol Microbiol*, **6**, 2065–72 (1992).
20. L Zhou, GP Xue, CG Orpin, GW Black, HJ Gilbert and GP Hazlewood, *Biochem J*, **297**, 359–64 (1994).
21. C Fanutti, T Ponyi, GW Black, GP Hazlewood and HJ Gilbert, *J Biol Chem*, **270**, 29314–22 (1995).
22. X Li, H Chen and L Ljungdahl, *Appl Environ Microbiol*, **63**, 4721–28 (1997).
23. S Raghothama, RY Eberhardt, P Simpson, D Wigelsworth, P White, GP Hazlewood, T Nagy, HJ Gilbert and MP Williamson, *Nat Struct Biol*, **8**, 775–78 (2001).
24. B Canard, T Garnier, B Saint-Joanis and ST Cole, *Mol Gen Genet*, **243**, 215–24 (1994).
25. T Shimizu, K Ohtani, H Hirakawa, K Ohshima, A Yamashita, T Shiba, N Ogasawara, M Hattori, S Kuhara and H Hayashi, *PNAS*, **99**, 996–1001 (2002).
26. O White, JA Eisen, JF Heidelberg, EK Hickey, JD Peterson, RJ Dodson, DH Haft, ML Gwinn, WC Nelson, DL Richardson, KS Moffat, H Qin, L Jiang, W Pamphile, M Crosby, M Shen, JJ Vamathevan, P Lam, L McDonald, T Utterback, C Zalewski, KS Makarova, L Aravind, MJ Daly, KW Minton, RD Fleischmann, KA Ketchum, KE Nelson, S Salzberg, HO Smith, JC Venter and CM Fraser, *Science*, **286**, 1571–77 (1999).
27. H Ogata, S Audic, P Renesto-Audiffren, P-E Fournier, V Barbe, D Samson, V Roux, P Cossart, J Weissenbach, J-M Claverie and D Raoult, *Science*, **293**, 2093–98 (2001).
28. WV Ng, SP Kennedy, GG Mahairas, B Berquist, M Pan, HD Shukla, SR Lasky, NS Baliga, V Thorsson, J Sbrogna, S Swartzell, D Weir, J Hall, TA Dahl, R Welti, YA Goo, B Leithauser, K Keller, R Cruz, MJ Danson, DW Hough, DG Maddocks, PE Jablonski, MP Krebs, CM Angevine, H Dale, TA Isenbarger, RF Peck, M Pohlschroder, JL Spudich, K-H Jung, M Alam, T Freitas, S Hou, CJ Daniels, PP Dennis, AD Omer, H Ebhardt, TM Lowe, P Liang, M Riley, L Hood and S DasSarma, *PNAS*, **97**, 12176–81 (2000).
29. U Deppenmeier, A Johann, T Hartsch, R Merkl, RA Schmitz, R Martinez-Arias, A Henne, A Wiezer, S Baeumer, C Jacobi, H Brueggemann, T Lienard, A Christmann, M Boemecke, S Steckel, A Bhattacharyya, A Lykidis, R Overbeek, H-P Klenk, RP Gunsalus, H-J Fritz and G Gottschalk, *J Mol Microbiol Biotechnol*, **4**, 453–61 (2002).
30. H-P Klenk, RA Clayton, J-F Tomb, O White, KE Nelson, KA Ketchum, RJ Dodson, M Gwinn, EK Hickey, JD Peterson, DL Richardson, AR Kerlavage, DE Graham, NC Kyrpides, RD Fleischmann, J Quackenbush, NH Lee, GG Sutton, S Gill, EF Kirkness, BA Dougherty, K McKenney, MD Adams, B Loftus, S Peterson, CI Reich, LK McNeil, JH Badger, A Glodek, L Zhou, R Overbeek, JD Gocayne, JF Weidman, L McDonald, T Utterback, MD Cotton, T Spriggs, P Artiach, BP Kaine,

SM Sykes, PW Sadow, KP D'Andrea, C Bowman, C Fujii, SA Garland, TM Mason, GJ Olsen, CM Fraser, HO Smith, CR Woese and JC Venter, *Nature*, **390**, 364–70 (1997).

31. EA Bayer, PM Coutinho and B Henrissat, *FEBS Lett*, **463**, 277–80 (1999).
32. B Lytle, C Myers, K Kruus and JH Wu, *J Bacteriol*, **178**, 1200–3 (1996).
33. E Leibovitz and P Beguin, *J Bacteriol*, **178**, 3077–84 (1996).
34. E Leibovitz, H Ohayon, P Gounon and P Beguin, *J Bacteriol*, **179**, 2519–23 (1997).
35. S Salamitou, O Raynaud, M Lemaire, M Coughlan, P Beguin and JP Aubert, *J Bacteriol*, **176**, 2822–27 (1994).
36. MT Rincon, SY Ding, SI McCrae, JC Martin, V Aurilia, R Lamed, Y Shoham, EA Bayer and HJ Flint, *J Bacteriol*, **185**, 703–13 (2003).
37. I Miras, F Schaeffer, P Beguin and PM Alzari, *Biochemistry*, **41**, 2115–19 (2002).
38. B Lytle and JH Wu, *J Bacteriol*, **180**, 6581–85 (1998).
39. S Chauvaux, P Beguin, JP Aubert, KM Bhat, LA Gow, TM Wood and A Bairoch, *Biochem J*, **265**, 261–65 (1990).
40. H Kawasaki and RH Kretsinger, *Protein Profile*, **2**, 297–490 (1995).
41. MR Nelson and WJ Chazin, *Biometals*, **11**, 297–318 (1998).
42. BW Schafer and CW Heizmann, *Trends Biochem Sci*, **21**, 134–40 (1996).
43. S Pages, A Belaich, JP Belaich, E Morag, R Lamed, Y Shoham and EA Bayer, *Proteins*, **29**, 517–27 (1997).
44. A Mechaly, S Yaron, R Lamed, HP Fierobe, A Belaich, JP Belaich, Y Shoham and EA Bayer, *Proteins*, **39**, 170–77 (2000).
45. BL Lytle, BF Volkman, WM Westler, MP Heckman and JH Wu, *J Mol Biol*, **307**, 745–53 (2001).
46. K Kruus, WK Wang, J Ching and JH Wu, *J Bacteriol*, **177**, 1641–44 (1995).
47. WK Wang, K Kruus and JH Wu, *J Bacteriol*, **175**, 1293–302 (1993).
48. BL Lytle, BF Volkman, WM Westler and JH Wu, *Arch Biochem Biophys*, **379**, 237–44 (2000).
49. HP Fierobe, S Pages, A Belaich, S Champ, D Lexa and JP Belaich, *Biochemistry*, **38**, 12822–32 (1999).
50. EA Johnson, M Sakajoh, G Halliwell, A Madia and AL Demain, *Appl Environ Microbiol*, **43**, 1125–32 (1982).
51. K Tokatlidis, S Salamitou, P Beguin, P Dhurjati and JP Aubert, *FEBS Lett*, **291**, 185–88 (1991).
52. F Schaeffer, M Matuschek, G Guglielmi, I Miras, PM Alzari and P Beguin, *Biochemistry*, **41**, 2106–14 (2002).
53. KS Akerfeldt, AN Coyne, RR Wilk, E Thulin and S Linse, *Biochemistry*, **35**, 3662–69 (1996).
54. BE Finn, J Kordel, E Thulin, P Sellers and S Forsen, *FEBS Lett*, **298**, 211–14 (1992).
55. GS Shaw, RS Hodges and BD Sykes, *Science*, **249**, 280–83 (1990).
56. AS Altieri, DP Hinton and RA Byrd, *J Am Chem Soc*, **117**, 7566–67 (1995).
57. DS Wishart and BD Sykes, *J Biomol NMR*, **4**, 171–80 (1994).
58. DS Wishart, BD Sykes and FM Richards, *Biochemistry*, **31**, 1647–51 (1992).
59. AG Palmer III, *Annu Rev Biophys Biomol Struct*, **30**, 129–55 (2001).
60. AM Mandel, M Akke and AGI Palmer, *J Mol Biol*, **246**, 144–63 (1995).
61. M Sastry, RR Ketchem, O Crescenzi, C Weber, MJ Lubienski, H Hidaka and WJ Chazin, *Structure*, **6**, 223–31 (1998).
62. H Kuboniwa, N Tjandra, S Grzesiek, H Ren, CB Klee and A Bax, *Nat Struct Biol*, **2**, 768–76 (1995).
63. NC Strynadka and MN James, *Annu Rev Biochem*, **58**, 951–98 (1989).
64. JS Park, Y Matano and RH Doi, *J Bacteriol*, **183**, 5431–35 (2001).
65. JJ Falke, SK Drake, AL Hazard and OB Peersen, *Q Rev Biophys*, **27**, 219–90 (1994).
66. HP Fierobe, A Mechaly, C Tardif, A Belaich, R Lamed, Y Shoham, JP Belaich and EA Bayer, *J Biol Chem*, **276**, 21257–61 (2001).
67. P Beguin, *Curr Opin Biotechnol*, **10**, 336–40 (1999).

Hemopexin domains

Hemopexin domains

F Xavier Gomis-Rüth

Institut de Biologia Molecular de Barcelona, C.S.I.C.; C/Jordi Girona, 18-26; Barcelona, Spain

FUNCTIONAL CLASS

Hemopexin: Iron metabolism and oxidative stress protection; Vitronectin: cell adhesion and protein–protein interactions; Matrix metalloproteinases: substrate binding, activation, and localization.

Hemopexin domains (HDs) are named after the mammalian serum glycoprotein hemopexin (HPEX) involved in the metabolism of heme (iron protoporphyrin IX) and in the protection against oxidative stress.[1-3] They are made up of several hemopexin-type repeats of about 35 to 45 residues[4] and are also found as C-terminal domains (CTDs) in matrix metalloproteinases (MMPs),[5] zinc-dependent hydrolases belonging to the metzincin clan of metalloproteinases (MPs).[6-8] In these MMPs, they deal with protein–substrate and protein–inhibitor interactions and in activation events. Both protein classes have been characterized structurally. Moreover, HDs are present within vitronectins, which are extracellular matrix glycoproteins.[9] In this case, similarity with HPEX has been inferred from sequence information,

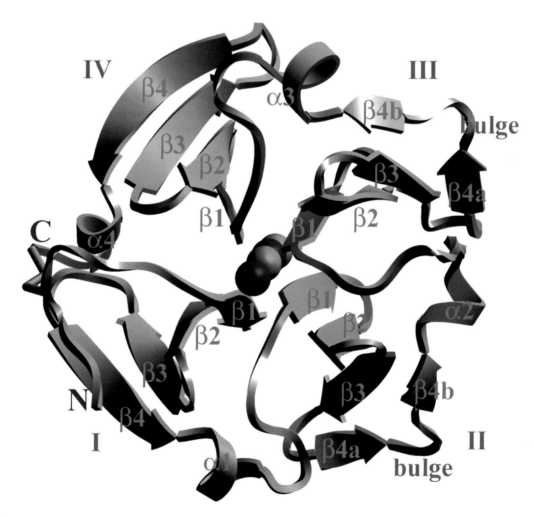

3D Structure Ribbon plot of the CTD of collagenase 3 (PDB 1PEX). View on the 'entry side' of the disc. The four blades are labeled (I to IV), as well as is each constituting β-strand (β1 to β4). The positions of the characteristic β-bulges in blades II and III are indicated, leading to the interruption of the corresponding strands β4, to render substrands β4a and β4b. The four ions localized in the central shaft (red, calcium cations; green, chloride anions) and the SS-bond tethering the MMP CTD (Cys278–Cys466, in yellow) are also shown.

as no 3D Structure has been reported to date. The presence of HDs in HPEXs, vitronectins, and in MMP CTDs suggests that these repeating peptide motifs have arisen by divergent evolution.[4]

OCCURRENCE

HPEX is a 57-kDa β_1-glycoprotein present in the serum of mammals, birds, and fish.[1] Together with transferrin and haptoglobin, it is one of the most abundant plasma proteins in mammals, after albumin, immunoglobulins, and plasma proteases.

Vitronectin (alias complement S-protein or serum spreading factor[4]) is a ubiquitous mammalian matrix glycoprotein, initially described in plasma, where it constitutes about 0.5% of the total protein content. It is present in various tissues and fluids, in both physiological and pathological conditions, such as seminal plasma, urine, amniotic fluid, cerebrospinal fluid, and bronchoalveolar lavage fluid.[10]

MMPs are metalloproteinases originally identified in vertebrates as the active principle involved in collagenolysis, as during tail resorption in the transition from the tadpole to the frog.[11] These zinc-dependent enzymes constitute a family within the metzincin clan of metalloendopeptidases[7,8] and have been subdivided into true collagenases, gelatinases, and stromelysins,[5,12] though other subdivisions have been proposed, for example, matrilysins, archetypal MMPs, convertase-activatable MMPs, gelatinases,[13] and so on.

BIOLOGICAL FUNCTION

Hemopexin

The circulating HPEX form is mainly produced by the parenchymal cells of the liver[3] and it acts as an acute-phase reactant. It belongs to the class-I-gene group of products – together with haptoglobin, α_1-acid glycoproteins, serum amyloid A1–A3, complement C3, and C-reactive protein, which are synthesized after an inflammatory event.[1,3,14–16] Therefore, serum levels of HPEX are monitored to reveal the severity of intravascular hemolysis. Altered HPEX serum or plasma levels are also associated with chronic neuromuscular diseases and acute intermittent porphyria.[17]

Heme is a lipophilic molecule of limited water solubility that intercalates into the lipid membrane and produces hydroxyl radicals. It may induce oxidative stress injury from hydrogen peroxide, oxidized lipopolysaccharide, and activated neutrophils. The main sources of extracellular heme are hemoglobin from lysed red blood cells, myoglobin, catalases, peroxidases, and cytochromes. Through its contribution to oxidative stress, heme has been associated with aging, stroke, atherosclerosis, and Alzheimer's disease.[18,19] Furthermore, heme levels in plasma are increased by hemolysis, and during pathologies like reperfusion and ischemia of kidney, heart, and brain in humans.[3] In particular, the kidney is most vulnerable to the toxic effects of hemolysis, and alterations of heme turnover in pathologic states impair the immune response.[20]

HPEX is involved in heme binding, though not through its attached glucides. In this way, heme is maintained in a water-soluble monomeric state. HPEX also contributes to heme degradation, thus protecting liver cells against heme-mediated oxidative stress and heme-bound iron loss.[1,3,21] In plasma, heme is first bound by albumin through its multiple low-affinity binding sites to render methemalbumin[22] and subsequently transferred to HPEX, which has a much higher affinity for it ($K_d < 1\,\text{pmol}\,\text{L}^{-1}$). The equimolar HPEX-heme complex undergoes a conformational change, enabling association with a cell-surface receptor on the hepatocyte plasma membrane. This ternary complex HPEX/heme/HPEX-receptor undergoes endocytosis, and antioxidant activities are boosted within the cell, probably (co)induced by iron regulatory proteins.[23] Heme is catabolized by heme oxygenase 1, which opens the porphyrin ring giving rise to biliverdin – further metabolized to bilirubin by biliverdin reductase[24] and carbon monoxide. Iron is rescued by ferritin, rendering the cation less available to cause deleterious reactions. Furthermore, intracellular protection is enhanced by the induction of metallothionein 1 synthesis and signaling pathways involving the N-terminal c-jun kinase and the antiapoptotic transcription factor NFκB.[25] Heme-depleted HPEX is recycled back to the extracellular compartment in a similar manner to transferrin. HPEX protects all cells against oxidation, especially those lacking HPEX receptors and those of the kidney, by confiscating the redox-active and toxic heme, probably in concert with haptoglobin.[3,20,26] Furthermore, HPEX confers 'nutritional immunity' by sequestering heme and iron from invading pathogens and preventing access for the latter to inhouse resources. HPEX receptors have been reported for *Haemophilus influenza*, proving such a role. HPEX is also associated with nitric oxide and carbon monoxide binding, thus protecting against NO-mediated toxicity, for example, after trauma and hemolysis.[17]

In addition to the liver, HPEX and its receptors are also expressed by cells in the placenta, ovary, peripheral neurons, and brain, as well as in retinal pigment epithelium and ganglion, that is, cells beyond the blood-brain and blood-retinal barriers.[3,27–29] In transected rat sciatic nerves, HPEX is expressed by fibroblasts, Schwann cells, and invading blood macrophages and is accumulated in the extracellular matrix (ECM). These findings suggest a further role for HPEX besides the multifunctional ones as the only heme transporter in plasma, as a player in iron homeostasis, bacteriostatic defense, nerve regeneration, and gene expression to promote cell survival. It may provide local protection for injured tissues against oxidative damage.[3]

Vitronectin

Reduced plasma levels have been reported for patients with severe liver failure, suggesting that this organ is the major site of vitronectin biosynthesis.[10,30] On the other hand, increased levels have been found in lung-disease patients, as well as in elective orthopedic surgery patients, and after stimulation with endotoxins or cytokines, implying that vitronectin, like hemopexin, is an acute-phase protein.[10,31,32] Immunohistochemical studies have revealed that vitronectin may be deposited in a series of normal human tissues, including the blood vessel wall. But it can also be found in sclerotic and necrotic tissues, like in hepatitis, rheumatoid arthritis, lung fibrosis, acute myocardial and kidney infarction, in plaques of atherosclerotic arteries, and in central nervous system disorders, so it can be used as a marker for tissue injury and necrosis.[10,33] Besides mammals, vitronectin-like molecules have been found in plants, algae, insects, and lower invertebrates.[34,35]

As an adhesive protein like fibrinogen, fibronectin, and von-Willebrand factor, vitronectin plays key roles in the attachment of cells to their surrounding ECM and may participate in the regulation of cell differentiation, proliferation, and morphogenesis, as observed during neuronal differentiation.[10,30,36,37] It mediates cell adhesion and spreading and facilitates cell proliferation on surfaces.[4] It can interact with platelets and the blood vessel wall and forms stable complexes with serpins and enzymes of the blood coagulation cascade, like thrombin.[38] The molecule also binds glycosaminoglycan and heparin. These interactions are mediated by a structural cell-attachment domain and confer on vitronectin a procoagulatory role.[10,39] However, other functional domains in the molecule confer a multifunctional role on vitronectin, as a regulator in fibrinolysis as well as in the immune system. When complement is activated, vitronectin is involved in the fluid-phase assembly of the terminal complement proteins, protecting neighboring cells from membrane damage by nascent complement complexes.[40] Furthermore, the molecule encompasses the integrin-binding footprint sequence R-G-D observed in disintegrins, enabling specific interaction with cognate vitronectin receptors.[41] Through the various interacting integrins, the versatile vitronectin molecule participates in signal transduction and gene regulation events, like those observed in defense mechanisms including wound healing, tissue repair, humoral and cellular immune system, and hemostasis.[42] The molecule can therefore function as a cross-linker between cells and the ECM through its two distinct binding sites.

These various activities are dependent on and regulated by the conformational states of the protein. Vitronectin is a conformationally labile molecule, subjected to changes to meet the requirements of the various interacting partners. Multimeric forms are recognized by several nonrelated surface receptors including integrins, urokinase receptor, and proteoglycans engaged in cell adhesion, migration, and invasion, and in processes related with tissue remodeling and bacterial tropism.[10,30] Together with its widespread distribution in the organism, these findings present vitronectin as a conceptual molecular link between cell adhesion, humoral immune response, and the hemostatic system, particularly at the blood vessel–wall interface.[39]

Matrix metalloproteinase C-terminal domain

Through the turnover of components of the vertebrate ECM, MMPs participate in homeostatic processes involving tissue resorption, remodeling, and repair, and also, when uncontrolled, in pathologies associated with indiscriminate tissue destruction, ranging from arthritis, fibrosis, and ulcers to cardiovascular disease and tumor neovascularization, invasion, and metastasis.[5,43] However, we have known for some years now that the roles of MMPs extend beyond tissue turnover. These are related to proteolytic processing or limited proteolysis and include substrates that transcend the original components of the ECM. In this way, MMPs participate in ectodomain shedding, zymogen activation, and inhibitor inactivation, being involved in apoptosis, intestinal defense-protein activation, and pathologies of the nervous system. In this line, these enzymes participate in atherosclerosis, Alzheimer's disease, multiple sclerosis, and apoptosis.[44,45] In accordance with such widespread roles, more than 30 different MMPs have been encountered in a number of tissues and organisms, which transcend vertebrates and range from archaebacteria to sea urchins, and from viruses to worms and plants.[8,46]

To perform all these specific functions, MMPs may display distinct domains. These include an ~20-residue signal peptide to enable secretion, a zymogenic ~80-residue pro-peptide, a 160 to 170-residue zinc- and calcium-dependent catalytic proteinase domain, and a CTD reminiscent of HPEX present in most MMPs (except for MMP-7, MMP-23, and plant and nematode MMPs[47]), among other domains, like fibrinogen type-II domains in gelatinases.[46,48] Most soluble secreted MMPs display their C-terminus at or just after the last cysteine residue within the CTD. The catalytic domain and the CTD are connected by a flexible linker or 'hinge' region of varying length and rich in proline residues, theoretically enabling separate movement of the two domains. The hinge regions range from 2 to 72 residues,[5] being 17 or 18 in classic collagenases[49] (MMPs-1, 8, and 13) and about 65 in membrane-type MMPs (MMP-15).

True collagenases (MMP-1, MMP-8, MMP-13, MMP-14, MMP-18, and MMP-22) cut through the triple helix of collagen at a single 'weak spot' – bonds Gly775-Ile776 and Gly775-Leu776 of the α_1 and α_2 chains of type I collagen – rendering characteristic one-fourth and three-fourth fragments.[2,50–52] Both the catalytic domain and the

CTD possess determinants essential for this specific activity, although the catalytic domains alone retain proteolytic activity toward other (peptide) substrates. On the other hand, chimeras between collagenase catalytic domains and CTDs of other MMPs, like stromelysins (MMP-3 and MMP-10), do not have triple-helicase activity.[53,54] In the latter case, however, the CTDs enable at least the binding of interstitial collagens (types I–III).[55] Besides the CTD, probably the gross conformation of its surface loops,[56] the sequence of the preceding flexible linker, though apparently not its length, also seems to be important for triple-helical collagenolysis.[54,57] These findings led to a model in which the native collagen triple helix would be trapped between the CTD and the catalytic domain of collagenases, resulting in initial proper orientation of the substrate and local disruption of the triple helix to provide access to the scissile bonds. Subsequently, peptide bond cleavage would occur, affecting one chain at a time.[50,52,58] The function of the proline-rich linker peptide that connects the catalytic and the hemopexin domains is not known, although its interaction with triple-helical collagen is hypothesized on the basis of molecular modeling.[46] In the case of gelatinases (MMP-2 and MMP-9), whose traditional substrates are gelatins and denatured collagens, a triple-helicase collagenolytic activity against native collagens of type IV, V, VII, X, and XI has been shown. In these cases, however, substrate binding does not appear to be mediated by the CTD, but probably by three characteristic fibronectin type-II repeats.[58]

Besides assisting catalytic domains during collagen binding, CTDs further interact with tissue inhibitors of metalloproteinases (TIMPs), the endogenous protein inhibitors of MMPs.[5,59] The inhibition of these enzymes involves the N-terminal domain of TIMPs and the MMP catalytic domain,[60] but initial docking between the TIMP and the MMP CTD moiety speeds up the rate of inhibition.[56] There is not only a stabilization site on the CTD for the TIMP inhibitory interaction with MMP-2 and MMP-9 but also a separate, high-affinity site that can bind TIMP-2, -3, and -4, but not TIMP-1.[58,61] Other CTD functions include zymogen activation. The hemopexin domain of MMP-2, for example, is required for the cell-surface activation of proMMP-2 by MT1-MMP, whose CTD must also dimerize during this process. Such an activation process may be inhibited by TIMP-2 through CTD binding, suggesting that similar regions of the hemopexin domain may be involved. In some cases, as with progelatinase B (proMMP-9), TIMP-CTD binding regulates the rate of activation of the zymogen, which also requires homodimerization through the CTD.[62] Interaction of the CTD with heparin was proposed to trigger the autolysis required for the complete proMMP-2 activation. Also, proMMP-13 was shown to be activated at the cell surface in a CTD-dependent manner.[63] Accordingly, MMPs can orchestrate their activation by the versatile use of their CTDs, in a tightly regulated way close to or at the cell surface.[56]

Finally, CTDs may be required to bind other proteins: sea-urchin MMP envelysin requires its CTD for substrate recognition during hatching of the fertilization envelope. Moreover, MMP-2 CTD has been reported to bind monocyte chemotactic protein 3, fetuin, fibronectin, and $\alpha_V\beta_3$ integrin, reminiscent of the function of vitronectin.[58]

AMINO ACID SEQUENCE INFORMATION

HPEXs comprise a single polypeptide chain of 430–440 residues.[64] The sequences display internal homology (about 25% identity), with two similar halves of about 200 residues organized in two HDs joined by an ~20-residue linker. This suggests duplication of an ancestral gene.[4] However, a detailed scrutiny reveals even a fourfold repetition of a core unit of about 40 residues within each HD, the so-called hemopexin-type repeats, followed by a proline-rich sequence. This suggests that a process of four elongative duplications of an ancestral exon leading to the common primordial gene for the pexin protein family has occurred.[1] In particular, human HPEX displays 439 amino acids – the gene also has an additional 23-residue signal peptide – and contains 20% carbohydrate,[65] being fully sialylated with four or five N-linked glycosylations. Its N-terminal threonine residue is O-glycosylated.[17] The mature rabbit ortholog, the only member of the family characterized at the 3D level (see below and Table 1) has 435 residues, preceded by a 25-residue signal peptide. A search for HPEXs in SwissProt/TrEMBL (www.expasy.ch) has revealed HPEX orthologues in humans (access code P02790), pig (P50828), rabbit (P20058), rat (P20059), mouse (Q91X72), and a partial sequence in the chicken (P20057). Further – probably more distant in evolution – HPEX-like sequences (complete or fragments) have been deposited for the rainbow trout (P79825; Q9DFF1), mouse (Q8K1U6), brown lemur (Q8MHV7, Q8MHV8, Q8MHV9, Q8MHW8, Q8MHY1, Q8MJT4, Q8MJT5, Q8MJT6, Q8MJT7, Q8MJT8), chicken (Q90WR3), human (Q9BS19), and the long-jawed mud sucker (Q9DFN1).

Vitronectin contains only seven hemopexin-type repeats within its collagen-binding domain, probably featuring two HDs, with the second one surrounding the basic heparin-binding site[9] and the interaction regions with type I collagen, serpins, and plasminogen.[10] Accordingly, incomplete gene duplication seems to have occurred, with the previously mentioned proline-rich linker found only once at the C-terminal end of the whole molecule. Therefore, duplication to render a molecule with two halves occurred independently and by a different mechanism than in HPEX.[4] At the vitronectin N-terminal part, a proteolytically susceptible cysteine-rich somatomedin B domain is found. Such domains are present in a number of proteins, including plasma cell membrane glycoprotein – with nucleotide pyrophosphate and alkaline

Table 1 Structures deposited with the Protein Data Bank (http://www.ebi.ac.uk/msd) displaying HDs

Name	PDB access code	Reference	Ions/solvent molecules in the channel at position[a]			
			1	2	3	4
Hemopexins						
Rabbit hemopexin C-terminal domain	1HXN	66	Na^+	Cl^-	Na^+	PO_4^{3-}
Rabbit hemopexin/heme complex (glycosyl.)	1QJS[b]	67	Na^+	Cl^-	Na^+	?
			Na^+	Cl^-	Na^+	PO_4^{3-}
			Na^+	Cl^-	Na^+	?
			Na^+	Cl^-	Na^+	PO_4^{3-}
Rabbit hemopexin/heme complex (deglycosyl.)	1QHU	67	Na^{+c}	Cl^-	Na^+	H_2O
			Na^{+c}	Cl^-	Na^+	PO_4^{3-}
MMP CTDs						
Porcine fibroblast collagenase (MMP-1)	1FBL	49	Ca^{2+}	H_2O	H_2O	H_2O
Human gelatinase A (MMP-2)	1GEN	68	Ca^{2+}	Cl^-	Na^+	H_2O
Human gelatinase A (MMP-2)	1RTG	69	Ca^{2+}	Cl^-	Ca^{2+}	H_2O
Human collagenase-3 (MMP-13)	1PEX	50	Ca^{2+}	Cl^-	Ca^{2+}	Cl^-
Human progelatinase A (proMMP-2)	1CK7	70	Ca^{2+}	Cl^-	Na^+	–
Human progelatinase A (proMMP-2)/TIMP-2	1GXD[b,d]	61	–	–	–	–
Human gelatinase B (MMP-9)	1ITV	71	Na^{+e}	H_2O	–	–
			–	H_2O	–	–

[a] The positions refer to the sequential order from the 'entry side' to the 'exit side' of the CTD disk (see Figure 5). Multiple entries for a structure refer to the different HPEX domains present in the asymmetric unit, in sequential order within one polypeptide chain first and then within subsequent chains.
[b] No solvent molecules are available for this entry.
[c] These ions are further coordinated by an additional ordered solvent molecule on the bulk-solvent side.
[d] The authors did not find any ions, as reported in the original publication.[61]
[e] Labeled as solvent 104 in the PDB entry.

phosphodiesterase I activities – and in placental protein 11 – with amidolytic activity. Human vitronectin is a 459-residue N-linked-carbohydrate glycoprotein, including a 19-residue signal peptide. Vitronectin molecules or precursor-form sequences have been reported for human (P04004), pig (P48819; P79272), mouse (P29788; Q8VII4), rat (Q62905), rabbit (P22458), and chicken (O12945), sharing more than 80% sequence homology. Further, similar complete or partial sequences from mouse (Q91X32; Q9QV98; Q9QVD2), guinea pig (Q9QVD0), rat (Q9QVD1), rabbit (Q9TRS8), sheep (Q9TRS4), goat (Q9TRS5), cattle (Q9TRS6), dog (Q9TRS7), rainbow trout (Q9DFD0), goose (Q9PS58), chicken (Q9PS59), slime mold (Q9TWP2; Q9TWP3; Q9TWP4; Q9TWX2), and common tobacco plant (P43643) have been deposited. No 3D structure for any vitronectin is available to date.

Regarding HDs belonging to the MMP CTDs, more than 160 sequences have been deposited for matrix metalloproteinases (see http://www.merops.ac.uk;[5]) from human, pig, Northern tree shrew, cattle, horse, goat, sheep, chicken, cat, dog, rat, mouse, rabbit, golden hamster, Chinese hamster, guinea pig, rainbow trout, carp, zebra fish, Japanese pufferfish, Japanese flounder, Atlantic salmon, cloudy catshark, Mexican axolotl, Japanese newt, African clawed frog, bull frog, sea urchins, fruitfly, African malaria mosquito, hydra, soybean, rice, thale cress, barrel medic, cucumber, nematodes, archaebacteria, uncultured crenarchaeote, bacteria, iridovirus, poxvirus, and granulovirus. Most of them display HPEX-like CTDs, and analysis of the sequences of representative MMPs reveals that they are between 184 and 220 residues in length, about half as long as HPEX sequences. Accordingly, MMP CTD sequences feature just one HD and possess an internal fourfold repeat.[47] It is noteworthy that almost all MMPs display a conserved disulfide bond cross-connecting the beginning with the end of the CTD. Substitution of these cysteines disrupts the CTD, impairing type I collagenolysis.[58]

Besides the previously mentioned biochemically and/or structurally characterized protein families, HDs might be harbored by other proteins, like a 65-kDa protein related with warm acclimation from the carp (Q8JIP8) and the goldfish (Q90310), toxic complex gene products (photopexins A and B) from *Photorhabdus luminescens* (Q9F4V3; Q9F4V4), mung-bean seed albumin (Q43680), and a human putative protein (Q9BS19).

PROTEIN PRODUCTION

HPEX is isolated from serum. Schematically, rabbit HPEX is obtained by precipitation of diluted rabbit serum with perchloric acid at 4 °C. The supernatant is neutralized with ammonium hydroxide, and 50% polyethylene glycol 3500 is added to a final concentration of 15%. After centrifugation, the pH of the solution is acidified and the concentration of polyethylene glycol is increased to 25%. A further centrifugation step is followed by the dissolution of the obtained pellet in acidified water. The solubilized protein is subjected to two anionic-exchange chromatography steps and finally lyophilized to be stored desiccated at 4 °C.[72]

Plasma vitronection is purified also by polyethylene glycol precipitation. Human plasma is enriched in barium chloride and polyethylene glycol 8000. The precipitate obtained is removed by centrifugation, and additional polyethylene glycol is added to the supernatant to a final concentration of 200 g L^{-1}. The precipitate thus obtained is dissolved in phosphate buffer and purified by anion-exchange and size-exclusion chromatography. Final heparin-sepharose and gel-filtration steps render purified vitronectin, which is stored as a 70% ammonium sulfate precipitate.[73]

MMP CTDs can be derived from the full-length MMPs by limited proteolysis within the linker region connecting the catalytic domain and the HD. MMPs are obtained by overexpression in mouse myeloma cells (MMP-2,[69] MMP-13,[50]) or in *Escherichia coli* as a fusion protein with β-galactosidase (MMP-1[74]). Only MMP-9 CTD was obtained directly cloning and overexpressing the HD.[71]

X-RAY STRUCTURE

HDs belong to the topological family of β-propeller domains, which encompasses proteins with number of blades ranging from four to eight (see http://scop.berkeley.edu/index.html[75]). HDs conform to all the reported proteins characterized by a fourfold propeller and show the overall shape of a disc or shallow cylinder. These proteins are made up of a succession of four structurally homologous hemopexin-type repeats of 35–45 residues, each repeat featuring a blade or β-leaflet, around a central pseudo-fourfold axis. This axis features a central funnel-like tunnel or shaft (see below), which traverses the molecule and in this way connects both flat disc surfaces (see Figure 1(b)). Accordingly, this β-propeller domain has been formed by modular tetramerization of an ancestral β-leaflet in a circular manner, such that the fourth blade packs against the first. Each of the four constituting blades (I to IV) of the propeller is made up of a four-stranded antiparallel β-sheet (strands β1 to β4) of simple up-and-down 'W' connectivity or β-leaflet topology[66] (Figure 1 and 3D Structure). The blades are arranged in such a way that the (sequentially and structurally) first inner strands β1 within each blade are located with radially directed hydrogen bonds to the following strands that run along the shaft of the propeller. The blades accumulate an internal twist on going from one strand to its outer neighbor, such that the outermost strand is nearly perpendicular to the innermost.[67] The direction of the β1 strands along the channel axis allows us to distinguish between an 'entry side' (origin of the strand) and an 'exit side' (end of the strand) of the thick disc or squat cylinder (Figure 1(b)). The resulting toroidal domain structure is tethered by an SS-bridge between the two terminal blades, which maintain domain integrity.[9] The central shafts serve as ion-binding sites in all HDs. This ion-binding property and the overall shape qualify these domains for a widespread use in macromolecular recognition, participating in protein–protein and protein–substrate interactions, though the fold is also versatile enough to enable binding of small molecules like heme.[9,66]

Rabbit hemopexin

The structure of rabbit hemopexin has been solved for its C-terminal domain alone[66] and for its full-length form, both glycosylated and deglycosylated, in complex with heme[67] (see Table 1). There are no significant differences between the crystal structures corresponding to both full-length forms and among the three C-terminal domains. As foreseeable from the amino acid sequence, HPEX is made up of two structurally homologous domains (see Figures 1 and 2) connected by a 20-residue linker (N-domain, residues 1–208, and C-domain, residues 228–435; numbering of HPEX according to PDB 1QHU and[67]). Overall, each domain is a disc of approximately 40 Å in diameter and 25 Å in height. Both domains have their N- and C-terminal tails on the same side of the propeller and show the internal pseudosymmetry characteristic of HDs. The constituting blades are packed against each other with predominantly hydrophobic face-to-face interactions. The first three β-strands and the loop-connecting strands β2 and β3 of each blade bear the highest conservation of all the leaflets,[66] whereas strands β4 are quite irregular and hardly contribute to the β-sheet hydrogen bonding pattern. In a given blade, residues 2, 4, and 6 of strand β1, residues 1, 3, and 5 of strand β2, and residue 2 of strand β3 contribute to the interface with the following blade, while residues 1,3, and 5 of strand β1, residues 2 and 4 of strand β2, and residue 3 of strand β3 shape the interface with the preceding leaflet. HPEX displays three *cis*-proline residues, two of them (Pro198 and Pro314, one in each domain) accounting for a bulge segment within one β4 strand, though in different leaflets. They affect blade IV of the N-domain and blade II of the C-domain (see Figure 2(a)). Three β-helices are encountered in the loops connecting leaflets I and II and II and III, as well as after the fourth leaflet, prior to the HD-characteristic disulfide bridge (Cys27–Cys208 in the N-domain and Cys230–Cys433 in the C-domain). The C-domain possesses a fourth helical segment (see below). Intradomain stabilization is further supported by the presence of long side chains, which better fill the leaflet interfaces. The central channel is mostly water solvated and funnel shaped, coated with hydrogen bond donors and acceptors provided by the

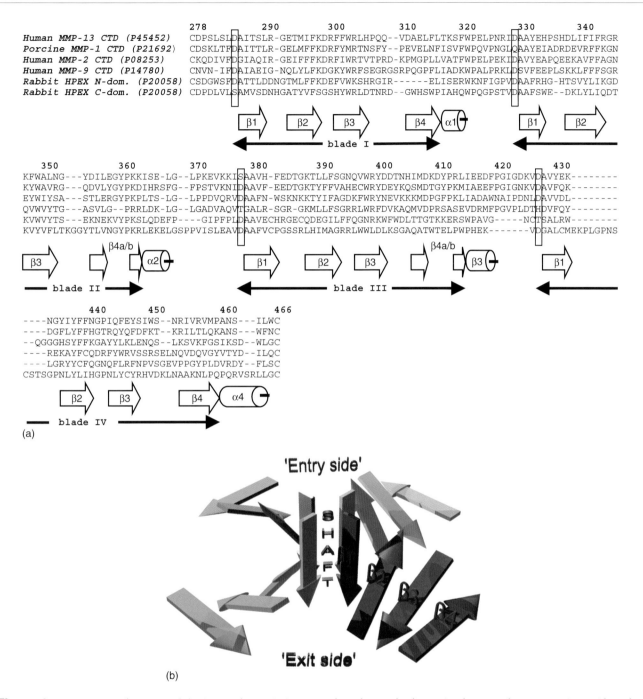

Figure 1 (a) Sequence alignment of the HDs, whose 3D Structures have been solved, running between the two cysteine residues that tether the domain architecture. The numbering, regular secondary structure elements (arrows for β-strands, cylinders for α-helices), and extension of each blade (defined from the first residue of strand β1 to the last of β4) correspond to the MMP-13 CTD structure,[50] which lacks residue number 367 to follow the numbering of MMP-1.[49] Boxes denote the four residues at the entrance of the funnel in each structure. Sequence access codes (SwissProt/TrEMBL) are indicated within parentheses. (b) Topology scheme of HDs, with each of the blades or leaflets (each in a different color and constituted by the antiparallel strands β1 to β4) arranged around the central pseudo-fourfold axis. The 'entry side', 'exit side', and central shaft are indicated.

blade's inner β-strands (see below). Three or four ions have been found in its interior. An acidic patch surrounds the tunnel entrance at the 'entry side' owing to conserved aspartate residues clustered around the tunnel opening in each domain (Figure 1), and this patch could play a role in binding the HPEX-receptor.[67]

Although the regular secondary structure elements within each blade superimpose quite well in both the N- and the C-domains, the connecting loops, comprising about half of the residues, vary greatly to provide the structural determinants for heme binding and for the dimerization surface (Figure 2(b)). Major differences are observed between loop

Hemopexin domains

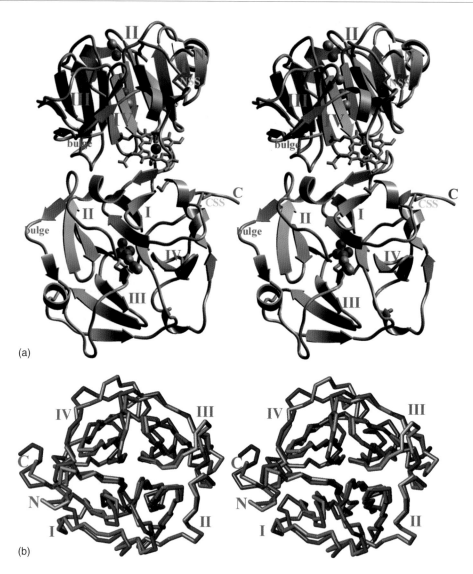

Figure 2 (a) Stereo view showing a ribbon plot of the structure of full-length deglycosylated hemopexin (PDB 1QHU) in complex with heme (green sticks). The ions/solvent molecules present in both shafts are displayed as spheres (magenta, sodium; cyan, chloride; iron, red; yellow, water; marine blue, phosphate). The disulfide bridges (yellow) and two apical proteinaceous iron-liganding histidines (gray) are also shown as sticks. The cystines maintaining the domain integrity are labeled (NSS, for N-domain SS-bond and CSS for C-domain SS-bond). Besides these, each domain has two further (distinct) disulfide bonds. The blades are labeled in roman numerals (I to IV), as are the positions of the *cis*-proline-mediated bulges and the N- and the C-terminus. (b) Superimposition of the N- (marine blue trace) and the C-domain (orange sticks) of rabbit hemopexin as a Cα-plot. Termini and blades are labeled. View of the 'entry side'.

β3β4 of blade I, which is 4 residues longer in the C-domain, and the same loop of blade II, defined in the C-domain but flexible in the N-domain. Also, the loop connecting blades II and III has a 4-residue bulge-like insertion in the C-domain on the 'entry side' of the disk and three residues more in loop β3β4 of blade III. Loop β1β2 of blade III is four residues longer in the N-domain, being folded backwards to interact with loop β3β4 of the C-domain in the full-length molecule. A helical segment is found between β3 and β4 of blade IV in the C-domain, not present in the corresponding part of the N-domain. As a result of this regular structural element, fewer interactions are observed within this blade between strands β3 and β4. The same effect is observed in blade I, in this case leading to an approach of this segment of the C-domain to loop β1β2 of blade I of the N-domain, thus contributing to the interface between the domains. By far, the greatest difference, however, is found at loop β1β2 of blade IV: no fewer than 12 extra residues are placed in the C-domain, accounting for a large protuberance on the 'exit side', which is not found in any other HD (Figures 2(b) and 3(b)). Besides the loops, the SS-bond pattern also diverges. In addition to the disulfide bond cross-linking each domain, two further bonds are found in both the N- and the C-domain, though at distinct positions: in the C-domain one bond connects the end of β1 of blade III and IV, at the funnel exit, and a second one anchors β1

Hemopexin domains

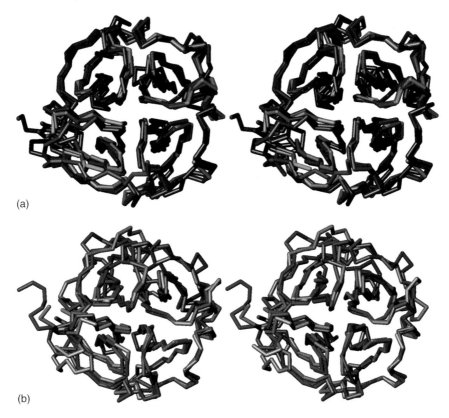

Figure 3 (a) Superimposition of the Cα-plot structures of MMP CTDs of porcine fibroblast collagenase (MMP-1; green), collagenase 3 (MMP-13; magenta), gelatinase A (MMP-2; violet) and B (MMP-9; red). View of the 'entry side' and orientation as in 3D Structure. The central ions in the shaft have been omitted for clarity. (b) Superimposition of MMP-13 CTD (magenta) with both domains of rabbit hemopexin (N-domain, marine blue; C-domain, orange). View of the 'entry side' and orientation as in 3D Structure.

with β2 and β3 of blade IV. In the N-domain, they connect strand β1 with loop β1β2 (within blade III) and the loop region preceding β1 with β2, both within blade IV.

The two domains are connected by a partially disordered linker (seven residues missing in PDB 1QHU) that ends in a small helical segment prior to the C-domain. They lock together at a 90° angle, edge (N-domain) to face (C-domain), and combine to bind a single heme molecule at their interface (Figures 2(a) and 4(a) ⑤). Whereas the N-domain is itself capable of heme binding, the isolated C-domain is not, but is engaged in HPEX-receptor binding.[76] The protein–domain interface, also formed by the ordered part of the connecting linker, delivers two histidine residues, His213 and His266, which trap the iron porphyrin ring through the two apical metal-coordinating positions (Figure 2(a)). The heme binding pocket is coated by a cluster of conserved aromatic residues, mainly provided by the N-domain, proposed to be required to preserve stability of the heme site. Besides this, another cluster of basic residues contributes to the binding of the heme propionate groups.[67]

Collagenase 3 (MMP-13) C-terminal domain

A series of MMP CTD structures have been solved to date, including those of MMP-1, MMP-2, MMP-9, and MMP-13 (see Table 1). Three correspond to molecules that also display the catalytic MMP domain, among others, and/or a TIMP-2 molecule,[49,61,70] and four correspond to isolated CTDs.[50,68,69,71] These MMP CTDs show sequence and structural similarity,[47,50] and so their structures can be exemplified by the discussion of the MMP-13 CTD[50] (see 3D Structure and Figures 1(a) and 3). Only MMP-9, probably due to the lack of ions in the central shaft, deviates somewhat from the common structure of the other MMP members (see below).

The MMP-13 CTD's shallow cylinder is approximately 45 Å in diameter and 30 Å in height, consistent with the values reported for MMP-1 and MMP-2[49,69] and displays the overall fourfold β-propeller architecture mentioned for HDs. As in the HPEX structure, the N- and the C-terminus are covalently linked (Cys278–Cys466) and located on the same side of the propeller, between blades I and IV. Again, as in HPEX, the structure is stabilized by the presence of increasingly bulkier residues on going from the center to the periphery and of internal hydrophobic clusters at the blade interfaces. Within each blade, strands β1 and β2 are linked by distorted β-hairpin loops of two (first three leaflets) or four (last blade) internal residues arranged around the 'exit side' of the tunnel, whereas 1,4-tight turns connect strands β2 and β3 and shape the 'entrance side'.

Hemopexin domains

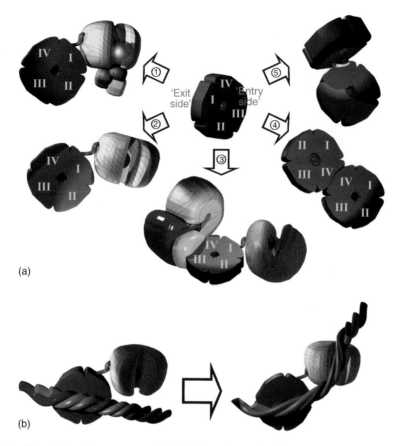

Figure 4 (a) HDs and their disposition with respect to additional domains and inhibitors as present in the different reported structures for MMP CTDs and in HPEX. The HD 'entry side' is characterized by the presence of a cation (Ca^{2+} or Na^+) at the tunnel entrance in most structures (red sphere embedded in the HD disc). HD blades are labeled with roman numerals. The relative disposition of the HD and the catalytic domain (yellow wooden shape with a violet zinc cation) is almost the same in ① progelatinase A (proMMP-2; PDB 1CK7; transparent prodomain and green fibronectin type-II repeats) and in ② porcine fibroblast collagenase (MMP-1; PDB 1FBL). ③ TIMP-2 (blue) interacts through its C-terminal domain (cyan) with the CTD in the proMMP-2/TIMP-2 complex (PDB 1GXD; prodomain and fibronectin domains not shown), while its N-terminal domain (dark blue) could block a catalytic domain (left yellow moiety) without steric clashes, as observed when superimposing the inhibitor moiety with the TIMP-2/MT1-MMP structure (PDB 1BUV). ④ Dimerization via the CTDs, as seen for MMP-9 CTD (PDB 1ITV), required for various functions including activation, involves the same CTD region as the interaction with TIMP C-terminal domain (see 3) and should be mutually exclusive. ⑤ In HPEX, the two HDs form an angle of about 90° between their 'exit sides' trapping the heme molecule (metallic ring) at their interface (see also Figure 2). (b) In the light of (a), a novel working mechanism for triple-helical collagenolysis could be thought of.

A structural peculiarity is a β-bulge structure, as found in HPEX (3D Structure and Figure 3(b)) which, however, here interrupts strands β4 of blades II and III, with four residues opposing one of the preceding strand β3. These bulges enable the outermost strand to keep in phase with the preceding one despite the overall sheet curvature. They are mediated by *cis*-proline residues (Pro361 and Pro412 for blades II and III respectively), which compensate the structural perturbations produced by the bulges. In the remaining two blades, these positions include a one-residue bulge in blade IV and a stronger curvature of β4 in leaflet I. In all blades, as in HPEX, helical segments looping around the periphery of the cylinder are encountered. In contrast to the plasma protein structure, however, all helices are placed in the four loops connecting the blades in MMP-13 (Figure 1(a) and 3D Structure).

Comparison with other MMP C-terminal domains

Comparison of MMP-13 CTD with the other collagenase CTD structure available (MMP-1) and those of (pro)gelatinases A and B, proMMP-2 and MMP-9, reveals, despite close overall structural similarity, that both collagenases are topologically more similar than both compared with gelatinases. In particular, around the first cysteine residue, the MMP-2 polypeptide chain reaches the entrance point of the first shaft strand in a wide loop, while both collagenases display multiple-turn structures. MMP-1, MMP-13, and MMP-2 exhibit the same β-bulges within β4 of blades II and III (see Figure 3(a)) caused by *cis*-proline residues, while MMP-9 has only the second bulge. The surface charge patterns confirm these findings, with greater similarity among collagenases than to gelatinase, in particular, at the 'entry side', positively charged in collagenases, in contrast to the more acidic MMP-2. The opposite happens on the 'exit side', mainly negative in the former and electropositive in the latter.[50] MMP-9, on the other hand, presents a rather uniform charge distribution pattern, with positive and negative patches all over the monomer surface.[71]

The most deviating structure among MMP CTDs is that of MMP-9. In this case, an asymmetric SDS-stable though not covalent (putatively functional) dimer is found within the crystal asymmetric unit. The dimerization mainly occurs through blade IV, with the axes of the HDs being roughly antiparallel (see Figure 4(a), ④). Upon comparison with the other CTDs, MMP-9 shows a significant repositioning of blades III and IV in both molecules of the dimer, predominantly along the propeller axis and leading to significant channel widening.[71] Furthermore, the relative arrangement of the leaflet blades is much more flexible than in the other CTDs, and so only the first three β-strands within each blade are superimposable with the other CTDs.[71]

Arrangement of the catalytic and C-terminal domains in MMPs: A novel mechanism for triple-helical collagenolysis?

Three structures have been reported for molecules comprising the catalytic MMP domain and the subsequent CTD, one for porcine fibroblast collagenase (MMP-1; PDB 1FBL) and two for human progelatinase A (proMMP-2; PDB 1CK7, 1GXD). Despite differences in the lengths of the 'hinge' region (18 residues between the last aromatic residue of the C-terminal helix of the catalytic domain and the first tethering CTD cysteine in MMP-1; 24 residues, partially disordered, in MMP-2) and enzyme specificity, the relative arrangement of the two domains is similar in all three structures (despite unique crystallographic environments), both in orientation and position. In all cases, the catalytic cleft runs almost parallel to the interface between blades I + II and III + IV, on the 'exit side' of the CTD (Figure 4(a), ① and ②). The main interactions with the catalytic domain affect the lateral cylinder surface around blade I. In particular, a clockwise rotation of just about 14° respectively around the CTD funnel axis (running from the 'entry side' to the 'exit side') and a subsequent translation of about 10 Å would place the MMP-1 catalytic domain on top of the corresponding domain of proMMP-2 in the different crystal forms reported, after optimal superimposition of their CTDs. This is important, as the relative disposition of the two domains was initially attributed to crystallographic packing in MMP-1, and free and independent motion due to the flexible linker was suggested.[2,50,58] Furthermore, the presence of the prodomain and the three fibronectin type-II repeats (in proMMP-2) is sterically compatible with the interaction via blade IV with a second CTD molecule (as observed in MMP-9 CTD[71] and postulated for proMMP-2 activation; see PDB 1ITV and Figure 4(a), ④). The spatial arrangement of the catalytic domain and the CTD is even compatible with the binding of the latter to TIMP-2 at its C-terminal domain (through blades III and IV; similar to clathrin ligand recognition[61]), thus preventing MMP dimerization via the CTDs,[58,77] as unrelated regions of the CTD disc are involved (PDB 1GXD and Figure 4(a), ③). (Pro)MMP-9 forms complexes with TIMP-1, (pro)MMP-1, or (pro)MMP-9, which are mutually exclusive, suggesting overlapping recognition sites within the CTD.[62,71,78] The proMMP-2/TIMP-2 complex is proposed to inhibit MT1-MMP via the N-terminal TIMP part[60,79] (Figure 4(a), ③) and to localize MMP-2 to the cell membrane where it is activated by membrane(-associated) proteinases.[58,61] A similar proMMP-9/TIMP-1/MMP-3 soluble complex has also been observed *in vitro*.[80]

All these findings suggest that binding to CTDs is a complex issue and the question arises as to whether the various domain arrangements found in the crystal structures discussed have biological significance. If the relative disposition of the catalytic and CTD moieties in the three distinct structures is not artifactual, a novel working mechanism for true collagenases against native collagen, which diverges from earlier models,[50,58] could be conceived: the CTD could interact with its 'exit side', putatively by means of the four protruding loops connecting strands β1 and β2 within each blade, with the substrate triple helix upstream or downstream of the 'weak spot' leading to the required partial unwinding or 'melting' of the triple helix. This is consistent with the differences in the charge pattern of the 'exit side' observed between collagenases and gelatinases (see above). In a second step, catalytic domain–mediated scission through each of the three loosened collagen chains would occur (Figure 4(b)). This would be in agreement with the extremely low turnover rate of native triple-helical collagens: merely 22 collagen molecules are cut through per molecule of collagenase per hour.[58]

CENTRAL SHAFT, METAL CONTENT, AND CALCIUM-BINDING SITE

The central shaft, coated by strands β1 of each leaflet, traverses the oblate ellipsoidal HDs along their short axis and has a width of about 4.5 to 5 Å (measured as interatomic distances) in its central part in hemopexin and in MMP-13 CTD. This channel is slightly funnel shaped because of the curvature of the β-strands that make it narrow at the 'entry side' (about 4–4.5 Å) and wide at its 'exit side' (about 7–9 Å) in both MMP-13 CTD and the C-domain of HPEX. The corresponding part in the HPEX N-domain is, interestingly, much narrower (4–5 Å) despite superimposable main-chain tracings, owing to the presence of several bulky charged side chains crowning the channel exit. In the center of the tunnel, the four β1 strands come closest together, so that packing requirements favor the presence of small side chains, mostly alanine residues. In HDs, several anions, cations, and solvent molecules are found in these shafts, and their funnel-like shape may support a mechanism to load the HDs with them from the more open 'exit side' of the channel. The function of these ions, however, is not clear. A stabilizing function for the whole domain has been suggested, in particular in light of the MMP-9 CTD structure,[71] which (besides one sodium ion, see below) lacks them and displays a flexible architecture and greatest deviations from the other reported structures (see above). However, this was not confirmed by a full-length structure of proMMP-2 (PDB 1GXD), which equally lacks ions but is rigid.[61]

Ions bind mainly at four positions (see Figure 5 and Table 1) and are mostly coordinated/bound in a planar manner by (1) the carbonyl oxygen atoms of residues at β-strand position R_1 of each of the four strands, that is, in most cases, the acidic residues acting as a gatekeeper at the tunnel entrance (see Fig. 1(a)); (2) the amide nitrogen atoms of positions R_3; (3) the main-chain oxygen atoms of positions R_3; and (4) the main-chain nitrogen atoms at positions R_5, albeit at a much greater distance. Accordingly, two potential binding environments for cations and two more for anions are found alternating in the shaft, though each of the reported structures displays distinct ion species or even just solvent molecules (Table 1). Among the structures analyzed, position 1, at the 'entry surface', seems to be the most important, being predominantly occupied by a calcium or sodium cation. Putatively, Ca^{2+} ions have been proposed to be required for structure stabilization, for supporting the appropriate distribution of hydrophobic residues at the blade interfaces and the ring-closing SS-bond, and for the binding of several ligands, among them heparin and fibronectin in the case of MMP-2.[68,81]

In HPEX C-domain,[66] the cation at position 1 (Figure 5 and Table 1) is either a Na^+ or an isoelectronic Mg^{2+}, coordinated by the four oxygen atoms at an average distance of 2.2 Å and a surface-located solvent molecule at 2.4 Å. The anion at position 2 was clearly identified as a chloride, surrounded by four amide groups at an average distance of 3.4 Å. It is further bound by the second cation at position 3 (at 2.8 Å). The latter is located in the center of the tunnel and bound by four carbonyl oxygens (at 2.3 Å) and, in the C-domain, by a phosphate found in position 4 at the wider end of the tunnel (at 2.9 Å). In the N-domain, this fourth position is occupied by a solvent molecule. The phosphate receives hydrogen bonds from the four R_5 amide nitrogens at an average distance of 3.5 Å (C-domain), while the solvent in the N-domain (2.6 Å from the Na^+ at position 3) is too far away from the R_5 nitrogen atoms (4.5–5.0 Å). This latter site 4 seems to have an enhanced affinity for larger anions in HPEX, at least in

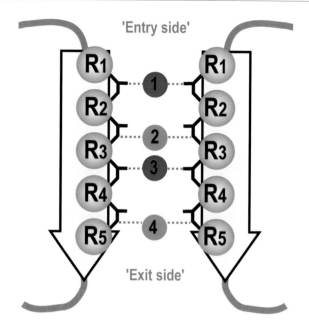

Figure 5 Schematic view of the central tunnel with the positions found to be occupied by ions in most of the distinct HDs, exemplified by MMP-13. Only two (opposed) of the four strands β1 are shown for clarity. Dashed gray lines denote hydrogen bonds.

the much wider C-domain, and also binds propionate and butyrate anions.

In MMP-13 CTD, two cations and two anions have been suggested for positions 1 to 4 on the basis of electron density, ligand sphere, and reasonable temperature factors after refinement, as well as their presence in the crystallization conditions.[50] They have been assigned Ca^{2+}, Cl^-, Ca^{2+}, and Cl^-, respectively (see Table 1). The first cation is coordinated by the R_1 carbonyl oxygens, 2.1 to 2.4 Å away, at typical binding distances for Ca^{2+}-oxygen bonds[82] (Figure 6). The first chloride, at position 2, is coordinated in the plane by the four R_2 nitrogen atoms at 3.3 to 3.7 Å and in one apical position by the calcium ion at position 3, at 2.9 Å. This latter cation displays a regular octahedral coordination sphere with the second apical position occupied by the chloride at position 4 (3.3 Å) and the four in-plane liganding atoms at 2.2 to 2.4 Å. Finally, the chloride at position 4 is already too far away to be bound by the R_5 amide nitrogen atoms (4.0–4.9 Å).

MMP-1 and MMP-2 CTDs share with MMP-13 the presence of a Ca^{2+} at position 1. In the collagenase, the coordinating carbonyl oxygens are at distances of 2.2 to 2.5 Å, while for the three coordinates deposited for MMP-2 with ions in the shaft these values are 2.4 to 2.5 Å (PDB 1RTG; one solvent molecule on the bulk-solvent side at 2.9 Å), 2.0 to 2.2 Å (PDB 1GEN; three solvent molecules on the bulk-solvent side at 2.0 to 2.3 Å rendering an overall slightly distorted octahedral + 1 coordination) and 2.5 to

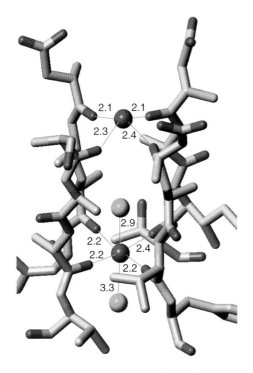

Figure 6 Calcium-binding sites of MMP HDs exemplified by the structure of MMP-13 CTD. The binding distances to the ligands are indicated. This protein presents (putatively) calcium cations (red spheres) in shaft positions 1 and 3 (see Figure 1). These cation sites are shared with and are almost identical with MMP-1 CTD (site 1) and MMP-2 CTD (sites 1 and 2). In the shaft, two further anions (putatively chloride ions) have been found.

2.8 Å (PDB 1CK7; no further ligands on the bulk-solvent side).[49,61,68–70] In MMP-9 CTD, a sodium cation is found at this position within one of the two molecules present in the asymmetric unit. This cation is bound by the four R_1 carbonyl oxygens at distances of 2.4 to 2.6 Å. No further anions or cations have been found in either of the two shafts in MMP-9. In particular, binding of the Na^+/Ca^{2+} ion at position 1 on the 'entry side' of the second molecule is prevented by a conserved aspartate residue (Asp663), mainly responsible for channel widening.[71] While MMP-1 displays solvent molecules at positions 2, 3, and 4 (the first two with rather short binding distances, though; 2.5 to 2.8 Å), there is some controversy about the MMP-2 structures. While in all cases a chloride anion was assigned to position 2 (in-plane ligands at 2.4 to 2.8 Å in PDB 1GEN, at 3.8 to 4.0 Å in PDB 1RTG, and at 3.6 to 4.3 Å in PDB 1CK7), the identity of the cation at position 4 is in doubt. Both a Ca^{2+} (at 3.1 Å in apical position from the chloride in position 3; PDB 1RTG) and a Na^+ (at 2.8 Å in PDB 1GEN or 3.2 Å in PDB 1CK7) have been described. The in-plane ligands for the three latter cases are, respectively, 2.6 to 2.7 Å, 2.0 to 2.4 Å, and 2.7 to 2.9 Å away. Among these three cations in position 3, the Ca^{2+} and the first Na^+ show a regular octahedral coordination sphere owing to the presence of a sixth (apical) ligand, occupying position 4. In both cases, a solvent has been described (at 2.4 Å and 2.3 Å respectively).

Finally, it may be mentioned that for one of the MMP-2 CTD structures,[68] a potential Zn^{2+}-binding site was suggested, which is not observed in other structures.[61,69,70] However, plasma mass spectrometry analyses did not reveal any significant zinc content in recombinant MMP-2 CTD.[81]

RELATIONSHIP IN THE FOLD WITH OTHER PROTEINS

HDs share the β-propeller or circular leaflet topology with twisted four-stranded antiparallel β-sheet blades, including, in some cases, inserted peripheral β-helices, and the overall disc shape with a number of other proteins of diverse functions of prokaryotic and (more commonly) eukaryotic origin. These proteins, however, vary in the order of the axial pseudosymmetry (see http://scop.berkeley.edu/index.html;[75]). It may be fivefold, as in the lectin tachylectin-2 (PDB 1TL2;[83]), whose central shaft is ion-depleted; and in α-L-arabinanase (PDB 1GYH;[84]), with chloride in the channel. Sixfold pseudosymmetry is encountered in the ion-depleted N-terminal domain of neuroaminidase (PDB 1EUT;[85]); in diisopropylfluorophosphatase (PDB 1E1A;[86]); and in phytase, alias myo-inositol-hexaphosphate 3-phosphohydrolase (PDB 2POO;[87]), the latter two proteins with calcium ions in the channel. The largest group has a pseudo-sevenfold axis. Examples are transducin, alias guanine–nucleotide binding protein G (PDB 1TBG;[88]); the WD40-domain proteins,[89] like the CTD of transcriptional repressor TUP1 (PDB 1ERJ;[90]); regulator of chromosome condensation RCC1 (PDB 1A12;[91]), a guanine–nucleotide-exchange factor for the nuclear Ras homologue Ran; methylamine dehydrogenase (PDB 2BBK;[92]); and the inserted propeller domain of prolyl oligopeptidase (1QFM;[93]). All these proteins display ion-depleted central channels, whereas β-lactamase inhibitor protein II (PDB 1JTD;[94]) harbors two calcium ions in the shaft. The tricorn protease protomer is of special interest as it has two such β-propeller domains, one with sixfold and a second with sevenfold pseudosymmetry, this latter with a slightly oblong shape, both being cation depleted (PDB 1K32;[95]). Finally, eightfold pseudoaxes are found in methanol dehydrogenase (PDB 4AAH;[96]), with a calcium ion bound by a pyrroloquinoline quinone in the funnel-like shaft, and the dimeric D_1 domain of cytochrome cd_1 nitrite reductase (PDB 1QKS;[97]). This structure has a heme D_1 prosthetic group (including an iron cation) in the shaft, which is not accessible to the bulk solvent.

Packing arguments suggest that the optimum symmetry for a β-propeller is sevenfold and that symmetries below sixfold are energetically unfavorable.[98] Accordingly, four-fold and fivefold β-propellers must incorporate additional elements to stabilize the circular arrangement, like the strictly conserved tethering disulfide bridge in HDs and large interfaces between leaflets, with many bulky side chains and extra residues inserted into the loops and helices connecting the blades. Also, the order of the propeller does not seem to affect the width of the central tunnel: the minimum diameter in HDs is approximately equal in sixfold and only slightly narrower than that in sevenfold domains.[98] Interestingly, propellers with orders higher than four do not show a systematic tunnel widening on their 'exit sides', a fact that might be related to the more abundant cross-linking ions in the shaft of fourfold propellers.

In most cases of higher-symmetry β-propellers, the blades are made up sequentially around the pseudosymmetry axis, with their four constituting β-strands with simple up-and-down connectivity, as observed in the fourfold HDs. The first strand is shaft-proximal and the last one the radial outermost. In some of the aforementioned cases, however, the N-terminal end of the polypeptide chain forms only the outermost strand β4 of blade I (as found in PDB 1TBG, 1ERJ, 1A12, 1EUT, 1QBI, and 1E1A) or just strands β2, β3, and β4 (PDB 2BBK) before entering blade II. Subsequently, the remaining blades are canonically formed by the up-and-down running chain. Finally, the remaining strands of blade I (β1, β2, and β3, or β1 respectively) are contributed by the C-terminal stretch of the polypeptide after the last blade. In other cases, this first blade is made up by two N-terminal strands and two C-terminal ones (PDB 1A12). This feature seems to enable ring closure and ring stabilization, in a sort of 'velcro' mechanism that compensates for the absence of the disulfide bond connecting the first and last blades of fourfold propellers. In other proteins, if the propeller constitutes a domain inserted in a complex structure, as in prolyl oligopeptidase and the tricorn protease, neither 'velcro' nor SS-bridges are observed. In these cases, the pertinence of the N- and C-termini to another domain provides the required structural stabilization.[75,93,95]

Accordingly, an axial arrangement of varying numbers of propeller-like blades constitutes the structural scaffold for proteins of major importance in cell biology, in particular, as mediators of protein–protein interactions.[67,89] These proteins participate in distinct functions and reactions as host defense, cellulose processing, connective-tissue turnover, heme, phosphate and sugar metabolism, organophosphate detoxification and protection against oxidative stress, transmembrane signaling, DNA repression, control of nucleo-cytoplasmic transport and the cell cycle, cell adhesion, humoral immune response and the hemostatic system, electron transport, maturation and degradation of peptide hormones and neuropeptides, antibiotic resistance, intracellular protein degradation, alcohol processing, and reduction of oxygen to water, among several others. This is an interesting example of divergent evolution adapted, from the functional point of view, to a number of tasks and, structurally, to render several different β-propeller orders.[4]

REFERENCES

1. U Muller-Eberhard, *Methods Enzymol*, **163**, 536–65 (1988).
2. W Bode, *Structure*, **3**, 527–30 (1995).
3. E Tolosano and F Altruda, *DNA Cell Biol*, **21**, 297–306 (2002).
4. D Jenne and KK Stanley, *Biochemistry*, **26**, 6735–42 (1987).
5. JF Woessner Jr and H Nagase, *Matrix Metalloproteinases and TIMPs*, Oxford University Press, New York (2000).
6. W Bode, FX Gomis-Rüth and W Stöcker, *FEBS Lett*, **331**, 134–40 (1993).
7. W Stöcker, F Grams, U Baumann, P Reinemer, FX Gomis-Rüth, DB McKay and W Bode, *Protein Sci*, **4**, 823–40 (1995).
8. FX Gomis-Rüth, *Mol Biotech*, **24**, 157–202 (2003).
9. LT Hunt, WC Barker and HR Chen, *Protein Seq Data Anal*, **1**, 21–26 (1989).
10. KT Preissner and D Seiffert, *Thromb Res*, **89**, 1–21 (1998).
11. J Gross and Y Nagai, *Proc Natl Acad Sci USA*, **54**, 1197–204 (1965).
12. H Nagase, AJ Barrett and JF Woessner Jr, *Matrix Suppl*, **1**, 421–24 (1992).
13. CM Overall and C López-Otín, *Nat Rev Cancer*, **2**, 657–72 (2002).
14. RJ Winzler, AW Devor, JW Mehl and IM Smyth, *J Clin Invest*, **27**, 609–16 (1948).
15. H Baumann and J Gauldie, *Immunol Today*, **15**, 74–80 (1994).
16. V Poli, *J Biol Chem*, **273**, 29279–82 (1998).
17. JR Delanghe and MR Langlois, *Clin Chim Acta*, **312**, 13–23 (2001).
18. BS Berlett and ER Stadtman, *J Biol Chem*, **272**, 20313–16 (1997).
19. MA Smith, K Hirai, K Hsiao, MA Pappolla, PL Harris, SL Siedlak, M Tabaton and G Perry, *J Neurochem*, **70**, 2212–15 (1998).
20. A Smith, *DNA Cell Biol*, **21**, 245–49 (2002).
21. M Juckett, Y Zheng, H Yuan, T Pastor, W Antholine, M Weber and G Vercellotti, *J Biol Chem*, **273**, 23388–97 (1998).
22. HF Bunn and JH Jandl, *J Biol Chem*, **243**, 465–75 (1968).
23. DM Davies, A Smith, U Muller-Eberhard and WT Morgan, *Biochem Biophys Res Commun*, **91**, 1504–11 (1979).
24. PJ Pereira, S Macedo-Ribeiro, A Párraga, R Pérez-Luque, O Cunningham, K Darcy, TJ Mantle and M Coll, *Nat Struct Biol*, **8**, 215–20 (2001).
25. JD Eskew, RM Vanacore, L Sung, PJ Morales and A Smith, *J Biol Chem*, **274**, 638–48 (1999).
26. JM Gutteridge, *Biochim Biophys Acta*, **917**, 219–23 (1987).
27. E Tolosano, MA Cutufia, E Hirsch, L Silengo and F Altruda, *Biochem Biophys Res Commun*, **218**, 694–703 (1996).
28. JP Swerts, C Soula, Y Sagot, MJ Guinaudy, JC Guillemot, P Ferrara, AM Duprat and P Cochard, *J Biol Chem*, **267**, 10596–600 (1992).
29. W Chen, H Lu, K Dutt, A Smith, DM Hunt and RC Hunt, *Exp Eye Res*, **67**, 83–93 (1998).
30. KT Preissner, *Annu Rev Cell Biol*, **7**, 275–310 (1991).
31. D Seiffert, M Geisterfer, J Gauldie, E Young and TJ Podor, *J Immunol*, **155**, 3180–85 (1995).
32. WR Pohl, MG Conlan, AB Thompson, RF Ertl, DJ Romberger, DF Mosher and SI Rennard, *Am Rev Respir Dis*, **143**, 1369–75 (1991).

33 D Seiffert, *Histol Histopathol*, **12**, 787–97 (1997).

34 K Miyazaki, T Hamano and M Hayashi, *Exp Cell Res*, **199**, 106–10 (1992).

35 N Nakashima, K Miyazaki, M Ishikawa, T Yatohgo, H Ogawa, H Uchibori, I Matsumoto, N Seno and M Hayashi, *Biochim Biophys Acta*, **1120**, 1–10 (1992).

36 BR Tomasini and DF Mosher, *Prog Hemost Thromb*, **10**, 269–305 (1991).

37 RJ Wechsler-Reya, *Trends Neurosci*, **24**, 680–82 (2001).

38 D Jenne, F Hugo and S Bhakdi, *Thromb Res*, **38**, 401–12 (1985).

39 KT Preissner, *Blut*, **59**, 419–31 (1989).

40 ER Podack, WP Kolb and HJ Müller-Eberhard, *J Immunol*, **119**, 2024–29 (1977).

41 R Pytela, MD Pierschbacher and E Ruoslahti, *Proc Natl Acad Sci USA*, **82**, 5766–70 (1985).

42 B Felding-Habermann and DA Cheresh, *Curr Opin Cell Biol*, **5**, 864–68 (1993).

43 CE Brinckerhoff and LM Matrisian, *Nat Rev Mol Cell Biol*, **3**, 207–14 (2002).

44 LJ McCawley and LM Matrisian, *Curr Opin Cell Biol*, **13**, 534–40 (2001).

45 VW Yong, C Power, P Forsyth and DR Edwards, *Nat Rev Neurosci*, **2**, 502–11 (2001).

46 H Nagase and JF Woessner Jr, *J Biol Chem*, **274**, 21491–94 (1999).

47 I Massova, LP Kotra, R Fridman and S Mobashery, *FASEB J*, **12**, 1075–95 (1998).

48 G Murphy, V Knäuper, S Cowell, R Hembry, H Stanton, G Butler, J Freije, AM Pendas and C López-Otín, *Ann N Y Acad Sci*, **878**, 25–39 (1999).

49 J Li, P Brick, MC O'Hare, T Skarzynski, LF Lloyd, VA Curry, IM Clark, HF Bigg, BL Hazleman, TE Cawston and DM Blow, *Structure*, **3**, 541–49 (1995).

50 FX Gomis-Rüth, U Gohlke, M Betz, V Knäuper, G Murphy, C López-Otín and W Bode, *J Mol Biol*, **264**, 556–66 (1996).

51 J Ottl, D Gabriel, G Murphy, V Knäuper, Y Tominaga, H Nagase, M Kröger, H Tschesche, W Bode and L Moroder, *Chem Biol*, **7**, 119–32 (2000).

52 JL Lauer-Fields, D Juska and GB Fields, *Biopolymers*, **66**, 19–32 (2002).

53 T Hirose, C Patterson, T Pourmotabbed, CL Mainardi and KA Hasty, *Proc Natl Acad Sci USA*, **90**, 2569–73 (1993).

54 V Knäuper, PML FX, SB Gomis-Rüth, A Lyons, AJP Docherty and G Murphy, *Eur J Biochem*, **268**, 1888–96 (2001).

55 JA Allan, RM Hembry, S Angal, JJ Reynolds and G Murphy, *J Cell Sci*, **99**, 789–95 (1991).

56 G Murphy and V Knäuper, *Matrix Biol*, **15**, 511–18 (1997).

57 SJ de Souza and R Brentani, *J Biol Chem*, **267**, 13763–67 (1992).

58 CM Overall, *Methods Mol Biol*, **151**, 79–120 (2001).

59 W Bode, C Fernández-Catalán, F Grams, FX Gomis-Rüth, H Nagase, H Tschesche and K Maskos, *Ann N Y Acad Sci*, **878**, 73–91 (1999).

60 FX Gomis-Rüth, K Maskos, M Betz, A Bergner, R Huber, K Suzuki, N Yoshida, H Nagase, K Brew, GP Bourenkov, H Bartunik and W Bode, *Nature*, 77–81 (1997).

61 E Morgunova, A Tuuttila, U Bergmann and K Tryggvason, *Proc Natl Acad Sci USA*, **99**, 7414–19 (2002).

62 GI Goldberg, A Strongin, IE Collier, LT Genrich and BL Marmer, *J Biol Chem*, **267**, 4583–91 (1992).

63 V Knäuper, L Bailey, JR Worley, P Soloway, ML Patterson and G Murphy, *FEBS Lett*, **532**, 127–30 (2002).

64 WT Morgan and A Smith, *J Biol Chem*, **259**, 12001–6 (1984).

65 N Takahashi, Y Takahashi and FW Putnam, *Proc Natl Acad Sci USA*, **82**, 73–77 (1985).

66 HR Faber, CR Groom, HM Baker, WT Morgan, A Smith and EN Baker, *Structure*, **3**, 551–59 (1995).

67 M Paoli, BF Anderson, HM Baker, WT Morgan, A Smith and EN Baker, *Nat Struct Biol*, **6**, 926–31 (1999).

68 AM Libson, AG Gittis, IE Collier, BL Marmer, GI Goldberg and EE Lattman, *Nat Struct Biol*, **2**, 938–42 (1995).

69 U Gohlke, FX Gomis-Rüth, T Crabbe, G Murphy, AJ Docherty and W Bode, *FEBS Lett*, **378**, 126–30 (1996).

70 E Morgunova, A Tuuttila, U Bergmann, M Isupov, Y Lindqvist, G Schneider and K Tryggvason, *Science*, **284**, 1667–70 (1999).

71 H Cha, E Kopetzki, R Huber, M Lanzendorfer and H Brandstetter, *J Mol Biol*, **320**, 1065–79 (2002).

72 W Morgan, P Muster, FM Tatum, SM Kao, J Alam and A Smith, *J Biol Chem*, **268**, 6256–62 (1993).

73 SV Bittorf, EC Williams and DF Mosher, *J Biol Chem*, **268**, 24838–46 (1993).

74 MC O'Hare, NJ Clarke and TE Cawston, *Gene*, **111**, 245–48 (1992).

75 V Fülöp and DT Jones, *Curr Opin Struct Biol*, **9**, 715–21 (1999).

76 WT Morgan, P Muster, FM Tatum, J McConnell, TP Conway, P Hensley and A Smith, *J Biol Chem*, **263**, 8220–25 (1988).

77 W Bode and K Maskos, *Methods Mol Biol*, **151**, 145–77 (2001).

78 V Knäuper, S Cowell, B Smith, C López-Otín, M O'Shea, H Morris, L Zardi and G Murphy, *J Biol Chem*, **272**, 7608–16 (1997).

79 C Fernández-Catalán, W Bode, R Huber, D Turk, JJ Calvete, A Lichte, H Tschesche and K Maskos, *EMBO J*, **17**, 5238–48 (1998).

80 H Kolkenbrock, D Orgel, A Hecker-Kia, J Zimmermann and N Ulbrich, *Biol Chem Hoppe-Seyler*, **376**, 495–500 (1995).

81 UM Wallon and CM Overall, *J Biol Chem*, **272**, 7473–81 (1997).

82 O Herzberg and MNG James, *Biochemistry*, **24**, 5298–302 (1985).

83 HG Beisel, S Kawabata, S Iwanaga, R Huber and W Bode, *EMBO J*, **18**, 2313–22 (1999).

84 D Nurizzo, JP Turkenburg, SJ Charnock, SM Roberts, EJ Dodson, VA McKie, EJ Taylor, HJ Gilbert and GJ Davies, *Nat Struct Biol*, **9**, 665–68 (2002).

85 A Gaskell, S Crennell and G Taylor, *Structure*, **3**, 1197–205 (1995).

86 EI Scharff, J Koepke, G Fritzsch, C Lücke and H Rüterjans, *Structure (Camb)*, **9**, 493–502 (2001).

87 NC Ha, BC Oh, S Shin, HJ Kim, TK Oh, YO Kim, KY Choi and BH Oh, *Nat Struct Biol*, **7**, 147–53 (2000).

88 J Sondek, A Bohm, DG Lambright, HE Hamm and PB Sigler, *Nature*, **379**, 369–74 (1996).

89 TF Smith, C Gaitatzes, K Saxena and EJ Neer, *Trends Biochem Sci*, **24**, 181–85 (1999).

90 ER Sprague, MJ Redd, AD Johnson and C Wolberger, *EMBO J*, **19**, 3016–27 (2000).

91 L Renault, N Nassar, I Vetter, J Becker, C Klebe, M Roth and A Wittinghofer, *Nature*, **392**, 97–101 (1998).

92 L Chen, M Doi, RC Durley, AY Chistoserdov, ME Lidstrom, VL Davidson and FS Mathews, *J Mol Biol*, **276**, 131–49 (1998).

93. V Fülöp, Z Böcskei and L Polgár, *Cell*, **94**, 161–70 (1998).
94. D Lim, HU Park, L De Castro, SG Kang, HS Lee, S Jensen, KJ Lee and NC Strynadka, *Nat Struct Biol*, **8**, 848–52 (2001).
95. H Brandstetter, JS Kim, M Groll and R Huber, *Nature*, **414**, 466–70 (2001).
96. Z Xia, W Dai, Y Zhang, SA White, GD Boyd and FS Mathews, *J Mol Biol*, **259**, 480–501 (1996).
97. V Fülöp, JW Moir, SJ Ferguson and J Hajdu, *Cell*, **81**, 369–77 (1995).
98. AG Murzin, *Proteins*, **14**, 191–201 (1992).

Annexins

Annexins: calcium binding proteins with unusual binding sites

Anja Rosengarth and Hartmut Luecke
Department of Molecular Biology & Biochemistry, University of California, Irvine, CA, USA

FUNCTIONAL CLASS

Calcium and phospholipid membrane-binding proteins.

OCCURRENCE

Annexins have been found in mammals, plants, fungi/molds, and protists. However, no annexin proteins are present in yeast.

BIOLOGICAL FUNCTION

The hallmark of the annexin protein family is the calcium-dependent binding to negatively charged phospholipid bilayers *in vitro* and *in vivo*. The exact biological role of annexins is still a matter of debate; however, they have been implicated in a number of membrane-related events according to their membrane activities. Among those are endocytosis, exocytosis, cytoskeleton linkages, and ion currents across membranes.[1-5] Calcium-independent binding of annexins to membranes at low pH has also been described; however, we decided to focus on the calcium- and membrane-binding properties of annexins in this review.[6-8]

AMINO ACID SEQUENCE INFORMATION

Numerous annexin proteins have been found in different organisms (mammals, fungi, plants), but in this review

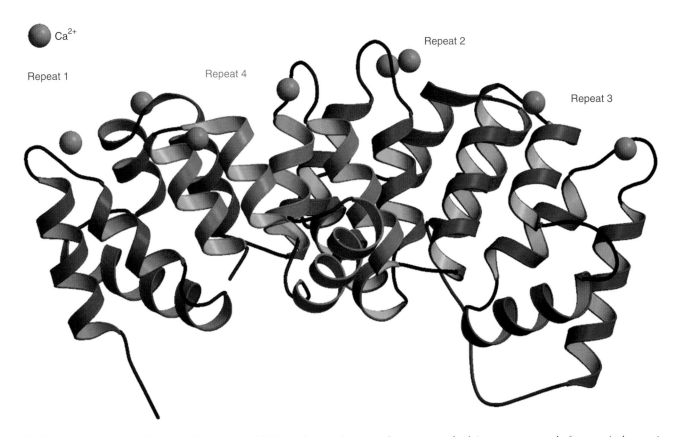

3D Structure Three-dimensional structure of full-length annexin A1 in the presence of calcium as an example for a typical annexin core-domain structure. The N-terminal domain is disordered in this case. Repeat 1 is shown in red, repeat 2 in green, repeat 3 in blue, and repeat 4 in purple. The calcium ions are depicted as orange spheres. Overall, eight calcium ions were located–four in AB loops, one in an AB′ site, and three in DE loops (PDB code: 1MCX).[60]

we would like to concentrate on mammalian annexins with complete sequences (for more information, see http://us.expasy.org/srs5bin/cgi-bin/wgetz).

Annexin A1: human: sequence from nucleic acid; 345 amino acids;[9]
pigeon (two isoforms): sequence from nucleic acid; 341 and 343 amino acids;[10,11]
bovine: sequence from nucleic acid; 346 amino acids;[12]
guinea pig: sequence from nucleic acid; 346 amino acids;[13]
mouse: sequence from nucleic acid; 345 amino acids;[14]
pig: sequence from nucleic acid; 346 amino acids;[15]
rabbit: sequence from nucleic acid; 346 amino acids;[16]
rat: sequence from nucleic acid; 345 amino acids;[17]
rodentia sp.: sequence from nucleic acid; 345 amino acids;[18]

Annexin A2: human: sequence from nucleic acid; 338 amino acids;[19]
bovine: sequence from nucleic acid; 338 amino acids;[20]
chicken: sequence from nucleic acid; 338 amino acids;[21]
mouse: sequence from nucleic acid; 338 amino acids;[22]
rat: sequence from nucleic acid; 338 amino acids;[23]

Annexin A3: human: sequence from nucleic acid; 323 amino acids;[24]
mouse: sequence from nucleic acid; 323 amino acids;[25]
rat: sequence from nucleic acid; 324 amino acids;[26]

Annexin A4: human: sequence from nucleic acid; 318 amino acids;[27]
bovine: sequence from nucleic acid; 318 amino acids;[28]
dog: sequence from nucleic acid; 318 amino acids;[29]
mouse: sequence from nucleic acid; 318 amino acids;[30]
rat: sequence from nucleic acid; 318 amino acids;[31]
pig: sequence from nucleic acid; 318 amino acids;[32]

Annexin A5: human: sequence from nucleic acid; 319 amino acids;[33]
bovine: sequence from nucleic acid; 320 amino acids;[34]
chicken: sequence from nucleic acid; 321 amino acids;[35]
mouse: sequence from nucleic acid; 319 amino acids;[36]
rat: sequence from nucleic acid; 318 amino acids;[37]

Annexin A6: human: sequence from nucleic acid; 672 amino acids;[38]
bovine: sequence from nucleic acid; 618 amino acids;[39]
chicken: sequence from nucleic acid; 671 amino acids;[40]
mouse: sequence from nucleic acid; 672 amino acids;[41]
rat: sequence from nucleic acid; 672 amino acids;[42]

Annexin A7: human: sequence from nucleic acid; 466 amino acids;[43]
mouse: sequence from nucleic acid; 463 amino acids;[44]

Annexin A8: human: sequence from nucleic acid; 327 amino acids;[45]
mouse: sequence from nucleic acid; 327 amino acids;[46]

Annexin A9: human: sequence from nucleic acid; 338 amino acids;[47]
mouse: sequence from nucleic acid; 338 amino acids;[48]

Annexin A10: human: sequence from nucleic acid; 324 amino acids;[49]
mouse: sequence from nucleic acid; 324 amino acids;[49]

Annexin A11: human: sequence from nucleic acid; 505 amino acids;[50]
bovine: sequence from nucleic acid; 503 amino acids;[51]
mouse: sequence from nucleic acid; 503 amino acids;[52]
rabit: sequence from nucleic acid; 503 amino acids;[53]

Annexin A13: human: sequence from nucleic acid; 315 amino acids; isoform A;[54]
dog: sequence from nucleic acid; 315 amino acids; isoform B;[55]

PROTEIN PRODUCTION, PURIFICATION, AND MOLECULAR CHARACTERIZATION

Annexins have been isolated directly from mammalian tissues and as recombinant proteins from *Escherichia coli*.

Two major protein-purification protocols for annexins have been established over the years. One utilizes the property of annexins to bind to phospholipid vesicles in a calcium-dependent manner. This protocol should only be applied for annexins whose N-terminal domains are not sensitive to proteolysis. The other one uses classical ion exchange chromatography in the absence of calcium and it is mainly used for annexins with N-terminal domains that are very sensitive for proteolytic cleavage. Working in the absence of calcium ensures that fewer proteases are activated.

1. Calcium-dependent binding to phospholipid vesicles[56]
Brain extract containing 80–85% phosphatidylserine (PS) is dissolved in chloroform/methanol (2:1) and dried under a stream of nitrogen. It is redissolved in ether and dried again, and liposome buffer is added (100 mM NaCl, 3 mM $MgCl_2$, 20 mM Tris, pH 8.0). This solution is sonicated for 15 min to obtain small unilamellar vesicles and the lysate of the E. coli cells that expressed the annexin protein is added to these vesicles. The solution is adjusted to an excess of 5 mM $CaCl_2$ and stirred for 30 min on ice.

The solution is centrifuged at 40 000 × g for 45 min (4 °C), the supernatant is removed, and the pellet is washed with the liposome buffer containing 5 mM $CaCl_2$. The solution is centrifuged again, the supernatant is removed, and the liposome buffer, containing 10 mM [ethylenebis(oxyethylenenitrilo)] tetraacetic acid (EGTA) to remove the bound annexin protein from the liposomes, is added. A centrifugation for 1 h at 50 000 × g (4 °C) follows and the supernatant is dialyzed against 20 mM Bis-Tris, pH 6.0, 0.02% (w/v) sodium azide. The dialysate is loaded on a diethylaminoethyl (DEAE) sepharose column and the protein is eluted with a linear gradient of sodium chloride (NaCl) (0–200 mM).

2. Ion exchange chromatography[57]
This protocol, which works well in the absence of calcium, was designed for annexins that are sensitive to proteolytic cleavage of the N-terminal domain in the presence of calcium.

The cell lysate from the E. coli cells that expressed the annexin protein is dialyzed against 10 mM imidazole pH 7.4, 10 mM NaCl, 1 mM EGTA, 1 mM sodium azide, and 1 mM phenylmethylsulfonyl fluoride (PMSF). The dialysate is applied to a DEAE-cellulose column and the flow-through is collected. If the pI of the respective annexin is around 7.0, it will be found in the flow-through. If the pI is much higher, the protein has to be eluted using a linear gradient of NaCl. The fractions that contain annexin protein are dialyzed against 50 mM MES/NaOH, pH 6.0, 1 mM EGTA and applied to a carboxymethyl (CM) cellulose column. The protein is eluted from the column using a linear NaCl gradient (0–250 mM). If the protein samples are not at least 95% pure, a gel filtration column should be used thereafter.

METAL CONTENT AND COFACTORS

Calcium binding to annexins has been studied using X-ray crystallography as well as biochemical and biophysical measurements.

Crystal structures

Twenty-nine X-ray structures of mammalian annexins in complex with calcium or without calcium have been deposited in the Protein Data Bank (PDB) to date. Among these are several mutant protein structures and structures of annexins in complex with ligands different from calcium. For simplicity, we only list the wild-type and truncated versions of the wild-type protein of mammalian annexins in the absence and presence of calcium:

Annexin A1: a truncated form of human annexin A1 (PDB code: 1AIN) and full-length porcine annexin A1 in the absence (PDB code: 1HM6) and in the presence of Ca^{2+} (PDB code: 1MCX)[58–60]
Annexin A2: truncated human annexin A2 (no PDB file deposited)[61]
Annexin A3: human annexin A3 (PDB code: 1AXN)[62]
Annexin A4: bovine annexin A4 (PDB code: 1AOW)[63]
Annexin A5: human, chicken, and rat annexin A5 (PDB codes: 1ALA, 1ANW, 1ANX, 1AVH, 1AVR, 2RAN)[64–68]
Annexin A6: bovine annexin A6 (PDB code: 1AVC)[69]

In these X-ray structures, the annexin A1 proteins have six to eight calcium ions bound, annexins A2 and A3 show five bound calcium ions, and annexin A5 can bind up to 10 calcium ions. Annexin A6, which contains two core domains, is able to bind six calcium ions (three per lobe).

Studies in solution

Interestingly, experiments concerning the calcium-binding equilibrium of annexins in solution have mostly been carried out in the presence of phospholipid structures. There are only a few publications that report calcium binding in the absence of phospholipid, and these focused on annexins A1, A2, A5, and A6. The goal of these experiments was to determine the binding constants for calcium ions using biochemical and biophysical techniques such as $^{45}Ca^{2+}$-equilibrium dialysis, a $^{45}Ca^{2+}$ gel filtration assay, or differential scanning calorimetry (DSC).

Annexin A1

Three different publications report the binding of calcium to annexin A1 in the presence and absence of phospholipid vesicles using $^{45}Ca^{2+}$-equilibrium dialysis: In 1987, Glenney et al. published a paper in which the authors reported that they could not detect calcium binding to annexin A1 for free calcium concentrations up to 40 μM. In the presence of PS vesicles, however, an apparent dissociation constant (K_d) of about $10\,\mu M$ was measured that corresponded to two calcium ions bound.[70] In the same year, Schlaepfer and Haigler also reported very weak binding of calcium in the absence of PS vesicles at micromolar calcium concentrations, but this binding increased in the presence of PS vesicles as evidenced by an apparent dissociation constant K_d of $75\,\mu M$. From a Scatchard analysis, the authors suggested that three to four calcium ions could bind to each annexin A1 molecule.[71] Finally, Ando et al. reported an apparent dissociation constant of $90\,\mu M$ in the presence of PS vesicles with four calcium ions bound in 1989.[72]

Rosengarth et al. determined the calcium-binding equilibrium of annexin A1 in the absence of phospholipid vesicles using differential scanning calorimetry. The authors reported a sequential binding mechanism, with two calcium ions bound in the first reaction and two to three more calcium ions in a second step. Half-maximal binding at $37\,°C$ occurs at $0.6\,mM$ for the first two calcium ions and at $44\,mM$ for the next two to three calcium ions.[73]

Annexin A2

Gerke and Weber determined calcium binding to the annexin A2 heterotetramer using a gel filtration assay involving $^{45}Ca^{2+}$. Comparable to the situation with annexin A1, the authors were not able to detect any calcium binding in the absence of PS vesicles.[74] In 1986, Glenney reported his results on the calcium binding of the annexin A2 heterotetramer using $^{45}Ca^{2+}$ equilibrium dialysis. In the absence of phospholipid vesicles, calcium binding was very weak, whereas the addition of PS vesicles resulted in increased affinity with a calcium concentration of $5\,\mu M$ at half-maximal binding.[75] A calcium-binding constant in the absence of lipids has not been determined yet.

Annexin A5

Schlaepfer et al. determined the calcium-binding properties of annexin A5 in the absence and presence of PS vesicles using $^{45}Ca^{2+}$-equilibrium dialysis. Half-maximal binding of calcium occurs at $101\,\mu M$ in the absence of acidic phospholipids and at $53\,\mu M$ in the presence of acidic phospholipids. Five calcium ions per annexin A5 molecules are reportedly bound.[76]

Annexin A6

Three different research groups published results on the calcium binding of annexin A6 in the 1980s. Owens and Crumpton used $^{45}Ca^{2+}$-equilibrium dialysis experiments in order to determine the apparent dissociation constant for calcium. In the absence of lipids, they determined the K_d to be $1.2\,\mu M$ with one calcium ion bound.[77] In 1986, Moore used the same technique and determined a K_d of $0.4\,\mu M$ with one calcium ion bound.[78] Finally, Mani and Kay utilized fluorescence spectroscopy for their calcium-binding experiments and reported an apparent K_d of $20\,\mu M$ with four calcium ions bound.[79]

All these measurements clearly show that annexins bind calcium ions only at high and most likely unphysiological calcium concentrations (annexin A6 seems to be the exception). However, the calcium-binding affinity is greatly increased in the presence of negatively charged phospholipids. Thus, calcium ions might only serve as a mediator or bridge between annexins and phospholipid headgroups, which could be the 'real' ligand for annexins.

In contrast to annexins, proteins of the EF hand family of calcium-binding proteins exhibit dissociation constants in the micromolar and nanomolar range, reflecting a higher affinity for calcium. Members of this protein family include S100 proteins, calmodulin, troponin C, parvalbumin, and many more. Calmodulin exhibits four calcium binding sites with calcium-binding constants at low ionic strength around 10^6 to $10^7\,M^{-1}$.[80] Another example, calbindinD9K, a member of the S100 protein family, binds two calcium ions with binding constants around 5×10^6 to $10^8\,M^{-1}$.[81] Parvalbumin, which also contains two calcium ions, exhibits calcium-binding constants around $10^9\,M^{-1}$.[82]

Activity test

Although annexins do not exhibit enzymatic activity, they interact strongly with negatively charged phospholipids in a calcium-dependent manner. Therefore, their 'activity' can be tested by calcium-dependent binding to phospholipid liposomes.

This assay can be performed according to the following protocol:[7]

To $10\,\mu g$ of annexin protein in $500\text{-}\mu L$ buffer containing $500\,\mu M\,Ca^{2+}$, phospholipid vesicles in a 1:1000 protein/lipid molar ratio are added. The phospholipid vesicles, which contain either 100% PS or mixtures of PS and phosphatidylcholine (PC), are prepared according to the Reeves/Dowben protocol.[83] The resulting solution is incubated at room temperature for $20\,min$ and centrifuged for $10\,min$ at about $14\,000\,rpm$ (Eppendorf benchtop centrifuge). The supernatant and the pellet are separated, and to each of the solutions, SDS-PAGE sample buffer is added. Then, these solutions are analyzed with an SDS-PAGE gel. If the annexin protein is 'active', it will be in the fraction containing the phospholipid pellet. A sample containing protein and lipids without calcium (plus a chelating agent such as EGTA) should serve as a control.

SPECTROSCOPY

In the case of annexin A2 and annexin A5, it has been shown that calcium and/or phospholipid binding can be monitored by tryptophan emission fluorescence spectroscopy.

Annexin A2

Human annexin A2[84] contains a single tryptophan residue at position 212, which is in the AB loop of the third repeat of the core domain. Thiel et al. have shown that upon calcium binding, the emission maximum of this tryptophan residue (excitation at 295 nm) is shifted from 321 nm to 311 nm ('blue shift'), whereas the emission intensity is unchanged.[84] This indicates a transition of the tryptophan side chain from a relatively hydrophilic environment in the absence of calcium to a more hydrophobic environment in the presence of calcium. Therefore, the tryptophan residue, which resides in a fairly hydrophobic environment, is probably in close proximity to the calcium binding site.

The authors also mutated residues in other calcium binding sites to tryptophan residues and measured the fluorescence emission of tryptophan as a function of calcium concentration. However, none of the other mutant proteins exhibit a blue shift in the tryptophan emission spectra upon calcium binding. Therefore, different calcium sites of annexins provide different environments.

Annexin A5

In the case of human annexin A5,[85,86] a conformational change upon calcium binding that is opposite to annexin A2 as judged by tryptophan fluorescence has been observed: The single tryptophan residue of annexin A5 also resides in the AB loop of repeat 3 of the core domain, but the fluorescence emission maximum of this tryptophan residue shifts from about 320 nm in the absence of calcium to 345 nm in the presence of calcium (excitation at 295 nm). In this case, the tryptophan side chain is transferred from a more hydrophobic into a hydrophilic environment, therefore resulting in a red shift of the emission maximum. Upon addition of phospholipid vesicles composed of PS/PC (1:3 molar ratio), the intensity of the emission spectrum increases four- to fivefold, but no shift in the maximum is observed.[85,86]

X-RAY STRUCTURES OF NATIVE ANNEXINS

Crystallization

Annexins crystallize best out of polyethyleneglycol (PEG) or ammonium sulfate (AS) with and without calcium. However, in the case of AS, one is only able to use up to 20 mM $CaCl_2$ in the stock solution owing to the limit solubility of $CaSO_4$. Examples of crystallization solutions resulting in annexin crystals are as follows:

Annexin A1: 8% PEG 4000, 0.05 cacodylate pH 6.5, 0.1 M Na(OAc), 10 mM $CaCl_2$, 20 mg mL^{-1} annexin A1 (final drop conditions) (PDB code: 1AIN).

Annexin A2: 20% PEG 6000, 0.6 M MES pH 6.0, 20 mM $CaCl_2$ (crystallization solution).

Annexin A3: 25% AS, 0.05 M Tris/HCl pH 7.5, 10 mM $CaCl_2$, 18 mg mL^{-1} annexin A3 (final drop condition) (PDB code: 1AXN).

Annexin A4: 1 M AS, 0.05 M Tris/HCL pH 7.3, 10 mg mL^{-1} annexin A4 (final drop condition) (PDB code: 1AOW).

Annexin A5: (a) 1.05 M AS, 10 mM Tris/HCl pH 8.0, 10 mM $CaCl_2$, 10 mg mL^{-1} chicken annexin A5 (final drop condition) (PDB code: 1ALA);

(b) 5% PEG 20 000, 50 mM Tris/maleate pH 6.4, 0.5 mM Ca^{2+}, 15 mg mL^{-1} annexin A5 (final drop condition) (PDB code: 1ANW);

(c) 1 M AS, 50 mM Tris/HCl pH 7–8, 10–60 mM Ca^{2+}, 15 mg mL^{-1} annexin A5 (final drop condition) (PDB code: 1ANX);

(d) 0.25 M AS, 0.01 M Tris-chloride, pH 8.5, 0–1 mM $CaCl_2$, 7 mg mL^{-1} annexin A5 (final drop condition) (PDB codes: 1AVH and 1AVR);

(e) 40–60% AS, 5–10 mM $CaCl_2$, 50 mM HEPES, 1 mM DTT, 1 mM sodium azide, 10–30 mg mL^{-1} annexin A5 (final drop condition) (PDB code: 2RAN).

Overall description of the atomic structure

Annexins are structurally divided into a conserved core domain and an N-terminal domain that is variable in sequence and length. The core domains, which exhibit a sequence homology of up to 80% among different annexins, are composed of four repeats (eight in the case of annexin A6), with five α-helices (named A to E) each. Four of these five α-helices (A, B, D, and E) form a coiled coil structure. The overall shape of the core is a slightly curved disk with loops connecting helices A and B and helices D and E located on the convex side of the disk. These loops harbor the calcium binding sites. The N-terminal domain is located on the concave side of the molecule (3D Structure), or in the case of annexin A1 in the absence of calcium, affiliated with the core in between the concave and the convex side.

Calcium site geometries

Annexins contain three different types of calcium binding sites: type II (AB loop), type III (DE loop) and AB′ sites.

1. Type II (AB loop) binding site

According to Huber *et al.*, three backbone carbonyl oxygens in the AB loop coordinate the calcium ion comprising the conserved sequence (M,L)-K-G-(A,L)-G-T. Additionally, the side chain of an acidic amino acid, about 39 residues downstream of the conserved sequence, coordinates in a bidentate fashion. The last two coordination sites complete a pentagonal bipyramidal coordination sphere and are provided by water molecules[87] (Figure 1). A coordination by eight oxygens in the type II binding site has also been observed in two annexin A5 structures: the calcium ion located in the AB loop of repeat 2 of rat annexin A5 in complex with glyceroethanolamine (PDB code: 1A4B) and the calcium ion in the AB loop of repeat 3 of human annexin A5 (PDB code: 1ANX).[66,88]

It should be noted that many annexin X-ray structures contain numerous empty, partially occupied, or high B-factor calcium sites, usually coupled with incomplete coordination shells.

2. Type III (DE loop) binding site

The coordination sphere of the calcium ion in this type of binding site is made up of two backbone carbonyl oxygens from the DE loop that is not part of a conserved sequence. A bidentate ligand of an acidic amino acid close in sequence and three water molecules complete the pentagonal bipyramidal coordination sphere of the calcium ion[58,68] (Figure 2).

3. AB′ site

The backbone carbonyl oxygen of a residue downstream of the conserved sequence for the AB loop, the side

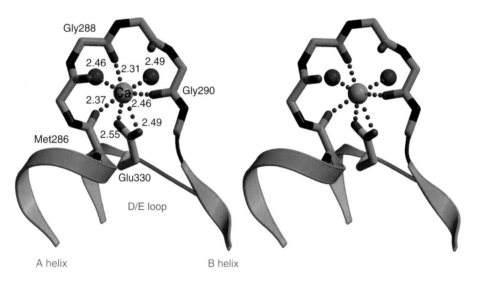

Figure 1 Typical type II calcium binding site. Stereo image of the coordination and Ca–O bond distances of the calcium ion located in the AB loop (type II binding site) of repeat 4 of full-length annexin A1. The calcium ion, shown as an orange sphere, is coordinated by the backbone carbonyl oxygens of residues Met286, Gly288, and Gly290. A bidentate carboxylate from residue Glu330 and two water molecules, shown as blue spheres, complete the pentagonal bipyramidal coordination sphere. The coordination is according to the canonical sequence (M,L)-K-G-(A, L)-G-T (PDB code: 1MCX).[56,60]

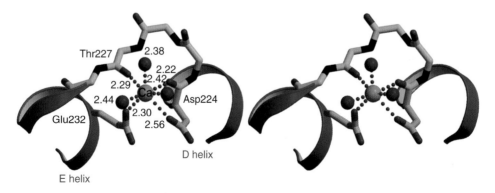

Figure 2 Typical type III calcium binding site. Stereo image of the coordination of the calcium ion in the DE loop (type III binding site) of repeat 3 of annexin A5. The backbone carbonyl oxygens of Asp224 and Thr227, OD2 of Asp224, OE2 of Glu232, and three water (blue spheres) molecules coordinate the calcium ion (orange sphere) in this pentagonal bipyramidal arrangement (PDB code: 1G5N).[89]

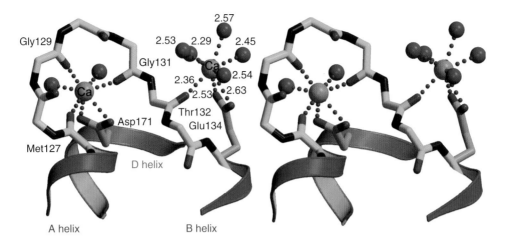

Figure 3 Typical AB' calcium binding site. Stereo image of the AB and AB' sites found in repeat 2 of full-length annexin A1: The calcium ion shown on the left is coordinated in the AB site and the one on the right is coordinated in the AB' site. Calcium ions are shown as orange spheres and water molecules are depicted as blue spheres. Coordination in the AB' site comprises the backbone carbonyl of Thr132, a bidentate ligand from Glu134, and five water molecules. Therefore, the calcium coordination in this AB' site is eight rather than seven (PDB code: 1MCX).[60] Octa-coordination has been observed for other calcium ions in annexin structures.[66,88]

chain of an acidic amino acid residue nearby, and water molecules make up this coordination sphere[60,88] (Figure 3).

Comparison with the EF hand calcium binding sites

EF hand calcium-binding proteins contain two types of calcium binding sites: the canonical and the pseudo EF hand. These binding sites, also called 'type I calcium binding sites', are also located in loops connecting two helices, but the coordination of the calcium ions is very different from that observed in annexins.[90]

The canonical EF hand is characterized by 12 amino acids in the loop in which residues 1, 3, 5, 7, 9, and 12 provide the seven oxygens for calcium coordination. By convention, the six coordinating positions are labeled by their approximate relative orientations centered around the calcium in a Cartesian coordinate system. Positions 1 (+X), 3 (+Y), and 5 (+Z) each provide one side-chain oxygen for coordination (most often an aspartate residue). Position 7 (−Y) is the only residue that coordinates via its backbone carbonyl oxygen. The residue at position 9 (−X) is usually a glutamate or glutamine residue whose side-chain oxygen coordinates a water molecule that in turn coordinates the calcium ion. The side chain of a highly conserved glutamate residue at position 12 completes the pentagonal bipyramidal coordination of the calcium ion by providing two oxygens as a bidentate ligand.[91]

The main difference in calcium coordination between annexins and EF hand proteins is the high number of water molecules coordinating the calcium ions in annexins. This could be the reason the calcium-binding constants are significantly lower for annexins. However, upon membrane binding, the water molecules in the coordination sphere of the calcium ions might be exchanged for oxygen atoms from the glycerol backbone of the phospholipids and the binding constants increase.

FUNCTIONAL ASPECTS

As mentioned above, annexins all bind to negatively charged phospholipids in the presence of calcium. The conserved core domain of the protein mediates this interaction. However, after having bound to a membrane, annexins are involved in different cellular functions that are most likely mediated by their N-terminal domain, which is variable in sequence and in length for each annexin. Here, we would like to discuss four members of the annexin protein family that exhibit different functions *in vitro* and most likely *in vivo*.

Annexin A1

Subcellular fractionation and immunochemical analysis of fixed cells have shown that annexin A1 is associated with early endosomes and multivesicular bodies.[15,92] The association with early endosomes is regulated by calcium and requires the complete N-terminal domain. A truncated version of annexin A1 lacking the first 26 amino acids no longer binds to early endosomes; instead, it is associated with late endosomes and mulitvesicular bodies.[15] These findings underscore the importance of the N-terminal domain in the regulation of annexin A1 function and a possible role of the protein in endocytosis. However, the exact mechanism of annexin A1-induced membrane aggregation is still unknown.

Additionally, annexin A1 has been shown to be a target of the epidermal growth factor receptor/kinase (EGF-R). However, phosphorylation at Tyr21 occurs only in the presence of calcium.[93] This is an interesting feature that we will discuss in the section 'Functional Derivatives'. EGF-R itself is located at multivesicular bodies, and therefore annexin A1 is also found in those lipid structures. There, annexin A1 is thought to play a role in the inward vesiculation of multivesicular bodies, a process that is regulated by the phosphorylation of annexin A1 by EGF-R.[92]

Studies concerning the subcellular distribution of annexin A1 in living cells have shown colocalization with early endosomes and the plasma membrane. The core of annexin A1 alone seems to bind to late endosomes,[94] which is in good agreement with the studies by Seemann et al.[15] Interestingly, the binding to cellular membranes is mostly mediated by one calcium binding site. A mutant protein lacking the calcium binding site in the AB loop of repeat 2 (all the other binding sites are intact) shows a cytosolic distribution because the membrane association is diminished.

The function of annexin A1 is regulated by phosphorylation, either by EGF-R or by other protein kinases.[95] Phosphorylation by protein kinase C in the presence of calcium and phospholipid vesicles resulted in a protein that required less calcium for lipid binding than the unphosphorylated form, but required more calcium to induce chromaffin granule aggregation. Unphosphorylated annexin A1 aggregated chromaffin granules with a half-maximal calcium concentration of 174 μM, whereas the phosphorylated isoform requires a half-maximal calcium concentration of 682 μM.[96]

In vitro, annexin A1 has been shown to promote membrane aggregation and even fusion of phospholipid vesicles and chromaffin granules. The N-terminal domain appears to play an important role in this process since a chimera comprising the N-terminal domain of annexin A1 fused to the core domain of annexin A5 induced membrane aggregation, whereas annexin A5 itself does not.[97-99]

The cellular distribution and therefore the function of annexin A1 is regulated by the calcium binding sites in the core domain, the N-terminal domain, and phosphorylation.

Annexin A2

An association with early endosomes has also been reported for annexin A2; however, this annexin has been reported to be part of a very interesting and important cellular process: fibrinolysis. In 1994, Hajjar et al. reported the discovery of an endothelial cell receptor for plasminogen/tissue plasminogen activator (t-PA), which they identified as annexin A2.[100] The authors even linked annexin A2 to acute bleeding in patients suffering from acute promyelotic leukemia and they reported a greater expression of annexin A2 on leukemic cells from patients with acute promyelotic leukemia than on other types of leukemic cells. These findings resulted in the conclusion that the abnormally high levels of annexin A2 expression on these cell types increase the production of plasmin and therefore lead to severe bleeding or other hemorrhagic complications in this form of leukemia.[101] A model suggests that annexin A2 could bind plasminogen and t-PA simultaneously and therefore facilitate the contact between those two proteins and eventually increase the rate of plasmin generation by t-PA. Plasmin in turn dissolves fibrinolytic clots.

Two different working theories have been published so far: Hajjar et al. reported the extracellular annexin A2 monomer to be the receptor for plasminogen and t-PA,[100-104] whereas Waisman's group favors the involvement of extracellular annexin A2 heterotetramers.[105-108] The heterotetramer contains two molecules of annexin A2 and a dimer of an S100 protein, S100A10 (also known as p11). S100A10 belongs to a family of EF hand-type calcium-binding proteins; however, the two EF hand loops in this protein are not able to bind calcium. Therefore, the binding of S100A10 to annexin A2 is calcium-independent,[109] unlike the binding of S100A11, another member of the S100 protein family that interacts with the N-terminal domain of annexin A1. The latter complex formation strictly depends on calcium binding to S100A11 and possibly annexin A1 and an N-acetylation of the first residue of the N-terminus of annexin A1.[110]

The interaction of annexin A2 with t-PA and plasminogen on endothelial cells seems to require calcium and negatively charged phospholipids.[103] However, the percentage of negatively charged phospholipids in the outer leaflet of the plasma membrane of different cell types is generally low in comparison to the percentage in the inner leaflet. Additionally, only 4% of the complete annexin A2 pool per cell has been found on the extracellular surface and, therefore, the majority of the protein is bound to negatively charged phospholipids in the inner leaflet of the membrane. Another problem arises from the fact that annexin A2 does not contain an export signal and the mechanism by which the protein is secreted to the extracellular space remains elusive. However, a report was published recently that showed that the insulin receptor and its signaling pathways might be involved in the secretion of annexin A2.[111]

Annexin A4

Like annexin A1, annexin A4 has also been shown to induce membrane aggregation of chromaffin granules in the presence of calcium.[112,113] The membrane aggregation seems to be induced by protein contacts between annexin

A4 molecules located on different membranes. Cryo-electron microscopy revealed a protein layer between two liposomes about 6 nm in thickness, which reflects a dimer, composed of two annexin A4 molecules that are about 3.5 nm in height.[114] Interestingly, an annexin A4 mutant that contains an aspartate rather then a threonine residue at position 6 does not promote membrane aggregation. The same phenomenon is observed for wild-type annexin A4, which is phosphorylated at Thr6. Phosphorylation or an amino acid exchange mimicking phosphorylation diminishes annexin A4–induced membrane aggregation. In general, phosphorylation of the N-terminal domain of annexins by tyrosine and serine/threonine kinases alters their properties.[95]

Annexin A4 has been described as an inhibitor of the calmodulin-dependent protein kinase II-activated chloride conductance in colonic T84 epithelial cells. The protein most likely inhibits anion conductance activation by preventing the calmodulin-dependent kinase II channel interaction by blocking the target of calmodulin-dependent kinase II, rather than by direct interaction with the enzyme.[115] Annexin A4 multimers lining the membrane could sterically hinder protein–protein interactions that are necessary for the phosphorylation of the target protein by the calmodulin-dependent kinase II. Phosphorylation of the target protein, however, would determine the chloride ion conductance in epithelial T84 cells. It has also been shown that the inhibition is regulated by the phosphorylation of annexin A4 and it has been proposed, in accordance with the structure of annexin A4, that annexin A4 might promote contact between the plasma membrane and vesicles that have been targeted at the site of chloride efflux. Phosphorylation would attenuate the cellular response by preventing further membrane–membrane interactions.[114]

Annexin A5

Annexin A5 is probably the most commonly known annexin because it is used as a marker for apoptotic cells. As with all other annexins, annexin A5 binds to negatively charged phospholipids in the presence of calcium, and this feature is used in a common apoptosis assay: Apoptotic cells have a high level of PS in the outer leaflet of the cell membrane, and if fluorescence-labeled annexin A5 is added to those cells, they can easily be detected. Annexin A5 also interacts with apoptotic cells *in vivo*, leading to the hypothesis that annexin A5 is involved in the regulation of coagulation and inflammation processes. Under physiological conditions, apoptotic cells will be removed from the body by phagocytes. However, pathological conditions yield a high level of apoptotic cells that may contribute to coagulation and inflammation due to the exposure of PS on the outer leaflet of the membrane. Annexin A5 could bind to those phospholipids and therefore would be able to regulate coagulation and inflammation caused by the high amount of apoptotic cells.[116]

The first annexin structure published was that of the human annexin A5 in 1990 by Huber *et al.*[117] No calcium ions could be located in this structure, but additional annexin A5 structures were solved, and it became evident that the loops between helices A and B and between helices D and E harbor calcium binding sites. The localization of calcium ions on the convex surface of the molecule raised the hypothesis that annexins might bind phospholipids by binding of calcium and phospholipid headgroups simultaneously. This binding mechanism would be accomplished by a so-called 'calcium-bridge', which was eventually verified through two structures of annexin A5 in complex with calcium and phospholipid analogs by Seaton and Co-workers in 1995.[88] In these structures, the calcium ion located in the AB loop of repeat 3 is coordinated by the backbone carbonyl oxygens of three residues in this loop, an acidic side chain, and the phosphoryl oxygen from the lipid analogue, thereby replacing the water molecule in the apical position. Protein ligands as well as oxygens of the glycerol backbone of the phospholipids most likely coordinate the calcium ions in the annexin A5 molecule when bound to the membrane.

FUNCTIONAL DERIVATIVES

In the previous section on 'Functional Aspects', we presented various functions and specific interactions of different annexins, which we will now discuss in light of the X-ray crystallographic studies.

Annexin A1

In 1993, Huber and Co-workers published the structure of annexin A1 lacking the N-terminal domain (amino acids 1–32), which was truncated during purification of the protein (PDB code: 1AIN). This structure reflected a typical annexin core domain structure.[58] Knowing that the N-terminal domain plays an important role in the membrane aggregation process, we decided to crystallize full-length annexin A1, initially in the absence of calcium, and later in the presence of calcium (PDB codes: 1HM6, 1MCX)[59,60] (Figure 4).

The structure of full-length annexin A1 in the absence of calcium shows the fold and position of the N-terminal domain very clearly. The first 26 amino acids form an α-helix with a kink at position 16. Amino acids 1 to 12 form an amphipathic helix that inserts into repeat 3 of the core domain, thereby replacing helix D of the core domain. Helix D in turn unfolds into an extended loop. In the structure published by Huber and Co-workers, helix D of repeat 3 is directly involved in the coordination of the calcium ion in the AB loop of this repeat. In the absence of calcium, the

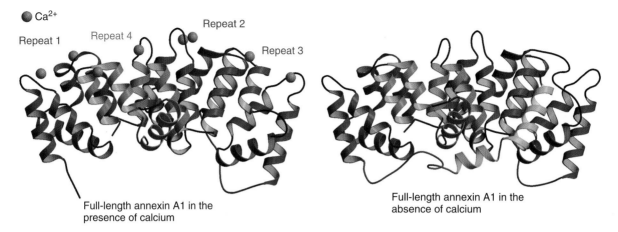

Figure 4 A comparison of the structures of full-length annexin A1 in the presence of calcium ions (on the left) and in the absence of calcium ions (on the right): The N-terminal domain is shown in yellow, repeat 1 in red, repeat 2 in green, repeat 3 in blue, repeat 4 in purple, and the calcium ions are depicted as orange spheres. In the absence of calcium, the first N-terminal helix is folded into repeat 3 of the core domain, thereby replacing helix D, which in turn unwinds into an extended loop that is folded over the N-terminal helix. In the presence of calcium, the N-terminal domain is expelled from the core domain and disordered, and helix D of repeat 3 is folded back into place for calcium binding (PDB codes: 1HM6, 1MCX).[59,60]

amphipathic helix of the N-terminal domain is buried in the core domain, and, therefore, this conformation might be the inactive form of the protein.

Upon calcium binding, the N-terminal domain is expelled from the core domain, and the sequence of helix D of repeat 3 folds back into place for calcium coordination, as shown in the structure of full-length annexin A1 in the presence of calcium. We hypothesized that the N-terminal amphipathic helix of annexin A1 (residues 1–12) might be able to interact with a second membrane after calcium-dependent binding to the first membrane, with concomitant expulsion of the N-terminal domain. This second binding event is thought to occur via hydrophobic interactions and would not be mediated directly by calcium ions. This would imply that the second lipid interaction would not be lipid-specific. In support of this model, it has been shown by other researchers that annexin A1 does contain two distinct membrane binding sites and that it can aggregate two membranes as a monomer. The secondary membrane binding site in this interaction was shown to be calcium-independent and lipid-unspecific.[98,118,119]

The three-dimensional structure of the amphipathic N-terminal helix of annexin A1 in the full-length structure is comparable to the one reported for the structure of just the N-terminal peptide of annexin A1 in complex with S100A11, a protein of the S100 family of EF hand calcium-binding proteins.[120] The N-acetylation of the first residue in the N-terminal helix of annexin A1 is required for the interaction with S100A11, and the complex formation is observed in the presence of calcium only (PDB code: 1QLS) (Figure 5).

It has been shown that annexin A1 targets S100A11 to early endosomes, but the physiological relevance for

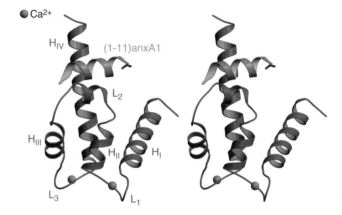

Figure 5 The annexin A1 N-terminal peptide comprising the first 11 residues in complex with S100A11 (PDB code: 1QLS). In this stereo image, the S100A11 monomer is shown in dark orange and the N-terminal annexin A1 peptide is shown in yellow. S100A11 comprises four helices (H) and three loops (L) marked H_I to H_{IV} and L_1 to L_3. Calcium ions (green spheres) are bound in loops 1 and 3. The annexin A1 N-terminal peptide interacts with L_2 and H_{IV}. Interestingly, the N-terminal peptide of annexin A1 needs to be acetylated in order to form a complex with S100A11. Therefore, we depicted the N-acetylation moiety in dark green.[110,120,121]

the complex formation and translocation is not yet understood.[110,121]

Annexin A2

The structure of an N-terminally truncated version of annexin A2 was published by Huber and Co-workers in 1996, but no PDB file is available. In this structure, the first residue with electron density is residue 33, so, as in the case of annexin A1, no information is available on the structure of the N-terminal domain. The calcium binding sites are located on the convex surface of the core domain, as described for annexin A5, and five calcium ions are bound. However, the electron density of the calcium ion in the AB loop of repeat 2 is not very well defined. The other calcium ions are coordinated in the AB loop of repeat 1, the AB loop of repeat 4, the DE loop of repeat 1, and the DE loop of repeat 3.[61]

Information on the structure of the N-terminal domain of annexin A2 in the absence of the core domain is available from the structure of the complex between the N-terminal peptide (residues 2 to 11) of annexin A2 and S100A10 (PDB code: 1BT6) (Figure 6). The first 11 residues of the N-terminal domain of annexin A2 form an α-helix comparable to the annexin A1 N-terminal helix.[122]

The structure of full-length annexin A2 in complex with S100A10 still needs to be determined in order to obtain information on the structural determinants of complex formation between these two proteins.

Annexin A4

Two X-ray structures of wild-type annexin A4 have been reported to date: a 2.3-Å structure from Sutton and Sprang in 1995 (PDB code: 1ANN)[124] and a 3.0-Å structure from Berni and Co-workers in 1998 (PDB code: 1AOW).[63]

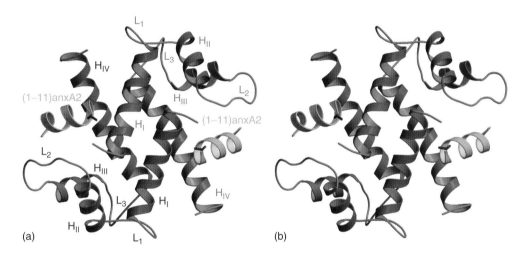

Figure 6 The annexin A2 N-terminal peptide comprising the first 11 residues in complex with S100A10 (PDB code: 1BT6).[122] In this stereo image, the S100A10 dimer is shown in dark orange and the annexin A2 N-terminal peptide is shown in yellow. The S100A10 monomer exhibits four helices (H) and three loops (L) marked H_I to H_{IV} and L_1 to L_3. No calcium ions are found in this structure because S100A10, in contrast to S100A11, does not contain calcium-binding motifs. The annexin A2 peptide interacts with L_2 and H_{IV}. Again, we colored the N-acetylation group that has been shown to be essential for complex formation in dark green.[123]

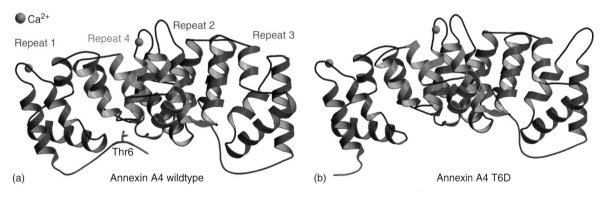

Figure 7 A comparison of the structures of (a) wild-type annexin A4 and (b) the T6D mutant protein: The N-terminal domain and repeat 1 are shown in red, repeat 2 in green, repeat 3 in blue, repeat 4 in purple, and the calcium ions are depicted as orange spheres. In the wild-type structure on the left, we also included the side chain of Thr6 to illustrate the close contact of this residue with the core. In fact, the side chains of residues Thr6, Val7, and Phe13 participate in hydrophobic interactions connecting the N-terminal loop to the core. In the T6D mutant structure, residues 1 to 9 are disordered; however, it is fairly obvious that the N-terminal domain is not in close contact with the core any longer. Residues 10 to 12 are pointing away from the core, whereas residue Phe13 remains at the same position as in the wild-type structure (PDB codes: 1ANN, 1I4A).[114,124]

The 3.0-Å structure is from an N-terminally truncated (des(1–9)) bovine annexin A4, whereas in the 2.3-Å structure, the first residues of the full-length protein are disordered. In the latter structure, residues 4 to 14 of the N-terminal domain fold into an irregular, extended structure resembling one β-strand. This strand runs along the concave surface of the molecule with residues Thr6, Val7, and Phe13 engaging in hydrophobic interactions with the core domain. The introduction of an aspartic acid residue at position 6 to mimic phosphorylation results in the loss of the membrane aggregation properties of annexin A4. The same is the case for wild-type annexin A4 phosphorylated at position Thr6. In the X-ray structure of the T6D mutant, the first nine amino acids are disordered, so the structural effect of the mutation is not clearly visible (PDB code: I4A). However, residues 10 to 12 point away from the concave side of the molecule (Figure 7). The introduction of a charge at position 6 might destroy the connection between the N-terminal domain and the core domain, resulting in the inability of annexin A4 to form protein–protein contacts that lead to membrane aggregation.

Phosphorylation also alters the membrane aggregation properties of annexin A1, whereas the binding to membranes is not impaired.[71,96] It has been shown that the phosphorylation of annexin A1 by protein kinase C, a serine/threonine kinase, and EGF-R, a tyrosine kinase, requires the presence of calcium and membranes. This

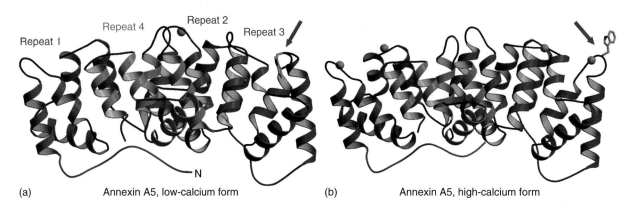

Figure 8 A comparison of the (a) low-calcium and (b) high-calcium forms of annexin A5. The color code is the same as in Figure 8. The Trp187 side chain in repeat 3 is buried in the low-calcium form (red arrow, (a)), whereas it is solvent exposed (red arrow, (b)) in the presence of high-calcium concentrations. This conformational change has also been confirmed by tryptophan fluorescence spectroscopy. Not only does the tryptophan side chain rearrange upon calcium binding but the length of the D-helix in repeat 3 increases as well. In the absence of calcium, the helix only contains one turn, whereas in the presence of calcium it shows two turns (PDB codes: 1AVH, 1ANX).[66,67,85,86]

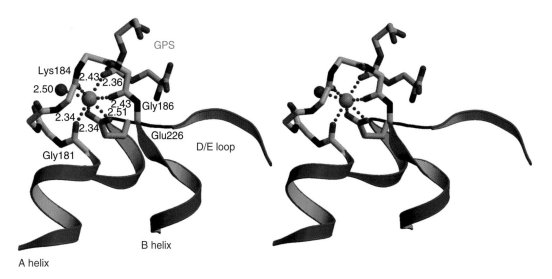

Figure 9 Stereo image of the calcium and 'lipid' binding site in human annexin A5 (PDB code: 1A8A). The phospholipid analog glycerophosphoserine (GPS) as well as a calcium ion is bound in the AB loop of repeat 3. The calcium ion (orange sphere) is coordinated by the backbone carbonyl oxygens of residues Gly181, Lys184, and Gly186, the bidentate carboxylate from Glu226, a water molecule (blue sphere) and the phosphoryl oxygen from GPS. Only one phospholipid analog per annexin A5 monomer has been found in the structure due to the crystal packing in the R3 space group.[88]

phenomenon makes perfect sense since residue Tyr21 is completely buried and residues Thr24 and Ser28 are partially buried in the absence of calcium.[59] Only in the presence of calcium and/or phospholipids will the N-terminal domain of annexin A1 be freely accessible, allowing the kinase to phosphorylate the respective amino acid.

Annexin A5

As mentioned earlier in this review, human annexin A5 was the first annexin structure to be solved. In the section 'Spectroscopy', we described that annexin A5 harbors one calcium binding site that is accessible at high calcium concentrations only. Tryptophan fluorescence spectroscopy revealed that the binding of a calcium ion in the AB loop of repeat 3 involves a large conformational change resulting in a change of solvent exposure of Trp187 from a more hydrophobic to a more hydrophilic environment.[66,67,85,86]

A comparison of the X-ray structures of the 'low-calcium' and the 'high-calcium' forms of annexin A5 (PDB codes 1AVH, 1AVR, and 1ANX) depicts the conformational changes in repeat 3 that had been expected on the basis of tryptophan fluorescence spectroscopy (Figure 8).

Structural studies concerning the lipid binding of annexins resulted in two structures of annexin A5 in complex with glycerophosphoserine and glycerophosphoethanolamine, which were used as model substances for phosphatidylserine and ethanolamine derivatives. In these two structures it was shown that the calcium ion in the AB loop of repeat 3 is coordinated by oxygen atoms from the protein and from the lipid analogue phosphoryl oxygen (Figure 9). Again, these structural studies confirmed the hypothesis that annexins bind to the biological membrane via a calcium bridge.[88]

REFERENCES

1 P Raynal and HB Pollard, *Biochim Biophys Acta*, **1197**, 63–93 (1994).

2 MA Swairjo and BA Seaton, *Annu Rev Biophys Biomol Struct*, **23**, 193–213 (1994).

3 S Liemann and R Huber, *Cell Mol life Sci*, **53**, 516–21 (1997).

4 V Gerke and SE Moss, *Biochim Biophys Act*, **1357**, 129–54 (1997).

5 V Gerke and SE Moss, *Physiol Rev*, **82**, 331–71 (2002).

6 R Langen, JM Isas, WL Hubbell and HT Haigler, *Proc Natl Acad Sci USA*, **95**, 14060–65 (1998).

7 JM Isas, JP Cartailler, Y Sokolov, DR Patel, R Langen, H Luecke, JE Hall and HT Haigler, *Biochemistry*, **39**, 3015–22 (2000).

8 A Rosengarth, A Wintergalen, HJ Galla, HJ Hinz and V Gerke, *FEBS Lett*, **438**, 279–84 (1998).

9 BP Wallner, RJ Mattaliano, C Hession, RL Cate, R Tizard, LK Sinclair, C Foeller, EP Chow, JL Browning, KL Ramachandran and RB Pepinsky, *Nature* **320**, 77–81 (1986).

10 ND Horseman, *Mol Endocrinol*, **3**, 773–79 (1989).

11 HT Haigler, JA Mangili, Y Gao, J Jones and ND Horseman, *J Biol Chem*, **267**, 19123–29 (1992).

12 JD Ernst, *Biochem J*, **289**, 539–42 (1993).

13 EF Sato, Y Tanaka and K Utsumi, *FEBS Lett*, **244**, 108–12 (1989).

14 T Sakata, S Iwagami, Y Tsuruta, R Suzuki, K Hojo, K Sato and H Teraoka, *Nucleic Acids Res*, **16**, 11818 (1988).

15 J Seemann, K Weber, M Osborn, RG Parton and V Gerke, *Mol Biol Cell*, **7**, 1359–74 (1996).

16 FHC Tsao, C Wen and J Hu, submitted to the EMBL/GenBank/DDBJ databases, GenBank code: 1052872 (1995).

17 M Tamaki, E Nakamura, C Nishikubo, T Sakata, M Shin and H Teraoka, *Nucleic Acids Res*, **15**, 7637 (1987).

18 A Robitzki, HC Schroeder, D Ugarkovic, M Gramzow, U Fritsche, R Batel and WEG Mueller, *Biochem J*, **271**, 415–20 (1990).

19 KS Huang, BP Wallner, RJ Mattaliano, R Tizard, C Burne, A Frey, C Hession, P McGray, LK Sinclair, EP Chow, JL Browning, KL Ramachandran, J Tang, JE Smart and RB Pepinsky, *Cell*, **46**, 191–99 (1986).

20 T Kristensen, CJM Saris, T Hunter, LJ Hicks, DJ Noonan, JR Glenney Jr and BF Tack, *Biochemistry*, **25**, 4497–503 (1986).

21 V Gerke and W Koch, *Nucleic Acids Res*, **18**, 4246 (1990).

22 CJM Saris, BF Tack, T Kristensen, JR Glenney Jr and T Hunter, *Cell*, **46**, 201–12 (1986).

23 T Ozakiand and S Sakiyama, *Oncogene*, **8**, 1707–10 (1993).

24 RB Pepinsky, R Tizard, RJ Mattaliano, LK Sinclair, GT Miller, LJ Browning, EP Chow, C Burne, KS Huang, D Pratt, L Wachter, C Hession, AZ Frey and BP Wallner, *J Biol Chem*, **263**, 10799–811 (1988).

25 MP Fernandez, NG Copeland, DJ Gilbert, NA Jenkins and RO Morgan, *Gene*, **207**, 43–51 (1998).

26 RB Pepinsky, R Tizard, RJ Mattaliano, LK Sinclair, GT Miller, JL Browning, EP Chow, C Burne, KS Huang, D Pratt, L Wachter, C Hession, AZ Frey and BP Wallner, *J Biol Chem*, **263**, 10799–811 (1988).

27 U Grundmann, E Amann, KJ Abel and HA Kuepper, *Behrin Inst Mitt*, **82**, 59–67 (1988).

28 HC Hamman, LC Gaffey, KR Lynch and CE Creutz, *Biochem Biophys Res Commun*, **156**, 660–67 (1988).

29 SL Fukuoka, submitted to the EMBL/GenBank/DDBJ databases, GenBank code: 540498 (1994).

30 CL Sable and J Shannon, submitted to the EMBL/GenBank/DDBJ databases, GenBank code: 1778312 (1997).

31 SL Fukuoka, submitted to the EMBL/GenBank/DDBJ databases, GenBank code: 21326828 (1994).

32 K Weber, N Johnsson, U Plessmann, PN Van, HD Soling, C Ampe and J Vandekerckhove, *EMBO J*, **6**, 1599–604 (1987).

33 U Grundmann, KJ Abel, H Bohn, H Loebermann, F Lottspeich and H Kuepper, *Proc Natl Acad Sci USA*, **85**, 3708–12 (1988).

34 MP Learmonth, SA Howell, ACM Harris, B Amess, Y Patel, L Giambanco, R Bianchi, G Pula, P Ceccarelli, R Donato, BN Green and A Aitken, *Biochim Biophys Acta*, **1160**, 76–83 (1992).

35 MP Fernandez, O Selmin, GR Martin, Y Yamada, M Pfaeffle, R Deutzmann, J Mollenhauer and K von der Mark, *J Biol Chem*, **263**, 5921–25 (1988).

36 ML Rodriguez-Garcia, CA Kozak, RO Morgan and MP Fernandez, *Genomics*, **31**, 151–57 (1996).

37 RB Pepinsky, R Tizard, RJ Mattaliano, LK Sinclair, GT Miller, JL Browning, EP Chow, C Burne, KS Huang, D Pratt, L Wachter,

C Hession, AZ Frey and BP Wallner, *J Biol Chem*, **263**, 10799–811 (1988).

38 MR Crompton, RJ Owens, NF Totty, SE Moss, MD Waterfield and MJ Crumpton, *EMBO J*, **7**, 21–27 (1988).

39 C Comera and CE Creutz, submitted to the EMBL/GenBank/DDBJ databases, GenBank code: 1842108 (1997).

40 X Cao, BR Genge, LN Wu, WR Buzzi, RM Showman and RE Wuthier, *Biochem Biophys Res Commun*, **197**, 556–61 (1993).

41 SE Moss, MR Crompton and MJ Crumpton, *Eur J Biochem*, **177**, 21–27 (1988).

42 H Fan, D Josic, YP Lim and W Reutter, *Eur J Biochem*, **230**, 741–51 (1995).

43 AL Burns, K Magendzo, A Shirvan, M Srivastava, E Rojas, MR Alijani and HB Pollard, *Proc Natl Acad Sci USA*, **86**, 3798–802 (1989).

44 ZY Zhang-Keck, AL Burns and HB Pollard, *Biochem J*, **289**, 735–41 (1993).

45 R Hauptmann, L Maurer-Fogy, E Krystek, G Bodo, H Andree and CPM Reutelingsperger, *Eur J Biochem*, **185**, 63–71 (1989).

46 MP Fernandez, NG Copeland, DJ Gilbert, NA Jenkins and RO Morgan, *Genome*, **9**, 8–14 (1998).

47 RO Morgan and MP Fernandez, *FEBS Lett*, **434**, 300–4 (1998).

48 MP Fernandez, NG Copeland, NA Jenkins, DJ Gilbert and RO Morgan, submitted to the EMBL/GenBank/DDBJ databases, GenBank code: 8745188 (2000).

49 RO Morgan, NA Jenkins, DJ Gilbert, NG Copeland, BR Balsara, JR Testa and MP Fernandez, *Genomics*, **15**, 40–49 (1999).

50 Y Misaki, GJ Pruijn, AW van der Kemp and WJ van Venrooij, *J Biol Chem*, **269**, 4240–46 (1994).

51 CA Towle and BV Treadwell, *J Biol Chem*, **267**, 5416–23 (1992).

52 MP Fernandez, NA Jenkins, DJ Gilbert, NG Copeland and RO Morgan, *Genomics*, **37**, 366–74 (1996).

53 H Tokumitsu, A Mizutani, MA Muramatsu, T Yokota, KI Arai and H Hidaka, *Biochem Biophys Res Commun*, **186**, 1227–35 (1992).

54 BM Wice and JL Gordon, *J Cell Biol*, **116**, 405–22 (1992).

55 K Fiedler, F Lafont, RG Parton and K Simons, *J Cell Biol*, **128**, 1043–53 (1995).

56 A Burger, R Berendes, D Voges, R Huber and P Demange, *FEBS Lett*, **329**, 25–28 (1993).

57 A Rosengarth, J Rosgen, HJ Hinz and V Gerke, *J Mol Biol*, **288**, 1013–25 (1999).

58 X Weng, H Luecke, IS Song, DS Kang, SH Kim and R Huber, *Protein Sci*, **2**, 448–58 (1993).

59 A Rosengarth, V Gerke and H Luecke, *J Mol Biol*, **306**, 489–98 (2001).

60 A Rosengarth and H Luecke, *J Mol Biol*, **326**, 1317–25 (2003).

61 A Burger, R Berendes, S Liemann, J Benz, A Hofmann, P Goettig, R Huber, V Gerke, C Thiel, J Roemisch and K Weber, *J Mol Biol*, **257**, 839–47 (1996).

62 B Favier-Perron, A Lewit-Bentley and F Russo-Marie, *Biochemistry*, **35**, 1740–44 (1996).

63 G Zanotti, G Malpeli, F Gliubich, C Folli, M Stoppini, L Olivi, A Savoia and R Berni, *Biochem J*, **329**, 101–6 (1998).

64 MC Bewley, CM Boustead, JH Walker, DA Waller and R Huber, *Biochemistry*, **32**, 3923–29 (1993).

65 A Lewit-Bentley, S Morera, R Huber and G Bodo, *Eur J Biochem*, **210**, 73–77 (1992).

66 J Sopkova, M Renouard and A Lewit-Bentley, *J Mol Biol*, **234**, 816–25 (1993).

67 R Huber, R Berendes, A Burger, M Schneider, A Karshikov, H Luecke, J Roemisch and E Paques, *J Mol Biol*, **223**, 683–704 (1992).

68 NO Concha, JF Head, MA Kaetzel, JR Dedman and BA Seaton, *Science*, **261**, 1321–24 (1993).

69 AJ Avila-Sakar, CE Creutz and RH Kretsinger, *Biochim Biophys Acta*, **1387**, 103–16 (1998).

70 JR Glenney Jr, B Tack and MA Powell, *J Cell Biol*, **104**, 503–11 (1987).

71 DD Schlaepfer and HT Haigler, *J Biol Chem*, **262**, 6931–37 (1987).

72 Y Ando, S Imamura, YM Hong, MK Owada, T Kakunaga and R Kannagi, *J Biol Chem*, **264**, 6948–55 (1989).

73 A Rosengarth, J Roesgen, HJ Hinz and V Gerke, *J Mol Biol*, **306**, 825–35 (2001).

74 V Gerke and K Weber, *EMBO J*, **3**, 227–33 (1984).

75 J Glenney, *J Biol Chem*, **261**, 7247–52 (1986).

76 DD Schlaepfer, T Mehlman, WH Burgess and HT Haigler, *Proc Natl Acad Sci USA*, **84**, 6078–82 (1987).

77 RJ Owens and MJ Crumpton, *Biochem J*, **219**, 309–16 (1984).

78 PB Moore, *Biochem J*, **238**, 49–54 (1986).

79 RS Mani and CM Kay, *Biochem J*, **259**, 799–804 (1989).

80 S Linse, A Helmersson and S Forsen, *J Biol Chem*, **266**, 8050–54 (1991).

81 S Linse, C Johansson, P Brodin, T Grundstrom, T Drakenberg and S Forsen, *Biochemistry*, **30**, 154–62 (1991).

82 HJ Moeschler, JJ Schaer and JA Cox, *Eur J Biochem*, **111**, 73–78 (1980).

83 JP Reeves and RM Dowben, *J Cell Physiol*, **73**, 49–60 (1969).

84 C Thiel, K Weber and V Gerke, *J Biol Chem*, **266**, 14732–39 (1991).

85 P Meers and T Mealy, *Biochemistry*, **32**, 5411–18 (1993).

86 J Sopkova, J Gallay, M Vincent, P Panoska and A Lewit-Bentley, *Biochemistry*, **33**, 4490–99 (1994).

87 R Huber, M Schneider, I Mayr, J Roemisch and EP Paques, *FEBS Lett*, **275**, 15–21 (1990).

88 MA Swairjo, NO Concha, MA Kaetzel, JR Dedman and BA Seaton, *Nat Struct Biol*, **2**, 968–74 (1995).

89 I Capila, MJ Hernaiz, YD Mo, TR Mealy, B Campos, JR Dedman, RL Linhardt and BA Seaton, *Structure*, **9**, 57–64 (2001).

90 JJ Falke, SK Drake, AL Hazard and OB Peersen, *Q Rev Biophys*, **27**, 219–90 (1994).

91 RR Biekofsky, SR Martin, JP Browne, PM Bayley and J Feeney, *Biochemistry*, **37**, 7617–29 (1998).

92 CE Futter, S Felder, J Schlessinger, A Ullrich and CR Hopkins, *J Cell Biol*, **120**, 77–83 (1993).

93 RA Fava and S Cohen, *J Biol Chem*, **259**, 2636–45 (1984).

94 U Rescher, N Zobiack and V Gerke, *J Cell Sci*, **113**, 3931–38 (2000).

95 B Rothhut, *Cell Mol Life Sci*, **53**, 522–26 (1997).

96 W Wang and CE Creutz, *Biochemistry*, **31**, 9934–39 (1992).

97 RA Blackwood and JD Ernst, *Biochem J*, **266**, 195–200 (1990).

98 P Meers, T Mealy, N Pavlotsky and AI Tauber, *Biochemistry*, **31**, 6372–82 (1992).

99 HAM Andree, GM Willems, R Hauptmann, I Maurer-Fogy, MC Stuart, WT Hermens, PM Frederik and CPM Reutelingsperger, *Biochemistry*, **32**, 4634–40 (1993).

100 KA Hajjar, AT Jacovina and J Chacko, *J Biol Chem*, **269**, 21191–97 (1994).

101 JS Menell, GM Cesarman, AT Jacovina, MA McLaughlin, EA Lev and KA Hajjar, *N Engl J Med*, **340**, 994–1004 (1999).

102 GM Cesarman, CA Guevara and KA Hajjar, *J Biol Chem*, **269**, 21198–203 (1994).

103 KA Hajjar, CA Guevara, E Lev, K Dowling and J Chacko, *J Biol Chem*, **271**, 21652–59 (1996).

104 KA Hajjar, L Mauri, AT Jacovina, F Zhong, UA Mirza, JC Padovan and BT Chait, *J Biol Chem*, **273**, 9987–93 (1998).

105 G Kassam, KS Choi, J Ghuman, HM Kang, SL Fitzpatrick, T Zackson, S Zackson, M Toba, A Shinomiya and DM Waisman, *J Biol Chem*, **273**, 4790–99 (1998).

106 KS Choi, J Ghuman, G Kassam, HM Kang, SL Fitzpatrick and DM Waisman, *Biochemistry*, **37**, 648–55 (1998).

107 G Kassam, BH Le, HS Choi, HM Kang, SL Fitzpatrick, P Louie and DM Waisman, *Biochemistry*, **37**, 16958–66 (1998).

108 SL Fitzpatrick, G Kassam, KS Choi, HM Kang, DK Fogg and DM Waisman, *Biochemistry*, **39**, 1021–28 (2000).

109 E Kube, T Becker, K Weber and V Gerke, *J Biol Chem*, **267**, 14175–82 (1992).

110 J Seemann, K Weber and V Gerke, *Biochem J*, **319**, 123–29 (1996).

111 WQ Zhao, GH Chen, H Chen, A Pascale, L Ravindranath, MJ Quon and DL Alkon, *J Biol Chem*, **278**, 4205–4215 (2003).

112 WJ Zaks and CE Creutz, *Biochim Biophys Acta*, **1029**, 149–60 (1990).

113 WJ Zaks and CE Creutz, *Biochemistry*, **30**, 9607–15 (1991).

114 MA Kaetzel, YD Mo, T Mealy, B Campos, W Bergsma-Schutter, A Brisson, JR Dedman and BA Seaton, *Biochemistry*, **40**, 4192–99 (2001).

115 HC Chan, MA Kaetzel, AL Gotter, JR Dedman and DJ Nelson, *J Biol Chem*, **269**, 32464–68 (1994).

116 CPM Reutelingsperger and WL van Heerde, *Cell Mol Life Sci*, **53**, 527–32 (1997).

117 R Huber, J Roemisch and EP Paques, *EMBO J*, **9**, 3867–74 (1990).

118 M de la Fuente and AV Parra, *Biochemistry*, **34**, 10393–99 (1995).

119 O Lambert, V Gerke, MF Bader, F Porte and AI Brisson, *J Mol Biol*, **272**, 42–55 (1997).

120 S Rety, D Osterloh, JP Arie, S Tabaries, J Seemann, F Russo-Marie, V Gerke and A Lewit-Bentley, *Structure*, **8**, 175–84 (2000).

121 J Seemann, K Weber and V Gerke, *FEBS Lett*, **413**, 185–90 (1997).

122 S Rety, J Sopkova, M Renouard, D Osterloh, V Gerke, S Tabaries, F Russo-Marie and A Lewit-Bentley, *Nat Struct Biol*, **6**, 89–94 (1999).

123 T Becker, K Weber and N Johnson, *EMBO J*, **9**, 4207–13 (1990).

124 RB Sutton and SR Sprang, in B Seaton (ed.), *Annexins: Molecular Structure to Cellular Function*, R. G. Landes Company, Austin, TX, pp 31–42 (1996).

Other Ca-proteins

Calcium pump (ATPase) of sarcoplasmic reticulum

Chikashi Toyoshima

Institute of Molecular and Cellular Biosciences, The University of Tokyo, Japan

FUNCTIONAL CLASS

Enzyme; EC 3.6.3.38; Ca^{2+}-transporting ATPase. Ca^{2+}-ATPase of skeletal muscle sarcoplasmic reticulum (Sarco(Endo)plasmic Reticulum Calcium ATPase 1, SERCA1) transports two Ca^{2+} ions from the cytoplasm to the lumen of SR against a concentration gradient, in exchange of two to three H^+ per ATP hydrolyzed. Plasma membrane Ca^{2+}-ATPase is thought to transport only one Ca^{2+} per ATP hydrolysis. Complete reverse reaction is possible under favorable conditions. Acetylphosphate can work as a substrate.

Reaction cycle

Reaction mechanism is usually interpreted on the basis of E1/E2 scheme (Figure 1,). In the E1 state, the Ca^{2+}

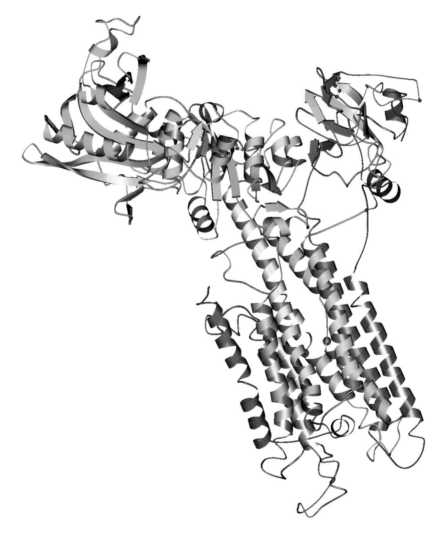

3D Structure Schematic representation of the Ca^{2+}-bound form of the calcium ion pump (Ca^{2+}-ATPase; SERCA1a) of the skeletal muscle sarcoplasmic reticulum, PDB code 1EUL. Two Ca^{2+} ions bound to the transmembrane high-affinity site are shown as purple spheres (also circled). Secondary structure was assigned with DSSP.[38]

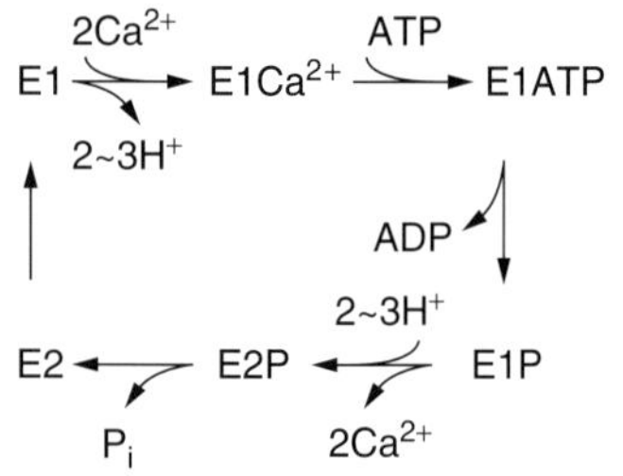

Calcium pump (ATPase) of sarcoplasmic reticulum

Figure 1 A simplified reaction scheme according to the E1/E2 model. Only the forward direction is shown.

binding sites have high affinity and are accessible from the cytoplasm. In the E2 state, the Ca^{2+} binding sites have low affinity and are accessible from the lumenal (extracellular) side. Hence, Ca^{2+}-ATPase is said to be a member of the E1/E2-type ATPase that includes Na^+K^+ ATPase and gastric H^+K^+-ATPase among others.[2] This type of ATPases is also called the P-type ATPase, because the enzyme becomes auto-phosphorylated during the reaction cycle. This is a unique feature of the P-type ATPase, distinct from F_1F_0 type, for example. Transfer of bound Ca^{2+} ions is thought to take place between E1P and E2P. These two phosphorylated states are distinguished by the sensitivity to ADP (i.e. ATP is synthesized from ADP in E1P but not in E2P). Each state of the reaction cycle can also be characterized by different susceptibilities to trypsin and proteinase K.[3]

OCCURRENCE

Ca^{2+}-ATPase is an integral membrane protein and is the most abundant (>60%) among the proteins in the SR membrane.

BIOLOGICAL FUNCTION

In muscle cells, Ca^{2+} ions stored in SR are released through Ca^{2+} release channels for contraction. The Ca^{2+} ions released have to be pumped back into SR to cause relaxation. Ca^{2+}-ATPase is responsible for this process and maintains 10^4-fold (0.1 μM vs. 1.5 mM) concentration gradient across the membrane. It is also recognized as an important source of body heat.

AMINO ACID SEQUENCE INFORMATION

SERCA1a is the adult form of the fast twitch skeletal muscle sarcoplasmic reticulum Ca^{2+}-ATPase (994 AA; Figure 2); SERCA1b is the neonatal form (1001 AA) having additional seven residues at the C-terminus.

SERCA2 is by far the most widespread of all SERCA isoforms and phylogenetically the oldest. SERCA2 has two splice variants and SERCA2a (997 AA) is the main isoform in the cardiac muscle and the slow twitch skeletal muscle.

Signature motif of the P-type ATPase superfamily can be written as D-K-T-G-T-[LIVM]-[TIS]. In the case of SERCA1a, this motif starts at Asp351 (Figure 2). This constitutes motif PS00154 in the PROSITE database. An extensive database of P-type ATPases is available at http://biobase.dk/~axe/Patbase.html. The phylogenic tree of the P-type ATPases is also available there. Various types of SERCA are reviewed by Møller et al.[2] and more recently by Wuytack et al.[4]

PROTEIN PRODUCTION, PURIFICATION, AND MOLECULAR CHARACTERIZATION

Production

The most common source is rabbit hind leg white muscle, in which only SERCA1a is present. Large-scale production of recombinant ATPase is still a challenging task. So far, the most successful system has been to use the adenovirus vector with COS-1 cells.[5] This system expresses SERCA as high as one-third of the native SR. Large-scale production using yeast has been reported by several groups but the expression level appears to be rather low.[6]

Purification

SR membrane (light SR fraction devoid of T-tubule containing Ca^{2+}-release channels) can be isolated by differential centrifugation (e.g. 7). This preparation contains usually $30\,mg\,ml^{-1}$ proteins and can be stored for a few months at $-80\,°C$ with 0.3 M sucrose. From 150 g of muscle, crude specimen of 500 mg can be routinely obtained. For purification of rabbit SR Ca^{2+}-ATPase, the membrane preparation is first dissolved with 2% octaethyleneglycol dodecylether ($C_{12}E_8$) and applied to affinity column chromatography using reactive red 120.[8] This dye appears to mimic the adenosine moiety of ATP. The enzyme can be eluted from the column with either ADP[9] or AMPPNP.[8]

Activity test

Enzyme-coupled ATPase assay that utilizes the absorption of NADH is widely used.[9,10] The method is applicable to both purified enzyme and native SR membrane, provided that sufficient amount of $C_{12}E_8$ or ionophore (A23187, Sigma) is present. Our typical value for the turnover number is $20.8\,s^{-1}$ at $25\,°C$ (in 123 mM KCl, 6.15 mM $MgCl_2$, 0.12 mM $CaCl_2$, 2 mM ATP, 0.1% $C_{12}E_8$, 61.5 mM MOPS, pH 7.0) for affinity-purified enzyme.

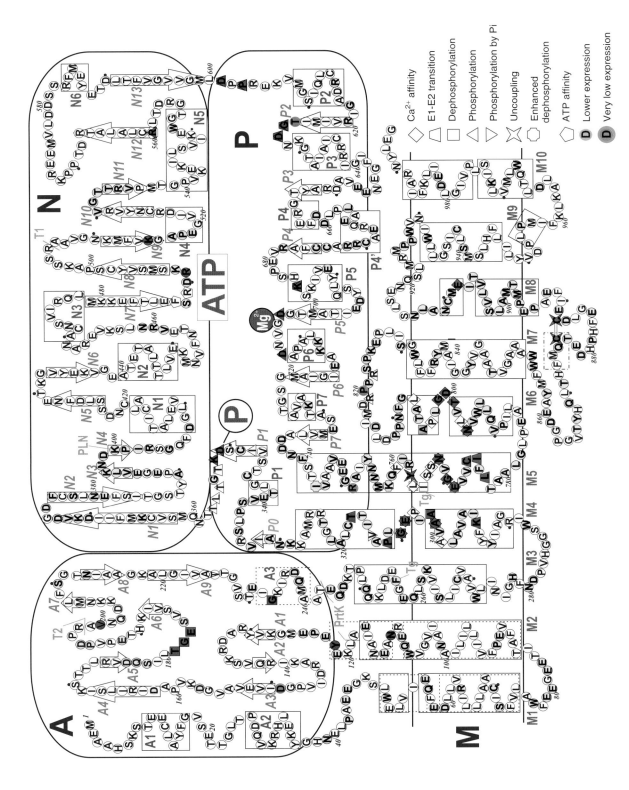

Figure 2 Two-dimensional diagram of the structure, the amino acid sequence, and the results of mutation experiments of the skeletal muscle Ca^{2+}-ATPase (SERCA1a). Secondary structure elements (boxes for helices; arrows for β-strands) follow those in the Ca^{2+}-bound form. T1 and T2 are trypsin digestion sites; PrtK, proteinase K digestion sites. Adapted from F Wuytack, L Raeymaekers and L Missiaen, *Cell Calcium*, **32**, 279–305 (2002).[4]

Inhibitors

Modulators of Ca^{2+}-ATPase have been reviewed by Kochegarov.[11] The best-known inhibitor is thapsigargin (TG), a sesquiterpene lactone from a plant that traps the pump in E2 state with a subnanomolar dissociation constant.[12] The mode of binding to SERCA is now determined.[13] Phospholamban (52 amino acids) and sarcolipin (31 amino acids) are endogenous inhibitors. They are related short transmembrane peptides and important regulators in cardiac muscle (for a recent review, see reference 4).

X-RAY STRUCTURE OF THE Ca^{2+}-ATPase (SERCA1a) FROM RABBIT SKELETAL MUSCLE

Crystallization

So far, SR Ca^{2+}-ATPase has been crystallized in five different states analogous to $E1Ca^{2+}$, E1ATP, E1P-ADP, E2P, and E2. In all cases, obtaining crystals required the addition of phospholipid (phosphatidylcholine) and could be accomplished only by the dialysis method. Crystallization conditions were published for two of them. Crystals in $E1Ca^{2+}$ were obtained by dialysis against a buffer containing 0.8 M sodium butyrate, 2.75 M glycerol, 10 mM $CaCl_2$, 3 mM $MgCl_2$, 2.5 mM NaN_3, 0.2 mM dithiothreitol, and 20 mM MES, pH 6.1.[14] Typical dimensions of the crystals were $100 \times 500 \times < 20\,\mu m$.

Because Ca^{2+}-ATPase is unstable without a high concentration of Ca^{2+}, for making crystals in E2 state, the enzyme had to be stabilized by thapsigargin. The crystals were obtained by dialyzing the solution containing $20\,\mu M$ Ca^{2+}-ATPase, 1 mM Ca^{2+}, $30\,\mu M$ thapsigargin, 3 mM EGTA, and $2\,mg\,mL^{-1}$ $C_{12}E_8$, against a buffer containing 2.75 M glycerol, 4% PEG 400, 3 mM $MgCl_2$, 2.5 mM NaN_3, $2\,\mu g\,mL^{-1}$ butylhydroxytoluene, 0.2 mM dithiothreitol, 0.1 mM EGTA, 20 mM MES, pH 6.1.[13] The crystals grew to the size $200 \times 200 \times 50\,\mu m$.

Crystals of SR Ca^{2+}-ATPase

So far, the structures of rabbit SR Ca^{2+}-ATPase have been determined to 2.6-Å resolution in the presence of Ca^{2+} ($E1Ca^{2+}$; PDB ID 1EUL) and to 3.1-Å resolution in the absence of Ca^{2+} and stabilized with TG (E2(TG); PDB ID 1IWO). $E1Ca^{2+}$ structure was re-refined by the author and is presented here. The $E1Ca^{2+}$ crystals belong to the space group C2 with unit cell parameters of $a = 166.0$ Å, $b = 64.4$ Å, $c = 147.1$ Å, $\beta = 98.0°$ (at 100 K), and contain one protein molecule in the asymmetric unit.[14] Two types of E2(TG) crystals with $P4_1$ or $P2_1$ symmetry were obtained. Those with $P4_1$ symmetry diffracted better and had unit cell dimensions of $a = b = 71.7$ Å, $c = 590.3$ Å, with two molecules in the asymmetric unit.[13]

Overall description of the structure

Ca^{2+}-ATPase is a tall molecule of about 150-Å high and 80-Å thick and comprises a large cytoplasmic headpiece, transmembrane domain made of 10 (M1–M10) α-helices, and small lumenal domain (Figure 2). The cytoplasmic headpiece consists of three domains, designated as A (actuator), N (nucleotide binding), and P (phosphorylation) domains. They are widely split in the presence of Ca^{2+} but gather to form a compact headpiece in the absence of Ca^{2+} (Figure 3). The A-domain is connected to M1–M3 helices, and the P-domain to M4 and M5 helices. The N-domain is a long insertion between two parts forming the P-domain. The M5 helix runs from the lumenal surface to an end of the phosphorylation domain and works as the 'spine' of the molecule. On the lumenal side, there are only short loops connecting transmembrane helices, except for the loop of ~40 residues connecting the M7 and M8 helices (Figure 2). The distance between the Ca^{2+} binding sites and the phosphorylation site is larger than 50 Å. For the phosphoryl transfer from ATP to Asp351 to take place, the N-domain has to approach the P-domain even closer than that observed in the E2(TG) form.

The orientations and positions of atomic models with respect to the lipid bilayer are illustrated in Figure 3, in which the models are positioned in the bilayer of DOPC generated by molecular dynamics calculation (Sugita, Y; Ikeguchi, M; and Toyoshima, C; unpublished result). Ca^{2+}-ATPase is reported to have full functionality when reconstituted in DOPC membrane.[15] These orientations and positions with respect to the membrane were determined from crystallographic constraints and from comparison with different crystal forms. The orientations were found to be the same as in the original crystals. The coordinates of the aligned models can be downloaded from the author's website (http://www.iam.u-tokyo.ac.jp/~StrBiol/models).

Organization of the transmembrane domain

As expected from the amino acid sequence, SERCA1 has 10 (M1–M10) transmembrane α-helices (Figure 2), two of which (M4 and M6) are partly unwound in both $E1Ca^{2+}$ and E2(TG) for efficient coordination of Ca^{2+} (Figure 4). M6 and M7 are far apart and are connected by a long cytosolic loop that runs along the bottom of the P-domain and interacts with M5 (Figure 3). The amino acid sequence is well conserved for M4 to M6 but not for M8 even within the members of closely related P-type ATPases, such as $Na^+ K^+$- and $H^+ K^+$-ATPases. M7–M10 helices are in fact lacking in bacterial type I P-type ATPases, and are

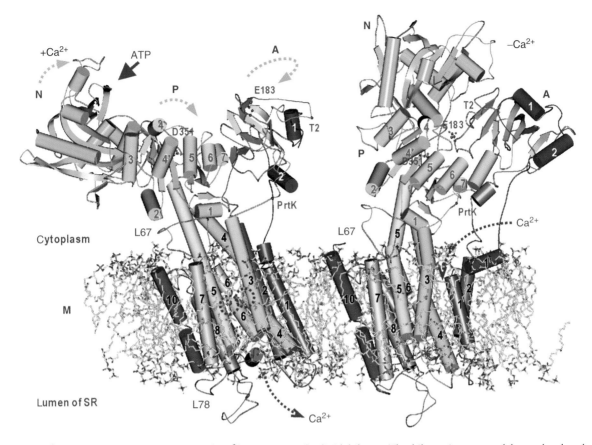

Figure 3 Ca^{2+}-bound and unbound forms of Ca^{2+}-ATPase in the lipid bilayer. The bilayer is generated by molecular dynamics simulation of dioleoylphosphatidylcholine (DOPC). M3 and M5 helices in the unbound form are approximated with 2 and 3 cylinders. The color changes gradually from the N-terminus (blue) to the C-terminus (red). The arrows indicate the directions of movements accompanying the dissociation of Ca^{2+}. Two bound Ca^{2+} are shown as purple spheres (circled). T2 trypsin digestion site and a proteinase K digestion site (PrtK) are also marked.

Figure 4 Organization of the transmembrane helices. Viewed from the cytoplasmic side normal to the membrane. Only the cytoplasmic half of each helix is shown as a cylinder. Violet, E1Ca^{2+}; light green, E2(TG). Two Ca^{2+} in the binding sites are shown as cyan spheres. M1' is the amphipathic helix lying on the membrane. Asp800 (M6) and Glu309 (M4) side chains are also shown.

apparently specialized for each subfamily.[2] Presumably reflecting this fact, the electron density map gives the impression that M7–M10 are segregated from M1–M6 (refer to Figure 3(a) of reference 14). Nevertheless, M5 and M7 are packed very tightly at G770 (M5), G841 (M7), and G845 (M7), forming a pivot for the bending of M5 (see later). The M7–M10 helices appear to work as a membrane anchor.

Details of the Ca^{2+}-binding sites

It is well established that SR Ca^{2+}-ATPase has two high-affinity transmembrane Ca^{2+}-binding sites and that the binding is cooperative. Nevertheless, there was much confusion about their locations, because the transmembrane Ca^{2+} binding sites do not appear to bind lanthanides.[16] Soaking of the crystals in a crystallization buffer containing

1 mM Tb for 90 h did not replace the bound Ca^{2+} in the high-affinity sites.[14] Nd ion binding to the N-domain has been reported.[17] Debates still exist about the presence of low-affinity Ca^{2+} binding sites, in particular, in the lumenal region. X-ray crystallography of the rabbit SERCA1a in 10 mM Ca^{2+} identified two binding sites in the transmembrane but none outside the membrane.[14] Valence search[18] showed no other peaks if the threshold was set higher than 1.6.

The two Ca^{2+} binding sites (I and II) are located side by side near the cytoplasmic surface of the lipid bilayer (Figure 3), with the site II ∼3 Å closer to the surface. Site I, the binding site for the first Ca^{2+}, is located at the center of the transmembrane domain (Figure 4) in a space surrounded by M5, M6, and M8 helices, and formed by Glu771 (M5), Thr799, Asp800 (M6), and Glu908 (M8) side-chain oxygen (Figure 5). Two water molecules also contribute to this site. M8 is located rather distally and the contribution of glutamic acid (Glu908) is not essential in that glutamine can substitute Glu908 to a large extent.[5] Any substitutions to other residues totally abolish the binding of Ca^{2+}.

Site II is nearly 'on' the M4 helix when viewed normal to the membrane, with the contribution of Asp800 (M6) and Asn768 (M5). (Figures 4–6). The M4 helix is partly unwound (between Ile307–Gly310). This part forms a γ-turn and provides three main-chain oxygen atoms to the coordination of Ca^{2+}. Glu309 provides two oxygens and caps the bound Ca^{2+} (Figure 5). This arrangement of oxygen atoms is reminiscent of the EF-hand motif (e.g. reference 19).

Thus, both site I and site II have 7 coordination but of different characteristics. Asp800, on the unwound part of M6, is the only residue that contributes to both sites (Figures 5 and 6). Even double mutations of Glu309 and Asn768 leave 50% Ca^{2+} binding, indicating that site II is the binding site for the second Ca^{2+}.[5]

Structure of the P-domain

The P-domain contains the residue of phosphorylation, Asp351. There are three critical aspartate (627, 703, and 707) residues clustered around the phosphorylation site

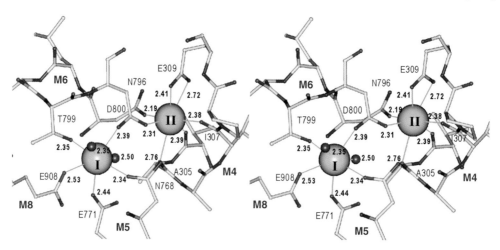

Figure 5 Details of the transmembrane Ca^{2+}-binding sites. Two Ca^{2+} in the binding sites are shown as cyan spheres and water molecules as small red spheres. The numbers show the distances between Ca^{2+} and coordinating oxygen atoms.

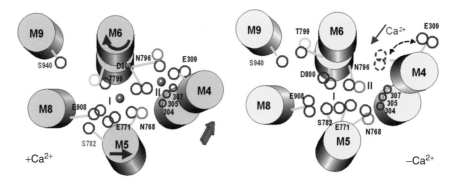

Figure 6 Schematic diagram of the Ca^{2+}-binding sites and the movement accompanying the dissociation of Ca^{2+}. Oxygen, nitrogen, and carbon atoms are represented by red, blue, and orange circles respectively. Ca^{2+} ions are supposed to enter into the binding cavity by the conformation change of the side chain of Glu309 as indicated by dotted circles. The arrows indicate the movements of the helices in the transition from $E1Ca^{2+} \rightarrow E2(TG)$.

(Figure 7), in addition to an absolutely conserved lysine residue (Lys684). These key residues are very well conserved throughout the haloacid dehalogenase (HAD) superfamily, but are also shared by the bacterial two-component response-regulator proteins, such as CheY and FixJ, which have folding patterns different from that of HAD.[20] The P-domain has a Rossmann fold, commonly found in nucleotide binding proteins, consisting of parallel β-sheet (7 strands in Ca^{2+}-ATPase) and associated short α-helices (Figure 7). The catalytic residue (i.e. Asp351 in Ca^{2+}-ATPase) is always at the C-terminal end of the first β-strand, which is connected to a long insertion, the N-domain (Figure 2). Thus, the P-domain is formed by two regions far apart in the amino acid sequence. This is why the P-type ATPase was thought to be an orphan in evolution for a long time.[21] This system requires Mg^{2+} for phosphorylation. Mg^{2+} is coordinated directly by Asp351, Asp703, and by Asp707 in the P-domain and residues in the A-domain through water molecules. It is not known if this Mg^{2+} is the same as that bound to ATP. Lys684 is particularly important for the binding of γ-phosphate of ATP.

Phosphoserine phosphatase (PSP) is structurally the best-studied member in the HAD superfamily. The atomic structures have been determined for five states in the reaction cycle at high resolution.[22] The atomic model superimposes very well with that of Ca^{2+}-ATPase. It is interesting that PSP has domains, though much smaller, corresponding to the A- and N-domains of Ca^{2+}-ATPase. Unfortunately, however, the conformation changes are relatively small and the amino acid sequence ends before M5 of Ca^{2+}-ATPase starts.

Structure of the N-domain

N-domain is the largest of the three cytoplasmic domains, consisting of residues approximately Asn359–Asp601 (Figure 2), and connected to the P-domain with two strands. These strands bear a β-sheet like hydrogen bonding pattern, presumably to allow large domain movements with a precise orientation.[23] In particular, consecutive prolines (Pro602–603) appear to serve as a guide that determines the orientation of the domain movement. The N-domain contains the binding site for the adenosine moiety of ATP.[14] Phe487 makes an aromatic adenine-ring interaction, which is a common feature in many ATP binding sites. Lys515, a critical residue, is located at one end of the binding cavity. This residue can be labeled specifically with FITC at alkaline pH and has been used for many spectroscopic studies.[24] The binding of the adenine ring appears predominantly hydrophobic (i.e. devoid of hydrogen bonds) and Lys515 and Glu442, which is also critical,[25] are important in lining the binding cavity.

Structure of the A-domain

A-domain is the smallest of the three cytoplasmic domains and consists of the N-terminal ∼50 residues that form two short α-helices and ∼110 residues between the M2 and M3 helices (Figure 2), which form a deformed jelly role structure. The A-domain contains a sequence motif [183]TGE, one of the signature sequences of the P-type ATPase. This sequence represents a loop that comes very close to the phosphorylation site Asp351, in the E2 and E2P states (Figure 3). Because this domain is directly connected to the M1–M3 helices and more indirectly to the M4–M6 helices, this domain is thought to act as the 'actuator' of the gates that regulate the binding and release of Ca^{2+} ions.

STRUCTURAL CHANGES ACCOMPANYING THE DISSOCIATION OF Ca^{2+}

As illustrated in Figure 3, Ca^{2+}-ATPase undergoes large structural changes upon the dissociation of Ca^{2+}.[13] Three cytoplasmic domains change their orientations and gather to form a compact headpiece. The P-domain inclines 30° with respect to the membrane and N-domain inclines 60° relative to the P-domain, whereas the A-domain rotates ∼110° horizontally. The structure of each domain, however, is hardly altered. In contrast, some of the transmembrane helices are bent or curved (M1, M3, and M5; Figure 3) or partially unwound (M2; Figure 8); M1–M6 helices undergo drastic rearrangements (Figure 4) that involve shifts normal to the membrane (M1–M4; Figure 8). Thus, the structural changes are very large and global; they are mechanically linked and coordinated by the P-domain.

M3–M5 helices are directly linked to the P-domain by hydrogen bonds (Figure 7). M3 is connected to the P1 helix at the bottom of the P-domain through a critical hydrogen bond involving Glu340.[26] The top part of M5

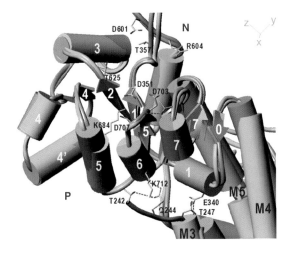

Figure 7 Organization of the P-domain and the linkage with the transmembrane helices.[13] Superimposition of the E1Ca^{2+} (violet) and E2(TG) (light green) forms fitted with the P-domain. The residues (in atom color) represent those in E2(TG).

Calcium pump (ATPase) of sarcoplasmic reticulum

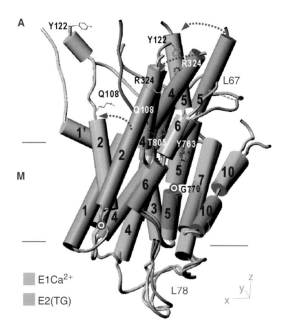

Figure 8 Rearrangement of transmembrane helices on the dissociation of Ca^{2+}. The models for $E1Ca^{2+}$ (violet) and E2(TG) (light green) are superimposed. The M5 helix lies along the plane of the paper. M8 and M9 are removed. Double circles show pivot positions for M2 and M5. Arrows indicate the directions of movements during the change from $E1Ca^{2+}$ to E2(TG). Orange broken line shows a critical hydrogen bond between the L67 loop and M5 helix. Adapted from C Toyoshima and H Nomura, Nature, 418, 605–11 (2002).[13]

is integrated into the Rossmann fold (Figure 7) and moves together with the P-domain as a single entity. M4 and M5 are 'clamped' by forming a short antiparallel β-sheet. M6 is also connected, though less directly, to the P-domain through L67 (Figure 3), which is in turn linked to M5 through a critical hydrogen bond (Figure 8). If the P-domain inclines, for instance, owing to the bending of M5, all these helices (M3–M6) will incline and generate movements that have components normal to the membrane (Figure 8). The distances of the movements depend on the distances from the pivoting point, located around Gly770 at the middle of the membrane (double circle in Figure 8). The lower part below Gly770 hardly moves. The shift is therefore small for M6 and large for M3 and M4; whole M3 and M4 helices move downwards in dissociation of Ca^{2+}, whereas M6 undergoes rather local changes around Asn796 to Asp800.

Ca^{2+} binding sites in the E2(TG) state

The most important movements of the transmembrane helices directly relevant to the dissociation of Ca^{2+} are (i) a shift of M4 towards the lumenal (extracellular) side by one turn of an α-helix (5.5 Å), (ii) bending of the upper part of M5 (above Gly770) towards M4, and (iii) nearly 90° rotation of the unwound part of M6.[13] As a result, profound reorganization of the binding residues takes place and the number of coordinating oxygen atoms decreases (Figure 6). For site I, this is due to the movement of Asn768 toward M4 caused by the bending of M5; for site II, replacement of Asp800 by Asn796 (i.e. rotation of M6) is critical. Then, why the large movement of M4 (and M3) toward the lumenal side is required is left unexplained.

Homology modeling of the cation binding sites of Na^+ K^+-ATPase[27] suggested that such movements are needed for countertransport. With the arrangements of residues observed with E2(TG) structure, it was straightforward to make two high-affinity K^+-binding sites, provided that Asn796 is replaced by asparagine following Na^+ K^+-ATPase sequence: the other coordinating residues are common to both ATPases. The key feature in the model is that the asparagine (Asn796 in Ca^{2+}-ATPase) is coordinated to both K^+, similar to Asp800 in coordination of 2 Ca^{2+}. Because Asn796 is located one turn below where Asp800 was, M4 must move downward to provide carbonyl groups for coordination of K^+. These rearrangements ensure that release of one type of cation coordinates with the binding of the other and allow the binding of ions of different radii with high affinity.

Entry pathway of Ca^{2+} to the binding sites

It is not obvious how Ca^{2+} reaches the transmembrane binding sites because of the lack of a vestibule similar to those found in ion channels.[28] However, evidence indicating the involvement of Glu309, the residue capping the site II Ca^{2+}, is accumulating. In E2(TG) structure, Glu309 points outside the binding cavity and toward the cytoplasm (Figure 6). The amphipathic short helix lying on the cytoplasmic surface (M1' in Figures 3 and 4) and the loop connecting the A-domain and M3 helix provide negatively charged residues around a path leading to Glu309.[13] This path is made by the large rotation of the A-domain and destroyed in $E1Ca^{2+}$. Thus, it is likely that conformation change of the Glu309 side chain delivers Ca^{2+} to the binding cavity. Glu309 has been proposed as the gating residue also from mutational studies.[29]

This idea agrees well with the cooperative binding of Ca^{2+}.[30] Site I is the binding site for the first Ca^{2+} and has a high affinity ($\sim$0.5 μM at pH 7.0).[31] Because Ca^{2+} reaches the site I through site II, site II has to have substantially lower affinity than site I. After the binding of Ca^{2+} to site I, site II may become a binding site with even higher affinity.[32] For this purpose, the region around Asp800 has to be flexible and fixed by the binding of the first Ca^{2+}.

Exit pathway from the binding sites

Where the exit of Ca^{2+} is located on the lumenal side is not obvious either. In $E1Ca^{2+}$, there are layers of hydrophobic residues below the binding cavity. Site I Ca^{2+} and the closest water molecule observed in the crystal structure (near the loop connecting M5 and M6) is distant by >12 Å in the direction normal to the membrane. In E2(TG), the loop connecting M3 and M4 comes closer to the L78 loop (Figure 3), sealing the access pathway at the very surface. Thus, in this aspect, the structure deviates from the classical E1/E2 models that assume lumenally opened binding sites.[1] However, in the normal reaction cycle, lumenal gate will open in E1P → E2P transition and close after the binding of counter ions (H^+ in this case; Figure 1). In the Mg^{2+}/F^- complex,[33] which is an analogue of E2P state, the position of the A-domain is different apparently to open the lumenal gate by moving the M1–M4 helices (Toyoshima, C; Nomura, H; unpublished observation).

Role of the closed configuration of the cytoplasmic domains in the E2 state

The closed configuration of the headpiece appears to be important in three aspects: (i) restriction of the delivery of ATP to the phosphorylation site; (ii) restriction of the thermal movements of transmembrane helices; (iii) positioning of the A-domain.

The reaction cycle of Ca^{2+}-ATPase (Figure 1) is regulated essentially by Ca^{2+} alone. ATP can bind to the enzyme even when Ca^{2+} is absent but, without Ca^{2+}, the reaction cycle cannot proceed. In the closed configuration, γ-phosphate of ATP comes close to but cannot reach the phosphorylation residue Asp351. It requires deeper inclination of the N-domain, or, the rotation of the A-domain to release the N-domain. This, in turn, requires the binding of Ca^{2+}, which will cause the shift of the transmembrane helices into different positions.

Because large-scale conformational changes shown in Figure 3 occur without any involvement of ATP or phosphate, such changes must be realized by thermal movements. In fact, in $Na^+ K^+$- and $H^+ K^+$-ATPases, transmembrane helices come out of the membrane in the absence of K^+ after proteolysis and mild heat treatment.[34,35] Also, SR vesicles with loaded Ca^{2+} show significant leakage[36] and thapsigargin can stop it. The closed configuration of the cytoplasmic domains will limit thermal movements of the transmembrane helices and, therefore, the leakage.

The rotation of the A-domain takes place in E1P → E2P[3] and must be another key event in the active transport. For example, the proteolytic cut of the link between the A-domain and M3 (Figures 2 and 3) results in complete inhibition of the ATPase activity.[37] Because the position of the A-domain appears to regulate the gates from and to the binding sites, it has to be fixed in the E2 conformation.

In summary, ion pumps use thermal movements very well and are integrated with the 'devices' to regulate such movements, with the P-domain as the coordinator. Animations illustrating these movements can be downloaded from the author's web site (http://www.iam.u-tokyo.ac.jp/StrBiol/animations).

REFERENCES

1. L de Meis and AL Vianna, *Annu Rev Biochem*, **48**, 275–92 (1979).
2. JV Møller, B Juul and M le Maire, *Biochim Biophys Acta*, **1286**, 1–51 (1996).
3. S Danko, K Yamasaki, T Daiho, H Suzuki and C Toyoshima, *FEBS Lett*, **505**, 129–35 (2001).
4. F Wuytack, L Raeymaekers and L Missiaen, *Cell Calcium*, **32**, 279–305 (2002).
5. Z Zhang, D Lewis, C Strock, G Inesi, M Nakasako, H Nomura and C Toyoshima, *Biochemistry*, **39**, 8758–67 (2000).
6. G Lenoir, T Menguy, F Corre, C Montigny, PA Pedersen, D Thines, M le Maire and P Falson, *Biochim Biophys Acta*, **1560**, 67–83 (2002).
7. S Eletr and G Inesi, *Biochim Biophys Acta*, **282**, 174–79 (1972).
8. RJ Coll and AJ Murphy, *J Biol Chem*, **259**, 14249–54 (1984).
9. DL Stokes and NM Green, *Biophys J*, **57**, 1–14 (1990).
10. KW Anderson, RJ Coll and AJ Murphy, *J Biol Chem*, **259**, 11487–90 (1984).
11. AA Kochegarov, *Exp Opin Ther Patents*, **11**, 825–59 (2001).
12. Y Sagara and G Inesi, *J Biol Chem*, **266**, 13503–6 (1991).
13. C Toyoshima and H Nomura, *Nature*, **418**, 605–11 (2002).
14. C Toyoshima, M Nakasako, H Nomura and H Ogawa, *Nature*, **405**, 647–55 (2000).
15. AG Lee, *Biochim Biophys Acta*, **1376**, 381–90 (1998).
16. F Henao, S Orlowski, Z Merah and P Champeil, *J Biol Chem*, **267**, 10302–12 (1992).
17. P Champeil, T Menguy, S Soulié, B Juul, AG de Gracia, F Rusconi, P Falson, L Denoroy, F Henao, M le Maire and JV Møller, *J Biol Chem*, **273**, 6619–31 (1998).
18. M Nayal and E Di Cera, *Proc Natl Acad Sci USA*, **91**, 817–21 (1994).
19. JP Glusker, *Adv Protein Chem*, **42**, 1–76 (1991).
20. LN Johnson and RJ Lewis, *Chem Rev*, **101**, 2209–42 (2001).
21. L Aravind, MY Galperin and EV Koonin, *Trends Biochem Sci*, **23**, 127–29 (1998).
22. W Wang, HS Cho, R Kim, J Jancarik, H Yokota, HH Nguyen, IV Grigoriev, DE Wemmer and SH Kim, *J Mol Biol*, **319**, 421–31 (2002).
23. S Hayward, *Proteins*, **36**, 425–35 (1999).
24. DJ Bigelow and G Inesi, *Biochim Biophys Acta*, **1113**, 323–38 (1992).
25. JD Clausen, DB McIntosh, B Vilsen, DG Woolley and JP Andersen, *J Biol Chem*, **278**, 20245–58 (2003).
26. Z Zhang, C Sumbilla, D Lewis, S Summers, MG Klein and G Inesi, *J Biol Chem*, **270**, 16283–90 (1995).
27. H Ogawa and C Toyoshima, *Proc Natl Acad Sci USA*, **99**, 15977–82 (2002).
28. DA Doyle, CJ Morais, RA Pfuetzner, A Kuo, JM Gulbis, SL Cohen, BT Chait and R MacKinnon, *Science*, **280**, 69–77 (1998).
29. B Vilsen and JP Andersen, *Biochemistry*, **37**, 10961–71 (1998).

30 G Inesi, M Kurzmack, C Coan and DE Lewis, *J Biol Chem*, **255**, 3025–31 (1980).
31 G Inesi, Z Zhang and D Lewis, *Biophys J*, **83**, 2327–32 (2002).
32 S Orlowski and P Champeil, *Biochemistry*, **30**, 352–61 (1991).
33 AJ Murphy and RJ Coll, *J Biol Chem*, **267**, 5229–35 (1992).
34 C Gatto, S Lutsenko, JM Shin, G Sachs and JH Kaplan, *J Biol Chem*, **274**, 13737–40 (1999).
35 S Lutsenko, R Anderko and JH Kaplan, *Proc Natl Acad Sci USA*, **92**, 7936–40 (1995).
36 G Inesi and L de Meis, *J Biol Chem*, **264**, 5929–36 (1989).
37 JV Møller, G Lenoir, C Marchand, C Montigny, M le Maire, C Toyoshima, BS Juul and P Champeil, *J Biol Chem*, **277**, 38647–59 (2002).
38 W Kabsch and C Sander, *Biopolymers*, **22**, 2577–2637 (1983).
39 PJ Kraulis, *J Appl Crystallogr*, **24**, 946–50 (1991).

Phospholipase A₂

Christian Betzel[†], Tej P Singh[‡], Dessislava Georgieva[†] and Nicolay Genov[§]

[†] University Hospital Hamburg-Eppendorf, Centre of Experimental Medicine, Institute of Biochemistry and Molecular Biology I, Hamburg, Germany
[‡] Department of Biophysics, All India Institute of Medical Sciences, New Delhi, India
[§] Institute of Organic Chemistry, Bulgarian Academy of Sciences, Sofia, Bulgaria

FUNCTIONAL CLASS

Enzyme; phosphatide *sn*-2 acylhydrolase (PLA$_2$); EC 3.1.1.4.

Other names: Phosphatidylcholine 2-acylhydrolase, Lecitinase *a*. Phosphatidase, Phosphatidolipase.

Cofactor: Calcium. With some exceptions (intracellular Ca^{2+}-independent phospholipolytic enzymes, iPLA$_2$s), PLA$_2$s are Ca^{2+}-dependent enzymes. The metal ion is bound to the so-called 'calcium-binding loop' and it stabilizes the oxyanion of the tetrahedral transition-state intermediate. PLA$_2$s catalyze stereospecifically the hydrolysis of the *sn*-2 ester bond of phospholipids (PL) releasing fatty acids and

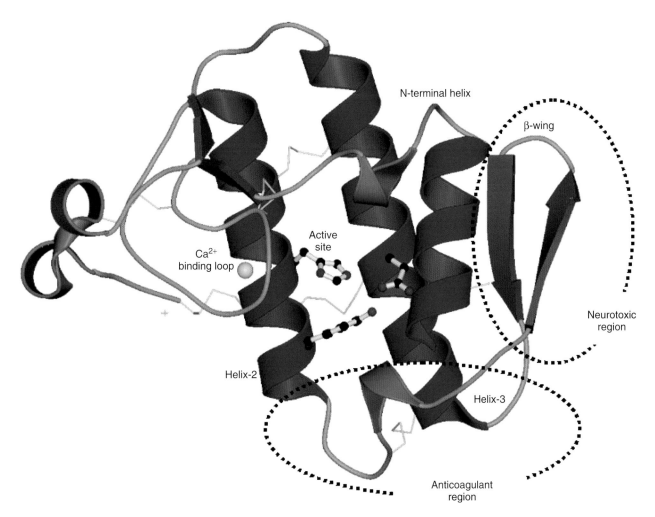

3D Structure 3D structure of the PLA$_2$ from *Agkistrodon halys pallas*, cartoon representation. The PDB code is 1PSJ. The active site residues are included as ball and stick models. The Ca^{2+} binding region and regions coupled with pharmacological functions are labeled. The figure is prepared with the programs MOLSCRIPT[41] and RASTER3D.[42]

lysophospholipids. These enzymes hydrolyze natural long fatty acid chain phospholipids such as phosphatidylcholine, choline plasmilogen, platelet aggregation factor (PAF), and PLs in biological membranes.

Reaction: Phosphatidylcholine + H_2O = 1-acylglycerophosphocholine + fatty acid. Phosphatidylcholine is the major structural phospholipid in the brain and it is also present in egg yolk. The liberated fatty acid residue is linoleic, palmitinic, oleic, or stearic, depending on the source of the substrate.

OCCURRENCE

Phospholipase A_2 enzymes are widely distributed in nature. They are present in high concentrations in snake[1] and insect[2] venoms, mammalian pancreas,[3,4] and synovial fluid.[5] PLA_2s are also present in lung, gastric mucosa, liver, spleen, stomach, alveolar macrophages, intestine, membranes, heart, placenta, and brain,[6] as well as in invertebrate animals (*Drosophila melanogaster*) and plants.[7]

BIOLOGICAL FUNCTION

PLA_2s play an important role in the PL digestion and metabolism, host defense, and signal transduction.[8] These enzymes catalyze the hydrolysis of cell membrane PLs leading to the production of lysophospholipids, which can induce tissue damage, and arachidonic acid – a precursor of eicosanoid mediators of inflammation: prostaglandins, leukotrienes, and tromboxanes.[9] In this way, PLA_2 is involved in chronic inflammatory diseases such as rheumatoid arthritis, asthma, platelet aggregation, and acute hypersensitivity.[10,11] Heparin-binding group IIA and V $sPLA_2$s (secreted PLA_2s) act as 'signaling' enzymes, connected with prostaglandin biosynthesis. Type X $sPLA_2$s are considerably less efficient for the biosynthetic pathway of the eicosanoids mentioned above.[12] Arachidonic acid is involved in superoxide generation and in the release of lysosomal enzyme.[13] Elevated PLA_2 activity has been found in the body fluids of patients with acute pancreatitis[14] and rheumatoid arthritis[15] including culture-positive septic arthritis and traumatized joints,[16] and in bronchoalveolar fluids from patients with adult respiratory distress syndrome[17] and patients with Crohn's disease[18] and from sickle cell patients suffering from acute chest syndrome.[19] In the last case, mainly group II PLA_2 have been identified.[20] PLA_2s are involved in a number of other inflammatory and infectious diseases, such as inflammatory bowel disease,[21] and injuries, especially in patients with abdominal trauma.[22] Protection of cell membranes from PLA_2s is of medical importance and can be used in the treatment of inflammatory processes. A large number of natural or synthetic compounds have been proposed as PLA_2 inhibitors and potential drugs for inflammatory diseases. However, there are important requirements that are not fulfilled in many cases: the compound has to inhibit different $sPLA_2$s, which play a major role in the production of arachidonic acid and do not enter the cell because it can block $cPLA_2$s (cytosolic PLA_2s) participating in the vital PL metabolism. A new approach for the control of inflammatory processes by the inhibition of $sPLA_2$s with cell-impermeable compounds, which do not affect the vital PL metabolism, has been proposed by Yedgar *et al.*[23]

Group IIA PLA_2s are effective antibacterial agents, especially against gram-positive bacteria.[24] The positive charge of the PLA_2 molecule facilitates the penetration into the anionic bacterial cell wall. Mammalian group IIA $sPLA_2$ have an affinity for anionic interfaces and exert a substrate specificity hydrolyzing preferentially anionic PLs, such as phosphatidylglycerol which is a component of the bacterial membranes.[25] At the same time, they bind weakly to the zwitterionic interfaces of the eukaryotic cell membranes.[26] Human group IIA PLA_2s are basic proteins with a pI > 10.5 and many positive charges on the interfacial binding surface.[27] The electrostatic charge is important for their preference for anionic interfaces. The specific hydrolysis of components of the bacterial membranes can explain the antibacterial effect of the mammalian $sPLA_2$s. Probably, these enzymes are part of the antibacterial defense of the body.

Snake venom PLA_2s ($svPLA_2$s) exert a cytotoxic activity that is not connected with catalysis.[25] Thus, Myotoxin II, a catalytically inactive Lys49 PLA_2 from the venom of *Bothrops asper*, displays bactericidal activity.[28] Heparin binds to this protein and inhibits its cytotoxicity. $sPLA_2$s also exhibit antiviral activity blocking the entry of HIV-virus into host cells.[29]

AMINO ACID SEQUENCE AND CLASSIFICATION

The complete or partial sequences of more than 200 PLA_2s have been determined so far. The main parts of these are summarized by Danse *et al.*[6] The predominant sequences are of enzymes from the snake venoms. $svPLA_2$s elaborated a specific scaffold of 118–133 amino acid residues. It is impressive how the relatively small PLA_2 molecule can express such significant diversity of pharmacological activities using this scaffold. A certain number of conserved residues are important for the organization of structural elements, common for these enzymes, and for the expression of the phospholipolytic activity. Specific regions of the polypeptide chains are responsible for the toxic activities.

Up to 15 different PLA_2s have been found in the venom of a single snake species.[7] The sequence homology ranges from 40 to 99% for the enzymes isolated from one species.[1] The PLA_2 isoenzymes in the venom of a single snake exert a variety of pharmacological effects, and it was supposed that these proteins evolved through gene duplication and accelerated evolution.[30] Kini and Chan[30] correlated the substitution rates of Group I/II svPLA$_2$ amino acid residues with their surface accessibility. These authors found that natural substitutions occur predominantly on the surface of the PLA$_2$ molecule: the fast-rate substitutions, which are 40 to 80% of the total number of changed amino acids, include residues on the surface of the globule while the considerably smaller slow rate substitutions are connected with 'buried' residues. These surface substitutions are important for the pharmacological activity because they alter the PLA$_2$ specificity for binding to target membranes, which results in new pharmacological effects.

The minimal level of identity between svPLA$_2$s is about 30%. However, group III enzymes share only 20% sequence homology with the Group I/II PLA$_2$s.[31] There are highly conserved residues and sequences, characteristic of Group I/II PLA$_2$s. They include the active site residues, His48, Asp99, Tyr52, and Tyr73; Ca^{2+}-binding loop (residues 26–34) and Asp49, which is the main ligand for calcium. The residues from the hydrophobic channel/substrate binding site are also well conserved, especially Leu2, Phe5, and Ile9.[32] A novel svPLA$_2$, highly homologous (61% identity) to the acidic PLA$_2$ from King cobra venom, has been isolated from the venom of the Malayan krait *Bungarus candidus*.[33] This enzyme should be classified as a Group IA PLA$_2$ because phospholipolytic enzymes from cobra and krait venoms belong to this subgroup. Surprisingly, the isolated PLA$_2$ contains a pancreatic loop, typical for Group IB mammalian pancreas enzymes. This is the first example of a Group IB PLA$_2$ isolated from krait venom, and it is also a demonstration that monomeric PLA$_2$s from both Group IA and IB can be expressed in the venom of this snake.[33] It was shown that human calcium-independent PLA$_2$ has a 90% sequence identity to iPLA$_2$s from hamster, rat, and mouse.[34] Several isoforms of the enzyme have been identified. Human iPLA$_2$ contains the so-called 'ankyrin' repeats, specific motifs involved in protein–protein interactions. The active form of the Group VI PLA$_2$ is a tetramer[35] and it is possible that these repeats enable the oligomerization of the iPLA$_2$ monomers.[36] However, the sequence of the 'ankyrin'–iPLA$_2$ can function as a negative regulator of the phospholipase A$_2$ activity precluding the formation of enzymatically active oligomers.[34] Recently, the first cDNA sequence of PLA$_2$ from the venom of *Trimeresurus jerdonii* has been reported.[37] This novel toxin was called 'jerdoxin'. It is an Asp49 enzyme and is very similar to viperid PLA$_2$s.

PLA$_2$s form a large superfamily of extracellular (secreted) or intracellular (cytosolic and Ca^{2+}-independent) enzymes. Secreted and cytosolic PLA$_2$s are Ca^{2+}-dependent hydrolases.[35,38] This family is rapidly growing and now includes several hundred members from different origin. Initially, PLA$_2$s were divided into two groups on the basis of their amino acid sequences and the pattern of disulfide bridges.[39] Thus, Group I contains PLA$_2$s from *Elapidae* and *Hydrophiidae* snake venoms, and from mammalian pancreas. PLA$_2$s of this group possess a specific disulfide bridge, Cys11–Cys80 and the so-called 'elapid' or 'pancreatic' loop (residues 55–67). Group IA PLA$_2$s have a specific elapid surface loop and the members of group IB possess a five–amino acid extension of the elapid loop (pancreatic loop). Group II consists of mammalian nonpancreatic PLA$_2$s and related enzymes from *Viperinae* and *Crotalinae* snake venoms. They have a C-terminal extension of 5–7 amino acid residues and cysteinyl residues at the C-terminus, and in position 50. The 'elapid' loop, as well as the two cysteines, Cys11 and Cys80, are missing in the members of Group II. Both groups of enzymes have six conserved disulfide bonds. On the basis of some specific features of the primary structure and the number of disulfide bridges, Group I PLA$_2$s were subdivided into 5 and Group II enzymes into 6 subgroups. The number of isolated and characterized PLA$_2$s increased rapidly and the concept for these enzymes as a short chain 13–15 kDa proteins changed after the discovery of 85-kDa cPLA$_2$ from macrophages.[35] Since then, several articles concerning the classification of PLA$_2$s appeared.[6,38,40] In a recent review,[40] these enzymes were classified into 11 groups, with a total of 23 subgroups, on the basis of four criteria: (i) the enzyme must catalyze the hydrolysis of the *sn*-2 ester bond of a natural phospholipid substrate; (ii) only proteins with known primary structures are included in the respective group; (iii) the members of the group are closely related proteins with a high degree of sequence homology. In the case of more than one PLA$_2$ gene in the species, each paralog is assigned a subgroup letter (e.g. IIA, IIB, IIC etc.); and (iv) catalytically active variants of the same gene are distinguished using Arabic numbers (e.g. VIA-1, VIA-2 etc.). Briefly, according to this classification, Groups I and II include the most abundant 13- to 15-kDa PLA$_2$s from different mammalian organs and fluids as well as from snake (*Viperinae and Crotalinae*) venoms. A representative three-dimensional structure is shown here.

Group III consists of 15- to 18-kDa PLA$_2$s from bee, lizard, and scorpion venoms. Their amino acid sequences differ considerably from those of the Group I and II enzymes; Group IV is formed by high molecular mass (61–114 kDa), cytosolic, phospholipolytic enzymes from human pancreas, liver, brain, heart, skeletal muscles, and rat kidney; Group V incorporates 14-kDa PLA$_2$s from mammalian heart and lung, and macrophages; Group VI contains 85-kDa iPLA$_2$s from macrophages, human B-lymphocytes, heart, and skeletal muscles; Group VII includes 45-kDa phospholipase A$_2$s from mammalian plasma; Group VIII contains 26-kDa enzymes from human brain; Group IX has at present only one member, a 14-kDa PLA$_2$ from the marine snail *Conus magus*; Group X contains 14-kDa enzymes from human spleen, thymus, and leukocytes, and finally, Group XI contains 12- to 13-kDa PLA$_2$s from plants. Groups I, II, III, V, IX, X, and XI

PLA$_2$s have a catalytic histidine in their active sites. The enzymes from groups IV, VI, VII, and VIII contain an active site serine, as serine proteinases. PLA$_2$s from groups I, II, V, and X are structurally related and have six conserved disulfide bonds, plus one to two additional S–S bridges.[40] For more details, see reference 40. In view of the increasing number of new PLA$_2$s from different sources, an improved classification and definition of novel groups and subgroups can be expected in the future.

PROTEIN PRODUCTION

PLA$_2$s are present in many organs and fluids of mammals and invertebrates. The vast majority of these enzymes have been isolated as extracellular proteins from mammalian pancreas, fluids, and snake, or bee venoms. Porcine[43] and bovine[44] pancreatic PLA$_2$s were cloned and overexpressed in *Escherichia coli*. Mammalian PLA$_2$ genes comprise four exons and three introns (group I PLA$_2$s),[45] or five exons and four introns (group II PLA$_2$s).[46] The snake venom glands synthesize large amounts of toxic phospholipolytic enzymes. In *Viperinae* snakes, PLA$_2$ genes form multigene families organized in five exons and four introns resulting from gene duplication.[6,47] In these multigene families, the introns are highly conserved. In contrast, there is no similarity between intron sequences of human and rat PLA$_2$ genes.[47]

It was shown[48] that the structural organization of both *Vipera ammodytes ammodytes* and *Vipera palestinae* PLA$_2$ multigene families is the same: five exons are separated by four introns. Genomic analysis demonstrated that the sequences of introns and the 5′ and 3′ flanking regions of all *Viperidae* PLA$_2$ genes are highly conserved. While exons 1, 2, 4, and 5 in the *V. palestinae* PLA$_2$ genes are highly homologous (the identity is more than 89%), the third exon shows a considerably lower degree of sequence identity – 71%. This means that a positive Darwinian selection is limited to exon 3, in contrast to other *Viperidae* PLA$_2$ genes. The *V. palestinae* PLA$_2$ genes are products of duplication and divergence of a single ancestral gene.[48] Recently, the genomic DNA sequences encoding venom PLA$_2$s from four *Viperinae* snakes, *Vipera aspis aspis*, *Vipera aspis zinnikeri*, *Vipera berus berus*, and a neurotoxic *Vipera aspis aspis* have been determined.[49] The authors sequenced five groups of genes, each corresponding to a different PLA$_2$. Single nucleotide differences between genes encoding the same PLA$_2$ as well as between genes encoding different PLA$_2$s have been observed in exons 3 and 5. Genes encoding ammodytins I1 and I2, described previously in *Vipera ammodytes ammodytes*, are also present in *Vipera aspis aspis* and *Vipera berus berus*. These data suggest that the population of neurotoxic *Vipera aspis aspis* snakes may have resulted from interbreeding between *Vipera aspis aspis* and *Vipera a. ammodytes*.

10 cDNAs encoding distinct PLA$_2$s have been cloned from the venom glands of *Callosellasma rhodostoma* (Malayan pitviper).[50] Two cDNAs result from the recombination of Lys49 and Asp49 PLA$_2$ genes. A basic PLA$_2$ homologue contains, unusual for the snake venom Lys49 PLA$_2$s, substitutions.

Purification of PLA$_2$s usually includes several chromatographic steps. The lyophilized crude material is dissolved in buffer and gel filtered. Further purification is achieved by FPL or HPL chromatography using highly effective columns (Mono S, Mono Q, RP-C8 etc.) and gradient systems.[37,51] PLA$_2$ has also been purified by direct HPL chromatography of a snake venom.[52]

PHARMACOLOGICAL ACTIVITIES

Snake venom PLA$_2$s display a number of pharmacological activities: neurotoxicity, cardiotoxicity, myotoxicity, antiplatelet, hypotensive, hemolytic, hemorrhagic, coagulant, anticoagulant, and edema-inducing effects.[53] These enzymes can exert their neurotoxicity through two different mechanisms:[54,55] (i) blockade of the transmission across the neuromuscular junctions of the breathing muscles (presynaptic toxins) or (ii) prevention of the binding of acetylcholine to its receptor (postsynaptic toxins). svPLA$_2$s affect hemostasis, a physiological process that includes clot formation and dissolution. Platelet activation is an important moment of this process and it is initiated by thrombin and arachidonic acid. sPLA$_2$s can inhibit platelet aggregation (anticoagulant activity) hydrolyzing PLs that act as cofactors of the prothrombinase complexes[56] that produce thrombin, an enzyme of critical importance for coagulation. On the other hand, some other sPLA$_2$s can hydrolyze PLs of the platelet membrane, releasing arachidonic acid, and/or platelet aggregation factor. In this way, they can induce aggregation.[57]

Basic PLA$_2$s with myotoxic activity have been isolated from the venom of *Crotalidae* snakes.[58] Myotoxins are proteins that damage the cells of skeletal muscle fibers.[59] Two isoforms have been isolated: 'Asp49' and 'Lys49' PLA$_2$s, the second one being catalytically inactive. Most probably, the two types of group II sPLA$_2$s exert their myotoxicity through different mechanisms.[60] iPLA$_2$s belong to the Groups VIA (mainly), IVC, VII, and VIII PLA$_2$s. Group VIA iPLA$_2$ were found in human tissues, especially in rat brain, and their major function is the phospholipid/membrane 'remodeling'.[36] In contrast to the sPLA$_2$s, these 85- to 88-kDa enzymes possess a catalytic serine.[61] Group VII and VIII iPLA$_2$s are PAF-acyl hydrolases and have a substrate specificity toward the platelet aggregation factor, which is an inflammatory mediator. iPLA$_2$s are also involved in some human diseases like myocardial ischemia,[62] and in phagocytosis.[63]

PHARMACOLOGICAL SITES

There is considerable progress in the localization of the pharmacologically important sites in svPLA$_2$s. Nevertheless, many aspects of the structure–pharmacological activity relationships in this group of enzymes remain unclear. Experimental results from different laboratories demonstrated that the toxic effects cannot be explained with the enzymatic activity only and that the catalytic and pharmacological sites do not overlap. Catalytically active svPLA$_2$s can be detoxified and enzymatically inactive representatives of this subfamily can induce pharmacological effects as myotoxicity, cardiotoxicity, and cytotoxicity. Not all svPLA$_2$s exert the whole spectrum of toxic effects and a single protein can exhibit different toxicities. The mechanism by which svPLA$_2$s exert their pharmacological effects can be dependent or independent of the phospholipid-hydrolyzing activity and includes binding of the enzyme to membrane phospholipids and/or to specific proteins on the cell surface.[64] The location of the pharmacological sites (indicated also in the 3D Structure) was studied by chemical modification of amino acid residues, specific antibodies, and comparison of structurally related toxic and nontoxic enzymes. Comparison of the sequences of three ammodytoxins from the venom of *Vipera ammodytes ammodytes*, which are structurally very similar but differ considerably in neurotoxicity, showed that the region between Tyr115 and Lys128 might be responsible for this type of pharmacological activity.[65] Immunological experiments with ammodytoxin A revealed the regions between residues 106–113 and 113–121 as sites of neurotoxicity.[66] The specific substructure called 'β-wing' (residues 76–81) as well as the C-terminal segment 119 to 125 and residues 6 and 12 have been identified as neurotoxic sites in the toxic PLA$_2$ (RV4) of the heterodimeric toxin from the venom of *Vipera russelli formosensis*.[67]

The region 115 to 129 of myotoxin II, a catalytically inactive Lys49 PLA$_2$ from the venom of *B. asper*, has been identified as the site responsible for the myotoxic, cytotoxic, edema-forming,[68] and bactericidal[69] activities. The antigenic structure of the region mentioned above is conserved among class II myotoxic PLA$_2$s.[69] The substitution of the three tyrosines in the peptide 115 to 129 by tryptophyl residues results in a drastic enhancement of its membrane-damaging activity. This peptide exerts myotoxicity, cytotoxicity, bactericidal action, and edema-forming activity. The substituted peptide reproduces the toxic effects of myotoxin II.[70] Synthetic peptide with a sequence identical to that of the 115 to 119 region of the Lys49 PLA$_2$ from the venom of *Agkistrodon piscivorus piscivorus* reproduces the enzyme's myotoxic effect *in vivo*.[60]

Asp23 and Arg30 participate in the phosphatidylcholine activator site of the group IA PLA$_2$ from the *Naja naja naja* (cobra) venom.[71,72] A 10- to 20-fold increase of the hydrolysis of phosphatidylethanolamine in the presence of phosphocholine-containing lipids has been observed.[73] Basic svPLA$_2$s exert strong hemolytic effect while neutral and acidic isoenzymes are devoid of such activity. A positively charged C-terminal 'face', including residues Lys111, Lys114, Lys115, Lys129, and Lys132 has been supposed to be responsible for the hemolytic activity of the basic PLA$_2$ from the venom of *Agkistrodon halys Pallas*.[74] Other positively charged region, located at the N-terminus of the PLA$_2$ polypeptide chain, has been proposed to be important for the degradation of bacterial phospholipids; that is, to be involved in the bactericidal activity.[75] The region consisting of residues 54–77 has been identified as an 'anticoagulant' site, which binds to anionic phospholipids.[76] Residues at positions 53 and 70 were also proposed to be important for this pharmacological activity.[74] It is not clear if there is a correlation between the anticoagulant and enzymatic activities of PLA$_2$s: the experimental data cannot answer this question definitely and the data published so far lead to controversial conclusions.

PLA$_2$ RECEPTORS AND OTHER PLA$_2$-BINDING PROTEINS

The absence of correlation between the pharmacological and enzymatic activities of svPLA$_2$s led to the suggestion that these hydrolases can bind to specific proteins in target tissues.[77] A number of membrane and soluble proteins have been identified as binding receptors for venom and mammalian sPLA$_2$s.[7] The isolation and characterization of these proteins demonstrated that PLA$_2$s exert their pharmacological activities through specific protein–protein interactions. Two structurally and pharmacologically distinct types of receptors, N-type and M-type, have been identified using radiolabeled svPLA$_2$s.[78–80] The N-type receptors are 18- to 85- kDa proteins present in brain membranes; they have a high affinity for neurotoxic sPLA$_2$s but low affinity for the nontoxic counterparts of these enzymes. This suggests that the receptors are involved in the PLA$_2$ neurotoxicity.[78,81,82] The M-type receptors are 180- to 200-kDa proteins found in various cell types, including skeletal muscle cells,[79] Swiss 3T3 and rat vascular cells.[83,84] In contrast to the N-type receptors, they bind with higher-affinity nontoxic sPLA$_2$s. M-type receptors have been cloned and their structure investigated.[80,85,86] The results demonstrated that they belong to the C-type lectins. Also, it was shown that amino acid residues, close to or from the Ca^{2+}-binding loop of sPLA$_2$, participate in the binding to the receptor,[87] which lead to inhibition of the enzyme's catalytic activity.[88] The carbohydrate moiety of an M-type glycoprotein receptor for porcine group IB sPLA$_2$ has been shown to be important for the binding to the enzyme.[89,90] Binding of the enzyme to its receptor initiates the production of inflammatory cytokines that leads to development of the endotoxic shock, a process connected with cell damage, tissue necrosis, and vascular disruption.[89] Both N- and M-type PLA$_2$ receptors are expressed in lung, liver, heart, and kidney.[7] The presence of receptors for venom PLA$_2$s in mammals raises the possibility that there are mammalian sPLA$_2$s, which can bind to venom PLA$_2$ receptors.[64] PLA$_2$ inhibitors (PLI) from the snake plasma can be grouped into three classes: (i) inhibitors with a carbohydrate recognition domain (CRD) of C-type lectins; (ii) PLI having a urokinase-type plasminogen activator receptor (U-PAR) like domain and (iii) leucine-rich α_2 glycoprotein (LRG) like domain that could be responsible for binding to basic PLA$_2$.[91,92] The C-type lectin-like venom proteins are particularly interesting because they affect the hemostatic system. The members of this family show a high-sequence homology to the animal C-type lectins, including the region of CRD.[93] The C-type lectin-like venom proteins are structurally homologous but functionally distinct and can have different effects on blood coagulation and platelet aggregation.[94] Recently, two representatives of the protein family mentioned above have been purified and characterized from the *Agkistrodon actus*[95] and *Trimeresurus mucrosquamatos*[94] venom. The first protein, named *Aa*ACP, has an anticoagulant activity inhibiting the factor Xa-induced plasma coagulation and most probably binds to the factor Xa in the prothrombinase complex. The second, called *TMVA*, is also a novel C-type lectin-like protein, which binds to platelet membrane proteins inducing platelet aggregation in a dose-dependent manner. Both are two-chain proteins, with a similar length of the polypeptide chain. PLA$_2$ inhibitor has been isolated from the serum of the nonvenomous snake *Python reticulatus*, termed PIP. This inhibitor binds the PLA$_2$ toxin of *Daboia russelli siamensis* in almost equimolar ratio and neutralizes the toxicity of various snake venoms and toxins and prevents the formation of edema in mice.[91] PIP contains a proline-rich cluster, which is absent in u-PAR-related proteins. This cluster may be involved in the binding of the inhibitor to PLA$_2$. Cytosolic PLA$_2\alpha$ is an enzyme expressed in most mammalian cells. It plays a critical role in the lipid mediator synthesis and regulates the arachidonic acid metabolism.[96] It was shown that cPLA$_2\alpha$ binds to vimetin, the major component of the intermediate filament.[97] This interaction is important for the cPLA$_2\alpha$ function during the process of the eicosanoid biosynthesis.

CATALYTIC MECHANISM

Initially, a mechanism of PLA$_2$ catalyzed reactions, similar to that of serine proteinases, has been proposed.[98] According to this mechanism, a water molecule, as a nucleophile, attacks the scissile bond. Later on, complexes of secreted Group I (Cobra venom;[99] PDB code: 1POB), Group II (human nonpancreatic;[5] PDB code: 1POE), and Group III (bee venom;[2] PDB code: 1POC) PLA$_2$s with the transition-state analogue L-1-O-octyl-2-heptylphosphonyl-*sn*-glycero-3-phosphoethanolamine [diC$_8$(2Ph)PE] as well as the complex between a porcine PLA$_2$ mutant and a substrate-derived analogue[100] (PDB code: 5P2P) have been used to study the mechanism of substrate binding and catalysis by X-ray crystallography. The transition-state analogue has been designed to simulate the tetrahedral intermediate formed during the hydrolysis of L-1,2-dioctanyl-*sn*-3-phosphatidyl ethanolamine.[5] The crystallographic models revealed specific interactions of the substrate analogue with the active site residues, Ca^{2+}, and conserved water molecules. It became clear that catalysis by PLA$_2$ is similar to that by serine proteinases, with some exceptions: absence of acyl-enzyme intermediate, differences in the 'catalytic' hydrogen bonds network, which in PLA$_2$ includes the phenolic hydroxyls of the conserved in the Group I/II enzymes Tyr52 and Tyr73 (in the bee-venom enzyme only the OH group of Tyr87, corresponding to the hydroxyl group of Tyr52, participates in the network) and the hydrogen-bonding pattern of the active site histidine.[31] The X-ray data[99] show that the acyl part of the *sn*-1 and *sn*-2 substituents of the inhibitor/transition state analogue lie in a hydrophobic channel made by the side chains of Ile9, Phe5, Leu2, Trp19, and Tyr69. Invariant or highly conserved amino acid residues form the channel.

Calcium plays an important role in the catalysis by secreted PLA$_2$s. It is essential for both the binding of the substrate and the proceeding of the catalytic process (Figure 1). The metal ion is hepta-coordinated in a pentagonal bipyramidal cage at the 'primary' binding site. The ligands forming the coordination sphere are the two carboxylate oxygens of Asp49, the backbone carbonyl oxygens of residues 28, 30, and 32 and two water molecules. The Ca^{2+} to ligand distances are between 2.29 and 2.69 Å[3] (Table 1; PDB code: 1MKT). At the active site of class I/II PLA$_2$s there is a 'catalytic network' including a hydrogen bond between the couple His48 and Asp99. A water molecule is hydrogen-bonded to the imidazole of His48.

The conserved Tyr52 and Tyr73 also participate in this network and their phenolic hydroxyls are hydrogen-bonded to OD1 and OD2 carboxylic atoms of Asp99 respectively (Figure 2). Abstraction of a proton from the

Table 1 Ligand to calcium distances in pancreatic bovine phospholipase A$_2^3$ (PDB code: 1MKT)

Coordination bond	Distance (Å)
Ca^{2+} – Tyr28 O	2.29
Ca^{2+} – Gly30 O	2.47
Ca^{2+} – Gly32 O	2.30
Ca^{2+} – Asp49 OD1	2.69
Ca^{2+} – Asp49 OD2	2.48
Ca^{2+} – Water1	2.43
Ca^{2+} – Water2	2.66

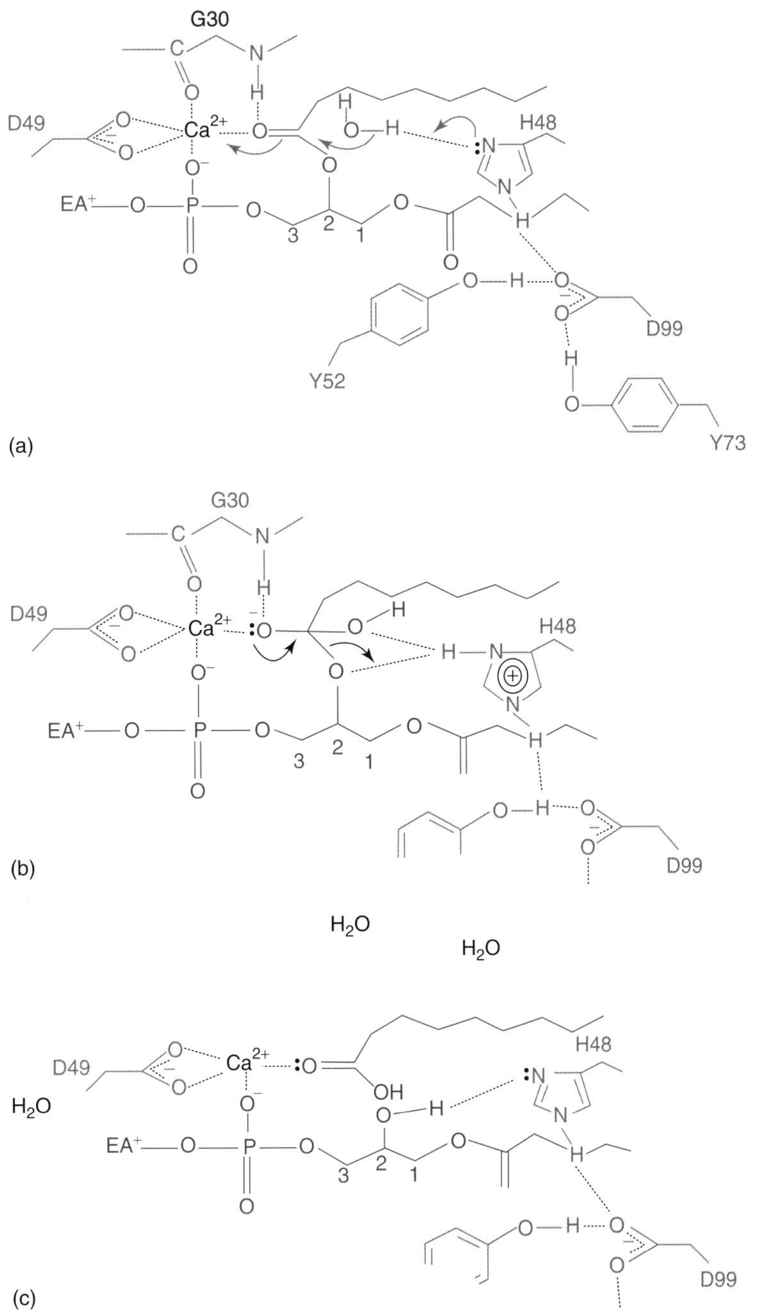

Figure 1 Scheme of the catalytic mechanism as proposed by Scott et al.[99,101] (a) showing the substrate bound in a productive mode; (b) the tetrahedral intermediate, and (c) the products in the active site to be replaced by water molecules after the product diffused out.

water molecule by His48 will initiate a nucleophilic attack on the scissile bond. The interaction between Asp99 and His48 will neutralize the positive charge on the imidazole group of the last residue. In the complex of the bee-venom PLA$_2$ with the transition-state analogue[2] (PDB code: 1POC), an oxygen atom of the phosphonate is hydrogen-bonded with the protonated ND1 of His48. The position of this oxygen atom is the same as that of the O atom of the fixed water molecule, which can act as the attacking nucleophile during catalysis. Comparison of the X-ray structures of the uninhibited and inhibited by a transition-state analogue PLA$_2$ from *Naja naja atra* (PDB code: 1POB) showed that the two water molecules, which are associated with the bound Ca^{2+} in the native enzyme, are displaced by the oxyanion of the substrate's tetrahedral intermediate and the nonbridging oxygen of the sn-3 phosphate.[101]

The oxyanion of the tetrahedral intermediate is stabilized by the calcium ion and the backbone N–H of Gly30. The nucleophile attacking the scissile bond is an 'activated'

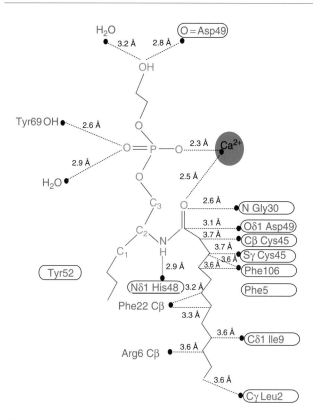

Figure 2 A scheme for the interaction of the substrate-derived inhibitor (R)-2-dodecanoyl-amino-1-hexanol-phosphoglycol with Ca^{2+} and the active site amino acid residues of porcine pancreatic PLA_2 mutant lacking the 'pancreatic loop' (residues 62–66), as proposed by Thunnissen et al.[100] (PDB code: 5P2P).

water molecule from which His48 extracts a proton. The extracted proton is subsequently transferred to the alkoxide-leaving group.

The vast majority of PLA_2s are Ca^{2+}-dependent enzymes. However, in several myotoxic PLA_2s from *Viperidae* snakes (species *Agkistrodon, Bothrops,* and *Trimeresurus*) the major ligand for Ca^{2+} binding, Asp49, is substituted by Lys49 (Figure 3).[102–105] The substitution makes the binding of Ca^{2+} to Lys49-PLA_2s impossible; these enzymes are catalytically inactive or their phospholipolytic activity is greatly reduced. It was shown[106] that the ε-NH_2 of Lys49 occupies the same site as the calcium cofactor. Nevertheless, the Lys49 PLA_2 from the venom of *Cerrophidion (Bothrops) godmani* damages membranes by a Ca^{2+}-independent mechanism.[107]

The crystal structure of a complex between Piratoxin II from the venom of *Bothrops pirajai* and a fatty acid demonstrated that Lys49 PLA_2 is the active enzyme in which the catalysis is interrupted at the stage of substrate release[109] (PDB: 1QLL). The apparent lack of activity of this enzyme has been explained by a failure of a product release.

INTERFACIAL CATALYSIS BY PLA_2

The activity of secreted PLA_2s toward micelles, membranes, and vesicles is several times higher than that on monomolecular dispersed substrates.[110] This phenomenon is known as 'interfacial activation'.[111] It was supposed that the lipolytic enzymes are activated after the adsorption at the water–lipid interface.[112,113] Further investigations on the enormously increasing number of PLA_2s from different origins demonstrated that the 'interfacial activation' is a characteristic feature of these enzymes. During the last decade, the mechanism of interfacial catalysis was investigated thoroughly. PLA_2s are 'interfacial' enzymes because they act on substrates at an organized interface to which they are associated. Catalysis by PLA_2 involves an initial binding of the enzyme to the lipid–water interface followed by the catalytic step. The efficiency of catalysis depends on the adsorption of the enzyme on the interface and the environment of the substrate. Electrostatic and nonelectrostatic forces are important for the binding, especially the interactions between positively charged groups from the enzyme's interfacial adsorption surface (IAS) and negatively charged phospholipids.[114] All types of aromatic residues, tryptophan, tyrosine, and phenylalanine are also important for the interfacial binding, especially in binding to zwitterionic membranes. Thus, Trp19, Trp61, and Phe64 of the secreted *Naja n. atra* PLA_2 are involved in membrane penetration, the last residue being fully inserted into the membrane during the interfacial binding, while the other two side chains are partially inserted.[115] Trp67 of mammalian $sPLA_2$ significantly contributes to the interfacial binding to zwitterionic vesicles.[116] However, it is difficult to obtain direct information about structural changes during the catalysis.[31]

The X-ray models of complexes between $sPLA_2$s and a transition-state analogue[99,101] (PDB codes: 1POB and 1POC) revealed an area enveloping the opening of the hydrophobic channel that forms the so-called 'interfacial adsorption surface'. This area consists of highly conserved hydrophobic residues, which should be 'buried' in the membrane to permit a transfer of the substrate and products of the reaction. IAS is responsible for the adsorption of the enzyme to the aggregated substrates and is also important for the efficiency of the interfacial catalysis proceeding at lipid–water interfaces. In the Group I/II PLA_2s, IAS includes the first two turns of the N-terminal helix, the side chain of the residue in position 19 and part of the Ca^{2+}-binding site. Some other residues can also participate and support the binding. Thus, the cationic side chains of Lys53 and Lys56 are important for the activation of bovine pancreatic PLA_2 during the interaction with anionic interface.[117] Mammalian (mouse and rat) $sPLA_2$s bind to cell surfaces via several positively charged residues located at the C-terminus of the enzyme polypeptide chain. Lys119 and Lys120 are especially important for the association.[118] For more detailed information about the interfacial catalysis by PLA_2, see references 31, 111, 119.

Phospholipase A$_2$

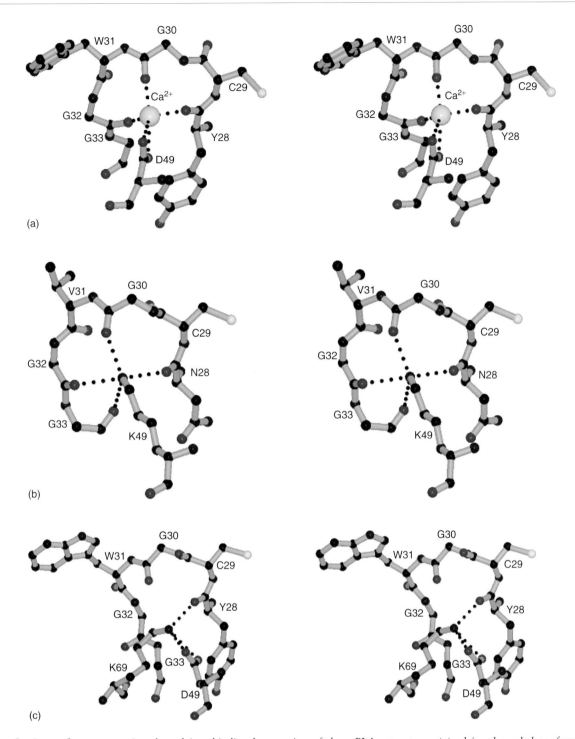

Figure 3 Stereo figure comparing the calcium binding loop region of three PLA$_2$ structures: (a) calcium-bounded conformation of bovine pancreatic PLA$_2^3$ (PDB code: 1MKT); (b) calcium-free conformation of the Vipoxin PLA$_2$[108,137] with the side chain of Lys69 of the Vipoxin inhibitor bound in the Ca^{2+} loop (PDB code: 1JLT), and (c) a Lys49 PLA$_2$ from cottonmouth snake[104,106] (PDB code: 1PPA). The direct interactions between the amino acids of the loop and the cations, Ca^{2+} or the ε-amino group in (a) and (c), are indicated as dotted lines.

MULTICHAIN PLA$_2$s

Multichain PLA$_2$s are complexes of several subunits in which at least one subunit possesses phospholipolytic activity. The number of subunits varies up to five and they are usually associated by noncovalent interactions. Exceptions to this rule are bungarotoxins in which the phospholipase A$_2$ subunit is linked by a disulfide bridge to a Kunitz-type proteinase inhibitor. Two types of bungarotoxins have been identified: α-type, which is a postsynaptic toxin blocking acetylcholine receptor on membranes of neuromuscular junctions, and β-neurotoxins, which bind to the presynaptic membranes and block the neurotransmitter release. Different isoforms of the β-bungarotoxin have been isolated and sequenced.[120] All β-bungarotoxins consist of a subunit of 120 amino acid residues (A-chain) and a considerably shorter polypeptide of 60 residues (B-chain), which are cross-linked by a disulfide bond. A-chains possess PLA_2 activity and are neurotoxic. B-chains are structurally homologous to mammalian proteinase inhibitors. However, these analogues of Kunitz-type inhibitors have no inhibitory effect on proteolytic enzymes.[121] A number of combinations of identical PLA_2s and different B-chains have been observed. For example, $β_1$- and $β_2$-bungarotoxins have identical toxic subunits but different B-chains.[121] The Kunitz module plays an important physiological role 'targeting' the PLA_2 subunit to the presynaptic membrane. The nonenzymatic subunit binds to the K+ channel receptor through a basic binding surface, which has been identified in $β_2$-bungarotoxin.[122] The binding of the Kunitz domain to the receptor should be important for the suitable orientation of the PLA_2 subunit toward the membrane and for the further phospholipolytic action of the enzyme because the PLA_2 active site and the ion channel binding region are located on the same site of the neurotoxin.[121] In this way, a nonspecific binding of the toxic enzyme is avoided and the phospholipolytic hydrolysis leads to a blockade of the neural transmission.

The dimeric svPLA_2s consist of two different polypeptide chains, and in rare cases they are homodimers.[123] Crotoxin from *Crotalus atrox* is an intensively studied representative of the heterodimeric snake venom toxins. It consists of two noncovalently bound polypeptide chains – one of them is toxic and the other is devoid of toxicity. The nontoxic subunit plays an important physiological role – it considerably potentiates the neurotoxic action of the pharmacologically active component. The complex is very stable in solution but dissociates in the presence of synaptic membranes. Only the toxic subunit binds to the membrane and the other remains in solution. The last subunit acts as a 'chaperon' preventing a nonspecific binding of the toxic chain.[123,124] The amino-terminal region of the *C. atrox* PLA_2 is important for catalysis and stabilization of the dimer in solution.[125]

An example of a 'synergistic' toxicity is the two-component toxin isolated from the venom of *V. palestinae*.[126] It consists of an acidic PLA_2 and enzymatically inactive basic protein for which there is no indication of forming a complex. The individual components are not toxic. However, the equimolar mixture of the two proteins is lethal. The mechanism of physiological action is not clear but is probably not neurotoxic.[126] The acidic component shares a high degree of sequence homology with acidic svPLA_2s or related proteins, forming stable neurotoxic complexes with basic phospholipolytic enzymes.

Several heterodimeric neurotoxic complexes have been isolated from the venom of *Viperinae* snakes inhabiting remote parts of the world: Vipoxin from the venom of the sand viper, *Vipera a. meridionalis*[127] (the central part of the Balkan peninsula, southeastern Europe), the complex RV-4/RV-7 from the Taiwan viper *Vipera russelli formosensis*[128] (Asia), and the toxin from *Vipera aspis zinnikeri*[129] (southwestern France). Surprisingly, all these neurotoxins, produced by the toxic glands of snakes living in different continents, were found to be closely related proteins. Each complex is formed by two oppositely charged, basic and acidic noncovalently bound 14-kDa subunits of the same length of the polypeptide chain, 122 residues (Figure 4).

The sequence homology between the respective subunits (basic-to-basic or acidic-to-acidic chains) is impressive: 92% for Vipoxin-RV-4/RV-7 and 97 to 98% for the *Vipera a. zinnikeri* complex-Vipoxin. The degree of identity between the basic and acidic components of each heterodimer is 62 to 65%. These dimeric complexes are very stable and can be dissociated only under highly acidic conditions and in the presence of a high concentration of urea. The complexes are formed through ionic, hydrophobic, and hydrogen bond interactions. The basic components of the neurotoxins are toxic Group IIA PLA_2 enzymes. The acidic subunit of Vipoxin, called 'inhibitor' (Inh), and that from the *Vipera a. zinnikeri* toxin, are PLA_2-like enzymatically inactive proteins. In both cases, the reason for the absence of enzymatic activity is the substitution of the catalytic His48 by glutamine.[129,131] The acidic component RV-7 of the neurotoxin from the venom of *Vipera r. formosensis* preserves the active site His48 and possesses a weak enzymatic activity – 100-fold lower than that of RV-4.[128] The complex formation reduces the phospholipolytic activity of the toxic Vipoxin PLA_2 and RV-4 from *Vipera r. formosensis* up to 40 to 60%. In both neurotoxins, the acidic subunits are nontoxic and act as natural

Figure 4 Surface potential of the complex Vipoxin with an underlying Cα-worm. The figure is prepared using the program GRASP.[130] The active PLA_2 is on the right and the Inh on the left side. The arrows are pointing toward the β-wing and the color code is according to the surface potential, with values indicated by the crossbar. The electrostatic differences between the two protein chains are obvious, especially in the region of the proposed toxic β-wing site.

inhibitors. However, in the case of RV-4/RV-7, the acidic component potentiates the neurotoxicity, while its counterpart in Vipoxin considerably reduces the pharmacological activity of the respective PLA$_2$.[128,132] The physiological function of RV-7 and that of the Vipoxin inhibitor is not clear. It can be supposed that they are important for the toxin–target membrane interaction and act as 'chaperons' preventing a nonspecific binding of the respective toxin to sites other than synaptic membranes. The two neurotoxic complexes from the venom of *Vipera a. meridionalis* and *Vipera r. formosensis* are examples of a modulation of the PLA$_2$ activity, which is of pharmacological interest and can be used for medicinal purposes.

X-RAY STRUCTURE

X-ray crystallography contributes enormously to the understanding of structure–function relationships in PLA$_2$s, especially for the enzyme–substrate complex formation, structure and mechanism of action of the active site, interactions with effectors (metal ions and low molecular mass substances), and the organization of pharmacological sites. These studies made it possible to evaluate the importance of the phospholipolytic enzymes for human diseases and revealed perspectives for structure-based drug design. The X-ray models revealed the hydrophobic nature of the substrate-binding 'channel', the organization of the catalytic site and the IAS, important for the PLA$_2$ binding to membranes, and for the interfacial catalysis. The first crystal structure of monomeric PLA$_2$, that from bovine pancreas, was published in 1978.[133] This structure provides important information about the shape and dimensions of the molecule as well as for the stereochemistry of the catalytic and metal-binding site, and for the interfacial surface, responsible for binding to the micelles and other aggregated substrates. Since that time, the X-ray structures of an increasing number of monomeric and dimeric PLA$_2$s have been determined[5,99,108,122,134–141] and the atomic coordinates are available from the Protein Data Bank. These studies revealed a conservation of structural elements, important for the function of the Group I/II PLA$_2$s: the N-terminal 12–14 residues α-helix, a pair of two nearly parallel α-helices, connected with two disulfide bridges, a β-wing structure (residues 74–85), which represents a short antiparallel β-sheet, and a loop (residues 25–33) binding Ca^{2+}. The metal ion is hepta-coordinated as it was described before. The α-helical content is approximately 50% of the whole polypeptide chain and the β-structure is only 10%. The active site residues are located on the two long α-helices, which are fixed by disulfide bridges in a distance of 10 Å, as shown in the 3D structure (PDB code: 1PSJ). This relatively 'rigid' structure ensures the conformational stability of the catalytic site. The X-ray studies showed that the Group I/II PLA$_2$s are 'rigid' enzymes, stabilized by seven disulfide bridges.

Comparison of the crystallographic structure of the Group I PLA$_2$ from bovine pancreas with that of its counterpart from the *Crotalus atrox* Group II enzyme[137] (PDB code: 1PP2) demonstrated a strong similarity of the backbone conformation, especially in the 'core' structure, formed by the three α-helices. The homologous segments, including 82 residues or 62% of the whole polypeptide chain, have been superimposed with an rms distance between the corresponding C$_\alpha$ atoms less than 1 Å.[137] The catalytic site and its hydrogen-bonding network are absolutely conserved. However, there are regions that differ in their backbone conformations. The β-wing of the Group II enzyme is more 'open' because the disulfide bridge Cys11–Cys77, connecting the β-structure to the 'core' region in the pancreatic Group I enzyme, is absent in its counterpart from the *C. atrox* venom. The crotaline venom enzyme also lacks the pancreatic/elapid loop typical for the mammalian/elapid venom Ca^{2+}-dependent PLA$_2$s, which creates significant conformational differences in the respective part of the polypeptide chain. The role of the 'pancreatic' loop has been studied by a combination of protein engineering and X-ray crystallography.[138] The removal of this surface loop from porcine pancreatic PLA$_2$ (PDB code: 5P2P) considerably enhances the activity toward zwitterionic micelles but decreases the activity on negatively charged substrates. Comparison of the crystallographic structures of the wild-type and mutant enzymes showed that the structure of the active site is preserved but structural changes in the region of the deletion have been observed. The deletion of the loop affects the binding site for aggregated substrates.[138] The bee-venom Group III PLA$_2$ preserves the functional substructures of the Group I/II enzymes but the overall architecture is different[2] (PDB code: 1POC). Also, the N-terminal part of the Group III enzyme does not form an α-helix and is not connected to the active site as it is in the Group I/II PLA$_2$s. The geometry of the Ca^{2+}-binding site is conserved among Group I/II/III enzymes.[2]

As noted in a preceding paragraph, Vipoxin is a representative of a group of heterodimeric neurotoxins found in the venom of *Viperidae* snakes inhabiting remote parts of the world. Characteristic feature of these toxins is that the acidic, nontoxic, and catalytically inactive subunit is highly homologous to the strongly toxic and enzymatically active basic PLA$_2$ of the complex. Usually, the acidic subunit partially inhibits the basic PLA$_2$. Most probably, the nontoxic protein is a product of evolution of the toxic enzyme. In this respect, the Vipoxin-type neurotoxins are examples of transformation of the toxic and enzymatic function into nontoxic and inhibitory ones. The modulation of the PLA$_2$ function is of great pharmacological interest.[140] The X-ray structure of Vipoxin, shown in Figure 3, has been determined at 2.0 Å[140] and refined to 1.4 Å[108] resolution (PDB code: 1JLT). The 3D model revealed that the complex formation, which results in a reduction of the enzymatic activity and toxicity of the basic PLA$_2$, is connected with a decrease of the accessible surface area of both subunits by ~1480 Å.^{2} The model explains the inhibitory effect of the acidic subunit (Inh), which partially shields the entrance to the active site of PLA$_2$. The complex is formed through ionic, hydrophobic, and hydrogen bond interactions. The two subunits, with a pI of 10.4 and 4.6, respectively, drastically differ in their charge. This creates a strong electrostatic attraction between the basic and acidic components. The two tryptophans of PLA$_2$ and the single indole group of Inh are located in the intersubunit region and participate in hydrophobic interactions. The complex is stabilized by interchain hydrogen bonds and a salt bridge between the major ligand of Ca^{2+}, Asp49 of the toxic enzyme, and Lys69 of the acidic subunit. In this case, the ε-NH$_3^+$ of Inh plays the role of the metal ion for neutralization of the Asp49 carboxylic group and stabilization of the Ca^{2+}-binding loop. This bond is observed in the absence of bound Ca^{2+} and probably represents a mechanism for self-stabilization of the neurotoxin when there is no calcium at the metal ion binding site. The X-ray data showed that the folding of the two Vipoxin subunits is very similar and that their polypeptide chains were superimposed with an rms difference in the C$_\alpha$ positions of 1.5 Å. Larger differences were observed in the surface region including residues 16–20, the β-wing, and C-terminus. The last two regions are involved in the putative toxicity site of the neurotoxic subunit. The hydrophobic channel is preserved in the acidic subunit and it can be supposed that fatty acid chain can occupy the substrate-binding site during the interaction with substrates or their analogues. However, the catalytic step cannot be realized because of the substitution of the active site His48 of PLA$_2$ by Gln48 in Inh, which results in the destruction of the catalytic machinery. X-ray studies of the structure of complexes between PLA$_2$ and substrate analogue/transition state analogue have been used as a strategy for investigating the mechanism of catalytic action of this enzyme.[2,5,99] The results were discussed in a preceding paragraph. Structural changes in PLA$_2$ molecule during the enzyme–substrate interaction have been studied by X-ray structure determination of a complex between bovine pancreatic PLA$_2$ and the substrate analogue N-dodecylphosphorylcholine.[142] The binding of this compound to the enzyme results in considerable changes in the native PLA$_2$ structure, especially in the N-terminal part of the polypeptide chain and in the region between residues 60–70. In the complex, the distance between the α-NH$_3^+$ group, which is important for the activity, and the OH group of Tyr52 is 1 Å shorter than that in the native protein. Structure-based drug design was also used as an approach for the development of anti-inflammatory agents. Thus, the complex between an acylamino analogue of phospholipids, named FPL67047XX,[143] and a human nonpancreatic secreted PLA$_2$ has been crystallized and the X-ray structure determined.[144] The inhibitor is highly potent toward the human enzyme and IC$_{50}$ = 1.3 × 10^{-8} M.[145] The 3D model revealed the contribution of different components to the tight binding of the inhibitor to the enzyme, that is, the structural basis of the high affinity of the synthetic compound for the enzyme active site.

Naturally occurring compounds, tolerant to living organisms, are perspective drugs for human diseases. Recently, the crystallographic structure of the complex between vitamin E and PLA$_2$ from the venom of *Daboia russelli pulchella* was analyzed at 1.8-Å resolution[146] (PDB code: 1JQ9). This is the first structural evidence for a specific inhibition of PLA$_2$ by α-tocopherol (Figure 5). Two PLA$_2$ molecules, A and B, held together through a long interface, were observed in the asymmetric unit; vitamin E is bound to one of them, molecule A. The enzyme exists as a dimer also in solution. The hydrocarbon chain of the inhibitor lies inside the hydrophobic channel of the enzyme and participates in hydrophobic interactions with the substrate binding site. An OH group of the aromatic part of vitamin E plays a key role in the enzyme–inhibitor complex formation interacting with the ND1 atom of the active site His48 and the OD1 of Asp49.

The orientation of Trp31 is important for the function of the PLA$_2$ dimer because it is located at the entrance of the substrate-binding site. The conformation of this

Figure 5 Binding of α-tocopherol at the active site of *Vipera russelli* phospholipase A$_2$[146] (PDB code: 1JQ9). The hydrocarbon chain of α-tocopherol fills the hydrophobic channel.

residue is different in the two subunits: in molecule A, which binds α-tocopherol, the side chain is oriented in a position suitable for binding, leaving the hydrophobic channel 'open'. However, in molecule B, the entrance to the hydrophobic channel is hindered by the side chain of Trp31, which prevents the binding of the inhibitor to the substrate-binding site. In this way, the X-ray data revealed important structure–function relationships in the vitamin E-PLA_2 complex and explains details of the mechanism of enzyme inhibition by this natural compound. The model can be used for a structure-based drug design of effective PLA_2 inhibitors.

ACKNOWLEDGEMENTS

The authors thank the Deutsche Forschungsgemeinschaft for financial support by the Project 436 BUL 113/115/01 & BE 1443/9-1. This project was further supported by a joint grant of the DAAD (Deutscher Akademischer Austauschdienst, Germany) and by a grant from the Department of Science and Technology (DST), New Delhi) in terms of the Project Based Personnel Exchange Program 2000.

REFERENCES

1 RM Kini, in RM Kini (ed.), *Venom Phospholipase A_2 Enzymes. Structure, Function and Mechanism*, John Wiley & Sons, Chichester, UK, pp 1–28 (1997).

2 DL Scott, Z Otwinowski, MH Gelb and PB Sigler, *Science*, **250**, 1563–66 (1990).

3 BW Dijkstra, KH Kalk, WGJ Hol and J Drenth, *J Mol Biol*, **147**, 97–123 (1981).

4 BW Dijkstra, R Renetseder, KH Kalk, WGJ Hol and J Drenth, *J Mol Biol*, **168**, 163–79 (1983).

5 DL Scott, SP White, JL Browning, JJ Rosa, MH Gelb and PB Sigler, *Science*, **254**, 1007–10 (1991).

6 JM Danse, S Gasparini and A Menez, in RM Kini (ed.), *Venom Phospholipase A_2 Enzymes. Structure, Function and Mechanism*, John Wiley & Sons, Chichester, UK, pp 29–72 (1997).

7 E Valentin and G Lambeau, *Biochim Biophys Acta*, **1488**, 59–70 (2000).

8 EA Dennis, SG Rhee, MM Billah and YA Hannun, *FASEB J*, **5**, 2068–77 (1991).

9 R Flower, *Trends Pharmacol Sci*, **2**, 186–88 (1981).

10 F Hirata and J Axelrod, *Science*, **209**, 1082–90 (1980).

11 P Vadas and W Pruzanski, *Lab Invest*, **4**, 391–404 (1986).

12 M Murakami, T Kambe, S Shimbara, K Higashino, K Hanasaki, H Arita, M Horiguchi, M Arita, H Arai and K Inoue, *J Biol Chem*, **274**, 31435–44 (1999).

13 W Pruzanski and P Vadas, *Immunol Today*, **12**, 143–46 (1991).

14 A Makela, B Sternby, T Kuusi, P Puolakkainen and T Schröder, *Scand J Gastroenterol*, **25**, 944–50 (1990).

15 P Vadas and W Pruzanski, *Adv Inflamm Res*, **7**, 52–59 (1984).

16 P Kortekangas, HT Aro and J Nevalainen, *Scand J Rheumatol*, **23**, 68–72 (1994).

17 DK Kim, T Fukuda, BT Thompson, B Cockrill, C Halles and JV Bonventre, *Am J Physiol*, **269**, 109–18 (1995).

18 G Olaison, R Sjödahl and C Tagesson, *Digestion*, **41**, 136–41 (1988).

19 LA Styles, GG Schalkwijk, AJ Aarsman, EP Vichinsky, BH Lubin and FA Kuypers, *Blood*, **87**, 2573–78 (1996).

20 AJ Aarsman, FW Neys, HA van der Helm, FA Kuypers and H van den Bosch, *Biochim Biophys Acta*, **1502**, 257–63 (2000).

21 T Minami, H Tojo, Y Shinomura, S Tarui and M Okamoto, *Gut*, **33**, 914–21 (1992).

22 W Uhl, M Büchler, TJ Nevalainen, A Deller and HG Beger, *J Trauma*, **30**, 1285–90 (1990).

23 S Yedgar, D Lichtenberg and E Schnitzer, *Biochim Biophys Acta*, **1488**, 182–87 (2000).

24 TJ Nevalainen, MM Haapamaki and JM Grönroos, *Biochim Biophys Acta*, **1488**, 83–90 (2000).

25 AG Buckland and DC Wilton, *Biochim Biophys Acta*, **1488**, 71–82 (2000).

26 T Bayburt, BZ Yu, HK Lin, J Browning, MK Jain and MH Gelb, *Biochemistry*, **32**, 573–82 (1993).

27 Y Snitko, RS Koduri, SK Han, R Othman, SF Baker, BJ Molini, DC Wilton, MH Gelb and WH Cho, *Biochemistry*, **36**, 14325–33 (1997).

28 L Paramo, B Lomonte, J Pizarro-Cerda, JA Bengoechea, JP Gorvel and E Moreno, *Eur J Biochem*, **253**, 452–61 (1998).

29 D Fenard, G Lambeau, E Valentin, JC Lefebre, M Lazdunski and A Doglio, *J Clin Invest*, **104**, 611–18 (1999).

30 I Nobuhisa, K Nakashima, M Desimaru, T Ogawa, Y Shimohigashi, Y Fukumaki, Y Sakaki, S Hattori, H Kihara and M Ohno, *Gene*, **172**, 267–72 (1996).

31 DL Scott, in RM Kini (ed.), *Venom Phospholipase A_2 Enzymes. Structure, Function and Mechanism*, John Wiley & Sons, Chichester, UK, pp 97–128 (1997).

32 C Betzel, N Genov, KR Rajashankar and TP Singh, *Cell Mol Life Sci*, **56**, 384–97 (1999).

33 I Ho Tsai, HY Hsu and YM Wang, *Toxicon*, **40**, 1363–67 (2002).

34 PKA Larsson, HE Classon and BP Kennedy, *J Biol Chem*, **273**, 207–14 (1998).

35 EJ Ackermann, ES Kempner and EA Dennis, *J Biol Chem*, **269**, 9227–33 (1994).

36 MV Winstead, J Balsinde and EA Dennis, *Biochim Biophys Acta*, **1488**, 28–39 (2000).

37 QM Lu, Y Jin, JF Wei, DS Li, SW Zhu, WY Wang and YL Xiong, *Toxicon*, **40**, 1313–19.

38 EA Dennis, *Trends Biochem Sci*, **22**, 1–2 (1997).

39 RL Heinrikson, ET Krueger and PS Keim, *J Biol Chem*, **252**, 4913–21 (1977).

40 DA Six and EA Dennis, *Biochim Biophys Acta*, **1488**, 1–19 (2000).

41 P Kraulis, *J Appl Crystallogr*, **24**, 946–50 (1991).

42 EA Merrit and MEP Murphy, *Acta Crystallogr*, **D50**, 869–73 (1994).

43 P de Geus, CJ van den Bergh, O Kuipers, HM Verheij, WPM Hoekstra and GH de Haas, *Nucleic Acids Res*, **15**, 3743–59 (1987).

44 JP Noel and MD Tsai, *J Cell Biochem*, **40**, 309–20 (1989).

45 B Kerfelec, KS La Forge, P Vasiloudes, A Puigsverver and GA Schelle, *Eur J Biochem*, **190**, 299–304 (1990).

46 C Kusunoki, S Satoh, M Koboyashi and M Niwa, *Biochim Biophys Acta*, **1087**, 95–97 (1990).

47. F Gubensek and D Kordis, in RM Kini (ed.), *Venom Phospholipase A₂ Enzymes. Structure, Function and Mechanism*, John Wiley & Sons, Chichester, UK, pp 29–72 (1997).
48. D Kordis, A Bdolah and F Gubensek, *Biochem Biophys Res Commun*, **251**, 613–19 (1998).
49. I Guillemin, C Bouchier, T Garrigues, A Wisner and V Choumet, *Eur J Biochem*, **270**, 2697–706 (2003).
50. I-H Tsai, Y-M Wang, L-C Au, T-P Ko, Y-H Chen and Y-F Chu, *Eur J Biochem*, **267**, 6684–91 (2000).
51. B Francis, JA Coffield, LL Simpson and II Kaiser, *Arch Biochem Biophys*, **318**, 481–88 (1995).
52. IH Tsai, YH Chen, YM Wang, MY Liau and PJ Lu, *Arch Biochem Biophys*, **387**, 257–64 (2001).
53. MZ Huang, P Gopalakrishnakone, MCM Chung and RM Kini, *Arch Biochem Biophys*, **338**, 150–56 (1997).
54. B Westerlund, P Nordlund, U Uhlin, D Eaker and H Eklund, *FEBS Lett*, **301**, 159–64 (1992).
55. JP Changeux, M Kasai and CY Lee, *Proc Natl Acad Sci USA*, **67**, 1241–47 (1970).
56. RM Kini and HJ Evans, *Toxicon*, **27**, 613–35 (1989).
57. S Braud, C Bon and A Wisner, *Biochemie*, **82**, 851–59 (2000).
58. JM Gutierrez and B Lomonte, *Toxicon*, **33**, 1405–24 (1995).
59. D Mebs and CL Ownby, *Pharmacol Ther*, **48**, 223–26 (1990).
60. CE Nunez, Y Angulo and B Lomonte, *Toxicon*, **39**, 1587–94 (2001).
61. K Conde-Frieboes, LJ Reynolds, Y Lio, M Hale, HH Wasserman and EA Dennis, *J Am Chem Soc*, **118**, 5519–25 (1996).
62. SL Hazen and RW Gross, *Circ Res*, **70**, 486–95 (1992).
63. MR Lennartz, JB Lefkowith, FA Bromley and EJ Brown, *J Leukoc Biol*, **54**, 389–98 (1993).
64. E Valentin and G Lambeau, *Biochemie*, **82**, 815–31 (2000).
65. I Krizaj, D Turk, A Ritonja and F Gubensek, *Biochim Biophys Acta*, **999**, 198–202 (1989).
66. F Gubensek, I Krizai and J Pungercar, in RM Kini (ed.), *Venom Phospholipase A₂ Enzymes. Structure, Function and Mechanism*, John Wiley & Sons, Chichester, UK, pp 245–68 (1997).
67. Y-M Wang, P-J Lu, C-L Ho and I-H Tsai, *Eur J Biochem*, **209**, 635–41 (1992).
68. B Lomonte, A Tarkowski, U Bagge and LA Hanson, *Biochem Pharmacol*, **47**, 1509–18 (1994).
69. L Calderon and B Lomonte, *Arch Biochem Biophys*, **358**, 343–50 (1998).
70. B Lomonte, J Pizarro-Cerda, Y Angulo, JP Gorvel and E Moreno, *Biochim Biophys Acta*, **1461**, 19–26 (1999).
71. BW Segelke, D Nguyen, R Chee, NH Xuong and EA Dennis, *J Mol Biol*, **279**, 223–29 (1998).
72. LJ Lefkowitz, RA Deems and EA Dennis, *Biochemistry*, **38**, 14174–84 (1999).
73. M Adamich, MF Roberts and EA Dennis, *Biochemistry*, **18**, 3308–14 (1979).
74. K Zhao, Y Zhou and Z Lin, *Toxicon*, **38**, 901–16 (2000).
75. J Weiss, G Wright, ACAPA Bekkers, CJ van den Bergh and HM Verheij, *J Biol Chem*, **266**, 4162–67 (1991).
76. HJ Evans and RM Kini, in RM Kini (ed.), *Venom Phospholipase A₂ Enzymes. Structure, Function and Mechanism*, John Wiley & Sons, Chichester, UK, pp 353–60 (1997).
77. E Valentin, RS Koduri, J-C Scimeca, G Carle, MH Gelb, ML Lazdunski and G Lambeau, *J Biol Chem*, **274**, 19152–60 (1999).
78. G Lambeau, J Barhanin, H Schweitz, J Qar and M Lazdunski, *J Biol Chem*, **264**, 11503–10 (1989).
79. G Lambeau, A Schmid-Alliana, M Lazdunski and J Barhanin, *J Biol Chem*, **265**, 9526–32 (1990).
80. G Lambeau, P Ancian, J Barhanin and M Lazdunski, *J Biol Chem*, **269**, 1575–78 (1994).
81. G Lambeau, M Lazdunski and J Barhanin, *Neurochem Res*, **16**, 651–58 (1991).
82. JP Nicolas, Y Lin, G Lambeau, F Ghomashchi, M Lazdunski and MH Gelb, *J Biol Chem*, **272**, 7173–80 (1997).
83. H Arita, K Hanasaki, T Nakano, S Oka and K Matsumoto, *J Biol Chem*, **266**, 19139–41 (1991).
84. K Hanasaki and H Arita, *J Biol Chem*, **267**, 6414–20 (1992).
85. J Ishizaki, K Hanasaki, K Higashino, J Kishino, N Kikuchi, O Ohara and H Arita, *J Biol Chem*, **269**, 5897–904 (1994).
86. P Ancian, G Lambeau, MG Mattei and M Lazdunski, *J Biol Chem*, **270**, 8963–70 (1995).
87. G Lambeau, P Ancian, JP Nicolas, S Beiboer, D Moinier, H Verhej and M Lazdunski, *J Biol Chem*, **270**, 5534–40 (1995).
88. P Ancian, G Lambeau and M Lazdunski, *Biochemistry*, **34**, 13146–51 (1995).
89. K Hanasaki and H Arita, *Arch Biochem Biophys*, **372**, 215–23 (1999).
90. H Fujita, K Kawamoto, K Hanasaki and H Arita, *Biochem Biophys Res Commun*, **209**, 293–99 (1995).
91. M-M Thwin, P Gopalakrishnakone, RM Kini, A Armugam and K Jeyaseelan, *Biochemistry*, **39**, 9604–11 (2000).
92. K Okumura, N Ohkura, S Inoue, K Ikeda and K Hayashi, *J Biol Chem*, **273**, 19469–75 (1998).
93. Y Sakurai, Y Fujimura, T Kokubo, K Imamura, T Kawasaki, M Handa, M Suzuki, T Matsui, K Titani and A Yoshioka, *Thromb Haemostasis*, **79**, 1199–207 (1998).
94. Q Wei, Q-M Lu, Y Jin, R Li, J-F Wei, W-Y Wang and Y-L Xiong, *Toxicon*, **40**, 1331–38 (2002).
95. A Tani, T Ogawa, T Nose, NN Nikandrov, M Deshimaru, T Chijiwa, C-C Chang, Y Fukumaki and M Ohno, *Toxicon*, **40**, 803–13 (2002).
96. CC Leslie, *J Biol Chem*, **272**, 16709–12 (1997).
97. Y Nakatani, T Tanioka, S Sunaga, M Murakami and I Kudo, *J Biol Chem*, **275**, 1161–68 (2000).
98. HM Verheij, JJ Volwerk, EHJM Jansen, WC Puijk, BW Dijkstra, J Drenth and GH de Haas, *Biochemistry*, **19**, 743–50 (1980).
99. SP White, DL Scott, Z Otwinowski, MH Gelb and PB Sigler, *Science*, **250**, 1560–63 (1990).
100. MMGM Thunnissen, AB Eiso, KH Kalk, J Drenth, BW Dijkstra, OP Kuipers, R Dijkman, GH de Haas and HM Verheij, *Nature*, **347**, 689–91 (1990).
101. DL Scott, SP White, Z Otwinowski, W Yuan, MH Gelb and PB Sigler, *Science*, **250**, 1541–46 (1990).
102. JM Marangone, G Merutka, W Cho, W Welches, FJ Kezdy and RL Heinrikson, *J Biol Chem*, **259**, 13839–43 (1984).
103. K Yoshizumi, SY Liu, T Miyata, S Saito, M Ohno, S Iwanaga and H Kihara, *Toxicon*, **28**, 43–54 (1990).
104. SY Liu, K Yoshizumi, N Oda, M Ohno, F Tokunaga, S Iwanaga and H Kihara, *J Biochem*, **107**, 400–8 (1990).
105. B Francis, JM Gutierrez, B Lomonte and II Kaiser, *Arch Biochem Biophys*, **284**, 352–59 (1991).
106. DR Holland, LL Clancy, SW Muchmore, TJ Ryde, HM Einspahr, BC Finzel, RL Heinrikson and KD Watenpaugh, *J Biol Chem*, **265**, 17649–56 (1990).
107. RK Arni, MRM Fontes, C Barberato, JM Gutierrez, C Diaz and RJ Ward, *Arch Biochem Biophys*, **366**, 177–82 (1999).

108 S Banumathi, KR Rajashankar, C Nötzel, B Aleksiev, TP Singh, N Genov and Ch Betzel, *Acta Crystallogr*, **D57**, 1552–59 (2001).

109 WH Lee, MT da Silva Giotto, S Marangoni, MH Toyama, I Polikarpov and RC Garratt, *Biochemistry*, **40**, 28–36 (2001).

110 W Yuan, DM Quinn, PB Sigler and MH Gelb, *Biochemistry*, **29**, 6082–94 (1990).

111 M Cygler and JD Schrag, *Methods Enzymol*, **284**, 3–7 (1997).

112 L Sarda and P Desnuelle, *Biochim Biophys Acta*, **30**, 513–19 (1958).

113 P Desnuelle, L Sarda and G Ailhard, *Biochim Biophys Acta*, **37**, 570–76 (1960).

114 MA Gelb, W Cho and DC Wilton, *Curr Opin Struct Biol*, **9**, 428–32 (1999).

115 M Sumandea, S Das, C Sumandea and W Cho, *Biochemistry*, **38**, 16290–97 (1999).

116 S Bezzine, RS Koduri, E Valentin, M Murakami, I Kudo, F Ghomashchi, M Sadilek, G Lambeau and MH Gelb, *J Biol Chem*, **275**, 3179–91 (2000).

117 J Rogers, B-Z Yu, M-D Tsai, OG Berg and MK Jain, *Biochemistry*, **37**, 9549–56 (1998).

118 M Murakami, Y Nakatani and I Kudo, *J Biol Chem*, **271**, 30041–51 (1996).

119 OG Berg, MH Gelb, MD Tsai and MK Jain, *Chem Rev*, **101**, 2613–53 (2001).

120 K Kondo, H Toda, K Narita and CY Lee, *J Biochem*, **91**, 1531–48 (1982).

121 K Kondo, H Toda, K Narita and C-Y Lee, *J Biochem*, **91**, 1519–30 (1982).

122 PD Kwong, NQ McDonald, PB Sigler and WA Hendrickson, *Structure*, **3**, 1109–19.

123 C Bon, in RM Kini (ed.), *Venom Phospholipase A$_2$ Enzymes. Structure, Function and Mechanism*, John Wiley & Sons, Chichester, UK, pp 269–86 (1997).

124 BJ Hawgood and JW Smith, *Br J Pharmacol*, **61**, 597–606 (1977).

125 A Randolph and RL Heinrikson, *J Biol Chem*, **257**, 2155–61 (1982).

126 I Krizai, A Bdolah, F Gubensek, P Bencina and J Pungercar, *Biochem Biophys Res Commun*, **227**, 374–79 (1996).

127 B Aleksiev and R Shipolini, *Hoppe-Seyler-Z Physiol Chem*, **352**, 1183–87 (1971).

128 Y-M Wang, P-J Lu, C-L Ho, I-H Tsai, *Eur J Biochem*, **209**, 635–41 (1992).

129 Y Komori, K Masuda, T Nikai and H Sugihara, *Arch Biochem Biophys*, **327**, 303–7 (1996).

130 A Nicholls, KA Sharp and B Honig, *Proteins*, **11**, 281–96 (1991).

131 I Mancheva, T Kleinschmidt, B Aleksiev and G Braunitzer, *Biol Chem Hoppe-Seyler*, **368**, 343–52 (1987).

132 B Aleksiev and B Chorbanov, *Toxicon*, **14**, 477–84 (1976).

133 BW Dijkstra, J Drenth, KH Kalk and PJ Wandermaelen, *J Mol Biol*, **124**, 53–60 (1978).

134 BW Dijkstra, J Drenth and KH Kalk, *Nature*, **289**, 604–6 (1981).

135 C Keith, DS Feldman, S Deganello, J Glick, KB Ward, EO Jones and PB Sigler, *J Biol Chem*, **256**, 8602–7 (1981).

136 S Brunie, J Bolin, D Gewirth and PB Sigler, *J Biol Chem*, **260**, 9742–49 (1985).

137 R Renetseder, S Brunie, BW Dijkstra, J Drenth and PB Sigler, *J Biol Chem*, **260**, 11627–34 (1985).

138 OP Kuipers, MMGM Thunnissen, P de Geus, BW Dijkstra, J Drenth, HM Verheij and GH de Haas, *Science*, **244**, 82–85 (1989).

139 JP Very, RW Schevitz, DK Clawson, JL Bobbit, ER Dow, G Gamboa, T Goodson Jr, RB Hermann, RM Kramer, DB McClure, ED Mihelich, JE Putnam, JD Sharp, DH Stark, C Teater, MW Warrick and ND Jones, *Nature*, **352**, 79–82 (1991).

140 M Perbandt, JC Wilson, S Eschenburg, I Mancheva, B Aleksiev, N Genov, P Willingmann, W Weber, TP Singh and Ch Betzel, *FEBS Lett*, **412**, 573–77 (1997).

141 L Gu, Z Wang, S Song, Y Shu and Z Lin, *Toxicon*, **40**, 917–22 (2002).

142 K Tomoo, H Ohishi, M Doi, T Ishida, M Inoue, K Ikeda and H Mizuno, *Biochem Biophys Res Commun*, **187**, 821–27 (1992).

143 HG Beaton, C Bennion, S Connoly, AR Cook, NP Gensmantel, C Hallam, K Hardy, B Hitchin, CG Jackson and DH Robinson, *J Med Chem*, **37**, 557–59 (1994).

144 S-S Cha, D Lee, J Adams, JT Kurdila, CS Jones, LA Marshall, B Bolognese, SS Abdel-Meguid and B-H Oh, *J Med Chem*, **39**, 3878–81 (1996).

145 P Elsbach and J Weiss, *Methods Enzymol*, **197**, 24–30 (1991).

146 V Chandra, J Jasti, P Kaur, Ch Betzel, A Srinivasan and TP Singh, *J Mol Biol*, **320**, 215–22 (2002).

Calsequestrin

ChulHee Kang
School of Molecular Biosciences, Washington State University, Pullman, WA, USA

FUNCTIONAL CLASS

Calsequestrin, a 40-kD protein, binds calcium ions with high capacity (40–50 mol Ca^{2+} mol^{-1} calsequestrin) but with moderate affinity ($K_d = 1$ mM) over the Ca^{2+} concentration range between 0.01 and 1 M and releases it with a high off-rate (10^6 s^{-1}).[1]

Calsequestrin, calreticulin (CRT), and a series of other acidic lumenal Ca^{2+} binding proteins provide a buffer for Ca^{2+}. Among these, calsequestrin is a major Ca^{2+} storage protein within the sarcoplasmic reticulum (SR) membranes of both cardiac and skeletal muscle.

The Ca^{2+} binding and dissociation mechanisms of calsequestrin are of crucial importance, but are not yet clearly understood. Ca^{2+} binding sites in calsequestrin are supposed to be very different from those in the Ca^{2+} pump, sarcoplasmic and endoplasmic reticulum calcium ATPases (SERCA), calmodulin (CaM), and troponinC (TnC). Calsequestrin sites need to be made and broken, but not over the low cytosolic Ca^{2+} concentration range or with the same stoichiometry and precision as those formed and subsequently disrupted in the Ca^{2+} pump or those that are intrinsic to the EF-hand structure.[2]

OCCURRENCE

High concentration of calsequestrin (up to 100 mg mL^{-1}) is present in the lumen of the junctional terminal cisternae

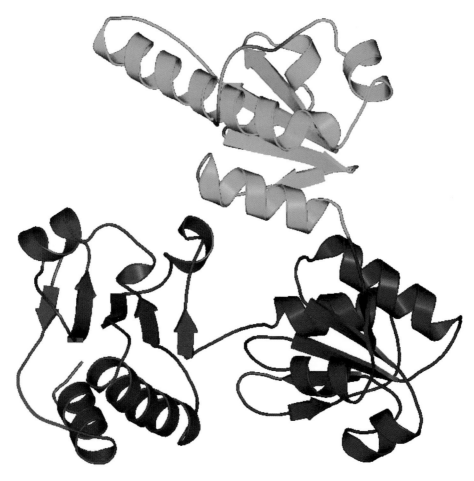

3D Structure Ribbon diagram showing the distribution of structural elements of rabbit skeletal calsequestrin (PDB:1A8Y). Domain I, II, and III are depicted as green, blue, and red respectively.

of the sarcoplasmic reticulum.[1] The SR membrane is a specialized elaboration of the endoplasmic reticulum (ER), and, like the ER, the SR serves as a calcium storage/release system for regulating calcium ion concentration in response to an electrical signal from the motor nerve. Calsequestrin does not contain the KDEL tetrapeptide ER/SR retrieval signal[3] and it is not uniformly distributed within the lumen of the SR membrane, but rather forms regular arrays that appear crystalline.[4–6] Calsequestrin, probably in a Ca^{2+}-bound form, is also physically bound at the calcium release channel, known as the ryanodine receptor (RyR). The details of how calsequestrin is associated with the junctional facing membrane and the RyR remains uncertain, although two additional proteins, junctin and triadin are implicated.[7–16] These two proteins interact with RyR in the junctional face region of the SR, and the network of interacting proteins assures that high concentrations of Ca^{2+} are stored very near the site of Ca^{2+} release.

Calsequestrin is also present in the ER vacuolar domains of some neurons and smooth muscles. The ER from other types of animal and plant cells contains very similar proteins that bind Ca^{2+} with a relatively high capacity and low affinity.

BIOLOGICAL FUNCTION

Calsequestrin regulates Ca^{2+} levels in the lumen of the SR

The regulation of calcium ion by the SR, in turn, controls the state of the actin–myosin fibrils, with release of Ca^{2+} from the SR bringing about muscle contraction and uptake of calcium by the SR bringing about relaxation. In this pump-storage release of calcium ion by the SR, not only does calsequestrin act as a Ca^{2+} buffer inside the SR, lowering free Ca^{2+} concentrations and thereby facilitating further uptake by the Ca^{2+}-ATPases, but also actively participates by localizing Ca^{2+} at the release site and regulating the amount of Ca^{2+} released through the RyR. Localization of calsequestrin at the lumenal face of the junctional SR, thereby sequestering and concentrating Ca^{2+} near the RyR, is thought to be essential for a short contraction/relaxation cycle that can be completed within a time interval of less than $100 \, \text{ms}^{-1}$. By this localization, diffusion time for Ca^{2+} release could be drastically reduced, since the diffusion time depends on the square of the distance.

Calsequestrin interacts with other proteins in the SR

Biochemical as well as electron microscopic data suggests that a complex of proteins may be involved in the release of Ca^{2+} at the junctional facing membrane of the SR.[17–20] A number of the protein components of this complex have been identified. These include RyR,[17,18] calsequestrin,[7,21–23] triadin, a putative anchoring protein that stabilizes calsequestrin at the junctional face membrane,[11,12,14,24,25] and junctin, a calsequestrin receptor protein.[7,10,13] Both junctin and triadin have been purified, cloned, and expressed.[10,13,25–27] Although junctin (26 kD) and triadin (30 kD) are the products of different genes, they exhibit significant similarities. Both have single-membrane spanning domains within short N-terminal sections followed by long C-terminal sections. The N-terminal regions of each protein are located in the cytoplasm and the C-terminal regions of each are located in the lumen of the SR. These C-terminal sections are very basic and show very high sequence similarity between species.[10,12,25,27] A potential calsequestrin binding domain in triadin was reported.[16,28] Junctin binds to both calsequestrin and to the RyR protein at the same time.[13] Recently, a direct and high-affinity interaction was demonstrated between calsequestrin and RyR,[29] even though opposite results were reported previously.[30]

These results, taken together, suggest that a quaternary protein complex may exist between junctin, triadin, calsequestrin, and the RyR, possibly allowing calsequestrin to sequester Ca^{2+} in the vicinity of the RyR during Ca^{2+} uptake and release.[28,31]

Related human disease

Both mutation and improper level of calsequestrin expression have been implicated to several diseases (references 32 and 33). Thyroid-associated ophthalmopathy (TAO) is a progressive orbital disorder associated with Grave's hyperthyroidism and, less often, with Hashimoto's thyroiditis. TAO results when autoantibodies react with orbital antigens and lead to exophthalmos and eye muscle inflammation. Calsequestrin was identified as one of the major antigens in these disorders.[34] Closely related Ca^{2+} storage and chaperone proteins such as calreticulin and protein disulfide isomerase (PDI) also play major immunological roles in the progression of hepatic disorders.[35,36]

Cardiac calsequestrin overexpression impairs calcium signaling in murine myocytes, leading to severe cardiac hypertrophy,[15,37,38] sporadic Ca^{2+} sparks,[39] severe forms of catecholaminergic polymorphic ventricular tachycardia,[40,41] systolic dysfunction,[42] depressed contractility in the mammalian heart and induction of a fetal gene expression program,[43] and premature death.[15,44] There is also emerging, although only suggestive, evidence that pathogenic mechanisms underlying age-related changes in the E–C coupling process are due to alteration in the interaction between RyR1 and calsequestrin.[45]

AMINO ACID SEQUENCE INFORMATION

Cardiac and skeletal muscle calsequestrin cDNAs have been sequenced and cloned from several species, including

Calsequestrin

Table 1 Negatively charged amino acids, Glu (E) and Asp (D) are depicted as blue and positively charged amino acids, Lys (K) and Arg (R) are depicted as red. Completely conserved amino acids among all the species are marked with stars

```
rabbit_cardiac     100.0%   EEGLNFPTYDGKDRVVSLSEKNFKQILKKY  DLLCLYYHAPVSADKVAQKQFQLKEIVLEL  VAQVLEHKEIGEFVMVDRKKEAKLAKKLGFD
dog_cardiac         93.6%   EEGLNFPTYDGKDRVVSLTEKNFKQVLKKY  DVLCLYYHESVSSDKVAQKQFQLKEIVLEL  VAQVLEHKDIGEFVMVDAKKEAKLAKKLGFD
human_cardiac       92.3%   EEGLNFPTYDGKDRVVSLSEKNFKQVLKKY  DLLCLYYHEPVSSDKVTPKQFQLKEIVLEL  VAQVLEHKRIGEFVMVDRKKEAKLAKKLGFD
mouse_cardiac       88.4%   QEGLNFPTYDGKDRVVSLSEKNLKQMLKRY  DVLCLYYHESVSSDKVSQKQFQLKEIVLEL  VAQVLEHKNIGEFVMVDSRKEARLAKRLGFS
chicken_skeletal    80.3%   EEGLNFPTYDGKDRVIDLNEKNYKHALKKY  DMLCLLFHEPVSSDRVSQKQFQMTEMVLEL  AAQVLEPRSIGFGMVDSKKDARLAKKLGLV
human_skeletal      67.7%   QEGLDFPEYDGVDRVINVNAKNYKNVFKKY  EVLALLYHEPPEDDKASQRQFEMEELILEL  AAQVLEDKGVGFGLVDSEKDAAVAKKLGLT
mouse_skel          68.2%   EDGLDFPEYDGVDRVINVNAKNYKNVFKKY  EVLALLYHEPPEDDKASQRQFEMEELILEL  AAQVLEDKGVGFGLVDSEKDAAVAKKLGLT
rabbit_skeletal     69.2%   EEGLDFPEYDGVDRVINVNAKNYKNVFKKY  EVLALLYHEPPEDDKASQRQFEMEELILEL  AAQVLEDKGVGFGLVDSEKDAAVAKKLGLT
frog_skeletal       63.3%   EDGLDFPEYDGEDRVIHISLKNYKAALLKKY EVLALLYHEPIGDDKASQRQFEMEELILEL  AAQVLEDKGVGFGLVDSEDDRAVAKKLGLD
                            ** ** *** ***          **   *   *    *      *         **         ** **     *****     ** **      * ** **

                                         120                             150                             180
rabbit_cardiac     100.0%   EEGSLYILKGDRTIEFDGEFAADVLVEFLL  DLIEDPVEIINSKLEVQAFERIEDHIKLIG  FFKSRDSEYYKAFEERAEHFQPYIKFFATE
dog_cardiac         93.6%   EEGSLYVLKGDRTIEFDGEFAADVLVEFLL  DLIEDPVEIINSKLEVQAFERIEDQIKLIG  FFKSEDSEYYKAFEERAEHFQPYIKFFATE
human_cardiac       92.3%   EEGSLYILKGDRTIEFDGEFAADVLVEFLL  DLIEDPVEIISSKLEVQAFERIEDYIKLIG  FFKSEDSEYYKAFEERAEHFQPYIKFFATE
mouse_cardiac       88.4%   EEGSLYVLKGDRTIEFDGEFAADVLVEFLL  DLIEDPVEIVNNKLEVQAFERIEDQTKLLG  FFKNEDSEYYKAFQERAEHFQPYIKFFATE
chicken_skeletal    80.3%   EEGSLYVFKEERLIEFDGELRTDVLVEFLL  DLLEDPVEIINSKLELQAFDQIDDEIKLIG  YFKGEDSEHYKAFEERAEHFQPYVKFFATE
human_skeletal      67.7%   EVDSMYVFKGDEVIEYDGEFSADTIVEFLL  DVLEDPVELIEGERELQAFENIEDEIKLIG  YFKSKDSEHYKAFEDARAEFHPYIPFFATE
mouse_skel          68.2%   EEDSVYVFKGDEVIEYDGEFSADTLVEFLL  DVLEDPVELIEGERELQAFENIEDEIKLIG  YFKSKDSEHYKAYEDARAEEFHPYIPFFATE
rabbit_skeletal     69.2%   EEDSIYVFKEKDEVIEYDGEFSADTLVEFLL DVLEDPVELIEGERELQAFENIEDEIKLIG  YFKSKDSEHYKAYEDARAEFHPYIPFFATE
frog_skeletal       63.3%   EESIYVFKDDEMIEYDGEFSADTLVEFLL   DVLEDPVEFIDGSHELARFENLDDEPKLIG  YFKNEDSEHYKAYEDARAEEFHPYIPFFATE
                            *   *    *     ** ***      *        *****        *  *****         *   **     *  **      ** ***   *** ** *****

                                         210                             240                             270
rabbit_cardiac     100.0%   DKGVAKKLSLKMNEVDFYEPFMDEPTPIPN  KPYTEEELVEEFVKEHQRPTLRRLRPEDMFE  TWEDDLNGIHIVPFAEKSDPDGYEFLEILK
dog_cardiac         93.6%   DKGVAKKLSLKMNEVDFYEPFMDEPIAIPD  KPYTEEELVEEFVKEHQRPTLRRLRPEEMFE  TWEDDLNGIHIVAFRAERSDPDGYEFLEILK
human_cardiac       92.3%   DKGVAKKLSLKMNEVDFYEPFMDEPIAIPN  KPYTEEELVEEFVKEHQRPTLRRLRPEEMFE  TWEDDLNGIHIVAFAEKSDPDGYEFLEILK
mouse_cardiac       88.4%   DKRAVAKKLSLKMNEVGFYEPFMDEPNVIPN KPYTEEELVEEFVKEHQRPTLRRLRPEDMFE  TWEDDLNGIHIVAFAESHPDGYEFLEILK
chicken_skeletal    80.3%   DKGVAKKLGLKMNEVEFYEPFMDEPVHIPD  KPYTEEELVEEFVKEHKRATLRKLRPEDMFE  TWEDDMEGIHIVAFAEEDDPDGFEFLEILK
human_skeletal      67.7%   DSKVAKKLTLKLNEIDFYEAFMEEPVTIPD  KPNSEEEIVNFVEEHRRSTLRKLKPESMYE   TWEDDLDGIHIVAFAEEEADPDGFEFLETLK
mouse_skel          68.2%   DSKVAKKLTLKLNEIDFYEAFMEEPVMTIPD KPNSEEEIVSFVEEHRRSTLRKLKPESMYE   TWEDDMLDGIHIVAFAEEEDPDGYEFLEILK
rabbit_skeletal     69.2%   DSKVAKKLTLKLNEIDFYEAFMEEPVTIPD  KPNSEEEIVNFVEEHRRSTLRKLKPESMYE   TWEDDMDGIHIVAFAEEDPDGYEFLEILK
frog_skeletal       63.3%   DAKVAKTLTLKLNEIDYYEPFHDEPITIPS  KPNSEKEIVDFLHQHKRPTLRKLRPDSMYE   TWEDDLNGIHIVAFAEEEDPDGYEFLQIIK
                            *   *** *** ** **    *  **              **    **   * ** *    *   *****  *****  ***   ***   **  *

                                         300                             330                             360
rabbit_cardiac     100.0%   QVARDNTDNPDLSIVVWIDPDDFPLLVAYWE  KTFKIDLFKPQIGVVNVTBADSVWMEIPDD  DDLPTAEELEDWIEDVLSGKINTEDDDNED
dog_cardiac         93.6%   QVARDNTDNPDLSIVVWIDPDDFPLLVAYWE  KTFKIDLFKPQIGVVNVTBADSVWMEIPDD  DDLPTAEELEDWIEDVLSGKINTEDDDNEE
human_cardiac       92.3%   QVARDNTDNPDLSILWIDPDDFPLLVAYWE   KTFKIDLFRPQIGVVNVTBADSVWMEIPDD  DDLPTAEELEDWIEDVLSGKINTEDDDNED
mouse_cardiac       88.4%   QVARDNTDNPDLSIVVWIDPDDFPLLVAYWE  KTFKIDLFKPQIGVVNVTBADSIWMEIPDD  DDLPTAEELEDWIEDVLSGKINTEDDD---
chicken_skeletal    80.3%   QVARDNTDNPDLSIVVWIDPDDFPLLITYWE  KTFKIDLFRPQIGIVNVTBADSVWMEIRDD  DDLPTAEELEDWIEDVLSGKINTEDDDDDD
human_skeletal      67.7%   AVAQDNTENPDLSIIWIDPDDFPLLVPYWE   KTFDIDLSAPQIGVVNVTBADSVWMEMDDE  EDLPSAEELEDWLEGEINTEDDDBDD
mouse_skel          68.2%   RVAQDNTENPDLSIIWIDPDDFPLLVPYWE   KTFDIDLSAPQIGVVNVTBADSVWMEMDNE  EDLPSADELEDWLEGEINTEDDDBDD
rabbit_skeletal     69.2%   SVAQDNTENPDLSIIWIDPDDFPLLVPYWE   KTFDIDLSAPQIGVVNVTBADSVWMEMDDE  EDLPSAEELEDWLEGEINTEDDDBED
frog_skeletal       63.3%   EVAEDNTDNPDLSIIWIDPDEDFPLLIPYWE  EKFGILSRPHIGVVNVTBADSVWMDMDDE   EDLPTVDELEDWIEDVLEGEVNTEDDDBDD
                            ** ***   ****** ****   ****       *   ** ****** ****                   ***  ***** **** *  ***** *

                                         390
rabbit_cardiac     100.0%   EDDDDDNDDDDDDDGN-SD--EEDNDDSDED  DE-----
dog_cardiac         93.6%   GDDGDDDEDDDDDGNNSD--EESNDDSDDD   DE-----
human_cardiac       92.3%   ------EDDDDDDDN---SD--EEDNDDSDDD DDE----
mouse_cardiac       88.4%   EDDDGDNDDDDDDDDDNDSEDNEDSDD      DDDDDE--
chicken_skeletal    80.3%   DDDDDDDDDDDDDD-------DDDDDDDDDD  DDDD----
human_skeletal      67.7%   DD----------------------         -------
mouse_skel          68.2%   DDDDDDDDDDD-------------         -------
rabbit_skeletal     69.2%   DDDDDDD------------------        -------
frog_skeletal       63.3%   DDDDDDDDDDDDDDDDDDDDDDDDDDDDDDDD DDDDDDDD
```

human, rabbit, dog, pig, mouse, rat, chicken, frog, and soil nematode.[21,22,43,46–50,51–62] In general, cardiac isoform of calsequestrin exhibits 60 to 70% sequence identity to the skeletal form, but its carboxy terminus extends to additional amino acids (Table 1).

STRUCTURE

Each calsequestrin molecule binds and releases large numbers of calcium ions during a contraction–relaxation cycle, estimated to be 40 to 50 ions per molecule. For the binding and release of that many divalent cations, calsequestrin needs to have a large number of negatively charged side chains. Consistent with this high-binding capacity, calsequestrin, for example, of rabbit skeletal muscle, has an isoelectric point near pH 3.7, a total of 110 carboxylate groups, and an excess of approximately 80 carboxyl side-chain groups over the sum of the positively charged ones[63] (Table 1). This large excess of negatively charged side chains is not uniformly distributed over the entire molecule. Rather, the amino terminal half of calsequestrin is slightly negative, whereas the residues in the carboxyl terminal half are more than 35% aspartic acids or glutamic acids. The most acidic and most variable sequence in different calsequestrins is the carboxy terminus, with the carboxy terminus of skeletal muscle calsequestrin containing 14 contiguous aspartic acids. Frog skeletal calsequestrin contains an even longer run of consecutive negative residues (45) at the carboxy terminus.[63] In general, the skeletal muscle calsequestrin has the twofold reduced Ca^{2+} binding capacity of cardiac calsequestrin despite very similar overall net-negative charge.[64]

Posttranslational modification

The posttranslational modifications for the calsequestrin have been reported. From the amino acid sequence, a glycosylation site and two potential phosphorylation sites were detected.[21] Three serine residues (Ser378, 382, 386) were shown to be partially phosphorylated in the purified canine cardiac isoform, whereas the rabbit fast-twitch isoform is phosphorylated on Thr373.[50] Both cardiac and skeletal muscle calsequestrins were phosphorylated by the casein kinase II. Recent mass spectrometry results show that significant amounts of calsequestrin contained glycan with only a single mannose residue, indicative of a novel postendoplasmic reticulum mannosidase activity.[65] The significance of these posttranslational modifications has not been fully understood, but unique glyco- and phosphoforms of calsequestrin might chart a complex cellular transport, with calsequestrin following trafficking pathways not present or not accessible to the same molecules in nonmuscle cells.[65]

Ionic strength induces folding of calsequestrin

The physicochemical properties, including Ca^{2+} binding affinity, of calsequestrin have been studied by tryptophan fluorescence,[66–69] circular dichroism (CD),[66,68–71] Raman spectroscopy,[68,72] NMR,[68] and proteolytic digestion.[7,57,69] It has been predicted that instead of a distinct Ca^{2+} binding site such as the EF-hand motif,[73] pairs of acid residues bind Ca^{2+} through the net charge density and Ca^{2+} binding is driven by the entropy gain from the liberation of many water molecules from the hydrated cations.[74,75]

The large negative charge of calsequestrin results in a significant tendency to be disordered, even at neutral pH. It has been reported that, when in low-salt environment (especially in the absence of calcium ion), calsequestrin is largely unfolded. Various monovalent, divalent, or trivalent ions cause calsequestrin to undergo substantial conformational changes. These changes are associated with an increased α-helical content and the internalization of tryptophans, as monitored by CD and intrinsic fluorescence, respectively.[7,56,67,68] They indicate that this protein is a mostly unfolded, random coil at low ionic strength (<0.01 mM), but folds into a compact structure as the concentration of ions is increased. Cations such as Zn^{2+}, Sr^{2+}, and Tb^{2+} bind to calsequestrin and cause changes analogous to those caused by Ca^{2+}.[66,67,76] Either proton (pH 6.0) or K^+, or both ions jointly, can also replace Ca^{2+} in eliciting the intrinsic fluorescence changes of calsequestrin;[77] thus, pH and K^+ changes *in vivo* are suggested to be an important regulator of calsequestrin function in connection with Ca^{2+}.[77,78] However, the mechanistic dependencies among Ca^{2+}, K^+, H^+, and calsequestrin are poorly understood and the role of ion-induced structural changes in calsequestrin remains to be elucidated.

In summary, calcium ions–induced folding could be as much the general effect of ionic strength on a highly charged protein, rather than the result of specific calcium interactions. It is also unlikely that there are specific Ca^{2+} sites, although the first few ions that bind at low Ca^{2+} concentrations may have some specificity.

High-capacity calcium binding leads to calsequestrin polymerization

Raising the concentration of Ca^{2+} further leads to concomitant low-affinity binding of large numbers of calcium ions and calcium-induced calsequestrin aggregation. Studies by Tanaka *et al.*[79] demonstrated that high-capacity Ca^{2+} binding by calsequestrin is established by the formation of Ca^{2+}/calsequestrin complexes at relatively high calsequestrin and Ca^{2+} concentrations. Under such conditions, two-thirds of the total bound Ca^{2+} is associated with Ca^{2+}/calsequestrin aggregates, while one-third is associated with the soluble form of calsequestrin. On the contrary, monovalent ion-induced folding does not lead to calsequestrin aggregation and precipitation and those Ca^{2+}/calsequestrin aggregates are even dissociated by K^+. Static laser light scattering and 3,3′-dithio-bis(sulfosuccinimidylproprionate) (DTSSP) cross-linking indicated that whole protein exhibited an initial Ca^{2+}-induced dimerization, followed by additional oligomerization as the Ca^{2+} concentration was raised or as the K^+ concentration was lowered.[64] Therefore, it has been suggested that K^+ might lower the affinity of Ca^{2+} for calsequestrin from both cardiac and skeletal muscle.[56,64,68,70,80]

During purification and storage in the presence of calcium ion, calsequestrin readily forms fibrils or needle-like crystals.[72,81,82] Likewise, electron microscopy reveals fibrous arrays in SR microsomes at the junctional membranes, and these arrays are believed to be calsequestrin in its calcium-bound form.[5,6,81] *In vivo* cross-linking studies indicate that most calsequestrin in the SR microsomes are involved in calsequestrin–calsequestrin complexes.[64,83] However, the low resolution of electron microscopy allows the possibility that other proteins are within these fibers. All these data suggest that calsequestrin is not only localized at the calcium-release channel but also forms calcium-induced linear polymers.[64] This calcium-induced polymerization likely contributes to the localization and function of calsequestrin, and, furthermore, the needle-like crystal probably reflects the morphology of the natural polymer.

X-RAY CRYSTAL STRUCTURE

Ca^{2+}-calsequestrin crystals were readily obtained in several laboratories.[79,81] However, these crystals often formed as needles that have not been amenable to X-ray analysis. Useful crystals were grown in the presence of 0.3 M K^+ together with a trace amount of Ca^{2+}.[31]

Calsequestrin is made up of three thioredoxin fold domains

The crystal structure of calsequestrin[31] shows that it is made up of three nearly identical tandem domains; each of

Calsequestrin

these has the thioredoxin protein fold[84] (3D Structure). Sequence analysis of this protein family prior to the structure determination gave no hint of the three-domain structure, since there is no indication of a significant repeat in the amino acid sequence.[21,22] The sequence similarities among the domains are not distinguishable from random by any method tested so far. Instead, the asymmetric charge distribution in calsequestrin led to the expectation that the N-terminal domain could fold into a typical globular protein, whereas the negatively charged C-terminus forms a largely disordered tail.[69] The last 130 residues in this protein are more negative than any stretch of comparable length of any protein previously deposited in the Protein Data Bank. The expectation that the C-terminal region would be disordered was at least partially realized. Residues 327–333, a polyanionic loop, and 348–368, the highly negative C-terminal end, were unobserved in the calsequestrin crystal structure, because of their structural disorder[31] (Figure 1).

Each of these thioredoxin domains is made up of a five-stranded β-sheet with two α-helices on both sides (Figure 2). These three domains interact to form an approximate disk-like shape with a radius of 35 to 45 Å and a thickness of 35 Å. Every domain has a hydrophobic core with acidic residues on the exterior, generating highly

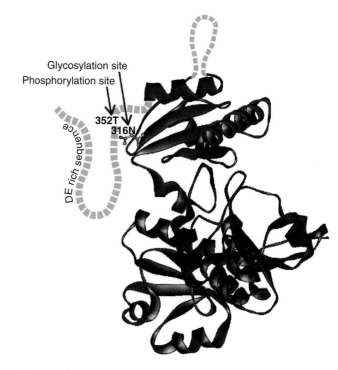

Figure 1 Schematic diagram of rabbit skeletal calsequestrin showing the sites for glycosylation and phosphorylation. Disordered regions are depicted in light blue color (PDB:1A8Y).

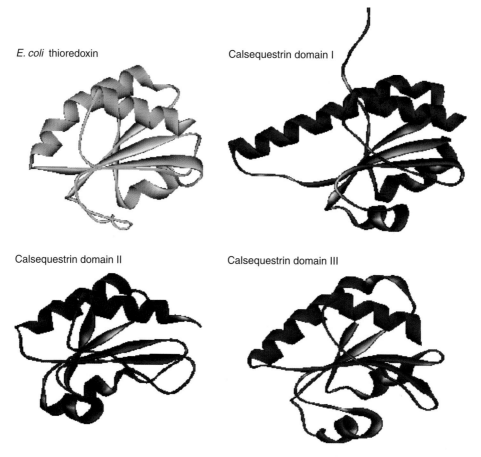

Figure 2 Comparison of *E. coli* thioredoxin structure (2TRX) with individual rabbit skeletal calsequestrin domain (1A8Y).

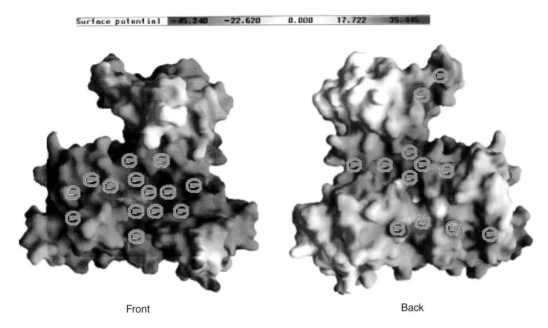

Figure 3 The molecular surface of calsequestrin (PDB: 1A8Y).[58] The front and back of the calsequestrin molecular surfaces show the electrostatic potential from $-54.24\,k_B T$ to $35.44\,k_B T$. Overall, both the surfaces show extreme negative values.

electronegative potential surfaces (Figure 3). The individual domains are connected by short sequences located interior to the domains themselves. These connecting loops and the secondary structural elements that fill the interdomain space contain mostly acidic residues, making the overall center of the protein hydrophilic rather than hydrophobic. Therefore, cations are required to stabilize the acidic center of three domains. Divalent cations, which can provide cross-bridging, might be more effective in this regard than monovalent cations.

All individual domains of calsequestrin contain a high net negative charge, ranging from about -13 to -32, and also a high aromatic amino acid composition, ranging from 9 to 13%. Statistical studies on peptide sequences suggest that charge imbalance is an important factor favoring the unfolded state and that high aromatic content is an important factor favoring the folded state.[85] Perhaps to offset the instability arising from the large net negative charges (-32) especially in domain III, the aromatic groups in that domain interact with each other in an edge-to-face manner, forming a 'herring-bone' chain like that in benzene crystals with seven aromatic groups organized as F-W-F-W-F-W-Y. Such interactions could provide exceptional stability[86] and thereby balance the destabilizing effects of the very high net negative charge of this domain.

A family of Ca^{2+} binding proteins in SR and ER has thioredoxin folds

Several Ca^{2+} binding proteins in the ER and SR lumens have been characterized, including PDI, CRT, Crp96, immunoglobulin binding protein (BiP or Grp78), ERp72, and ER calcistorin.[87] Regardless of their other functions, all these ER (and SR) proteins have the secondary characteristic of binding Ca^{2+} as the price to pay for existing within a high Ca^{2+} milieu, thereby contributing to overall Ca^{2+} storage and homeostasis.[88] Crp96, PDI, calcistorin, and calreticulin bind 10, 19, 20, and 25 Ca^{2+} ions respectively.[89] Thus, ER (or SR) lumen proteins may all be dual or multifunctional, with one of the functions being the buffering of Ca^{2+}. Taking this one step further, to coordinate function and Ca^{2+} levels, there is likely to be mutual regulation between the functions of these proteins and Ca^{2+} binding by them.

From sequence analysis, several of the multifunctional ER-, SR-calcium storage proteins were predicted to contain one or two thioredoxin domains linked in tandem with calsequestrin-like domains,[89] which now, owing to the crystal structure of calsequestrin, are realized to contain even more repeats of the thioredoxin fold together with other structural motifs.

Calsequestrin monomer stacks into continuous, linear polymers

The few positive charges in calsequestrin are mostly involved in salt bridges between domains within a monomer. In crystal lattice, the monomers stack with alternate orientations forming a ribbon-like polymer. These alternating orientations define two distinct twofold axes that are perpendicular to the long axis of the ribbon, leading to two types of packing interfaces within the polymer of $70 \sim 90$ Å in diameter[31] (Figure 4). The two dimer interfaces involve burying large surface areas and

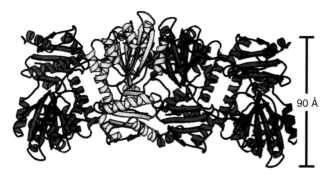

Figure 4 Diagram illustrating the interaction of the four calsequestrin molecules. Front-to-front interaction: green-red, blue-yellow (left two and right two); back-to-back interaction: red-yellow (central two).

very intricate interactions, both of which are unlike typical crystal contacts. Not only does the polymer in the crystal lattice have the linear morphology inferred for the physiologically relevant aggregation of calsequestrin, but the two contacts that stabilize this polymer also have structural details that would lend themselves to control by Ca^{2+} binding.

The first type of interface, front-to-front interface, involves the fitting of the convex globule from domain II on one subunit into a concave depression on the other, and this interface also involves 'arm exchange or domain swapping' between the extended amino terminal ends of the two adjacent molecules. Each extended arm comprises 10 N-terminal residues from domain I. This extended arm binds along a groove between two β-strands of domain II in the neighbor (Figure 4). Also, the front-to-front interface contains a number of negatively charged groups.[31] The second type of interface, back-to-back interface, involves bringing together two jawlike openings that lie between domains I and III, thus creating a substantial cavity within this interface. The last observed residue at the two carboxyl terminal ends lie within this cavity; each monomer contains an additional polyanionic tail that is unobserved because of protein disorder;[31] it is unclear how many of these 20 additional disordered residues reside within the cavity and how many extend to the exterior of this structural feature.

On two sides of this cavity formed by a back-to-back interaction, a symmetry-related pair of helix–helix interface contains a pair of lysine/glutamate salt bridges and a glutamate from one helix that associates with the amino terminal (e.g. the positive dipole) end of its neighbor; it is as if a side chain from one helix acts as an N-cap residue for its neighbor. The sequences and overall folding motifs of this specific helix–helix interaction show high similarity to those of the interaction between CaM (or troponinC) and its target peptide, the dibasic hydrophobic (DBH) sequence.[90–94]

Trifluoperazine (TFP), an antipsychotic drug, affects calsequestrin structure and function. Studies with peptide fragments of CaM led to the conclusion that TFP, which is a potent inhibitor of CaM, binds to a particular sequence on CaM (TFP binding site), thereby inhibiting interactions between this site on CaM and a DBH site on the target enzyme.[95–97] Because calsequestrin contains two helix-forming sequences similar to the TFP binding and DBH sites, it was suggested that TFP binding might interfere with an interaction between these two sites and thereby alter the function of calsequestrin.[21] Addition of TFP to the solution of calsequestrin leads to inhibition of Ca^{2+}-induced protein aggregation. That is, at Ca^{2+} concentration levels that cause 60% of calsequestrin to be precipitated, TFP lowers the amount of calsequestrin precipitate below about 2%.[69] The amount of Ca^{2+} in the precipitate shows a sharp drop as TFP concentration increases. Addition of TFP before Ca^{2+} causes calsequestrin to remain susceptible to protease digestion. Taken together, these results show that folding, aggregation, and Ca^{2+} binding are all coordinated events and TFP simultaneously inhibits calsequestrin aggregation and high-capacity calcium binding.[69] A significant uncertainty is whether these coordinated events are relevant to calsequestrin function *in vivo*. Determination of the crystal structure of calsequestrin[31] revealed a trimer of the thioredoxin fold, a structure motif that often provides the platform for small molecule binding. In this regard, the inhibition of folding, aggregation, and precipitation by TFP is especially interesting because all these features suggest the possibility of the regulation of Ca^{2+} binding by certain ligands.

Important structural features for calcium binding and release

The polymerization or matrix formation of calsequestrin has been hypothesized as a regulation mechanism of Ca^{2+} fluxes,[14,31,88,98] and a couple of different structural models for the polymerization have been presented.[31,99] Strong cooperative Ca^{2+} binding accompanies the polymerization of calsequestrin, and this association of calsequestrin monomers to form macrostructures has a character similar to condensation or crystallization. As mentioned earlier, the crystal lattice contains a polymeric organization that could account for the observed polymeric structures *in vivo*. The interactions are more extensive and more complex than those typically observed in crystal contacts; rather, these interfaces are similar to the interfaces in proteins that function as oligomers. Raman spectroscopy showed that the calsequestrin structure in the crystals is extremely similar to its structure in solution.[72] Dimer formation is a key feature in the overall crystal structure of calsequestrin.[31] In both 'front-to-front' and 'back-to-back' contacts, electronegative pockets are formed, resulting in the production of a linear polymer through extensive fitting between the two monomer proteins. For these reasons, the polymer is likely to be the functional form of calsequestrin, and several observations support the physiological relevance of this polymer. First, at the ion and protein concentrations (100 mg mL^{-1}) within the SR,[4] it seems likely that calsequestrin remains polymerized during a contraction/relaxation cycle. Each polymer provides a highly acidic and extended surface onto which the calcium ion can be adsorbed. The attractive forces exerted by such a surface would have a longer range than those from an isolated molecule. A sparingly soluble ion such as Ca^{2+} would tend to spread over this surface of the calsequestrin polymer, forming a readily exchangeable film. An array of such polymers would create long, narrow, negatively charged solvent channels leading to the calcium-release channel. One-dimensional diffusion of Ca^{2+} along the surface of the polymer[88] and restriction of the solvent to long, narrow tubes would work together to speed up the diffusion of calcium from its binding sites in calsequestrin to its release channel. Second, the lack of fixed structure of the calcium binding sites means that binding affinities would be low, and diffusion-limited on-rates and off-rates would be as fast as possible. The use of Ca^{2+} as a cross-linker rather than as a tightly bound form free of H$_2$O speeds up dissociation. Thus Ca^{2+} diffusion from calsequestrin to the Ca^{2+} release channel is likely to involve surface diffusion, a more rapid process than diffusion through liquid.[2,88,100] Comparison of the 3D Structure of rabbit skeletal muscle calsequestrin with a homology model of canine cardiac calsequestrin from the point of view of the coupled Ca^{2+}-binding and polymerization mechanism leads to a possible explanation for the twofold reduced Ca^{2+} binding capacity of cardiac calsequestrin despite very similar overall net negative charge for the two molecules.[64]

Thioredoxin fold in calsequestrin and hypothetical ligand

Thioredoxin-like folds often bind ligands or have a redox site at a specific locus.[101,102] In spite of the considerable variability in size, the number of β-strands, and their strand order, the positions of the binding site of proteins belonging to this type of fold (open α/β-sheet motif) were successfully predicted from topology diagrams.[102] In the case of thioredoxins and other thiol-redox proteins, the redox-active site is located at the same topologically predicted site. From the topology diagram and these considerations, a total of three potential active sites (or binding sites) can be predicted in each calsequestrin molecule. These sites occur at chain reversals, which generate a crevice defined by the edge of one β-sheet and the carboxy ends of the adjacent β2 and β3 strands.[102] Not only are these sites predicted to be binding sites from the topology diagrams but they also look like flavin or nucleotide binding sites in the 3D Structure, that is, these sites form hydrophobic grooves that are bounded by exposed hydrophilic side chains, with considerable structural similarity to known nucleotide and flavin binding sites in other open α/β-sheet structures. Given the observed interaction with TFP, calsequestrin can bind planar, hydrophobic molecules, and such ligand binding can evidently regulate aggregation and Ca^{2+} binding. PDI, CRT, and several other ER-, SR-calcium binding proteins apparently contain thioredoxin domains and are also high-capacity Ca^{2+} binding proteins. PDI and CRT are known to bind ATP.[103,104] Overall, these observations are consistent with the crystal structure of calsequestrin, indicating that calsequestrin has a structural motif that is commonly involved in flavin or nucleotide binding.

Doxorubicin (former generic name, Adriamycin) and daunorubicin (Daunomycin) are two widely used and highly effective anticancer drugs (anthracycline derivatives). The biological properties of these cytotoxic DNA intercalators involve ternary interactions of drug–DNA complexes with DNA binding factors.[105,106] Unfortunately, these drugs produce severe cardiotoxicity, often irreversibly damaging the heart, which limits their therapeutic potential.[107] Even though details of the molecular basis of this side effect have not been determined, evidence has been reported of interactions between these anthracycline molecules and the components of the SR Ca^{2+} release complex and that anthracyclines can stimulate calcium release from isolated SR.[108–110] Anthracyclines and their metabolites were shown to persist in heart tissue and other muscle tissues owing to the propensity of muscle to accumulate these types of small molecules.[107,110,111] The fact that calsequestrin is the most abundant protein in the heart and that it contains a potential binding site that modulates Ca^{2+} levels is reason enough to speculate the nonspecific and damaging interaction of calsequestrin with anthracyclines or their metabolites. In fact, the major side effects of the TFP, the antipsychotic drug, are muscle-related symptoms including palpitation.

Summary: current model for calsequestrin structure/function relationships

Integrating all the information on calsequestrin presented above leads to the scheme in Figure 5 with the following specific steps:

- *At low ionic strength, calsequestrin is partially or completely unfolded*: In Figure 5, UF refers to the unfolded form (random coil) observed at low ionic strength. Because the domain surfaces are largely polar and have a large negative charge, folding of calsequestrin is ionic-strength dependent.
- *Higher ionic strengths are needed to completely fold calsequestrin*: As the ionic strength is raised to about 150 mM, UF converts first to the intermediate form, IF$_1$, which is characterized by the folding of domain I and II, but with domain III still unfolded. Folding of domain III converts IF$_1$ to the second intermediate form, IF$_2$ at ionic strengths above 350 mM.[64]

Calsequestrin

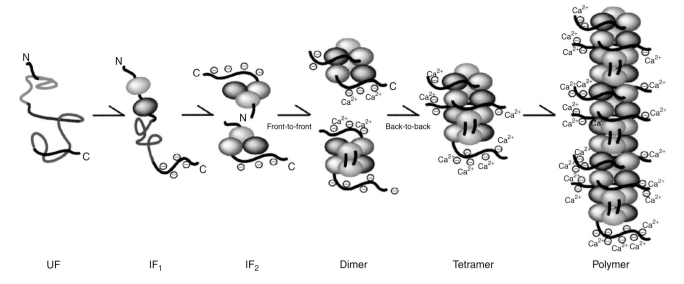

Figure 5 Schematic diagram of the conformations of calsequestrin at different Ca^{2+} (and other cation) concentrations in sarcoplasmic reticulum. UF (unfolded): unfolded calsequestrin under nonphysiological condition, IF_1 (folding intermediate 1): partially folded intermediate, IF_2 (folding intermediate 2): appropriate ionic strength and calcium level (0.01–1 mM Ca^{2+}) are able to promote the collapse of the three domains into one compact structure. Front-to-front dimer: dimer formation through interactions of the N-terminal arm exchange. Back-to-back interaction: Ca^{2+} chelation by the two C-terminals in the dimer interface at high Ca^{2+} concentration.

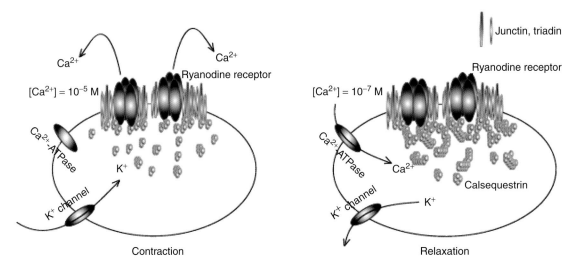

Figure 6 Schematic diagram of the conformations of calsequestrin at different Ca^{2+} concentrations in junctional sarcoplasmic reticulum, as suggested by the crystal structure. Calcium uptake and release might also be modulated by other counterions, and the SR membrane is known to be permeable for other cations Na^+, H^+, and Mg^{2+}, and the anion Cl^-.[112]

- *Divalent cations induce the formation of linear polymers*: Without divalent cations, the folding stops at the monomeric state. With divalent cations, calsequestrin polymerization is induced by the formation of two different dimerization interfaces, front-to-front and back-to-back.[31,64] Since both these contacts involve cross-bridging as acidic groups are brought into proximity, on-rates are likely to be close to the diffusion limit ($\sim 10^9\,M^{-1}\,s^{-1}$) and off-rates to be approximately $10^6\,s^{-1}$ for a binding constant of $10^3\,M^{-1}$(2). On the basis of the much greater negative charge of the back-to-back interface, it is likely that the front-to-front-type dimer forms ahead of the back-to-back-type dimer.

Current hypothetical model for calcium binding and release in the SR is as follows (Figure 6) but should be further tested.

- *Reduced diffusion dimensionality accelerates Ca^{2+} release*: The negative charges on the calsequestrin polymer surface would accelerate Ca^{2+} release by reducing its dimensionality of diffusion. Therefore, calsequestrin polymer located near the release channel can act as a Ca^{2+} wire using a huge Ca^{2+} gradient ($\sim 10^4$) maintained between the cytosolic and the lumenal spaces. Acceleration of diffusion by dimensionality reduction was suggested previously for another system.[113] The highly charged disordered tails are proposed to contribute to these approximately one-dimensional diffusion pathways along the calsequestrin polymer surface.

- *Binding calsequestrin to junctin and/or triadin by 'arm exchange' or other means*: The distribution of calsequestrin is uneven within the lumen with a markedly high concentration in close proximity to the RyR,[98,114] possibly by calsequestrin-anchoring proteins such as triadin and junctin.
- *Concomitant calsequestrin aggregation and high-capacity Ca^{2+} binding*: As the Ca^{2+} levels increase and monovalent ion levels (H^+, K^+, Na^+) decrease owing to the charge valence in the SR lumen, there is a concomitant formation of calsequestrin polymers or aggregates and high-capacity Ca^{2+} binding.

ACKNOWLEDGEMENTS

Original research described in this review was supported by grants to CHK from the National Science Foundation (NSF), the American Heart Association (AHA), and the Murdock Charitable Trust.

REFERENCES

1. DH MacLennan and P Wong, *Proc Natl Acad Sci USA*, **68**, 1231–35 (1971).
2. DH MacLennan, N Abu-Abed and C Kang, *J Mol Cell Cardiol*, **34**, 897–918 (2002).
3. J Meldolesi and T Pozzan, *Trends Biochem Sci*, **23**, 10–14 (1998).
4. AV Somlyo, H Gonzalez-Serratos, H Shuman, G McClellan and AP Somlyo, *J Cell Biol*, **90**, 577–94 (1981).
5. C Franzini-Armstrong, L Kenney and E Varriano-Marston, *J Cell Biol*, **105**, 49–56 (1987).
6. A Saito, S Seiler, A Chu and S Fleischer, *J Cell Biol*, **99**, 875–85 (1984).
7. R Mitchell, H Simmerman and LR Jones, *J Biol Chem*, **263**, 1376–81 (1988).
8. E Damiani and A Margreth, *Biochem Biophys Res Commun*, **172**, 1253–59 (1990).
9. J Collins, A Tarcsafalvi and N Ikemoto, *Biochem Biophys Res Commun*, **167**, 189–93 (1990).
10. LR Jones, L Zhang, K Sanborn, A Jorgensen and J Kelley, *J Biol Chem*, **270**, 30787–96 (1995).
11. W Guo and KP Campbell, *J Biol Chem*, **270**, 9027–30 (1995).
12. W Guo, A Jorgensen, LR Jones and KP Campbell, *J Biol Chem*, **271**, 458–65 (1996).
13. L Zhang, J Kelley, G Schmeisser, Y Kobayashi and LR Jones, *J Biol Chem*, **272**, 23389–97 (1997).
14. M Ohkura, K Furukawa, H Fujimori, A Kuruma, S Kawano, M Hiraoka, A Kuniyasu, H Nakayama and Y Ohizumi, *Biochemistry*, **37**, 12987–93 (1998).
15. LR Jones, Y Suzuki, W Wang, Y Kobayashi, V Ramesh, C Franzini-Armstrong, L Cleemann and M Morad, *J Clin Invest*, **101**, 1385–93 (1998).
16. D Shin, J Ma and D Kim, *FEBS Lett*, **486**, 178–82 (2000).
17. C Franzini-Armstrong and A Jorgensen, *Annu Rev Physiol*, **56**, 509–34 (1994).
18. M Inui and S Fleischer, *Methods Enzymol*, **157**, 490–505 (1988).
19. E Johnson and J Sommer, *J Cell Biol*, **33**, 103–29 (1967).
20. G Meissner, *Annu Rev Physiol*, **56**, 485–508 (1994).
21. L Fliegel, M Ohnishi, M Carpenter, V Khanna, R Reithmeier and DH MacLennan, *Proc Natl Acad Sci USA*, **84**, 1167–71 (1987).
22. B Scott, H Simmerman, J Collins, B Nadal-Ginard and LR Jones, *J Biol Chem*, **263**, 8958–64 (1988).
23. T Kawasaki and M Kasai, *Biochem Biophys Res Commun*, **199**, 1120–27 (1994).
24. C Knudson, KK Stang, AO Jorgensen and KP Campbell, *J Biol Chem*, **268**, 12637–45 (1993).
25. C Knudson, KK Stang, CR Moomaw, CA Slaughter and KP Campbell, *J Biol Chem*, **268**, 12646–54 (1993).
26. A Caswell, NR Brandt, JP Brunschwig and S Purkerson, *Biochemistry*, **30**, 7507–13 (1991).
27. G Wetzel, S Ding and F Chen, *Mol Genet Metab*, **69**, 252–58 (2000).
28. Y Kobayashi, B Alseikhan and LR Jones, *J Biol Chem*, **275**, 17639–46 (2000).
29. A Herzog, C Szegedi, I Jona, FW Herberg and M Varsanyi, *FEBS Lett*, **472**, 73–77 (2000).
30. B Murray and K Ohlendieck, *FEBS Lett*, **429**, 317–22 (1998).
31. S Wang, W Trumble, H Liao, C Wesson, A Dunker and C Kang, *Nat Struct Biol*, **5**, 476–83 (1998).
32. M Berchtold, H Brinkmeier and M Muntener, *Physiol Rev*, **80**, 1215–65 (2000).
33. DH MacLennan, *Eur J Biochem*, **267**, 5291–97 (2000).
34. K Gunji, S Kubota, C Stolarski, S Wengrowicz, J Kennerdell and J Wall, *Autoimmunity*, **29**, 1–9 (1999).
35. W Kreisel, A Siegel, A Bahler, C Spamer, E Schiltz, M Kist, G Seilnacht, R Klein, PA Berg and C Heilmann, *Scand J Gastroenterol*, **34**, 623–28 (1999).
36. S Nagayama, T Yokoi, H Tanaka, Y Kawaguchi, T Shirasaka and T Kamataki, *J Toxicol Sci*, **19**, 163–69 (1994).
37. B Linck, P Boknik, S Huke, U Kirchhefer, J Knapp, H Luss, FU Muller, J Neumann, Z Tanriseven, U Vahlensieck, HA Baba, LR Jones, KD Philipson and W Schmitz, *J Pharmacol Exp Ther*, **294**, 648–57 (2000).
38. B Knollmann, BE Knollmann-Ritschel, NJ Weissman, LR Jones and M Morad, *J Physiol*, **525**, 483–98 (2000).
39. W Wang, L Cleemann, LR Jones and M Morad, *J Physiol*, **524**, 399–414 (2000).
40. H Lahat, E Pras, T Olender, N Avidan, E Ben-Asher, O Man, E Levy-Nissenbaum, A Khoury, A Lorber, B Goldman, D Lancet and M Eldar, *Am J Hum Genet*, **69**, 1378–84 (2001).
41. A Postma, I Denjoy, T Hoorntje, J Lupoglazoff, A Da Costa, P Sebillon, M Mannens, A Wilde and P Guicheney, *Circ Res*, **91**, 21–26 (2002).
42. L Stull, R Matteo, W Sweet, D Damron and C Schomisch Moravec, *J Mol Cell Cardiol*, **33**, 449–60 (2001).
43. Y Sato, DG Ferguson, H Sako, GW Dorn 2nd, VJ Kadambi, A Yatani, BD Hoit, RA Walsh and EG Kranias, *J Biol Chem*, **273**, 28470–77 (1998).
44. M Cho, A Rapacciuolo, WJ Koch, Y Kobayashi, LR Jones and HA Rockman, *J Biol Chem*, **274**, 22251–56 (1999).

45 A Margreth, E Damiani and E Bortoloso, *Acta Physiol Scand*, **167**, 331–38 (1999).

46 A Zarain-Herzberg, L Fliegel and DH MacLennan, *J Biol Chem*, **263**, 4807–12 (1988).

47 D Clegg, JC Helder, BC Hann, DE Hall and LF Reichardt, *J Cell Biol*, **107**, 699–705 (1988).

48 J Fujii, H Willard and DH MacLennan, *Somat Cell Mol Genet*, **16**, 185–89 (1990).

49 E Choi and D Clegg, *Dev Biol*, **142**, 169–77 (1990).

50 S Cala and LR Jones, *J Biol Chem*, **266**, 391–98 (1991).

51 M Arai, N Alpert and M Periasamy, *Gene*, **109**, 275–79 (1991).

52 S Treves, B Vilsen, P Chiozzi, J Andersen and F Zorzato, *Biochem J*, **283**, 767–72 (1992).

53 K Otsu, J Fujii, M Periasamy, M Difilippantonio, M Uppender, DC Ward and DH MacLennan, *Genomics*, **17**, 507–9 (1993).

54 KW Park, JH Goo, HS Chung, H Kim, DH Kim and WJ Park, *Gene*, **217**, 25–30 (1998).

55 MM Rodriguez, CH Chen, BL Smith and D Mochly-Rosen, *FEBS Lett*, **454**, 240–46 (1999).

56 J Slupsky, M Ohnishi, M Carpenter and R Reithmeier, *Biochemistry*, **26**, 6539–44 (1987).

57 M Ohnishi and R Reithmeier, *Biochemistry*, **26**, 7458–65 (1987).

58 SL Hamilton, MJ Hawkes, K Brush, R Cook, RJ Chang and HM Smilowitz, *Biochemistry*, **28**, 7820–28 (1989).

59 PJ Yazaki, S Salvatori, RA Sabbadini and AS Dahms, *Biochem Biophys Res Commun*, **166**, 898–903 (1990).

60 AG McLeod, AC Shen, KP Campbell, M Michalak and AO Jorgensen, *Circ Res*, **69**, 344–59 (1991).

61 A Knoll, A Stratil, G Reiner, LJ Peelman, M Van Poucke and H Geldermann, *Anim Genet*, **33**, 390–92 (2002).

62 JH Cho, YS Oh, KW Park, J Yu, KY Choi, JY Shin, DH Kim, WJ Park, T Hamada, H Kagawa, EB Maryon, J Bandyopadhyay and J Ahnn, *J Cell Sci*, **113**, 3947–58 (2000).

63 K Yano and A Zarain-Herzberg, *Mol Cell Biochem*, **135**, 61–70 (1994).

64 H Park, S Wu, AK Dunker and C Kang, *J Biol Chem*, **278**, 16176–82 (2003).

65 J O'Brian, ML Ram, A Kiarash and SE Cala, *J Biol Chem*, **277**, 37154–60 (2002).

66 N Ikemoto, B Nagy, G Bhatnagar and J Gergely, *J Biol Chem*, **249**, 2357–65 (1974).

67 T Ostwald, DH MacLennan and K Dorrington, *J Biol Chem*, **249**, 5867–71 (1974).

68 B Aaron, K Oikawa, R Reithmeier and B Sykes, *J Biol Chem*, **259**, 11876–81 (1984).

69 Z He, AK Dunker, CR Wesson and WR Trumble, *J Biol Chem*, **268**, 24635–41 (1993).

70 N Ikemoto, G Bhatnagar, B Nagy and J Gergely, *J Biol Chem*, **247**, 7835–37 (1972).

71 B Cozens and R Reithmeier, *J Biol Chem*, **259**, 6248–52 (1984).

72 R Williams and T Beeler, *J Biol Chem*, **261**, 12408–13 (1986).

73 N Strynadka and M James, *Annu Rev Biochem*, **58**, 951–98 (1989).

74 D Wright, J Holloway and C Reilley, *Anal Chem*, **37**, 884–912 (1965).

75 K Krause, M Milos, Y Luan-Rilliet, D Lew and J Cox, *J Biol Chem*, **266**, 9453–59 (1991).

76 A Jorgensen, A Shen, KP Campbell and DH MacLennan, *J Cell Biol*, **97**, 1573–81 (1983).

77 C Hidalgo, P Donoso and P Rodriguez, *Biophys J*, **71**, 2130–37 (1996).

78 P Donoso, M Beltran and C Hidalgo, *Biochemistry*, **35**, 13419–25 (1996).

79 M Tanaka, T Ozawa, A Maurer, J Cortese and S Fleischer, *Arch Biochem Biophys*, **251**, 369–78 (1986).

80 DH MacLennan, *J Biol Chem*, **249**, 980–84 (1974).

81 A Maurer, M Tanaka, T Ozawa and S Fleischer, *Proc Natl Acad Sci USA*, **82**, 4036–40 (1985).

82 K Hayakawa, L Swenson, S Baksh, Y Wei, M Michalak and ZS Derewenda, *J Mol Biol*, **235**, 357–60 (1994).

83 P Maguire, FN Briggs, NJ Lennon and K Ohlendieck, *Biochem Biophys Res Commun*, **240**, 721–27 (1997).

84 A Holmgren, B Soderberg, H Eklund and C Branden, *Proc Natl Acad Sci USA*, **72**, 2305–9 (1975).

85 Q Xie, G Arnold, P Romero, Z Obradovic, E Garner and A Dunker, *Genome Inf*, **9**, 193–200 (1998).

86 S Burley and G Petsko, *Science*, **229**, 23–28 (1985).

87 D Ferrari and H Soling, *Biochem J*, **339**, 1–10 (1999).

88 DH MacLennan and RA Reithmeier, *Nat Struct Biol*, **5**, 409–11 (1998).

89 M Michalak, P Mariani and M Opas, *Biochem Cell Biol*, **76**, 779–85 (1998).

90 J Gariepy and R Hodges, *FEBS Lett*, **160**, 1–6 (1983).

91 M Payne, YL Fong, T Ono, RJ Colbran, BE Kemp, TR Soderling and AR Means, *J Biol Chem*, **263**, 7190–95 (1988).

92 P Kelly, R Weinberger and M Waxham, *Proc Natl Acad Sci USA*, **85**, 4991–95 (1988).

93 R Colbran, YL Fong, CM Schworer and TR Soderling, *J Biol Chem*, **263**, 18145–51 (1988).

94 D Blumenthal, K Takio, AM Edelman, H Charbonneau, K Titani, KA Walsh and EG Krebs, *Proc Natl Acad Sci USA*, **82**, 3187–91 (1985).

95 W Cook, L Walter and M Walter, *Biochemistry*, **33**, 15259–65 (1994).

96 R Levin and B Weiss, *Mol Pharmacol*, **13**, 690–97 (1977).

97 P Cachia, J Van Eyk, RH Ingraham, WD McCubbin, CM Kay and RS Hodges, *Biochemistry*, **25**, 3553–62 (1986).

98 C Szegedi, S Sarkozi, A Herzog, I Jona and M Varsanyi, *Biochem J*, **337**, 19–22 (1999).

99 G Gatti, S Trifari, N Mesaeli, J Parker, M Michalak and J Meldolesi, *J Cell Biol*, **154**, 525–34 (2001).

100 R Williams, *Biochim Biophys Acta*, **505**, 1–44 (1978).

101 S Katti, D LeMaster and H Eklund, *J Mol Biol*, **212**, 167–84 (1990).

102 C Branden, *Q Rev Biophys*, **13**, 317–38 (1980).

103 S Nigam, AL Goldberg, S Ho, MF Rohde, KT Bush and M Sherman, *J Biol Chem*, **269**, 1744–49 (1994).

104 E Corbett, KM Michalak, K Oikawa, S Johnson, ID Campbell, P Eggleton, C Kay and M Michalak, *J Biol Chem*, **275**, 27177–85 (2000).

105 Y Pommier, RE Schwartz, LA Zwelling and KW Kohn, *Biochemistry*, **24**, 6406–10 (1985).

106 I Berger, L Su, J Spitzner, C Kang, T Burke and A Rich, *Nucleic Acids Res*, **23**, 4488–94 (1995).

107 R Olson, PS Mushlin, DE Brenner, S Fleischer, BJ Cusack, BK Chang and R Boucek Jr, *Proc Natl Acad Sci USA*, **85**, 3585–89 (1988).

108 F Zorzato, G Salviati, T Facchinetti and P Volpe, *J Biol Chem*, **260**, 7349–55 (1985).

109 I Pessah, EL Durie, MJ Schiedt and I Zimanyi, *Mol Pharmacol*, **37**, 503–14 (1990).

110 P Mushlin, BJ Cusack, RJ Boucek Jr, T Andrejuk, X Li and RD Olson, *Br J Pharmacol*, **110**, 975–982 (1993).

111 D Stewart, D Grewaal, RM Green, N Mikhael, R Goel, VA Montpetit and MD Redmond, *Anticancer Res*, **13**, 1945–52 (1993).

112 R Fink and C Veigel, *Acta Physiol Scand*, **156**, 387–96 (1996).

113 P von Hippel and O Berg, *J Biol Chem*, **264**, 675–78 (1989).

114 C Franzini-Armstrong and F Protasi, *Physiol Rev*, **77**, 699–729 (1997).

C-type animal lectins

William I Weis

Departments of Structural Biology and of Molecular & Cellular Physiology, Stanford University School of Medicine, Stanford, CA, USA

FUNCTIONAL CLASS

Carbohydrate binding proteins.

OCCURRENCE

C-type lectins are the founding members and a subgroup of the C-type lectin-like superfamily.[1] Sequences of C-type lectin-like domains (CTLDs) have been identified throughout the animal kingdom. In many cases, CTLDs have been identified solely from sequence data but have not yet been characterized at the protein or functional levels.

Each CTLD-containing protein can be classified into a subgroup according to its domain organization and how the sequence of the CTLD compares to other CTLDs.[1] On the basis of structural data, the presence of residues essential for forming the characteristic Ca^{2+} and carbohydrate-binding site can be used to predict carbohydrate-binding activity from sequence data. Ca^{2+}-dependent lectin activity has been experimentally demonstrated for some members of groups I (proteoglycans), II (hepatic lectins and several macrophage receptors), III (collectins), IV (selectins), and VI (mannose receptor family). This chapter focuses on those family members with demonstrated Ca^{2+}-dependent carbohydrate-binding activity. In these proteins, the CTLD is a carbohydrate-recognition domain (CRD).

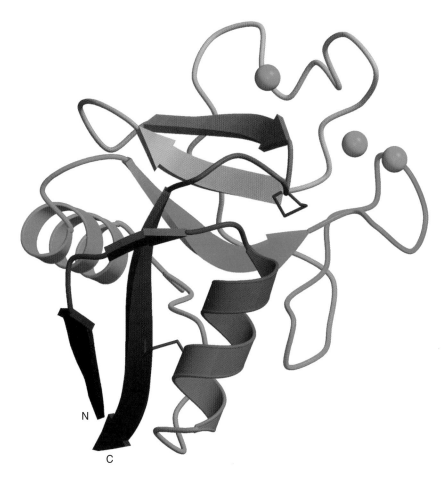

3D Structure Ribbon diagram of the CRD from rat MBP-A. The color changes from blue to red as it traverses the sequence. Calcium ions are shown as spheres. Disulfide bonds are shown in pink. PDB code 2MSB; prepared with MOLSCRIPT[92] and RASTER3D.[93]

BIOLOGICAL FUNCTION

C-type lectins function in the extracellular milieu. Many are integral membrane proteins that serve as cell-surface receptors, and others are secreted proteins. The most well-understood members of the family are those that function in the immune system.[2] The collectins, which include serum mannose binding proteins (MBPs) and pulmonary surfactant proteins, are found in vertebrates and recognize carbohydrate structures characteristic of pathogenic surfaces. Binding to pathogens triggers complement-mediated killing of these organisms. The mannose receptor (MR) found on the surface of macrophages serves in pathogen clearance, including the clearance of by-products of inflammation such as released lysosomal enzymes and collagen fragments.[3]

C-type lectins also play important roles in cell adhesion. The selectins mediate the initial contacts between circulating leukocytes and the vascular endothelium.[2] The intial weak interaction causes the leukocytes to roll along the endothelium. This triggers cytokine production and subsequent upregulation of integrins; the now firmly adherent leukocytes migrate through the endothelial wall into the lymphatic system. DC-SIGN is believed to be involved in the initial low-affinity contacts formed between antigen-presenting dendritic cells and T cells in the secondary lymphoid organs that allow the latter to scan the dendritic cell surface for appropriate peptide-MHC complexes.[4] DC-SIGN may also mediate the rolling of circulating dendritic cells along the vasculature.[5] DC-SIGN and the related DC-SIGNR, which is expressed on certain endothelial cells, also serve as receptors for HIV and other viruses. It is thought that binding of HIV to the dendritic cell surface at primary sites of virus exposure, such as the mucosal epithelium of the genital tract, allows the virus to be transported to the secondary lymphoid tissues, where it can infect T cells.[6–9]

Mammalian asialoglycoprotein receptors and their avian counterparts are present on the surface of hepatocytes.[10] These proteins bind to circulating complex N-linked carbohydrates whose terminal sialic acid (mammals) or sialic acid and galactose residues (birds) have been removed. The exposed terminal galactose or N-acetylglucosamine residues are recognized by the receptor, resulting in endocytosis of the glycoprotein. The physiological role of this glycoprotein clearance is poorly understood. There are a number of cell surface C-type lectins present in macrophages, but the roles of these proteins are not known. Also unknown is the biological function of C-type lectin activity in some proteoglycans.[11]

AMINO ACID SEQUENCE INFORMATION

The presence of at least one C-type CRD is the defining characteristic of the family. The CRD is found in all primary structure orientations, that is, it can be the N- or C-terminal domain of a multidomain protein, or it can be flanked on both sides by other domains.

A database of C-type lectin sequences can be found at http://ctld.glycob.ox.ac.uk/ctld/. Statistical search profiles for CTLDs are available in all major databases, including Prosite (PS00615 and PS50041), Pfam (PF00059), InterPro (IPR001304), and SMART (SM00034). The sequences and SwissProt identification numbers for the examples discussed in this chapter are as follows:

Human MBP (hMBP)	P11226
Rat MBP-A (MBP-A)	NP036731
Rat MBP-C (MBP-C)	NP073195
Human pulmonary surfactant protein-D (SP-D)	NP003010
Human mannose receptor (MR)	P22897
Human hepatic lectin-1 (HHL-1)	P07306
Rat hepatic lectin-1 (RHL-1)	P02706
Chicken hepatic lectin (CHL)	P02707
Human E-selectin	P16581
Human P-selectin	P16109
Human dendritic cell-specific ICAM-3 grabbing nonintegrin (DC-SIGN)	Q9NNX6
Human DC-SIGN-related receptor (DC-SIGNR)	Q9H2X3
Tunicate lectin TC14	BAB16304

PROTEIN PRODUCTION, PURIFICATION, AND MOLECULAR CHARACTERIZATION

Several family members have been purified to homogeneity from their natural sources, for example, the hepatic lectins[12,13] and the collectins.[14–22] The key step in purification is affinity chromatography on a matrix derivatized with sugars[23] in the presence of Ca^{2+}.

The CRDs, or CRDs with one or more other domains of the full C-type lectin, have been produced in large quantities as recombinant proteins for biochemical and structural analysis. Production in *Escherichia coli* has been achieved using periplasmic[11,24–26] and cytoplasmic[27] expression systems. Small amounts of correctly folded material are produced directly in these systems, but efficient production has usually entailed denaturation/renaturation procedures to achieve refolding. Generally, the protein is denatured in the presence of β-mercaptoethanol or dithiothreitol to reduce disulfide bonds, which are likely to be scrambled. This is followed by removal of denaturant by dialysis in the presence of air to facilitate formation of the correct disulfide bonds. Application to an affinity column in the presence of Ca^{2+} insures purification of functional material; elution with a Ca^{2+} chelator provides Ca^{2+}- and sugar-free material for subsequent studies.

Bacterial expression has been the most successful with CRDs that do not have N-linked glycosylation sites, presumably because the eukaryotic machinery that uses the glycosylation/chaperone machinery[28] is not critical for proper folding in these cases. For other C-type lectins, expression in eukaryotic cells, including the yeast *Pichia pastoris*,[29] the baculovirus/SF9 system,[30,31] Chinese hamster ovary (CHO) cells,[32–35] and mouse myeloma cells,[36] has been successful. In some cases, like the selectins, the carbohydrate ligand is too expensive for preparation of affinity columns; in these cases, affinity chromatography using antibodies[33] or purification tags such as the Fc portion of an antibody[35] or polyhistidine can be employed.

Each C-type CRD contains two conserved disulfide bonds, and some family members have additional disulfides. N-linked glycosylation sites are present in many cases, but these are not conserved. The CRD has one absolutely conserved Ca^{2+} site, and many family members have a second Ca^{2+} site. The Ca^{2+} affinity of these proteins is fairly weak (K_d values typically 10–1000 μM).

The binding of Ca^{2+} to C-type CRDs has been characterized in a number of ways. In the absence of Ca^{2+}, the C-type CRD becomes sensitive to proteases, indicating that Ca^{2+} is essential for maintaining the structure of the domain. Measurement of resistance to proteolysis as a function of Ca^{2+} can therefore be used to obtain an effective equilibrium constant K_{Ca} as well as the stoichiometry of the interaction.[25,37,38] Likewise, measurement of binding to sugar ligands as a function of Ca^{2+} concentration (see next section) can be used to obtain K_{Ca} and stoichiometry.[25,37,39] The close agreement in stoichiometry and K_{Ca} values determined by these methods emphasizes the essential nature of Ca^{2+} in maintaining a structure competent for binding to sugar ligands. Structural data reveal that a highly conserved tryptophan residue lies directly underneath the conserved Ca^{2+} site, allowing changes in tryptophan fluorescence as a function of Ca^{2+} to be used for measurement of Ca^{2+} affinity.[40,41] Fluorescence resonant energy transfer from Tb^{3+} to the conserved tryptophan has also been employed to study the interaction of this ion with MBP-C.[40]

The ability of other ions to replace Ca^{2+} structurally and functionally has been investigated for several C-type lectins. The effectiveness of group IIa (alkaline earth) ions in protecting CHL from proteolysis falls in the order $Sr^{2+} > Ca^{2+} > Ba^{2+}$; Mg^{2+} does not afford significant protection.[37] The failure of Mg^{2+} to protect CHL from proteolysis is consistent with its inability to support sugar binding by the rabbit homolog.[12] Similar experiments performed on E-selectin showed that Ba^{2+} and Sr^{2+} are fivefold more effective than Ca^{2+}, with Mg^{2+} again showing little effect.[38] Curiously, Ba^{2+} completely and Sr^{2+} partially inhibit E-selectin function, as assessed in a cell-based adhesion assay.[38] Trivalent lanthanide ions also substitute functionally and structurally for Ca^{2+} in MBPs.[25,40,42] These ions bind much more tightly to the CRD than Ca^{2+}.

Many C-type lectins are oligomers. In general, the affinity of single CRDs for carbohydrate ligands is weak, but strong binding can be achieved by the clustering of multiple binding sites in an oligomeric structure. The CRD itself can dimerize in some instances, whereas in others a different domain of the protein provides the oligomerization activity.[1] For example, a trimeric coiled-coil of α-helices precedes the CRD in the collectin subgroup.

ACTIVITY AND INHIBITION TESTS

Like most lectins, C-type CRDs bind rather weakly to monovalent carbohydrate ligands, with K_d values typically in the 50 μM to 10 mM range. High avidity interactions are frequently achieved through oligomerization and multivalent binding.[43] Several different ligand-binding assays have been used to characterize binding affinity and specificity. For inexpensive sugar ligands, the most straightforward qualitative method is affinity chromatography, where sepharose is conjugated with mannose, galactose, N-acetylglucosamine, fucose, and so on, or larger polysaccharides such as mannan or those present on glycan-rich proteins like invertase.[12,23,44] Application of the protein to the column in the presence of Ca^{2+} results in retention or, in some cases, retardation on the column. Elution with Ca^{2+} chelator such as ethylenediaminetetraacetic acid (EDTA) confirms the Ca^{2+} dependence of the interaction. More quantitative data come from a solid-phase assay in which the protein is coated onto plastic wells in the presence of Ca^{2+}. Radiolabeled bovine serum albumin (BSA) that has been previously derivatized with a simple sugar serves as a highly avid, multivalent ligand. After incubation with this neoglycoprotein, the well is washed to remove unbound ligand and counted.[45] Different sugars can be attached to BSA, allowing comparison of selectivities. The Ca^{2+} dependence of the interaction can also be obtained in this assay by adding increasing amounts of chelator. Solution-based assays that employ similar radiolabeled neoglycoconjugates have also been described.[12,46]

The solid-phase format is convenient for measuring inhibition constants K_i. Radiolabeled sugar-BSA is bound to the immobilized CRD. After washing away unbound radiolabeled ligand, a nonradioactive competing ligand is added in increasing concentrations. After washing, the remaining radioactivity is counted, thereby generating an inhibition curve. A similar assay has been used for selectins, in which inhibition of adhesion of cells expressing the selectin to target cells expressing the carbohydrate ligand can be measured in the presence of increasing concentrations of test ligand.

The weak affinity and rapid exchange of ligand on the protein has allowed direct measurement of dissociation constants by NMR spectroscopy. In these experiments, structural data indicated the presence of aromatic residues near the binding site. Resonances from these residues could be assigned and followed during the titration.[39,47,48] In other experiments, the sugar resonances themselves could be followed.[39,48] In either case, a plot of line shift as a function of sugar concentration provides a simple binding isotherm that yields absolute dissociation constants K_d. Importantly, in these experiments the relative K_d values for the interaction of several sugar ligands with MBP-A agree with relative K_i values obtained in the solid-phase assay, indicating that relative binding affinities can be obtained from the latter assay, which is faster and requires less protein. K_d values for binding to cells have also been obtained from Scatchard analysis (e.g. reference 49).

X-RAY AND NMR STRUCTURES

Crystal structures of several C-type CRDs are known: MBPs (rat MBP-A and MBP-C[50,51] and the single human protein,[52]) E- and P-selectin,[35,53] human hepatic lectin-1 (asialoglycoprotein receptor),[54] pulmonary surfactant D,[29] CRD-4 of the mannose receptor,[55] DC-SIGN and its close relative DC-SIGNR,[56] and TC14, a galactose-binding lectin from a tunicate.[27] Crystallization conditions have generally involved polyethylene glycol (PEG), and Ca^{2+}; the prevalence of PEG likely reflects the fact that many common crystallization salts form precipitates with Ca^{2+}. The relatively weak Ca^{2+} affinity of these proteins (10–1000 μM) implies that the solubility product of these precipitates is sufficiently low to prevent full occupancy.

In addition to carbohydrate-binding family members, structures of several CTLDs have been determined by X-ray crystallography or NMR: lithostathine,[57] tetranectin,[58,59] type II antifreeze protein from sea raven,[60] a group of snake venom proteins,[61–65] and receptors found on natural killer cells (reviewed in reference 66), and a related protein encoded by the Epstein–Barr virus.[67] C-type lectin-like folds have also been found in the hyaluronan-binding link module,[68] the angiogenesis inhibitor endostatin,[69,70] and domains of the bacterial adhesion proteins intimin[71] and invasin.[72] The sequences of these proteins, however, are unrelated to those of C-type lectin-like domains, suggesting that they may have independently evolved to a similar fold.[1]

The fold of the C-type CRD is illustrated by rat MBP-A[50,73] (Figure 1(a)). The domain is approximately 40 × 25 × 25 Å. The structure starts as a β-strand, followed by a 10-residue α-helix, an extended stretch of 10 residues, and a second α-helix. A short turn after this helix leads into a second β-strand that turns sharply at a conserved glycine. The backbone then enters a region of about 45 amino acids lacking regular secondary structure, featuring two loops, an extended stretch, and two more loops. After the fourth loop, the backbone forms two antiparallel β-strands that end in a tight turn. The conserved cysteine residues at the start of β3 and in the turn following β4 form a disulfide bond. The structure enters a short loop, and finishes with a β-strand that pairs in an antiparallel orientation with the first amino terminal strand. The β5 strand also contains a conserved cysteine that forms a disulfide bond with a conserved cysteine in the first α-helix. The pairing of the

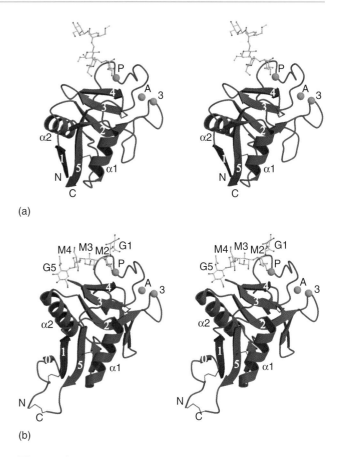

Figure 1 The C-type CRD fold. (a) Stereo view of ribbon diagram of the rat MBP-A CRD bound to a high-mannose oligosaccharide (PDB ID 2MSB). The Ca^{2+} proteins are shown as cyan spheres, marked 'P' for principal site, 'A' for auxiliary site, and '3' for the third site. The two α-helices are indicated, and the five β-strands are numbered. Disulfide bonds are shown in pink. The oligosaccharide is shown in yellow. (b) Stereo view of the DC-SIGN CRD is shown bound to a pentasaccharide (PDB ID 1K9I). Coloring and labeling are as in (a). Residues in the pentasaccharide are G, N-acetylglucosamine, and M, mannose. The extra β-strand at the beginning of this long-form CRD is designated β0. Reprinted with permission from H Feinberg, DA Mitchell, K Drickamer and WI Weis, *Science*, **294**, 2163–66 (2001).[56] Copyright 2001 American Association for the Advancement of Science.

first and last β-strands places the beginning and end of the CRD next to one another. This arrangement explains the presence of C-type CRD sequences in a variety of primary structure orientations. The CRD is located at either the amino- or carboxy-terminal end of many C-type lectins, whereas in others the CRDs are internal, flanked on both sides by various protein modules. This 'loopout' topology implies that for membrane-bound C-type lectins, the CRD can adopt the same orientation with respect to the cell surface whether it is at the amino or carboxy terminus.

Many CTLDs have an additional N-terminal β-strand (β0) that pairs with β1. In this subset of CTLDs, the polypeptide preceding β0 contains a conserved cysteine that forms a disulfide bond with another conserved cysteine (Figure 1(b)). The presence of these cysteines allows classification of CTLD sequences into 'long' or 'short' forms.[1] MBP-A is an example of a short form C-type CRD, and the DC-SIGN CRD is an example of the long form. Various family members have additional disulfide bonds (e.g. DC-SIGN; Figure 1(b)), but these are not conserved.

Apart from the short- and long-form classes, the principal difference amongst the known C-type CRD structures is the size and orientation of the second α-helix. An example of this difference is illustrated in Figure 1(b), where α2 of DC-SIGN is one turn longer and angled upwards relative to that of MBP-A. In this case, the DC-SIGN helix forms an essential part of an extended carbohydrate-binding site.[56]

The structure of the MBP-A CRD contains three calcium ions, designated sites 1, 2, and 3 for historical reasons. MBP site 2 is conserved in all C-type lectins, and is referred to as the *principal Ca^{2+} site*. Sugar binding occurs at the principal Ca^{2+} site. MBP site 1 is found in many family members, and is referred to as the *auxiliary Ca^{2+} site*. MBP site 3 has also been found in the crystal structures of MBP-A,[73,74] pulmonary surfactant protein-D,[29] and DC-SIGN and DC-SIGNR.[56] Most of the coordination ligands at this site are water molecules, and the two protein ligands are also shared with the auxiliary site. Binding curves of resistance to proteolysis or sugar binding as a function of Ca^{2+} binding are best fit by a model that is second order in Ca^{2+} concentration,[25] so it is likely that this site is an artifact of the very high Ca^{2+} concentrations used in the crystallization. Indeed, this site is absent in the closely related MBP-C(51) and in MBP-A crystallized with lanthanide ions instead of Ca^{2+}.[50,75]

The structure of the irregular loops that constitute the Ca^{2+}-binding end of the molecule is largely determined by the presence of Ca^{2+}, as shown by structures of MBP-A lacking the principal Ca^{2+}, and apo-MBP-C.[42] The selectins are examples of C-type lectins that contain only the principal Ca^{2+} site. In these proteins, the extended loops that form the auxiliary site in MBP are shortened. In the case of nonlectin family members that lack either Ca^{2+} site, the elaborate loops that form the two sites in MBPs are significantly shortened.

METAL SITE GEOMETRY

Crystal structures of several Ca^{2+}-bound C-type lectins define the coordination geometry of both sites. In the refined structures, the Ca^{2+}–oxygen coordination distances vary between about 2.3 and 2.6 Å (Table 1). These numbers must be interpreted cautiously, however, because at subatomic resolutions typical of protein structures, distance restraints are imposed during refinement to prevent the coordination ligands from being drawn into the strong electron density of the bound Ca^{2+}. Analysis of small molecule databases indicates that the distance between

Table 1 Ca^{2+} coordination distances in MBP-A. The water-bound principal site is shown in Figure 2(a), and the distances are taken from chain A of PDB entry 1KWT. The mannose-bound site is shown in Figure 2(b), and is from chain B of PDB entry 2MSB. The auxiliary and third sites are shown in Figure 2(c), and are also from chain B of PDB entry 2MSB

Residue	Atom	Distance from Ca^{2+} (Å)
(a) Principal site, water-bound		
Glu185	OE2	2.6
Asn187	OD1	2.4
Glu193	OE2	2.4
Asn205	OD1	2.6
Asp206	O	2.5
Asp206	OD2	2.3
Wat338	O	2.5
Wat340	O	2.9
(b) Principal site, mannose-bound		
Glu185	OE1	2.7
Asn187	OD1	2.5
Glu193	OE1	2.3
Asn205	OD1	2.4
Asp206	O	2.5
Asp206	OD1	2.4
Man9	O3	2.5
Man9	O4	2.5
(c) Auxiliary site		
Asp161	OD1	2.7
Asp161	OD2	2.5
Glu165	OE1	2.5
Glu165	OE2	2.6
Asp188	OD1	2.5
Glu193	O	2.4
Asp194	OD1	2.4
Wat10	O	2.4
(d) Third site		
Glu165	OE1	2.3
Asp194	OD1	2.6
Asp194	OD2	2.5
Wat19	O	2.4
Wat23	O	2.4
Wat28	O	2.2
Wat30	O	2.3

Ca^{2+} and carbonyl or carboxylate ligands is typically 2.4 Å; bidentate carboxylate ligands, hydroxyl groups, and water molecules tend to display a slightly longer coordination distance.[76,77] In our laboratory, we use a soft van der Waals repulsive potential set to optimize the distance between Ca^{2+} and oxygen atoms at 2.4 Å. No other restraints such as oxygen–Ca^{2+}–oxygen angles are imposed. In the descriptions that follow, the residues refer to rat MBP-A unless otherwise noted.

The principal Ca^{2+} site coordination geometry is an almost ideal pentagonal bipyramidal arrangement, with

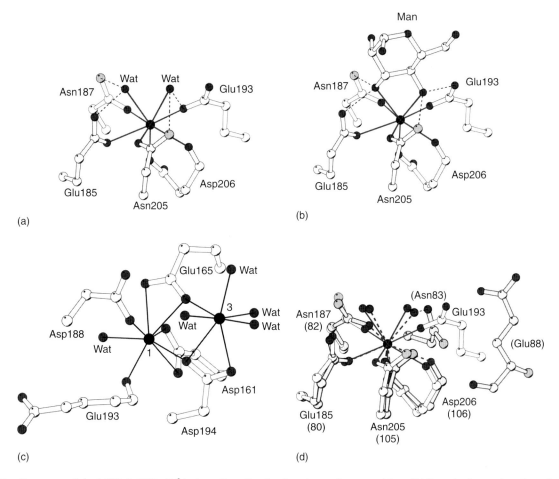

Figure 2 Geometry of the MBP-A CRD Ca^{2+} sites. Coordination bonds are shown as thin solid lines, hydrogen bonds as dashed lines. Oxygen and nitrogen atoms are shown in black and gray respectively, and the Ca^{2+} is shown as a larger black sphere. Coordination bond distances for panels (a), (b), and (c) are provided in Table 1. (a) The principal site, with water molecules occupying the seventh and eighth coordination positions (PDB ID 1KWT). (b) The principal site is bound to mannose (PDB ID 2MSB). Note the displacement of water molecules by the 3- and 4-OH groups of mannose. (c) The auxiliary and third sites (PDB ID 2MSB). (d) Overlay of the principal MBP-A (PDB ID 2MSB) site (white bonds) with that of E-selectin (gray bonds) (PDB ID 1ESL). Selectin residue numbers are shown in parentheses. In this panel, the thick dashed lines represent the Ca^{2+} coordination bonds in the selectin structure. Note the replacement of E-selectin Glu88 with Asn83 in the coordination sphere. Part (a) reprinted from KK-S Ng, AR Kolatkar, S Park-Snyder, H Feinberg, DA Clark, K Drickamer and WI Weis, *Journal of Biological Chemistry*, **277**, 16088–95 (2002)[78] with permission from the Journal of Biological Chemistry; parts (b) and (c) from WI Weis, K Drickamer and WA Hendrickson, *Nature*, **360**, 127–134 (1992)[73] with permission from Nature.

eightfold coordination (Figure 2(a) and (b); Table 1). The pentagonal plane is formed by the side chain carboxylate or carbonyl oxygen atoms of Glu185, Asn187, Glu193, and Asn205, and the main chain carbonyl oxygen of Asp206. A side chain carboxylate oxygen of Asp206 forms one apex, and the other apical position is bisected by two ligands to form an eight-coordinate complex. These ligands are either two water molecules (Figure 2(a)), or vicinal hydroxyl groups from a bound sugar (Figure 2(b)).

The auxiliary Ca^{2+} site also displays pentagonal bipyramidal geometry with eightfold coordination (Figure 2(c); Table 1). A bent pentagonal plane is formed from the two carboxylate oxygen atoms of Glu165, a carboxylate oxygen of Asp194, the main chain carbonyl oxygen of Glu193, and a water molecule. A carboxylate oxygen from Asp188 forms one apex. The second apical position is bisected by the two side chain carboxylate oxygen atoms of Asp161.

The principal and auxiliary sites are structurally coupled. The side chain of Glu193 is a coordination ligand in the principal site, and its backbone carbonyl oxygen is a ligand in the auxiliary site (Figure 2(c)). Moreover, the water molecule that serves as a coordination ligand in the auxiliary site forms a hydrogen bond with Asp206, a principal site coordination ligand. Thus, it is not surprising that the absence of Ca^{2+} from one site can affect the geometry of the other. Indeed, the structures of several C-type lectins that lack the auxiliary Ca^{2+} site reveal a slightly different coordination scheme in the principal site from that described for MBP-A. In E- and P-selectin, for example, residues equivalent to MBP-A Glu185, Asn187,

Asn205, and Asp206 are conserved and occupy the same coordination positions.[35,53] However, selectin residue Glu88, which is equivalent in sequence to MBP-A Glu193, is swung out of the site (Figure 2(d)). The coordination ligand normally provided by this residue is instead Asn83, the residue equivalent to the MBP-A auxiliary site ligand Asp188. A similar scheme is observed in TC14, but in this case the coordination scheme is further altered: a serine occupies the position equivalent to MBP-A Asn187, and the equivalent coordination position is missing.[27]

The absence of the auxiliary Ca^{2+} in selectins and TC14 probably changes the energetic balance of coordination such that the asparagine is a more favorable ligand. The energetic differences must be small, however, since in the complex of P-selectin with a P-selectin glycoprotein ligand-1 (PSGL-1)-derived glycopeptide,[35] the loop from Asn83 through Asp89 moves, resulting in Glu88 moving into the principal Ca^{2+} site and performing coordination and hydrogen bond interactions similar to those of MBP-A Glu193.

As noted above, a third and likely adventitious Ca^{2+} site is formed near the auxiliary site. This seven-coordinated site displays a pentagonal bipyramidal geometry (Figure 2(c); Table 1). The two apical positions are occupied by two water molecules. The pentagonal plane comprises two other water molecules, a carboxylate oxygen of Glu165, and both Asp194 carboxylate oxygen atoms. Note that both carboxylate oxygen atoms of Glu165 ligate the auxiliary Ca^{2+} and one is also a ligand for the third Ca^{2+}, whereas the two carboxylate oxygen atoms of Asp194 ligate the third Ca^{2+} and one is also a ligand in the auxiliary site.

COMPLEX STRUCTURES AND BINDING SPECIFICITY

The crystal structures of several C-type lectins bound to carbohydrate ligands have been reported: rat MBPs bound to a variety of ligands,[51,73,78] DC-SIGN and DC-SIGNR bound to a pentasaccharide,[56] TC14 bound to galactose,[27] and E- and P-selectins bound to both carbohydrate and glycoprotein ligands.[35] In addition, site-directed mutagenesis has been used to introduce alternative specificities into rat MBP-A in order to mimic other family members,[41,48,79,80] and structures of these mutants bound to sugar ligands have also been reported.[81–84]

The essential feature of C-type lectin–carbohydrate interactions is direct ligation of the principal calcium ion by vicinal OH groups of a pyranose ring. In MBPs, the 3- and 4-OH groups of mannose replace the two water molecules that form the seventh and eighth coordination positions at one apex of the pentagonal bipyramid (Figure 2(b)). The site is constructed to exploit the full noncovalent bonding potential of the sugar OH groups. An OH group has two lone pairs of electrons and a proton available for noncovalent interactions. Each of the sugar OH groups contributes one lone pair of electrons to a calcium coordination bond. The other lone pair serves as a hydrogen bond acceptor from the NH_2 functionalities of two asparagine residues, and the proton of each OH is a hydrogen bond donor to the oxygen atoms of acidic side chains. The four residues that form hydrogen bonds with these sugar hydroxyls are also direct coordination ligands. Thus, sugar binding results in an intimately linked ternary complex of protein, sugar, and Ca^{2+}. The need for all hydrogen bond partners was shown by the mutation MBP-A Asn187 → Asp, which preserves Ca^{2+} binding but ablates sugar binding by removing an essential donor group.[73]

For MBPs, Ca^{2+} ligation appears to provide the majority of the binding energy. Only a few additional van der Waals contacts are observed in the crystal structures, and mutagenesis has shown that only one, a contact between a pyranose ring carbon and Cβ of His189 (MBP-A), is energetically significant.[47] MBPs recognize a variety of pathogenic cell surfaces, and the paucity of direct interactions between MBPs and sugar likely reflects the broad specificity required of these proteins. MBPs bind to any pyranose with vicinal OH groups in the same stereochemical arrangement as the 3- and 4-OH of mannose, including glucose, N-acetylglucosamine, and fucose. MBPs will also bind to noncarbohydrate compounds with this arrangement of OH groups, such as myo-inositol.[85] Indeed, in a crystal of rat MBP-A, a molecule of glycerol, which was used as a cryoprotectant, binds at the principal Ca^{2+} site with vicinal OH groups occupying the seventh and eighth coordination positions.[74] Structures of MBPs bound to N-acetylglucosamine and fucose confirms the basic coordination scheme shown in Figure 2(b); in each case there are at most two additional unique contacts.[51,78]

Apart from displacement of the two water molecules bound at the Ca^{2+}, no structural changes occur in MBPs upon ligand binding. NMR analysis of TC14 confirms that the binding site is preformed and rigid in the absence of sugar.[86] The one significant change upon ligand binding observed to date is in the interaction of P-selectin with a PSGL-1 glycopeptide, where Glu88 moves into the principal coordination site in place of Asn83 and forms a hydrogen bond with the 2-OH of fucose.[35]

The sugar OH groups that form the seventh and eighth coordination positions of the principal Ca^{2+} are related by a local twofold symmetry axis. Given the sparsity of contacts with the rest of the protein, one might imagine that carbohydrates would bind in either orientation, that is, with the 3-OH on the Glu185/Asn187 side of the site (Figure 2(b)) or the twofold related configuration with the 4-OH on the Glu185/Asn187 side. However, a variety of crystal structures in the 1.7 to 2.0 Å resolution range have provided no evidence for mixtures of orientations in the site.[51,78] In the case of MBP-C, all ligands tested, including mono-, di-, and trisaccharides, and some larger ligands, bind in the same orientation. In contrast, monosaccharides and α(1,6)-linked glycosides bind to MBP-A in the same orientation as observed in MBP-C, but α(1,2)- and α(1,3)-linked glycosides bind in the opposite orientation.[78] Mutagenesis suggests that His189 of MBP-A, which lies next to the binding site and interacts with bound sugars, is responsible for this difference, which is likely due to steric exclusion of some conformations of α(1,2)- or α(1,3)-linked carbohydrates.[78] However, the basis of the preferred binding of monosaccharides in one orientation, which at first glance would appear to be entropically disfavored, remains unclear.

In broad terms, C-type lectins recognize D-mannose, D-glucose, L-fucose and their derivatives (mannose-type ligands), or D-galactose and its derivatives. The difference between these two sets of ligands lies in the stereochemistry of adjacent hydroxyl groups: in mannose-type ligands, the hydroxyl groups equivalent to the 3- and 4-OH of mannose are equatorial, whereas in galactose the 3-OH is equatorial and the 4-OH is axial. Examination of C-type lectin sequences reveals that residues equivalent to MBP-A Glu193, Asn205, and Asp206 (Figure 2(b)) are highly conserved regardless of sugar specificity. In contrast, MBP-A positions 185 and 187 are glutamic acid and asparagine in mannose-binding members of the family, whereas these positions are replaced by glutamic acid and aspartic acid in most galactose-binding C-type lectins. Site-directed mutagenesis of MBP-A Glu185/Asn187 to Gln185/Asp187 (termed the 'QPD' mutant, as these two residues flank a highly conserved proline) results in preferential binding to galactose.[79] However, the absolute affinity of this mutant toward galactose is weak. A tryptophan residue equivalent to MBP-A His189 is highly conserved in galactose-binding C-type lectins. When introduced into the QPD mutant, galactose affinity comparable to the galactose-binding asialoglycoprotein receptor is obtained (QPDW).[48] Introduction of glycine-rich loop immediately following the tryptophan in RHL provides discrimination against mannose (QPDWG mutant).

The crystal structures of the QPDWG and related mutants bound to galactose and N-acetylgalactosamine (GalNAc) reveal that the 3- and 4-OH groups form a similar set of Ca^{2+} coordination and hydrogen bonds with the principal Ca^{2+} site as observed in the wild-type MBPs[81–83] (Figure 3(a)). However, the different stereochemistry of galactose places its ring in a different spatial position. In particular, one face of the galactose ring is relatively hydrophobic, and this face packs on the conserved tryptophan (Figure 3(a)). The discrimination provided by the glycine-rich loop is explained by its positioning the tryptophan for optimal binding to galactose, while preventing the tryptophan from moving out of the way to accommodate mannose, whose ring is more 'upright' or parallel to the Ca^{2+} coordination apex than that of galactose.[81] Thus, galactose binding still employs the basic Ca^{2+} coordination and hydrogen-bonding scheme found in the mannose binding proteins, but in this case, it appears that this interaction is inherently weaker, since additional binding energy provided by the interaction with the tryptophan side chain is required.

The importance of the galactose–tryptophan interaction is highlighted by the structure of the natural galactose-binding C-type lectin TC14 bound to galactose.[27] This protein has a divergent set of Ca^{2+} ligands relative to other C-type lectins, and one of the hydrogen bond acceptors for the sugar OH groups is a water molecule rather than a Ca^{2+} ligand. The galactose is bound in an inverted orientation relative to the QPD mutant, which correlates with the observation that the arrangement of hydrogen bond donors and acceptors is opposite to that of the QPD site. Moreover, a tryptophan is located on the opposite side of the Ca^{2+} site relative to that of RHL, and it stacks on the aliphatic face of the bound galactose. Thus, despite a highly divergent set of ligands, the chemistry of the binding interaction is preserved.

The similarity of binding chemistry of mannose-type ligands to the EPN site or galactose to the QPD site raises the question of why this small difference, which essentially exchanges the position of hydrogen bond donors and acceptors on one side of the site, gives rise to this specificity difference. It should be noted that in the case of the QPD site, the discrimination is not absolute: in the absence of constraints of the tryptophan position, which in its galactose-binding position sterically excludes mannose, mannose binding is observed, albeit weaker than galactose.[48] The EPN site, however, does not seem to accommodate galactose. It appears that the discrimination lies in the detailed electrostatic and steric arrangements of the binding sites that accommodate the equatorial or axial 4-OH group,[87,88] but a full explanation has not been achieved.

Many C-type lectins display more restricted carbohydrate specificity than MBPs. In these cases, crystal structures have shown that the basic Ca^{2+} coordination schemes described above are preserved, but additional interactions are made with surrounding regions of the protein. The nature of these extended binding sites is quite variable. Rat hepatic lectin-1 (RHL-1) displays a 60-fold preference for GalNAc over galactose, and site-directed mutagenesis demonstrates that His256 is essential for this difference.[26] Introduction of this residue into the MBP-A mutant QPDWG background provided a ninefold preference for GalNAc over Gal, and the crystal structure of this mutant bound to GalNAc revealed a van der Waals contact between the methyl group of the acetamido moiety and the distal carbon of the imidazole ring.[82] Increased discrimination was achieved by the addition of several other conserved residues from RHL that had been shown to contribute further to the GalNAc specificity of RHL-1.[26] The structure of this mutant bound to GalNAc reveals a similar contact between the methyl group and the imidazole ring; the other required residues appear to stabilize the precise

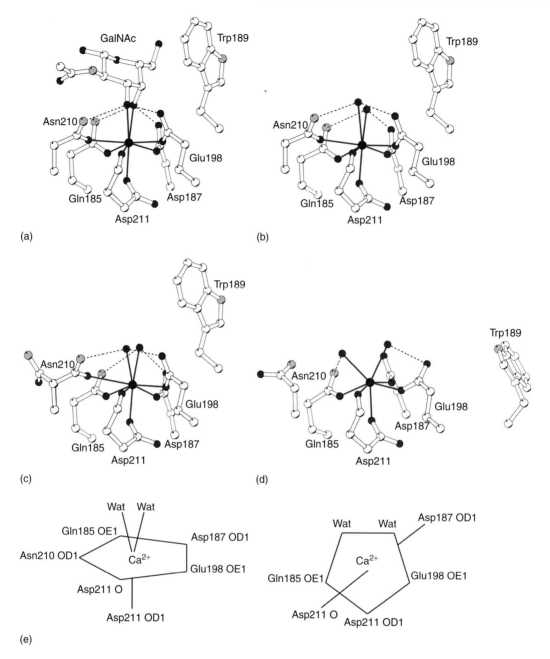

Figure 3 Galactose binding by C-type lectins. The structure of the GalNAc-specific mutant of MBP-A[83] is shown. Atom and bond types are indicated as in Figure 2. Owing to the insertion of a five-residue loop, residues Glu198, Asn210, and Asp211 correspond to the wild-type MBP-A residues 193, 205, and 206. (a) GalNAc bound to the principal site (PDB ID 1FIH). The view is rotated approximately 90° about the vertical relative to Figure 2(b). (b) Copy 1 of the unliganded site (PDB ID 1FIF), showing the binding-competent form in which the structure is exactly the same as that bound to GalNAc apart from two water molecules occupying the same positions as the 3- and 4-OH groups of GalNAc. (c) Copy 2, showing the change in Asn210 rotamer. The alternate conformation of Asn210 is occupied in 50% of the molecules. (d) Copy 3, showing more extensive rearrangement of the site (see text for description). (e) Schematic diagram of the difference in principal site geometry between copy 1 and copy 3. Reprinted from H Feinberg, D Torgerson, K Drickamer, and WI Weis (2000), *J Biol Chem*, **275**, 35176–84[83] with permission of the Journal of Biological Chemistry.

position of the histidine, and also provide an intricate hydrogen-bonding network that includes a water-mediated bond between the carbonyl oxygen of the acetamido group and an acid.[83] Importantly, the positions of these residues in the MBP-A chimera correspond precisely to those in the unliganded wild-type human hepatic lectin-1,[54] indicating that the interactions observed in the chimera likely exist in the native receptor.

RHL-1 is an example in which several residues on the surface create a subsite for a substituent on the pyranose ring that is bound in the principal Ca^{2+} site. In other cases, the specificity is for oligosaccharides rather than for single sugars, and the binding sites are correspondingly more extended. DC-SIGN and DC-SIGNR show selectivity for high-mannose N-linked oligosaccharides,[56,89] and have the characteristic EPN motif in the principal Ca^{2+} site. Structures of a pentasaccharide bound to these proteins revealed the typical binding of mannose at the principal Ca^{2+} site, with the other portions of the pentasaccharide also interacting specifically with the surrounding surface.[56] In particular, the second α-helix of the CRD is long and tilts upwards toward the principal Ca^{2+} site to form, along with the loop between β-strands 3 and 4, a continuous surface that interacts with several of the sugars (Figure 1(b)).

E- and P-selectin specifically recognize derivatives of the Lewis^x epitope Galβ(1,4)[Fucα(1,3)]GlcNAc. Structures of these proteins bound to sialyl-Lewis^x (sLe^x; NeuNAcα(2,3)Galβ(1,4)[Fucα(1,3)]GlcNAc) and a glycopeptide derived from PSGL-1, a high-affinity ligand of P-selectin that displays sLe^x, reveal that the fucose moiety binds at the principal Ca^{2+} site by direct ligation.[35] Although this is a 'mannose-type' site with the conserved EPN sequence, it was surprising to see that the 3- and 4-OH of L-fucose coordinate the Ca^{2+}, rather than the 2- and 3-OH groups that are the stereochemical equivalents of the 3- and 4-OH of D-mannose. In this case, the equivalent of MBP-A Glu193 is replaced by an asparagine (Asn83), whose oxygen provides a coordination ligand. However, the Asn83 amide group cannot be a hydrogen bond acceptor, and instead forms a hydrogen bond with a water molecule that in turn fulfills this role. The basis of this unusual fucose binding mode is not clear, but it is probably due to the favorable energy of the interactions made by other portions of the carbohydrate with the selectin surface. In the interaction with sLe^x, the galactose and sialic acid residues interact with residues in the β3–β4 loop. In the case of PSGL-1 binding to P-selectin, additional affinity and specificity is provided by interactions between two sulfotyrosine residues in PSGL-1 and other portions of the P-selectin polypeptide.[35] The examples of DC-SIGN and the selectins show that the loops between α2 and β2, and between β3 and β4, are quite variable in sequence and structure, and are used to create binding sites for other portions of the ligand.

FUNCTIONAL ASPECTS

Specific interactions of C-type lectins with carbohydrate ligands govern several kinds of cell surface recognition phenomena.[2,10] MBPs display a broad specificity for mannose-type ligands, concordant with the need to recognize a variety of pathogenic cell surfaces. These ligands bind to MBP with dissociation constants on the order of 1 mM; multivalent interactions provide high avidity binding. It is essential that MBPs trigger cell killing only of pathogens and not host cells. Selective, high-avidity binding to pathogens is achieved in the placement of binding sites on the MBP trimer. The structure of a trimeric fragment of MBP containing the coiled-coil α-helical trimerization and lectin domains reveals that the principal Ca^{2+} and carbohydrate-binding sites are 53 Å (rat MBP-A(74)) or 45 Å (human MBP(52)) apart. Spectroscopic and modelbuilding studies have shown that the terminal mannose residues in vertebrate high-mannose oligosaccharides are about 20 to 30 Å apart.[46,90] Thus, the binding sites in the MBP trimer are too far apart to interact multivalently with host oligosaccharides; in fact, common high-mannose vertebrate oligosaccharides bind with affinities comparable to simple mannosides.[85] The wide spacing of sites presumably allows multivalent binding to the dense, repetitive arrays of sugars present on pathogenic cell surfaces.[74] It is likely that a similar situation holds in the case of the mannose receptor. In this case, however, multivalency is achieved by having multiple CRDs in a single polypeptide chain.[2]

Some other family members recognize carbohydrate ligands with somewhat higher specificity and affinity than MBPs. In these cases, the extended binding sites present on the surface near the principal Ca^{2+} site provide the interactions. For example, the selectins have a binding surface optimized for interaction with Le^x derivatives. Curiously, these proteins do not bind to isolated fucose, despite the interaction of the fucose moiety with the Ca^{2+} site. The basis for the discrimination against the monosaccharide remains unclear, but is perhaps related to the unusual interaction of the 3- and 4-OH groups of fucose with Ca^{2+} described above, and/or the lack of a contact equivalent to that between mannose and the Cβ of His189 of MBP-A.

Selectins function in the initial interaction of circulating leukocytes with the vascular endothelium, in which the leukocytes initially roll along the vascular surface at reduced speed prior to becoming firmly adherent. During rolling, contacts are repeatedly made and broken, and the relatively low affinity and rapid on/off kinetics characteristic of protein–carbohydrate interactions likely has a role in this process.

CHANGES ACCOMPANYING REMOVAL OF Ca^{2+} AND THEIR FUNCTIONAL IMPLICATIONS

C-type lectins function in the extracellular milieu, where the Ca^{2+} concentration is in the 1 to 5 mM range and the pH is slightly basic. These proteins display relatively weak Ca^{2+} affinities ($K_d \sim 10–1000\,\mu M$), so the sites are likely to be saturated or close to saturated under normal conditions. The relatively weak affinity for Ca^{2+} is exploited by those C-type lectins that serve as endocytic receptors. After binding to a carbohydrate ligand at the cell surface, the receptor–ligand complex is endocytosed. In the acidic environment of the endosome, the ligand is released, and the receptor is sorted from the ligand and recycled to the cell surface for another round of binding. Therefore, an essential aspect of these C-type lectins is the coupling of sugar binding, Ca^{2+} binding, and pH.

The involvement of oxygen ligands, some from carboxylate groups and all with partial negative charge, would be expected to make Ca^{2+} affinity sensitive to pH. Indeed, the Ca^{2+} dependence of sugar binding and proteolytic sensitivity change with pH.[37,91] For example, in MBP-A, at 1 mM Ca^{2+} half-maximal sugar binding occurs at pH 5.0, which likely reflects direct titration of one or more Ca^{2+} ligands.[91] In addition to the immediate ligation shell, various C-type lectins have additional nearby titratable groups whose pK_a is matched to a pH appropriate for the endosome in which the ligand must be released. Protonation of one or more groups produces changes in the Ca^{2+} coordination shell, such that the precise arrangement required for sugar binding is disrupted, resulting in release of the sugar. Thus, RHL-1, which unlike MBP-A functions as an endocytic receptor, half-maximal sugar binding occurs at pH 7.1 at 1 mM Ca^{2+}. In this case, a key histidine residue controls the pH sensitivity.[91]

The same histidine that is required for the pH-dependent sugar and Ca^{2+} binding of RHL-1 also confers selective binding to GalNAc versus Gal, as described above. The crystal structure of the mutant of MBP-A containing this histidine and other residues required for positioning it for binding to GalNAc was determined in the absence of sugar.[83] The crystals contain three crystallographically independent molecules. In one copy, the principal site has exactly the same structure as observed when bound to GalNAc, except that water molecules replace the 3- and 4-OH groups of the sugar (Figure 3(a) and (b)). This structure is competent to bind to a sugar ligand. In the second copy, the site is similar, except that Asn210 is observed to be swung out of the site in 50% of the molecules (Figure 3(c)). In its normal, sugar-binding conformation (Figure 3(a) and (b)), the NH_2 group of Asn210 is hydrogen bonded to the essential histidine. If the histidine was to be protonated, this bond would have to be broken, and since the histidine is locked into place by other interactions, it is likely that the observed change in Asn210 rotamer would have to occur. In the third copy, Asn210 is similarly swung out of the site in all molecules, and is replaced by a water molecule in the Ca^{2+} coordination shell (Figure 3(d)). In addition, the $\chi 3$ angle of Glu198 changes by ~75°. Collectively, these changes alter the coordination geometry from eightfold to sevenfold, and in addition, the plane of the pentagonal bipyramid rotates 90° (Figure 3(e)). The observed structural rearrangements remove hydrogen bond donors and acceptors from the positions needed to interact with the bound sugar, suggesting that they are plausible intermediates for the pathway of ligand release in the endosome.[83]

Changes accompanying the complete loss of one or more Ca^{2+} have been visualized in crystal structures of rat MBP-A bound to a single lanthanide ion in the auxiliary site, and rat MBP-C crystallized in the absence of Ca^{2+} (apo-MBP-C)[42] (Figure 4). The apo-MBP-C crystals contain four independent copies of the CRD in the asymmetric unit, providing multiple independent views. Removal of the principal Ca^{2+} results in rearrangements of loops 3 and 4. A conserved proline (residue 186 of MBP-A, 191 of MBP-C) separates these loops. This proline adopts the *cis* peptide conformation in the presence of Ca^{2+}, which turns the backbone in such a way as to allow both the flanking glutamic acid and asparagine residues to serve as Ca^{2+} ligands (Figure 4(a) and (b)). In two copies of apo-MBP-C, the proline remains in the *cis* conformation, but the Ca^{2+} site is rearranged, with the rotamers of several Ca^{2+} ligands altered and swung out of the site (Figure 4(c) and (d)). These changes likely reflect electrostatic repulsion of full and partial negative charges that would be neutralized by the Ca^{2+}. In another copy of apo-MBP-C and in the one-ion MBP-A, the proline is isomerized to the *trans* conformation, resulting in large changes in the surrounding loops (Figure 4(e) and (f)). (This region is disordered in the fourth copy of the apo-MBP-C structure.) Kinetic measurements of the conformational changes accompanying removal of Ca^{2+} demonstrated that the proline isomerization occurs in solution; at equilibrium, the principal site proline residue is *trans* in 80% of the apo-MBP-C molecules, similar to the distribution of *cis* and *trans* isomers found in proline-containing peptides.[40] Because *cis*–*trans* proline isomerization occurs on a timescale of minutes, loss of the principal Ca^{2+} could lock many of the molecules in a *trans* proline conformation unable to bind the ligand.[40] The relevance of such a kinetic trap is unclear, however, since it is not known whether this site is completely lost during endocytosis in true endocytic receptors.

Structural rearrangements also occur in the auxiliary site in apo-MBP-C.[42] This is of interest given the coupling between the auxiliary and principal Ca^{2+} sites. The possible effects of changes in the auxiliary site on sugar binding were illustrated by the crystal structure of the mannose receptor CRD-4, which crystallized with its principal site occupied, but the auxiliary site empty, because of the formation of a domain-swapped dimer in the crystals.[55] In this case, the glutamate equivalent to MBP-A Glu193 was swung out of the site and replaced by an asparagine, similar to the situation described above for the selectins and other family members lacking the auxiliary site. Loss of the glutamate alters the arrangement of hydrogen bond donors and acceptors in the site, which appears to render it unable to bind mannose; no mannose was observed in this crystal structure despite its presence at 100 mM in the crystallization drop.[55] It was proposed that the auxiliary site could serve as a pH sensor for ligand release in the endosome in this case.

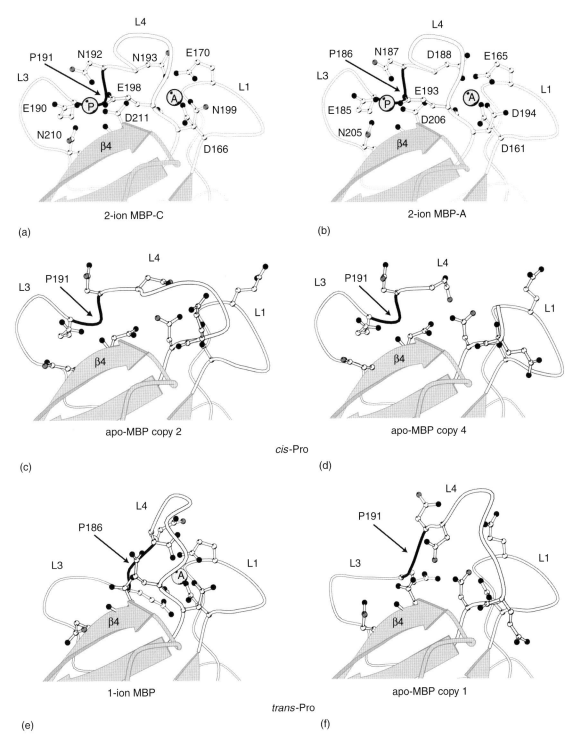

Figure 4 Conformational changes upon loss of Ca^{2+} from the C-type CRD. *cis*-Pro is present in the principal site of (a) MBP-C (PDB ID 1RDO) and (b) MBP-A (PDB ID 2MSB), and (c,d) two copies of apo-MBP-C (PDB ID 1BV4). (e) *trans*-Pro is present in a one-ion form of MBP-A (PDB ID 1BUU) and (f) another copy of apo-MBP-C. The backbone trace for loops 1 to 4 (designated L1–L4; residues 165–199 of MBP-C and 160–194 of MBP-A) is drawn in white. The remainder of the CRD is drawn in light gray, with arrows representing β-strands. The Ca^{2+} are marked 'P' for principal site and 'A' for auxiliary site, all of the residues that serve as Ca^{2+} ligands are drawn in ball-and-stick representation, and are labeled in panels A and B. Carbon, oxygen, nitrogen, and Ca^{2+} are shown as white, black, gray, and large white spheres respectively. The arrow points to the backbone (black) of the conserved proline residue in the principal Ca^{2+} site. Reprinted with permission from KK-S Ng, S Park-Snyder and WI Weis, *Biochemistry*, 37, 17965–76.[42] Copyright (1998), American Chemical Society.

ACKNOWLEDGEMENTS

I thank Kurt Drickamer for comments on the manuscript, and Hadar Feinberg for comments and preparation of the figures. Studies of C-type lectins in the author's laboratory have been supported by grant GM50565 from the National Institutes of Health.

REFERENCES

1. K Drickamer, *Curr Opin Struct Biol*, **9**, 585–90 (1999).
2. WI Weis, ME Taylor and K Drickamer, *Immunol Rev*, **163**, 19–34 (1998).
3. SJ Lee, S Evers, D Roeder, AF Parlow, J Risteli, L Risteli, YC Lee, T Feizi, H Langen and MC Nussenzweig, *Science*, **295**, 1898–1901 (2002).
4. TBH Geijtenbeek, R Torensma, SJ van Vliet, GCF van Duijnhoven, GJ Adema, Y van Kooyk and CG Figdor, *Cell*, **100**, 575–85 (2000).
5. TBH Geijtenbeek, DJEB Krooshoop, DA Bleijs, SJ van Vliet, GCF van Duijnhoven, V Grabovsky, R Alon, CG Figdor and Y van Kooyk, *Nat Immunol*, **1**, 353–57 (2000).
6. TBH Geijtenbeek, DS Kwon, R Torensma, SJ van Vliet, GCF van Duijnhoven, J Middel, ILMHA Cornelissen, HSLM Nottet, VN KewalRamani, DR Littman, CG Figdor and Y van Kooyk, *Cell*, **100**, 587–97 (2000).
7. S Pöhlmann, F Baribaud, B Lee, GJ Leslie, MD Sanchez, K Hiebenthal-Millow, J Münch, F Kirchhoff and RW Doms, *J Virol*, **75**, 4664–72 (2001).
8. S Pöhlmann, EJ Soilleux, F Baribaud, GJ Leslie, LS Morris, J Trowsdale, B Lee, N Coleman and RW Doms, *Proc Natl Acad Sci USA*, **98**, 2670–75 (2001).
9. AA Bashirova, TBH Geijtenbeek, GCF van Duijnhoven, SJ van Vliet, JBG Eilering, MP Martin, L Wu, TD Martin, N Viebig, PA Knolle, VN KewalRamani, Y van Kooyk and M Carrington, *J Exp Med*, **193**, 671–78 (2001).
10. K Drickamer and ME Taylor, *Annu Rev Cell Biol*, **9**, 237–64 (1993).
11. S Saleque, N Ruiz and K Drickamer, *Glycobiology*, **3**, 185–90 (1993).
12. RL Hudgin, WEJ Pricer, G Ashwell, RJ Stockert and RG Morell, *J Biol Chem*, **249**, 5536–43 (1974).
13. T Kawasaki and G Ashwell, *J Biol Chem*, **252**, 6536–43 (1977).
14. Y Mizuno, Y Kozutsumi, T Kawasaki and I Yamashina, *J Biol Chem*, **256**, 4247–52 (1981).
15. R Townsend and P Stahl, *Biochem J*, **194**, 209–14 (1981).
16. J Wild, D Robinson and B Winchester, *Biochem J*, **210**, 167–74 (1983).
17. Y Maynard and JU Baenziger, *J Biol Chem*, **257**, 3788–94 (1982).
18. N Kawasaki, T Kawasaki and I Yamashina, *J Biochem (Tokyo)*, **94**, 937–47 (1983).
19. A Persson, D Chang and E Crouch, *J Biol Chem*, **265**, 5755–60 (1990).
20. S Hawgood, BJ Benson and RLJ Hamilton, *Biochemistry*, **24**, 184–90 (1985).
21. K Drickamer, MS Dordal and L Reynolds, *J Biol Chem*, **261**, 6878–86 (1986).
22. U Holmskov, B Teisner, AC Willis, KBM Reid and JC Jensenius, *J Biol Chem*, **268**, 10120–25 (1993).
23. N Fornstedt and J Porath, *FEBS Lett*, **57**, 187–91 (1975).
24. RA Childs, T Feizi, C-T Yuen, K Drickamer and MS Quesenberry, *J Biol Chem*, **265**, 20770–77 (1990).
25. WI Weis, GV Crichlow, HMK Murthy, WA Hendrickson and K Drickamer, *J Biol Chem*, **266**, 20678–86 (1991).
26. ST Iobst and K Drickamer, *J Biol Chem*, **271**, 6686–93 (1996).
27. SF Poget, GB Legge, MR Proctor, PJG Butler, M Bycroft and RL Williams, *J Mol Biol*, **290**, 867–79 (1999).
28. L Ellgaard and A Helenius, *Curr Opin Cell Biol*, **13**, 431–37 (2001).
29. K Håkansson, NK Lim, H-J Hoppe and KBM Reid, *Structure*, **7**, 255–64 (1999).
30. ME Taylor and K Drickamer, *J Biol Chem*, **268**, 399–404 (1993).
31. FX McCormack, HM Calvert, PA Watson, DL Smith, RJ Mason and DR Voelker, *J Biol Chem*, **269**, 5833–41 (1994).
32. M Kuhlman, K Joiner and RAB Ezekowitz, *J Exp Med*, **169**, 1733–45 (1989).
33. SH Li, DK Burns, JM Rumberger, DH Presky, VL Wilkinson, M Anostario, BA Wolitzky, CR Norton, PC Familletti, KJ Kim, AL Goldstein, DC Cox and K-S Huang, *J Biol Chem*, **269**, 4431–37 (1994).
34. DZ Simpson, PG Hitchen, EL Elmhirst and ME Taylor, *Biochem J*, **343**, 403–11 (1999).
35. WS Somers, J Tang, GD Shaw and RT Camphausen, *Cell*, **103**, 467–79 (2000).
36. JE Schweinle, M Nishiyasu, TQ Ding, K Sastry, SD Gillies and RAB Ezekowitz, *J Biol Chem*, **268**, 364–70 (1993).
37. JA Loeb and K Drickamer, *J Biol Chem*, **263**, 9752–60 (1988).
38. M Anostario and K-S Huang, *J Biol Chem*, **270**, 8138–44 (1995).
39. NP Mullin, KT Hall and ME Taylor, *J Biol Chem*, **269**, 28405–13 (1994).
40. KK-S Ng and WI Weis, *Biochemistry*, **37**, 17977–89 (1998).
41. S Bouyain, S Rushton and K Drickamer, *Glycobiology*, **11**, 989–96 (2001).
42. KK-S Ng, S Park-Snyder and WI Weis, *Biochemistry*, **37**, 17965–76 (1998).
43. WI Weis and K Drickamer, *Annu Rev Biochem*, **65**, 441–73 (1996).
44. D Torgersen, NP Mullin and K Drickamer, *J Biol Chem*, **273**, 6254–61 (1998).
45. MS Quesenberry and K Drickamer, *J Biol Chem*, **267**, 10831–41 (1992).
46. RT Lee, Y Ichikawa, T Kawasaki, K Drickamer and YC Lee, *Arch Biochem Biophys*, **299**, 129–36 (1992).
47. ST Iobst, MR Wormald, WI Weis, RA Dwek and K Drickamer, *J Biol Chem*, **269**, 15505–11 (1994).
48. ST Iobst and K Drickamer, *J Biol Chem*, **269**, 15512–19 (1994).
49. KG Rice, OA Weisz, T Barthel, RT Lee and YC Lee, *J Biol Chem*, **265**, 18429–34 (1990).
50. WI Weis, R Kahn, R Fourme, K Drickamer and WA Hendrickson, *Science*, **254**, 1608–15 (1991).
51. KK-S Ng, K Drickamer and WI Weis, *J Biol Chem*, **271**, 663–74 (1996).
52. S Sheriff, CY Chang and RAB Ezekowitz, *Nat Struct Biol*, **1**, 789–94 (1994).
53. BJ Graves, RL Crowther, C Chandran, JM Rumberger, S Li, K-S Huang, DH Presky, PC Familletti, BA Wolitzky and DK Burns, *Nature*, **367**, 532–38 (1994).
54. M Meier, MD Bider, VN Malashkevich, M Spiess and P Burkhard, *J Mol Biol*, **300**, 857–65 (2000).

55 H Feinberg, S Park-Snyder, AR Kolatkar, CT Heise, ME Taylor and WI Weis, *J Biol Chem*, **275**, 21539–48 (2000).

56 H Feinberg, DA Mitchell, K Drickamer and WI Weis, *Science*, **294**, 2163–66 (2001).

57 JA Bertrand, D Pignol, J-P Bernard, J-M Verdier, J-C Dagorn and JC Fontecilla-Camps, *EMBO J*, **15**, 2678–84 (1996).

58 BB Nielsen, JS Kastrup, H Rasmussen, TL Holtet, JH Graversen, M Etzerodt, HC Thogersen and IK Larsen, *FEBS Lett*, **412**, 388–96 (1997).

59 JS Kastrup, BB Nielsen, H Rasmussen, TL Holtet, JH Graversen, M Etzerodt, HC Thøgersen and IK Larsen, *Acta Crystrallogr, Sect D*, **54**, 757–66 (1998).

60 W Gronwald, MC Loewen, B Lix, AJ Baugulis, FD Sönnichsen, PL Davies and BD Sykes, *Biochemistry*, **37**, 4712–21 (1998).

61 H Mizuno, Z Fujimoto, M Koizumi, H Kano, H Atoda and T Morita, *Nat Struct Biol*, **4**, 438–41 (1997).

62 H Mizuno, Z Fujimoto, M Koizumi, H Kano, H Atoda and T Morita, *J Mol Biol*, **289**, 103–12 (1999).

63 K Fukuda, H Mizuno, H Atoda and T Morita, *Biochemistry*, **2000**, 1915–23 (2000).

64 S Hirotsu, H Mizuno, K Fukuda, MC Qi, T Matsui, J Hamako, T Morita and K Titani, *Biochemistry*, **40**, 13592–97 (2001).

65 H Mizuno, Z Fujimoto, H Atoda and T Morita, *Proc Natl Acad Sci USA*, **98**, 7230–34 (2001).

66 K Natarajan, N Dimasi, J Wang, RA Mariuzza and DH Margulies, *Annu Rev Immunol*, **20**, 853–85 (2002).

67 MM Mullen, KM Haan, R Longnecker and TS Jardetzky, *Mol Cell*, **9**, 375–85 (2002).

68 D Kohda, CJ Morton, AA Parkar, H Hatanaka, FM Inagaki, ID Campbell and AJ Day, *Cell*, **86**, 767–75 (1996).

69 E Hohenester, T Sasaki, JR Olsen and R Timpl, *EMBO J*, **17**, 1656–64 (1998).

70 Y-H Ding, K Javaherian, K-M Lo, R Chopra, T Boehm, J Lanciotti, BA Harris, Y Li, R Shapiro, E Hohenester, R Timpl, J Folkman and DC Wiley, *Proc Natl Acad Sci USA*, **95**, 10443–48 (1998).

71 G Kelly, S Prasannan, S Daniell, K Fleming, G Frankel, G Dougan, I Connerton and S Matthews, *Nat Struct Biol*, **6**, 313–18 (1999).

72 ZA Hamburger, MS Brown, RR Isberg and PJ Bjorkman, *Science*, **286**, 291–95 (1999).

73 WI Weis, K Drickamer and WA Hendrickson, *Nature*, **360**, 127–34 (1992).

74 WI Weis and K Drickamer, *Structure*, **2**, 1227–40 (1994).

75 FT Burling, WI Weis, KM Flaherty and AT Brünger, *Science*, **271**, 72–77 (1996).

76 H Einspahr and CE Bugg, *Acta Crystallogr*, **B36**, 264–71 (1980).

77 H Einspahr and CE Bugg, *Acta Crystallogr*, **B37**, 1044–52 (1981).

78 KK-S Ng, AR Kolatkar, S Park-Snyder, H Feinberg, DA Clark, K Drickamer and WI Weis, *J Biol Chem*, **277**, 16088–95 (2002).

79 K Drickamer, *Nature*, **360**, 183–86 (1992).

80 O Blanck, ST Iobst, C Gabel and K Drickamer, *J Biol Chem*, **271**, 7289–92 (1996).

81 AR Kolatkar and WI Weis, *J Biol Chem*, **271**, 6679–85 (1996).

82 AR Kolatkar, AK Leung, R Isecke, R Brossmer, K Drickamer and WI Weis, *J Biol Chem*, **273**, 19502–8 (1998).

83 H Feinberg, D Torgerson, K Drickamer and WI Weis, *J Biol Chem*, **275**, 35176–84 (2000).

84 KK-S Ng and WI Weis, *Biochemistry*, **36**, 979–88 (1997).

85 RT Lee, Y Ichikawa, M Fay, K Drickamer, M-C Shao and YC Lee, *J Biol Chem*, **266**, 4810–15 (1991).

86 SF Poget, SMV Freund, MJ Howard and M Bycroft, *Biochemistry*, **40**, 10966–72 (2001).

87 S Elgavish and B Shaanan, *Trends Biochem Sci*, **22**, 462–67 (1997).

88 S Elgavish and B Shaanan, *J Mol Biol*, **277**, 917–32 (1998).

89 DA Mitchell, AJ Fadden and K Drickamer, *J Biol Chem*, **276**, 28939–45 (2001).

90 YC Lee, RT Lee, K Rice, Y Ichikawa and T-C Wong, *Pure Appl Chem*, **63**, 499–506 (1991).

91 S Wragg and K Drickamer, *J Biol Chem*, **274**, 35400–6 (1999).

92 PJ Kraulis, *J Appl Crystallogr*, **24**, 946–50 (1991).

93 EA Merritt and DJ Bacon, *Methods Enzymol*, **277**, 505–24 (1997).

Structural calcium (trypsin, subtilisin)

Gary L Gilliland and Alexey Teplyakov

Center for Advanced Research in Biotechnology of the University of Maryland, Biotechnology Institute, and the National Institute of Standards and Technology, Rockville, MD, USA

FUNCTIONAL CLASS

Trypsin: enzyme; hydrolase; serine endopeptidase; EC 3.4.21.4.

Trypsin-like serine proteases are all endopeptidases, have a catalytic triad consisting of histidine, aspartic acid, and serine in that order in the sequence, and have a common double β-barrel fold.[1,2] Their evolution from a single ancestral protein is generally accepted.[3] Trypsin is characterized by the substrate specificity toward positively charged lysine and arginine side chains.

Subtilisin: enzyme; hydrolase; serine endopeptidase; EC 3.4.21.62.

Subtilisins are serine proteases that contain the serine, histidine, and aspartic acid catalytic triad that has evolved independently from the chymotrypsin and trypsin family, and have a common α/β/α fold with a seven-stranded, central parallel β-sheet with a left-handed crossover between strands 2 and 3. The enzymes catalyze the hydrolysis of protein peptide bonds. As a rule, they have broad specificity, but they also have a preference for large non-β-branched hydrophobic residues at the P1 position.[4,5] This

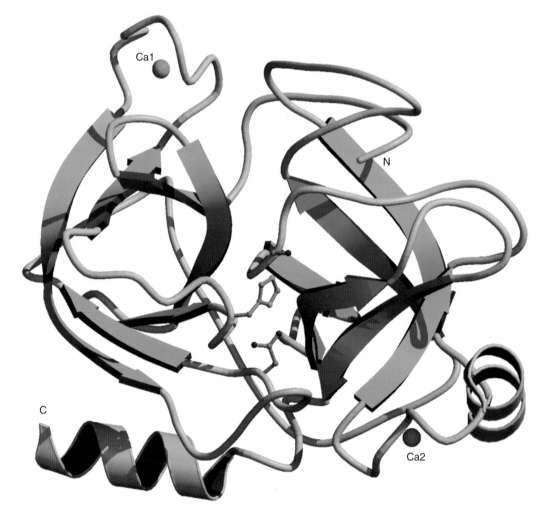

3D Structure: Trypsin Schematic representation of the structure of bovine pancreatic trypsin PDB code 5PTP.[15] The catalytic triad is shown as ball-and-stick model, and Ca1 is the primary site observed in trypsin and elastase. Ca2 is the location of the site observed in *S. griseus* trypsin (PDB code 1SGT).[27] Prepared with the programs MOLSCRIPT[61] and RASTER3D.[62]

family of enzymes is characterized by the presence of one or more bound calcium ions with varying affinities that are not involved in catalysis but help in stabilizing the structure.[6]

OCCURRENCE

Trypsin Trypsin is one of several digestive enzymes secreted into the intestine of animals. It is found in all animals, including mammals, birds, insects, fish, and crustaceans.[7] In bovine pancreatic secretions, it represents approximately 15% of the digestive enzymes.[8] Trypsin was also found in some microorganisms from taxon *Streptomyces* and in fungi.

Subtilisin Subtilisin BPN' is found in *Bacillus amyloliquefaciens*, and similar enzymes are found in other *Bascillus* species variants such as *Bacillus subtilis*. A large number of homologous enzymes with similar structure and function are found in other species of bacteria, and a number of eukaryotic homologs with diverse functions such as prohormone convertases are known. Only the bacterial subtilisins will be discussed further.

BIOLOGICAL FUNCTION

Trypsin Trypsin is synthesized as a pre-proenzyme by the acinar cells of the pancreas and is stored as the proenzyme trypsinogen in secretory granules. Following its release into the gut, trypsinogen is activated by enterokinase or by trypsin itself. Once activated, trypsin cleaves and activates other pancreatic serine protease zymogens, procarboxypeptidases, and prolipases. Their inactivation and degradation is also regulated by proteolysis. Trypsin also contributes to the digestion of the consumed protein.

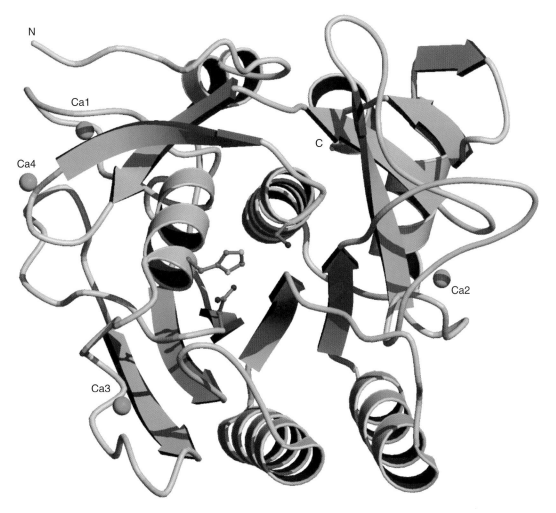

3D Structure: Subtilisin BPN' Schematic representation of the structure of *B. amyloliquefaciens* subtilisin BPN' (PDB code 2ST1).[30] Indicated in the figure are four unique calcium binding sites that are found in subtilisins. Two sites, Ca1 and Ca2, are found in subtilisin BPN'; two of the calcium binding sites Ca1 and Ca3 are found in thermitase with site Ca2 occupied by sodium (PDB code 1THM);[47] and three of the calcium binding sites, Ca1, Ca2, and Ca4, are occupied in *Bacillus* Ak.1 protease with Ca2 occupied by a sodium ion (PDB code 1DBI).[31] Prepared with the programs MOLSCRIPT[61] and RASTER3D.[62]

Structural calcium (trypsin, subtilisin)

An R117H mutation in the human cationic trypsin prevents autolysis at this site and may cause hereditary pancreatitis.[9] The stability of the mutant enzyme could result in increased proteolytic activity that could damage the pancreas.

Subtilisin Bacterial subtilisins are expressed as a pre-proenzyme[10] which is then secreted into the extracellular matrix and, thus, they serve a nutritive role for the parent organism.[11] The 77-amino acid residue proenzyme domain is required for the initial folding.[12]

AMINO ACID SEQUENCE INFORMATION

Trypsin Amino acid sequences of trypsin from a wide range of species, including vertebrates, insects, crustaceans, fungi, and bacteria, are now available from sequence databases. Multiple isoforms are present in most species. The sequence identity between the anionic and cationic forms of bovine trypsin is 74%. The following list includes the species for which the three-dimensional information is also available. Most functional and structural studies have been carried out on bovine trypsin, which will be mostly referred to in the text if not otherwise stated. The traditionally used numbering system of the residues for all trypsin-like serine proteases follows that of chymotrypsin.

- *Bos taurus*, 243 amino acids (AA), cationic isoform, SWISS-PROT: P00760,[13] PDB: 5PTP.[14,15]
- *Rattus norvegicus*, 246 AA, anionic isoform, SWISS-PROT: P00763 sequence from nucleic acid,[16] PDB: 1TRM.[17]
- *Sus scrofa*, 231 AA, SWISS-PROT: P00761,[18] PDB: 1MCT.[19,20]
- *Homo sapiens*, 247 AA, cationic isoform, SWISS-PROT: P07477,[21] PDB: 1TRN.[22]
- *Salmo salar*, anionic isofrom, SWISS-PROT: P35031,[23] PDB: 1BIT.[24]
- *Fusarium oxysporum*, 248 AA, SWISS-PROT: P35049,[25] PDB: 1TRY.[25]
- *Streptomyces griseus*, 259 AA, SWISS-PROT: P00775,[26] PDB: 1SGT.[27]
- *Streptomyces erythraeus*, 227 AA, SWISS-PROT: P24664.[28]

Subtilisin Subtilisins comprise a large family of enzymes that are present in a wide variety of organisms. Sequence information is provided only for the bacterial enzymes for which three-dimensional structures have been determined.

- *Bacillus amyloliquefaciens*, precursor, 382 amino acids (AA), sequence of the isolated gene,[29] SWISS PROT: P00782 PDB: 2ST1.[30]
- *Bacillus* Ak.1, precursor, 401 AA, SWISS-PROT: Q45670.[126]
- *Bacillus alcalophilus*, 269 AA, SWISS-PROT: P27693 PDB: 1AH2.[127]
- *Bacillus lentus*, 269 AA, SWISS-PROT: P29600,[33] PDB: 1GCI.[34]
- *Bacillus licheniformis*, precursor, 379 AA, SWISS-PROT: P00780, sequence of the isolated gene;[35] and 274 AA, SWISS-PROT: P00781,[36] PDB: 1SCA.[37]
- *Bacillus pumilus*, 275 AA, SWISS-PROT: P07518,[38] PDB: 1MEE.[39]
- *Bacillus* novo sp. MN-32, (kumamolisin) SWISS-PROT: Q8RR56,[128] 384 AA, PDB: 1GT9.[40]
- *Bacillus subtilis*, precursor, 381 AA, SWISS-PROT: P04189, sequence of the isolated gene,[41] PDB: 1SJC.[42]
- *Pseudomonas* sp. 101, (carboxyl proteinase), 269 AA, SWISS-PROT: P42790,[43] PDB: 1GA6.[44]
- *Pseudomonas* sp. KSM-K16, (Protease M), 269 AA, SWISS-PROT: Q99405,[45] PDB: 1MPT.[45]
- *Thermoactinomyces vulgaris*, (thermitase), 279 AA, SWISS-PROT P04072,[46] PDB: 1THM.[47]
- *Tritirachium album limber*, (proteinase K), 384 AA, SWISS-PROT: P06873,[48] PDB: 1IC6.[49]

PROTEIN PRODUCTION, PURIFICATION, AND MOLECULAR CHARACTERIZATION

Trypsin Pancreatic trypsin is commercially available from Sigma, Boehringer Mannheim, Worthington, and Fluka. Recombinant trypsin has been expressed in many different systems, including *Escherichia coli*, yeast, and *Pichia pastoris*. Purification typically involves affinity chromatography on immobilized benzamidine or aprotinin. Literature references to purification procedures for different trypsins can be found on the website of BRENDA (http://www.brenda.uni-koeln.de/php/result_flat.php3?ec no=3.4.21.4).

The inactive precursor of trypsin, trypsinogen, is activated by removal of the N-terminal hexapeptide Val10-Asp-Asp-Asp-Asp-Lys15 to yield single-chain β-trypsin. Subsequent limited autolysis, first at Lys145 (α-trypsin) and then at Arg117 and Lys61, produces other active forms having two or more peptide chains bound by six disulfide bonds.

Subtilisin A variety of approaches have been used to produce, purify, and characterize members of the subtilisin family. A description of examples of the methods for subtilisin BPN′ are presented here. The gene for subtilisin BPN′ from *B. amyloliquefaciens* has been cloned, and it can be expressed at high levels from the promoter sequences in *B. subtilis*.[29,50] The secretion of the enzyme from the cells at high levels (up to 200 mg L^{-1}) facilitates the purification from the fermentation broth.[51] A number of purification protocols have been published.[51-53] A relatively simple recent procedure uses acetone precipitation after removal of the cells followed by two chromatography steps using DE52 and Poros HP 20 columns.[51]

ACTIVITY AND INHIBITION TESTS

Trypsin Trypsin exists as a monomer of 23 to 25 kDa molecular weight. The pI may vary significantly between the isoforms of the protein, for example, pI = 8.2 for cationic bovine trypsin, and pI = 4.7 for the anionic form. The pH optimum is approximately 8, and varies slightly with species. The presence of Ca^{2+} is required for maximum activity and stability of the enzyme. Under these conditions, the catalytic efficiency (k_{cat}/K_m) of protein (such as bovine serum albumin or insulin B chain) cleavage by trypsin is about 10 to 30 $min^{-1} \mu M^{-1}$.[54] The catalytic efficiency for synthetic substrates is much higher, for example, $K_m = 3 \mu M$, $k_{cat} = 1600 min^{-1}$ for benzoyl-Arg ethyl ester cleavage by bovine α-trypsin.[55] The temperature optimum for bovine trypsin is 45 °C.[56]

Subtilisin Assays are typically performed with the chromogenic substrate Suc-Ala-Ala-Pro-Phe-p-nitroanilide as described by Delmar *et al.*[57] Salt concentrations of 100 mM KCl and 2 mM $CaCl_2$ are often included for stability.[58] Depending on the enzyme variant or species, the values of k_{cat} range from 30 to 200 s^{-1}, values for K_m range from 0.13 to 1.2 μM, and values for k_{cat}/K_m range from 10^5 to $10^6 M^{-1} s^{-1}$.[59]

THREE-DIMENSIONAL STRUCTURES

Overall description of the structure

Trypsin Chymotrypsin was the first member of the superfamily of trypsin-like serine proteases, for which the crystal structure was determined.[2] The structure of bovine trypsin was determined a few years later,[14] and appeared to be very similar to that of chymotrypsin. It has a two-domain architecture, with the active site cleft between the domains (3D Structure: Trypsin). Each domain consists of a 6-stranded β-barrel with a 'Greek key' topology. It has been suggested that the two-domain structure is the result of an ancient gene duplication and fusion event.[60]

Subsequent structure determination of trypsinogen and comparison to the trypsin structure revealed the mechanism of the enzyme activation.[63,64] Cleavage of the propeptide allows the new N-terminal amino group to reach the invariant Asp194 at the bottom of a deep pocket, inducing the conformational changes that lead to the formation of the oxyanion hole and the substrate binding sites.[65]

Subtilisin Of the three crystal forms, the crystal structure of the C2 form was the first to be determined at 2.5-Å resolution[66] followed by that of the $P2_1$ form at 2.8 Å.[67] Much later, the structure of the $P2_12_12_1$ orthorhombic crystal form was reported.[30] Because subtilisin BPN′ has been the subject of numerous protein engineering efforts, structures at higher resolution have been determined, for example, at 1.6 Å for the C2 in which the crystals were obtained with slightly different conditions[68] and at 1.75 Å for the $P2_1$ crystal form.[69] Structures of subtilisins from other species have also been determined sometimes as enzyme–inhibitor complexes: subtilisin Carlsberg from *B. lichenformis*, the eglin c complex at 1.2-Å resolution;[70] subtilisin DY from *B. lichenformis* at 1.75-Å resolution;[71] subtilisin BL or savinase from *B. lentus* at 0.78-Å resolution;[34] subtilisin E and the propeptide from *B. subtilis* at 2.0-Å resolution;[42] protease M from *Bacillus* sp.KSM-K16 at 2.4-Å resolution;[45] and subtilisin from *Bacillus* Ak.1 at 1.8-Å resolution.[31] In addition, numerous crystal structures of variants of this enzyme have been published (for a review of protein engineering efforts, see Bryan[72]). The NMR structure of a subtilisin, serine protease PB92, from *B. alcalophilus* has also been reported.[32]

Wright *et al.*[66] were the first to provide a general description of the structure of subtilisin BPN′. The monomeric enzyme is roughly heart-shaped with a diameter of ∼42 Å (3D Structure: Subtilisin BPN′). Unlike many enzymes, no pronounced cleft is present on the surface. The fold has been classified as a doubly wound parallel α/β structure.[73] The active site is located on the flat surface of the hemisphere. The enzyme secondary structure is dominated by a central β-sheet comprised of seven parallel β-strands. This is bound on both sides by parallel α-helices. On one side, a bundle of 7α-helices are packed against the central β-sheet. Five of these helices are approximately antiparallel to the β-strands. On the other side of the β-sheet are two α-helices also running antiparallel to the β-strands. A left-handed βαβ crossover is found connecting β-strands 2 and 3. Additional pairs of antiparallel β-strands can also be identified that are either independent of or associated with the central β-sheet.

Trypsin-bound calcium ions

Calcium does not play any catalytic role in trypsin-like serine proteases, neither is it involved in substrate binding. However, the stabilizing role of Ca^{2+} against thermal denaturation and proteolytic degradation in many of these enzymes is well documented.[74] It has been suggested that the requirement for Ca^{2+} ensures that the enzymes are not active in the cytoplasm where the Ca^{2+} concentration is low.[75]

There are two Ca^{2+} binding sites in trypsinogen. One of them is related to the activation propeptide, which has an unusual, though highly conserved, sequence and plays a dual role in the activation process. It should be a good substrate for enterokinase, and a poor, but nonetheless most favorable, tryptic cleavage site on the trypsinogen molecule. Binding of Ca^{2+} at this site ($K_d = 16$ mM) is essential for complete and efficient activation of trypsinogen.[76] The effect of binding is to improve the substrate character of the propeptide to favor tryptic hydrolysis of the Lys15–Ile16 bond. Upon Ca^{2+} binding, K_m decreases by a factor of 3, while k_{cat} remains unchanged.[77] Studies on synthetic peptides similar to the propeptide have shown that tryptic hydrolysis of peptides containing only one aspartate residue does not depend on Ca^{2+}, while two and more aspartate residues promote Ca^{2+} dependence.[77] It was suggested that Ca^{2+} binds to the aspartate residues of the propeptide. However, this Ca^{2+} site has never been observed crystallographically because the entire propeptide is disordered in trypsinogen crystals.[63,64] It seems that the loosely structured activation hexapeptide is protected against trypsin more by its negative charge than by tertiary structure. Binding of Ca^{2+} to the propeptide neutralizes the negative charge and thus promotes the binding and hydrolysis by trypsin.

The other Ca^{2+} ion remains bound to the protein molecule after trypsinogen activation. This site has a higher affinity for Ca^{2+}, with a K_d of 0.6 mM,[77] and thus is considered a primary site. The calcium ion at this site stabilizes the protein toward thermal denaturation and autolysis.[76] The primary Ca^{2+} binding site was first identified in bovine trypsin by Bode and Schwager.[78] Later, it was also observed in the structures of trypsinogen.[63,64]

The Ca^{2+} binding loop is formed by residues 70 to 80. It is located at the periphery of the molecule, 20 Å from the catalytic center (3D Structure: Trypsin). Ca^{2+} is octahedrally coordinated to the carboxylic groups of Glu70 and Glu80, the main-chain carbonyl groups of residues 72 and 75, and two water molecules, which are in turn bound to Glu70 and Glu77 (Figure 1). The distances between Ca^{2+} and its oxygen ligands are in the range 2.21 to 2.34 Å as measured in the high-resolution crystal structure.[15]

Some differences in the coordination bond lengths have been reported in the early studies of the trypsinogen structure. Although calcium was excluded from all trypsinogen crystallization solutions to avoid autoactivation, the primary site was found to be fully occupied by a Ca^{2+} ion in the 1.9-Å structure of Kossiakoff et al.,[64] obtained from 30% ethanol at pH 7.5. In the structure determined by Fehlhammer et al.,[63] at 1.8-Å resolution from 1.5 M $MgSO_4$ at pH 6.9, the Ca–O distances were as short as 2.0 to 2.1 Å, which is significantly below normal values. The electron-density peak of about eight electrons was interpreted as incomplete occupation of the site. It was suggested that about half of the protein molecules were Ca^{2+}-deficient. Such heterogeneity would cause the effect of shorter Ca–O distances when refined as a single-conformation atomic model. This conclusion is in agreement with the relatively high-temperature factors of

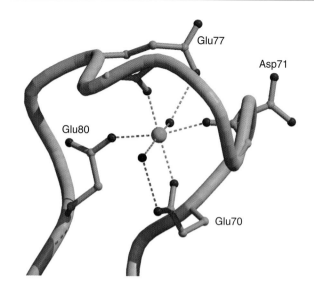

Figure 1 Calcium binding site found in trypsin (PDB code 5PTP).[15] Prepared with the programs MOLSCRIPT[61] and RASTER3D.[62]

the Ca^{2+} ligands. A similar picture is observed in trypsin crystals obtained at pH 5, when protonation of glutamate ligands reduces the affinity for Ca^{2+}.[63] After soaking the trypsinogen crystals in 30% PEG 2000 solution containing 0.1 M $CaCl_2$, full occupancy of the Ca^{2+} site was restored, and the coordination distances were normal. The difference in Ca^{2+} occupancies in the two trypsinogen structures is probably due to different procedures of protein purification and different crystallization conditions (though the crystals were isomorphous).

A direct comparison of the conformational changes accompanying binding of Ca^{2+} was available in the crystal structure of the Atlantic salmon (*S. salar*) trypsin.[79] There are four crystallographically independent protein molecules, and only one of them contains Ca^{2+}. The lack of calcium destabilizes not only the binding region but also the surrounding loops. The molecules without Ca^{2+} seem to be more flexible.

While all crystal structures of pancreatic trypsins in several crystal forms indicated only one Ca^{2+} binding site in the loop 70 to 80, somewhat contradictory data were obtained by using lanthanide ions as probes in NMR[80] and fluorescence energy-transfer[81] experiments. The distance between the paramagnetic Gd^{3+} ion in place of Ca^{2+} and the phosphorus of diisopropyl fluorophosphate (DIP) attached to the catalytic Ser195 was estimated to be about 10 Å. This led to the proposal that the Ca^{2+} binding site of trypsin is comprised of the side chains of Asp194 and Ser190, and the carbonyl group of Ser139, and raised the possibility of different protein conformations in solution and in the crystals. Epstein et al.[82] argued that because of the large amounts of Ca^{2+} present in the sample the primary binding site may have been saturated with Ca^{2+}, and the distances measured referred to weaker binding sites occupied by Gd^{3+}. Subsequent NMR relaxation measurements[83] produced the Gd^{3+} to ^{31}P distance of 21 Å, which is consistent with the crystallographic model.

The effect of Ca^{2+} on the conformational stability of trypsin and trypsinogen against thermal and chemical denaturation was studied at different pHs using UV spectroscopy and ^1H NMR.[84] At pH 5.8, Ca^{2+} enhances the conformational stability of trypsinogen (as measured by the free energy change) by $1.8 \, kcal \, mol^{-1}$, and the temperature of denaturation by 2.6 K. At higher pH, the stabilization effect becomes more profound. At pH 8.3, binding of Ca^{2+} triples the stability of trypsinogen, when compared to pH 5.8. At the same time, binding of Ca^{2+} at pH 8.3 is almost an order of magnitude weaker than at pH 6.6, where it reaches maximum. Although binding of the Ca^{2+} ion does not induce any major conformational change in the protein, it probably induces both local and long-range small structural changes, qualitatively similar to those observed in proteinase K.[85] A large water network extends from the calcium site to the autolysis loop 142 to 151 and Asp194. Conformation of the Ca^{2+} loop, which is apparently sensitive to the presence of the metal, may affect the orientation and dynamics of the autolysis loop.[86] Substantial differences in the NMR spectra following binding of Ca^{2+} were interpreted as a long-range rearrangement involving Trp141 and the His40–Asp194 ion pair, which is considered crucial in the trypsinogen-activation process. In a similar experiment on the urea denaturation of trypsin immobilized on glass beads (to eliminate the possibility of autolysis), Ca^{2+} decreased the denaturation rate significantly and also accelerated the rate of renaturation of denatured trypsin.[87] Stabilization of trypsin by Ca^{2+} was not accompanied by changes in enzymatic activity or in ligand-binding properties.

The activation effect of Ca^{2+} was observed in a series of experiments carried out at above the temperature optimum T_{opt} (45 °C).[88] High (millimolar) concentrations of Ca^{2+} enhanced the proteolytic activity of trypsin, while in the absence of Ca^{2+} and at low (micromolar) Ca^{2+} concentrations, the activity dropped quickly as the temperature raised to 60 °C. The result was interpreted as an indication of an altered, Ca-stabilized form of trypsin with the enhanced enzymatic activity. The Arrhenius activation-energy values derived from these data were $8.2 \, kcal \, mol^{-1}$ for the low, and $12.8 \, kcal \, mol^{-1}$ for the high-temperature forms. In the absence of Ca^{2+}, a value of $8.2 \, kcal \, mol^{-1}$ was obtained. Optical rotatory dispersion spectra showed significant conformational rearrangements above the T_{opt}, suggesting the formation of a more compact structure with a possible increase in the helical content. The existence of this form of trypsin was never confirmed by X-ray or NMR structural studies.

The involvement of Ca^{2+} in the stabilization of the autolysis sites Lys61–Ser62 and Arg117–Val118 was demonstrated in the mutagenesis studies on rat trypsin.[89] Both single mutants K61N and R117N, and particularly the double mutant K61N/R117N, showed significantly increased stability against autolysis and decreased sensitivity to Ca^{2+}.

Calcium sites in trypsin-related proteases

Trypsin-like serine proteases constitute one of the largest enzyme families, both in terms of the number of sequenced proteins and in the number of different peptidase activities. Over 50 different proteases have been structurally characterized to date according to the SCOP database.[90] These are mostly mammalian enzymes including digestive proteases and proteins of the blood coagulation cascade. Prokaryotic peptidases are largely represented by the components of *S. griseus* pronase, a mixture of extracellular enzymes.

The Ca^{2+}-related properties of trypsin-like serine proteases vary significantly between the members of the family. Even closely related proteins differ in their dependence on Ca^{2+}. In contrast to mammalian trypsins, no Ca^{2+} binding sites are present in trypsins from the fungus *F. oxysporum*[25], larvae *Tineola bisselliella*,[91] and *S. erythraeus*.[92] On the other hand, *S. griseus* trypsin has one Ca^{2+} binding site,[27] although different from the primary trypsin site (see below). In most cases, the lack of Ca^{2+} binding can be explained by the amino acid replacements in the respective fragments of the protein sequence. However, the answer to the question of what are the alternative stabilization mechanisms in these Ca^{2+}-independent proteases is not so trivial.

Chymotrypsin Chymotrypsin (EC 3.4.21.1) is characterized by the preference to substrates with aromatic amino acids in the P1 position. Although studies of chymotrypsin in solution[93–95] suggested a high-affinity Ca^{2+} binding site, no indication of such a site was found in the crystal structures of chymotrypsin determined at different pHs (from 4.5 to 7.8) and under different crystallization conditions. The conformation of the primary site loop 70 to 80 is very similar to that in trypsin, but the replacement of Glu80 with a hydrophobic amino acid (isoleucine in most species) apparently precludes binding of Ca^{2+} at this site in chymotrypsin.

Elastase Elastase is defined by its ability to hydrolyze insoluble elastin fibers. There are two genetically distinct porcine elastases called I (EC 3.4.21.36) and II (EC 3.4.21.71). Ca^{2+} binds to elastase I with high affinity ($K_m = 45 \, \mu M$ for porcine enzyme), but does not influence the catalytic properties of the enzyme. Terbium luminescence and fluorine NMR studies have indicated that the Ca^{2+} binding site involves Glu70 and Glu80.[96]

The crystal structure of porcine elastase I has been determined in free and inhibited states.[97] The Ca^{2+} binding

Structural calcium (trypsin, subtilisin)

site is formed by the loop 70 to 80, with the carboxyl groups of Glu70, Asp77, and Glu80, carbonyl groups 72 and 75, and a water molecule at the top of an octahedron. The geometry is close to a pentagonal bipyramid with the carboxylate of Gly70 being a bidentate ligand in the equatorial plane that also includes Asp77, carbonyl 75, and a water molecule. Compared to trypsin, the carboxyl oxygen of Asp77 replaces a water molecule in the coordination sphere of Ca^{2+} in elastase. This aspartic acid is conserved in the sequences of mammalian elastases and is replaced by a glutamic acid in the fish enzymes, where it also directly coordinates Ca^{2+}.[98] In trypsin, the equivalent Glu77 is involved in Ca^{2+} binding not directly, but through a water molecule (Figure 1). The difference may be due to the presence of the negatively charged aspartic acid in position 71 in trypsin that causes mutual repulsion of Gly77 and Asp71. In all elastases I, position 71 is occupied by a histidine.

Although the physiological metal in elastase I is Ca^{2+}, it was noted[99] that most crystal structures in the PDB probably contain Na^+ in the binding loop that was misinterpreted as Ca^{2+}. The standard crystallization conditions include 0.1 M sodium acetate buffer pH 5.1 and 0.2 M Na_2SO_4, which favor binding of Na^+, as was demonstrated by using anomalous diffraction.[99] Ca^{2+} occupies the site when crystallization solution is complemented with 5 mM $CaCl_2$ in sodium citrate buffer pH 5.1. Comparison of the two structures revealed the same geometry for both metals, and slightly shorter metal–oxygen distances for Ca^{2+} (2.25–2.38 Å as compared to 2.30–2.43 Å for Na^+). The binding of different metals does not cause any significant structural rearrangement, which is consistent with the fact that neither Na^+ nor Ca^{2+} has a major influence on the catalytic activity of the enzyme.[100] It should be noted that elastase II does not bind Ca^{2+} because of the highly conserved basic amino acid in position 70.

Besides the pancreatic elastase, many animals have leukocyte elastase (EC 3.4.21.37) that cleaves not only elastin but also different types of collagen, fibronectin, immunoglobulins G and M, and a number of coagulation proteins thereby modulating their activity.[101] Despite the sequence and structural similarity to pancreatic elastases, human leukocyte elastase does not bind Ca^{2+}, which probably reflects different physiological roles and regulation of these two elastases.

Blood coagulation factors Blood coagulation factors represent a large group of trypsin-like proteases that are also Ca^{2+}-dependent.[102] However, the Ca^{2+} binding sites are not on the protease domain of these enzymes, but are associated to the other domains, namely, the Gla domain and the EGF domains. The presence of a lysine residue at position 70 in thrombin precludes binding of cations in the 'Ca loop'. The cation observed in several crystal structures of thrombin in the loop 221 to 224 of the protease domain and described as Ca^{2+}, is likely to be Na^+, as can be judged from the coordination geometry and from comparison to other thrombin structures.

Bacterial trypsin-like proteases The soil bacterium *S. griseus* produces several serine proteases that belong to the trypsin structural superfamily. One of them is particularly similar to pancreatic trypsin sharing 30% sequence identity and a strong preference for substrates with basic P1 residues. This bacterial trypsin (SGT) possesses a calcium site, although different from that in pancreatic trypsin as the key glutamate residues in positions 70 and 80 are not conserved in SGT. The crystal structure of SGT[27] revealed the Ca^{2+} ion bound in the loop 165 to 180. Given the twofold internal symmetry of the trypsin molecule, the site is located in a topologically symmetrical position with respect to the primary Ca^{2+} site in the loop 70 to 80 (3D Structure: Trypsin). The loop of residues 165–180 provides three out of four protein ligands. The site is composed of the side chains of Asp165 (bidentate coordination) and Glu230, the carbonyl groups of Ala177 and Glu180, and two water molecules. The coordination geometry is pentagonal bipyramidal, which is typical for calcium complexes.[103] The distances to ligands are in the range 2.24–2.52 Å. Somewhat unusual for the calcium sites in proteins is that the site is not built from one short stretch of the polypeptide, but includes a distal ligand, Glu230.

S. griseus proteases A (SGA) and B (SGB) have broad substrate specificity with some P1 preference to aromatic amino acids and act best on extended peptides. Both have a large propeptide of about 80 residues. The overall structure is similar to trypsin, with some of the largest differences being around the 'Ca loop', which is 10 residues shorter and does not bind Ca^{2+}. The 'truncated' N-terminus resides near the 'Ca loop' rather than in the other domain, not far from the extended C-terminus. These structural differences result in the formation of a cation binding site in the loop between residues 120 and 121, where six residues (120A to 120F) are inserted as compared to chymotrypsin. The coordination sphere of the cation (Figure 2) involves the C-terminal carboxyl group (bidentate ligand), the carboxyl group of Asp120F, the carbonyl groups of residues 120 and 121, and two water molecules. The pentagonal bipyramidal geometry and the distances (2.3–2.5 Å) suggest that the cation is Ca^{2+}.[103] The binding site is observed in all structures of *Streptomyces* proteases solved to date, namely, in SGA,[104] SGB,[105] *S. griseus* protease E, also known as glutamyl endopeptidase II (EC 3.4.21.80),[106] and protease from *Streptomyces fradiae* (EC 3.4.21.81).[107] It should be noted, however, that the cation was interpreted as Ca^{2+} only in SGB,[108] while it was modeled as a water molecule (with very low B-factors) in the other structures. The fact that some of these crystals were obtained from calcium-free solutions indicates that the Ca^{2+} ion observed in the

Structural calcium (trypsin, subtilisin)

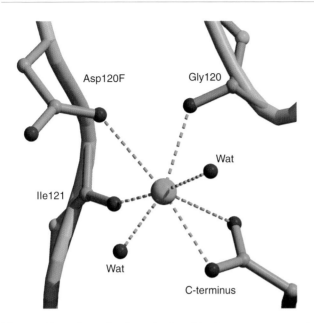

Figure 2 Calcium binding site found in *S. griseus* protease A (PDB code 3SGA).[109] Prepared with the programs MOLSCRIPT[61] and RASTER3D.[62]

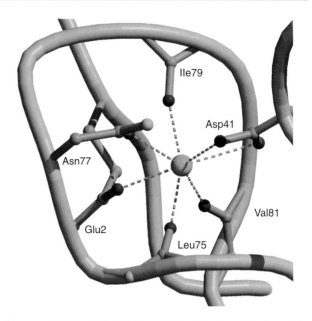

Figure 3 The high-affinity calcium binding site found in subtilisin BPN′ (PDB code 2ST1).[30] Prepared with the programs MOLSCRIPT[61] and RASTER3D.[62]

structure is an integral part of the protein likely related to its function. Given that both termini of the protein chain are involved in Ca^{2+} binding (C-terminal directly, N-terminal through a hydrogen bond to the Ca^{2+} binding loop), a possible role of this cation may be to stabilize and protect the structure from exopeptidases. On the other hand, the Ca^{2+} binding site is located in the same region of the molecule where the primary Ca^{2+} site is observed in trypsin, and therefore it may play a role of protecting the protein from autolytic degradation typical for other trypsin-like serine proteases.

Subtilisin-bound calcium ions

Subtilisin BPN′ has two calcium binding sites, A and B, a high-affinity and low-affinity binding site, respectively. The binding energy of calcium ions to specific sites in the tertiary structure contributes to the enzyme stability and increases the unfolding activation energy. The presence of calcium is also known to inhibit autolysis.[110] Both sites have been found in all *Bacillus* enzymes determined to date. The high-affinity calcium binding site is associated with a loop near the N-terminus of the helix containing the active site residue His64. The calcium ion at the high-affinity site is encircled with the backbone carbonyl oxygen atoms of Leu75, Ile79, and Val81 and the side-chain oxygen of Asn77 is shown in Figure 3. Three additional oxygen atoms from the side chains of Glu2 and Asp41 coordinate the cation from either side of the loop. The distances of the ligands to the calcium ions range from 2.3 to 2.4 Å.[30] A calcium-free apoenzyme can be produced in which the calcium binding affinity can be determined using fluorescence spectroscopy and microcalorimetry.[111] The binding parameters from these measurements were $\Delta H = -11\,kcal\,mol^{-1}$ and $K_a = 7 \times 10^6\,M^{-1}$ at 25 °C. The binding free energy is primarily enthalpic with a value of 9.3 kcal mol^{-1}.

The low-affinity calcium binding site, located in a narrow crevice more than 30 Å away from the high-affinity binding site, has a dramatically different coordination geometry – a distorted pentagonal bipyramid. The calcium ion ligands shown in Figure 4 include the peptide carbonyl oxygen atoms of Gly169, Tyr171, Val174, and Glu195, and two water molecules. The carboxylate oxygen atoms of the Asp197 side chain bind directly with the calcium ion. The distances of the ligands to the calcium ions are longer than those found associated with the high-affinity site, ranging from 2.7 to 3.1 Å.[30] Because of discrepancies in the structure of this site, a systematic structural investigation was undertaken that verified that this site can accommodate calcium or a monovalent cation such as Na^+ or K^+ depending on the concentration of calcium.[112] In the presence of sodium, the location of the site moves by 2.7 Å. The ligand pattern is different in that it includes the carbonyl oxygen atoms of Gly169, Tyr171, Val174, and Glu195, a side-chain carboxylate oxygen of Asp197, and two water molecules. Recently, using a variant missing the high-affinity A site, binding affinity for the low-affinity calcium binding site have been determined. At 65 °C, $K_a = 100\,M^{-1}$.[51]

Calcium sites in subtilisin-related proteases

Thermitase The 1.4-Å resolution structure of thermitase from *T. vulgaris*, a subtilisin homolog, revealed two calcium

Structural calcium (trypsin, subtilisin)

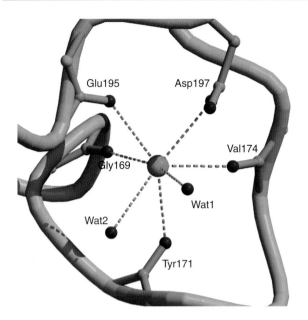

Figure 4 The low-affinity calcium binding site found in subtilisin BPN' (PDB code 2ST1).[30] Wat1 (#365 in the 2STI coordinates) is axial. Prepared with the programs MOLSCRIPT[61] and RASTER3D.[62]

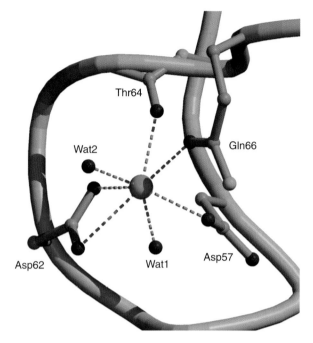

Figure 5 The second calcium binding site in thermitase (PDB code 1THM).[47] Wat2 (#370 in the 1THM coordinates) is the axial ligand. Prepared with the programs MOLSCRIPT[61] and RASTER3D.[62]

binding sites.[47] The first was equivalent to the high-affinity site found in subtilisins BPN', but the second was a site not seen previously. The site is located near one edge of the central β-pleated sheet in a loop of residues 57 to 66 (Figure 5). The geometry of binding is pentagonal bipyramidal with the apical ligands Asp57 being the carboxylic oxygen and the water molecule Wat2. The distances of the ligands to the calcium ions range from 2.25 to 2.64 Å. A third potential calcium binding site was also discovered corresponding to the low-affinity site observed for subtilisin BPN'. A full calcium occupancy at this site in thermitase was achieved when crystallization medium was complemented with 100 mM $CaCl_2$.[113] The calcium ions at each of the three sites have no impact on specificity, and all are implicated only in the role of stabilizing the enzyme structure.

Proteinase K The first crystals of fungal enzyme Proteinase K from *Tritirachium album* reported diffracted to 1.4-Å resolution.[114] The crystal structure was solved and reported at 3.3-Å resolution.[115] From the structure, it was recognized that the enzyme was related structurally to subtilisin BPN'. At this time, the amino acid sequence of proteinase K had not been determined. This structural data helped establish the subtilisin family of proteases that are distinctly different from the enzymes in the trypsin family. It was not until the structure was determined at 1.5-Å resolution[116,117] that the identification of two bound calcium site ions was reported.[116] Recently, the structure of the enzyme has been reported at 0.98 Å, providing new details of the interactions of the calcium ions with the protein.[49] As shown in Figure 6,

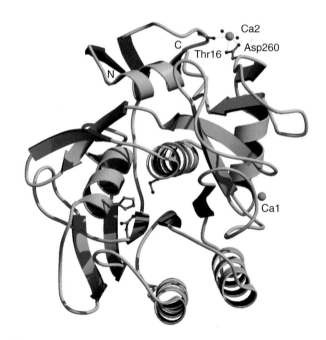

Figure 6 Proteinase K fold (PDB code 2PRK)[117] with the catalytic triad and two calcium sites. Ligands of Ca1 (bidentate Asp260, carbonyl Thr16, and two waters) are shown. Prepared with the programs MOLSCRIPT[61] and RASTER3D.[62]

the two calcium ions, Ca1 and Ca2, are found associated with bridging loops near the surface of the protein. The Ca1 ion shown in Figure 7 is eightfold coordinated with the ligands forming a pentagonal bipyramid. The four water

Structural calcium (trypsin, subtilisin)

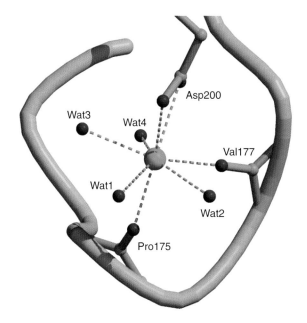

Figure 7 The first calcium binding site in proteinase K (PDB code 2PRK).[117] Prepared with the programs MOLSCRIPT[61] and RASTER3D.[62]

molecules, 301, 303, 304, and 729 (corresponding to Wat1 to Wat4 in the Figure), and the carbonyl oxygen atom of Val177 lie in a plane between the two apical ligands, the carbonyl oxygen of Pro175 and the carboxylate oxygens of Asp200. This site is equivalent to the low-affinity binding site found in subtilisin BPN'. The binding of Ca2 ion shown in Figure 6 is less well defined. The 1.5-Å structure identified only six ligands[116] but the recent high-resolution structure identified eight, again in a pentagonal bipyramid arrangement. The backbone carbonyl oxygen atoms of Thr16 and a water molecule, 683, occupy the apical positions. The ligands on the equatorial plane include four water molecules, 305, 581, 589, and 655, and one of the carboxylate oxygen atoms of Asp260. The role of calcium in stability and in preventing autolysis was established by Bajorath et al.[118]

Bacillus Ak.1 protease The structure of this enzyme revealed three calcium binding sites.[31] One corresponds to the high-affinity site of subtilisin BPN', and the second site corresponds to the second site observed in thermitase. The third site is unique to the *Bacillus* Ak.1 protease. The ligand interactions with the ion form a pentagonal bipyramidal arrangement. The ligands include an oxygen atom of the side-chain carboxyl group of Asp50, both oxygen atoms of the side-chain carboxyl group of Glu83, the carbonyl oxygen atom of Pro47 and three water molecules. The distances between the oxygen atoms and the calcium ion vary from 2.2 to 2.6 Å. Calcium binding sites 1 and 3 are closely linked through interactions with Glu83. The carbonyl oxygen atom of Glu83 is one of the ligands of calcium binding site 1. It should be mentioned that a sodium binding site was also discovered that corresponds to the weak binding equivalent binding site found in both subtilisin BPN' and thermitase. It was speculated that calcium site 3 and its interaction with calcium site 1 may be cooperative in stabilizing the protease against thermal denaturation, and thus the third site is a major factor in this protease's stability at elevated temperatures.

Serine–Carboxyl proteinases Recently, the structures of two serine–carboxyl proteinases from *Pseudomonas* sp. 101 and *Bacillus* novosp. MN-32 (kumamolysin) have been reported.[40,44,119] These enzymes have a subtilisin fold with approximately 20% sequence identity, and the active site histidine is replaced with a glutamate; hence the name serine–carboxyl proteinase. Only a single, common calcium binding site has been identified in both enzyme structures that is quite different from the earlier reported calcium binding sites in other subtilisins. In the *Pseudomonas* enzyme, the ligands are in an octahedral arrangement with a single carboxylate oxygen atom from both Asp328 and Asp348 at the two apical positions and the peptide carbonyl groups of Val329, Gly344, and Gly346 along with a single water molecule (401) at the equatorial positions[119] (Figure 8). The bond distances of the ligands range from 2.3 to 2.5 Å. In the *Bacillus* structure, analogous ligands are found bound to the calcium ion: carboxylate oxygen atoms from Asp316 and Asp338 are at the two apical positions and the peptide carbonyl oxygens of Ile317, Gly334, and Gly336 along with a single water molecule (552) are at the equatorial positions.[40] The calcium ion stabilizes the

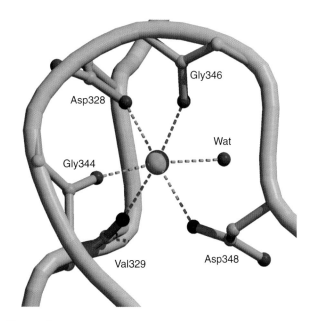

Figure 8 Calcium binding site in serine carboxyl-protease (PDB code 1GA4).[44] Prepared with the programs MOLSCRIPT[61] and RASTER3D.[62]

loop structure in the C-terminal region of the molecule composed of residues 328–353. Before it was realized that this protease family had a novel catalytic triad and believing that it might be an acid proteinase, site-directed mutagenesis studies discovered that replacement of Asp328 in the *Pseudomonas* structure eliminated both catalytic and autoprocessing activities.[120] Thus, the importance of bound calcium has been established for this family of enzymes, although its role has not been clearly defined at this point.

CALCIUM SITE PROTEIN ENGINEERING

Trypsin   Trypsin has served as a model system to test the design of metal binding sites for specific functions. Successful examples are the incorporation of the divalent cation sites to reversibly inhibit the catalytic activity,[121] and to change substrate specificity toward histidine at the P2′ position.[122]

Subtilisin   Subtilisin has been the subject of numerous protein engineering efforts that have been directed at changing many of its properties. During the course of these studies, mutations for more than 50% of the 275 amino acid residues have been reported.[72] One focus area has been to change the dependence upon Ca^{2+} for stability. The high-affinity A site has been removed by deleting residues 75–83 (Δ75–73).[111] The structure of a subtilisin variant containing this mutation revealed that the omission of this loop, which extends out from a helix containing residues 63–85, results in an uninterrupted helix.[123] The folding rate of the deletion mutant is 10^4 times faster than that observed for the wild-type enzyme folded in 0.1 M KP_i at neutral pH, but the unfolding rates are very similar.[111] Using a process of directed evolution, a series of mutations were introduced that increased the stability of subtilisin by a 1000-fold in the presence of strong chelating conditions.[124]

Subtilisin BL has also been the subject of protein engineering efforts, in this case to introduce the third calcium binding site of thermitase to increase the enzyme's thermal stability.[125] This involved site-directed mutagensis efforts to introduce the thermitase amino acid sequence for residues 50–60 and residue 92. Unfortunately, the enzyme undergoes autolysis, cleaving the protein in the modified loop region between amino acid residues 54 and 55. The X-ray crystal structure revealed the modified loop in the active site of a molecule in a symmetry-related position.

REFERENCES

1   AJ Barrett, ND Rawlings and JF Woessner, *Handbook of Proteolytic Enzymes*, Academic Press, San Diego (1998).

2   BW Matthews, PB Sigler, R Henderson and DM Blow, *Nature*, **214**, 652–56 (1967).

3   D Hewett-Emmett, J Czelusniak and M Goodman, *Ann N Y Acad Sci*, **370**, 511–27 (1981).

4   DA Estell, TP Graycar, JV Miller, DB Powers, JP Burnier, PG Ng and JA Wells, *Science*, **233**, 659–63 (1986).

5   H Gron, M Meldal and K Breddam, *Biochemistry*, **31**, 6011–18 (1992).

6   G Voordouw, C Milo and RS Roche, *Biochemistry*, **15**, 3716–24 (1976).

7   KA Walsh, *Methods Enzymol*, **19**, 41–63 (1970).

8   P Keller, E Cohen and H Neurath, *J Biol Chem*, **233**, 344–49 (1958).

9   DC Whitcomb, MC Gorry, RA Preston, W Furey, MJ Sossenheimer, CD Ulrich, SP Martin, LK Gates Jr, ST Amann, PP Toskes, R Liddle, K McGrath, G Uomo, JC Post and GD Ehrlich, *Nat Genet*, **14**, 141–45 (1996).

10 SD Power, RM Adams and JA Wells, *Proc Natl Acad Sci USA*, **83**, 3096–3100 (1986).

11 MY Yang, E Ferrari and DJ Henner, *J Bacteriol*, **160**, 15–21 (1984).

12 S Strausberg, P Alexander, L Wang, F Schwartz and P Bryan, *Biochemistry*, **32**, 8112–19 (1993).

13 O Mikes, V Holeysovsky, V Tomasek and F Sorm, *Biochem Biophys Res Commun*, **24**, 346–52 (1966).

14 RM Stroud, LM Kay and RE Dickerson, *Cold Spring Harbor Symp Quant Biol*, **36**, 125–40 (1972).

15 JS Finer-Moore, AA Kossiakoff, JH Hurley, T Earnest and RM Stroud, *Proteins*, **12**, 203–22 (1992).

16 CS Craik, QL Choo, GH Swift, C Quinto, RJ MacDonald and WJ Rutter, *J Biol Chem*, **259**, 14255–64 (1984).

17 S Sprang, T Standing, RJ Fletterick, RM Stroud, J Finer-Moore, NH Xuong, R Hamlin, WJ Rutter and CS Craik, *Science*, **237**, 905–9 (1987).

18 MA Hermodson, LH Ericsson, H Neurath and KA Walsh, *Biochemistry*, **12**, 3146–53 (1973).

19 Q Huang, S Liu, Y Tang, F Zeng and R Qian, *FEBS Lett*, **297**, 143–146 (1992).

20 RM Sweet, HT Wright, J Janin, CH Chothia and DM Blow, *Biochemistry*, **13**, 4212–28 (1974).

21 M Emi, Y Nakamura, M Ogawa, T Yamamoto, T Nishide, T Mori and K Matsubara, *Gene*, **41**, 305–10 (1986).

22 C Gaboriaud, L Serre, O Guy-Crotte, E Forest and JC Fontecilla-Camps, *J Mol Biol*, **259**, 995–1010 (1996).

23 R Male, JB Lorens, AO Smalas and KR Torrissen, *Eur J Biochem*, **232**, 677–85 (1995).

24 AO Smalas and A Hordvik, *Acta Crystallogr, Sect D*, **49**, 318–30 (1993).

25 WR Rypniewski, S Hastrup, C Betzel, M Dauter, Z Dauter, G Papendorf, S Branner and KS Wilson, *Protein Eng*, **6**, 341–48 (1993).

26 JC Kim, SH Cha, ST Jeong, SK Oh and SM Byun, *Biochem Biophys Res Commun*, **181**, 707–13 (1991).

27 RJ Read and MN James, *J Mol Biol*, **200**, 523–51 (1988).

28 T Yamane, M Kobuke, H Tsutsui, T Toida, A Suzuki, T Ashida, Y Kawata and F Sakiyama, *J Biochem (Tokyo)*, **110**, 945–50 (1991).

29 N Vasantha, LD Thompson, C Rhodes, C Banner, J Nagle and D Filpula, *J Bacteriol*, **159**, 811–19 (1984).

30 R Bott, M Ultsch, A Kossiakoff, T Graycar, B Katz and S Power, *J Biol Chem*, **263**, 7895–7906 (1988).

31 CA Smith, HS Toogood, HM Baker, RM Daniel and EN Baker, *J Mol Biol*, **294**, 1027–40 (1999).
32 JR Martin, FA Mulder, Y Karimi-Nejad, J van der Zwan, M Mariani, D Schipper and R Boelens, *Structure*, **5**, 521–32 (1997).
33 C Betzel, S Klupsch, G Papendorf, S Hastrup, S Branner and KS Wilson, *J Mol Biol*, **223**, 427–45 (1992).
34 P Kuhn, M Knapp, SM Soltis, G Ganshaw, M Thoene and R Bott, *Biochemistry*, **37**, 13446–52 (1998).
35 M Jacobs, M Eliasson, M Uhlen and JI Flock, *Nucleic Acids Res*, **13**, 8913–26 (1985).
36 P Nedkov, W Oberthur and G Braunitzer, *Hoppe-Seyler's Z Physiol Chem*, **364**, 1537–40 (1983).
37 PA Fitzpatrick, AC Steinmetz, D Ringe and AM Klibanov, *Proc Natl Acad Sci USA*, **90**, 8653–57 (1993).
38 I Svendsen, N Genov and K Idakieva, *FEBS Lett*, **196**, 228–32 (1986).
39 Z Dauter, C Betzel, N Genov, N Pipon and KS Wilson, *Acta Crystallogr, Sect B*, **47**, 707–30 (1991).
40 M Comellas-Bigler, P Fuentes-Prior, K Maskos, R Huber, H Oyama, K Uchida, BM Dunn, K Oda and W Bode, *Structure (Camb)*, **10**, 865–76 (2002).
41 ML Stahl and E Ferrari, *J Bacteriol*, **158**, 411–18 (1984).
42 SC Jain, U Shinde, Y Li, M Inouye and HM Berman, *J Mol Biol*, **284**, 137–44 (1998).
43 K Oda, T Takahashi, Y Tokuda, Y Shibano and S Takahashi, *J Biol Chem*, **269**, 26518–24 (1994).
44 A Wlodawer, M Li, Z Dauter, A Gustchina, K Uchida, H Oyama, BM Dunn and K Oda, *Nat Struct Biol*, **8**, 442–46 (2001).
45 T Yamane, T Kani, T Hatanaka, A Suzuki, T Ashida, T Kobayashi, S Ito and O Yamashita, *Acta Crystallogr, Sect D*, **51**, 199–206 (1995).
46 B Meloun, M Baudys, V Kostka, G Hausdorf, C Frommel and WE Hohne, *FEBS Lett*, **183**, 195–200 (1985).
47 AV Teplyakov, IP Kuranova, EH Harutyunyan, BK Vainshtein, C Frommel, WE Hohne and KS Wilson, *J Mol Biol*, **214**, 261–79 (1990).
48 K-D Jany, G Lederer and B Mayer, *Biol Chem Hoppe-Seyler*, **367**, 87 (1986).
49 C Betzel, S Gourinath, P Kumar, P Kaur, M Perbandt, S Eschenburg and TP Singh, *Biochemistry*, **40**, 3080–88 (2001).
50 JA Wells, E Ferrari, DJ Henner and EY Chen, *Nucleic Acids Res*, **11**, 7911–25 (1983).
51 PA Alexander, B Ruan and PN Bryan, *Biochemistry*, **40**, 10634–39 (2001).
52 MD Ballinger, J Tom and JA Wells, *Biochemistry*, **34**, 13312–19 (1995).
53 M Rheinnecker, G Baker, J Eder and AR Fersht, *Biochemistry*, **32**, 1199–1203 (1993).
54 W Rick, *Methods Enzymol Anal*, **1**, 1052–63 (1974).
55 G Foucault, F Seydoux and J Yon, *Eur J Biochem*, **47**, 295–3022 (1974).
56 R Venkatesh and PV Sundaram, *Protein Eng*, **11**, 691–98 (1998).
57 EG DelMar, C Largman, JW Brodrick and MC Geokas, *Anal Biochem*, **99**, 316–20 (1979).
58 M Philipp and ML Bender, *Mol Cell Biochem*, **51**, 5–32 (1983).
59 MD Ballinger, JA Wells, in AJ Barret (ed.), *Handbook of Proteolytic Enzymes*, Academic Press, San Diego, pp 289–94 (1999).
60 AM Lesk and WD Fordham, *J Mol Biol*, **258**, 501–37 (1996).
61 PJ Kraulis, *J Appl Crystallogr*, **24**, 946–50 (1991).
62 EA Merritt and DJ Bacon, *Methods Enzymol*, **277**, 505–24 (1997).
63 H Fehlhammer, W Bode and R Huber, *J Mol Biol*, **111**, 415–38 (1977).
64 AA Kossiakoff, JL Chambers, LM Kay and RM Stroud, *Biochemistry*, **16**, 654–64 (1977).
65 R Huber and W Bode, *Acc Chem Res*, **11**, 114–22 (1978).
66 CS Wright, RA Alden and J Kraut, *Nature*, **221**, 235–42 (1969).
67 J Drenth, WG Hol, JN Jansonius and R Koekoek, *Eur J Biochem*, **26**, 177–81 (1972).
68 T Gallagher, J Oliver, R Bott, C Betzel and GL Gilliland, *Acta Crystallogr, Sect D*, **52**, 1125–35 (1996).
69 GL Gilliland, DT Gallagher, P Alexander and P Bryan, *Adv Exp Med Biol*, **379**, 159–69 (1996).
70 W Bode, E Papamokos and D Musil, *Eur J Biochem*, **166**, 673–92 (1987).
71 S Eschenburg, N Genov, K Peters, S Fittkau, S Stoeva, KS Wilson and C Betzel, *Eur J Biochem*, **257**, 309–18 (1998).
72 P Bryan, *Biochim Biophys Acta*, **1543**, 203–22 (2000).
73 JS Richardson, *Adv Protein Chem*, **34**, 167–339 (1981).
74 RB Martin, in H Siegel (ed.), *Metal Ions in Biological Systems*, Vol. 17, Marcel Dekker, New York, pp 1–49 (1984).
75 RH Kretsinger, *Int Rev Cytol*, **46**, 323–93 (1976).
76 JH Northrop, M Kunitz and RM Herriot, *Crystalline Enzymes*, 2nd edn, Columbia University Press, New York (1948).
77 JP Abita, M Delaage and M Lazdunski, *Eur J Biochem*, **8**, 314–24 (1969).
78 W Bode and P Schwager, *J Mol Biol*, **98**, 693–717 (1975).
79 HK Schroeder, NP Willassen and AO Smalas, *Acta Crystallogr, Sect D*, **54**, 780–98 (1998).
80 F Abbott, DW Darnall and ER Birnbaum, *Biochem Biophys Res Commun*, **65**, 241–47 (1975).
81 DW Darnall, F Abbott, JE Gomez and ER Birnbaum, *Biochemistry*, **15**, 5017–23 (1976).
82 M Epstein, J Reuben and A Levitzki, *Biochemistry*, **16**, 2449–57 (1977).
83 F Adebodun and F Jordan, *Biochemistry*, **28**, 7524–31 (1989).
84 G Bulaj and J Otlewski, *J Mol Biol*, **247**, 701–16 (1995).
85 J Bajorath, S Raghunathan, W Hinrichs and W Saenger, *Nature*, **337**, 481–84 (1989).
86 HD Bartunik, LJ Summers and HH Bartsch, *J Mol Biol*, **210**, 813–28 (1989).
87 JE Gomez, ER Birnbaum, GP Royer and DW Darnall, *Biochim Biophys Acta*, **495**, 177–82 (1977).
88 T Sipos and JR Merkel, *Biochemistry*, **9**, 2766–75 (1970).
89 E Varallyay, G Pal, A Patthy, L Szilagyi and L Graf, *Biochem Biophys Res Commun*, **243**, 56–60 (1998).
90 AG Murzin, SE Brenner, T Hubbard and C Chothia, *J Mol Biol*, **247**, 536–40 (1995).
91 CW Ward, *Biochim Biophys Acta*, **391**, 201–11 (1975).
92 T Yamane, A Iwasaki, A Suzuki, T Ashida and Y Kawata, *J Biochem (Tokyo)*, **118**, 882–94 (1995).
93 F Friedberg and S Bose, *Biochemistry*, **8**, 2564–67 (1969).
94 ER Birnbaum, F Abbott, JE Gomez and DW Darnall, *Arch Biochem Biophys*, **179**, 469–76 (1977).
95 J De Jersey, RS Lahue and RB Martin, *Arch Biochem Biophys*, **205**, 536–42 (1980).
96 JL Dimicoli and J Bieth, *Biochemistry*, **16**, 5532–37 (1977).

Structural calcium (trypsin, subtilisin)

97. EF Meyer, R Radhakrishnan, GM Cole and LG Presta, *J Mol Biol*, **189**, 533–39 (1986).
98. GI Berglund, NP Willassen, A Hordvik and AO Smalas, *Acta Crystallogr, Sect D*, **51**, 925–37 (1995).
99. MS Weiss, S Panjikar, E Nowak and PA Tucker, *Acta Crystallogr, Sect D*, **58**, 1407–12 (2002).
100. BS Hartley, DM Hotton, in PD Boyer (ed.), *The Enzymes*, 3rd edn, Academic Press, New York, pp 323–73 (1971).
101. JG Bieth, in RP Mecham (ed.), *Biology of Extracellular Matrix*, Vol. 1, Academic Press, New York, pp 217–320 (1986).
102. EW Davie, K Fujikawa and W Kisiel, *Biochemistry*, **30**, 10363–70 (1991).
103. H Einspahr, CE Bugg, in H Sigel (ed.), *Metal Ions in Biological Systems*, Vol. 17, Marcel Dekker, New York, pp 51–97 (1984).
104. GD Brayer, LT Delbaere and MN James, *J Mol Biol*, **124**, 261–83 (1978).
105. LT Delbaere, GD Brayer and MN James, *Can J Biochem*, **57**, 135–44 (1979).
106. VL Nienaber, K Breddam and JJ Birktoft, *Biochemistry*, **32**, 11469–75 (1993).
107. K Kitadokoro, H Tsuzuki, H Okamoto and T Sato, *Eur J Biochem*, **224**, 735–42 (1994).
108. HM Greenblatt, CA Ryan and MN James, *J Mol Biol*, **205**, 201–28 (1989).
109. MN James, AR Sielecki, GD Brayer, LT Delbaere and CA Bauer, *J Mol Biol*, **144**, 43–88 (1980).
110. HG Brittain, FS Richardson and RB Martin, *J Am Chem Soc*, **98**, 8255–60 (1976).
111. P Bryan, P Alexander, S Strausberg, F Schwartz, L Wang, G Gilliland and DT Gallagher, *Biochemistry*, **31**, 4937–45 (1992).
112. MW Pantoliano, M Whitlow, JF Wood, ML Rollence, BC Finzel, GL Gilliland, TL Poulos and PN Bryan, *Biochemistry*, **27**, 8311–17 (1988).
113. P Gros, M Fujinaga, BW Dijkstra, KH Kalk and WG Hol, *Acta Crystallogr, Sect B*, **45**, 488–99 (1989).
114. JK Dattagupta, T Fujiwara, EV Grishin, K Lindner, PC Manor, NJ Pieniazek, R Saenger and D Suck, *J Mol Biol*, **97**, 267–71 (1975).
115. A Pahler, A Banerjee, JK Dattagupta, T Fujiwara, K Lindner, GP Pal, D Suck, G Weber and W Saenger, *EMBO J*, **3**, 1311–14 (1984).
116. C Betzel, GP Pal and W Saenger, *Eur J Biochem*, **178**, 155–71 (1988).
117. C Betzel, GP Pal and W Saenger, *Acta Crystallogr, Sect B*, **44**, 163–72 (1988).
118. J Bajorath, W Hinrichs and W Saenger, *Eur J Biochem*, **176**, 441–47 (1988).
119. A Wlodawer, M Li, A Gustchina, Z Dauter, K Uchida, H Oyama, NE Goldfarb, BM Dunn and K Oda, *Biochemistry*, **40**, 15602–11 (2001).
120. H Oyama, S Abe, S Ushiyama, S Takahashi and K Oda, *J Biol Chem*, **274**, 27815–22 (1999).
121. JN Higaki, BL Haymore, S Chen, RJ Fletterick and CS Craik, *Biochemistry*, **29**, 8582–86 (1990).
122. WS Willett, SA Gillmor, JJ Perona, RJ Fletterick and CS Craik, *Biochemistry*, **34**, 2172–80 (1995).
123. T Gallagher, P Bryan and GL Gilliland, *Proteins*, **16**, 205–13 (1993).
124. SL Strausberg, PA Alexander, DT Gallagher, GL Gilliland, BL Barnett and PN Bryan, *Biotechnology (N Y)*, **13**, 669–73 (1995).
125. C Paech, DW Goddette, T Christianson and CR Wilson, *Adv Exp Med Biol*, **379**, 257–68 (1996).
126. B Maciver, RH McHale, DJ Saul and PL Bergquist, *Appl Environ Microbiol*, **60**, 3981–88 (1994).
127. JC van der Laan, G Gerritse, LJM Mulleners, RA van der Hoek and WJ Quax, *Appl Environ Microbiol*, **57**, 901–9 (1991).
128. H Oyama, T Hamada, S Ogasawara, K Uchida, S Murao, BB Beyer, BM Dunn and K Oda *J Biochem*, **131**, 757–65 (2002).

The superfamily of Cadherins: calcium-dependent cell adhesion receptors

Thomas Ahrens, Jörg Stetefeld, Daniel Häussinger and Jürgen Engel
Department of Biophysical Chemistry and Structural Biology, Biozentrum, University of Basel, Basel, Switzerland

FUNCTIONAL CLASS

The superfamily of cadherins (calcium-dependent adherent proteins)[1] is classified into six subfamilies that differ in domain composition, genomic structure, and phylogenetic analysis.[2,3] These subfamilies of cadherins are the 'classical' or type I cadherins (E-, P-, N-, R- and C-cadherin), 'atypical' or type II cadherins (e.g. VE–cadherin), desmocollins, desmogleins, protocadherins, and Flamingo cadherins.[2] Cadherins are transmembrane receptors, which mediate homophilic cell–cell adhesion.[4] The adhesive function of cadherins is critically dependent on calcium binding of the extracellular domain and connection of the cytoplasmatic domain to the actin cytoskeleton or the interfilament system. In addition to the adhesive function, cadherins are important signaling molecules, which transmit extracellular signals through the cytoplasmatic domain and associated proteins to the nucleus.

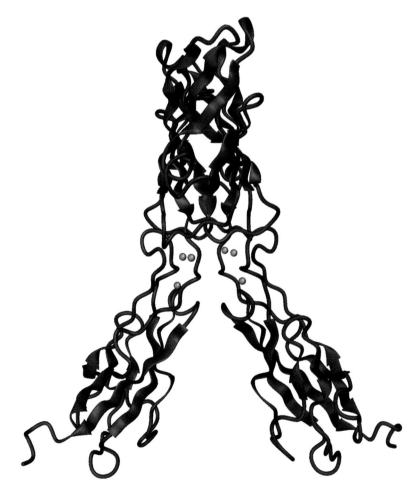

3D Structure   Schematic representation of the structure of murine E-cadherin CAD1–2, PDB code: 1 FF5. Two E-cadherin molecules (colored red and blue) per asymmetric unit are arranged in an intertwisted X-shaped form. Three calcium ions (yellow spheres) per molecule are bound in the interdomain sections. The figure was prepared using DINO.[78]

OCCURRENCE

Cadherins are expressed in vertebrates as well as in invertebrates like tunicates[5] and nematodes.[3] Owing to sequence similarities in the cytoplasmic domains of *human*, *Drosophila melanogaster*, and *Caenorhabditis elegans* cadherins, it can be assumed that proteins common to human and to one or both of the two organisms were probably already present in the earliest metazoan.[3] The cytoplasmic domain, however, is associated with different extracellular domains. The five-domain type I cadherins may therefore represent a special development of chordates. To date, around 100 different cadherins have been identified in the human and murine genome.[2,6,7]

BIOLOGICAL FUNCTION

Cadherins are tissue specifically expressed in multicellular organisms. This might reflect the need of multicellular organisms for many types of highly coordinated intercellular connections.[8] Cadherins primarily mediate homophilic cell–cell adhesion during tissue morphogenesis and maintenance, but heterophilic interactions, for example, with integrins are also observed.[9,10] E-cadherin, for example, which is a type I cadherin expressed in adherens junctions of epithelial cells, exerts cell adhesion in a two-step process. In the first step, E-cadherin molecules on the surface of one cell dimerize and, in the second, these E-cadherin dimers interact with dimers located on an adjacent cell.[11] Cadherins are crucially involved in early mouse development during compaction of morulae and blastocysts,[12–14] in the formation and maintenance of the epithelial junctional complex,[15] in epithelial cell polarity,[16] and in cellular shape changes during embryogenesis in *C. elegans*.[17,18] In addition, cadherins play a pivotal role in neural morphogenesis,[19,20] for example, for axon guidance,[21] formation of synaptic junctions,[22] or in motor neuron pool sorting.[23] The recent discovery of a gene cluster coding for ∼50 so-called protocadherins gives rise to an enormous number of possible combinations of differently expressed cadherins in the central nervous system (CNS).[7,24] Assuming homophilic interactions, these cadherins could be responsible for formation of neuronal circuits or the modulation of synaptic structures.[25]

Cadherins are anchored to the cytoskeleton through intracellular adapter and signaling proteins.[26] These are important for cell adhesion[27] and epithelial cell polarity,[16] and are involved in signaling processes.[28–30]

AMINO ACID COMPOSITION

Type I and type II cadherins are transmembrane proteins composed of an extracellular domain, a transmembrane region, and a cytosolic domain. The extracellular domain consists of modular arranged individual folding units that are called cadherin domains (CAD).[31,32] Each CAD (Figure 1) consists of around 110 amino acids and contains conserved peptide sequences (LDRE, DXXD, and DXD), which are involved in binding of calcium ions (Figure 1, black boxes).

For detailed phylogenetic analysis, classification, and amino acid sequence comparisons of the cadherin superfamily, see references 2 and 3. Accession numbers for SWISS-PROT and PDB files of cadherins used in this review are given below:

- E-cadherin (*Mus musculus*): 884 amino acids; Swiss-Prot: P09803[33]
  PDB codes: 1SUH[34]
  1EDH[35]
  1FF5[36]
- N-cadherin (*Mus musculus*): 906 amino acids; Swiss-Prot: P15116[37]
  PDB codes: 1NCG, 1NCH, 1NCI[38]
  1NCJ[39]
- C-cadherin (*Xenopus laevis*): 880 amino acids; Swiss-Prot: P33148[40]
  PDB code: 1L3W[41]
- VE-cadherin (*Mus musculus*): 783 amino acids; Swiss-Prot: P55284[42]

PROTEIN PRODUCTION, PURIFICATION, AND CHARACTERIZATION

Studies on the full-length extracellular domain of E-cadherin were initially performed on tissue extracted material.[43] Most recent work employed recombinantly in eukaryotic cell lines expressed protein. One suitable cell line is the human embryonal kidney cell line 293 (EBNA), which produces properly glycosylated protein. After cDNA transfection, the proteins are secreted into the cell-culture supernatant and are purified by DEAE-cellulose and Superose 12 column chromatography.[11] Oligomerized cadherin ectodomains were purified with the same procedure or were expressed as C-terminal 6xHis- or Strep-tag fusion proteins and purified by affinity chromatography.[11,44]

For the purpose of solving the crystal structure of the full extracellular domain of C-cadherin, the protein was expressed as a C-terminal 6xHis-tag fusion protein in Chinese hamster ovary (CHO) cells and purified from cell culture supernatants by nickel-affinity chromatography.[45]

Pairs of domains 1 and 2 of E-cadherin, which contain no glycosylation sites were expressed in *E.coli* strain BL21 (DE3) as C-terminal 6xHis-tag fusion proteins. Purification by nickel-affinity chromatography and cleavage of the 6xHis-tag was performed as described.[46] Uniformly ^{15}N or $^{15}N/^{13}C$-labeled E-cadherin for NMR studies was prepared

The superfamily of Cadherins: calcium-dependent cell adhesion receptors

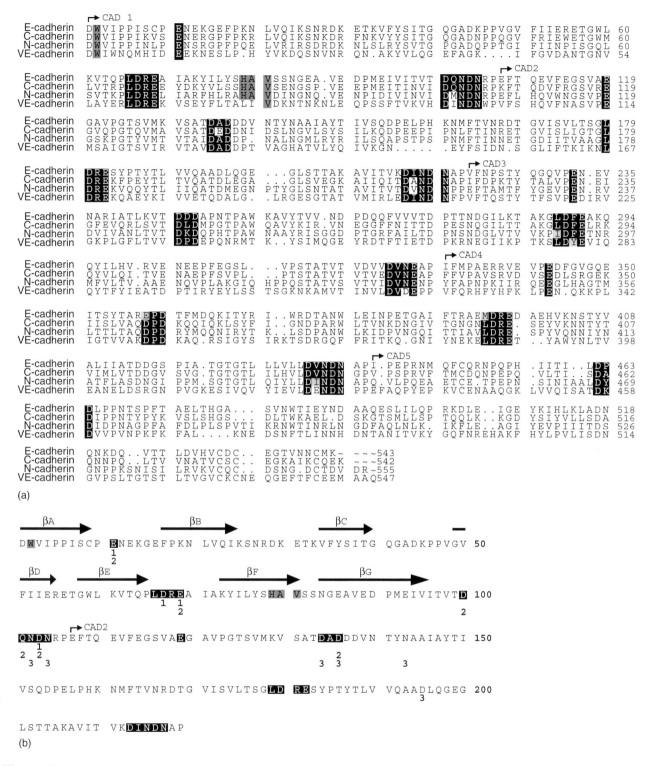

Figure 1 (a) Amino acid sequence alignment of the extracellular domains of murine E-cadherin, *Xenopus laevis* C-cadherin, murine N-cadherin, and murine VE-cadherin. Arrows indicate the individual cadherin domain boundaries. Conserved calcium binding motifs are highlighted in each domain by a black background. Residues labeled orange are proposed to be involved in adhesive contacts of cadherins. (b) Calcium coordination in E-cadherin CAD1–2. Arrows indicate regions of β-strand formation. Numbers 1, 2, and 3 indicate residues that coordinate calcium ions Ca1, Ca2, and Ca3 respectively. Ca1–3 are defined in Figure 4.

HANDBOOK OF METALLOPROTEINS 733

in deuterated form (>85% ^{2}H) by using modified M9 minimal media prepared in ^{2}H$_2$O with ^{15}N-NH$_4$Cl as the sole nitrogen and ^{13}C-glucose as the sole carbon source.[47]

CADHERINS AND DISEASE

Cadherins have been implicated in various diseases. E-cadherin is the entry point for infection by the human pathogen *Listeria monocytogenes*,[48] the cause of listeriosis. Another gram-positive bacterium, *Shigella flexneri*, uses cadherins for the spread between epithelial cells.[49]

Loss or mutations of cadherins can be correlated with invasive behavior of tumor cells.[50] In transgenic mice, loss of E-cadherin is associated with a transition from adenoma to invasive carcinoma.[51] In patients suffering from lobular breast cancer, mutations in the E-cadherin gene have been reported.[52] In familial gastric cancer, germline mutations of E-cadherin have been observed.[53] Among these mutations amino acid changes in calcium binding motifs of the extracellular domain were detected.[54]

FUNCTIONAL ASPECTS OF CALCIUM BINDING TO CADHERINS

Calcium triggers cadherin-based cell adhesion

Initial studies on the function of cadherins showed their decisive role in compaction of cells in early development.[12,13] Calcium promotes the compaction of embryonal cells mediated by E-cadherin (formerly also known as uvomorulin) and trypsin digestion of the protein is prevented in the presence of calcium.[55] With the onset of cDNA cloning and recombinant expression of E-cadherin in cells,[33,56] the molecular mechanisms of calcium-dependent cell adhesion was investigated in more detail. In cell-aggregation assays with transfected cell lines, the specificity and calcium dependence of cadherin interactions was first demonstrated.[57] Owing to predictions of possible calcium binding motifs in the extracellular domain of E-cadherin,[56] site-directed mutagenesis was applied to analyze changes in calcium binding motifs.[58] Mutation of Asp134 in the DXD motif of CAD2 of murine E-cadherin (Figure 1(b)) into alanine or lysine changed trypsin sensitivity and completely abolished the adhesive function. The tumor-associated mutation of a DXXD motif in CAD2 of E-cadherin also results in markedly reduced homophilic cell adhesion.[59]

Affinity of calcium binding and induction of conformational changes

Consequences of calcium binding to the purified extracellular fragment of E-cadherin were further analyzed by electron microscopy, circular dichroism, and spectroscopic titrations.[60] For the extracellular domain, electron microscopy revealed a reversible conformational change from a rodlike shape toward a collapsed globular structure upon depletion of calcium (Figure 2(a)). At 2 mM calcium concentration, 22-nm long, curved rods were observed (Figure 2(b)).

When the secondary structure was analyzed by circular dichroism spectroscopy, the spectra indicated a large fraction of β-structure in the presence of calcium. A minimum of the mean molar ellipticity was observed at 215 nm. This minimum changed to 208 nm and the ellipticity values decreased at 5 mM ethylenediaminetetraacetic acid (EDTA) indicating changes in the secondary structure toward a

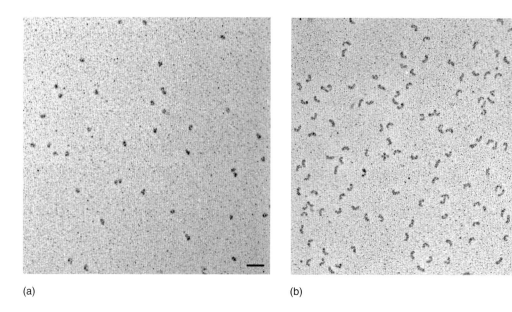

Figure 2 Comparison of the full length purified extracellular domain of native E-cadherin (a) in the absence and (b) in the presence of calcium by electron microscopy after rotary shadowing. Bars represent 50 nm.

random coil conformation. In a recent publication using NMR spectroscopy, these changes were localized to the interdomain section of E-cadherin domains, not interfering with the β-barrel structure of individual cadherin domains.[47] The K_d values of calcium binding to the whole extracellular domain obtained by titrations of the circular dichroic and fluorescence signal were 45 μM and 150 μM respectively.[60] Equilibrium dialysis performed to analyze calcium binding to the entire extracellular domain of E-cadherin resulted in a K_d of 30 μM and the total number of calcium ions bound was found to be 9 per molecule.[46] Determination of calcium binding affinities was also extended to domain pairs of cadherin.[46,61] Calcium binding to a CAD1–2 comprising fragment of E-cadherin by continuous flow dialysis revealed high cooperativity (Hill coefficient of 2.4), which indicates three binding sites.[61] An apparent K_d for calcium binding of 20 and 23 μM was determined by flow dialysis and tryptophane fluorescence spectroscopy.[61] This, however, is in contrast to an average dissociation constant of 460 μM for calcium binding calculated from equilibrium dialysis of a CAD1–2 fragment of E-cadherin.[46] Here, the fit of the data yielded two calcium binding sites with a K_d of 330 μM and a third binding site with a much higher K_d of 2000 μM.

It follows from the above survey of K_d values that most calcium binding sites in cadherins are saturated in the extracellular space where the free calcium concentration is about 1 mM. This value was derived from measurements of calcium levels in blood where the concentration usually is tightly regulated to 1.2 mM by calcium sensing receptors.[62] Local or temporary changes of calcium concentrations in the extracellular space are, however, possible.[63,64] Variable saturation of the low-affinity site in the interdomain section of CAD1 and 2 may have consequences for cadherin function[65] (see also the section on calcium and adhesive interactions).

CRYSTALLOGRAPHIC STUDIES ON CALCIUM BINDING TO CADHERINS

Several high-resolution structures of cadherins have been solved, which clarified the molecular details of calcium coordination.[34–36,38,39,41] In general, the fold of a cadherin domain consists of seven β-strands (A, B, C, D, E, F and G) where the N- and C-termini are located at opposite ends of one domain (Figure 3). The β-barrel topology is similar to that of variable and constant domains of the immunoglobulin (Ig) family.[66]

The prediction of four calcium binding pockets in the interdomain sections of the extracellular domain of type I cadherins was proven recently in a report on the crystal structure of the full extracellular domain of C-cadherin (Figure 3).[41] In total, 12 calcium ions were bound to the protein, 3 per binding pocket. The curved shape of the C-cadherin ectodomain in the crystal resembles the shape

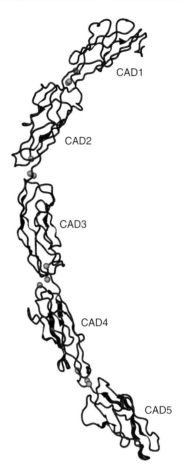

Figure 3 Schematic representation of the structure of the full ectodomain of C-cadherin, PDB code: 1L3W. Calcium ions are labeled as yellow spheres.

of E-cadherin ectodomains in electron microscopic pictures (Figure 2(b)).

In the solution structure of the N-terminal CAD1 of E-cadherin[34] and in the crystal structure of CAD1 of N-cadherin,[38] no calcium binding was observed, but from Yb^{3+} and UO_2^{2+} binding, the presence of partial coordination sites was deduced.[38] The molecular details of calcium binding to E-cadherin were first described in the crystal structure of E-cadherin 1–2.[35] In the crystal, two molecules per asymmetric unit dimerize in a calcium-dependent manner and are arranged in an intertwisted X-shaped form. In Figure 4, the crystallographic dimer of ECAD1–2 is shown.[35,36] Note that in contrast to Figure 3, dimers are shown in Figure 4. In the case of ECAD1–2, two molecules with contacts in the calcium binding region were found in the unit cell. It is still unknown whether this contact is physiological or induced by crystal forces. The calcium binding pocket of each molecule in the dimer is formed by the interplay of binding motifs in two adjacent cadherin domains. The calcium ions bound to the linker region of CAD1 and 2 of murine E-cadherin are labeled Ca1–3. Side chains of amino acids participating in calcium binding are indicated in a ball and stick representation (Figure 4(a)).

The superfamily of Cadherins: calcium-dependent cell adhesion receptors

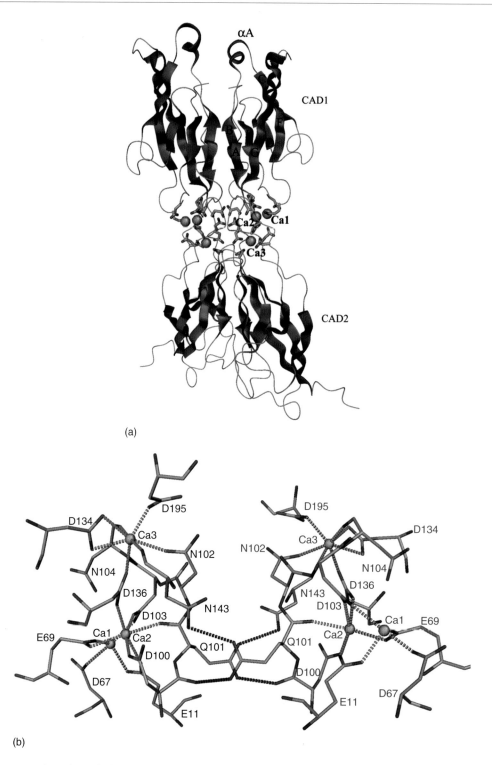

(a)

(b)

Figure 4 (a) Amino acid residues of adjacent cadherin domains participate in calcium coordination. Side chains of calcium-coordinating residues of E-cadherin are shown in a ball and stick representation. PDB code: 1FF5; (b) Details of the calcium binding region of E-cadherin CAD1–2, PDB code: 1EDH. Calcium ions (Ca1, Ca2, and Ca3) are shown as yellow spheres. Direct hydrogen bonds formed between amino acids in the dimer interface are shown as red dashed lines, calcium–ligand bonds are shown as light blue dashed lines.

A detailed view of the Ca^{2+} binding sites is shown in Figure 4(b).[35] Ca3 is coordinated by seven oxygen atoms resulting in a pentagonal bipyramidal coordination geometry. Ca2 and Ca1 are each coordinated by six oxygen atoms. Two oxygen atoms from water molecules, which coordinate Ca1, are not shown. The calcium–ligand distances for each molecule in the crystallographic dimer are listed in Table 1 and reflect typical distances seen for

Table 1 Calcium–ligand distances in the crystal structure of E-cadherin 1–2[a]

Atom 1	Atom 2	Distance (Å) A[b]	B
Ca1	OE1 69[c]	2.28	2.23
Ca1	OD1 67	2.35	2.13
Ca1	OD2 103	2.23	2.45
Ca1	OE1 11	2.32	2.31
Ca2	OD1 136	2.25	2.33
Ca2	OD1 103	2.16	2.41
Ca2	OE2 11	2.41	2.35
Ca2	OD1 100	2.32	2.30
Ca2	OE2 69	2.48	2.53
Ca2	O 101	2.43	2.47
Ca3	OD2 134	2.49	2.44
Ca3	OD1 134	2.56	2.54
Ca3	OD2 195	2.31	2.41
Ca3	OD1 102	2.44	2.44
Ca3	O 143	2.40	2.36
Ca3	OD2 136	2.42	2.44
Ca3	O 104	2.35	2.24

[a] The data are from Nagar et al. (35)
[b] A and B refer to the left and right molecule of the E-cadherin 1–2 dimer in Figure 4(b).
[c] Zucchini AO numbering.

protein–calcium complexes. The interactions involve OE1 11, OD1 67, OE1 69 and OD2 103 for Ca1, OE2 11, OE2 69, OD1 100, O 101, OD1 103 and OD1 136 for Ca2 as well as OD1 102, O 104, OD1 134, OD2 134, OD2 136, O 143 and OD2 195 for Ca3.

Rigid contacts between adjacent domains are stabilized by interactions of OD1 102 and OD2 103/OD1 103, which bind Ca3 and Ca1/Ca2, respectively. OD1 and OD2 134 are coordinating Ca3, thus explaining the dramatic effect on cell adhesion after mutation of this residue to alanine.[58] The amino acid residues involved in coordination of calcium ions Ca 1–3 of E-cadherin are also highlighted in Figure 1(b). Altogether, the calcium-mediated interactions account for a high proportion of noncovalent interactions between adjacent cadherin domains leading to the rodlike structure of the extracellular domain (Figures 2 and 3).

CALCIUM AND LATERAL (cis) DIMERIZATION OF CADHERINS

Lateral dimerization (cis dimerization) of cadherin molecules on the cell surface precedes the adhesive contact.[45] Lateral dimers interact with dimers on adjacent cells to form the adhesive cell–cell contact (trans association). Although there is no doubt about the importance of calcium for the adhesive contact, the role of calcium for lateral dimerization is controversial. There are reports on the calcium-independent formation of lateral dimers.[45,67,68] The intrinsic problem of this hypothesis is the lack of a calcium-free state at physiological conditions. In the absence of calcium, lateral dimerization can hardly be detected by electron microscopy because cadherins appear in a globular collapsed state.[60]

Other workers report on a strong calcium dependence of cis dimerization. Evidence for a calcium-dependent dimerization of E-cadherin CAD1–2 has been derived from analytical ultracentrifugation experiments.[46,61] Calcium-dependent lateral dimers of E-cadherin have been detected by cross-linking of E-cadherin at the cell surface.[69] In the crystal structure of E-cadherin, CAD1–2 direct interactions of amino acids in the dimer interface were observed (Figure 4(b), red dashed lines).[35] As mentioned, there is no evidence that the contact near the calcium binding regions seen in Figure 4 is a physiological lateral contact. Different contact regions were proposed on the basis of NMR experiments (see below) and crystallographic evidences.[41]

Occupation of Ca^{2+} binding sites in interdomain sections of cadherins is clearly a prerequisite for lateral dimerization probably because of conformational changes in the protein. Cadherin interactions are modulated by a combination of different binding affinities for calcium and local changes in cadherin–protein concentration at the cell surface. It has been postulated that the binding sites for Ca2 and Ca3 represent the high affinity (K_d of 330 µM), and the binding site for Ca1 represent the low-affinity (K_d of 2 mM) sites of the calcium binding pocket of E-cadherin.[46] Ca1 is the only one exposed to the surface of the molecule and coordination of Ca1 is not involved in stabilizing interactions in the lateral dimer interface.

HOMOASSOCIATION OF E-CADHERIN ANALYZED BY NMR SPECTROSCOPY

NMR spectroscopy at different calcium and protein concentrations was used to investigate the calcium-dependent homoassociation of E-cadherin.[47] ^1H-^{15}N hetronuclear single quantum coherence (HSQC) spectra recorded in the absence of calcium at 40 µM and 0.6 mM E-cadherin CAD1–2 protein concentration were almost indistinguishable. Without calcium, E-cadherin remains monomeric for concentrations up to several millimolar, which supports the calcium dependence of lateral cis dimerization. After adding calcium to the protein chemical shift, the line width changes as well as the changes in ^{15}N relaxation times indicate protein dimerization above 100 µM, which is in agreement with reported dimerization constants of 0.08 to 0.17 mM. At a protein concentration of 40 µM and 1 mM free calcium concentration, the protein is still in a monomeric state and chemical shift changes mainly involve the calcium binding pocket in the interdomain region of CAD1 and CAD2 (Figure 5(a)). Chemical shift changes are mainly observed for amino acids, which were shown to bind calcium in crystal structures. The relative orientation of the two domains resembles an elongated rather than

The superfamily of Cadherins: calcium-dependent cell adhesion receptors

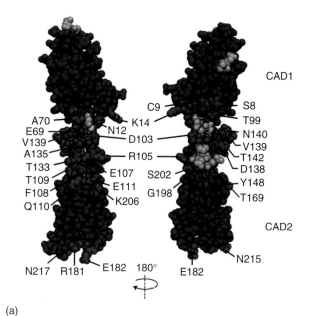

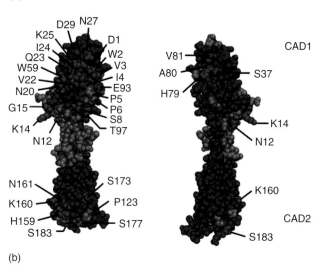

Figure 5 Calcium binding and dimerization of cadherins analyzed by NMR spectroscopy. (a) Chemical shift differences induced by the binding of calcium to E-cadherin CAD1–2 (40 μM). Atoms colored in red correspond to residues for which averaged chemical shift differences between calcium-free and calcium-bound state or for which intensity changes in the ^{1}H-^{15}N HSQC spectra were observed. Nonaffected residues are labeled in blue, calcium ions are labeled in yellow. (b) Chemical shift differences observed for amino acids in E-cadherin CAD1–2 when changing the protein concentration from 40 μM (monomeric state) to 0.6 mM (dimeric state) at 1 mM calcium concentration. Chemical shift differences were color-coded onto the crystal structure PDB code 1FF5. [Reprinted from *J Mol Biol*, **324**, D Haussinger, T Ahrens, HJ Sass, O Pertz, J Engel and S Grzesiek, Calcium-dependent homoassociation of E-cadherin by NMR spectroscopy: changes in mobility, conformation and mapping of contact regions, 823–39 (2002) with permission of Elsevier[47]].

a collapsed structure and the transition to an elongated structure after calcium binding is independent of homoassociation.

E-cadherin CAD1–2 associates at 0.6 mM protein concentration in the presence of calcium (Figure 5(b)). The relative orientation of the two domains in each CAD1–2 molecule remains unchanged, but the NMR data differ from crystal structures regarding the alignment of the molecules in the dimeric state. On the basis of alignment and diffusion tensor data, a much more parallel alignment of the molecules in the dimer is observed. The residues involved in this lateral (*cis*) interaction cluster around Trp2 on only one side of CAD1. In addition, a second smaller, nonsymmetric contact around residue Ala80 of CAD1 and a region at the C-terminal end of CAD2 centered around Lys160 was detected. This second contact was also described to constitute the lateral dimerization interface of C-cadherin.[41]

CALCIUM AND THE ADHESIVE (*trans*) INTERACTION OF CADHERINS

Although a direct contribution of calcium ions and calcium-coordinating amino acids in lateral (*cis*) contacts of cadherins has been observed,[35] the main consequence of calcium binding is an overall rigidification resulting in a curved rodlike shape of the extracellular domain. Calcium binding restrains the positions of adhesive sites of the molecule to those that are important for the formation of adhesive cell–cell contacts. These adhesive sites seem to be located distant from the calcium binding pocket. The N-terminal located Trp2 of classical cadherins has been implicated to be important for the adhesive interaction.[36,39,41] A region around a conserved sequence in the βF-strand in CAD1 of type I cadherins (His-Ala-Val) has been proposed to determine the specificity and adhesive surface.[70,71] To mimic the high surface density in adherens junctions on the cell surface, ectodomains of cadherins have been oligomerized by fusing them to oligomerization domains of cartilage oligomeric matrix protein (COMP),[11] the Fc-part of IgGs[72–74] and the coiled-coil domain of matrillin 1 (CMP).[75] Through oligomerization, the intrinsic concentration of ectodomains can be increased to ~2 mM. Analysis of oligomerized extracellular domains of cadherins by electron microscopy showed a calcium-dependent rigidification and subsequent association of ectodomains to ring-like structures, which represent lateral dimers (Figure 6).[11,36,75]

Two rings frequently associate to double ring structures or 'associated rings'. These may represent adhesive (*trans*) contacts of cadherins and have been observed for most members of the type I subfamily of cadherins and VE-cadherin.[11,75] From the crystal structures and the electron microscopy (EM) studies, it can be concluded that the N-terminal domains CAD1 and 2 are mainly mediating lateral and adhesive interactions of cadherins. Notably, the regions determining homophilic specificity of cadherins were also localized to the N-terminal domain.[76]

A model for the role of calcium in cadherin-mediated adhesion is depicted in Figure 7.

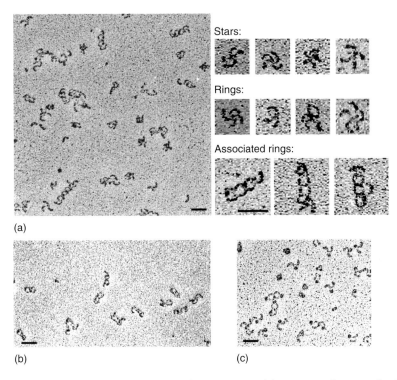

Figure 6 Electron microscopy of oligomerized cadherin ectodomains: A model system analyzing cadherin interactions. (a) Electron micrographs of E-cadherin CAD1–5-COMPcc; (b) VE-cadherin CAD1–5-CMPcc; (c) N-cadherin-Fc fusion proteins in the presence of calcium are shown. In the right panel of (a), galleries of magnified molecules representing stars, ring-like and associated rings of E-cadherin-COMPcc are depicted. Bars represent 50 nm. [Reprinted from *J Mol Biol*, **325**, T Ahrens, M Lambert, O Pertz, T Sasaki, T Schulthess, RM Mege, R Timpl and J Engel, Homoassociation of VE-cadherin follows a mechanism common to 'classical' cadherins, 733–42, (2003) with permission from Elsevier.[75] Reproduced from O Pertz, D Bozic, AW Koch, C Fauser, A Brancaccio and J Engel, A new crystal structure, Ca^{2+} dependence and mutational analysis reveal molecular details of E-cadherin homoassociation, *Embo J*, (1999) **18**, 1738–47 by permission of Oxford University Press[36]].

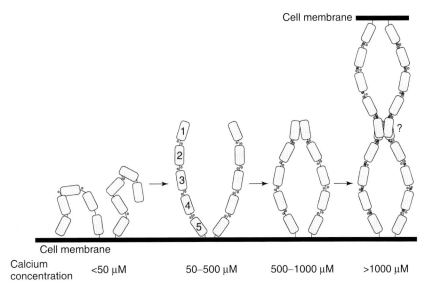

Figure 7 Model of the calcium-dependent adhesive function of cadherins. At low calcium concentrations (<50 μM), the ectodomain appears as a collapsed structure. At calcium concentrations ranging from 50 μM to 1000 μM, the rodlike structure of the ectodomain is stabilized and lateral (*cis*) dimerization of cadherins clustered at the cell surface is observed. At high calcium concentrations (>1000 μM), adhesive (*trans*) contacts of dimers are established. Cadherin domains 1 to 5 are shown as gray blocks and calcium ions as green (high-affinity binding sites) and red spheres (low-affinity binding sites). [Reproduced from O Pertz, D Bozic, AW Koch, C Fauser, A Brancaccio and J Engel, A new crystal structure, Ca^{2+} dependence and mutational analysis reveal molecular details of E-cadherin homoassociation, *Embo J*, (1999) **18**, 1738–47 by permission of Oxford University Press[36]].

Upon calcium binding, cadherin ectodomains are rigidified, resulting in a curved, rodlike shape of the molecules. Following calcium binding and local increase of cadherin molecules, for example in adherens junctions, lateral (*cis*) dimerization is induced. When the low-affinity calcium binding sites are occupied, *cis* dimers interact to form the adhesive *trans* contact. Trp2 of the N-terminal domain of type I cadherins is critically involved in this process; however, the exact mechanism is still under investigation. Regarding the role of calcium in cadherin mediated cell–cell adhesion, the key to stable adhesion would lie in the collective assembly of properly oriented structures at high surface density.[77]

ACKNOWLEDGEMENT

This work was supported by the Swiss National Science Foundation (Grant 31-49281.96 to J.E.).

REFERENCES

1. C Yoshida-Noro, N Suzuki and M Takeichi, *Dev Biol*, **101**, 19–27 (1984).
2. F Nollet, P Kools and F van Roy, *J Mol Biol*, **299**, 551–72 (2000).
3. E Hill, ID Broadbent, C Chothia and J Pettitt, *J Mol Biol*, **305**, 1011–24 (2001).
4. BM Gumbiner, *Cell*, **84**, 345–57 (1996).
5. L Levi, J Douek, M Osman, TC Bosch and B Rinkevich, *Gene*, **200**, 117–23 (1997).
6. BD Angst, C Marcozzi and AI Magee, *J Cell Sci*, **114**, 629–41 (2001).
7. T Yagi and M Takeichi, *Genes Dev*, **14**, 1169–80 (2000).
8. RO Hynes and Q Zhao, *J Cell Biol*, **150**, F89–F96 (2000).
9. KL Cepek, SK Shaw, CM Parker, GJ Russell, JS Morrow, DL Rimm and MB Brenner, *Nature*, **372**, 190–93 (1994).
10. JD Whittard, SE Craig, AP Mould, A Koch, O Pertz, J Engel and MJ Humphries, *Matrix Biol*, **21**, 525–32 (2002).
11. A Tomschy, C Fauser, R Landwehr and J Engel, *EMBO J*, **15**, 3507–14 (1996).
12. R Kemler, C Babinet, H Eisen and F Jacob, *Proc Natl Acad Sci USA*, **74**, 4449–52 (1977).
13. F Hyafil, D Morello, C Babinet and F Jacob, *Cell*, **21**, 927–34 (1980).
14. D Vestweber and R Kemler, *Exp Cell Res*, **152**, 169–78 (1984).
15. B Gumbiner, B Stevenson and A Grimaldi, *J Cell Biol*, **107**, 1575–87 (1988).
16. H McNeill, M Ozawa, R Kemler and WJ Nelson, *Cell*, **62**, 309–16 (1990).
17. J Pettitt, WB Wood and RH Plasterk, *Development*, **122**, 4149–57 (1996).
18. M Costa, W Raich, C Agbunag, B Leung, J Hardin and JR Priess, *J Cell Biol*, **141**, 297–308 (1998).
19. K Hatta, S Takagi, H Fujisawa and M Takeichi, *Dev Biol*, **120**, 215–27 (1987).
20. M Takeichi, *Curr Opin Cell Biol*, **7**, 619–27 (1995).
21. M Matsunaga, K Hatta, A Nagafuchi and M Takeichi, *Nature*, **334**, 62–64 (1988).
22. AM Fannon and DR Colman, *Neuron*, **17**, 423–34 (1996).
23. SR Price, NV De Marco Garcia, B Ranscht and TM Jessell, *Cell*, **109**, 205–16 (2002).
24. Q Wu and T Maniatis, *Cell*, **97**, 779–90 (1999).
25. M Frank and R Kemler, *Curr Opin Cell Biol*, **14**, 557–62 (2002).
26. M Ozawa, H Baribault and R Kemler, *EMBO J*, **8**, 1711–17 (1989).
27. R Kemler, *Trends Genet*, **9**, 317–21 (1993).
28. J Behrens, JP von Kries, M Kuhl, L Bruhn, D Wedlich, R Grosschedl and W Birchmeier, *Nature*, **382**, 638–42 (1996).
29. S Kuroda, M Fukata, M Nakagawa, K Fujii, T Nakamura, T Ookubo, I Izawa, T Nagase, N Nomura, H Tani, I Shoji, Y Matsuura, S Yonehara and K Kaibuchi, *Science*, **281**, 832–35 (1998).
30. N Kohmura, K Senzaki, S Hamada, N Kai, R Yasuda, M Watanabe, H Ishii, M Yasuda, M Mishina and T Yagi, *Neuron*, **20**, 1137–51 (1998).
31. J Schultz, RR Copley, T Doerks, CP Ponting and P Bork, *Nucleic Acids Res*, **28**, 231–34 (2000).
32. P Bork, AK Downing, B Kieffer and ID Campbell, *Q Rev Biophys*, **29**, 119–67 (1996).
33. A Nagafuchi, Y Shirayoshi, K Okazaki, K Yasuda and M Takeichi, *Nature*, **329**, 341–43 (1987).
34. M Overduin, TS Harvey, S Bagby, KI Tong, P Yau, M Takeichi and M Ikura, *Science*, **267**, 386–89 (1995).
35. B Nagar, M Overduin, M Ikura and JM Rini, *Nature*, **380**, 360–64 (1996).
36. O Pertz, D Bozic, AW Koch, C Fauser, A Brancaccio and J Engel, *EMBO J*, **18**, 1738–47 (1999).
37. S Miyatani, K Shimamura, M Hatta, A Nagafuchi, A Nose, M Matsunaga, K Hatta and M Takeichi, *Science*, **245**, 631–35 (1989).
38. L Shapiro, AM Fannon, PD Kwong, A Thompson, MS Lehmann, G Grubel, JF Legrand, J Als-Nielsen, DR Colman and WA Hendrickson, *Nature*, **374**, 327–37 (1995).
39. K Tamura, WS Shan, WA Hendrickson, DR Colman and L Shapiro, *Neuron*, **20**, 1153–63 (1998).
40. D Ginsberg, D DeSimone and B Geiger, *Development*, **111**, 315–25 (1991).
41. TJ Boggon, J Murray, S Chappuis-Flament, E Wong, BM Gumbiner and L Shapiro, *Science*, **296**, 1308–13 (2002).
42. G Breier, F Breviario, L Caveda, R Berthier, H Schnurch, U Gotsch, D Vestweber, W Risau and E Dejana, *Blood*, **87**, 630–41 (1996).
43. D Vestweber and R Kemler, *EMBO J*, **4**, 3393–98 (1985).
44. T Ahrens, O Pertz, D Haussinger, C Fauser, T Schulthess and J Engel, *J Biol Chem*, **277**, 19455–60 (2002).
45. WM Brieher, AS Yap and BM Gumbiner, *J Cell Biol*, **135**, 487–96 (1996).
46. AW Koch, S Pokutta, A Lustig and J Engel, *Biochemistry*, **36**, 7697–705 (1997).
47. D Haussinger, T Ahrens, HJ Sass, O Pertz, J Engel and S Grzesiek, *J Mol Biol*, **324**, 823–39 (2002).
48. J Mengaud, H Ohayon, P Gounon, RM Mege and P Cossart, *Cell*, **84**, 923–32 (1996).
49. PJ Sansonetti, J Mounier, MC Prevost and RM Mege, *Cell*, **76**, 829–39 (1994).
50. J Behrens, MM Mareel, FM Van Roy and W Birchmeier, *J Cell Biol*, **108**, 2435–47 (1989).
51. AK Perl, P Wilgenbus, U Dahl, H Semb and G Christofori, *Nature*, **392**, 190–93 (1998).

52 G Berx, AM Cleton-Jansen, F Nollet, WJ de Leeuw, M van de Vijver, C Cornelisse and F van Roy, *EMBO J*, **14**, 6107–15 (1995).

53 P Guilford, J Hopkins, J Harraway, M McLeod, N McLeod, P Harawira, H Taite, R Scoular, A Miller and AE Reeve, *Nature*, **392**, 402–5 (1998).

54 G Berx, KF Becker, H Hofler and F van Roy, *Hum Mutat*, **12**, 226–37 (1998).

55 F Hyafil, C Babinet and F Jacob, *Cell*, **26**, 447–54 (1981).

56 M Ringwald, R Schuh, D Vestweber, H Eistetter, F Lottspeich, J Engel, R Dolz, F Jahnig, J Epplen, S Mayer, C Muller and R Kemler, *EMBO J*, **6**, 3647–53 (1987).

57 A Nose, A Nagafuchi and M Takeichi, *Cell*, **54**, 993–1001 (1988).

58 M Ozawa, J Engel and R Kemler, *Cell*, **63**, 1033–38 (1990).

59 G Handschuh, B Luber, P Hutzler, H Hofler and KF Becker, *J Mol Biol*, **314**, 445–54 (2001).

60 S Pokutta, K Herrenknecht, R Kemler and J Engel, *Eur J Biochem*, **223**, 1019–26 (1994).

61 JR Alattia, JB Ames, T Porumb, KI Tong, YM Heng, P Ottensmeyer, CM Kay and M Ikura, *FEBS Lett*, **417**, 405–8 (1997).

62 EM Brown, G Gamba, D Riccardi, M Lombardi, R Butters, O Kifor, A Sun, MA Hediger, J Lytton and SC Hebert, *Nature*, **366**, 575–80 (1993).

63 P Maurer, E Hohenester and J Engel, *Curr Opin Cell Biol*, **8**, 609–17 (1996).

64 EM Brown, PM Vassilev and SC Hebert, *Cell*, **83**, 679–82 (1995).

65 S Chakrabarty, V Radjendirane, H Appelman and J Varani, *Cancer Res*, **63**, 67–71 (2003).

66 C Chothia and EY Jones, *Annu Rev Biochem*, **66**, 823–62 (1997).

67 NA Chitaev and SM Troyanovsky, *J Cell Biol*, **142**, 837–46 (1998).

68 J Klingelhofer, OY Laur, RB Troyanovsky and SM Troyanovsky, *Mol Cell Biol*, **22**, 7449–58 (2002).

69 H Takeda, Y Shimoyama, A Nagafuchi and S Hirohashi, *Nat Struct Biol*, **6**, 310–12 (1999).

70 OW Blaschuk, R Sullivan, S David and Y Pouliot, *Dev Biol*, **139**, 227–29 (1990).

71 E Williams, G Williams, BJ Gour, OW Blaschuk and P Doherty, *J Biol Chem*, **275**, 4007–12 (2000).

72 M Lambert, F Padilla and RM Mege, *J Cell Sci*, **113**, 2207–19 (2000).

73 W Baumgartner, P Hinterdorfer, W Ness, A Raab, D Vestweber, H Schindler and D Drenckhahn, *Proc Natl Acad Sci USA*, **97**, 4005–10 (2000).

74 E Corps, C Carter, P Karecla, T Ahrens, P Evans and P Kilshaw, *J Biol Chem*, **276**, 30862–70 (2001).

75 T Ahrens, M Lambert, O Pertz, T Sasaki, T Schulthess, RM Mege, R Timpl and J Engel, *J Mol Biol*, **325**, 733–42 (2003).

76 A Nose, K Tsuji and M Takeichi, *Cell*, **61**, 147–55 (1990).

77 JR Alattia, H Kurokawa and M Ikura, *Cell Mol Life Sci*, **55**, 359–67 (1999).

78 A Phillipsen, *DINO: Visualizing Structural Biology*, http://www.dino3d.org (2002).

Metal-dependent type II restriction endonucleases

Éva Scheuring Vanamee and Aneel K Aggarwal

Structural Biology Program, Mount Sinai School of Medicine, NY, USA

FUNCTIONAL CLASS

Enzyme; hydrolases; endodeoxyribonucleases producing 5′-phosphomonoesters; EC 3.1.21.4; type II site-specific deoxyribonucleases.

Restriction endonucleases are phosphodiesterases that bind double-stranded DNA with high specificity and cleave the DNA to yield 5′-phosphate and 3′-hydroxyl groups as products.

OCCURRENCE

Restriction endonucleases (REases) with their corresponding methyltransferases (Mtases) form the restriction–modification (R–M) systems of bacteria. R–M systems are ubiquitous in bacteria; only a few are known not to contain any R–M gene. The recently sequenced *Helicobacter pylori* strain 26695, for instance, reveals more than a dozen R–M systems, though less than 30% are functional.[1]

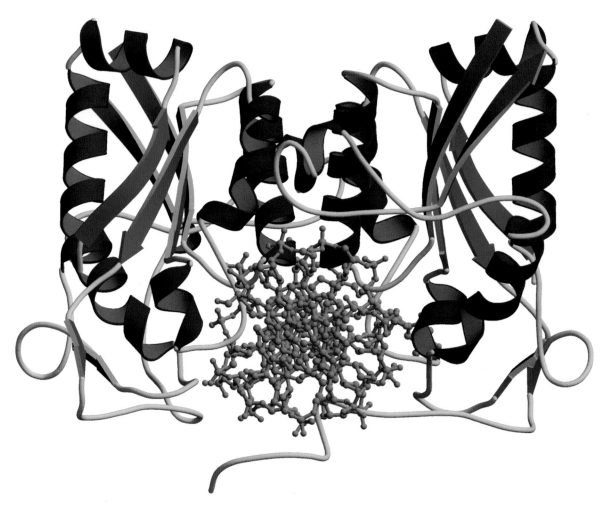

3D Structure Schematic representation of the type IIP restriction endonuclease *Bam*HI in complex with its cognate DNA (PDB code: 2BAM). Produced with the programs MOLSCRIPT[64] and RASTER3D.[65]

BIOLOGICAL FUNCTION

The endonuclease and methyltransferase enzymes have opposing functions: the REase cleaves DNA specifically, while the corresponding Mtase adds a methyl group to each strand within the recognition sequence by which it prevents the REase from binding and cleaving the DNA. Methylation of one strand alone is usually sufficient to prevent cleavage by the REase. It is believed that R–M systems serve as the host defense of bacteria against DNA infection. The invading DNA is cleaved by the REase, while the host DNA is protected from self-cleavage by the Mtase. According to another view, R–M systems act primarily as selfish gene elements.[2] This notion is strengthened by the fact that once established in a genome, R–M systems cannot be removed easily. The genes that code for R–M systems are under tight regulation to avoid self-destruction by the highly lethal endonuclease. The REase gene is only activated after all the recognition sites within the cell's own DNA are methylated by the corresponding Mtase.

R–M systems have been classified into three large families on the basis of cofactor requirements, position of the cleavage site with respect to the recognition site, and subunit composition[3] (a new family, type IV, has been proposed recently to include those systems that cleave only methylated DNA as their substrate).[4] The function of type I and type III R–M systems depend on the hydrolysis of ATP or GTP. The type I systems are the most complex with both restriction and modification functions carried on the same enzyme. Type I restriction endonucleases recognize an asymmetric sequence and cleave thousands of base pairs away from it. Type III systems carry the restriction and modification functions on two separate subunits. The type III restriction enzymes also recognize an asymmetric sequence but cleave 25 to 27 base pairs to one side of the recognition sequence. For a review of nucleoside triphosphate-dependent restriction enzymes, see Reference 5.

The type II REases cleave within or close to their recognition site, and instead of NTP require Mg^{2+} as a cofactor for the cleavage reaction. To date, more than 3500 type II REases have been discovered.[6] They share little or no sequence homology and show remarkable specificity. A single change within the recognition sequence results in a millionfold decrease in cleavage activity.[7] Owing to their high specificity, type II REases have become indispensable tools in molecular biology and serve as excellent model systems for studying site-specific protein–DNA interactions. Here we focus on the structure-function of metal-dependent type II REases, for which both structural and biochemical data are available.

AMINO ACID SEQUENCE INFORMATION

- R. *Bam*HI from *Bacillus amyloliquefaciens* H: 213 amino acid (AA) residues.[8]
- R. *Bgl*I from *Bacillus subtilis*: 299 AA.[9]
- R. *Bgl*II from *Bacillus subtilis*: 223 AA.[10]
- R. *Bse*634I from *Geobacillus stearothermophilus*: 293 AA[11] – isoschizomer of *Cfr*10I.
- R. *Bso*BI from *Geobacillus stearothermophilus*: 323 AA.[12]
- R. *Cfr*10I from *Citrobacter freundii*: 285 AA.[13]
- R. *Eco*RI from *Escherichia coli*: 277 AA.[14]
- R. *Eco*RV from *Escherichia coli*: 245 AA.[15]
- R. *Fok*I from *Flavobacterium okeanokoites*: 583 AA.[16]
- R. *Hinc*II from *Haemophilus influenzae* 258 AA.[17]
- R. *Mun*I from *Mycoplasma species*: 202 AA.[18]
- R. *Nae*I from *Lechevalieria aerocolonigenes*: 317 AA.[19]
- R. *Ngo*MIV from *Neisseria gonorrhea* M: 286 AA.[20]
- R. *Pvu*II from *Proteus vulgaris*: 157 AA.[21,22]

PROTEIN PRODUCTION, PURIFICATION, AND MOLECULAR CHARACTERIZATION

To avoid cleavage of the host genome by the REase, the endonuclease gene is expressed in the presence of the Mtase gene. An intein-based expression system has been developed for large-scale production and single-step purification of REases.[23] The REase gene is cloned into an *E. coli* expression vector to create a target protein–intein–chitin binding domain (CBD) fusion. The expression of the fusion construct is controlled by an IPTG-inducible T7 promoter. The three-part fusion protein is bound to chitin beads and the REase is released from the intein-CBD tag after induced on-column cleavage by 1,4-dithiothreitol (DTT).

ENZYME ACTIVITY ASSAY

REase activity is assayed on λ-DNA. One unit of activity is defined as the amount of enzyme required to completely digest 1 µg of substrate DNA in a total reaction volume of 50 µL in 1 h using the appropriate buffer. The reaction is stopped by heat inactivation of the enzyme. The reaction mixture is then run on a DNA gel to evaluate the amount of product DNA generated.

METAL CONTENT AND COFACTOR REQUIREMENT

Type II REases require only divalent metals as cofactor for catalysis with the exception of *Bfi*I, a metal-independent REase that does not require any cofactor at all.[24] The natural cofactor is Mg^{2+} that can be replaced by other divalent cations such as Mn^{2+}, Co^{2+}, Zn^{2+}, and Cd^{2+} with the exception of Ca^{2+}. Ca^{2+} is an effective inhibitor of restriction enzymes as well as nucleases such as *E. coli* DNA polymerase I and the *Tetrahymena* ribozyme.[25–27] The larger ionic radius of

Ca^{2+}, its electronegativity, and coordination geometry are likely contributing factors for its inhibitory effects. Ca^{2+} readily binds in place of Mg^{2+} at near-identical sites, and so the inhibitory effects of Ca^{2+} are rather subtle and are concentrated over the transition state rather than in the initial binding to the ground state.[28,29] Recent calculations for *Bam*HI show that the bond-breaking induced by the nucleophile has a substantially higher energy barrier (~20 kcal mol^{-1}) when magnesium is replaced by calcium, and the difference is attributed to the fact that Ca^{2+} tends to over-coordinate, which destabilizes the transition state.[29] This is likely to be the case for other restriction enzymes. It is interesting to note that while Ca^{2+} plays an inhibitory role in restriction endonuclease–catalyzed phosphodiester cleavage, other phosphodiesterases such as deoxyribonuclease I and staphylococcal nuclease[30] employ Ca^{2+} for catalysis. This indicates that the protein environment around the catalytic site has a crucial role in controlling metal selectivity. Since Ca^{2+} can replace Mg^{2+} without promoting cleavage in REases, it is often used in crystallization studies to visualize the enzyme before catalysis. In addition to catalysis, the divalent metal plays an important role in site-specific DNA binding and in allosteric activation.[31]

X-RAY STRUCTURAL STUDIES

Crystallization

REases and their DNA co-crystals can be crystallized from a variety of conditions, including polyethylene glycol (PEG), 1,5-methyl-pentene-diol (MPD), and even salts such as ammonium sulfate and ammonium acetate. In the co-crystals of DNA-bound complexes, the DNAs often align to form pseudo-continuous helices.

Tertiary Structure

The three-dimensional structures of more than a dozen type II endonucleases including *Eco*RI,[32,33] *Eco*RV,[34–36] *Bam*HI,[28,37–40] *Pvu*II,[41,42] *Fok*I,[43,44] *Cfr*10I,[13] *Bgl*I,[9] *Bgl*II,[45,46] *Bso*BI,[47] *Nae*I,[48,49] *Ngo*MIV,[50] *Bse*634I[11] (an isoschizomer of *Cfr*10I), *Mun*I,[51] and *Hinc*II[52] (Table 1) have now been determined. All except *Cfr*10I and *Bse*634I have been determined in complex with their cognate DNA sites. The enzymes share little or no sequence homology (except the isoschizomers) but they all consist of a central β-sheet that is flanked by α-helices on both sides. Figure 1 shows the ribbon diagram of four representative members of type II REases: (a) the monomer of the classical dimeric type IIP REase *Bam*HI, (b) the cleavage domain of the monomeric type IIS REase *Fok*I, (c) the monomer of tetrameric *Ngo*MIV, a member of the type IIF family, and (d) the Endo domain of the type IIE enzyme *Nae*I. Interestingly, a similar α/β core is also present in other DNA-acting enzymes such as λ-exonuclease,[53,54] *Mut*H,[55] Vsr endonuclease,[56,57] and TnsA[58] from the Tn7 transposase. In the common core, only three β-strands are absolutely conserved.[49] Two of these strands contain the amino acid residues directly involved in catalysis. The similarity at the tertiary structure level is strongest between endonucleases that share a similar cleavage pattern, such as between *Bam*HI and *Eco*RI, which cleave DNA to leave four-base 5′ overhangs, or between *Eco*RV, *Pvu*II, which cleave DNA to produce blunt ends.[59] An exception is *Bgl*I that has a similar fold to *Eco*RV, *Pvu*II, and *Nae*I but cleaves DNA to leave 3′ overhangs.[9] Overall, the similarity reflects constraints in positioning of the active sites: 17–19 Å apart to produce four-base 5′ overhangs and 2 Å to produce blunt ends.[39] The active sites occur at one end of the central β-sheet and contain at least three superimposable residues that are critical for catalysis (see Table 2). Two of these residues are acidic, while the third residue is usually a lysine, except in *Bam*HI, which has a glutamate,[37] and in *Bgl*II, which has a glutamine,[45] in the same position. The active site geometry with the metal binding sites will be discussed in more detail later.

Quaternary structure

While type II REases show remarkable similarities at the tertiary structure level, the quaternary structures show striking differences underlying their functional diversity. They can form monomers, dimers, or tetramers in solution and have the ability to bind a single recognition site or two DNA sites simultaneously. To highlight the similarities and differences of REases at the quaternary structure level, we will describe examples of a classical homodimeric REase *Bam*HI, a monomeric type IIS REase *Fok*I, a tetrameric type IIF enzyme *Ngo*MIV, and finally a type IIE enzyme *Nae*I.

*Bam*HI: a type IIP REase

The classical type II (or according to a new nomenclature,[4] type IIP) REases are homodimeric enzymes that recognize a 4–8 base-pair palindromic sequence and cleave within that recognition sequence. Owing to their importance in recombinant DNA technologies, the type IIP enzymes are the best studied type II REases. Nine out of the 14 REases with 3D Structures – *Bam*HI, *Bgl*I, *Bgl*II, *Bso*BI, *Eco*RI, *Eco*RV, *Hinc*II, *Mun*I, and *Pvu*II – are type IIP enzymes.

*Bam*HI is a type IIP REase that recognizes the palindromic sequence 5′-GGATCC-3′ and cleaves after the first nucleotide, leaving four-base 5′ overhangs. The subunit structure consists of a large six-stranded mixed β-sheet, which is sandwiched on both sides by α-helices (Figure 1(a)).[37] Strands β3, β4, and β5 are antiparallel and

Table 1 Single crystal X-ray studies on type II restriction endonucleases[a]

Name	Description	Oligonucleotides used for co-crystallization	Number and nature of metal bound	Space group	Resolution (Å)	PDB code	Reference
BamHI	Apoenzyme	N/A	No metal	C2	1.95	1BAM	37
BamHI	Specific complex	5'-TAT**GGATCC**ATA-3'	No metal	$P2_12_12_1$	2.20	1BHM	39
BamHI	Nonspecific complex	5'-AT**GAATCC**ATA-3'	No metal	$P2_12_12_1$	1.90	1ESG	40
BamHI	Prereactive complex with Ca^{2+}	5'-TAT**GGATCC**ATA-3'	2 Ca^{2+} in the R subunit only	$P2_12_12_1$	2.00	2BAM	28
BamHI	Postreactive complex with Mn^{2+}	5'-TAT**GGATCC**ATA-3'	2 Mn^{2+} in the R subunit only	$P2_12_12_1$	1.80	3BAM	28
BglI	Specific complex	5'-ATC**GCC**TAATA**GGC**GAT-3'	2 Ca^{2+}	$C222_1$	2.20	1DMU	9
BglII	Apoenzyme	N/A	No metal	$P3_221$	2.30	1ES8	46
BglII	Specific complex	5' TATTAT**AGATCT**ATAA-3' 3'-AATA**TCTAGA**TATTAT-5'	1 Ca^{2+}	$P2_12_12_1$	1.50	1DFM	45
Bse634I	Apoenzyme	N/A	No metal	$P2_12_12$	2.17	1KNV	11
BsoBI	Specific complex	5'-TATA**CTCGAG**TAT-3' 3'-TAT**GAGCTC**ATAT-5'	No metal	$P2_12_12_1$	1.70	1DC1	47
Cfr10I	Apoenzyme	N/A	No metal	I222	2.15	1CFR	13
EcoRI	Apoenzyme	N/A	No metal	C2	3.0	1QC9	–
EcoRI	Specific complex	5'-TCGC**GAATTC**GCG-3 5'-GCG**CTTAAG**CGCT-3'	No metal	$P32_1$	2.70	1ERI	33
EcoRI	Postreactive complex with Mn^{2+}	5'-TCGC**GAATTC**GCG-3 5'-GCG**CTTAAG**CGCT-3'	1 Mn^{2+}	$P32_1$	2.50	1QPS	–
EcoRI	Methylated DNA	5'-CGC**GA(m6)ATTC**GCG-3'	No metal	$P2_12_12_1$	2.00	4DNB	61
EcoRV	Apoenzyme	N/A	No metal	$P2_12_12_1$	2.50	1RVE	34
EcoRV	Specific complex	5'-GG**GATATC**CC-3'	No metal	$C222_1$	3.00	4RVE	34
EcoRV	Noncognate complex	5'-C**GAGCTC**GAGAGCTCG-3'	No metal	$P2_1$	3.00	2RVE	34
EcoRV	Prereactive complex	5'-AAA**GATATC**TT-3' 3'-TT**CTATAG**AAA-5'	2 Mg^{2+}	P_1	2.10	1RVB	36
EcoRV	Product complex	5'-AAA**GATATC**TT-3' 3'-TT**CTATAG**AAA-5'	4 Mg^{2+}	P_1	2.10	1RVC	36
FokI	Apoenzyme	N/A	No metal	$P2_12_12_1$	2.30	2FOK	43
FokI	Specific complex	5'-TC**GGATG**ATAACGCTAGTCA-3' 3'-G**CCTAC**TATTGCGATCTGTA-5'	No metal	$P2_1$	2.80	1FOK	44
HincII	Specific complex	5'-CCG**GTCGAC**CGG-3'	No metal	$I2_12_12_1$	2.60	1KC6	52
MunI	Specific complex D83A mutant	5'-GC**CAATTG**GC-3'	Ca^{2+} soak but no metal binding found	$P2_12_12_1$	1.70	1D02	51
NaeI	Apoenzyme	N/A	No metal	$P2_1$	2.30	1EV7	49
NaeI	Specific complex	5'-TGCCAC**GCCGGC**GTGGC-3' 3'-CGGTG**CGGCCG**CACCGT-5'	No metal	$P2_12_12_1$	2.50	1IAW	48
NgoMIV	Postreactive complex	5'-TGC**GCCGGC**GC-3' 3'-CG**CGGCCG**CGT-5'	8 Mg^{2+} (2 per monomer)	$P2_12_12_1$	1.60	1FIU	50
PvuII	Apoenzyme	N/A	No metal	$P2_12_12$	2.40	1PVU	42
PvuII	Specific complex	5'-TGAC**CAGCTG**GTC-3' 3'-CTG**GTCGAC**CAGT-5'	No metal	$P2_12_12_1$	2.80	1PVI	62
PvuII	Specific complex, cross-linked	5'-TGAC**CAGCTG**GTC-3' 3'-CTG**GTCGAC**CAGT-5'	2 Ca^{2+}	$P2_12_12_1$	2.50	1F0O	63

[a] This is not a complete list of all structures. In particular, for PvuII, EcoRI, and EcoRV several other structures have been solved.

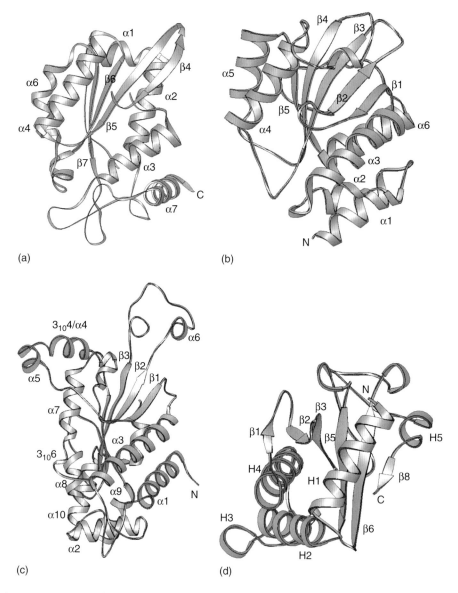

Figure 1 Schematic representation with secondary structure assignment of (a) *Bam*HI monomer (PDB code: 1BAM), (b) the cleavage domain of *Fok*I (1FOK), (c) *Ngo*MIV monomer (1FIU), and (d) the Endo domain of *Nae*I (1IAW). Produced with the program RIBBONS[60]

form a β-meander; strands β5, β6, and β7 are parallel and resemble a Rossman fold, with α4 and α6 acting as the crossover helices. The dimer interface is formed primarily by helices α4 and α6, which pair with the corresponding helices from the symmetry-related subunit to form a parallel four-helix bundle. The dimerization interface is similar in the other dimeric REases. The catalytic residues of Asp94, Glu111, and Glu113 are clustered at one end of the β-meander, close to the scissile phosphate group. The *Bam*HI cognate DNA complex (Figure 2) displays extensive protein–DNA interactions, with side-chain and main-chain atoms and tightly bound water molecules all contributing toward recognition of the *Bam*HI recognition sequence.[39] Interactions occur both in the major and minor grooves. In the major groove, every hydrogen bond donor and acceptor group takes part in direct or water-mediated hydrogen bonds with the protein. This complementarity at the protein–DNA interface ensures that only the *Bam*HI recognition sequence can make all the necessary interactions. In contrast, *Bam*HI forms a loose nonspecific complex with the 5′-G_AATCC-3′ sequence that differs only by a single base pair (underlined) from the cognate sequence.[40] All interactions of the protein with the DNA bases are lost throughout the whole recognition sequence, and not just the substituted base pair. This highlights the remarkable selectivity with which REases recognize their specific DNA sites. While type IIP restriction endonucleases bind as homodimers to a single DNA site, there are increasing examples of modifications of this theme. In particular, a growing number of type II REases have now been shown to act on two distant DNA sites.[66] The type IIE enzymes were shown to require a second

Table 2 Summary of restriction endonucleases for which 3D Structure is available

Name	Recognition sequence	Catalytic residues
Type IIP Reases: homodimers recognizing palindromic sequences		
BamHI	5'-G/GATCC-3'	Asp94, Glu111, Glu113
BglI	5'-GCCN$_4$/NGGC-3'	Asp116, Asp142, Lys144
BglII	5'-A/GATCT-3'	Asp84, Glu93, Gln95
BsoBI	5'-C/YCGRG-3'[a]	Asp212, Glu240, Lys242
EcoRI	5'-G/AATTC-3'	Asp91, Glu111, Lys113
EcoRV	5'-GAT/ATC-3'	Asp74, Asp90, Lys92
HincII	5'-GTY/RAC-3'	Asp114, Asp127, Lys129[b]
MunI	5'-C/AATTG-3'	Asp83, Glu98, Lys100
PvuII	5'-CAG/CTG-3'	Asp58, Glu68, Lys70
Type IIE Reases: two recognition sites are required but only one is cleaved; the other site serves as an allosteric activator		
NaeI	5'-GCC/GGC-3'	Asp86, Glu95, Lys97
Type IIF: two recognition sites are cleaved concertedly by a homotetrameric enzyme		
Bse634I	5'-R/CCGGY-3'	Asp146, Glu212, Lys198
Cfr10I	5'-R/CCGGY-3'	Asp134, Glu204, Lys190
NgoMIV	5'-G/CCGGC-3'	Asp140, Glu201, Lys187
Type IIS: asymmetric recognition site with cleavage occurring at a defined distance		
FokI	5'-GGATGN$_9$/-3' 3'-CCTACN$_{13}$/-5'	Asp450, Asp467, Lys469

[a] R = A, G and Y = T, C.
[b] N Horton, personal communication.

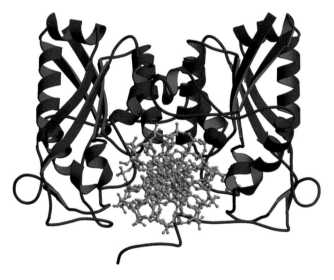

Figure 2 Overall structure of the type IIP dimeric endonuclease BamHI (PDB code: 3BAM); the L subunit is shown in blue and the R subunit in magenta. The metal binding sites are represented by gray spheres. Produced with the programs MOLSCRIPT[64] and RASTER3D.[65]

DNA site for allosteric activation,[67] with two members of this class, EcoRII[68] and NaeI,[69] known to function as dimers. Other enzymes such as the type IIF enzymes SfiI,[70] SgrA1,[71] Cfr101,[72] and NgoMIV[50] have now been shown to function as homotetramers and to mediate interactions between two recognition sequences. There is growing evidence that restriction enzymes requiring two recognition sites are related to other DNA-modifying enzymes that bring distant DNA sites together. The type IIE enzyme EcoRII, for instance, shows sequence homology to the integrase family of recombinases,[73] while NaeI has been shown to have topoisomerase and recombinase activities.[74] SfiI shows a close relationship to a family of recombinases that simultaneously catalyze a four-strand DNA break,[70] and a FokI-like fold has been described in TnsA, one of the two proteins of the Tn7 transposase that mediates the release of a transposon.[58]

FokI: a type IIS REase

Members of the type IIS family of endonucleases, unlike the type IIP enzymes, recognize a nonpalindromic recognition site and cleave at a defined distance away from it. The type IIS REase FokI, for instance, cleaves DNA nonspecifically at a fixed distance of 9 and 13 nucleotides downstream of its 5'-GGATG-3' recognition sequence.[75] FokI has a modular structure consisting of an N-terminal DNA recognition domain and a C-terminal cleavage domain.[76] The bipartite nature of FokI has made it an excellent candidate for the design of hybrid endonucleases with novel sequence specificities via the attachment of the cleavage domain of FokI to DNA binding domains of transcription factors such as zinc fingers and homeodomains.[77–82] As anticipated, the FokI-DNA structure[44] (Figure 3), determined with a 20-mer oligonucleotide in the absence of divalent metals, reveals two domains: an N-terminal DNA recognition domain (372 AA) and a C-terminal cleavage domain (195 AA) joined by a flexible linker (16 AA). The recognition domain is made of three smaller subdomains (D1, D2, and D3) that are unrelated to the conserved REase fold but are instead related to the helix-turn-helix-containing DNA binding domain of the catabolite gene activator protein CAP.[83] The CAP core has been embellished in the first two subdomains, whereas in the third subdomain it has been co-opted for protein–protein interactions. The cleavage domain, on the other hand, is structurally similar to a BamHI monomer, despite the fact that the two enzymes display limited sequence homology. Residues Asp450, Asp467, and Lys469 within the cleavage domain overlap with the catalytic residues Asp94, Glu111, and Glu113 of BamHI. Mutational studies had previously shown that replacing any of these three residues in FokI results in the complete loss of cleavage activity on both strands, suggesting the existence of a single catalytic center.[84] The most surprising feature of the structure is that the cleavage domain is packed alongside the recognition domain and hence is positioned away from the site of cleavage. Although unexpected at first, this sequestration

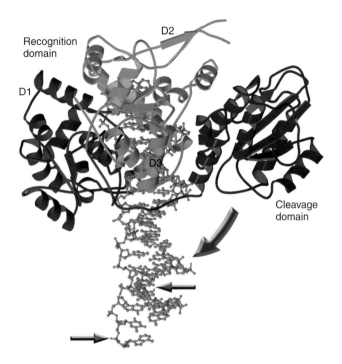

Figure 3 The overall structure of the type IIS endonuclease FokI (PDB code: 1FOK). The recognition domain consists of three smaller subdomains D1 (magenta), D2 (green), and D3 (light blue). The cleavage domain is shown in dark blue and the linker segment connecting it to the recognition domain is shown in red. Small arrows indicate the sites of cleavage and the large arrow conveys the motion of the cleavage domain toward the cleavage site. Produced with the programs MOLSCRIPT[64] and RASTER3D.[65]

FokI molecule. The recognition domain of the second FokI molecule could either remain free, be accommodated on the same DNA molecule, or alternatively bind to another DNA molecule containing the appropriate recognition sequence. Recent biochemical and biophysical evidence points to a requirement for a specific DNA and divalent metal in the formation of an active complex.[31] This two-site model offers a two-level control mechanism to prevent nonspecific cleavage. First, the specific binding of two FokI molecules to two DNA sites is required in order to release the cleavage domains and, second, dimerization is required to properly align the two catalytic domains for double-strand DNA cleavage. This mechanism is likely to be true for other monomeric type IIS enzymes.

NgoMIV: a type IIF REase

NgoMIV is a type IIF REase that recognizes the palindromic hexanucleotide sequence 5'-GCCGGC-3' and cleaves after the 5'-G on each strand to produce four-base 5' staggered ends. The crystal structure of the enzyme in complex with a cleaved 11 bp oligonucleotide (Figure 4) shows that NgoMIV, like other members of the type IIF family, is tetrameric and binds two copies of its recognition sequence.[50] In the tetramer, two primary dimers (comprising subunits A and B, and C and D respectively) are arranged back to back and the two oligonucleotides are

of the cleavage domain is consistent with footprinting results that reveal very weak protection at the cleavage site in the absence of divalent metals.[85] Together, these structural and biochemical observations raise the question of how a monomeric FokI manages to cleave both DNA strands. Two mechanisms appear plausible: (i) the cleavage domain first cuts one strand of the DNA and then flips over to cut the other strand, or (ii) two FokI molecules dimerize on the DNA through their cleavage domains and cut both strands simultaneously. Evidence for the second mechanism comes from the crystal structure of the apoenzyme[43] that reveals a dimer as well as from kinetic studies that support a cooperative second-order reaction mechanism.[86] Interestingly, dimerization in the apoenzyme structure was found to be mediated by two helices that are the exact analogues of the helices that mediate dimerization in BamHI.[43] Indeed, mutations of two residues (Asp483 and Arg487) that form a salt bridge in the apoenzyme dimer resulted in the loss of cleavage activity.[86] On the basis of these results, the following mechanism of DNA cleavage has been proposed:[43] First, a FokI molecule interacts with its target recognition site, resulting in the dissociation of the cleavage domain from its sequestered state. Second, an efficient DNA cleavage follows dimerization of the first FokI molecule with the cleavage domain of a second

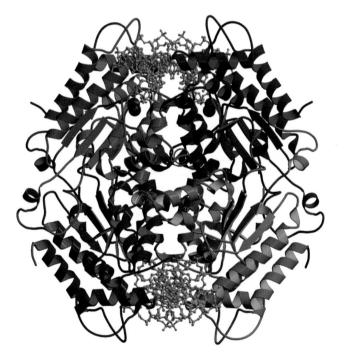

Figure 4 The overall structure of the tetrameric type IIF endonuclease NgoMIV (PDB code: 1FIU). The Mg^{2+} ions are represented as gray spheres. Produced with the programs MOLSCRIPT[64] and RASTER3D.[65]

bound on opposite sides of the tetramer with their helical axes tilted at a 60° angle.

The core of the NgoMIV monomer and its primary dimers are structurally similar to that of dimeric restriction enzymes that produce four-base 5′ overhangs (BamHI, EcoRI, and MunI). However, a loop region – mainly involved in the tetramerization interface – extends 25 Å out of the core of each monomer, which is not present in the dimeric REases. Dimerization occurs primarily by contacts between helices $3_{10}6/\alpha7$. These helices deserve special mention because, in addition to participating in dimerization, they are also involved in tetramerization and DNA interactions. The tetrameric assembly of NgoMIV is fixed by side-by-side contacts (between subunits A/D and B/C) and cross contacts (between subunits A × C and B × D) between primary dimers. The most extensive contacts in the NgoMIV tetramer, between subunits A/D and B/C, are made by the tetramerization loop (residues 147–176, including helix 6) that spans across the neighboring monomer. In total, the contacts between subunits at the tetramerization interface comprise eight salt bridges and numerous hydrophobic interactions. Accessible surface area calculations indicate that a large monomer surface area of 4200 Å^2 is buried upon tetramerization. The contribution of individual intersubunit contacts to the totally buried surface area varies greatly and the surface of each monomer buried during tetramerization exceeds the surface area buried at the dimer interface by almost three times. Tetramerization can also be detected in classical type II enzymes under special circumstances. In BamHI, tetramerization occurs at micromolar concentration at lower than 0.3 M salt concentration.[87] This suggests that the tetrameric form is a natural evolutionary step from the more simple homodimeric form.

DNA contacts in NgoMIV are made primarily from the major groove. The central 5′-CCGG-3′ part of the NgoMIV recognition sequence is contacted by residues located on a short stretch of amino acids with the sequence 191RSDR. The sequence region 194RPDR is present at the structurally equivalent positions of the Cfr10I restriction enzyme, which recognize the Pu/CCGGPy sequence. In SsoII, biochemical evidence suggests that the **RXXR** motif plays an important role in DNA recognition.[88] In addition, several other REases that recognize adjacent GG sequences contain this **RXXR** motif. Thus, NgoMIV, Cfr10I, and other REases with the **RXXR** motif may utilize a similar mechanism to interact with the GG bases.[88]

NaeI: a type IIE enzyme

NaeI is a remarkable REase with both endonuclease and topoisomerase activities.[74] It recognizes the same DNA sequence as NgoMIV, 5′-GCCGGC-3′, but cleaves in the middle to generate blunt-ended products. NaeI is a homodimer but it must bind two DNA recognition sequences in order to cleave one DNA sequence, with the second DNA acting as an allosteric activator.[89,90] 'Activator' and 'substrate' DNAs bind to two independent domains on NaeI: the Endo and Topo domains. The presence of these two independent DNA binding domains suggest an enzyme more complex than is needed for simple monofunctional cleavage of DNA. Strikingly, single Leu43 to lysine mutation converts NaeI from an endonuclease to topoisomerase/recombinase, suggesting that NaeI is a bridge between these common protein families.[91,92]

The structure of the NaeI–DNA complex (Figure 5) marks the first example of novel recognition of two copies of the same DNA sequence by two different sets of amino acid residues, in two different structural motifs, in a single polypeptide.[48] NaeI in complex with DNA (17-mer) has the same overall fold as that of NaeI without DNA.[49] However, two loops, Ser191–Asn194 and Asn254–Val258, that are disordered in the apoenzyme structure are well ordered in the complex, and residues Asn254–Leu262 form a new α-helix in the complex. Each NaeI monomer consists of two structural domains: an N-terminal Endo domain (residues 1–162) and a C-terminal Topo domain (residues 172–313). The domains are linked by a relatively extended loop (residues 163–171). Dimerization is mediated primarily by

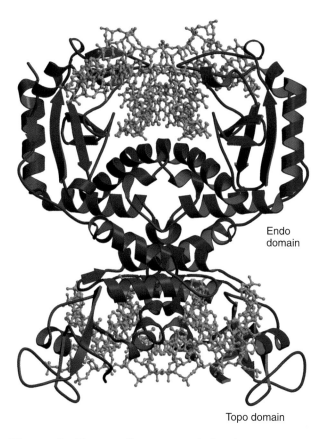

Figure 5 The overall structure of the dimeric type IIE endonuclease NaeI (PDB code: 1IAW). The substrate DNA binds in the Endo domain and the activator DNA in the Topo domain. Produced with the programs MOLSCRIPT[64] and RASTER3D.[65]

the Endo domain, which has the characteristic α/β fold of type II endonucleases. The cleft between the two Endo domains is the site of DNA binding with the catalytic residues, Glu70, Asp86, Asp95, and Lys97. While the *Nae*I Endo domain has the same basic structural fold as *Ngo*MIV and also recognizes 5'-GCCGGC-3', it lacks the **RXXR** motif present in *Ngo*MIV and uses different hydrogen bond pairs for recognition.[48] This is an example of the same structural fold recognizing the same DNA sequence in multiple ways. The Topo domain contains a CAP DNA binding motif observed in many transcription factors,[93] DNA processing proteins, such as DNA topoisomerases IA and II,[94] as well as the type IIS REase *Fok*I.[44] The two Topo domains in the dimer form a second cleft, which binds the second 'activator' DNA molecule.

Conformational changes upon DNA binding

Endonucleases such as *Bam*HI, *Eco*RI, and *Ngo*MIV that cleave the DNA to leave 5' overhanging ends approach the DNA from the major grove, while REases such as *Eco*RV, *Pvu*II, *Mun*I, and *Nae*I that cleave DNA to leave blunt ends approach the DNA from the minor groove. The structure of *Bgl*I is the only example of an REase that cleaves DNA to leave 3' overhanging ends, and it approaches DNA from the minor groove. Both protein and DNA undergo conformational changes upon binding, but there is no clear correlation between DNA distortion and the mode of binding or the cleavage pattern. For example, a 20° to 50° bend in the central bases of the DNA has been found with enzymes *Eco*RI, *Eco*RV, *Bgl*I, *Nae*I, and *Mun*I. The DNA is not significantly bent, however, upon binding *Bam*HI, *Pvu*II, or *Ngo*MIV. Conformational changes in enzymes include rotation and translation of subunits, and folding of disordered regions. However, in *Bam*HI the most striking conformational change is the unraveling of the carboxy-terminal α-helices to form partially disordered 'arms'.[39] The arm from one subunit fits into the minor groove, while the arm from the symmetry-related subunit follows the DNA sugar-phosphate backbone. In *Bgl*II, the two subunits undergo a large scissor-like motion to capture the DNA,[46] as opposed to the 'tong-like' motion in *Bam*HI. Moreover, the catalytic active site residues are sequestered in the *Bgl*II apoenzyme by a large 'lever-like' motion of the residues nearby. In *Fok*I, notable changes occur in the protein–protein interface of the recognition and cleavage domains. Disruptions of the hydrogen-bonding network between the two domains widens the interface, possibly readying the cleavage domain to dissociate from the recognition domain, a step necessary to initiate cleavage.[43] In *Nae*I, a comparison between free and DNA complexed forms reveals dramatic rearrangements of the dimer upon DNA binding.[48] In the DNA complex, the two Topo domains move 16 Å closer to each other, bringing the recognition helices in contact with the DNA major groove.

In contrast, the Endo domains move 8 Å away from each other, thus widening the DNA binding cleft. The substrate-binding cleft of the Endo domain is not wide enough to accommodate the DNA substrate in the apoenzyme form. Binding of DNA to the Topo domains not only narrows the effector-binding cleft but also opens the active site in the substrate-binding cleft in the Endo domain. Thus, the Topo domain serves as an allosteric activator for the binding and cleavage of DNA in the Endo domain.

METAL SITE GEOMETRIES IN THE CATALYTIC SITE

The active sites of *Eco*RI, *Eco*RV, *Bam*HI, *Pvu*II, *Cfr*10I, *Bse*634I, *Bgl*I, *Bgl*II, *Ngo*MIV, *Fok*I, *Mun*I, and *Nae*I are all structurally similar, containing overlapping residues that occur at one end of a central β-sheet. The residues follow the weak consensus **Asp-X_n-Glu/Asp/Ser-Z-Lys/Glu/Gln**, where n varies from 9 to 54 residues and Z is usually a hydrophobic residue. The first two residues of the consensus, corresponding to Asp94 and Glu111 in *Bam*HI, are generally acidic except for the type IIF enzymes *Cfr*10I, *Ngo*MIV, and *Bse*634I, all of which have a serine in place of the second acidic residue in their primary sequence. However, in all three of these enzymes, a glutamate residue from a distal helix takes the place of the second catalytic residue. In the case of *Ngo*MIV, Glu201 overlaps spatially with Glu111 of *Bam*HI.[50] Thus, the first two catalytic residues in all known REase structures are acidic. The third residue is a lysine in all of the endonucleases except *Bam*HI, which has a glutamate (Glu113), and *Bgl*II, which has a glutamine (Gln95). This difference is critical: there is a severe loss of cleavage activity when Glu113 in *Bam*HI is substituted by a lysine residue,[95] or conversely, when Lys92 in *Eco*RV[96] or Lys113 in *Eco*RI[97] is substituted by a glutamate residue. Similarly, there is a loss of cleavage activity when Glu113 in *Bam*HI is substituted by the glutamine in *Bgl*II (I. Schildkraut, personal communication). A fourth acidic residue often precedes the consensus in some of the enzymes (e.g. Glu77 in *Bam*HI, Glu45 in *Eco*RV, and Glu70 in *Ngo*MIV), but it shows poor superposition. Moreover, in *Eco*RI, *Bgl*II, *Pvu*II, and *Mun*I, there is no equivalent of the fourth negatively charged catalytic residue.

In *Bam*HI, two metal binding sites (A and B) were determined on the basis of the pre- and postreactive states of the enzyme obtained with Ca^{2+} and Mn^{2+} respectively (Figure 6).[28] The Ca^{2+} and Mn^{2+} binding sites are nearly identical, showing that despite its inhibitory role, Ca^{2+} readily binds in place of the cofactor in the ground state of the enzyme. Site A is between Asp94, Glu111, and the scissile phosphate, while site B is next to the cleavable bond in a pocket formed by the side chains of Asp94 and Glu77, and the distance between the two metals is ~4.0 Å. The metal binding sites in *Bam*HI are superimposable on the metal sites found in the *Ngo*MIV-product complex,

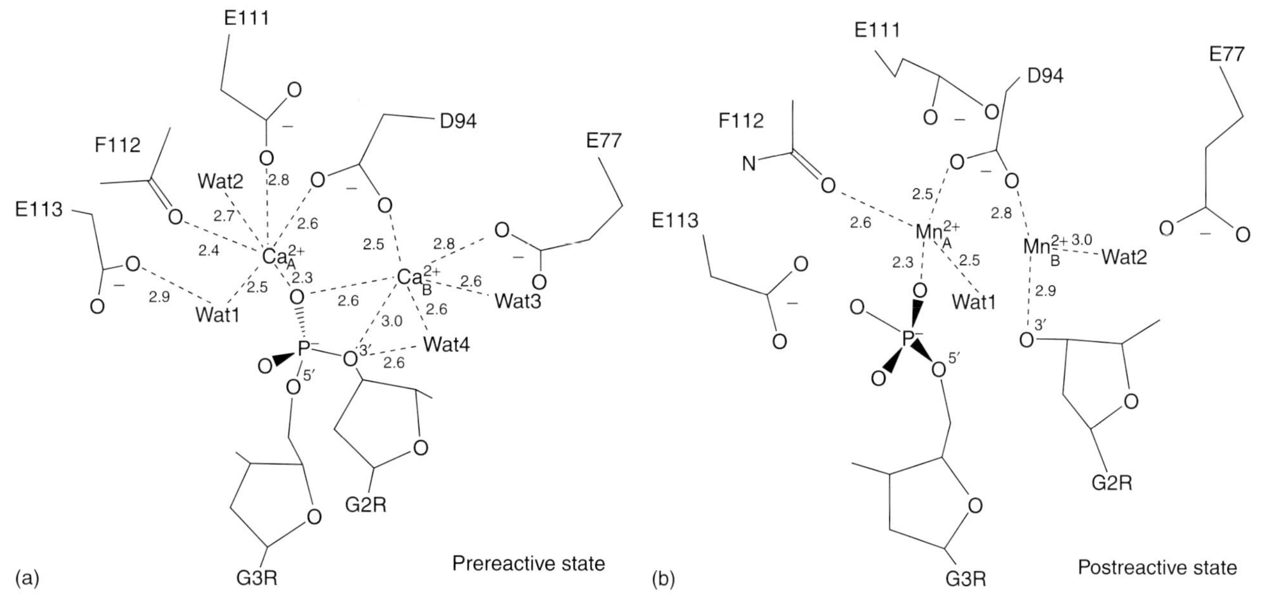

Figure 6 (a) The prereactive state of *Bam*HI captured in the presence of Ca^{2+}. Both metal ions are coordinated to six ligands in octahedral arrangements. Wat1 to Wat4 refers to water molecules within the active site. The distance between Ca$_A^{2+}$ and Ca$_B^{2+}$ is 4.3 Å. (b) The postreactive state of *Bam*HI captured in the presence of Mn^{2+}. The two Mn^{2+} ions are disposed similarly to the Ca^{2+} ions although they are slightly closer (3.9 Å) and have looser coordination spheres as well. The overall rmsd between the Cα positions of the pre- and postreactive complexes is only 0.33 Å. The main difference is in the location of the scissile phosphate group. After cleavage, it is displaced by 2.5 Å, bringing it within the vicinity of Glu113.

where metal A is coordinated by Asp140 and Glu201 and metal B is coordinated by Asp140 and Glu70 and the two metal sites are 3.8 Å away from each other.[50] Two metals have also been found in the active sites of *Bgl*I,[9] *Pvu*II,[63] and *Eco*RV.[36] In *Eco*RI (1QPS, unpublished) and *Bgl*II,[45] only a single metal binding site has been identified that corresponds to the position of metal A in *Bam*HI. Figure 7 shows the active sites of *Bam*HI, *Ngo*MIV, and *Bgl*II (the distances are listed in Table 3). In the *Mun*I-D83A mutant-DNA complex, no metal has been found despite the fact that the crystals were soaked in Ca^{2+}.[51] Asp83 of *Mun*I is structurally equivalent of Asp94 in *Bam*HI, and the lack of metal binding in the *Mun*I catalytic mutant structure highlights the importance of the first acidic residue in metal binding. In the *Bam*HI prereactive complex,[28] two water molecules are noteworthy: Wat1 – coordinated to Ca^{2+}A (2.5 Å) and Glu113 (2.9 Å), and 3.2 Å from the scissile phosphodiester group – is positioned ideally as the attacking nucleophile, maintaining in-line orientation with the cleavable P-O3' bond. On the other side of the scissile phosphodiester group, Wat4, coordinated to Ca^{2+} B, forms a hydrogen bond (2.6 Å) with the O3', possibly acting as a general acid in donating a proton to the leaving group. In the crystal structures of the other REase–DNA complexes, water molecules in positions equivalent to Wat1 and Wat4 have also been observed.

In summary, all REase structures solved to date share the following features in their catalytic sites:

- The catalytic site is located at one end of the conserved central β-sheet.
- There are two conserved acidic residues – an aspartic acid and a glutamic acid/aspartic acid.
- There is a divalent metal ion at identical or a similar position to metal A in *Bam*HI. This metal is coordinated by the two conserved acidic residues.
- There are two water molecules. The first water molecule (Wat1 in *Bam*HI) is hydrogen bonded to metal A and is approximately in-line with the scissile phosphate bond. The second water molecule (Wat4 in *Bam*HI) is in close proximity to the 3'-phosphate leaving group.

The variability occurs in

- the identity of the third catalytic residue – Glu in *Bam*HI, Gln in *Bgl*II, and Lys in all others;
- the number of metal ions.

The common elements described above indicate that the type II REases might employ a similar cleavage mechanism.

CATALYTIC MECHANISM

Type II REases – with the exception of *Bfi*I[24] – require only Mg^{2+} as a cofactor to catalyze the hydrolysis of DNA phosphodiesters. The reaction occurs through an S$_N$2 mechanism, with an in-line displacement of the 3'-hydroxyl group and an inversion of configuration of the 5'-phosphate

Table 3 Metal–ligand and metal–metal distances from Figure 7. Labels used in Figure 7 are in parenthesis

Atom 1	Atom 2	Distance (Å)
BamHI (PDB: 3BAM)		
Mn 901 (MnA)	OE2 94	2.5
MnA	O 112	2.6
MnA	O2P G1 (G3R)	2.3
MnA	O HOH1 (1)	2.5
MnA	Mn 902 (MnB)	3.9
MnB	OE1 94	2.8
MnB	O3′ G4 (G2R)	2.9
MnB	O HOH2 (2)	2.9
NgoMIV (PDB: 1FIU)		
Mg 5555 (MgA)	O HOH221 (1)	2.2
MgA	O HOH38 (2)	2.2
MgA	O2P C5 (C2)	2.1
MgA	O ACY 2001 (Ac)	2.3
MgA	O 186	2.2
MgA	OD2 140	2.2
MgA	Mg 4444 (MgB)	3.8
MgB	OXT ACY 2001 (Ac)	2.1
MgB	O2P C5 (C2)	2.3
MgB	OD1 140	2.2
MgB	O HOH22 (3)	2.2
MgB	O HOH199 (4)	2.3
MgB	O HOH30 (5)	2.3
BglII (PDB: 1D2I)		
Mg 601 (Mg)	O 94	2.3
Mg	OD2 84	2.4
Mg	O1P G8	2.4
Mg	O HOH24 (NW)	2.7
Mg	O HOH131 (2)	2.4
Mg	O HOH57 (3)	2.4

group. The important elements of the reaction are an activated water molecule to carry out the nucleophilic attack, a Lewis acid to stabilize the negatively charged transition state, and a general acid to donate a proton to the leaving O3′ atom. Several catalytic mechanisms of DNA cleavage have been proposed, differing mainly in the number and the role of the metal in catalysis.[88] A two-metal mechanism has been proposed for BamHI[28] similar to that proposed for the 3′-5′-exonuclease domain of the E. coli DNA polymerase I.[39] According to the mechanism, metal A activates the attacking water molecule, while metal B stabilizes the buildup of negative charge on the leaving O3′ atom. At the same time, both metals (acting as Lewis acids) help stabilize the pentacovalent transition state. The positions of the Ca^{2+} and Mn^{2+} ions in the pre- and postreactive complexes conform to the classical description of a two-metal mechanism, as discussed by Beese and Steitz for E. coli DNA polymerase I.[26] The two metals lie on a line that is parallel to the apical direction of the trigonal bipyramid geometry that would be expected at the scissile phosphodiester during the transition state. The distance between the two metals (~4 Å) correlates with the anticipated distance of 3.8 Å between the apical oxygens in the transition state. Thus, the ions are in positions to stabilize the 'entering' and 'leaving' oxygens at the apical positions, and to reduce the energy in forming the 90° O–P–O bond angles between the apical and the equatorial oxygens.

The metals in the active site of the NgoMIV-product complex overlap with the metals of the BamHI-postreactive complex.[50] This is despite the fact that NgoMIV contains a lysine as the third catalytic residue instead of Glu113 in BamHI, and its second catalytic residue Glu201 shows only spatial but not sequential alignment. A two-metal mechanism has also been postulated for PvuII,[63] but the crystallographically determined metal positions in that case are different from BamHI. Two metal sites in EcoRV are similar to that of PvuII and a two-metal mechanism proposed for EcoRV[98] closely resembles that of PvuII.

Can BamHI, NgoMIV, and the other type II REases share a common cleavage mechanism? The conserved α/β core and the overlapping catalytic sites argue in favor of a similar cleavage mechanism. The most striking difference between BamHI, BglII, and the other type II enzymes

Figure 7 Stereo view of the active sites around the scissile phosphate bond. (a) The active site of the BamHI-postreactive complex determined in the presence of Mn^{2+} (PDB code 3BAM). Metal A is coordinated by the OE2 of Asp94 (2.5 Å), the carbonyl oxygen of Phe112 (2.6 Å), the O2 atom of the cleaved phosphate group (2.3 Å), and a water molecule (2.5 Å). Metal B is coordinated by the OE1 of Asp94 (2.8 Å), the O3′ of Gua 2R (2.9 Å), and a water molecule (2.9 Å). The metal-to-metal distance is 3.9 Å. (b) The active site of the NgoMIV postreactive complex determined in the presence of Mg^{2+} (PDB code 1FIU). The three active site residues of NgoMIV Asp140, Glu201, and Lys187 are the equivalent of Asp94, Glu111, and Glu113 of BamHI. Two Mg^{2+} ions exhibit octahedral coordination. The O2P of the 5′ phosphate, the carboxylate of Asp140, and the acetate molecule (from the reservoir of the crystallization buffer) contribute to the coordination of both Mg^{2+} ions. The remaining three ligands of Mg^{2+} ion A are the backbone oxygen of Cys186 and water molecules 1 and 2. Water molecules 3 to 5 complete the octahedral coordination of Mg^{2+} ion B. The ligand oxygen atoms are within 2.2–2.3 Å of Mg^{2+} ion A, and within 2.1–2.4 Å of Mg^{2+} ion B. The metal-to-metal distance is 3.8 Å. (c) The active site of the BglII prereactive complex (PDB code 1D2I). Residues Asp84, Glu93, and Gln95 correspond to the active site residues Asp94, Glu111, and Glu113 of BamHI. The octahedrally coordinated cation is possibly a Mg^{2+}. Five of its six ligands have distances ranging from 2.3–2.4 Å; the sixth ligand (the proposed nucleophile labeled NW) is 2.7 Å. Produced with the programs MOLSCRIPT[64] and RASTER3D.[65]

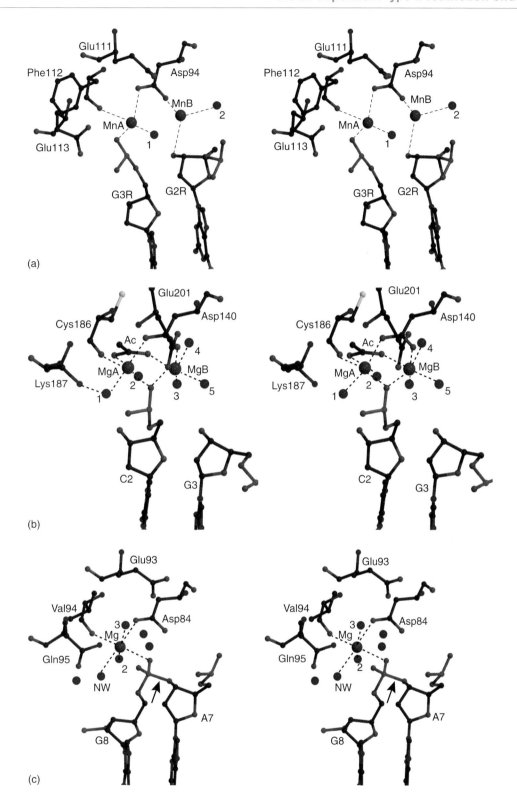

Figure 7

is the identity of the third catalytic residue: a glutamic acid in *Bam*HI, a glutamine in *Bgl*II, and a lysine in the others. Both glutamic acid[28] and lysine,[50] as well as the neighboring phosphate group in a substrate-assisted catalytic mechanism, have been postulated as a possible

general base.[99–101] At least for *Bam*HI, recent calculations confirm the two-metal mechanism and favor the transfer of a hydroxide from bulk solution rather than its generation at the active site by a general base.[29,102] In addition, a substrate-assisted mechanism for *Bam*HI is even less likely than a general base mechanism. This is in agreement with earlier calculations for DNA Polymerase I.[103] A lysine residue is less likely to act as a general base than Glu113 of BamHI; moreover, Gln95 in BglII cannot function as a general base. A possible exception is BsoBI, in which His253 could act as a general base analogous to BfiI that also contains a catalytic histidine and requires no divalent metal for catalysis. Overall, without the requirement for a general base, a more unified cleavage mechanism of type II REases can emerge. Calculations have also shown that the role of the second metal is less important.[102] A single metal can coordinate the nucleophile, and through active site reorganization can also stabilize the negatively charged pentavalent transition state as was postulated for staphylococcal nuclease that utilizes a single metal cleavage mechanism.[104] The less stringent requirement for a second catalytic metal could explain the discrepancies found among type II REases in the number of metals in their active sites. BamHI, EcoRV, BglI, PvuII, and NgoMIV contain two metals, while EcoRI and BglII appear to contain one metal. The equivalent of metal A in BamHI is the only metal that has been seen in all REase structures and it is stabilized by two acidic residues conserved in all type II REases. Besides its role in the activation of the nucleophile, the metal may also act as Lewis acid to stabilize the pentacovalent transition state, in a common mechanism of REase cleavage.

ACKNOWLEDGEMENTS

The authors gratefully acknowledge financial support from the National Institute of Health (grant GM44006 to Aneel K Aggarwal and GM20015 to Éva S Vanamee).

REFERENCES

1. LF Lin, J Posfai, RJ Roberts and H Kong, *Proc Natl Acad Sci USA*, **98**, 2740–5 (2001).
2. I Kobayashi, A Nobusato, N Kobayashi-Takahashi and I Uchiyama, *Curr Opin Genet Dev*, **9**, 649–56 (1999).
3. GG Wilson and NE Murray, *Annu Rev Genet*, **25**, 585–627 (1991).
4. RJ Roberts, M Belfort, T Bestor, AS Bhagwat, TA Bickle, J Bitinaite, RM Blumenthal, S Degtyarev, DT Dryden, K Dybvig, K Firman, ES Gromova, RI Gumport, SE Halford, S Hattman, J Heitman, DP Hornby, A Janulaitis, A Jeltsch, J Josephsen, A Kiss, TR Klaenhammer, I Kobayashi, H Kong, DH Kruger, S Lacks, MG Marinus, M Miyahara, RD Morgan, NE Murray, V Nagaraja, A Piekarowicz, A Pingoud, E Raleigh, DN Rao, N Reich, VE Repin, EU Selker, PC Shaw, DC Stein, BL Stoddard, W Szybalski, TA Trautner, JL Van Etten, JM Vitor, GG Wilson and SY Xu, *Nucleic Acids Res*, **31**, 1805–12 (2003).
5. DT Dryden, NE Murray and DN Rao, *Nucleic Acids Res*, **29**, 3728–41 (2001).
6. RJ Roberts, T Vincze, J Posfai and D Macelis, *Nucleic Acids Res*, **31**, 418–20 (2003).
7. RJ Roberts and SE Halford, in RJ Roberts (ed.), *Nucleases*, Cold Spring Harbor Laboratory Press, Cold Spring Harbor, New York (1993).
8. JE Brooks, PD Nathan, D Landry, LA Sznyter, P Waite-Rees, CL Ives, LS Moran, BE Slatko and JS Benner, *Nucleic Acids Res*, **19**, 841–50 (1991).
9. M Newman, K Lunnen, G Wilson, J Greci, I Schildkraut and SE Phillips, *EMBO J*, **17**, 5466–76 (1998).
10. BP Anton, DF Heiter, JS Benner, EJ Hess, L Greenough, LS Moran, BE Slatko and JE Brooks, *Gene*, **187**, 19–27 (1997).
11. S Grazulis, M Deibert, R Rimseliene, R Skirgaila, G Sasnauskas, A Lagunavicius, V Repin, C Urbanke, R Huber and V Siksnys, *Nucleic Acids Res*, **30**, 876–85 (2002).
12. H Ruan, KD Lunnen, ME Scott, LS Moran, BE Slatko, JJ Pelletier, EJ Hess, J Benner II, GG Wilson and SY Xu, *Mol Gen Genet*, **252**, 695–99 (1996).
13. D Bozic, S Grazulis, V Siksnys and R Huber, *J Mol Biol*, **255**, 176–86 (1996).
14. AK Newman, RA Rubin, SH Kim and P Modrich, *J Biol Chem*, **256**, 2131–39 (1981).
15. L Bougueleret, M Schwarzstein, A Tsugita and M Zabeau, *Nucleic Acids Res*, **12**, 3659–76 (1984).
16. K Kita, H Kotani, H Sugisaki and M Takanami, *J Biol Chem*, **264**, 5751–56 (1989).
17. H Ito, A Sadaoka, H Kotani, N Hiraoka and T Nakamura, *Nucleic Acids Res*, **18**, 3903–11 (1990).
18. V Siksnys, N Zareckaja, R Vaisvila, A Timinskas, P Stakenas, V Butkus and A Janulaitis, *Gene*, **142**, 1–8 (1994).
19. CH Taron, EM Van Cott, GG Wilson, LS Moran, BE Slatko, LJ Hornstra, JS Benner, RB Kucera and EP Guthrie, *Gene*, **155**, 19–25 (1995).
20. DC Stein, R Chien and HS Seifert, *J Bacteriol*, **174**, 4899–906 (1992).
21. TR Gingeras, L Greenough, I Schildkraut and RJ Roberts, *Nucleic Acids Res*, **9**, 4525–36 (1981).
22. A Athanasiadis, M Gregoriu, D Thanos, M Kokkinidis and J Papamatheakis, *Nucleic Acids Res*, **18**, 6434 (1990).
23. S Chong, FB Mersha, DG Comb, ME Scott, D Landry, LM Vence, FB Perler, J Benner, RB Kucera, CA Hirvonen, JJ Pelletier, H Paulus and MQ Xu, *Gene*, **192**, 271–81 (1997).
24. A Lagunavicius, G Sasnauskas, SE Halford and V Siksnys, *J Mol Biol*, **326**, 1051–64 (2003).
25. IB Vipond, GS Baldwin and SE Halford, *Biochemistry*, **34**, 697–704 (1995).
26. LS Beese and TA Steitz, *EMBO J*, **10**, 25–33 (1991).
27. CA Grosshans and TR Cech, *Biochemistry*, **28**, 6888–94 (1989).
28. H Viadiu and AK Aggarwal, *Nat Struct Biol*, **5**, 910–16 (1998).
29. T Mordasini, A Curioni and W Andreoni, *J Biol Chem*, **278**, 4381–84 (2003).
30. PW Tucker, EE Hazen Jr and FA Cotton, *Mol Cell Biochem*, **22**, 67–77 (1978).
31. ES Vanamee, S Santagata and AK Aggarwal, *J Mol Biol*, **309**, 69–78 (2001).
32. JA McClarin, CA Frederick, BC Wang, P Greene, HW Boyer, J Grable and JM Rosenberg, *Science*, **234**, 1526–41 (1986).
33. YC Kim, JC Grable, R Love, PJ Greene and JM Rosenberg, *Science*, **249**, 1307–9 (1990).
34. FK Winkler, DW Banner, C Oefner, D Tsernoglou, RS Brown, SP Heathman, RK Bryan, PD Martin, K Petratos and KS Wilson, *EMBO J*, **12**, 1781–95 (1993).

35 MP Thomas, RL Brady, SE Halford, RB Sessions and GS Baldwin, *Nucleic Acids Res*, **27**, 3438–45 (1999).

36 D Kostrewa and FK Winkler, *Biochemistry*, **34**, 683–96 (1995).

37 M Newman, T Strzelecka, LF Dorner, I Schildkraut and AK Aggarwal, *Structure*, **2**, 439–52 (1994).

38 M Newman, T Strzelecka, LF Dorner, I Schildkraut and AK Aggarwal, *Nature*, **368**, 660–64 (1994).

39 M Newman, T Strzelecka, LF Dorner, I Schildkraut and AK Aggarwal, *Science*, **269**, 656–63 (1995).

40 H Viadiu and AK Aggarwal, *Mol Cell*, **5**, 889–95 (2000).

41 X Cheng, K Balendiran, I Schildkraut and JE Anderson, *EMBO J*, **13**, 3927–35 (1994).

42 A Athanasiadis, M Vlassi, D Kotsifaki, PA Tucker, KS Wilson and M Kokkinidis, *Nat Struct Biol*, **1**, 469–75 (1994).

43 DA Wah, J Bitinaite, I Schildkraut and AK Aggarwal, *Proc Natl Acad Sci USA*, **95**, 10564–69 (1998).

44 DA Wah, JA Hirsch, LF Dorner, I Schildkraut and AK Aggarwal, *Nature*, **388**, 97–100 (1997).

45 CM Lukacs, R Kucera, I Schildkraut and AK Aggarwal, *Nat Struct Biol*, **7**, 134–40 (2000).

46 CM Lukacs, R Kucera, I Schildkraut and AK Aggarwal, *Nat Struct Biol*, **8**, 126–30 (2001).

47 MJ van der Woerd, JJ Pelletier, S Xu and AM Friedman, *Structure (Camb)*, **9**, 133–44 (2001).

48 Q Huai, JD Colandene, MD Topal and H Ke, *Nat Struct Biol*, **8**, 665–69 (2001).

49 Q Huai, JD Colandene, Y Chen, F Luo, Y Zhao, MD Topal and H Ke, *EMBO J*, **19**, 3110–18 (2000).

50 M Deibert, S Grazulis, G Sasnauskas, V Siksnys and R Huber, *Nat Struct Biol*, **7**, 792–99 (2000).

51 M Deibert, S Grazulis, A Janulaitis, V Siksnys and R Huber, *EMBO J*, **18**, 5805–16 (1999).

52 NC Horton, LF Dorner and JJ Perona, *Nat Struct Biol*, **9**, 42–47 (2002).

53 R Kovall and BW Matthews, *Science*, **277**, 1824–27 (1997).

54 RA Kovall and BW Matthews, *Proc Natl Acad Sci USA*, **95**, 7893–97 (1998).

55 C Ban and W Yang, *EMBO J*, **17**, 1526–34 (1998).

56 SE Tsutakawa, H Jingami and K Morikawa, *Cell*, **99**, 615–23 (1999).

57 SE Tsutakawa, T Muto, T Kawate, H Jingami, N Kunishima, M Ariyoshi, D Kohda, M Nakagawa and K Morikawa, *Mol Cell*, **3**, 621–28 (1999).

58 AB Hickman, Y Li, SV Mathew, EW May, NL Craig and F Dyda, *Mol Cell*, **5**, 1025–34 (2000).

59 AK Aggarwal, *Curr Opin Struct Biol*, **5**, 11–19 (1995).

60 M Carson, *Methods Enzymol*, **277**, 493–505 (1997).

61 CA Frederick, GJ Quigley, GA van der Marel, JH van Boom, AH Wang and A Rich, *J Biol Chem*, **263**, 17872–79 (1988).

62 X Cheng, K Balendiran, I Schildkraut and JE Anderson, *Gene*, **157**, 139–40 (1995).

63 JR Horton and X Cheng, *J Mol Biol*, **300**, 1049–56 (2000).

64 P Kraulis, *Acta Crystallogr, Sect D: Biol Crystallogr*, **24**, 946–50 (1991).

65 EA Merritt and DJ Bacon, *Methods Enzymol*, **277**, 505–24 (1997).

66 SE Halford, DT Bilcock, NP Stanford, SA Williams, SE Milsom, NA Gormley, MA Watson, AJ Bath, ML Embleton, DM Gowers, LE Daniels, SH Parry and MD Szczelkun, *Biochem Soc Trans*, **27**, 696–99 (1999).

67 DH Kruger, GJ Barcak, M Reuter and HO Smith, *Nucleic Acids Res*, **16**, 3997–4008 (1988).

68 M Mucke, R Lurz, P Mackeldanz, J Behlke, DH Kruger and M Reuter, *J Biol Chem*, **275**, 30631–37 (2000).

69 BK Baxter and MD Topal, *Biochemistry*, **32**, 8291–98 (1993).

70 LM Wentzell, TJ Nobbs and SE Halford, *J Mol Biol*, **248**, 581–95 (1995).

71 DT Bilcock, LE Daniels, AJ Bath and SE Halford, *J Biol Chem*, **274**, 36379–86 (1999).

72 V Siksnys, R Skirgaila, G Sasnauskas, C Urbanke, D Cherny, S Grazulis and R Huber, *J Mol Biol*, **291**, 1105–18 (1999).

73 MD Topal and M Conrad, *Nucleic Acids Res*, **21**, 2599–603 (1993).

74 K Jo and MD Topal, *Science*, **267**, 1817–20 (1995).

75 H Sugisaki and S Kanazawa, *Gene*, **16**, 73–78 (1981).

76 L Li, LP Wu and S Chandrasegaran, *Proc Natl Acad Sci USA*, **89**, 4275–79 (1992).

77 W Szybalski, *Gene*, **40**, 169–73 (1985).

78 YG Kim, J Smith, M Durgesha and S Chandrasegaran, *Biol Chem*, **379**, 489–95 (1998).

79 YG Kim, Y Shi, JM Berg and S Chandrasegaran, *Gene*, **203**, 43–49 (1997).

80 SC Kim, AJ Podhajska and W Szybalski, *Science*, **240**, 504–6 (1988).

81 YG Kim, L Li and S Chandrasegaran, *J Biol Chem*, **269**, 31978–82 (1994).

82 L Li and S Chandrasegaran, *Proc Natl Acad Sci USA*, **90**, 2764–68 (1993).

83 SC Schultz, GC Shields and TA Steitz, *Science*, **253**, 1001–7 (1991).

84 DS Waugh and RT Sauer, *Proc Natl Acad Sci USA*, **90**, 9596–600 (1993).

85 A Yonezawa and Y Sugiura, *Biochim Biophys Acta*, **1219**, 369–79 (1994).

86 J Bitinaite, DA Wah, AK Aggarwal and I Schildkraut, *Proc Natl Acad Sci USA*, **95**, 10570–75 (1998).

87 G Nardone and JG Chirikjian, *Gene Amplif Anal*, **5**, 147–84 (1987).

88 A Pingoud and A Jeltsch, *Nucleic Acids Res*, **29**, 3705–27 (2001).

89 M Conrad and MD Topal, *Proc Natl Acad Sci USA*, **86**, 9707–11 (1989).

90 CC Yang and MD Topal, *Biochemistry*, **31**, 9657–64 (1992).

91 K Jo and MD Topal, *Nucleic Acids Res*, **24**, 4171–75 (1996).

92 K Jo and MD Topal, *Biochemistry*, **35**, 10014–18 (1996).

93 CO Pabo and RT Sauer, *Annu Rev Biochem*, **61**, 1053–95 (1992).

94 JM Berger, D Fass, JC Wang and SC Harrison, *Proc Natl Acad Sci USA*, **95**, 7876–81 (1998).

95 LF Dorner and I Schildkraut, *Nucleic Acids Res*, **22**, 1068–74 (1994).

96 U Selent, T Ruter, E Kohler, M Liedtke, V Thielking, J Alves, T Oelgeschlager, H Wolfes, F Peters and A Pingoud, *Biochemistry*, **31**, 4808–15 (1992).

97 G Grabowski, A Jeltsch, H Wolfes, G Maass and J Alves, *Gene*, **157**, 113–18 (1995).

98 GS Baldwin, RB Sessions, SG Erskine and SE Halford, *J Mol Biol*, **288**, 87–103 (1999).

99. A Jeltsch, J Alves, H Wolfes, G Maass and A Pingoud, *Proc Natl Acad Sci USA*, **90**, 8499–503 (1993).

100. A Jeltsch, M Pleckaityte, U Selent, H Wolfes, V Siksnys and A Pingoud, *Gene*, **157**, 157–62 (1995).

101. NC Horton, KJ Newberry and JJ Perona, *Proc Natl Acad Sci USA*, **95**, 13489–94 (1998).

102. M Fuxreiter and R Osman, *Biochemistry*, **40**, 15017–23 (2001).

103. M Fothergill, MF Goodman, J Petruska and A Warshel, *J Am Chem Soc*, **117**, 11619–27 (1995).

104. J Aqvist and A Warshel, *Biochemistry*, **28**, 4680–89 (1989).

List of Contributors

Aneel K Aggarwal
Structural Biology Program, Mount Sinai School of Medicine, NY, USA

Thomas Ahrens
Department of Biophysical Chemistry and Structural Biology, Biozentrum, University of Basel, Basel, Switzerland

Ian M Armitage
Department of Biochemistry, Molecular Biology and Biophysics, University of Minnesota, Minneapolis, MN, USA

David S Auld
Center for Biochemical and Biophysical Sciences and Medicine and Department of Pathology, Harvard Medical School, Cambridge, MA, USA

Francesc X Aviles
Departament de Bioquímica i Biologia Molecular and Institut de Biotecnologia i Biomedicina, Universitat Autònoma de Barcelona, Spain

Ulrich Baumann
Departement für Chemie und Biochemie, University of Berne, Switzerland

Jozef Van Beeumen
Laboratory of Protein Biochemistry and Protein Engineering, Gent University, K.L. Ledeganckstraat, Belgium

Christian Betzel
University Hospital Hamburg-Eppendorf, Centre of Experimental Medicine, Institute of Biochemistry and Molecular Biology I, Hamburg, Germany

Klaus Bister
Institute of Biochemistry, University of Innsbruck, Austria

Wolfram Bode
Max-Planck-Institut für Biochemie, Abteilung für Strukturforschung, Martinsried, Germany

Rolf Boelens
Department of NMR Spectroscopy, Bijvoet Center for Biomolecular Research, Utrecht University, Utrecht, The Netherlands

Emma J Boswell
Biochemistry Department, University of Oxford, Oxford, OX1 3QU, England

Mark A Brown
Lund University, University Hospital, Malmö, Sweden

Robert L Campbell
Department of Biochemistry, Queen's University, Kingston, Ontario, Canada

Eila S Cedergren-Zeppezauer
Biochemistry, Center for Chemistry and Chemical Engineering, Lund University, Lund, Sweden

Paulette Charlier
Centre d'Ingénierie des Protéines, Université de Liège, Institut de Chimie B6, Sart Tilman, Belgium

Jonathan B Cooper
School of Biological Sciences, University of Southampton, UK

Senena Corbalán-García
Departmento de Bioquímica y Biología Molecular (A), Facultad de Veterinaria, Universidad de Murcia, Murcia, Spain

Miroslaw Cygler
Biotechnology Research Institute, National Research Council of Canada, Montreal, Quebec, Canada

Glenn E Dale
Morphochem AG, Basel, Switzerland

J H David Wu
Department of Chemical Engineering, University of Rochester, Rochester, NY, USA

Peter L Davies
Department of Biochemistry, Queen's University, Kingston, Ontario, Canada

Srikripa Devarakonda
Department of Pharmacology, University of Virginia, Charlottesville, USA
and
Interdisciplinary Program in Biophysics, University of Virginia, Charlottesville, USA

Otto Dideberg
LCM, Institut de Biologie Structurale, Rue Jules Horowitz, Grenoble-Cedex1, France

List of Contributors

Cyril Dominguez
Department of NMR Spectroscopy, Bijvoet Center for Biomolecular Research, Utrecht University, Utrecht, The Netherlands

A Kristina Downing
Biochemistry Department, University of Oxford, Oxford, OX1 3QU, England

David M Duda
Department of Biochemistry and Molecular Biology, University of Florida, Gainesville, FL, USA

Jürgen Engel
Department of Biophysical Chemistry and Structural Biology, Biozentrum, University of Basel, Basel, Switzerland

Peter T Erskine
School of Biological Sciences, University of Southampton, UK

Robert Etges
IBFB Pharma GmbH, Leipzig, Germany

Ignacio Fita
Departamento de Biología Estructural, IBMB (C.S.I.C.), Jordi Girona, Barcelona, Spain

Paula MD Fitzgerald
Merck Research Laboratories, NJ, USA

Gert E Folkers
Department of NMR Spectroscopy, Bijvoet Center for Biomolecular Research, Utrecht University, Utrecht, The Netherlands

Lloyd D Fricker
Department of Molecular Pharmacology, Albert Einstein College of Medicine, NY, USA

Günter Fritz
Mathematisch-Naturwissenschaftliche Sektion, Fachbereich Biologie, Universität Konstanz, Konstanz, Germany
and
Department of Pediatrics, Division of Clinical Chemistry and Biochemistry, University of Zürich, Zürich, Switzerland

Jean-Marie Frère
Centre for Protein Engineering, Liège University, Institut de Chimie B6, Sart Tilman, Belgium

Stéphane M Gagné
Université Laval, Québec, Canada

Nicolay Genov
Institute of Organic Chemistry, Bulgarian Academy of Sciences, Sofia, Bulgaria

Dessislava Georgieva
University Hospital Hamburg-Eppendorf, Centre of Experimental Medicine, Institute of Biochemistry and Molecular Biology I, Hamburg, Germany

Gary L Gilliland
Center for Advanced Research in Biotechnology of the University of Maryland, Biotechnology Institute, and the National Institute of Standards and Technology, Rockville, MD, USA

Colette Goffin
Centre for Protein Engineering, Liège University, Institut de Chimie B6, Sart Tilman, Belgium

Juan Carmelo Gómez-Fernández
Departmento de Bioquímica y Biología Molecular (A), Facultad de Veterinaria, Universidad de Murcia, Murcia, Spain

F Xavier Gomis-Rüth
Institut de Biologia Molecular de Barcelona, C.S.I.C.; C/Jordi Girona, 18–26; 08034 Barcelona, Spain

Claus W Heizmann
Department of Pediatrics, Division of Clinical Chemistry and Biochemistry, University of Zürich, Zürich, Switzerland

Osnat Herzberg
Center for Advanced Research in Biotechnology, University of Maryland Biotechnology Institute, MD, USA

Daniel Häussinger
Department of Biophysical Chemistry and Structural Biology, Biozentrum, University of Basel, Basel, Switzerland

Erhard Hohenester
Department of Biological Sciences, Imperial College, London, UK

Mitsuhiko Ikura
Department of Medical Biophysics, University of Toronto and Division of Molecular & Structural Biology, Ontario Cancer Institute, Canada

Tina M Iverson
Imperial College of Science, Technology and Medicine, Division of Biomedical Sciences, Wolfson Laboratories, London, SW7 2AZ, UK

List of Contributors

ChulHee Kang
School of Molecular Biosciences, Washington State University, Pullman, WA, USA

Evan R Kantrowitz
Department of Chemistry, Boston College, Merkert Chemistry Center, Chestnut Hill, MA, USA

Hiroshi Kawasaki
Division of Plant Genetic Engineering, Kihara Institute for Biological Research, Yokohama City University, Yokohama, Japan

Caroline Kisker
Department of Pharmacological Sciences, Center for Structural Biology, State University of New York at Stony Brook, Stony Brook, NY, USA

Robert Konrat
Institute of Theoretical Chemistry and Molecular Structural Biology, University of Vienna, Austria

Georg Kontaxis
Institute of Theoretical Chemistry and Molecular Structural Biology, University of Vienna, Austria

Robert H Kretsinger
Department of Biology, University of Virginia, Charlottesville, VA, USA

Nyoman D Kurniawan
Biochemistry Department, University of Oxford, Oxford, OX1 3QU, England

Tatiana G Kutateladze
Department of Pharmacology, University of Colorado Health Sciences Center, Denver, Colorado, USA

John H Laity
Division of Cell Biology & Biophysics, School of Biological Sciences, University of Missouri-Kansas City, Kansas City, Missouri, USA

William N Lipscomb
Department of Chemistry and Chemical Biology, Harvard University, Cambridge, USA

Hartmut Luecke
Department of Molecular Biology & Biochemistry, University of California, Irvine, CA, USA

Betsy L Lytle
Department of Biochemistry, Medical College of Wisconsin, Milwaukee, WI, USA
and
Department of Chemical Engineering, University of Rochester, Rochester, NY, USA

Wolfgang Maret
Center for Biochemical and Biophysical Sciences and Medicine, Harvard Medical School, Cambridge, MA, USA

Klaus Maskos
Max-Planck-Institut für Biochemie, Abteilung für Strukturforschung, Martinsried bei München, Germany

Brian W Matthews
Howard Hughes Medical Institute, Institute of Molecular Biology, University of Oregon, Eugene, OR, USA

Robert McKenna
Department of Biochemistry and Molecular Biology, University of Florida, Gainesville, FL, USA

Rob Meijers
Dana Farber Cancer Institute, Harvard Medical School, Boston, MA, USA

Peter Metcalf
School of Biological Sciences, University of Auckland, New Zealand

Satoshi Mitsuhashi
Kyowa Hakko Kogyo Co., Ltd, Tokyo, Japan

Tudor Moldoveanu
Department of Biochemistry, Queen's University, Kingston, Ontario, Canada

Susumu Nakayama
Department of Biochemistry, Nagasaki University Graduate School of Biomedical Sciences, Sakamoto, Nagasaki, Japan

Herbert Nar
Boehringer Ingelheim Pharma, Biberach, Germany

Wendy F Ochoa
Departamento de Biología Estructural, IBMB (C.S.I.C.), Jordi Girona, Barcelona, Spain

Christian Oefner
Morphochem AG, Basel, Switzerland

Michael Overduin
Department of Pharmacology, University of Colorado Health Sciences Center, Denver, CO, USA

Fraydoon Rastinejad
Department of Pharmacology, University of Virginia, Charlottesville, USA Interdisciplinary Program in Biophysics, University of Virginia, Charlottesville, USA
and
Department of Biochemistry and Molecular Genetics, University of Virginia, Charlottesville, USA

List of Contributors

Mahadev Ravi Kiran
Department of Pharmacology, University of Colorado Health Sciences Center, Denver, CO, USA

Han Remaut
Laboratory of Protein Biochemistry and Protein Engineering, Gent University, K.L. Ledeganckstraat, Belgium

Josep Rizo
University of Texas Southwestern Medical Center at Dallas, TX, USA

Christophe Romier
Institut de Génétique et de Biologie Moléculaire et Cellulaire, CNRS/INSERM/ULP, Illkirch, France

Anja Rosengarth
Department of Molecular Biology & Biochemistry, University of California, Irvine, CA, USA

Maria A Schumacher
Department of Biochemistry and Molecular Biology, Oregon Health and Science University, Portland, OR, USA

Tej P Singh
Department of Biophysics, All India Institute of Medical Sciences, New Delhi, India

G David Smith
The Hospital for Sick Children, Toronto, Ontario Hauptman-Woodward Medical Research Institute Inc., Buffalo, NY, USA

Johan Stenflo
Lund University, University Hospital, Malmö, Sweden

Jörg Stetefeld
Department of Biophysical Chemistry and Structural Biology, Biozentrum, University of Basel, Basel, Switzerland

Walter Stöcker
Institut für Zoophysiologie, Westfälische Wilhelms-Universität Münster, Germany

Norbert Sträter
Biotechnologisch-Biomedizinisches Zentrum der Universität Leipzig, Leipzig, Germany

Dietrich Suck
European Molecular Biology Laboratory, Heidelberg, Germany

Alexey Teplyakov
Center for Advanced Research in Biotechnology of the University of Maryland, Biotechnology Institute, and the National Institute of Standards and Technology, Rockville, MD, USA

Rupert Timpl
Max-Planck-Institut für Biochemie, Martinsried, Germany

Chikashi Toyoshima
Institute of Molecular and Cellular Biosciences, The University of Tokyo, Japan

Tomitake Tsukihara
Institute for Protein Research, Osaka University, Osaka, Japan

Éva Scheuring Vanamee
Structural Biology Program, Mount Sinai School of Medicine, NY, USA

Josep Vendrell
Departament de Bioquímica i Biologia Molecular and Institut de Biotecnologia i Biomedicina, Universitat Autònoma de Barcelona, Spain

Nuria Verdaguer
Departamento de Biología Estructural, IBMB (C.S.I.C.), Jordi Girona, Barcelona, Spain

Anne Volbeda
Institut de Biologie Structurale J.P. Ebel, CEA-CNRS-UJF, Grenoble, France

Brian F Volkman
Department of Biochemistry, Medical College of Wisconsin, Milwaukee, WI, USA

William I Weis
Departments of Structural Biology and of Molecular & Cellular Physiology, Stanford University School of Medicine, Stanford, CA, USA

Jean-Pierre Wery
Concurrent Pharmaceuticals, Inc., West Center Office Drive, Fort Washington, PA, USA

Eiki Yamashita
Institute for Protein Research, Osaka University, Osaka, Japan

Kyoko L Yap
Department of Medical Biophysics, University of Toronto and Division of Molecular & Structural Biology, Ontario Cancer Institute, Canada

Irene Yiallouros
Institut für Zoophysiologie, Westfälische Wilhelms-Universität Münster, Germany

Klaus Zangger
Institute of Chemistry, Organic and Bioorganic Chemistry, University of Graz, A-8010 Graz, Austria

Hong Zhang
University of Texas, Southwestern Medical Center at Dallas, TX, USA

PDB-Code Listing

1QRM Carbonic Anhydrase from Methanosarcina Thermophila; *T. M. Iverson, B. E. Alber, C. Kisker, J. G. Ferry, D. C. Rees*

1A12 Regulator of Chromosome Condensation (Rcc1) of Human; *L. Renault, N. Nassar, I. Vetter, J. Becker, M. Roth, A. Wittinghofer*

1A16 Aminopeptidase P from *E. Coli* with the Inhibitor Pro-Leu; *M. C. J. Wilce, C. S. Bond, P. E. Lilley, N. E. Dixon, H. C. Freeman, J. M. Guss*

1A1F Dsnr (Zif268 Variant) Zinc Finger-DNA Complex (Gacc Site); *M. Elrod-Erickson, T. E. Benson, C. O. Pabo*

1A1G Dsnr (Zif268 Variant) Zinc Finger-DNA Complex (Gcgt Site); *M. Elrod-Erickson, T. E. Benson, C. O. Pabo*

1A1H Qgsr (Zif268 Variant) Zinc Finger-DNA Complex (Gcac Site); *M. Elrod-Erickson, T. E. Benson, C. O. Pabo*

1A1I Radr (Zif268 Variant) Zinc Finger-DNA Complex (Gcac Site); *M. Elrod-Erickson, T. E. Benson, C. O. Pabo*

1A1J Radr (Zif268 Variant) Zinc Finger-DNA Complex (Gcgt Site); *M. Elrod-Erickson, T. E. Benson, C. O. Pabo*

1A1K Radr (Zif268 Variant) Zinc Finger-DNA Complex (Gacc Site); *M. Elrod-Erickson, T. E. Benson, C. O. Pabo*

1A1L Zif268 Zinc Finger-DNA Complex (Gcac Site); *M. Elrod-Erickson, T. E. Benson, C. O. Pabo*

1A25 C2 Domain from Protein Kinase C (β); *R. B. Sutton, S. R. Sprang*

1A4B Azurin Mutant with Met 121 Replaced by His, pH 6.5 Crystal Form, Data Collected at −180 Degrees Celsius; *A. Messerschmidt, L. Prade*

1A5T Crystal Structure of the Prime Subunit of the Clamp-Loader Complex of *Escherichia Coli* DNA Polymerase III; *B. Guenther, R. Onrust, A. Sali, M. O'Donnell, J. Kuriyan*

1A6Y Crystal Structure of Reverb-DNA Binding Complex; *Q. Zhao, S. Khorasanizadeh, F. Rastinejad*

1A71 Ternary Complex of an Active Site Double Mutant of Horse Liver Alcohol Dehydrogenase; *T. D. Colby, B. J. Bahnson, J. K. Chin, J. P. Klinman, B. M. Goldstein*

1A72 An Active-Site Double Mutant (Phe93→Trp, Val203→Ala) of Horse Liver Alcohol Dehydrogenase in complex with the Isosteric Nad Analog Cpad; *T. D. Colby, B. J. Bahnson, J. K. Chin, J. P. Klinman, B. M. Goldstein*

1A73 Intron-Encoded Endonuclease I-PpoI Complexed with DNA; *K. E. Flick, R. J. Monnat Junior, B. L. Stoddard*

1A7I Amino-Terminal Lim Domain from Quail Cysteine and Glycine-Rich Protein, NMR, Minimized Average Structure; *G. Kontaxis, R. Konrat, B. Kraeutler, R. Weiskirchen, K. Bister*

1A7T Metallo-β-Lactamase with Mes; *P. M. D. Fitzgerald, J. K. Wu, J. H. Toney*

1A8A Rat Annexin V Complexed with Glycerophosphoserine; *M. A. Swairjo, N. O. Concha, M. A. Kaetzel, J. R. Dedman, B. A. Seaton*

1A8H Methionyl-tRNA Synthetase from Thermus Thermophilus; *I. Sugiura, O. Nureki, Y. Ugaji, S. Kuwabara, B. Lober, R. Giege, D. Moras, S. Yokoyama, M. Konno*

1A8Y Crystal Structure of Calsequestrin from Rabbit Skeletal Muscle Sarcoplasmic Reticulum at 2.4 A Resolution; *S. Wang, W. R. Trumble, H. Liao, C. R. Wesson, A. K. Dunker, C. Kang*

1AAY Zif268 Zinc Finger-DNA Complex; *M. Elrod-Erickson, M. A. Rould, C. O. Pabo*

1ADB Crystallographic Studies of Isosteric NAD Analogues Bound to Alcohol Dehydrogenase: Specificity and Substrate Binding in Two Ternary Complexes; *H. Li, W. A. Hallows, J. S. Punzi, K. W. Pankiewicz, K. A. Watanabe, B. M. Goldstein*

1ADC Crystallographic Studies of Isosteric NAD Analogues Bound to Alcohol Dehydrogenase: Specificity and Substrate Binding in Two Ternary Complexes; *H. Li, W. A. Hallows, J. S. Punzi, K. W. Pankiewicz, K. A. Watanabe, B. M. Goldstein*

1ADF Crystallographic Studies of Two Alcohol Dehydrogenase-Bound Analogues of Thiazole-4-carboxamide adenine dinucleotide (TAD), the Active Anabolite of the Antitumor Agent Tiazofurin; *H. Li, W. A. Hallows, J. S. Punzi, V. E. Marquez, H. L. Carrell, K. W. Pankiewicz, K. A. Watanabe, B. M. Goldstein*

1ADG Crystallographic Studies of Two Alcohol Dehydrogenase-Bound Analogues of Thiazole-4-carboxamide adenine dinucleotide (TAD), the Active Anabolite of the Antitumor Agent Tiazofurin; *H. Li, W. A. Hallows, J. S. Punzi, V. E. Marquez, H. L. Carrell, K. W. Pankiewicz, K. A. Watanabe, B. M. Goldstein*

1AF0 Serratia Protease in Complex with Inhibitor; *U. Baumann*

1AGN X-ray Structure of Human Alcohol Dehydrogenase; *T. D. Hurley, P. Xie*

1AH2 Serine Protease Pb92 from Bacillus Alcalophilus, NMR, 18 Structures; *R. Boelens, D. Schipper, J. R. Martin, Y. Karimi-Nejad, F. Mulder, J. V. D. Zwan, M. Mariani*

1AH7 Phospholipase C from Bacillus Cereus; *R. Greaves*

1AIN Crystal Structure of Human Annexin I at 2.5-A Resolution; *S.-H. Kim*

1AJ5 Calpain Domain Vi Apo; *M. Cygler, P. Grochulski, H. Blanchard*

1AK0 P1 Nuclease in Complex with a Substrate Analog; *C. Romier, D. Suck*

1AKL Alkaline Protease from Pseudomonas Aeruginosa Ifo3080; *H. Miyatake, Y. Hata, T. Fujii, K. Hamada, K. Morihara, Y. Katsube*

1ALA Structure of Chicken Annexin V at 2.25-A Resolution; *D. A. Waller, M. C. Bewley, R. Huber*

1ALK Reaction Mechanism of Alkaline Phosphatase Based on Crystal Structures. Two-metal Ion Catalysis; *E. E. Kim, W. Wyckoff*

1ALV Calcium Bound Domain Vi of Porcine Calpain; *S. V. L. Narayana, G. Lin, D. Chattopadhyay, M. Maki*

PDB-Code Listing

1ALW Inhibitor and Calcium Bound Domain Vi of Porcine Calpain; *S. V. L. Narayana, G. Lin*

1AMP Crystal Structure of Aeromonas Proteolytica Aminopeptidase: A Prototypical Member of the Co-Catalytic Zinc Enzyme Family; *B. Chevrier, C. Schalk, H. D'Orchymont, J. M. Rondeau, D. Moras, C. Tarnus*

1ANN Annexin IV; *R. B. Sutton, S. R. Sprang*

1ANU Cohesin-2 Domain of the Cellulosome from Clostridium Thermocellum; *L. J. W. Shimon, S. Yaron, Y. Shoham, R. Lamed, E. Morag, E. A. Bayer, F. Frolow*

1ANW The Effect of Metal Binding on the Structure of Annexin V and Implications for Membrane Binding; *A. Lewit-Bentley, S. Morera, R. Huber, G. Bodo*

1ANX The Crystal Structure of a New High-calcium Form of Annexin V; *J. Sopkova, M. Renouard, A. Lewit-Bentley*

1AOH Single Cohesin Domain from the Scaffolding Protein Cipa of the Clostridium Thermocellum Cellulosome; *P. M. Alzari, G. Tavares*

1AOW Annexin IV; *G. Zanotti, G. Malpeli, F. Gliubich, C. Folli, M. Stoppini, L. Olivi, A. Savoia, R. Berni*

1AP4 Regulatory Domain of Human Cardiac Troponin C in the Calcium-saturated State, NMR, 40 Structures; *M. X. Li, L. Spyracopoulos, S. K. Sia, S. M. Gagne, M. Chandra, R. J. Solaro, B. D. Sykes*

1APO Three-dimensional Structure of the Apo form of the N-terminal EGF-like Module of Blood Coagulation Factor X as Determined by NMR Spectroscopy and Simulated Folding; *M. Ullner, M. Selander, E. Persson, J. Stenflo, T. Drakenberg, O. Teleman*

1APQ Structure of the EGF-like Module of Human C1R, NMR, 19 Structures; *B. Bersch, J.-F. Hernandez, D. Marion, G. J. Arlaud*

1ARD Structures of DNA-binding Mutant Zinc Finger Domains: Implications for DNA Binding; *R. C. Hoffman, R. X. Xu, S. J. Horvath, J. R. Herriott, R. E. Klevit*

1ARE Structures of DNA-binding Mutant Zinc Finger Domains: Implications for DNA Binding; *R. C. Hoffman, R. X. Xu, S. J. Horvath, J. R. Herriott, R. E. Klevit*

1ARF Structures of DNA-binding Mutant Zinc Finger Domains: Implications for DNA Binding; *R. C. Hoffman, R. X. Xu, S. J. Horvath, J. R. Herriott, R. E. Klevit*

1AST Structure of Astacin and Implications for Activation of Astacins and Zinc-ligation of Collagenases; *W. Bode, F. X. Gomis-Rueth, W. Stoecker*

1AT1 Crystal Structures of Phosphonoacetamide Ligated T and Phosphonoacetamide and Malonate Ligated R States of Aspartate Carbamoyltransferase at 2.8-A Resolution and Neutral pH; *J. E. Gouaux, W. N. Lipscomb*

1AUI Human Calcineurin Heterodimer; *C. R. Kissinger, H. E. Parge, D. R. Knighton, L. A. Pelletier, C. T. Lewis, A. Tempczyk, J. E. Villafranca*

1AUT Human Activated Protein C; *T. Mather, V. Oganessyan, P. Hof, W. Bode, R. Huber, S. Foundling, C. Esmon*

1AVC Bovine Annexin Vi (Calcium-Bound); *A. J. Avila-Sakar, C. E. Creutz, R. H. Kretsinger*

1AVH Crystal and Molecular Structure of Human Annexin V after Refinement. Implications for Structure, Membrane Binding and Ion Channel Formation of the Annexin Family of Proteins; *R. Huber, R. Berendes, A. Burger, M. Schneider, A. Karshikov, H. Luecke, J. Roemisch, E. Paques*

1AVN Human Carbonic Anhydrase II Complexed with the Histamine Activator; *F. Briganti, S. Mangani, P. Orioli, A. Scozzafava, G. Vernaglione, C. T. Supuran*

1AVR Crystal and Molecular Structure of Human Annexin V after Refinement. Implications for Structure, Membrane Binding and Ion Channel Formation of the Annexin Family of Proteins; *R. Huber, R. Berendes, A. Burger, M. Schneider, A. Karshikov, H. Luecke, J. Roemisch, E. Paques*

1AVS X-ray Crystallographic Study of Calcium-saturated N-terminal Domain of Troponin C; *N. C. J. Strynadka, M. N. G. James*

1AW5 Aminolevulinate Dehydratase from Saccharomyces Cerevisiae; *P. T. Erskine, J. B. Cooper, S. P. Wood*

1AXE Crystal Structure of the Active-site Mutant Phe93→Trp of Horse Liver Alcohol Dehydrogenase in Complex with Nad and Inhibitor Trifluoroethanol; *T. D. Colby, J. K. Chin, B. M. Goldstein*

1AXG Crystal Structure of the Val203→Ala Mutant of Liver Alcohol Dehydrogenase Complexed with Cofactor Nad and Inhibitor Trifluoroethanol Solved to 2.5 Angstrom Resolution; *T. D. Colby, J. K. Chin, B. J. Bahnson, B. M. Goldstein, J. P. Klinman*

1AXN The High-resolution Crystal Structure of Human Annexin III Shows Subtle Differences with Annexin V; *B. Favier-Perron, A. Lewit-Bentley, F. Russo-Marie*

1AZ9 Aminopeptidase P from E. Coli; *M. C. J. Wilce, C. S. Bond, P. E. Lilley, N. E. Dixon, H. C. Freeman, J. M. Guss*

1B3D Stromelysin-1; *L. Chen, T. J. Rydel, C. M. Dunaway, S. Pikul, K. M. Dunham, F. Gu, B. L. Barnett*

1B4E X-ray Structure of 5-Aminolevulinic Acid Dehydratase Complexed with the Inhibitor Levulinic Acid; *P. T. Erskine, J. B. Cooper, G. Lewis, P. Spencer, S. P. Wood, P. M. Shoolingin-Jordan*

1B4K High Resolution Crystal Structure of A Mg2-dependent 5-Aminolevulinic Acid Dehydratase; *N. Frankenberg, D. Jahn, D. W. Heinz*

1B59 Complex of Human Methionine Aminopeptidase-2 Complexed with Ovalicin; *S. Liu, J. C. Clardy*

1B66 6-Pyruvoyl Tetrahydropterin Synthase; *T. Ploom, B. Thoeny, J. Yim, S. Lee, H. Nar, W. Leimbacher, R. Huber, J. Richardson, G. Auerbach*

1B6Z 6-Pyruvoyl Tetrahydropterin Synthase; *T. Ploom, B. Thoeny, J. Yim, S. Lee, H. Nar, W. Leimbacher, R. Huber, J. Richardson, G. Auerbach*

1B8J Alkaline Phosphatase Complexed with Vanadate; *K. M. Holtz, B. Stec, E. R. Kantrowitz*

1BAM Structure of Restriction Endonuclease Bamhi Phased at 1.95 A Resolution by MAD Analysis; *A. K. Aggarwal, M. Newman*

1BBO High-resolution Solution Structure of the Double Cys2His2 Zinc Finger from the Human Enhancer Binding Protein MBP-1; *G. M. Clore, J. G. Omichinski, A. M. Gronenborn*

1BC2 Zn-dependent Metallo-beta-lactamase from Bacillus Cereus; *S. M. Fabiane, B. J. Sutton*

PDB-Code Listing

1BF9	N-terminal EGF-like Domain from Human Factor Vii, NMR, 23 Structures; *A. Muranyi, B. E. Finn, G. P. Gippert, S. Forsen, J. Stenflo, T. Drakenberg*
1BGY	Cytochrome Bc1 Complex from Bovine; *S. Iwata, J. W. Lee, K. Okada, J. K. Lee, M. Iwata, S. Ramaswamy, B. K. Jap*
1BHI	Structure of Transactivation Domain of Cre-Bp1/Atf-2, NMR, 20 Structures; *A. Nagadoi, K. Nakazawa, H. Uda, T. Maekawa, S. Ishii, Y. Nishimura*
1BIT	Structure of Anionic Salmon Trypsin in A 2Nd Crystal Form; *G. I. Berglund*
1BLL	X-ray Crystallographic Determination of the Structure of Bovine Lens Leucine Aminopeptidase Complexed with Amastatin: Formulation of a Catalytic Mechanism Featuring a Gem-diolate Transition State; *H. Kim, W. N. Lipscomb*
1BM6	Solution Structure of the Catalytic Domain of Human Stromelysin-1 Complexed to a Potent Non-peptidic Inhibitor, NMR, 20 Structures; *Y. Li, X. Zhang, R. Melton, V. Ganu, N. C. Gonnella*
1BMC	Structure of a Zinc Metallo-beta-lactamase from Bacillus Cereus; *A. Carfi, S. Pares, E. Duee, O. Dideberg*
1BMI	Metallo-beta-lactamase; *A. Carfi, E. Duee, O. Dideberg*
1BMO	Bm-40, Fs/Ec Domain Pair; *E. Hohenester, P. Maurer, R. Timpl*
1BN4	Carbonic Anhydrase II Inhibitor; *P. A. Boriack-Sjodin, S. Zeitlin, D. W. Christianson*
1BN5	Human Methionine Aminopeptidase 2; *S. Liu, J. Widom, C. W. Kemp, C. M. Crews, J. C. Clardy*
1BOA	Human Methionine Aminopeptidase 2 Complexed with Angiogenesis Inhibitor Fumagillin; *S. Liu, J. Widom, C. W. Kemp, C. M. Crews, J. C. Clardy*
1BOR	Transcription Factor Pml, a Proto-oncoprotein, NMR, 1 Representative Structure at pH 7.5, 30 °C, in the Presence of Zinc; *K. L. B. Borden, P. S. Freemont*
1BP6	Thymidylate Synthase R23I, R179T Double Mutant; *R. J. Morse, J. S. Finer-Moore, R. M. Stroud*
1BQQ	Crystal Structure of the Mt1-Mmp–Timp-2 Complex; *C. Fernandez-Catalan, W. Bode, R. Huber, D. Turk, J. J. Calvete, A. Lichte, H. Tschesche, K. Maskos*
1BT6	P11 (S100A10), Ligand of Annexin II in Complex with Annexin II N-terminus; *S. Rety, J. Sopkova, M. Renouard, D. Osterloh, V. Gerke, F. Russo-Marie, A. Lewit-Bentley*
1BTK	pH Domain and Btk Motif from Bruton'S Tyrosine Kinase Mutant R28C; *M. Hyvonen, M. Saraste*
1BTO	Horse Liver Alcohol Dehydrogenase Complexed to Nadh and (1S,3R)3-Butylthiolane 1-Oxide; *S. Ramaswamy, B. V. Plapp*
1BUU	One Ho3+ Form of Rat Mannose-Binding Protein A; *K. K.-S. Ng, S. Park-Snyder, W. I. Weis*
1BUV	Crystal Structure of the Mt1-Mmp-Timp-2 Complex; *C. Fernandez-Catalan, W. Bode, R. Huber, D. Turk, J. J. Calvete, A. Lichte, H. Tschesche, K. Maskos*
1BV4	Apo-Mannose-binding Protein-C; *K. K.-S. Ng, W. I. Weis*
1BVT	Metallo-beta-lactamase from Bacillus Cereus 569/H/9; *A. Carfi, E. Duee, O. Dideberg*
1BY4	Structure and Mechanism of the Homodimeric Assembly of the Rxr on DNA; *Q. Zhao, S. A. Chasse, S. Devarakonda, M. L. Sierk, B. Ahvazi, P. B. Sigler, F. Rastinejad*
1BYN	Solution Structure of the Calcium-bound First C2-Domain of Synaptotagmin I; *X. Shao, I. Fernandez, T. C. Sudhof, J. Rizo*
1C07	Structure of the Third Eps15 Homology Domain of Human Eps15; *J. L. Enmon, T. De Beer, M. Overduin*
1C23	E. Coli Methionine Aminopeptidase: Methionine Phosphonate Complex; *W. T. Lowther, Y. Zhang, P. B. Sampson, J. F. Honek, B. W. Matthews*
1CA1	Alpha-toxin from Clostridium Perfringens; *C. E. Naylor, A. K. Basak, R. W. Titball*
1CCF	How an Epidermal Growth Factor (EGF)-like Domain Binds Calcium. High Resolution NMR Structure of the Calcium Form of the NH2-terminal EGF-like Domain in Coagulation Factor X; *M. Selander-Sunnerhagen, M. Ullner, M. Persson, O. Teleman, J. Stenflo, T. Drakenberg*
1CDL	Target Enzyme Recognition by Calmodulin: 2.4 A Structure of a Calmodulin-peptide Complex; *W. E. Meador, F. A. Quiocho*
1CDO	Alcohol Dehydrogenase (E.C. 1.1.1.1) (Ee Isozyme) Complexed with Nicotinamide Adenine Dinucleotide (Nad), and Zinc; *S. Ramaswamy, H. Eklund*
1CEJ	Solution Structure of an EGF Module Pair from the Plasmodium Falciparum Merozoite Surface Protein 1; *W. D. Morgan, B. Birdsall, T. A. Frenkiel, M. G. Gradwell, P. A. Burghaus, S. E. H. Syed, C. Uthaipibull, A. A. Holder, J. Feeney*
1CF1	Arrestin from Bovine Rod Outer Segments; *J. A. Hirsch, C. Schubert, V. V. Gurevich, P. B. Sigler*
1CG2	Carboxypeptidase G2; *S. Rowsell, R. A. Pauptit, A. D. Tucker, R. G. Melton, D. M. Blow, P. Brick*
1CGE	Crystal Structures of Recombinant 19-kDa Human Fibroblast Collagenase Complexed to Itself; *B. Lovejoy, A. M. Hassell, M. A. Luther, D. Weigl, S. R. Jordan*
1CGL	Structure of the Catalytic Domain of Fibroblast Collagenase Complexed with an Inhibitor; *B. Lovejoy, A. Cleasby, A. M. Hassell, K. Longley, M. A. Luther, D. Weigl, G. Mcgeehan, A. B. Mcelroy, D. Drewry, M. H. Lambert, S. R. Jordan*
1CHC	Structure of the C3HC4 Domain by 1H-nuclear Magnetic Resonance Spectroscopy. A New Structural Class of Zinc-finger; *P. N. Barlow, R. D. Everett, B. Luisi*
1CIT	DNA-binding Mechanism of the Monomeric Orphan Nuclear Receptor Ngfi-B; *G. Meinke, P. B. Sigler*
1CJY	Human Cytosolic Phospholipase A2; *A. Dessen, J. Tang, H. Schmidt, M. Stahl, J. D. Clark, J. Seehra, W. S. Somers*
1CK7	Gelatinase A (Full-length); *E. Morgunova, A. Tuuttila, U. Bergmann, M. Isupov, Y. Lindqvist, G. Schneider, K. Tryggvason*
1CKX	Cystic Fibrosis Transmembrane Conductance Regulator: Solution Structures of Peptides Based on the Phe508 Region, the Most Common Site of Disease-causing -F508 Mutation; *M. A. Massiah, Y. H. Ko, P. L. Pedersen, A. S. Mildvan*
1CLL	Calmodulin Structure Refined at 1.7 A Resolution; *R. Chattopadhyaya, F. A. Quiocho*
1CTD	Determination of the Solution Structure of a Synthetic Two-site Calcium-binding Homodimeric Protein Domain by NMR Spectroscopy; *G. S. Shaw, B. D. Sykes*

PDB-Code Listing

1CTR   Drug Binding by Calmodulin: Crystal Structure of a Calmodulin-trifluoperazine Complex; *W. J. Cook, L. J. Walter, M. R. Walter*

1CXV   Structure of Recombinant Mouse Collagenase-3 (Mmp-13); *I. Botos, E. Meyer, S. M. Swanson, V. Lemaitre, Y. Eeckhout, E. F. Meyer*

1CXX   Mutant R122A of Quail Cysteine and Glycine-rich Protein, NMR, Minimized Structure; *K. Kloiber, R. Weiskirchen, B. Kraeutler, K. Bister, R. Konrat*

1D0Q   Structure of the Zinc-Binding Domain of Bacillus Stearothermophilus DNA Primase; *H. Pan, D. B. Wigley*

1D1S   Wild-type Human Sigma (Class IV) Alcohol Dehydrogenase; *P. T. Xie, T. D. Hurley*

1D1T   Mutant of Human sigma Alcohol Dehydrogenase with Leucine at Position 141; *P. T. Xie, T. D. Hurley*

1D2I   Crystal Structure of Restriction Endonuclease Bglii Complexed with DNA 16-Mer; *C. M. Lukacs, R. Kucera, I. Schildkraut, A. K. Aggarwal*

1D8D   Co-crystal Structure of Rat Protein Farnesyltransferase Complexed with a K-Ras4B Peptide Substrate and Fpp Analog at 2.0 A Resolution; *S. B. Long, P. J. Casey, L. S. Beese*

1DAN   Complex of Active Site Inhibited Human Blood Coagulation Factor Viia with Human Recombinant Soluble Tissue Factor; *D. W. Banner*

1DAQ   Solution Structure of The Type I Dockerin Domain from the Clostridium Thermocellum Cellulosome (Minimized Average Structure); *B. L. Lytle, B. F. Volkman, W. M. Westler, M. P. Heckman, J. H. D. Wu*

1DAV   Solution Structure of the Type I Dockerin Domain from the Clostridium Thermocellum Cellulosome (20 Structures); *B. L. Lytle, B. F. Volkman, W. M. Westler, M. P. Heckman, J. H. D. Wu*

1DBI   Crystal Structure of a Thermostable Serine Protease; *C. A. Smith, H. S. Toogood, H. M. Baker, R. M. Daniel, E. N. Baker*

1DCE   Crystal Structure of Rab Geranylgeranyltransferase from Rat Brain; *H. Zhang, M. C. Seabra, H. Deisenhofer*

1DD6   Imp-1 Metallo Beta-lactamase from Pseudomonas Aeruginosa in Complex with a Mercaptocarboxylate Inhibitor; *N. O. Concha, C. A. Janson, P. Rowling, S. Pearson, C. A. Cheever, B. P. Clarke, C. Lewis, M. Galleni, J. M. Frere, D. J. Payne, J. H. Bateson, S. S. Abdel-Meguid*

1DDZ   X-ray Structure of a Beta-carbonic Anhydrase from the Red Alga, Porphyridium Purpureum R-1; *S. Mitsuhashi, T. Mizushima, E. Yamashita, S. Miyachi, T. Tsukihara*

1DE5   L-rhamnose Isomerase; *I. P. Korndorfer, W. D. Fessner, B. W. Matthews*

1DEH   Crystallization of Human beta-1 Alcohol Dehydrogenase (15 Mg/Ml) in 50 Mm Sodium Phosphate (pH 7.5), 2.0 Mm Nad+ and 1 Mm 4-Iodopyrazole at 25 Oc, 13% (W/V) Peg 8000; *T. D. Hurley, G. J. Davis*

1DF0   Crystal Structure of M-Calpain; *C. M. Hosfield, J. S. Elce, P. L. Davies, Z. Jia*

1DFO   Crystal Structure at 2.4 Angstrom Resolution of E. Coli Serine Hydroxymethyltransferase in Complex with Glycine and 5-Formyl Tetrahydrofolate; *J. N. Scarsdale, G. Radaev, G. Kazanina, V. Schirch, H. T. Wright*

1DFS   Solution Structure of the Alpha-domain of Mouse Metallothionein-1; *K. Zangger, G. Oz, J. D. Otvos, I. M. Armitage*

1DFT   Solution Structure of the Beta-domain of Mouse Metallothionein-1; *K. Zangger, G. Oz, J. D. Otvos, I. M. Armitage*

1DGS   Crystal Structure of Nad+-dependent DNA Ligase from T. Filiformis; *J. Y. Lee, C. Chang, H. K. Song, S. T. Kwon, S. W. Suh*

1DJY   Phosphoinositide-specific Phospholipase C-delta-1 from Rat Complexed with Inositol-2,4,5-Trisphosphate; *L.-O. Essen, O. Perisic, R. L. Williams*

1DKV   Crystal Structure of Human M-calpain Form II; *S. Strobl, C. Fernandez-Catalan, M. Braun, R. Huber, H. Masumoto, K. Nakagawa, A. Irie, H. Sorimachi, G. Bourenkow, H. Bartunik, K. Suzuki, W. Bode*

1DMC   Three-dimensional Solution Structure of Callinectes Sapidus Metallothionein-1 Determined by Homonuclear and Heteronuclear Magnetic Resonance Spectroscopy; *S. S. Narula, M. Brouwer, Y. Hua, I. M. Armitage*

1DME   Three-dimensional Solution Structure of Callinectes Sapidus Metallothionein-1 Determined by Homonuclear and Heteronuclear Magnetic Resonance Spectroscopy; *S. S. Narula, M. Brouwer, Y. Hua, I. M. Armitage*

1DMO   Calmodulin, NMR, 30 Structures; *M. Zhang, T. Tanaka, M. Ikura*

1DMT   Structure of Human Neutral Endopeptidase Complexed with Phosphoramidon; *C. Oefner, A. D'Arcy, M. Hennig, F. K. Winkler, G. E. Dale*

1DSY   C2 Domain from Protein Kinase C (alpha) Complexed with Ca2+ and Phosphatidylserine; *N. Verdaguer, S. Corbalan-Garcia, W. F. Ochoa, I. Fita, J. C. Gomez-Fernandez*

1DSZ   Structure of the Rxr/Rar DNA-binding Domain Heterodimer in Complex with the Retinoic Acid Response Element Dr1; *F. Rastinejad, T. Wagner, Q. Zhao, S. Khorasanizadeh*

1DT7   Solution Structure of the C-terminal Negative Regulatory Domain of P53 in a Complex with Ca2+-bound S100B(Bb); *R. R. Rustandi, D. M. Baldisseri, D. J. Weber*

1DVI   Calpain Domain Vi with Calcium Bound; *M. Cygler, H. Blanchard, P. Grochulski*

1DVP   Crystal Structure of the Vhs and Fyve Tandem Domains of Hrs, a Protein Involved in Membrane Trafficking and Signal Transduction; *Y. Mao, A. Nickitenko, X. Duan, T. E. Lloyd, M. N. Wu, H. Bellen, F. A. Quiocho*

1DX5   Crystal Structure of the Thrombin-thrombomodulin Complex; *P. Fuentes-Prior, Y. Iwanaga, R. Huber, R. Pagila, G. Rumennik, M. Seto, J. Morser, D. R. Light, W. Bode*

1E0E   N-terminal Zinc-binding Hhcc Domain of HIV-2 Integrase; *A. P. A. M. Eijkelenboom, F. M. I. Van Den Ent, R. H. A. Plasterk, R. Kaptein, R. Boelens*

1E1A   Crystal Structure of Dfpase from Loligo Vulgaris; *J. Koepke, E. I. Scharff, G. Fritzsch, C. Luecke, H. Rueterjans*

1E3E   Mouse Class II Alcohol Dehydrogenase Complex with Nadh; *S. Svensson, J. O. Hoeoeg, G. Schneider, T. Sandalova*

1E3I	Mouse Class II Alcohol Dehydrogenase Complex with Nadh; *S. Svensson, J. O. Hoeoeg, G. Schneider, T. Sandalova*	1ESO	Monomeric Cu, Zn Superoxide Dismutase from *Escherichia Coli*; *A. Pesce, C. Capasso, A. Battistoni, S. Folcarelli, G. Rotilio, A. Desideri, M. Bolognesi*
1E3J	Ketose Reductase (Sorbitol Dehydrogenase) from Silverleaf Whitefly; *M. J. Banfield, M. E. Salvucci, E. N. Baker, C. A. Smith*	1EUL	Crystal Structure of Calcium ATPase with Two Bound Calcium Ions; *M. Nakasako, C. Toyoshima, H. Nomura, H. Ogawa*
1E3L	P47H Mutant of Mouse Class II Alcohol Dehydrogenase Complex with Nadh; *S. Svensson, J. O. Hoog, G. Schneider, T. Sandalova*	1EUT	Sialidase, Large 68Kd Form, Complexed with Galactose; *A. Gaskell, S. J. Crennell, G. L. Taylor*
1E4U	N-terminal Ring Finger Domain of Human Not-4; *H. Hanzawa, M. J. De Ruwe, T. K. Albert, P. C. Van Der Vliet, H. T. Timmers, R. Boelens*	1EV6	Structure of the Monoclinic Form of The M-Cresol/Insulin R6 Hexamer; *G. D. Smith, E. Ciszak, L. A. Magrum, W. A. Pangborn, R. H. Blessing*
1EB3	Yeast 5-Aminolaevulinic Acid Dehydratase 4,7-Dioxosebacic Acid Complex; *P. T. Erskine, L. Coates, R. Newbold, A. A. Brindley, F. Stauffer, S. P. Wood, M. J. Warren, J. B. Cooper, P. M. Shoolingin-Jordan, R. Neier*	1EVX	Apo Crystal Structure of The Homing Endonuclease, I-PpoI; *E. A. Galburt, M. S. Jurica, B. S. Chevalier, D. Erho, B. L. Stoddard*
		1EW2	Crystal Structure of a Human Phosphatase; *M. H. Le Du, T. Stigbrand, M. J. Taussig, A. Menez, E. A. Stura*
1ED8	Structure of *E. Coli* Alkaline Phosphatase ; *B. Stec, K. M. Holtz, E. R. Kantrowitz*	1EWC	Crystal Structure of Zn2+ Loaded Staphylococcal Enterotoxin H; *M. Hakansson, K. Petersson, H. Nilsson, G. Forsberg, P. Bjork, P. Antonsson, A. Svensson*
1ED9	Structure of *E. Coli* Alkaline Phosphatase without the Inorganic Phosphate at 1.75 A Resolution; *B. Stec, K. M. Holtz, E. R. Kantrowitz*		
1EDH	E-cadherin Domains 1 and 2 in Complex with Calcium; *B. Nagar, M. Overduin, M. Ikura, J. M. Rini*	1EWW	Solution Structure of Spruce Budworm Antifreeze Protein at 30 Degrees Celsius; *S. P. Graether, M. J. Kuiper, S. M. Gagne, V. K. Walker, Z. Jia, B. D. Sykes, P. L. Davies*
1EDM	Epidermal Growth Factor-like Domain from Human Factor Ix; *Z. Rao, P. Handford, M. Mayhew, V. Knott, G. G. Brownlee, D. Stuart*		
1EE2	The Structure of Steroid-active Alcohol Dehydrogenase at 1.54 A Resolution; *H. W. Adolph*	1EYW	Three-dimensional Structure of the Zinc-containing Phosphotriesterase with Bound Substrate Analog Triethylphosphate; *H. M. Holden, M. M. Benning, F. M. Raushel, S.-B. Hong*
1EE8	Crystal Structure of Mutm (Fpg) Protein from Thermus Thermophilus Hb8; *M. Sugahara, T. Mikawa, T. Kumasaka, M. Yamamoto, R. Kato, K. Fukuyama, Y. Inoue, S. Kuramitsu*	1F2I	Cocrystal Structure of Selected Zinc Finger Dimer Bound to DNA; *B. S. Wang, R. A. Grant, C. O. Pabo*
		1F4L	Crystal Structure of the *E. Coli* Methionyl-tRNA Synthetase Complexed with Methionine; *L. Serre, G. Verdon, T. Chonowski, N. Hervouet, C. Zelwer*
1EF4	Solution Structure of the Essential RNA Polymerase Subunit Rpb10 from Methanobacterium Thermoautotrophicum; *C. D. Mackereth, C. H. Arrowsmith, A. M. Edwards, L. P. Mcintosh*	1F4O	Crystal Structure of Grancalcin with Bound Calcium; *J. Jia, Q. Han, N. Borregaard, K. Lollike, M. Cygler*
		1F4Q	Crystal Structure of Apo Grancalcin; *J. Jia, Q. Han, N. Borregaard, K. Lollike, M. Cygler*
1EH2	Structure of the Second Eps15 Homology Domain of Human Eps15, NMR, 20 Structures; *T. De Beer, R. E. Carter, K. E. Lobel-Rice, A. Sorkin, M. Overduin*	1F7M	The First EGF-like Domain from Human Blood Coagulation Fvii, NMR, Minimized Average Structure; *Y.-H. Kao, G. F. Lee, Y. Wang, M. A. Starovasnik, R. F. Kelley, M. W. Spellman, L. Lerner*
1EKJ	The X-ray Crystallographic Structure of Beta Carbonic Anhydrase from the C3 Dicot Pisum Sativum; *M. S. Kimber, E. F. Pai*	1F8H	Structure of the Second Eps15 Homology Domain of Human Eps15 in Complex with Ptgssstnpfr; *T. De Beer, A. N. Hoofnagle, J. L. Enmon, R. C. Bowers, M. Yamabhai, B. K. Kay, M. Overduin*
1EMN	NMR Study of a Pair of Fibrillin Ca2+ Binding Epidermal Growth Factor-like Domains, Minimized Average Structure; *A. K. Downing, I. D. Campbell, P. A. Handford*		
		1FAK	Human Tissue Factor Complexed with Coagulation Factor Viia Inhibited with a Bpti-mutant; *E. Zhang, R. St Charles, A. Tulinsky*
1EN7	Endonuclease Vii (Endovii) from Phage T4; *H. C. A. Raaijmakers, O. Vix, I. Toro, D. Suck*		
1EQW	Crystal Structure of Salmonella Typhimurium Cu, Zn Superoxide Dismutase; *A. Pesce, A. Battistoni, M. E. Stroppolo, F. Polizio, M. Nardini, J. S. Kroll, P. R. Langford, P. O'Neill, M. Sette, A. Desideri, M. Bolognesi*	1FAR	Raf-1 Cysteine Rich Domain, NMR, Minimized Average Structure; *H. R. Mott, S. L. Campbell*
		1FB1	Crystal Structure of Human GTP Cyclohydrolase I; *G. Auerbach, A. Herrmann, A. Bracher, G. Bader, M. Gutlich, M. Fischer, M. Neukamm, H. Nar, M. Garrido-Franco, J. Richardson, R. Huber, A. Bacher*
1ERJ	Crystal Structure of the C-terminal Wd40 Domain of Tup1; *E. R. Sprague, M. J. Redd, A. D. Johnson, C. Wolberger*	1FBL	Structure of Full-length Porcine Synovial Collagenase Reveals a C-terminal Domain Containing a Calcium-linked, Four-bladed Beta-propeller; *J. Li, P. Brick, D. M. Blow*
1ESD	The Molecular Mechanism of Enantiorecognition by Esterases; *Y. Wei, J. L. Schottel, U. Derewenda, L. Swenson, S. Patkar, Z. S. Derewenda*		
		1FBV	Structure of a Cbl-Ubch7 Complex: Ring Domain Function in Ubiquitin-protein Ligases; *N. Zheng, P. Wang, P. D. Jeffrey, N. P. Pavletich*
1ESL	Insight into E-selectin/Ligand Interaction from the Crystal Structure and Mutagenesis of the lec/EGF Domains; *B. J. Graves, R. L. Crowther*	1FBX	Crystal Structure of Zinc-containing *E. Coli* GTP Cyclohydrolase I; *G. Auerbach, A. Herrmann,*

PDB-Code Listing

	A. Bracher, A Bader, M. Gutlich, M. Fischer, M. Neukamm, H. Nar, M. Garrido-Franco, J. Richardson, R. Huber, A. Bacher
1FF5	Structure of E-cadherin Double Domain; *O. Pertz, D. Bozic, A. W. Koch, C. Fauser, A. Brancaccio, J. Engel*
1FFM	The First Egf-like Domain from Human Blood Coagulation Fvii (Fucosylated at Ser-60), NMR, Minimized Average Structure; *Y.-H. Kao, G. F. Lee, Y. Wang, M. A. Starovasnik, R. F. Kelley, M. W. Spellman, L. Lerner*
1FFY	Insights into Editing from an Ile-tRNA Synthetase Structure with tRNA(Ile) and Mupirocin; *L. F. Silvian, J. Wang, T. A. Steitz*
1FI6	Solution Structure of the Reps1 Eh Domain; *S. Kim, J. D. Baleja*
1FIF	N-acetylgalactosamine-selective Mutant of Mannose-binding Protein-A (Qpdwg-Hdrpy); *H. Feinberg, D. Torgersen, K. Drickamer, W. I. Weis*
1FIH	N-acetylgalactosamine Binding Mutant of Mannose-binding Protein A (Qpdwg-Hdrpy), Complex with N-acetylgalactosamine; *H. Feinberg, D. Torgerson, K. Drickamer, W. I. Weis*
1FIU	Tetrameric Restriction Endonuclease Ngomiv in Complex with Cleaved DNA; *M. Deibert, S. Grazulis, G. Sasnauskas, V. Siksnys, R. Huber*
1FJM	Protein Serine/Threonine Phosphatase-1 (alpha Isoform, Type I) Complexed with Microcystin-Lr Toxin; *J. Goldberg, A. C. Nairn, J. Kuriyan*
1FME	Solution Structure of Fsd-Ey, a Novel Peptide Assuming a Beta-beta-alpha Fold; *C. A. Sarisky, S. L. Mayo*
1FNO	Peptidase T (Tripeptidase); *K. Hakansson, C. G. Miller*
1FOK	Structure of Restriction Endonuclease Foki Bound to DNA; *D. A. Wah, J. A. Hirsch, L. F. Dorner, I. Schildkraut, A. K. Aggarwal*
1FPO	Hsc20 (Hscb), A J-type Co-chaperone from *E. Coli*; *J. R. Cupp-Vickery, L. E. Vickery*
1FSD	Full Sequence Design 1 (Fsd-1) of Beta-beta-alpha Motif, NMR, 41 Structures; *B. I. Dahiyat, S. L. Mayo*
1FSV	Full Sequence Design 1 (Fsd-1) of Beta-beta-alpha Motif, NMR, Minimized Average Structure; *B. I. Dahiyat, S. L. Mayo*
1FT1	Crystal Structure of Protein Farnesyltransferase at 2.25 Angstroms Resolution; *L. S. Beese, L.-W. Park*
1FWY	Crystal Structure of N-acetylglucosamine 1-Phosphate Uridyltransferase Bound to Udp-Glcnac; *K. Brown, F. Pompeo, S. Dixon, D. Mengin-Lecreulx, C. Cambillau, Y. Bourne*
1G0E	Site-specific Mutant (His64 Replaced with Ala) of Human Carbonic Anhydrase II Complexed with 4-Methylimidazole; *D. Duda, R. Mckenna*
1G1K	Cohesin Module from the Cellulosome of Clostridium Cellulolyticum; *S. Spinelli, H.-P. Fierobe, A. Belaich, J.-P. Belaich, B. Henrissat, C. Cambillau*
1G25	Solution Structure of the N-terminal Domain of the Human Tfiih Mat1 Subunit; *V. Gervais, E. Wasielewski, D. Busso, A. Poterszman, J. M. Egly, J. C. Thierry, B. Kieffer*
1G2D	Structure of a Cys2His2 Zinc Finger/Tata Box Complex (Clone #2); *S. A. Wolfe, R. A. Grant, M. Elrod-Erickson, C. O. Pabo*
1G2F	Structure of a Cys2His2 Zinc Finger/Tata Box Complex (Tatazf; Clone #6); *S. A. Wolfe, R. A. Grant, M. Elrod-Erickson, C. O. Pabo*
1G4Y	1.60 A Crystal Structure of the Gating Domain from Small Conductance Potassium Channel Complexed with Calcium-calmodulin; *M. A. Schumacher, A. Rivard, H. P. Bachinger, J. P. Adelman*
1G5B	Bacteriophage Lambda Ser/Thr Protein Phosphatase; *W. C. Voegtli, D. J. White, N. J. Reiter, F. Rusnak, A. C. Rosenzweig*
1G5C	Crystal Structure of the 'Cab' Type Beta Class Carbonic Anhydrase from Methanobacterium Thermoautotrophicum; *P. Strop, K. S. Smith, T. M. Iverson, J. G. Ferry, D. C. Rees*
1G5N	Annexin V Complex with Heparin Oligosaccharides; *I. Capila, M. J. Heraiz, Y. D. Mo, T. R. Mealy, B. Campos, J. R. Dedman, R. J. Linhardt, B. A. Seaton*
1G71	Crystal Structure of Pyrococcus Furiosus DNA Primase; *M. A. Augustin, R. Huber, J. T. Kaiser*
1G7A	1.2 A Structure of T3R3 Human Insulin at 100 K; *G. D. Smith, W. A. Pangborn, R. H. Blessing*
1G7B	1.3 A Structure of T3R3 Human Insulin at 100 K; *G. D. Smith, W. A. Pangborn, R. H. Blessing*
1GA4	Crystal Structure Analysis of Pscp (Pseudomonas Serine-Carboxyl Proteinase) Complexed with Inhibitor Pseudoiodotyrostatin (This Enzyme Renamed 'Sedolisin' In 2003); *A. Wlodawer, M. Li, Z. Dauter, A. Gustchina, K. Uchida*
1GA5	Crystal Structure of the Orphan Nuclear Receptor Rev-Erb(alpha) DNA-binding Domain Bound to its Cognate Response Element; *M. L. Sierk, Q. Zhao, F. Rastinejad*
1GA6	Crystal Structure Analysis of Pscp (Pseudomonas Serine-Carboxyl Proteinase) Complexed with a Fragment of Tyrostatin (This Enzyme Renamed 'Sedolisin' In 2003); *A. Wlodawer, M. Li, Z. Dauter, A. Gustchina, K. Uchida*
1GAX	Crystal Structure of Thermus Thermophilus Valyl-tRNA Synthetase Complexed with tRNA(Val) and Valyl-adenylate Analogue; *S. Fukai, O. Nureki, S. Sekine, A. Shimada, J. Tao, D. G. Vassylyev, S. Yokoyama*
1GCI	The 0.78 Angstroms Structure of a Serine Protease – Bacillus Lentus Subtilisin; *R. Bott, P. Kuhn*
1GEN	C-terminal Domain of Gelatinase A; *A. M. Libson, A. G. Gittis, I. E. Collier, B. L. Marmer, G. G. Goldberg, E. E. Lattman*
1GJP	Schiff-base Complex of Yeast 5-Aminolaevulinic Acid Dehydratase with 4-Oxosebacic Acid; *P. T. Erskine, L. Coates, R. Newbold, A. A. Brindley, S. P. Wood, M. J. Warren, J. B. Cooper, P. M. Shoolingin-Jordan, R. Neier*
1GJY	The X-ray Structure of the Sorcin Calcium Binding Domain (Scbd) Provides Insight into the Phosphorylation and Calcium Dependent Processess; *A. Ilari, K. A. Johnson, V. Nastopoulos, D. Tsernoglou, E. Chiancone*
1GKC	Mmp9-inhibitor Complex; *S. Rowsell, R. A. Pauptit*
1GKD	Mmp9 Active Site Mutant-inhibitor Complex; *S. Rowsell, R. A. Pauptit*
1GKP	D-hydantoinase (Dihydropyrimidinase) from Thermus Sp. In Space Group C2221; *J. Abendroth, K. Niefind, D. Schomburg*
1GKR	L-hydantoinase (Dihydropyrimidinase) from Arthrobacter Aurescens; *J. Abendroth, K. Niefind, D. Schomburg*

1GLU	Crystallographic Analysis of the Interaction of the Glucocorticoid Receptor with DNA; *B. F. Luisi, W. X. Xu, P. B. Sigler*
1GO7	The Metzincin'S Methionine: Prtc M226C-E189K Double Mutant; *T. Hege*
1GO8	The Metzincin'S Methionine: Prtc M226L Mutant; *U. Baumann*
1GT9	High Resolution Crystal Structure of a Thermostable Serine-carboxyl Type Proteinase, Kumamolisin (Kscp); *M. Comellas-Bigler, P. Fuentes-Prior, K. Maskos, R. Huber, H. Oyama, K. Uchida, B. M. Dunn, K. Oda, W. Bode*
1GTQ	6-Pyruvoyl Tetrahydropterin Synthase; *H. Nar, R. Huber, C. W. Heizmann, B. Thoeny, D. Buergisser*
1GUP	Structure of Nucleotidyltransferase Complexed with Udp-Galactose; *J. B. Thoden, I. Rayment, H. Holden*
1GXD	Prommp-2/Timp-2 Complex; *E. Morgunova, A. Tuuttila, U. Bergmann, K. Tryggvason*
1GYG	R32 Closed Form of Alpha-toxin from Clostridium Perfringens Strain Cer89L43; *A. K. Basak, J. T. Eaton, R. W. Titball*
1GYH	Structure of D158A Cellvibrio Cellulosa Alpha-L-arabinanase Mutant; *D. Nurizzo, J. P. Turkenburg, S. J. Charnock, S. M. Roberts, E. J. Dodson, V. A. Mckie, E. J. Taylor, H. J. Gilbert, G. J. Davies*
1GYT	E. Coli Aminopeptidase A (Pepa); *N. Straeter*
1GZG	Complex of a Mg2-Dependent Porphobilinogen Synthase from Pseudomonas Aeruginosa (Mutant D139N) with 5-Fluorolevulinic Acid; *F. Frere, W.-D. Schubert, F. Stauffer, N. Frankenberg, R. Neier, D. Jahn, D. W. Heinz*
1H7N	Schiff-base Complex of Yeast 5-Aminolaevulinic Acid Dehydratase with Laevulinic Acid at 1.6 Å Resolution; *P. T. Erskine, R. Newbold, A. A. Brindley, S. P. Wood, P. M. Shoolingin-Jordan, M. J. Warren, J. B. Cooper*
1H7O	Schiff-base Complex of Yeast 5-Aminolaevulinic Acid Dehydratase with 5-Aminolaevulinic Acid at 1.7 Å Resolution; *P. T. Erskine, R. Newbold, A. A. Brindley, S. P. Wood, P. M. Shoolingin-Jordan, M. J. Warren, J. B. Cooper*
1H7P	Schiff-base Complex of Yeast 5-Aminolaevulinic Acid Dehydratase with 4-Keto-5-Amino-Hexanoic (Kah) at 1.64 Å Resolution; *P. T. Erskine, R. Newbold, A. A. Brindley, S. P. Wood, P. M. Shoolingin-Jordan, M. J. Warren, J. B. Cooper*
1H7R	Schiff-Base Complex of Yeast 5-Aminolaevulinic Acid Dehydratase with Succinylacetone at 2.0 Å Resolution; *P. T. Erskine, R. Newbold, A. A. Brindley, S. P. Wood, P. M. Shoolingin-Jordan, M. J. Warren, J. B. Cooper*
1HC7	Prolyl-tRNA Synthetase from Thermus Thermophilus; *A. Yaremchuk, M. Tukalo, S. Cusack*
1HCP	DNA Recognition by the Estrogen-receptor – from Solution to the Crystal; *J. W. R. Schwabe, D. Rhodes, D. Neuhaus*
1HCQ	The Crystal Structure of the Estrogen Receptor DNA-binding Domain Bound to DNA: How Receptors Discriminate Between their Response Elements; *J. W. R. Schwabe, L. Chapman, J. T. Finch, D. Rhodes*
1HCW	23-residue Designed Metal-free Peptide Based on the Zinc Finger Domains, NMR, 35 Structures; *B. Imperiali, M. Struthers, R. P. Cheng*
1HDX	Structures of Three Human Beta-alcohol Dehydrogenase Variants. Correlations with their Functional Differences; *T. D. Hurley, L. M. Amzel*
1HDY	Structures of Three Human Beta-alcohol Dehydrogenase Variants. Correlations with their Functional Differences; *T. D. Hurley, L. M. Amzel*
1HDZ	Structures of Three Human Beta-alcohol Dehydrogenase Variants. Correlations with their Functional Differences; *T. D. Hurley, L. M. Amzel*
1HET	Atomic X-ray Structure of Liver Alcohol Dehydrogenase Containing a Hydroxide Adduct to Nadh; *R. Meijers, R. J. Morris, H. W. Adolph, A. Merli, V. S. Lamzin, E. S. Cedergen-Zeppezauer*
1HEU	Atomic X-ray Structure of Liver Alcohol Dehydrogenase Containing Cadmium and a Hydroxide Adduct to Nadh; *R. Meijers, R. J. Morris, H. W. Adolph, A. Merli, V. S. Lamzin, E. S. Cedergen-Zeppezauer*
1HFC	1.56 Å Structure of Mature Truncated Human Fibroblast Collagenase; *J. C. Spurlino, D. L. Smith*
1HFS	Crystal Structure of the Catalytic Domain of Human Fibroblast Stromelysin-1 Inhibited with the N-Carboxy-Alkyl Inhibitor L-764,004; *J. W. Becker*
1HI9	Zn-dependent D-aminopeptidase Dppa from Bacillus Subtilis, a Self-compartmentalizing Protease; *H. Remaut, C. Bompard-Gilles, C. Goffin, J. M. Frere, J. Van Beeumen*
1HJ7	NMR Study of a Pair of Ldl Receptor Ca2+ Binding Epidermal Growth Factor-like Domains, 20 Structures; *S. Saha, P. A. Handford, I. D. Campbell, A. K. Downing*
1HJK	Alkaline Phosphatase Mutant H331Q; *J. E. Murphy, B. Stec, L. Ma, E. R. Kantrowitz*
1HLD	Structures of Horse Liver Alcohol Dehydrogenase Complexed with NAD+ and Substituted Benzyl Alcohols; *S. Ramaswamy, H. Eklund, B. V. Plapp*
1HM6	X-ray Structure of Full-length Annexin 1; *A. Rosengarth, V. Gerke, H. Luecke*
1HO5	5′-Nucleotidase (E. Coli) in Complex with Adenosine and Phosphate; *T. Knoefel, N. Straeter*
1HP1	5′-Nucleotidase (Open Form) Complex with ATP; *T. Knoefel, N. Straeter*
1HPU	5′-Nucleotidase (Closed Form), Complex with Ampcp; *T. Knoefel, N. Straeter*
1HQM	Crystal Structure of Thermus Aquaticus Core RNA Polymerase-Includes Complete Structure with Side-chains (Except For Disordered Regions)-Further Refined from Original Deposition-Contains Additional Sequence Information; *L. Minakhin, S. Bhagat, A. Brunning, E. A. Campbell, S. A. Darst, R. H. Ebright, K. Severinov*
1HQV	Structure of Apoptosis-linked Protein Alg-2; *J. Jia, S. Tarabykina, C. Hansen, M. Berchtold, M. Cygler*
1HR8	Yeast Mitochondrial Processing Peptidase Beta-E73Q Mutant Complexed with Cytochrome C Oxidase IV Signal Peptide; *A. B. Taylor, B. S. Smith, S. Kitada, K. Kojima, H. Miyaura, Z. Otwinowski, A. Ito, J. Deisenhofer*
1HRA	The Solution Structure of the Human Retinoic Acid Receptor-DNA-binding Domain; *R. M. A. Knegtel, M. Katahira, J. G. Schilthuis, A. M. J. J. Bonvin, R. Boelens, D. Eib, P. T. Van Der Saag, R. Kaptein*
1HSO	Human Alpha-alcohol Dehydrogenase (Adh1A); *M. S. Niederhut, B. J. Gibbons, S. Perez-Miller, T. D. Hurley*

PDB-Code Listing

1HSZ Human Beta-1 Alcohol Dehydrogenase (Adh1B*1); *M. S. Niederhut, B. J. Gibbons, S. Perez-Miller, T. D. Hurley*

1HT0 Human Gamma-2 Alcohol Dehydrogenase; *M. S. Niederhut, B. J. Gibbons, S. Perez-Miller, T. D. Hurley*

1HTB Crystallization of Human Beta-3 Alcohol Dehydrogenase (10 Mg/Ml) in 100 Mm Sodium Phosphate (pH 7.5), 7.5 Mm Nad+ and 1 Mm 4-Iodopyrazole at 25 °C; *T. D. Hurley, G. J. Davis*

1HTO Crystallographic Structure of a Relaxed Glutamine Synthetase from Mycobacterium Tuberculosis; *H. S. Gill, D. Eisenberg*

1HV5 Crystal Structure of the Stromelysin-3 (Mmp-11) Catalytic Domain Complexed with a Phosphinic Inhibitor; *A. L. Gall, M. Ruff, R. Kannan, P. Cuniasse, A. Yiotakis, V. Dive, M. C. Rio, P. Basset, D. Moras*

1HXN 1.8 A crystal structure of the C-terminal Domain of Rabbit Serum Haemopexin; *H. R. Faber, E. N. Baker*

1HYI Solution Structure of The Eea1 Fyve Domain Complexed with Inositol 1,3-Bisphosphate; *T. Kutateladze, M. Overduin*

1HYT Redetermination and Refinement of the Complex of Benzylsuccinic Acid with Thermolysin and its Relation to the Complex with Carboxypeptidase A; *A. C. Hausrath, B. W. Matthews*

1I0U Solution Structure and Backbone Dynamics of a Concatemer of Egf-Homology Modules of The Human Low Density Lipoprotein Receptor; *N. D. Kurniawan, K. Aliabadizadeh, I. M. Brereton, P. A. Kroon, R. Smith*

1I4A Crystal Structure of Phosphorylation-mimicking Mutant T6D of Annexin IV; *M. A. Kaetzel, Y. D. Mo, T. R. Mealy, B. Campos, W. Bergsma-Schutter, A. Brisson, J. R. Dedman, B. A. Seaton*

1I6N 1.8 A Crystal Structure of Ioli Protein with a Binding Zinc atom; *R. Zhang, I. Dementiva, F. Collart, E. Quaiterandall, A. Joachimiak, R. Alkire, N. Maltsev, O. Korolev, L. Dieckman*

1I6P Crystal Structure of E. Coli Beta-carbonic Anhydrase (Ecca); *J. D. Cronk, J. A. Endrizzi, M. R. Cronk, J. W. O'Neill, K. Y. J. Zhang*

1I8J Crystal Structure of Porphobilinogen Synthase Complexed with the Inhibitor 4,7-Dioxosebacic Acid; *J. Kervinen, E. K. Jaffe, F. Stauffer, R. Neier, A. Wlodawer, A. Zdanov*

1IAA Crystal Structures, Spectroscopic Features, and Catalytic Properties of Cobalt(II), Copper(II), Nickel(II), and Mercury(II) Derivatives of the Zinc Endopeptidase Astacin. A Correlation of Structure and Proteolytic Activity; *F.-X. Gomis-Rueth, W. Stoecker, W. Bode*

1IAB Crystal Structures, Spectroscopic Features, and Catalytic Properties of Cobalt(II), Copper(II), Nickel(II), and Mercury(II) Derivatives of the Zinc Endopeptidase Astacin. A Correlation of Structure and Proteolytic Activity; *F.-X. Gomis-Rueth, W. Stoecker, W. Bode*

1IAC Refined 1.8 A X-ray Crystal Structure of Astacin, a Zinc-endopeptidase from the Crayfish Astacus Astacus L. Structure determination, Refinement, Molecular Structure and Comparison with Thermolysin; *F.-X. Gomis-Rueth, W. Stoecker, W. Bode*

1IAE Crystal Structures, Spectroscopic Features, and Catalytic Properties of Cobalt(II), Copper(II), Nickel(II), and Mercury(II) Derivatives of the Zinc Endopeptidase Astacin. A Correlation of Structure and Proteolytic Activity; *F. Grams, W. Stoecker, W. Bode*

1IAW Crystal Structure of Naei Complexed with 17Mer DNA; *Q. Huai, J. D. Colandene, M. D. Topal, H. Ke*

1IC6 Structure of a Serine Protease Proteinase K from Tritirachium Album Limber at 0.98 A Resolution; *C. Betzel, S. Gourinath, P. Kumar, P. Kaur, M. Perbandt, S. Eschenburg, T. P. Singh*

1ILE Isoleucyl-tRNA Synthetase; *O. Nureki, D. G. Vassylyev, M. Tateno, A. Shimada, T. Nakama, S. Fukai, M. Konno, P. Schimmel, S. Yokoyama*

1IOU Solution Structure of Ykt6P (1-140); *H. Tochio, M. M. K. Tsui, D. K. Banfield, M. Zhang*

1IQ3 Solution Structure of The Eps15 Homology Domain of a Human Pob1; *S. Koshiba, T. Kigawa, J. Iwahara, A. Kikuchi, S. Yokoyama*

1IS8 Crystal Structure of Rat Gtpchi/Gfrp Stimulatory Complex Plus Zn; *N. Maita, K. Okada, K. Hatakeyama, T. Hakoshima*

1ITU Human Renal Dipeptidase Complexed with Cilastatin; *Y. Nitanai, Y. Satow, H. Adachi, M. Tsujimoto*

1ITV Dimeric Form of the Haemopexin Domain of Mmp9; *H. Cha, E. Kopetzki, R. Huber, M. Lanzendoerfer, H. Brandstetter*

1IWO Crystal Structure of the Sr Ca2+-ATPase in the Absence of Ca2+; *C. Toyoshima, H. Nomura*

1IXA The Three-dimensional Structure of the First EGF-like Module of Human Factor IX: Comparison with EGF and TGF-alpha; *M. Baron, D. G. Norman, T. S. Harvey, P. A. Hanford, M. Mayhew, A. G. D. Tse, G. G. Brownlee, I. D. C. Campbell*

1J5L NMR Structure of the Isolated Beta$_C$ Domain of Lobster Metallothionein-1; *A. Munoz, F. H. Forsterling, C. F. Shaw III, D. H. Petering*

1J5M Solution Structure of the Synthetic 113Cd$_3$ beta$_N$ Domain of Lobster Metallothionein-1; *A. Munoz, F. H. Forsterling, C. F. Shaw III, D. H. Petering*

1J79 Molecular Structure of Dihydroorotase: A Paradigm for Catalysis Through the Use of a Binuclear Metal Center; *J. B. Thoden, G. N. Phillips Jr., T. M. Neal, F. M. Raushel, H. M. Holden*

1J7O Solution Structure of Calcium-calmodulin N-terminal Domain; *J. J. Chou, C. B. Klee, A. Bax*

1J7P Solution Structure of Calcium-calmodulin C-terminal Domain; *J. J. Chou, C. B. Klee, A. Bax*

1JAN Complex of Pro-leu-gly-hydroxylamine with the Catalytic Domain of Matrix Metallo Proteinase-8 (Phe79 Form); *P. Reinemer, F. Grams, R. Huber, T. Kleine, S. Schnierer, M. Pieper, H. Tschesche, W. Bode*

1JAO Complex of 3-Mercapto-2-benzylpropanoyl-ala-gly-Nh2 with the Catalytic Domain of Matrix Metallo Proteinase-8 (Met80 Form); *F. Grams, P. Reinemer, J. C. Powers, T. Kleine, M. Piper, H. Tschesche, R. Huber, W. Bode*

1JAP Complex of Pro-leu-gly-hydroxylamine with the Catalytic Domain of Matrix Metallo Proteinase-8 (Met80 Form); *W. Bode, P. Reinemer, R. Huber, T. Kleine, S. Schnierer, H. Tschesche*

1JAQ Complex of 1-Hydroxylamine-2-isobutylmalonyl-ala-gly-Nh2 with the Catalytic Domain of Matrix Metallo Proteinase-8 (Met80 Form); *F. Grams,*

	P. Reinemer, J. C. Powers, T. Kleine, M. Piper, H. Tschesche, R. Huber, W. Bode
1JI9	Solution Structure of the Alpha-domain of Mouse Metallothionein-3; G. Oz, K. Zangger, I. M. Armitage
1JIW	Crystal Structure of the Apr-aprin Complex; T. Hege, R. E. Feltzer, R. D. Gray, U. Baumann
1JIZ	Crystal Structure Analysis of Human Macrophage Elastase Mmp-12; H. Nar, K. Werle, M. M. T. Bauer, H. Dollinger, B. Jung
1JJD	NMR Structure of the Cyanobacterial Metallothionein Smta; P. J. Sadler, N. J. Robinson
1JJE	Imp-1 Metallo Beta-lactamase from Pseudomonas Aeruginosa in Complex with a Biaryl Succinic Acid Inhibitor (11); P. M. D. Fitzgerald, N. Sharma
1JK1	Zif268 D20A Mutant Bound to Wt DNA Site; J. C. Miller, C. O. Pabo
1JK2	Zif268 D20A Mutant Bound to the Gct DNA Site; J. C. Miller, C. O. Pabo
1JK3	Crystal Structure of Human Mmp-12 (Macrophage Elastase) at True Atomic Resolution; R. Lang, A. Kocourek, M. Braun, H. Tschesche, R. Huber, W. Bode, K. Maskos
1JLT	Vipoxin Complex; S. Banumathi, K. R. Rajashankar, C. Notzel, B. Aleksiev, T. P. Singh, N. Genov, C. Betzel
1JM7	Solution Structure of the Brca1/Bard1 Ring-domain Heterodimer; P. S. Brzovic, P. Rajagopal, D. W. Hoyt, M.-C. King, R. E. Klevit
1JOC	Eea1 Homodimer of C-terminal Fyve Domain Bound to Inositol 1,3-Diphosphate; J. J. Dumas, E. Merithew, D. Rajamani, S. Hayes, D. Lawe, S. Corvera, D. G. Lambright
1JQ9	Crystal Structure of a Complex Formed between Phospholipase A2 from Daboia Russelli Pulchella and a Designed Pentapeptide Phe-leu-ser-tyr-lys at 1.8 Resolution; V. Chandra, J. Jasti, P. Kaur, S. Dey, C. Betzel, T. P. Singh
1JQB	Alcohol Dehydrogenase from Clostridium Beijerinckii: Crystal Structure of Mutant with Enhanced Thermal Stability; I. Levin, F. Frolow, O. Bogin, M. Peretz, Y. Hacham, Y. Burstein
1JR3	Crystal Structure of the Processivity Clamp Loader Gamma-complex of E. Coli DNA Polymerase III; D. Jeruzalmi, M. O'Donnell, J. Kuriyan
1JTD	Crystal Structure of Beta-lactamase Inhibitor Protein-II in Complex with Tem-1 Beta-lactamase; D. C. Lim, H. U. Park, L. De Castro, S. G. Kang, H. S. Lee, S. Jensen, K. J. Lee, N. C. J. Strynadka
1JU9	Horse Liver Alcohol Dehydrogenase Val292Ser Mutant; J. K. Rubach, S. Ramaswamy, B. V. Plapp
1JUO	Crystal Structure of Calcium-free Human Sorcin: A Member of the Penta-Ef-hand Protein Family; X. Xie
1JVB	Alcohol Dehydrogenase from the Archaeon Sulfolobus Solfataricus; L. Esposito, F. Sica, A. Zagari, L. Mazzarella
1JWH	Crystal Structure of Human Protein Kinase Ck2 Holoenzyme; K. Niefind, B. Guerra, I. Ermakowa, O. G. Issinger
1K07	Native Fez-1 Metallo-beta-lactamase from Legionella Gormanii; I. Garcia-Saez, P. S. Mercuri, R. Kahn, C. Papamicael, J. M. Frere, M. Galleni, O. Dideberg
1K1D	Crystal Structure of D-hydantoinase; Y. H. Cheon, H. S. Kim, K. H. Han, J. Abendroth, K. Niefind, D. Schomburg, J. Wang, Y. Kim
1K32	Crystal Structure of the Tricorn Protease; H. Brandstetter, J.-S. Kim, M. Groll, R. Huber
1K3W	Crystal Structure of a Trapped Reaction Intermediate of the DNA Repair Enzyme Endonuclease Viii with DNA; G. Golan, D. O. Zharkov, R. Gilboa, A. S. Fernandes, J. H. Kycia, S. E. Gerchman, R. A. Rieger, A. P. Grollman, G. Shoham
1K5W	Three-dimensional Structure of the Synaptotagmin 1 C2B-domain: Synaptotagmin 1 as a Phospholipid Binding Machine; I. Fernandez, D. Arac, J. Ubach, S. H. Gerber, O. Shin, Y. Gao, R. G. W. Anderson, T. C. Sudhof, J. Rizo
1K7G	Prtc from Erwinia Chrysanthemi; T. Hege, U. Baumann
1K7I	Prtc from Erwinia Chrysanthemi: Y228F Mutant; U. Baumann, T. Hege
1K7Q	Prtc from Erwinia Chrysanthemi: E189A Mutant; U. Baumann, T. Hege
1K82	Crystal Structure of E. Coli Formamidopyrimidine-DNA Glycosylase (Fpg) Covalently Trapped with DNA; R. Gilboa, D. O. Zharkov, G. Golan, A. S. Fernandes, S. E. Gerchman, E. Matz, J. H. Kycia, A. P. Grollman, G. Shoham
1K83	Crystal Structure of Yeast RNA Polymerase II Complexed with the Inhibitor Alpha Amanitin; D. A. Bushnell, P. Cramer, R. D. Kornberg
1K8U	Crystal Structure of Calcium-free (Or Apo) Human S100A6; Cys3Met Mutant (Selenomethionine Derivative); L. R. Otterbein, R. Dominguez
1K93	Crystal Structure of Edema Factor Complexed with Calmodulin; C. L. Drum, S.-Z. Yan, J. Bard, Y.-Q. Shen, D. Lu, S. Soelaiman, Z. Grabarek, A. Bohm, W.-J. Tang
1K94	Crystal Structure of Des(1-52)Grancalcin with Bound Calcium; J. Jia, N. Borregaard, K. Lollike, M. Cygler
1K95	Crystal Structure of Des(1-52)Grancalcin with Bound Calcium; J. Jia, N. Borregaard, K. Lollike, M Cygler
1K96	Crystal Structure of Calcium Bound Human S100A6; L. R. Otterbein, R. Dominguez
1K9I	Complex of Dc-Sign and Glcnac2Man3; H. Feinberg, D. A. Mitchell, K. Drickamer, W. I. Weis
1KAP	Three-dimensional Structure of the Alkaline Protease of Pseudomonas Aeruginosa: A Two-domain Protein with a Calcium Binding Parallel Beta Roll Motif; U. Baumann, S. Wu, K. M. Flaherty, D. B. Mckay
1KB2	Crystal Structure of Vdr DNA-binding Domain Bound to Mouse Osteopontin (Spp) Response Element; P. L. Shaffer, D. T. Gewirth
1KBC	Human Neutrophil Collagenase Catalytic Domain in Complex with an Inhibitor; M. Betz, F. X. Gomis-Rueth, W. Bode
1KBP	Kidney Bean Purple Acid Phosphatase; T. Klabunde, N. Strater, B. Krebs
1KEV	Structure of Nadp-dependent Alcohol Dehydrogenase; Y. Korkhin, F. Frolow
1KFU	Crystal Structure of Human M-calpain Form II; S. Strobl, C. Fernandez-Catalan, M. Braun, R. Huber, H. Masumoto, K. Nakagawa, A. Irie, H. Sorimachi, G. Bourenkow, H. Bartunik, K. Suzuki, W. Bode
1KFV	Crystal Structure of Lactococcus Lactis Formamidopyrimidine DNA Glycosylase (Alias Fpg Or Mutm)

PDB-Code Listing

	Non Covalently Bound To An Ap Site Containing DNA; L. Serre, K. Pereira De Jesus, S. Boiteux, C. Zelwer, B. Castaing
1KFX	Crystal Structure of Human M-calpain Form I; S. Strobl, C. Fernandez-Catalan, M. Braun, R. Huber, H. Masumoto, K. Nakagawa, A. Irie, H. Sorimachi, G. Bourenkow, H. Bartunik, K. Suzuki, W. Bode
1KHK	E. Coli Alkaline Phosphatase Mutant (D153Hd330N); M. H. Le Du, C. Lamoure, B. H. Muller, O. V. Bulgakov, E. Lajeunesse, A. Menez, J. C. Boulain
1KHO	Crystal Structure Analysis of Clostridium Perfringens - toxin Isolated from Avian Strain Swcp; N. Justin, D. S. Moss, R. W. Titball, A. K. Basak
1KJO	Thermolysin Complexed with Z-L-Threonine (Benzyloxycarbonyl-L-Threonine); M. Senda, T. Senda, S. Kidokoro
1KLF	Fimh Adhesin-Fimc Chaperone Complex with D-Mannose; C. S. Hung, J. Bouckaert
1KLO	Crystal Structure of Three Consecutive Laminin-type Epidermal Growth Factor-like (Le) Modules of Laminin Gamma-1 Chain Harboring the Nidogen Binding Site; J. Stetefeld, U. Mayer, R. Timpl, R. Huber
1KLR	NMR Structure of the Zfy-6T[Y10F] Zinc Finger; M. J. Lachenmann, J. E. Ladbury, N. B. Phillips, N. Narayana, X. Qian, M. A. Weiss
1KOL	Crystal Structure of Formaldehyde Dehydrogenase; N. Tanaka, Y. Kusakabe, K. Ito, T. Yoshimoto, K. T. Nakamura
1KQA	Galactoside Acetyltransferase in Complex with Coenzyme A; X.-G. Wang, L. R. Olsen, S. L. Roderick
1KSO	Crystal Structure of Apo S100A3; P. R. E Mittl, G. Fritz, D. F. Sargent, T. J. Richmond, C. W. Heizmann, M. G. Grutter
1KU0	Structure of the Bacillus Stearothermophilus L1 Lipase; S.-T. Jeong, H.-K. Kim, S.-J. Kim, S.-W. Chi, J.-G. Pan, T.-K. Oh, S.-E. Ryu
1KW7	Methionine Core Mutant of T4 Lysozyme; N. C. Gassner, W. A. Baase, B. H. Mooers, R. D. Busam, L. H. Weaver, J. D. Lindstrom, M. L. Quillin, B. W. Matthews
1KWT	Rat Mannose Binding Protein A (Native, Mpd); K. K. S. Ng, A. R. Kolatkar, S. Park-Snyder, H. Feinberg, D. A. Clark, K. Drickamer, W. I. Weis
1KXR	Crystal Structure of Calcium-bound Protease Core of Calpain I; T. Moldoveanu, C. M. Hosfield, D. Lim, J. S. Elce, Z. Jia, P. L. Davies
1KZP	Protein Farnesyltransferase Complexed with a Farnesylated K-Ras4B Peptide Product; S. B. Long, P. J. Casey, L. S. Beese
1L1T	Mutm (Fpg) Bound to Abasic-site Containing DNA; J. C. Fromme, G. L. Verdine
1L3W	C-cadherin Ectodomain; T. J. Boggon, J. Murray, S. Chappuis-Flament, E. Wong, B. M. Gumbiner, L. Shapiro
1L3Y	Integrin Egf-Like Module 3 from the Beta-2 Subunit; N. Beglova, S. C. Blacklow, J. Takagi, T. A. Springer
1L6J	Crystal Structure of Human Matrix Metalloproteinase Mmp9 (Gelatinase B); P. A. Elkins, Y. S. Ho, W. W. Smith, C. A. Janson, K. J. D'Alessio, M. S. Mcqueney, M. D. Cummings, A. M. Romanic
1L6S	Crystal Structure of Porphobilinogen Synthase Complexed with the Inhibitor 4,7-Dioxosebacic Acid; E. K. Jaffe, J. Kervinen, J. Martins, F. Stauffer, R. Neier, A. Wlodawer, A. Zdanov
1L6Y	Crystal Structure of Porphobilinogen Synthase Complexed with the Inhibitor 4-Oxosebacic Acid; E. K. Jaffe, J. Kervinen, J. Martins, F. Stauffer, R. Neier, A. Wlodawer, A. Zdanov
1L8D	Rad50 Coiled-Coil Zn Hook; K. P. Hopfner, J. Tainer
1LAM	Leucine Aminopeptidase (Unligated); N. Straeter, W. N. Lipscomb
1LAN	Leucine Aminopeptidase Complex with L-Leucinal; N. Straeter, W. N. Lipscomb
1LAT	Glucocorticoid Receptor Mutant (DNA Binding Domain) Complex with Non-cognate DNA; D. T. Gewirth, P. B. Sigler
1LBU	Hydrolase Metallo (Zn) Dd-peptidase; J. P. Wery, P. Charlier, O. Dideberg
1LCP	Bovine Lens Leucine Aminopeptidase Complexed with L-leucine Phosphonic Acid; N. Straeter, W. N. Lipscomb
1LDD	Structure of the Cul1-Rbx1-Skp1-F Boxskp2 Scf Ubiquitin Ligase Complex; N. Zheng, B. A. Schulman, L. Song, J. J. Miller, P. D. Jeffrey, P. Wang, C. Chu, D. M. Koepp, S. J. Elledge, M. Pagano, R. C. Conaway, J. W. Conaway, J. W. Harper, N. P. Pavletich
1LDE	Horse Liver Alcohol Dehydrogenase Complexed to Nadh and N-formyl Piperdine; S. Ramaswamy, B. V. Plapp
1LDJ	Structure of the Cul1-Rbx1-Skp1-F Boxskp2 Scf Ubiquitin Ligase Complex; N. Zheng, B. A. Schulman, L. Song, J. J. Miller, P. D. Jeffrey, P. Wang, C. Chu, D. M. Koepp, S. J. Elledge, M. Pagano, R. C. Conaway, J. W. Conaway, J. W. Harper, N. P. Pavletich
1LDY	Horse Liver Alcohol Dehydrogenase Complexed to Nadh and Cyclohexyl Formamide (Cxf); S. Ramaswamy, B. V. Plapp
1LFW	Crystal Structure of Pepv; D. Jozic, G. Bourenkow, H. Bartunik, H. Scholze, V. Dive, B. Henrich, R. Huber, W. Bode, K. Maskos
1LML	Leishmanolysin; E. Schlagenhauf, R. Etges, P. Metcalf
1LNA	Structural Analysis of Zinc Substitutions in the Active Site of Thermolysin; D. R. Holland, A. C. Hausrath, D. Juers, B. W. Matthews
1LOK	The 1.20 Angstrom Resolution Crystal Structure of the Aminopeptidase from Aeromonas Proteolytica Complexed with Tris: A Tale of Buffer Inhibition; W. T. Desmarais, D. L. Bienvenue, K. P. Bzymek, R. C. Holz, G. A. Petsko, D. Ringe
1LXA	Udp N-acetylglucosamine Acyltransferase; S. L. Roderick
1LXF	Structure of the Regulatory N-domain of Human Cardiac Troponin C in Complex with Human Cardiac Troponin-I(147–163) and Bepridil; X. Wang, M. X. Li, B. D. Sykes
1M6H	Human Glutathione-dependent Formaldehyde Dehydrogenase; P. C. Sanghani, H. Robinson, W. F. Bosron, T. D. Hurley
1M6W	Binary Complex of Human Glutathione-dependent Formaldehyde Dehydrogenase and 12-Hydroxydodecanoic Acid; P. C. Sanghani, H. Robinson, W. F. Bosron, T. D. Hurley

1M8N	Choristoneura Fumiferana (Spruce Budworm) Antifreeze Protein Isoform 501; *E. K. Leinala, P. L. Davies, Z. Jia*	1MRT	Conformation of [Cd7]-metallothionein-2 from Rat Liver in Aqueous Solution Determined by Nuclear Magnetic Resonance Spectroscopy; *W. Braun, P. Schultze, E. Woergoetter, G. Wagner, M. Vasak, J. H. R. Kaegi, K. Wuthrich*
1MA0	Ternary Complex of Human Glutathione-dependent Formaldehyde Dehydrogenase with Nad+ and Dodecanoic Acid; *P. C. Sanghani, H. Robinson, W. F. Bosron, T. D. Hurley*	1MWN	Solution NMR Structure of S100B Bound to The High-affinity Target Peptide Trtk-12; *K. G. Inman, R. Yang, R. R. Rustandi, K. E. Miller, D. M. Baldisseri, D. J. Weber*
1MAT	Structure of the Cobalt-dependent Methionine Aminopeptidase from *Escherichia Coli*: A New Type of Proteolytic Enzyme; *S. L. Roderick, B. W. Matthews*	1MXE	Structure of the Complex of Calmodulin with the Target Sequence of Camki; *J. A. Clapperton, S. R. Martin, S. J. Smerdon, S. J. Gamblin, P. M. Bayley*
1MC5	Ternary Complex of Human Glutathione-dependent Formaldehyde Dehydrogenase with S-(Hydroxymethyl)Glutathione and Nadh; *P. C. Sanghani, W. F. Bosron, T. D. Hurley*	1NCG	Structural Basis of Cell-cell Adhesion by Cadherins; *L. Shapiro, A. M. Fannon, P. D. Kwong, A. Thompson, M. S. Lehmann, G. Grubel, J.-F. Legrand, J. Als-Nielsen, D. R. Colman, W. A. Hendrickson*
1MCT	Refined 1.6 A Resolution Crystal Structure of the Complex Formed between Porcine Beta-trypsin and MCTI-A, a Trypsin Inhibitor of the Squash Family. Detailed Comparison with Bovine Beta-trypsin and its Complex; *Q. Huang, S. Liu, Y. Tang*	1NCS	NMR Study of Swi5 Zinc Finger Domain 1; *R. N. Dutnall, D. Neuhaus, D. Rhodes*
1MCX	Structure of Full-length Annexin A1 in the Presence of Calcium; *H. Luecke, A. Rosengarth*	1NFG	Structure of D-hydantoinase; *Z. Xu, Y. Yang, W. Jiang, E. Arnold, J. Ding*
1MEE	Complex between the Subtilisin from a Mesophilic Bacterium and the Leech Inhibitor Eglin-C; *Z. Dauter, C. Betzel, K. S. Wilson*	1NL1	Bovine Prothrombin Fragment 1. In Complex with Calcium Ion; *M. Huang, G. Huang, B. Furie, B. Seaton, B.C. Furie*
1MEY	Crystal Structure of a Designed Zinc Finger Protein Bound to DNA; *C. A. Kim, J. M. Berg*	1NWD	Solution Structure of Ca2+/Calmodulin Bound to the C-terminal Domain of Petunia Glutamate Decarboxylase; *K. L. Yap, T. Yuan, T. K. Mal, H. J. Vogel, M. Ikura*
1MGO	Horse Liver Alcohol Dehydrogenase Phe93Ala Mutant; *J. K. Rubach, B. V. Plapp*		
1MGX	Coagulation Factor, Mg(II), NMR, 7 Structures (Backbone Atoms Only); *S. J. Freedman, B. C. Furie, B. Furie, J. D. Baleja*	1OCC	Structure of Bovine Heart Cytochrome C Oxidase at the Fully Oxidized State; *T. Tsukihara, H. Aoyama, E. Yamashita, T. Tomizaki, H. Yamaguchi, K. Shinzawa-Itoh, R. Nakashima, R. Yaono, S. Yoshikawa*
1MHU	The Three-dimensional Structure of Human [113Cd7]; *W. Braun, B. A. Messerle, A. Schaeffer, M. Vasak, J. H. R. Kaegi, K. Wuthrich*	1ONW	Crystal Structure of Isoaspartyl Dipeptidase from *E. Coli*; *J. B. Thoden, R. Marti-Arbona, F. M. Raushel, H. M. Holden*
1MKT	Carboxylic Ester Hydrolase, 1.72 Angstrom Trigonal Form of the Bovine Recombinant Pla2 Enzyme; *M. Sundaralingam*	1PAA	Structure of a Histidine-X4-histidine Zinc Finger Domain: Insights into ADR1-UAS1 Protein-DNA Recognition; *B. E. Bernstein, R. C. Hoffman, S. J. Horvath, J. R. Herriott, R. E. Klevit*
1MMB	Complex of Bb94 with the Catalytic Domain of Matrix Metalloproteinase-8; *W. Bode, F. Grams*		
1MMP	Matrilysin Complexed with Carboxylate Inhibitor; *M. F. Browner, W. W. Smith, A. L. Castelhano*	1PED	Bacterial Secondary Alcohol Dehydrogenase (Apo-form); *Y. Korkhin, F. Frolow*
1MMQ	Matrilysin Complexed with Hydroxamate Inhibitor; *M. F. Browner, W. W. Smith, A. L. Castelhano*	1PEX	Collagenase-3 (Mmp-13) C-terminal Hemopexin-like Domain; *F. X. Gomis-Ruth, U. Gohlke, M. Betz, V. Knauper, G. Murphy, C. Lopez-Otin, W. Bode*
1MMR	Matrilysin Complexed with Sulfodiimine Inhibitor; *M. F. Browner, W. W. Smith, A. L. Castelhano*	1PFX	Porcine Factor Ixa; *H. Brandstetter, M. Bauer, R. Huber, P. Lollar, W. Bode*
1MNC	Structure of Human Neutrophil Collagenase Reveals Large S1' Specificity Pocket; *T. Stams, J. C. Spurlino, D. L. Smith, B. Rubin*	1PM0	Replication of a Cis-syn Thymine Dimer at Atomic Resolution; *H. Ling, F. Boudsocq, B. Plosky, R. Woodgate, W. Yang*
1MOO	Site Specific Mutant (H64A) of Human Carbonic Anhydrase II at High Resolution; *D. M. Duda, L. Govindasamy, M. Agbandje-Mckenna, C. K. Tu, D. N. Silverman, R. Mckenna*	1POB	Crystal Structure of Cobra-venom Phospholipase A2 in a Complex with a Transition-state Analogue; *S. P. White, D. L. Scott, Z. Otwinowski, P. B. Sigler*
1MPT	Structure of a New Alkaline Serine-protease (M-Protease) from Bacillus Sp Ksm-K16; *T. Yamane, T. Kani, T. Hatanaka, A. Suzuki, T. Ashida, T. Kobayashi, S. Ito, O. Yamashita*	1POC	Crystal Structure of Bee-venom Phospholipase A2 in a Complex with a Transition-state Analogue; *D. L. Scott, Z. Otwinowski, P. B. Sigler*
1MQ1	Ca2+-S100B-Trtk-12 Complex; *K. A. Mcclintock, G. S. Shaw*	1POE	Structures of Free and Inhibited Human Secretory Phospholipase A2 from Inflammatory Exudate; *D. L. Scott, S. P. White, P. B. Sigler*
1MRB	Three-dimensional Structure of Rabbit Liver [Cd7]metallothionein-2a in Aqueous Solution Determined by Nuclear Magnetic Resonance; *W. Braun, A. Arseniev, P. Schultze, E. Woergoetter, G. Wagner, M. Vasak, J. H. R. Kaegi, K. Wuthrich*	1PON	Site III – Site IV Troponin C Heterodimer, NMR; *G. S. Shaw, B. D. Sykes*
		1PP2	The Refined Crystal Structure of Dimeric Phospholipase A2 at 2.5 A. Access to a Shielded Catalytic center; *S. Brunie, P. B. Sigler*

PDB-Code Listing

1PPA	The Crystal Structure of a Lysine 49 Phospholipase A2 from the Venom of the Cottonmouth Snake at 2.0-A Resolution; *D. R. Holland, L. L. Clancy, S. W. Muchmore, T. J. Rydel, H. M. Einspahr, B. C. Finzel, R. L. Heinrikson, K. D. Watenpaugh*
1PSJ	Acidic Phospholipase A2 from Agkistrodon Halys Pallas; *X. Q. Wang, Z. J. Lin*
1PTQ	Protein Kinase C Delta Cys2 Domain; *G. Zhang, J. H. Hurley*
1Q2L	Crystal Structure of Pitrilysin; *Maskos, K., Jozic, D.*
1QBI	Soluble Quinoprotein Glucose Dehydrogenase from Acinetobacter Calcoaceticus; *A. Oubrie, H. J. Rozeboom, K. H. Kalk, J. A Duine, B. W. Dijkstra*
1QF8	Truncated Form of Casein Kinase II Beta Subunit (2-182) from Homo Sapiens; *L. Chantalat, D. Leroy, O. Filhol, A. Nueda, M. J. Benitez, E. Chambaz, C. Cochet, O. Dideberg*
1QFC	Structure of Rat Purple Acid Phosphatase; *J. Uppenberg, F. Lindqvist, C. Svensson, B. Ek-Rylander, G. Andersson*
1QH3	Human Glyoxalase II with Cacodylate and Acetate Ions Present in the Active Site; *A. D. Cameron, M. Ridderstrom, B. Olin, B. Mannervik*
1QH5	Human Glyoxalase II with S-(N-Hydroxy-N-Bromophenylcarbamoyl)Glutathione; *A. D. Cameron, M. Ridderstrom, B. Olin, B. Mannervik*
1QHU	Mammalian Blood Serum Haemopexin Deglycosylated and in Complex with its Ligand Haem; *M. Paoli, H. M. Baker, W. T. Morgan, A. Smith, E. N. Baker*
1QHW	Purple Acid Phosphatase from Rat Bone; *Y. Lindqvist, E. Johansson, H. Kaija, P. Vihko, G. Schneider*
1QIB	Crystal Structure of Gelatinase A Catalytic Domain; *V. Dhanaraj, M. G. Williams, Q.-Z. Ye, F. Molina, L. L. Johnson, D. F. Ortwine, A. Pavlovsky, J. R. Rubin, R. W. Skeean, A. D. White, C. Humblet, D. J. Hupe, T. L. Blundell*
1QJI	Structure of Astacin with a Transition-state Analogue Inhibitor; *F. Grams, W. Bode, W. Stocker*
1QJJ	Structure of Astacin with a Hydroxamic Acid Inhibitor; *F. Grams, W. Bode, W. Stocker*
1QJK	Metallothionein Mta from Sea Urchin (Alpha Domain); *R. Riek, B. Precheur, Y. Wang, E. A. Mackay, G. Wider, P. Guntert, A. Liu, J. H. R. Kaegi, K. Wuthrich*
1QJL	Metallothionein Mta from Sea Urchin (Beta Domain); *R. Riek, B. Precheur, Y. Wang, E. A. Mackay, G. Wider, P. Guntert, A. Liu, J. H. R. Kaegi, K. Wuthrich*
1QJT	Solution Structure of the Apo Eh1 Domain of Mouse Epidermal Growth Factor Receptor Substrate 15, Eps15; *B. Whitehead, M. Tessari, A. Carotenuto, P. M. Van Bergen En Henegouwen, G. W. Vuister*
1QKS	Cytochrome Cd1 Nitrite Reductase, Oxidised Form; *V. Fulop*
1QLH	Horse Liver Alcohol Dehydrogenase Complexed to Nad Double Mutant of Gly 293 Ala and Pro 295 Thr; *S. Ramaswamy, B. V. Plapp*
1QLI	Quail Cysteine and Glycine-rich Protein, NMR, Minimized Average Structure; *R. Konrat, R. Weiskirchen, B. Kraeutler, K. Bister*
1QLJ	Horse Liver Alcohol Dehydrogenase Apo Enzyme Double Mutant of Gly 293 Ala and Pro 295 Thr; *S. Ramaswamy, B. V. Plapp*
1QLL	Piratoxin-II (Prtx-II) – A K49 Pla2 from Bothrops Pirajai; *W.-H. Lee, I. Polikarpov*
1QLS	S100C (S100A11), or Calgizzarin, in Complex with Annexin I N-Terminus; *S. Rety, J. Sopkova, M. Renouard, D. Osterloh, V. Gerke, F. Russo-Marie, A. Lewit-Bentley*
1QQ9	Streptomyces Griseus Aminopeptidase Complexed with Methionine; *R. Gilboa, H. M. Greenblatt, M. Perach, A. Spungin-Bialik, U. Lessel, D. Schomburg, S. Blumberg, G. Shoham*
1QQO	E175S Mutant of Bovine 70 Kilodalton Heat Shock Protein; *E. R. Johnson, D. B. Mckay*
1QQT	Methionyl-tRNA Synthetase from *Escherichia Coli*; *Y. Mechulam, E. Schmitt, L. Maveyraud, C. Zelwer, O. Nureki, S. Yokoyama, M. Konno, S. Blanquet*
1QRE	A Closer Look at the Active Site of Gamma-carbonic Anhydrases: High Resolution Crystallographic Studies of the Carbonic Anhydrase from Methanosarcina Thermophila; *T. M. Iverson, B. E. Alber, C. Kisker, J. G. Ferry, D. C. Rees*
1QRF	A Closer Look at the Active Site of Gamma-carbonic Anhydrases: High Resolution Crystallographic Studies of the Carbonic Anhydrase from Methanosarcina Thermophila; *T. M. Iverson, B. E. Alber, C. Kisker, J. G. Ferry, D. C. Rees*
1QRG	A Closer Look at the Active Site of Gamma-carbonic Anhydrases: High Resolution Crystallographic Studies of the Carbonic Anhydrase from Methanosarcina Thermophila; *T. M. Iverson, B. E. Alber, C. Kisker, J. G. Ferry, D. C. Rees*
1QRL	A Closer Look at the Active Site of Gamma-carbonic Anhydrases: High Resolution Crystallographic Studies of the Carbonic Anhydrase from Methanosarcina Thermophila; *T. M. Iverson, B. E. Alber, C. Kisker, J. G. Ferry, D. C. Rees*
1QTW	High-resolution Crystal Structure of the *Escherichia Coli* DNA Repair Enzyme Endonuclease IV; *D. J. Hosfield, Y. Guan, B. J. Haas, R. P. Cunningham, J. A. Tainer*
1QU2	Insights into Editing from an Ile-tRNA Synthetase Structure with tRNA(Ile) and Mupirocin; *L. F. Silvian, J. Wang, T. A. Steitz*
1QYP	Thermococcus Celer Rpb9, NMR, 25 Structures; *B. Wang, D. N. M. Jones, B. P. Kaine, M. A. Weiss*
1RDO	Mannose-binding Protein, Subtilisin Digest Fragment; *K. K.-S. Ng, K. Drickamer, W. I. Weis*
1RGD	Structure Refinement of the Glucocorticoid Receptor-DNA Binding Domain from NMR Data by Relaxation Matrix Calculations; *M. A. A. Van Tilborg, A. M. J. J. Bonvin, K. Hard, A. Davis, B. Maler, R. Boelens, K. R. Yamamoto, R. Kaptein*
1RMD	Rag1 Dimerization Domain; *S. F. Bellon, K. K. Rodgers, D. G. Schatz, J. E. Coleman, T. A. Steitz*
1RSY	Structure of the First C2 Domain of Synaptotagmin I: A Novel Ca2+/Phospholipid-binding Fold; *R. B. Sutton, S. R. Sprang*
1RTG	C-terminal Domain (Haemopexin-like Domain) of Human Matrix Metalloproteinase-2; *U. Gohlke, W. Bode*
1RXR	High Resolution Solution Structure of the Retinoid X Receptor DNA Binding Domain, NMR, 20 Structure; *S. M. A. Holmbeck, M. P. Foster, D. R. Casimiro, D. S. Sem, H. J. Dyson, P. E. Wright*

1SAT	Crystal Structure of the 50 kDa Metallo Protease from Serratia Marcescens; *U. Baumann*
1SCA	Enzyme Crystal Structure in a Neat Organic Solvent; *P. A. Fitzpatrick, A. C. U. Steinmetz, D. Ringe, A. M. Klibanov*
1SDY	Structure Solution and Molecular Dynamics Refinement of the Yeast Cu, Zn Enzyme Superoxide Dismutase; *K. Djinovic, G. Gatti, A. Coda, L. Antolini, G. Pelosi, A. Desideri, M. Falconi, F. Marmocchi, G. Rotilio, M. Bolognesi*
1SGT	Refined Crystal Structure of Streptomyces Griseus Trypsin at 1.7 A Resolution; *R. J. Read, M. N. G. James*
1SIC	Refined Crystal Structure of the Complex of Subtilisin BPN' and Streptomyces Subtilisin Inhibitor at 1.8 A Resolution; *Y. Mitsui, Y. Takeuchi, S. Hirono, H. Akagawa, K. T. Nakamura*
1SLM	Crystal Structure of Fibroblast Stromelysin-1: The C-Truncated Human Proenzyme; *J. W. Becker*
1SLN	Crystal Structure of the Catalytic Domain of Human Fibroblast Stromelysin-1 Inhibited with the N-carboxy-alkyl Inhibitor L-702,842; *J. W. Becker*
1SMG	Calcium-bound E41A Mutant of the N-domain of Chicken Troponin C, NMR, 40 Structures; *S. M. Gagne, M. X. Li, B. D. Sykes*
1SML	Metallo Beta Lactamase L1 from Stenotrophomonas Maltophilia; *J. H. Ullah, T. R. Walsh, I. A. Taylor, D. C. Emery, C. S. Verma, S. J. Gamblin, J. Spencer*
1SMP	Crystal Structure of a Complex between Serratia Marcescens Metallo-protease and an Inhibitor from Erwinia Chrysanthemi; *U. Baumann, M. Bauer, S. Letoffe, P. Delepelaire, C. Wandersman*
1SP1	NMR Structure of a Zinc Finger Domain from Transcription Factor Sp1F3, Minimized Average Structure; *V. A. Narayan, R. W. Kriwacki, J. P. Caradonna*
1SP2	NMR Structure of a Zinc Finger Domain from Transcription Factor Sp1F2, Minimized Average Structure; *V. A. Narayan, R. W. Kriwacki, J. P. Caradonna*
1SPD	Amyotrophic Lateral Sclerosis and Structural Defects in Cu, Zn Superoxide Dismutase; *H. E. Parge, J. A. Tainer*
1SPY	Regulatory Domain of Human Cardiac Troponin C in the Calcium-free State, NMR, 40 Structures; *L. Spyracopoulos, M. X. Li, S. K. Sia, S. M. Gagne, M. Chandra, R. J. Solaro, B. D. Sykes*
1SRD	Three-dimensional Structure of Cu, Zn-superoxide Dismutase from Spinach at 2.0 A Resolution; *Y. Kitagawa, Y. Katsube*
1SRP	Structural Analysis of Serratia Protease; *K. Hamada, H. Hiramatsu, Y. Katsuya, Y. Hata, Y. Katsube*
1SUH	Amino-terminal Domain of Epithelial Cadherin in the Calcium Bound State, NMR, 20 Structures; *M. Overduin, K. I. Tong, C. M. Kay, M. Ikura*
1TBG	Beta-gamma Dimer of the Heterotrimeric G-protein Transducin; *J. S. Sondek, A. Bohm, D. G. Lambright, H. E. Hamm, P. B. Sigler*
1TBN	NMR Structure of a Protein Kinase C-G Phorbol-Binding Domain, Minimized Average Structure; *R. X. Xu, T. Pawelczyk, T. Xia, S. C. Brown*
1TCF	Crystal Structure of Calcium-Saturated Rabbit Skeletal Troponin C; *J. Soman, G. N. Phillips Junior*
1TCO	Ternary Complex of a Calcineurin A Fragment, Calcineurin B, Fkbp12 and the Immunosuppressant Drug Fk506 (Tacrolimus); *J. P. Griffith, J. L. Kim, E. E. Kim, M. D. Sintchak, J. A. Thomson, M. J. Fitzgibbon, M. A. Fleming, P. R. Caron, K. Hsiao, M. A. Navia*
1TEH	Structure of Human Liver Chichi Alcohol Dehydrogenase (A Glutathione-dependent Formaldehyde Dehydrogenase); *Z.-N. Yang, T. D. Hurley*
1TF3	Tfiiia Finger 1-3 Bound to DNA, NMR, 22 Structures; *M. P. Foster, D. S. Wuttke, I. Radhakrishnan, D. A. Case, J. M. Gottesfeld, P. E. Wright*
1TF6	Co-crystal Structure of Xenopus Tfiiia Zinc Finger Domain Bound to the 5S Ribosomal RNA Gene Internal Control Region; *R. T. Nolte, R. M. Conlin, S. C. Harrison, R. S. Brown*
1THJ	Carbonic Anhydrase from Methanosarcina; *C. Kisker, H. Schindelin, D. C. Rees*
1THM	Crystal Structure of Thermitase at 1.4 A Resolution; *A. V. Teplyakov, I. P. Kuranova, E. H. Harutyunyan*
1TL2	Tachylectin-2 from Tachypleus Tridentatus (Japanese Horseshoe Crab); *H.-G. Beisel, S. Kawabata, S. Iwanaga, R. Huber, W. Bode*
1TLP	Crystallographic Structural Analysis of Phosphoramidates as Inhibitors and Transition-state Analogs of Thermolysin; *D. E. Tronrud, A. F. Monzingo, B. W. Matthews*
1TMN	Binding of N-carboxymethyl Dipeptide Inhibitors to Thermolysin Determined by X-ray Crystallography: A Novel Class of Transition-state Analogues for Zinc peptidases; *A. F. Monzingo, B. W. Matthews*
1TN4	Four Calcium Tnc; *M. L. Love, R. Dominguez, A. Houdusse, C. Cohen*
1TNP	Structures of the Troponin C Regulatory Domains in the Apo and Calcium-saturated States; *S. M. Gagne, B. D. Sykes*
1TNQ	Structures of the Troponin C Regulatory Domains in the Apo and Calcium-saturated States; *S. M. Gagne, B. D. Sykes*
1TNW	NMR Solution Structure of Calcium-saturated Skeletal Muscle Troponin C; *C. M. Slupsky, B. D. Sykes*
1TOP	Structure of Chicken Skeletal-muscle Troponin-C at 1.78 Angstrom Resolution; *M. Sundaralingam*
1TRM	The Three-dimensional Structure of Asn102 Mutant of Trypsin: Role of Asp102 in Serine Protease Catalysis; *S. Sprang, T. Standing, R. J. Fletterick*
1TRN	Crystal Structure of Human Trypsin 1: Unexpected Phosphorylation of Tyr151; *C. Gaboriaud, J. C. Fontecilla-Camps*
1TRY	Structure of Inhibited Trypsin from Fusarium-Oxysporum at 1.55-Angstrom; *W. R. Rypniewski, C. Dambmann, C. Von Der Osten, M. Dauter, K. S. Wilson*
1TRZ	Crystallographic Evidence for Dual Coordination Around Zinc in the T3R3 Human Insulin Hexamer; *E. Ciszak, G. D. Smith*
1UBD	Co-crystal Structure of Human Yy1 Zinc Finger Domain Bound to the Adeno-associated Virus P5 Initiator Element; *H. B. Houbaviy, A. Usheva, T. Shenk, S. K. Burley*
1UEA	Mmp-3/Timp-1 Complex; *W. Bode, K. Maskos, F.-X. Gomis-Rueth, H. Nagase*
1UMS	Stromelysin-1 Catalytic Domain with Hydrophobic Inhibitor Bound, pH 7.0, 32Oc, 20 Mm CaCl2,

	15% Acetonitrile; NMR Ensemble of 20 Structures; *S. R. Van Doren, A. V. Kurochkin, W. Hu, E. R. P. Zuiderweg*
1UMT	Stromelysin-1 Catalytic Domain with Hydrophobic Inhibitor Bound, pH 7.0, 32Oc, 20 Mm CaCl2, 15% Acetonitrile; NMR Average of 20 Structures Minimized with Restraints; *S. R. Van Doren, A. V. Kurochkin, W. Hu, E. R. P. Zuiderweg*
1USH	5′-Nucleotidase from E. Coli; *T. Knofel, N. Strater*
1UTE	Pig Purple Acid Phosphatase Complexed with Phosphate; *L. W. Guddat, A. Mcalpine, D. Hume, S. Hamilton, J. De Jersey, J. L. Martin*
1VFY	Phosphatidylinositol-3-Phosphate Binding Fyve Domain of Vps27P Protein from Saccharomyces Cerevisiae; *J. H. Hurley, S. Misra*
1VHH	A Potential Catalytic Site Revealed by the 1.7-A Crystal Structure of the Amino-terminal Signalling Domain of Sonic Hedgehog; *T. M. T. Hall, J. A. Porter, P. A. Beachy, D. J. Leahy*
1VSR	Very Short Patch Repair (Vsr) Endonuclease from Escherichia Coli; *S. E. Tsutakawa, T. Muto, H. Jingami, N. Kunishima, M. Ariyoshi, D. Kohda, M. Nakagawa, K. Morikawa*
1WDC	Scallop Myosin Regulatory Domain; *A. Houdusse, C. Cohen*
1WHE	Coagulation Factor, NMR, 20 Structures; *M. Sunnerhagen, G. A. Olah, J. Stenflo, S. Forsen, T. Drakenberg, J. Trewhella*
1WHF	Coagulation Factor, NMR, 15 Structures; *M. Sunnerhagen, G. A. Olah, J. Stenflo, S. Forsen, T. Drakenberg, J. Trewhella*
1WJA	Solution Structure of the N-Terminal Zn Binding Domain of HIV-1 Integrase (D Form), NMR, Regularized Mean Structure; *G. M. Clore, M. Cai, M. Caffrey, A. M. Gronenborn*
1XGM	Methionine Aminopeptidase from Hyperthermophile Pyrococcus Furiosus; *T. H. Tahirov, T. Tsukihara*
1XGS	Methionine Aminopeptidase from Hyperthermophile Pyrococcus Furiosus; *T. H. Tahirov, T. Tsukihara*
1XJO	Structure of Aminopeptidase; *H. M. Greenblatt, D. Barra, S. Blumberg, G. Shoham*
1XSO	Three-dimensional Structure of Xenopus Laevis Cu, Zn Superoxide Dismutase b Determined by X-ray Crystallography at 1.5 angstrom Resolution; *K. Djinovic Carugo, A. Coda, A. Battistoni, M. T. Carri, F. Policelli, A. Desideri, G. Rotilio, K. S. Wilson, M. Bolognesi*
1YAI	X-ray Structure of a Bacterial Copper, Zinc Superoxide Dismutase; *Y. Bourne, S. M. Redford, T. P. Lo, J. A. Tainer, E. D. Getzoff*
1YKF	Nadp-dependent Alcohol Dehydrogenase from Thermoanaerobium Brockii; *Y. Korkhin, F. Frolow*
1YSO	Yeast Cu, Zn Superoxide Dismutase with the Reduced Bridge Broken; *H. E. Parge, B. R. Crane, J. Tsang, J. A. Tainer*
1YUI	Solution NMR Structure of the Gaga Factor/DNA Complex, Regularized Mean Structure; *G. M. Clore, J. G. Omichinski, A. M. Gronenborn*
1YUJ	Solution NMR Structure of the Gaga Factor/DNA Complex, 50 Structures; *G. M. Clore, J. G. Omichinski, A. M. Gronenborn*
1ZAA	Zinc finger-DNA Recognition: Crystal Structure of a Zif268-DNA Complex at 2.1 A; *N. P. Pavletich, C. O. Pabo*
1ZFD	Swi5 Zinc Finger Domain 2, NMR, 45 Structures; *D. Neuhaus, Y. Nakaseko, J. W. R. Schwabe, D. Rhodes, A. Klug*
1ZIP	Bacillus Stearothermophilus Adenylate Kinase; *M. B. Berry, G. N. Phillips Junior*
1ZNB	Metallo-beta-lactamase; *N. O. Concha, O. Herzberg*
1ZNF	Three-dimensional Solution Structure of a Single Zinc Finger DNA-binding Domain; *M. S. Lee, G. P. Gippert, K. V. Soman, D. A. Case, P. E. Wright*
1ZNI	Insulin; *M. G. W. Turkenburg, J. L. Whittingham, G. G. Dodson, E. J. Dodson, B. Xiao, G. A. Bentley*
1ZNM	A Zinc Finger with an Artificial Beta-turn, Original Sequence taken from the Third Zinc Finger Domain of the Human Transcriptional Repressor Protein Yy1 (Ying and Yang 1, A Transcription Factor), NMR, 34 Structures; *J. H. Viles, S. U. Patel, J. B. O. Mitchell, C. M. Moody, D. E. Justice, J. Uppenbrink, P. M. Doyle, C. J. Harris, P. J. Sadler, J. M. Thornton*
2ADR	Adr1 DNA-binding Domain from Saccharomyces Cerevisiae, NMR, 25 Structures; *P. M. Bowers, R. E. Klevit*
2ANH	Alkaline Phosphatase (D153H); *J. E. Murphy, T. T. Tibbitts, E. R. Kantrowitz*
2APS	Cu/Zn Superoxide Dismutase from Actinobacillus Pleuropneumoniae; *K. T. Forest, P. R. Langford, J. S. Kroll, E. D. Getzoff*
2BAM	Restriction Endonuclease Bamhi Complex with DNA and Calcium Ions (Pre-reactive Complex); *H. Viadiu, A. K. Aggarwal*
2BBK	Refined Crystal Structure of Methylamine Dehydrogenase from Paracoccus Denitrificans at 1.75 A Resolution; *L. Chen, F. S. Mathews*
2BBM	Solution Structure of a Calmodulin-target Peptide Complex by Multidimensional NMR; *G. M. Clore, A. Bax, M. Ikura, A. M. Gronenborn*
2CBA	Structure of Native and Apo Carbonic Anhydrase II and Structure of Some of its Anion-ligand Complexes; *K. Hakansson, M. Carlsson, L. A. Svensson, A. Liljas*
2CTN	Structure of Calcium-saturated Cardiac Troponin C, NMR, 30 Structures; *S. K. Sia, M. X. Li, L. Spyracopoulos, S. M. Gagne, W. Liu, J. A. Putkey, B. D. Sykes*
2DRP	The Crystal Structure of a Two Zinc-finger Peptide Reveals an Extension to the Rules for Zinc-finger/DNA Recognition; *L. Fairall, J. W. R. Schwabe, L. Chapman, J. T. Finch, D. Rhodes*
2GLI	Five-finger Gli-DNA Complex; *N. P. Pavletich, C. O. Pabo*
2HRV	2A Cysteine Proteinase from Human Rhinovirus 2; *J. F. W. Petersen, M. M. Cherney, H.-D. Liebig, T. Skern, E. Kuechler, M. N. G. James*
2MAT	E. Coli Methionine Aminopeptidase at 1.9 angstrom Resolution; *W. T. Lowther, A. M. Orville, D. T. Madden, S. Lim, D. H. Rich, B. W. Matthews*
2MHU	Three-dimensional Structure of Human [113Cd7]metallothionein-2 in Solution Determined by Nuclear Magnetic Resonance Spectroscopy; *W. Braun, B. A. Messerle, A. Schaeffer, M. Vasak, J. H. R. Kaegi, K. Wuthrich*
2MRB	Three-dimensional Structure of Rabbit Liver [Cd7]metal-lothionein-2a in Aqueous Solution Determined by Nuclear Magnetic Resonance;

2MRT   W. Braun, A. Arseniev, P. Schultze, E. Woergoetter, G. Wagner, M. Vasak, J. H. R. Kaegi, K. Wuthrich

2MRT   Conformation of Cd-7 Metallothionein-2 from Rat Liver in Aqueous Solution Determined by Nuclear Magnetic Resonance Spectroscopy; W. Braun, P. Schultze, E. Woergoetter, G. Wagner, M. Vasak, J. H. R. Kaegi, K. Wuthrich

2MSB   Structure of a C-type Mannose-binding Protein Complexed with an Oligosaccharide; W. I. Weis, K. Drickamer, W. A. Hendrickson

2NLL   Retinoid X Receptor-thyroid Hormone Receptor DNA-binding Domain Heterodimer Bound to Thyroid Response Element DNA; F. Rastinejad, T. Perlmann, R. M. Evans, P. B. Sigler

2OHX   Refined Crystal Structure of Liver Alcohol Dehydrogenase-Nadh Complex at 1.8 angstroms Resolution; S. Al-Karadaghi, E. S. Cedergren-Zeppezauer

2OXI   Refined Crystal Structure of Cu-substituted Alcohol Dehydrogenase at 2.1 angstroms Resolution; S. Al-Karadaghi, E. S. Cedergren-Zeppezauer

2PF2   The Ca2+ Ion and Membrane Binding Structure of the Gla Domain of Ca-prothrombin Fragment 1; M. Soriano-Garcia, K. Padmanabhan, A. M. de Vos, A. Tulinsky

2POO   Thermostable Phytase in Fully Calcium Loaded State; N.-C. Ha, B.-H. Oh

2PRK   Synchrotron X-ray Data Collection and Restrained Least-squares Refinement of the Crystal Structure of Proteinase K at 1.5 A Resolution; C. Betzel, G. P. Pal, W. Saenger

2PSR   Human Psoriasin (S100A7) Ca2+ and Zn2+ Bound Form (Crystal Form II); D. E. Brodersen, J. Nyborg, M. Kjeldgaard

2RAN   Rat Annexin V Crystal Structure: Ca(2+)-induced Conformational Changes; N. O. Concha, J. F. Head, M. A. Kaetzel, J. R. Dedman, B. A. Seaton

2SCP   Structure of a Sarcoplasmic Calcium-binding Protein from Nereis Diversicolor Refined at 2.0 A Resolution; W. J. Cook, S. Vijay-Kumar

2SOD   Determination and Analysis of the 2 A-structure of Copper, Zinc Superoxide Dismutase; J. A. Tainer, E. D. Getzoff, J. S. Richardson, D. C. Richardson

2SPT   Differences in the Metal Ion Structure between Sr- and Ca-prothrombin Fragment 1; A. Tulinsky

2SRT   Catalytic Domain of Human Stromelysin-1 at pH 5.5 and 40Oc Complexed with Inhibitor; P. R. Gooley, J. F. O'Connell

2ST1   The Three-dimensional Structure of Bacillus Amyloliquefaciens Subtilisin at 1.8 A and an Analysis of the Structural Consequences of Peroxide Inactivation; R. Bott

2TCI   X-ray Crystallographic Studies on Hexameric Insulins in the Presence of Helix-stabilizing Agents, Thiocyanate, Methylparaben, and Phenol; J. L. Whittingham, E. J. Dodson, P. C. E. Moody, G. G. Dodson

2TCL   Structure of the Catalytic Domain of Human Fibroblast Collagenase Complexed with an Inhibitor; F. K. Winkler, N. Borkakoti, A. D'Arcy

2TMN   Crystallographic Structural Analysis of Phosphoramidates as Inhibitors and Transition-state Analogs of Thermolysin; D. E. Tronrud, A. F. Monzingo, B. W. Matthews

2USH   5'-Nucleotidase from E. Coli; T. Knofel, N. Strater

2XAT   Complex of the Hexapeptide Xenobiotic Acetyltransferase with Chloramphenicol and Desulfo-coenzyme A; T. W. Beaman, M. Sugantino, S. L. Roderick

3AYK   Catalytic Fragment of Human Fibroblast Collagenase Complexed with Cgs-27023A, NMR, Minimized Average Structure; R. Powers, F. J. Moy

3BAM   Restriction Endonuclease Bamhi Complex with DNA and Manganese Ions (Post-reactive Complex); H. Viadiu, A. K. Aggarwal

3BTO   Horse Liver Alcohol Dehydrogenase Complexed to Nadh and (1S,3S)3-Butylthiolane 1-Oxide; S. Ramaswamy, B. V. Plapp

3CLN   Structure of Calmodulin Refined at 2.2 A Resolution; Y. S. Babu, C. E. Bugg, W. J. Cook

3CTN   Structure of Calcium-saturated Cardiac Troponin C, NMR, 30 Structures; S. K. Sia, M. X. Li, L. Spyracopoulos, S. M. Gagne, W. Liu, J. A. Putkey, B. D. Sykes

3EGF   Solution Structure of Murine Epidermal Growth Factor Determined by NMR Spectroscopy and Refined by Energy Minimization with Restraints; G. T. Montelione, K. Wuthrich, H. A. Scheraga

3MAT   E. Coli Methionine Aminopeptidase Transition-state Inhibitor Complex; W. T. Lowther, A. M. Orville, D. T. Madden, S. Lim, D. H. Rich, B. W. Matthews

3PAL   Ionic Interactions with Parvalbumins. Crystal Structure Determination of Pike 4.10 Parvalbumin in Four Different Ionic Environments; J. P. Declercq, B. Tinant, J. Parello, J. Rambaud

3RPB   The C2B-domain of Rabphilin: Structural Variations in a Janus-faced Domain; J. Ubach, J. Garcia, M. P. Nittler, T. C. Sudhof, J. Rizo

3SGA   Structures of Product and Inhibitor Complexes of Streptomyces Griseus Protease A at 1.8 A Resolution. A Model for Serine Protease Catalysis; A. R. Sielecki, M. N. G. James

3TDT   Complex of Tetrahydrodipicolinate N-succinyltransferase with 2-Amino-6-Oxopimelate and Coenzyme A; T. W. Beaman, J. S. Blanchard, S. L. Roderick

3TMN   The Binding of L-valyl-L-tryptophan to Crystalline Thermolysin Illustrates the Mode of Interaction of a Product of Peptide Hydrolysis; H. M. Holden, B. W. Matthews

3ZNF   High-resolution Three-dimensional Structure of a Single Zinc Finger from a Human Enhancer Binding Protein in Solution; Zinc Finger DNA Binding Domain

456C   Crystal Structure of Collagenase-3 (Mmp-13) Complexed to a Diphenyl-ether Sulphone Based Hydroxamic Acid; B. Lovejoy, A. Welch, S. Carr, C. Luong, C. Broka, R. T. Hendricks, J. Campbell, K. Walker, R. Martin, H. Van Wart, M. F. Browner

4AAH   Methanol Dehydrogenase from Methylophilus W3A1; F. S. Mathews, Z.-X. Xia

4INS   The Structure of 2Zn Pig Insulin Crystals at 1.5 A Resolution; G. G. Dodson, E. J. Dodson, D. C. Hodgkin, N. W. Isaacs, M. Vijayan

4KBP   Kidney Bean Purple Acid Phosphatase; T. Klabunde, N. Strater, B. Krebs

4MT2   Comparison of the NMR Solution Structure and the X-ray Crystal Structure of Rat Metallothionein-2; A. H. Robbins, C. D. Stout

4TLN   Binding of Hydroxamic Acid Inhibitors to Crystalline Thermolysin Suggests a Pentacoordinate Zinc

PDB-Code Listing

	Intermediate in Catalysis; *B. W. Matthews, M. A. Holmes*	6ADH	Structure of a Triclinic Ternary Complex of Horse Liver Alcohol Dehydrogenase at 2.9 A Resolution; *H. Eklund*
4TMN	Slow- and Fast-binding Inhibitors of Thermolysin Display Different Modes of Binding: Crystallographic Analysis of Extended Phosphonamidate Transition-state Analogues; *H. M. Holden, D. E. Tronrud, A. F. Monzingo, L. H. Weaver, B. W. Matthews*	6TMN	Structures of Two Thermolysin-inhibitor Complexes that Differ by a Single Hydrogen Bond; *D. E. Tronrud, H. M. Holden, B. W. Matthews*
		7INS	Structure of Porcine Insulin Cocrystallized with Clupeine Z; *P. Balschmidt, F. B. Hansen, E. Dodson, G. Dodson, F. Korber*
4ZNF	High-resolution Three-dimensional Structure of a Single Zinc Finger from a Human Enhancer Binding Protein in Solution; *A. M. Gronenborn, G. M. Clore, J. G. Omichinski*	7TLN	Structural Analysis of the Inhibition of Thermolysin by an Active-site-directed Irreversible Inhibitor; *B. W. Matthews, M. A. Holmes, D. E. Tronrud*
5CPA	Refined Crystal Structure of Carboxypeptidase A at 1.54 A Resolution; *W. N. Lipscomb*	7ZNF	Alternating Zinc Fingers in the Human Male Associated Protein ZFY: 2D NMR Structure of an Even Finger and Implications for 'Jumping-linker' DNA Recognition; *M. Kochoyan, H. T. Keutmann, M. A. Weiss*
5P2P	X-ray Structure of Phospholipase A2 Complexed with a Substrate-derived Inhibitor; *B. W. Dijkstra, M. M. G. M. Thunnissen, K. H. Kalk, J. Drenth*		
5PTP	Structure of Hydrolase (Serine Proteinase); *R. M. Stroud, J. Finer-Moore*	830C	Collagenase-3 (Mmp-13) Complexed to a Sulphone-based Hydroxamic Acid; *B. Lovejoy, A. Welch, S. Carr, C. Luong, C. Broka, R. T. Hendricks, J. Campbell, K. Walker, R. Martin, H. Van Wart, M. F. Browner*
5TLN	Binding of Hydroxamic Acid Inhibitors to Crystalline Thermolysin Suggests a Pentacoordinate Zinc Intermediate in Catalysis; *B. W. Matthews, M. A. Holmes*		
5TMN	Slow- and Fast-binding Inhibitors of Thermolysin Display Different Modes of Binding: Crystallographic Analysis of Extended Phosphonamidate Transition-state Analogues; *H. M. Holden, D. E. Tronrud, A. F. Monzingo, L. H. Weaver, B. W. Matthews*	8ADH	Interdomain Motion in Liver Alcohol Dehydrogenase. Structural and Energetic Analysis of the Hinge Bending Mode; *T. A. Jones, H. Eklund*
		9PAP	Structure of Papain Refined at 1.65 A Resolution; *I. G. Kamphuis, J. Drenth*
5TNC	Refined Crystal Structure of Troponin C from Turkey Skeletal Muscle at 2.0 A Resolution; *O. Herzberg, M. N. G. James*	1CGF	Crystal Structures of Recombinant 19-kDa Human Fibroblast Collagenase Complexed to Itself; *B. Lovejoy, A. M. Hassell, M. A. Luther, D. Weigl, S. R. Jordan*
5ZNF	Alternating Zinc Fingers in the Human Male Associated Protein ZFY: 2D NMR Structure of an Even Finger and Implications for 'Jumping-linker' DNA Recognition; *M. Kochoyan, H. T. Keutmann, M. A. Weiss*	I4A	Crystal Structure of Phosphorylation-mimicking Mutant T6D of Annexin IV; *M. A. Kaetzel, Y. D. Mo, T. R. Mealy, B. Campos, W. Bergsma-Schutter, A. Brisson, J. R. Dedman, B. A. Seaton*

Index

α-class carbonic anhydrases (α-CA) 249–263, *249, 254, 256, 257, 259, 260*
   inhibition constants 279–280
α-tocopherol 688–689, *688*
α$_2$-macroglobulin 192
   astacin 119
2Zn insulin 368
aAC *see* anthrax adenylyl cyclase
AAP *see* Aeromonas proteolytica
aceticlastic pathway 271–272, *272*
Acetivibrio cellulolyticus 618, *620*
activation factor-1 (AF-1) 326
actomyosin 460
acute chest syndrome 678
acute intermittent poryphria 632
acute pancreatitis 678
ADAMs, astacin 116, 120
adenylate kinase 407
ADHs *see* alcohol dehydrogenases
ADR1 319–320
Aeromonas
   A. hydrophila, metallo β-lactamases 219
   A. proteolytica (AAP)
      [D]-aminopeptidase DppA 212, *213*
      cocatalytic zinc sites 425–427
      leucine aminopeptidase 206
AF-1 *see* activation factor-1
afterhyperpolarization (AHP), calmodulin 473, 482, 484
Agkistrodon 677, 681, 682, 684
AHP *see* afterhyperpolarization
AID *see* autoinhibitory domain
ALA *see* 5-aminolaevulinic acid
[D]-Ala-[D]-Ala carboxypeptidase (DDC) 164–175, *164–173*
ALAD *see* 5-aminolaevulinic acid dehydratase
alcohol dehydrogenases (ADHs) 5–33
   ligand complexes 9–11
   phylogenies 6–7
   structural diagrams *5, 13, 16–19, 21, 23, 24, 26*
   structural zinc sites *404, 406*
ALG-2, penta EF-hand proteins 516, 517, 519, 520–526, *523, 525, 526*
alkaline phosphatases (AP) 71–82, *71, 73, 74, 75, 77, 79*
   two–metal ion mechanism 59
allosterism
   insulin 368–369
   metal binding sites 289–290, *289, 290*
Alzheimer's disease
   hemopexin 632
   metallothioneins 354
   neprilysin 105
   pitrilysins 194
   6-pyruvol-tetrahydropterin synthase 297
Amadori rearrangements 235, 243
amastatin 200–201, *203, 204*
amidohydrolases *418–419*, 424–425
5-aminolaevulinic acid (ALA) 283–284
5-aminolaevulinic acid dehydratase (ALAD) 283–295, *283, 287–290, 293*

p-aminomethylbenzenesulfonamide (*p*AMBS) 253
[D]-aminopeptidase DppA 208–214, *208, 210–213*
aminopeptidases *419*, 425–426
AMP, 5'-nucleotidase 62–69, *65, 66, 69*
Anas platyrhynchos 178
anchoring protein receptor for activated C-kinases (RACK) 590
angiogensis 96
annexins 649–663
   annexin A1 *649*, 650–653, *655*, 655–656, 657–658, *658*
   annexin A2 650–653, *656*, 658–659, *659*
   annexin A4 650, 651, 653, 656–657, *659*, 659–661
   annexin A5 650–653, *654*, 657, *660, 661*
   annexin A6 650–652
anthrax adenylyl cyclase (aAC) 451, *454*
AP *see* alkaline phosphatases
apamin 482–484
apo cbEGFs *557, 558*
apo-Gla-domain 581, *581*
apo-m-calpain 493–494, *495*, 495, 498
apo-MBP-C 714
AR-JP *see* autosomal recessive juvenile parkinsonism
Arabidopsis
   A. thaliana
      cocatalytic zinc sites 428
      Eps 15 homology domain 542
      FYVE domain 391, 392
      γ-class carbonic anhydrases 273
arachidonic acid 678
Archaeglobus fulgidus 67
Argobacterium tumefaciens 308–309
aspartate carbamoyl transferase 403, 407
Aspergillus
   A. oryzae, nuclease P1 52
   calmodulin 472
astacin *116*, 116–129, *120–126*
   matrix metalloproteinases 144
Astacus astacus 116, 117, 119
ATF-2 315
atherosclerosis, hemopexin 632
ATPase, [D]-aminopeptidase DppA 213
autoinhibitory domain (AID) 448, 453
autosomal recessive juvenile parkinsonism (AR-JP) 348
Axl1, pitrilysins 194–195

β-class carbonic anhydrases (β-CA) *264, 266, 267*, 264–269
B-DNA, nuclease P1 56
β-hydroxylation, calcium-binding EGF-like domains 555–557
β-lactamases 217–234, *217, 222–226*
   cocatalytic zinc sites *420*, 428–429
β-lactams, [D]-Ala-[D]-Ala carboxypeptidase 174
Bacillus
   B. anthracis, calmodulin 484
   B. cereus
      cocatalytic zinc sites 423, 428–429
      metallo β-lactamases 217–232, *224*
      neprilysin 111
      nuclease P1 54, 55

Index

Bacillus (continued)
  B. fragilis, metallo β-lactamases 217–232, *217*, *222–224*
  B. methanolicus, [D]-aminopeptidase DppA 209
  B. pumilus, [D]-aminopeptidase DppA 208
  B. sphaericus, carboxypeptidases 178
  B. subtilis
    alkaline phosphatases 80–81
    [D]-aminopeptidase DppA 208–214
    cocatalytic zinc sites 427
    protein interface zinc sites 438
  B. thermoproteolyticus, thermolysin 85, 86
  Bacillus Ak.1 proteinase 727
  subtilisin 719–721, 727
bacitracin 192
bacterial trypsin-like proteases 724–725
Bam HI 742, 744–747, *746*, *747*, 750–751, *752*, *753–754*
barium, C-type lectin-like domains 706
base recognition, nuclease P1 58–59
basement membrane protein BM-40 509–515, *509*, *511–514*
batimastat 135, 138, *140*
Bemisia argentifolii 8
benzylpenicillin
  [D]-Ala-[D]-Ala carboxypeptidase 174
  metallo β-lactamases 227–229
bestatin 200–201, *203*, 204, *204*
BH$_4$ *see* tetrahydrobiopterin
bicarbonate, leucine aminopeptidase 202–203, 205
biphenyl tetrazole inhibitors 225
blood coagulation factors 724
BM-40 *see* basement membrane protein
Boophilus microplus 63, 67
Bos taurus
  carboxypeptidases 178–179
  leucine aminopeptidase 200
  5'-nucleotidase 63
  S100 proteins 529
Bothrops asper 678
Botulinum Neurotoxin Type B protease 93
BRCA1 protein 347
bungarotoxins 685–686
Bungarus candidus 679
Burton's tyrosine kinase 407

C-cadherin 731, *732*, 733, 735, *735*
c-Cbl-Ubch7 complex 345–347, *346*
C-terminal domains (CTDs) 631–634, *631*, *637*, *639*, *639–643*, *640*, *643*
C-type lectin-like domains (CTLDs) 704–717, *704*, *707*, *712*, *715*
C2-domain proteins 599–613, *599*, *602*, *605*, *608*, *609*
C2-like domains, membrane binding 587–598, *587*, *589*, *590*, *594–596*
C2H2 *see* Cys$_2$His$_2$
CA *see* carbonic anhydrases
CaaX motif 37, 39, 43–44
cadherins 731–741, *739*
  C-cadherin 731, *732*, 733, 735, *735*
  desmocollins 731
  desmogleins 731
  E-cadherin 731, *731*, 733, 734–738, *736*, *738*, 739
  Flamingo cadherins 731
  N-cadherin 731, *732*, 733, 735
  P-cadherin 731

  protocadherins 731
  R-cadherin 731
  VE-cadherin 732, *733*, 739
cadmium
  alkaline phosphatases 73
  cocatalytic zinc sites 424
  matrix metalloproteinases 140
  medium-chain dehydrogenases/reductases 6, 20
  metallo β-lactamases 220–221
  metallothioneins 353, 362–363
  pitrilysins/inverzincins 191
  RING domain proteins 342
  serralysin 155
  thermolysin 92
  type II restriction endonucleases 743
Caenorhabditis elegans
  basement membrane protein BM-40 510
  C2-domain proteins 607
  cadherins 732
  calmodulin 472, 474
  Eps 15 homology domain 542
  FYVE domain 391
  LIM domain proteins 378, 380
  neprilysin 105
  nuclear hormone receptors 325
  protein prenyltransferases 38
calbindin 532, *533*
calcineurin
  calmodulin 453
  5'-nucleotidase 63
calcium
  annexins 649–663
  astacin 127
  basement membrane protein BM-40 509–515
  C-type lectin-like domains 704–717
  C2-domain proteins 599–613
  cadherins 731–741
  calmodulin 447–458, 471–488
  calpain 489–500
  calsequestrin 692–703
  carboxypeptidases 176–189
  dockerin domains 617–628
  Eps 15 homology domain 541–550
  Gla-domain 573–583
  hemopexin domains 631–646
  inverzincins 191
  matrix metalloproteinases 139–140
  membrane binding C2-like domains 587–598
  parvalbumin 501–508
  penta EF-hand proteins 516–528
  phospholipase A$_2$ 677–691
  pitrilysins 191
  S100 proteins 529–540
  SERCA1 667–676
  serralysin 152–153, *152*, *153*
  subtilisin 718–721, 725–728
  troponin C 459–470
  trypsin 718–725, 728
  type II restriction endonucleases 743–744, 750, 751, 753
calcium activated potassium channels (SK channels) 471–488, 486
calcium binding regions (CBRs) 593–595
calcium-binding EGF-like domains (cbEGFs) 553–570, *553*, *556*, *557*, *559–566*

Index

calcium-binding loop 677
calcium-bridges 657
calcium-dependent adherent proteins *see* cadherins
Callinectes sapidus, metallothioneins 353–354, 360–361
calmodulin (CaM) 447–458, *447, 449–451, 454, 455*
   annexins 652
   calcium activated potassium channels 471–488, *471, 478–483, 485, 506*
   calsequestrin 692, 698
   carbonic anhydrases 270–281, *274, 275, 277, 281*
calpains 489–500, *489–491, 494–496*
   penta EF-hand proteins 517, 518, 520, 522, *524*
calreticulin (CRT) 692, 693
calsequestrin 692–703, *692, 696–698, 700*
Cam *see* calmodulin
cancer
   PFT inhibitors 46–47
   protein prenyltransferases 38, 46–47
canonical EF-hand
   annexins 655
   Eps 15 homology domain 547
   penta EF-hand proteins 522
   S100 proteins *529*, 531, 534–535, *535, 536*
carbapenems 217–218, 227–229
carbohydrate recognition domains (CRDs)
   C-type lectin-like domains 704, *705–708, 707, 709, 715*
   phospholipase A_2 682
carbon monoxide dehydrogenase (CODH) 271–272
carbonic anhydrase activators 258–259
carbonic anhydrase inhibitors 256–257
carbonic anhydrases (CA)
   α-class 249–263, *249, 254, 260,* 279–280
      binding sites *256, 257, 259*
   β-class 264–269, *264, 266, 267*
   γ-class 270–282, *270, 274, 275, 277, 281*
carboxypeptidases (CP) 176–189, *176, 180–182, 184, 187*
cardiac isoform of troponin C (cTnC) 459, 461, 463, 467–469, *468*
cartilage oligomeric matrix protein (COMP) 738
casein kinase II (CK2) 408
catalytic domains (CAT) 135–137
cbEGFs *see* calcium-binding EGF-like domains
CBRs *see* calcium binding regions
cellobiohydrolase (CeIS) 622–624
cellulosome 617, 618
cephaloridine 227–229
cephalosporins
   [D]-Ala-[D]-Ala carboxypeptidase 174
   metallo β-lactamases 217–218, 227–229
cephalothin 174
CGS 31447, neprilysin *108*, 110–113, *110, 112*
CH *see* cyclohydrolases
chemical rescue 259–260
Chionoecetes opilio 116, 117
Chlorobium 297
chloroplast processing peptidase (CPE) 195
cholesterol metabolism 553
chronic neuromuscular diseases 632
chymotrypsin 723
CK2 *see* casein kinase II
Clostridium
   C. beijerinckii 6
   C. perfringens
      cocatalytic zinc sites 423

   membrane binding C2-like domains 588
   nuclease P1 54
   C. thermocellum, dockerin domains 618–620, 623, 625–626
   dockerin domains 618, 619
clotrimazole 455
CNG *see* cyclic nucleotide gated
cobalt
   α-class carbonic anhydrases 257–258
   alkaline phosphatases 73
   astacin *122*, 126, 127
   cocatalytic zinc sites 426
   γ-class carbonic anhydrases 274–275, 278–279
   leucine aminopeptidase 200
   LIM domain proteins 383
   medium-chain dehydrogenases/reductases 6, 20
   metallo β-lactamases 220
   methionine aminopeptidases 95–103
   5′-nucleotidase 64
   pitrilysins/inverzincins 191
   6-pyruvol-tetrahydropterin synthase 299
   RING domain proteins 342
   serralysin 155
   thermolysin 92
   type II restriction endonucleases 743
cocatalytic zinc sites 416–431, *417–420, 422, 427*
   amidohydrolases *418–419,* 424–425
   aminopeptidases *419,* 425–426
   β-lactamases *420,* 428–429
   glyoxalase II *419,* 428
   hydrolases *417–420, 420–429*
   methionine aminopeptidases *419,* 426–427
   oxidoreductases *417,* 420
   phosphatases *417–418,* 420–424
   phosphotriesterases *418*
   purple acid phosphatases *418,* 424
   superoxide dismutases 416, *417, 420*
   UDP-sugar hydrolases *418,* 424
CODH *see* carbon monoxide dehydrogenase
cohesins, dockerin domains 617–620, 625–627, *626*
coiled coil regions 397
collagenases 130
   MMP-13 C-terminal domain 639–643
collectins 704, *705*
COMP *see* cartilage oligomeric matrix protein
constitutive, tethering interactions 481–482
Conus
   C. magus, phospholipase A_2 679
   Gla-domain 573
copper
   astacin *122*, 126, 127–128
   cocatalytic zinc sites 416
   LIM domain proteins 383
   matrix metalloproteinases 140
   metallothioneins 362–363
   nuclease P1 60
   pitrilysins/inverzincins 191
   S100 proteins *529,* 531, 538
   serralysin 155
Core proteins 195–196, *196*
CP *see* carboxypeptidases
CPE *see* chloroplast processing peptidase
cPLA2 *see* phospholipase A2
CRDs *see* carbohydrate recognition domains
Crohn's disease, phospholipase A_2 678

Index

Crotalus, phospholipase A_2 679, 686, 687
Crotoxin, phospholipase A_2 686
CRP *see* cysteine-rich proteins
CRT *see* calreticulin
CTDs *see* C-terminal domains
CTLDs *see* C-type lectin-like domains
cTnC *see* cardiac isoform of troponin C
cyanobacterial Smta 355, 361–362
cyclic nucleotide gated (CNG) channels 485
cyclohydrolases (CH) 235–245, *235*, *240–243*
Cys_2His_2 zinc finger proteins 307–323, *307*
    ADR1 319–320
    ATF-2 315
    design/selection 321
    DNA bound *311*, 315–321
    GAGA factor 319, *319*
    glioblastoma oncogene 317, 321
    SW15 finger 1 314–315, *314*
    TFIIIA 318–319
    TGEKP linker sequence 308, 320–321, *320*
    tramtrack 317–318, 321
    Xfin finger 31 310–312, *311*
    Ying-Yang protein 318
    ZFY finger 6 313–314, *313*
    Zif268 315–317, *315*, *316*, 321
cysteine switches 132, 142
cysteine-rich proteins (CRP) 379–380, *382*, 382, *384*, *385*, 386–387
cytochrome c oxidase 406–407
cytokines 705
cytoskeleton linkages 649
cytosol 471

Daboia
    D. russelli pulchella 688
    D. russelli siamensis 682
N-dansylated oligopeptides 121
DAPS *see* 1,2-diacetyl-*sn*-phosphatidyl-[L]-serine
dAp(S)dA, nuclease P1 57, 58
daunorubicin 699
DBD *see* DNA binding domain
DBH *see* dibasic hydrophobic
DCPA *see* 1,2-dicaproyl-*sn*-phosphatidic acid
DDC *see* [D]-Ala-[D]-Ala carboxypeptidase
dehydrogenases *404*, 406
des-penta insulin 368
desmocollins 731
desmogleins 731
1,2-diacetyl-*sn*-phosphatidyl-[L]-serine (DAPS) 594–596, *594*, *595*
diaryl succinic acid inhibitors 225–226, *226*
dibasic hydrophobic (DBH) sequence 698
1,2-dicaproyl-*sn*-phosphatidic acid (DCPA) 594–596, *594*
Dictyostelium discodeum 516
7,8-dihydroneopterin aldolase 303
dihydroneopterin triphosphate (H_2NTP) 235, 237, 296, 299, 300–301
dioleoylphosphatidylcholine (DOPC) 671
4,7-dioxosebacic acid 291–293
Discopyge ommata 63
dithiophosphorylated oligonucleotides 58
DNA binding domain (DBD) 326–334, *327*, *330*, *332–334*, *333*

DNA bound Cys_2His_2 zinc finger proteins *311*, 315–321
DNA-dependent RNA polymerases (RNAP) 408
DNA ligase 411
DNA response elements (DREs) 324, 328–329
DNA transferases 408–409
dockerin domains 617–628, *617*, *623*, *624*, *626*
domain pairs 561–563, *562*, *563*
DOPC *see* dioleoylphosphatidylcholine
doxorubicin 699
DppA gene 208–214
DREs *see* DNA response elements
Drosophila melanogaster
    astacin 117
    basement membrane protein BM-40 510
    cadherins 732
    calcium-binding EGF-like domains 556
    calmodulin 448, 472
    carboxypeptidases 178
    Cys_2His_2 zinc finger proteins 308–309, 317–318, 319
    Eps 15 homology domain 542
    FYVE domain 391–393
    neprilysin 105
    nuclear hormone receptors 325
    protein prenyltransferases 38

E-cadherin *731*, 731, *733*, 734–738, *736*, *738*, *739*
E2(TG) state, SERCA1 674
E41A-sNTnC 466–468, *467*
EBIO *see* 1-ethyl-2-benzimidazolinone
EC *see* extracellular calcium
ECE *see* endothelin-converting enzymes
EEA1 *see* endosome antigen 1
EF-hand calcium binding proteins 502–503, *504*, *506*
    basement membrane protein BM-40 509–515
    C2-domain proteins 599–613
    calmodulin 447–458, 471–488
    calpain 489–500
    Eps 15 homology domain 541–550
    gating domains 471–488
    parvalbumin 501–508
    penta EF-hand proteins 516–528
    S100 proteins 529–540
    troponin C 459–470
EGF *see* epidermal growth factor
EGFs *see* epidermal growth factor-like domains
EH *see* Eps 15 homology
eicosanoids 678
elastase 723–724
endocytosis 649
endonucleases 410
endopeptidases
    subtilisin 718–721, 725–728
    trypsin 718–725, 728
endosome antigen 1 (EEA1) 390, 391, 394–398, *395–397*
endostatin 707
endothelial cell protein C receptor (EPCR) 579–580
endothelin-converting enzymes (ECE) 105
endotoxins 436
enkephalinase *see* neprilysin
Enterococcus
    E. faecium 167–169
    E. gallinarum 168–169, *168*
EPCR *see* endothelial cell protein C receptor

Index

epidermal growth factor (EGF) 549
epidermal growth factor-like domains (EGFs) 553–570, *553, 556, 557, 559–566*
Eps 15 homology (EH) domain 541–550, *541, 544–546, 548*
  S100 proteins 538
Epstein–Barr virus 707
Equus caballus 9–10
Erwinia 148–149, *150,* 154–155
Escherichia coli
  α-class carbonic anhydrases 253
  alkaline phosphatases 71–82, *71, 73–75, 77, 79*
  5-aminolaevulinic acid dehydratase 283–294, *283, 287–290, 293*
  [D]-aminopeptidase DppA 209
  annexins 650–651
  β-class carbonic anhydrases 264–269, *267*
  basement membrane protein BM-40 510
  C-type lectin-like domains 705
  calcium-binding EGF-like domains 555
  calmodulin 448
  calpain 490, 493, 497
  calsequestrin 696
  cocatalytic zinc sites 421–425
  Cys$_2$His$_2$ zinc finger proteins 308–309
  Eps 15 homology domain 543
  FYVE domain 392
  γ-class carbonic anhydrases 273, 274
  guanosine triphosphate cyclohydrolase I 235–241, *235, 240–243*
  inverzincins 190–198
  leucine aminopeptidase 200, 203
  medium-chain dehydrogenases/reductases 12
  membrane binding C2-like domains 591
  metallo β-lactamases 218
  metallothioneins 355
  methionine aminopeptidases 95–96
  nuclear hormone receptors 329
  5′-nucleotidase 62, 63, 67–69
  pitrilysins 190–198
  protein interface zinc sites 437, 438
  protein prenyltransferases 39
  6-pyruvol-tetrahydropterin synthase 297
  S100 proteins 531
  serralysin 149
  structural zinc sites 403–412
ESTs *see* expressed sequence tags
ethanol 8
1-ethyl-2-benzimidazolinone (EBIO) 484, *486*
exocytosis 649
expressed sequence tags (ESTs) 517–518
extracellular calcium (EC) binding domain 509–514, *514*

familial hypercholesterolemia 567
farnesoid X receptor (FXR) 325
farnesyl diphosphate (FPP) 39–45, *43, 44*
FDH *see* formaldehyde dehydrogenase
FENS-1 protein 398
fetal alcohol syndrome 8
fibrillins 554, *555,* 559, *560,* 564–565, 567
fibrinolysis 553
fibronectin 560
Flamingo cadherins 731
flexible Glu 19–21

Fluoribacter gormanii 219, 221, *225*
*Fok*I 744, *746,* 747–748, *748,* 750–751
follistatin-like (FS) domain 509–514, *514*
formaldehyde dehydrogenase (FDH) 6, 12, 20, *21,* 26
formamidopyrimidine-DNA glycosylase MutM (Fpg) family *405,* 410
Fpg *see* formamidopyrimidine-DNA glycosylase MutM
FPP *see* farnesyl diphosphate
free RING domains 343–344
Fringe, calcium-binding EGF-like domains 556
FS *see* follistatin-like
fumagillin 96, *98,* 100
functional bipartism 475, 477–480, *480*
FXR *see* farnesoid X receptor
FYVE domain 390–399, *390, 393–397*
  RING domain proteins 341, *341,* 344, 344–345
γ-class carbonic anhydrases 270–282, *270, 274, 275, 277, 281*

Gadus callarias 506–507
GAG *see* glycosaminoglycan
GAGA factor 319, *319*
galactose-1-phosphate uridylyltransferase 409
Gallus gallus
  calmodulin 474
  nuclear hormone receptors 325
  S100 proteins 529
gating domains 471–488
'gating' loops 18
gating triggers 480–481
GDH *see* glucose dehydrogenase
gelatinases 130
GEMSA *see* guanidinoethylmercaptosuccinic acid
geranylgeranyl diphosphate (GGPP) 39, 42–43
GFRP *see* GTP cyclohydrolase feedback regulatory protein
GGPP *see* geranylgeranyl diphosphate
GIF *see* growth inhibitory factors
GIPL3 *see* glycoinositolphospholipid
Gla-domains 573–583, *573, 575, 578, 579, 580*
  C2-domain proteins 599–613
glioblastoma oncogene (GLI) 317, 321
glucose dehydrogenase (GDH) 5, 6, 13
glutamate decarboxylase *454*
glutamates 432–433
glycoinositolphospholipid (GIPL) 158
glycophosphatidylinositol (GPI) anchors
  leishmanolysin 158
  membrane–anchoring signal 133
  5′-nucleotidase 63
glycosaminoglycan (GAG) modification *514*
glyoxalase II *419,* 428
gp63 *see* leishmanolysin
GPI *see* glycophosphatidylinositol
grancalcin 516–518, 520–522
Grave's hyperthyroidism 693
Grotthus diffusion 255
growth inhibitory factors (GIF) 353–354
GTP *see* guanosine 5′-triphosphate
GTP-CH-I *see* guanosine triphosphate cyclohydrolase I
guanidinoethylmercaptosuccinic acid (GEMSA) 184, *184*
guanosine 5′-triphosphate cyclohydrolase feedback regulatory protein (GFRP) 235, 237–238, *242*
guanosine 5′-triphosphate (GTP) 297

Index

guanosine triphosphate cyclohydrolase I (GTP-CH-I) 235–245, *235, 240, 241, 242, 243*

H_2NTP *see* dihydroneopterin triphosphate
HAD *see* haloacid dehalogenase
Haemophilus influenzae
   hemopexin 632
   5′-nucleotidase 67
haloacid dehalogenase (HAD) 673
Halobacterium halobium 80–81
Hashimoto's thyroiditis 693
HB-EGF *see* heparin-binding epidermal growth factor–like growth factor
hCA II *see* human carbonic anhydrase II
HD *see* homeodomains
HDs *see* hemopexin domains
heavy metal detoxification 354
Helicoverpa armigera 178–180, 187
Heliobacter pylori 742
helix-loop-helix (HLH) motif 476
hemicellulose 617
hemopexin domains (HDs) 631–646, *631, 635, 637–640*
hemopexin-like (PEX) domains 133, 135, 142
hemophilia B 567
heparin-binding epidermal growth factor–like growth factor (HB-EGF) 194
heparin-binding group IIA 678
Herzberg, Moult and James (HMJ) model 461–462, 466
histamine 258–259
histone lysine methyl transferases 412
HIV integrase 411
HLH *see* helix-loop-helix
HMJ *see* Herzberg, Moult and James model
holo cbEGFs 557, *558*
Homarus americanus 353–354, 355, 360–361
homeobox (HOX) 378
homeodomains (HD) 378–381
Homo sapiens
   α-class carbonic anhydrases 250–251
   astacin 117, *118*
   basement membrane protein BM-40 510
   cadherins 732
   calmodulin 474
   calpain 489–499
   carboxypeptidases 178–179
   Cys_2His_2 zinc finger proteins 308–309
   Eps 15 homology domain 542
   FYVE domain 390–398
   Gla-domain 576
   matrix metalloproteinases 131–133
   medium-chain dehydrogenases/reductases *10, 11*
   metallothioneins 354, 358–360
   neprilysin 105, 109–114
   nuclear hormone receptors 325–326, *325–326*
   5′-nucleotidase 63
   S100 proteins 529
homophilic celle–cell adhesion, cadherins 731–740
HOX *see* homeobox
HPEX *see* hemopexin
human carbonic anhydrase II (hCA II) 249, 250–260, *256, 257, 259, 260*
hydride transfer 5–6, 18, 28–30
hydrolases
   cocatalytic zinc sites *417–420*, 420–429
   ester bond active
      alkaline phosphatases 71–82
      nuclease P1 51–61
      5′-nucleotidase 62–70
   non-peptidic C-N bond active
      guanosine triphosphate cyclohydrolase I 235–245
      metallo β-lactamases 217–234
   peptidic C-N bond active
      [D]-Ala-[D]-Ala carboxypeptidase 164–175
      [D]-aminopeptidase DppA 208–214
      astacin 116–129
      carboxypeptidases 176–189
      inverzincins 190–198
      leishmanolysin 157–163
      leucine aminopeptidase 199–207
      matrix metalloproteinases 130–147
      methionine aminopeptidases 95–103
      neprilysin 104–115
      pitrilysins 190–198
      serralysin 148–156
      thermolysin 85–94
   structural zinc sites *405*, 409–411
hydrophobic anchor points 481
4-hydroxynonenal 8
hyperphenylalaninemia
   guanosine triphosphate cyclohydrolase I 236
   6-pyruvol-tetrahydropterin synthase 297, 298

I-EGFs *see* integrin epidermal growth factor-like domains
IBS *see* inflammatory bowel disease
IDE *see* insulin-degrading enzyme
IEEHV *see* immediate early equine herpesvirus
IK *see* intermediate conductance potassium channels
immediate early equine herpesvirus (IEEHV) 342, 343–345, *344*
inflammatory bowel disease (IBS) 678
insulin 367, 367–377
   allosterism 368–369
   protein interface zinc sites 433
   R_6 structures 373–376, *374, 375*
   $T_3R_3^f$ structures 369–373, *371–373*
   T_6 structures 368–370
insulin-degrading enzyme (IDE) 194
insulysin *see* insulin-degrading enzyme
integrin epidermal growth factor-like domains (I-EGFs) 559, 559–560
interface zinc sites 432–441
interferons 436
intermediate conductance potassium channels (IK) 473
Intersectin 543
intimin 707
invasin 707
inverzincins *190*, 190–198
iron
   astacin 127
   cocatalytic zinc sites 416, 424, 426
   γ-class carbonic anhydrases 274–275
   hemopexin domains 631–646
   methionine aminopeptidases 97
   5′-nucleotidase 67
   thermolysin 92
isomerases *405*, 411
isoprene diphosphate binding sites 42–43

junctin 700–701

KAH *see* 4-keto-5-aminolaevulinic acid
KELL, neprilysin 105
4-keto-5-aminolaevulinic acid (KAH) 291–292
ketose reductase (KR) 6
killer cell Ig-like receptors (KIR) 436
Klebsiella pneumoniae 218
Knorr-type condensations *284*
KR *see* ketose reductase
KTS linkers 308

L-EGFs *see* laminin epidermal growth factor-like domains
L1 lipase 410
LA *see* laevulinic acid
Lactobacillus delbrueckii 427
laevulinic acid (LA) 289, 291–292
laminin epidermal growth factor-like domains (L-EGFs) *559*, 559–560
lanthanides 507
LAP *see* leucine aminopeptidase
LBD *see* ligand binding domain
LCI *see* leech carboxypeptidase inhibitor
LDLR *see* low-density lipoprotein receptors
lectins 704–717
leech carboxypeptidase inhibitor (LCI) 183, 186–187, *187*
Leishmania 157–163
　L. major 158, 159
　L. pifanoi, nuclease P1 52
leishmanolysin 157–163, *157*, *160–162*
Lepidosiren paradoxus 529
[L]-leucinal 203, *203*, *204*
leucine aminopeptidase (LAP) 199–207, *199*, *202–204*
[L]-leucine phosphonic acid 203
leucine-rich α₂ glycoprotein (LRG) like domains 682
leukemia, annexins 656
leukocytes 705
leukotrienes 678
ligand binding domain (LBD) 326–331, *327*
ligases *405*, 411–412
LIM domain proteins *378*, 378–389
　cysteine-rich proteins 379–380, *382*, *382*, *384*, *385*, *386–387*
　RING domain proteins *341*, *341*, 344–345, *344*
　rubredoxin 'full knuckles' 383–385
LIM-homeodomain (LIM-HD) 378–381
LIM-only proteins (LMO) 378–380
linkers, Cys₂His₂ zinc finger proteins 308, 320–321, *320*
lipid modification 588
lipophosphoglycans (LPG) 158
Listeria
　L. innocua 209
　L. monocytogenes 734
lithosthine 707
liver X receptor (LXR) 325
LMO *see* LIM-only proteins
low-density lipoprotein receptors (LDLR) 554, 560, 561, 563, 565
LPG *see* lipophosphoglycans
LRG *see* leucine-rich α₂ glycoprotein
lumen regulation 693
Lutzomyia logipalpis 63

LXR *see* liver X receptor
lyases
　α-class carbonic anhydrases 249–263
　5-aminolaevulinic acid dehydratase 283–295
　β-class carbonic anhydrases 264–269
　Cys₂His₂ zinc finger proteins 307–323
　γ-class carbonic anhydrases 270–282
　6-pyruvol-tetrahydropterin synthase 296–304

m-calpain 489–500, *489–491*, *494–496*
maC₂H₂ *see* multi-adjacent-fingered Cys₂His₂
magnesium
　alkaline phosphatases 71, 73, 74
　5-aminolaevulinic acid dehydratase 288–290
　C-type lectin-like domains 706
　C₂-domain proteins 607
　calsequestrin 700
　cocatalytic zinc sites 416, 421–423
　dockerin domains 620
　Gla-domain 577, 581–582
　guanosine triphosphate cyclohydrolase I 237
　leucine aminopeptidase 200
　nuclease P1 59
　5′-nucleotidase 64
　parvalbumin 505, 507
　pitrilysins/inverzincins 191
　6-pyruvol-tetrahydropterin synthase 299
　S100 proteins 532
　troponin C 459
　type II restriction endonucleases 743–744, 751–753
major histocompatibility complex (MHC) 200
major surface proteinase (MSP) *see* leishmanolysin
manganese
　astacin 127
　cocatalytic zinc sites 424, 426
　inverzincins 191
　leucine aminopeptidase 200
　matrix metalloproteinases 140
　methionine aminopeptidases 97
　5′-nucleotidase 64–65, *65*
　pitrilysins 191
　serralysin 155
　thermolysin 92
　type II restriction endonucleases 743, 750, 751, 753
mannose binding proteins (MBPs) 705–714, *708*, *709*, *712*, *715*
mannose receptor (MR) 705
MARCKS *see* myristorylated alanine-rich C kinase substrate
Marfan syndrome 566, 567
matrix metalloproteinase inhibitors (MMPIs) 135
matrix metalloproteinases (MMPs) 130–147, *132*, *136*
　[D]-Ala-[D]-Ala carboxypeptidase 166–172, *168*, *169*
　astacin 116, 120
　hemopexin domains 631–634, *631*, *637*, *639–643*, *639*, *640*, *643*
　leishmanolysin 159, 162
　MMP-2 *130*, *141*
　MMP-8 *136*, *137*
　MMP-12 *138*, *139*
　MMP-TIMP complexes 131, 142, *143*
　structural zinc sites *405*, 409–410
MBPs *see* mannose binding proteins

Index

medium-chain dehydrogenases/reductases (MDRs) 5–33, *5, 9–11, 16, 19, 20*
membrane binding C2-like domains *see* C2-like domains, membrane binding
membrane interaction modules 390–399
membrane traffic 599–613
meprin 122
mercaptocarboxylate inhibitors 225–226, *226*
mercury
  α-class carbonic anhydrases 255–256
  astacin *122, 126, 127*
5-mercury-cytidine complex 57
MES *see* morpholineethanesulfone
Mesorhizobium loti 54
metal exchange 155
metal-dependent type II restriction endonucleases 742–756, *745, 746*
  Bam HI 742, 744–747, *746, 747,* 750–751, *752,* 753–754
  Fok I 744, *746,* 747–748, *748,* 750–751
  Nae I 744, *746,* 749–751, *749*
  Ngo MIV 744, *746,* 748–749, *748,* 750–751, 753–754
metallo β-lactamases 217–234, *217, 222–226*
metallocarboxypeptidases *see* carboxypeptidases
metallophosphoesterases 63
metallothioneins (MT) 353–364, *353, 358–360, 362*
MetAPs *see* methionine aminopeptidases
Methanobacterium thermoautotrophicum 265–269, 485
Methanosarcina thermophila 270–276
methionine 155
methionine aminopeptidases (MetAPs) 95–103, *95, 97, 98*
  cocatalytic zinc sites *419,* 426–427
methionine phosphonate 99, 100, *101*
methionine phosphonite 99, 100, *101*
methyl-laudanosine 483
4-methylimidazole (4-MI) 259–260, *259, 260*
methyltransferases (Mtases) 742
metzincins
  hemopexin domains 631
  leishmanolysin 157–163, *160*
  matrix metalloproteinases 130–147
  serralysin 149, 155
MHC *see* major histocompatibility complex
4-MI *see* 4-methylimidazole
Michaelis complexes
  alkaline phosphatases 79
  cocatalytic zinc sites 426
  metallo β-lactamases 227, 230, 231
  5'-nucleotidase 66
Micrococcus xanthus 165, *168,* 168–169
mitochondrial processing peptidases (MPPs) 195–197, *196, 197*
MLCK *see* myosin light chain kinase
MMPIs *see* matrix metalloproteinase inhibitors
MMPs *see* matrix metalloproteinases
molecular complementation 76
morpholineethanesulfone (MES) 221
MPPs *see* mitochondrial processing peptidases
MR *see* mannose receptor
MSP *see* leishmanolysin
MT *see* metallothioneins
Mtases *see* methyltransferases
MTMR *see* myotubularin-related proteins
multi-adjacent-fingered Cys$_2$His$_2$ (maC2H2) 308–309
multichain PLA$_2$s 685–687
multiple valency, FYVE domain 397–398

murine sonic hedgehog (SHH) 168–172, *168, 171, 172*
Mus
  M. muscularis
    α-class carbonic anhydrases 251, 252
    Cys$_2$His$_2$ zinc finger proteins 308–309
  M. musculus
    basement membrane protein BM-40 510
    cadherins 732
    calmodulin 474
    carboxypeptidases 178
    medium-chain dehydrogenases/reductases 10–11
    metallothioneins 354, 358–360
    neprilysin 105
    nuclear hormone receptors 325
    5'-nucleotidase 63
    S100 proteins 529
myosin light chain kinase (MLCK) 448, 450–451, 484
myosins 453, *454*
myotubularin-related proteins (MTMR) 392, 397
myristorylated alanine-rich C kinase substrate (MARCKS) 451

N-arginine dibasic convertase (NRD convertase) 194
N-cadherin 731, 732, *733,* 735
NAD(H)/NADP(H) 5–6, 13–15, 17–19, 24–30
*Nae*I 744, *746,* 749, 749–751
Naja naja naja 681
Nardilysin *see* N-arginine dibasic convertase
native tropomyosin 460
Nell-1/Nell-2 proteins 553
neprilysin (NEP) 104–115, *104, 109, 110, 112, 113*
nerve growth factor (NGF) 436–437
nerve growth factor–induced-B (NGFI-B) 325
neuromuscular diseases 632
neutral endopeptidase (NEP) *see* neprilysin
NGF *see* nerve growth factor
NGFI-B *see* nerve growth factor–induced-B
*Ngo*MIV 744, *746,* 748–749, *748,* 750–751, 753–754
nickel
  astacin *122, 126, 127*
  cocatalytic zinc sites 426
  matrix metalloproteinases 140
  medium-chain dehydrogenases/reductases 6
  methionine aminopeptidases 97
  serralysin 155
nitric oxide synthase 438
nitroanilides 121
nitrocefin 227–229, 231
Notch-1 proteins 554, 556, 564–565, 567
nprM gene 86
NRD convertase *see* N-arginine dibasic convertase
5'-NT *see* 5'-nucleotidase
nuclear hormone receptors 324, 324–337
  DNA binding domain 326–334, *327, 330, 332–334, 333*
  DNA response elements 324, 328–329
  ligand binding domain 326–331, *327*
  receptor hinges 335–336
  thyroid hormone receptor 324, 325, 331, 334
  vitamin D receptor 325, *331, 334*
  zinc finger motifs 328
nuclease P1 51–61, *51, 54, 56, 57, 60*
5'-nucleotidase (5'-NT) 62–70, *62, 65, 66, 68, 69*
oncomodulin 506, *507*

Oncorhynchus mykiss 474
ordered bi-bi mechanism 24–25
Oryctolagus cuniculus 354, 358–360
osteonectin *see* basement membrane protein BM-40
osteopetrosis 250
ovalicin 98
oxidoreductases
    cocatalytic zinc sites *417*, *420*
    medium-chain dehydrogenases/reductases 5–33
    structural zinc sites *404*, *406–407*

P family see phospholipase C
P-cadherin 731
p53 gene, S100 proteins 537, *537*
p53-Mdm2 complex 347–348
paired homeobox (PAX) 378
*p*AMBS *see p*-aminomethylbenzenesulfonamide
pantophobiacs 475
papains 489–500
PAPs *see* purple acid phosphatases
parallel β-rolls 151–153, *152*
Paramecium
    calmodulin 472
    P. tetraurelia, calmodulin 448
Parkin protein 348
Parkinson's disease
    6-pyruvol-tetrahydropterin synthase 297
    RING domain proteins 347, 348
particularly interesting new cysteine–histidine protein (PINCH) 379–381
parvalbumins 501–508, *505*
    annexins 652
    penta EF-hand proteins 520
PAX *see* paired homeobox
PBG *see* porphobilinogen
PCI *see* protein carboxypeptidase inhibitors
PCPs *see* procarboxypeptidases
PDE *see* phosphodiesterases
PDI *see* protein disulfide isomerase
PEF *see* penta-EF-hand
PEFLINS *see* penta EF-hand proteins
penicillins
    [D]-Ala-[D]-Ala carboxypeptidase 174
    metallo β-lactamases 217–218, 227–229
Penicillium citrinum
    cocatalytic zinc sites 423
    nuclease P1 51, 52
penta EF-hand (PEF) domains 490, 491, 498–499
penta EF-hand proteins (PEFLINS) 501, *502–503*, 504, 516, 516–528, *522*
    ALG-2 516, 517, 519, 520–526, *523*, *525*, *526*
    calpains 517, 518, 520, 522, *524*
    EF1 hand motif *522*, 522–523, *523*
    EF2 hand motif *522*, *523*
    EF3 hand motif *522*, *523*
    EF4 hand motif *522*, *523*, *523*
    EF5 hand motif *522*, *523–524*
    grancalcin 516, 517, 518, 520–522, *524*
    parvalbumins 520
    peflin 517–518, 519, 522, *524*
    sorcin 517, 518–519, 520, *522*
peptidoglycan binding domain (PGBD) 166–167

peroxisome proliferator-activated receptors (PPAR) 325
PEX *see* hemopexin-like
PEX gene 105
PFT *see* protein farnesyltransferase
PGBD *see* peptidoglycan binding domain
PGGT-I *see* protein geranylgeranyltransferase type I
PGGT-II *see* protein geranylgeranyltransferase type II
PH *see* pleckstrin homology
PHD domain 341, *341*, 344, *344–345*
1,10-phenanthroline 121, 125
phosphatases *417–418*, *420–424*
phosphatide *sn*-2 acylhydrolase *see* phospholipase A_2
phosphatidylinositol 3-phosphate (PtdIns[3]P) 390, 391–392, 394–398
phosphocalmodulin 449
phosphodiesterases (PDE)
    calmodulin 448
    nuclease P1 51–61
phospholamban 670
phospholipase A_2 inhibitors (PLI) 682
phospholipase A_2 (PLA$_2$) 677–691, *677*, *685*, *688*
    membrane binding C2-like domains 587, 591, *592*, *596*
phospholipase C (PLC) 587
phospholipase C-δ1 (PLC-δ1) 603, 606
phospholipid binding 609, 611
phospholipid membrane-binding proteins 649–663
phosphomonoesterase *see* alkaline phosphatases
3′-phosphomonoesterases 51–61
phosphonates 99, 100, *101*
phosphonites 99, 100, *101*
phosphoramidon 87, 89
    neprilysin *108*, *109*, 110–113, *110*, *112*
phosphorylase kinase *454*
phosphoserine phosphatase (PSP) 673
phosphoseryl intermediates 74–75, *75*, 77–80
phosphotriesterases *418*
piccolo, C2-domain proteins 600, 604, 610–611
Pichia pastoris
    C-type lectin-like domains 706
    carboxypeptidases 179
PINCH *see* particularly interesting new cysteine–histidine protein
Ping-Pong mechanism 25
Piratoxin II 684
Pisum sativum 265–269, *267*
'pita-bread' fold *95*, 98
pitrilysins *190*, 190–198, *192*
PKCs *see* protein kinase C isoforms
PLA$_2$ *see* phospholipase A_2
plasminogen 656
PLC *see* phospholipase C
PLC-δ1 *see* phospholipase C-δ1
pleckstrin homology (PH) 391
PLG-NHOH, astacin 123, *124*
PLI *see* phospholipase A_2 inhibitors
polypeptide folds 276
porphobilinogen (PBG) 283–284, 293
Porphyridium purpureum 264, 264–269, *266*, *267*
post-translational modifications 555–557, 565, 694–695
potassium, basement membrane protein BM-40 512
potassium channels, calmodulin 471–488
PPAR *see* peroxisome proliferator-activated receptors
PPs *see* protein phosphatases
PPT *see* protein prenyltransferases
PRGP *see* protein-rich Gla protein

Index

pro-astacin activation 117
pro-meprin 117, *118*
procarboxypeptidases (PCPs) 177
promastigote surface proteinase (PSP) *see* leishmanolysin
prostaglandins 678
protein C
    calcium-binding EGF-like domains *563*
    Gla-domain *575*
protein carboxypeptidase inhibitors (PCI) 186–187, *187*
protein disulfide isomerase (PDI) 693
protein farnesyltransferase (PFT) *37*, 37–47, *42–44*
protein geranylgeranyltransferase type I (PGGT-I) 37–39
protein geranylgeranyltransferase type II (PGGT-II) *37*, 37–43, *41*, *43*, 45–46
protein hormones 367, 367–377
protein interface zinc sites 403, 432–441, *433–435*
protein kinase C isoforms (PKCs) 587, 591, 592, *594–596*
protein kinases *404–405*, 407–408
protein phosphatases (PPs) 63, 66, 67, *67–69*
protein phosphorylation 588
protein prenyltransferases (PPT) 37–48
protein S
    calcium-binding EGF-like domains 564–565
    Gla-domain *574*
protein tramtrack (TTK) 317–318, 321
protein-rich Gla protein (PRGP) 574
proteinase K 726, *726–727*, *727*
proteolysis
    C-type lectin-like domains 708
    calcium-binding EGF-like domains 567
    calpain 489–490, 494
prothrombin fragments 578–579, *578–580*
protocadherins 731
proton transfer 255
Pseudomonas
    P. aeruginosa
        5-aminolaevulinic acid dehydratase 289–290
        γ-class carbonic anhydrases 273
        metallo β-lactamases 218, 221, 224, 226
    P. diminuta, cocatalytic zinc sites 424, *425*, 427
    serralysin 148–149, *150*, *154*
    subtilisin 727–728
PSP *see* leishmanolysin; phosphoserine phosphatase
PtdIns[3]P *see* phosphatidylinositol 3-phosphate
pteridines, synthesis 236, *236*
PTPS *see* 6-pyruvol-tetrahydropterin synthase
purple acid phosphatases (PAPs)
    cocatalytic zinc sites *418*, 424
    5′-nucleotidase 63, 66, *67*, 67, 68
pylol dehydrogenases 406
Pyrococcus
    P. abyssi, [D]-aminopeptidase DppA 209
    P. furiosis, methionine aminopeptidases 95
6-pyruvol-tetrahydropterin synthase (PTPS) 296–304, *296*, *300–302*
    guanosine triphosphate cyclohydrolase I 239, 242, *243*
Python reticulatus 682

R-cadherin 731
R_6 insulin 373–376, *374*, *375*
RA *see* rheumatoid arthritis
Rab escort proteins (REP) 39, 40
rabbit hemopexin 636–639

Rabenosyn-5 398
RabGGT *see* PGGT-II
rabphilin
    C2-domain proteins *600*, 600–604
    membrane binding C2-like domains 587–588, *589*, *590–591*
RACK *see* anchoring protein receptor for activated C-kinases
RAGE *see* receptor for advanced glycation endproduct
Ralstonia solanacearum 209
Rana catesbeiana 529
random bi-bi mechanism 25
RAR *see* retinoic acid receptors
Rattus
    calpain domain VI 520
    R. norvegicus
        calmodulin 474, 476–482
        calpain 490–499
        neprilysin 105
        5′-nucleotidase 63
        S100 proteins 529
    R. rattus, metallothioneins 354, 358–360
    rat hepatic lectin-1 711–713
reactive oxygen species (ROS) 354
really interesting new gene (*RING1*) 338–350
REases *see* restriction endonucleases
receptor for advanced glycation endproduct (RAGE) 531
receptor hinges 335–336
REP *see* Rab escort proteins
Reps1 543, *544*
respiratory distress syndrome 678
restriction endonucleases (REases) 742–756, *745*, *746*
    Bam HI *742*, 744–747, *746*, *747*, 750–751, *752*, 753–754
    Fok I 744, *746*, 747–748, *748*, 750–751
    Nae I 744, *746*, 749, 749–751
    Ngo MIV 744, *746*, *748*, 748–751, 753–754
restriction–modification (R–M) systems 742
retinoic acid 8, 26
retinoic acid receptors (RAR) 325
retinoid X receptor (RXR) *324*, 325
rhamnose isomerase 411
rheumatoid arthritis (RA) 678
rhinovirus-2 proteinase 410–411
RHL-1 (rat hepatic lectin-1) 711–713
RING domain proteins 338–350, *338*, *341–344*
    protein interface zinc sites 439
RNAP *see* DNA-dependent RNA polymerases
ROS *see* reactive oxygen species
Rossmann folds 674
RTX-toxins 149
rubredoxin 'full knuckles' 383–385
RXR *see* retinoid X receptor

S family *see* synaptotagmins
S100 proteins 529–540
    amino acid sequence *532–534*
    annexins 656, 658, *659*
    biological functions *530*
    structural diagrams *529*, *535–537*
SA *see* succinylacetone
Saccharomyces cerevisiae
    calcium-binding EGF-like domains 555
    calmodulin 448, 450, 472

carboxypeptidases 178
Cys$_2$His$_2$ zinc finger proteins 308–310, 314, 319–320
Eps 15 homology domain 542
FYVE domain 391, 393
mitochondrial processing peptidases 195–197
pitrilysins 195–197
Sags *see* superantigens
sAHP *see* slow afterhyperpolarization
Salmonella typhimurium
   cocatalytic zinc sites 427
   methionine aminopeptidases 95
   5′-nucleotidase 63
salt bridges
   astacin 123
   calpain 497, *497*
SARA *see* Smad anchor for receptor activation
sarcolipin 670
sarcoplasmic and endoplasmic reticulum calcium ATPases (SERCA) 667, 667–676, *669*, *671–674*, 692
sarcoplasmic reticulum (SR) 692–701
scaffoldins 617–620, 625–627
SCF ubiquitin ligase complex 347
SEH *see* staphylococcal enterotoxin H
selectins 707, 709, 710, 713
separated-paired-fingered Cys$_2$His$_2$ (spC2H2) 308–309
SERCA *see* sarcoplasmic and endoplasmic reticulum calcium ATPases
serine endopeptidases 718–730
serine-carboxyl proteinases 727, *727*–728
serine/threonine protein phosphatases (PPs) 63, 66, 67, *67*
serralysins *148*, 148–156, *152–154*
   astacin 124, 128
Serratia marcescens
   metallo β-lactamases 218
   serralysin 148–151, *150*
SF-1 *see* steroidogenic factor-1
Shaker channels 472, 485
SHH *see* sonic hedgehog
Shigella flexneri 734
silver, metallothioneins 362–363
single-strand specificity 56
site-directed mutagenesis 227–230
SK channels *see* small conductance calcium activated potassium channels
skeletal isoform of troponin C (sTnC) 459, *459*, 461–468, *464–466*
slow afterhyperpolarization (sAHP) 473, 482, 484
Smad anchor for receptor activation (SARA) 392
small conductance calcium activated potassium channels (SK channels) 471–488, *486*
small GTPase regulation 588
snake venom phospholipase A$_2$s (svPLA$_2$s) 678–689
SNAP-25, C2-domain proteins 607–609
SNAREs *see* soluble N-ethylmaleimide sensitive factor attachment protein receptors
sNEP *see* soluble extracellular domain of human NEP
SODs *see* superoxide dismutases
soluble extracellular domain of human NEP (sNEP) *109*, 109–114
soluble N-ethylmaleimide sensitive factor attachment protein receptors (SNAREs) 600, 607–610
sonic hedgehog (SHH) 168, 168–172, *171*, *172*
sorcin (soluble resistance-related calcium binding protein) 517–520, 522

spacers
   protein interface zinc sites 432
   receptor hinges 335
   structural zinc sites 403
SPARC *see* basement membrane protein BM-40
spC2H2 *see* separated-paired-fingered Cys$_2$His$_2$
SPP *see* chloroplast processing peptidase
squash trypsin inhibitor 432–433
SR *see* sarcoplasmic reticulum
staphylococcal enterotoxin H (SEH) 435, *435*
Staphylococcus
   S. aureus, protein interface zinc sites 433–434, 438
   S. pyogenes, protein interface zinc sites 433–434
steroidogenic factor-1 (SF-1) 325
sTnC *see* skeletal isoform of troponin C
Strenotrophomonas maltophilia 217–232, *217*, *223*, *225*
Streptomyces
   S. albus G, [D]-Ala-[D]-Ala carboxypeptidase 164–175
   S. griseus
      [D]-aminopeptidase DppA 212
      leucine aminopeptidase 206
   S. lividans
      [D]-Ala-[D]-Ala carboxypeptidase 165
      calmodulin 473
   S. tanashiensis, neprilysin 109
   trypsin 719, 720, 724–725
stromelysins 130
Strongylocentrotus purpuratus 354, 360
strontium
   C-type lectin-like domains 706
   Gla-domain 577, 579, 580
structural calcium 718–730
   subtilisin 718–721, *719*, 725–728, *725*, *726*
   trypsin 718–725, *718*, *722*, *725*, *728*
structural zinc sites 403–415
   formamidopyrimidine-DNA glycosylase MutM family *405*
   hydrolases *405*, 409–411
   isomerases *405*, 411
   ligases *405*, 411–412
   oxidoreductases *404*, 406–407
   transferases *404–405*, 407–409
substrate specificity 58–59
subtilisin 718–721, *719*, 725–728, *725*, *726*
succinylacetone (SA) 291–292
sulfate 257–258
Sulfolobus solfataricus
   medium-chain dehydrogenases/reductases 6
   protein interface zinc sites 437
sulphonamide inhibitors 250, 253, 256–257
superantigens (Sags) 433–436, *434*
superoxide dismutases (SODs) 416, *417*, 420
Sus scrofa
   calmodulin 474
   S100 proteins 529
svPLA$_2$s *see* snake venom phospholipase A$_2$s
SW15 finger 1 314–315, *314*
synaptobrevin II 93
synaptotagmins
   C2-domain proteins *599*, 600–611, *600*, *602*, *605*, *608*, *609*
   membrane binding C2-like domains *587*, *589*, *590*, *592–594*, *596*
Synechococcus 355, 361–362
syntaxin 607, *608*

$T_3R_3^f$ insulin 369–373, *371–373*
T_6 insulin 368, 369, *370*
tamoxifen 455
TAO *see* thyroid-associated ophthalmopathy
tC$_2$H$_2$ *see* triple-fingered Cys$_2$His$_2$
testicans 514
tetrahydrobiopterin (BH$_4$) 236, 297, 299
tetranectin 707
TFIIIA 318–319
TFP *see* trifluoperazine
TGEKP linker sequence 308, *320*, 320–321
thapsigargin 670, 675
thermitase 725–726, *726*, 728
Thermoactinomyces vulgaris 178, 180
thermolysin (TLN) 85–94, *85*, *87*
    inhibitor complexes 87–91, *88–90*
    matrix metalloproteinases 144
    neprilysin 108, 111, *113*, 113–114
Thermoplasma acidophilum 6
Thermosynechococcus elongatus 272, *273*
Thermotoga maritima 80–81
thioredoxin fold domains 695–697, *696*, *699*
thiorphan *108*, *110*, 110–113, *112*
thrombin 560, *561*
thrombomodulin EGF domains 560, *561*
thromboxanes 678
thyroglobulin-like (TY) domain *514*
thyroid-associated ophthalmopathy (TAO) 693
thyroid hormone receptor (TR) *324*, *325*, *331*, *334*
thyroid peroxidase 554
TIM-barrel folds *283*, *285*, *286*
tissue inhibitors of metalloproteinases (TIMPs)
    astacin 119, *120*
    hemopexin domains 634
    MMP-TIMP complexes 131, 142, *143*
TLN *see* thermolysin
TMG *see* transmembrane Gla protein
TnC *see* troponin C
TNP-470 98
TNPFR ligand *541*, *548*
TR *see* thyroid hormone receptor
TR$_1$C 463–464
tramtrack (TTK) 317–318, 321
transcription factors
    Cys$_2$His$_2$ zinc finger proteins 307–323
    nuclear hormone receptors 324–337
    RING domain proteins 338–348
transferases
    protein prenyltransferases 37–48
    structural zinc sites *404–405*, *407–409*
transmembrane Gla protein (TMG) 574
transphosphorylation 71
triadin 700–701
trifluoperazine (TFP)
    calmodulin *455*, 455–456
    calsequestrin 698, *699*
Trimeresurus 679, 682
triple fingered Cys$_2$His$_2$ (tC2H2) 308–309
triple helical collagenolysis 641
tRNA synthetases *405*, 411
troponin C (TnC) *459*, 459–470, *462*, *463*
    calsequestrin 692, 698
    parvalbumin *505*, 505–506, *507*

Trypanosoma 38
    leishmanolysin 158
    T. brucei, protein prenyltransferases 47
trypsin *718*, 718–725, *722*, *725*, *728*
TTK *see* tramtrack
tubular acidosis 250
two–metal ion mechanism 59
TY *see* thyroglobulin-like
type I tyrosineamia 284
type II restriction endonucleases 742–756, *745*, *746*
    Bam HI *742*, 744–747, *746*, *747*, 750–751, *752*, *753–754*
    Fok I 744, *746*, 747–748, *748*, 750–751
    Nae I 744, *746*, 749–751, *749*
    Ngo MIV 744, *746*, *748*, 748–749, 750–751, *753–754*
tyrosine kinase Lck 438
tyrosineamia (type I) 284

U-PAR *see* urokinase-type plasminogen activator receptors
ubiquitination
    membrane binding C2-like domains 588
    RING domain proteins 338–340, *339*, *345*, *346*
UDP-sugar hydrolases *418*, 424
uridine diphosphate *N*-acetylglucosamine *O*-acyltransferase 276
urokinase-type plasminogen activator receptors (U-PAR) 682

vanadate
    alkaline phosphatases 77, *78*
    cocatalytic zinc sites 422–423
vancomycin 164, *167*, 168–172, *170–172*
variable ligands 19, *19*
VDR *see* vitamin D receptor
VE-cadherin *732*, *733*, *739*
vesicular transport 588
Vibrio parahaemolyticus 63
Vipera 679–681, 684, 686–688
Vipoxin 686–688
vitamin D receptor (VDR) *325*, *331*, *334*
vitamin K-dependent proteins 573–583, *575*
vitronectin 631–635

Williams–Beuren syndrome transcription factor (WSTF) 345

Xenopus laevis
    basement membrane protein BM-40 510, 514
    cadherins 732
    Cys$_2$His$_2$ zinc finger proteins 308–310
Xfin finger 31 310–312, *311*

Ying-Yang protein (YY1) 318

ZFY finger 6 *313*, 313–314
Zif268 315–317, *315*, *316*, 321
zinc
    α-class carbonic anhydrases 249–263
    [D]-Ala-[D]-Ala carboxypeptidase 164–175
    alkaline phosphatases 71–82
    5-aminolaevulinic acid dehydratase 283–295
    [D]-aminopeptidase DppA 208–214

astacin 116–129
β-class carbonic anhydrases 264–269
bridges 438, *438*
carboxypeptidases 176–189
clusters 412, *412*
cocatalytic zinc sites 416–431
FYVE domain 390–399
γ-class carbonic anhydrases 270–282
guanosine triphosphate cyclohydrolase I 235–245
hook motif 437–438, *437*
insulin 367–377
inverzincins 190–198
leishmanolysin 157–163
leucine aminopeptidase 199–207
LIM domain proteins 378–389
matrix metalloproteinases 130–147
medium-chain dehydrogenases/reductases 5–33
metallo β-lactamases 217–234
metallothioneins 353–364
methionine aminopeptidases 95–103
neprilysin 104–115
nuclease P1 51–61
5′-nucleotidase 62–70
pitrilysins 190–198
protein interface sites 432–441
protein prenyltransferases 37–48
6-pyruvol-tetrahydropterin synthase 296–304
S100 proteins 529–540
serralysin 148–156
structural zinc sites 403–415
sulfur interactions 437–439
supplements 436
thermolysin 85–94
type II restriction endonucleases 743
zinc-fingers
  Cys_2His_2 *307*, 307–323
    ADR1 319–320
    ATF-2 315
    design/selection 321
    DNA bound *311*, 315–321
    GAGA factor 319, *319*
    glioblastoma oncogene 317, 321
    SW15 finger 1 *314*, 314–315
    TFIIIA 318–319
    TGEKP linker sequence 308, *320*, 320–321
    tramtrack 317–318, 321
    Xfin finger 31 310–312, *311*
    Ying-Yang protein 318
    ZFY finger 6 *313*, 313–314
    Zif268 *315*, 315–317, *316*, 321
  nuclear hormone receptors 324–337
  RING domain proteins 338–350
Zinkov 192
zymogens
  astacin 117
  carboxypeptidases 182–183, 185–186
  matrix metalloproteinases 141–142
  serralysin 149
Zyxins 380–381

Contents of Volumes 1 and 2

Volume 1

IRON 1

Heme Proteins: Oxygen Storage and Oxygen Transport Proteins 3

Myoglobin 5
George N Philips Jr.

Hemoglobin 16
Massimo Paoli and Kiyoshi Nagai

Heme Proteins: Cytochromes Section 31

Mitochondrial cytochrome c 33
Lucia Banci and Michael Assfalg

Cytochrome c' 44
Maria João Romão and Margarida Archer

Cytochrome c_2 55
Kunio Miki and Satoshi Sogabe

Cytochrome c_{551} 69
Francesca Cutruzzolà, Marzia Arese and Maurizio Brunori

Cytochrome c_{553} 80
Atsushi Nakagawa

Cytochrome c_6 87
Wolfgang Reuter and Georg Wiegand

Cytochrome c_4 100
Niels H Andersen, Hans EM Christensen, Gitte Iversen, Allan Nørgaard, Christina Scharnagl, Marianne H Thuesen and Jens Ulstrup

Cytochrome c_7 110
Lucia Banci and Michael Assfalg

Photosynthetic reaction centers of purple bacteria 119
C Roy D Lancaster and Hartmut Michel

Cytochrome c_{554} 136
Tina M Iverson, Michael P Hendrich, David M Arciero, Alan B Hooper and Douglas C Rees

Nine-heme cytochrome c 147
Maria Arménia Carrondo, Cláudio M Soares and Pedro M Matias

b-Type cytochrome electron carriers: cytochromes b_{562} and b_5, and flavocytochrome b_2 159
F Scott Mathews

Cytochrome f 172
Glenda M Soriano, Janet L Smith and William A Cramer

Cytochrome f/plastocyanin complex 182
M Ubbink

Heme Proteins: Cytochrome Peroxidases 193

Horseradish peroxidase 195
Michael Gajhede

Myeloperoxidase 211
Roger E Fenna

Contents of Volumes 1 and 2

Arthromyces peroxidase 222
Keiichi Fukuyama

Chloroperoxidase 233
Munirathinam Sundaramoorthy

Prostaglandin endoperoxidase H_2 synthases-1 and -2 245
R Michael Garavito

Heme Proteins: Cytochrome P-450 265

Cytochrome P450 267
Huiying Li

Heme Proteins: Oxygenases 283

Inducible nitric oxide synthase 285
Robin J Rosenfeld, John A Tainer and Elizabeth D Getzoff

Endothelial nitric oxide synthase 300
CS Raman

Heme oxygenase 317
David J Schuller

Heme Proteins: Oxidoreductases 329

Bacterial cytochrome *c* oxidase 331
Aimo Kannt and Hartmut Michel

Mitochondrial cytochrome *c* oxidase 348
Shinya Yoshikawa, Kyoko Shinzawa-Itoh, Eiki Yamashita and Tomitake Tsukihara

ba_3-Cytochrome c oxidase from Thermus thermophilus 363
Manuel E Than and Tewfik Soulimane

Succinate:quinone oxidoreductases 379
C Roy D Lancaster

Cytochrome bc_1 complex 402
Thomas A Link

Cytochrome cd_1 nitrite reductase 424
James WA Allen, Stuart J Ferguson and Vilmos Fülöp

Cytochrome c nitrite reductase 440
Oliver Einsle

Hydroxylamine oxidoreductase 454
Noriyuki Igarashi and Nobuo Tanaka

Sulfite reductase hemoprotein 471
M Elizabeth Stroupe and Elizabeth D Getzoff

Heme-catalases 486
Maria J Maté, Garib Murshudov, Jerónimo Bravo, William Melik-Adamyan, Peter C Loewen and Ignacio Fita

Non-Heme Proteins: Iron-Sulfur Clusters 503

Rubredoxin 505
Jacques Meyer and Jean-Marc Moulis

Fe–S Rieske center 518
Thomas A Link

The [2Fe–2S] ferredoxins 532
Giuliana Zanetti, Claudia Binda and Alessandro Aliverti

Ferredoxins containing one [4Fe–4S] center 543
Keiichi Fukuyama

Ferredoxins containing one [3Fe–4S] cluster. Desulfovibrio gigas ferredoxin II – solution structure	553
José JG Moura, Anjos L Macedo, Brian J Goodfellow and Isabel Moura	
Ferredoxins containing two different Fe/S centers of the forms [4Fe–4S] and [3Fe–4S]	560
Charles D Stout	
The 2[4Fe–4S] ferredoxins	574
Larry C Sieker and Elinor T Adman	
Hybrid cluster protein	593
Susan Bailey, Serena J Cooper, WF Hagen, AF Arendsen and Peter F Lindley	
High potential iron sulfur proteins	602
Charles W Carter Jr.	

Non-Heme Proteins: Mononuclear Iron Proteins 611

Naphthalene 1,2-dioxygenase	613
S Ramaswamy	
Protocatechuate 3,4-dioxygenase	622
Douglas H Ohlendorf and Matthew W Vetting	
2,3-Dihydroxybiphenyl 1,2-dioxygenase	632
Jeffrey T Bolin and Lindsay D Eltis	
Deacetoxycephalosporin C synthase	643
Inger Andersson, Anke C Terwisscha van Scheltinga, Graziella Ranghino and Karin Valegård	
Phthalate dioxygenase reductase	652
Martha L Ludwig, David P Ballou and Louis Noodleman	
Fe superoxide dismutase	668
Anne-Frances Miller	

Volume 2

IRON continued 683

Non-Heme Proteins: Dinuclear Iron Proteins 685

Hemerythrin	685
Ronald E Stenkamp	
Ribonucleotide reductase	699
Hans Eklund	
Methane monooxygenase hydroxylase	712
Douglas A Whittingotn and Stephen J Lippard	
Δ^9 Stearoyl-acyl carrier protein desaturase	725
Ylva Lindqvist	
Iron-only hydrogenases	738
Brian J Lemon and John W Peters	
Purple acid phosphatase	752
Andreas Vogel, Fridrich Spener and Bernt Krebs	

Non-Heme Proteins: Iron Storage 769

Ferritin	769
Elizabeth C Theil	
Cytochrome b_1 – bacterioferritin	782
Felix Frolow and Aaron Joseph Kalb (Gilboa)	

Contents of Volumes 1 and 2

Non-Heme Proteins: Iron Transport 791

Transferrins 793
Peter F Lindley

Lactoferrin 812
Clyde A Smith

The ferric hydroxamate uptake receptor FhuA and related TonB-dependent transporters in the outer membrane of gam-negative bacteria 834
Andrew D Fereguson, James W Coulton, Kay Diederichs and Wolfram Welte

Iron-dependent regulators 850
Michael D Feese, Ehmke Pohl, Randall K Holmes and Wim GJ Hol

NICKEL 865

Urease 867
Robert P Hausinger and P Andrew Karplus

Nickel-iron hydrogenases 880
Michel Frey, Juan C Fontecilla-Camps and Anne Volbeda

Methyl-coenzyme M reductase 897
Wolfgang Grabarse, Seigo Shima, Felix Mahlert, Evert C Duin, Rudolf K Thauer and Ulrich Ermler

Peptide deformylase 915
Andreas Becker and Wolfgang Kabsch

Diphtheria toxin repressor: metal ion mediated control of transcription 929
Dagmar Ringe, Andre White, Shuyan Chen and John R Murphy

MANGANESE 939

Manganese superoxide dismutase 941
M Elizabeth Stroupe, Michael DiDonato and John A Tainer

Arginase 952
Maria C Bewley and John M Flanagan

Concanavalin A 963
A Joseph Kalb (Gilboa) and John R Helliwell

Aminopeptidase P 973
J Mitchell Guss, and Hans C Freeman

COBALT 981

Glutamate mutase 983
Christoph Kratky and Karl Gruber

Methylmalonyl CoA mutase 995
Karl Gruber and Christoph Kratky

Cobalamin-dependent methionine synthase 1010
Karl Gruber and Christoph Kratky

MOLYBDENUM/TUNGSTEN 1023

Nitrogenase 1025
Benedikt Schmid, Hsiu-Ju Chiu, Vijay Ramakrishnan, James B Howard and Douglas C Rees

Aldehyde oxidoreductase (MOP) 1037
Maria João Romão and José JG Moura

Dimethylsulfoxide reductase 1048
Hung-Kei Li and Hermann Schindelin

Trimethylamine N-oxide reductase 1063
Chantal Iobbi-Nivol, Richard Haser, Vincent Méjean and Mirijam Czjzek

Dissimilatory nitrate reductase 1075
Maria Joaão Romão, João Miguel Dias and Isabel Moura

Formaldehyde ferredoxin oxidoreductase 1086
Roopali Roy, Ish K Dhawan, Michael K Johnson, Douglas C Rees and Michael WW Adams

Aldehyde ferredoxin oxidoreductase 1097
Roopali Roy, Ish K Dhawan, Michael K Johnson, Douglas C Rees and Michael WW Adams

Formate dehydrogenase H 1109
Peter D Sun, Jeffrey C Boyington and Thressa C Stadtman

Sulfite oxidase 1121
Caroline Kisker

CO dehydrogenase 1136
Holger Dobbek, Lothar Gremer, Ortwin Meyer and Robert Huber

COPPER 1149

Cupredoxins (Type-1 Copper Proteins) 1151

Plastocyanin 1153
Hans C Freeman and J Mitchell Guss

Azurin and azurin mutants 1170
U. Kolczak, C. Dennison, A. Messerschmidt and G. W. Canters

Pseudoazurin 1195
Elinor T Adman

Amicyanin and complexes of amicyanin with methalamine dehydrogenase and cytochrome c_{551i} 1203
F Scott Mathews

Cucumber basic protein 1215
J Mitchell Guss and Hans C Freeman

Stellacyanin, a member of the phytocyanin family of plant proteins 1219
Aram Migran Nersissian, Peter John hart and Joan Selverstone Valentine

Rusticyanin 1235
Menachem Shoham

Type-2 Copper Enzymes 1243

Prokaryotic copper amine oxidases 1245
Michael J McPherson, Mark R Parsons and Carrie M Wilmot

Eukaryotic copper amine oxidases 1258
Diana L Wertz and Judith P Klinman

Galactose oxidase 1272
Michael J McPherson, Mark R Parsons, R Kate Spooner and Carrie M Wilmot

Copper-zinc superoxide dismutase in prokaryotes and eukaryotes 1284
Domenico Bordo, Alessandra Pesce, Martino Bolognesi, Maria Elena Stroppolo, Mattia Falconi and Alessandro Desideri

Binuclear Copper: Type-3 Copper Enzymes 1301

Hemocyanins from arthropods and molluscs 1303
Karen A Magnus

Catechol oxidase 1319
Christoph Eicken, Carsten Gerdemann and Bernt Krebs

Contents of Volumes 1 and 2

Binuclear Copper: CuA Copper 1331
Binuclear copper A 1333
Peter MH Kroneck

Multicopper Enzymes 1343
Ascorbate oxidase 1345
Albrecht Messerschmidt

Laccase 1359
Gideon J Davies and Valérie Ducros

Ceruloplasmin 1369
Peter F. Lindley

Copper nitrite reductase 1381
Elinor T Adman and Michael EP Murphy

Copper Storage and Transport 1391
Structure of the fourth metal-binding domain from the Menkes copper-transporting ATPase 1393
Wayne J Fairbrother

Yeast copper metallothionein 1405
Cynthia W Peterson

VANADIUM 1415

Vanadium haloperoxidases 1417
Ron Wever and Wieger Hemrika

List of Contributors 1429

PDB code list 1437

Index 1449